17th D A S C

The AIAA/IEEE/SAE Digital Avionics Systems Conference

Bellevue, WA

Oct. 31 – Nov. 7, 1998

Proceedings

Volume I

IEEE Catalog Number 98CH36267
ISBN 0-7803-5086-3 (Softbound Edition)
ISBN 0-7803-5087-1 (Casebound Edition)
ISBN 0-7803-5088-X (Microfiche Edition)

Prepress Production By:

ISOGEN International Corp.
2760 N. Academy Blvd, Suite 301
Colorado Springs, CO 80917
719 572-1706
sgml@isogen.com

Printed in USA by:

Lucent Technologies
Customer Information Center
2833 N. Franklin Road
Indianapolis, IN 80917

Welcome to the Emerald City

On behalf of all citizens of Seattle, I would like to welcome you to the beautiful Emerald City. World-renowned for its people and beauty — our state is also home to Boeing, Paccar and Microsoft. These companies are world leaders in their field and build products addressed by the DASC. While we need to instill a greater confidence in our community and must work together as partners, I commend DASC in organizing a world leading conference as you bring together the high technology industry and governmental organizations to address electronically-enabled systems of all vehicle types. We hope you enjoy your stay and make new connections. Again, welcome.

The Honorable Paul Schell, Mayor of Seattle

Chairman's Message

Welcome to the 17th DASC. Our modern world requires continuous learning and networking among professionals to stay competitive. In few areas is this as true as it is for electronically-based systems in vehicles. Daily, we all rely upon the safety and functionality of systems in our cars and airplanes and depend upon the security resulting from our military's readiness. Consumers also now demand more entertainment and communications from their vehicles. Our 17th DASC team has assembled over 250 papers and 32 tutorials to bring you the latest business realities, technology advances and systems integration skills in these areas. We have a number of programs and networking events to allow you to make lasting connections with other professionals, while you enjoy the hospitality of the great Northwest.

Paul E. Gartz, The Boeing Company

DASC History and Current Scope

The DASC began in 1975 as a visionary initiative to bring together multiple professional societies to address newly emerging directions in vehicle, digital electronic systems. Initially, this was in on-board aviation and space vehicle systems called avionics, hence the namesake of the "Digital Avionics Systems Conference". From the earliest conferences, the DASC has been committed to bridging the gap between commercial and defense sectors that has become so important in recent years.

Today, digital, electronically-based systems are essential to all aspects of all modern vehicles. The 17th DASC has again taken a visionary step to bring together all the involved industry and governmental organizations to address electronically-enabled systems in all vehicle types (air, space and ground), including the latest, large-scale FAA, DOT and DoD transportation and integration initiatives.

In cars, these systems enable modern fuel efficiencies, deploy air bags, provide audio and now video entertainment, offer cellular communications, assure anti-lock braking and provide your mechanic with vehicle diagnostic information. In commercial airplanes, these systems allow pilots to plan their flights, automatically control cabin air pressure and temperature, reconfigure electrical power systems in case of failures, assure safe, automatic landings in poor visibility, provide the latest in sophisticated In-Flight Entertainment (IFE) Systems and record and transmit maintenance-related information. In military vehicles, electronically based systems lie at the heart of the Gulf War victory, control the same vehicle systems as on commercial aircraft and cars, provide similar maintenance support, provide battlefield communications, manage stores and fire weapons.

The deployment of these systems over the past two decades has resulted in a world fleet of automobiles, airplanes and defense vehicles that can now be linked together to manage transportation and the military battlefield situation at higher, broader levels...so called "Systems-of-Systems". These include the FAA's Air Traffic Management System, the DOT's Intelligent Transportation System (ITS) and the military's modern Information Battlefield and integrated US Transportation Command.

DASC Track Concept

Ten Areas of Interest

The DASC addresses the interaction of five previously separate industries that are beginning to work together in vehicle systems: aerospace and automotive vehicles with telecommunications, PCs and entertainment. Modern development of these vehicles requires close cooperation among customer, business, governmental and technical perspectives to assure all needs are met in a timely fashion. Furthermore, the technical development of a vehicle and its systems require close cooperation between program design functions andthe technical skill and process base of the company that supports these programs.

To meet these needs and to bring order to our industry dialog, the 17th DASC has been organized into interest areas. We call them "Tracks". There are three categories: Customer, Engineering Disciplines and Design. By analogy these can be thought of as a company's Market and Customer Service organizations that understands the customer, the Education, Skill Resource and Process arms that maintain technical competency and the Programs that design the vehicles. These tracks organize all three major DASC programs: Tutorials, Technical and Exhibits.

Customer Track

Any successful system design begins and ends with the customer. In this track, the end paying/using customers review their operational needs for and in-service experiences with vehicle systems. Economic realities and values that govern business success are given prime emphasis. We also promote a "Virtual Customer Track" whose goal is to encourage all authors to discuss customer and business considerations in every DASC paper. (See Track A)

Engineering Discipline Tracks

These tracks address the competencies required for Systems Integration, the key goal in systems development. Included are the processes, standards, technologies and facilities needed for each of the four major systems integration disciplines: Systems Engineering, Software Engineering, Hardware Engineering and Human/Machine Engineering. Authors and papers are mixed in the sessions from all industries (aerospace, automotive, telecommunications, commercial and defense) to achieve maximum cross-fertilization and advancement of each of these disciplines. (See Tracks B - E)

System Design Tracks

Through design description and case studies these tracks address system architectures, designs and lessons learned for each of the specific industries and vehicle types the DASC addresses. The goal here is to concentrate knowledge within an industry. (See Tracks F- I)

Customer

A. Customer, Business & Operations
Jean Valentine, ARINC

Discipline

B. Systems Engineering
Doug Chappelle, Boeing

C. Software Engineering
Marc Pitarys, USAF
Steven Miller, Rockwell Collins

D. Hardware Engineering
Jeff Volp, Draper Lab

E. Human Engineering
Amy Pritchett, Georgia Tech

Design

F. Air Traffic Management
Alex Smith, Rannoch

G. Aircraft Systems
Erv Ulbrich, Boeing
Steve Groethe, Honeywell

H. Spacecraft Systems
Mike Jahan, JPL
Dave Barnett, Lockheed Martin

I. Automotive Systems
Wayne Devereux, Veridian
Paul Skalny, Nat'l Automotive Center

J. Information Systems
Tom Brown, Draper Lab

Program Chairs

Deputy Conference Chair: Jose Bolanos - Boeing Commercial Airplane Group

The DASC is co-sponsored by IEEE, AIAA and SAE to bring you a broad representation of electronic, aerospace and automotive professional interests. We have arranged a series of programs to meet your needs. Participate in any one of them independent of the others or come to all to absorb the full scope of what the DASC has to offer. I want to especially encourage you to attend the tutorials which provide in depth learning in Systems Integration, Design and Business skills even for those not in aerospace or automotive fields.

Tutorials Program Co-Chairs: John Mari - Lockheed Martin, Ken Waichunas – Analogy

Our tutorial program has been formulated to provide you high value learning in each of the DASC track subject areas. The presenters were sought as practicing experts in their fields that are not only knowledgeable in technical aspects of their subjects but also bring you their wisdom of implementation realities and give you "behind the scene" insight into the issues.

Technical Program Chair: Steve Runo - Scitor

The heart of the DASC has always been the technical program. This year, over 250 papers in 40 technical sessions offer opportunities to discover the latest in vehicle system designs and advances in development methodologies. The 17th DASC technical program provides a unique opportunity to sample innovation and meet associates in related, yet different industries and disciplines. Two panel sessions feature timely discussions of crowded transportation lanes and the limits of human automation.

Exhibits Program Chair: Paul Kostek - Tek Sci

The DASC exhibits are designed to give you a chance to see new technology first hand and talk with the companies developing it. Exhibitors will include leaders from the aerospace, automotive and military sectors. Exhibitors will offer rotating presentations of introducing their company and products of interest to you. This year the DASC will also offer tool suppliers to provide you a look at the world class tools system development professionals need to accomplish the twin goals of technical quality and business efficiency.

Social Program/Banquets Chair: Judy Makue - Northrup Grumman

Those of you who have attended a DASC before realize the importance (and fun) of our traditional Thursday night social event. Here we all have a chance to relax and meet each other in a more casual environment. This year we have a special outing that will maximize your exposure to the best of the Northwest. We will take a boat across the Puget Sound and have dinner and entertainment at Kiana Lodge.

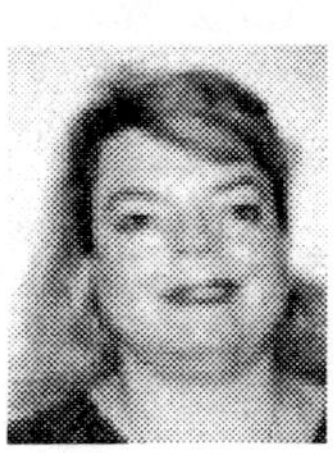

Spouse Program Chair: Mary Bolanos - State of Washington

Often our spouses miss the opportunity to travel with us for both enjoyment and to learn what we do in our careers. The DASC has felt strongly that spouses should be an important part of our program. We will offer a comprehensive series of outings to introduce you to the Seattle environment, its history, its beauty and its shopping. Spouses have made lasting friendships at the DASC and even helped their DASC-attending partners to connect with each other. We can also help you with considerable Pacific Northwest and Seattle tour information...come early!

17th DASC
Presentations

Track A - Customer & Business

Session A1: Executive Perspectives

1	Supplier's Perspective on Commercial Air Transport Avionics Business Practices	Mr. Gust M. Tsikalas
2	PROCESS.... TECHNOLOGY...TEAMWORK: Ingredients for Success in an Era of Change (Not Published)	Captain Larrie Cable
3	Optimizing Intermodal Cargo Movements (Not Published)	Andrea Riniker
4	Strategic Decision Making Increases Shareholder Value (Not Published)	Mr. Nick Pudar

Session A2: Maintenance Operations

1	Integrating the Aircraft with the Airline the need for improved 'open', industry standard business systems in the aircraft operational environment (Not Published)	Mr. Ian Briscoe
2	A Diagnostic Decision Support Tool (Not Published)	Mr. Karl Rein-Weston
3	Remote Aircraft Maintenance (Not Published)	Mr. Christian Orlowski
4	The Evolution of Central Maintenance Systems	Mr. Robert J. Manelski
5	Loop Resistance Tester: A Non-Intrusive Method To Measure Connector And Shield Resistance	Mr. Erik Godo
6	Application of Causal Networks for Advance Maintenance	Mr. Chuck Sitter
7	Electronic Maintenance Task Support System (Not Published)	Mr. William Ahl

Session A3: Economic Analysis

1	Cost Impacts of Real-Time Non-Intrusive (RTNI) Monitoring Technology To Real-Time Embedded Systems	Mr. Daniel Ferens
2	Total Ownership Cost and Decision Support in Business Management	Mr. Ellis Hitt
3	Technology Obsolescence Impact on Total Cost	Mr. Ellis Hitt
4	Correct Decisions Based On Uncertain Requirements and Imperfect Information Within Product Partnerships - A Contradiction?	Dr. Michael Kokes
5	Managing Long-Lived Information Assets - Emerging Standards and Tools can $ave BIG $$	Mr. Kent Taylor
6	Cost Benefit Analysis of the Combined WAAS/LAAS System	Mr. Alex Smith

Track B Systems Engineering

Session B1: Processes & Methods

1	The House of CE+ - A Systemic Framework for the Proper Selection of Practices	Mr. Ernst Fricke
2	Drawing Roadmaps To The Future: The Need For An Integrated Systems Engineering Approach	Dr. Ir. Eric Theunissen
3	Process Mapping Of Concurrent Engineering Processes For Integrated Systems	Mr. Ernst Fricke
4	Towards a Logical Basis for Systems Engineering	Manfred Broy
5	Model-Based Systems Engineering Of Automotive Systems	Mr. Jim Long
6	Streamlining the System Engineering Process for Developing Vehicle Management System Software for Air, Ground, and Space Applications (Not Published)	Mr. Peter Ross
7	The Impact of Acquisition Reform on System Engineering: Selected Examples (Not Published)	Mr. Mel M. Cutler

Session B2: Systems Architecture

1	Tactical Aircraft Moving Map Cabability (TAMMAC) (Not Published)	Mr. Ervin Pagel
2	Managing Multi-Platform Sensor Systems	Mr. Kurt Himmelreich
3	An Interactive Tool Suite for Real-time Application Development on Heterogeneous High Performance Computing Platforms	Dr. Minesh Patel Ph.D.
4	The Generic Unified Systems Engineering Metamodel	Mr. George Caple
5	Architecting Mobile Command and Control Communications (Not Published)	Mr. Jeff Lankford

17th DASC
Presentations

Track B Systems Engineering (Continued)

Session B3: Safety, Certification & Standards

	Title	Presenter
1	Current Government and Industry Developments in the Area of System Safety Assessment	Mr. Brett Portwood
2	Overview of EIA 632: Processes for Engineering a System	Mr. James N. Martin
3	Implementation of EIA-632 - Process Compliance or Imbedded Infrastructure?	Mr. Richard M. Harwell
4	An Overview of the Systems Engineering Capability Model EIA/IS 731	Mr. Don E. Barber
5	Comparison Of The Number Of 'Shalls' In Quality System Standards	Mr. William Walker
6	Low-cost Integrated Display Certification	Uma Satyen

Session B4: Test & Simulation

	Title	Presenter
1	BITE Is Not The Answer(But What Is The Question?)	Mr. Ronald Carson Ph.D.
2	The Propulsion Instrumentation Working Group Current Status Report (Not Published)	Mr. Robert Anderson
3	The Application Of Function Reconfiguration For Digital Fly-By-Wire Flight Control System	Yao Yiping
4	Processor Design and Implementation for Real-time Testing of Embedded Systems	Ms. Gail Walters
5	Meeting Military System Test Requirements with the Usage of COTS Products	Mr. Stephen Pizzica
6	An Overview of the Electromagnetic Compatibility Requirements in RTCA Standard DO-160D, Environmental Conditions and Test Procedures for Airborne Equipment	Mr. Erik Borgstrom

Session B5: System Tools

	Title	Presenter
1	Life Extension of USAF Fighter Aircraft: Lessons Learned from F-22 Subsystem Durability Life Tests	Mr. Stefan Glista
2	SpecTRM: A CAD System for Digital Automation	Ms. Nancy Leveson
3	Overview of TRIZ Technology for Innovation (Not Published)	Zinovy Royzen
4	Integrated Interface Definition Tool Development - IFTrac	Mr. Keith Ryan
5	OpEMCSS: A Graphical Discrete Event Simulation Library	Mr. John R. Clymer
6	Rapid Evolution Of All Your Systems: Your Problem Or Opportunity?	Mr. Jack Ring

Track C - Software Engineering

Session C1: Advanced Software Concepts

	Title	Presenter
1	Scheduling Paths - Routing Aircraft Or Communication Packets A Fast Parallel Concurrent Shortest Path Algorithm	Mr. David Lawson
2	Compiler Optimization and its impact on Development of Real-Time Systems	Mr. Kevin Tucker
3	Lean Software Development: Is It Feasible ?	Sowmyan Raman
4	Cost Effectiveness Of Formal Methods In The Development Of Avionics Systems At AEROSPATIALE	Mr. Francois Pilarski
5	Dynamic Scheduling for Avionics Mission Computing	Mr. David L. Levine
6	Software Process and Early Project Planning: Issues, Insights, and Lessons Learned	Mr. Jogeswari P. Sarkar

17th DASC
Presentations

Track C - Software Engineering (Continued)

Session C2: Test & Certification

1	Analyzing, Specifying and Verifying a Fault-Diagnosis-Isolation System (Not Published)	Gao Jianping
2	Low and Non-Intrusive Software Instrumentation - A Survey of Methods	Mr. Richard Fryer
3	Implementation and Testing of International Space Station Flight Software for the Pointing and Support Portion of GN&C	Mr. Douglas Yazell
4	Using Software Capability Maturity Model on Certification Projects	Ms. Leanna Rierson
5	A Comparison of Fly-Fix-Fly Testing to the Software Development Facility Testing Approach	Mr. Charles Satterthwaite
6	Methodologies for Predictable Real-Time System Design	Srini Vasan

Session C3: Software Architecture

1	Simulator, Workstation, and Data Fusion Components for Onboard/Offboard Multi-Target Multi-Sensor Fusion	Ms. Angela Pawlowski
2	Resource Allocation for Distributed Computing	Mr. Russell L. Carter
3	Fitting the Pieces Together: System/Software Analysis and Code Integration Using MetaH	Mr. Jonathan Kruegar
4	Selecting a Programming Language for your Project	Mr. David Naiditch
5	A Distributed Computing Environment for Signal Processing and Systems Health Monitoring	Mr. Larry Freudinger
6	Results of the Open System Development Initiative (OSDI) at NAWCWPNS	Ms. Debra Borden

Session C4: Software Modeling & Analysis

1	A Formal Methods Approach to the Analysis of Mode Confusion	Mr. Rick Butler
2	Club Med Avionics Systems At Motel 6 Costs: Win/Win Development Strategies	Mr. James Sutton
3	Experiences from Specifying the TCAS II Requirements using RSML (Requirements State Machine Language)	Dr. Mats P.E. Heimdahl
4	SCR: A Practical Method for Requirements Specification	Dr. Constance L. Heitmeyer
5	Using Models for Test Generation and Analysis	Dr. Mark R. Blackburn Ph.D.
6	Crafting Avionics Display Applications in Java	Dr. Roger K. Shultz Ph.D.

Track D - Hardware Engineering

Session D1: Sensor Systems

1	Microfabricated Chemical Sensor Systems For Safety And Emission Control Applications	Mr. Gary Hunter
2	Air-Cooled Flight Worthy Eye-Safe Coherent Lidar Transceiver	Mr. Mark Poling
3	Microfabricated Thin-Film Physical Sensors For Hostile Environment	Jih-Fen Lei
4	Real-Time Digital Signal Processing Enables A Reduced-Cost Atmospheric Remote Sensorcraft	Mr. Michael Grant
5	Reliability Flight Tests On A Fiber Optic Temperature Sensor	Ms. Margaret Tuma
6	Enhanced Recognition And Sensing Ladar (ERASER) Flight Demonstration (Not Published)	Mr. Robert Zumrick
7	Air Data Sensor Failure Detection System	Drew Houck

17th DASC
Presentations

Track D - Hardware Engineering (Continued)

Session D2: Processors & Commercial-Off-The-Shelf Systems

Session D3: Displays & Optoelectronics

Session D4: EMI HIRF [1] Electromagnetic Phenomenology

Session D5: EMI HIRF [2] Testing Effects And Test Facilities

17th DASC
Presentations

Track E - Human/Machine Engineering

Session E1: Methods & Measurements

1	A Process for Creating a Human-Centered System	Dr. Sheryl L. Chappell Ph.D.
2	Human Performance Considerations in Aircraft Certification (Not Published)	Dr. Elizabeth Lyall Ph.D.
3	A Pilot-Centered Autoflight System Concept	Mr. Victor Riley
4	Making the Cockpit a Human Friendly Environment: A Customer's Perspective	Mr. Edward A. Rafacz Jr.
5	Future Flight Deck Design - Emerging Safety Issues (Not Published)	Dr. Ratan Khatwa
6	Real-Time Operational Evaluations Using Advanced Flight Simulators	A.E. Dillard

Session E2: Automation & Decision Aiding

1	An Algorithmic Method for Predicting Pilot-Mode Interaction Difficulties	Denis Javaux
2	Analyzing Mode Transition Matrices	Sanjay Vakil
3	Human Factors Issues of Flight Deck Automation	Mr. Kenneth H. Funk II
4	Operator Directed Common Conceptual Models for Advanced Aircraft Automation	Sanjay Vakil
5	Some Formal Aspects of Human-Machine Interaction (Not Published)	Dr. Asaf Degani Ph.D.
6	Improving Pilot Conformance to Alerting System Commands	Ms. Amy Pritchett
7	Task Engagement: A Means to Situation Awareness	Mr. Paul C. Schutte

Session E3: Traffic Information Displays

1	CDTI Feature Evaluation for Enhancement of Oceanic Operations	Mr. Baltazar O. Olmos
2	Cognitive Factors In Aviation Display Design	Mr. Christopher D. Wickens
3	An Operational Implementation of CDTI (Not Published)	Mr. Bob Hilb
4	Pilots Spatial Mental Models for Memory of Heading and Altitude	Ms. Susanne Delzell
5	Shared Situation Awareness in the Flight Deck-ATC System	Professor John Hansman Jr.
6	Investigation of Cockpit Display of Traffic Information as an Aid for More Efficient Air Traffic Management Procedures	Ms. Amy Pritchett

Session E4: Advanced Displays

1	Experimental Study of Helmet-Mounted Display Symbology for Terrain Avoidance During Low-Level Maneuvers	Mr. James K. Kuchar
2	Advanced Flight Deck for Next Generation Aircraft	Mr. Robert Faerber
3	Projected Flight Path Displays and Controlled Flight into Terrain Accidents	Mr. Barry Strauch
4	A Display Concept for Staying Ahead of the Airplane	Mr. Eric N. Johnson
5	The Concept of 'Brain Mapping' for Flight Safety Avionics (Not Published)	Mr. Ivan Y. Burdun
6	Spatial Terrain Displays: Promises And Potential Pitfalls	Dr. Ir. Eric Theunissen

Session E5: Navigation & Guidance Displays

1	Synthetic Vision Flight Tests For Curved Approach And Landing	Dr. Gottfried Sachs
2	GPS User-Interface Design Problems	Mr. Kevin W. Williams Ph.D.
3	An International Usability Survey of GPS Avionics Equipment	Mr. Kurt M. Joseph
4	The 4-D Display	Dr.-Ing. Harro Von Viebahn
5	4D Flight Guidance Displays, A Gate to Gate Solution	Pro. Dr. -Ing Wolfgang Kubbat
6	Preliminary Flight Tests of Pseudo-Attitude Control Using Single Antenna GPS Sensing	Mr. Richard P. Kornfeld

A. Customer and Business Track

Customer requirements, operations and economics increasingly drive the design and application of avionics. This track focuses on the business considerations that are an integral element of the modern avionics design process.

Track Chair: Jean Valentine, ARINC

Session A1: Executive Perspectives

Chair: Jean Valentine, ARINC

	Title	Author
1	Supplier's Perspective on Commercial Air Transport Avionics Business Practices	Mr. Gust M. Tsikalas
2	PROCESS.... TECHNOLOGY...TEAMWORK: Ingredients for Success in an Era of Change (Not Published)	Captain Larrie Cable
3	Optimizing Intermodal Cargo Movements (Not Published)	Andrea Riniker
4	Strategic Decision Making Increases Shareholder Value (Not Published)	Mr. Nick Pudar

Session A2: Maintenance Operations

Chairs: Ian Briscoe, Penny & Giles; Phil Burke

	Title	Author
1	Integrating the Aircraft with the Airline the need for improved 'open', industry standard business systems in the aircraft operational environment (Not Published)	Mr. Ian Briscoe
2	A Diagnostic Decision Support Tool (Not Published)	Mr. Karl Rein-Weston
3	Remote Aircraft Maintenance (Not Published)	Mr. Christian Orlowski
4	The Evolution of Central Maintenance Systems	Mr. Robert J. Manelski
5	Loop Resistance Tester: A Non-Intrusive Method To Measure Connector And Shield Resistance	Mr. Erik Godo
6	Application of Causal Networks for Advance Maintenance	Mr. Chuck Sitter
7	Electronic Maintenance Task Support System (Not Published)	Mr. William Ahl

Session A3: Economic Analysis

Chair: Ellis Hitt, Battelle; Eric Haapala, Air Economics Group, Inc.

	Title	Author
1	Cost Impacts of Real-Time Non-Intrusive (RTNI) Monitoring Technology To Real-Time Embedded Systems	Mr. Daniel Ferens
2	Total Ownership Cost and Decision Support in Business Management	Mr. Ellis Hitt
3	Technology Obsolescence Impact on Total Cost	Mr. Ellis Hitt
4	Correct Decisions Based On Uncertain Requirements and Imperfect Information Within Product Partnerships - A Contradiction?	Dr. Michael Kokes
5	Managing Long-Lived Information Assets - Emerging Standards and Tools can $ave BIG $$	Mr. Kent Taylor
6	Cost Benefit Analysis of the Combined WAAS/LAAS System	Mr. Alex Smith

Track A – Customer and Business

SUPPLIER'S PERSPECTIVE ON COMMERCIAL AIR TRANSPORT AVIONIC'S BUSINESS PRACTICES

Gust Tsikalas, *Honeywell, Phoenix, Arizona*

Summary

Shift to the Bottom Line

I'd like to begin by casting your thoughts for a moment back to the period of the late 1980s and early 90s.

Airlines found themselves in a precarious double-whammy position: overcapitalized and unable to make a profit. They couldn't fill the seats of the airplanes they *had*, let alone the airplanes they had on *order*. Because of this, a shift in buying habits occurred at the airlines as the bottom line became the primary driver of decision-making. This has driven the industry ever since. Unfortunately, many of the business practices that have been ingrained in the industry have not shifted accordingly.

And that's exactly what I'd like to talk about in my brief allotted time.

Here's a quick example: Many value added functions are provided through software, but while the customers, in the best case situation, are willing to pay the nonrecurring cost to develop this software — and possibly allow a small profit, they seem not willing to enter into a discussion of its true value.

This adherence to cost-based pricing as opposed to value-based pricing is a serious threat to the delicate supplier-customer balance.

Total Quality Management

This is not to say that cost reduction initiatives don't need to continue — I can't emphasize enough that we can't let up in this area. Concepts like Total Quality Management have proven successful at both reducing costs and improving overall quality. These kinds of quality initiatives have had a very positive impact on the industry and have led to gains for all.

Business Model

What is needed is a business model that allows suppliers to benefit from its investments – to receive value, where they add value. This is important as the industry has many needs and the creativity to identify solutions to these pressing needs is required if we are to move forward in an effective manner. The solution process is bounded by the business risks to a given business practice. The question of what is the return on my investment has to be answered. At the industry forums, frustration is often verbalized that we are moving too slow. The technology is there; perhaps it is the business practices that are holding us back.

Environment

Airlines

Now, the major airlines are turning record profits and are more disciplined in maintaining their bottom lines. There is pressure to privatize in Europe and, globally, major alliances and acquisitions are occurring all the time. The outlook is fairly solid if there is a cautious management of capacity.

In fact, airlines are keeping older planes longer to fill capacity needs rather than buying all new—again, as a measure of caution.

OEMs

There are fewer OEMs today than a few years ago due to consolidation and airline buy-

0-7803-5086-3 /98/$10.00 ©1998 IEEE

ing preferences. In essence, Airbus and Boeing represent a duopoly in the commercial air transport arena. The OEMs have been developing derivative airplanes with a focus on more integration and value-added technology. The consolidations have led to stronger buying power, and this, coupled with market share strategies has resulted in lower acquisition prices and commitments over the full life cycle.

Regulations

Everybody complains about the bureaucracy of "regulations," but, of course, we all know that they're a necessary and important element of the overall safety process. Nevertheless, regulations having to do with development processes do increase cost and cycle time—with no appropriate counterbalance. In addition, avionics firms carry the responsibility of being knowledgeable and compliant with the regulations in our business. These deal with safety of flight for the travelling public and cannot be underestimated. This is no small task, as it in fact constitutes an entry criterion for this business. With the aging aircraft fleet, more effort is required, as well as prolonged liability exposure.

Economics

Pricing pressures have pretty much settled what part of an aircraft's purchase price an OEM can afford to allocate to avionics, putting pressure on acquisition costs. This in turn results in OEMs looking for areas where they can reduce costs, including working with suppliers to reduce *their* prices.

The OEMs are also looking to reduce their inventory by having suppliers participate in Just In Time processes. These initiatives have driven the suppliers to embrace Total Quality Management concepts, striving to eliminate waste and improve quality. This has led to process improvements and production cycle time reductions to bring costs in line with industry affordability.

This has also meant that suppliers have had to revisit their *own* supplier management strategies to be consistent with a changing industry. Differences in COTS parts, for example, and how to deal with increasing obsolescence issues are more and more demanding attention.

In addition, the need for airlines to lower their cost of ownership has resulted in avionics suppliers having to take on additional life cycle requirements. Problem is, the additional *value* provided by addressing cost of ownership has been real, but no premium has been received in return for this value. The Boeing 777 improved cost of ownership by 35%, resulting in lower need for spares. But there's still dissatisfaction in spares prices (although fewer are required).

Suppliers must continue to put a strong emphasis on cost reduction (acquisition and life cycle), but will also need to move into areas that provide better profit opportunities. With the large numbers of suppliers chasing a smaller business, it is expected that avionics firms will go the way of the airlines and OEMs and move towards broader system/product offerings through acquisitions/mergers /alliances. The trend has already started with alliances being formed to meet specific product needs, but this will need to accelerate. The industry is global, and more global solutions will need to be formed.

Supplier's Role

More and more, customers want to do business with a smaller, *preferred* set of suppliers. The criteria to be on these lists vary, but to be considered there are two things that a customer wants:

A supplier who's *dependable*

A supplier that has overall low cost.

Performance is more critical than ever. Suppliers are being driven to improve their performance; delivering better products, at less

0-7803-5086-3 /98/$10.00 ©1998 IEEE

cost and with shorter production cycle times. Reliance on the supplier to have the *right* parts in the *right* place at the *right* time with the *right* quality and the *right* functionality is essential for the OEMs and airlines to meet their obligations.

Without a doubt, suppliers recognize their obligation to reduce costs and cycle times while improving quality and delivery performance. This has been healthy to the industry and is something we all need to continue to do. In fact if a supplier is *unable* to do this, then he will not be able to satisfy his customer's needs for dependable, low cost products—and thus find himself out of business.

But expectations should be reasonable. Not all requirements can be downflowed to all sub-suppliers. High volume commercial suppliers have shorter product lives and different business models. The days of "mil spec" parts are for the most part gone, and the parts cost benefits have already been reaped. Added costs are being absorbed by the avionics firm in the form of more intensive supplier management, parts screening, more rapid obsolescence, compensating for the portions of the purchase agreements that cannot be downflowed and increased costs due to having to multiple source critical components. Cost reduction, cycle time improvements and quality improvements should benefit all parties. Currently, OEMs expect these economic benefits to be passed along to them in full, without consideration of added costs.

In addition, when a supplier chooses to make an investment, he has the right to not only recoup the investment, but to make a profit on it. In order for the various avionics firms to fill the needs of the industry, they need to know how the industry will let them make money.

Once *that* is understood then the avionics firms will be able to make intelligent decisions on their investments and provide solutions more in line with industry needs.

Business Practices

Traditional

The traditional way of making money in the world of commercial avionics can be summarized as follows:

1. Perform the development work—and get paid for it,
2. Sell equipment to OEMs,
3. Sell equipment to airlines,
4. Repair and overhaul

Thus, avionics manufacturers sold hardware and subsequently earned revenue from the repair and overhaul of that hardware. What a customer was willing to pay was a price based on the cost of the development work and the cost of the hardware. The cost paid undergoes heavy scrutinizing during detailed fact-findings. In a value-based system residing in a risk-sharing environment, this overhead would not be necessary.

Current Business Practice

Business practices have been established around this traditional model and long-standing agreements are permeated with it. As the new emphasis to the bottom line has been introduced, agreements have been modified where beneficial to an OEM's or an airline's bottom line—*but the previous obligations remain.*

In short, the rules have changed.

A supplier is expected to invest his own money for development as a risk-sharing partner in the best case, but more often, as a risk-*taking* partner. In effect, he is told: "Share risk in the development, but not in the revenue—except as increased volume."

Even *that* risk sharing opportunity is negated when OEMs go to dual sourcing. Prices are negotiated to low flat rates at virtually no margin with the expectation that the avionics supplier will obtain margin by improving his processes and that of his suppliers. Even then, the avionics firm may be subjected to renegotiations at the whim of the OEM.

Dual sourcing practices have the further difficulty of having to compete at both the OEM and the airline. This may have some short term advantages to the OEMs and airlines, but it is difficult to establish a business case for the avionics manufacturer as a standalone item.

The airlines are no longer willing to offset OEM prices and are establishing their own economic models to determine price. This, coupled with higher reliability of products means lower spares revenue and margins.

In addition, the OEMs and airlines have fallen into a practice of late payments, which help their finances but put a further burden on the avionics manufacturer.

On the repair and overhaul end of the business, the cost of capital is increasing for the more sophisticated equipment and, through agreements, test equipment is sometimes provided to airlines. This not only ups the cost of doing business, but also makes it easier for airlines to enter and compete for the same business.

In the area of Buyer Furnished Equipment, or BFE, the market is commodity-based where low cost prevails. Market entry requires a supplier with the ability to meet specifications and have a reliable product, but competition is based on the lowest price. This means that the product cost has to be low and its margins not only have to cover initial investments, but reinvestments when the product is in the field. For example, whenever the FAA mandates changes for TCAS, the OEMs expect reimbursement for their certification costs and the airlines expect the change at no charge. Even the buying patterns of the airlines may not be conducive for receiving best value. Airlines typically buy by ATA chapter (for example ATA Chapter 34 – Navigation or ATA Chapter 23 - Communication), which may preclude integrated solutions.

Obsolescence

Obsolescence is increasing due to the increased reliance on Commercial Off The Shelf Software, COTS, and the fact that older airplanes are not being retired as fast. COTS, by definition, means a shorter product life than required by our industry; hence, some of the parts in the older airplanes are getting harder to obtain, especially when those airplanes are out of production.

This problem is aggravated by the fact that customers expect prices to go down over time. In fact, though, the point of obsolescence *reverses* that cost curve. Older parts actually cause the cost curve to go *up* at a point in time; they don't go down forever.

The industry needs to recognize these economics and allow more retrofit solutions for dealing with obsolescence. Agreements may need to be made for fixed periods of time. Recognition of the higher piece part cost of older parts and the option of more retrofit solutions needs to take place. *Not* dealing with obsolescence is not an option, but avionics firms cannot sustain the greater financial burden of resolving them.

Purchase Agreements

Purchase agreements need to be written in a way that allows the avionics company to satisfy the airline customer's cost of ownership requirements so that we can all have an advantage.

Current practices typically mandate the price of spares as a ratio to original equipment prices without factoring in other latitudes such as reliability.

What is needed, therefore, is method of receiving credit for having increasing reliable products.

0-7803-5086-3 /98/$10.00 ©1998 IEEE

Initial attempts to market "power by the hour" never gained favor. Issues ranged from airline cashflow to supplier capital investment. These obstacles need to be overcome, and the flexibility for more innovative solutions needs to be available.

New Technology

In today's environment, new technology must have an absolute financial justification. There are some terrific technologies that have been developed and applied and have tremendous upside. Yet, the products that contain this new technology have not been received as might be expected. This can be attributed to the fact that new products need to satisfy three marketing criteria.

Not only must there be a "need" … and a "desire" … for the new product, but the "want" needs to be established.

SATCOM is a good example. SATCOM is an excellent new technology offering gate to gate communication. There was a "desire" by the airlines to increase revenue by providing services in the passenger compartment using this technology; there was a need for this technology in the CNS/ATM environment; but the problem is in trying to quantify the return.

The "want" needs to be established in terms of problems solved in the cockpit and the value of gate to gate communications. If you look at the total picture, then the "want" should drive new business practices which could include major shifts and expansions in the current build, ship and collect mentality. How the avionics firm offers its products and services should dramatically change, including receiving value for software and providing value added services for the products provided.

Product Opportunities

Talking about system trends for a moment will underscore the industry need for change. There are many product opportunities in today's environment. Here are a few of the more obvious ones:

- Emphasis on systems that will answer the safety needs of tomorrow;
- Systems that facilitate efficient use of airspace
- High integrity systems that answer airplane mission needs of tomorrow: 18 hour flight segments; flight control and visual systems for supersonic flight
- Improved man-machine interface

Business practices need to change so that avionics firms can make the business decisions necessary to find solutions for these pressing needs.

Conclusion

Squeeze Play

Let me recount the factors that are leading to what avionics manufacturers perceive as a kind of "squeeze play" on their business.

First: Airlines have increased their power over suppliers through alliances and acquisitions

Second: OEMs have increased their power over suppliers in the new duopoly.

This leaves a large number of suppliers competing for the same business with still other companies entering for niche opportunities.

On top of this, increasing system reliability is cutting into aftermarket sales.

And aftermarket sales are further declining due to cost of ownership improvements that have been built into newer products.

Finally, avionics firms are losing power over their own suppliers with the shift to Commercial Off the Shelf parts and being a minor customer to those suppliers. This also

0-7803-5086-3 /98/$10.00 ©1998 IEEE

makes it difficult to "flow down" the more severe requirements required by our industry.

These factors are making it exceedingly difficult to find the profit zone.

Business Practices Need to Change

Here's the long and short of it in a single sentence: The industry needs to change.

Hardware is becoming a commodity. The *value* is in software and integration. And our customers need to be able to *see* the benefit in software and integration or they won't *want* it.

This is no easy task. The benefits of FANS, for example, vary widely depending on the issue. If a supplier is late in providing FANS functionality, then a customer is inclined to overvalue its benefit. But during a negotiation, the customer may tend to *undervalue* its benefit.

Regardless, we need to shift from cost based pricing to value based pricing.

Pricing of FANS as cost-based were established by the OEMs and the investments by the avionics firms are questionable as to whether they will pay off under that model.

Features are either of benefit to the customer or they are not. These features for the most part reside in software and integrated solutions. Whenever customers assign value to a product, then the value assigned should be in line with what is desired. Value should be assigned by benefit, rather than cost-based elements such as weight and digital bytes. Otherwise, the industry will be misaligned and it will be difficult to make the investments required on the right set of products and services that indeed fulfill the real needs, rather than some historically derived measures, that could drive the wrong decisions.

Highly integrated systems have fewer parts, weigh less, take up less room, consume lower power, are more reliable, have lower software overhead per function, provide for fault tolerance, aid in the dispatch of the airplane and require less spares. These are all important attributes on the surface, but yet, they do not provide added value to the avionics firm— merely more cost to deal with over the life cycle without the traditional methods of recouping investments and support costs.

Profit Zone

If the industry is not willing to pay for the value added by integrated products and software functionality, then the avionics firms need to somehow recoup the investments on future commitments or not make them under the current set of practices. This would lead to a different set of commitments based on where an avionics firm can make money.

The aviation industry needs to determine where an avionics firm can make a profit and allow the avionics firms to satisfy this very understandable, stockholder-driven need. The traditional model is gone and a new one needs to be established.

0-7803-5086-3 /98/$10.00 ©1998 IEEE

Abstract

Central Maintenance systems (CMS) have become a common element of modern aircraft designs that feature digital systems. The need for such reporting systems is warranted by a need to troubleshoot and return to service digital systems within which many complex and often subtle interrelationships exist. By the very nature of electronics, they do not lend themselves to traditional troubleshooting techniques and therefore require a new set of tools. This article explains both the philosophical and practical aspects of the Central Maintenance systems in existence today. The intent is to clarify and summarize a complex subject.

History

The earliest attempts to centralize fault information from more than one unit were seen in autopilot systems. Examples are the MD-80 Status Test Panel, the A300 Auto Flight System, and the B757/767 Maintenance Control and Display Panel. While these concepts offered a marked improvement, technology and cost factors limited their capability. The late 1980's saw the first generation of Central Maintenance Systems which provide airplane wide diagnostic coverage.

General Benefits

When contrasted to a federated BITE architecture, the key advantages provided by any Central Maintenance System are:

- Standardized English language
- Central location for fault information
- Fault storage and retrieval
- Central location for ground test initiation
- Ground test warning and inhibits
- Central data source for disk drive, printer and ACARS

Industry Standards

Before considering the details and features of Central Maintenance systems, it is important to understand that there are two fundamental CMS architectures in use; Central Fault Display System (CFDS) and Onboard Maintenance System (OMS).

CFDS (Centralized Fault Display System)

A CFDS collects BITE data from all or most systems on the airplane and provides a Central place to display this data. This system is outlined in ARINC report 604 which provides a philosophical vision for a Central Maintenance system. ARINC report 604 was originally published in 1985. It is important to be aware that when this report was published, Boeing and Airbus had each already embarked on mutually exclusive ways of achieving the intent of report 604. Boeing was working on the 7J7 and a centralized architecture, while Airbus was developing the A320 with a distributed architecture. The centralized architecture was not formally included in the ARINC 604 report until 1988 with the addition of appendix 7.

Distributed vs. Centralized

Distributed Architecture

The distributed classification is derived from the fact that the intelligence (BITE) of the system resides solely in the member systems and is thus *distributed* among these units. As such, when the user requests fault information from a specific system or LRU, the CFDS simply relinquishes control of the display over to a member system. While this implementation provides the conveniences mentioned previously, it does not accomplish the task of sorting out complex intersystem relationships and dependencies often encountered when troubleshooting modern avionics. In addition, correlation of individual system BITE messages to FDE's and other flight crew indications is predominately a manual task which relies heavily upon the use of manuals and/or a highly educated user. In order to gain more insight as to the root cause of a given flight deck indication, the user must interrogate each system via the interactive mode. The interactive mode is available only for those systems that are connected to the maintenance system via ARINC 429. Because the BITE data is developed by vendors who are often not the airframe OEM, this implementation can lead to inconsistent content and format i.e. the same fault type may appear very different from one system to another. To a single user who is troubleshooting different systems on the same airframe, this can be very misleading. Transport category aircraft that use this architecture include the MD-11, B737 next generation series, A320, A330 and A340.

Centralized Architecture

The Centralized implementation includes a Central Maintenance Computer (CMC) that collects BITE data from the same systems. However, this scheme requires that the member systems be designed to interface with a CMC. While adding to the suite of avionics, and thus introducing fault scenarios of its own, the benefits of a CMC have proved to outweigh these negative aspects. By providing a processing device which has access to all member system data in structured formats, the centralized architecture provides three key additional services via the CMC. These services are:

- Fault correlation to Flight Deck Effects
- Cascade effect removal
- Fault Consolidation

These attributes are vital to troubleshooting efforts that involve systems which incorporate such features as; multi-user databuses, Integrated Modular Avionics (IMA) architectures and Fault Tolerance.

Transport category aircraft which employ a centralized architecture include the B747-400 and the B777. Note: The A330/340 have implemented a hybrid architecture that allows for limited fault correlation to FDE's. The A330/340 uses a time based correlation scheme which is discussed later.

OMS (Onboard Maintenance System)

ARINC report 624 Defines a Onboard Maintenance System (OMS) which includes:

- Centralized architecture and the aforementioned attributes
- Integration of BITE and troubleshooting manuals
- Onboard Maintenance Documentation (OMD)
- Integration of BITE and Airplane Condition Monitoring (ACMS)

OMD is yet largely undefined in terms of an industry standard. As such, implementation on a large scale has not yet occurred. This slow progression is largely attributable to the many legacy systems that are not compatible with the envisioned implementation of OMD. In brief, OMD requires publishing systems and document type definitions (DTD's) that necessitate a substantial investment in infrastructure.

Aircraft Condition Monitoring Systems (ACMS) allow for more effective failure management through the techniques of trend monitoring. Such techniques are widely used in the area of turbine engine maintenance but are underutilized in other areas in spite of the availability of data and analysis programs.

The B777 is the only implementation that meets and/or is provisioned for the full intent of ARINC report 624.

Users and Presentation

General

There is only one set of BITE logic in any given LRU or system. BITE is BITE. This logic does however serve two very different audiences; the flight crew and the maintenance crew.

Flight Crew

The flight crew must be informed of failures that effect the continuation of the flight and the status of the airplane as it pertains to maintenance release. For these needs, messages derived from the BITE data are shown on the forward display where they may be viewed by the flight crew from their seated position. Messages related to the continuation of flight are available to the flight crew at all times without overt action. Those messages pertaining to maintenance release are optimally geared to cater to the use of the MEL and are not generally required to be viewed by the flight crew after engine start. These messages are intended for use by both the flight and maintenance personnel to indicate what system or functions are inoperative. These messages are not intended to provide detailed troubleshooting information, rather they are provided as a high level indication of the status of systems and functions. These messages are used to indicate the failure of items that effect airworthiness. Such messages are not germane to items such as entertainment systems or coffee makers. It is advantageous to provide a cross reference of these messages to the individual MEL items. Examples of such message systems are:

- Engine Indicating and Crew Alerting System (EICAS) - Boeing

- Electronic Centralized Aircraft Monitoring (ECAM) - Airbus

Maintenance

From the same BITE information that is used to generate the messages described above, there must also come detailed troubleshooting (diagnostic) information for exclusive use by maintenance personnel. This troubleshooting information is supplied by either front panel BITE or via a Central Maintenance system. This information must be specific enough to allow the user to pinpoint the root cause of a failure.

Regulatory Implications

At this point it is important to consider one of the prime objectives of both ARINC 604 and 624. As industry participants in the development of these standards, the operators made it clear that any maintenance system must be implemented such that it is *NOT* required for flight. This requirement is understandable from a practical viewpoint in that any information shown by the maintenance system is diagnostic in nature. Diagnostics are not required once a valid maintenance release is attained. A valid release is evidence that everything pertaining to airworthiness has been dispositioned by either repair or deferral under an approved MEL. Without a need for further repair, there is a pursuant lack of need for diagnosis, or diagnostic information. This is no different than front panel BITE. The problem is that it is more easily identifiable when the maintenance system is inoperative as opposed to knowing when a single LRU front panel BITE is inoperative. The underlying principle; Diagnostic information shown on a maintenance system is not germane to the flight crew "during the course of their normal duties" and should not be required for flight. There are two distinct positions regarding the use of this information for establishing airworthiness. The FAA has established that deferral of an item under an approved MEL must not be predicated on information attained solely from a maintenance system. As such, Boeing has taken the position that there are no periodic maintenance requirements that mandate review and correction of maintenance system messages. While use of a maintenance system indication may be the most efficient method of accomplishing a required task, there is nothing that requires a specific task to be accomplished using that indication. The Airbus implementation requires corrective action of certain maintenance system messages at a specified interval. Both systems are listed on the respective MMEL.

The Primary User - Line Maintenance

From the previous discourse, it is clear that the primary user of the maintenance information is a line mechanic. During the development of any maintenance system, it is of vital importance that this premise by kept in the forefront. It is very easy to overindulge in the ability to display endless amounts of data; *very little of which may be of real value to the primary user*. A designer must remember that the primary objective of the system is to facilitate faster fault resolution. The ARINC reports provide a suggested data structure to prioritize the information in relation to primary and secondary users.

Secondary Users

Secondary users include extended maintenance, engineering and manufacturing personnel. Extended maintenance uses features such as dataloading, configuration reports and rigging in an environment where time is less crucial. Engineering can use a CMS to troubleshoot a persistent problem using features such as data bus monitoring and ACMS. Modern manufacturing operations integrate Automatic Test Equipment (ATE) and the CMS to reduce production cycle time.

Display

The ARINC reports specify a Maintenance Access Terminal (MAT) for display of CMS data. There are presently two implementations. The most common is the Multi Purpose Control Display Unit (MCDU). The MCDU is typically 3.5 inches in size and controlled by a set of keys located on either side of the display. There is a keypad which allows for alpha numeric entries. This is available in both black and white and color versions. The second is a full page sized MAT which is color and is controlled using a track ball cursor and an external keyboard. This device also incorporates a disk drive and hard drive for storage of large quantities of data. In addition, the ARINC reports specify Portable Maintenance Access Terminals which provide the same functionality as the MAT but are self contained and can be plugged into the airplane in various locations.

Fault Processing

Fault Consolidation

Consolidation is the reduction of faults with a common root cause to a single fault. For example, if five LRU's report a failure of the same component, consolidation logic would provide a means to indict the faulty component via a single message as opposed to five separate messages. It is important to note that these five messages would all indicate the same component is at fault. This differs from cascade effects.

Cascade Effects

Cascade effects or "ripple effects" are also single failures that are reported by several external (to the faulty system) monitors. The primary difference is that cascade effects do not indict the same root cause. A cascade effect is one that becomes more prevalent in integrated architectures where there are multiple users of a single data source. For example, a multi-user data bus may allow over one hundred devices to simultaneously communicate on a single twisted pair. In this scenario it is simple to envision a case where the loss of one piece of data could be reported by many users. However, that would merely be a case of consolidation. Cascade effect removal comes into play when the first user of the missing data is unable to perform its function, which is in turn recognized and reported by a third party. In such cases a central maintenance computer uses a list of effected signals to sort through these relationships and interdependencies to present the root cause in the form of a single message. In modern IMA

(Integrated Modular Avionics) architectures where several modules share common I/O data it is easy to recognize the impact to the user if forced to sort out the subtleties of these effects manually. These latter relationships are sometimes referred to as effected signals. That is, if a member system needs data (a signal) from another system in order to perform its intended function, it can be said that this source signal causes an effect on the output of the member system using that signal. These relationships can be several layers deep.

Processing Methodologies

There are two approaches presently in use to provide consolidation and cascade effect removal. One system uses a rule based method while the other uses a model.

Rule Based

Rule based systems use a series of logic equations as the primary means of performing cascade effect and consolidation logic. This involves the arduous task of manually creating lists of effected signals by determining their relationships and interdependencies. These lists are then used as the basis of the aforementioned equations. A engineer must also create a list of which maintenance messages can correlate to which FDE's. While this approach is effective and yields a high fidelity solution, it is difficult to maintain due to fact that in large architectures the same signal my appear in many equations. This also results in large regression test efforts whenever the signals are changed. In a rule based system, fault correlation to FDE's is accomplished using a timing window scheme. In this methodology, when a FDE is received, the CMC looks in the correlation list to see if any of the maintenance messages that are presently active are able to be correlated to the active FDE. There is a limited amount of time after the FDE is set that correlation is allowed to occur.

Model Based

In a model based architecture each member system engineer uses a common tool to enter the effected signals for the respective system into a common database shared by all member systems. Similar to the rule based architecture, a member system engineer must also enter a list of which flight deck effects can correlate to which maintenance messages.

Using this common database as a platform, integration tools can be built to exactly model the effected signal lists and create "chains" which model these relationships. Each time a fault is reported to the CMC the chain is followed until a single conclusion is reached. That single conclusion is shown to the user as a maintenance message. This same method is used to map fault correlation to flight deck effects. The drawback of the present implementation of this architecture is that these chains are static much like the logic equations of the rule based architecture. The ideal implementation of this concept is one that dynamically responds to a new fault. The negative effects of the present implementation are limited to systems which employ fault tolerance and are subsequently allowed to operate for long periods of time with latent faults. Because the chains are static in the sense that time of occurrence is not considered when fault correlation is accomplished, older faults which satisfy the relationship established via the chain will be correlated. This is opposed to a situation where the user is presented with the single most recent fault that precipitated the FDE. In the present implementation of a model based system it is still necessary to employ rule based equations for utilities such as global inhibits for ground test initiation and fault storage.

Conclusion

There are many ways to implement a Central Maintenance System within the present industry standards. With the continued trend toward integration of multiple functions onto common hardware and the expanded use of software, more effective diagnostics will likewise be in demand.

Centralized takes this one step further. MMO is one step further than class three fault which must be fixed at the next check. If no internal fault is " correlated to the event" than time based "correlation" is made to the first "identifier" in the list. or the "type 1" systems. Avionics status mixes class one two and three together thus mixing MEL and non-MEL issues together since the class three reports are not required to be fixed until the next maintenance check.

LOOP RESISTANCE TESTER: A NON-INTRUSIVE METHOD TO MEASURE CONNECTOR AND SHIELD RESISTANCE

Erik L. Godo, The Boeing Company, Seattle, Washington
Bruce Van Deventer, The Boeing Company, Seattle, Washington

Abstract

The Loop Resistance Tester is a portable electronic instrument for measuring the resistance of electronic cable shielding as installed in aircraft, without requiring the cables to be disconnected. It comprises three elements: couplers, joint probes, and an instrument assembly. The couplers are used to measure the loop resistance formed by the shielding in the cable under test and airplane structure. The joint probes are used to measure the resistance of joints in the cable (cable-to-backshell, connector-to-bracket) to isolate bad joints if the loop measurement is out of tolerance. The algorithms and circuitry the Loop Resistance Tester uses to measure the loop and joint resistances are described.

Shielding Difficult To Test

Today several Electromagnetic Interference/High Intensity Radiated Fields (EMI/HIRF) and lightning threats exist for aerospace electronics. Electrical wiring shielding is used to reduce EMI/HIRF and lightning effects and to ensure that they do not cause hard failures in the electronics of the system and eventual system failure. Some benefit can be taken because of the shielding of the structure, but this is usually not enough to obviate the need for shielding in the cable.

In a typical aerospace application, the shielding of a cable is terminated inside the connectors, and then the connector connects the shield to structure on both ends of the cable. The shield can be viewed as a conduit which periodically connects to structure, and through which the signal and return wires of the system are routed. See Figure 1. In general the effectiveness of the shield in protecting the core wires is a function of the loop resistance; the lower the resistance, the better the shielding.

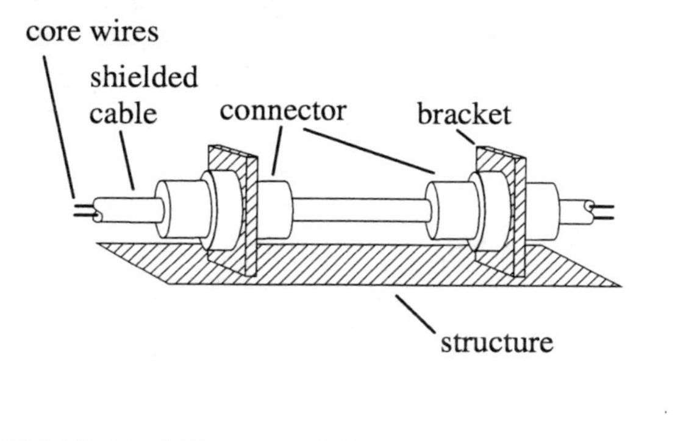

Figure 1. Shielding And Internal Wires

Aerospace environments are detrimental to materials and bonds of the shielding. High levels of moisture, salt, and extreme temperatures are among conditions that create corrosion problems. If certainty of the shielding effectiveness is needed, it must be tested.

This is usually done by using a bonding meter, disconnecting the cable under question, and measuring the resistance of the shield from the disconnected end to the structure connection. This is a costly test because once the connections for the system have been broken, a full system test must be run to make sure all the connections are still good after reconnecting the connectors.

Loop Resistance Tester

Because of the above problems, Boeing has developed a new instrument, the Loop Resistance Tester (LRT). Similar equipment does exist, but that equipment is not designed for testing aerospace wiring shielding, and does not have all the capability of the Loop Resistance Tester (LRT). The LRT is shown in Figure 2.

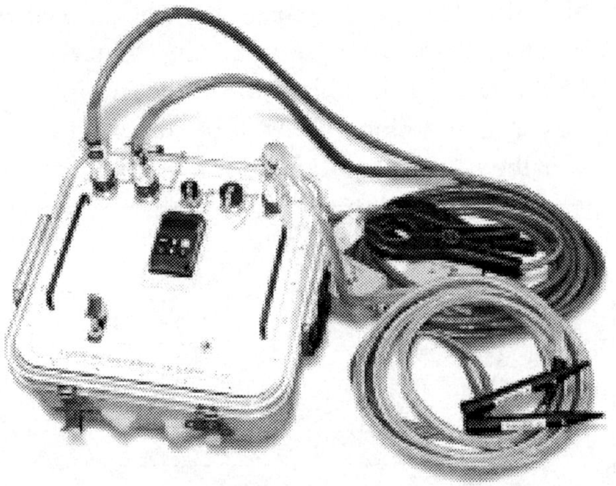

Figure 2. The Loop Resistance Tester

Shield Loop And Coupler Is A Transformer

An ideal transformer will transform impedance from the input of the transformer to the output proportional to the square of the turns ratio. The output voltage is the input voltage times the turns ratio, while the output current is the input current divided by the turns ratio. Therefore the impedance on the input of the transformer is multiplied by the turns ratio squared when viewed at the output of the transformer.

The shield shown in Figure 1 can be viewed as a conductive loop, going through the structure, connector, cable shield, connector, and back to structure. By connecting a coupler to this loop, the coupler forms the output winding and magnetic core of the transformer, while the shield loop forms the input winding of the transformer. By forcing a voltage on the coupler winding, and then measuring the current, the impedance of the loop can be found (voltage divided by current, divided by the turns ratio squared). This is the basic method of operation of the LRT. A voltage is forced around the loop, the current is measured, and the impedance of the loop is found.

Two Couplers Instead Of One

Note that this is only if the transformer is ideal, or that the coupler is ideal. To keep the coupler operating as close to ideal conditions as possible, the LRT uses two couplers. One coupler is used to drive the voltage around the loop (drive coupler). The second coupler is used to sense the current flowing in the loop (sense coupler). See Figure 3.

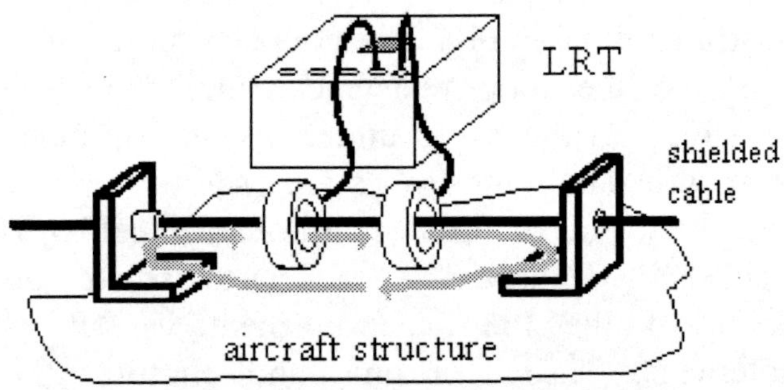

Figure 3. Two Couplers On A Shield Loop

By using two couplers, the current sensing coupler can be used with a current-to-voltage converter that keeps the voltage across the coupler near zero, so the coupler core will not become saturated. Also, with two couplers, a wiring bundle with multiple shielded cables can be tested by connecting the drive coupler across all the cables, and then moving the sense coupler from cable to cable to measure the current flowing in each cable.

Loop Troubleshooting

Once a shield is found to have a larger than allowable value, how is the problem found and fixed? This can be done by removing the cable in question, and testing it with a bonding meter. But by removing it, the corrosion may be cut through and the cable appears now to be perfectly fine, and the problem area in the cable cannot be found. The cable may be returned to service, only to put the system in danger in the field as it continues to corrode. What is needed is some way to zero in on the troublesome connection while the cable is still installed. This would also help in those cases where the problem can be repaired without removing the cable, because many cables are difficult and costly to remove and replace.

Joint Mode

To troubleshoot a bad loop, the LRT has the joint mode where instead of measuring the loop voltage and the loop current, it measures the voltage across a joint (joint voltage) and loop current. The joint voltage is measured by a separate pair of joint probes which are directly connected across the joint of interest. Given the joint voltage and loop current, the joint impedance can be measured in the same way that the loop impedance was measured. By measuring the joints in the loop, the bad joint connection can be found without having to remove the cable.

Sense Winding

Instead of measuring the drive voltage and dividing by the turns ratio to get the loop voltage, the LRT uses an extra winding to measure the voltage forced around the loop. This is shown in Figure 4.

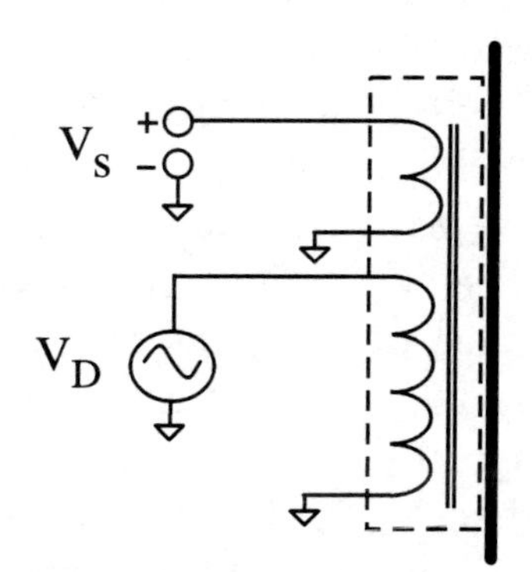

Figure 4. Extra Sense Winding

Figure 4 shows the view of the coupler and its main winding and the core of the coupler and a portion of the shield loop. The extra winding is called the sense winding. The sense winding is a low number of turns and is wound to approximate the coupling of the main winding into the shield loop. In the LRT the main winding has 1000 turns while the sense winding has 10 turns.

By measuring the voltage on the sense winding, V_S, the exact coupling of the coupler can be compensated for. For example, if the coupler faces have a slight air gap, the voltage coupled to the loop will decrease, but the voltage coupled to the sense winding will also decrease. The drive voltage change will not be proportional to the loop (and sense) voltages because the air gap produces a leakage inductance with a voltage drop. By using the sense winding to measure the voltage forced around the loop, the effect of the coupling on the loop voltage can be cancelled out.

Sensing Of Coupler Closure

The sense winding also allows the LRT to sense if the coupler jaws are closed adequately to complete a reading. From our earlier discussion about transformers, we know that the output of an ideal transformer is the input voltage times the turns ratio. If we measure the input voltage to the coupler, V_D, and measure the output of the sense winding, V_S, we can determine how close to an ideal transformer the coupler is functioning. As mentioned before, the turns ratio of the main winding to the sense winding is 100 turns. If V_S is less than 1/100 V_D, the coupler is degraded. This could occur if something is stuck between the jaws of the coupler and the coupler doesn't close completely, or if the coupler is being placed in a location that is hard to see and the coupler is closed on a bolt or some other piece of hardware. With the sense winding the LRT can detect these cases and flag the operator before a bad reading is given. The method of implementing this function is shown in Figure 5 below.

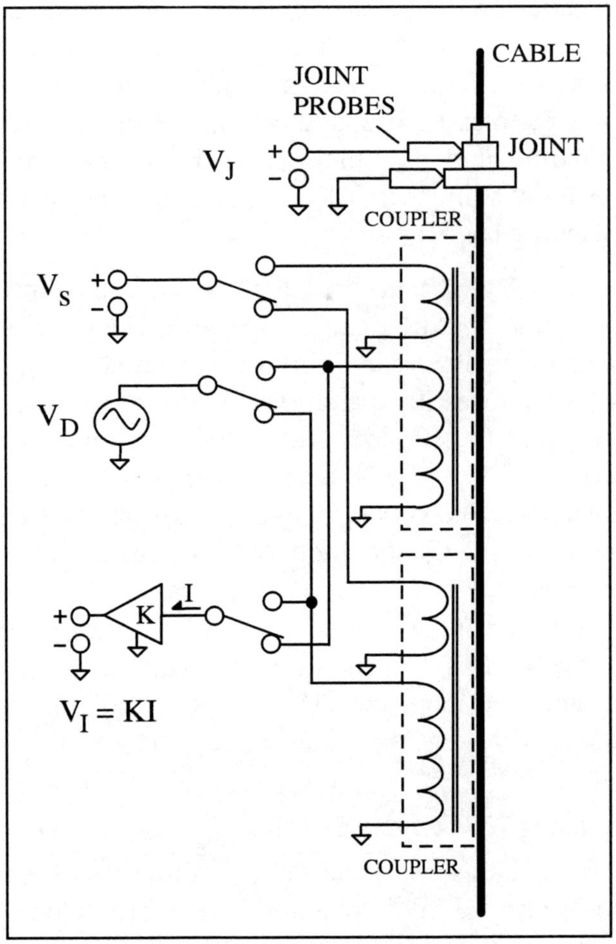

Figure 5. Coupler Closure Apparatus

At the top of the picture are the joint probes, with the voltage across a joint as V_J. Next we have the two couplers, the drive and sense coupler. Both couplers are identical. A series of switches switch the circuitry between the drive coupler and the sense coupler. V_D is the drive voltage, V_S is the sense voltage from the voltage sensing winding, and V_I is the voltage output of the current-to-voltage converter, which represents the current in the shield loop. By switching the drive voltage and voltage sensing circuitry to both the drive and sense couplers and measuring V_D and V_S, the closure check can be done on both couplers with a minimum of circuitry.

Sensing of Joint Probe Connection

A different sensing method is used to detect that the joint probes are connected to the joint before a reading is taken in joint mode. Because joints in general are very low impedance, usually around 1 milliohm, this can be used to detect when the joint probes are connected. The method is shown in Figure 6.

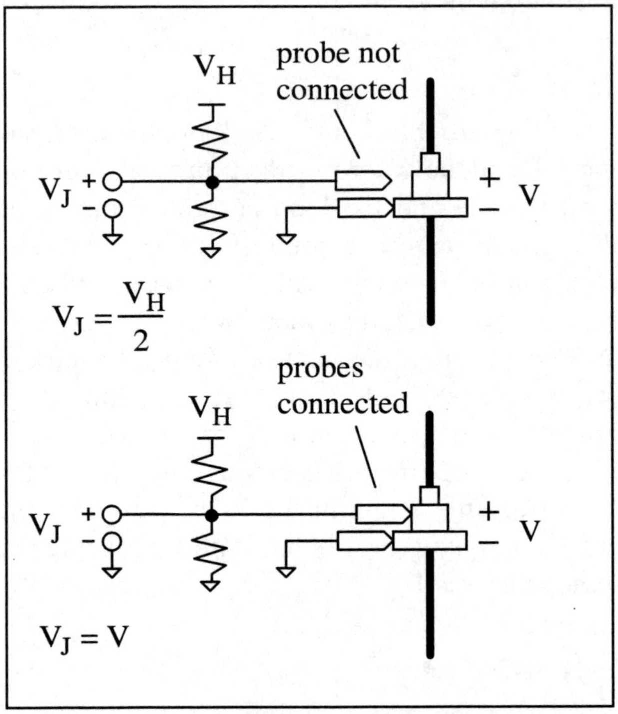

Figure 6. Joint Probe Connection Sensing

A large source impedance voltage is forced on the joint probes such that V_J is $V_H/2$ when the joint probes are not connected. The resistances shown in the figure are approximately one Mohm, greater than 10^6 times a typical joint. V_H is a few volts, which is a very large voltage compared to the typical voltage across a joint. Once the joint probes are connected across the joint, the much lower source impedance of the joint over-drives the V_H source, and the actual measured voltage is the joint voltage, V_J. Since the joint voltage is much lower than $V_H/2$, the change in voltage from $V_H/2$ to V_J can be detected and the LRT can determine if the joint probes are connected to the joint before taking a measurement.

Note that figure 6 does not show the series resistance such that there is no low-impedance path between the LRT circuitry and the loop being tested.

Frequency Of Operation

During development of the tester concept, several different frequencies were used to measure the resistance of the loop. A low frequency is needed to minimize the effects of the loop inductance, but not too low such that it cannot be coupled effectively using off-the-shelf couplers. A low frequency also provides good skin-depth penetration such that measurements made with the LRT will agree reasonably well with DC resistance measurements. 200Hz was picked because it did not coincide with the 50Hz, 60Hz, and 400Hz frequencies typically used for power.

This 200Hz signal is generated by the microprocessor as an adjustable amplitude square wave, which is then filtered to produce a clean sine-wave drive signal.

Magnetics

The couplers used in the LRT are modified Fluke 80i-600a couplers. They are designed for use at 60Hz, and can be considered ideal under the frequencies and currents of LRT operation. These couplers were picked also because of their hinged mechanical design that ensures that both faces of the coupler mate with minimum air gaps.

Safety

An important feature of the LRT is that it can be used in certain hazardous locations. This is made possible by adding series resistance to the couplers. This resistance limits the worst-case fault current that can flow in the coupler and hence the energy that can be stored inductively in a coupler to a safe value. This series resistance also limits the inductively stored energy in the loop being tested. Since this resistance is about 1000 ohms, the same resistance reflected into the loop is about 1 milliohm, so the safety feature does not start limiting loop current significantly unless the loop resistance is below a few milliohms. There is no adverse effect on accuracy because the loop voltage is being measured by the sense winding and not the input to the main winding. An added feature of this resistance is that the range of currents that need to be measured is automatically restricted for low-impedance loops.

The Measurement Algorithm

The LRT uses digital sampling of the loop current and loop (or joint) voltage to obtain the loop or joint resistance. This is shown in Figure 7.

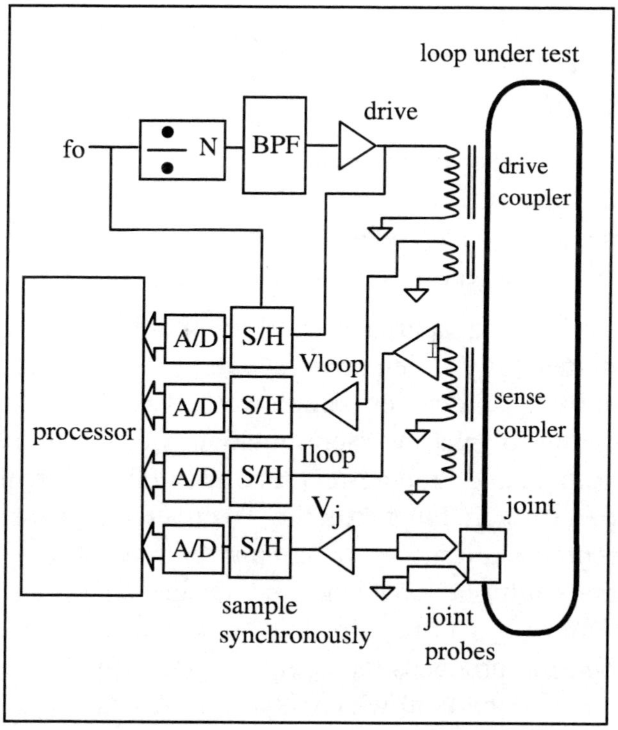

Figure 7. LRT System Block Diagram

The digital approach was favored over analog demodulation of the signals for increased flexibility and because errors introduced by the digital process are repeatable (there is no extra component drift or tolerances to factor in).

Approximately 4,000 samples of current and voltage are obtained at a rate of 128 samples per 200 Hz cycle. Two precision 16 bit A/D converters are used such that the voltage and current samples are synchronous. This helps reject common-mode noise sources because they will be removed when voltage is divided by current.

There are several possible "demodulation" techniques that could be used to turn the time series samples of voltage and current into resistance. A power-based approach can be used as shown below:

$$\boldsymbol{R} = \frac{\sum_{i=1}^{n} I_{loop}(i)V(i)}{\sum_{i=1}^{n} I_{loop}^{2}(i)}$$

n is the number of samples and $V(i)$ and $I_{loop}(i)$ are the loop voltage and current time series.

The disadvantage of this approach is that it has low numerical stability as I_{loop} becomes small, which is the case as the loop resistance gets large.

The approach used in the LRT is to numerically "demodulate" the voltage and current time series samples $V(i)$ and $I_{loop}(i)$ to create a phasor representation of the loop voltage and loop current, and then take the real part of their complex ratio. To create the phasor representation the voltage and current samples are multiplied by sine and cosine terms and summed. The ratio of the voltage and current phasors produces the complex loop impedance, the real part of which is the loop resistance. This is expressed mathematically below

$$\vec{\mathbf{Z}} = \frac{\sum_{i=1}^{n} V(i)\sin(i\omega T/N) + j\sum_{i=1}^{n} V(i)\cos(i\omega T/N)}{\sum_{i=1}^{n} I_{loop}(i)\sin(i\omega T/N) + j\sum_{i=1}^{n} I_{loop}(i)\cos(i\omega T/N)}$$

j is the square root of negative 1, N is 128 samples/period, T is the sampling period, and ω is the 200 Hz excitation signal in radians/sec.

The A/D conversion sampling frequency is guaranteed to be 128 times the AC excitation frequency because the 200 Hz drive signal is divided down from the A/D conversion control signal. This would otherwise produce serious errors in the data as the sampling time increased or decreased.

When measuring a joint, the joint voltage is treated in the same manner as the loop voltage was when measuring a loop. There should be no *a priori* relationship between the phase of the joint voltage and the loop current; however, doing the full calculation rejects quadrature noise sources - noise or stray signals which are capacitively or inductively coupled. Therefore, better joint accuracy is obtained this way than by discarding the joint phase information.

There are several variations on this approach that may yield improved results. One improvement may be digital filtering of the voltage and current data to remove non-200Hz components, or windowing the data before demodulating it.

Conclusion

The Loop Resistance Tester has capabilities that are beneficial to testing cable shielding where it is costly to test using normal methods. The joint mode of the LRT allows isolation of the bad joint before removal of the cable. The LRT senses if the couplers are closed and if the joint probes have good contact before completing a measurement. An improved version of the Loop Resistance Tester is currently under development. For more information see the Boeing website (www.boeing.com) and search on Loop Resistance Tester.

ADVANCED MAINTENANCE USING CAUSAL NETWORKS

Charles J. Sitter, Rockwell Collins, Cedar Rapids, Iowa

Gregory M. Provan, Rockwell Science Center, Thousand Oaks, California

Abstract

We present a flexible, powerful, easily maintainable diagnostics system that integrates diagnostic development functions into product development tools and processes.

Challenges of Developing Diagnostic Systems for Complex Applications

There exists a wide variety of modeling and inference methods for diagnostics, including expert systems, neural networks, and model-based approaches. All of these approaches can represent and analyze the same artifact but from different viewpoints.

Complex systems, such as commercial aircraft, represent a challenge both in modeling and inference. Next generation improvements in the performance of diagnostic systems will require, at the very least, higher fidelity models, guarantees on model consistency, and reasoning algorithms that are less susceptible to state-space explosion problems.

Moreover, analyses of the diagnostic models for complex systems must be correct and exhaustive. Incomplete analyses using probabilistic arguments become progressively more tenuous for complex systems and unacceptable for safety-critical applications.

Powerful and generic modeling paradigms and algorithms that seamlessly serve many analytic needs are key ingredients for developing the next generation diagnostic systems. Other critical aspects of models are capturing aspects of temporal and dynamic systems, and the ability to scale to complex and non-deterministic scenarios.

To address these challenges in the design of diagnostic systems for complex applications, we propose a model-based solution that integrates two state-of-the-art technologies: (1) causal network diagnostic modeling and inference algorithms and (2) distributed computing.

We argue that these two technologies can synergistically provide:

- The ability to model a fault using current design data and thus troubleshoot and repair a failure either (a) not previously identified or (b) with no existing maintenance procedures.
- The ability to provide diagnostics feedback to design engineering during the design and integration process.
- A significantly enhanced ability to scale to large complex applications while maintaining accuracy.
- The ability to guarantee computational requirements, *a priori,* based on model fidelity.

We organize this paper as follows. The following section summarizes the diagnostic representation, causal networks [1], and introduces a simple example of causal network modeling. The next section summarizes Rockwell Collin's avionics diagnostic system, briefly describing each major component of the system. The final sections provide conclusions and references.

0-7803-5086-3 /98/$10.00 ©1998 IEEE

Diagnostic Representation

This section introduces an example that we use to describe our approach, and the representation that we adopt for modeling, causal networks [1].

A causal network is a graph-based representation that is used for diagnosing failures of a system. Causal networks provide probabilistic, order of magnitude, and symbolic representations. They also have predictable scaling properties for the eventual embedded runtime diagnostic system.

A causal network encodes the causal relations of a system; e.g., in Figure 1, the transmitter in Tx causes Receivers Rx1 through Rx4 to receive data. A causal network is thus a good way of representing the flow of data in such systems, and hence reasoning about the root causes of data flow; e.g., Tx is the root cause of data being received by Rx1 through Rx4. Figure 2 shows the causal network structure for this simple example. Conversely, if data is not flowing as intended, the causal network can help track down the reason for the "breakdowns" of correct data flow; e.g., Tx could be the root cause of data being absent at Rx1 through Rx4. We call this diagnosing the faults in the system.

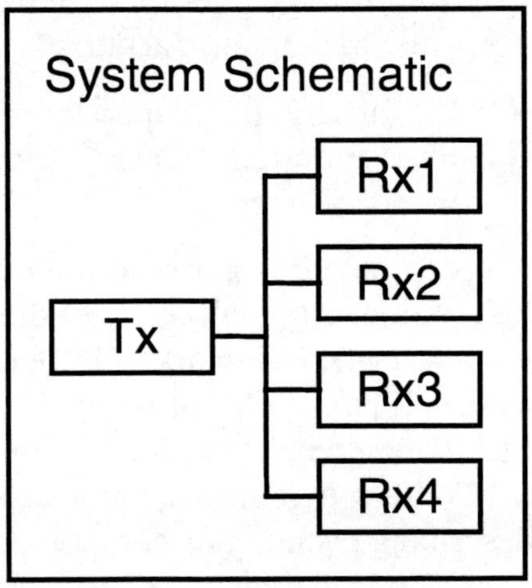

Figure 1: Simple Avionics Example

More formally, a causal network specifies the causal relationships among a set *V* of variables by encoding each variable in *V* with a node, and encoding the causal influence of V_1 on V_2 by a directed arc from V_1 to V_2. Hence a causal network is a directed graph (*N*,*A*) of nodes *N* and directed arcs *A*.

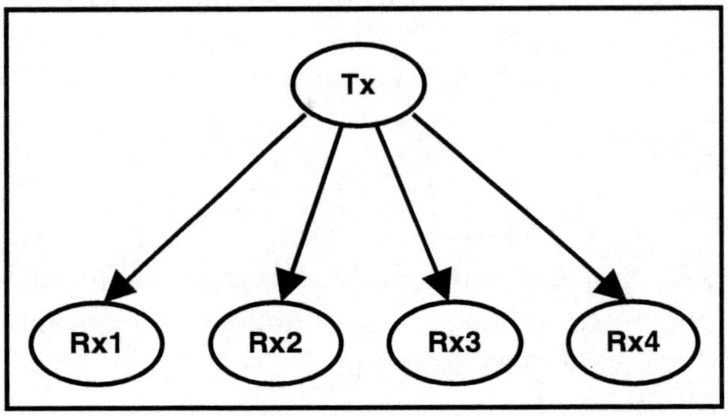

Figure 2: Causal Network Structure for Simple Avionics Example

The example we have just described covers only hardware, but note that the causal network modeling language allows functional specifications of both software and hardware.

Causal Network Specification Language

To specify a causal network model we need to define:

1. the variables in the model, which represent the components, e.g., Tx, Rx1, Rx2, Rx3, Rx4 from Figure 1;
2. the values for the variables, which represent the data that flows through various parts of the system;
3. the assumables, which are the variables that describe the operating characteristics of the components, such as OK or broken;
4. the quantifications, which relate the variables and assumables;
5. the evidence variables, which are the variables that can be observed;
6. the weights for the assumables, which specify the likelihood or relative ranking of the assumable fault modes.

The component modes characterize the ways the component works under an exhaustive set of scenarios. Rx1 has associated assumables, OK and broken, and a quantification as follows: if OK, Rx1 receives data from Tx, but if broken receives no data. In propositional logic, we might write this as:

- IF (Rx1-mode = OK) AND (Tx = transmit) THEN (Rx1 = receive)
- IF (Rx1-mode = broken) OR (Tx = not-transmit) THEN (Rx1 = not-receive).

Figure 3 shows the causal network model with the assumables and quantification defined.

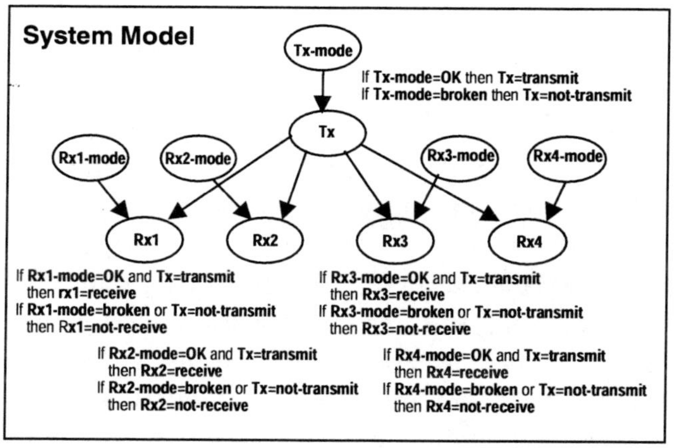

Figure 3: Causal Network for Simple Avionics Example

As described later in the paper, we have implemented a graphical user interface (GUI) that allows a user to specify a system using a simple set of components that are connected based on data flows. For each component, the user needs to specify only the inputs and outputs, the functional relationships between the inputs and outputs, and the operating modes of the component.

Note that the causal network representation can describe not only deterministic systems as in the above example, but also stochastic systems (using probabilities and order-of-magnitude probabilities (OMP) [1,2]), and discrete-event systems [4].

Rockwell Avionics Diagnostic System

Rockwell has developed a distributed maintenance architecture that is based on causal networks and open system standards, and provides the capability to significantly reduce the costs of development and of upgrading and extending the onboard maintenance system (OMS) of the future. The Rockwell avionics diagnostic system has several characteristics that, taken together, distinguish it from competing systems. These key characteristics include:

1. an avionics-oriented user interface for model development, simulation and diagnostic testing;
2. guarantees of diagnostic correctness and completeness given the models;
3. a compiler that allows the user to embed a real-time diagnostic system on-board the aircraft;
4. the deployment of the diagnostic system within a distributed, open architecture.

This section now describes the overall architecture of this diagnostic system.

Diagnostic System Architecture

The general architecture of the tool set, is shown in Figure 4. Diagnostic system development and deployment is divided into two phases: off-line model generation and compilation, and on-line diagnostic evaluation.

The *off-line* phase entails developing and analyzing a model, and then compiling that model so that it can be efficiently evaluated by the on-board avionics system. Model development takes place using a graphical user interface called EASIER (Enhanced Analysis and Simulation Integrated Environment from Rockwell), and compilation is based around a causal network inference system called

CNETS. EASIER supports development of primitive blocks and subsequent interactive model development and simulation for avionics applications. The output of EASIER is a causal network on which inference is performed by CNETS. Models generated in EASIER can be output to a diagnostic compiler in CNETS to generate embeddable diagnostic code.

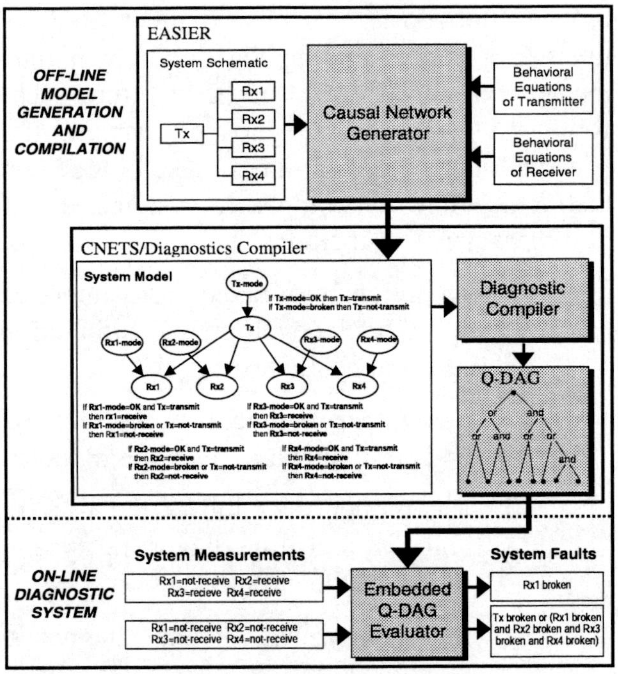

Figure 4: Diagnostic System Architecture: On-Line vs. Off-Line Subsystems

The *on-line* component of this tool-set is a real-time evaluator/interpreter that operates on the compiled diagnostic model or Q-DAG (Query Directed Acyclic Graph) to generate diagnostic results from a set of input variables. A Q-DAG is a compiled representation of the causal network, and is an expression that is optimized for real-time embedded diagnostic inference.

It is important to note that the underlying diagnostic tool, CNETS, as well as the associated compilation system, is a completely general system that can perform domain-independent causal network inference. The other component, EASIER, was designed specifically for the avionics domain, although the concepts of the EASIER interface can be applied to other domains (e.g., communications, digital circuits).

A second important point to note is that the EASIER/CNETS system is meant to hide the details of diagnostic modeling and inference so that engineers can use the system without having to acquire significant new skills. The EASIER/CNETS system has been explicitly tailored to avionics engineers, presenting to these engineers a front end with which they will be familiar. This approach unlocks the potential of causal networks to avionics applications.

Distributed Diagnostics Architecture

For aircraft onboard maintenance system (OMS) applications for the 21st century, the OMS must move to "open systems" and use "industry standards" in an attempt to get away from the proprietary "black box" implementations. The model-based approach is critical to allow model updating and re-use, and to ensure that different types of models can be stored in a single database, thereby leveraging common information. Further, it is widely recognized that to take advantage of current and future technology, the traditional federated avionics system application approach of the past is being replaced with an advanced, multiprocessing, distributed model that provides for future flexibility and expansion. Part of this strategy is to develop a "Distributed Diagnostic Architecture" that does not require specialized systems to be incorporated in the aircraft system to perform the diagnostic function.

Rockwell has defined a distributed maintenance architecture, based on open system standards that can significantly reduce the development costs, as well as the costs for upgrading and extending the OMS of the future. The application of Rockwell's next generation OMS divides the user interface, data storage/retrieval, diagnostic engine, and OMS application code. Separation of these components provides a system design approach that allows flexible applications for the diagnostic reasoning methods, the data collection and concentration methods, and the maintenance display methods.

Additional benefits of this architecture are improved responsiveness, throughput, and extensibility. Using a distributed system allows for functions to reside on multiple processors within the avionics system. The well-defined object interface allows the CPU loading to be shared over a network, as well as to dynamically change the distribution of tasks for better load sharing and redundancy management. Individual components can simply be upgraded without affecting the other components. If needed, additional functional objects can be added to this open system to allow unplanned options to be incorporated with less risk and cost than previous designs could provide.

A distributed architecture is composed of object-oriented components that are freely distributed across various computers in a network. The key to this distributed environment is developing standardized methods for brokering and communicating objects throughout the network. This is the responsibility of an Object Request Broker (ORB). The two major competing standards are the Common Object Request Broker Architecture (CORBA), developed by the Object Management Group (OMG), and the Distributed Component Object Model (DCOM), developed by Microsoft Corporation. These models effectively separate an object's interface from its implementation by using an Interface Definition Language (IDL). The IDL provides a standardized representation of an object and its methods and attributes, which remain consistent regardless of the language or platform implementation. By using standard IDL interfaces, CORBA and DCOM make distributed programming simple, allowing developers to treat and use any remote object as if it were local to the user.

OMG has also developed an Internet Inter-ORB Protocol (IIOP) standard that guarantees CORBA interoperability over the Internet. IIOP is a transport alternative to the popular HTTP that incorporates an object-oriented communications paradigm instead of the current page-oriented one. The CORBA IIOP approach allows an ORB from one manufacturer to have full interaction with an ORB from another manufacturer without any intervention or special programming.

In order to reuse system information and models (e.g., aircraft system interconnects, data dictionaries) between different organizations (e.g., system engineering, maintainability, reliability, hardware, software) or different tools, the different organizations must either share a common repository or their tools must be able to easily exchange data. Most tool vendors have proprietary formats and databases underneath their tools, so it is virtually impossible to get these tools to share data without some standard data interchange format, such as the CASE Data Interchange Format (CDIF).

CDIF defines a single architecture for exchanging data between modeling tools and repositories, and defines the interfaces of the components to implement this architecture. CDIF also defines a standard way of moving data between tools, without the need for customized interfaces. Each tool only needs one import and one export interface to be able to communicate with other CDIF-conformant tools.

Whether a single repository should be used for storing the data or if each tool should store its own data is still being debated. Our view is that a single common repository should be used to provide a single source for configuration management, version control, model synchronization, etc. Without this single source of control, sophisticated mechanisms must be developed to make sure that a model change in one tool is automatically reflected in all other tools.

Taking into account these standards, a new distributed architecture, shown in Figure 5, is defined for the current development system.

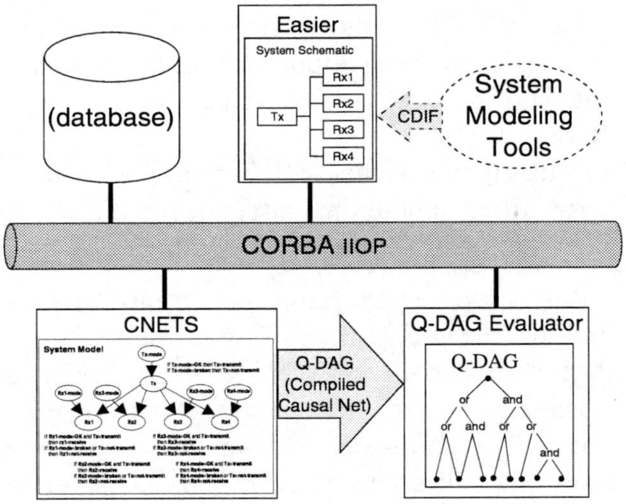

Figure 5: Architecture for Distributed Diagnostic Development System

Similarly, the embedded system, shown in Figure 6, can be distributed over three separate components that communicate over a network using CORBA technology: a data collector, a diagnostic engine (Q-DAG evaluator), and a presenter. EASIER generates the loadable tables necessary for the embeddable components. The *data collector* would monitor the raw input data, perform any filtering on that data, and pass state changes to the diagnostic engine. The *diagnostic engine* would update the Q-DAG based on these state changes and generate the latest diagnosis based on the current state of the system. These results would then be sent (either automatically or on demand) to the *presenter*, which would display the ambiguity group(s) to the appropriate user.

The distributed environment also allows for multiple data collectors and/or presenters, if desired. Separating these components provides a system design that allows flexible applications for the diagnostic reasoning methods, the data collection and concentration mechanisms, and the maintenance display platform. Additional benefits of this architecture are improved responsiveness, throughput, upgradability, and extensibility.

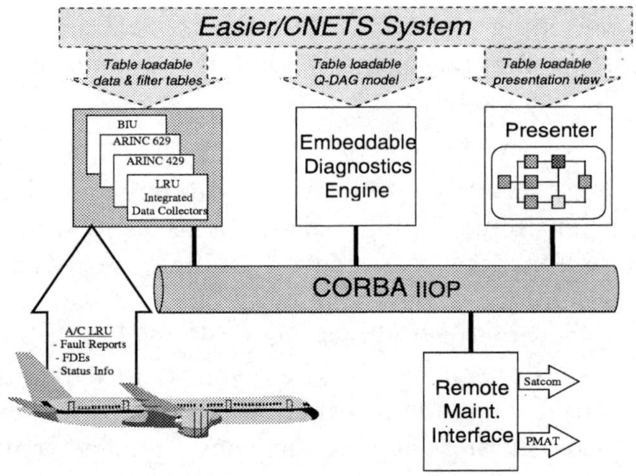

Figure 6: Architecture for Distributed Diagnostic Run-Time System

Diagnostic System Components

This section summarizes the major modules in the diagnostic system, the EASIER interface and the CNETS inference system and compiler.

EASIER User Interface

Diagnostic models developed in EASIER/CNETS are viewed as a collection of interconnected primitives. These primitives, or diagnostic model building blocks, are customized to achieve the appropriate diagnostic "view" for an application. Only information relevant to this diagnostic view (failure characteristics, system variables) is modeled (e.g., general functional characteristics not related to diagnostics are not modeled). Depending on the problem specifications, a primitive could be a logic gate, a bus, or a complex LRU.

Once a set of primitives is identified for a diagnostic problem, causal network structures are compiled and stored for these primitives. These primitives are then available in the EASIER graphical interface—users develop a system model by placing instances of these primitives on the screen and connecting them to define a complete system. The tool then generates a causal network system model from the specification of these primitives and their interconnections.

The behavior equations (which map into causal network properties) defined for each primitive building block can be simple or complex. Flexible equations are used to capture this behavior. CNETS supports relationships based on both multi-valued propositional logic and conditional probabilities for equations, and OMP or probabilistic weights for modeling component reliabilities.

Diagnostic models developed in EASIER can be debugged and refined through EASIER's simulation utility. EASIER allows the user to simulate normal and abnormal conditions in the system behavior. System users can also simulate and refine the model by changing the state of individual components. In the system, the colors of the blocks indicate the state of the block, e.g., OK or broken. The fully interactive environment provides immediate feedback (e.g., highlighting of affected pathways, changing of color-coded states) to gain understanding of model behavior. In the airplane-level diagnostics domain, EASIER's simulation capabilities provide mechanisms to solve cascade effect removal, fault isolation, fault consolidation, and flight deck effect correlation in one integrated visualization environment.

Causal Networks Inference System (CNETS)

CNETS is a system for representing and reasoning with causal networks, which include probabilistic and symbolic causal networks. CNETS allows the user to create a model by specifying a causal structure and a quantification of this structure using any of the methods described earlier.

Off-Line Inference

The chosen method of quantification depends on the application domain, the reasoning task, and the available information about the system being modeled. Independent of the causal network used, the user will typically engage in the following type of scenario with CNETS: First, the user asserts available evidence about the modeled system. Second, the user makes queries about the behavior of the system, typically simulation, prediction and/or diagnostic queries.

CNETS implements a number of inference algorithms that can be used to process such queries. The fundamental class of algorithms used by CNETS transforms the input causal network into a form that facilitates the particular type of query-processing. The primary algorithm used is a modified version of the clique-tree algorithm [3]. If the queries are probabilistic, then an algorithm like the clique-tree algorithm can be used directly. CNETS also supports a symbolic form of this algorithm for answering symbolic queries for avionics systems and for discrete-event systems [4].

CNETS Compiler for On-Line Inference

CNETS has a compiler that allows the user to generate a run-time version of the system model by using a compiler [2] in conjunction with an algorithm like [3]. Using this novel compilation technique, CNETS is able to compile a causal network into a Boolean expression, called a Query Directed Acyclic Graph, or Q-DAG, that contains all possible diagnoses for the complete system, given inputs for the observable variables of the system model.

The Q-DAG provides the same complete diagnostic information that is available with the original causal network, given that we specify which variables are used as input values and which variables are used for diagnostic conclusions. The main benefit of this compiled approach is an embeddable model that, together with a very simple evaluator, requires minimal computer processing resources and can be ported to other platforms and languages with relative ease.

Causal Network Guarantees

One of the key features of the causal network approach is that, for diagnostic inference, it provides a number of guarantees. First, the causal network approach provides soundness and completeness guarantees based on the model [1]. In other words, if any diagnosis is computable from the model, CNETS is guaranteed to compute it. This guarantee allows users, when developing models, to test the BIT (built-in test), BITE (built-in test equipment) and diagnostic coverage.

Second, it can provide complexity guarantees in terms of the model parameters. The structure of the network is the most critical aspect that governs the complexity of inference, and this is why causal networks are one of several classes of techniques termed "structure-based" approaches. For example, in the causal graph the maximum number of parent nodes of any node is one of the most significant parameters. Because all significant complexity parameters are ultimately derivable from the system structure, the structure can be manipulated to obtain acceptable inference complexity without altering the diagnostic coverage. Approaches that are not structure-based do not have this capability to alter inference complexity without altering the diagnostic coverage.

Beyond the complexity guarantees, the causal network approach has been optimized in a number of ways. This approach uses a focusing mechanism to focus inference on only the most likely diagnoses [1]; in addition, it makes use of observations to prune and to decompose the system (and hence reduce overall inference complexity) [5].

Conclusions and future work

This paper has summarized a new approach to designing an advanced maintenance system that provides important capabilities both for system design and for on-board diagnostics. This maintenance system provides a graphical user interface that has been tailored for avionics and is designed for ease of use by engineers. This visualization allows engineers to design systems using a schematic-oriented, system-level framework with which they are familiar. The human computer interface (HCI) is connected to the other system software components using CORBA middleware. The on-board diagnostics system can be compiled automatically from the models created by the engineers through the user interface. Together these features constitute a maintenance system that significantly advances the state-of-the-art.

We intend to extend this approach in a number of ways. One important extension is to implement prognostics and remaining functionality algorithms, such that we can predict future faults and define remaining vehicle functionality given the predicted or existing faults, respectively.

References

[1] A. Darwiche: "New Advances In Structured-Based Diagnosis: A Method For Compiling Devices," Proc. Of The 8th International Workshop On Principles Of Diagnosis (Dx97)

[2] Adnan Darwiche and Gregory Provan, "Query-DAGs: A Practical Paradigm for Implementing Belief-Network Inference," *Journal of AI Research*, 1996.

[3] V. Jensen, S. L. Lauritzen, and K. G. Olesen, "Bayesian Updating in Recursive Graphical Models by Local Computation," *Computational Statistics Quarterly*, 4:269--282, 1990.

[4] Cassandras, C. G., *Discrete Event Systems*, IRWIN Inc. 1993.

[5] Adnan Darwiche and Gregory Provan, "The Effect of Observations on the Complexity of Model-based Diagnosis," *Proc. AAAI, pp. 91-94, 1997.*

COST IMPACTS OF REAL-TIME NON-INTRUSIVE (RTNI) MONITORING TECHNOLOGY TO REAL-TIME EMBEDDED SYSTEMS

Michael D. Lewis, SA-ALC/TIS, San Antonio, TX
Daniel V. Ferens, Air Force Research Laboratory, Dayton, OH

ABSTRACT

The use of RTNI monitoring has had an impact on life cycle costs of existing programs through a reduction in debug time. Other areas in which RTNI monitoring can provide potential benefits to future programs are through the use of increased dynamic testing and the sharing of test time among more engineers. There are a number of areas in which software life cycle costs are impacted by various cost drivers.

To determine which areas were affected by the use of RTNI monitoring, a panel of expert users of RTNI monitoring was created using a form of the Nominal Group Technique methodology to achieve group consensus. The group concluded that the areas described above were most important to their programs. However, merely using RTNI monitoring alone may not have an impact on future programs; management commitment to integrate RTNI monitoring into test programs is also necessary for success.

INTRODUCTION

Software costs are continuing to rise in the Department of Defense (DoD) and in commercial agencies [1]. Much of the cost of developing software is due to testing the software. For example, Boehm attests that testing can account for 50% of software development costs [2]. Furthermore, testing affects the costs of supporting, or maintaining computer software. According to a Government Accounting Office GAO) report [3], software support costs can account for 70% of software life cycle costs. About 14% of the support costs can be attributed to correcting software errors missed during developmental testing. Improvements in testing, therefore, can significantly reduce software life cycle costs.

Much of the software developed for the DoD environment is for use in real-time embedded systems. These are systems in which computer hardware and software are embedded into a larger system [4]. Embedded systems often contain an amalgamation of hardware and software that is difficult to test. They are often dependent on various data feeds that have timing constraints imposed by real-world processes [5]. Integration and testing of embedded systems is challenging since feeds must be serviced in real-time without affecting software operation.

RTNI monitoring, which is a relatively new technology, sends signals through embedded systems and monitors results at several locations across the processors to pinpoint specific problem areas. Hardware problems are identified with tools such as oscilloscopes or frequency generators, while software problems can be identified through techniques such as static walk-throughs of code or dynamic debugging tools. A strength of RTNI monitoring lies in its ability to detect software defects using monitors attached to the system at

0-7803-5086-3 /98/$10.00 ©1998 IEEE

various points. Bugs are detected during real time, and the tester does not have to examine output and track errors to the source [6].

One limitation of RTNI monitoring is that there is no definitive proof that RTNI monitoring will save more than it costs to implement. However, there are potential cost-benefits. This paper discusses research by Capt Michael D.Lewis [7] performed at the Air Force Institute of Technology (AFIT) in 1997 to determine the benefits of using RTNI monitoring for DoD programs.

BACKGROUND

Information is now presented to set the stage for the AFIT study. Four areas are addressed: software cost drivers, types of software testing, the role of RTNI in software testing, and previous RTNI cost-benefit studies.

Software Cost Drivers

According to Ould and Unwin, a design is feasible if life cycle benefits exceed life cycle costs [8]. Since these costs are not known until after a program is complete, cost estimation methods must be used to predict life cycle costs in the interim. Various software cost models have been developed to estimate software development and support costs early in a program.

According to Royer, models contain "cost drivers" which are those factors that contribute to total software cost [9]. Different models contain different cost drivers with different weights or degrees on influence on costs. Since RTNI monitoring is a method of testing, it will most strongly affect testing cost drivers in cost models and methods.

For test cost drivers, Royer suggests that program size and program specification and design are strong drivers. Beizer, however, suggests that size is not a strong driver; he lists other drivers including testing schedule [10]. Schedule (and, hence, cost) can be reduced by increasing defect detection rate, which is a feature of RTNI monitoring. Brooks states that designing programs with fewer defects can greatly reduce support costs.[11]. In the SEER-SEM model, for example, increased testing rigor raises software development costs, but use of modern tools and practices (such as RTNI monitoring) will lower both development and support costs [12].

Types of Software Testing

Two basic, high level forms of testing are static and dynamic testing. Static testing, according to Daich, is evaluating software without executing the software [6]. Examples of static testing methods include desk checking, walk-throughs, and inspections. Static testing alone can not find faults in the operating environment. Dynamic testing, according to Daich, is executing a piece of software to determine if the software functions as expected [6]. Dynamic testing discovers defects in the operating environment, but is only as effective as the data used.

There are also four phases of software testing: unit, integration, system, and qualification testing According to Ould and Unwin, unit testing is testing a module to insure it correctly implements design and is ready to be integrated into a system or subsystem [8]. Integration testing, according to Daich, determines how individual modules perform together as an integrated entity [6]. System testing insures that the combination of hardware and software works together to satisfy system requirements, and qualification testing, also called acceptance testing, verifies that the system meets the needs of the user [6].

Glass also describes two phases of testing for real-time systems, host and target testing [4]. Host testing is done only on the host (or development) computer, and primarily involves eliminating coding errors. Target testing is testing on the computer on which the software will execute in the operational environment, and often involves real-time simulation.

Role of RTNI In Software Testing

Instrumentation for RTNI monitoring must meet the requirements of being both "real-time" and "non-intrusive". Real-time processes use real-world time as a parameter, and are dependent on time as well as other inputs [5]. RTNI instrumentation, according to Fryer, must be able to operate throughout the full speed gamut of the system being tested [13]. Intrusive systems allow software to be modified to allow monitoring of data elements. Therefore, non-intrusive instrumentation, according to Daich, must be able to monitor internal software data elements without modifying the software in any way [6].

RTNI monitoring is useful in dynamic testing, where it provides an analysis of the software as it operates on the actual hardware and in the real-time sequence of events. It does not compare inputs and outputs, but detects errors as they occur, which decreases testing time. Also, because RTNI monitoring is non-intrusive, modifications to the software do not have to be tracked or corrected.

RTNI monitoring is useful during any of the four testing phases; however, Glass implies that it is most useful during the target testing phase [4]. Here, typical testing tools only allow for a post-mortem analysis of failures. RTNI monitoring, however, allows the testers to monitor execution at several points and detect errors as they occur rather than waiting until after a test case has failed [4].

Previous RTNI Cost-Benefit Studies

Two studies conducted by ITCN Corporation showed that use of RTNI monitoring can result in substantial cost savings. The first, conducted for the F-18 Software Support Activity in China Lake, CA, showed that RTNI saved about $2.6 million [14]. The savings were due to faster resolution of software trouble reports and less flight testing delays. The second, conducted for integration of new circuit devices into theF-111 at Sacramento, CA. The results of the use of RTNI monitoring, according to the study [15], were that the system integration phase was reduced from 17 months to 12 months, and a savings of $200,000 was realized.

Unfortunately, the results of the two studies have not been independently verified. Furthermore, according to Ferens and Christensen, the accuracy of software cost estimating models for DoD projects is questionable [16]. Even if the testing effects of RTNI monitoring are known and can be accurately input to cost models, the costs and benefits of using RTNI monitoring may still be hard to quantify.

STUDY METHODOLOGY

The methodology used by Lewis in his RTNI monitoring study will now be explained. This will include the cost methods used, the interview instrument, and the use of the Nominal Group Technique (NGT). It should be noted that, since there is no verified historical data to show the savings from use of RTNI monitoring on completed programs, a structured interview technique was used to assess potential savings from RTNI monitoring.

Cost Estimating Method

According to Boehm, there are seven cost estimating methods [2]; however, only three were examined for this study. These were algorithmic (or parametric) models, estimation by analogy, and expert judgment estimating. Neither algorithmic models nor estimating by analogy were appropriate for this study. Algorithmic models were judged inappropriate because they do not readily accommodate new techniques (such as RTNI monitoring), and because the cost driver and size inputs must be accurate, which was difficult here. Analogy techniques were also deemed inappropriate because data that would show the effects of different tools on costs was not available.

The method chosen for this study was the expert judgment method. According to Boehm, this method has the advantage of being useful for new techniques (such as RTNI monitoring), but is subject to experts' bias [2]. To overcome this bias, a structured interview was used and several expert opinions were solicited.

Interview Instrument

A technique known as "snowball sampling" was used to find suitable participants for the study. According to Cooper and Emory, snowball sampling consists of identifying a group that has a particular characteristic and, in turn, identifies others with the same characteristic [17]. For this study, the characteristic is "users of RTNI technology".

For this study, the initial list came from customer lists of companies that market RTNI instrumentation. These customers provided additional names. Of course, the snowball sampling technique will not usually identify the entire population, so this study did not identify all RTNI users.
(Only six were available for further participation, so a structured interview was used instead of a survey, which is often used for larger groups.)

NGT Description and Analysis

The initial design for the structured interview was the use of the NGT. According to Zimmerman and Muraski, advantages of the NGT are that it avoids domination by certain individuals, and it allows conversion of responses to numerical data [18]. Usually, NGT requires all respondents to be in a conference room at the same time; however, since respondents were geographically dispersed, telephone and electronic mail interviews were used.

The NGT consists of the following seven stages:

- Define the question under study.
- Select participants.
- Begin the session.
- List nominations of responses.
- Discuss and clarify items, and eliminate duplicates.

- Select individual items.
- Rank selected items.

The responses and ratings are then compiled, and statistical analysis performed on each item. The results of the NGT show what items the group feels are important, and the weights the participants assigned to each item. The use and results from the NGT for this study are now discussed.

FINDINGS

The results of the seven stages of the NGT technique for this study are now summarized. A more detailed analysis of the results is presented where it is appropriate.

Stage 1: Definition of Question

The question defined was, "In what areas do you see the use of RTNI monitoring having an impact on life cycle costs, whether positive (decreasing cost) or negative (increasing cost)". This question limited responses to RTNI monitoring, allowed for positive and negative responses, and resolved ambiguity as to what constituted a positive or negative response.

Stage 2: Selection of Participants

Initial searches through DoD found six programs that currently use RTNI. It was found that NASA also uses RTNI in developing deep space probes. In the end, six RTNI users from six programs agreed to participate in the study. Three were Air Force personnel, two were Navy personnel, and one person was from NASA. The programs were diverse; two used RTNI during development and four used RTNI during support.

Stage 3: Beginning the Session

The participants were provided the research question and asked to respond within 48 hours. All participants responded within one week. The responses varied, but contained some common areas.

Stage 4: List nominations

The responses were listed by similar areas. These are:

- Testing costs.
- Complexity.
- Software size.
- Reliability.
- Maintenance costs.
- Hardware procurement costs.
- Training costs.

Within each area, respondents listed several items or ideas. For example, in the area of testing costs, two items were "debugging time reduced" and "testing time reduced".

Stage 5: Discussion

Capt Lewis, who acted as administrator to insure all comments and responses were anonymous, provided each participant a copy of the items for discussion. One week's time was allowed for responses. Unfortunately, responses were paltry because some respondents were not in town and others felt no need for further discussion. Still, enough information was gathered to generate a final list.

Stage 6: Selection of individual items

As a result of Stages 4 and 5, a list of 12 items was generated for further analysis. These items are as follows:

A. Reduction of debug time.
B. Sharing of test time among engineers.
C. Reduction of regression test time.

D. Additional testing data that was not available before.
E. Increased amount of dynamic testing.
F. Reduction in test case development time.
G. Increase in testing reliability.
H. Reduction in time needed for integration.
I. Increase in training cost.
J. Increase in overall system reliability.
K. Increase in hardware procurement cost.
L. Reduction in maintenance cost.

Of the 12 items identified by the participants, ten were positive; they are expected to result in cost reduction. Only two areas, increase in training cost (I) and increase in (RTNI monitoring) hardware procurement cost (K), were negative, or expected to result in increased costs.

Stage 7: Ranking Selected Items

Participants were asked to rank the items by selecting the "Top 5" that were most important. They were then asked to rank order their Top 5 items in descending order by assigning a rating of "5" for most important to "1" for least important.

Table 1 shows the results of the rankings by each participant. The letters for the items correspond with the list above, and the numerical ratings are from 5 (most important) to 1 (least important) of the top five items selected by each participant. Table 2 shows, in descending order, the number of participants that selected each item (frequency) along with the percentage of respondents that selected the item. This information will be used in the analysis of results.

Table 1. Ranking of Items by Participants

Item	Respondent					
	1	*2*	*3*	*4*	*5*	*6*
A	4	5	4	5	5	5
B	5			3		2
C						
D		4				
E		3	1	4	3	
F	2		5	2		
G		2		1	5	4
H	3					
I		1				
J			2		4	3
K						1
L	1		3		1	

Table 2. Items ordered by selection frequency

Item	Frequency	Percentage
A	6	100
E	4	67
G	4	67
B	3	50
F	3	50
J	3	50
L	3	50
D	1	17
H	1	17
I	1	17
K	1	17
C	0	0

Analysis of NGT Results

Table 3 shows the statistical average ranking for each item (Average) for the 1-5 rankings. The average is an indication of the relative importance of that item within the Top 5 for each respondent. This average does *not* include instances where the item was not in the top five (i.e., empty cells in Table 1), so it can be misleading. For example, Item D has an average rating of 4, but the item was only selected by one person as a "Top 5" item.

A better measure of the item's importance is a weight, which combines the average rating and the percentage of those selecting the item as a Top 5 item. The weight for each item, as shown in Table 3, is a product of the average, percentage (from Table 2) divided by 100, and a value of "6" for the number of participants. Table 3 lists the items in decreasing order of weight.

Table 3. Items ordered by weight

Item	Average	Percentage	Weight
A	4.67	100	28
E	2.75	67	11
B	3.33	50	10
G	2.25	67	9
F	3	50	9
K	3	50	9
L	1.67	50	5
D	4	17	4
H	3	17	3
I	1	17	1
K	1	17	1
C	0	0	0

According to the composite expert opinion of the group, Item A, reduction of debug time, is the most significant benefit of RTNI monitoring. As shown in Table 1, all six respondents ranked it among their Top 5, and ranked it first or second (rating 5 or 4) in importance. Other areas receiving high weights are: increased dynamic testing (Item E), sharing of test time among engineers (Item B), testing reliability increased (Item G), reduction in test case development time (Item F), and overall system reliability increased (Item J). It is noteworthy that the two negative items (I and K) are near the bottom of the weighted rankings.

CONCLUSIONS

The study indicates that the use of RTNI monitoring can significantly reduce life cycle costs of embedded systems. The major area of impact is reduction of debugging time. Using RTNI monitoring, an engineer will not have to spend long periods of time analyzing code for bugs, but can see the exact location where an error occurs or a calculation fails during execution. There were several other areas judged to be significant cost reducers, including increased amount of dynamic testing and greater sharing of test data among engineers.

A limitation of this study is that it did not provide a method for quantifying cost savings associated with RTNI monitoring. In order to be effective, RTNI monitoring must have proper implementation and commitment by management. The managers often have to be shown cost-benefits before approving a program such as RTNI monitoring. Still, with proper support, RTNI monitoring should be considered for real-time systems because of the many potential benefits.

REFERENCES

1. Mosemann, Lloyd K., II, *Guidelines for Successful Acquisition and Management of Software-Intensive Systems*, Hill AFB, UT, Software Technology Support Center, June, 1996.

2. Boehm, Barry W., *Software Engineering Economics*, Englewood Cliffs, NJ, Prentice-Hall, 1981.

3. Government Accounting Office, *Test and Evaluation: DoD Has Been Slow in Improving Testing of Software-Intensive Systems*, GAO-NSIAD-93-198, 1993.

4. Glass, Robert L., "The Lost World of Software Debugging and Testing", *Communications of the ACM*, Vol. 3, No. 5, pp. 264-271 (May, 1980).

5. Plattner, Bernard, "Real-Time Execution Monitoring", *IEEE Transactions on Software Engineering*, Vol. SE-10, No. 6, pp. 756-764 (November, 1984).

6. Daich, Gregory T., *et al*, *Software Test Technologies Report*, Hill AFB, UT, Software Technology Support Center, August, 1994.

7. Lewis, Michael D., Captain, *A Comparative Study and Estimation of the Life-Cycle Cost Impact of Application of Real-Time Non-Intrusive Monitoring Technology to Real-Time Embedded Systems* Air Force Institute of Technology, 1997 (GSS/LAS/97D-2).

8. Ould, Martyn A., and Charles Unwin, *Testing in Software Development*, University Press, Cambridge, UK, 1986.

9. Royer, Thomas C., *Software Testing Management*, Englewood Cliffs, NJ, Prentice-Hall, 1993.

10. Beizer, Boris, *Software Systems Testing and Quality Assurance*, New York, Van Nostrand Reinhold, 1984.

11. Brooks, Frederick P., *The Mythical Man-Month (20th Anniversary Edition)*, Reading, MA, Addison-Wesley, 1995.

12. *SEER-SEM User's Manual, Release 4.5*, El Segundo, CA, GA SEER Technologies, December, 1996.

13. Fryer, R.E., *et al*, "Real-Time Non-Intrusive Monitoring and Embedded Simulation", *Special Proceedings of the Society of Computer Summer Simulation Conference*, July, 1994.

14. *C-TAC RTNI Case Study #1*, Centerville, OH, ITCN, 1990.

15. *C-TAC RTNI Case Study #2*, Centerville, OH, ITCN, 1990.

16. Ferens, Daniel V., and Dr. David S. Christensen, "Calibrating Software Cost Models to Department of Defense Databases", *Proceedings of the Joint SCEA/ISPA International Conference*, Toronto, June, 1998, pp. 1282-1297.

17. Cooper, Donald C., and C. William Emory, *Business Research Methods*, Chicago, Irwin, 1995.

18. Zimmerman, Donald E., and Michel Lynn Muraski, *The Elements of Information Gathering*, Phoenix, Oryx Press, 1995.

TOTAL OWNERSHIP COST USE IN MANAGEMENT

Ellis F. Hitt, Battelle, Columbus, OH

Abstract

Total ownership cost (TOC) is being adopted by government and industry to manage business processes. Total ownership cost provides managers actual costs and the causes of these costs. Actual costs are often visible to management after some rollup from lower levels has taken place. This rollup process often obscures the causes or cost drivers and leaves the manager without the information needed to drive down costs. Knowledge of the high cost drivers is the first step in driving down costs. The second step is the process of defining alternatives in sufficient detail that the investment and operating and support costs can be determined considering technical, schedule, and cost risk. The third step is the computation of the return on investment, payback period, and cost savings/growth associated with each alternative. This paper describes the use of total ownership cost and the decision support tools in government and industry.

Introduction

Total Ownership Cost (TOC) involves determination of all actual costs associated with a business or government agency. Your point of view of TOC depends on whether you're the customer or the supplier to the customer. The customer's objectives include minimizing TOC. The supplier's objective is to maximize profit. Two significant factors influencing both customer and supplier are the market size and the demand generated by the customer for the product offered by the supplier. Market size in turn influences the number of suppliers competing to satisfy the demand for the product. The aerospace industry is a supplier to the military and air transportation market. The market for new aircraft is predominately commercial. Only a very small number of military aircraft are being produced each year. The primary demand generated by the Department of Defense is reprocurement of spare parts to maintain existing aircraft, or modernization of existing aircraft systems to enhance safety, add capability or reduce TOC. This demand will decrease as older aircraft are phased out even if replaced one for one with new aircraft since the new aircraft should have lower equipment failure rates due to use of newer technology.

Variables Driving TOC

The USAF and other Department of Defense organization are using TOC to manage future expenditures to reduce Total Ownership Cost for weapon systems under development, and operating and support (O&S) costs of existing weapon systems. TOC includes research, development, test and evaluation (RDT&E), acquisition, O&S, and disposal costs.

The variables driving TOC have similarities and differences for government and different types of industry. TOC applies to government (federal, state, local), industries (automobile, aerospace, airlines, electronics, information, etc.) and individual businesses. For customers such as the USAF, TOC is driven by parameters such as those given in Table 1.

0-7803-5086-3 /98/$10.00 ©1998 IEEE

Table 1. Parameters Driving USAF TOC

- Mission type-airlift, surveillance, air combat, space, training
  - Number of sorties and flying hours required to perform missions
  - Number of launches, etc.
- Constraints
  - Available fleet size by mission type
  - Budget allocation
    - Operation and maintenance funds
    - Military personnel funds
  - Reliability of equipment

Operations generate demands on personnel, fuel, and the elements of the infrastructure including the supply system. The demand for items from suppliers is driven by the operating hours, failure rates of equipment, and the inventory of items in supply that are available for installation as replacements for items that fail.

For aerospace suppliers such as avionics companies, TOC is driven by parameters such as those given in Table 2.

Table 2. Parameters Driving Avionics Supplier TOC

- Market size
- Labor-wages, retirement benefits, medical benefits, training
- Utilities-electricity, gas, water, sewer
- Facilities-acquisition, maintenance, depreciation
- Equipment-procurement and maintenance, depreciation
- Materials used in production
- Taxes
- Profit

Comparison of Tables 1 and 2 reveals that both the customer and supplier TOC include labor, facilities, equipment, and material. The differences include a supplier's inclusion of depreciation, taxes, and profit. The link between the customer's TOC and the supplier's TOC is the demand generated by the customer for items from the supplier. The supplier has little control over the market size, but may be able to capture a larger share of the existing market by reducing TOC, which in turn improves the supplier's competitive position against other suppliers who do not reduce their TOC. A supplier may also attempt to acquire all other suppliers until there is only one supplier. While this may seem a feasible strategy, it may not be economically feasible as the market size decreases when the return on available assets could be higher by producing commercial products and abandoning the military market. Stockholders may be better served by a supplier quitting the military market. Where does this leave the military? One approach, should this happen, is to return to organic design and flexible manufacturing. Another is to acquire services from existing commercial providers.

Managing using TOC requires an understanding of the business processes driving TOC and an examination of alternative processes to see if they can drive down the cost of the high cost processes. Managing using TOC involves making decisions whether TOC can be reduced through improved productivity, to invest in improving an asset or to dispose of an asset, or to acquire the needed mission capability through new business approaches such as lease versus buy. Investing in an asset having limited capability, or remaining useful life is not wise if a new asset can be leased or purchased for a lower cost per operating hour at a good return on investment (ROI). TOC also involves knowing where assets are located, and the value of those assets. As the USAF mission changes, and the aircraft fleet grows smaller, the USAF examines shifting roles and aircraft between the active duty, reserve, and Air National Guard organizations.

Total Ownership Cost Determination

Determination of total ownership cost requires collection of all actual cost data involved with the operation of a government agency or business. Many agencies and businesses set up a number of appropriations and attempt to track their costs against each appropriation. Often the appropriations are for a department and not broken down into the chart of accounts used to collect costs. This approach results in loss of management visibility into how funds are actual spent. The actual costs must be organized to permit determination of the expenditures by operating units with the costs logically grouped according to a chart of accounts. This chart of accounts is dependent upon the type of business or government agency

Government agencies are not required to make a profit, whereas an industry must make a profit if it is to survive. Many types of taxes paid by industry are not included in government accounts. A manufacturing company must know the cost of producing and marketing a product in order to establish pricing. The airline industry takes the total ownership cost and converts it into a cost per seat mile based on the number of seat and miles flown.

The Department of Defense maps cost into elements defined by the Cost Analysis Improvement Group (CAIG). These are then used analyzed to rank the CAIG elements in terms of metrics such as Cost per Flying Hour (CPFH). This permits comparison of costs across similar weapon systems operated at different bases by major commands (MAJCOMs). Assigned personnel, their level of training, and all the variables associated with their cost are analyzed using various metrics such as total cost, CPFH, and cost per aircraft (CPA). Items consumed such as fuel, materials, etc. are tracked in terms of quantities and costs which are then expressed in terms of CPFH and CPA. The time required for maintenance processes and the labor per maintenance action are also computed for use in determination whether process changes will reduce TOC. Similar metrics apply whether government or contractor personnel perform the maintenance. Costs not directly attributable to a specific weapon system such as maintenance of buildings, infrastructure, medical costs, etc. are analyzed to permit comparison across operating locations using metrics such as cost per assigned personnel.

Managing Using TOC

Management decisions are made using decision support tools that present actual costs and future planned expenditures. USAF decision support tools are hosted on USAF Web sites with access limited to authorized users. These tools present a user selectable view of actual (prior years and year to date (YTD)) costs (the baseline), and future costs (ABIDES, forecast from the baseline using planned operating scenarios with user selectable options of planned modifications implemented or not implemented, or costs forecast considering technology obsolescence).

The TOC data provides visibility into multi-year obligations and actual expenditures. The metrics implemented use multiyear inventory and flying hour (FH) data to convert total costs to the cost per flying (CPFH) and cost per aircraft (CPA) by year. Use of these metrics provides data for comparison of planned appropriations in ABIDES with forecasts of future costs based on these values, the forecast inventory of aircraft and flying hours. Significant differences between

ABIDES data and forecast future costs alerts managers that management action is required to minimize these differences. This could involve reducing the flying hour program or increasing the appropriations.

TOC provides visibility into cost drivers and provides graphical indications of cost elements by MDS and MAJCOM Wing. Cost elements that deviate significantly from the average for an MDS operated by different wings in a command, or across multiple commands, require AF managers' attention. This needs to be examined over a period of 2-3 years to determine if the cause is expenditures catching up to prior year obligations, or whether the deviations from the average require detailed analysis to identify the cause of the deviation.

Analysis of cost elements such as contractor logistics support (CLS) indicates that CLS significantly drives up O&S costs in terms of CPFH and CPA when compared to organic support. There are no available data in USAF databases that permit allocation of CLS costs to cost elements such as labor, training, software maintenance, travel, and transportation cost elements. Without these data, the analyst is limited to comparing the CPFH for CLS with comparable cost elements that would be included under CLS.

Reducing Total Ownership Cost

Comparison of like cost element metrics indicates to USAF managers costs that require more detailed analysis, and management action. All high costs are not due to hardware. Changes in policies, processes, and procedures can also drive down costs. The TOC information indicates units having CPFH and CPA lower than the average. Analysis of contributors to these below average CPFH and CPA can identify improved policies, processes, and procedures used by these units that, if adopted by other units, can result in lower costs. HQ USAF and MAJCOMs can implement training to institute these improved policies, processes, and procedures in units whose costs are significantly higher than the average.

The tools also include a process for evaluation of alternatives to high cost hardware items comparing each alternative to the baseline data for the item it modifies or replaces. This process requires input of the number of aircraft requiring this proposed modification, the budget requirements (RDT&E and acquisition) by FY for the alternative, and computes the O&S payback period and return on investment (ROI) of the alternative relative to the baseline. To maximize ROI and savings, the development and production time period should be minimized to get the high cost item out of the inventory as soon as possible.

To assist the tool user in definition of candidate alternatives, key descriptors of high cost items can be loaded into an Internet web search engine that can return candidate web site connections containing matching data on alternatives. The user accesses these sites, views the data, and may download the data. The user must assess whether these candidates are form, fit, function, interface (F^3I) replacements or require modification to meet the interface constraints (weight, power, cooling, data flow).

A manager uses the TOC information by examining the sorted list of high-cost drivers to determine if the required solution is in the budget. The user then assesses whether this is a common need whose solution can replace multiple high cost NSNs across multiple weapon systems or is a unique need involving one of a few NSNs on a single MDS. With the diminishing demand, common solutions must be considered in all analyses.

The current management processes tend to result in MDS-unique solutions since each USAF system program office deals with a single weapon system and does not look at other weapon systems to determine if they have a similar need. As a result, a duplication of RDT&E expenditure occurs if each Program Office develops a unique solution when a common solution would have been more cost effective. The OEMs encourage and benefit from this unique solution, because competition for the replacement is nonexistent, and each OEM gets RDT&E funding, and often ends up with a proprietary product. If the OEM conducts the competition for the replacement LRU or subsystem, the original supplier that delivered the original high cost of ownership system to be replaced has an inside track because of teaming agreements, with the result being a flow of RDT&E and acquisition funds to the same vendor that delivered the original "problem" item. USAF management now has the data and tools to examine other approaches to fixing the "problem" item. The key descriptors in the high cost database provide the capability for managers to look at all high cost items having the same key descriptors. This permits identifying all weapon systems having a need for a potentially common item. The challenge is for USAF and the MAJCOMs to work toward a common solution when feasible. This includes common reusable software.

If the search for an alternative returns a limited number of candidates, this may indicate a need for investment in a significant RDT&E effort. Since TOC captures and forecasts future costs, the analysis process should determine whether the need with limited or no available alternative should be funded as a 6.2, or 6.3 science and technology (S&T) program or whether the technology is sufficiently mature to proceed to a 6.4 program with the effort managed as a common solution or a weapon system-unique solution.

The user should analyze more than one alternative to make sure the most cost-effective is selected for inclusion in the budgeting process. The user of the alternative analysis tool, typically a MAJCOM, Program Office, or Product Group Manager, provides the recommended program information to the next office up the budgeting process chain for funding consideration. Having all managers providing inputs for budget funding use the same tool aids the MAJCOMs and HQ USAF since it assures the same process was used in generating the budget submission data, e.g. the same method was used for computing payback period and ROI.

Programs requesting funding that promises to drive down high costs and satisfy a priority mission need would probably have a higher probability of funding than one having the same savings and ROI but not meeting a priority mission need in a time of limited funding for new starts.

Summary

Substantial progress has been made in defining Air Force high O&S cost items. Until the investment is made to develop, produce, and install reliable products to replace/modify high O&S cost items, substantial long-term savings will not occur. This is the next step to avoid continued high O&S costs, and have money available for future force modernization. Affordability investments should not only drive down costs, but also provide increased capability whenever technically compatible and financially feasible.

TECHNOLOGY OBSOLESCENCE (TO) IMPACT ON FUTURE COSTS

Ellis F. Hitt, Battelle, Columbus, OH
Jerry Schmidt, Battelle, Columbus, OH

Abstract

Products become obsolete for a number of reasons. A product may be made obsolete by advances in technology and the lack of a viable market for the few replacements required for failures that occur. A product may become noncompetitive because the manufacturer failed to keep costs in control relative to the costs of their competitors making a similar product. Technological advances often contribute to obsolescence. Another factor is diminishing manufacturing sources; mergers of companies result in the new entity deciding that the resources required to manufacture a product having few buyers could be put to better use manufacturing a new product with higher profit.

Obsolete technology impacts government and business in many ways. It impacts their cost of doing business and hence profits, as well as their day-to-day operations. Management needs to be aware of the impact of technology obsolescence on all aspects of their business, and factor it into their decision processes. Many of the avionics systems installed use obsolete technology, but not all parts of an avionics subsystem, or line replaceable units have high total ownership cost. This paper addresses a method of assessing technology obsolescence, and the tools to determine the impact of technology obsolescence and new technology on business management.

Many complex systems, particularly major aircraft and military weapon systems, often have life spans far in excess of several life spans of many of the supporting components. As consumer use of technology accelerates, new products are introduced at ever increasing frequencies. Understanding and dealing with these market forces will separate the very successful program managers from those who constantly work only today's problems.

Introduction

The 1940s, 50s, and 60s were the hay-day of military equipment development and production. Many aircraft were required to inflict the damage required to overcome a determined adversary. The advent of precision weapons, in the 1970s, greatly increased the cost of delivery systems and, concurrently, reduced the number of weapon systems required to inflict the desired damage.

The military, during that time frame, was the clear leader in science and technology development and employment; it fueled many new developments and was often the first and/or only user of the newly developed technology. The development of a very robust consumer electronics product line in the late 70s and early 80s started to change the dynamics of the entire

0-7803-5086-3 /98/$10.00

marketplace. Products developed by the military were rapidly assimilated into commercial products. As the market grew, profits from the consumer products far exceeded the profits from the military and the products were changed to gain market share. As more manufacturers entered the market, competition increased, and the rate of change accelerated.

The military, in the 80s, faced a dual dilemma, 1) a share of the market that was becoming smaller every year, and 2) technology that was evolving because of forces beyond their control. By the early 90s, the military share of the United States electronics market, measured in dollars spent, was less than five percent and shrinking. Manufacturers were increasingly leaving the military market, or modifying their products developed for the military, for the more lucrative commercial market. Retaining the equipment, software, and technical talent required to support the military over decades became less and less a viable option for many producers. Often the first hint the military receives that products can no longer be procured is when they try to purchase a product and receive no bids.

As the military market continues to shrink, more and more products are being added to the list of items that are simply too expensive to continue in the military inventory. New analysis techniques are required to insure we can continue to operate systems over decades and keep the costs to operate these systems at an affordable level. Technology obsolescence is a major consideration in today's military.

Technology's Impact

Technology impacts RDT&E, acquisition, and future operating and support (O&S) costs. O&S costs are a function of a number of variables including failure rate of an item, the operating hours and the cost to repair the item including overhead, and even the weight of the item. In the United States Air Force, the primary variables that determine equipment O&S are the labor costs and the depot level reparable (DLR) costs. Both are impacted by technology. Improved design processes that minimize repair labor can significantly impact O&S costs. DLR costs include labor, materials, depot overhead, transportation, etc. costs. Decision-makers need to determine whether to continue to maintain the current product (existing technology which may be state-of-the-art, becoming obsolete, or is obsolete) or insert new technology. This decision is driven by an economic analysis such as that depicted in Figure 1 that needs to be updated yearly. This analysis needs to factor the current product's technology into forecasts of the future recurring costs for the current product. The current product has two major components that drive its costs; the cost to supply (CTS) new units from the vendor(s) and the cost to repair failed units already fielded. In the ideal world, new units from a supplier in a follow-on buy would have improved components and hence lower failure rates than the currently fielded units, as well as lower costs. Unfortunately, this is often not the case. In fact, the costs are often higher due to the supplier producing fewer units with the same equipment and facilities used to produce the original buy.

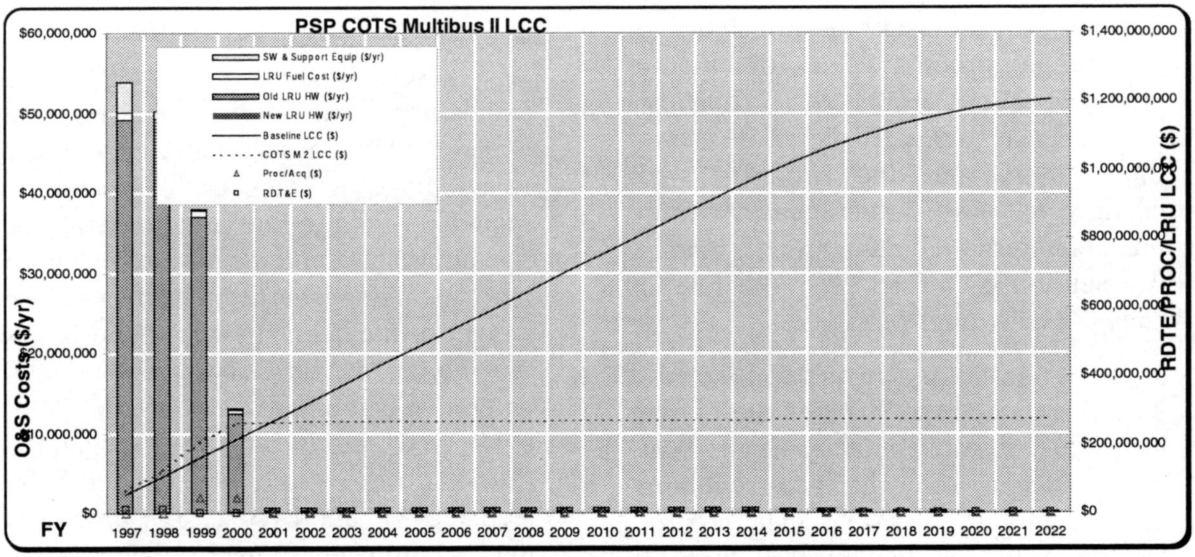

Figure 1. Example of economic analysis

The CTS for a new unit of current inventory products includes set-up time, material, labor, overhead, and profit. Set-up time is often very expensive when the tools have been stored and the factory floor space is in use for other production products. The overhead includes the fixed costs of the facility and machines required to manufacture the new units. The vendor is in business to make a profit. That means the set-up costs and fixed costs must be recovered in the unit price. Smaller quantities mean the fixed costs per unit must be higher. Material cost is another consideration in establishing the unit price is the material. If the suppliers of the material are amortizing their fixed costs over a small number of units, the result is a multiplier effect on the end item.

The manufacturer may have committed the facilities and equipment to produce another product that results in more profit. Stopping that production may be a detriment to the corporate profit picture and may result in the vendor declining to bid for the limited procurement. Decisions, made for the corporate health (translated – bottom line profit), can result in fewer vendors, one vendor, or no vendors of a current product. This may be considered *economic obsolescence*. The world is full of products that did not gain enough market share relative to other uses of resources to remain in production.

Planned obsolescence is also a factor that is market driven. In order to maintain or gain market share, companies have increased the rate at which they introduce new technology with a corresponding shorter product life for the current product. Intel has demonstrated a unique capability to engineer major product improvements and release those products to market every 18 to 24 months. The price of the current product is reduced when the new product is introduced to extend the life of the older product, but the commercial market quickly assimilates the new product and the older product is quickly phased out as the gross margin decreases with subsequent price decreases.

Technology Obsolescence Impact on Future Costs

Technology obsolescence impact on future O&S costs is a function of the technology used in the components of a mission design series (MDS) in the USAF and the demand for these components. An approach to factoring technology obsolescence into the forecast O&S costs is to develop cost factors that are technology dependent for each MDS. Technology used in an MDS is dependent on the years required to design the MDS and the years they were produced, the state of the art of the technology used in the different subsystems, line and shop replaceable units, and the individual components. Table 1 presents data for some typical MDS still in service or soon to enter service. Note that the time period from start of design to the beginning of production is increasing.

Table 1. MDS Technology Age

MDS	Design Period	Production Period
A/OA-10A	1970-1974	1975-1982
AC130H	1951-1973	1969 as AC-130E
AC130U	1987-1989	1989-1992
B-1B	1969-1984	1984-1988
B-2	1985-1989	1988-1998
B-52H	1952-1960	1960-1961
C-5A	1964-1966	1966-1970
C-5B	1975-1984	1983-1989
C-9A	1964-1967	1967-1973
C-17	1986-1991	1991-1998
C-130E	1955-1961	1961-1972
C-130H	1970-1973	1973-1996
C-141B	1958-1963	1963-1967
E-3B	1970-1975	1971-1979
E-3C	1975-1980	1980-1983
E-4B	1970-1978	1973-1975
E-8C	1986-1997	1992-1998
F-15C/D	1969-1979	1978-1986
F-15E	1980-1985	1986-1992
F-16A/B	1972-1978	1978-1987
F-16C/D	1981-1985	1983-1994
F-22	1988-1999	2000-

The dates of these aircraft procurements paint a very vivid story. We will see many of these aircraft well into the next century. The commercial new product cycle for many technologies, including the rapidly changing electronic component industry, is two years or less. The support for those components extends for about ten years. With aircraft that now can expect to see a service life of 40 to 90 years, the obsolescence problem is clearly evident.

The last entry in Table 1 for the F-22 illustrates the acquisition problem the USAF faces. The time from design to production keeps stretching out to the point that many of the technologies used during the design are obsolete before production starts. In the case of the F-22, production will start more than a decade after many of the design decisions were made and components selected. Many of the F-22 parts are obsolete before production will be started.

Using the national stock number (NSN) and Standard Reporting Designator (SRD) from the Air Force's standard base supply system (SBSS) for each MDS, along with cataloging data from another system, it is possible to identify the technology used in the various subsystems and systems of the MDS. A search of published data on technology obsolescence and price as a function of time since the technology first entered production, and the number of vendors of that technology permits development of an empirical relationship between a technology's age and cost based on demand. The forecaster analyzes the past data characterizing a time series of costs as a function of a number of variables and extrapolates the pattern into the future in order to prepare a forecast. All forecasts involve some uncertainty. Inclusion of an irregular component in the forecast is a common practice to represent unexplained or unpredictable variations in the data. The forecast is also dependent on the accuracy in the prediction of all of the other components of a time series. The greater the detail, the less uncertainty in the cost forecast. An example of the technology types used in a subsystem of an MDS is given in Table 2. A flight control (FC) system (FCS) includes sensors, processors, databuses, control sticks, FCS mode select, and actuators to move the control surfaces.

Table 2. Examples of Technology Types used in Flight Control

Subsystem	Line Replaceable Unit	Shop Replaceable Unit	Technology Type
Attitude Heading Ref	Rate Gyro Package	Rate Gyro	Inertial Sensor
Attitude Heading Ref	Accelerometer Assy	Normal Accelerometer	Inertial Sensor
Flight Control Computer	Digital Computer	Circuit Card Assembly	Microprocessor
Flight Control Computer	Digital Computer	Circuit Card Assembly	Memory
FC Panel Assembly	Mode Select Panel	Logic Circuit Card	TTL logic
FC Actuator	Rudder Actuator	Hydraulic Piston	Hydro-mechanical
Leading Edge Flaps	Power Drive Unit Assy	LE Flap Hydraulic Motor	Motor

The costs can be characterized in terms of unit price, or depot level reparable (DLR) cost (expressed in total $, cost per flying hour (CPFH), or cost per aircraft (CPA) by fiscal year (FY). Actual costs are for the current time. The predicted costs consider the demand for a replacement part that is a function of the flying hour program and the respective TO factors for future years. As the hours flown continue to decrease, due to both decreases in the numbers of aircraft and the number of hours flown each year by each aircraft, the demand for replacement parts grows smaller since the suppliers' costs must be spread across fewer units. Consequently, the cost of each new buy of spare parts will increase. Table 3 presents a notional example in constant dollars (note then year $) of the growth in DLR CPFH for some of the subsystems on a typical USAF aircraft. Note that the future is not predicted.

Table 3. DLR CPFH (Constant $)

Installed Equipment	FY94	FY95	FY96	FY97	FY98	FY99	FY00	FY01	FY02
Altimeter	$0.24	$0.27	$0.26	$0.36					
Horizontal Situation Ind.	$6.16	$6.53	$6.76	$7.12					
Central Computer	$10.98	$11.24	$11.76	$12.38					
UHF Radio	$1.79	$1.98	$1.96	$2.18					
HF Radio	$2.44	$2.54	$2.73	$2.93					
Inertial Navigation Unit	$70.47	$75.33	$78.26	$83.54					
Brakes	$8.33	$8.42	$8.55	$8.69					
Tire	$0.13	$0.13	$0.14	$0.14					

Available technology changes as a function of time. Available avionics technology changes more rapidly than structural or landing gear technology. There are different cost acceleration factors as a function of time since the technology was originally produced over a specific time period. Acceleration factors are influenced by introduction of new technology. New technology production often displaces older technology production on a line. As demand for new technology products increases, more lines are converted from the older product whose demand is decreasing because the new product has a price/performance advantage.

The challenge is to couple the forecast of the costs of maintaining the existing technology that is, or will soon be, obsolete with an economic analysis tool. This tool will give the decision maker the information needed to determine when to budget for replacement of obsolete technology with newer technology which has a larger number of users, and is easily upgraded when it becomes obsolete.

This requires forecasting the cost of staying with the obsolete technology as well as the investment required to develop, acquire, and make operational the new technology item or system. This is an optimization problem with a number of variables. The items causing high cost may be one or two or many line replaceable units in the case of avionics. If the decision maker decides to examine alternatives ranging from replacing a circuit card assembly to the replacement of many of the line replaceable units, the forecast of technology obsolescence must cover the future costs of all items that may be replaced or modified to reduce total ownership cost. The most accurate approach is to forecast the future cost of each item and then to add up the future O&S costs of all items which then becomes the baseline. The technology obsolescence factor is a function of time for each item with a numerical range starting at 1 for the current FY and increasing up to maybe 20 or 100 in terms of a multiplier for constant dollars unit price.

New technology drives obsolescence as does diminishing manufacturing sources. Examples of technology obsolescence are the replacement of the rotating mass gyro with ring-laser or fiber-optic gyros. Another is the use of flat panel displays to replace cathode ray tubes. The number bought each time also drives unit cost. Smaller quantities often result in higher unit prices. The analysis of the alternative must take into account the number of aircraft to be retrofit and the phase out of the item with high TOC. The item with high TOC should be removed from the inventory as soon as possible to eliminate the costs associated with that item. Spreading production over a long time period to keep a factory producing that item invariably results in high TOC than producing the item in a short time period and shifting production to another item. Avionics manufacturers should be capable of reconfiguring their production lines to manufacture items for both the military and commercial avionics markets, and consumer electronics.

The investment costs and future O&S costs of the alternatives are forecast using a similar technique. The USAF uses a standard economic analysis technique [1] to apply the same discount factors to the existing and proposed alternative. The return on investment, payback period, and life cycle cost savings are computed to assist the decider in whether to continue with the present item, or request funding to replace the high cost item.

Mergers of companies and discontinuing certain product lines cause diminishing manufacturing sources. Other factors include a vendor deciding to quit a product because his return-on-investment (ROI) is too low. He may be able to build another profitable product on the same production line.

The composite of all of the above factors enter into the determination by stock number of the impact of technology obsolescence and diminishing manufacturing sources on future costs. Figure 2 presents a notional representation of future costs considering technology obsolescence compared with future costs without consideration of technology obsolescence for specific items. The P-3X forecast is an example of the impact of inserting next technology to replace an obsolete item.

Figure 2. Impact of technology obsolescence on future costs

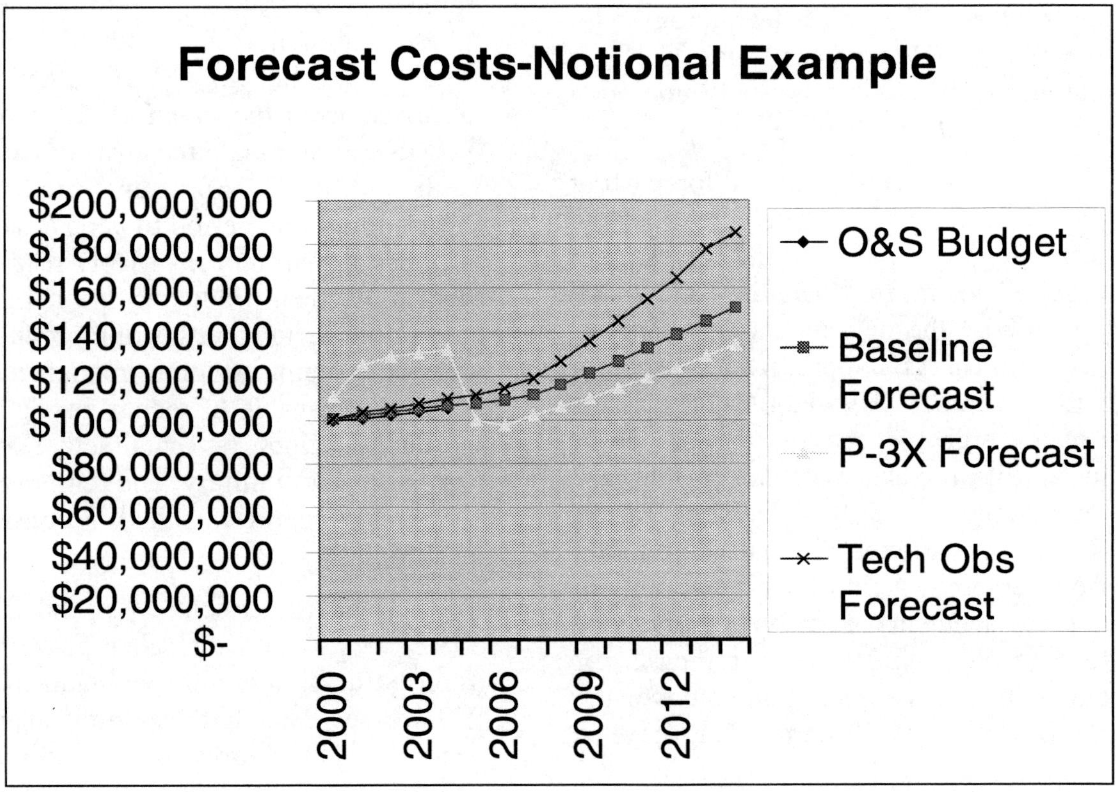

Summary

This paper described the need for considering technology obsolescence in the decision making process during the acquisition and sustainment of major military and commercial systems. Diminishing demand and the availability of new technologies accelerates technology obsolescence. An understanding of these factors will be critical in future years for both government and industry.

References

[1] Economic Analysis, AFMAN 65-506, 1 July 1995.

CORRECT DECISIONS BASED ON UNCERTAIN REQUIREMENTS AND IMPERFECT INFORMATION WITHIN PRODUCT PARTNERSHIPS – A CONTRADICTION ?

Günter Hertel, Daimler-Benz AG, Research and Technology, Frankfurt am Main, Germany

Thomas Jahns, Massachusetts Institute of Technology, Cambridge, MA

Michael Kokes, Daimler-Benz AG, Research and Technology, Frankfurt am Main, Germany

Thomas Neff, Daimler-Benz AG, Research and Technology, Frankfurt am Main, Germany

Isaac Trefz, Massachusetts Institute of Technology, Cambridge, MA

Wilfried Virt, Daimler-Benz AG, Research and Technology, Frankfurt am Main, Germany

1. Challenges for Automobile Manufacturers

The automotive industry is currently in a phase of upheaval, which has manifested itself in the form of far-reaching structural change. This change is being unleashed and accelerated by worldwide overcapacity, globalization, corporate mergers, and shorter product life cycles. In reaction to the altered competitive environment, manufacturers have withdrawn to their core competencies and significantly reduced the range of their production and development capabilities. Single sourcing and integration of third party parts and systems into production processes as well as the increasing participation of system suppliers in development efforts have led to a much more complex relationship between end producers and their system suppliers. Today, an automobile manufacturer may produce less than 40% of a given vehicle in-house [1]. An enterprise wide, process oriented product development process is necessary in order to ensure the competitiveness of system manufacturers under these conditions.

Companies are interlaced with value chains, whose configuration and coordination determine a company's competitiveness. An enterprise's value chains incorporate themselves into complex value networks extending from parts manufacturers through component and system suppliers all the way to the OEM [2]. The goal of value chain optimization is an efficient consumer response (ECR) [3]. ECR means, that the OEM must be in a position to recognize the changing demands of customers early on and meet these demands in a prompt and targeted manner. This however requires proactive decision making, which takes into account customer needs and willingness to pay for certain products or features.

As shown in Figure 1.1, final product

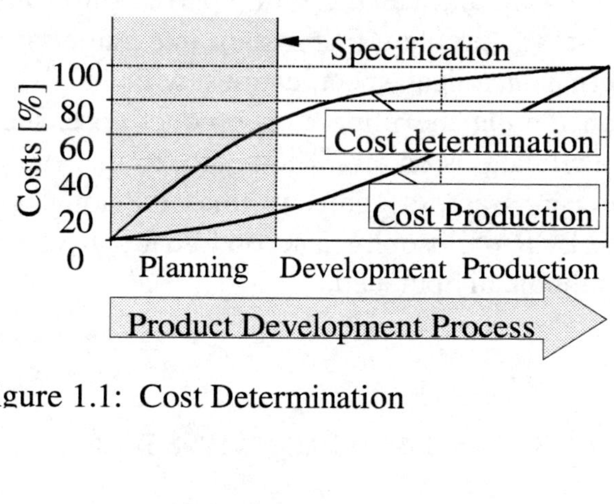

Figure 1.1: Cost Determination

0-7803-5086-3 /98/$10.00 ©1998 IEEE

characteristics are strongly influenced in early product planning phases. A product's technical specifications determine 70% of its total life cycle costs.

Two basic problems are in evidence here: The more a product specification meets anticipated customer wishes in an early phase, the more sucessful the product will be. In other words, if future customer desires are being poorly anticipated, the product will be developed for a nonexistent market. The AIM-Model with its phases "find-communicate..." [4], originally developed for use within the service sector, is helpful for the implementation of a product planning process that takes customer needs into account.

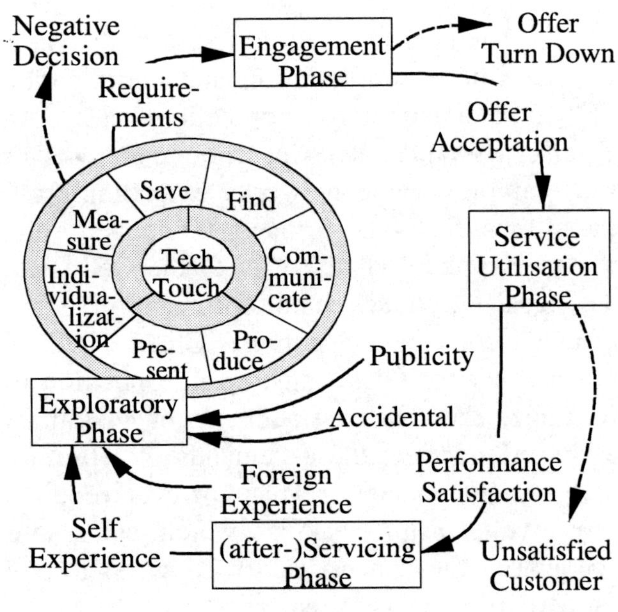

Figure 1.2: 4-Phase AIM-Model

Within value creation partnerships there exists the problem that system integrators often define technical specifications without always considering the impact on product costs these specifications entail. The ability to make early decisions across company lines is, along with the ECR philosophy, a second goal that will be methodically pursued.

2. Incomplete Information within the Product Planning Process

It is clearly apparent that the general problem of decision-making in the presence of incomplete information is connected to the realization of the two goals mentioned in the previous section. Henceforth, the adjectives "incomplete" and "uncertain" will be used for the term incomplete information. It is characteristic of early product planning and development phases that a large portion of required information regarding product requirements, technical specifications, and other external data can only be estimated or is incomplete. In order to process this incomplete information a methodology has been developed, which describes all data as beta random variables [5]. The beta distribution provides a good approximation to other distribution functions (normal, Weibull, uniform, etc.) [6,7].

Figure 2.1 shows a typical beta density function (solid line) and the corresponding distribution function (dotted line). The gray shaded area represents a 70% confidence interval.

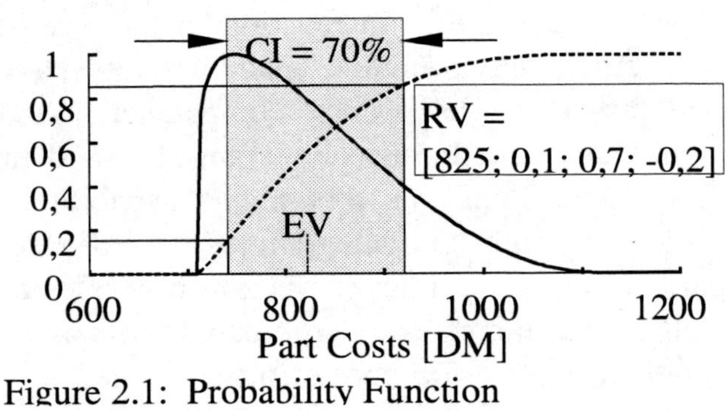

Figure 2.1: Probability Function

For clarity, the maximum value of all density functions has been normalized to one in all figures. Each density function is described by its expected value EV, coefficient of variation VC (relative error), skew SK (deviation from a symmetric density function), and kurtosis KU. These four parameters can be lumped together to form the following vector, RV:

$$RV=\left[EV, VC, SK, KU\right]$$

Two effects greatly determine the quality of the results: the compression and expansion effects. As more and more independent random variables are added together, the relative error of the result decreases (Figure 2.2) [8, 9, 10].

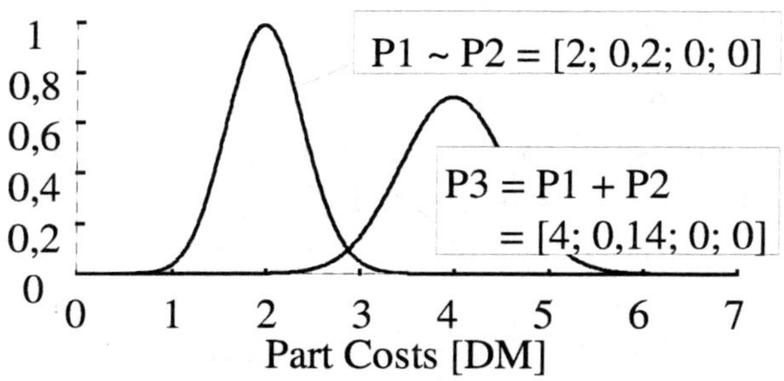

Figure 2.2: Compression Effect due to Addition of 2 Random Variables

In the case shown above, the relative error is reduced from 20% down to 14%. The validity of product models can be verified and influenced by the number of submodels during the modelling phase (Figure 2.3).

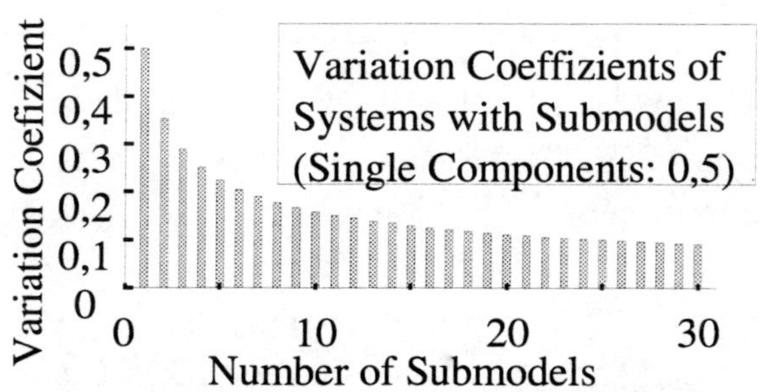

Figure 2.3: Compression of Uncertainty due to Addition of n Random Variables

Division of a model into ten submodels is sufficient to reduce the relative error from 50% (single components) to 16% (in relation to the entire system). In contrast to the compression effect, which increases the quality of results, multiplication increases the relative error of results (Figure 2.4).

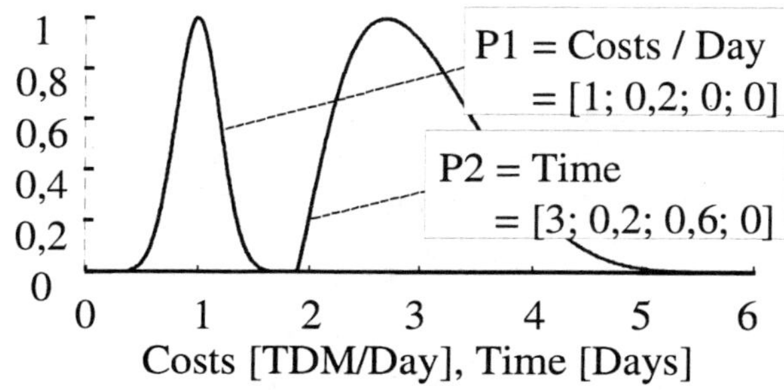

Figure 2.4: Expansion Effect due to Multiplication of 2 Random Variables (Input)

The example in Figure 2.4 shows the imprecise input data used to determine the development costs, while Figure 2.5 shows the resulting density function.

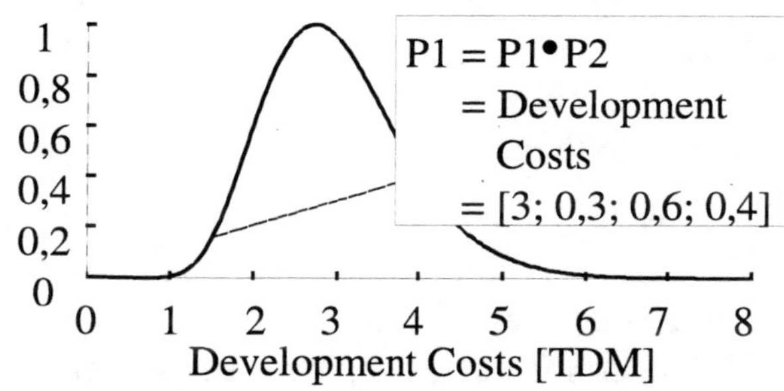

Figure 2.5: Expansion Effect due to Multiplication of 2 Random Variables (Output)

The variation coefficients of both input density functions are 0.2, while the variation coefficient of their product is 0.3. Both effects consequently determine the quality of results and must be taken into consideration when modelling products. The incorporation of uncertainty into product models will be illustrated through the example of one electronic component that is used in trucks.

3. Modelling of an Intelligent Switch Module

Component and product models were developed for the conception of new electronic structures to be used in trucks. Figure 3.1

shows one such component, an intelligent

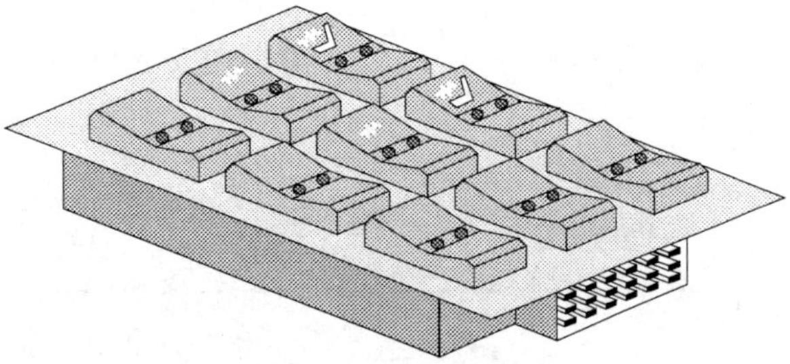

Figure 3.1: Intelligent Switch-Box switch module.

This intelligent switch module is described by the following parameters: number, maximum current rating, and type of digital outputs, number of analog and digital inputs, number and type of switches as well as the number of total units and production schedule. In early product planning phases, all of these parameters can be represented as random variables.

Within the product model there are submodels for high side outputs, low side outputs, microprocessors, CAN bus drivers, platinum, connectors, the power supply, buttons and chasis. All of these submodels take into account manufacturing, material, mounting, development and tooling costs as well as overhead costs. Model development precedes product and structural cost analyses. In this manner, elements with the greatest influence on product costs can be identified. Model development is then continued on the basis of these elements.

Figures 3.2 and 3.3 show examples of the cost structure of one box based on two different production volumes.

Each figure shows the expected value and minimum and maximum values calculated from the coefficient of variation at a confidence interval of 70%. Development and tooling costs are included in the total cost along with material and manufacturing costs. Sensitivity analyses can be accomplished using parameter variation (e.g. production volume). Each cost category can be further subdivided as the need arises. Figure 3.4 shows the structure of the initial costs.

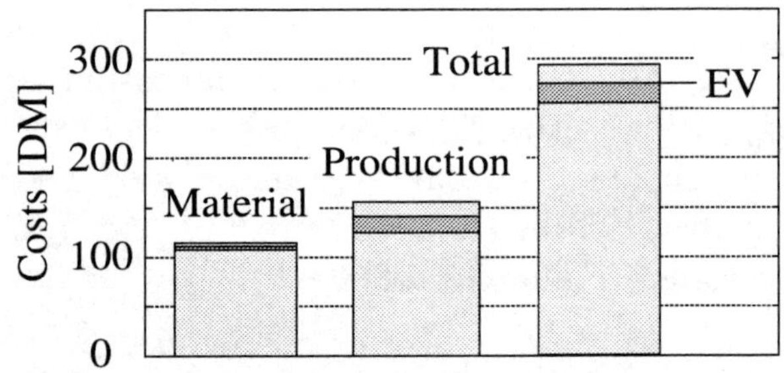

Figure 3.2: Cost Structure at 10000 Units

A transparent representation of cost structures is conducive to creating a product plan that gives due consideration to costs and forms the basis for effective target cost management. This approach supports an interdisciplinary design process as models created in the product planning department must be generated and cultivated in conjunction with the pre- and series development departments.

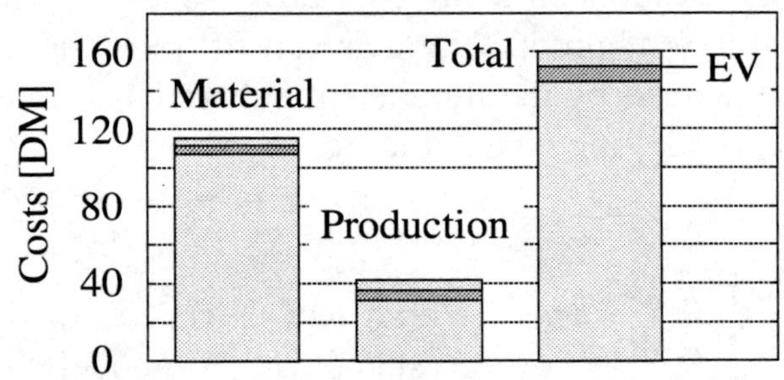

Figure 3.3: Cost Structure at 50000 Units

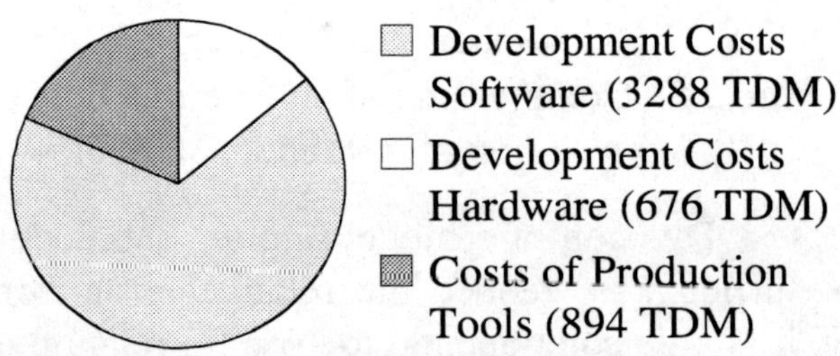

Figure 3.4: Structure of Initial Costs

0-7803-5086-3 /98/$10.00 ©1998 IEEE

However, direct representation of product costs is important in the product planning process. Various concepts may be quickly and efficiently assessed in this way, leading to significant savings in time, since cost optimization can be undertaken early and later changes for the sake of savings in cost can be avoided.

4. Profit Optimization of Accessories in the Planning Phase

With the aid of the product models introduced thus far, extensive analyses are possible, which account for customers' willingness to pay for a given product. Our objective is to maximize profits through an optimized pricing scheme and the proper concept choice. This is explained in more detail below using the example of an accessory (e.g. a sunroof). Figure 4.1 shows the results of a fictitious survey on the customers' willingness to pay for such a product and the transformation of these results into a continuous random variable.

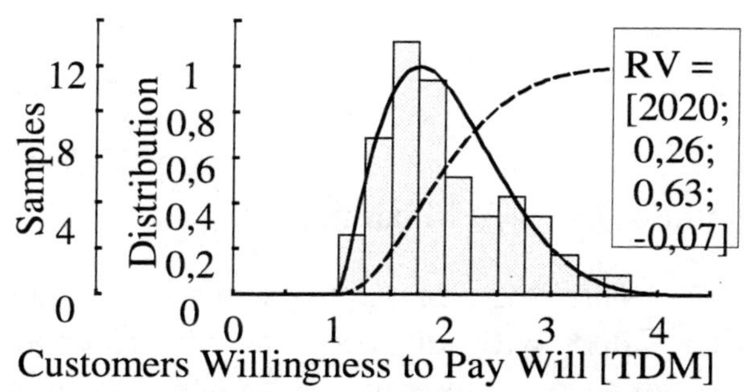

Figure 4.1: Cusomers Willingness to Pay

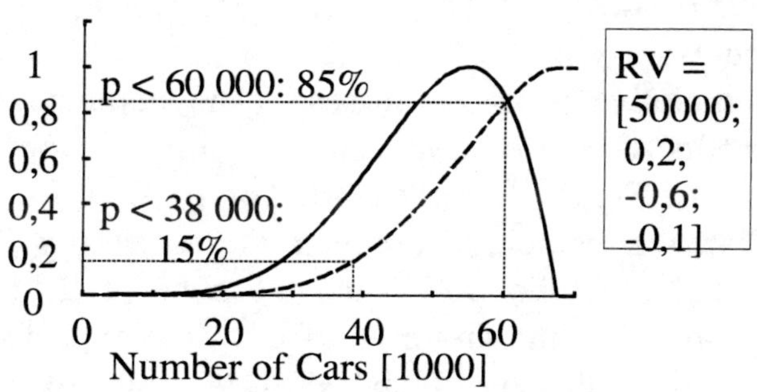

Figure 4.2: Probability of Cars Sold

Alongside the customers' willingness to pay a prognosis about the overall production volume of cars is necessary. This serves as a basis for the portion of the initial costs that must be added to the cost of each individual car. Figure 4.2 shows a typical such prognosis.

Two different concepts are under consideration for use. While concept I is a fairly conventional system characterized by low initial costs, concept II is an innovative concept with considerably higher initial costs. The initial costs of both concepts are shown in Figure 4.3.

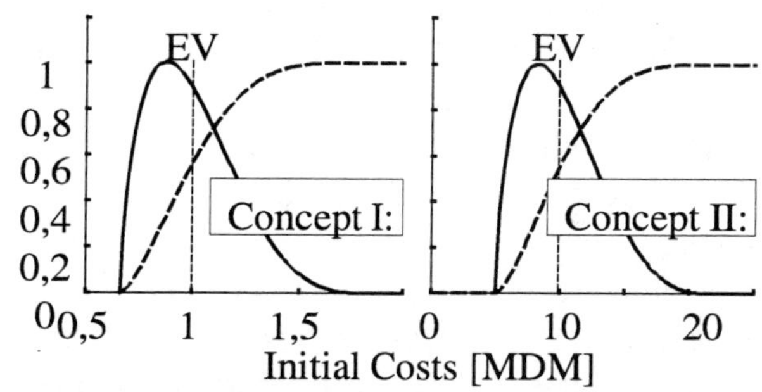

Figure 4.3: Initial Costs of Two Concepts

The higher initial costs of conept II are compensated for by lower manufacturing costs. It is implied that both concepts exhibit equal functionality. Thus, a consumer would be willing to pay the same amount for both realizations.

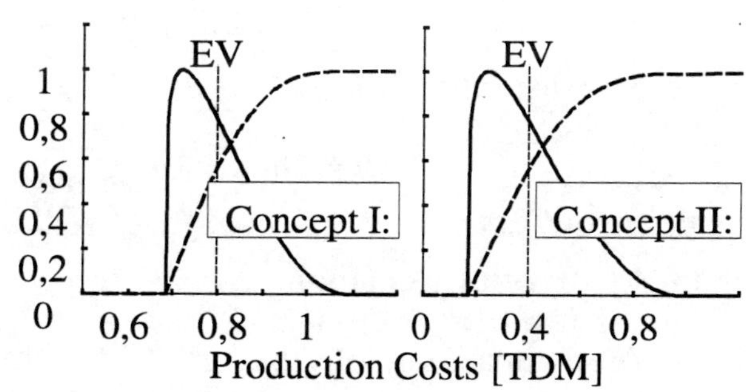

Figure 4.4: Production Costs of Two Concepts

As can be seen in Figure 4.4, the manufacturing costs of concept II ($VK_{II} = 0.3$) were chosen to be much less precise than those of concept I in order to represent the greater uncertainty inherent in concept II. This uncertainty carries over into the cost per car, which is displayed in Figure 4.5 as a function of the total production volume.

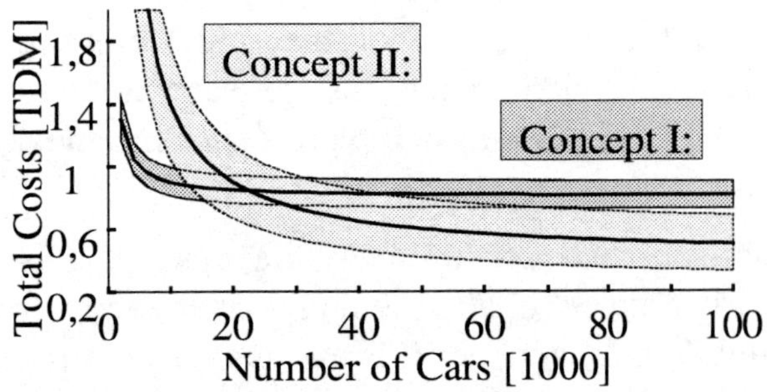

Figure 4.5: Total Costs over Number of Cars

The share of total cars into which an accessory is installed is however determined by consumer willingness to pay for that accessory. It is assumed that the effect of the accessory on the total production volume is negligible. The percentage of cars into which the accessory is installed and the profit on each car is determined by an optimized pricing scheme. Since these two effects oppose each other an approach that takes both into account is necessary. Figure 4.6 illustrates this calculation scheme.

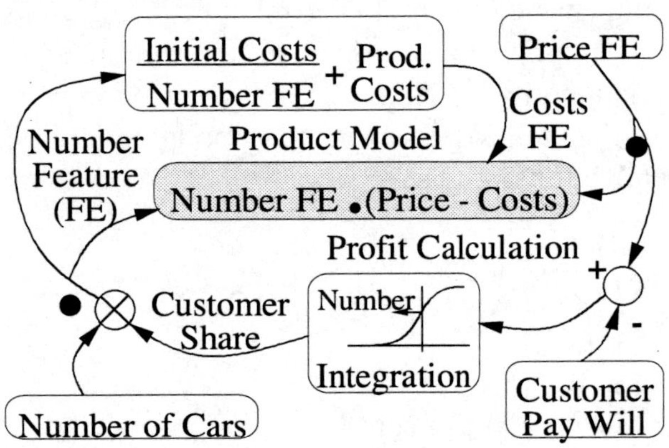

Figure 4.6: Profit Calculation

A comparison of the price of the accessory with the willingness of the consumer to pay for it and integration yields the percentage of cars containing the accessory, which in turn has an effect on the production costs of the accessory.

We however wish to maximize the overall profit made, which is determined by the number of units sold and the profit per unit.

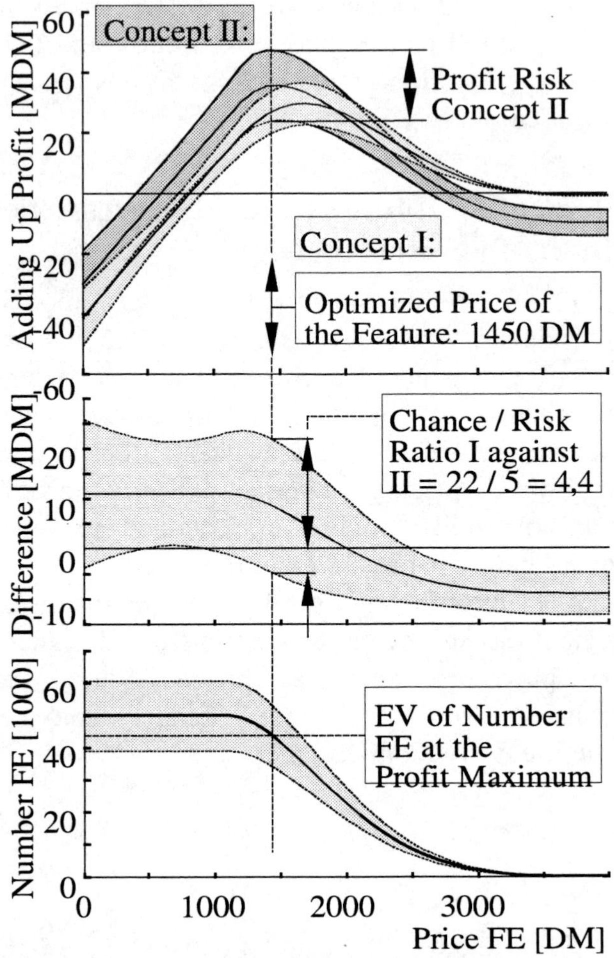

Figure 4.7: Profit Maximization

The share of vehicles out of the total (FE) fitted with an accessory is calculated through a comparison of consumer willingness to pay a given price for that accessory (see Figure 4.7). This share decreases with increasing price. At the same time, the cumulative profit increases. The cumulative profits of both concepts are shown in the upper part of Figure 4.7. At a price of 1450 DM the share of vehicles in which the accessory is installed reaches 88%, and the concept II has reached its maximum profit. The expected value of this profit is 35 DM with likely minimum and maximum values of 23 DM and 47 DM respectively (confidence interval of

70%) Concept I attains a lower profit with a higher price.

Subtraction of the two cumulative profit curves yields the difference in profit between the two concepts show in the middle of Figure 4.7. The chance/risk relationship of the concepts can be derived by examining the regions outside of the 70% confidence interval. Figure 4.7 shows this parameter to have a value of 4.4. Thus, concept II is clearly the more advantageous of the two concepts. Naturally, this assertion was obtained by examing the problem from only one perspective and must be complemented by strategic considerations related to market positioning.

5. Conclusions and Outlook

Early and effective integration of subsuppliers during the early phases of the product development process has become a strategic factor key to a manufacturer's competitiveness. It will therefore be important in the future to act upon subsuppliers' innovations in a significantly more timely manner than is practiced today [11]. The realization of this task will render the sphere of development crucially important. Not only the technical functionality of a future product is defined in the early phases of the product development process. Through specification of technical parameters a product's cost as well as indirect and direct costs associated with a product's production may be determined.

The transformation of this strategy into reality requires the ability to work with incomplete information, which, when coupled with accurate mathematical and modelling methods, can contribute to making the goal of higher profits a reality.

Through the example of a vehicle accessory, it was shown that with the help of probability based assessment techniques that are able to model the kind of incomplete input data often present in early planning stages, well founded decisions can be reached. Timely recognition of risks and of requirement to act as well as transparent and accountable decision making that takes customer wishes into account are the primary goals that will be pursued in order that economically robust products are created.

6. References

[1] Bullinger, Hans-Jörg; Thaler, Klaus: Vom Teilefertiger zum Wertschöpfungspartner. FB/IE 42 (1993), Heft 1, S. 24-27.

[2] Porter, Michael E.: The Competive Advantage of Nations. London: Macmillan, 1991.

[3] Swoboda, Bernhard: Wertschöpfungspartnerschaften in der Konsumgüterwirtschaft. WiSt 26 (1997), Heft 9, S. 449-454.

[4] Hertel, Günter; Fischer, Klaus et al.: Qualitätssicherung in Dienstleistungsprozessen. FQS-DGQ-Schriftenreihe 08-04. Frankfurt/Main, Berlin, u.a.: Beuth-Verlag, 1995.

[5] McNichols, G.R.: On the Treatment of Uncertainty in Parametric Costing. University of Pennsylvania: Diss., 1976.

[6] Johnson, Norman L.; Kotz, Samuel; Balakrishnan, N.: Continuous Univariate Distributions, Vol. 2. New York: John Wiley & Sons, 1995.

[7] Springer, Melvin D.: The Algebra of Random Variables. New York, Toronto: John Wiley & Sons, 1979.

[8] Hertel, Günter; Fischer, Klaus: Theoretische Grundlagen für die strategische Planung von Qualitäts-zielen im Dienstleistungsbereich. Forschungszwischenbericht. AIF-Förderthema 4245-D, März 1993.

0-7803-5086-3 /98/$10.00 ©1998 IEEE

[9] Fischer, Klaus; Hertel, Günter: Bedienungsmodelle im Transportwesen. Theoretische Grundlagen und Anwendungen. Berlin: transpress, 1990.

[10] Hertel, Günter: Verkehrssystemtheorie. Europäische Verkehrsentwicklung und Verkehrssystemtechnik. Dresden: Fachtagung "Verkehrsingenieur – Planung und Gestaltung der Bahn." 11./12. November 1993, TU Dresden. Vortragsband, S. 3.1-1 bis 3.1-37.

[11] Rudnitzki, Johannes; Deiß, Frank: Direktvergabe statt Konzeptwettbewerb. In: Automobil-Produktion 11 (1998), Heft 2, S. 38 –40.

0-7803-5086-3 /98/$10.00

MANAGING LONG-LIVED INFORMATION ASSETS ... EMERGING STANDARDS AND TOOLS CAN $AVE BIG $$.

Kent M. Taylor, ISOGEN International Corp, Colorado Springs, CO

Abstract

The digital revolution has enabled rapid advances in product/systems capability and functionality that can only be described as *astounding*. Corresponding advances in the complexity of systems and even individual components is equally astounding, and occasionally *confounding*. Add to this the ability to fundamentally change the behavior of software-driven products and/or systems in near real time, and you have rapidly *compounding* growth in the amount of information that is needed to support today's products and systems over their long lifespans.

Effective management, distribution, and use of this continually growing stockpile of information is becoming a significant effort, and a plethora of tools have been introduced to address various aspects of the information management challenge. While the individual tools may streamline the variety of information creation, storage, change, management, production, and distribution processes, they also have the potential to compound the overall problem ... especially in the support of complex, long life cycle products. Major enterprise-level inefficiencies can be introduced if individual information handling systems cannot communicate *perfectly* with each other. And major disruptions can occur over the life of a product when information handling systems are replaced or upgraded ... and legacy data is found to be incompatible with the new system.

The Problem

It is generally agreed that the technical information is being generated at an exponentially increasing rate, as illustrated in Figure 1, below.

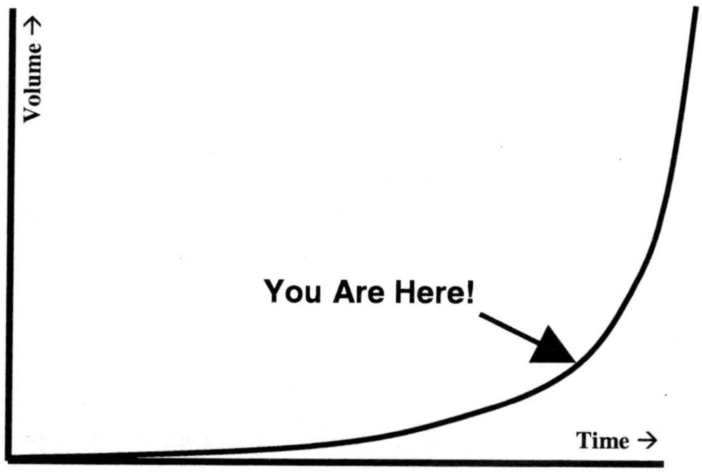

Figure 1 – Growth in Technical Information

The growth rate appears to be relatively independent of industry, with key drivers being

- Shorter Product Life Cycles (Technology)
- Software Driven Products (Easily Upgradable)
- More Suppliers (Global economy)
- Easier to Create (Computer-based Publishing Tools)

We can debate the relative position of the "you are here" marker, but it is my contention that we are at or near the knee of the curve; the real "information explosion" is just around the corner. While many enabling technologies are in place, I believe that what many have been calling the "information explosion" will barely qualify as the "fuse" when we look back from ten years out.

Today's situation feels like an explosion largely because we are either using old tools and concepts to manage and distribute the rapidly increasing volume of information ... or, we are applying new tools to manage information the old way.

0-7803-5086-3 /98/$10.00 ©1998 IEEE

Most of the volume increase depicted in Figure 1 comes from the automation and/or computerization of existing processes. It's not coincidental that the shape of the information volume curve looks a lot like the shape of computer chip performance curves; incremental increases in Workstation performance has contributed directly to increased ability to create/compile information. But we're reaching the point of diminishing returns by applying new technology to "Business as Usual" processes ... we need to look elsewhere for the next order-of-magnitude improvements.

Locally Optimized Solutions

Information is created, stored, and used in a variety of ways, by a number of different organizations within a typical business enterprise. For any given product,

- Engineering develops requirements and specifications
- Marketing/Product Management develops product plans
- Tech Pubs develops "User Documents"
- Sales develops advertising collateral
- "Everyone" develops administrative information.
- And so on for other types of organizations and the information they develop.
-

Good News: Over the last five to ten years, each of these functional organizations has discovered the joy of word processing, computer graphics, and/or desktop publishing ... and greatly improved their personal and departmental productivity. Many have also discovered alternate (other than paper) means of storing and distributing information.

Bad News: What should be a smooth pipeline of information from initial product definition through final User Documents, is now full of roadblocks. The tools used by Engineering are not compatible with those used by Marketing, which are not compatible with those used by Tech Pubs ... After all the information outputs from each of these departments *looks* different. They "*have to*" use different tools to get the desired output, despite the fact the fact that much of the content may be very similar, or even identical.

Summary: While each of the departments is efficient/productive when viewed independently, the efficiency of the Enterprise suffers as a result of the multiple translations/transformations that must take place at organizational boundaries.

"Obvious" Enterprise Solutions

Translation Programs: Because word processing and desktop publishing systems (and their users) are focused on how the *output looks,* and because there are multiple ways to make printed output *look* identical, the "obvious" solution of an automated/software translation rarely exceeds 80% accuracy. And a human being must then examine 100% of the output to find the 20% errors. Bottom line – it is frequently more effective to rekey or rewrite information that already exists in another department.

Make Everyone Use (fill in your favorite Word Processor): Remember why different departments have already chosen different tools ... they each have different *output* requirements. Sales brochures *look* a lot different than product specifications, which look different than Operators' Manuals ... Tools that are effective at creating jazzy brochures aren't real effective for creating specs or multi-volume User Manuals, and vice-versa.

This "solution" also leaves you with a legacy problem ... all of the active existing documents must be converted to the system of choice ... at 80% accuracy. And they will likely need to be converted again, when a new system comes along that promises a 20% productivity improvement (or the old system becomes obsolete/unsupported).

0-7803-5086-3 /98/$10.00 ©1998 IEEE

Object-Oriented Solutions

Given the right mindset, text and graphic objects can be created, stored, and manipulated in much the same way as software. And, just as the software development community has migrated from large linear programs to modular programs to object-oriented techniques, content creators are moving from large linear documents to modular documents, to object-oriented techniques.

A number of progressive companies are rethinking their strategies for managing their information assets, and reengineering their information development, storage/management, and distribution processes to take advantage of emerging information standards and tools.

Emerging Standards

There are a series of ISO standards that deal specifically with text, tabular, graphical, and multimedia information objects, and methodologies for describing them *independent of their presentation/formatting.* In other words, these standards focus on the logical structure and nature of content objects, and their relationships to one another.

This paper addresses the primary structural standard, ISO 8879-1987 *Standard Generalized Markup Language (SGML).* ISO 8879 defines structural and semantic rules for the descriptive markup of human readable text. Note once again, that the rules *do not address presentation/format of the text!* Also note that once the rules have been established, text files can be *parsed* to ensure that they conform to the rules.

This approach enables a new systems-level view separating Authoring/Editing, Storage/Management, and Presentation/Distribution processes, as illustrated in Figure 2.

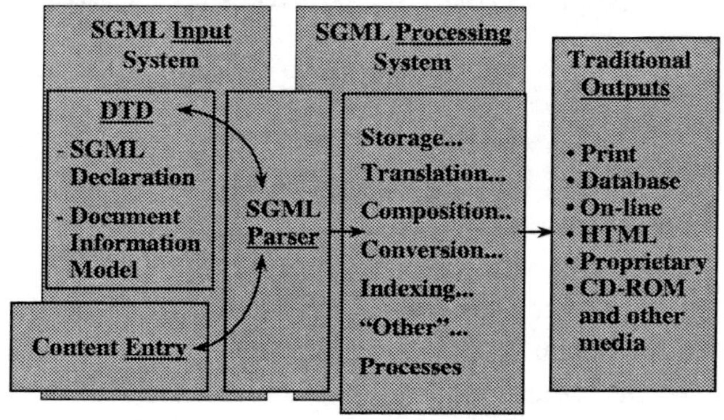

Figure 2 – SGML Systems-Level View

This view of the world results in features that can be extremely attractive in life-cycle management processes for information supporting large, complex systems:

- Authoring/editing tools can enforce structural and semantic rules through interactive, guided editing and/or after the fact parsing/reporting … all the way from initial Engineering Requirements through final Customer-deliverable documents
- Parsed Information can be interchanged easily across organizational and tool boundaries
- Parsing of text files against established, consistent rules enables accurate processing, translation, and/or transformation of text files
- Parsed files can be automatically formatted for multiple outputs (eg, paper, CD-ROM, Web, etc)
- Authors can focus on the quality of their content (versus format)
- Configuration management/control can be effectively applied at a more granular level
- Descriptively marked-up content can serve multiple business needs (eg, informational needs, control needs, ordering/inventory control)
- Information can be effectively reused (boiler plate, common procedures, and descriptions, admonitions, etc)
- Information can be effectively repurposed, customized, translated/localized

Sounds Good ... What's the Catch?

Catch #1 – Significant Process Change

SGML-based information management enables a basic paradigm shift ... away from the familiar/comfortable world of document-oriented single-step publishing where the author simultaneously addresses both content and format issues, toward a new object-oriented world where authors are concerned primarily with content and logical structure at a much more granular level. Figure 3 describes the conceptual shift that authors will face.

Single-step publishing....

1) Content + Format = Document

is replaced by multi-step processing....

content development (authoring) first...
1) Content + Description = Polished Info Object
followed by the formatted output of choice (publishing)....
2) Info Objects + Format = Document

Figure 3 – Difference in Value Added by Authors

This seemingly minor change in viewpoint can be traumatic for authors; after a lifetime of controlling both content and form of the final deliverable, authors will have to adjust to dealing almost exclusively with content. They may never even see many of the final products built from their individual information components. (Note: This "catch" can be overcome with appropriate doses of training and communication.)

Catch #2 - Significant up-front costs.

An effective SGML application requires a good deal of analysis, planning, systems engineering, and training of all participants, in addition to hardware and software costs.

Catch #3 – Legacy Conversion

The legacy problem still exists, and still has the problems outlined earlier. However, once converted to consistent, parsable SGML, content can easily be used by any conforming tool (authoring, editing, production, presentation), or translated for use by non-conforming tools and systems.

So, Why Bother?

In general, there are 4 overwhelming reasons to implement an SGML solution.

Reason #1 -- Customer Demand

SGML applications are gathering wide acceptance as Interchange Standards in many industries, including

- Aerospace
- Automotive
- Defense
- Telecommunications
- Semiconductor Manufacturing
- Pharmaceuticals

If your trading partner(s) demand that product information be delivered in an industry standard exchange format, it's probably a good idea to provide it ...

Reason #2 – Customer Demand

Alternatively, if you have several large customers, all of whom demand *different* interchange formats, an internal SGML application is also likely to be the most cost-effective solution. Likewise if different market segments require different information products (eg, paper, CD-ROM, WWW, etc.), customized or translated/localized information products.

Reason #3 – Breakdowns in Current Systems and Processes.

As large-scale systems become more sophisticated, complex, and software driven, the information required to operate, administer, maintain, and repair them has grown exponentially. Document sets of tens or even hundreds of thousands of pages are not unusual in aerospace, defense, telecommunications, and other industries. Requirements for the simultaneous release of product and supporting

0-7803-5086-3 /98/$10.00 ©1998 IEEE

User Manuals in 20 or more custom versions, in as many languages is not unusual in Consumer-oriented industries. Requirements for "built-in" electronic "manuals" are becoming commonplace in everything from space vehicles to household appliances.

Old publishing systems and processes designed to put ink on paper just aren't capable of keeping up with today's requirements for the "first release" ... much less managing the on-going changes and updates over the life of the product.

Reason #4 – Cost, Quality, and Time-to-Market Improvements

If you are in a market niche where your Customers aren't looking for higher quality, lower prices, quicker time to market, and/or greater flexibility/customization, you may skip this section.

However, if you're among the other 99%, you may be interested to know that *serious* SGML applications typically pay back in 2 - 3 years, while significantly reducing cycle times and improving quality. I stress the word *serious* because the exceptional savings require a serious commitment to taking advantage of the paradigm shifts enabled by a well thought out SGML application. A "business as usual" SGML application generally results in (incremental) "savings as usual ..."

Show Me ...

SGML applications have been implemented by Fortune 500 companies in many industries, generally for one or more of the four reasons cited above. While Customer demand or impending breakdowns may have provided the initial motivation for exploring SGML-based solutions, implementation approval and funding has generally been based on expected financial benefit and/or market advantage.

And SGML applications have been demonstrated to deliver – big time – on both accounts.

We've worked primarily with Technical Publications Departments of companies both large and small, across many different industries, to migrate to SGML-based systems and processes, and have seen authoring productivity improvements of 10% to 50%, and production interval reductions from months to hours. Add to this the added bonus of noticeable improvements in the quality of the content, while enabling the production of innovative new information products, and the front-end *cost* starts to look more and more like a *wise investment.*

It is our expectation that even greater benefits will be realized by SGML applications at the enterprise level, addressing (at least) Hardware/Software Engineering, Technical Publications, Technical Training, and Customer Technical Support organizations.

Want to Know More?

Independent Case Studies

$GML – The Billion Dollar Secret, Chet Ensign, © 1997 by Prentice Hall PTR, Upper Saddle River, NJ. Presents five detailed case studies of large companies that have implemented major SGML systems. Includes a management-oriented overview of SGML.

Comprehensive SGML/XML Bibliography

The most comprehensive SGML/XML bibliography in the known universe is maintained by Robin Cover, at:

http://www.oasis-open.org/cover/

Additional information my be found at the host site,

http://www.oasis-open.org

COST BENEFIT ANALYSIS OF THE COMBINED WAAS/LAAS SYSTEM

David M. Gustafson, Alexander E. Smith, Rick Cassell
Rannoch Corporation
Alexandria, Virginia

Abstract

The Federal Aviation Administration (FAA) plans to replace current terrestrial navigation systems with a satellite-based system called the Global Navigation Satellite System (GNSS). The two main systems planned for navigation are the Wide Area Augmentation System (WAAS) and the Local Area Augmentation System (LAAS). Both systems rely on the Global Positioning System (GPS) satellite constellation, augmented by a combination of additional geostationary satellites and ground reference signals. These augmentations are required to meet aviation requirements for service availability, integrity, continuity, and accuracy. Together, WAAS and LAAS will provide seamless navigation coverage throughout the National Airspace System (NAS). WAAS will provide en route and terminal guidance throughout the NAS and Category I landing guidance at most airports. LAAS will be used exclusively to provide Category II and III landing guidance at all airports, and Category I where it is not provided by WAAS.

This paper presents the approach and methodology used in developing a combined cost benefit analysis of both systems. Previous analyses addressed separate and independent implementations of each system. The combined analysis is designed to capture all costs and benefits attributable to both systems, ensuring a comprehensive and fair assessment. The analysis attributes costs and benefits to air carriers and other aircraft operators, passengers of air carriers, airport operators, and the FAA. Cost avoidance is included through the decommissioning of the current terrestrial navigation infrastructure.

This paper contains only the understanding and views of the authors and is not intended to reflect the official position or views of the U.S. Government or Federal Aviation Administration.

Introduction to WAAS and LAAS

FAA is implementing satellite-based navigation technology throughout the NAS. In 1989, the Department of Defense (DoD) launched the first production series of GPS satellites. This effort was the initial step in revolutionizing the way we find our location on earth. The 24-satellite system, consisting of 6 orbits of 4 satellites per orbit, was declared operational by the DoD in December 1993. A similar declaration was made by the FAA regarding civil operational status in February 1994. The basic GPS signal is currently being used as a primary means of navigation over oceanic areas and a supplemental navigation aid for en route and non-precision approaches. Basic GPS service cannot meet all worldwide civil aviation requirements, especially those for precision approaches and landings. Civil aviation requires augmentation to improve GPS accuracy, integrity, continuity, and availability. WAAS (Figure 1), is designed to meet navigation and landing needs down

0-7803-5086-3 /98/$10.00 ©1998 IEEE

to or near the lowest Category (CAT) I decision height of 200 feet. Where the WAAS is unable to meet these needs, LAAS is required. In addition, LAAS is required to meet the more stringent needs for CAT II/III services. Outside of the aforementioned requirements, the LAAS also provides an all-weather surface navigation capability.

LAAS (Figure 2) will replace existing instrument landing systems (ILS) and expand functionally equivalent service to new airports and runways. WAAS will replace existing en route systems and extend precision approach coverage for landing operations.

Figure 1. WAAS Architecture

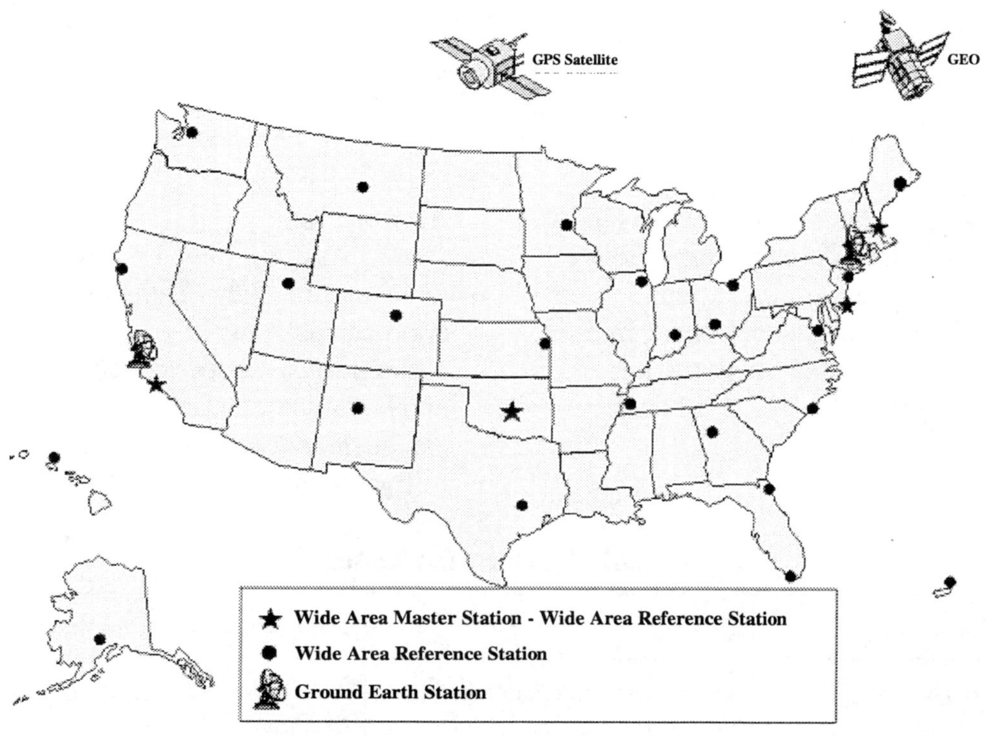

FAA Cost Benefit Analysis

This cost benefit analysis was performed for FAA under a GSA contract. Rannoch assembled an expert team composed of a WAAS/LAAS ground equipment manufacturer and a WAAS/LAAS avionics manufacturer. Additional information was gathered from the industry at large. The scope of Rannoch Corporation's WAAS/LAAS cost benefit analysis was to evaluate the incremental costs and benefits of using WAAS and LAAS to provide user services now delivered by other systems.[1] The FAA relies on analyses such as this to evaluate investment decisions (Figure 3)—thereby satisfying the investment analysis segment of its Acquisition Management System (AMS). Rannoch's corporate process for these analyses is outlined in Figure 4. Investment analysis is performed once FAA mission analyses have shown a shortcoming

[1] WAAS/LAAS Cost Benefit Analysis, CBA for a Combined WAAS/LAAS Architecture, Rannoch Corporation, GSA FEDSIM Contract No. GSOOK-96-AJD-1008, April 30, 1998.

Figure 2. LAAS Architecture

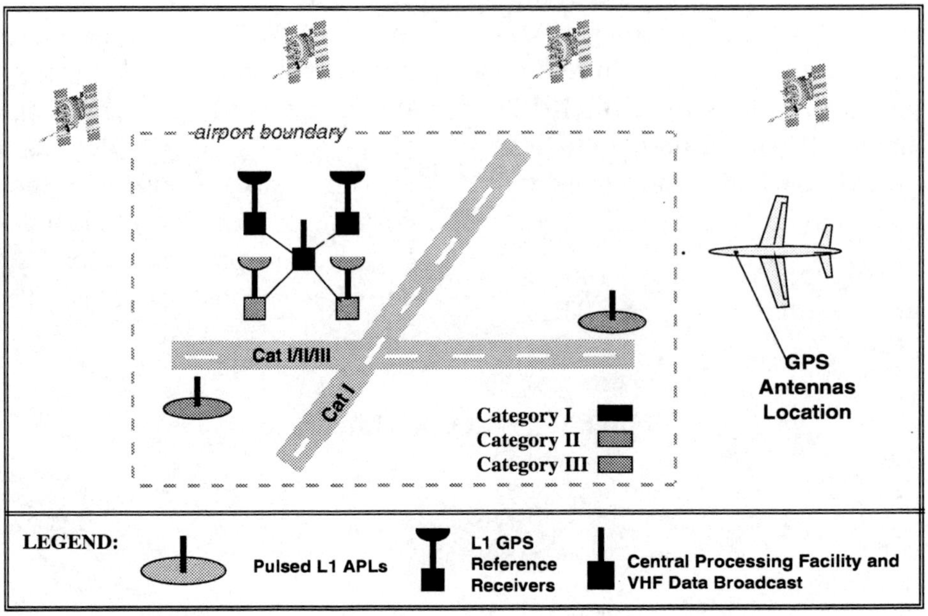

in agency assets and the ability to satisfy existing and future demand for services. The analysis is used to document system capability requirements; perform market research, identify alternatives, and to determine the cost, schedule and benefits of the system.[2]

Assumptions

Assumptions are key to the analysis. One of the key elements to Rannoch's approach is to gain early agreement of the assumptions, scope of costs and benefits, and implementation schedule. Early establishment of these foundation elements mitigates the risk of costly revisions toward the end of the analysis process.

Table 1. Aircraft Classes

Class	Description	Representative Examples
Air Carrier	Turbine aircraft operating under FAA Part 121 rules.	Airbus, Boeing, Lockheed, McDonnell Douglas
Air Taxi/Commuter /High-End General Aviation	Aircraft operating under FAA Part 135. High end GA aircraft (Part 91) were limited to turbines.	Beechcraft, British Aerospace, Cessna, Dornier, Embraer, Falcon Jet, Gulfstream
Mid to Low-End General Aviation	Non-turbine aircraft (with electrical systems) operating under part 91.	Cessna, Beechcraft, Mooney, Piper, Socata

It is also important to manage the inclusion of assumptions so as to arrive at a

2 Federal Aviation Administration Acquisition Management System, Federal Aviation Administration, April 1, 1996.

Figure 3. FAA Acquisition Process

defensible analysis. Benefit assumptions include those relating to transition/timing and those corresponding to efficiency. Transition/timing benefits are a result of converting from existing systems to WAAS/LAAS. These benefits include not having to support the current ground system infrastructure (cost avoidance). Efficiency benefits are those incremental benefits made possible by WAAS/LAAS direct flight path capabilities.

Cost assumptions are grouped into three general categories: ground equipment, avionics, and the space segment. Ground equipment cost assumptions are directly related to the transition and implementation schedules of WAAS/LAAS. Avionics assumptions were driven by the size of the fleet within three aircraft classes (Table 1), and the implementation schedules summarized in Table 2.

Data and Data Sources

Specifically contributing to the credibility of the cost and benefit values, was the selection of data sources. Rannoch performed the analysis with several team members. Involved were manufacturers of ground and avionics systems for WAAS and LAAS. As likely manufacturers of WAAS and LAAS avionics and ground systems, these participants were in the best position to supply life cycle acquisition and operations costs.

WAAS and LAAS received credit for benefits when they were to be deployed and in service. WAAS and LAAS incremental costs were positive while redundant systems remained in place to provide the same services, then negative as WAAS and LAAS replace existing systems.

User Costs

The basic method of costing the avionics equipment was to obtain price quotes from potential vendors for their actual planned products. That way, the vendor took on the task of estimating the market size, their share of it, and to amortize non-recurring costs. In this way, system pricing included the non-recurring as well as the per-unit recurring portion of the equipment costs and profit. Most avionics

Table 2. LAAS/WAAS Avionics Equipage Assumptions

Air Carrier	Will equip with WAAS/LAAS over the period 1998-2007; 100% of the aircraft will be equipped with CAT II operational capability in 2007.
Air Carrier	Will equip with LAAS over the period 1998-2007; 50% of the aircraft will be equipped with CAT III operational capability in 2007.
Air Taxi/ Commuter/High End General Aviation	Will equip with WAAS/LAAS over the period 1998-2007; 100% of the aircraft will be equipped with CAT I operational capability in 2007.
Air Taxi/ Commuter/High-End General Aviation	Will equip with LAAS over the period 1998-2007; 7% of the total aircraft will be equipped with CAT II operational capability in 2007 (based on 25% of the air taxi aircraft equipping with CAT II).
Mid/Low-End General Aviation	Will equip with WAAS only over the period 1998-2011; 80% of the aircraft will be equipped with CAT I operational capability in 2011 (similar to today's fleet with 20% not equipped with basic navigation equipment).

vendors plan to incorporate WAAS and LAAS capability in a single integrated box such as an MMR (Multi-Mode Receiver) or card with a single price. Based on information from the manufacturers, it was assumed that WAAS precision-approach and LAAS capability will appear on the market in the same time frame.

Avionics equipment costs for each of the four aircraft classes were affected by the per aircraft avionics cost for LAAS and WAAS, fleet size, and the timing for conversion from ILS to GPS. It was assumed that in a LAAS environment air carriers would not equip solely with CAT I capability and mid/low-end general aviation will not equip at all with LAAS. It was also assumed that no air taxi/commuter/high-end general aviation users will equip for CAT III.

MMR air carriers and air taxi/commuter/high-end general aviation were to equip with both LAAS and WAAS beginning in 1998. Their respective fleets would be entirely retrofitted to LAAS/WAAS capability over a 10-year period ending in 2007. Avionics costs beyond 2007 accounted for projected growth in the fleet. Mid/low end general aviation would not equip with LAAS. WAAS equipage was to begin in 1998 in new aircraft and in 2007 as retrofits, as ILS began to be decommissioned. This class of aircraft was to be completely retrofitted with WAAS by 2011 once ILS was completely decommissioned.

Provider Costs

Costs for LAAS and WAAS ground systems were categorized into four major groupings: acquisition, operations and maintenance (O&M), program/system engineering and decommissioning costs related to ILS/VOR/NDB. Acquisition costs included those costs associated with government and contractor system development, purchasing the equipment and the installation and commissioning of that equipment. LAAS acquisition costs were based on information provided by a potential equipment supplier. WAAS acquisition costs were composed of anticipated contractor costs as part of a current FAA WAAS contract, plus those costs associated with the FAA's settlement with the original contractor.

In general terms, O&M costs included: hardware life cycle maintenance, software life cycle maintenance, and replacement costs. These costs include elements such as hardware and software maintenance and hardware replacement every five years. For WAAS, O&M costs included the same general elements as LAAS. WAAS O&M costs included government operations, various satellite communication support costs, and terrestrial communication support.

Program/system engineering costs for LAAS included various government operations such as facilities and engineering, miscellaneous O&M costs, logistics and flight inspection. WAAS program/system engineering costs included FAA facilities and engineering costs such as NAS implementation, leasing terrestrial communications, orbit determination and satellite support.

Decommissioning costs were based exclusively on a combined LAAS/WAAS system where all ILS (CAT I, II, and III), VOR and NDB's are removed from service. The costs associated with deactivating these navaids were classified as decommissioning costs.

User Benefits

Aircraft operators and their passengers were collectively referred to as the users of NAS. Inclement weather disrupts aircraft operations, both landings and departures. Disruptions to landing operations are especially critical for both safety and efficiency. Precision approach capabilities provided by WAAS and LAAS reduce the number and severity of disruptions that would otherwise be encountered by users. Efficiency benefits were found as a result of reduced disruptions to flight operations. Disruption occurs when there is an interruption to a planned or optimum flight path. It consists of delay, diversion or cancellation.

The benefits computed for WAAS and LAAS included benefits for all flight disruption savings starting the year after WAAS or LAAS ground services were to have been installed, certified, and authorized for operations. Efficiency benefits were computed based on the number of flights that encounter weather that would have disrupted their approach if they had not been equipped to continue their landing operation (to a compatibly equipped airport). Computations were split into two steps: computing the number of eligible flights affected and applying the cost savings factors that pertain to each type of user. The consequences for affected flights were produced by weighting cancellations, diversions, and delays by the expected percentage of occurrence over a large sample of disruption events.

The analysis involved annual computations of flights and passengers affected by potential disruptions. Disruptions have costs for both aircraft operators and passengers. Flight efficiency benefits were reported for four types of users (air carriers, air taxi/commuters, general aviation, and passengers).

A WAAS/LAAS system enables aircraft to fly a more direct flight path than the current navigation systems. For the en-route phase of flight it was assumed that the routes would be restructured, resulting in more direct paths. The resulting savings were estimated and assumed to be an average of one minute of flight time. For the approach phase of flight, the current ILS requires approximately five miles of stable, straight-in controlled flight prior to landing.

LAAS can allow both shorter turns to final approach and curved approaches.

Figure 4. Rannoch Cost-Benefit Analysis Process

Process Elements	***Rannoch Management Role***	***Study Risk Areas***
DEFINE OBJECTIVE · Cost Benefit Analysis · User Investment Analysis	· Attempt to manage/calibrate client and user expectations for a defensible (not necessarily conservative) study, where all of the benefits in the numerator are individually defensible.	· Reaching for unlikely benefits will undermine study credibility.
DEFINE SCOPE · Operational Assumptions · Specify System · Define Target Environment · Identify Effect Mechanisms · Identify Data Sources · Define Impact on Environment	· Assess level of data availability. · Identify areas where additional data is required. · Develop brief plan/cost estimate with alternatives showing cost vs. data fidelity. · Gain acceptance of study assumptions with key organizations and study stockholders through participation. 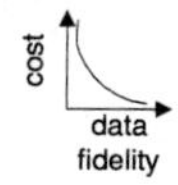	· Data gaps impact accuracy of study. · Safety systems present the risk that conventional calculations of benefits, quanitified in dollar terms, will not exceed costs. · Correct definition and agreement on major study assumptions at early stage is key. Late changes often require more extensive rework.
PERFORM DEMAND/ DEPLOYMENT ANALYSIS · Specify Time Horizon · Clarify Economic Incentives · Specify Detailed Assumptions · Allow for Uncertain Implementation · Realistic De-Commissioning	· Ensure that study addresses the issues for each target group. · Service providers like to see the overall costs and benefits. · Users want to see benefits and costs, such as a reduction in direct operating costs and life cycle cost of ownership of a new system.	· In many studies, results are presented at a macro-economic level. This has the risk of not being meaningful, understandable or useful to the target audience. The result is a shelved study that is not given appropriate consideration. · Demand uncertainty is usually a high-risk area.
COST/BENEFIT ANALYSIS · Review Data Sources · Develop Computational Models	· Ensure benefits are exclusive to those systems under evaluation. · To ensure that correct system acquisition cost data is used, industry should provide or review system cost estimates.	· Avoid capturing benefits attributable to other systems. · There are risks associated with the fidelity of cost information, particularly system acquisition costs.
INTEGRATE RESULTS AND SENSITIVITY ANALYSIS · Draft Numerical Results · Cross-Validate to Other Studies · Submit to Expert Peer Review	· Be an "honest broker" of study information. · Some scenarios may have low ratios, especially those with unfavorable user equipage scenarios. 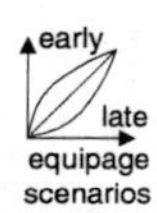	· Study credibility can be undermined by emphasizing only the most favorably results.
PREPARE REPORT · Briefings · Draft · Internal Review · Sponsor Review · Sensitivity Analyses · Recommendations · Net Present Value	· Consider presenting combined benefit cost analysis, such as those cases where systems are complemented (e.g. LAAS/WAAS). This ensures that the same benefits are not double-counted. In the final report, all costs and benefits should be traceable to their appropriate sources.	· The risk at this stage is in losing the reader by presenting results in a macro sense, e.g. benefits in $ billions per year.

In the most likely case, it was assumed that final approach could be shortened by approximately two miles or one minute on final approach.

Safety benefits would be realized from a reduction in the number of aircraft landing accidents. This results from the availability of WAAS at runways that currently do not have an ILS installed. Most of the benefits resulted from the use of WAAS for precision approaches at airports without ILS on any runways. At airports where a single ILS is already installed, a small benefit was realized due to precision approaches being available at all runway ends. Therefore all of these safety benefits were due to WAAS. LAAS provided no benefit because it does not improve safety over existing ILSs.

These numbers were the basis for computations in current year, or constant year dollars. Discounting factors applied to these annual numbers, according to Office of Management and Budget guidelines, provided the basis for conversion into present value dollars.

Provider Benefits

The implementation of a combined LAAS/WAAS navigation system creates the potential to avoid certain future costs. Specifically, ground system support costs pertaining to the current NAS system including ILS, VOR and NDB can be avoided. As these systems are decommissioned, anticipated replacement and O&M costs for the decommissioned sites is realized as a distinct benefit. All current navaids (ILS, VOR, DME, TACAN, NDB and LORAN C) will be decommissioned by 2010.

This number was calculated by first determining the per aircraft savings of avionics equipage of LAAS/WAAS over ILS/NAV. The per aircraft savings was then multiplied by the number of aircraft entering the respective segment of the U.S. fleet beginning in 2011—the year ILS and VOR are to be phased out. Dual equipage is no longer required after that date. There is no credit given for cost avoidance during the period of dual equipage. Air carriers would eventually benefit the most as GLS systems are the standard equipage (and much lower cost than current avionics) when they become available in 2012.

Conclusions

It is very difficult to "sell" safety with many aviation programs. Improved safety is substantial due to additional vertical guidance, but it is small in comparison to other benefit categories. The majority of the benefits impacted users over the provider. The benefit category contributing the most to WAAS/LAAS benefits was direct flight paths. This benefit was due to the enabling of a more efficient route structure (both en-route and terminal area). The safety benefit resulted from the availability of precision approach with WAAS at runways that are currently without a precision approach capability.

This cost-benefit analysis determined that the WAAS/LAAS implementation costs were roughly equally divided between the provider (FAA) and users. On the provider side, the greatest costs would be the program and system engineering efforts for the ground systems, followed by the acquisition of said systems. For users, large acquisition costs and maintenance of these new LAAS/WAAS avionics are the price for reaping large efficiency benefits. Decommissioning costs and avionics certification are comparatively less significant costs for providers and users, respectively.

B. Systems Engineering Track

Modern market forces, whether defense, space, or automotive, are driving a systematic optimization of the development process. The study and implementation of these best practices embodies systems engineering and hence is the key to many organizations future. This track explores a number of recent developments and offers several tutorials to augment understanding.

Track Chair: Doug Chappelle, Boeing

Session B1: Processes & Methods

Chair: Leo Little, Sverdrup Technology, Inc.

1	The House of CE+ - A Systemic Framework for the Proper Selection of Practices	Mr. Ernst Fricke
2	Drawing Roadmaps To The Future: The Need For An Integrated Systems Engineering Approach	Dr. Ir. Eric Theunissen
3	Process Mapping Of Concurrent Engineering Processes For Integrated Systems	Mr. Ernst Fricke
4	Towards a Logical Basis for Systems Engineering	Manfred Broy
5	Model-Based Systems Engineering Of Automotive Systems	Mr. Jim Long
6	Streamlining the System Engineering Process for Developing Vehicle Management System Software for Air, Ground, and Space Applications (Not Published)	Mr. Peter Ross
7	The Impact of Acquisition Reform on System Engineering: Selected Examples (Not Published)	Mr. Mel M. Cutler

Session B2: Systems Architecture

Chair: Stephen Kuehle

1	Tactical Aircraft Moving Map Cabability (TAMMAC)	Mr. Ervin Pagel
2	Managing Multi-Platform Sensor Systems	Mr. Kurt Himmelreich
3	An Interactive Tool Suite for Real-time Application Development on Heterogeneous High Performance Computing Platforms	Dr. Minesh Patel Ph.D.
4	The Generic Unified Systems Engineering Metamodel	Mr. George Caple
5	Architecting Mobile Command and Control Communications	Mr. Jeff Lankford

Session B3: Safety, Certification & Standards

Chair: Sharon Crockett, NASA Langley Research Center

1	Current Government and Industry Developments in the Area of System Safety Assessment	Mr. Brett Portwood
2	Overview of EIA 632: Processes for Engineering a System	Mr. James N. Martin
3	Implementation of EIA-632 - Process Compliance or Imbedded Infrastructure?	Mr. Richard M. Harwell
4	An Overview of the Systems Engineering Capability Model EIA/IS 731	Mr. Don E. Barber
5	Comparison Of The Number Of 'Shalls' In Quality System Standards	Mr. William Walker
6	Low-cost Integrated Display Certification	Uma Satyen

Track B Systems Engineering (Continued)

Session B4: Test & Simulation

Chair: Frank Doehrenberg, Allied Signal

Session B5: System Tools

Chair: Jeff Lankford, Northrup Grumman Corp.

THE HOUSE OF CE+ — A SYSTEMATIC FRAMEWORK FOR SELECTING THE "RIGHT" PRACTICES

Herbert Negele, Ernst Fricke, and Stefan Wenzel
Fachgebiet Raumfahrttechnik (Institute of Astronautics), Technische Universität München
Boltzmannstr. 15, 85748 Garching, Germany

Abstract

To be successful in a highly dynamic environment, with increased global competition, shorter life cycles, more demanding customers/users, and more complex products, different industries should take advantage of learning from each other's problems and solutions. For example, practices like QFD, Taguchi, CAX, Robust Design, or DFX that are applied (successfully?) in daily practice within automotive industries, could be used more efficiently in the space domain. On the other hand, commercial companies can learn from aerospace industries' traditional strengths, in the areas of requirements engineering, modeling and simulation, or systems engineering. A variety of potential synergies exists that should be exploited to the greatest possible extent. But what method or tool should we use, and when?

For an effective application of methods and tools in product development, a comprehensive framework is needed, which supports the selection, coordination, and assessment according to domain- and company-specific needs and constraints. Unfortunately, tools and methods often are used because it is popular to use them, but not because they ideally support the solution of specific problems. Therefore, in this paper a systematic approach is proposed on how methods and tools can be selected for a successful product development system, independent of the industrial sector of a company, but dependent on its problems, market constraints, and declared goals.

0-7803-5086-3 /98/$10.00 ©1998 IEEE

A structured model, the House of CE+ (CE+ means concurrent engineering and more), was derived from the analysis of different product development processes at commercial companies in Germany, England, and the US (Wenzel & Bauch 1996). In the House of CE+, the interdependencies and relations between goals and problems, essentials/ processes, and practices can be structured. It is possible to visualize in a matrix, how and to what extent the practices (HOWs) contribute to the realization of certain essentials (WHATs), and which essentials support which goals or solve which problems (WHYs).

1. Introduction

All kind of markets have been changing quickly in the last years, including space, automotive, and other markets. This trend is influenced by political situations, as well as global interrelation of national markets, and increasing competition. Each market sector has its own combination of problems and constraints, which are assumed to be unique to that market. Nevertheless, due to ongoing changes in these markets, problems formerly typical for the space sector become also typical for the automotive sector and strengths needed to survive in the automotive market becomes neccessary for space companies. A successful engineering of systems for these markets requires effective and efficient solutions and overcoming specific problems and constraints.

Comparing space and automotive industries, they have two major features in common: both develop complex and hybrid products and they are integrators of subsystems drawing on a broad range of technologies (ISU 1997). But space and automotive markets differ in many aspects, resulting in differing market problems and constraints. For example, typical for the automotive industry is a highly competitive, purely commercial, global market with an immense number of different customers. Therefore, the automotive industry developed processes, methods, and tools in order to know its markets and customers, and their needs. The automotive industry generally knows WHAT to do to meet "fuzzy" customer needs and also HOW to do it. Now, with the increasing commercialization of space markets, space companies have to tackle problems that the automotive industry is already used to.

On the other hand, cars are getting more and more complex and hybridized. The number of electrical and electronic parts in a car is increasing and will increase much more in the future, especially in the areas of guidance assistance, entertainment systems, electronical control of engine or brakes, and other safety related functions. Automotive development is following a trend here akin to the direction of satellite development, which was also experienced by airplane development already (Gartz 1997). Managing this increasing complexity and hybridization and the related processes is a problem where automotive industry may learn from space industry.

2. Solutions, Methods and Tools

2.1 Overcoming Problems with the Right Solutions

A wide variety of different methods and tools can be found in automotive as well as in space industries. They are supposed to support the achievement of the objectives and to provide solutions for the problems of the respective development processes. Methods describe **HOW** we should do what we are doing and tools are means (mostly, but not necessarily, computer-based) that should help us to accomplish the HOWs faster, better, and cheaper. Therefore, a tool without an adequate method and environment is rarely of much help, whereas methods can't be applied very efficiently without adequate tools. An example is, a method to enhance the product development process (or specific parts of it) could be to conduct product assessments with the help of proper metrics. A tool that collects and updates all relevant data needed to quantify the metrics could be used to implement the method more efficiently.

The following extract of a collection of more than 50 methods and (best) practices (Wenzel & Bauch 1996) shows the variety of solutions applied in product development. Depending on the specific problems, constraints, goals, essentials, processes, and environment the right mix of these (and other) methods/ tools has to be chosen und implemented.

- Agreeable Environment
- Common Data Base
- Computer Aided "X" (CAX)
- Collocation (physical or virtual)
- Customer-Satisfaction-Index
- Design for "X" (DFX)
- Design of Experiments
- Failure Mode and Effects Analysis
- House of CE+
- Integrated Product Teams
- KANO Model
- Metrics
- Multi-Project Resource Plan
- Process Modelling and Simulation
- Product Value Matrix
- Quality Function Deployment
- Rapid Prototyping
- System Modelling and Simulation
- Taguchi Methods
- Target Costing
- TRIZ
- ZOPH-Model

2.2 Inefficient Use of Methods and Tools

Unfortunately, the best methods and tools lose impact when they are not used effectively and efficiently. This insufficiency is generally a result of the usage of *too many*, *wrong* or *not embedded and adjusted* means (Wenzel & Bauch 1996). On the other hand, the benefits connected with new methods and tools, such as machines, computers, QFD, CAD, or Rapid Prototyping are usually multitudinous. They are often seen as the universal remedy for all kind of difficulties. Benefits expected are for example:

- The "reduction" of complexity
- Cost and time savings, higher productivity
- Better quality
- Reproducibility
- Higher flexibility
- Higher accuracy
- Early verification

But the expected benefits usually do not come true without an early investment of time, money and resources (Figure 1). The pay back for these new means is similar to new products. Money has to be invested to buy new means or the appropriate knowledge. But with the acquisition of the tools, only the first step is done. The means have to be adapted to the specific requirements of the users and people have to get acquainted with new processes and rules. This takes mental setup time and attention. This time is often not honestly considered, rather by "sunshine-commitments". The maximum benefit is highly connected with the rate of acquaintance. The more the users are familiar with the methods and tools, the higher is acceptance and efficiency, and the lower are mental barriers and setup times.

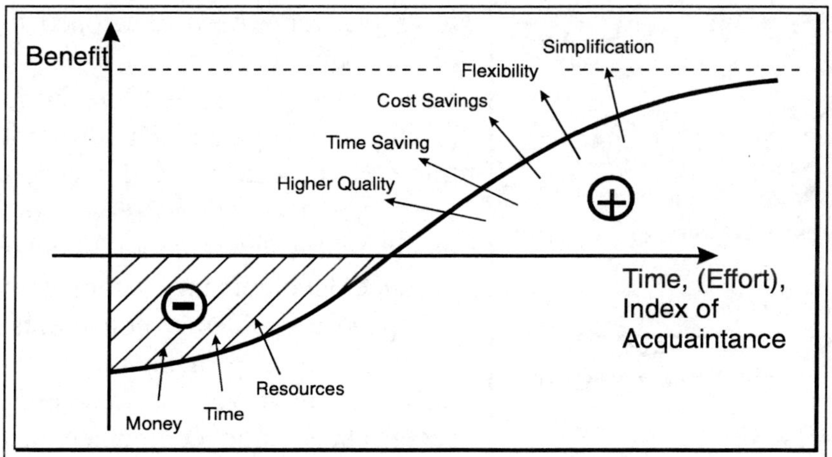

Figure 1: Tools & Methods: Investment and Benefit

The problem can be clearly illustrated with an example many people have to deal with privately or at work. It's the difficulty we have with the continuous upgrading of application programs for personal computers. Some perceived benefits associated with a new word processing program are: better user-ergonomics ('just a mouse click'), faster processing, better connectivity to other applications, and more features (e.g. integrated graphics program). But, in order to make a profit from these improvements, the user has to upgrade to a more powerful system and invest money for buying the program. Before (s)he is able to improve the processing time, (s)he has to install the program and invest learning-time to get acquainted with the new program. Thus, there is a setback in efficiency, although *this aspect is to be improved.* This kind of paradox is qualitatively visualized in Figure 2. After a learning- or acquaintance-time, depending on previous experience and the need to learn, the investment pays back and overall efficiency improves.

If the capacity of an individual or an organization is assumed to be constant, the total amount of its activities can be categorized (Figure 2):

- "Effective Performance", which supplies the required results (e.g. creation of a drawing),
- "Supportive Activities", which are necessary to achieve the effective performance (e.g. looking through current drawings on micro film), and
- "Blind Performance", which is useless for the fulfillment of the tasks (e.g. waiting for drawings until they are mailed).

New means typically aim at improving the ratio between these kind of activities. But, with the introduction of a new method or tool, the blind performance does not necessarily decline at the beginning. Rather, the effort to get acquainted with the new processes reduces the effective performance and increases the support activities. After a certain time the users get more and more familiar with the new tools and the ratio between effective and blind performance improves: The "Break Even Point" is reached when the effective performance is on the former level. From now on, the tools and methods reduce the blind-performance up to a certain degree of saturation (Figure 2).

Often the actual benefits are not as drastic as expected, as people are reluctant or get tired of methods/tools. Reasons for these obstacles are:

- **Wrong methods and tools:** Methods and tools are not universal remedies though they often are promoted as such. They only can serve as a support to better carry out the right processes in the proper environment.
- **Competing and overlapping**: In many companies several different means are applied to similar tasks. This does not have to be a drawback necessarily, since it can enhance flexibility and reinforce decision-making. On the other hand, it can lead to competition between organizational units using and promoting different methods and tools, and result in reduced efficiency.
- **Not embedded, adjusted, and integrated:** Means should be learned by doing. Consequently, they should be adjusted to the specific situation and environment of the company, in order to avoid unnecessary ballast. Another problem, especially with computer-based tools, is the lack of integration and interoperability, e.g. data exchange between different tools is not possible.
- **Wrong environment:** In many companies the introduction of new methods and tools is seen as a major threat to people who will have to change some of their habits. This kind of antagonistic atmosphere kills many promising attempts to introduce new solutions.
- **Not user-oriented:** Supporting means often are developed without closely involving the later users of these means. Therefore, they tend to be too theoretical, sophisticated, or difficult to learn and use.
- **Not accepted:** Methods and tools which are not accepted will not be used. They need to be promoted by management (but adjusted by the user) and embedded into a supportive framework, such as processes and rules.
- **Too many:** A host of tools and methods means too much effort for learning, supporting, and maintaining all the different means. Less is more and a few well-known and utilized means are certainly better than a lot of unused.

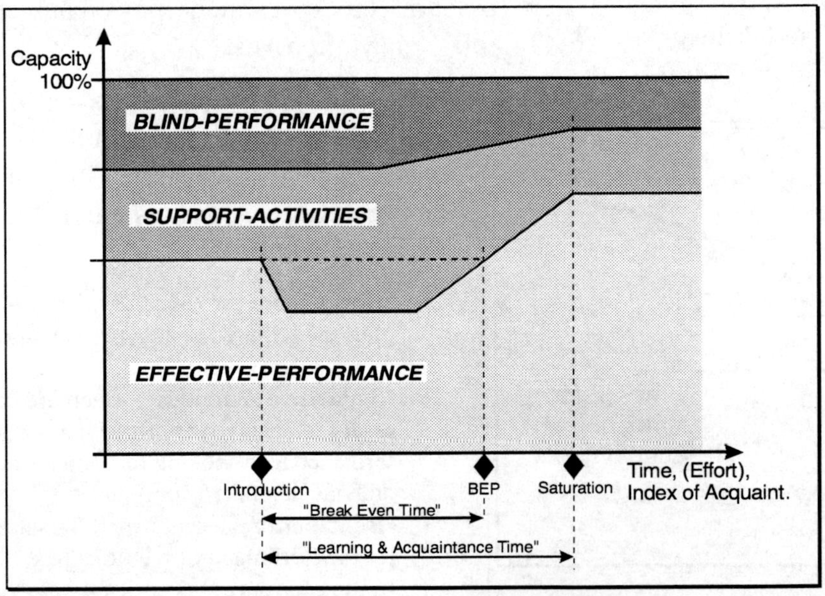

Figure 2: Setup Time for New Tools and Methods (qualitative)

3. The House of CE+

There are a lot of different solutions, practices, methods, and tools that have been developed to overcome problems and to enhance effectivity and efficiency in product development. The difficult thing is to select and implement the *right mix of solutions* and means that fit best into a particular product development system with its specific problems, constraints, goals, processes, organization, products, and environment. It was observed that some companies just saw that certain methods and /or tools (e.g., a specific simulation tool) were successfully applied in other companies or departments, and then thought using these methods/tools would solve their problems as well. Just taking over solutions from others without looking at the special situation and adapting them to specific needs rarely will yield great results. In order to support a systematic selection and implementation of the right solutions and practices the **House of CE+** is proposed.

3.1 Fundamental Structure

The House of CE+ (Concurrent Engineering and more) was developed as a methodical approach to structure the interdependencies and relations between goals, problems, essentials, methods, and tools for the specific use in practice (Wenzel et al. 1997, Negele et al. 1997, Negele & Fricke 1997). In analogy to the House of Quality (Hauser & Clausing 1988), it is possible to visualize in a matrix, how and to what extent, the practices, methods, and tools (HOWs) contribute to the realization of certain essentials (WHATs) and which essentials are related to certain goals or problems (WHYs). Figure 3 provides an impression of how this matrix looks like in principle and gives examples for possible relationships. In the matrix, HOWs are connected with WHATs by a small spanner symbol, whereas WHYs and WHATs are interrelated with dots.

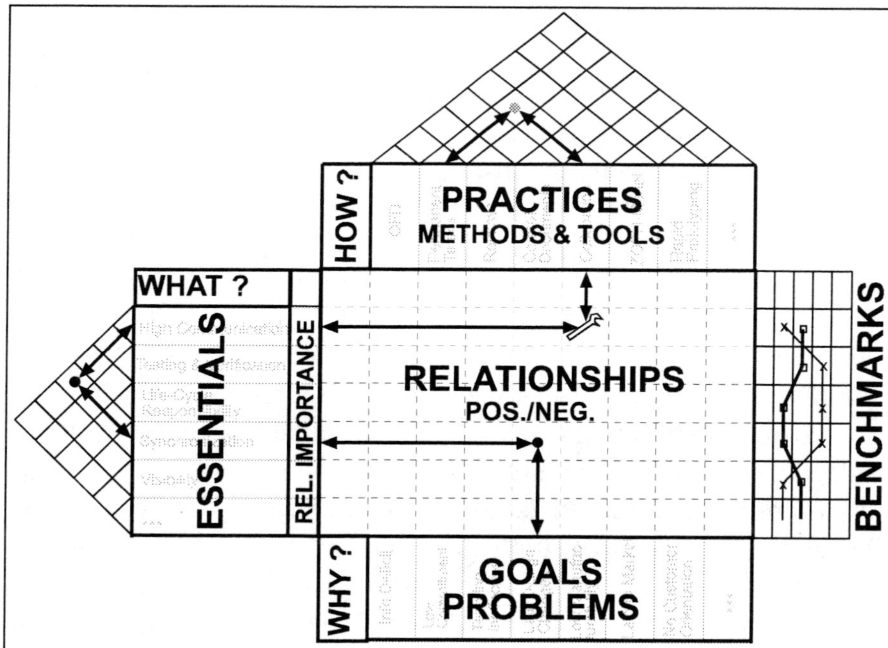

Figure 3. Structure of the House of CE+

Depending on industry or line of business, adequate essentials or processes can be defined and assigned a relative importance for the realization of the goals set, or the elimination of the problems occurring during the development process. After the definition of the most important essentials, proactive or corrective action can be initiated by the implementation of selected practices. These practices enable the essentials and should be "hands-on" methods or tools for the use in day-to-day development work. Also, the specific environment in which practices are to be implemented, with associated constraints have to be regarded (e.g., tools already used, available resources, software and hardware issues, skills of concerned people).

In order to visualize how strong (e.g., high–medium–low) each essential and practice affects another (positive/negative), they should also be related among themselves. This aspect is exemplarily shown in the roof of the house and in the triangle on the left. In a similar way this can be done for problems (problem net), goals (goal net), or problems versus goals in the "basement" It is important to carry out these steps, as the combination of certain essentials or practices can promote or sometimes hinder the impact expected. Thus, a specific mix of the "right" practices can be put together for each project. In order to improve the development system continuously, its performance can be measured along with the fulfillent of the most important essentials (Negele et al. 1997). This benchmark is placed on the right side of the house.

3.2 WHATs – Essentials for Successful Product Development

Companies have to excel in several key aspects — Quality and Cost to market, Effectiveness and Efficiency, Time to market and Flexibility — if they want to be and remain economically successful. In order to reach this objective a set of 17 **essentials** has been defined that describe basic success factors (**WHATs**) for new product development (Wenzel & Bauch 1996). The essentials are quite neutral and they should guide the user rather independently from the product, company or environment. The essentials are an "answer" to the goals and problems typically found in today's product development environments (Clark & Fujimoto 1991, Wheelright & Clark 1992, Clausing 1994, Negele & Fricke 1997). With the realization of the essentials, the crucial aspects of Concurrent Engineering are taken into account. The essentials are listed below. A short description of each essential is given in Wenzel et al. (1997) or Negele et al. (1997):

- **High Communication Density**
- **Testing & Verification**
- **Big Loops: "Lessons learned"**
- **Education & Training**
- **Life Cycle Responsibility**
- **2D-Integration**
- **Synchronization**
- **Clear Goals, Rules and Roles**
- **Early and Fast Check of Profitability and Risk**
- **Cost, Time and Resource Visibility**
- **Metrics in the Loop**
- **Self-accountable, cross-functional Product Teams**
- **Technology Carrier**
- **Balancing Technology and Product Development**
- **Clear Supplier Strategy**
- **Robust Design and Customization**
- **Customer Focus**

3.3 Further Potentials

The essentials listed in Sector 3.2 are one attempt to describe the WHATs for successful product development. But, the generic structure of the House of CE+ can be applied to varying fields, e.g. business processes other than product development, like production or maintenance, and at different levels, e.g. company-level, project-level, or subproject-level. With adjusted essentials (processes) the expansible framework also can be employed for the "big picture" Systems Engineering or Project Management ("House of SE" or "House of PM") as well as for a more focused view at specific aspects, e.g. Change Management. For a "House of SE", e.g. the 19 different focus areas of the Systems Engineering Capability Model SECM (EIA/IS 731-1 SECM) could be used as WHATs.

They are organized in three categories (Widmann 1998): *Systems Engineering Technical Category* (Define Stakeholder and System Level Requirements, Define Technical Problem, Define Solution, Assess and Select, Integrate System, Verify System, Validate System), *Systems Engineering Management Category* (Plan and Organize, Monitor and Control, Integrate Disciplines, Coordinate with Suppliers, Manage Risk, Manage Data, Manage Configurations, Ensure Quality), and *Systems Engineering Environment Category* (Define and Improve the Systems Engineering Process, Manage Competency, Manage Technology, Manage Systems Engineering Support Environment). These focus areas can be mapped to problems and goals concerning Systems Engineering, and to methods and tools that support them.

Similar to the QFD method, several linked matrices or houses can be used to make visualization and application easier (Figure 4). Furthermore, a time dimension can be added. In different phases of, e.g., product development different problems/goals are dominant and corresponding methods and tools will have to be used. Specific houses can be established for the main phases of the product development process, thus focusing process capability assessment. To avoid dozens of different matrices, all information can be kept in a software tool that provides criteria (hierarchy level, phase, focused process, etc.) for flexibly filtering out and visualizing the matrices and information one

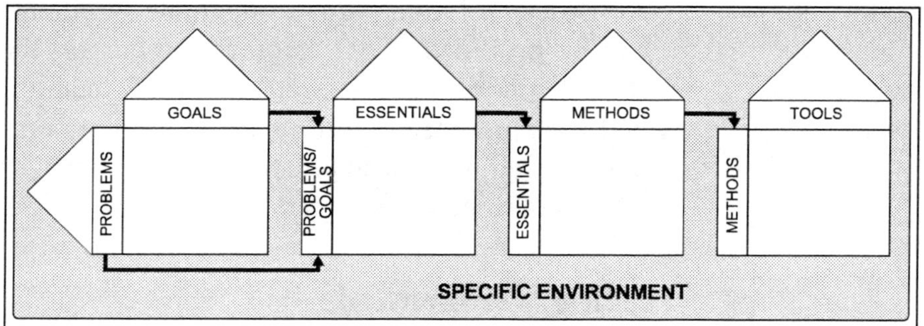

Figure 4. Linked Houses

needs.

4. Application of the House of CE+

The practical application of the House of CE+ is illustrated by means of a simplified example. For an easier understanding the information contained in the example House of CE+ is split up into two steps (Figures 5 and 6). In the basement of the House in Figure 5 several problems found in a given product development are listed. In our example, two of the main issues were:

- Over-engineering, leading to increased product costs,
- Insufficient customer satisfaction, leading to lower sales volumes than planned.

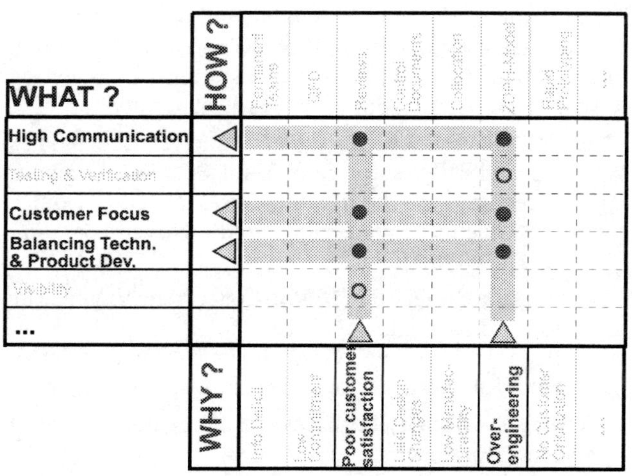

Figure 5: Step 1 – Tracing WHY-WHAT-Relations

These problems provide the WHYs for corrective action. By tracing the WHY-WHAT-relations (dots) in Figure 5 we can find the areas that should be focused on in order to improve the development system, from the list of WHATs. Here, this trace would lead to (using the 17 essentials listed above):

- High communication
- Balancing technology and product development
- Customer focus
- ...

Knowing WHAT has to be done in order to overcome these problems, in a second step, the HOWs can be identified. This includes the practices, methods, and tools that support an practical implementation of the related WHATs.

This second step is shown in Figure 6. By tracing the WHAT-HOW-relations (spanners) we find a practice or method that supports all of the identified essentials is Quality Function Deployment (QFD). There are, of course, many other possibilities to overcome poor customer satisfaction and over-engineering, but here we just want to explain the application of the House of CE+ in principle. Also, in addition, the specific environment has to be taken into consideration, e.g., which organizational structures already exist in the company, which methods and tools are already used, which information technology (hardware and software) is in use or available, human and political factors, and others.

In daily practice, with the House of CE+ a project manager is able to filter from the "complete" row of practices the specific solutions which best solve the problems with regard to a balance of effort/cost and benefit. Furthermore, it is not sufficient to just select and implement any or all of the practices found in related columns. Mutual dependencies and impacts of methods and tools have to be considered in order to get a coordinated mix of practices. For example, collocation and permanent teams have a positive impact on each other.

Hence, it would be recommendable to implement both at the same time. This coordination of methods and tools is supported by the information contained in the "roof" of the House of CE+ (see Figure 3).

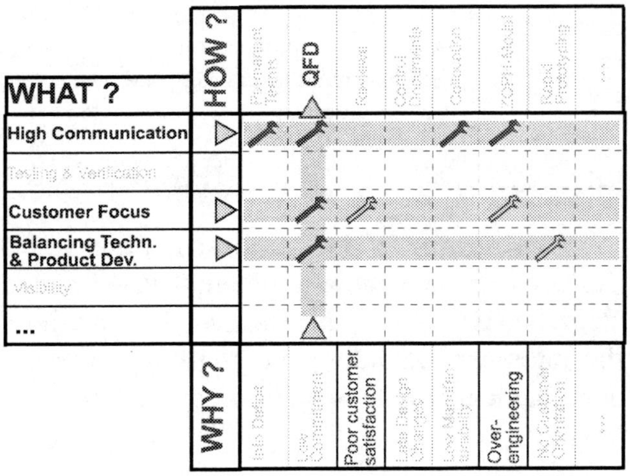

Figure 6: Step 2 – Tracing WHAT-HOW-relations

Conclusions

It was shown that many different problems can hinder an effective and efficient usage of practices, methods, and tools in system development. In order to successfully support product development, it is necessary to have appropriate answers to these problems and other given constraints. These answers have to show **what** to do and **how** to do it. Therefore, the House of CE+, a comprehensive and systematic approach was presented to enable the structuring and tracing of interdependencies and relations between the problems and goals of a company (or for a certain process, etc.), necessary essentials which show what to do, and possible methods and tools which support these essentials. Applying this approach helps to find the right mix of methods and tools for a company's specific mix of problems and goals.

In order to identify synergies and potential for learning from each other, different Houses can be compared for different industry sectors, e.g., automotive and space industry. Thus, a systematic approach to transferring well-tried solutions for comparable problems can be provided. Next steps will focus on the integration of cost-effectiveness of methods and tools into the House of CE+. Also, a computer-based tool implementing the House of CE+ is currently developed.

Acknowledgements

We would like to thank Valerie Gundrum for proof-reading and her valuable comments.

References

Clark, K. B.; Fujimoto, T. (1991): Product Development Performance: Strategy, Organization, and Management in the World Auto Industry, Harvard Business School Press, Boston, Mass.

Clausing, D. (1994): Total Quality Development: A Step-By-Step Guide to World-Class Concurrent Engineering, ASME Press.

Gartz, P.E. (1997): Commercial Systems Development in a Changed World, IEEE Transactions on Aerospace and Electronic Systems, 33, 2, 632-636.

Hauser, J.R.; Clausing, D. (1988): The House of Quality, Harvard Business Review 66, No. 3, pp. 63-73.

ISU (1997): Technology Transfer-Bridging Space and Society, Design Project Report of the Summer Session Program '97, International Space University, Houston.

Negele, H.; Fricke, E.; Wenzel, S. (1997): The House of CE+ — A Systematic Approach To Comprehensive Engineering of Product Development Systems, Advances in Concurrent Engineering, Proceedings of CE97: Fourth International Conference On Concurrent Engineering: Research and Applications, Oakland University.

Negele, H.; Fricke, E. (1997): Interrelating WHYs, WHATs, and HOWs for Successful Systems Engineering. Proceedings of the 1st Joint ESA/INCOSE Conference on Systems Engineering, 11-13 November 1997, ESTEC, The Netherlands.

Wenzel, S.; Bauch, T. (1996): Concurrent Engineering and More – Essentials for Successful Product Development from International Case Studies, Masters Thesis. Fachgebiet Raumfahrttechnik. Techn. Univ. München.

Wenzel, S.; Bauch, T.; Fricke, E.; Negele, H. (1997): Concurrent Engineering and More – A Systematic Approach to Successful Product Development, Proceedings of the 7th International Symposium of INCOSE.

Wheelwright, S.; Clark, K. (1992): Revolutionizing Product Development – Quantum Leaps in Speed, Efficiency, and Quality; New York.

Widmann, E.R. (1998): Key Features of the "Merged" EIA/IS 731-1 Systems Engineering Capability Model. Proceedings of the 8th International Symposium of INCOSE.

DRAWING ROADMAPS TO THE FUTURE: THE NEED FOR AN INTEGRATED SYSTEMS ENGINEERING APPROACH

D. van Willigen, E. Theunissen, P.B. Ober,
R.C. Meijer, A.W.S. Helwig, and G.W.A Offermans

Delft University of Technology, P.O. Box 5031, 2600 GA Delft, The Netherlands

Introduction

Given the continuing growth of air transport, the demand for an increase in airspace capacity is likely to continue. The situation in the Netherlands clearly demonstrates that as a consequence of future increases in airspace capacity the environmental constraints will become more and more important. It also shows that if these new constraints are not dealt with in a serious way, illogical and sometimes unfair constraints are imposed on the airport and the airlines. An example is the limit in the Netherlands on the number of passengers (44 million) and cargo (3.3 million tons) imposed in 1995. In 1997 the Netherlands' Government decided that this limit will be lifted.

Although it is generally agreed that developments in the area of communication, navigation, and surveillance (CNS) combined with new operational procedures have the potential to allow a growth in air transport while at the same time reducing the relative burden on the environment, there is certainly no consensus on how this should be achieved. Tulder (1996) provides examples of the operational benefits which would result from smaller separation standards made possible by increased navigation performance and dependent surveillance. To be able to determine the potential to increase capacity from the possibilities offered by new technology, a roadmap must be defined which outlines how the capabilities will be exploited and also describes how the issues of safety, economy and environmental burden are influenced. Based on this roadmap, airlines can invest in systems to benefit from new infrastructures. Today's rapid advances in technology sometimes cause an unjustified exaggerated optimism, which in turn may obscure fundamental problems. Significant extra costs can result from wrong assumptions regarding the availability of certain capabilities, since this may require additional system modifications and upgrades of systems which were planned to be replaced. Therefore, it is important to make an up-front research investment in order to have a realistic estimate about possibilities and constraints of future navigation systems as a function of time. Results from research into advanced navigation systems can be used for a more accurate specification of roadmaps, which rely on anticipated navigation technology. This paper discusses why the complexity of today's navigation systems requires an integrated-systems-engineering approach, and why prototyping is considered an essential element of the research into navigation systems. It shows how the need for an integrated systems approach determines the research roadmap used by the radionavigation group of Delft University of Technology. Some examples in which the concept of commercial technology insertion has been used extensively in the realization of prototypes are discussed.

0-7803-5086-3 /98/$10.00 ©1998 IEEE

Roadmaps

Roadmaps to the future must be defined at various levels. To start with, the government defines a roadmap outlining the potential for future growth combined with the constraints following from environmental considerations. In the Netherlands, a roadmap outlining the future of the air transport is described in the government's white paper on Schiphol, the airport of Amsterdam (Anonymous, 1996).

In addition, the authorities providing the ground infrastructure define a roadmap describing the infrastructure, which will be available as a function of time. Based on these two basic roadmaps, airlines can define roadmaps outlining the required functionality needed to optimally benefit from the available infrastructure and maximize operational benefits within the constraints imposed by the government and aviation authorities. It is clear that decisions with respect to the creation of a future infrastructure should be based on a roadmap of the anticipated use and on the available technology.

To define a roadmap, an understanding of the possibilities and constraints is required. A distinction must be made between technology-induced constraints, and constraints imposed by the authorities. In general, the amount of technology-induced constraints reduces as time advances. However, a single wrong assumption can result in huge penalties.

For example, in the 1970's it was concluded that the instrument landing system (ILS) would not suffice to support all future needs, and therefore new precision approach and landing systems have been developed. The microwave landing system (MLS) was selected as the replacement for ILS, and a roadmap was defined for the introduction of MLS in 1985. Due to development delays, the planned introduction was postponed to 1995. ILS was foreseen to be phased out by 1998. In the mid 1980s, proponents of the global positioning system (GPS) were suggesting the use of GPS for civil aviation. In that case a single system could be used worldwide for all phases of flight. As the advantages would be tremendous, ICAO decided in 1995 to drop the MLS mandate. Unfortunately, risks and problems of using GPS, although identified, were not really taken serious. The (unjustified) "GPS can solve all our navigation needs" optimism caused airports to delay decisions with respect to the installation of MLS, and as a result airlines delayed decisions to install on-board MLS equipment. As predicted years earlier by the research community, vulnerability to jamming proved to be a major bottleneck. Finally, the FAA officially acknowledged this vulnerability and the large impact of the non-availability risk of GPS in 1997, which led to the requirement of a terrestrial backup navigation system.

In a description of Boeing's system method for CNS/ATM development and integration, Tulder (1996) points out '*Early and accurate identification of these requirements for new systems or upgrades is key to provide the airlines an additional revenue generation capability*'. He continues with '*In the effort to improve the current air transportation system the tendency is to focus more attention on new technology solutions rather than on critical system performance requirements*'. The focus on the impressive accuracy provided by the new GPS technology is a striking example. In the discussion about GPS for landings and also for en-route, the jamming susceptibility issue, certainly a critical system performance requirement, did not receive sufficient attention.

Defining a roadmap requires predictions. Unfortunately, the accuracy of such predictions reduces with an increasing time horizon. An important challenge for a typical University community is to identify potential weak links in future operational concepts, and to point these out to everyone involved in the definition of roadmaps, which rely on such operational concepts.

An ISE approach

'Integrated systems engineering is concerned with the systematic development of large-scale systems by means of using integrative knowledge across several disciplines' (Johannsen, 1994). Successful integrated-systems design requires the seamless fusion of domain specific knowledge from the encompassing disciplines into a single product. It is evident that here an integrated-systems-engineering (ISE) approach is also needed to define a roadmap towards the objective of an increase in overall capacity and safety, while reducing the relative cost and environmental burdens. This roadmap should address the potential of new technologies as a function of time. When defining a roadmap, it is unrealistic to assume that all elements that are required for a new operational concept can be introduced at a single point in time. Some of the elements needed for a new operational concept may have a long time-to-market as a result of extremely high safety requirements. Therefore, an evolutionary approach is usually needed.

The systems development process can be divided into a number of stages. Sage (1992) identifies the following:

Table 1. Systems development process

1	Requirements and specifications
2	Preliminary conceptual design
3	Logical design and architectural specifications
4	Detailed design and testing
5	Operational implementation
6	Evaluation and modification
7	Operational deployment

University research roadmaps

As stated by Gartz (1997), a total system advances only as far as its weakest link allows. As indicated earlier, an important challenge for the University community is to identify potential weak links in future operational concepts and to point these out to groups involved in the definition of roadmaps that rely on such operational concepts. The goal of the research of the Delft radionavigation group can be summarized as *'To explore/develop new/better navigation technologies, in which 'better' can be translated into aspects of capacity, safety, economy, and environment'*.

Typically, the university's research time-horizon is longer than that of industry. Therefore, research will mostly be limited to the first four stages of the systems development process. To be able to determine the potential of new ideas, it is important that there is also an understanding of constraints which follow from the last three stages in the systems development process: operational implementation, evaluation and modification, and operational deployment.

Thus, in the definition of a university research roadmap, factors related to all seven stages of Table 1 should be taken into account. In the research into navigation systems, prototyping is used to reach stage 4. As a result of the longer time-horizon and the often less detailed requirements, the frequency of changes in the first 3 stages is relatively high. The required flexibility of prototypes poses specific requirements on the implementation process. This justifies a significant up-front investment in the flexibility of the research infrastructure. Although rapid prototyping methods offer flexibility through a reduction in implementation time, they should ensure that modifications are made at the specification level instead of at the implementation level.

A navigation system can be divided into a positioning system, a data communication system, a system containing the navigation

0-7803-5086-3 /98/$10.00 ©1998 IEEE

functionality, a navigation database, and a man-machine interface (Fig. 1).

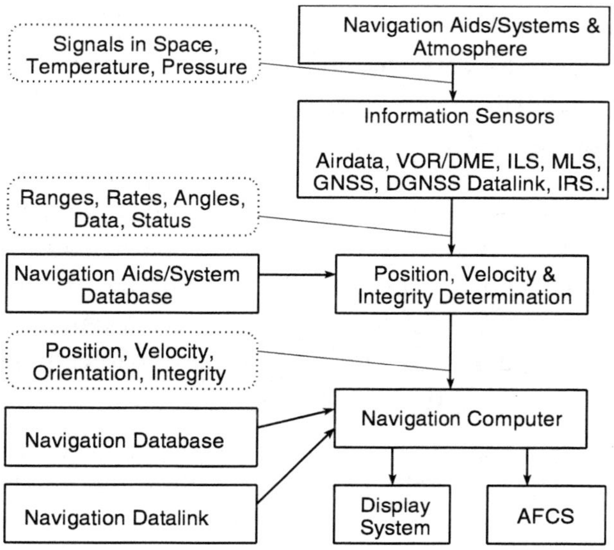

Figure 1. Overview of a navigation system.

This classification is the basis for the different types of research performed in the radionavigation group. The current research topics comprise:

- pilot vehicle interfaces for navigation, guidance, and control
- electronic positioning systems
- datalinks
- navigation systems (combination of previous three plus functionality for navigation)

To perform research up to the level of implementing prototypes of new concepts for position determination, datalinks, and pilot-vehicle interfaces, system integration issues must be addressed. To support an efficient integration, the research infrastructures for the just mentioned three disciplines should be developed with system integration as a requirement. Here the standards offered through the use of COTS are a great help. The research roadmap should contain both the research topics and a path describing the development of the necessary infrastructure.

Research infrastructure

In order to be able to perform and stay involved in research into navigation systems up to the level of developing and testing prototypes, a highly flexible research infrastructure is needed. One of the typical characteristics of research oriented systems is the high frequency of changes in the functional requirements during their entire lifecycle. Furthermore, research oriented systems are characterized by highly interactive hardware and software. In the design of such systems it is of crucial importance to integrate software requirements with hardware design synthesis. Due to the rapid increases in performance of computer hardware this is not easy to fulfill. Plant (1993) refers to the effect of the rapidity of technology changes in the area of information technology as the 'ΔT effect'. He states that '*Unless management and developers control their software processes through an understanding of the impact that hardware, software, practical and theoretical developments have upon it, these technology changes may have a serious and detrimental effect upon the developers software creation process*'.

Research oriented systems should be able to keep up with advances in technology during their entire lifecycle. Due to the rapid increase in hardware performance, the lifecycle of the hardware is typically shorter than that of the implemented functionality. Therefore, the software that provides the functionality should be designed such that upgrading hardware does not require major re-engineering or programming. In other words, the software lifecycle should be extended beyond the hardware lifecycle. As a result it is necessary to design hardware with the objective of upgradeability and scalability, and software with the objective of rehostability in mind.

Results

In order to perform research up to the level of developing, implementing, and integrating prototypes of subsystems within the often tight budgetary constraints of the university community, the concept of commercial technology insertion (CTI) is used extensively. For the display target systems used for the research into pilot-vehicle interfaces, a generic architecture had to be developed which can be used to implement current and expected future concepts. Fig. 2 presents an overview of the systems involved in the presentation of navigation data. This figure shows the various locations at which the data presentation system will interface with the other systems and provides a common frame of reference for the earlier mentioned research topics.

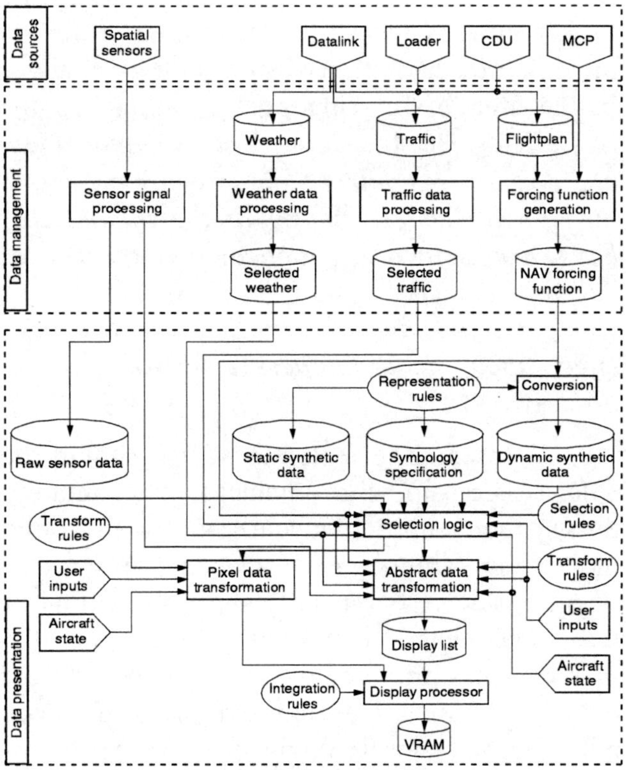

Figure 2. Overview of the systems involved in the presentation of navigation data.

Based on the desired generic architecture for the datapresentation, a display system architecture was developed and implemented. This development took place in 1990 and 1991. The first prototype has been used from 1991 to 1995 in Delft University's moving base flight simulator. The second prototype was used in Delft's Cessna Citation II for in-flight tests of advanced display concepts. To efficiently deal with the high frequency of changes in the functional requirements, which are typical in a research environment, software development methods relying on automatic code generation have been implemented (Theunissen, 1994). The use of a high-level specification language for the autocode generation also enables easy re-use of MMI components.

The software architecture that was developed in 1990 had the interface to the graphics API encapsulated in a bridge pattern. This allowed an easy migration to various new target systems. Until 1998, the display hardware has been upgraded two times. The first time was in 1994 and included the addition of high speed shading processors, resulting in a target system of approximately $20k. The second time took place in 1997, and the architecture was changed to accommodate VOODOO™ graphics processors. The current target system, with a hardware cost below $3k, can generate 3-D flight displays with integrated terrain data in real-time and supports many of today's graphics capabilities such as anti-aliasing, z-buffering, transparency, texture mapping and tri-linear filtering.

The following examples discuss how the concept of commercial technology insertion has been used in the realization and integration of prototypes for various research projects performed by the radionavigation group of Delft University of Technology.

Example 1: Integrated positioning and pilot vehicle interfaces

For in flight testing of a new tunnel-in-the-sky display developed by the radionavigation group, the earlier mentioned display target system was integrated with a prototype of a positioning system providing D-GPS data (Theunissen, 1995). This system was also developed by the radionavigation group to validate the possibility of uplinking the differential corrections through the auxiliary dataword channel of a modified MLS ground station. This concept, partly as a research contract with the CAA in the United Kingdom, was investigated in the context of the multi-mode integrated approach system (MIAS) research (Breeuwer, 1995). Recently, this option has received renewed attention (Marsh, 1998). For the development of the MIAS system hardware, extensive use was made of commercial of the shelf components.

Example 2: Eurofix

In 1989 Delft University proposed Eurofix, an integrated radio-positioning system based on DGPS and Loran-C. By additionally time modulating the Loran-C signals, a long-range data communication channel is achieved that broadcasts DGPS information to the users (Fig. 3).

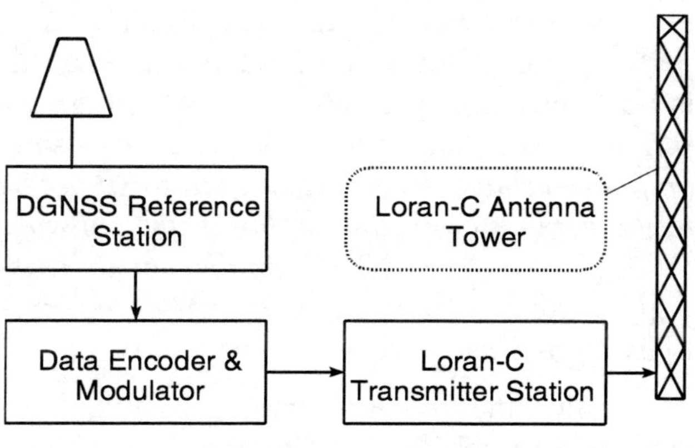

Figure 3. Eurofix transmitter station.

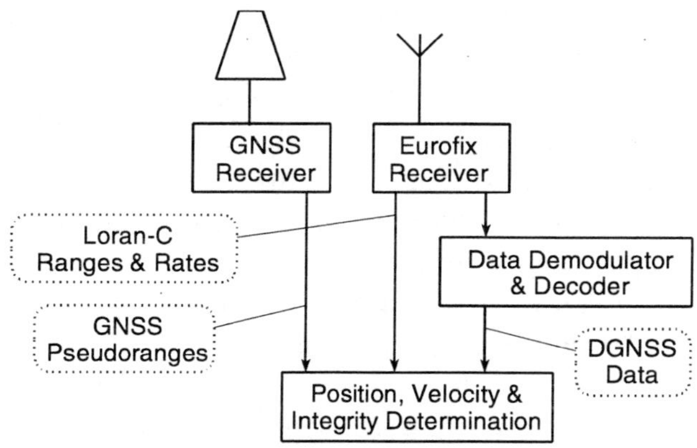

Figure 4. Eurofix receiver block diagram for integrated navigation.

In the beginning of this research it was not allowed to perform tests real-life Loran-C transmitters. So, a very realistic simulation technique has been implemented. Instead of adding time modulation on the very precise time-controlled transmissions, the sampling clock of the Eurofix receiver in the laboratory has been time modulated. In this way an extensive test phase could be executed without interrupting existing Loran-C operations. So, while taking the last three phases of Table 1 into account, all first four phases could efficiently be carried out. On the bases of these simulation tests, the German DoT and the Steering Committee of NELS (Northwest European Loran System) gave in November 1996 the green light to implement Eurofix on the Loran-C transmitter station at Sylt (Germany). On request of the DoT, the US Coast Guard executed Eurofix tests on the Wildwood Loran-C station in April 1998. Both real-life tests fully confirmed the simulation results. The development of prototype hardware for the evaluation of the Eurofix concept also relied on the use of COTS that proved to be a cost and labor efficient approach.

During the development of the Eurofix system, one of the research objectives was the investigation of the complexity of algorithms (resulting in demands for processing power) that were needed for a Eurofix receiver. For flexibility reasons, it was decided that the receiver should be designed according to the software radio concept with digitizing of the signal at the RF-level. Because of this evolutionary character of the receiver software, it was decided to use a highly scalable and flexible hardware platform. The choice of PC-based hardware was logical, given the short learning phase and the plethora of hardware, compilers, performance, as well as mathematical and debugging tools available for this platform. During the development, the actual hardware as well as the compiler have changed for various reasons; the changes to the software that needed to be made were minor.

Another milestone in the Eurofix development was the installation of a Differential GPS reference station at the location of Sylt, Germany. Many of the considerations for the Eurofix receiver applied here as well, resulting in the choice for a PC-platform. A high quality off-the-shelf GPS receiver on a PC-card was chosen as a reference receiver, thereby minimizing difficulties in interfacing with the rest of the software. In this case, a ruggedized PC with a watchdog-timer was chosen to provide an escape in case of computer lock-ups. The software design was such that a modem interface for remote maintenance was easily incorporated. Since the installation, the reference station has been operating flawlessly for over 16 months without the need for any on-site maintenance.

Again, the applied ISE approach made the step to a real implementation of Eurofix relatively easy and efficient. Further, it was shown that by first analyzing the navigation requirements just minor modifications of the existing stations were needed to achieve a large added value to the Loran-C system. Eurofix can provide both DGPS and a backup navigation service to the user that is not possible with any other system currently available (Fig. 4).

Future challenges

The last fifteen years have demonstrated an enormous increase in the performance of onboard computers. According to Moore's law (Intel), performance doubles every 18 months. However, the increase in performance is not only used to speed up processes, but also to add functionality. This increases complexity, especially of the avionics. Ulbrich (1996) summarizes the importance of avionics as '*Avionics is running the aircraft today. It is involved in flight controls, the flight management, the vehicle management, the cockpit, the communication/navigation/identification, the external sensors, the data management, and various military functions*'. In the cockpit, all systems come together and the pilot-vehicle interface must keep the pilot in the loop while '*avionics is running the aircraft*'. There are numerous studies that illustrate the serious consequences resulting from problems in the pilot-vehicle interface. Beevis (1987) already pointed out the need for an integrated systems approach with these kinds of systems. He stated: '*An approach which combines the interaction of hardware, software and human functions is made especially necessary by the impact of advanced technology on the roles of human operators and maintainers, on the man-machine interface, and on the system development process itself*'. Whereas in the past limitations in performance might have been a bottleneck, this is no longer the case. Wilkie et al. (1995) describe the situation as follows: '*The speed limit on future vehicle development cycles will not be in the domain of hardware, but in the software that must be developed to implement increasingly complex algorithms and provide connectivity to the various vehicle subsystems*'.

For the research into navigation systems, a major challenge is to cope with the increasing complexity by selecting and/or developing methods to easily specify and implement different types of functionality.

Conclusions

In order to have a realistic estimate about possibilities and constraints of future navigation systems as a function of time it is important to make an up-front research investment.

Results from research into foreseeable bottlenecks of advanced navigation systems can be used for a more accurate specification of roadmaps that rely on such anticipated technology.

In order to gain sufficient feedback and have a common frame of reference with industry, research must often be performed up to the level of implementing prototypes of new concepts.

When the research is multidisciplinary, the supporting infrastructures should be developed with overall system integration as a main requirement.

Experience has shown that today's COTS components provide the performance and flexibility needed to implement prototypes of advanced navigation systems in a cost-efficient way.

The selected software design paradigm has proven that it is easy to extend the lifecycle of software for research oriented systems beyond that of the hardware lifecycle.

The next challenge is to select/develop methods that allow researchers to efficiently cope with the higher complexity resulting from the continuous increases in functionality of navigation systems.

References

Anonymous (1996). *The Government White Paper on Schiphol's Future*, http://www.schiphol.nl/press/p96b.htm.

Breeuwer, E.J. et al (1995). 'Design, Test and Evaluation of a CAT-IIIB C-Band Data Link', *Proc. of the 1995 Int. Symp. On Precision Approach and Automatic Landing,* February 21-24, Braunschweig, Germany.

Gartz, P.E. (1998). 'Does Aerospace Technology have a Future in Aerospace'. *IEEE Spectrum*, Jan. 1998.

Johannsen, G. (1994). 'Integrated Systems Engineering: The Challenging Cross-Discipline', *Proceedings of the IFAC Conference on Integrated Systems Engineering*, pp. 1-10, Baden-Baden, Germany.

Marsh, G. (1998). 'The Global Positioning System: A View from Europe', *Avionics Magazine*, June, pp. 34-37.

Plant, R.T. (1993), 'Software Development Process Model for Aerospace Systems', *Proc. of the 9th AIAA Computing in Aerospace Conference, pp. 883-890.*

Sage, A.P. (1992), *Systems Engineering,* Wiley, NY.

Theunissen, E. (1994). 'The Development of the DELPHINS Display Design System', *Proceedings of the 5th International Training and Equipment (ITEC) Conference*, pp. 583-588, The Hague, The Netherlands.

Theunissen, E. (1995). 'In-Flight Application of 3-D Guidance Displays: Problems and Solutions', *Proceedings of the 1995 IFAC Conference,* June 27-29, Cambridge, MA, USA.

Tulder, P.A. (1996). 'System Method for Communications Navigations Surveillance/Air Traffic Management (CNS/ATM) Development and Integration', *Proceedings of the 15th Digital Avionics Systems Conference,* pp. 3-10.

Ulbrich, E. (1996). 'The Coming Avionics Utility', *Proceedings of the 15th Digital Avionics Systems Conference*, pp. 293-299.

Wilkie, B.F., Frank, R., and Suchyta, J. (1995). 'Silicon or Software: The Foundation of Automotive Electronics', *IEEE Vehicular*

Technology Society News, August 1995, pp. 17-20.

MODELING OF CONCURRENT ENGINEERING PROCESSES FOR INTEGRATED SYSTEMS DEVELOPMENT

Ernst Fricke, Herbert Negele, Lutz Schrepfer
Fachgebiet Raumfahrttechnik (Institute of Astronautics), Technische Universität München
Boltzmannstr. 15, 85748 Garching, Germany

Alfred Dick, Bernd Gebhard, Nicole Härtlein
BMW AG, 80788 München, Germany

Abstract

For the reengineering of development processes and for its continuous improvement, the development process has to be modeled and documented. A process map is the basis for the way how a system will be or has been designed. The standard methods and tools for process modeling are mostly used for business process modeling and redesign. Due to the special nature of concurrent engineering processes, a method for process modeling has to be able to support and easily map the high interconnectivity between the process of different engineering disciplines over all hierarchies. The dominant elements in a model for concurrent engineering processes are the informational relations and flows between the different processes. These build the basis for an information-based engineering process.

The analysis of models and textual documentation of the engineering processes in an automotive company revealed that different departments used quite different methods and tools for generating their process descriptions. This resulted in very heterogeneous process models with different degrees of detail that could not be interrelated and coordinated. Therefore, a single method for mapping and interconnecting all the engineering processes of all different disciplines was agreed upon, which describes inputs and outputs (e.g. informational objects) for every process. This supported the idea of client-supplier relations between different processes.

A detailed market study was conducted, in which about 30 tools for business process modeling (e.g. ARIS, Bonapart, IDEF, Facets, etc.) were analyzed and evaluated by a variety of criteria. None of these tools met the requirements of the project partner, not only concerning the method, but also concerning supporting factors like easy and fast input of data, multiuser and database support. Another important factor was the fact, that the process model also should be the basis for project planning and scheduling. Hence, a smooth data flow from process mapping to project planning and scheduling should be possible.

Therefore, in order to meet all requirements a special process modeling tool was developed which is based on the same database as the project planning tool already in use. The paper describes the reasons for an engineering process driven modeling method and its use, as well as the rationales for developing a new software tool. Also, lessons learned from this approach are described.

1. Introduction

Nowadays systems are becoming more and more complex and integrated. Therefore, also the development process, particularly in the time of concurrent engineering, is becoming

0-7803-5086-3 /98/$10.00 ©1998 IEEE

a highly interconnected network of processes concurrently involving many different disciplines. For the automotive industry a main reason for this is the strong increase of electrical/electronical components. E.g., guidance assistance, entertainment systems, electronical control of engine or brakes, and many other (also more and more safety related) functions are added.

Due to the increased overlapping of the development processes the amount of information that has to be exchanged for an effective and efficient way of working has strongly increased, too. In order to reduce development cycle times and costs, the development processes were reengineered in the last years and new organizational concepts and structures were implemented (e.g., team-oriented organizations with Integrated Product Teams) (Clark & Fujimoto 1991).

It was observed in several companies that process reengineering often meant to identify and optimize several different process chains. Often this optimization is done separately for each of these chains, as if they were independent of each other. This corresponds to observations of psychologists: people tend to think in causal *chains*, in particular when they have to tackle complex problems (Dörner, 1989). As a result, interactions and interfaces between different chains were ignored ending up in an mediocre (or even poor) overall development process. Reality cannot be described adequately by isolated, sequential and one-dimensional chains. It rather resembles a network (*process net*), with many interrelated processes and process chains (Negele et al., 1997).

Applied to the field of product development, besides the processes, additional (interrelated) aspects have to be taken into account to get a comprehensive view of this system (e.g., customer and user needs, requirements and goals, products, people, resources, organizations). Systemic methods and tools can help to manage these complex systems successfully by enabling a holistic analysis, modeling, and examination of all relevant elements and their interrelationships. Such a method was proposed as the "ZOPH Model", a comprehensive systems modeling approach embracing, structuring, modeling, and interrelating information essential for product development systems (Negele et al., 1997; Negele, 1998). It structures all the information relevant to a given development system by using four different system types that form the abbreviation "**ZOPH**", which is derived from the first characters of the german terms

- "**Z**ielsystem" (goal system),
- "**O**bjektsystem" (product system),
- "**P**rozeßsystem" (process system), and
- "**H**andlungssystem" (agent system).

Here, we want to focus on the development processes (process system) and how they can be modeled in order to meet the requirements arising from their specific characteristics, taking into account a concurrent engineering environment with decentrally acting teams and individuals.

2. Modeling Development Processes

2.1 Why Modeling Development Processes?

First, we have to answer the question, why modeling development processes is necessary and worth the effort. Several points can be put forward here (Figure 1):

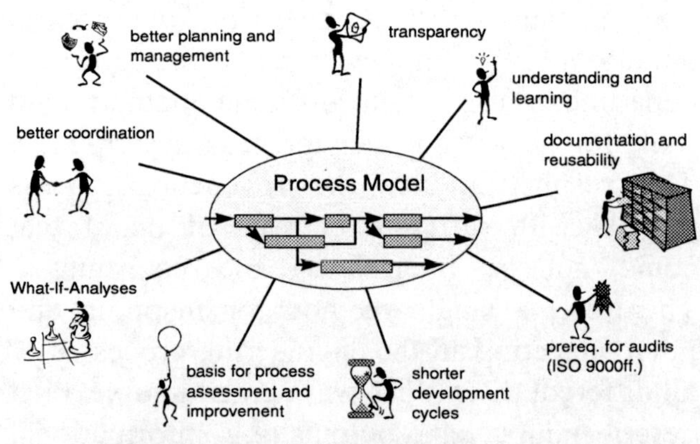

Figure 1: Why Modeling Processes?

- **Transparency**: A process model helps people to get an overview, to understand what part they play in the game, and to see who is doing what and when. This is all the more necessary when processes are changing very often (e.g., due to reengineering efforts).
- **Understanding and Learning:** A transparent process model supports and communicates understanding of complex processes and their interactions and dependencies within the organization. It also provides an excellent learning aid for employees that are new or have changed jobs.
- **Coordination:** In the course of the development many process interfaces (flows of information, material, money, etc.) have to be coordinated in content and time. A consistent process model promotes better communication (people talk about the same things) and allows early planning of future interactions.
- **Better planning and management:** By enabling transparency and early coordination a modeled process represents a sound basis for detailed planning and easier management of the actual development project.
- **Documentation and Reusability:** Process models are a kind of documentation that (or parts of it) can be easily reused as a starting point or "building blocks" in subsequent development projects.
- **Prerequisites for Audits:** In order to achieve certification (e.g., compliant with the ISO 9000 Series of Standards, esp. ISO 9001) a documented process and evidence that the process is performed as documented are required. A process model that is used (or "lived") by all people involved in the development activity can provide both.
- **What-if Analyses:** A process model can be used to conduct what-if analyses to determine the effects of process changes. Moreover, process simulation capabilities can be built upon the model.
- **Basis for Process Assessment and Improvement:** Only if you know what you are doing (which can be described in a process model), you can assess how good you are doing it and use it as a basis for improvement.
- **Shorter Development Cycles:** One main reason for process modeling is the achievement of shorter development times. Process models are the starting point for process reengineering and optimization activities.

These arguments for modeling (development) processes are, at the same time, the goals that have to be achieved by process modeling activities. In order to describe processes in the product development context adequately, we first have to have a closer look at the specific characteristics of this kind of processes.

2.2 Characteristics of Development Processes

Today's development processes differ significantly from other business processes, like production, logistics, or supply-chain processes. While these rather resemble sequential process chains that are performed several times in a very similar or almost identical way, development processes are rather like **process nets** (see above), with processes that are highly interconnected, including feedback-loops and interactions on different levels (Figure 2).

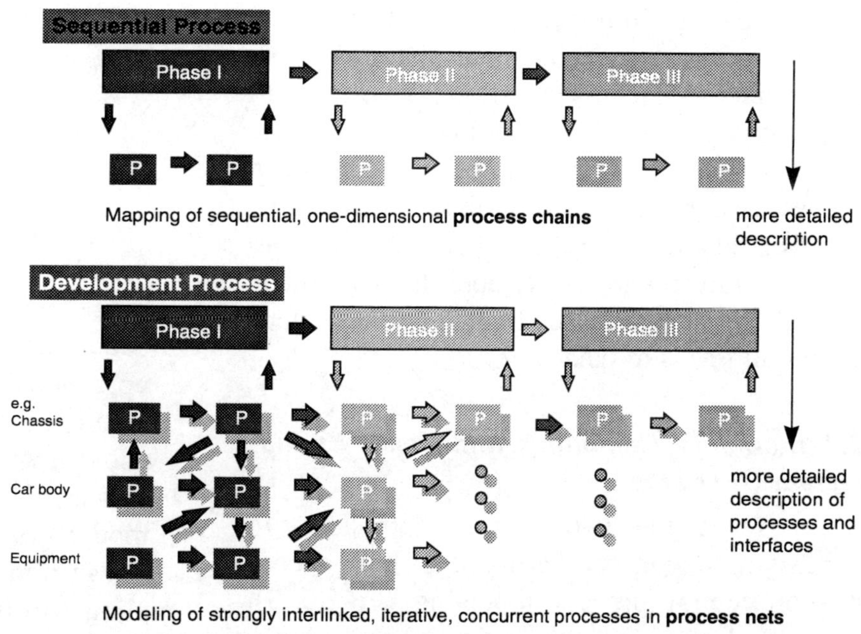

Figure 2: Development Processes vs. Other ("Sequential") Processes

Typical characteristics that can be used to describe development processes are, e.g. (Negele 1998):

- creative and innovative
- dynamic
- interdisciplinary
- strongly interrelated

- strongly parallel
- iterative
- communication intensive
- anticipatory
- planning intensive
- uncertain
- risky

Many development activities have an unique and intuitive character that is very difficult to capture in a model. Many decisions have to be made anticipatory relying on assumptions. Because of the newness of many development tasks acquired information often is tainted with uncertainty. Usually, development processes are treated as what we called "sequential" processes. There seems to be a lack of understanding for the different character of development processes. Negele (1998) describes a comprehensive approach to modeling development processes (rsp. systems) in more detail.

Another important issue that has to be tackled is how to adequately structure the process model. Most methods (e.g., SADT/IDEF) use a strict hierarchical, top-down approach to model processes where interrelations between processes can only be described on the same level of detail. In practice, there are manifold logical, informational, temporal, etc. interactions between the different process steps across all levels of detail.

2.3 Boundary Conditions and Some Fundamental Questions

In general, two different approaches to process modeling can be distinguished: decentral or central process modeling (Figure 3). With the latter approach, typically, some process specialists collect all relevant information for a top-down structured process model. Advantages of this approach are: modelers can be specially trained for their job and for supporting tools; they are capable of generating high sophisticated models (high "information density"); and, since the number of modelers is quite small, the modeled information should be quite consistent concerning content and degree of detail.

A fundamentally different approach is to let all people involved in development work decentrally on the process model. The number of modelers or users of a corresponding process modeling tool can amount to several hundreds of persons. The advantages of this approach are: if all people are using the model on a regular basis, the information contained is up-to-date; since the modelers know their processes best, the model is likely to be quite realistic; the effort for updating the process model is limited, because it is shared by many individuals and loss of information can be avoided.

usage:	central	decentral
users:	specialists	"everybody"
number of users:	few	many
approach:	top-down	bottom-up
info consistency:	high	limited
info "density":	high	limited
model up-to-date?:	limited	more likely
model realistic?:	limited	more likely
updating effort?:	high	limited

Figure 3: Comparison of Process Modeling Approaches

As always, the best alternative had to be found in a compromise of both approaches. Since several unsuccessful attempts already had been made at the project partner's site, to build up a detailed process model centrally, a combined top-down and bottom-up approach was chosen. A centrally generated and coordinated master plan and a common, top-down process model structure provide the basis for a practical integration of the distributed, bottom-up modeling efforts.

Also, for a reengineering project, engineers had started to model their processes with a quite simple input-process-output (IPO) logic, describing what they are doing (P), what they need to do it (I), and what they produce (O). The output of one process can be used as input by other processes. These output-input relations represent the interactions between processes. This supports the idea of an information-based system development, where the structure of the information flow defines the structure of the development process (Fricke & Negele 1997, Gartz 1997).

The IPO description was done in ordinary MS Word forms everybody could generate. The problem was that no possibility existed to support the coordination of output-input links. Also, in daily practice, many different methods and tools for process modeling were used (CAD-tools, spreadsheets, presentation or word processing software, project scheduling tools, etc.). The usage of different tools for capturing, visualizing, and analyzing process information does not necessarily have to be a problem, if a common method for process modeling were used and generated data were interchangeable. Unfortunately, this was not the case. Therefore, the need for a common tool that supports modeling, planning, and coordination of processes and their interactions became evident.

Several boundary conditions have to be taken into account here. Since the work load of the engineers generally is very high, it is crucial that they have an operational need and benefit of applying such a tool. For sure, it has to be very user-friendly and easy to use. Moreover, engineers have to overcome the conviction that everything they are doing is unique and therefore can't be modeled. Also, they have to be persuaded that their processes should be documented for the reasons mentioned above and that they don't become replaceable by doing so. Fricke (1998) describes principles and methods for realizing such an user-centered approach.

Other important issues were the amount of training necessary, serviceability, and maintainability of the tools. Also, the visualization capabilities for an adequate representation of the process net were considered to be fundamental. As process plans should later be used in the ongoing project for project scheduling, a data interface to a project scheduling tool should be possible. This requires that correspondent information of both methods can be mapped on each other.

3. Analysis of Existing Methods and Tools

The requirements for selecting a method and tool for modeling development processes, according to the strategy of decentral process modeling, were based on the simple Input-Process-Output description. Additionally, the assignment of process owners, customers (of outputs), and suppliers (of inputs) should be enabled. Also, rough temporal information for process duration and output-input interactions has to be added.

First, seven goals were set for the evaluation of the process modeling tools:

1. Support of IPO method with flexible hierarchy
2. Interactions modeled by input-output relations
3. Interface to project scheduling
4. Consistency and reusability of generated data
5. Visualization with filters
6. High user-acceptance and minimal necessary training
7. Fast implementation of the tool

Based on these goals the requirements for the modeling of concurrent engineering processes, and therefore for a supporting tool, were defined. For each single requirement its contribution to the goals was specified and determined if it is a shall or should requirement. A market analysis was conducted to evaluate which process modeling tools (COTS) support the defined set of requirements.

The criteria by which the tools were evaluated were structured in 5 classes: Requirements concerning

- **Modeling method**
- **Data storage**
- **User-acceptance**
- **Information technology environment**
- **Others**

Requirements concerning the modeling method:

The requirements in this class included

the capability of modeling the elements of the IPO method, e.g. inputs, outputs, processes, connectivity of processes by inputs/outputs, process owners, flexible hierarchy.

Requirements concerning data storage:

These requirements were all related to the tool's possibilities for storing data of the modeled processes and its interfaces to other tools, like use of a database, API, data interfaces.

Requirements concerning user-acceptance:

These included requirements for the ease of use of the tool as provided by different functionality and adaptability. Examples are: information filters, visualization of process net, definable layouts, intuitivity.

Requirements concerning IT-environment:

Here, for example, supported platforms, client/server operation, multi-user access and intranet capabilities were used as requirements.

Other requirements:

These included requirements for acquisition costs, supplier assessment, and usage at the project partner's site.

More than 30 commercially available tools which are or could be used for process planning were evaluated. They were separated into three different groups:

1. **Graphical visualization tools**
2. **Project planning tools**
3. **Process modeling tools (for sequential processes)**

Some tools like Visio, are more or less graphical visualization tools, which have no possibility to analyze modeled data automatically. Other tools are more project scheduling tools than process modeling tools. The third group are the typical process modeling tools like ARIS, which were originally designed for more sequential processes, like manufacturing and logistics processes. A subgroup of them consisted more or less of simulation tools. Most of the process modeling tools didn't support data storage in a database, which was a shall-criterion. Those which stored model data in a database, didn't support the IPO-method and the requirements for a flexible hierarchy. E.g. all SADT/IDEF based modeling tools required a strict process hierarchy.

Additionally, only a few tools supported a connection to project scheduling tools, which was also a crucial point. No existing tool seemed to satisfy the needs derived from modeling concurrent engineering processes for integrated systems. There seems to be a lack in understanding the special characteristics of these development processes, which might be a hint why many reengineering activities of engineering processes in different industries failed.

Therefore, the decision was made to develop a new tool for the input-process-output modeling method, serving the needs derived from concurrent, highly-interactive development processes. Additionally it will be integrated with the project scheduling tool in use in that way, that the modeled processes can be mapped on the method used in the scheduling tool and vice versa. Additionally the graphical user interface should look similar to the format of the word documents already in use for process modeling, as it turned out to be the best way for the engineers to model or document their processes. For sure, additional functions, that can be made available in a computer-based tool, should be implemented, like drag and drop for connecting inputs and outputs.

4. Brief Description of Developed Method and Tool

In Figure 4 the proposed graphical user interface of the tool supporting the IPO method is shown. Major elements are the navigator, the process plans (with optional details) and the connector.

Process plans are the core element of the tool. The graphical user interface design was kept very similar to the MS Word forms formerly used, being divided into three columns which represent inputs, processes and outputs. Additional attributes, e.g. duration, responsibility, and type, of each process can be captured and displayed in a separate detail area.

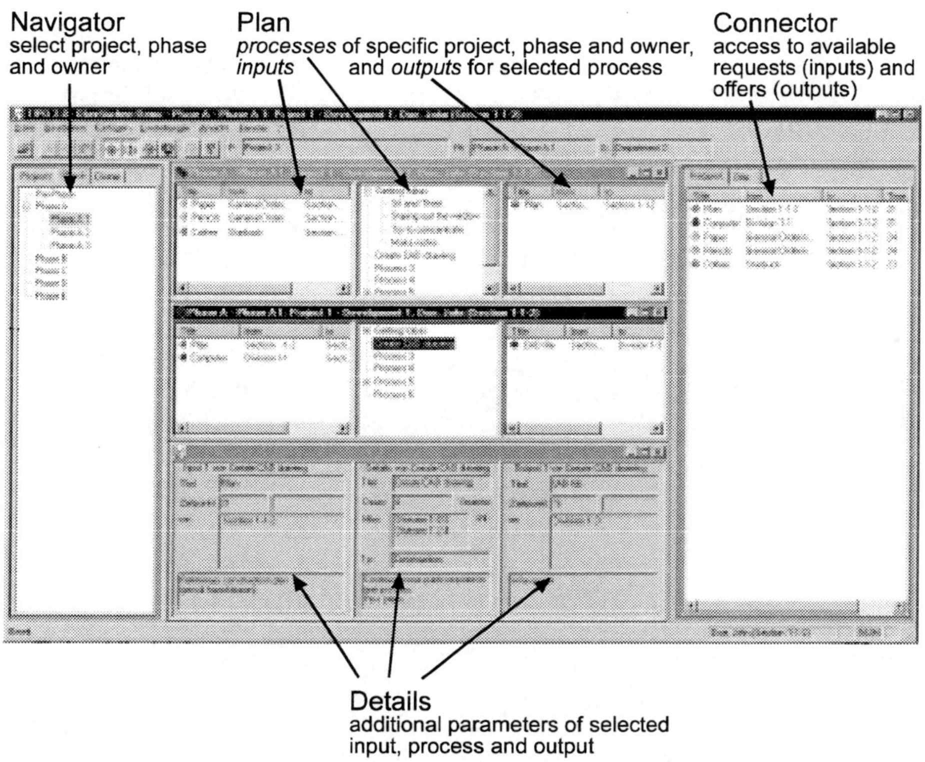

Figure 4: Prototype for the graphical user interface

The navigator provides an overview of the existing projects, phases, and owners. After selecting one of each item, the according process plan can be opened.

Corresponding to the IPO method, the processes are interlinked with each other exclusively by output-input relations. Therefore, the connector supports linking the output of one process as an input of another process, and vice versa. This is done by simply dragging an available offer (output) and dropping it at the input column of a process plan. Existing requests (inputs) are linked to processes in the same manner. This interlinking contains a coordination procedure to ensure that both process owners agree on contents and time of delivery.

In addition to the list view shown in Figure 4, a graphical representation of the process net (boxes and arrows) can be generated from the captured data.

In a multi-user environment, this approach helps to establish a highly interconnected process net very quickly. This is also the basis for further analyses of dependencies and future improvements of the development process.

Conclusions

One of the main reasons why engineering processes have to be modeled and documented are the ongoing efforts in all industries for reengineering their development processes. Otherwise, a process would be optimized that is not understood, communicated, or transparent to the engineers.

Concurrent engineering development processes of integrated systems differ significantly from other, sequential business processes. They are rather like process nets, including manifold interrelations, feedback-loops and interaction on different hierarchy levels. Also, they are planning intensive, uncertain, and communication intensive.

Two different approaches can be distinguished for process modeling: decentral (everyone) and central (specialists). The advantage of a mixed bottom-up and top-down, but strongly decentral, approach was shown. To make it work in daily practice, a simple process modeling method had to be chosen. Every single engineer has to have an operational benefit from modeling his processes, which is supported by using these modeled processes later in his project scheduling, too.

An analysis of the market showed, that no tool seems to satisfy the needs deriving from modeling concurrent engineering processes of integrated systems. There seems to be a lack in understanding the special characteristics of these development processes, which might be a hint why many reengineering activities of engineering processes in different industries failed.

A brief description of a method and tool which supports the needs derived for modeling concurrent engineering processes was given.

Future papers will describe the modeling method, the developed tool and lessons learned from the practical application of a prototype in more detail.

References

Clark, K. B.; Fujimoto, T. (1991): Product Development Performance: Strategy, Organization, and Management in the World Auto Industry, Harvard Business School Press, Boston, Mass.

Dörner, D. (1989): Die Logik des Mißlingens. Rowohlt, Reinbeck.

Fricke, E.; Negele, H. (1997): Informationsbasiertes Systems Engineering zur Erhaltung der Wettbewerbsfähigkeit der Luft- und Raumfahrtindustrie, Proceedings: Deutscher Luft- und Raumfahrtkongress, München.

Fricke, E. (1998): Der Änderungsprozeß als Grundlage einer nutzerzentrierten Systementwicklung. PhD thesis (submitted), München.

Gartz, P.E. (1997): Commercial Systems Development in a Changed World, IEEE Transactions on Aerospace and Electronic Systems, 33, 2, 632-636.

Negele, H.; Fricke, E.; Igenbergs, E. (1997): ZOPH — A Systemic Approach to the Modeling of Product Development Systems, Proceedings of the 7th International Symposium of INCOSE.

Negele, H. (1998): Systemtechnische Methodik zur ganzheitlichen Modellierung am Beispiel der integrierten Produktentwicklung, PhD thesis (submitted), Technische Universität München.

List of Analyzed Tools

Attached is the list of the analyzed tools. For sure, it is not complete and there are more tools on the market. Additionally, it has to be emphasized, that for other process modeling needs than described in this paper most of the tools might be appropriate.

- ABC Flowcharter
- Ablauf-Profi 3.5
- ARIS
- Autoplan II
- Bonapart
- BPWin
- CA-Super-Project
- Enterprise Modeller
- Facets
- Hyper ABC Power
- IDEF (Design/IDEF)
- ITHINK (Analyst 5.0)
- KAISHA Modeler Pro
- MODELER/2000
- oCTAVe
- Optima
- ORACLE PROCESS MODELLER
- OROS+Link Engine
- Powersim
- PrimaVera
- ProCAP/ProSim
- Process Builder (Analyst Edition)
- Process Charter
- Process Wise Integrator
- Quesheet
- ReThink
- SA/BPR, PBLink
- SES Workbench/Object Bench
- SIM Process
- Visio
- WITNESS

0-7803-5086-3 /98/$10.00 ©1998 IEEE

SYSTEMS ENGINEERING NEEDS A FORMAL BASIS

Manfred Broy, Technische Universität München, D-80290 Munich, Germany

Peter Scholz, Technische Universität München, D-80290 Munich, Germany

Abstract

Today, more and more electronic parts of automobiles or aircrafts are realized as software, often distributed on a network of high-performance processors that are embedded in the car or airplane. In the systematic development of distributed interactive system we distinguish the following views: the interface view, the data view, the distribution view, and the process view. Each of these views is helpful and has its place in the development and design process of systems. We show how to formalize these different views by logical means. The development of a system is carried through several levels of abstraction. We also demonstrate how to formalize the typical steps in the development process. In particular we may identify three directions of development: refinement within one level of abstraction, transition from one level of abstraction to another, incremental development by enlarging the functionality. We introduce refinement relations to capture these three dimensions of the development space. We give verification conditions for these refinement steps. In this way, a logical basis for the development of systems is created.

1 Introduction

The number of electronic control units (ECUs) in the avionics sector increases continuously. ECUs are used to guarantee an optimum of pilot and passenger security, flying comfort, and operational convenience. Behind these units is the more and more complex software which controls many important functions of an aircraft. The development of this software has to be completely and faultlessly finished at latest with the start of serial production. Beta-tests which are well-known from the software industry cannot be used in this context: pilots and passengers cannot serve as test persons. Thererfore, it is important to use a formal, mathematical basis for software development.

For practical issues, the scaling up of a development method is decisive. Here, modularity is a key concept. For a discipline of modular system development firmly based onto a scientific theory we need a clear notion of components and ways to manipulate and to compose them. In this paper, we introduce a mathematical model of a component with the following characteristics: a component is interactive; it is connected with its environment by named and typed channels; it receives input messages from its environment on its input channels and generates output messages to it on its output channels; a component can be nondeterministic, this means that for a given input history there may exist several output histories that the component may produce; the interaction between the component and its environment takes place in a global time frame.

Throughout this paper we work exclusively with discrete time. Discrete time is a sufficient model for most of the typical applications for information processing systems. For an extension of our model to continuous time see [9]. Based on the ideas of an interactive component we define composition techniques. We basically introduce only one form of composition, *namely parallel composition with feedback*. This form of composition allows us to model *concurrent execution* and *interaction* of components within a network. We briefly show that other forms of

0-7803-5086-3 /98/$10.00 ©1998 IEEE

composition can be introduced as special cases of parallel composition with feedback. For the systematic stepwise development of components we introduce the concept of *refinement*. By refinement we develop a given component in a stepwise manner. We study three refinement relations namely *property refinement*, *glass box refinement*, and *interaction refinement*. We claim that these notions of refinement are sufficient for a systematic top down system development. Our approach is *compositional*. This means that a refinement step for a composed system is obtained by refinement steps for its components. As a consequence, global reasoning about the system can be structured into local reasoning about the components. Compositionality relates to *modularity* in systems engineering. The main issue of this paper is the relational version of the stream processing approach as developed at the Technische Universität München (under the keyword FOCUS, see [5], [6]).

This paper is organized as follows. We begin with an informal introduction to the concept of interactive components. This concept is based on communication histories called streams that are introduced in section 2. Additionaly, a mathematical notion of component is introduced and operators for composing components into distributed systems are defined. In section 3 we introduce three notions of refinement to develop systems and show the compositionality of these notions.

2 Components

We introduce the mathematical notion of a component and on this basis a concept of component specification. A component specification is given by a description of the syntactic interface and a logical formula that relates input and output histories.

The notion of component is essential in systems and software engineering. Especially in software engineering a lot of work is devoted to the concept of *software architecture* and to the idea of *componentware*. Componentware is a catchword in software engineering (see [8]) for a development method where software systems are composed from given components such that main parts of the systems do not have to be reimplemented every time again but can be obtained by new configurations of existing software solutions. A key issue for such an approach are well designed *software architectures*. Software architectures mainly can be described as distributed systems, composed of components. For this, a clean and clear concept of a component is needed. In software engineering literature the following informal definition of a component can be found: a *component* is a physical encapsulation of related services according to a published specification. According to this definition we work with the idea of a component which encapsulates a local state or a distributed architecture. We provide a logical way to write a specification of component services. We will relate these notions to glass box views, to the derived black box views, and to component specifications. A powerful semantic concept of a component interface is an essential ingredient for the key issues in system development: well structured system (software and hardware) construction, system architecture, systematic systems engineering.

In the following we introduce a mathematical concept of a component. We show how basic notions of development such as specification and refinement of components can be based on this concept.

2.1 Syntactic and Semantic Interfaces

A *stream* is a finite or infinite sequence of messages or actions. Streams are used to represent communication histories of channels or histories of activities. Let M be a given set of messages. A stream over the set M then is a finite or an infinite sequence of elements from M. We use the following notation: M^* denotes the set of finite sequences over M with the *empty* sequence $\langle\rangle$, M^∞ denotes the set of infinite sequences over M (which can be represented by mappings $N^+ \to M$, where N^+ denotes the positive natural numbers). A stream is an element of the set M^ω, which is defined by $M^\omega = M^* \cup M^\infty$. On streams we specify the prefix ordering for $x, y \in M^\omega$ by the formula

$$x \sqsubseteq y \equiv \exists\, z \in M^\omega: x\hat{\ }z = y$$

Here $x\hat{\ }z$ denotes the concatenation of the stream x and the stream z. If x is infinite then $x\hat{\ }z = x$. Throughout this paper we do not work with the simple concept of a stream as introduced so far but find it more appropriate to work with *timed streams*. A timed stream represents an infinite history of communications over a channel or an infinite history of activities that are carried out in a discrete time frame. The discrete time frame represents time as an infinite chain of time intervals of equal length. In each time interval a finite number of messages can be communicated or a finite number of actions can be executed. Therefore we represent a communication history of a system model with such a discrete time frame by an infinite sequence of finite sequences of messages or actions. By $(M^*)^\infty$ we denote the set of timed streams. The k-th sequence s.k in a timed stream $s \in (M^*)^\infty$ represents the sequence of messages exchanged on the channel in the k-th time interval or the sequence of actions executed in the k-th time interval. A timed stream over M is an infinite sequence of finite sequences of elements from M.

In the following, we use streams exclusively to model the communication histories of sequential communication media called channels. In general, in a system several communication streams occur. Therefore, we work with *channels* to identify the individual communication streams. Hence, in our approach, a channel is just an identifier that is related to a stream in every execution of the system. We work with typed channels. Let a set S of *sorts* or *types* be given. By C we denote a set of typed channels. We assume that we are given a type assignment for the channels in the set C:

$$\text{type}: C \to S$$

Given a set C of typed channels a *channel valuation* (let M be the set of all messages and $[\![T]\!]$ the set of elements of type T) is an element of the set $\vec{C}$ defined as follows:

$$\vec{C} = \{x: C \to (M^*)^\infty: \\ \forall\, c \in C: x.c \in ([\![\text{type}(c)]\!]^*)^\infty \}$$

A channel valuation $x \in \vec{C}$ associates a stream of elements of type type(c) with each channel $c \in C$. The operators of streams induce operators on valuations by pointwise application.

Given a set of typed input channels I and a set of typed output channels O we introduce the notion of a *syntactic interface* of a component:

(I, O)	syntactic interface,
I	set of typed input channels, and
O	set of typed output channels.

In addition to the syntactic interface we need a concept for describing the *behavior* of a component. We work with a very simple and straightforward notion of behavior. A behavior is a *relation between input histories and output histories*. Input histories are represented by valuations of the input channels and output histories are represented by the valuations of output channels. We represent the black box behavior of a component by the set valued function F over the semantic interface $(\vec{I}, \vec{O})$:

$$F: \vec{I} \to \wp(\vec{O})$$

Given $x \in \vec{I}$, by F.x we denote the set of all output histories which a component with behavior F may produce on the input x. Of course, a set valued function, as well-known, is isomorphic to a relation. We prefer set-valued functions to emphasize the roles of input and output. We call the function F an *I/O-function*.

Using logical means, an I/O-function F can be described by a logical formula relating input channels with output channels.

Syntactically, such a formula uses typed channels as identifiers for streams.

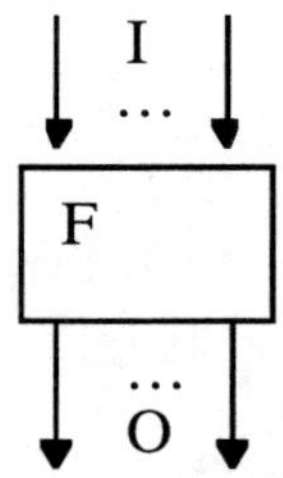

Figure 1. Graphical Representation of a Component F with Input Channels I and Output Channels O

A specification of a component provides the following information: its syntactic interface, describing how the component is connected to its environment and its behavior by specifying a formula Φ relating input and output channel valuations. This way we obtain a specification technique that gives us a very powerful method to describe components. By the specifying formula of a specification of an I/O-function we may prove properties about the specified function.

We use the following notation for a component F to refer to the constituents of its syntactic interface: In(F), the set of input channels I, and Out(F), the set of output channels O. In the following we introduce some basic notions for I/O-functions. An I/O-function

$$F: \vec{I} \rightarrow \wp(\vec{O})$$

(1) is called *properly timed*, if for all times $k \in \mathbb{N}$ we have

$$x{\downarrow}k = z{\downarrow}k \Rightarrow F(x){\downarrow}k = F(z){\downarrow}k,$$

(2) *time guarded* (or *causal*), if for all times $k \in \mathbb{N}$ we have

$$x{\downarrow}k = z{\downarrow}k \Rightarrow F(x){\downarrow}k{+}1 = F(z){\downarrow}k{+}1,$$

(3) *partial*, if $F(x) = \varnothing$ for some $x \in \vec{I}$ and *total* otherwise, and *realizable*, if for a time guarded function $f: \vec{I} \rightarrow \vec{O}$, for all $x \in \vec{I}$:

$$f.x \in F.x$$

Time-guardedness is a very basic notion. It formalizes the asymmetry between input and output. For time independent systems, time guardedness has a strong relationship to *prefix monotonicity* for nontimed channels. See [5] for a functional approach to system modeling without notions of time.

We do not require that an I/O-function described by a specification has always all the properties introduced above. We are much more liberal. We may add such properties to specifications freely whenever appropriate and therefore deal with all kinds of specifications of I/O-functions that do not have these properties.

A special case of I/O-functions are partial functions which are functions that for certain input histories may have an empty set of output histories. An extreme case is a function that maps every input history onto an empty set. Such functions are not very interesting when used for modeling the requirements for the implementation, since an implementation shows at least one output for each input. However, partial functions may be interesting as intermediate steps in the specification process, since based on these functions we can construct other functions that are not partial and more adequate for composition and implementation (see section 3).

2.2 Composition

In this section we introduce a notion of *composition* for components. We prefer to introduce only one very general form of composition and later define a number of other composition forms as special cases.

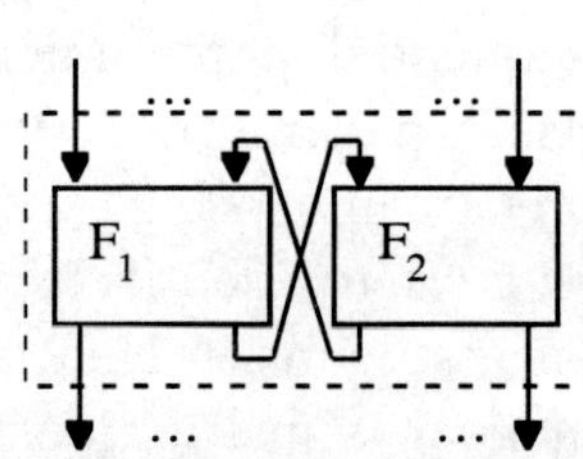

Figure 2. Parallel Composition with Feedback

Given I/O-functions F_i with i=1,2 and $O_1 \cap O_2 = \varnothing$, $I_1 \cap I_2 = \varnothing$:

$$F_i : \vec{I}_i \rightarrow \wp(\vec{O}_i)$$

we define the parallel composition with feedback by the I/O-function

$$F_1 \otimes F_2 : \vec{I} \rightarrow \wp(\vec{O})$$

where the syntactic interface is specified by $I = (I_1 \cup I_2) \backslash (O_1 \cup O_2)$, $O = (O_1 \cup O_2) \backslash (I_1 \cup I_2)$. The resulting function is specified by the following equation (with $y \in \vec{C}$, where C is an abbreviation for $I_1 \cup I_2 \cup O_1 \cup O_2$):

$$(F_1 \otimes F_2).x = \{ y|O : y|I = x|I \wedge y|O_1 \in F_1(y|I_1) \wedge y|O_2 \in F_2(y|I_2) \}$$

Here y denotes a valuation of all the channels of F_1 and F_2. By $y|C$ we denote the restriction of the valuation y to the channels in C. For this form of composition we can prove the following facts by rather straightforward proofs:

(1) if the F_i are *time guarded* for $i = 1,2$, so is $F_1 \otimes F_2$,

(2) if the F_i are *realizable* for $i = 1,2$, so is $F_1 \otimes F_2$.

If the F_i are total and properly timed for $i = 1, 2$, we cannot conclude, however, that the function $F_1 \otimes F_2$ is total. This shows that the composition works only in a modular way for well-chosen subclasses of specifications. Further forms of composition that can be defined (we do not give formal definitions for them, since these are quite straightforward):

- *feedback without hiding:* μF

  let $F: \vec{I} \rightarrow \wp(\vec{O})$, then we define: $\mu F: \vec{J} \rightarrow \wp(\vec{O})$ where $J = I \backslash O$ by the equation (here we assume $y \in \vec{C}$ where $C = I \cup O$):

  $$(\mu F).x = \{y|O: y|I = x|I \wedge y|O \in F(y|I)\}$$

- *parallel composition:* $F_1 \parallel F_2$

  only defined if $(I_1 \cup I_2) \cap (O_1 \cup O_2) = \varnothing$ then we have $F_1 \parallel F_2 = F_1 \otimes F_2$

- *sequential composition:* $F_1 ; F_2$

Sequential composition of the components F_1 and F_2 requires $O_1 = Out(F_1) = In(F_2) = I_2$; it is denoted by

$$F_1 ; F_2$$

In the special case where $O_1 = I_2 = (O_1 \cup O_2) \cap (I_1 \cup I_2)$ we can reduce sequential composition to parallel composition with feedback as follows:

$$F_1 ; F_2 = F_1 \otimes F_2$$

Further composition operators are hiding and renaming of channels.

3 Refinement

Refinement relations (see [7]) are the key to formalize development steps (see [3]) and the development process. We work with the following basic ideas of refinement:

- *property refinement* (enhancing requirements): allows us to add properties to a specification,
- *glass box refinement* (designing implementations): allows us to decompose a component into a distributed system or to give a state transition description for a component specification,
- *interaction refinement* (relating levels of abstraction): allows us to change the granularity of the interaction, the number and types of the channels of a component (see [4]).

We claim that these notions of refinement are sufficient to describe all the steps needed in the idealistic view of a strict hierarchical top down system development. The three refinement concepts mentioned above are explained in detail in the following.

3.1 Property Refinement

Property refinement allows us to replace an I/O-function by one with additional properties. A behavior

$$F: \vec{I} \rightarrow \wp(\vec{O})$$

is refined by a behavior

$$\hat{F} : \vec{I} \rightarrow \wp(\vec{O})$$

if $\forall x \in \vec{I}: \hat{F}(x) \subseteq F(x)$. Obviously, property refinement is a partial order. A property refinement is a basic refinement step as it is needed in requirements engineering. In the process of requirements engineering, typically the overall services of a system are specified. This, in general, is done by requiring more and more sophisticated properties for components until a desired behavior is specified.

3.2 Glass Box Refinement

Glass box refinement is a classical concept of refinement that we need and use in the design phase. In the design phase we typically decompose a system with a specified black box behavior into a distributed system architecture or we represent this behavior by a state transition machine. In other words, a glass box refinement is a special case of a property refinement of the form

$$F_1 \otimes F_2 \otimes \ldots \otimes F_n \subseteq F$$

Accordingly, a glass box refinement is a special case of property refinement where the refining component has a special syntactic form. In the case of a glass box refinement that transforms a component into a network.

In the case of the design of an architecture, by the decomposition we are fixing the basic implementation structure of a system. These components have to be specified and we have to prove that their composition leads to a system with the required functionality. The components $F_1, \ldots, F_n$ can be hierarchically decomposed again into a distributed architecture, until a granularity of components is obtained which should not be further decomposed into a distributed system but realized by an interactive state machine.

As explained, in a glass box refinement we replace a component by a design which is given by a network of components $F_1 \otimes F_2 \otimes \ldots \otimes F_n$. Of course, a glass box refinement can be applied to each of the components F_j in a network of components. This form of iterated glass box refinement leads to a hierarchical, top down refinement method.

3.3 Interaction Refinement

Interaction refinement is the refinement notion that we need for modeling development steps between levels of abstraction. Interaction refinement allows us to change the number and names of input and output channels and the granularity of the messages on the channels of a component. An *interaction refinement* requires two functions

$$A: \vec{C}' \rightarrow \wp(\vec{C}) \quad R: \vec{C} \rightarrow \wp(\vec{C}')$$

that relate the abstract with the concrete level of a development step leading from one level of abstraction to the other. Given an abstract history $x \in \vec{C}$ each $y \in R(x)$ denotes a concrete history representing x. Calculating a representation for a given abstract history and then its abstraction yields the old abstract history again. This is expressed by the requirement:

$$R ; A = Id$$

where Id denotes the identity relation. A is called the *abstraction* and R is called the *representation*. R and A are called a *refinement pair*. Interaction refinement allows us to refine components, given appropriate refinement pairs for the input and output channels. The idea of a interaction refinement is visualized in Fig. 3.

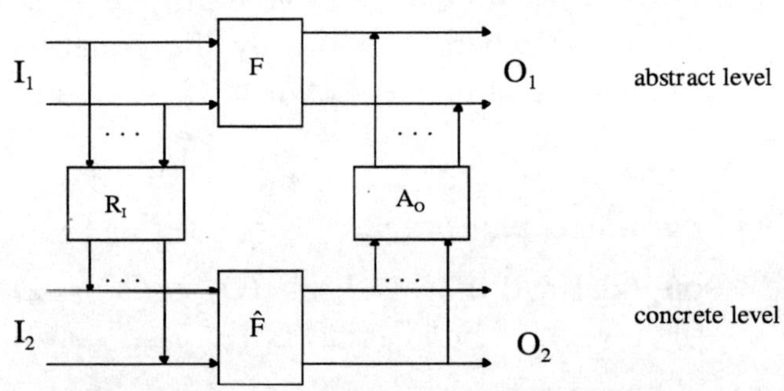

Figure 3. Interface Interaction Refinement (*U-simulation*)

Given interaction refinements

$$A_I : \overset{\perp}{I}_2 \rightarrow \wp(\overset{\perp}{I}_1) \qquad R_I : \overset{\perp}{I}_1 \rightarrow \wp(\overset{\perp}{I}_2)$$

$$A_O : \overset{\perp}{O}_2 \rightarrow \wp(\overset{\perp}{O}_1) \qquad R_O : \overset{\perp}{O}_1 \rightarrow \wp(\overset{\perp}{O}_2)$$

for the input and output channels we call the I/O-function

$$\hat{F} : \overset{\perp}{I}_2 \rightarrow \wp(\overset{\perp}{O}_2)$$

an *interaction refinement* of

$$F: \overset{\perp}{I}_1 \rightarrow \wp(\overset{\perp}{O}_1)$$

if one of the following propositions holds:

U^{-1}-Simulation: $\hat{F} \subseteq A_I ; F ; R_O$

Downward Simulation: $R_I ; \hat{F} \subseteq F ; R_O$

Upward Simulation: $\hat{F} ; A_O \subseteq A_I ; F$

U-Simulation: $R_I ; \hat{F} ; A_O \subseteq F$

These are different versions of useful relations between levels of abstractions. A more detailed discussion can be found in [7].

4. Tool Support

The CASE tool AutoFocus [1], which has been developed at the Technische Universität München, spans the bridge between the two difficult-to-integrate worlds of practical systems engineering and its logical framework presented here. As a consequence, AutoFocus provides an infrastructure for systems engineering with practice-oriented description techniques while also giving a formal basis for them. The first protoype of our tool allows to specify a system graphically in different views. With System Structure Diagrams (SSDs) we model the architecture of the system under development, including all communi-cation channels. The behavior of a system can be described by State Transition Diagrams (STDs), and a scenario-oriented view can be given by Extended Event Traces (EETs), a MSC-like language.

The special feature of AutoFocus is that all system views are based on the same formal basis. Thus, different system parts specified in different languages can be integrated seamlessly. This is the best proof that our compositional approach pays. Hence, each part can be modelled in the description technique that is most appropriate for this system part. Furthermore, it is possible to capture the system's requirements with EETs and then translate or, in other words, *refine* the result of this development phase into STDs to further develop the system. This way, AutoFocus supports an incremental development process.

AutoFocus also provides static consistency checks to verify type correctness of different system views. In the near future, we will also implement dynamic checks to verify the semantic consistence of these views. For instance, this enables us to prove automatically whether a state transition diagram is consistent with the corresponding EET. Great store has been set by open interfaces, that is, our CASE tool provides an interface for arbitrary, user-defined properties that can be given in first order logic.

A further component of AutoFocus is its simulation component, named „SimCenter". It permits simulating executable system parts by means of Java code.

Due to the open interfaces of AutoFocus, further tools, like those for visualizing system runs, can be linked up with it. Currently, we are extending our tool not only by automatic verification techniques such as Model Checking or Testing but also by abstraction techniques that enable the formal, automatic verification of systems with infinte state spaces. Abstraction is just the inversion of refinement and again the key role of a formal basis for a CASE tool becomes clear.

5. Conclusion

What we have presented in the previous chapters is a comprehensive method for a system and software development which supports all steps of a hierarchical stepwise refinement development method. It is compositional and therefore supports all modularity requirements that are generally needed. What we have presented is a method that provides the following ingredients:

- a mathematical notion of a syntactic and semantic interface of a component,
- a formal specification notation and method,
- a precise notion of composition,
- a mathematical notion of refinement and development,
- a compositional development method,
- a flexible concept of software architecture,
- concepts of time and the refinement of time (see [9]).

What we did not mention throughout the paper are concepts that are also available and helpful from a more practical point of view including combination with tables and diagrams.

The simplicity of our results is a direct consequence of the specific choice of our semantic model. The introduction of time makes the model robust and expressive. The fact that communication histories are explicitly included allows us to avoid all kinds of complications like prophecies or stuttering and leads to an abstract and relational view of systems.

Of course, what we have presented is just the scientific kernel of the method. More pragmatic ways to describe specifications are needed. These more pragmatic specifications can be found in the work done in the SysLab-Project (see [2]) at the Technical University of Munich. For extensive explanations of the use of state transition diagrams, data flow diagrams, and message sequence charts as well as several versions of data structure diagrams we refer to this work.

References

1. F. Huber, B. Schätz, G. Einert: Consistent Graphical Specification of Distributed Systems. In: J. Fitzgerald, C. B. Jones, P. Lucas (ed.): FME '97: 4th International Symposium of Formal Methods Europe, Lecture Notes in Computer Science 1313, 1997, 122 -141
2. R. Breu, R. Grosu, Franz Huber, B. Rumpe, W. Schwerin: Towards a Precise Semantics for Object-Oriented Modeling Techniques. In: H. Kilov, B. Rumpe (eds.): Proceedings ECOOP'97 Workshop on Precise Semantics for Object-Oriented Modeling Techniques, 1997, Also: Technische Universität München, Institut für Informatik, TUM-I9725, 1997
3. M. Broy, B. Möller, P. Pepper, M. Wirsing: Algebraic Implementations Preserve Program Correctness. Science of Computer Programming 8 (1986), 1 - 19
4. M. Broy: Compositional Refinement of Interactive Systems. Digital Systems Research Center, SRC Report 89, July 1992, To appear in JACM
5. M. Broy, F. Dederichs, C. Dendorfer, M. Fuchs, T. F. Gritzner, R. Weber: The Design of Distributed Systems - An Introduction to Focus. Technische Universität München, Institut für Informatik, Sonderforschungsbereich 342: Methoden und Werkzeuge für die Nutzung paralleler Architekturen TUM-I9202, January 1992
6. M. Broy, F. Dederichs, C. Dendorfer, M. Fuchs, T. F. Gritzner, R. Weber: Summary of Case Studies in FOCUS - A Design Method for Distributed Systems. Technische Universität München, Institut für Informatik, Sonderforschungsbereich 342: Methoden und Werkzeuge für die Nutzung paralleler Architekturen TUM-I9203, January 1992
7. M. Broy: Interaction Refinement – The Easy Way. In: M. Broy (ed.): Program Design Calculi. Springer NATO ASI Series, Series F: Computer and System Sciences, Vol. 118, 1993
8. M. Broy: Towards a Mathematical Concept of a Component and its Use. First Components' User Conference, Munich 1996. Revised version in: Software-Concepts and Tools 18, 1997, 137 - 148
9. M. Broy: Refinement of Time. M. Bertran, Th. Rus (eds.): Transformation-Based Reactive System Development. ARTS'97, Mallorca 1997. Lecture Notes in Computer Science 1231, 1997, 44 - 63, To appear in TCS

MODEL-BASED SYSTEMS ENGINEERING OF AUTOMOTIVE SYSTEMS

Gerard H. Fisher, Vitech Corporation, Vienna, VA

Abstract

For as long as systems engineers have been around, people have thought of them as "paper pushers." The primary products of systems engineers all seem to be documents such, as specifications and trade-off studies. Communicating the requirements to hardware and software designers with documents is fraught with problems. By using a model-based approach to systems engineering, there is no ambiguity. The model fully defines the functional behavior, inputs and outputs, and the physical architecture, as well as the performance and resource requirements. This method provides a unified, consistent and traceable design.

This paper describes the design of the Automotive Personal Assistance System (APAS) exploiting the features of the Global Positioning System, which is similar to the Cadillac OnStar System. It includes the entire systems engineering life cycle, from source requirements analysis, through behavior analysis and physical architecture, to verification and validation. The behavior model can be fully executed to verify its correctness long before money is wasted building prototypes. The use of the model-based approach will aid in the reduction of product cost and the increase in product quality.

Introduction

Systems engineers (SE) have been prolific producers of proscriptive prose. The SE's most prominent products are documents; the heavier, the better. Systems engineers perform many tasks, but always convey their results in the form of a document (e.g., system specification, unit specification, trade-off study report, etc.). When the SE passes a specification to the hardware or software engineers, there is a tremendous opportunity for miscommunication. Because of the multiple meanings of English words, it is easy to misunderstand the author's intent. Furthermore, words are not the only requirements. Requirements may also come in the form of relationships, attributes and diagrams. It is very difficult to perfectly coordinate all of these items. Using an object-oriented model-based tool, such as CORE®, for system development will greatly aid the systems engineer in this task.

This paper discusses the mythical design of a automotive system to aid the driver through the use of Global Positioning System (GPS) data. Cadillac offers such a system called OnStar. There are also several "after market" versions available.

Originating Requirements

Figure 1 illustrates the simple requirements for the APAS in the form of a context diagram. As the requirements are analyzed, it is useful to draw such a diagram to help understand the overall context for the system to be designed. The large ellipse represents the system of interest. The smaller ellipses show the external interfaces to our system. Key inputs and outputs are shown as labeled arrows.

The APAS is required to perform the following:

- Receive and process Global Positioning System data.
- Alert monitoring station of automobile accident

0-7803-5086-3 /98/$10.00 ©1998 IEEE

- Remotely unlock doors at owner's request
- Monitor & assess its own performance
- Retain automobile position history for 24 hours.

In this example, the customer came back later with some additional requirements (ECP #1), which added the following:

- Alert monitoring station of auto theft and current position
- Assist owner in finding car in parking lot by sounding horn and flashing lights.

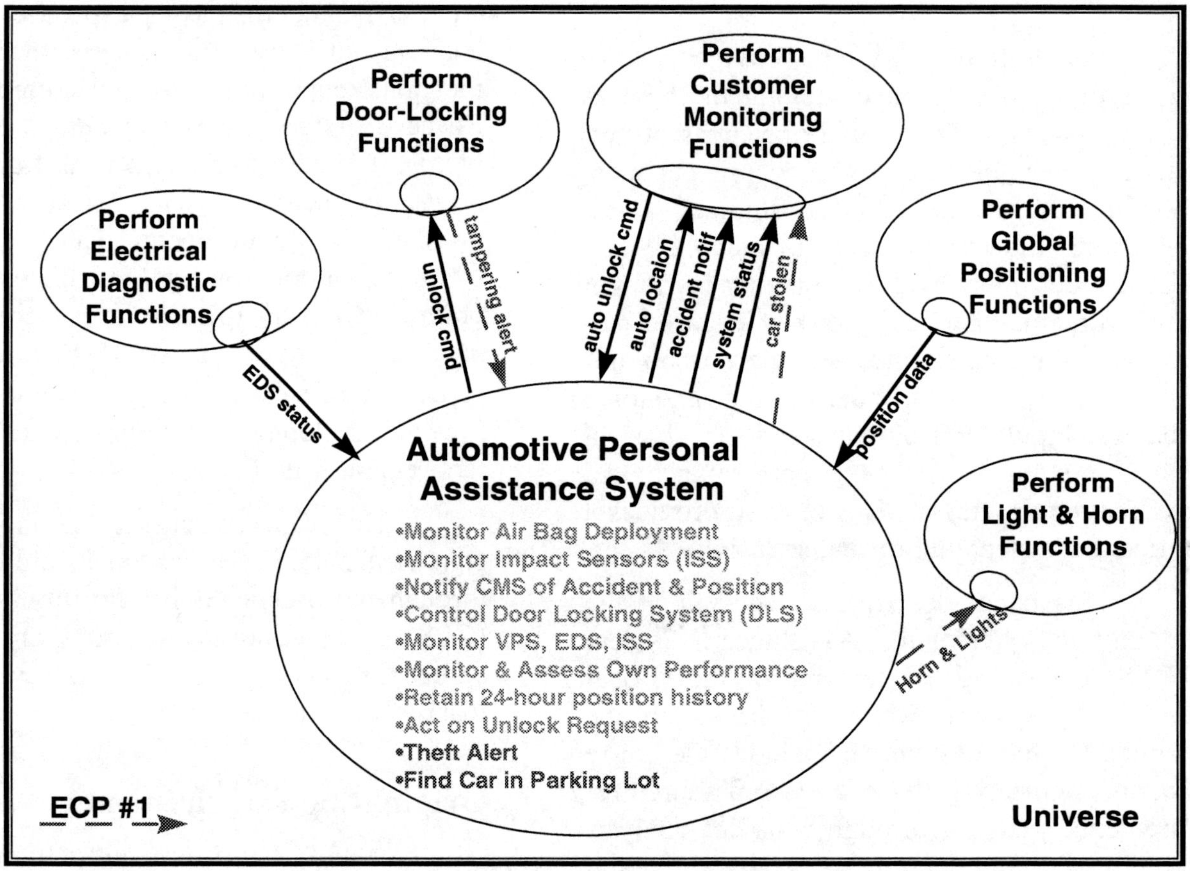

Figure 1. APAS Context Diagram.

These requirements are analyzed by the systems engineer and extracted. Compound requirements are broken into simple statements. Surely some of the requirements will be ambiguous. Those ambiguities are usually handled by creating an action item or *issue*.

The issue describes the problem associated with the originating requirement and assigns a severity, ownership and a due date. These ambiguities are perhaps the major impediment to progress in the initial design phase. Customer TBDs (to be determined) and ambiguous requirements will stretch the schedule and add cost to a project. Once sufficient information is obtained, the issue should document the assumptions made, the alternatives considered and the eventual decision, along with the usual configuration control data. By keeping the customer informed of these issues and diligently tracking them to conclusion, one can minimize the impact. Figure 2 illustrates the requirements hierarchy for the basic re-

0-7803-5086-3 /98/$10.00 ©1998 IEEE

quirements for APAS. It traces the requirements from their source specification through high-level compound requirements to the leaf-level, implementable and verifiable requirements.

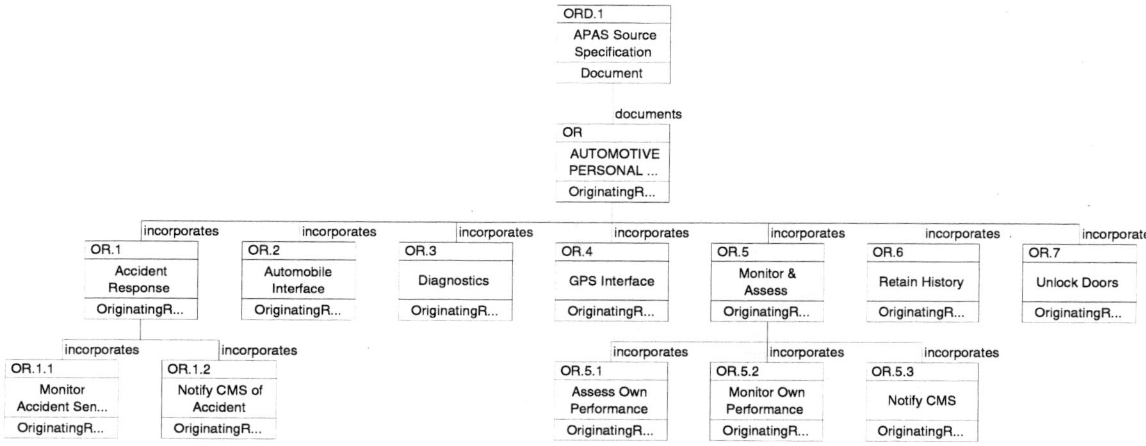

Figure 2. Requirements Hierarchy.

Behavior Modeling

Once all requirements have been extracted, analyzed and understood, it is time to postulate one or more functional models that would fulfill these requirements. All requirements should trace to a function. If you end up with functions that do not trace to a requirement, the function may be superfluous. If, on the other hand, you find requirements without a corresponding function, your model is incomplete. The models should be constructed in layers. The top layer should be similar to the context diagram described above.

An example of such a high-level model is shown in Figure 3, in the form of an enhanced function flow block diagram (E-FFBD). The system under design is "Operate APAS." The other functions shown represent the external interfaces to APAS. These include the GPS interface, the door lock system in the car, the automobile's electronic diagnostic system (EDS), automobile's horn & light systems, and the customer monitoring station (CMS) that supports calls for help. The *and* construct indicates that all functions are operating in parallel. The functions have two types of inputs and outputs shown. Those with only a source or a destination are called *data stores*. The data is assumed to be available whenever it is needed. The other type of data element is called a *data trigger*. A data trigger is generated by one function and sent to one or more other functions. The functions requiring the data trigger cannot begin until it arrives. This feature permits the E-FFBD to be simulated.

The use of a discrete event simulator is very helpful in verifying the logical correctness of the design. Whenever possible, a systems engineer should postulate two or more behavior models and trade them off to select the best design. The simulator is one of the tools that will aid in these trade-offs. However, this type of trade-off requires the combination of economic, performance and subjective data to arrive a single figure of merit. Another technique that would be useful for this task is the analytic hierarchy process (AHP).

0-7803-5086-3 /98/$10.00 ©1998 IEEE

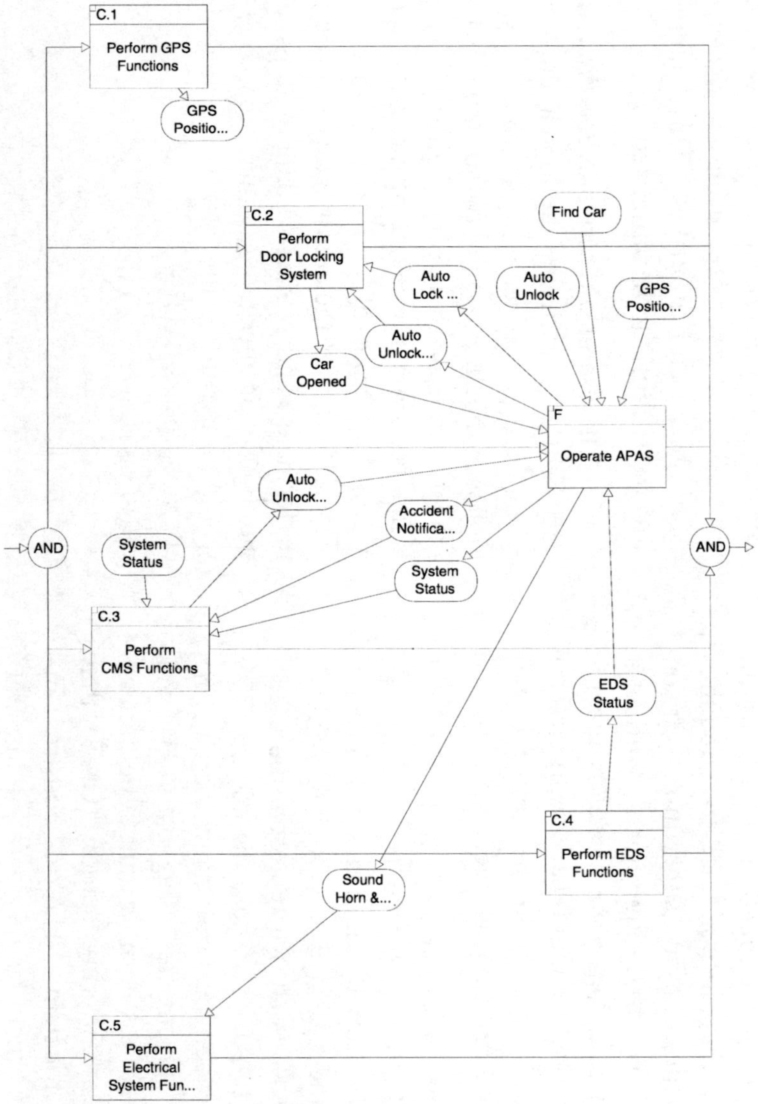

Figure 3. Functional Context Diagram.

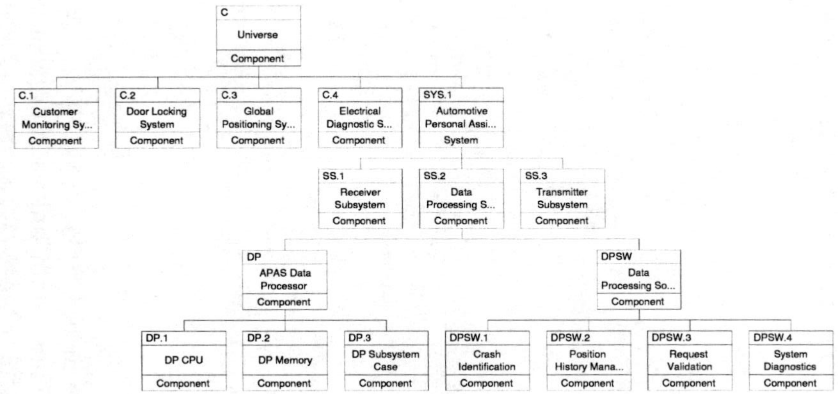

Figure 4. Physical Design Hierarchy.

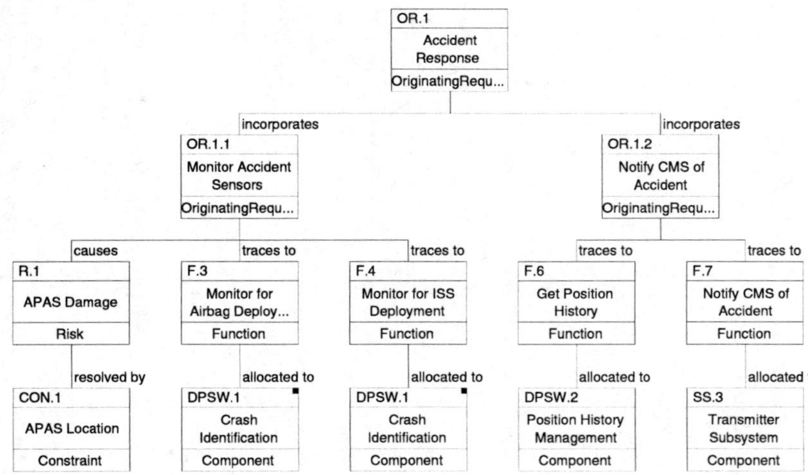

Figure 5. Traceability from Requirements to Components.

0-7803-5086-3 /98/$10.00 ©1998 IEEE

AHP is a form of multi-attribute decision analysis, developed by Dr. Thomas Saaty in the early 1970's. AHP includes development of a hierarchical model of the criteria to be used in a decision and then develops a set of relative weights using a pairwise comparison method. The SE might make a series of Monte Carlo runs for each proposed behavior model, measuring task times, wait times, queue sizes, etc. These results, along with other measures and estimates, could be plugged into an AHP model to determine the most compliant design.

Physical Design

Once a preferred behavior is selected, the SE can begin developing several physical designs. As part of the physical design, the systems engineer will allocate functions to hardware, software, and humans. These are the three types of physical components in a system. The way in which these three function together is the *system* or *process*. In a similar fashion to the process above, these designs should be traded off to select the preferred design. If one were to give the same requirements and behavior model to three experienced systems engineers, the results would be three unique system designs. There are always some obvious choices for software implementation or hardware implementation, but some require serious trade-offs. This could be another application for the AHP. Some of the key parameters one considers in physical design trade-offs are cost (both initial and recurring), power consumption, performance, reliability, maintainability, supportability, technical risk, etc. Figure 4 illustrates the physical design hierarchy for APAS. The interfacing systems are numbered C.1 through C.4. Our system, SYS.1, has three major subsystems: receiver, transmitter and data processing. The data processing subsystem is comprised of both hardware (the data processor) and software (the data processing software). Each of these can then be broken down to as low a level as necessary. However, the systems engineers should cease design when they have allocated function to the specialty (HW or SW). Let the specialty engineer make the design and implementation choices below that level. This is not to imply that the specialty engineers should use a different method. They can and should stay with the same model-based approach throughout their design phase. When the hardware engineers are ready to design VHDL (the hardware language used to specify design) and the software engineers are ready to code, they can just export the design from the tool. There are several tools that interface with VHDL and code generators. As a final check of your design, Figure 5 shows you a trace from specific requirements through functions to physical components. The inverse would ensure that all physical components had a reason for being in the design. If they could not be traced back to a specific requirement, the component might be unnecessary.

0-7803-5086-3 /98/$10.00 ©1998 IEEE

Verification and Validation

During the requirements analysis phase, a key step is the assurance that compliance with a requirement can be proven. If not, the requirement is not valid and the issue should be resolved with your customer. If the requirement can be verified, then a high-level *verification requirement* should be generated during the requirements analysis phase. You have though it out and it is a good idea to record it while it is fresh in your mind. For instance, if a requirement specified a 95% isolation of all hardware faults to a single circuit board, you might specify the verification requirement "verify 95% fault isolation to a single circuit board by doing 100% fault insertion on all input / output pins on the board." This would give you enough information to be able to estimate the cost of the effort. When detailed test planning is done, that verification requirement would be expanded to describe a *verification event*, assign a responsible individual, develop a test plan and procedure, and identify the test configuration needed to perform such a test.

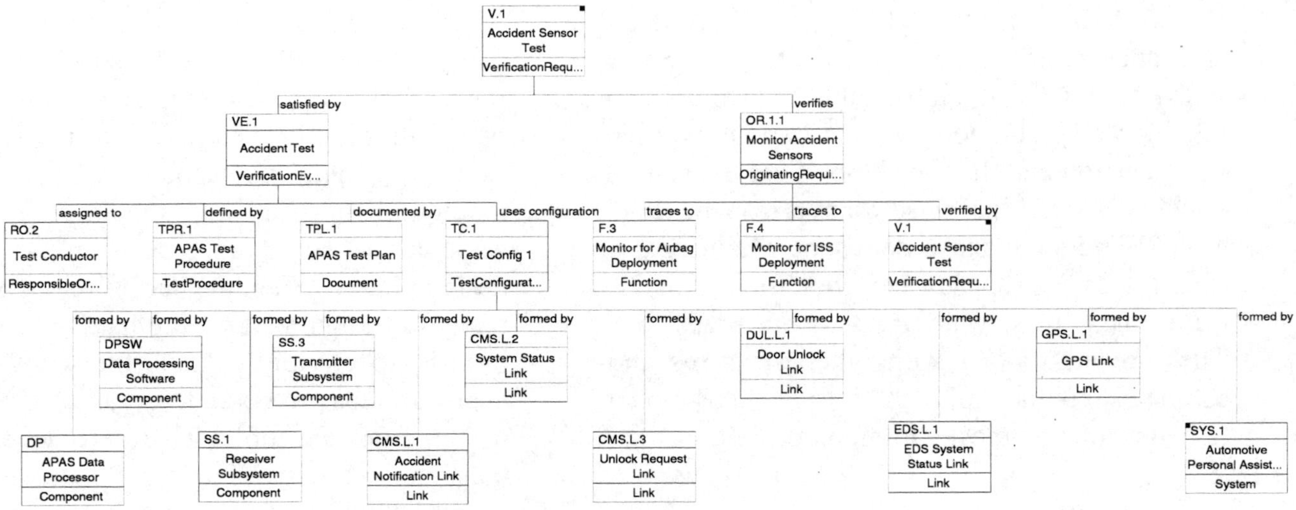

Figure 6. Verification Hierarchy.

Figure 6 illustrates the verification hierarchy for the APAS Accident Sensor Test. In addition to the information described above, it also identifies the requirements and functions to be verified. This becomes especially useful in large systems, where test resources are at a premium. Being able to identify the equipment needed will permit scheduling of tests to avoid conflicts.

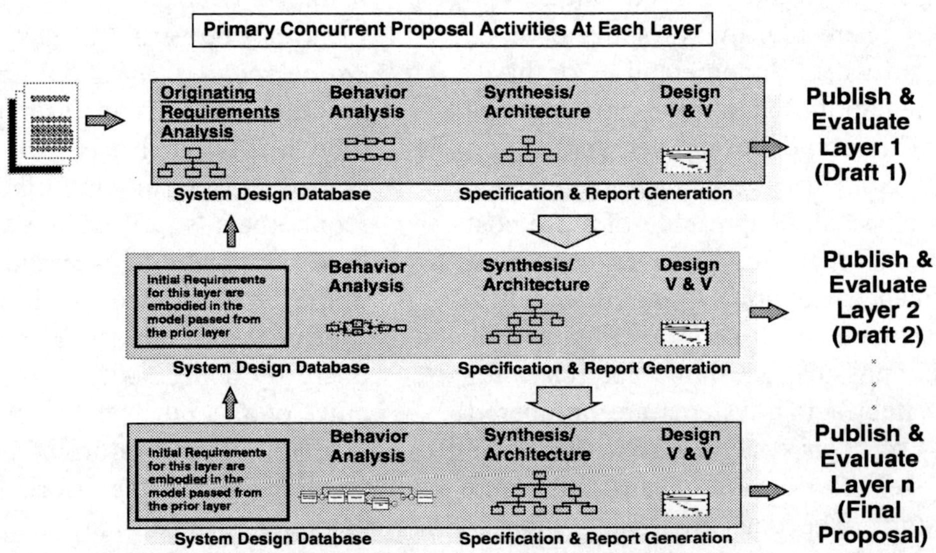

Figure 7. The Onion Model.

0-7803-5086-3 /98/$10.00 ©1998 IEEE

The Onion Model

Figure 7 illustrates the recommended method for doing model-based design. First extract high-level requirements and develop a high-level model for behavior. Then develop a corresponding high-level physical design and verification and validation (V&V) plans. Trace requirements to the model, design, and V&V to ensure all requirements were addressed. This is a complete design and would greatly assist your team in understanding and reviewing its completeness. Then proceed to the next level of abstraction in the requirements, model, design, and V&V. This process should be continued until all requirements are addressed.

Summary

The model-based approach to systems engineering offers clear advantages over the traditional paper-driven approach. The most obvious is that the database containing the model becomes the "corporate memory" for the project. Systems engineers change jobs often and take their project knowledge with them. With all rationale and decisions in the database, the knowledge stays with the project. Documents get lost and one is never sure whether the copy you have is the most up to date. This is not a problem with the model-driven approach.

The model also permits "what-if" analysis later. After the system is fielded, the customer might come to you and ask for the impact of making some upgrades to the system. With an up-to-date model, such an analysis is almost trivial.

Biography

Gerard H. Fisher is a Member of the Technical Staff at the Vitech Corporation, in Vienna, Va. He provides consulting services on the application of CORE® to support systems engineering, decision analysis, product development and program management. Prior to joining Vitech, Mr. Fisher was the corporate Director for Enterprise Process Improvement, for CACI. For 25 years, Mr. Fisher was a Senior Systems Engineer at IBM Federal Systems (and Loral after the acquisition) in Manassas, Virginia. He was involved in systems engineering on the Advanced Ballistic Missile Defense System and several high-performance signal processors for federal customers. He chairs the International Council on Systems Engineering (INCOSE) System Reengineering Working Group and is Technical Chair for the INCOSE WMA Chapter. He received his B.S. in Physics from St. Bonaventure University and M.S. in Systems Engineering from Virginia Tech.

0-7803-5086-3 /98/$10.00 ©1998 IEEE

TACTICAL AIRCRAFT MOVING MAP CAPABILTY (TAMMAC)

Using Commercial-Off-the-Shelf (COTS) Technology to Solve Common Military Aircraft Avionics Obsolescence Problems

Ervin Pagel, The Boeing Company, St. Louis, MO

Abstract

Many military aircraft have on-board computers that generate moving map video for display in the cockpit. Many of these same aircraft have computers that record aircraft data in flight. The F/A-18 Hornet and the AV-8B Harrier, manufactured by Boeing (formerly McDonnell Douglas Aerospace (MDA)), have both computers. However, the computers that perform these tasks on the Hornet and Harrier have become obsolete. They were designed in the late 1970's, built in the early 1980's, and contain several electronic components within them that can no longer be procured.

In 1995, MDA teamed with the Naval Air System Command's Air Combat Electronics or ACE group, to design and develop the TAMMAC system to replace these computers on the F/A-18 and AV-8B as well as be the common digital map/data recording system for all Navy aircraft.

This paper describes the system engineering process used to develop the TAMMAC system and how this process helped maximize the application of COTS technology within TAMMAC while maintaining performance in the military aircraft environmental conditions.

Program Background

Consumer electronics technology is evolving at such a fast pace that by the time you purchase a computer, bring it home, and learn how to use it, a new one is available in the store that is better, cheaper, and faster. This same problem is magnified within the defense electronics industry because the same technology is being used but it takes several years to purchase a new avionics computer system. By the time the system gets defined and built, a better, cheaper, and faster solution exists. This problem was addressed on the TAMMAC development by adopting new acquisition rules available to the Navy to try and reduce the obsolescence problems by using as many COTS components as possible. By maximizing use of COTS and giving suppliers lifetime responsibility of the circuitry inside the computer, the suppliers will have an easier time inserting new technology into their designs throughout the life of the system.

Digital Map Computer Obsolescence

The F/A-18 and AV-8B aircraft have a Digital Video Map Set or DVMS that generates digital map video for display in the cockpit. This computer, like many others, has many parts obsolescence problems.

The most visible is the Write-Once-Read-Many (WORM) optical disk used to store digital map data within the unit. The foreign supplier has stopped production of the disks. The Navy has purchased a limited number of these disks, but it is not sufficient to meet future needs.

Other obsolescence problems exist. The DMVS does not have enough processor throughput or storage capacity to support growing demands. The digital map database used by the DVMS is unique and the Navy wants to discontinue support for it. The Navy

0-7803-5086-3 /98/$10.00 ©1998 IEEE

wants to use the standard Department of Defense digital map database for all activities, but the DVMS can not use this database without costly preprocessing.

Data Transfer Computer Obsolescence

The F/A-18 and AV-8B aircraft also have a Digital Storage Set or DSS that is used to transfer data between the aircraft and ground processing stations. This computer is also facing parts obsolescence. The most critical DSS problem is its limited memory capacity. The DSS only contains less than one Megabyte of memory to store data recorded in flight. As a result, the amount of flight time that can be recorded becomes a limitation that is not acceptable to the Navy. Replacing the DVMS described above requires the data transfer computer to support a new method to load digital map data into the host aircraft. The DSS does not support this new concept.

Lack of Funding for New Developments

A result of shrinking defense budgets is not being able to fix all obsolescence problems when they are identified. Problems that are less critical are given lower priorities and some less than optimal computer systems remain in use longer. For years the F/A-18 and AV-8B Navy program offices and MDA have unsuccessfully tried to establish a development program to replace the systems described above, but funding was not available. However, the Navy's Air Combat Electronics (ACE) group's group was funded to develop a digital map system very similar to the one required by the F/A-18 and AV-8B.

The ACE group's charter is to provide avionics that are common to several Navy aircraft. The benefits of developing common equipment called Government Furnished Equipment or GFE result when the needs of several Navy programs can be pooled together to gain economies of scale. However, the F/A-18 and AV-8B were reluctant to use GFE for this application because they would have to relinquish control of the equipment's development and they have had some problems with GFE developments in the past.

The TAMMAC Solution

After several weeks of meetings, the F/A-18 and AV-8B agreed to use GFE on their aircraft - provided it contained both digital map *and* data transfer functionality. The ACE group also adopted a suggestion from the F/A-18 and AV-8B programs to use MDA as a partner in this GFE development. MDA has successfully developed new systems for the F/A-18 and AV-8B before, so the Navy program offices believed if the ACE group worked with MDA on this development, the probability of the program's success would increase. Also, the ACE group needed customers to maintain their development budgets so they agreed to add the data transfer functions to their requirements for this development. The Navy later approved MDA as the sole source developer for this system and the TAMMAC program was formally established.

Since TAMMAC was to be developed as GFE, it was also planned to be part of the program to update the H-1 helicopter's cockpit and replace the current digital map and data loading systems on the V-22. In total, approximately 2,500 production TAMMAC Systems will be built.

Program Goals

The goals of the TAMMAC program were formally documented in the Navy's Operational Requirements Document or ORD. Some of the goals are listed below:

- Replace F/A-18 AV-8B obsolete systems
- Fit into DVMS and DSS mechanical envelopes
- Install new system in the F/A-18 E/F aircraft in December of 2000
- Reduce unit count from three to two
- Reduce cost
- Reduce weight
- Increase reliability

- Adopt a two-level maintenance concept
- Increase performance
- Establish new digital map loading concept
- Reduce/eliminate use of military specifications
- Reduce/eliminate use of military standards
- Maximize use of COTS technology

Many of these goals are standard for a new development program, but trying to accomplish all of them at the same time was very challenging. In order to have an opportunity to succeed, everyone working on the program needed to know what their role was and to have their tasks clearly defined.

Program Structure

New Working Relationships

The ACE group had never had worked directly with an aircraft contractor before because they historically contracted directly to avionics box suppliers. For this development, MDA would be the prime contractor. The Navy's F/A-18 and AV-8B program offices were accustomed to working directly with MDA on avionics box developments. For this development they were going to have to work through the ACE group to formally communicate with MDA. A TAMMAC Integrated Program Team (IPT) was created to define how these organizations and the suppliers would work together.

Organization

A lead organization was identified for each one of the major program tasks and each member of the team was assigned to a task. The Navy took responsibility to lead efforts for non-MDA (UH-1Y, AH-1Z, and MV-22) aircraft integration activities, program test and evaluation, and logistics. MDA was given the lead for subsystem development (digital map, data transfer, and high-speed interface), mission planning software development, and MDA aircraft (F/A-18 and AV-8B) integration activities.

Communication Protocol

Establishing communication protocol was another important step in defining the program's structure. Formal contracts were to be placed between the ACE group and MDA and between MDA and the avionics suppliers. The ACE group also had informal agreements between themselves and aircraft program offices. Primary points of contact for formal communications (contractual changes in scope) and informal communication (requests for information) were established. Communication protocol was emphasized through team training prior to contract award and subsequently became a "team norm." It was critical to the program's success that everyone followed the protocol. For example, preventing the Navy's aircraft program offices from directing changes in scope to one of the avionics suppliers without going through the ACE group and MDA was necessary.

Program Phases

The TAMMAC program was divided into five phases the following phases to ensure a step by step process was used to flow down program goals into subsystem or unit requirements. The phases include:

- Requirements Study
- Integrated Product Definition
- Engineering and Manufacturing Development
- Low Rate Initial Production
- Full Rate Production

Phase 1 - Requirements Study

This phase was used to merge preliminary common digital map requirements and trade study results compiled by the ACE group with the specific F/A-18 and AV-8B digital map and data transfer requirements developed by MDA. The new concept for loading digital map data into the host aircraft was also refined. MDA continued investigating use of Personal Computer Memory Card International Association (PCMCIA) standards or "PC Cards" to transfer data between the aircraft and the ground processing stations. The product of merging and refining the requirements was a collection of specifications that now defined common digital map and data transfer computers that could be used on both fixed wing and rotary wing Navy aircraft. A functional requirements document was generated to show how ORD requirements were being allocated into the various equipment performance specifications. The following specification tree was determined:

- Operational Requirements Document (ORD)
  - TAMMAC System Specification
    - Digital Map System (DMS)
    - Advanced Memory Unit (AMU)
    - Flight Worthy PC Card
    - Non-Flight Worthy PC Card
    - High-Speed Interface Cable

Because MDA uses subcontractors for developing avionics, the study phase also funded initial source selection. Fifteen companies sent a "request for information" asked for preliminary design proposals for the DMS and AMU. The list of companies to receive the request included potential digital map suppliers identified by the ACE group and MDA and potential data transfer equipment suppliers identified by MDA. The Navy and MDA collaborated on the criteria for selecting the suppliers that would continue working into the next phase. After evaluation of fourteen responses five suppliers were selected.

Another product of this phase was a program plan that defined the communication protocol, roles, responsibilities and organizational structure described earlier.

Phase 2 - Integrated Product Definition or IPD

MDA began using IPD in the early 1990's as a method to improve the avionics procurement process. Prior to IPD, MDA would send out a request for proposal that included technical, data, management, and contractual requirements. The responses were evaluated with limited supplier interaction (teleconferences) to resolve issues. The winning supplier would be selected based solely on the proposal submitted.

The same requirements are generated using IPD, but at least two (sometimes three) suppliers are subjected to a three to six month competition where they have an opportunity to explain their design solution. During this period several meetings are held with the each supplier where they are encouraged to challenge, question, and propose changes to the requirements. Technical and business representatives from MDA and the Navy attend each meeting to support one-on-one meetings to refine requirements. During IPD, MDA is responsible for ensuring that the minimum baseline requirements are maintained. For the TAMMAC program, two candidate suppliers were selected for the AMU IPD and three for the DMS IPD. Although the process was very labor intensive, it resulted in higher quality specifications that were better understood by all three organizations.

An example of some of the technical issues studied during the TAMMAC IPD phase included:

- PC Card performance in military environment
- PC Card benchmarking
- AMU and DMS Processor benchmarking
- AMU door and PC Card receptacle mechanization
- Magnetic hard disk drive performance under military environment
- Use of flow through heat exchangers on circuit cards
- Programmable DMS output video
- Mezzanine circuit card applications for the DMS

At the conclusion of IPD, suppliers submit their final "buy-to package." This package includes the following elements:

- Technical Specifications
- Management Requirements / Statement of Work
- Data List
- Schedule
- Purchase Order (Pricing)
- Contract Terms and Conditions

Using criteria established by MDA and approved by the Navy prior to IPD, MDA evaluates and scores each supplier's performance during IPD and their final buy-to package. The winner is selected and the components of the buy-to package shown above become the actual contract placed between MDA and the winning supplier.

Phase 3 - Engineering and Manufacturing Development (EMD)

In parallel to IPD, the Navy and MDA established the elements for the TAMMAC EMD contract. MDA was awarded a contract for TAMMAC EMD from the Navy's ACE group on 20 December 1996. It contained five major components:

- Development of the TAMMAC System (DMS, AMU, PC Cards, High-speed Interface)
- Integration of the TAMMAC System onto the F/A-18 aircraft
- Integration of the TAMMAC System onto the AV-8B aircraft
- Support the integration of the TAMMAC System onto non-MDA aircraft (AH-1Y, UH-1Z, & MV-22)
- Development of Tactical Automated Mission Planning System (TAMPS) software for TAMMAC

Boeing subsequently awarded subcontracts in January 1997 for the TAMMAC system components. Harris Corporation in Melbourne, FL was selected to develop the Digital Map System and Smiths Industries in Grand Rapids, MI was selected to develop the Advanced Memory Unit.

Although a traditional development program was established, the team continued to use non-traditional ways to achieve some of the milestones listed below.

- Systems Requirements Reviews,
- Preliminary Design Reviews
- Critical Design Reviews
- Prototype DMS and AMU Deliveries
- Pre-production DMS and AMU Deliveries
- DMS and AMU Qualification Testing
- DMS and AMU Reliability Testing
- Delivery of Mission Planning Software
- Development of F/A-18 and AV-8B Aircraft Software Modifications
- Integration Testing at MDA
- Aircraft Modifications to Accommodate TAMMAC
- Support Flight Testing at China Lake and Pax River
- Production Readiness Reviews
- Functional and Physical Configuration Audits
- Support of System Technical Evaluation
- Support of System Operational Evaluation

For example, the design reviews were divided into segments and were accomplished incrementally. This allowed portions of the design to evolve on their own pace. Some design components within the units were adopted from other programs so they were much more mature than the new "clean sheet" designs created specifically for TAMMAC. The critical design reviews for some of the mature elements were completed concurrent to the preliminary design review for the other newer elements. Hardware development also lagged software development so the design reviews for those components were separated and accomplished independently. The suppliers, to tie all the increments together, presented DMS and AMU design summaries. A "System" design review was also conducted by MDA to describe how system requirements were being met.

One of the lessons learned from the TAMMAC program development was the suppliers had to verify COTS component documentation prior to establishing final subsystem designs. All COTS component design parameters identified in the data sheets had to be benchmarked through testing.

Phase 4 - Low Rate Initial Production (LRIP)

LRIP will include producing up to 250 TAMMAC systems for installation into the F/A-18 and AV-8B aircraft. The Navy will contract through MDA to procure LRIP systems. A five-year warranty will be placed on each DMS delivered and a five-year AMU maintenance agreement will be activated concurrent with delivery of the first production AMU.

Phase 5 - Full Rate Production (FRP)

FRP will be used to procure approximately 2,500 production systems. The Navy ACE group will procure the TAMMAC equipment directly for Harris Corporation and Smiths Industries during this phase.

Separate contracts will be placed by the Navy's program offices (e.g. F/A-18 and AV-8B) to prepare the host aircraft for receipt of TAMMAC. In the F/A-18 and AV-8B's case, a high-speed interface cable and mission computer software modification are prerequisites for TAMMAC installation.

Description of TAMMAC System

The TAMMAC System consists of an Advanced Memory Unit (AMU) and a Digital Map System (DMS) connected together via a High Speed Interface (HSI) Cable Assembly. The diagram below shows how the TAMMAC system is to be incorporated into the host aircraft's architecture.

TAMMAC Block Diagram

Advanced Memory Unit (AMU)

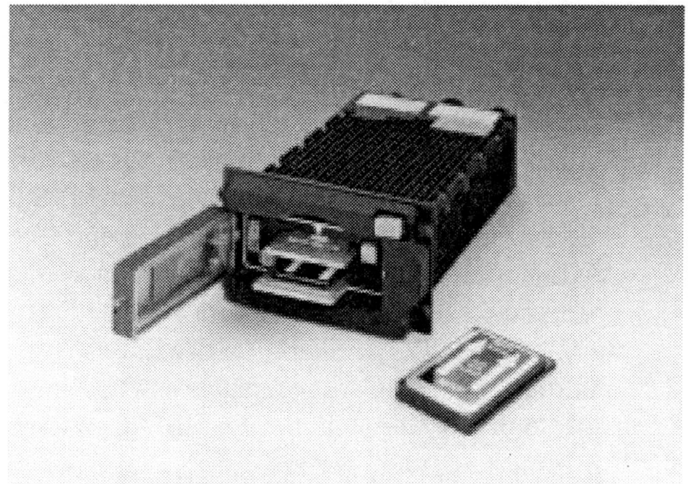

The TAMMAC AMU consists of the AMU itself and two types of PC Cards. It is designed as a ‘drop-in’ replacement for the DSS described earlier or the Mission Data Loader - another data transfer computer used on many military aircraft. The AMU provides reliable mass memory, data transfer and recording functions that can be used in flight under extreme environmental conditions. It features a commercial Power PC processor, dual PC Card receptacles, a MIL-STD-1553 interface, and an ANSI-STD Fiber Channel-based high-speed interface. The AMU uses removable non-volatile memory PC Cards for data transfer between the aircraft and the ground stations. The body of the AMU containing the processor, the 1553 interface circuitry, the high-speed interface circuitry, and the power supply remains in the host aircraft. The AMU can be used with the TAMMAC DMS or as a stand-alone data loader/recorder. The AMU has enough processing power to support hosting up to eight user-defined programs or mini-applications in addition to its normal functions. These mini-applications are run as background tasks and can be used to support various non-time-critical processing functions. For example the AV-8B aircraft uses one of these mini-applications to calculate aircraft fatigue-tracking data in flight.

Personal Computer Memory Card International Association (PCMCIA) "PC" Cards

Two different ground-processing systems are used to support most Navy aircraft - the Tactical Automated Mission Planning System or TAMPS and the Automated Maintenance Environment or AME. TAMPS is used by the pilots for mission planning and AME is used by maintenance personnel to process aircraft maintenance data collected during flight. TAMPS and AME will be modified with PC Card receptacles so that they capable of working with the TAMMAC system.

The AMU will contain two different sized PC Card receptacles to support use of three different kinds of PC cards (Type I, Type II, and Type III). PC Cards have the same dimensions as a credit card, except they vary in thickness depending on their memory technology and capacity. Type I and Type II cards use Flash EPROM and Type III use either Flash EPROM or magnetic disks. The memory densities available for TAMMAC in production will depend on the commercial PC Card market. The Type II and Type III PC Card maximum memory capacities available today are 280 and 520 Megabytes, respectively.

Since the AMU is being designed with two PC Card receptacles, it provides the capability for the host aircraft to separate mission and maintenance data. This allows operational (pilots) and maintenance personnel to have their own PC Card.

Mission Card

The pilots will primarily use the Mission Card for loading initialization data (e.g. flight plan, radio frequencies, targets, etc.) into the aircraft. They can also use it to record tactical data during flight. The Mission Card will be inserted into the AMU's Mission Card receptacle. The Mission Card and its receptacle will support transferring classified data. The Mission Card receptacle will be designed to accept either a Type I or Type II PC Card. The Mission Card will be a Type II PC Card that is qualified for use in flight.

Maintenance Card

Maintenance personnel will use the Maintenance Card for recording maintenance data collected during flights. It will be inserted into the AMU's Maintenance Card receptacle. Only unclassified data will be transferred to the Maintenance Card. The Maintenance Card receptacle will accept a Type I, Type II or Type III PC Card. The Maintenance Card will also be a Type II PC Card that is qualified for use in flight.

Map Card

Maintenance personnel will use the Map Card to load digital map data into the aircraft. Map loading will only occur when the aircraft is on the ground. The Map Card will be inserted into the AMU's Maintenance Card receptacle. Since this card will always be used on the ground, it will be a Type III PC Card that will not be qualified for flight use. This designation increases the number of commercial sources available to the Navy for the buying the card.

Digital Map System (DMS)

The TAMMAC DMS will be a "drop-in" replacement for the current DVMS. The TAMMAC DMS contains map processing and digital map data storage within a single unit - unlike its predecessor (DVMS) that needed two units to provide these functions. It was designed with an open architecture with standard circuit card interfaces. It uses a ruggedized magnetic disk drive to store the digital map data and can operate using 115 VAC or 28 VDC source power.

TAMMAC DMS Features

The basic function of the TAMMAC DMS is to generate moving map video images with symbol overlays for display in the cockpit. The primary map data source used by the pilots are aeronautical charts (CADRG data). This data consists of map images of various scales and is used to generate various scales of plan (or bird's eye) views of moving map video. The DMS can generate other moving map images derived from Digital Terrain Elevation Data (DTED), Controlled Imagery Base (CIB) data, or digital bit maps called Data Frames. All of these databases, except for Data Frames are standard map products produced by NIMA. The DMS can display any of these images with north-up, heading-up, or track-up references.

The TAMMAC DMS can generate symbol overlays that are either geographically referenced to the moving map or referenced to the display screen. An example of a geographically referenced symbol is a threat and an example of a screen-referenced symbol is a pushbutton legend.

The TAMMAC DMS can produce other overlays on the map images that are dynamic or change with the aircraft's flight. Dynamic threat intervisibility is a translucent tinting over the map image that shows the pilot where his aircraft is vulnerable to surrounding threats. This overlay is generated based on the characteristics of threat data loaded prior to the mission or received in real time over the 1553 interface. Another dynamically generated tinted overlay is elevation banding. It provides the pilot with a quick-look situation awareness tool. Terrain that is higher than the host aircraft's current altitude is tinted one color (e.g. red) and terrain that is within a programmable distance below the aircraft's current altitude is tinted another color (e.g. yellow). The elevation bands change in width as the aircraft changes altitude.

The TAMMAC DMS also uses DTED to calculate line of sight information. Symbol overlays can be displayed on the map to show if the aircraft has optical line of sight to a target or what altitude is required to achieve optical line of sight.

TAMMAC Map Data Loading

The method used to load digital map data into the F/A-18 and AV-8B aircraft will change drastically with the incorporation of the TAMMAC system. New map theaters will be generated on TAMPS for the TAMMAC system. Map data is selected and loaded onto the Map Cards using on TAMPS. The Map Cards are brought to the aircraft and inserted into the TAMMAC AMU. The AMU extracts the map data from the Map Card and transfers it to the DMS using the HSI Cable Assembly. This process provides a much more flexible capability for the Navy. It currently takes approximately 2-3 weeks to order a new map theater on an optical disk for the DVMS. With TAMMAC, the squadron will have the capability to create their own map theaters using TAMPS in a few hours.

TAMMAC High Speed Interface (HSI) Cable Assembly

An HSI Cable Assembly connects the DMS to the AMU. The HSI Cable consists of a modified commercial quad cable with MIL-STD-38999, Series 3 connectors on each end. The cable and the connectors were designed to support data transfer rates up to 1 Gigabits per second over a sixty-foot cable length. W.L. Gore, Inc. was selected as the sole source supplier of the cable during TAMMAC EMD.

MANAGING MULTI-PLATFORM SENSOR SYSTEMS

Kurt A. Himmelreich, Raytheon Company, McKinney, Texas

Introduction

The task of developing a stand-alone sensor system for multiple aircraft is limited primarily to mechanical and interface considerations. Oh, the good old days...

Raytheon's Terrain Following/Terrain Avoidance (TF/TA) radar systems are designed primarily to provide terrain following control cues and terrain avoidance information to the pilot. Satisfying this functionality goes beyond the simple provision of the radar's presentation of terrain. Aircraft-dependent and situational factors must be continuously managed and processed to ensure top performance and safety-of-flight. Herein lies the challenge in developing and adapting today's sensor systems for multiple, modern aircraft.

Background

Over the years, the TF/TA functionality in Raytheon Terrain Following Radars (TFRs) has been adapted to many platforms and coupled with additional radar modes. The result is a Multi-Mode Radar (MMR) that provides a low-level operational capability at night, in adverse weather and in high-threat environments for a wide variety of aircraft.

There exists a high degree of commonality between the LANTIRN (Low-Altitude Navigation and Targeting Infrared System for Night) AN/APN-237A TFR subsystem and its MMR derivatives. Five of the six Weapons Replaceable Assemblies (WRAs) in the MMR are essentially LANTIRN designs. Only the Radar Interface Unit (RIU) is unique to the MMR.

This commonality thread is maintained from LANTIRN through the AN/APQ-174B and C versions and, to the maximum extent possible, in the AN/APQ-186. Utilizing this approach gives the MMR the benefit of combat-proven control algorithms and highly reliable designs to provide a high-confidence system with high user acceptability.

The MMR system configuration consists of a Radar Pod Assembly, the RIU, a Pressurization Unit, and the Radar Pod Mount as shown in Figure 1. The WRAs mounted on the Pod Assembly are the Antenna/Gimbal, the Transmitter, the Receiver/Exciter, and the Power Supply. The following pages address the history of the system and the development approach that has helped make this line of radars successful on multiple, airborne platforms.

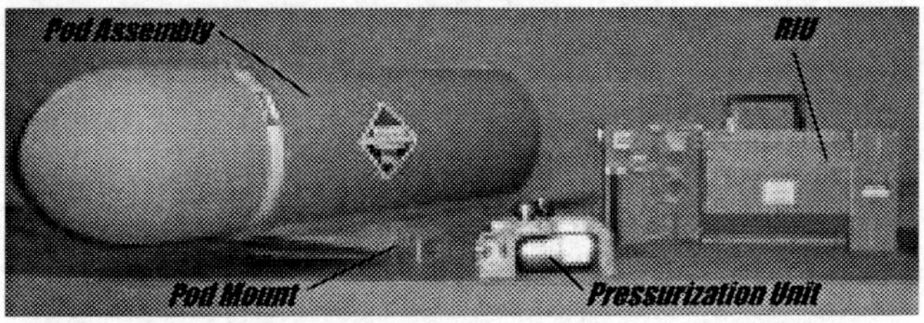

Figure 1. AN/APQ-186 Multi-Mode Radar

Evolution

To be successful for an extended time and on multiple platforms, the Raytheon line of TFRs was required to adapt not only to the various aircraft, but also to the evolving mission needs of the user. Developing a system to meet this challenge required the use of similar or common hardware and software in each "new" system. Maintaining this commonality amongst the systems is important for more than just the associated development costs. Perhaps more than any other airborne radar, of paramount importance in TFRs is user confidence. If the pilots don't trust the TFR, it's just another sensor. Thus, in addition to

0-7803-5086-3 /98/$10.00 ©1998 IEEE

Life Cycle Cost (LCC) benefits, commonality in TFRs saves time and money in development costs, flight testing, and user-acclimation.

Creating a system that could be easily adapted to fit various airborne platforms required considerable effort to identify the potential problems associated with installation and integration on multiple existing aircraft. Terms like scalability, open architecture, and modular design are commonplace today. However, when the LANTIRN development was started, TFRs where generally custom designs for each aircraft and application. While the original LANTIRN design was intended only for fixed-wing aircraft such as the F-16 and A-10, the use of standard interfaces and flexible software made the base design adaptable to rotorcraft and the tiltrotor.

Key Development Considerations

As with any airborne system, size was a key concern in the design. In airborne radars, it is even more significant because of the aerodynamics associated with the exposed surfaces. To minimize space requirements, LANTIRN was designed for Ku-band operation. This choice of frequency provided a good compromise between the small size needed to fit in an aircraft's nose or external pod versus the increasing atmospheric/rain attenuation of radar signals as the frequency increases. Like some of its predecessors, LANTIRN also featured a flat-plate antenna to reduce the antenna's physical size.

Developing a system for multiple applications requires a high degree of flexibility. Thus, the key radar parameters in the MMR were put under software control. This was crucial to allow the addition of new operational modes in future applications.

Do to the flight-safety considerations associated with terrain following, extensive flight testing is required to qualify the algorithms. To minimize this expensive test time, simulation techniques were developed which allow efficient optimization of the algorithms for various environments. The thorough design and high accuracy of these simulations reduce the need for algorithm adjustments during the flight test phase.

While TF was the primary purpose of the LANTIRN system, growth capabilities were designed into the system to allow modifications to meet the needs of future customers and platforms. As a result, each new Raytheon MMR system has been able to adapt to new requirements while maintaining a high level of commonality with its predecessors.

Mission Requirements

In addition to the examination of hardware changes, the evolution of this line of radars may be illustrated by examining the high level functionality that has been added. A look at the radar modes in Figure 2 for the latest MMR, the AN/APQ-186, illustrates the significance of the changes made since the LANTIRN TFR. The AN/APN-237A TFR subsystem supported only the TF mode (and Standby) with two TF submodes. Without sacrificing the benefits of commonality, various navigation and cross-scan (interleaved) modes have been added to both the AN/APQ-174 and the AN/APQ-186 systems. The addition of these new modes and submodes provides the pilot with the situational awareness required for new missions, thereby allowing the continued success of this line of radars.

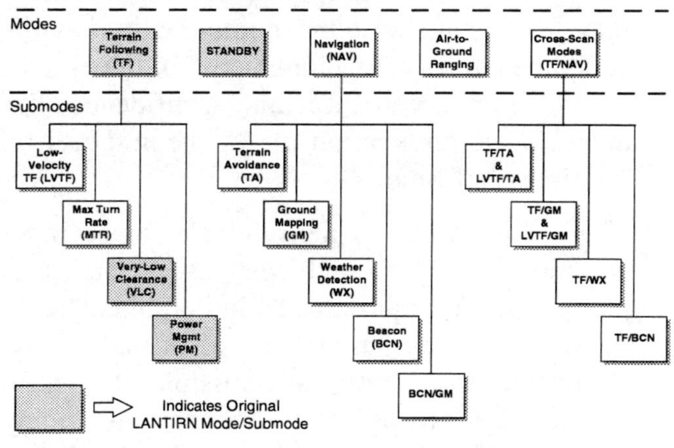

Figure 2. AN/APQ-186 System Operational Modes

0-7803-5086-3 /98/$10.00 ©1998 IEEE

While some new modes and submodes required hardware modifications, the majority were accomplished via software controlled parameters and algorithms. The common enablers that allowed the addition of new modes in the LANTIRN system design include:

- Available processing power,
- Software-controlled radar parameters,
- Standard interfaces with available bandwidth,
- Flexible (software-controlled) scan patterns, and
- Available margin in the radar timeline.

The addition of new operational modes was accomplished without the usual increase in size. With improved technology, several modes were added to the AN/APQ-186 for the CV-22 while reducing the overall size of the MMR system. This was primarily the result of the redesigned RIU, in which the weight was cut in half while doubling the potential interface bandwidth, tripling the memory capacity, and increasing the processing power potential by almost ten times!

System Engineering

Systems engineering may be viewed as the process of translating user needs into a product that satisfies those needs. In today's military environment, common hardware is one of the primary needs. Financial analysts can quickly point out the advantages of reduced development costs, inventory and the like which can be realized by employing common parts. For some systems, commonality has the added benefit of user familiarity. In the case of TFRs, user acceptance and confidence has aided in the application of Raytheon TFRs on multiple platforms.

Regardless of the rationale for utilizing common hardware and systems on multiple platforms, the requirements and architecture to facilitate success must be identified early. While it may not be possible to fully characterize all potential applications, many steps can be taken to maximize the likelihood of success. The following sections address some of the unique challenges associated with the development of a versatile TFR, along with many of the techniques and approaches used to ensure success on multiple, airborne platforms.

Modern Challenges

The development of a basic airborne sensor (e.g., a basic weather radar) for multiple platforms would likely focus on mechanical structures and a common interface bus. With the sophisticated avionics in many of today's aircraft, it seems that few tasks are so simple.

The challenges in developing TFRs for multiple platforms are made even more difficult when the overall TF functionality is considered. Below, Figure 3 illustrates the closed-loop interaction between the aircrew and avionics systems for TF operation. While each platform will have greatly varied characteristics for aircraft dynamics, the development process is also required to consider the various interface differences and aircrew performance. While platform differences necessitate variations of the TF algorithms and some interface structures, success on multiple platforms hinges on flexibility and commonality.

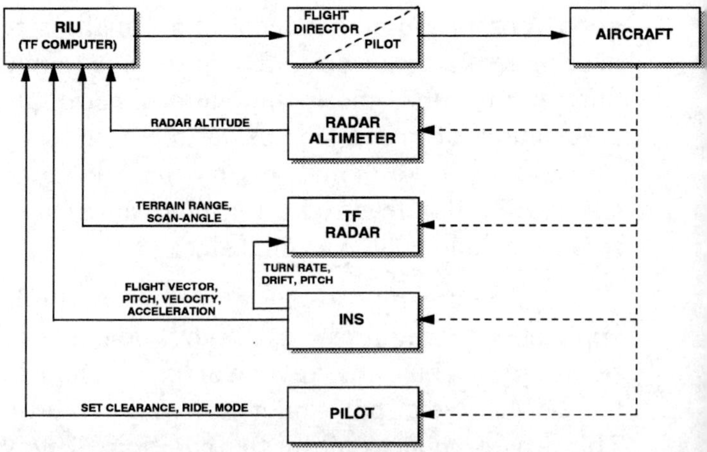

Figure 3. Typical TF Avionics Interaction

Hardware and Mechanical Concerns

The LANTIRN TFR called for the navigation pod (housing the TFR) to be suspended below the aircraft. In rotorcraft, the

0-7803-5086-3 /98/$10.00 ©1998 IEEE

MMR pod is often mounted in the nose of the aircraft. Thus, it was clear that considerable effort must be put into minimizing the physical size if the design was to be successful on multiple platforms. Ku-band operation was selected to reduce the size of Radio Frequency (RF) components (primarily the antenna), and considerable restraints were put on the configuration of the electronics and support structures to maintain a 12-inch diameter pod.

In the front-mounted configuration of rotorcraft and the CV-22, the pod is largely enclosed by the nose cover or other aircraft structure, limiting access to WRAs. Thus, mechanical structures and procedures were necessary to allow removal and installation of the pod without compromising the antenna boresight, which would result in poor TF performance and, in the extreme, unsafe flight. To facilitate compatibility with multiple platforms, these mounting and alignment structures are common across all applicable aircraft, for a given MMR type, to allow the pod to be interchangeable from one aircraft type to another while maintaining boresight.

Along with the benefits of commonality comes an increased obsolescence challenge. An electronics system that is designed to successfully meet the needs of the user for an extended period will require a concerted effort to keep up with the times. In addition to a formal parts obsolescence effort which continually examines the designs, the Raytheon TFR systems have successfully mitigated many obsolescence issues by using backwards-compatible designs for system upgrades. Thus, when a board, for example, is redesigned for new functionality, it maintains compatibility with "older" systems such that it replaces boards that house obsolete components.

Interaction and Interfaces

The use of proprietary busses and non-standard interfaces requires custom circuit designs and causes difficulties when trying to adapt avionics into an existing airframe. To maintain platform flexibility, the MMR systems were designed to use interfaces which are common in military aircraft. Specifically, MMR system control and data input/output is handled via a MIL-STD-1553B data bus. The 1553 bus allows the structure of messages to be adapted to many installation configurations. Video is output in RS-170/RS-343 standard analog video format, and power for the TFR is derived from the standard 400 Hz aircraft supply. The use of these common interfaces minimizes risk, cost, and integration time when fitting the MMRs on new platforms.

Aircraft Dynamics

The LANTIRN-based TFRs were designed for use on a wide variety of aircraft. However, as aircraft and aircrew performance are directly related to the terrain following operation, each platform requires custom algorithms. The cost (both time and money) of developing and verifying new algorithms from scratch would prohibit applications in many platforms. To avoid such costs and minimize development risks, the algorithms were developed in a modular fashion such that airframe-dependent parameters and algorithms could be altered without compromising much of the proven functionality in the code. This approach makes the Raytheon MMRs more appealing as the TF algorithms are proven on multiple platforms, and common simulation techniques allow the efficient modification of those algorithms for new designs and upgrades.

Flexibility and Software Control

Developing a system that will have flexibility in its application generally requires that it also have flexibility in its design. Developing a radar for numerous airborne platforms necessitated a modular hardware architecture and software control of key radar parameters. This was needed not only to allow installation in multiple aircraft, but also to facilitate the addition of new operational modes to meet the growing requirements of the user.

The LANTIRN TFR and its derivatives were architected with distinct boundaries between primary functions. This modular architecture allows upgrades and modifications to WRAs, without compromising the proven designs in other areas. In the case of the development of the AN/APQ-174 from the AN/APQ-168, it also allowed the RIU to be separated from the pod without a complete electrical redesign. The RIU separation was needed to minimize location-critical equipment in the CV-22 tiltrotor and other rotorcraft.

Designing the TFRs with software control of the radar parameters may have been the most significant contributor to multi-platform success. While the early fixed-wing applications required only TF capability, newer rotorcraft and tiltrotor applications required greater functionality for Special Operations Forces (SOF) and other users. The addition of specialized modes (for new applications) would not have been feasible without software control of the antenna, thresholds, power levels, pulse repetition frequency (PRF), etc.

In addition to ensuring that software had control over the internal operation, it was also necessary to make the software tolerant of different platforms. As TF requires adaptation of algorithms for each vehicle, the software must refine system operation in each case to permit installation in multiple aircraft. For a given system, software for all supported platforms is resident in EEPROM (Electrically Erasable Programmable Read Only Memory) in the RIU. This allows that RIU, for example, to be moved from one aircraft type to another, provided that each utilize the same MMR, without installing new software. The platform type is identified by the RIU through the 1553B bus at startup, and the appropriate mission software is loaded accordingly.

User Confidence

While "confidence" cannot be designed in, it is an important part of success in modern avionics, especially in the military. The successful application of a TFR system on multiple platforms would not be possible without the pilot's trust. Because of the safety issues associated with terrain following, LANTIRN-based TFRs feature continuous built-in-test (BIT) while the system is in operation and an interruptive BIT for further checkout. These extensive test features, along with a highly reliable system, aid in developing the confidence required for continued success.

Although confidence must be earned, a little future-proofing goes a long way in maintaining it. The manufacturers of modern avionics systems need to be prepared to meet the future needs of the aircrew. While largely a communications issue, the system must be designed for relatively easy adaptation to new requirements and platforms.

AN/APQ-186 Features

The AN/APQ-186, being developed for the United States Special Operations Command (USSOCOM) CV-22, is the latest of the LANTIRN-based MMRs. Several new modes and features have been added to fulfill customer needs. The AN/APQ-186 exhibits the flexible architecture and control from its predecessors, along with growth capabilities that position it for strong consideration on even more aircraft. The following paragraphs detail features of the AN/APQ-186 that highlight the focus on multi-platform applications. While some of the features discussed below are relatively simple, the key to success lies in long-term planning. Each development phase must be viewed and designed as a stepping stone to new applications.

Mechanical Structures

The modular architecture of the LANTIRN TFRs allowed the RIU to be separated from the Pod Assembly in the AN/APQ-174 and AN/APQ-186 MMRs. This allowed the RIU to be located in less critical areas and shortened the Pod Assembly by approximately 15 inches to ease installation

0-7803-5086-3 /98/$10.00 ©1998 IEEE

requirements in various airframes. While the choice of Ku-band operation allowed relatively small RF components, the boresighting requirements for safe TF operation necessitated special mounting hardware.

A guided Pod Mount fixture was developed which, once properly installed, allows the Pod Assembly to be removed and reinstalled without realigning the boresight. This fixture provides the Antenna/Gimbal with the references necessary to measure terrain angles with the precision required for the various operating modes. The use of this Pod Mount on different airframes also permits the radar Pod Assemblies to be interchanged from one aircraft type to another. This commonality feature reduces inventory and special equipment requirements for the user.

To facilitate maintenance and repair, the pod mounting structure is also equipped with slides that allow the Pod Assembly to be extended out of the surrounding structure of the aircraft so that the WRAs are accessible. This minimizes the need for special hoists to support the radar for routine maintenance or inspection.

System Interfaces and Communication

In today's complex avionics environment, there are few standalone systems. Various aircraft systems are interdependent and in constant communication. While standalone operation would certainly simplify a design, interfaces with multiple systems (see Figure 3) are required for a TF system. Thus, a standard interface structure is the best method to simplify integration on multiple platforms.

In the AN/APQ-186, the entire Pod Assembly is under the control of the RIU. This eliminates the need for the Pod Assembly to communicate with other aircraft systems. The AN/APQ-186 RIU is equipped with dual redundant MIL-STD-1553B data busses. The 1553 bus is common in many aircraft and provides a message structure which is adaptable to many installation configurations.

Versatility is also achieved by providing analog outputs in a standard format, where possible. For example, the original LANTIRN TFR utilized a radar video output in RS-170 format. Newer MMRs still support RS-170, but are also capable of the RS-343 format.

The definition and control of interfaces is one of the most underrated tasks in product development. If done well, it goes unnoticed. In the Raytheon MMRs, care is taken to maintain a common interface control document (ICD) that is applicable to multiple platforms. While there may be some differences in the platforms, the MMR system remains common.

Hardware Commonality

In the MMRs, common hardware has been a key focus. The ability to maintain high commonality comes from a convenient physical design coupled with a robust architecture. Despite the addition of numerous new modes, three of the six WRAs in the CV-22 MMR system maintain an estimated 90% hardware commonality, or better, with the fixed-wing LANTIRN systems. The commonality for the Receiver/Exciter and Antenna/Gimbal WRAs are estimated at 83% and 68%, respectively. Only the RIU WRA is virtually a new design.

The redesign of the RIU was primarily the result of the need for additional processing power to support new requirements. Digital Signal Processors (DSPs) were incorporated into the RIU as they exhibit high growth potential by delivering processing power far more efficiently than the original processors.

The installation and removal of the AN/APQ-174C RIU in the CV-22 aircraft was a difficult task. Consequently, when redesigning the AN/APQ-186 RIU, electrical and mechanical design changes were made that resulted in a smaller, lighter, faster, and more flexible system. The design also eliminated parts obsolescence problems related to eight boards which were removed or replaced.

Algorithm Development

The AN/APQ-174 and AN/APQ-186 MMRs utilize a modified version of the versatile ADLAT (Advanced Low Altitude Techniques) terrain following algorithms. The modified ADLAT algorithms provide safe terrain following operation and dynamic compensation for aircraft performance. These algorithms have been adapted for the MH-60K, MH-47E, and the CV-22. Utilizing ADLAT algorithms for each aircraft type has allowed extensive closed-loop simulation and testing software to be developed and used to create and characterize the platform-specific control algorithms for TF. This same methodology can be used to easily adapt the algorithms to other platforms while minimizing expensive and time-consuming flight-test time.

Architecture / Software Control

As stated earlier, designing for software control of key radar parameters may have been the most significant element of multi-platform success. As each platform often has a different mission, new modes were needed for new applications. Flexible software control allowed the new modes to be incorporated without significant hardware changes, thereby reducing the expensive flight testing required to re-qualify new hardware used for TF. Some of the software controlled parameters are listed below, along with a brief example of the benefits to the radar system. Refer to Figure 2 for the AN/APQ-186 mode tree.

- **Transmit Frequency** – Permits interference avoidance and improved detection of certain targets.
- **Transmit Power** – Allows low-power modes (e.g., the Power Management TF submode).
- **Pulsewidth** - Enables improved detection for long-range submodes, such as Ground Mapping.
- **Pulse Repetition Frequency** - Allows range ambiguities to be avoided in the various modes.
- **Receiver Bandwidth** – Permits optimized matching to the expected return pulse in a given mode.
- **Receiver Thresholds** - Allows control of the false-alarm rate and detection probability for each mode.
- **Antenna Positioning** - Allows the use of modes which require external antenna control.
- **Scan Pattern/Timeline** – Permits the scan pattern to be customized to suit the needs of each mode, including cross-scan modes that combine (interleave) TF and NAV data/scans.
- **Polarization** – Optimizes operation in clear and adverse weather depending on the mode/conditions.
- **M-of-N Filtering** - Provides a mechanism to control false alarms by requiring the correlation of returns to any of 16 possible settings (mode dependant).

Configuration Management

Configuration Management often refers to an organization or infrastructure which exists to ensure that the correct components and software are utilized in the manufactured product. In addition to such an infrastructure, a design with multiple applications should foolproof the potential problems associated with, in this case, use on multiple aircraft.

Raytheon MMR systems are designed such that there are no hardware differences when a given MMR type (e.g., an AN/APQ-186) is installed on multiple aircraft. Thus, all platform-dependent functionality is controlled by software.

To avoid the possibility of mixing aircraft types in the software, completely separate loads are maintained. EEPROM in the RIU contains the software for all supported platforms. This allows the RIU to be interchanged between aircraft without reprogramming the EEPROM. At system startup, the aircraft type is identified through the MIL-STD 1553B bus, and the appropriate load is downloaded from EEPROM. To avoid errors, the aircraft type is verified periodically through the normal built-in-test cycle.

Growth Potential

While there are certainly markets where a "one size fits all" product can do well, military avionics are often called upon to perform beyond their original requirements. Systems that can not adapt to these new expectations will not be as well positioned for installation on additional airframes. As this is especially true for Special Operations Forces, the AN/APQ-186 system was designed with an eye toward the future. While much of the front-end radar components remain largely common with the LANTIRN systems, the processing center (the RIU) was redesigned with considerable enhancements. The prime item development specification for this system called for limiting memory utilization to 50% and processor loading to 50%. Raytheon went a few steps further.

The new AN/APQ-186 MMR computer configuration (Figure 4) includes:

- A single TMS320C40 microprocessor Circuit Card Assembly (CCA) which replaces 10 TMS9989 microprocessor CCAs of the previous MMR and LANTIRN configurations
- A combined Microprocessor Interface (MPI) CCA which provides bus arbitration and system control for both the radar and the processor bus
- A single EEPROM CCA which replaces the two EEPROM CCAs in previous configurations
- Expansion slots to provide for the addition of up to three microprocessor CCAs (for as much as four times the processing power) and up to two Global Memory CCAs (to expand the memory capacity by up to 24 Mbytes) for future growth
- A new 1553B Interface CCA with expanded memory and two complete dual redundant MIL-STD-1553B channels.

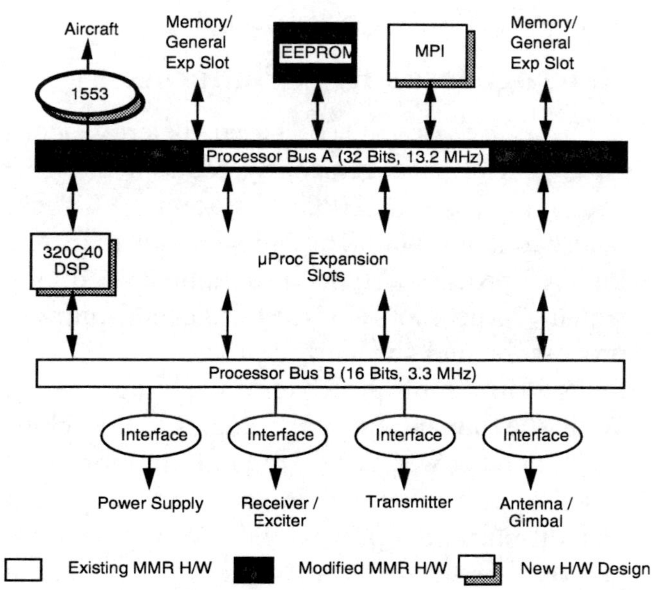

Figure 4. AN/APQ-186 RIU Processing Architecture

A Glimpse of the Future

The multi-platform approaches and techniques of the past have not been forgotten. Raytheon continues to lay the foundation for further applications and upgrades of this radar family. One of the ever-increasing needs for Special Operations Forces is stealth. The flexibility of Raytheon MMR systems creates a wealth of opportunities to minimize emissions and increase the likelihood of mission success. The AN/APQ-186 RIU was designed with such upgrades in mind. There are additional slots available to upgrade the processing power by four times and the memory by up to 24 megabytes. Thus, the AN/APQ-186 is poised to incorporate new capabilities without hardware redesign.

With the increased use of highly-flexible, multi-purpose radars integrated in tactical aircraft (e.g., the Active-Element Scanned Array system on the USAF F-22), there comes further challenges for the special-purpose sensors such as TFRs. While the special-purpose sensors may become less common in new development platforms, the reality of budget cuts, and the resulting extended operational life of today's aircraft, further enhance the opportunities for system upgrades and applications in existing airframes. Benefiting from these opportunities requires a system designed with flexibility beyond that necessary for a single aircraft. In hindsight, many of the design approaches used for the base LANTIRN system and its derivatives may seem like common sense. In aggregate, however, maintaining the focus on flexibility required considerable insight and dedication to avoid trading LANTIRN's versatile architecture for short-term simplifications. The result was a LANTIRN TFR that not only excelled on fixed-wing aircraft, but also spawned new systems which continue to meet the needs of the customer.

An Interactive Tool Suite for Real-time Application Development on Heterogeneous High Performance Computing Platforms

Minesh I. Patel; Honeywell Space Systems; Commercial Systems Operations; Clearwater, Florida

Abstract

With the emergence of inexpensive commercial off-the-shelf (COTS) parts, heterogeneous multi-processor HPC platforms have now become more affordable. However, the effort required in developing real-time applications that require high-performance and high input/output bandwidth for the HPC systems is still difficult. There are no software tools that allow an engineer to develop and field applications efficiently on the HPCs. This paper describes Honeywell's Multi-domain Embedded System Architect tool suite (MESA). The tool suite allows an engineer to rapidly develop an application on a COTS based heterogeneous HPC platform, effectively reducing the development time. To illustrate the cost benefit and the user friendliness of MESA[1], the mapping of the bistatic radar application on Honeywell's HPC will be presented.

Introduction

Many Military and Industrial systems require real-time, high-performance and high input/output bandwidth performance. Such applications include radar, signal and image processing, computer vision, pattern recognition, real time controls and optimization. The complexities of high performance computing (HPC) resources have made it difficult to port and fully implement the various applications. With the availability of inexpensive HPC systems based on commercial hardware, the high demands of military and industrial applications can be met. However, the potential benefit of using high performance parallel hardware is offset by the effort required to develop the application. A set of user friendly tools would offer the engineer ways to use the computing resources for application development. By tuning processes, improving application efficiency and throughput, and automatic mapping, partitioning and source code generation, the engineer can improve productivity and turn around time, and lower development cost.

This paper describes the Mult-domain Embedded System Architect (MESA), a suite of system and application development tools developed and released by Honeywell for use with a family of heterogeneous platforms. This paper will first provide a brief overview on the benefits of HPC platforms. Next, a description of the MESA suite of tools will be given. Finally, the steps required by an engineer to develop and port a radar processing application to a target HPC will be presented.

Heterogeneous HPC Platforms

The use of COTS based heterogeneous HPC platforms significantly reduces the cost over custom HPC systems. These heterogeneous platforms allow more than one kind of processor type (programmable digital signal processors (DSP), general-purpose processors, and specialized processing chips) to be used in a multiprocessor computing system. With the advancing technologies, DSP chips have more power, less size, and are more cost efficient at vector and image processing than general-purpose processors. Most general-purpose processors are becoming better at scalar and super scalar processing. Specialized processors (FFT, data compression, neural network, or FPGAs) offer an order of

[1] D. Bhatt, et.al., "Multi-Domain Embedded System Architect (MESA) Methodology and Processing Overview, Honeywell Technical Report, August 1997.

0-7803-5086-3 /98/$10.00 ©1998 IEEE

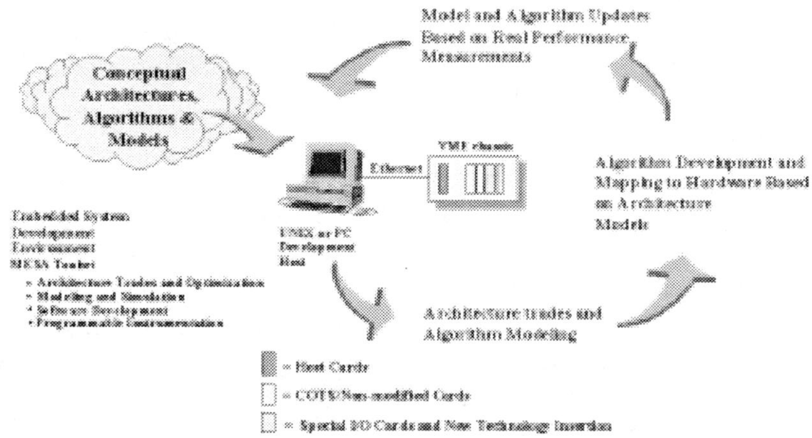

Figure 1. Life Cycle Development

magnitude throughput improvement for specific algorithms compared to program based processors. The end result of the heterogeneous system is a high performance multi-processor platform that is scalable and inexpensive.

With a parallel heterogeneous environment, system designers can choose the most efficient and cost effective processor(s) for the type of computations or algorithm. A heterogeneous configuration that can use different processors optimizes the data flow throughput, increases extensibility and scalability, and leads to fewer overall hardware components (i.e. processors, memory, etc). Consequently, the smaller system is less expensive because it requires less hardware, and it becomes easier to maintain, package, and provide power and cooling. Another benefit is that the heterogeneous HPC environment reduces the life-cycle costs by providing a standard platform for upgrades and enhancements. This allows the rapid insertion of new computing elements. But, the advances in technology and extensibility ease results in increased HPC complexity and application development costs.

Multi-domain Embedded System Architect (MESA)

Honeywell has developed an integrated tool suite for system design called the Multi-domain Embedded System Architect (MEA). The tool suite provides complete lifecycle development through an integrated combination of tools potentially reducing the development costs and source lines of code in half. Figure 1 graphically depicts a typical system development lifecycle. The MESA approach to embedded application development is to bring together under a common GUI, a set of collaborating applications or tool designed specifically for each phase of a system's development lifecycle. MESA consists of the visual design tool, the architecture trades and optimization tool, the performance modeling and simulation tool, and the programmable instrumentation tool. The following will describe the uses of MESA in each life cycle phase.

Requirements Phase

At the beginning of a system's development cycle, the requirements of the application, system and hardware architecture are entered through MESA's interactive GUI environment, the Visual Design Tool[2]. Software and hardware co-design can be performed using the design tool's two editors, the system editor and the hardware editor. The system editor is used to build a graphical view or model of the application by connecting functional or

[2] Visual Design Tool Product and Data Sheet Specifications Report, Report Number 1, Honeywell Inc., 1997.

0-7803-5086-3 /98/$10.00 ©1998 IEEE

behavioral blocks (hierarchical) in a data flow manner through user defined or COTS functional libraries. In the hardware editor, the hardware architecture is built hierarchically from the processor all the way up to the system level. After the designs have been built, the user can specify other requirements such as the levels of functional parallelism and synchronization, latency, period, throughput, scheduling, FLOPS(floating point operations), KSLOC(thousands of source lines of code).

When selecting candidate architectures, the architecture trades and optimization tool of MESA[3] can be used to compare different target hardware architectures in terms of which one best meets the application's overall requirements. The application and various target hardware requirements and specifications can be fed into the trades component of MESA where a study can be performed using criteria such as throughput, latency, cost, size, weight, and power. Integrated into the trade tool is a genetic optimization and Monte-Carlo simulation environment, which iteratively cycles through the criteria parameters and selects the optimal solution. Additionally, CoCoMo models that are incorporated into MESA are used to estimate software development. Once target architectures have been selected, and given the computation and communication requirements of the functions, the optimization portion of the tool can suggest an optimal mapping and partitioning of the application functions onto the hardware components.

Design Phase

Once a set of candidate architectures is established, the performance modeling and simulation tool of MESA[4] allows VHDL (VHSIC (Very High-Speed Integrated Circuits) Hardware Description Language) simulations. The simulations are done by automatically synthesizing VHDL behavioral models from the system design. The models can be simulated on any of the commercially available VHDL simulators. Operating systems, communication libraries, and domain specific run-time layers can all be modeled and simulated, thus allowing the engineer to verify that the performance, timing, and latency requirements are being met. Simulation results can be used to refine the design, ensuring development of an optimal system. In addition to custom models, a library of generic model templates for components such as processors, operating systems, communication systems, etc. is provided.

Implementation Phase

After selecting the hardware architecture, the engineer can further refine the hardware and system designs that together represent the total system architecture. All primitive and hierarchical blocks are stored on design "shelves." Replacing current components with those stored on the shelf can refine system/software and hardware design. Items on the hardware shelf include workstations, other embedded computers, CPU chips, memory, ASICs, FPGAs, etc. The application and system designs can be refined using the software shelf items such as other COTS functional or user defined blocks. The entire software development environment integrates COTS-supplied components (compilers and run-time system, and libraries), along with custom, user-supplied software and hardware components (application code, libraries, etc.). Combining elements from the hardware shelf, the software shelf, and trade information, the engineer can construct an executable which maps software components onto hardware resources. The MESA software code generator will then construct the glue-code necessary to connect the various components making up the application executable. Build information (makefiles) is generated and the components are compiled and linked using the third-party tools.

[3] Architecture Trades and Optimization Tool Product and Data Sheet Specification Report, Report Number 2, Honeywell Inc., 1997.

[4] Performance Modeling and Simulation Tool Product and Data Specification Report, Report Number 3, Honeywell Inc., 1997.

0-7803-5086-3 /98/$10.00 ©1998 IEEE

In addition to the manual approach to application partitioning and mapping described above, the MESA toolset provides domain knowledge for parallel and data flow designs to automate the application development process. Using the hardware information captured in the hardware shelf, the engineer can specify the multiple levels of parallelism for component functions, as well as synchronization information. This information, knowledge that is normally isolated in an expert user's mind, is captured in the MESA tool's knowledge base of design optimizations to be shared among all users and developers. The MESA tools can then automate the software-to-hardware mapping and glue-code generation. The benefit is a major reduction in application development time and cost. By simply capturing the parallelism, interconnection scheme, and hardware mapping of library functions, many applications can be easily developed.

Integration and Validation Phase

The final tool in the MESA toolset is the real-time performance monitoring and visualization tool[5]. Using the software GUI component interface, instrumentation "hooks" or probes are placed within the hardware and software diagrams. Upon execution, the hooks provide the performance monitoring information, which drives the instrumentation displays. The displays can show software and hardware architecture usage and hot spots; profiles and timelines of the application's functional executions, DMA, and I/O overheads; Output data in 2D, 3D or tabular formats. The instrumentation facility provides real-time, as well as post-mortem analysis of the execution results.

Radar Processing Application

Bistatic radar is a target detection system that contains a single transmitter and multiple receivers[6]. The application is divided into a front-end and back-end. The application receives two channels or data dwells, the sum and reference channel data each consisting of 16x2048 complex data sets. The front-end takes the data and performs a fix to floating point conversion, channel correlation by normalizing the reference channel and performing an FFT operation on the normalized reference channel and the sum channels. Then a complex vector multiply is performed on the results. An inverse FFT is performed on the product. This is followed by a matrix transpose, an FFT, and another matrix transpose. The last three operations correspond to Doppler processing. The back-end further processes the results of the front-end data. The port of the application's front-end will be done using MESA.

The data arrives to the system application every 50 milliseconds and the system hardware must be able to process the data at that rate or faster. The HPC platform consists of a PC/NT with an Alacron FTSHARC accelerator board. The accelerator board contains eight 40MHz 21060 SHARC DSP processors each with 256Kbytes of IRAM (Internal RAM) and 1M bytes VRAM (Video RAM), an Intel i860 control processor, and 8M bytes of DRAM. The board is supplied with Alacron's optimized DSP functional libraries and software.

The process of mapping an application to a target platform is entirely done through the various GUI interfaces that are provided by MESA. The mapping process consisting of 8 steps: (i) select an application and target hardware, (ii) develop custom functions if required by the application, and (iii) choose a desired parallelism type and parallelize and partition the application. Step (iv) create the application and hardware description on the system and hardware editors, (v) generate and compile the source code, and execute.

The application development begins with the first step, identify the application and target platform. For this example, it is the Bistatic

[5] Programmable Instrumentation Product and Data Specifications Report, Report Number 4, Honeywell Inc., 1997.

[6] "HPC for Bistatics Program," Honeywell Inc, SaSSO Technical Report, July 11, 1997, pp. 1-68.

0-7803-5086-3 /98/$10.00 ©1998 IEEE

radar application. The target platform was chosen to be the FTSHARC accelerator board. Since the Alacron hardware was available, a trades study was not done. The second step is to develop custom libraries if necessary. MESA allows the user to insert custom written functions into the library. Because the Bistatic application requires 10 functions of which 2 are custom, the user can insert the source code for each custom function. The functions then become a part of the functional library.

Step three, choose a parallelism type and parallelize and partition the application. There are a variety of parallel architectures and methodologies. They can be classified into SIMD (Single Instruction Multiple Data), MIMD (Multiple Instruction Multiple Data), SPMD (Single Program Multiple Data), by the information exchange method i.e. shared memory or distributed memory (message passing), and functional and data parallelism.

the lowest level, most processors take full advantage of functional parallelism to perform more than one instruction per clock cycle. For example, in an ideal situation, an Intel i860 processor can compute three instructions per clock cycle. However, this architecture is not generally controllable by the user. The user can gain control at a higher level by executing independent program blocks in parallel to take advantage of multiple processors. The traditional divide and conquer approach requires that the problem be divisible into chunks. The chunks are then assigned to different processors, and letting each processor work on its portion. Each function in the Bistatic radar application can be mapped to the single i860 processor or stripped (divided) over the 1-8 SHARCS on the accelerator board. For example, the data coming into an FFT function can be partitioned over the 8 SHARCS where the FFT function is executed using ***n*** SHARCS

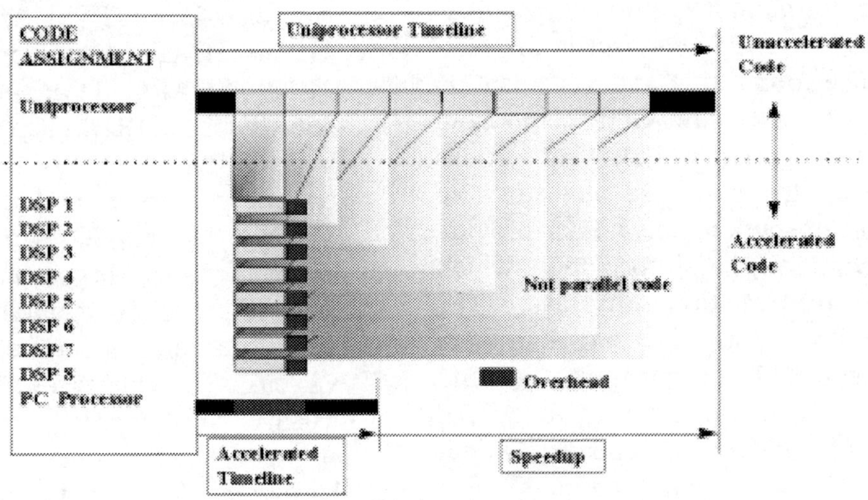

Figure 2. Application Acceleration

The method used in this example is functional or algorithmic parallelism, which executes different functions or algorithms simultaneously as shown in Figure 2. The selection of functional parallelism does not limit the user from choosing other types of parallelism. Figure 2 shows the speed up achieved over a uni-processor implementation when parts of an application are mapped to different processors (functional parallellism). At

on different parts of the input data dwell. The Bistatic application can be broken down into 10 functions as shown in Figure 3.

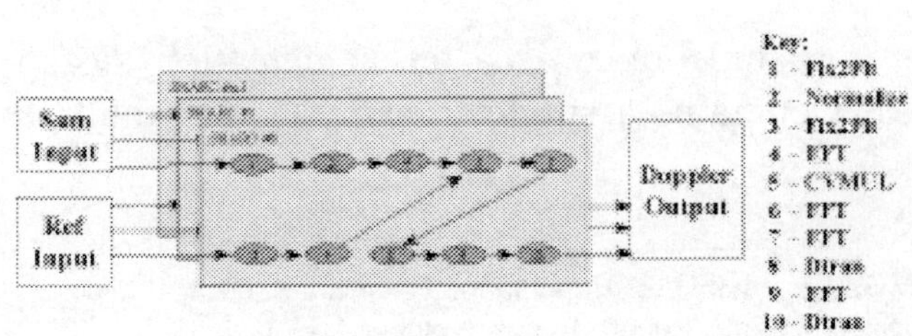

0-7803-5086-3 /98/$10.00 ©1998 IEEE

Figure 3. Bistatic Application Decomposition

The figure shows the input and output requirements, the listed functions and the flow of data between the functions. The input data consists of two channels each a 16 by 2048 complex matrix (one dwell) arriving every 50 mS. The output is a processed complex matrix of size 16 by 2048, outputting every 50 mS. The function list contains 8 functions from the COTS library and 2 user written custom functions (Normalize vector and Dtran(matrix transpose)). The Fix2Flt function converts a fixed-point number to a floating-point number. The Normalize function normalizes entries in the matrix. The FFT is a Fast Fourier Transform function and CVMUL is a complex vector multiply function.

Creating the application and hardware description on the system and hardware editors is done in step four. The co-design environment of the visual design tool allows the user to enter the application design using simple point and click mouse operations. The software and hardware editors for the Bistatic Application are shown in Figure 4. The system editor is shown in the upper left corner of Figure 4. It consists of the functional diagram of the application with the 10 identified functions. The application is built by selecting functions from the system design shelf that is populated with the functions from the COTS and user defined library. Simply clicking, cutting, and pasting can quickly model an application. It should be noted that the libraries are error free. Thus when creating an application using the system builder and auto source code generating feature, there is no code debug time.

The GUI window to the upper right corner of Figure 4 shows the property inspector. Through the property inspector, the user can set such functional properties as the functional threading, mapping to hardware resource, creating composite blocks (these blocks are a composite of many functions in a hierarchical top down decomposition), setting the type of buffering between functions, and setting function loading and scheduling.

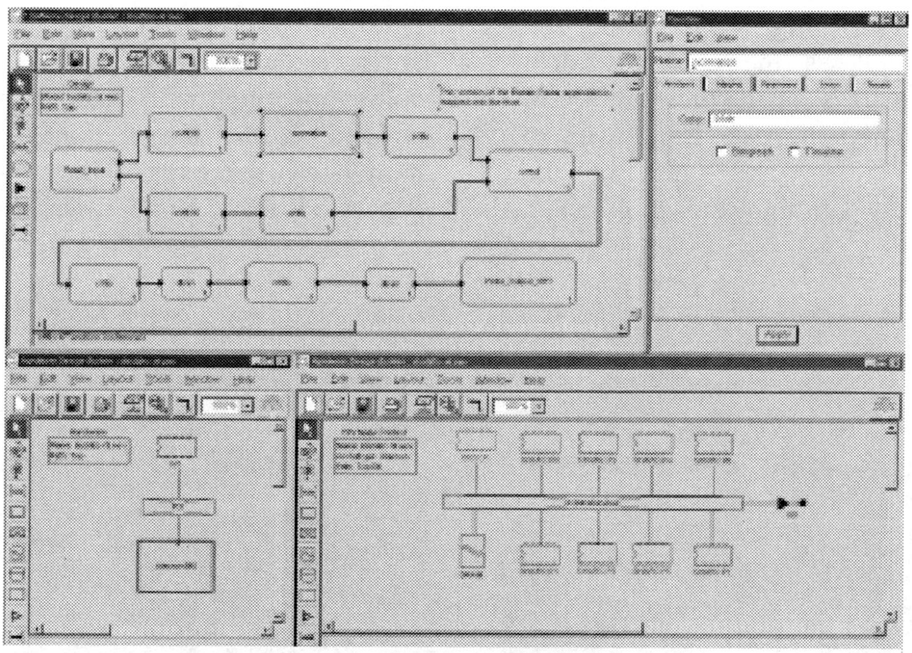

Figure 4. Illustration of Editors

The hardware editor is shown in the lower left corner of Figure 4. The user can build the target platform by selecting components from the hardware design shelf that is populated with hardware primitives such as processors, networks, and memory. These components can

0-7803-5086-3 /98/$10.00 ©1998 IEEE

be grouped in a hierarchical fashion creating larger systems. The platform for the Bistatic application consists of a PCI bus connecting an Alacron accelerator block and PC/NT workstation. The Alacron block can be decomposed into the window shown in the lower right of Figure 4. The figure shows the Alacron FTSHARC board with a single i860 control processor, 8 Mbytes of DRAM and the 8, 21060 SHARC processors each connected to VRAM across a SHARC bus. A PCI bus interconnect is used to connect components together. Using the software and hardware editors, instrumentation hooks can be placed within the respective editors for retrieving data and information. Such information consists of but not limited to processor utilization, functional efficiency, hot spot identification, and timeline and profiling information. The design tool also allows the user to select various types of displays for the presentation of the data. Additionally, the user is allowed to specify headings, titles, and units of the displays on the fly through the tool. Such instrumentation displays include 2-d and 3-d graphs as shown in Figure 5.

The figure shows two displays for the Bistatic application. The bar graph display shows the processor utilization. The different gray scale shades show how efficient the functions are per processor, overall processor utilization, the I/O overhead, memory usage, idle times, and busy wait times. The second display is a 3-d plot of the output complex 16 by 2048 data cube. The plot shows all the peaks and valleys and is color-coded by magnitude. The headings, titles, and units are derived from settings within the design tool and can be changed at design time by the user without reconfiguring the code for the displays.

In step five, the application source code is automatically generated and compiled for execution on the target platform. After the user has defined the software and hardware architecture, the design tool can automatically generate the source code for the target platform. The source code consists of such attributes as vendor specific software for communicating with the board, parallelism defined in the application, and function to hardware mappings. The source code is generated using a hardware/vendor specific

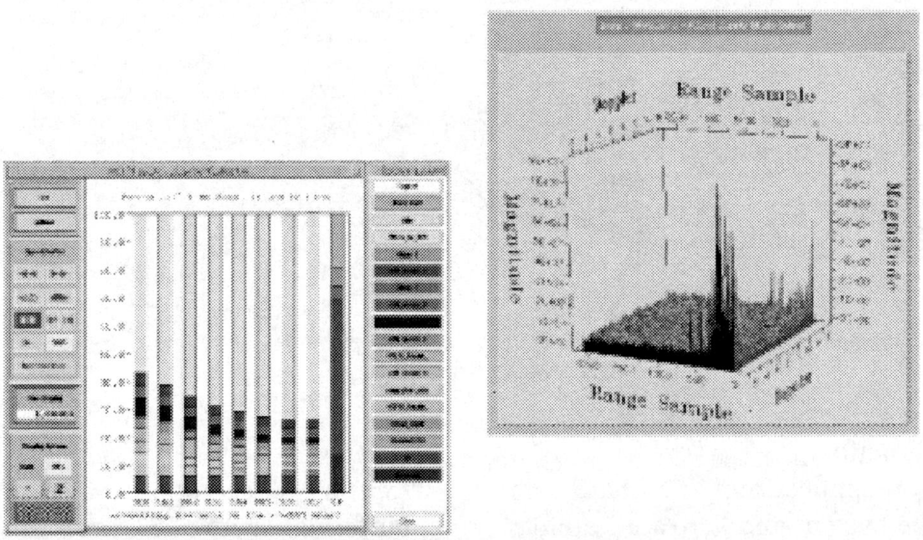

Figure 5. Typical Displays for Bistatic Radar

0-7803-5086-3 /98/$10.00 ©1998 IEEE

code generator. The code generator is built based on a scripting language from Honeywell Technology Center[7] called Alter Script.

The generated source code can be compiled using the COTS compiler provided by the vendor of the target hardware. The design tool can intelligently select the correct compiler to use based on the information provided by the system and hardware editors. After compilation of the source code, it can be executed on the target platform. The results of the Bistatic execution are shown in Figure 6.

The figure shows several results from the execution. The center block shows the optimal compute metrics which were generated through graph and table are generated as shown in Figure 8. The graph shows that optimal performance was achieved by using functional parallelism with 5 SHARCS with an end-to-end latency of 117 milliseconds. As more or fewer SHARCs were employed, the times increased due to compute and I/O bound restrictions. The hand coding of the application for the optimal number given in the figure required 4 months debugging and coding time, whereby using MESA, the time required was less than 2 days using the simple point and click development structure. It was observed that the optimal hand coded implementation was close in latency and used less hardware. The intent of the

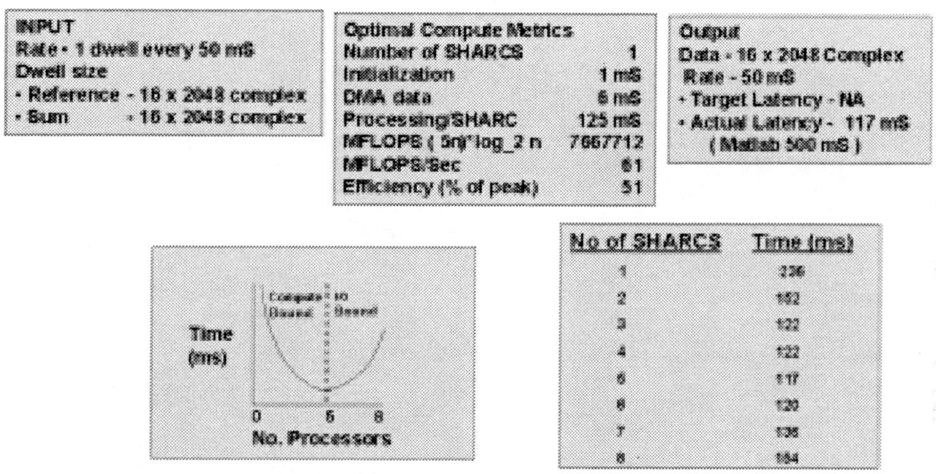

Figure 6. Results of the Bistatic Radar Execution

a hand coding of the application to a single SHARC. The two blocks at the bottom show an analysis done when mapping the application with MESA where 8 simulations were done. The first execution required that the input data cube is sent to only one SHARC and the entire applications execution resides on that SHARC. Each subsequent simulation required the data cube and application execution to be partitioned over additional processors. After the executions, the Time (end-to-end latency for an application). vs. Number (No.) of processors simulations was to show the rapid development time when compared to hand coding an application to the target hardware. In order to approach the optimal solution generated by hand coding, the user could try different architectural configurations and parallelism structures.

Conclusions

The MESA tool suite provides a powerful graphical and interactive interface for the creation of executable systems and applications based on customer defines specifications with fewer errors and an order of magnitude

---

[7] "DoME 5.0b: Alter Programmer's Reference Manual," Honeywell Technology Center, Minneapolis, MN, 1997.

0-7803-5086-3 /98/$10.00 ©1998 IEEE

reduction in development time. This paper has presented Honeywell's MESA tools suite. MESA's ease of use was demonstrated by the step by step description of the radar application development to the HPC hardware. In addition, other application are currently being studied such as computer vision and image processing (Global Yankee-Shrapnel Detection), synthetic aperture radar (SAR), space time adaptive processing (STAP), automatic target recognition (ATR), and target tracking.

0-7803-5086-3 /98/$10.00 ©1998 IEEE

GENERIC UNIFIED SYSTEMS ENGINEERING METAMODEL

SYSTEMS DESIGN of A SINGLE SYSTEM

Author : George Caple, Principal Systems Engineer

GEC Marconi Avionics, Radar & Countermeasures Systems

Stanmore, England

Abstract

Focusing on "key issues" without losing sight of the whole picture means being able to see the whole picture in the first place. This paper focuses on a part of systems engineering which, unfortunately, can be greatly obscured by attention to nominally more important issues such as finance, programme, hardware design, software design, safety planning, reliability planning, EMC test planning, etc., etc., & yet which, being diminished, by ignorance, lack of belief, management support, & training, will in turn impact severely upon those greater issues.

The focus is on : ***Systems Design***.

The choice is : ***Systems Engineering or Systems Design,*** as the central & major focus of any project within any domain.

Problem Scenario

Problems with current, & past single system projects highlight the need to recognise & more importantly, implement, ***disciplined*** Systems ***Design*** methods from concept onward.

Only correctly structured Systems ***Design*** methods can enable designers to create the design, with traceability, rationale, & compliance with the Customer's Requirement. Even more important is the need to move effectively from bid phase into contractual phase. (unpublished presentation material available)

"Design", simply in accordance with a "Project Management Programme" based on previous experience can not suffice. Circuit design, software design & project management are not sufficient.

One may argue, therefore, that there is a tool missing from the project management toolbox. *Systems Design.*

Systems ***Engineering*** is :--

Everything that you can think of that should be done, on a project, & within the encompassing business sphere, exemplified by, for example, IEEE P1220 :--

IMMENSE !

The author would suggest that, to resolve, & manage this complexity, it is necessary to throw light upon :--

Systems ***Design,***

& thereby illuminate & define systems design as the core of systems creation, with Project Management, all forms of planning & all supporting roles & disciplines then falling into their true,

0-7803-5086-3 /98/$10.00 ©1998 IEEE

relative, supporting positions. Tasks, processes & products will be seen clearly as such.

Systems ***Design***

What is systems design ? [domain independent]

Unfortunately thought of by many as a small element, & somewhat detached from the realities of systems engineering (getting requirements right, optimising solutions, prototype build, cutting metal & hacking code), **Systems *Design*** is the heart of creating systems, & indeed producing bid material in order to obtain a contract. Further, at the heart of **systems design** is **STRUCTURE** as presented to **INCOSE (UK)** in June 1997 as the Generic Unified Systems Engineering Metamodel. [.G.U.S.E.M.] [also in an exploratory presentation at Bragbury End, Stevenage : INCOSE (UK) SEPDC June 1995]

Systems **Design** is :--

"The creation of the full & complete design definitive, manufacture definitive, & test definitive, **structured** Data Repository (Information Model), as in G.U.S.E.M., in accordance with repeatable, teachable, developable processes", (Process Models) aided by, & supporting as necessary, circuit design, microwave design, mechanical design, software design, & other disciplines, in accordance with the project programme and plans." [Caple]

This accords to a degree with the Boeing "Common, Shared Database" [Gartz] , & particularly York University. [Vickers, Smith, J]

The Structured, rigorous, Data Repository in turn applies a **discipline** to the **DESIGN** "product & process" combinations, while leaving more than sufficient creative "Design Space" for the design engineers. (products ranging from specifications to hardware & software)

Systems Design must be Science

Based, not Process or List Based .

There must be structure & there must be exact rules for a Systems *Design* model, internally for its structure, its processes, & the people & material within it, & externally in regard of the containing environment. (Caple : INCOSE, June 1997 : Generic Unified Systems Engineering Metamodel : Rules of the Repository, the Processes, & the Containing Environment) It must bring together, & integrate, Project Management & Systems *DESIGN,* within the containing Systems *Engineering.* This must be achieved for the single system, (THIS PAPER) the "system of systems", (paper in preparation) & the complete platform, (paper in preparation) at least.

Systems Design must be consistently methodical,, definitive, logical, & repeatable. One must, as Derek Hitchins has written, manage complexity in process & product, & reduce entropy in regard of both. Systems Design method must be rediscovered, & recorded for use, development, & teaching.

How can Systems Design be Achieved ?

It is proposed that the application of the **Generic Unified Systems Engineering Metamodel (G.U.S.E.M.)** satisfies the question & achieves the objective, by identifying Systems *Design* within Systems *Engineering,* & maintaining the relationships between them, & their interdependencies. This is greatly assisted by the use of Soft Systems Methodology [S.S.M.] (Checkland, Lancaster) applied within all of the process areas. Relevant Project Management methods may then be applied intelligently. [see IEE Transactions on Aerospace &

Electronic Systems Vol. 33, No. 2 April 1997, page 584, 4) Myron Kayton.]

This paper does not attempt the analysis of the containing environment through the medium of S.S.M.

Systems Design within Systems Engineering

For those who have not met G.U.S.E.M.

G.U.S.E.M. is shown below in its simplest "Logo" format. Its 3 domains, intended to impose upon design, a repeatable teachable developable **discipline,** are :--

1. The Data Repository bi-cone, containing in the top half, structured, integrated, leveled, & grouped , definitive data, from Operational Requirements, through Customer Requirement, system ***design (adjective) requirements,*** black box, module, mechanical, electrical & software design into the manufacturing datapack at the centre, & with all test & proving definitive data in the inverted cone. No time line exists.

2. The processes, shown simply as 3 elipses, which are used to guide the ***design (verb)*** the system at each level & within each level in groups, & create the definitives [Eg. Design Requirements] demanded by the data repository. In reality there are a number of interconnected & decomposed processes acting together at each level of the repository, as will be shown later. No time line exists.

3. The programme, or PERT domain, shown as a straight, simple time line. All tasks carried out within the processes in order to create the definitives for the Data Repository appear in the programme as resourced tasks with durations. However, there is no absolute sequence, nor an absolute top down movement. ***Completion*** of the definitive ***design*** must be top down. (Best illustrated by an "Animated Sequence", Caple 1997, unpublished)

It is above all the structured Data Repository (Information Model) which provides the **discipline** & systems design focus upon which the directly related subservient guiding processes are founded, the whole of which is converted into a programme with ***design*** definitive outputs (deliverables), from Operational Requirement to Manufacturing Datapack.

GENERIC UNIFIED SYSTEMS ENGINEERING METAMODEL

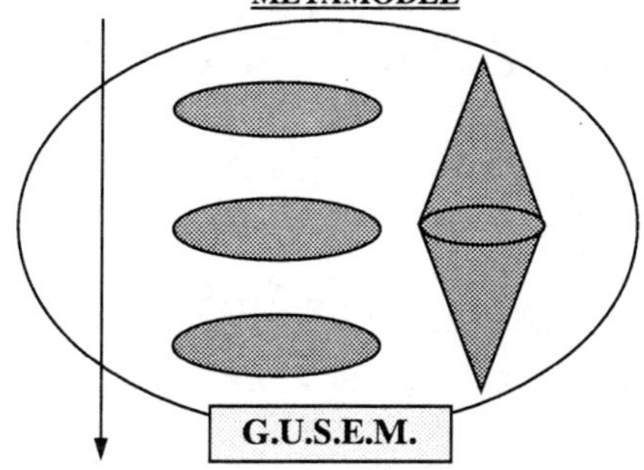

Programme / PERT : Processes : Data Repository

For this paper, only the top definitive semi-cone is addressed, as this is where definitive ***DESIGN*** resides, in the form of ***design*** requirements, circuit design data, mechanical design data, etc., all being classed as *definitive.* Plans such as those for safety, reliability etc. are task *supportive*, within the relevant processes. It should be agreed that we can not build & deliver a *plan.*

The Repository structure must include a number of architectural levels in own right, normally block schematic, or other diagrams A view of the repository is shown below. [Single system]

It should be noted that the **Repository** serves from concept, onward into disposal, applying a most beneficial **discipline.** It is the core discipline of

G.U.S.E.M. & is greatly expandable & decomposable.

[DRS = DESIGN REQUIREMENT]

In the following paragraphs only selected levels of G.U.S.E.M. are addressed by means of S.S.M., to save space.

The Essential Structure of the Repository

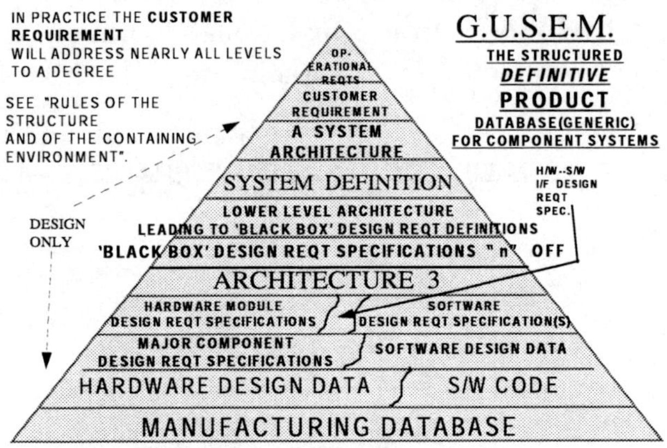

THIS IS THE SPECIFICATION TREE

This is also an information model, with rules.

System Design Level

Creation of System Level Design Requirement Specifications

Analysis of the SYSTEM DEFINITION level process(es) in accordance with Checkland (Eg. Soft Systems Methodology in Action, pages 35, 36, 37, & Fig. 2.9) needed to create the Repository content, results in the following :--

<u>Root Definition</u>

"A systems design business-owned & manned process, in which Systems Design Engineers create the System Definition Level Definitives of the Data Repository from the Customer Requirements, via Architecture 1, by applying training & experience in a disciplined, organised manner, for further work by Systems Design Engineers at the Architecture 2 & LRI levels."

This may be expressed usefully in Checkland's C.A.T.W.O.E., by means of which, applied to the System Definition Level of the model, we can deduce :--

System Design, System Definition Level

Customer : Systems Design Engineers responsible for Architecture 2, & LRI DRS's

Actors : Systems Design Engineers supported by all relevant disciplines

Transformation : Customer Requirement via Architecture 1 into System Definition Level Design Requirements & Interface definitions, inclusive of MMI.

World view : disciplined, organised, reliable, improvable, repeatable method. CMM compatible.

Owner : Any systems design business

Environmental constraints : Those of the containing business domain.

Of the above, "Transformation" may be seen to be central within GUSEM, as indicated in the following Figure. See SYSTEM DEFINITION ARCHITECT [TRANSFORMATION MASTER] One of a number of System Design "Roles"

The diagram below shows the satellite ring of cooperative processes devoted to System level ***design,*** in accordance with proven, generic set of ***Design*** Requirement templates to create the definitive data for the Repository.

0-7803-5086-3 /98/$10.00 ©1998 IEEE

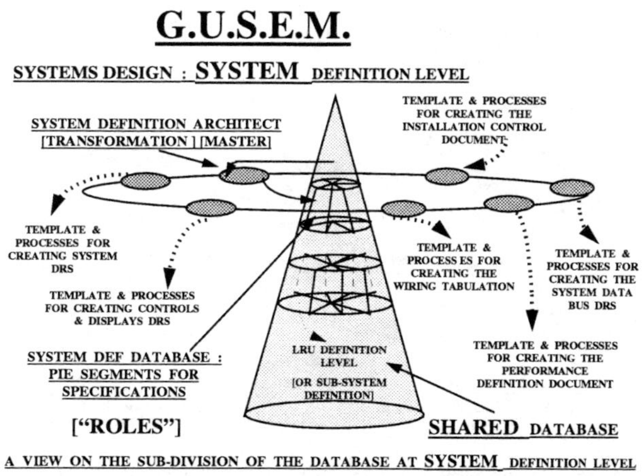

It is under control of the System Level Architect. Whatever form the ***processes*** may be expressed in, they are subservient to the relevant specification template within which is an implicit process known to the suitably skilled. Nevertheless the processes should be determined & set down to aid memory & training, & "process development" (CMM !).

The DRS Templates & processes encompass safety, EMC, reliability, human factors, system interfaces (mechanical, electrical, RF etc), & all relevant plans at SYSTEM level. (refer to "Endless lists." Caple, INCOSE (UK) 1997). In accordance with the rules of GUSEM, all plans are classed as "Supportive" to the creation of *Definitive* ***design*** data for the Repository. Each process has to be resourced in accordance with the needs of each specific project.

System Level Intranet

This view (below) of the System Level ***design*** satellite ring is indicative of process communications via a local I.T. network, & the general form of data to be agreed, modified or otherwise worked upon

. Project Management of the "level" is given its own process. It is suggested that, as with the use of RDD, there should be only one person able to change the definitive data within the Repository, but all concerned have read access. (Configuration management, & change control)

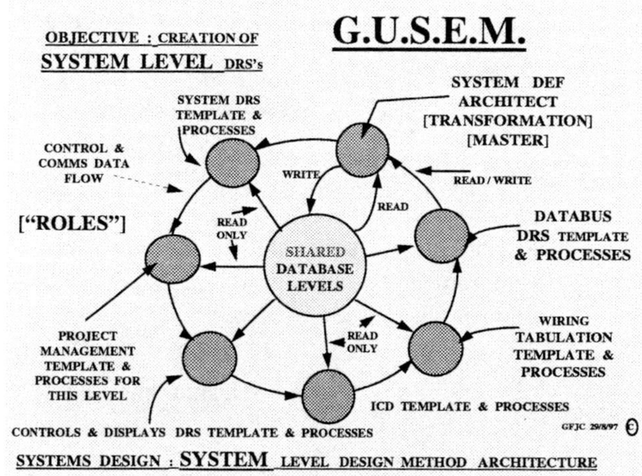

Quoting from the Proceedings of the RAeS, Jan. 1998, one may also describe the satellite level as containing a cross - functional team, in regard of which there are design team practices, plus I.T. & Data Management, which equate to a structured approach to product (specification within the Repository) development. (design). [this paper's author's words in brackets]

LRU (LRI) Level

Creation of LRI Design Requirement Specifications

Root Definition

"A systems design business-owned & manned process, in which Systems Design Engineers create the LRI Level Definitives of the Data Repository from the System Level Requirements, via Architecture 2, by applying training & experience in a disciplined, organised manner, for further work by Systems Design Engineers, hardware Design, & Software Design Engineers, at the Architecture 3, board / module, & software design requirements levels."

Single LRI (as opposed to "n")

Customer : the target LRI Manager

Actor : the responsible Systems Engineer

Transformation : Repository System Definition Level via Architecture 2 into the target LRI DRS.

World view : complete & accurate Design Requirement Specification is created, with all relevant disciplines involved.

Owner : any Systems Design business

Environmental constraints : Those of the containing business domain

In the definition above, a Systems ***Design*** "Actor" transforms higher level Repository definitive data into lower level definitive data for his "Customer" (who is involved actively) The Product of this Process is the ***Design*** Requirement Specification, to a standard template, for a single LRU (LRI) which is given to the LRU Manager & team. They carry out circuit, microwave, mechanical etc ***design*** against the DRS, & respond also to EMC, safety, reliability etc requirements, for which they prepare action plans, or respond to input plans. It will be observed that the Repository contains a second architectural level, which comprises LRI schematics & other definitive drawings, created as part of the "Tranformation" process : System Level to LRI Level.

Below, we have the model (minus PERT) with the LRI Definition Level shown. It employs a single template for a common LRI ***Design*** Requirement Specification, with related processes for a number of LRI DRS's.

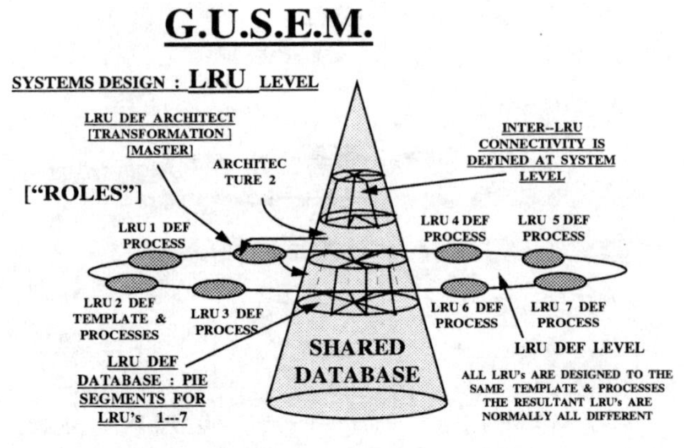

A VIEW ON THE SUB-DIVISION OF THE DATABASE AT LRU DEFINITION LEVEL

Below, we have the Satellite ring shown in more detail, indicating a number of DRS's being produced to the single template. Checkland P34 Fig. 2.7. is relevant. Other material is available to define the data & plans in circulation.

LRI Level Intranet

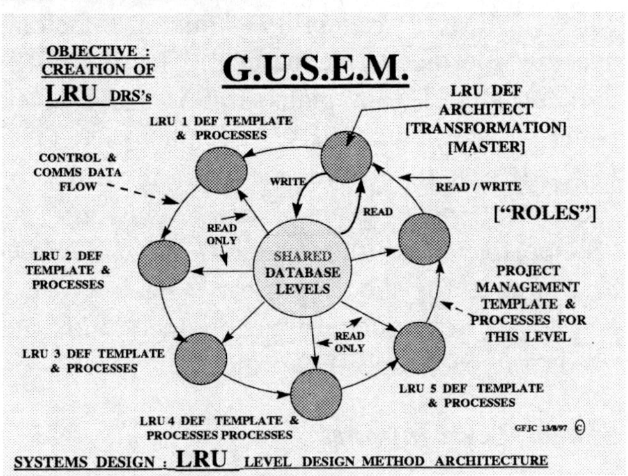

Sub-processes are drawn, & described below.

Creation of LRI Design Requirement Specifications

The diagram below is indicative of sub-processes **(Checkland Soft Systems definable)** beneath, or decomposed from the transformation process devoted to the transformation into the multiple LRI level of the Repository. Sub-processes relate to the creation of ***design*** requirements for safety, EMC, reliability, function, performance, mechanical, maintainability, self-test, & other requirements, as described above, in accordance with a common LRI ***Design*** Requirement Specification template.

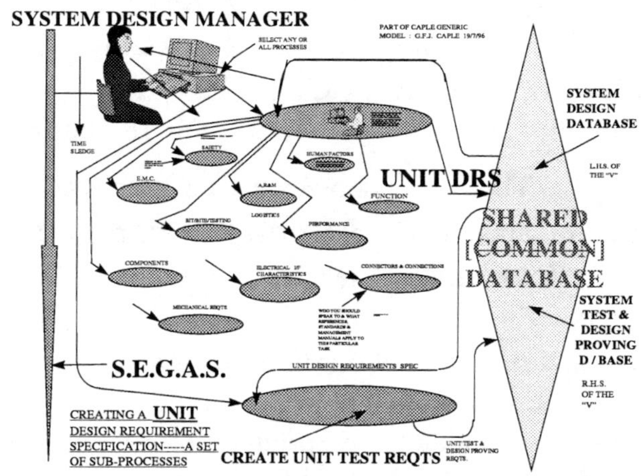

All specification data, whether paper or electronic is lodged within the LRI DRS level of the Repository.

Relationships with RDD Design Guide

In this context, the RDD guide talks in terms of "Evaluate Unit design status.", & "Define / redefine tasks."

Further, in regard of the "Architect" in the Figure above, one may state : "By virtue of her systems ***design*** experience, coupled with a structured design / manufacture / design proving data Repository, the System Design manager will "automatically" carry out work comparable with the RDD processes of the "Design Guide for Project A, Fig. 2." It would be sensible & advantageous nevertheless, to write down for her the essentials of the required "supporting" processes, & in RDD terms provide "Admin. Support" [Log task turnover etc.] These would be entirely subsidiary to the "Product / Process Pairs" of G.U.S.E.M."

[Each process must be be devoted to the creation of a product for entry within the Repository, or supportive, or executive, of a definitive product creative process] A definitive design product is a specification or drawing, A plan may be definitive, but NOT of the ***design***.

Sub-Assembly, Module, & Board Level

Creating board & software design requirements

At the level of the diagram below, the initiative passes from the systems designers to the circuit, microwave, mechanical, & other ***design*** engineers, & the directly appropriate processes.

G.U.S.E.M.

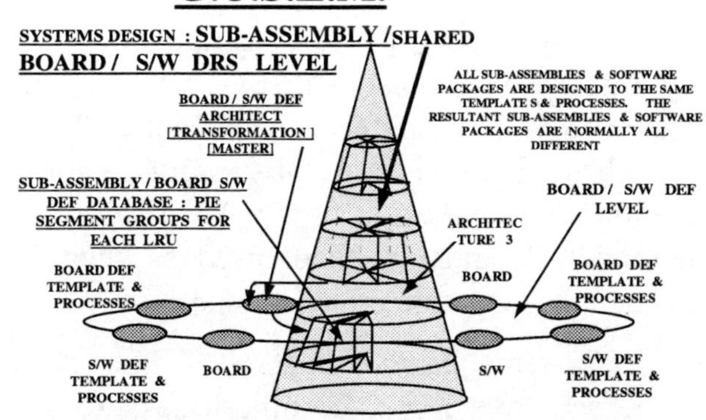

A VIEW ON THE SUB-DIVISION OF THE DATABASE AT LRU SUB-ASSEMBLY / BOARD & SOFTWARE DESIGN REQUIREMENTS DEFINITION LEVEL

Below this level (not shown) we have circuit design, mechanical & software design, with their own process / product combinations. The conclusion is the "definitive" Manufacturing Datapack. (see unpublished presentation material)

System Solution / Selection

Optimisation of Alternative Designs

G.U.S.E.M.

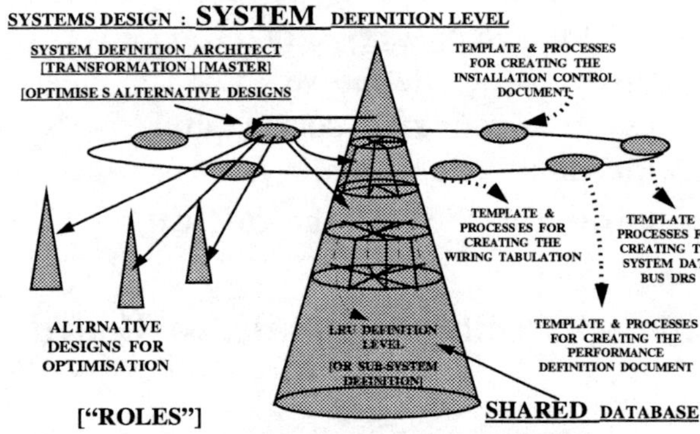

A VIEW ON THE SUB-DIVISION OF THE DATABASE AT SYSTEM DEFINITION LEVEL

Within the diagram above, one may observe that while optimising system solutions, one is inevitably creating a number of partially populated Repositories, & analysing their content, in order to make an informed choice.

Conclusions

This paper has demonstrated that systems, & other *design* disciplines, are the core of Systems Engineering. Plans & generalised methods are subservient to the creation of the *design* at any level within the model, & must be recognised as such through the structure of the model.

Systems *design* must NOT be submerged, overcome, or otherwise set aside, either accidentally, or through ignorance,. Normal management disciplines must remain, but must recognise, make room for, plan & resource, the full range of systems *design.* products & processes.

Without a model such as G.U.S.E.M. this will not be accomplished, to the great detriment of systems engineering.

There will only be a robust, repeatable ***systems design discipline*** if the engineering society is willing to accept the constraint of a model with rules, practices & processes centred on systems ***design***. This will, by its very nature, bring about the integration of all other disciplines, including project management, primarily through the influence of the structure & content of the *domain independent* definitive data Repository. Systems ***Engineering*** will benefit from recognising Systems ***Design,*** resourcing & managing it correctly, from start to finish, top to bottom

There are choices to be made, across all domains. Let's put ***design - complete, & integrated -*** back into systems engineering by means of a suitable model, ***designed*** for the purpose, & clarify the whole business. Think G.U.S.E.M. Analyse your own methods & procedures by means of the disciplined G.U.S.E.M. approach.

Choice

(i) Choose to stay with an unstructured unmodelled reductionist amalgum of plans, analyses, requirements capture, requirements analysis, requirements synthesis, risk assessment, integrated logistics support, compliancy matrices, circuit design, software design, EMC & physical requirements, functional decomposition, functional analysis, manufacture, test, V & V, trials, acceptance, design certification, programme, costings, cost-effectivenes, metrics, deliverables, computer modelling, & reviews. Believe that systems design has been taken care of in the bid phase by means of the Technical Proposal, the Compliancy Matrix, the Test & Evaluation Master Plan, & Product Descriptions.

OR

(ii) Choose to go for structured Systems Design in accordance with the model described above, as the core, focus, & key to Systems Engineering. Recognise the ammount of work required to produce the necessaty definitive Data Repository content. Resource that work, & plan it into the programme **flexibly.** Realise that everything else (Plans etc.) is there to ensure that the **design** is right. GET THE DESIGN RIGHT & include integrated plans, etc. Recognise this as the key to the bid phase also. Apply a good model consistently, & improve it with practice, as will be required by the CMM.

Litmus Tests (D.K. Hitchins)

Proof of G.U.S.E.M.'s validity by means of the above will have to be the subject of a separate paper.

0-7803-5086-3 /98/$10.00 ©1998 IEEE

CURRENT GOVERNMENT AND INDUSTRY DEVELOPMENTS IN THE AREA OF SYSTEM SAFETY ASSESSMENT

Brett E. Portwood, Federal Aviation Administration, Lakewood, California

Introduction

The use of increasingly more complex systems and higher levels of system integration on transport aircraft systems and equipment has created new safety and certification challenges. Recent and current initiatives are providing aviation industry and regulatory authorities technical guidance in a manner that helps meet their mutual objectives and responsibilities in the area of system safety. The proposed new amendment to §/JAR 25.1309 and its associated guidance material which references two new industry standards, Society of Automotive Engineers (SAE) Aerospace Recommended Practice (ARP) 4754, "Certification Considerations of Highly Integrated Complex Systems" and SAE ARP 4761, "Guidelines and Methods of Performing the Safety Assessment Process on Civil Airborne Systems and Equipment", represents a collective government and industry effort to pursue solutions to current challenges in the area of system safety.

Rulemaking Activities

There are current efforts by working groups of the Aviation Rulemaking Advisory Committee (ARAC)[1] to revise and harmonize §/JAR 25.1309 and §/JAR 25.901 and the associated advisory information. A significant part of the activity is to align the requirements of §/JAR 25.1309 to powerplant installations.

The ARAC Powerplant Installations Harmonization Working Group is proposing changes to §/JAR 25.901(c) which would be revised to establish a general requirement for a powerplant safety assessment and would apply the requirements of §/JAR 25.1309 to powerplant installations, except, these requirements would not apply to:

- The effects of an engine case rupture
- Uncontained engine rotor failure
- Propeller debris release

In addition, the ARAC Powerplant Installations Harmonization Working Group has developed a proposed new associated common §/JAR Advisory Circular/Advisory Material Joint, Safety Assessment of Powerplant Installations, No. 25.901(c).

The ARAC Systems Design and Analysis Harmonization Working Group is proposing changes to §25.1301 and §25.1309, a new §25.1310, and a common §/JAR Advisory Circular/Advisory Material Joint for §/JAR 25.1309.

The proposed regulatory changes include:

- A revised §25.1301 which would

[1] By notice in the Federal Register (58 FR 13819, March 15, 1993), the FAA chartered a working group of industry and government specialists from Europe, the United States, and Canada.

0-7803-5086-3 /98/$10.00 ©1998 IEEE

permit systems which do not affect safety to be installed, even if they do not perform their intended function.

- Revisions to §25.1309:
    - Require safety related systems to function properly under any operational or environmental condition approved for the airplane.
    - Require that catastrophic failure conditions be extremely improbable and not result from a single failure.
    - Require that Hazardous failure conditions be extremely remote.
    - Require that Major failure conditions be remote.
    - Explicitly identify by section the regulations to which §25.1309 does not apply.
    - Move the analysis requirements of the existing §25.1309 from the rule to the Advisory Circular and extend the analysis requirements to include design errors.

- Establish a new section §25.1310 to contain the electrical power requirements now found in §25.1309

The following tables depict the relationship between the probability and severity of failure condition as defined in the new rules and associated advisory material:

	NO SAFETY EFFECT
Effect on Airplane	No effect on operational capabilities or safety
Effect on occupants	Inconvenience for Passengers
Effect on flight crew	No effect on flight crew
Qualitative probability	No Probability Requirement
Quantitative Probability	No Probability Requirement

	MINOR
Effect on Airplane	Slight reduction in functional capabilities or safety margins
Effect on occupants	Physical discomfort for passengers
Effect on flight crew	Slight increase in workload or use of emergency procedures
Qualitative probability	Probable
Quantitative Probability	$< 1.0 \times 10^{-3}$ per flight hour

	MAJOR
Effect on Airplane	Significant reduction in functional capabilities or safety margins
Effect on occupants	Physical distress for passengers, possibly including injuries
Effect on	Physical discomfort or a significant

flight crew	increase in workload
Qualitative probability	Remote
Quantitative Probability	< 1.0 x 10^{-5} per flight hour

	HAZARDOUS
Effect on Airplane	Large reduction in functional capabilities or safety margins
Effect on occupants	Serious or fatal injury to a small number of occupants
Effect on flight crew	Physical distress or excessive workload impairs ability to perform tasks
Qualitative probability	Extremely Remote
Quantitative Probability	< 1.0 x 10^{-7} per flight hour

	CATASTROPHIC
Effect on Airplane	Hull Loss
Effect on occupants	Multiple Fatalities
Effect on flight crew	Fatalities or Incapacitation
Qualitative probability	Extremely Improbable
Quantitative Probability	< 1.0 x 10^{-9} per flight hour

The ARAC Systems Design and Analysis Harmonization Working Group held its last meeting on April 23, 1998. The FAA intends to publish a Notice of Proposed Rule Making for comment in late 1998.

Industry Documents

History of aircraft systems reveals many accidents and incidents which came about through losses or violations of what was thought to be sufficient redundancy. One purpose of a safety assessment is to identify for removal all potential redundancy violators, such as single point failures, latent failures, too high probability combinations of failures, and installation induced failures. Therefore, we need a comprehensive system safety assessment approach that will address these types of failures.

Two of the safety assessment industry guidance documents that has been in circulation since the late 1970's and throughout the 1980s are SAE ARP 926A, "Fault/Failure Analysis Procedure" and SAE ARP 1834, "Fault/Failure Analysis for Digital Systems and Equipment". Although the two ARP's, authored by the SAE S-18 Committee, were useful tutorials for conducting Failure Modes and Effects Analyses (FMEA) and Fault Tree Analyses (FTA), they did not address safety exclusively in a comprehensive manner. The documents have become outdated with respect to current industry trends since they do not address aircraft level analysis, common mode analysis, or the concept of a Preliminary System Safety Assessment (PSSA).

Safety assessment guidance that adequately addressed complex integrated systems was lacking and motivated the SAE S-18 Committee in 1991 to generate a comprehensive "how to" safety assessment guideline that would supersede ARP 926A and ARP 1834 for the purposes of system safety assessment. Coinciding with the S-18 activity, another SAE committee, "Systems Integration and Requirements Taskforce (SIRT), was developing industry guidance which addresses systems issues associated with complex integrated airborne systems. Both committees eventually coordinated their efforts with SIRT focusing on the systems engineering process and S-18 covering the safety assessment process. In late 1996, both committees released and published their respective guidance documents.

SAE ARP 4754

SAE ARP 4754, "Certification Considerations for Highly-Integrated or Complex Systems", describes the systems engineering process and describes the interfaces with the certification process. The guidance also depicts the requirements allocation and capture process with emphasis on requirements validation and verification. The safety assessment process is also described at a high level and references ARP 4761 for detailed descriptions of the various safety assessment tools. Much time and effort was spent during the development of ARP 4754 to describe how design assurance levels are determined based on architectural considerations such as functional dissimilarity and monitoring.

In particular, the document recognizes that a parallel, functionally dissimilar, multiple channel architecture may provide protection from both random physical failures and anomalies due to design errors. Furthermore, in those cases, the document recognizes that it may be possible for the design assurance levels for the individual channels to be selected below that associated with the top level failure condition.

SAE ARP 4761

ARP 4761, "Guidelines and Methods of Performing the Safety Assessment Process on Civil Airborne Systems and Equipment", describes in detail the system safety assessment process. The process is defined as three sub-processes:

1. Functional Hazard Assessment (FHA), which established the top level safety requirements or a system.
2. Preliminary System Safety Assessment (PSSA), which establishes the lower-level safety requirements of the system.
3. System Safety Assessment (SSA), which is a systematic, comprehensive evaluation of the implemented system to be certified to show that the qualitative and quantitative safety requirements as defined in the FHA and PSSA have been met.

ARP 4761 also provides tutorials for each of the following major safety assessment tools:

- Functional Hazard Assessment
- Fault Tree Analysis (FTA)
- Markov Analysis (MA)
- Failure Modes and Effects Analysis (FMEA)
- Failure Modes and Effects Summary (FMES)
- Zonal Safety Analysis (ZSA)
- Particular Risks Analysis (PRA)
- Common Mode Analysis (CMA)

SAE ARP 4761- New Concepts

Traditionally, Functional Hazard Assessments have been conducted at the system level. For federated architectures where systems are relatively independent from other systems, this approach seems reasonable. However, with today's levels of integration where aircraft level functions may be distributed across multiple integrated systems, analysis at the aircraft level might be more appropriate. ARP 4761 describes the FHA process at both the aircraft and the system level.

Both ARP 4761 and ARP 4754 introduce the concept of a Preliminary System Safety Assessment (PSSA). This assessment conducted early in the development program would minimize certification risks by providing the certification authority and applicant up-front agreement on the basic system architecture concept. The PSSA looks at a given proposed architecture and starting with the FHA, allocates safety requirements to the lower levels. The PSSA shows how item failures combine to lead to the considered failure condition, identifies requirements for event independence verification, and shows that quantitative and qualitative requirements can be met.

ARP 4761 further refines the treatment of common-cause faults. The regulatory advisory material[2] indicates the safety analysis must give special attention to analyzing common-cause faults. Independence claims made in the Fault Tree Analysis and Failure Modes and Effects Analysis are verified such that common cause events are precluded for catastrophic failure conditions and must be within the assigned probability budget for hazardous failure conditions. Common-cause analysis as defined in ARP 4761 is subdivided into three types of analysis:

- Zonal Safety Analysis
- Particular Risks Analysis
- Common Mode Analysis

Zonal Safety Analysis is performed on each geographic zone of the aircraft with the objective of ensuring that the equipment installation meets the safety requirements with respect to basic installation, interference between systems, and maintenance error potential.

Particular Risks Analysis addresses those events or influences which are outside the system of concern, such as fire, wheel rim release, or bird strike, which may violate failure independence claims.

Common Mode Analysis is performed to verify that ANDed events in the Fault Tree Analysis are independent in the actual implementation.

In ARP 4761, Fault Tree Analysis formulas are provided in an attempt to standardize the way in which fault trees are normalized. For example, probability per flight hour is based on the probability calculations of the failure condition on a per flight basis divided by the average flight time for the particular model aircraft. In addition, exposure time for latent failures is defined with standard formulas as are other cases of monitored failures with imperfect monitor coverage.

An example of one of the more controversial fault tree calculations that was clarified in ARP 4761 is depicted in Figure 1 which represents a fault tree calculation when two items cause a loss of a function and one of the items can fail latent but the other cannot fail latent. In this case, Item #1 can fail at any point between when it is checked (time=0) and when it is next checked (time=T). Item #2 is known to be operating at the start of each flight and never fails latent. The order of failure does not matter. In this case there is a difference between the average probability of the top event per flight and the probability of the top event for the worst case flight. Consider that there are n flights of t_f hours such that $T=nt_f$.

[2] AC/AMJ 25.1309

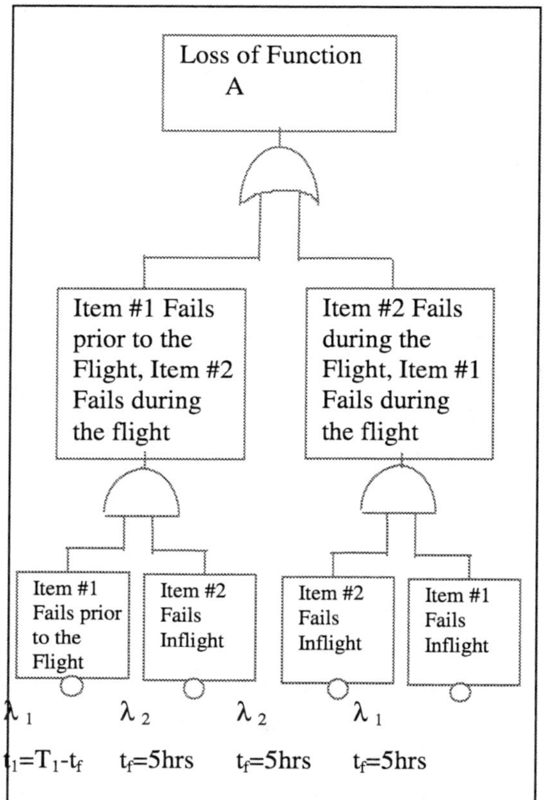

Figure 1

Using the approximation of $P_f = \lambda t$ for small λt, gives the following:

The average probability per flight equals the sum of the probabilities for each flight, divided by n, the number of flights in the latency period.

$$P_f\,(\text{average}) = 1/n * \Sigma i\ \lambda_1 \lambda_2 t_f^2 \text{ for } i = 1 \text{ to } n$$

$$= (\lambda_1 \lambda_2 t_f^2)/n * \Sigma i \text{ for } i = 1 \text{ to } n$$

$$= (\lambda_1 \lambda_2 t_f^2)/n * (n*(n+1))/2$$

$$= 1/2 * \lambda_1 \lambda_2 t_f (n t_f + t_f)$$

$$= 1/2 * \lambda_1 \lambda_2 t_f (T + t_f)$$

The factor of 1/2 appears as a result of calculating the average probability that the function will fail on any single flight within the latency period and not from using a mean exposure time of T/2.

Conclusions

The proposed new regulations and industry guidance in the area of transport aircraft system safety assessment provide baseline safety objectives, along with a defined design discipline and development structure required to ensure that complex or highly integrated system safety and operational requirements can be fully substantiated. This should provide the aircraft manufacturers, equipment suppliers, and certification authorities a common understanding of the fundamental safety assessment issues to be addressed in the certification of modern transport aircraft.

OVERVIEW OF THE EIA 632 STANDARD: PROCESSES FOR ENGINEERING A SYSTEM

James N. Martin, Raytheon Systems Company, Plano, TX

Abstract. The EIA 632 standard has been developed that describes the "processes for engineering a system." This standard evolved from EIA/IS 632—the interim standard that described a systems engineering process. This paper will describe the structure of the standard, elements of the process, and key concepts such as stakeholder requirements, enabling products, building blocks, and development layers.

INTRODUCTION

In April 1995, the G47 Systems Engineering Committee of the Electronic Industries Association (EIA) chartered a working group to convert the interim standard EIA/IS 632 into a full standard. This full standard has been developed over the last three years and is planned to be released in August 1998.

The interim standard, EIA/IS 632, was titled "Systems Engineering." The full standard was expanded in scope to include *all the technical processes for engineering a system.* It is intended to be a higher level abstraction of the activities and tasks found in the IS version plus those other technical activities and tasks deemed to be essential to the engineering of a system.

This paper describes the elements of the process, and key concepts. The intended purpose is to give the reader of the standard some background in its development and to help other standards activities in developing their own standard. (Since this standard was not released at the time of this paper submittal, some differences may be found between this paper and the released version.)

There is a separate paper that describes the evolution from an interim standard to the full standard [Martin 1998]. This standard is intended to be a "top tier" standard for the *processes essential to engineering a system.* It is expected that there will be second and third tier standards that define specific practices related to certain disciplines (e.g., systems engineering, electrical engineering, software engineering) and industry domains (e.g., aircraft, automotive, pharmaceutical, building and highway construction).

It is important to understand several things that are *not* covered by this standard:

a) Does *not* define what "systems engineering" is;
b) Does *not* define what a "systems engineer" is supposed to do; and
c) Does *not* define what a "systems engineering organization" is supposed to do.

STRUCTURE OF THE STANDARD

Clause 1	*Scope*
Clause 2	*Normative references*
Clause 3	*Definitions and acronyms*
Clause 4	*Requirements*
Clause 5	*Application guidance*
Clause 6	*Application context*
Annex A	*Glossary*
Annex B	*Enterprise-based life cycle*
Annex C	*Process tasks*
Annex D	*Planning documents*
Annex E	*Technical reviews*
Annex F	*Unprecedented and precedented development*

ROLE OF EIA 632

Implementation of the requirements of EIA 632 are intended to be through establishment of enterprise policies and procedures that define the requirements for application and improvement of the adopted processes from the standard.

0-7803-5086-3 /98/$10.00 ©1998 IEEE

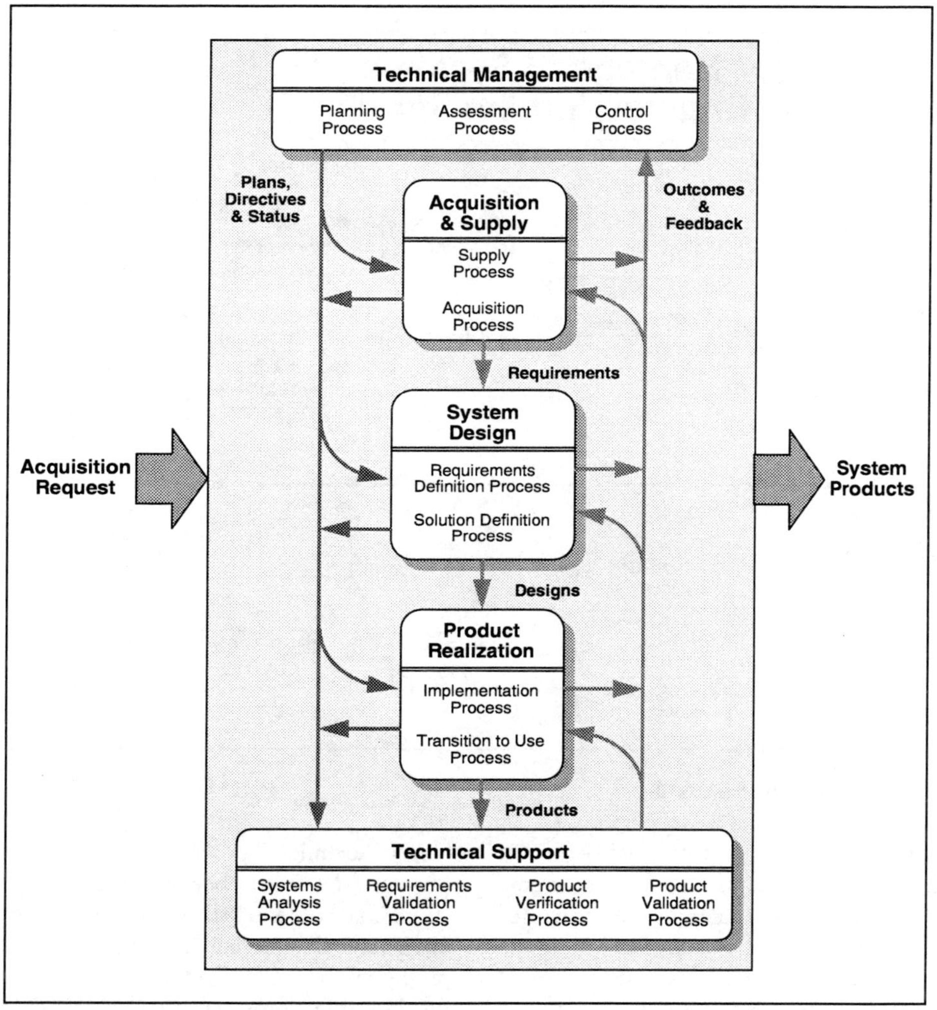

THE PROCESSES

The figure to the left shows the processes described in EIA 632 and their relationship to one another. Each enterprise will determine which of these processes are implemented by systems engineering personnel.

PROCESS HIERARCHY

The processes for engineering a system are grouped into the five categories as shown below. This grouping was made for ease of organizing the standard and is not a required structure for process implementation.

Traditional systems engineering is most often considered to include two of these processes: Requirements Definition and Systems Analysis. Often, Planning and Assessment are included in what is called systems engineering management.

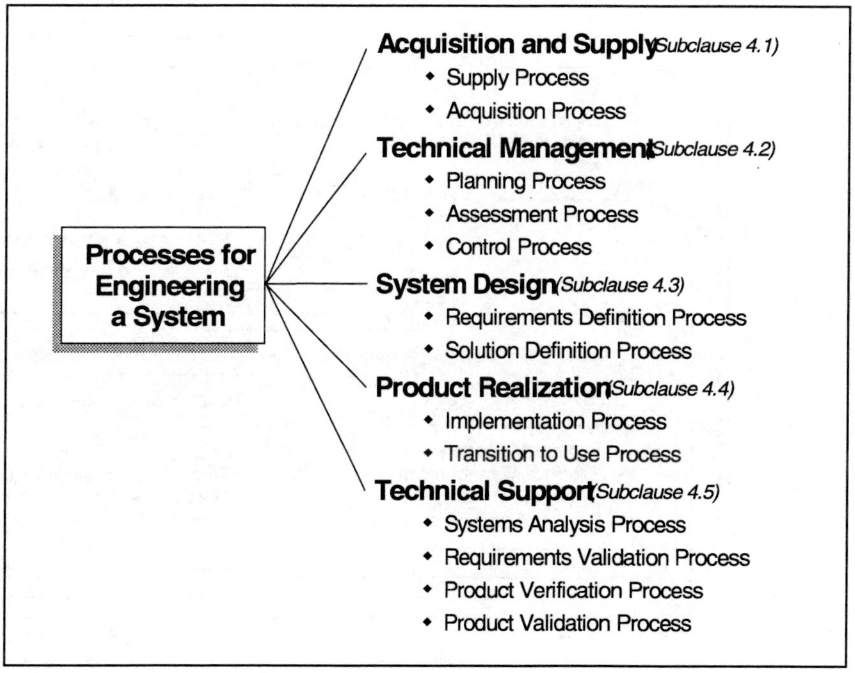

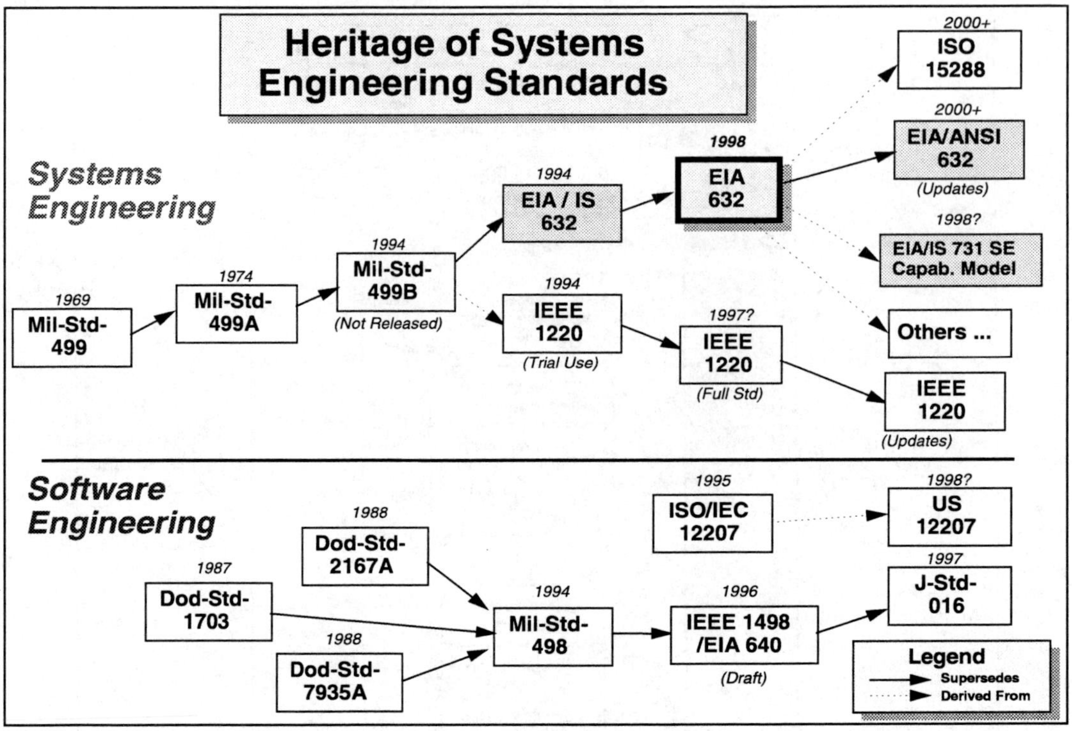

HERITAGE OF EIA 632

The figure above shows the relationship between EIA 632 and other standards on systems engineering. Some of the key software engineering standards are shown for comparison since there has been an intimate relationship between the development of both types of standards. There has been much activity recently in unifying the processes contained in each.

PROJECT CONTEXT

These "technical" processes fit into a larger context of a project (see figure below), and the project resides in some sort of enterprise, which in turn resides in an environment external to the enterprise. There are processes in the project and within the enterprise (but outside the project) that significantly affect the successful implementation of the technical processes.

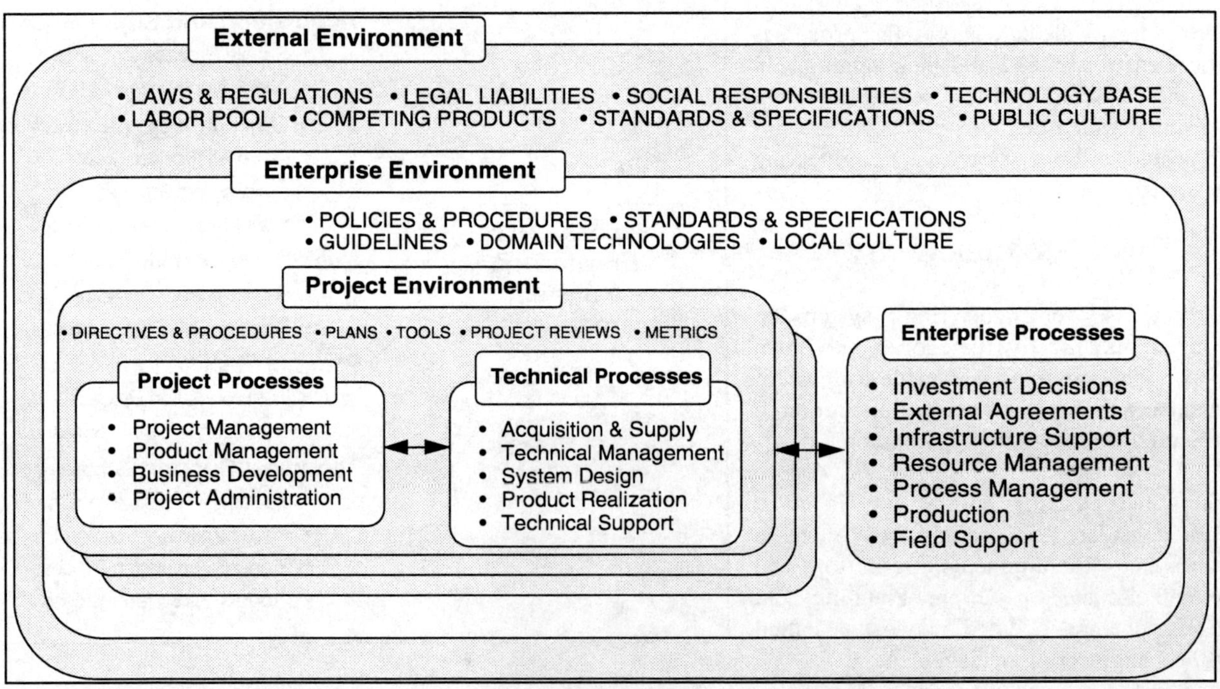

KEY CONCEPTS

To understand the processes as described in this standard, it is essential to understand the distinct use of certain terms and the conceptual models that underlie each process. Some of the key terms are: system, product, verification, and validation. Some of the key concepts are: building block, end products, associated processes, and development layers.

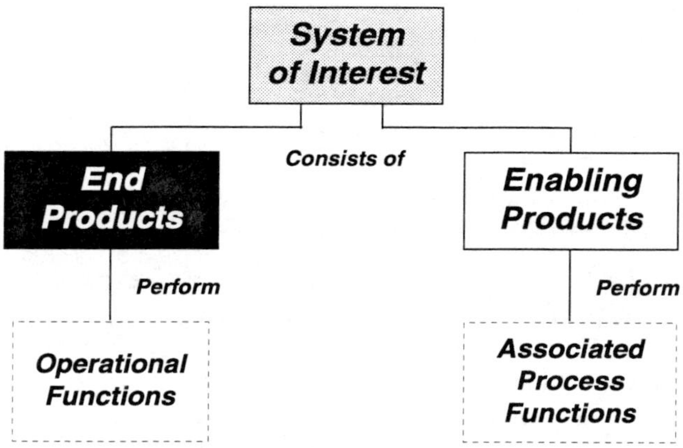

What is a system? The term "system" is commonly used to mean the set of hardware and software components that are developed and delivered to a customer. This standard uses this term in a broader sense in two aspects.

First, the system that needs to be developed consists of not only the "operations product" (that which is delivered to the customer and used by a user), but also the enabling products associated with that operations product. The operations product consists of one or more end products (so called since these are the elements of the system that "end up" in the hands of the ultimate user). The associated processes are performed using enabling products that "enable" the end products to be put into service, kept in service, and retired from service.

Second, the end products that need to be developed often go beyond merely the hardware and software involved. There are also people, facilities, data, materials, services, and techniques. See figure below.

This is not intended to be an exhaustive list of the "basic" product types since these will vary depending on the particular business or technology domain. For example, in the television industry, "media" is certainly one of the system elements that constitute the overall system that is developed. "CBS News" might be considered the system, for example, with end products like:

a) cameras, monitors (hardware)
b) schedule management tools, video compression algorithms (software)
c) camera operators, news anchor (personnel)
d) studio, broadcasting tower (facilities)
e) script, program guide (data)
f) pictures, stories (materials)
g) airplane transportation, telephone (services)
h) presentation method, editing procedures (techniques)
i) airwaves, worldwide web (media)

Note that any or all of these end products could be "off the shelf" (OTS). But some of them may need to be conceived, designed, and implemented. Even if one of these items is truly OTS, it may still need some enabling

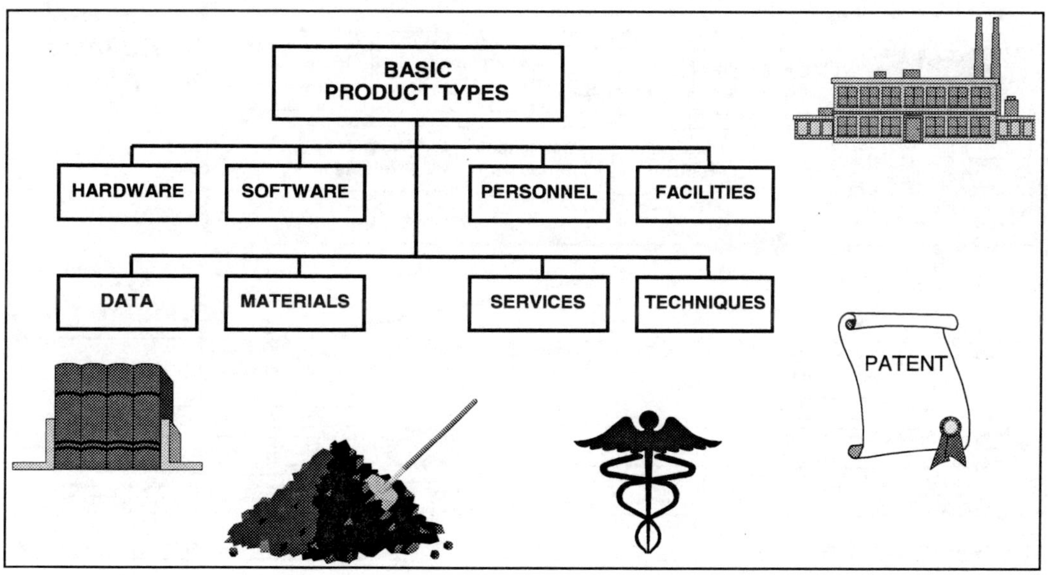

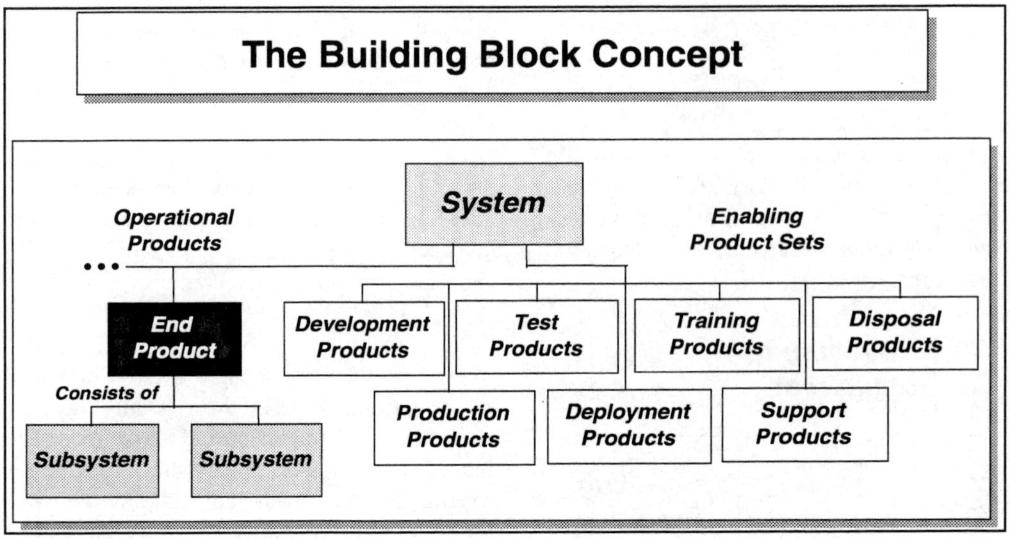

product to allow effective use of that item. For example, even if you can use existing editing procedures, you may need to develop a training program to train the editors.

BUILDING BLOCK

As we can see from the description above of the "CBS News" system, the non-hardware/software items may be crucial to successful realization of the whole system. You may also need to develop the associated processes along with the relevant enabling products. If we tie all these elements together, we can illustrate this using the so-called "building block" shown to the left. There are seven associated processes related to these sets of enabling products. These processes are used at various points in the life of a product (sometimes called "system life cycle elements"). Hence, the use of the building block concept is intended to help the developer in ensuring that the full life cycle of the end products is properly considered.

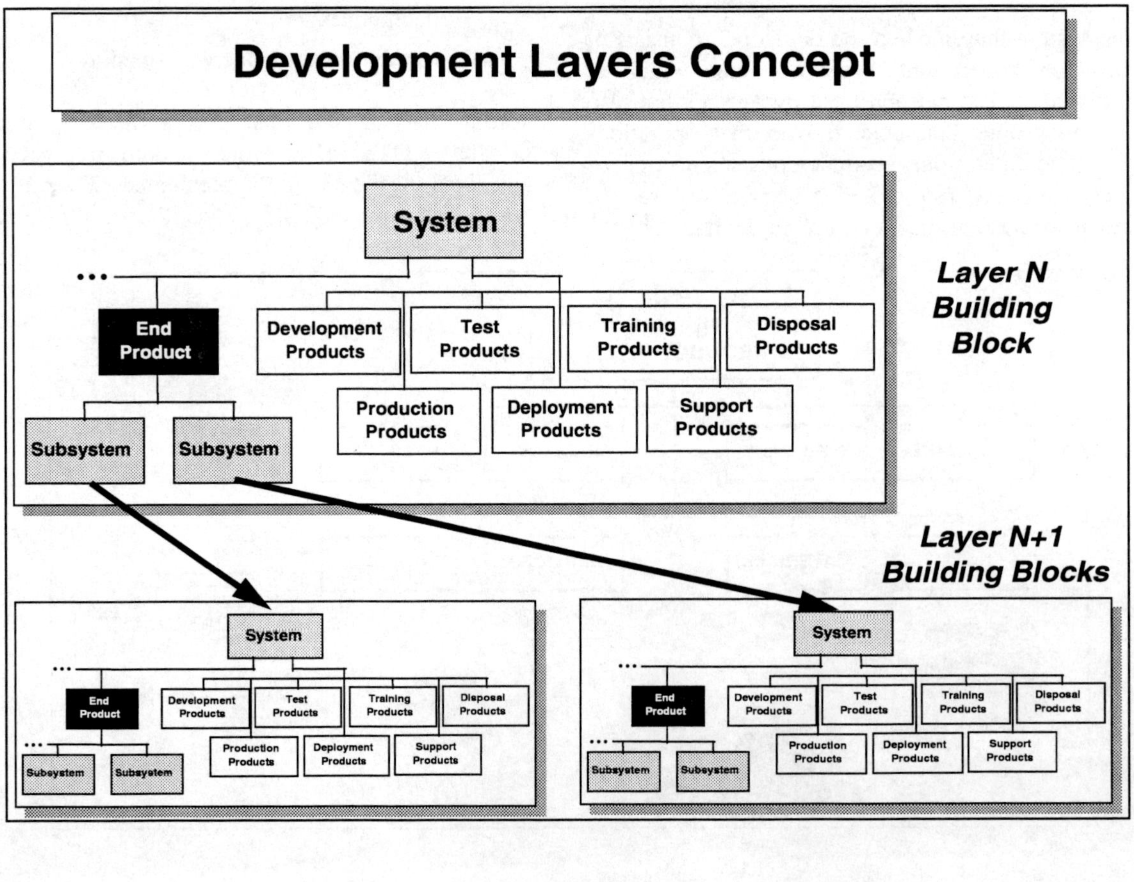

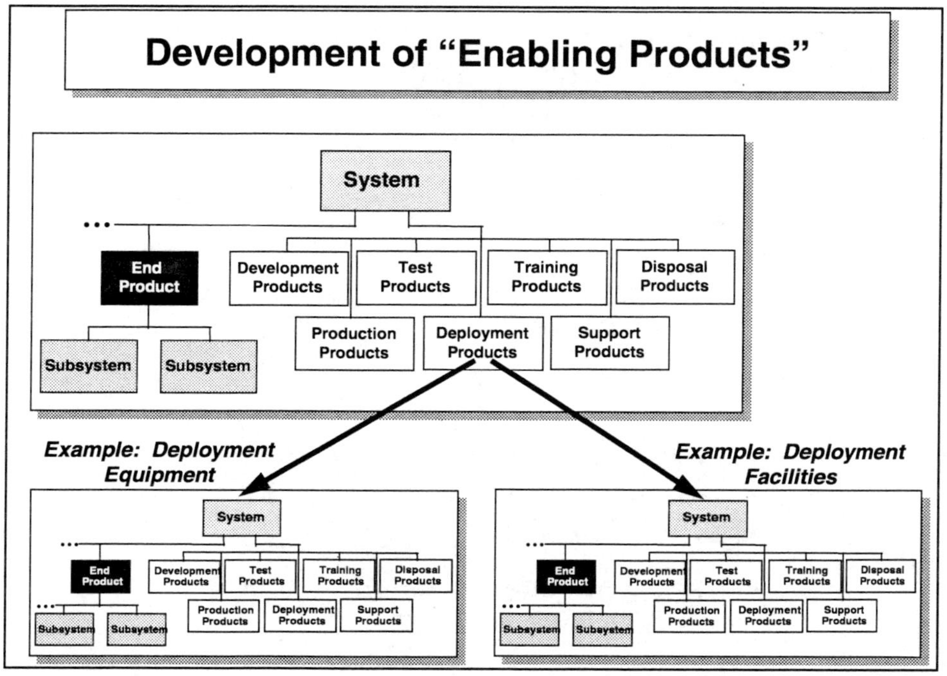

Note that each end product can consist of subsystems. Each of these subsystems may need its own development. The building block can be used at the various "development layers" of the total system. These development layers are illustrated in the figure below.

DEVELOPMENT OF ENABLING PRODUCTS

As I mentioned above, the enabling products may need to be developed also. For each associated process, there could be enabling products that either exist already or that need some degree of development. The figure above shows how enabling products related to the deployment process have their own building blocks.

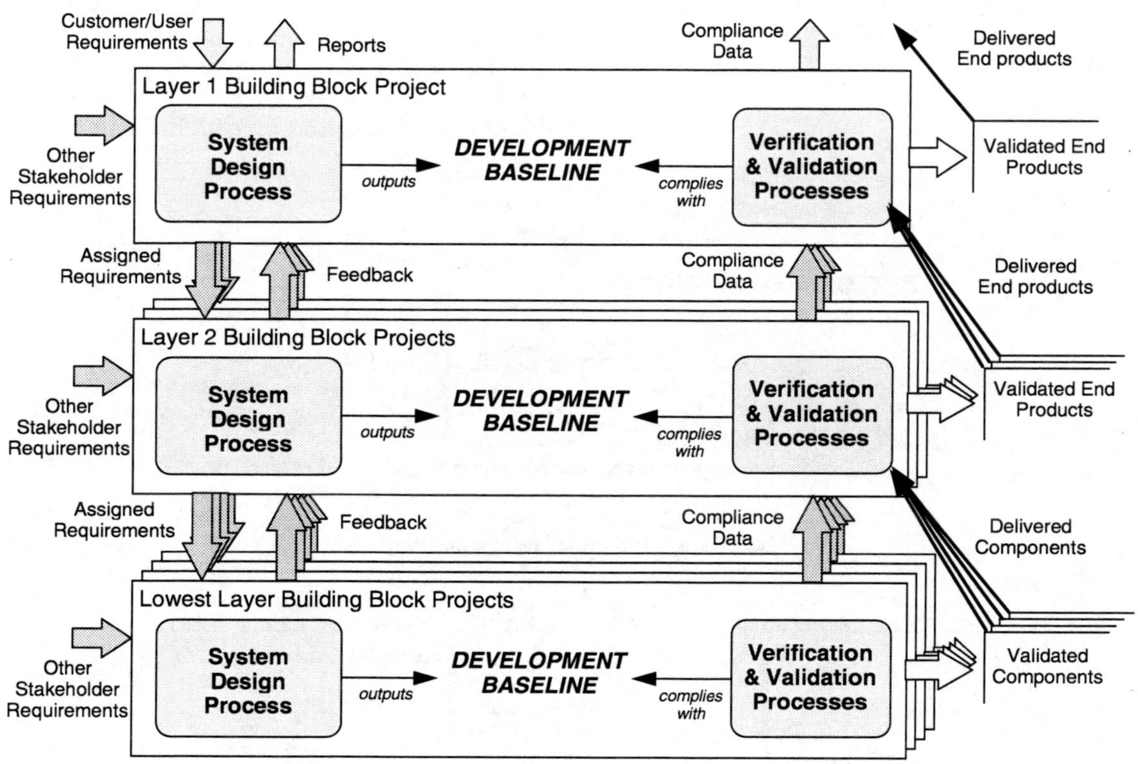

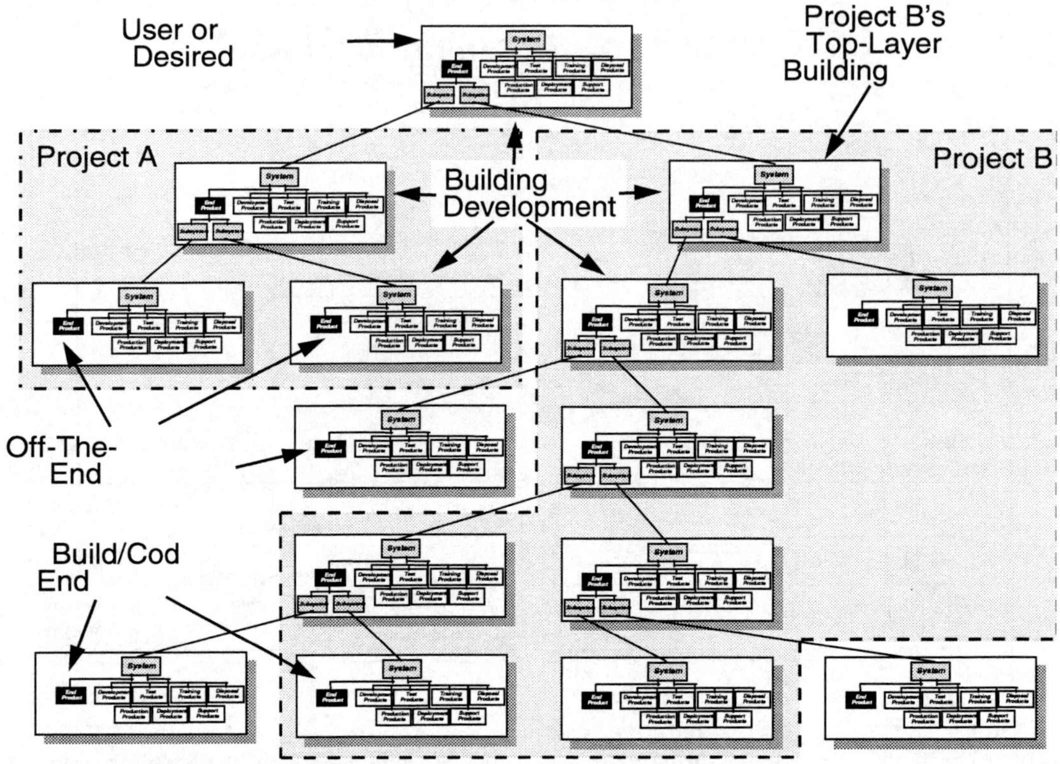

RELATIONSHIP BETWEEN THE BUILDING BLOCKS AND THE PROCESSES

The building blocks relevant to a particular system development can be "stacked" into a System Breakdown Structure (SBS). The System Design Process and Verification and Validation Processes have a special relationship to the SBS as shown in the figure below.

HIERARCHY OF BUILDING BLOCKS

Typically a project is developing a system of such a size that it will require more than one building block to complete the development. The figure above shows an example of how several building blocks would be related to each other within a project. Notice also how an "adjacent" Project A has building blocks that might have interfaces to the systems under development by Project B. There also may be building blocks above

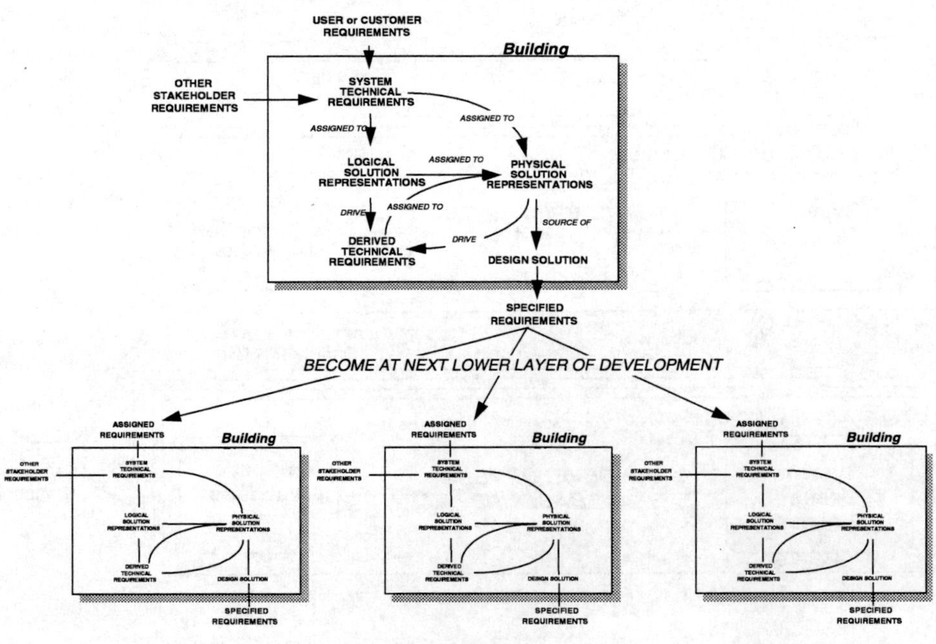

and below a project.

REQUIREMENTS

Requirements are especially useful for engineering a system, especially when the system consists of several building blocks. Requirements act as a kind of "decoupling mechanism" between the building blocks. They are the "terms and conditions" between the layers of development. The figure below shows the different types of requirements dealt with by the processes described in EIA 632.

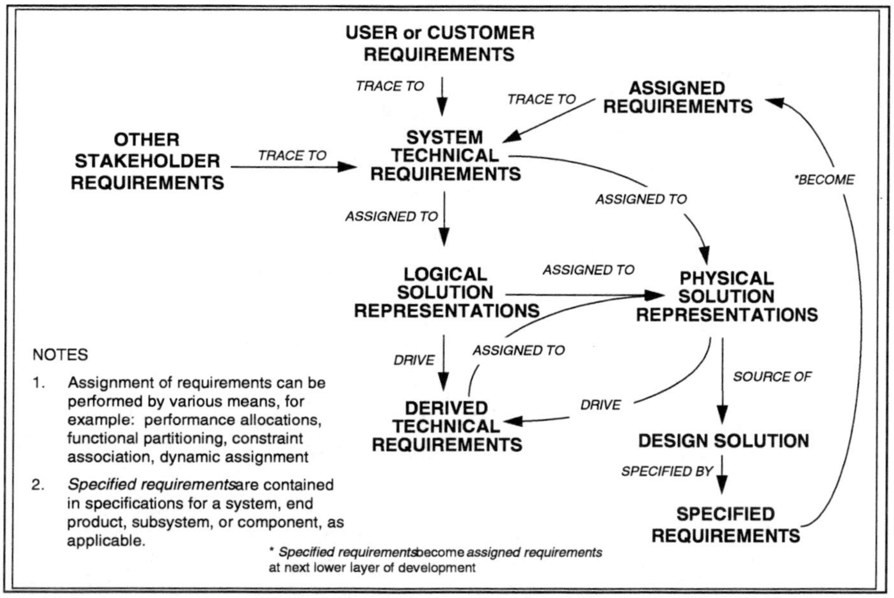

Requirements in general do not come from a particular customer. And even when they do, as is often the case in government-sponsored developments, they do not represent the full spectrum of the requirements that will eventually need to be defined for development of the products of a system.

User Requirements. User requirements are often non-technical in nature and usually conflict with each other. These need to be translated into "technical requirements" that are in the vernacular of the engineers on a project and are specific to the domain of the technology to be used in the system.

Customer Requirements. Customers will often translate their perceived user requirements into a set of requirements to be given to a development project. These also, like the user requirements, are often not technical enough and usually conflict with each other. Moreover, a particular project may have more than one customer for its products.

Stakeholder Requirements. In defining the technical requirements for the system, one must consider that there are other stakeholders for that system. For example, the manufacturing organization within your company may desire that you utilize their existing processes, tools, facilities, etc. There are constraints associated with these "enabling products." Hence, you must identify the stakeholder requirements that drive the development.

Requirements Decomposition. Once the system technical requirements are defined, further requirements are often "derived" by analyzing both the logical and physical aspects of the system. Some of these derived requirements, plus whatever system technical requirements are relevant, are "specified" for the system, its end products, and possibly subsystems of those end products.

Given these specified requirements for an item, the process starts all over again for items needing further development. In other words, another layer of building blocks may be required to develop an item far enough such that the bottom-most item can be procured from a supplier. This is illustrated in the figure above.

SUMMARY

The purpose of this paper is to give a high-level overview of the processes for engineering a system described in EIA 632. Further details can be found in the full publication of the standard.

REFERENCES

EIA/IS 632, "Systems Engineering." 1994.

EIA 632, "Processes for Engineering a System." To be published in 1998.

Martin, James N., "Evolution of EIA 632 from an Interim Standard to a Full Standard." INCOSE 1998 Symposium.

BIOGRAPHY

James Martin is a systems engineer at Raytheon Systems Company working on airborne and satellite communications systems. He is leader of the working group responsible for developing EIA 632, a US national standard that defines the processes for engineering a system. He previously worked at AT&T Bell Labs on wireless telecommunications products and underwater fiber optic transmission products. He has a book published by CRC Press called "Systems Engineering Guidebook".

IMPLEMENTATION OF EIA-632 - PROCESS COMPLIANCE OR EMBEDDED INFRASTRUCTURE?

Richard M. Harwell, SYSTEM Perspectives, Alpharetta, Georgia

Abstract

An effective implementation of EIA-632 ("Processes for Engineering a System") cannot rely on just sending the product team through an accelerated process course and claiming compliance. Implementation begins with senior management conviction that the enterprise needs an integrated infrastructure to provide the overall guidance and support to projects, and that projects adopt viable process requirements necessary to meet their customer's needs and expectations within this infrastructure. These, then become the ingrained and inherent resources to be used by the project team in developing products that meet stakeholder needs and expectations the first time around!

The Typical Implementation

The following scenario is well known in industry:

- A new process standard has just been issued
- A customer strongly indicates compliance as an essential in the development of its desired system
- A process specialist quickly extracts the "key" aspects of the standard
- All members of the enterprise crowd into a large theater to attend an overview of these key aspects of the standard
- All are told they are to follow the new standard in performance of their assigned tasks

Unfortunately, this scenario is typical of the approach companies take in applying new process standards. And when the new processes do not appear to deliver as anticipated, the post-mortems often fail to identify this as a contributing reason.

Influencing Factors

The typical approach is the result of a number of factors including:

- Cosmetic adjustment of company processes and procedures to demonstrate the meeting of a customer-directed process requirement
- Cost of supporting infrastructure
- Time and cost to integrate new processes with existing company policies and procedures
- Lack of senior management embracing the new process as the way to do business

In each case, the company's original culture and methods of doing business remain relatively unchanged, yet appearing to demonstrate compliance with the governing process requirements. Any of these indicators sends a signal to the project team that it is acceptable to continue as usual – without adverse consequences. However, new processes are often specified by the customer (or process specialists) because doing business "as usual" has been deemed non-responsive to the marketplace.

Short-Term vs. Long-Term Needs

It is easier to be influenced by the factors noted above than to consciously develop a new or modified infrastructure with each new process document, or to modify the company's culture that may have driven a number of "wins" in the past. But the current way of doing business may not enable a company to meet today's complex challenges

0-7803-5086-3 /98/$10.00 ©1998 IEEE

and today's informed stakeholder base. Like it or not, today's customer is a partner in the development process.

EIA 632 Implementation Approach

EIA 632 focuses on the entire development process, wherever it might fit into an enterprise's life cycle. It cites only thirty-three requirements, each with representative tasks and expected outcomes. These requirements are encapsulated into five process groups as shown in Figure 1.

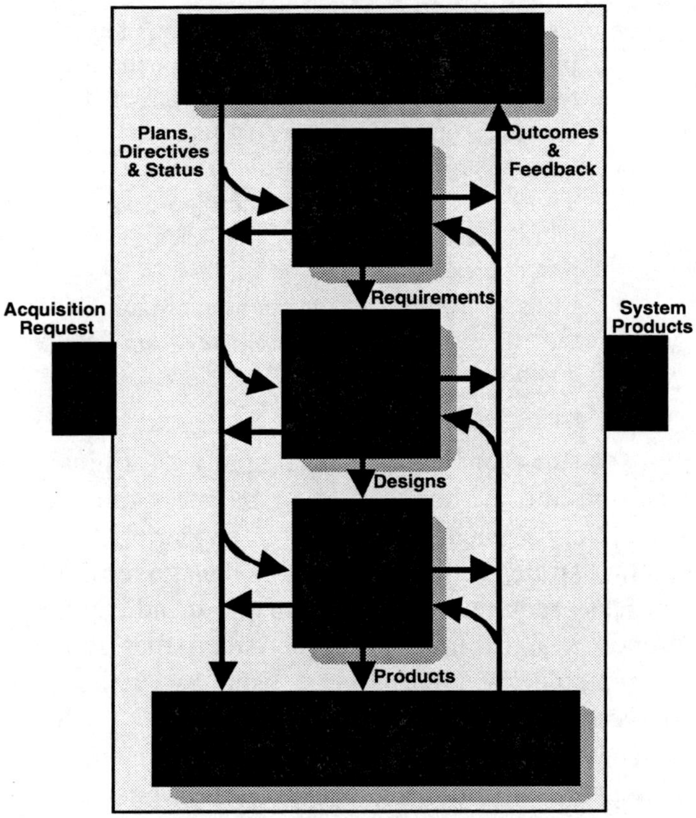

Figure 1. EIA 632 Processes[1]

Unlike many other standards, EIA 632 does not specify required methods for implementing the thirty-three process requirements. In fact, it states that a developer should:

- determine which of the specified processes apply to their enterprise
- determine which of the process requirements apply for the selected processes
- establish implementation policies and procedures for the selected processes
- define tasks for each of the selected requirements
- establish implementation methods to support the defined tasks
- select appropriate tools to enable the implementation methods

Responsive To The Needs Of The User

Just as an initial set of stakeholder requirements might result in a number of different designs – all compliant with the requirements set – EIA 632's set of process requirements enables the user to determine how they will be applied. For a small enterprise (developing a simple commercial product), the tasks, methods, and toolsets may be straightforward and the planning effort relatively simple. For a major enterprise; the policies, procedures, tasks, methods, and toolsets may be complex; and the planning effort substantial.

Note that planning is an essential element of both.

Textbook Versus Reality

Effective processes are essential to long-term viability. However, their application more often than not fails to reflect reality; i.e., the way we think and work. In short, they are often "textbook solutions" that look good on paper, but are too cumbersome to work in practice.

This occurs because most process training is based on abstract, sequential process elements. The typical engineer does not necessarily think in a rigid, sequential manner. Analysis of a problem or a situation usually occurs in a series of starts, jumps, and plateaus – the analyst examining the collected information at each point for patterns and

[1] From Standard EIA 632, "Processes For Engineering A System", Figure 4.0, copyrighted by Electronic Industries Alliance and used with permission

conflicts until a logical solution is established. Lacking correlation between process and reality, the engineer often abandons the new process tasks and proceeds with business as usual.

It is practically impossible for accelerated process briefings to overcome this inherent conflict. Most training of this nature is accomplished in as little as a few hours, but occasionally may take as much as two or three days. Such training is usually limited to a series of lectures and "vu-graph" process charts, with little correlation to the individual needs of each unit. This enables the training material to be sufficiently generic so an entire branch and all its varied disciplines can be covered under one training package.

The generic approach is essentially an accepted method because:

- It takes little time to prepare the training materials
- An expert capable of communicating ideas to a group is not required
- Training time is kept to a minimum
- Senior management only needs to know that a "box was checked" to show compliance to customer or regulatory requirements

The students dutifully place the process training manuals in their bookshelves for the customer to see, but rarely ever again opens them for use in their work. Why should they? As noted in the introduction to this paper, by the time they look up the information, try to determine its applicability to the current problem and then understand how to apply it, events have moved on to new concerns.

The Best Shortcut Is No Shortcut!

Does this, then, mean that the development of improved processes is a waste of time and money. Absolutely not! However, taking shortcuts in the implementation of an improved or new process *is a definite waste of time and money.*

So, then, how should a new process standard be implemented?

Process Standard Implementation – Not Reengineering

It is essential to recognize that integrating a new process standard into an enterprise's practices differs from reengineering the enterprise. The former is the adjustment of an existing process system to meet new requirements. The latter is the recreation of the enterprise into a new integrated unit.

Know And Understand The Enterprise's Operating Environment

No enterprise operates in a vacuum. It is significantly affected by the external environment; i.e., laws and regulations (which differ for each industry), legal liabilities, social responsibilities, technology base, labor pool and relationship with unions, competing products, and public culture and perception. Within these external constraints, each enterprise must develop its own local culture, which is influenced by significant stakeholders and domain technologies. From these foundations, an enterprise develops the guiding principles governing performance, policies and procedures, and from these, the standards and requirements for day to day operations. These need not be volumes of procedural books, but a simple statement of principles with guidelines for those tasks not intuitive to a skilled and trained staff.

Integrating A New Process Standard At The Enterprise Level

Unfortunately, determination of the impact on the enterprise is rarely accomplished in the implementation of a new process standard. Yet failure to redefine this infrastructure is often a major reason new processes fail to achieve the stated goals. Any integration of a new process standard must build on this foundation. Key questions to aid in achieving this include:

- What is the current culture?
- What processes make it work?
- Does it work? (Understand the subtleties)
- Does the new process standard improve on the existing processes? Or is it at odds with the current processes?
- Does the new process standard add value?
- What are the significant impacts of implementation on the enterprise?
- Can the existing processes be adapted in an integrated manner with the new process requirements?
- Is the enterprise capable of making the needed adjustments in a timely and cost-effective manner?

Use the new standard as a wake-up call to challenge the status quo. That is, resist the desire to state that changes are not necessary since "it has always been done this way!" Recognize that market conditions may have changed, and that an enterprise's way of doing business may need to change as well.

All units in the enterprise should participate in developing responses to these questions, and ultimately, to the implementation of changes resulting from these answers. This includes senior management, project teams, operations units, and supporting units. In this manner, all members of the enterprise become owners in the changed process culture. As necessary, the principal developers of the changes should meet with each enterprise unit to review and discuss the changes in the method of doing business as it affects the enterprise as a whole, and as it affects the individual units.

A key component of the enterprise level effort is determining the need for, and the providing of, new (or modified) resources necessary to implement the changed process elements. Absent this; all the assessments, plans, changes in policies and practices, and implementation meetings will be for naught. Nothing sends a signal to enterprise teams that changes are simply cosmetic faster than a lack of resources to implement them.

Project Teams Implement Process Standards Within The Enterprise Infrastructure

Once the new enterprise culture has been defined, disseminated, and understood by all units; each project needs to tailor implementation to its unique needs. Key steps in achieving this include identifying:

- The specific process requirements applicable to the individual project
- The specific tasks necessary to implement each identified requirement
- The results expected from the accomplishment of each task, and the criteria by which to determine if that result has, in fact, been achieved
- The specific methods for implementing each task
- The tools needed in support of implementing each task
- The additional resources needed to accomplish the identified tasks

Rather than an intrusion into a project's tight schedule, these efforts should be considered as a key element of the initial project or engineering planning. (Should the new process standard be levied during a project's life cycle, then this effort would be considered as part of the replanning needed to meet the new customer/marketplace requirements.) As with the enterprise, all units of the project should be involved in this planning to ensure process ownership and understanding at the project level. As part of this effort, it is essential that projects establish clearly defined responsibilities and boundaries for implementation of each process related task.

Again, as with the enterprise, the principal developers of the changes should meet with each project unit to review and discuss the changes in the method of doing business as it affects the project as a whole, and as it affects the individual project teams.

EIA 632's System Approach

EIA 632 states that application of its requirements is first dependent upon the user establishing enterprise policies and procedures that define the enterprise requirements for project implementation of the adopted EIA 632 processes. Then a project implements the adopted process requirements within the framework of the enterprise policies and procedures. This relationship is shown in Figure 2.

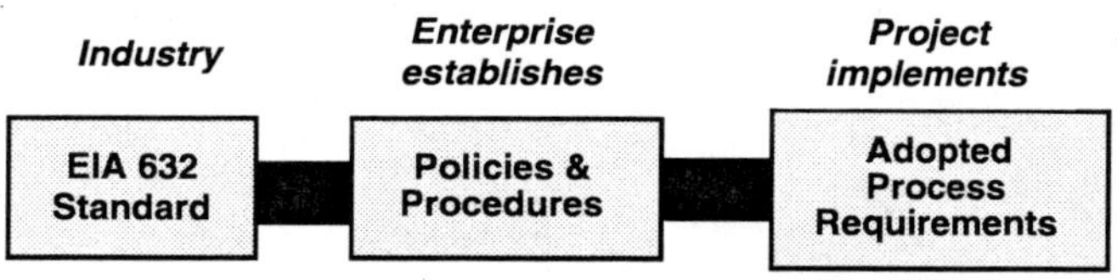

Figure 2. EIA 632 Relationship To A Development Project[2]

Clause 5 of EIA 632 describes the application context for the thirty-three process requirements contained in the Standard.

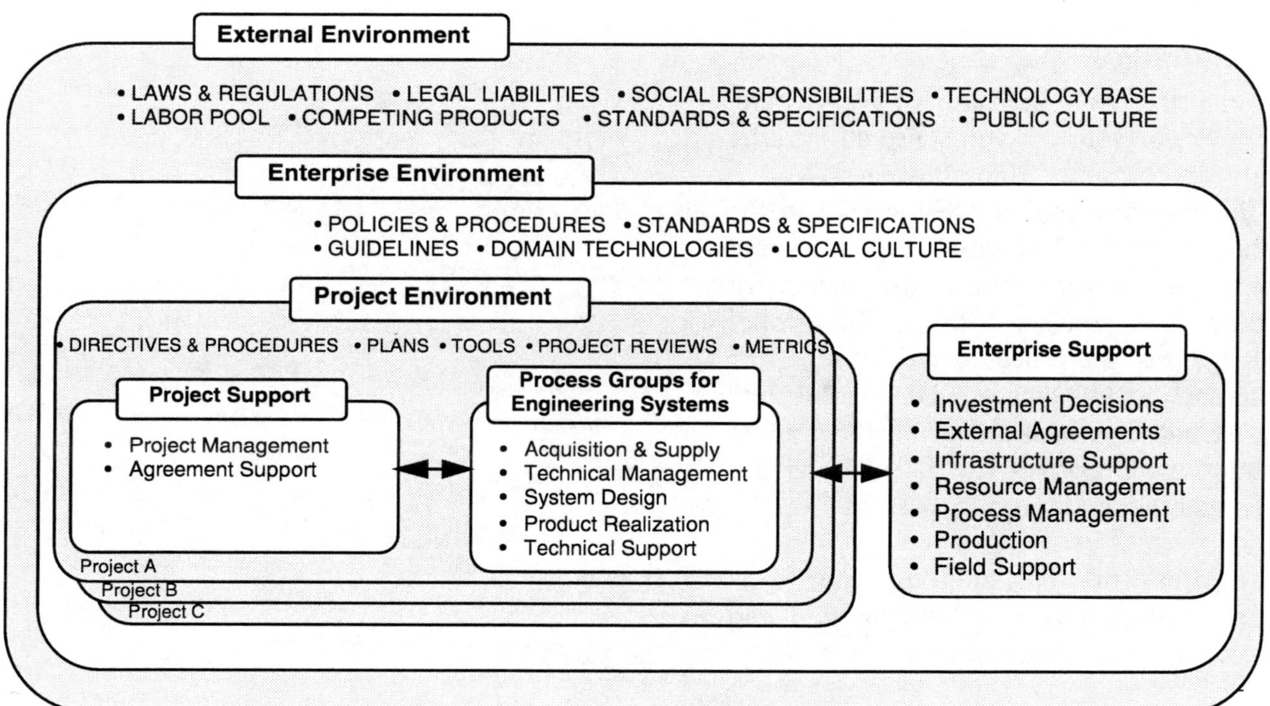

Figure 3. EIA 632 Application Context[3]

As shown in Figure 3, EIA 632 identifies the external enterprise and project factors that have the potential to affect, or to be affected by, project interfaces.

EIA 632 states that the enterprise is the context in which the process requirements " . . . are intended to be adopted, directed, and improved." Specifically, it identifies a number of support functions expected to be provided to a project by the governing enterprise, noting that the availability and adequacy of these supporting functions determine:

- The existence of the project
- Schedule of project tasks
- Capability to satisfy an established agreement with another enterprise
- The availability of personnel having the skills and knowledge to complete project responsibilities

EIA 632 does not levy requirements on the enterprise, but does clearly denote the need for enterprise support of its requirements.

[2] Ibid., Figure 1.3

[3] Ibid., Figure 5.0

Similarly, EIA 632 states that projects provide the context in which a system is engineered using the EIA 632 process requirements to satisfy agreements. As with the enterprise, this Standard identifies project support functions necessary to determine:

- Tasks and scope of the processes for engineering a system
- The tools, equipment, and metrics to be used
- The reporting and management review requirements

Process Compliance Through Embedded Infrastructure

Unlike most process standards, EIA 632 recognizes that the most effective compliance is achieved through the establishment of an infrastructure which integrates the requirements of the Standard into an enterprise's way of doing business – and provides insight into the context in which this should be achieved.

An Afterthought

The street is dark, shadows everywhere, and as you near your car you sense an intruder stealthily approaching from the rear. Not to worry though - for you've been trained in the martial arts process. Knowingly, you whip out your process standard, check the index for "nighttime rear attack", turn to the proper section, and read the process steps for repelling the intruder. Of course, the martial arts process standard is later found on the ground adjacent to your body - the intruder having moved onto other endeavors!

Meeting the challenge of change is much more productive when preparation and understanding precedes adventure.

An Overview of the Systems Engineering Capability Model EIA/IS 731

by
Don E. Barber
Chairman of the Capability Assessment Working Group
don.barber@cas.honeywell.com

Abstract

As the world marketplace continues to demand "better, faster and cheaper," organizations are searching for new tools to assist them in meeting this need. Systems Engineering is at the heart of product development and improving the performance of this discipline is key to organizational success. The Systems Engineering Capability Model (SECM) is a tool that organizations can use to evaluate the capability of their current Systems Engineering process. The SECM also provides a framework that can be used as a guide for developing or improving a structured Systems Engineering process.

1. Introduction

In the early 1990s, organizations were beginning to see the benefits of using various capability models to assess their software development process. As these benefits were realized, the need to improve other aspects of the product development process became apparent. Individuals and organizations began to investigate methods to extend the assessment capability into the Systems Engineering discipline. In 1992, the International Council on Systems Engineering (INCOSE) formed the Capability Assessment Working Group with a purpose of developing a method of assessing Systems Engineering Capability. During the same period, several companies provided resources to form the Enterprise Process Improvement Collaboration (EPIC) with a similar purpose. INCOSE released the first version of their System Engineering Capability Assessment Model (SECAM) in 1994 and EPIC released the Systems Engineering Capability Maturity Model (SE-CMM) in the same year. Both of these models have gone through several revisions and have been used by a wide range of companies. In 1996, the Electronics Industry Alliance (EIA) joined with INCOSE and EPIC to merge the SECAM and the SE-CMM into a single model that captures the essence of both. The EIA 731 Working Group was formed with representatives from INCOSE, EPIC, and EIA member companies.

The merged model was titled the "Systems Engineering Capability Model" (SECM) and was assigned the number EIA/IS 731 by the Electronic Industry Alliance. EIA/IS 731 has two parts:
  Part 1: The Systems Engineering Capability Model
  Part 2: The SECM Appraisal Method

An initial draft of the model was released in August 1997 and a review and comment cycle was conducted through December of 1997. In all, the EIA 731 Working Group collected over 1300 comments during the review cycle. The EIA 731 Working Group incorporated over 800 of the changes and over 250 were deferred to a later date. Approximately 300 of the comments were rejected.

In May of 1998, EIA 731was proposed for acceptance as an Interim Standard. Interim status is sought until experience is reported from companies piloting the model and the appraisal method. SECM successfully passed the ballot process and has been accepted as an Interim Standard. After some experience is gained through application, the model will be updated to incorporate lessons learned and will then be submitted for vote as a full standard.

2. Model Architecture

0-7803-5086-3 /98/$10.00 ©1998 IEEE

Like the SECAM and the SE-CMM, EIA/IS 731 is a continuous capability model. A continuous model contrasts with a staged model in that a staged model awards a single score based on satisfying a selected group of focus areas, while a continuous model provides a rating for each individual Focus Area. The SECM is composed of 19 Focus Areas that are grouped into three categories and each Focus Area is rated on a scale of 0 to 5. The 19 Focus Areas are intended to provide complete coverage of the process contained in EIA 632, *Processes for Engineering a System*, and to be consistent with IEEE STD 1220-1994, *IEEE Trial-Use Standard for Application and Management of the Systems Engineering Process.*

2.1 Categories and Focus Areas

The SECM model contains three categories, with each category representing a natural grouping of Focus Areas. The Technical category is made up of seven Focus Areas that address the 'engineering' aspects of Systems Engineering. The SE Management category contains eight Focus Areas that deal with planning, tracking, information & data management, and quality assurance. The SE Environment category has four Focus Areas that address organizational activities that enable Systems Engineering and align the SE process with the business objectives.

Technical Category
- FA 1.1 Define Stakeholder and System Level Requirements
- FA 1.2 Define Technical Problem
- FA 1.3 Define Solution
- FA 1.4 Assess and Select
- FA 1.5 Integrate System
- FA 1.6 Verify System
- FA 1.7 Validate System

Systems Engineering Management Category
- FA 2.1 Plan and Organize
- FA 2.2 Monitor and Control
- FA 2.3 Integrate Disciplines
- FA 2.4 Coordinate with Suppliers
- FA 2.5 Manage Risk
- FA 2.6 Manage Data
- FA 2.7 Manage Configurations
- FA 2.8 Ensure Quality

Systems Engineering Environment Category
- FA 3.1 Define and Improve the Systems Engineering Process
- FA 3.2 Manage Competency
- FA 3.3 Manage Technology
- FA 3.4 Manage Systems Engineering Support Environment

2.2 Focus Area Elements

To assist the user, each Focus Area is subdivided into themes. The themes are a collection of related Specific Practices that help the user interpret the intent of the Focus Area and the differences between capability levels within the Focus Area. Capability within a Focus Area is determined by evaluating Specific Practices, Generic Practices, and Generic Attributes.

Specific Practices

Specific Practices define activities that are essential to accomplishing the purpose of a Focus Area or that helps to accomplish the purpose of the Focus Area more effectively or efficiently. They are unique to a Focus Area and each is associated with a capability level.

Generic Practices

Generic Practices define activities that are essential to accomplishing all Focus Areas. Generic Practices enhance the capability to perform the Specific Practices. Generic Practices are grouped by capability level, except for Level 1, that does not have any Generic Practices assigned.

Generic Attributes

Generic Attributes are designed to assess the effectiveness of a process being used and to assess the value of products produced by a process. Generic Attributes are defined for each level of capability.

2.3 Capability Levels

Capability levels are defined based on widely observed plateaus of performance that organizations typically achieve as they strive to improve their business processes and systems engineering activities. They can be used as goals, guidance, or benchmarks to assist an organization in structuring its improvement efforts. Process attributes address the manner in which the practices of each Focus Area are performed. Non-process attributes are indicative of the judgment used in executing the Focus Area practices, the effectiveness of the work, and the value of the work and work products. The SECM has six defined capability levels. The combination of performance in conducting the Specific Practices, Generic Practices, and Generic Attributes relative to a Focus Areas indicates the capability level. The lowest level is "Level 0," and the highest is "Level 5." Characteristics of Level 5 represent the highest, or optimal, level of performance.

Level 0 - Initial
There is a general failure to perform the specific practices in the process and there are no easily identifiable work products. This is the default level.

Level 1 - Performed
Specific Practices of the Focus Area are performed informally and performance is not rigorously planned and tracked. Execution is dependent upon individual knowledge. There are work products and records to testify that practices are accomplished.

Level 2 - Managed
The activities of the Focus Area are planned and tracked. The process is documented in some manner. Work products are reviewed for adequacy and corrective actions are taken when needed. The work products are placed under version control.

Level 3 - Defined
A standard process is documented and used within the organization. Performance of practices is done or tailored with respect to this standard. Information related to the use of the defined process is gathered and is used in planning and managing the standard process.

Level 4 - Measured
Measurable quality goals are established for the standard process. Detailed measures of the performance of the standard process are collected and analyzed to identify potential improvements. There is a quantitative understanding of the standard process and an improved ability to predict project performance. Causes of defects are tracked back to process deficiencies.

Level 5 - Optimized
Process effectiveness goals are established based upon business goals of the organization. The process undergoes continuous refinement based on quantitative feedback.

3. SECM Appraisals

A SECM appraisal is an assessment of a business-area's current systems process by a trained team of experienced engineers. The appraisal is based on responses to the appraisal questionnaires, in-depth discussion with project managers and practitioners, review of key projects, and the appraisal team's knowledge and experience. Figure 3-1 highlights the key steps in the appraisal process. The objectives of the appraisal process are:

- to understand your organization's systems engineering processes
- to understand your organization's willingness and ability to improve
- to identify key areas for improvement
- to facilitate your organization's initiation of actions to improve

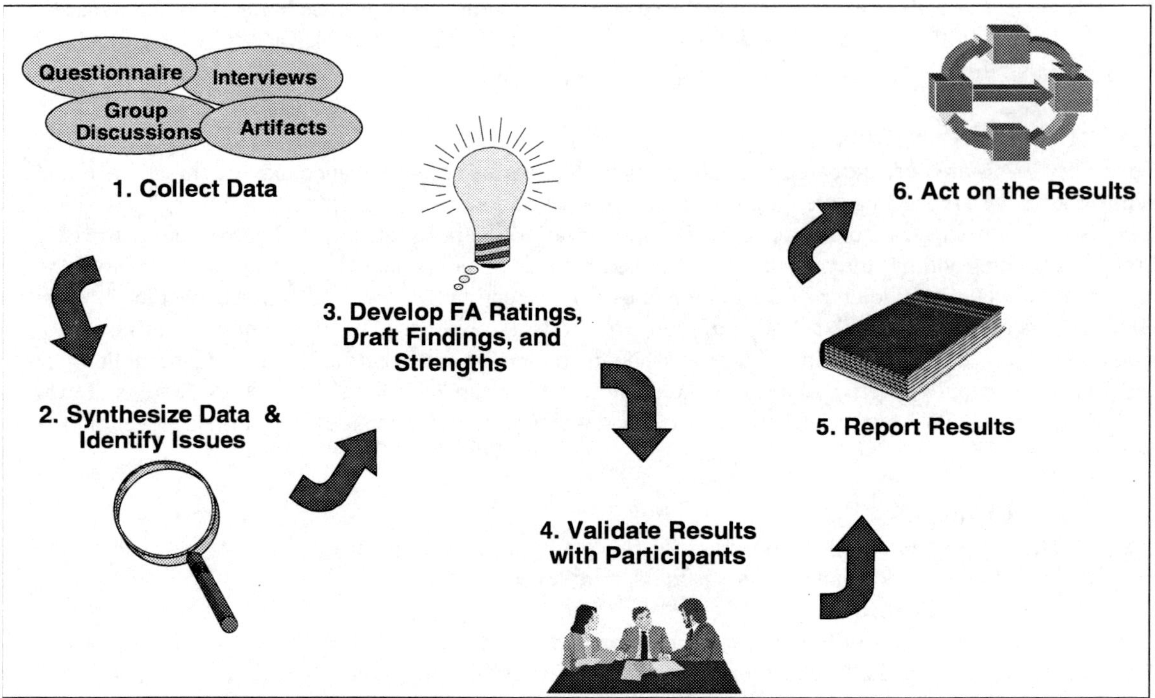

Figure 3-1 SECM Appraisal Process

3.1 Conceptual Approach to an Appraisal

Each appraisal consists of three distinct phases. The preparation phase begins as much as six months prior to the actual on-site phase and involves all the activities necessary to set up the on-site period. These activities include: enlisting management support, identifying and educating participants, developing an Appraisal Plan, and administering the questionnaire.

The on-site phase is normally conducted two to three weeks after the questionnaire data is collected. This gives the appraisal team an opportunity to review the questionnaire responses and identify areas of focus for the interview sessions. The on-site phase typically lasts four days. During the on-site phase the appraisal team will conduct interviews, review project artifacts, develop draft scores and findings, and validate these findings with the participants. In most cases the appraisal team also provides a summary briefing to management.

Following the on-site phase the appraisal team generates a Findings and Recommendations Report. This report is provided to the appraisal sponsor and is usually reviewed with senior management. After the report is documented the appraisal team assists the participants with development of a Process Improvement Action Plan.

3.2 Developing Findings and Ratings

The appraisal team analyzes all the data collected during the preparation and on-site phases to determine Findings. The appraisal team attempts to distill the data into set of key actions that are appropriate for the organization based on the demonstrated capability level. The Findings identify strengths in the organization's process as well as areas for improvement. These Findings are the key element of the process improvement effort.

The appraisal team assigns a capability level to each Focus Area based on the data collected. Each rating is assigned based on the three elements discussed in section 2.2. In general, Specific Practices count for half the score, Generic Practices count for one-quarter of the score, and Generic Attributes count for one-quarter of the score for each Focus Area.

The Specific Practices for a level of a Focus Area are considered to be met if:

- there are 4 or fewer Specific Practices and all are met
- there are 5-7 Specific Practices and all but 1 are met
- there are 8 or more Specific Practices and all but 2 are met

The Generic Practices for a level of a Focus Area are considered to be met only if all the Generic Practices at that level are satisfied.
The Generic Attribute Scoring is determined by a numerical calculation based on the evaluation of value and effectiveness.

Figure 3-2 represents a sample scoring template used by the appraisal team to determine ratings for the Focus Areas and Figure 3-3 represents a sample scoring profile that would be presented as part of the final report.

Technical Category	Level 1 Specific Practices are performed	Results are at least of marginal utility	Level 2 Specific Practices are performed	Level 2 Generic Practices are performed	Results are at least of adequate utility	Level 3 Specific Practices are performed	Level 3 Generic Practices performed	Results are of at least significant utility	Level 4 Specific Practices are performed	Level 4 Generic Practices performed	Results are at least of measurably significant utility	Level 5 Specific Practices are performed	Level 5 Generic Practices performed	Results are of optimum utility
1.1 Define Stakeholder & System Level Reqts														
1.2 Define Technical Problem														
1.3 Define Solution														
1.4 Assess and Select														
1.5 Integrate system														
1.6 Verify System														
1.7 Validate System														
	Level 1		Level 2			Level 3			Level 4			Level 5		

Figure 3-2 Sample Scoring Template

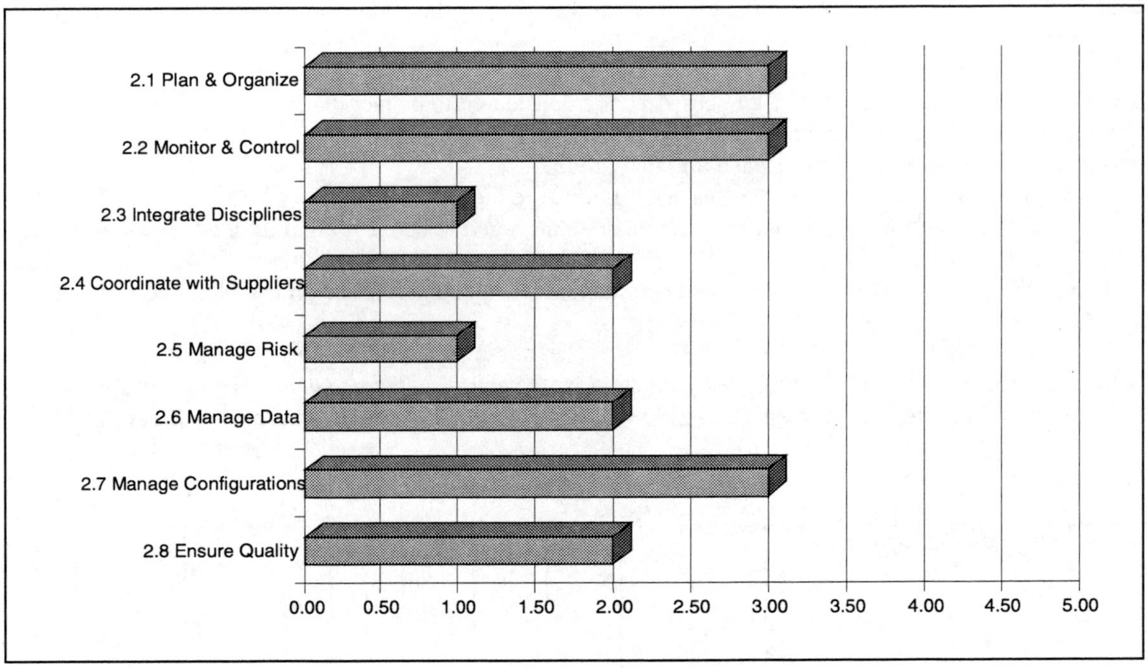

Figure 3-3 Sample Scoring Profile

4. Benefits of a Performing a SECM Appraisal

There are many reasons for using EIA/IS 731 within your organization. If your organization is currently using SECAM or SE-CMM the similarity in structure and content should allow for a smooth transition to EIA/IS 731. If your organization is new to the concept of systems engineering assessment, EIA/IS 731 will allow you to establish an initial baseline and form a plan for improvement. Both the participating organizations and the EIA 731 Working Group will benefit from use of the Interim Standard.

Participating organizations can use the SECM to:

- gain an understanding of the organization's current SE strengths, and key areas for improvement
- provide a baseline for measuring future improvements
- provide a means for obtaining sponsorship and support of action within the organization
- provide a framework for defining corrective actions
- provide a valuable learning experience for the participants

When participating organizations coordinate with the EIA 731 Working Group, the appraisal can be used to assist in validating the model and provides data for improvement of EIA/IS 731. This coordination also provides data to improve and standardize the appraisal process.

5. Future Direction

EPIC and INCOSE have committed to fully support EIA/IS 731 and no longer support their respective models. SECAM will no longer be distributed for use by INCOSE and the Capability Assessment Working Group (CAWG) will no longer support SECAM assessments. The CAWG will continue to serve as the INCOSE focal point for supporting EIA/IS 731. As experience is gained through application of this interim standard, changes will be proposed and incorporated at the direction of the EIA. When sufficient experience has been gained the updated SECM will be proposed as a full standard.

Along with the work being performed through the EIA, the Office of the Secretary of Defense (OSD), the National Defense Industry Association (NDIA) and the Software Engineering Institute (SEI) are collaborating on a project to integrate Systems, Software and Integrated Product Development (IPD) maturity models. Their goal is to produce a common framework that adds a degree of consistency to all the appraisals and to reduce the redundancy between the various models. The current plan calls for a prototype to be released based on:

- SEI SW CMM version 2
- EIA/IS 731 version 1
- IPD CMM version 0.9

Future revisions to EIA/IS 731 (or upgrade to a full Standard from an IS) will likely be influenced by the relative success of this capability model integration effort.

References

EIA, *Processes for Engineering a System,* EIA 632 (draft), Electronics Industry Association, Washington DC, 1997.

EIA, *System Engineering Capability Model*, EIA/IS 731 (draft), Electronics Industry Association, Washington DC, 1998.

IEEE, *Trial Use Standard for Application and Management of the Systems Engineering Process*, P1220, Institute of Electrical and Electronic Engineers, 1994

INCOSE, Systems Engineering Capability Assessment Model, CAWG-1996-01-1.50, International Council On Systems Engineering, Seattle WA, 1996.

SEI, *A System Engineering Capability Maturity Model*, Version 1.1, CMU/SEI-95-MM-003, Software Engineering Institute, Carnegie Mellon University, Pittsburg PA, 1995.

Don Barber is one of the authors of the Systems Engineering Capability Model, EIA 731. As chairman of the INCOSE Capability Assessment Working Group, Mr. Barber was directly involved with the development of Version 1.5 of the INCOSE SECAM and its assessment method. He has performed numerous systems engineering assessments across a wide range of products. He continues to facilitate assessments and provide informal assessor training. Mr. Barber is currently a Technical Staff Engineer at Honeywell Inc. responsible for process improvement. He is chairman of Honeywell's corporate Systems Engineering Committee and represents Honeywell on the INCOSE Corporate Advisory Board. Mr. Barber received his BS in Mechanical Engineering from the U.S. Naval Academy and a Masters in Engineering Management from Washington University.

WHY A QUALITY SYSTEM AND SPC?

William A. Walker, AlliedSignal Inc., Redmond, WA.

Why a Quality System?

Let us look at why companies are installing a company wide Quality System that meets their customers needs. Or has some company executive finally seen the rewards that are obtained by installing a Quality System and doing SPC (Statistical Process Control)?

Quality Systems and SPC are HOT items in the worldwide arena of both domestic and international trade. Whether we provide a service or produce a product for our customers, they want it right, on time and at the price agreed to. No longer can we produce 125 items and ship the best 100 items to meet schedule only. Our customers are demanding better product, on time and at a competitive price. (Product can be hardware, software, services, process materials or any combination of these.) If we cannot perform to our customer requirements, our customers will find someone who can. Don't think that just because we have the cheapest price that we will always get the contract. Our customers have developed supplier-rating systems that not only include on- time deliveries, but also include rejection tag costs, response to corrective action requests and other factors.

Today we will explore the different domestic and international Quality Standards, what they mean and what it takes to be compliant or registered to a Quality System Standard. The Quality System Standards that we must be aware of and the requirements of each and how they differ will be analyzed. The Quality System Standards are:

AS 9000	**Aerospace Document**
D1-9000® Revision "A"	Boeing Advanced Quality System®
EN 9000-1	Quality Systems-Model for Quality Assurance in Design, Developı Production, Installation and Servicing for Aerospace (European)
ISO 9001-1994 Development,	Quality System-Model for Quality Assurance in Design, Production, Installation, and Servicing
ISO 9002-1994 Installation,	Quality System-Model for Quality Assurance in Production, and Servicing.
QS 9000	Automotive Quality System (GM, Ford & Chrysler developed)

Which above standard or standards are your customers contractually requiring that you have a documented Quality System to? I said standards because some companies are registered to more then one Quality System Standard.

0-7803-5086-3 /98/$10.00 ©1998 IEEE

Registration is accomplished by hiring an independent third party called a registrar and paying a fee of several thousands dollars. Contract with a registrar is usually a three-year contract. Registration has more meaning and recognition then just saying that your company is compliant.

Boeing does not accept third party registration to the above Quality System Standards and will do their own audit of your facility to see if you are compliant to D1-9000®. The Boeing document contains ISO 9002-1994 requirements plus 74 additional supplier shalls, wills, and musts. Shall, will, and must indicate a mandatory requirement. The other Quality System Standards above only discuss that a supplier shall do this or shall do that. The significant thing to note here is that the other Quality Standards tell what has to be done and you decide how to do it that fits your organization. Also remember that all the Quality Standards are for implementation of a Quality System and are not for product approval.

The ISO 9001 Quality System Standard contains 20 elements, 4.1-4.20. A comparison of the Quality System Standards by element numbers can be found on Attachment "A". QS 9000 has over 224 supplier shalls and 3rd edition is now available. (The word supplier refers to us as we are the supplier of product to our customers.) Also all the above Quality System Standards contain ISO 9001 or ISO 9002 completely within. Now the scary part. Companies have based their Quality System level 1, level 2 documentation, Audit checklists, training materials, contract terms and conditions etc. on the current 20 elements. The ISO 9001 year 2000 proposed revision has completely renumbered the elements and no old element numbers are used in the new revision. This is not a cost-effective idea nor is it value-added. The cost to current ISO 9001/9002, AS 9000, D1-9000®, EN 9000-1, and QS 9000 compliant or registered companies world-wide will run into millions to billions of dollars to change their documented Quality System. It is not to late to let the members of TC 176 know that you object to this proposed change before it costs your company an unnecessary expense. Please convey to TC 176 your objection(s) and how this renumbering of the element numbers will affect your organization, whether you will continue or drop your registration and/or compliance, and any other information you wish to convey. I have been told that they, TC 176, have to respond to letters and e-mail received in regard to questions about ISO 9001. So here is your opportunity to save your company thousands of dollars by stopping this non-value added change to the all the Quality System Standards shown above.

Why SPC?

Control charts were first introduced by Western Electric in 1924 by Walter Shewhert. After World War II the use started again but did not last long because it was easy to sell items to people because the items they were purchasing were items they did not have during the war. So since there was no value-added or contractual requirements, why do control charts. Well here we are in the 90's and control charts are a way of life. A way to stay in business. A way to stay competitive, a way to satisfy our customers.

SPC (Statistical Process Control) is a method to analyze processes and implement process improvements. Processes are analyzed by taking a sample of the process and using control charts and a Pareto Diagram to help with evaluation. More and more contracts by our customers require SPC and a thing called Cpk. Cpk is a method that will tell us the facts about the process and the number of defects per million. Cpk is called process performance and this index number is what is called out in contracts. For example: A Cpk of 1.00 would mean that the process may produce 2,7000 defects per million of produced product. Cpk of 1.33 equals 63 defects per million and a Cpk of 2.0 equals 3.4 defects per million. Most companies are asking their suppliers for product that has Cpk of 1.33 or 1.67. If you are a Six Sigma company you will ask your suppliers for Cpk of 2.0. If you do not meet the Cpk index number when sampling the process, then instead of sampling the process you will be back to 100% inspection of the product produced by process. You will also be asked to provide positive corrective action and tell the customer what you have done to meet the Cpk contractual requirement and prevent the process from not meeting the Cpk index number in the future.

Another SPC term is Cp which tells us the process potential, sometimes called process capability. If we know the Cp we can tell if we have a chance to meet the Cpk. Example: If Cpk contract requirement is 2.0 and the Cp is 1.6, we know that we can not meet the Cpk requirement. When determining the Cpk we divide the Normal Curve in two parts and calculate the Cpk upper half of the normal curve and Cpk lower half. If the Cpk lower half was calculated as 2.8 and Cpk of upper half at 1.2, the true Cpk that we tell our customers is 1.2, the smaller of the two calculated Cpk numbers. The Cpk can not exceed the Cp. What you have just learned is Process Capability. Time permitting we will do a simple Cp and Cpk calculation during my presentation.

Now that you understand Process Capability, you need to determine the Method of distribution. Do you want your suppliers to send in the Control Charts and data used t o determine Cpk with the product or store the information for 7 years? I want the Control Charts, Histogram and data sent in with the product and will discuss my reasons during the presentation.

So why do we do SPC? Besides the customer contractual requirements for SPC and Cpk, we also have to worry about JIT (Just In Time) deliveries and making a profit. SPC can help companies identify out of control processes, analyze process variation, determine the accuracy of equipment, measure processes and determine improvement made or needed, plus other areas of concern.

I can share success stores about the use of SPC and failures because companies have not implemented SPC unless it was a contractual requirement. Where 125 parts were made to be able to ship 100 was no longer necessary because the operator made process adjustments as he/she made the control charts and plotted the process in "real time". Real time means that while the process was producing product, samples were taken and the control charts were constructed and analyzed while the process was making the part, not after the production order had been completed.

Summary

Whether the customer places a contractual requirement for your company to be compliant or registered to one or more of the Quality System Standards above, it is necessary that it be accomplished <u>now</u>. Why wait, especially if your competition is in the process or has already done it. It is a great marketing tool to show both your current customers and potential customers that your company has a documented Quality System. The

0-7803-5086-3 /98/$10.00 ©1998 IEEE

journey to registration usually takes 18 months and requires someone, with management support, to devote full time to this project. Installation of a Quality System has also saved companies money, reduced variation, reduced employee turnover, and increased profits.

SPC is a tool that will also save companies money, increase production, eliminate waste (muda), and provide information for management to make decisions. Do you know what the contractual Cpk requirement is from your suppliers? More and more companies are now requiring that their suppliers do SPC and provide control chart information on the processes that are used to produce product. If you don't currently do SPC you need to start the training and start shipping product on time and to the contractual Cpk.

If your company does not have a Quality System and are not doing SPC, you are going to find the amount of contracts that you can bid on and win to be very limited. Don't wait until you have a customer that you need to win the contract to stay in business and you can not win because you were not prepared.

0-7803-5086-3 /98/$10.00 ©1998 IEEE

ATTACHMENT "A"

COMPARISON OF THE NUMBER OF "SHALLS" IN QUALITY SYSTEM STANDARDS
ISO 9001 vs AS 9000 vs prEN 9000-1 vs D1-9000 REV. "A" BQS (Section 1)

Paragraph Number	ISO 9001 SHALLS	AS 9000* SHALLS	EN 9000-1* SHALLS	D1-9000 BQS ONLY* SHALLS	D1-9000 BQS ONLY* MUST
4.1 Mgmt Resp	9	1	1	4	2
4.2 Q System	9	5	5	1	0
4.3 Contract Rev	5	0	3	0	0
4.4 Design	19	1	20	1	0
4.5 Doc & Data	7	3	0	3	0
4.6 Purchasing	10	5	7	9	0
4.7 Cust Product	2	0	0	0	0
4.8 ID & Trace	3	0	2	0	0
4.9 Process Cntrl	8	6	20	8	0
4.10 Insp & Test	15	3	9	11	2
4.11 I, M & T E	9	1	2	0	0
4.12 I & T Status	2	1	3	10	0
4.13 Nonconform	7	4	6	5	0
4.14 C & PA	5	0	2	2	0
4.15 H S P P& D	9	0	2	3	0
4.16 Q Records	7	1	0	2	0
4.17 Q Audits	6	0	1	5	4
4.18 Training	3	0	1	1	0
4.19 Servicing	1	1	1	0	0
4.20 Statistics	2	2	0	1	0
Totals	138	34*	85*	66* (66+8=74)	8*

***The "SHALLS" & "MUST" shown in these columns are in addition to what is already required in ISO 9001.**

ATTACHMENT "A" (CONT'D)

COMPARISON OF THE NUMBER OF "SHALLS" IN QUALITY SYSTEM STANDARDS ISO 9001 vs AS 9000 vs prEN 9000-1 vs D1-9000 REV. "A" BQS (Section 1)

The European Association of Aerospace Industries (AECMA) is in the process of reviewing prEN 9000-1 which is the European version of AS 9000. The document is not released but is scheduled for release by the end of 1998. PR means "Prestandard" and has been drawn up under the responsibility of AECMA (Association Europeenne des Constructeurs de Material Aerospatial). After examination and signature of the AECMA Standard Checking Centre (NPS) and formal agreement of the Official Services of the Member countries it will be submitted as a draft European Standard to CEN (European Committee for Standardization) for formal vote.

The comparison that I did shows the additional requirements beyond ISO 9001 for AS 9000, prEN 9000-1, and D1-9000 Revision "A" BQS (Basic Quality System) which is Section 1. Section 2 of the D1-9000 is all SPC and Advanced Quality System Requirements and has 52 Shalls, 31 Must, 1 Will and 47 Should. For Boeing a "Shall, Will and Must" are a mandatory requirement. A "Should" indicates a preferred approach. This means that the supplier may use an alternative approach, but must be able to show that the alternate approach meets the intent of the Boeing Standard.

Please note that where the Boeing BQS has added only 74 mandatory requirements, the prEN 9000-1 document has added 85 mandatory requirements. Actually there are more requirements as 4.1.4.2 added items e), f), g) & h) to an already existing shall requirements a) through d). A "shall" can add one or more mandatory requirements.

prEN 9000-1 has added 20 "Shalls" to element 4.4 Design Control and 4.9 Process Control. In third place is 4.10 Inspection and Testing with 9 Shalls, then 4.6 Purchasing with 7 Shalls and 4.13 Control of Nonconforming Product with 6 Shalls.

Included in the material that I received is information that an Audit questionnaire on EN 9000-1 is available. Also that the committee is working on EN 9000-2 Standard for Repair Stations and EN 9000-3 Standard for Distributors/ Stockists.

(Comparison information prepared by Bill Walker, AlliedSignal AT&R, Redmond, WA. Any Questions or input is requested. Phone is 425-885-8514, e-mail: william.walker@alliedsignal.com.)

LOW-COST INTEGRATED DISPLAY CERTIFICATION

Uma D. Satyen, The MITRE Corporation (CAASD), McLean, VA
Andrew D. Zeitlin, The MITRE Corporation (CAASD), McLean, VA
David J. Lubkowski, The MITRE Corporation (CAASD), McLean, VA

Abstract

FAA certification requirements and processes appear to have disproportionate impact on overall system costs and production schedules for General Aviation avionics. In particular, we are investigating the effects of certification on integrated cockpit display systems that provide weather, traffic, navigation or terrain information to the pilot.

This work is aimed at fostering a more practical and cost-effective certification process. Industry should benefit by being able to leverage commercially available products for airborne use, and the ability to price their systems to appeal to the large GA market. FAA and National Airspace System users should benefit from implementation of streamlined certification processes, and from rapid, widespread system equipage by General Aviation, allowing potentially earlier transition to modern ATC systems. General Aviation should benefit by acquiring systems that are lower in cost and which provide useful information and services to the pilot. Lower costs will encourage greater numbers of General Aviation owners to equip. Overall safety of flight will improve with widespread use of this added information.

This paper documents the results of this project to date; additional results of interviews and the ensuing analysis will be presented at DASC.

Introduction

The MITRE Corporation is a not-for-profit organization that operates two Federally Funded Research and Development Centers (FFRDCs): one working with the Department of Defense; the other, the Center for Advanced Aviation System Development (CAASD), working with the FAA. MITRE develops systems engineering solutions to address a variety of technological issues; we do not manufacture products for retail sale. The information we receive from governments and industry is kept private and confidential. We regularly engage in non-disclosure agreements with members of industry in order to consider the most accurate information and to develop recommendations that are based on facts.

Project Purpose

The MITRE Corporation is sponsoring an independent research project concerning the effect that FAA certification requirements and processes have on overall system costs and production schedules for General Aviation avionics. In particular, we are investigating the effects of certification on integrated cockpit display systems that provide weather, traffic, navigation or terrain information to the pilot.

This project is an outgrowth of a previous MITRE Sponsored Research (MSR) project in which a Universal Access Transceiver and Airborne Research Prototype were developed and demonstrated for General Aviation use. The cockpit display showed weather, traffic and navigation information. The system cost was dominated by the certified display, which seemed quite costly, compared with ordinary computer displays (Liquid Crystal Displays (LCDs) and Cathode Ray Tubes (CRTs)). It seemed that the steps necessary to qualify for certification may be the reason such airborne displays are so costly. The "Low-cost Integrated Display Certification" MSR project was created to determine if this was the case, and to investigate potential means by which the overall system costs could be reduced through synergistic improvements to the design and certification processes.

Goals of the Project

The focus of this project concerns low-end General Aviation display systems, rather than systems for higher-end business jets and commercial airlines. The system should be affordable and provide pilots with two or three of the most needed airborne applications. Graphical color displays are highly desirable for presentation of information.

The project objectives are to:

- Investigate current computer display technology, its cost, and its applicability to airborne systems
- Investigate and recommend possible system architectures that would promote low cost and modularity, and permit easy upgrade and certification, in terms of both hardware and software. Review these with the FAA and with avionics system manufacturers regarding which might be easiest to develop and certify.
- Review and evaluate the display certification process required by the FAA and used by system manufacturers. Investigate current development practices and allocation of costs in developing and deploying new avionics systems. Meet with FAA regulatory authorities to review results. Recommend changes to streamline the certification process. Suggest ways in which widely available technologies can be leveraged for avionics applications (e.g. use of COTS software and hardware).

0-7803-5086-3 /98/$10.00 ©1998 IEEE

Expected Benefits

The project is aimed at developing recommendations for practical and cost-effective steps in the certification process. Industry would benefit by being able to leverage commercially available products for airborne use, and the ability to price their systems to appeal to the large GA market. FAA would benefit with implementation of streamlined certification processes, and from rapid, widespread system equipage by General Aviation, allowing potentially earlier transition to modern ATC systems. General Aviation would benefit with systems that are lower in cost and which provide useful information and services to the pilot. Overall safety of flight will improve with use of this added information.

Survey Participants

We met with representatives from various facets of product development, including requirements definition specialists, systems engineers, software QA specialists, marketing representatives. We also talked with the FAA National Resource Specialists (NRS), Small Aircraft Directorate, Area Certification Office (ACO), and Designated Engineering Representatives (DER). We have developed a questionnaire that we used as a guideline for the interview. Each interview took between three to four hours.

Survey questions were designed to assess current certification requirements for avionics displays and associated processing units, and review current practices of display manufacturers for applicability to those requirements.

This paper documents the results of this project to date; additional results of interviews and the ensuing analysis will be presented at DASC.

Survey Results and Recommendations

Survey results were examined to recommend ways in which manufacturers can improve their development processes to overcome certification difficulties, as well as ways in which the certification process can be streamlined or relaxed to reduce cost and time.

Recommendation 1:

Use Avionics Computer Resource (ACR) architecture to facilitate technology advances

Introduction of new technology into avionics has been a problem with regard to the time and expense required for certification.

The use of partitioned architectures is increasingly prevalent in order to isolate the more critical functions that need the highest levels of protection.

Upgrading and retrofit lead to major certification problems since current systems are not built as true open systems. This is not an immediate problem for the General Aviation fleet. However, as the industry envisions implementing situation awareness instruments that accommodate free flight in phases, it would be most cost effective to be able to upgrade or retrofit equipment built as an open system.

RTCA Special Committee 182 and EUROCAE Working Group 48 are jointly developing Minimum Operational Performance Standards (MOPS) for an Avionics Computer Resource (ACR). The ACR will be a generic resource to perform multiple aircraft functions, that will be loaded with partitioned software applications at the time of the aircraft-specific avionics system design. These functions could easily be upgraded, or new functions could be added, at any time. This approach aims to reduce the development and modification costs of avionics by improving portability and streamlining the hardware and software qualification (certification) process. Different customers could install custom suites of avionics functions without the need for total re-certification. Related avionics functions could be integrated to a higher degree than ever before. The ACR manages integrity of its partitions using a health and safety monitor.

Recommendation 2:

Clarify the certification process

Many of the participants noted the need for a clear and common understanding of the certification process by applicants and FAA. The need for education in regard to intra- and interagency responsibilities within the FAA was noted. Although the responsibilities of the various offices have been spelled out, in practice the process of certification may have to more explicitly consider how an avionics instrument is to be used. This factor traditionally has fallen under the jurisdiction of Flight Standards. The FAA's Integrated Product Teams do not appear to be working as intended in all cases: key personnel with certification expertise have too many demands on their time to permit their participation in every relevant activity. Further, there seems to be a steep learning curve for the rules regarding certification of avionics. There are several venues available for education of ACOs and DERs. Because of the widespread demand for skilled personnel in the software industry, FAA is finding it hard to hire and retain qualified staff. With high turnover rates and the steep learning curve, the increased use of new staff is perceived as further reducing the quality of service to the certification applicant. Additionally, the level of scrutiny used on a project by the DERs is not always adequate because of schedule and cost limitations. Such situations tend to undermine the ACO's confidence in the DER process.

There have been many instances where an applicant may follow the published TSO rules, yet these rules are deemed inadequate by the ACO for safety reasons. Applicants need to address the bigger picture of the aircraft environment in which they are installing the equipment.

Both the FAA and the applicants noted that application of rules on service history for previously developed software was not always consistent. The other areas of uneven application are the determination of whether a change is major or minor, which greatly affects the required documentation, and software tool qualification.

Some rules are considered ambiguous and open to interpretation; it is hard to recognize when an interpretation is violating the intent of policy.

Many of the FAA rules were scripted following incidents or accidents. The rules live on even when the technology that caused that rule has long been obsolete. It is recommended that the FAA rules be reviewed for currency and technological appropriateness.

Recommendation 3:

Keep requirements simple

Avionics must not detract from the pilot's fundamental need to fly the aircraft. Instruments have become more complex to use, as technology has facilitated manufacturers striving for market advantage by providing complex features. When the avionics are built with the pilot in mind, unduly complex requirements should be avoided. Such instruments also would keep the testing and certification processes simpler. However, instead of complexity, manufacturers would have to promote quality (e.g., resolution of the display, size of the screen, weight and volume) to gain a market advantage.

An example is in the area of computer-human interaction. Frequently, an instrument is failed during flight testing because it is found impossible to use as intended. There are many considerations that go into the design of controls and operational modes including the pilot workload and ease of understanding what is presented on the display. These considerations are better understood by involving pilots in the design. This would lead to a better chance of success in certification and in marketing.

Recommendation 4:

Work to foster cooperation and trust

Human elements of interaction between the applicant and the FAA (i.e., personalities) sometimes make or break the certification process. A professional environment needs to be established in which differences of opinion may be discussed, and applicants need not fear punishment or reprisal for appealing a decision. Moreover, incentives should be provided for the ACO to help applicants to succeed.

Applicants also desired that more responsibility be delegated to DERs.

Recommendation 5:

Work effectively under business and legal constraints

It was reported that ACOs take great care not to disclose applicants' methods to others. It would be desirable to train ACOs how

0-7803-5086-3 /98/$10.00 ©1998 IEEE

they may give applicants helpful guidance while they avoid divulging proprietary information.

Recommendation 6:

Introduce effective practices

There is a large amount of objective evidence that is required by regulations. Many companies are finding it to be very expensive to furnish all the detail specified in DO-178. Some users have successfully provided Executive Summaries, with details available upon FAA request. Others have assembled details on a CD-ROM with hypertext provided to support data retrieval.

Applicants seek to reduce repeat inspections by ACOs and MIDOs. They should be provided guidance to help them minimize this expense.

A good practice would involve aircraft DERs and the software DERs in an integrated team so that both the minimum requirements and the aircraft environment issues are addressed.

Applicants feel that the FAA is not fast enough in reviewing the certification papers, leading to significant delays in product release time. The FAA needs more ACOs and DERs to cope with the workload. Further, there are no rules specifying paperwork turnaround time.

Hesitation to sign off on new technology may have been caused by lack of a precedent and the FAA's tendency to be conservative. This problem may be mitigated with continuing education for the ACO and DERs.

Applicants are requested by the ACO to rewrite their documents if the documents are not exactly in the same form as indicated by the rules. They feel it is more productive to be concerned about the contents rather than the form. The ACO should indicate where the exact form is important.

Both the FAA and the applicants note that every project is started up as a brand new project without gleaning lessons learned from the previous projects. The same mistakes are repeatedly made and corrected at great expense. It is recommended that lessons learned be applied from one project to the next.

Both the FAA and the applicants preferred standardized interfaces to accommodate choice in the market place. In such systems, applications may be written by any vendor. Such competition will create innovation and lower price.

The FAA noted that developing new technology in consortia is the cheapest way to build and certify new technology. Such systems have caused many participants to review different perspectives to the problem. The process also allows an easier way for the FAA to become familiar with these systems.

RTCA has provided a forum in Task Force 4 to improve certification. Manufacturers are invited to come up with better ways of certification and influence the process. This platform may also be used for sharing war stories. Such activities would help the manufacturers to learn from each other's experiences.

Some of the FAA is of the opinion that by certifying companies as having a DO-178B (compliant) Process, there is no need to re-inspect this company's process at every certification. This streamlining would help the FAA to focus its attention on product verification.

Conclusion

We are continuing to interview avionics and airframe manufacturers, General Aviation industry groups, and additional FAA experts. We plan to conduct interviews and sample new technology offerings at the Oshkosh, WI airshow. Additional interviews are scheduled in Seattle and at FAA Headquarters. New results

0-7803-5086-3 /98/$10.00 ©1998 IEEE

from these visits will be included in the briefing to DASC. Our final report also will be published in October.

0-7803-5086-3 /98/$10.00 ©1998 IEEE

BITE IS NOT THE ANSWER (BUT WHAT IS THE QUESTION?)

Ronald S. Carson, The Boeing Company, Seattle, WA

Abstract

Complicated avionics systems increasingly rely on built-in-test equipment (BITE) to perform fault detection and aid in fault isolation. Recent experience with the large quantity of avionics in the passenger cabins of commercial airlines (Velocci 1997) indicates that (a) reliability is generally inadequate, and (b) maintenance regularly results in a no faults found ratio (NFF) of 50%. The expectation that adding BITE will improve reliability and reduce NFF has not been realized for these large cabin systems.

Reducing NFF requires (1) verifying that a failure has actually occurred prior to taking action, and then (2) correctly replacing the failed unit on the first maintenance action. This in turn depends on the correct isolation of failure by whatever methods are available.

We propose a methodology for performing line maintenance that is less dependent on the scope of built-in-test than is traditionally thought to be necessary, and, when followed scrupulously, results in low NFF rates. The method flows out of the failure modes and effects analysis (FMEA), by identifying and correlating the effects of failure (including annunciated test results and observables such as loss or degradation of function). By working "backward" from the failure effects using the analysis, the most-probable source of failure can be identified.

We analyze an extensive cabin system with audio and video distribution, and show that the NFF rate is relatively insensitive to the actual amount of BITE, and very sensitive to following the recommended procedures for isolating a failure and replacing suspected units.

The question is not, "How do I find out what is failed?", which is the focus of BITE, but is, "What must be done to restore full functionality?"

Introduction: What's the Problem?

Cabin equipment avionics can be treated much differently from flight-related avionics for the simple reason that failures are not generally likely to affect airplane safety. As in any system, overall reliability is the key ingredient in achieving satisfaction. A large in-flight entertainment (IFE) system consisting of perhaps several thousand separate units of electronics demands reliability in each unit that generally surpasses the state-of-the art. Because of these two factors the ability to quickly isolate failures for maintenance actions is critical in maintaining system functionality.

Reliability and seat availability for IFE systems are increasingly seen as key elements in achieving passenger satisfaction (IFE 30 March 1998). The cabin equipment can be treated as "revenue critical" because the performance of the entertainment system can contribute to long-term customer loyalty (or customer avoidance).

Is BITE the Answer?

The "traditional" method for identifying and isolating failures has been based on built-in-test equipment (BITE), which performs some level of passive monitoring in combination with active testing to determine the health of the system. The amount of BITE is generally expressed in terms of "coverage":

0-7803-5086-3 /98/$10.00 ©1998 IEEE

the fraction of failures detected and annunciated by the BITE, weighted by failure rate. It is generally thought that a high number (approaching 1.0) for BITE coverage will equate to accurate maintenance actions (reduced no-faults found).

The sad truth is that the NFF rate appears to be independent of the level of BITE. Recent service data (AWST, 11 May 1998) indicates that, for example, "of 2,100 removals of avionics units with BITE capability, about 1,500 came back from the shop with no fault found." Although no statement is given regarding the expected level of BITE coverage, this works out to a staggering NFF value of 71%! What is more compelling is that, "of the NFF units, mechanics had not bothered to run a BITE test on 55%....", indicating a failure to adhere to the minimum procedural requirements for troubleshooting.

Is More BITE Really Needed?

Yet, BITE is still considered to be a capability of which more is needed: "In-flight entertainment hardware manufacturers must provide airlines with better built-in-test equipment (BITE)." (IFE, 11 May 1998). The stated goal is for the airlines to have "BITE systems that will measure, record and report system performance and availability." (Smith, 1998).

From an engineering perspective we are keenly interested in the details of any failure: what fails, how often, and why (root cause). This kind of information enables us to improve the system reliability and maintenance performance over time. As engineers we also want the annunciated fault to indicate as much information as possible regarding the failure.

From a maintenance perspective, however, we want to display "only the information the end user (line mechanic) needs" (AWST, 11 May 1998). But what is that "information" and how do we determine it and present it?

Is Better BITE Needed?

One solution is thought to be in "standardizing" the BITE messages and displays in a "universal, plain English format" (AWST, 11 May 1998). But "standardizing" the messages fails to address the content: "What *is* the information needed to accurately *maintain* the airplane, regardless of the cause of failure?" It is this question that we address herein.

If the line mechanic could confidently effect replacements in a predetermined manner that was shown, over time, to be accurate (*i.e.*, minimizes NFF), then we would be providing the necessary information. This information is not, however, strictly related to the level of BITE coverage and reporting.

What Else Might be Useful?

We show that the level of BITE in terms of faults detected as a percentage of faults occurring is less important in identifying the correct order of replacement than simply adhering to a defined maintenance procedure, using whatever information is available in addition to the BITE data. For it is the *totality* of information that identifies the root cause of failure. Having identified the failure source, one can then take the correct maintenance action (generally, remove and replace) to return the system to full functionality.

This totality of information includes not only that available from BITE (*i.e.*, test results), but the information the cabin crew and line mechanic can determine on their own. For of what use is it to annunciate a "television broken" test report while the crew or mechanic is watching TV? This is a simple example of a "false positive" or "nuisance" fault that leads to loss of confidence. The television may indeed by "broken" (fails self-test), but the failure doesn't cause a loss of function.

We analyze an extensive cabin system with audio and video. Field-service data associated with the cabin system is compared

0-7803-5086-3 /98/$10.00 ©1998 IEEE

with predictions of BITE coverage. It is found that the predicted no-faults found rate is relatively insensitive to the level of BITE coverage. Discrepancies with field-service data are attributed to failures to implement the defined maintenance procedures.

Possible Solutions

Increase BITE (decrease reliability).

One might be tempted to think that simply increasing the amount or effectiveness of BITE would have an immediate payoff in either improving reliability or reducing NFF. However, any increase in the quantity of equipment (test or otherwise) adds components which use additional power, creating more heat. For most avionics systems, more equipment also means more lines of executing software, with more opportunities for the discovery of unwanted "features".

System reliability can be improved by a combination of (a) reducing the number of components in the design, (b) improving the reliability of those components that remain in the design, or (c) reducing the environmental stress on the component (lower temperature and vibration). Certainly, adding more components, such as BITE, will only degrade the system reliability compared with not having them. Improving the reliability of those components remaining is a special discipline in itself.

Key contributors to component reliability include (a) operating in an ambient environment which minimizes environmental swings (temperature, vibration), (b) designing with large design margins for both temperature, vibration, and other effects, and (c) determining and mitigating failure modes for specific components rather than simply accepting "random failure" as a necessary consequence of avionics systems (AWST 1997, Rassaian, 1998). Boeing has experienced MTBFs significantly exceeding predictions when conservative thermal analyses are used for the electronics.

How effective is BITE if it is ignored?

Reducing or eliminating NFF depends on infrequently or never removing a unit that is not failed. This in turn depends on correctly identifying the source of failure for any loss of function. So, how do we identify the source of any failure? Equivalently, what is "the information the end user (line mechanic) needs" to maintain the airplane (AWST, 11 May 1998)?

As indicated above, unacceptable levels of NFF can be experienced even in the presence of high-coverage BITE if the BITE is not used or is ignored. Without delving into psychology too deeply, ignoring the BITE system or the data it produces can be related to an assessment by the user that the information is unreliable, *i.e.*, fails to correctly identify the failed unit and associated maintenance action. Indeed, in the example cited above where 55% of the removals had no BITE testing, 29% had "passed" their BITE tests, "and 16% had incorrectly indicated they were not functioning properly" (AWST 11 May 1998).

How Does BITE Get Respect?

An element of reducing NFF must therefore be the reliability of the *information* provided to the line mechanic. For if the information is ignored or disbelieved, it is of negative value, since the developer has paid for its development, and the airline has paid for the equipment, but it is only using maintenance time non-productively since the information is not influencing the maintenance actions.

So, by what method could we improve the reliability of the information available to the line mechanic? How do we correctly determine exactly which of many possible units should be identified as the cause of failure during a loss of function? If the cause of failure can be correctly identified with high confidence, and the mechanic follows-through

0-7803-5086-3 /98/$10.00 ©1998 IEEE

by replacing the identified unit, we have done as much as is possible to reduce NFF.

Since we are concerned with maintenance actions in the presence of failures, it is sensible to begin with the failure analysis of the system.

Failure Analysis

Creating maintenance procedures based on the failure analysis is a technique developed at The Boeing Company, and used in two recent programs (777 Cabin Management System, Phased Array Communication Antenna System). The objective is to provide the maintainer with a level of information equivalent to that available to the engineer, synthesized in a manner useful to his/her activity (line fault isolation and repair).

The key is to determine and document the system responses in the presence of failure (more on multiple failures later). Thus, for any postulated failure mode in the system, the failure analysis identifies the system responses at the level at which the maintainer interacts with the system. Such responses include functionality changes as well as test results.

"Functional failure analysis" is an area of growing interest (Carson 1996, Bowles, 1998, Spangler 1999), because we can correlate failure with loss of function beginning with the design phase of system development. For example, having lots of BITE to monitor elements of hardware that don't contribute to system functionality is clearly a waste of money, even though it may provide information of interest to the engineer. Conversely, we can use functional failure analysis to help select cost-effective fault monitoring provisions (most functional coverage for least cost).

We perform a failure analysis for the end item with the goal of using the information to create fault isolation procedures. We are concerned not only with the local effect for the end item being analyzed, but we are also concerned with the manifestation of failure at the system level. This is the so-called, "end-level" or "system-level effect".

Such system level effects should relate to the mission or functions of the system being analyzed. In cabin systems, end-level effects will include loss of audio, loss of video, loss of lights or control of lights, loss or degradation of passenger address or cabin interphone systems. The scope of the end-level effect is also important in identifying the root cause of the failure.

For example, "loss of entertainment audio" as a system effect must be further refined. For if the audio is lost airplane-wide, the cause is most likely at the audio source (tape player, etc.) or in the distribution system, rather than at all of the seats. To differentiate between these possibilities, one would need to know whether all audio programs were lost (focus on the distribution system), or only one program was lost (tape player).

Based on these very simple observables, we can perform a high degree of fault isolation in the airplane cabin, mainly because the cabin is full of "observables" and "observers".

The addition of BITE increases the fault isolation capability. BITE results will generally be categorized as either being isolated to a single unit (self-detected or imputed as an "internal fault"), or being within a limited "ambiguity group".

When all of these indications in the presence of failure (test results and observables) are documented, we have a substantial portion of what is usually in the failure analysis or BITE coverage analysis. If we now "turn the process around," we attempt to infer the source of failure given the presence of certain indications.

Mathematically, for the failure consequence analysis we write:

$$F \rightarrow R,$$

where F is the failure antecedent (what failed),

0-7803-5086-3 /98/$10.00 ©1998 IEEE

and R is the result or consequence (end-level effect).

The problem we wish to solve, is, how to determine F give some set **R** of observables, $\mathbf{R}=\{R_1, R_2, R_3,\ldots R_n\}$. Obviously, it cannot be assumed that $R \rightarrow F$ (*i.e.*, the inverse of any syllogism is not necessarily true), since there may be many failures that produce the same result.

Linking Failure to Action

The method for isolating the source of failure is two-fold. First, the analysis examines the consequences of any failure for more than one operating or test mode. While not always the case, the consequences of many failure modes which are identical in one system operating mode turn out to be different in another operating mode. Test results may differ for the same observable, or different observables may correlate with different test results, depending on the source of failure. This is no different than the use of "test vectors" to ensure that a set of tests provides the required level of completeness for any system being tested. Test vectors relate input conditions with output results and observations to distinguish non-equivalent faults (Kohavi 1978).

The second aspect of the method is to examine the likelihood of failure for multiple failure modes with otherwise identical consequences. If we can determine that one replaceable unit has a much higher failure rate than other units which have failure modes with identical consequences, then we prioritize the order of replacement to reflect this data. This gives a quantitative basis for prioritizing the replacement, and also allows us to predict the no-fault-found rate.

For if the procedures are scrupulously followed and the analysis is correct, the only units removed with NFF should be those which are part of a multi-unit ambiguity group and which were not failed but were listed ahead of a failed unit.

0-7803-5086-3 /98/$10.00 ©1998 IEEE

As noted above, Boeing has developed an automated tool which, beginning with the failure modes and effects analysis, links the consequence(s) of failure to the source of failure, and prioritizes the resulting ambiguity groups for producing fault isolation procedures.

The result of the analysis is a maintenance procedure that:

1. Begins with what the line mechanic will observe, either from a test mode or using eyes, ears, nose, etc.
2. States what test or observation is to be made in order to compare failure mode consequences across different operating states.
3. States the possible outcomes of the recommended tests or observations, and directs additional tests or observations as appropriate.
4. Recommends a list of units which should be replaced which is both complete (contains all LRUs which could possibly cause the observations), and is prioritized (lists the possibilities in order of decreasing failure rate).

Two implementations are possible. The first is the traditional "Fault Response Manual/ Fault Isolation Manual" or FRM/FIM. This captures the observation (test result) as an index to the recommended maintenance action. While complete in theory, it suffers for two reasons. First, the FIM is generally not based on the fault analysis we discussed above. Rather, it is based on an examination of engineering data and connectedness, rather than an analysis of response to failure. Therefore, the procedures themselves may be incomplete or even incorrect. Second, the FIM is rarely maintained at the location where the maintenance is performed, because of its bulk (hundreds of pages in several volumes). Once the initial observations are made, the mechanic is sometimes forced to literally drive miles to obtain the relevant FIM pages.

The second possible implementation is to provide all the necessary data at the location where the maintenance action is to take place. For airplanes, that means "onboard". Given, however, the large quantity of data, maintaining ever-changing paper copies in multiple volumes leads to a logistics nightmare for the users. Boeing has begun providing its fault isolation and other manuals on laptop computers as Portable Maintenance Aid Terminals (PMAT).

An alternative to actually creating and maintaining the equivalent of the procedural, paper-based FIM, even when hosted on a portable computer, is to provide all of the fault analysis as a database on a portable computer. If the line mechanic were able to enter his/her observations, the database can sort by these indications, recommend operating modes to investigate, and finally narrow the list of possibilities to a prioritized list, as described above. In effect, the mechanic would be able to conduct the fault isolation analysis in real time using his/her observations as inputs to the mathematical tool which performs the correlation of the indications to failure modes. This alternate method doesn't require any BITE at all, since the allowable inputs would be any observation, whether or not from a test result. Alternative prioritization schemes might be cost of spares, ease of replacement, availability of spares, but these are difficult to predict in a manner consistent for multiple users.

Relating NFF to Expected Replacements

The Boeing Company uses "expected replacements" to quantify the degree of confidence in our fault isolation design and analyses. The average quantity of expected replacements for any given maintenance action is expressed as the number of units replaced in an ambiguity group when weighted in increasing order of failure rate, FR_i, averaged over all ambiguity groups. Given an ambiguity group of n_j LRUs, with individual failure rate, FR_{ij}, the expected replacements for this j^{th} ambiguity group is:

$$ER_j = \frac{\sum_{i=1}^{n_j} i * FR_{ij}}{\sum_{i=1}^{n_j} FR_{ij}}.$$

One can easily see that ER = 1 whenever there is only one LRU in an ambiguity group. Conversely, as n_j gets larger and the failure rate FR*ij* is larger for large *i*, ER will increase, indicating that, on average, substantially more than one LRU must be replaced to "find" the failed unit.

Over all ambiguity groups, *A*, the expected replacements for the system being analyzed is then:

$$ER = \sum_{j=1}^{A} \frac{\sum_{i=1}^{n_j} i * FR_{ij}}{\sum_{i=1}^{n_j} FR_{ij}}$$

In contrast, "no faults found" is the fraction of units removed that are not failed. Currently, this averages approximately 50% or more for the airline cabin systems. While it is easy to determine NFF in practice (measuring the results on returned units), the *a priori* calculation is somewhat tedious, but follows the methodology of ER to a certain extent.

NFF is the sum over all ambiguity group elements and ambiguity groups of the probability that the next unit removed is not failed, given that all prior elements of the ambiguity group have been removed and are not failed. For a simple, two-element ambiguity group, ER = 1* FR_1 + 2 *FR_2. For this ambiguity group, NFF = (FR_2)/(FR_1 + FR_2) [the probability that the first unit removed is not failed]. Solving for NFF in terms of ER yields, NFF = ER-1.

One common aspect of these calculations is the notion of prioritized replacement. The calculations cannot be

0-7803-5086-3 /98/$10.00 ©1998 IEEE

performed if the order of replacement is random. Given the data quoted above, (71% NFF rate) with 55% of LRUs untested) it is not unreasonable to speculate that no disciplined procedure is being followed. Line mechanics are possibly (a) simultaneously removing all units in an ambiguity group, or (b) removing all possible units for which spares are available, or (c) a combination of (a) and (b). What must be lacking is a disciplined maintenance procedure.

Cabin Systems Analysis

Using our ideas of (1) differentiation of ambiguity group elements using multiple operating modes and (2) prioritized replacement based on failure rate of identical failure mode effects, we have performed analyses on two different cabin systems.

We have analyzed a large passenger services system with audio and video distribution, but without the head-end audio/video sources or the televisions and audio outputs. This system comprises 337 LRUs (of which roughly 250 are of 3 types), with a system MTBCF of 962 hours. Two analyses are considered in table 1. "1a" is the system with all BITE active. "1b" is the system with most BITE deactivated in the analysis. "Actual" results are only available in the first case.

Table 1. System Performance

System	BITE Coverage *	NFF (calculated / actual)	ER (calculated)
1a	92.9%	0.052/ 0.32	1.05
1b	23.8%	0.087/NA**	1.08
* % critical faults detected ** NA = not available			

What is most striking about the results is that, for system 1b, "turning off" most of the BITE in the analysis results in only a slight drop in the calculated fault isolation performance (slight increase in ER and expected NFF). In other words, when the consequences of failure are reduced in the analysis to ONLY descriptions of loss of function, there is still enough information available to produce fault isolation procedures with a theoretical NFF of less than 10%.

In addition, the actual NFF rate for system 1a for calendar year 1997 field service data is 32% of 618 units returned. This is much greater than the calculated values, and much higher than would be expected given the very extensive BITE coverage.

All of the calculations assume that the maintainer is following a *disciplined maintenance procedure* which is faithful to the actual system behavior in the presence of failure. Thus, it is the fidelity of the maintenance procedure and the discipline of the mechanic which are the strongest predictors of NFF, NOT the level of BITE coverage.

What accounts for the dramatic failure of BITE to produce low NFF rates? As discussed in the introduction, the tendency in airplane maintenance is to ignore the BITE results, rendering the calculated BITE effectiveness, "ineffective". BITE will only be effective if the mechanic is provided procedures that are trustworthy and then follows them scrupulously

But as table 1 also indicates, producing effective fault isolation procedures does not require a high level of BITE coverage because of the large quantity of observables in a cabin IFE system. It may well be that cabin maintenance is currently operating on the basis of observables, where the mechanics are mainly using their experience in maintaining the cabin and ignoring the fault isolation procedures. There seems no other way to rationalize the NFF experience for the 777 Cabin Systems.

Effect of Concurrent Multiple Failures

The presence of concurrent failures can

0-7803-5086-3 /98/$10.00 ©1998 IEEE

confuse the maintainer, since he/she is never certain whether the list of indications is attributable to exactly one or more than one failure mode. Most failure analyses are performed with the ground rule that only one failure is active at any given moment. Safety analyses look for combinations of failures that may produce unwanted consequences in order to calculate the probability of such combinations occurring.

If one is to use maintenance procedures based on fault analyses which assume only a single failure, then the maintenance approach must seek to ensure that any single failure is resolved (failure cause identified) as soon as the indication occurs, even if the actual maintenance is deferred. The accumulation of multiple failures will render virtually any fault isolation procedure ineffective as the mechanic struggles to sort out the possible relationships among numerous indications.

Summary and Conclusions

We have shown that the goal of reducing NFF rates is less sensitive to the level of BITE coverage than has been assumed in the airline industry. NFF can be reduced using disciplined procedures that are based on the system failure analysis, not assumptions about how the system will behave in the presence of failure. The level of on-airplane BITE coverage is then a choice that can be made to rely less on observables, or take advantage of BITE that is available from factory testing, but it is not necessarily a system requirement for the sake of maintenance.

The line mechanic cares less about "what's broken" than about what to do. Fault isolation procedures based on failure analysis provide that information.

References

Bowles, John B., "The New SAE FMECA Standard," *Proceedings of the Annual Reliability and Maintainability Symposium*, January 1998, pp. 48-53.

Carson, Ronald S., "Designing for Failure: Anomaly Identification and Treatment in System Requirements Analysis," *Proceedings of INCOSE 1996*, Boston, MA.

----- "Better BITE is top of airlines' wish list," *In-Flight Entertainment International*, Volume 4, Number 9, 11 May 1998, pp. 3ff.

----- "IFE should be treated as flight critical system," *In-Flight Entertainment International*, Volume 4, Number 8, 30 March 1998, pp. 3ff.

Kohavi, Zvi, *Switching and Finite Automata Theory*, McGraw-Hill Book Company, New York (1978), pp. 223ff.

----- "Program Rigor Key to IFE Reliability," *Aviation Week & Space Technology,* Vol. 147, No. 20, 17 November 1997, page 49.

Rassaian, Mostafa, and Lee, Jung-Chuan, "Thermomechanical Multi-Domain Stress Modeling for Fatigue Analysis of Electronic Packaging," 1998 IEEE Aerospace Conference Proceedings, paper 4.703, March 23, 1998.

Smith, Mark, quoted in "Better BITE is top of airlines' wish list," *In-Flight Entertainment International*, Volume 4, Number 9, 11 May 1998, page 3.

Spangler, C.S., "Equivalence Relations within the Failure Mode and Effects Analysis", to be published, *Proceedings of the Annual Reliability and Maintainability Symposium*, January 1999.

----- "Standardizing BITE", *Aviation Week & Space Technology*, Vol. 148, No. 19, May 11, 1998, page 17.

Velocci, Anthony L., "IFE Vendors Grapple With Technology Issues," *Aviation Week & Space Technology*, Vol. 147, No. 20, November 17, 1997, pp. 45-46.

0-7803-5086-3 /98/$10.00 ©1998 IEEE

THE APPLICATION OF FUNCTION RECONFIGURATION FOR DIGITAL FLY-BY-WIRE FLIGHT CONTROL SYSTEM

Pei Ying
No.303 Department of Automatic control
Beijing University of Aeronautics and Astronautics, Beijing P.R.China
Li Peiqiong Yao Yiping
No.303 Department of Automatic control
Beijing University of Aeronautics and Astronautics, Beijing P.R.China

Abstract

This paper is tend to describe a method of function reconfiguration, dealing with the failure of helms. When a helm of the aircraft is fail, how to reconfigure the remained normal helms to balance the aircraft again is an important problem. This paper introduce a method to solve this problem, and simulate the airplane with computer. In the Fly-By-Wire flight control system, we can intercalate a control mixer, which reassigns the control dictate to helms, between the control law and the actuator. Normally, it is an identify matrix; when a failure occurs, an algorithm will be used to calculate the matrix, so as to the control law have the same effect as the state before failure. There are two algorithms to two fail modes of helm: turn to middle and lock. As an illustration example, the function reconfiguration method is exercised on airplane, it is shown that with the helm turning to middle fail mode, the effect of using the reconfiguration algorithm is satisfactory. In addition it is also shown via simulation that with lock fail mode, the using of this reconfiguration method is feasible when failed helm locks within the limited angle.

¢ñ Instruction

Reliability, viability and fault tolerance are required in air war in the future, redundancy system and fault tolerant system are effective measurement to win it. In both systems, three functions have been achieved: fault detect, fault isolation and reconfiguration. With the development of digital computer and analytic redundancy technique, fault can be detected and located in time. Reconfiguration is the process of reassigning resource after fault, there are two type of resource in normal system: hardware resource and functional resource, so there are two methods of reconfiguration: redundancy reconfiguration and functional reconfiguration.

Traditionally, the implementation approach used to provide the desired reliability has been the use of multiple hardware with attendant selection logic, namely redundancy reconfiguration. When a failure appears, the failed hardware is replaced by spare in good condition. However, there are penalties in component cost, volume, power consumption and weight for every duplication for the sake of redundancy. By compounding the redundancy and selection logic, the apparent gain in reliability is soon lost due to increased failure rate associated with the higher parts count. Thus, most systems of today do not exceed quadruple implementation.

In some systems, there are no spare for component, such as the helm of aircraft, once a failure appears, no spare replace it. In order to solve this problem, a new idea called functional reconfiguration is put forward.

The main idea of function reconfiguration is that there is relativity in function among components, components in good condition offer function redundancy for failed component. For example, normally the stabilizer of aircraft control pitch, flap control

0-7803-5086-3 /98/$10.00

roll, when stabilizer fails, flap can deflect to the same direction, thus, pitch can be finished by flap. This helm, which offers redundancy in function, is called functional redundancy. It realized by changing control law of aircraft. With the advent of present digital computer technology, it has become feasible to utilize real-time digital computation for recalculate control law, therefore, our purpose is to investigate the possibility of increasing the functional reliability of flight control system by using functional reconfiguration.

There are three sections in this paper. At first, function reconfiguration is instructed, at the second section, the practical design method— pseudo-inverse method is illustrated in details, at last, we exercised this method on an airplane, simulating it by computer, and displaying the result.

In redundancy system, there are spare components, when failure appears, system can turn to the backup to recover, but for other system, there is no spare, such as helm of airplane, when it fails, no backup replace it. In order to solve this problem, a new idea is put forward, it is called function reconfiguration. The main idea of function reconfiguration is to use the components which are correlated with failedt component of function. When fail appears, control law will be changed, the function of failure component is reassigned to the remain normal helms.

hardware or software, when failure occurs, backup component can be interposed in system to replace fail component, and function of system is recover. But in some system, such as helm of aircraft, there is only a suit of hardware, once a helm fails, there are no backup to realize

its function. In order to solve this problem, helms of aircraft are thought multifunction, although helm of aircraft is one suit of hardware, different combination among helms would produce different control torque. such as if stabilizer move to the same direction, they control pitch, if they deflect differently, they control roll, so do flap, and we call it function redundancy. This concept have been used in flight control system, such as YF-16.

We can conclude from table 1 that the combination of different helm provide functional redundancy for failed helm, it is called multifunctional helm.

Function reconfiguration is realized by control law reconfiguration. For aircraft, the control law of flight control has been designed before, and may hardly be changed during flight. A method called pseudo-inverse is used to solve this problem. Based on the matrix theory, it uses a matrix to reassign the direct of control law(fig 1). It is characterised by simple design and feasible of real-time application.

Practically, a control mixer is intercalate between the control law and actuator; it is an

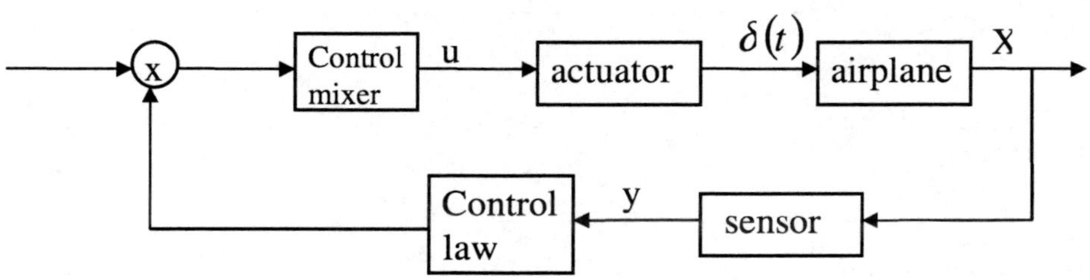

Fig 1 control system with control mixer

¢ò Function Reconfiguration

In redundancy system, there are spares for

identify matrix normally, and the actuator

0-7803-5086-3 /98/$10.00 ©1998 IEEE

executes the direct of control law, when a failure occurs, an algorithm will be used to calculate the matrix to reassign the failed helm function to other helms in good condition, so as to the control law have the same effect as the state before failure.

¢ó Pseudo-inverse Method

Pseudo-inverse method is a simple and effective reconfiguration method, put out by Razza, it means a control effective reassigned method based on matrix pseudo-inverse. Directions to helms are reassigned to balance system according to fault detection by computer. Pseudo-inverse method intercalate a control mixer between the control law and the actuator(fig 2). Normally it is an identify matrix, once a failure occurs, control mixer is changed to corresponding gain matrix calculated before.

The effect of helm failure is to change the input of airplane, the main idea of pseudo-inverse is to assure that the input of airplane after failure is the same as that before failure. Consider the linear-invariant system describing the aircraft:

$$\dot{X}(t) = AX(t) + B\delta(t) \quad (1)$$

where $X(t) \in R_n, \delta(t) \in R_m$, and A, B are real constant matrices of compatible order. In addition, when the dynamic process of actuator is ignores, it can also be described as follow:

$$\delta(t) = Ku(t) \quad (2)$$

where $u(t) \in R_m$, it is direct of control law. K is control mixer.

It is assumed that dynamic characteristic of aircraft is invariable, we can obtain the respond vector by input:

$$\dot{X}_u(t) = BKu(t) \quad (3)$$

for system in good condition:

$$\dot{X}_u(t) = B_0 K_0 u(t) \quad (4)$$

where $B_0 = B$, and K_0 is an identify matrix. The control matrix of system after helm fail is B_I, whose column corresponding to failed helm is replaced by zero column vector, so after a helm failed formula (4) can be replaced by:

$$\dot{X}_u(t) = B_I K_I u(t) \quad (5)$$

where $K_I \in R_{(m-1)\times m}$, it can be calculated by pseudo-inverse.

For helms, there are two fail modes: turn to middle and lock. When a failure occurs, the failed helm does not execute direct and turn to middle without deflecting. On the contrary, the failed helm does not turn to middle, but stay on the location where helm get at the time of the lock failure mode appears. In the next section,

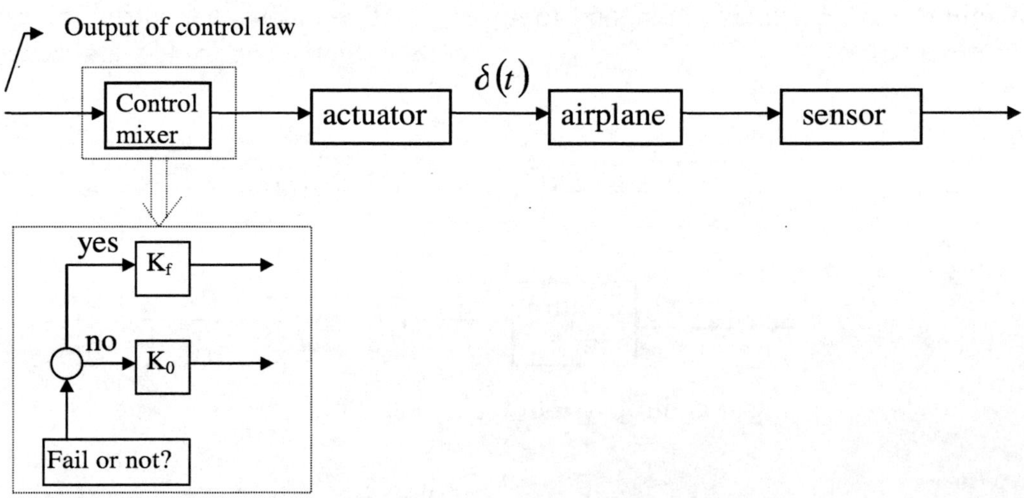

fig 2 concept of control mixer

how to deal with these two fail mode is considered.

1. mode 1: turn to middle

When this failure occurs, there is not' unexpected effect by the failed helm, that is, the failed helm offers no control torque. In order to ensure the dynamic characteristic is invariable, the following requirement must be satisfied:

$$B_0 K_0 = B_I K_I \qquad (6)$$

it is shown that when (6) is satisfied, the respond vector of normal system and that of failed system is the same. Based on the matrix theory, K_I can be solved as follow:

1) $m = n$ B_I is odd matrix

$$K_I = B_I^{-1} B_0 K_0 \qquad (7)$$

2) $m < n$ the order of B matrix is m

$$K_I = \left(B_I^T B_I\right)^{-1} B_I^T B_0 K_0 \qquad (8)$$

3) $m > n$

$$K_I = B_I^T \left(B_I B_I^T\right)^{-1} B_0 K_0 \qquad (9)$$

2.mode 2: lock

The effect of this fail mode is different from above mode, when this failure occurs, the failed helm offers unexpected torque. Now, it need to offset the unexpected torque. As we all know, normal system state equation is described that:

$$\dot{X}(t) = AX(t) + Bu(t) \qquad (10)$$

when some helm locks in a fixed angle, (10) can be replaced by:

$$\dot{X}(t) = AX(t) + B_1 u_1(t) + B_2 F \qquad (11)$$

where $u_1(t)$ is the control vector by deleting the failed helm, $u_1(t) \in R_{m-1}$, and B_1 is the matrix with column corresponding to the failed helm deleted, B_2 is vector corresponding to failed helm in B matrix, F is the lock location of failed helm. For a failed system, control law can be predigested as follow:

$$u_1(t) = -KX(t) + TF \qquad (12)$$

and we can get the close-loop system:

$$\dot{X}(t) = \left(A - B_1 K\right) X(t) + B_1 T + B_2 F \qquad (13)$$

for the purpose that the unexpective torque should be offset, namely:

$$B_1 T + B_2 F = 0 \qquad (14)$$

based on the pseudo-inverse theory, control mixer T can be solved:

$$T = -\left(B_1^T B_1\right)^{-1} B_1^T B_2 \qquad (15)$$

In engineering application, the control mixer is calculated during design process. For mode 1, calculation is simple, because the failed helm to middle, we need not consider the effect brought by the failure. For a special failure, such as the failure of left stabilizer, the control mixer is unique. But for mode 2, the location of failed helm will effect the control torque. It is not practical to pre-calculate the control mixer for every location of every helm, this work is cockamamie and skimble-skamble. The practical method is that the deflecting angle area of helm should be divided in several parts, the key angle has been found, in a part of area, only the mixer of the key angle is calculated, when failure appears, control system can get the vicinalest control mixer to reconfigure. The magnitude of the part is decided by the reliability requirement, the little of the part, the more successful the reconfiguration.

¢ô Design of Control Mixer and Simulation

The helm of traditional aircraft is stabilizer, flap and helm, controlling pitch, roll and yaw respectively. Normally stabilizer deflect to the same direction to offer pitch torque, and ailerons deflect to different direct to offer roll torque, ailerons associate with helm to control yaw. Lengthways and lateral is independent, any helms failure will affect the control effect of it. Now we think, they are independent, namely, stabilizers may deflect to different direction, and flap may deflect to the same direction.

The proposed reconfiguration method was exercised on the real aircraft. The state equation is in the form of (1), where the state X is comprised by attack angle, pitch rate, sideslip angle, roll rate, and yaw rate as denoted by $X = [\alpha \;\; \omega_Z \;\; \beta \;\; \omega_x \;\; \omega_y]^T$, the control vector $\delta(t)$ is comprised by δ_{zr} (right stabilizer deflection), δ_{zl} (left stabilizer deflection), δ_{xr} (right aileron deflection), δ_{xl} (left aileron deflection), δ_y (rudder deflection).

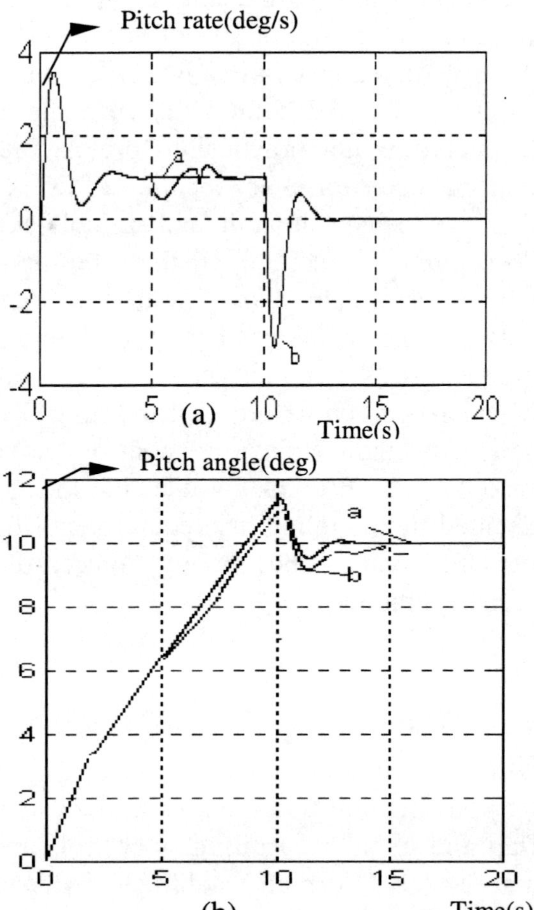

fig 3 responding line of mode 1

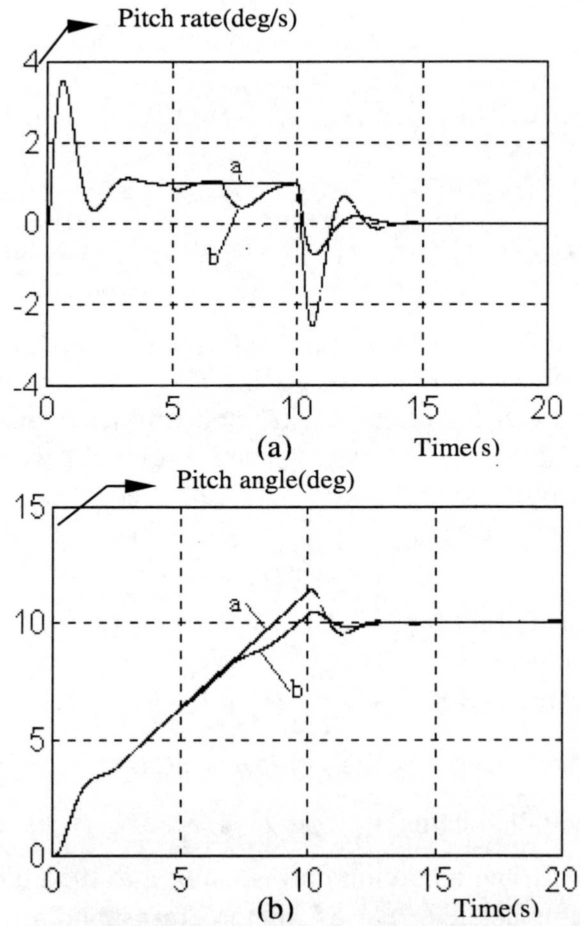

fig 4 responding line of mode 2

a-----system performance without failure

b-----system performance with failure

For normally flight condition, the A and B are given by:

$$A = \begin{bmatrix} -0.7688 & 1 & 0 & 0 & 0 \\ -1.6282 & -1.1874 & 0 & 0 & 0 \\ 0 & 0 & -0.2430 & 0 & 1 \\ 0 & 0 & -29.93 & -2.24 & -1.0693 \\ 0 & 0 & -4.7270 & -0.0379 & -0.3703 \end{bmatrix}$$

$$B = \begin{bmatrix} -0.0795 & -0.0795 & -0.0241 & -0.0241 & 0 \\ -4.7315 & -4.7315 & -1.1125 & -1.1125 & 0 \\ 0.0019 & -0.0019 & 0.0026 & -0.0026 & -0.0399 \\ -15.77 & 15.77 & -11.944 & 11.944 & -8.003 \\ 0.2404 & -0.2404 & 0.1833 & -0.1833 & -2.403 \end{bmatrix}$$

for the purpose of simulating a helm failure, it is assumed that the left stabilizer has failed, only right stabilizer control lengthways, consider two fail modes: 1) turn to middle and 2) lock, we have different design for control mixer.

For the first fail mode:

$$K_I = \begin{bmatrix} 1 & 0.9911 & 0 & 0 & 0 \\ 0 & 0 & 0 & 0 & 0 \\ 0 & -1.2954 & 1 & 0 & 0 \\ 0 & 1.3334 & 0 & 1 & 0 \\ 0 & 0 & 0 & 0 & 1 \end{bmatrix}$$

for the second mode:

$$T = [-0.9911 \quad 1.2594 \quad 1.3334 \quad 0]^T$$

The results of simulation are illustrated as fig 3 and fig 4.

The responding line is hole process of system performance. For system performance with failure, system is normal in time 0s to 5s, the failure appears at the time of 5 second and the time for fault detection is 2 second, that is, from moment of 7 second, the reconfiguration have affected system.

It illustrates in fig 3.a that after reconfiguration scheme has worked, the pitch rate can trance the normal system quickly, but in fig 3.b, there are static difference in pitch angle because of the integration of pitch rate during fault detection, and it may be solved with a new lengthways command after system is steady. Fig 4 is the same with fig 3. But for the lock mode, the reconfiguration result has something to do with the lock angle of helm, with a great deal of examples, the result is satisfied when the failure helm locks with 10 degrees.

IV Conclusion

Based on the simulation results it has been shown that the use of pseudo-inverse method for reconfiguration is a viable method for promote the function reliability of the aircraft. For fail mode1, the reconfiguration result is satisfied, although there are static difference in angle, this static difference can be eliminated by control system; but for fail mode 2, there are satisfied result when lock angle within a limitation.

Reference

1. Wang Xingren, Fault Tolerance Control System,", the Third Research Laboratory of BUAA, 1988
2. Li Qing and Guo Suofeng, Analysis and Simulation of Control Mixer Concept for Reconfigurable Aircraft, Transaction of Nanjing University of Aeronautic & Astronautics Vol.12 No.2 1995
3. Pei Ying, Study of Reconfiguration application theory of system", Master paper of BUAA, 1998.3

This paper is funded by National Defence Science & Technology Foundation of China.

PROCESSOR DESIGN AND IMPLEMENATION FOR REAL-TIME TESTING OF EMBEDDED SYSTEMS

G. Walters, E. King, R. Kessinger, CPU Tech, Pleasanton, CA 94588
R. Fryer, ATAC, Ridgecrest, CA 93555

Abstract

As more complex devices with higher levels of integration are inserted into real-time systems, traditional testing methods are becoming obsolete. The most difficult obstacle to thorough testing of real time embedded systems is the lack of visibility into the operations of processing elements while application software is executing. It is now possible to design and implement processors for embedded applications that are binary compatible with commercial instruction sets and have specific features for visibility to facilitate the test, debug, and maintenance of real-time processing systems. These features include Real Time Non-intrusive Instrumentation (RTNI) and Behavioral Verification Technology™ (BVT™) and do not interfere in any way in the operation of the system.

RTNI increases developer productivity by enabling direct observation of processor operation during system development, support and maintenance. BVT is used to automatically test the correct functional behavior of the integrated hardware and software against the system specification. The combination of RTNI and BVT significantly reduces system validation time, risk and cost, while increasing the coverage and assurance level. These features can be implemented in processors that are very high performance, low power, commercial grade, or radiation-hardened. Application of this approach is underway in the development of processors for military and commercial applications.

Background

Historically, DOD-designed embedded processors contained in avionics systems have incorporated 'test hooks' to facilitate the debugging of the system and software. Specifically, the Air Force's MIL-STD-1750A family[1] and the Navy's AN/AYK-14 family[2] are two examples of processors that contain these features. The processors were typically used in the implementation of the most complex and critical functions on the aircraft.

Former Secretary of Defense William Perry's Commercial-Off-The-Shelf (COTS) directive and the demand for increased processor throughput has resulted in the insertion of high performance COTS microprocessors in complex, real-time embedded systems. Since most of the software that executes on COTS microprocessors is not real-time, and high performance real-time applications tend to be low volume, microprocessor suppliers have not included extensive test hooks, such as Real-Time Non-intrusive Instrumentation (RTNI) that support real-time software applications. Some suppliers provide performance monitoring

---

[1] J. Weber and W. Cannon, "Real-Time, Non-Intrusive Software Testing: The Technique for the Future, " *SAE International Congress & Exposition*, Feb 28 1994. pp. 167-177 and JIAWG, 1989. "User Console Interface Specification," Joint Integrated Avionics Working Group Specification J88-N1F, R. Fryer, Ed. (Dec).

[2] Control Data Corporation. 1990. "Feasibility Study Results to Implement a Real Time Non-Intrusive Monitor into the AN/AYK-14(v) VHSIC Processor Module." NWC Contract N60530-89-C-0355. (Jan 23).

0-7803-5086-3 /98/$10.00 ©1998 IEEE

registers[3], but these do not enable real-time operations to be traced. Other proposed approaches required an external embedded test processor[4] that can be difficult to employ if space is tight. None of the approaches solve the problem of how to non-intrusively monitor real-time what is occurring on the microprocessor chip and how to adequately test the system.

Real Time Non-intrusive Instrumentation (RTNI)

Although some benefit can be obtained by testing each software section in non-real-time, bugs often do not appear until the system has been fully integrated and is running at speed. As stated by Da Silva and Sim at GO DSP[5], "Real-time execution causes event timing and the CPU profile to change, since it is affected by real-time interrupts, real-time task-switching, and inter-task dependencies."

As microprocessors become more complex and systems on silicon becomes closer to a reality, real-time visibility into the internal operation decreases and in many cases is completely blocked. Even common features such as on-chip caches and wide buses contribute to the problem. While the microprocessor is running out of the cache, there are no bus cycles at the system bus, so a traditional bus-based debugging system is completely blind to what the microprocessor is doing. With the wide buses, the microprocessor fetches multiple instructions or operands when it does a memory access. Bus-based debugging systems are unable to locate instruction or operand boundaries within the block of memory accessed. Clearly, bus-based debugging systems are incapable of adequately controlling and debugging modern microprocessors. This forces the firmware /software developer to intrusively trap real time code.

To solve these problems, CPU Tech has incorporated RTNI into its microprocessor designs. This work is based on previous work sponsored by the Navy[6] and the Air Force[7]. RTNI moves the debugging system onto the microprocessor where it has complete visibility into what the microprocessor is doing every cycle. In order to be effective, as noted by Fryer, Naval Air Warfare Center Weapons Division, "The instrumentation mechanisms must be transparent to the behavior of the software. Transparency has typically been deemed adequate if the flow of addresses in program execution is identical and if the time relationships of all software detectable events are equivalent (i.e. interrupts, sequence of procedures run, time spent in a process, etc.)."[8] The effect, as depicted in Figure 1, is as if all of the key points were accessible through a giant multiplexer.

[3] *MPC750 RISC Microprocessor User's Manual*, Motorola document MPC750UM/AD, August 1997, Chapter 11, pp. 11-1 through 11-12.

[4] K. Petersén, B. Magnhagen and H. Lindén, "Embedded Test Processors (RTP)," *Baltic Electronic Conference*, 1996.

[5] G. Da Silva, E. Sim "DSP Code Development Tools: Features for Modern Application Needs," white paper.

[6] R.E. Fryer, G. Vansteenkiste, and J. Van Campenhout. "Real Time Non-Intrusive Monitoring and Embedded System Simulation," in *Special Proceedings on Embedded Systems; of the Society for Computer Simulation Summer Simulation Conference, (July 1994)* and R.E. Fryer, 1973. "The memory bus monitor," In *Proceedings of the AFIPS National Computer Conference*, V42. (New York, N.Y.) 42, 75~79.

[7] W.J. Cannon, M. T. Michael, and D. D. Beeson. 1992. "Real Time, Non-Intrusive Instrumentation of Reduced Instruction Set Computer (RISC) Microprocessors," in *Proceedings of the National Aerospace and Electronics Conference (NAECON)*, (Dayton, OH). IEEE (Piscataway, N.J.), 550~557.

[8] R.E. Fryer, G. Vansteenkiste, and J. Van Campenhout. "The 'Instrumentation Memory', A Measuring Feature for Simulation of Real Time Embedded Software," in *Proceedings of the 1995 Society for Computer Simulation European Multiconference*, (Prague, CR), July 1994.

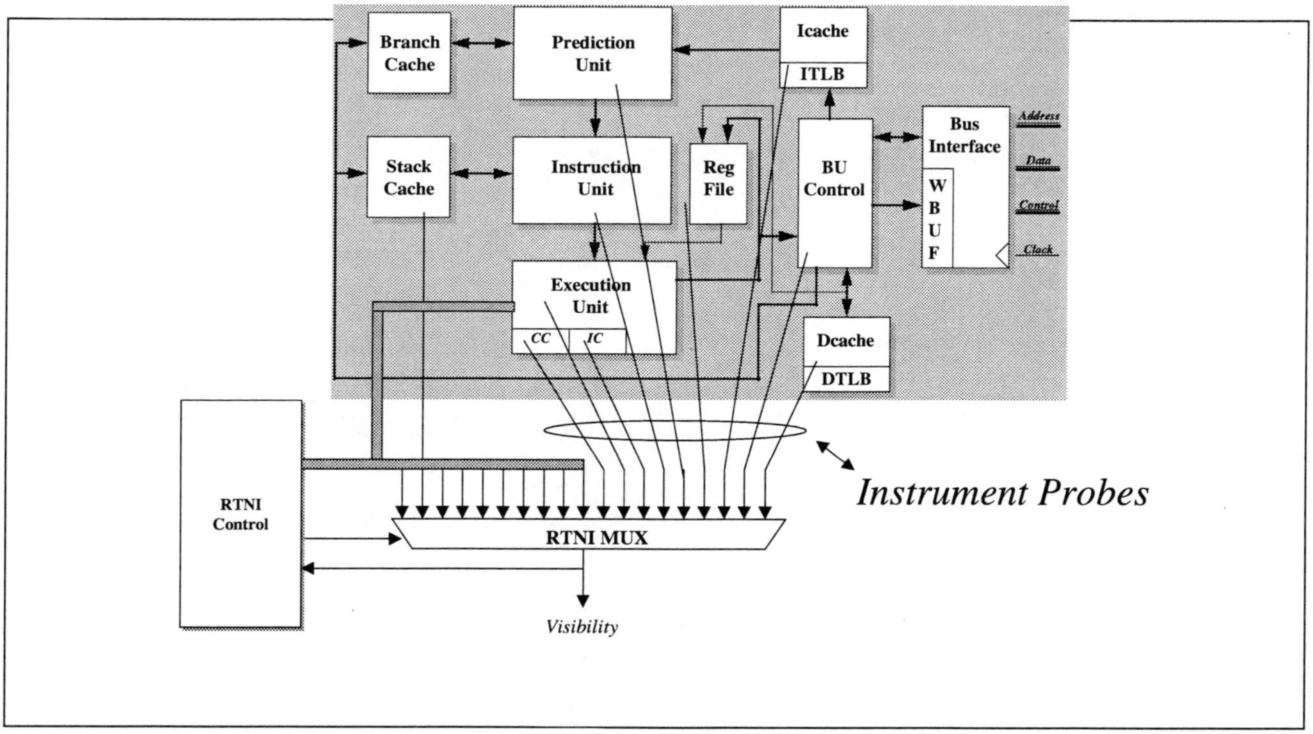

Figure 1. Representation of RTNI

External RTNI pins provides a means to transmit RTNI data off the microprocessor. In essence, parts of the traditional external debug monitor, those that control the chip's behavior, have been incorporated onto the microprocessor. To minimize the impact on the number of pins, the RTNI shares the same pins as the Built-In Self Test (BIST) and JTAG[9]. Since these functions are mutually exclusive, no conflicts arise from this arrangement.

RTNI supports the following debug functions:

- Trace Stop/Start on Event
- Single Step
- Software/Hardware Breakpoints
- Inspect/Change Memory
- Up to (16) Hardware Traps
- Timers/Event Counters
- Event Timing
- Inspect/Change Registers
- Inspect/Change I/O Ports
- Reset/Configure Performance Monitor
- Enter/Exit Console Mode
- Reset RTNI
- Run/Stop

As shown in Figure 2, the System Debug Port (SDP), external to the microprocessor, is comprised of the RTNI interface, and an Ethernet controller. The RTNI interface controls storage of RTNI information into the trace buffer and communicates RTNI control and data information with the host debugging station. Placing RTNI on chip and SDP on the microprocessor board means there are no other boards that need to be plugged into the system in order to debug it.

[9] IEEE Std. 1149.1-1990; IEEE Standard Test Access Port and Boundary-Scan Architecture.

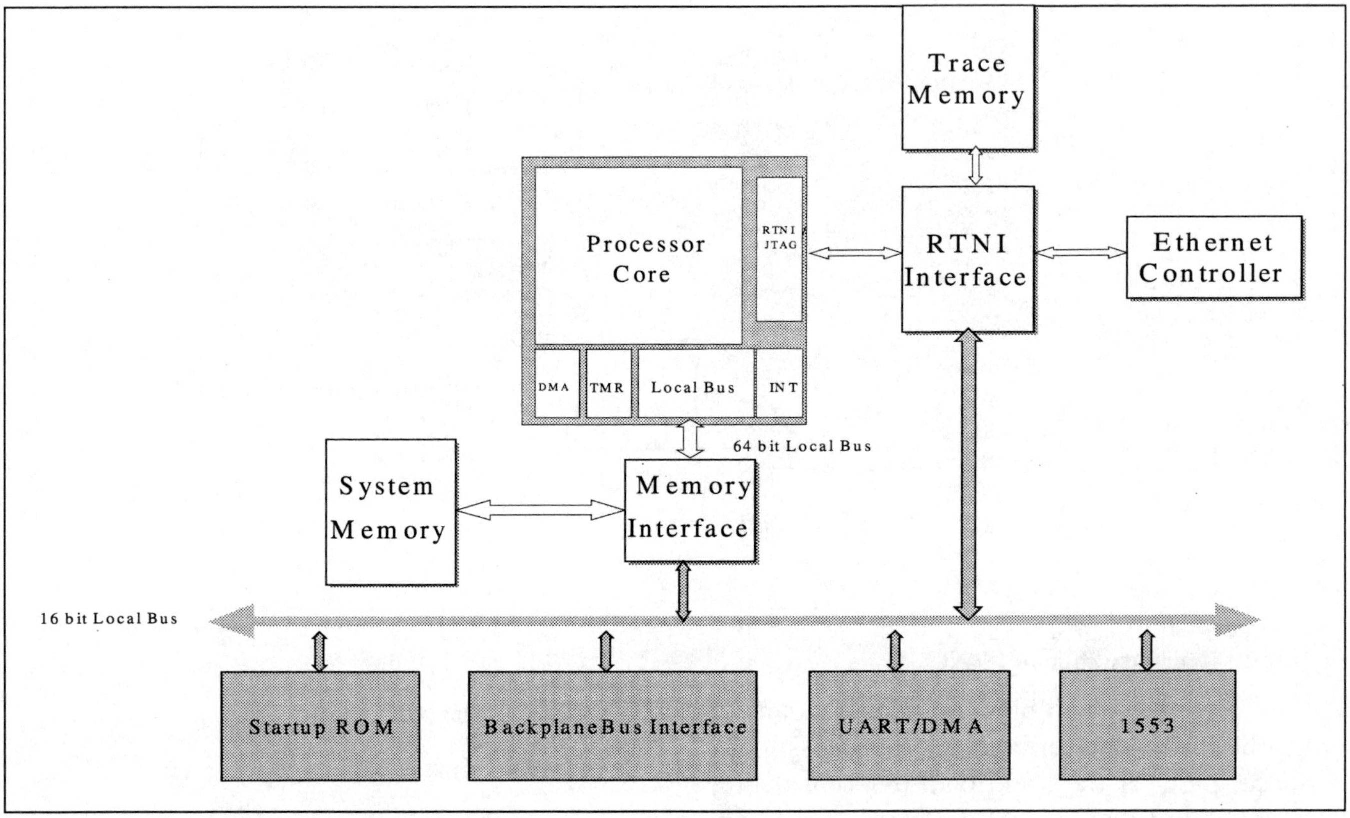

Figure 2. System Block Diagram

The RTNI interface chip is comprised of an ASIC/FPGA. The interface itself is made up of an 8-bit command bus, 32-bit data bus, and 14-bit status bus. The nineteen commands support the functions listed above. The data bus is used to:

- Specify the register number for read/write register commands
- Specify the address for read/write memory/IO commands
- Specify the data value for write commands
- Provide the data value for read commands
- Provide the trace data when tracing is enabled

The status bus is used to indicate:

- RTNI can accept a new command
- Current RTNI command complete
- Current state of the microprocessor
- Valid trace data
- First word of multi-word trace packet
- Hit/Miss status for the internal breakpoint registers

The breakpoint registers are loaded using the 'Write Register' RTNI command. Individual breakpoints are enabled or disabled through a mask register. Once a breakpoint is encountered, it can either stop the processor execution, or start/stop tracing. The amount of trace data is limited solely by the amount of

Trace Memory contained on the board. The chaining of the breakpoints is accomplished through the RTNI interface logic.

As implemented by CPU Tech, RTNI readily supports industry standard symbolic debuggers through an Application Programming Interface (API). It can also interface to simulators and other tools.

RTNI has been implemented as a general-purpose diagnostic port architecture. The architecture is open, scalable and consistent with commercial development tools, resulting in a low power, highly integrated solution for high-end embedded applications. It is incremental and compatible to the extensive R&D efforts in the industry for next generation high-end embedded systems.

Behavioral Verification Technology™ (BVT™)

While a key element, especially for software debug, RTNI is only part of the solution for testing embedded systems. RTNI provides the means to test, but BVT provides the actual diagnostics tests.

Embedded microprocessors are growing exponentially more complex with each new generation. Traditional methods of verification, which rely on manual, ad-hoc testing using existing applications software, are inadequate to guarantee that all of the functions of a microprocessor and its embedded systems are actually working properly.[10] Existing operating systems and application software do not utilize all of the functionality of these microprocessors. Worse yet, with ad-hoc methods it is unknown which functions are being tested and which are not. Future software is also at risk. Embedded systems may contain design flaws which are not exposed by today's software, but will cause catastrophic errors in future software programs.

Testing systems using exhaustive pseudo-random testing is not feasible. To perform and validate all combinations of possible instructions, modes, addresses and data patterns would take longer than the lifetime of the sun for even a simple microprocessor. The ability to properly validate all of the system's functions is the major problem faced by embedded system developers.

BVT approaches verification through specification-based behavioral diagnostic tests. The structured diagnostics are architected to systematically test the functions of the microprocessor and the system.

BVT embodies the following characteristics:

- It is automated, checking against expected results thereby eliminating the uncertainty of ad-hoc, random testing.
- It is based on a measurable and enforceable specification. (Claiming compatibility to an operating environment is not measurable.)
- It provides comprehensive specification coverage. It tests the defined capabilities of the processor or system, not just the ones in common usage.
- It is applicable throughout the development process as well as on the finished product.
- It integrates new tests as the specification evolves or as the developer requires.
- It is an efficient diagnostic and reporting tool, which directly identifies any problems that it detects so one knows exactly what does and does not work.

[10] M. Scheitrum, A. Smith, "Behavioral Validation and Its Application to Pentium Class Processors," *PCI Developers' Conference*, 1995.

- It provides repeatable and predictable test results.
- It is easy to use.

These features enable processor manufacturers to quickly identify design flaws and validate revisions to the design.

BVT consists of an extensive set of software programs. Each program is comprised of a set of self-diagnosing tests. The expected results of each test are embodied in the test program itself. These pre-engineered expected results are created in a semi-automated method which utilizes custom development tools. As the software program performs diagnostic tests on the unit under test, any discrepancy between actual and expected results are immediately reported. All specified behaviors of each function are thoroughly tested. Each test case is composed of code that sets up the initial state of the units, performs the function, and then compares the actual results against the expected results. Each test checks that proper results are stored in registers, condition codes, memory, etc. In addition, tests are performed to detect any undesired side-effects of the function.

The validation suite is organized in a fashion such that the simplest functions are tested first. The testing of these functions does not require more complicated functions to be operational. Once a function has been verified, it can then be used to verify more complex functions. This "bootstrapping" structure allows the designer to implement and debug the design in an incremental fashion, rather than requiring him or her to implement complex functions prior to testing simple operations. While BVT exercises a large number of different test cases, it is structured such that any specific test case can be isolated and exercised individually. The modularity enhances the productivity of the development team.

The basic organization of BVT is comprised of three categories: basic function tests, corner case tests, and sequence tests. An exceptional validation suite must provide comprehensive tests in each of these areas.

Basic Function Tests

Basic function tests provide coverage of every operation in all available modes. Basic tests focus on each operation or function individually. These tests verify that the fundamentals of an operation are working properly --- the proper result is calculated, registers and memory are updated correctly, and condition codes are set properly. Basic function tests also verify that unwanted side-effects do not exist.

Corner Case Tests

Corner case testing involves validating the boundary conditions and exception cases of a function. Many corner cases can be tested in an individual fashion, and this is done throughout the validation suite. Boundary conditions may involve certain operations that cause, for example, underflow or overflow in floating point operations, or that cause address calculations to cross memory boundaries. Exception cases are also tested in this section. Of particular concern is the proper prioritization of multiple exceptions that can occur in a complex system.

Sequence Tests

Individual testing of functions ensures that each function operates properly in a stand-alone manner. Yet even if all functions work correctly individually, errors could occur when multiple functions are executed sequentially or concurrently. This is due to dependencies between instructions for register values, memory contents, and flag settings. Additional tests are necessary in the suite to verify that sequences of multiple instructions interact properly.

Embedded systems have numerous functions, from simple to extremely complex, and all functions can be validated in the manner described above. The process is applicable to all functional components, including embedded controllers. BVT can be applied during the design and after the system is built, including maintaining the system after it is fielded.

Applications

The concepts embodied in RTNI and BVT are in use today. RTNI has been implemented in a MIL-STD-1750A microprocessor, called the CPU-1750A, that provides over 12 DAIS MIPS sustained at 60 MHz. The CPU-1750A is currently being inserted in an Air Force avionics system. The RTNI is enabling the replacement of the legacy software development environment with a modern environment that has more capabilities. The effort will be completed before the end of 1999.

RTNI is also being implemented in an x86-compatible microprocessor, called the MS1, being developed in conjunction with the Navy.[11] This effort will also be completed before the end of 1999. The RTNI interface for the MS1 will be the same as for the CPU-1750A.

BVT has been applied to the testing of microprocessors and systems since 1992. It has demonstrated that it provides more test coverage than other approaches such as vector, application and/or random testing. BVT has been licensed to major microprocessor companies.

Conclusions

Processors are the most complex building blocks ever made by man. Transistor counts in processors are doubling every two to three years. Because of the fact that functional complexity can double with only an incremental increase in transistor count, the functional complexity as well as probability of design errors is growing at an exponential rate. Traditional methods of testing cannot adequately verify the correctness of these devices. While COTS microprocessors contain some additional features to support software debug, they typically are either not documented or not supported.[12]

RTNI is now available that can keep up with the single cycle execution of modern COTS processors. It can be standardized across multiple processor types, including Digital Signal Processors. The ability to visualize real-time data allows developers to isolate where and why a particular real-time bug occurs. Otherwise, it can be a frustrating and time-consuming task to debug real-time software. Bus-based debugging systems are incapable of adequately controlling and debugging modern microprocessors. RTNI re-establishes this debugging capability.

BVT, which is based on the systematic verification of the specification, is significantly more effective in finding, identifying, correcting, and re-testing bugs than verification approaches based on using only system level applications. It can be created to validate specifications for both software and hardware. BVT provides a thorough, effective solution to the problem of validating the functionality of increasingly complex processors and embedded systems.

---

[11] E. King, G. Walters, "Dual-Mode Processor Technology for Legacy System Renewal," in *Digest of Papers from the Government Microcircuit Applications Conference (GOMAC)*, (Arlington, VA), Mar 16-19, 1998, pp. 324-327.

[12] R. Fryer, "Low and Non-Intrusive Software Instrumentation, A Survey of Requirements and Methods," *IEEE 17th Digital Avionics Systems Conference*, Oct 31-Nov 6, 1998 (these proceedings).

References

[1] Weber, J. and Cannon, W., "Real-Time, Non-Intrusive Software Testing: The Technique for the Future," *SAE International Congress & Exposition*, Feb 28 1994. pp. 167-177.

[2] JIAWG, 1989. "User Console Interface Specification," Joint Integrated Avionics Working Group Specification J88-N1F, R. Fryer, Ed. (Dec).

[3] Control Data Corporation. 1990. "Feasibility Study Results to Implement a Real Time Non-Intrusive Monitor into the AN/AYK-14(v) VHSIC Processor Module." NWC Contract N60530-89-C-0355. (Jan 23).

[4] *MPC750 RISC Microprocessor User's Manual*, Motorola document MPC750UM/AD, August 1997, Chapter 11, pp. 11-1 through 11-12.

[5] Petersén, K., B. Magnhagen, H. Lindén, "Embedded Test Processors (RTP)," *Baltic Electronic Conference*, 1996.

[6] Da Silva, G., E. Sim "DSP Code Development Tools: Features for Modern Application Needs," white paper.

[7] Fryer, R. E., G. Vansteenkiste, and J. Van Campenhout. "Real Time Non-Intrusive Monitoring and Embedded System Simulation," in *Special Proceedings on Embedded Systems; of the Society for Computer Simulation Summer Simulation Conference, (July 1994).*

[8] Fryer, R. E., 1973. "The memory bus monitor," In *Proceedings of the AFIPS National Computer Conference*, V42. (New York, N.Y.) 42, 75~79.

[9] Cannon, W.J., M. T. Michael, and D. D. Beeson. 1992. "Real Time, Non-Intrusive Instrumentation of Reduced Instruction Set Computer (RISC) Microprocessors." In *Proceedings of the National Aerospace and Electronics Conference (NAECON),* (Dayton, OH). IEEE Piscataway, N.J., 550~557.

[10] Fryer, R. E., G. Vansteenkiste, and J. Van Campenhout. "The 'Instrumentation Memory', A Measuring Feature for Simulation of Real Time Embedded Software," in *Proceedings of the 1995 Society for Computer Simulation European Multiconference*, Prague, CR, July 1994.

[11] IEEE Std. 1149.1-1990; IEEE Standard Test Access Port and Boundary-Scan Architecture.

[9] Hetzel, W., ed. *Program Test Methods*. Englewood Cliffs, J.JK.: Prentice-Hall, 1973.

[10] Demillo, R.A. et. al. *Software Testing and Evaluation*. Benjamin Cummings, Menlo Park, CA 1987 pp20-21,

[11] Gupta, N. K., R. E. Seviora, "An Expert System Approach to Real Time System Debugging," *IEEE Conference on Artificial Intelligence Applications*, December 5-7 1984.

[12] Helmbold, D. and D. Luckham. 1985. "Debugging Ada Tasking Programs." *IEEE Software*, 2 no. 2 (Mar): 47~57.

[37] Weiss, R. 1991. "Debugging Support Goes On-Chip.", *Electronic Engineering Times* (May 20): 49~53.

[38] "Debug tools – not an option," editorial in *Electronic Engineering Times,* April 16, 1990.

[40] Walters, G., E. King, R. Kessinger, "Processor Design and Implementation for Real-time Testing of Embedded Systems," *IEEE 17th Digital Avionics Systems Conference*, Oct 31-Nov 6, 1998 (these proceedings).

MEETING MILITARY SYSTEM TEST REQUIREMENTS WITH THE USAGE OF COTS PRODUCTS

Stephen V. Pizzica, Raytheon Systems Company, El Segundo, CA

Abstract

The Perry Initiative in 1994 provided a top-down directive for the defense industry to use the benefits of Commercial Off The Shelf (COTS) technology to effect major cost savings in military programs.

While such military product cost reduction advantages have become well publicized, less well known is the concomitant problem of meeting military system test requirements with equipment that has been designed for commercial applications.

This paper describes and addresses the test problems caused by the insertion of COTS products into military systems. A systematic approach is presented that provides a general methodology for meeting military system test requirements with COTS product usage. This process maximizes reuse of existing COTS supplier tests, and gives guidelines for determining and closing performance gaps regarding military requirements .

Some specific techniques for assessing the quality of existing COTS product tests are described. Also discussed is how to fulfill the need for a library of test improvement techniques that can be applied to closing COTS product test performance gaps to meet military test requirements. Finally, military field supportability implications of COTS product usage are considered.

Introduction

In 1994, the Perry Initiative ushered in the era of Acquisition Reform in the defense industry. This was based on the potential of deriving the twin benefits of utilizing commercial sector technological advances without paying for development costs. In many instances, commercial technology with its shorter product life cycles has been outpacing the rate of defense advances.

The perspective of this paper is that of a defense contractor required to provide an electronics system or subsystem to a military customer. In the new COTS environment, the contractor's development process is evolving from being totally an in-house process to becoming a system integrator of COTS products. Figure 1 contrasts the old and new development processes. The scope of this paper will be to focus on the test aspects of COTS product usage in military systems.

Impact of COTS on Military Test

The non-developmental advantage of using COTS now prevents the test function from specifying that military test capabilities be incorporated into designs, as was done in the past with in-house designs. This problem presents a new challenge to meeting military test requirements.

0-7803-5086-3 /98/$10.00 ©1998 IEEE

Benefits of COTS in Military Systems

Before covering COTS test issues, the overall motivation for military exploitation of COTS products will be reviewed. The economic and technological benefits of COTS utilization are summarized as follows.

- Development cost savings reduce non-recurring costs
- Commercial economies of scale reduce system recurring costs
- Off-the-shelf product usage reduces system development cycle time
- Proven common industry standards allow open system architecture to be used
- Commercial sector technological advances can be directly incorporated
- Debugged products reduce developmental risks

COTS Test Problems in Military Systems

The pre-COTS in-house military product development process had two major advantages regarding test implementation.

1. Having complete control over the design allowed insertion of military Design-For-Testability features.
2. There was total access to all design information.

Conversely, in the COTS environment, lack of these advantages present significant test problems, as described:

1. COTS equipment is not designed per military testability/supportability requirements, including fault detection and isolation as specified in:
   - MIL-STD-2165 Testability Program for Electronic Systems and Equipments
   - MIL-STD-2084 General Requirements for Maintainability of Avionic and Electronic Systems and Equipment
2. There is a lack of rights/access to proprietary COTS design data.
   - Not paying for development costs implies no rights to proprietary design information.
   - Can't make detailed circuit model for test simulations/fault grade.
3. There are problems in using existing COTS supplier tests.
   - Test thoroughness is often unknown, even by supplier.
   - Available COTS supplier tests must be evaluated
   - Tests must be improved/added to close any test performance gaps

General COTS Test Methodology

The position is taken that the optimization of a COTS military test implementation strategy must be driven by two overriding factors. The first is that military test requirements must be met. The second is that the cost savings of the COTS reuse philosophy must be extended into maximum reuse of existing COTS supplier test capabilities.

Based on this supposition, the following general COTS test implementation methodology is presented below.

1. Define all military project test requirements.

2. Evaluate all available COTS supplier tests and test-related information.

3 Determine all test performance deficiencies vs. military requirements

4. List all test upgrades that are needed to meet project test requirements.

5. Assemble a list of test improvement techniques that can be applied to COTS products.

6. Correlate needed test upgrades with the best applicable test method/tools.

7. Implement these optimum test solutions.

Guidelines in Reuse of COTS Supplier Tests

There are two main phases in optimally applying existing available supplier tests to COTS products used in military systems.

The first is assessing the quality of available COTS supplier tests.

The second is determining the performance gap between the existing test capabilities and the military system test requirements.

Assessing Quality of Available COTS Supplier Tests

In order to extend the cost saving benefits of off-the-shelf product usage to testing, available COTS supplier test capabilities should be applied towards meeting project test requirements.

The first step is to ascertain from the supplier all available tests that can be used. These should be divided into two broad categories. The first category is the built-in-self-tests that typically reside on COTS modules or ASICS. The second is off-line diagnostic tests.

Built-in-self-tests can be directly reused at the desired vertical test level (i.e. system, unit or module). Selected off-line diagnostics can be converted and stored in on-board firmware as additions to the built-in-tests (BIT). They can also be reused on an off-line basis for production unit level testing.

All supplier test methods should be determined for each type of circuitry to be tested. This analysis should include a determination of the design-for-test (DFT) techniques used. The most common methods to investigate are:

- JTAG IEEE 1149.1 boundary scan methods
- CPU test methods
- Memory test methods
- How ASIC's are tested
- I/O and bus test methods
- Any analog test methods
- Any timing tests
- Functional vs. structural test methods

Supplier test quality information should be obtained to the extent available. The most important test quality metric is the fault grade (percent fault detection), along with the types of faults graded, such as:

- Stuck at one
- Stuck at zero
- Open pins
- Adjacent pin shorts
- Memory cell faults
- Timing tests

Any fault isolation capabilities, such as boundary scan on module I/O, should also be ascertained.

If available, the best measure of fault grade is that provided by a computer fault simulation. However in practice, COTS suppliers typically do not do costly test fault simulations or use expensive module level ATE (Automatic Test Equipment), as has been the practice in military programs. COTS modules are not normally designed with the test access required for military module level ATE.

0-7803-5086-3 /98/$10.00 ©1998 IEEE

COTS module suppliers generally make use of BIT for module go/no-go testing. Rather than do computer fault simulations, they tend to verify their BIT thoroughness by analysis and/or by some type of physical fault insertion. If deficiencies are found, additional iterative BIT tests are generated as necessary.

Determine Test Performance Gap

Any test performance gaps between existing COTS test capabilities and project test requirements must be analyzed and quantified before test upgrading methods can be prescribed. The process below is offered as a guideline for determining such performance gaps.

1. Define all contractual project test requirements specified by the customer.

2. Define any manufacturing, system integration, and customer test requirements.

3. Define all applicable and available off-the-shelf supplier tests and test related data.
   - Determine test quality
   - Fault detection percentage for what fault types
   - Any relevant isolation capability
   - Determine credibility of test quality claims
   - How quantified (fault simulation, rigorous analysis, manual fault insertion, engineering judgment)

4. Determine/list all test deficiencies vs. project test needs.
   - Fault detection
   - Fault isolation
   - Any ATE compatibility requirements
   - Any environmental testing requirements

5. Decide if any exceptions can/should be taken to project test requirements. Under Acquisition Reform, it is often permissible to trade off a large cost savings for a small performance decrement.

6. At this point a test upgrade plan needs to be generated with the objective of finding/matching corrective test methods with specific COTS test performance deficiencies. The next section provides some examples of test methods applicable to COTS test improvement.

COTS Test Improvement Methods

Some specific methods are given below of ways to improve existing COTS tests. Within the scope of this paper it is not possible to provide an exhaustive list of techniques that covers all test situations. The purpose of this section is to provide examples of the types of techniques that a military system integrator of COTS equipment needs to assemble in order to create a test improvement library. Different integrator companies/organizations would obviously have different sets of test improvement methods dependent on their test requirements and types of equipment to be tested.

Use of JTAG IEEE 1149.1 Boundary Scan

A major common denominator across both military and commercial test applications is the IEEE 1149.1 standard.

COTS modules using components with 1149.1 access can have test vectors inserted serially to augment the test of these components. This can be done for either built-in-tests, or as off-line tests used in acceptance testing of units comprised of COTS modules.

Of special interest is the usage of boundary scan tests to detect shorts and opens at the I/O pins of 1149.1 components. This is important since military fault detection requirements typically specify that the percentage detection of node shorts/opens be in the high 90's. Boundary scan test algorithms exist which provide analytical verification of 1149.1 component I/O node fault coverage.

The tri-stating feature of boundary scan can be exploited to provide better fault isolation between modules that have 1149.1 at their I/O.

Augmentation of Built-In-Tests

An inventory should be made of any unused firmware memory space that can be used to store additional built-in-tests. These tests can be targeted to provide such test improvements as:

a. More thorough functional tests of complex components
b. Tests of memory cells
c. Tests of component I/O shorts and opens
d. Addition of applicable available COTS supplier off-line tests, or subsets thereof
e. Storage of signatures, requiring very little memory, for checksum type tests

Empirical Quantification of COTS Test Thoroughness

Where the thoroughness of existing COTS built-in-tests is unknown, the percent detection coverage can be determined by physical insertion of faults at each COTS module node. Where it is impractical to do this for every node, a random sample of nodes can be used to provide the percent fault coverage within a statistically determined confidence level.

Information on areas of the module lacking in test coverage can also be extracted. Any such areas with test voids can thus be more efficiently targeted with additional tests.

Use of COTS-Based Test Equipment

Extensive STE (Special Test Equipment) is needed for acceptance test of military systems. This STE can also be comprised of COTS equipment. There is an existing industry standard, called VXI, where a wide variety of test interface cards are commercially available. COTS cost saving benefits can thus also be applied to test equipment.

Reusable Built-In-Tests for Common Test Functions

Significant test software savings can be made by maximizing the reusability of common built-in-tests. This takes advantage of the fact that many test functions are generic in nature and can thus be reused across projects.

In principle, test routines could be written in a common high level language, such as C, and be recompiled and reused, with minor custom tailoring, in modules/units using a CPU with a C compiler. Examples of these generic test functions fall into two categories. The first is test overhead functions such as: BIT initiation, test comparisons, results reporting, and fault logging. The other is testing components/circuitry with applicable common algorithms such as RAM march tests.

Fault Filtering

It is becoming more common for military customers to require suppression of false alarm failure indications, since this causes expensive unneeded repair actions. These can be the result of transients or other soft failures. These faults can be filtered and distinguished from hard failures by repeating the built-in-test and rejecting single fault occurrences.

Creating a Library of COTS Test Data Bases

It is useful for a company integrating COTS-based military systems to create its own library of COTS test information. Especially in the large corporations resulting from defense consolidations, this would greatly reduce duplications of research and test methods development. A suggested list of test data bases is presented as follows:

1. COTS supplier product data
2. Test literature including test textbooks and periodicals
3. Proceedings from conferences covering test, such as ITC, DAC
4. Project lessons learned
5. List of subject matter experts
6. Data base of reusable test methods
7. Data base of any proprietary corporate test innovations
8. Records of any corporate test user groups

Supportability Implications

In military programs, the test and supportability functions are closely related. In order to ensure meeting customer field support needs in the COTS environment, it is important that these two functions work closely together. The projected impact of COTS test on supportability is outlined as follows.

1. Customer needs to stock spare modules instead of having depot test/repair, since COTS modules are not designed for military ATE

2. Need service agreements with COTS suppliers, including need to select optimum service plan using logistics/life cycle cost analyses to evaluate alternatives

3. Need logistics plan for spare module replacement that is coordinated with the customer and COTS supplier

4. Low cost modules need repair vs. discard trade-offs

5. Need to get accurate MTBF data from supplier and/or do own MTBF analysis, including environmental factors such as temperature/vibration

6. There is a new design-for-maintainability paradigm: Must design for replacement due to frequent COTS upgrades

7. Must have plans for COTS obsolescence problem, such as:
   - Plan for life-time quantity buys
   - Obtain contingency data rights and rights to produce
   - Plan for replacement with upgraded COTS product

8. Need to adjust military equipment LCC (Life Cycle Cost) models and LSA (Logistics Support Analyses) to cover fielded COTS equipment

Conclusions

It is critical that military system test requirements be met while deriving the advantages of implementing COTS products into military systems. Test considerations should be an important attribute in evaluating COTS product alternatives as applied to military system usage.

It has been demonstrated that a methodology can be applied to evaluating and closing COTS test performance gaps regarding military test requirements.

The bottom line is that reuse of existing COTS test capabilities should be considered to be part of the COTS development cost savings. Meeting military test requirements thus becomes an add-on rather than start-over expense.

Resources

Annual Conference Proceedings

1. International Test Conference (ITC)
2. Design Automation Conference (DAC)
3. AutoTestCon

Useful Websites

1. Defense Acquisition Reform Homepage www. acq.osd.mil/ar
2. Acquisition Reform Network www.arnet.gov
3. Aegis Communicator of Acquisition Reform aegis.dynsys.com

An Overview of the EMC Requirements in RTCA/DO-160D

Erik J. Borgstrom
TÜV Product Service, Inc.
1775 Old Hwy. 8 NW, Suite 104
New Brighton, MN 55112

Abstract: RTCA/DO-160D, *Environmental Conditions and Test Procedures for Airborne Equipment*, defines a series of minimum standard environmental test conditions (categories) and applicable test procedures for airborne equipment [1]. The purpose of these tests is to provide a laboratory means of determining the performance characteristics of airborne equipment in environmental conditions representative of those which may be encountered in airborne operation of the equipment.

This paper will provide an overview of the Electromagnetic Compatibility requirements in RTCA/DO-160D. Future directions and changes to DO-160D will also be discussed from the author's perspective as a member of RTCA/Special Committee 135, which is responsible for producing DO-160D.

INTRODUCTION

RTCA, Inc. is a private, not-for-profit organization that addresses requirements and technical concepts for aviation. RTCA's main activity is to develop and publish recommended standards and guidance documents that focus on the application of electronics technology to implement new or modified concepts and to satisfy related requirements.

The main focus for EMC within RTCA is in Special Committee 135 (SC-135), which deals with Environmental Conditions and Test Procedures. In May of 1997, SC-135 approved the Final Draft of DO-160D, and in July of 1997, RTCA/DO-160D was approved for publication. DO-160D supersedes the previous version, DO-160C (published in 1989) and it's associated Changes 1, 2, and 3 [2].

The work of SC-135 was coordinated with the European Organisation for Civil Aviation Electronics (EUROCAE) Working Groups 14, 31, and 33. EUROCAE concurs with RTCA on the environmental conditions and test procedures set forth in DO-160D and, when approved by EUROCAE, the document will be identified jointly as RTCA DO-160D/EUROCAE ED-14D.

DO-160D has 25 Sections and 3 Appendices. Sections 1 - 3 cover *Purpose and Applicability*, *Definitions of Terms*, and *Conditions of Tests*. Sections 4 -14 and Section 24 cover environmental conditions such as *Temperature and Altitude*, *Vibration*, *Sand and Dust*, and *Icing*. Sections 15 - 23 and Section 25 cover EMC, and it is these sections that will be discussed in this paper, with particular emphasis placed on the differences between DO-160D and the previous version, DO-160C.

SECTION 15: MAGNETIC EFFECT

This test is performed to determine how much the Equipment Under Test (EUT) will deflect a compass needle.

A standard compass that has a large enough dial to read one degree of needle deflection is the only test equipment required. The EUT is simply moved closer to the compass on an East-West line until one degree of deflection away from magnetic North is observed. The separation distance is then measured and the "Equipment Class" is determined.

Equipment Classes

There are four Equipment Classes (A, B, C, and Z) that apply to separation distances of 30 to 300 centimeters.

Changes from the Last Revision

None.

SECTION 16: POWER INPUT

The tests in this section are performed to determine that the EUT can operate as required during all of the different conditions of AC and DC supply variations that occur during aircraft operation. These tests are comparable with the requirements in Mil-Std-704D and cover conditions that range from normal to emergency [3].

Normal conditions include the typical and expected fluctuation of voltage, frequency, ripple, etc., that would be found during normal operation of the aircraft and it's AC and DC power sources.

Emergency conditions occur when the primary power systems are unable to provide sufficient or proper power, and operation on emergency backup power is required. Emergency power usually has limited capability, resulting in significantly lower quality (voltage, ripple, frequency, etc.) AC and DC power being provided to the equipment.

DC Input tests

On DC inputs, there are tests that cover:

0-7803-5086-3 /98/$10.00 ©1998 IEEE

- Steady-state over- and under-voltage conditions
- Ripple voltage
- Momentary power interruption
- Momentary sags and surges

AC Input tests

AC inputs are subjected to the following tests:

- Steady-state over- and under-voltage conditions
- Steady-state over- and under-frequency conditions
- Steady-state phase unbalance (three-phase power)
- Voltage and frequency modulation
- Momentary power interruption
- Momentary sags and surges

Equipment Categories

There are four Equipment Categories (A, B, C, and Z) that indicate the type of power used by the equipment and the type of AC and/or DC power source that the equipment is compatible with.

Changes from the Last Revision (DO-160C)

A very few changes in test levels make some of the categories a bit more severe than the previous version of DO-160. For example, in the Abnormal Surge Voltage (DC) test, the surge voltage requirement for category B is 60 VDC for 100 milliseconds and 40 VDC for one second. In DO-160C, the requirement was 46.3 VDC for 100 milliseconds and 37.8 VDC for one second.

Future Changes and Directions

There have been discussions about adding a requirement to measure and set limits for Total Harmonic Distortion (THD) or Harmonic Currents. This has been brought up by some aircraft manufacturers who say that the proliferation of new equipment attached to the AC power bus (e.g. seat-back video systems using non-linear supplies) is causing harmonic distortion in the AC power distribution system. They want a THD and/or harmonic current limit for aircraft electronics similar to those used in Europe (EN61000-3-2) [4].

SECTION 17: VOLTAGE SPIKE

This test determines whether the EUT can operate as required during and/or after voltage spikes are applied to the AC and/or DC power input(s). Any method of generating the spike may be used, provided that the pulse produced has a duration of at least 10 microseconds, a risetime of less than 2 microseconds, and a source impedance of 50 ohms. A minimum of 50 voltage spikes are applied within 1 minute. This test is very similar to Mil-Std-462 test method CS06, except for the source impedance of the generator, which is 5 ohms for CS06 [5].

Equipment Categories

There are two Equipment Categories. The Category B test level is twice the AC (rms) and/or DC line voltage (or 200 volts, whichever is less). The Category A test level is 600 volts.

Changes from the Last Revision

DO-160C had two different voltage spike waveforms, where DO-160D has only one. There are no changes in the categories and test levels, but the section no longer includes diagrams for building the generators to produce the pulses. Overall, the section has been greatly simplified and is much easier to use.

SECTION 18: AUDIO FREQUENCY CONDUCTED SUSCEPTIBILITY - POWER INPUTS

This test is performed to determine that the EUT will operate as specified when audio frequency interference is applied to the AC and/or DC power input. The test setup and procedure are nearly identical to Mil-Std 462D test method CS101, with the only difference being the actual test level and frequency range [6]. The audio frequency interference is transformer coupled onto the power input lead, and the rms voltage level of the interference signal is measured across the power input and return leads. Test levels are up to 5% of the nominal input voltage, and the frequency range is 10 Hz to 150 kHz.

Equipment Categories

There are four Equipment Categories (A, B, E, and Z) that indicate the type of power used by the equipment and the type of AC and/or DC power source that the equipment is compatible with.

Changes from the Last Revision

Overall, the section has been simplified and is easier to use, but there are no changes in the requirements.

Future Changes and Directions

As mentioned above, harmonic distortion of the AC power available on the aircraft has become a concern of the aircraft mnufacturers. As a result, there has been discussion about increasing the test level on the AC power input to 8% of the nominal AC input voltage.

SECTION 19: INDUCED SIGNAL SUSCEPTIBILITY

The tests in this section are performed to determine that the EUT can operate as required when the equipment and interconnecting cables are subjected to audio frequency electric fields, magnetic fields and transient voltage spikes. These tests are similar to the "Chattering Relay" and power frequency magnetic fields test in Mil-Std-462 test method RS02, except that RS02 has the radiating wire wrapped around the EUT, and Section 19 uses a straight wire radiator for magnetic fields induced into equipment.

Unless otherwise specified, interconnecting cables shall be at least 3.3 meters in length for these tests.

The test levels for the interconnecting cable tests are determined by the length of wire that is exposed to the radiating wire. For the induced spikes test, the exposed length is either 1.2 or 3.0 meters, with the amplitude of the Inductive

Switching Transients applied to the radiating wire being at least 600 Vpp.

For the magnetic and electric fields induced into cables, the test level is defined as the product of the length of interconnecting cable that is exposed to the radiating wire and the rms voltage or current applied to the wire This test level is given as "volts x meters" (V-m), or "amps x meters" (A-m). For example, category Z requires an electric field test level of 1800 V-m, which is typically obtained by exposing 3 meters of cable to a radiating wire with 600 volts rms applied to it. If less than 3 meters of cable is exposed to the radiating wire (due to space restrictions, for example), the voltage applied to the wire must be increased so that the test level of 1800 V-m is achieved. The exception to this requirement is when the actual length of the cable in the final installation is known to be less than 3 meters. In this case, the test level may be reduced in proportion to the ratio of the reduced coupling length.

The test level for magnetic fields induced into equipment for all categories is 20 amps rms at 400 Hz.

Equipment Categories

There are four Equipment Categories (A, B, C, and Z) that indicate the tests performed and severity level of the tests. Category B has no requirement for electric or magnetic fields induced into cables. Equipment category C has a requirement of 120 A-m for magnetic fields, and 5400 V-m for electric fields.

Changes from the Last Revision

The only real change is the addition of equipment category C.

Future Changes and Directions

There has been some discussion about replacing the induced spikes test (chattering relay) with a method that is more repeatable and has tighter requirements on the voltage waveform. Two methods that have been discussed are EN61000-4-4 and Mil-Std-462D test method CS115 [7].

SECTION 20: RADIO FREQUENCY SUSCEPTIBILITY (RADIATED AND CONDUCTED)

These tests are performed to determine that equipment will operate as specified when the EUT and it's interconnecting cables are exposed to Radio Frequency interference. CW, square wave AM, and pulse modulated RF signals are required, and for some categories, it is suggested to gate the modulated signal on and off at a slow (1-3 Hz) rate. A Line Impedance Stabilization Network (LISN) must be inserted in series with each power lead and ungrounded power return lead, with a 10 uF capacitor connected between the power input of the LISN and the ground plane. Unless otherwise specified, interconnecting cables shall be at least 3.3 meters in length, and power leads will be no more than 1 meter in length for these tests.

Background

This section was by far the most discussed and the most controversial of all the sections in DO-160D. In fact, section 20 as it is now, is really an interim procedure, put into DO-160D so that it could be published without further delay. This was done with the intention to release a new version of section 20 as a change notice within 18 months after DO-160D was published. This allowed the important changes and updates in other sections to be put into use, while more discussions continued on the nature of the aircraft RF environment, and what the RF susceptibility test levels and procedures should be.

Due to the interim nature of section 20 as it exists today, it will be covered rather briefly here, with only the parts that will survive the change notice being addressed. Instead, more time will be spent on what section 20 will be after the upcoming change notice.

Conducted Susceptibility

The RF conducted susceptibility test procedure is similar to Mil-Std-462D test method CS114. RF interference is coupled into the EUT interconnecting cables and power leads using an injection probe that is calibrated to the required test level prior to performing the test. The calibration is performed in a 50 ohm fixture. The amount of RF power applied to the injection probe that is required to achieve the specified RF current in the fixture is recorded for each test frequency. This calibration table, showing RF power required at a given frequency, is then used during the actual test. The amount of RF current induced into the cable or lead under test is also monitored during the test. The test frequency range is 10 kHz to 400 MHz.

Radiated Susceptibility

The RF radiated susceptibility test procedure is similar to Mil-Std-462D test method RS103. The EUT and it's interconnecting cables and power leads are exposed to RF radiated fields in the frequency range of 100 MHz to 18 GHz.

There are two RF radiated susceptibility test methods allowed in section 20. The first uses a standard semi-anechoic chamber as in Mil-Std-462D test method RS103. Calibration of the RF field prior to placement of the EUT is required. The power needed to achieve the specified test level is recorded for each antenna used. RF absorber is required, and the minimum performance of the absorber is specified. A minimum antenna distance of 1 meter is specified, and multiple antenna positions are required when the beamwidth of the antenna does not totally cover the system.

The second method uses a mode-stirred chamber, which requires a power density calibration prior to use. Chamber Q, time constant, and stirring ratio verification are also required.

Table 1. Proposed Radiated Susceptibility Test Levels

Frequency	Category F		Category E		Category D		Category C		Category B		Category A	
Range	CW	Pulse	CW	Pulse	CW	Pulse	CW	Pulse	CW	Pulse	CW	Pulse

Frequency Range	Category F		Category E		Category D		Category C		Category B		Category A	
	CW	Pulse	CW	Pulse	CW	Pulse	CW	Pulse	CW	Pulse	CW	Pulse
100 - 200 MHz	200		100		50		25		13		7	
200 - 400 MHz	200		100		50		25		13		7	
400 - 700 MHz	200	730	50	700	25	350	12.5	175	7	88	4	45
700 - 1000 MHz	240	1400	100	700	50	350	25	175	13	88	7	45
1 - 2 GHz	250	5000	200	2000	100	1000	50	500	25	250	13	125
2 - 4 GHz	490	6000	200	3000	100	1500	50	750	25	375	13	390
4 - 6 GHz	400	7200	200	3000	100	1500	50	750	25	375	13	190
6 - 8 GHz	170	1100	200	1000	100	500	50	250	25	125	13	63
8 - 12 GHz	330	5000	200	3000	150	1500	75	750	38	375	19	190
12 - 18 GHz	330	2000	200	2000	100	1000	50	500	25	250	13	125

Equipment Categories

In this interim section 20, there are 6 Equipment Categories for conducted susceptibility, and 8 Equipment Categories for radiated susceptibility. These categories indicate the severity level of the tests performed, and/or the type of modulation used. Category S is the least severe at 1 V/m up to 2 GHz, and category P is the most severe at 600 V/m up to 18 GHz.

Changes from the Last Revision

Since this is an interim procedure, the changes were not as sweeping as originally planned, but they are still significant. The biggest change is the addition of the mode-stirred chamber method for radiated susceptibility. The other changes of note are the addition of equipment categories S and P, and the performance requirements for the RF absorber material used in the semi-anechoic chamber radiated susceptibility method.

Future Changes and Directions

As mentioned above, there are significant changes in the works for this section, and a change notice is expected to be issued early in 1999. The list below details the changes to section 20 that are *expected* at the time this paper is being written.

- The elimination of all existing equipment categories, with the possible exception of category S.
- The addition of 7 new equipment categories (A thru F, and Z) for both radiated and conducted susceptibility.
- Category Z will be used to indicate tests at levels or modulations other than those specified in section 20.
- Pulse modulation will be required for all radiated susceptibility categories, with the possible exception of category S.
- Pulse widths of 1 and 4 microseconds will be required for radiated susceptibility tests with categories A thru F.
- The maximum conducted susceptibility test level will be increased to 500 mA for category F.
- The radiated susceptibility test levels will be increased significantly, to as high as 7200 V/m for category F (pulse modulated).
- The mode-stirred chamber radiated susceptibility test method will be updated to use "state-of-the-art" procedures for calibration and verification of the RF field produced in the chamber.

The conducted susceptibility test method will require that the RF current induced in the cable or lead under test be equal to or greater than the specified test level current for the equipment category selected. In addition, the maximum amount of forward RF power applied to the injection probe will be limited to 6 db above the power recorded during the injection probe calibration procedure for the test frequency.

SECTION 21: EMISSION OF RADIO FREQUENCY

The tests in this section are performed to determine that the EUT does not emit Radio Frequency interference in excess of the specified limits. Conducted RF emissions appearing on interconnecting cables and power leads are measured. Radiated RF emissions from the EUT, interconnecting cables, and power leads are also measured. Measurements must be made with an instrument using a peak detector, and with IF bandwidths as specified in the procedure for the frequency range being scanned. A LISN must be inserted in series with each power lead and ungrounded power return lead, with a 10 uF capacitor connected between the power input of the LISN and the ground plane. Unless otherwise specified, interconnecting cables shall be at least 3.3 meters in length, and power leads will be no more than 1 meter in length for these tests. Ambient emission levels should be at least 6 dB below the applicable limit, and must be measured and recorded if any signals are found to be within 3 dB of the applicable limit.

Conducted Emissions

Conducted RF currents on interconnecting cables and power leads are measured with a clamp-on current probe. The probe is positioned 5 centimeters from the EUT and measurements are made over the frequency range of 150 kHz to 30 MHz.

Radiated Emissions

Radiated RF fields are measured with linearly or circularly polarized antennae over the frequency range of 2 MHz to 6 GHz. Multiple antenna positions may be required, based on the beamwidth of the measurement antennae used and the layout of the system under test. Multiple EUT orientations may also be required, depending on the design of the EUT and the location of possible emission sources such as displays, connectors, vents, covers, etc.

Equipment Categories

There are four Equipment Categories (B, L, M, and H) that indicate the location of the equipment and the separation between the equipment and aircraft antennae. In general, the closer the equipment is to an aircraft antenna, and the more it approaches a "direct view" of an aircraft antenna, the tighter the emissions limits.

Changes from the Last Revision

The list below is a summary of the changes.

- Equipment categories A and Z have been eliminated, and categories L, M, and H have been added.
- Separate broadband and narrowband measurements and limits have been eliminated and replaced by measurements made with a peak detector and specified IF bandwidths.
- All conducted emissions measurements must be made with a current probe. LISN measurements are no longer allowed.
- The frequency range for radiated emissions measurements is now 2 MHz to 6 GHz.
- The radiated emissions limits for new categories M and H have several deep "notches" in them that correspond to aircraft RF sensors and receivers operating frequencies.
- Performance requirements for the RF absorber material used in the semi-anechoic chamber are specified.

SECTION 22: LIGHTNING INDUCED TRANSIENT SUSCEPTIBILITY

These tests determine whether the EUT can operate as specified during and/or after various lightning induced transient waveforms are injected into connector pins, interconnecting cables, and power leads. The three test methods used are pin injection, cable induction, and ground injection. The pin injection method is normally used to show damage tolerance, while the cable induction and ground injection methods are normally used to show upset tolerance. In each test method, a minimum of ten transients of each polarity are required for each waveform. For the cable induction and ground injection tests, a LISN must be inserted in series with each power lead and ungrounded power return lead, with a 10 uF capacitor connected between the power input of the LISN and the ground plane. Unless otherwise specified, interconnecting cables shall be at least 3.3 meters in length, and power leads will be no more than 1 meter in length for these tests.

Pin Injection

During pin injection testing, the EUT is normally unpowered or minimally powered, so that the circuits being tested are biased as they would be in normal operation. The test level is defined as an open circuit voltage (Voc) with a specified source impedance from the generator. For example, waveform 3, test level 2 specifies Voc as 250 volts, with a short circuit current (Isc) of 10 amps. The ratio of Voc to Isc yields a generator source impedance requirement of 25 ohms.

The generator is adjusted to produce waveform 3 with these specified characteristics, and the transient waveform is then applied directly to the interface pins. After all the pins have been tested, the EUT is evaluating to determine if it's performance has been degraded.

Cable Induction

This test method uses an injection probe to induced the transient waveforms into interconnecting cables and power leads. For each waveform, either a voltage or current test level is given, along with a current or voltage limit. For example, waveform 2, test level 3 specifies a voltage test level (VT) of 300 volts, and current limit (IL) of 600 amps. This means that during the test, the generator level is increased until the peak voltage measured on a single turn monitor loop placed thru the injection probe reaches 300 volts, **or** the monitored induced current in the cable or lead reaches the 600 amp limit.

Ground Injection

This test method is very similar to the cable induction method, except that the transient waveform is applied between the EUT case and the ground plane, after the EUT has been isolated from the ground plane by lifting all local grounds and returns, and insulating the case from the ground plane. The injected transient is thus forced into the cable shields and any other return path it can find back to the ground plane. The voltage developed between the EUT and the ground plane is monitored during the test, as well as the current through the cable(s) under test. The test levels are specified and determined in the same way as the cable induction method.

Equipment Categories

There are 4 pin injection waveforms grouped together in 2 waveform sets, and there are 6 cable bundle waveforms grouped together in 4 waveform sets. There are 5 test levels for all of the waveforms. Category designations consist of four characters that describe the pin and cable test waveform sets and test levels. For example, category A3E3 indicates pin waveform set A at test level 3, and cable waveform set E at test level 3.

A designator of Z in any part of the category description is used to indicate a non-standard test waveform or level, and a designator of X is used to indicate no test is required.

Future Changes and Directions

There have been discussions about adding multiple burst and multiple stroke test methods and levels to this section, but this will not happen until SAE committee AE-4L completes it's work in this area.

SECTION 23: LIGHTNING DIRECT EFFECTS

The tests in this section are performed to determine the ability of externally mounted electrical and electronic equipment to withstand the direct effects of a severe lightning strike. The equipment will not normally be powered during the test, and these tests usually cause damage (sometimes spectacular damage) to the EUT. High voltage and/or high current tests at levels of thousands of kVolts and/or hundreds of kAmps are required.

Equipment Categories

There are four Equipment Categories (1A, 1B, 2A, 2B, and 3) that indicate the nature and the severity level of the tests performed. These categories correspond to the aircraft "lightning zones" where the equipment is to be located in it's final installation. There is also a special F suffix, which, when added to one of the equipment categories described above, indicates that the equipment can be used in "fuel vapor zones" on the aircraft.

Changes from the Last Revision

None but a few additional explanations of the test procedures and levels.

SECTION 25: ELECTROSTATIC DISCHARGE (ESD)

This test determines whether the EUT can operate as specified during and after being subjected to an electrostatic air discharge event. The test procedure is similar to EN61000-4-2, except that the EUT is bonded to the ground plane and only air discharge is specified [8]. The generator and discharge tip used have exactly the same specifications as the ESD simulator in EN61000-4-2. Test points are chosen based on their accessibility to personnel, then 10 positive and 10 negative polarity discharges at 15 kV are applied to each one.

Equipment Categories

There is only one category (A), which indicates a test level of 15 kV.

Changes from the Last Revision

Section 25 is brand new for this revision and is the first ESD requirement in the history of DO-160.

SUMMARY

RTCA/DO-160, and it's European twin, EUROCAE/ED-14, are truly the world standards for Electromagnetic Compatibility requirements for aircraft electronic equipment. This latest revision contains more severe test levels than the previous version, and one new section covering ESD.

Section 20 had the most significant changes, even though it is really an interim procedure. This version of section 20 will be used until the change notice that will increase test levels even more is released, which is expected to be in early 1999.

The test levels, requirements, and procedures are intended to reflect the "state-of-the-art" in aviation technology and EMC testing methodology.

Since both aviation technology and EMC testing methodology are evolving at a rapid rate, work is continuing on future change notices and on the next revision (DO-160E).

REFERENCES

[1] RTCA/DO-160D, "Environmental Conditions and Test Procedures for Airborne Equipment," RTCA, Incorporated, July 29, 1997.

[2] RTCA/DO-160C, "Environmental Conditions and Test Procedures for Airborne Equipment," RTCA, Incorporated, Decmber 4, 1989.

[3] Mil-Std-704D, "Aircraft Electric Power Characteristics" US Dept. of Defense, September 30, 1980.

[4] EN61000-3-2, "Electromagnetic compatibility (EMC), Part 3: Limits, Section 2: Limits for harmonic current emissions (equipment input current ≤ 16 A per phase," CENELEC, April 1995.

[5] Mil-Std-462, "Electromagnetic Interference Characteristics, Measurement of," US Dept. of Defense, July 31, 1967

[6] Mil-Std-462D, "Measurement of Electromagnetic Interference Characteristics, " US Dept. of Defense, January 11, 1993.

[7] EN61000-4-4, "Electromagnetic compatibility (EMC), Part 4: Testing and measurement techniques, Section 4: Electrical fast transient/burst immunity, Basic EMC Publication" CENELEC, March 1995.

[8] EN61000-4-2, "Electromagnetic compatibility (EMC), Part 4: Testing and measurement techniques, Section 2: Electrostatic discharge immunity test, Basic EMC Publication" CENELEC, March 1995..

LESSONS LEARNED FROM THE F-22 AVIONICS INTEGRITY PROGRAM

Stefan Glista, Lockheed Martin Aeronautical Systems

Abstract

The F-22 is contractually required to be significantly more reliable than the F-15C/D. Prior to the Gulf War, the F-15 C/D's Mean Time Between Maintenance (MTBM) total was 0.6 hours[1]. The F-22's MTBM (total) requirement is 3.0 hours. This required improvement in maintenance demand is based on improvements in equipment reliability. This high reliability must be achieved in spite of increased functional requirements, subsystem complexity and exposure to more severe induced operating environments.

The F-22 Team's approach to designing reliability into subsystem hardware is called the F-22 Subsystem Integrity Program. The goals of the program are to increase reliability, campaign effectiveness and safety. These goals will be attained through a rigorous application of engineering principals including life analyses and realistic environmental testing. As part of the Subsystem Integrity Program, the F-22 Team sponsored industry wide supplier conferences to share design, analysis and test approaches and results.

The purpose of this paper is to summarize the top level management and technical lessons learned in the implementation of the F-22's Subsystem Integrity Program. The two most important management lessons learned are 1) schedule - the disciplined implementation of the integrity program was key to keeping flight certification process on schedule and 2) cost - reliable subsystem components reduce initial acquisition costs (initial spares) as well as customer life cycle costs. A key technical lesson learned is that many Mil-Std tests are not adequate in severity or duration to achieve the cost and schedule benefits described.

Background

As discussed, the F-22's improvement in reliability must be achieved even as the functional performance requirements 1) drive the design to increased complexity, and 2) drive the induced operating environments to be more severe than in previous fighter aircraft.

Design Complexity

The F-22's functional performance requirements have driven electronic designs to increase packaging density, operating speeds and therefore to increase cooling requirements. Complexities include 1) Surface Mount Technology, 2) Liquid Flow Thru Cooling, and 3) large (2 in x 4 in) Multi-Chip-Modules (MCMs). In addition, digital electronics are susceptibility to data loss from microsecond intermittent events. Microsecond intermittent events may be caused by connector fretting, solder joint cracks, plated through hole cracks, and dendritic growth between ever closer compliant pins.

Severe Environments

The F-22's operational environments are more severe than previous fighter aircraft's. The ability to access equipment is a problem on any fighter aircraft. One approach used is to locate equipment in weapons bays and wheel wells; however, these bays have severe vibration, acoustic, temperature and corrosion environments. Another approach is to design

[1] NOR 86-235 High Reliability Fighter Concept Investigation, Terry E. Gaumer

0-7803-5086-3 /98/$10.00 ©1998 IEEE

equipment to extremely high reliability requirements to minimize access frequency.

Implementation Philosophy

The basic philosophy is that high equipment reliability can be attained through a disciplined application of rigorous engineering principals during each phase of the development process. In other words, improvements to the product can only be achieved by improving key aspects of the development process.

The five subsystem integrity tasks define the development process for subsystem equipment. The tasks are defined as:

- Design Information and Development Planning
- Design Analysis and Development Tests
- Subsystem and Component Testing
- System Ground and Flight Testing
- Production Quality Control and Life Management.

The F22's Subsystem Integrity Program is organized to establish synergy between management concepts championed by E. Deming[2] and the analytical and test technologies developed for the Aircraft Structural Integrity Program (ASIP) program. For example, the tailoring of the integrity program plan by each supplier was documented with the use of fishbone diagrams (recommended by Deming) which listed all potential causes of equipment failures. The cause and effect diagrams were then used to construct tabular analysis, test and product quality control plans that addressed each alpha numerically categorized failure mechanism. Examples of fishbone diagrams and tabular plans were used extensively in training, brainstorming and contract negotiations with over 1000 supplier and prime contractor personnel. The following is a brief discussion of the first three integrity tasks as well as the lessons learned in each phase.

(I) Design Information and Development Planning

The objective of Task I is to understand the general design requirements and develop a plan that will result in a disciplined development of specific requirements and criteria for materials and process selection and design of systems. The creation of the Weapon Systems Integrity Master Plan (WSIMP) and Environmental Criteria Document (ECD) during task I coincides with the description of:

- Design Service Life and Design Usage
- Design Environments
- Component Criticality Classification
- Engineering Design Criteria
- Product Integrity Control Plans

A component's criticality is used to determine the design criteria, the amount of analysis, testing, quality control and maintenance requirements.

Durability Design Criteria Probabilities

Component design analysis and test criteria are established based on the probability of all the equipment items in the fleet completing a service life without failure.

The F-22's durability margins and resulting probabilities of failure are consistent with the FAA's guidance of 10^{-7} probability of failure for a given safety critical function. The probability of 10^{-7} is defined as "extremely remote" and an "acceptable numerical probability range" for "fatal injury to a small number of occupants" [3]. The F-22 margins and resulting probabilities were calculated as end of life probabilities versus constant failure rate probabilities.

[2] W. Edwards Deming, Out of the Crisis, MIT Center for Advanced Engineering Study, 1982.

[3] JAR 25, Joint Aviation Requirements, change 14, 27 May 1994 , page 3-F-13.

Combine control point probabilities in series.

If an equipment item is required to meet a high reliability requirement, then each durability control point of the item must meet high reliability requirements. If a computer has 384 highly stressed control points, then each control point must be designed to more than 16 lifetimes for the computer to achieve a low probability of failure. For example, if a computer has 12 circuit cards and each circuit card has 4 large hybrids with 8 critical -- highly loaded corner lead wires, then the computer has 384 durability control points. If the lead wires' solder joints are designed for a 16 lifetime margin, then the probability of the computer failing due to a solder joint failure is 3.84×10^{-5} per hour. In other words, the probability of failure for each joint (1×10^{-7}) is combined in series.

Similarly, if an aircraft has 40 safety critical parts that are each designed for a 16 lifetime margin (1×10^{-7} POF), then the probability of failure of a safety critical item is approximately 4×10^{-6} per hour at the end of the 20 year service life.

Criticality Control Logic

As previously mentioned, the F-22 classified its equipment as Safety, Mission or Durability Critical. The same concept should be applied to analysis and test margins of aging aircraft. The purpose of classifying equipment is to require more margin in analysis and testing, manufacturing controls, and field traceability.

(II) Design Analysis and Development Tests

The objective of Task II is to effectively characterize, through analysis and testing, the response of the subsystem design for the environment in which the subsystem must operate. Durability analysis combines deterministic and probabilistic approaches to model (and hopefully mitigate) time dependent failure mechanisms. Examples of time dependent failure mechanisms analyzed include low cycle fatigue (which includes fatigue caused by inertial loads, hinge moments, thermal cycling environments etc.), and high cycle fatigue (which includes vibration).

The primary elements of Task II are:

- Design Analysis
- Development/Materials Tests
- Durability Analysis
- Damage Tolerance Analysis

Equipment Improvement

Thermal cycling fatigue life of electronics was improved through 1) Coefficient of thermal expansion (CTE) matching, 2) Omega and other strain relieving lead wire designs for large devices, 3) Plated Thru hole improvements - nickel plating, 4) Rigid Flex Wrap Around replaced with co-fired ceramic substrates. Vibration fatigue life of electronics was improved through 1) frequency separation of circuit boards from chassis and shelves, 2) minimized relative motion to reduce edge card connector wear, and 3) use of multi-contact edge card connectors to eliminate intermittent failures.

(III) Subsystem Testing

The goal of Task III is to verify the durability and performance capability of items over the full F-22 life. This includes verification of the analysis as well as the identification of durability modes not identified in analysis and development tests. Verification tests may be conducted on individual parts, in simulated system installation environments, or during system ground and flight tests - as appropriate.

The primary elements of Task III are:

- Safety of Flight Testing
- Qualification Testing

F-22 Qualification Test Lessons Learned.

The F-22 Subsystem Integrity Program required that suppliers tailor analysis and test requirements using comprehensive fish bone diagrams which listed all potential causes of equipment failures. The prime contractor rigorously developed test requirements equivalent to the damage from the realistic environment. In comparison to F-22 test requirement, the vibration and corrosion tests specified by MIL-STD-810[4] and RTC/DO-160 [5] are inadequate both in severity or duration or both. The realistic vibration and corrosion environments are much more severe than previous equipment tests required. The greatest risks appear to be for 1) equipment mounted on or near engines, 2) corrosion susceptible materials, 3) items that were incorrectly tested with respect to installation transmissibility, and 4) items susceptible to wear.

Vibration Test Improvements

The typical vibration test for MIL-STD-810 or RTC/DO-160 is only one or two hours long. The typical F-22 vibration test is 20 hours long. The longer F-22 test length is due to differences in the usage analysis and the amount of test compression. The usage equation, used for F-22 usage analysis is derived from measured data – the measured data shows that the vibration level (Wo) is almost proportional to dynamic pressure (Q) ($Wo = kQ^{1.2}$). Mil-STD-810's equation to estimate vibration assumes a squared relationship $Wo = kQ^2$. Starting with the maximum expected or measured vibration level, the estimates of all other vibration levels throughout the mission profile analysis, using MIL-STD-180's equation, will result in a significant under estimate of the vibration levels at lower dynamic pressures.

[4] RTCA/DO-160C, 4 December 1989, Environmental Conditions and Test Procedures for Airborne Equipment

[5] MIL-STD-810E, 14 July 1989, Environmental Test Methods and Engineering Guidelines

In addition to longer test times, the F-22 program required forced displacement testing of circuit card connectors (0.002 inches rms for over 30 million cycles). The forced displacement tests were designed to account for the number of wear cycles caused by the vibration environment. The test compression technique listed in MIL-STD-810 is based on equivalent fatigue damage and does not account for wear damage. If the test compression technique is based on wear, the length of the test increases by a factor of 10.

Finally, the typical vibration test program from MIL-STD-810 or RTCDO-160 does not include transmissibility effects. The prime contractors measure aircraft vibration with accelerometers attached to "primary structure" such as bulkheads. Electronics are typically mounted on shelves or other secondary structure. Care was taken to separate the frequency of the circuit cards from the resonance of the secondary structure. For example, if a shelf, loaded with various electronics, responds at 200 Hz; the F-22 program's transmissibility analysis would require that the supplier stiffen the circuit cards up to 400 Hz to minimize damage. Installation transmissibility effects typically increased the random vibration level at low frequencies (less than 200 Hz) by a factor of 100. Failure to account for transmissibility effects may result in early solder joint and lead wire fatigue as well as connector wear.

Engine Vibration Environments

Engine vibration levels are based on a worst case envelope that includes assumptions about out of balance tolerances - and aging. The engine vibration levels may increase over the life of the engine, but will stay within the specified envelope. The historical test approach of 10 to 30 minute dwells at an endurance condition may have been adequate to prevent failures for engine mounted equipment -- when the engines are relatively new. The increase in vibration levels between a new engine and an aging engine may result in a factor of five increase in high cycle fatigue stress levels. In addition, wear tests are not as easily accelerated as fatigue tests. Vibration tests, accelerated based on fatigue slope factors, are significantly over accelerated - shortened - with respect to wear test slope factors [6,7]. Many aspects of aircraft equipment items are susceptible to early wear such as connectors, wiring, and seals. The F-22 program requires that at least 30 million cycles of vibration damage be completed for equipment with engine or inlet influenced vibration environments.

Thermal Cycling Test Improvements

The F-22 program requires approximately ten times more thermal cycles for qualification, than are typically required in reliability test programs. The F-22 program's typical durability life test requires from 500 to 1500 thermal cycles on one unit, whereas a typical reliability program would complete about 50 to 150 cycles.

Corrosion Test Improvements

The F-22 program requires approximately ten times more corrosion test time, than is typically required in traditional salt fog test programs. The F-22 program's salt fog test requires is 500 hours long, whereas the MIL-STD-810 and RTC/DO-160 salt fog test is only 48 hours long.

TAAF versus Durability Life Testing of Electronics

Reliability testing of aircraft electronics was historically based on simulated mission test requirements (typically on three units) and a Test, Analyze And Fix (TAAF) philosophy. Reliability analysis and test approaches assume that the failure rate is a Weibull distribution with Beta = 1. In other words, failures are assumed to occur "randomly" in time and thus the failure rate remains constant. This is not a valid assumption when equipment nears its wear out phase or Beta >1 condition. One of the goals of the reliability tests is to demonstrate that the equipment meets the Mean Time Between Failure (MTBF) requirement. A typical MTBF for a complex electronic system ranges from 200 to 2000 hours. Traditional TAAF tests are similar to the F-22's full scale test program, except that as a minimum one F-22 unit was tested to at least one lifetime. The design analysis included consideration of the damaging effects on electronics from thermal fatigue. F-22 developed analysis could be used to assess the effects of inactive storage.

The Business Case for Improved Testing

The improvements in the test approach for the F-22 were developed from fishbone analysis of known weaknesses in currently fielded equipment. F-22 suppliers demonstrated (by analysis and test) that conventionally designed electronic equipment items will experience wear out far sooner than the F-22 required 20 year service life. It is cost effective for the USAF to require improvements in analysis and testing of avionics. The cost to extend the service life for the F-15E's airframe from 8,000 to 16,000 hours is estimated at $200,000 per Aircraft per year [8]. The FY95 cost to maintain the F-15E's electronics is estimated at $2,880,000 per aircraft per year [9]. In addition, the low reliability of electronics results in accelerating redesigns due to the unavailability of spare parts. Multiple hardware and software

---

[6] Lockheed Martin Publication, F-22 Weapon System Master Plan

[7] David S. Steinberg, Vibration Analysis for Electronic Equipment, Wiley, 1988

[8] Ellis F. Hitt, Batelle, "Avionics Cost of Ownership" IEEE 1997

[9] John W. Lincoln, ASC, Ronald A. Mellliere, McDonnell Douglas Corp., "Economic Life Determination for a Military Aircraft"

configurations are created to deal with obsolete parts. These multiple configurations require unique spares, support equipment and training. Longer tests would have been required to verify the assumption that the electronics failure rate (and the resulting support costs) will remain constant over a ten or twenty year life.

Conclusion:

This paper offers a summary of information that the F-22 program adds to development and qualification of electronic equipment. Additional data available from the F-22 program include the F-22 Weapon System Integrity Master Plan[10], the F-22 Environmental Criteria Document [11], and the hundreds of analysis and test reports of suppliers equipment.

The historical military standards for testing subsystem equipment were reviewed to assess the risk of service life extension on aging equipment. The comparison of the FAA and MIL-STD tests to the current F-22 Subsystem Integrity Program requirements showed significant weaknesses in the duty cycling tests required for redundant - durability critical equipment. Previous papers have documented that there is less comparative risk for single - point failure - safety critical equipment [12]. It was demonstrated that the current military and FAA test requirements are not of sufficient duration or levels to demonstrate that durability critical subsystem equipment will meet an 8,000 hour, 20 year service life.

The greatest risks appear to be for 1) equipment mounted on or near engines, 2) corrosion susceptible material, 3) items that were incorrectly tested with respect to installation transmissibility, and 4) items susceptible to wear.

The true accomplishments of the integrity process can only be measured in the reliability of fielded equipment. The overall progress and value of the integrity process can, however, be measured by how many of our suppliers have adopted integrity design, analysis and test concepts in other business lines beyond the F-22.

The team's successes are many. We have changed the way that we specify the environment, criticality, design criteria, analysis, and test requirements. We have improved industry's analysis and test methods. We have changed the way we review designs. We have changed the way we envision new designs. We have added discipline to the subsystem development process. The disciplines with improved processes and interfaces with other disciplines include system engineering, configuration management, aerodynamic performance, vibration, thermodynamics, maintainability, stress, fatigue, fracture, corrosion control, system safety, and reliability.

---

[10] Lockheed Martin Publication, F-22 Weapon System Master Plan

[11] Lockheed Martin Publication, F-22 Environmental Criteria Document

[12] Stefan Glista, "Life Extension of USAF Fighter Aircraft: Lessons Learned from F-22 Subsystem Durability Life Tests" – Internal Lockheed Publication

SpecTRM: A CAD SYSTEM FOR DIGITAL AUTOMATION

Nancy G. Leveson, Massachusetts Institute of Technology, Cambridge, MA
Jon Damon Reese, Safeware Engineering Corporation, Seattle, WA
Mats P.E. Heimdahl, University of Minnesota, Minneapolis, MN

Introduction

In the system engineering of complex systems that include digital automation, the most vexing and potentially costly problems arise in the early stages of development. Few adequate tools exist to assist in developing system requirements and architectures and translating the system requirements to software requirements. Serious unsolved problems also exist at the other end of the lifecycle in changing or upgrading automated control tasks without introducing errors. In addition, these two system development phases present the most serious and unsolved problems in certification and hazard analysis.

SpecTRM-RL (Specification Tools and Requirements Methodology) is a CAD system for digital automation. It is not intended to replace engineers, but instead to use the latest in research ideas to assist engineers in managing the requirements, design, and evolution process. SpecTRM emphasizes:

- Finding errors early in development so they can be fixed with the lowest cost and impact on the system design.
- Tracing not only requirements but design rationale (including safety constraints) throughout the system design and documentation.
- Building required system properties into the design from the beginning rather than emphasizing assessment at the end of the development process when effective response is limited and costly.
- Supporting the construction of families of related systems and the reuse of the early parts of the system development process.

Complex systems cannot be built successfully without the interaction of multiple disciplines—the most challenging systems include electromechanical components, computers, and humans. While we train engineers to be experts in individual fields, these complex heterogeneous systems require knowledge across engineering disciplines. The introduction of integrated product teams or other approaches to organizing projects so specialists work together have helped to some degree, but the problem goes deeper than simply management structures. SpecTRM attempts to provide bridges between diverse groups of system designers and builders. It does this by providing modeling, specification, and analysis tools (1) to ease communication and coordinated design of components and interfaces and (2) to provide seamless transitions and mappings between the various development and maintenance stages.

SpecTRM was designed to act as a workbench for teams of engineers to enhance communication by using common models and analysis tools that execute on the models (see Figure 1). At the heart of the system development process using SpecTRM is an executable model of the system components. Our modeling language (SpecTRM-RL) emphasizes readability and requires little training. At the same time, it has a formal foundation, which allows formal analysis on the models built. In addition, visualization tools are used to assist in understanding the models and to provide the most appropriate views of the system for the task being performed. Visualization tools are also included to provide animations of the output of the model as it executes.

0-7803-5086-3 /98/$10.00 ©1998 IEEE

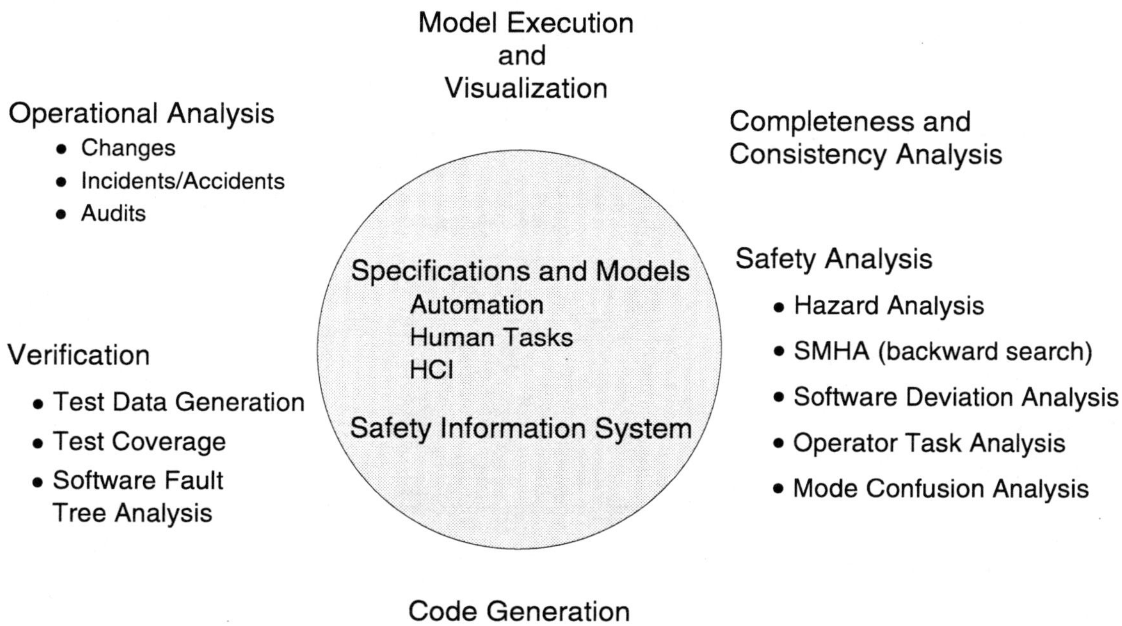

Figure 1. The SpecTRM Systems Engineering Workbench for Digital Systems.

Tools are also included to generate or assist in generating various types of analyses such as fault trees, hazard analysis, human error analyses, robustness analysis, and consistency and completeness analysis. We plan to add tools for timing analysis, reliability analysis, test data generation, blackbox test coverage analysis, and code generation. A safety information system is included to assist in maintaining hazard logs and audit trails. But instead of being separate from the development process, hazard and safety information is tightly integrated into the environment in which safety-related decisions are made.

SpecTRM is written entirely in Java and is supported on most Windows, Macintosh, and UNIX platforms. The architecture is based on a client-server paradigm where the client side can be launched from Java-enabled Web browsers or as a stand-alone, thus supporting both individual and shared use of SpecTRM design and development artifacts (such as specifications, models, safety information system contents, and modeling results). SpecTRM supports revision control, encryption, and access control.

SpecTRM is an example of technology transfer from the university to industry. Safeware Engineering Corporation is building commercial quality tools to support SpecTRM and will also act as a conduit from university research to continually upgrade the SpecTRM CAD system with additional tools and analysis capabilities as they are invented. Parts of SpecTRM have been used experimentally on real systems including a flight management system and TCAS II. NASA has contracted to use this methodology in the safety assessment of the upgrades they are building for the U.S. Air Traffic Control System.

The rest of this paper describes some of the unique features of SpecTRM that differentiate it from other system engineering tools and environments.

Intent Specifications

Many of the goals stated above are achieved through our unique specification methodology. Intent specifications are a new way of structuring system specifications that is based on research in the fundamental principles

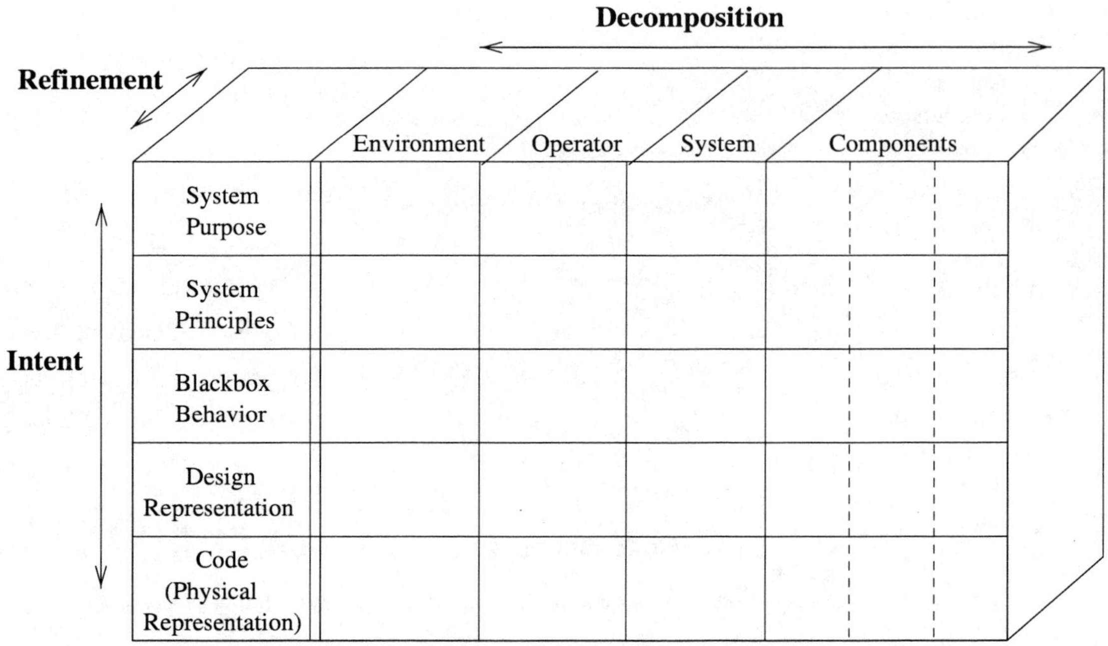

Figure 2. The form of a SpecTRM-RL specification.

of problem-solving and abstraction that humans use to make complex tasks intellectually manageable. The problems in performing system engineering and software engineering activities are rooted in complexity and intellectual manageability. Psychologists have found that complexity itself is not a problem if humans are presented with meaningful information in a coherent, structured context. "People don't mind dealing with complexity if they have some way of controlling or handling it … if a person is allowed to structure a complex situation according to his perceptual and conceptual needs" [3].

One approach found to be effective in dealing with complexity is the use of hierarchical abstraction—that is, structuring the situation such that the problem solver can transfer the problem to a different level of abstraction. Most specifications use decomposition and refinement to provide hierarchical abstraction, where each level of the usual specification can be thought of as providing "what" information while the next lower level describes "how." Intent specifications add a type of hierarchical abstraction based on goals or purpose and thus organize the levels of the specification not only in terms of what and how but also in terms of "why." By organizing the specification in this way, higher-level purpose or intent for design decisions can be immediately determined.

Systems and software are continually changing and evolving; they must be designed to be changeable and the specifications must support evolution without compromising the confidence in the properties that were initially verified. One requirement for correct and safe system evolution is knowing the rationale behind a design. Unfortunately, the reasons why something was done a certain way are often not recorded. Maintainers of a system may not know why a particular design feature was included and inadvertently remove it while making a change. Recording all the design rationale is difficult and may not be effective if the necessary information is difficult to find when a change is made. Intent specifications record and link the rationale behind design decisions ("why" information) directly in the system specification so that it is easy to find and use.

System and software specifications in SpecTRM are organized along three abstraction dimensions: intent, refinement, and decomposition (see Figure 2). The vertical dimension specifies the level of intent at which the problem is being considered, i.e., the language or model that is currently being used. The decomposition and refinement dimensions allow users to change their focus of attention to more or less detailed views within each level or model. The information at each level is fully linked to related information at the levels above and below it.

The intent dimension has five levels of abstraction. The highest level of the specification contains the overall goals and safety constraints. Some of the information here is generated through the preliminary hazard analysis process.

The next lower level contains the underlying scientific principles upon which the design at the lower levels is based and through which the goals and constraints at the highest level are satisfied. Models and specifications at this level may be subjected to operations research and other types of system analyses to evaluate alternative designs (such as various aircraft spacing and routing alternatives for an ATC system) with respect to the higher-level goals and constraints. The third level contains a blackbox functional behavior model of the system components. These models are executable and formally analyzable. The fourth and fifth levels contain design and implementation information.

Because each level is mapped to the appropriate parts of the intent levels above and below it, traceability of not only requirements but also design rationale and design decisions is provided from high-level system goals and constraints down to code (or physical form if the function is implemented in hardware) and vice versa.

Each level of an intent specification supports a different type of reasoning about the system and represents a different model of the same system. The model at each level is described in terms of a different set of attributes or language. Level 1 (System Purpose) assists system engineers in their reasoning about system-level properties such as goals, constraints, hazards, priorities, and tradeoffs. The second level allows engineers to reason about the system in terms of the physical principles and laws upon which the design is based. The third level enhances reasoning about the logical design of the system as a whole and the interactions between components as well as the functional state without being distracted by implementation issues. The lowest two levels provide the information necessary to reason about individual component design and implementation issues. The mappings between levels provide the relational information that allows reasoning across the hierarchical levels and tracing from high-level requirements down to implementation and vice versa.

The intent information represents the design rationale upon which the specification is based and thus design rationale is integrated directly into the specification. Each level also contains information about underlying *assumptions* upon which the design and validation is based. Assumptions are especially important during analysis of operations, such as safety audits. When conditions change such that the assumptions are no longer true, then a new safety analysis should be triggered.

Because the separation of human factors and the design of the human-computer interface from the main system and component design can lead to serious deficiencies in each, we have attempted to integrate both types of specifications at each level and across levels. Interface specifications and specifications of important aspects of the system environment are also integrated into the intent specification. Finally, each level of the intent specification includes a specification of the requirements for and results of the verification and validation activities at each level.

Safety Information System

Setting up a comprehensive and usable safety information system can be time consuming and costly, but such a system is crucial to the success of safety efforts and resources invested in it are well spent. Studies of organizations and accidents have shown that an effective safety information system ranks second only to top management concern about safety in discriminating between safe and unsafe companies matched on other variables [1, 2].

Control of any activity requires information—a crucial aspect of any management system is the feedback of information on which to base decisions. Information can be used to describe, to diagnose, to compare, to evaluate, and to improve. For example, a safety information system can provide information necessary (1) to identify and control hazards and to improve designs and standards, (2) to evaluate the effectiveness of proposed or implemented safety controls, (3) to evaluate the implications of suggested changes for their safety impact, (4) to compare models and risk assessments with actual behavior, and (5) to detect trends and deviations that presage an accident.

Documenting and tracking hazards and their resolution are basic requirements for any effective safety program. A complete hazard log and audit trail will show what was done and how and why safety decisions were made during system development, operational use, maintenance and evolution. The safety information system should also contain the most recent update of the System Safety Program Plan and the status of all the activities included in the plan, results of hazard analyses, incident and accident information including corrective action, trend analysis data, etc. Interfaces with various project databases, such as system and software configuration and control, should be well defined.

One of the most difficult aspects of maintaining a safety information system is dissemination of information in a useful form. If information is not presented to decision makers in a meaningful way, use of and learning from the data is inhibited. The information needs to be presented in a form that people can learn from, apply to their daily jobs, and use throughout the life cycle of projects. Also, the method of presenting the data should be adaptable to the cognitive styles and models of the users. SpecTRM tightly integrates the safety information system into the process and engineering environment in which safety-related decisions are made. The support provided for complete traceability from hazard analysis to design and implementation is valuable in both designing safety into the system and ensuring that safety is not compromised during operational use and system evolution.

Modeling Language

Building models is not easy and will (and should) only be done if there is enough payoff from the process. We believe that formal modeling can provide the most assistance at the system design phase where functionality is allocated to individual components. It is at this phase where tools are most needed to assist in design validation and tradeoff decisions. Models of the allocated component behavior can be executed and both formally and informally analyzed to ensure that required system functionality has been incorporated, the system design and architecture exhibits various desirable properties such as fault tolerance and safety, human—machine interactions have been appropriately designed, and tradeoffs between conflicting goals are resolved adequately. The formal modeling language in SpecTRM, called SpecTRM-RL (SpecTRM Requirements Language) is specifically designed to assist at this phase of system development. It is used in Level-3 of an Intent Specification.

SpecTRM-RL reflects lessons we learned from our use of an earlier language, called RSML, that we designed for the FAA to specify the requirements for TCAS II, an aircraft collision avoidance system[1]. Like many of the currently popular specification languages, SpecTRM-RL uses a state machine as its underlying formal model. And like those languages, SpecTRM-RL is executable and has an associated suite of visualization and analysis tools. However, SpecTRM-RL is unique in several ways.

The first is the way that SpecTRM-RL deals with complexity. The complexity of our new systems is starting to overwhelm our current tools and cognitive capabilities. Engineers are attempting to build systems where the interactions between components cannot be thoroughly planned, understood, anticipated, and guarded against. Digital automation is exacerbating the problems by allowing greater levels of coupling with more integrated, multi-loop control and large numbers of dynamically interacting components. Besides allowing and sometimes encouraging new levels of complexity, computers introduce new types of failures modes not handled well by traditional approaches to designing for reliability and safety. In addition, computers introduce new types of problems in the interactions between components, including the interaction between humans and automation. Increased complexity and coupling is making it difficult for the designers to consider all the potential system states or for operators to handle all normal and abnormal situations and disturbances safely and effectively.

SpecTRM-RL has been designed to assist with intellectual manageability, both in terms of the difficulty inherent in building models of complex systems and in the design and engineering of complex systems themselves. Models are completely blackbox, that is, they describe component behavior only in terms of outputs and the inputs that stimulate or trigger those outputs: The model does not include any information about the internal design of the components themselves, only their externally visible behavior. The overall system behavior is described by the combined behavior of the components, and the system design is modeled in terms of these component behavior models and the interactions and interfaces between the components.

Blackbox modeling allows separation of concerns so that the review and analysis of the specified blackbox functional behavior of a system component can be separated from the review and analysis of the internal design and implementation of the component. We believe that black box functional specifications are closer to the mental model of the engineer evaluating the system architecture than specifications that are complicated by the addition of internal component design decisions. Readability and reviewability will be enhanced by using languages that allow building models that are semantically close to the reviewer's mental model of the system. That is, the *semantic distance* between the model in the expert's mind and the specification model should be minimized. Using blackbox models reduces this distance compared to models that also include component design information.

SpecTRM-RL has been tailored to support the specification of requirements for reactive and, in particular, control systems. For example, the language supports specifying complex systems in terms of operational modes. Modes are abstractions that allow categorizing classes of system behaviors and therefore enhance the understanding of complex system behavior. However, the mode structure itself can become complex and some mode-related behavior of automated systems can lead to mode confusion on the part of users. By including the specification of component behavior in terms of modes, SpecTRM-RL encourages the design of simplified mode structures and also allows mode-related analyses, such as mode confusion analysis (see below).

[1] Another paper at this conference describes those lessons learned. See "Experiences From Specifying the TCAS II Requirements Using RSML".

In our experience using a state-machine modeling language (such as RSML) for modeling the TCAS II system requirements, we found that some features of these languages are error-prone. For example, the semantics of internally broadcast events were difficult for reviewers to understand and led to different interpretations of specified behavior by different readers of the models. SpecTRM-RL substitutes less ambiguous and complex modeling facilities. The semantically complex features of general-purpose modeling languages are not needed when building blackbox models only.

In specifying the TCAS II system requirements, we also learned a lot about readability and understandability of formal models—out specification was reviewed by a large number of people having varied backgrounds and knowledge, such as computer scientists, aeronautical engineers, and pilots. Ideally, specification languages should be both formally analyzable and readable without graduate-level mathematical training.

Readability is critical. While automated tools are helpful and may even be necessary to analyze some aspects of large and complex models, our experience in using these models for industrial projects is that the most important errors will be found by expert (human) review. The analysis tools we build and the mathematical theories upon which they are based cannot possibly incorporate all the domain-specific knowledge (e.g., for TCAS this knowledge includes FAA rules and procedures, basic aeronautical engineering, human factors and cognitive psychology) required to find subtle safety-critical flaws in a complex system design.

On the other hand, the complexity of these systems leads to the need to use formal models and automated assistance to support human navigation and understanding of the models and system specifications and to perform automated analysis where possible. Note, however, that any potential design flaws found by automated tools will need to be evaluated by humans. Thus readability and understandability of the models is a requirement for human processing of the mathematical analysis results.

We solve this dilemma in SpecTRM-RL by providing both a readable specification language we believe is semantically close to a designer's mental model of the system as well as a formal modeling language that underlies the more readable specification language. The automated analysis tools are based on the underlying formal model. In SpecTRM-RL, this formal model is called the Requirements State Machine (RSM), which is a form of Mealy automaton. Only the most intuitive aspects of this formal model need be understood by readers or even builders of SpecTRM-RL specifications, and these features can be described in a few minutes.

Flawed understanding of requirements is a major cause of software-related accidents. Incomplete specifications play a significant role in these misunderstandings. Previously, we defined a set of criteria describing what is needed for a specification of a process-control system to be sufficiently complete with respect to safety. Some of these criteria are derived from mathematical aspects of the formal model underlying the specification language. Others are related to lessons learned from accidents and incidents. Many are related to human-computer interaction. A few of the criteria, namely those related to mathematical completeness, can be checked by our automated tools (and such checking for mathematical completeness is provided for most commercial and research modeling languages). Our other criteria, many of which are derived from basic control engineering concepts or from experience, have been used primarily in checklist form in industry. In designing SpecTRM-RL, we included many of the required features in the language syntax so they can either be checked by a parser or can easily be checked manually.

Analysis Tools

SpecTRM includes a suite of tools (Figure 1) that (1) assist in building, using, and maintaining the specifications and models, (2) perform various types of automated analysis on SpecTRM-RL models, and (3) perform or assist in performing various system development activities.

The intent specification editor is used to create, change, and use intent specifications. Tools also assist in building SpecTRM-RL models and visualizations. We are planning to include facilities to provide the analyst with various views of the models, both during model development (design) and evaluation (execution).

The consistency and completeness tool can be used to identify some types of incompleteness (specifically, mathematical completeness) and nondeterminism in SpecTRM-RL models. The tool detects all conditions for which no behavior has been specified and identifies parts of the model with inconsistent and nondeterministic behavior (different behaviors under the same conditions). The specification can be analyzed in small pieces and need not be finished, which means the requirements model can be analyzed incrementally as it is being constructed.

A set of safety analysis tools is also provided for requirements and hazard analysis. These tools include support for backward analysis (starting from a hazardous state and identifying predecessor states to determine how and if the hazard can be reached), forward robustness analysis (to evaluate the operation of the software in an imperfect environment), and some forms of human task and mode confusion analysis (to identify features of the required software behavior that may lead to critical human errors). Additional analysis tools are planned but will not be part of the initial release of the tool set.

Finally, SpecTRM will include tools to assist in later phases of the system and software engineering process including (1) verification (test data generation from the SpecTRM-RL models, blackbox test coverage analysis, and software fault tree analysis), (2) code generation, and (3) operational analysis (e.g., tracing the effect of proposed changes on safety).

Summary

SpecTRM is a system engineering environment to support modeling and analysis during requirements generation, functional decomposition and tradeoff analysis, subsystem specification, implementation, verification, and system maintenance and evolution. A general goal is to build bridges among disciplines by providing integrated specifications and modeling tools that can be used by system engineers, software engineers, hardware engineers, and human factors experts. We also hope to provide seamless transitions and mappings between the various system development and maintenance stages.

Because many automated real-time systems have safety-critical aspects, SpecTRM provides support for hazard analysis and building safety into the design. The safety information and activities on a project are integrated into the development and decision making environment.

Beta testing of the basic components of SpecTRM will begin in Fall 1998. The first full release of the tool set is planned for June 1999.

References

[1] U. Kjellan. Deviations and Feedback Control of Accidents. in J. Rasmussen, K. Duncan, and J. Leplat (eds.), New Technology and Human Error, pages 143-156, John Wiley & Sonds, New York, 1987.

[2] N.G. Leveson. Safeware: System Safety and Computers. Addison-Wesley Publishing Company, 1995.

[3] J.R. Newman. Extensions of Human Capability Through Information Processing and Display Systems. Technical Report SP-2560, System Development Corporation, 1966.

INTEGRATED INTERFACE DEFINITION TOOL DEVELOPMENT: INTERFACE TRAC

Keith Ryan, Boeing Information, Space and Defense Systems, Kent, Washington

Abstract

The system engineering discipline has begun to mature in recent years and several companies have addressed the need for requirement tools. The second generation tools available today have improved their interfaces and capabilities a great deal. However, they still remain relatively rigid in their application of system engineering and rarely do they address all facets of engineering systems.

Programs that existed without system engineering as we know it today (such as E-3 AWACS) frequently have difficulty in adapting to the available tools. When this happens, there are several choices. Obviously, the system engineer can continue on with the methodology developed over years of use. The system could be reengineered, essentially scrapping the existing methodology and building tools a new. Or, a progressive evolution can be initiated.

This papers explores the first tentative steps of evolving internal interface management automation, using Microsoft Access®, on the NATO Mid Term AWACS Program. This tool set automates avionics interface modeling and documentation needed to support interface identification, definition and control. The NATO Mid Term Program is a mid-life technology update and insertion program for the 17 NATO E-3A AWACS aircraft.

Introduction

The Boeing E-3 Airborne Warning and Control System (AWACS) is a highly complex system that contains hundreds of electronic devices integrated into a single airframe. Considerable effort is spent in developing the racks, consoles and wiring, not to mention software, that bring all these pieces together in a package that will satisfy the system requirements. Because air vehicle accommodations, mission avionics and software elements are designed concurrently, interface definition and control is a critical engineering function that helps ensure that the entire system is fully integrated when it first comes together. Boeing has traditionally used Interface Control Drawings (ICDs) and Interface Design Definitions (IDDs) to define and control those interfaces.

ICDs can be labor intensive. The very nature of trying to control a portion of the design prior to completion goes against the nature of most designers in general as it constrains the solution space. Thus developing and controlling interface designs that enable concurrent engineering can require a great deal of personal intervention just to get the equipment designers attention. Current automation tools, with the exception of some software development environments, are not focused on interface management leaving the engineering team to either develop and reconcile the data set by hand, or develop their own tools and processes to automate this task.

Boeing has begun an effort to upgrade the computer, navigation, and communication systems on NATO's fleet of 17 Boeing E-3 aircraft. Particular emphasis is placed on using non-developmental equipment, especially Commercial-Off-The-Shelf (COTS) equipment, to reduce both non-recurring and recurring costs and to expedite the overall EMD so that it can be completed within approximately three years of contract award, including flight testing. Because of the short development time and some of the unique problems the use of COTS in a military aircraft presents, particular emphasis has been placed on interfaces, and the Boeing NATO-Mid Term Team has developed and used some automation tools and techniques to facilitate that effort.

Interface Analysis

One of the first tasks the systems integration team at Boeing faced the identification of the interfaces. It is, however, not enough to identify the interface, one must also determine how that interface will be defined and controlled so that no changes to the interface design can be made without the concurrence of all affected parties.

Intersection Interface Identification

The system integration team began by using a matrix listing equipment and software items, grouped by major subsystem, along each axis and identifying the intersecting nodes where interfaces existed. This was a derivative of the N-squared approach defined in "System Integration", by Jeffrey O. Grady, and other system engineering guides. [i]

	Item 1	Item 2	Item 3	Item 4	Item 5
Item 1		X			X
Item 2	X		X	X	X
Item 3		X			
Item 4		X			
Item 5	X	X			

Figure 1, Identifying Interfaces Using Intersections

This met with limited success. Symmetry had to be maintained by hand, and was error prone. Furthermore, because of the shear magnitude of equipment items, the matrix began to outgrow its usefulness. Finally, there was a desire to more fully characterize the nature of the interface and define the controlling documentation for each interface.

Data Driven Intersection Analysis

To overcome these problems, the systems integration team developed a relational database using Microsoft Access® to support interface identification and analysis. The first goal was to replicate the matrix that had been developed and guarantee that symmetry would always be maintained. A 'point and shoot' user interface allowed the user to identify the items that interfaced with the selected target and those that did not. The user could connect and disconnect items simply selecting an item on the "interfaces with" or the "doesn't interface with" list and executing the appropriate command. Figure 2 shows that user interface.

[i] Jeffery O. Grady, System Integration (Boca Raton, Florida: CRC Press, 1994) pp. 185-192.

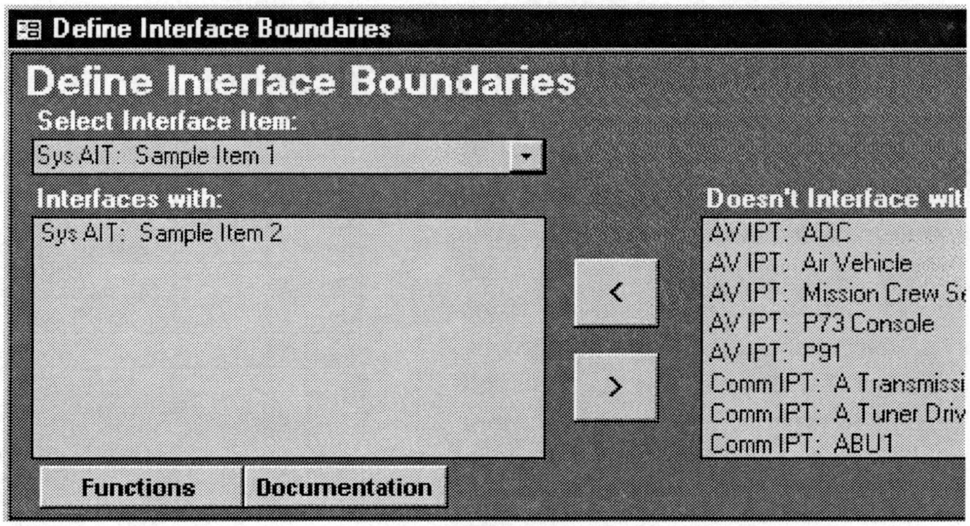

Figure 2, Identifying Interfacing Items

Once the capability to identify which items interfaced with each other was developed, it was extended to characterize the nature of the interface. A table of interface 'functions' was used and linked to the intersecting nodes, allowing the integration team to define how the items interfaced. Structure, power, analog signals and digital bus sub-systems were some of the more common characterizations of the interface. A single intersection between two items could have many interfaces under this model.

Characterizing the interfaces was not sufficient to ensure that the interface would be appropriately defined and controlled. To support this goal, one or more controls (ICDs or other appropriate documentation) were assigned to each characterization. The documentation often was an ICD, but not always. IDDs, configuration item specifications, vendor specifications and commercial standards were cited where appropriate as the controlling documentation. Note that ideally an interface is defined by a single controlling document. However there are occasions where an interface may have to be documented more than once, and the tools and processes put in place allow for relatively easy identification for all controlling documentation for a given interface.

Through this process, and tool, the integration team developed a high level model of the system interfaces. The product teams assisted in creating the data required to complete the model. In some cases whole subsystems were treated as a single entity. In other cases, individual configuration items were uniquely identified.

Developing the detailed model served two purposes. First, it provided confidence that the interfaces were well understood and appropriate interface controls would be applied. Second, looking at the model from the perspective of the interface documentation allowed the engineers to identify what interfaces should be covered in any given interface document or drawing. This proved useful in the initial documentation development and later on during reviews and audits as some ICDs covered multiple interfaces that needed to be tracked. These ICDs typically contain integration information regarding an entire subsystem and all its constituent equipment items. Capturing the ICD scope proved somewhat less useful for other interface documentation that was more focused, potentially defining but a single interface.

Interface Documentation

The Plan

One of the areas that the NATO Mid-Term Integration Team focused on was the interface documentation process, specifically the physical interfaces. The initial goal was to do away with traditional ICDs and create a data driven process that would allow all designers on the program to tap into a central data repository of information. Standard template reports compiled into documents would the serve as more formal documentation. Additionally, the configuration item drawings would capture the same information as an output of the database, and present it as requirements to the subcontractor developing the hardware.

After studying the problem we concluded their are three fundamental data 'types' that we were concerned with. The first was the mechanical, or mating interface. This has been traditionally represented with drawings, and drawings are superior in communicating the spatial relationships between the equipment items and the cabinets, consoles and shelves that they mate with. The integration team made no attempt to capture this information in the database, instead relying on mechanical drawing computer aided design tools to capture the physical mating, feature location and envelope interface data.

The second data type was textual data. In the past full formal requirements was the norm in these ICDs. As the ICDs of concern were primarily internal documents, one goal was to simplify the presentation of the requirement by capturing the key characteristic in a single field or value to make it easy to find the pertinent information. Other fields were also used to capture the other required information to complete the 'requirement'. We developed a table of these requirements and linked each to an equipment item

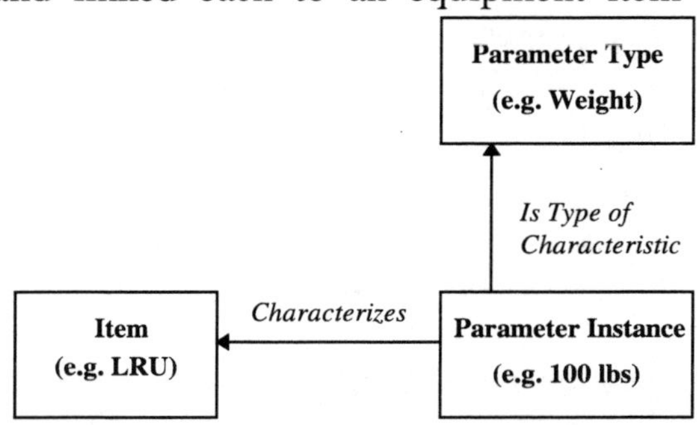

Figure 3, Characterizing Equipment Items

and the category of parameter that the entry defined as shown in Figure 3.

This relationship allowed the users to find all the parameters associated with an LRU, or all the parameters of a given type. Examples of categories are weight, power and cooling requirements. This can be an extremely useful tool when dealing with specialists who are responsible for a narrow group of requirements across all the items being integrated into the system. The specialists can filter on the characterizations of interest across a wide range of interfacing items described in the database.

In addition to the field for the target value, or primary characterization of the parameter, the table structure also contains fields for upper and lower tolerances (or maximum and minimum values, as appropriate), actual values, and textual information. Separate entries for tolerances were used in an attempt to force the designers to define the tolerances rather than specify a point value. The actual value will be filled in when the system is developed. Capturing actual values will provide a much-needed mechanism for capturing information that is difficult to produce in a systematic fashion today. Currently the designers can only look at the limiting conditions for each interface parameter to assume the absolute worst case in terms of tolerance stack-up. The textual information allows the users to use more formal requirements should they be necessary, or to add clarifying information.

The last data type is connectivity data. This includes full connector characterization and how the pins of one connector are wired to the pins of one or more connectors of another box, or even the same box if needed. The goal was to automate the building of complex wiring data sets that typically represent a large fraction of the physical ICDs. This information documents how the boxes are to be for those responsible for designing and building the cabling used in the system.

Connectivity definition has two discrete tasks. Before describing connections between two boxes, the connectors and pins on the boxes must first be characterized. Therefore the first task was to develop the schema and supporting tools to define the connectors for each equipment item. Figure 4 shows an approximation of the schema used for this purpose. Each pin relates to a connector, which in turn relates to a single equipment item. Additional information describing interface functions, such as primary power or a specific data bus, and sub-functions, such as Highs and Returns, characterized the pins and allowed the integration team facilitated the use consistent terminology. From this information set consistent tables describing every connector in the system can easily be generated.

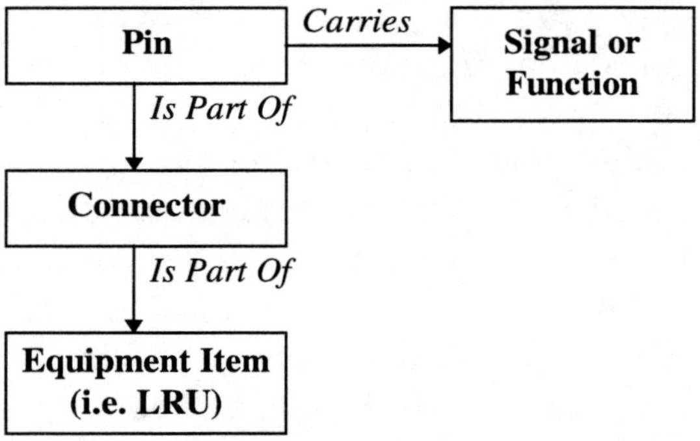

Figure 4, Defining Electrical Connectors

After fully characterizing the connectors, true connectivity could be modeled and output in tables. Because a given equipment item may be used multiple times within the system, each instance, or occurrence, of an LRU within the airframe was uniquely identified. The task then was to establish a complex relationship between the primary equipment item instance, pin, and the connecting equipment item instance(s) and pin(s). Figure 5 illustrates the schema used. The relationships were implemented using a table of 'nodes' that each combination of pin, equipment item and instance can 'connect' to. When two or more item /instance / pin combinations are related to the same node, a connection is modeled. The use of a form driven system hides the underlying queries and routines that make establishing and maintaining these relationships possible.

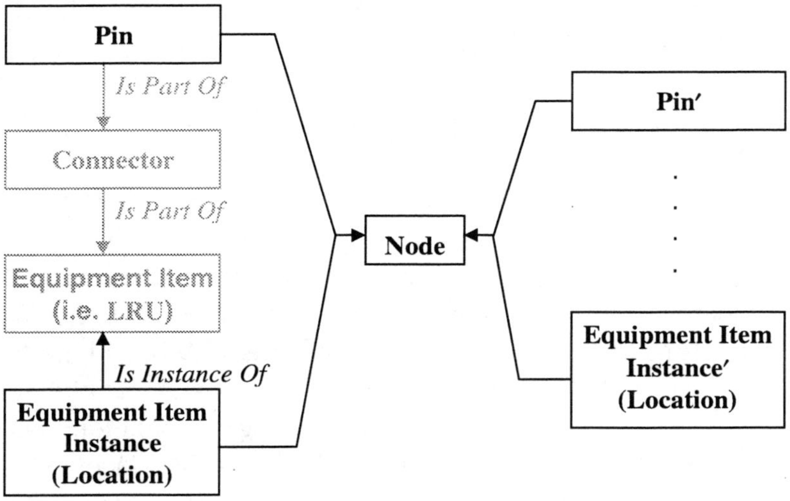

Figure 5, Capturing Connections Between Pins

The Reality

The effort to automate the interface documentation process has admittedly had only limited success. Several problems were associated with this approach.

The goal of eliminating the ICDs all together was summarily dismissed by the avionics equipment engineers and management, perhaps wisely so in hindsight. Instead of placing the interface requirements in the configuration drawings, the configuration item drawings referenced formal ICDs that would call out the interface details. While this was accommodated through automation, it presented a setback in that it created an emphasis on developing formal documentation rather than on engineering the interface details.

The software required to automate the interface documentation process has been a rapid development effort, and sometimes lagged the needs of the engineers developing the ICDs. The team encountered typical problems associated with implementing automation tools, including training. As the tool set was developed concurrently with the engineering data, no formal training and documentation was available making the transition to automation difficult at best. Furthermore, there was general resistance in abandoning formal 'shall statements' for a more structured approach to documenting the data with just the performance characteristics.

The drive to model the connectivity information came at the expense of connector diagrams. Unfortunately no mechanism was put in place to capture complex shielding and over-braiding requirements. This meant that in some cases the connector diagrams were generated in addition to the tabular connectivity data. In other cases the more labor-intensive connector diagrams were used for those individual engineers that expressed preferences for them.

Finally, as this was a legacy system, many of the interfaces had been previously defined in ICDs. The ICDs for unchanging equipment was generally left alone. New equipment that was to use existing interface designs had existing documentation to use as a starting point. Engineers responsible for developing the ICDs were reluctant to utilize a new system to create the updated ICDs when the documentation most of the needed documentation already existed. Thus much of the program did not utilize the system. This had two consequences. Often the characterization of equipment items, instances, connectors and pins at the other end of a cable were not always defined causing extra work to capture this information. Furthermore, because not every equipment item was fully characterized in the database, the specialists had limited use for the database.

Interface Problem Management

It would be remiss not to reflect on one of the other, and more successful, areas of interface management that has been automated on the NATO Mid Term program. Interface problem identification and tracking can consume a lot of resources managing the generated data. Identified problems are documented, assigned a tracking ID, linked to specific ICDs, and actions leading to the closure of the problem are assigned both to teams and individuals representing those teams. These actions to close out the open issues are tracked to completion. This effort requires regular status reports of the identified problems and actions sorted and filtered by affected ICD(s), team, individual, dates and status. Automating this process is relatively easy and provides a high return on investment by allowing the engineering staff to focus on the problems, instead of focusing on problem accounting.

Conclusions

Despite all the problems we encountered their are some positives in this experience. First and foremost, interface planning was greatly enhanced by automating the interface identification and documentation assignment effort. We developed, fielded and utilized the system for automating the interface documentation process; Several ICDs were successfully developed using the system. There are also formative plans to populate the tool with data that was not entered due to some of the problems cited above, providing the integration team with the means to quickly locate data describing the equipment items in the system. For future applications, having an existing automation system and plan for handling interfaces prior to the start of engineering development may make the difference between experiencing cost and schedule overruns, or not.

Finally, the success a program has using automated interface management, like many other system engineering automation tools, depends upon program or product maturity. New systems under development provide an opportunity to add value and reduce costs by applying new tools, methods and/or techniques across a whole program. More mature systems, with existing engineering in place, get less overall value from such tools and techniques. And where the system is sufficiently defined, no or negative value is provided as the cost of implementing automated interface management exceeds its benefits.

Keith Ryan is a Systems Engineer at the Information, Space & Defense Systems at the Boeing Company in Kent, Washington. He has provided system integration support for space transportation, commercial aircraft, systems, and is currently assigned to support airborne surveillance systems.

OpEMCSS: A GRAPHICAL DISCRETE EVENT SIMULATION LIBRARY

John R. Clymer
Applied Research Center for Systems Science
California State University Fullerton
Fullerton, CA

Abstract

A graphical discrete event simulation library is described that works with EXTEND (Imagine That Inc), an inexpensive and easy to use simulation software package. This graphical simulation library is called Operational Evaluation Modeling for Context-Sensitive Systems (OpEMCSS). Operational Evaluation Modeling (OpEM) is a graphical language that explicitly describes interacting concurrent processes. Context-Sensitive Systems (CSS) is a systems theory, based on general finite state machines, that can assist a system designer in understanding and evaluating systems. OpEMCSS facilitates the modeling of systems operation, functional flow, and architecture as they affect systems effectiveness. The parallel process view of systems is discussed in terms of CSS theory. A simulation of a space factory system that is built using some of OpEMCSS blocks is discussed as an example.

Introduction

Simulation has become, in the last few years, a mandatory part of the systems engineering process, especially as applied to the engineering of complex-adaptive systems. Large military, manufacturing, and transportation systems are too complex to be designed and evaluated without the use of simulation. Business organizations are currently being re-engineered, and simulation is a natural tool to design and evaluate them.

Some graphical simulation views in use today are queuing systems, Petri nets, and functional flow models.

- In a queuing system model, transactions flow through a network of queues and servers. This view does not explicitly describe any interactions among separate queuing processes other than the flow of transactions. It is difficult, thinking in terms of queuing systems alone, to imagine transaction flow in one queuing process depending on what is happening in another queuing process. For example, the EXTEND Discrete Event Simulation (DES) library is also based on queuing processes. To apply the EXTEND DES library to model process interactions in surface ship combat, attribute sharing and process synchronization blocks had to be added to the library.
- In a Petri net model, tokens flow from place to place as defined by transitions. What tokens signify can be as varied as the change of state in a process or data flow. This view is based on general finite-state machine theory. One of the early goals of Petri net developers was to be able to predict mathematically, model characteristics such as consistency and deadlock. Since then, the Petri net model has been expanded to model process timing, and what tokens can represent has also been expanded by allowing tokens to have "Color." With these additions, the Petri net modeling approach has lost the ability to predict things mathematically and has evolved

0-7803-5086-3 /98/$10.00 ©1998 IEEE

from an analytical to a simulation based method.

- In a functional flow model, data flows in a network of functional blocks. Each block receives data from other blocks, transforms this data into output data which is then sent out to other blocks. The functional flow model is derived from the traditional mechanistic view of dynamic systems where differential equations describe the functional transformations

OpEMCSS is based on general finite state machine theory, and it is able to model the effects of the operational, functional, and architectural aspects of a system, represented as time sequences of system state transitions. State transitions are organized into processes that transform data and send messages to each other. The OpEM view is similar to Petri nets in that it is based on finite state machines, but it is a more general model.

The parallel process view of systems is discussed in terms of general finite-state machine theory, and it is shown how context-sensitive state transitions can model system operation and functional flow. A simulation of a space factory system that is built using some of the OpEMCSS blocks is discussed as an example.

Parallel Process

A parallel process is defined as the collection of all possible sequences of system states and events. Each system state dimension has a value that identifies a condition of the system for a period of time. A condition could be the function currently being performed by a process or subsystem data. Events mark points in time where a change of system state occurs.

System state dimensions of a parallel process consist of the discrete state of each parallel subprocess plus the current value of each state variable. The discrete states represent periods of time where either: (1) system functions are being performed by resources or (2) the process is waiting for a specified logical condition to be satisfied before it can continue. The state variables represent other process conditions such as process control and data flow.

A general finite-state machine is defined as a set of states (a state space) and transition rules between these states. Let $(D_iD_j...D_k)$ be dimensions of system state and events $E_aE_b...E_n$ indicate value changes in some of these dimensions, respectively. A system transition is described by the rule:

$$(D_i\ D_j...D_k) \rightarrow E_aE_b...E_n\ (D'_i\ D'_j...D'_k)$$

This rule indicates a transition from system state dimensions $(D_iD_j...D_k)$ to new values $(D'_iD'_j...D'_k)$ occurs after event chain $E_aE_b...E_n$ is executed. Each event E_i is associated with an individual sequential process, and it may cause a change in one or more state dimensions. An event occurs in zero time; a state may persist for non-zero duration. The transition rule notation is commonly used in computation theory [12] to define languages or system behavior as strings of symbols.

The operational view, describing changes of process state during a period of time, and the functional flow view, describing flow of information from one subsystem to another in a network, can both be described by transition rules.

- A transition for a single sequential process that depends on state dimensions of two or more concurrent processes and results in a change in discrete state. A process is permitted to adapt its behavior based on knowledge from other processes. The OpEMCSS alternate action block models this interaction.

0-7803-5086-3 /98/$10.00 ©1998 IEEE

- A synchronized transition for two or more concurrent processes that depends on discrete state dimensions of each concurrent process and results in a change of discrete state for each process. Processes can coordinate begin and end of tasks using the OpEMCSS split action and assemble event blocks to model this interaction.
- Functional flow transition rules that can depend on state variable dimensions (knowledge or data) for one or more processes and result in a change in one or more state variable dimensions for one or more processes. These rules model the functional transformations of data and data-flows found in the functional-flow model. OpEMCSS context-sensitive, message, local, and global event action blocks model this interaction.

Because of the split action and assemble event pairs in a system process model, process instances can come into and out of existence as a function of time. For example, the mission execution process in the space factory model has a variable number of instances active at any given time. Such variable instantiation of processes is similar to the Object-Modeling Technique (OMT) where instances of object classes are created and deleted as the program executes [11].

Model Development

Model development should follow the following steps:

1. Define the system to be evaluated, describe system scenario or context. This step produces a context diagram.
2. Define mission of system, measures of effectiveness. This step defines system effectiveness and performance parameters to be estimated by the model.
3. Define objectives for study of the system. What questions a model must answer often dictates model details.
4. List tasks that must be performed to achieve each mission objective. What must the system do in order to achieve mission objectives?
5. Subdivide tasks into periods of time represented by states where: (a) a task is being performed by a resource and (b) a task is waiting for a logical condition to be satisfied.
6. Group purely sequential tasks into the same processes and tasks that can be performed concurrently into separate processes. Always allow for maximum permissible parallelism: tasks are sequential only if they must be done that way.
7. Develop timelines of concurrent states that describe system operation and assist visualization. Show process states on the timeline and indicate when one process sends data to another process.
8. Develop a directed graph model diagram on the computer screen. Start with tasks and wait states then add state variables to model interactions and physical details. The model is visualized from the timeline then generalized for all cases.
9. Generate event-state traces or timelines to verify and validate the model and modify the directed graph model diagram until satisfied.

A systems engineer begins model development by extracting knowledge about required system operation and control from the customer. One way to do this is to involve the customer in the process of model development. Involvement can either be direct or indirect through a review process. The customer, assisted by the systems engineer, can understand the OpEM primitives that form a directed graph model diagram of system operation.

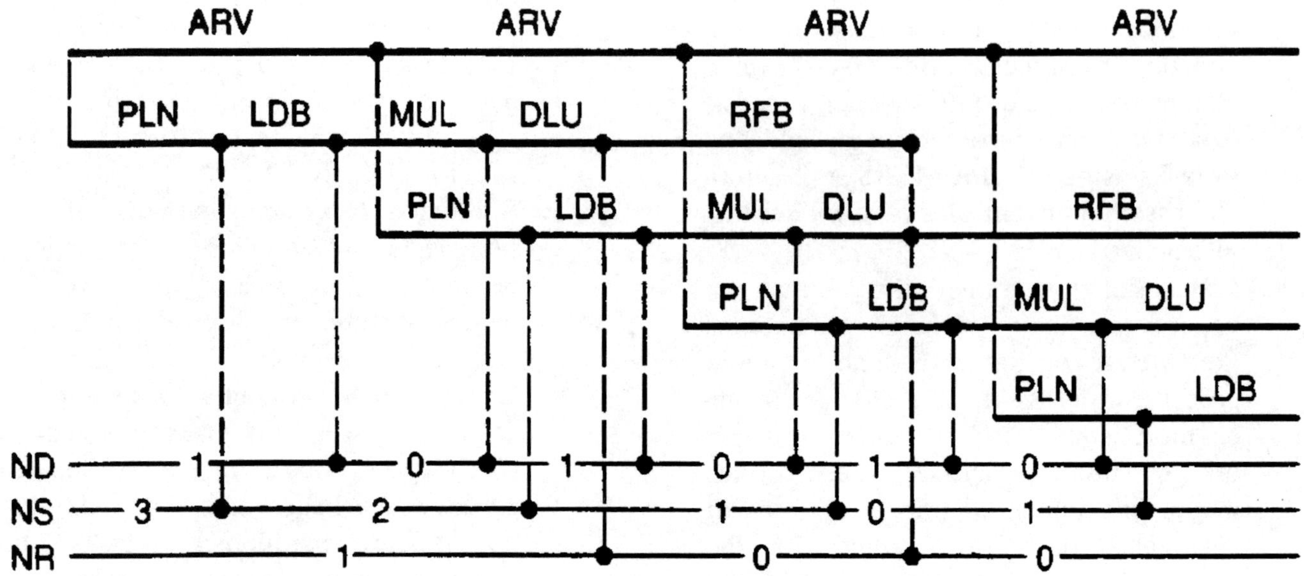

Figure 1. A Timeline of Shuttle Mission Operation

These primitives describe simply and concisely the essential simulation programming concepts required for description of a parallel process. Thus, the systems engineer can specify the directed graph model diagram on the computer screen through a dialog with the customer. This dialog can be either direct or indirect, and it can be done without either person understanding any of the complex, underlying simulation-programming details of the OpEMCSS library.

OpEMCSS library blocks are summarized by categories:

- Begin Event, End Event, and Evolutionary Algorithm blocks that define a system process instance (EXTEND calls these "runs");
- Split Action and Assemble Event that define the begin and end of concurrent processes;
- Global Reaction Time Event, Reaction Time Event, and Wait Until Event that model the time spent in a discrete state;
- Alternate Action, Context-Sensitive Event Action, Event Action, Global Event Action, Initialize Event Action, Input Event Action, Local Event Action, and Message Event Action that perform event actions; and
- Executive Block, that sequences events in simulated time.

A simple example can assist understanding the OpEM approach to model development and concomitant system specification.

Suppose a system consists of a space station factory in low earth orbit having (ND) docks at the space station, (NS) shuttle vehicles, and (NR) refurbishing facilities for the shuttles. A shuttle mission consists of: (1) mission planning - PLN; (2) load cargo aboard shuttle and boost shuttle to low earth orbit, arriving at the space station - LDB; (3) move to dock, unload, and load cargo - MUL; and (4) deorbit, land, and unload cargo - DLU. When a shuttle finishes a mission, it must be

0-7803-5086-3 /98/$10.00 ©1998 IEEE

refurbished - RFB before it can be assigned to another mission.

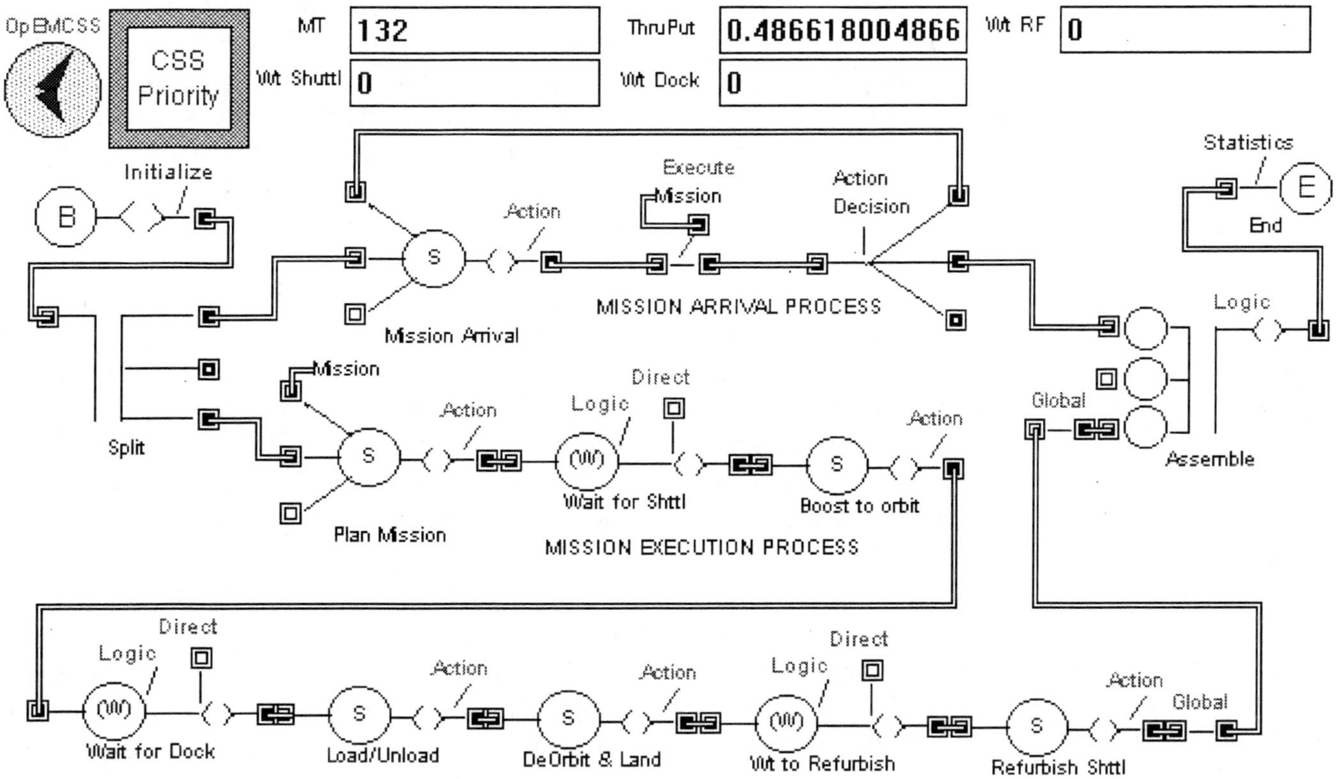

Figure 2. OpEMCSS Directed Graph Model.

A mission is generated every 48 hours. Mission times are PLN - 24, LDB - 18, MUL - 24, DLU - 18, and RFB - 48 hours. If no waiting is required, a shuttle mission process takes 132 hours to complete. The problem is to determine the number of each resource ND, NS, and NR needed to keep up with the demand for shuttle missions.

System operation can be visualized as two processes in parallel; a mission generator process and a mission execution process instance for each shuttle mission in the system. A timeline, shown in figure 1, shows how the system operates. The vertical dotted lines indicate process interactions and data flow. Each mission generated is connected to the beginning of an instance of a shuttle mission execution process. When mission execution is complete, refurbishment follows. From the timeline one can see that one dock (ND = 1), three shuttles (NS = 3), and one refurbishment center (NR = 1) are sufficient to handle the missions generated when all reaction times are constant.

A directed graph model diagram of this system is shown in figure 2 on the next page. The top process is the mission arrival process. The bottom process is the shuttle mission execution process.

At the very top of the figure are shown the mission time, throughput, and the three average wait times. Note the mission time is 132 hours as predicted above for constant reaction times.

A Begin block followed by a Split block connects to the mission arrival process and the

0-7803-5086-3 /98/$10.00 ©1998 IEEE

mission execution process. These two parallel processes connect to an Assemble Event block followed by an End block. The logic that is used to synchronize the two parallel processes is specified in the Assemble Event dialog. These four blocks define the overall structure of the simulation model and allow one or more replications of the model to be run.

The first OpEMCSS block in the mission arrival process is a Reaction Time Event block. This block models mission interarrival times. When a mission arrival event occurs, the Event Action Block creates a new shuttle mission process instance and sends it to the first block of the shuttle mission execution process which is mission planning-PLN. An Alternate Action Block in the mission arrival process decides whether to loop back for another mission arrival or to transition to the Assemble Event Block. This decision is based on the number of missions that have arrived.

The first block of the shuttle mission execution process is a Reaction Time Event Block that models mission planning -PLN. When mission planning is complete, the next state of the process is wait for shuttle-WTS. If the number of shuttles currently available is greater than zero, the process transitions to the boost to orbit state-LDB. Otherwise, the process instance waits until a shuttle becomes available.

Each shuttle mission execution process instance has reaction time states representing plan mission (PLN), boost to orbit (LDB), load/unload (MUL), deorbit and land (DLU), and refurbish shuttle (RFB) discussed previously.

Each of these states has a specified reaction time representing execution of a system task. Each wait state (wait for shuttle - WTS, wait for dock - WTD, and wait for refurbishment - WTR) occurs whenever its associated resource is not available. For each wait state a "P operation" is used to decide if the process will continue now or wait. For a P operation, if the state variable is positive, it is decremented and the process continues. If the state variable is zero, the process waits. A "V operation" increments a state variable when a resource is released by a process.

An interesting result observed from running this model is that if all reaction times are constant, then no waiting time occurs. The state timing is perfect for the resources available. However, if the mission interarrival time is gamma distributed, the random variation in the interarrival times causes shuttle and refurbishment waiting time to occur.

Summary and Conclusions

The Operational Evaluation Modeling (OpEM) graphical discrete event simulation language [1,2] was developed to describe the context-sensitive transitions required for complex systems. The challenge, in developing the graphical version of this language, was to apply lessons learned from experience obtained from the OpEM Pascal Simulation programming system [1-9], used during the past fifteen years. This system, though flexible in terms of data structures and event action code that can be created, is not easy to program.

A major challenge, for which experience with OpEM Pascal simulation was applied, is in determining a primary set of event action blocks sufficient to model any system of interest efficiently. To solve this problem, OpEMCSS event action blocks were developed to provide the capability necessary to model a set of benchmark models. When the number of benchmark models increases, other event action blocks may be needed.

Some currently planned expansions of OpEMCSS are as follows.

- Subsystem motion blocks are being added to OpEMCSS such that each subsystem can be located in a three dimensional (X, Y, Z) space. Each subsystem can be associated with a collection of process instances, and it can move in the space relative to other subsystems. When one subsystem comes within a specified range (or angle) of another, a process instance can be notified. This process instance can execute an appropriate response for the subsystem. The subsystem motion blocks allow continuous motion and subsystem interaction to occur based on discrete events. This is much faster than a time-step model of motion.
- An evolutionary fuzzy logic transformation block is being considered in which each block has process state variables as inputs. This block generates an output based on fuzzy sets and fuzzy rules, and allows process decision making to evolve, discovering the most effective rules to control the system.

See "http//ecs.fullerton.edu/~jclymer/" to download the OpEMCSS library and benchmark models.

The EXTEND simulation programming system, the foundation upon which OpEMCSS is built, provides an excellent (and inexpensive) platform from which a more sophisticated simulation language can be developed. It is a relatively high level language to begin with and already provides the capability for graphical interaction. Its proprietary C-like language makes it very flexible and creation of new types of blocks relatively easy.

References

[1] Clymer, J. R., Systems Analysis Using Simulation and Markov Models, Englewood Cliffs, NJ: Prentice-Hall Inc, 1990.

[2] Clymer, J. R., Corey, P.D., and N. Nili, "Operational Evaluation Modeling", IN Simulation, San Diego, CA: The Society of Computer Simulation International, December 1990 issue, pages 261-270.

[3] Clymer, J. R., "System Design Using OpEM Inductive/Adaptive Expert System Controller", IN IASTED International Journal of Modeling & Simulation, Volume 10, Number 4, 1990, pages 129-136.

[4] Clymer, J. R., Corey, P. D., and J. Gardner, "Discrete Event Fuzzy Airport Control", IN IEEE Transactions on Systems, Man, and Cybernetics, Volume 22, Number 2, March-April 1992, pages 343-351.

[5] Clymer, J. R., D. J. Cheng, and D. Hernandez, "Induction of Decision Making Rules for Context Sensitive Systems", IN Simulation, San Diego, CA: The Society of Computer Simulation International, September 1992 issue, Volume 59, Number 3, pages 198-206.

[6] Clymer, J. R., P. D. Corey, and H. Bandukwala, "Induction of Classification Rules for a Sonar System", IN Simulation, San Diego, CA: The Society of Computer Simulation International, April 1994 issue, Volume 62, Number 4, pages 256-267.

[7] Clymer, J. R., "Induction of Fuzzy Rules for Air Traffic Control", IN Proceedings-1995 IEEE International Conference on Systems, Man, and Cybernetics, Vancouver, British Columbia, Canada, October 22-25, 1995, pages 1495-1502.

[8] Clymer, J. R., "Expansionist/Context-Sensitive Methodology: Engineering of Complex Adaptive Systems", IN IEEE Transactions on Aerospace and Electronic Systems, April 1997 issue, Volume 33, Number 2, pages 686-695.

[9] Corey, P. D., and J. R. Clymer, "Discrete Event Simulation of Object Movement and Interactions", IN Simulation, San Diego, CA: The Society of Computer Simulation International, March 1991 issue, pages 167-174.

[10] Pritsker, A.A.B., Introduction to Simulation and SLAM II, New York, NY: John Wiley and Sons, 1986.

[11] Rumbaugh, J. et al, Object-oriented Modeling and Design, Englewood Cliffs, NJ: Prentice-Hall Inc., 1991.

[12] Yeh, R. T., Applied Computation Theory: Analysis, Design, and Modeling, Englewood Cliffs, NJ: Prentice-hall, 1976.

Author Biography

JOHN R. CLYMER is a professor of electrical engineering at California State University Fullerton (CSUF) and consults in the area of systems engineering, simulation, and artificial intelligence. In addition to consulting, he presents intensive short courses at various locations around the United States and abroad. His teaching assignments have included computer engineering, system control, continuous systems simulation, operational analysis and DES simulation, optimization and mathematical programming, and artificial intelligence (fuzzy logic and control, neural networks, and expert systems). Dr. Clymer's current research interests are focused in the area of intelligent, complex adaptive systems, applying integrated simulation, artificial intelligence, and evolutionary programming methods to study such systems. He is a founding member of the Applied Research Center for Systems Science at CSUF.

0-7803-5086-3 /98/$10.00 ©1998 IEEE

RAPID EVOLUTION OF ALL YOUR SYSTEMS – PROBLEM OR OPPORTUNITY?

Jack Ring, Innovation Management, Scottsdale, Arizona
Ernst Fricke, Institute, of Astronautics, Technical University of Munich, Germany

Abstract

Engineering practices for land, sea, air and spacecraft have been driven by different forces and priorities. These forces and priorities are now converging and alert leaders recognize that engineering practices must converge, as well. Ideally the engineering profession will seek an intelligent convergence; one that preserves the current practices that are future-sufficient while innovating new practices where necessary. All engineers across these industries have much to learn from one another[1]. However, the degree to which we converge the upstream, systems engineering practices will be the major determinant as to whether all other engineering disciplines will collaborate -- or collide.

Systems engineering practitioners and managers must respond to two kinds of challenges. Firstly, how to model and specify increasingly encompassing "systems of systems" in a way that proactively orchestrates their conjoint evolution? Secondly, how to manage change -- profitably -- which amounts to executing the classic ECP > ECN > ECO cycle with ten-fold more ramifications in each change while taking only one-tenth the time for change processing and decision.

This paper explores the problem, identifies key, future-sufficient practices and describes the concepts, technologies and behaviors that will enable Change Proficiency.

[1] Gartz, Paul Ebner Does Aerospace Technology Have a Future in Aerospace? November 26, 1997. Incorporated in IEEE Spectrum Magazine Technology Update Issue; January 1998

Drivers and Attractors

Engineering a whole solution.

Future-sufficient engineering practices will be responsive to Market, Regulatory, Technology and Corporate drivers. Such drivers include; mass customization; increasing product scope and complexity; much shorter cycle times; cost containment of controls, software and electronics technologies; technology half-lives less than product market windows; IPPD initiatives; regulatory demands and Value Chain alliances.

The implications are severe. In automotive sector, for example, the numbers of electronic components is increasing and electronic systems are getting more complex, bringing more software which, in turn, brings many changes that are not easy to control. Additionally, because the interconnectivity of all electronic systems is very high, the task of accurately estimating change impacts is one of the larger engineering problems, especially because the technology "half-life" is shorter than the model year development cycle.

The focus must shift from engineering products to engineering solutions. It is no longer sufficient to focus only on the design of the deliverable system. Engineering must design, as one integrated solution, the deliverable system, the maintenance and support system, the operator training (including simulation) system, the test system, the production system, the distribution logistics system, the marketing and selling system, and even the aftermarket customization system, Further, a design must provide for a series of versions that can be developed and released over time.

The Mars Pathfinder was designed from a total mission or solution viewpoint and

0-7803-5086-3 /98/$10.00 ©1998 IEEE

included the ability to upload new software after landing. This turned out to be a mission saving capability when the operators recognized malfunction conditions, figured out work arounds and uploaded revised software in a matter of minutes. In contrast, the automobile industry spends tens of millions of dollars updating their customer's powertrain computer PROMs.

As a member of a system of systems

Delivered systems must be designed not only to meet customer needs, but increasingly, to meet the requirements and constraints of other systems that share its operational context.

The air transport context includes aircraft, regulatory compliance verification systems, airports, weather forecasting systems, air traffic control systems, navigation systems, ground control systems, airlines (operations, maintenance and support systems), passengers and luggage, infotainment, and hospitality.

Similarly, the ground transport context includes automobiles, energy replenishment systems, traffic management systems, trip navigation systems, theft recovery systems, highway systems, regulatory compliance verification systems, maintenance systems, and passengers, as well as infotainment and other hospitality systems.

Each a higher-order system

The focus must shift to the design of higher order systems, also variously called complex, adaptive systems or goal-seeking systems.

System performance envelopes require that we design systems that operate consistently closer the "edge of chaos" than in the past. Space, capital, time, and natural resources are insufficient for the traffic demands and the market is becoming more demanding regarding pollution of all kinds. Aircraft must fly faster while closer together. Automobiles must travel faster than the anticipated actual 10 MPH on Los Angeles freeways or Tokyo expressways or the 7 MPH in Manila[2]. Emissions and noise, already reduced several fold, must be reduced still farther.

The other side of this coin is that systems must be evolvable after their initial delivery. Users (or maintenance service providers) must be able to adapt their respective systems, easily, to changes in their context and changes in user operating needs (within reasonable bounds).

Because performance and flexibility can be natural enemies, at least at one level of engineering awareness, future designs must accommodate both. For example, each system must include both order-maintaining elements and change integrity elements in addition to its primary mission elements.

Model-based engineering is key

A system model is an abstract depiction of the key entities and relationships in the system. A good system model depicts the entities and relationships in the system context along with those in system content and depicts the mesh between the context and content networks[3]. In the good old days (yesterday) engineering documentation was produced to document the design decisions and communicate the design to others who then helped further design and materialize the system. Today, the age of big screen, high resolution workstations and object technology makes documentation secondary. Rather, each participant can view the model from a number of different viewpoints and even watch an emulation or simulation of the model's response to various stimuli[4]. A system specification may be thought of as one view of this model. Likewise, the System Test Spec is

---

[2] Transportation's perennial Problems, W. Wayt Gibbs, Scientific America, Special Issue, The Future of Transportation, October97, pg 54

[3] Wymore, A. Wayne, Model-based Systems Engineering, CRC Press, 1993

[4] Taylor, David, Business Engineering with Object Technology, Wiley, 1992

another view, the Operational Availability estimates another and so forth.

Implications on Engineering Practice

Power still equals work per unit time

Although solution systems will get bigger and increasingly interoperable we cannot be less precise in our designs. For example, suppose we are managing a million dollar business to the nearest dollar, or to one part per million. If we take on the job of managing a billion dollar business we cannot be successful by continuing to manage to one part in a million because that would mean one thousand dollar-size errors. No, we still have to manage to the nearest dollar. Similarly, as we maintain the granularity of decision while designing systems of ever-larger scope, the volume of considerations and decisions must increase. Ten-fold increases are clearly possible. Simultaneously, project cycle time must be decreased by a factor of 10. Doing more work in less time means the system that performs engineering must be able to generate more "horsepower" and apply it to the problem.

The name of the game is agility

According to the Agility Forum, Lehigh, U., agility is "the ability to thrive in an environment of unforeseen change." An agile system is designed to be easily evolved. For example, the ten principles for the design of agile manufacturing and business systems as listed in Table 1 are applicable to DASC engineering practices, as well[5].

Integrating structure and process

Because the number of interrelationships increase as the square of the number of elements in a system, the challenge of managing an efficient, consistent set of interfaces grows even faster than the demand for power. Knowing the interface between your part and the next is no longer sufficient. Rather, you have to look beyond the immediate structural interface to many more considerations and ramifications. This means that process and structure must be integrated in several dimensions:

- The product structure with the product production process.
- The product gestation process (the engineering system) with the enterprise roles, responsibilities and competencies structure.
- The production system gestation process (industrial engineering) with the enterprise roles, responsibilities and competencies structure as well as the product structure and product gestation process.
- The test system structure with its production process.
- The training system structure with its production process.
- The business processes with the product structure and organization structure and the production and gestation processes.
- The business gestation process (management of change or the learning organization) with the business processes and the organization structure.

For years management sought appropriate structure (functional vs. project vs. matrix) as the key to success. More recently management practice has turned to process and process-orientation has paid off for many but has also demonstrated disappointing limits. This is no surprise to systems thinkers who know that both structure and process must be harmonized. The future-sufficient response will integrate process and structure to produce an N-way relational map -- not a hierarchy but a web of entities and relationships.

---

[5] Dove, R.K, "Tools for Analyzing and Constructing Agile Capabilities", Agility Forum, PA96-01, Jan 1996

Table 1: Agile System Design Principles
(Paradigm Shift International, www.parshift.com)

Self-Contained Units	System composed of distinct, separable, self-sufficient units not tightly coupled.
Plug Compatibility	System units share common interaction and interface standards, and are easily inserted or removed.
Flexible Capacity	Allows unrestricted unit populations that allow large increases and decreases in total system capacity or responsiveness.
Unit Redundancy	Duplicate unit types, capabilities and quantities to provide capacity fluctuation options and fault tolerance.
Facilitated Re-Use	Unit inventory management, modification tools, and designated maintenance responsibilities.
Non-hierarchical interaction	Direct negotiation, communication, and interaction among system units.
Distributed Control	Units respond to objectives, decisions made at point of knowledge, data retained locally but accessible globally.
Deferred Commitment	Relationships are transient, binding is postponed until immediately necessary.
Self-Organizing Relationships	Dynamic unit alliances and scheduling, open bidding, and other self-adapting behaviors.
Evolving Standards	Evolving, open system framework capable of accommodating legacy, common and completely new units.

The flow of design through this web will not be simple and sequential. Instead, as a design decision is made the ramifications will spread across the web like the pattern of neuron firings in the human brain. Likewise, design changes to either structure or process will appear almost randomly and spawn asynchronous flows that overlap and often are mutually conflicting. In fact, the turbulence of change will morph the integrated structures and processes web, itself. A future-sufficient web must be agile – with integrity.

Pursuit

Engineering practitioners must continuously adapt the engineering system to accommodate changing requirements, technologies, risk mitigation schemes, regulations, etc. This means that the engineering team will need more maneuverability as well as more horsepower. To cope with such change the engineering practice must take the form of a "design completion pursuit system" that accommodates the implications of increasingly interconnected customer systems and increasingly interdependent design and production systems

Future-sufficient Change Proficiency

The goal is 100X improvement.

- In volume and rate of change -- Even if development cycles are compressed considerably with IPT, IPPD, etc., the incidence of in-process changes will not diminish. Rather, the urgency for decision will increase and the need for team-wide change awareness will be come more critical.
- In change Cycle Time -- Anecdotal evidence of change management practices indicates "change decision cycle times" of weeks and months. For example; the integrated change process for Space Station was set to 55 working days. Not very responsive

but evidently the change performance limit. Similarly, the Software Productivity Group suggests that changes wreck software engineering projects if the change rate exceeds 2% of function points per month.

- In change decision accuracy --

On larger scale projects (more than 15 people) change decision accuracy can be expected to be in the range of only 80%, meaning 20% of changes cause unforeseen problems that beget even more changes. Recently a large Internet provider updated their software and the whole system crashed. Apparently they didn't check the compatibility of old and new modules before attempting to evolve the system. We cannot allow these kinds of decision errors when evolving our future integrated transportation system.

For fast, accurate changes it is necessary to easily find out the impacts of a change on the system. For example, if an Integrated Product Team wants to change something it must decide whether the change is a local decision or whether collaboration with another team is required, or whether a higher Integration Team has to take the decision. But if all changes are passed to a higher team an information overload may ensue along with the risk that an important and seriously impacting change may be delayed. Accordingly, each team needs a way to accurately assess the potential impacts of a change.

A higher order view of change must be taken.

Configuration and Change Control principles and concepts are still valid as is the Change Control Board, as orchestrator and controller. However, the current practice of change management is not future-sufficient. These ideas must be applied at a higher level than just product end items. Change must be managed throughout the web of structures and processes. More is required than just administration and coordination of change. And a strict measure of Change Effectiveness must be applied -- similar to the measures for project management, namely, mitigation of risk and convergence of conformance to quality, cycle time and cost requirements.

A new change proficiency process is required

The better and more efficient we become in implementing appropriate changes, the smaller is the gap between the decision about product functionality and the product delivery.

Strategies for coping with change have been identified and described;[6]

- Prevention,
- Front-loading,
- Selection,
- Efficient Implementation and
- Learning

These should be pursued, but still, numerous in-process changes will have to be accommodated.

Prevention and early discovery of changes must be augmented with effective selection and efficient implementation. This will be easier if we build-in the agility to accommodate change, not only in the solution systems but also in our product gestation systems.

Assessment of Real Value: Change ideas, both real and anticipated, are encouraged, then strongly filtered. Filtering must be done on a Risk-discounted, Value Added basis and only the best change ideas allowed to survive.

Decision Scheduling: The evaluation process accommodates multiple pending, overlapping changes and avoids "race" conditions, deadlocks and deadly embraces typical of an asynchronous, distributed process. Improving the cycle time, accuracy and completeness of;

- assessing the change impact in terms of scope and risk
- assessing the Risk-discounted Value of the proposed change.

---

[6] Fricke, E., No Innovation Process without Changes, but..., Proceedings of the 7th Annual Symposium, INCOSE, 1998

- interpreting the change down to the level of implementation (specs, designs, process instructions, edits, test cases, etc., and
- integrating the change into on-going activities
- measuring the actual value
- learning the characteristics of the Change Proficiency process, itself, and improving the process in both logistical and forecasting dimensions.

First In First Out scheduling cannot be used. Approved changes will be combined, sorted and restructured so that implementation can be packaged into a logical sequence regardless of the original organization of the change requests.

But we've always done it this way --

New behaviors will be required. While some conservatism must be sustained an acceptance of change and even a proactive anticipation of change will be beneficial. Also, although we tend to think of processing ECP's about the product, we will need to process ECP's about technology choices, processes, relationships and decision authority structures. We all know of product improvements that were a great idea but did not survive the NIH syndrome. In tomorrow's demanding markets, those who withhold better products in favor of internal political advantage will annihilate themselves as their business perishes.

Model-based Change is Key

Current practice uses documents as the point of change control and manifestation. This complicates Change Proficiency because any given system characteristic or attribute is likely to appear in dozens of places throughout the engineering documentation tree. With model-based system engineering and design, there is only one master reference[7]. The various documents simply become renderings of the design model.

The model-based process must include provisions for:

- handling external demands for change,
- handling internal demands for change,
- making it easy for participants to articulate change ideas, both real and anticipated,
- identifying the baseline assumed in the proposed change,
- identifying the primary impacts of the proposed change,
- identifying collateral impacts of the proposed change,
- quantifying the risk factors in making the change (including the current and anticipated state of the system model and project health),
- making a unit decision in the context of several outstanding change proposals and yet-to-be-implemented changes,
- confirming the anticipated Risk-discounted, Value as the change implementation details become available,
- prioritizing and scheduling changes.
- confirming absence of unforeseen impacts,
- learning regarding the Change Management process.

New tools are required

Although a tool cannot solve any problem sans an intelligent user, the existence of appropriate tools can encourage better thinking and behaviors. For example, paraphrasing one frustrated process manager, "Does anyone have a knowledge repository for a large organization? I have yet to find a tool that combines ease of use by process users with ease of maintenance by process developers. One tool makes it easy for a user to plunk out a

[7] Burke, Miles, White Paper: Introduction to Demand Modeling. Also, White Paper, Introduction to Integrated Product and Process Change Management. See www.mba.com.

project plan from a process, but it is hell to maintain the process and impossible to control and track changes to process steps. Another tool is much better at tracking changes at the object level, and lets you keep documents in their native format, which makes maintenance and configuration management much easier, but the product is not being actively marketed and supported."

Imagine what his frustration level will be when he grows to the level of integrating multiple processes and structures.

The information and decision automation chosen for the model-based change proficiency system must support the CCB, the individual engineers responsible for various chunks of the system, and the project engineers in predicting time, cost, risk, etc. of proposed changes. Furthermore, the system must span time -- from proposal to Last Spare Part (20 - 40 Years), space -- multiple workstations (10 to 50,000) in multiple companies (2 to 200), and media -- text, graphics and animation, including real time video. Finally, the system must contain and manage terrabytes to petabytes of data. Many data elements with average more than 10 relations to other elements.

ConOps for a change proficiency tool

The ideal change proficiency tool will:

1. Hold a multipartite, multiversion model

   To adequately represent the system design model materials, the tool must provide for[8]:

   - Registration: the data that identifies the model.
   - Revision: the current version and a reconstructable history of all previous versions.
   - Elements: objects which encapsulate individual facts expressed in the model.
   - Compositions: the structural relationships among the Elements.
   - References: the pointers to related models, compositions or elements
   - Renderings: allow incorporation of multimedia elements such as documents, pictures, film clips, voice and video. Keep documents in their native format making maintenance and configuration management easier. Captured images shall be associated with an Object and with Process Steps. These shall be subjected to revision and composition effectivity. Captured object (product, process, resource, etc.) definitions can be "rendered" as WORD documents which themselves are revision controlled in the system. The latter means that the specific content (by revision) of anything printed can be reproduced at any time from the object definition
2. Self-execute a change management process including evaluation, approval and implementation planning.. Allow processes for ECP, ECN, and ECO interacting with CCB to be configured.
3. Perform revision control, content effectivity, and composition effectivity. Also, Find and Where Used. Also, Create, Clone, Compose, and Compare.
4. Track changes at the object level
5. Generate diagrams from the process structure.
6. Publish and Subscribe to the web
7. Allow for the expression of rules which can be used to cross-check each of the components, so that the process is internally correct and sufficient
8. Provide extensive editing definition as the model and specifications are captured. Check, as a function of revision, the consistency of the specific combination of captured elements. Allow requests, as a function of revision release, to check the

[8] Burke, Miles, White Paper: Introduction to Demand Modeling. Also, White Paper, Introduction to Integrated Product and Process Change Management. See www.mba.com.

model and documentation content against manufacturing and other capabilities (that themselves are managed as revision controlled specifications).

9. And even generate project plans from the process, including effort, duration, and cost estimates.

The value of change proficiency

According to Advance Manufacturing Research Newsletter, July 1998, "Manufacturing companies typically take weeks or months to get new product changes into production." In most companies a month delay in market introduction can mean financial losses exceeding the product development budget. Furthermore, we all have heard horror stories of how a seemingly innocuous change blossomed into a series of costly impacts. These experiences tend to make engineers suppress change even when there is great potential value. We need to understand that change is not the problem. How we handle change, our change proficiency behavior, is the problem. And the value is in reducing the time and cost to bring products to market.

Feasibility

Model-based design is gathering momentum and will prevail in five years (at least among the survivors). Paper-based design representation is being replaced with visualization workstations. Tools for representing non-hierarchical structures and non-linear process behaviors are becoming available. Advances beyond object oriented thinking and design are in hand and enable the kinds of constructs needed for n-way, situational integration of structures and processes.

The Time to Act is NOW.

It will take two to five years to become future-sufficient. Start now.

Place your demands on your Engineering Automation vendors, Enterprise Information vendors, and Knowledge Management vendors and enlist the muscle of your industry associations and professional associations. A casual survey at the 1997 PDM/EDAM Conference[9] revealed no evidence that any vendor is developing a tool nor supporting the change proficiency process described herein. Interested users must speak out.

Also, back at work, acknowledge the kind of engineer who;

- parses the problem,
- identifies the system of systems framework, the role of your module and the interrelations with the other modules
- builds or facilitates evolving models of your systems, both deliverable and internal,
- sustains the understanding of the systems throughout the project time,
- helps judge the value and risk of changes and
- helps converge the thousands of design tradeoffs.

Whether you call them systems engineers, systems managers, or "hey, you" get clear on your needs for such people and make sure you have, or are developing, that unique competency.

[9] Kalthoff International, www.kalthoff.com

C. Software Engineering Track

This year's Software Engineering Track reflects on the establishment of software as the dominant intellectual and economic component within embedded systems. The Track will provide examples illustrating how the embedded systems community has decisively entered the Software Age. In this new age, the Software Engineer determines the architecture of embedded systems and respective interactions of its software components. Examples will be given in the Track's Architecture Session. In addition, this Track will address methods for assuring software product quality in the Test and Certification Session. The Software Modeling and Analysis Session examines new methods and techniques for rapid construction and analysis of complex designs. Finally, the Advanced Concepts Session introduces promising new approaches and designs that reach beyond the current state of practice. In all, this Track will offer examples of how Software Engineers, who apply intellect to hardware commodities, can create highly effective and affordable software systems.

Track Chairs: Mark Pitarys, USAF
Dr. Stephen Miller, Rockwell-Collins

Session C1: Advanced Software Concepts

Chair: Jonathan Preston, Lockheed-Martin

1	Scheduling Paths - Routing Aircraft Or Communication Packets A Fast Parallel Concurrent Shortest Path Algorithm	Mr. David Lawson
2	Compiler Optimization and its impact on Development of Real-Time Systems	Mr. Kevin Tucker
3	Lean Software Development: Is It Feasible ?	Sowmyan Raman
4	Cost Effectiveness Of Formal Methods In The Development Of Avionics Systems At AEROSPATIALE	Mr. Francois Pilarski
5	Dynamic Scheduling for Avionics Mission Computing	Mr. David L. Levine
6	Software Process and Early Project Planning: Issues, Insights, and Lessons Learned	Mr. Jogeswari P. Sarkar

Session C2: Test & Certification

Chair: Richard Fryer, Naval Air Weapons Center

1	Analyzing, Specifying and Verifying a Fault-Diagnosis-Isolation System (Not Published)	Gao Jianping
2	Low and Non-Intrusive Software Instrumentation - A Survey of Methods	Mr. Richard Fryer
3	Implementation and Testing of International Space Station Flight Software for the Pointing and Support Portion of GN&C	Mr. Douglas Yazell
4	Using Software Capability Maturity Model on Certification Projects	Ms. Leanna Rierson
5	A Comparison of Fly-Fix-Fly Testing to the Software Development Facility Testing Approach	Mr. Charles Satterthwaite
6	Methodologies for Predictable Real-Time System Design	Srini Vasan

Track C - Software Engineering (Continued)

Session C3: Software Architecture

Chair: Kenneth Littlejohn

No.	Title	Presenter
1	Simulator, Workstation, and Data Fusion Components for Onboard/Offboard Multi-Target Multi-Sensor Fusion	Ms. Angela Pawlowski
2	Resource Allocation for Distributed Computing	Mr. Russell L. Carter
3	Fitting the Pieces Together: System/Software Analysis and Code Integration Using MetaH	Mr. Jonathan Kruegar
4	Selecting a Programming Language for your Project	Mr. David Naiditch
5	A Distributed Computing Environment for Signal Processing and Systems Health Monitoring	Mr. Larry Freudinger
6	Results of the Open System Development Initiative (OSDI) at NAWCWPNS	Ms. Debra Borden

Session C4: Software Modeling & Analysis

Chair: Dr. Stephen P. Miller, Rockwell International

No.	Title	Presenter
1	A Formal Methods Approach to the Analysis of Mode Confusion	Mr. Rick Butler
2	Club Med Avionics Systems At Motel 6 Costs: Win/Win Development Strategies	Mr. James Sutton
3	Experiences from Specifying the TCAS II Requirements using RSML (Requirements State Machine Language)	Dr. Mats P.E. Heimdahl
4	SCR: A Practical Method for Requirements Specification	Dr. Constance L. Heitmeyer
5	Using Models for Test Generation and Analysis	Dr. Mark R. Blackburn Ph.D.
6	Crafting Avionics Display Applications in Java	Dr. Roger K. Shultz Ph.D.

CONCURRENT PATH SELECTION AND REAL-TIME APPLICATIONS:

AIRCRAFT ROUTING AND TASK SCHEDULING

David I. Lawson, Honeywell CSO, Houston, TX

Abstract

A method of finding the shortest concurrent path on a graph for p concurrent players, $p \geq 1$, is presented. Real-time applications, the routing of aircraft and the scheduling of collections of tasks are then discussed and a parallel Bellman shortest path algorithm is described which finds the shortest path from node r to node s on a graph G. Using this algorithm and the proper configuration of processors the shortest path on a graph with k nodes can be found in O(log n) time for $2^{(m-1)}+1 < k < (n=2^m)+1$. This is faster than other algorithms currently available within the professional literature. To find concurrent paths is a new problem which has particular application for the best use of resources. The algorithm can be adapted to schedule a wide variety of concurrent tasks.

Introduction

Finding the shortest path on a graph is a classical problem, and there are a great many shortest path algorithms with many applications. The applications include the routing of aircraft, the routing of freight, and the scheduling of tasks, and the intent is always to minimize a cost applied to the path. Modern day applications are dynamic, with players entering very quickly, often at the same time. The intent of the application can then become to find the shortest path for two or more players concurrently. This problem has a surprising twist.

Concurrent Processing

There are problems that arise when finding two concurrent shortest paths. The two paths P and Q are traced simultaneously across the graph. How is the best pair <P,Q> of concurrent paths to be found?

There is a serial solution, which fails. To demonstrate that this is so, first find the best path P, and then the best path Q. The examples that follow demonstrate that this method will not always yield the optimal solution.

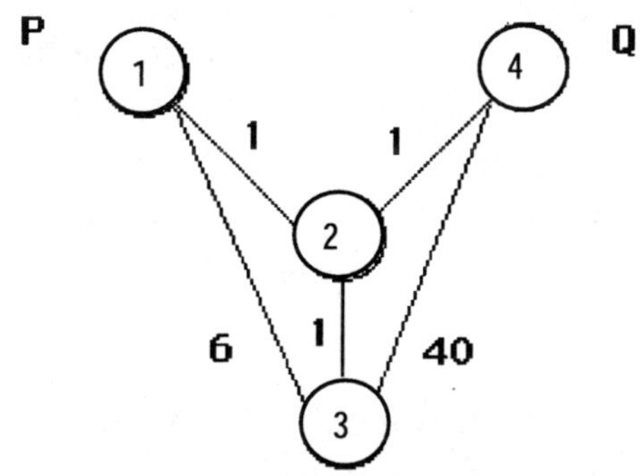

Figure 1. Example 1.

P will travel from node 1 to node 3. Q will travel from node 4 to node 2. P and Q cannot occupy the same node at the same time, but they can travel along the same arc concurrently.

Example 1

This is an example that shows that the order of choice makes a difference. In other words, there are cases when to first find the best path p_1 for P and then the best path q_1 for Q yields an answer which differs from the answer

0-7803-5086-3 /98/$10.00 ©1998 IEEE

found when one first finds the best path q_2 for and then the best path p_2 for P.

Let total = Len(best path for P) + Len(best path for Q). The following is a demonstration showing that total is different if a) one first finds a best path for P and then a best path for Q, or b) one first finds a best path for Q and then a best path for P.

If the path for P is found first the best pair of paths are <1,2,3> for P and <4,3,2> for Q. Len(<1,2,3>) = 2 and len(<4,3,2>) = 41 for a total length of 43. If the path for Q is found first the best pair of paths are <4,2> for Q and <1,3> for P. Len(<4,2>)= 1 and len(<1,3>)=6 giving a total length of 7.

For concurrent paths, finding the best path for one participant and then the best path for the other may not yield the correct answer. The reason: as demonstrated, the order of choice determines the outcome. Perhaps surprisingly, the minimum path <p,q> can not always be found by first choosing a path for P and the choosing a path for Q, or vice versa. The next example will demonstrate this fact.

Example 2

An example in which the best pair of paths cannot be found by first choosing the best path for one player and then the best path for the other, no matter which play is chosen first.

If a path for player P is chosen first then player P uses the path < 1, 2, 3, 5> with a length of 3, and the player Q is forced to use the path < 4, 6, 5> with a length of 210. The total length of the two paths is then 213.

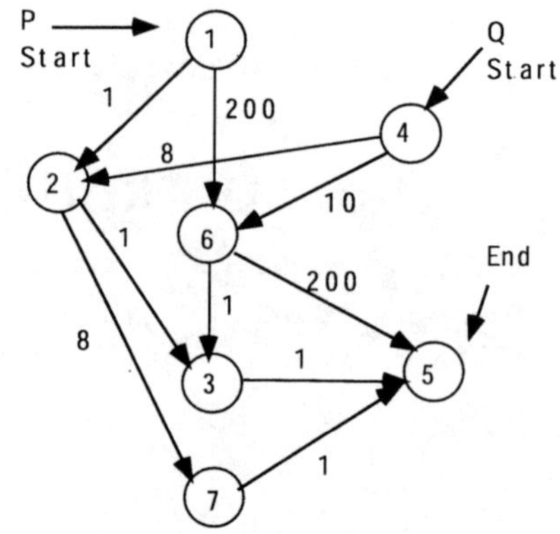

Figure 2. Example 2.

Let P start at node 1. Let Q start at node 4. Let both end at node 5. Don't let either player occupy the same node at the same time (although they can arrive at the end node at the same time).

If a path for the player Q is chosen first, then Q will use the path < 4, 2, 3, 5 > with a length of 10. The player P is then is forced to use < 1, 6, 5> with a length of 400. The total combined length for the two is then 410.

The shortest concurrent path for the two players is strikingly different. This path will have: p using < 1, 2, 7, 5> length = 10, q using < 4, 6, 3, 5> length = 12 for a total length of p+q = 22.

The Concurrent Architecture

If the graph G has k nodes, the two player model will use pairs < r, s> for r and s nodes on G. If an arc connects r_1 to r_2 and s_1 to s_2 then (and only then) an arc will connect < r_1, s_1> to < r_2, s_2>. The length from <r_1, s_1> to < r_2, s_2> will be defined to be dist(r_1, r_2) + dist(s_1, s_2), dist(i,j) is the length of the arc from node i to node j on the graph G. The model will consist of k^2 pairs of nodes connected by the arcs described. The model will thus be a graph m(G) derived in this fashion from the graph G. The concurrent shortest path problem for the graph G then becomes the problem of finding the shortest path on the graph m(G).

0-7803-5086-3 /98/$10.00 ©1998 IEEE

The Model

The model m(G) for c concurrent processes will be a graph with nodes of the form $< r_1, r_2, \ldots, r_c>$, for each r_j a node on G, j = 1,...,c. $< r_1, r_2, \ldots, r_c>$ will have an arc to $< s_1, s_2, \ldots, s_c>$ if and only if there is an arc from r_j to s_j, j = 1,...,c. $< r_1, r_2, \ldots, r_c>$ is a permissible candidate, as the node r_h and r_i do not have to be distinct if h /= i. dist.($< r_1, r_2, \ldots, r_c>$, $<s_1, s_2, \ldots, s_c>$) is defined to be $\Sigma_{i=1..c}dist(r_i,s_i)$. The model will consist of at most k^c c-tuples of nodes connected by the arcs described.

Modeling Concurrency

'Concurrent' means 'at the same time'. The problem presented by 'concurrent activity' is the clash of the use of resources. Resources must be shared. The same resources cannot be used at the same time by different entities. To solve concurrent sharing of resources using a shortest path algorithm requires that 'concurrence' (or time) be captured on a graph in some fashion. One method follows.

Building the temporal graph

Modeling concurrence can present a problem. Begin with a graph G with n nodes. If $<c_1, c_2, .., c_n>$ is a node of c(G) only if ($c_r \neq c_s$) for all pairs r and s, then no path can have two players occupy the same node concurrently. Underlying this statement is the assumption that $<c_1, c_2, .., c_n>$ are the nodes occupied by players $p_1, p_2,..p_n$ on the kth **hop** of each player, as each traverses the graph. The word 'hop' is in boldface because it emphasizes that the 'concurrence is a 'hop' concurrence, when some (perhaps most) applications are concerned with 'time' concurrence, not 'hop' concurrence.

The 'time quantized' model is introduced in the next section to represent 'time concurrence'. A special model has been introduced, because if one uses c(G), and finds that at node n, the node has been visited at a previous hop, but at the same time, then BOTH PATHS ARE TERMINATED. BUT, the path that has been to the node at an earlier hop must be terminated and A NEW PATH MUST BE CALCULATED, FROM THAT POINT OF TERMINATION.

Rather than recalculate paths from earlier hops we have chosen another approach which synchronizes hops and time. We call this graph the TIME QUANTIZED GRAPH, or just the quantized graph, and discuss its construction and utilization in the next section.

Building the time quantized graph: time conservation

For the concurrent graph to accurately reflect/model concurrence, the path of two tasks that collide must be discontinued. It is from the point of collision that the paths must be recalculated. If one task has gone on to continue to develop a path, then that calculation must be thrown away, and calculation must begin again.

To avoid recalculation we introduce another concept, the point of which is to synchronize hops and time. In this model two paths will reach a node at the same time, if and only if they reach the node on the same hop. To do this effectively it is required, in the model we present, that tasks be 'nearly homogeneous in size'. In other words, we require that tasks be the same size, or approximately the same size. If one task is several orders of magnitude bigger than another then another method of scheduling the tasks should be found. The reason for this restriction should become apparent as the model is developed.

The size(time) quantization graph.

Begin by quantizing the graph. The meaning of this: Given k tasks, $t_1, t_2, ..t_k$, let the time to complete task t_j be s_j*q, or approximately so. This requires that some number q, the quantization value, be identified that will make this possible. As an example: task 1 = size1, task 2 = size2 = 2*size1, and task 3 = size3= 4*size 1. Let M = the maximum of the s_j. In the example M = 4 and q = size1. The quantization graph will be a graph, as before. But, as a path is transversed the time to travel the path will be represented by the path length. To transform a graph G into a quantized graph q(G), spacer nodes will be inserted into the graph G. Before describing how a spacer node will be inserted look at the following example.

path1= <p_1, p_2, s, p_3, R>

path2= <s_1, s_2, s_3, w_1, R>

The s and s_i represent nodes connected to the next node in the path by an arc of length 0. In this case, the quantization value of w_1 is 4, the quantization value of p_1 and p_2 is 1, and p_3 is 2. When a player has finished task w_1 the time will be 4q. when a player has finished the path <p_1, p_2, p_3> the time elapsed will be 4q. Both players will reach the task R simultaneously. Thus, path selection on the grid will, be time consistent. The time of a path will be geometrically captured in the time-quantized graph.

The rule for insertion of the spacer nodes: If node x is directly connected to node w, and the size quantization of w is d, eg. size of w = dq, then insert d-1 spacer nodes s_1, s_2, ..$s_{(d-1)}$ nodes between node x and node w, and let the length of the arc from s_{d-1} to w_1 be the length of the (former) arc from x to w, and let the length of the other new arcs = 0.

The time-quantized graph correctly represents the original cost function, AND it maintains concurrence (it creates a relationship between time elapsed and the number of hops). e.g. the representation on the quantization graph conserves time, time is conserved (not lost).

Let n and m be two nodes on the graph G, and let an arc connect n and m. In addition let the time associated with the node n be k*q.

Proposition 1. On the graph q(G), the time-quantized graph, the number of hops from n to m = quantization value of the path from n to m.

proof. This is by construction of the time-quantized graph. If a node n has an associated time k*q, then the quantization value of n is k, and n is followed by k-1 spacer nodes. It then takes k hops to move from node n to node m.

Proposition 2. Let S and R be nodes in the graph G. Let paths i and j on the graph q(G) begin at the same node S, and then subsequently arrive at a new node R Then, paths i and j arrive at node R at the same time if and only if they arrive at node R on the same hop.

proof. This follows from Proposition 1.

Application 1: Aircraft Routing

There are two different problems that could both be called "routing." One is trajectory optimization to minimize the cost of getting from point A to point B, based on distance traveled, winds, congestion delays, etc. Airlines do this, it is usually called “flight planning” and most of the approaches are proprietary.

The other problem is traffic management, which is essentially network flow with propagation delays. This consists largely of local methods for smoothing out the flow and negotiating conflicts. This is a real-time operation, planes travel within an airspace, and move from there to another airspace or land. There are often many aircraft moving into a local airspace concurrently. Figure 3 is an example of what can happen when a mistake in traffic management occurs.

0-7803-5086-3 /98/$10.00 ©1998 IEEE

Controller sends jetliner within 200 feet of plane
NASHVILLE, Tenn. (AP) - A mistake by an air traffic controller sent a Delta Air Lines jet within 200 feet of a private plane earlier this month near the Nashville airport. Houston Chronicle, Friday, June 26, 1998

Figure 3. A Routing Mistake

A newspaper headline and first paragraph as they appeared in the Houston Chronicle this year.

Both problems are difficult. The local problem, which we address here is difficult because aircraft appear in an airspace concurrently and in real-time and air traffic control depends not only on the traffic within the airspace, but conditions on the ground as well. The costs that must be considered as aircraft are routed through the local airspace include the cost of accidents, fuel costs, penalties paid to passengers and more. In addition, local methods for smoothing out the flow must include ground traffic, because this too impacts when and where an aircraft can land. Ground traffic consists of airplanes queued for take-off and airplanes that have landed but must wait for terminal space

Example 2 demonstrates that network throughput can be increased if a concurrent shortest path algorithm is used. Can Example 2 or a similar problem arise during everyday operation? One would expect that it could. The problem arose because the best free path for one player forced the second into a much less desirable situation. Infact, local operations do run well, and there have been many years of experience developing smooth flow procedures. BUT when accidents, storms, and/or closure of traffic lanes on the ground occur, they can cause ripples into the flow pattern in the air. There are many more than two players involved, and optimal choices, even good choices, may not be easily made. One would suspect that flow control needs to include concurrent path planning.

Concurrent path planning can increase network throughput, we suspect that there are cases where throughput can be increased substantially in some cases. We do not have data to substantiate this claim. It is at present a conjecture.

Application 2: Scheduling a Collection of Tasks

A collection of tasks can be modeled as a graph G. Each task can be represented as a node on the graph If R and S are two nodes on the graph, and task R can be followed by task S, then a directed arc should point from R to S. The arc length should be the length of time to complete task S. Then, using a time-quantized graph q(G), as described above, a shortest path algorithm can find the 'best' concurrent assignment and order of the tasks.

The shortest path algorithm used should make it impossible for a task to be completed twice. If there are d concurrent participants then a node on the concurrent graph will include $<c_1,..,c_d>$ tasks. After k hops, k*d tasks will be complete. It will then be easy to determine when or if all of the tasks have been completed.

For purposes of clarity, if two tasks must be performed sequentially, collapse them to one task. Call such a graph an INDEPENDENT graph.

Theorem. If an independent time-quantized graph is completely connected, then a path can be found through each of the nodes of G. Thus, all of the tasks can be completed.

Proof. Suppose the longest path for one player $p =<n_1,..., n_q>$ that can be formed on G is only q nodes long. In addition, suppose that N, the number of node in G is > q. Then there is a node w which is not in the path p. G is completely connected, so here is an arc connecting nq to w. Thus p+w is a path on G which is longer than p. This proof , which is just for one player, can be easily extended to k concurrent players for k>1.

This may be too restrictive in some cases. Scheduling tasks can become quite sophisticated as constraints are applied to the problem. As an example, it may be the case that some participants can do some tasks, but not others. To model such a situation, let node B be connected only to tasks that participant b can complete. Then connect each of these tasks to themselves. Participant B will then be connected to all of the tasks he can participate in. As the concurrent path is built, do not allow participant b to travel to an unauthorized node. This can be modeled by representing this constraint in the model c(G). Alter the construction of c(G) in the following fashion: If b represents the kth component ck of the element $<c_1, c_2,..,c_k,..c_n>$ of c(G), then connect ck only to elements that b is allowed to do. If the time-quantized graph is used, the best concurrent path can be found.

Real-Time Applications

For an application to be solved in real-time the size of the problem must not overpower the solution. The parallel implementation of Bellman's algorithm that is presented in the following section is a parallel solution and as such it can be scaled up by adding additional hardware. Then the solution speed will be of O(log n) for a graph of size k, for $2^{(m-1)}+1 < k < (n=2^m)+1$.

A Parallel Implementation of Bellman's Algorithm

This algorithm is presented in complete detail in [Lawson, 1998]. The architecture for the parallel implementation of Bellman's algorithm, pictured in Figure 4, is used to find the shortest path on a graph G of k nodes. Let $n=2^m$ for m the least integer such that $k<=2^m$. The hardware will consist of a grid of nodes with k rows and n+1 columns. A node r in column i is connected to a node s in column i+1 if node r is connected to node s by an arc in the graph G. Figure 4 is a picture of the k x (n+1) grid.

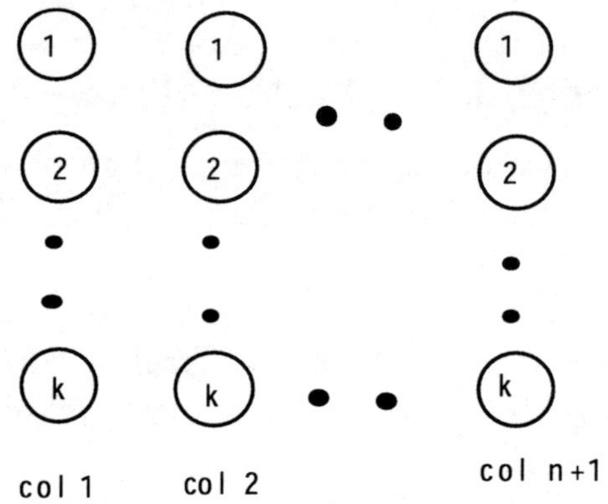

Figure 4. The network architecture.

The network architecture of the rows and columns used by the parallel algorithm's hardware.

The Parallel Shortest Path Algorithm

A brief description of the solution follows. This is a parallel implementation of Bellman's algorithm. The algorithm has two parts. First the shortest distance is found. Then a path with that distance is found. The implementation of the parallel algorithm uses the network pictured in Figure 4.

STEP 1. Find the minimum 2-hop path lengths

Step 1.1. Use each node k in column 3 to calculate the minimum distances dist(1,k), for node 1 in col 1 and node k in col 3. The distances dist(1,k) must be sent to the node 1 in each column. In particular, each node 1 should have a memory location available to hold this value once it has been computed. Dist(i,k) = min{dist(i,j) + dist(j,k)} for i in column 1, j in column 2 and k in column 3. Dist(i,k) is computed at node k by finding the minimum in parallel. To find the minimum of k inputs in parallel requires k processors and uses a parallel algorithm. The minimum distance is the length of the path $<1, x_1, k>$ for some node x_1 in column 2. Save the node x_1 at node k so that the x_1 can be recovered when the shortest path is generated. The node x_1 may not be unique. If it is not unique then just pick one. The node x_1 will be used to build the path with the shortest distance from the start node to the finish node.

Step 1.2. Use the nodes in column 4 to calculate the minimum distances dist(2,k), for node 2 in col 2 and all k in col 4. Step 1.2 occurs at the same time as step 1.1. Note: the minimum distance dist(2,k), for node 2 in col 2 and k in col 4 equals the minimum distance dist(2,k), for node 2 in col 1 and k in col 3. The distances are computed using columns 2 and 4 so that the operation of finding the distance from node 2 to nodes k can occur in parallel with the computation of the distances from node 1 to the nodes k. The distances dist(2,k) must be sent to the node 2 in each column. In particular, each node 2 should have a memory location available to hold this value once it has been computed.

Step 1.j. At the same time as the other calculations in Step 1 occur use the nodes in column j+2 to calculate the minimum distances dist(j,k), for node j in column j and all k in column j+2 mod g, for g = number of nodes in the graph G. The minimum $dist(j,k) = \min\{ dist(<j, x_j>) + dist<x_j, k>) : x_j$ a node in column j+1}. After computing the minimum distances, broadcast the results. The distances dist(j,k) must be sent to each node j in each column. In particular, each node j should have a memory location available to hold this value once it has been computed. The result is that in step 1 each column of the kx(n+1) array is involved in the computation. Nodes in column 3 compute the minimum inputs from node 1 to each node j. Nodes in column 4 compute the minimum inputs from node 2 to each node j. Nodes in column k compute the minimum inputs from node k-2 to each node j.

Collapse the Network

After completing step 1 remove the columns 2, 4, .., $(2^m)/2$. Form arc lengths on the new network by connecting node j on column r to node k in column r+1 if there has been a two-hop distance between the two nodes which has been computed in step 1. After connecting node j in column r to node k in column r+1, give it the arc length computed in step 1. The network formed after the completion of step 1 has $(2^m/2)+1$ columns.

STEP k.

Repeat step 1 on the network formed after the collapse following the completion of step k-1. The network formed after the completion of step k has $(2^m / 2^k)+1 = 2^{(m-k)}+1$ columns.

Theorem. The parallel algorithm finds the shortest distance between nodes i and j for each i and j provided a path exists between the two.

Proof. The theorem is true for 1-hop paths. Suppose, as the induction hypothesis, that the theorem is true for n-hop paths. Let $<p_1, p_2, .., p_{(n+2)/2}, .., p_{n+1}>$ be the shortest path returned by the algorithm that connects nodes p1 and pn+1. Suppose there is another path $<p_1, r_2, ..., r_{(n+2)/2}, .. p_{n+1}>$ that is shorter. Then by the induction hypothesis the distance of path $<p_1, .. r_{(n+2)/2}>$ is the shortest from node p_1 to node $r_{(n+2)/2}$. The distance of the path from node $r_{(n+2)/2}$ to node $p_{(n+1)}$ is the shortest from node $r_{(n+2)/2}$ to node $p_{(n+1)}$. The length of the path $< p_1, r_2, ..., r_{(n+2)/2}, .. p_{(n+1)} >$ is the sum of these distances. This length is presented as an input to node $p_{(n+1)}$. But, the minimum of these inputs is the length of the path $<p_1, p_2, ..., p_{(n+2)/2}, .., p_{(n+1)}>$. Thus, the length of the path $<p_1, r_2, ..., r_{(n+2)/2}, .. p_{(n+1)}>$ cannot be shorter and the theorem is proven.

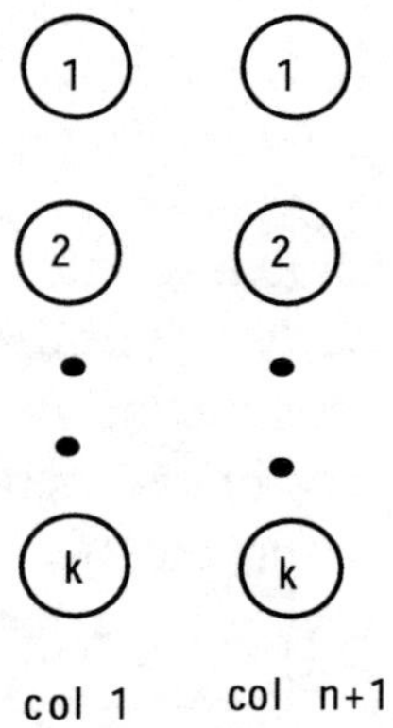

Figure 5. When the algorithm terminates.

The algorithm terminates with a two column Logical grid, column 1 and column n+1. Node R column 1 has a Logical Connection to node S column n+1 exactly when a path exists from node R to node S. The length of the Logical Connection is computed by Step Log(n), and that length is the shortest distance from node R to node S. After $O(\log_2 n)$ steps the network consists of just 2 columns: columns 1 and n+1.

Recovering the Shortest Path

Unwind the algorithm by working backward. See [Lawson, 1998] for details.

Conclusion

We have shown by example that the concurrent choice of paths requires a special concurrent method if the 'best/minimal' path is to be found. We have also demonstrated one such method, the creation and use of the concurrent graph c(G) from the original graph G. The time-quantized graph q(G) is also introduced. Two applications for concurrent path selection using c(q(G)) are examined: the routing of aircraft and he scheduling of tasks.

If the concurrent choice of paths is to be made in real-time then a parallel algorithm is well suited for the task. The parallel Bellman algorithm presented above is valuable because it is faster than other shortest path algorithms currently available in the literature.

References

Lawson, David, I.: Real-Time Concurrent Path Selection for the Movement of Multiple Robots on a Factory Floor. The Proceedings of the International Symposium on Mobile Control in Robotics (ISMCR '98), Prague, The Czech Republic, June 8-12, pp. 321-326, 1998.

0-7803-5086-3 /98/$10.00 ©1998 IEEE

COMPILER OPTIMIZATION AND ITS IMPACT ON DEVELOPMENT OF REAL-TIME SYSTEMS

Kevin Tucker, DDC-I, Inc., Phoenix, AZ

Eli Solomon, Northrop Grumman, Electronic Sensors & Systems Sector (ES^3), Baltimore, MD

Kenneth Littlejohn, United States Air Force, Air Force Research Laboratory, WPAFB, OH

Abstract

Real-time systems are increasing in size and complexity. Software developers seek to maximize the performance, as well as minimize the memory requirements in order to meet the constraints of these systems. Traditionally, software developers have overcome these constraints by using an optimizing compiler. However, optimization carries a hidden price. Many Commercial-Off-The-Shelf (COTS) debuggers are unable to work effectively with optimized applications, because the one-to-one mapping of source to object code is often disturbed by optimization. While optimizing, a compiler is free to reorder or interleave the code for statements, so long as the semantics of the program are preserved. This causes two major problems for a debugger. First, selecting a representative instruction to be used when setting a breakpoint on a source-level statement becomes difficult. Second, at a given breakpoint, the value of a variable may differ from what it would be in the non-optimized version. This condition is known as non-currency. Each of these problems becomes a serious issue if a debugger is used to debug the deliverable code or to perform verification testing. This paper presents an in-depth study of the effects of loop-invariant optimization on debugging user applications. The paper will show how to significantly improve the performance of the debugger, with respect to breakpoints and non-current variables, when operating on code affected by this optimization. A summary of the research done, including specific implementation issues and lessons learned while enhancing the DDC-I SPARCStation Solaris native Ada compiler system, is presented.

Introduction

The inability of debuggers to effectively handle optimized code imposes a real burden on the application developer. A bus-analyzer and assembly listings are often used to track down an obscure bug in the optimized, deliverable code. Most debuggers do not completely circumvent optimized code; rather, each debugger makes its best attempt at dealing with the optimized code. The problem with this approach is that the debugger does not communicate to the user its inability to correctly complete the requested operation. Control flow in code generated by an optimizing compiler is often different from the control flow that a user would expect from the source code. Consequently, a non-cognizant debugger might present the values of non-current variables without any warning to the user, or set a breakpoint at an incorrect location, or fail to set all relevant breakpoints; these are some of the scenarios which could

0-7803-5086-3 /98/$10.00 ©1998 IEEE

lead the user astray. These scenarios are being addressed under the Debugging Highly Optimized Ada with Code Motion (DHACM) program. The goal of the DHACM project is to significantly improve the performance of the debugger when operating on optimized code. Loop-invariant optimization, where the compiler hoists out of the loop, statements whose results are not dependent upon the loop-variable, was chosen as the focus of this investigation. The rationale behind this is that it clearly demonstrates both major problems caused by compiler optimization, viz. determining the location of breakpoints, and non-currency of variables. A method of determining the correct location of breakpoints, for a loop-invariant statement, is provided. Furthermore, using the example program given in Figure 1, the effect of loop-invariant optimization on the currency of program variables at a breakpoint, is demonstrated by using different zones delineated by comments.

Setting Breakpoints

Normally, when a breakpoint is set on a source code statement, it is placed at a point prior to the execution of the action described by the statement. This point is termed the "semantic" breakpoint by Zelleweger[1]. At the semantic breakpoint, the user can examine, and possibly change, the state of the system prior to the execution of the statement. This breakpoint is contrasted with the "syntactic" breakpoint, which is at the location of the statement in the source code. For non-optimized code, these locations are the same, but for loop-invariant and other code-motion optimizations, they are separate and distinct. For example, in Figure 1, the syntactic breakpoint for the loop-invariant statement remains inside the loop, whereas the semantic breakpoint has moved above the loop, to an area known as the loop "pre-header" (Zone 2). The pre-header identifies the area after the loop has been initialized, and before the iteration portion of the loop has been entered. Those debuggers that are unaware of code-motion optimizations having been performed by the compiler, will set a breakpoint only at the syntactic position, a point at which it is too late for appropriate user action. In order to correctly set a breakpoint on a source code statement affected by loop-invariant optimization, the debugger must locate and set breaks at both positions. This ensures that the breakpoint will be hit both prior to executing the statement, and also each time through the loop, as the user would expect. However, this approach has the problem of setting one too many breakpoints. In a loop with N iterations, N+1 breakpoints will be hit (the extra breakpoint is at the semantic location). A potential remedy for this problem is the use of a conditional breakpoint at the syntactic location that is not activated on the last iteration.

To improve the setting of breakpoints, the compiler was enhanced to insert, in the debugger tables, additional line information that marks the segment of code as belonging to its original line, together with an indication of the optimization that caused this sudden switch. In addition, another line marker is output to mark the end of the optimized region, returning the current line to the next sequential line following the pre-header. This is explained in more detail in Tucker[2]. While recognition of the semantic breakpoint is necessary to correctly handle loop-invariant optimized code, it is not sufficient. The problem of non-current variables becomes self-evident when execution is stopped at the semantic breakpoint.

[1] Zelleweger, Polle-Trescott: Interactive Source-level Debugging For Optimized Programs, Ph.D. Dissertation, University Of California, Berkeley. May 1984.

[2] Tucker, Kevin: Debugging Highly-Optimized Ada With Code Motion (DHACM), Proceedings Tri-Ada '97. November 1997.

Non-currency Problems

During the testing and verification of the operation of real-time systems, debuggers are frequently used, in conjunction with simulators or in-circuit emulators, to set breakpoints in critical areas of the software, and to examine the values of variables. Consequently, it is imperative that accurate information about the value of a variable, be displayed by the debugger. If this requirement cannot be precisely fulfilled by the debugger, then the current value, together with a warning, must be presented to the user, to alert the user of the situation. To illustrate this condition, the program fragment in Figure 1 has been annotated to mark five different zones of interest.

```
--Prior to the loop (Zone 1)

X := 1;

for I in 0..10

--Loop pre-header (Zone 2)

loop

--Between the pre-header and the original position of the hoisted
--statement (Zone 3)

   X:= X+I;

--Original position of hoisted statement (Zone 4)

   F:=G+H;

--After the original position of the hoisted statement (Zone 5)

   A(I):=G;

end loop;
```

Figure 1. Example Program with Annotated Zones.

These zones identify the occurrence of different non-currency conditions, and the user will see a difference in the way affected variables are displayed and assigned. For example, in Zone 2, a variable may be non-current because an expected assignment has not yet taken place. In Zone 3, assignment to a variable may be inhibited because that variable is involved in a statement which has been affected by loop-invariant optimization. In Zone 5, the inhibition upon assignment, placed in Zone 3, could be lifted. The special characteristics of each zone are explained in this section.

Prior To The Loop (Zone 1)

The location of the boundary between Zone 1 and Zone 2 was selected so that it would include the determination on whether the loop is to be entered or bypassed. This check is necessary in order to assure that the semantics of the program remain unchanged. Aside from this, no special handling for non-current variables is required here.

Loop Pre-header (Zone 2)

This is the area to which loop-invariant statements are moved. A unique feature of Zone 2 is that it is guaranteed to be executed only once, so that the debugger can determine with certainty if a given variable is non-current.

When execution is stopped here (as a result of encountering a semantic breakpoint for a loop-invariant statement), it will appear to the user as if prior statements in the loop (from Zone 3) have already been executed, when in fact they have not. Therefore, display of any variable (such as X in Figure 1) modified by a statement in Zone 3, must be accompanied by a warning, stating that the expected assignment has not yet occurred.

To implement this warning, the debugger must receive an indication of the condition when it is determining the value of the variable. In the DDC-I system, this indication is stored in a data structure known as the Storage Interpreter (SI) tables. These tables are generated by the compiler, for every basic block affected by optimization, and contain information about the dynamic properties of program variables. Furthermore, these tables are used by the debugger when looking up a variable, and the information in these tables indicates if a variable has been moved to a register, or has been replaced by a constant, or is to be protected against debugger modification. To allow the debugger to detect non-current variables, these tables were enhanced to include an indication of optimization. Once the loop-invariant optimization phase has completed, all statements between the loop pre-header and the original position of the loop-invariant statements (Zone 3) are inspected, and an optimization mark is added to the SI information for all the variables that are modified in that region.

Between The Loop Pre-Header And The Original Position Of The Hoisted Statement (Zone 3)

Here, the non-currency issues surround variables involved in the statements that have been hoisted out of the loop. In non-optimized code, when a breakpoint is reached in this zone, none of the loop-invariant statements has been executed. Because of the non-sequential execution created by optimization, special treatment is required for variables both read and modified by the loop-invariant statements.

From the user's point of view, the loop-invariant statements have yet to execute. Therefore, any debugger modifications to variables used by these statements, (such as G and H in Figure 1), should have an effect on the result of the loop-invariant statements. Since this will not happen, the user must be notified that debugger assignments to these variables will not have their intended effect. As a result, the debugger has two options. One is to prevent the assignment, with an appropriate error message. The other is to allow the assignment, with an appropriate warning message. While either method is acceptable, the first option was undertaken for this project, in order to remain consistent with the existing treatment of variables that are involved in common sub-expression elimination.

The situation is similar for variables that are modified by a loop-invariant statement (such as F in Figure 1). The user can reasonably expect that debugger modification to a variable, which is the result of a loop-invariant statement, will be overwritten in the next iteration of the loop. Since this is not the case, the debugger faces the same options as for referenced variables. The option undertaken for the DHACM project was to prevent the assignment.

A further non-currency condition occurs when a variable modified by a loop-invariant statement is displayed. As explained above, the user does not expect the loop-invariant statements to have been executed yet. Consequently, the user would not expect any variables modified by such statement to have received their new values yet. However, due to optimization, those variables have in fact been updated. This deviation from expected behavior must be flagged by the debugger with a warning message to the user (such as the display of the variable F in Figure 1).

It is interesting to note that the last condition is manifested only until the loop has executed once. Beyond that point, the user will expect that the loop-invariant statement has been executed, and the debugger should not need to warn about non-currency of the modified variables, any longer. However, the debugger does not currently track execution history, and has no way of knowing how many times the loop has executed, so the warning message, stating that the variable in question "may" be non-current, is presented during execution of the entire loop. This is in contrast to the situation in Zone 2, where the debugger can state with confidence when a variable is non-current. Although it may be possible to determine if the loop is past its first iteration by comparing the current value of the loop index against its bounds, this capability was not implemented for the DHACM project.

Implementation of this enhancement involves improving the compiler to output additional information during the optimization phase. Specifically, for all loop-invariant statements, the compiler will output one of two types of Storage Interpreter (SI) information for each variable involved in a loop-invariant statement. For right-hand-side, or referenced, variables, the compiler emits a SI mark with a type of Loop-Invariant-Operand, and for left-hand-side, or modified variables, the compiler emits a SI mark with a type of Loop-Invariant-Result. The reason for the differentiation lies in the debugger's interpretation of the SI marks. When the debugger is asked to display a variable that carries a mark of Loop-Invariant-Result, this signifies that this variable was involved in an assignment that has been hoisted, and as such is non-current. Therefore, the debugger should inform the user of this when the variable is displayed. A variable that carries a mark of Loop-Invariant-Operand was only referenced, but not modified, in the loop-invariant statement, so display requires no special action. However, modification is an issue. Modification of a variable that has been marked as a Loop-Invariant-Operand will not have the expected result, and should not be allowed. An attempt at modification will cause an error message to this effect to be displayed.

Original Location Of The Hoisted Statement (Zone 4)

This region is where the syntactic breakpoint is located. However, no special action for non-current variables is performed in this region.

After The Hoisted Statement (Zone 5)

In this area, the inhibitions on assignment placed upon the variables in the loop-invariant statement could be relaxed in order to allow changes to the involved variables to have an effect on subsequent loop statements. An example of this is the assignment of G to A(I) in Figure 1. However, because the statements are still inside the loop, the user will still expect that changes to these variables will affect the result of the loop-invariant calculation on the next iteration; therefore, the locks are kept in place. If the debugger could discern when this is the last iteration of the loop, the debugger could safely release the locks. Again, although this determination could be made by comparing the current value of the loop index against its bounds, this capability was not implemented for the DHACM project.

Efficiency Issues

Improving debugger response to optimization should not significantly increase the size of the executable, or degrade the performance of the debugger. On the DDC-I system, all debugging information is kept in the Ada library, so the size of the executable remains constant, whether debugging information is present or not. The size of the debugging information stored in the library has increased slightly. It is estimated to be one 32-bit word per source-code statement for the improved handling of breakpoints, and one 32-bit word for each variable involved in a statement that has been moved due to loop-invariant optimization.

The speed of handling breakpoints has not changed significantly, since the debugger needs to do only one additional check to see if the statement has been affected by optimization, and if so, search for the additional breakpoint. For non-currency checking, the speed of the debugger has changed only minimally, since the required checks were already being performed. The minimal degradation in debugger response time is attributable to the additional SI marks that are processed.

Concluding Remarks

This paper has examined the problems faced by developers when loop-invariant compiler optimization has impacted the generated code, including issues with the location of breakpoints, and the non-currency of variables. Using an example program, the regions of code that are affected by this optimization have been clearly outlined, and the non-currency issues associated with each region have been enumerated. Suggestions for improving compiler and debugger technology have been presented in a general way, suitable for use in any language and any compiler implementation. The technology that was explored and demonstrated by the DHACM project, and the methods suggested by this paper, have since been transitioned into the design of the next-generation of DDC-I products, which include Ada95, C, and C++ compilation systems. In addition, problems found with the chosen approach, which is a basis for further research, have been identified.

Acknowledgments

This project is sponsored by the Air Force Research Laboratory, Wright-Patterson Air Force Base, Ohio, under contract F33615-95-C-1648 to Northrop Grumman, Electronic Sensors and Systems Sector (ES^3), Baltimore, Maryland, with DDC-I, Inc. as a subcontractor. Acknowledgments and thanks are extended to Karl Rehmer and Jackie Benton of DDC-I, and to Jon Squire of Northrop Grumman ES^3, for their work on the project.

Further References

Wismuller, Roland. *Debugging of Globally Optimized Programs using Data Flow Analysis*. ACM SIGPLAN Notices, Vol. 29, No 6, June 1994.

Ali-Reza, Adl-Tabatabai, and Gross. *Detection and recovery of Endangered Variables Caused by Instruction Scheduling*. ACM SIGPlan Notices, Vol. 28, No 6, June 1993.

Ali-Reza, Adl-Tabatabai, and Gross. *Source-Level Debugging of Scalar Optimized Code*. ACM SIGPlan Notices, Vol. 31, No 5, May 1996.

LEAN SOFTWARE DEVELOPMENT: IS IT FEASIBLE ?

Sowmyan Raman, The Boeing Co, Seattle, WA 98124.

ABSTRACT

The Lean Aerospace Initiative originally began as Lean Aircraft Initiative (LAI) in the summer of 1992 at Massachusetts Institute of Technology (MIT). The refashioning of Lean Aircraft Initiative to Lean Aerospace Initiative (LAI) took place in November 1997 to reflect the addition of Space Sector. A consortium of Air Force and 21 defense firms including the Boeing Co and the erstwhile McDonnell Douglas was originally formed to investigate how the "Lean Principles" that were applied in the automobile industry particularly in Toyota could be adopted in the Aerospace industry including Electronics and Avionics segments [1].

It is no secret that the next millennium will be digital. Software will not only play an important role in the Aerospace systems but almost in every aspect of our life It is the dream of every enterprise to have software developed cheaper, faster, and better. During the last several years many technologies, methodologies and software languages and development tools have emerged as the answers to the above dream. We are yet to accomplish that dream

Modern aircraft is nothing but a bunch of "flying" computers connected together and it is just "Electronics in motion" or flying electronics. It is estimated that 50% of the development costs for any modern airplane is attributed to systems and a substantial portion to software development.

Toyota is accepted not only as a leader in the automobile industry but as a world class manufacturing company which is very successful in adopting lean principles which they call as Toyota Production System (TPS) [2]. This paper will briefly describe what is "Lean Thinking" and explore the feasibility of applying these principles to software development. It will include discussions on the emerging software development methods and Capability Maturity Model (CMM)[9] of the Software Engineering Institute (SEI) which gives us a framework. It will attempt to answer whether these principles of Lean can to applied to Software Development.

What Is Lean ?

Lean is all about getting the right things to the right place at the right time the first time while minimizing waste and being open to change. The production system that was pioneered after world war II by the Toyota company for the manufacture of automobiles was labeled as "Lean Production" by the authors of the landmark book: The Machine That Changed the World [2]. This book was the direct result of the study by The Massachusetts Institute of Technology 's five million, five year effort, on the future of The Automobile. Lean Aircraft Initiative was formed to extend

0-7803-5086-3 /98/$10.00 ©1998 IEEE

the Lean production paradigm and knowledge-base from this study. A key goal was to articulate, define and apply concepts, principles and practices of Lean production to defense aircraft industry initially and aerospace industry in general latter, through quantitative bench-marking of best industrial practices.

The word *muda* a Japanese word is the center of this whole Lean revolution. *Muda* means "waste" specifically any activity that absorbs resources but outputs no value. Taiichi Ohno (1912-1990) , an executive of Toyota who developed and implemented Toyota Production System (now known as Lean) was horrified at different forms of *muda* which made him develop methods to eliminate them [3]. Some of them he identified are: Mistakes that are subsequently rectified, unwanted production and inventory, processes which are not really needed, unnecessary movement of people/things, waiting, and products and services which do not meet the needs of the customer. Some of the above are obviously more applicable to hardware environment rather than software.

What are the basic concepts of Lean Thinking. These are: Value, the Value Stream, Flow, Pull, Perfection. Let me describe these concepts[4].

Value: Value can only be described by the ultimate customer. It is useful when expressed in terms of specific product which meets customer's needs at a price at a specified time. In the context of aircraft industry Phil Condit, Chairman and CEO of The Boeing Co, has recently said " the first 75 years were defined by constant push for performance. The aerospace industry is going to change to value driven industry" [5]. Airlines want "value" as perceived by their ultimate customers, rather than advanced and exotic technology or super performance.

Value Stream: is a set of actions required to bring a value product or service to the customer. An aerospace example would be a titanium ingot made 10 times the weight of the finished part. By redesigning the ingot to suit several parts , unnecessary forging and machining that added no value were eliminated. By making supply chain transparent to all the players the value stream can be strengthened.

Flow: Once the unnecessary steps are eliminated the remaining value creating steps need to be made to flow. Instead of organizing functionally, development or production activities should be organized as continuous flow with responsibilities for complete processes or sub-processes. This type of arrangement eliminates sub-optimization of any functional group which does not contribute to the ultimate output.

Pull: The production system should be developed such that the customer pulls the product. Every down stream process should pull from its upstream process. This way wait and inventory in between processes can be avoided.

Perfection: This is a principle based on the concept that there is no end to the process of reducing effort, time, space, cost and mistakes. Basically this is the Japanese concept of *Kaizen*[7] or continuous improvement. Always try to be better , cheaper and faster and there is nothing like one has reached the perfection.

Software Development - Today

Professor Niklaus Wirth laments that almost all software have grown "fat". He contributes this tendency to growing hardware performance and the unnecessary complexity built in the software. Obviously "fat" software needs "fat" software development [6].

The causes for complexity are: (1) software vendors adopt almost any features that users may want and (2) monolithic design with all conceivable features. Complex problems require complex solutions and resulting complex and "fat" software. Complexity is also viewed as power by many. Vendors use the strategy of customer dependence rather than customer education which is more profitable.

Sophisticated software are complex and good solutions are developed by iterative and time-consuming thought processes. In view of time constraints software inadequacies are corrected by quickly conceived additions increasing the complexity and size of the software. Prof.Wirth recommends good engineering by gradual stepwise refinement of the product which we understand as continuous improvement or *kaizen* [7].

Lean Thinking in Software Development

Let us see whether the basic Lean principles indicated earlier can be applied to software development.

Value: value is what the customer perceives as value or useful to him. In the mass market software industry Microsoft has been very successful. According to Michael Cusumano, Microsoft and Bill Gates value making money and they believe that delivering value to the customers only will make the most money [8]. They do not value technology for the technology's sake. Most of the successful software technologies Microsoft adopted were invented elsewhere including Windows.

There are enough stories about completed software that were not usable by the customers. The value for the customer from his point of view, need to be understood at the stage of Requirements development. SEI Capability Maturity Model [9] includes Requirements Management in level 2 which is one of the early stages of Capability Maturity. Usability by the customer is very important as that contributes to his value.

The question that arises is how can we extract the "value" as perceived by the customer. Rapid Prototyping is one of the processes that may help here. By rapidly prototyping and reviewing with the customer one can develop the customer's requirements and his value. The erstwhile McDonnell Douglas Aerospace has used RAPIDS process very successfully. Microsoft uses flexible incremental approach by incrementally changing evolving features targeting specific customer activities and values. The "spiral" software development model as well as "stacked spiral model" use the concept of prototyping [19]. One way to achieve speedy prototyping is to use "software components". Object-Oriented technology plays an important role here.

Quality Function Development (QFD) is a team-based technique that helps in identifying and translating customer requirements [10]. QFD has been used successfully in hardware environments and its application in software is not yet dominant. Many defense programs like The Joint Advanced Strike Technology is employing QFD. QFD is based on the popular article "House Of Quality" appeared in the Harvard Business Review [11]

First and foremost is to understand clearly what the customer wants and what he considers as "value".

Value Stream: is a series of actions for developing a value product for the customer from order to delivery. By identifying the stream and those activities which actually add value, it would be possible to eliminate activities that do not contribute any value. Gordon Bethune the Chairman and CEO of the Continental Airlines in his recently published book: From Worst to First [12], cites how even after improving the on-time arrivals of the flights, the customer's satisfaction was poor due to the delays in getting their baggage. It is important that the whole "value stream" has to be improved to add value to the product.

In software development it is not just enough to understand customers requirements but is equally important that the design is developed meticulously using good engineering by gradual stepwise refinement of products. Sloppy design to save time will deliver a product with less value during the life of the product. A good design architecture and a Programming language based on object oriented technology will ensure value to the customer.

By identifying the stream and the activities that actually add value it would be possible to eliminate steps which do not contribute to the value for the customer. CMM helps in providing a framework. In SEI CMM level 2 the emphasis is to get the processes under control and establish repeatability. Once the processes are defined and repeatable focus on the process (level 3) and process improvement (level 5) are possible. By defining the process one can eliminate those processes that do not add value[9]

Flow: The third concept in Lean Thinking is Flow and we should make value flow. Flow is to make every activity in the product development take place continuously with no stoppage, no back flows and no scrap. Translating this to software development at Microsoft would be syn-and-stabilize approach according to Cusumano. In this approach developers activities as individuals and as members of teams working on different features are continuously coordinated as project proceeds. Microsoft people refer to this as "nightly build" or "Zero-defect" process[8]. The term "build" refers to integrating partially completed or finished pieces of software during development process to see what functions work. This structure, while allowing people enough flexibility to be creative and evolve product's details in stages, results in coordination of individual engineers and teams. Developers are required to test features with customers and refine their design during development. Scrapping of efforts can be avoided by putting in place processes that ensure continuous product development as described above.

Pull: Pull is an interesting concept in Lean. It means that no one upstream should produce a good or service until the customer downstream asks for it. This has been the basis for "Just In Time" (JIT) inventory concept in a manufacturing environment. In software environment JIT may not be applicable but it is needless to emphasis that only what is required by the customer should be built in the software. The software community has a tendency to include features which they think are "nice" for the customer to have. The result is "fat" and unnecessarily complex software.

Perfection: This last concept in Lean Thinking is to continuously move towards perfection or *Kaizen*. This is quite opposite to the reengineering concept that has emerged past few years. SEI CMM emphasizes continuously improving the process and defect prevention as key process areas at the highest level 5. There are two aspects in perfection. The first is eliminate defects while developing the product. The second, is improve the processes so that all "waste" can be removed. Michael Hammer the author of "reengineering" has recently conceded in his current book "Beyond Reengineering" [13] that reengineering has failed since process focus was not integrated in the reengineering efforts. Process Management defined as one of the Key Process Area under SEI CMM level 5 is very essential for moving towards perfection. By establishing appropriate processes defects in the software development can be avoided. Component-Based software development enables to use already proved components thus ensuring defect free [14].

Lean Software Development

The literature survey has produced only two articles on Lean Software Development. The work of Alexander C.Hou under the auspices of Lean Aircraft Initiative is an exhaustive paper encompassing Lean Hardware and Software System Development[15]. The author has described and discussed five methodologies : (1) The Rapid Development Process, (2) The GriTech rapid development process, (3) Ptolemy-supported hardware and software co-design (4) The RASP (Rapid Prototyping of Application Specific Signal Processors) design methodology and (5) Clean room software engineering. He has recommended study of ARPA/Tri-service RASPP program, ARPA STARS (Software Technology for Adaptable, Reliable Systems), and F22 Avionics development case study. Prof Wirth has argued for Lean Software and has not discussed Lean Software Development . He has however recommends simple object-oriented programming language Oberon instead of C++ which is becoming a very popular language.

Japanese Experiences

Michael Cusumano has described in detail the Japanese experiences of improving Software development through what he calls as "Software Factories" [16]. Software Factory is a concept or a philosophy that some software can be produced in a manner more akin to engineering and manufacturing than craft practices. Hitachi, Toshiba, NEC, and Fujitsu have successfully adopted this concept. So if one accepts the Software Factory Philosophy then application of Lean principles can also be accepted easily.

Even though the Software Factory was

Even though the software factory was originally proposed in US it did not live very long in US. However US Department Of Defense had subsequently developed the concept of software development " component-by-component" rather than "instruction-by-instruction". This was called "mega-Programming" [17]. Emergence of Object-Oriented programming languages rekindled interest in "re-use" of software components and there is considerable interest now in the industry for Component Based Software Development. Software leaders like Microsoft have developed Class libraries for object oriented languages which promote re-use.

The "re-use" help in achieving in improved quality and reduced flow time and cost since already developed, tested and proven components are used. Reuse is a very dominant principle under Lean.

Japanese have been using Software Factory concept and continuously improving the software processes. These have helped many Japanese companies to reach SEI CMM level 5. Fujitsu for example has software groups functioning at CMM level 5 and they do certify some of their software "Zero defect". This is based on "perfection" principle of Lean. Many Japanese firms attribute their higher quality and productivity to software reuse but they also define reuse as systematic way of integrating components that have been certified as "zero defect" [18].

Flexible Software manufacturing

Software Factory concept introduced the factory model to software development. Object-Oriented technologies promoted reusability to a great extent.

Paul Bassett comments that the most difficult problem in software reuse is not technical rather it is establishing a manufacturing culture[19]. The craftsman's mentality of distrusting what others do has been the main source of obstacle in implementing aggressive reuse in software development. He argues that to manufacture software (not develop) managers must support development processes, infrastructure and most importantly culture. Development processes should include processes for developing reusable components and for developing systems. The infrastructure must align with a reuse strategy. Both execution and construction standards may be the answers to this problem. Culture is a very dominant in changing any processes and software processes are no exception. Most of the problems in the direction of software manufacturing are people related.

It is claimed that by using Object-oriented reusability technologies based on Framework have achieved 70% reduction in schedule and 84% in costs [19]. Both Flow time and cost reductions are the basic goals of Lean thinking.

Lean Enterprise Model

Before a Lean Software Development model is suggested, a brief review of the Lean Enterprise Model as developed by Lean Aircraft Initiative is worthwhile.

Lean Enterprise model of LAI is based on the meta-principles: Responsiveness to change and Waste Minimization. The enterprise principles consist of Right thing at Right Place, Right Time, and in Right quantity/quality, effective relationships within the Value Stream, Continuous Improvement, and optimal First Delivered Unit Quality. This model suggests enterprise level metrics like flow time, stakeholders satisfaction, resource utilization and quality yield. This model also recommends the following 12 overarching practices:

(1) Identification and optimization of enterprise flow

(2) Assuring Seamless information flow

(3) Optimization of capability and utilization of people

(4) Decisions at lowest possible level,

(5) Integrated product and process development

(6) Mutual trust and commitment

(7) Continuous focus on the customer

(8) Lean leadership at all levels

(9) Challenging of existing processes

(10) Nurture a learning environment

(11) Ensure process capability and maturation

(12) Maximize stability in changing environment.

The above apply to any enterprise including those which are involved in the development of software.

Lean Software Development Model

While it is true that software development is very different from normal hardware product development and production the Lean principles can still be applied. There is always an argument that Software development is a creative work as such not amiable to hardware development practices which I believe, is not true. Lean thinking is applicable to any activity or processes including software development processes.

My first suggestion is that Software Factories model based on reusability of software components should be adopted. Industrializing software development is the first step.

There are several technologies emerging. IEEE has a Reuse Steering Committee (RSC) with a scope related to the analysis, design, implementation, verification, validation, documentation and maintenance of reusable software assets, as well as their supporting infrastructure in the creation and maintenance of new software systems. RSC will ensure coordination of standards related to software reuse and their general harmonization with the more general body of software engineering standards. Emerging programming languages particularly those based on Object-Orientation like C++, Java, Visual Basic, and Ada to mention a few facilitate reusability.

Concurrent Engineering has been effectively used to reduce the flow time of product development in the hardware world. Software product development can benefit by using this concept. Software development concurrency often involves the use of system prototypes that allow the software architect to build a skeletal system and try it out in the real world environment. The "spiral" software development model as well as "stacked spiral " model use this concept of prototyping [17]. Prototyping helps not only to resolve design issues but also the most important software requirements issues.

SEI's CMM[9] should be adopted to move the organization to higher levels with emphasis on the processes.

Continuous Improvement and "perfection" concept of Lean can be followed by using what Wirth calls as "stepwise refinement" or "synch-and-stabilize" approach used at Microsoft. Microsoft also assigns the responsibility of ensuring that the "build" is not broken by the individual's software component or code fragments to the individual developer. Thus the waste introduced in the build by defective code is eliminated.

Testing is a very important and also very expensive phase in the software development. In Japan quality control or quality assurance is defined as the development, design, manufacture and service of products that will satisfy the customer's needs and at the lowest possible cost. The customer's satisfaction with product quality is an end in itself. Inspection by specialized inspectors have been minimized in Lean factories. Inspectors whose activities are outside the manufacturing process perform operations with no value added and thus impact productivity and cost. The responsibility for the quality of the product lies with the producer. This would be a radical thinking in software development. In the current scenario extensive testing is done at various levels, unit test, Integration test, functional test and final tests by the end user (Alpha and Beta tests). There is scope to adopt Lean thinking here by delegating the responsibility to the developer or the integrator for quality. This is an area which requires some investigation. It is also evident that if software components already produced are used testing will be minimum. We understand that as the use and age of the software component increases its quality also does.

Conclusion

I have discussed above what is meant by Lean and the various software methodologies/technologies that contribute to Lean thinking in software development. The next steps in the software community is to move from a craft to an engineering discipline, industrialize the software development utilizing factory model and Lean Thinking. Reusability, rapid prototyping, spiral model, object-oriented technologies, component-based software development , concurrent engineering, quality function deployment are some of the concepts that will help in the direction of Lean

The question whether Lean Software Development is Feasible can easily be answered with "yes".

References

(1) Lean Air II , Winter 1998, A Publication of the Lean Aerospace Initiative, Massachusetts Institute of Technology, Vol 5, No 3.

(2) J.P.Womack, D.T.Jones, and D.Roos, The Machine That Changed The World, Simon & Schuster, 1990.

(3) Taichii Ohno, Toyota Production System, Productivity Press, 1988.

(4) J.P.Womack, and D.T.Jones, Lean Thinking, Simon & Schuster, 1996.

(5) "Condit, Stonecipher Speak Out", Boeing News, an internal Boeing company magazine, June 12, 1998.

(6) N.Wirth, "A Plea for Lean Software", IEEE Computer, Vol 28, No 2; February 1995.

(7) M.Imai, Kaizen, The Key to Japan's Competitive Success, McGraw-Hill Publishing Co, New York, 1986.

(8) M.A.Cusumano, and R.W.Selby, Microsoft Secrets, The Free Press, 1995

(9) M.C.Paulk, W.Curtis, M.B.Chrissis, and C.V.Weber, Capability Maturity Model for Software, Version 1.1, CMU/SEI-93-TR-2, Software Engineering Insitute, Carnegie Mellon University, Pittsburgh, PA 15213, 1993.

(10) Y.Akao, Quality Function Deployment, Integrating Customer Requirements into Product Design, Productivity Press, 1988.

(11) J.R.Hauser and D.P. Clausing, "House of Quality", Harvard Busines Review, May/June 1988.

(12) B.Gordon, From Worst to First: Behind the Scenes of Continental's Remarkable Comeback-A Flighgt Plan for Success, John Wiley & Sons, 1998.

(13) M.Hammer, Beyond Re-Engineering: how the process-centered organization is changing our work and our lives, Harper Business, 1996.

(14) A.W.Brown, Component-Based Software Engineering, Selected papers from the Software Engineering, IEEE Computer Society Press, 1996.

(15) A.C.Hou, "Toward Lean Hardware/Software System Development: Evaluation of Selected Complex Electronic System Development Methodologies, Report-Lean 95-01, Lean Aircraft Initiative, Massachusetts Institute of Technology, Cambridge, 1995.

(16) M.A.Cusumano, Japanese Software Factories, Oxford University Press, 1991.

(17) Draft Department of Defense Software Technology Strategy, Office of The Director of Defense Research and Engineering, Washington DC, 1991

(18) J.Vu, "Boeing Scientist Tours Japan", unpublished Boeing internal document.

(19) P.Bassett, "Managing for Flexible Software manufacturing", IEEE Computer, July 1998

COST EFFECTIVENESS OF FORMAL METHODS IN THE DEVELOPMENT OF AVIONICS SYSTEMS AT AEROSPATIALE

François PILARSKI Design and specification methods - Systems Department
Aerospatiale, Toulouse, France

Summary

AEROSPATIALE has been using formal methods since the early 80's to develop part of the embedded systems, and more specifically most of the avionics systems. During the last two decades, such an approach has demonstrated significant cost savings. This paper presents some aspects of this approach and focuses more precisely on the way it has helped in reducing embedded system development and maintenance costs.

This first section of this paper is devoted to a general description of the overall context and it gives some elements of history. Basically this section is focused on the tool SCADE and its predecessors (SAO) which implements the main, though not only, formal method used at AEROSPATIALE.

The second section summarizes the benefits AEROSPATIALE has found in using SAO first, and now SCADE, and constitutes an attempt to synthesize the properties and characteristics that make a formal method efficient (at AEROSPATIALE).

In this second section the cost aspects are emphasized instead of safety aspects. Indeed when speaking about formal methods one generally thinks of proof techniques in order to ensure "a complete and exhaustive" validation. It is obvious that these techniques, when applicable, are useful but it is a fact that aircraft (Airbus A320, A340, for example) have already been certified, and are now flying safely every day, thanks to classical techniques such as RAMS analysis, ground tests, flight tests, ...

The third section summarizes some aspects of formal methods, which have not been extensively used yet but which are likely to induce extra cost reduction, namely proof techniques.

As a conclusion it is recalled that formal methods have been demonstrated to be cost effective at AEROSPATIALE, especially SAO and now SCADE. In addition same extra cost saving is expected, thanks to the industrial maturity of proof techniques.

1. introduction

Many of the computers on board AIRBUS aircraft are specified using formal or semi-formal methods. This is not the case for all embedded computers but one can notice an increase in the number of such formally specified computers from A/C programme to A/C programme.

AEROSPATIALE, who specifies many of the A/C systems, definitely believes that such an investment, in terms of the designer training needed to use formal methods and appropriate tools is profitable particularly because these methods provide automatic code generation capability, that is to say :

- possible use of simulation as early as the specification phase
- quick debugging of the specification and very short modification cycle during the test phases.

Considering also that the automatic code generators, which provide the embedded code, have been "qualified" (that is to say known as producing a code strictly corresponding to the specification itself), the benefit becomes obvious. Indeed the unit test phase of the embedded software has been significantly shortened.

For the time being no proof techniques have been used at AEROSPATIALE except for prospective experimentation. Indeed such techniques have to be demonstrated to be mature enough to be used for industrial programmes.

0-7803-5086-3 /98/$10.00 ©1998 IEEE

In any case it is a fact that such techniques are not necessary to demonstrate the required safety level in the aeronautics world : indeed nowadays aircraft are flying safely everyday without the use of these techniques. Nevertheless, proof techniques could be used as alternatives to existing validation techniques and it is likely that they will induce extra cost savings in certain cases. The problem that has to be faced now is to define where and when such techniques can be efficiently used with the whole validation process.

At AEROSPATIALE the SAO method (SAO stands for Computer Aided Specification) is mainly used, but some other formal methods have also been used, SDL for example. The experience is twofold :

- one single formal method used all along the project : this has been basically the case for the existing AIRBUS programmes,
- more than one method used simultaneously to specify different aspects of the same system, each formalism being chosen because it is particularly suitable for a specific problem. Indeed this multi-formalism approach implies coherency problems that have been mastered until now.

In fact the SAO "language" is not really a single language but composed of different dialects, even if all these dialects have a common basis.

These different SAO variants do not favor an optimization of the benefits of formal methods. So AEROSPATIALE has engaged in a rationalization approach around SAO, within a partnership with Schneider Electric and Verilog, which has finally led to the marketing of the toolkit SCADE, now available on the market.

2. overall context

2.1. scope

Computer aids in aircraft concern flight control, guidance, navigation and air traffic management. These functions are shown on the figure below.

Flight control consists in controlling the attitude of the aircraft around its center of gravity.

Guidance consists in slaving the position of the aircraft to a given trajectory.

Navigation consists in defining trajectories.

At present, there is no computerized link between air traffic management and navigation; this situation should change in the years to come.

Figure 1. A simplified aircraft

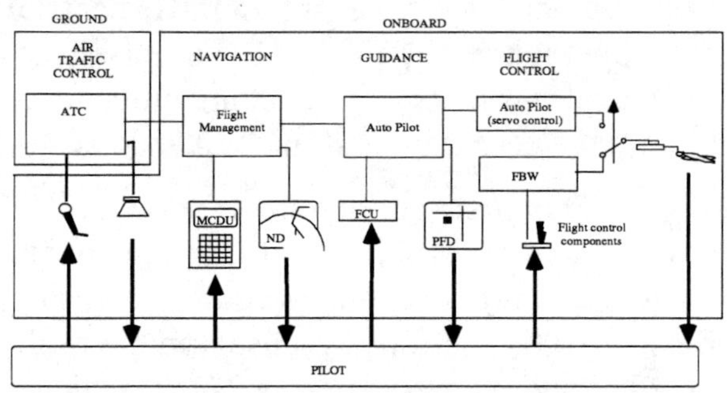

The SAO specifications mainly concern flight control and guidance and, recently, a part of navigation.

In order to better describe the use and benefits of SAO, we will give below the characteristic principles and the concepts of the elements on the previous figure which are specified, at least in part, with SAO.

FBW: Fly By Wire. These are all the flight control components, computers, links, servocontrols, etc. placed between the pilot and the control surfaces. These are elements that are especially critical from a safety viewpoint and require monitored and redundant architectures to tolerate failures.

The aim of the FBW functions is to provide the crew with intuitive, easy and repetitive flight control by harmoniously combining the stability and manoeuvrability requirements. In addition, they can include so-called protection functions intended, on the one hand, to prevent unwanted excursions outside the normal flight envelope and, on the other, if the need is felt, to facilitate safe flight at the limits of this envelope.

MCDU: the MCDU is an alphanumerical console (keyboard and screen) allowing the pilot to give orders, mainly to the flight management system, and to receive information such as flight plan, runway heading, etc.

Screens. They display information, which previously was given by mechanical equipment, such as speed, altitude, vertical speed, set speed values (to be reached, not to be exceeded), etc., and navigation information. In addition, certain screens display system status and warning information (for example: engine, landing gear, cabin, etc.). All these screens can fail: there are therefore a certain number of rules for switching information from one screen to another in case of failure. Also, the pilots have the possibility of selecting the display of information on such or such a screen.

The emergence of the Future Air Navigation System has been taken into consideration by AEROSPATIALE with the introduction of specific equipment in charge of managing the data link with the onground stations and transmitting information to relevant onboard systems. This equipment is known as ATSU and its links with other systems are shown in the figure below. As protocols were concerned at first and because some of them were already specified using SDL, this automata description formal language was chosen to specify the core part of the ATSU. The use of SDL follows the same rules as SAO and leads to comparable benefits.

Figure 2. A solution for FANS

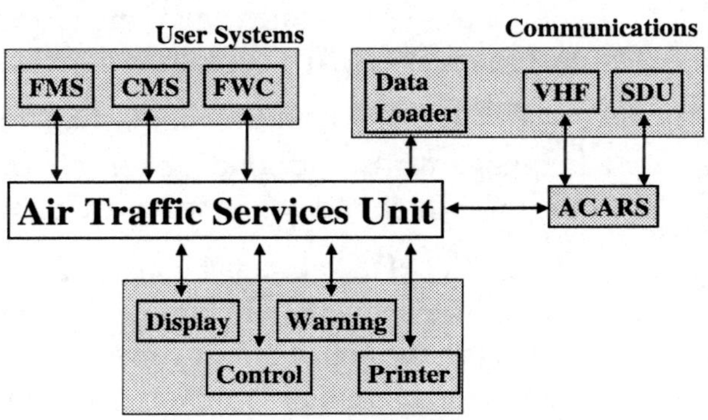

To sum up, formal methods are used at AEROSPATIALE to specify:

- Using SAO :
  - slave control loop-type processing operations,
  - reconfiguration and mode engagement and change logics,
  - interactions with the pilot in alphanumerical mode.
- Using SDL (within the FANS context) :
  - Protocols (ground / onboard and interequipment)
  - "Bounding" part of the Air Traffic Control applications.

It has to be noted that despite many benefits detailed hereafter, formal methods are not systematically used: indeed the use of such methods is not free of charge and has to be carefully assessed with respect to the type of application.

2.2. a typical process

Let us consider for instance the case of the flight control laws.

First of all the designer uses a tool of the MATLAB type to define the flight control laws which are looped up with the aircraft model during an initial validation stage thanks to a simulation of the assembly. This tool allows the law designer / automatic control specialist to conduct conventional stability, robustness and performance studies using a linear aircraft model (flight point by flight point tangent with the complete model).

The detailed specification is then "expressed" into SAO and validated by simulation with suitable tools (real-time) which then integrate a complete aircraft model including the non-linearities of the aerodynamics or of the systems. This aircraft model is a single model gradually tuned over time and used for all other applications (other aircraft systems, other simulations, etc.): this is THE aircraft reference model. The law model (in SAO), which is established on this occasion, enters, at this time, into an iterative process leading it to define THE flight control law specification.

The designer then expresses in SAO the specification of all law switching operations and information exchanges between redundant calculation units, monitoring, etc. and integrates all the specifications. A validation phase of this integrated set of SAO specifications is then started using simulation tools, this time not real time tools, mainly (today) on account of the limited performances of the machines. However, SAO specification integration goes as far as "exceeding" the boundaries of the equipment. The simulation means ("multi-equipment") therefore become a "virtual system integration bench".

Lastly, the SAO specifications are delivered to the equipment manufacturers concerned who code them, automatically for the major part, and integrate them into the equipment : establishing links with the hardware and the basic software services. The equipment will then be subjected to integration tests on the benches and will finally be installed on the aircraft for flight tests.

For a new standard, the specification follows the same process with, in addition, many non-regression tests before being supplied to the equipment manufacturer.

2.3. elements of history

The use of SAO at AEROSPATIALE dates back to around 1980. At that time, computers were analogue and the basic need was to establish unambiguous dialogue with the equipment manufacturers. As the main problem posed by flying aids was a slave control problem, the dialogue language was naturally oriented towards a block diagram type language suited to automatic applications.

The original language therefore had two basic characteristics:

- it was a graphics language,
- it was a "speciality" language.

We would add that it is not a computer language and therefore the sheets are not explicitly complete: a non-negligible part of the information is implicit and related to the speciality context. This greatly simplifies the expressing of the specification.

The advent of embedded data processing, first with the AIRBUS A310 then with the AIRBUS A320, allowed greater utilisation flexibility in particular with automatic code generation possibilities. The language therefore changed somewhat to be more rigorous in expression and towards a certain standardization of the component library, which is the elements of the language. We must add to this evolution, related to the technology, a natural evolution due to wider use and therefore more accurate formalization concepts, etc. SAO was born.

As far as the AIRBUS A340 programme is concerned, it was especially marked by a generalization of automatic coding for the majority of equipment while maintaining the coding for embedded equipment / coding for simulation duality.

But the use of SAO remained limited to the specification of system status management and automatic application problems. Prospective studies then allowed the investigation of variants adapted, on the one hand, to the specification of alphanumerical dialogues with the pilot (typically the MCDU) and, on the other, the specification of data processing operations (typically the Flight Management System).

These studies especially showed that the scope of SAO could be extended but that this course, which consisted in introducing new library elements and code generation directives, made the specification less simple and especially less readable.

On the strength of this experience, AEROSPATIALE, within a partnership with SCHNEIDER ELECTRIC and VERILOG, defined the SCADE environment, which completes SAO while remaining close to the designers' needs. To sum up, we can say that:

SCADE is based on the LUSTRE language whilst introducing complementary notions which today are not in LUSTRE.

SCADE remains however a language suited for designers who are not computer experts in the sense that it conserves the three main characteristics of the original language invented by the designers themselves; that is, it is a speciality-oriented language, it is graphical and especially remains simple as it allows purely data processing constraints to be masked (scheduling, explicit exhaustivity of information flows, etc.) by high-level and abstract concepts (declarative language, hidden flows, strong typing, etc.).

Finally such a success story led AEROSPATIALE to look for other formal languages when necessary to address different types of specification problems. This led to the use of SDL within the FANS context to address generally speaking automata specification.

3. benefits

Looking back at the previous generation aircraft, namely the A310, it is a fact that the first benefit in formalizing the communication language was to ease the mutual understanding of the actors. Its certainly difficult to assess precisely the impact of such an effect but it can be observed that the number of coding errors has been reduced by a factor of ten.

Because we do believe the overall validation process is sure enough to detect safety critical errors at least, and indeed it has been demonstrated as such, one cannot find any effect on safety. On the contrary, impacts on costs are obvious. Because cost assessments within the project management were not as precise as nowadays, and because the coding errors were detected all along the validation chain, and finally because the equipment itself was not so complex as today's equipment, it was not possible to quantify the induced cost savings.

The second effect induced by the formalization of the language was the possibility to automatically generate code. Here again two different effects have to be considered the first one is the impact on simulation and the second one the impact on the embedded software development itself.

a) Impacts on simulation.

It is a fact that for a long time one has been looking for means to debug at least and validate the specification itself. For example the certification of category III autoland has been using simulation for a long time, long before the emergence of formal methods. The point was to demonstrate the representativeness of the simulated models with respect to the actual aircraft. So models have "always" been used to simulate systems and related environment. At first these models were manually coded as the corresponding embedded software: this activity was expensive, lasted some time, and was definitely specific to simulation. The use of such an approach was rather rare because of its cost. Generally speaking this implied validating the specification on the real equipment. The formalization of the specification allowed the simulation code to be generated automatically, and so the simulation of the specification to be generalized at almost no cost. As this was not generally the case the simulations of specification only changed the development process rather than having a direct impact on the cost.

Nevertheless, without changing the validation process, always based on the final equipment, this new development process increased the specification quality as it permitted the designer himself to "close the loop" and directly test his specification. As an indirect effect, the quantity of specification problems was reduced.

b) Impacts on embedded code

Because of the formalization of the specification it was possible to generate the embedded code automatically. In fact, part of the embedded code still had to be coded manually. As a direct effect the time-to-code was significantly reduced and so was the equipment cost. At the same time the idea of qualifying the embedded code generator emerged, which as a direct effect made the unit tests unnecessary. Globally the reduction of the time-to-code and the elimination of the unit tests induced cost reduction by a factor of at least two despite the necessary investment in terms of code generator development and qualification.

The indirect effects induced by the early simulation of the specification and the better quality of the specification have not been measured precisely. But it induced a reduction in the modification cycle by a factor of 3 to 4.

These points are summarized in the figure below.

Figure 3. Effectiveness of methods

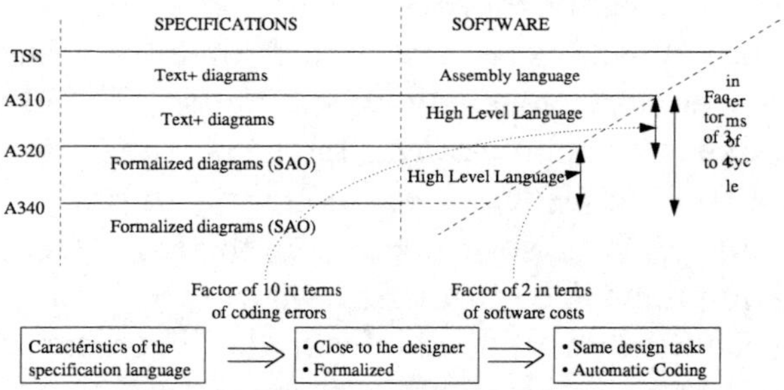

Finally it is important to notice that the validation process is unchanged with or without the formalization of the specification, the development process only is impacted.

Two reasons lead to this fact:

- it would be expensive and maybe too difficult to qualify the simulation code generator in order to guarantee the identity of the simulation code with respect to the embedded code. Even if possible this has not yet been done, as far as we know.
- in any case the simulated specification needs complementary models to be run. Particularly the aerodynamic models (generally speaking the aircraft models) are not modeled with formal methods. In addition what is formalized is only the application part of the systems: Operating systems, Hardware, and related software are not modelled.

Finally, even if it is formalized, the specification is not demonstrated as having strictly the same behavior as the embedded software.

Concerning the language itself, it is also important to note that the generalization of formalization is efficient because the language is accepted and shared between all AEROSPATIALE participants in the avionics field of activity. This is certainly one of the keys to success. For example another experience with SDL has not been demonstrated to be so efficient at the beginning because the language, though graphic too, is less easy to understand at first. People have to be trained first.

As we saw above, SAO is a language common to the various participants in the system design process. Indeed, at all levels: flight tests, integration tests, automatic control specialists who design the laws, system architecture specialists or even equipment manufacturers, all those involved refer to the SAO sheets: in addition to being a specification language, it is also a communication language.

The formal aspect of the language therefore allows:

- the establishing of an unambiguous dialogue. But, each aircraft programme and even each system adds small differences, complements and variants, etc. depending on its specificities: the language is not truly the only one today.
- the generation of an executable code reflecting the specification and at the same time an early validation thanks to the simulation.
- the generation of the major part of the embedded code with good guarantee of similarity to what was validated by simulation.
- and finally the elimination of coding errors.

4. expected extra cost savings

As described above only, the development process has already been impacted by formal methods. We do believe that this part of the overall process can be cost-optimized, for example using a more general language than SAO, unifying the existing dialects and developing a single (qualified) code generator. Indeed this is what has already been done and what has lead to the SCADE environment, which is now currently being used for new projects.

It is more difficult to optimize the validation process, but this is certainly a source of cost savings. Two ways of cost improvement can be found:

- as shown before it could be advantageous to directly use the work done when debugging the specification to formally validate it : this implies demonstrating first of all the representativeness of the simulation and to do so it implies qualifying the simulation code generator, or rehosting the "real" software in a simulated hardware environment, for example.

- another way is to investigate one particular benefit of formal methods : namely proof techniques.

It is obvious that none of the existing validation means is able to cover, by itself, all the aspects necessary to certify an aircraft. This is why a validation strategy is defined (and formalized into a validation plan) to organize the validation activities and the use of the related means. Todays proof techniques have not yet been industrially used within the aeronautical context as far as we know.

Like other means, it is clear that proof techniques will not cover, themselves all the aspects either. Nevertheless it is likely that they could help to solve some problems more efficiently than existing techniques.

For example, demonstrating that a given state transition will "never" occur (in fact will occur with a probability less than 10-9 for example) necessitates a large quantity of test cases using simulation, but the same could be easy to demonstrate within the considered hypothesis.

It is important to note that in this example, but this fact is general, choosing one or the other technique implies "exchanging" probability for hypotheses (more precisely a certain expression of the hypotheses): neither of these two techniques is perfect!

To go further in the analysis, when speaking of proof techniques, one generally means 3 different types of results:

- proof techniques make it possible to verify that a given property assumed to be satisfied by one expression is still satisfied by the translation of the expression into another one. Typically the translation can be a refinement or a compilation.
- proof techniques make it possible to validate the fact that a given formal expression satisfies a given property. Typically this is an alternative to tests.
- proof techniques are used to generate test cases to be executed in order to evaluate whether a given property is satisfied by an existing implementation of a specification. Typically this can be used as acceptance test.

If these techniques were able to demonstrate an industrial level of maturity they could certainly efficiently help the validation process.

5. conclusions

SAO is the SUCCESSFUL use of a formal method. This is a clear demonstration of the applicability of formal methods in the industrial world.

It is successful as the language is shared by all participants: design office, flight tests, product support, etc. and even beyond the boundaries of the companies as certain equipment manufacturers share this language. We could think that this appropriation of the language is due to what is intrinsic to the designers who initialized, then accepted, the changes as they improved the matching of the language to their preoccupations.

It is also successful through the benefits that AEROSPATIALE has drawn from the formal aspect in terms of:

- validation of the specification and no longer of the code, or even of the equipment (thanks to simulation),
- elimination of coding errors,
- modification cycle (thanks to embedded code generation),
- easy dialogue with all those involved in the process (including the equipment manufacturers).

These benefits have been evaluated, though not too precisely, as part of an experience feedback from previous AIRBUS programmes. It has been shown that the use of formal methods has reduced:

- coding errors by a factor of 10, firstly only because of the formalization of the specification. Then coding errors have been completely eliminated thanks to the use of qualified code generators.
- the modification cycle by a factor of 3 to 4
- software costs by a factor of at least 2

Another experience with the SDL language and associated tools seems to show similar results, though it is not yet finalized.

Use of SAO within the scope of previous aircraft programmes or prospective studies has shown the need for a change in the language towards greater formalism and to high-level concepts. This new SAO+ language allows a new step to be made towards improved efficiency thanks to:

0-7803-5086-3 /98/$10.00 ©1998 IEEE

- the disappearance of dialogues, therefore a simpler automatic code generation
- simulation which is more representative as it is strictly in compliance with the single semantics of the language,
- better control of complexity thanks to the fact that the language takes time aspects into account (scheduling), processing hierarchy, data typing, etc..

The implementation of SCADE, a totally formal method, will thus allow progress to be made along other promising channels such as proof techniques, generation of pertinent tests, etc.

DYNAMIC SCHEDULING STRATEGIES FOR AVIONICS MISSION COMPUTING

David L. Levine, Christopher D. Gill, and Douglas C. Schmidt
{levine,cdgill,schmidt}@cs.wustl.edu
Department of Computer Science, Washington University, St. Louis, MO 63130, USA *

Abstract

Avionics mission computing systems have traditionally been scheduled statically. Static scheduling provides assurance of schedulability prior to run-time and can be implemented with low run-time overhead. However, static scheduling handles non-periodic processing inefficiently, and treats invocation-to-invocation variations in resource requirements inflexibly. As a consequence, processing resources are underutilized and the resulting systems are hard to adapt to meet worst-case processing requirements.

Dynamic scheduling has the potential to offer relief from some of the restrictions imposed by strict static scheduling approaches. Potential benefits of dynamic scheduling include better tolerance for variations in activities, more flexible prioritization, and better CPU utilization in the presence of non-periodic activities. However, the cost of these benefits is expected to be higher run-time scheduling overhead and additional application development complexity. This report reviews the implications of these tradeoffs for avionics mission computing systems and presents experimental results obtained using the Maximum Urgency First dynamic scheduling algorithm.

1 Introduction

1.1 Motivation

Supporting the quality of service (QoS) demands of next-generation real-time applications requires object-oriented (OO) middleware that is flexible, efficient, predictable, and convenient to program. Applications with deterministic real-time requirements, such as avionics mission computing systems, impose severe constraints on the design and implementation of real-time OO middleware. Avionics mission computing applications manage sensors and operator displays, navigate the aircraft's course, and control weapon release.

Middleware for avionics mission computing must support applications with both deterministic and statistical real-time QoS requirements. Support for deterministic real-time requirements is necessary for mission computing tasks that must meet all their deadlines, *e.g.*, weapon solutions and navigation. Support for statistical real-time requirements is desirable for tasks such as built-in-test and low-priority display queues, which can tolerate minor fluctuations in scheduling and reliability guarantees, but nonetheless require QoS support.

1.2 Design and Implementation Challenges

Figure 1 illustrates the architecture of an avionics mission computing application developed at Boeing [1] using OO middleware components and services based on the Object Management Group's Common Object Request Broker Architecture (CORBA) [2]. CORBA Object Request Brokers (ORBs) allow clients to invoke operations on target object implementations without concern for where the object resides, what language the object is written in, the OS/hardware platform, or the type of communication protocols and networks used to interconnect distributed objects [3]. To achieve these benefits for avionics applications, however, requires

*This work was supported in part by Boeing, DARPA contract 9701516, and NSF grant NCR-9628218.

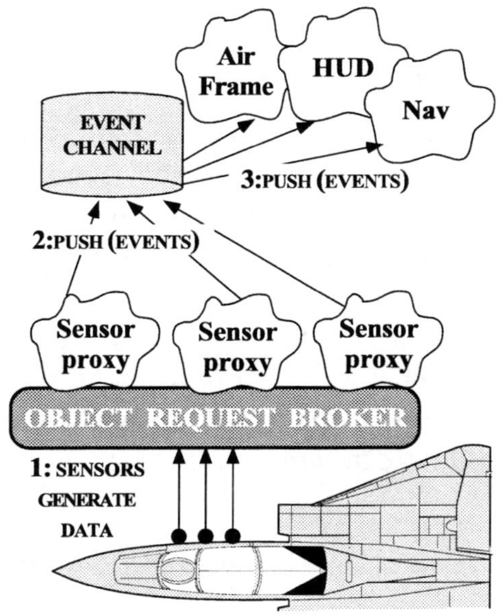

Figure 1: Example Avionics Mission Computing Application

the resolution of the following design and implementation challenges:

Scheduling assurance prior to run-time: In avionics applications, the consequences of missing a critical deadline at run-time can be catastrophic. For example, failure to process an input from the pilot within the necessary time frame could be disastrous, especially in critical situations such as air-to-air engagement or weapons release. Therefore, it is essential to validate that all critical processing deadlines will be met *prior* to run-time.

Severe resource limitations: Processing must be minimized due to limited resource availability, such as weight and power consumption restrictions. A consequence of using static, off-line scheduling is that worst-case processing requirements drive the schedule. Therefore, resource allocation and scheduling must always accommodate the worst case, even in non-worst case scenarios.

Distributed Processing: In complex avionics systems, mission processing must be distributed over several physical processors and computations on separate processors must communicate effectively. Clients running on one processor must be able to invoke operations on servants in other processors. Likewise, the allocation of operations to processors should be flexible, *e.g.*, it should be transparent whether a given operation resides on the same processor as the client that invokes it.

Testability: Avionics software is complex, critical, and long-lived. Maintenance is particularly problematic and expensive [4]. A large percentage of software maintenance involves testing. Current scheduling approaches are validated by extensive testing, which is tedious and non-comprehensive. Thus, analytical assurance is essential to help reduce validation costs by focusing the requisite testing on the most strategic system components.

Adaptability across product families: Current avionics applications are custom-built for a specific product family. Development and testing costs can be reduced if large, common portions can be factored out. In addition, validation and certification of components can be shared across product families, potentially reducing development time and effort.

The remainder of this paper is organized as follows: Section 2 reviews the drawbacks of off-line, static scheduling and introduces the dynamic scheduling strategy we are evaluating, *Maximum Urgency First* (MUF) [5]. Section 3 presents experimental results showing the cost of dynamic scheduling. Section 4 presents concluding remarks.

2 Dynamic Scheduling Strategies

This section describes the limitations of purely static scheduling and outlines the potential benefits of applying dynamic scheduling. We also evaluate the limitations of purely dynamic scheduling strategies. This evaluation motivates the hybrid static/dynamic MUF scheduling approach for CORBA operations used by TAO's real-time scheduling service (described in [6]).

2.1 Limitations of Static Scheduling

Many hard real-time systems have traditionally been scheduled statically using rate monotonic scheduling (RMS). Static scheduling provides schedulability assurance prior to run-time and can be implemented with low run-time overhead [6]. However, static scheduling has these disadvantages:

Inefficient handling of non-periodic processing: Static scheduling treats aperiodic processing as if it was periodic, *i.e.*, occurring at its maximum possible rate. Resources are allocated to aperiodic operations either directly or through a sporadic server[1] to reduce latency. In typical operation, however, aperiodic processing may not occur at its maximum possible rate. One example is interrupts, which potentially may occur very frequently, but often do not.

Unfortunately, with static scheduling, resources must be allocated pessimistically and scheduled under the assumption that interrupts occur at the maximum rate. When they do not, utilization is effectively reduced because unused resources cannot be reallocated.

Utilization phasing penalty for non-harmonic periods: In statically scheduled systems, achievable utilization can be reduced if the periods of all operations are not harmonically related. Operations are harmonically related if their periods are integer multiples of one another. When periods are not harmonic, the phasing of the operations produces unscheduled gaps of time. This reduces the maximum schedulable percentage of the CPU, *i.e.*, the *schedulable bound*, to below unity.

Inflexible handling of invocation-to-invocation variation in resource requirements: Because priorities cannot be changed easily[2] at run-time, allocations must be based on worst-case conditions. Thus, if an operation usually requires 5 msec of CPU time, but under certain conditions requires 8 msec, static scheduling analysis must assume that 8 msec will be required for every invocation. Again, utilization is effectively penalized because the resource will be idle for 3 msec in the usual case.

[1] A sporadic server [7] reserves a portion of the schedule to allocate to aperiodic events when they arrive.

[2] Priorities can be changed via mode changes, but that is too coarse to capture invocation-to-invocation variations in the resource requirements of complex applications.

In general, static scheduling limits the adaptability of systems to changing conditions and changing configurations. In addition, static scheduling compromises resource utilization in order to guarantee access to resources at run-time. To overcome the limitations of static scheduling, therefore, we are investigating the use of dynamic strategies to schedule CORBA operations for applications with real-time QoS requirements.

2.2 *Synopsis of Scheduling Terminology*

Precise terminology is necessary to discuss and evaluate static, dynamic, and hybrid scheduling strategies in terms of their priority assignment and dispatching characteristics. Figure 2 shows the relationships between key terms, defined below.

RT_Operation and RT_Info: In TAO, an `RT_Operation` is a scheduled CORBA operation [6]. In this paper, we use *operation* interchangeably with `RT_Operation`. An `RT_Info struct` is associated with each operation and contains its QoS parameters. The `RT_Info` structure contains the following operation characteristics described in [6].

- **Criticality:** Criticality is an application-supplied value that indicates the significance of a CORBA operation's completion prior to its deadline. Higher criticality should be assigned to an operation that incurs greater cost to the application if it fails to complete execution before its deadline. Some scheduling strategies, such as MUF, take criticality into consideration, so that more critical operations are given priority over less critical ones.

- **Worst Case Execution Time:** This is the longest time it can take to execute a single dispatch of the operation.

- **Period:** Period is the interval between dispatches of an operation.

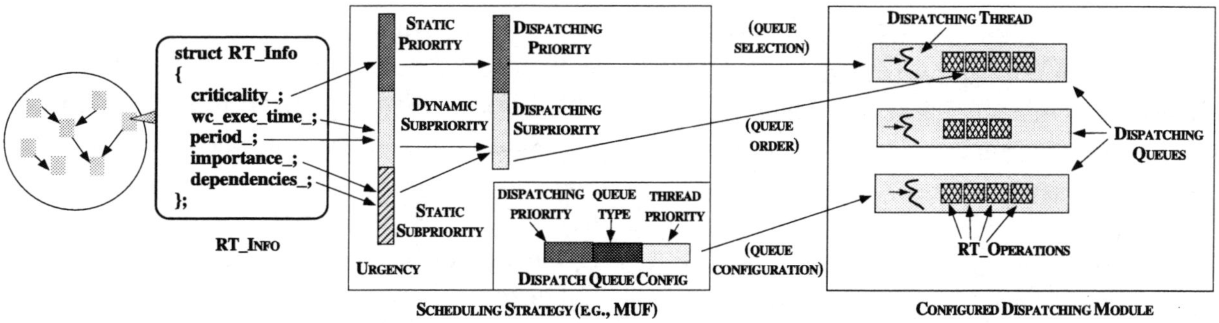

Figure 2: Relationships Between Operation, Scheduling, and Dispatching Terms

- **Importance:** Importance is a lesser indication of a CORBA operation's significance. Like its criticality, an operation's importance value is supplied by the application. Importance is used as a tie-breaker to distinguish between operations that otherwise would have identical priority.

- **Dependencies:** An operation depends on another operation if it is invoked only via a flow of control from the other operation.

Scheduling Strategy: A scheduling strategy (1) takes the information provided by an operation's `RT_Info`, (2) assigns urgency to the operation, (3) maps urgency into dispatching priority and dispatching subpriority values for the operation, and (4) provides dispatching queue configuration information so that each operation can be dispatched according to its assigned dispatching priority and dispatching subpriority.

- **Urgency:** Urgency [8] is an ordered tuple consisting of (1) static priority, (2) dynamic subpriority, and (3) static subpriority. Static priority is the highest ranking priority component in the urgency tuple, then dynamic subpriority, and last static subpriority. Figure 2 illustrates these relationships.

- **Static priority:** Static priority assignment establishes a fixed number of priority partitions into which all operations must fall. The number of static priority partitions is established off-line. An operation's static priority value is often determined off-line, though the value assigned a particular dispatch of the operation could vary at run-time, depending on the scheduling strategy.

- **Dynamic subpriority:** Dynamic subpriority is a value generated and used at run-time to order operations *within* a static priority level, according to the run-time and static characteristics of each operation. For example, a subpriority based on nearest deadline must be computed dynamically.

- **Static subpriority:** Static subpriority values are determined prior to run-time. Static subpriority acts as a tie-breaker when both static priority and dynamic subpriority are equal.

- **Dispatching priority:** An operation's dispatching priority corresponds to the real-time priority of the thread in which it will be dispatched. Operations with higher dispatching priorities are dispatched in threads with higher real-time priorities.

- **Dispatching subpriority:** Dispatching subpriority is used to order operations within a dispatching priority level. Operations with higher dispatching subpriority are dispatched ahead of operations with the same dispatching priority but lower dispatching subpriority.

- **Queue Configuration:** A separate queue must be configured for each distinct dispatching priority. The scheduling strategy assigns each queue a dispatching type (static, deadline, or laxity), a dispatching priority, and a thread priority.

Dispatching Module: The dispatching module (1) constructs the appropriate type of queue for each dispatching priority, and (2) sets each dispatching thread's priority to the value provided by the scheduling strategy.

2.3 *Survey of Dynamic Scheduling Strategies*

Several other forms of scheduling exist beyond RMS. For instance, Earliest Deadline First (EDF) scheduling assigns higher priorities to operations with closer deadlines. EDF is commonly used for dynamic scheduling because it permits run-time modification of rates and priorities. In contrast, static techniques like RMS require fixed rates and priorities.

Dynamic scheduling does not suffer from the drawbacks described in Section 2.1. If these drawbacks can be alleviated without incurring too much overhead or non-determinism, dynamic scheduling can be beneficial for avionics applications. However, many dynamic scheduling strategies do not offer the *a priori* guarantees of static scheduling. For instance, purely dynamically scheduled systems can behave non-deterministically under heavy loads. Therefore, operations that are critical to an application may miss their deadlines because they were (1) delayed by non-critical operations or (2) delayed by an excessive number of critical operations, *e.g.*, if admission control of dynamically generated operations is not performed.

The remainder of this section reviews several strategies for dynamic and hybrid static/dynamic scheduling. These include purely dynamic strategies such as EDF and MLF, and hybrid approachs such as MUF and two-level scheduling.

2.3.1 *Purely Dynamic Scheduling Strategies*

Earliest Deadline First (EDF): EDF [9, 10] is a dynamic scheduling algorithm that orders dispatches[3] of operations based on time-to-deadline. Operation executions with closer deadlines are dispatched before those with more distant deadlines. The EDF scheduling algorithm is invoked whenever a dispatch of an operation is requested. The new dispatch may or may not preempt the currently executing operation, depending on the implementation strategy.

[3] A *dispatch* is a particular execution of an *operation*.

A key limitation of EDF is that an operation with the earliest deadline is dispatched whether or not there is sufficient time remaining to complete its execution prior to the deadline. Therefore, the fact that an operation cannot meet its deadline will not be detected until *after* the deadline has passed. If the operation is dispatched even though it cannot complete its execution prior to the deadline, the operation consumes CPU time that could otherwise be allocated to other operations. If the result of the operation is only useful to the application prior to the deadline, then the entire time consumed by the operation is essentially wasted.

Minimum Laxity First (MLF): MLF [8] refines the EDF strategy by taking into account operation execution time. It dispatches the operation whose *laxity* is least. Laxity is defined as the time-to-deadline minus the remaining execution time.

Using MLF, it is possible to detect that an operation will not meet its deadline prior to the deadline itself. If this occurs, a scheduler can reevaluate the operation prior to allocating the CPU for the remaining computation time. For example, one strategy is to simply drop the operation whose laxity is not sufficient to meet its deadline. This strategy could decrease the chance that subsequent operations will miss their deadlines, especially if the system is transiently overloaded.

Evaluation of EDF and MLF:

- **Advantages:** From a scheduling perspective, the main advantage of EDF and MLF is that they overcome the utilization limitations of RMS. In particular, the utilization phasing penalty described in Section 2.1 that can occur in RMS is not a factor. This is because EDF and MLF prioritize operations according to their dynamic run-time characteristics. They handle harmonic and non-harmonic periods comparably, and respond flexibly to invocation-to-invocation variations in resource requirements, allowing CPU time one operation does not use to be reallocated to other operations. Thus, they can produce schedules that are optimal in terms of CPU utilization [9]. Moreover, both EDF and MLF can dispatch operations within a single static priority level and do not prioritize operations by rate [9, 8].

• **Disadvantages:** Purely dynamic scheduling approaches like MLF and EDF potentially relieve the utilization limitations of the static RMS approach. However, they have a higher cost to evaluate the scheduling algorithm at run-time. In addition, these purely dynamic scheduling strategies offer no control over *which* operations will miss their deadlines if the schedulable bound is exceeded. As operations are added to the schedule to achieve higher utilization, the margin of safety for *all* operations decreases. Therefore, the risk of missing a deadline increases for every operation as the system become overloaded.

2.3.2 Maximum Urgency First

The Maximum Urgency First (MUF) [8] scheduling algorithm supports both the deterministic rigor of the static RMS scheduling approach and the flexibility of dynamic scheduling approaches such as EDF and MLF. RMS assigns all priority components statically and EDF/MLF assign all priority components dynamically. In contrast, MUF can assign both static *and* dynamic priority components. The hybrid priority assignment in MUF overcomes the drawbacks of the individual scheduling algorithms by combining techniques from each. The remainder of this section describes how each characteristic of the MUF scheduling algorithm helps to achieve this.

Criticality: In MUF, operations with higher *criticality* are assigned to higher static priority levels. Assigning static priorities according to criticality prevents operations critical to the application from being preempted by non-critical operations.

Ordering operations by application-defined criticality reflects a subtle and fundamental shift in the notion of priority assignment. RMS, EDF, and MLF (1) exhibit a rigid mapping from empirical operation characteristics to a single priority value, and (2) offer scheduling guarantees only for all operations or none. In contrast, MUF (1) gives applications the ability to distinguish operations arbitrarily, (2) can offer scheduling guarantees for a critical *subset* of the entire set of operations.

Dynamic Subpriority: At the instant of evaluation, dynamic subpriority in MUF is a function of the the laxity of an operation.

Assigning dynamic subpriorities according to laxity (1) offers higher utilization of the CPU than static approaches, and (2) allows deadline failures can be detected *before* they actually occur, except when an operation that would otherwise meet its deadline is preempted by a higher criticality operation. Moreover, MUF can apply various types of error handling policies when deadlines are missed [8]. For example, if an operation has negative laxity prior to being dispatched, it can be demoted in the priority queue, so an operation that can still meet its deadline can be dispatched instead.

Static Subpriority: In MUF, *static subpriority* is a static, application-specific, optional priority. Static subpriority has lower precedence than either criticality or dynamic subpriority. It is used to order the dispatches of operations that have the same criticality and the same dynamic subpriority. Assigning a unique static subpriority to each operation that has the same criticality ensures a total dispatching ordering of operations at run-time, for any operation laxity values having the same criticality. A total dispatching ordering ensures that for a given arrival pattern of operation requests, the dispatching order will always be the same. This in turn improves the reliability and testability of the system.

The variant of MUF used in TAO's strategized scheduling service enforces a complete dispatching ordering by providing an `importance` field in the TAO `RT_Info` CORBA operation QoS description `struct` [6]. TAO's scheduling service uses importance, as well as a topological ordering of operations, to assign a unique static subpriority for each operation within a given criticality level.

2.3.3 *Hybrid Approaches*

Hybrid static and dynamic approaches may be used to combine the benefits of both. Multi-level scheduling integrates different approaches at different scheduling *levels*. One example is two-level hierarchical scheduling, which allows real-time applications to coexist with non-real-time applications in an open OS environment [11]. Another is standardized in the ARINC Avionics Application Software Standard Interface (APEX) for Integrated Modular Avionics (IMA) [12]. One level consists of *partitions*, which are executed cyclically and scheduled statically and off-line. Within each partition, application *processes* are scheduled using potentially more flexible approaches.

Each task in a partition is characterized statically by period (for periodic tasks), deadline within the period, and worst-case execution time. Aperiodic tasks are supported; Audsley and Wellings offer an analysis approach assuming minimum arrival time for aperiodic task periods [13]. TAO used this same approach initially to handle aperiodic tasks with rate monotonic scheduling and analysis.

APEX Partitions are scheduled cyclically. Each partition is characterized statically by parameters including criticality level, period, and duration. Therefore, a straightforward static scheduling approach can be used.

The APEX approach provides static schedulability analysis and fault tolerance across partitions. However, it suffers from the drawbacks of static scheduling described in Section 2.1. In particular, it is not clear how APEX can appreciably improve resource utilization when compared to conventional static scheduling approaches. For instance, jitter may be high when the the period of a task is not a multiple of its partition's period [13]. In that case, the task could become ready to run at a time when another partition was executing, and therefore would have to wait for its partition's activation.

3 Dynamic Scheduling Overhead

To assess the run-time cost of dynamic scheduling, we used an experimental setup based on TAO's Event Channel [14]. It consisted of a single high-priority supplier/consumer pair, and a varied number of low-priority event suppliers and consumers. We measured the latency in event delivery between the high-priority supplier and consumer. This latency included the time required for the TAO run-time scheduler to satisfy the Event Channel dispatch module scheduling request.

The test was run on a Sun Ultra 30 in the RealTime scheduling class, with a single 300 MHz UltraSPARC CPU, for two different scheduling strategies. The static scheduling strategy used off-line RMS and table lookup at runtime. The dynamic strategy used MUF, and therefore required an additional laxity calculation at runtime.

As shown in Figure 3, there appears to be a small (up to 10 percent) overhead for dynamic scheduling.

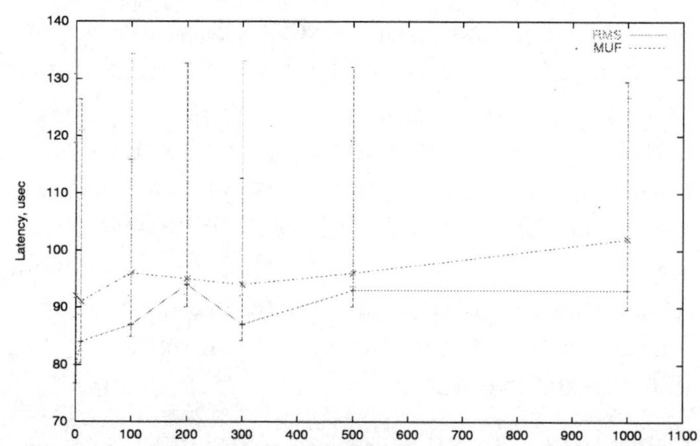

Figure 3: The Cost of Dynamic Scheduling

4 Concluding Remarks

As mission system computing evolves to address increasingly complex user requirements, run-time variation and non-periodic activities will become increasingly common. In addition, required system flexibility, in support of changing mission conditions, will no longer be achievable with static modal behavior. Conventional static approaches to scheduling mission computer activities will not support these increasingly complex behaviors.

Dynamic scheduling approaches offer potential solutions to these increasingly complex requirements, but at some run-time cost. The contribution of this work has been the implementation and analysis of selected dynamic scheduling algorithms within a strategized scheduling framework. From this work, we have concluded that the Maximum Urgency First scheduling algorithm has desirable properties for use within a mission computing application and acceptable run-time overhead.

In future work, we plan to measure the run-time cost of dynamic scheduling in TAO's scheduling service under a variety of of operational conditions. Further work will also explore scheduling in distributed systems. Dynamic scheduling appears to be a prerequisite for distributed system scheduling, due to the loose coupling between operations on separate processors.

5 Acknowledgments

This work was funded in part by Boeing. We gratefully acknowledge the support and direction of the Boeing Principal Investigator, Bryan Doerr.

References

[1] T. H. Harrison, D. L. Levine, and D. C. Schmidt, "The Design and Performance of a Real-time CORBA Event Service," in *Proceedings of OOPSLA '97*, (Atlanta, GA), ACM, October 1997.

[2] Object Management Group, *The Common Object Request Broker: Architecture and Specification*, 2.2 ed., Feb. 1998.

[3] S. Vinoski, "CORBA: Integrating Diverse Applications Within Distributed Heterogeneous Environments," *IEEE Communications Magazine*, vol. 14, February 1997.

[4] J. R. Newport, *Avionics Systems Design*. Boca Raton, Florida: CRC Press, 1994.

[5] D. B. Stewart, D. E. Schmitz, and P. K. Khosla, "Implementing Real-Time Robotic Systems using CHIMERA II," in *Proceedings of 1990 IEEE International Conference on Robotics and Automation*, (Cincinnatti, OH), 1992.

[6] D. C. Schmidt, D. L. Levine, and S. Mungee, "The Design and Performance of Real-Time Object Request Brokers," *Computer Communications*, vol. 21, pp. 294–324, Apr. 1998.

[7] J. P. Lehoczky, L. Sha, and J. K. Strosnider, "Enhanced Aperiodic Scheduling in Hard Real-Time Environments," in *Proceedings of the IEEE Real-Time Systems Symposium*, IEEE Computer Society Press, 1987.

[8] D. B. Stewart and P. K. Khosla, "Real-Time Scheduling of Sensor-Based Control Systems," in *Real-Time Programming* (W. Halang and K. Ramamritham, eds.), Tarrytown, NY: Pergamon Press, 1992.

[9] C. Liu and J. Layland, "Scheduling Algorithms for Multiprogramming in a Hard-Real-Time Environment," *JACM*, vol. 20, pp. 46–61, January 1973.

[10] M. H. Klein, T. Ralya, B. Pollak, R. Obenza, and M. G. Harbour, *A Practitioner's Handbook for Real-Time Analysis: Guide to Rate Monotonic Analysis for Real-Time Systems*. Norwell, Massachusetts: Kluwer Academic Publishers, 1993.

[11] Z. Deng and J. W.-S. Liu, "Scheduling Real-Time Applications in an Open Environment," in *Proceedings of the 18th IEEE Real-Time Systems Symposium*, IEEE Computer Society Press, Dec. 1997.

[12] ARINC Incorporated, Annapolis, Maryland, USA, *Document No. 653: Avionics Application Software Standard Inteface (Draft 15)*, Jan. 1997.

[13] N. Audsley and A. Wellings, "Analysing APEX Applications," in *Proceedings of the 16th Real-Time Systems Symposium*, pp. 39–44, Dec. 1996.

[14] T. H. Harrison, C. O'Ryan, D. Levine, and D. C. Schmidt, "The Design and Performance of a Real-time CORBA Event Service," *submitted to the Journal on Selected Areas in Communications special issue on Service Enabling Platforms for Networked Multimedia Systems*, 1998.

SOFTWARE PROCESS AND EARLY PROJECT PLANNING: ISSUES, INSIGHTS AND LESSONS LEARNED

Jogeswari P. Sarkar, Lockheed Martin Tactical Aircraft Systems, Fort Worth, Texas

Abstract

A matured software process is important, particularly for large or long-lived systems. It can help to ensure quality and reduce maintenance cost. Software Processes based upon Software Engineering Institute Capability Maturity Model will head off many problems and leads to efficient, steady-state development. However, the business acquisition phase omits many important process considerations. When this happens, it is hard to reverse and virtually impossible to optimize the eventual contract and process. Several categories of breakdown can occur in such areas as communication, coordination, timely execution, training, capital planning and procurement, and roles and role transitions. These above can be considered as management challenges, and are solvable. This paper contains examples of problems arising, and presents some ideas on an enlightened business acquisition model that avoids process problems during program execution.

Introduction

With the new concept of efficient business organization taking effect, the Programs enjoy greater autonomy. Integrated Product Development (IPD) teams have near-full authority to act in a manner that seems best for the team to deliver its products, on schedule and at the least cost. Naturally, there are situations that tempt the teams to bypass process steps and save cost and schedule in the near-term. In that case, for software intensive projects, the software integrators must assume the responsibility of ensuring a consistent project/process integration.

For software development, the Software Engineering Institute Capability Maturity Model lays down the goals, commitments and activities that are required for project success to attain long-term organizational software maturity. Company standard processes are developed within that framework. Attainment of higher levels of maturity (SEI Level 4 or 5) show a commitment to excellence, and may become a significant factor in future government contracts. Mature processes help to ensure quality products that reduce the cost of maintenance through the life of the software (which often measures in tens of years in the defense industry).

Application of a disciplined process in early project planning is at least as important -if not more so - as it is during software development. The long-term implications of early planning on cost and schedule force us to carefully explore all possible interfaces and achieve all required early coordination. This is done to assure that mistakes are not made at the start which the software development team must suffer at the end.

0-7803-5086-3 /98/$10.00 ©1998 IEEE

Problem Characterization

The software process essentially starts with new business acquisition phase and ends with customer delivery of the product. In reality, the business acquisition activities are often confined to a few marketing and contracts-type people with occasional help from a professional software person. This software person (or Software Point-of-Contact - POC) is assigned either ad-hoc by the Program or by the Engineering functional center on an availability basis. Often the result is that the Software POC is unable to follow the standard planning process for lack of training; or, the person is under stress to deliver at a quick turn-around, or there are activities that need to be accomplished before a formal plan could be established.

The company standard software process manual puts all the burden of planning and development on the Software Product Manager (SPM). The planning document that the SPM develops and follows is the Software Development Plan (SDP). The SDP, once approved, becomes 'the holy book' for the software developers and Software Quality Assurance. Unfortunately, this SDP frequently does not become ready for review even in its early draft form until several weeks or months after the go-ahead of the project. This is particularly true for projects that are medium to big size with development span time of 18 - 30 months. The early planning phase for software (Figure 1), which begins during business acquisition and ends sometime after go-ahead, is a period of vigorous activity when the Software POC essentially must act as an SPM. The real SPM assignment is unlikely to have happened until a program becomes firm.

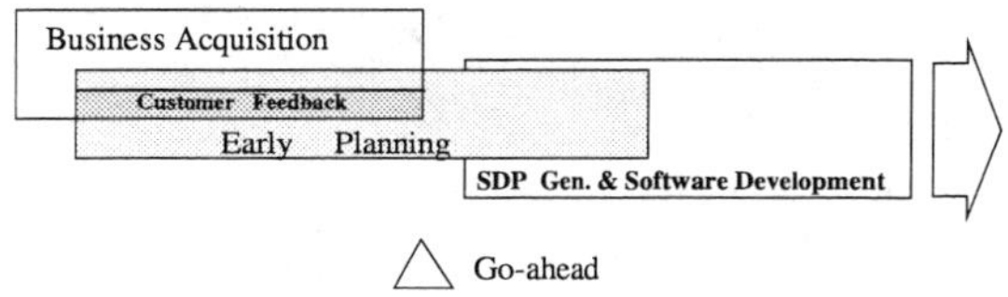

Figure 1. Early Planning Envelope

Therefore in the software early planning phase, the Software POC must do a myriad of complex interfacing, pre-planning and coordinating that could have a significant implication on the success of the program. These interfaces are usually confined within the company (Figure 2).

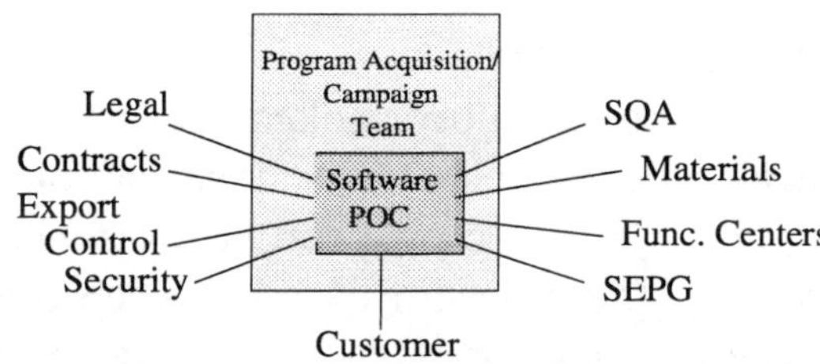

Figure 2. Early Planning Interfaces

While the Memorandum of Understanding (MOU) process is going on, the Software POC has activities to work on in parallel, and must feed the MOU team with the updates. A typical list (Figure 3) includes activities that very often take long and thoughtful deliberations with the interfacing groups. The Software POC's experience and expertise play a big role in arriving at the ultimate best decisions for the company.

Early Planning Checklist	
Project size Estimation	Reuse
Make/Buy plan	Supplier Software Capability
Supplier Source List	Technology Integration
Legal Limitation	Workshare
Export Restriction	Risk Analysis
Co-development	Deliverables
Company Strategic Plan Integ.	Facility requirements
Development Model Selection	Special QA plan
Programming Language	Standard Process Application
GFE/ CFE	Project Management Tool
Customer Participation	

Figure 3. Early Planning Checklist

Early Planning Issues

A discussion of some of the major checklist items would reveal their significance on the downstream planning of a project:

Legal Limitation

An often-overlooked area is the legal. Many companies have a long heritage of legacy software which undergo modifications and updates to generate new products. The user (or reuser) of such legacy software may not be aware of who originally funded the development of the software or who has the proprietary rights. There may have been tools developed and subsequently improved by using part customer funding, part internal funding and part by a second customer funding. In those types of cases, making a decision on the assumption that the software can be made part of a delivered product for a new contract could create a legal problem about violation of some customers' proprietary rights.

Since software never dies unless you kill it, there is continuous improvement made on some software, particularly in-house tools. Appropriate caution should be applied to determine the company proprietary rights on the available software before making any commitment on it.

Export Restriction

Government restriction on exports can be thrusted from various Departments. Statutory regulations as well as special restrictions could apply. All restrictions in exportability must be clarified before signing a contract with a foreign customer. DoD (Department of Defense) may allow a technology for export since it is now an old technology that is in existence for many years in defense and is being replaced by newer technology. However, DoC (Department of Commerce) may not allow the same technology for export since it is feared that the technology can be successfully applied by the foreign country in commercial field and eventually pose a threat to the current, favorable commercial trade.

An good approach is to write a technology transfer control plan (TTCP) in very detail and unambiguous words and get that approved by all the appropriate government agencies before the contract. [Only that way we will know whether the flight control software, for example, is as exportable as the avionics software.] The language used in the plan document should be simple so that its interpretation is easy. If necessary, the trade names may be used instead of generic names to assert specificity when applying restriction on tools. A procured software may contain internal code that is not exportable. In this type of cases, early-on TTCP interpretation and making sure that the procured software complies with the restriction would save cost and valuable schedule by avoiding unnecessary

0-7803-5086-3 /98/$10.00 ©1998 IEEE

coordination effort with the release authority.

Customer Participation

Participation by the customer in the developmental process can be helpful in achieving success by acquiring a common understanding of the problems and "solving together". Programs start with a jointly-agreed statement of work that contains overall requirements of the contract and the system. The system requirements are then developed and approved by the customer. The customer signs a large document without really internalizing much of it. However, the expectations of the customer grow as the development progresses, primarily because the customer becomes more and more familiar with the internals of the system through reviews and briefings. This expectation growth can become a major impediment if not checked.

Before agreeing to any customer participation, the depth and breadth of participation as well as the nature of participation need to be thoroughly examined. While participation during system development is helpful in order to restrict the expectation growth later, the same would most certainly be a hindrance during development and testing. It is not far-fetching that the customer likes to monitor or may even want to do intelligence-gathering activities in the name of participation. In this situation, reviews /walkthroughs will have to be carefully planned in order to check unrestricted access to proprietary data. For a foreign customer to be on sight, a separate facility arrangement may have to be constructed with restricted security access in order to comply with export regulations and prevent unwarranted visits.

Technology Integration

Although insertion of new technology may go along the line of company or department strategic planning, the integration plan for a new technology in a contracted program need to be addressed more on the basis of cost and schedule rather than on its technical merit. A technology may have been validated in a small IRAD (Internal Research and Development) or CRAD (Contracted Research and Development) environment, but introducing the same in a production environment certainly needs a better validation in a demo or prototype project environment. Also, procurement of a new hardware, for example, a custom-designed multi-chip module, may turn out to be a major problem since there are not enough suppliers in the market yet. If we have to 'create' a supplier, then cost and schedule slip may become unmanageable.

Many new technologies should not introduced together since risks multiply. Introduction of only one (or very few) new technology into the rest proven technologies is a prudent planning.

Workshare

Workshare is decided at the MOU and clarified in precise terms in the contract. It is generally defined as a percentage of total work estimated in dollars. The collaborating companies try to leverage the share upon their technology strength among other things or, may even be, upon their political strength in international deals. Since the funding of the program is prone to fluctuate based upon either the government spending or the market demand or the dollar-foreign currency exchange ratio, the task definitions and the planning should be sufficiently modular in structure. This would allow an easy transfering a portion of the workshare to the partner company, should an imbalance show up in mid-term, without causing any major impact to the overall plan. For cost-plus programs, however, this problem does not exist.

The tasks that are clearly coupled in terms of technical interdependency should be the workshare on the same side. If the Operational Flight Program (OFP) of a core avionics system is a task, then the Software Test Station for that OFP should also be a task for the same company. If due to imbalance the workshare needs to tilt on the other side, it is better to give away the back-end task rather than the front-end one. Test Station Development is to be sacrificed rather than the OFP development since OFP development is much more independent and does not require intensive programmatic interfacing. The final delivery of the product should take place at the owner's facility and not at the other company's facility or in another country. The transportation of the product, putting it all together and making it work in other's facility can become a major headache, sometimes simply due to logistic reasons.

Crtical Decision Making

Many important decisions are taken at the MOU negotiations or in early contract negotiations, most of which are critical because these generally have long-term, far-reaching implications. Changing the MOU or contract mid-stream in an on-going program is often impossible. Figure 4 lists some areas where management is challenged to display far-sightedness decisions.

Organization/ Group	Critical Decisions
Legal	Reuse of copy-right code
Contracts	Workshare, Co-development
Export Control	Code exportability, Sanitization
Security	Classified Code handling
Quality Assurance	Customer-requested Special Provisions
Customer	Customer involvement in Development
SEPG	Adaptability of Standard Development Process, Use of Company software database
Functional Center	Technology Integration, Tools

Figure 4. Critical Decision Coord. Requirements

Some decisions taken at the early planning stage must be communicated to the customer since the customer has to plan their activities accordingly. Most significantly, almost all the decisions will have some impact on the customer funding which in turn has impact on the workshare.

As we can see, the Software POC has a grave responsibility upfront to make things right for the program (and so for the company). How does that person do everything right? Or, is it even possible for the Software POC to proceed rightly? The Software POC is not an SPM and probably did not have the training in the software process. So, the Software POC-followed procedures are likely to be unreliable.

The standard software process manual essentially starts with an SPM and a Software Development Plan (SDP). Since none of them exists at the early planning phase, the software POC will be at liberty to act and behave as he/she thinks fit. Thus, recognizably there is a void.

Solutions Approach

In order to take care of the void, the areas that need to be defined are:

- Assignment of the Software POC and SPM
- Organizational relationships for SPM
- Early planning tool establishment
- Transition to software development phase

A discussion of the essential processes follows:

Timing the assignment of the software POC and the SPM

The Software POC needs to be selected by the Engineering functional center, who keeps a list of experienced and trained engineers suited for the job. A foreign campaign team should need an individual who preferably has some knowledge of the country, culture and foreign program exposure. Managerial experience in software product management or software subcontract management should be a primary requirement. This is because the Software POC will be interfacing with many departments or groups at mid or senior management level in order to obtain effective and timely inputs in the overall coordination and make key decisions.

If and when the campaign is announced as a full program, an SPM needs to be assigned almost immediately. Transitioning of the Software POC into the SPM role is definitely an excellent idea, the feasibility of which, however, would depend on the organization structure, the individual's own performance and training, Program/ functional organization acceptability and other factors.

Organizational relationship model for the SPM

In a typical model, the Program organization assigns an Integrated Product Team (IPT) for each product area. The Leader (IPTL) of each IPT reports to a manager in the overall Program Management. The SPM is responsible for the software portion of the IPT product and reports to the IPTL. The SPM activities in this structure do not have a direct visibility to the Program Management.

Typically, the software projects consume a large share of funding. The interface complexities are enormous. There are many limitations (or government restrictions for international programs) and company process compliances. Direct visibility of the software issues and risks to the Program Management on a regular basis would allow problems to be addressed in a timely manner. The SPM should have a direct or near-direct access to the Program Management through weekly meetings, reviews and special presentations.

The appointment of the SPM should be through a joint involvement of the Program Management and the cognizant Engineering functional center. The functional center makes the selection and the Program reviews and approves it. The Program then makes the announcement with the organization framework for the SPM to function. The ideal model should look like Figure 5.

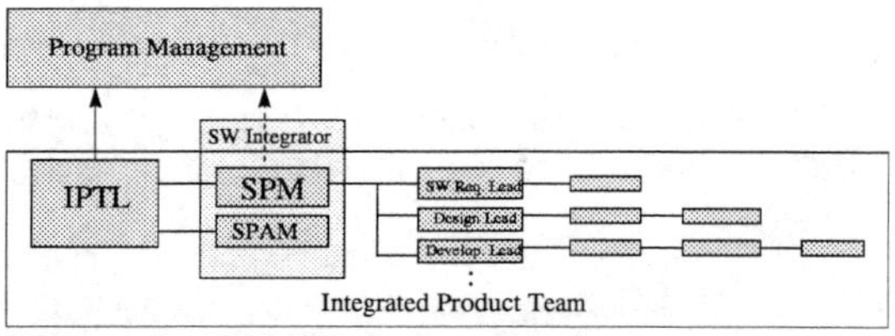

Figure 5. SPM Relationship in Org.

The SPM's direct supervisor is the IPTL. Management visibility to all SPM activities is provided through the IPTL on a regular basis. The SPM obtains project/process integration oversight from the Software Integrator for the business area/Program.

Checklists used as Early Planning Tools

All activities to be accomplished in the early planning phase need to be itemized in a checklist, much like the Phase Activity Planning (PAP) for Business Acquisition & Planning (BA&P) . The start date and the finish date for each listed item should be assigned with an assignee name. Most of the activities, however, will have to be assigned to the Software POC (or the SPM) himself/herself. Some good examples of checklist items are the items that have been discussed earlier (see Early Planning Issues).

Replanning of the start and finish dates for the activities is not unusual since the funding level at the business acquisition/ campaign level often varies and many tasks are dependent on external inputs without a good assessment of the projected completion dates. However, with dependency established among the list items, an activity network can be created. Management of the activity network could be enormously difficult because of the uncertain nature of timely task completion. Sometimes, a periodical checklist review may be the only viable way to status the activities.

Conceptually a software planning maturity index can be designed similar to a company software maturity index. However, in the absence of explicit early planning activities and processes in the Software CMM, the maturity index will have to be designed with a set of carefully selected objective criteria.

Making the Transition to the Development phase

SDP generation begins sometime after the program go-ahead. Early planning results need to be assimilated into the SDP. If the same person (as the software POC) takes over as the SPM, then seamless planning would be expected. However, with the early planning materials documented and process managed, a good transition into the software development phase is expected.

Summary Recommendation

A successful program must start with a successful early planning. The essential management steps that need to be ensured are:

- Appoint an experienced senior software manager (Software POC) in charge very early in the program in its campaign phase
- Ensure that the manager has been trained in company standard software process
- Have an early planning checklist made with all possible peculiar (ever imaginable) issues included
- Address methodically each issue to a well-coordinated resolution
- Plan for a smooth, fairly long transition of responsibility from the Software POC to an SPM

References

[1] Humphrey, W. S., *Managing the Software Process,* Reading, MA: Addison-Wesley Publishing Company, 1990

[2] Humphrey, W. S., *A Discipline for Software Engineering,* Reading, MA: Addison-Wesley Publishing Company, 1995

[3] Hermanson, S. E., "Software Supportability: A Managers Perspctive", *AIAA/NASA Symposium on the Maintainability of Aerospace Systems,* 26-27 July 1989

[4] Basili, V. R., H. D. Rombach, "The TAME Project: Towards Improvement-Oriented Software Environments", *IEEE Transactions on Software Engineering*, Vol. 14, No. 6, June 1988

[5] Paulk, M. C., Bill Curtis, Mary Beth Chrissis, Charles V. Weber, "Software Project Planning, a key process area for Level 2: repeatable", *Capability Maturity Model for Software,* Version 1.1

[6] *Lockheed Martin Tactical Aircraft Systems (LMTAS) Software Development Process Manual, PM-4001A*

[7] Safford III, E. S., *"Improving Software Development Through the Proper Implementation of a Standard Software Process: Case Study and Lessons Learned", AIAA/IEEE Digital Avionics Systems Conference,* 12th DASC

Acknowledgment

The author is grateful to Ed Safford for review of this article and suggestions made from his past experiences which were closely coupled with the author's.

LOW AND NON-INTRUSIVE SOFTWARE INSTRUMENTATION

A SURVEY OF REQUIREMENTS AND METHODS

R. Fryer, ATAC, Ridgecrest, CA 93555

Abstract

The Commercial Off The Shelf (COTS) initiatives reduce access to some of the traditional methods of customizing systems for better testing. Testing of embedded systems has often been aided by providing special access to test data. In mission critical and high value real time embedded systems (DOD, nuclear, telphony, transportation, etc.) special 'hooks' have been provided to assist software and system testing. A broad set of requirements with emphasis on real time embedded systems is provided, followed by some of the past and current methods of meeting these requirements.

Some emphasis is placed on the payoff for instrumenting embedded systems and some possible ways to move forward given the limitations of present COTS processors.

Background

In the industry as a whole, software and software-based system test methods focus largely on aspects of software that might be termed 'static.' Static testing assesses the correctness of the program specification; proofs of correctness, symbolic testing, functional testing and other aspects that might be determined from the software without executing it in its target environment [1].

These methods are especially useful in embedded systems as it is often quite difficult to provision a test harness for embedded systems that faithfully simulate the external behavior (data sources and sinks) required for correct or 'effective' testing.

To improve the testability of DOD's embedded avionics systems, embedded processors have in many cases been enhanced with software 'test hooks' to facilitate performance measurements and to aid software or system debugging. 'System' debugging here addresses problems that arise from interaction of hardware and software components. In recent avionics systems, RTNI – Real Time Non-Intrusive methods have been used within the Mil-Std 1750 family (developed primarily by the Air Force but widely used in all services) [2, 3, 4] and in all existing members of the Navy's AYK-14 standard computer family [5]. Similarly in other fields (e.g. Telecom) where high availability is crucial, extensive special hardware has been designed into systems [6].

In the scientific computer market (which has emphasized parallel systems for many years), system performance assessment and debugging has also been a focus area, usually calling for the development of custom hardware.

The emphasis on use of COTS in embedded systems and specifically in vehicles with human life related issues calls for an assessment of the testability of these systems.

General purpose tooling in the target area of commercial 'embedded computing' is exceptional in terms of both system development and software *and* system test tools. The commercial embedded market has typically focused on equipment, process and vehicle control functions. These needs have usually been well met at the 'low performance'

0-7803-5086-3 /98/$10.00 ©1998 IEEE

end of the computing spectrum. Capabilities for high performance (desktop, scientific and server) class processors have not enjoyed similar support [7]. It is these processors that have the capabilities required for DOD avionics applications.

This survey is prepared to summarize the many approaches taken in solving the unique testing problems that arise in high performance embedded systems (for both uniprocessor and multiprocessor systems). Practitioners from the system test community continue to evaluate adequacy of these embedded systems designs in the area of 'testability'. It is hoped that this may also stimulate dialog regarding new approaches to testing high performance COTS based processors in mission critical embedded systems.

REQUIREMENTS

This section sets the stage for describing and comparing various instrumentation approaches. For the purpose of this survey, requirements are divided between software and system ***Testing*** requirements and requirements to assess the ***Performance*** of the system. Testing requirements will only address tests that of necessity cannot be done using conventional methods (static, etc.) that are shared by non-embedded systems.

Testing

Testing is usually carried out to assure that system behavior is correct under defined test conditions. It is often also conducted to determine the performance limits of a system and as well to determine if failure modes are acceptable under conditions that fall out of the expected envelope (i.e. fail-soft systems).

In years past, 'testing' and 'debugging' were often thought of as similar or equivalent processes. In recent years, opinions have varied on whether testing is intended to find the (expected) errors in a system [8] or to demonstrate their absence (in the domain of test cases) [9]. Either way, both white box and black box testing [10] may in real time systems call for special instrumentation.

Debugging

It has been recognized that 'debugging' real time systems is one of the most challenging tasks facing the embedded systems developer [11, 12]. Witschorik [13] emphasized that debugging might be categorized into 'non-interfering,' 'moderately-interfering,' and 'interfering.' He clarified that to conduct 'debugging' examination of the code behavior, that state transitions needed to be recorded *adequate to replay (preferably both backward and forward) the behavior of the program in the vicinity of suspected problems.*

Conventional (commercial) debuggers have improved in capability substantially over the last few decades, incorporating support for real-time operating systems and multiple execution threads. However, even the most capable commercial source level debugger requires control over the target system execution sequencing to carry out its functions [14]. Further, these are *not* designed to operate with multiple monitoring sources (Tabak defines 'Multiprocessor' as a system composed of CPUs, memory modules, Input/Output (I/O) units, and an interconnection network [15]). For this reason, 'multiprocessors' rather than 'Distributed Real Time Systems' [16] is preferred in this report.

Hardware Software Interactions

Software that operates as expected in a unit test or simulation environment may not work correctly when it is combined with the target hardware. This can occur because the hardware itself may not operate as expected (hardware documentation is often incomplete in terms of software interfaces; hardware changes are not always reflected in changes to software interface documentation, etc.).

Timing Problems

Hardware / Software interaction problems are most evident due to implicit timing requirements that were not known or understood at software design time (e.g. interrupt service routine, bus sharing between DMA transfers and other subsystems such as processor execution, and other hardware interface constraints). The ideal test equipment facility will provide for testing of these low level and time sensitive interfaces without impacting the behavior of the software / hardware interface. This implies a high degree of non-intrusiveness.

Software Layer Interactions

In a modern system design there may be a number of 'layers' of software from low level board interface software up to the application code. These 'layers' are usually developed and supported by as many vendors. Documentation and testing approaches will be equally varied, and the challenges of identifying problems that exist between layers (that is, at the software interfaces) have caused the collapse of some projects.

Real Time Scheduling

Embedded systems are now embracing modern software engineering concepts. New designs use features in POSIX[1] (such as multi-threading), Distrubuted Component Object Models (DCOM) and a Common Object Request Broker Architecture (CORBA) components. The interaction between the system elements and the operating system task scheduler becomes a great deal more complex than in most prior embedded systems.

However, relatively straight-forward uni-processor systems were already notorious for suffering from scheduling problems. The high visibility of this problem area has led to a large body of work [17]. Application of Rate Monotonic Scheduling (RMS) theory however requires excellent measurement capability for code execution in the target environment (see 'Performance Monitoring' below).

Multiprocessor Systems

Testing and to an even greater degree debugging of multiprocessor based systems dictates some of the groundrules for instrumentation. The interdependence of tasks running in multiprocessor systems effectively prohibits the use of an interactive debugger running separately on each element of the system (processor, bus, etc.). The interaction of such systems with the system clock, with 'real world' time and in the absence of a digital global time that is synchronized in all sytstems with 'real' time dictate that tests not intrude on behavior of specific sub-elements of the system [18, 19].

Embedded multiprocessor systems are inherently non-deterministic and test repeatability (at least precise execution repetition) is not possible. Finally, timing constraints within these systems prohibit intrusive monitoring [20] – or at least intrusion that could change the ordering of events within the system.

Performance Assessment & Monitoring

Modeling software as part of a detailed hardware/software co-design calls for a detailed model of software behavior [21, 22]. A persistent experience in embedded system design is that interaction of software and hardware components are sources of many surprises to system designers and implementers, an experience that indicates the need to carefully measure and validate these models.

Non-Intrusive (and in some cases, Low-Intrusion) monitoring capabilities can provide the needed data to model software systems construction and both verify performance & predict the results of enhancements [23]. Typical parameters that will be required for useful modeling include:

[1] IEEE Std P1003.4b Draft 8. 1993b. *Portable Operating System Interface (POSIX)*

- Task (or process) initiation times, lifetimes & invocation rates
- Delay (between software critical events) and distribution of delays
- Signaling Distributions
- Software processing delays (Queue arrival and departure times).
- Branching Ratios or Alternate Action ratios
- Frequency & distribution of interrupts (usually highly mode specific)
- Message Frequency & Size
- Coordination delays between multiple processors
- I/O block distributions and sizes
- Distribution of values of selected variables

The major difference between these broad goals for instrumentation and those in the Testing category is that code tracing (detailed execution history) is needed in addition to the needs described for performance modeling.

Monitoring Measurement Detailed Requirements

Tracing

Tracing program execution or variable behavior over time is the most challenging requirement of an instrumentation system. In previous system designs, data access to support tracing was usually designed into the processor to provide the needed data. The challenge was finding a way to store enough execution history information (or state history information) to reconstruct execution history. The usual method of managing the volume of this data is by using Complex Triggering to control log activation and termination.

In modern commercial RISC processors other methods must be found. Cache line fetches coupled with foreknowledge of program contents can provide for 'Execution Replay' to determine, for example, which path must have been taken at each branch [24].

The application domain may include process to process messages that may remain entirely internal to a processor cache during an execution invocation. As on-chip cache memories increase in size, this may become a major inhibitor for this method.

Code Coverage

Code coverage is usually accomplished by keeping statistics on how often each instruction and/or data location are 'touched' during a test sequence. This may be modified to 'written,' 'executed,' or 'read.'

Task (or process) initiation times, lifetimes & invocation rates:

If instruction tracing can be derived as described above, then task lifetimes and invocation rates may be directly determined from the resulting log, as an external timebase can be used. Precise initiation times depends on the length of time that a cache line fetch containing the initial instruction of a task awaits execution. Therefore, some signal must be passed to the outside that coincides with the completion of such an instruction.

Delay (between events) and distribution of delays

Signaling Distributions

Queue arrival and departure times.

&

Branching Ratios, Comparison of Alternate Actions

These measurements may be directly made from a time-accurate tracing mechanism.

Frequency of interrupts, distribution of interrupts

Most interrupt requests, other than internal processor clocks and faults, are visible external to the processor. (Some internally generated interrupt requests and interrupt service mechanisms can be derived from a successful tracing implementation.) To derive correct initiation time of internally generated interrupts/traps will require special signaling unless 'execution replay' along with tracing can reveal that information.

Message Frequency & Size

I/O block distributions and sizes

Distribution of values of selected variables

Coordination delays between multiple processors

A successful monitoring program – especially for a multiprocessor system, will demand concurrent monitoring of I/O ports and the interprocessor network to extract this data.

Bus Performance

Connection networks for multiprocessor systems are an intense area of development. Much work has been done predicting the performance of actual and hypothetical system busses. Once a system has been developed, measurement of actual bus use and the interaction of the processors with the bus becomes a fertile area for system improvement.

Summary of Testing Requirements

The primary requirements that can be derived from these test goals may be categorized as *Tracing* (Real-Time Execution Monitoring [25]), *Complex Triggering* [26] and *Timing* capabilities. Timing needs are usually satisfied by time-tagging traces and events detected by complex triggers. Multiprocessor systems create complexities here too when the timebase for each processor uses a different clock and the system design may call for asynchronous operation.

Variable access histories are also frequently needed for both debugging and for system performance confirmation. 'Complex Triggering' is the notion of allowing a tester to use various assertions that should (or might be) be true of the system behavior to constrain the collection of behavior logs and to determine when (or if) the assertion is satisfied. This can reduce the bandwidth of data logged and narrow the search space to isolate problem causes.

SOFTWARE AND SYSTEM INSTRUMENTATION APPROACHES

Successful methods for meeting these requirements in real-time systems fall into *non* and *minimally* intrusive approaches. Non-intrusive approaches require specialized hardware to be added somewhere in the target system, though a COTS processor does not necessarily need to be modified. Minimal intrusion is an acceptable approach as long as the monitoring system does not, when activated, change the timing of events or execution pattern of the application. Non-intrusive means that there is no 'visible' memory or execution footprint of the monitoring system – even when it is active.

Uniprocessor Systems

At the 'lowest' level, a number of vendors have offered a 'background debug monitor' on some embedded processors. High perfomance processors such as the Power PC desktop family (60X and 7XX) offer a 'JTAG'[2] port as an alternative. This port provides a low level access point to the compute engine on-chip and can provide a number of functions that could be invisible to execution of the machine code.

While this can be an asset in a uniprocessor system, the time intrusion appears to be milliseconds on processors of interest so that it is of very restricted use for the purposes in this survey. Further, the maximum bandwidth of this bus is perhaps two orders of magnitude below the value needed to carry critical trace event data.

In addition to JTAG support, the Intel pentium family contain a number of registers that fall outside normal application use. These seem to offer interesting capabilities for software tracing and test support with very low intrusion characteristics but are not documented for access by end users[3].

The Motorola PowerPC 750 does contain a significant number of performance monitoring registers with related event detection logic [27]. The primary use of this would appear to be to tune compilers and examine L1 and L2 cache behavior, and to date

[2] IEEE 1149.1-1990 *Standard Test Access Port and Boundary Scan Architectecture.*

[3] `http://www.x86.org/`

none of the commercial tool vendors support these registers.

Some embedded processors support low and even non-intrusive tracing fully compliant with the goals cited here [7]. However, those providing internal visibility fall at the low capability end of the performance spectrum (max system clock of 20 MHz as of this writing).

Hybrid instrumentation systems combine some hardware and software elements to extend instrumentation capabilities [28]. In many cases this calls for adding custom logic on the computer board that cooperates with the performance of a COTS processor.

The cost penalty may seem high using hybrid techniques. For example, the CodeTEST product that instruments source code for cooperation with external monitors may bost 10% of system throughput [29]. And, for some approaches, external (to the CPU) hardware must still be designed into the system.

Figure 1 below demonstrates a method of solving uniprocessor instrumentation by resorting to a custom version of a COTS architecture. The 'Instrumentation Memory' augments a CPU core with additional logic that carries out software test and performance instrumentation capabilities without any impact on visible memory or throughput, but with a potentially high impact on the silicon footprint [30]. Such an approach might become very attractive should the industry be able to standardize on low level functions that were commonly needed in embedded systems.

Finally, many mechanisms have been devised to determine the actual behavior of a machine extracted from a fairly limited set of observable events [31]. These methods have also been extended to multiprocessor systems. Based on the 'instant replay' [24] technology but also incorporate the timing constraints particular to embedded systems.

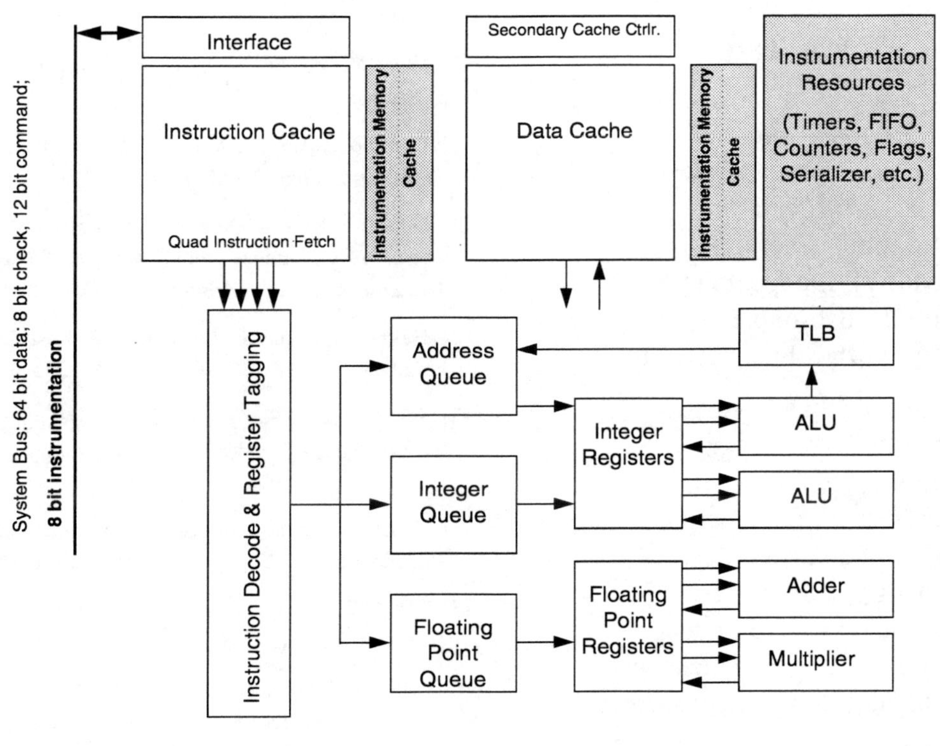

Figure 1 – An 'Instrumentation Memory' can provide internal tracing and condition testing

Multiprocessor Systems

In terms of debugging alone, several authors have explored the problems of multiprocessor systems [32]. The approach taken uses minimally intrusive Operating System (OS) support calls to examine system events, and provides aids in examining system trace logs. However the processors used were not RISC based and tracing was feasible.

'Deterministic replay' is a term widely used in the distributed processing research community to describe the approach to reconstruct system behavior from system instrumentation [33]. This approach focuses on isolating the events that must be observed and the critical data needed in conjunction with this data to reconstruct the behavior in a manner that minimizes ambiguities [34].

The concept of the hybrid monitor has been extended into this realm also. Hybrid approaches referenced for Uniprocessors have been used in multiprocessors as well [35].

Conclusions: Design For Test

As restated by Maples at Lawrence Berkeley Lab [36], "*There are only two difficulties in constructing a general purpose multiprocessor system – hardware and software*." On-chip support is not only occasionally provided [37], it appears to be inevitable [38] to provide affordable multiprocessor systems with robust software.

A number of approaches have been proposed *and* implemented [39, 40] for RISC systems. The problems appear not to be technological but rather lack of an adequate demand from the embedded software quality and software engineering communities. Perhaps this is due to the hardware intensive nature of the best solutions. Nevertheless, as system testers, we must not leave the customer at risk because it's 'hard.'

REFERENCES

[1] Demillo, R. et. al. 1987, *Software Testing and Evaluation*, 1987, Benjamin Cummings, Menlo Park, CA.

[2] Weber, J. and Cannon, W., "Real-Time, Non-Intrusive Software Testing: The Technique for the Future," *SAE International Congress & Exposition*, Feb 28 1994. pp. 167-177.

[3] "VHSIC 1750A Module Software Development Support Interfaces," Texas Instruments, Inc. September 1988.

[4] JIAWG, 1989. "User Console Interface Specification," Joint Integrated Avionics Working Group Specification J88-N1F, R. Fryer, Ed. (Dec).

[5] Control Data Corporation. 1990. "Feasibility Study Results to Implement a Real Time Non-Intrusive Monitor into the AN/AYK-14(v) VHSIC Processor Module." NWC Contract N60530-89-C-0355. (Jan 23).

[6] Holy, Z., D. J. Lewis, T. J. Todd, "Non-Intrusive Tool for Software Testing and Debugging: An Actual Industrial Application Example," *Proceedings of the IEEE Software Maintenance Conference*, 1990.

[7] Neugass, H. "An Inside Look at Chip Debugging," *Insight Journal*, V3 #2 1998 pp18-21.

[8] Myers, G. J. *The Art of Software Testing*. New York: John Wiley, 1979.

[9] Hetzel, W., ed. *Program Test Methods*. Englewood Cliffs, J.JK.: Prentice-Hall, 1973.

[10] Demillo, R.A. et. al. *Software Testing and Evaluation*. Benjamin Cummings, Menlo Park, CA 1987 pp20-21,

[11] Gupta, N. K., R. E. Seviora, "An Expert System Approach to Real Time System Debugging," *IEEE Conference on Artificial Intelligence Applications*, December 5-7 1984.

[12] Helmbold, D. and D. Luckham. 1985. "Debugging Ada Tasking Programs." *IEEE Software*, 2 no. 2 (Mar): 47~57.

[13] Witschorik, C. A. "The Real-Time Debugging Monitor for the Bell System 1A Processor," *Software Practice and Experience*, V 13, 1983, pp 727-743.

[14] Weiss, R. "Emerging tools help designers debug 32 –bit embedded systems," *Computer Design*, July 1995.

[15] Tabak, D. *Multiprocessors*. Prentice Hall, Englewood Cliffs, NJ, 1990, p11.

[16] Tsai, J.P., Fang, K.Y., and Chen, H.Y., 1990. "A Noninvasive Architecture to Monitor Real-Time Distributed Systems," *IEEE COMPUTER*, (March) 11-23.

[17] Sha, L., R. Rajkumar, and J. Lehoczky. 1990. "Priority Inheritance Protocols: An Approach to Real-Time Synchronization," *IEEE Transactions on Computers*, (Sept) pp. 1175-1185.

[18] Lamport, L. "Time, Clocks, and the Ordering of Events in a Distributed System," *Comm. ACM*, Vol. 21, No. 7, July 1978, pp. 17-27.

[19] Raju, S. C. V., R. Rajkumar, F. Jahanian, "Timing Constraints Monitoring in Distributed Real-Time Systems. *IEEE Proceedings of Real-Time Systems Suymposium*, 1992, pp. 57-67.

[20] Tsai, J., S. Yang, *Monitoring and Debugging of Distributed Real-Time Systems*, IEEE Computer Society Press, Los Alamitos, CA., 1995. pp. 7-9.

[21] Kumar, S. ; R. Klenke,; J. Aylor; B. Johnson; R. Williams; R. Waxman, "A Unified System Level Modeling Design Environment," In *Proceedings of the 1st Annual RASSP Conference*, (Crystal City, VA August 15-18, 1994) 114-123.

[22] Leshem, G. 1992. "Simulation of Real-Time Embedded Systems." *ACM Simulation Digest*, 22 no. 2 (Fall): 35~41.

[23] Fryer, R. E., G. Vansteenkiste, and J. Van Campenhout. "Real Time Non-Intrusive Monitoring and Embedded System Simulation," in *Special Proceedings on Embedded Systems; of the Society for Computer Simulation Summer Simulation Conference, (July 1994).*

[24] LeBlanc, T.J. and J.M. Mellor-Crummey, "Debugging Parallel Programs with Instant Replay," *IEEE Transactions on Computers*, Vol. C-36, No. 4, Apr. 1987

[25] Plattner, B. 1984. "Real Time Execution Monitoring." *IEEE Transactions on Software Engineering* SE-10 no. 6 (Nov) 756~763.

[26] Fryer, R. E., 1973. "The memory bus monitor," In *Proceedings of the AFIPS National Computer Conference*, V42. (New York, N.Y.) 42, 75~79.

[27] *MPC750 RISC Microprocessor User's Manual*, Motorola document MPC750UM/AD, August 1997, Chapter 11, pp. 11-1 through 11-12.

[28] Carpenter, R. 1989. "Performance Measurement Instrumentation at NBS." In *Proceedings of the Workshop on Instrumentation for Future Parallel Computing Systems* (Sante Fe, New Mexico, May) ACM Press, New York, New York, 159~184.

[29] Ganssle, J. "The Challenges of Real-Time Programming," *Embedded Systems Programming*, Vol. 11, No. 7, July 1998. pp. 20-26.

[30] Fryer, R. "A Core Architecture Feature for System Software Test and Instrumentation in Real Time Embedded Systems," preprint.

[31] Tsai, J. K. Fang, H. Chen and Y. Bi, "A Noninteference Monitoring and Replay Mechanism for Real-Time Software Testing and Debugging, " *IEEE Transactions on Software Engineering*, Vol. 16, No. 8, Aug. 1990, pp. 897-915.

[32] Timmerman, M., F. Gielen, P. Lambrix, *ACM/ONR Workshop on Parallel and Distributed Debugging*, May 17-18, 1993, San Diego, CA. pp. 151-157.

[33] McDowell, C. and D. Helmbold, "Debugging Concurrent Programs," *ACM Computing Surveys*, Vol. 21, No. 4, Dec. 1989, pp. 593-622.

[34] LeBlanc, R. and A. Robbins, "Event-Driven Monitoring of Distributed Programs, " *IEEE Conference on Distributed Computing Systems*, Denver, CO. May 13-17 1985, pp. 515-522.

[35] Mink, A. R. Carpenter, G. Nacht, and J. Roberts, "Multiprocessor Performance-Measurement Instrumentation," *IEEE Computer*, (September 1990), pp. 63-75.

[36] Maples, C. "Analyzing Software Performance in a Multiprocessor Environment," *IEEE Software*, July 1985. pp. 50-63.

[37] Weiss, R. 1991. "Debugging Support Goes On-Chip.", *Electronic Engineering Times* (May 20): 49~53.

[38] "Debug tools – not an option," editorial in *Electronic Engineering Times,* April 16, 1990.

[39] "Cannon, W.J., M. T. Michael, and D. D. Beeson. 1992. "Real Time, Non-Intrusive Instrumentation of Reduced Instruction Set Computer (RISC) Microprocessors." In *Proceedings of the National Aerospace and Electronics Conference (NAECON),* (Dayton, OH). IEEE Piscataway, N.J., 550~557.

[40] Walters, G., E. King, R. Kessinger, "Processor Design and Implementation for Real-time Testing of Embedded Systems," *IEEE 17th Digital Avionics Systems Conference*, Oct 31-Nov 6, 1998 (these proceedings).

IMPLEMENTATION AND TESTING OF INTERNATIONAL SPACE STATION FLIGHT SOFTWARE FOR THE POINTING AND SUPPORT PORTION OF GUIDANCE, NAVIGATION, AND CONTROL

Douglas J. Yazell, Honeywell Space Systems/Commercial Systems Operation, Houston, Texas

Implementation

Boeing engineers designed and implemented these algorithms using MOODS (Multibody On-Orbit Dynamics Simulation), a Boeing/NASA Ada non-realtime simulation running on a DEC Alpha platform in Boeing Tower 2 in Houston, Texas. The Ada flight software was then written in MatrixX superblocks on Hewlett Packard computers, using the MatrixX autocode capability to generate the Ada code.

After Flight Stage 5A of the International Space Station (ISS), the Pointing and Support (P&S) subsystem is activated and provides the Guidance, Navigation, and Control (GN&C) interface for most of the other U.S. subsystems. P&S supplies state vector (ISS position and velocity), attitude, and attitude rate data to other ISS systems, as well as

- Targets for the solar array beta joints
- Solar line of sight vector and rise and set times for two orbits
- The line of sight data for two TDRSs (Tracking and Data Relay Satellites) and TDRS rise and set times for one orbit
- ISS mass property data, including adjustments for the location and mass properties of dynamic objects such as the Mobile Servicing System (MSS) and the Space Station Remote Manipulator System (SSRMS)
- GPS time to the Command and Control (C&C) MDM to synchronize timing in all MDMs.

The MatrixX user interface is a set of graphical elements including blocks and wires, along with menus which allow creation and placement of these elements, etc. Blocks and wires typically fill a page (a superblock, which can be viewed on screen in one window and printed on one page). Many of the superblocks contain other superblocks. Wires show the I/O variables for each block and the data flow within blocks, including elements such as summers and multipliers. Additional I/O interfaces are handled with global or internal "read from" and "write to" blocks.

One of the drawbacks of the MatrixX tool is that an equation that is relatively simple to write by hand, in a word processor, or in hand coded Ada can become a large tangle of blocks and wires when implemented in these superblocks.

While both pros and cons exist with this MatrixX tool, the GN&C community has decided not to use it for the next major update (Flight Stage 8A), using hand-coded Ada in its place.

Testing

The selloff of the flight software is done with Formal Qualification Testing (FQT), where the performance of the software is compared to requirements (pointing accuracy, for example). The flight software is first run in the MDM in the Honeywell MDM Application Test Environment (MATE), then the flight software inputs are used to drive the flight software models in MOODS, which generates a signature for comparison to the flight software outputs.

Additional testing defined by Boeing is called informal, or development testing. This

0-7803-5086-3 /98/$10.00 ©1998 IEEE

includes unit tests, CSC tests, CSCI (scenario) tests, and Hardware/Software Integration (HSI) tests. This paper focuses on the P&S CSC tests for Flight Stage 5A.

CSC Definition

For these tests, P&S is defined to consist of six CSCs:

1. Solar Ephemeris
2. Generic Data Computation
3. Rise/Set Times (Solar and TDRS)
4. Mass Properties
5. Target Rotary Joints (for sun pointing)
6. TDRS Pointing

Table 1 below summarizes the approximate number of I/O variables for each CSC, where the CSC ID numbers correspond to the list above. Some of the indirect outputs are repeats of data written out to other locations, either as direct or indirect outputs.

Table 1. CSC Input/Output Summary

CSC ID	Number of Inputs		Number of Outputs	
	Direct	Indirect	Direct	Indirect
1	0	13	0	7
2	23	20	40	56
3	8	39	0	25
4	0	87	7	15
5	9	2	0	7
6	21	38	0	43

Test Purposes

Test purposes were to:

- Demonstrate required functionality
- Exercise all component entry points
- Exercise all control flow branches, exception handling, and decision points down to the unit level
- Stress the component

Test Case Definition

Inputs were varied over the expected operational range and slightly outside those limits.

Pass/Fail Criteria

Each CSC was allowed to pass or fail as a unit. Goals (not requirements) were set for comparisons with signatures, such as

- 10 matching significant digits when comparing floating point numbers
- 10 digits after the decimal when comparing components of unit vectors and quaternions
- one second when comparing rise and set times

Test Execution

The first step in testing each CSC was to find corresponding inputs and outputs in MOODS and in MatrixX. Significant work was often needed to modify MOODS to set up this correspondence. Also, MOODS does not model data quality indicators, so they were tested separately, as discussed below.

Test cases were run in MOODS by Honeywell engineers, saving the inputs and outputs of the flight software models. These outputs were the signatures. The input and output files were then sent to Boeing engineers in Huntington Beach, California, where the inputs drove the MatrixX-generated Ada flight software on a Hewlett Packard computer.

Data quality indicators were tested with the flight software running on the Hewlett Packard computer without MOODS signatures, since signatures were not needed to ensure correct results. In general, past values instead of current values of variables are output when the data quality indicators fail.

All test-related files are kept under Configuration Management (CM). These files include descriptions of test procedures. Procedures which allow the MOODS results to be rerun and reproduced at any time were created using the DEC Alpha VMS Code Management utilities. This involves recreating the state of MOODS at the time of the original test, since frequent updates change the MOODS tool.

Test Results

All six CSCs were eventually accepted as valid and were retested and passed again with various updates of the flight software.

Solar Ephemeris CSC Test Results

This CSC passed after we learned that the flight software used a value for a constant (time of J2000 epoch) which was different from the value used in MOODS. The differences were determined to be acceptable, though the MOODS constant was more accurate. The value used in the flight software is less accurate due to the following approximations:

- The speed of light is 3 x 10^8 m/s instead of the smaller, more accurate value used in MOODS (2.997 924 58 x 10^8 m/s)
- The difference between Terrestrial Dynamic Time and GPS time is 51.184 seconds instead of the more accurate value of 51.184 381 7 used in MOODS

These approximations are documented in the Technical Description Document for the PG-1 Guidance, Navigation, and Control System (MDC 95H0223, Revision D, 17 January 1997).

Generic Data Computation CSC Test Results

This CSC passed after we learned that the computation of two delta time values in the flight software differed from the MOODS signatures. MOODS uses a time utility which carries more accuracy than the flight software (nanosecond accuracy vs. millisecond accuracy). When two very large and similar numbers (time in seconds since GPS epoch) are subtracted in the flight software to obtain a delta time, the result is usually different enough from the signature to cause some results to mismatch. In all cases, the mismatches were judged to be acceptable, even if matching to less than 10 significant digits, etc. MOODS values for delta times were also used in the flight software for selected test cases to show that the results were improved.

A source of error in the comparisons involved vector calculations which create a unit vector and change the coordinates from one coordinate system to another. The order of these two operations affects the outcome. MOODS was changed to match the flight software.

Rise/Set Time Determination CSC Test Results

In addition to the above problem with delta time calculations, an error related to outputting rise and set data was found and corrected. The error was in the logic path executed when the "iterations exceeded" flag was set to "true". In this case, an incomplete set of rise and set events was output. This was changed so that the past values for these events were output. All test results passed.

Table 2 lists partial results for comparisons of solar event times and TDRS event times. Cases 5 and 8 are shown in the table since they contain the worst case results. After considerable work by the test team, satisfactory results were obtained. The largest absolute difference is 0.6 second for test case number 5, output pin number 10, which is solar event time #4. The pass/fail limit for these absolute differences in event times is one second, since the convergence tolerance in the integrator is 0.5 second. This tolerance allows for up to a 1.0 second difference between the flight software and MOODS.

0-7803-5086-3 /98/$10.00 ©1998 IEEE

Table 2. Rise and Set Event Times Test Results

Test Case	Output Pin	Description	Absolute Difference (Pass/Fail Limit: 1 second)
5	7	Solar Event 1	0.00
	8	Solar Event 2	0.20
	9	Solar Event 3	0.00
	10	Solar Event 4	**0.60**
	20	TDRS Event 1	0.40
	21	TDRS Event 2	0.30
	22	TDRS Event 3	0.40
	23	TDRS Event 4	0.10
8	7	Solar Event 1	0.00
	8	Solar Event 2	0.00
	9	Solar Event 3	0.00
	10	Solar Event 4	0.01
	20	TDRS Event 1	0.00
	21	TDRS Event 2	**0.48**
	22	TDRS Event 3	0.00
	23	TDRS Event 4	0.00

Figure 1 displays the data from Table 2 (test cases #5 and #8) and similar data from other test cases.

Mass Properties CSC Test Results

All results were acceptable.

Target Rotary Joints CSC Test Results

Two changes were made to the flight software after the first tests were completed, and the tests were repeated with the new version. All test results were judged acceptable.

The first change was to add logic to the time tag output so that the past value was output if the related solar LOS (Line of Sight) data quality flag was failed.

The second change was to add the two beta gimbal biases to the outputs. These ground-uplinked biases can be used for correcting the solar panel pointing calculations for structural misalignments.

TDRS Pointing CSC Test Results

Test results were acceptable. The delta time problem reported above appears here, also, but the errors were again small enough to be tolerated.

An error related to data output in the presence of a failed quality indicator was found and corrected. When the TDRS Line of Sight (LOS) quality flag was set to invalid, the TDRS satellite IDs were wrongly outputting current values instead of past values.

Acknowledgments

A hardworking team is responsible for the good results with this task. Thomas Gederberg/Honeywell deserves special recognition. Among many who deserve mention, I wish to acknowledge Tom Russell, Sheldon Lee, David Anderson, and Roger Corson of Boeing, Jack Justus of AAI, and Bill Graham of Dynacs.

Figure 3.4-1 Test Results from Cases 1, 2, 4, 5, 7, and 8
Compared to the 1.0-Second Pass/Fail Limit (Data Shown in Table 3.4-1)

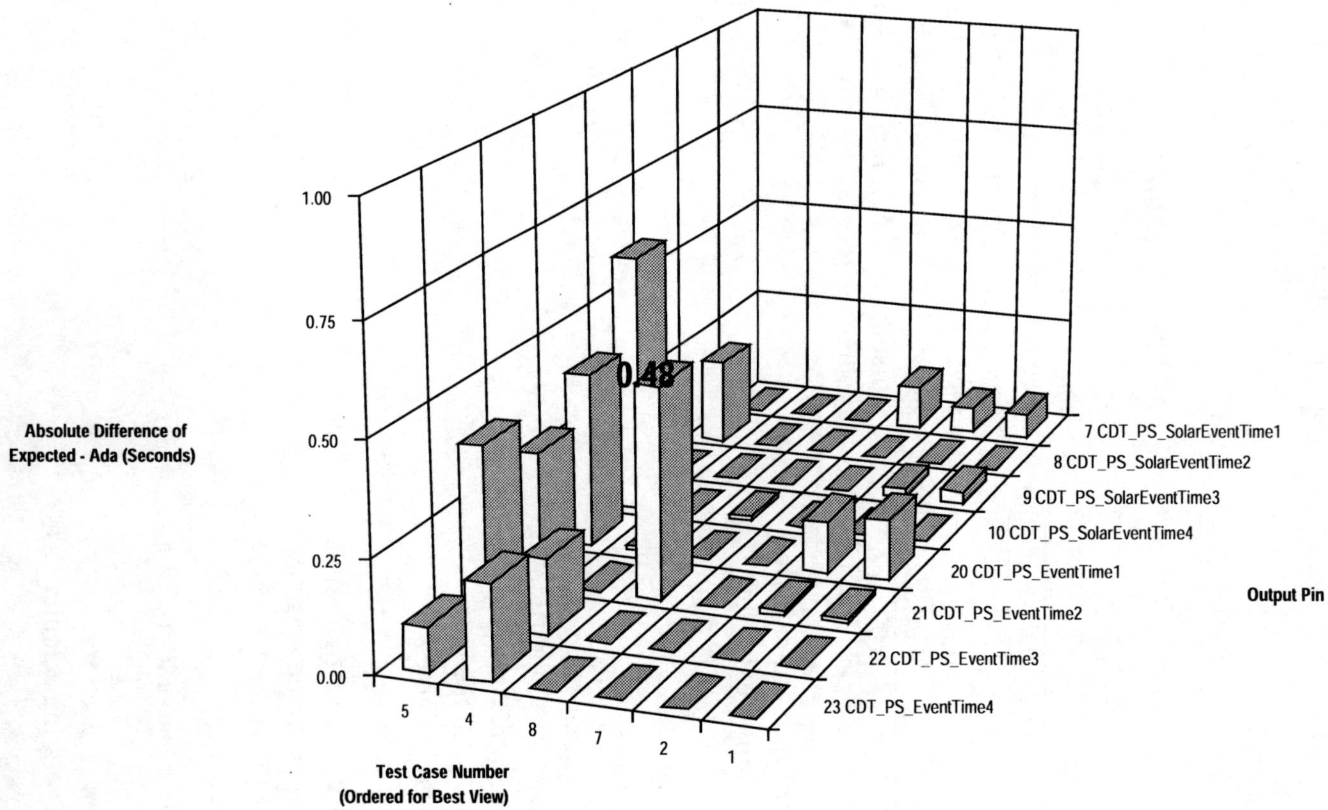

Figure 1. Absolute Differences of Event Times from Table 2 and Similar Data from Other Cases (Labels Unchanged from Original Test Report)

0-7803-5086-3 /98/$10.00 ©1998 IEEE

USING THE SOFTWARE CAPABILITY MATURITY MODEL FOR CERTIFICATION PROJECTS

Leanna K. Rierson, Federal Aviation Administration, Washington, D.C.

Abstract

The purpose of this paper is to explore the use of the Software Engineering Institute's Software Capability Maturity Model (SW-CMM) on civil aviation projects. The paper will examine SW-CMM and RTCA DO-178B/EUROCAE ED-12B by considering the basic concepts of each standard, keys to successful integration of the standards, and benefits of integrating the two standards.

Introduction

Throughout the 1990s the Software Capability Maturity Model (SW-CMM) has emerged as a yardstick for measuring software process maturity. The SW-CMM was developed at Carnegie Mellon by the Software Engineering Institute. It soon became clear that software process maturity was insufficient without looking further into the systems engineering process and the acquisition life cycle. This led to the development of the Systems Engineering Capability Maturity Model (SE-CMM) and the Systems Acquisition Capability Maturity Model (SA-CMM). The SW-CMM, SE-CMM, and SA-CMM form the foundation for a number of other Capability Maturity Models (CMMs) being developed to meet the specialized needs of the industry. For example, the Federal Aviation Administration's (FAA) Research and Acquisition Organization recently released the Integrated Capability Maturity Model (known as FAA-iCMM) to meet the software, systems, and acquisition needs of the FAA for acquiring software intensive systems. Likewise, a Security Systems CMM was developed for the specialized needs of developing secure systems.

The SW-CMM is based on good, common-sense software engineering practices and provides a measuring process for companies to baseline and improve their processes. These quality attributes give incentive for many avionics developers to implement the SW-CMM into their organizations today. Additionally, many acquirers of avionics systems now require vendors to meet specific SW-CMM levels prior to contract award.

Avionics projects for civil aviation must meet the regulations of the FAA to receive certification or authorization. For the software aspects of certification, the FAA's Advisory Circular 20-115B recognizes RTCA DO-178B, "Software Considerations in Airborne Systems and Equipment Certification", as an acceptable means of compliance for the evaluation of software in airborne systems. The European Community recognizes European Organization for Civil Aviation Equipment (EUROCAE) document ED-12B, which is identical to DO-178B. DO-178B/ED-12B contains a set of objectives that are based on the level of safety required by the system.

As described above, many developers of avionics systems must meet both the SW-CMM and DO-178B/ED-12B. This has led to many questions and concerns regarding the relationship of these two assurance standards. This paper will address the three most often asked questions:

- Are DO-178B/ED-12B and SW-CMM compatible?
- Can SW-CMM be used instead of DO-178B/ED-12B?

0-7803-5086-3 /98/$10.00 ©1998 IEEE

- How can a company apply both SW-CMM and DO-178B/ED-12B?

To effectively integrate these two standards for civil certification projects, this paper first provides an overview of both SW-CMM and DO-178B/ED-12B. Next, the paper compares SW-CMM and DO-178B/ED-12B and provides suggestions for integration of the two standards. The paper ends with a summary of the benefits of using SW-CMM on certification projects.

It should be noted that this paper is not the official FAA position. The author is a FAA employee and the paper is intended to be consistent with FAA policy; however, it has not been coordinated through the FAA's approving officials and merely represents the opinions of the author.

Overview of SW-CMM

The SW-CMM framework was started in 1986 by Software Engineering Institute (SEI) with assistance from MITRE Corporation. In 1991, version 1.0 was released and began to be used by the software community. Version 1.0 was revised by the software community in 1991 and 1992—leading to the release of SW-CMM version 1.1 in early 1993. Version 1.1 is currently being used world-wide. SW-CMM provides a model that leads to software process improvement (Paulk, viii).

The SW-CMM categorizes the overall company process maturity into five levels of maturity. For the purpose of SW-CMM, a "software process can be defined as a set of activities, methods, practices, and transformations that people use to develop and maintain software and associated products (e.g., project plans, design documents, code, test cases, and user manuals)" (Paulk, 3). The maturity levels, 1 to 5, indicate the overall effectiveness of the company's software engineering practices (Pressman, 27). Each increasing level is based on achieving the attributes of the previous low levels. The five levels are describe as follows (Paulk, 8-9):

- **Level 1: Initial** – The software process is characterized as ad hoc, and occasionally even chaotic. Few processes are defined, and success depends on individual effort.
- **Level 2: Repeatable** – The project management processes are established to track cost, schedule, and functionality. The necessary process discipline is in place to repeat earlier successes on project with similar applications.
- **Level 3: Defined** – The software processes for both management and engineering activities is documented, standardized, and integrated into a standard software process for organization. All projects use an approved, tailored version of the organization's standard software process for developing and maintaining software.
- **Level 4: Managed** – Detailed measures of the software process and product quality are collected. Both the software process and products are quantitatively understood and controlled.
- **Level 5: Optimizing** – Continuous process improvement is enabled by quantitative feedback from the process and from piloting innovative ideas and technologies.

Each maturity level has associated key process areas (KPAs) that describe the software engineering attributes that must be present to satisfy that particular level (see Table 1). As the maturity level increases, quality and productivity increase and risk of unsuccessful and unpredictable projects decreases. Within each KPA there are goals/objectives that must be achieved to satisfy the KPA. Each KPA is rated using the components described below:

- **Activities** – Tasks to be completed to successfully achieve the KPA.
- **Commitments** – Organizational requirements imposed to assure the activities are carried out.
- **Abilities** – Things in place to enable organization to meet commitments.
- **Measurement of Implementation** – Process of monitoring activities.

- **Verification of Implementation** – Process of assuring that the KPA is being properly achieved.

Table 1. Key Process Areas For SW-CMM

Maturity Levels	Key Process Areas
5 – Optimizing (focus on continuous improvement)	• Process Change Management • Technology Innovation • Defect Prevention
4 – Managed (focus on product & process quality)	• Quality Management • Process Measurement & Analysis
3 – Defined (focus on engineering process)	• Peer Reviews • Intergroup Coordination • Software Product Engineering • Integrated Software Management • Training Program • Organization Process Definition • Organization Process Focus
2 – Repeatable (focus on project management)	• Software Configuration Management • Software Quality Assurance • Software Sub-contract Management • Software Project Tracking & Oversight • Software Project Planning • Software Requirements Management
1 – Initial (focus on individual)	• None

Each KPA is defined by a set of key practices (such as policies, procedures, and activities) that must be in place before the KPA is achieved.

This section has provided a very high overview of the SW-CMM process. There are numerous sources to assist in deeper study of the subject, including the SW-CMM document and the SEI web-site (http://www.sei.cmu.edu).

Overview of DO-178B/ED-12B

DO-178/ED-12 was first developed by the international civil aviation community in 1982. It was revised in 1985 to add more detail. In 1992, DO-178B/ED-12B was completed and has become the software "standard" for airborne software in civil aviation products. The DO-178/ED-12 document and all of its revisions were sponsored by RTCA and EUROCAE, with the involvement of aviation, software, and certification experts from across the world.

DO-178B/ED-12B focuses on the software aspects of system development. As part of the systems engineering task, a system safety assessment must be performed before DO-178B/ED-12B can be applied to the software development effort. A system safety assessment is a process to identify the hazards, failure conditions leading to these hazards, and the effects of mitigation strategies. The safety assessment task determines a software level based upon the contribution of the software to the potential failure conditions defined in the system safety assessment process. The five software levels, A to E, are summarized in Table 2 (DO-178B, page 7).

These software levels define differing degrees of rigor. Annex A in DO-178B/ED-12B lists the objectives that must be met for each specific software level. These software levels define a number of desirable attributes for the software development and verification processes. The differences in rigor are determined by the number of objectives which need to be satisfied, whether a specific objective is satisfied with independence, and the formality of configuration control of the software data produced during development. For example, the number of objectives for each software level is listed below:

- Level A: 66 objectives
- Level B: 65 objectives
- Level C: 58 objectives
- Level D: 28 objectives
- Level E: 0 objectives

DO-178B/ED-12B is divided into development activities and integral processes. The development activities include planning, requirements, design, code, and integration. The integral processes include verification, configuration management, quality assurance, and certification liaison. The integral processes are overlaid on each of the development activities (i.e., verification, configuration management, quality assurance, and certification liaison are applied to each development activity).

Table 2. DO-178B/ED-12B Software Levels

Failure Condition Category	Description	SW Level
Catastrophic	Failure conditions which would prevent continued safe flight and landing of the aircraft.	A
Hazardous	Failure condition which would reduce the capability of the aircraft or the ability of the crew to cope with adverse operation conditions to the extent that there would be: (1) a large reduction in safety margins or functional capabilities, (2) physical distress or higher workload such that the flight crew could not be relied on to perform their tasks accurately or completely, or (3) adverse effects on occupants including serious or potential fatal injuries to a small number of occupants.	B
Major	Failure conditions which would reduce the capability of the aircraft or the ability of the crew to cope with adverse operation conditions to the extent that there would be, for example, a significant reduction in safety margins or functional capabilities, as significant increase in crew workload or in conditions impairing crew efficiency, or discomfort to occupants, possibly including injuries.	C
Minor	Failure conditions which would not significantly reduce aircraft safety, and which would involve crew actions that are well within their capabilities.	D
No Effect	Failure conditions which do not affect the operational capability of the aircraft or increase crew workload.	E

The objectives of DO-178B/ED-12B are listed in Annex A and are organized around the development activities and integral processes previously described. There are ten tables in Annex A with objectives—the subject of each table is listed below:

- Table A-1: Software Planning Process
- Table A-2: Software Development Processes
- Table A-3: Verification of Outputs of Software Requirements Process
- Table A-4: Verification of Outputs of Software Design Process
- Table A-5: Verification of Outputs of Software Coding & Integration Processes
- Table A-6: Testing of Outputs of Integration Process
- Table A-7: Verification of Verification Process Results
- Table A-8: Software Configuration Management Process
- Table A-9: Software Quality Assurance Process
- Table A-10: Certification Liaison Process

Table A-4 objective 1 is used in Figure 1 to illustrate the Annex A table layout and structure. The first set of columns contains information about the DO-178B/ED-12B **objectives**: objective number, description, and reference to DO-178B/ED-12B paragraph where that objective is further detailed. The next set of columns with headers A, B, C, D show the applicability of that particular objective to the software level. For example, objective 1 is applicable for levels A, B, and C; however, it does not need to be satisfied for software level D. If the circle indicating applicability is filled in, then that objective must be satisfied with independence. The next series of columns describe the **outputs** produced as evidence that the objective is satisfied. The "Description" column lists where that data is found. The "Ref." Column identifies the paragraph within Chapter 11 of DO-178B/ED-12B that details the attributes of that software data. The last 4 columns correlate the rigor of configuration management of the particular output with the associated software level. Control category 1 requires more configuration management activities than control category 2. For instance, control category 1 requires problem reporting and change control, where as control category 2 requires only change control.

Assessment to DO-178B/ED-12B is performed through on-site reviews and/or desk-top (data) reviews by FAA personnel, Designated Engineering Representatives, and/or software developer's team members. The assessment evaluates the data to determine if the objectives listed in Annex A of DO-178B/ED-12B are met. In June of 1998, the FAA released a job aid entitled, "Conducting Software Review Prior to Certification". The job aid outlines a process for assuring compliance to the objectives of DO-178B/ED-

12B. The job aid is available electronically and is designed to be tailored to meet the specific needs of the evaluator or project.

Figure 1. Portion of Table A-4 in DO-178B

	Objective		Applicability by SW Level				Output		Control Category by SW level			
	Description	Ref.	A	B	C	D	Description	Ref.	A	B	C	D
1	Low-level requirements comply with high-level requirements.	6.3.2a	●	●	○		Software Verification Results	11.14	2	2	2	

This section has provided a very high level overview of DO-178B/ED-12B. More information may be obtained by reading DO-178B/ED-12B itself, by participating in related RTCA and EUROCAE activities, and by reviewing the FAA job aid.

Comparison of SW-CMM and DO-178B/ED-12B

Although both SW-CMM and DO-178B/ED-12B are assurance standards, their primary focus is somewhat different. The SW-CMM looks not only at the development processes but also at their management and refinement. The better defined, managed, and improved a process is, the better the software quality and the greater the likelihood that the development is within cost and schedule.

The focus of DO-178B/ED-12B is solely on design assurance. Specific DO-178B/ED-12B objectives are identified to provide the required assurance by criticality level. For level A and B, specific classes of software development errors are targeted for removal from code through stringent verification activities. Although management, cost, and schedule considerations may be important to developers, this is not the purview of FAA certification authorities. FAA authorities are concerned solely in the design assurance aspects of software development.

Acquisition organizations are also interested in quality software; however, they must also consider managerial issues such as cost and schedule. To mitigate known risks many organizations require their suppliers to be SW-CMM assessed to a specific level.

Because avionics developers are striving for design assurance, certification, and efficient processes, many are applying both SW-CMM and DO-178B/ED-12B. This has led to the question: "Can SW-CMM be used as a substitute for DO-178B/ED-12B?" (i.e., can SW-CMM be considered an alternate means of compliance to DO-178B/ED-12B?). While there are many similarities between DO-178B/ED-12B and SW-CMM, there are also a number of significant differences that make it clear that SW-CMM is not an acceptable alternate means of compliance to DO-178B/ED-12B—especially for the more safety critical systems. Even when SW-CMM is used on a project, the objectives of DO-178B/ED-12B must be met for the specific software level.

Together SW-CMM and DO-178B/ED-12B create a quality software process. Companies that have SW-CMM level 2 or higher are more able to apply DO-178B/ED-12B efficiently and across product lines. Some software experts believe that it requires at least a level 2 or 3 maturity in order to apply DO-178B/ED-12B effectively; i.e., DO-178B/ED-12B is intended for mature organizational use. This has led to some difficulties in applying DO-178B/ED-12B, as most software development organizations are level 1. Aviation manufacturers can benefit by the combined use of SW-CMM and DO-178B/ED-12B for certification projects. Unfortunately, many companies use excessive resources by trying to apply DO-178B/ED-12B and SW-CMM separately, to the exclusion of one or the other. The best approach is to integrate the SW-CMM and DO-178B/ED-12B processes.

Many companies have a difficult time integrating the SW-CMM and DO-178B/ED-12B processes successfully. Below are a few items that should be carefully considered when

0-7803-5086-3 /98/$10.00 ©1998 IEEE

striving to integrate. The items are not in any particular order.

1. Evaluate current processes.

If the company already has software processes in place, it is beneficial to examine those processes in order to determine if change is needed. Some things to consider are:

- Do the current processes meet DO-178B/ED-12B objectives for the appropriate level?
- Has the process in place been evaluated on a project by the FAA?
- Are the current processes consistent across projects and product lines?
- Has the company performed an internal SW-CMM assessment?
- Have the processes been evaluated by an independent SW-CMM evaluator/assessor?
- Do the current processes meet the desired SW-CMM level?

If the answer is "no" to any or all of these questions, there will likely need to be some revamping of the company processes. In most cases, the change in processes will simultaneously benefit the company's ability to meet SW-CMM and DO-178B/ED-12B.

2. Perform a mapping between DO-178B/ED-12B and SW-CMM.

It is nearly impossible to perform a general comparison between the DO-178B/ED-12B and SW-CMM compliance, because both are designed to be flexible for company implementation. Therefore, the comparison should be done by the company for the particular implementation. It is beneficial to perform a mapping. Starting with either DO-178B/ED-12B or the SW-CMM, map the existing company processes against the selected assurance standard. This will result in a list of evidences to support a claim as meeting either a key process area from the SW-CMM or an objective from DO-178B/ED-12B, depending on the starting point. The remaining assurance standard can then be mapped to expose any remaining deficiencies. The mapping should be performed by a team of company experts; it might even be beneficial to bring in external experts. The team should include experts in SW-CMM, DO-178B/ED-12B, each product line, current company processes, and other areas as needed.

The FAA recently performed a mapping between the FAA-iCMM and DO-178B/ED-12B to determine the compatibility of the two assurance documents for acquiring ground systems. FAA-iCMM combines SEI's SW-CMM, SA-CMM, and SE-CMM; therefore, the scope is slightly different than a strict SW-CMM to DO-178B/ED-12B comparison. However, the trends that were discovered in this mapping are applicable to those who apply SW-CMM as well.

The DO-178B/ED-12B to FAA-iCMM was performed by a team of DO-178B/ED-12B experts and FAA-iCMM experts. Since the FAA's Research and Acquisition organization is already applying FAA-iCMM to projects, the effort was a one-way mapping to determine what additional effort would be required to implement DO-178B/ED-12B on the FAA projects. The mapping occurred by listing the DO-178B/ED-12B objectives in one column and listing the differences in the second column. Below are some of the major differences between DO-178B/ED-12B and FAA-iCMM discovered during the mapping are listed:

- DO-178B/ED-12B is more specific in what is required for planning.
- DO-178B/ED-12B is more specific in what standards are required.
- DO-178B/ED-12B specifies "object code" in a number of cases—this is not so for FAA-iCMM.
- DO-178B/ED-12B requires integration on the target computer; FAA-iCMM does not specify target computer.
- Partitioning and protection have different connotations between DO-178B/ED-12B and FAA-iCMM. FAA-iCMM has a more traditional meaning—DO-178B/ED-12B focuses on safety partitioning and protection.

- FAA-iCMM does not specifically mention normal and robustness testing for each specific software requirement.
- Statement coverage, decision coverage, and modified condition/decision coverage are not specifically called out in FAA-iCMM.
- FAA-iCMM does not address the certification liaison process or certification specific documents.
- DO-178B/ED-12B control categories are not addressed in FAA-iCMM.
- FAA-iCMM does not address the exercise of archived data.
- FAA-iCMM does not address the protection against unauthorized changes.
- Transition criteria is not explicitly mentioned in FAA-iCMM.

In general, DO-178B/ED-12B is targeted more at the software for a specific product being used in safety-related airborne applications. Whereas the FAA-iCMM was written for all types of systems and deals more at an organization and management level.

The above listing of DO-178B/ED-12B to FAA-iCMM differences is not extensive and only provides an example of the types of things to be considered when performing a company-specific mapping. In almost all cases, the requirements of the two assurance standards are not contradictory but are complimentary. Generally, the mapping results in only minor changes or extensions to already existing company processes.

3. Seek certification authority input.

If a company is establishing new processes, it is advisable to get FAA input as early in the process as possible to ensure that any interpretations of DO-178B/ED-12B are in accordance with any regulatory guidance. This becomes more important as new technology and non-traditional methods are employed. Too often certification authorities are not consulted until the process is already implemented. It is much easier to change a process during the planning stages than after it has been implemented.

Both FAA and industry are striving to have early communication on processes and projects that affect certification outcomes. The early partnering of FAA and industry help to avoid surprises and high risk changes late in the program.

4. Strive for consistency across product lines.

Many companies have excellent processes in one product line, such as the flight controls; however, the processes for other product lines, such as displays, are weak and chaotic. The inconsistency of applying DO-178B/ED-12B across product lines has been a great concern for the FAA. SW-CMM promotes the cross-product line consistency.

When implementing new company-wide processes, representatives from all product lines should be involved. The benefit is the easy transition of personnel from one product line to another.

5. Seek outside help, as needed.

Establishing new processes is a tremendous investment. Unfortunately, it is often difficult for company representatives to see their own processes objectively. In many cases, it is beneficial to seek the help of an independent consultant to provide an objective perspective. Independent consultants also tend to have experience with a number of companies and have seen what does and does not work.

6. Strive to maintain and improve processes.

Once a good process is established, it is essential to strive for maintenance and improvement of that process. As people change jobs and new technologies are introduced, it is easy to slip back into bad habits or processes.

When the processes are set up, there should be a process in place for continuous improvement and for addressing problems as they arise.

Benefits of Using SW-CMM on Certification Projects

There are several benefits of applying the SW-CMM to certification projects already requiring DO-178B/ED-12B. The major benefits are described below:

Cost Benefits

DO-178B/ED-12B does not consider whether a process is efficient; it is only interested in whether it meets the objectives. Issues such as training, management, metrics, and improvement strategies that are found in SW-CMM help predict and reduce the cost of a particular DO-178B/ED-12B development. For example, in many certification projects, the test aircraft is used as a test bed for software debugging. This is not the most cost-effective approach. In an organization with well-defined processes, the software bugs are likely to be addressed prior to installation on the aircraft.

Additionally, SW-CMM encourages developers to learn from previous problems faced, created, and resolved from one project to the next. SW-CMM helps developers increase their long term memory and become less reliant on tribal knowledge. By avoiding repeating the same errors over and over, cost is reduced.

Schedule Benefits

When an aircraft misses its certification date, both the aircraft manufacturer and the responsible avionics developers are penalized for each day the certification is delayed. The lack of management of schedule has led many avionics manufacturers to use high priced consultants and test houses late in the cycle. As expensive as this approach is, it is certainly not as bad as the penalty they would incur if they are the system that holds up certification.

With this in mind, predictability in terms of schedule is essential. The more managed and improved a process is, the more predicable the schedule. SW-CMM promotes predictability.

FAA Confidence

The FAA is currently developing criteria to determine the level of FAA involvement required for software approvals. At the beginning of each project, the FAA will assess both applicant and developer. Assessment criteria will include but not be limited to such things as:

- company history
- company certification experience
- software processes across product lines
- designee support
- company experience with DO-178B/ED-12B

Companies that have higher levels of SW-CMM in place are likely to receive more favorable ratings on the assessment, and hence minimize the level of FAA involvement in the project. This is a significant benefit to both the FAA and industry workloads.

Summary

Higher levels of SW-CMM approval can improve the overall software approval on certification projects. SW-CMM must be considered in conjunction with DO-178B/ED-12B and applied simultaneously. This paper has provided some ideas for successful integration of SW-CMM and DO-178B/ED-12B. Successful application of SW-CMM on certification projects may lead to cost savings, schedule reductions, and higher confidence from the FAA.

References

Paulk, Mark C. Report: "Capability Maturity Model for Software, Version 1.1", dated February, 1993.

Pressman, Roger S. Software Engineering: A Practitioner's Approach. McGraw Hill, 1997.

RTCA, document DO-178B/ED-12B, "Software Considerations in Airborne Systems and Equipment Certification", dated December 1, 1992.

A COMPARISON OF FLY-FIX-FLY TESTING TO THE SOFTWARE DEVELOPMENT FACILITY TESTING APPROACH

Charles Satterthwaite, AFRL/IFTA, Wright-Patterson AFB, Ohio
Gary Miyahara, Raytheon, Los Angeles, California
Steve Tomashefsky, Raytheon, Los Angeles, California
Ivan Harnage, WR-ALC/LYSF, Robins AFB, Georgia

Abstract

New and Legacy Aircraft Weapon Systems depend on an increasingly complex verification and validation process to decrease the risks in operation, to increase mission success, and to provide flexibility for rapidly changing mission requirements. In this process, embedded software is a key ingredient for changing, recording, proving, and preserving the usefulness of the weapon system. The increased use of software is placing an increased demand for innovation to capture and observe the impacts of weapon system changes.

The heavily relied upon Fly-Fix-Fly testing approach to verification and validation will always have its place, but is losing its effectiveness to affordably, thoroughly, responsively assure the weapon system changes.This paper will provide a comparison of Fly-Fix-Fly Testing to the Software Development Facility Testing approach.

Introduction

The concepts of verifying design requirements and validating the desired functionality of a system are not new. When man first became innovative, ideas were considered worthy by their labor savings. Revisions were made by trial and error. As these innovations improved, machines replaced men and animals. New issues of safety and efficiency became important. These new ideas required more formal techniques of proof for public acceptance. Government regulation and industry standards became common ways of tracking this new public concern.

In the area of aviation, innovation drives a continuing evolution of design, material, and system performance. The techniques for verifying and validating these new ideas are also relatively new and innovative. Wind tunnels, stress testing, anechoic chambers, and flight simulators, to name a few, are relatively recent techniques for monitoring and proving aerospace innovation.

The ultimate proof is still by trial and error, and for aircraft, that proof is called flight test (Fly-Fix-Fly). Until a pilot can put the innovation through the rigors of flying in the environment and conditions that this system change will face, it is non-flight certified. For much of aviation's history, and for many aircraft systems today, flight test has been the only way of verifying design requirements and validating the desired functionality of the change. Experienced pilots and instrumented aircraft can provide specific information by flying the system change against its expected result while monitoring the aircraft for performance gains or losses. Through a series of pre-arranged flight tests, modifications (or upgrades) to aircraft are approved (or disapproved) for operational use.

0-7803-5086-3 /98/$10.00 ©1998 IEEE

Flight-testing will remain a viable means for verification and validation, because of its comprehensive ability to operate all the aircraft functionality in the dynamic environment. In addition, the experience of flight crews offers invaluable information to the understanding of the effects of changes to aircraft. But flight-testing has its drawbacks: it is increasingly expensive. Tests that require numerous hours of flying are increasingly unaffordable. Flight tests don't allow frequent iterations of small events. Tests that require hundreds of mode switching or reconfigurations are unavailable in flight test. Flight tests set up times are prohibitively long. These set ups require trained pilots, instrumented aircraft, available and proper range conditions, and the whole modified aircraft. Flight-testing is a system level test, and many times inappropriate for unit or sub-system testing.

The Software Development Facility (SDF) is an alternative to some Flight-testing. The SDF is built to develop and maintain aircraft unit and sub-system software. A major sub-system of many aircraft, and of most modern fighters, is its radar. A large portion of aircraft maintenance and upgrade complexity centers on the system's radar software. The SDF provides real and simulated environments to stimulate sub-systems as well as full-time monitoring capabilities to record and analyze unit level and sub-system testing. Set up time in a SDF is short, and can be configured to automatically iterate numerous times through a test. The drawback of an SDF is that it can't exercise the entire aircraft in its real environment. It depends heavily on simulations. Real avionics components are often in short supply, so simulations and emulations, are depended upon with less then real-time fidelity. This paper will address some of the issues of flight-testing and SDF testing. It will make some comparisons of these techniques, highlight advantages and disadvantages of each, and then recommend where each might be improved upon through research and development projects. The ultimate goal of this paper is to improve the effectiveness of aircraft verification and validation by pointing out the strengths of existing testing practices and suggesting where researchers might develop new techniques.

The Fly-Fix-Fly Approach To Verifying & Validating Operational Flight Programs

The Fly-Fix-Fly approach to verifying and validating OFPs is generally defined as an approach where the ground-based tests of the embedded software are largely limited to component level tests and modest tests of the integrated system, with the bulk of the verification and validation of the integrated system being performed through tests in the actual flight environment of the aircraft. While this definition is not formally recognized, the term is also generally applied to approaches where a spiral development methodology is applied with considerable program resources dedicated to the iterative identification of performance problems in the flight test environment.

While the current practicality of extreme examples of this approach is questioned, it has historically been applied to aircraft weapons systems where it must be recognized that the final validation of the system must eventually be proven in the real world flight environment. Factors that lead to Fly-Fix-Fly approaches to program development include: programs where little or no capability exists for laboratory-based systems level tests, programs where the qualification requirements of the system are perceived to be almost entirely testable at the component level, as well as programs with low risk system validation requirements allocated to the flight environment. The costs and program risks of the iterative flight tests of a Fly-Fix-Fly approach must be evaluated against the potential costs of investment in laboratory test capabilities that reduce flight test requirements [6,7,9].

The Software Development Facility Approach To Verifying & Validating Operational Flight Programs

The SDF approach is defined in this paper as an approach that provides for system and software integration and test capabilities in a laboratory environment While it is still recognized that the system must eventually be validated in the flight test environment, this approach strives to minimize the need for flight tests and more generally overall program costs, by identifying and fixing design and implementation problems as early as possible in the development process [6,8,9].

Though much attention has been given to the engineering design tools and component level test capabilities of the SDF, this paper focuses on the trade-off between the integrated software and

Table 1. **SDF Positive Aspects**

SDF Positive Aspects
Fast Fix - Hardware and Software
Rapid Retest
Data Extraction & Fast Analysis
Change Scenarios Rapidly
Change System Configuration
New Hardware/Software
Add/remove Hardware
Available More Often
Simulate Impossible Conditions
Do Unit, Module & Integrated Test
Software Experts on Hand
Pre-Production Hardware Tests
Better Fidelity/Repeatability
Lower Cost per "Flight"

system verification and validation that can be performed in the laboratory as opposed to that accomplished during flight test. Key factors influencing the application of laboratory test include: the fidelity required of laboratory simulations of the system environment, the availability or cost of laboratory test support equipment, the potential need for a dedicated avionics system for the laboratory, and the costs of operating and maintaining the lab facility [6,7,9].

A Comparison Of The Approaches

Tables 1-4 list the positive and negative aspects of the SDF and Aircraft (Fly-Fix-Fly) approaches to verifying and validating OFPs. These observations are from the perspective of Radar OFP developers, maintainers, and researchers.

The immediate observation is that the SDF offers flexibility for testing options, while the Aircraft approach provides realistic feedback. For complete confidence, both flexibility and realism are required. The trick is to combine the best aspects of both approaches.

Table 1 is the positive aspects of using the SDF to verify and validate OFPs. The striking advantages are flexibility, accessibility, lower cost, and speed. The SDF is incredibly flexible because of the availability of spare and redundant hardware and software resources. These resources can be quickly configured and initiated for different levels of testing. The SDF is designed to provide great accessibility to individual units and sub-systems as well as integrating various combinations of these together. The unit, sub-system, and integrated test costs, on an individual basis, are greatly reduced. This is because they can be carried out repeatedly without further set-up expenses and they can be differentiated to the specific required test, rather then require the complete weapon system. The SDF provides a means to quickly examine changes to units, sub-systems, and integrated systems by providing on-line access to previous versions of the OFP and its components as well as on-line analysis tools.

Table 2 is the positive aspects of using Aircraft to verify and validate OFPs. The main advantage is the realism of flight-testing. There will never be a substitute for exercising the real system in its real environment. In this type of testing, every aspect of operational usage can be exercised. Aircraft vibration, temperature extremes, random selection of sub-systems activation, live threat scenarios, and aircrew response time occur in real-time. Instrumented test aircraft provide the most accurate verification and validation data.

Table 2. **Aircraft Positive Aspects**

Real World
All Aircraft Systems Involved
Full Systems Integration
Real Scenarios
Pilot Expertise
Realistic Targets
Multi-Aircraft Scenarios
Real Munitions

Table 3 is the negative aspects of using the SDF to verify and validate OFPs. Chief obstacles to the SDF approach are the various physical constraints and the initial set-up costs. For radar applications, it is very difficult to provide an environment where an antenna can operate in three-degrees-of-freedom and fully radiate. The best scenarios to date are tower mounted units overlooking uninhabited ranges. Many of the inputs needed to stimulate system responses must be simulated. These include flight inputs, threats, sub-system interactions, and aircrew interfaces. The initial costs, which are often high, include the laboratory facilities, computer resources, simulators, specialized support equipment, and the skilled technicians. It usually requires several years to build up a support capability. The skills and experience necessary to man this support capability must be provided as early as possible with as much of the SDF available as possible.

Table 3. **SDF Negative Aspects**

Fidelity
Simulations
Hardware - A/C Cables, etc.
Antenna Placement
Radiation Restrictions
Physical Constraints
Gyros
Antenna Roll
Initial Cost

0-7803-5086-3 /98/$10.00 ©1998 IEEE

Table 4 is the negative aspects of using Aircraft to verify and validate OFPs. The major disadvantages of the Aircraft approach is the scheduling of resources to perform meaningful flight tests. Aircraft flight testing is very expensive, so every flight test hour must be designed to provide the maximum information. Flight test aircraft are heavily instrumented to record the results of the flight test. Often times this instrumentation is removed after flight testing is complete, and installed on the next available operational aircraft, when the

Table 4. **Aircraft Negative Aspects**

Availability of Aircraft
Knowledgeable Personnel
Scheduling
Weather Problems
Cost
Aircraft
Range
Personnel

commanders can spare them. Flight

testing requires specialized testing ranges and the provision of the technicians and proper test conditions. Multiple iterations of flight test conditions are impractical. Flight tests usually are designed around broad scenarios which exercise as much of the weapon system functionality as possible in through a scripted sortie. Repetitive passes are almost impossible to duplicate.

Offsetting Software Development Facility Overhead Costs

At a first look, the up-front costs of an SDF are prohibitive. These costs include the facilities, specialized equipment, dedicated experts, and the access of operational assets. Commanders, who often make comparisons by available aircraft, might see an SDF costing 4 or more fighters up front with 2 more fighters for annual maintenance. The commander also sees an SDF as competing for valuable spare parts, which are often in short supply.

The problem with minimizing up-front costs to maximize available number of aircraft is that it ignores a much larger cost, the over-all life-cycle cost. Aircraft modifications, obsolescence, mission changes, and weapon system life extensions inflate this cost. It is well documented that over 70% of the life cycle costs occur in operation and maintenance. It is poorly documented that acquisition programs exclude operation and maintenance in their life-cycle planning exercises [2,4].

In the life-cycle view, the commander would see that there are actually more aircraft available over time by incorporating operational and maintenance design requirements. Many of these design requirements can be met and exceeded by an SDF.

The issue remains that the up-front investment for an SDF is hard to justify. The next 3 sections offer some brief discussions on how some of these cost issues could be reduced, or at least softened.

Multi-Weapon System Usage

The SDF becomes more attractive as a support option when the costs are shared between 2 or more weapon systems. When some portion of the multiple aircraft's systems are common, it becomes attractive to jointly support those areas. The challenge is to offer the multiple weapon systems full coverage of their individual functional requirements, while sharing the same resources. The key issues are resource scheduling and configuration management. If these issues can be adequately addressed, there exists an opportunity for a multi-SDF [6,7,9].

Open Architectures for Adaptation to Evolving and Breakthrough Technology

When technical innovation offers cost savings, it is advantageous to fully implement it. Innovations in computer processing power and storage are occurring daily. Verification and validation techniques are continually being improved. Information management and accessibility are making dramatic improvements. Unfortunately, many weapon systems and their support equipment were built with peculiar requirements. These requirements often were based on proprietary one-of-a-kind designs. To repair, upgrade, or replace these requirements takes significant resources and justification. Major upgrades are infrequent, expensive, and slow to implement.

An open architecture avoids the pitfalls of peculiar requirements. The architectural designer is unconstrained by the unique requirement. The unique requirement is analyzed against future constraints and alternate solutions. With an open architecture SDF, improvements can be implemented, as they become available, thus expanding the support capability of the SDF. With the flexibility of the open architecture SDF, repairs, upgrades, and replacements become routine.

Writing Off Sunken Costs of Archaic and Inefficient Support Equipment

Sunken costs are those expenditures that have been made in the past for equipment, processes, and training, which currently have a declining or negative value. In SDFs, a typical sunken cost is a main frame computer being used to support maintenance and development processes. The cost of maintaining this equipment is high because of its unique functionality and limited application domain. The cost of upgrading or replacing the support equipment can also be perceived to be prohibitively high unless new support processes are also considered along with the incorporation of more open support equipment architectures that mitigate future upgrade costs. With computer and software innovations today, turn-around times are much shorter. A proper cost-benefit-analysis would show that, multiple upgrades of support equipment over the life of a weapon system would outweigh the buy-once-philosophy. New software intensive systems are made to be flexible in design and functionality. Any support capability dependency, which freezes this flexibility, limits the future usefulness of the weapon system. If the SDF is a museum of antique development and maintenance capability, its usefulness is increasingly limited. But, if the SDF is frequently modified to take advantage of evolving technologies, it can remain potent as a verification and validation capability for years to come.

A Futuristic Fly-Fix-Fly Approach, The Aircraft Weapon System As Its Own Software Development Facility

If the following conditions could be accommodated by advancements in technology, a futuristic weapon system could approach being its own SDF. These conditions include:

a. Increased computer memory size, power, and available space.

b. Availability of complex software and firmware.

c. Ability of aircrew/maintenance crew to interpret and relay the data/results of flight-tests.

d. Ability to highly modify flight/support software.

e. A capability which allows dynamic reconfiguration of flight assets for mission and for testing.

f. Fully instrumented operational aircraft.

The engineering of a futuristic weapon system usually starts out with many good intentions, but with sub-optimal implementation. Much of the blame for this is the lack of decision maker's understanding of the over-all cost (life-cycle costs) of a weapon system. Many active system today were designed to last 20-30 years but have been extended to live 50 or even 100 years. Most of the original mission requirements of active systems change several times over its lifetime. The problem is that the original design requirements are not allowed to address this dynamic life. Instead they focus on the narrow shorter life span and mission designs to minimize front-end costs. Modifications and extensions are added incrementally at inflated costs, to offset the higher front-end costs of newer weapon systems.

The proper way to engineer a weapon system, and minimize its life-cycle costs is to take into account an extended lifetime and a much varied mission capability. The maintenance and development costs of the extended lifetime are now front-end costs, and are included in the weapon system design. The design is highly adaptable, and the over-all weapon system is readily tested with on-board and off-board verification and validation techniques.

The futuristic extension of this design is the increased autonomy of the individual weapon system for verification and validation. The weapon system monitors itself, and gives self-diagnostic reports to the air crew and maintenance crews. The weapons system utilizes a high degree of fault tolerance and dynamic reconfiguration (with diagnostics) to adapt to mission changes, battle damage, or system failure. The weapon system has libraries of system upgrades it implements during ingress and egress periods of its mission to maximize its current task. The complete weapon system sortie is recorded with diagnostics for the aircrew and maintenance-crew debriefings. New technical innovations utilize these weapon systems for proof of concept, and thus provide a standardization path for technical transition into the weapon system itself.

With this futuristic design, the Fly-Fix-Fly approach to verification and validation becomes increasingly attractive. Testing information is provided interactively from each operational weapon system. System and sub-system upgrades can be switched in and out dynamically during the course of every flight. As the system ages, it can be optimized for alternative missions, with an onboard capability to evaluate its new role [1,2,3,4,5].

Summary & Conclusions

With the increase in weapon system complexity comes a need to increasingly verify and validate lifecycle requirements. Those requirements include the operational and maintenance aspects of the lifecycle. They also include predictions and preparations for life extensions.

The continued comparison of ground based SDF approaches and Fly-Fix-Fly approaches should help both techniques evolve their unique capabilities. The comparisons will also show where the approaches can compliment each other. Improved modeling and simulation techniques will offer immediate advantages to the SDF, but will also become a reliable technique during flight test. Improvements in computer and flight recorder size, speed, and functionality will make operational aircraft more available for traditional based type testing.

From an intuitive perspective, OFP developers and maintainers know that an SDF offers great advantages over flight test. Changes to an OFP often require minute adjustments that are easily performed and verified with a high-end personal computer and a flight simulator. During an OFP Block-cycle period of 1-2 years, there are numerous changes made. The labor and cost savings of many of these changes often goes undocumented. The type of information and saving provided by these changes could not be performed in flight test. The challenge to these techniques is to document the value of these small changes. In order to perform this documentation, the alternatives with value, such as flight test hours, must be determined and recorded.

Fly-Fix-Fly approaches are not necessarily bad; they just need to be improved upon. Flying a blind system change is probably considered foolish by most. The over-all system functionality, especially flight critical dimensions, better be reliable before risking one's life. But this technique can offer invaluable information when the exercised system can be properly evaluated. An experienced flight crew and an instrumented aircraft are a potent combination for evaluating the performance ceilings of a weapon system.

In comparing these approaches, it is desirable to use the combination that maximizes the supportability of weapon system. The complexity of the weapon system defines the details of test required. The details are provided by the capabilities offered by each approach. Real-time guided weapon delivery and scoring lends it to Flight Testing techniques. Sub-system debugging is best done in the SDF.

Research and development of verification and validation techniques for futuristic OFPs needs to offer significant improvements in both the SDF approach as well as the Fly-Fix-Fly approach. Additionally, work should be performed quantifying the value of individual techniques and integrating the information gained from each technique. Increasingly, the approaches should be complimentary.

References

(1) Satterthwaite, C.P., Getting A Handle On Designing For Avionics Software Supportability and Maintainability, NAECON, May 1992, DTIC A726 362

(2) Satterthwaite, C.P., The Maintenance of Operational Flight Programs, DASC, Oct. 1992, DTIC A276 363

(3) Satterthwaite, C.P., Testing Operational Flight Programs, AGARD, May 1993, DTIC A276 361

(4) Satterthwaite, C.P., A Feedback Model For The Software Lifecycle Focused At The Operational And Maintenance Phase Of The Cycle, TFWGCON 95, Presented June 1995.

(5) Loyall, J., Satterthwaite, C.P., The Science of Testing - Managing Test Complexity, International Test and Evaluation Association Symposium, 14-18 October 1996

(6) Miyahara, G., Satterthwaite, C.P., The Software Development Facility as A Multi-Platform Support Environment, International Test and Evaluation Association Symposium, 14-18 October 1996

(7) Miyahara, G., Satterthwaite, C.P., Improvements to Support the Verification & Validation of Air-to-Air Ground Radar Modes in a Shared Facility, JAWSSS 97, 9-13 June 1997.

(8) Loyall, J., Mathison, S., Satterthwaite, C.P., Impact Analysis and Change Management for Avionics Software, NAECON 97, July 1997.

(9) Miyahara, G., Satterthwaite, C.P., A Multi-Platform Support Environment, Co-authored with NAECON 97, July 1997.

A METHODOLOGY FOR PREDICTABLE REAL-TIME SYSTEM DESIGN

Srini Vasan, TimeSys Corporation, Pittsburgh, Pennsylvania

Ragunathan (Raj) Rajkumar, Department of Computer Science, Carnegie Mellon University, Pittsburgh, Pennsylvania

Abstract

Timing predictability and analyzability are key requirements of real-time systems. As a result, a temporal model of the system must be used during the requirements, design and implementation phases. The temporal model is useful in representing latencies and delays associated with scheduling and data transmissions in distributed systems and for modeling various processor and network elements. Such a temporal model is directly applicable to all phases of system development: (a) the methodical capture of end-to-end predictability and performance requirements during the requirements phase, (b) the structuring of tasks, critical sections and shared data during the design phase (c) system profiling and performance optimizations during the implementation phase.

This paper discusses a state-of-the art methodology, a widely used temporal model called the Rate-Monotonic Analysis (RMA) and TimeWiz™, a commercially available integrated solution that allows uniform modeling, analysis and simulation of processing and networking components used in distributed real-time systems.

The Benefits of Modeling

Modeling techniques are used in computer system design for a variety of reasons. Models present a higher abstraction of the system implementation enabling easier communication among members of the design, development and verification teams. For example, object modeling such as entity-relationship techniques are widely used in object-oriented systems, especially in business applications. State modeling is used to simplify the understanding of telecommunications applications, where the status of a session moves among multiple states.

Temporal Modeling

In hard real-time systems, such as avionics and nuclear plant systems, missed deadlines can be catastrophic, leading to loss of life and/or property. For example, a timing bug on the $175 million Mars Pathfinder threatened the whole mission [2]. The inability to react quickly to a temperature alarm in a nuclear power plant can lead to significant damage. In soft real-time systems, system fidelity and predictability can also be compromised by missed deadlines. For example, a video conference in a military command and control context may have soft deadlines, but a pre-defined quality of service must be guaranteed to be delivered for the system to satisfy its users. Hence, the design of such systems needs to deal with an appropriate mapping of system functionality to available resources in such a way that the system timing constraints are always satisfied. Thus, a key requirement of real-time systems is timing *predictability*.

The objective of temporal modeling of real-time systems is to present a consistent view of the system design such that its timing predictability can be readily determined using efficient and proven scientific techniques. These techniques include

0-7803-5086-3 /98/$10.00 ©1998 IEEE

- mathematical methods based upon Rate-Monotonic Analysis (RMA) [1] for analyzing the worst-case timing behavior of the system, and
- discrete-event simulation for determining the average-case behavior of the system.

In the rest of this paper, we will focus upon a methodology that can be used to

- capture both the hardware and software characteristics of distributed real-time systems,
- analyze and predict the worst-case timing behavior of the system using RMA, and
- study the average-case behavior of the system using discrete-event simulation techniques.

Finally, we will discuss a commercial product named TimeWiz™ that supports this methodology.

Processor Model

Modern real-time systems tend to exhibit relatively higher degrees of functionality and use real-time operating systems to deal with inputs and outputs to the physical world, application-level concurrency, synchronization among applications. As a result, timing predictability, analysis and simulation of these systems must address the following primitives and mechanisms:

1. Concurrent threads
2. Multiple priorities
3. Interrupts
4. Critical Sections
5. Shared Data
6. Synchronization
7. Context Swaps

In other words, the processor temporal model must be able to represent, model and analyze the above primitives.

The temporal model that we will consider is based upon the popular Rate-Monotonic Analysis (RMA) framework developed at Carnegie Mellon University and the Software Engineering Institute [1]. RMA represents a set of scientific principles by which real-time systems can be modeled and analyzed using mathematical equations. For example, the ability of a periodic event to satisfy *all* its deadlines in the presence of other events and logically shared resources can be tested using a single mathematical inequality. The resource demands imposed by each event occurrence are fed as inputs to the inequality, which if satisfied guarantees that timing constraints will always be satisfied.

The primary entities in the model consist of "Resources", "Actions" and "Events". We describe each of these entities in detail below.

Resources

Resources represent elementary objects on which *actions* execute, or objects that are required during their execution. A resource can be a *physical* hardware resource (a processor, network element, backplane, etc.) or a *logical* resource (a buffer, semaphore, or shared memory).

Table 1 lists some key properties relevant to the RMA temporal model:

Table 1. Properties of Physical Resources

Physical Resource Properties	
Normalized Speed	Speed of this processor relative to an arbitrary benchmark.
Scheduling Policy	An enumerated list, e.g., Rate Monotonic, Deadline Monotonic, HKL, or Fixed Priority.
Data Sharing Policy	An enumerated list, e.g., Priority Inheritance,

Physical Resource Properties	
	Priority Ceiling, Highest Locker Priority, Interrupt Masking, FIFO.
Context Switch Time	Amount of time taken by the OS to perform a task switch
Schedulable	Usually a result of mathematical analysis, a Boolean indicating if the processor is schedulable (1 = schedulable).
Utilization	Fraction indicating the processor usage. Note that while utilization > 1 means the resource is not *schedulable*, utilization < 1 *does not* mean that the resource is *schedulable*. In other words, even at very low utilization levels, deadlines can be missed.
Logical Resources	List of available logical resources for this resource.

Actions

Actions represent a single, sequential, executable segment of code that requires one physical resource and zero, one or more logical resources. Actions also need to be *triggered* through an Event as part of the *response* to that Event. The subsequent *Event* section defines these terms. More than one event may trigger the same Action as part of its response.

An action that is not bound to a physical resource does not *execute* in the system. Also, an Action, which is not in the response of any Event, does not execute, even if it is bound to a Physical Resource.

Table 2 lists some of the key properties of actions that are relevant to this temporal model.

Table 2. Properties of Actions

Action Properties	
Normalized Execution Time	Represents the worst-case execution time of the action on a fictitious CPU of unit speed.
Physical Resource	Represent the processing element on which the action executes.
Logical Resource List	Represents the logical resource(s), if any needed during the execution of the action.
Priority Class	An enumerated e.g., Application, Kernel, Hardware Interrupt.
Priority Value	Represents the priority of the action.
Jitter	Specifies the jitter value associated with the execution time.
Worst Completion	Usually, an output of a mathematical analysis, represents the worst case completion time, given the scheduling decisions in the presence of multiple actions and shared resources. Note that this completion time can be 'unbounded'.
Utilization	Fraction indicating processor usage of this action.

Events

Events represent a periodic (or aperiodic) scheduling of a sequence of actions on a single resource. On a single processor resource, these represent the 'arrival' or 'ready state' of a thread or process; they may also represent the arrival of an interrupt, in response to which a sequence of actions is initiated.

Table 3. Properties of Events

Event Properties	
Arrival Type	Either Periodic or Sporadic
Period	Inter-arrival interval for a periodic event sequence.
Response	Represents a sequence of actions that execute in response to the event.
Deadline	Time by which the entire response to the event must be completed.
Utilization	Ratio of *Total Execution Time* and *Period.*
Worst Completion	Usually a result of mathematical analysis, this represents the time required to complete all actions in the response of this event. Note that the completion time may be 'unbounded'.
Blocking Time	Total amount of blocking time for the sequence of actions. If there are logical resource requirements for any of the actions in the response, this may be automatically calculated based upon analysis; or an estimate can be provided.

Relationships among Objects

Figures 1 through 3 illustrate the relationships among resources, actions and events.

Figure 1 represents a resource or, hardware configuration diagram, which illustrates that a physical resource may have several logical resource(s) associated with it.

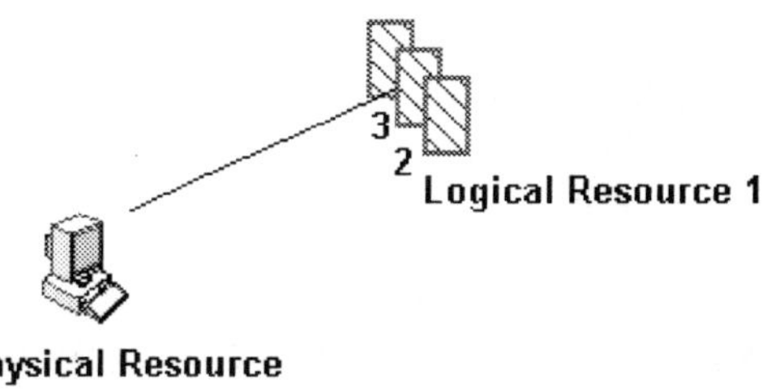

Figure 1. Hardware Configuration Diagram

Figure 2 represents a software configuration showing an event that initiates a response triggering the execution of a sequence of three actions. The event can be *periodic* or *aperiodic* in nature. Typically, a timing constraint called a *deadline* is associated with an event. In other words, when this event occurs, its corresponding sequence of actions must be completed within the event's deadline.

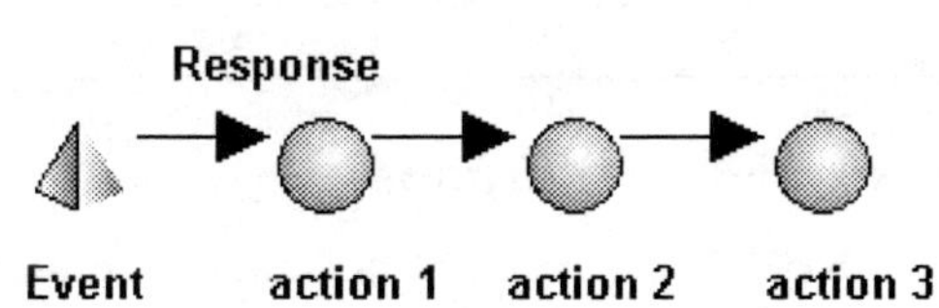

Figure 2. Software Configuration Diagram

The Actions "action1", "action2" and "action3" in Figure 2 may be dynamically bound to a resource such as the Physical Resource (CPU) in Figure 1. In other words, an action can be bound to at most one Physical Resource and is said to execute on that resource. Zero or more Logical Resource(s) may also be specified for consumption by an Action.

System Development Process

The temporal model described in the previous section can be applied to all phases of the Waterfall and Spiral models of software development. Due to space limitations, we shall illustrate this by applying the temporal model only to the waterfall model of software development.

The traditional waterfall model of the software development process consists of the phases indicated in Figure 3.

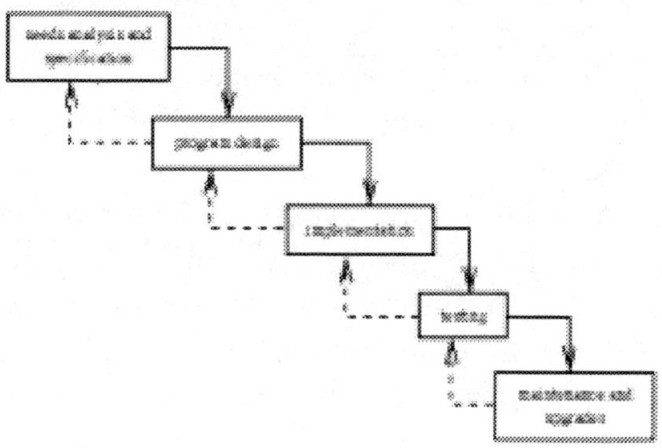

Figure 3. Waterfall Model

Analysis and Specification

Real-time systems have end-to-end timing constraints or performance requirements. In cases such as missile defense systems, fly-by-wire avionics, air traffic control and medical systems, it is easy to visualize life-threatening situations if software tasks are unable to complete execution of their functions on time or components fail. In automated manufacturing, assembly and refining plants, the inability to meet timing constraints can result to lost quality and damaged assets if not injury to operators.

In the model discussed above, such timing requirements translate to deadlines and periods. System temporal models can be built at this stage and the parameters documented. The methodical capture of timing requirements is enforced by the use of temporal models.

Program Design

Key software design issues in real-time systems include the following:

- the assignment of appropriate priorities for threads,
- setting the scheduling policy for the operating platform,
- the design of shared resource accesses and their placements within the threads,
- setting the data access policies for shared resources,
- an assessment of spare capacity that must be left behind for future growth vs. required resources to satisfy current deliverables,
- guaranteeing predictability of timing behavior,
- meeting deadlines in the worst case, and
- the synchronization of multiple data streams (e.g., audio, video, data)

The RMA temporal model can be used to represent, model and perform a comprehensive worst-case analysis to provide answers to the above issues as follows:

- Priority assignment can be modeled using action priorities and predictability automatically analyzed.
- Resource scheduling policies and data sharing policies can be interactively set and their impact can be studied.
- Alternative data sharing designs can be represented and analyzed.
- Synchronization scenarios may be represented using logical resources and

analyzed to determine the impact of problems such as *priority inversion* [3]. Priority inversion is said to occur when a higher priority thread is forced to wait for a lower priority thread. Priority inversion can lead to subtle but damaging timing problems which cannot be easily reproduced [2].

Implementation and Testing

Profiling techniques are often used during implementation and testing phases of system development to analyze the time response characteristics of the system. Most scenarios resulting in the worst-case timing behavior cannot be usually tested easily. Temporal models play an important role in this phase of the system development. Multiple branches executed during a system execution can be analyzed (often using automation tools) to extract the worst-case time for the various branches. A mathematical analysis can then be performed taking into consideration the worst-case task phasing and context swap scenario.

The worst-case time response arrived at using such temporal model can be pessimistic but *must* be acceptable if failure of the system cannot be tolerated. Since the temporal model captures the characteristics of both the hardware and software characteristics of the system, average-case studies can be also be performed using simulation techniques to simulate the temporal model.

Maintenance Upgrades

Real-time systems rarely stay the same after they are implemented. Changes are often required during the system lifetime. System predictability and performance requirements should continue to be satisfied through all such changes.

Temporal models, once built, present a cost-effective way to certify the timing predictability in the presence of maintenance changes.

Network Model

The temporal model outlined above for processors can be extended to network elements supporting predictable transport protocols. The properties of a network element would need to be appropriately modified to represent the physical characteristics of the network medium and the logical characteristics of the network protocol. For example, consider CANbus, a Controller Area Network which is used in manufacturing and automotive systems. A CANbus resource would need properties to support header lengths, network speed; CANbus actions would need properties to support the message size (or number of octets), message IDs (which represent the priority of messages).

The RMA temporal model can be used with some modification to represent, analyze and track the latencies and communication delays (as opposed to processing delays, in processor-based nodes) associated with CANbus network elements. In fact, this can be done for the token-ring, full-duplex ethernet switched systems, ATM, FDDI and Fibre-Channel systems besides CANbus and PCI busses.

Distributed Systems

Distributed systems consist of several processors, networks and intercommunicating subsystems. The temporal model for single processors can be extended to distributed systems by considering each physical element in the system to be *resource* and each entity consuming some portion of a physical resource to be an *action*. The worst-case latency and transmission delays of dataflow paths can be modeled by viewing it as a sum of multiple paths, each of which is confined to a single processor or network elements.

Additional considerations in distributed systems exist for external events and the manner in which they are handled by the various elements in the systems. Specifically, there are distinct paths through system elements, hardware and software, that constitute the processing of that external event. This can be modeled as a set of asynchronous concurrent tasks executing on different physical resources. External events with end-to-end timing constraints that span multiple physical resources can be modeled as a special type of Event. These events called *Tracers* do not necessarily cause an execution sequence of actions, but may represent data that flows through multiple actions. By studying this dataflow and the delays that it encounters, we can gain insight into the latencies and transmission or processing delays associated with the handling of the external event. The dataflow path traced by these Tracer events is called a *Tracer Path.*

Similar to the event properties described above, arrival patterns and deadlines can also be associated with Tracers. Since the tracer events represent dataflows through actions in the system, the worst-case system time response to tracer events can be calculated simply as the sum of worst-case response to each constituent action in the corresponding tracer path. This result can be used throughout the system design process as outlined in the previous section to ensure that the system always meets the response requirements of all external events.

TimeWiz™: A Commercial Solution

TimeWiz™ from TimeSys Corporation (www.timesys.com) is an integrated design environment for distributed real-time systems. TimeWiz™ supports real-time system capture, the analysis of the worst-case timing behavior of a real-time system, and the study of the average-case behavior of the system using discrete-event simulation. TimeWiz™ is commercially available, and runs on Windows NT and Windows 95. An example screen-shot of its highly visual and user-friendly interface is provided in Figure 4.

In TimeWiz™, the hardware and software configurations of a distributed real-time system can both be represented hierarchically and visually using standard or custom-defined palettes containing visual objects with standard and customizable property lists. Software objects expected in the system can then be bound to hardware resources in the distributed real-time system using interactive drag-and-drop interfaces or automatically using intelligent resource allocation algorithms.

Custom analysis and simulation modules can be developed and seamlessly integrated into the TimeWiz environment using a built-in plug-and-play interface. Detailed time-line information can be generated to provide the user with detailed sequences of execution that facilitate detailed visualization and debugging. In addition, visual and customizable charts and reports can be generated to provide continual documentation of system characteristics. Put together, system architects, designers, developers and integrators can view system predictability and performance during the requirements, development, implementation and verification stages of the project.

Concluding Remarks

Temporal modeling is an effective approach that can be used during all phases of real-time system design, development, implementation and certification to understand, analyze and improve timing characteristics of real-time systems such as avionics systems. Proven, established representation and analysis techniques such as Rate-Monotonic Analysis (RMA) can be used to guarantee system predictability and performance.

Temporal modeling is complementary to and therefore can be used along with other modeling techniques such as state modeling and object modeling.

The availability of TimeWiz™, a commercial off-the-shelf (COTS) software tool, automates this process and greatly enhances the enterprise-wide adaptation of temporal modeling technologies.

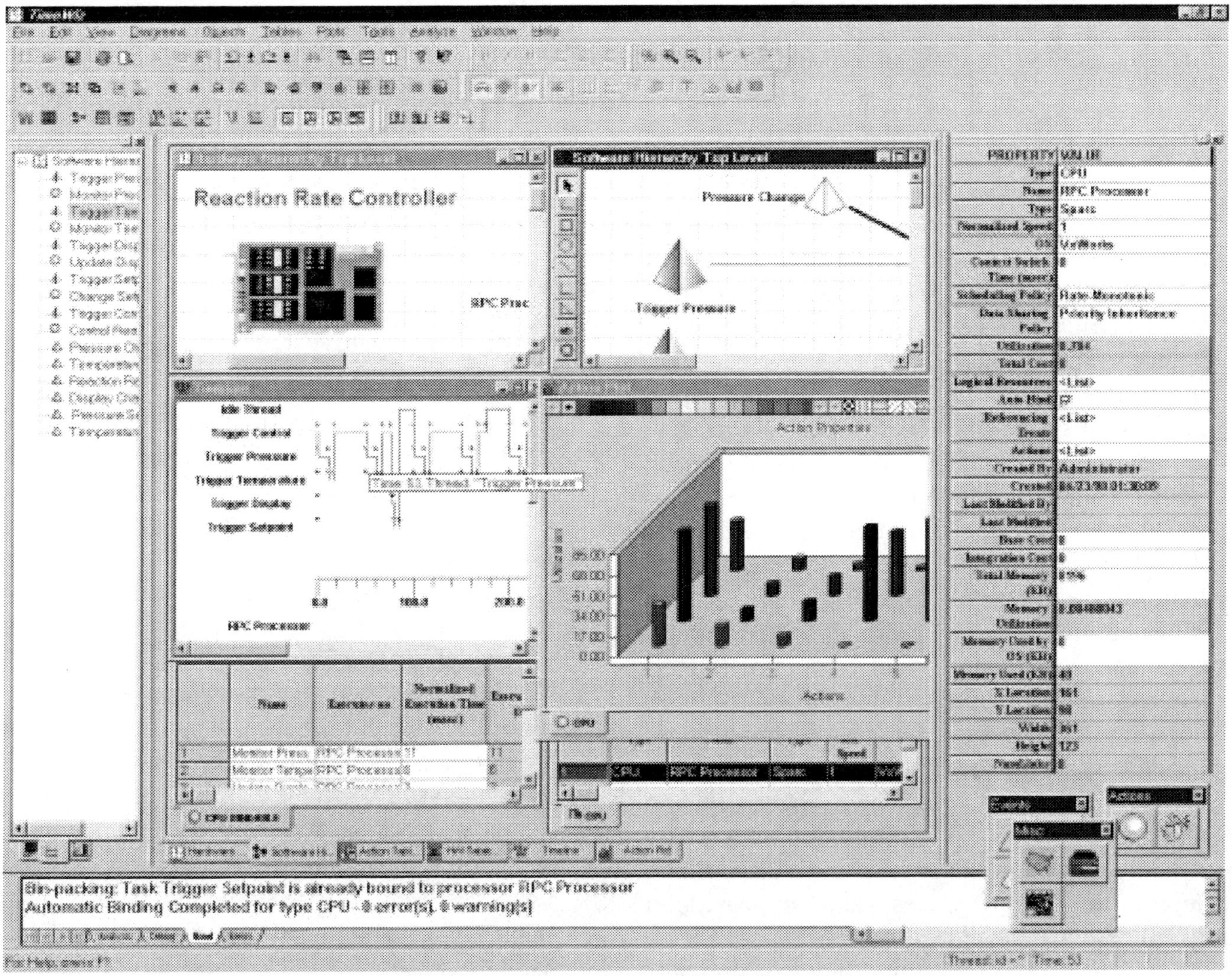

Figure 4. A Sample Screen Image of TimeWiz (see www.timesys.com), a commercially available tool for temporal modeling of real-time systems. TimeWiz is an integrated design environment for all phases of real-time system development and design.

References

[1] Mark H. Klein, Thomas Ralya, Bill Pollak, Ray Obenza and Michael Gonzalez Harbour, "*A Practitioner's Handbook for Real-Time Analysis: Guide to Rate Monotonic Analysis for Real-Time Systems*", Kluwer Academic Publishers, 1993.

[2] M. Jones, "*What Happened on Mars?*", Posting on Internet Bboards, December 5, 1997.

[3] L. Sha, R. Rajkumar and J. P. Lehoczky, "Synchronization in Real-Time Systems", IEEE Transactions on Computers, 1990.

SIMULATOR, WORKSTATION, AND DATA FUSION COMPONENTS FOR ONBOARD/OFFBOARD MULTI-TARGET MULTI-SENSOR FUSION

Angela M. Pawlowski, Lockheed Martin Advanced Technology Laboratories, Camden, NJ

Peter M. Gerken, Lockheed Martin Advanced Technology Laboratories, Camden, NJ

Abstract

The Lockheed Martin Advanced Technology Laboratories (ATL), participating in the Army's largest Advanced Technology Demonstration, has developed a real-time, onboard/offboard multi-sensor, multi-target data fusion system that operates on 14 different sources of information. As the Data Fusion software developer, ATL was created two additional software products in order to effectively design, build, measure, and demonstrate the capabilities of the embeddable information fusion system. A user-friendly, graphical Data Fusion Input Simulator (DFIS) was required to facilitate building multi-target, multi-sensor scenarios for fusion stimulation. In addition, since the embedded fusion system has no graphical output, a Data Fusion Workstation (DFWS) was needed to display metrics and output data within each algorithmic step of the fusion process and to provide a fusion demonstration facility.

The Data Fusion Input Simulator replicates the performance of 14 onboard and offboard data sources to provide tracks, contact reports, group tracks, and line-of-bearing data from the following systems: JSTARS, E2C, AWACS, FAADSC2I, TADIXS-B/TRAP, TIBS, TRIXS, digital intelligence, IR, RF, EO, laser warning, millimeter wave radar, and the Battlefield Combat ID (IFF) System. These systems may report on a variety of land, rotorcraft, fixed wing, and missile entities. A user-friendly graphical interface permits engineers to specify the ground-truth paths of the participating entities and a wide selection of kinematic and classification parameters. Then, by specifying and placing sensors and sources of information, time-tagged data streams of sensor output per entity are recorded for input to the Data Fusion System.

The Data Fusion System is a collection of hybrid, hard real-time algorithms that correlate and fuse the information from the diverse list of sensors named above. Data Fusion manages up to 200 tracks from sources that report up to ten times a second. A fully managed, correlated and fused trackfile is exported from Data Fusion once a second, although all algorithmic processing is nominally accomplished ten times a second, permitting 10 Hz output for critical tracks.

To view the inner workings of the embeddable Data Fusion System, a Data Fusion Workstation was created to execute the algorithms in total, as well as one at a time. Designed to be non-obtrusive to the real-time fusion system, its graphical user interface displays track and other output from each processing step. Execution is possible in both near-real-time (full system) and non-real-time (stepwise through the algorithms). Fusion system metrics collected during runtime are displayed by the Workstation for an excellent performance demonstration.

All software is object-oriented, written in C++, and developed on a Silicon Graphics platform. Due to the transferable nature of this fusion technology, these products have been transitioned to other tactical domains.

Introduction

The Data Fusion component of the Army's largest ATD, Rotorcraft Pilot's Associate, is responsible for correlating and fusing 14 types of non-homogeneous onboard and offboard sensor data.

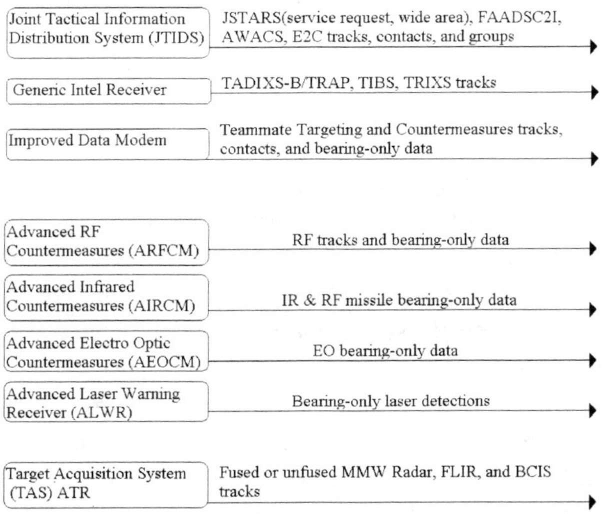

Figure 1 Onboard/Offboard Sensor Types

The fusion end-product - a trackfile of consolidated, merged data - is exported to battlefield assessment reasoning functions that aide the pilot and co-pilot on the tactical rotorcraft platform. ATL engineers have developed customized data fusion algorithms for this system that perform with acceptably low error rates in dense, highly maneuverable targeting scenarios that include ground, air, and missile entities.

Data Fusion within the Rotorcraft Pilot's Associate (RPA) is an embedded function, and, as such, required engineering verification and validation prior to integration. Like many embedded functions within large programs, Data Fusion has no direct graphical interface. While Data Fusion operates on a portion of the same data that stimulates the overall RPA system, it was not trivial to construct fusion-specific tests with the comprehensive RPA stimulation system that were controlled at a sufficiently low level to enable analysis and monitoring of internal variables during embedded execution. Therefore, to accomplish rigorous engineering test and analysis, LM ATL constructed two additional software systems: the Data Fusion Input Simulator and the Data Fusion Workstation. The Data Fusion Input Simulator provides a user-friendly, quick way to construct ground truth tracking scenarios to stimulate the Data Fusion System. The Data Fusion Workstation is a graphical user interface to the Data Fusion System algorithms, designed to unobtrusively permit insight into fusion algorithm execution and performance. The Data Fusion System, Data Fusion Input Simulator, and Data Fusion Workstation will be detailed in the following sections.

The Data Fusion Simulator (DFIS)

Sensor reports on a variety of targets and scenarios were needed to test the functionality and performance of the Data Fusion system. The Data Fusion Input Simulator (DFIS) was developed as a graphical engineering tool that allows these scenarios to be easily created and modified by engineers.

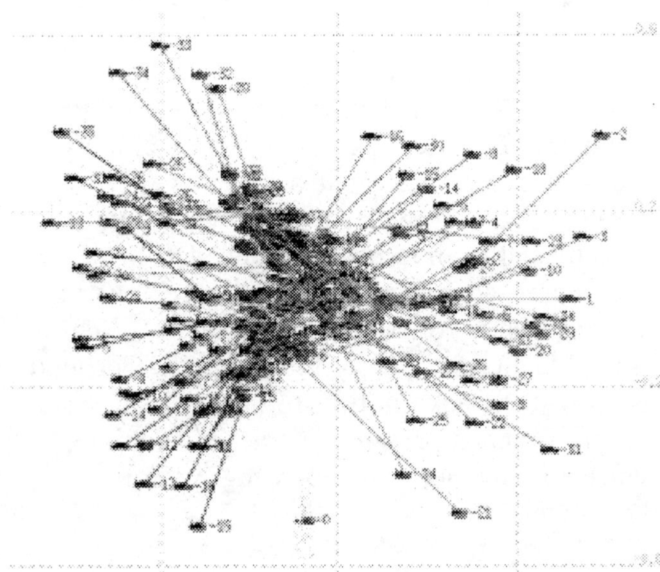

Figure 2 Data Fusion Input Simulator Tactical Situation Window

Figure 2 shows a portion of the graphical user interface for DFIS. The Tactical Situation window shows a view above the current scenario. This display uses custom XWindows and Motif graphics libraries previously developed at ATL. The libraries include support for a grid based coordinate system, displaying ground and air entity icons, selection of a specific icon, moving icons, individual pixel manipulation, centering the display on all of the entities, and the ability to zoom into and out of a boxed region. These features make it very easy to create a complicated scenario with a large number of track entities and sensors. The scenario shown in Figure 2 has 35 tracks entering a notional mountain pass and then dispersing again.

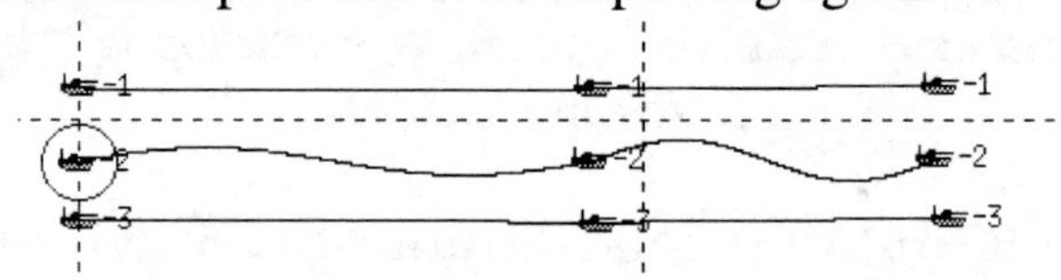

Figure 3 Waypoints for Track Entities

Track Entities

A scenario consists of a set of track entities and sensors. The paths to be traveled by the track entities are represented by a connected set of waypoints; each represented by an icon, as shown in Figure 3. An easy to use graphical interface permits the user to essentially drag and drop these icons to form the path. The user specifies a set of kinematic and classification attributes for each waypoint. The points in-between the waypoints are automatically interpolated by DFIS at a maximum rate of ten times per second, using a chosen linear or sinusoidal pattern. These interpolated positions represent the ground truth of the track entities. In this fashion, track entities can represent a tracked vehicle, a wheeled vehicle, a fixed wing vehicle or a helicopter.

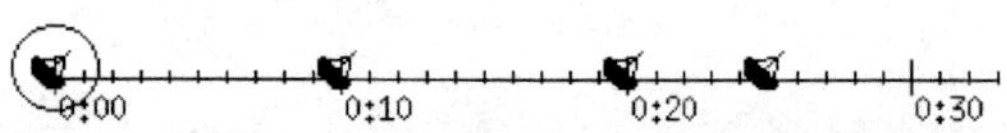

Figure 4 Waypoints for a Sensor

Sensors

Sensor behavior is similarly selected by specifying a set of parameters for a chosen set of waypoints across time as shown in Figure 4. The sensor types span a diverse spectrum, including onboard targeting sensors, countermeasures, offboard intelligence and teammate data. A sensor can be placed at a specific position or can be linked directly to a track entity. The sensor can be set to report tracks, contact reports, line-of-bearing, or group tracks.

Sensor parameters can be set that determine its performance during the periods of time distinguished by the sensor waypoints. For example, some of these parameters are applied uniformly between waypoints, like the range and search angle parameters, which determine how far away the sensor can see the various entities. Error rates determine the accuracy of the sensor when reporting a position. Still other parameters are computed for a sensor per track entity on a percentage basis, like probability of detection, which determines how often the sensor sees an entity and percentage of correct classification, which determines how often the entity's classification is correctly reported.

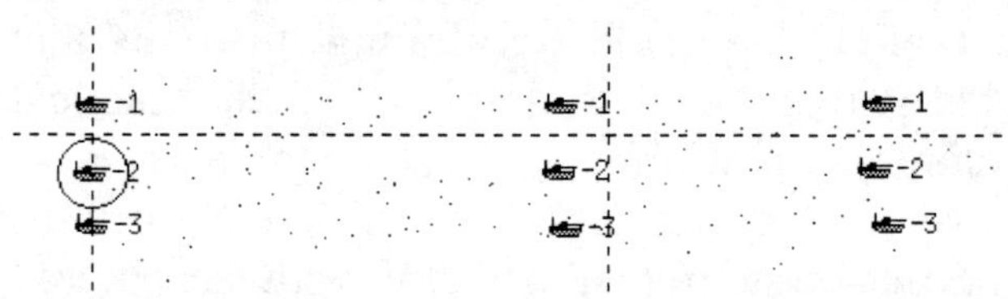

Figure 5 Sensor Reports for Track Entities

Sensor Reports

Figure 5 shows the simulated sensor reports generated by DFIS for the ground truth shown in Figure 3. These reports are determined by user specified entity and sensor parameters. The probability of a sensor's ability to detect an entity at a given time can be set in the entity's waypoint for each type of sensor. This allows a scenario to simulate various effects for a specific entity/sensor combination. For example, a low probability of detection could be attributed to bad weather conditions or rough terrain, and these conditions may be simulated by DFIS entity/sensor specifications. Simulated sensors

0-7803-5086-3 /98/$10.00 ©1998 IEEE

also have a general probability of detection parameter that determines their overall performance. A low probability of detection could be attributed to a faulty sensor or bad weather conditions. If a sensor detects an entity, then the entity's position is determined by the positional accuracy of the sensor.

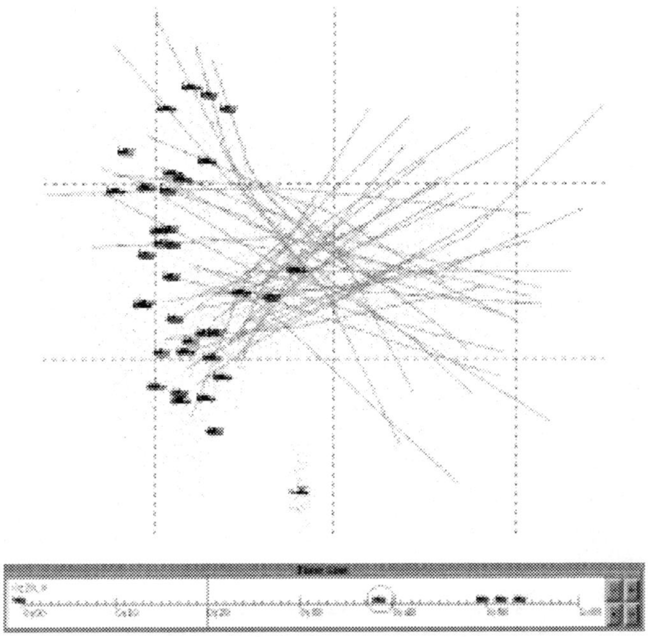

Figure 6 Scenario at Time 0:20

Various overlays can be toggled on and off to help view different parts of the scenario. Figure 6 shows the same display as shown in Figure 2 with the ground truth and icon overlays turned off, so only the sensor reports are displayed in the Tactical Situation window.

The Sliding Timeline window, below the Tactical Situation Window in Figure 6 has also been activated in this example, and displays an entity's or sensor's waypoints linearly across time. This window permits the user to drag the vertical slider bar across the timeline, thus causing the icons in the Tactical Situation Window to move forward in time. Its effects are seen in Figure 7. This function allows a visual pre-play of the scenario as it unfolds in time.

DFIS writes the generated sensor reports, like those shown in Figures 6 and 7, to a binary file that meets the Data Fusion System input interface specification. The sensor reports are arranged in time-ordered fashion by sensor type in this file. Minimal effort is required to modify the format of this file for use by other software systems.

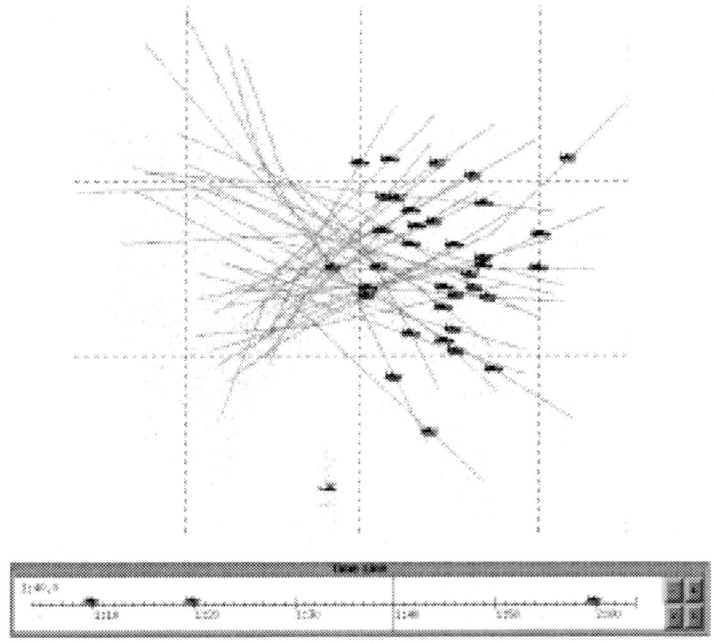

Figure 7 Scenario at Time 1:40

The Data Fusion System

The purpose of the Data Fusion System within the Rotorcraft Pilot's Associate is to correlate and fuse tracks from a wide variety of sensors onboard and offboard the helicopter to enhance situational awareness and to provide better quality input data to the onboard decision aiding functions. This process is accomplished by a set of algorithms that manipulate kinematic and classification data as shown in Figure 8.

As required, this system fuses sensor data present in highly maneuverable, dense tracking environments, and adheres to strict real-time requirements. Up to 200 targets are reported on at a maximum input rate of 10 Hz while Data Fusion produces a fused trackfile at a rate of 1 Hz. Versions of the software algorithms that accomplish Data Fusion, as well as the Data Fusion Input Simulator and the

Data Fusion Workstation, currently execute on both Silicon Graphics and Sun Sparcstation platforms. In the Silicon Graphics configuration, one processor is dedicated to Data Fusion input and output tasks and another is dedicated to the customized data fusion algorithms. In the Sparcstation configuration, all algorithms execute on a single processor. Additionally, the Data Fusion algorithms have been ported to the Real Time Symmetric MultiProcessor (RTSMP), a flight-ready system built by Lockheed Martin Federal Systems and Concurrent Computer Corporation.

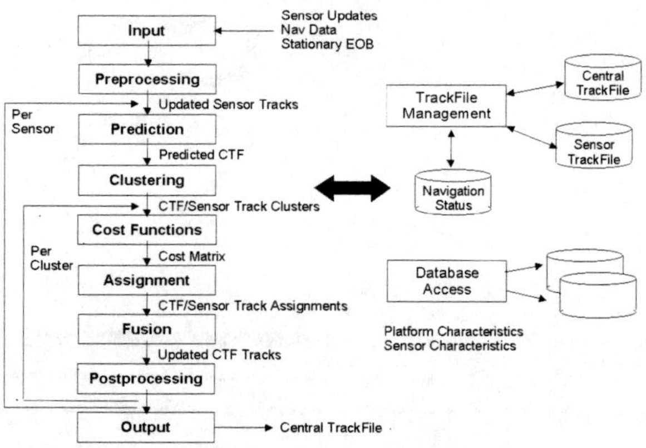

Figure 8 Functional Flow Diagram

Kinematic Data vs. Classification Data

One early premise from which the Data Fusion System evolved was the notion that the correctness of track correlation could be improved by employing the available target classification data, not just the kinematic data, in the track comparison process.

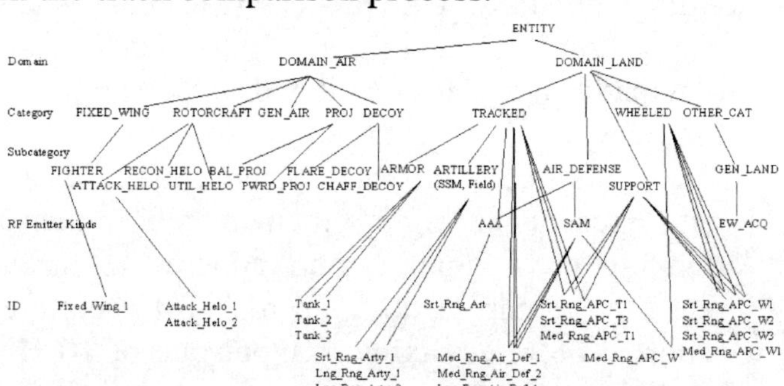

Figure 9 Air/Ground Classification Taxonomy Graph

This premise lead to the development of two very different types of algorithms: those which manipulated kinematic data, such as position and velocity, and those which manipulated a taxonomy of possible classification values as partially illustrated in Figure 9. Both types of processing form the algorithms discussed below.

Custom Algorithms

It was determined that conventional algorithms for this data fusion problem were in some cases either too slow or too inaccurate to meet the system processing and timing requirements. Therefore, hybrid algorithms were developed to perform the fusion-related functions that support the central function in the fusion system: the Assignment function. Since the Assignment function is critical to any fusion problem, we opted to utilize a proven, off-the-shelf algorithm called JVC, which had much success in the Canadian frigate program.[1] These fusion algorithms execute in a linear fashion, as shown in the process abstraction in Figure 10. The circles represent sensor reports and the squares represent tracks from the Central Trackfile (CTF).

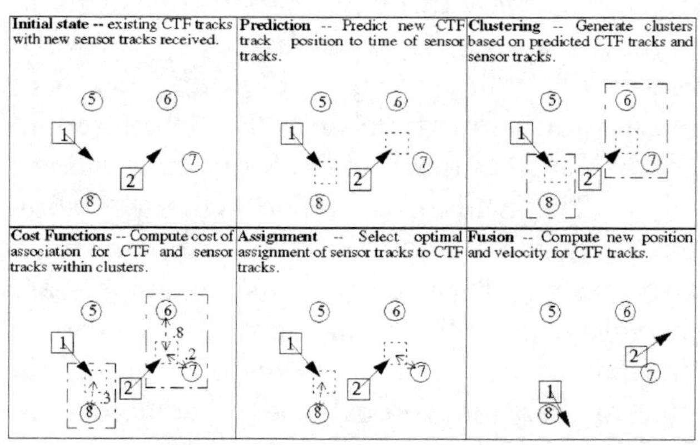

Figure 10 Process Abstraction

The set of Data Fusion System algorithms shown in Figure 10 are described in more detail below.

[1] Malkoff, Dr. Donald, "Evaluation of the Jonker-Volgenant-Castanon (JVC) Assignment Algorithm for Track Association," SPIE 11th Annual International Symposium on Aerospace/ Defense Sensing and Control, Orlando, Florida, April 1997.

Preprocessing

The Preprocessing functions attempt to derive track classification information when it is not included in an input sensor report. This algorithm potentially creates more data than what is given from the sensors and therefore improves track comparisons by permitting more data points on which to reason. For instance, based on a track's velocity, we can infer via a rule-based process, that it is behaving more like a helicopter than a ground vehicle.

Prediction

The prediction algorithm extrapolates position and error based on an adaptable alpha-beta-gamma filter that weights the input contributions according to their relative error ellipses. This algorithm quickly predicts highly maneuverable target positions and positional error with an accuracy and reliability that can not be accomplished using traditional Kalman Filtering.

Clustering

The clustering algorithm executes in linear time, separating the tracks into clusters to decompose the fusion processing into smaller bundles. While an intuitively useful concept, clustering is often not performed in real time fusion systems because the execution time of most approaches to clustering increases faster than the number of tracks, but the linear order of this algorithm makes it fast enough to be used and reduces overall assignment time by a factor of 10.

Cost Functions

The cost functions calculate a measurement of similarity between each pairing of tracks in a cluster based on kinematic and classification data. The use of classification has been very helpful in the fusion process as it expands the number of factors contributing to a track pairing's overall cost.

Assignment

The only off-the-shelf algorithm employed in the fusion process is utilized for assignment. It is the Jonker-Volgenant-Castanon (JVC) algorithm, constructed from the Auction and Munkres algorithms, and provides the speed and accuracy needed to assign sensor tracks to the tracks in the central track file.

Fusion

The fusion functions merge the kinematic and classification data from sensor tracks with the corresponding data from the assigned central track file tracks. Separate functions fuse kinematic and error ellipse data, and a third type merges classification data in a unique modified Bayesian evidence accumulation scheme[2] for the possible track classifications shown in Figure 9.

Postprocessing

During the previous algorithm steps, even tracks without classification may have under gone kinematic modification. If classification of a track has not been determined, then a classification derivation function is again employed in this postprocessing step. This function permits up to three classification hypotheses to be generated and managed for use by external situational awareness functions.

The Data Fusion System was extensively tested under a variety of operational and engineering scenarios. The results from the tests determined the following conclusions:

- RPA data fusion contributes to a pilot's ability to assess a tactical situation
- a scenario's complexity can be directly related to fusion performance
- fusion performance is consistent across different scenarios, relative to scenario complexity

[2] Hoffman, Martin, "Multi-Sensor Track Classification in Rotorcraft Pilot's Associate Data Fusion," American Helicopter Society 53rd Annual Forum, Virginia Beach, Virginia, April 29 – May 1, 1997.

- use of classification data enhances data fusion performance
- the RPA data fusion system performs with less than 1% assignment error in comprehensive laboratory tests.

The Data Fusion Workstation

The Data Fusion System was designed to work without a graphical user interface because it needs to perform in an embedded, real time environment. To enable engineering and demonstration of this system in a stand-alone fashion, the Data Fusion Workstation was designed to display the output of each of the major fusion functions, as well as the metrics collected for system analysis. A window displaying the central trackfile and the incoming sensor tracks to be fused is shown in Figure 11. A sample of the graphical output for each fusion function is shown below.

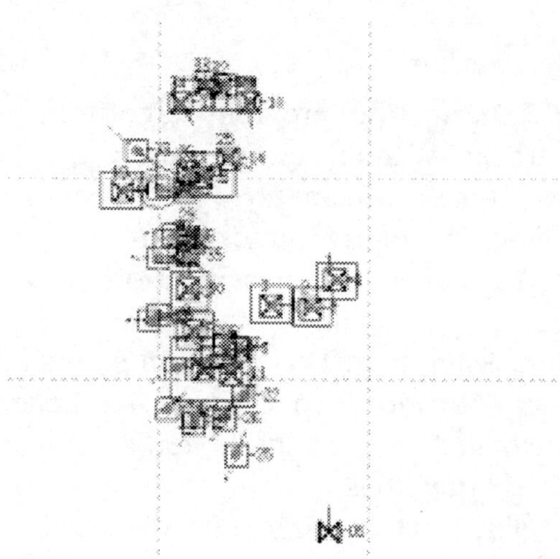

Figure 11 Data Fusion Workstation

Prediction

If the Data Fusion System were paused after the Prediction function had executed, the window shown in Figure 12 would be visible. The square boxes designate tracks in the central track file. The diamond shapes around each central trackfile track represent major and minor axes of the associated error ellipse. The X's represent a sensor report received by the data fusion system, while the diamonds around each X represent the sensor report's error ellipse. A heavy dot represents the predicted position of the track.

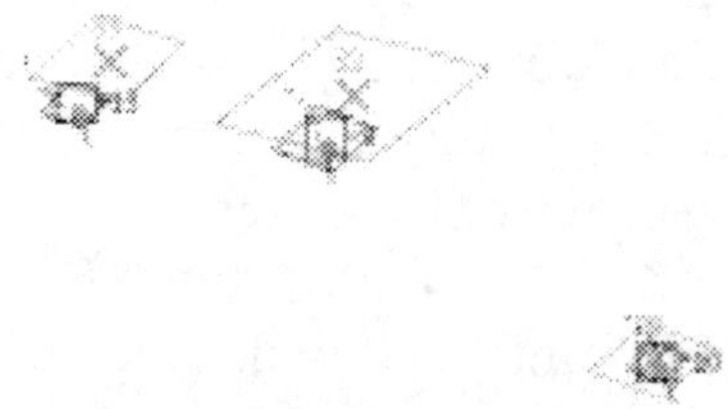

Figure 12 Data Fusion Workstation window: Prediction

Clustering

Pausing the Data Fusion System after the Clustering function, would cause the window in Figure 13 to be visible. The large boxes around the tracks show which sensor tracks and central trackfile tracks belong to a cluster. In this example, sensor reports for tracks 32 and 33 belong in a cluster separate from the sensor report for track 28. The clustering algorithm determines which sensor tracks should be compared to which central trackfile tracks, so in this example, the sensor track 28 will not be compared with central trackfile track 9 or 13, because they are in a different cluster.

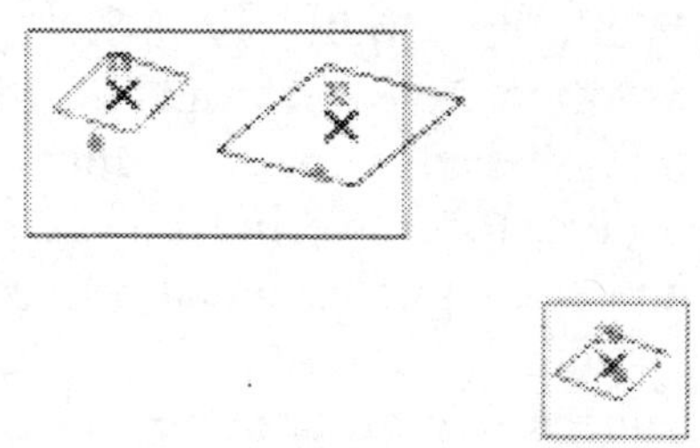

Figure 13 Data Fusion Workstation window: Clustering

Cost Functions

The individual cost functions that contribute to the measures of similarity result in a single value. Matrices of these values are the input to the JVC Assignment algorithm. It is these matrices, with sensor reports across the horizontal axis and central trackfile tracks across the vertical axis, that are displayed on the Data Fusion Workstation, if the Data Fusion System is paused after the Cost Functions, as shown in Figure 14.

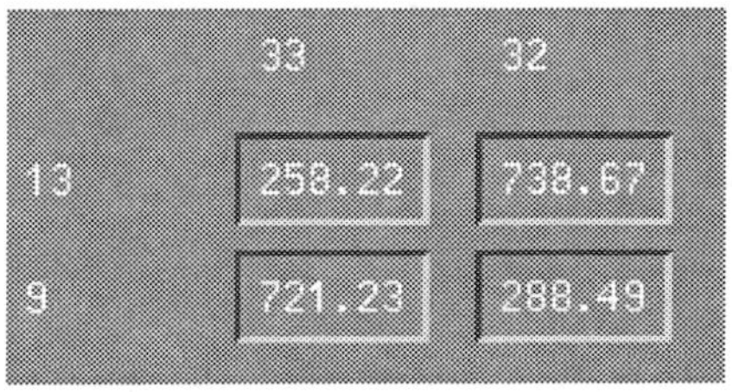

Figure 14 Data Fusion Workstation window: Cost Functions

Assignment

The Assignment function uses the results of the cost function to make near-optimal matches of the sensor tracks with the central trackfile tracks. For this example, the lower numbers represent higher similarity. The matches are displayed on the Data Fusion Workstation as highlighted cells in the cost matrix if the Data Fusion System is paused after the Assignment function, as shown in Figure 15.

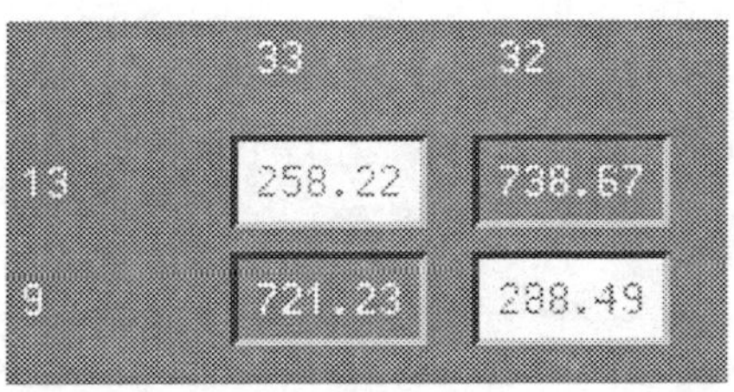

Figure 15 Data Fusion Workstation: Assignment

Fusion

When the Fusion functions execute, the graphical result on the Data Fusion Workstation is a merged central trackfile position and error ellipse, as shown in Figure 16. Rather than stepping through each fusion function, it is possible to select only new sensor inputs to be displayed and the resultant updates to the central trackfile. In this mode, the Data Fusion Workstation demonstrates the execution of the Data Fusion System operating in near-real time.

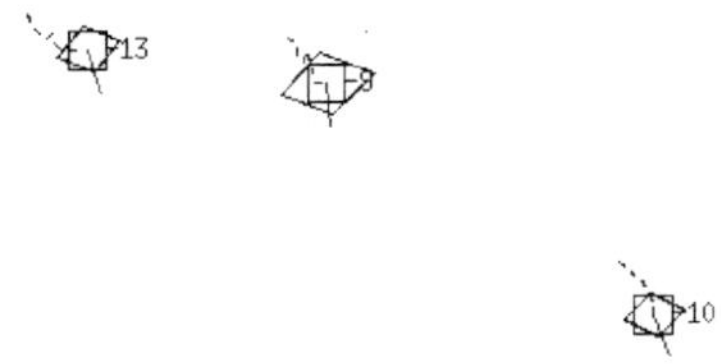

Figure 16 Data Fusion Workstation window: Fusion

Summary

All three components of the data fusion software were written in C++, with the graphical components also employing XWindows and Motif. Although all components were initially developed on a Silicon Graphics Challenge, they have also been ported to Sun Sparcstations. The Data Fusion Input Simulator has also been transitioned into two other programs: the Electronic Warfare Input Simulator (EWIS) and the Ballistic Missile Scenario Generator (BMSG), for use in four other projects at ATL. Currently, ATL has delivered six software builds on the RPA program, with a system flight demonstration scheduled for July 1998. The Data Fusion System algorithms and supporting components are transitionable to a variety of adjacent tactical domains due to their object-oriented design and the requirements they meet for the development of most correlation and fusion problems.

RESOURCE ALLOCATION IN A DISTRIBUTED COMPUTING ENVIRONMENT

Russell L. Carter, Conceptual Systems & Software, Chino Valley, AZ, rcarter@consys.com

David St. Louis, Conceptual Systems & Software, Chino Valley, AZ, stlouis@consys.com

Ed P. Andert Jr. , Conceptual Systems & Software, Chino Valley, AZ, andert@consys.com

Abstract

Advanced transportation systems can be expected to increasingly employ modern software development methodologies for the implementation of their operational capabilities. This is demanded by the shift to open architecture hardware and software solutions in order to meet stringent low-cost objectives. The Common Object Request Broker Architecture (CORBA) is an important maturing standard offering interoperability among software components and portability across hardware platforms.

While CORBA can provide needed interoperability and portability, the question arises how to manage distributed CORBA clients and servers in a fault tolerant manner. Future mission critical and distributed software deployments require effective resource allocation. Critical resource allocation issues that need to be addressed include component migration, scheduling, and load balancing across dynamic computational resources. A management system that addresses these issues in a uniform and cost-effective way is needed.

This paper describes an integrated approach to optimal resource allocation and fault tolerance in a distributed computing environment. The approach leverages a commodity-off-the-shelf (COTS) resource allocation system to provide the infrastructure for the specific resource allocation and fault tolerance enhancements required to host a reliable CORBA applications on heterogeneous networked workstations.

Distributed Workload Components

Modern distributed computing environments commonly process workloads consisting of a variety of application architectures. These architectures may include a mix of one or more of ordinary sequential processes, CORBA (or other middleware) applications, and parallel applications.

The Common Object Request Broker Architecture (CORBA) [1] is a popular standardized architecture for developing distributed object-orientated applications. Communication is performed through an intermediary, the Object Request Broker (ORB). For the purpose of optimal resource allocation the architecture of the applications running in the distributed computing environment is of interest.

A CORBA application typically has a client-server architecture. One or more clients interact with one or more servers, with the ORB providing the underlying transport. Modules provided with the ORB act as the intermediate repository for object interface information, directory, authentication, and other essential services. The ORB along with the collection of services is generically called the middleware environment.

Implementations of parallel applications commonly exhibit a structure analogous to standard client-server architecture. A master program initiates, then spawns one or more slaves. These slaves generally run on different systems in the distributed environment in order to maximize computational resources for the

0-7803-5086-3 /98/$10.00 ©1998 IEEE

program. Slave processes correspond to CORBA clients and the master program is the server. In more complex applications, heterogeneous modules may alternate the master/slave functionality. Similarly, distributed CORBA applications may have a more complex architecture, where many of the components expose both client and server interfaces.

Parallel applications running on networks of workstations (NOW) may be hosted in a middleware environment called the Parallel Environment (PE). Implementations of PEs available commercially or from the research community include Parallel Virtual Machine (PVM) [2] and Message Passing Interface (MPI) [3]. There are a variety of approaches, but each PE implementation typically handles directory services, format translations, slave startup and management, etc. in direct analogy to the functionality provided by CORBA middleware.

Resource Allocation for CORBA Components

Distributed applications run on collections of networked computers, called clusters or networks of workstations (NOW). Computers are generally multiuser; they run more than one application at a time. In order for a networked application to have acceptable performance, the amount of work (applications/CORBA objects) given to individual computers in the cluster must be matched to the capability of the computer. If the distributed application runs on a large enough number of computers, the situation can arise where some of the computers are given little work, while others are given too much. This situation is nearly guaranteed to occur without special metering software, and results in poor use of costly computers and networking infrastructure. The special metering software goes by several names, depending on the type of metering that is done, but can be generically called resource allocation software.

0-7803-5086-3 /98/$10.00 ©1998 IEEE

Figure 1 indicates the situation in a heterogeneous cluster of workstations running different operating systems, with initially similar workloads.

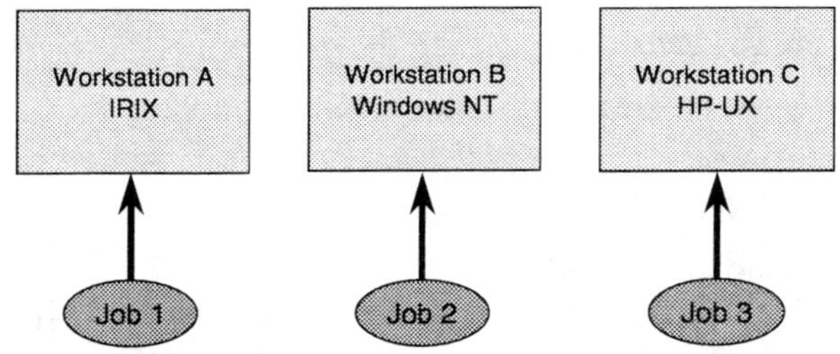

Figure 1: Initial Load on Cluster

At first the load is balanced, but after time, in the absence of a cluster-wide system to allocate resources, the load will become unbalanced, with some workstations overwhelmed with work while other workstations sit idle, as illustrated in Figure 2. Here Workstation A has three jobs running, while two jobs are swapped out. Workstation B has one job running with one job swapped out. Workstation C is idle. Workstation A has a load normalized to 150%, while Workstation C is nearly idle at 1% load. The heavy load on Workstation A precludes efficient processing, and the cluster as a whole is will have poor throughput.

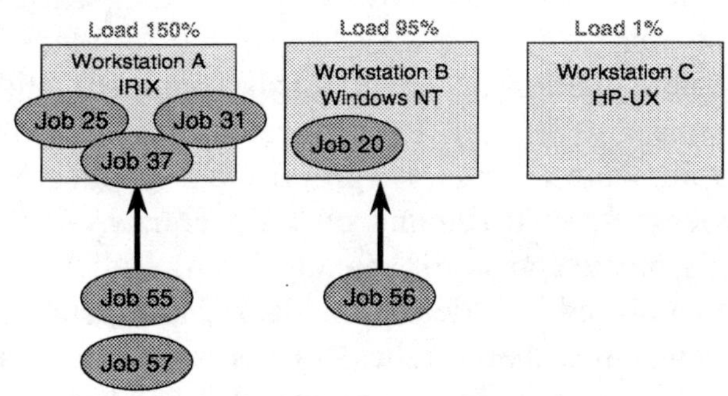

Figure 2: Load at Later Time

In order to address loading and other management issues pertaining to efficient use of a cluster the resource allocation system needs to implement policies differentiating the priorities of classes of work, and provide other ancillary services as needed including authentication, fault tolerance, and resource limits.

Transaction Processing Monitors

Resource allocation packages are available that meter access to particular CORBA objects, but only for narrowly specialized applications. The metered CORBA object typically is a database server, in which case the resource allocation package is called a transaction processing (TP) monitor. On Line Transaction Processing (OLTP) and distributed client-server databases are target applications. Examples of commercially available TP-Monitors include Tuxedo from Novell, Encina from Transarc, and CICS from IBM. Figure 3 provides a schematic view of a client-server implementation with a TP-Monitor providing flow control for incoming client service requests via a message queue.

Replicated Servers

A new approach suitable for some specialized client-server applications uses multicast group communication among a collection of processes to deliver fault tolerance and limited resource allocation. The method is termed virtual synchrony [4], and various research groups have produced implementations, c.f. Horus [5], Totem [6], and Transys [7]. A replicated server is implemented by a group of replicated processes, each running on a different system. This process group is managed by middleware and may add or delete members, for instance when a member crashes. Clients see only a single service. Since the middleware handles the client to service interactions it is usually straightforward to add basic resource allocation functionality to the service such as load balancing among the process group members. In the simplest case, each client request is decomposed into a number of parallel actions within the group.

Similarly, fault tolerance is achieved for the distributed services by running the same executable image on as many machines as is desirable to give the appropriate degree of fault tolerance. For example, if a process group consists of five processes providing a service, the service remains available in the presence of up to four group member failures.

The cited virtual synchrony toolkits provide support for replicated processes, whereas applications need support for standardized CORBA objects. CORBA support via proprietary extensions is achieved by layering an object adapter over the basic group communication structure to map process groups to object groups in Orbix+Isis [8], Electra [9], and Eternal [10].

Cluster Management Systems

Resource allocation packages are available that meter many types of (non-CORBA) applications running on a wide variety of heterogeneous clusters. These packages are known as Cluster Management Systems (CMS). The parallel computing approach has been to use batch queuing systems to provide essential resource allocation services for a wide variety of environments, including mainframe, supercomputer, massively parallel processor, and networks of workstations (NOW). Available CMS implementations displaying varying degrees of functionality for managing distributed computing environments, and specifically heterogeneous networks of workstations (NOW).

Tradeoffs

Commonly, the collection of available computers on which the target workload is to be run consists of various models with dissimilar capabilities. At the present time no resource allocation packages are available that meter multiple generic CORBA objects on heterogeneous clusters of computers. Each of the three existing schemes for resource

0-7803-5086-3 /98/$10.00 ©1998 IEEE

allocation, TP-Monitors, Replicated Servers, and CMS, differ in their capabilities for resource allocation. A TP-Monitor manages one or possibly several dedicated, distributed, client-server software applications. The target TP-Monitor architecture is generally many clients accessing a single (abstract) server. A TP-Monitor does not provide a facility to optimally allocate and schedule independent parallel applications on a NOW. In particular, it does not manage the system resources clients may use, just access to the managed server object.

Similarly, replicated servers cannot automatically preempt an actively replicated process group based on scheduling priorities assigned to an incoming job stream, so that a higher priority application may take over the system to complete a time critical computation. As with TP-Monitors, replicated servers do not manage the system resources used by clients. Rather, they focus solely on management of access to and uptime of the managed (replicated) server object. In addition, the replication technology increases hardware and software system costs linearly with the number of replicated nodes. Also, there are serious unresolved issues [12] with the implementation of CORBA object groups which simultaneously provide client and server functionality. An effort to address these issues produced an Object Group Service [13]; however, a prototype implementation displayed overheads of about a factor of ten [12]. These issues imply that the appropriate target domain for replicated servers are highly available, load-balanced process-oriented servers that require little or no persistent state [14].

CMS is designed specifically to manage all user level application requests for resources from a NOW The difference is fundamental: a TP-Monitor and Replicated Servers manage the internal resources of specifically targeted server objects, while CMS provides external resource management of all jobs requesting system resources from a specific NOW. In fact, there is no fundamental inhibitor to running a TP-Monitor as a part of a particular (possibly dedicated) client-server application in a NOW managed by CMS. For this reason, a CMS is the appropriate choice for managing NOW resources when the workload does not consist solely of a limited number of dedicated client-server applications, optimal resource allocation strategies are to be implemented, and cluster wide load-leveling and component fault tolerance is desired.

Resource Allocation in Practice

Networks of workstations (NOW) are an increasingly popular architecture for running distributed workloads. A large component of the popularity derives from the potential price/performance advantage commodity workstation clusters have compared to other competing technologies for processing distributed application workloads. In order to achieve an advantage, the NOW must be efficiently utilized. Techniques for efficiently processing workloads on a NOW architecture usually seek to minimize idle resources in existing workstation networks.

Workloads are generally composed of sequential jobs and distributed applications. The sequential jobs may be the result of interactive use of individual workstations or may be the result of exploiting the multitasking capability of workstations to run additional (usually non-interactive) sequential jobs through load leveling techniques across a network of unsaturated processors. Distributed jobs may be run in a dedicated NOW or may be run in a mixed environment with interactive, sequentially load-shared jobs. Due to the variety of ways the resources of a NOW may be allocated, it is essential that the CMS provide the flexibility to implement local resource allocation and scheduling schemes encompassing all of the required site-specific modes of operation of the NOW.

Generally, the target for optimal resource allocation is to maximize throughput of the job stream through the system. In the

0-7803-5086-3 /98/$10.00 ©1998 IEEE

simplest case, for a system composed of a network of workstations, this is roughly equivalent to maximizing utilization of individual workstations. Site specific policy decisions may reduce utilization on individual nodes; a site may wish to minimize interference with interactive response, and hence forbid non-interactive jobs from using possibly lightly loaded interactive nodes.

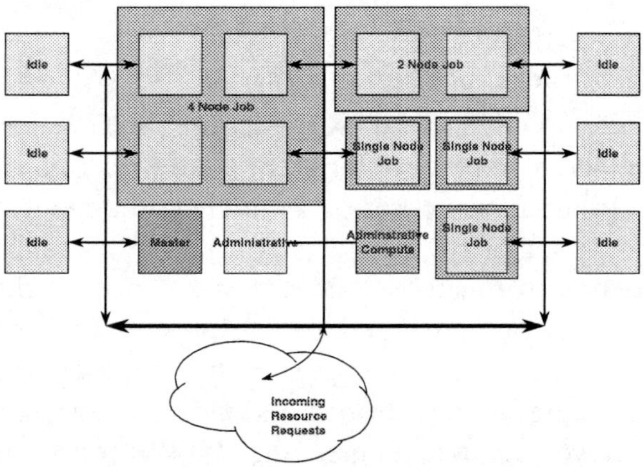

Figure 3: Lightly loaded NOW processing 1, 2, and 4 node jobs

Figure 3 provides a schematic of a NOW processing a workload consisting of jobs requiring one or more nodes. In this case, the resources are not completely allocated; some nodes are idle. Jobs which need more than the total number of idle nodes must wait for running jobs to finish until enough nodes are available to allocate the request.

In practice, optimal scheduling in a NOW running a variety of distributed CORBA and/or parallel jobs is strongly dependent on site specific policy, and therefore a mandatory requirement is that the cluster management system accommodate the variety of policies required to implement efficient processing of the site specific workload.

Application Checkpointing on a NOW

Checkpointing refers to the periodic generation of a restart file from a running application. If the process dies, whether by intention or by failure, the process may then be restarted from the most recent checkpoint file and run successfully to completion. Therefore, the restart file must contain all of the program state necessary to resume computation. For certain kinds of applications, the amount of state can be considerable.

A checkpoint facility is essential to achieving high levels of utilization on all of the nodes of a NOW. It provides the ability to implement site specific allocation policies. For instance, if a job requires more nodes than are currently available in the NOW, jobs with lower priority may be checkpointed so that resources are freed for the higher priority large resource job. Similarly, if a component has mission-critical or real-time requirements, appropriate resources may be obtained this way.

Fault Tolerance for CORBA Clients

For a workload running on a NOW the need for fault tolerance varies according to the requirements of the jobs comprising the workload. "Mission-critical" jobs require the highest level of fault tolerance. The CMS for such a NOW must be able to ensure that CORBA client components are handled appropriately in the event that one or more nodes become unavailable from any of a variety of events. Such events may include renewed interactive access to a node assigned exclusive interactive access, software or hardware failure, etc. Jobs assigned less critical priorities should be handled as required by site policy.

Implementation of fault-tolerance also relies on checkpoint facilities. Consider a job running on a workstation, generating periodic user checkpoints. The CMS generates periodic "heartbeat" probes to determine the health of the system. In the event of a failure, for example a network interface, the heartbeat test fails and an appropriately configured CMS responds by restarting the appropriate jobs on another system in the cluster with compatible resources. The restarted job is first placed in

0-7803-5086-3 /98/$10.00 ©1998 IEEE

the queue for the target resource. When the resource becomes available the job begins executing, then reads the restart file, at which point it may continue normally.

Dynamic Load Balancing

The load on individual nodes may vary significantly over time in a NOW supporting a distributed CORBA environment. This occurs despite the best efforts of the resource allocation system to schedule incoming jobs. Typical reasons for the load variance include unpredictable runtimes due to inaccurate forecast or unexpected exits due to exceptions, drained queues, etc. For sequential load-shared client object components, the existence of a checkpoint facility allows for the implementation of dynamic load balancing. Checkpoints for designated job classes are generated regularly and stored on persistent storage.

If a higher priority job enters the workload and needs lower priority checkpointed jobs may be terminated. The lower priority job may then be migrated by the CMS to another node with appropriate resources. If no such node exists, the lower priority job is requeued on the initial system and waits until the higher priority job completes. Migration is performed similarly to the response to system faults. The job is first queued on the target system and then restarted from the latest checkpoint.

CMS To CORBA Integration

The work we report here determined the interfaces required to support fault tolerant migration of CORBA client components. The Server and client components need to be migrated to available nodes in the event of one or multiple failures of nodes in the NOW. The CMS detects failure of nodes and failure of managed client and/or server CORBA components. Failure detection of a system hardware or software component is determined through the use of a heartbeat function. If a heartbeat test fails, jobs running on the affect node(s) are marked for restart. Assuming appropriate resources are available elsewhere in the cluster, a job may then be continued.

Integrating CORBA applications into the CMS checkpointing environment requires a method of communicating a unique identifier between the CORBA application and the checkpointing environment. Our approach used a system specific identification number (jid) to uniquely specify each particular job. This number is unique to a job throughout its active life. In particular, the CMS assigns the same jid to restarted jobs. A CORBA application managed by the CMS may access the current jid by reading an environment variable. This identifier may then be used to generate checkpoint filenames, etc.

The remaining functionality required to interface CORBA applications to the CMS checkpoint and restart facilities is the method by which a CORBA application is terminated when the job must be migrated. The method for terminating a CORBA application from the CMS is to implement user checkpointing command. This command is run with the same environment as the initial job. The command is also a CORBA client, which contacts the CheckPoint Service object with the appropriate parameters. Since the original CORBA application (which it is desired to terminate) is also a client of the CheckPoint Service object, that terminating command may communicate to the appropriate client using the passed environment variables, which can then perform appropriate cleanup actions and terminate.

Demonstrated Functionality

A prototype was implemented demonstrating and optimal resource allocation and fault tolerance for CORBA clients managed by a CMS. The demonstration was hosted on a Silicon Graphics workstation running the IRIX 6.2 operating system and a Hewlett Packard PA-RISC workstation running the HP-UX 10.10 operating systems. Both

0-7803-5086-3 /98/$10.00 ©1998 IEEE

computer systems hosted the Orbix 2.0 CORBA ORB [15] and the CODINE 4.02b Cluster Management System [16].

First, heterogeneous CORBA job migration as a component of optimal resource allocation was demonstrated. A prioritized set of job queues was configured on each of the workstations. Jobs consisting of CORBA clients subscribing to the Checkpoint Object Service were submitted to the lower priority queues on the HP system. While these were still active, higher priority jobs were submitted. As required, the lower priority CORBA objects were automatically terminated, migrated, and restarted without loss of state to the SGI system when a higher priority job began execution.

Finally, fault tolerance for CORBA clients was demonstrated. A pool of "compatible" queues was configured on the SGI and HP systems. CORBA clients subscribing to the Checkpoint Object Service were submitted to queues on the SGI system. Subsequently, the network was disconnected from the SGI system. The CMS detected the failure of the system, and automatically restarted without loss of state the active CORBA client jobs on the HP system.

Future Work

The approach elaborated in this paper provides a solution for implementing optimal resource allocation and fault tolerance for CORBA clients operating in heterogeneous environments. Functionality required for a complete solution would extend these capabilities to CORBA servers as well. Possible approaches include integrating the replicated servers approach into the CMS environment, or re-implementation of the comprehensive capabilities now provided by COTS CMS into a native CORBA service. The Information Manager [17] and Piranha [18] efforts take the second approach. Piranha is targeted to Electra [9] compliant servers, only. TAO [19] seeks to provide static and dynamic scheduling for predefined real-time CORBA workloads. Each of the approaches requires significant study to determine the potential issues to be uncovered by deployment in operational environments characterized by heterogeneous application architectures.

Conclusion

Given the increasing penetration of middleware solutions into distributed computing environments, provisions for managing diverse workloads consisting of a variety of application architectures are required. The approach investigated here provides a cost-effective, COTS-based approach to addressing traditional resource allocation issues for CORBA clients operating in a heterogeneous environment.

References

[1] OMG. *CORBA 2.0 Specification*, Technical Document PTC/96-03-04, Object Management Group, http://www.omg.org.

[2] V. Sunderam, PVM: *A Framework for Parallel Distributed Computing*, Concurrency: Practice and Experience, 2(4):315-339, December 1990.

[3] *Message Passing Interface Standard Version 1.1*, Message Passing Interface Forum, June 1995.

[4] K. Birman and R. van Renesse. *Reliable Distributed Computing with the Isis Toolkit*, IEEE Computer Society Press, 1993.

[5] R. Van Renesse, K.P. Birman, and S. Maffeis. *Horus: A Flexible Group Communication System*. Commun. ACM 39, 4 (Apr. 1996).

[6] L.E. Moser, P.M. Melliar-Smith, D.A. Agarwal, R.K. Budhia, and C. A. Lingley-Papadopoulos. *Totem: A Fault-Tolerant Multicast Group Communication System*, Comm. ACM 39, 4 (April 1996).

[7] D. Dolev and D Malki. *The Transis Approach to High Availability Cluster Com-*

0-7803-5086-3 /98/$10.00 ©1998 IEEE

munication, Commun. ACM 39, 4 (April 1996).

[8] Iona Technologies PLC and Isis Distributed Systems. *An Introduction to Orbix+Isis*. 1994

[9] Silvano Maffeis. *Run-Time Support for Object-Oriented Distributed Programming*. PhD thesis, University of Zurich, 1995.

[10] P. Narasimhan, L. Moser, and M. Melliar-Smith. *Exploiting the Internet Inter-ORB Protocol to Provide CORBA with Fault-Tolerance*, USENIX Conference on Object-Oriented Technologies and Systems, 1997.

[11] M. A. Baker, Geoffrey C. Fox and Hon W. Yau, *A Review of Commercial and Research Cluster Management Software*, NHSE Review, 12 June, 1996.

[12] Rachid Guerraoui, Pascal Felber, Benoit Garbinato, and Karim Masouni, *System Support for Object Groups*, Technical Report – May 1998, Operating Systems Laboratory, Swiss Federal Institute of Technology of Lausanne CH-1015 Lausanne, Switzerland.

[13] P. Felber, B. Garbinato, and R. Guerraoui. *The Design of a CORBA Group Communication Service*, IEEE Symp. on Reliable Distributed Systems, 1996

[14] Silvano Maffeis and Douglas Schmidt. *Constructing Reliable Distributed Communication Systems with CORBA*, IEEE Comm. Magazine, Vol 14 No. 2, April 1997.

[15] Iona Technologies PLC. *Orbix 2.0*. http://www.iona.com

[16] Genias Software GmbH, *CODINE*, http://www.genias.de.

[17] Lockheed Martin Federal Systems. *Information Manager*, Owego, NY http://ww.owego.com/awacs/ovrview_x.html

[18] Silvano Maffeis. *Piranha: A CORBA Tool for High Availability*, IEEE Computer, April 1997.

[19] Christopher D. Gill, David L. Levine, and Douglas C. Schmidt. *Evaluating Strategies for Real-Time CORBA Dynamic Scheduling*, submitted to the 19th IEE Real-Time Systems Symposium, Madrid, Spain, Dec. 2-4, 1998.

0-7803-5086-3 /98/$10.00 ©1998 IEEE

FITTING THE PIECES TOGETHER: SYSTEM/SOFTWARE ANALYSIS AND CODE INTEGRATION USING METAH

Jonathan W. Krueger, Honeywell Technology Center, Minneapolis, MN
Steve Vestal, Honeywell Technology Center, Minneapolis, MN
Bruce Lewis, US Army AMCOM, Huntsville, AL

Introduction[1]

Embedded systems software integration is in desperate need of improved architectural methodologies and tools. Traditional approaches are running up against accelerating trends that increase complexity. More functionality is being crammed into vehicles: on-board maintenance and diagnostics, multimedia communications, situation awareness, mission simulation and training, higher levels of cockpit automation, greater autonomy and less dependence on central traffic management, etc. Software applications previously deployed at arm's length on federated boxes are coalescing onto "stock" multi-processing hardware. Engineers from a wide variety of specialties are thrust together to wrangle out decisions with little means of assessing their impact.

Architecture Description Languages (ADLs) are beginning to address some of these issues. An ADL represents the system as a set of interconnected components subject to an overall execution model. An ADL aimed at a specific domain can give designers a way to check their system against particular properties required by their application. Formal ADLs can also be transformed automatically into executable "glue" code that ties the components together according to the specified connections and execution model.

MetaH is an ADL and toolset originally developed to meet the strict requirements of flight control and avionics, including hard real-time, safety, security, fault-tolerance and multi-processing. However, an increasingly broad range of applications shares these requirements.

A systems developer launches a project using MetaH by first establishing (possibly reusing) an evolvable architecture. High level, complexity-hiding abstractions decompose gradually down to processes, data flows, events and hardware descriptions. This first step by itself is often very valuable. It captures in a formal way what is often left sparsely and ambiguously documented.

The process level of the specification can be incrementally populated with functional code, some of which may come from other domain-specific code generators like SystemBuild, SimuLink and ACSL. The system is tuned by tool-assisted analysis, then compiled, loaded and executed on the hardware. The toolset collects timing data from the system, provides signal injection, and provides easy re-tuning based on the target hardware results.

MetaH provides a simple wrapper approach to interfacing code from a variety of sources including application code generators, legacy code, and new, manually developed code. Application engineers can use other tools specialized for parts of the application, then system engineers can wrap them for use with MetaH and rapidly integrate them onto the embedded hardware.

[1] This work has been supported by DARPA, ONR and AMCOM under contracts N00014-91-C-0195 and DAAH01-97-C-R233.

0-7803-5086-3 /98/$10.00 ©1998 IEEE

MetaH minimizes the impact of evolving hardware and software by explicitly representing key software/hardware interfaces in a unified way across platforms, and by automatically customizing a layer of the executive according to the application-specific architecture. Examples of architectural change accommodated by MetaH are process to processor binding, process and processor scaling, communication, processor type, bus type, fault recovery strategy, process rates, event communication, I/O ports, I/O rates, safety dependencies, process isolation and level of control, etc. MetaH greatly simplifies the effort to get on the embedded hardware and to evolve the hardware and software. MetaH reduces the V&V required when components change by limiting change impact, and significantly reducing the cost of updating glue code.

The MetaH Language

The essential concepts of MetaH are *components* and *connections*. Each component has a set of *attributes*, an *interface* and zero or more *implementations*. Connections link components together to form an *architecture*. MetaH specifications can also refer to a series of components called a *path*.

MetaH establishes a strict definition for what constitutes an interface. An interface is a set of *visible* elements associated with a particular component. The interface elements are usually "ports" for communicating data or events, but there are several possible classes of interface elements in MetaH.

An implementation specifies the behavior for an interface through the use of other components and connections between them. It is possible to declare more than one implementation for each of these kinds of components, as shown in Figure 1.

Of the various classes of components, *ports*, *types* and *events* are primitive and do not themselves have interfaces or implementations; all others can have both interfaces and implementations. Ports represent the supply of, or demand for data. Each port has an associated type (e.g. integer, array, record) and can be connected only to ports of the same type. Events are similar to ports but convey only predetermined signals.

MetaH contains language elements that support the description of both software and hardware components. It also has a means of hierarchically decomposing descriptions of both types. At bottom, the concrete software and hardware MetaH components correspond to actual objects found in real-time, embedded systems.

A concrete software component, or code module, is any separately compiled unit in a traditional programming language like Ada, e.g., a function or package. One or more software components are contained in a source file. Software component implementations can be produced by a wide variety of methods, including code generators and traditional manual development.

Concrete software components can be grouped together to form more abstract components such as processes, modes and macros. A process groups subprograms together into a thread of control. A process is also the basic unit of security and fault containment, memory protection and compute time enforcement. A mode groups processes together into an active set (only one mode can be active at a time). A macro similarly groups processes together but macros can execute in parallel.

The concrete hardware components in MetaH reflect the typical features of real-time embedded systems: processors, channels, memories, devices, etc. Hardware components can be grouped together into systems.

Hardware and software components are grouped together at the highest level of MetaH abstraction: the application.

0-7803-5086-3 /98/$10.00 ©1998 IEEE

All components and their interface elements have attributes that specify values used in analysis and code generation. Some attributes of a component are inherited by all of its implementation parts. For example, the Binding attribute on a mode has no meaning for the mode itself, but is instead inherited by all components, subcomponents, etc. within that mode. Some attributes require values that are time or space measurements.

When building a multi-processor embedded application, each process must ultimately be bound to a particular processor. Each data transfer connection must ultimately be routed through some sequence of channels and processors. Each subprogram, package or monitor (semaphore-protected data structure) that is shared between processes must ultimately be bound to a particular memory that can be accessed by the processors involved. The language allows an explicit specification of these bindings if desired. It is also possible to specify that an object must be bound to a particular class of hardware object or to a member of a designated set of objects, where the binder tool makes its selection from among the allowed objects. The binder will automatically select a binding within the specified constraints, where this binding attempts to achieve both feasible processor scheduling and feasible channel communication scheduling.

Analysis

Three forms of analysis have so far been defined for MetaH specification: schedulability, reliability and safety/security.

Schedulability

The schedulability analysis tool allows the system architect to determine whether an application can be feasibly scheduled (all processes can be dispatched at their specified periods or event arrival rates and complete by their specified deadlines). Schedulability analysis is currently performed each time a MetaH specification is processed by the MetaH tool. The tool also produces sensitivity analysis information that shows how individual process, monitor, package or subprogram execution times can be varied while maintaining feasibility (or how much execution times need to be reduced to achieve a feasible schedule).

Schedulability analysis is based on the compute path and source execution time attributes given in the MetaH specification. This means that the schedulability analysis results are only as good as those attribute values.

The timing analyzer allows the execution time of a process to be modeled using an execution path through components of a process. Restart times may optionally be included as an addition to the normal compute times. Absolute worst-case analysis requires that both maximum restart and maximum compute paths be used in the compute time model.

A separate schedulability analysis is performed for each mode of operation, and two different kinds of reports are written for each mode.

A *processor timing* report is produced for each processor that has active processes in the mode. The report lists the processes in order of descending criticality. Schedulability analysis results are produced for each process and for each component that appears along the compute path specified for that process.

For each mode in which communication between two different processors or a processor and a device occurs, a channel timing report will be also produced. This report lists every channel used in the system during that mode and, for each channel, gives a profile of the communication loading that occurs at the various application periods.

Figure 2 gives an example of a processor report produced by the schedulability analysis tool for a mode.

Processor timing, application exam1, mode M2, processor I80960MC

Module	Period	Budget	Critical	Maximum %	Margin %	Util
M2_kernel	25000	1000	6944	22400	95.5	4.0
_kernel		1000	6944	22400	95.5	4.0
P3	100000	4700	32639	92760	94.9	4.7
KERNEL						
AWAIT_DISPATC		1000	6944	11700	91.5	1.4
_self		1000	6944	89060	98.9	1.0
compute		2500	17361	90560	97.2	2.5
_kernel		200	1389	21600	99.1	0.2
P4	500000	14800	102776	455100	96.7	3.0
KERNEL						
AWAIT_DISPATC		1000	6944	11700	91.5	1.4
_self		1000	6944	221150	99.5	0.4
comp1		2500	17361	222650	98.9	1.0
comp2		2500	17361	222650	98.9	1.0
_kernel		400	2778	21800	98.2	0.2

Total utilization = 11.7%
Breakdown utilization = 81.0%
Critical scaling factor = 6.94
Processor schedule is feasible

Times are reported in microseconds
Nominal compute times were used in the compute paths.

Figure 2. Example Schedulability Analysis Report

Processor schedulability analysis uses the so-called exact characterization for schedulability. However, there are three cases where the current analysis may be more conservative (i.e., the analysis may state that a mode might not be feasibly scheduled under all possible circumstances when in fact it will be feasibly scheduled). The first case occurs when a process is given both a pre-period deadline and a high criticality. Such processes may be transformed in a way that results in an overly conservative schedulability model. The second case occurs when a process accesses two or more shared monitors. We currently model a process as being blocked by the sum of all such monitor compute times, but in most cases the maximum blocking time is the largest compute time among the monitors that it accesses. The third case concerns aperiodic processes which MetaH handles using a deferrable server approach. This is more conservative than, e.g., period-enforced and slack-scheduled approaches. Future versions of the toolset may generate and solve more accurate schedulability models in these three cases.

In addition to the processor schedulability analysis, the tool also produces, for a multi-processor system, an approximate channel schedulability analysis based on the total number of bytes moved over each channel at each particular periodicity in each mode. Synchronization overheads and delays are currently not included in the timing model (which makes the reported results optimistic), and possible parallelism during bi-directional communication over a channel is currently not included (which makes the reported results pessimistic).[2]

The report lists, for each channel, the processes that send data over the reported channel in that mode. With each process is a list of the ports of that process that send data over the channel. Processes are listed in order of smallest period (highest rate) to greatest period (lowest rate).

A total is given for all processes whose periods are less than or equal to that of the given period. The overall form of the report for a given channel is a list of processes from small to large period, with "total at period" analysis values that give the cumulative channel loading due to all processes up to a given period. Finally, the grand total is given for the cumulative channel loading due to all processes at all periods that communicate over that channel in the given mode.

Reliability

The reliability modeling and analysis tools[3] allow the system architect to determine the probability of failure of a fault-tolerant system that is subject to randomly arriving fault events. The exact set of faults, errors, and the response of various objects to faults can be specified by the system architect. The reliability analysis will derive two kinds of models from user-supplied error modeling input.

---

[2] Future versions of the MetaH timing analyzer will include, in addition to the summary values given here, estimates of total actual communication times that take into account parallelism and synchronization overheads and delays among the processors connected to a channel.

[3] Not all features of the MetaH ADL are supported in the current version. Missing are: support for mode changes in fault-tolerant systems, and capabilities to reduce the size of the (often large) state space size.

1. A stochastic, concurrent process model. There is an intuitive and well-defined mapping between objects in this model, objects in the MetaH specification, and sets of implementation code modules and hardware entities.
2. A Markov chain model. This model may be submitted to tools that solve Markov chain reliability models (this file can be directly input by the NASA SURE, PAVE and PAWS tools, for example).

The information needed to produce and solve a reliability model comes from two places in a MetaH specification: error models and reliability-specific attributes. Error models define kinds of faults, errors and error behaviors. Attributes of objects can be set to make specific choices for specific objects, including fault event rates, propagation rates, error paths, and error-masking protocols (e.g., voting).

The reliability modeling features of MetaH have been designed so that it is possible to perform analyses with different error model implementations without making any other changes to the specification. This makes it easy to evaluate system reliability for different error models, or to perform initial reliability analysis (which can be time-consuming) with a simple error model then switch to a more detailed error model to obtain a more precise analysis. Figure 3 shows an example of an error model.

The reliability modeler initially generates a stochastic concurrent process reliability model (where "concurrent process" means finite state machine). Each processor, memory, channel, device, process, monitor, package, subprogram and port connection is represented by a finite state machine in this model, where the particular state machine for each object is the error paths selected for that object. A Markov model is then generated as the reachable state space of this set of concurrent error state machines, up to death states.

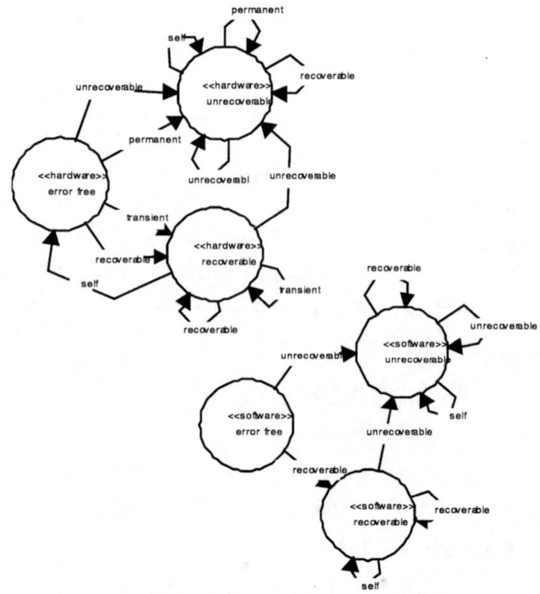

Figure 3. Example Error Propagation Model

Several tools exist to analyze a Markov chain model. They will typically allow you to specify the mission duration, solve for the probability of being in each particular application error state, and specify various parameters of the solution algorithm. The exact analysis method and available results depend on which tool is used.

Safety/Security

The safety/security modeling and analysis tool[4] enables the system architect to determine if a specification is consistent with a stated set of safety/security attributes. Safety/security analysis insures that specified patterns of connections, data sharing, software/hardware binding and scheduling do not allow objects to interact in ways that violate a set of safety or security classifications for the individual objects.

Objects in a specification can be assigned a *safety level* that ranges from `A0` (highest safety level) to `Z9` (lowest level of safety). A specification is unsafe if one object can affect in any way the proper operation of a

[4] In the current version of the MetaH toolset, only safety analysis is performed.

second object when the second object has a higher security level than the first. A successfully completed safety analysis will ensure that there are no mechanisms by which a defect in a lower-safety-level object could possibly affect the proper operation of a higher-safety-level object.

Safety levels can be assigned to hardware as well as software objects. A hardware object is considered to affect all software objects hosted on it for the purposes of safety/security analysis. The security level attribute[5] is a list of user-specified secure data classes, e.g. RED, BLACK. One possible policy to check is that no object can receive any information from an object for which it does not have an appropriate safety level.

Inheritance of safety/security attribute values is more restricted than normal inheritance of attribute values. Assignment of safety/security attributes for high-level mode, macro, system and application objects cannot be used to increase the safety level or security rights of any contained source object or hardware entity components. Safety/security attributes are considered inherent properties of an implementation object that can only be changed in the description of that object. For example, the safety level of a subprogram can only be raised in the MetaH description of the subprogram source object itself, which should be done only when the associated code modules are certified to a higher level.

The safety/security attributes for components of software objects and hardware entities must be consistent with the attributes of the containing objects. For example, a subprogram cannot be assigned a lower safety level than that of the process containing it.

Code Generation and Build

The procedure used to generate an executable image from a MetaH specification can be broken into two steps using two tools. The MetaH compiler transforms MetaH specifications into source code and compiler and linker directives. The application build tool (MakeH) uses information produced by the MetaH compiler to compile, link and, possibly, load images for the designated target hardware system (one for each processor). Figure 4 shows how the various code segments fit together to form a load image.

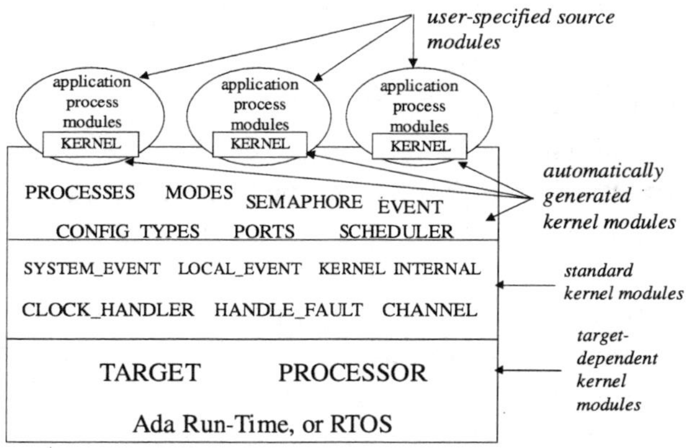

Figure 4. Layers of a MetaH-Integrated Application

Portions of the kernel on each target processor are automatically generated for each individual application. The generated kernel declarations, tables and subprograms control the resource allocations, scheduling, synchronization, modes, and event and data communication for an application. The process is fully automatic; it requires no user intervention.

The design schema for the generated executive is based on preemptive, fixed priority scheduling theory. The code generator derives priority, period transformation, and dispatch and time slice refill information from process attributes in the MetaH specification. Port connections and process timing attributes become scheduled data movement code. Code is also generated to vector events for the dispatching of aperiodic processes and for triggering mode changes. Mode changes themselves are managed by more generated code.

---

[5] Not currently implemented.

Once an application has been completely specified and the MetaH compiler has successfully completed its code generation, then MakeH can be run to build the executable images. MakeH does this fully automatically.

MakeH directs the compilation and linking of each process and the kernel into a virtual address space, and it combines these virtual address spaces to form the final physical executable image for each processor. The final executable image reflects the pattern of memory sharing or allocation needed for that application (e.g. shared monitor objects).

The MakeH tool directs the build process incrementally, using information about files that have been updated by the MetaH tool since the last build. The MetaH tool keeps such changes to the minimum. This makes it practical to use the tools in an iterative fashion even with large applications.

The MakeH tool uses a library of scripts and makefiles, catalogued according to the names of processor interfaces and implementations. This is because different processors use different language cross-development toolsets. New scripts and makefiles usually need to be added to this library when new kinds of processors are targeted.

Related Work

MetaH is similar to a number of other systems such as Durra [1], Maruti [2], ABE [3], LileAnna [4] and Onika/Chimera [5] in its use of an architectural specification language for composing source modules written in a traditional programming language. It is similar to Real-Time Euclid [6], Maruti, SARTOR and Onika/Chimera in its integration of real-time schedulability theory and analysis with automated code generation or composition. Modes and macros provide capabilities to specify hierarchical system-level state transitions that are comparable to aspects of Durra, Statemate [7] and Modechart [8]. MetaH represents a selection, integration and extension of a variety of basic technologies in an attempt to meet the requirements of a range of real-time, fault-tolerant, securely-partitioned, multi-processor applications.

Conclusion

In this paper we have described an approach to systems development that allows a variety of domain-specific tools and subsystems to be integrated into a complex system, and closely couples architectural specification, formal modeling, analysis and implementation. The initial MetaH toolset focuses on hard real-time, fault-tolerant, securely partitioned, multi-processor on-board avionics GN&C systems. However, we feel the general approach is broadly applicable, and we are pursuing research in areas that would extend the language and tools to more widely distributed systems hosting more complex real-time workloads. For example, we are currently upgrading our aperiodic scheduling and analysis to utilize emerging slack scheduling techniques. We have recently implemented support for undelayed message passing between processes.[6] Work is also underway to extend MetaH to better accommodate legacy software and to enhance portability and reusability through the use of open systems standards.

The MetaH approach has been reasonably successful for systems that primarily perform hard real-time periodic processing. For example, it has been used at AMCOM to develop a tightly coupled, multi-processor implementation of real-time missile flight code, using the original i80960 Tartan implementation of MetaH. In order to support the execution of the flight code in a simulated environment, a six degree-of-freedom simulation of the missile physics was also

[6] When a message goes from process A to process B, MetaH normally delivers this message value at the next dispatch of B that followed the completion of A. An undelayed message requires A to always be scheduled before B.

developed. The 6-DOF simulation was designed as five MetaH processes, and the flight code was designed as ten MetaH processes. AMCOM estimated 2X first-use savings over traditional methods and tools. Overall application size and performance were comparable to those of the original, handcrafted system. This approach is now being applied during the modeling of the avionics for an advanced helicopter.

MetaH is serving as the foundation for a new standard Avionics Architecture Description Language under the SAE AS5 committee, with sponsorship from Open Systems Joint Task Force (OSJTF). Contact Bruce Lewis (lewis@sed.redstone.army.mil) for more information or if you are interested in participating.

References

[1] Mario Barbacci, Charles Weinstock, Dennis Doubleday, Michael Gardner and Randall Lichota, "Durra: A Structure Description Language for Developing Distributed Applications," *Software Engineering Journal,* March 1993.

[2] M. Saksena, James da Silva and Ashok Agrawala, "Design and Implementation of Maruti-II," in *Principles of Real-Time Systems*, Sang Son Ed., Prentice Hall, 1994.

[3] Frederick Hayes-Roth, James E. Davidson, Lee D. Erman and Jay S. Lark, "Frameworks for Developing Intelligent Systems," *IEEE Expert*, June 1991.

[4] Will Tracz, "LILEANNA: A Parameterized Programming Language," *Proceedings of the Second International Workshop on Software Reuse*, March 1993.

[5] Matthew W. Gertz, David B. Stewart and Pradeep K. Khosla, "A Human-Machine Interface for Distributed Virtual Laboratories," *IEEE Robotics and Automation Magazine*, December 1994.

[6] Alexander Stoyenko, "A Schedulability Analyzer for Real-Time Euclid," *Proceedings of the Real-Time Systems Symposium*, December 1987.

[7] David Harel, Hagi Lachover, Amnon Naamad, Amir Pnueli, Michal Politi, Rivi Sherman, Aharon Shtull-Trauring and Mark Trakhtenbrot, "STATEMATE: A Working Environment for the Development of Complex Reactive Systems," *IEEE Transactions on Software Engineering*, April 1990.

[8] Farnam Jahanian and Alysius K. Mok, "Modechart: A Specification Language for Real-Time Systems," *IEEE Transactions on Software Engineering*, v20 n12, December 1994.

SELECTING A PROGRAMMING LANGUAGE FOR YOUR PROJECT

David Naiditch, Raytheon Systems Company, El Segundo, California

Introduction

In 1997, the mandate was dropped that the Ada language had to be used in all DOD projects. Since then, software managers have had more freedom to choose which programming language to use on their DOD projects. With this new freedom has come new responsibilities. Too often, discussions about what language to select break down into emotional appeals for one's favorite language. Instead, a fair and rational process should be established that bases the choice of a programming language on issues related to customer needs and business goals. In particular, for large projects, a language should be selected according to objective evaluation criteria, a system of priorities, and an overall scoring algorithm. This selection process should be done by those having a working knowledge of the languages under consideration. The winning language that emerges from such a processes might not be the language initially considered as the obvious best choice. For projects that are too small to afford the time and effort needed to evaluate different languages, the choice of language might be made at the product line level or higher.

For most new DOD software projects that need a general purpose high-level language, Ada, C, or C++ are the main contenders, with increasing attention paid to the new language celebrity, Java. (In this paper, "Ada" refers to both Ada 83 and Ada 95.) Furthermore, multilanguage systems seems to be getting increasingly popular. As a result, companies such as Rational, Aonix, and Green Hills offer integrated multilanguage programming environments. (In response to this trend, Ada 95 has added several new pragmas [compiler directives] and special packages [Interface and its child packages] that enable Ada 95 code to easily interface with other languages.)

As Table 1 shows, there are many important differences between Ada 95, Ada 83, C++, C, and Java. The comments in the right column of the table provide some rationale for the rankings. A full discussion of each category in Table 1 is beyond the scope of this brief paper, but some issues are discussed subsequently when I cover common misconceptions.

Although each programming language has its own strengths and weaknesses, the reason for choosing a particular language may ultimately be based on factors having nothing to do with the technical merits of the language itself. (This is yet another reason that the Ada mandate was dropped.) Such factors may include the availability of compilers for the host/target, the maturity and efficiency of available compilers, the availability of programmers who already know the language, whether the language can easily interface with existing systems such as graphical user interfaces or data bases, the existence of legacy software written in a specific language, or how well the language fits in with adopted commercial-off-the-shelf (COTS) products.

A few common misconceptions that may affect the choice of a programming language are listed below.

0-7803-5086-3 /98/$10.00 ©1998 IEEE

Misconception 1: The government dropped the Ada mandate because Ada is a dying language.

When, in 1997, the Honorable Emmett Paige, Jr., Assistant Secretary of Defense for Command, Control, Communications and Intelligence, dropped the Ada mandate, he made it very clear that this action should not be seen as the government's lack of commitment to Ada. He felt that the mandate became a "contentious point of resentment," and that Ada was fully capable of standing on its engineering merits.

In fact, Ada is very unlikely to go away in the foreseeable future. The DOD has a tremendous investment in Ada. Over 50 million lines of Ada code have been developed for warfighting systems. Many of these systems will be used well into the next millennium. Furthermore, Ada will continue to be chosen for its reliability and lower development/maintenance cost, both of which are so important for large, long-lived software systems (see misconception 11).

Misconception 2: C is the most efficient language since it is closer to assembly code.

C is not necessarily the most efficient language. When runtime checks are turned off, Ada code may run about as fast as C. (Standard C does not perform runtime checks.) Java, however, usually runs considerably slower because it makes extensive use of dynamic (heap allocated) data structures and has automatic garbage collection, and because the generated bytecode is typically interpreted. (The last problem can be mitigated if a "just-in-time" [JIT] compiler, or better yet, a native Java compiler, is available.) Of course, efficiency can significantly vary from one implementation of a language to another. Furthermore, different compilers are optimized for different language features.

Misconception 3: Very terse source code results in very efficient object code.

The only guarantee is that very terse code results in unreadable and therefore unmaintainable code. Do not assume, for example, that this C code

```
if ((A = B) == C) C += 2;
```

executes any faster than

```
A = B;
if (A == C) C = C + 2;
```

Interestingly, Java's syntax is a somewhat cleaned-up version of C++'s syntax. For instance, Java eliminates the possibility of confusing an assignment with a test for equality in a conditional. Thus, Java rejects this code

```
if (A = B)
```

However, Java embraces C's error-prone syntax for the switch (case) statement, where a "break" can be accidentally omitted.

Misconception 4: Ease of coding is a major advantage.

Ease of coding is not a major advantage compared with ease of reading. For complex code that must be maintained over many years, maintenance costs typically run 70% to 90% of the software life cycle costs. Because one must be able to read code in order to maintain it, the extra effort taken in developing readable code amply pays off. And don't forget that the annoyance of excess key strokes is no excuse for writing unreadable code, not even for two-finger typists.

Misconception 5: Software should be easy to compile.

Software should *not* necessarily be easy to compile. As many bugs as possible should be caught at compile time, since the sooner a bug is caught, the easier and cheaper it is to fix.

Ada, for example, catches many errors at compile time (for example, calling subprograms with the wrong type or number of parameters) that other languages catch at runtime, or not at all. As a result, Ada code is harder to compile than, for example, C code, but once compiled, many potential errors have been eliminated.

Misconception 6: Java is a great language because the same code can run on Windows 95, Solaris, UNIX, Macintosh, or just about any other platform.

Java's platform independence is wonderful (although a bit overstated). However, this independence derives not so much from the Java language itself, but from the ability of the Java compiler to generate Java bytecode that can be interpreted on any machine with a Java-enabled web browser such as Netscape or Internet Explorer. Note that Java bytecode does not have to be generated by a Java compiler. For instance, Intermetrics and Aonix offer Ada 95 compilers that generate Java bytecode.

Although Java's portability does not have that much to do with the language itself, Java does address portability through predefined numeric types with specified sizes and formats. Thus, type "int" is always a 32-bit signed integer type, whereas in C and C++, int can be of any size. If any of Java's numeric types are not supported by the target architecture, then that type must be implemented in software, which could significantly degrade performance (see misperception 2). In Ada, a programmer can select a numeric type by range and/or precision rather than only by name, which guarantees that the numeric type has the range and precision that is required.

Misconception 7: Using operating system (OS) services for concurrent programming is just as good as (if not better than) using special language constructs.

Unlike Ada and Java, C and C++ do not provide constructs for writing concurrent code (code that can do more than one thing at once). To write concurrent software, C and C++ programmers often rely on OS services such as semaphores that are nonportable, low level, error prone, and difficult to debug. For example, with semaphores, there is a risk of locking (seizing) an item and then forgetting to unlock (release) it when one is finished with it so that other tasks can access the item. In contrast, Ada 95 allows protected operations to be defined that automatically provide mutually exclusive access to a protected item. This feature eliminates the risk of having an unbalanced lock/unlock. (For hard real-time programs that cannot even afford the overhead of protected types, Ada 95 offers portable semaphore services.)

Misconception 8: Procedural-based programming has been replaced by object-oriented programming.

Although the programming world has largely gone object oriented, applications do exist (for example, embedded real-time or scientific applications) where procedural-based programming seems more natural. In Ada 95, the language features that support object-oriented programming are distinct and can be used separately. Therefore, Ada 95 programmers are free to select those object-oriented features they want to use without being forced into fully embracing the object-oriented methodology. In contrast, Java only supports object-oriented programming. Java cannot be used for procedural-based programming. (In Java, even exceptions and threads are defined in object-oriented terms.) C++ is like Ada 95 in that it does not force programmers to embrace object-oriented programming. However, object-oriented features of C++ are not as distinct as those of Ada 95. For instance, data encapsulation, which is needed for creating abstract data types (called private types in Ada), cannot be performed without introducing a class, which is a central feature of object-oriented programming.

Misconception 9: Java is a small and simple language that is easy to learn.

Even though Java is a small and simple language, it has some subtle features and includes a huge library of over 150 classes and interfaces. It takes considerable effort to determine which classes to use and how best to use them. Programmers accustomed to large languages such as Ada and C++ often find that Java is too small. For example, Java is missing local constants, method parameter modes, enumeration types, records/structures, operator overloading, variant records/unions, explicit pointers, fixed-point types, unsigned integer types, range constraints, the goto statement, and generics/templates. Some of these features were undoubtedly omitted in the interest of simplicity, portability, or security, but not all of these omissions are easy to justify.

In general, for small languages to be effectively used, programmers often need to write code that simulates a missing language feature, or to rely on tools or practices that are external to the language itself. For instance, Ada defines a basic library manager that prevents an executable module from ever being built from out-of-date components. Other languages obtain this capability through the use of a Make utility and the careful use of naming conventions. However, as a result, portability is lost and complexity increased, since the proper use of such languages depends on nonstandardized utilities and practices that exist outside the language. Thus language complexity should not be considered in isolation, but as part of the software engineering environment.

Misconception 10: Computer languages should give programmers the freedom to do what they want to do.

This libertarian view is also held by those who resent the government forcing them to wear a motorcycle helmet. Without a doubt, Ada and, to some extent, Java, are "paternalistic" languages that attempt to protect programmers from themselves and other programmers. In a world where all drivers and programmers are perfectly rational and sober, such paternalism may seem unnecessary. However, on the freeways populated by maniacs, and in the real software development world where mind-bogglingly complex programs are being maintained over many years by gaggles of programmers, such optimism is unwise.

Consider, for example, the use of pointers. In C and C++, pointers can point willy-nilly to any address. Not surprisingly, the misuse of pointers is responsible for the majority of C and C++ programming errors. Often pointers end up pointing to the wrong data or to items that no longer exist. Java takes a Draconian approach to this problem of pointer abuse by not allowing programmers to explicitly declare pointers! In Ada 95, pointers are flexible, but also quite safe. Pointers can point to heap allocated objects (as in Ada 83), to stack allocated objects, and to subprograms. Checks are automatically made (mostly at compile time) to guarantee that (in the absence of "unchecked programming") a pointer does not point to an object of the wrong type, or point to a stack allocated object or subprogram that no longer exists.

Misconception 11: In terms of life cycle costs, it does not matter what programming language is chosen.

Various studies have shown that the choice of programming language does impact live cycle costs. Such claims, however, are difficult to substantiate. Part of the difficulty is that large projects are never done in parallel with different languages, but with all other factors held constant. Perhaps the study that best approaches this ideal was done by Dr. Stephen Zeigler of Verdix (now Rational Software Corporation). Dr. Zeigler based his study on records kept at Verdix documenting all software changes. Although these records were not intended to be used to compare the development costs of Ada and C, they turned out to be very useful because few factors varied besides the language being used: The same 60 programmers were developing code in both Ada and C, the same work environment was used, the same debugging tools, same editors, same testing tools, and the same design methodology. Most of these programmers had master degrees in computer science, and the more experienced programmers tended to work more in C. When first hired, 75% of the programmers knew C, while only 25% knew Ada. Despite the bias in C's favor, Dr. Zeigler showed that the cost of coding in Ada is about half the cost of coding in C. This difference in cost resulted because code written in Ada contained 70% less bugs discovered before product delivery and 90% less bugs discovered after product delivery. (As discussed under misconception 5, the later a bug is caught, the more expensive it is to fix.) Dr. Zeigler expects the cost of C to be even more pronounced in other organizations, because Verdix had the advantage of understanding common C problems and enforcing thorough software testing procedures. Interestingly, C++ bugs were running even higher than C bugs. Dr. Zeigler's paper, "Comparing Development Costs of C and Ada," can be obtained at http://sw-eng.falls-church.va.us/AdaIC/docs/reports.

Table 1. Language Comparisons

Feature	Ada95	Ada83	C++	C	Java	Comments
Built-in support for concurrency (multithreading)	Excellent	Very Good	None	None	Good	For C or C++, concurrent programming requires calls to nonportable and low-level operating system (OS) services.
Full support of object-oriented programming	Yes	No	Yes	No	Yes	Ada 83 has most, but not all object-oriented features.
Full support for procedural-based programming	Yes	Yes	Yes	Yes	No	To use Java, object-oriented programming must be fully embraced.
Support of modern software engineering principles	Excellent	Excellent	Average	Poor	Good	Principles include strong typing, information hiding, localization, modularity. Ada is unique in strong type checking of integer types and array index types.
Completeness of language features	Excellent	Good	Very Good	Poor	Poor	Java lacks many common language features, such as enumeration types and records/structures. Ada 83 lacks pointers to subprograms and stack-allocated objects and certain OOP features. C++ lacks concurrency features.
Portability	Very Good	Very Good	Average	Good	Excellent	Java's portability derives mostly from the Java virtual machine, not from the language itself. C++ implementations still do not fully conform to the recently introduced ISO standard.
Life cycle costs	Excellent	Excellent	Poor	Poor	Good	Historic data show that, line per line, Ada code is about half the cost of C or C++ code (see reference to Zeigler's article).
Safety and reliability	Excellent	Excellent	Poor	Poor	Very Good	Ada catches many bugs at compile time and then others at runtime. Ada is the most strongly typed language and guards against dangling pointers.
Ease of reuse	Excellent	Excellent	Excellent	Poor	Good	Neither Java nor C have generics/templates. C's macro expander is a poor substitute.
Support for inter-language communication	Very Good	Poor	None	None	None	Ada 95 provides several pragmas (compiler directives) and packages for this purpose.
Language support	Good	Fair	Good	Very Good	Excellent	Many Ada vendors have transitioned to Ada 95 and consider Ada 83 to be obsolete. Ada host and target compilers are readily available for mainstream microprocessors, but may be unavailable for special microprocessors.

A DISTRIBUTED COMPUTING ENVIRONMENT FOR SIGNAL PROCESSING AND SYSTEMS HEALTH MONITORING

Lawrence C. Freudinger,[*] NASA Dryden Flight Research Center, Edwards, California
Matthew J. Miller, Creare, Inc., Hanover, New Hampshire
Karl Kiefer, Invocon, Inc., Conroe, Texas

Abstract

The evolution of systems health monitoring and signal processing is toward on-line automated techniques that improve performance, cost effectiveness, and safety associated with operating modern machines. The evolution of computing is toward distributed networks of dissimilar applications communicating through standard interfaces and tools. This papers gives an overview of a scaleable collaborative computing environment that routes real-time and non-real-time data via transaction control protocol/Internet protocol (TCP/IP). A primary enabling technology is an object-oriented, Java™-based, software data server and data management tool. A network composed of one or more of these data servers provides storage, data stream parsing and merging, and compatibility among similar and dissimilar applications running anywhere on the network. Connectivity to mobile platforms is accomplished through a wireless spread spectrum approach. Applications to flight research and aviation safety are discussed.

Introduction

The United States decided in 1996 that an order-of-magnitude reduction in aircraft accident rates over a 20-yr period is required to keep the global aviation system efficient, safe, and affordable. This challenge is difficult, given that most of the aircraft expected to be flying in 2016 already exist. To exacerbate the anticipated aging fleet and fleet retrofit problems, the volume of air traffic may triple over the same period.[1]

The role of NASA in the aviation safety challenge is to close technology gaps and lower the risk of adopting promising technologies. Information technology (IT), system health monitoring, modeling, and simulation tools are important components that must be invented, matured, or applied. The envisioned solution includes integration of all possible data sources in a massively decentralized computing environment—a decentralized decision-making environment—to identify and address risk-increasing events before the events cascade into accidents.

A critical component of future aviation safety is the data-sharing environment on which the IT components will be built. This data-sharing infrastructure is the common denominator that will make large-scale system integration affordable and manageable over time. The infrastructure will include distributed databases and distributed components (hardware, software, and humans). Furthermore, the flow of information necessarily includes real-time data. The time interval inferred by "real-time" can vary from

---

[*] l.freudinger@dfrc.nasa.gov. *This paper is a declared work of the U. S. Government and is not subject to copyright protection in the United States. Use of trade names or names of manufacturers in this document does not constitute an official endorsement of such products or manufacturers, either expressed or implied, by the National Aeronautics and Space Administration.*

microseconds to minutes, depending on the application. Weather data, flightpath information, and vehicle health monitoring alarms are examples of real-time data types relevant to aviation safety. Lower-level infrastructure that will support aviation safety can be thought of as a general framework applicable to many automation or collaborative computing tasks.

This paper discusses two ongoing projects at NASA Dryden Flight Research Center (Edwards, California) to develop enabling technologies for an affordable, real-time, distributed decision-making environment. The first project focuses on a software solution for managing real-time data streams over transaction control protocol/Internet protocol (TCP/IP) networks. The second project focuses on connecting aircraft and other mobile platforms to standard networks. Applications of interest include aviation safety and aircraft flight test, but solutions are designed with general engineering problems in mind. Examples of specific accomplishments are provided.

Data Management Software

A commercial software tool for managing streams of data in a distributed collaborative environment is currently under development.* Originally intended to support distributed signal processing during aircraft flight test operations, the implementation is very general. Called the Ring Buffered Network Bus (RBNB)™ (Creare, Inc., Hanover, New Hampshire), this section describes its essential features.

Description

The RBNB™ is a software data server that functions as a virtual network for arbitrary channels of information. Incoming data are stored in random access memory and optionally on disk for seamless transition between archived and most-current-value information. Multiple RBNB™ servers can communicate with each other to scale performance locally or across wide area networks. Built on top of TCP/IP and written in Java™ (Sun Microsystems Inc., Palo Alto, California) language, the RBNB™ software serves as an intermediary between data sources and destinations. The RBNB™ can be thought of as a cross-platform shared-memory server with features similar to data acquisition systems. Applications use the RBNB™ network as the destination or source of real-time and recorded information.[2] In short, the RBNB™ is a tool to move any data into any display or algorithm anywhere on a network.

Figure 1 shows a schematic of the

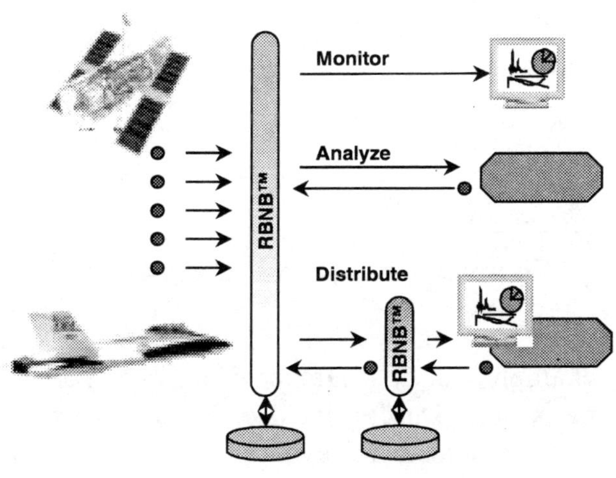

Figure 1. The RBNB™ Concept

RBNB™ concept. In this figure, the larger RBNB™ suggests a local primary server providing data to displays, to and from on-line algorithms, and to remote sites. The smaller RBNB™ represents a secondary site serving users local to that site.

* Funded by NASA Dryden, Contract No. NAS4-98008 (patent pending).

Architecture

The core components of the RBNB™ architecture are the software objects that manage input and output data streams. On the input side, a ring buffer object (RBO) is a data server and archive manager for a particular local or remote source. The "ring" part of an RBO refers to a circular buffer that overwrites the oldest data, if necessary. The RBO manages frames of multiple data channels, each channel composed of one or more data elements (fig. 2). One RBO exists for each data source (for example, each incoming arrow in figure 1). Each frame is time-stamped for synchronization with other frames. Successive frames within an RBO typically have the same

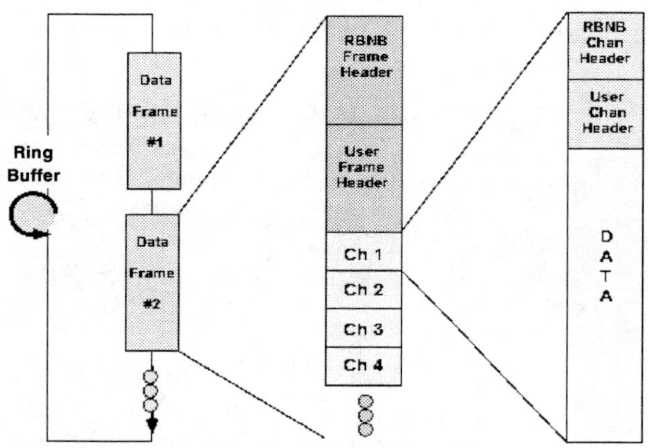

Figure 2. RBO Data Frame Structure

channel map, but dynamic channel maps (for example, varying number of channels) are also permitted. Descriptive user header information can be added to evolve structure and add meaning to the data. User header "templates" organize information for data types, applications, or user groups. The adaptive RBO data structure is a key feature of the RBNB™ (fig. 2).

On the output side, a network bus object (NBO) handles requests for data and provides stream parsing and channel merging from multiple RBO sources. The NBOs can access local or remote RBOs. Access to remote RBOs uses routing or mirroring capabilities between multiple servers. An RBNB™ control object (RCO) creates and manages input and output connections.

Application developers are provided with interface libraries, written in "C" and Java™ languages. The Java™ library, like the RBNB™ itself, is platform independent. These libraries permit applications to manage multiple connections to multiple servers and, in general, permit applications to function as data sinks, data sources, or both. Development of a format conversion template to facilitate communications with common object request broker architecture (CORBA)* objects is in progress. This template is being implemented using the Java Beans™ (Sun Microsystems, Inc.) component model and offers an alternative approach for using CORBA and ActiveX™ (Microsoft Corporation, Redmond, Washington)† in aviation applications.[3] The RBNB™ is a degenerate object request broker (ORB) specialized for connection-oriented data management.‡

Figure 3 shows a network topology for distributed computing and collaborative applications. This topology is a hybrid client-server concept, exploiting the use of RBNBs™ as both clients and servers in order to maximize flexibility in managing processing requirements, connections, and bandwidth.

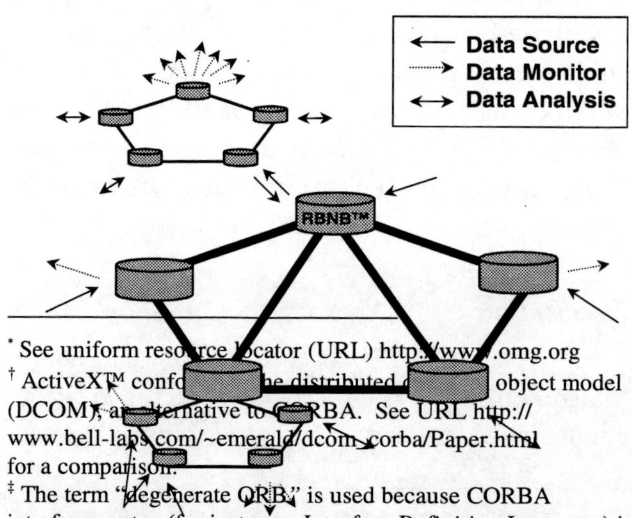

Figure 3. Virtual Network Topology

* See uniform resource locator (URL) http://www.omg.org

† ActiveX™ conforms to the distributed component object model (DCOM), an alternative to CORBA. See URL http://www.bell-labs.com/~emerald/dcom_corba/Paper.html for a comparison.

‡ The term "degenerate ORB" is used because CORBA interface syntax (for instance, *Interface Definition Language*) is not required and CORBA services are not provided.

0-7803-5086-3 /98/$10.00 ©1998 IEEE

Performance

Recent performance measurements indicate aggregate throughput on a Sun Microsystems, Inc. Ultra5™ workstation to peak at approximately 12 MB/sec in streaming mode. Over 10BaseT networks, performance is restricted to approximately 1 MB/sec (network bottleneck). Meaningful benchmark measurements for network servers like RBNB™ need to be developed and are being addressed. The RBNB™ code involves substantial process threading, so the use of multiprocessor computers improves performance. In addition, through inter-RBNB™ routing and mirroring features, multiple RBNBs™ across several computers can be configured as one logical server. This feature is useful for performance scaling.

Security Features

An RBNB™ owner can restrict access to information, data, and commands by configuring authorizations on a client-by-client basis. Internal UNIX-like user name and password or external authentication servers are supported. For example, a particular user may only have access to a single channel of data from a particular data source. Alternatively, administrators may have full access to commands for server maintenance, yet have no way of seeing or rerouting secured data.

The RBNB™ does not perform any encryption services; however, the user can manage encrypted data streams using the RBNB™. Several firewall configuration scenarios and security augmentation tools are being considered.

Development Status

Commercial introduction of the RBNB™ product is scheduled for 1998. Nine task groups (table 1) encompass ongoing efforts to mature the technology.

These efforts include agreements with multiple third parties. In some cases, the focus is on integration with existing commercial products or more special-purpose tools. In other cases the focus is on future hardware and software products. Detailed discussion of each task is beyond the scope of this paper, but four example applications provide a sample of accomplishments to date.

Table 1. RBNB™ Development Task Areas

Task No.	Task Description
1	Flight Research Applications
2	Third-Party Software Interfaces
3	Third-Party Hardware Interfaces
4	CORBA Compatibility
5	Real-time Performance Optimization
6	Audio-Visual Data Server Applications
7	Databases; On-line Analytical Processing
8	Aviation Safety Applications
9	Technology Assessment

Example 1: Distributed FFT Analyzer

A simple distributed computing application helps to illustrate the use of a single RBNB™ server for connecting algorithms (task 2) across a wide area network. Figure 4 shows a Java™ applet screen image showing a time history and the fast Fourier transform (FFT) of that time history. The time history shown is the most current data for which an FFT was available. The time history is acquired on the east coast of the United States (New Hampshire), and the FFT is calculated on the west coast (Southern California). The applet runs anywhere on the network, talking in this case to an RBNB™ server on the east coast.

0-7803-5086-3 /98/$10.00 ©1998 IEEE

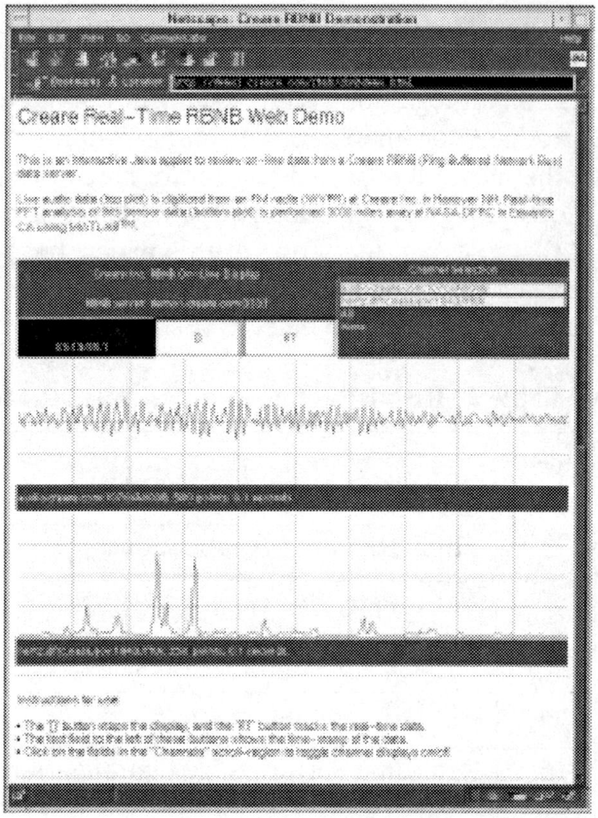

Figure 4. Distributed FFT Analyzer

Note the use of heterogeneous components: LabVIEW™ on a Macintosh™ (Apple Computer, Cupertino, California) acquiring the time history, Matlab™ on HP-UX™ (Hewlett Packard, Inc., Palo Alto California) calculating the FFT, RBNB™ server running on Windows NT™ (Microsoft Inc.), and Java™ applets running from Navigator™ (Netscape, Inc., Mountain View,

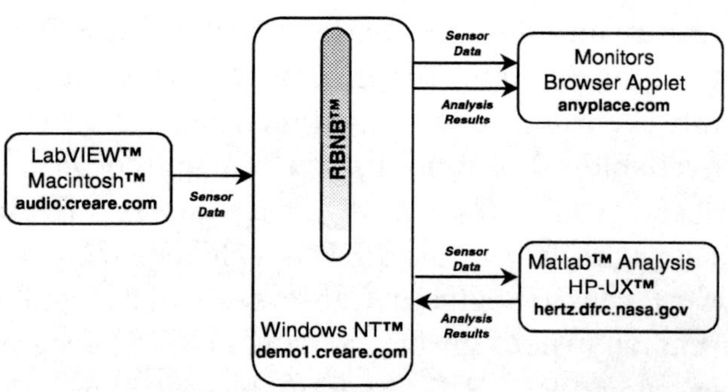

Figure 5. Distributed FFT Analyzer Schematic

California) or Explorer™ (Microsoft, Inc.). Figure 5 shows a schematic of these four processes.

Example 2: On-Line Flight Data Processing

The RBNB™ is being used to reduce costs and improve productivity for an ongoing flight research program at NASA Dryden (task 1). The Adaptive Performance Optimization (APO) program uses an L-1011 aircraft.[4] The research goal is to perform drag minimization by adjusting control surface positions. Aircraft modifications to support the APO program include additional sensors and installation of an onboard research engineering test station (RETS). All aspects of the APO program are conducted from the RETS.

The empirical determination of optimal control surface positions is a nonlinear optimization problem. Initial flight tests included data recording, but data analysis was performed postflight using Matlab™. Deferment of analysis until after the flight is costly when the analysis indicates poor maneuver conditions or otherwise suggests a more productive use of flight time. Implementation of the RBNB™ in the RETS allowed use of Matlab™ within the RETS system and the optimization is now being performed on-line, eliminating a need for posttest data analysis in most cases.

Figure 6 shows a screen image of a Java™ application used for managing the connections to an RBNB™. Figure 7 shows a schematic of the corresponding data flow. The particular configuration shown is an early simulation of the APO flight implementation. Icons distinguish the various objects: circles represent input streams (RBOs); triangles represent output streams (NBOs); and a rectangle designates a single control object. At the lowest level within an RBO are individual channels of information.

0-7803-5086-3 /98/$10.00 ©1998 IEEE

Figure 6 shows three input streams: two from the simulated RETS, and one from the machine used for Matlab™ analysis. The default naming scheme of *<hostname>:<port>* is shown, but the generating source is free to use a more meaningful naming scheme.

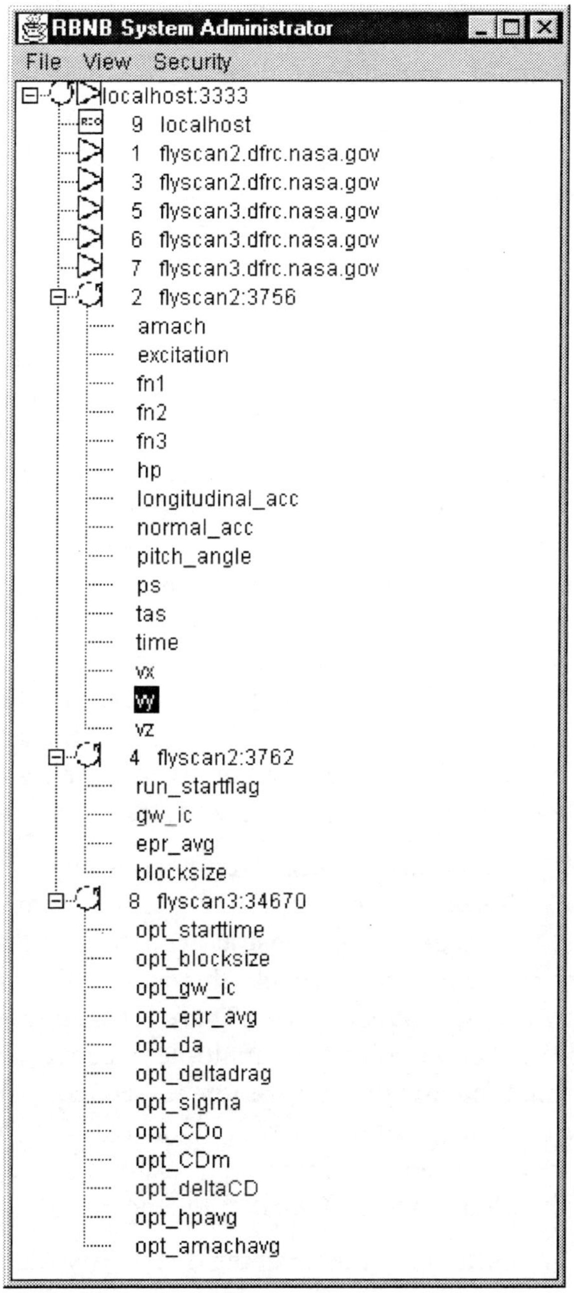

Figure 6. RBNB™ Connection Map, L-1011/APO

The first RBO, flyscan2:3756, contains 15 channels of 50 sample/sec sensor data. Although not indicated in the figure, these RBO frames actually contain 0.25 sec of data with 4 updates/sec. The second RBO, flyscan2:3762, also comes from the RETS. This stream, however, is updated occasionally during a flight maneuver and contains analysis initialization, analysis control, and flight card information.

The on-line Matlab™ algorithm uses a specific parameter (run_startflag) as a run trigger for the analysis sequence and uses the remaining parameters to control algorithm flow (fig. 7). The Matlab™ optimization code runs in a loop with successively larger blocks of incoming data (time since trigger) until a reset command is received. Each iteration includes an update of the third RBO, which is a results database. Meanwhile, the test engineer observes the analysis and makes a real-time decision on the quality of the run. The essential parameter (opt_da) from each acceptable run is passed to the control surface actuator.

Example 3: Flight Test Data Server

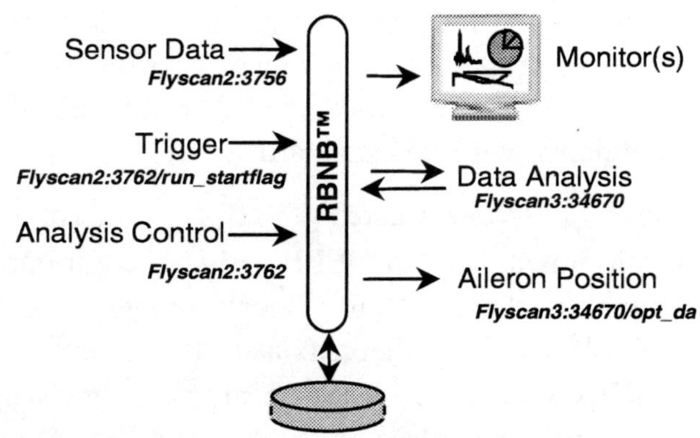

Figure 7. L-1011/APO RBNB™ Data Flow

The historical driving force that led to the RBNB™ concept was the desire to improve and automate hazardous flight flutter testing of aircraft. Interfaces to telemetry systems are required for processing flutter test data on-line. For this reason, the RBNB™ is being integrated into both a commercial telemetry front-end

0-7803-5086-3 /98/$10.00 ©1998 IEEE

processor and the currently unique flight test infrastructure at NASA Dryden (tasks 1–3, 8). These prototype efforts have been demonstrated and are being evaluated and refined.[*]

In addition, work has begun on merging audio and video streams (task 6) with telemetry data so that extending these other important data types to remote sites will be possible. This work includes high-performance wavelet-based compression.[†]

Example 4: Smart Sensor Network

Construction of very large sensor networks will require convenient ways to integrate sensor subarrays. This requirement is the premise of IEEE P1451, a proposed standard that focuses on unifying the fragmented sensor and actuator markets by evolving plug-and-play capability for future networks of "smart" transducers.[5] The standard involves a common object model for transducers and establishes a common interface that facilitates the creation of networked transducers (task 3). This proposed standard is relevant to larger integration efforts within the transportation industries. Other techniques also exist to migrate existing sensor configurations to a TCP/IP data link that do not include compliance with this standard.[6]

The RBNB™-related effort involves development of an IEEE P1451–compliant sensor package (four accelerometers and network interface) that acts as a data source for an RBNB™. In this case, a proxy utility will configure connections to the network interface, making integration simple and straightforward. Data fusion of IEEE P1451–compliant sensors with noncompliant sensor types is possible using the RBNB™ proxy server (for sensors already interfaced to TCP/IP) or by using the API to implement an RBNB™-aware network interface.

Mobile Platform Connectivity

The RBNB™ establishes a powerful collaborative environment across TCP/IP networks. If TCP/IP networks were extended to aircraft and other mobile platforms, then maximum potential of distributed computing technology (not just RBNB™-based) could be achievable in an affordable manner. Before distributed computing can be extended to mobile units, the underlying network itself must first be extended with wireless technology.

An approach for establishing wireless networks amongst several aircraft and between an aircraft and the ground involves "spread spectrum" radio communications. This concept is being explored via a project dubbed AIRNET.[‡] The goal of AIRNET is to demonstrate affordable vehicle health monitoring and conditioned-based maintenance through data links that are at least as fast as mainstream terrestrial modem technology. The in-flight demonstration will include an Internet web site where live general aviation aircraft data can be viewed and analyzed (task 8).

For this project, compatibility with IEEE 802.11 is not a requirement[§] because AIRNET seeks to demonstrate capabilities that are not supported under the new wireless networking standard. These capabilities include code division multiple access for optimal bandwidth utilization, discovery and use of nearby aircraft as routers or relays to the nearest ground station or satellite, and application in the 800-MHz frequency band.

Although the ongoing feasibility study includes all applicable frequency bands, the

---

[*] These efforts are supported under SEWP contract No. 11600 and NASA Contract No. E-02965D.

[†] NASA Contract No. E-02982D.

[‡] Aircraft Integrity Radio Network Evaluation Technology funded by NASA Dryden, SBIR Contract No. NAS4-98020.

[§] See URL http://www.wlana.com for a discussion of IEEE 802.11.

0-7803-5086-3 /98/$10.00 ©1998 IEEE

800-MHz region is of interest because the Federal Communications Commission recently permitted experimental use (frequency reuse) of the 800-MHz band specifically for airborne cellular communications. Air-to-ground data links could be demonstrated using established 800-MHz infrastructure with minimal hardware upgrades,* thus eliminating what could otherwise be prohibitive integration costs for demonstration of AIRNET technologies.

Concluding Remarks

Internet protocol is a common denominator from which large-scale fleet management solutions for the transportation industries can be built. The transaction control protocol/Internet protocol (TCP/IP) layer separates the physical network from the logical network, allowing evolution of network hardware and application software to progress independent of each other. The desired evolution toward software connectivity that is manageable over time for large enterprises such as global transportation fleets will benefit from additional independence between a data sharing infrastructure and the algorithms or components which process that data.

The Ring Buffered Network Bus (RBNB)™ is a Java™- and TCP/IP–based real-time data sharing, archival, playback, and information management architecture that offers software developers an easy way to move information among any set of processes running on one or more computers over a network. The RBNB™ is useful for applications and data sources that may or may not yet be compatible with each other. The RBNB™ provides flexible data handling for large-scale integration of heterogeneous applications. Using this approach, powerful health management applications for machines and electronics can be quickly implemented, and the ability to make those applications components of a larger collaborative effort will be straightforward.

Finally, the introduction of TCP/IP as a communication protocol amongst ground, air, and space vehicles is viewed as a next step toward integrating the transportation industries into the growing world of high-performance object-oriented distributed computing. The AIRNET concept contributes to the networking solution by applying spread spectrum techniques in a manner consistent with the needs of future aviation safety environments.

References

[1]Huettner, Charles. H., *Toward a Safer 21st Century: Aviation Safety Research Baseline and Future Challenges*, NASA NP-1997-12-231-HQ, 1996.

[2]Miller, Matthew J., Lawrence C. Freudinger, Ian A. Brown, and William R. Baschnagel, "On-Line Test and Measurement in a Distributed Collaborative Environment," *Proceedings of the 16th International Modal Analysis Conference*, vol. 2, Feb. 1998, pp. 1313–1319.

[3]Pedersen, Robert, "Object Request Broker Software Technology: Applications in an Advanced Open Systems Avionics Architecture," *Proceedings of the 16th Digital Avionics Systems Conference*, Oct. 1997, vol. 1, pp. 5.2-18–5.2-25.

[4]Gilyard, Glenn, *In-Flight Transport Performance Optimization: An Experimental Flight Research Program and an Operational Scenario*, NASA TM-97-206229, 1997.

[5]Institute of Electrical and Electronics Engineers (IEEE), "Draft Standard for a Smart Transducer Interface for Sensors and Actuators: Network Capable Application Processor (NCAP) Information Model," P1451.1, D1.83, Dec. 1996.

[6]Lutz, Thomas A., "Using TCP/IP as an Instrument Interface," *Sensors*, vol. 15, no. 7, July 1998, pp. 43–46.

* See, for example, URL http://www.aircell.com.

0-7803-5086-3 /98/$10.00 ©1998 IEEE

AN OPEN SYSTEM DEVELOPMENT INITIATIVE

David Breitigam, Naval Air Warfare Center, Weapons Division, China Lake, CA

Debra L. Borden, Naval Air Warfare Center, Weapons Division, China Lake, CA

Abstract

With the Fiscal Year 1997 software maintenance release, the AV-8B aircraft mission computer and stores management computer hardware and software reached its operational limit. To maintain the war-fighting readiness of the aircraft, this limitation had to be addressed. The selected approach is to replace the existing AYK-14 systems with VME-based PowerPC single board computers. The F/A-18 E/F aircraft is planning an upgrade of a similar nature. This hardware architecture change requires the conversion of the existing AYK-14 assembly language operational flight program (OFP) to a high order language (HOL). To facilitate the transition, the Open System Development Initiative (OSDI) project was established. The following discussion will identify the open systems tools and methods employed and the resulting architectural and design patterns in the prototype application. It will also discuss some of the specific lessons learned with the tools, methods and project.

Background

The Naval Air Warfare Center, Weapons Division (NAWCWPNS) has been actively involved in the maintenance of Operational Flight Programs (OFPs) for avionics systems such as the F/A-18 A/B and AV-8B Day Attack Mission Computers, Stores Management Computers, and Display Computers. Since these systems were fielded in the mid-1980s, they were developed using late 1970s software tools and native assembly language. Today these systems continue to be maintained using the same tools and language. The avionics software engineers who have maintained these systems have a strong background in the application domain. As these legacy systems are converted to 1990s technology, the same software engineers will have to be trained in the newer software development technologies and tools.

The OSDI provides an environment wherein F/A-18 and AV-8B avionics software engineers and managers develop knowledge of emerging software technologies. Each OSDI member joined the team as an expert in an area of the application domain. Re-training domain experts in the newer software development environment was believed to be more cost-effective than training a software engineer familiar with the newer technologies in the application domain.

Technology training included object-oriented analysis and design (OOAD) methodology, pattern based avionics software architectures, OOAD program management, and the C++ language. Training on the use of commercial tools included Wind River System's Tornado development environment, and VxWorks Real-time Operating System (RTOS). Rational Software Corporation provided on-site consultation in the use of their Rose visual modeling tool and OOAD methodology as well as formal classes in Rose and the Unified Modeling Language (UML).

The technology and supporting tools were used to create a prototype open system avionics architecture for the AV-8B mission computer. The experience gained is used to critically evaluate the application of these methods and tools to the AV-8B and F/A-18 upgrade programs. Members of the OSDI team serve as

0-7803-5086-3 /98/$10.00 ©1998 IEEE

technology consultants on the AV-8B and F/A-18 upgrade programs.

Software Architecture

In the legacy military avionics computers, the architectural layering of flight software is a design investment with negative return. Even in relaxed layered architectures, much of the data is passed through intermediate layers for various transformations and de-multiplexing. Because of this data intensive approach, correct layer and class size is difficult to achieve to balance reusability and maintainability against complexity and performance. Even in the best designs, layering of software is costly in performance and until now, flight processors did not have the performance to handle it.

Rationale for Layering

Layers are not new to software engineering. Perhaps the best known example of a layered architecture is the OSI 7-layer model. In this ISO standard, each layer performs a specific role in the domain of network communication. Each layer uses only the services of the next layer below. By adopting this layered approach, the standards bodies are able to specify standard program interfaces that a developer can use to develop applications or components that conform to the standard. Such an interface provides a dual benefit to the architecture.

Maintainability

With a well-specified mature interface providing layer access, changes in implementation will not spread into other layers, since the interface does not change. Thus the interface acts as a firewall against the chain-reaction of change that may occur due to seemingly minor changes to the software.

Reusability

Since a mature interface specification is stable, software components and applications that conform to the specification are interchangeable, promoting software re-use. If an interface remains stable through multiple applications it may have a chance at industry-wide acceptance as a standard.

Relaxed Layered Software Architecture

The OSDI project chose a relaxed layered architecture for implementation of an aircraft flight program. Such a system retains the benefits of layers while shedding some of the performance problems associated with a strict layered architecture. In the relaxed layered architecture, layers are allowed to access *any* layer below, rather than only the lower adjacent layer. Such a system removes the useless transport of data by a layer that only occurs to make the services of a lower layer available to an upper non-adjacent layer.

Such a relaxing of the layered architecture reduces the maintainability benefit. For example, if the navigation model in the software depends on a feature set specific to a certain brand of inertial navigation system (INS), and the standard interface for INS does not support those features, then the navigation model must bypass the interface to use the extended feature set. Bypassing the interface is the equivalent of relaxing the layering. The navigation model is bypassing a logical INS device presented through a standard interface. Instead the model is reaching to a lower layer and accessing vender specific INS features. Later, if the INS is changed on such a program, the effect of the change will directly cause changes in the navigational model of the system. Such effects are extremely difficult to avoid, especially in the case where the requirements have been written to display unique physical avionics data in the cockpit. Such requirements cause multiple layer impacts when the hardware is changed.

Layer Inheritance

In a layered system, where lower layers implement interface, it may become necessary for an upper interface to specify the implementation. In functional designs this was implemented with a callback; the upper layer

0-7803-5086-3 /98/$10.00 ©1998 IEEE

passes the address of a call for the lower layer to execute as part of the interface implementation. Thus implementations could be switched during the execution and the layering of the system was not violated.

With object-oriented design the callback method may be abandoned. The upper layer inherits a lower layer class interface and re-implements one or more methods of the class. This is called layer inheritance. When an upper layer instantiates the class as part of the lower layer system, effectively the lower layer is calling methods that were implemented in an upper layer, again allowing up-calls without violating the layering of the system. For example, the logical device layer may implement a class interface that contains generic abstract methods for rendering any avionics display. Such a class implements the rendering through specific calls, over an avionics bus, to an actual avionics display. An upper level class in the cockpit layer will inherit the interface and override the method, Draw() to draw a *specific* display, without knowledge of any specific avionics hardware. Such a specific display may be instantiated as part of the logical display. When the logical device is notified that the specific display is current, it issues the Draw() message without knowledge of the display implementation specified in the cockpit layer.

OSDI Layered Architecture

The OSDI architecture is a relaxed layered architecture. With this architecture OSDI has endeavored to maintain high performance and simplicity without degrading the potential maintainability, extensibility and longevity that is important to open systems in the military. Eight major layers of the OSDI architecture are discussed below. Each of the top layers fills a specific role within the application. Each of the

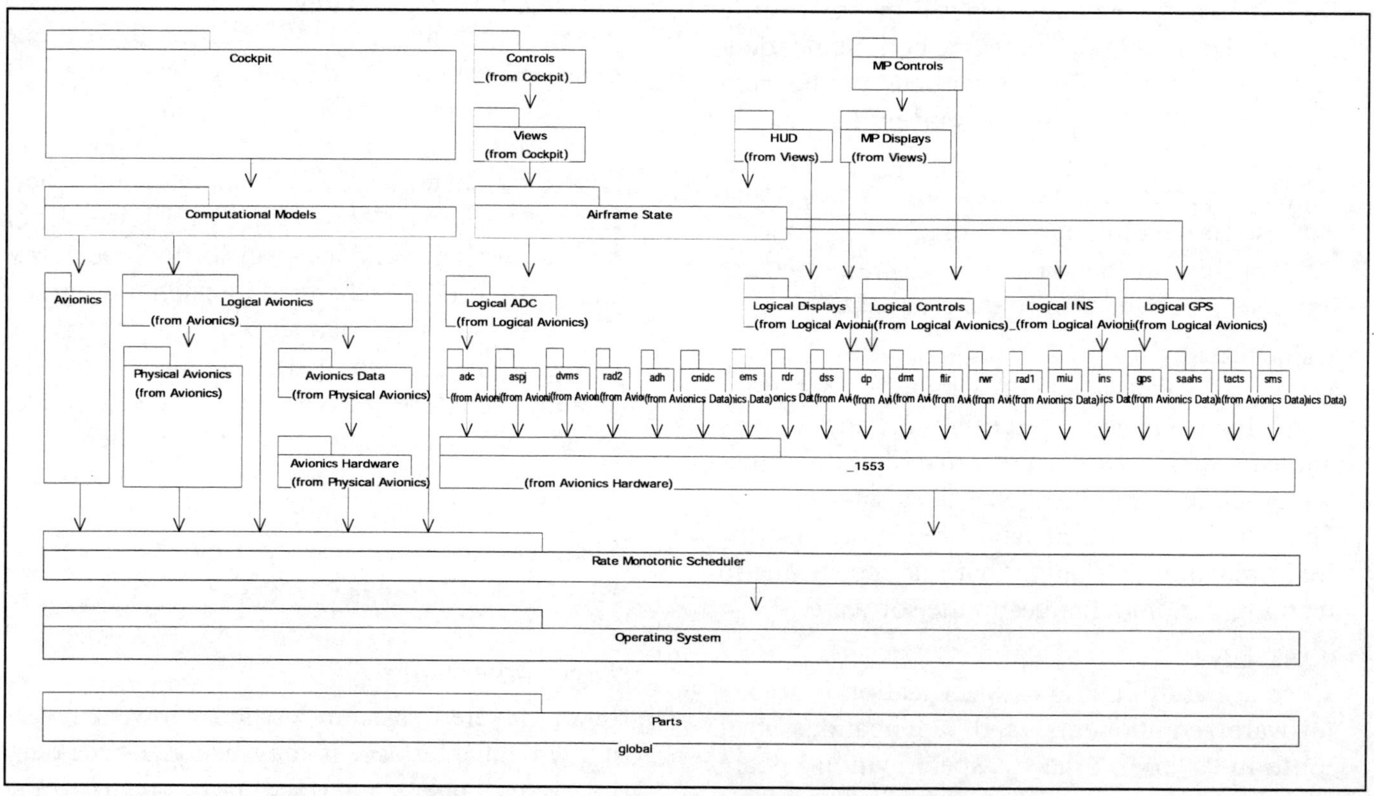

Figure 1. OSDI Iteration 5 Demonstration Architecture

layers may be further decomposed into sub-layers, packages and classes that form the complete architecture. [Fig. 1]

Board Support Layer

The board support package (BSP) is an adaptive layer supplied by the board manufacturer. The primary role of the layer is to provide a common interface to the hardware for the real-time operating system (RTOS). OSDI used the BSP supplied by DY-4 systems with the PowerPC 603e board. The BSP is designed specifically to support the VxWorks RTOS.

DY-4 also supplied as part of the BSP a device driver for the MIL-STD-1553B mezzanine boards. OSDI had problems with the supplied driver because it performed its own scheduling that was incompatible with the scheduling already provided by the OSDI architecture. Rather than simply ignoring the enhanced features, OSDI was forced to modify the driver, removing the portions of the driver that were clock-based in order to achieve 1553 messaging via the OSDI OFP scheduler. This provides an interesting example of a commercial software component that was not suitable for use in a specific architectural design, because it performed roles that encroached on roles already performed by pre-existing critical components.

Real-time Operating System (RTOS) Layer

For the RTOS layer, OSDI used the VxWorks RTOS from WindRiver Systems. The VxWorks system provided some POSIX.1 compliant calls. However, since the interface was not object oriented, OSDI provided an object adapter sub-layer for the specific RTOS objects used by the flight program. The adapter not only provided an object-oriented interface, but also provided an abstraction of some specific RTOS objects. The abstraction could facilitate a future migration to a different RTOS if required.

Real-time Scheduler Layer

The flight program required a typical Rate-monotonic cyclical scheduler. Using layer inheritance the OSDI project created a design whereby classes (representing tasks) conforming to a standard interface could be scheduled to execute at a specified rate. Also their execution could be dependent on the completion of other like interface conforming classes. All scheduling could be performed at run-time. However, no scheduling occurred after the scheduler was started so that the system could be analyzed using rate-monotonic analysis. Such a uniform treatment of tasks could allow for a flexible level of runtime instrumentation for task timing and external observation of the running system.

Physical Device Layer

The flight program supports 20 specific remote terminals (RTs) over the MIL-STD-1553B buses. Each RT communicates via a certain number of fixed-format messages. Each message format is specified in an interface control document for the bus. OSDI created a custom tool to convert the interface control document to C++ class specifications. OSDI also was able to reverse engineer the generated code into the Rational Rose design tool. The role of each class was to pack or unpack (depending on the direction of the data) a specific message data set and convert the values, when appropriate, to machine representations (floating point, radians, signed and unsigned integers, and Boolean values). Each value was represented as an inline class method.

Logical Device Layer

The role of the logical device layer is to hide the physical source of data (RT and message) and create logical groupings of data. For example, in the AV-8B, the mission computer communicates with many other aircraft subsystems via a single display processor (DP). These sub-systems are various cockpit displays and controls. All this data is represented at the physical layer with a single RT category (see DP in Fig. 1) and its associated messages. At the logical device layer, the display computer data is broken up into multiple devices. In the Harrier aircraft this is the multi-purpose color displays (MPCD), the up-front control (UFC), the Options Display Unit (ODU), the Hand On Throttle And Stick (HOTAS) and other miscellaneous data sources. Furthermore we sub-bundle data into logical groupings. The most common kind of sub-grouping is the value/validity pair; a data value bundled with its validity bit. At the physical level the data and its associated validity may appear as separate attributes of a message. At the logical level they are retrieved as one class type. When the value is retrieved from the logical layer, the client must check the validity to assure the data may be used in a calculation.

Computational Model Layer

The computational models comprise the core of the flight program. Computational models perform four roles.

- Retrieve input from logical devices or other computational models
- Calculate the state of the aircraft and its environment.
- Provide the state of the aircraft and its environment as class methods available to peer models and the cockpit.
- Output the state of the aircraft and its environment to logical devices.

Cockpit Displays Layer

Ideally cockpit displays would provide the pilot with views of the computational models, adhering to a strict model-view-controller architecture. In reality actual requirements dictate that pilots are provided views down into the physical device layer, involving at least some part of each layer in between. Thus, from the display layer, the whole system (all lower layers) becomes the "model." This was the greatest driving force in typical OFP requirements that caused OSDI to adopt a relaxed layer approach.

Cockpit Controls Layer

Cockpit controls are often interpreted in the context of the current display-state. Thus, we find the controls at our top-most layer. The classes in the control layer handle control events arriving from the logical device layer, triggering the appropriate adjustments in the display layer, and lower layers that are clients of these events.

Design Patterns

Much recent progress in object oriented software techniques has been driven by the field of patterns, that is, the communication of generalized design solutions gleaned from repeated experience. This is an efficient method of communicating about software designs borrowed from the field of building architecture. The OSDI project has used the demonstration OFP to explore the application of popular software architectural and design patterns to an object-oriented real-time avionics system. This section discusses four sample patterns that proved useful in the solution of multiple design problems.

Adapter Pattern

OSDI used an adapter pattern to adapt the vendor specific functional RTOS interface to the upper layers of the OFP [Fig 2]. This served to make the RTOS interface object-oriented; as well as limit the changes required to port the OFP to a different RTOS. If the OSDI RTOS interface is correctly designed, the implementation of the adapter objects could be changed to accommodate a different OS, and no changes to the software would be required outside of the RTOS layer. Since different and future RTOS interfaces were unknown, the

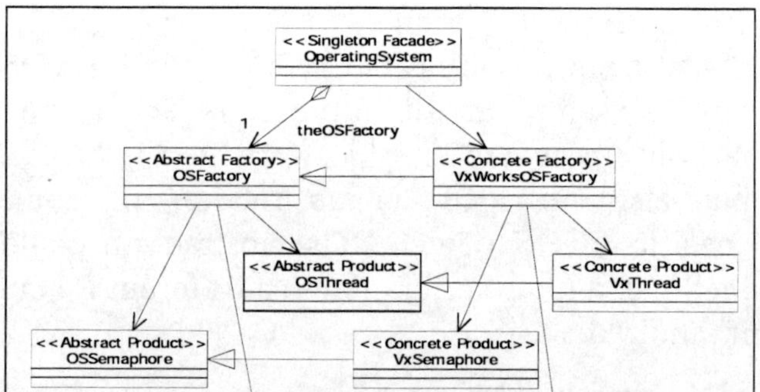

Figure 2. RTOS object Factory

OSDI demonstration RTOS interface is immature, and it is expected that such an interface would undergo future refinement to accommodate other RTOS implementations. Only after successfully porting the layer to many RTOS implementations would such an interface be considered mature.

Abstract Factory Pattern

The OSDI OFP defined a family of RTOS objects to be available in the RTOS adapter layer. The Abstract Factory pattern simply provided the interface for creating the family of RTOS objects, without specifying their concrete classes [Fig 2]. OSDI defined an interface on an abstract factory base class to create the various components of the RTOS class family. The interface returned references to the abstract RTOS class types of the family. OSDI then created a factory implementation for the VxWorks RTOS that conformed to the base class interface. The factory implementation actually returned VxWorks implementations of the objects. The abstract factory assured that the rest of the OFP was dependent on neither the instantiation nor the implementation of the RTOS classes. In the future, a different system

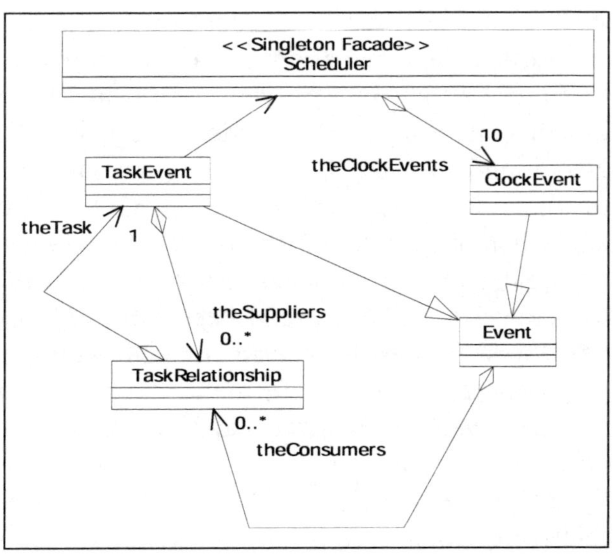

Figure 3. Scheduler Facade

could be configured with a different factory that produces RTOS objects for a different brand of RTOS. The combining of the object family into one interface will prevent future OFP designers from mixing RTOS objects from different implementations.

Façade Pattern

An object-oriented design solution is often achieved through the clustering of several objects within a localized area. The façade pattern is used to provide a unified interface to the object cluster, providing an interface definition to the outside user. For example, the rate monotonic scheduler is composed of clocks, events and tasks combined in a complex way [Fig 3]. There is however a very limited number of things that the rest of the OFP must know about the scheduler in order to use it. OSDI created a scheduler class that provides the interfaces to use the scheduler objects. The façade provides a simplified interface to the scheduler and hides implementation details that the rest of the OFP need not know. OSDI found the use of singleton facades to be universal throughout the OFP, often forming the layer interfaces.

Singleton Pattern

The singleton pattern is used extensively throughout the OSDI OFP for the design of a single or a limited number of object classes. When a class was instantiated as one object or a limited number of objects, OSDI used the singleton pattern to assure its limited cardinality as well as making it accessible in a standardized way. Essentially the singleton pattern allows the definition of classes that do not form the aggregation or composition of any other class. The class contains the static instances of itself which can be accessed through a static method rather than finding the owner through a possibly deep traversal of a specific object ownership structure.

Single interface facades to single clusters of objects were a common use of the singleton

0-7803-5086-3 /98/$10.00 ©1998 IEEE

pattern [Fig 3]. But single real-world objects such as a specific remote terminal or bus message also worked well with the pattern.

Singleton objects instantiate themselves at the first access of their instance. For real-time usage of the pattern, OSDI found that some singleton objects had to be specifically accessed at system startup. This assured that instantiation of objects did not occur after the scheduler had started, creating inconsistent runtimes in early scheduler cycles.

Lessons Learned

Design Methodology and Tool

OSDI used the Rational Rose/C++ visual modeling tool with the Unified Modeling Language (UML) to create software designs.

OSDI chose to use the code generation capabilities of the tools to generate C++ code. Since code generation is limited to class structure, the class diagram, which communicates the class structure, is the primary source of code generation. OSDI used primarily class and interaction diagrams to communicate the design.

As the class diagrams evolved, the interaction diagrams could become outdated but the generated code stayed in sync with the classes. Engineers had to manually update interaction diagrams to keep fidelity with the code.

The component diagram also impacted code generation. In early iterations it was used to map eight character (MS-DOS) filenames to classes for an older compiler. In later iterations it was abandoned in favor of long filenames and was used only for the rare time when multiple classes were combined in one file.

State diagrams are very useful for communicating real-time event driven designs. Because they suffer from the same code sync problems as interaction diagrams, software engineers avoided them. OSDI did not use them much as they should have.

Rose includes a scripting language that gives the user access to the software design. OSDI used a customized script to generate Microsoft Word documentation for design peer reviews.

Rose supports configuration of the code generation property sets to define a custom specification of what the design generates. In early iterations OSDI adopted a variety of sets to support various portions of the design. The variety created confusion about what the design actually generated and thus decreased the fidelity between the design and the code. In later iterations the sets were decreased to just one. With the single set, customization was often still required, but it was always small and relative to a single baseline set.

A big advantage of the Rose product is the round trip engineering of source code. There is no reason for class diagrams and code to be out of sync. OSDI created a tool to produce classes from the 1553 interface control document (ICD) tables. The generated classes were reverse-engineered into the design.

Object Oriented Metrics

OSDI used the McCabe Visual Reverse Engineering Tools to analyze both conventional source code features such as complexity, as well as object oriented measures. The tool supports an overwhelming variety of metrics and reports that, by volume, are unsuitable for use in a streamlined peer review process. OSDI also found the tool lacked sufficient guidance on the relative importance of the OO metrics and when or how to use them. OSDI developed a single page spreadsheet report that distilled the metrics. The spreadsheet allowed the developers to easily experiment with weighing the importance of each metric. The set of metrics was normalized and combined to produce an overall quality value for each class. When sorted, the classes with the lowest relative quality were prioritized for inspections by the developers. By refining the weights, OSDI hopes to predict and prevent rework.

Iterative Approach

OSDI used a phased iterative approach where each iteration is the design and construction of a testable executable. Sets of requirements were allocated to each iteration with the goal of placing high-risk items in early iterations. OSDI experienced difficulty in achieving executable status in early iterations because of the large amount of external data and hardware interfaces that had to be built to support the application. The amount of time given to this task was severely underestimated. OSDI found that the iterative approach to object-oriented embedded systems differed from desktop applications. It was a bottom-up hardware-centric approach and required a significant investment of time to build up to the user interface in the highest layers of the architecture.

Training

The OSDI project found that the new tools and methodologies discouraged some software engineers. The amount of learning required to use the tools and construct the code was, in some cases, overwhelming. Of the eight software engineers involved with OSDI, one engineer resigned so that he could work on a project that still uses the older technologies.

OSDI found training the most beneficial when it could be applied immediately and could be customized to the problem domain. The technology consultants from Rational customized their training for OSDI and this accelerated project progress.

Summary

The military effort to adopt open systems is driven by the promise of lower cost, shortened schedules and product longevity. The tools and methodologies that OSDI has demonstrated are currently the choice of commercial industry and, although they are somewhat immature, we believe they currently have the best chance at longevity. They are viable solutions for building real-time embedded avionics systems. However, OSDI participants with no previous exposure struggled to learn the new concepts. Preliminary OSDI experience indicates that achieving proficiency in these tools and methods takes longer and is more difficult than first expected. We believe that a careful approach to layered object-oriented OFP architecture, with emphasis on simplicity and standard interfaces, will yield some of the whole life-cycle savings that the open systems approach promises. The OSDI project would welcome a teaming with industry to further explore the use of these methods for the real-time embedded avionics application.

References

Booch, Grady, *Object-oriented Analysis and Design with Applications*, Addison-Wesley Publishing Company, Inc., Menlo Park, CA, 1994. ISBN 0-8053-5340-2.

Booch, Grady, *Object Solutions Managing the Object-Oriented Project*, Addison-Wesley Publishing Company, Inc., Menlo Park, CA, 1996. ISBN 0-8053-0594-7.

Buschmann, Frank, et. al., *Pattern-Oriented Software Architecture: A System of Patterns*, John Wiley & Sons, West Sussex, England, 1996. ISBN 0-471-95869-7.

Douglass, Bruce Powel, *Real-Time UML: Developing Efficient Objects for Embedded Systems*, Addison-Wesley, 1998. ISBN 0-201-32579-9.

Gamma, Erich, et. al., *Design Patterns: Elements of Reusable Object-oriented Software*, Addison Wesley Longman, Inc., 1994. ISBN 0-201-63361-2.

Acknowledgments

The authors wish to express their sincere appreciation for the contributions made to this project by the OSDI team members: Dale Christensen, Fernando Escobar, Jeff Fisher, Lee Lucas, John T. Paul, Chris Rickets, Linda Shelton, and John Veazey.

A FORMAL METHODS APPROACH TO THE ANALYSIS OF MODE CONFUSION

Ricky W. Butler, NASA Langley Research Center, Hampton, Virginia
Steven P. Miller, Rockwell Collins, Cedar Rapids, Iowa
James N. Potts, Rockwell Collins, Cedar Rapids, Iowa
Victor A. Carreno, NASA Langley Research Center, Hampton, Virginia

Introduction

The goal of the new NASA Aviation Safety Program (AvSP) is to reduce the civil aviation fatal accident rate by 80% in ten years and 90% in twenty years. This program is being driven by the accident data with a focus on the most recent history. Pilot error is the most commonly cited cause for fatal accidents (up to 70%) and obviously must be given major consideration in this program. While the greatest source of pilot error is the loss of "situation awareness", mode confusion is increasingly becoming a major contributor as well. The January 30, 1995 issue of Aviation Week lists 184 incidents and accidents involving mode awareness including the Bangalore A320 crash 2/14/90, the Strasbourg A320 crash 1/20/92, the Mulhouse-Habsheim A320 crash 6/26/88, and the Toulouse A330 crash 6/30/94 [2].

These incidents and accidents reveal that pilots sometimes become confused about what the cockpit automation is doing. Consequently, human factors research is an obvious investment area. However, even a cursory look at the accident data reveals that the mode confusion problem is much deeper than just training deficiencies and a lack of human-oriented design. This is readily acknowledged by human factors experts. For example, Charles E. Billings, writes in Aviation Automation: The Search for a Human-Centered Approach (pg 144) [1]:

> *...today's flight management systems are "mode rich" and it is often difficult for pilots to keep track of them ... The second problem, which is related to the first involves lack of understanding by pilots of the system's internal architecture and logic, and therefore a lack of understanding of what the machine is doing, and why, and what it is going to do next.*

Similarly, Sarter and Woods write [7]:

> *What is needed is a better understanding of how the machine operates, not just how to operate the machine.*

It seems that further progress in human factors must come through a deeper scrutiny of the internals of the automation. It is in this arena that formal methods can contribute. Formal methods refers to the use of techniques from logic and discrete mathematics in the specification, design, and verification of computer systems, both hardware and software. The fundamental goal of formal methods is to capture requirements, designs and implementations in a mathematically based model that can be analyzed in a rigorous manner. Research in formal methods is aimed at automating this analysis as much as possible. By capturing the internal behavior of a flight deck in a rigorous and detailed formal model, the dark corners of a design can be analyzed.

This paper will explore how formal models and analyses can be used to help eliminate mode confusion from flight deck designs and at the same time increase our confidence in the safety of the implementation. The paper is based upon interim results from a new project involving NASA Langley and Rockwell Collins in applying formal methods to a realistic business jet Flight Guidance System (FGS).

The Targeted Flight Guidance System

A *Flight Guidance System* (FGS) is a component of the overall *Flight Control System* (FCS) (see Figure 1). The FGS compares the measured state of an aircraft (position, speed, and attitude) to the desired state and generates pitch and roll guidance commands to minimize the difference between the measured and desired state. When engaged, the *Autopilot* (AP) translates these commands into movement of the aircraft's control surfaces necessary to achieve the commanded changes about the lateral and vertical axes. An FGS can be further broken down into the *mode logic* and the *flight control laws*. The mode logic accepts commands from the flight crew, the *Flight Management System* (FMS), and information about the current state of the aircraft to determine which system modes are *active*. The active modes in turn determine which flight control laws are used to generate the pitch and roll guidance commands. The active lateral and vertical modes are displayed (annunciated) to the flight crew on the *Flight Director* (FD), a portion of the Electronic Flight Instrumentation System (EFIS). The magnitude and direction of the lateral (roll) and vertical (pitch) commands generated by the FGS are also displayed on the EFIS as *guidance cues*.

The specification of the Flight Guidance System used in this project was developed at Collins as part of a project to investigate different methods of modeling requirements [5]. While it is a simplified composite of several actual Flight Guidance Systems, and does not describe an actual aircraft in service, it is complex enough to serve as a realistic example [4].

Goals of Formal Modeling

Moving new technology into practice is always more difficult than the creation of that technology, especially in the case of software development. Over the years a multitude of new software development methodologies have been produced, yet few of them have been accepted

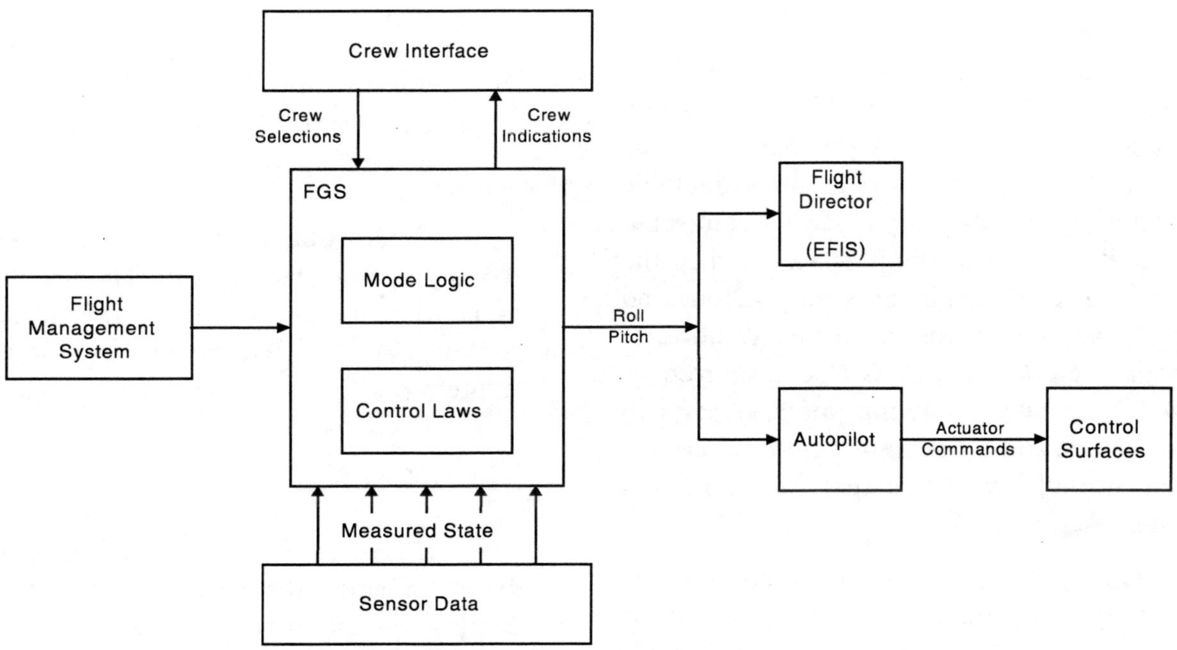

Figure 1 - Flight Control System Overview

by industry. A major goal of this project was not only to show how formal methods could be used to remove mode confusion from a flight guidance system and to discover design flaws, but to do so in a way that would be accepted by industry.

Companies such as Rockwell Collins typically build variations of the same products over and over. Increasingly, these companies are looking for strategies that support the systematic reuse of common artifacts. One such approach is Product Family Engineering, also known as Domain Engineering. Central to the Product Family approach is the development of a domain architecture consisting of those requirements, design, implementation, and verification artifacts that are common to all members of the family and the variations of these artifacts that are supported by the domain. Prior to this project, Collins had conducted a Commonality Analysis [9] of the FGS mode logic described in [4] and developed a tentative product family architecture. Consequently, an important goal of this project was to build on that work and develop a formal model consistent with that architecture.

Another central goal of the project was to make the mode logic accessible to pilots and experts in human factors. To achieve this, the mode logic is also specified as an executable ObjecTime model [8]. This model is connected to a mock-up of the Flight Deck so that the model can be executed by pressing buttons and turning dials on the mock-up. A visualization of the mode logic (Figure 3) is also displayed as the model executes, allowing pilots, experts in human factors, and the design engineers to relate the behavior of the automation to the human computer interface.

Yet another goal was to be able to formally analyze the model for various forms of consistency, completeness, safety properties, and properties related to human factors. To achieve this, the ObjecTime model was manually translated into the PVS specification language [6]. The desired properties could then be analyzed with this model using the PVS theorem prover. This overall strategy is illustrated in Figure 2.

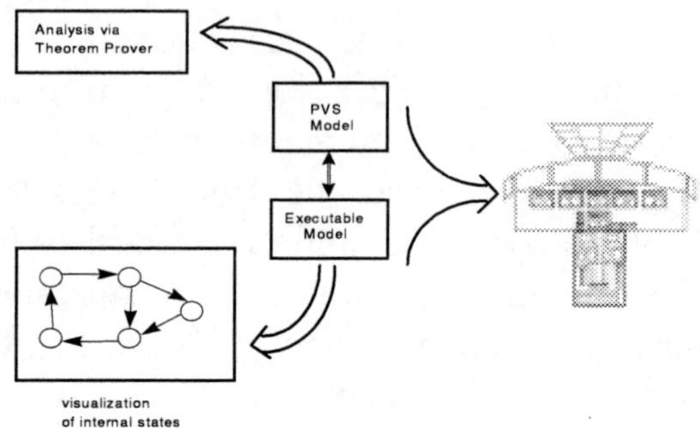

Figure 2 - Overall Strategy

As a result, there were several goals that affected the style of the formal models. In particular, the formal models had to be:

1. suitable for mathematical analysis of their potential for mode confusion
2. consistent with the executable ObjecTime model
3. conceptually simple enough to display during pilot training
4. consistent with the product family architecture

Interestingly, we found that (1) and (4) worked against each other. The proofs became more difficult as the model was structured to achieve (4). This will be discussed in detail in a future report.

Mode Confusion

Mode confusion can be traced to at least three fundamental sources: (1) opacity (i.e., poor display of automation state), (2) complexity (i.e., unnecessarily complex automation), and (3) incorrect mental model (i.e., flight crew misunderstands the behavior of the automation). Traditional human factors research has concentrated on (1), and significant progress has been made. However, mitigation of mode confusion will require addressing problem sources (2) and (3) as well. Towards this end, our approach uses two complementary strategies based upon a formal model:

Visualization Create a clear, executable formal model of the automation that is easily understood by flight crew and use it to drive a flight deck mockup from the formal model.

Analysis Conduct mathematical analysis of the model.

It is hoped that this approach will (1) force designers to commit to a clear conceptual model of the automation, (2) facilitate discussion between designers, human factors experts, and the flight crew, (3) enhance the training process by direct exposure to an accurate mental model of the automation, and (4) through analysis, uncover characteristics of the automation that historically have been a source of mode confusion.

Model Visualization

Development of a flight deck around an executable formal model enables several innovative strategies for pilot training. First, the executable model can be used in place of a rapid prototype for early life cycle discussions with pilots. These discussions can be focused on the

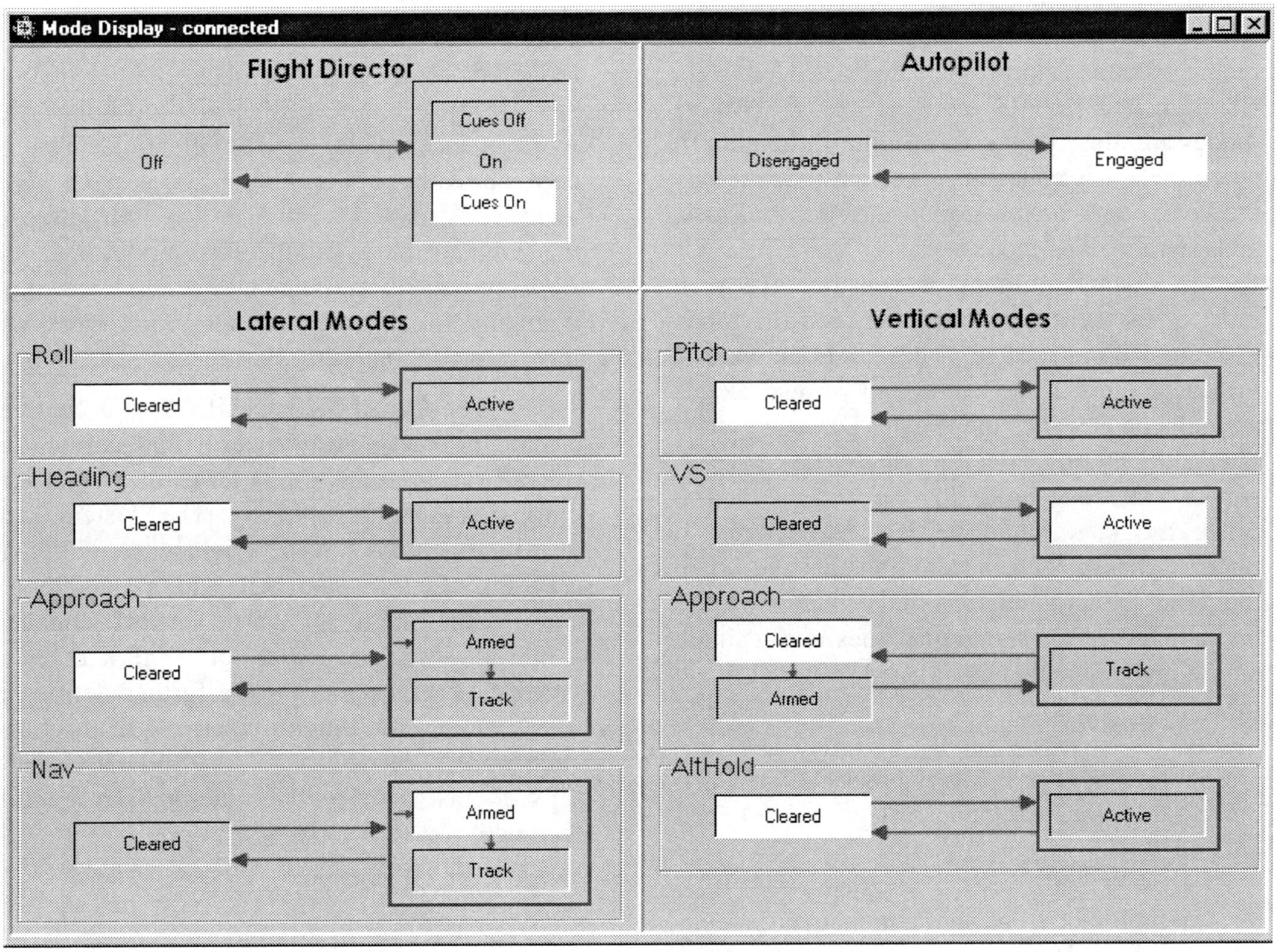

Figure 3 - Visualization of the FGS Modes

identification of model complexities that are confusing to the pilots. Second, during training the flight deck can be augmented with an additional display that directly exposes the internal structure of the automation and its dynamic changes. This display will not be present in the cockpit of an operational aircraft. It will be used exclusively to help the pilot form an accurate model of what the automation is doing.

An example of the visual display of the mode logic is shown in Figure 3. The state of the Flight Director (FD), Autopilot (AP), and each of the lateral and vertical modes are modeled as small, tightly synchronized finite state machines. In Figure 3, the FD is *On* with the guidance cues displayed; the AP is *Engaged*; lateral Roll, Heading, and Approach modes are *Cleared*; lateral NAV mode is *Armed*; vertical modes Pitch, Approach, and AltHold are *Cleared*; and the VS mode is *Active*. Active modes are those that actually control the aircraft when the AP is engaged. These are indicated by the heavy dark boxes around the *Active, Track,* and lateral *Armed* modes.

A small number of constraints govern most of the synchronization between the mode machines. For example, if the Flight Director is turned on (so that the lateral and vertical modes are annunciated on the Flight Director), then one, and only one, lateral mode can be active. A similar constraint holds for the vertical modes. Since the lateral and vertical modes are so tightly synchronized, a common mistake when modeling the mode logic is to try to combine the lateral and vertical mode machines into a single lateral mode machine and a single vertical mode machine. Besides violating the modularity needed to support a family of products, combining the modes in this way breaks down when other modes that are more loosely synchronized are added.

This visual model of the automation is connected to a simulation of the cockpit and can be executed by pressing buttons and turning dials on the mockup. In this way, pilots, experts in human factors, and the designers can easily relate the behavior of the automation with the human computer interface

Model Analysis

In a new paper entitled "Analyzing Software Specifications for Mode Confusion Potential" [3], Nancy Leveson, et. al., identify six categories of design that have historically been a source of mode confusion: (1) inputs interpreted differently in different modes, (2) indirect mode changes, (3) behavior that is different in different modes, (4) operator authority limits, (5) unintended sides effects, (6) lack of appropriate feedback. The critical question here is whether these categories can be understood well enough that they can be mathematically characterized. If so, the formal models can be analyzed against these characterizations. Although this work is incomplete, we can illustrate the concept on categories (2) and (3) above. The analysis is made possible by the translation of the ObjectTime visualization model into the PVS specification language. This was done manually for this project, but future work will look into automatic translation.[1]

The formal PVS specification is centered around a "next_state" function that defines the overall system transition in terms of several synchronized state machines. The system state vector includes four fields: LATERAL, VERTICAL, FD, and AP which contain the state of the lateral guidance, vertical guidance, Flight Director and Autopilot. The lateral and vertical guidance models are further defined in terms of several synchronized mode machines such as PITCH, ROLL, NAV, and HDG. Details of this model will be included in a future technical report.

---

[1] Alternatively one could drive the visualization from the PVS specification. The ultimate goal is to use one model for specification, training, analysis, and implementation.

Indirect Mode Changes

The first problem is formally defining what constitutes an indirect mode change. Let's begin by defining it as a mode change that occurs when there has been no crew input:

```
Indirect_Mode_Change?(s,e): bool =
  NOT Crew_input?(e) AND Mode_Change?(s,e)

No_Indirect_Mode_Change: LEMMA
   Valid_State?(s) IMPLIES
         NOT Indirect_Mode_Change?(s,e)
```

We then seek to prove the "false" lemma above using GRIND, a brute force proof strategy that works well on lemmas that do not involve quantification.[2] The resulting unproved sequents elaborate the conditions where indirect mode changes occur. For example,

```
{-1}    Overspeed_Event?(e!1)
{-2}    OFF?(mode(FD(s!1)))
{-3}    s!1  WITH [FD := FD(s!1) WITH [mode := CUES],
            LATERAL := LATERAL(s!1) WITH
                        [ROLL := (# mode := ACTIVE #)],
            VERTICAL := VERTICAL(s!1) WITH
                        [PITCH := (# mode := ACTIVE #)]]
        = NS
{-4}    Valid_State(s!1)
   |-------
{1}     mode(PITCH(VERTICAL(s!1))) =
        mode(PITCH(VERTICAL(NS)))
```

The situations where indirect mode changes occur are clear from the negatively labeled formulas in each sequent. We see that an indirect mode change occurs when the overspeed event occurs and the Flight Director is off. This event turns on the Flight Director and places the system into modes ROLL and PITCH.[3]

Discovering Inconsistent Behavior

Precisely defining the concept of inconsistent behavior is nontrivial and likely to be a long term endeavor. With no pretense of fully capturing the notion of inconsistent behavior, we offer the following as simple examples to illustrate the concept:

- Button pushes that are ignored in some modes but not others
- Button pushes that act like toggles in some modes but not others

We define an "ignored command" as one in which there is a crew input and there is no mode change. We seek to prove that this never happens:

```
No_Ignored_Crew_Inputs: LEMMA
    Valid_State(s) AND Crew_Input?(e) IMPLIES
        NOT Mode_Change?(s,e)
```

The result of the failed proof attempt is a set of sequents similar to the following:

```
{-1} VS_Pitch_Wheel_Changed?(e!1)
{-2} CUES?(mode(FD(s!1)))
{-3} TRACK?(mode(NAV(LATERAL(s!1))))
{-4} ACTIVE?(mode(VS(VERTICAL(s!1))))
|-------
{1} ACTIVE?(mode(ROLL(LATERAL(s!1))))
{2} ACTIVE?(mode(HDG(LATERAL(s!1))))
```

The negatively labeled formulas in the sequent clearly elaborate the case where an input is ignored, i.e., when the VS/Pitch Wheel is changed and the Flight Director is displaying CUES and the active lateral mode is ROLL and the active vertical mode is PITCH. In this way, PVS is used to perform a state exploration to discover all conditions where the lemma is false, i.e., all situations in which a crew input is ignored.

We can determine whether the HDG Switch acts like a toggle by seeking to prove the following lemma which asserts that the HDG mode toggles between CLEARED and ACTIVE whenever the HDG switch is pressed.

```
HDG_Toggle?: LEMMA
   HDG(LATERAL(next_state(s, HDG_Switch_Hit))) /=
      HDG(LATERAL(s))
```

This lemma is easily proved for the current model. Of course, this may not remain true as more modes are added and the mode logic becomes more complex.

[2] As the model has grown larger, we have had to develop more effective custom proof strategies to keep the proof times at manageable durations.

[3] PITCH mode is selected because the model is still under construction. In the final model, Flight Level Change (FLC) mode will be selected as the active vertical mode when overspeed occurs.

Safety Analysis

Of particular importance to the analysis of safety-critical systems is the fact that formal methods provides a way to investigate all of the behaviors of a model. In other words, they can explore whether a property is true over its entire input space. And total exploration of the entire input space is the only way to gain assurance that catastrophic failure does not lie hidden among the vast number of possible behaviors. Simulation and testing simply cannot accomplish this in practical amounts of time. Although no accident in civil aviation has been blamed directly on a bug in the software, many serious incidents have occurred and are occurring with increasing frequency. Billings writes (pg. 149) [1]: *It must be noted that automation also makes apparently random, unpredictable "errors" (e.g. the flap lockup at Hong Kong, 1994).* He then offers six examples where this has occurred.

A key advantage of a formal model is that it can be mathematically analyzed to insure that key safety properties are not violated. For example, one would seek to prove

```
GO_AROUND_SAFETY_PROP: LEMMA
    Valid_state(s) IMPLIES
        LATERAL(next_state(s,e) = GA    IFF
        VERTICAL(next_state(s,e) = GA
```

In other words, the lateral mode will only be in Go Around if the vertical mode is in Go Around and vice versa for all possible inputs and reachable states. This lemma is proved for all possible inputs, so conceptually this is equivalent to exhaustive testing[4]. Another example is:

```
At_Least_One_Lateral_Mode_Active(s): bool =
    ON?(FD(s)) IFF
        At_Least_One_Mode_Active(LATERAL(s))

ALOLMA: LEMMA
At_Least_One_Lateral_Mode_Active(s) IMPLIES
At_Least_One_Lateral_Mode_Active(next_state(s,e))
```

[4] Note. This is not a claim for perfection because we have no guarantee that all of the needed properties have been elaborated and that they have all been stated correctly.

This property must be proved as a consequence of modeling the lateral modes as a set of parallel state machines.

Currently we have identified over 50 key functional properties. Three are shown here:

```
AP_TURNS_ON: LEMMA
    NOT Engaged?(AP(s))  AND
    Engaged?(AP(next_state(s, e)))
        IMPLIES
            AP_Engage_Switch_Pressed?(e)  AND
            Disconnect_Bar_Up?(AP(s))

FD_TURNS_OFF: LEMMA
    On?(FD(s))  AND
    NOT On?(FD(fast_next_state(s, e)))
        IMPLIES
            FD_Switch_Hit?(e) AND
            NOT Engaged?(AP(s)) AND
            NOT Overspeed?(FD(s))

HDG_SELECTED: LEMMA
    NOT ACTIVE?(mode(HDG(LATERAL(s))))
    AND HDG_Switch_Hit?(e)
        IMPLIES
ACTIVE?(mode(HDG(LATERAL(next_state(s,e)))))
```

It is interesting that the need to prove certain key properties has led us to discover even more key functional properties than had been cited in [4] and [5]. This has led us to speculate about the possibility of a more abstract specification in which these properties are explicitly stated, rather than being implicit as in the current operational model.

Future Work

Finally, because the current flight deck has grown incrementally over two decades, the result has not been a coherent, well integrated means of controlling an aircraft. Tony Lambregts, the FAA National Resource Specialist for Automated Controls writes "... the great majority of automation deficiencies and unnecessary complexities are the result of bad formulation of requirements and adhering too long to outdated technologies, engineering design concepts and processes." In particular he argues that much of the complexity of the flight deck derives from the independent design of the autopilot and autothrottle. Thus, it seems essential that a multidisciplinary approach using (1) integrated control laws (i.e. combining autopilot and autothrottle), (2) formal models, and (3) human factors will be necessary to fully solve the mode confusion problem. It is also clear that the formal models must be extended to include information about the control laws in order to enable a full analysis of systems that would result from such a multidisciplinary development. The crash of an Airbus A330-322 in Toulouse, France on 6/30/1994, highlights this need. During a test flight of simulated engine failure, an unexpected mode transition to altitude acquisition (ALT*) occurred. Pitch protection was not provided in ALT* mode, although it was present in all of the other modes. Detection of inconsistent behavior such as this will necessitate the elaboration of the basic properties of the control laws in the model in addition to the mode structure.

Conclusions

Many of the proposed solutions to other accident categories in the NASA AvSP program involve the use of new automation. It seems likely that as failures in other accident categories are reduced, the problems associated with automation will increase in significance. Use of formal methods can

- aid the pilot in training through direct display of internal states
- reveal "dark corners" and non-intuitive interactions through analysis
- discover properties of the design that have a potential for mode confusion
- discover design errors early in life cycle

References

[1] Charles E. Billings. Aviation Automation: The Search for a Human Centered Approach. Lawrence Erlbaum Associates, Inc., Mahwah, NJ, 1997.

[2] Dan Hughes and Michael Dornheim, Automated Cockpits: Who's in Charge?, Aviation Week & Space Technology, January 30-February 6, 1995

[3] Nancy Leveson, et al, Analyzing Software Specifications for Mode Confusion Potential, 1997.

[4] Steven P. Miller and Karl F. Hoech, Specifying the mode logic of a flight guidance system in CoRE, Rockwell Technical Report WP97-2011, Rockwell Collins, November 1997.

[5] Steven P. Miller, Specifying the Mode Logic of a Flight Guidance System in CoRE and SCR, in Proceedings of the Second Workshop on Formal Methods in Software Practice (FMSP'98), pg. 44-53, Clearwater Beach, Florida, March 4-5, 1998.

[6] Sam Owre, John Rushby, Natarajan Shankar, and Friedrich von Henke. Formal Verification for Fault-tolerant Architectures: Prolegomena to the Design of PVS. IEEE Transactions on Software Engineering, 21(2):107-125, Feb. 1995.

[7] N.B. Sarter and D.D. Woods. Decomposing Automation: Autonomy, Authority, Observability and Perceived Animacy. First Automation Technology and Human Performance Conference, April 1994.

[8] Bran Selic, G. Gullekson, and P. Ward, *Real-Time Object-Oriented Modeling*, John Wiley & Sons, 1994.

[9] David M. Weiss, Defining Families: The Commonality Analysis, Lucent Technologies Bell Laboratories, 1000 E. Warrenville Rd, Napierville, IL, 60566, 1997.

"CLUB MED"™ AVIONICS AT "MOTEL 6"™ COSTS: WIN/WIN DEVELOPMENT STRATEGIES

James Sutton
Boeing North American

Entering "The Lobby"

Where would *you* spend your dream vacation? For many of us, a trip to a top-quality Caribbean resort like "Club Med" would suit quite nicely. Accompanied by your favorite person...sitting on beach chairs in the sun, watching the waves and the people...sipping the drink of your choice...getting up to snorkel or play golf—*you* decide...checking out crafts made by the locals...and finishing with unlimited quantities of the finest foods on the planet...ahhh!

Coming back to earth after such a stay in heaven would be tough enough...but 30 days after you re-entered your front door, reality would *really* crash in: Your credit-card bill in the morning mail! You *knew* you should have gone to the nearest State Park and stayed at the Motel 6. It would have been neat and clean and plenty good enough; but nooo, *you* had to play at "Saudi Royalty!" Now you'll be making atonement for your extravagance with every extra penny you earn for the next five years!

Avionics system development offers us a similar dilemma. You know how you'd like to develop systems...highest possible integrity, lowest failure rates, best ergonomics.

You also know what will price you out of the market and abandon the business to your competitors: "bleeding edge" efforts to extract a theoretical last "scoshe" from your product.

What we really need is a "Club Med" experience on a survivable "Motel 6" budget.

Of course, everybody tells us this is crazy...heaping on helpful truisms like "there's no such thing as a free lunch" or "if it sounds too good to be true, it is."

Sometimes "everybody" is wrong.

People go for centuries "being right" about "common knowledge"...then everyone changes. For decades people knew that outer space was filled with something called "ether" (Einstein eloquently defended it in a 1920 speech). This knowledge worked reasonably well until experiments early in this century conclusively proved ether nonexistent.

For many years people were positive that light was a wave and electrons were particles, based on the most obvious behaviors of each...until scientists like de Broglie, Schroedinger, and Born proved that light could be particles and electrons waves (for awhile Einstein had a hard time with that one, too).

In each case, the results depended completely upon how one defined the circumstances around the observation.

People "know" that avionics is *either* high-integrity *or* low-cost, but never *both.* And this "knowledge" has shown itself true in project after project. Anyone would be foolish to question this wisdom, right?

As with knowing of ether or electrons, the answer depends upon the circumstances of the question. Circumstances have recently been found where avionics development can be *both* high integrity and low cost. The circumstances that produce such results come from alterations to the nature and characteristics of the avionics product development process. The results are a project history of reduced costs, accelerated developments (e.g., "lean engineering"), and greatly minimized defects.

0-7803-5086-3 /98/$10.00 ©1998 IEEE

Concierge

It's time for a look around our *affordable* "Club Med." What does our approach look like?

First, it allows us to develop avionics systems with all the following characteristics:

- High Integrity and Safety
- Easily Adaptable and Reusable
- Good Functional Performance
- Satisfying to the Customer
- Low Cost
- Low Risk

While times are extremely good for many of our companies, challenges continue to grow from every quarter. We can take our market positions less for granted than at any time in the history of avionics.

These product characteristics are traditionally thought of as mutually exclusive. When goals are conflicting, the only way to resolve them is often a change of paradigm. The Swiss discovered this truth: long the makers of the most accurate timepieces on earth, they lost almost all market share to the Japanese, when the Japanese decided not to compete with the Swiss spring-based actions but rather to build clocks using quartz electronics.

Quartz electronics gave the Japanese a way to produce a watch that was simultaneously:

- More Accurate
- More Durable
- Less Expensive

The Japanese didn't beat the Swiss at their game; the Japanese changed the game to one the Swiss didn't know how to play!

The current avionics systems development "game" is getting old and worn and vulnerable…it *needs* to be replaced by new game that isn't subject to the same limitations.

Such a game has been found. It was used on the Lockheed-Martin C-130J program to produce a Mission Computer system whose software, though safety critical, was both developed more productively than standard industry costs for *non*-safety critical software, and achieved an error/defect rate of just 1/10 that of typical safety-critical software. Additionally, turn-around times of just days were achieved from safety-critical requirements changes to new flight-ready software load on aircraft. The MoD and the FAA both expressed the highest levels of satisfaction and admiration for the processes used and results obtained.

Similar principles are now being used on avionics development for the Boeing Brimstone Missile and Launcher avionics. Early results are also very promising.

The Restaurant

Just as the restaurant is the heart of every resort hotel, there is a guiding principle at the heart of the high-integrity/low-cost process. This principle is "coherence," and it will either eat our lunch, or set the table.

"Coherence" is a term from laser engineering. It implies that, to the greatest degree possible, every photon emitted will strike a multiplier device and cause many additional photons to be emitted. In other words, emitted photons are almost all used to increase the laser's output. Few photons are wasted on other than the laser's eventual goal: to put as much light on the target as possible.

"Coherence" in the engineering process means that almost all efforts at each stage of the development process "hit home" on efforts at a later stage…few efforts are lost to no discernable effect on the end goal of the development: producing an avionics system with the characteristics we discussed earlier (see Figure 1).

0-7803-5086-3 /98/$10.00 ©1998 IEEE

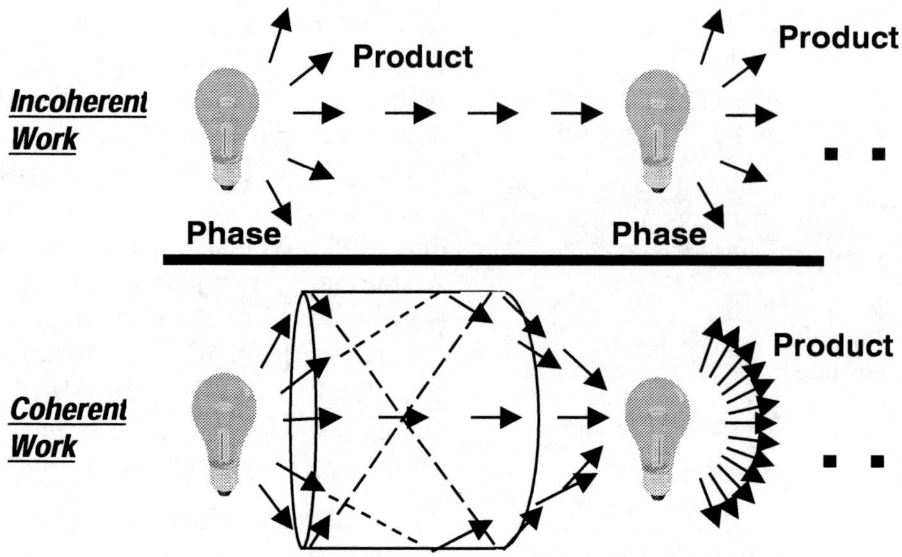

Figure 1: Incoherent vs. Coherent Work

The Kitchen

Behind every great restaurant there's an exceptional kitchen. It is hidden to the happy diners, but it is the key to the quality of the dining experience.

The "kitchen" of the exceptional avionics development process is equipped with "cooking appliances:" principles that enable or limit what can be achieved.

- Everything Justified on Business (ROI) Basis
- Process is Customer-Driven
- Development Based on Integrated Product Line
- "Correctness by Construction"
- Verification-Driven Development
- Continuous Improvement

The "cooks," or avionics developers, can use these appliances because they have the necessary experience and management support to do so. That support includes:

- Use of Mathematics (to Make Verifiability Possible)
- Development of Sound Requirements
- Collection of Quantified Customer Information

Without mathematics, there is arguably no engineering. This seems obvious to most engineering disciplines. However, because avionics are becoming increasingly dependent on software to implement functionality and achieve integration, the nature of the software engineering discipline is increasingly dominating the characteristics of the overall avionics development lifecycle.

And software, as currently practiced, is unique among engineering disciplines by being almost completely non-mathematical.

Of course, software is highly mathematical in the sense that code consists of "1"s and "0"s, and these taken together form numbers. This is a very shallow sense of mathematics, however.

Software is almost completely non-mathematical in the sense of using underlying laws and principles for specifying requirements and the criteria for validating the final system's successful performance of those requirements. Neither the software requirements (or often, consequently, the overall system requirements) nor the resulting software component of the system is amenable to any truly mathematical treatment.

Without quantified information about the customer, how can the development effort be efficiently directed to achieving arguably its highest goal: the customer's satisfaction (especially if you broaden the definition of "customer" to include corporate customers of your development organization: those who evaluate you based on your achievement of corporate goals like profitability or building of long-term business)?

One must know who your customers are (as we've seen, this can be a much larger list than simply "the buyer"), the relative importance (numerically) of each customer to the project's success, what each customer wants and how dearly they want it (quantitatively) and finally, a net balancing out of all customer desires into a single list that can be used to help make project management and engineering decisions.

What's *not* in the kitchen is almost as important as what is there. There's no need in the avionics development process for:

- "Silver-Bullet" Methods
- Extreme Levels of Automation
- "Edge of Reality" Engineering Environments

They simply do not enhance the "coherence" of the development: its ability to achieve its most important goals at minimum cost.

"Kitchen Tools"

Finally, avionics development requires use of specific tools. A wok would hardly be helpful if all one had in their hand was a measuring cup. The right tools make it possible to use the "appliances" of the previous section to their highest potential.

The "tools" of high-integrity/low-cost avionics development are:

- Formal Methods (i.e., Mathematical Systems)
- Domain Engineering ("Product Line Engineering")
- "Customer Voice"
- Appropriate Process

Formal Methods are a key to better avionics for less cost. Formal methods include mathematical systems such as Z, VDM, SCR, and CoRE.

Formal methods have very attractive characteristics. They are rigorous: they constrain development to greater "coherence" and fewer wasteful departures. They are inherently verifiable by applying the rules of the underlying mathematical systems.

Formal methods have some equally undesirable traits. Most of them require expensive and project-delaying training. Their greatest costs come at the time of project definition…so if the project's requirements are likely to be fluid for a significant part of the product lifecycle, they are unaffordable. And formal methods do not "scale up" well: they can only currently be used efficiently on small pieces of functionality.

Domain engineering, including practices defined by DARPA and the Software Productivity Consortium (SPC), have strengths and weaknesses which almost perfectly complement those of formal methods.

Domain engineering restricts the propagation of changes throughout a system by focusing that system's architecture on the functions and elements least likely to change over the product's lifecycle. This type of organization minimizes the exposure of formally-defined system elements to change…and thus to unaffordable modification.

Domain engineering also makes "correctness by construction" practical, by segmenting a system into much smaller, but inherently modifiable, pieces which can be implemented formally and thus with known correctness.

The "Customer Voice" is brought into the development process using techniques such as "Quality Function Deployment" (QFD), and the "7 Management and Planning" tools (7MP). The resulting quantitative customer-driven goals can then drive all the engineering and development done using formal methods and domain engineering.

Finally, the process model for the avionics development should be chosen carefully. How often has the tale been told of the project planned as a linear waterfall, only to discover that the customer as of project initiation had the wrong idea of what they wanted? A little "QFD" would have revealed this, and the appropriate (say, iterative or spiral) process chosen to accommodate the inevitable requirements cycles to come.

"Checkout"

If all of these fundamentals are observed, history has shown that one can have a very effective and successful development program.

We're all successful only if we consistently provide "win/win" development experiences for customers and developers alike.

We achieve a "win/win" result through crafting a "coherent" product-development lifecycle: one that is customer-focused, business-justified (as demonstrated by ROI), and operating within an integrated product-line context.

To achieve that coherence we use mathematics—in process, methods, and achieving an understanding of our customers—and we use domain-driven development.

In this environment, everyone benefits: our business, our customers, the general public, and the project staff.

These principles have resulted in a major avionics program (the Lockheed-Martin C-130J) achieving better-than-industry productivity and unprecedented integrity for such a large commercial system (250,000+ Ada lines of code). The results were so positive that several of the principles have been adopted by Boeing for its Brimstone missile program.

Consistent application of such a "win/win" process can only translate into long-term competitive advantages...and many more trips for us to the avionics "Club Med," at our special "Motel 6" development rates!

Biographical

James Sutton is on the system engineering team for Boeing North American's Brimstone missile system development. He also serves as its lead software methodologist and software safety criticality engineer. He previously served as lead methodologist and architect on the mission software for Lockheed Martin's C-130J program, as well as safety critical methods liaison with US and international regulatory agencies, reuse IRAD principal investigator, and reuse lead engineer for the F-22. He has authored a college textbook entitled "Power Programming" for Prentice Hall, and has presented and/ or published for numerous conferences including NAECON, Tri-Ada, ERA Avionics Conference (U.K.), Ada-Europe, IEEE DASC, and Compass (safety critical software).

EXPERIENCES FROM SPECIFYING THE TCAS II REQUIREMENTS USING RSML

Mats P.E. Heimdahl, University of Minnesota, Minneapolis, MN
Nancy G. Leveson, Massachusetts Institute of Technology, Cambridge, MA
Jon Damon Reese, Safeware Engineering Corporation, Seattle, WA

Introduction

TCAS II (Traffic alert and Collision Avoidance System II) is an avionics system required on all commercial aircraft with more than 30 passengers. In 1990, FAA deemed the Minimal Operational Performance Standard (MOPS) for TCAS II, expressed in plain English and low-level pseudocode, unacceptable as a basis for government certification. To correct this problem, a high-level software requirements specification (SRS) was developed by reverse engineering the pseudocode and capturing the required behavior using RSML (Requirements State Machine Language), a requirements language based on hierarchical communicating finite state machines [1]. The Irvine Safety Research Group, in cooperation with industrial and government representatives, specified the requirements between 1990 and 1992 [1].

In this paper we provide an overview of this project and the specification technique we used. In the following section we discuss the TCAS II project. Next, we cover some desirable properties of a high-level specification language and provide an overview of RSML. Finally, we share lessons learned and outline current developments.

The System—TCAS II

In 1981, the FAA decided to develop and implement TCAS II. TCAS II is an airborne device that functions independently of the ground-based air traffic control system to provide collision avoidance protection for a broad spectrum of aircraft types (commercial aircraft and larger commuter and business aircraft). To avoid threatening aircraft, TCAS II alerts the pilot of nearby traffic (traffic advisories) and, if necessary, provides recommended escape maneuvers (resolution advisories) in a vertical direction. In 1989, the FAA required that TCAS II be installed on commercial aircraft with more than 30 seats by December 1993 and on commercial aircraft with 10 to 30 seats by 1995.

At this time, the TCAS II requirements were defined by the Minimal Operational Performance Standard (MOPS) document. The MOPS was expressed using a combination of English and approximately 7,000 lines of low-level pseudocode.

Because of perceived deficiencies in this document (discussed in the next section) and the difficulty of FAA certification without high-level system or software requirements, an effort was begun in 1990 to provide such a high-level requirements document for TCAS II.

The Original Specification; The MOPS

The MOPS was expressed using both English and pseudocode, the English is used to comment and explain the code. The pseudocode is a low-level language (called `E`) containing only simple data types, arithmetic expressions, if statements, loop statements, and subroutines. All variables are global: There are no local variables and formal parameters but there are language constructs to indicate which variables are used and modified by a subroutine (few subroutines actually use this feature in the TCAS specification). An example of a routine

0-7803-5086-3 /98/$10.00 ©1998 IEEE

can be seen in Figure 1. The only complex data structure allowed is a "group" that provides for grouping related variables into a "data structure", that is, giving them a group name. In summary, the pseudocode lacks many desirable features of a modern programming language and, more importantly, most desirable features of a high-level specification language.

```
PROCESS Ground_level_estimation;

  IF (O.OOGROUND EQ $TRUE)
    THEN G.ZGROUND = G.ZOWN;
  ELSEIF (G.RADAROUT GT P.RADARLOST)
    THEN G.ZGROUND = -P.ZLARGE;
  ELSEIF (G.RADAROUT GT 0)
    THEN <do not update ZGROUND>}
  ELSEIF (G.RADAROUT EQ -P.ZLARGE)
    THEN
      IF (O.ZRADAR LT P.KNOWGROL)
        THEN G.ZGROUND = G.ZOWN -
O.ZRADAR;
        ELSE; <ground level unchanged>
  OTHERWISE
    IF (O.ZRADAR GT P.KNOWGROH)
      THEN G.ZGROUND = -P.ZLARGE;
      ELSE G.ZGROUND = G.ZOWN - O.ZRADAR;

END Ground_level_estimation;
```

Figure 1. A sample routine from the MOPS

The Participants

To provide a proper requirements specification, a committee of industry and government representatives (RTCA Special Committee 147, or SC-147) was formed and began to develop an English language specification. Participants in this committee included airframe and avionics manufacturers, air traffic controllers, airline representatives, pilots, and FAA representatives. At this time, Dr. Nancy Leveson was looking for a real world system she could use as a testbed for her research in safety-critical systems. Because of its size and complexity, TCAS provided a challenging experimental application of formal specification and analysis methods to a real system.

Initially, the Irvine Safety Research group developed a requirements document purely for research purposes, but it was later adopted by RTCA Special Committee 147 for their specification of the TCAS system requirements.

The requirements were expressed using RSML. The group developed RSML concurrently with the TCAS II requirements, and the experiences and feedback gathered during this effort was continually used to refine and improve the language. A brief description of RSML can be found in the next section. A detailed discussion about the reasons for developing RSML is included in [1]. In short, no other language satisfied our demands for readability and usability combined with formality.

The Solution – RSML

During the initial stages of the project, we reviewed many common specification methods. Unfortunately, we could not find any language and method that satisfied our needs.

The first step in designing a specification language or modeling method is to determine goals and criteria for the language. This section describes general design criteria for such a requirements specification language as well as a short overview of the language used to specify TCAS—RSML.

The Goals of RSML

We identified several criteria that were important with respect to our goals and that we believe apply in general to this type of specification language (a partial list is shown in Figure 2).

- Black-box
- Minimal
- Semantically simple
- Coherent, consistent, and concise
- Unambiguous and formal
- Readable, reviewable, and usable by application experts and developers
- Flexible notations
- Readability given priority over writability
- User needs given priority over personal preferences

Figure 2. Design criteria for the language.

0-7803-5086-3 /98/$10.00 ©1998 IEEE

The first criterion is that the language specifies blackbox behavior of the software only and does not include internal design information in the specification.

Two other criteria are minimality and simplicity. Minimality implies that the specification should contain only the information needed by the developers and analysts. Otherwise, time is wasted in specifying things that are not used. Many of the popular real-time requirements specification languages include facilities that are not strictly necessary. The problem with the "kitchen sink" approach is that the specification language becomes unnecessarily complex and the specification process becomes unnecessarily tedious and time-consuming.

Related to the minimality and simplicity criteria are coherency, consistency, and conciseness. Other specification languages for reactive systems, e.g., Statemate [2], Hatley/Pirbhai [3], and Ward/Mellor [4] include a variety of diverse models, some of which are not formally defined. Our goal was to specify all the required information using one formally defined modeling language based on one underlying state-machine model. We also wanted our language to represent information as economically as possible while still maintaining readability.

Because of our goal to provide a safety analysis of the specification, the language must be unambiguous and the underlying model must have a mathematical foundation. At the same time, the requirements specification must be readable, reviewable, and usable. In some respects, these criteria may be conflicting but it is possible to satisfy both. The specification must be unambiguous and translatable into a formal foundation, but it need not itself include arcane mathematical symbols that are unfamiliar to the application experts and software developers. We spent considerable time and energy developing a notation that was readable yet maintained the underlying formal state-machine model. This notation has graphical, symbolic, and tabular aspects depending on which was best for specifying a particular type of information [5].

Because readability and writability are often conflicting goals, we chose readability in cases where a conflict existed: The added investment in constructing the requirements specification pays off in terms of discovering more requirements-level errors.

As mentioned earlier, we received continual industry feedback on our specification effort. One of the advantages of the feedback was to help us overcome our individual preferences. When devising the specification language, we usually had ourselves in mind as the user. However, our familiarity with certain notations, especially mathematical notations such as predicate calculus, hid their weaknesses. Our first attempts at devising our language, therefore, were failures: the notation was clear to us but not to others. The feedback from a diverse group of users helped us to evaluate the evolving specification language more objectively. The resulting language as well as specification was by all measures very successful.

RSML Overview

RSML is a state-based requirements specification language suitable for the specification of reactive systems. RSML includes several features developed by Harel for Statecharts [2, 6], for example, superstates, AND decomposition, broadcast communication, and conditional connectives. In addition, RSML has some unique syntactic and semantic features that were developed to enhance readability, reviewability, analyzability, and the ability to handle complex systems. A complete description of RSML is provided in [1]. This section contains only a brief overview of the language.

A simple finite-state machine is composed of *states* connected by *transitions* (see Figure 3). *Default* or start states are

0-7803-5086-3 /98/$10.00 ©1998 IEEE

signified by states where a connecting transition has no source. In Figure 3, state *Within-Limits* is the start state. Transitions define how to get from one state to another.

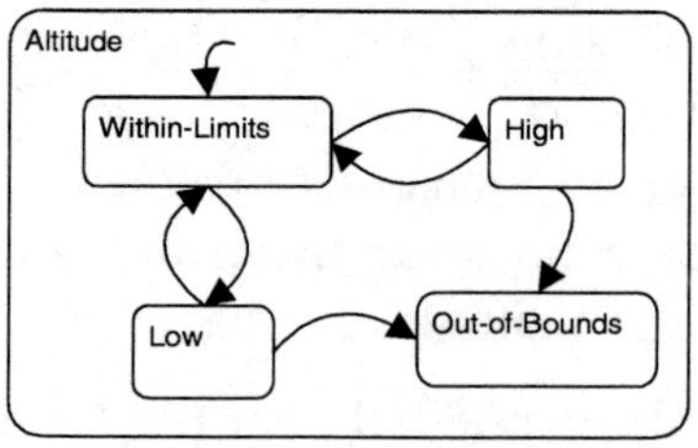

Figure 3. A Basic State Machine Example.

Superstates

In RSML (and Statecharts), states may be grouped into *superstates* (see Figure 4). Such groupings reduce the number of transitions by allowing transitions to and from the superstate rather than requiring explicit transitions to and from all of the grouped states (*substates*). Superstates can be entered in two ways. First, the transition to the superstate may end at the superstate's border (transition T_1 in Figure 4). In this case, a default state must be specified within the superstate. In the example, state *Climb* is entered upon taking transition T_1. Alternatively, the transition may be made to a particular state inside the superstate (transition T_2 in Figure 4). The same superstate may have transitions ending at the border and at any number of the inner states. The superstate may be exited in two ways (transitions T_3 and T_4 in Figure 4). Analogous to transitions into the superstate, transitions out of the superstate may originate from the border or from an inner state.

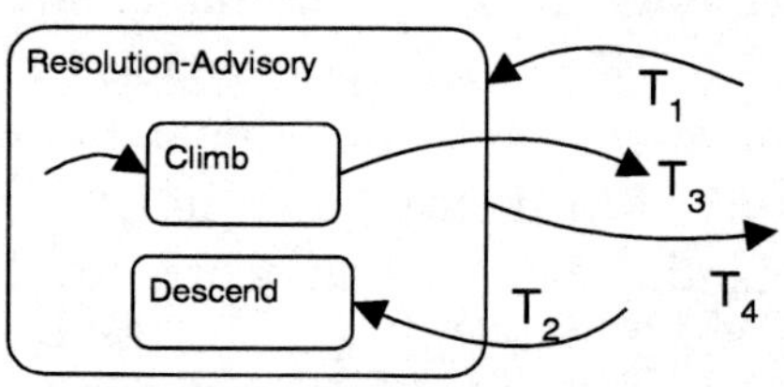

Figure 4. A Superstate Example.

AND Decomposition

One of the most important innovations in Statecharts is what Harel calls the *parallel state*[1], which contains two or more state machines separated by dashed borders (Figure 5). When the parallel state *S* is entered, *each* of the state machines *A, B, C,* and *D* within it is entered. All state machines are exited when *any* transition is taken out of the parallel state. The use of parallel states greatly reduces the size of the specification.

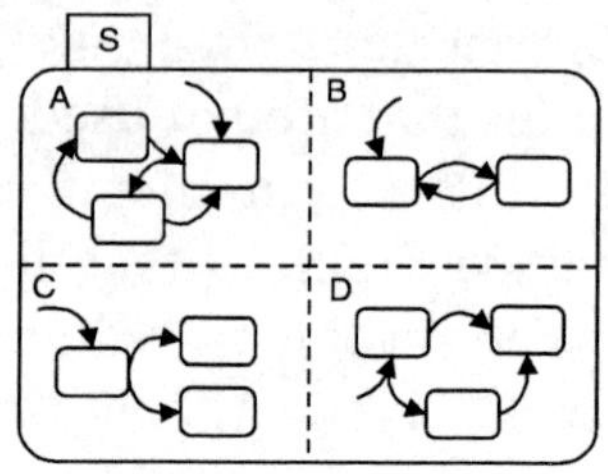

Figure 5. The Parallel State.

Naturally, the TCAS II specification is significantly more complex than the small examples above. To illustrate, we have included a part of the TCAS II specification.

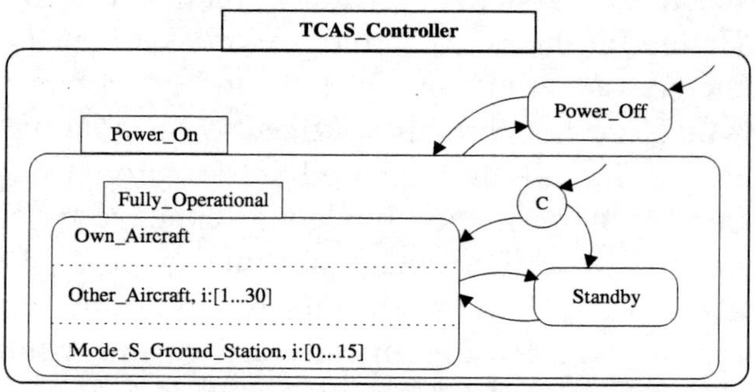

Figure 6. The Highest Level of TCAS.

The highest level TCAS state machine is shown in Figure 6. At this level, TCAS is either on or off; if it is on, it may be either fully operational or in standby mode. In the case of the TCAS logic, the states of three types of process components are modeled: our own aircraft, other aircraft, and mode-S ground radar stations. Each of the three subcomponents of TCAS is elaborated in more detailed RSML models.

[1] Parallel states are also known as "orthogonal products"', "product states"', and "AND states"'.

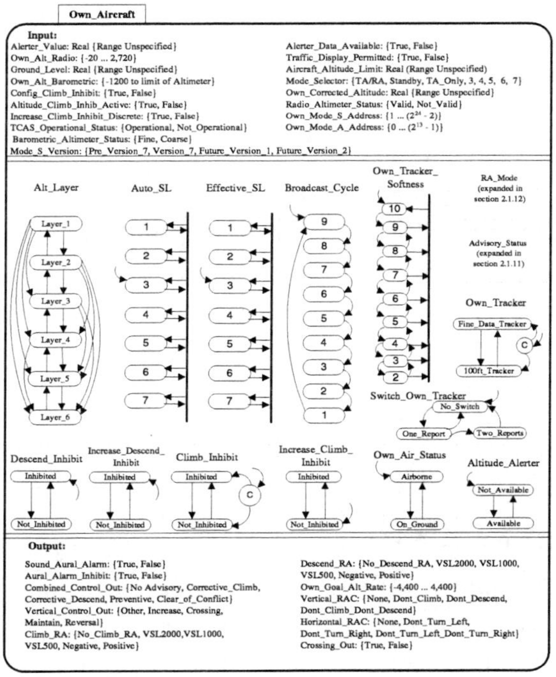

Figure 7. The Model of Own Aircraft

Figure 7 shows the expanded Own-Aircraft portion of the TCAS model. The top portion of the diagram lists variables that represent inputs to TCAS from the sensors that provide information about the state of Own-Aircraft. The bottom portion of the diagram lists variables that represent outputs to TCAS actuators. The middle portion of the diagram represents the parts of the derived Own-Aircraft state necessary for the evaluation of the TCAS control function. Again, for a detailed discussion the reader is referred to [1].

Transition Definitions

Transition definitions in RSML contain five parts: (1) the identification (the source and destination of the transition), (2) the location, (3) the triggering event, (4) the guarding condition, and (5) the output action. The identification, location, and triggering event are the only required parts.

Transitions are taken upon the occurrence of the *trigger event*, provided that the guarding condition is true. The *guarding condition* defines preconditions on the transition and is specified using AND/OR tables, described below. *Output actions* identify events that are generated when the transition is taken. These newly generated events may now trigger transitions elsewhere in the state machine. This *event propagation mechanism* is used to sequence and synchronize the execution of the parallel state machines in the model.

AND/OR Tables

Many state-based languages use standard logic notation to describe the guarding conditions on the transitions [6,8]. Our TCAS external reviewers, however, did not find this notation natural or reviewable. Instead, we decided to use a tabular representation of disjunctive normal form that we call AND/OR tables (see Figure 8 for a transition from the TCAS II requirements).

The far-left column (the wide column) of the AND/OR table lists the logical phrases in the condition. Each of the other columns represents a conjunction of those phrases and contains the logical values of the expressions (a column denotes the logical AND of the phrases in the wide column). A column evaluates to true if all of its elements are true. A dot denotes "don't care". The collection of columns represents a disjunction (indicated by the OR above the columns). Thus, if one of the columns is true, then the guarding condition (the table) evaluates to true.

Experience using RSML

We used this language to develop an experimental specification of TCAS II. This initial specification was well-liked by the RTCA TCAS II Committee. The members felt that adopting a formal approach would help to accurately capture the requirements and enable a shorter time to certification. The RSML language has been subsequently formally adopted for the official requirements specification.

Location: Own_Aircraft$_{s\text{-}40}$ ⊳ Auto_SL$_{s\text{-}89}$
Trigger Event: Descend_Inhibit_Evaluated_Event$_{e\text{-}682}$
Condition:

AND				OR	
	Effective_SL$_{s\text{-}97}$ **in one of** {1, 2}	.	.	.	T
	Effective_SL$_{s\text{-}97}$ **in state** 3	.	.	T	.
	Own_Alt_Radio$_{v\text{-}43}$ ≤ 900 ft$_{(ZSL3TO2)}$	.	.	T	.
	Climb_Descend_Inhibit$_{m\text{-}370}$	.	T	F	.
	Own_Air_Status$_{s\text{-}141}$ **in state** Airborne	F	T	T	T
	Traffic_Display_Permitted$_{v\text{-}50}$	T	.	.	.
	Radio_Altimeter_Status$_{v\text{-}46}$ = Valid	.	.	T	T
	Own_Alt_Radio$_{v\text{-}43}$ < 1,100 ft$_{(ZSL2TO3)}$	.	.	.	T

Output Action: Auto_SL_Evaluated_Event$_{e\text{-}682}$

Figure 8. A transition definition from TCAS II.

After SC-147 decided to adopt the RSML specification, we received continual feedback on our effort from the full SC-147 and from a smaller ad-hoc Requirements Working Group chartered with the development of the requirements.

In March 1992, the document was delivered to SC-147 for verification. The RSML specification contained approximately 47 input and output variables, 140 states, and 170 transitions. The specification has since then undergone an extensive verification and validation effort, and is currently being maintained and extended by the Rannoch Corporation.

Lessons Learned

During the course of the project we learned several lessons, some of which are summarized in this section. The lessons generally fell into two categories (1) general lessons regarding requirements specification and (2) lessons related to the use of formal specification languages.

General Lessons Learned

Keeping the design out is hard: Initially, we had difficulty abstracting away from the design. Even when we did not look at the pseudocode, we found it difficult in the beginning to eliminate functional decomposition and flowchart-like logic, i.e., to specify the problem without trying to solve it. With practice we became better at omitting design information, but the struggle never entirely abated. The very low level of the pseudocode also made the process of abstraction more difficult as many purely implementation features, such as flags, were used extensively in the pseudocode. After the specification of the CAS logic was completed, an independent verification and validation was performed to compare the pseudocode specification and the RSML specification. The verifiers experienced the same problems that we did, and a large number of identified discrepancies between the pseudocode and the RSML specification resulted in no change to the RSML specification because they merely represented design peculiarities of the pseudocode and not requirements.

Understanding intent is difficult: Although it may be a function of the particular system we were working on, we found it impossible to derive the requirements specification strictly from the pseudocode and an accompanying English language description. Although the basic information was all there, the *intent* was missing. The rationale for the various system design tradeoffs was never recorded. Therefore, distinguishing between requirements and artifacts of the implementation was not possible

0-7803-5086-3 /98/$10.00 ©1998 IEEE

in all cases. An audit trail of decisions and the reasons why decisions were made is absolutely essential. This was not done for TCAS over the 15 years of its development.

We should have started from scratch: The final requirements specification model would have been different and much simpler if we had been starting from scratch. Because the TCAS pseudocode specification had evolved over a period of more than 15 years, the current version contains more complexity than is necessary. This is a common maintenance dilemma, and TCAS was no exception. When changes are made to design or code without backing up all the way to requirements, such problems arise and increase as time passes. For TCAS, the highest-level specification *was* the pseudocode.

Because of the necessity of building a requirements specification that matched the TCAS systems actually in use (which were certified against the pseudocode specification), our resulting model was more complicated than necessary, included more than the minimum required behavior, and was harder to understand than was strictly necessary. This was frustrating as we first built a nice, simple model and found that we had to complicate it for no better reason than that it had to match some errors or poor design decisions in the pseudocode. We believe that if a blackbox behavioral model of our type had been built originally, not only would the final specification be simpler and more understandable, but making changes without introducing errors or unnecessarily complicating the resulting requirements also would have been simplified.

Language Lessons Learned

Formal requirements are feasible: One result of this effort was a demonstration that formal specifications can be applied to complex, reactive systems and that such specifications can be readable and reviewable by application experts with a minimal knowledge of mathematics and computer science. Formal specifications are clearly usable if their design takes into consideration the training and backgrounds of those who are to read and review the specification. Some engineers working with us on the TCAS specification reported that they liked the AND/OR table description of the transition condition because it resembles the logic tables that they are used to using and that the state machines and logic tables fit the way they think about systems.

Readability, simplicity, and usability are extremely important: Reviews of our document for correctness by users during development made clear that specifications should include graphical, symbolic, tabular and textual notation, depending on the type of information being conveyed. For example, the graphical state machines were a great help during reviews for finding certain types of errors as were the tables for finding other errors. A language that contains only graphics or only tables or only symbolic strings is probably less useful than one in which different notational techniques are used to communicate different types of information.

Easy to be blind to the users needs: Although formal specification languages obviously have to be defined in an unambiguous and mathematical way, the syntax itself does not have to contain obscure mathematical symbols that are familiar and comfortable to neither the application expert nor the implementor of the system. Currently, formal specification languages are designed primarily by mathematicians who use a notation with which they are comfortable, but which is foreign to those who must use the language. One solution is to train hardware and software engineers to think like mathematicians while our alternative solution is to provide languages that allow the user to think about the system in the way that they have been trained in their discipline. We hypothesize that providing a model of a system that is closer to the mental model that the reviewer and implementor have of the system will aid in finding errors in the specification itself and reduce the numbers of errors that are introduced when implementing the specification. This hypothesis, of course, still needs to be experimentally validated, although our experience provides some convincing anecdotal support.

Summary and Recent Developments

This paper has discussed an approach to specifying system requirements for real-time, reactive systems, some criteria that should be used in designing a language for such requirements, and some lessons learned while writing a system requirements specification for an aircraft collision avoidance system.

Since the completion of the project we have made advances in many areas, two of which are of particular interest to the practicing engineer: (1) Specification and traceability of intent and design rationale and (2) an improved specification language and modeling methodology.

Intent Specifications

Leveson has developed a new way of structuring specifications called *Intent Specifications* [8]. Instead of the usual hierarchical abstraction based on *what* and *how*, Intent Specifications abstract on intent or *why*. Because each level of the specification is mapped to the appropriate parts of the intent levels above and below it, traceability of design rationale and design decisions is provided from system-level requirements and constraints down to code (or physical form if the function is implemented in hardware) and vice versa.

Intent specifications integrate formal and informal aspects of system and software development. The structure is designed to facilitate the tracing of system level requirements and constraints into the design and assist in the assurance of various system properties—such as safety, security, and survivability—in the initial design as well as reduce the costs of implementing changes and reanalysis when the system is changed, as it inevitably will be. Leveson has applied these ideas to specifying control systems with shared human and computer control and to assuring safety in these systems. Leveson and Reese then demonstrated the approach's feasibility by creating a complete (800-page) intent specification (from high-level system goals down through code) for TCAS II.

SpecTRM

After the completion of the TCAS II project our method and language have evolved. A commercial version of the approach is named SpecTRM (pronounced "spectrum") [8]. The name stands for Specification Tools and Requirements Methodology. The specification language in the SpecTRM approach is called SpecTRM-RL (RL for Requirement Language).

SpecTRM-RL is our new language based on the lessons learned from the experiences specifying TCAS as well as subsequent large case studies. SpecTRM-RL improves on usability, readability, and analyzability over RSML. It supports a complete system engineering methodology, is integrated with intent specifications, and extensive tool support will be commercially available shortly.

References

[1] N.G. Leveson, M.P.E. Heimdahl, H.Hildreth, and J.D. Reese. Requirements specification for process-control systems. *IEEE Transactions on Software Engineering*, vol-20, no-9,September 1994.

[2] D. Harel, H. Lachover, A. Naamad, A. Pnueli, M. Politi, R. Sherman, A. Shtull-Trauring, and M. Trakhtenbrot. Statemate: A working environment for the development of complex reactive systems. *IEEE Transactions on Software Engineering*, vol-16, no-4, April 1990.

[3] D. Hatley and I. Pirbhai. *Strategies for Real Time System Specification*. Dorset House Publishing, 1987.

[4] P. Ward and S. Mellor. *Structured Development for Real-Time Systems*. Yourdon Press, 1985.

[5] M. Fitter and T.R.G. Green. When do diagrams make good computer languages? *International Journal on Man-Machine Studies*, no-11, 1979.

[6] D. Harel. Statecharts: A visual formalism for complex systems. *Science of Computer Programming*, 8:231—274, 1987.

[7] C.L. Heitmeyer, R. Jeffords, and B.L. Labaw. Consistency checking of SCR-style requirements specifications. *ACM Transactions on Software Engineering and Methodology*, vol-5(3):231—261, July 1996.

[8] SpecTRM: A CAD System for Digital Automation. *Proceedings of the 17th Digital Avionics Systems Conference (DASC).* Seattle, WA, 1998.

SCR: A PRACTICAL METHOD FOR REQUIREMENTS SPECIFICATION

Constance Heitmeyer, Naval Research Laboratory, Washington, DC

Abstract

A controversial issue in the formal methods research community is the degree to which mathematical sophistication and theorem proving skills should be needed to apply a formal method. A premise of this paper is that formal methods research has produced several techniques with potential utility in practical software development, but that mathematical sophistication and theorem proving skills should not be prerequisites for using these techniques. In the paper, several attributes needed to make a formal method useful in practice are described. These attributes include user-friendly notation, automated (i.e., push-button) analysis, and easy to understand feedback. To illustrate the attributes of a practical formal method, a formal method for requirements specification called SCR (Software Cost Reduction) is introduced.

Formal Methods in Practice: Current Status

During the last decade, researchers have proposed numerous *formal methods* for developing computer systems. These include formal specification languages and formal analysis techniques, such as model checkers and mechanical theorem provers. One area in which formal methods have already had a major impact is hardware design. Not only are companies such as Intel beginning to use model checking, along with simulation, as a standard technique for detecting design errors, in addition, some companies are developing their own in-house model checkers. Moreover, a number of model checkers customized for hardware design have become available commercially.

In contrast, the use of formal methods in practical software development is rare. A significant barrier is the widespread perception among software developers that formal notations and formal analysis techniques are difficult to understand and apply. Many software developers also express serious doubts about the scalability and cost-effectiveness of formal methods.

Another reason for the minimal impact of formal methods in software development is the absence in most software development processes of two features common in hardware design. First, hardware designers routinely use one of a small group of languages, e.g., Verilog or VHDL, to specify their designs. In contrast, precise specification and design languages are rarely used in software development.

Second, at hardware companies, integrating a formal method, such as a model checker, into the design process is relatively easy because other tools, such as simulators and code synthesis tools, are already a standard part of the design process. In software development, in contrast, no standard engineering process exists. For example, although "object-oriented design" (OOD) has become quite popular in recent years, OOD is still largely informal. Moreover, despite the growing availability of commercial CASE (Computer-Aided Software Engineering) tools, software developers rarely use CASE tools during the early stages of software development when such tools (and formal methods) have the greatest benefits.

While formal methods have significant potential for reducing software development

0-7803-5086-3 /98/$10.00 ©1998 IEEE

costs and increasing software quality, the above barriers must be overcome before these benefits can be realized. Below, I recommend several improvements needed before formal methods can be useful in practical software development. I then introduce the SCR requirements method, an example of a "practical" formal method useful in software development.

Toward *Practical* Formal Methods

Described below are four improvements needed to make current formal methods more useful in software development. While most improvements can be achieved by better engineering, in some cases, additional research is needed.

Minimize Effort and Expertise Needed to Apply the Method

To be useful in practice, formal methods must be convenient and easy to use. Currently, most software developers are reluctant to use formal methods because they find the learning curve too steep and the effort required to apply the method too great. In many cases, deciding not to use a formal method is rational. The time and effort required to learn about and apply a formal method may not be worth the information and insight provided by the method; in some cases, the effort could be better spent applying another method, such as simulation. Suggested below are three ways in which the difficulty of learning and applying a formal method can be reduced.

- **Offer a language that software developers find "natural".** To the extent feasible, a language syntax and semantics familiar to the software practitioner should be supported. The language should have an explicitly defined formal semantics and should scale. Specifications in the language should be translated automatically into the language of a formal analysis tool, such as a model checker.
- **Make formal analysis as automatic as possible.** To the extent feasible, analysis should be "push-button", that is, the user should be able to invoke an analysis technique with the mere push of a button.
- **Provide good feedback.** When formal analysis exposes an error, the user should be provided with easy-to-understand feedback useful in correcting the error.

Provide a Suite of Analysis Tools

Because different tools detect different kinds of errors, a "complete" suite of tools, carefully integrated to work together, should be available. Many hardware designers are already using suites of tools, such as simulators, model checkers, equivalence checkers, and code synthesis tools. One benefit of a suite of tools is that properties shown to hold with one tool may simplify analysis with a second tool. For example, demonstrating that the specified behavior is deterministic can simplify later analysis with a model checker.

Integrate the Method into the User's Software Development Process

To the extent feasible, formal methods should be integrated into the existing user design process. Techniques for exploiting formal methods in object-oriented software design and in software development processes which use semiformal languages, such as Statecharts, should also be explored. How formal methods can be integrated into the user's design process should be described explicitly.

Provide a Powerful, Customizable Simulation Capability

Many researchers underestimate the value of simulation in exposing defects in software specifications. Unlike formal techniques which check the specification for properties of interest, simulation helps the user *validate* the specification. By using the simulator to symbolically execute the system, the user can ensure that the specified behavior captures his intent. One promising approach to selling formal methods is to build customized simulator front-ends, tailored to particular application domains. Such customized simulators can serve as system "prototypes", useful in demonstrating and analyzing the required system behavior prior to coding.

The SCR Requirements Method

The SCR (Software Cost Reduction) requirements method is a formal method based on tables for specifying the requirements of safety-critical software systems. Originally formulated in 1978 by NRL (Naval Research Laboratory) researchers to document the Operational Flight Program (OFP) requirements of the US Navy's A-7 aircraft, SCR has been used in practice by numerous organizations, including Grumann, Ontario Hydro, Bell Laboratories, and Lockheed, to specify software requirements.

The SCR method uses a tabular notation to specify requirements. Underlying the tabular notation is a state machine model. Specifications based on tables are relatively easy for software practitioners to understand and to produce. In addition, tables provide a precise, unambiguous basis for communication among practitioners and a natural organization for independent construction, review, modification, and analysis of parts of a large specification. Finally, tabular notations scale. Evidence of the scalability of tabular specifications has been demonstrated by engineers at Lockheed, who used SCR-style tables to specify the complete requirements of the C-130J OFP, a program containing over 230K lines of Ada.

The SCR Toolset

Introduced in 1995, SCR* is an integrated suite of tools supporting the SCR requirements method. The toolset includes a *specification editor* for creating a requirements specification, a *dependency graph browser* for displaying the variable dependencies in the specification, a *consistency checker* for detecting errors such as type errors and missing cases, a *simulator* for validating the specification, and a *model checker* for checking application properties. Currently, more than 50 commercial, academic, and government institutions in the US, Canada, UK, and Germany, are experimenting with SCR*.

Specification Editor

To create, modify, or display a requirements specification, the user invokes the specification editor. Each SCR specification is organized into dictionaries and tables. The dictionaries define the static information in the specification, such as the user-defined types and the names and values of variables and constants. The tables specify how the variables change in response to input events. One important class of tables specifies the values of the system outputs.

Dependency Graph Browser

Understanding the relationship between different parts of a large specification can be difficult. To address this problem, the Dependency Graph Browser represents the dependencies among the variables in a given SCR specification as a directed graph. By examining this graph, a user can detect errors, such as undefined variables and circular definitions. The user can also use the DGB to display and extract parts of the dependency graph, e.g., the subgraph containing all variables upon which a selected controlled variable depends.

Consistency Checker

The consistency checker detects syntax and type errors, variable name discrepancies, missing cases, unwanted nondeterminism, and circular definitions. When an error is detected, the consistency checker provides detailed feedback to facilitate error correction by displaying the table (or dictionary) containing the error and highlighting the erroneous entries. It also provides a "counterexample", i.e., an example that demonstrates the error. A form of static analysis, consistency checking is usually less expensive computationally than model checking. In developing an SCR specification, the user normally invokes the consistency checker first and postpones more expensive analysis, such as model checking, until later. Exploiting the special properties guaranteed by consistency checking (e.g., determinism) can make later analyses more efficient.

Simulator

To validate a specification, the user can run the simulator and analyze the results to ensure that the specification captures the intended behavior. Additionally, the user can define properties believed to be true of the required behavior and, using simulation, execute a series of scenarios to determine if any violate the properties.

The simulator supports the construction of front-ends, tailored to particular application domains. One example is a customized front-end for pilots to use in evaluating an attack aircraft specification (see **Figure 1**). Rather than clicking on variable names, entering values for them, and seeing the results of simulation presented as variable values, a pilot clicks on visual representations of cockpit controls and views the results on a simulated cockpit display. This front-end allows the pilot to move out of the world of requirements specification and into the world of attack aircraft, where he is the expert. Such an interface facilitates validation of the specification.

Model Checker

Recently, the explicit state model checker Spin was integrated into SCR*. After using the SCR tools to develop a requirements specification, a developer can automatically translate the specification into *Promela*, the language of Spin, and then invoke Spin within the toolset to check that the specification satisfies properties of interest. The user can use the simulator to demonstrate and validate any property violation detected by Spin.

The number of reachable states in a state machine model of real-world software is usually very large, sometimes infinite. To make model checking practical, three push-button techniques have been developed, which derive sound abstractions from SCR specifications. The methods are practical: none requires ingenuity on the user's part, and each derives a smaller, more abstract specification automatically. For example, prior to invoking Spin to check a weapons system specification for a safety property, we used our abstraction methods to automatically reduce the number of variables in the specification from over 250 to 55 and to replace several real-valued variables with finite-valued variables. Without such reductions in the number of variables and in the type sets of the real-valued variables, model checking would be infeasible.

Practical Use of the SCR Tools

To date, the SCR tools have been applied in three pilot projects external to NRL. In the first, researchers at NASA's IV &V Facility used SCR* to detect missing cases and instances of nondeterminism in the prose requirements specification of software for the International Space Station. In the second project, engineers at Rockwell-Collins used the tools to expose 24 errors, many of them serious, in the requirements specification of an example flight guidance system. Of the detected errors, a third were uncovered by entering the specification into the toolset, a third in running the consistency checker, and the remaining third in executing the specification with the simulator. In a third project, researchers at the JPL (Jet Propulsion Laboratory) used SCR* to analyze specifications of two components of NASA's Deep Space-1 spacecraft for errors.

In a fourth pilot project, NRL applied the SCR tools to a sizable contractor-produced requirements specification of the Weapons Control Panel (WCP) for a safety-critical US military system. The tools uncovered numerous errors in the contractor specification, including a safety violation which could result in the malfunction of a weapon. Translating the contractor specification into the SCR tabular notation, using SCR* to detect specification errors, and building a working prototype of the WCP required only one person-month, thus demonstrating the utility and cost-effectiveness of the SCR method.

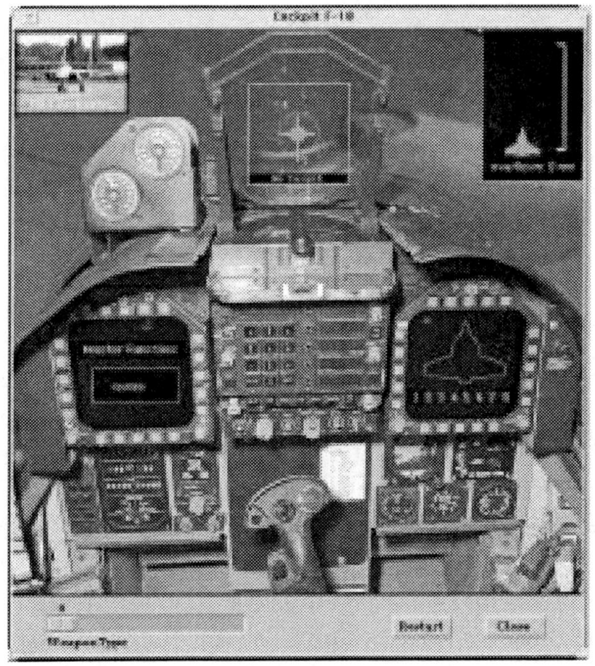

Figure 1. Customized Simulator Front-End for an Aircraft Specification

Conclusions

The SCR tools can be distinguished in three major ways from commercial tools and other research tools. First, unlike most commercial tools for requirements specification, SCR* has a solid mathematical foundation, thus allowing mathematically sound analyses, such as consistency checking and model checking, unsupported by current CASE tools. Second, the SCR tools, unlike most research tools, have a well-designed user interface, are integrated to work together, and provide detailed feedback when errors are detected to facilitate their correction. Finally, users of SCR* can do considerable analysis *without* interaction with application experts or formal methods researchers, thereby providing formal methods usage at low cost.

0-7803-5086-3 /98/$10.00 ©1998 IEEE

USING MODELS FOR TEST GENERATION AND ANALYSIS

Mark R. Blackburn, Software Productivity Consortium, Herndon, Virginia

Introduction

Systems are increasing in complexity. More systems perform mission-critical functions, and dependability requirements such as safety, reliability, availability, and security are vital to the users of these systems. The competitive marketplace is forcing companies to define or adopt new approaches to reduce the time-to-market as well as the development cost of these critical systems. Much focus has been placed on front-end development efforts, not realizing that testing accounts for 40 to 75 percent of the lifetime development and maintenance costs [Bei83; GW94]. Testing is traditionally performed at the end of development, but market-driven schedules often force organizations to release products before they are adequately tested. The long-term effect is increased warranty costs due to products' poor reliability and poor quality.

Model-based development tools are increasing in use because they provide tangible benefits by supporting simulation and code generation, in addition to the traditional design and analysis activities. These tools help users develop requirement and design models of target systems. Certain tools are based on formal models, and the underlying models are represented using specification languages. Such formal specifications provide a basis for test case generation. However, the underlying development models are generally not represented in a form that supports automatic test case generation. The key challenge is to translate development-oriented modeling languages into a form that is suitable for automated test vector generation, specification-based test coverage analysis, requirement-to-test traceability, and design-to-test traceability.

Using Models for Testing and Analysis

Figure 1 illustrates a conceptual view for using models to support test generation and analysis. Models and their associated tools typically provide various views of the system under development. When modeling tools are based on precise semantics, user models can also support:

- ***Test Vector Generation***. A test vector includes inputs, expected outputs, and an association with the specification from which it was derived
- ***Static Analysis***. Typically used to determine if there are contradictions in the specification
- ***Dynamic Analysis***. Analysis based on execution of the model.

Modeling tools are beginning to support simulation and code generation. Simulation of a model can help developers assess the correctness of the model with respect to the user requirements; however, it can be time consuming to develop simulation data required for thorough dynamic analysis. Automatically generated test vectors can provide a cost-effective way to exercise a model in a simulator using the boundary values associated with the constraints of a model specification; it is at the boundaries where model anomalies are typically discovered. In addition, these same test vectors can also be used to test the code in a host or target environment.

0-7803-5086-3 /98/$10.00

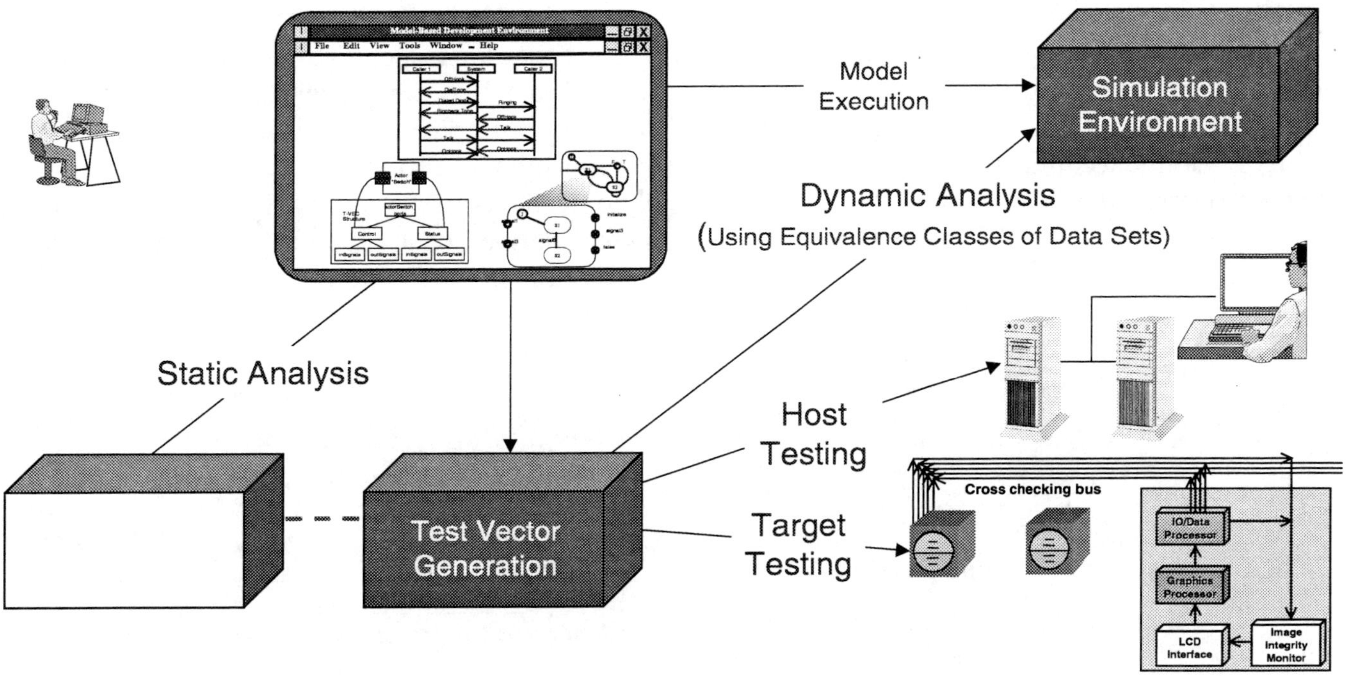

Figure 1. Using Models for Test Generation and Analysis

Scope

This paper describes the use of automated test generation and analysis from specification models. Through the integration of commercial off-the-shelf (COTS) model-development and test generation tools, a process has been developed that eliminates most of the traditional testing activities. This approach has been demonstrated to identify many types of specification errors prior to any implementation. This paper is based on the experiences in developing two model translators [BF98; Bla98] supporting:

- Software Cost Reduction (SCR) [Hen80]/Consortium Requirements Engineering Method (CoRE) [SPC93] for modeling requirements
- Real-Time Object-Oriented Modeling (ROOM) [SGW94] method for analysis and design

For each respective method and associated tool, the translators produce a specification that is used by the T-VEC tool system to generate test vectors and perform specification-based test coverage analysis. The model transformation process is briefly described using a specification example. The paper summarizes the results of applying the process and tools to industrial applications.

Models and Specifications

Formal specifications provide simple abstract descriptions of the required behaviors describing what the software should do. Because formal specifications have, in the past, been considered difficult to use, they have not been widely used. Recent advances in visual model-based development tools provide the basis for developing formal specifications while hiding the formalism.

It has been commonly accepted that formal specifications provide a basis for test case generation. Goodenough and Gerhart may have been the first to claim that testing based only on a program implementation is fundamentally flawed [GG75]. Gourlay developed a mathematical framework for specification-based testing [Gou83]. Figure 2 graphically represents Gourlay's mathematical framework for testing and the key relationships between specifications, tests, and programs. Given a **specification** that describes the requirements for some system, there is one or more **programs** that implement the specification. **Tests** are derived from the specification; if every test executed by a program computes the appropriate expected results (i.e., passes every test), there is some level of confidence that the program satisfies the specification.

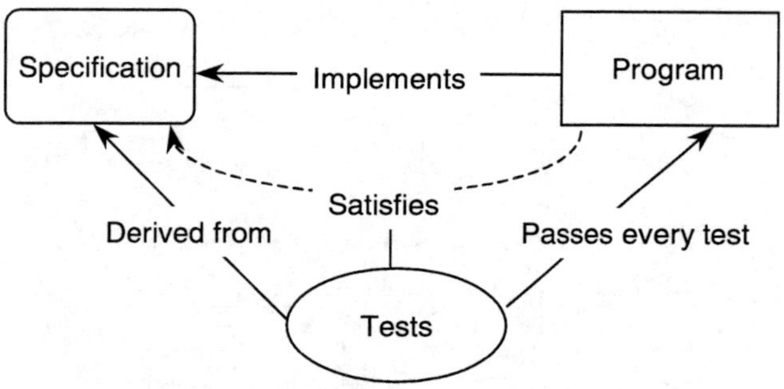

Figure 2. Testing Model and Relationships

In Figure 2, the specification symbol (i.e., rounded rectangle) is generically used to represent requirement, design, or test specifications. Certain specification languages have tool support that helps in developing complete and consistent specifications. Such tools provide the syntactic and semantic rigor that is required for transforming specifications into a form suitable for test vector generation. Model-based specification methods that support functional, state transition, and event-based techniques are increasing in popularity and use because the tool support has helped make them easier to use.[1]

A **model-based specification** approach constructs an abstract model of the system states and characterizes how a state is changed by abstract or concrete operations (paraphrased from Cohen et al. and Cook et al. [CHJ88; CGDDTK96]). Operations in the system are specified by defining the state changes or events that affect the model using existing mathematical constructs like sets or functions. **State transitions** define relationships between sequences of states based on conditions of the system state. **Event specifications** define certain conditions related to a change in the system state.

A **test specification model** is defined by a set of test specification elements, as shown in Figure 3. A **test specification element** is an input-to-output relation and an associated constraint defined by a conjunction (i.e., logically ANDed) of Boolean-valued relations that define constraints on the inputs associated with the input-to-output relation.

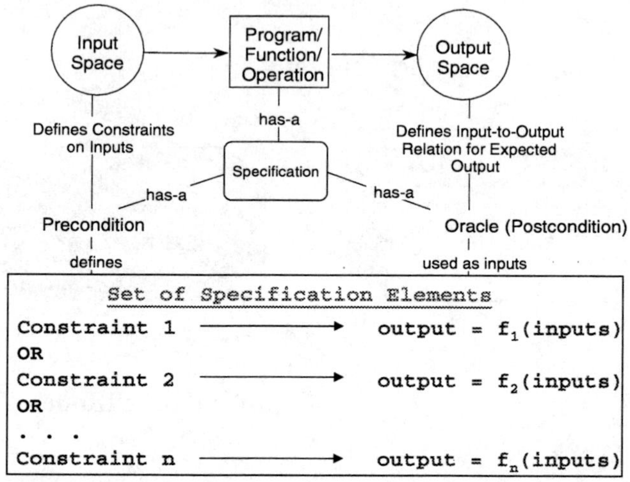

Figure 3. Representation of Test Specification Model

Given a specification element, a **test vector** is a set of test inputs values derived from the constraint, and an expected output value derived from the input-to-output relation with respect to the test input values [BB96]. Informally, from a test generation perspective, a specification is **satisfiable** if at least one test vector exists for every specification element [BBF97].

Model Transformation

Model transformations are typically required to transform model-based specifications into a form to support test generation. Hierons describes rewriting rules for Z specifications [Hie97] to support test case generation, but does not address specifications composed of combinations of specification techniques, particularly specifications composed using event specification techniques. In general, model transformation to support tool interoperability is an important area of investigation [Gil97].

[1] Zave and Jackson [ZJ97] identify potential implementation bias of model-oriented techniques but support the claim that model-oriented techniques are gaining in popularity.

Blackburn [Bla98] describes a tool-based approach for transforming a model-based specification into a form that supports test vector generation. The model-based specification supports composition using function, state, and event specifications. A translator implements rules for transforming SCR model specifications into a language used by the T-VEC test vector generation tool. The development of the prototype translator and evaluation environment helped identify shortcomings in the rules described in prior work that was presented at the 1997 Computer Assurance Conference [BBF97].

Similar model transformation efforts, not described in this paper, were performed for the ROOM method using the ObjecTime Developer toolset as part of the validation environment [BF98].

Evaluation Environment

Figure 4 identifies generic tool types that are related to the elements of the test model shown in Figure 2. Such tools use or produce the three primary types of system artifacts (i.e., specifications, programs, and tests). A specific instance of this model was created to support the model transformation approach using the SCR tool (referred to as SCR* - pronounced SCR star) as the source for model-based specifications and the T-VEC tool system as the tool that supports test generation and specification-based coverage analysis.

SCR*, developed by the Naval Research Laboratory, supports modeling and analysis of requirement specifications using a formal modeling language (i.e., a language with well-defined syntax and semantics).

T-VEC, developed by T-VEC Technologies, Inc. supports:

- ***Test Vector Generation.*** A test vector generator produces test vectors from test specifications.
- ***Specification-Based Coverage Analysis.*** This tool analyzes the transformed specification to determine whether all specification elements have a corresponding test vector. This is the mechanism used to assess satisfiability of the transformed specifications.

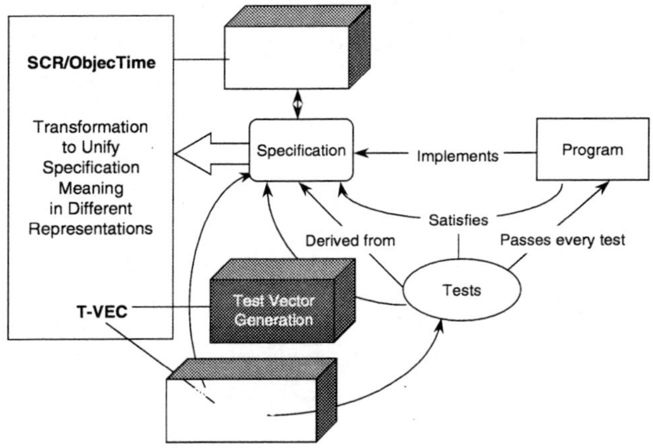

Figure 4. Tools of the Evaluation Environment

Applications and Results

The remainder of this paper describes a simple example to illustrate the use of this approach for model analysis and testing. Consider the example shown in Figure 5 of an electronic regulator. The requirements for the regulator are:

- When the temperature reaches the High zone (i.e., 180 degress), the valve opens.
- The amount the valve opens is a function of the temperature from 120 degrees (closed) up to 300 degrees (fully open).
- Once the valve is open, it remains open until the temperature reaches the Low zone (i.e., 120 degrees).

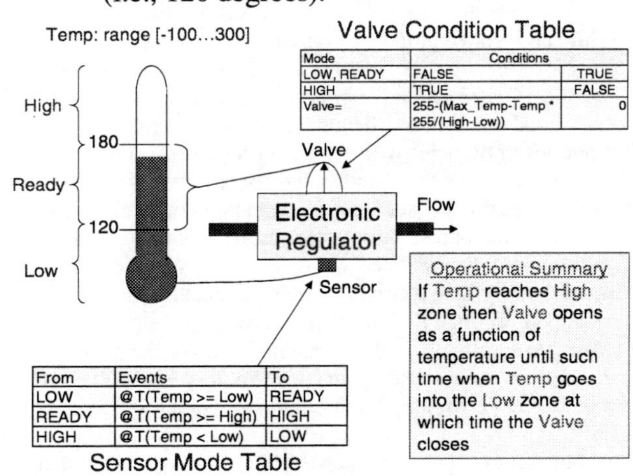

Figure 5. Example of Electronic Regulator

The specification is described in the SCR tabular notation. Heitmeyer et al. [HJL96] describes the SCR method. The specification is defined in two parts. The **first part of the specification** defines the relationships between the temperature and the associated modes that relate to the temperature zones. This is referred to as the Sensor Mode Table shown in Figure 5. The system can be in one of three modes: LOW, READY, and HIGH. At the time when the temperature becomes greater than the constant Low (i.e., 120 degrees), the system transitions into the mode READY. The formal expansion of the event is:

```
@T(Temp >= Low) means:

if the previous value of Temp
denoted NOT(_Temp >= Low) and the
new value of Temp >= Low then the
event is true and the mode
transitions from LOW to READY
```

Table 1 shows the translated meaning for each event specification of the Sensor Mode Table. For each constraint, there is a minimal set of tests as shown in Table 2. The T-VEC test generation system uses a test selection heuristic based on domain testing theory where low-bound and high-bound values are selected for each constraint.[2] For example, the first test selects the low-bound value for the previous state value of _TEMP[3] (-100), which is less than the constant Low, and selects a value of 120 for the next state value of TEMP. For the high-bound selection, the value of 119 (i.e., one less that the constant Low) is selected for _TEMP, and 179 for TEMP (i.e., one less than the constant High).

Table 1. Relationship of Translated Constraints

Events	Translation
@T(Temp >= Low)	(Temp >= Low) AND (_Temp < Low)
@T(Temp >= High)	(Temp >= High) AND (_Temp < High)
@T(Temp < Low)	(Temp < Low) AND (_Temp >= Low)

Table 2. Tests for Each Translated Constraint

	Output	Inputs		
Translation	Sensor	_Sensor	Temp	_Temp
(Temp >= Low) AND	READY	LOW	120	-100
(_Temp < Low)	READY	LOW	179	119
(Temp >= High) AND	HIGH	READY	180	120
(_Temp < High)	HIGH	READY	300	179
(Temp < Low) AND	LOW	HIGH	-100	180
(_Temp >= Low)	LOW	HIGH	119	300

The **second part of the specification** defines the constraints and functions for the Value Condition Table shown in Figure 5. This table depends on the Sensor Mode Table. The Valve Condition Table is interpreted as follows:

```
if Sensor mode = High then
  Valve = 255-(Max_Temp-Temp
          * 255/(High-Low))
else if Sensor mode = LOW
     or Sensor mode = READY then
  Valve = 0
endif
```

Each SCR output variable and associated function map to a T-VEC functional relationship of an output variable with respect to the constraints on the input variables. The SCR model does not necessarily define a system state strictly in terms of constraints on the input variables as is required for T-VEC. For example, the Sensor mode is defined in terms of a mode transition table. This results in table dependencies as illustrated in Figure 6. The mode variables and the associated table relations must be transformed into constraints on the input variables.

---

[2] White and Cohen proposed domain testing theory as a strategy to select test points to reveal domain errors [WC80]. Their theory is based on the premise that if there is no coincidental correctness, then test cases that localize the boundaries of domains with arbitrarily high precision are sufficient to test all the points in the domain. When there is a strong correlation between the specification constraints and implementation paths, the selected test data should uncover computation and domain errors. As defined by Howden and refined later by Zeil, a computation error occurs when the correct path through the program is taken, but the output is incorrect due to faults in the computation along the path. A domain error occurs when an incorrect output is generated due to executing the wrong path through a program [How76; Zei89].

[3] An underbar (_) precedes the variable name to indicate that the variable represents the previous state variable before the event versus the next state variable after the event.

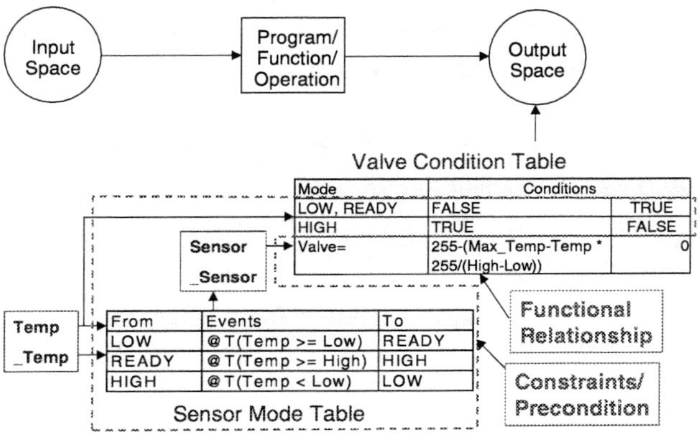

Figure 6. Dependency Relationship

Figure 6 provides a perspective of the example SCR specification represented in a form that is compatible with the test model shown in Figure 3. The constraint for the Valve Condition Table includes the conditions of the Valve table and the Sensor Mode transition table. This means that the constraint:

```
(Temp >= High) AND (_Temp < High)
```

must be satisfied (i.e., the Sensor mode is HIGH) as a requirement for the value to be computed using the functional relationship:

```
1)      255-(Max_Temp-Temp
2)      * 255/(High-Low))
```

In general, mode transition tables can have dependencies on other terms and modes. Events for modes and terms create the need to identify the previous and next state variable dependencies. As shown in Figure 6, the Sensor mode table depends on both the previous and next state input value of Temp; similarly the condition table Valve depends on the previous and next state variables Temp and Sensor.

A test specification requires the constraints of a specification to be defined strictly in terms of the input and output variables. A model-based approach defines states that are relations of inputs, terms, or state variables (e.g., Sensor, _Sensor). This allows the constraint/precondition and functional relationship (defined in terms of a Condition Table) to be defined as a relation on inputs, states, or terms. This approach typically simplifies the task of specifying behavior, but it is the key reason why a model transformation process is required.

Static Analysis

Static analysis helps determine whether there are contradictions in the model without executing the model. Contradictions exist if constraints cannot be satisfied. This is typically the most common problem, especially when the dependencies of specifications become large. This example is a simple 2-level dependency problem, but typical systems can have 10 or more dependency levels. It is also possible to identify functional relationships that specify values that are inconsistent with the domain of the output variables. These are analogous to computation errors in the code.

Consider the function to compute the Valve function. The requirements are that as the temperature reaches the maximum temperature (i.e., 300 degrees), the valve should be completely opened, and when the value reaches the constant Low (i.e., 120 degrees), the valve should be closed. Electronically controlled devices typically use some type of digital value to represent a fully open valve (in this case 255 – an 8-bit unsigned integer), and the value should be 0 when the valve is closed. It is common for implementors to make errors in scaled arithmetic conversions. To illustrate this point, the computation has two errors.

Figure 7 shows a sample test vector that has identified a problem in the computation. A warning is appended to the expected output because the computation is out of range. This is typically an indication that there is a computation error in the specification or that there are missing constraints on the inputs. The original expression (line 1 of the functional relationship under Figure 6) is missing parentheses around Max_Temp-Temp. In line 2, the subtraction should be Max_Temp-Low rather than High-Low. The correct computation is as follow:

```
255-((Max_Temp-Temp) *
255/(Max_Temp-Low))
```

Identifying this type of problem is time consuming. In addition, it is well known that identification and removal of errors in the implementation or integration phase is much more costly than it is during the requirements phase.

Indicates output out of range

```
Valve<<1>>,  RP__1<<1>>
OUTPUT
  Valve FLOAT 32 720.0 {0.00..255.0} "WARNING: VALUE OUTSIDE of EXPECTED RANGE"
INPUTS
  Sensor    ENUMERATION '32  3 HIGH  {LOW .. HIGH}
  Temp      INTEGER      32  180  {-100 .. 300}
  _Sensor   ENUMERATION  32  2 READY  {LOW .. HIGH}
  _Temp     INTEGER      32  120  {-100 .. 300}
JUSTIFICATION {
  SOLUTION : 1
  STATE_SPACE_SCAN : OFF
  SWITCHES : LEAST_RECENT, LOW_BOUND, SINGLE, OPPOSITE
  DCP : 1
    Valve, Valve_FR__1, cv_Valve_RP__1, Valve_RP__1, Valve_RP__0,
    Sensor_LS, Sensor<<2>>, Sensor_FR__2, Sensor_RP__2,
    Sensor_RP__0, Sensor_Valve_2
}
```

Figure 7. Internal Form of Test Vector With Warning

Figure 8 provides a summary of a minimal set of test values for the translated condition table for Valve. In this figure, the associated test selection mode (i.e., LOW_BOUND, HIGH_BOUND) is also shown.

Mode	Conditions		Output Valve	Input Sensor	_Sensor	Temp	_Temp	Test Selection Mode
HIGH	TRUE	FALSE	85.00	HIGH	READY	180	120	LOW_BOUND
			255.00	HIGH	READY	300	179	HIGH_BOUND
LOW, READY	FALSE	TRUE	0.00	LOW	HIGH	-100	180	LOW_BOUND
			0.00	LOW	HIGH	119	300	HIGH_BOUND
			0.00	READY	LOW	120	-100	LOW_BOUND
			0.00	READY	LOW	179	119	HIGH_BOUND

Figure 8. Test Vectors for Valve Condition Table

Sample Results

Table 3 shows some sample results on the application of this approach to other systems. Each specification originally had one or more specification problems or anomalies. As seen in Figure 5, the electronic regulator problem is very small (two tables, five functional relationship, two constraints, and maximum depth of two table dependencies). A flight guidance system is a real-world industrial problem [Mil98]; it has 78 tables, 423 functional relationships, 7,349 constraints, and a maximum dependency depth of 12. The results on this project are planned for publication in the next year.

Table 3. Sample Results Statistic

System/Projects	Condition Table	Event Table	Mode Table	Functional Relationship	Constraint	Level
Temperature regulator	1		1	5	6	2
Safety injection	1	1	1	10	68	3
Electronic flight instrumentation system	37	5	0	88	389	3
Elevator system	10	6	0	38	90	3
Flight guidance system	49	15	14	423	7349	12

Summary

Software testing will play a role in the development of software systems for some time to come. Although testing can account for 40 to 75 percent of the lifetime development and maintenance costs, the results summarized in this paper provide promising evidence that the use of test automation to support the manually intensive test generation and model-based analysis is feasible and practical.

There is a great need to demonstrate and integrate new and advanced technologies. This paper describes an environment developed to validate the use of model-based translators on real-world applications. The environment integrates model-based development tools with a specification-based test vector generator and specification-based coverage analyzer.

As modeling tools and associated methodologies continue to evolve, these results provide the basis for building translators for other modeling tools. This allows new tooling technology to be integrated with existing tools and has the indirect effects of reducing the cost and time of specialized training and tool expenditures.

The ability to integrate front-end development tools with back-end testing tools fosters the use of model-development tools, and such tools can significantly reduce the maintenance phase of a product, which typically consumes 70 percent of the product life cycle. Maintenance typically requires minimal development effort but typically large efforts in testing. Because the original developers usually are not available to assist in maintenance and evolution efforts, test automation can significantly minimize reverification efforts because the designer's requirement and design knowledge is captured in model specifications.

References

[BB96] Blackburn, M.R., R.D. Busser, T-VEC: A Tool for Developing Critical System. In Proceeding of the Eleventh International Conference on Computer Assurance, Gaithersburg, Maryland, pages 237-249, June, 1996.

[BBF97] Blackburn, M.R., R.D. Busser, J.S. Fontaine, Automatic Generation of Test Vectors for SCR-Style Specifications, In Proceeding of the 12th Annual Conference on Computer Assurance, Gaithersburg, Maryland, pages 54-67, June, 1997.

[Bei83] Beizer, B. Software Testing Techniques, New York, New York: Van Nostrand Reinhold, 1983.

[BF98] Blackburn, M.R., J.S. Fontaine, Specification Transformation to Support Automated Testing, TR SPC-97036-MC, Version 02.00.01, Software Productivity Consortium, March 1998.

[Bla98] Blackburn, M.R., Specification Transformation and Semantic Expansion to Support Automated Testing, Ph.D. Dissertation, George Mason University, 1998.

[CGDDTK96] Cooke, D., A. Gates, E. Demirors, O. Demirors, M. Tankik, B. Kramer, Languages for the Specification of Software, Journal of Systems Software, 32:269-308, 1996.

[CHJ88] Cohen, B., W. T. Harwood, M.I. Jackson, The Specification of Complex System, Addison-Wesley, Great Britain, 1988.

[GG75] Goodenough, J. B., S. L. Gerhart, Toward a Theory of Test Data Selection, IEEE Transactions on Software Engineering, 1(2):156-173, 1975.

[Gil97] Gill, D. H., Formal Methods for Software Evolution, Solicitation, Defense Advanced Research Projects Agency, BAA 98-10, November, 1997.

[Gou83] Gourlay, J.S., Introduction to the Formal Treatment of Testing, Software Validation. Proceeding of the Symposium on Software Validation, 1983.

[GW94] Ghiassi, M., K.I.S. Woldman, Dual Programming Approach to Software Testing, Software Quality Journal, 3:45-58, 1994.

[Hen80] Heninger, K., Specifying Software Requirements for Complex Systems: New Techniques and Their Application, IEEE Transactions on Software Engineering, 6(1):2-13, 1980.

[Hie97] Hierons, R. M. Testing from a Z Specification, Journal of Software Testing, Verification and Reliability, 7:19-33, 1997.

[HJL96] Heitmeyer, C., R. Jeffords, B. Labaw, Automated Consistency Checking of Requirements Specifications. ACM TOSEM, 5(3):231-261, 1996.

[How76] Howden, W.E., Reliability of the Path Analysis Testing Strategy, IEEE Transactions on Software Engineering, 2(9):208-215, 1976.

[Mil98] Miller, S., Specifying the Mode Logic of a Flight Guidance System in CoRE and SCR. Accepted to the Second Workshop on Formal Methods in Software Practice (FMSP'98), Clearwater Beach, Florida, March, 1998.

[SGW94] Selic, B., G. Gullekson, P.T. Ward, Real-Time Object-Oriented Modeling. New York, New York: John Wiley & Sons, Inc, 1994.

[SPC93] Software Productivity Consortium, Consortium Requirements Engineering Guidebook, SPC-92060-CMC, version 01.00.09. Herndon, Virginia, 1993.

[WC80] White, L.J., E.I. Cohen, A Domain Strategy for Computer Program Testing. IEEE Transactions on Software Engineering, 6(3):247-257,May, 1980.

[Zei89] Zeil, S.J., Perturbation Techniques for Detecting Domain Errors, IEEE Transactions on Software Engineering, 15(6):737-746, 1989.

[ZJ97] Zave, P., M. Jackson, Four Dark Corners of Requirements Engineering, ACM Transactions on Software Engineering and Methodology, 6(1):1-30, 1997.

CRAFTING AVIONICS DISPLAY APPLICATIONS IN JAVA™

Roger K. Shultz, Rockwell Collins, Cedar Rapids, Iowa
David S. Hardin, Rockwell Collins, Cedar Rapids, Iowa

Introduction

Java has emerged as a set of compelling technologies for the development and deployment of object-oriented applications. The safety features of Java, coupled with its simplified object model and language support for multithreading, has much appeal in the avionics domain. However, concerns over the size, speed and determinism of off-the-shelf Java environments have prompted caution in the real-time embedded community.

Rockwell Collins has developed hardware, software, and tools to make real-time embedded Java applications feasible. We have developed the world's first direct-execution Java microprocessor, the JEM1™. JEM1, the first in the JEM™ family of low-cost, low-power embedded Java CPU's, features direct execution of the entire Java Virtual Machine instruction set [1]. The JEM tool chain leverages off-the-shelf compiler and debugger technology developed for the broader Java market. The final step in the JEM build process is a unique static linker that accepts standard format Java class files and produces very small runtime memory images for dedicated, embedded Java applications.

We have also shown that one can create traditional statically allocated avionics applications in Java. In particular, a JEM/Java avionics display prototype showed performance and memory space improvement when compared to the same application in Ada [2] which executed on a typical processor board.

™ JEM and JEM1 are trademarks of Rockwell. Java and all Java-based marks are trademarks or registered trademarks of Sun Microsystems, Inc. in the United States and other countries.

Java

Sun Microsystems first unveiled Java in 1995, after several years of internal development. The basic Java technologies include the following.

- The Java object-oriented programming language
- The Java Virtual Machine, a platform-independent instruction set for mobile code
- The Java "applet" model, providing a safety and security "sandbox" for mobile code
- A large number of Application Programmatic Interfaces (API's) allowing Java code to call on just about any imaginable system service

The Java Programming Language

The Java programming language is an object-oriented language with syntax similar to C++, but with much simpler semantics. Java features single inheritance, no explicit pointer manipulation, and an Ada-like exception model. Type casts in Java are closely scrutinized, and memory deallocation is implicit. In addition to these safety and security features, Java also embraces multithreading, with the "synchronized" keyword and language-defined priority-preemptive thread model.

The Java Virtual Machine

Java compilers translate Java source into instructions for the Java Virtual Machine (JVM). The JVM is a stack-based 32-bit architecture, with a number of unique instructions, such as virtual method invocation with lock detection. The Java Virtual Machine supports a variety of basic data types, including

0-7803-5086-3 /98/$10.00 ©1998 IEEE

32- and 64-bit integers, 16-bit Unicode characters, and 32- and 64-bit IEEE floating point types.

In a typical Java runtime environment, portable applications or applets compiled into Java Virtual Machine bytecodes are interpreted by a software JVM implementation, or Just-In-Time (JIT) compiled to native machine code. This approach has been quite successful in the desktop and server worlds, where the increased productivity of the Java language, along with its "Write Once, Run Anywhere" promise has fueled a flurry of software development.

The JEM Microprocessor Family

While the portability of Java applications and applets holds great appeal, neither interpretation nor JIT compilation is attractive for real-time embedded use. Thus, in 1996, Rockwell Collins began to develop a direct execution Java microprocessor family, called JEM. The instruction set for JEM includes all the Java Virtual Machine bytecodes; as such, no Java interpreter of Just-In-Time (JIT) compiler is required. JIT's require a good deal of RAM to function effectively, something an embedded system can ill afford. Worse, JIT's introduce unpredictability into the run time of embedded applications, which is very undesirable. By contrast, the space-efficient JEM instruction set can more than halve the amount of code space required for a typical application, and has been designed for predictable Java bytecode execution.

The JEM1 microprocessor, the first in the JEM family of low-cost, low-power CPU's, features direct execution of the entire Java Virtual Machine instruction set, including object-oriented method invocation and IEEE floating point. The JEM instruction set architecture was designed with an embedded systems focus, and thus maximizes the amount of runtime data that can be ROM'ed, and eliminates runtime instruction stream modification. JEM directly supports the Java thread model in hardware, yielding extremely fast thread switching times. JEM also defines a set of 'extended' instructions for physical hardware interfacing and other systems programming tasks; these extended instructions are not available to dynamically downloaded Java applets.

The JEM1 core is only 6 square millimeters in 0.5 micron CMOS. The JEM1 core dissipates only 60 mW at 3.3V and 50 MHz. JEM1 is intended mainly for use as a microcontroller core in applications such as telecommunications, navigation, and 'smart cards', although the first JEM1 device is a standalone CPU. The JEM1 core includes an interrupt controller and two programmable timers with prescaler. JEM1 supports 32, 16, or 8 bit external data buses, features two power-saving modes, and provides an IEEE 1149.1 (JTAG) test interface.

Deterministic Execution Support

One of the most significant issues in adopting Java for embedded systems is the non-determinism introduced by garbage collection and runtime class resolution. We have addressed the garbage collection problem by providing hardware support for multiple logical Java Virtual Machines, each with potentially different space and time management policies, on the same CPU. Hard real-time JVM's on a JEM CPU allocate dynamic memory only at initialization time, and do not attempt to free it via garbage collection. Other 'applet' JVM's, which can be dynamically downloaded, can use garbage collection (if needed) to manage the heap space allocated to them by the system. An applet, however, cannot exceed its initial heap allocation. The system also sets up an applet's time allocation, and does not allow the applet to exceed its time budget.

Most Java implementations rely on runtime class resolution, but this is not necessary if all the necessary classes are present at build time. Thus, we have directed the development of a unique set of software for the creation of embedded Java systems. A static linker computes the class dependencies for a standard format Java class file, and creates JEM RAM and ROM memory images only for those classes that are needed. The static linker can also eliminate unused methods, fields, and constants, resulting in significantly reduced memory requirements (often by a factor of 10) for dedicated Java applications. Java's safety and security orientation, particularly the avoidance of pointer manipulation, helps make this build-time analysis possible.

JEM Runtime: No Assembly Required

The Rockwell Java CPU team decided early on that JEM should be able to be programmed completely in Java, including substitutes for the many 'native methods' [1] in the Java runtime system. To that end, the static linker supports the substitution of certain method invocations with JEM instructions. With this approach, there is no need for an assembler, and developers can create Java implementations for the JEM instruction classes (e.g. physical memory access) in a host simulation environment. The result is that the entire JEM runtime is written in Java, thus easing both testing and maintenance.

JEM/Java Display Prototype

The display prototype was implemented in Java and benchmarked on a JEM1 against the Ada version running on a Collins Advanced Architecture Microprocessor-5 (AAMP5) [3].

The Collins Common Graphics Engine Interface (CGEI) is a library of graphical operations applications call to draw objects on an LCD panel. The CGEI consists of a well-documented object oriented design. In order for it to be implemented in Ada83, the class hierarchy was flattened because Ada does not support inheritance. The Java CGEI closely follows the original design.

This benchmark application consists of avionics graphics. Collins Air Transport Division provided frequency of occurrence data for strings, characters, lines, and shapes. The data is representative of avionics display products and was used to generate the test application.

The Ada version was timed on the Graphics Processor Module containing an AAMP 5. The Java version was timed on a JEM1 board.

Each application was characterized by the following metrics:

- Benchmark identification
- Benchmark description
- Language
- Compilation tools and versions
- Compilation options
- Host platforms
- Target platforms
- Display update rate
- Application source size
- Library source size
- Application object code
- Executable code sizes (i.e. RAM, ROM, Stack, Heap, Libraries, Constants)

Java Configuration

The Java compiler was the Sun JDK *javac* version 1.1.5. with options -O The following table describes our Java tools.

Linker:	Intermetrics SLINK V3.4
JEM Runtime	1.12
PC JVM	Borland JBuilder v1.0
Unix JVM	Sun JDK java version 1.1.5
Target	JEM1r0
External Clock Speed	50Mhz
Internal Clock Speed	25Mhz
Bus Speed	25Mhz
Wait States	0

Ada Configuration

The Ada compiler was the ADACS version 7.6 with compiler options for optimization and run-time checks disabled. The following table describes our Ada tools.

Linker:	ADACS version 7.6(CAPS Linker)
Target:	AAMP5vr1
External Clock Speed	20Mhz
Internal Clock Speed	20Mhz
Bus Speed:	20Mhz
Wait States:	0

Speed Comparisons

To determine speed, the refresh time was captured for the JEM1 and AAMP5. Each refresh involves drawing all of the text and graphics on the display, and executing the display swap command.

For the optimized configurations listed:

PROCESSOR	REFRESH	UPDATE RATE
JEM1:	8.8ms	113.6 Hz
AAMP5:	14.3ms	69.7 Hz

Java/JEM Memory Requirements

The significant source lines of code include non-blank and non-comment lines.

COMPONENT	SOURCE LINES
CGEI	2355
Test Application	692

The Java ROM space contains Java byte codes, class tables, and constant data.

ROM AREA	SIZE (BYTES)
CGEI :	13882
Test Application :	7450
Java Libraries :	15570
JEM1 runtime :	9410
TOTAL	46312

The Java RAM space contains class variables in data RAM, and object variables and object data in the heap RAM.

RAM AREA	SIZE (BYTES)
Objects (Heap)	11168
Stack	908
Data	700
TOTAL	12776

Ada/AAMP Memory Requirements

COMPONENT	SOURCE LINES
CGEI	3637
Test Application	941

The Ada ROM space contains AAMP5 instructions, and constant data.

ROM AREA	SIZE (BYTES)
CGEI	13372
Test Application	10486
Ada Libraries	1958
RTOS	14372
TOTAL	40188

The Ada RAM space contains return addresses and local variables in the application stack RAM, arrays and records in the data stack RAM, and variables in the data RAM.

RAM AREA	SIZE (BYTES)
Application Stack	2000
Data Stack:	6000
Data :	684
RTOS	3626
TOTAL:	12310

JEM1 Vs AAMP5 Conclusions

As shown in Figure 5, the display refresh rate was used as the measure of throughput. The JEM1 was 63% faster than the AAMP5 for this application.

The JEM1 was similar in memory requirements to the AAMP5 in some aspects. In other aspects, it required more memory. Figure 6. shows that the CGEI and the test application code space requirements were similar. This is an important comparison because if the application is large, this size will dominate the ROM or Flash EEPROM requirements. The AAMP5 instruction set is typically 2-1 more efficient in code space than a 486. It appears that Java byte codes are as efficient as AAMP5 instructions for this application.

The total RAM requirement was similar for the two processors but in different places. As figure 6 shows, the Java version required more RAM for heap. This is because in Java, most variables are stored in objects on the Heap. In contrast, Ada83 is not object oriented and stores most if it's data on the stack.

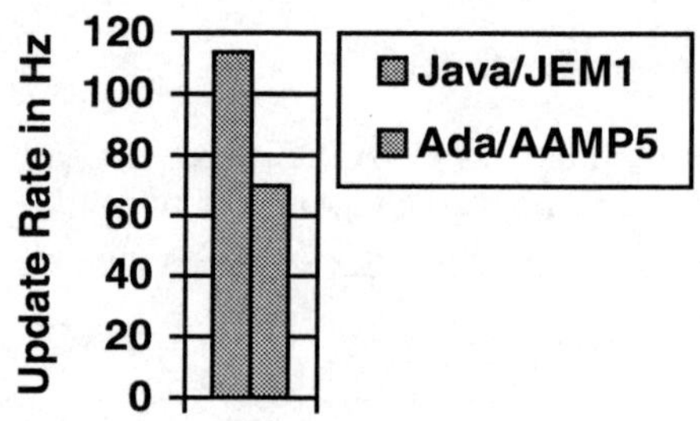

Figure 5. Ada and Java Performance

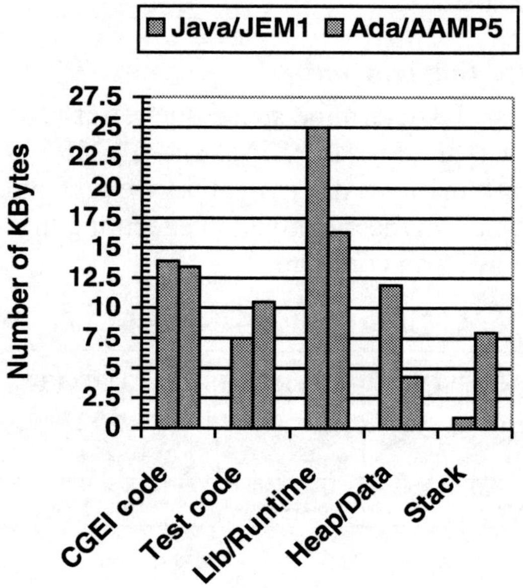

Figure 6. RAM use comparisons

References

[1] Lindholm, Tim and Frank Yellin. The Java Virtual Machine Specification. Addison-Weseley, Reading, Massachusetts, 1997.

[2] Reference Manual for the Ada Programming Language. ANSI/MIL-STD-1815 A, United States Department of Defense, 1983.

[3] Best, David W., Charles E. Kress, Nick M. Mykris, Jeffrey D. Russell, and William J. Smith. An Advanced-Architecture CMOS/SOS Microprocessor. IEEE Micro, V2, N3, August 1982.

D. Hardware Engineering Track

In today's world we are constantly dealing with the economic pressures driving higher performance for lower cost. As engineers we constantly hear demands for faster, smaller, lighter, lower power, and cheaper. Fortunately, the rapid evolution of electronic technology enables us to progress down this path. This track will focus on several of the important issues facing electronic hardware engineers today.

Track Chair: Jeff Volp – Draper Labs

Session D1: Sensor Systems

Chairs: Rich Richmond, USAF Wright Lab; Nick Tomljanovich, MITRE; Neil Barbour, AIAA Sensor Systems

1	Microfabricated Chemical Sensor Systems For Safety And Emission Control Applications	Mr. Gary Hunter
2	Air-Cooled Flight Worthy Eye-Safe Coherent Lidar Transceiver	Mr. Mark Poling
3	Microfabricated Thin-Film Physical Sensors For Hostile Environment	Jih-Fen Lei
4	Real-Time Digital Signal Processing Enables A Reduced-Cost Atmospheric Remote Sensorcraft	Mr. Michael Grant
5	Reliability Flight Tests On A Fiber Optic Temperature Sensor	Ms. Margaret Tuma
6	Enhanced Recognition And Sensing Ladar (ERASER) Flight Demonstration (Not Published)	Mr. Robert Zumrick
7	Air Data Sensor Failure Detection System	Drew Houck

Session D2: Processors & Commercial-Off-The-Shelf Systems

Chairs: Larry Hart, Mission Computers and Processors; Jay Prizant, Draper Lab

1	The Mini-ACE Family of MIL-STD-1553 Terminals	Mr. Mike Glass
2	Keeping COTS-Based Systems Current Through The Use Of Programmable Electronics	Dr. Gary L. Fitzhugh
3	Flight Certification in the Rapidly Evolving COTS Processing Environment (Not Published)	Mr. Larry Hart
4	Ultrareliable Real-Time Control Systems -- Future Trends	Mr. Robert C. Hammett
5	High Speed Communicator for Real Time Fault Tolerant Systems	Mr. Jay Prizant
6	Evolving Avionics Systems from Federated to Distributed Architectures	Dr. Delbert L. Swanson Ph.D.
7	The Impact of COTS on a Ground Support System Procurement	Mr. Peter Lister
8	Keeping COTS-Based Systems Current Through the Use of Programmable Electronics (Not Published)	Gary Fitzugh
9	Building Standards Based COTS Multiprocessor Computer Systems for Space Around a High Speed Serial Bus Network	Mr. Joseph Marshall

Track D - Hardware Engineering (Continued)

Session D3: Displays & Optoelectronics

Chairs: Becky Morgan; Ted Riffle, Battelle Labs

1	Solving the Automotive Customers Needs with Optical Systems: Defense Diversification (Not Published)	Mr. Phil Parr-Burman
2	Development of Display Technology for the Next Century	Mr. Jerry Schmidt
3	A Review Of The Fly-By-Light Optical Aileron Trim Flight Demonstration System	Mr. John Todd
4	Integrated F-16 / A-10 Advanced Display Processor	Mr. Ted Riffle
5	Virtual Retinal Display Technology	Mr. Thomas M. Lippert

Session D4: EMI HIRF [1] Electromagnetic Phenomenology

Chairs: Bill Larson, FAA – NASA Ames; Gerry Fuller, Veda, Inc; Wayne Devereux, Veridian

1	Assessment of Aircraft Internal EMI using Electromagnetic Analytical Modeling Codes	Mr. R. Wayne Devereux
2	Use of Computational Electromagnetics for Inter and Intra System Aircraft Validation	Dr. Paul A. Johnson
3	Statistical Results for the UHF Data From the NASA EME Flight Tests (Not Published)	Mr. Stephen Scearce
4	The Influence of Modulated Signal Risetime in Flight Electronics Rdiated Immunity Testing with a Mode-Stirred Chamber (Not Published)	Jay Ely
5	Mode Stirred vs Anechoic Chamber Testing for EMI Susceptibility: A Comparison of Results (Not Published)	Mr. Karl J. Moeller
6	Coupling & Vulnerability of CNI Systems to an On Board Source	Mr. Gerald Fuller
7	Investigations into a System for the Detection and Location of Potentially Harmful Radiation from Portable Electronics Carried on Board Aircraft	Dr. Albert D. Helfrick

Session D5: EMI HIRF [2] Testing Effects And Test Facilities

Chairs: Bill Larson, FAA – NASA Ames; Gerry Fuller, Veda, Inc; Wayne Devereux, Veridian

1	Failure Tolerance of Sensors in Flight Control Systems in Electro-Magnetic Environment	B.C. Chang
2	EMI Models for Controller Analysis and Synthesis (Not Published)	W. S. Gray
3	Recent Reverberation Chamber Research and Implementation for Vehicle and Component Level EMC Testing	Galen Koepke
4	Evaluation of Honeywell Recoverable Computer (RCS) in Presence of Electromagnetic Environments	Mahyar Malekpour
5	Design, Construction, Computational EM Modeling, and Characterization of an Aircraft-Sized Reverberation Chamber and Stirrer	Dr. Paul A. Johnson
6	Determination of Upper Destructive Limit for HIRF Experiments on a Fault Tolerant Flight Control Computer (Not Published)	Ms. Laura J. Smith
7	An Experimental Study of the Effect of Placing an Aircraft Inside a Large Welded Zinc-Plated Steel Electromagnetic Reverberation Chamber:	Dr. Paul A. Johnson

Microfabricated Chemical Sensors for Safety and Emission Control Applications

Gary W. Hunter, Philip G. Neudeck, Liang-Yu Chen, NASA Lewis Research Center, Cleveland, OH, USA 44135*

Dak Knight, Cortez III/NASA Lewis Research Center, Cleveland, OH, USA 44135

C. C. Liu and Q. H. Wu, Electronics Design Center, Case Western Reserve University Cleveland, OH, USA 44106

** National Research Council Research Associate*

ABSTRACT

Chemical sensor technology is being developed for leak detection, emission monitoring, and fire safety applications. The development of these sensors is based on progress in two types of technology: 1) Micromachining and microfabrication (MEMS-based) technology to fabricate miniaturized sensors. 2) The development of high temperature semiconductors, especially silicon carbide. Using these technologies, sensors to measure hydrogen, hydrocarbons, nitrogen oxides, carbon monoxide, oxygen, and carbon dioxide are being developed. A description is given of each sensor type and its present stage of development. It is concluded that microfabricated sensor technology has significant potential for use in a range of aerospace applications.

INTRODUCTION

Aerospace applications require the development of chemical sensors with capabilities beyond those of commercially available sensors. Three areas of particular interest are leak detection, emission monitoring, and fire detection. In launch vehicle safety applications, the detection of hydrogen leaks is important for applications involved with, for example, operation of the Space Shuttle. In 1990, the leaks on the Space Shuttle while on the launch pad temporarily grounded the fleet until the leak source could be identified. In response to the hydrogen leak problems, NASA endeavored to improve propellant leak detection capabilities during assembly, pre-launch operations, and flight. In particular, efforts were made to develop an automated hydrogen leak detection system using point-contact hydrogen sensors. However, no commercial sensors existed at that time that operated satisfactorily in this and other space related applications. For example, commercially available sensors often needed oxygen or depended upon moisture to operate.[1] Such sensors did not meet the needs of this application where hydrogen detection was necessary in inert environments or upon exposure to cryogenic environments.[2] Thus the development of new types of sensors was necessary.

The development of a new class of sensors is also necessary for the monitoring of emissions from aircraft engines. The control of emissions from aircraft engines is an important component of the development of the next generation of these engines. The ability to monitor the type and quantity of emissions being generated by an engine is important in not only controlling those emissions but also in determining the status of the engine. Ideally, an array of sensors placed in the emission stream close to the engine could provide information on the gases being emitted by the engine. However, there are very few sensors available commercially which are able to measure the compo-

0-7803-5086-3 /98/$10.00 ©1998 IEEE

nents of the emissions of an engine in-situ. The harsh conditions and high temperatures inherent near the reaction chamber of the engine render most sensors inoperable.

In fire safety applications, fire detection equipment presently used in the cargo holds of many commercial aircraft relies on the detection of smoke. Although highly developed, these sensors are subject to false alarms. These false alarms may be caused by a number of sources including: changes in humidity, condensation on the fire detector surface, contamination from animals, plants, or other contents of the cargo bay.[3] A second method of fire detection to complement existing techniques, such as the measurement of chemical species indicative of a fire, will help reduce false alarms and improve aircraft safety. Although, many chemical species are indicative of a fire, two species of interest are carbon monoxide and carbon dioxide.[4]

In order to meet the needs of these applications, a new generation of sensor technology must be developed. This paper discusses point-contact sensor technology being developed to address these needs. The development of these sensors is based on progress in two types of technology: 1) Micromachining and microfabrication technology (MEMS-based technology) to fabricate miniaturized sensors, and 2) The development of high temperature semiconductors, especially silicon carbide, to provide electrical components and sensors operable at the temperatures of interest. Using these technologies, point-contact sensors are being developed to measure hydrogen (H_2), hydrocarbons (C_xH_y), nitrogen oxides (NO_x), carbon monoxide (CO), carbon dioxide (CO_2) and oxygen (O_2). A description is given of each sensor type and its present stage of development. The silicon (Si) based hydrogen sensor is at a relatively mature stage of development while the state of development of the other sensors ranges from the proof of concept level to prototype stage. The number of dual-use commercial applications makes this general area of sensor development a field of significant interest.

SENSOR DEVELOPMENT

Hydrogen Sensor Technology

One component of the sensor development program at NASA Lewis Research Center (LeRC) and Case Western Reserve University (CWRU) involves the development of palladium (Pd) alloy Schottky diodes on silicon (Si) substrates. A Schottky diode is composed of a metal in contact with a semiconductor (MS) or a metal in contact with a very thin insulator on a semiconductor (MIS). For gas sensing applications, the metal is often a catalytic film. The advantage of a Schottky diode sensing structure in gas sensing applications is its high sensitivity: factors of over 1000 change in reverse current for 0.5% H_2.[5] Further, this type of sensor does not need oxygen for operation.

The sensor structure is shown in Figure 1. The structure includes a Pd alloy Schottky diode, a temperature detector, and a heater all incorporated in the same chip. The sensor is fabricated using a n-type silicon wafer on which approximately 50 Å of SiO_2 is thermally grown in the sensor region. The heater and temperature detector are platinum resistors. Gold leads are applied by thermal compression bonding and the sensor is mounted on a TO5 header or on a ceramic flat package. The surface area of the Schottky diode is 6.1×10^{-3} cm^2 and the sensor dimensions are approximately 2.2 mm on a side. The response of the Schottky diodes was determined by measuring the diode's reverse current.

A hydrogen sensor using palladium chrome (PdCr) as the H_2 sensitive alloy is presently under development for use on the NASA experimental vehicle, the X-33, in its hydrogen leak detection system.[6] Palladium chrome was chosen due its ability to detect H_2 up to 100%. A resistor, whose resistance is dependent on the H_2 concentration, has been added to the sensor design to expand the detection range of the sensor (Figure 1); a Schottky diode provides sensitive detection of low concentrations of H_2 while the resistor (also made of PdCr) provides sensitivity up to 100% H_2. A complete system with hardware and software included with the sensor is included on the X-33 safety system. This complete system is also scheduled for a demonstration flight on the Shuttle for the fall of 1998.[6]

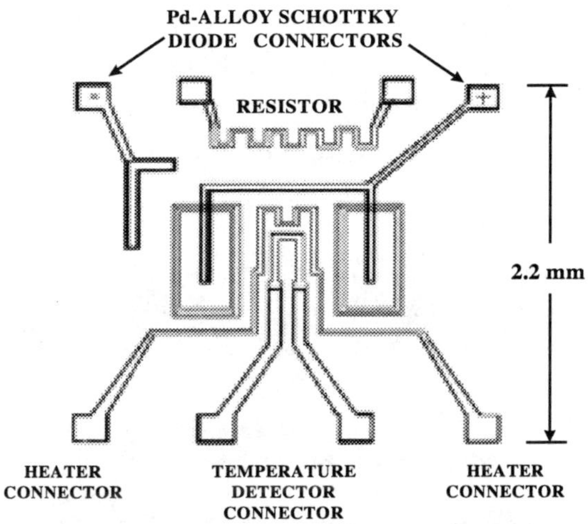

Figure 1. Schematic diagram of the Schottky Diode Hydrogen Sensor. The Pd alloy Schottky diode resides symmetrically on either side of a heater and temperature detector. The resistor expands the sensitivity range of the sensor package to 100% hydrogen.

Hydrocarbon Detection

The development of hydrocarbon sensors for use in harsh environments has centered on the development of a stable SiC-based Schottky diode. The advantage of SiC over Si is its ability to operate as a semiconductor at temperatures in excess of 600°C. This allows SiC-based gas sensors to operate at temperature high enough to allow the detection of hydrocarbons.[7] The first Schottky diode structure developed at NASA LeRC was Pd on SiC MS structures (Pd/SiC). Direct contact between the catalytic metal and the semiconductor allows changes in the catalytic metal to have maximum effect on the semiconductor. Studies of this baseline system help determine limits of diode sensitivity, potential material interactions between Pd and SiC, and whether a barrier layer between the Pd and SiC is necessary for long-term sensor stability. The sensor detects C_xH_y in inert or oxygen containing environments. The details of this work are reviewed elsewhere.[7]

Figure 2 illustrates the advantage of SiC over Si in C_xH_y sensing applications. Shown is the zero bias capacitive response of a Pd/SiC Schottky diode to one hydrocarbon (propylene) over a range of temperatures. The sensor temperature is increased from 100°C to 400°C in steps of 100°C and the response of the sensor is observed. At a given temperature, the sensor is exposed to air for 20 minutes, N_2 for 20 minutes, 360 ppm of propylene in N_2 for 20 minutes, N_2 for 10 minutes, and then air for 10 minutes.

The magnitude of sensor response to 360 ppm propylene depends strongly on the operating temperature. A sensor operating temperature of 100°C is too low for propylene to dissociate on the Pd surface, so the device does not respond at all. The three other curves for 200°C, 300°C, and 400°C show that elevating the temperature increases the sensor's response to propylene. The presence of propylene can be detected at any of these higher temperatures with 200°C being the minimum operating temperature determined in this study. Since the standard long-term operating temperature of Si is usually below 200°C, these results demonstrate the significant advantages of using SiC rather than Si in C_xH_y gas sensing applications.

However, the Pd/SiC sensor response is affected by extended high temperature heating.

Prolonged heating at 425°C has been shown to change the sensor properties and to decrease sensor sensitivity. The reason for this change in diode properties is likely due to reactions between the Pd and SiC at the interface upon heating.

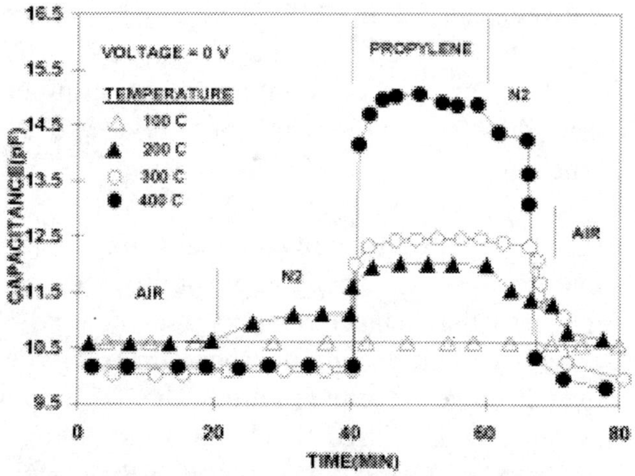

Figure 2. The temperature dependence of the zero bias capacitance to various gas mixtures. The response to propylene is seen to be strongly temperature dependent.

Efforts have been underway to stabilize the sensor structure for long-term, high temperature operation. Two new structures have been demonstrated which improve the stability of the Pd-based Schottky diode structure over that of Pd/SiC.[8] These structures involve 1) The use an alloy as the catalytic metal. 2) The use of a MIS structure where the "insulator" stabilizes the structure and is also reactive to the gases of interest. Preliminary testing of both of these structures suggests significantly improved stability and response over Pd/SiC.

Nitrogen Oxide and Carbon Monoxide Detection

A microfabricated and micromachined Si-based structure with a chemically sensitive resistor is being used to measure NO_x and CO detection. The Si is used as a platform on which the structure necessary for the sensor is fabricated. This sensor structure, shown in Figure 3, includes a temperature detector, heater, and sensing element. The microfabrication process allows the sensor to be small in size with low heat loss and minimal energy consumption. Energy consumption is further reduced by etching out the backside of the Si wafer so that the sensor components (temperature detector, heater, and sensing element) are over a diaphragm region. This minimizes the thermal mass of the sensing area thereby decreasing power consumption for heating and decreasing the time for thermal equilibrium. The temperature detector and heater are doped into the Si substrate for operation over a wide temperature range. The electrode material is Pt and the sensor dimensions are approximately 300 microns on a side with a height of 250 microns.

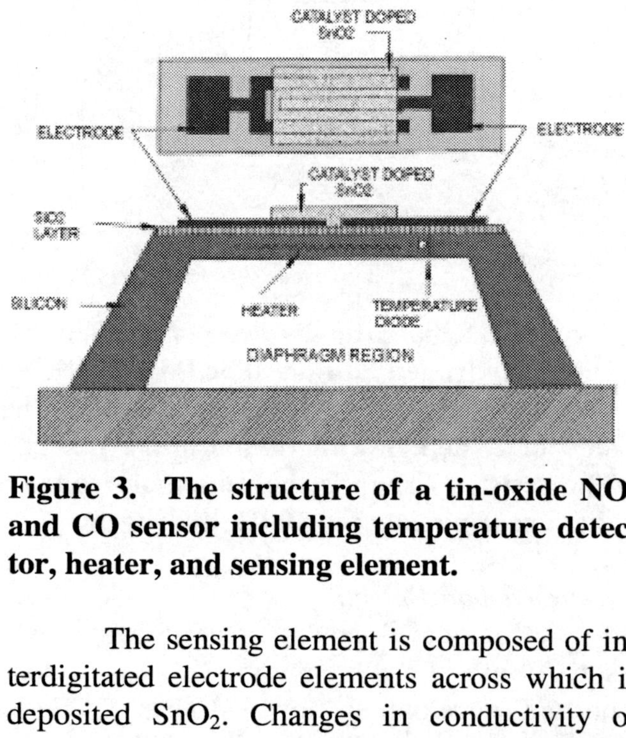

Figure 3. The structure of a tin-oxide NO_x and CO sensor including temperature detector, heater, and sensing element.

The sensing element is composed of interdigitated electrode elements across which is deposited SnO_2. Changes in conductivity of doped SnO_2 across the interdigitated electrodes is measured and correlated to NO_x concentration. A major component of this development work is to stabilize the SnO_2 for long-term, high temperature operation. Drift in the properties of SnO_2 with long term heating due to grain boundary annealing have been previously noted.[9-10] This drift results in changes in the sensor output with time and reduces sensor sensitivity. In order to stabilize the SnO_2 structure for long term operation, the fabrication of nanocrystalline SnO_2 is being investigated. Nanocrystalline materials have several inherent advantages over conventionally fabricated materials including increased stability and sensitivity at high temperature.[11-12]

Several prototypes of these devices using nanocrystalline SnO_2 have been fabricated and evaluated. The sensitivity of the SnO_2 to NO_2 at 360°C is shown in Figure 4. The sensor is shown to be able to detect NO_2 down to the 5 ppm level. The sensor tends to saturate at higher NO_2 concentrations but the highest level of sensitivity is in the lower ppm range which is the region of interest. Similar results are seen with the detection of CO.

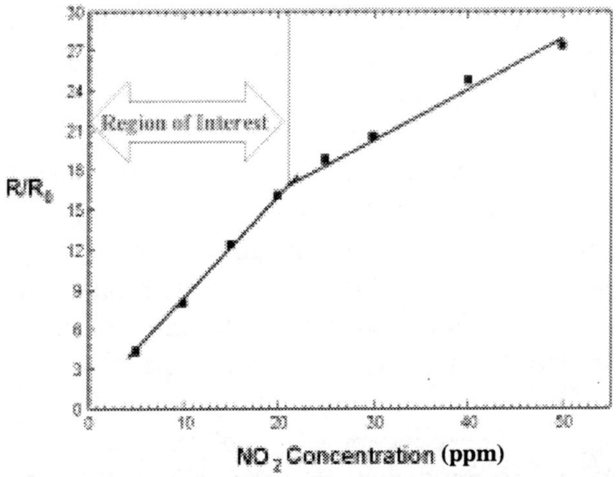

Figure 4. The response of a tin-oxide sensor to a range of NO_2 concentrations at 360°C.

Oxygen Detection

The development of a microfabricated O_2 sensor has been initiated for safety purposes in aerospace applications but significant applications also exist in the area of aeronautics emission control. Commercially available O_2 sensors are typically electrochemical cells using zirconium dioxide (ZrO_2) as a solid electrolyte and Pt as the anode and cathode.[13] The anode is exposed to a reference gas (usually air) while the cathode is exposed to the gas to be detected. Zirconium dioxide becomes an ionic conductor of $O^{=}$ at temperatures of 600°C and above. This property of ZrO_2 to ionically conduct O_2 means that the electrochemical potential of the cell can be used to measure the ambient oxygen concentration at high temperatures. However, operation of these commercially available sensors in this potentiometric mode limits the range of oxygen detection. Further, the current manufacturing procedure of this sensor is relatively labor intensive and costly resulting in a complete sensor package with a power consumption on the order of several watts.

The objective of this research is to develop a zirconium dioxide solid electrolyte O_2 sensor using microfabrication and micromachining techniques. The combination of a fuel with O_2 produces a hazardous environment. Thus, the measurement of both the fuel and O_2 simultaneously is useful in leak detection applications. Also, the presence of O_2 often affects the response of H_2, C_xH_y, and NO_x sensors. An accurate measurement of the O_2 concentration will help quantify the response of other sensors in environments where the O_2 concentration is varying. Therefore, the combination of an O_2 sensor with other microfabricated gas sensors is envisioned to optimize the ability to detect leaks and monitor emissions.

A schematic of the sensor design is shown in Figure 5. As discussed in the NO_x detection section above, microfabricating the sensor components onto a micromachined diaphragm region allows the sensor to be small in size and have decreased energy consumption and time for thermal equilibrium. When operated in the amperometric mode, the current of this cell is a linear function of the ambient O_2 concentration. This linear response to oxygen concentration significantly increases the O_2 detection range of the sensor. A chamber structure with a well-defined orifice is micromachined to cover the sensing area. This orifice provides a pathway to control oxygen diffusion which is important in amperometric measurements. This orifice also protects the integrity of the sensing electrode from impinging particles. Preliminary testing of the complete O_2 sensor has been accomplished and further improvements on the design are planned.

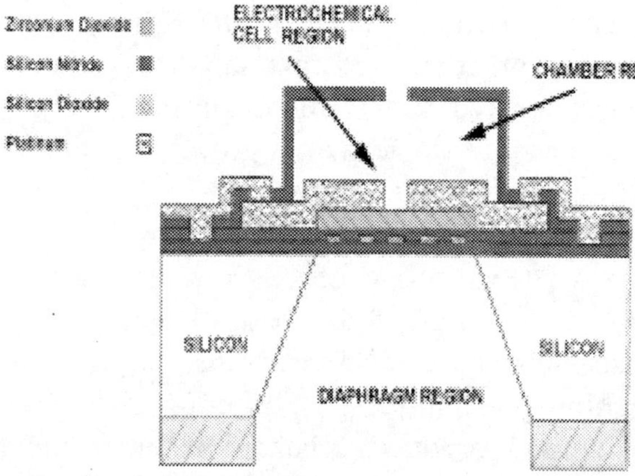

Figure 5. The structure of a microfabricated amperometric oxygen sensor. The dimensions of this sensor are comparable to that of the NO_x sensor shown in Figure 4.

CO2 Detection

The detection of CO_2 is, like the detection of oxygen, based on the use of a solid electrolyte. The significant difference from the O_2 sensor will be the use of NASICON (sodium super ionic conductor) as the solid electrolyte. NASICON is an ionic conductor composed of $Na_3Zr_2Si_2PO_{12}$ which has previously been shown to be sensitive for CO_2 detection. The preparation of the NASICON will be performed using a sol-gel technique. The sensor structure will be similar to that of Figure 5: a microfabricated, miniaturized sensor structure which can be incorporated with other sensors such as the CO sensor. A combined CO_2/CO sensor is of interest not only for fire safety applications but for combustion monitoring applications as well.

High Temperature Electronic Nose Concept

The successful development of the individual high temperature sensors discussed above will allow the formation a sensor array which will allow the detection of a number of gases on a single chip. For example, the formation of an array of the sensors discussed in this paper will detect H_2, C_xH_y, NO_x, CO, O_2 and CO_2. These gases will be detected using three different platforms: a Schottky diode, a resistor, and an electrochemical cell.

Development of such a micro-fabricated gas sensor array operable at high temperatures and high flow rates would be a dramatic step in allowing the monitoring and control of emissions produced by an aeronautic engine. This gas sensor array would, in effect, be a *high temperature* electronic nose and be able to detect a variety of gases of interest. Several of these arrays could be placed around the exit of the engine exhaust to monitor the emissions produced by the engine. The signals produced by this nose could be used to determine the constituents of the emission stream and this information then used to control those emissions. The microfabrication of these sensors is necessary: a conventional bulky system would add weight to the aircraft and impede the flow gases leaving the engine.

The concept of an electronic nose has been in existence for a number of years.[14] Commercial electronic noses presently exist and there are a number of efforts to develop other electronic noses. However, these electronic noses depend significantly on the use of polymers and other lower temperature materials to detect the gases of interest. These polymers are generally unstable above 400°C and thus would not be appropriate for use in harsh engine environments. Thus, a separate development effort is necessary for a high temperature electronic nose.

COMMERCIAL APPLICATIONS

The gas sensors being developed at NASA LeRC and CWRU are meant for aeronautics and aerospace applications but can be used in a variety of commercial applications as well. For example, hydrogen sensors using palladium silver that were initially developed for application on the launch pad of the space shuttle. These sensors can be applied to an automotive application. GenCorp Aerojet Corporation, in conjunction with NASA Marshall Space Flight Center, has developed hardware and software to monitor and control the NASA LeRC/CWRU sensors. The system can be customized to fit the user's needs, e.g., to monitor and display the condition of the tank of a natural gas vehicle. Several of these systems have been purchased for use on the Ford Motor Company assembly line for natural gas vehicles (NGV). This complete system received a 1995 R&D 100 Award as one of the 100 most significant inventions of that year.[5]

Likewise, the high temperature H_2, C_xH_y, NO_x, and O_2 sensors are being developed for aeronautical applications but can be applied in commercial applications. For example, the conditions in an aeronautic engine are similar to those of an automotive engine. Thus, sensors that work in aeronautic engine applications may be operable in automotive engine applications. Other possible applications include catalytic reactor monitoring, alarms for high-temperature pressure vessels, and chemical plant processing.

SUMMARY

The needs of aerospace applications require the development of sensors with capabilities beyond those of commercial sensors. These requirements include operation in harsh environments, high sensitivity, and minimal size weight and power consumption. Sensor technology is being developed to address these requirements using microfabrication and micromachining technology as well as SiC semiconductor technology. Several types of sensors have been described all of which have aeronautic and space applications. Some of the sensor designs are relatively mature while the development of others is ongoing. The combination of these technologies may allow the development of a high temperature electronic nose to provide complex chemical analysis in harsh environments.

Sensors designed for aeronautic and space applications also have significant commercial applications. One example is the use of the hydrogen sensor in automotive applications. This application involved a sensor not completely developed for a space application but with excellent properties for use in the automotive application. Further, given the similarities of aeronautic engine environments to those of automotive engine environments and chemical process monitoring, sensors developed for aeronautic applications also have a wide range of applicability. Although each application is different and the sensor needs to be tailored for that environment, the base technology being developed for aeronautic and space applications can have significant impact on a range of fields.

ACKNOWLEDGMENTS

The authors would like to acknowledge the contributions of Dr. J. Lei, Dr. H. Will, and Dr. D. Larkin of NASA LeRC, Dr. D. Makel of Makel Engineering, and the technical assistance of D. Androjna and C. Salupo of Cortez/NASA LeRC. L. Chen is a NRC Fellow.

REFERENCES

1. Hunter, Gary W.; “A Survey and Analysis of Commercially Available Hydrogen Sensors”, NASA Technical Memorandum 105878, November 1992.

2. Hunter, Gary W.; "A Survey and Analysis of Experimental Hydrogen Sensors", NASA Technical Memorandum 106300, October 1992.

3. “Proceedings of the 1995 Workshop on Fire Detection Research”, W. L. Grosshandler, Editor, June, 1995, Gaithersburg, MD, NISTIR 5700.

4. Grosshandler, W. L.; “A Review of Measurements and Candidate Signatures for Early Fire Detection,” NISTIR 555, Nat. Inst. Of Stand. And Tech., Gaithersburg, MD, January 1995.

5. Hunter, G. W.; Neudeck, P. G.; Chen, L.-Y.; Liu, C. C.; and Wu, Q.H.; Makel, D. B.; and Jansa, E.; "A Hydrogen Leak Detection System for Aerospace and Commercial Applications," 31st IAA/ASME/SAE/ASEE Joint Propulsion Conference and Exhibit, San Diego, CA, July 10-12,1995 Tech. Rep AIAA-95-2645, 1995.

6. Hunter, G. W.; Neudeck, P. G.; Chen, L.-Y.; Makel, D. B, Liu, C. C.; and Wu, Q.H. and "A Hazardous Gas Detection System for Aerospace and Commercial Applications," 34th IAA/ASME/SAE/ASEE Joint Propulsion Conference and Exhibit, Cleveland, OH July 1998, Tech. Rep AIAA-98-3614,1998.

7. Chen, L.-Y.; Hunter, G. W.; Neudeck, P. G.; Knight, D.; Liu, C. C.; and Wu, Q. H.; “SiC Based Gas Sensors”, in Proceedings of the Third International Symposium on Ceramic Sensors, Anderson, H.U.; Liu, M.; and Yamazoe, N., ed., Electrochemical Society Inc., pp. 92-105, 1996.

8. Hunter, G. W.; Neudeck, P. G.; Chen, L.-Y.; Knight, D.; Liu, C. C.; and Wu, Q. H.; “SiC-Based Schottky Diode Gas Sensors”, Proceedings of International Conference on SiC and Related Materials, Stockholm, Sweden, Sep. 1-7, 1997.

9. Ogawa, H.; Nishikawa, M.; and Abe, A.; “Hall Measurement Studies and an Electrical Conduction Model of Tin Oxide Ultrafine Particle Films”, Journal of Applied Physics, vol. 53, pp. 4448-4455, 1988.

10. Xu, C.; Tamaki, J.; Miura, N.; and Yamazoe, N.; “Grain Size Effects on Gas Sensitivity of Porous SnO_2-Based Elements, Sensors and Actuators, B, vol. 3, pp. 147-155, 1991.

11. Yoo, D. J.; Tamaki, J.; Park, S. J.; Miura, N.; and Yamazoe, N.; “Effects of Thickness and Calcination Temperature on Tin Dioxide Sol-Derived Thin-Film Sensor", Journal of Electrochemical Society, vol. 142, pp. L105-L107,1995.

12. Vogel, R.; Hoyer, P.; and Weller, H.; “Quantum-Sized PbS, CdS, Ag_2S, Sb_2S_3, and Bi_2S_3 Particles as Sensitizers for Various Nanoporous Wide-Bandgap Semiconductors”, Journal of Physical Chemistry, vol. 98, pp. 3183-3188, 1994.

13. Logothetis, E. M.; "Automotive Oxygen Sensors", Chemical Sensor Technology, Vol. 3, N. Yamazoe, ed., Kodansha Ltd., pp. 89-104, 1991.

14. Gardner, Julian W.; Bartlett, Philip N.; “A Brief History Of Electronic Noses.” Sensors and Actuators, B, 18(1-3), 211-20,1994.

AIR COOLED FLIGHT WORTHY EYE-SAFE COHERENT LIDAR TRANSCEIVER

Hal R. Bagley, Coherent Technologies, Inc. Lafayette, Colorado
Stephen M. Hannon, Coherent Technologies, Inc. Lafayette, Colorado
Mark A. Poling, Coherent Technologies, Inc. Lafayette, Colorado
Dane L. Schnal, Coherent Technologies, Inc. Lafayette, Colorado

Abstract

Coherent Lidar systems based on solid-state laser technology and operating at eye-safe wavelengths have rapidly evolved as useful sensors for several ground-based and airborne wind-related remote sensing applications. Until now, these systems did not lend themselves to many airborne applications due the constraints of their cost, size, weight and a large, power-consuming chiller required to keep the system temperature stabilized. In recent and ongoing engineering developments as part of the Laser Wind and Hazard Profiler (LWiHP) program, we have succeeded in producing a convective-air-cooled lidar transceiver with a factor of four reduction in size, weight, parts count and subsequent cost reductions. This effort was specifically aimed at installing lidar systems on both military helicopter (the Apache AH-64D) and high performance fighter aircraft (the F/A-18 and the Joint Strike Fighter) for several different wind sensing and obstacle detection applications. This paper describes the engineering development that enables these technology advances.

1.0 Introduction

Solid-state lidar systems have been proven to be very accurate for a variety of wind related remote sensing applications. Coherent Technologies, Inc. (CTI) demonstrated the first coherent systems based, on Nd:YAG lasers operating at 1 μm and Tm,Ho:YAG lasers operating at 2 μm (an eye-safe wavelength). These early systems were used for many successful demonstrations of the technology. Examples range from a 1 J, 10 Hz Nd:YAG system, used at Kennedy Space Center in 1991 to demonstrate wind profiling to and altitude of 27 km during the launch of the Space Shuttle, and demonstrations of wake vortex detection and tracking at Stapleton International airport using a 30 mJ, 5 Hz, 2 μm system in 1993. There have been many airborne wind measurement programs using functional prototypes as well. These include the measurement of winds from military air transport aircraft for a precision air cargo drop application, ballistics delivery from the C-130 Gunship, optical air data from a C-130, and an ongoing program which recently took the first successful measurement of clear air turbulence ahead of an aircraft with sufficient warning time (30 to 50 seconds) to take necessary precautions to avoid injury.

The main challenge to practical implementation of these systems into airborne applications is the size and weight of both the transceiver (which contains the transmit and receive laser, detectors, seeding laser and optics) and the ancillary equipment required to operate the lidar system. By the nature of the lidar design, there are very tight requirements for the intracavity optical stability of the various mirrors and lenses that make up the system. The laser transmit power is reduced if portions of the optical train become misaligned. Each optical component has different requirements for this pointing stability, but experience shows that component-to-component misalignments of 20 microradians or more result in a

0-7803-5086-3 /98/$10.00 ©1998 IEEE

significant drop in laser performance. For this reason, previous metal (aluminum) lidar systems were maintained at a constant temperature; to prevent temperature excursions or gradients from causing the supporting structure to move. This technique worked well for validating the technology, as the temperature of the bench was maintained to ± 0.5 deg by an internal heat exchanger and an industrial chiller. The size of the transceiver and its associated chiller using this type of thermal stability was limited to a lower bound of 8+ cubic feet and 120 pounds. While tolerable for demonstrations, practical implementation for military and commercial applications requires significant reductions in size and weight.

2.0 Program Requirements

The Laser Wind and Hazard Profiler (LWiHP) program is a joint U.S. Government-Industry program comprised of the U.S. Air Force and the Defense Advanced Projects Agency (DARPA) on the government side and a consortium of Honeywell, CTI, McDonnell Douglas (Boeing) and Lightwave Electronics on the industry side. The LWiHP program is aimed at meeting the requirements for two distinct applications. The first application is for optical air data sensing aboard military helicopters. Air data is measured beneath the rotors, as well as outside the influence of the rotors, to provide airspeed information used by the onboard ballistics fire control computer. The second application is aboard high performance military fixed wing aircraft for accurate, high rate update, three dimensional airspeed measurements. Existing pitot-tube-based air data instruments increase the aircraft's radar cross section, require fairly significant life cycle maintenance and calibration, and tend to be inaccurate during high angle of attack maneuvers. CTI has been responsible for developing the transceivers for incorporation into the Honeywell system; therefore, this paper is focused on the size and weight improvements recognized in the transceiver. The Manufacturing Technology (ManTech) Directorate at Wright Patterson AFB has also contributed to the maturation of certain components which were subsequently incorporated into the transceiver.

The two applications addressed by the LWiHP program restricted the lidar transceiver design to the following physical requirements:

- A total sensor volume of less than 2 cubic feet; this eliminated the need for a chiller.
- No reliance on a liquid cooling loop of any kind; this meant that only conditioned or ram air cooling could be used.
- Operation in an ambient temperature environment ranging from -40 to +50 °C.
- Both environments had moderate to severe vibration requirements under which the system had to operate.
- Total system weight less than 40 lbs; this flowed down to a combined transceiver, telescope and scanner weight requirement of less than 15 lbs.

The mission requirements themselves forced a very flexible design: the helicopter air data application transmitter power requirements are optimized for low Pulse Repetition Frequency (PRF) operation and a fairly low average power coherent lidar (0.1-0.5 Watts, average power). The fixed wing air data application demands a higher average power sensor due to the fact that the sensor must operate at high altitudes, where the atmospheric aerosol concentrations are significantly reduced from their levels in the planetary boundary layer. The CTI design approach taken to meet the LWiHP transceiver performance requirements is described in the next section.

3.0 General Technical Approach

As can be seen by reviewing the requirements, the LWiHP program dictated that a new lidar transceiver design approach be taken. In order to meet the size requirement, not only would the transceiver (the Transmit & Receive Laser portion of the system) have to be significantly reduced in size, but the chiller would also have to be eliminated. This required that the transceiver (hereafter referred to as the "optical bench" or just "bench") needed to operate without temperature stabilization and still maintain optical alignment within 20 microradians. Additionally, the tight mounting locations also dictated that a forced air cooling technique be used. The moderate to severe vibration environments (10-15 g RMS) required the bench to be very stiff so vibration inputs would not cause any structural damage and dynamic displacements would be low. Lastly, the higher power fixed wing optical air data system (2.0 mJ pulse energy, 480 Hz PRF) required a design that could deal with the heat of the additional power generators (laser diodes, thermal electric coolers) needed to achieve the required laser power output.

These constraints lead to the LWiHP design concept: namely a low or zero Coefficient of Thermal Expansion (CTE) bench that was stiff as well as lightweight. The heat producing components (laser diodes and laser cavity elements) would have their heat pumped directly to an external, air cooled, finned heat exchanger via thermal electric coolers (TEC's) and high conductance graphite heat straps. In addition, the size requirements dictated that the telescope and scanner be compact; specifically, they needed to reside within the transceiver's enclosure. The overall concept is that, by using near-zero CTE graphite bench material, coupled with directly attached heat removal sources for the power dissipating components, the optics would not experience any CTE-induced misalignments as the bench temperature varies over the ambient ranges. Graphite composites are ideally suited to this application since they possess a high stiffness-to-weight ratio and the tailorability to attain near-zero CTE. A CTE on the order of one percent that of aluminum would allow operation over a temperature range of +/-50 °C. By going to a non-temperature-stabilized structure, the optics' mounts, which hold the mirrors and lenses in an athermalized condition, would have to be redesigned.

4.0 Specific Technical Approach

In order to take advantage of the low CTE optical bench structure, new optical mounts and components were developed. Low Expansion Invar was chosen as the best candidate for use in these sensitive components. With the optical mount material chosen, the specifics of the graphite could be generated. In addition to requiring a low CTE, cost minimization and compatibility with the Invar used for the optical components were also necessary. A choice fiber with approximately +0.25 parts per million per Kelvin (PPM/K) CTE for in-plane directions. These values were ideal, with the CTE being very low, but still positive (negative CTEs are possible and common with graphite composites) for a reasonable match with the approximate one PPM/K of the mating component's Invar material. The material also provided a reasonable cost, a 13 Million PSI modulus, and a low density. New products on the market have since provided an equivalent resin which lowers cost and which CTI intends to use on future composite structure efforts. A significant amount of time was spent in the development and testing of the fixed wing air data scanner and the optical mounts to ensure compatibility with the graphite composite bench and size/weight constraints. Both are described below.

4.1 Fixed Wing Air Data Scanner Design

The fixed wing air data sensor design required three or more lines of sight, each visited in a stop-and-stare fashion in order to maintain velocity precision at the target 80 Hz vector velocity update rate. The three beams are required to derive the vector velocity from the radial velocity measurements made outside the influence of the aircraft. The baseline PRF for the fixed wing air data application was 480 Hz. A compact three-telescope configuration was selected to meet these requirements. Each telescope utilized a refractive wedge to provide the correct pointing vector. The transmit laser was switched between telescopes with a continuously rotating switching mechanism. The switching mechanism comprises a stacked set of three planar optics, as shown in Figure 1. Each of the three optics is segmented into four "blades" or "facets." When mounted at a 45 degree angle of incidence (as shown) the combination presents 12 apertures to the beam with each complete revolution in a repeating pattern of three. The first pulse reflects off of the first optic, the second pulse reflects off of the second optic, and the third pulse reflects off of the third optic, and the sequence repeats. With the optic assembly rotating at 40 Hz, this provides 480 apertures each second, matching the laser PRF. Because the optics are staggered by virtue of being stacked on one another, the outgoing pulses are arranged in three parallel beams. This allows pairs of fold mirrors to "pick off" the beams and translate them to the inputs of the three Gallilean output beam-expander/telescopes.

Each of the switching optic's 12 apertures is wide enough to allow the return pulse, as well as the outgoing pulse, to strike while accounting for motor control and optic fabrication tolerances. The laser pulse triggering is slaved to the rotating scanner. An additional field director position feedback device was included in the motor for this use and provides for one degree position accuracy.

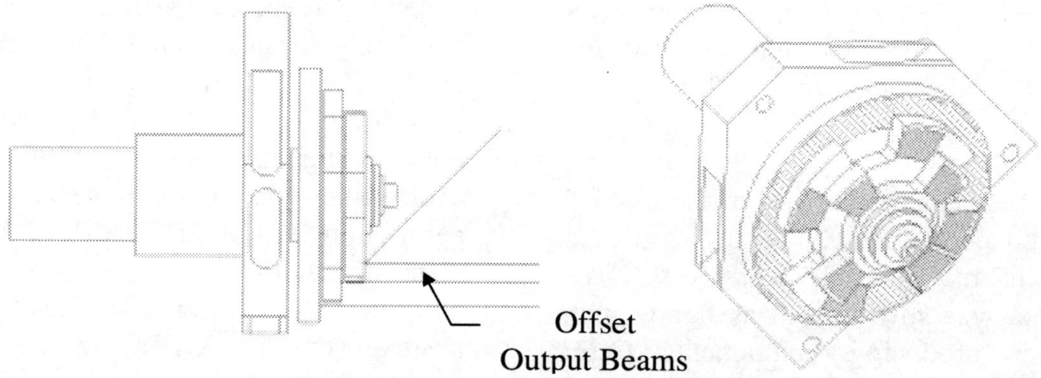

Figure 1. Switching Mechanism

Because the laser triggering uncertainties are less than the optic rotational position uncertainty, and because the actual PRF (nominally 480 Hz) is not critical, this approach provides the easiest and most robust implementation.

The simplest arrangement was for a single optic assembly mounted directly on a high precision motor. The motor was procured from Astro Instruments in Florida. This approach does require that the three optic planes be very parallel and that each of the 12 apertures be flat and parallel as well. The three optics were fabricated at MRC in Florida and mounted one precise hub with the mating surfaces being lapped and the assembly being tested for surface figure and alignment by interferometer.

4.2 Optical Mount Design

The optic mounts underwent significant development with a number of design iterations leading to the final configuration described here. The design, which finally proved to perform the best over temperature, incorporated a few unique features. Because space was an issue, the optic is bonded to an Invar mount, rather than utilizing clips and retainers. This, in itself, presented one of the biggest challenges for the program. A typical optic assembly is shown in the Figure 2. Stability testing, using a theodolite and finite element structural analysis, led to the final design where a square optic was bonded into a close fitting forked mount with a specific geometry. Optic movements on the order of 40 microradians were recorded when the temperature was reduced from room temperature to -40 °C. The bond sizes and locations are by a controlled proprietary configuration and process.

In addition to the optic pointing, optical figure was important. Analysis and experimentation led to a specific epoxy which provided a robust and stable bond, while minimizing distortion of the optic. A mount was also tested for surface figure performance by placing it front of an interferometer. The mount was then heated ± 15 °C from room temperature and the surface figure measured. The performance was excellent, with no more than 0.005 waves RMS change measured from the initial 0.011 waves at room temperature.

4.3 Optical Bench Design

Extensive analysis was performed on the transceiver optical bench assembly. The result was a first natural frequency prediction in the range of 750 Hz. The very high stiffness is advantageous for operating lasers in vibrational environments and is only obtained with high stiffness-to-weight ratio materials such as the graphite used on LWiHP, with very thick and resultantly heavy structures made from common

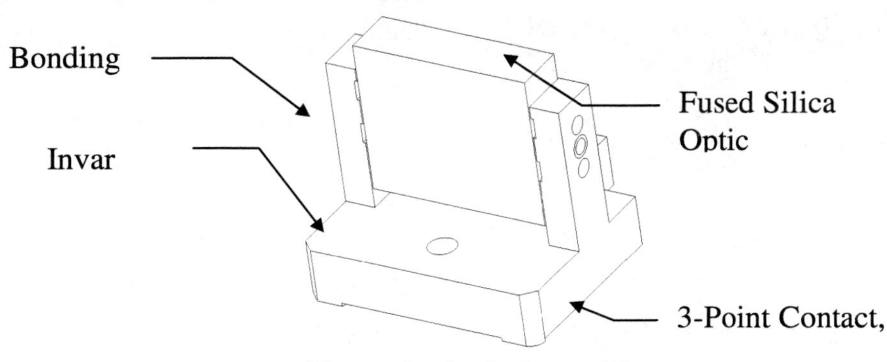

Figure 2. Optic Assembly

0-7803-5086-3 /98/$10.00 ©1998 IEEE

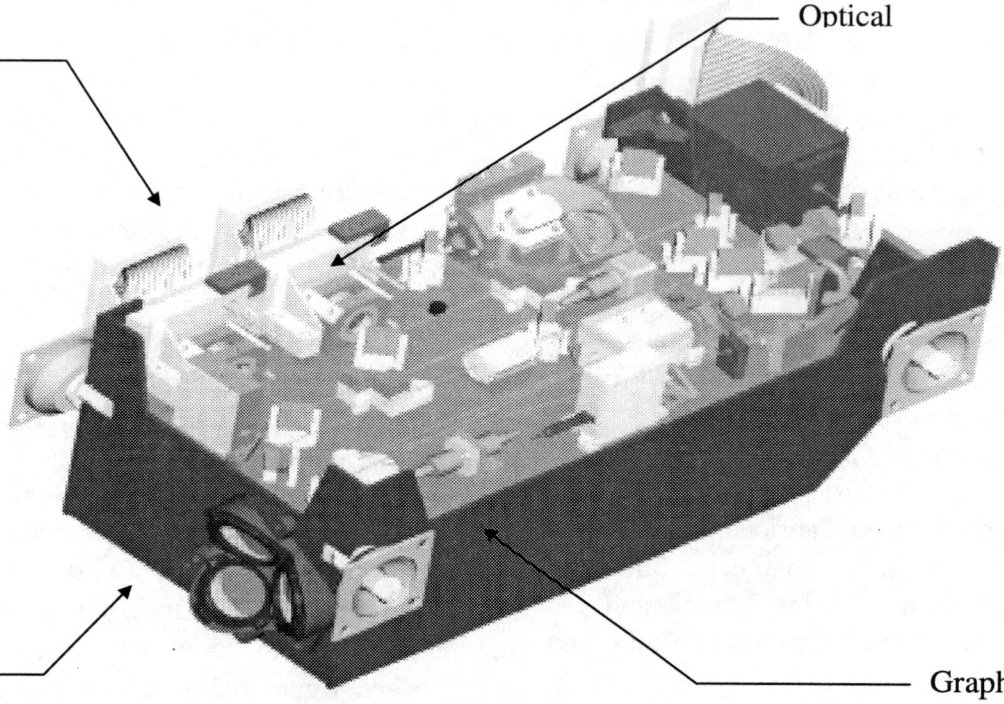

Figure 3. LWiHP Transceiver

metallic materials, or very expensive materials such as beryllium, which is difficult to process. The final design of the LWiHP transceiver can be seen in Figure 3.

5.0 Results to Date

The LWiHP transceiver has been built and tested. The results to date show that the first design iteration performed as expected. Initial tests were conducted to determine the performance of the bench over the environments of vibration and thermal excursions. The vibration testing was performed to a 10 grms environmental which simulated one of the high speed lighter mounting locations, per MIL-STD-810. The system performed very well; showing no loss in power during the vibration. The thermal testing was performed from -40 to +50 °C. These tests were performed with the bench temperature soaked in an environmental chamber. The measure of importance was the percentage of peak average laser power measured (average laser power measured at temperature divided by average laser power at ambient temperature). The performance goal was 75% (less than a 25% loss in power). The initial results are shown in Figure 4.

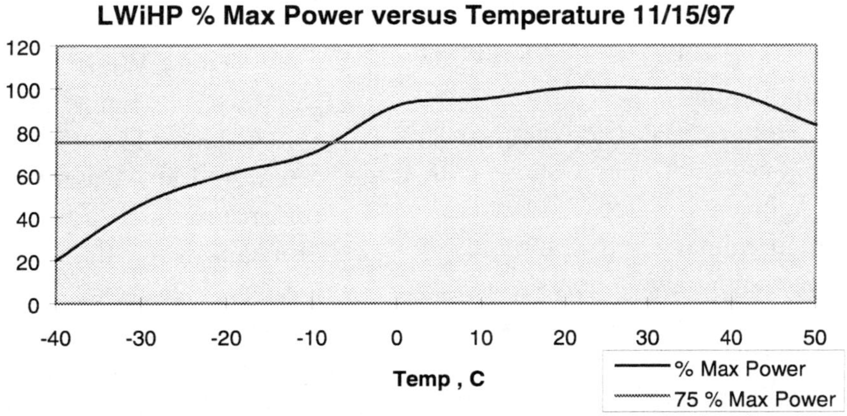

Figure 4. Maximum Power Versus Temperature Graph

The fixed wing air data version of the LWiHP sensor was tested at a field site near Boulder, Colorado in late July, 1998. During this test, the transceiver performed well and transmitted 1 mJ pulses with a PRF of 400 Hz. The transmit/receive beam diameter utilized for these tests was only 1.55 cm (0.6 inches). The system, because of new customer interest, was used to demonstrate long-range wire detection as well as range-resolved wind sensing. The LWiHP transceiver successfully detected and discriminated wires at 540 meters. These wires ranged in diameter from 3/32" to 1/4".

A sample wind measurement is shown in the Figure 5. The left hand panel shows the Range-Resolved Doppler spectrum, with the horizontal axis extending from -25 MHz (or -25 m/sec) to +25 MHz (or +25 m/sec). The vertical axis extends from a few hundred meters to 4.2 km. A hillside target intercepts the laser beam at a range of 3.5 km. The aerosol (wind return) is evident all the way to the target. The right hand panel shows the resultant radial wind velocity estimates as a function of range (range is now the horizontal axis). The winds were fairly light and variable, due to the afternoon mixed layer turbulence at the test site.

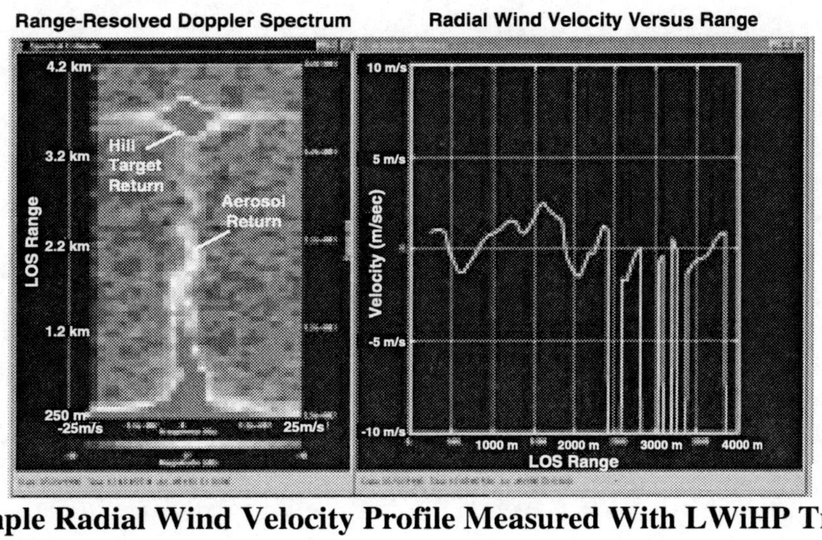

Figure 5. Sample Radial Wind Velocity Profile Measured With LWiHP Transceiver

6.0 Summary

Coherent lidar systems, operating at various wavelengths, have been operational for many years. This activity has demonstrated that coherent lidar systems can now be engineered for *practical* implementation in military and commercial applications. Size and weight goals can be met without diminishing perform in the harsh rotary wing and fixed wing environments. While achieving these goals, parts count and assembly techniques were developed to significantly reduce future production costs. The production costs will be further addressed by the remainder of the ManTech program and subsequent EMD programs.

Acknowledgments

The LWiHP Program is funded by Dr. Mark McHenry, DARPA, and managed by Mr. Dan Thompson at the Air Force Research Lab, Dayton, Ohio. The LWiHP consortium is led by Honeywell, Inc, Minneapolis, MN, with team members Boeing Helicopter, Mesa AZ, Boeing Aerospace, St. Louis, MO, and Lightwave Electronics, Mountain View, CA., who were responsible for developing the master oscillator component. The ManTech contract, in process at CTI is, managed by Mr. Walt Spaulding, AFRL, Dayton, OH. The team's continued support allows the technology and engineering advances to be realized.

MICROFABRICATED THIN-FILM PHYSICAL SENSORS FOR HOSTILE ENVIRONMENT

Jih-Fen Lei, Army Research Lab, NASA Lewis Research Center, Cleveland, Ohio
Herbert A. Will and Lisa C. Martin, NASA Lewis Research Center, Cleveland, Ohio

Abstract

Advanced microfabricated thin film sensors are being developed to provide accurate temperature, heat flux and strain measurements for use in hostile propulsion environments. These sensors are sputter deposited and microfabricated directly onto the test articles with no additional bonding agent. The thickness of the sensors is only a few micrometers, which creates minimal disturbance of the gas flow over the test surface. Thus thin film sensors have the advantage over conventional wire-based sensors by providing minimally intrusive measurement and having a faster response. These microfabricated thin film sensors are being developed for characterization of advanced materials and structures in hostile, high-temperature environments, and for validation of design codes. This paper presents the advances of microfabricated physical sensor technologies developed at NASA Lewis Research Center: thermocouples, heat-flux gages and resistance strain gages. The fabrication techniques of these thin film sensors are described. Sensor demonstrations on a variety of engine materials in several hostile, high-temperature test environments are presented. The advantages and limitations of thin film sensor technology are also discussed.

Introduction

In the development of advanced propulsion systems where new materials and components are being developed to increase operating temperatures and efficiencies and to reduce noise, size and weight, there is a need to develop in-parallel advanced sensor technologies. These advanced sensors are required to characterize advanced materials and structures in hostile propulsion environments and to provide experimental verification of computational design models. Microfabricated thin film sensors are fabricated using vapor deposition and lithography techniques. They can be fabricated directly onto the test surface where the measurements are needed; or they can be batch fabricated, diced and then packaged into multiple devices. These sensors in general have thicknesses on the order of a few micrometers (μm) which are much thinner than the commonly used sensor wires. They have fast response times (in milliseconds), add negligible mass to the test surface, and create minimal disturbance of the gas flow over the surface. Consequently, microfabricated thin film sensors have minimal impact on the thermal, heat load, strain, and vibration patterns that exist in the operating environment and provide a minimally intrusive means of accurate measurements.

This paper presents three advanced microfabricated sensor technologies that are being developed at the NASA Lewis Research Center (LeRC): thermocouples, heat flux gages, and strain gages. Sensor preparation techniques and procedures, including thin film sputter-deposition, electron beam evaporation, sensor patterning, and lead wire attachment techniques, are discussed. The characteristics of the sensors and their test results on various materials including superalloys, ceramics and advanced ceramic matrix composites and in various environments are also presented.

0-7803-5086-3 /98/$10.00 ©1998 IEEE

Microfabricated Sensor Systems

The fabrication of a microfabricated thin film sensor system is completed in a clean room to minimize possible contamination. For a sensor of micron size, contamination or particles could react with the films at high temperatures and possibly change the properties of the sensor system. The fabrication also requires sophisticated and expensive equipment including thin film sputter-deposition and evaporation vacuum systems, wire-bonding systems, wet and dry etching systems, photolithography processes facilities, and surface profiler.

The microfabrication techniques enable batch fabrication. When a large quality of a sensor system is needed, an array of sensors can be fabricated simultaneously on a substrate, which is then diced and packaged into multiple devices. The maximum size of the components that can be directly instrumented with thin film sensors is limited by the size of the vacuum chamber available for thin film deposition. The largest rig at LeRC can accommodate parts up to 80 x 80 x 80 cm^3. The area of the parts that can be coated at any one time is limited by the size of the sputter targets (<20 cm in diameter) or evaporation sources. Several depositions may be necessary when a large area of the component needs to be instrumented.

The fabrication procedures of a microfabricated sensor system depend heavily on the substrate materials and surface roughness. The required layers between the sensor and substrate material need to be tailored to the substrate material to ensure good adhesion, sufficient resistance-to-ground and no chemical interaction between the sensor and the substrate material[1-5]. For those applications that require protection from oxidation or corrosion, a protective overcoat is deposited onto the sensor, Fig. 1.

Basecoat and Overcoat

To successfully insulate the sensor from the substrate or to protect the sensor from oxidation, corrosion or contamination, the coating should have no pinholes or cracks that can allow interdiffusion or interactions between the substrate and sensor. This is achieved by fabricating a dense, thick aluminum oxide layer with the substrate heated during sputter-deposition or evaporation[1,6]. Alumina was chosen as both the basecoat and overcoat for the sensor systems because it is stable and has high electrical resistivity at high temperatures. Substrate heating increases the mobility of the depositing atoms and minimizes the thermal stresses in a thick film. The thickness of the deposited oxide layer ranges from 2 μm to 8 μm depending upon the sensor system.

Thermocouples

The work at LeRC to develop and apply thin film thermocouples (TFTCs) has concentrated on platinum-13% rhodium (Pt13Rh) vs. platinum (Pt) materials due to high temperature capability and stability. The thin film thermocouples can be patterned simply with stenciled shadow masks during two separate sputter-deposition processes: one for the Pt element and one for the Pt13Rh element. The substrate is heated to approximately 250-300°C during the Pt and Pt13Rh thin film deposition process. Oxygen is introduced into the sputter chamber during the first few minutes of deposition to enhance film adhesion[3-5].

The width and length of the sensor can be designed to meet various operational requirements. For example, a thin film thermocouple can be extended from the hot zone of the test area to the cold zone before lead wires are attached. This is to minimize the possibility of sensor failure due to lead wire detachment. When the gaging area is limited and a fine sensor pattern is required, a more complicated photolithography lift-off technique is then used for patterning the sensor.

0-7803-5086-3 /98/$10.00 ©1998 IEEE

Heat Flux Gage

There are numerous methods used to measure heat flux. An easy method that can be implemented with microfabricated thin film technology uses two thermocouple temperature sensors overcoated with two different thicknesses of an insulating material[7-8]. The heat flux measured is proportional to the temperature difference measured by these two thermocouples. Using a thermopile can enhance the output signal of this sensor system; the number of thermocouple pairs multiplies the signal, Fig. 2.

The thermocouples are fabricated with the photolithography lift-off technique to obtain a fine pattern with junctions at the points of the circular zigzag pattern. The change in the insulation layer thickness is fabricated by first depositing an insulating layer on the entire surface (but not on the contact pads). Then a second thermal insulation layer is deposited only on the inside ring of the thermocouple junctions. The thermopile consists of 40 Pt-Pt13Rh thin film thermocouple pairs. Aluminum oxide is the insulating layer and has thicknesses over the outer and inner rings of the thermocouples of 2 microns and 4 microns, respectively. There are three thin film lead pads connected to the sensors so as to obtain heat flux as well as surface temperature simultaneously[8].

Resistance Strain Gage

The size and shape of a resistance strain gages can be varied to meet different testing requirements. A gage can be made so small that it is aligned with fibers in a composite material to determine the strain variation along various fiber orientations. The gage can also be made big enough to determine an average strain level in a large area. This is easily accomplished by using microfabrication techniques. The thinner the film is, the smaller the gage can be. However, the oxidation resistance of a thin small gage will be compromised and an oxidation protection coating will be required[9-10].

The strain gages developed at LeRC are based on a palladium-13wt%chromium (hereafter, PdCr) high temperature alloy. This alloy, in bulk form, is structurally stable up to at least 1100 °C and its apparent strain versus temperature characteristic is linear, repeatable and insensitive to heating and cooling rates[11-13]. The maximum operating temperature of the gages, however, heavily depends on the purity of the PdCr sputtering target[1, 13]. A dynamic strain gage made of PdCr is prepared first by a sputter deposition of PdCr thin film followed by patterning using photolithography and chemical etching with ferric chloride ($FeCl_3$). For static strain applications, a temperature compensator element of platinum (Pt) is included to minimize the temperature effect on the resistance change of the PdCr. The Pt film is deposited and patterned first to be around the PdCr gage grid and is connected to the adjacent arm of a Wheatstone bridge circuit to minimize the temperature effect[14] (Fig. 3).

Lead Wire Attachment

Fine leadwires of 75μm diameter or smaller are attached to the thin films via a parallel gap welding process. The other end of the leadwires is spot welded to heavy trunk wires, which connect to the measurement circuits. Pt13Rh and Pt wires are used for Pt13Rh and Pt thin film thermocouple elements, respectively. Pt fine leadwires and Hoskin 875 trunk wires are used for strain gage. The leadwires are secured onto the substrates by means of spot welding, flame-spraying[15], high temperature ceramic cement, or ceramic joining[16]. The preferred technique depends on the substrate material.

Sensor Characteristics and Testing

Thermocouples

Thin film thermocouples (TFTCs) have been applied to various materials with either flat or

0-7803-5086-3 /98/$10.00 ©1998 IEEE

curved surfaces. These include nickel-based superalloys, ceramics, and ceramic matrix composite materials. TFTCs on a nickel-based superalloy have been tested up to 1100 °C in conditions similar to, though not as severe as, the Space Shuttle Main Engine (SSME) high-pressure fuel turbopump environment[4-5]. Sensors tested on flat panels at low-pressure conditions proved to be highly adherent and durable. The lifetime of TFTCs on SSME turbine blades tested under high pressure, high temperature conditions were limited. This was mainly attributed to poor contact between the thin films and leadwires[5]. TFTCs instrumented on ceramic materials were evaluated in ceramic tube air furnaces under steady state and thermal cycling up to 1000 to 1500 °C for times up to 150 hours at ambient pressures[2]. No sensor failures occurred during this total test time. The drift rates of TFTCs at temperatures greater than 1250 °C, however, increase rapidly as test temperature increases. TFTCs on silicon nitride and mullite also exhibited good durability, surviving a 20-cycle test with heating rates up to 2500 °C/sec [2].

The most recent applications of TFTCs have been on ceramic matrix composite (CMC) materials designed for advanced turbine engine applications. TFTCs have been instrumented on different CMC materials (such as hydridopolysilazane fiber/silicon carbide matrix (HPZ/SiC), SiC/SiC, alumina/alumina, and carbon/SiC) of various shapes (such as a hoop, combustion liner, plate, and cylinder), Fig 4. These instrumented CMC parts were tested in either a combustion rig that operates on jet fuel at pressures of 0.7 to 2 MPa (100-300 psi) or a Mach 0.3 atmospheric burner rig to 1100 °C. The eventual failure mechanism for the TFTCs was again due to the poor contacts between the thin film and the leadwires as well as poor leadwire attachment on the CMC substrate material[1,3]. The improvement of lead wire attachment techniques is being addressed.

Heat-Flux Gage

Prototypes of several heat flux sensors have been constructed and tested[8], Fig 5. Sensors have been fabricated on silicon, aluminum oxide, silicon nitride, MARM-200, and Inconel-100 alloy to provide both surface temperature and heat flux information with minimal disturbance of gas flows. Most of the sensors were fabricated on flat 2.54 x 5.08 cm (1"x2") coupons and tested in the range of 0.1 Mw/m^2 to 1.5 Mw/m^2 in a heat flux calibration high-energy arc lamp facility. The sensors survived several temperature cycles to 920 °C[7-8].

The fabrication of a microfabricated thin film heat flux gage requires many photographic patterning steps which makes its fabrication on a curved surface extremely difficult. A demonstration of fabrication feasibility on the curved surface of a SSME turbine blade was accomplished with the use of a flexible mylar film as the photo mask[1]. More work still needs to be done to develop better ways of masking the blade during fabrication and on methods for aligning the patterns.

Resistance Strain gages

The PdCr based thin film strain gages (TFSG) have been fabricated and tested on various substrate systems including nickel-based superalloys, silicon nitride (Si_3N_4), aluminum oxide (Al_2O_3) ceramics, and SiC/SiC ceramic matrix composites (CMC). The strain sensitivity (or gage factor) of a PdCr strain gage is approximately 1.8 and 1.4 at room temperature and 1000°C, respectively. The apparent strain of a PdCr dynamic strain gage of thickness greater than 5 μm is stable and repeatable between thermal cycles to 1100 °C[13]. The stability of a PdCr gage at high temperature and its reproducibility between thermal cycles, however, decrease as the thickness or the purity of the gage decreases. The maximum operating temperature of the gage depends on the gage thickness and purity [9-13].

This advanced PdCr dynamic strain gage has been applied to an advanced silicon nitride turbine blade that is utilized in the IHPTET environment. The gage survived tests up to a maximum speed of 42500 rpm as well as fatigue tests under 2000 microstrain, up to 1000 °C for a million cycles[1]. The characteristics of a PdCr strain gage under higher frequency loading are being further evaluated for high cycle fatigue applications.

The gage factor of a PdCr/Pt compensated strain gage is approximately 1.3 at room temperature, which decreases approximately 22% to 1100°C[13]. The strain response of this PdCr based static gage is stable and linear to the applied mechanical loads even at 1050 °C. The TFSGs have been evaluated in a heat flux calibrator with heating rates up to 1100 °C/sec under heat fluxes from about 0.1 to 2 MW/m. No sensor failures occurred even after the sensors were subjected to the maximum heat flux.

Most recent applications of TFSGs, like TFTCs, have been concentrated on ceramic matrix composite (CMC) materials, Fig. 6. Instrumented CMC flat components were tested in either a combustion rig that operates on jet fuel at pressures of 0.7 to 2 MPa (100-300 psi) or a Mach 0.3 atmospheric burner rig. The maximum lifetime of a TFSG operating on a CMC material up to 1100 °C was approximately 35 hours. The main failure mechanism for the TFSGs was again due to poor lead wire attachment. Thin Film strain gage applications on CMCs has been limited to flat components since its fabrication technique on large and complex curved surfaces are still being investigated.

The benefit of batch fabrication presented by the microfabrication techniques has been demonstrated with strain gage. Arrays of PdCr strain gages were fabricated on both sides of a wafer as shown in Fig. 7. The wafer was diced and fabricated into 40 multiple probes for turbulence vibration measurements.

Summary and Future Work

Measurement techniques for propulsion systems are being advanced through the development of microfabricated thin film sensor technologies. These thin film sensors have the advantage of providing minimally intrusive measurements and fast response times in harsh high temperature environments. Sensor fabrication, procedures need to be tailored to the substrate materials. Thin film thermocouples, heat flux gages and strain gages have proven to be applicable to measurements on various materials ranging from superalloys, ceramics and ceramic matrix composites. Data has been obtained in furnace testing, under high heat flux conditions, and in harsh engine environments.

Future work will be concentrated on the improvements in sensor reliability and durability in high temperature, high pressure, and high velocity gas stream environments. Improvements in lead wire attachment techniques to both the thin films and the substrate materials are being addressed. Gaging techniques on curved surfaces is also an area in need of further investigation. In addition, the next generations of microfabricated sensor systems, which provide integrated multi-functions, are being developed.

Acknowledgments

The authors would like to acknowledge the invaluable assistance of Chuck Blaha, Drago Androjna of Gilcrest/NASA LeRC and Gerald A. Danzey of NASA LeRC for the fabrication of the test specimens.

References

1. Lei, J. -F., Martin, L.C., Will, H. A., "Advance in Thin Film Sensor Technologies

for Engine Applications", NASA TM 107418, 1997.

2. Holanda, R., "Development of Thin Film Thermocouples on Ceramic Materials for advanced Propulsion System Applications", 7th International Symposium on Temperature: Its Measurement and Control in Science and Industry, Toronto, Canada, April. 1992.

3. Martin, L. C. and Will, H. A., "Testing of Thin Film Thermocouples on Ceramics Matrix Composites in a Burner Rig", NASA CP-10146, Paper 39, 1995.

4. Martin, L. C., and Holanda, R., "Applications of Thin Film Thermocouples for Surface Temperature Measurements", NASA TM-106714, 1994.

5. Grant, H. P., Przybyszewski, J. S. and Claing, R. G., "Turbine Blade Temperature Measurements Using Thin Film Temperature Sensors", PWA-5604-31, NASA CR-165201, 1981.

6. Kayser, P., Godefroy, J. C., and Leca, L., "High Temperature Thin Film Strain Gages", Eurosensors VI Conference Proceeding, October 1992.

7. Bhatt, H., Zeller, M., and Will, H., "Novel Thin-Film Heat Flux Sensors", National Aerospace Planes Conference, Orlando, Florida, Dec 1-4 1992.

8. Will, H and Bhatt H, "Status of Thin Film Heat Flux Sensors", 1994 Conference on Advanced Earth-to-Orbit Propulsion Technology, NASA CP-3282, vol. 1, p.127-132, 1994.

9. Niska, R. H., and Gregory, O.," Thin Film Palladium Chrome Dynamic Strain Gages on Silicon Nitride Components", Ceramic Sensors symposium, ECS Meeting, Abstract no. 840, May 1998.

10. Gregory, O.J., Dyer, S. E. and Cooke, J. D., "Overcoats for the Improved Performance of PdCr High Temperature Thin film Strain Gages", submitted to Thin Solid Films, 1998.

11. Grant, H. P., Przybyzewski, J. S., Anderson, W. L. and Claing, R. G., "Thin Film Strain Gage Development Program", NASA CR-174707, 1983.

12. Hulse, C. O., Bailey, R. S., Grant, H.P., and Przybyzewski, J. S., "Development of a High Temperature Thin Film Static Strain Gage", NASA CP-2493, 1987.

13. Lei, J. -F., "High Temperature Thin film Strain Gage", British Society for Strain Measurement 1995 Annual Conference Proceeding, p.13-16, 1995.

14. Lei, J. -F, Englund, D. R., and Croom, C., "The Temperature Compensation Technique for a PdCr Resistance Strain Gage", SEM Fall Conference for Structural Testing Technology at High Temperature Proceeding, P. 190-196, 1991.

15. Lei, J. -F., Cuy, M. D., and Wnuk, S. P., "Attachment of Free Filament Thermocouples for Temperature Measurements on CMC", NASA TM 107488, June 1997.

16. Kiser, D., Shish J., Lei J. -F., and Martin, L.C., "Novel Approach for Positioning Sensor Lead Wires on SiC-Based Monolithic Ceramic and FRCMC Subcomponents Curved Surfaces" NASA TM, to be published.

Figure 1. Schematic diagram of a thin film sensor system.

Figure 2. A microfabricated thin film heat flux gage.

Figure 3. PdCr/Pt temperature compensated strain gage for static strain measurements.

Figure 4. Thin film thermocouples fabricated on a CMC hoop.

Figure 5. A probe instrumented with a microfabrication heat flux sensor.

Fig. 6 Thin film strain gages and thermocouples fabricated on CMC panels.

Figure 7. Batch fabrication of PdCr resistance strain gages.

0-7803-5086-3 /98/$10.00 ©1998 IEEE

REAL-TIME DIGITAL SIGNAL PROCESSING ENABLES A REDUCED-COST ATMOSPHERIC REMOTE SENSORCRAFT

Michael S. Grant, NASA-Langley Research Center, Hampton, Virginia

Stephen P. Sandford, NASA-Langley Research Center, Hampton, Virginia

1. Abstract

NASA's Langley Research Center (LaRC) is sponsoring the development of a revolutionary new concept for remote sensing from space, which has been termed a sensorcraft. Sensorcraft designates a space flight remote sensing system that does not distinguish between the instrument, which performs a remote measurement, and the spacecraft bus, which provides the operational resources needed by the instrument in orbit. The small size and integrated nature of the design of both the instrument and spacecraft bus can provide a 10X reduction in estimated mission costs for atmospheric remote sensing.

The Gas and Aerosol Monitoring Sensorcraft, GAMS, will measure vertical distributions of ozone, water vapor, aerosols, temperature, and pressure in the atmosphere from the mid-troposphere through the stratosphere. A key feature of GAMS is the use of extensive onboard data processing to reduce the sensorcraft size and weight, and thereby reduce total mission cost. The GAMS real-time digital signal processing (DSP) data processing algorithm reduces the instrument's raw data volume by a factor of 30. Combined with post event processing that achieves a similar reduction, the total GAMS data volume is reduced by a factor of approximately 1000 from that of the current generation of space-based remote sensors. The numerical techniques used in the real-time algorithm, including the fundamental assumptions made about the measurement error in the data and its relationship to the processing will be discussed.

2. Introduction

The Earth Observing System (EOS) currently being deployed by the National Aeronautics and Space Administration is centered around three large orbiting facilities each of which includes a number of complex, well-calibrated remote sensing instruments. Each mission required hundreds of millions of dollars and many years to design, build, test and launch. This approach to Earth science, although it has many advantages, will not be acceptable in the current environment of reduced federal spending. If adequate space-based Earth science measurements are to continue, new approaches must be conceived. Fortunately, new understanding and improved models of the Earth system afforded by the first phase of the complex EOS, should allow more optimized, less expensive sensing systems to be designed.

One effort, which seeks to provide long term monitoring capability of the upper troposphere and the stratosphere, is the Gas and Aerosol Monitoring Sensorcraft or GAMS. A primary goal of the GAMS development is a factor of ten reduction in total mission cost. A key element in the proposed sensor system is the use of on-board processing. This paper describes how on-board processing, specifically a real-time algorithm, enables the GAMS remote sensing concept to be realized.

3. Background

The Gas and Aerosol Monitoring Sensorcraft development has been a highly integrated, concurrent engineering activity that

0-7803-5086-3 /98/$10.00 ©1998 IEEE

began with the goal of reducing total mission costs by an order of magnitude. The primary total mission cost drivers are ground operations and data analysis, the launch vehicle, and spacecraft integration costs. It was determined that the major design objectives driving mission costs are mission lifetime and allowable risk. A major objective of the GAMS design was to reduce total sensorcraft mass, since this is a key estimator of mission cost. Reducing mass reduced total mission cost until the total payload mass was low enough to be suitable for a flight opportunity on many different launch vehicles. However, reducing the mass beyond this point actually *increased* total mission cost since the ballistic coefficient of the resulting satellite was reduced, which in turn reduced the lifetime of the sensor in its intended orbit.

In order to achieve the science measurement objectives and to attack the largest mission costs in the areas of design and calibration, the technique of solar occultation was chosen. This technique removes the need for highly sensitive detectors and for complex calibration schemes. Solar occultation uses the sun as a source while the spectrometer measures the modulation of the solar spectrum by the atmosphere. Measurements are made during sunrise and sunset as seen by the orbiting sensor.

It was also determined that an integrated design of the sensor (spectrometer) and the traditional spacecraft bus functions allowed an optimum reduction of total mission costs. The term sensorcraft is used to indicate the integration of these traditionally separated functions. An example of this integration is the use of a single computer to control both the spectrometer and the satellite operation functions such as power distribution and communications. A second example is the attitude control system, which uses data from the spectrometer as input for its control algorithm.

The use of onboard science data processing in GAMS, to greatly reduce the data volume, results in significant savings in satellite mass and ground operations and data analysis costs. The onboard data processing is described below.

4. Real-Time, Onboard Digital Signal Processing (DSP): A Key Enabling Technology for the Sensorcraft:

4.1 DSP Algorithm Introduction

Traditional instruments, capable of generating detailed science information, are large and download only raw data. This paradigm is ideal for risk free data analysis, but maximizes the instrument's resource requirements and necessitates an extensive ground operations effort, both of which contribute to increased mission costs.

A key enabling technology to the sensorcraft concept is the use of onboard data processing. The resulting reduced data volume allows a significantly reduced data transmission bandwidth requirement since the amount of data transmitted to the ground per orbit is relatively small. This allows a smaller antenna and lower average power, each of which reduces mass. More significantly, it reduces ground operations and data analysis costs.

To our knowledge, all previous remote atmospheric sensing instruments have transmitted only unprocessed data to the ground. There is a long history of atmospheric science data reduction from solar occultation instruments. This gives high confidence that ground data processing algorithms can be used inflight, and that the original unprocessed data can be discarded after it is used, without compromising mission success.

The GAMS onboard processing consists of both real-time processing (during the solar occultation event) and post-event processing. The post-event processing requirements are currently being formalized by the GAMS Principle Investigator (PI). The real-time portion of the processing is addressed in this paper.

0-7803-5086-3 /98/$10.00 ©1998 IEEE

The real-time algorithm is described by Figure 1. The input is the total absorption spectrum of a section of the atmosphere; each section being that portion of the atmosphere through which the instrument views the sun. The calculated coefficients for spectra of individual gas species are the output. These output coefficients are weighting factors

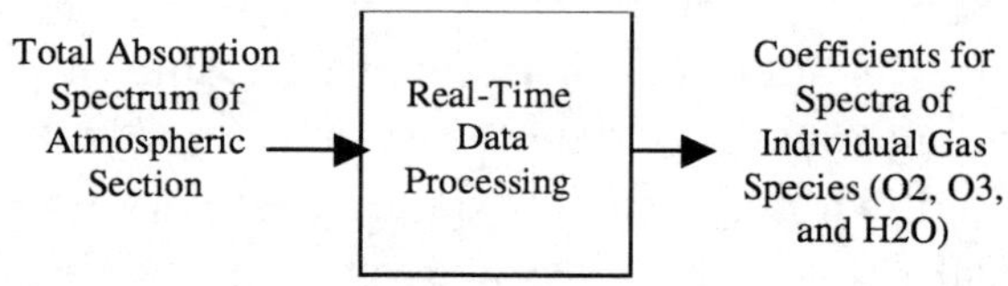

Fig. 1 Real-Time Data Processing Diagram

for the basis vectors associated with a given atmospheric constituent. The real-time algorithm uses Multiple Linear Regression (MLR) to fit a measured spectrum to a primary set of 13 basis vectors. This reduces the original 501-word (16-bits/word) absorption spectrum (or simply 'spectrum') to only 13 coefficients, each one word. The remaining 501-word residual is then reduced by polynomial curve fitting to 4 coefficients. Thus, a 501-word measured spectrum is converted to only 17 words for onboard data storage and eventual transmission to the ground.

Figure 2 shows the fit (a spectrum reconstructed from the computed coefficients and the basis vectors), and the three atmospheric components that are extracted from a nominal 4 km spectrum of optical depth data. An optical depth spectrum is essentially a scaled, inverted version of the original absorption spectrum.

4.2 Multiple Linear Regression and GAMS Spectral Component Extraction

Multiple Linear Regression is an algorithm that is used in the overall GAMS real-time

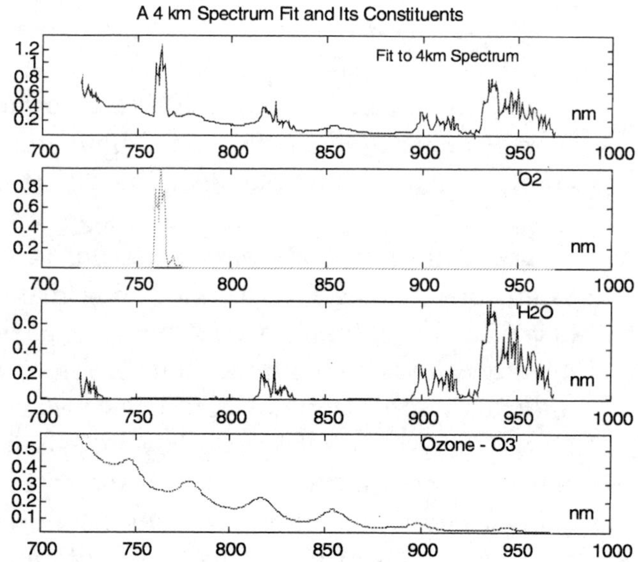

Fig. 2 A 4 km Spectrum Fit and Its Constituents: O2, H2O, and O3

processing algorithm, to obtain spectral information about three gas species: O2, H2O, and O3 (Ozone). MLR is used to extract the desired information from the complicated data set representing a given total atmospheric absorption spectrum. Each spectrum is measured by the GAMS instrument CCD and data sampling electronics, and is then processed in various ways to produce the desired science data output. There are 13 vectors in the GAMS "primary" basis. These 13 vectors are used a priori to derive the other sets of basis vectors needed for the real-time processing. Since MLR is the primary task of the GAMS real-time processing, some fundamentals related to this mathematical technique will now be discussed.

With MLR, several independent variables model (or mathematically describe) a single dependent variable. For GAMS, the independent variables are the various basis vectors. The dependent variable is the 501-

0-7803-5086-3 /98/$10.00 ©1998 IEEE

point vector of "raw" science data for a single absorption spectrum.

The general matrix equation that is solved is:

$$\mathbf{Ac} = \mathbf{y} \qquad \text{(eqn 1)},$$

where, in GAMS, **A** is any one of the sets of basis vectors. (**Bold** type is used to denote either a vector or a matrix.) Each basis, **A**, is formed by arranging a set of basis vectors in columns. Vector **c** is the set of coefficients for which the solution is sought, and vector **y** is the spectrum being modeled or "fit" to a given basis.

Since there are more equations (501) than unknowns (at most 13) the system of equations is overdetermined, and a "least-squares" approach (i.e. MLR) was chosen to obtain a "best" fit. With this approach, the system is solved so as to minimize the sum of all the squared differences between the actual data, **y**(x), and the "model" of the data which is represented by:

$$\mathbf{y}(x; \mathbf{a}_1 .. \mathbf{a}_M) = c_0 + c_1 \mathbf{a}_1 + c_2 \mathbf{a}_2 + \ldots + c_M \mathbf{a}_M \cong \mathbf{y}(x) \quad \text{(eqn 2)},$$

where $\mathbf{a}_1 .. \mathbf{a}_M$ are the columns/basis vectors in matrix **A** and $c_1 .. c_M$ are elements of vector **c**. Thus the residual is minimized where:

$$\text{residual} = \sum_{i=1}^{N} [\mathbf{y}(x)_i - \mathbf{y}(x_i; \mathbf{a}_1 .. \mathbf{a}_M)]^2 \qquad \text{(eqn 3)},$$

where we fit N data points to M basis vectors.

For simplicity, the data, $\mathbf{y}(x)_i$, may be represented by $\mathbf{y}_i$, where the measured data vector, **y**, is understood to be a function of x and the index, i, is used to denote individual **y**-data points.

0-7803-5086-3 /98/$10.00 ©1998 IEEE

4.3 Fundamental Assumptions Related to The GAMS Measurement Errors

Several fundamental assumptions are made about the measurement errors in the GAMS data, related to the least-squares estimation approach. The measurement error for each data point, $\mathbf{y}_i$, is comprised of a random (or statistical) and a systematic component. Any systematic error, which unlike random error cannot be "averaged out," is assumed to be negligible. Systematic errors will be minimized during system design. It is also assumed that the measurement errors for data points in **y** are independently random, i.e. the error for point $\mathbf{y}_i$ is not dependent on the error of any another **y**-data point. In addition, the measurement errors for any data point, $\mathbf{y}_i$, are assumed to have zero mean, and to be normally (Gaussian) distributed. Further, the standard deviation, σ, of the statistical measurement error is assumed to be the same for all data points.

Note that if the individual standard deviations, σ_i, for every **y**-data point were known, then "chi-squared" (χ^2) fitting would be used, where the quantity:

$$\chi^2 = \sum_{i=1}^{N} [(\mathbf{y}_i - \mathbf{y}(x_i; \mathbf{a}_1 .. \mathbf{a}_M)) / \sigma_i]^2 \qquad \text{(eqn 4)}$$

is minimized. However for cases like GAMS where the standard deviations of the measurement errors, σ_i, are not known, they are assumed to have the same value, which is an arbitrary constant; generally 1. This in turn assumes that the set of basis vectors chosen will provide an acceptably good fit to the data.

4.4 Numerical Techniques in the C-code Real-Time Algorithm Implementations

The real-time algorithm was first developed in MATLAB, before being converted to

C-code running on a desktop PC, and finally being ported to a Texas Instruments 320C40 DSP-based development system. MATLAB and other mathematics manipulation software packages generally use numerical analysis techniques which are (correctly) designed to maximize the accuracy of the solution. Thus the MATLAB algorithm results were used as the baseline against which the C-code implementations were compared.

The tradeoff for the high accuracy achieved by mathematics software such as MATLAB, is increased calculation time and memory usage. Because of the real-time processing constraints on GAMS, careful consideration was given to the numerical methods used to perform the MLR in the C-code, compared to those used in the MATLAB version, in order to maintain both a high degree of accuracy and minimal processing time. The numerical methods used to solve the MLR, and the reasons each method is used at various points in the data processing, will now be discussed.

4.4.1 Numerical Methods Used To Solve the Multiple Linear Regression

The two methods used to solve the various overdetermined systems of equations that occur in the processing are: 1) Lower-Upper Decomposition (LUD) with backsubstitution to solve the so-called normal equation (eqn 5, below), and 2) Single Value Decomposition (SVD) to solve the basic matrix equation (eqn 1) directly.

Reference 2 provides a derivation of the normal equation:

$$(\mathbf{A}^T \cdot \mathbf{A}) \cdot \mathbf{c} = (\mathbf{A}^T \cdot \mathbf{y}) \qquad \text{(eqn 5).}$$

A solution for **c** from this equation can be subject to round-off error. To help minimize this error and thus maintain numerical accuracy, matrix **A** and vector **y** are put into "corrected" form as per Reference 1, prior to solving for **c** with LUD. This is done by subtracting the mean of **y** from every element in **y**, and by subtracting the mean of every column of **A** from the elements in the respective column.

The C-code module that solves the normal equation with LUD, was obtained from Reference 2. To use this code, the terms of equation 5 are combined such that:

$$\alpha = (\mathbf{A}^T \cdot \mathbf{A}) \qquad \text{(eqn 5a)}$$

and

$$\beta = (\mathbf{A}^T \cdot \mathbf{y}) \qquad \text{(eqn 5b).}$$

The normal equation (eqn 5) then becomes:

$$\alpha \cdot \mathbf{c} = \beta \qquad \text{(eqn 6),}$$

where the corrected forms of $\mathbf{A}^T$ and **y** are used.

Note that an advantage of this method over others such as Gauss-Jordan elimination is that once the LU decomposition of the quantity, α, has been performed, different sets of coefficients can be calculated using the same LUD of α with different β's.

To minimize processing time in GAMS during a solar occultation event, the LU decompositions are calculated for the α-matrices associated with each basis, **A**, prior to the actual sunrise or sunset measurement event. As each new spectrum, **y**, is measured, a new β-quantity is calculated and the coefficients are found in real-time (current goal is 5 msec. or less for each new spectrum).

Solving for the constant term, c_0, completes the normal equation/LUD solution, and is found by:

$$c_0 = \bar{\mathbf{y}} - \mathbf{c}^T \cdot \bar{\mathbf{A}} \qquad \text{(eqn 7)}$$

where **y**-bar is the mean of **y**, $\mathbf{c}^T$ is the transpose of the coefficient vector, **c**, and **A**-bar is

0-7803-5086-3 /98/$10.00 ©1998 IEEE

the vector of means for each column (basis vector) in **A**, as derived in Reference 1.

The SVD method was required for cases where the normal equation/LUD method resulted in significant round-off error. This is a method that solves the basic matrix equation (eqn 1) directly. One consequence of using the SVD method is that it requires more code space than does LUD.

4.4.2 C-code Implementations of Processing Block Diagram

In the GAMS real-time algorithm C-code implementations to date (PC and DSP system versions), the normal equation/LUD was generally used, with SVD used when round-off error associated with the LUD method became significant. This section will address where specific numerical techniques are used in the processing. The real-time processing block diagram is given in Figure 3.

Block C of Figure 3 computes the coefficients for the "water" section, **wS**, of **S**, in the range 880 - 970 nm. The model for **wS** is:

$$\mathbf{wS} = \mathbf{a[H_2O]} + \mathbf{b[H_2O]}^{-1} + c\lambda \qquad \text{(eqn C)},$$

and combining terms yields:

$$\mathbf{wB} * \mathbf{wcoefs} = \mathbf{wS} \qquad \text{(eqn C1)}$$

which is solved for **wcoefs** by generating the normal equation and by using LUD with back-substitution as described in section 4.4.1. In this case the results were acceptably close to the MATLAB results (matching to within 0.5%).

In Block D the fit of the water information is removed from the spectrum, and in Block E a polynomial fit to the remaining spectral information is performed. (Note that the narrow, higher-frequency oxygen (O_2) feature from 759 to 768 nm is "snipped" out of **S'** prior to calculating the polynomial fit, so the fit to the mostly "low-frequency" information is not influenced by this feature.)

The Block E processing is a case where the system of equations required the use of SVD to obtain results acceptably close to the MATLAB solution. In order to calculate the polynomial coefficients, a basis, **Bnlam**, was constructed (a priori) of 501-element vectors where each column is a power of normalized lambda, **n**λ, values:

$$[\ \mathbf{n}\lambda^0\ \mathbf{n}\lambda^1\ \mathbf{n}\lambda^2\ \mathbf{n}\lambda^3\ ] = \mathbf{Bnlam}\,.$$

From Figure 2, note that the λ-variable is our "**x**" variable, which is wavelength of the incident energy over the range 720 – 970 nm in 0.5 nm increments. The vector **n**λ is λ normalized to 1 to reduce the size of the numbers in this basis, and thus reduce the potential for round-off error.

The resulting "normalized" polynomial coefficients, $\mathbf{pc_{norm}}$, were initially determined as in previous processing blocks using LUD to solve the normal equation for this system, generated from the matrix equation:

$$\mathbf{Bnlam} * \mathbf{pc_{norm}} = \mathbf{S'_{snip}} \qquad \text{(eqn E)}.$$

A method was derived to calculate the desired "de-normalized" polynomial coefficients, **pc,** from the normalized coefficients. However, this normalization technique alone was not enough to prevent the apparent round-off errors from the LUD/normal equation solution, which resulted in significant differences from the MATLAB results.

SVD was thus used to avoid the significant round-off errors that occurred with LUD in this case. However, not only does SVD require a larger code segment than does LUD, it should also be noted that some discretion is required with SVD concerning where to set the limit of singular values. Space does not allow for a discussion of these considerations, which are fully addressed in Reference 2.

In Block "G", coefficients for the fit to **S''** are again able to be accurately determined by solving the normal equation of the system with LUD and backsubstitution.

In Block "J" the coefficients, $\mathbf{pc_{resid}}$, for a 3rd-order polynomial fit to the residual are determined using SVD, similar to Block E; again to avoid excessive round-off error.

5. Conclusions

A key feature of the GAMS sensorcraft is the use of extensive onboard data processing to reduce the sensorcraft size and weight, and thereby reduce total mission cost. The GAMS real-time data processing algorithm was first developed in MATLAB and then converted to C-code running on a desktop PC, before being ported to a Texas Instruments 320C40 DSP-based system. Fundamental assumptions were made about the GAMS data measurement error in relation to the proper use of the Multiple Linear Regression or "least-squares" technique. By appropriately using both Lower-Upper Decomposition (LUD) and Single Value Decomposition (SVD) in the C-code implementations, excessive round-off error was avoided, and results were obtained which match the MATLAB solutions to 0.5% or better.

6. Acknowledgements

We acknowledge Dr. Joseph M. Zawodny, the GAMS Principle Investigator, for his pioneering work advocating the use of extensive onboard science data processing for atmospheric remote sensing.

7. References

1. Weisberg, S. (1980), *Applied Linear Regression*, John Wiley & Sons, New York.

2. Press, W. H, Teukolsky, S. A., Vetterling, W. T., Flannery, B. P. (1992*) Numerical Recipes in C*, 2nd edition, Cambridge, New York.

3. Zawodny, J. M. (1997), *GAMS Real-Time Processing Block Diagram*, NASA Langley Research Center, Hampton, VA.

4. The Mathworks Inc. (1992), *MATLAB Reference Guide*, 24 Prime Park Way, Natick, Mass., 01760.

5. Hines, W. W. and Montgomery, D. C. (1990), *Probability and Statistics in Engineering and Management Science*, 3rd edition, John Wiley & Sons, New York.

6. Koch, M., Sanford, S. and Zawodny, J., *Next Generation Remote Sensing: Gas and Aerosol Monitoring Sensorcraft (GAMS)*, Multispectral Imaging for Terrestrial Applications II, SPIE, International Society for Optical Engineering, vol. 3119, July 30-31, 1997, Ca.

7. Texas Instruments Inc. (1996), *TMS320C4x User's Guide*, Houston, Texas.

0-7803-5086-3 /98/$10.00 ©1998 IEEE

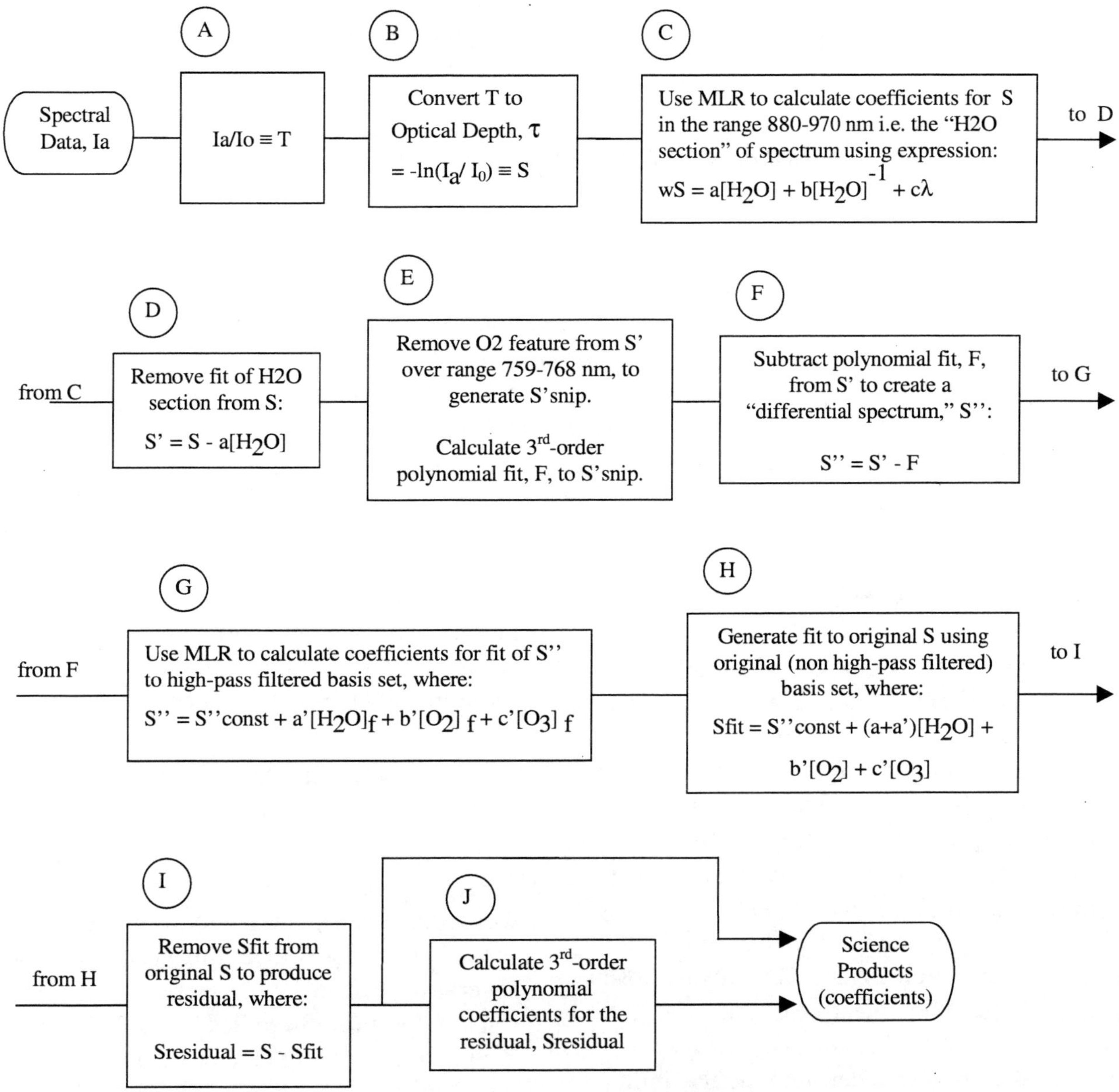

Figure 3. GAMS Real-Time Data Processing Algorithm Block Diagram

FLIGHT TESTS ON A FIBER OPTIC TEMPERATURE SENSOR

Margaret L. Tuma
NASA Lewis Research Center, Cleveland, OH

Takeo Sawatari
Sentec Corporation, Walled Lake, MI

Yuping Lin
Sentec Corporation, Walled Lake, MI

Kristie A. Elam
Gilcrest Electric/NASA Lewis Research Center, Cleveland, OH

Abstract

For aircraft engine control, one key parameter to detect on an airplane is the exhaust gas temperature (EGT). Presently, thermocouples are used to perform this measurement. These electrical sensors perform adequately; however, fully utilizing the benefits of optical sensors requires replacing electrical architectures with optical architectures. Part of this requires replacing electrical sensors with optical sensors, such as the EGT sensor chosen for these tests. The objective of the development and testing of this prototype sensor system was to determine the feasibility of operating an optical sensor in a hostile aircraft environment. The fiber optic sensor system was developed to measure temperatures from 20°C to 600°C in an aircraft environment and was utilized to monitor the EGT of an OV-10D aircraft engine. The sensor has successfully flown over 50 hours and proven to be immune to surface deterioration of the optical element (located inside the sensor head) and able to withstand and operate in normal and sustained severe flight conditions where forces on the airplane exceeded 4 g's. Potential commercial uses for this sensor include monitoring temperature for aeropropulsion system control, military vehicle and naval engine control, conventional and nuclear power plant monitoring, and industrial plant monitoring where EMI issues are critical.

Introduction

For future supersonic and hypersonic aircraft, there is a need for high performance as well as safe sensors for aircraft engine control. Fiber optic sensors meet this need and have several benefits over conventional electronic sensors for aircraft applications. These benefits include immunity to electromagnetic interference, larger bandwidth, as well as providing a safety advantage due to the lack of a sparking hazard. Perhaps the largest potential benefit of optical sensors and a fly by light architecture is the greatly reduced system weight compared to electrical systems. In contrast, conventional electronic sensors such as thermocouples, thermistors, and bi-metal type devices are susceptible to electromagnetic interference, require shielding, and possess a potential fire hazard due to sparking. Presently, no single optical sensor is capable of detecting temperature over the 20°C to 600°C range. There are several sensors which operate at low

0-7803-5086-3 /98/$10.00 ©1998 IEEE

temperatures (<300°C) and blackbody radiation sensors are only useful in detecting temperatures greater than 600°C whereas this single sensor is capable of detecting temperature over the entire range. This paper describes a sensor that was developed to cover the entire temperature range and withstand up to 4 g's of acceleration.

This sensor system wavelength encodes changes in temperature onto the spectrum of a white light source. This wavelength encoding is achieved by utilizing a Fabry-Perot interferometer as the transducer. Previous Fabry-Perot type temperature sensors have utilized a thin film coating as the etalon,[1-2] which limits the temperature range to approximately 300°C due to surface adhesion problems. Other Fabry-Perot temperature sensors have utilized sapphire fiber to detect higher temperatures, but are mainly single-mode devices[3-4] which are more difficult to couple light into.

Theory

A broadband fiber optic temperature sensor system has been constructed utilizing a Fabry-Perot interferometer as the sensing mechanism.[5] The sensor, shown in Figure 1, consists of an optical cavity, d, between a metal reflector and the end of an optical sapphire fiber. At room temperature, d is 1 μm. Light enters the optical sapphire fiber via the SMA connector. A small portion of the incident light (~4%) is reflected at the distal end of the sapphire fiber at the sapphire/air interface. The remaining light propagates through the air gap and is incident upon the metal reflector. Light reflected off the air/metal interface (~90%) re-enters the fiber and propagates back to the SMA optical connector. These two reflected beams interact to produce interference fringes. As the sensed temperature varies, the distance between the metal reflector and the fiber, d, varies due to thermal expansion and contraction of the sensor housing and sapphire fiber. This property is exploited to deduce temperature by interrogating the sensor with broadband light and analyzing the modulated spectrum.[5] As the temperature increases, the number of fringes in this reflected spectrum increases.

The sensor is wavelength modulated as opposed to intensity modulated. Hence, it is expected to be immune from power fluctuations of the source, fiber bends, changes in the reflectivity of the metal reflector due to oxidation, and changes in the spectrometer's sensitivity.

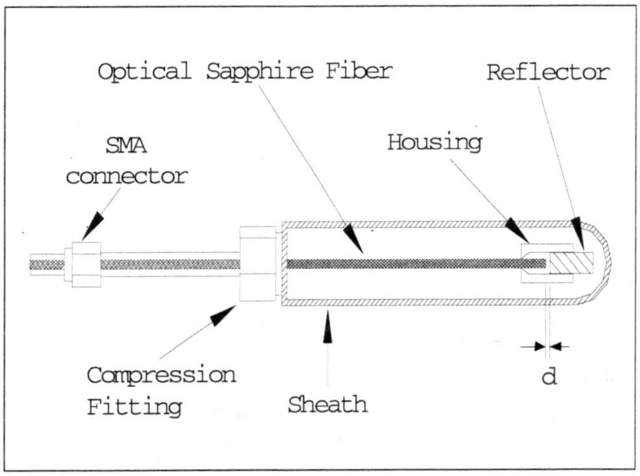

Figure 1. Schematic Drawing of Sensor Head

Experimental Setup

The fiber optic sensor, shown in Figure 1, was fabricated by Sentec Corporation.[6] The prototype sensor assembly is approximately 6" long with an outer diameter of 0.19". Light enters the sensor via an SMA optical connector, located on the left side of Figure 1, and propagates down the sapphire optical fiber, located inside the inconel sheath. The opposite end of the sapphire fiber is optically polished and inserted into a housing in close proximity to the reflector. High temperature cement was used to bond the sapphire fiber to the housing, which was found sufficient for preliminary testing. However, another bonding technique will be utilized in future work.

As shown in Figure 2, a type K thermocouple is attached directly to the sensor by three spot-welded metal tie wraps. Due to its lack of rigidity, the thermocouple requires visual inspection and replacement. As shown in Figure 2, the mounting plate is affixed to the shroud, with the sensing portion penetrating into the exhaust region. The maximum temperature observed was 600°C.

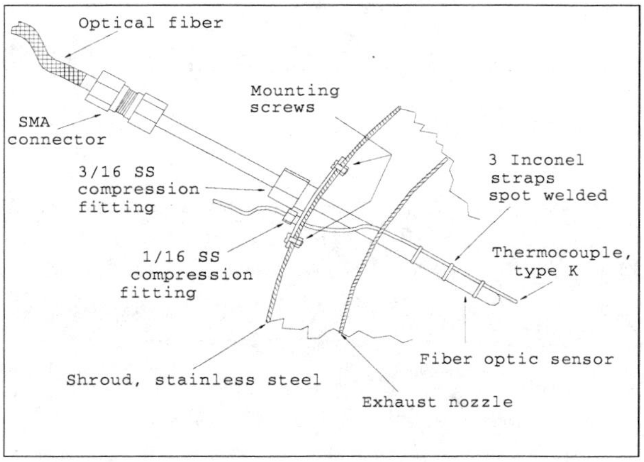

Figure 2. Cross Section Schematic of Sensor Mount

The sensor system, shown in Figure 3, consists of a tungsten halogen white light source, a 300 micron core optical 2 x 1 coupler, an SMA connector, 16 feet of 320 μm core multimode optical cable, the fiber optic sensor, 3 type K thermocouples, a spectrometer card which detects optical wavelengths from 500-1000 nm, an eight channel thermocouple card, and a Pentium computer.

The white light source exhibits a nearly Gaussian profile of intensity vs. wavelength from 300-1900 nm with its peak at 1100 nm. As illustrated in Figure 3, output from the white light source is coupled into one arm of a 2 x 1 optical coupler. This light enters the sensor's optical sapphire fiber via an SMA optical connector.

Referring again to Figure 3, light reflected from the metal surface is coupled out of the sensor via the SMA connector and travels through the multimode fiber into the 2 x 1 coupler. This light continues into the opposite coupler leg which directs approximately 50% of the reflected light to the SMA connector located on a computer plug-in spectrometer card. The data-acquisition Pentium computer and white light source were located in the cargo area of the plane. Care was taken to minimize the number of bends in the optical fiber and maximize each bend radius.

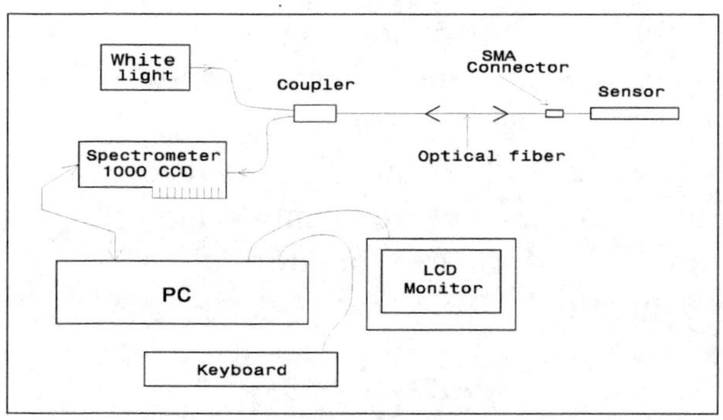

Figure 3. Fiber Optic Temperature Sensor System

Using a single fiber to interrogate the sensor and transmit the modulated signal provides system simplicity and reduced production cost. The modulated light is incident upon a CCD array located in the spectrometer card which converts the optical spectral data into an electrical signal. This 1024 element CCD linear array is responsive over the 500 to 1000 nm range and is capable of recording a single spectrum in a few milliseconds.

Due to the vibration of the plane, it was determined that a hard disk was likely to fail during flight tests. Therefore, data was saved to a floppy disk. This limited the data acquisition rate to 1 spectral reading every 20 seconds, due to the limited storage capacity of the floppy disk. This did not adversely impact the sensor system's performance because the sensor's response time is 60 seconds due to its thermal mass.

In the exhaust region, the maximum temperature observed was approximately 600°C. Three type K thermocouples interfaced to the data acquisition computer were used to monitor the sensor system. As previously shown in Figure 2, the first thermocouple is attached directly to the sensor head to validate its performance. The second thermocouple was attached to the outside of the exhaust duct to monitor the heat transfer from the sensor patch hole. For health monitoring purposes it was necessary to observe the SMA connector temperature because failure would occur at temperatures exceeding 125°C. Therefore, the third thermocouple was welded to the base of the SMA connector.

During the initial ground tests, the sensor was attached to the exhaust duct via the 4 mounting screws on the mounting plate shown in Figure 2. However, the sensor was found to vibrate excessively and in preliminary tests, a spot weld on the sensor housing weakened and eventually failed. Therefore, it was determined that (1) the sensor portion external to the shroud required a brace and (2) the spot-weld required reinforcement. To dampen the vibration, plastic tubing was placed around the sensor/fiber connection. To further dampen vibration at this connection, a brace was used to support a portion of the multimode optical fiber. To reinforce the weld, a bead of nickel was placed over the original spot weld and then laser-welded which greatly improved the strength and durability of the sensor housing.

Sensor Calibration

A calibration was required and performed prior to the flight tests. Ideally, the sensor would be installed and calibrated in its operating environment, however, in this case it was not possible to enclose the sensor in a furnace while installed in the airplane's exhaust.

A diagram of the calibration furnace is shown in Figure 4. An automated calibration was performed from room temperature to 600°C with the furnace temperature increasing at a rate of approximately 0.16°C/minute. A total of 117 reflected spectra were taken every 35 minutes. This corresponded to roughly 5°C increments in sensor temperature, based upon the output of the thermocouple attached to the sensor. Each of the 117 spectra was saved to a computer data file to be used during the flight tests.

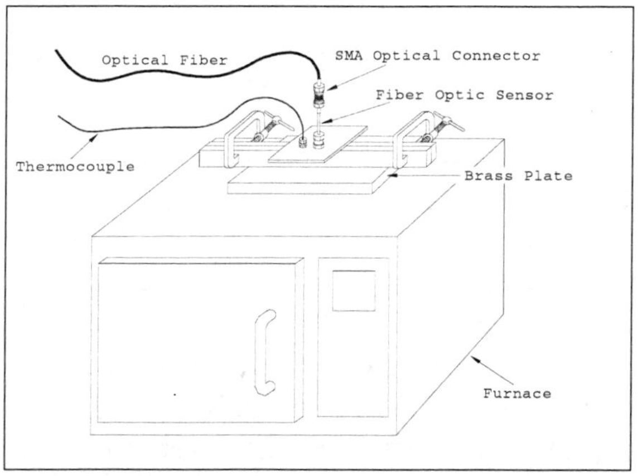

Figure 4. Configuration of Calibration Furnace

Data Acquisition - Procedure

The temperature of the exhaust duct was nearly 20°C prior to each day's initial flight. Approximately 1 hour prior to engine start-up, the optical SMA connectors were cleaned and reconnected using a small amount of index matching fluid. At the same time, the light source was powered and allowed to warm up in order to reduce fluctuations. After the 1 hour warm-up period, a reference spectrum was taken.

Just prior to engine start-up, external power was removed from the plane. Without power to the light source and computer, it was not possible to monitor the exhaust temperature during start-up. Approximately one minute after engine start-up, aircraft power was initiated. Once this occurred, the sensor system began acquiring optical spectra and calculating the exhaust temperature every 20 seconds.

Sensor performance was monitored by a flight test engineer during each test. The most significant problem encountered during 4 particular flights was loosening of the SMA connector. A commercial version of this sensor would transfer the optical connection to a more stable location and/or utilize a more rugged connection.

Data Reduction

Spectral data is normalized by dividing the raw reflected spectra by the reference spectrum, taken prior to each flight. Therefore, at room temperature, the normalized spectrum is nearly a straight line. As the sensed temperature increases, the normalized spectra become sinusoidal in appearance with increasing numbers of fringes. Each sensed temperature is associated with a particular reflected spectral curve. Figure 5 illustrates a typical high temperature reflected spectrum.

Temperature is deduced by comparing the normalized spectrum to each of the 117 calibration curves, which are stored in a look-up table. The computer algorithm calculates the minimum squared difference between the current fringe pattern and the set of calibration curves. However, the temperature corresponding to this minimum can only be determined to within the 5°C calibration increment. In order to more precisely determine the temperature, a parabolic approximation is used: a parabolic curve fits three points centered at the minimum and the minimum of this parabolic curve is defined as the measured temperature. This procedure improves the resolution to at least ± 2.5°C.

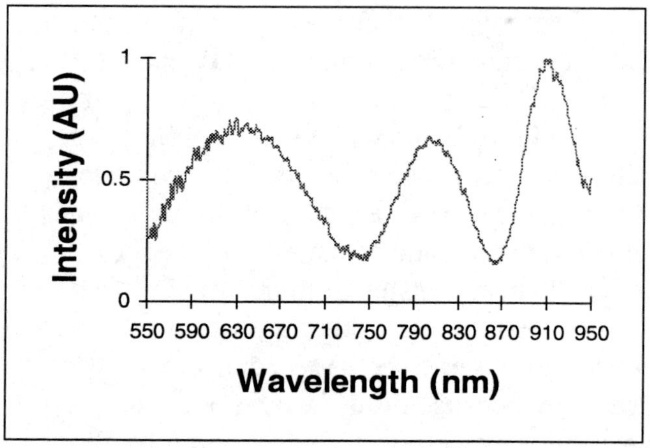

Figure 5. Spectrum Reflected from Sensor Exposed to 500°C

Results

Over 50 hours of flight tests were successfully completed consisting of 40 flights. The flight profiles for the first 20 flight test hours were straight and level. After 20 hours, the flight profiles were much more severe, exposing the sensor to forces exceeding 4 g's when performing maneuvers. These maneuvers consisted of loops, aileron rolls, Cuban Eights, Barrel Rolls, hard turns with 90 degree angle of bank, and touch and go landings.

Calculated temperature vs. thermocouple readings for a few flight profiles are shown in Figures 6-8. Figure 6 illustrates the temperature results for a typical straight and level flight profile. Figure 7 illustrates the temperature results for a flight profile with varied engine power levels, and Figure 8 for a flight where maneuvers were performed. Based on these flight test results, the fiber optic sensor accurately detected the exhaust gas temperature.

The average difference between the thermocouple output and the sensor output and standard deviations of the three flights are given in Table 1. The straight and level flight profile produced the most accurate results because the sensor was in near equilibrium throughout the flight. Flight test profiles with varied engine power settings produced less accurate temperature results, while the flight profiles that included maneuvers yielded the highest errors. The time constant, τ, was approximately 60 seconds. The larger deviation in the non-straight flight profiles is most likely caused by the slow response time of the sensor.

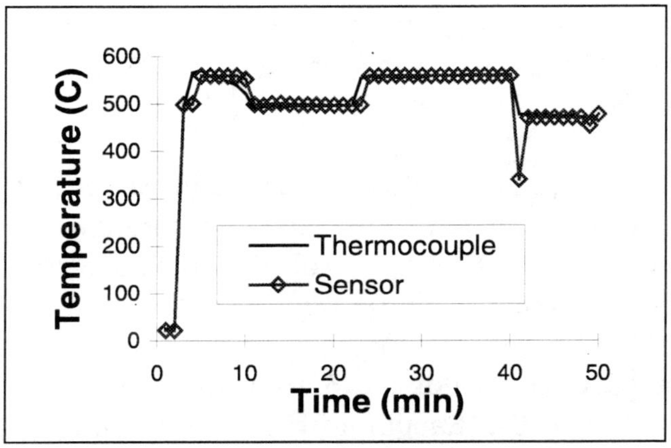

Figure 6. Flight Test Results for a Straight and Level Flight

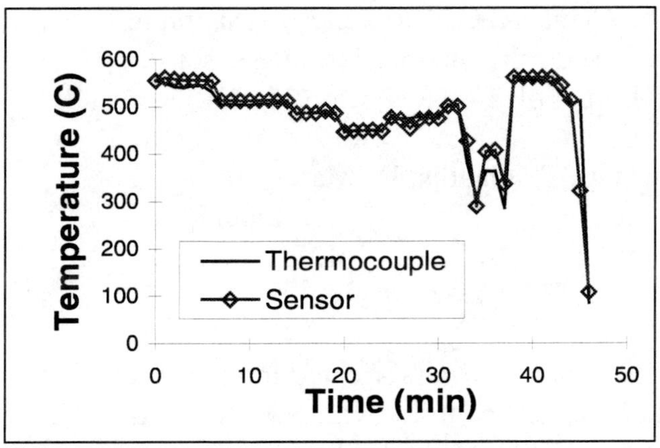

Figure 7. Flight Test Results for a Flight with Varied Engine Power Settings

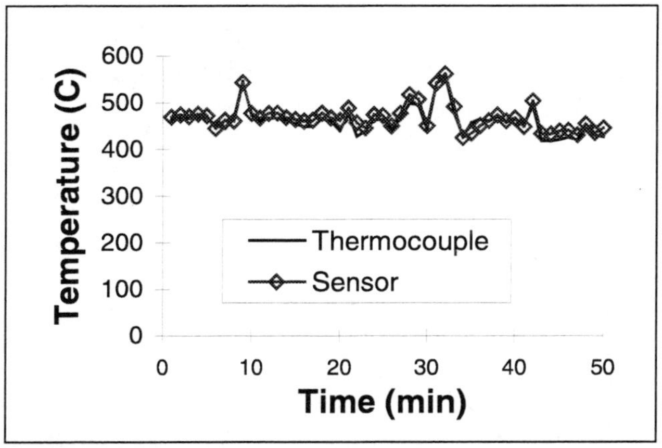

Figure 8. Flight Test Results for a Flight with Maneuvers

Table 1. Statistical Analysis of Three Flights

Average T Difference	Standard Deviation	Flight Profile
4.5°C	9.3°C	Straight
7.5°C	12.8°C	Varied
9.8°C	19.8°C	Maneuver

Improvements

Based upon these initial flight tests, improvements to the sensor have been identified and are being incorporated into a new sensor. The most important change is reducing the sensor's thermal mass in order to improve its response to temperature change. This new sensor will be installed in a Learjet engine exhaust in place of an existing thermocouple.

Heat flow calculations were performed using the new sensor geometry to determine the theoretical response time of the new sensor. As shown in Figure 9, the theoretical response time was reduced by 50% to roughly 30 seconds.

To improve the sensor's performance and reliability, the cement bond between the sapphire fiber and the sensor housing was replaced by a gold braze. Flight tests will be performed on this sensor to confirm the heat flow analysis and test the new sensor's performance and reliability.

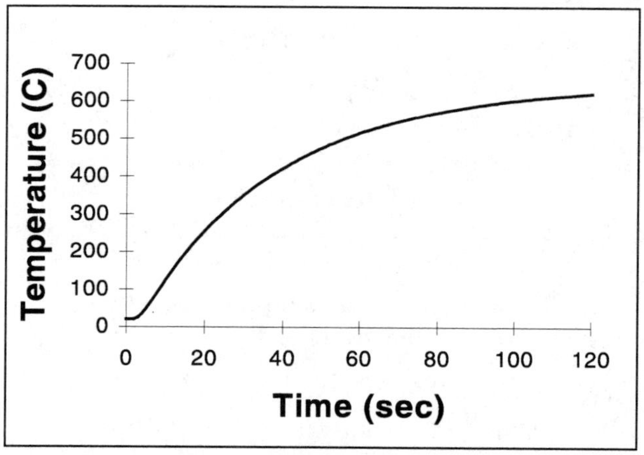

Figure 9. Results of Heat Flow Calculation

Summary

Ground and flight tests performed on an OV-10D airplane verified the prototype sensor system's operation and demonstrated the feasibility of utilizing such a system in the harsh environment of an airplane exhaust. For a prototype system, the sensor performance exceeded expectations. The sensor system successfully operated over 50 flight hours which consisted of straight and level flights in addition to varied engine power settings, maneuvers, and touch and go landings. These tests proved that it is feasible to operate such a fiber optic sensor system in the hostile environment of an aircraft engine exhaust.

Acknowledgments

The authors wish to thank Rich Ranaudo, William Rieke, and Larry Schultz for piloting the OV-10D airplane during the sensor flight tests, Bud Schutte for maintaining the airplane and installing the sensor system, Mike Perez for assisting in the sensor installation, Ed Emery and Steve Plaskon for coordinating the installation of the associated electronic components, Mike Ernst for design of the sensor vibration-damping mount, George Saad for welding the thermocouples to the sensor and reinforcing the sensor welds, Jih-Fen Lei for the use of her programmable furnace, and Gus Fralick for performing the heat transfer calculation.

References

[1]Beheim, G., Sotomayor, J.L., Tuma, M.L., and Tabib-Azar, M., "Fiber-Optic Temperature Sensor Using Laser Annealed Silicon Film," SPIE Proceeding, Vol. 2291, pp. 92-98, 1994.

[2]Schultheis, L. Amstutz, H., Kaufmann, M., "Fiber-optic temperature sensing with ultrathin silicon etalons," Opt. Lett, Vol. 13, No. 9, pp. 782-784, 1988.

[3]Wang, A., Gollapudi, S., Murphy, K. A., May, R., Claus, R. O., "Sapphire-fiber-based intrinsic Fabry-Perot interferometer," Op. Lett. Vol. 17, No. 14, pp. 1021-1023, 1992

[4]Wang, A., Gollapudi, S., May, R. G., Murphy, K. A., Claus, R. O., "Advances in sapphire-fiber-based intrinsic interferometric sensors," Opt. Lett. Vol. 17, No. 21, pp.1544-1546, 1992.

[5]Sawatari, T., Gaubis, P., Mattes, B., Charnetski, C., "Fiber Optic Temperature Sensor," Invited Paper, SPIE Proceeding, Vol. 2686, pp. 132-145, 1996.

[6]Sentec Corporation fabricated the sensor as part of an SDB contract with NASA Lewis Research Center (NAS3-27202).

AIR DATA SENSOR FAILURE DETECTION

Drew Houck, Boeing Commercial Airplane Group, Seattle, Washington
Les Atlas, Dept. of Electrical Engineering, University of Washington, Seattle, Washington

Introduction

A review of safety event report summaries from world wide sources reveals repeated instances of anomalous air data events resulting from blocked or disconnected pitot-static probes and ports over the last 15 years. Two aircraft losses in 1996 included obstructed pitot static probes or ports as contributing factors.[1,2]

This paper describes initial work to develop and validate signal processing techniques that could be applied to a practical and reliable system for detecting pitot-static sensor failures, both on the ground, prior to committing to takeoff, as well as inflight. The candidate approaches require no new hardware and could be implemented with software modifications to aircraft currently being produced and as a retrofit for many airplanes already in service.

System Description

Background

Air data driven instruments are among the most basic displays on modern aircraft. These instruments measure air pressure, static and total, inducted through probes and ports on the exterior surface of the airplane and display the data as pressure altitude and airspeed.

The altimeter detects static air pressure changes, measured at static ports, and maps it to pressure altitude using the equation for a standard atmosphere.[3]

$$P_a = P_{a_{SL}}\left(1 - 6.8735e^{-6}H_c\right)^{5.2561}$$

H_c = Pressure Altitude
P_a = Atmospheric Pressure
P_{aSL} = Standard Sea Level Atmospheric Pressure

Airspeed indicators uses both static and total pressure to provide the pilot with an indication of velocity through the air mass.

Calibrated airspeed, V_c, indicators are simple to design and construct, easy to calibrate and sufficiently accurate. In general, the aircraft stall speed, landing speed and handling characteristics are proportional to V_c for a given gross weight.

$$V_c = \sqrt{\frac{2\gamma}{\gamma - 1}\frac{P_{a_{SL}}}{\rho_{SL}}\left[\left(\frac{q_c}{P_{a_{SL}}} + 1\right)^{\frac{\gamma-1}{\gamma}} - 1\right]}$$

V_c = Calibrated Airspeed
γ = Ratio of specific heats
P_{asl} = Standard Sea Level Atmospheric Pressure
ρ_{SL} = Standard Sea Level Air Density
q_c = $P_T - P_S$
P_T = Total Pressure
P_S = Static Pressure

These equations are adequate for this discussion with the additional stipulation that P_s is assumed equal to P_a.

Since both static and dynamic pressure are used to calculate airspeed, failure of either

---

[1] NTSB Report No: A-96-20, 11/21/97
[2] NTSB Report No: A-96-17, 9/8/97

[3] USAF TPS, "Pitot Static Flight Testing", pp 3.1-3.33, May 1979

0-7803-5086-3 /98/$10.00 ©1998 IEEE

parameter will result in an erroneous V_c output. If an airplane takes off with trapped static pressure, V_c may appear normal during the ground roll, but the airspeed error will increase with increasing altitude during the initial climb-out. (Figure 1) A partially blocked static port may be in error during an altitude change, slowly correct to an accurate pressure while the aircraft is in level flight and then again become erroneous with the next altitude change. (Figure 2) The airspeed error will be fast or slow depending on the whether the plane is climbing or descending.

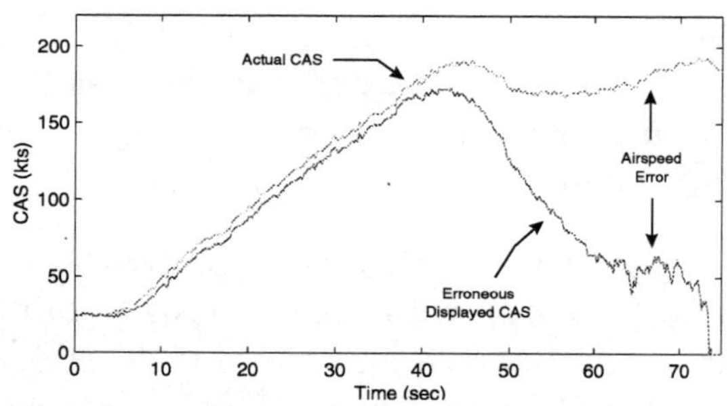

Figure 1 - Static Pressure Port Failure; Takeoff and Initial Climb

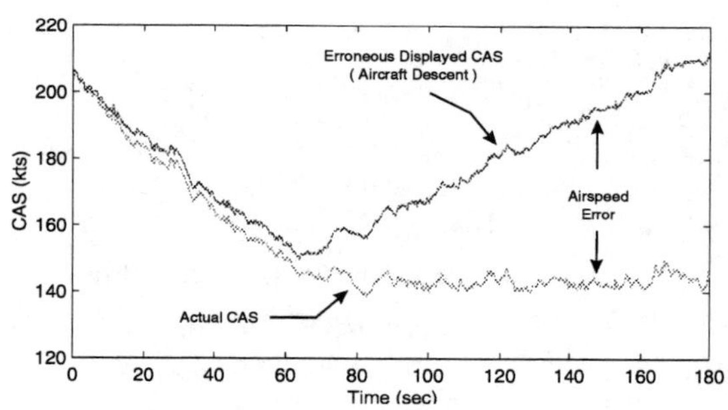

Figure 2 - Static Pressure Port Failure; Descent

In addition to flight deck displays, other aircraft systems, including the autopilot, can be adversely affected by pitot-static faults. An aircraft descending with partially blocked static ports could have autothrottles attempt to capure an incorrectly low airspeed. In this situation the pilot could be confronted with low speed warnings which are independent of pitot-static inputs (angle-of-attack) while observing a high and apparently increasing airspeed.

Air Data Sensors

Air data sensor probes and ports are simple devices with no moving parts and electrical power provided only for anti-icing. Current airplanes have no direct method of determining if the probes themselves are functioning properly. Reliability is improved by providing redundant sensors on each airplane.

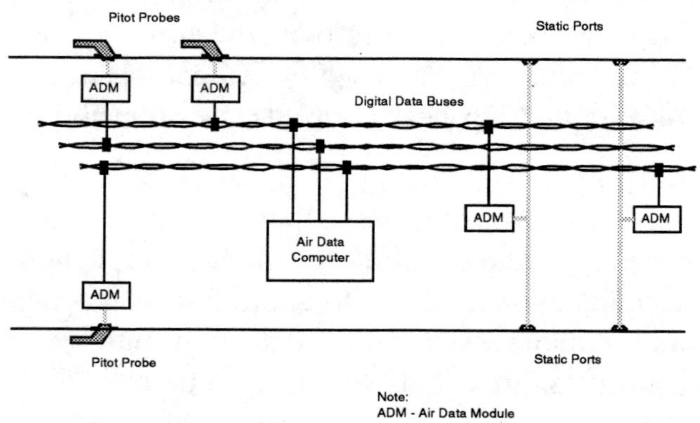

Figure 3- Pitot-Static Sensor Schematic

This approach implicitly assumes that sensor failures are independent. In general however, if the conditions that contribute to obstructed sensor ports (icing, ash, bugs, maintenance actions) are present at one sensor they are likely to be present at all the plane's ports and/or probes. This lack of sensor independence reduces the overall system dependability. The use of multiple sensors also implies that the crew will be able to differentiate a failed sensor from a properly operating one. A failed sensor on the ground may be impossible for the crew to detect by simply observing their instruments. Airborne failures can be insidious and confusing.

Given the past history of pitot static instrument failures and the potential for serious consequences, this study was intended to investigate real time methods of detecting

0-7803-5086-3 /98/$10.00 ©1998 IEEE

failed sensors and alerting the crew in a timely manner. For this study a failed system was specifically defined as an obstructed probe or port orifice.

Failure Detection

Design Requirements

This study defined the following design requirements as necessary to ensure an operationally viable detection system.

1. The ability to detect failures while still on the ground, prior to committing to a takeoff. A more stringent requirement for future investigation is failure detection prior to aircraft movement on the ground.
2. No hardware modifications or additions. This is required to improve the ability to retrofit airplanes already in service and reduce costs. Consistent with current aircraft avionics architectures, allowable computational complexity was limited to what would normally be available in a modern display processor. This requirement also limits the signal processing rate to 20 Hz.
3. A low false alarm probability. Any system that is plagued by false alarms will not be trusted by aircrews, could be economically unfeasible and might be uncertifiable.[4]
4. A high failure detection rate. This is of course the basic purpose of the detection system. The potential consequences of a failure are sufficiently serious that a missed alert is very undesirable.

Design Proposition

The study's proposition was the assumption that post flight analysis of a failed pitot-static sensor system using operationally available onboard data would be sufficient to confirm or identify the failure. If this failure identification could be accomplished after the flight, would it then also be possible to develop a viable, real time algorithm that could provide the flight crew with a timely, reliable fault alert during the flight?

Pitot-Static Pressure Signals

The initial design step collected and analyzed actual commercial aircraft static and total pressure data to identify any unique characteristics that might provide a useful signature of failed sensors. The data was obtained from Boeing 737 and 777 aircraft. Thirty-eight test sequences were evaluated, reflecting a variety of flight and atmospheric conditions. The majority of the data included a takeoff sequence. On-ground identification of a failed sensor prior to takeoff was considered the most essential and difficult element of the design requirements.

Actual failure data, represented by an obstructed pitot probe or static port, was not readily available. The expense of a dedicated flight test to obtain this data was beyond the scope of this preliminary inquiry. A failure was synthesized by obtaining data from a parked aircraft on a very calm day. This appeared to adequately simulate the expected failure condition where no external air pressure changes excite the pressure sensor. It did not address however any possible effects of sensor vibration.

The initial analysis revealed the following observations:

1. If any air mass perturbations are present, either from atmospheric conditions or aircraft movement, an operating systems had a higher de-meaned energy content than an obstructed ones. Both were essentially white noise. (Figure 4). It is be difficult to define a decision threshold however since a normal background noise level for one sensor may be a failed threshold for another.

---

[4] FAA, "AC 25-1309-1A", 8.f, May 21, 1988

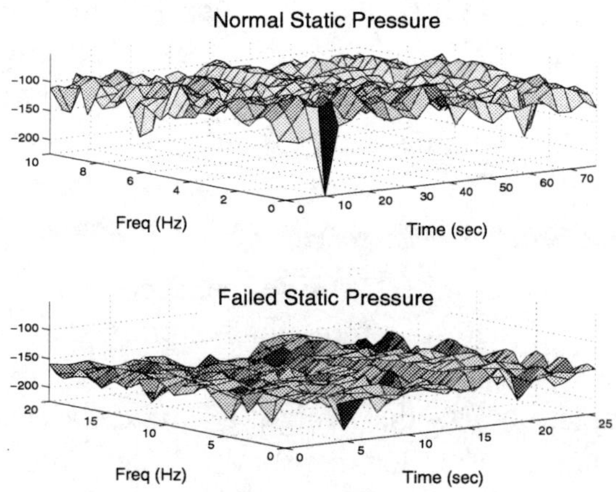

Figure 4 - Static Pressure Time-Frequency Spectrums

2. An indication of an operable sensor, static or pitot, is an appreciable pressure slope or time-derivative. However, a constant pressure is not necessarily an indication of a failed sensor. (Figure 5)

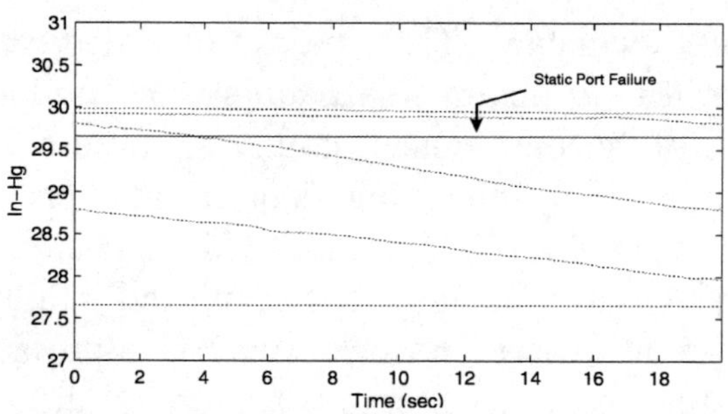

Figure 5 - Static Pressure Sequences

3. While static pressure sensors generally measure variations due to vertical aircraft movement, the static pressure data also showed that an accelerating aircraft, even at constant elevation, exhibits a non-zero static pressure derivative or slope. (Figure 6 and Figure 7).

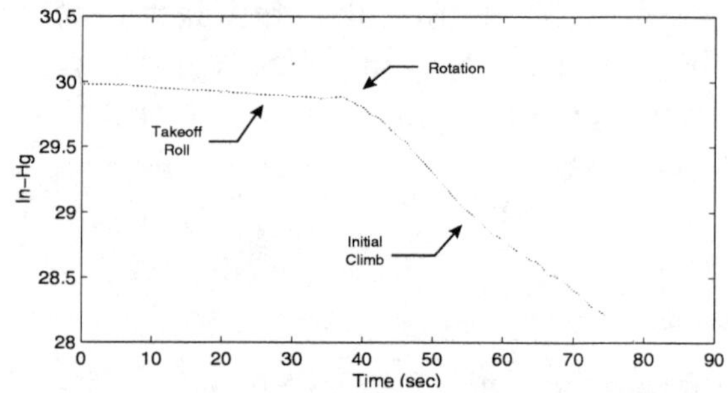

Figure 6 - Takeoff and Initial Climb Static Pressure.

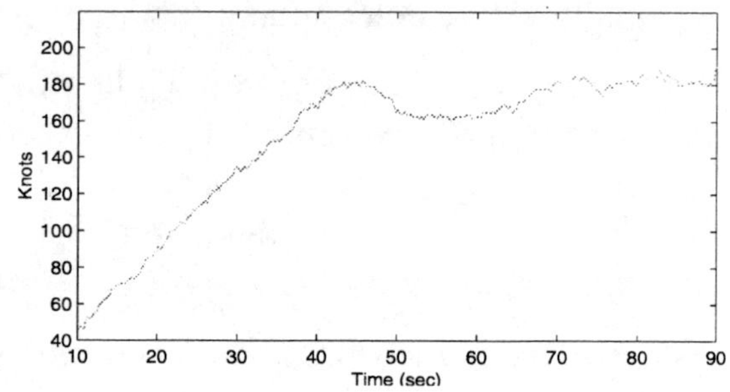

Figure 7 - Takeoff and Initial Climb Airspeed (V_c)

Differentiator Filter

The data suggests a potentially reliable indicator of an unobstructed static port is a non-zero static pressure signal derivative. When pressure altitude changes, static pressure will also normally change. That the static pressure consistently changes when the airplane is accelerating at a constant runway altitude is not so obvious. As the airstream experiences an increase in velocity moving around the aircraft fuselage, the total airstream energy is unchanged. This increase in kinetic energy is balanced by a decrease in potential energy, represented by static pressure. Referencing Euler's Equation:

$$P_T = P_S + \frac{1}{2}\rho V^2$$

The relatively simple implementation selected to take advantage of these P_s slopes is summarized in the following steps.

1. Compute the static pressure signal differential.

2. Compare that differential against a static pressure model to determine if it matches a normal or failed signal signature.

A simple, efficient method of obtaining derivative samples of a band limited signal is a discrete-time differentiator with frequency response

$$H_{diff}\left(e^{jw}\right) = \frac{1}{T} jw \quad ; \quad -\pi < w < \pi$$

The high frequency components of the differentiator output are not important to the application and it is necessary only to ensure that the lower frequency portion of the filter accurately represents the desired response. The filter response illustrated in Figure 8 meets the requirement for a smoothed low frequency differentiator.

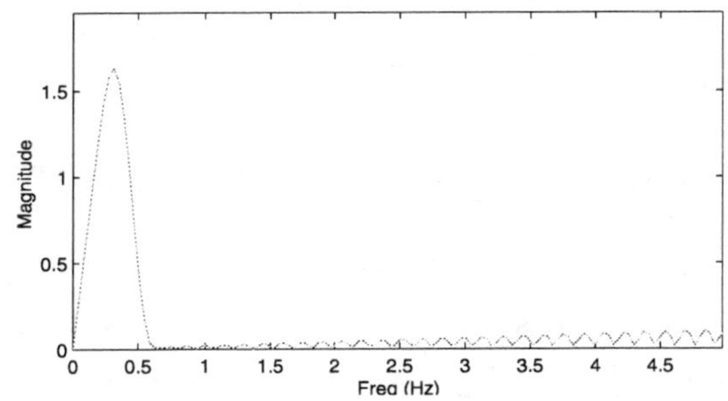

Figure 8 - Smoothed Differentiator Response

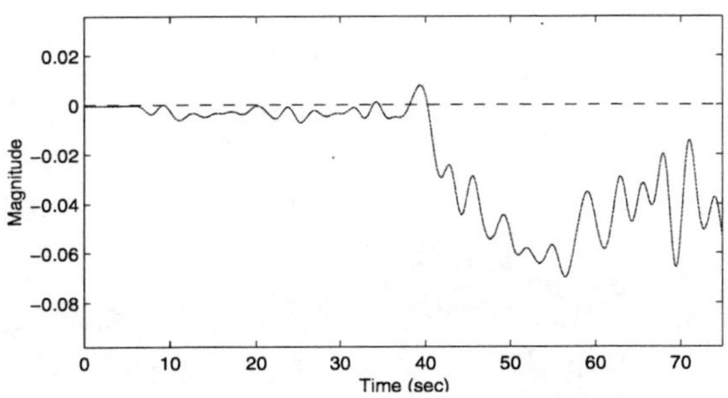

Figure 9 - Smoothed Differentiator Output

Correlation

Armed with the derivative of the static pressure, the next step was to determine if it represented the output of a clear or obstructed port. This determination can not reliably be made without correlating or comparing the static pressure derivative with another parameter since, as stated before, a failed system will always have a zero or minimal low frequency differential, but an operating system may under some conditions also exhibit that same characteristic. Several possible approaches were considered.

1. To identify which of two static systems has failed, monitor the mean value difference of both systems. If the difference exceeds a preset threshold, compare the pressure derivative of both static signals and condemn the lower derivative value. The rational for this approach is that if two system values on the same airplane differ appreciably, one has failed and the most likely candidate is the signal that is not changing.

2. To identify one or more static sensor failures, select a correlating parameter that is independent of the pitot-static system. If the pressure derivative is below a defined decision threshold and the correlating parameter is above a defined correlation threshold, then declare the static pressure port failed. This is the approach that was investigated.

The task is to find a parameter or parameters that can be used to determine when the static pressure derivative should normally be above some threshold. The correlating parameter must be readily available, independently reliable and meet all the initial design requirements previously defined, especially the ability to determine pitot static system health prior to committing to takeoff. Inertial accelerations, vertical and longitudinal, are parameters that appear to meet these prerequisites. The longitudinal inertial acceleration, either obtained directly or by differentiating ground speed, V_g, allows correlation with the static pressure change resulting from the Bernoulli effect discussed earlier. Total pressure was another possible

0-7803-5086-3 /98/$10.00 ©1998 IEEE

correlating parameter. A constantly changing total pressure is indicative of either a changing airspeed (dynamic energy component) or altitude (potential energy component). All these effects normally result in a static pressure change.

Static Pressure Model

Multiple models may be used to cover the different sources of static pressure derivatives. The takeoff case is the most difficult and will be covered in some detail. Other conditions could be handled in an analogous fashion.

Figure 10 is a plot of the static pressure differential for a typical takeoff generated using the filter technique just discussed. Figure 11 is a two-dimensional histogram representing 6500 takeoff data points obtained from 25 separate takeoffs. These plots suggest a correlation between d(Ps)/d(t) and d(Vg)/d(t). The heavy concentration of data in the 3-4 kt/sec regime is a function of aircraft performance characteristics and crew procedures.

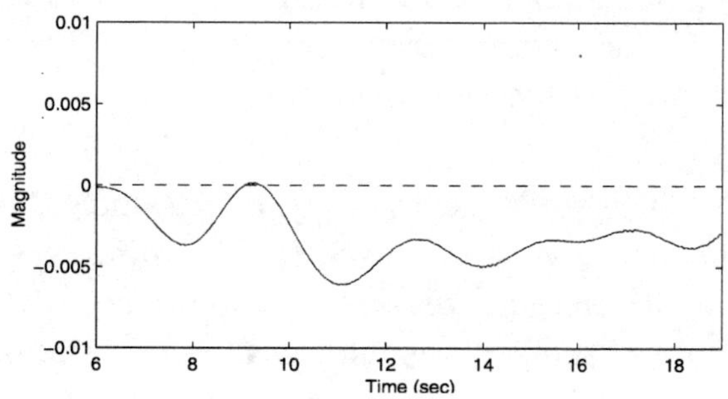

Figure 10 - Takeoff Static Pressure Differential

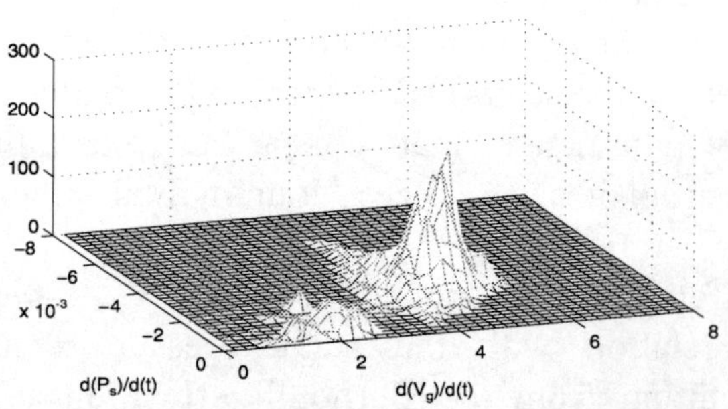

Figure 11 - Static Pressure Differential Histogram

This flight test data was used to develop a static pressure model for an operating system. The histogram was divided into ground speed rate bins and a gaussian distribution computed for each bin to obtain a mean and standard deviation. An example bin histogram gaussian distribution statistics is illustrated in Figure 12.

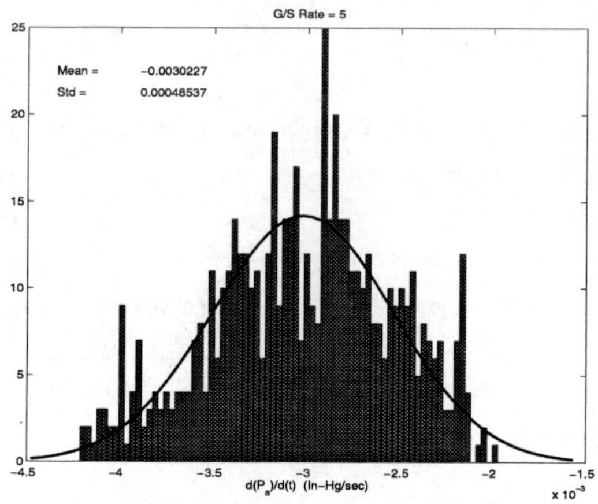

Figure 12 - V_g Rate Bin Statistics

Using the statistical summary generated from the all the processed data, a Probability Density Function (PDF) model of $d(P_s)/d(t)$ vs $d(V_c)/d(t)$ was constructed. This model matched the generally expected result predicted by the Bernoulli equation. The static pressure derivative approaches zero for $d(Vg)/d(t) = 0$ and increases linearly with increasing ground speed derivative.

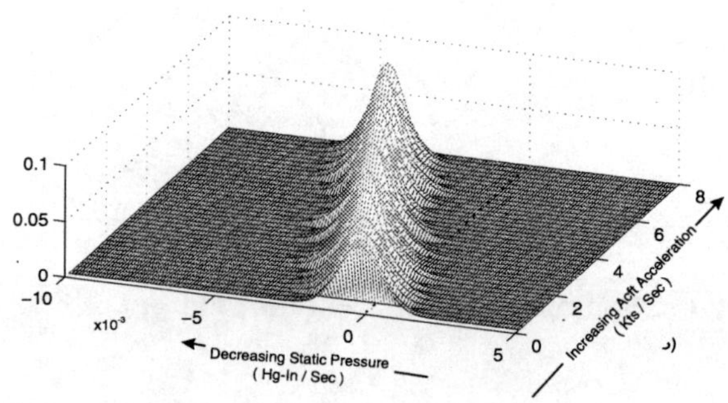

Figure 13- Normal Static Pressure PDF Model

In a fashion similar to the normal

operating case, a PDF model was also constructed for the failed sensor case. The failure model is represented by a constant zero mean $d(P_s)/d(t)$ for all $d(V_g)/d(t)$.

These two individual models combine to define the system model for the normal and failure cases. As expected at low acceleration rates, the normal static pressure differential is also low and difficult or impossible to differentiate from the failure condition. As the acceleration increases however the ability to identify a failed or normal systems improves. (Figure 14)

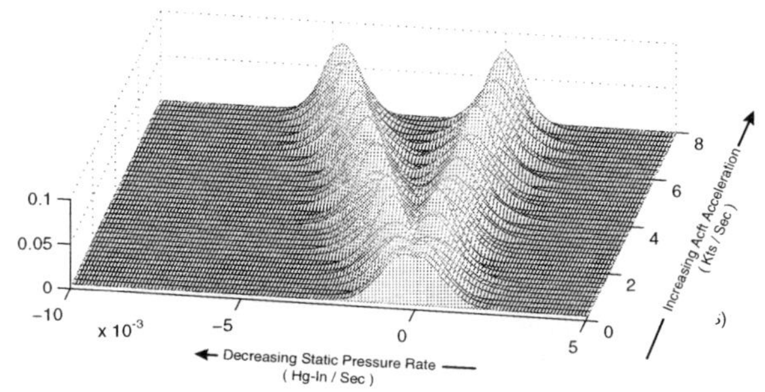

Figure 14 - Normal/Fail Static Pressure PDF Model

Decision Boundary

Given an efficient, linear method of determining the static pressure rate and correlating ground speed rate, a remaining issue is determination of the optimum decision boundary for declaring a sensor failure.

In this case we are concerned with 4 possibilities.

	Definition	Condition
$P_{(D|F)}$	Probability of detecting a failure given a failure	Alert
$P_{(ND|NF)}$	Probability of no detected failure given no failure	Normal Operation
$P_{(D|NF)}$	Probability of detecting a failure given no failure	Nuisance Alert
$P_{(ND|F)}$	Probability of no detected failure given a failure	Missed Alert

The approach to this problem is analogous to the two hypothesis detection performance problem for a binary, gaussian PDF signal. (Figure 15)

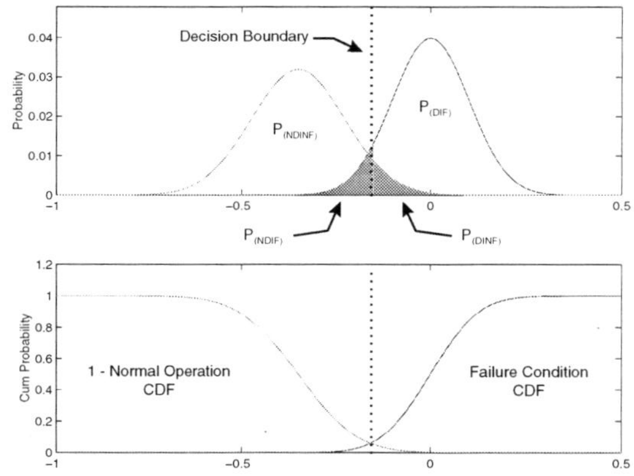

Figure 15- Static Pressure Model Decision Boundary

Detection performance increases with distance between the means of the two PDF's. For the static sensor problem the distance is zero for $d(V_g)/d(t) = 0$ (assuming no other sensor excitation sources) and increases linearly with increasing acceleration.

Figure 16 is a receiver operating characteristic (ROC) for $P_{(D|NF)}$ vs $P_{(ND|F)}$ with increasing values of $d(V_g)/d(t)$. Like the communication problem, an important case is the desire to minimize the total probability of error. For this application however, a very low probability of false alarm is a primary system requirement. Crews must have confidence that a detected failure is very likely to be indicative of an actual system failure. The actual probably of a failure is unknown, but estimated to be very low. The possibly significant consequences of a failure, not the high likelihood of one, make this a potentially beneficial application.

0-7803-5086-3 /98/$10.00 ©1998 IEEE

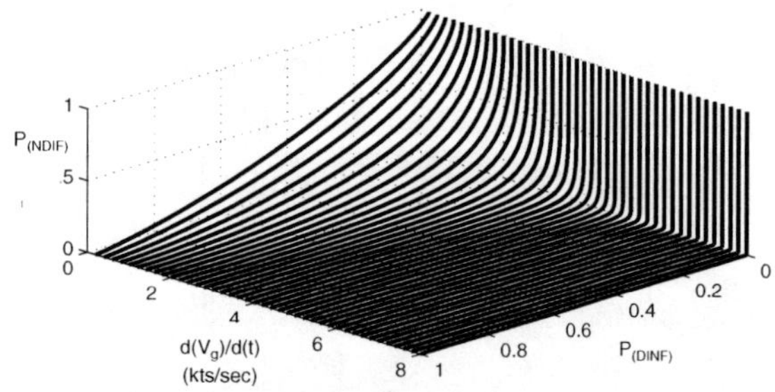

Figure 16-Two Dimensional $P_{(D|NF)}$ vs $P_{(ND|F)}$ ROC

The probability of a false alarm can be decreased by making a system failure decision based upon multiple independent samples. If the samples are separated by at least the filter buffer length, sample independence appears to be a reasonable assumption. For a typical takeoff, about 400 static pressure samples are available for analysis. To ensure an analysis is complete prior to the takeoff decision speed (V_1), the pilot should be alerted to a system failure within the first 15 seconds of the takeoff. Assuming 3 independent samples are obtained and that a system failure is not declared unless all three are individual failures, a false alarm rate of 10^{-6} is obtainable. The probability of detecting a failure if a failure exists is about 97%.

The system can be further tuned and improved by taking more independent samples over a longer period of time, using more sophisticated strike-counter schemes to make the system failure decision and implementing multiple models to cover other failure modes.

Conclusions

This project identified a potentially viable real time processing technique for detecting common pitot-static sensor failures. Additional effort to ensure adequate model fidelity and robustness for specific aircraft over a wide range of operating conditions is necessary to validate the system utility in an operational environment.

Additionally, there may be other approaches to this problem that could be investigated. Angle of attack indicators is one example. Synthesized altitude and airspeed sources, completely independent of pitot-static systems, have also been the subject of initial evaluations.

THE MINI-ACE PLUS FAMILY OF MIL-STD-1553 TERMINALS

Mike S. Glass, ILC Data Device Corporation, Bohemia, N.Y.

Abstract

The Advanced Communications Engine (ACE) family of integrated dual redundant MIL-STD-1553 RT (Remote Terminal) and BC/RT/MT (Bus Controller/Remote Terminal/Bus Monitor) terminals with enhanced functions has been upgraded. The Mini-ACE Plus series addresses many of the key issues in today's market; specifically, the demand for COTS/MOTS, reduced cost and size, and requirements for MIL-STD-1760 stores applications. This paper reviews the features and the technology that were used to achieve the cost and functional improvements, and size reduction.

The MINI-ACE Plus is a functional equivalent to the initial generation ACE family, but in a package that is half the size. The MINI-ACE Plus is presently the **smallest production RT/BC/MT terminal with 64K words of RAM on the market at one square inch.** ILC DATA DEVICE CORPORATION (DDC) was able to accomplish this task by utilizing state-of-the-art monolithic fabrication technology, along with a custom designed ceramic co-fired package. The resultant MCM design supports both through-hole and surface mount applications.

This paper will discuss the processes and the technology used in the development of the MINI-ACE Plus MCM. This includes the issues and advantages of migrating to co-fired ceramic packaging, along with the evolution of the analog transceiver and digital protocol monolithic chips used in the Mini-ACE Plus terminals. This includes a functional overview of the ACE terminals. In addition, the paper discusses new features, involving an option for 64K X 16 RAM, an RT boot-up option, and additional clock frequencies. Finally, there is a discussion of software-related issues, including compatibility to previous generation terminals, and the availability of libraries that are portable to different hardware and software platforms.

Introduction

Current demands for military components entail a multiplicity of increasingly exacting requirements. These include the needs for lower size, lower power, and lower cost (e.g., COTS and MOTS); plus demands for increased performance in terms of offloading the host processor, while maintaining software compatibility for code written for legacy applications.

The subject of this paper is an extension of ILC Data Device Corporation's (DDC's) Advanced Communication Engine (ACE) MIL-STD-1553 terminals. In line with its continuing leadership role of supplying a full range of products to the MIL-STD-1553 community, DDC offers the Mini-ACE and Mini-ACE Plus families of RT and BC/RT/Monitor terminals.

These offerings fulfill a multiplicity of needs for users of 1553 components and boards. The cornerstone elements of these

0-7803-5086-3 /98/$10.00 ©1998 IEEE

products include a monolithic transceiver chip and a new digital multiprotocol BC/RT/MT chip. The components are packaged in co-fired ceramic. In addition, with respect to DDC's previous generations of BUS-61553 AIM-HY, BUS-61559 AIM-HY'er, and BU/65170 and BU-61580 ACE series hybrids, they provide software compatibility, plus a number of enhancements.

RT-only, BC/RT/Monitor Versions

The Mini-ACE terminals comprise a family of complete integrated interfaces between a host processor and a MIL-STD-1553 bus. The Mini-ACE series of multi-chip hybrids (MCMs) provide RT-only or BC/RT/MT functions in an identical 1.0 square inch (1.0 X 1.0 X 0.155 (max)) ceramic package footprint.

The Mini-ACE RT and BC/RT/MT terminals include dual transceiver, protocol, and 4K X 16 of internal shared RAM. As illustrated in Figure 1, the Mini-ACE Plus BC/RT/MT terminal includes an additional 64K X 16 RAM chip. The Mini-ACE terminals are available in the choice of 81-pin PGA for through-hole applications, or the choice of 72-pin flatpack or gull lead packages for surface mount applications. The Mini-ACE hybrids include eight pins on the package floor, to allow for active trimming of the two transceivers.

Figure 1. Mini-ACE Plus Photo

Protocol Chip

The Mini-ACE (Plus) protocol chip is fabricated in 0.8 ? CMOS. This provides complete backwards compatibility to the previous generation protocol chip, which is fabricated in 1.2 ? CMOS, while reducing power and enabling a factor of nearly two in reduction of overall real estate.

Transceivers

The transceivers in the Mini-ACE are completely monolithic, requiring no additional components, except for decoupling capacitors. The Mini-ACE terminals consume very low power, with a typical dissipation under 0.5 watts at idle, and 1.7 watts at 100% duty cycle.

The Mini-ACEs family is powered from a single +5V power supply. In addition to eliminating the need for power supplies for other voltage(s), the use of a 5V transceiver entails the use of a step-up, rather than step-down isolation transformer. This increases the terminal's inherent input impedance, reducing the demands on system layout in meeting the MIL-STD-1553 impedance requirement.

Functional Overview

Figure 2 is a simplified block diagram of the Mini-ACE RT and BC/RT/MT, along with the Mini-ACE Plus BC/RT/MT MIL-STD-1553 terminals.

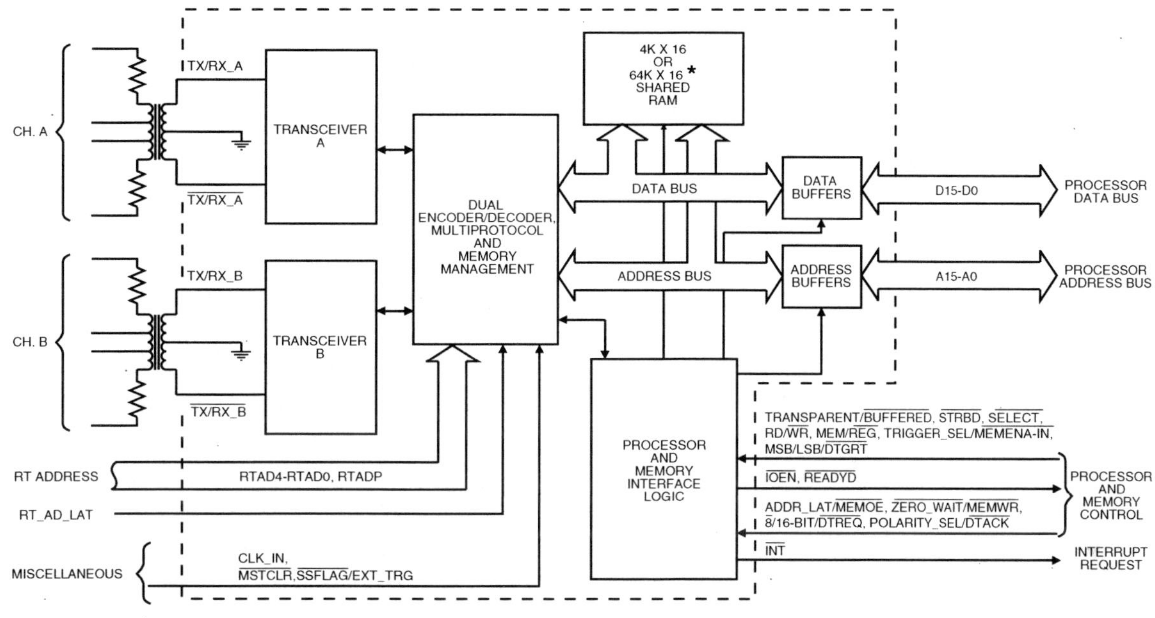

Figure 2. Mini-ACE Block Diagram

Host Interface

As a means of minimizing P.C. board space and "glue" logic, the Mini-ACE and Mini-ACE Plus terminals provide a high degree of flexibility in interfacing to a host processor and internal/external buffer RAM. The interface options include 8-bit or 16-bit buffered modes to 4K (or 64K) of internal RAM, 16-bit transparent and DMA configurations, and the capability to easily interface to processors that do not have an "acknowledge" input signal to enable wait states (e.g., microcontroller ports). The transparent and DMA configurations allow the use of up to 64K X 16 of external RAM.

Software Compatibility

The advanced functional architecture of the Mini-ACE terminals provides software compatibility to DDC's previous generation ACE, AIM-HY'er and AIM series hybrids. This is a key consideration in allowing the re-use of software code written for legacy systems.

Multiprotocol Support

The Mini-ACE components provide complete multiprotocol support of MIL-

0-7803-5086-3 /98/$10.00 ©1998 IEEE

STD-1553A/B Notice 2 and STANAG 3838. The Mini-ACEs incorporate a myriad of architectural features supporting flexible operation while offloading the host processor. For BC mode, the Mini-ACE and Mini-ACE Plus include capabilities for programmable intermessage gaps, as well as automatic retries and frame repetition.

RT, MT Architecture

For RT mode, the Mini-ACE supports either hardwired, latchable, or software programmable RT address, programmable illegalization and busy, and the choice of single message, double buffering and circular buffering modes, programmable by T/R bit, broadcast, and subaddress.

For monitor mode, the Mini-ACE and Mini-ACE Plus implement a true message monitor or combined RT/monitor mode with message selection based on RT address, T/R* bit and subaddress.

Interrupts, Clock Input

A wide array of maskable interrupts are provided for each of the three modes. The Mini-ACE hybrids may be operated with a choice of four clock input frequencies: 16 MHz, 12 MHz, 10 MHz, or 20 MHz.

The Mini-ACE and Mini-ACE Plus terminals are manufactured in DDC's DOD-STD-1772 facility. 1772 processing is required in order to screen MCMs or hybrids to 883B. An intermediate level, ensuring operation over the full military temperature range of -55 to +125? C. but without QCI testing, as well as units with limited temperature ranges of -40 to +85? C, and 0 to 70? C, are also available.

Enhanced Functionality

With respect to the previous generation ACE hybrids, the Mini-ACE family adds functionality in a number of areas. These include the choice of 10 and 20 MHz clock input frequencies, in addition to the options of 16 and 12 Mhz for the previous generation. In many system designs, this eliminates the need for an additional clock oscillator.

Two optional features provide support for MIL-STD-1760 mission stores applications. One is an option to automatically latch the RT address internally following power turn-on. Another is the capability to automatically initialize into RT mode, with the Busy status word bit set to a value of logic "1". These features enable operation in scenarios where the host processor has a boot-up/initialization time of several seconds, but in which the RT must meet the 1760 requirements of responding (with Busy) within 150 ms following power turn-on, prior to responding with a full 1760-compliant message within 500 ms.

The Mini-ACE Plus version (Figure 1) provides a BC/RT/MT terminal with 64K words of internal shared RAM. For bus controller mode, this allows for the storage of a large number of multi-message frames. For the Mini-ACE's message monitor mode, this enables the storage of large amounts of bus traffic, while reducing the frequency with which the host processor needs to read the monitor command and data stacks.

Ceramic Packaging

All versions of the Mini-ACE are

packaged in co-fired ceramic packages. With this technique, the interconnect signal layers are embedded into the base of the package. The use of co-fired ceramic offers several advantages over Kovar packages, which were used with earlier generation products:

- **Reduced Weight.** In comparison to Kovar, ceramic (90% Al_2O_3) provides a 30% reduction in weight.

- **Hermeticity.** In a Kovar package, the glass beads enclosing the package leads are subject to cracking from mishandling of the part, lead forming, and placement on a board. This cracking can compromise the package hermeticity, since once a glass bead is cracked, moisture can leak into a circuit, resulting in contamination and corrosion of the circuit.

  With a ceramic package, the leads are brazed directly to the body of the package, providing built-in strain relief located at the mounting/braze joint. In addition to eliminating the problems associated with the glass beads, the brazed leads are much more rigid, allowing for ease of flat package lead forming and placement of the hybrid onto the PC board. For example, the Mini-ACE hybrids are available with a surface mount package with formed gull leads.

- **Manufacturability.** The use of ceramic eliminates several steps that are necessary parts of the manufacturing process for Kovar hybrids. These steps include the fabrication and attachment of dielectric layers, a substrate and a preform.

- **Thermal Resistivity.** The thermal conductivity of ceramic is greater than that of Kovar (0.46 vs 0.40 Watts/?C-inch). In addition, eliminating the various layers in the package (see previous paragraph) produces even greater reductions in the hybrids' thermal resistance. The net result is a reduction in the junction-to-case thermal resistance of about 80%, and an improvement (reduction) in the junction-to-case temperature difference of approximately 10 ?C. These improvements in thermal design result in a lower junction temperature for the chips in a hybrid, leading to an increase in the part's reliability.

- **Lower Package Profile.** As the result of eliminating several layers from the package, there is a minimum reduction in the vertical stack of 0.040 inches. This allows for a reduction in overall package height. As a result, the maximum height for the Mini-ACE hybrids is 0.155 inches.

- **Circuit Density.** The co-firing of the package with the circuit embedded in the base of the package allows for tighter design rules for hybrid layout. This allows the mounting of more die in a smaller surface area, with the end result being a reduction in overall real estate.

Software

The BUS-69080 C software library provides comprehensive support of the Mini-ACE and Mini-ACE Plus series hybrids. The library, developed in Microsoft Visual C++ 4.0, comprises a suite of C function calls that is portable to any hardware/software platform. The software library serves to offload a great deal of low-level tasks from the application programmer.

The BUS-69080 series includes drivers and menu software for use on PCs running DOS, Windows 3.1, Windows 95, and Windows NT. For Windows 95 and NT, system information for up to four Mini-ACE (Plus) devices may be stored in the Windows registry.

Portability

The libraries incorporate a small set of low-level interface routines for reading and writing the Mini-ACE's registers and memory, along with function calls for servicing interrupts, including support of user-supplied interrupt service routines. As a result, the libraries may be ported to any number of hardware platforms and embedded systems, for use with various operating systems and compilers. This enables the library to provide a versatile tool for the development of software drivers and/or applications in many environments.

Memory Allocation

For each mode of operation - BC, RT, and MT - the library operates under an open/access/close model, in which areas of RAM are autonomously allocated and de-allocated by means of low-level routines. While these functions may be invoked directly by an application, in general their operation is transparent to the application programmer. The library's memory manager module performs autonomous allocation of shared memory for stacks and data blocks. This provides a high degree of flexibility for sizing various data structures.

The memory management functions make use of handles consisting of starting addresses and sizes of memory blocks, along with status information delineating whether particular areas of shared RAM are unused, used, or protected. For each mode, there are functions to transfer data between shared RAM data blocks and buffers in host memory. In addition, there are functions to access consolidated data structures providing both message status information, as well as 1553 message words.

Mode-Specific Software

For BC mode, in addition to memory allocation, the library's lower level routines perform tasks such as automatically formatting the BC Control words and Command words. There are mid-level functions for sending and accessing individual messages, as well as functionality supporting a higher level of abstraction for automatically formatting minor frames. Other functions initiate the processing of single or repeated message frames.

For RT mode, there is high level operation for configuring and utilizing the Mini-ACE's single-buffer, double-buffer, and circular buffered subaddress memory management schemes. This includes methods for accessing synchronously and asynchronously received message data. There is also a mechanism provided for automatically reading and accessing the most recently received message. In addition, there is high-level support of subaddress illegalization and use of the busy bit, enhanced mode code handling, along with functions allowing for accessing user programmable status and BIT words.

For the message monitor mode, functions are provided for programming the command and data stack sizes, programming of the monitor "filter" table (which addresses/T-R/subaddresses to monitor), along with high-level tools that decode monitored messages, and transfer status information and message words to host RAM in a consolidated stack data structure.

Conclusion

The current generation of Mini-ACE and Mini-ACE Plus components provide MIL-STD-1553 RT and BC/RT/MT functionality in 1.0 inch square packages, with a maximum height of 0.155 inches. These hybrids incorporate 5V-only monolithic transceivers and a new 0.8 ? protocol chip. They provide functionality, including software compatibility to the previous generation ACE, while incorporating a flexible host interface, BC retries and automatic frame timing, flexible RT buffering, a selective message monitor, and a comprehensive interrupt architecture.

New features include additional clock options, an RT auto-boot option, and an option for 64K of internal RAM. The use of co-fired ceramic packages provides advantages in weight, hermeticity, manufacturability, thermal design, height, and circuit density. There is software available to support the Mini-ACE, in the form of a C library that is portable to any hardware/software platform. The software library performs memory allocation and de-allocation, while offloading a great deal of low-level tasks from the application programmer.

REFERENCES

(1) Fioriglio, J., *Advantages of Ceramic Packaging,* Application note for ILC Data Device Corporation, Bohemia, N.Y., 1996.

(2) Friedman, S., The ACE Family of MIL-STD-1553 Terminals Gets Smaller, *DASC Proceedings for AIAA/IEEE Digital Avionics Systems Conference, Volume I,* American Institute of Aeronautics and Astronautics, and IEEE, New York, October, 1997.

Keeping COTS-Based Systems Current through the use of Programmable Electronics

Dr. Gary L. Fitzhugh, VisiCom Laboratories, Inc.
225 Reinekers Lane, Suite 280, Alexandria, Virginia 22314-2875

The Pros and Cons of COTS

On June 29, 1994 Dr. Perry signed a memorandum, entitled Specifications & Standards - A New Way of Doing Business, that changed the way the Department of Defense (DoD) thinks about the procurement of new weapon systems. This memorandum paved the way for acquisition reform within the DoD, and with it, the infusion of commercial technology into government systems. This change in focus was accompanied by an increasing reliance on technology developments in the commercial sector. In fact, commercial technology is the driving force behind acquisition reform. These developments are neither paid for, nor will they be controlled, by the DoD or defense industry.

The DoD share of the technology market has been sliding for the past two decades. DoD's share of the worldwide semiconductor market has fallen from 17 percent in 1975, to about 1 percent in 1997. It was quite apparent that the DoD could not afford to be unique. The question is, can it afford COTS? Many government specifications were canceled, and scores of program managers began to build and modify their systems using COTS technology. The use of COTS technology has impacted nearly every government system, from older "legacy" systems to new start programs. While several success stories have emerged, and millions of (up front) dollars have been saved, there is another side to this story.

In today's economic environment of reduced DoD budgets, a number of defense industry trade journals have capitalized on the increased applications of COTS procurement strategies, some praising its benefits while others point out potential risks and possible disasters. Credibility of either claim is usually dependent on the specifics of a particular procurement. But one fact cannot be avoided. Operations and Support (O&S) costs of COTS based military systems are now the challenge. The O&S cost for a complex government electronics system is typically more than 60 percent of the total budget for that system. In the past, this cost covered a largely organic infrastructure to support the system. The use of COTS is changing the make-up of this O&S infrastructure.

An inherent advantage of COTS is the allowed and necessary deviation from older regulations that evolved from purely developmental programs. However, while providing the flexibility to adopt a commercial product to meet a requirement, risks associated with loss of DoD control are also introduced. Thus, advantages of rapid expanding technology, product identification and pricing, reduced procurement time, performance and reliability may be offset by lack of full logistic support, little (if any) control over future product modifications, incompatible repair parts, premature obsolescence, and inaccessible design details.

Perhaps the biggest change the DoD must anticipate is in the design process.

0-7803-5086-3 /98/$10.00 ©1998 IEEE

Widespread use of COTS will require an entirely new mindset for defense industry design engineers. In the past, virtually everything defense electronics did carried military specifications. Defense industry came to rely on, and perhaps, in many ways, found comfort in building to the MILSPEC requirements down to the component level. In many cases, the paradigm in today's defense industry design strategy is "best commercial manufacturing practices (BCMP)". For VisiCom, this means that while we may not use a specification or standard, there are portions of those documents contained in the BCMP.

The challenge today is determining how much and what is required to ensure our customer, the DoD, is getting the best product possible. We must understand what exists in the commercial marketplace, how well that will fit our requirements, and if they're not a complete answer, how we can adapt them with minimal change. We must also understand how the COTS can be interfaced, ruggedized, packaged, assembled, tested, and logistically supported to meet DoD operational needs. The focus clearly shifts to reuse of existing designs rather than new custom designs.

One of the ways that VisiCom Laboratories has tried to meet this challenge is through the use of Reconfigurable Logic Engines and a process called Rapid Retargeting.

Rapid Retargeting

Rapid Retargeting is a process developed to address the critical obsolescence problem currently being realized by the Military Services, as well as many commercial electronic applications.

Rapid Retargeting makes it possible to continuously upgrade the electronic device technology within a fielded system in a manner that is totally transparent to the operation of the system. It is believed that this process, when applied, will totally obviate digital microelectronic obsolescence. With the recent advances made in programmable analog arrays and programmable digital signal processing, this program could do the same in the analog and mixed signal domain.

Since the life cycle of a system tends to be much longer than that of the underlying semiconductor device technologies employed by the system, Rapid Retargeting was developed to enable users of electronic systems to protect their investment by providing a path for technology upgrade. This is currently accomplished at the electronic module and subsystem level.

Aging semiconductor technologies and diminishing sources of supply threaten system maintainability long before the system itself becomes obsolete. Rapid Retargeting is a design process which uses a collection of sophisticated analysis, simulation, and modeling tools to transform an existing electronic module from the fielded system into a new module with identical form, fit, and function with respect to the target module. The retargeted module, however, is implemented using the most current semiconductor devices available at the time. In addition, the resulting functional representation of the module in software makes subsequent hardware modifications and upgrades much easier. Furthermore, since the current proven design of the target module is the basis for the new module, the design cycle is much shorter than it would be if the design were to be started from scratch.

Hardware retargeting is typically done when existing hardware becomes obsolete or there is some value (such as a new system development) in having descriptive models of hardware functions. Reasons to develop hardware description models include the desire to reuse functions across several designs, to change the mechanical form factor of a board, or to allow for planned system upgrades. The use of descriptive models greatly decreases the design cycle and lowers the risk of upgrading a complex electronic system.

At VisiCom the hardware retargeting effort, for both our COTS line of products and custom legacy systems, begins by extracting the functionality of the target hardware and capturing it in a hardware description language (HDL). The resulting HDL models are simulated and compared with the original hardware for verification. Once verified, these models are ported to a new hardware design. Subsequently, parts obsolescence is no longer an issue, since HDL models can be rehosted whenever new technology becomes available. Once rehosting is complete, a packaging engineer then decides which packaging and interconnect technology is required for the desired level of environmental protection.

Elements of Rapid Retargeting

The elements of Rapid Retargeting used at VisiCom are graphically depicted in figure 1-1.

Hardware Description Language.

VHDL is a language used to describe digital electronic systems. It arose out of the United States Government's Very High Speed Integrated Circuits (VHSIC) program, initiated in 1980. In the course of this program, it became clear that there was a need for a standard language for describing the structure and function of integrated circuits (ICs). Hence the VHDL was developed, and subsequently adopted as a standard by the Institute of Electrical and Electronic Engineers (IEEE).

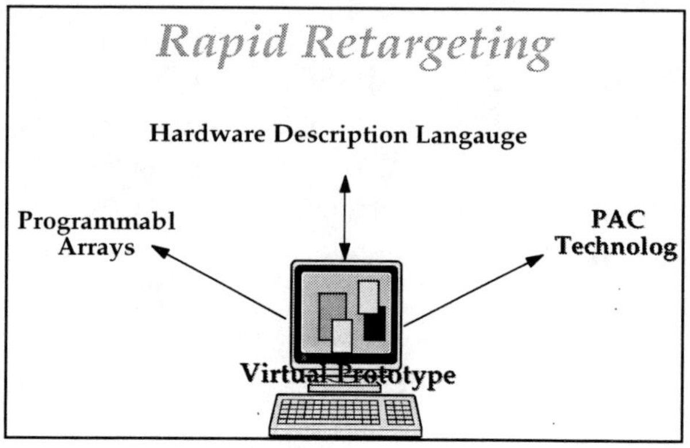

Figure 1-1

VHDL is designed to fill a number of needs in the design process. First, it allows description of the structure of a design. In other words, how it is decomposed into sub-designs, and how those sub-designs are interconnected. Second, it allows the specification of the function of designs using familiar programming language forms. Third, as a result, it allows a design to be simulated before being manufactured, so that designers can quickly compare alternatives and test for correctness without the delay and expense of hardware prototyping.

Through the use of today's Computer Aided Engineering (CAE) tools, it is possible to build detailed and accurate simulation models of electronic sub-system and system level. Until recently, such detailed simulation models have been restricted to board level complexities. Models of complete systems have lacked detail and accuracy due to the lack of

0-7803-5086-3 /98/$10.00

computer performance required to simulate highly complex systems in a timely fashion.

Today's desktop computer work-stations are far more powerful than the Cray supercomputer of yesterday, and it is now possible to simulate complete systems very accurately. Also, with the advent of hardware description languages, writing software models of hardware makes the modeling process more efficient and relatively painless.

VHDL allows an engineer to define the functionality of hardware in one of two ways:

Behavioral: A logical representation of the hardware behavior, independent of how the hardware implements the function. This is very good for general simulation allowing the engineer to determine what is required of the part being modeled. Behavioral models are easier to write than structural models.

Structural: An HDL description of the hardware behavior based on the actual physical implementation. This type of model is generally used as a precursor to creating new hardware. Structural models can be used as input to CAE tools that converts them to hardware data used by silicon chip manufacturers. Structural models can also be used as simulation models, and generally are treated as such to verify their content prior to conversion to hardware.

Through the use of these tools, it is possible to create a simulation model of a typical digital system that will actually execute the software designed for the original hardware. Such a system would be a virtual system (prototype) since it can operate as the original, but only exists as a computer model. There are a number of advantages to having such a virtual system. These include:

- Hardware development and debugging support
- Obsolescence solutions support
- Hardware migration validation

During the development cycle of hardware, it has become more and more difficult to predict the final operational characteristics of any particular piece of hardware prior to hardware fabrication. Through the use of these modeling techniques it is possible to better predict the characteristics, and determine if the hardware will actually operate correctly prior to assembly. Such an ability supports a more efficient and faster development cycle.

The ability to define hardware in a descriptive model also makes changes during development more efficient. Since the function of the hardware is defined logically and in a modular fashion, the representation of the function is easier to interpret. This again improves development efficiency.

If a functional block within the virtual system has been modeled in detail, it is possible to re-engineer the functional block to remove obsolete parts. The re-engineering effort can be evaluated in the virtual system to validate the changes made prior to fabricating physical hardware. Such a validation effort would reduce development time and costs.

If the virtual system is architected in an appropriate form, it is possible to expand the capability and to use such a system to engineer systems migration. Models of each of the functional blocks can be systematically replaced within the virtual system, and their performance and interaction with the other blocks in the system can be validated. Both the effects on overall hardware and software performance and requirements can be evaluated should there be any.

This approach would assist in the migration of the existing avionics systems to the next generation architecture. A model would exist of the legacy system, and as new subsystems in the new architecture were developed, their integration could be evaluated and validated.

VHDL-AMS is the language currently being defined by the IEEE 1076.1 Working Group, one of the working groups of the VHDL Analysis and Standardization Group of the Design Automation Standards Committee of the Computer Society of the IEEE. VHDL-AMS intends to extend the VHDL language to support the description and simulation of continuous time behavior (commonly called analog) and mixed continuous/discrete time behavior (commonly called analog/digital).

Since VHDL-AMS is not a new language, but an extension of an existing one, the development and definition of VHDL-AMS can build on the syntax and semantics of its parent VHDL.

Figure 1-2 is a graphic depiction of the Rapid Retargeting process.

Programmable Arrays

Programmable devices, digital, analog, and mixed logic, no longer mean just a simple sum-of-products array of limited function and size. Devices are now available that extend to all aspects of technology, price, size, and performance. While the more simple devices are still available, newer, more mature architectures have appeared that are providing much higher functional capabilities than had been possible with user-configurable devices.

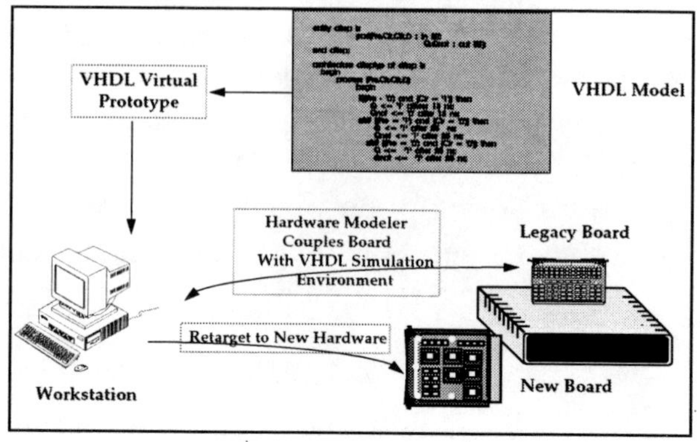

Figure 1-2

Programmable arrays are general-purpose combinational or sequential digital components whose ultimate functions are determined by the designer. They leave the manufacturer in an unprogrammed state. The configurations of internal switches are fixed after the particular logic function for the array has been prepared and checked using a computer-aided design package appropriate for the array family used. Arrays belong to the family of application-specific integrated circuits (ASICs).

Programmable arrays are manufactured today in most digital integrated circuit technologies principally: fuses, antifuses, floating-gate MOSFETs and RAM cells. The floating-gate devices can usually be erased and reprogrammed while the RAM-based devices can be re-configured dynamically. Most arrays have to be programmed with the aid of a special programmer, a unit that is able to deliver the appropriate sequences of programming pulses which configure the array's switches in the pattern specified by the user. Some arrays can be programmed by sending a data stream to the device in its application environment. This is called "in-system"

0-7803-5086-3 /98/$10.00 ©1998 IEEE

programming. A graphic depiction of the synthesis or array partitioning is shown in figure 1-3.

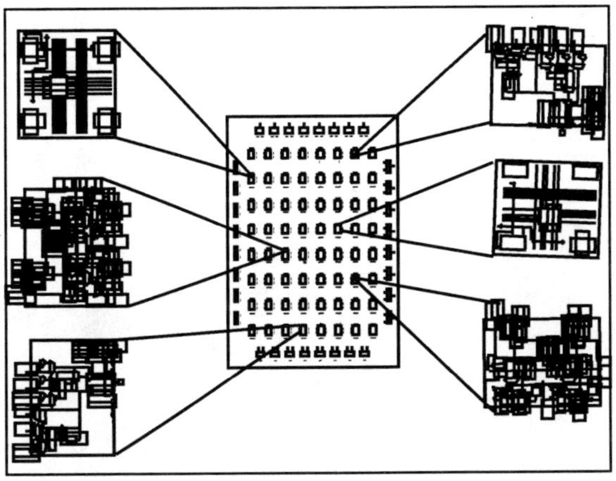

Figure 1-3

Field Programmable Gate Arrays (FPGAs)

Over the past ten years FPGAs have revolutionized the way many systems are designed by providing a low-cost, fast-turnaround implementation alternative. They represent a new method of designing custom ICs. Unlike conventional gate arrays, FPGAs can be user-programmed, like RAM or ROM, depending on the type of chip. They represent the logical development from less complex programmable logic devices, but have a much greater logic capacity and offer greater flexibility.

Field programmable gate arrays combine the integration of an ASIC with the flexibility of user-programmable logic. FPGAs present the user with basic cells and interconnect resources, which serve as the building blocks for design implementation. Users specify their design with a schematic or hardware description language. This design is then converted into a vendor-specific format by software, with the components of the design mapped onto the basic cells of the FPGA. Once the design has been successfully simulated, interconnect resources are then programmed by the user, typically with a special device programmer.

FPGAs are extremely useful for Rapid Retargeting. Designs can be described either with schematic layout tools or using synthesis from a hardware description language model, such as VHDL.

Analog and Mixed Signal Arrays

Until recently programmable electronic devices were limited to the digital domain. With the advances in technology there is now programmable analog technology as well. Two new devices have recently been made available in the analog domain: the Field Programmable Analog Array (FPAA) and the Field Programmable Mixed Array (FPMA).

The FPAA/FPMA cell is designed to be employed according to the same principles as FPGA cells. This is the principle that each cell should implement the set of the most primitive analog or mixed functions that can reliably be described, without referring to individual components

PAC Technology™.

PAC technology is a proven semiconductor device packaging and interconnection process that permits the use of the most advanced chips in severe environments. In this configuration, hermetic ceramic packages with high-density, square arrays of I/O terminal pads are directly attached to a ceramic, copper-thick-film multilayer interconnect board (MIB) by means of conventional surface mount technology, high-power, high-speed chips compared with alternative packaging technologies. This outstanding thermal management permits PAC technology to out-perform alternative technologies in either harsh environments, or where conduction-cooling is required.

Rapid Retargeting Applications

VisiCom has completed several Rapid Retargeting efforts directed at the custom military system market and have several underway directed at both the COTS and custom market segments.

COTS Applications

NTDS 4003

This effort is being implemented in a 6U VME NTDS input/output board. Its predecessor is used by the Navy in several operational weapon systems. Figure 1-4 is a photo of the current NTDS 4002 COTS Module.

Figure 1-4

***Vortex*.**

Vortex is an affordable, high-performance product combining a Color Frame Grabber, Accelerated SVGA Display Controller, Video Output and Real-Time Image Processing. This single board solution is based on the state-of-the-art S3 Virga GX2 2D/3D graphics accelerator. Vortex is implemented in a PCI form factor and acts as a PCI Bus Master with a burst transfer rate of up to 107 megabytes per Second. The unique architecture of the Vortex is built around a Xilinx-based 4000 FPGA RLE, which provides fast, reconfigurable image processing. The Vortex module is shown in figure 1-5.

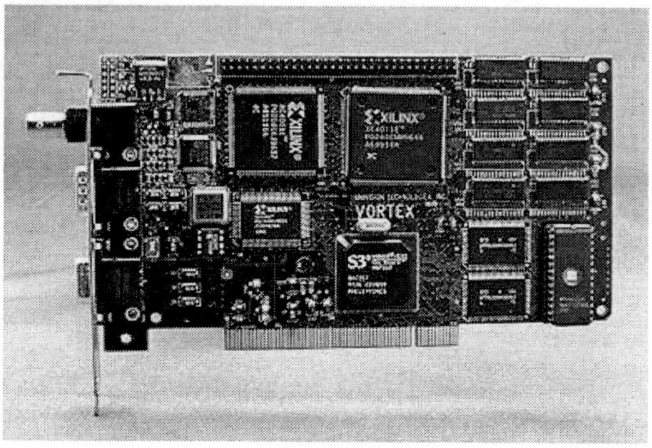

Figure 1-5

Sceptre

Sceptre is a generic, open systems architecture, Reconfigurable Logic Engine that can be adapted to many different functions and several form factors. The 6U VME implementation of Sceptre is show in figure 1-6.

Custom Applications

Universal Standard Electronic Modules

With the cooperation of the Navy, VisiCom has successfully rehosted several of the Standard Electronic Modules (SEMs) into common hardware components to produce the Universal Standard Electronic Module (UniSEM). UniSEM is a universal hardware design that can replace 26, or more, of the keycodes in the AN/UYK-44 system. UniSEM is manufactured using a single module design that employs Field Programmable Gate Arrays. The SEM is "personalized" to replace a specific board type by programming the FPGA. See figure 1-7.

Figure 1-6

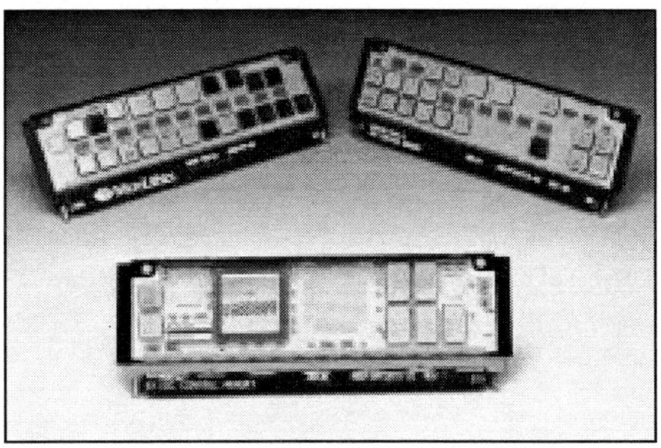

Figure 1-7

Generic SEM (GSEM)

GSEM was developed as a prototype digital module, through the use of VHDL and appropriate FPGAs, which is capable of duplicating the functional and parametric performance of over 200 Format A SEMs. A photo of GSEM is shown in figure 1-8

Target Acquisition System (TAS)

This is a recent effort in which the Navy has used a Reconfigurable Logic Engine approach to solve an obsolescence problem. To date, ten highly complex modules have been replaced with a single design. A total of sixty-four modules can be emulated by a single RLE design. TAS is show in figure 1-9.

Figure 1-8

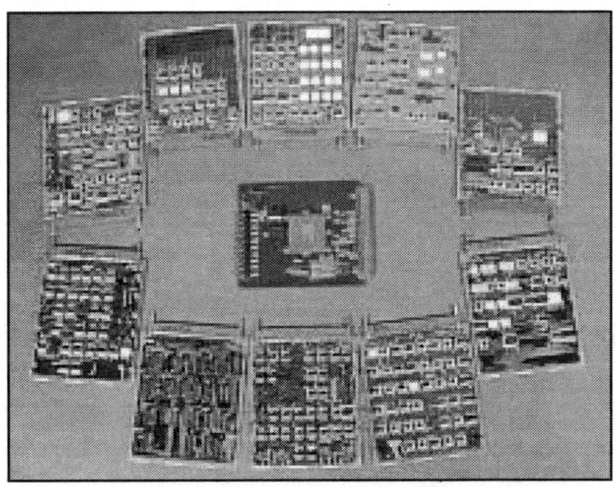

Figure 1-9

ULTRA-RELIABLE REAL-TIME CONTROL SYSTEMS - FUTURE TRENDS

Robert C. Hammett, Senior Member of the Technical Staff, C. S. Draper Laboratory, Cambridge MA

Introduction

The last few decades have been revolutionary times in the development of control systems. Mechanical, hydraulic, and pneumatic control devices such as flyweight speed governors and integrating pistons and diaphragms have been replaced by electronic computers, sensors, and electrically operated actuation devices. Using complex real-time software algorithms, these controls can optimize the performance and efficiency of the system being controlled, and allow it to operate with minimal human supervision. This equipment has leveraged the huge computer electronics industry to provide these benefits at reasonable cost and with good reliability. Yet despite these benefits, designers of these systems must remain very cautious of the failure modes of these electronic controls. Unlike the mechanical devices they replaced, electronic components usually fail without warning and will rarely provide degraded operation after failure that is sufficient for the system to "limp home." Using redundant computers, sensors, and other components, it is possible to make these electronic controls completely dependable, but these systems are very costly. Only critical applications such as the Fly-By-Wire(FBW) flight controls used on commercial transport and military aircraft that demand the highest levels of reliability and are relatively insensitive to cost have been able to fully exploit the control advantages of electronics. Applying ultra-reliable but affordable controls to smaller and less expensive vehicles and other subsystems would have many benefits: Their designers could reduce weight by using dependable controls to alleviate stress in lightweight structures too fragile to survive worst-case design load; dependable automatic controls could routinely pilot the vehicle and operate systems with reaction times beyond human capabilities; human operators that serve as back-up control systems can be eliminated and complex controls that enhance operating efficiencies can be depended on, reducing the need for safety fuel reserves. Examples of systems that could benefit from sophisticated, ultra-reliable, but affordable controls are: unmanned aircraft and spacecraft that can safely over-fly populated areas; autonomous spacecraft requiring minimal control from earth, but providing safety critical navigation or communications functions; marine ships that replace crew members with automation; drive-by-wire, auto-piloted ground vehicles that carry passengers and rely on sophisticated controls for safety. The challenge is to provide these ultra-reliable systems at an affordable cost.

Abstract

Today's aircraft use ultra-reliable real-time controls for demanding functions such as Fly-By-Wire (FBW) flight control. Future aircraft, spacecraft and other vehicles will require greater use of these types of controls for functions that currently are allowed to fail, fail to degraded operation, or require human intervention in response to failure. Fully automated and autonomous functions will require ultra-reliable control. But ultra-reliable systems are very expensive to design and require large amounts of onboard equipment. This paper will discuss how the use of low-cost sensors with digital outputs, digitally commanded fault-tolerant actuation devices

0-7803-5086-3 /98/$10.00 ©1998 IEEE

and interconnecting networks of low-cost data buses offer the promise of more affordable ultra-reliable systems. Specific technologies and concepts to be discussed include low-cost automotive and industrial data buses, "smart" actuation devices with integral fault masking capabilities, management of redundant sensors, and the fault detection and diagnosis of the data network. The advantages of integrating the control and distribution of electrical power with the control system will be illustrated. The design, installation, and upgrade flexibility benefits provided by an all-digital and shared network approach will be presented. The economic benefits of systems that can operate following failure and without immediate repair will be reviewed. The inherent ability of these redundant systems to provide effective built-in-test and self-diagnostics capabilities will be described. The challenges associated with developing ultra-reliable software for these systems and the difficulties associated with exhaustive verification testing will be presented as will additional development hurdles that must be overcome.

Future Control Systems Must Be Sophisticated but Dependable

As systems become more highly automated, autonomous, and totally dependent on their automatic controls, more control functions must remain fully operational for the vehicle or system to complete its mission, and do so safely. Vehicles and subsystems that today use simple controls will utilize digital controls to realize efficiency, weight and automation benefits, and will be totally inoperative without them. At the same time that the need for dependability is increasing, the level of control sophistication is also rising. New systems will use controls requiring more sensors and actuation devices to optimize efficiency and performance. Complex new sensor inputs, such as machine vision for obstacle avoidance, may be used and will perform critical functions. Sensors will be added to provide the automatic health monitoring and fault diagnostic capabilities that are in demand to reduce the maintenance costs. Highly sophisticated control and decision making algorithms will reside in ever faster control computers to manage these systems. Affordable, ultra-reliable control systems will find widespread application in future systems.

I/O Circuitry is the Key to Affordable Ultra-Reliable Systems

Consumer demand for ever faster personal computers, portable laptop computers, and handheld equipment has made miniaturized, rugged, and low power microprocessor components available at low cost. Using these components, affordable Fault-Tolerant Processors (FTP) can be developed that will form the core of an affordable, ultra-reliable system. [reference (1)] But the FTP is only a fraction of the system. To realize affordable, highly dependable systems, a simple and cost-effective means to sense many different measurements and provide sensor redundancy for fault tolerance is needed. These systems will also require affordable, compact, lightweight and fault-tolerant means to operate control actuation devices.

The designers of existing FBW flight controls have already taken steps to manage I/O complexity. These designers soon realized that building large, highly redundant systems using all analog electronic signaling was impractical. **Figure 1** illustrates a single-channel control system using all analog signaling and a centralized digital computer. Using this approach, long wire runs may be required to connect sensors and actuators to the control computer. To make this system ultra-reliable, the entire system must be replicated several times to form redundant channels. Making each analog signal redundant results in many independent wiring paths requiring separate connectors and introducing problems of signal accuracy and electro-magnetic interference.

0-7803-5086-3 /98/$10.00

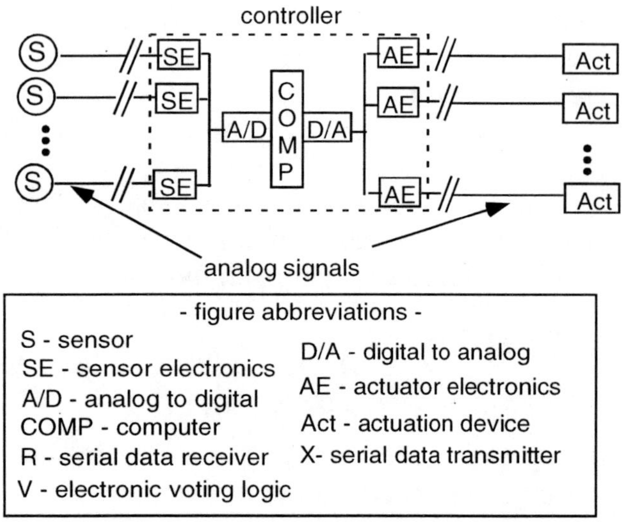

Figure 1. Control System Using All Analog I/O Signals

To overcome these problems, data multiplexing was introduced. **Figure 2** illustrates how multiplexing units and serial data buses are used to consolidate or multiplex many sensor and actuator command signals onto common wiring and transmit them to/from the central computer.

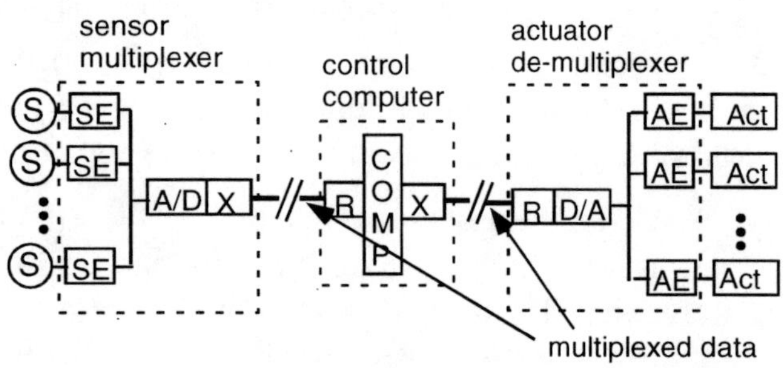

Figure 2. Control System Using Data Multiplexing

These systems use data multiplexer units located as near as possible to the sensors and actuators to minimize analog wire lengths. The data multiplexer amplifies and transforms the sensor analog signals into digital data and vice versa. Today's FBW systems use this approach, replicating the data multiplexers into multiple channels to make the system redundant and ultra-reliable. Data multiplexing has helped reduce the amount of wiring and eliminate analog signal problems, but it also provides a standard interface between equipment from different manufacturers. But even with this data multiplexing, redundant systems remain very expensive and complex. The multiplexer and demultiplexer electronic units are expensive, heavy, require power and must be made redundant. Extensive wiring is still needed to connect the redundant analog sensors and actuation equipment to the data multiplexing units. Automatic fault detection and localization of failures in the remaining analog circuitry is difficult, limiting the effectiveness of built-in-test and system health monitoring. And given the relatively high cost of each I/O signal, even the most extravagant of today's FBW systems minimize redundancy by only applying it where demanded by safety. The software is complicated by the need to manage mixed levels of redundant I/O within one system.

These systems remain too expensive and complex for all but the most critical of functions. The key to affordable, ultra-reliable systems will be to further reduce the size, weight, and cost of the I/O electronics.

Fault Masking Actuation Devices Also Increase Cost

The need for fault-tolerant actuation devices is another contributor to the high cost of ultra-reliable systems. These systems need actuation devices that are themselves fault tolerant, providing a means to reject a faulty command from any one channel of electronics. Today's FBW controls typically use specialized fault masking actuation devices as the final safeguard or "voting plane" to prevent system failure. As an example, a force sum voting hydraulic actuator, as illustrated in **figure 3**, are used for critical functions such as to control the position of an aircraft aerodynamic control surface like an aileron or elevator. These devices serve to translate the multiple, redundant electrical command signals into a hydraulic actuator position that represents the consensus of these signals. In this way, the effect of an electrical failure does not result in incorrect motion of the surface and a loss of control. This is accomplished by a small three (or more) piston hydraulic actuator that sums the hydraulic forces from three electro-hydraulic servo valves to position a control valve. This valve in turn applies hydraulic pressure to one side of the actuation cylinder under control to move it to the desired position.

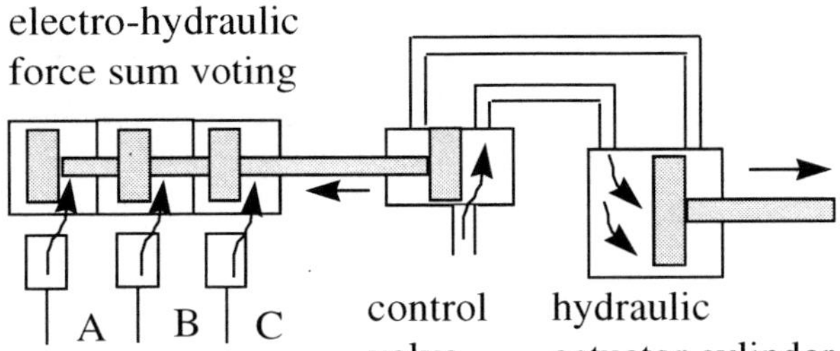

Figure 3. Voting Hydraulic Actuator

Bringing the benefits of ultra-reliable controls to a wider range of systems will require that affordable alternatives to these specialized voting actuators be developed. One promising approach is to minimize the complexity of the actuator itself, using redundant electric voting circuits instead of hydraulics. This also conforms to the current trend toward greater use of low-maintenance electrical actuation in favor of hydraulic actuation. Another type of actuation voting is also needed for ultra-reliable control systems. On or off (discrete) output types such as energizing solenoid valves, turning on or off sources of electrical power, firing pyrotechnic separation and ignition devices, and operating electric motors also require fault-tolerance. The Space Shuttle orbiter provides an example of the need for an ultra-reliable electrical actuation device. As part of the shuttle launch to orbit sequence, the avionics system is required to control the firing of the pyrotechnic devices that ignite the Solid Rocket Boosters (SRBs) and later control separation of the SRBs and external fuel tank. Either a failure to fire, or premature firing, would have catastrophic consequences for the shuttle. To ensure dependable operation, the shuttle uses special Master Event Controller (MEC) electronic units to vote on the consensus of computer commands to initiate these critical events. [reference (2)] Conceptually, the MEC uses logic similar to the simple discrete voting logic illustrated below, in **figure 4**.

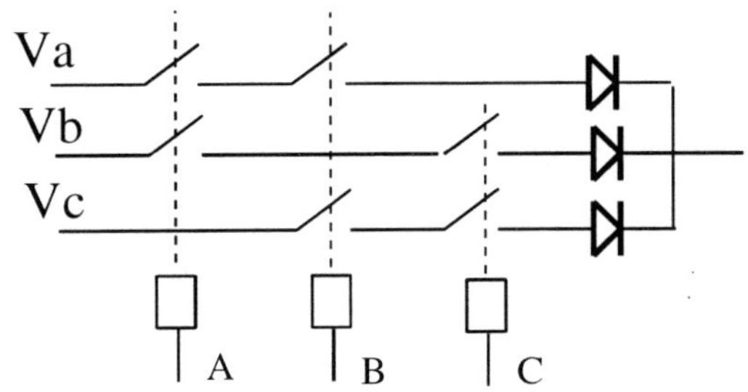

Figure 4. Discrete Electronic Voting

Briefly, the discrete electronic voter uses electronically controlled series switches operated by channels A, B and C of the control system to switch power source Va, Vb or Vc to diode "or" logic. Should any two of the three channels issue commands to close the switches, the output will be energized (fired). No channel acting alone can fire or prevent the firing of the output. These types of devices also insure that failure of one source of electrical power, Va, Vb or Vc does not result in a loss of the function. Any approach directed towards making ultra-reliable systems affordable must not lose sight of the importance of these output voting devices and costs associated with them.

Applying Automotive And Industrial Electronics to Ultra-Reliable Controls

A promising approach for reducing the cost of I/O for affordable but dependable systems can be found in components being introduced into automobiles and industrial control systems. These industries have made a large investment towards making data multiplexing available for cost sensitive applications. Two popular approaches are the Controller Access Network or CAN bus and the Echelon Corporation LonWorks bus. The CAN bus, originally developed by Robert Bosch GmbH for the automotive market, has also been adapted to industrial control applications. The LonWorks bus has been primarily applied to industrial controls. Either bus, and other similar buses, offer great potential for developing affordable, dependable controls. In today's automobile and industrial controls, these buses are applied primarily to allow to allow two control computers to communicate. For future systems, the advantage that the automotive / industrial approach offers is that the data multiplexing electronics become small, inexpensive, rugged and low enough in power to be embedded directly into the sensor or actuation device. Ideally, a single "chip" is integrated directly with the sensor or actuator, allow that device to directly interface to a shared I/O data bus without need for analog signal connections as shown in **figure 5**. Further miniaturization and cost reductions could be realized by integrating the data multiplexing electronics with single chip sensor or by combining them with Micro Electro Mechanical Systems (MEMS) sensors.

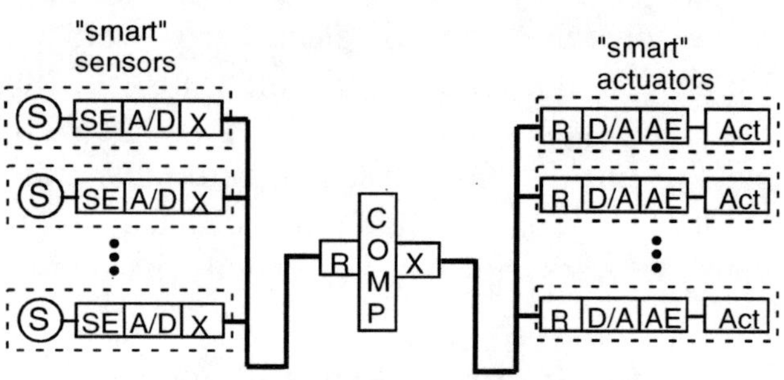

Figure 5 - Control System Using Automotive / Industrial I/O Bus

Using these "smart" sensors and actuators that directly produce digital outputs, data multiplexing units are eliminated, along with the analog signals and associated wiring. The CAN and LonWorks buses both offer data rates as high as 1 million bits per second, allowing potentially hundreds of sensors and/or actuators to share the same bus wires. Using this I/O bus approach, the cost and complexity associated with the redundant sensors and multiple input, voting actuation devices needed for ultra-reliable systems is made manageable. **Figure 6** illustrates an ultra-reliable, redundant system architecture constructed using the I/O bus approach.

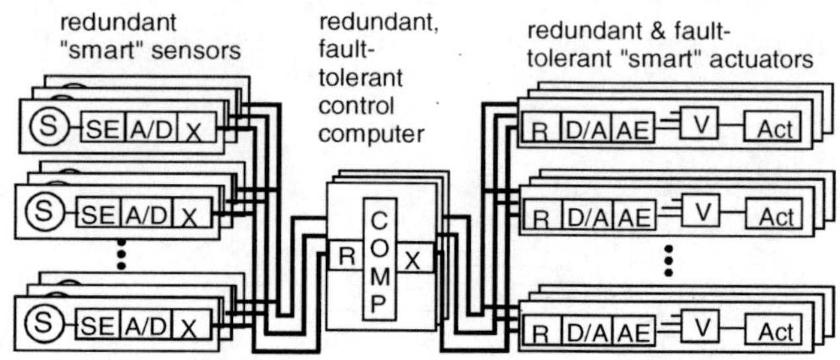

Figure 6 - Ultra-Reliable System Architecture Using the I/O Bus Approach

For the actuation devices, electronic circuits that provide the voting function are also embedded into the actuator, interfacing directly to the automotive / industrial demultiplexer electronics. At Draper Laboratory, experimental testing of a breadboard system using the CAN bus and integral electrical voting planes has demonstrated these concepts. Low-cost multiplexing and integral electrical voting plane architectures holds great promise for making ultra-reliable systems affordable. There are also many other benefits of this approach.

Flexibility of the I/O Bus Approach

The design of the electronic control system has a significant impact on the cost and schedule associated with developing new systems. The design effort is significantly increased if ultra-reliability is a goal, owing to redundancy requirements and attention to common mode failure concerns. In today's FBW systems, the addition of a new sensor or actuator, or the relocation of equipment can substantially impact the design of wiring and the number and location of data multiplexing units. Changing the allocation of analog sensor signals and actuation power signals that are bundled together in wiring requires a re-evaluation of electromagnetic interference and grounding issues. Spare I/O channels must be allocated in anticipation of growth or change. Exceeding the anticipated growth can require major redesign. Software must be designed or redesigned with knowledge of the physical path between computers and I/O devices. A significant development advantage can be obtained by using an I/O bus approach. With a method to easily connect sensors and actuators to the bus, sensors and actuation devices can be added or moved easily, with few restrictions on growth. The use of digital signals minimizes concerns about signal interference. Software can be written to interface with sensors and actuators by logical address only, independent of physical location. The system can easily grow, without weight, power or size penalties for carrying spare I/O capacity.

Affordable Hardware Redundancy Simplifies Redundancy Management

The development of Fault Detection, Isolation, and Reconfiguration (FDIR) software for ultra-reliable systems is difficult and expensive. Somewhat counter-intuitively, much of this difficulty stems from too little, rather than too much, hardware redundancy. The process of detecting sensor faults from three redundant sensors is straightforward. The well established approach of middle value selection and majority signal voting can be applied, and the designer can be confident in the ability to detect and recover from even subtle sensors failures. The problem becomes much more difficult when less then three redundant sensors are used: sensors and electronics must be designed to provide easily recognizable failure signals; to increase assurance that all failures are detected, complex mathematical models are often included into the real-time software to provide a comparison to the sensor readings. Similarly, for outputs, utilizing triple redundant command signals with a hardware voting plane at the actuator is very effective for detecting and masking the effects of failures. When only two outputs are available, some form of an active/standby arrangement is needed, with complex and critical software needed to determine when switchover from the active output to the back-up is required to accommodate failure. The use of the I/O bus approach makes a straightforward triple redundant with output voting architecture small and affordable. This eliminates the incentive to use less redundancy that results in complex redundancy management software.

Ultra-Reliable System Architectures Offer Inherent Health Monitoring and Diagnostic Capabilities

The need for I/O redundancy to allow the system to fail operational is obvious. An additional benefit associated with redundant sensing and actuation is that it can greatly enhance the performance of automated Vehicle Health Monitoring (VHM). The availability of redundant measurements improves the ability to detect failures, minimizes false alarm fault reporting, improves the ability to automatically isolate faults to a single unit, and eliminates ambiguities in deciding if a failure is mechanical or electrical. [reference (3)] Effective VHM can reduce maintenance costs and ensures that back-up redundancy is ready to operate when needed.

Ultra-Reliability Allows For Deferral of Maintenance

The ability of an ultra-reliable system to provide uninterrupted operation following failure not only provides the high operating reliability needed by autonomous systems, but can greatly simplify maintenance logistics. Instead of requiring immediate, unscheduled maintenance to restore the system, repair personnel have the option to postpone the repair until parts or service personnel are available, the system has arrived at an appropriate repair depot, or the system is shut down for other routine scheduled maintenance. The ability to operate following failure is made possible by a combination of redundancy and confidence in the ability to detect, correctly isolate, and contain the effects of a failure. Affordable redundancy brings this benefit to more systems.

Benefits of Integrated Electrical Power Control

Because of concerns for common mode failures and the high cost of providing fault-tolerant switching of electrical power, previous systems have avoided integrating the control of electric power with other control functions. In systems such as the Space Shuttle, the control of electric power is performed manually by the crew. Placing control of electric power under the control system computer offers many advantages: system power-up and shut down can be automated; electric power load monitoring and management can be fully automated; electrical power can be automatically rerouted using power bus cross ties in response to failures; automatic cutoff of electric power can be used as a means of securing failed equipment to provide safe operation following failures and power wiring and manually operated switches and circuit breaker panel hardware are reduced. The redundant I/O bus approach, complemented by other advances in low-cost power switching devices, makes it practical to provide affordable, fault-tolerant power switching and control, allowing power control to be fully integrated with other control system functions.

Software Errors Will Remain a Problem

Even with an ultra-reliable hardware architecture, software errors remain as a common mode of failure. Techniques such as N-versions of dissimilar software in each channel can be applied, but this approach is expensive and introduces new concerns. For now, the widely accepted approach appears to be a tightly controlled software design process and exhaustive verification to eradicate software errors. As previously described, the I/O bus architecture can simplify the redundancy management software, at least provide some simplification to the verification of these software functions. Hopefully, new techniques for software development or formal methods for verification will be developed that provide a better solution.

Realization of Affordable, Ultra-Reliable Controls

The I/O bus architecture approach described is very promising, but would be difficult to produce today. Few sensors are available that can directly interface to a low cost bus. Those that are were designed for industrial applications where weight, power or rugged design are not as great of a concern. For some applications, the temperature range that some sensors and actuators must operate in is too severe for off-the-shelf automotive or industrial parts. There is limited consensus on bus standards that would open the door for greater availability of off-the-shelf sensors and actuators. Few industrial control applications demand ultra-reliability and as a result, off-the shelf fault-tolerant actuators are not available. But it seems likely that demand for these types of systems will prevail. Parallel developments in single chip sensors, high temperature electronics, MEMS and power switching semiconductors will all contribute towards making the approach practical. Visionary developers will build these systems, at first using custom designed sensor and actuation components. Later, these custom components will evolve into off-the shelf equipment, with the development of standards to allow for interchangeable components from different manufacturers. The availability of completely dependable but affordable control systems will free system designers to use controls to extract greater performance and efficiencies from tomorrow's systems.

Summary

The case has been made that future systems will require controls that are as sophisticated and dependable as today's FBW systems, but must be much smaller, lighter, lower in power and less expensive. The difficulty of meeting this need with today's system architectures has been described. An approach that eliminates analog signaling and separate multiplexing units by embedding data multiplexing devices directly into sensors and actuators has been described. The benefits that this provides both in terms increased reliability and indirect benefits of increased flexibility during development, improved fault diagnostic capability, simplified redundancy management software and integral control of electric power were described. The need for visionary developers to take the first steps toward building such systems was discussed.

References

(1) High Speed Communicator for Real Time Fault Tolerant Systems, Prizant, J., Digital Avionics Systems Conference, Seattle, WA, November 1998

(2) Space Shuttle Avionics Systems, Hanaway, J., Morehead, R., NASA publication SP-504, TL3025.H36

(3) Hammett, R.C., "Seawolf Ship Control Performance Monitoring Provides Fault Tolerance and Simplified Maintenance," ASNE Intelligent Ship Symposium, Philadelphia, PA, November 1996.

HIGH SPEED COMMUNICATOR FOR FAULT TOLERANT SYSTEMS

Jay Prizant, The Charles Stark Draper Laboratory, Cambridge, MA

1.0 Overview

Critical missions require ultra high reliability systems. Quad-redundant fault-tolerant computers are a rigorously proven solution to this ultra high reliability requirement. In the past, custom-built computers, in which the redundant computers are synchronized in lock step, were built to provide the solution. The quad-redundant ship control computer used in the Seawolf Submarine[1] is an example of such a system delivered within the last 5 years. This Fault-Tolerant Computer, designed by the Charles Stark Draper Laboratory (CSDL), uses custom designed modules (68020 based) in a non-standard backplane. Inter-channel communication is facilitated with byte-wide discrete I/O at 1 Mbyte/sec.

Custom computer architectures are increasingly undesirable due to their expense, lack of COTS processor boards and backplanes, and lack of easy technology insertion (such as upgrading to faster processors).

In any fault-tolerant computer, the communication between redundant channels necessary to achieve rigorous fault tolerance is a bottleneck, imposing a severe performance penalty for the added fault tolerance.

This paper will present the design for a communicator that allows four COTS single-board computers, each in a standard backplane such as VMEbus or CompactPCI, to be networked together to facilitate a quad-redundant fault-tolerant computer. The communicator features very high bandwidth, high utilization, and low latency, resulting in minimal performance penalty for interprocessor communication. The communicator uses off-the-shelf Gigabit Link Modules (designed for use in the Fibre Channel physical layer) as high-bandwidth inter-channel data links. The data links are used to implement a time-division-multiplexed two-round data exchange, providing the fundamental source congruency operation required for rigorous fault tolerance[2]. The communicator is called a "Carousel" because its message scheduling can be thought of as messages moving in and out of the slots of a rotating slide projector carousel.

2.0 Carousel Architecture

The Carousel Communicator is an evolutionary step whose precursor is the Fault Tolerant Parallel Processor (FTPP) developed by CSDL[3]. The FTPP architecture is currently being designed into the Flight Critical Computer of NASA's X38 vehicle by CSDL. Section 3.0 will discuss some differences between the Carousel Architecture and the FTPP Communicator, called a Network Element.

2.1 Topology

The Carousel architecture is based on a quad-redundant computer in which the computing channels are connected via a mesh network. Figure 1 shows the mesh network interconnecting the four channels. The four communicators are fully connected and operate

[1] Seawolf Submarine Ship Control System: A Case Study of a Fault-Tolerant Design, Naval Engineers Journal, January 1994

[2] Reaching Agreement in the Presence of Faults, Journal of the Association for Computing Machinery, Vol 27, No. 2, April 1980, pp228-234

[3] Harper R.E. and J.H. Lala, "Fault-tolerant Parallel Processor", Journal of Guidance,Control, and Dynamics, vol. 14, no. 3, (1991), pp554-563.

0-7803-5086-3 /98/$10.00 ©1998 IEEE

in tight synchrony to perform message exchanges. The inter-channel links are fiber optic links operating at 1.06 gigabits/sec. Each channel contains a standard backplane (such as VME), one or more processor boards in each channel, I/O boards as needed, and the Carousel Communicator. The as-built prototype supports one processor per channel. Section 3.0 discusses enhancements that could support multi-processing, allowing more than one processor per channel. All boards are COTS, except the communicator board. The Gigabit Link Modules on each communicator are COTS.

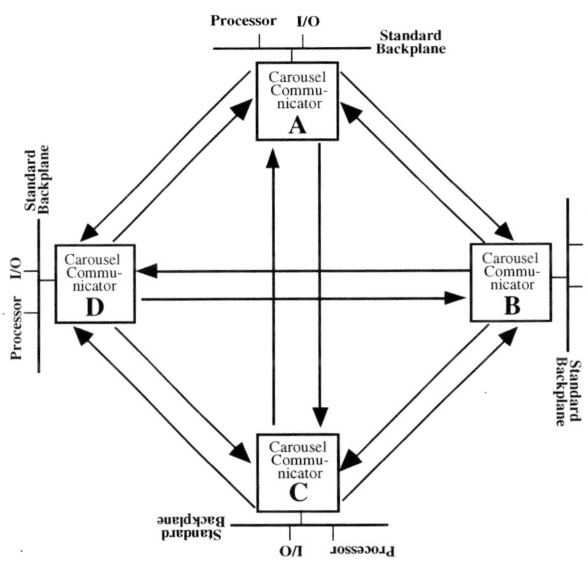

Figure 1. Mesh Network

The Carousel Communicator is an ultra high-speed data pipe engine distributing data from one channel to all the others implementing a reliable fault-tolerant mechanism known as a two round exchange. The processors in each channel are executing the same program. In order for the redundant computers to work together as a system and not diverge, all channels must receive and process the same data sets. This data may represent sensor data, or the result of calculations. If this data were distributed on a bus, such as Ethernet, or with a simple point-to-point flow, then a transient error or degradation could cause one channel to receive or interpret a piece of data different than the other channels. Even if a Cyclic Redundancy Code (CRC) were inserted to detect an error, a retry protocol would have to be added which would reduce throughput and add complexity. One application requiring a very high bandwidth data exchange might be memory realignment to reintegrate a channel back into a quad that has had a transient failure, or to integrate a cold spare. Another application requiring a high speed data exchange might require continuous high data rates, such as digitized video. The Carousel Communicator allows each channel to have a communication bandwidth approaching 1/4 (for a quad-redundant system) of the total link data rate.

2.2 Two Round Messages

Two-round messages, also known as source congruency messages, can contain channel-specific information such as the value of a processor clock. The Communicator performs two rounds of exchange of this message, allowing each channel to compare its own copy of the message to that received by the other channels. Figure 2 and Figure 3 show a two-round exchange.

Figure 2 shows that in the first round, channel A transmits its single source data (x) on all its links to the other three channels. On the second round, each channel, except the original channel that sourced the message, reflects the message out on the links. This is shown in Figure 3. The received second round is then voted at each channel. As shown in the diagram, channel D, for example, receives reflections from channels C and B. The three pieces of data that are voted in channel D are the reflections from B and C along with the saved value that D originally received from A.

These exchanges and votes ensure that all channels will receive and interpret the same data, even if any one link is broken.

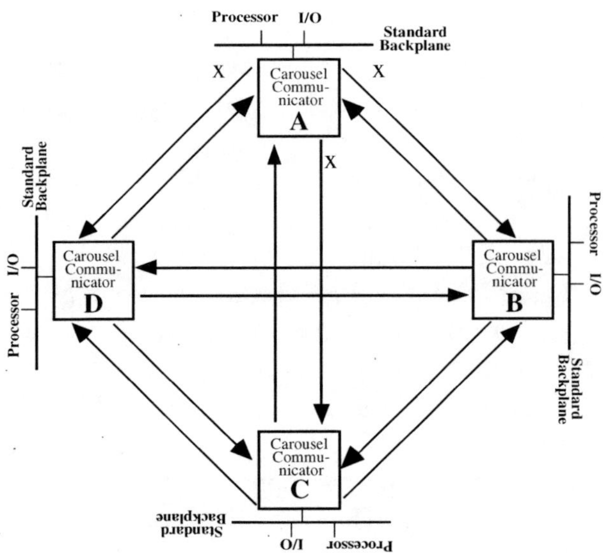

Figure 2. 1st Round of 2-Round Exchange

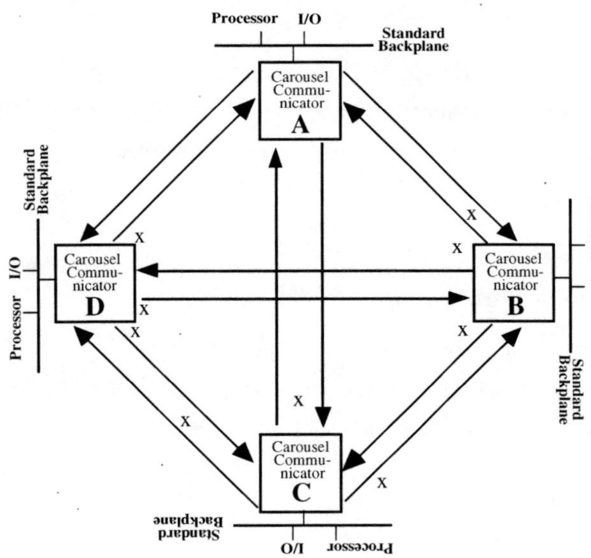

Figure 3. 2nd Round of 2-Round Exchange

2.3 Data Flow

The data exchange is composed entirely of source congruent two-round exchanges. At each clock tick, a 32-bit word is exchanged on every link. The clock ticks are broken up into repeating 4 tick frames. Each frame consists of an A-tick, a B-tick, a C-tick and a D-tick. Each channel is only allowed to send a word during its own tick. This is true whether the word is being sent in the first round, or as a reflection in the second round. This means that all data occupying an A-tick, for example, was sourced by channel A, either representing the first round being sent by channel A or reflecting the data that originally came from channel A.

Four Redundant channels A,B,C,D are shown in the Figure 1 along with the communications links between them. Table 1 shows the data flow for 9 successive ticks. In Table 1, there is a row for every link. For example AB means the link that sends data from A to B. The columns from left to right represent time, with each column representing a time tick in which a word is exchanged. The data contents are in the form of a letter and a number, i.e., A0. The letter represents the channel that sourced the word, and the number is the sequential word number. For example, word 1 from channel A is sent from channel A to all the other channels during the leftmost tick in the table. This is the first round of the exchange. 4 ticks later[4] this word is reflected from channels B, C, and D. This is the second round of the exchange. During the second round, channel A sends out its next word, word 2. The data exchanges in the table are continuous and are independent of data exchange requests by the processors in each channel. If a processor wants to source a message, then it will write the word to its communicator, and the communicator will put the data out onto the links when its slot is next available in the time-division-multiplexed pattern shown.

We can think of this time-division-multiplexed data flow just described as the second lowest layer of a layered model, analogous but not identical to the Open System

[4] Actually the number of ticks later that the word is reflected is 4n, where n is around 4. This is due to the pipelining of the ENDEC and the logic.

Interconnect Reference Model (ISO) seven layer model. The bottom layer is the physical access layer. This includes the electrical characteristics of the links, and the 8b/10b encoding and decoding of data. This is described further in section 2.5. The second-lowest layer, which can be called the Carousel layer, comprises the two-round exchange primitives within the 4-tick time division multiplexed data flow described. The Carousel Communicator that is currently built at CSDL implements the bottom 3 layers. The third-lowest layer, called the message layer, groups words into messages from 1 to 31 words long. Each word is sent through a two-round exchange by the Carousel layer, and then the message layer assembles the received words into messages and puts each assembled message into Dual-Port RAM to be read in VMEspace by the processor. The message layer also puts an info word into each message buffer. This info word contains the word count for the message, user fields that the processor can use for control or message number info, and vote error information called syndromes. Figure 4 shows a possible snapshot of a data flow pattern on a link. Each channel can source a 32-bit word every 4 ticks, giving it 1/4 of the total link bandwidth. A different channel can source a message every tick, making very efficient use the network. The last word of every message is a control word signifying to the message layer that the end of the message has arrived. This word is called the "footer," and can also contain control information.

Time →

Link	Tick	Tick	Tick	Tick	Tick	Tick	Tick	Tick	Tick
AB	A1	B0	C0	D0	A2	B1	C1	D1	A3
AC	A1	B0	C0	D0	A2	B1	C1	D1	A3
AD	A1	B0	C0	D0	A2	B1	C1	D1	A3
BA	A0	B1	C0	D0	A1	B2	C1	D1	A2
BC	A0	B1	C0	D0	A1	B2	C1	D1	A2
BD	A0	B1	C0	D0	A1	B2	C1	D1	A2
CA	A0	B0	C1	D0	A1	B1	C2	D1	A2
CB	A0	B0	C1	D0	A1	B1	C2	D1	A2
CD	A0	B0	C1	D0	A1	B1	C2	D1	A2
DA	A0	B0	C0	D1	A1	B1	C1	D2	A2
DB	A0	B0	C0	D1	A1	B1	C1	D2	A2
DC	A0	B0	C0	D1	A1	B1	C1	D2	A2

Table 1. Data Flow

The voted assembled messages are placed into one of four input buffer areas in the memory map of the processor. Each buffer represents one of four possible sources for the messages. These buffers are in a relative format, and are "myself," "left," "middle," and "right." The "myself" buffer contains messages that originated in this channel. The "left" buffer contains messages that originated in the channel to the left of this one, etc. For each of these buffer types, the as-built Carousel Communicator has a ring of 32 input buffers, allowing 32 input messages to arrive before the processor reads them. If the ring is full and a new message arrives, then the new message is lost and an overflow bit is set.

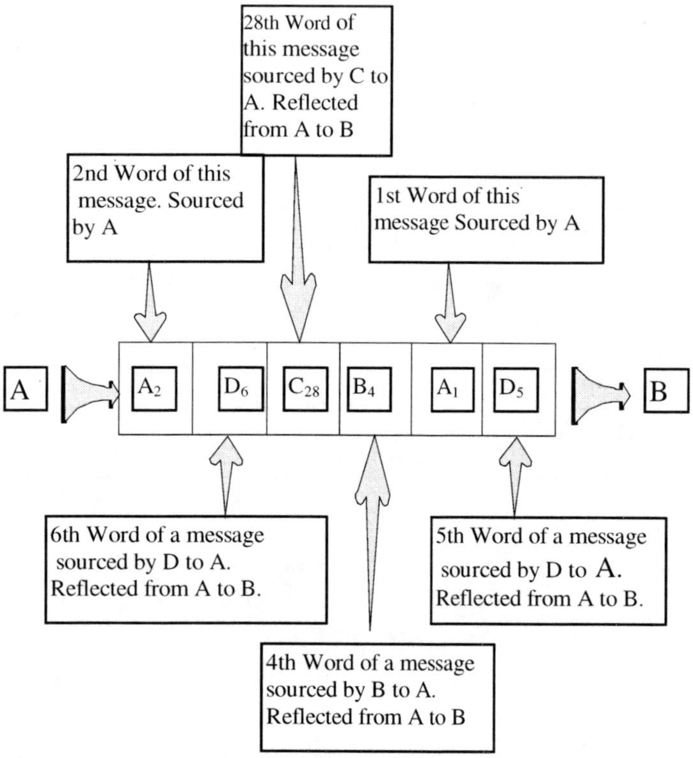

Figure 4. Link Snapshot

2.4 Clock Synchronization

The COTS processors in each channel are not clock synchronized, however the four communicators must be clock synchronized in order to ensure that the data word sets being voted together arrive within a bounded skew (with about one tick allowed skew). Every 26 MHz clock tick, new 32-bit data words arrive at the communicators' three input channels. At every new tick, this data represents a different channel's sourced data and must be routed to the appropriate voter. The system complexity would quickly blow up if data had to be stored waiting for the appropriate fellow words of data with the same tick and the same message word count number to arrive. If the reflected data words are to be sent on an A'tick, then each channel's A'tick must occur at the same time. If not, then the received 2nd round words to be voted would not arrive together. A common clock cannot be routed to all the communicators, as that would violate fault containment, another requirement for rigorous fault tolerance. Fault containment ensures that a single fault does not propagate to more than one channel. This requirement leads to each channel having separate power supplies and electrical isolation. The solution to this problem is as follows: Each channel has its own oscillator and generates its own copy of a fault-tolerant clock (FTC), which is a pulse that occurs every 64 frames. Remember that a frame consists of a 4-tick A,B,C,D sequence. After powerup, each channel's FTC is at an arbitrary phase with respect to each other. For the data exchange engine to work correctly, the FTC edges in each channel must line up with each other within a bounded skew. In addition, once they do line up, there must be some adjustment mechanism to compensate for the drift of each channels oscillator with respect to each other. The mechanism for synchronizing FTC edges to each other involves each channel receiving the FTC clocks from the other channels and voting the edges with an algorithm that determines whether its own FTC period needs to slow down or speed up. The FTC rising edge is embedded as a special command word in the data stream. Each receiver reconstructs the incoming FTC pulses from the other channels on the existing links. Figure 5 shows a data stream with the FTC clock embedded. The adjustment that a channel uses to move its FTC clock involves deciding whether to have 0, 1, or 2 optional clock idle ticks during the adjustment phase. The FTC clock voting algorithm involves detecting the second incoming rising and falling edges and creating a medium bound signal that the channel tries to align itself to by adjusting its own FTC (myftc) one clock cycle at a time. Although the algorithm seems complicated, it is a better solution than using dedicated links for the FTCs, which double the number of fiber links. Figure 6 shows a representation of the FTC algorithm.

The FTC rising edge is also used as a reference to determine whether an incoming data word is an A-tick, B-tick, C-tick, or D-tick word, and what frame number the word belongs to. The first tick after the FTC edge is always an A-tick of frame number 0. Frame numbers are used to determine when to reflect a given word to insure that the reflections to be voted will all arrive at a channel together.

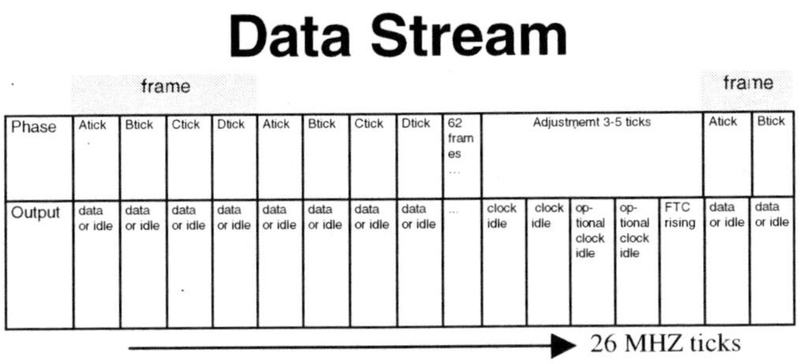

Figure 5. Data Stream

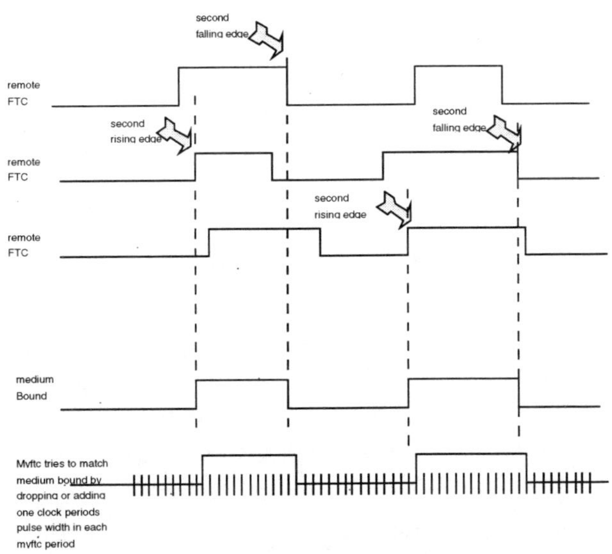

Figure 6. Fault Tolerant Clock Voting/Synchronization

2.5 Link Discussion

The inter-channel links use fiber optic Gigabit Link Modules (GLMs). These modules conform to GLM standard FCSI-301-Revision 1.0. They provide serial transmission at 1.06 gigabits/sec. These modules are normally used as the physical layer devices for Fibre Channel. A separate IC, called an ENDEC, is used to provide 8b/10b optical encoding and decoding. The encoding translates each byte into a 10-bit pattern. The 10-bit pattern always has multiple data transitions and ensures clock recovery at the receiver by providing DC balance. The ENDEC also can encode various command words into the data stream. This is possible because codes not needed for data representation are available as control/command characters. These commands are used to embed an FTC (Fault Tolerant Clock) that allows all channels to vote clock edges and synchronize to a common clock, even though all channel have independent oscillators. Idle words, when there is no data, and clock idle words, used for FTC adjustment, are also embedded as Fibre Channel-like command words. Command characters are also used in word framing. The full Fibre Channel FC-1 layer protocol is not used. The lower section of the Fibre Channel FC-1 layer that is being used is the 8b/10 encoding, and the control/command character usage. The upper part of the Fibre Channel layer, used for link management and message framing, is not used, but is replaced with the protocol described.

2.6 Physical Implementation

Figure 7 shows a representation of the 6U VME board implementation. At the front are the three Gigabit Link Modules (GLMs). These provide the fiber optic links to the other three channels. In a relative representation, these are referred to as the "right", "middle", and "left" links. Each GLM has an ENDEC associated with it. A Dual-Port RAM (DPRAM) is shown which provides the buffer storage. There is a transmit buffer and the four receive buffers described in section 2.3. To give the reader an idea of the amount of logic required, the Field Programmable Gate Array (FPGA) used is an Altera 10k100 SRAM based device with 100,000 equivalent gates. About 75% of the device is used.

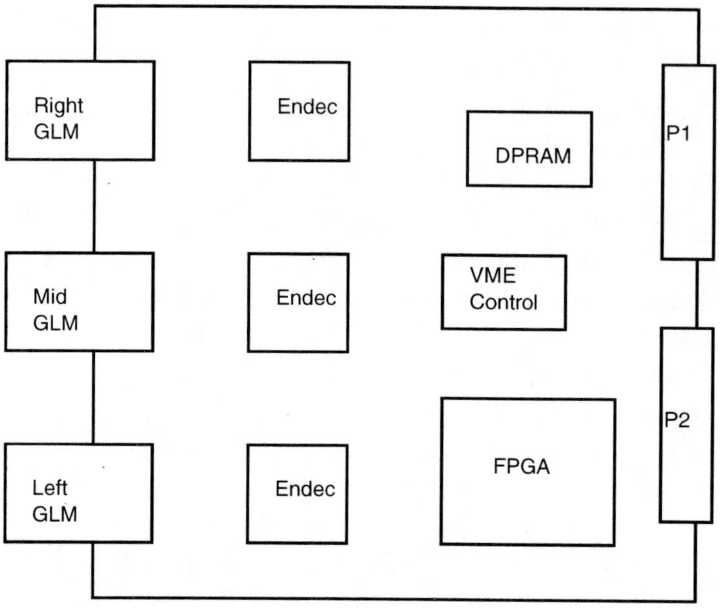

Figure 7. 6U VME Board

3.0 Upper Layers

The Carousel Communicator as currently constituted implements the bottom three layers as described in section 2.3. A layer above these three would provide various services. One service could be a voted one-round exchange. If all redundant members of the quad have identical data that they want to compare, they could exchange and vote the data in one round of data exchange, if this was supported. The Seawolf Ship Control Computer and the FTPP have a one-round voted exchange primitive. The Carousel Communicator does not have a one-round exchange primitive, at least in its lowest three layers. This capability was given up in order to allow the high network utilization and low overhead of the elegant and simple Carousel scheduling described. In the FTPP, additional overhead is incured in order to accommodate one-round exchanges in addition to two-round exchanges. Each data exchange round must be preceded by a round of exchanges whose purpose is for all communicators to reach consensus on what type of exchange is to be performed next. In the Carousel, there is no need for such a consensus, since only two round exchanges are used and the A-, B-, C- and D-tick assignments provide consensus. However, a one-round voted exchange is possible with the Carousel by implementing it with a layer-four function This could be accomplished by having each channel source a two-round exchange. The four copies of the data resulting from the two-round exchanges can then be voted in software, or by an additional hardware layer added to the communicator for this purpose. Another possible upper-layer function might be providing communication between multiple computers, with each computer being of arbitrary redundancy. There would still be four communicators in a mesh configuration, but there could be multiple processor boards in each channel on each backplane. This is the configuration of the CSDL FTPP. For a Carousel implementation, the footer at the end of each message would contain fields specifying which processor sourced the message. The "which processor" specified here would not specify which processor in a redundant group, but rather, on an orthogonal axis, which parallel computer, that in itself can be composed of redundant processors. The control fields do not have to specify which member or channel of a redundant group sourced the message because the receiver knows which tick the message arrived in, which specifies the channel that sourced the message.

4.0 Conclusion

A layered architecture for a communicator was presented which allow four COTS processor boards to be networked together to produce a quad-redundant Fault Tolerant Computer. The communicator features a high-bandwidth data exchange making data reliably congruent across the four channels.

0-7803-5086-3 /98/$10.00 ©1998 IEEE

EVOLVING AVIONICS SYSTEMS FROM FEDERATED TO DISTRIBUTED ARCHITECTURES[1]

Delbert L. Swanson, General Dynamics Information Systems, Minneapolis, MN

Abstract

Typical avionics systems today are structured in a federated architecture. Upgrading such systems has proven unwieldy and expensive, even as the advance of technology and increased functional requirements have made upgrades more imperative. Of particular concern is the systemic integration required for information superiority through multi-source data fusion.

To improve this situation, the Maritime Avionics Subsystems and Technology Scalable Open Architecture Project (MAST SOAP) has designed a distributed avionics architecture. It is characterized by general-purpose hardware resources linked by an extensible communications network, an Object Request Broker infrastructure, and commercially available components with open interface standards.

This paper outlines our approach to legacy system upgrades in the context of an open distributed architecture, presents the lessons we have learned that are applicable to such a system evolution, and discusses the performance implications involved in the use of such an architecture for an avionics system.

Introduction

Within operational federated architectures, the subsystems are typically encapsulated in special-purpose hardware elements tightly tied to their input and output devices, and are integrated with one another by conventions of data writing and reading over predefined interconnects. The hardware has often been specially designed for the functions it executes, and the software has been specially designed to execute on this hardware.

Upgrading such systems is problematic and costly. Components have become obsolete and deficient. Design and development of special purpose replacement elements is too costly. Yet the introduction of commercially available technology is hampered by the architectural structure, which is usually not compatible with modern open interface standards, and not conducive to modern principles of software system design.

General Dynamics Information Systems is in the third year of MAST SOAP, under contract to NAVAIR 4.5. We have designed a distributed avionics architecture, and are evaluating the architecture by iteratively including components, applications, and upgrade methodologies within it. Efforts to this point have concentrated on the functional and performance characteristics of CORBA and Fibre Channel in support of real time graphics and display, signal processing, and off-board data acquisition and integration. The current phase of the project incorporates data management and information fusion applications into the architecture, and investigates issues related to fault tolerance and dynamic reconfiguration.

[1] The work described in this paper was sponsored by the Office of Naval Research and NAVAIR 4.0T, under Advanced Technology Support Program contract F04606-95-D-0068, DMEA/MEDI.

Objectives

The over-arching objective of the project is to design, prototype, and evaluate an architectural model of an avionics system that can support all the traditional computing requirements of military aircraft, and be flexible enough to be easily upgraded and to accommodate the introduction of advanced mission capabilities. Examples of such advanced capabilities include advanced graphics with data fusion, mission planning updated from off-board communications, and expanded sensor and signal processing.

This project is particularly directed toward the support of legacy system upgrades. While we hope and expect that the architecture will be suitable for new program development as well, any trade-offs will be decided in favor of approaches that better support existing systems that are likely to be upgraded in equipment, software, or capabilities.

The intention of the project is that by doing advanced architectural work in a prototyping mode, we can reduce the risk for systems in this domain which are about to be upgraded, or are planned to be upgraded. Within a low-cost environment, we can alert full-scale development programs to beware of designs, development approaches, processes, and components that pose cost, schedule, or functional risks. Likewise, we will point program managers to emerging technologies and system approaches that have a high likelihood of success within such programs, since they have been proven in a prototype environment.

In accord with the objective to support legacy systems, the architectural structure and services are intended to require only minimal change for legacy components to be integrated into the system. This is in recognition of the economic reality that systems are more likely to be improved by evolutionary change than reengineered in a single step. Further, it is intended that the isolation of system elements within strict interfaces will result in the streamlining of the recertification process for incrementally upgraded avionics systems.

Approach

The approach to our project's objectives is best seen in terms of the architectural choice and the implementation choices.

Architectural Choice

The encompassing technological approach of the project is to frame the avionics functionality within a distributed architecture. The structure of the system and the underlying system software support functional integration of multiple components for the cooperative solution of problems. The computational resources for a specific function are not necessarily confined to a single processing node. Further, the processing of the components is independent of the system morphology: the functions are logically independent of location.

This contrasts to most legacy systems, which assume a defined static location for each functional unit, and the processing of a function is confined to a single computing node. The advantages of the distributed architecture over a federated one include added flexibility of design and implementation, easier portability, easier insertion of functional upgrades, and easier introduction of dynamic reconfiguration for fault recovery or reallocation of system resources for changing situations.

The chief software support for this distributed architecture is its object oriented design. This departs from traditional systems, which traditionally have used functional decomposition of the system design. Since we build our evaluation models from existing components (see below, Incremental Evaluation) this approach has required the building of object oriented wrappers around existing components.

In addition to its generic appeal derived from its support of robust software design with modularity and encapsulation, object orientation scales readily to the distributed realm. We have chosen the Common Object Request Broker Architecture (CORBA) as our integrative approach. Our use of CORBA was reported at this conference last year.[2]

On the hardware side, our approach has been to build this distributed architecture with a minimal number of types of generic components. Avionics programs toward which the MAST SOAP architecture is aimed are typified by the use of special-purpose components, often specifically developed to fulfill requirements particularly defined for the platform.

The intention is to add flexibility of system structure for system configuration and dynamic configuration. An added benefit of no small significance is the cost reduction due to the minimization of special purpose processors. To this point, we have constructed our architecture exclusively of generic commercially available components. The components are all available (or soon will be available) in ruggedized versions, and we expect them to be suitable for introduction into operational avionics systems with minimal adaptation.

Figure 1 is a schematic diagram of our vision of an avionics architecture. It illustrates the generic nature of the system elements.

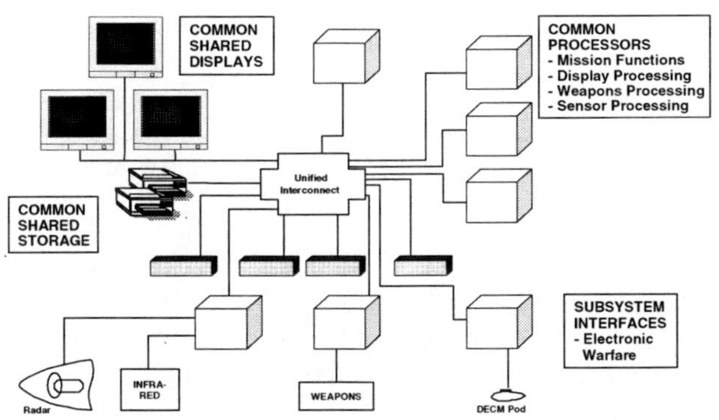

Figure 1. Generic Architecture Schema

Implementation Choices

Important elements of our implementation choices are the components chosen for the architecture demonstrations, the iterative methods of evaluation, and the Protocol Adaptation Devices used as an upgrade strategy.

Component Choices

Intrinsic to our approach is the use and evaluation of commercial off-the-shelf (COTS) components. Within this realm, we give preference in our choices to components and subsystems that are most available or have most promise of being made available with qualifications suitable for avionics applications. A strict requirement is that they must conform to open systems standard interfaces.

While the use of open systems standards has not provided the magic, effortless integration and interchangeability promised in the most optimistic literature, we have been helped by the use of open interface standards, and expect that their use will benefit legacy upgrade programs.

The key open standards used in our architectural demonstrations to this point include:

- CORBA, the Common Object Request Broker Architecture
- Fibre Channel Physical and Signaling Interface
- The OpenGL© Graphics System
- POSIX, the Portable Operating System Interface
- The VME Bus Standard

Incremental Evaluation

The architectural evaluations of this project are carried out with realistic prototypes. Rather than using artificially created elements, we populate the architecture with existing software components. A system is then built from these elements, exercised, and evaluated for functional and performance adequacy. The system is now in its fourth incremental iteration.

In anticipation of each incremental upgrade, the project leadership selects candidate applications for inclusion into the architecture. This selection is based on the perceived direction of technology upgrades of existing systems, and some understanding of the availability of applications for use in this sort of project.

[2] Robert Pedersen, "Object Request Broker Software Technology: Applications in an Advanced Open Systems Avionics Architecture," *Reflections to the Future*, Proceedings of the 16th AIAA/IEEE Digital Avionics Systems Conference, 1997, Volume I, pp. 5.2-18 – 25.

0-7803-5086-3 /98/$10.00 ©1998 IEEE

Among the candidates, selections are made based on a combination of practical and technical criteria. The application must be available for our use, and suitable for us to integrate into the system. This usually means that source code is available, in Ada, C, or C++, and the application must be of a suitable size for experimentation. Its interfaces must be well defined. The application should not depend upon exotic, overly expensive, or difficult to obtain hardware or software. We thus develop a pool of real possibilities.

From these possibilities, we apply technical criteria for the selection of included components. Full programs are preferred over prototypes. Preference is given to applications that are expected to exercise or stress some interesting aspect of the architecture, that add functional gaps in the existing evaluation system, that complement currently existing functions, and that represent advanced functions being considered for introduction into avionics platforms.

Depending upon their size and expected complexity of integration, two or three applications are introduced into the architecture at each increment.

Protocol Adaptation Devices

One of the problems faced by the upgrade of legacy systems is how to integrate existing components that depend on older technology into an upgrading system. For the instances when the upgrading technology is an interconnect component, we have introduced the notion of a Protocol Adaptation Device (PAD).

A PAD is a piece of hardware or software, or combination of both, that bridges the gap between old and new technology. It allows the legacy component to continue working with its traditional interfaces, and transforms the information into a form acceptable to the functional replacement. At a suitable time, the legacy component may be replaced by a newer one, which will interface directly with the newer technology.

Progress

MAST SOAP is in its third year, as a multiple organizational, multiple site activity. General Dynamics Information Systems is the lead team member, with Boeing Corporation as the major subcontractor. Additional technical work has been contributed by NAWC at Patuxent River. Cambridge Research Associates has a companion research contract, in Advanced Visualization and Data Fusion, and has cooperated in this effort under the auspices of that contract.

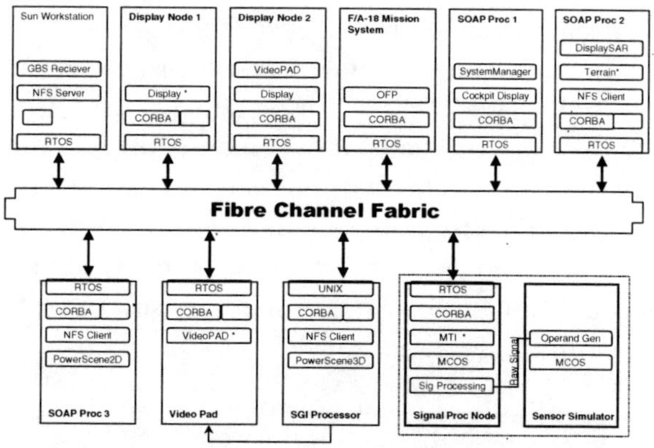

Figure 2. Notional Representation of Current Architectural Instantiation

As Figure 2 illustrates, to this point we have established a foundational architecture with integrated resources available to the applications of the system. We have integrated advanced graphics, display functionality, and signal processing elements into the system. We have begun experimentation with Real Time CORBA, fault tolerance, and dynamic reconfiguration. System upgrades are demonstrated semi-annually, with functional analysis and performance evaluations.

Infrastructure

The physical skeleton of our instance of the distributed architecture is a switched Fibre Channel network. This network connects multiple VME-based and PCI-based embedded chassis, as well as Sun and Silicon Graphics workstations. For demonstration purposes, the system is integrated with an F/A-18 flight and cockpit simulator, which executes version 11-C of the Operation Flight Program of that aircraft.

The software infrastructure is based on the CORBA standard. The programming support environment and the operating systems provide a multiple language system. The operating systems are COTS, and are POSIX compliant. For the purposes of ease of integration, we have developed a core of common interface and usage conventions, documented in an Interface Control Document.

Common Resources

In line with the distribution approach of the project, we are populating the architecture with functional resources, which are available via standard interfaces to any of the application functions that require such services. This is a step in disjoining the tight coupling of functional processing with ancillary capabilities.

The first such resource instantiated within our architecture was display processing. We populated two nodes (for illustration purposes) with graphics processing capabilities and OpenGLX libraries. For purposes of experimentation, and in accommodation to limited windowing capabilities in the software that was ready to hand, we made the OpenGLX services available to distributed functions via CORBA interfaces. By means of appropriate surrogate libraries, legacy applications are able to use this system for remote display without code change.

More recently, we have been introducing a general capability for data acquisition from off-board sources, associated with common mass memory and data base management functions. This has been demonstrated with image updates from a Global Broadcast Service (GBS) source for graphics processing updates. We expect it to have ready application in other scenarios as well, for example data fusion and mission planning updates.

Graphics Processing

We have introduced several graphics generation applications to the architecture. Most of these have their OpenGL commands transported to either of the display resources. One (PowerScene) is connected to a display node via a Video PAD, since it uses OpenGLX commands not yet available in our library.

The Cockpit Display Object generates a set of standard F/A-18 navigation display formats for presentation on the Display Node. It generates normal navigational information based on data it receives periodically from the F/A-18 OFP Object. Selection of the display format and Display Node is controlled by CORBA messages from a centralized System Manager.

The source code used to generate the display formats is ANSI C, automatically generated by the Visual Applications Builder (VAPS) tool. The display formats are output in the form of OpenGL calls to the server on the Display Nodes. All parameters to and from the Cockpit Display are defined by CORBA IDL.

PlanView is map presentation tool developed by Cambridge Research Associates, now embedded as a function of PowerScene, described just below. It uses the same database of geographic information, and displays a two dimensional map, which can move relative to the position of the aircraft. For MAST, this tool has been separately re-hosted on an embedded processor, under the control of a real time operating system. Its OpenGL output has served as a stress load on the infrastructure.

PowerScene® is a visualization tool developed by Cambridge Research Associates. It displays terrain imagery in real-time using high resolution imagery, digital terrain elevation data, and digital maps. It accepts digitized elevation models and constructs 3-D Perspective Scenes. The most common source of this data is the Defense Mapping Agency's (DMA) Digital Terrain Elevation Data (DTED). PowerScene mates imagery with the elevation data to produce a photo specific database of an area.

PowerScene was developed to use Silicon Graphics computers in several different configurations, generally including a main

® PowerScene is a registered trademark of Cambridge Research Associates

multiple processor computer with a Reality Engine to store and render elevation and imagery data into a 3-D Perspective Scene. In cooperation with MAST SOAP, it is being revised and rehosted on an embedded target, to make it amenable to embedded avionics use.

Signal Processing

SOAP aims to accommodate signal processing requirements for future avionics applications, guided by two functional ideals. We are introducing shared, scalable digital signal processing resources within the architecture, for multiple signal processing roles. We are pushing the signal processing as close to the sensor as interconnect technology, digital processing, and shared resources allow.

Our current prototype incorporates a Moving Target Indicator application, developed to showcase a heterogeneous signal processor subsystem. It integrates multiple signal processors with data processing elements, fed by remotely generated radar signals. It provides evaluation opportunities for the accommodation of heterogeneity within the system, as well as high-bandwidth communications streams.

Advanced Distribution

The latest iteration of growth of the architecture has included ventures into real time versions of CORBA and fault tolerance with dynamic reconfiguration.

While we have demonstrated the flexibility and modularity advantages of the CORBA standard, we never expected that it would be adequate for real time avionics systems in its current version. As a result, we have been tracking the progress of the minimal and real time CORBA work by the OMG.

Now that the structures and features of the prospective real time standard are firming up, and research versions of real time implementations are becoming available, we have started to rebuild our architecture with such an implementation. We are particularly interested in the priority schema and the streaming communications features. As of this writing, we are not far enough along in our experimentation to provide specific experience in this area.

One of the guiding principles of the SOAP architecture was to provide a framework flexible enough that the system could be fault tolerant without an overburdening of spare resources. To this point, our strategy is to define three fault recovery modes. The most time-critical functions will be redundant for fastest recovery. Other functions can live with the delay of a re-load and restart recovery. Some of either of these modes are likely to require a checkpoint capability. Our current iteration of the system includes first order implementations of the first two modes.

Interim Results

Our activities to this point are encouraging, from a functional point of view. That is, it seems that an architecture of the type we have designed and prototyped can provide suitable resources to support a migration path for federated avionics systems in their upgrades into modern, commercially based systems.

In our development, experimentation, reconfigurations, and incremental upgrades, the architecture has demonstrated that it offers advantages of flexibility, generality, commercial availability, and an easing of the problems associated with continuous system upgrades.

Issues of system performance, however, are of some concern. In the computing industry, we have come to expect that gains in design abstraction, in interface and protocol standards, and in programming ease and reliability are accompanied by a performance cost. We are making efforts within our project to minimize that performance penalty for the advantages we provide. But as of this writing, it is not yet clear whether some of our architectural decisions, or some of the underlying infrastructure, extract an inordinate performance penalty for their advantages.

Lessons Learned

In the performance of this project, we have learned many lessons that are likely to be applicable to development efforts intending to use COTS and open systems standards. The high-level lessons are summarized here:

Although the use of COTS hardware and software components dramatically decreases the time and effort involved in over-all system design and development, it will have a much smaller (perhaps negligible) effect on integration. This seems to follow from some of the items outlined below, and from the fact that interfaces among components with a relatively small market space have not been honed to the "plug and play" level.

Integration and customization effort should be planned for when specifying the use of COTS for military embedded systems. COTS *technologies* offer immediate advantage in a military system. *Components* that have been developed in response to commercial market forces will likely need modification.

Even on a relatively small project such as ours, we discovered complex interactions in our choices of standards, components, and products. The selection of any given equipment, software, or standard can severely limit performance, flexibility and component availability. A part of a system design should be a System Component Correlation Matrix to aid in understanding the constraints resulting from design alternatives.

We discovered later than we should have, the impact on performance of configuration parameters (and of configuration variables we wanted to change but could not), and the amount of effort that can be spent trying to optimize configuration variables that are unwieldy to change. The configurability of components should be a key selection factor for inclusion in a system.

Some of the critical technologies of a project are likely to fall in the category of the esoteric. It is important to identify these, and obtain the key necessary resources early. These resources include equipment, development tools, and human resources. Contingency plans should also be developed for coping with esoteric problems that arise during program execution.

The nature of our project prompted us to select technologies and products that are not yet mature or market proven. This caused us enough frustration that we have become convinced that full-scale development programs should use mature technologies, or technologies proven through validation programs.

There are advantages and penalties associated with the use of CORBA. With reference to a particular system, a trade-off study should be performed to determine if the flexibility is worth the overhead. Particular attention should be paid to real-time characteristics.

Performance Issues

Three sorts of performance issues have been encountered by the MAST SOAP activities. Two have relatively straightforward approaches to overcome. The third is more troublesome.

The first performance issue entails the unavailability within a form factor suitable for our target avionics domain, of technology that is readily available in the commercial marketplace. A particular example of this is graphics acceleration hardware. It is to be hoped that this problem will improve within the near future, as technology advances, and newer generations infiltrate the embedded, ruggedized domains.

The second issue involves applications that are dependent on the performance of processing architectures that have not yet been integrated into a resource-limited, real time embedded environment such as we target. Within our project, advanced graphics, signal processing, and the coming challenges of information fusion are examples of this issue. Technology advance is not expected to resolve this issue soon, but the incorporation of support for such architectures, and innovative approaches for application design make this seem to be a tractable problem.

The more difficult issue threatens some of the foundational precepts of our architecture. Our distributed approach depends on high effective bandwidth and low latencies among remote nodes. Established communications networks and protocols would seem, on the surface, to meet this demand. But our experience parallels other published results[3] that, within environments such as we operate, the software and processor overhead at the origins and destinations of the communication paths radically reduces the bandwidth promised by such advanced communications technology primarily targeted to other domains.

We are currently seeking alternatives to overcome these barriers. Incremental improvements can be made in drivers and firmware. A CORBA streaming interface holds promise. Hardware acceleration has been proposed. The path is not yet apparent. But we sincerely hope that we can present demonstrable progress at the next DASC.

Future Directions

The remainder of this project will be executed in accordance with two major thrusts. First, we will complete and consolidate our current efforts, particularly in the incorporation and evaluation of real time CORBA, and its use for fault tolerance and dynamic reconfiguration. Second, we will devise support for significantly more intense data and information fusion, and the associated merging of on-board and off-board information sources and mission planning and execution.

---

[2] See, particularly, Douglas C. Schmidt, Tim Harrison, and Ehab Al-Share, "Object-Oriented Components for High-Speed Network Programming", First Conference on Object Oriented Technologies, USENIX, Monterey, CA, 1995; Anirudlha Gokhale and Douglas C. Schmidt, "Measuring the Performance of Communication Middleware on High-Speed Networks, Proceedings of the SIGCOMM Conference, Stanford University, 1006; and the CORBA Comparison Project at Charles University, Prague, available at http://nenya.ms.mff.cuni.cz.

THE IMPACT OF COTS ON A GROUND SUPPORT SYSTEM PROCUREMENT

Peter M Lister, Aerosystems International Ltd., Yeovil, Somerset, England

1. Abstract

This paper investigates some of the benefits and draw-backs associated with the application of COTS hardware and software. Previous papers have identified that COTS requires a disciplined engineering approach to achieve good results [1]. COTS issues are discussed against the background of an ongoing project to supply a field deployable Ground Support System (GSS). Many of the issues have not yet been completely resolved, however the author believes that the insights gained so far on this project will be of value to others.

The system is being developed to provide mobile support to a military helicopter. The system will be deployed in a number of vehicles and provide both Mission Planning and Maintenance Data facilities on separate workstations. The workstations are based on COTS PC components that are packaged to meet the rugged environment. A COTS operating system (Windows NT 4) and COTS software utilities are employed to host 'bespoke' software applications. The software applications, whilst being specifically developed for the programme, are based on previously developed applications used for other aircraft types.

This deceptively simple design concept has raised a host of issues to be resolved, including the following:

- Performing system requirements and functional analyses taking account of COTS capabilities, whether they limit or enhance the user requirement;
- Meeting safety and security requirements;
- Meeting environmental constraints (temperature, EMC etc.) with COTS hardware modules;
- Coping with the limitations of COTS documentation;
- Future proofing the system design.
- Supporting the system in use given a fast moving technology base;

This paper proposes some practical approaches to these and other issues, but there is much still to be learned.

2. Requirements and Functional Analyses

2.1 Establishing the System Requirements

The first challenge with a COTS based procurement is ensuring that the requirements agreed with the customer are consistent with the use of COTS equipment [2]. The system that forms the background of this paper was defined at the outset as being a COTS procurement. The customer had a preferred COTS infrastructure based on a PC running a Windows NT operating system.

The customer had prepared an extensive, and somewhat over-prescriptive, requirement specification for the system. Our first job was to generate a definitive Product Specification. This customer specified document was interpreted as two SSSs (System/Segment Specifications), prepared in accordance with MIL-STD-2167A, supported by some other documents that responded to specific customer documentation requirements.

[1] References 1 and 2 address this assertion in more detail.

[2] Reference 3 expands on the reasons for this approach (Lesson 2)

0-7803-5086-3 /98/$10.00 ©1998 IEEE

It soon became apparent that some of the requirements, whilst being perfectly reasonable for a bespoke system, could not be easily achieved with the COTS solution. Some of these requirement short-falls would have caused problems with any COTS 'families' whilst others were specific problems for a PC/Windows NT solution.

A prime example of such a requirement was as follows:

"Each level of assembly should have sufficient test points to allow defect location to the next lower level of assembly without the need for further dismantling."

This requirement would have been an impossible selection constraint for COTS modules. In any case, we were proposing a 'First and Fourth Line' maintenance policy, so the testability features of at module level was not an issue for the customer. Also in most cases repair at module level would be by replacement, and so deep analysis of the fault would be inappropriate. On the basis of this reasoning, and the fact that the requirement was a 'should' rather than a 'shall', it was agreed that this requirement would be waived.

Some may have difficulties with taking such a pragmatic approach to the requirements. Our point of view was that the customer's prime requirement was for an 'open system' that conformed to a defined infrastructure. Implementation specific requirements could reasonably be argued out if they were not compatible with this prime requirement. The reasonableness of the approach was in some ways academic because the development was being carried out to a fixed price that was based on a COTS procurement. There was no budget (either in terms of time or cost) to pursue bespoke designs.

The success of this approach with our customer hinged on two key points:

- The customer was very keen on procuring 'Open Systems' and did not wish to dilute the COTS approach;
- The use of COTS will deliver additional useful functionality to the customer.

In the event there were very few of the customer's original requirements that were completely rejected. None of these requirements impacted the fundamental capability of the system to perform its task. In most cases requirements were modified in the Product Specification to be compatible with the use of COTS. All the changes to requirements were documented and discussed with the customer. Those requirements that were completely rejected were documented in a Null Requirements Report that forms part of the Product Specification. It should be noted that this process was both slow and arduous. Formal agreement to the Product Specification was not achieved until the hardware and software reached the detailed design phase.

2.2 *Defining the System Functions*

The author consulted Reference 1 for assistance, and found that it described a rigorous approach to COTS selection. Although technically rigorous this did not fit the situation well – the COTS infrastructure was mandated by the customer and there were few major selection decisions to be made. There were however many detailed trades that had to be made.

The solution that we adopted was to build a prototype system and in effect verify the things that it could do and the things that it couldn't [3]. This illustrates a major benefit of using COTS; it is possible to 'try-before-you-buy'. For a cost equivalent to a few man weeks of labour we were able to procure a PC that was functionally representative of our proposed design. We also had access to prototypes of both software applications, and were able to run these to gain an early indication of performance on our selected hardware.

With the information gained from the initial prototyping exercises in our minds, we

[3] Reference 3 proposes that 'hands-on evaluation' is an essential part of COTS product selection.

held a number of Design Conferences. The lead engineers from each area of the design (mission planning, maintenance management, hardware, system software) and relevant specialists (human factors, security, networking etc.) met to discuss critical areas of the design. Where there were options that could not be resolved at the meeting, individuals were actioned to undertake specific analyses to identify the most appropriate solution.

A typical example was deciding the physical architecture of the work station. There were a number of potentially conflicting requirements – such as the need for a large high-resolution screen and LRUs capable of being a 'one-man-lift' (that is less than 20 Kg in weight). The Design Conference 'brainstormed' the potential arrangements and rated their benefits. The hardware Lead Engineer then performed a preliminary design analysis of the most preferred options to confirm their properties.

The Design Conference approach proved to be extremely useful. Discussing issues with all concerned lead to a better understanding of the problems and the potential solutions. The least preferred options could often be discarded quite quickly, and the key selection criteria for the most preferred options identified to minimise the amount of work to assess the trade off. This meant that when someone was tasked to investigate the details of an option, the terms of reference of the investigation were clear. This made it easier and quicker to home in on the ultimate solution. In the example of the hardware architecture, the Design Conference identified the preferred solution from an operational viewpoint. The only reason why this would not be adopted was if the weight of any LRU exceeded the 20Kg limit, and this therefore was the question posed to the hardware LE.

The prototype processor has been modified several times in order to confirm the feasibility and performance of alternative configurations. The prototype has also been used to try out various COTS software packages that were believed to be of benefit to the design. We have made great use of the prototypes to obtain customer 'buy-in' to our design solutions. It is much easier to show what the solution is than tell somebody about it. All of our formal design reviews at system and hardware/software level have featured prototype demonstrations. This approach has helped to mitigate the risk (exposed in the previous section) associated with gaining approval of the Product Specification.

The customer now has a clear picture of what the system is going to look like when it is finished. There should be no unpleasant surprises when we seek customer acceptance.

3. Safety and Security

The GSS has to meet safety and security requirements. The security requirements have so far proved the easiest to manage.

The Windows NT 4 operating system, with a third party security extension package, is already being evaluated. The only challenge that we faced was that of ensuring the security level necessary for the GSS did not exceed the level at which the operating system was being evaluated. There are clearly risks that have to be considered; for example should there be any problem with the evaluation then we would have to revisit this area and seek alternative solutions.

The safety requirements for the GSS stem from the fact the data analysed by or uploaded from the system can impact on aircraft safety. A safety analysis had to be performed and coupled into the safety analysis performed for the aircraft itself. Because the GSS is operated 'off-line', the only potential hazards result from undetected faults. If the machine crashes or fails in an obvious way, there has at least been a warning that a fault has occurred and that data should be checked before proceeding.

The analysis confirmed the adequacy of an initial assumption that software should be developed in accordance with RTCA DO-178A level 2 as requested by the customer. This requires a rigorous and traceable software development and test process, but falls short of calling for static code analysis or similar activities.

Safety and security are areas where using COTS can be a problem simply because it may not be feasible to design a way out of a fundamental constraint. For example, it is not possible to perform a full safety analysis of a commercial operating system because the end user cannot easily obtain the source code and design records. (Attempting to analyse millions of lines of code that can be combined in an almost infinite variety of ways would also be prohibitively expensive. Neither is it obvious what one would do if the analysis identified any problems.)

4. Meeting Environmental Constraints

Environmental constraints include temperature, humidity, vibration, EMC, dust, drip proofing etc. Within reason, these can be met by careful packaging of the COTS components. Low temperature requirements can be countered with heaters. Vibration and shock can be survived with additional mechanical restraints and anti-vibration mountings. EMC standards can be met with carefully designed screening. Dust and drip proofing can be achieved with careful mechanical design.

There are however practical limits, beyond which the cost and complexity of the packaging outweighs the benefits of using readily available COTS. For the GSS the unit cost of the hardware is approximately 6 times the cost of the COTS elements. A proportion of this cost is in the assembly, which can become rather intricate if EMC filters are required.

The greatest challenges are the high end of the temperature spectrum and the most stringent EMC classes. But the best advice is to ensure that the hardware supplier (or designer) is consulted regarding the feasibility of meeting environmental requirements, before committing to them in the contract. It is also a good idea to see whether the hardware supplier has met similar requirements before. Also make sure that the performance has been tested and not just 'designed to meet'.

5. Coping With COTS Documentation

The title of this section was chosen carefully, because one of the major limitations of COTS is that in many cases the documentation is at best sketchy or at worst almost non-existent. ILS led procurements, which are currently fashionable for military customers, demand enormous amounts of design data. In some cases it is possible to create appropriate documentation for COTS, but this is expensive and may be of little value. The only real solution is to ensure that your customer is aware of the limitations in this area.

This is one of the reasons why a first and fourth line maintenance policy has been proposed for the GSS. This ensures that the user is only concerned with documentation at the LRU level, and at this level the design is under our control.

In software areas it is important to ensure that where COTS packages are used they are either fully configured by/with the application or are used in such a peripheral way that 'undocumented-features' are unlikely to cause problems. The set-up and features that are used (and tested) should be documented so that the user is clear about the scope of the 'developed' system. Users are likely to try and exploit other features and, unless the scope of the system has been identified, the developer could end up fixing and supporting features that were never intended to be used.

6. Future Proofing the System

The total development time for a system such as the GSS is in the order of 2 years. This development period is extended by the time taken to integrate the GSS with the aircraft it supports, and the further time taken to bring the entire weapon system (aircraft plus support) up to operational readiness. There is plenty of time for the user of the system to realise that necessary features have been omitted from their initial specification.

It is essential consider future needs for any system, however COTS has the additional problem that it is not possible to be certain what type of equipment will be available in four or five years time. Of course one can look back at the past, but it is questionable how this helps predict an unknown future. There are however certain key features of current PC architecture that will hopefully stand the test of time.

6.1 Buses and Backplanes

For the GSS, the key future proofing design decision was the implementation of a passive back plane, containing a mixture of ISA and PCI slots. Provided that these two backplane standards withstand the test of time, then in theory it will be possible to replace the entire complement of internal cards without impacting the bespoke (mainly mechanical) design aspects of the Processor Unit. As mentioned earlier a considerable part of the unit cost of the GSS (and indeed the development cost) is the environmental protection provided by the bespoke packaging and associated support systems.

A SCSI expansion bus is ported to the outside of the Processor Unit. This will allow future growth in an additional chassis if required.

6.2 Choice of Processor

During the development period of the GSS there were several changes in the readily available processor types. The first prototype system used a Pentium Pro, which offered excellent performance over the standard P5 device. The big question was, would the 'Pro' last the course until the time came in mid 1998 to procure the production quantities. This date was set because to the need to build production standard prototypes and perform qualification tests. There is no point in purchasing production quantities of COTS modules if it becomes necessary to reselect to meet the qualification tests.

The original choice was brought into question by the emergence of the Pentium II. Would it oust the Pentium Pro, which seemed to have only a limited market? This indeed seems to be the case. For our application the Pentium II had two main drawbacks, firstly it has a higher power consumption than the P5 and secondly it is a physically larger device. These features would have increased the size of the Processor Unit (possibly exceeding the weight limit), increased power consumption (not a good idea for a system to be powered from a vehicle battery source), and would have caused packaging problems (the profile of the Pentium II could have reduced the available card slots in the passive backplane).

The final choice was the 233 MHz P5, which seemed to offer the best balance between performance and power consumption. There are indications from some recent catalogues for industrial PCs that we may have selected wisely. These catalogues, which incidentally still offer 386 and 486 processors, offer the 233 MHz P5 as the 'standard' Pentium. This suggests that the device may be around for some time.

6.3 Memory and Disk Storage

The best advice in this area is to fit as much as you can, and test with the maximum installed capacity that can be supported by the design. Benchmarking tests were performed on prototypes of both main applications to determine what the worst case requirement might be. As the development proceeded, there were changes mooted by the customer that resulted in a re-assessment of the memory and disk needs.

As mentioned earlier, the cost of the COTS components is relatively small part of the overall unit cost. So for example, the delta cost of moving from a 4 Gbyte hard disk to a 9 Gbyte hard disk is small compared to the cost of having to retrofit and re-test the hardware at a later date.

Fitting the largest memory modules available ensures that any spare memory slots will provide a real pay-off if the memory capacity has to be expanded. There is also the risk that smaller memory modules will become unobtainable, and thus require a complete re-population of the processor memory.

6.4 Crystal Ball Gazing

Ultimately any decisions related to the future-proofing of a design are based on guess-work. As very small users of COTS hardware we cannot influence developments, and so our only option is to try and predict what is likely to happen [4].

It could be argued that there is no point in trying to 'second guess' the market. In the short to medium term it is essential, because there are very real risks that re-development might be required if a selected product goes obsolete before the production parts buy. Money and time invested in qualifying COTS based equipment can be wasted if a product becomes obsolete before you have secured an adequate supply.

In the long term, the only approach is to keep abreast of developments and be prepared to exploit whatever emerges. The rate at which new products are developed is the 'double-edged sword' of COTS – the shelf will be full of wonderful products, but they will not be exactly what you expected.

7. Supporting the System

As previously mentioned, a first and fourth line maintenance policy was proposed for the GSS. This decouples the 'internal' COTS elements from the user, and provides greater freedom for the manufacturer to replace the COTS elements with functionally equivalent items without impacting the user support environment.

This is fine in theory, but in practice, replacing any of the COTS elements could require re-qualification of the hardware. Given the short update cycles of many COTS items it becomes necessary to stockpile modules for future repairs if we are to avoid re-qualifying the system every time a module becomes obsolete. A life-time buy is impractical for a system that could be in service for 20 to 30 years because it is impossible to predict the future spares utilisation with any degree of confidence over that long period of time.

When considering solutions to this problem it was realised that it would be negating one of the real benefits of COTS to even attempt to maintain the GSS in its original configuration for an extended period. The users' needs are almost certain to evolve and the cost of hardware with superior performance will drop. The concept that we have evolved to balance these two extremes is called Planned Technology Insertion (PTI).

The essence of PTI is that target date is selected when a system upgrade would be appropriate, and that date is built into the Support Plan. A limited one-time buy of COTS components that are likely to become obsolete during this period can then be purchased, to ensure that the system as originally configured and qualified can be maintained until the PTI. Usage of the one-time buy spares can be monitored, and if usage is higher than anticipated the PTI date could be brought

[4] Reference 3 advocates planning and budgeting for 'frequent and uncontrollable COTS product end-of-life and update events'.

forward. Similarly if the spares usage is lower than planned then it may be possible to extend the PTI period.

The PTI will be a mini-development programme. The scope of the PTI will depend on the current technology status, future technology developments and most importantly, a review of customer needs and observations. It is expected that there will have been software improvements or package upgrades carried out prior to the PTI, but it would be prudent to consider whether an all-round software upgrade is required. For example the latest hardware specification may be a precursor to adopting the latest operating system upgrade.

At the end of the PTI the customer will have an upgraded machine, probably looking externally the same as the old one. Functionally it would behave much the same as the earlier model, apart from customer led changes. It would have better performance, the latest software packages (allowing compatibility with more recently developed systems). A new supply of COTS spares would be held at fourth line to keep the system operational until the next PTI.

8. Conclusions

COTS is here to stay, if only because bespoke system development is perceived by potential acquirers to be too costly and time consuming to develop. Development of a COTS based system represents a new challenge to the system designer; how to exploit the considerable benefits of COTS components whilst avoiding the many pitfalls.

Hopefully, this paper has demonstrated that COTS requires a different design approach to the traditional 'bespoke' systems engineering. It is necessary to consider the options and make pragmatic decisions that balance what is available with the customers' needs to achieve a viable and useful product.

Beware of those who believe that it is possible to take short cuts on the systems engineering process 'because it's COTS'. If anything, more care is required with COTS, because there are very few avenues open if it fails to meet a fundamental customer need. The opposite is also true, there is no benefit in performing a 'top-down' analysis of requirements if COTS is to be used. The developer and the user have to be prepared to trade requirements that are not consistent with a COTS approach.

The requirements that are eventually agreed need to be properly documented. COTS can aid the agreement of requirements because it is possible to prototype rapidly and cheaply and demonstrate what can be achieved and what can't. Good requirements are essential to scope COTS features that are specifically needed by the customer. This places at least some constraints on the boundaries of qualification testing.

Both developers and customers still have much to learn regarding the procurement and operation of COTS in military systems. Documenting and discussing the lessons learnt will hopefully stimulate the learning process so that in future we can exploit the benefits of COTS more fully and suffer fewer of the disadvantages

9. References

1. COTS/NDI Assessment and Selection Methodology, Dinesh Verma et al, published in the proceedings of the 6th Annual INCOSE Symposium, Boston, July 7-11 1996.
2. A Coherent Engineering Process for COTS Acquisition and Integration, Jerry Nolte and David Newbern, published in the proceedings of the 5th Annual NCOSE Symposium, St Louis July 22-26 1995.
3. COTS Integration and Support Model, Carolyn K Waund, published in the proceedings of the 5th Annual NCOSE Symposium, St Louis July 22-26 1995.

BUILDING STANDARDS BASED COTS MULTIPROCESSOR COMPUTER SYSTEMS FOR SPACE AROUND A HIGH SPEED SERIAL BUS NETWORK

Joseph R. Marshall, Lockheed Martin Federal Systems, Manassas, Virginia

Introduction

This paper summarizes work done for the Naval Research Lab under various contracts as well as synergistic internal research and development.

In 1995, Loral Federal Systems (now Lockheed Martin Federal Systems) in Manassas, Virginia was deeply involved in transferring and radiation-hardening the RISC Single Chip processor of the IBM RS/6000 family for use in space processor systems. The Mars Pathfinder success in mid-1996 was the first in a continuing set of successful insertions of this high performance processor (the RAD6000) into space applications. Most of these applications have been in the Command and Data Handling functions of a spacecraft and have only required a single processor (or a processor and a backup) to perform the needed processing.

At this same time, a group in Manassas was looking into higher performance processing requirements where as many as a dozen RAD6000s would work together to perform payload processing and other onboard processing tasks on a satellite. This work led to an alternate subsystem architecture to the common parallel bus used to interconnect elements in a typical subsystem. After an comprehensive trade study, a multiplexed set of serial busses based on an emerging IEEE standard (IEEE 1394-1995) was chosen as the most cost effective and scalable architecture for future multiprocessor payloads.

A radiation-hardened Application Specific Integrated Circuit (ASIC) was developed to implement a bridge to this interface from the RAD6000 and its support chip. This chip was fabricated, tested and demonstrated in two similar multiprocessor systems. This paper describes the trades that led to this chip, its features and performance. It describes the device drivers developed for running under the VxWorks operating system and the performance that was measured. The paper will describe a loosely coupled thirteen-processor system interconnected by these chips that was developed. It will describe how IEEE 1394's updates being worked out in standards bodies this year may be applied to future space systems requiring both higher performance and more remote interconnections between processors such as the Power PC.

Space Multiprocessor Architectures

Most Command and Data Handling (C&DH) and other single processor applications are able to use a parallel bus. Current popular busses for space usage include the Peripheral Component Interface (PCI), VME, and the Advanced Spaceborne Computer Module (ASCM) Module Bus, a bus developed for the ASCM program sponsored by Phillips Research Site of the Air Force Research Lab. A parallel bus usually provides the greatest bandwidth, as the processor is the main user of the interface as it uses the bus to send and receive data, commands and telemetry. The parallel bus slows down when too many peers require the bus at the same time as arbitration and waiting for the bus introduces unacceptable delays.

The natural way to solve this is to provide more interfaces between processors, each with their own dedicated bandwidth. If parallel busses are used, this may quickly exceed the available pin counts between processors or (especially in space) the amount of power

0-7803-5086-3 /98/$10.00 ©1998 IEEE

available. Thus if more than two or three interfaces are needed, an architectural switch to the use of serial busses may be necessary to minimize the pinouts and power required. Serial busses typically have lower bandwidths than parallel busses due to their smaller number of connections. Also, arbitration delays are often more significant with a serial bus due to the fact that data and arbitration events share the same medium and require longer messages to minimize arbitration times from significantly impacting performance. Currently, there are various serial busses such as MIL-STD-1553B, Universal Serial Bus (USB), and IEEE 1394 as well as many simple non-standard serial busses using proprietary protocols.

Serial interfaces allow connections to be made between a processor and all of its peers with a separate connection between each data source and sync. Serial interfaces by themselves may be overwhelmed if the number of peers grows much beyond eight or ten. It is at this point that many system architects move to a switching network between serial interfaces. This reduces the number of ports a processor must have and places most of the complexity in the switching fabric. This will usually lead to the greatest bandwidth but the extra switching fabric components will add to the system power. Asynchronous Transfer Mode (ATM) and Crossbar switches such as Myrinet are examples of this switching fabric.

The Lockheed Martin Federal Systems, Manassas architects searched for a solution that consumed less power and provided almost as much bandwidth as a switching fabric. The concept of a multiplexed set of serial busses was evaluated and found to be the best power / performance compromise. Figure 1 shows how this concept was attached between processors. Each processor has one or more channels of multiplexed busses physically wired to it. These are wired to all processors in the subsystem. For each channel, a processor is able to choose which of the busses it will operate on. Thus any two processors may choose one of the busses to communicate over. If only those two processors choose that bus, then they have the full bandwidth of that bus available for their communication. The bus operates like a single bus for those two but arbitration delays are minimized. A bus may also be used as a broadcast source for more than one receiver. For N processors, N/2 pairs of processors may each use one of the busses for a given channel effectively multiplying the bandwidth of the serial interconnect by N/2. For example, if there are twelve processors connected by at least six busses of a multiplexed 50 Mbps serial interface, the bandwidth potential is 6 x 50 or 300 Mbps per channel.

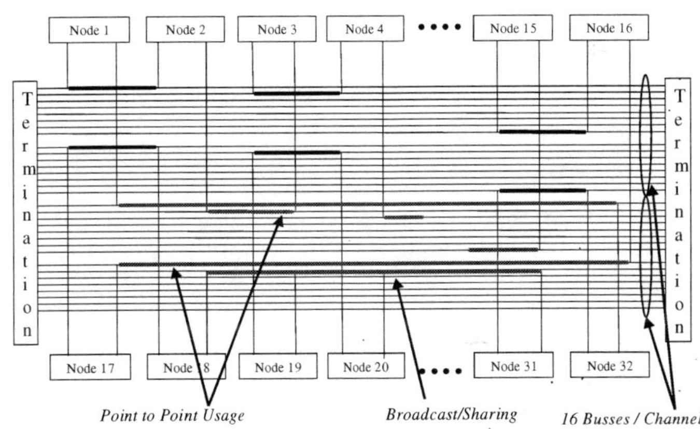

Figure 1. Multiplexed Serial Bus Interconnect

This approach also allows for reconfiguration of the channels to match the application. For fault tolerance, multiple channels could be used. Or if half (or less) of the available busses were used by the processors, then the other half of the busses could be used as a redundant set. In Lockheed Martin Federal Systems, Manassas subsystems, one of the channels was dedicated as a setup and status bus and shared between all processors. This allowed peer to peer dynamic setup. It may also be used by one master processor to broadcast configuration changes to all other processors. Growth was permitted up to the number of busses implemented in a single channel and it was also possible to bridge busses between groups of processing elements.

PLASMA Chip

The PCI Local Architecture Serial Media Access or PLASMA chip was developed to provide the multiplexed serial interfaces. A block diagram of the chip is shown in Figure 2. The chip's main function is to provide two channels of multiplexed IEEE 1394 backpanel busses to any PCI device. The PCI Bus is the main interface to the RAD6000 processor through its support chip and thus this provided the RAD6000 direct access to the PLASMA's two channels. The PCI interface in the current PLASMA chip was designed to PCI version 2.0 and uses 3.3V signaling on the 32 bit 33 MHz PCI Bus. Two independent IEEE 1394 channels share the single PCI interface. The PLASMA is powered off a single 3.3V voltage with a 2.1V reference voltage provided for the integrated Space Bus Transceiver Logic (S-BTL) transceivers.

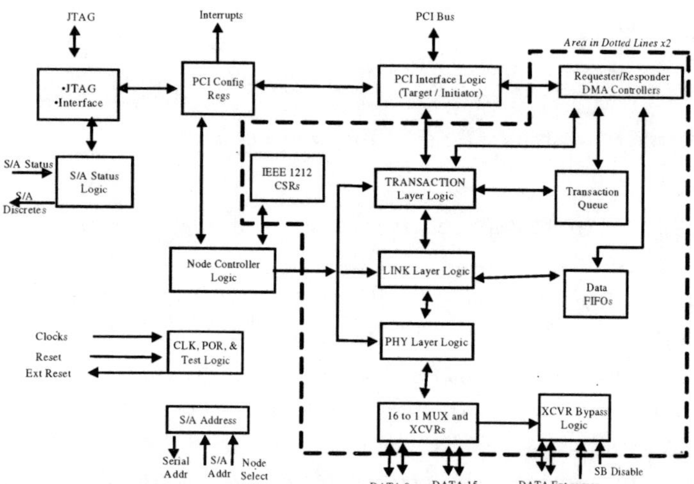

Figure 2. PLASMA Chip Block Diagram

In most general-purpose 1394 devices on the market today, the various sections of the interface are split between functions. As shown in Figure 2, this is not the case for PLASMA. In order to save on power and processing overhead, the transaction, node control, link, physical and transceivers are all integrated into the single 12.5 mm ½ micron CMOS device. All elements of the 1394 interface are fully duplicated for each channel in the device. Additionally, DMA controllers are integrated into each of these channels allowing the RAD6000 to setup dual independent transfers each up to four megabytes in length.

Other notable features in the PLASMA implementation are a set of semaphore registers that are controllable using lock transactions on the IEEE 1394 bus, a set of discrete I/O outputs that may be used to control external devices, and a JTAG interface that can control self test of the chip and provide access to certain critical status registers. Significant fault tolerant detection and recovery mechanisms are implemented in the PLASMA including hot-bit state machines and duplicated critical registers with self-checking. Cyclic redundancy checks or parity is maintained on all datapaths. The chip was designed using SEU–hardened latches which reduce the likelihood of a fault occurring in the first place and these mechanisms are then used as part of a system fault tolerance strategy.

From the address space point of view, PLASMA contains a wide variety of options to meet the various uses foreseen for this interface. A PIO access by the RAD6000 to certain reconfigurable address spaces on the I/O or memory space maps directly out onto the IEEE 1394 bus to a subset address space of the device on the bus. The IEEE 1394 bus' 64-bit address space allows all the remote PCI Busses (each requiring 8 GB of space) on the 31 possible devices to be flatly mapped in one address space. Each DMA controller allows full addressability to this flat 64-bit space. Additionally, all internal registers of the PLASMA are accessible from the PCI bus or from either 1394 channel.

The processor initiating a transfer controls all of this addressability. When the target processor wants to control where the data it receives is placed, it usually must buffer it and then move the data as it is received from the controlling processor. To improve performance in these situations, a set of remap registers are implemented for each channel and a 32 KB block of address space is set aside for each device. Any data written to this remote block is "remapped" by PLASMA logic upon receipt and stored in the new location. The addresses for the remapping is set by the target processor. When a block of data hits the last block in the remap space, an interrupt is activated to the target processor so that it can change the remap for the next 32 KB block. Since data is moved in 256 byte packets on the 1394 backpanel interface, there is sufficient time between messages for the remap region to be retargeted.

Support Software and its Development Environment

As initially demonstrated by the Mars Pathfinder mission, VxWorks is used as the COTS Real Time Operating System for RAD6000 applications. To utilize the PLASMA, a device driver was written providing an Application Programmer Interface (API) to the PLASMA. The API is scaleable to handle one to five PLASMAs per RAD6000 processor, a limitation imposed by the support chip managing the PCI Bus. Figure 3 shows the software structure of the PLASMA drivers. The VxWorks I/O system utilizes the PLASMA API to control the PLASMA devices. It was apparent early that some form of messaging layer would make it much easier to utilize the multiplexed 1394 interfaces in an efficient manner. Therefore, as part of the device driver effort, a messaging protocol was also developed that utilized the PLASMA's semaphores to share resources and pass messages between processors. The application layer may call elements from the messaging layer, VxWorks I/O System, or the PLASMA driver in order to perform needed operations.

The JTAG interface on the PLASMA provides an alternate path from the RAD6000 during initialization or error handling. Initialization code running in the processor may direct the PLASMA to run self test or initialize certain critical registers. Many card designs tie unused redundant interfaces together so that card wrap tests may be performed. Software initially developed to verify the PLASMA design was later utilized as part of this diagnostic code.

Software is normally developed using the Multi system. This supports Ada or C as well as other language front ends. Software developed for the RAD6000 may be easily retargeted to the PowerPC since the architectures of these processors is very similar.

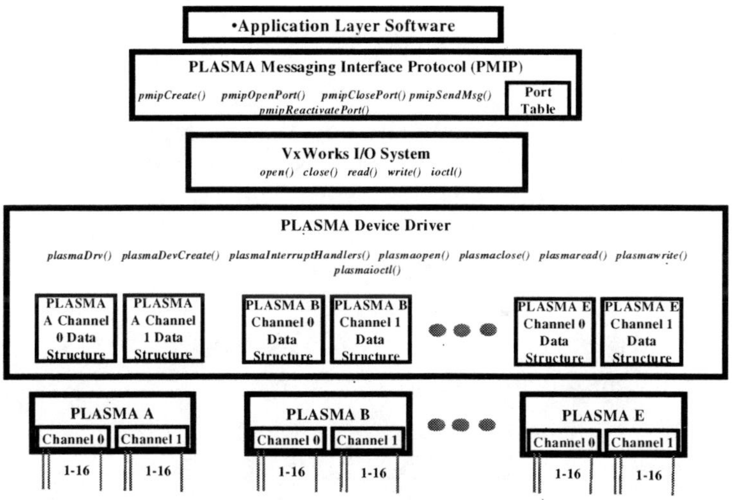

Figure 3. PLASMA Driver Architecture

Space Multiple Processor System

An initial brassboard system to checkout the PLASMA 1394 interface was built in early 1996 and named the High Performance Serial Bus (HPSB) subsystem. A backpanel supporting twelve processor cards of a common implementation were designed containing the RAD6000 and its support chip as well as three PLASMA chips and 4 MB of SRAM each.

Three of these cards were built, tested and characterized in the HPSB backpanel.

A second brassboard system representing the ultimate target application was built in late 1996. This solid state compressive recorder (SSCR) has 13 processor slots, each with same components of the HPSB processors. Eleven of these slots also contained 320 MB of DRAM. The other two slots had additional spacecraft interfaces instead of the DRAM. A fully populated SSCR provides over 400 Dhrystone MIPS, 3.5 GB of DRAM, in a box weighing less than 71 pounds and drawing less than 175 W. Each slot has its own separate power control so that power could be removed from spare cards or any that weren't needed in order to conserve power..

Both subsystems were integrated and used to debug the ASICs (including the PLASMA) that provided the various functions. Initial capabilities of the subsystems were demonstrated in late 1996 and then fully demonstrated in the second quarter of 1997. This included a performance measurement benchmark and two demonstration applications written on top of the PLASMA device drivers and communication layers.

Demonstrations and their Results

Figure 4 shows the timeline of the PLASMA Device Driver Performance Measurement application. The device driver is invoked through a function call at the left end. A short time later (61 microseconds median), the device driver completes all the setup of registers in the PLASMA and activates the DMA Write Operation. This causes a 4088 byte block of data to be transferred from one PLASMA across the IEEE 1394 backpanel to another PLASMA on another board. The transfer occurs using 256 byte blocks that take advantage of the double buffering in the PLASMA to move the data at a median rate of 40.42 Mbps over a 49.152 Mbps link. When the DMA is complete and the last acknowledgment has been received from the target PLASMA, an interrupt is generated to the RAD6000. This is handled by VxWorks and transfers control to the interrupt handler in the device driver. This marks the measured end of the transfer. It then takes a median time of 119.5 microseconds before the interrupt handler returns from the function call that initially activated the DMA request and marks the end of the activity from the initiating processor point of view.

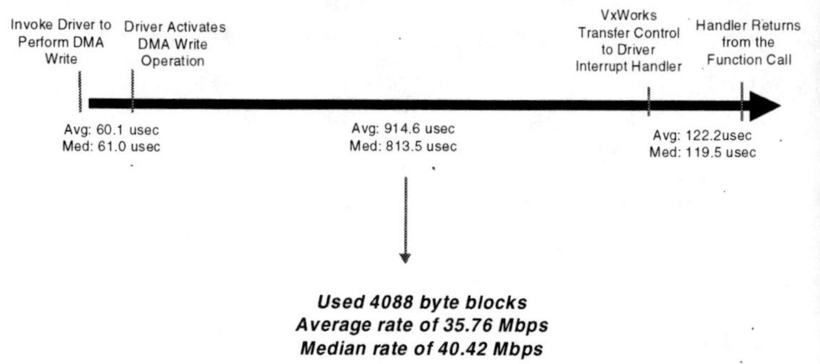

Figure 4. PLASMA Driver Performance Timeline

This was measured using the real time clock in the RAD6000, which runs with a resolution of 813 nanoseconds. Thus it was only possible to measure at points where an application program could query the real time clock (thus excluding any measurements within VxWorks). If a 1394 bus analyzer with sufficient storage had been available at that time, the end of the transfer could have been pulled in to the actual assertion of the interrupt, which would be less (and thus faster) than the measured values. We took 10 separate measurement runs through this application of which nine were very close in value and one was lower than the median. Thus the average is weighted lower and is probably close to worst case while the median is closer to the typical performance. This throughput measurement used 4088 byte blocks due to an early error (subsequently corrected) in the RAD6000 support chip. As the block sizes that are transferred become larger, the times at the beginning and end of the transfers remain constant and approach 0 in their contribution to the overall performance, leaving the center time segment showing the maximum performance.

A functional demonstration utilizing all the channels between all the processors in both common bus and point to point connection schemes was written, debugged and demonstrated. It utilized many of the elements of the performance demonstration as well as the various verification test software routines. Data was provided into one of the processors through a test port and extracted from all three processors through UART ports on the processor boards. Figure 5 shows the software demonstration application structure running in each processor in more detail.

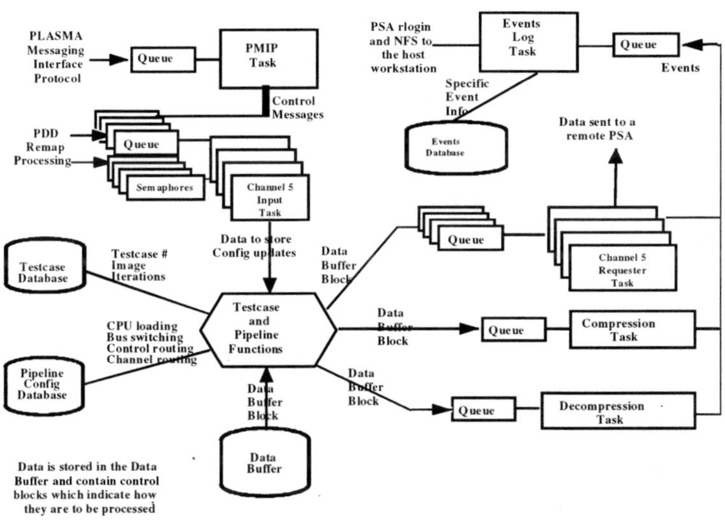

Figure 5. HPSB System Demonstration Software Structure

The final demonstration of the HPSB system added COTS IEEE 1394 hardware to the mix. A Sony IEEE 1394 video camera was purchased. This camera had a single 1394 cable port for power, setup and control. Since the PLASMA/HPSB system was designed only to interface to the IEEE 1394 backpanel, a 1394 PCI Mezzanine Card (PMC) from Technobox was also purchased and installed in one of the PMC connectors on one of the HPSB processor boards. Software drivers were written for these COTS products so that they could be added to the demonstration. An MPEG-2 encoding algorithm was adapted and coded for the RAD6000. The setup and dataflow of this demonstration is shown in Figure 6. When the application started, the RAD6000 would set up the 1394 camera to send pictures at 100 Mbps. The RAD6000 would receive this data using the 1394 cable board attached to its PMC port. It would then transfer the data to a second processor that would encode a small number of frames. This data would then be pulled out of that processor's UART port and stored in a workstation. That workstation would use standard MPEG decoding routines to display the images on its screen. This was demonstrated to various groups and individuals throughout the summer of 1997. A final step was moving the camera, PMC board and demonstration code to a group of COTS PowerPC boards and using it there. The backpanel in that case was VME. As such, this was a pathfinder and demonstration vehicle for the various capabilities of IEEE 1394 for space and how one might mix and match cable and backpanel capabilities in the same subsystem.

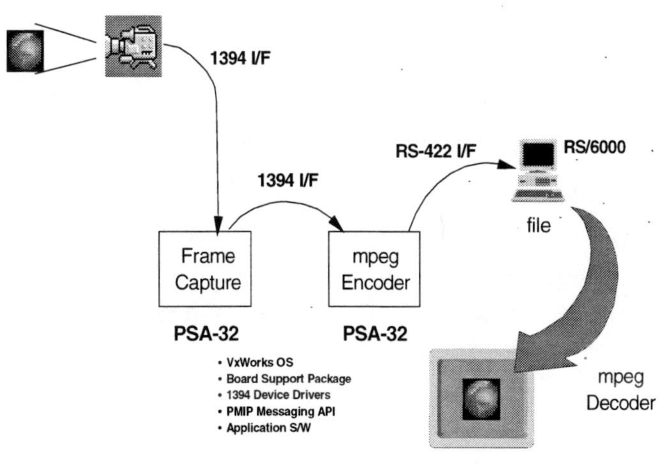

Figure 6. 1394 Cable and Backpanel Demonstration Diagram

Expanding 1394's Use in Space

Lockheed Martin Federal System, Manassas' use of IEEE 1394 in space applications has to date been centered around our multiplexed backpanel implementation. The best COTS 1394 Backpanel parts available today are only rated at double our basic bus speed, around 100 Mbps. While this is quite sufficient for a system based around the RAD6000 with its 35 Dhrystone MIPS of performance, it becomes a bottleneck as we look to the future and the next generation of processors, such as the PowerPC or clusters of Digital Signal Processors (DSP). Just as importantly, almost all of the current COTS attention is focused on the cable implementation of the IEEE 1394 standard.

The cable implementation of IEEE 1394 is currently defined to run at one of three speeds: just under 100, 200 or 400 Mbps. It allows for point to point four wire (or six wire with power) connections between devices with each device having one to sixteen ports for these connections. Up to 63 devices may be connected on a single interface structure with no more than 16 hops between any two devices. In a maximum system, each connection may be no more than 4.5 meters in length. This is used to guarantee arbitration latency – if one has a system with fewer nodes or hops, one may use longer cables between devices as long as the overall arbitration latency remains below the specified threshold. Also, there may be no loops in a single interface structure. This is because a hardware tree-identify scheme is used to assign addresses and roles after any reset, connection or disconnection of a device to or from the interface structure and this mechanism cannot reach a conclusion if there is a loop in the network. Using the IEEE 1394 cable interface and various multiport devices, one may build a variety of interconnect structures between devices that include a daisy chain, tree, mesh or hybrids of these three.

With these capabilities and the growth and popularity (and thus availability of tools, designs, and support) of this standard in the commercial world, there is significant interest in utilizing the IEEE 1394 cable in space. The most obvious usage would be to make this the new standard interface between boxes or subsystems in a spacecraft providing at least two orders of magnitude increase over established interfaces like MIL-STD-1553B. Other proposed uses include replacing the backpanel or internal board busses with IEEE 1394 on a spacecraft processor assembly and using this interface to connect to large numbers of remote command and telemetry devices throughout a spacecraft. Such applications favor tree and daisy chain configurations where a second parallel interface (such as the two channels on the PLASMA chip) would be present for redundancy and error recovery.

Utilizing a mesh topology with four channels per device would allow the IEEE 1394 cable to be utilized in large arrays of processors where connections beyond the nearest neighbor could be handled through the 1394's large 64 bit address space totally in hardware. Each four-channel IEEE 1394 device could contain hardware bridging logic for transferring data to nodes not directly connected together mimicking a switching fabric. Groups of DSPs and PowerPCs, which are expected to be part of many payload applications in the future, could easily be mapped onto such a topology greatly increasing application onboard throughput. Figure 7 shows how these various topologies may be interconnected in various systems. By identifying, architecting, implementing and inserting the IEEE 1394 building blocks needed, Lockheed Martin Federal Systems, Manassas is working with potential users to make IEEE 1394 in space a reality.

0-7803-5086-3 /98/$10.00 ©1998 IEEE

Summary

The PLASMA chip represented a major leap forward in bringing commercial standards to space applications and to increasing performance in multiprocessor applications while lowering power per Mbps. Through an initial design focused on a multiplexed backpanel bus structure, sustained API callable performance of 40 Mbps per two wire bus was measured and demonstrated in a variety of applications in two different subsystems.

The IEEE 1394 standard continues to grow in popularity and capabilities and should find a responsive set of users in the space community. Lockheed Martin Federal Systems, Manassas will be involved in bringing the commercial cable version of IEEE 1394 to space through upgrades and application of its PLASMA chip and other designs to the processors of today and tomorrow.

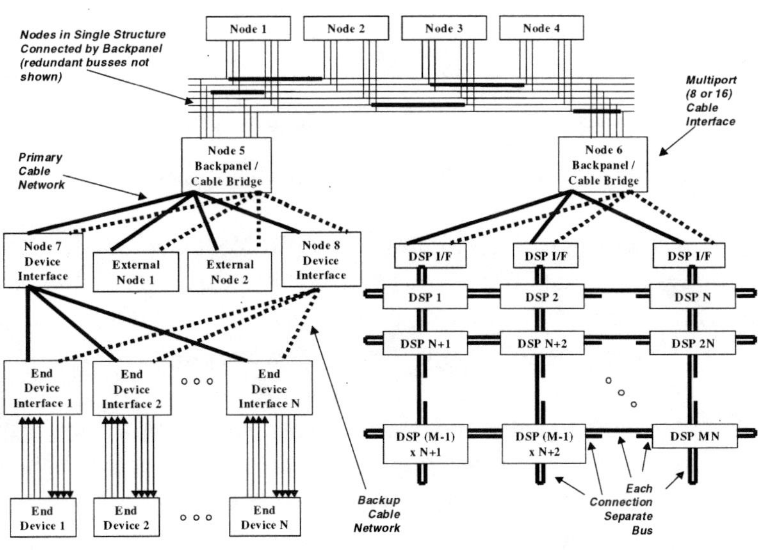

Figure 7. 1394 Topologies

0-7803-5086-3 /98/$10.00 ©1998 IEEE

DEVELOPMENT OF DISPLAY TECHNOLOGY FOR THE NEXT CENTURY

Jerry D. Schmidt, Battelle Memorial Institute, Columbus, Ohio

Introduction

The need for new technology displays is evident in almost every military weapon system. Many of the cathode ray tubes (CRTs) built during the past forty years are no longer procurable at a reasonable price, if procurable at all. Brookhaven National Laboratory and Battelle Memorial Institute are combining technologies from the two organizations to promote a new military display system. That display system, which is being commercialized by Inflection Point Technologies, Inc., will have wide commercial application. This display system will solve many current display problems and, at the same time, demonstrate the technology that will be required for the next generation of military systems. This year, Brookhaven and Battelle will provide technologies critical to the fielding of the first engineering models of a Polyplanar Optic Display unit to replace the 6"x8" CRTs in the B-52 aircraft. Brookhaven has demonstrated the viability of this technology in a breadboard model and demonstrated that it will fit into the current B-52 display casting. Battelle will add microscreen coatings to the surface of the displays that will eliminate all reflections and enable excellent daylight readability. Battelle will also automate the building of the display screens using plastic sheets. The plastic screens will be added to the B-52 program in future years. The joint effort of these two major research and development organizations will, over the next five years, provide the required system engineering and build preproduction models for wrap-around cockpit daylight readable displays.

The Polyplanar Optic Display

Dr. James Veligdan, a research scientist at the Brookhaven National Laboratories, developed the first internal projection display using sheets of glass with a refractive index of n, and black cladding with a reflective index of less than n. Dr Veligdan demonstrated that light, when input into the thin pane of glass with cladding, would travel the length of the glass sheet and emerge where the glass had been cut at an angle and frosted. Figure 1 shows

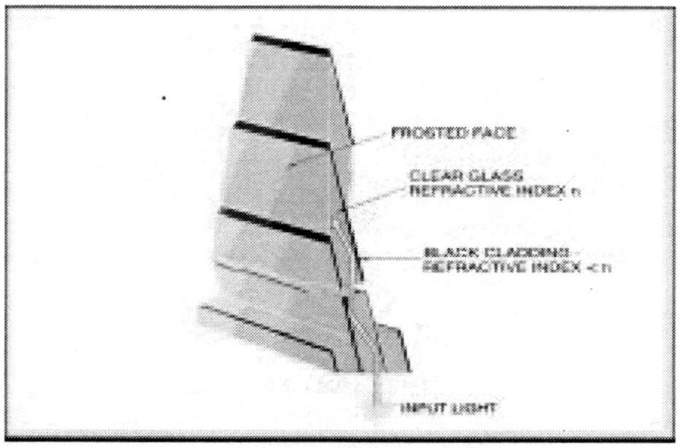

Figure 1. Principle of Polyplanar Optics

a section of the planar waveguide structure where each core sheet is surrounded by a low refractive index cladding. When light is injected at the bottom of the waveguide, it travels up the core sheet, without loss, by Total Internal Reflection (TIR). As the light reaches the top of each sheet, it encounters a beveled surface that has been treated to scatter the light out to a viewer. One of the most unique aspects of this screen is the apparent blackness one sees on the face. The display appears black even though the viewer is looking at transparent glass or plastic. Each sheet of glass could be made longer than the adjoining sheet, and the sheets could be cut on a shallow angle. This allowed a very thin set of many glass sheets to reestablish one dimension of a projected picture which had been collapsed in that dimension via optics. The Polyplanar Optic Display is thus neither front nor rear projection. This technology is called internal projection where the image is projected internally into an optical waveguide device. Figure 2 shows a schematic of both the screen and the image projection device. On the right side of Figure 2 is shown the preferred embodiment of the display where the light engine is packaged at the top rear of the display.

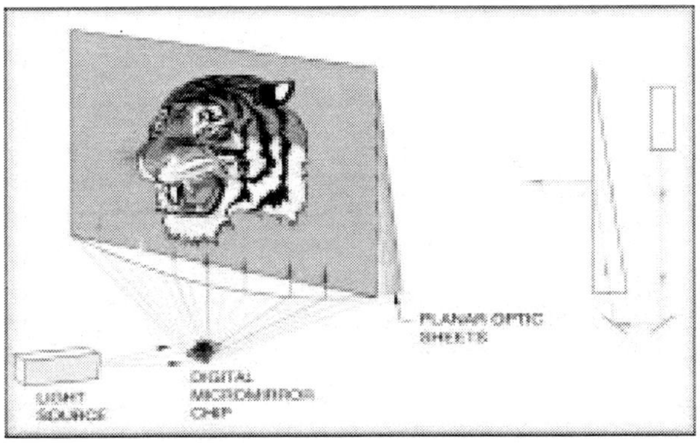

Figure 2. Polyplanar Optic Display (POD)

This technique allowed the building of projection displays that could be thin enough to hang on walls yet large enough to cover the wall. The total thickness of the display, including the light path down the back and the projection device, should be no more than four inches. Note that the projector could be from any of a number of light sources. The laboratory models built to date have used the Digital Light Processing (DLPTM) System; these are products of Texas Instruments, Incorporated and are widely used in computer projection display systems. The light from the DLPTM is focused, via optics, on the correct sheet of the glass to present the reformatted display to the viewer.

Dr. Veligdan correctly reasoned that the same display could be built using plastic instead of glass. By using plastic, a display could be made that would weigh only about as much as a sheet of Plexiglas.

Dr. Veligdan reasons that projection systems will continue to be the best method to produce large (one meter and larger) displays. However, current technologies, in order to produce a large display, require considerable volume. The Polyplanar Optic Display (POD) is a novel type of projection display screen, which for the first time, makes it possible to produce a large projection system that is self-contained and only inches thick. In addition, this display screen is matte black in appearance, allowing it to be used in high ambient light conditions. This makes the screen ideal for cockpit applications. The POD screen is also interactive and can be remotely controlled via an infrared optical pointer resulting in mouse-like control of the display. Furthermore, this display need not be flat, allowing it to be curved to wrap around a viewer. The display may also be flexible.

Since the POD screen will eventually be made approximately 1 inch thick, at its widest part, there is ample room at the top rear for many types of light engines. Thus, the entire display can be made to have a depth of only 4 inches with only minor modifications to existing engines such as Texas Instruments' DLPTM. Other light engines may be smaller, such as the 2-inch thick PDLC engine from Digital Reflection, Inc. The novel POD screen technology can be interfaced with virtually any light engine or laser projector, even those still being developed. It is very compatible with the drive for open systems and the insertion of the latest technology for product improvements.

The low index cladding is deliberately made to be black. The light, which enters at the bottom of the POD, is collimated such that it all enters and traverses the waveguide within the acceptance angle or numerical aperture of the device. However, most ambient room light is not within the numerical aperture of the screen and this ambient light is absorbed by the black cladding. We are, therefore, able to produce high contrast images in high ambient light.

Figure 3 shows a schematic of the interactive nature of the POD. The POD screen allows a user to interface and interact with the display by using an optical pointer. As a user aims the optical pointer at the display, a cursor appears on the screen just like the cursor on a computer. A remote viewer can now interact with the display by simply pointing and clicking.

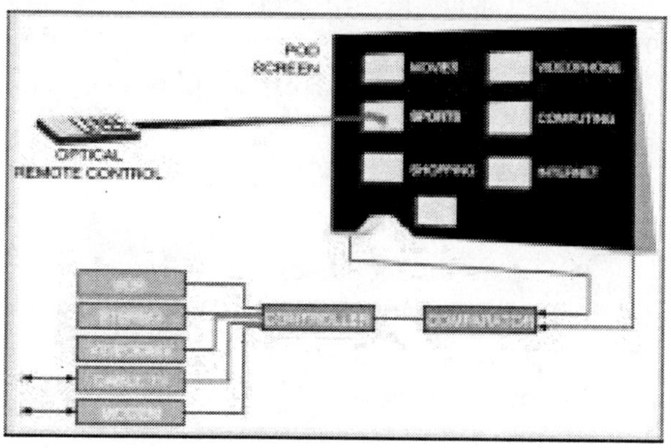

Figure 3. Interactive Display

In operation, the interactivity is possible because of the two-way light directionality of the display. To sense the light that is emanating from a user's remote pointer, a photo-diode array is placed on or near the bottom (entrance face) of the display. If we assume a 1:1 correspondence between the photo-diode and the planar sheets, the vertical position of the optical pointer signal can be determined simply by observing which sheet the pointer light is coming into. In other words, we look for the presence of a signal on one or more of the diodes. To determine horizontal position, we use a comparator to measure the relative intensity (left to right) of the optical pointer signal. This optical pointer would preferably be an infrared laser diode with a known modulation to distinguish it from ambient light.

Thus, the POD screen technology can provide the following:

1. Large Area Displays - which are only inches thick
2. Interactive Displays - via remote optical pointer
3. Curved Displays - where entire instrument panel can be a POD display
4. Flexible Displays - conceptually possible when using cross-linked co-polymer poly-acrylate
5. High Resolution Displays - possible when the number of layers (waveguides) is increased
6. High Contrast Displays - the viewing face is black

The goals of our technical approach are to:

1. Advance the technology to produce a display suitable for military use in all cockpit applications.
2. Perform the engineering required to show the risks, benefits and costs of the POD technology when compared with all competing display technologies.
3. Work with IPT, Inc. to develop the manufacturing path necessary to commercialize the POD such that the military will have the ability to purchase commercial off-the-shelf

technology for command and control applications.

No Reflection Coatings and Light beam Shaping

Dr. Richard Ridgway and Mr. Paul Andrus, research scientists at Battelle Memorial Institute, Columbus, Ohio, have developed a microstructured laminate that completely eliminates reflections and allows light emitted from a surface to be shaped and directed for optimum viewing. This technology, when applied to a POD, provided several improvements. There was 1) a marked improvement in contrast, 2) a redirection of light to an angle more suitable for viewing the POD, and 3) a reduction in the specular reflectance of the front face (i.e., a reduction in glint).

The microstructured material is presently produced using a heat-embossing process to form sharp ridges in thin layers of polyurethane. The ridges are about 160 micrometers in height and about 100 micrometers peak-to-peak. A scanning electron microscope (SEM) photograph of the microstructure is shown in Figure 4. To form the silicone rubber mold, we first form the microstructure in the photoresist using a focused laser beam and then reproduce the microstructure in the silicone rubber using standard replication processes.

Figure 4. SEM photograph of Ridged Microstructures

When the microstructured polyurethane is placed in intimate contact with an image (photographic or display), the coating significantly reduces the specular reflection (glare) from lights or the sun without significantly degrading the image. In addition, it enhances contrast by making black areas look blacker.

We applied the microstructured coating to a portion of several PODs. Figure 5 shows a photograph of the POD with the clear microstructured coating applied to the left half. Figure 6 shows the

Figure 5. Photograph of Polyplanar Display

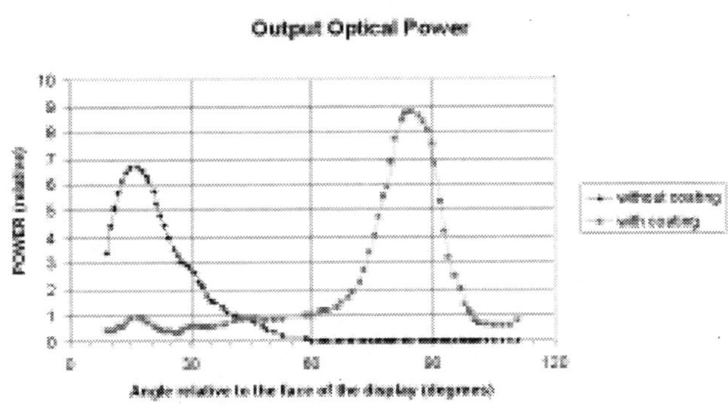

Figure 6. Experimental evaluation of Polyplanar Display with and without Microstructured Coating

experimental results of the laser-based measurement. Without the microstructured coating, the output light exits the face of the display at an angle of 15 degrees from the face (or 75 degrees off normal). With the microstructured coating, the light exits at an angle of 85 degrees from the face (or 5 degrees off normal), which is a much more desirable angle for a display. Dr. Ridgway noted that this could be enhanced and corrected to 90 degrees by slightly changing the angles on the coating material.

Polymer MultiLayer (PML) Manufacturing Technology

Mr. John D. Affinito, a senior research scientist at Battelle's Richland, Washington, facility, has developed and patented proprietary processes for high speed vacuum deposition of polymer layers with appropriate coatings to build complete multilayer, thin film, polymer sheets, and pigment coatings. This process will alloy the high-speed production of super high quality sheets of multilayer material for the POD program.

Figure 7 shows the current PML web coater being used to build the multilayer material. The polymer material is added continuously as the inner drum rotates. The black cladding, required for the POD display is deposited on the plastic through a vacuum

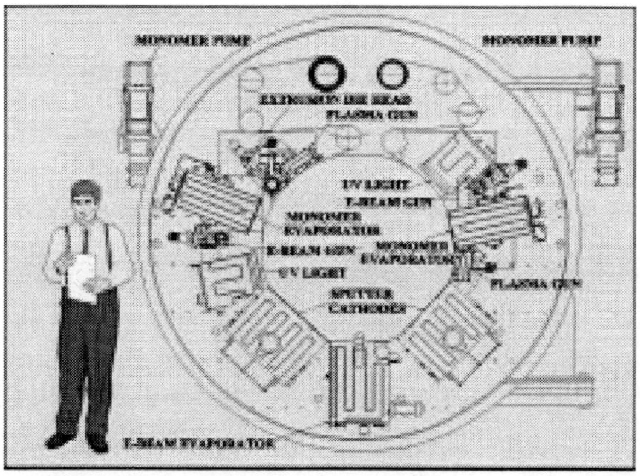

Figure 7. Layout of the PML Web Coater

deposition process. The thickness of the layers can be controlled and various sizes can be built to determine the optimum thickness for maximum light passage and minimum losses inside the plastic.

Figure 8 shows the results of the coating process. Note the smooth surface (lower left) once the coating has been applied. This smoothness provides for very high quality light paths, which transmit undistorted images through the material.

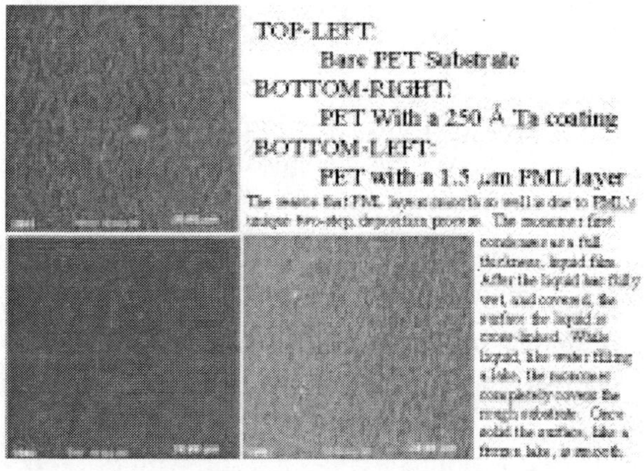

Figure 8. PML Coatings

Future High Technology Displays

Dr. Darrell Hopper, a senior display scientist for the Air Force Research Laboratory, has stated on many occasions that future high technology fighter cockpits will require 4-6 million pixels of digital information to be displayed. This information must be presented to the pilot in all weather conditions, including full sunlight.

Battelle and Brookhaven are uniquely positioned to provide this capability. Battelle has built and the Air Force is now procuring digital displays with 1280x1024 pixel resolution through the Common Large Area Display System (CLADS). These displays are currently being procured for the AWACS aircraft to replace obsolete CRTs. One of the technologies being considered for the

AWACS display is the Texas Instruments Digital Micromirrow Device (DMD). Using Battelle's PML material built on a 56-inch diameter drum, displays can be crafted that will show all information 28 inches from the pilot (human factors engineers want it 25 to 31 inches from the crew). Tiling four of the 1280x1024-pixel DMDs (or another projection technology) onto the curved screen would provide more than 5.2 million pixels of information to the cockpit, well within the required 4-6 million pixel range required (see Figure 9). Some software development will be required to tile the four projection devices.

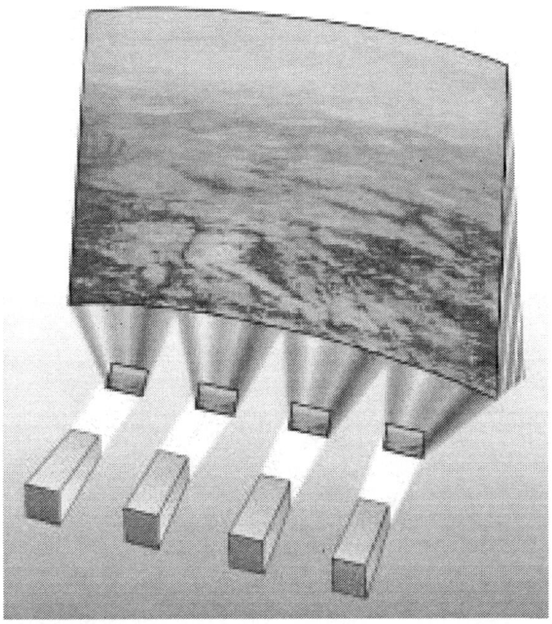

Figure 9. Future Cockpits

Real-time information in the cockpit (RTIC) requirements are forcing the display community to respond with new technology that can clearly display large amounts of information from many sources. Battelle and Brookhaven have developed and demonstrated innovative solutions to the challenges facing these new technology displays. These innovations are key to ensuring that next generation displays can satisfy the RTIC requirement. We are now capable of providing a screen that is black, has no reflections, has the pixel count required, and can even be interactive. The challenges that remain are development of process engineering for production and commercialization. Over the next couple of years we simple have to finish the engineering required to tie it all together in a working model. Use of this and other advanced technologies in future advanced aircraft will ensure we are the dominant technological military power well into the next century.

Acknowledgements

Dr. James T. Veligdan of the Brookhaven National Laboratories on Long Island, Dr. Richard W. Ridgway from the Battelle Memorial Institute in Columbus, Ohio, and Mr. John Affinito, a research scientist at Battelle's Northwest Division in Richland, Washington, provided all of the viewgraphs and many of the words that you have read in this article. I extracted the words from various notes and memos that they have written to me as we have attempted to glue the various parts of this effort together. I stand in awe of the accomplishments of these gentlemen. They are providing the leadership for the displays that you may all be viewing in your living rooms in a few years.

References

1. L. Beiser and J. T. Veligdan, "Ten Inch Planer Optic Display," Proceedings to The International Society for Optical Engineering (SPIE), Vol. 2734, April 1996.
2. Leonard DeSanto and Cyrus Biscardi, "Polyplaner Optical Display Electronics," Proceedings of The International Society for Optical Engineering (SPIE), Vol. 3057, April 1997.

Author Contact Information

Jerry D. Schmidt
Battelle
5100 Springfield Pike, Suite 210
Dayton, Ohio 45431-1231
(937) 258-6717
(937) 254-8583 Fax
schmidtj@battelle.com

Author Bio

Mr. Schmidt in a Senior Marketing Manager in Battelle's Government Market Sector Group. He has been involved in many Battelle display initiatives including the Common Large Area Display System (CLADS) currently being procured for the AWACS and proposed for the Joint Stars, ABCCC, and Navy E-2C. Prior to joining Battelle, Mr. Schmidt spent 33 years as a civilian employee of the Air Force. For several of those years he was the senior civilian engineer for the Air Force Logistics Command. Mr. Schmidt holds patents in both radar and electronic countermeasures. He is a graduate of the Georgia Institute of Technology with a B.S. in Electrical Engineering and the George Washington University with a M.S. in International Affairs.

A REVIEW OF THE FLY-BY-LIGHT OPTICAL AILERON TRIM FLIGHT DEMONSTRATION SYSTEM

John R. Todd, The Boeing Company, Long Beach, California, USA

ABSTRACT

In January of 1998 Boeing conducted successful flight testing of a prototype fly-by-light/power-by-wire concept demonstration system performing a flight control function. This paper discusses the demonstration system and the concepts it was designed to illustrate. Additionally, the aircraft installation and flight test effort will be overviewed.

INTRODUCTION

In 1994 a team companies led by McDonnell Douglas began a DARPA dual use Technology Reinvestment Program to develop and validate a new generation of fiber optic and other advanced flight control hardware. This program was designated the Fly-by-light Advanced Systems Hardware (FLASH) program. The FLASH program was administered by Wright Labs (now called Air Force Research Labs) and included participation from NASA, the Navy and Army. All the final ground demonstrations for the FLASH program were completed in 1996. The Fly-by-light Optical Aileron Trim (FLOAT) was the final demonstration and only flight test conducted on the FLASH program. On January 26, 1998 the FLOAT system made its inaugural flight on a Boeing MD-90 operated by the Flight Test organization of Douglas Products Division. Figure 1 shows the MD-90 on its first takeoff with the FLOAT system.

Figure 1. FLOAT system first flight.

BACKGROUND

The FLOAT flight demonstration system was designed to demonstrate and validate in flight fiber optic interconnect and sensor components as well as the basic concepts of a distributed flight control system.

For the MD-90 flight demonstration the FLOAT system replaced the existing cable and pulley mechanical aileron trim system with a fly-by-light/power-by-wire distributed control system. Figure 2 shows the basic installation and major features of the FLOAT system on the MD-90 flight test aircraft.

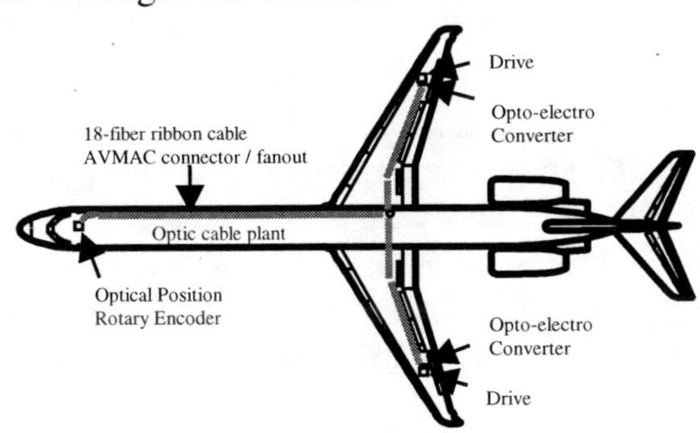

Figure 2. General installation of the FLOAT flight test system.

0-7803-5086-3 /98/$10.00

FLOAT System Configuration

The FLOAT system was designed to duplicate the operation and response of the production MD-90 mechanical aileron trim system.

The FLOAT system was comprised of an Allied Signal 12 bit digital optical rotary position for trim command input, a fiber optic cable harness using both 18 fiber ribbon cable and single fiber cables, 2 remote actuator controllers and 2 GEC electromechanical actuators. Actuator position feedback was also provided to the pilot on a bargraph display. Figure 3 shows the FLOAT system configuration for the MD-90 flight demonstration.

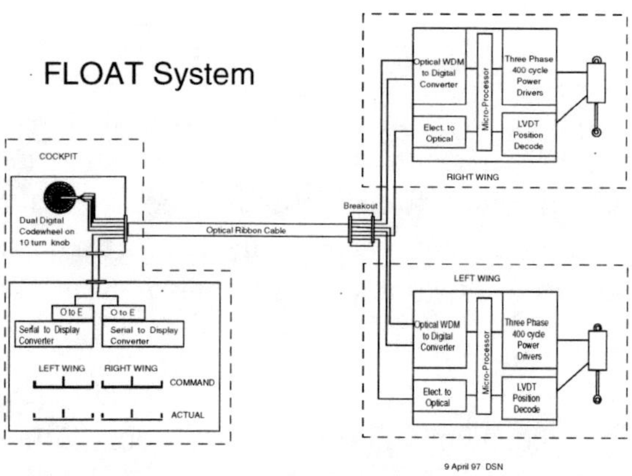

Figure 3. FLOAT system configuration.

There were several unique aspects the FLOAT flight demonstration system. The system itself was designed to be cost competitive with both mechanical and fly-by-wire trim systems. All of the electronics used were industrial grade off the shelf components and the controllers themselves helped define the configuration and capabilities for these type units in future production distributed systems. Additionally, the FLOAT flight demonstration system included the first use of the Avionics Multi-fiber Array Connector (AVMAC) 18 fiber ribbon cable interconnect system.

FLOAT System Integration

By mid-1996 all the hardware for the FLOAT flight demonstration system had been completed and tested. Simulator test flights had established that even in the event of a simultaneous asymmetric hardover of both trim actuators, the flight test aircraft was controllable over its entire flight regime within acceptable flight test limits, so no special safety of flight testing was required. For the flight test installation, electrical power disconnect switches were wired in the cockpit to remove power to the controllers and actuators in the event of irregular behavior.

The FLOAT electronic controllers did incorporate inline monitoring and failure detection on the fiber optic command sensor as well as monitoring the commanded position versus the ram position feedback. In the event of anomalies from the fiber optic sensor or a disagree between the commanded and feedback position, the controller would remove power to its respective actuator and thus locking it in place. A maximum acceptable turn rate limit for the 10 tun trim wheel was also incorporated to disengage the system if the trim wheel was rotated too rapidly to full deflection in either direction. The system was designed to reengage itself after a partial or full disconnect on the next valid and different command input after a momentary timeout.

FLOAT Installation & Flight Test

The FLOAT flight demonstration system completed its integration and testing in the fall of 1996 and preparations were begun for installation and flight demonstration on the MD-90 flight test aircraft. The system was installed on the aircraft in January 1998. Figure 4 shows the installation of the aileron trim electromechanical actuator (EMA) and electronic controller on one wing of the MD-90.

0-7803-5086-3 /98/$10.00 ©1998 IEEE

Figure 4. EMA and controller installation.

The MD-90 FLOAT flight test system was installed in several stages during normal aircraft downtime for modification and checkout of other flight test systems. The installation was conducted in stages to assure the proper operation of all the major hardware before hooking it all together. This was done to facilitate any troubleshooting of the system on the aircraft. No problems were encountered installing the various pieces of hardware or the fiber optic data bus interconnect system. During the final integration and checkout of the overall system, one of the electromechanical actuators was miswired to the 400hz three phase electrical power, but this problem was corrected with no harm to the system.

Once the FLOAT system installation was completed, it underwent a fully operational checkout on the aircraft. This was followed by review and signoff by the flight safety review board. Final pilot and crew briefings were held and the FLOAT system was ready for its maiden flight.

On January 26, 1998 the MD-90 flight test aircraft moved onto the flight ramp at Long Beach Airport in California. Final checkout of the FLOAT system was conducted in preparation for flight. At 11:02 Pacific Standard Time the MD-90 flight test aircraft lifted off from the runway with the FLOAT system operational. The first flight of the FLOAT system lasted three and a half hours and included multiple takeoffs and landings from Palmdale airport. On the following day a similar duration and flight profile was conducted. During both flights the FLOAT prototype fly-by-light/power-by-wire system worked flawlessly.

INTEGRATED F-16 AND A-10 ADVANCED DISPLAY PROCESSOR

Ted Riffle, Battelle, Columbus, Ohio

Colonel Chris Mears, HQ USAF/REOR, Washington DC

Abstract

The Air Force Reserve Command (AFRC) and Air National Guard (ANG) are seeking to increase the capability of both their A-10 and F-16 aircraft by implementing a flexible and upgradable Advanced Display Processor (ADP) and color Multi-Function Display (MFD) into each of the aircraft's avionics architectures. An engineering feasibility study was conducted to evaluate the feasibility of incorporating common hardware and software for both of the platforms. This study defined an approach for incorporation of an ADP/color MFD into the aircraft. On the F-16 the ADP will replace the XPDG electronics unit and a color MFD will replace the current monochrome CRTs in the cockpit. The A-10 upgrade will be accomplished by replacing the current Armament Control Panel (ACP) and associated stores interface electronics with a software configurable ADP and color MFD. These enhancements are needed to meet the diverse and continually expanding mission requirements of AFRC/ANG A-10 and F-16 aircraft for the next few decades. This paper presents the results of the engineering feasibility study.

AFRC/ANG Requirements

AF/REOR is the Requirements and Integration Division for the AFRC. This office is responsible for engineering analysis, development, and procurement for all avionics upgrades for AFRC. For A-10 and F-16 programs AF/REOR and the Air National Guard Directorate for Acquisition (ANG/AQ) work closely to assure that Guard and Reserve forces have the capability required to go to war when called upon. One of the joint AFRC/ANG programs is the Situational Awareness Data Link (SADL). SADL is a low cost program for the A-10 and F-16 aircraft. It uses off the shelf Enhanced Position Location Reporting System (EPLRS) radios with the express purpose of saving soldiers and airmen's lives by preventing fratricide and enhancing situational awareness while providing accurate, albeit a limited, combat ID capability. The system is secure, jam resistant, and has a low probability of intercept. It provides fighter-to-fighter, fighter-to-ground, and ground-to-fighter data link and combat ID capability and takes advantage of the developing Army and Marine Corps digitized battlefield [1]. Another important AFRC/ANG joint program is the Precision Attack Targeting System (PATS). PATS is intended to increase the F-16 Block 25/30/32 precision attack capability through the incorporation of a non-developmental targeting pod.

These and other programs illustrate the continual need for more capability and flexibility for less dollars on AFRC/ANG aircraft, particularly in the area of core computer processing and display technology upgrades. In the cockpit information is useless to the pilot unless it can be processed and displayed in a way which makes sense to him. Thus, high resolution color displays and expanded display processing capabilities will be critical to the success of precision attack targeting and situational awareness in the A-10, F-16 and other aircraft. Due to overall shrinking budgets, the Reserves/Guard require systems with much lower

0-7803-5086-3 /98/$10.00 ©1998 IEEE

life cycle costs. To achieve this, we are taking the lead in multi-platform use of common avionics hardware and software.

Technical Approach

Limited aircraft upkeep budgets do not allow wholesale gutting of airplane avionics systems and replacement with all new equipment. Thus, the technical approach for implementing common hardware/software architecture between the F-16 and A-10 aircraft must be consistent with the fiscal realities of aircraft operating and support (O&S) budgets. This means that any upgrade added to the aircraft must have immediate payback in terms of O&S dollar reductions. An incremental approach where high cost of ownership items are replaced with more reliable and maintainable (and thus lower cost of ownership) is the logical solution. Thus, the technical approach chosen for the F-16 and A-10 is to replace existing systems on each aircraft with form fit functional replacement systems with improved R&M (and performance) features. These new systems will then be leveraged to provide an enhanced pilot vehicle interface (PVI) and greatly expanded processing capability. This combination of R&M and performance improvements provides a cost effective growth capability for each of the aircraft.

F-16 Advanced Display Processor Implementation

In the case of the AFRC/ANG F-16s, the approach is to replace the existing Programmable Display Generator (PDG) and monochrome CRT based Multi-Function Display (MFD). Each of these items has significant performance limitations which prevent the AFRC/ANG from implementing desired functionality enhancements (especially with regard to SADL).

Figure 1 shows the relevant components of the F-16 system architecture. The PDG is replaced with a modular COTS based Advanced Display Processor and the monochrome MFD is replaced with a color AMLCD based MFD. This implementation is described in detail in reference 2. With this implementation, the increased processing capability of the ADP can be utilized to implement the desired features of SADL which were not possible with the PDG implementation. Studies have also been accomplished which show that R&M problems with the APG-68 fire control radar (FCR) which are currently contributing to high cost of ownership on the F-16 can be reduced significantly with this architecture [3]. This is accomplished by eliminating the programmable signal processor (PSP) line replaceable unit (LRU) and absorbing its functionality into the ADP. The PSP is one LRU of the APG-68 FCR subsystem.

As described in reference 2, this implementation of the F-16 ADP uses a set of hardware and software modules to implement the desired functionality. It is these hardware and software modules which will be common between the F-16 and A-10. In essence the concept is to establish a "library" of modules which can then be used in different combinations to meet the current and future requirements for each aircraft application. Some interest has also been expressed in having an entire Operational Flight Program (OFP) commonality between different aircraft (i.e., one OFP that is indistinguishable between the A-10 and F-16) [4]. This goal, while extremely ambitious, begins to become a reasonable objective if the common hardware/software module approach is first adopted on both aircraft.

Another key feature of the F-16 ADP implementation is the use of standard and open interfaces. Internal (to the ADP) standard interfaces is what makes modularity of the hardware and software possible. External interfaces also need to be open and standard. As shown in Figure 1, the interface to each MFD is standard RGB video, RS-422 serial communications and DC power (either 270 VDC or 28

0-7803-5086-3 /98/$10.00 ©1998 IEEE

VDC). Each of these interfaces is very well defined and widely used. Thus, the MFD

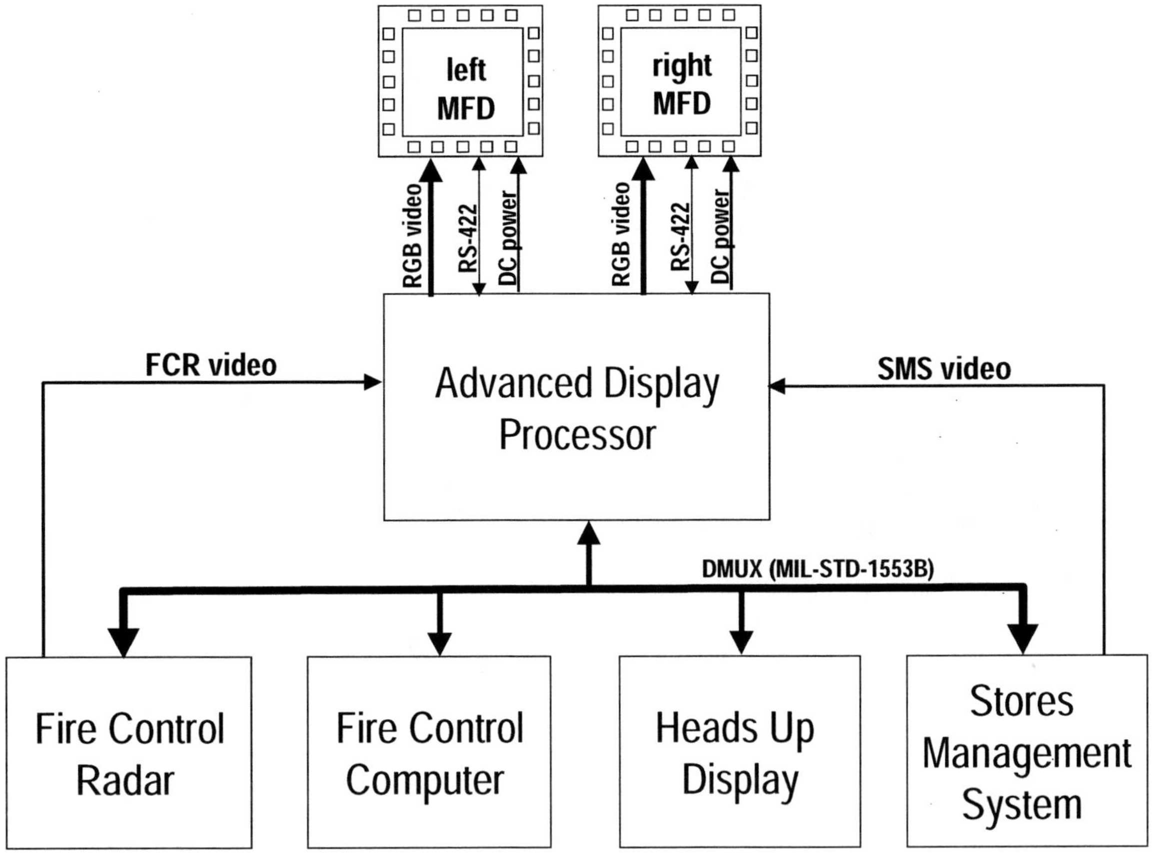

Figure 1. F-16 ADP Implementation Architecture

implementation (from the ADP perspective) is quite straight forward and flexible which promotes multi-vendor and multi-technology choices during MFD procurement.

A-10 Advanced Display Processor Implementation

To meet expanding mission requirements, AFRC/ANG planners desire to incorporate a multi-function display into the A-10 aircraft. The concept explored for this study was to replace the current armament control panel (ACP) in the cockpit with an MFD. Figure 2A shows a picture of an A-10 cockpit.

The ACP is the control panel at the lower left of the front console (just above the pilot's left knee). Figure 2B shows a diagram of an un-installed ACP. As Figure 2B shows, the ACP is an old style control panel with discrete and dedicated switches, knobs and LED indicators. These controls/indicators are used by the pilot to set-up parameters for stores control and delivery. Thus, any replacement of the ACP with an MFD would obviously need to replicate the functionality of the existing discrete controls/indicators.

To develop an integration strategy for replacing the ACP with an MFD, we first reviewed the architecture of the existing A-10 armament control system (ACS). Figure 3 shows the A-10 ACS architecture. Besides the ACP in the cockpit, the A-10 ACS consists of two other type of components: the Interstation Control Unit (ICU) and eleven station control units (SCU). The ICU is remotely mounted behind access panel F14. It provides an intermediary control/junction box between the ACP and each of the SCUs. The 11 store stations each have an SCU which provides the direct connection to the stores. As Figure 3 shows, there are several discrete interfaces to various other aircraft subsystems.

Figure 2A. A-10 Cockpit

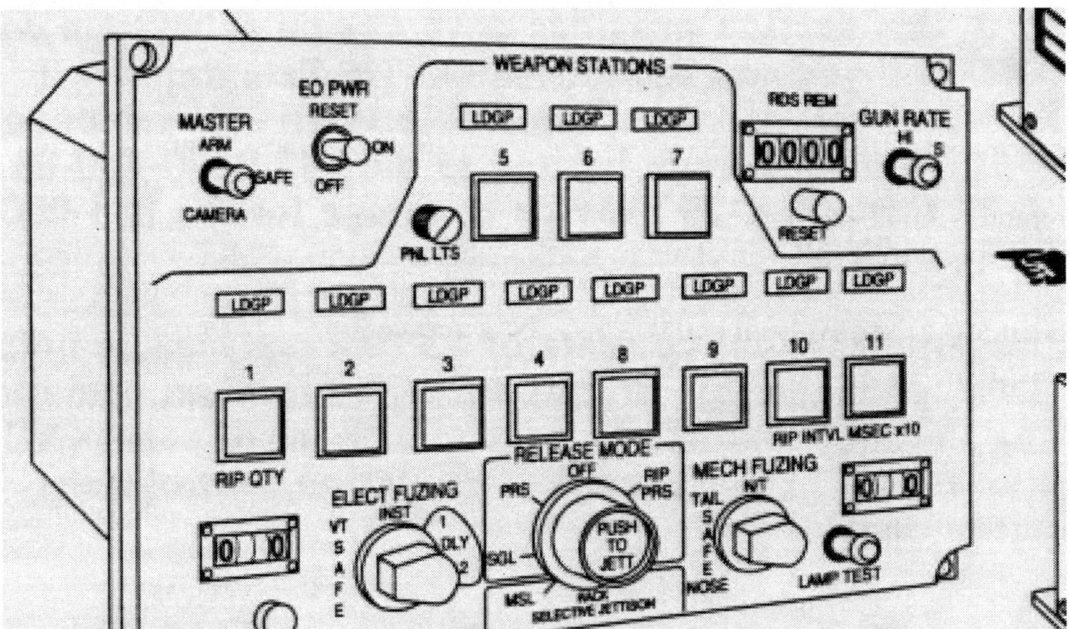

Figure 2B. A-10 Armament Control Panel

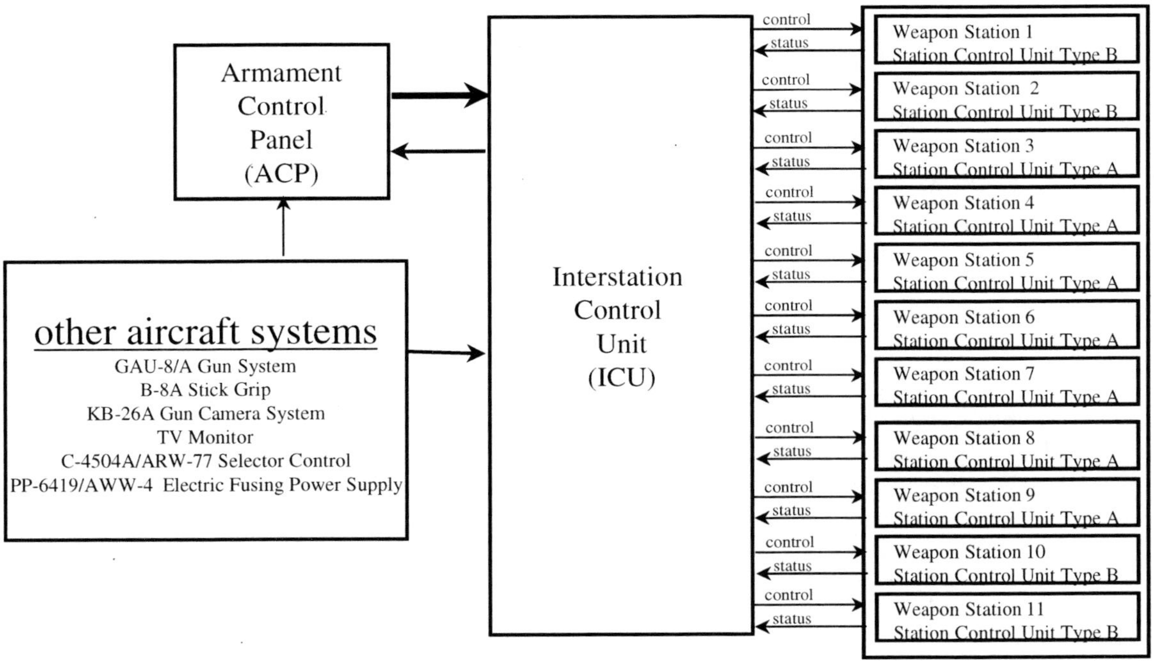

Figure 3. A-10 Armament Control System Architecture

Figure 4 shows the A-10 ACS system from a functional perspective. As the figure indicates, there is a large block of ACS logic which is distributed between the ACP and the ICU.

Each of the nine items indicated in this block corresponds to a circuit card assembly (CCA) with two of the CCAs in the ACP and seven in the ICU. This distribution of logic (between the ACP and ICU) complicates the integration of an MFD to replace the ACP since there is no a distinct standard interface division between the physically separated logic cards. Thus, an approach of replacing the ACP without modifying the ICU would have to replicate all of these special purpose logic interfaces. To maintain compatibility with the ADP architecture (as established for the F-16) this would be undesirable. Figure 5 shows the chosen approach for replacing the ACP with an MFD.

In this approach the eleven CCAs have been eliminated. The physical space (in the ICU) previously occupied by the CCAs is now used to host an ADP Graphics/Data Processor Module (GDPM) and a special ACS logic card. The majority of functionality previously implemented in the eleven CCAs will be implemented in software on the high throughput GDPM processor. Logic which cannot be implemented in software (due to special timing constraints) will be implemented on the ACS logic card. The interface to the MFD is RGB video and RS-422 serial data stream (consistent with the established F-16 ADP architecture). An interactive user interface which replicates the discrete switches and indicators (of the old ACP) will be displayed on the MFD with pilot inputs via the bezel switches. The bezel switch activations are translated to the "new ICU" via the RS-422 serial data stream where there are processed by software running on the GDPM. In this approach, the interface from the ICU to each of the store stations (SCUs) will remain unchanged. Also, existing wires between the ICU and ACP will be utilized (if possible) for the "new ICU" to MFD interface. Thus, wiring change requirements to the aircraft will be minimized.

The approach described above (Figure 5) provides a form fit function replacement for the ACP and ICU on the A-10 aircraft. Thus, all current functionality is maintained and cost

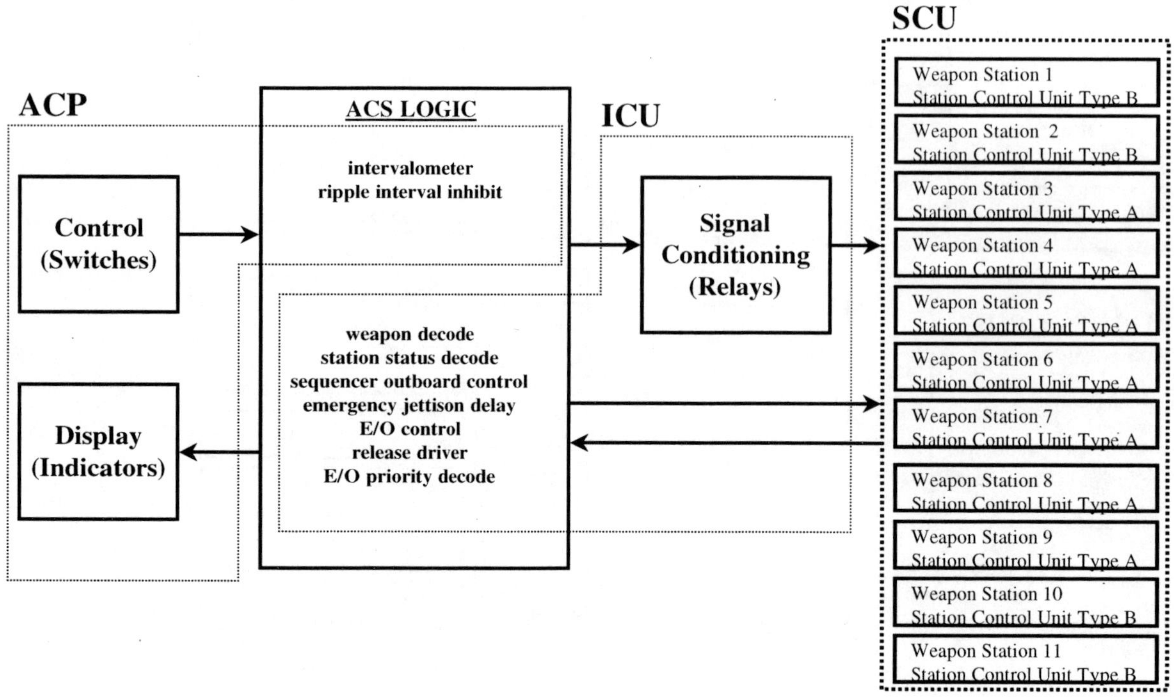

Figure 4. A-10 ACS Functional Diagram

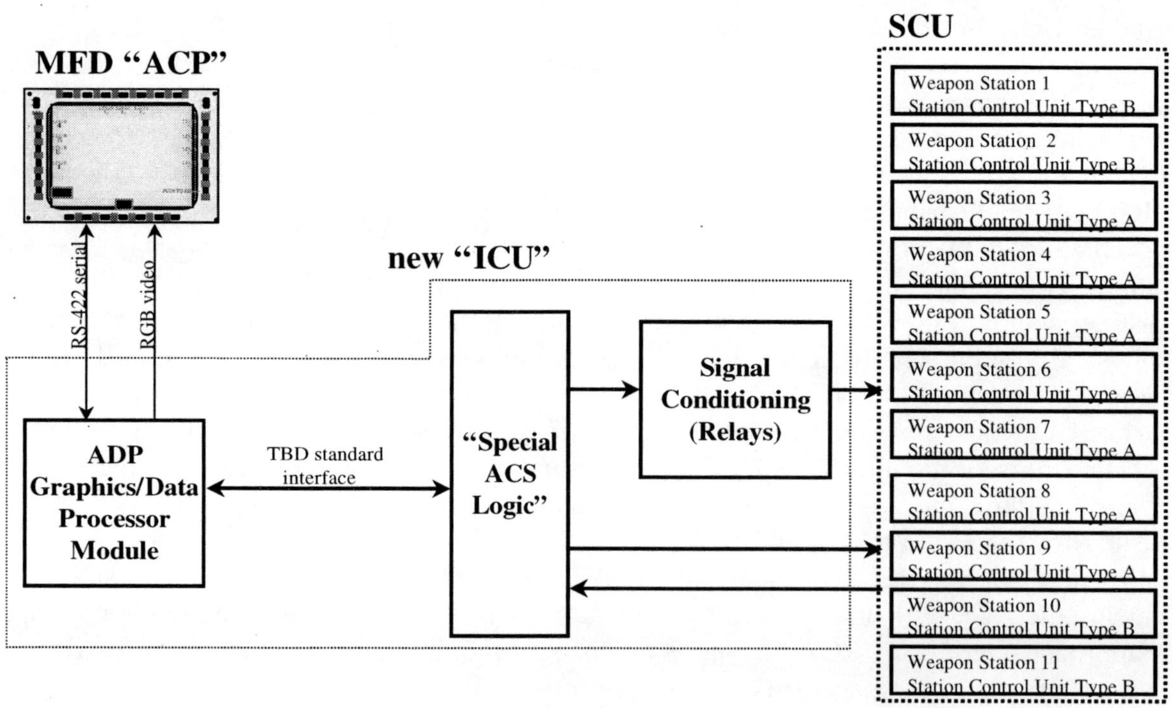

Figure 5. A-10 MFD/ADP Architecture

of ownership is reduced (due to R&M improvements). However, this approach has the added advantage of introducing a modular COTS based system architecture to the A-10. This system architecture will be expandable to include additional capabilities beyond F3 replacement of the ACP. Some possibilities include integration of Maverick and/or other sensor video to the larger (than the existing Maverick display) MFD. Also, this architecture supports one or multiple MIL-STD-1553B avionics buses (either as a remote terminal or bus controller) and MIL-STD-1760 (Smart Weapons Interface standard). Introduction of a flexible display/processing architecture such as this will be key to the future success of the A-10 aircraft in meeting the changing mission roles and requirements as discussed previously in this paper.

Summary/Conclusions

The Air Force Reserves and Air National Guard must make use of innovative strategies to maintain aircraft mission readiness in a shrinking defense budget environment. One such strategy is through the use of common systems across multiple aircraft types. This paper presents an approach to upgrade the PVI interface and mission computing capabilities for the A-10 and F-16 aircraft. The approach is to use an existing Advanced Display Processor system architecture with key hardware and software components common to both aircraft. This approach provides a cost effective and supportable growth path to accommodate future mission requirements.

References

[1] ANG/AQ web site, http://www.ang.af.mil/hqang/aq/F-16.html

[2] Riffle, Ted, "Open Systems Architecture and Commercial Off The Shelf Technology Applied to Retrofit Military Display Subsystems", AIAA/IEEE 15th Digital Avionics Systems Conference Proceedings, October 27-31, 1996.

[3] "Trade Study for Reliability and Maintainability Improvements to the AN/APG-68 Radar System", report prepared by Battelle for WR-ALC/LY and OO-ALC/LFMR under contract number F04606-95-D-0018-RJ01, March 20, 1997.

[4] Presentation/discussion session led by Colonel Doug Marlowe (ACC/DRG), June 9, 1998, AATC Test Center, Tucson, Arizona.

VIRTUAL RETINAL DISPLAY TECHNOLOGY

Microvision; Seattle, Washington

Abstract

All the technological developments required to achieve useful operation of miniature, laser scanned, virtual retinal displays have been demonstrated. Prototype devices operating at VGA resolutions (640 x 480 pixels) have been delivered to military and commercial customers for several years, and Microvision's Virtual Retinal Display™ (VRD™) is now poised for explosive performance growth. SVGA resolution (800 x 600 pixels) was demonstrated in March 1998. A functional head- or helmet-mountable 1280 x 1024 pixel, full color, biocular VRD system is scheduled for completion in late 1998, with a digital high definition television (HDTV) version scheduled for delivery in early 1999. This paper describes the VRD's fundamental operating principles (such as synthetic vision information display) and some potential applications of VRD technology.

Fundamental Operating Principles

The VRD is display technology that is fundamentally different from other, existing display technologies[1]. The VRD uses a modulated, low power beam of light to "paint" an image directly onto the retina of the viewer's eye (in a raster scan pattern similar to that in a conventional television set), and it creates a high-resolution, full-motion, full-color image without the use of screens. In certain applications, an image appears in the viewer's field of vision as if the viewer were only an arm's length distant from a high quality video screen. In other applications, the VRD can superimpose an image on the viewer's field of vision (augmented vision), enabling the viewer to see data or other information within the context of his or her natural surroundings. Regardless of how the VRD is used, the viewer perceives a high resolution, bright image.

A typical VRD system has three units:

1. System Unit. The System Unit system contains the drive electronics and a light source module.
2. Interconnect cable. The interconnect cable carries optical fibers and several electrical conductors between the System Unit and the Display Unit.
3. Display Unit. The Display Unit is head-mounted or helmet-mounted. It contains the scanner assembly, the pupil expander, and the viewer optics.

Figure 1 illustrates the five major components in a typical VRD system.

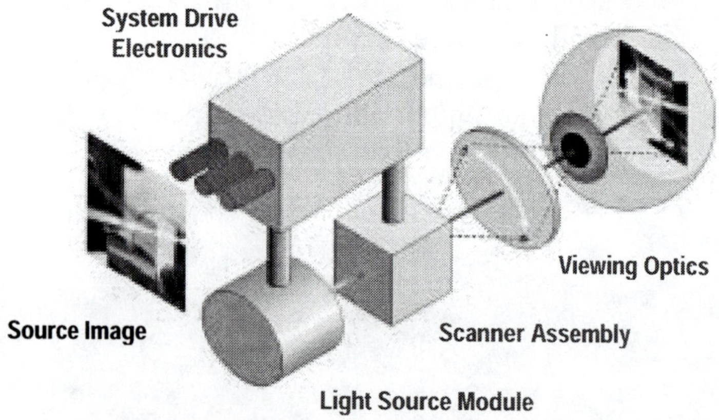

Figure 1 — *Virtual Retinal Display; Simplified Functional Diagram.*

[1] AMLCD (active matrix liquid crystal display); AMEL (active matrix electro-luminescent); FLCD (ferro-electric liquid crystal display); LCD (liquid crystal display); CRT (cathode ray tube); FED (field emission display).

0-7803-5086-3 /98/$10.00 ©1998 IEEE

1. System drive electronics. The system drive electronics receive and process an incoming video signal, provide image compensation, and control image display.
2. Light source module. The light source module contains laser light sources, light modulators, and a color combiner. The laser light sources are red, green, and blue (RGB) or a combination of the three. The light modulators modulate the light from each of the laser light sources. Modulation is controlled by signals from the system drive electronics. The color combiner combines the individual modulated red, green, and blue beams to produce a single beam of white light.
3. Scanner assembly. The scanner assembly contains two scanning mirrors. One scanning mirror sweeps the beam of laser light (from the light source module) horizontally at a high frequency and in a raster pattern. A second scanning mirror sweeps the beam of laser light vertically, but at a much lower frequency. Synchronizing signals generated by the system drive electronics control both scanning mirrors. Position feedback signals from both mirrors provide the drive electronics with instantaneous information on exact mirror position.
4. Pupil expander. The exit pupil expander (not shown in Figure 1) is an optical device that increases the viewer's field of view. The raster image created by the horizontal and vertical scanners passes through the pupil expander and to the viewer optics.
5. Viewer optics. The viewer optics relay the scanned raster image to the oculars worn by the user.

The VRD for Synthetic Vision Information Systems

The proliferation of innovative imaging sensor and display technologies argues for a reformulation of digital synthetic/augmented vision systems. The operational objective for these systems is to provide range resolution and target identification performances equal to or better than those achieved with normal unaided vision, but under Instrument Meteorological Conditions.

A synthetic vision system display must serve three functions. First, the display must be designed for maximum information transfer to the operator. This means that the image attributes of brightness, contrast (shades of gray), spatial resolution, and color must be engineered into a head- or helmet-mounted display (HMD) of low mass and high utility. Color fully engages the human visual system to detect and identify targets, permit display from imaging sensors outputting false-color information, support color encoding of symbolic information, and enhance legibility of information displayed against ambient luminances. Second, the spatial orientation of the operator's head must be tracked in order to align the real and virtual worlds. Many differing technologies exist, but when possible, a videometric approach is recommended due to high accuracy, high update rate, and environmental robustness. Finally, eye-tracking capability is required. Eye-tracking permits a 'look-to-activate', or 'look-to-shoot' virtual environment where the user's eye functions as a hands-free 'mouse'. Simple, inexpensive approaches are now becoming available with the accuracy and update rate required. Microvision will be embedding commercial off-the-shelf (COTS) head- and eye-tracking capability into its visually-coupled HMD systems.

Traditionally, the miniature, militarized, monochrome-green cathode-ray tube (CRT) has been the image source of choice for HMDs, but its performance limitations, high-voltage requirement, and cost have spurred major efforts to develop four basic categories of "CRT-replacement" technology: emissive, transmissive, reflective, and scanning. All of these technologies for CRT-replacement promise reduced package sizes, weights, voltage requirements, and costs, and each has its performance strengths and weaknesses. But only the scanning type, exemplified by Microvision's VRD, is proving through rapid advancement to provide the high image quality of a monochrome display, but in full color, and at levels of spatial resolution and display brightness *unmatched by any of the alternatives*.

The VRD's superiority lies in its "loosely-coupled" architecture. Coherent light sources of any color(s) or power(s) are modulated with an image signal and combined as a full-color pixel 'stream' into a single optical fiber conduit. The fiber may run a considerable distance to a head-worn, 15-gram scanner assembly which is unlike a matrix-addressed display in that there is no pixel structure or "yield" issue, and for which the only resolution limits are imposed by optical beam diffraction, scatter, and effective acousto-optic modulation (AOM) frequency. As light sources, AOMs, fiber-couplers, scanning technologies, or viewing optics are improved, the VRD advances.

For instance, in late 1998 Microvision will deliver a monochrome green binocular HMD system to the Army with 1355 horizontal pixels per eye and 1470 foot-Lamberts displayed raster luminance at the eye. Microvision has modularized the system design to permit delivery of a full-color 1280 x 1024-pixel biocular head-mounted system only 3 months later, as well as an HDTV (1920 x 1080 pixels) HMD to the Air Force by the year 2000. Even higher spatial resolutions are feasible, placing the VRD well beyond the capabilities of competing miniature display technologies. Clearly, no other miniature display image source in existence is supported by either the physics or the engineering which characterize the VRD.

Potential Applications

Because the versatility of VRD technology makes it suitable for a broad range of applications and a wide variety of potential products, Microvision believes that markets for personal displays incorporating the VRD are very large in scale and scope, with the potential to sustain rapid growth.

Microvision divides the marketplace for its technology into five major categories:

- Defense and public safety
- Healthcare
- Industrial
- Wireless communications
- Consumer electronics

Defense and Public Safety

Because battlefield success increasingly turns on technology advantage, many nations are pursuing technologies that can enhance combat effectiveness and survivability while slowing steadily increasing costs.

As a part of this effort, military leaders have focused on increasing battlefield "situational awareness" by providing visual information to participants in air, ground, and sea operations. Participants wearing devices capable of displaying critical information are considered more likely to survive and to prevail in military operations. Such "wearable" display systems must be obtainable at reasonable cost, consume very little power, and provide a high quality image without encumbering the user or blocking vision.

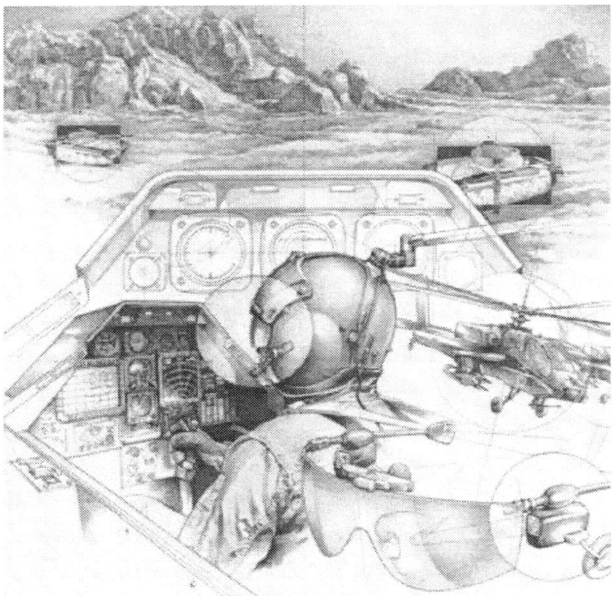

The suitability and cost of currently available personal display systems, commonly miniaturized CRTs and flat panel displays, have been compromised by trade-offs and limitations in image quality, brightness, power consumption and system bulk.

However, opportunities abound in the military marketplace for a display system that overcomes these limitations. As defense agencies worldwide equip new aircraft with more sophisticated information systems and seek ways to extend the life and enhance the combat effectiveness of existing aircraft, display systems will play an important role.

Major market opportunities exist for:

Advanced helmet-mounted display systems for jet aircraft and helicopter crew systems. More than 75,000 combat aircraft are already in service worldwide, representing a substantial retrofit market for new display technology. The Teal Group estimates that 2,846 additional jet fighter aircraft will be built between 1997-2005 worldwide, a market worth approximately $110 billion, and that 7,139 helicopters valued at approximately $38 billion will be built over that same period.

Aircraft simulation and training systems. Microvision estimates that the worldwide market for such systems (including newly built units and retrofits) will exceed 24,000 units over the next five years. Displays designed for these systems are expected to range in price from $25,000 to $100,000.

Other military applications. Microvision also anticipates a future opportunity to sell advanced display systems for ground vehicles and even ground troops. More than 100,000 ground vehicles and 23.4 million ground troops are in service worldwide, representing substantial market potential.

Public Safety. Many products designed and built for military purposes will be attractive to the public safety markets, which include law enforcement, fire fighting, emergency service and other applications.

Healthcare

Widespread use of imaging techniques in medical diagnostics has made healthcare professionals increasingly dependent upon high quality visual displays for the analysis and interpretation of diagnostic images. Such sophisticated techniques as ultrasound, nuclear medical imaging, computed tomography and magnetic resonance imaging are critical sources of information-rich electronic images used by surgeons.

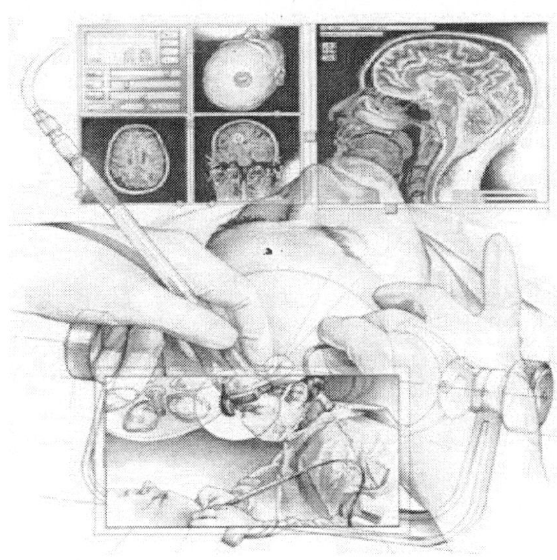

In the emerging field of "image-guided surgery," complex visual data is displayed to assist the physician during a surgical procedure. However, current systems require surgeons to refer to images on inconveniently located CRT monitors and to instruments on the periphery of the operating theater, away from the work at hand.

A more desirable solution is to present high contrast, high resolution and full color images in the surgeon's field of view, where he may refer to them at any time during the procedure without looking away from his work. Using VRD technology, an image can be superimposed over the target area on the patient's body to help guide the surgeon's hands through the procedure.

Image-guided surgery would be especially useful during delicate procedures involving the brain and spine. This technique will represent a market approaching $10 billion by the year 2000, according to industry analysts.

The increasing number and complexity of surgeries using minimally invasive techniques, such as laparoscopy and endoscopy, indicates that a significant opportunity exists to improve the success rate of these techniques by improving the surgeon's hand-eye coordination. Laparascopic procedures will comprise almost half of general surgeries by the year 2000, according to the Society of Laparoendoscopic Surgeons. According to Frost & Sullivan, U.S. sales of endoscopic equipment, currently growing at a 17.5% compound annual rate, will reach $1.8 billion by 1999.

Industrial

Manufacturing and maintenance professionals in a wide range of industries have identified a means to achieve productivity gains and increased performance by using portable displays offering augmented vision to guide and support complex manual procedures. Augmented vision allows workers wearing personal displays to maintain full visual contact with their work while viewing pertinent information such as blueprints, instructions and diagnostic information. Several target industries and their needs include:

Manufacturing. Manufacturing workers would benefit from consultimg blueprints and instructions often as they perform manual tasks. Production timelines could be substantially reduced using devices that give assembly operators a "heads-up" view of their plans and instructions while working.

Maintenance and Repair: Field maintenance workers such as aircraft mechanics and photocopy machine technicians could use real-time portable visual displays in order to consult digitized repair manuals that guide their work.

Construction. Supervisors would benefit from the ability to visually check, share and coordinate project blueprints, especially on large, complex construction sites.

Inventory Control. Warehouse workers could use a portable, body-wearable display to locate and identify items wherever they may be stored within an expansive facility. Due to their inherent limitations, flat panel displays have failed to meet the requirements for portable personal displays that can deliver augmented (see-through) vision capabilities. A major shortcoming of screen technologies is a lack of adequate image brightness under high ambient light conditions such as exist on a sunny day. In its augmented vision mode, VRD technology provides the industrial user with bright, crisp images under any light condition.

Wireless Communications

Manufacturers of wireless communication products have identified a need for users to view the same kinds of information now available on their desktops, such as e-mail, web pages, faxes, and other graphical documents.

A new generation of "personal data communicator" products represents the latest advance toward an emerging all-digital wireless communications environment in which voice communications will be complemented with high resolution, still images and even full-motion videoconferencing. Such devices must include a display that presents a desktop-size image to the viewer but adds little to its size, power requirements or cost. VRD technology is uniquely capable of meeting the demanding combination of performance requirements for such go-anywhere access.

The burgeoning popularity of data communications and the possibility of handheld remote access to the Internet and other business networks have created substantial opportunity for new products that can provide such capabilities.

Industry analysts forecast that demand for cellular telephones will reach 350 million units worldwide by the year 2000, up from 80 million units in 1997. Wireless data revenues of $3.12 billion expected in 1997 could more than triple in the same period, according to Frost & Sullivan.

Consumer Electronics

Increasingly powerful graphics image processors are becoming available at lower and lower costs. As a result, simulations for professional, educational, and entertainment uses offer an unprecedented level of visual realism. The experience offered by such powerful new interactive media products could be enhanced through the addition of wearable stereoscopic displays that engage the user with vivid 3-d images over a wide field of view. Such displays must be very reasonable in cost, should provide a high quality image, and should be comfortable for extended wear.

In the past, products that have attempted to meet this need have provided too little performance at too great a cost. Microvision believes that VRD technology can enable the development of gaming peripherals that would be lightweight and affordable, and that provide life-like realism to interactive gaming and educational products.

The current popularity of personal computer- and console-based video games indicates a robust market for such peripheral devices. Total sales in the category in 1996 were estimated at $5 billion. Dataquest estimates that 10 million console-based game units will be sold worldwide during 1997. In its first six months of availability, the Nintendo 64 sold more than four million units.

ASSESSMENT OF AIRCRAFT INTERNAL EMI USING ELECTROMAGNETIC ANALYTICAL MODELING CODES

R. Wayne Devereux, Veridian, Veda Operations, Lexington Park, Maryland
Dr. Bruce Archambeault, Veridian, Veda Operations, Durham, North Carolina
Gerald L. Fuller, Veridian, Veda Operations, Mariposa, California

Introduction

Determining the response of a wide variety of aircraft types and avionics installations to onboard electromagnetic interference (EMI) is extremely difficult and very costly. Numerical electromagnetic modeling offers a valuable solution to separate the complex system level EMI test problem into smaller pieces to gain knowledge about coupling inside complex cavities with multiple apertures, antenna factors near walls, windows, seats and ceilings, etc. Modeling can also aid in analyzing EMI responses without major impact to in-service aircraft schedules, aid in post-test data analysis, help plan future tests, and to reduce costly iterative testing of new mitigation techniques.

This paper addresses the basic problem of modeling an aircraft's cabin or cargo bay compartments and onboard electromagnetic sources to determine the amount of Radio Frequency (RF) leakage through various apertures and the resultant interference levels received by internal avionics and external aircraft Communications, Navigation and Identification (CNI) antennas. The standard compartment apertures include cockpit and passenger cabin windows, doors and emergency exits or avionics and cargo compartment access doors, apertures and wiring ducts. This is a very complex problem to model since the frequencies of interest cover the range of well below 100 MHz to above 3 GHz. The aircraft cabin and cargo bays are often multiple wavelengths and resonance effects must be included in the simulation. Internal 'features' such as absorption details of seats, people, and the aircraft's interior materials, etc. can be included as required.

Problem Analysis

Veridian's Veda Operations, the Federal Aviation Administration (FAA), and other organizations have been researching the threat posed to commercial aircraft by onboard EMI transmitters located in the passenger cabin and the cargo holds [1, 2, 3, 4]. These transmitters may be in the form of unintentional use of Portable Electronic Devices (PEDs) or even intentional RF threat sources from hostile passengers. Part of this investigation has been to examine different electromagnetic modeling codes to support research and experiment data analysis of RF vulnerabilities and coupling modes of aircraft CNI systems and flight control avionics [5, 6]. A variety of numerical modeling techniques and available modeling codes were researched and considered to accomplish this modeling task. Because each modeling technique has weaknesses, ongoing research is exploring a hybrid modeling approach to accurately solve these complex problems and provide a cost effective means of analyzing and comparing various mitigation techniques and minimize the need for iterative experimentation.

Objective of Modeling, As A First Step

Numerical and analytic modeling can be useful to help understand the phenomenon displayed in the data. Both the normative behavior and the seeming irregularities can be supported. Perhaps more important is the use of a data validated model to apply these results to conditions beyond those of the tests, especially to aid in evaluating the test data when the test aircraft is not longer available and unusual data characteristics are needing to be further analyzed.

0-7803-5086-3 /98/$10.00 ©1998 IEEE

When considering the RF paths from a cabin mounted interference source to a victim aircraft antenna, the geometric layout of the cylinder size, windows and source and victim antenna locations must first be determined to assess the emissions through the primary apertures (windows). Figure 1 shows a generic sketch of a cylinder, window and antenna model. The frequency of concern, source antenna types, source position within the cabin, window size and location relative to the source, resonance modes within the cavity, and victim antenna type and location will dramatically impact the results. A variety of source and window locations must be used to determine the worst-case emissions through the windows. This was initially bounded using experimental data already acquired using cavity mode-stirring for optimum worst-case emissions data [7]. Once these window emissions are known they can be used to determine the amount of interference received by different antennas at different locations on the external skin of the aircraft. External wings, tail structures, etc. can be included as necessary.

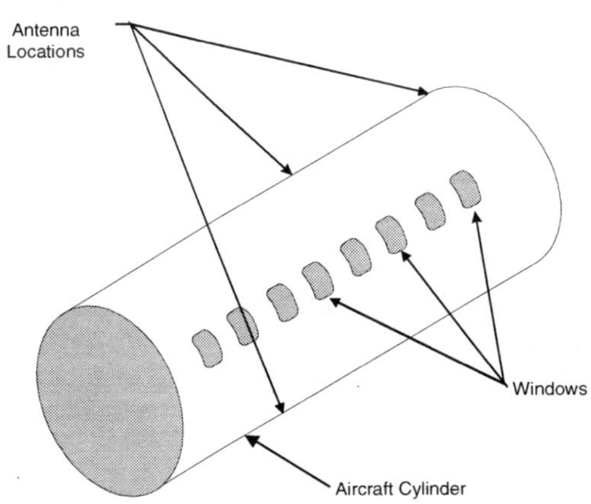

Figure 1 Aircraft Cylinder Sketch

This paper shows several modeling results using a sample of passenger cabin RF source locations and different polarized antennas over a broad range of frequencies. Several figures show results of relative interference field strength present at different top mounted external aircraft antennas from sources located in different standard windows. Also, the aperture effects an emergency exit door, test antennas near a curved surface, and various sizes of aircraft are shown.

Modeling and Simulation Strategy

While modeling techniques and specific software tools are much more powerful than in the past, they are still not capable of modeling the effects of EMI sources to an entire aircraft over all frequencies of interest. Since the initial focus has been on RF sources coupling to the CNI antennas on the outside of the aircraft, and the source is assumed to be on the inside of the aircraft, then both inside and outside of the aircraft must be modeled. This is a considerable challenge to the modeling tools and to the potential computer resources required to run the simulations. Even with the use of super computers, the frequency range must be severely limited when modeling entire aircraft the size of today's commercial aircraft.

Regardless of the limitations of the modeling techniques and computer resources, some information about the source effectiveness and potential ways to reduce the threat effects is still desired. Limited models can be developed where the important portions of the entire aircraft are included in the model. Care must be taken to ensure the important portions are not omitted, but this can be easily checked through an iterative process where additional parts of the aircraft are included with no discernible change in the results. Early model results, no matter how generic, should always be checked against real data to determine whether some key aspects of the model, such as infrequent and forgotten apertures, are important.

One of the greatest strengths of using modeling and simulation is the ability to separate the various effects and analyze them each in turn. A number of different specific effects were analyzed using modeling.

Another major strength of simulation and molding is the ability to remove physical test and measurement effects that add confusion to the analysis. For example, a dipole antenna was used as the source antenna during the measurements. At VHF frequencies, this antenna was too long to physically fit close to the window, without touching the metal floor and curved ceiling. The

0-7803-5086-3 /98/$10.00 ©1998 IEEE

antenna was physically moved away from the window to provide several inches of clearance from the floor and ceiling. The effects of the floor and ceiling on the antenna are known to be dramatic, yet are unavoidable in real-world measurements. However, these effects can be easily analyzed using modeling, as shown later.

Modeling RF Path Loss from Window to Antenna

One example of a reduced size model was used to determine the RF path loss between windows in the passenger compartment and the various VHF and UHF receive antennas on the outside of an aircraft. The aircraft body was modeled as a cylinder with windows along the side, and antennas on the top and bottom, while maintaining the relative distances between them, shown in Figure 1. For this set of models, the windows were considered to be the source of the RF energy, so the model's domain was only on the outside of the aircraft. For this problem, both Method of Moments (MoM) and Finite Difference Time Domain (FDTD) codes were considered as potential solution techniques. While both modeling techniques were used, only the FDTD models will be discussed here.

The FDTD computational domain was broken into 10 cm cubed cells. This would provide accurate results at frequencies up to about 300 MHz (using the 1/10th Lambda rule for cell size), and so would include both the VHF and UHF frequency range of interest. The aircraft cylinder was modeled using the FDTD cells in a stairstep arrangement. The source was placed in each window, in turn, while the amount of field received by the antenna (located on the top or bottom of the cylinder) was recorded. Both horizontal and vertical polarizations of the source were modeled. Figure 2 shows one of the FDTD models where a 70 x 190 x 70 cell computational domain was used. A source (vertical or horizontal) was placed in each window in turn, and the field strength at the various antenna locations recorded.

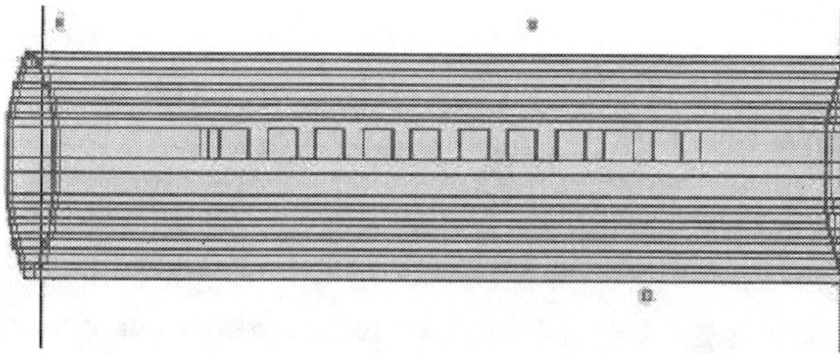

Figure 2 FDTD Model Example for External Path Loss from Various Windows to Receive Antennas

Figures 3 and 4 show the vertical and horizontal results for one of the VHF antennas, and Figures 5 and 6 show the vertical and horizontal results for a UHF antenna. The results in Figure 3 show a significant increase in received signal strength when the VHF source was placed in window #7. This window position corresponds to the location of the antenna on the top of the cylinder. Depending upon which window is illuminated, as much as a 15 dB difference in the received signal strength was observed. Figure 4 shows the same source, but turned horizontally. Since the receive antenna is vertically polarized, it does not receive as much signal from a horizontally polarized source, and in fact, the closest window (#7) received the least amount of signal.

Similar results are shown in Figures 5 and 6 for the UHF case. One interesting result was the range of signal strengths received depending upon the frequency. As much as a 6-7 dB variation over the UHF frequency range was observed (in addition to the proximity to window #7 effects).

Modeling the Effect of Emergency Exit Windows

While a clear trend was apparent from the previous modeling, a trend which showed that the closer the window to the antenna (for the same polarization) resulted in high received levels, post-test analysis of measurement data had shown some anomalies when compared to the model results. The VHF and UHF measurement data indicated that the windows with emergency exits are the sources of the highest received energy, rather than simply the closest window.

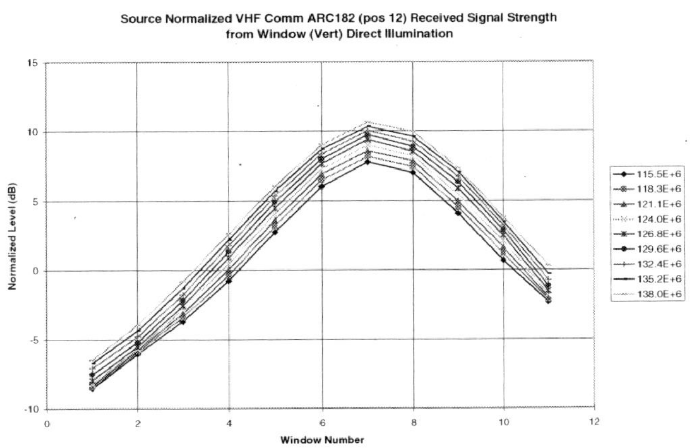

Figure 3. Example of VHF Antenna Received Signal Strength from Vertically Polarized Source

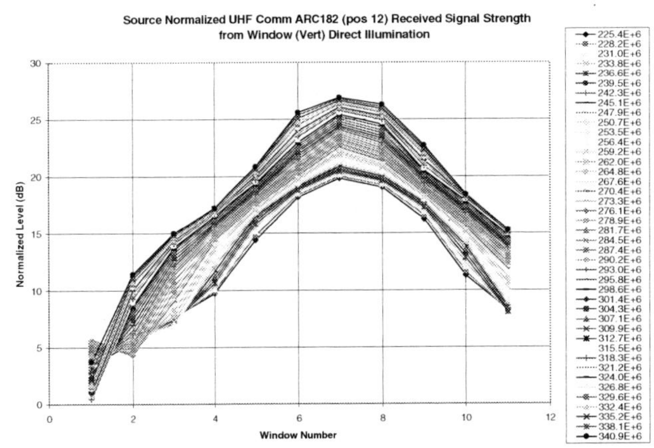

Figure 5. Example of UHF Antenna Received Signal Strength from Vertically Polarized Source

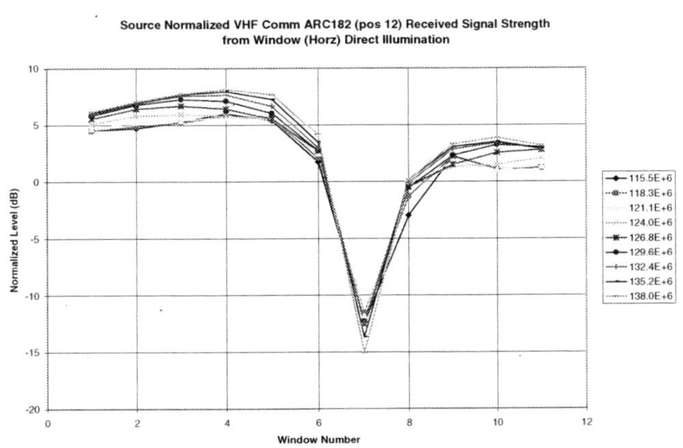

Figure 4. Example of VHF Antenna Received Signal Strength from Horizontally Polarized Source

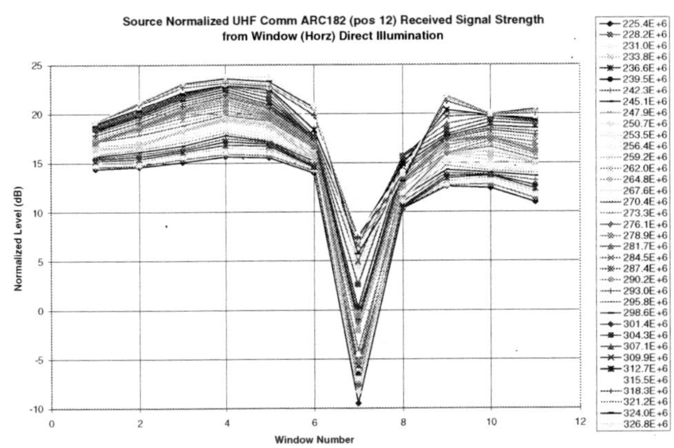

Figure 6. Example of UHF Antenna Received Signal Strength from Horizontally Polarized Source

To understand this disagreement between model and measured results, it's necessary to examine how the measurements were made originally. In this case, a dipole antenna was placed inside the aircraft directly in front of the window being illuminated. The antenna was positioned so that it was as close to the window as the curve of the aircraft body allowed. In the case of the emergency exit windows, this meant the entire emergency exit was illuminated, and if the emergency exit door was poorly bonded to the aircraft skin, the EMI leakage would be dramatically increased. The original experiment data analysis had not accounted for the emergency exit doors located at specific windows, and was initially interpreted as unstable. A new report being issued contains the benefit of this modeling knowledge and accounts for the corrected emergency exit door data [4].

The original data was interpreted as unstable when common window data sets were compared. This data problem was further analyzed with a model to better understand the response. A simple window opening with the transmit antenna behind it, was modeled initially and plotted, but did not agree with experimental data. The emergency exit was then modeled as a larger opening in the cylinder, with a metal plate in this opening, and a window in the metal plate. Figure 7 shows the increase in received signal strength at the antenna location due to the emergency exit. Note the plotted differences at the 100 MHz and VHF frequency range. These new results explained the original difference in the simplified model with no emergency exits and the experimental measurements, and helped in the overall understanding, and analysis of the measured results from actual aircraft cavities.

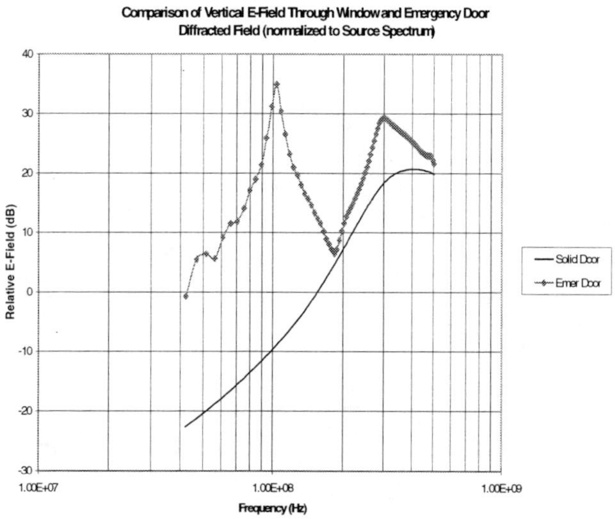

Figure 7 Relative Difference in Received Signal Strength With Emergency Exit

This model was created to help understand the effects of the emergency exit illumination, as compared to illuminating a solid panel with a window. FDTD was used since the model required the characterization of the metal shield (with different sized openings) and MoM does not allow the RF currents to be different on either side of a shield (as must exist in the real-world). For this model, the FDTD cell size was set to 1 cm, and only a portion of the cylinder was included in the model. The relative location of the receive antennas (relative to the emergency exit) was included.

Near-by Window Leakage

While the path loss between the windows and the receive antennas was characterized previously, during measurements the transmit antenna was positioned slightly back from the window itself, due to the length of the antenna, and the curve of the metal wall/ceiling. Therefore, while most of the transmitted RF energy was radiated out through the 'intended' window, some amount of the transmitted RF energy was leaked through adjacent windows too. A separate FDTD model was created to identify the amount of this leakage through other windows, so that the total effect could be calculated (using superposition). Figure 8 shows an example of the FDTD model.

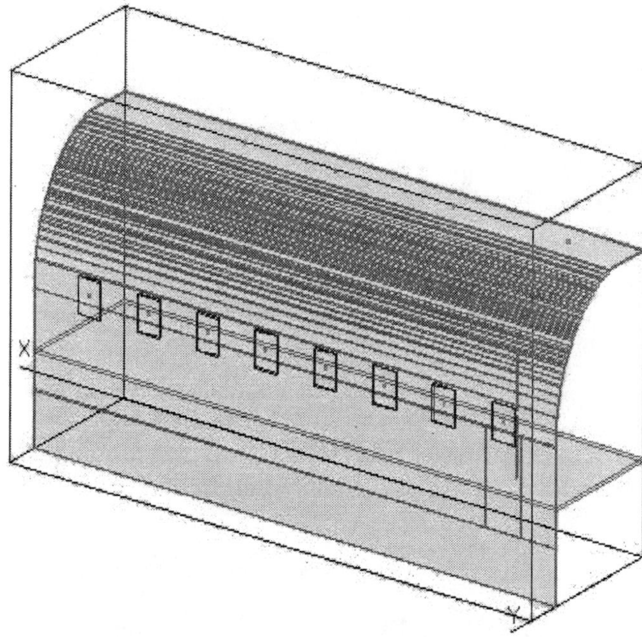

Figure 8 FDTD Model for Adjacent Window Leakage

The amount of E-field in each window was determined for both vertical and horizontal polarizations. Figure 9 through 13 shows the relative electric fields in the various windows for the vertical and horizontal polarizations of the different VHF, UHF and Glideslope (GLS) frequency ranges used.

Effect of Nearby Metal Walls/Floor on Dipole Antenna

As stated earlier, the amount of electric field created from the source dipole antennas would be modified due to the close proximity of the metal ceiling and floor. The antenna was set up so that in the worst case, the ends of the dipole were about three inches away from the metal ceiling and floor. The previous FDTD model was modified so that the effect of the close proximity could be included. This allowed the creation of an accurate near-field antenna factor for the dipole in this configuration. Figure 14 shows the antenna factors for the dipole alone in free space, and then with the ceiling and floor present. In all cases, the 'receive point' was at the location of the window, approximately 13 inches from the dipole's center at VHF, 3 inches at UHF, and in the window at GLS frequencies.

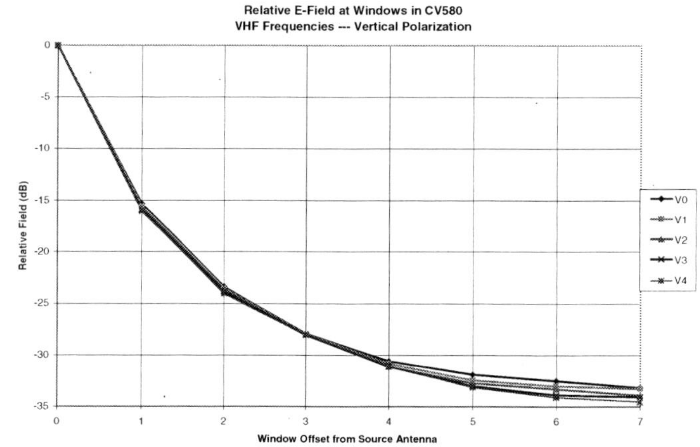

Figure 9 Relative E-field at Adjacent CV580 Windows VHF Frequencies - Vertical Polarization

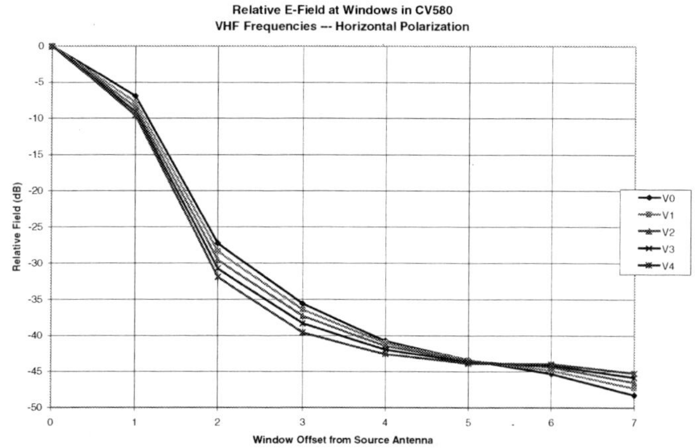

Figure 10 Relative E-field at Adjacent CV580 Windows VHF Frequencies - Horz Polarization

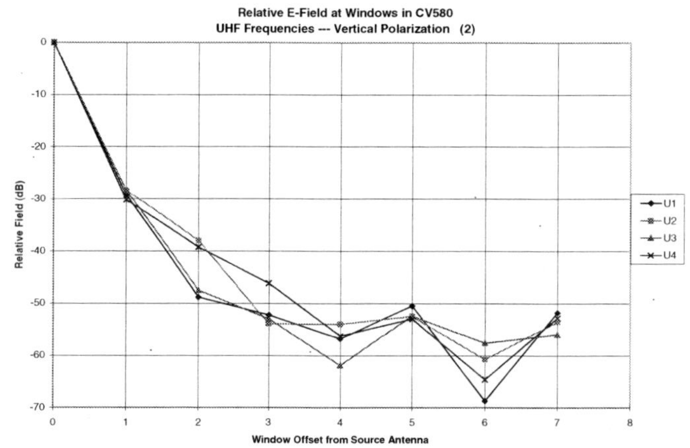

Figure 11 Relative E-field at Adjacent CV580 Windows UHF Frequencies - Vertical Polarization

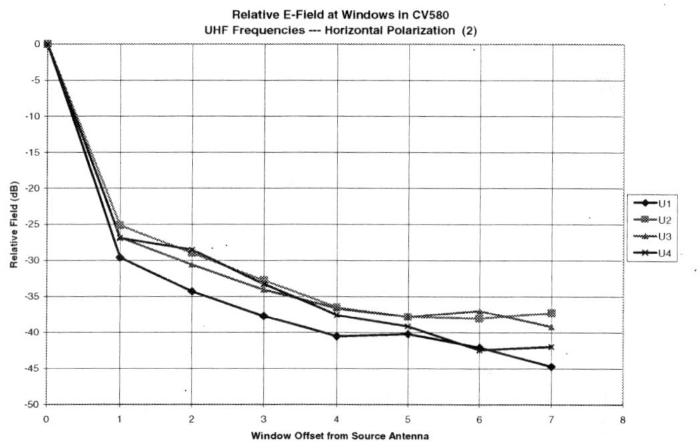

Figure 12 Relative E-field at Adjacent CV580 Windows UHF Frequencies - Horz Polarization

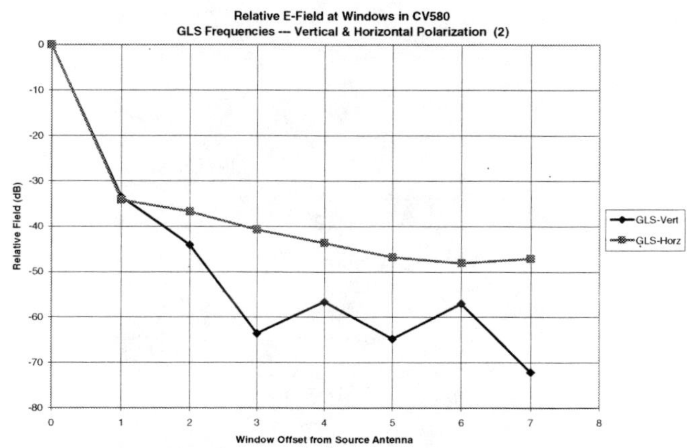

Figure 13 Relative E-field at Adjacent CV580 Windows Glideslope Frequencies

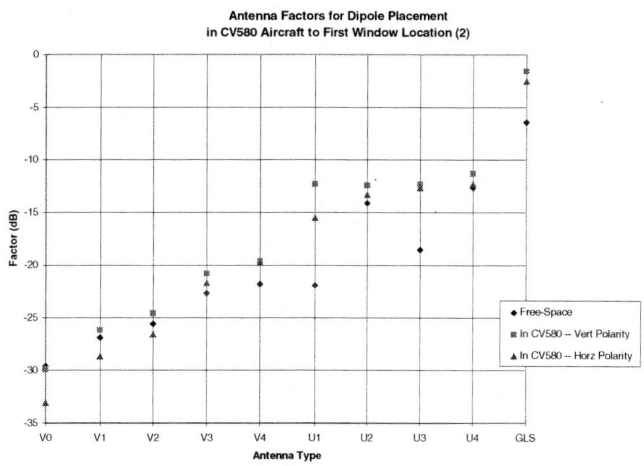

Figure 14 Comparison of Free-Space Near-Field Dipole Antenna Factors with Metal Ceiling/Floor in Close Proximity

Diffraction Effects On External Path Loss for Various Sized Aircraft

The main propagation of fields through the windows are not directed to the antennas located on the top and bottom of the aircraft. Rather, any fields present at the top/bottom receive antennas are due to diffraction around the external curved surface of the aircraft. This diffraction-path-loss would change with the size of different aircraft. Therefore, a FDTD model was created to determine the amount of change due to aircraft size on this external diffraction path loss. Figure 15 shows the FDTD model where the source is located in the center of a window, and the receive antenna is directly above the window.

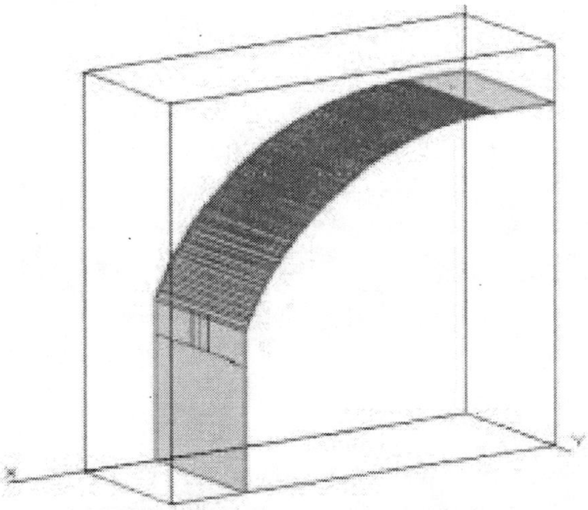

Figure 15 FDTD Model Example for Various Aircraft Body Sizes for External Path Loss Difference

Similar models were created for the CV580, the narrow body, the standard body, the wide body (747), and intermediate body (767) aircraft styles. The CV580 and narrow body aircraft are almost the same size diameter cylinders, while the standard body, wide body, and intermediate body aircraft are also almost the same size diameter cylinders. Figure 16 shows the results of the different sized aircraft, each behaving with a maximum amplitude variation over the frequency range of 10 dB on the same size cylinder. This Figure indicates that RF path loss data measured on smaller scale aircraft is only 15 to 22 dB less at VHF and UHF and only 10 to 12 dB less at microwave frequencies then data for the largest of aircraft cylinder diameters.

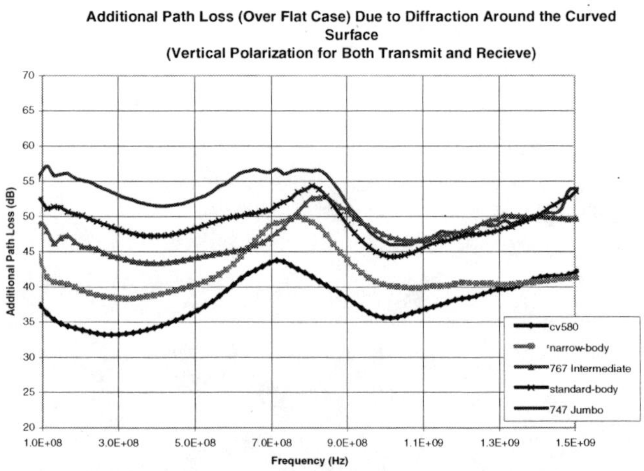

Figure 16 Diffraction Path Loss for Various Size Aircraft Bodies (Vertical Polarization)

Movies for Modeling Results Analysis

The analysis of complex models and wave propagation through apertures is sometimes very difficult to interpret from general data plots and graphs. The FDTD code used to evaluate these modeling effects has an animation viewer to allow complex wave propagation and amplitude response to be naturally analyzed in visual format over time. Figures 17 and 18 show two views of a FDTD model of a cylinder with windows and a floor. The dynamic wave propagation can be seen using the animation by clicking on the figure when the hand appears.

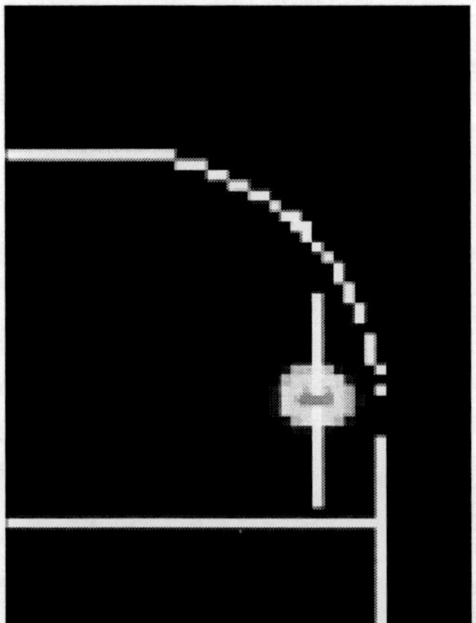

Figure 17 FDTD Movie Viewer Output for Side View of Aircraft Cylinder with Vertical Source in Windows

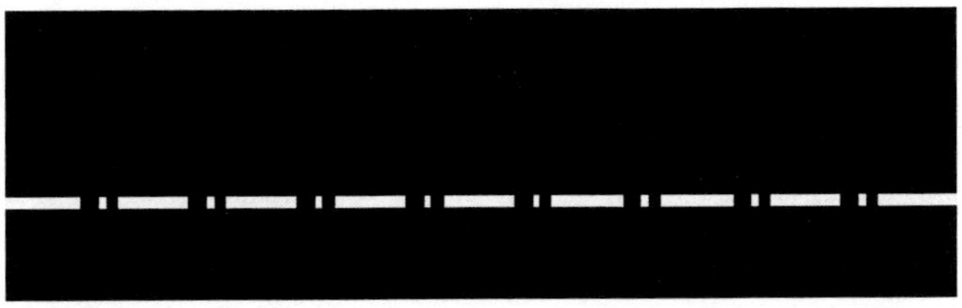

Figure 18 FDTD Movie Viewer Output for Top View of Aircraft Cylinder with Vertical Source in Windows

Conclusions

Unfortunately, there is no one modeling technique that can be used to completely model this problem. The best solution is to use a hybrid technique with both the FDTD and MoM techniques. FDTD is used to model the inside of the aircraft cabin, and find the amount of electric field escaping through a window, cargo door seal, or other fuselage aperture, given a source position and field strength inside the cabin or cargo bay. MoM techniques are best for solving the internal cabin cable harness currents by using the cavity field level results from FDTD techniques. The requirement to solve for high frequencies drive the need for electrically small FDTD cells and MoM segments, which result in computer memory and processing time requirements that are often excessive. The Concurrent Complementary Absorbing Boundary Condition (CC-ABC) dramatically reduces the required white space domain (which largely contributes to the excessive memory problems) and allows more accurate modeling [8]. The CC-ABC method should be implemented into existing and future FDTD codes.

This paper has shown the usefulness of electromagnetic modeling for helping determine and analyze the various types of potential RF coupling paths to aircraft navigation and communications antennas from internal cavities with apertures. Both MoM and FDTD were used for modeling different effects on the aircraft. While no practical models of the entire aircraft are possible with today's technology over the entire frequency range of interest (above 1 GHz), portions of the aircraft can be conveniently modeled, and individual effects analyzed individually. An extremely important point was demonstrated by using modeling to simulate individual aspects of the over all problem. This allowed individual effects to be analyzed separately, and even physical measurement effects can be removed from the overall data analysis. This allows much more focused analysis than is possible with only measured data.

Acknowledgment

This research has been performed under the FAA Aviation Research Grant Program, Grant No. 97-G-001. We would like to thank Mr. Ray Schillinger of the FAA William J. Hughes Technical Center, Code AAR-520, for his support throughout this research effort.

References

1. R. W. Devereux and R. Schillinger, *Electromagnetic Coupling Paths Evaluated From Inside A Commercial Aircraft*, 15th Digital Avionics Systems Conference (DASC) Proceedings, Oct. 1996
2. R. W. Devereux, G. L. Fuller, R. Schillinger, *Further Analysis of the CV-580 and B-727 Aircraft RF Coupling Experiment Data*, 16th DASC Proceedings, Oct. 1997
3. R. W. Devereux, G. L. Fuller, R. Schillinger, *Electromagnetic Susceptibility of Installed Avionics*, 16th DASC Proceedings, Oct. 1997
4. R. W. Devereux and G. L. Fuller, *Coupling and Path Loss from Internal Source to CNI Antennas*, 17th DASC Proceedings, Nov. 1998
5. R. W. Devereux, B. Archambeault, G. L. Fuller, *Assessment Of Analytical Codes For Use In Modeling Aircraft Onboard EMI Threats*, 16th DASC Proceedings, Oct. 1997
6. R. W. Devereux, B. Archambeault, G. L. Fuller, *Electromagnetic Analytical Modeling Used to Aid Aircraft EMI Data Analysis*, 98 International Symposium on Electromagnetic Compatibility (EMC), Aug. 1998
7. *CV-580 RF Coupling Validation Experiment Report*, Veridian Report Number 79689-96U/P30041, Nov. 15, 1996
8. O. M. Ramahi, *Complementary Boundary Operators for Wave Propagation Problems*, Journal of Computational Physics, Volume 133, 1997, pages 113-128

AN EXPERIMENTAL STUDY OF THE PLACEMENT AN AIRCRAFT INSIDE A LARGE WELDED ZINC-PLATED STEEL ELECTROMAGNETIC REVERBERATION CHAMBER

Paul A. Johnson and Kevin R. Goldsmith, Defence Science and Technology Organisation, PO Box 1500, SA 5108, Australia

Introduction

Electromagnetic Compatibility (EMC) and full EM Interference (EMI) testing are extremely important for electronic equipment at a component, system and platform level. This is true for both the civilian and military markets. For the latter it is important that the manufacturer must ensure the original military equipment performs in some stated environment without interfering with other equipment, and the Australian Defence Force (ADF) must ensure that a subsequent modification remains EMC compliant.

Australia is a consistent modifier of platforms, owing to its unique geography, location and climate, and the ADF is vulnerable to modification-induced EMI problems. It is therefore of particular importance for aircraft that no such problems exist, as the consequences can be catastrophic or expensive. Whilst there are no recorded aircraft losses due to EMI in Australia, it cannot be said they have not happened because there is often no physical evidence left by EMI. Indeed, there is great concern worldwide about EMI related aircraft problems amongst civil and military operators, with the rise of internally carried consumer electronics (eg. laptops and mobile phones), external sources (radar, radio, TV and communications emitters), the increasing digitalisation of aircraft electronics and the advent of fly-by-wire control systems.

Standard EMI and EMC testing facilities include Open Area Test Sites (OATS), Anechoic Chambers, Transverse Electromagnetic Mode (TEM) Cells and Shielded rooms. In addition, Electromagnetic Reverberation Chambers (RC) are rapidly gaining acceptance as an alternative, though their operation is imperfectly understood. They have many advantages, such as high efficiency (a large electric field is achievable for a modest RF input power), a uniform, isotropic and randomly polarised field, and highly predictable statistical behaviour.

DSTO has built one of the largest Reverberation Chambers in the world at its Salisbury site. A description of its design, commissioning and characterisation can be found elsewhere in these proceedings [1].

Recently, a military aircraft (Blackhawk S-70A-9 helicopter) was temporarily installed into the chamber. It is believed that this is the first time a platform of this type has been studied in an RC. Using low-level fields only, the chamber was re-characterised, and the effect of the aircraft on the chamber deduced. In this paper, we describe the results of this work, in particular the derivation of the loading effect of this aircraft upon the chamber. This information can now be used to estimate what RF power is required to achieve full threat–level fields inside an aircraft-filled chamber.

0-7803-5086-3 /98/$10.00 ©1998 IEEE

Description of the Chamber

The DSTO Chamber was constructed inside an existing large building (an old diesel engine workshop) on the Salisbury site. Its dimensions are given in Table 1. It was constructed from steel sheets of thickness 2 mm, with a 90 micron zinc coating, and the seams welded throughout the entire chamber. The chamber has two personal access doors, and two large doors which form one entire wall, enabling large test objects to be introduced.

At one end of the chamber a large tuner assembly is mounted. The tuner is required for exciting modes in the chamber. A central shaft runs floor to ceiling, driven by a gearbox and induction motor on the roof, and attached to the shaft are four large (4m x 4m) paddles of various opening and rotation angles.

Chamber Parameters	
Length, L	20.85 m
Width, W	10.89 m
Height, H	6.0 m
Volume, V	1361 m^3
Surface Area, S	835 m^2

Table 1: Reverberation Chamber Parameters

The chamber is more fully described elsewhere [1].

Experimental Method

Installation of Aircraft into Chamber

The aircraft landed at approximately 1.30pm on a Thursday, on a hard standing area adjacent to the chamber building. The ground crew were already standing by. The four main rotors were removed, and then the aircraft pushed by a ground-cart and by hand into the chamber. The aircraft cleared the doorway in the main building only by 20cm each side, and the doorway lintel on top only by rotating the tail rotor very carefully! By the time the main chamber doors were sealed up, it was approximately 4.30pm. Only three hours were therefore required to install the aircraft into the chamber. The aircraft was available to us for the weekend, and left on the Monday. A similar amount of time was required to remove the aircraft to the point of its departure. In future, we believe an aircraft could be in and out the same day within a couple of days, having been subjected to a modification test or periodic retest, which is important when aircraft are removed from service.

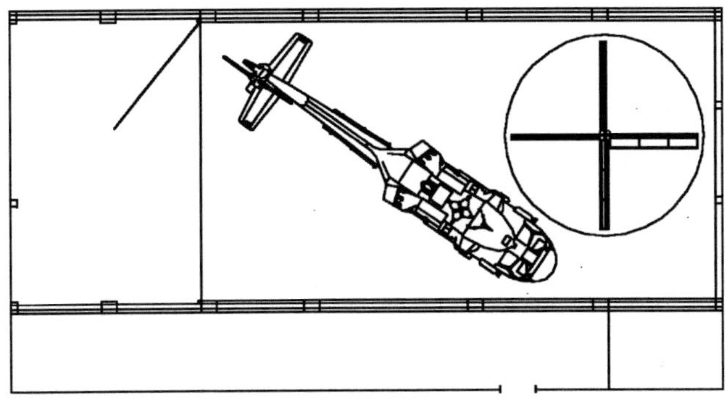

Figure 1: Scale Drawing of location of Aircraft Inside Chamber

Figure 2: Aircraft installed into chamber, with main chamber doors still open.

The location of the aircraft inside the chamber building is shown to scale in Figure 1, and pictorially in Figure 2. It is evident that there is not much clearance between the aircraft and the walls or paddle. For this reason, the stabilator was in the folded position for this experiment.

Characterisation of Chamber

The method used to characterise the chamber was the same for both the empty and aircraft-filled case. The signal source was a HP series 71209A Spectrum Analyser with Tracking Generator up to 18 GHz, with output power set to –10dBm. The signal was fed to a Power Meter via a directional coupler, as well as the chamber, for forward power measurement, P_{in}. A series of approximately 35 discrete frequencies were used between 26 MHz and 18 GHz. Mode stirring was employed, where for a given discrete frequency, the paddle wheel was allowed to turn continuously rotated at 1 RPM, and the signal recorded for one full rotation.

Below 1 GHz two EMCO 3143 and 3141 biconilogs were used as transmit and receive antennas, and above this frequency two EMCO 3115 Double-ridged horns were used. When in use, the TX biconilog was placed facing the upper corner behind the paddle, and the RX biconilog was placed facing the opposite upper corner. In the higher frequency range, the horns were both placed in the same corner nearest the instrumentation (cable losses would otherwise be severe at the highest frequencies). The TX horn faced the upper corner, and the RX horn towards the centre of the chamber. See Figure 3 for antenna locations.

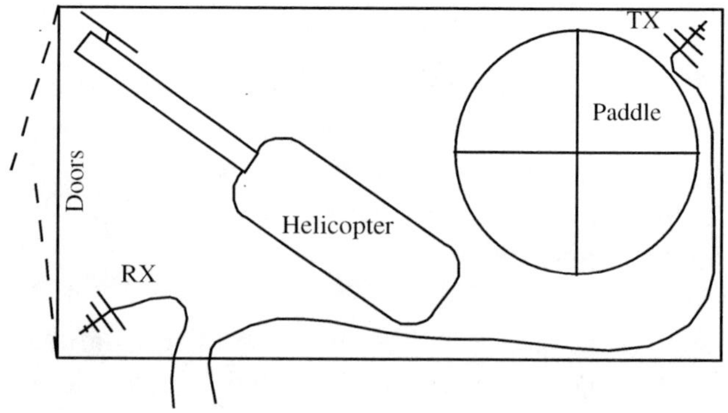

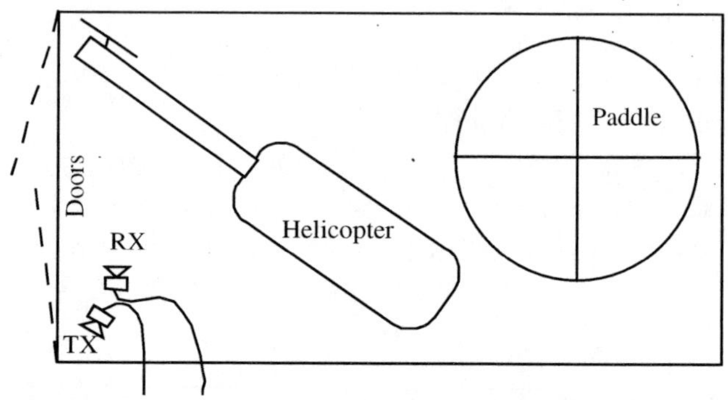

Figure 3: Schematic layout of chamber for the two frequency ranges: below 1GHz (top) and above 1GHz (bottom). This is identical to empty-chamber layout.

The equipment set-up was identical for both the empty and aircraft-filled chamber (eg: cables, directional couplers, pre-amplifiers, antennas etc). The cable losses were measured at the same frequencies used for the characterisation.

Results

When mode–stirring, for a particular frequency the power received from the antenna is recorded by setting the trace time on the spectrum analyser to 60 sec, in order to capture a full turn of the paddle wheel. The whole data acquisition process is automated, and a file of 1001 data points is saved, along with a description file containing run information. The paddle position resolution is therefore 0.36 deg. A sample signal plot is given in Figure 4.

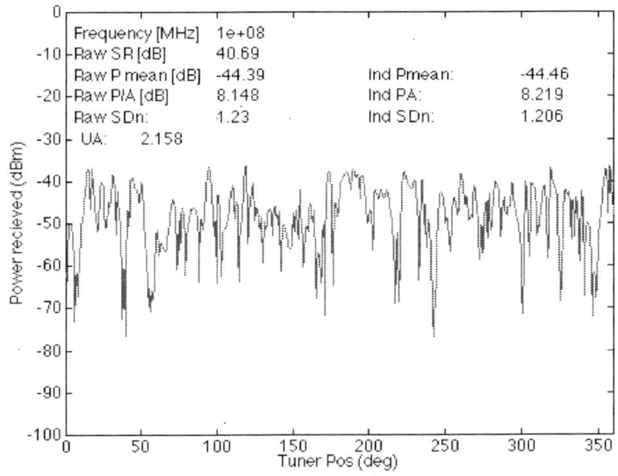

Figure 4: An example plot of signal against paddle position, for a frequency of 100 MHz.

For each plot, the following information is derived: the stirring ratio (SR), which is the maximum to minimum signal (conventionally, it is accepted that proper operation of the chamber requires at least 20dB [2]); average received power $<P_{rec}>$, (which must be a linearised mean); peak–to–average ratio (PA), the ratio of the maximum to the mean, (which should theoretically be about 7-8dB, depending on the number of independent data points); normalised standard deviation (σ_n), a linearised standard deviation normalised to the mean (which should be unity).

It is important to obtain stastically sufficient independent data points for the analysis to be valid. The field in the chamber changes significantly only when the paddle has turned by an equal to what we term the "uncorrelation angle" (θ_u). The θ_u is frequency dependent, and is determined for each frequency by correlating the signal plot with itself, shifting the plot one paddle position at a time until the correlation coefficient drops to 1/e at the uncorrelation angle. An example of the correlation function is given in Figure 5, where the plot in Figure 4 becomes uncorrelated after 6 paddle steps, or 2.1 deg. Having determined θ_u, every independent point is sampled (every sixth point in this example), to determine corrected versions of <Prec>, PA, and σ_n.

If the chamber Electromagnetic Environment (EME) is fully random, the Probability Distribution Function (PDF) of the E field at a given point should follow that of a χ^2 (Chi-squared) distribution with 2 degrees of freedom (χ^2_2). Representing this distribution in cumulative form makes arbitrary binning of data unnecessary, and so the Cumulative Distribution Function (CDF) was obtained for each data set. An example is given in Figure 6. The correlation coefficient, γ_{CDF}, of the distribution compared with the theoretical one is also calculated, together with the Kolmogorov-Smirnov probability [3] (p_{KS}) that the CDF is drawn from the theoretical distribution.

In the past it has been empirically determined that the EME should be fully random below that frequency above which there exists at least 60 modes [2]. For this chamber, that critical frequency is 60 MHz.

The quantities SR, PA, σ_n, θ_u, γ_{CDF} and p_{KS} are given in Figure 7 – Figure 12 respectively, for both the empty and the aircraft–filled chamber. In figures comparing the empty and loaded chamber, circles are used for the empty chamber, and crosses for the aircraft–filled chamber.

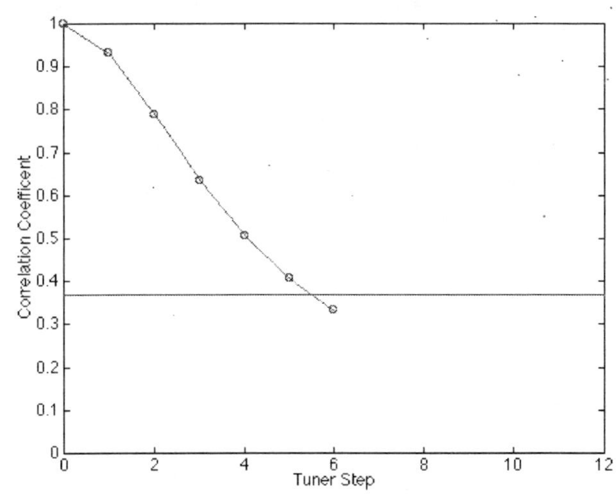

Figure 5: An example of the correlation coefficient varying with paddle position.

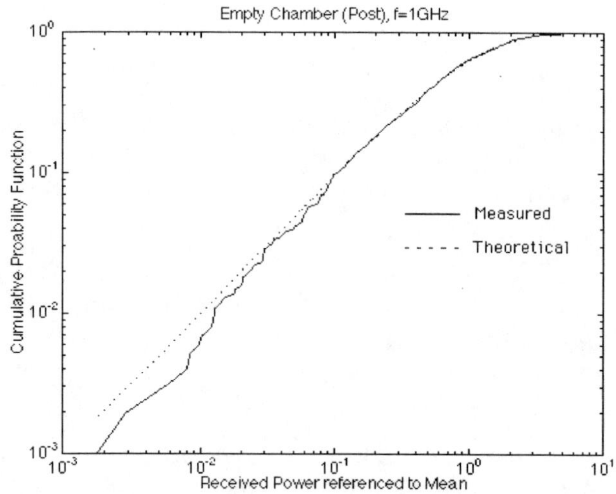

Figure 6: Cumulative Probability Function of the Received Power Referenced to the Mean (Empty Chamber at 1GHz)

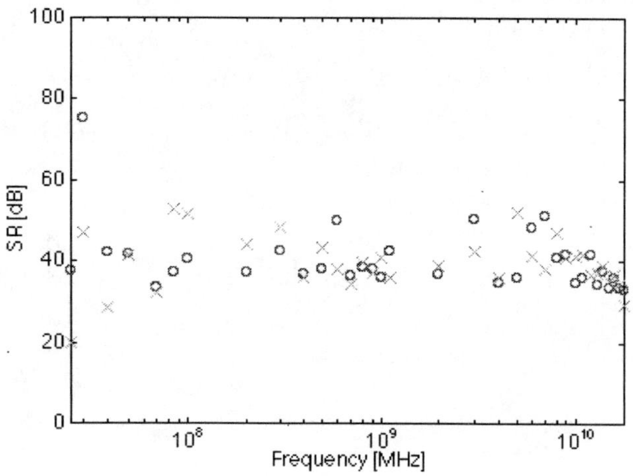

Figure 7: Stirring Ratio. (o = empty chamber, x = aircraft–filled chamber).

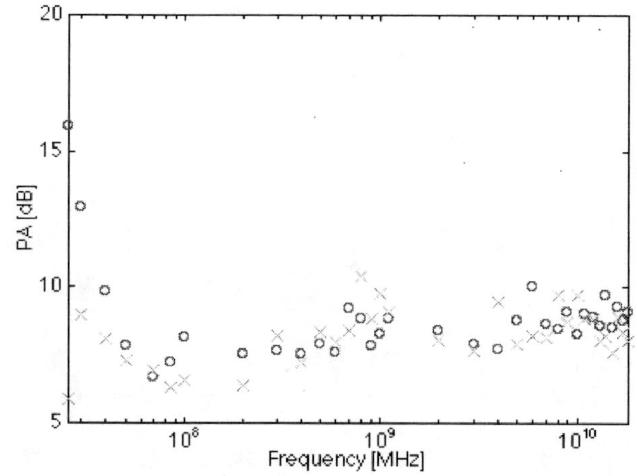

Figure 8: Peak to Average Ratio. Symbols as in Figure 7.

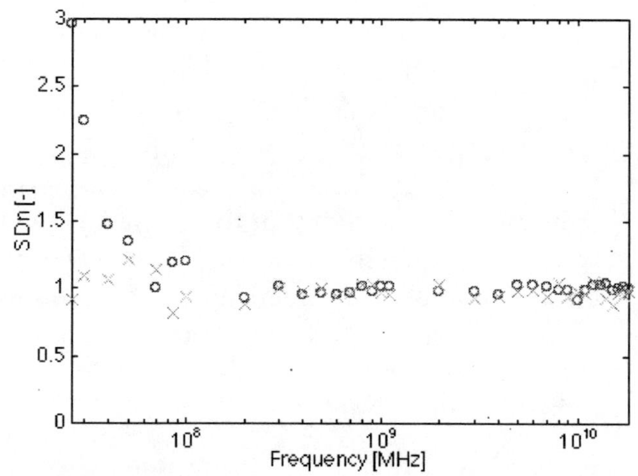

Figure 9: Standard Deviation, normalised to mean. Symbols as in Figure 7.

0-7803-5086-3 /98/$10.00 ©1998 IEEE

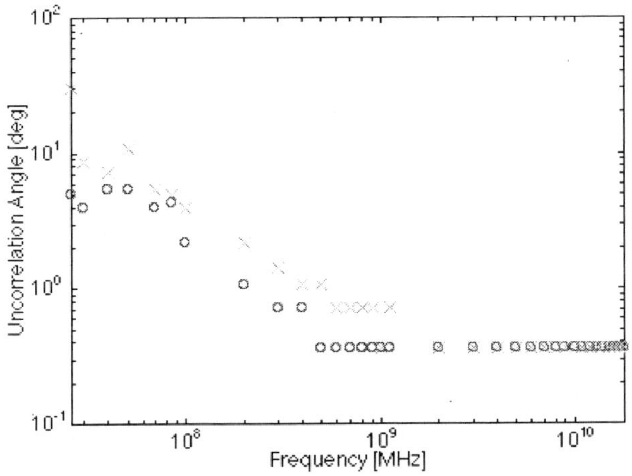

Figure 10: Uncorrelation Angle. Symbols as in Figure 7.

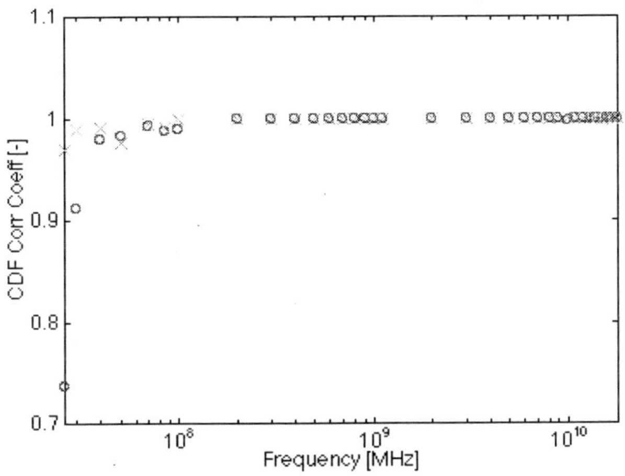

Figure 11: Correlation Coefficient of the Cumulative Probability Distribution Function compared with theory. Symbols as in Figure 7.

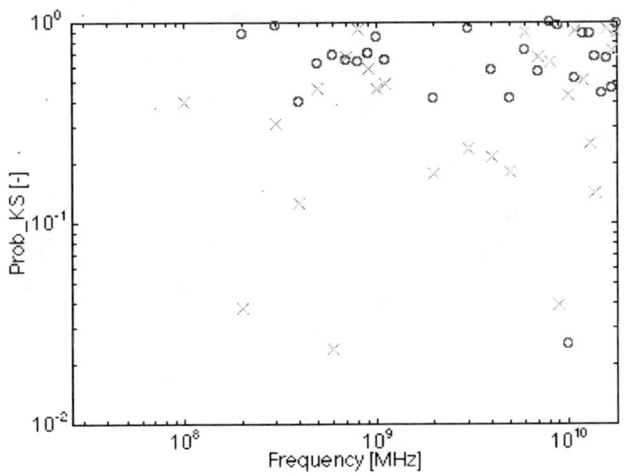

(a)

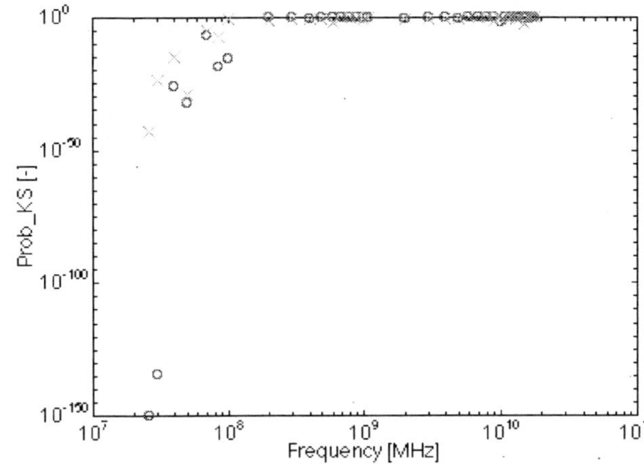

(b)

Figure 12: The probability that the CDF is drawn from the theoretical distibrution ie. That the EME is fully random (a) in fine detail (b) overall.

The Q of the chamber comprises dominant components due to the antenna, Q_{ant}, and due to the walls of the chamber, Q_{wall}, where

$$Q_{ant} = (16\pi^2/\eta m c^3)\ V\ f^3$$

$$Q_{wall} = 3(V/S)\ (f/MHz)^{0.5}\ (\sigma_{con}/\mu_r)^{0.5}$$

and the total expected Q_{theo} of the chamber, is given by

$$1/Q_{theo} = 1/Q_{ant} + 1/Q_{wall}$$

For the loaded chamber, an additional component is due to the aircraft itself, Q_{ac}, causing the Q to fall by an amount ΔQ. The determination of this parameter is the chief aim of this paper.

The measured Q of the chamber can be calculated from

$$Q=(16\pi^2/c^3)\ V\ f^3\ (\langle P_{rec}\rangle/P_{in})$$

The theoretical Q for an empty chamber, and the measured Q for empty and loaded chamber is given in Figure 13, and the drop in Q caused by the aircraft, ΔQ, is shown in Figure 14. An earlier measurement of the empty chamber is given in the same Figure.

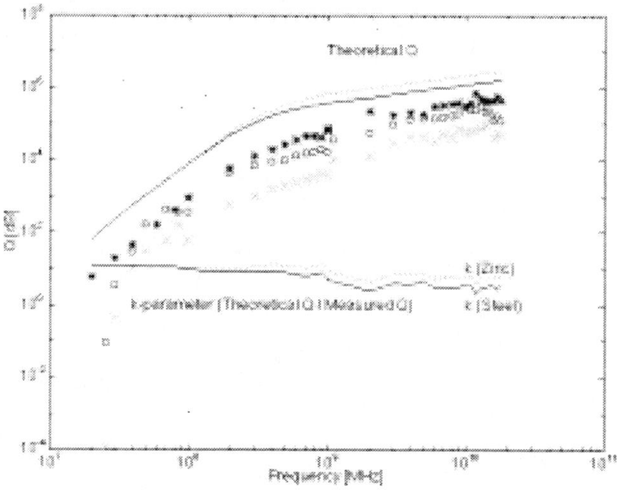

Figure 13: Q of chamber. (o = empty chamber, x = aircraft–filled chamber, * = previously derived data).

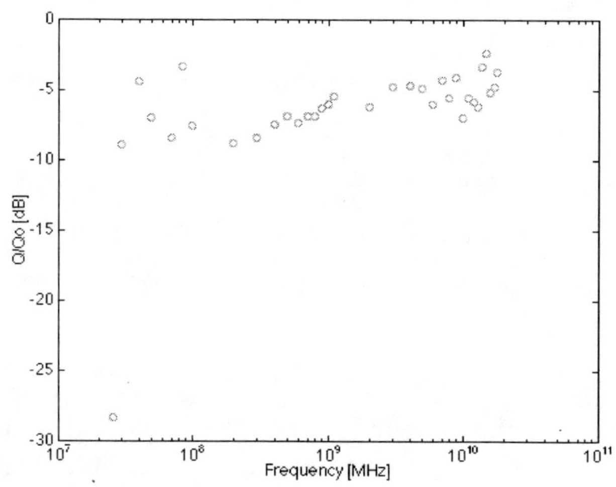

Figure 14: Change in Q caused by presence of aircraft, ΔQ

Finally, the expected E-field achievable with 1W of input power is given in Figure 15.

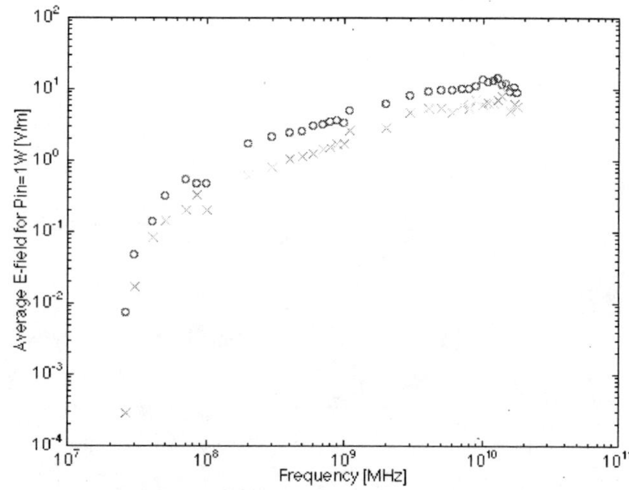

Figure 15: Electric Field achievable for 1W of input power. Symbols as in Figure 7.

Discussion

There appears to be little difference in the statistical parameters between the empty and loaded chambers. The major difference is in the change of chamber Q, due to the loading caused by the reflecting and absorbing materials which comprise the structure of the aircraft.

Statistical Properties

The Stirring Ratio (SR), is between 20dB and 50dB for the whole frequency range for both cases. See Figure 7. This is a requirement, but not sufficient it itself for the chamber to perform correctly.

The normalised standard deviation (σ_n) agrees very well above 100MHz with the theoretical value of unity in both cases. See **Figure 9**. Below this frequency, there is some scatter in the values for the aircraft–filled chamber and an interesting departure towards 2–3 for the empty chamber.

The correlation coefficient of the CDF compared with the theoretical value for a fully randomised EME, γ_{CDF}, is unity across the frequency range for the aircraft–filled chamber, dropping only to 0.97 at 26MHz. See Figure 6. For the empty chamber, however, unity is only

0-7803-5086-3 /98/$10.00 ©1998 IEEE

reached above 40MHz, actually falling to 0.7 at 26MHz.

The probability that the EME is random, given in Figure 12, suggests a good agreement above 100 MHz and, but rapidly decreasing probability below this frequency. It is important to realise the difference between the parameters γ_{CDF} and p_{KS}. With large amounts of independent data, it is possible to say that two distributions are very similar (say γ_{CDF} =0.99), but that the probability that they are exactly the same is very low (say pKS = 10^{-5}). The two variables provide complementary information.

Taken together with the standard deviation data, the data suggests that at frequencies below 100MHz, chamber EME is actually statistically better with the aircraft in place: although the Q is less, the relationships between Peak and Average field etc. are more determinate. This could be due to the aircraft causing more reflections inside the chamber, preventing direct coupling between the TX and RX antenna. widening the bandwidth of the Q to included more modes at the low frequencies, effectively adding more low frequency modes to the chamber. In both cases the EME is random down to frequencies below the empirically determined 60 MHz. Conversely, at frequencies above 100MHz, p_{KS} implies that the data are better behaved in the empty chamber (Figure 12a).

The uncorrelation angle θ_u varies smoothly from 5 deg for the empty chamber, and 10–20 deg for the loaded chamber, to 0.36 deg at higher frequencies. See Figure 10. This latter value represents the resolution of the sampled paddle position (the intrinsic paddle positional accuracy is actually 0.1 deg, but the spectrum analyser stores only 1001 samples). At lower frequencies, it is interesting that successive paddle positions appear less correlated for the loaded chamber compared with the empty chamber, even though the field is more random when averaged over a full turn of the paddle wheel.

The peak-to-average ratio, PA is in the range 6–10 dB for the whole frequency range for the aircraft–filled chamber. See Figure 8. The theoretical value of PA depends on the number of independent samples: for N=1000 it is 8.5 dB, whilst for N=20 it is only 5 dB. The measured values seem to agree well with the theoretical values. For the empty chamber, there is similar agreement except at the lowest frequencies, where the chamber EME is not random.

The Q of the Chamber

It became evident after the experiment was completed that a power divider had been used in the low frequency range rather than a directional coupler. This had the effect that instead of reading forward power only, the total power (forward and reverse) was sampled by the power meter. At the lower frequencies, where the VSWR becomes bad, this causes the chamber Q to appear to roll-off. The earlier results from a set-up that used a directional coupler do not show this roll-off. Fortunately, this error should not greatly effect the *difference* in Q between the empty and loaded chamber, only its *absolute* value.

Other than this roll-off, there is good agreement between the old and new sets of empty chamber Q.

If we compare the old empty Q data with that expected from theory, by calculating

$$k=Q_{theory}/Q_{measured}$$

we find that the so-called "k-parameter" varies from a factor of 10 at lowest frequencies, to about 3–5 at the highest frequencies, the latter depending on whether a theoretical chamber of steel or zinc is considered. Better agreement is found with a steel chamber, which seems reasonable, since the chamber walls contain far more steel than zinc. This range of k compares favourably with other chambers [4].

The difference in Q between the empty and loaded chamber is probably the most important result of this paper. It shows that the presence of the aircraft loads the chamber by an amount of about 10dB at 26MHz, rising smoothly, with some scatter, to 5dB at 18GHz. This is encouraging, as it might be expected that a modern military aircraft, comprising lots of composite and plastic would load the chamber more than this amount. This makes it feasible to produce high electric fields inside the chamber, even in the presence of an aircraft, with affordable amounts of RF power.

Conclusions

The EME inside a large EM reverberation chamber has been characterised, both for an empty chamber, and also with an aircraft–filled chamber.

For the empty chamber, the EME appears to be random down to about 40 MHz, which is better than expected. The Q is within a factor 3–10 lower than expected from theory, but this compares favourably with other chambers.

For the chamber loaded with an aircraft, the EME appears to be random all the way down to 26 MHz, much better than expected, and surprisingly better than the empty chamber. Conversely, the paddle appears more effective at lower frequencies in the empty chamber. It is not obvious why this should be the case. The Q in the aircraft–filled chamber is in the range 5–10 dB lower, depending on frequency.

Since the loading of the chamber by the aircraft is not severe, high field strengths will be achievable for modest power input.

References

[1] Goldsmith K. and Johnson P.A. "Design, Construction, Computational EM Modeling, and Characterization of an Aircraft–Sized Reverberation Chamber and Stirrer", these Proceedings.

[2] Crawford M .L. and Koepke G.H., National Bureau of Standards Report, Technical Note 1092 (1986)
[3] Press W. H. et al., "Numerical Recipes", pub. Cambridge, pp. 472-475 (1986)
[4] Hatfield M.O., Freyer G.J and Slocum M.B., Naval Surface Warfare Center Report NSWCDD/TR-96/149 (1996)

COUPLING AND VULNERBILITY OF CNI SYSTEMS TO AN ONBOARD RF SOURCE

R. Wayne Devereux, Veridian, Veda Operations, Lexington Park, Maryland
Gerald L. Fuller, Veridian, Veda Operations, Mariposa, California

Abstract

The Federal Aviation Administration (FAA) Technical Center and Veridian's Veda Operations have investigated avionics' response to onboard intentional electromagnetic interference (EMI). Many other aerospace organizations have previously performed applicable research in areas similar to this goal, however, several basic questions about coupling remained unanswered. This paper presents further refined analysis and conclusions from data gathered during testing and modeling of a CV-580 and B-727 aircraft. Coupling paths and path loss information between an intentional internal radio frequency (RF) source and aircraft receivers and antennas at frequencies between VHF and Global Positioning System (GPS) are provided.

Background

The FAA's Office of Aviation Security R&D at the Williams J. Hughes Technical Center and Veridian's Veda Operations have been investigating the coupling mechanism and path losses from an internal source to Communications, Navigation and Identification (CNI) receiving systems and to flight control and data handling avionics systems. The preliminary results of Phase I work done in 1996, was reported on at the 15th Digital Avionics Systems Conference (DASC) by Ray Schillinger of the FAA and Wayne Devereux of Veda Incorporated [ref. 1]. Phase II experimental results were presented by Schillinger, Devereux and Fuller at the 16th DASC [ref. 2].

The initial experiments were conducted using multiple aircraft receive antennas installed aboard the FAA's CV-580 aircraft. Test frequencies included VOR starting from 100 MHz and ran to GPS frequencies of 1.6 GHz. Additional testing at GPS frequencies was conducted on the FAA's B-727. This same FAA B-727 was also used to measure path losses from forward and aft cargo compartments to the main avionics bay starting from 100 MHz and running to 1.6 GHz. These experiments confirmed that the primary RF path between a source in the passenger cabin to an external fuselage mounted antenna was out the windows (passenger or cockpit depending on the antenna location) rather than internal coupling to the aircraft's receiver wiring or connectors. Coupling from sources inside cargo bays to receivers was through the door seams.

Objectives

The following objectives were derived from recommendations developed from the previously reported experimental and modeling work:

1) Understand the anomalous coupling path losses discovered in the previous experiments

2) Explore how this path loss data could be extended to the "in-service" aircraft of the widebody and jumbo jet class of aircraft so that fleet susceptibility could be assessed

Introduction

This paper presents the analysis from further definitive experiments aimed at these questions. A companion paper explores the modeling effort and its relationship to the data analysis. It represents the extension of our results to estimating path losses for the "in service" aircraft.

0-7803-5086-3 /98/$10.00 ©1998 IEEE

Experimentation

CNI System Path Loss Anomalies

The analysis of path loss data has shown three anomalies, namely:

1) The VHF Comm #1 had minimal path losses; almost 10 dB less than other VHF systems. This data did not fit with the ensemble of results from 100 to 1600 MHz
2) The relationship of the minimum path loss to receiver antenna location shows a generalized pattern forming. However, the relationship also exhibited a strong unexplained cyclic pattern at VHF frequencies.
3) The minimum path loss variation with the receive antenna placement was not as large as expected. Particularly at GPS frequencies. There also was only an 8 dB change when placing the source antenna at the extreme locations on the tail, nose and fuselage of the B-727.
4) Of lesser concern was that the DME systems exhibited a strong secondary coupling path internal to the aircraft.

Experiments were conducted to explore these issues:

Cargo Bay to Avionics Path Losses

An opportunity arose for Mr. Fuller to conduct limited path loss tests on a SAS B-767 aircraft in Copenhagen, Denmark in June 98. Path loss tests from the fore and aft cargo bays to the main avionics bay have been made and then compared to those performed on the FAA B-727 and presented at last year's DASC. Surprisingly, similar results were obtained on these two aircraft. A full report on the SAS B-767 aircraft will be presented after the final data analysis is complete.

Most Recent CNI System Path Loss Analysis

The path losses to each antenna system as a function of window number are shown in chart form in Table 1. By highlighting the windows having the first, second and third least path losses a strong preference emerged at windows 3, 5, and 11. This effect was most pronounced at VHF and less so with increasing frequency.

The CV-580 aircraft has emergency exits at starboard windows 3, 5 and 11. These emergency exits were about 20 inches (in.) wide by 28 in. high. The passenger windows were 15 in. x 16 in. and were centered horizontally and slightly above center in the vertical. The emergency exit seams could not be examined in detail, so we assumed a non-conducting rubber gasket pressure seal with a residual gap between metal surfaces of about ¼ in. to ½ in.. These seams behave quite differently than the nearly square (or circular) windows. The seams are highly polarization sensitive and have a sharper frequency response than the round hole. The VHF Comm #1 antenna was located just ahead of window # 1 and presented the opportunity at VHF for a stronger source out the horizontal seam from windows # 3 and 5. The seam is also closer to the antenna which results in a stronger signal.

To explore the net effect of those emergency exit seams on the VHF Comm #1 receiver, the path loss vs. window number plots were compared for three VHF frequencies with and without windows 3, 5 and 11 as seen in Figures 1a and 1b. Without windows 3, 5 and 11 the data are much smoother and fit a clear pattern that makes logical sense. All of the path loss data for each system was plotted in histogram form and cumulative distributions made to highlight the impact of windows 3, 5 and 11's emergency exit seams on minimum path loss. Figures 2a and 2b shows a sample of these data plots at VHF with and without the emergency exits. It can be seen that at VHF the emergency exit window data seriously skews the distributions. Figure 3 shows a similar comparison of emergency exits with UHF cumulative distribution data plots.

Table 1

Samples of Window to Window Path Loss Variation

		STARBOARD WINDOW NUMBERS										
FREQ.BAND - SYSTEM	**POL**	**1**	**2**	**3**	**4**	**5**	**6**	**7**	**8**	**9**	**10**	**11**
V0/VOR	HORZ	51	47	44	48	**42**	51	52	44	49	46	42
V0/VOR	VERT	56	56	52	54	54	62	57	55	52	48	**44**
V1 - VHF COM #1	HORZ	48	44	43	52	**35**	46	44	48	56	48	39
V1 - VHF COM #1	VERT	48	50	**46**	60	56	56	56	58	60	61	56
V2 - VHF COM #1	HORZ	40	45	35	48	**30**	40	38	40	46	42	33
V2 - VHF COM #1	VERT	45	46	**40**	46	48	48	50	52	50	46	47
V3 - VHF COM #1	HORZ	40	48	38	44	**30**	42	46	39	46	44	38
V3 - VHF COM #1	VERT	48	47	**41**	51	51	49	54	54	54	51	51
U1 - UHF PROJECT	HORZ	56	54	60	51	46	48	49	48	49	54	**42**
U1 - UHF PROJECT	VERT	55	55	57	51	45	53	48	41	41	45	**41**
U2 - UHF PROJECT	HORZ	56	58	60	58	**49**	58	52	50	56	52	51
U2 - UHF PROJECT	VERT	59	64	62	59	58	53	50	48	46	**44**	53
U4 - UHF PROJECT	HORZ	57	67	58	62	52	54	52	52	57	51	**49**
U4 - UHF PROJECT	VERT	60	56	57	57	57	50	51	51	53	53	**47**
G1 - GLIDESLOPE	HORZ	57	57	65	57	**54**	58	65	62	62	62	54
G1 - GLIDESLOPE	VERT	63	66	**58**	62	72	70	68	68	68	70	76
D1 - DME #1	HORZ	**51**	55	51	67	66	67	71	71	71	67	65
D1 - DME #1	VERT	**55**	**45**	53	63	65	71	77	63	69	73	69
D2 - DME #1	HORZ	**47**	57	53	57	55	59	57	55	48	63	55
D2 - DME #1	VERT	**52**	**47**	55	65	67	63	67	65	61	67	69
D3 - DME #1	HORZ	**48**	59	55	63	55	68	65	65	63	58	61
D3 - DME #1	VERT	**45**	60	58	65	68	80	75	68	72	65	70

NOTES:

1. All Window #'s are Starboard – Windows 3, 5 and 11 are in Emergency Exits.
2. Numbers in Squares are the Minimum Path Loss for that window at that frequency range.
3. Specific Frequencies Measured for V0, V1, V2, V3, U1, U2, U3, U4, G1, D1, D2 and D3 can be found in Reference 3.

At UHF the cumulative distribution is not skewed significantly but individual points still show a preference of minimum path loss for the emergency exit windows. At microwave the effect of these exit windows is negligible. This is in keeping with the resonant frequency behaviors of the seams and window apertures. The seams have a lower frequency of resonance than the windows but by UHF frequencies the window transmission dominates.

Two efforts were conducted to better understand the propagation pattern of signals from window and seam apertures to the antennas placed at different locations on the top centerline of the aircraft. First, surface path loss mapping experiments were conducted at VHF, UHF and microwave. This test was primarily aimed at understanding why moving the GPS antenna from nose to tail had so little effect. Second, numerical electromagnetic modeling was undertaken at VHF and UHF.

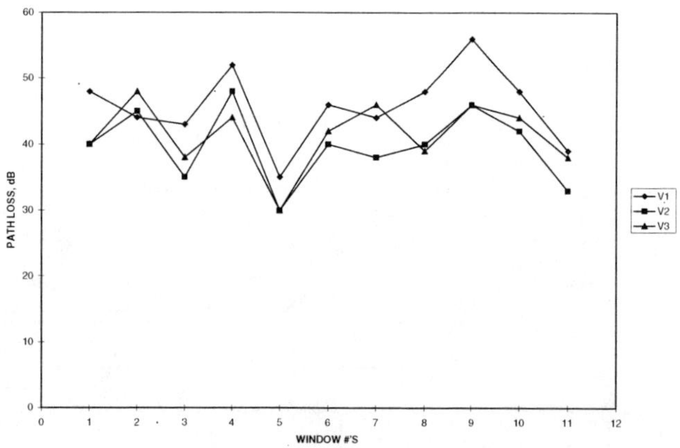

Figure 1a - VHF Com#1, Horizontal Polarization With All Windows

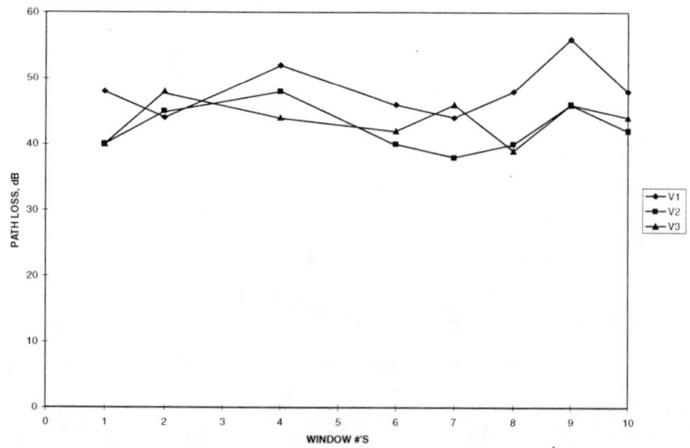

Figure 1b - VHF Com #1, Horizontal Polarization Without Windows 3, 5 and 11

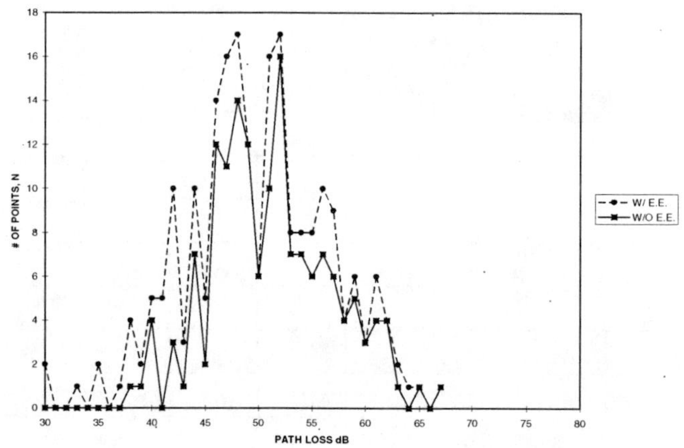

Figure 2a - VHF Histogram Distribution

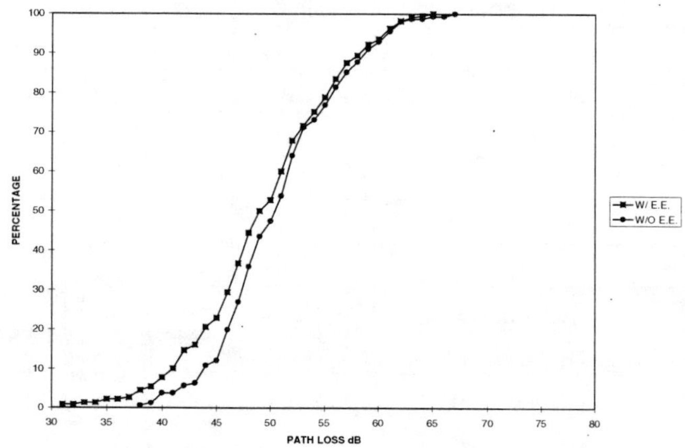

Figure 2b - VHF Cumulative Distribution

Measurements were made at points along the aircraft top centerline, the window line, and at a line midway between. The illuminating tuned dipole antennas were positioned along these lines from the cockpit region back to window 3. The results for two VHF frequencies, V1 and V3, are shown above each physical measurement point in Figure 4. An inspection of the values on the figures immediately shows that the path loss on a path around the circumference is much larger than on a path down the long dimension of the cylinder. Modeling of this effect was done and is presented in the companion paper [ref. 4]. The data were reduced to produce gradients of path loss in dB per foot along circumferential and longitudinal paths. These data are shown in Figure 5.

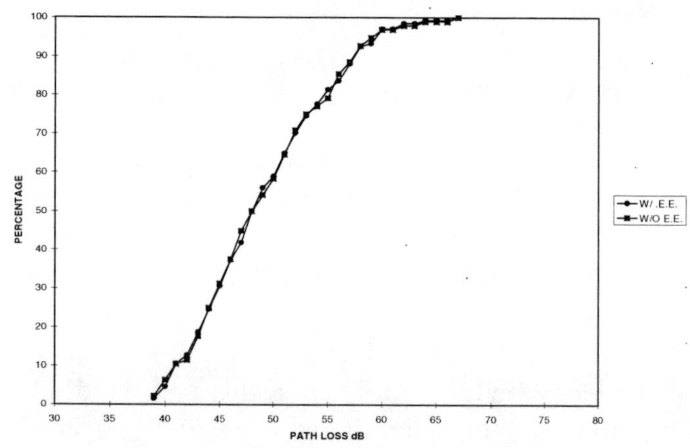

Figure 3 - UHF Cumulative Distribution

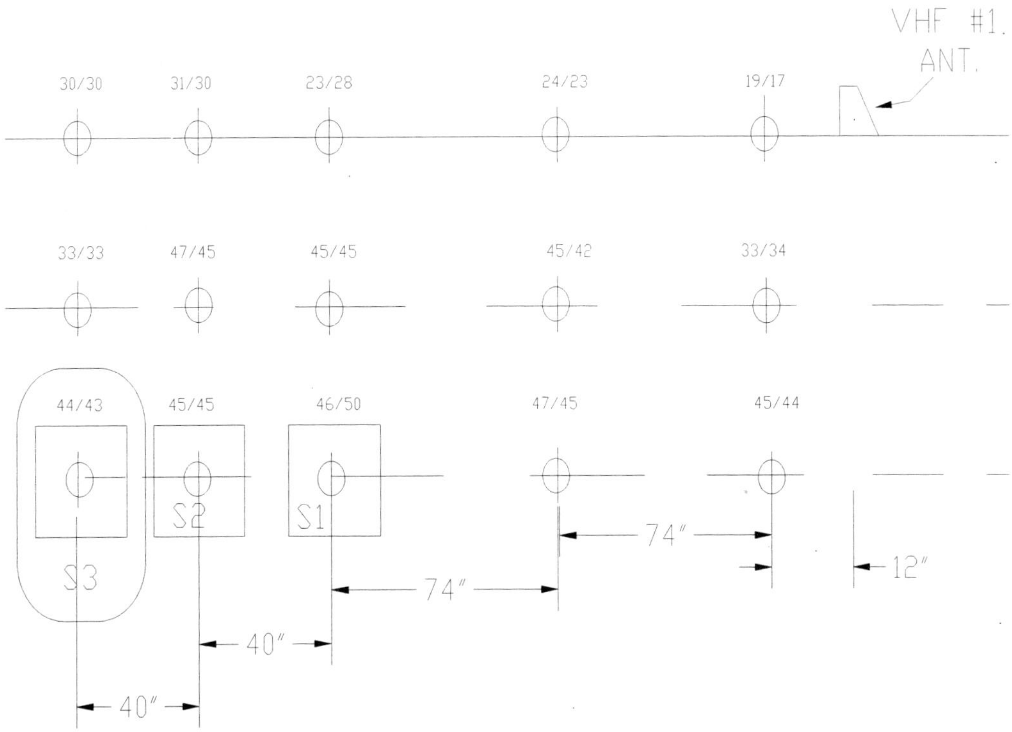

Figure 4 - VHF External Path Loss Measured on Curved Cylindrical Surface

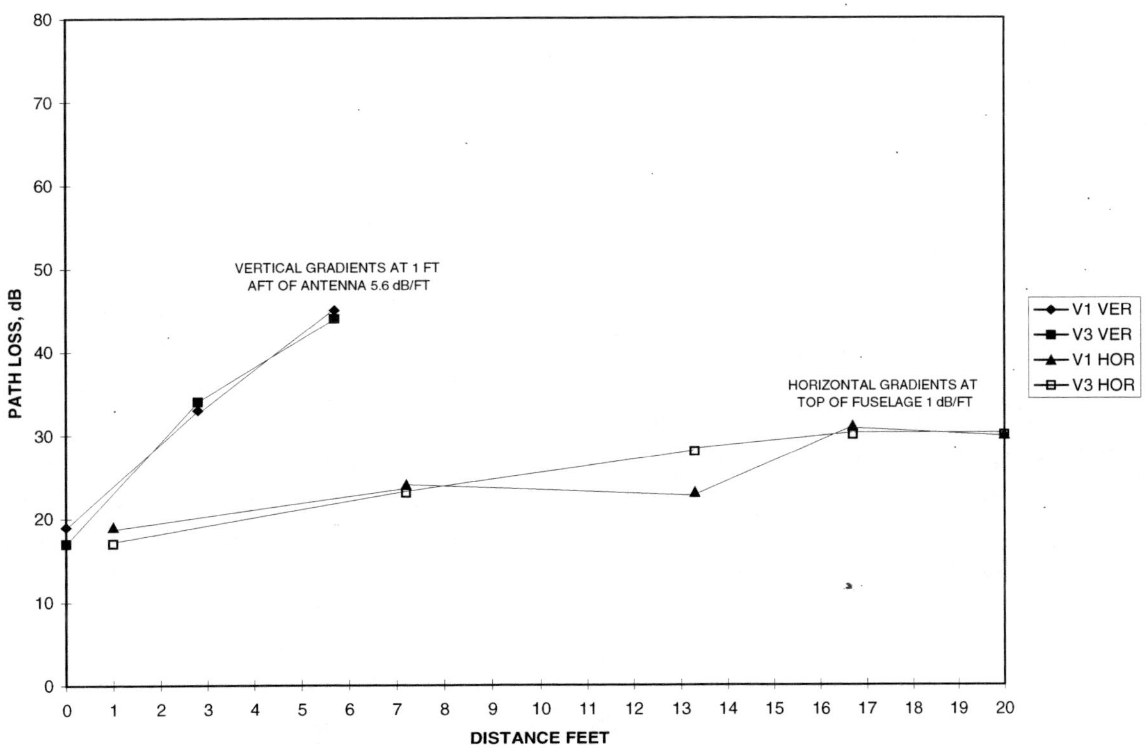

Figure 5 - VHF #1 Vertical Polarization Gradients

There is a consistent factor of about 5 times more loss (in dB) around the circumference than along the long dimension of the cylinder. Since the Communications and Navigation antennas are placed on the top or bottom centerlines of the aircraft the minimum loss is defined by the aircraft diameter and the window to top centerline distance. Since commercial passenger aircraft are designed to have the windows at eye level and the passengers hips at the widest part of the cylinder, the window to top centerline grows in proportion to diameter.

Commercial airline diameters are from 10 feet (ft.) for regional commuter aircraft, 12 ft. for standard narrow bodies, about 18 ft. for standard wide bodies and 21 ft. for the B-747. The B-767 is about 15 ft. in diameter. This variation in diameter is about a factor of 2. The variation from the B-727 at 12 ft. to the largest jet at 21 ft. results in a window to top center line change from 7 ft. to about 14 ft.. By linear extrapolation this might cause a 35 dB increase in path loss at GPS, based on a gradient of 5 dB/foot. This is probably an over estimate of increased loss since the radius of curvature affects the gradient, the less curved B-747 would have less gradient than measured on the B-727. Modeling results can be found in companion papers by Devereux, Fuller and Archambeault [ref. 4 and 5].

Fields present at the top/bottom Communications and Navigation antennas from an internal RF source are generally due to coupling through the window apertures and wave diffraction around the external curved surface of the aircraft. A Finite-Difference Time Domain (FDTD) model was created to determine and bound the amount of path loss change due to aircraft size on this external diffraction path loss[ref. 5]. In the FDTD model, the RF source was located in the center of a window and the receive antenna was directly above the window on the top of the cylinder.

These FDTD models were created for the CV-580 scale fuselage, the narrow body, the standard body, the wide body (747), and intermediate body (767) aircraft styles. Figure 6 shows the results of the different sized aircraft, each behaving with a maximum amplitude variation over the frequency range of 10 dB on the same size cylinder. This Figure indicates that RF path loss data measured on smaller scale aircraft is only 15 to 22 dB less at VHF and UHF and only 10 to 12 dB less at microwave frequencies then data for the largest of aircraft cylinder diameters.

In any case, path losses will be larger for the bigger diameter aircraft than measured on the FAA's CV-580 and B-727 aircraft. The modest additional path loss increase experienced by moving the antenna along the top centerline results can be seen to be due to the gradient in the longitudinal direction being small which is dominated by the much larger path loss gradient effect of propagation around the cylinder circumference.

DME Antenna Leakage

The path determination test performed during Phase I of the research Grant showed the DME antenna's least path separation was 5 to 12 dB between the out the window path and the internal coupling path. A detailed test of internal path loss was conducted by moving the source antenna to various locations around the antenna and its cables inside the aircraft. It was found that at the underside of the antenna a strong signal, only 5 to 12 dB below that induced out the window was induced into the antenna. It was determined that the leakage was not through the cable or connector. Instead, the DME antenna mounting plate and bolts were allowing a strong signal to leak through the plate and bolts and directly couple to the antennas elements. We did not assess the technical or operational impact of this leakage.

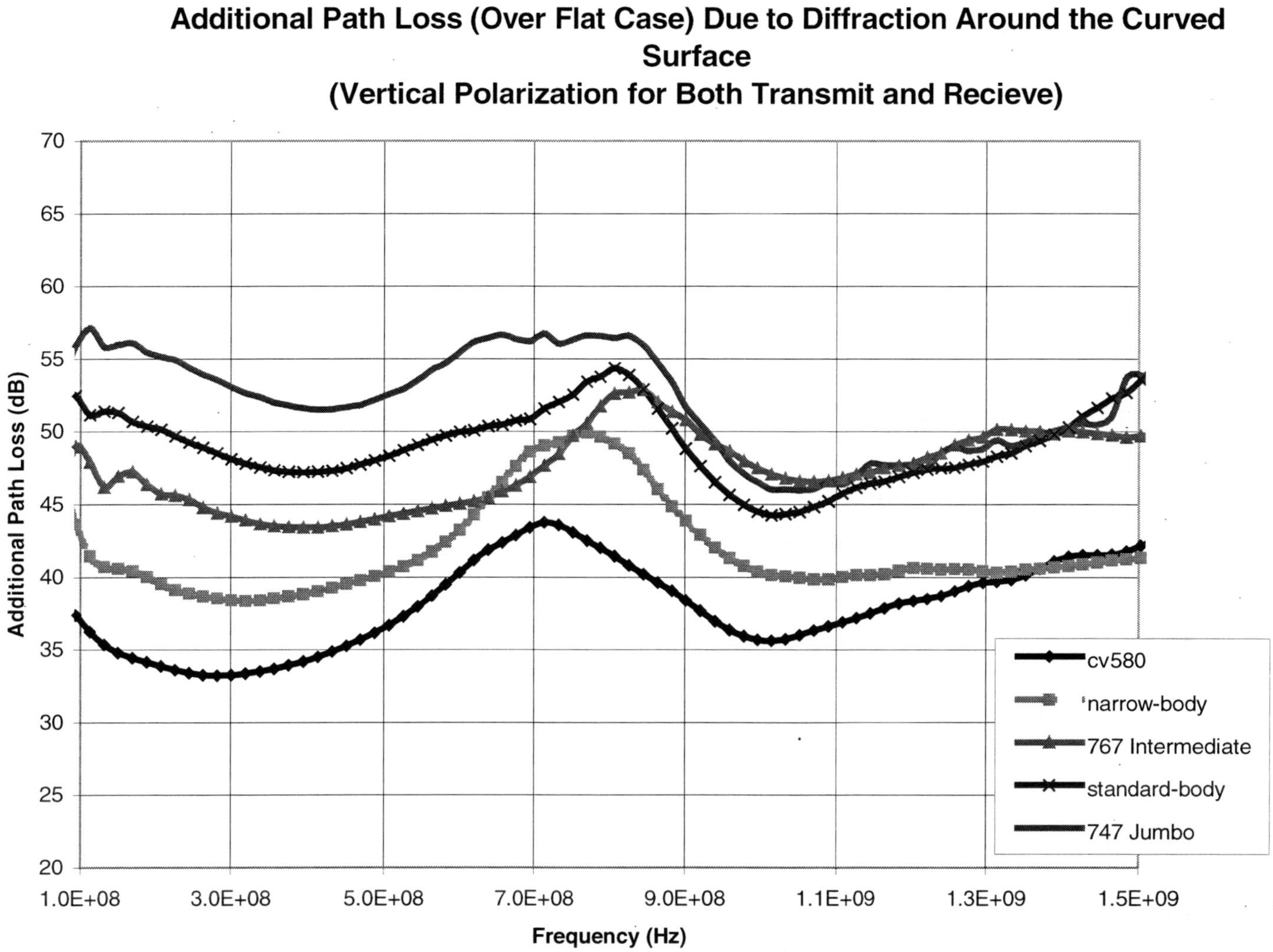

Figure 6 - Diffraction Path Loss for Various Size Aircraft Bodies (Vertical Polarization)

Conclusion

The dual discoveries that emergency exit seams (and presumably passenger and cargo bay door seams) allow low frequency VHF signals to be greater than those out the passenger windows, and that the wave/signal propagates with much greater loss around the aircraft's circumference than down its long dimension have important implications for estimating the minimum path losses for wide body and jumbo jets.

The larger diameter aircraft will have greater path loss, in general, due to the longer paths around the circumference. However, since the larger aircraft have smaller windows but larger door seams than the CV-580 test aircraft the minimum path loss, at VHF, will be less than expected for larger aircraft due solely to the distance factor. Modeling has played a major role in our understanding of path loss in separating those factors out and in estimating the changes in path loss due to seam apertures and window diameter sizes.

Since this modeling process is only as good as our ability to account for physical features of the aircraft, it is of first priority and importance to conduct validation experiments of all sizes of modern "in-service" aircraft as soon as possible to determine the range/limits of path losses to CNI system antennas for these operational aircraft.

The significance of the threat from passenger cabin sources depends critically on aircraft size. Whether or not the increased path loss will change the susceptibility conclusions depends on the amount of minimum path loss. This can only be determined by a carefully conducted test of "in-service" aircraft. Likewise detailed path loss tests from the fore and aft cargo bays to the avionics bay is vitally needed to assess flight control and data handling system susceptibility to a cargo based source for larger "in-service" aircraft.

We have evaluated modeling codes, developed models and shown their importance in understanding aircraft coupling and in estimating path loss for variations of aircraft antenna configurations [ref. 6]. It is equally important to use "in-service" aircraft data to numerically validate and calibrate these models so that various mitigation options can be evaluated quickly and inexpensively using these models.

Acknowledgment

This research has been performed under the FAA Aviation Research Grant Program, Grant No. 97-G-001. We would like to thank Mr. Ray Schillinger of the FAA William J. Hughes Technical Center, Code AAR-520, for his support throughout this research effort.

References

1. R. W. Devereux and R. Schillinger, *Electromagnetic Coupling Paths Evaluated From Inside a Commercial Aircraft,* 15th Digital Avionics Systems Conference (DASC) Proceedings, October, 1996
2. R. W. Devereux, G. L. Fuller, R. Schillinger, *Further Analysis of the CV-580 and B-727 Aircraft RF Coupling Experimental Data,* 16th DASC Proceedings, October, 1997
3. *CV-580 RF Coupling Validation Experiment Report*, Veda Report Number 79689-96U/P30041, November 15, 1996
4. R. W. Devereux and B. Archambeault, *Electromagnetic Analytical Modeling Used to Aid Aircraft EMI Data Analysis*, 1998 IEEE International Symposium on Electromagnetic Compatibility (EMC), August 1998
5. R. W. Devereux, B. Archambeault, G. L. Fuller, *Assessment Of Aircraft Internal EMI Using Electromagnetic Analytical Modeling Codes*, 17th DASC Proceedings, November 1998
6. R. W. Devereux, B. Archambeault, G. L. Fuller, *Assessment Of Analytical Codes For Use In Modeling Aircraft Onboard EMI Threats*, 16th DASC Proceedings, October 1997

INVESTIGATIONS INTO A SYSTEM FOR THE DETECTION AND LOCATION OF POTENTIALLY HARMFUL RADIATION FROM PORTABLE ELECTRONICS CARRIED ONBOARD AIRCRAFT

Albert Helfrick, Embry-Riddle Aeronautical University, Daytona Beach, Florida
Anthony Wilson, Federal Aviation Administration, William J. Hughes Technical Center, Atlantic City International Airport, N.J.

Abstract

The problem of interference from portable electronic devices, PEDs, to avionics systems has been known for more than 30 years. This problem has been periodically studied but other than identifying the airlines as the responsible party for policing the situation, no hardware has ever been designed to combat the problem. The latest report on the subject from the RTCA, DO-233 suggested a hardware detection system as a possible safeguard against future problems. This paper describes the preliminary design of such a detection system.

Background

Radio frequency interference, RFI, is a problem as old as radio. The first RF interference problem was engine ignition systems that, in the early days of aircraft radio, was ranked near the top of the list of difficult hurdles. Noise from the spark plugs rendered radio receivers installed in aircraft for communications or navigation, ineffective. RFI did not disappear with the fully shielded ignition system. The new sources of trouble were the aircraft's own electronics equipment. This prompted the development of standards and test procedures and standards organizations such as the Radio Technical Commission for Aeronautics. Through testing and controlling all of the systems in an aircraft, compatibility is insured.

A new player in the RFI game arrived in the late 1950s; the portable electronic device, or PED. The transistor, then some thirteen years old, was making portable radios affordable and popular. Long before airlines offered sophisticated onboard entertainment systems, passengers toted their own. The RFI compatibility specifications and required tests in existence at the time had no bearing on the new interference. The RTCA Executive Committee formed a special committee, SC88 to investigate the problem. In 1963 the report of SC88, DO-119 was published.

The problem of PED interference was recognized by the first RTCA investigation, but was determined to be not of a serious nature. The FAA responded to the report by making a ruling that the operator was responsible for insuring no harmful interference would occur.

In 1983 the first portable computers were beginning to appear. The air carriers, responsible for insuring safety, categorically prohibited passengers from using these new portable computers. Computer manufacturers suggested that there was no need to have such a strict requirement and the FAA requested the RTCA to form another committee to investigate the problem.

The first meeting of SC-156 was held in December of 1983 and the terms of reference were modified to include a minimum operational performance standard, MOPS. SC-156 produced the document, DO-199, *Potential Interference to Aircraft Electronic Equipment*

From Devices Carried Aboard.

In May 1992, the FCC, again, requested the RTCA to revisit the problems of portable electronic devices carried aboard aircraft. PEDs are proliferating at a tremendous rate with the appearance of a broad range of portable computers, video camcorders, and electronic games. Complexity was growing and digital clock frequencies were rising.

The special committee tasked with the PED problem, SC-177, finished its work and published a final report in late 1996. The report document, DO-233, concludes that the problem of interference from PEDs in large aircraft is not an immediate threat but should be monitored. A number of recommendations were proposed including research into the feasibility of a PED radiation detector system. It is this recommendation that is addressed by this paper.

The design of a detection system would be an important adjunct for allowing continued or increased use of PEDs which is the desire of the traveling public. If a serious incident should ever be attributed to a PED, one solution to the elimination of further dangers is the total cessation of PED use aboard aircraft. Aircraft equipped with PED radiation detectors could possibly be exempt from increased PED restrictions should a PED-induced incident occur.

Introduction

A system is desired that can detect and locate RF radiators within the cabin of an aircraft. This system should be capable of detecting such radiators with a high level of probability, and, more importantly, not produce an unacceptable level of false alarms. At this stage of development, there are no paradigms for the accuracy of the position indication of the radiator or the maximum level of false alarms. To satisfy the need for a design goal, the accuracy of location is plus or minus one row of seats in a conventional air-transport aircraft, and the probability of false alarm as one per million flight hours.

There are a number of issues that must be considered in the design of a system of this nature, such as the type and location of RF sensors, the method of alarming, the display technology and cost and weight considerations. Specifically, there are important issues such as the method of spectrum surveillance, discrimination of radiators and background noise and the algorithms for generating an alarm.

Without a doubt, the most difficult part of this system design is the effect of the background radiation or noise. If it were not for noise and interference, detecting RF radiation would be a simple task. The background noise level in an aircraft is troublesome for a number of reasons. They are:

The actual background radiation environment is not known.

In spite of all the work that has been done relative to RF compatibility, there are very little data that will definitively characterize the background RF environment in a modern air transport aircraft. Measurements that are made of one aircraft type do not represent the levels in another airframe. Furthermore, even between similar airframes, the electronic equipment installed therein can be significantly different. There are DO-160 specifications that will define the levels that avionics must endure. But these DO-160 specifications exclude on-channel signals. There are studies and evidence to show the background radiation is complex, intense, and variable. There are very little data as to the actual background radiation levels available for, perhaps, a specific aircraft.

Aircraft systems include transmitters.

This is one part of the background radiation environment that is known. Transmission on specific frequencies such as

1090 MHz is expected. Other transmissions in the VHF communications band are also expected. The difficulty is twofold. The first difficulty is a transmission by an aircraft transmitter is not a reason for an alarm, while transmission from a PED carried aboard is cause for an alarm. This requires the PED detector to discriminate between legitimate onboard transmitters and PEDs. The second problem associated with onboard transmitters is the level of radiation is very high which will require the detection system to have a wide dynamic range.

The vulnerable spectrum covers a very wide range of frequencies.

The lowest frequencies currently in use in aircraft are lightning detectors centered in the 50 kHz region. Above that is LORAN-C and the spectrum continues up to X band for airborne weather radar. There are relatively broad gaps in which no airborne system would be vulnerable. On the other hand, there is at least one radio-based avionics systems operating in the LF, MF, HF, VHF, UHF, and SHF frequency ranges.

Much of the cabin environment is due to signal ingress.

Because the aircraft is a mobile platform, the environment outside the cabin is extremely variable and is a major factor in the internal environment. Although, in most cases, the cabin is surrounded by a conducting structure, there are plenty of egress/ingress points, such as windows, that couple in a significant amount of the external RF fields.

Design Considerations

Item three is one area where compromises may be made to improve the efficacy of the PED detection system. PED detection can be concentrated to specific frequencies that are particularly vulnerable. The frequency capability of the PED detector will cover most of the spectrum used for avionics. As it was pointed out, radio-based avionics systems are well distributed throughout the radio spectrum. Signals in the LF, MF and HF range tend to be very strong and since PEDs are poor radiators in this portion of the spectrum, a detection system may skip this part of the spectrum. Furthermore, avionics systems operating in this frequency range such as LORAN-C, lightning detection, and non-directional beacons, NDBs, are seldom used aboard air transport category aircraft. HF communication is used for transoceanic flight but was considered to be a low risk system because the interference is heard by an operator. In the case of HF ACARS, transmissions are not involved with safety of flight. Only select portions of this frequency range will be analyzed, but, the hardware will cover a total frequency range from 50 MHz to 2 GHz.

Detection algorithms will concentrate on vulnerable frequencies or bands and, thus, it is necessary to identify the vulnerable frequency ranges. It is clear that certain frequencies are particularly vulnerable such as GPS L1 and L2 at 1575.42 and 1227.6 MHz respectively. In other cases, limited bands of frequencies are vulnerable such as 329 MHz to 335 MHz, which covers the glide slope frequency range and 108.10 to 112.95 MHz, which covers the localizer.

The glide slope frequency range covers 4 MHz, the localizer 5 MHz and GPS represents two discrete frequencies. Of the entire 50 MHz to 2 GHz frequency range covered by the hardware, only 9 MHz constitutes the most vulnerable frequencies. Even if the remainder of the VHF navigation band was included, the range of vulnerable frequencies would be 14 MHz. It is clear that if only vulnerable frequencies were monitored, the false alarm rate due to background could be reduced. Furthermore, the level of radiation will be weighted based on the typical signal levels and the sensitivity of the affected system. The detection system must be frequency agile in order to concentrate on the vulnerable frequency ranges.

One very useful input, when available, would be the active VOR, localizer, ADF etc. frequencies. If it is known what frequencies are currently in use, this information may be used to narrow the monitored frequencies and even further reduce false alarms. Aircraft with abundant use of standard ARINC data busses will permit this level of sophistication.

Simply detecting a signal will not provide sufficient reason to provide an alarm, even if the signal has been detected at one of the vulnerable frequencies. Significant signal analysis will be required to insure reliable alarms. As a simple example, a pulse of energy occurring on a glide slope or localizer frequency may not constitute an alarm condition. The bandwidth of the localizer and glide slope is below 200 Hz. Occasional narrow pulses are not likely to cause serious problems in the highly-damped narrow bandwidth localizer and glide slope systems. One signal analysis that will be employed is the ability to learn. The background radiation can, literally, be learned. As an example, since the PED problem has been identified as not being an immediate high threat by SC 177, a PED detector installed in an aircraft literally has the time to learn its own environment. Thus, when installed in an aircraft, the first 100 flight hours could be used to establish a background on the environment. The alarm thresholds may start at high levels and be adjusted as the system gathers background data. There are other situations where the alarm would be disabled, such as on the ground, where the system may continue to gather background data. Essentially, the system will alarm on situations that are significantly out of the ordinary.

Antennas

As it was discussed previously, the detection system hardware is to cover as much of the RF spectrum as practical. The hardware should be able to accommodate the spectrum from 50 MHz to at least 2 GHz. This frequency range can be handled reasonably by readily-available cost-effective hardware. One area where hardware is not off-the-shelf is in the area of sensors.

Preliminary Laboratory Prototype of Dipole Sensor

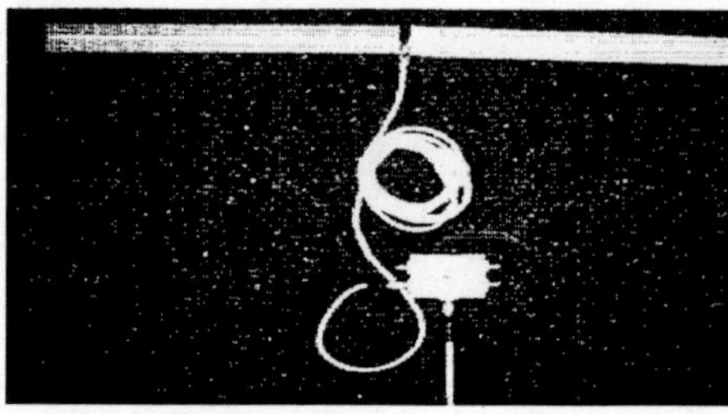

Figure 1

There were a number of potential sensor technologies that were considered. The first was leaky coax. This choice represented a very inexpensive method with a large number of products available and a well-known technology. Leaky coax is usually used for distributing signals into otherwise RF-free areas such as for cellular telephones. There were a number of reasons that this technique would not be preferable. Two major reasons were; 1. Available leaky coax was rather large and heavy, and 2, the leaky coax did not lend itself to a method of signal location.

A second consideration was a planar antenna applied to the aircraft window. This has some merit in that the most serious egress point for radiation from the cabin to the outside and to the avionics radio antennas would also be the site of the sensor. A number of conductive films were investigated and methods of providing cable attachment were considered. Although there were advantages to this sensor, the cost of retrofitting each window in an aircraft, 144 in a B 747, is high.

The calculated sensitivity of the detection system for the planar window sensor ranged from +33 dBμv/m for a monopole at 100 MHz, and a best case sensitivity of +2 dBμv/m at

300 MHz for a solid sheet similar to a patch antenna. The sensitivity is calculated on the proposed RF distribution system and detection algorithms and are for purposes of comparison.

The third sensor is an impedance-loaded dipole. This sensor has the advantage that it is thin and may be mounted at strategic locations that will provide good location ability. The

B-737 PSU/ Fillets

Figure 2

dipoles may be manufactured with adhesive backing that may allow the sensors to be applied at locations within the cabin without high retrofit costs. An example of this antenna is shown in Figure 1.

A dipole can be broad banded through loading. Sensitivity can be improved, at the expense of size by making longer dipoles. The length of the dipole is limited by physical constraints of the possible mounting areas.

Several mounting areas were considered; seat backs, arm rests and the overhead passenger service unit or PSU. An example of a PSU is shown in Figure 2. The coverage areas were simulated as shown in Figure 3 and the PSU allowed for the least expensive installation without serious compromise of sensitivity. Two lengths of loaded dipoles were considered. A 0.35 meter dipole and, a slightly longer, 0.50 meter version were analyzed. Both dipoles will fit in the normal PSU.

The sensitivity of loaded dipoles are shown in the chart below. As before, the sensitivity is based on performance in the proposed system.

Antenna	Sensitivity at 100 MHz	Sensitivity at 300 MHz
0.50 m dipole linear	17 dBμv/m	11 dBμv/m
0.35 m dipole linear	22 dBμv/m	17 dBμv/m
0.35 m with end loading	11 dBμv/m	6 dBμv/m
0.5m conical	13 dBμv/m	8 dBμv/m

The location determination will be made relative to magnitude and not rely on any sort of phase measurement. It was decided that maintaining any degree of phase coherence within the cabin of the aircraft would be virtually impossible. Therefore, detection algorithms will be dependent on signal strength and time.

The overhead PSU was selected as the best site for the dipole sensors. For a large air transport aircraft as many as 64 sensors would be required. In order to reduce the cost of retrofit and weight, four antennas are time multiplexed on a miniature coaxial cable. This is accomplished by use of low-loss, broadband GaAs MMIC switches. In addition, to preserve system noise figure, preamplifiers are used following each MMIC switch.

Detection System

The 50 – 2000 MHz spectrum is converted to a 10 MHz IF using a frequency agile local oscillator. The local oscillator heterodynes a DDS and a phase-locked loop to provide a local oscillator range of 40 – 1990 MHz, as show in Figure 4.

After down conversion, the resulting output is sampled at a rate of 25 to 30 MSPS. A FFT is performed on the resulting samples.

Four antennas are shared by the down conversion, sampling and FFT. In a large configuration, 16 down converters and samplers will be required.

The detection algorithms include the use of an optimum detector in white noise and subjecting each frequency block to a whitening process. The detection algorithm will measure the total power in each of the N whitened sub-bands. If any sub-band shows a power that exceeds a lower limit threshold, M additional measurements are made and if the total energy exceeds a higher threshold, detection is indicated.

The whitening process is adjustable and makes use of any a-priori knowledge, such as the VHF communications frequency currently in use. The current VHF navigation frequencies, when available, may be used to improve the detection algorithm by dynamically adjusting the detection threshold.

When an aircraft has an ARINC bus available for communications frequencies, this may be used to discriminate against the aircraft's own communications transmitters. Other onboard transmissions are expected such as transponders and DMEs. These transmissions can be categorically discriminated against because of their pulse characteristics. However, these onboard transmitters can cause overload and intermodulation products.

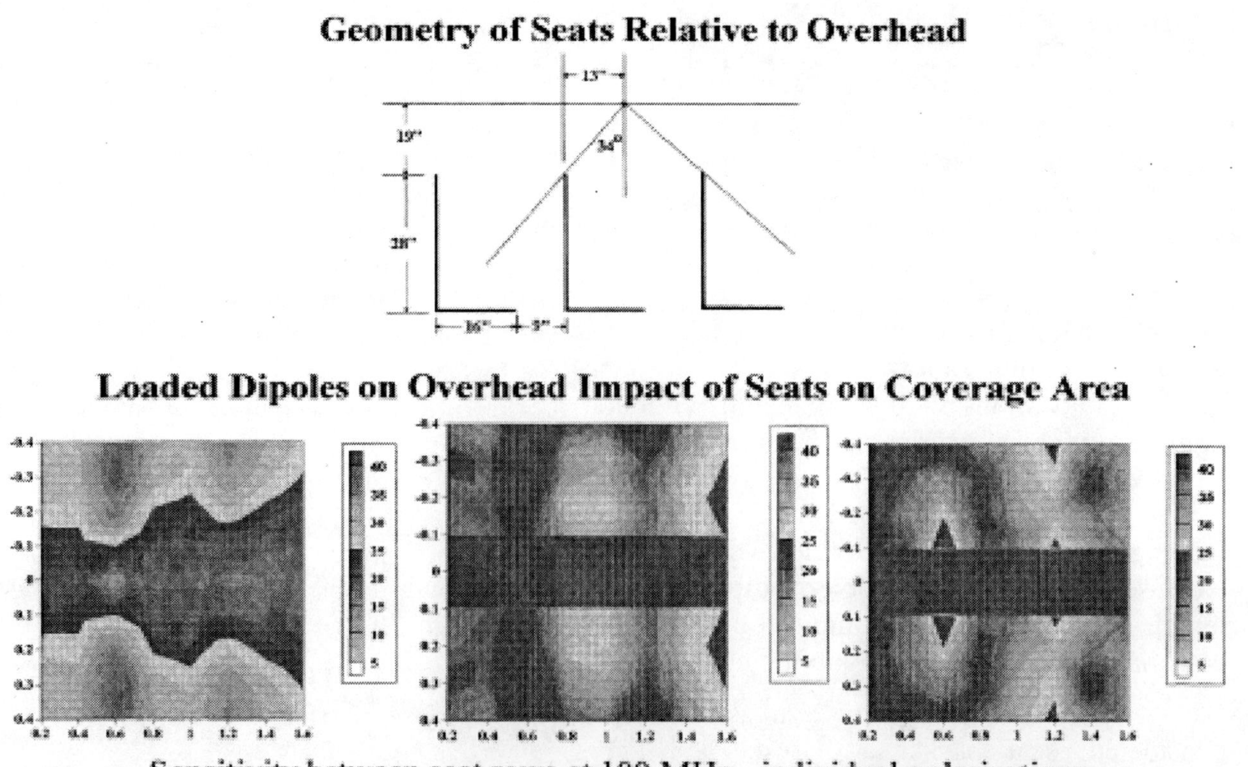

Sensitivity between seat rows at 100 MHz - individual polarizations

Figure 3

PED Detector Project: Major Subsystems

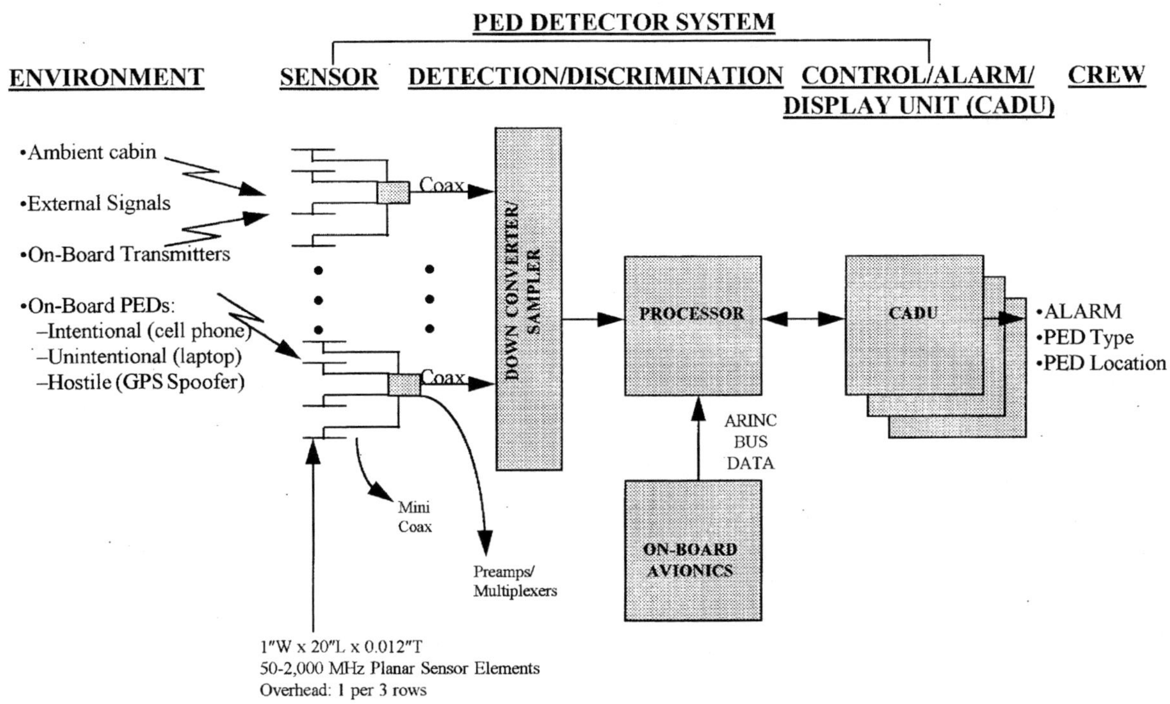

Figure 4

A method of discriminating onboard transmitters from the aircraft's own communications transmitters can be implemented with software. These transmitters are in the VHF communications band and include both voice and data, (ACARS or future VDL). Any transmission that emanates from outside the cabin should not produce an alarm. This would include the aircraft's own transmitters and radiation from high powered broadcast stations, particularly TV and FM. When a potentially interfering signal enters from outside the cabin, the field intensity will exceed threshold levels from many antennas and over a wide area of the cabin. Another possible tool to discriminate from high powered signals from outside the cabin is to provide an external antenna. The external radio frequency spectrum may be compared to the internal RF spectrum and those signals that are stronger external to the cabin than within the cabin will not cause an alarm.

For the case of PED intentional radiators, the field intensity levels are orders of magnitude above the typical PED. Any field intensity that exceeds about +80 dBμv/m exceeds the threshold for an alarm when the aircraft is airborne. The issuance of an alarm on the ground may be inhibited due to the fact that ground personnel often use hand-held two-way radios from within the aircraft. The criterion for an alarm from an intentional radiator after the aircraft has been pushed back from the loading ramp is based strictly on amplitude and does not depend on frequency. Therefore, whitening and dynamic thresholding are not employed for very strong signals. However, the threshold level must be set to be greater than any background radiation from on-board transmitters.

Although the most likely radiator is a narrow band signal from a PED or a broadband signal from a digital PED, it should not be forgotten that deliberate, or hostile, jammers may be encountered. The detection algorithms must be capable of detecting both spread spectrum and frequency hopping radiators.

Future Work

The next phase of this project will involve actual design and build of hardware to be evaluated. It is expected that the first installed systems will not actually cause alarms but evaluate the background radiation. The PED problem has caused three special committees to be formed and the lack of data has been a common deficiency for each committee. The first application of the PED detector is to gather much needed data concerning the background levels and the occurrence of potentially harmful PED radiation. After this additional data are generated, alarm thresholds may be set and the effectiveness of locating PEDs may be evaluated. Finally, an entire fleet may be outfitted with the detector. Since the conclusion of SC-177 was the problem of PED interference does not constitute an immediate threat, this gradual phase in will be very effective in developing a system that is free of false alarms, yet be very effective.

This project was funded by the FAA under an SBIR awarded to MegaWave Corporation, Boylston Mass.

Failure Tolerance of Sensors in Flight Control Systems in Electro-Magnetic Environment[*]

Jon Suh, B.C. Chang, A.C. Lau
Department of Mechanical Engineering and Mechanics
Drexel University, Philadelphia, PA. 19104

Abstract

High Intensity Radiated Field (HIRF) conditions may cause sensor and/or controller failure in flight control system. In this paper, a flight dynamics imitation scheme is proposed to establish the failure tolerance of sensors in Electro-Magnetic Environment (EME).

1. Introduction

The flight control systems in newer generations of civil aircraft use electronic flight control or Fly-By-Wire (FBW) technology [1], and the safety issue of electronic failure caused by Electro-Magnetic Environment (EME) has been of much concern. The increasing use of composites in aircraft bodies makes them more susceptible to such EME disturbances. The sensors are influenced by EME when there is lightning, High Intensity Radiated Field (HIRF) associated with radar and radio frequency transmitters, and/or nuclear electro-magnetic pulses generated by military ground equipment [2-4]. According to a study, even a small radio frequency emissions from carry-on electronic devices affect avionic systems as well [5].

Electronic sensors form an important sub-system of the FBW and other avionic systems. The sensor malfunction induced by EME can affect flight stability and operational performance. In a typical longitudinal flight control system, the sensors measure the output of the plant and convert them into electrical signals, which in turn are sent to compare with the reference inputs that represent the desired aircraft speed and angle of attack. The difference of the reference inputs and the measured outputs are used as the inputs to the controller.

The controller is an electronic system or a computer designed to achieve closed-loop stability and desired performance. But during the Electro-Magnetic (EM) interference, the sensor output signal may be so contaminated that the incorrect measured values would mislead the system to instability. Thus, in order to guarantee robust stability and robust performance, i.e., a safe and comfortable flight, the control system needs to be improved so that the closed-loop system remains stable and keeps states within the range of desired performance.

To ensure stability and a desired performance during HIRF interference, aircraft could be insulated from EM induced disturbances with special material. However, it is not feasible in term of affordability and compatibility for the existing airplanes. It is very costly, and the space needed for implementation may not be available. On the contrary, the proposed dynamics imitation scheme can be easily implemented on the existing airplanes to improve their stability with respect to EME disturbances. The basic concept of the dynamics imitation scheme is to construct a model which imitates the plant dynamics to generate a substitute measurement during sensor failure. The imitators can be shielded from HIRF interference and can be added onto the existing flight control systems without major reconfiguration and design changes. The cost of such an exercise would be minimal making it

[*] This research was supported by NASA Langley Research Center under Contract NCC-1-278.

affordable. In this paper, a plant dynamics imitator is augmented to the controller in order to improve the robustness of the flight control system during the period of sensor failure.

2. Sensor Failure

During the EM interference the sensor signals are corrupted, producing unreliable measurement signals. These measurements may mislead the controller resulting in unsatisfactory system response; therefore, in such a case they should be discarded.

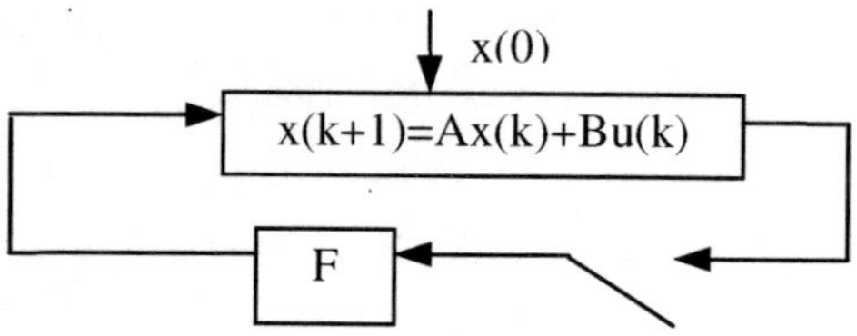

Figure 1. System dynamics with sensor failure.

For the purpose of illustration, consider a simple system model [6] shown in Figure 1, in which A=1.2, B=1.8, F=-.35, and the initial condition x_0=1. Normally the switch is closed; it is open whenever sensor failure occurs. The responses for both cases are shown in Figure 2. The simulation with feedback converges while the break in the closed loop makes the response divergent as the system is deprived of signals for error correction.

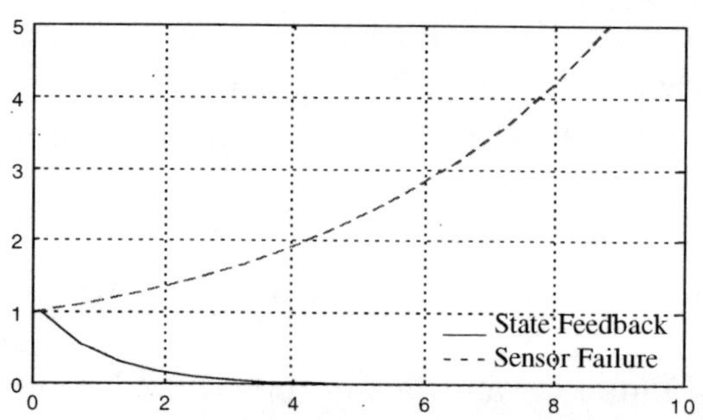

Figure 2. Responses with/without the feedback

3. Dynamics Imitation

Now again, consider the system experiencing EM interference as above. The sensor fails and the system will lose its closed loop stability as shown in Figure 2. However, if the control system can somehow imitate the dynamics of the real system by estimating the states, it is possible to regain stability of the system by forming a closed loop imitation. First, let's consider an ideal case. Assume that an estimated model is created exactly the same as the real one; i.e., $\hat{A}=A$, $\hat{B}=B$, $\hat{C}=C$, and z_0=x_0 as shown in Figure 3. Then the response of the system with imitated feedback is the same as that of the original with the state feedback. Hence, the stability and performance may remain intact under the influence of EME if such a plant dynamics imitator could be designed.

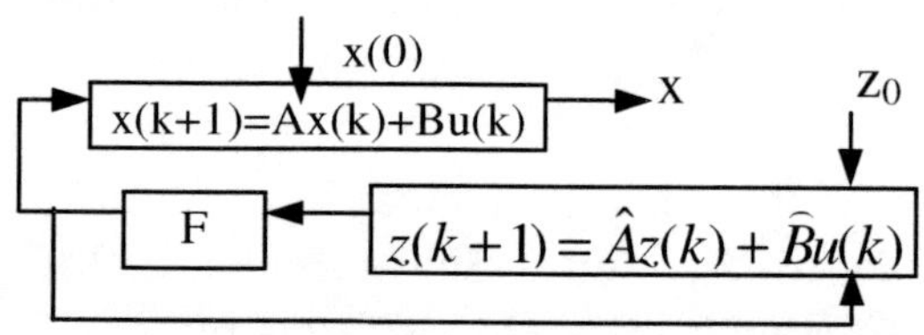

Figure 3. State imitation during sensor failure

Obviously in the real world, the imitator output will not be exactly the same as that of the dynamic system. However, the proposed state imitation scheme greatly enhances the survival opportunity of an aircraft during the period of sensor failure caused by the EME, even if the imitation is not perfect.

4. A Longitudinal Flight Control Example

In the following, the proposed state imitation is applied to a longitudinal flight control problem. Here, a linearized longitudinal aircraft dynamics model from Blakelock [7] with actuator dynamics is considered. The actuator transfer function is defined as [-1/(s+10)], and for simulation purpose, a disturbance model for HIRF interference is created to represent possible disturbances and noises.

The aircraft dynamics is represented in state space form as,

$$\dot{x}_a = Ax_a + B\delta_e$$
$$y = Cx_a$$

where the state vector x_a is $[u\ \alpha\ \theta\ q]^T$ in which u stands for the Forward Velocity, α the Angle of Attack, θ the Pitch Angle, and q the Pitch Rate. The input and output of the dynamic system are the elevator deflection angle δ_e and the Pitch angle θ, respectively.

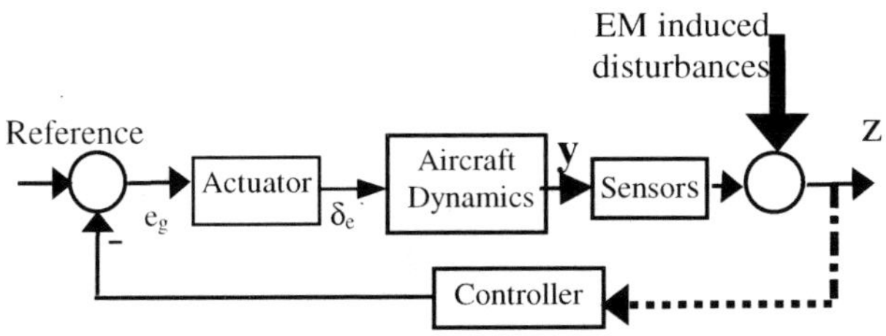

Figure 4. Typical Flight Dynamics in EME.

As shown in Figure 4, the actuator is located before the aircraft dynamics block and its dynamics is described by the following state equation.

$$\dot{\delta}_e = -10\delta_e - e_g$$

The actuator and the aircraft dynamics can be combined and represented by the following composite dynamic equations.

$$\begin{bmatrix} \dot{x}_a \\ \dot{\delta}_e \end{bmatrix} = \begin{bmatrix} A & B \\ 0 & -10 \end{bmatrix} \begin{bmatrix} x_a \\ \delta_e \end{bmatrix} + \begin{bmatrix} 0 \\ -1 \end{bmatrix} e_g$$

$$y = \begin{bmatrix} C & 0 \end{bmatrix} \begin{bmatrix} x_a \\ \delta_e \end{bmatrix}$$

or

$$\dot{x} = A_1 x + B_1 e_g$$
$$y = C_1 x$$

where the new state vector is,

$$x = [u \quad \alpha \quad \theta \quad q \quad \delta_e]^T$$

Based on this model, an observer-based controller is designed to achieve closed loop stability and some desired performance. The detailed design procedure is given in the Appendix.

5. Simulations and Results

For the closed-loop system with the controller designed in the Appendix, a step signal is applied to the reference input in order to change the pitch angle from the 0 initial condition to 5 degree. Under normal operation, the pitch angle response rises up from 0 with a very slight overshoot, and then converges to 5 degree as shown in Figure 6.

Again, the same simulation is repeated with an addition that an EM interference is introduced at the 150 seconds mark as shown in Figure 5. The EM disturbances are generated by summing several random numbers. The magnitudes of these disturbances are approximately 40 times of the reference input.

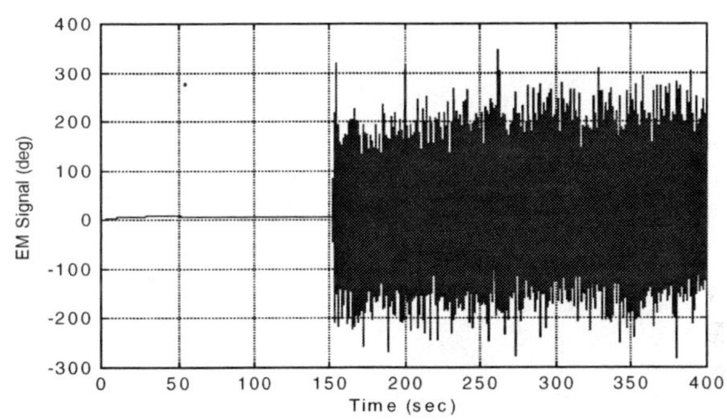

Figure 5. EM induced signals

Under the influence of EM interference, the pitch angle response starts to fluctuate and diverge as shown in Figure 6. Such response would mean a catastrophe in actual flight.

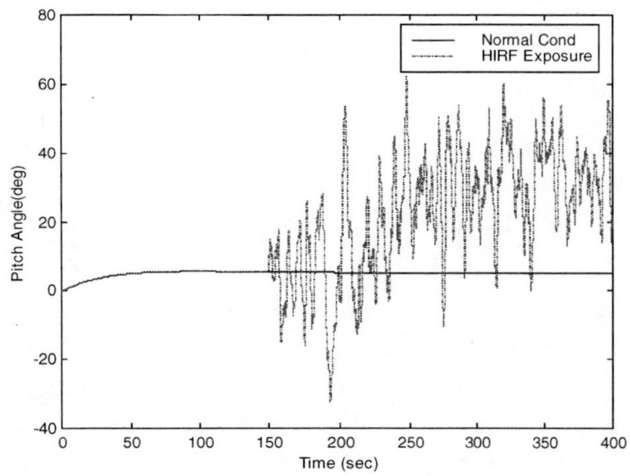

Figure 6. Pitch Angle responses with/without EM induced disturbances.

In order to prevent any such a catastrophe, a flight dynamics imitation scheme is employed. As shown in Figure 7, a flight imitator is constructed based on the plant model to generate a substitute measurement whenever sensor failure occurs. If the perfect plant model is available, the substituted measurement would be as good as the real one, and the pitch response would be the same as that shown in Figure 6 under normal condition. In reality, the flight imitator dynamics $(\hat{A}_1, \hat{B}_1, \hat{C}_1)$ is not the same as the plant dynamics (A_1, B_1, C_1). In the following simulation, we assume that $\hat{A}_1 = 1.1A_1$, $\hat{B}_1 = 1.1B_1$, $\hat{C}_1 = C_1$.

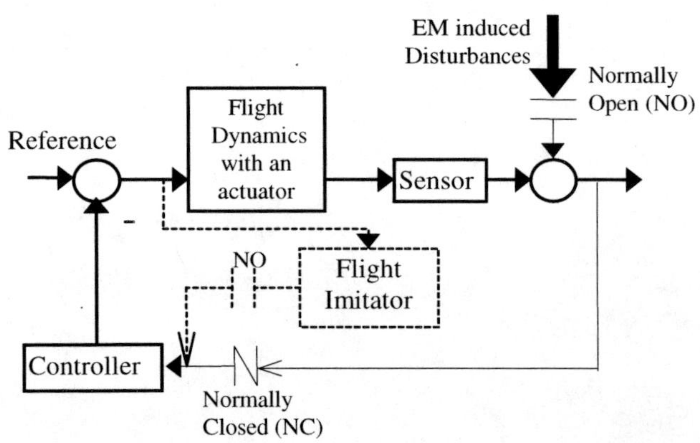

Figure 7. Modified Controller schematic to compensate EM induced problem

The design schematic to do the simulation is shown in the Figure 7. In the block diagram there are two types of switches: Normally Open (NO) and Normally Closed (NC). Under normal condition the sensor information is provided through NC; but when EM interference is detected, NO closes while NC opens. Thus, the controller uses the imitated dynamics signal instead of the false signals from the sensor under HIRF interference. As before, the reference input is a step function to command the pitch angle, changing from its 0 initial position to 5 degrees, and the EM interference starts at the 150 second mark as shown in Figure 5. The pitch angle response of the closed-loop system with flight imitator is shown in Figure 8.

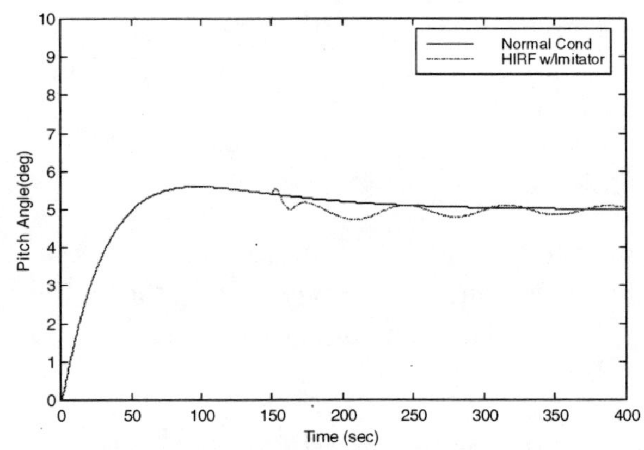

Figure 8. Pitch Angle responses of Aircraft with an imitator.

The results in Figure 8 show that the pitch angle response is much better than the diverging one shown in Figure 6. There are still a few, small deviations of the pitch angle due to difference between aircraft model and imitated dynamics. If the imitation could be modeled much closer to the actual dynamics, the response would be like the one under normal condition. It is seen that even with 10% perturbations in the imitator model, the system retains stability and stays very close to the desired response.

6. Conclusion

In this paper, a longitudinal flight control problem is used to demonstrate that a sensor failure could lead to a catastrophe and how it can be avoided by implementing the proposed flight dynamics imitation scheme in the system. It is verified that the proposed approach helps sustain stability of aircraft during the sensor failure.

References

1. D. Parry, "Fly-by-Wire," *Miltronics*, 1990, pp. 39-41.
2. C. M. Belcastro. "Monitoring Functional Integrity in Critical Control Computers Subjected to Electromagnetic Disturbances", *Proceedings of the 13th World Congress International Federation of Automatic Control,* San Francisco CA, Jun., 1996.
3. C. M. Belcastro. "Closed-Loop Hirf Experiments Performed on a Fault Tolerant flight Control Computer." *Proceedings of the 14th Digital Avionics Systems Conference*, Cambridge MA, Nov., 1995.
4. C. M. Belcastro, R. F. Fischl, "Upset Detection for Closed-Loop Laboratory HIRF Testing of Fault Tolerant Aircraft Control Computers," *Proceedings of the 14th Digital Avionics Systems Conference*, Cambridge MA, Nov., 1995.
5. Perry & Geppert, "Do portable electronics endanger flight?", IEEE Spectrum, Sept., 1996.
6. B. C. Chang,. "Robustness Issues of flight Control In High Intensity Radiated Field", *Proceedings of the American Control Conference*, Philadelphia PA, Jun. 1998.
7. John H. Blakelock., "*Automatic Control of Aircraft and Missiles.*", 2nd edition, A Willey-Interscience publication., 1991.

Appendix

The state-space representation of the linearized longitudinal aircraft dynamics is derived from the equations (eq. 1-59) in Blakelock. The flight model parameters are computed using the aerodynamic data for a F-94A, column 4 which aircraft is in cruise at 12,000 ft.

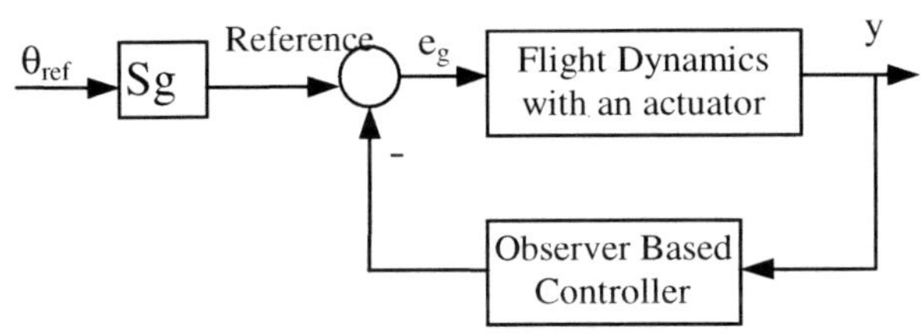

Figure 9. Plant Model with Observer based Controller.

OBSERVER BASED CONTROLLER

It is found that the controllability and the observability matrices of the system to be controlled are of full rank. Hence, the regulator gain (Rg) and the observer gain (Og) are chosen to place the poles as in the following table.

Open Loop Poles	Regulator Poles	Observer Poles
-0.8662+1.8643i	-10.0000	-100.0000
-0.8662- 1.8643i	-0.9000	-0.4200
-0.0031+0.0882i	-0.0210	-0.4400
-0.0031- 0.0882i	-0.0220	-18.0000
-10.0000	-0.8000	-16.0000

SET POINT CONTROL

After obtaining the observer-based controller, the set point gain (Sg) shown in

Figure 9 is computed so that the pitch angle response y will correspond to the command θ_{ref}.

NEW ELECTRIC FIELD EXPRESSIONS FOR EMC TESTING IN A REVERBERATION CHAMBER[1]

Galen H. Koepke, National Institute of Standards and Technology, Boulder, CO
John M. Ladbury, National Institute of Standards and Technology, Boulder, CO

Abstract

The statistical models used in reverberation chamber analysis have demonstrated good agreement with measured data from well-designed chambers. We use the models to derive several corrections to the field prediction formulas quoted in earlier publications. In this paper we will present a brief summary of these statistical models and the impact on earlier estimations of fields in the chamber. We will demonstrate how these statistical models are applied to actual measurements in reverberation chambers. This work has a direct impact on the estimate of chamber fields to which a device-under-test or vehicle is exposed during electromagnetic vulnerability tests.

Introduction

The cornerstone for the accurate interpretation of reverberation chamber data is to use the proper mathematical relationships for estimating the electromagnetic field from a limited number of measurements. Well-founded theoretical models have been developed and compared to an extensive amount of measurement data [2-17]. These models take different approaches to solving for the fields in an electrically large complex cavity. The first approach involves the expansion of the fields in terms of generalized expressions for the eigenmodes of an arbitrary cavity [4, 5, 6, 15] whose coefficients are random variables. An alternative approach formulated by Hill [14, 15] describes the fields as combinations of a large number of plane-waves with random coefficients (phase, polarization, incident angle). Fortunately, these diverse derivations all result in the same statistical model for fields in a reverberation chamber.

Models of the underlying probability distributions for the field quantities in a 'well-stirred' reverberation chamber provide several distinct advantages in the analysis of EMC measurements performed in such a facility [16, 17]. The knowledge of the distributions allows us to

derive relationships for fields quantities not available using electromagnetic theory alone,
estimate one quantity from measurements of another (such as the maximum E field from measurements of average received power),
estimate measurement uncertainties more accurately, and

[1] Contribution of the National Institute of Standards and Technology, not subject to copyright in the United States.

optimize the number of measurements required to complete the test.

For the following discussion we assume the measurements are power received from an antenna (placed at some arbitrary location in the chamber) at a single frequency and sampled at a sufficient number of paddle positions. Also, the frequency is high enough that the chamber is reverberant. The analysis for frequencies where the mode density is low, for a small number of samples, or for other techniques such as frequency stirring will not be considered in this paper.

The Origin of the Specifics

Relationships have been established using electromagnetic theory and simple descriptive statistics (such as sample mean) to estimate the average squared magnitude of the electric field (rectangular component or total) from measurements of power received from an arbitrary antenna in a reverberation chamber. Hill [15] has shown that the average received power is the product of the scalar power density, the effective area of an isotropic antenna and a polarization mismatch factor of ½:

$$\langle P_{rec} \rangle = \frac{1}{2} \frac{\langle |E_T|^2 \rangle}{\eta} \frac{\lambda^2}{4\pi}. \quad (1)$$

Where $\langle |E_T|^2 \rangle / \eta$ is the scalar power density expressed in terms of the total RMS electric field E_T and the wave impedance of free space $\eta = 120\pi$. The term $\lambda^2/4\pi$ is the effective area of an isotropic antenna, where λ is the wavelength. We use the symbols $\langle\ \rangle$ to indicate the average (mean) or expected value of the quantity. NBS Technical Note 1092 [1] used this relationship without the factor of ½; however recent measurements [16] and theory [15] support including it. Solving for the average squared magnitude of the total electric field, we have

$$\langle |E_T|^2 \rangle = \frac{8\pi\eta \langle P_{rec} \rangle}{\lambda^2}. \quad (2)$$

Three arbitrary, linearly polarized orthogonal (rectangular) electric field components E_1, E_2, E_3 are related to the total electric field by

$$|E_T|^2 = |E_1|^2 + |E_2|^2 + |E_3|^2. \quad (3)$$

It can be argued [15, 17] that each component is independent and has the same probability distribution and expected or average value. We can then express equation (3) as

$$\langle |E_T|^2 \rangle = 3 \langle |E_R|^2 \rangle. \quad (4)$$

The subscript R indicates any one of the three rectangular components. The expression for the average squared magnitude of a rectangular component of electric field is simply one third of equation (2), or

$$\langle |E_R|^2 \rangle = \frac{8\pi\eta \langle P_{rec} \rangle}{3\lambda^2}. \quad (5)$$

If we assume that received power and the squared magnitude of a rectangular electric field component can be described by the same family of distributions [17], then (5) can be used to estimate the maximum squared magnitude of electric field from the maximum measured received power. In addition, since the square root of the maximum is equal to the maximum of the square roots for any set of samples, we can take the square root of (5) as applied to the maximums. The maximum magnitude of a rectangular electric field component can now be estimated from the square root of the maximum received power [17].

$$\lceil |E_R| \rceil = \sqrt{\frac{8\pi\eta \lceil P_{rec} \rceil}{3\lambda^2}} = \sqrt{\frac{320\pi^2 \lceil P_{rec} \rceil}{\lambda^2}} = \frac{8\pi}{\lambda}\sqrt{5 \lceil P_{rec} \rceil}. \quad (6)$$

The symbols $\lceil\ \rceil$ indicate the maximum observed value. This relationship can also be applied to the minimum of $|E_R|$ using similar rationale. However, attempting to apply the square root of (5) to the averages can introduce errors due to the fact that, in general, the mean value of a random variable is not necessarily equal to the square root of the mean of the squared variable. In other words,

$$\langle X \rangle \neq \sqrt{\langle X^2 \rangle}. \quad (7)$$

The situation with the total electric field is even more limited. Since we cannot expect that the distribution for total electric field squared is the same as received power (or the rectangular electric field squared) a similar estimation for maximum $|E_T|$ as that given in (6) for $|E_R|$ would be incorrect. Also, even though (4) is valid, the extreme values of $|E_R|$ and $|E_T|$ (maximum and minimum) are not necessarily related by $\sqrt{3}$ for the same reason [17].

Statistics and the Missing Link

At this point, we are unable to extend these expressions to relate measurements of received power from an antenna in the reverberating chamber to the remaining field parameters. However, if we were to ignore a rigorous statistical analysis of (2) and (5) and apply these relationships to estimate the average or maximum total electric field or the average rectangular electric field, the errors may not be large enough to cause serious problems in a typical EMC test. The work presented in [1] did not have the benefit of the recent developments in the theoretical models and consequently used (2), (4), and (5) to estimate the total and rectangular electric field. This approach overestimates the fields by about 1 dB or less for average values and as much as 4.77 dB for the maximum [17] (not including the factor of ½ mentioned earlier). While this is not trivial, it can be absorbed into the large (5 to 10 dB or more) estimated uncertainties given in [1]. On the other hand, a recent estimate of measurement uncertainty for received power indicated, under certain conditions, values of about 1 dB [16]. These errors would then be significant. Fortunately, analysis of the underlying probability distributions can be used to derive more accurate relationships.

In a complex cavity the real and imaginary parts (or alternatively amplitude and phase) of each component of the electric field at any point in the cavity result from the combination of a large number of random eigenvectors, rays, or plane waves [4, 5, 6, 15, 17]. These constituents are argued to be independent and identically distributed random variables. The central limit theorem predicts that the combination of these random variables is best described by the normal or Gaussian probability density function [18]. Also, N independent and normally distributed random variables yield either the *chi statistic with N degrees of freedom (dof)* if combined as the square root of the sum of squares or the *chi-square statistic with N dof* when combined as the sum of the squares [18]. Hence, a single rectangular electric field component $|E_R|$ composed of two independent normal random variables (real and imaginary) is distributed as a *chi with 2 dof* and $|E_R|^2$ as a *chi-square with 2 dof*. The total electric field $|E_T|$ is composed of six components (three orthogonal rectangular components each with a real and imaginary part) and is distributed as a *chi with 6 dof*. The distribution of $|E_T|^2$ is *chi-square with 6 dof*. The power received from any linear antenna P_{rec} is proportional to some combination of the squared magnitude of rectangular electric field components $|E_R|^2$ over the aperture of the antenna and is described by the same distribution, *chi-square with 2 dof*. There is considerable measurement data supporting this model [11, 16]. We will assume it is valid and use the characteristics of the distributions to relate the electric field to received power. A detailed compilation of these characteristics can be found in [16] or [17].

As an example, we will examine how the characteristics (mean, variance, distribution of extreme values, etc.) of the *chi* and *chi-square* are used to bridge the gap between received power measurements and the rectangular components of electric field. A similar derivation could be performed for the total electric field in (2).

A comparison of the mean or expected value of each distribution can provide a solution to the square root of (5) and overcome the limitation noted in (7) [16, 17]. Since we are confident $|E_R|^2$ is distributed as a *chi-square with 2 dof* we know that $\langle |E_R|^2 \rangle = 2\sigma^2$ (where σ^2 is the variance of the contributing normal distributions). Also, $|E_R|$ is *chi with 2 dof* and $\langle |E_R| \rangle = \sigma\sqrt{\pi/2}$. We can now show that

$$\langle |E_R| \rangle = \sqrt{\langle |E_R|^2 \rangle} \left\{ \frac{\sqrt{\pi}}{2} \right\}. \quad (8)$$

Substituting in (5) we obtain

$$\begin{aligned} \langle |E_R| \rangle &= \left(\sqrt{\frac{8\pi\eta \ \langle P_{rec} \rangle}{3\lambda^2}} \right) \left(\frac{\sqrt{\pi}}{2} \right) \\ &= \frac{4\pi}{\lambda} \sqrt{5\pi \ \langle P_{rec} \rangle}. \end{aligned} \quad (9)$$

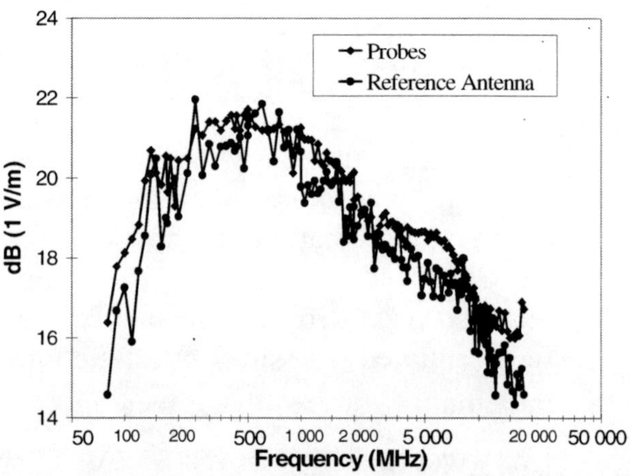

Figure 1. Average Electric Field.

Figure 1 is the average rectangular electric field measured directly by an array of 30 small probes compared to the same quantity estimated from received power measurements using (9). The agreement is within the estimated measurement uncertainty of about ±1 dB for the probes and about ±1.5 dB for average received power [16].

We now have an expression for the average rectangular electric field in the chamber. However, there seems to be (unofficial) consensus in the EMC vulnerability testing community that devices are really being stressed to the maximum rectangular electric field rather than the average [13]. The most direct (and least accurate) method to arrive at the maximum electric field is to measure the received power at a large number of paddle positions, take the largest sample, and use (6) to calculate maximum $|E_R|$. A more efficient (and accurate) approach is to measure the average received power to some desired uncertainty and calculate the maximum electric field. We could use (9) to calculate the average electric field, determine the relationship of the maximum to the average field for a given number of samples, and estimate the maximum electric field. The maximum-to-average field relationship, however, is involved and best evaluated numerically [16,17]. An alternative approach, with relatively simple approximate formulas, is to first estimate the maximum received power and then calculate the electric field. Lehman [5, 13] noted that the ratio of the maximum to average received power is essentially the mean of the probability density function for the maximums of many trials (each drawn from N samples of the *chi-square with 2 dof*) to the mean of the *chi-square with 2 dof*. This "Max-to-Ave" ratio for N samples of received power, shown within the brackets in (10), can be used to estimate the maximum received power from the average of N measurements.

$$\lceil P'_{rec} \rceil \approx \left\{ 0.577 + \ln(N) + \frac{1}{2N} \right\} \langle P_{rec} \rangle. \quad (10)$$

The estimate for maximum received power $\lceil P'_{rec} \rceil$ in (10) is, by the above definition, the mean value of many maximums. When we use that value in (6) to determine the maximum rectangular electric field we have a problem with the square root as noted in (7). Fortunately, we can derive a simple factor from the distributions and calculate the maximum electric field [17]:

$$\lceil |E_R| \rceil = \frac{8\pi}{\lambda} \sqrt{5 \lceil P'_{rec} \rceil} \left\{ \frac{\Gamma(N + 1/2)}{\sqrt{N}\ \Gamma(N)} \right\}. \quad (11)$$

The gamma (factorial) functions can be easily evaluated for typical values of N. Also, this correction factor approaches unity as N becomes large and may be unnecessary for N larger than about 50.

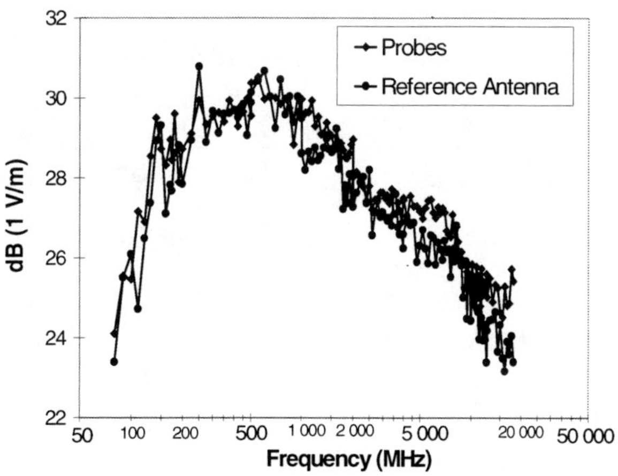

Figure 2. Maximum electric field

We show in Figure 2 a comparison of the average of the maximum rectangular electric field from each of 30 small probes and that quantity estimated from received power. In this example N was 225. The uncertainties are similar to Figure 1, since we are again using the average received power. The majority of the data are in very good agreement.

Conclusions

We have presented new expressions to estimate the rectangular components of electric field from measurements of power received from an antenna in a reverberation chamber. Also, we have argued that it is more reliable and accurate to measure the received power at a limited number of paddle positions, take the average, and calculate the maximum electric fields rather than to use the maximum received power directly. This observation should motivate chamber operators to reevaluate procedures and formulas proposed in [1] in light of these derivations.

References

[1] M.L. Crawford and G.H. Koepke, "Design, evaluation, and use of a reverberation chamber for performing electromagnetic susceptibility / vulnerability measurements," Nat. Bur. Stands. (U.S.) Tech. Note 1092, April 1986.

[2] R.E. Richardson, "Mode-stirred chamber calibration factor, relaxation time, and scaling laws," IEEE Trans. Instrum. Meas., vol. 34, No. 4, pp. 573-580, December 1985.

[3] J.M. Dunn, "Local, high-frequency analysis of the fields in a mode-stirred chamber," IEEE Trans. Electromag. Compat., vol. 32, pp. 53-58, 1990.

[4] J.G. Kostas and B. Boverie, "Statistical model for a mode-stirred chamber," IEEE Trans. Electromag. Compat., vol. 33, pp. 366-370, 1991.

[5] T.H. Lehman, "Statistics of MSCs," presented at the Anechoic Chamber and Reverberation Chamber Operators Group Meeting, Naval Surface Warfare Center, Dahlgren, VA, November, 1992.

[6] T.H. Lehman, "A statistical theory of electromagnetic fields in complex cavities," USAF Phillips Laboratory, Interaction Note 494, May 1993.

[7] R.H. Price, H.T. Davis, and E.P. Wenaas, "Determination of the statistical distribution of electromagnetic-field amplitudes in complex cavities," Phys. Rev. E, vol. 48, pp. 4716-4729, 1993.

[8] D.A. Hill, "Spatial correlation function for fields in a reverberation chamber," IEEE Trans. Electromag. Compat., Vol. 37, No. 1, February, 1995.

[9] D.A. Hill, "A reflection coefficient derivation for the Q of a reverberation chamber," IEEE Trans. Electromag. Compat., vol. 38, pp. 591-592, 1996.

[10] P. Corona, G. Ferrara, and M. Migliaccio, "Reverberation chambers as sources of stochastic electromagnetic fields," IEEE Trans. Electromag. Compat., vol. 38, pp. 348-356, 1996.

[11] G.J. Freyer, M.O. Hatfield, D.M. Johnson, and M.B. Slocum, "Comparison of measured and theoretical statistical parameters of complex cavities," IEEE Int. Symp. Electromag. Compat., Santa Clara, CA, August 1996.

[12] A.K. Mitra and T.F. Trost, "Statistical simulations and measurements inside a microwave reverberation chamber," IEEE Int. Symp. Electromag. Compat., Austin, Texas, August,1997.

[13] T.H. Lehman and G.J. Freyer, "Characterization of the maximum test level in a reverberation chamber," IEEE Int. Symp. Electromag. Compat., Austin, Texas, August,1997.

[14] D.A. Hill, "Plane-wave, integral representation for fields in reverberation chambers," IEEE Trans. Electromag. Compat., to be published.

[15] D.A. Hill, "Electromagnetic theory of reverberation chambers," Natl. Inst. Stand. Technol. Technical Note, in preparation.

[16] J.M. Ladbury, G.H. Koepke, and D.G. Camell, "Evaluation of the NASA Langley Research Center mode-stirred chamber facility," Natl. Inst. Stand. Technol. Technical Note, to be published.

[17] J.M. Ladbury, "Statistical theory of reverberation chambers," Natl. Inst. Stand. Technol. Technical Note, in preparation.

[18] A. Papoulis, "Probability, Random Variables, and Stochastic Processes," Second Edition. New York: McGraw-Hill; 1984.

EVALUATION OF HONEYWELL RECOVERABLE COMPUTER SYSTEM (RCS) IN PRESENCE OF ELECTROMAGNETIC EFFECTS

Mahyar Malekpour, NASA Langley Research Center, Hampton, VA

Abstract

The design and development of a Closed-Loop System to study and evaluate the performance of the Honeywell Recoverable Computer System (RCS) in electromagnetic environments (EME) is presented. The development of a Windows-based software package to handle the time critical communication of data and commands between the RCS and flight simulation code in real-time, while meeting the stringent hard deadlines is also presented. The performance results of the RCS while exercising flight control laws under ideal conditions as well as in the presence of electromagnetic fields is also discussed.

Introduction

The problem of verifying the integrity of control computers in adverse as well as nominal operating environments is a key issue in the development, validation, certification, and operation of critical control systems for advanced aircraft. An adverse operating environment of particular concern relative to validation and certification of critical systems is caused by electromagnetic disturbances. Sources of electromagnetic disturbances include lightning, High Intensity Radiated Fields (HIRF) caused by RF transmitters and radars, portable electronics devices carried onto the airplane, and electromagnetic incompatibilities of equipment installed on the aircraft [1].

Soft faults in digital avionics have traditionally been manually corrected. More recently, architecture design measures for the automatic correction of soft faults have begun to be developed. It is perceived that significant benefits can be gained through soft fault protection measures designed into the basic system mechanization. Architecture designs with soft fault protection provide the ability to tolerate disruption of either input/output data or internal computation. Fault clearing and computation recovery must be rapid enough to be "transparent" relative to functional operation and flight deck effect [2].

RCS

An example of an architectural soft fault protection philosophy in the design of computing platforms is the Aircraft Information Management System (AIMS) used on Boeing 777 aircraft. All computing and I/O management resources are lock-step compared on a processor cycle-by-cycle basis. In this approach, if a soft or hard fault event occurs, the processor is interrupted and service handlers take control and no data can be exported [2].

The first stage of a prototype computing platform for fast recovery from soft faults has been developed for NASA. This prototype includes patented technology for transparent soft fault recovery that had never been built or tested. These concepts were implemented and integrated into the basic lock-step processing module of the AIMS architecture. This prototype was delivered to NASA in 1997 for evaluation at the Langley Research Center (LaRC) system integration facility. A general illustration of a computing platform with rapid (transparent) recovery elements is depicted in Figures 1 and 2.

0-7803-5086-3 /98/$10.00 ©1998 IEEE

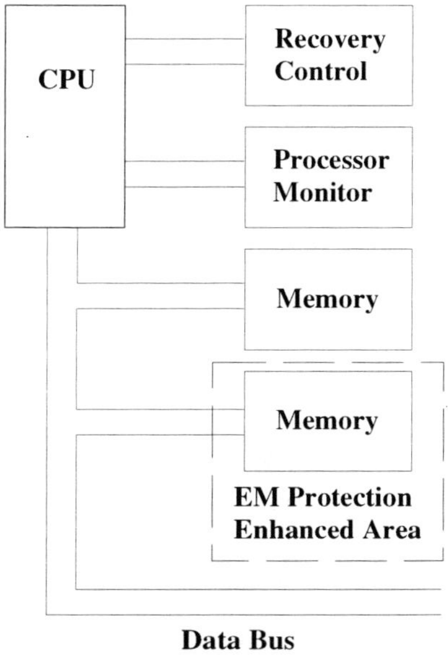

Figure 1. Digital Processor with Elements for Rapid Recovery

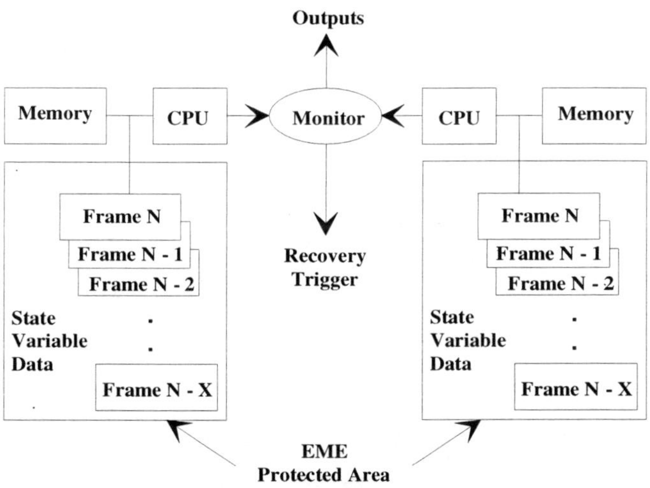

Figure 2. Rapid Recovery Upset Detection and Recovery Flow

Such a platform is intended to provide soft fault recovery that would be virtually software independent and transparent to a system function. As shown in Figure 2, the rapid recovery concept involves protecting and recovering the significant state variables of a system function. An aircraft autoland (arm, capture and track) function is being used for evaluation of the rapid recovery concept. Since this is the only function being implemented in the application program software, the partitioning portion of the platform SAFEbus technology was not employed. Communication with the LaRC host system integration facility is intended to be accomplished primarily through an optical 429 bus structure. LaRC has identified the platform integrated with the host facility as the RCS [2].

System Architecture

The proposed Closed-Loop architecture consists of the RCS, a VME-based 429-optic card for conversion of data from electrical to optical and vice versa, a flight simulation host processor, and a PC for the development and display of data as shown in Figure 3.

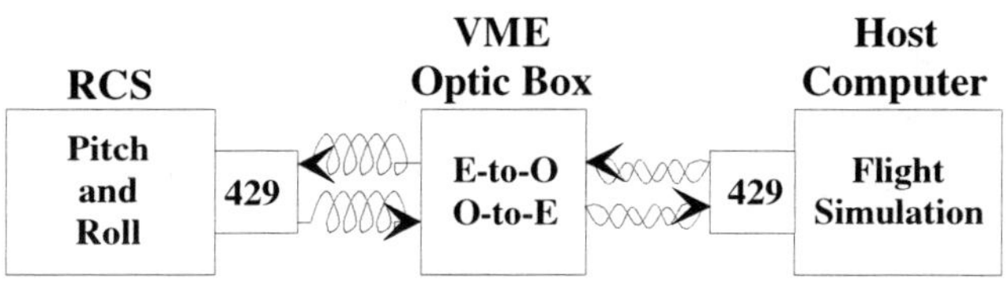

Figure 3. Closed-Loop System.

Two platforms were considered. The first alternative is a VME-based system, Figure 4, where the flight simulation host is a VME-based processor that communicates with the development processor, a PC, via a secondary link, an RS232 or an Ethernet.

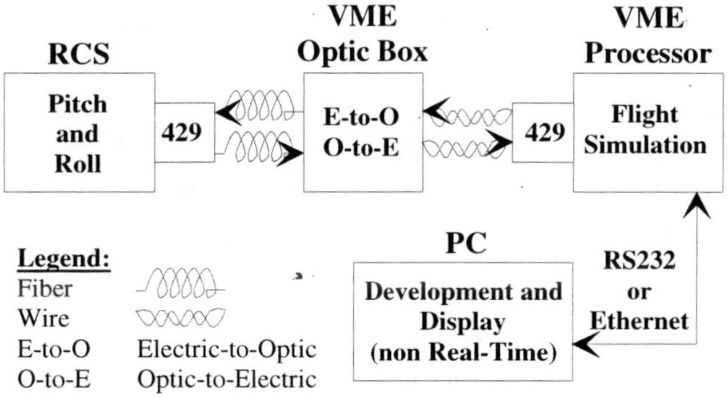

Figure 4. VME-Based System.

The second alternative bypasses the VME-based processor and implies implementation as well as operation of the flight simulation on one or two PCs, Figure 5.

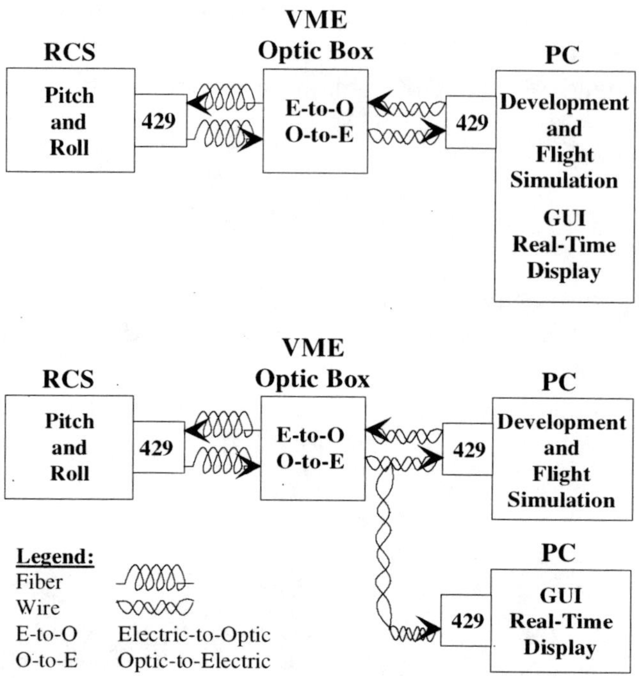

Figure 5. PC-Based Systems.

PC or VME?

A VME-based system consists of expensive hardware and software compared to a PC-based system. Although there are extensive and specialized software packages available for the VME-based systems, there are steep learning curves associated with them. However, with in-house knowledge of C/C++ and Windows Development software, the PC-based system imposed only a minimal learning curve. Regardless of the choice of architecture, the only new element that needed to be developed was a 429-interface package.

The VME-based systems were expected to be able to handle real-time operation of the flight simulation. However, real-time display of data would still have to be done through a secondary network and on a PC. As a result, managing the critical timing issues between the two computers was not a trivial task. Since available PCs are very fast and powerful and relatively inexpensive, it seemed practical to attempt to perform the flight simulation operation, as well as collection and display of data on one PC. The challenge, therefore, was proving the feasibility of managing the system and real-time display of data on a PC.

Software Development

A Graphical User Interface (GUI) package was developed for Windows 95 using C/C++ languages. This modularized and user-friendly software consists of 429-interface and flight simulation code. It manages communication between the RCS and the flight simulation code and displays the data in real-time while meeting timing requirements of the RCS, Figure 6. This software consists of 11 windows that depict all Closed-Loop System activities. Roll and Aileron command (DELAC), Pitch and Elevator command (DELEC), and Altitude (ALT) and Radio Altitude (HR-ALT) are displayed in three different windows to provide the user with visual correlation between the RCS inputs and their corresponding output commands. The lateral position (YCG), throttle (THROT), and yaw are also displayed separately. In addition to an error monitoring window, there are four other widows that display the discrete values reported by RCS. As a result, this software monitors all activities of this Closed-Loop System. This real-time monitoring capability is essential for testing the system in the presence of RF.

To prevent damaging the RCS, RF and field strength are gradually increased until a region of susceptibility is visually detected. During test runs, the airplane attitude as well as RCS status information, i.e., 429 discrete words, are displayed in real-time in their corresponding windows. Upon upset detection, RF is turned off for the remainder of the flight while system activities are continuously monitored and data

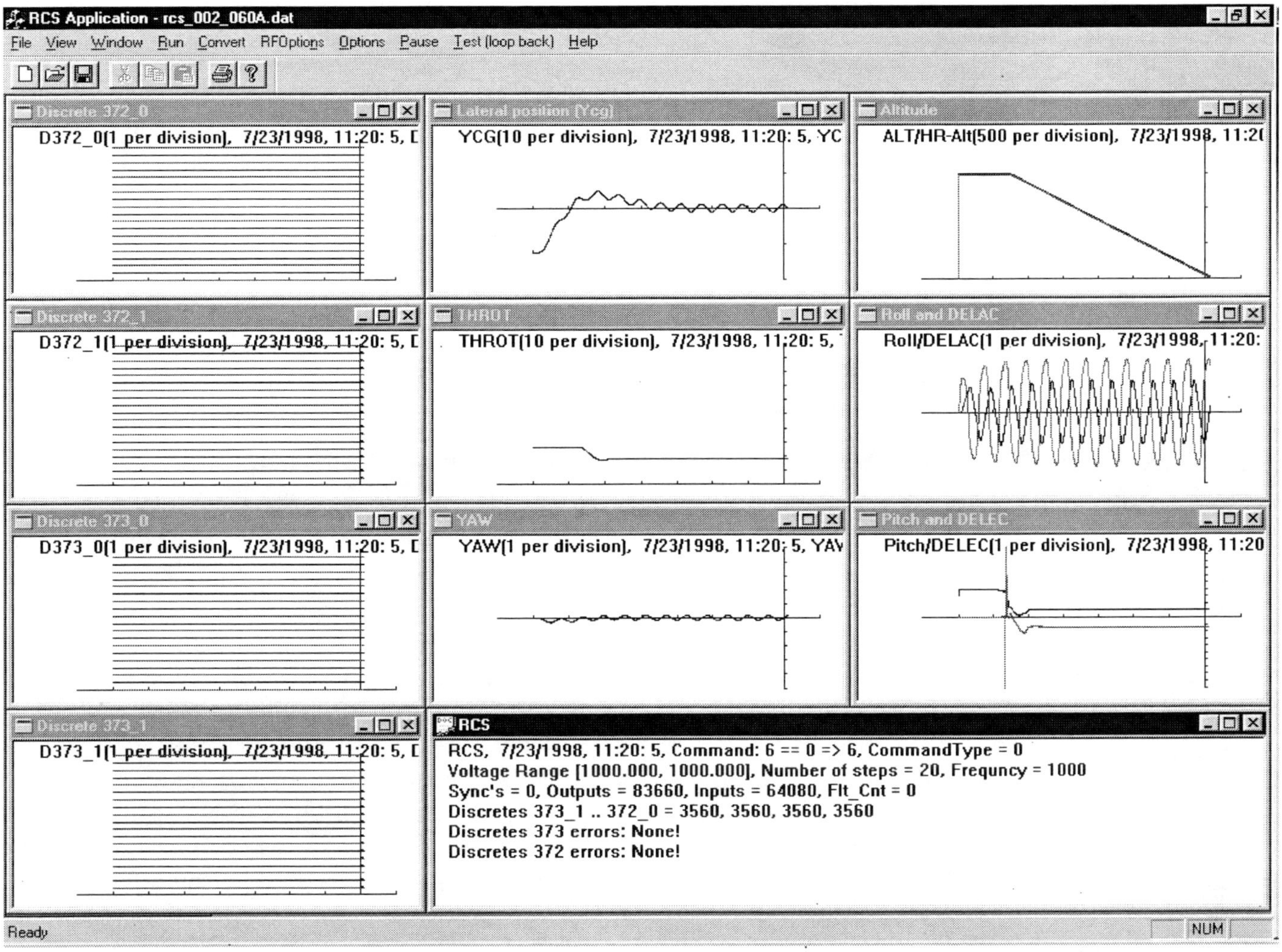

Figure 6. RCS-Win Software.

collected. Through this user-friendly software, test conditions can be specified and recorded along with the data for future references.

Data Management

The data along with the test conditions such as the RCS control flags, field strength range, and initial frequency are stored on the hard disk. Each file is approximately 1.5 Mbyte. A 1 Gbyte removable hard drive and a CD-ROM writer are used for archival storage. The collected data are in ARINC format [3] in order to minimize the storage requirements. The user interface software is then used to convert these data files to the desired format for post analysis purposes.

Tests

Tests for the RCS are conducted in two phases. Phase one consists of testing under ideal conditions where EME is absent. Phase two consists of testing in the HIRF chambers in the presence of EME. The specific goals are to characterize the RCS functionality and the RCS upset recovery scheme, to verify control laws and flight simulation integrity, and to assess RCS performance under various conditions.

Phase one was necessary in order to resolve discrepancies in the Interface Control Document (ICD) and to debug the system as a whole. Completion of phase one also led to the detection of errors in the implementation of the flight control codes in the RCS, Figure 7.

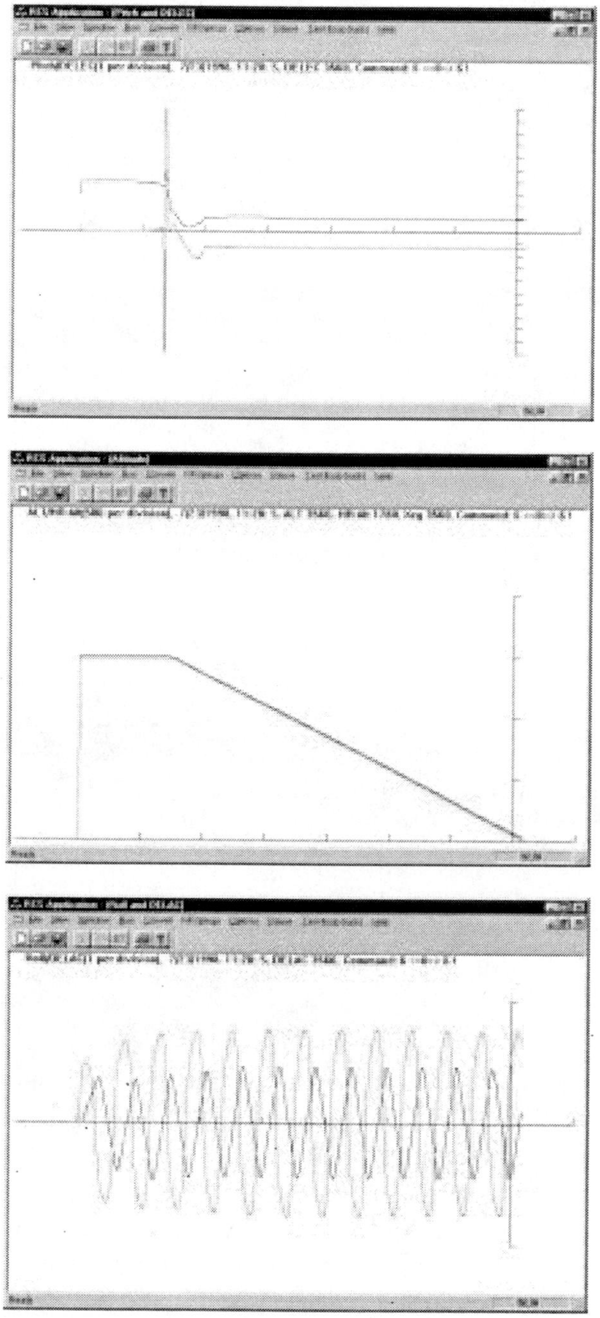

Figure 7. RCS 737 Autoland, no EME.

As shown in the Pitch and DELEC window of Figure 7, there is a spike in the DELEC command at the glide slope engagement as reported by the RCS. Although this discrepancy is not detrimental to the operation of this Closed-Loop System, it is currently being resolved.

In Figure 7, Pitch and Elevator command, Altitude and Radio Altitude, and Roll and Aileron command are displayed in three different windows, respectively, to provide visual correlation between the RCS inputs and their corresponding outputs. The solid lines are for RCS inputs and the dotted lines for RCS outputs.

Phase two of the test is currently in progress and to date, two tests have been completed.

Description of Experiments

The radiation environment of the HIRF Laboratory is the reverberation chamber. Within the chamber are both transmitting and receiving antennas as well as any sensors that might be needed for a specific test. The components of the signal generation and measurement instrumentation consist of a synthesized sweeper for signal generation, a network analyzer, a spectrum analyzer, an oscilloscope, and a high-power amplifier [4, 5]. These devices are software controlled from within the HIRF Laboratory. The advantage of performing tests in mode-stirred chambers is that the equipment is subjected to fields at all angles of incidence simultaneously. Therefore, the worst case conditions are represented during the test. The reverberation chambers within the HIRF Laboratory provide near-homogeneous randomized electromagnetic fields and make it possible to place the entire target unit in the test environment.

During the test, the RCS is placed inside one of the mode-stirred chambers of the HIRF Laboratory and is interfaced to the flight simulation computer. Electrical isolation is achieved via fiber optic cables.

While testing the RCS in the HIRF chamber, RF and/or field strength are gradually increased until a region of susceptibility is detected. To prevent damaging the RCS and upon upset detection, RF is turned off for the remainder of the test. The test is then resumed with finer increments of frequencies and/or field strength around the susceptibility region.

Closed-Loop HIRF Data

The RCS was exposed to RF continuously; however the field strength was ramped up from minimum to maximum in 10-second intervals while keeping the frequency constant. The same process was repeated for all frequencies. Table 1 is an example of the test conditions.

Table 1. Test Conditions with Continuous HIRF Exposure.

Frequency (MHz)	Min Field (V/m)	Max Field (V/m)
200	200	1000
250	200	1000
…	…	…
2000	200	1000

To date, no repeatable upsets have been detected for the frequencies and power levels shown in Table 1. Besides Continuous Wave (CW), RCS has been exposed to Square Wave (SQW) with all modulations per DO-160 specification for all of the frequency and field strength ranges specified in Table 1. Preliminary data from 175 flights indicate that the RCS is resilient to the EME.

Upon disabling the recovery capability of the RCS and subjecting the RCS to the same set of tests, RCS seemed somewhat more vulnerable to RF. Figure 8 shows a second spike in the Pitch and DELEC window that is a result of RF induced upset. As evident in the Altitude window, this upset caused the airplane to crash. Out of 165 flights, two upsets were observed, but they were not repeatable.

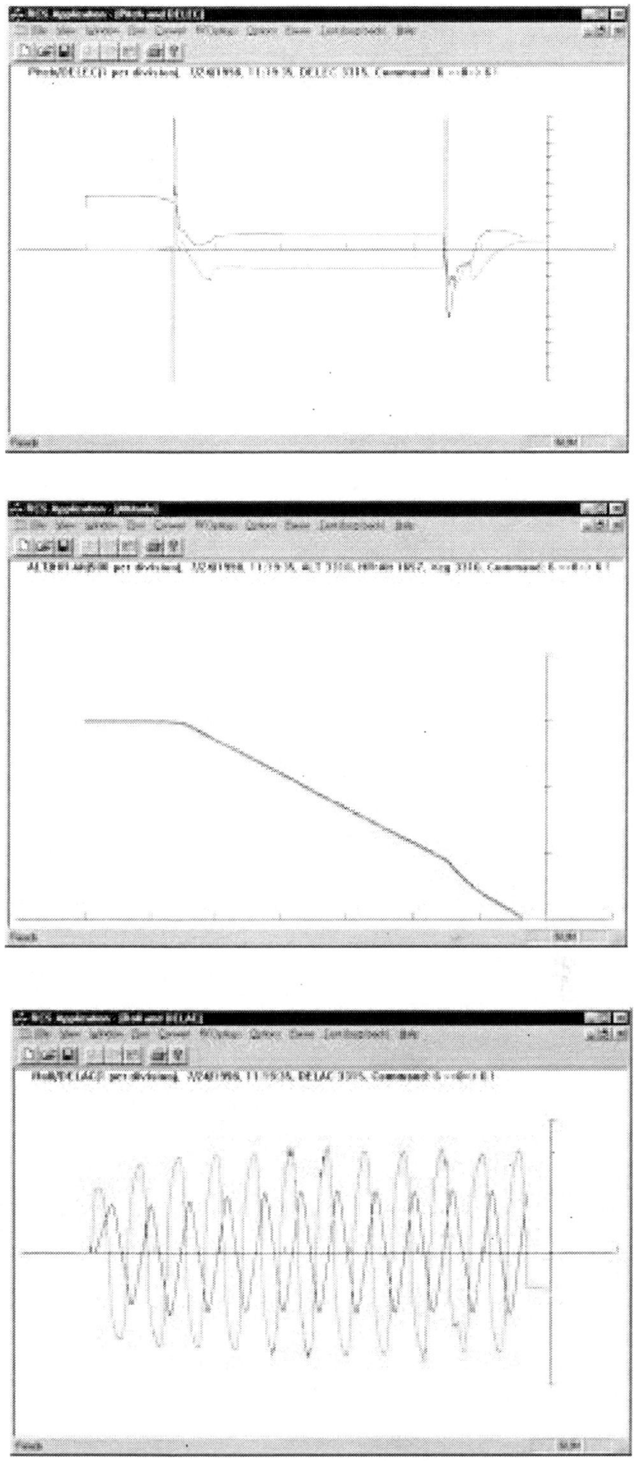

Figure 8. RCS 737 Autoland with Continuous HIRF Exposure and RCS Recovery Scheme Disabled.

Future Test Plans

Preliminary test results indicate resiliency of the RCS to the EME at frequencies and power levels used in tests. Since following the DO-160 specification will require considerable amount of time in order to find susceptible frequencies and power levels, another approach will be considered. The next test of RCS will be conducted in the Semi-Anechoic chamber of the HIRF Laboratory. In this chamber emission tests are expected to result in detection of some of the susceptibility regions of the RCS. The test will continue in the reverberation chambers to verify and characterize the predicted susceptibility regions. In order to fully characterize the performance capabilities of the RCS to recover from upsets, some of the RF shielding will have to be removed to increase the number of upsets that occur during the tests.

Summary

The design and development of a Closed-Loop System to study and evaluate the performance of the advanced RCS architecture when subjected to EME is presented. The integration of this hardware proved to be a major undertaking. During this process a few bugs in the 737-control law implementation of the RCS were found that are being resolved. In support of this Closed-Loop System, a Windows-based software package was developed to handle the time critical communication of data and commands between the RCS and flight simulation code, in real-time while meeting the stringent hard deadlines. This package consists of an ARINC429 bus driver, flight simulation code and GUI-based displays of the key elements of the control laws, as well as the airplane attitude. As a result, this package enables the researchers to monitor all activities of the airplane during the flight in real time. The real-time capability of this package is crucial during the EME testing of the hardware.

References

[1] Celeste M. Belcastro, "Closed-Loop HIRF Experiments Performed on a Fault Tolerant Flight Control Computer," 16th AIAA/IEEE DASC, Irvine, CA, October 1997.

[2] R.F. Hess, "Computing Platform Architectures for Robust Operation in the Presence of Lightning and Other Electromagnetic Threats," 16th AIAA/IEEE DASC, Irvine, CA, October 1997.

[3] ARINC Specification 429P1-15, September 1, 1995.

[4] Daniel M. Koppen, "Open-Loop HIRF Experiments Performed on a Fault Tolerant Flight Control Computer," 16th AIAA/IEEE DASC, Irvine, CA, October 1997.

[5] Rueben A. Williams, "The NASA High Integrity Radiated Fields Laboratory," 16th AIAA/IEEE DASC, Irvine, CA, October 1997.

DESIGN, CONSTRUCTION, COMPUTATIONAL EM MODELLING, AND CHARACTERISATION OF AN AIRCRAFT SIZED REVERBERATION CHAMBER AND STIRRER

Kevin R Goldsmith, Defence Science and Technology Organisation, Salisbury, South Australia
Paul A Johnson, Defence Science and Technology Organisation, Salisbury, South Australia

Introduction

Conventional methods for assessing the Electromagnetic Vulnerability (EMV) of whole aircraft require an open area test site or a large anechoic chamber. Chamber–based methods have the advantage of not causing interference to other users of the electromagnetic spectrum. However, the opportunity cost of an anechoic chamber is too high for the Australian Defence Force, other Defence Forces and Industries.

A pivotal event in the development of the reverberation chamber as an alternative to an anechoic chamber was the publication of [1]. The published work of Hatfield, Freyer [2,3,4, 5] and others has encouraged the Defence Science and Technology Organisation to conduct a research program to investigate the performance of reverberation chambers in establishing military aircraft EMV.

Critical Design Parameters

The construction of an aircraft–sized chamber for research purposes offers special challenges, mostly financial. The most critical design parameters are the chamber dimensions, mode tuner size and wall material.

Chamber dimensions

The research aircraft available to DSTO was an Aeromacchi 326H jet trainer. The aircraft has a complete avionics suite, engine and fatigued airframe making it ideal for the purposes of this work. The Macchi, like most aircraft, has a wingspan approximately equal to its fuselage length. The wingspan defines the minimum width and length of the chamber.

The major properties of a mode stirred chamber are achieved by changes in the boundary conditions of the standing waves in the chamber, thereby changing the spatial locations of the nodes. Moving a reflecting object, the tuner, inside the chamber changes the boundary conditions. A single large tuner, capable of operating at low frequencies, should have a diameter that is a significant portion of two dimensions of the chamber, normally height and width. The tuner must have clearance from the walls and performs better if mounted in a corner. The aggregation of these facts drives a chamber length to width ratio of about 2:1.

The height of aircraft from ground to vertical tail generally is approximately half of the fuselage length. This additional fact gives a ratio of chamber length to width to height of 4:2:1. These dimensions are not consistent with the ratio of dimensions of current research and equipment level chambers. It was not known whether this ratio of dimension would produce the required field distributions without changes to current concepts for the mode tuner and to methods to drive the chamber.

The nominal dimensions for the DSTO chamber (21m by 11m by 6m) were selected to be consistent with the 4:2:1 ratio and the existing building in which the chamber was to be constructed. The finished internal size is 20.85m by 10.89m by 6.0m.

0-7803-5086-3 /98/$10.00 ©1998 IEEE

Mode Tuner Design

Ladbury et al [6]and private discussions Carter [7] have indicated that the use of a large Mode Tuner, particularly when mounted in the corner, offered the best solution. However, this was at the cost of a large structure and some loss of useable chamber space. At the time of the design, the use of both mode stirring and mode tuning was contemplated, the former mode requiring constant rotation, the latter mode requiring a large number of tuner steps.

The design requirement was therefore for a tuner which was a significant portion of the width and height of the chamber, and capable of being stepped at 0.5 degree steps or better. This indicated a diameter for the tuner of approximately 8m giving a diameter–to–width ratio of 0.75. The height of the tuner was determined by a self imposed Occupational Health and Safety Requirement to have it above head height. Roof clearance was decided on using a different rationale, (ie about 0.5 m for a working clearance), resulting in a nominal height of 3.5 m. The similarity in the individual paddle dimensions was not considered significant nor has it been investigated.

A tuner of this size requires significant mechanical design effort, due to its weight, moment of inertia, and the small step size. An estimate of the distance the outside edge moved in a 0.5 degree step revealed a distance of 35 mm or a wavelength at 8.6 GHz. This frequency was considered too low to generate repeatable results at the highest frequency to be used in the chamber (18GHz). A compromise between mechanical design difficulties and step size was reached giving a 0.1 degree step or 7mm at the edge or a wavelength at 43 GHz.

Conventional tuners are made from polystyrene covered in thick aluminium foil or from aluminium plate. Neither of these designs is suitable for a tuner of this size and step size due to mechanical rigidity. A search for a commercial material made from a sandwich of aluminium and polystyrene lead to the use of cold–room wall material, 'Bondor'. Bondor is a sandwich of 0.6 mm light zinc–coated sheet steel with a full surface glued polystyrene core. Bondor is available in various thicknesses, and 75mm was chosen for its structural performance. Bondor is available in continuous lengths of width 1.2 m. This width met the requirement for a tuner diameter of 8 m by using four panels of 3 strips wide.

The drive system for the tuner was designed at DSTO. Backlash in the reduction gearbox was a critical parameter for the design due to the 0.1 degree step size. A design solution using a commercial variable speed AC motor driving a low backlash 30:1 gearbox and final 2 stage toothed belt reduction was chosen.

The motor, gearbox and controller provided microprocessor based closed loop control for speed and position. The two stage toothed belt reduction (6.8:1 by 2) provided a zero backlash high torque reduction at reasonable cost. The maximum speed of the tuner was chosen to be 1 rpm, which coincides to approximately three times the resolution of the tuner for every sample of the spectrum analyser. A shaft encoder located directly on the tuner shaft provides the tuner position sensor for the operator and data recording.

Chamber Wall Material

The chamber walls are required to be highly reflective. However, a chamber that is too reflective will also have a very sharp Q at each of the basic resonant frequencies of the chamber. Crawford and Koepke [1] recommended the use of galvanised sheet steel as a cost-effective option. Under Australian Occupational Health and Safety Laws, galvanised materials require special and expensive procedures for welding, eliminating this choice of material. Zincalum a 90 micron zinc plated steel was chosen as it is weldable, was expected to have a similar surface reflection coefficient to zinc with the permeability of the steel providing the reflection (and shielding) at low frequencies.

The choice of wall thickness was based on the need for a considerable number of skin depths at the lowest frequency, the need for weldability and cost. Two millimetres was the cost-effective choice.

Predicted Electromagnetic Characteristics

Prior to the finalisation of the chamber design, the expected characteristics were predicted. This included estimating the resonant frequencies and their spacing, chamber Q, and tuner performance.

Chamber Resonant Frequencies

The resonant frequencies for the chamber are given by

$$f_{MHz} = 150\sqrt{[(a/l)^2 + (b/w)^2 + (c/h)^2]}$$

where f is the frequency in MHz
l is the length in metres
w is the width in metres
h is the height in metres
a,b,c are positive integers.

Hence, the lowest resonance is at 15.44 MHz and the frequency for the sixth mode, accepted as the lowest useable frequency, occurs at 60.1 MHz.

A key factor in successful chamber performance is the spacing of the resonant frequencies. Evenly spaced and dense modes create better performance. Crawford and Keopke [1] demonstrated the ratio of the dimensions determined the density of the resonances, and that integer relationships are to be avoided. The dimensions of the DSTO chamber are constrained by the building in which it is installed. However the ratios of 20.85:10.89, 20.85:6 and 10.89:6 are not ideal as they too close to an integer. Nevertheless the ratios were considered to be suitable, as the ratios of the squares were reasonable.

Chamber Quality Factor (Q)

The Q of the chamber comprises dominant components due to the antenna, Q_{ant}, and due to the walls of the chamber, Q_{wall}, where

$$Q_{ant} = (16\pi^2/\eta m c^3)\ V f^3$$

$$Q_{wall} = 3(V/S)\ (f/MHz)^{0.5}\ (\sigma_{con}/\mu_r)^{0.5}$$

and the total expected Q_{theo} of the chamber, is given by

$$1/Q_{theo} = 1/Q_{ant} + 1/Q_{wall}$$

The Q that is achievable in practice is expected to be 10 to 15 times lower than this, based on a large number of chambers of various sizes [15]. This 'k' factor is unknown for a particular design, being dependant on the difference between the actual wall surface conductivity and the theoretical values used and the number of antennas within it. Figure 1 gives the Q factor prediction for the DSTO chamber for zinc and steel walls for a single antenna.

Figure 1 Predicted and Measured Chamber Q

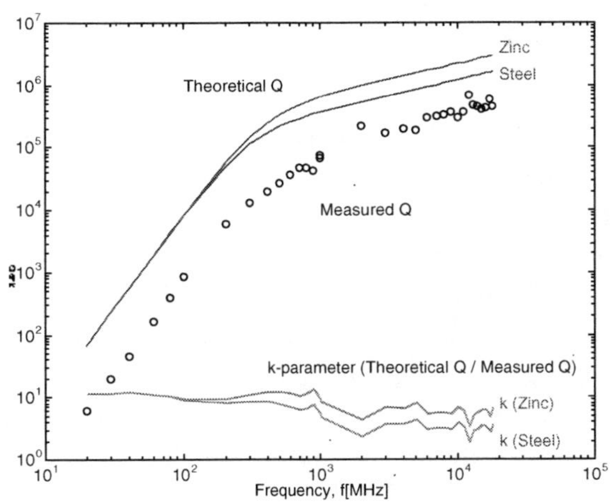

The Q also determines the electric field magnitude achievable for a given input power. This relationship is given by

$$E_{av} \sim 134\ \{Q\ P_{in}/V f_{MHz}\}^{0.5}$$

This analysis is used to verify that the chamber is likely to develop sufficient field strength for an affordable level of RF power.

0-7803-5086-3 /98/$10.00 ©1998 IEEE

The Q of a chamber drops as test objects of increasing size and absorption characteristics are introduced. Previously unknown for aircraft, this drop is measured and described in Johnson and Goldsmith [8].

Tuner Performance

The design of tuners is largely empirical for example it is not known if the bends in the tuner panels are required for good performance, and if they are what angles are best. The angles chosen for the DSTO chamber are the same as used by Carter in his large chamber, nominal 160, 140, 120 and 100 degrees internal angles. Johnson and Goldsmith [9] predicted the properties of the DSTO tuner using Computational Electromagnetics (CEM) after its design but before construction. The CEM analysis was performed using GEMACS (General Electromagnetic code for the Analysis of Complex Systems), a software code obtained from the USAF [10]. Three techniques are available, MOM (Method-of-Moments, for lower frequencies) [11], GTD (Geometrical Theory of Diffraction, principally for higher frequencies) [12] and FD (Finite Differences, mainly interior problems).

The chamber was represented with a 14–plate GTD model, six plates for the chamber walls, and eight for the stirrer, shown schematically in Figure 2. A source of RF radiation is placed about 1 m from the upper corner nearest the stirrer. The E-field is calculated at the mid point of the chamber for each of 360 positions of the stirrer ie. for every degree in stirrer azimuth. The computational time to obtain this data *for one particular frequency* is about 4 hours on a Silicon Graphics Infinite Reality R10000 computer.

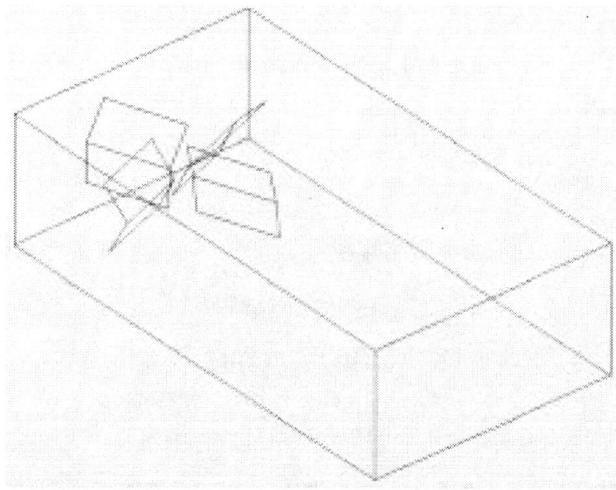

FIGURE 2: CEM 14-plate GTD representation of chamber

Lehman [13, 14] predicted that if the chamber has a gaussian field distribution for each of its two polarisations and three direction components, the field measured by a typical receiving antenna should have a Chi Squared distribution with two degrees of freedom. The Chi squared distribution is characterised by a single random variable, the mean, with a normalised standard (SDn) deviation of one. Crawford and Keopke [1] found that the stirring ratio (SR) should be at least 20 dB and the peak to average (PA) should be 7 to 8 dB for adequate chamber performance.

The simulation was performed for 23 separate frequencies, and for each the SR, PA, SDn and average power density was calculated. Additionally, the number of tuner steps required to generate an uncorrelated SR plot was calculated. An example SR plot is given in Fig 3 together with the equivalent correlation curve in Fig 4.

The PA, SR and SDn are also shown in Figures 5 and 6 for the 23 separate frequencies used.

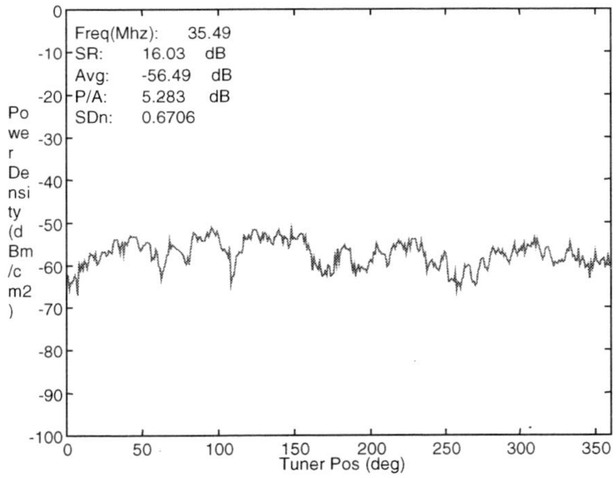

FIGURE 3: Computed Power Density v Tuner Position at 35 MHz

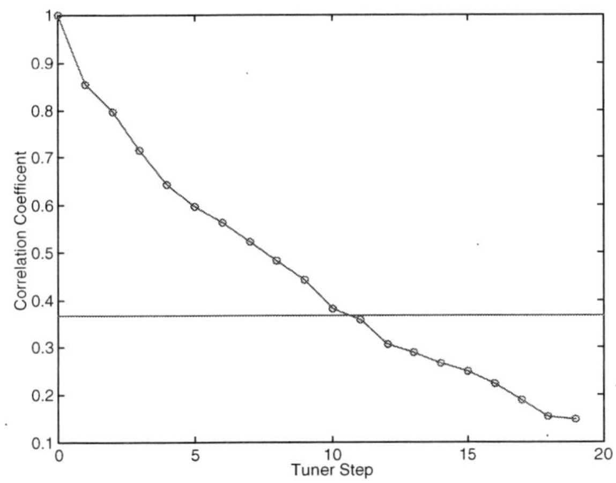

FIGURE 4: Computed Correlation Curve for Stirrer at 35MHz

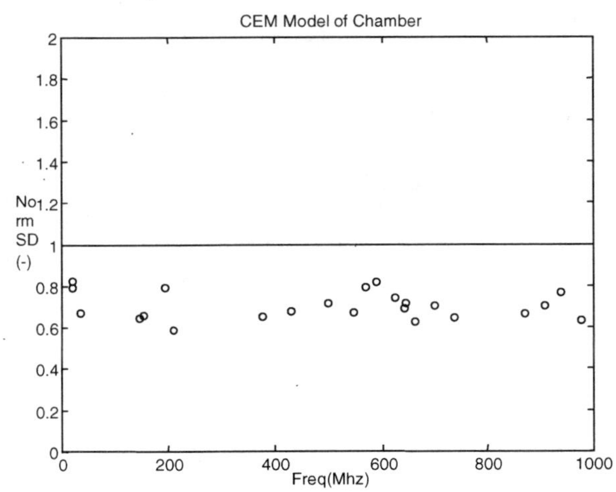

FIGURE 5: Computed Normalised Standard Deviation as a function of frequency

The computed results as shown in figures 3, 4 and 5, although low, indicated that the chamber may perform at a frequency well below the empirical '60 modes' frequency [1] of 64 MHz. Analysis conducted at this and other frequencies confirmed the design was suitable.

Measured Chamber Performance

The completed chamber is shown in figure 6. Measurement of the chamber performance has been undertaken on a number of occasions, to verify the expected high repeatability of the chamber characteristics. The repeatability is excellent with changes between measurements using different cables, antenna placements and power levels being about 1 to 2 dB.

Mode Tuner Drive System

The mode tuner drive system has proven to be excellent. The tuner has been stepped in 3 degree increments for a full revolution, with the maximum indicated error of 0.1 degree. The cumulative error for a full revolution was less than 0.1 degree. The highest rotational speed is 1.08 rpm.

FIGURE 6: Completed Mode Stirred Chamber

Chamber Q

Initial chamber measurements were taken using a HP series 71209A Spectrum Analyser with Tracking Generator from 0–18 GHz, using only -10dBm of input power. Below 1GHz, two EMCO Biconilog Antennae were used, and above 1GHz, two Double–ridged horns. Cable losses for both setups were obtained for all frequencies used. The tuner was set into continuous motion at 1 RPM, and a SR plot obtained using a sweep time of 60 sec

The measured chamber Q is shown in figure 1. The difference between the theoretical and measured Q performance is consistent with other chambers in the database kept by Hatfield and Freyer [15].

Statical Parameters

Sample SR plots are shown in figures 7,8,9,10 and compare well with the predicted and expected SR, PA and SDn.

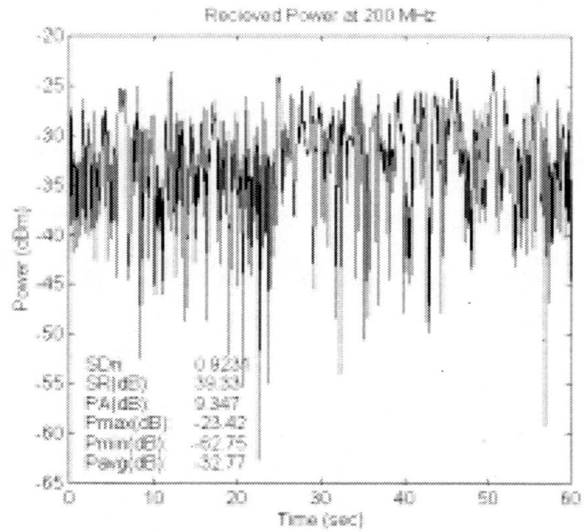

FIGURE 8 Received Power vs Time at 200 MHz

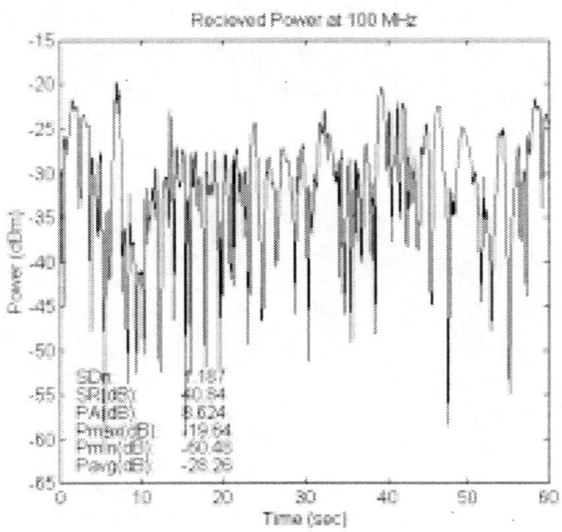

FIGURE 7 Received Power vs Time at 100 MHz

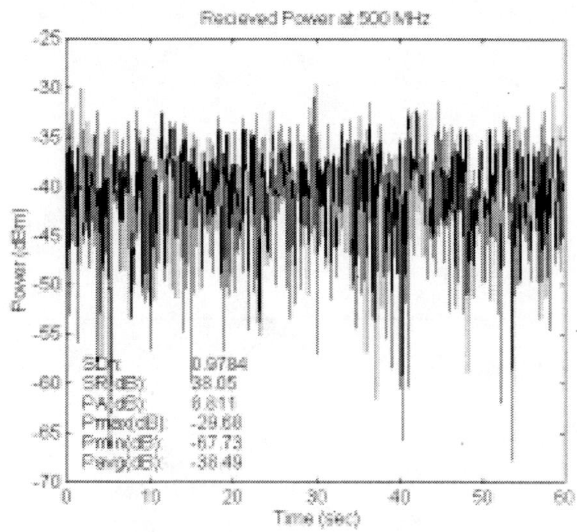

FIGURE 9 Received Power vs Time at 500 MHz

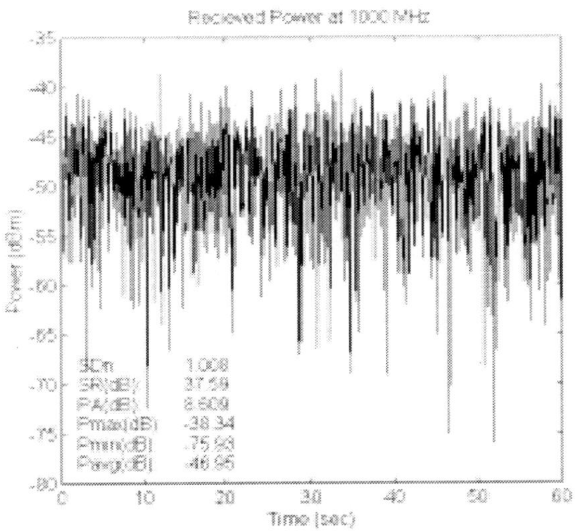

FIGURE 10 Received Power vs Time at 1000 MHz

Conclusions

The design of mode stirred chambers is largely empirical, however it is relatively simple. The empirical work of Crawford and Keopke has been well substantiated in this work and it remains an important reference. The work of Johnson and Goldsmith shows that design confidence can be achieved with rudimentary computational methods and broad assumptions. However, computational methods still need considerable work to achieve accurate results.

A key factor in Mode Stirred chamber design is its robust scalability. This chamber has been designed with aircraft friendly dimensions resulting in significantly different ratios of those dimensions compared with other designs. It has been constructed of inexpensive materials using common building industry methods. All of these factors have not significantly degraded its performance.

The use of a single large tuner consumes a significant portion of the chamber volume, but is effective in providing a useable mode stirred environment to frequencies below the current empirical expectation. The design of a large tuner presents difficulties with the step size and the maximum frequency within the chamber. This tuner is close to the practical limit, based on step size, dynamic structural, and drive system considerations.

The drive system designed for this chamber and its use of toothed belt reduction has proved successful.

Bibliography

1 Crawford M.L. and Koepke G.H., "Design, Evaluation and Use of a Reverberation Chamber for Performing Electromagnetic Susceptibility/Vulnerability Measurements", *NBS Technical Note* 1092 (1986)

2 Hatfield, M and Freyer, G, Summary of Test Techniques Available for Radiated Immunity Testing, Third Australiasian Instrumentation and Measurement Conference, Adelaide, Australia, April 1994

3 Hatfield, M and Freyer, G, Typical Results of radiated Immunity Testing Using Reverberation Chambers, Third Australiasian Instrumentation and Measurement Conference, Adelaide, Australia, April 1994

4 Hatfield, M and Freyer, G, Radiated Immunity Testing of Aircraft Systems Using Reverberation Chambers, Third Australiasian Instrumentation and Measurement Conference, Adelaide, Australia, April 1994

5 Hatfield, M and Freyer, G, Attributes of a Reverberation Chamber For Radiated Immunity Testing, Third Australiasian Instrumentation and Measurement Conference, Adelaide, Australia, April 1994

6 Improvements in the CW Evaluation of Mode-Stirred Chambers, Ladbury, J ,Keopke, G, Camell, D, National Institute of Standards and Technology

7 Discussion between K Goldsmith and N Carter, Under AAMOUR Topic 120, December 1997, Defence Experimental and research Agency, Farnborough, UK

8 Johnson, P and Goldsmith, K, An Experimental Study of the placing of an Aircraft Inside a Large Zinc-Plated Reverberation Chamber, 17th Digital Avionics Systems Conference, Bellevue Washington, 1998

9 Johnson, P, and Goldsmith, K Computational Electromagnetic Modelling of an Aircraft-Sized Reverberation Chamber, EMAC Adelaide, Australia, July 1998

10 A Users View of GEMACS, Johnson P.A. and Goldsmith K.R., DSTO Technical Note DSTO-TN-0100 (1997)

11 Harrington R.F., "Matrix Methods for Fields Problems", Proceedings of the IEEE, 55, 136-149 (Feb 1967).

12 Kouyoumjian R.G., "The Geometrical Theory of Diffraction and its Applications" in Numerical and Asymptotic Techniques in Electromagnetics (ed. Mittra), Springer-Verlag, New York (1975)

13 Lehman, T, Statistics of MSCs, presented at the Anechoic Chamber and Reverberation Chamber Operators Group Meeting, Naval Surface Warfare Centre, Dahlgren, VA, November 1992

14 Lehman, T, Statistical Theory of Electromagnetic Fields in Complex Cavities, Interaction Notes, Note 494, USAF Phillips Laboratory, May 1993

15 Freyer, G, et al, Verification of Fields Applied To An EUT IN A Reverberation Chamber Using Statistical Theory, IEEE EMC Symposium, August 1998.

AN EXPERIMENTAL STUDY OF THE PLACEMENT AN AIRCRAFT INSIDE A LARGE WELDED ZINC-PLATED STEEL ELECTROMAGNETIC REVERBERATION CHAMBER

Paul A. Johnson and Kevin R. Goldsmith, Defence Science and Technology Organisation, PO Box 1500, SA 5108, Australia

Introduction

Electromagnetic Compatibility (EMC) and full EM Interference (EMI) testing are extremely important for electronic equipment at a component, system and platform level. This is true for both the civilian and military markets. For the latter it is important that the manufacturer must ensure the original military equipment performs in some stated environment without interfering with other equipment, and the Australian Defence Force (ADF) must ensure that a subsequent modification remains EMC compliant.

Australia is a consistent modifier of platforms, owing to its unique geography, location and climate, and the ADF is vulnerable to modification-induced EMI problems. It is therefore of particular importance for aircraft that no such problems exist, as the consequences can be catastrophic or expensive. Whilst there are no recorded aircraft losses due to EMI in Australia, it cannot be said they have not happened because there is often no physical evidence left by EMI. Indeed, there is great concern worldwide about EMI related aircraft problems amongst civil and military operators, with the rise of internally carried consumer electronics (eg. laptops and mobile phones), external sources (radar, radio, TV and communications emitters), the increasing digitalisation of aircraft electronics and the advent of fly-by-wire control systems.

Standard EMI and EMC testing facilities include Open Area Test Sites (OATS), Anechoic Chambers, Transverse Electromagnetic Mode (TEM) Cells and Shielded rooms. In addition, Electromagnetic Reverberation Chambers (RC) are rapidly gaining acceptance as an alternative, though their operation is imperfectly understood. They have many advantages, such as high efficiency (a large electric field is achievable for a modest RF input power), a uniform, isotropic and randomly polarised field, and highly predictable statistical behaviour.

DSTO has built one of the largest Reverberation Chambers in the world at its Salisbury site. A description of its design, commissioning and characterisation can be found elsewhere in these proceedings [1].

Recently, a military aircraft (Blackhawk S-70A-9 helicopter) was temporarily installed into the chamber. It is believed that this is the first time a platform of this type has been studied in an RC. Using low-level fields only, the chamber was re-characterised, and the effect of the aircraft on the chamber deduced. In this paper, we describe the results of this work, in particular the derivation of the loading effect of this aircraft upon the chamber. This information can now be used to estimate what RF power is required to achieve full threat–level fields inside an aircraft-filled chamber.

0-7803-5086-3 /98/$10.00 ©1998 IEEE

Description of the Chamber

The DSTO Chamber was constructed inside an existing large building (an old diesel engine workshop) on the Salisbury site. Its dimensions are given in Table 1. It was constructed from steel sheets of thickness 2 mm, with a 90 micron zinc coating, and the seams welded throughout the entire chamber. The chamber has two personal access doors, and two large doors which form one entire wall, enabling large test objects to be introduced.

At one end of the chamber a large tuner assembly is mounted. The tuner is required for exciting modes in the chamber. A central shaft runs floor to ceiling, driven by a gearbox and induction motor on the roof, and attached to the shaft are four large (4m x 4m) paddles of various opening and rotation angles.

Chamber Parameters	
Length, L	20.85 m
Width, W	10.89 m
Height, H	6.0 m
Volume, V	1361 m^3
Surface Area, S	835 m^2

Table 1: Reverberation Chamber Parameters

The chamber is more fully described elsewhere [1].

Experimental Method

Installation of Aircraft into Chamber

The aircraft landed at approximately 1.30pm on a Thursday, on a hard standing area adjacent to the chamber building. The ground crew were already standing by. The four main rotors were removed, and then the aircraft pushed by a ground-cart and by hand into the chamber. The aircraft cleared the doorway in the main building only by 20cm each side, and the doorway lintel on top only by rotating the tail rotor very carefully! By the time the main chamber doors were sealed up, it was approximately 4.30pm. Only three hours were therefore required to install the aircraft into the chamber. The aircraft was available to us for the weekend, and left on the Monday. A similar amount of time was required to remove the aircraft to the point of its departure. In future, we believe an aircraft could be in and out the same day within a couple of days, having been subjected to a modification test or periodic retest, which is important when aircraft are removed from service.

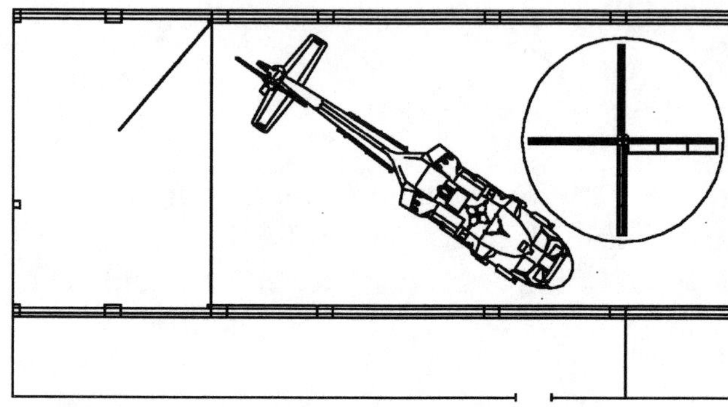

Figure 1: Scale Drawing of location of Aircraft Inside Chamber

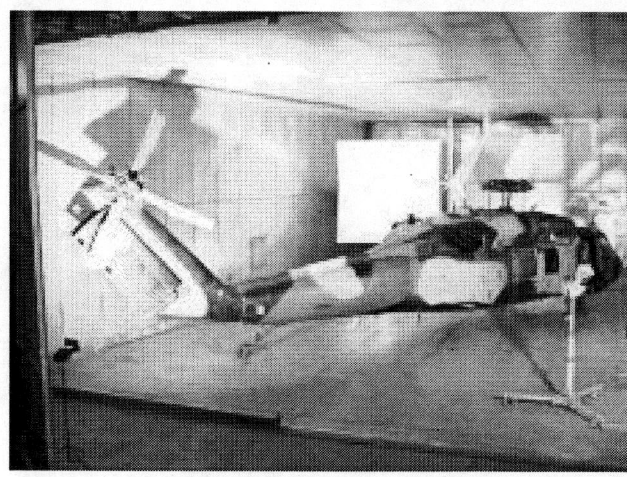

Figure 2: Aircraft installed into chamber, with main chamber doors still open.

The location of the aircraft inside the chamber building is shown to scale in Figure 1, and pictorially in Figure 2. It is evident that there is not much clearance between the aircraft and the walls or paddle. For this reason, the stabilator was in the folded position for this experiment.

Characterisation of Chamber

The method used to characterise the chamber was the same for both the empty and aircraft-filled case. The signal source was a HP series 71209A Spectrum Analyser with Tracking Generator up to 18 GHz, with output power set to –10dBm. The signal was fed to a Power Meter via a directional coupler, as well as the chamber, for forward power measurement, P_{in}. A series of approximately 35 discrete frequencies were used between 26 MHz and 18 GHz. Mode stirring was employed, where for a given discrete frequency, the paddle wheel was allowed to turn continuously rotated at 1 RPM, and the signal recorded for one full rotation.

Below 1 GHz two EMCO 3143 and 3141 biconilogs were used as transmit and receive antennas, and above this frequency two EMCO 3115 Double-ridged horns were used. When in use, the TX biconilog was placed facing the upper corner behind the paddle, and the RX biconilog was placed facing the opposite upper corner. In the higher frequency range, the horns were both placed in the same corner nearest the instrumentation (cable losses would otherwise be severe at the highest frequencies). The TX horn faced the upper corner, and the RX horn towards the centre of the chamber. See Figure 3 for antenna locations.

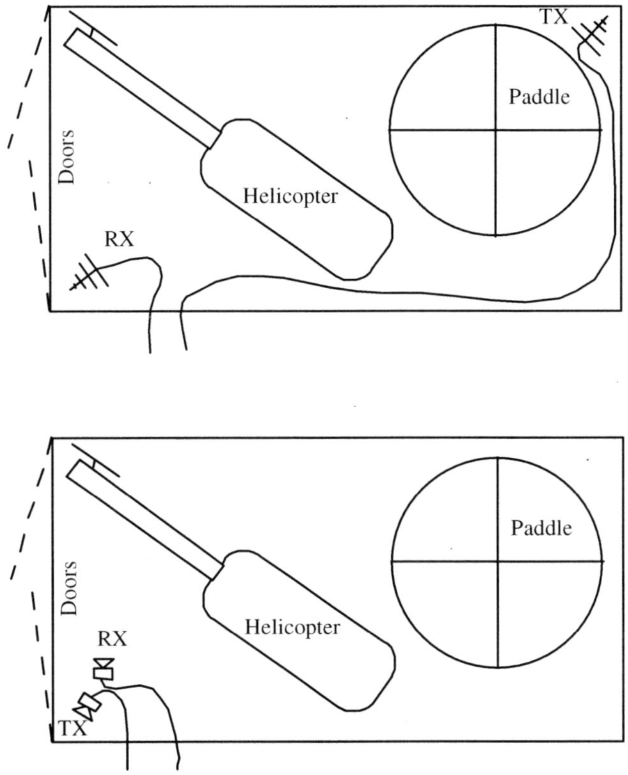

Figure 3: Schematic layout of chamber for the two frequency ranges: below 1GHz (top) and above 1GHz (bottom). This is identical to empty-chamber layout.

The equipment set-up was identical for both the empty and aircraft-filled chamber (eg: cables, directional couplers, pre-amplifiers, antennas etc). The cable losses were measured at the same frequencies used for the characterisation.

Results

When mode–stirring, for a particular frequency the power received from the antenna is recorded by setting the trace time on the spectrum analyser to 60 sec, in order to capture a full turn of the paddle wheel. The whole data acquisition process is automated, and a file of 1001 data points is saved, along with a description file containing run information. The paddle position resolution is therefore 0.36 deg. A sample signal plot is given in Figure 4.

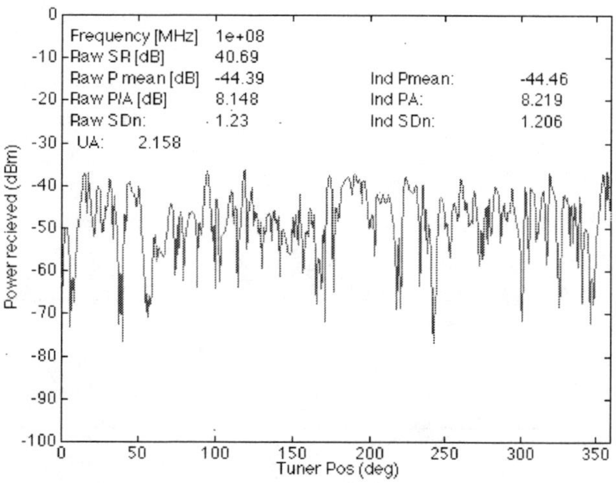

Figure 4: An example plot of signal against paddle position, for a frequency of 100 MHz.

For each plot, the following information is derived: the stirring ratio (SR), which is the maximum to minimum signal (conventionally, it is accepted that proper operation of the chamber requires at least 20dB [2]); average received power $\langle P_{rec} \rangle$, (which must be a linearised mean); peak–to–average ratio (PA), the ratio of the maximum to the mean, (which should theoretically be about 7-8dB, depending on the number of independent data points); normalised standard deviation (σ_n), a linearised standard deviation normalised to the mean (which should be unity).

It is important to obtain stastically sufficient independent data points for the analysis to be valid. The field in the chamber changes significantly only when the paddle has turned by an equal to what we term the "uncorrelation angle" (θ_u). The θ_u is frequency dependent, and is determined for each frequency by correlating the signal plot with itself, shifting the plot one paddle position at a time until the correlation coefficient drops to 1/e at the uncorrelation angle. An example of the correlation function is given in Figure 5, where the plot in Figure 4 becomes uncorrelated after 6 paddle steps, or 2.1 deg. Having determined θ_u, every independent point is sampled (every sixth point in this example), to determine corrected versions of <Prec>, PA, and σ_n.

If the chamber Electromagnetic Environment (EME) is fully random, the Probability Distribution Function (PDF) of the E field at a given point should follow that of a χ^2 (Chi-squared) distribution with 2 degrees of freedom ($\chi^2{}_2$). Representing this distribution in cumulative form makes arbitrary binning of data unnecessary, and so the Cumulative Distribution Function (CDF) was obtained for each data set. An example is given in Figure 6. The correlation coefficient, γ_{CDF}, of the distribution compared with the theoretical one is also calculated, together with the Kolmogorov-Smirnov probability [3] (p_{KS}) that the CDF is drawn from the theoretical distribution.

In the past it has been empirically determined that the EME should be fully random below that frequency above which there exists at least 60 modes [2]. For this chamber, that critical frequency is 60 MHz.

The quantities SR, PA, σ_n, θ_u, γ_{CDF} and p_{KS} are given in Figure 7 – Figure 12 respectively, for both the empty and the aircraft–filled chamber. In figures comparing the empty and loaded chamber, circles are used for the empty chamber, and crosses for the aircraft–filled chamber.

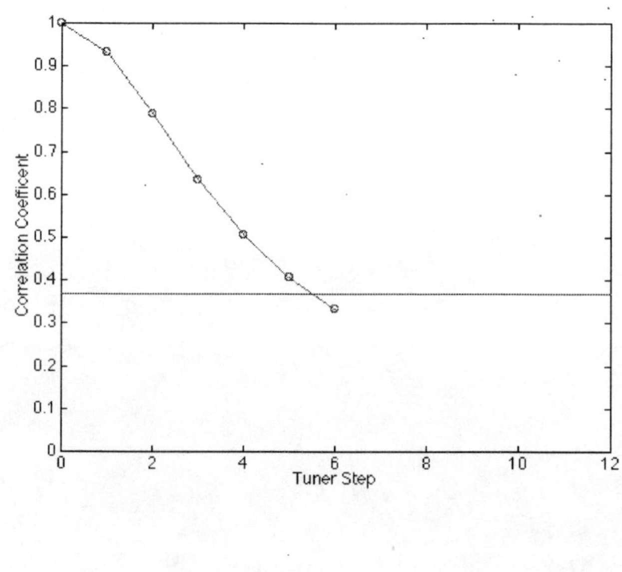

Figure 5: An example of the correlation coefficient varying with paddle position.

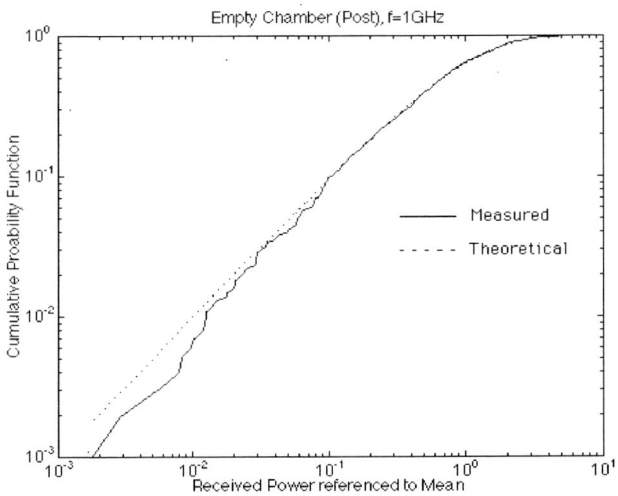

Figure 6: Cumulative Probability Function of the Received Power Referenced to the Mean (Empty Chamber at 1GHz)

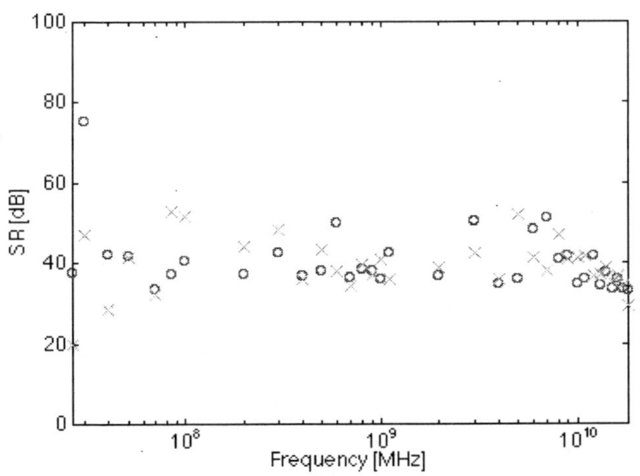

Figure 7: Stirring Ratio. (o = empty chamber, x = aircraft–filled chamber).

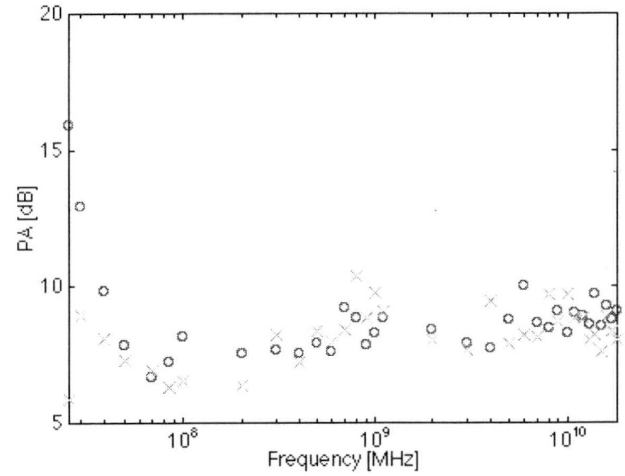

Figure 8: Peak to Average Ratio. Symbols as in Figure 7.

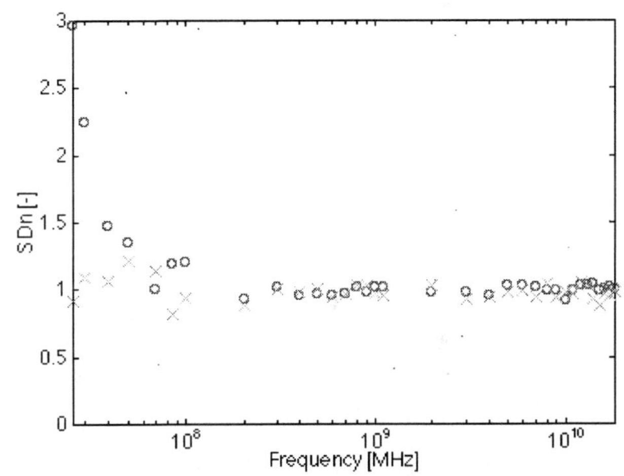

Figure 9: Standard Deviation, normalised to mean. Symbols as in Figure 7.

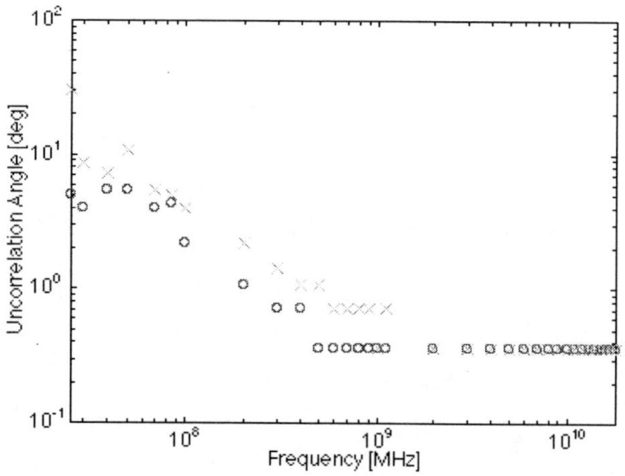

Figure 10: **Uncorrelation Angle. Symbols as in Figure 7.**

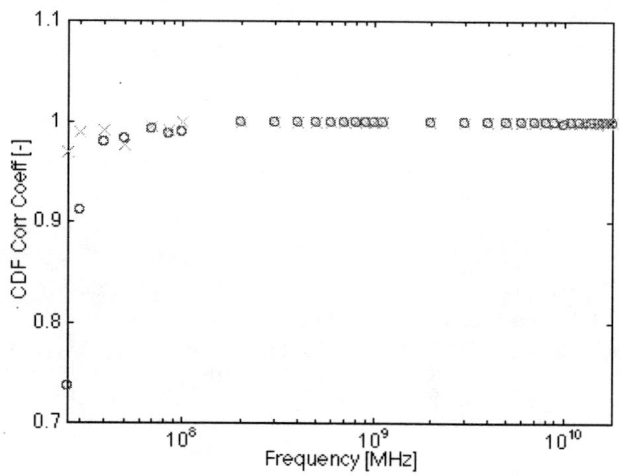

Figure 11: Correlation Coefficient of the Cumulative Probability Distribution Function compared with theory. Symbols as in Figure 7.

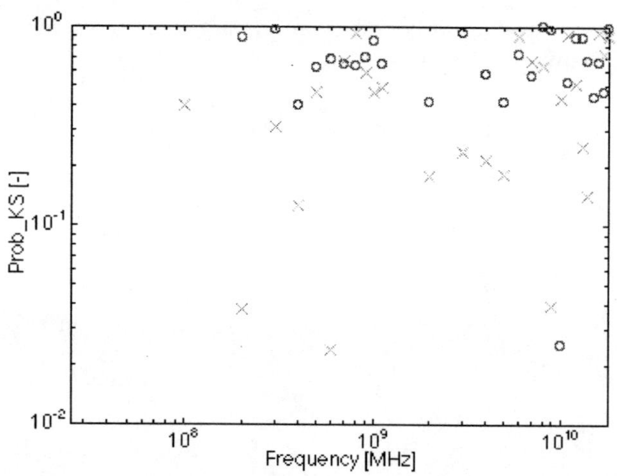

(a)

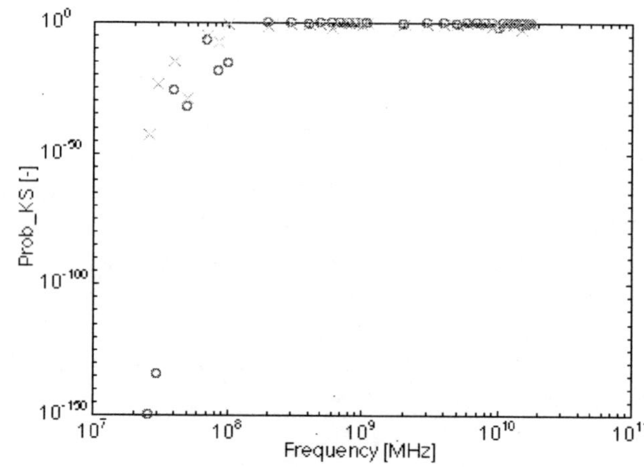

(b)

Figure 12: The probability that the CDF is drawn from the theoretical distibrution ie. That the EME is fully random (a) in fine detail (b) overall.

The Q of the chamber comprises dominant components due to the antenna, Q_{ant}, and due to the walls of the chamber, Q_{wall}, where

$$Q_{ant} = (16\pi^2/\eta mc^3)\ V\ f^3$$

$$Q_{wall} = 3(V/S)\ (f/MHz)^{0.5}\ (\sigma_{con}/\mu_r)^{0.5}$$

and the total expected Q_{theo} of the chamber, is given by

$$1/\ Q_{theo} = 1/Q_{ant} + 1/Q_{wall}$$

For the loaded chamber, an additional component is due to the aircraft itself, Q_{ac}, causing the Q to fall by an amount ΔQ. The determination of this parameter is the chief aim of this paper.

The measured Q of the chamber can be calculated from

$$Q=(16\pi^2/c^3)\ V\ f^3\ (<P_{rec}>/P_{in})$$

The theoretical Q for an empty chamber, and the measured Q for empty and loaded chamber is given in Figure 13, and the drop in Q caused by the aircraft, ΔQ, is shown in Figure 14. An earlier measurement of the empty chamber is given in the same Figure.

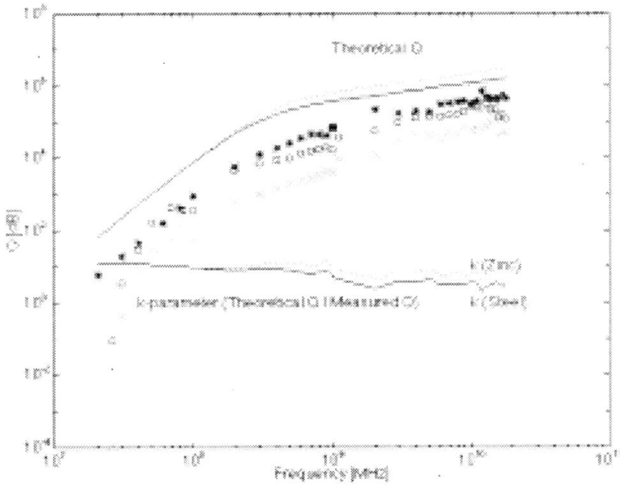

Figure 13: **Q of chamber. (o = empty chamber, x = aircraft–filled chamber, * = previously derived data).**

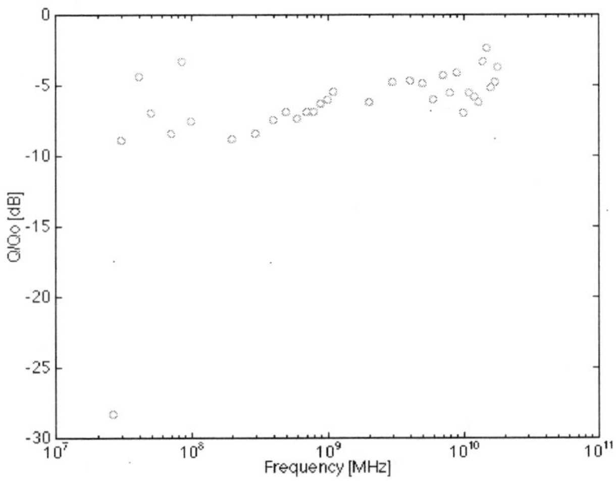

Figure 14: **Change in Q caused by presence of aircraft, ΔQ**

Finally, the expected E-field achievable with 1W of input power is given in Figure 15.

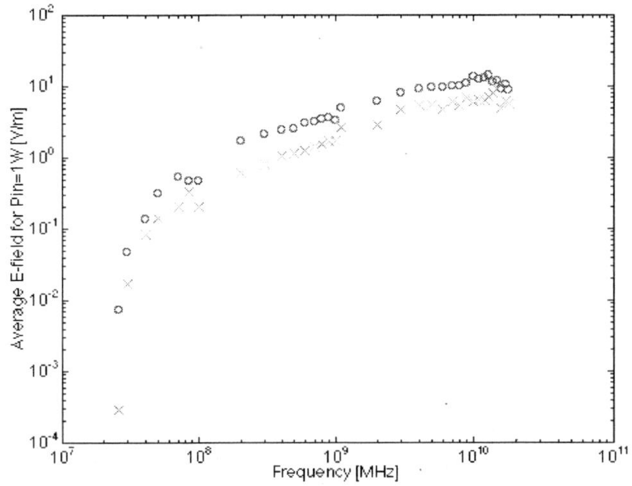

Figure 15: Electric Field achievable for 1W of input power. Symbols as in Figure 7.

Discussion

There appears to be little difference in the statistical parameters between the empty and loaded chambers. The major difference is in the change of chamber Q, due to the loading caused by the reflecting and absorbing materials which comprise the structure of the aircraft.

Statistical Properties

The Stirring Ratio (SR), is between 20dB and 50dB for the whole frequency range for both cases. See Figure 7. This is a requirement, but not sufficient it itself for the chamber to perform correctly.

The normalised standard deviation (σ_n) agrees very well above 100MHz with the theoretical value of unity in both cases. See **Figure 9**. Below this frequency, there is some scatter in the values for the aircraft–filled chamber and an interesting departure towards 2–3 for the empty chamber.

The correlation coefficient of the CDF compared with the theoretical value for a fully randomised EME, γ_{CDF}, is unity across the frequency range for the aircraft–filled chamber, dropping only to 0.97 at 26MHz. See Figure 6. For the empty chamber, however, unity is only reached above 40MHz, actually falling to 0.7 at 26MHz.

The probability that the EME is random, given in Figure 12, suggests a good agreement above 100 MHz and, but rapidly decreasing probability below this frequency. It is important to realise the difference between the parameters γ_{CDF} and p_{KS}. With large amounts of independent data, it is possible to say that two distributions are very similar (say $\gamma_{CDF} = 0.99$), but that the probability that they are exactly the same is very low (say $pKS = 10^{-5}$). The two variables provide complementary information.

Taken together with the standard deviation data, the data suggests that at frequencies below 100MHz, chamber EME is actually statistically better with the aircraft in place: although the Q is less, the relationships between Peak and Average field etc. are more determinate. This could be due to the aircraft causing more reflections inside the chamber, preventing direct coupling between the TX and RX antenna. widening the bandwidth of the Q to included more modes at the low frequencies, effectively adding more low frequency modes to the chamber. In both cases the EME is random down to frequencies below the empirically determined 60 MHz. Conversely, at frequencies above 100MHz, p_{KS} implies that the data are better behaved in the empty chamber (Figure 12a).

The uncorrelation angle θ_u varies smoothly from 5 deg for the empty chamber, and 10–20 deg for the loaded chamber, to 0.36 deg at higher frequencies. See Figure 10. This latter value represents the resolution of the sampled paddle position (the intrinsic paddle positional accuracy is actually 0.1 deg, but the spectrum analyser stores only 1001 samples). At lower frequencies, it is interesting that successive paddle positions appear less correlated for the loaded chamber compared with the empty chamber, even though the field is more random when averaged over a full turn of the paddle wheel.

The peak-to-average ratio, PA is in the range 6–10 dB for the whole frequency range for the aircraft–filled chamber. See Figure 8. The theoretical value of PA depends on the number of independent samples: for N=1000 it is 8.5 dB, whilst for N=20 it is only 5 dB. The measured values seem to agree well with the theoretical values. For the empty chamber, there is similar agreement except at the lowest frequencies, where the chamber EME is not random.

The Q of the Chamber

It became evident after the experiment was completed that a power divider had been used in the low frequency range rather than a directional coupler. This had the effect that instead of reading forward power only, the total power (forward and reverse) was sampled by the power meter. At the lower frequencies, where the VSWR becomes bad, this causes the chamber Q to appear to roll-off. The earlier results from a set-up that used a directional coupler do not show this roll-off. Fortunately, this error should not greatly effect the *difference* in Q between the empty and loaded chamber, only its *absolute* value.

Other than this roll-off, there is good agreement between the old and new sets of empty chamber Q.

If we compare the old empty Q data with that expected from theory, by calculating

$$k = Q_{theory}/Q_{measured}$$

we find that the so-called “k-parameter” varies from a factor of 10 at lowest frequencies, to about 3–5 at the highest frequencies, the latter depending on whether a theoretical chamber of steel or zinc is considered. Better agreement is found with a steel chamber, which seems reasonable, since the chamber walls contain far more steel than zinc. This range of k compares favourably with other chambers [4].

0-7803-5086-3 /98/$10.00 ©1998 IEEE

The difference in Q between the empty and loaded chamber is probably the most important result of this paper. It shows that the presence of the aircraft loads the chamber by an amount of about 10dB at 26MHz, rising smoothly, with some scatter, to 5dB at 18GHz. This is encouraging, as it might be expected that a modern military aircraft, comprising lots of composite and plastic would load the chamber more than this amount. This makes it feasible to produce high electric fields inside the chamber, even in the presence of an aircraft, with affordable amounts of RF power.

Conclusions

The EME inside a large EM reverberation chamber has been characterised, both for an empty chamber, and also with an aircraft–filled chamber.

For the empty chamber, the EME appears to be random down to about 40 MHz, which is better than expected. The Q is within a factor 3–10 lower than expected from theory, but this compares favourably with other chambers.

For the chamber loaded with an aircraft, the EME appears to be random all the way down to 26 MHz, much better than expected, and surprisingly better than the empty chamber. Conversely, the paddle appears more effective at lower frequencies in the empty chamber. It is not obvious why this should be the case. The Q in the aircraft–filled chamber is in the range 5–10 dB lower, depending on frequency.

Since the loading of the chamber by the aircraft is not severe, high field strengths will be achievable for modest power input.

References

[1] Goldsmith K. and Johnson P.A. "Design, Construction, Computational EM Modeling, and Characterization of an Aircraft–Sized Reverberation Chamber and Stirrer", these Proceedings.

[2] Crawford M .L. and Koepke G.H., National Bureau of Standards Report, Technical Note 1092 (1986)
[3] Press W. H. et al., "Numerical Recipes", pub. Cambridge, pp. 472-475 (1986)
[4] Hatfield M.O., Freyer G.J and Slocum M.B., Naval Surface Warfare Center Report NSWCDD/TR-96/149 (1996)

0-7803-5086-3 /98/$10.00 ©1998 IEEE

E. Discipline Track – Human/Machine Engineering

This track will provide a forum for discussion of the requirements of the humans who will ultimately use avionics and electronic systems. Sessions will span diverse domains, ranging from air traffic control and aircraft automation to cockpit, spacecraft and ground vehicle displays. The applied nature of this conference provides an opportunity to present research and development work that has implications for current design practices and operations.

Track Chair: Ms. Amy Pritchett, Georgia Tech

Track E - Human/Machine Engineering
Session E1: Methods & Measurements

Chairs: Dr Kathy Abbot, FAA National Resource Specialist; Amy Pritchett

1	A Process for Creating a Human-Centered System	Dr. Sheryl L. Chappell Ph.D.
2	Human Performance Considerations in Aircraft Certification (Not Published)	Dr. Elizabeth Lyall Ph.D.
3	A Pilot-Centered Autoflight System Concept	Mr. Victor Riley
4	Making the Cockpit a Human Friendly Environment: A Customer's Perspective	Mr. Edward A. Rafacz Jr.
5	Future Flight Deck Design - Emerging Safety Issues (Not Published)	Dr. Ratan Khatwa
6	Real-Time Operational Evaluations Using Advanced Flight Simulators	A.E. Dillard

Session E2: Automation & Decision Aiding

Chair: Paul Schutte, NASA Langley Research Center

1	An Algorithmic Method for Predicting Pilot-Mode Interaction Difficulties	Denis Javaux
2	Analyzing Mode Transition Matrices	Sanjay Vakil
3	Human Factors Issues of Flight Deck Automation	Mr. Kenneth H. Funk II
4	Operator Directed Common Conceptual Models for Advanced Aircraft Automation	Sanjay Vakil
5	Some Formal Aspects of Human-Machine Interaction (Not Published)	Dr. Asaf Degani Ph.D.
6	Improving Pilot Conformance to Alerting System Commands	Ms. Amy Pritchett
7	Task Engagement: A Means to Situation Awareness	Mr. Paul C. Schutte

Session E3: Traffic Information Displays

Chair: Anand Mundra, MITRE

1	CDTI Feature Evaluation for Enhancement of Oceanic Operations	Mr. Baltazar O. Olmos
2	Cognitive Factors In Aviation Display Design	Mr. Christopher D. Wickens
3	An Operational Implementation of CDTI (Not Published)	Mr. Bob Hilb
4	Pilots Spatial Mental Models for Memory of Heading and Altitude	Ms. Susanne Delzell
5	Shared Situation Awareness in the Flight Deck-ATC System	Professor John Hansman Jr.
6	Investigation of Cockpit Display of Traffic Information as an Aid for More Efficient Air Traffic Management Procedures	Ms. Amy Pritchett

Track E - Human/Machine Engineering (Continued)

Session E4: Advanced Displays

Chair: George P. Boucek, Boeing Commercial Airplane Group

1	Experimental Study of Helmet-Mounted Display Symbology for Terrain Avoidance During Low-Level Maneuvers	Mr. James K. Kuchar
2	Advanced Flight Deck for Next Generation Aircraft	Mr. Robert Faerber
3	Projected Flight Path Displays and Controlled Flight into Terrain Accidents	Mr. Barry Strauch
4	A Display Concept for Staying Ahead of the Airplane	Mr. Eric N. Johnson
5	The Concept of 'Brain Mapping' for Flight Safety Avionics (Not Published)	Mr. Ivan Y. Burdun
6	Spatial Terrain Displays: Promises And Potential Pitfalls	Dr. Ir. Eric Theunissen

Session E5: Navigation & Guidance Displays

Chair: George P. Boucek, Boeing Commercial Airplane Group

1	Synthetic Vision Flight Tests For Curved Approach And Landing	Dr. Gottfried Sachs
2	GPS User-Interface Design Problems	Mr. Kevin W. Williams Ph.D.
3	An International Usability Survey of GPS Avionics Equipment	Mr. Kurt M. Joseph
4	The 4-D Display	Dr.-Ing. Harro Von Viebahn
5	4D Flight Guidance Displays, A Gate to Gate Solution	Pro. Dr. -Ing Wolfgang Kubbat
6	Preliminary Flight Tests of Pseudo-Attitude Control Using Single Antenna GPS Sensing	Mr. Richard P. Kornfeld

CERTIFICATION OF USABILITY:
A PROCESS FOR CREATING A HUMAN-CENTERED SYSTEM

Sheryl L. Chappell, Delta Technology, Atlanta, GA

Abstract

If a system or procedure is being developed that will be used by humans, the design should be human-centered. Human-centered design capitalizes on and accommodates human skills in perception, attention, and cognition, while minimizing the opportunities for and effects of human error. This paper describes a mature practical process for accomplishing this goal of human-centered design. Usability is the foundation of this process.

The first step in creating a human-centered system is to determine who the operators are, how, where, and what they are using the system for now, and what they would like to use it for in the future. The next step is to set relevant and realistic usability goals for the user interface of the new system. These goals include the time to accomplish the task and error tolerance.

The final step is to perform usability testing, starting with a prototype. Many usability issues will become apparent by creating scenarios that 1) exercise a representative sample, if not all, of the operator functions and 2) provide a realistic operational environment for testing. By collecting and analyzing usability data, specific interface features can be evaluated and any mismatch between the design and the operational use will be revealed. Through iterative design improvements the final product will be easy to use and difficult to use incorrectly.

Introduction

The reasons for designing and producing highly usable systems are obvious. When the design centers around the user, the product can enjoy a broad acceptance and a smooth integration into operation. Product liabilities are minimized and training is reduced. Designing a usable system is accomplished by making the user the central focus in all aspects of the interface design. This process includes defining user requirements, designing a user-friendly interface, and testing and validating the usability of the product.

While these steps may not be universally followed in developing aeronautical systems and have not been common practice in the certification process, they are a mature science in the development of computer systems. Software and hardware product usability are highly correlated with acceptance and revenue. Therefore, the usability engineer has become an integral part of the development team in the computer industry. The author draws from both the computer industry and the aeronautical human factors research discipline to create a practical guide for system developers. (For a thorough text on the issues of Human Factors Certification see Wise, Hopkin and Garland, 1994.)

Creating a good human-centered design is not done with smoke and mirrors, but rather by following a clear, predictable process. This paper will provide a step-by-step procedure for each phase of the development of user-centered systems. Note that while this paper discusses the design and development of a system, these same steps may be applied to the design and development of a procedure or service.

User Requirements Definition

The following steps are a general recipe for creating a usable system. These steps may

0-7803-5086-3 /98/$10.00

be customized for the specific product being developed. For example, updates or revisions to existing systems may not require all of these steps.

There are several things that need to be considered before designing a human-centered system. These considerations include understanding *who* the users are, understanding *what* activities they do in their jobs, understanding *how* they accomplish these activities, and examining *where* they do their jobs. Each of these issues will be explored.

Operator Profiles: Who

There are two steps to answering the 'who' question. First the user interface designer must determine the roles and responsibilities of the target operators and any additional operators. The second step is to collect and analyze operator demographics. Of special importance is the amount of training on the use of any existing systems. For example, if the system being developed requires the use of a pointing device, it is easy to determine what percentage of the target population has previous experience with each type of pointing device. Matching the new product with users' experience will reduce the requirements for training and improve user acceptance.

Operational Task Analysis: What and How

After determining who the users are, the next important step in the initial user interface design is determining what activities users do in their jobs and how they do them. This is determined by conducting an operational task analysis. The task analysis will provide the insight into what the users do and how they do their jobs. Steps in conducting a task analysis include the following: 1) determine the various operators' goals, 2) analyze how the operators now accomplish their goals, 3) ascertain effects of concurrent activities, interruptions, and use of additional systems, 4) document job aids that are used in conjunction with the systems, including hard-copy reports, manuals, and cheat sheets, 5) determine how operators communicate with others regarding the use of the system, e.g., input from a nearby coworker, 6) determine current and desired output requirements of the system, including performance standards, and 7) reduce operators' tasks to the core elements and actions, i.e., what information is used and what action needs to be taken upon each information element. Note that the abstraction of core elements from the current tasks eliminates the effects of the current interface on the process of accomplishing operator goals.

It's important in conducting the task analysis to not overlook other activities that are perhaps orthogonal to the primary tasks. A testimony to the lack of this consideration is the fact that aircraft systems have often been damaged by spilled beverages, particularly on the center console. Providing room in the cockpit for food and beverages to be accessed and consumed must be part of the design.

Environmental Analysis: Where

The environmental analysis will assess operators' physical work environment, including location, lighting conditions, noise, workstation layout, shared workstations and methods of communicating with others. These considerations will guide the design of a system that integrates well in the work environment.

Establish Usability Objectives

Set relevant and realistic usability goals for the user interface of the new system. Determine 'how good is good enough'. These goals must be agreed upon by the management and the operators. Factors to consider include: time to accomplish each action and task, error tolerance, and training requirements.

For example, if the system will be used by both novices and experts, one objective may be that novices will be able to perform tasks as quickly as experts after performing a task five times. As another example, if another system is being used already, and the new system will replace it, one usability objective may be that users will be 20% more productive with the new system.

Note that in the development phase of the project, practical implications of the design may require a reevaluation of these goals. In the extreme case, certain design features necessary to accomplish these goals may be impossible to incorporate due to development costs, product pricing, or other pragmatic considerations.

User Interface Design

To design a usable system, the designer must follow user interface design principles, adhere to user interface standards and base the interface on an appropriate user metaphor. Each of these key ingredients is described in this section.

User Interface Design Principles

Human-centered interface design principles have been developed that take advantage of human skills in perception, attention, and cognition, while minimizing the opportunities and effects of human error. These design principles have evolved as systems have become more flexible. Operator sophistication has also evolved.

There are many texts that describe in detail user interface principles (e.g., Laurel, 1992 and Nielsen, 1993). This paper will only briefly list several principles that must be followed in the development of the user interface.

Consistency

The interface must be consistent in all aspects including: appearance, terminology, functionality, and navigation. As if this is not challenging enough, the system should also strive to be consistent with all other systems in use by the operator. A simple example of a violation of the consistency principle is the ACARS touch-panel alphabetic keyboard in some transport aircraft that uses a different alpha-numeric key layout than the keyboard on the control display unit.

Feedback

Effective feedback is appropriate, informative, and timely. It should not be presented only when an error situation has occurred, but instead should continuously inform users about what the system is doing. Feedback can be visual and/or auditory.

Users need to remain informed about: 1) where they are within the user interface, 2) the current state of the system, 3) the current state of data being acted upon, and 4) changes in the state of the system or data being acted upon.

Providing feedback to the users allows them to see how the actions they take affect the behavior of the system. It helps users to understand what happened as a result of actions taken and whether or not these actions were successful. In the event of error, feedback informs users what steps need to be taken to correct the problem.

Forgiveness

Forgiveness is the extent to which an application tolerates common and unavoidable human error and allows for easy recovery from any undesired actions taken by the user. Forgiveness is an important principle of user interface design. The user should feel confident that the system is robust enough to handle any kind of input, including errors.

Forgiveness qualities include: 1) allowing only appropriate commands or choices, 2) providing 'undo/redo' and 'cancel' functions, 3) requiring users to take explicit 'enter' actions to initiate processing of data, 4) providing "are you sure" messages for commands that may have destructive consequences, 5) providing clearly marked 'emergency exits' when system functions are chosen by mistake, and 6) avoiding modes, whenever possible.

Simplicity

A fairly common mistake in interface design is to try to provide all the functionality that any user could possibly ever want or need. The result is often a very complex interface that is overwhelming and confusing to the user. It is better to provide less functionality, that will be used, than so much functionality that the interface actually inhibits use of the system.

Present exactly the information the user needs, at exactly the time and place where it is needed. Use defaults, whenever possible. Eliminate unnecessary or extra steps for navigating through the system. Consolidate multiple page displays, whenever possible. Layer the interface so that novices can gradually use more advanced and sophisticated functions.

Familiarity

Familiarity is the principle of using facts that are already known by the user to facilitate usage of the new system. If there is a great deal of carry-over of user interface behaviors from other systems, training is minimized. Guidelines for familiarity include: 1) make information objects and operations accessible in a sequence that matches the way users will most effectively and productively do things, 2) provide familiar metaphors, wherever possible, and 3) avoid technical jargon and unfamiliar concepts and representations.

User Control

A usable and effective user interface puts the user in charge of the system, thereby allowing the user to initiate all actions, rather than being controlled and directed by the application. User control of both the system and the interface allows users to learn more easily how the system and interface work. User control also provides a one-to-one correspondence between an action taken by a user and the system's response to that action.

Flexibility

Flexibility is the extent to which a user interface accommodates for variations in user skill and preferences. Flexible user interfaces are especially important for user tasks that are inherently unstructured, since they allow for user tailoring and user choices. Furthermore, flexible interfaces allow more user control over a system. To ensure that user interfaces are flexible, developers should: 1) enable users to tailor frequent actions and take shortcuts. 2) cater to both inexperienced and experienced users, 3) allow users to personalize aspects of the interface to suit their skill level and/or preferences, 4) provide accelerator mnemonics for expert users, and 5) provide abbreviations, special keys, hidden commands, and macro facilities to knowledgeable users.

Users will soon gain a sense of mastery and will feel encouraged to explore the system as long as the user interface is familiar, simple, consistent, flexible, and forgiving.

User Interface Standards

What Standards Are

User interface standards are necessary to build an efficient user interface. The application of standards results in decreased training and support time and increased usability and user satisfaction. Many interface standards exist, including the US military standard (MIL-STD-1472), the NASA standard (NASA-STD-3000/Vol.1/Rev.A), the FAA Aircraft Alerting Systems Design Guidelines (DOT/FAA/RD-81/38,I), and standards for computer systems (e.g., Smith and Mosier, 1986).

Standards are the first element in designing a user-centered system. One can compare designing an interface to designing a house. The standards are the building codes and materials. Standards provide at least a partial template for the interface. Following the standards ensures that the product will meet some minimum specifications.

0-7803-5086-3 /98/$10.00 ©1998 IEEE

What Standards Are Not

Having the right materials is only the first step. In order to build a good house, a design plan is needed, the proper specialists are required (electricians, plumbers, and carpenters), and the last ingredient is input from the buyers. Without all of these elements, the final result may very well be a poorly constructed house that may be unsafe to live in, regardless of how good the building materials are. Another possibility is to wind up with a well-constructed house that no one wishes to purchase because it does not meet any buyers' needs. For example, the house may have a kitchen that is not accessible, except from the outside. The same holds true for designing and building a good interface. Following the proper standards is a necessary, but not sufficient, step in creating a usable interface.

Creating the Human Interface

The design principles and standards provide a framework upon which to design a human-centered interface. The actual design can take on an infinite variety of characteristics, as paintings constructed from the same palette can have an infinite variety of appearances.

Operator Functions

The primary consideration in the design of the human interface is function allocation, the definition of the roles of the system and the roles of the human (i.e., who does what). This issue has been addressed by many for a long time (e.g., Fitts' List (1962)) and there are no all-encompassing answers. It is important to question all functions. Redefining the task and the roles may greatly improve the overall human-machine performance. Small initiatives in functionality can make a big difference in usability; for example, if data reside in the system, the human should not reenter those data. This will reduce errors and increase task productivity.

Operator Metaphors

There is usually a variety of useful metaphors for any system, however one metaphor may emerge as the desired solution due to transfer from an existing system. The designer must determine the appropriate operator metaphor for the over-all design and the interface elements. For example, when the Traffic-alert and Collision Avoidance System (TCAS) was being designed for aircraft, the pilots on the design team created a traffic display that used a plan-view to show the relative location of the traffic. This information was displayed on the radar screen, which used the plan-view presentation and was familiar to pilots. The information for the maneuver required to avoid traffic was displayed on the vertical speed indicator. Logically, an up arrow was displayed to tell the pilot to climb and a down arrow indicated a descent was required for traffic avoidance. This design will be discussed further in the next section on usability testing.

Prototype Development

A pivotal step in the design process is to create a user interface prototype. Prototypes are developed with increasing fidelity and decreasing numbers as the design process progresses.

The number and level of functionality of the prototypes will vary across products. The more expensive, critical and widely used the product, the more prototyping is needed.

For a simple system the first prototypes will usually consist of several designs created as static pictures or screen shots. User input is sought and usability ratings are usually definitive in narrowing down these user interface options to a small number, generally from one to four. The next step is likely to be to develop the prototype(s) with limited functionality to create a dynamic model of the most highly rated user interface concepts. The degree of functionality and fidelity will vary with the system, but prototyping tools are providing increasing functionality with decreasing development time and cost. In the extreme, the prototypes may be a fully-functioning aircraft, such as those created for military contracts.

The prototypes are then tested for usability. These tests are described in the next section.

Usability Testing

Usability testing is the final and most important step in designing a human-centered system. There are no substitutes or short-cuts for proper usability tests. Some of the best designs were proven to be unusable in the usability tests.

In the example described earlier, the Traffic-alert and Collision Avoidance System, the maneuver displays were designed by the project test pilots and tested extensively in actual Boeing 727 flight encounters with intruder aircraft. When the usability tests were conducted in a Boeing 727 simulator the red arrows on the vertical speed indicator were responded to incorrectly by approximately half of the airline pilot participants. That is, when the system told them to climb to avoid colliding with the other aircraft, half of the pilots descended. Their rationale was that the red up arrow was indicating they should not climb. The vertical speed display was changed before production to indicate the unsafe and required vertical speeds. In usability testing of the new displays the error rate was reduced to less than 2%. The result is the highly successful TCAS system we have today.

Output from Usability Tests

Two types of data are collected in usability studies: objective performance data and subjective ratings. Both types of data are important for the assessment of the prototype and later the product.

The objective performance metrics of the usability tests come from the usability goals set earlier. For example, if the system logic relies on a five-second operator response to a warning, the response times measured in the usability tests should all be faster than five seconds.

Using our TCAS example again, the human performance data drove the design of the system. The mean pilot response time derived in simulator testing plus two standard deviations was used as the system criterion for issuing a second, opposite direction command (e.g., climb changing to descent).

Audio and video recordings of the users' actions may be useful in quantifying the results of the usability tests. This is a cumbersome and intrusive medium for data collection, but in some cases there are no alternatives for documenting usability issues. In one study (unpublished), video recordings were used to determine the time to recover from a user error and the error frequency. These data justified the costs for system improvements.

Subjective data are also important for determining the participants' assessment of the system's usability. Ratings of the navigation through the system, the over-all look and feel, the accessibility of important features and data, etc. are good metrics for determining when the usability is acceptable.

The product of the usability tests is the list of usability problems and virtues. Solutions to the problems may be obvious, or several alternatives may be tested to determine the best 'fix'. It is also important to note what user interface features are used with ease. These can become the template against which the modifications are made to the system.

Designing the Usability Tests

Proper design and conduct of the usability tests will ensure that the findings are accurate and sufficient. Several good references on this topic are Dumas & Redish, 1994 and Rubin, 1994. Some of the issues in the design of usability tests are described briefly here.

Considerations in Scenario Design

There are many considerations in the design of test scenarios. The scenarios need to be representative of real-world situations that exercise a sample, if not all, of the operator functions and exercise the full capacity of the system and the user. The scenarios should be conducted as though the user was in the full operational environment, to the extent possible. All attempts should be made to provide a realistic context, including the presence of all supporting equipment, manuals, and personnel, the occurrence of unforeseen events and interruptions, and time pressure, as appropriate.

Realistic workload is a key element in scenario design. Both too much and too little workload affect the usability of system. Therefore the full range of workload should be represented in the usability tests. This can be accomplished by ensuring that the users conduct all necessary concurrent tasks, including tasks not related to the primary task, e.g., documenting the operation, or as mentioned above, eating and drinking. All administrative activities should be part of the scenarios, e.g., in the flight environment, checklists, flight logs, and communications with air traffic control, the cabin and the company.

Another topic to be addressed in the design of scenarios is the failure of the system and errors by the user. Failure modes get increasingly complex as systems get more sophisticated. The recognition of and recovery from failures in the system or errors by the user must be investigated in the usability testing.

Fidelity of the Test Environment

With each stage of the development of the system, the fidelity of the usability test environment increases. As mentioned in the section on prototyping above, the first tests can be done with static sketches. The final testing may be done in the full operational environment. The appropriate level of fidelity is critical to obtaining valid results from a usability test. No golden rules exist for how much fidelity is required. If fidelity is added and the results do not change, then it was unnecessary. A relevant study of the level of simulator fidelity needed to evaluate cockpit traffic displays was conducted by the author (Chappell & Kreifeldt, 1982).

Training for Usability Tests

It is important that the participants' training for the usability tests is representative of the minimal training expected in operational use. For example, if some groups of users will only receive a manual and no formal training, then the participants for the usability tests should only receive this extent of training. Naturally the participants must be naive to the product design and functionality prior to receiving this training. Project subject matter experts cannot be participants in usability tests.

Iterative Design and Testing

The best method for developing human-centered systems is to iteratively design and test the user interface prototypes. Lessons learned from the usability tests are incorporated in the prototype and tested further. As the system evolves, the new functionality is tested for usability and any interactions with previous functions. This eliminates the possibility of fixing one problem and unknowingly creating a worse problem.

Final Implementation Usability Testing

The final testing will ideally occur in the operational environment. For example, when testing an aircraft system, the final testing should occur in flight with test missions designed to exercise all the capabilities of the product. The pilot participants should be operationally qualified and not familiar with the product. If the previous steps have been carried out, this test is merely a formality, a validation. No new findings should emerge in these final tests.

The product may then be put into limited operational use. Citing the TCAS example again, after the simulation testing, three airlines installed the prototype systems in one or two aircraft that were being used in revenue service by line pilots. After several months of operational use, some improvements were deemed appropriate, prior to the full implementation of the final system. This testing is cost-effective and leads to an end product that is void of any major usability problems.

How much is enough testing? It is impossible to create every situation in which the system will be used during the usability testing. When a number of successive tests yields no new findings, one can feel reasonably confident that the usability is fairly well understood. Those having experience in conducting usability tests on similar systems can provide guidance in determining when the testing is sufficient.

Operational Follow-up

Finally the development is complete and the product rolls out for operational use. However, the usability engineer's job is not over. Follow-up data collection of usability issues in the operation of the system is vital to ensuring that the product is truly usable in the intended operational environment, and in the unforeseen environments in which the system ends up being used. When the training on the system is long forgotten and the operators invent uses the designers never dreamed of, a new batch of usability issues may emerge.

This paper has described a process for creating human-centered systems and procedures. Following these steps through certification will ensure that the product will be usable and resistant to, yet tolerant of human error.

References

Dumas, J. S. and Redish, J. C. (1994). A Practical Guide to Usability Testing. Norwood, NJ: Ablex Publishing Corporation.

Chappell, S. L. & Kreifeldt, J. G. (1982). Facility requirements for cockpit traffic display research. Proceedings of the Annual Meeting of the Human Factors Society. Santa Monica, CA: Human Factors Society.

Fitts, P. M. (1962). Functions of men in complex systems. Aerospace Engineering, 21(1), 34-39.

Laurel, B. (1992). The Art of Human-Computer Interface Design. New York: Addison-Wesley. (ISBN: 0-20-151797-3)

Nielsen, J. (1993). Usability Engineering. Boston: Academic Press. (ISBN: 0-12-518406-9)

Rubin, J. (1994). Handbook of Usability Testing: How to Plan, Design, and Conduct Effective Tests. New York: John Wiley and Sons. (ISBN: 0-471-59403-2)

Smith, S. L. & Mosier, J. N., (1986). Guidelines for Designing User Interface Software. Technical report for the MITRE Corp.

Wise, J. A., Hopkin, V. D., & Garland, D. J. (1994). Human Factors Certification of Advanced Aviation Technologies. Daytona Beach, FL: Embry-Riddle Aeronautical University Press. (ISBN: 1-884099-03-3)

A PILOT-CENTERED AUTOFLIGHT SYSTEM CONCEPT

Victor Riley, Bob DeMers, Chris Misiak, Brian Schmalz

Honeywell Technology Center

Minneapolis, MN

Introduction

Ease of use has become a primary concern in avionics recently. Pilot training time represents a significant cost to aircraft operators, and the importance of human factors considerations in safety has been highlighted by the FAA [1], who notes that human error is a cause in roughly 70% of aircraft accidents. These concerns have led to a new focus on human centered design, which seeks to define system functionality based on the user's perspective rather than that of the designer. We have used the principles of user centered design to develop a new autoflight system concept that integrates the autopilot and flight management system functions into a single user interface that most pilots can learn to use within about fifteen minutes. This is in contrast to the weeks required to learn a typical flight management system.

User Centered Design

Human centered design has often been used as an updated term for human factors, but we believe there is a valuable distinction to be made. Many designers consider human factors to be concerned with the design of the displays and controls, while the underlying functionality is solely determined by the functional engineering requirements of the system. This leads to the traditional development process shown in Figure 1 [2].

In contrast, user centered design begins with the premise that how the system works, rather than what it looks like to the user, is the most common source of usability problems. For example, to use a flight management system (FMS), the pilot must make a series of decisions:

1. Which device should I use to execute the clearance (FMS or autopilot)?

2. If FMS, which FMS function is best suited to the clearance requirements?

3. Where do I find that function in the page architecture?

4. How do I use the function (how do I start, what do I do next, etc.)?

5. What data entry format does the function require?

These questions have to do with how the FMS works (its functionality and functional logic) rather than how the user interface is designed, and the pilot must learn the answers to these questions for every FMS function. This is why learning to use the FMS can take so long.

This suggests that the major benefits of user centered design are to be gained by addressing the level of system functionality and functional logic, rather than waiting to solve problems at the interface. In the traditional FMS, the functions are defined according to how the engineer thinks about managing the flight, and the interface must span the gap between how the engineer thinks and how the pilot thinks. Because that gap is too large for the interface to span completely, training is required so the pilot can think about the system like the engineer does. In a user centered design, the functionality and functional logic are defined according to how the user thinks

0-7803-5086-3 /98/$10.00 ©1998 IEEE

about the task. The interface is just a natural extension of the user-centered functionality.

To reflect this, the motto of our project has been that "we want to make the system work the way the pilot thinks, so we don't have to train the pilot to think the way the system works." A user-centered design process to accomplish this is shown in Figure 2 [2].

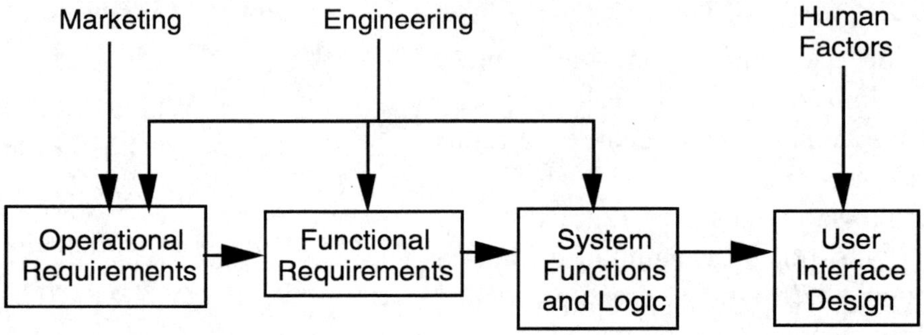

Figure 1: In the traditional process, human factors are addressed only during the user interface design, after system functions and logic have already been defined.

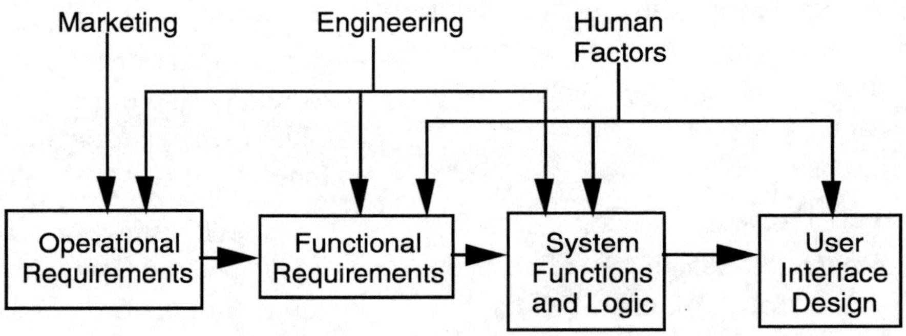

Figure 2: In a more human-centered process, human factors considerations are applied to define functional requirements to support crew task and decision needs, system functions and logic to match their task requirements, and the user interface design.

The key to developing pilot-centered functionality is to employ an appropriate metaphor. To illustrate the effect of an appropriate metaphor on usability, consider the desktop metaphor used in the personal computer industry. Basically, the graphical user interface is easy to use because it takes advantage of what the user already knows about managing paper documents and uses that as a metaphor for operating system functions. This enables new users to transfer or copy files, delete files, open applications, and perform other operating system functions without requiring training or a visit to the users' manual. To the extent the on-screen world works the way the real world does, most users already know how to perform those functions for which there is a metaphorical counterpart. Users require additional help only where the metaphor breaks down or doesn't apply. Within the bounds of the metaphor, users find the system "intuitive" because it works the way they expect it to based on their existing knowledge.

Note that the use of an appropriate metaphor separates out functional complexity from conceptual complexity. DOS is **functionally** less complex than Windows, but Windows is **conceptually** less complex than DOS for most users. Note also that the use of the graphical user interface reduces the need for the typical user to think like a computer scientist in order to use a computer. Our challenge for avionics is similar: to reduce the need for the pilot to think like an avionics engineer simply to use the avionics. However, the desktop metaphor is probably not the appropriate metaphor for the flight deck, because it does not provide a natural framework for organizing the functions related to flight path management. Therefore, our first challenge is to find the appropriate metaphor for the flight deck.

The Cockpit Control Language

The Cockpit Control Language (CCL) is a conceptual framework for defining and organizing flight deck functions in a way that pilots already understand. Because every instrument-rated pilot understands air traffic control clearances, and because the range of possible clearances contains the range of functionality required for flight path definition, the CCL uses the clearance as a metaphor for the system's operational logic. This enables the pilot to enter commands into the system in the same order they are presented in the clearance. This in turn eliminates the need for extensive training and for memorizing system functions, how to access them, what input procedure to follow, and so forth.

To define the required functionality, we began with a list of clearances allowed in today's environment, then extended the list to account for Free Flight, CNS/ATM, and other anticipated changes in the air traffic control environment. We then expressed the essential intent of each clearance in a consistent, reduced format; for example, several clearance types essentially mean "change your altitude". These essential intent statements contained the necessary elements of an efficient control language; we then classified the elements into "actions" and "targets", where targets were the physical parameters that define the flight path (airspeeds, altitudes, waypoints, navigation aids, even the entire flight plan) and actions defined what the pilot could do to the targets (fly **to** a target, fly away **from** a target, stay **above** or **below** a target, do something **at** a target, etc.). Essentially, the process of interacting with the system is one of invoking actions and targets in the same order they are heard in the clearance. A command to the autoflight system is given in the form of a text string containing all the actions and targets called for in the clearance.

Our initial user interface prototype for a CCL-based system is shown in Figure 3. It consists of a candidate glareshield controller, a navigation display, and a primary flight display. We included a variety of interface devices to illustrate that the same logic applies whether the pilot is manipulating knobs, buttons, and switches as on the glareshield controller or graphic objects as on the navigation display, or entering in command strings with a keyboard, speech recognition, or data link. The input devices are no longer directly linked to autoflight functions. Instead, they each generate an appropriate command language text string. Since the same logic applies to all devices and methods of entry, the pilot can mix entry types within a given clearance based on workload, convenience, head up time, and other considerations.

To date, we have shown the prototype to around one hundred and fifty pilots. In our demonstrations, we ask the pilot to program the system, starting with a simple altitude clearance. We follow with intercepting a jetway to the flight plan and intercepting a jetway at a waypoint. After illustrating how the concept works with these examples, we then give the pilot a required time of arrival over a waypoint. Most pilots can enter this clearance with little or no further instruction; those that still require help on this clearance, however, can usually enter the following clearance, a holding pattern around an unpublished waypoint, on their own. In general, pilots can learn to use the system within fifteen minutes. The ability of these pilots to enter in the holding clearance on their own is particularly instructive because it can be a very difficult clearance for even experienced FMS users to enter into conventional systems.

To illustrate how the system works, consider a simple altitude change, such as “climb and maintain FL 220”. In current aircraft, the pilot usually turns an altitude select knob and engages the appropriate mode to start the climb. With the CCL, the physical action is exactly the same. But because altitudes are targets and work the same way all other targets do, the CCL adds a command string “TO FL220” when the pilot turns the knob; when the pilot presses GO to tell the aircraft to follow the command, the autoflight system interprets the target (FL220) and the implied action (TO). The fact that all targets comply with the same underlying rules means that the pilot can combine them with any actions that make syntactic and semantic sense. Thus, when we later tell the pilot to “cross ABQ between 10:30 and 10:35 Zulu”, he or she knows to enter the time constraint the same way an airspeed or altitude constraint would be entered. This allows the system to be extended to encompass future capabilities with no additional training.

In addition to the interface, we use a syntactic parser to assess the syntactic legality of command strings and provide shortcuts wherever possible. This greatly reduces the number of keystrokes or inputs required to execute a clearance. For example, when the pilot is asked to “intercept J64 at Farmington”, he or she can select the jetway graphically on the navigation display, then select the waypoint, and the parser will dynamically generate a list of possible relationships between the jetway and the waypoint (the pilot can intercept the jetway **before**, **after**, **at**, or **until** Farmington, or **between** Farmington and some other waypoint). Or, when given the clearance “hold at 20 miles southwest of Farmington”, the pilot need only enter **hold 20 southwest Farmington**, and the parser will add in the additional modifiers to make the complete command string clear (“HOLD AT 20 MILES SOUTHWEST OF FARMINGTON”).

Conclusion

The Cockpit Control Language offers a pilot-centered interface to the autoflight system based on knowledge that the pilot already has. Once the pilot understands how his or her existing knowledge is being used as a metaphor for the system’s operational logic, the system is intuitive and easily learned. This demonstrates the value of user centered design for developing systems that require minimal training.

References

[1] U. S. Department of Transportation, Federal Aviation Administration. (1990). The national plan for aviation human factors.

[2] Riley, V. (1996). “What Avionics Engineers Should Know About Pilots and Automation.” IEEE Aerospace and Electronic Systems Magazine, 11(5), 3-8.

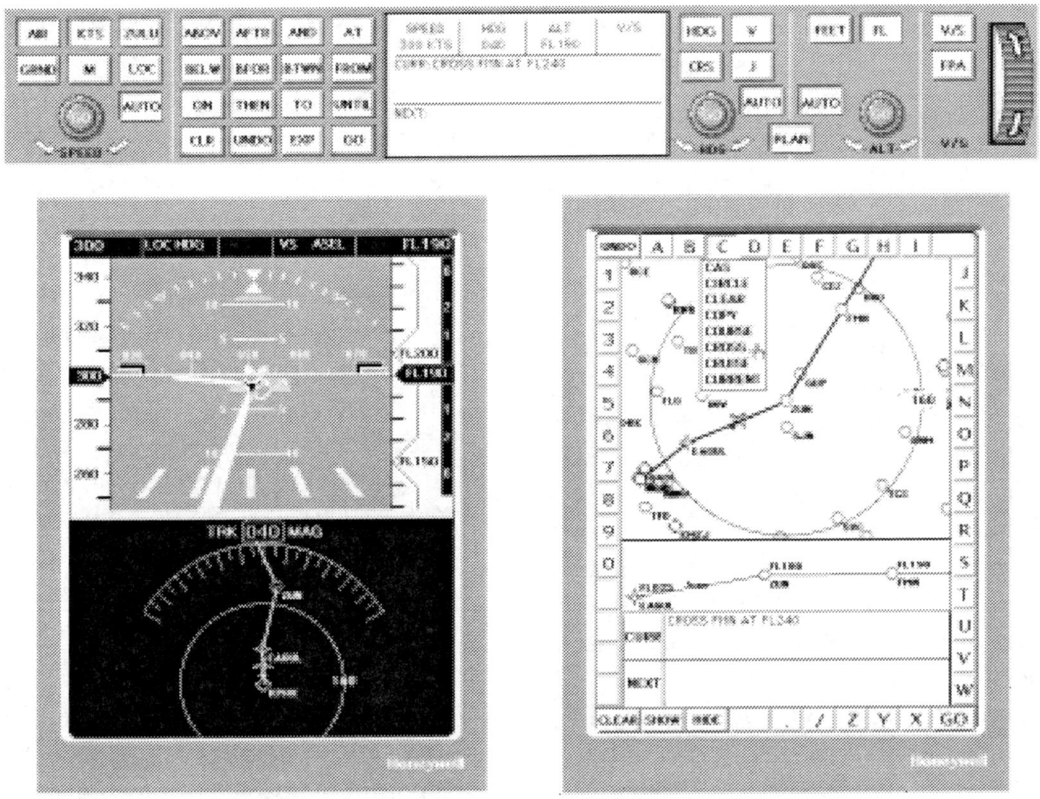

Figure 3: The prototype Cockpit Control Language interface

MAKING THE COCKPIT A HUMAN FRIENDLY ENVIRONMENT: A CUSTOMER'S PERSPECTIVE

Edward A. Rafacz Jr.

Introduction

Those who do not study history are doomed to repeat it. In the Aviation Industry, it means we kill people. Those who were the first to put information in the cockpit through the use of analog gauges had no standards to guide them. The result, pilots were more apt to get disorientated in bad weather conditions. It was not until an Industry Standard was developed, in this case putting flight instruments in a consistent placement, that pilots could fly safely in instrument conditions in almost any aircraft.

We need to learn from these lessons. If we do not, the Airline Industry may be pushed even faster toward a fate of one major aircraft hull loss per week. Careful design and implementation of Electronic Information Displays could also be the key to elimination of such a ghastly statistic. By striving for human factors excellence, and more importantly, system user input before designs are created, the electronics and software industry could help the Federal Aviation Administration get closer to its goal of zero accidents.

Present Day Environments

Present day cockpits are a conglomeration of keyboards, touch screens, cathode ray tubes, as well as switches, buttons, liquid crystal displays, and warning lights. Most aircraft differ from airline fleet to airline fleet because one airline will try to standardize its aircraft fleet to meet its needs. From an airline perspective, it my seem that standardization issues are contained. Unfortunately, that is not the case in today's airline industry. As more people want to fly and aircraft manufacturers have difficulty with productions, finding aircraft becomes more and more difficult.

The end result of our present situation is airlines purchase aircraft from companies with financial difficulty. Since standardization of cockpit information systems is not a reality, the information may be displayed differently in a purchased aircraft in such a benign way that it my not be deemed necessary to be changed. Such a change might be having an aircraft fuel gauges read in liters when the crew has always used gallons. The problem is aggravated by the fact so may airlines are trying to standardize their fleets by only having as few types of aircraft on the property. It is ironic, but by not thinking through standardization, airlines sell the different types and purchase new and used aircraft of a generic type. To cover interment inconsistencies, airlines create large sections in their manuals on differences that cover several those benign differences.

0-7803-5086-3 /98/$10.00 ©1998 IEEE

Present Equipment

Focusing attention to electronic displays in the current cockpit, we can see that a single cockpit can be comprised of several keyboards and screens placed in locations were the only criterion is where an open space exists. The present solution to supplying information to the pilot appears to be the additions of keyboards and screens which add to the weight and complexities of aircraft systems. There seems to be a disregard for the fact that cockpits already have a way to display information. The software becomes key.

Not only do displays and keyboards vary from aircraft manufacture to manufacturer, but within an aircraft type. One company, Airbus Industries, is addressing this problem head on by designing their whole fleets' cockpits the same. This is a step in the right direction, but even more fundamental topics need to be studied.

The two keyboards that I am familiar with both have different design philosophy. Although each goes in alphabetical order as opposed to a typewriter keyboard, numbers and punctuation are in different areas. One of the keyboards happens to be a touch screen which becomes difficult to use in turbulent conditions.

Display screens also seem to be un-standardized in their placement and their display. They may be placed in areas that pull the pilots attention away from their first and most important job, flying the aircraft. The size of the display may also be too small for practical use. The small display combine with turbulence to produce a useless light that interferes more than it helps the user.

Organization of information needs attention in the modern cockpit. The data required by the pilot may be hidden under some obscure menu with no cognitive path to find it. With all development of search engines, it is amazing that no manufacturer has approached this topic. And once one system's software has been mastered, there is most likely no way for that navigational knowledge to be transposed to another system's software organization.

The irony is that every new technology requires its own display and software. Once some company understands that it can just focus on software development since all the hardware is in the cockpit, it can sell a homogenous system that will not add to the weight of the aircraft. It will also require less down time to install since a minimal amount of maintenance is needed. Just think of the competitive advantage a company would have it if used one computer for its Word, Excel, PowerPoint, and Exchange programs while the competition used a different computer for each program.

Future

The future for informational displays in the cockpit is very promising. Systems being tested right now will not only increase the efficiency of the transportation system, but will add to its safety. New approaches in the use of graphic displays will convey more information in a less confusing format.

Future Air Navigation System, or FANS is using a data link to connect the

0-7803-5086-3 /98/$10.00 ©1998 IEEE

controller to the pilot. From its primary tests, pilots have found that the display needs to move to an area so as not to disrupt aircraft operations. Some envision the use of voice recognition to eliminate the keyboard all together. The implications of this technology could produce solutions to safety difficulties of pilots removing their attention from fight information.

Enhanced Ground Proximity Warning system, also known as EGPWS, is a tool that is displaying information graphically that was once done with numbers. Instead of reading your distance above the terrain as a single number, a color depiction is shown in front of the pilot. Not only does this informational display produce better situational awareness, but it also depicts an escape solution. Pilots surveyed in Aero Information Magazine, a web publication at www. safepic. com, thought very highly of the system. They seem to like the ability of the display to advise them of situations before it becomes a problem.

Heads Up Displays, or HUDs, is a unique way to disseminate flight information to the pilot. Not only does it display some information as cues, but it displays this information on the windshield. The elegance of this solution rests in the fact that transitioning from instrument to visual conditions is extremely easy. Currently, a pilot processes information from several instruments below the windshield, then moves attention up to the windshield to see the runway. HUDs make it possible to look through the information and see the runway in order to land.

Other efficiencies are acquired through the fact that much of the equipment needed to do an approach is carried onboard the aircraft. The onboard equipment is easier and cheaper to maintain than the traditional equipment used. Other residual benefits are reduced tire and brake ware due to more accurate approach information. More accurate information translates into more judicious use of the runway allowing the pilot to need less braking.

Another concept is 4-D Navigation Instrumentation. Presently a research project involving VDO-L Frankfurt and Lufthansa Frankfurt is in progress at the Technical University Darmstadt. This displays a color picture on the pilots Primary Flight Display along with flight cues. One of the innovative flight cues is a flight path display that shows a corridor to navigate. This visual depiction of the flight path combines a huge amount of information to the user in an easy to understand format. Transition from instrument to visual conditions seems to be almost as easy as the HUD because a picture of the outside world is produced. These two philosophies, real world views with cues, and pictures with numerical and cue data, will do battle for the best navigation system.

In the future of aircraft manual, picture maps may be the protocol. Pictures that allow to user to point and click are very intuitive to use. Combined with standard indexes and search engines that are found on the Internet, the pilot will have a powerful tool to drill down into the database. Other documents, such as the Aviation Information Manual, company manuals, and safety bulletins, could use hyperlinked words, search engines and general category tabs to retrieve data.

Remember that the informational revolution is still in its infancy and we pilots are not tied into any of these concepts. We do have a few guidelines. The solution must focus on high resolution, reliable access, and easy organization. Basically Star Trek but more refined.

Resolution has to be exceptional to the user. It needs to handle lighting conditions from extremely bright to full dark. Other factors such as a completely smoke filled cockpit, lightning flashes, and turbulence need to be addressed. Size of the screen and size of information are also key in the success of the next generation displays.

Reliability must be 100%. Just as the need for triple redundancy with navigation devices, there needs to be ways to guarantee that data is received. Once the data is received, it must be of the highest integrity. If it is to use electricity, it needs to be functional for a complete electrical failure until the aircraft lands.

Organization of the information must be of the simplest and intuitive form available. In an emergency situation when second are critical, the information must be accessible. It must also deliver information in a non-distracting way.

Are we asking for a great deal? You bet. We are in an era where companies are cutting budgets and looking at information systems to reach this goal. Airlines are looking at way to push more aircraft into the skies without decreasing safety. Information delivered correctly, like the standardized navigation instrument layout, can achieve a solution for both problems.

REAL-TIME OPERATIONAL EVALUATIONS USING ADVANCED FLIGHT SIMULATORS

A.E. Dillard
Flight Technologies and Procedures Division
Federal Aviation Administration
Oklahoma City, OK

ABSTRACT

The increased speed with which new technologies are being introduced into the modern aviation operating environment has made it necessary to find new ways of evaluating certification, human factors, operational and safety issues. We no longer have the luxury of an extended development program, followed by an evolutionary period of products maturing into more complex forms, with an extended useful life. Modern technology delivers fully formed products to the marketplace with rapid wide distribution and, in many cases, a limited operating life due to forced obsolescence caused by new advances and technologies. Aviation has always been a technology leader, and this hasn't changed, so the introduction of new communication, navigation, surveillance and display technology is moving forward at a fast pace. Aviation is also a very competitive business, and maximum benefit comes from the early implementation of innovative new products and applications.

While some time elements of the product life cycle have changed, critical requirements for validating safety, reliability and system integrity in civil aviation have not. The process of operationally integrating a new technology into an existing, highly complex, costly and potentially hazardous domain, such as airports and aircraft cockpits, demands an exhaustive evaluation of their effects on the existing system, while maintaining safety and performance standards, support logistics and affordability. To shorten the time required for equipment and procedural development, and operational implementation, the use of simulation has grown in importance. Simulation can consist of virtual modeling on a computer workstation, part task devices with actual system hardware and software, or full-mission man-in-the-loop simulators with visual systems and motion. All have their place in the process, and all play a roll in shortening development time and cost. We will be looking at the use of full-mission simulators for piloted operational evaluations.

USE OF MODERN FLIGHT SIMULATORS

Flight Crew Training

The most widespread use of modern flight simulators is for flight crew training. The reason for this is the lower cost, safety for men and equipment and more efficient use of resources. All of the major air carriers have extensive simulator complexes, such as those of American at DFW, Delta in Atlanta, Alaska in Seattle and United in Denver. New simulators that meet FAA Level C and D approval criteria can cost $15 million or more.

0-7803-5086-3 /98/$10.00 ©1998 IEEE

Use of Level D devices can enable virtually a zero aircraft flight time transition for experienced pilots between aircraft types. Operating a large aircraft simulator costs approximately 10% of what it costs to operate an aircraft of the same type per hour, you don't endanger men and equipment, and the actual aircraft stay in revenue service. An added advantage is that the daily operating service time for simulators can be twice that of aircraft. In addition to the bottom line advantages, you get much more efficient transfer of learning in a simulator as compared to the aircraft. The possible faults and abnormal operational exercises is unlimited, and it is possible to train to proficiency to accommodate the differing needs of different students.

Evaluation Programs

Simulators have been used for research for years by a number of organizations for human factors, procedural development, equipment development, accident investigation and design work. One of the best known, of course, is NASA, but the airlines now operate several hundred approved simulators all over the world. The aircraft manufacturers also build prototype simulators for new aircraft types for early crew training and to support design activities. Many colleges and universities maintain research devices in their laboratories, such as those of Arizona State Williams Campus and Georgia Tech.

A more recent development has been the use of FAA approved training simulators for operational evaluation test programs. Because of the constraints imposed on approved training simulators by the certifying officials, most operators are reluctant to do extensive changes or modifications to the equipment. We have learned in recent years, however, that you can indeed make changes and modify the simulators, within these constraints, for evaluation programs, but you must allow provisions to return to the approved training software and hardware configurations when needed. Since most of the new systems are software based, you can load training software programs and return to training status in a short time. Working within the approval constraints, most simulators can be returned to an approved status within 30 minutes if changes are software only, but can require up to four hours or more if hardware must be replaced and checked out.

Increased computer capacity and speed, coupled with more disk storage, and the capability to network simulators for data collection and joint exercises have further increased capabilities. There are research situations, of course, where it is necessary to keep changes or modifications in place for extended periods. In this case, approved training simulators may not be appropriate to use.

FAA APPROVED SIMULATORS

Advanced Simulator Program

The advanced simulator program administered by the Federal Aviation Administration for air carriers and commercial training centers is based on requirements set forth in Advisory Circulars (AC) AC 120-40, Airplane Simulator Qualification, and AC 120-45, Airplane Flight Training Device Qualification. The AC's give specific

0-7803-5086-3 /98/$10.00 ©1998 IEEE

criteria required to obtain and maintain approval on commercial simulators to be used for flight crew training. Key elements of the approval process include aerodynamic database fidelity, aircraft system fidelity, and realistic motion and visual systems performance, including through-put.

The fidelity of the simulators is established by comparison of simulator test data to actual aircraft flight test data and systems design requirements. In addition to the original approval, all commercial simulators must be rechecked a minimum of twice annually over the operational life of the equipment to maintain approval. There are generally 100 to 150 operational tests that are included in an approval test guide to check performance. The specific number can vary, depending on the aircraft type. More complex systems can add more tests. These tests are all repeated in the course of a year, and performed randomly in an order specified by the evaluator. Failure to meet any of the test requirements can mean loss of training credit approval. Each simulator is built to match, by performance and configuration, a specific aircraft, identified by aircraft tail number. The military is now in the process of developing their own approval process. The Department of Defense, (DoD) system will inevitably be more complex because of the sophisticated on-board tactical systems involved.

Typically, any kind of regulatory effort is looked upon as burdensome. The simulator approval process, however, was jointly developed by the FAA and industry as a way to ensure continued safety, while saving industry enormous amounts of time and money. The gains in quality, reliability, time and money saved have convinced most operators of the value of the program.

Non-Approved Simulators

One of the primary reasons for establishing the advanced simulator approval program was to improve fidelity, improve maintenance programs and ensure currency with existing aircraft configurations and performance in simulator equipment. Non-approved simulators do not meet the same stringent oversight requirements, so are limited in training credit allowed by the authorities.

The fidelity levels achieved and quality maintained varies widely between operators on non-approved equipment. Software programs are often changed considerably from the original loads. Similar problems can occur with cockpit hardware. Not all non-approved simulators suffer from these problems, of course, but it is difficult for an operator to maintain software and hardware configuration control without a formal process in place.

Many of the non-approved devices are in use by educational establishments and smaller training centers who find it difficult to justify the cost of maintaining certification status. NASA and some educational institutions, such as Embry-Riddle, have obtained FAA Level C or D approval on newer devices to receive added credits for their use and to ensure fidelity. The general aviation (G/A) community has also been slower to move into the approved simulator arena. In many cases for G/A training, an actual aircraft can be much cheaper than an advanced simulator. But as costs come down on simulator equipment, the market is expanding. Some operators, such as Flight Safety, have sizable simulator operations for G/A training.

Advantages and Disadvantages of Approved Simulators

The most obvious benefit from using the approval process is the documented fidelity, as compared to the actual aircraft, that leads to the additional training credit allowed. This eliminates most of the aircraft flying time required. System performance is assured within defined limits leading to more consistent results. There are normally built-in diagnostics which allow one to easily check performance and integrity to ensure continued fidelity.

The approval process has built-in provisions for tracking changes and modifications, and maintaining configuration control. As practiced, this is not as onerous as you might expect. In most cases changes can be evaluated on-line before being permanently installed for use. Most hardware changes are driven by update programs to the aircraft and do not effect flight performance, so have limited impact on the approval process beyond initial evaluation after installation. Research specific hardware can be installed for use and then removed to return to training status.

The credibility of using approved simulators for top-end man-in-the-loop evaluations can not be ignored. The time saved in system development, procedural development, training requirements assessment and risk assessment on safety issues is difficult to achieve in any other way.

Disadvantages of the approval process include cost and time requirements. The manpower needed to maintain daily checks and repairs can be considerable. Likewise with spare parts and in-house technical talent. To justify the overhead of maintaining an approved simulator program, an operator must get maximum utilization from the equipment. There are situations with smaller operators where the burden of the approval process is not needed, nor desired. In most of these cases they buy simulator time from larger operators.

EVALUATION PROGRAMS

Evaluation program using flight simulators fall into five categories: systems specific evaluations, human factors, procedural development, accident investigations and safety assessment. Often times an evaluation will cut across several, if not all, of these categories. In all cases it is essential that qualified pilots are used. It works out better in most situations if the test subjects are currently active pilots in the aircraft cockpit type being flown. These requirements can vary, depending on the evaluation program, but must be established by the test director. The pilot qualifications and other evaluation particulars should be specified in a formal test plan prepared before testing commences. The plan should also contain a clear definition of success and failure, if appropriate for the evaluation. Any evaluation program must have comment sheets and questionnaires to be completed by observers and test subjects to capture observations of the participants.

Systems Specific Evaluations

This category includes things such as airport lighting, global positioning systems, multi-function displays, data link, cockpit display of traffic information and head-up-displays. The simulator gives you the ability to add a complete set of weather effects, including various fog models, contaminated runways and runway visual ranges. In the case of targeted integrated avionics evaluations, the program manager should design a separate method of validating systems functionality since the systems are not looked at as closely as other simulator factors. Likewise, system modeling and integration must be validated as a separate effort for new systems to ensure accuracy. Remember, the original design took only the existing operational systems into account when it was built.

Human Factors Studies

This includes such things as Laser visual interference, pilot performance issues, all aspects of display symbology and runway approach light patterns. Again, a full range of weather effects can be introduced along with any new navigational aides, display content, crew workload and airport modeling. The possibilities of using modern high fidelity simulators for human factors operational evaluation programs is virtually unlimited. There are human factors issues associated with virtually all evaluation programs, and any new equipment introduced into the cockpit has extensive human factors issues associated with it. When you add extensive integration with other systems and new procedures, it gets even more complicated. Many times these concerns can be addressed in an overall program plan, but it can require a separate targeted effort. Use of the simulator to address these issues is one of the few ways you can look at them in a realistic operational environment. Given that pilots are highly trained and skilled individuals, having a realistic environment is even more important.

Procedural Development

This work normally involves looking at the introduction of operations using new runway configuration, such as closely spaced parallel runways, or the introduction of new equipment into the cockpit requiring new or modified procedures, such as GPS or CDTI. Airspace infrastructure, such as airports, navigation aides, air traffic and weather can be included in the process before systems are actually fielded.

Accident Investigations

Almost any accident that occurs today can be replicated on an advanced simulator if there is sufficient data recovered from the aircraft data recorder or other sources. Some aircraft do not have the data recorders, which can make the job much more difficult. Examples of this type of simulation includes windshear induced accidents and mechanical failures.

Safety Assessments

The use of advance simulators to evaluate safety related phenomena is growing rapidly. Such things as the effects of varying edge light spacing at LaGuardia in New York, windshear and wake vortex effects during closely spaced in-trail or parallel approaches are examples.

***Airport Infrastructure** modeling*

The FAA is responsible for evaluating new, or proposed, airport design and lighting issues. Some of the programs recently evaluated using flight simulators included the evaluation of approach light patterns, high speed runway exit designs, land and hold short stop bars (LAHSO), surface movement and guidance system (SMGCS) plan for the new Denver Airport, contaminated runways and foam bed arresting pads on runway overruns.

All of these programs were evaluated in an approved commercial flight simulator regularly used for training modified to support the specific program being evaluated. Most of the programs, such as the arresting foam bed, LAHSO, SMGCS and approach lights, involved extensive visual modeling. They also required exhaustive model validation of the surface friction for contaminated runways, fog and slant range RVR against known scientific data. In most cases the work can be accomplished with in-house resources. If the resources are not on staff, there are competent contractors available that can support similar activities.

DISTRIBUTED INTERACTIVE SIMULATION (DIS)

Current DIS Applications

A step up from using a single simulator is to create a network of simulators operating as individuals within their own separate domains, but dynamically connected to transfer data real-time between cockpits, and, possibly, with a central control center via modem and phone lines. Data collection can be done at the individual sites and/or at the control center. The possibilities for interactive operations are practically unlimited. This process has proven extremely useful for parallel runway operations and data link prototyping. Other near term uses include cockpit display of traffic information and converging approaches.

DoD DIS

DoD apparently envisions this approach as possibly being used widely for future training and gaming exercises. The use of simulators in this manner will eliminate a great deal of expense, and just as the civilian world found out, these cost savings are substantial. Of course, the more complex the system, the more expensive it becomes to operate and maintain. Most of our applications require 30 or 40 data variables being sent at 10 times a second over commercial phone lines. The protocol being developed by the military is much larger and will use secure dedicated satellite channels or land lines for massive data transfer requirements.

Issues With Using DIS

Some of the concerns with using DIS for evaluation programs include equipment security, identifying fidelity levels of participants, cost and special expertise needed to implement protocols. Most of the hardware for civilian application can be reasonably inexpensive. For instance, our current efforts include use of a 486 or Pentium computer (PC) for a simulator interface and a standard modem for data transfer. The software modules used in the simulators are relatively simple and serve as input/output files. The programming for the PC is not too elaborate, but generally must be reworked for each program, since most have different data requirements. Voice communications can also be a problem. We most often use standard phone lines tied into the simulator communications system, but you can also use a multiplexing modem on a dedicated line with voice on separate channels.

OTHER APPROACHES TO SIMULATION

Earlier I mentioned that there are a number of approaches to simulation including virtual and constructive simulations of varying degrees of fidelity, realism and complexity. In any constructive simulation you will be dealing with a physical representation of systems components or cockpit layout usually using actual aircraft hardware and software. These all have a valid role to play in program development. In fact, many part task limited simulation programs targeting specific systems, such as GPS, TCAS and CDTI, are an essential step towards final design and full integration that's needed in a full-mission simulator or aircraft.

Virtual Simulation

The latest simulation technique is referred to as virtual simulation. A virtual cockpit is a representation of a single device with operational logic and graphical representations of cockpit components and controls. It allows a designer to dynamically interact with functional components and to evaluate and refine concepts within a short period of time. These models can then be integrated with other virtual models and functionally integrated prior to actually constructing a hardware based system.

This has it's place in the overall process of program design, but it also has it's limitations. Through virtual simulation a designer can model, animate, operate and integrate a new system, such as a display, into a virtual cockpit on a computer. The skills required, and the process itself, are extremely complex, but the value of the model is only as good as the designers input. It is essential that any flight operational evaluation ultimately be based on a pilot centered design with a qualified pilot at the controls and operating in a realistic environment. Even if you have a pilot "flying" the virtual model on a computer, it is still not a realistic environment. Virtual simulation is at its most valuable when used in the conceptual design stages, evaluating proposed changes and for quick looks at "what if" scenarios with clearly defined variables.

It is possible today to link constructive simulators, such as approved pilot training devices, with virtual simulators where the constructive cockpit reflects the current technology or procedure, and the virtual simulator can be set up to reflect proposed design changes. Another use, and one used often in our technical evaluations, is to take actual flight test data from piloted approved simulators or aircraft, install it on a virtual aircraft model and run fast-time simulations for risk assessment. These virtual aircraft models are very complex, in fact similar to the sophistication of advanced simulator programs, representing specific aircraft models and performance. This enables us to get the data we need in spite of the limited availability of simulator and test pilot resources.

The value of virtual modeling for engineering design work and systems integration is without question . The results speak for themselves, but we need to remember that it is essential that we strive to keep real people in the process. Any tool is at its most valuable when used properly.

CONCLUSION

The sophistication and fidelity of modern flight simulators has enabled their application as evaluation tools to a level not possible a few years ago. The validation process used to approve advanced training simulators has provided the credibility necessary to apply the use of the equipment to an ever expanding range of programs. Lower level and non-approved devices can also be used in a similar manner, but the test director must provide for separate validation for the device or systems being used. A definitive simulator evaluation program must include accurate visual presentation, weather effects, navigation data bases and avionics displays. Every evaluation program must have a formal test plan completed prior to commencement of testing. Data analysis should include a thorough review of subjective comments from test participants and observers.

It is essential that operational evaluation programs for procedures and new systems for aircraft use appropriately qualified test subjects and pilots should be part of the program design team. Engineers seldom are able to accurately construct equipment without qualified pilot input. The situational awareness needed to operate a modern , complex aircraft is greater than ever. Many new systems coming to market from small, high-tech, entrapreneurial companies were developed without an adequate understanding of the human factors issues in the cockpit. In many cases they have a new product to sell and aviation looks like a good market. The use of virtual simulation is a valuable tool for the development and design of new systems or procedures, but in most cases final decisions should not be made until qualified flight crews are tested in a realistic environment.

AN ALGORITHMIC METHOD FOR PREDICTING PILOT-MODE INTERACTION DIFFICULTIES

Denis Javaux, Work Psychology Dpt, University of Liège, Belgium, d.javaux@ulg.ac.be

Introduction

This paper sketches a method for predicting pilot-autopilot interaction difficulties on modern airliners. It is a continuation to a series of other papers dedicated to the study of the cognitive complexity of pilot-mode interaction (Javaux, 1997; and in particular Javaux, 1998). A model is proposed, and a method for predicting the difficulties is derived. The impact of the structural and functional properties of the autopilot on the difficulties is investigated in particular. This research is conducted and validated on the B737-EFIS.

Modes and Pilot-Mode Interaction

The notion of *mode* is fundamental to the understanding of pilot-automation difficulties on highly-computerized aircraft. A mode is a basic function of the automation that achieves a specific task on the behalf of the pilots. On the B737-EFIS, the ALT HOLD (altitude hold) mode is for example in charge of maintaining the aircraft at a constant altitude. A mode can be either disengaged (inactive), armed (prepared) or engaged (active). A *mode transition* occurs when the engagement status of a mode is changed.

Commanded and Uncommanded Mode Transitions

Mode transitions are usually dissociated into commanded and uncommanded mode transitions. To understand *commanded mode transitions*, let's consider the following example. It involves a Boeing B737-EFIS flying at a constant altitude.

The aircraft is stabilized with the ALT HOLD (altitude hold) mode engaged at the altitude selected on the MCP (mode control panel) of 6000 ft. The ATC (air traffic control) requests a descent to 4000 ft.

To comply with the request, the pilot decides to use a vertical mode known as LVL CHG (level change). He first specifies the new target altitude by rotating the altitude selector on the mode control panel and then pushes the LVL CHG switch on the same panel. These two actions trigger the transition from ALT HOLD to LVL CHG. This is a *commanded mode transition*: it occurs because some commands have been issued by the pilot. A third action consists in verifying the engagement of LVL CHG (fig. 1). The actions to execute for triggering the transition are simple in this case. More complex action sequences can however be observed with other modes or in other contexts. Some sequences require the verification of several preconditions and conditions before the actions can be undertaken. Difficulties with commanded mode transitions may occur, usually because the pilot has a simplified mental model of the transitions (e.g. when some preconditions or conditions are unknown). This may happen with rare or complex mode transitions.

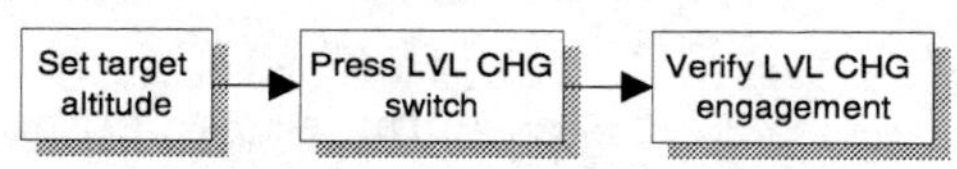

Figure. 1. the actions for the LVL CHG commanded mode transition

To understand *uncommanded mode transitions*, let's consider what happens on the B737-EFIS after the LVL CHG mode has been engaged in the previous example.

The aircraft is now descending in LVL CHG towards its new target altitude of 4000 ft. The distance

0-7803-5086-3 /98/$10.00 ©1998 IEEE

to the target altitude is reducing at a constant rate (vertical speed is constant).

The pilot here assesses the situation and detects that a mode transition is likely to occur: since the autopilot is in LVL CHG and the distance to the target altitude is reducing, the ALT ACQ (altitude acquisition) mode will be automatically substituted to LVL CHG when the aircraft is close to the target altitude of 4000 ft. This is an uncommanded mode transition: it occurs independently of any intervention by the pilot. All the pilot has to do is monitoring the distance to the target altitude and verifying the occurrence of the transition to ALT ACQ. The sequence of actions to execute is simple (fig. 2), but more complex action patterns can be observed when more conditions or more complex conditions are involved. A major difficulty with uncommanded mode transitions consists in detecting precursor situations. This may be difficult with rare uncommanded mode transitions such as mode reversions (their triggering conditions are not well known) or with transitions that involve the detection of events with low feedback. Pilots may therefore be surprised by the automation, either because something unexpected has occurred or because something expected has not occurred (Palmer, 1995).

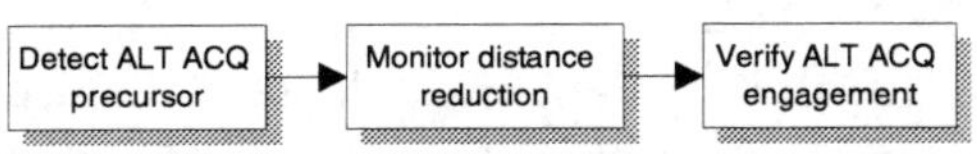

Figure. 2. the actions for the ALT ACQ uncommanded mode transition

In both cases (commanded and uncommanded mode transitions), the pilot has to produce, execute and verify a sequence of actions. A thesis of this paper is that the cognitive complexity of these operations is strongly influenced by the functional properties of the mode transitions.

Representing the Functional Properties of Mode Transitions

Mode transitions obey to rather precise rules. The formalism used in this paper to represent these rules is a variation on state transition diagrams. A detailed description can be found in Javaux (1998). The representation shows (fig. 3) several attributes of a rule: its preconditions, conditions and terminal state as well as the complex temporal relationships between conditions. Attributes such as rule frequencies, condition frequencies and feedback are of particular concern for the cognitive model described below. The frequency of a rule is the frequency with which it is experienced by the pilot in operational conditions. Rule frequencies range from very low to very high. Conditions also have frequencies. Condition (or conditional) frequency measures the frequency with which the condition is true when it is evaluated by the automated system (for triggering the transition). Values also range from very low to very high. Conditions that are almost always true when evaluated have very high conditional frequencies. The salience of the feedback on the condition is another attribute of a condition. Two types of value are dissociated: p-feedback is the salience of the feedback when the condition is true and n-feedback is the salience of the feedback when the condition is false. Another attribute (not depicted in the figure 3) is dependency to the pilot: some conditions require an action by the pilot to change their status while others are autonomous and can change independently of any actions.

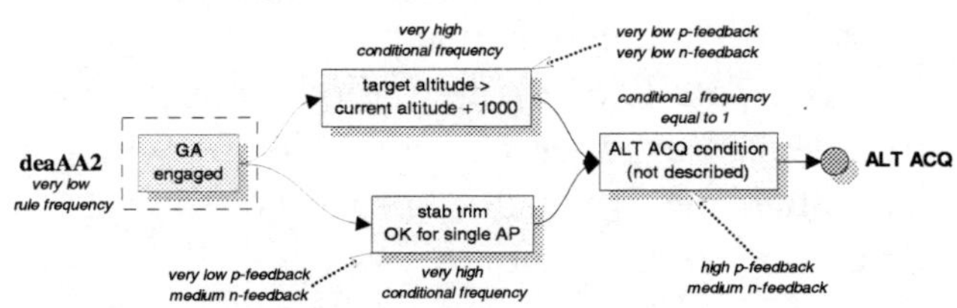

fig. 3. the rule for the uncommanded mode transition from GA to ALT ACQ

Pilot-mode interaction difficulties

Pilot-mode interaction difficulties have been invoked as a causal factor in many recent incidents and accidents. Pilots are not always able to assess easily which modes are active, what they are doing and why they are doing it as well as what they are going to do next,...

0-7803-5086-3 /98/$10.00 ©1998 IEEE

Pilot-mode interaction problems have therefore been studied in great details by several researchers (Sarter & Woods, 1992; Degani, 1996, Johnson & Pritchett, 1995; Vakil & Hansman, 1995,...). Many ways of improving things are also explored (Hutchins, 1993; Vakil & Hansman, 1995; Prevot, 1998; Riley & al, 1998). This paper is also an attempt to improve the current state of the art.

A Cognitive Model for Predicting Pilot-Mode Interaction Difficulties

Predicting the difficulties the pilots encounter with commanded and uncommanded mode transitions requires an understanding of the mechanisms by which the difficulties are generated. What are the cognitive processes and representations involved in pilot-mode interaction ? How are they organized, how are they used to produce the actions – and the errors – we observe on the flight deck.

The model presented below is a possible answer. It is extremely specific since it addresses the production of the actions related to commanded and uncommanded mode transitions. Its architecture however is general and it could therefore probably be used for studying the other procedural skills involved in the cockpit (e.g. the pilot-mode interaction sub-tasks described in Javaux, 1998). The model is an evolution of the rather simple and crude cognitive model described in Javaux (1998). The full details of the new model are provided in a separate paper (Javaux, 1998b) which is available on demand (requests should be addressed to the author).

The Architecture of the Model

The global architecture of the model is similar to the architecture of ACT-R developed at Carnegie Mellon University by John Anderson and his colleagues. ACT-R (Anderson, 1993) is a cognitive architecture which has been rather successful at explaining and simulating a whole bunch of psychological phenomenon and effects (e.g. the learning and performance of procedural skills). A common feature between ACT-R and our model is the usage of two main long-term memory stores: a declarative memory and a procedural memory. Both models also resort to a single short-term memory store: the working memory. Both models are hybrid systems because they combine symbolic processing with activation propagation mechanisms.

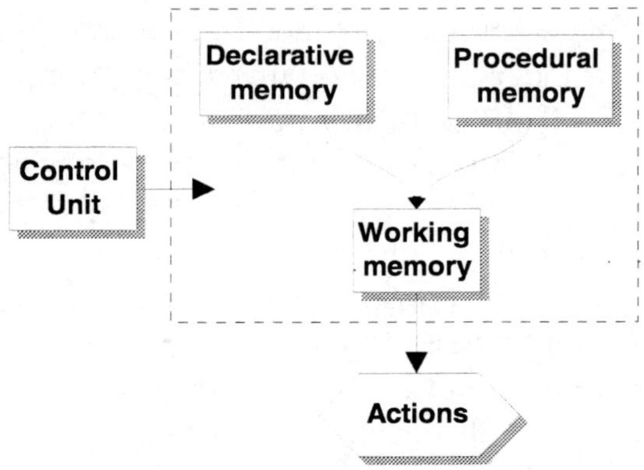

Figure. 4. the general architecture of the model

The Main Components of the Model

Declarative memory is a long-term memory (LTM) store. It contains the declarative knowledge related to the rules that govern mode transitions. Each rule is stored separately as a whole unit (cfr. the notion of chunk in ACT-R), by means of an oriented graph similar to the one in figure 3. Each rule and each node within a rule exhibits a specific activation level. The activation level corresponds to the 'strength' of the memory trace for the corresponding element. Activation levels are dynamic.

Procedural memory is another long-term memory (LTM) store. It contains the procedural knowledge related to the mode transitions. The basic data type in procedural memory is the procedure. A procedure is a template for producing actions. The same type of procedures is used for managing commanded and uncommanded mode transitions. While declarative memory is exhaustive, procedural memory is selective:

only a small set of procedures is stored in procedural memory. They correspond to the most frequently used action plans. Procedures are stored as an organized set of production rules (fig. 5). A production rule is constituted of preconditions and postconditions of the form $<(cond_1, cond_2, \ldots cond_n) \Rightarrow (action_1, action_2, \ldots action_m)>$. Each procedure and each production rule presents a specific activation level. The activation-related mechanisms found in declarative memory also apply here. Procedures are indexed (or addressed) in procedural memory by pairs of the form <mode transition, context>.

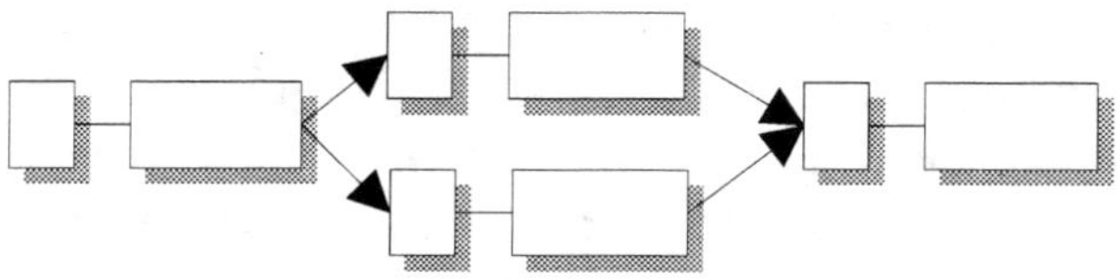

Figure 5. An example of procedure

Working memory, the third type of memory in the model is a short-term memory store: information will stay here for a limited amount of time. The role of working memory (WM) is important in the model as in most cognitive architectures. It is the place where most of the cognitive processing takes place. The capacity of working memory is limited. While the material in declarative and procedural memory is very specific, working memory can hold various types of material and it must be considered as a general purpose memory. Working memory is therefore shared with processes that do not resort to the management of mode transitions. This makes the processing of information in WM sensitive to external interference (e.g. during plan execution when responding to an ATC request).

The *control unit* is an essential component of the model. Its role is to coordinate the interaction between the different processes and data structures (or components) in the model. The control unit organizes the three mechanisms that produce the action plans (see below). It is also responsible for the circulation of information between the memory stores. The control unit is strongly coupled with working memory where most of the processing occurs. Like working memory it is a general purpose unit involved in multiple cognitive processes. As a consequence, it also is sensible to external interference.

As demonstrated by the two examples in the figures 1 and 2, the output of the model must be a *series of actions* the pilot executes, either for triggering a commanded mode transition or for predicting and monitoring an uncommanded mode transition. These actions correspond to the instantiation of an action plan present in working memory. Action plans are usually rather simple and linear as in the two examples above but more complex plans may exist, especially when more complex declarative rules are involved or when the rule chaining mechanism (see below) is invoked. Production rules of the form $<(cond_1, cond_2, \ldots cond_n) \Rightarrow (action_1, action_2, \ldots action_m)>$ have been chosen for representing action plans because they allow to implement elegantly many complex features. Production rules are also a basic component of procedures in procedural memory. Procedures and plans are similar in nature. Procedures should be considered as frequent plans.

The Production and Execution of an Action Plan

Commanded and uncommanded mode transitions are treated in similar ways in the model because they rely on the same kind of data structures in declarative and procedural memory. Differences appear when the working memory and control unit are concerned. The reader should refer to the examples in figures 1 and 2 to improve his or her understanding of the demonstration.

With *commanded mode transitions*, all starts with an intention and a context. The pilot wants to modify the engagement status of some modes (e.g. engaging LVL CHG in place of ALT HOLD) in the current context. According to the model, three mechanisms contribute to the *production of an action plan* for a given mode transition and a given context.

Retrieval in procedural memory. If one of the procedures available in procedural memory matches the pilot's intention (a mode transition) and the current context, the procedure is selected and transferred into working memory.

Contextual instantiation. If a procedure is not available in procedural memory, the pilot has to produce an action plan with a more complex and slower cognitive mechanism. It involves the transformation of a rule in declarative memory into an action plan. The detail of the process is described in Javaux (1998). The first step consists in selecting a rule that achieves the intended mode transition and transferring it to working memory. The conditions that are already true in the current context are then removed from the rule (contextual instantiation). An action plan for achieving each of the remaining false conditions is then generated. This last operation is usually simple and straightforward because single conditions can generally be achieved by single actions (e.g. pressing the ALT HOLD switch will achieve the 'ALT HOLD switch pressed' condition). Plan generation can be more complex when the conditions interact or when a single condition cannot be achieved by a single action. A particular case is when the achievement of a condition involves another mode transition. Rules must be *chained* (Javaux, 1998) and this leads to a hierarchical plan. The contextual instantiation mechanism is dependent upon the limited capacity of working memory (this can be a major concern in case of rule chaining, Javaux, 1998). Moreover, since working memory is a general purpose device, contextual instantiation is subject to possible external interference.

Problem solving and high-level reasoning. The third plan generation mechanism is invoked when the two other mechanisms fail. This happens in particular if the representation of the rule in declarative memory is weak, uncertain, incomplete or even non existent. The pilot has to resort to higher cognitive processes for bringing high level knowledge into the plan generation process. Mental models, abstractions, analogies,... are used here to build a more complete representation of the rule in working memory. Once completed, the representation is used in connection with the instantiation mechanism for producing an action plan.

According to the model, things are slightly different with *uncommanded mode transitions*. The role of the second and third mechanisms is less important here. This is because the search for potential precursors must be (nearly) permanent. The search is achieved by the control unit and working memory, by matching environmental information with the preconditions in the rules (in declarative memory) or in the procedures (in procedural memory). This continuous assessment of the current context can only be achieved on an automatic level of cognitive processing in the medium to high cognitive workload situations where most of the uncommanded mode transitions occur. A variation of the first plan generation mechanism is used in these situations. This means that the pilot will not be able to anticipate the rarest uncommanded mode transitions, because they are not represented in procedural memory. The second and third mechanism, while theoretically able to support the detection and the anticipation of uncommanded mode transitions cannot be used in these contexts, because of the high cognitive workload. These mechanisms can be involved at a latter stage, for providing explanations and supporting problem-solving in the situations where automation surprises occur.

The activation of the three mechanisms, either for commanded or uncommanded mode transitions, is scheduled by the control unit. One notes how cognitive complexity and response times increase as one steps from the

0-7803-5086-3 /98/$10.00 ©1998 IEEE

simpler to the more complex plan generation mechanisms

The *execution of an action plan* is also coordinated by the control unit. Action plans are normally executed after they have been obtained in working memory. In some cases, the execution will start during the plan generation. This will occur in particular to overcome the limitations of working memory (e.g. when the second and third plan generation mechanisms are invoked). Rarely practiced mode transitions involving rule chaining will typically lead to early execution. Executing a plan means executing its actions. Serialization of actions occurs when no specific order is specified.

Errors and Difficulties

The limited space in this paper does not allow to fully develop the consequences of the model for explaining and predicting errors and difficulties. Only the most important sources and mechanisms of errors are addressed here. The interested reader should consult Javaux (1998b) for complementary information.

Frequential simplification. This is an error mechanism described in Javaux (1998). It involves the transformation of the rules in declarative memory, thanks to the effects of rule and condition frequencies. Rule frequency impacts the strength of the trace for the corresponding rule. When rule frequency is low, the trace (activation level) of the rule is low, and its internal preconditions or conditions are more prone to be affected by external effects (e.g. frequential and inferential simplifications, mental models,...). Rules that are frequently used are more robust regarding the effects of these mechanisms. Frequential simplification itself is determined by the differential values of the condition frequencies in the rule. Javaux (1998) gives an example where the two internal conditions in the rule deaAA2 in the figure 3 are dropped from the representation in declarative memory because they are usually true when evaluated by the automated system (their conditional frequencies are very high). The rule stored in declarative memory is a simplification of the correct rule, hence the name of the error mechanism.

Inferential simplification. This error mechanism is also described in Javaux (1998). It is particularly influential when the trace of the rules in declarative memory is weak. Errors appear here in the form of distortions or simplifications applied to the rules already present in declarative memory. Errors appear because the human mind shows a natural tendency at making abstractions and searching for regularities and high-level patterns. Abstractions are high-level data structures that express the similarities that exist between the rules in declarative memory. Abstractions can be generated (or infered) by different means (e.g. finding syntactical similarities at the level of the preconditions or conditions within the rules, finding semantic similarities, for example when two rules trigger the same kind of aircraft behavior,...). Abstractions usually take the form of parametric rules, that is rules with variables (see Javaux, 1998 for an example). Substituting the variables with specific values produces an instance whose effects can be either constructive or detrimental on the rules in declarative memory. Some instances are fully correct (i.e. rules that actually belong to the functional specification of the autopilot) and they therefore reinforce the trace of the rules already present in declarative memory. Some instances however are incorrect and lead to a distortion of the representation of the rules (especially if their memory trace is weak). The conditions that are not compatible with an abstract rule will be removed (their trace will be weakened). New conditions will also be added to rules to which they do not belong.

Slips of action (Norman, 1988) represent a very frequent category of errors. They fit quite well with the model, mostly because of its hybrid nature (i.e. dynamic activation associated with the elements in the different memory stores). Capture errors (a sequence of actions is produced in place of the correct one) will occur when the pilot selects the wrong procedure in procedural memory or the wrong functional rule in declarative memory. This may happen because of the distribution of activation levels in the two memory stores. Frequency and recency effects are responsible for these errors. Omission or loss-of-activation errors (one or more planned actions are not executed) will occur when the activation level of the action plan in working memory is low and an external event disturbs its execution (e.g. when the control unit and the working memory have to get involved in another task). Description errors (actions performed on the wrong object) will also occur at the execution stage, mostly because of poor man-machine interface design.

The frequential and inferential error mechanisms, because they affect the information at an early stage of processing will influence all the later stages, in particular when the contextual instantiation mechanism is invoked. The action plans produced will be incorrect in some contexts. This will also lead to the storage of procedures that are simplifications of the correct forms (procedures are 'learned' from the plans manipulated in working memory, see Javaux, 1998b).

A Computerized Implementation of the Model

The model presented above constitutes a general architecture for studying the kind of procedural skills involved in the management of commanded and uncommanded mode transitions. The model relies on a series of assumptions. Many of them are plausible in regard to other theoretical constructs or architectures in cognitive psychology and cognitive science.

According to the model, three main factors determine pilot performance when dealing with commanded or uncommanded mode transitions: the functional characteristics of the modes, the frequencies of occurrence of the different rules and conditions, and the characteristics of the man-machine interface. Any pilot flying a specific aircraft in a specific operational context is immersed into a particular configuration of these three factors (rules, frequencies, interface). An algorithmic and computerized specification of the model should help investigating the relationship between a given configuration of factors and pilot performance. It will generate predictions in terms of behavior, response times and error patterns, and it will also allow to test existing configurations. Vakil & Hansman (1998) discuss four main strategies for improving pilot-automation interaction: procedure changes, training changes, improved feedback and automation modification. A computerized implementation of the model could probably contribute to each level of improvement.

Producing a computerized implementation of the model involves solving both the algorithmic and the implementation level problems (Anderson, 1993). A fully detailed algorithm must be defined to specify how the data structures are actually coded in the model and how they are processed. The algorithm must be implemented in a computer language. Choices must be made at both levels. The computerized model should therefore remain flexible, in order to test alternative algorithmic and implemented versions of the model (e.g. by tuning some of its parameters). The specification of the model at the algorithmic level is described in Javaux (1998b). The specification at the implementation level is currently under development (November 1998).

Acknowledgment

This research is carried out within the framework of a Belgian Scientific Policy program. The author would like to thank the people at the Sabena's flight training department where the research is performed.

References

Anderson, J.R. (1993). *Rules of the Mind.* Lawrence Erlbaum Associates Publishers, Hillsdale, NJ.

Degani, A. (1996). *Modeling Human-Machine Systems: On Modes, Error, and Patterns of Interaction.* Ph. D. thesis, Georgia Institute of Technology.

Hutchins, E. (1993). An Integrated Mode Management Display. Paper presented at the *NASA-Ames Mode Meeting*. NASA-Ames Research Center: Moffett Field, CA.

Javaux, D. (1997). Measuring Cognitive Complexity in Glass-Cockpits. A Generic Framework and its Application to Autopilots and their Modes. In R. Jensen & L. Rakovan (eds*), Proceedings of the Ninth Symposium on Aviation Psychology*, Columbus, Ohio.

Javaux, D. (1998). Explaining Sarter & Woods' Classical Results. The Cognitive Complexity of Pilot-Autopilot Interaction on the Boeing 737-EFIS. In N. Leveson & C. Johnson (eds), *Proceedings of the 2nd Workshop on Human Error, Safety and Systems Development*, April 1-2, Seattle, WA.

Javaux, D. (1998b). Predicting Pilot-Mode Interaction Difficulties. A Cognitive Model. Internal Report, Work Psychology Dpt, University of Liège, Belgium.

Johnson, E.N. & Pritchett, A.R. (1995). Experimental Study of Vertical Flight Path Mode Awareness. *ASL-93-3*, Department of Aeronautics and Astronautics, MIT: Cambridge.

Norman, D.A. (1988). *The Psychology of Everyday Things*. Basic Books.

Palmer, E. (1995). Oops, "It Didn't Arm." A Case Study of Two Automation Surprises. In R. S. Jensen & L.A. Rakovan (Eds.), *Proceedings of the Eighth International Symposium on Aviation Psychology*, pp. 227-232, April 24-27, Columbus: Ohio.

Prevot, T. (1998). A Display for Managing the Vertical Flight Path – An Appropriate Task with Inappropriate Feedback. In G. Boy, C. Graeber & J.M. Robert (Eds), *Proceedings of the International Conference on Human-Computer Interaction in Aeronautics 1998*, May 27-29, Montreal, Canada, Editions de l'Ecole Polytechnique de Montréal.

Riley, V., DeMers, B., Misiak, C. & Schmalz, B. (1998). The Cockpit Control Language: A Pilot-Centered Avionics Interface. In G. Boy, C. Graeber & J.M. Robert (Eds), *Proceedings of the International Conference on Human-Computer Interaction in Aeronautics 1998*, May 27-29, Montreal, Canada, Editions de l'Ecole Polytechnique de Montréal.

Sarter, N. & Woods, D.D. (1992). Pilot Interaction with Cockpit Automation. I. Operational Experiences with the Flight Management System. *International Journal of Aviation Psychology*, 2, 303-321.

Vakil, S.S., Hansman, J., Midkiff, A. & Vanek, T. (1995). Feedback Mechanism to Improve Mode Awareness in Advanced Autoflight Systems. In R.S. Jensen & L.A. Rakovan (Eds), *Proceedings of the 8th International Symposium on Aviation Psychology*, Ohio State University, Columbus, Ohio.

Vakil, S.S. and Hansman, J. (1998). Applying a Common Conceptual Model Approach to Vertical Navigation Automation. In N. Leveson & C. Johnson (eds), *Proceedings of the 2nd Workshop on Human Error, Safety and Systems Development*, April 1-2, Seattle, WA.

UNDERSTANDING AUTOMATION STRUCTURE BY ANALYZING MODE TRANSITION MATRICES

Sanjay S. Vakil, Massachusetts Institute of Technology, Cambridge, MA
R. John Hansman, Massachusetts Institute of Technology, Cambridge, MA

Abstract

Mode transition matrices are an extension of standard state diagrams and transition matrices. These matrices can be used as an organizational tool for the analysis of automation system composed of multiple modes. This paper will explore how these matrices may provide insight into underlying automation structure and be used in the process of human centered system design.

Motivation

The increased levels of automation in current generation aircraft have lead to a new set of human factors related concerns which are suspect in a large number of incidents. These include a crash caused by autopilot disconnect confusion in an A300 at Nagoya,[1] incorrect mode selection in an A320 at Strasbourg,[2] overspeed problems with the B757[1] and several others. These problems have raised the efforts of the design and human factors communities and, more recently, were the focus of the FAA Task Force Report on Interfaces Between Flightcrews and Modern Flight Deck Systems.[3] In addition, these concerns are also applicable to other highly automated arenas: air traffic control, process control plants and other highly automated vehicles.

Several researchers have suggested that the complexity in the modal structure of aircraft automation systems may be a contributing factor to the accidents and incidents which have occured.[4] While modes have come under criticism, they are essential to allow the functionality which is expected. A single "operating mode" is likely to be unable to cover all of the possible required functionality required since it must deal with differing demands. Modes allow an increase in capability by partitioning the possible behaviours of the automation to allow an operator to only actively recall and monitor a subset of possible behaviours.

In the context of this work, a mode is a set of unambiguously defined automation behaviours. Modal automation can be considered to consist of a set of modes linked by a set of transitions determined by a set of transition criteria. Transitions consist of both a starting and an ending mode and the criteria which will initiate the transition. These transitions can be caused by various factors, including intervention by the human operator (pressing a button), environmental changes (winds), and due to specific conditions being met (reaching a waypoint or speed limit).

In aircraft automation, each individual mode can be modeled by Control Block Diagrams, which capture the dynamic behaviour of a single mode. However, the global behaviour of the system, including the transitions between modes is difficult to capture in this notation. A previous ASRS review found that transitions between modes were a significant cause of problematic incidents.[5] Of the incident reviews related to automation, 20% were categorized as "mode transition problems". Mode transition diagrams and their associated matrices show promise in capturing

0-7803-5086-3 /98/$10.00 ©1998 IEEE

mode transition dynamics, and providing an organizational tool for mode research.

Mode Transition Matrices

Mode transition matrices are a tool used to represent modal automation and are based on the formalism of Finite State Machines (FSMs). FSMs are a standard tool used in the field of computer science to describe and design complex systems, including flight automation. FSMs consist of a set of states, a set of transitions between states and the criteria which cause transitions to occur.

Modal automation systems can be represented using the same notation which is used for Finite State Machines. In this nomenclature, the states correspond to the modes of the automation and transitions move between modes. The analogous diagrams are termed mode diagrams and the matrices which are derived are mode transition matrices.

Two representations are used. The first is a "state transition diagram" or "state diagram" as show in Figure 1 for a simple autoflight system. The transition criteria have been removed to simplify the diagram. The circles denote the possible states, and the directed arcs represent the possible transitions. There are a total of six states or modes in this diagram: Horizontal Autopilot Off (HOFF), Localizer Track (LOC), Vertical Autopilot off (VOFF), Heading Track (HDG) Vertical Speed (VS) and Glideslope Track (GS).

In this example, HOFF can transition to LOC or HDG, but not to VOFF, GS or VS. VOFF can be transitioned to from VS or GS and so on.

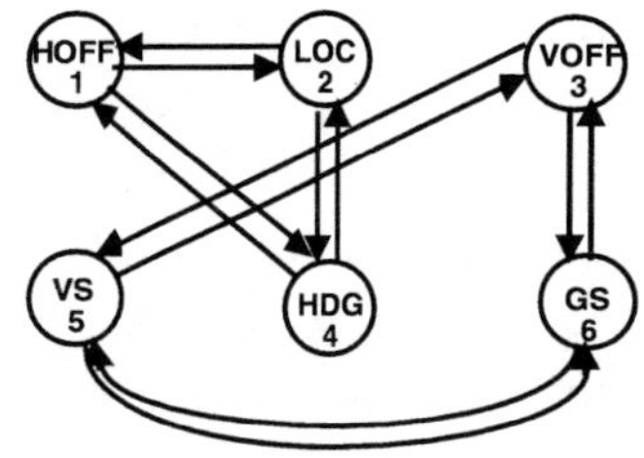

Figure 1: Simple Autoflight System

The autoflight system can also be described by a state transition table or matrix. Figure 2 shows the matrix associated with State Diagram A. Each row i is the set of transitions which leave state i. The column j in row i has an entry corresponding to whether a transition exists from state i to state j. If the state transition matrix is some matrix T, then T_{ij} is equal to one if a transition exists between states i and j.

Mode	HOFF	LOC	VOFF	HDG	VS	GS
HOFF		1		1		
LOC	1			1		
VOFF					1	1
HDG	1	1				
VS			1			1
GS			1		1	

Figure 2: Transition Matrix A

Allowable-transition Matrices

Allowable mode transition matrices have matrix elements corresponding to whether a transition is possible or allowed between two different modes, as shown in Figure 2. When creating the mode transition matrix, the correct ordering of modes may not be clear. In Figure 2, there is little apparent structure to the system. It is possible to reorder the matrix to show underlying structure to the system. In order to maintain the topology of the system, rows and columns must be reordered in tandem.

It is possible to reorder a matrix with k different modes in k! possible different ways. Since the number of possible permutations is so large, it is necessary to have a basis on which to determine the most useful configuration. The evaluation method which has been developed is to minimize the weighted sum of the squares of the distances of transitions from the diagonal axis of the matrix. This distance can be described as the difference between the row and column indices of the transition. The weighting term is the number of different transitions between the set of modes, T_{ij}. Equation 1 shows the function which is minimized. Transitions which do not change modes, namely those which lie directly on the diagonal, do not contribute to the sum.

This technique places each mode as close as possible to the other modes to which it can transition, grouping modes together which have transitions between them. This minimizes the width of the diagonal which bounds the existing transitions and thereby maximizes the areas which have no transitions.

$$f(k) = \sum_{i=1}^{k} \sum_{j=1}^{k} T_{ij} \cdot (i-j)^2 \qquad \textbf{Eq. 1}$$

This technique has been applied to the simple example matrix shown in Figure 2. The reordered matrix is shown in Figure 3. Two distinct areas are apparent and the top left and bottom right quadrants are completely empty. Modes HOFF, LOC and HDG can be grouped together in a set which does not interact with modes VOFF, VS and GS. If interactions were to occur, they would appear in the top-right and bottom-left quadrants of the matrix to allow transitions the lateral modes and the vertical. In independent systems, it is often the case that the modes are meant to be used in parallel, simultaneously, and so should be designed to be independent.

Mode	HOFF	LOC	HDG	VOFF	VS	GS
HOFF		1	1			
LOC	1		1			
HDG	1	1				
VOFF					1	1
VS				1		1
GS				1	1	

Figure 3: Reordered Matrix of Simple Autoflight System

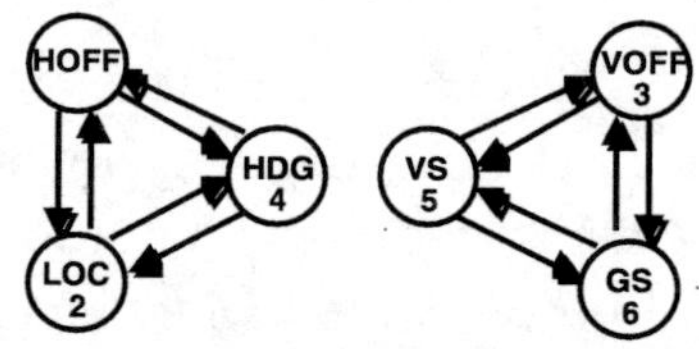

Figure 4: Reordered Diagram of Simple Autoflight System

The mode diagram can then be redrawn, as shown in Figure 4. This system can be seen to consist of two completely independent subsystems which cannot transition between each other such as the vertical and horizontal autopilots of an aircraft. This reduces the complexity and order of the system by reducing the interactions which an operator would have to internalize. These two controls may be implemented as logically and/or physically separate systems.

In some cases, it may be useful to examine the converse of the allowable-transition matrix.. By doing so, the criteria for not allowing access to specific behaviours can be examined. This may be useful in situations where a mode change could lead to a dangerous aircraft configuration.

Frequency-based Transition Matrices

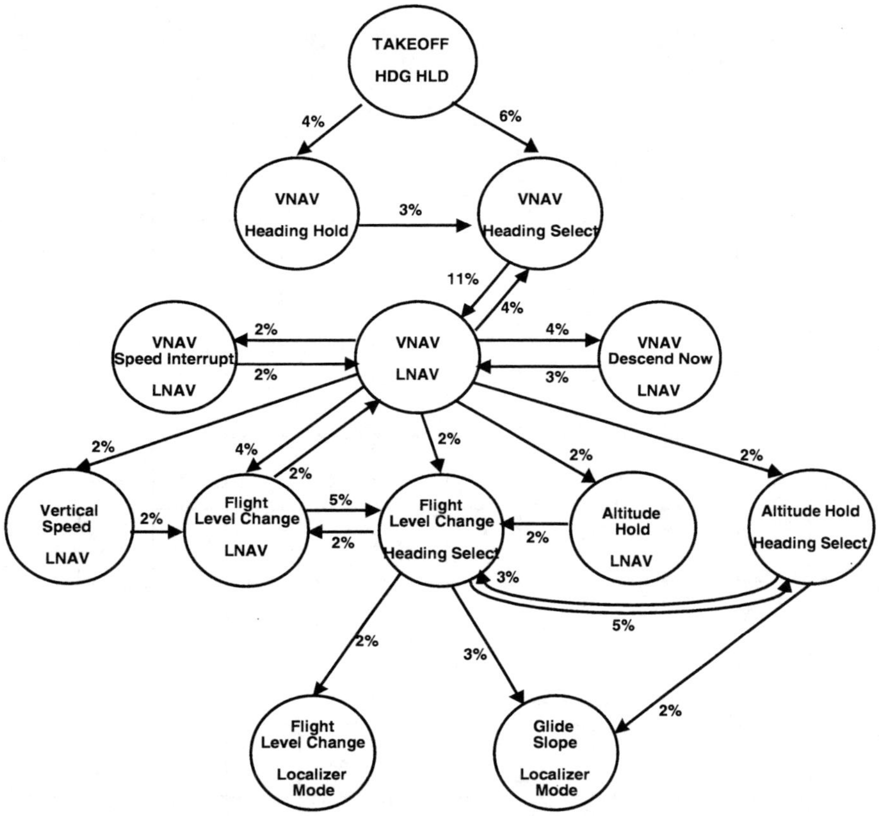

Figure 5: Observed Autoflight System Frequency Transition Matrix

Other aspects of mode transitions, such as the absolute or relative frequency of transitions may also be investigated by populating the mode transition matrix with different values. This is an extension from the use of Finite State Machines in computer science, where the matrices are typically populated with the allowable transitions between states.

Figure 5 is a mode transition diagram adapted from work by Degani (1994) observing mode transitions in operational, revenue-earning flights of a Boeing 757.[6] As opposed to the earlier example, the aircraft autoflight modes include both vertical and lateral behaviours. Mode transitions which were observed with a frequency of less than 2% were not included on the diagram.

The mode transition matrix based on Degani's work associated with Figure 5 is shown in Figure 6. There are several insights to be gained from the matrix. The first is that certain modes were only observed to transition to one other mode: VNAV-Heading Hold was only observed to transition to VNAV-Heading Select. This does not imply that this is the only allowable transition from VNAV-Heading Hold, but does highlight the use of modes operationally. Since the matrix is based on a finite set of data, all possible transitions were unlikely to be observed. However, by looking at the frequency data, the nature of the mode transition matrix can be seen to be dependent on the operating policies and training characteristics of the particular airline. If data across airlines can be examined, a nominal usage pattern of mode transitions may be apparent.

	Takeoff Heading Hold	VNAV Heading Hold	VNAV Heading Select	VNAV Spd Int LNAV	VNAV LNAV	VNAV Des Now LNAV	Vertical Speed LNAV	Flight Level Ch LNAV	Flight Level Ch Heading Select	Altitude Hold LNAV	Altitude Hold Heading Select	Flight Level Ch Localizer	Glide Slope Localizer
Takeoff Heading Hold		4%	6%										
VNAV Heading Hold			3%										
VNAV Heading Select					11%								
VNAV Spd Int LNAV					2%								
VNAV LNAV			4%	2%		4%	2%	4%	2%	2%	2%		
VNAV Des Now LNAV					3%								
Vertical Speed LNAV								2%					
Flight Level Ch LNAV					2%				5%				
Flight Level Ch Heading Select								2%			5%	2%	3%
Altitude Hold LNAV									2%				
Altitude Hold Heading Select									3%				2%
Flight Level Ch Localizer													
Glide Slope Localizer													

Figure 6: Observed Autoflight System Absolute Frequency Transition Matrix

Other modes have no observed exit transitions at all. The Flight Level Change-Localizer and Glide Slope-Localizer Modes are terminal modes, as can be seen from the empty row in the transition matrix. While there are allowable transitions from these modes, usually to a Go Around Mode, they were not observed.

Large blank areas can be observed in the matrix, corresponding to sets of transitions which were not observed. In many instances, this is due to certain modes being used during certain stages of flight. An obvious example is Takeoff-Heading Hold which only used during takeoff and climbout and not later in flight.

Columns and rows which are highly populated represent the modes in which the aircraft automation spends the bulk of its time. In this data VNAV-LNAV has a large number of transitions both in and out. In particular, note the most frequent transition between VNAV-Heading Select and VNAV-LNAV. This is due to vector based air traffic control transitioning to flying a predetermined flightplan.

0-7803-5086-3 /98/$10.00 ©1998 IEEE

	Takeoff Heading Hold	VNAV Heading Hold	VNAV Heading Select	VNAV Spd Int LNAV	VNAV LNAV	VNAV Des Now LNAV	Vertical Speed LNAV	Flight Level Ch LNAV	Flight Level Ch Heading Select	Altitude Hold LNAV	Altitude Hold Heading Select	Flight Level Ch Localizer	Glide Slope Localizer
Takeoff Heading Hold		40%	60%										
VNAV Heading Hold			100%										
VNAV Heading Select					100%								
VNAV Spd Int LNAV					100%								
VNAV LNAV			18%	9%		18%	9%	18%	9%	9%	9%		
VNAV Des Now LNAV					100%								
Vertical Speed LNAV								100%					
Flight Level Ch LNAV					29%				71%				
Flight Level Ch Heading Select								17%			42%	17%	25%
Altitude Hold LNAV									100%				
Altitude Hold Heading Select									60%				40%
Flight Level Ch Localizer													
Glide Slope Localizer													

Figure 7: Observed Autoflight System Relative Exit Frequency Transition Matrix

Relative versus Absolute Frequency Matrices

The matrix shown in Figure 6 is an example of an absolute frequency matrix. Each of the elements in the matrix is normalized relative to all observed transitions. Another way of examining the same data is to normalize the frequency of the transitions by the transitions out of a specific mode. Figure 7 shows the data normalized in this manner.

It is clear from this matrix, that a large number of modes only have a single transition. Others have sets of transitions with varying likelihoods. Another way of looking at the data would be to normalize by columns rather than the rows, which will show the relative frequencies of the entry transitions into modes.

Further Uses of Transition Matrices

It may be possible to use transition matrices to investigate other behaviours, such as the types of transitions that occur in automation. There are three categories of transitions in most autoflight systems: commanded transitions, automatic uncommanded transitions and armed conditional transitions. Each of these types of transitions can be shown on separate mode transition matrices for comparative purposes to examine the behaviour of the automation.

Another potential use of transition matrices is to highlight the set of mode transitions which are used infrequently by examining the difference between a normalized allowable matrix and a frequency matrix. The difference of these two matrices will be a measure of the infrequency of transitions: those which can occur, but do not appear during normal operation.

If these transitions are determined to no longer be necessary, this information can be used as guidance in the simplification of subsequent generations of automation systems. It may also be the case that the transitions are necessary (such as transitions to a Go Around Mode), but are operationally infrequent. Additional training can be brought to bear on these transitions.

Conclusion

Flight automation systems are becoming more complex with time and large portion of this complexity is due to their modal nature and the manner in which transitions between modes occur. Mode transition matrices provide an organizational framework in which to examine mode transitions in a methodical manner. The matrices may also provide insight into the underlying structure of automation systems.

Several types of mode transitions matrices have been outlined. The manner in which different types of matrices may be used in conjunction has also been discussed.

Acknowledgments

This work was supported by NASA Grants NAG-1-1581 and NAG-1-1857 and by FAA and NASA Grant FAA-95-G-017.

References

1. Mecham, M (1994). Autopilot Go-Around Key to CAL Crash *Aviation Week & Space Technology*. May 9, 1994.

2. Hughes, D. (1995). Accidents Direct Focus on Cockpit Automation, *Aviation Week & Space Technology*. January 30, 1995.

3. 1996 FAA Report on the Interfaces Between Flightcrews and Modern Flight Deck Systems

4. Sarter, N. (1994). Pilot Interaction with Cockpit Automation II: An Experimental Study of Pilots' Model and Awareness of Flight Management Systems, *The International Journal of Aviation Psychology 1*, no. 1: 1-28.

5. Vakil, S., Hansman, R.J., and Midkiff, A., "Feedback Mechanisms to Improve Mode Awareness in Advanced AutoFlight Systems," Eighth International Symposium on Aviation Psychology, April 1995.

6. Degani, A. (1994). Modelling Human-Machine Errors: on Modes, Error and Patterns of Interaction. Doctoral Thesis. Atlanta, GA: Georgia Institute of Technology.

HUMAN FACTORS ISSUES OF FLIGHT DECK AUTOMATION

Ken Funk, Oregon State University, Corvallis, OR

Beth Lyall, Research Integrations, Inc., Tempe, AZ

Introduction

It is widely acknowledged that commercial transport aircraft automation has improved the economy of airliners, and that accident rates for advanced technology commercial aircraft are lower than those of comparable conventional technology aircraft. Nevertheless, criticism of the human factors of modern flight deck automation is common among pilots and other aviation professionals (e.g., Billings, 1997).

Until recently, though, there was no comprehensive list of flight deck automation human factors issues, much less a comprehensive summary of evidence (data and other reasonably objective information) related to those issues. As a result, avionics designers have been somewhat in the dark about the usability of the equipment they have created to help improve the efficiency and safety of commercial air transportation.

This paper summarizes a study we conducted to identify and compile evidence related to flight deck automation human factors issues. To be useful to those individuals directly involved in the development of future automation systems, the paper focuses primarily on those issue related to automation design. More complete descriptions of the study are in a paper (Funk, Lyall, Wilson, Vint, Niemczyk, Suroteguh, and Owen, forthcoming) and a website (http://flightdeck.ie.orst.edu/).

Objectives

The objectives of our study were to:

1. develop a comprehensive list of flight deck automation human factors issues;
2. compile a large body of existing data and other evidence related to those issues; and
3. disseminate the list of issues and supporting data to the aviation research, development, manufacturing, operational, and regulatory communities.

Phase 1

To identify flight deck automation issues, in Phase 1 of our study we compiled a list of possible problems with, or concerns about, flight deck automation, as expressed by pilots, scientists, engineers, and flight safety experts. We reviewed 960 source documents, including papers and articles from the scientific literature as well as the trade and popular press, accident reports, incident reports, and questionnaires filled out by pilots and others. We also conducted our own survey of 128 pilots and others experienced with flight deck automation, asking them to describe their concerns. In these source documents, we found 2,428 specific citations and classified them into 114 possible problems and concerns. Although these problems and concerns covered everything from pilot training to company policies and procedures, many were related to automation design, such as the following.

- The pilot-automation interface may be poorly designed with respect to human factors considerations, possibly resulting in poor pilot performance or pilot dissatisfaction.
- Keyboard alphanumeric data entry may be prone to errors, which may adversely affect safety.

0-7803-5086-3 /98/$10.00 ©1998 IEEE

- Automation may be too complex in that it may consist of many interrelated components and may operate under many different modes. This makes automation difficult for pilots to understand and use safely.

In Phase 1 we merely enumerated possible problems and concerns and did not verify that these citations were justified in any way. That we addressed in Phase 2.

Phase 2

In Phase 2 we updated and refined the list of possible problems and concerns identified in Phase 1 and reviewed a wide variety of sources to locate and record evidence (data and other information) related to them. Because an *issue* is a point of disagreement or debate, we generally refer to these possible problems and concerns in the remainder of this paper as *flight deck automation issues,* or just *issues*. We refer to the statement associated with each issue (e.g., the three statements listed above) as the *issue statement*.

General Methodology and Results

The sources we reviewed for evidence included aircraft accident reports, documents describing aircraft incident report studies, and documents describing scientific experiments, surveys and other studies. We also conducted a survey of individuals with broad expertise related to human factors and flight deck automation. We reviewed these sources for data and other objective information related to an issue. For each instance of this evidence (i.e., each piece of relevant data or information) we recorded an excerpt from the source document and assessed the extent to which it supported one side of the issue or the other, assigning a numeric strength rating between -5 and +5. We assigned a positive strength rating to evidence supporting that side of the issue suggested by its issue statement (supportive evidence) and a negative strength rating to evidence supporting the other side (contradictory evidence). We recorded each excerpt, the associated issue, the assigned strength, and other related information in a database, which is now accessible from our website.

The overall process yielded a list of 92 flight deck automation issues and 742 instances of evidence related to those issues, of which 520 instances of evidence are supportive and 222 instances are contradictory. Sixty-eight issues are design-related. There are 556 instances of evidence for these design-related issues, of which 391 are supportive and 165 are contradictory. The remainder of this section summarizes the methodology and results for each specific part of Phase 2.

Expert Evidence

As a first step in meeting our second objective (see above) we conducted a survey of flight deck automation experts to develop data to be used as evidence. We asked forty-seven individuals to participate in the survey based on their broad research or performance experience with flight deck automation. Thirty completed surveys were returned. The participants included pilots of several automated aircraft types, university researchers, airline management pilots, industry designers and researchers, and government regulators and researchers.

The survey requested general demographics information then presented 114 statements, one for each of the problems and concerns identified in Phase 1. Each statement was presented as an unqualified assertion. For example, the statement for the first Phase 1 problem/concern listed above was as follows: *The pilot-automation interface is poorly designed with respect to human factors considerations, resulting in poor pilot performance or pilot dissatisfaction.* For each statement, the respondent was asked to rate their level of agreement that the assertion was true on a scale from 1 (strongly disagree) to 5 (strongly agree). There was also a place to mark "cannot address" for each statement. In addition to the agreement rating, each statement was rated for criticality to flight safety from 1 (not critical) to 5 (extremely critical). Additionally, there was a place for the respondents to indicate the type of information upon which they based their ratings for each issue (e.g., research data, personal experience).

We summarized the responses for each issue by recording the mean expert agreement rating and the mean expert criticality rating. Table 1 presents the results for three design-related issues.

We summarized the responses for each statement and recorded that information as an evidence excerpt in our database. We rated evidence in support of an issue statement with a strength rating based on the percentage of responses of "agree" and "strongly agree" combined, and we recorded contradictory evidence based on the percentage of responses indicating "disagree" or "strongly disagree." These strength ratings, ranging from -5 to +5, were consistent with the strength ratings we assigned to evidence obtained from other surveys we reviewed in Phase 2 (see below).

Accident Report Evidence

From the literature, comments from our experts, news media, and other sources, we identified 34 aircraft accident reports we thought might contain evidence related to the flight deck automation issues. We were able to obtain 20 of these reports, which we reviewed, looking for statements by the investigating board identifying one or more of the flight deck automation issues as contributing to the accident. We recorded excerpts and assigned strengths based on how closely the investigating board's statements about their findings corresponded to an existing flight deck automation issue and the extent to which the board suggested that the issue contributed to the accident.

Table 1. Expert ratings of three design-related flight deck automation issues. See text for rating interpretations.

issue number& issue statement	mean ratings agree-ment	criti-cality
issue165: Cultural differences may not be adequately considered in automation design, training, certification, and operations. If they are not considered, they may have resulting effects on performance and how automation is used.	4.4	3.9
issue102: The attentional demands of pilot-automation interaction may significantly interfere with performance of safety-critical tasks. (e.g., "head-down time", distractions, etc.)	4.3	4.0
issue100: Automation design may not be guided by a philosophy that gives adequate attention to the proper role and function of the human and to human capabilities and limitations. This may compromise system effectiveness and safety.	4.0	4.0

We found evidence related to flight deck automation issues in 17 of the 20 accident reports we reviewed. Table 2 presents the number of instances of supportive (+) evidence found in accident reports for three design-related issues. Due to the nature of accident reports, we found no contradictory evidence in them.

Incident Report Study Evidence

We reviewed eight studies of Aviation Safety Reporting System (ASRS) incident reports, including one we conducted ourselves. In each of the incident studies we reviewed, the investigators selected a set of incident reports from the larger ASRS database based on study-specific criteria, then reviewed the report narratives for information identifying and/or describing automation-related issues. We reviewed the investigators' summaries and conclusions in search of evidence for the flight deck automation issues identified earlier in our study. We recorded excerpts from these summaries and rated the strength of this evidence based on the percentage of incident reports reviewed that supported or contradicted the statement of the issue. Table 3 presents the number of instances of supportive (+) evidence found in incident report studies for three design-related issues. Due to the nature of incident reports, we found no contradictory evidence in incident report studies.

Experiment Evidence

We reviewed documentation from 18 flight simulator experiments and found evidence related to flight deck automation issues in 15 of them. We recorded excerpts, assigning strength ratings based on the methodology and type of subjects used and the type of evidence yielded by the experiment. For example, in experiments in which a specific hypothesis was tested by collecting data from subjects performing an experimental task, we considered the type of subjects used (i.e., whether they were trained pilots or not) and the nature of the task (i.e., how close it was to line flying) in assigning strengths. Table 4 presents the number of instances of supportive (+) and contradictory (-) evidence found in experimental studies for three design-related issues.

Table 2. Summary of accident evidence for three representative design-related flight deck automation issues.

issue number& issue statement	+ evidence
issue105: Pilots' understanding of automation may be inadequate for the performance of their duties.	6
issue092: Displays may not be designed for adequate visibility, legibility, and readability, possibly resulting in misinterpretation of display information.	3
issue099: Not enough information may be displayed for pilots to adequately perform their tasks.	2

Table 3. Summary of incident report study evidence for three representative design-related flight deck automation issues.

issue number& issue statement	+ evidence
issue112: Procedures for data entry and programming automation may be unclear, overly difficult, complex, and time consuming. This may cause errors and delays that may lead to unsafe conditions.	3
issue102: The attentional demands of pilot-automation interaction may significantly interfere with performance of safety-critical tasks. (e.g., "head-down time", distractions, etc.)	2
issue108: Automation may perform in ways that are unintended, unexpected, and perhaps unexplainable by pilots, possibly creating confusion, increasing pilot workload to compensate, and sometimes leading to unsafe conditions.	2

Table 4. Summary of experiment evidence for three representative design-related flight deck automation issues.

issue number& issue statement	evidence +	evidence -
issue095: Pilots may not be able to tell what mode or state the automation is in, how it is configured, what it is doing, and how it will behave. This may lead to reduced situation awareness and errors.	2	0
issue038: Display layout in automated flight decks may change the traditional instrument scan pattern, possibly leading to loss of skills which may be needed upon transitioning to conventional aircraft.	2	1
issue079: Automation may increase overall pilot workload, or increase pilot workload at high workload times and reduce pilot workload at low workload times, possibly resulting in excess workload and/or boredom.	0	7

Survey Evidence

We reviewed the results of 25 surveys (including our own -- see above) conducted to obtain opinions on flight deck automation from pilots, scientists, and other aviation professionals. As in our own study, most of these surveys asked respondents to rate their level of agreement with various assertions about automation. As we recorded this evidence, we assigned strengths based on the proportion of responses supportive of or contradictory to each issue. Table 5 presents the number of instances of supportive (+) and contradictory (-) evidence for three design-related issues.

Other Evidence

In addition to the studies described above, we also reviewed studies involving the direct observation of pilots in line operations and in line-oriented flight training (i.e., training simulator) sessions.

Table 5. Summary of survey evidence for three representative design-related flight deck automation issues.

issue number& issue statement	evidence +	evidence -
issue121: Automation design may not take into consideration the operational knowledge of pilots. This may lead to designs that are counter-intuitive to pilots, possibly increasing pilot workload and the opportunity for error.	4	3
issue046: Pilots may lack confidence in automation due to their experience (or lack thereof) with it. This may result in a failure to use automation when it should be used.	12	12
issue166: Company policies and procedures for the use of automation may be inappropriate or inadequate in some circumstances, possibly compelling pilots to use automation when they prefer not to and/or leading to pilot confusion or frustration.	4	4

Meta-Analysis

In our study we compiled a large amount of evidence on flight deck automation issues from a large number of sources. In an attempt to interpret this information, we conducted an analysis of the evidence. The objectives of this meta-analysis were to summarize the evidence collected in Phase 2 to identify those issues that are problems in need of solutions, those issues that do not appear to be problems, and those issues which require more research.

We ranked the issues based on several criteria, including the following:

- number of citations -- the number of (possibly unsubstantiated) citations of each issue as a possible problem or concern, as found in Phase 1;
- expert agreement rating -- the mean agreement rating given to each issue by the experts in our Phase 2 expert survey (that is, the extent to which the experts agreed with the statement suggested by the issue statement);
- expert criticality rating -- the mean criticality rating given to each issue by the experts in our Phase 2 expert survey (that is, how critical to safety the experts felt the issue was); and
- sum of strengths -- the sum of all evidence strength ratings for the issue (that is, a total "weight" of evidence on both sides of the issue).

Recall that we gave each piece of evidence a strength rating: +5 for evidence strongly supporting that side of an issue suggested by its issue statement, -5 for evidence strongly contradicting that side of an issue suggested by its issue statement, and values between those extremes for intermediate evidence. Considering only the overall "weight" of evidence on both sides of an issue, the sum of strength ratings from all of its evidence, the three design-related issues with the greatest overall supportive evidence are presented in Table 6.

The three design-related issues with the greatest overall contradictory evidence are presented in Table 7.

To form a composite ranking based on multiple criteria, for each issue we summed its ranks for each of the four criteria described above, then produced a "meta-ranking" by sorting the issues in increasing order of this rank sum. The three design-related issues ranking highest in multiple criteria are presented in Table 8.

Table 6. Design-related flight deck automation issues ranking high in supportive evidence.

rank	issue number& issue statement	sum of evidence strengths
1	issue105: Pilots may not understand the structure and function of automation or the interaction of automation devices well enough to safely perform their duties.	+63
2	issue083: The behavior of automation devices -- what they are doing now and what they will do in the future based upon pilot input or other factors -- may not be apparent to pilots, possibly resulting in reduced pilot awareness of automation behavior and goals.	+35
4	issue092: Displays (including aural warnings and other auditory displays), display formats, and display elements may not be designed for detectability, discriminability, and interpretability. This may cause important information to be missed or misinterpreted.	+32

Conclusions and Recommendations

In reviewing these results, the reader should keep in mind several limitations of our overall approach to evidence collection. First, in spite of our broad approach and our best efforts to identify all relevant evidence, there is certainly some that we missed. Second, the nature of some of the issues and the nature of the sources we reviewed to discover evidence may have reduced the opportunity for obtaining contradictory evidence.

Table 7. Design-related flight deck automation issues ranking high in contradictory evidence.

rank	issue number& issue statement	sum of evidence strengths
2	issue079: Automation may increase overall pilot workload, or increase pilot workload at high workload times and reduce pilot workload at low workload times, possibly resulting in excess workload and/or boredom.	-11
4	issue139: The presence of automation may reduce inter-pilot communication, possibly resulting in less sharing of information.	-9
5	issue156: Automation may induce fatigue, possibly leading to poor pilot performance.	-4

General Conclusions and Recommendations

Nevertheless, it is still possible to draw some general conclusions and recommendations from our findings. We consider those issues with the greatest overall supportive evidence (including those in Table 6) and especially those issues ranking highest in multiple criteria (Table 8) as problems which require solutions. Scarce resources should not be expended on further efforts to show that they are problems -- the evidence already exists. Now the resources should be applied to developing solutions.

We believe that the issues with the greatest overall contradictory evidence (including those in Table 7) do not represent significant problems, and resources would be better used in solving other problems (see above) or further exploring unresolved issues.

Table 8. Design-related flight deck automation issues ranking high in multiple criteria.

meta-rank	issue number& issue statement
1	issue102: The attentional demands of pilot-automation interaction may significantly interfere with performance of safety-critical tasks. (e.g., "head-down time", distractions, etc.)
2	issue108: Automation may perform in ways that are unintended, unexpected, and perhaps unexplainable by pilots, possibly creating confusion, increasing pilot workload to compensate, and sometimes leading to unsafe conditions.
4	issue025: It may be difficult to detect, diagnose, and evaluate the consequences of automation failures (errors and malfunctions), especially when behavior seems 'reasonable', possibly resulting in faulty or prolonged decision making.

Recommendations for Automation Designers

The design community must be the ones to implement solutions to the problems we have identified. We recommend that designers use our database to help identify the issues that they may want to address as they modify existing designs and develop new designs. As tradeoffs are made during the design process, the highest-ranking issues may be considered priorities, including those presented in Tables 6 and 8.

The research community has begun identifying possible solutions to some of these problems and we recommend that designers consider them. In particular, Sarter (1998) and her colleagues are exploring new ways to capture the flightcrew's attention and give them feedback through peripheral visual and tactile cues. This work may lead to future solutions addressing many issues, including 025, 083, 092, and 102 found in Tables 6 and 8. All of these issues relate to the ways in which the automation communicates with the flight crew.

We also identified several issues related to the way the flight crew communicates with the automation, including issues 083 (Table 6), 102, 108 (Table 8), and 112 (Table 3). Riley and his colleagues (1998) have been addressing these issues through their development of a new pilot interface with the flight management system. Their approach produces an interface based upon the flight crew's knowledge of flight control and how the airplane must react to requests from air traffic control and others.

Many of the issues are strongly related to pilots' understanding of automation, including issues 105, 083 (Table 6), 102, 108, and 025 (Table 8). Sherry and his colleagues (1995) are investigating a method of facilitating this understanding by adopting, early in the development process, a common functional representation that can serve as a basis not only for defining automation functional requirements and specifications, but also for designing automation interfaces, flight deck procedures, manuals, and training materials.

The experimental approaches described above may lead to solutions to the problems specifically mentioned, as well as those related to the more general issues of design philosophy (issue100 in Table 1) and having operator knowledge contributed to the design process (issue121 in Table 5), especially when those solutions are pursued in the context of a rigorous systems integration process (e.g., SAE, 1995). We recommend that designers give these and related approaches due consideration.

We acknowledge that developing solutions to the automation problems we have identified is not a trivial task -- there are significant technical, economic, and even political challenges to face. However, we hope that the findings summarized here and presented more completely in our website (http://flightdeck.ie.orst.edu/) will help designers create a new generation of flight deck automation that is not only functionally effective, but more usable by pilots.

Acknowledgment

This research was funded by the Federal Aviation Administration Office of the Chief Scientific and Technical Advisor for Human Factors (AAR-100). John Zalenchak, Tom McCloy, and Eleana Edens were our technical monitors. We gratefully acknowledge their support.

References

Billings, C.E. (1997). *Aviation Automation: The Search for a Human-Centered Approach*. Mahwah, NJ: Lawrence Erlbaum Associates.

Funk, K., B. Lyall, J. Wilson, R. Vint, M. Niemczyk, C. Suroteguh, and G. Owen (forthcoming). Flight deck automation issues, *International Journal of Aviation Psychology*.

Riley, V., B. DeMers, C. Misiak, and B. Schmalz (1998). The cockpit control language: a pilot-centered avionics interface, in *Proceedings of HCI-Aero 98*, 27-29 May 1998, Montreal, Quebec, Canada.

SAE (1995). *Aerospace Recommended Practice: Pilot -System Integration*, ARP 4033, Warrendale, PA: SAE International.

Sarter, N.B. (1998). Multisensory feedback in support of human-automation coordination in highly dynamic event-driven worlds. Presentation at the Third Automation Technology and Human Performance Conference, 25-28 March 1998, Norfolk, Virginia.

Sherry, L., D. Youssefi, and C.H. Hynes, (1995). *A Formalism For the Specification Of Operationally Embedded Reactive Avionics Systems* (Honeywell publication no. C69-5350-001). Phoenix, AZ.

OPERATOR DIRECTED COMMON CONCEPTUAL MODELS

FOR ADVANCED AIRCRAFT AUTOMATION

Sanjay S. Vakil, Massachusetts Institute of Technology, Cambridge, MA

R. John Hansman, Massachusetts Institute of Technology, Cambridge, MA

Abstract

The incorporation of advanced flight automation systems into modern aircraft has resulted an increase in errors in interaction with aircraft automation. A prior study identified the lack of a simple, consistent model of the automation as a contributing component and found pilots creating their own ad-hoc models of aircraft automation.

The design of new, complex automation system requires a shift from the current developmental process to one which addresses the needs of the operator explicitly. The Operator Directed Process is an attempt to enforce the creation of a high level system models suitable for operator use and training materials.

Introduction

The incorporation of advanced flight automation systems into modern aircraft has resulted in a new set of human factors related concerns. This set of problems has led to a large set of incidents, including accidents of an A300 at Nagoya,[1] an A320 accident at Strasbourg,[2] and incidents with an A310 at Orly,[3] overspeed problems in the B757.[1] They have also raised sufficient concerns to warrant government investigation in the form of the 1996 FAA Report on the Interfaces Between Flightcrews and Modern Flight Deck Systems.[4]

Modern flight automation systems have evolved from simpler versions through the incorporation of additional functionality designed to improve flexibility and provide new features such as automatic flight planning, the inclusion of envelope protection modes, fuel optimal descent profile calculation and a proliferation of warning and safety mechanisms.

One of the goals intended to be met by highly automated aircraft was a reduction in the aircraft flight automation issues attributed to flight crew error. This goal has not been attained and flight crew error remains the dominant cause of aircraft incidents.[4] Numerous incidents have been reported where there exists some confusion between the pilots' expectations of the AFS and what the system is actually doing.[5]

An analysis was made of reports from the Aviation Safety Reporting System (ASRS), a volunteer mechanism for documenting problems in flight operations.[6] Four years (1990-4) worth of reports were searched for automation-related keywords and returning a total of 300 reports. This set was then reviewed by a set of pilots and a total of 184 reports were categorized as Mode Awareness Problems. The results were further categorized by both the flight domain effected (vertical/speed or horizontal) and the type of problem reported.

Figure 1 shows the results of this report, noting in particular the preponderance of reports of Programming Errors, Mode Transition Problems and Insufficient Understanding of Automation. This indicates

0-7803-5086-3 /98/$10.00 ©1998 IEEE

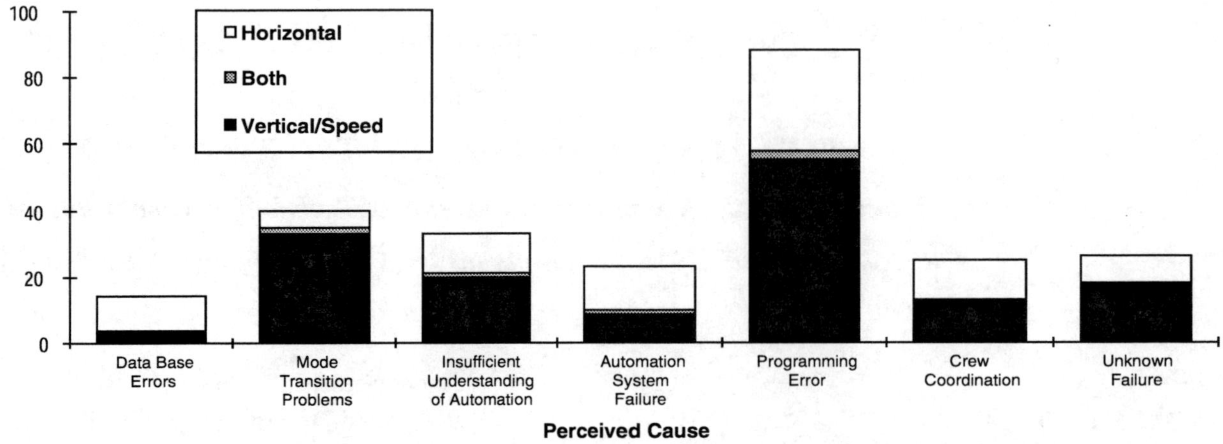

Figure 1: Results of ASRS Review from 1990-4

the need for mechanisms to improve operator understanding of complex systems.

This paper is organized into three sections. First, the need for a change in the development process of automation systems will be motivated by an examination of the existing pilot training literature and experience with the current Flight Management Systems in modern aircraft. Next, an Operator Directed Process will be described in the context of system design and certification. Finally, an experimental means to evaluate the efficacy of such a process will be presented.

Motivation

In a study of mode awareness problems in current flight automation, a review was conducted of the existing flight automation systems of seven aircraft (B757, B767, B747-400, A320, A300, MD-11, F-100).[6]

Absence of Consistent Models of Automation

Based on available training literature, focused interviews with line pilots and check airmen and contact with avionics manufacturers, a consistent global model does not exist. In the absence of a simple, consistent and communicable model of flight automation, pilots appear to develop their own ad-hoc mental models which may not accurately reflect the operation of the automation.[7] Ad-hoc mental models developed in an experiential manner by pilots have several potential problems. One of the key issues is that the training of these systems is based on a simplified rule-based, operational model, with little causality or connection with the structure of the underlying system.

It is hypothesized that operators develop mental models based on interaction with the system and experience.[8] It is expected that the actual mental models used by the pilots are more sophisticated than those put forward during training and a function of their individual pilot backgrounds.

Since these models are created independently by individual pilots, specific ad-hoc models may not be accurate. In addition, the fact that ad-hoc models are created during nominal operations (where the vast majority of pilot experience occurs), they may not hold (or may even be a liability) in emergency situations.

As an example, consider the accident which occured at Nagoya on an Airbus 300 when the crew inadvertently put the aircraft into a go-around mode. The autopilot on the aircraft can be disengaged by forcing forward on the yoke with sufficient force. However, the automation is not completely disengaged and maintains control over the trim of the horizontal stabilizer. The autopilot is programmed to trim the horizontal stabilizer to the flight path of the engaged mode. In the case of the go-around mode, the autopilot trims the horizontal stabilizer to a nose up configuration to allow the aircraft to climb. The crew, believing that they had disengaged go-around, pushed the the aircraft nose down to continue the approach. When the crew decided to abort the approach, the aircraft pitched up violently due to the combination of horizontal stabilizer trim and elevator input, stalled and crashed.

Another example occurs in advanced Boeing aircraft. In general, transitions between modes can be initiated by pressing buttons on the Mode Control Panel which select the new mode: pressing the ALT HLD button initiates the Altitude Hold mode. However, the Approach Mode cannot be transitioned out of without switching off the autopilot entirely. Attempting to engage a new more via the Mode Control Panel is ignored. While this prevents the accidental engagement of a spurious mode during a critical flight segment, the accidental engagement of Approach can lead to serious confusion since the automation will suddenly appear to be nonresponsive to switching modes.

Need for Automation Models

It is hypothesized that operators of modern automation systems need a consistent global representation of automation.[8, 10, 11] This model is necessary as a basis on which to build robust mental models for operational use. Such a model is also critical to allow the understanding of automation behaviour in non-nominal operations.

Compared to other automation systems, clear mental models of flight automation systems, which operate in a time critical, are particularly important. In current aircraft automation, the pilot is given final authority and responsibility over aircraft safety. In order to effectively and appropriately intervene, especially in emergency situations, the pilot must be able to understand the behaviour of the automation. It may be the case that a consideration in aircraft automation design should be the level of complexity which the operator is expected to supervise. While the underlying implementation may be more complex, the automation with which the operator interacts must be designed to allow the construction of a robust mental representation.

Existing Automation Systems

Current aircraft automation systems have evolved from more basic systems through the incorporation of additional functionality. The evolutionary development of these systems and increased capabilities have resulted in a complex structure of operating modes. This process has been termed "an entropic growth of complexity". This entropic growth has resulted in the lack of a consistent global model of the automation.

A side effect of this incremental growth is that mode proliferation is becoming a concern. This proliferation is due to a variety of factors, including attempts at product differentiation by manufacturers and the maintenance of previous modes as safety fallbacks. There can also be an inability to remove older modes, since they may be used by specific air carriers. A contributing factor may also be a reluctance by system designers to update older modes since this will result in additional certification costs.

Possible Solutions to Automation Problems

Three possible solutions to the problems causes by incremental system development are to change operational procedures and training, enhance feedback to the flightcrew and ultimately to design a more operator consistent structure to automation

Changing Procedures and Training

To deal with problem found during aircraft introduction, near term, and relatively inexpensive, solutions support procedural fixes and training modifications. Procedures are regularly updated with changes designed to work around problems with existing automation. On a larger scale, there is work underway on an advanced Vertical Navigation (VNAV) trainer[14]. This tool is designed to show pilots the underlying complexity of the VNAV system and the implications of a particular set of mode choices.

The shortest term solution to automation problems is to publicly detail the concern and change operational procedures to avoid these situations. An example is the warnings published by Airbus aircraft on flap overspeed concerns after an incident at the Orly airport.[3]

Enhance Feedback to Flight Crew

Mid-term solutions consist of enhancing feedback in the cockpit to prevent confusion and to allow pilots the ability to accurately predict what automation will do next. This may involve the installation of new displays, or at the least a modification of existing display software, causing this solution to be more expensive and longer term. Work has been undertaken in the design and evaluation of an Electronic Vertical Situation Display[13] and in the creation of a new mode annunciation displays, such as the Immediate Mode Management Interface.[15]

Operator Consistent Structure in Automation

The most comprehensive solution is to develop operator consistent automation that is less vulnerable to problems. The costs and development times involved make this unlikely to occur as retrofits to existing aircraft. More likely, this solution will only be undertaken when new functionality in the cockpit requires redesign in future aircraft. Research is underway to determine how to design new systems, including the Operator Directed Process detailed in this paper.

Changing Development Processes

The existing development processes for flight automation were developed in an era where computing power was at a premium and the capabilities of computing systems limited. The shift to a new development process is justified based on the current flexibility, capability and complexity afforded by modern processors and software systems.

One of the unique elements of software engineering is that there are few physically imposed constraints. In conventional engineering the complexity and size of a design is constrained by physical limitations, such as requiring to be light enough to fly, or requiring to be simple enough to maintain. Software engineering does not have these limitations in the conventional sense, since computing power, memory and storage continue to increase at exponential rates. In the absence of physical constraints, software systems can become excessively complex with little apparent penalty. One of the goals of the Operator Directed Process is to constrain the complexity of these systems to a level which human operators can internalize and understand while maintaining the desired functionality.

An additional concern with existing development processes is the lack of a high level model of the automation system. Current practices create a set of low level specifications that are verified as being fulfilled in the certification process. Unfortunately, these specifications have been shown to be incomplete in existing systems, such as the Traffic Collision and Avoidance System.[16] The low level specifications are also difficult to trace to high level system goals to verify the flow down to implementation details. High level functional representations of automation that can be understood by operators will aid in high level validation.

Operator Directed Process

In order to support the development and certification of complex automation systems that consider flight crew operational understanding, the use of an Operator Directed Process is proposed. Concepts similar in nature to the Operator Directed Process have been put forward previously, such as User Centered Design[8], Knowledge-based Interface Design[10] and others[11].

The Operator Directed Process is shown in Figure 2. The major difference in this process is that the training material is the source of the system specification rather than vice versa. This contrasts with existing development cycles that use training material to document system design. The intent is to develop a less error-prone and more understandable system by requiring consonance between the training material and the software, and by limiting the complexity of the system through the imposition of a model for the operator.

The next sections describe each of the stages in the Operator Directed Process.

Functional Analysis

The first stage of this process is to determine the functionality that the automation system requires. This analysis needs to be based on the existing environment in which the automation must function and the anticipated operational and procedural usage of the automation.

Automation Model

An Automation Model for use by the operator is derived based on the functional analysis and input from current operators and expert users. The Automation Model is a representation of the system which is complete

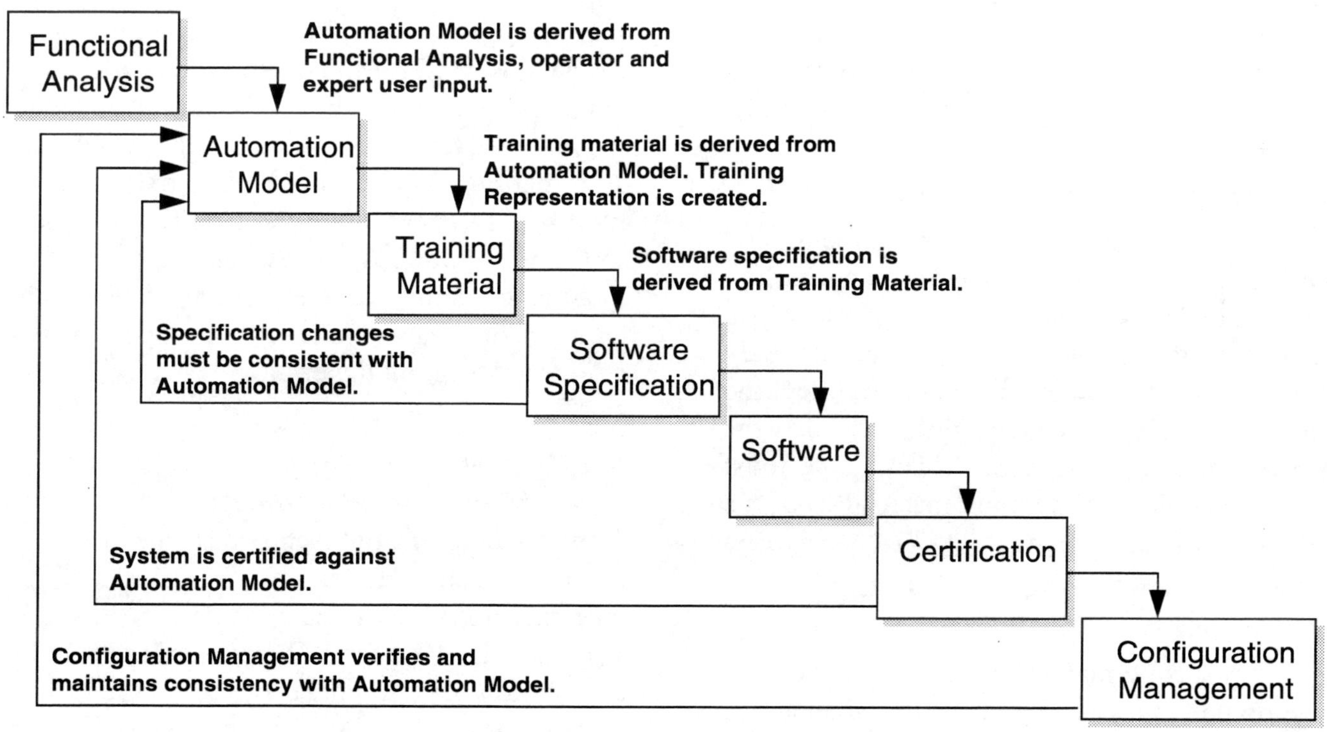

Figure 2: Operator Directed Process

at a level appropriate for its operation. This implies that additional detail which does not have an impact on operation should be excluded to simplify the model. This model is intended to be a high level, but complete, description of the system which captures the philosophical and design goals which lead to specific design criteria at more detailed levels.

The intent is that by enforcing the creation of such a model, the complexity of the final system will be constrained to be able to be used effectively by the operators. If, at this stage, the model for the operator is deemed to be excessively complex in preliminary testing, the functional analysis may need to be reexamined.

Training Material

The training material which is presented to the operators is derived directly from the model for the operators. In this way, the interpretation and translation is minimized and the operator has access to as accurate as description of the model of the system as possible.

Software Specification and Development

By generating the specifications and software from training material, an explicit constraint is developed which can serve to guide system development. In the unlikely event that the specification cannot be reflect the training material accurately, the model for the operator must be amended. This change must be reflected in the training material and the software specification rederived.

Certification

A second major departure from existing development process involves the manner in which certification is undertaken. Rather than only certifying that the software conforms to its specifications, as is done in the RTCA document entitled "Software Considerations in Airborne Systems and Equipment Certification" (DO-178B)[9], in the Operator Directed process, the final system is compared to the original model created for operators. By doing so, the levels of interpretation and translation which have been traversed in order to design the system can be determined to be correct.

Certifying to a known Automation Model may also have advantages in identifying the exceptional cases of the automation. Since the Automation Model is constrained to a simple form communicable to the operators, exceptional behaviours and modes are likely to be highlighted. Examining these cases may allow certification officials to focus on the more problematic issues.

Configuration Management

Many of the problems seen with existing aircraft automation are due to modifications and enhancements which add complexity to the original, simpler, system. By maintaining consistency with the Automation Model designed for the operator, this additional complexity may be better controlled. There may, however, be instances in which the Automation Model must be extended or modified to encapsulate the changes necessary to the system. This process should be handled in a manner similar to the original design by the creation of a new model for the operator and associated training material.

Experimental Evaluation

While these concepts have been recognized as useful techniques, the have not been validated empirically. Research is currently underway to develop an experiment to validate the Operator Directed Process for automation development by comparing it to existing, less constraining processes.

Representations for Automation Models

An important issue in the development of Automation Models is to identify the representational form which is most effective at conveying the behaviour of the system. In Table 1, a set of forms is presented.

It is unlikely that the full behaviour of a complex automation system can be captured effectively in a single representation. However, particular forms are amenable to the description of behaviour of specific elements of the system. As an example, Control Block Diagrams can effectively model the dynamic behaviour of a single mode, but cannot capture the transitions between modes.

Table 1: Subset of Possible Representations

Control Block Diagrams
Procedural Constructs
State Charts
Mode Transition Matrices
Analogical Descriptions
Anthropomorphic Descriptions
Linguistic Descriptions
Petri Nets
Finite State Representations
Explanatory Descriptions

In contrast, transitions between modes can be captured via Petri Nets, State Charts, Mode Transition Matrices or other Finite State representations.[12] Procedural, Linguistic, Analogical, Anthropomorphic and Explanatory representations are effective at conveying the highest level of structure to the system. These representations must be used with care, however, since an operator may make inappropriate analogical extrapolations and expect the system to behave differently than designed.

Conclusion

Interactions between pilot and the automation systems in modern aircraft are still subject to errors. A contributing factor to these errors is the lack of a simple, consistent modem of the automation. The Operator Directed Process is an attempt to redefine the design process to incorporate the development of an automation model suitable for presentation to the operators and as the basis for the development of training material. By constraining the design process to implementations which can be used effectively by operators at the basis of a robust mental representation, it is expected that less error-prone automation systems will be developed. An experimental evaluation of the Operator Directed Process is currently being designed.

Acknowledgments

This work was supported by NASA Grants NAG-1-1581 and NAG-1-1857 and by FAA and NASA Grant FAA-95-G-017.

References

1. Mecham, M (1994). Autopilot Go-Around Key to CAL Crash *Aviation Week & Space Technology.* May 9, 1994.

2. Hughes, D. (1995). Accidents Direct Focus on Cockpit Automation, *Aviation Week & Space Technology.* January 30, 1995.

3. Sparaco, P. (1994). A310 Inquiry Target FCU, *Aviation Week & Space Technology.* October 10, 1994.

4. 1996 FAA Report on the Interfaces Between Flightcrews and Modern Flight Deck Systems

5. Sarter, N. (1994). Pilot Interaction with Cockpit Automation II: An Experimental Study of Pilots' Model and Awareness of Flight Management Systems, *The International Journal of Aviation Psychology 1*, no. 1: 1-28.

6. Vakil, S., Hansman, R.J., and Midkiff, A., "Feedback Mechanisms to Improve Mode Awareness in Advanced AutoFlight Systems," Eighth International Symposium on Aviation Psychology, April 1995.

7. Vakil, S., Hansman, R.J., and Midkiff, A., "Impact of Vertical Situation Information on Vertical Mode Awareness in Advanced AutoFlight Systems," 14th AIAA/IEEE Digital Avionics Systems Conference, November 1995.

8. Norman, D. (1989). The Design of Everyday Things. New York, NY: Basic Books.

9. RTCA Document DO-178B (1992), "DO-178B Software Considerations in Airborne Systems and Equipment Certification".

10. Mark, , William. (1986). User Centered System Design. Ed. Norman D., Draper, S. Hillsdale, NJ: Lawrence Erlbaum Associates.

11. Denning , S. (1980). On Learning how to Predict, *Proceedings of the IEEE*, vol. 68 no. 9: 1099-1103.

12. Vakil, S., Hansman, R.J., "Understanding Automation Structure by Analyzing Mode Transition Matrices," 17th Digital Avionics Systems Conference, October 1998.

13. Vakil, S. (1996). Development and Evaluation of an Electronic Vertical Situation Display. MIT Aeronautical Systems Laboratory Report ASL-96-2

14. Chappell, A. R. & Mitchell, C. M. (1997). Intelligent tutoring systems for 'trained novices': Building expertise in mode management. Proceedings of the 1997 International Symposium on Aviation Psychology. Columbus, OH.

15. Hutchins, E. (1996). Immediate Mode Management Interface Final Report. NASA Ames Research Center.

16. Leveson, N.G., Palmer, E. (1997). Designing Automation to Reduce Operator Errors. Proceedings of Systems, Man, and Cybernetics Conference.

EFFECTS OF DISPLAYS ON PILOT CONFORMANCE TO AUTOMATIC ALERTS DURING CLOSELY-SPACED PARALLEL APPROACHES

Balázs Vándor and Amy R. Pritchett

Abstract

This article examines the effects of displays on pilot conformance to automatic alerts. Closely-Spaced Parallel Approaches were used as a case study. An experiment was conducted to examine subject reactions to potential collisions in different alert and display conditions. Results indicate that the use of both automatic alerts and traffic displays had a significant effect on when subjects decided the situation to be hazardous. Recommendations for alerting system design and realistic procedures for alerting system use will be given.

Introduction

Nonconformance to Alerting System Commands

Traditional alerting systems have an informative role, serving to draw the pilot's attention to a comparatively clear-cut condition. In contrast, newer "executive" alerting systems display alerts and commanded actions with the assumption that the pilot will carry out the actions as specified. That is, not only does the system determine if there is a problem, but it also takes over the decision-making and resolution-selection tasks.

The assumption that pilots will conform to alerting system commands is not always valid. Studies of currently operational alerting systems have identified a significant number of non-conformance situations. (Ciemier et al, 1993, Vallauri, 1997, DeCelles, 1991)

Non-conformance might be acceptable, or even desired, if the pilot decides that the situation is safe based on additional information that is not available to the alerting system. However, many executive alerting systems are intended for conditions where pilots may not have enough resources to use high performance decision strategies. In these cases, non-conformance may reduce the performance benefit expected from the automatic commands, and does not allow the pilots to be relieved of the workload of the decision-making tasks.

The information shown on situation displays is theorized to affect nonconformance. Displays indicating how the alerting system logic operates and when is an alerting threshold surpassed might improve pilot perception of the alerting system, possibly increasing conformance to the automatic alerts. A display where an explicitly displayed criterion supports the automatic alerts is said to be in *consonance* with the automatic alerts. If the information display implies a different criterion for reaction from the automatic alerts, than the automatic alerts and the display are in *dissonance* with each other. This work will explicitly examine the effects of consonance, dissonance, and other automatic alert-display interactions on subject reactions.

Closely-Spaced Parallel Approaches (CSPA's) were chosen as a case study to gain insights into nonconformance issues. CSPA's involve a simultaneous approach and landing to adjacent parallel runways by two aircraft in instrumental meteorological conditions (IMC).

0-7803-5086-3 /98/$10.00 ©1998 IEEE

CSPA's were chosen due to the complexity and time-criticality of the collision avoidance task. Additionally, the task of collision avoidance during CSPA's has qualities that have been found to make it susceptible to pilot non-conformance to automatically displayed alerts and commanded avoidance maneuvers.

Experimental Design

In this part-task flight simulator experiment, 45 subjects supervised Closely-Spaced Parallel Approaches and were instructed to react when the aircraft on the parallel approach appeared to be on a collision course that endangered the own aircraft. Both the automatic alerts and the display used one of three possible conditions: one not based on any alerting criterion, one based on the "NTZ criterion", and one based on the "MIT criterion". The NTZ criterion represents a simple, low-performance alerting logic consistent with subject reactions in a previous experiment. The MIT criterion uses a more complex, high-performance alerting strategy that takes into account several parameters of the dynamics of the two aircraft and has been developed specifically for CSPA's. (Carpenter & Kuchar, 1996)

Three display types were used in this experiment, shown in Figure 1. The left-most, called the Basic Display, shows the two aircraft on the parallel approach path without displaying either of the alerting criteria. The center display, called the NTZ display, contains all the information from the Basic Display, but also depicts the "danger zone" that forms the basis of this alerting criterion. The third display is called the MIT display, also based on the Basic Display, but with an additional "danger zone" representing the MIT criterion. As this criterion changes with the parameters of the dynamics of the two aircraft, the size and shape of this zone also changes as this information is updated.

Similarly, there were three automatic alert conditions used in the experiment. These were the conditions of No Alert, NTZ Alert, and MIT Alert. The first one was the condition without an automatic alert, while the second and third were based on the NTZ and the MIT criteria respectively.

The unpaired-by-displays test matrix of the nine combinations is illustrated in Figure 2. Nine combinations of alerting system and display conditions were tested, resulting from three alert conditions and three display conditions. The different alert and display conditions resulted in cells that contained alert effects only, display effects only, consonance effects, and dissonance effects.

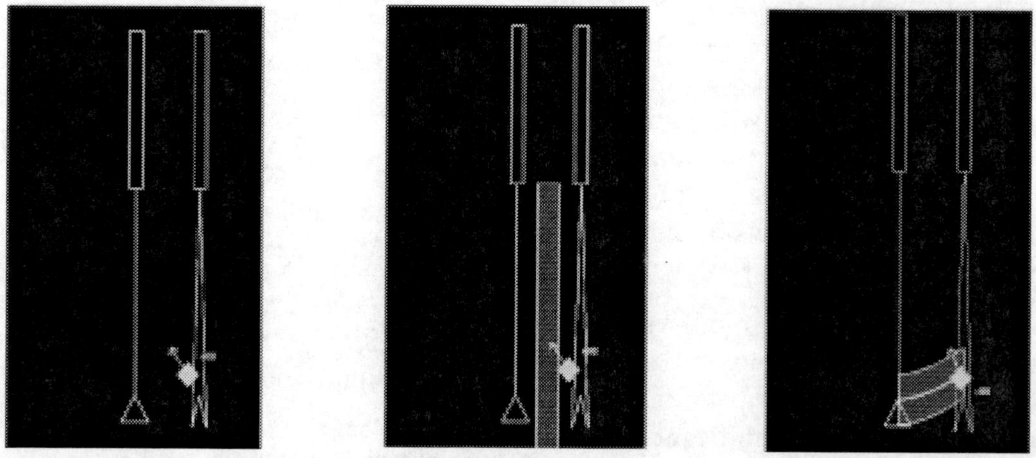

Figure 1: The Basic Display, Display of the NTZ Criterion and the Display of the MIT Criterion

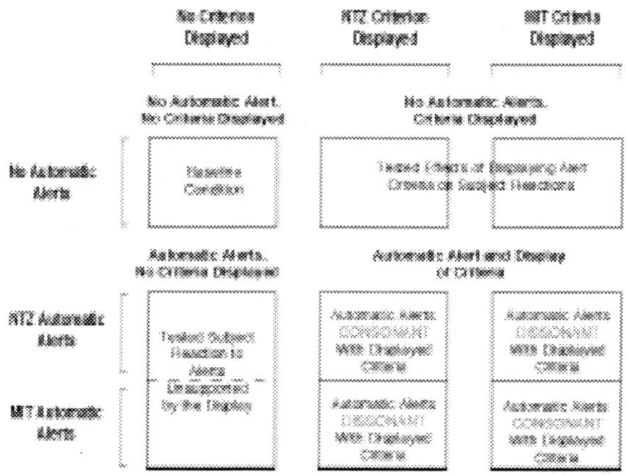

Figure 2: Test Matrix of the Experiment

Each subject completed a total of 36 runs using one of the three display types. These runs were divided equally among three scenario types and one type of placebo. The three scenarios were a low-convergence blunder, a high-convergence blunder, and a high-convergence missed approach intercept by the other aircraft.

The briefing used for this experiment emphasized that subjects should evaluate the situation to decide when the potential collision hazard, in their judgement, requires an action. The subjects were not required to react to an intruding aircraft when it entered the “danger zone” on the display or when the automatic alert was given. They were told that the alerting criteria used for the basis of the displays and automatic alerts may or may not be correct. This was done to approximate real-life reactions, where pilots may not trust an automatic alert unconditionally.

Measurements included state data at relevant times during the scenarios and performance measures for each subject. The state data was then combined into more informative combinatorial measurements, comparing the time of subject reaction to the triggering of the MIT criteria and the NTZ criteria.

Results

Altogether, 1215 datapoints collected found. For specific analysis of their effects, the data was subdivided by display conditions, by alert conditions, and by consonance and dissonance. Several statistical tests were used to evaluate statistical significance. First, planned comparisons were performed using linear combinations that were thought to be significant based on the hypotheses. Second, one-way ANOVA’s were performed on the data to point out all significant differences. Last, Tukey multiple comparisons were performed on datasets found significantly different by the ANOVA, to point out which of the examined conditions are different and by how much. For the sake of brevity, the measure of the average time between the triggering of the MIT criterion and the subject reaction will be used as an example that illustrates the types of effects found.

Display Effects

The data was first examined to ascertain whether subject reactions would be influenced by the explicit display of an alert criterion to have reactions more consistent with that criterion.

Figure 3 illustrates the average time between the triggering of the MIT criterion and the subject reaction, subdivided by display conditions, regardless of what kind of automatic alert was used. According to a one-way ANOVA and Tukey tests, there are statistically significant differences in the data. This is one of the more informative measures, as it measures the difference between subjects’ reactions and a high-performance alerting strategy.

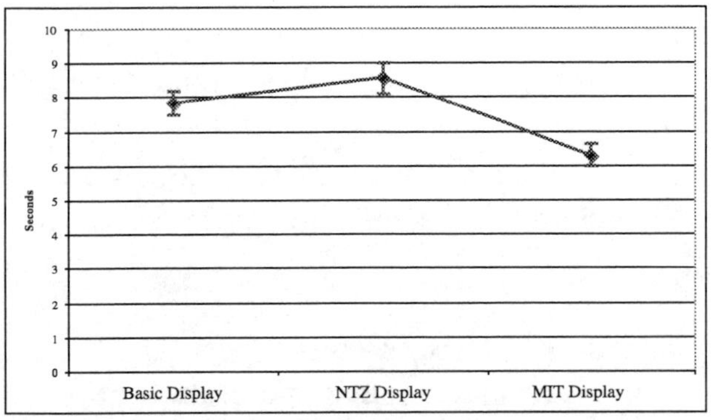

Figure 3: Time between Subject Reaction and the Triggering of the MIT Criterion for the Three Display Conditions

Planned comparisons found that the MIT Display condition had reactions, on average, 2.26 seconds closer to the triggering of the MIT criterion than the NTZ Display condition. When examining the subset of the data in the conditions without automatic alerts, this difference is 3.97 seconds. Using a Tukey multiple comparison, the MIT Display condition was found have closer subject reactions to the triggering of the MIT criterion than the Basic Display condition (a 95% confidence interval for the difference of the means is 0.35 to 2.75 seconds) and than the NTZ Display condition (a 95% confidence interval for the difference of the means is 1.06 to 3.46 seconds). No significant difference was found between the Basic Display and the NTZ Display conditions.

This measure illustrates that the display of the MIT criterion influenced subject reactions to be closer to the triggering of this criterion. A similar trend can be observed in other measures, such as time between when the NTZ criterion triggered and when the subject reacted, lateral separation, and miss distance achieved after the avoidance maneuver.

The display of the MIT criterion therefore was found to have beneficial effect on subject reaction. Subjects who saw this display had reactions closer to the triggering of the MIT criterion, and there is also a corresponding increase in aircraft separation at the time of the reaction and in the miss distance achieved after the avoidance maneuver.

The display of the NTZ criterion, on the other hand, did not show a consistent pattern regarding subject reactions and therefore aircraft separation. In some cases it had no effect on subject reactions. In other cases, it brought reactions closer to the triggering of the NTZ criterion. In a few cases, it had the negative effect of delaying critical subject reaction.

Alert Effects

Statistics identical to the analysis by display types was performed with the data divided by types of automatic alert. It was examined whether subject reactions will be influenced by automatic alerts to have reactions more consistent with the criterion used. The data is divided by automatic alert types, regardless of what kind of display was used.

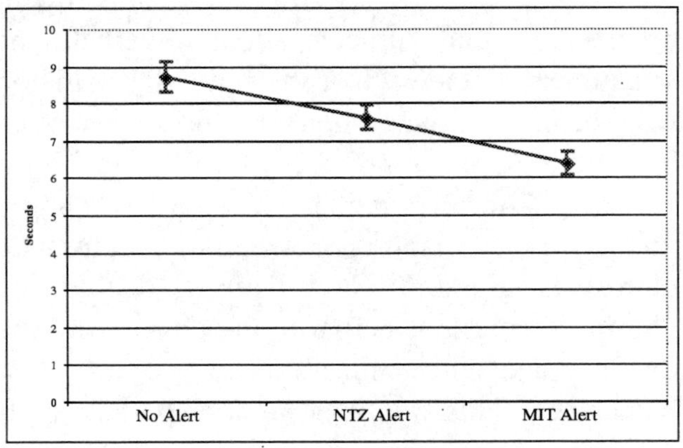

Figure 4: Time between Subject Reaction and the Triggering of the MIT Criterion for the Three Alert Conditions

Figure 4 illustrates the average time between the triggering of the MIT criterion and the subject reaction. Planned comparisons found that the combined conditions with automatic alerts had, on average, 1.70 seconds difference in times than the No Alert condition. When the NTZ and MIT Alert conditions were compared, the latter was found to have, on average, a 1.20-second lower value. Using a

Tukey multiple comparison, the MIT Alert condition was found have subject reactions closer to the triggering of the MIT criterion than the No Alert condition (a 95% confidence interval for the difference of the means is 1.12 to 3.47 seconds) and than the NTZ Alert condition (a 95% confidence interval for the difference of the means is 0.01 to 2.40 seconds). No significant difference was found between the No Alert and the NTZ Alert conditions.

This measure shows that using an automatic alert reduced the time between the triggering of the MIT criterion and subject reaction. Moreover, the automatic alert based on the MIT criterion had a greater effect on subject reaction than the ones based on the NTZ criterion. Therefore, an automatic alert based on a high-performance alerting criterion will bring reactions closer to this criterion than other or no automatic alerts.

Examination of the time between the automatic alert going off and the subject reacting revealed that subjects reacted sooner after an NTZ automatic alert then after an MIT automatic alert. This is not surprising given that subjects tended to react late and the NTZ criterion triggered later in the more dangerous high-convergence scenarios. These results underscore the earlier finding that found the NTZ criterion is more consistent with the subjects' unassisted reactions.

Subjects rated the timeliness of the automatic alert after each run. In all comparisons, alerts based on the MIT criterion are rated as earlier than the ones based on the NTZ criterion. This is likely due to the fact that for all subjects, two out of three scenarios were the more dangerous high-convergence scenarios. In these subjects tended to react late, and alerts based on the MIT criterion were triggered earlier than ones based on the NTZ criterion.

Automatic alerts had a weaker effect on pilot performance than displays. The automatic alert based on the MIT criterion has beneficial effects on subject reaction, as subjects who heard this alert had reactions closer to the triggering of the MIT criterion. Also, subjects reacted closer to an automatic alert based on the NTZ criterion then to one based on the MIT criterion.

Consonance and Dissonance Effects

To examine the effects of dissonant automatic alerts, statistics identical to the analysis by display types were performed. However, consonance and dissonance data was only considered from conditions where both an automatic alert was used and a criterion was displayed. The data was divided into the two conditions of consonance and two conditions of dissonance. It was hypothesized that consonance between display of criteria and automatic alerts would shift the pilot's alerting strategy to be more consistent with the alerts. Correspondingly, it was hypothesized that dissonance would have the opposite effects.

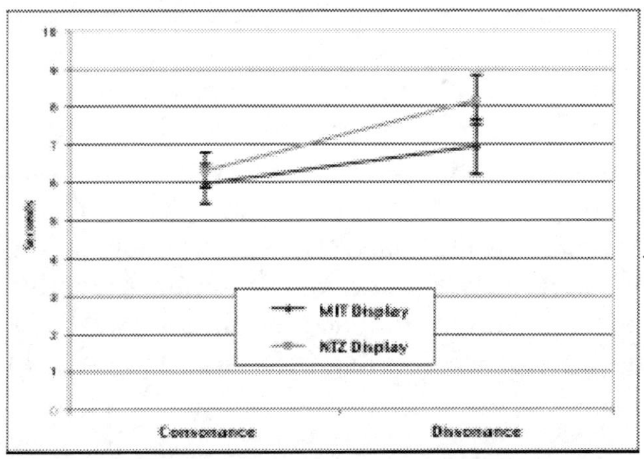

Figure 5: Time between Subject Reaction and the Triggering of the MIT Criterion for the consonance and the dissonance conditions

The means and standard error bars for the measure of the time between subject reaction and the triggering of the MIT alert are illustrated in Figure 5. Planned comparisons found that Consonance condition were on average 1.41 seconds closer to the triggering of the MIT criterion than the Dissonance conditions. No difference was found within the

0-7803-5086-3 /98/$10.00 ©1998 IEEE

two cases of Consonance and within the two cases of Dissonance. A Tukey multiple comparison found the MIT Consonance condition to have a smaller difference between automatic alert and subject reaction time than the dissonance condition *(NTZ Display and MIT Alert)*, a difference significant at the $p < 0.05$ level). Both of these conditions used an automatic alert based on the MIT criterion, but the former used a consonant display that reinforced the automatic alert. The latter used a display dissonant with the automatic alerts, resulting in delayed subject reactions. In fact, this dissonance condition showed the lowest performance of the four conditions in every significant measure, most significantly in the measure of the time between subject reaction and given alert.

These results suggest that consonance can influence reactions to be closer to the triggering of a high-performance criterion to a larger extent than dissonance. However, these results were not observed consistently throughout the measurements. For example, it would be expected that the consonance condition, with automatic alerts and displays based on the same criterion, would result in stronger effects than the individual effects from automatic alerts or from displays. This was not the case for the measures of Lateral Separation, the time between the MIT criterion triggering and subject reaction, and the time between the automatic alert going off and the subject reaction.

Discussion

Measuring Alerting System Effectiveness

Before discussing the implications of these results, it is useful to consider the difficulty in testing alerting system effectiveness. In experimental situations subjects' reactions can be biased by several factors. All such experiments are subject to expectation and learning effects. Subjects expect to hear an automatic alert in such an experiment, and after a few scenarios, they might learn approximately when the alerts will go off. These effects prime the subjects to react sooner after an automatic alert, while in real-life situations an element of surprise is present and the pilot might try to understand and evaluate the automatic alert. Conversely, in real life the perception of a higher-risk situation might make pilots more attentive and/or conservative. As the magnitude of these effects is not known, such experiments can compare different alerting systems, and can provide a rough estimation of the reaction times, but can not provide the exact reaction times that will be found in actual operations.

In real-life situations, pilots might not exactly follow the procedures for an alerting system, but might instead evaluate the situation. For this reason the subjects were not told to react as soon as the alert went off or the intruder entered the protected zone on the display. Instead, this experiment encouraged subjects to use their best judgement when deciding on an alert with the hope that this would best mimic real-life reactions.

For comparison, a follow-up experiment was conducted to measure the reaction time it takes for the subjects to press the alert button if they are briefed to react as soon as the automatic alert triggers or the intruding aircraft enters the protected zone on the display. It was found that in this case the reaction times were an order of magnitude shorter. They did not vary with alert or scenario types, and varied with display types by a statistically significant, but functionally small amount. This establishes that all the difference between reaction times between the scenarios or alert types can be attributed to other sources, such as analyzing the situation.

Thus, alerting system effectiveness can not be measured with experiments that request the subjects to react exactly according to procedures, as there is nothing to guarantee that pilots will react accordingly in real-life situations. To minimize experimental effects, subjects must be encouraged to behave as they normally would in such a situation.

Conclusion

Recommendations for Alerting System Design

The alerting systems described in this article are automated systems where the assessment of the presence and magnitude of a hazard may not be similar between the pilot and the alerting system. In such a case the pilot may attempt to verify the alert and then to reconcile it with the own understanding of the situation. To be effective, such alerting systems might also incorporate displays that give a visual representation of the situation, in addition to automatic alerts. This can help with both the verifying and the reconciling processes.

The amount of information and the way it is presented on a display can have strong effects on behavior. The information needed for many tasks in the cockpit is distributed across the external information displays and the internal minds of the pilots. The more of the information is presented externally, the more efficient the display. (Zhang, 1996)

As part of the display of higher-level information, displays should communicate the basis for the alerting logic, guiding the pilot's internal mental model of the situation and driving appropriate behavior. For example, the display of the MIT criterion provided the synthesized information from the alerting system's logic and the threshold used by the system to alert.

When an automatic alert is given, pilots might elect to verify it either by examining the alerting system or the environment to find reasons for the alert. If the results of this verification process do not make sense of the alerting system's commands, the pilot may try to reconcile the alert and his or her own judgement. The findings of this experiment reinforce that a display consonant with the automatic alerts might simplify the verification process and make this reconciliation self-apparent.

During the design of alerting systems, procedures are typically assumed to serve the needs of the system, and might result to be at odds with the expected human behavior. A particular system may require a very short reaction time and specific actions by the pilot to function properly. The procedure will call for these actions and reaction time to achieve the desired performance. This is typically acceptable for systems where there is little potential for disagreement about the presence and magnitude of a hazard.

Procedures required of the operator using safety-critical systems must be compatible with the expected range of human behavior in the given situation. Human-machine system specialists should work together with the engineers to develop a system that is likely to be used as intended, with procedures that take into account the range of human behavior and human limitations.

Recommendations for Closely-Spaced Parallel Approaches

The main concern regarding CSPA's is the possibility of late pilot reactions, which can compromise the effectiveness of the alerting system. Executing CSPA's represents a challenging task, with no simple solution at hand.

Several experiments in collision detection and avoidance in CSPA's have found a variety of alerting behavior. (Pritchett, 1995; Pritchett, 1996; Kockzo, 1996) The different results are possibly due to experimental effects, such as different training and briefing effects. To get consistent results in different experiments, all experimental factors must be carefully controlled.

It is possible that experimental effects contributed to the results showing low feasibility of CSPA's. A series of higher fidelity experiments would be needed to prove effective any system for this type of operation.

Acknowledgements

This work was partly supported by NASA Research Grant 2-1146, with Dr. Roger Remington serving as technical monitor. The authors would also like to thank all the subjects for the time they volunteered to participate in the experiment.

References

Carpenter, C. & Kuchar, J.K. "A Probability based Collision Alerting Logic for Closely-Spaced Parallel Approaches", *Proceedings of the AIAA 35th Aerospace Science Meeting*, January 1997

Ciemier, C.E., Gambarani, G.P., Sheridan, E.H., Stangle, S.C., Stead, R.P. & Tillotson, D.H. "The Results of the TCAS II Transition Program (TTP)", The ARINC Corporation, April 1993

DeCelles, J.L. "The Delayed GPWS Syndrome", Aviation Research and Education Foundation, Herndon VA, July 1991

Kockzo, S. "Co-ordinated Parallel Runway Approaches", NASA Contractor Report CR201661, October 1996

Asari, K., Carpenter, B.D., Hansman, R.J., Kuchar, J.K., Pritchett, A. "Issues in Airborne Systems for Closely-Spaced Parallel Runway Operations", *Proceedings of the 14th AIAA/IEEE Digital Avionics Systems Conference*, Cambridge, MA, November 1995

Hansman, R.J., Pritchett, A. "Experimental Study of Collision Detection Schema Used by Pilots During Closely Spaced Parallel Approaches", MIT Aeronautical Systems Laboratory Report, ASL-96-1, January 1996

Vallauri, E. "Suivi de la mise en oeuvre du TCAS II en France en 1996", Report CENA/RT97016, Toulouse, France, June 1997

Zhang, J. "Distributed Representation as a Principle for the Analysis of the Cockpit Information Displays", The International Journal of Aviation Psychology, 7(2), 105-121, 1996

POTENTIAL NEAR TERM ENHANCEMENTS TO OCEANIC OPERATIONS WITH A COCKPIT DISPLAY OF TRAFFIC INFORMATION

James J. Cieplak, Baltazar O. Olmos, Anand Mundra

The MITRE Corporation/Center for Advanced Aviation System Development, McLean, VA

Abstract

The oceanic domain may be an ideal venue for introducing the Cockpit Display of Traffic Information (CDTI) into Air traffic Management because of the need for improved efficiencies, the more benign traffic geometries, and a somewhat lower pilot workload compared to other domains. This paper describes two potential CDTI-assisted oceanic procedures and documents results of a simulation study designed to determine CDTI features required for these procedures.

Introduction

In oceanic airspace where radar surveillance is not available, controllers rely mainly on position reports by pilots for determining the location of aircraft. To ensure that the aircraft remain safely separated, Air Traffic Control (ATC) non-radar procedures require relatively large separations of up to 100 nautical miles (nmi) laterally (between tracks) and 10–15 minutes along-track, depending on the procedures in use. These large separation distances can quickly "use up" the available airspace and cause "crowding" on the track system, seriously constraining the ability of ATC to approve user requests, such as desired en route altitudes. The operational procedures described in this paper offer some methods by which clearances for more efficient altitudes may be issued more frequently with the use of a CDTI based on surveillance from an Automatic Dependent Surveillance-Broadcast (ADS-B) system.[1] The procedures described and studied in this paper require no ground infrastructure investment and only small changes to current ATC procedures.

CDTI-Assisted Oceanic Procedures

In 1994, the FAA authorized operational trials of an in-trail climb (ITC) procedure (Figure 1) in the Oakland and Anchorage Flight Information Regions using the Traffic Alert and Collision Avoidance System (TCAS) II traffic display [1]. In 1996, these operational trials were expanded to include an in-trail descent (ITD) procedure; they were also expanded to include any airline that demonstrates compliance with appropriate requirements.

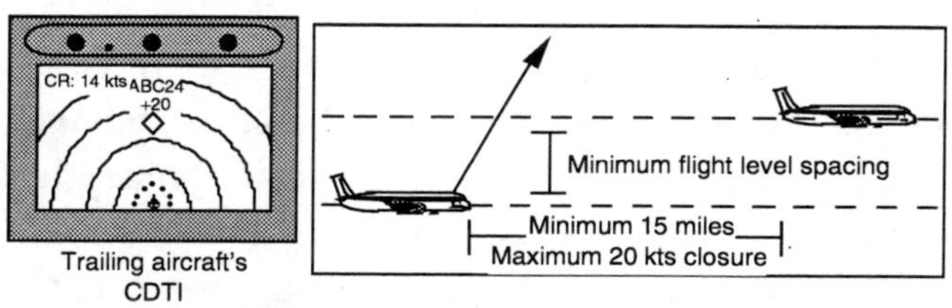

Figure 1. In-Trail Climb

In spite of the relative success of the ITC operational trials [2], it must be acknowledged that the TCAS II traffic display has numerous limitations with respect to its use in the ITC/ITD procedures. For example, the

---

[1] ADS-B is a surveillance concept that involves each equipped aircraft to periodically broadcast state information (e.g., position, altitude, velocity vector) that could be used by other proximate aircraft or ground facilities.

0-7803-5086-3 /98/$10.00 ©1998 IEEE

procedure requires the trailing aircraft to positively identify the lead aircraft in a cumbersome manner through the use of a transponder squawk-standby procedure. It also requires the determination of closure rate through voice communication. The TCAS II system also has a limited range and may see proximate targets out to 20 to 40 nmi in the forward direction depending upon the installation. Also, the ITC procedure incorporates certain distance safety buffers due partly to the inadequate availability characteristics of the current traffic display. These limitations in the current ITC procedure can be directly addressed with an ADS-B based CDTI.

CDTI-Assisted ITC/ITD Procedures

The first proposed CDTI oceanic procedure enhancement is based directly on the TCAS ITC/ITD, but uses CDTI capabilities to simplify the operations. The CDTI is used to satisfy the initial requirements of positively identifying the aircraft of interest by direct reference to the display instead of the transponder squawk-standby procedure. It is also used to establish that the range is not less than 15 nmi and that the closure rate is not more than 20 knots by direct readout of that information on the CDTI. An ADS-B CDTI installation could also significantly enhance surveillance coverage by providing reception ranges of up to 120 nmi, including the aft quadrant. This would provide a proportionately greater likelihood of the applicability of the procedure, as well as enable lead climb and descent procedures. (See Figure 2 for the simulation study pilot procedures checklist.)

CDTI-Assisted Climb or Descent to Co-Altitude Procedures

The climb or descent to co-altitude procedures builds on the CDTI-assisted ITC/ITD procedures and enables a leading or trailing aircraft to climb or descend to the same altitude as that of the target aircraft. The procedures are based on the TCAS ITC and a currently authorized procedure in Anchorage Air Route Traffic Control Center that utilizes Data Measuring Equipment (DME)/area navigation (RNAV) reports [3]. The current Anchorage procedure reduces longitudinal separation to 30 nmi between aircraft if distance is verified by DME reports or radar observation, or to at least 40 nmi for aircraft reporting RNAV positions. Application of DME/RNAV separation is authorized for up to 75 minutes without maintaining direct pilot/controller communications. Aircraft involved are also assigned mach numbers in accordance with Order 7110.65, paragraph 8-3-3, Mach Number Technique. The CDTI-based procedure simply substitutes a CDTI distance measurement reported to ATC in place of requiring ATC to obtain DME reports from both the in-trail and lead aircraft (Figure 3).

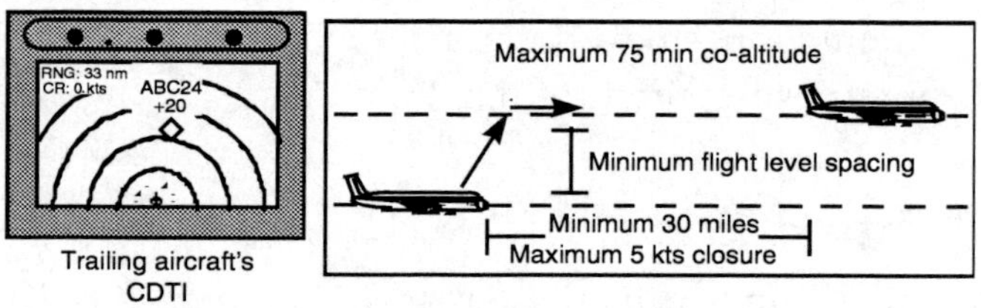

Figure 3. ITC to Co-Altitude

Pilot tasks for the climb or descent to co-altitude procedures are basically the same as for the CDTI enhanced ITC/ITD, except that the initial parameters for the procedure are slightly different. The proposed parameters for the minimum distance between the aircraft before the initiation of the procedure is 30 nmi, and the closure rate is required not to exceed 5 knots. It was also stated in this procedure that if distance to traffic decreases to less than 30 nmi, the pilot may adjust airspeed/mach number, within ATC clearance tolerances (10 kts or 0.02 Mach), to bring the initial requirements into conformity. It should be noted that slightly adjusting the airspeed/mach

These procedures are intended to be used in non-radar environments to enable a trailing (leading) aircraft to climb or descend through the altitude of a lead (in-trail) aircraft operating on the same track. The requesting aircraft must be at the minimum flight level spacing relative to the lead (in-trail) aircraft. ATC retains separation responsibility throughout this procedure; however, it is based on the pilot reporting an accurate initial separation distance measurement. Once above or below the leading (trailing) aircraft by the required vertical separation, operations are in accordance with standard oceanic procedures and aircraft cannot return to the vacated altitude without an ATC clearance.

- Procedures shall be reviewed prior to use.
- Procedures are only approved for use in designated airspace and between participating airlines.
- Procedures are authorized for day and night, visual and instrument meteorological conditions.
- Pilots must remain at their duty stations during the entire procedure.
- Procedures shall be conducted using the autopilot.
- Traffic display must be used to determine call-sign, bearing, distance, and ground speed closure of lead (in-trail) aircraft.
- Procedures are not authorized if the displayed target is intermittent prior to the procedure (i.e., target must be steady for at least 60 seconds).
- TCAS Resolution Advisory (RA) policies remain in effect.

Initial Requirements:

IDENTIFY LEAD (IN-TRAIL) AIRCRAFT **CALL-SIGN**

TARGET BEARING ___ °

- Must be within ±30° off nose (tail)

DISTANCE TO TRAFFIC ___ **NMI**

- Minimum 15 nmi

DETERMINE CLOSURE GROUNDSPEED ___ **KNOTS**

- Maximum allowable closure rate is 20 knots.

ATC Clearance:

ESTABLISH VHF VOICE COMMUNICATIONS WITH LEAD (IN--TRAIL) AIRCRAFT

- On common VHF, establish voice communications with lead (in-trail) aircraft, state intentions, and request notification of any change in speed, heading, or altitude.

REQUEST IN-TRAIL (LEAD) CLIMB OR DESCENT CLEARANCE FROM ATC

- Use exact terminology when requesting this procedure from ATC. Identify traffic by call sign and separation distance; use phrase "in-trail (lead) climb" or "in-trail (lead) descent." "<Facility>, <own call sign> is ____ miles in-trail (in-front) of <lead (in-trail) call sign>, request in-trail (lead) climb (descent) to FL ____."

WHILE WAITING FOR CLEARANCE, MONITOR INITIAL REQUIREMENTS

When Clearance Received:

INITIATE CLIMB OR DESCENT TO ASSIGNED FLIGHT LEVEL

- Advice lead (in-trail) of clearance.
- If unable to sustain 300 fpm climb rate while below the lead (in-trail) aircraft, an abnormal condition exists. This should be dealt with as unable to comply with ATC clearance.
- Descent rate of at least 1,000 fpm is required for the in-trail (lead) descent procedure.
- Distance to traffic may decrease to less than 15 nmi once climbing or descending.
- Climb or descent may be continued, after initiated, even if the target drops off the traffic display.
- Monitoring of the traffic display throughout the procedure is advised.

If Unable to Comply with ATC Clearance:

- Request an amended in-trail (lead) climb or in-trail (lead) descent clearance from ATC for a new altitude. (Note: Initial requirements must be met and trailing (leading) aircraft must be at or within minimum flight level spacing relative to the lead (in-trail) aircraft.)

"<Facility>, <own call sign> unable <requested FL>. We are ____ miles in-trail (in-front) of <lead (in-trail) call sign>, request in-trail (lead) descent (or in-trail (lead) climb) and return to FL____."

- If the in-trail (lead) climb or in-trail (lead) descent clearance cannot be completed or is inappropriate, then execute the standard oceanic contingency procedures contained in the *(specific reference must be made to standard company oceanic contingency procedures in appropriate flight manuals, e.g., Pacific Ocean High / Low Enroute Charts, P(H / L) 1 / 2 or P(H / L) 3 / 4).*

ADVISE LEAD (IN-TRAIL) AIRCRAFT

When Established at New Flight Level:

ADVISE ATC

ADVISE LEAD (IN-TRAIL) AIRCRAFT

FILL OUT DEBRIEF FORM

- Upon landing please return completed form to *(appropriate individual)*.

Figure 2. Pilot Checklist for CDTI-Assisted Climb or Descent Procedures (In-Trail or Lead)

number, even with an assigned airspeed/mach number, is consistent with current operations. Lastly, monitoring of the CDTI throughout the procedure was not mandatory (separation is assured through the initial requirements); however, it was advised.

Simulation Study

A simulation study was conducted in an effort to identify appropriate CDTI features for these procedures in the oceanic domain. A secondary objective of the simulations was an initial determination of the feasibility of these procedures. Eight (8) line pilots participated in this evaluation with an average total flight time of 11100 hours (~580 hours in the past year). Subject experience with a TCAS traffic display averaged 3400 hours.

Simulation Environment

The CDTI platform consisted of an integrated Electronic Flight Instrumentation System (EFIS) display modeled after a 747-400 'glass cockpit' navigation display. All CDTI features, along with standard TCAS symbology, were overlaid onto the navigation display (Figure 4).

The simulation test bed consisted of a generic mid-fidelity transport cockpit with an out-the-window view, a controller station, and simulated traffic representing oceanic operations on the San Francisco to Honolulu R464 oceanic track. Subject pilots operated as pilot flying and a confederate experimenter performed the duties of pilot-not-flying (e.g., communications, checklists, call-outs, etc.). A voice communication line was provided to allow emulated communications with an ARINC operator (including communication delays), and direct aircraft-to-aircraft communications for coordination during the procedures (as is done in current oceanic operations over a common VHF frequency). Weather, Flight Management

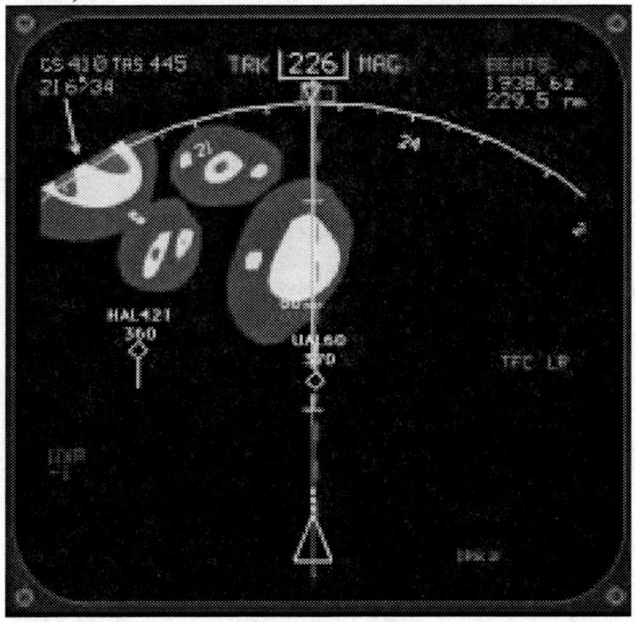

Figure 4. 747-400 Navigation Display with Weather and Traffic (Flight ID and Ground Track Vectors on all Targets)

System (FMS) route information, and traffic were overlaid on the navigation display. Targets and weather appearing on the navigation display could be correlated with visible traffic and weather (e.g., thunderstorm) in the out-the-window view at appropriate ranges. Variable winds through different flight levels were also simulated and affected aircraft performance.

CDTI Features

The CDTI features identified for detailed examination consisted of the following (refer to Figures 4 and 5 and also see [4] for a more detailed description of the features):

Target Selection and Highlighting

A push button panel was developed as an illustration of a simple pilot interface for the selection and removal of CDTI features. An initial button press displays Flight IDs (i.e., call signs) for all aircraft on the CDTI (Figure 4). A subsequent press begins the target selection process in the order of increasing target range with respect to own-ship. The highlighting

0-7803-5086-3 /98/$10.00 ©1998 IEEE

symbol was composed of a green diamond shape which outlined the TCAS target symbol (Figure 5). A reset button to remove all CDTI features was also provided.

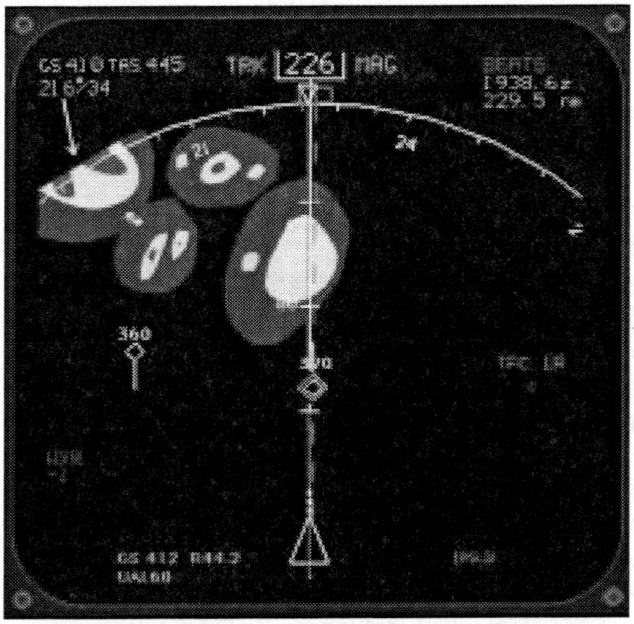

Figure 5. Selected Target with Ground Speed, Range, and Ground Track Vectors

Target Flight ID

Subjects could display Flight IDs in two ways. First, IDs for all targets could be displayed through the initial press of the selection button. Second, when a specific target was selected through a subsequent press, all IDs were removed and the call sign for the selected target was displayed in the lower left-hand corner of the CDTI (Figure 5).

Target Ground Track Vectors

The ground track vector symbology consisted of a line extending from each target icon in the direction of their predicted trajectories. The end of the vector represented the predicted position of the aircraft at a constant (but parameter adjustable) future time. For the current study, the vector depicted the target's predicted position over the ground in 60 seconds. Turn rate information was not incorporated into the target vector.

Target Speed Cues

A speed cue for the target was displayed in the lower left-hand corner of the CDTI (Figure 5). Three types of speed cues were available:

Alphanumeric Ground Speed consisted of a three digit display of the selected target's ground speed in knots. The ground speed display was supplemented with the acronym 'GS' to denote a ground speed representation.

Alphanumeric Closure Rate consisted of a three digit display of the target closure rate in knots. This would be displayed in the same location as the GS display in Figure 5. The closure rate display was supplemented with the acronym 'CR' to denote a closure rate representation.

The *Differential Ground Speed* cue was developed in light of results from a previous experiment [4]. Differential ground speed equals ownship ground speed minus target ground speed independent of the encounter geometry (e.g., head-on, in-trail, etc.). This would be displayed in the same location as the GS display in Figure 5, and have an associated acronym 'DGS'.

Target Alphanumeric Range

Due to the large map display ranges (i.e., >20 nmi scale) that are likely to be used within the oceanic environment, a more accurate display of target range may be needed. As a result, pilots were provided with an alphanumeric display of range to the selected target, located to the right of the ground speed display in Figure 5. This was quantized to 0.1 nmi.

Experimental Design

Three speed cues were comparatively examined against a CDTI with and without an alphanumeric display of range. Subjects flew two scenarios with each speed cue condition for a total of six scenarios. Three scenarios simulated enhanced climb or descent procedures (in-trail or lead) and three simulated enhanced climb or descent to co-altitude procedures (in-trail or lead) with the target (Table 1).

Table 1. Scenario Design

	1 In-Trail Climb (Long rng nominal)	2 Lead Climb (Late contingency during climb loose climb performance)	3 In-Trail Descent (Stressed)	4 Co-Altitude In-Trail Climb (Stressed)	5 Co-Altitude Lead-Descent (Long rng nominal)	6 Co-Altitude In-Trail Climb (Late contingency lead turn off track 30 right for weather (wx) @ co-altitude
Scenario Set-up	R464 50 nmi east of BAKON, flight planned for FL370, max altitude (alt) for weight (wt) is FL390, wx up ahead forecasted south of us, traffic ahead at FL350	R464 80 nmi east of BAKON, flight planned for FL370, max alt for wt is FL380 - just asked ATC for a climb, but was told unable due to traffic 3 min in-trail at FL350, but FL370 available thru lead-climb	R464 15 nmi east of BANDY (2/3rds of way to HNL), AOC reports expect mod/severe turb past BANDY, & better winds at FL350, wants us down to FL350, traffic ahead at FL370 but FL350 avail thru ITD	R464 80 nmi east of BAKON, flight planned for and optimum for FL350, available with co-altitude ITC	R464 40 nmi east of BILLO (about 1/2 way to HNL), we can expect mod chop past BILLO so we requested FL350, ATC said unable due to traffic 8 min in trail, FL350 available thru co-altitude LD	R464 10 nmi east of BILLO, going direct to BEATS, flight planned for FL370 and still heavy for FL380, traffic ahead at FL370 might be available thru ITC to co-altitude, also wx up ahead
Starting Alt Requested Alt	FL330 FL370	FL330 FL370	FL390 FL350	FL330 FL350	FL370 FL350	FL350 FL370
Initial Separation (nmi)	45	-18	15.1	33	-60	42
Winds (kts)	FL330 285° 30 FL350 290° 36 FL370 300° 40	FL330 235° 20 FL350 240° 28 FL370 245° 60	FL350 245° 24 FL370 250° 42 FL390 260° 50	FL330 285° 30 FL350 290° 36 FL370 300° 40	FL330 285° 30 FL350 290° 36 FL370 300° 40	FL330 225° 25 FL350 230° 35 FL370 240° 35

Each pilot received two hours of training to become familiar with the simulator and the CDTI features. Following training, subjects flew the six experimental scenarios, which took between two and three hours. Pilots were tasked to conduct a series of enhanced climb and descent procedures and were briefed that the CDTI information was available as an enhancement to oceanic operations. During the scenarios typical oceanic flying tasks, such as position reporting and weather deviation, were required.

Since the oceanic domain is a fairly benign environment (i.e., aircraft are established on tracks with low closing velocities and large separation distances), the data collected was primarily subjective in nature. Additional quantitative data (e.g., distance and closure rate) was collected to support procedure development; however, it is not reported here.

Experiment Results

Overall Feature Rank Order

When all scenarios were completed, pilots rank-ordered the CDTI features that were available during the oceanic scenarios. The ranking data was submitted to a Friedman two-analysis of variance by ranks, revealing a significant effect for CDTI feature (Chi-Square = 24.5, $p < .0001$). A Bonferroni analysis for all pair-wise comparisons revealed Target Selection and Highlighting and Flight ID were ranked significantly higher ($p < .05$) than the three speed cues and the alphanumeric range display (Figure 6).

0-7803-5086-3 /98/$10.00 ©1998 IEEE

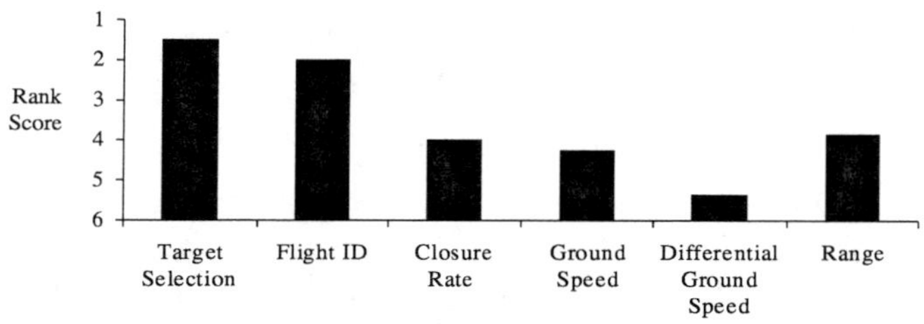

Figure 6. Mean Rank Order of CDTI Features from 1 (Most Useful) to 6 (Least Useful)

Speed Cue Rank Order

Pilots rank-ordered the three speed cues that were available during the oceanic scenarios. The ranking data was submitted to a Friedman test revealing no significant difference amongst the mean speed cue rankings ($p > .1$). However, pilot preference for the differential ground speed cue was much lower than the alphanumeric displays of ground speed and closure (Figure 7).

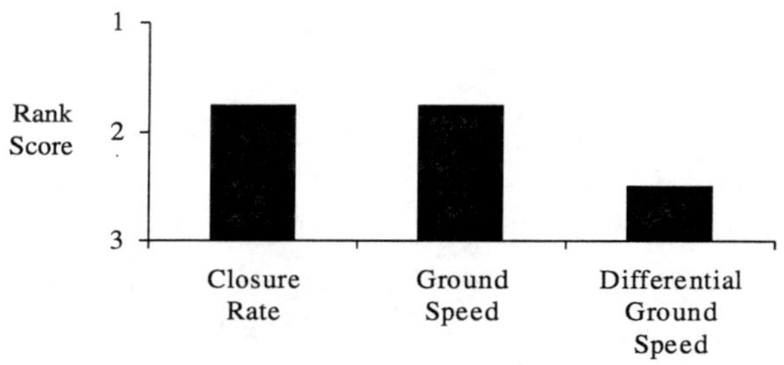

Figure 7. Mean Speed Cue Rank Order from 1 (Most Useful) to 3 (Least Useful)

Range Determination

Given it was unclear whether an alphanumeric readout of range (quantized to 0.1 nmi) to the selected target would be necessary for the various climb and descent procedures, a comparison was made between subject range assessments based on using the traffic display with and without a range readout (Table 2). The error in actual range versus reported range to ATC when using the range readout was a constant 0.3 nmi, whether at short or long ranges. This was due to the pilot rounding-off to the nearest nmi, when reporting the range to ATC. During the short range scenarios (i.e., near minimum separation distances), the average error when using the traffic display without a range readout was 0.8 nmi, and for long range scenarios it was 2.5 nmi. It is arguably more important to be precise with range estimates when near minimum separation standards; the use of the range readout provided this. In addition, the pilots were able to keep a precise range estimate to the target using the range readout, even at longer ranges. This could have the added benefit of allowing the pilot to keep the display set to a higher range, say for navigation purposes, while still maintaining a precise awareness of range to target.

Table 2. Pilot's Reported Range Estimate Error

Scenarios (Table 1)	With Range Readout (nmi)	Without Range Readout (nmi)
Short-Range (Scen 2,3,4)	0.3	0.8
Long-Range (Scen 1,5,6)	0.3	2.5
All	0.3	1.8

Subjective Comments

Pilots were queried about the utility of the CDTI feature set for the enhanced oceanic procedures. Subjective comments are reported in Table 3.

Table 3. Pilot Subjective Comments

1	"It was helpful to know which airplanes were around us and the speed and direction they were moving."
2	"It helped, knowing what the other guy is doing/experiencing. This always factors in to my decision to change altitude, speed, etc."
3	"It definitely helped performance—gave me a better feel for what other aircraft were actually doing. This should have a positive effect on fuel consumption by being able to change altitudes easier."
4	"It definitely helped, if the key is separation, and I believe it is, determining position and performance of target aircraft is a basic requirement."

5	"It helped, I had a higher confidence level in determining the required parameters for the climb/descent procedures. The additional information provided was well worth the interface/clutter cost."
6	"It helps you feel much more comfortable with the reduced separation standards. You can keep an eye on the other aircraft and vice-versa. I think this could be a valuable tool in helping airlines increase the number of flights overseas."
7	"It provided information not currently available, enabling more efficient use of the airspace safely."
8	"I was able to better operate with the flow of other traffic."

Next Steps

These simulations evaluated potential CDTI features for the proposed oceanic procedures. They did not address operational implementation issues such as controller acceptance, controller workload, or controller/pilot interaction. These latter issues must be addressed with appropriate future simulation studies or analyses. Also, the proposed procedures were based on the existing ITC procedure, and in particular the fact that separation responsibility was maintained by ATC. The specific separation standards proposed were similarly based on existing procedures. The actual design will need to take into account not only the existing standards cited, but also any other relevant oceanic and ICAO standards.

Conclusion

In this research we examined a basic CDTI feature set to support two near-term oceanic applications—enhanced climb and descent procedures (in-trail or lead) and enhanced climb and descent to co-altitude procedures (in-trail or lead). As a whole, pilots felt that CDTI greatly augmented their conduct of the proposed oceanic procedures. A review of the oceanic subjective feature rankings revealed a strong preference for the Target Selection and Flight ID features. The speed cues and alphanumeric range display were equally favored, with the exception of the differential ground speed display. In general, pilots felt the differential ground speed display was not very useful and should not be implemented on future displays.

However, some care should be exercised when interpreting these results. It should be noted that the current experiments only partially addressed the range of possible CDTI applications that are candidates for implementation. Indeed, the ADS-B MASPS [5] identifies nearly 80 CDTI applications, including farther term air-to-air conflict probing and resolution that could ultimately facilitate a mature free flight state. The implementation of additional CDTI applications may place other requirements on the CDTI that must be resolved in a wider context of information selection. Also, this simulation study focused on information content of the CDTI capabilities, and it is assumed that high fidelity simulations and field testing must be conducted to validate these findings for operational use.

Nevertheless, these evaluations demonstrated that a fairly simple set of CDTI features may enable certain potential near-term enhancements to oceanic procedures. One attractive factor in these proposed procedures is that they should require no ground infrastructure enhancements (i.e., they would depend on the air-to-air exchange of data made possible through ADS-B and relatively small changes to current ATC procedures).

References

[1] Cieplak, J. J., A. D. Mundra, K. W. Peppard, and F. Sweeney, November 1994, *Extension of Distance-Based Non-Radar Separation Rules to an Oceanic In-Trail Climb Procedure,* Proceedings of the 9th Annual International Oceanic Airspace Conference.

[2] Cieplak, J. J., A. D. Mundra, and K. W. Peppard, September 1995, *Operational Trials of the In-Trail Climb Procedure: Interim Results,*

MTR 95W0000121, The MITRE Corporation, McLean, VA.

[3] Federal Aviation Administration (1995) *Memorandum: Renewal of Waiver 93-01-140 to Order 7110.65, Air Traffic Control, Paragraph 6-1, Distance Measuring Equipment (DME)*, Speck, L. L., December 1995, Washington, DC.

[4] Olmos, B. O., et al., September 1998, *Evaluation of Near-Term Applications for ADS-B/CDTI Implementation*, SAE/AIAA 1998 World Aviation Congress, Anaheim, CA.

[5] RTCA. (1998). Minimum Aviation System Performance Standards (MASPS) for Automatic Dependent Surveillance Broadcast (ADS-B), DO242, Washington, DC.

COGNITIVE FACTORS IN AVIATION DISPLAY DESIGN

Christopher D. Wickens, University of Illinois, Aviation Research Laboratory, Savoy, Illinois
Steve Fadden, Lockheed Martin Aeronautical Systems Company, Atlanta, Georgia
David Merwin, Boeing Commercial Airplane Group, Seattle, Washington
Patricia May Ververs, Honeywell, Inc., Minneapolis, Minnesota

Introduction

Three-dimensional display technology is rapidly reaching the necessary levels of speed and robustness to serve the aircraft cockpit in a number of different navigational tasks. Indeed since the pilot's world is that of a three-dimensional volume of airspace, such technology would seem to be of considerable utility. This paper will review the research that we have done in our laboratory regarding the strengths, and sometimes hidden weaknesses of that 3D technology for a variety of tasks involving navigational awareness. In introducing this research we distinguish two general categories of such awareness: **flight path awareness** or guidance, simply refers to the pilot's knowledge of the current and future trajectory requirements. **Hazard Awareness** includes awareness of any hazards (traffic, weather, terrain, wake vortices) within some proximity of the aircraft. This may of course consist of hazards on or near the forward flight path; but it also includes what we refer to as **global awareness**, incorporating any hazards within the 360 degree sphere around the aircraft. In our discussion, we also refer to "three-dimensional displays" as any display that capitalizes upon the natural depth cues that humans use to navigate and understand the world, in order to create a sense of three dimensionality or depth within a displayed environment (Wickens, Todd, & Seidler, 1989). We do not consider any particular cue, such as stereoscopic viewing, to be the defining characteristics of a "3D display" (although it is the case that stereo appears to be one of three cues, along with occlusion and motion parallax, that conveys the strongest sense of depth; Wickens et al., 1989).

A critical issue addressed in this paper concerns the **frame of reference** by which the airspace is presented to the pilot (Wickens, 1998). As shown in Figure 1, this can be represented by a camera or display viewpoint, positioned in a number of locations and orientations, relative to the pilot's eyes. In the left panel (a) is shown an immersed perspective or egocentric view which essentially mimics the pilots vision. In panel b, the geometric field of view (McGreevy & Ellis, 1986) is expanded to provide a broader region of coverage. In panel c, the viewpoint is pulled away, as if "tethered" behind and above ownship. In panel d, representing the classic map viewpoint, this tether is oriented with a -90° elevation angle, as if looking straight down on own ship. To provide altitude information, this 2D rendering is coupled with a side view "profile" display to produce a **coplanar** suite.

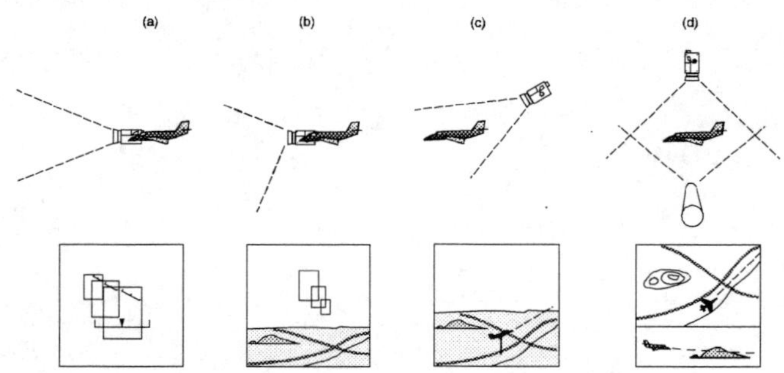

Figure 1. Four different frames of references for depicting aircraft navigation information shown beneath each are associated performance costs.

The Tunnel Display: Effects of Immersion

A Head-Down Display

Since the earlier work of Adams (1982) and Jensen (1981), avionics designers have considered the concept of a tunnel, pathway or "highway in the sky", presenting a 3D forward looking perspective of the required flight path, and future (predicted) aircraft position, from an immersed frame of reference typical of that shown in panel A (e.g., Jensen, 1981; Nataupsky & Crittendon, 1988; Theunissen, 1996; Grunwald, 1984). The particular rendering shown in

0-7803-5086-3 /98/$10.00

Figure 2, developed by Wickens, Haskell, and Harte (1988; Haskell & Wickens, 1993), was evaluated in a series of simulated flight trials and compared against a **coplanar** counterpart, for flight path awareness. This, and other evaluations (Reising, Liggett, Solz, & Hartsock, 1995; Flohr & Huisman, 1997; Grunwald, 1984; Regal & Whittington, 1995; Wickens & Prevett, 1995; Olmos, Wickens, & Chudy, 1997b), have revealed the distinct benefits of the three-dimensional tunnel or pathway rendering for flight path control. In depth analysis performed by Haskell and Wickens (1993) revealed that the specific advantage of this display over its co-planar counterpart, which presented the identical predictive information but separated into three orthogonal planes referenced to the aircraft flightpath, was the fact that the tunnel and predictor integrated current and future lateral and vertical command and status information into a single panel. No visual scanning was required to capture both axes, and thus the display integration of these two primary axes of flight was compatible (Wickens & Carswell, 1995) with their "cognitive integration" necessary for flight path control. This integration is necessary given the natural coupling of lateral and vertical axes on most aircraft. Additional benefits are associated with the compatible coincidence of the axis of viewing rotation and control rotation, and with the high display gain associated with the display when the aircraft rotates.

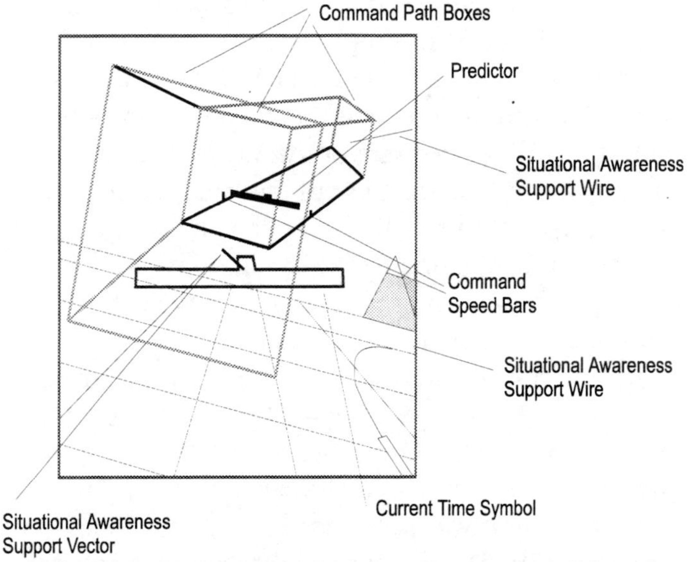

Figure 2. Tunnel in the sky display used head down by Wickens, Haskell, and Harte (1988) and Haskell and Wickens (1993).

A Head-Up Display

The distinct benefits of the tunnel display in a head down location, suggest that such an immersed perspective might offer benefits on a head-up display as well (see cell B of Table 1). Indeed the 3D depiction of a runway outline, a simplified immersed display, has become a standard feature of many commercial HUDs designed to support aircraft landing. The representation of a flight path velocity vector in the forward view, is also an example of immersed or ego-referenced symbology. Such symbology in HUDs has been referred to as **conformal** in that it conforms to features of the outside world (Weintraub & Ensing, 1992). However it should be noted that presentation of any information head up (cell A or B of Table 1), invites the possibility that HUD imagery may obscure the view of critical hazards on the forward flight path, and that the processing of that imagery itself may be degraded by the view of imagery in the far domain beyond the aircraft (terrain or clouds).

Table 1. Display Comparisons

Display Location		Imagery: Nonconformal	Imagery: Conformal (e.g., tunnel)
	HUD	A	B
	Down	C	D

In our laboratory, we have examined these costs in terms of the attentional costs of focusing attention on one domain (instruments or world), while ignoring visual information emanating from the other domain (Wickens, 1997; Ververs & Wickens, 1998; Martin-Emerson & Wickens, 1997). Against these well documented costs, are placed the real benefits of reduced visual scanning and reduced attention switching, resulting when symbology is displayed head up. Indeed our comparisons of head-up versus head-down presentation of identically formatted instrumentation, generally reveals that the costs of scanning (head down) dominate the costs of the clutter from overlapping imagery (head up), yielding an overall benefit to head-up presentation for most tasks (Fadden, Ververs, & Wickens, 1998), although clutter costs can still be well documented through careful analysis and experimental manipulation (Ververs & Wickens, 1998). However from the perspective of the current paper, most of the clutter costs are eliminated, and the advantages of head-up presentation (over head down presentation of identical imagery) become most pronounced, when the imagery is configured conformally, in 3D perspective, so as to overlay and move in synchrony with its counterpart in the far domain (cell B vs. cell D). This appears to be particularly true when runway outlines are used, either in landing (Wickens & Long, 1995) or in taxiing (Foyle, McCann, & Shelden, 1995; Lasswell & Wickens, 1995). The advantages of conformal symbology head up are conferred not only on flight path control, but also on the tasks of focusing attention on both near and far domain events (e.g., airspeed monitoring, traffic monitoring; Fadden et al., 1998).

Recently, capitalizing upon the success of the 3D tunnel in its head down location (cell D), we have examined its benefits head up (cell B; Fadden & Wickens, 1997; May Ververs, 1998; Fadden, 1998; see Figure 3), as the tunnel provides a more compelling sense of three dimensionality than does the mere runway overlay and flight path vector. Indeed the concept of a tunnel HUD underlies some versions of the future general aviation cockpit, envisioned under the AGATE program. Our evaluations suggest that the advantage to flight path control of the tunnel over conventional symbology when both are presented head up (cell B vs. cell A), are just as strong as they are when the two are compared head down (cell D vs. cell C; Ververs, 1998). However when a tunnel is compared between its head-down and head-up location (cell B vs. cell D), no differences in flight path performance appear to emerge (Fadden & Wickens, 1997; Fadden, 1998). Instead, lateral and vertical deviations (from an ILS path or a curved MLS path) are so small in **both** conditions, as to be nearly irreducible. Furthermore the ability to focus attention on the far domain is helped by the head-up location of the tunnel (cell B vs. cell D; Fadden & Wickens, 1997). That is, the reduced scanning outweighs the costs of clutter, and focused attention is somewhat helped in the head-up location by use of the tunnel, rather than the conventional ILS guidance information (cell B vs. cell A; Ververs, 1998).

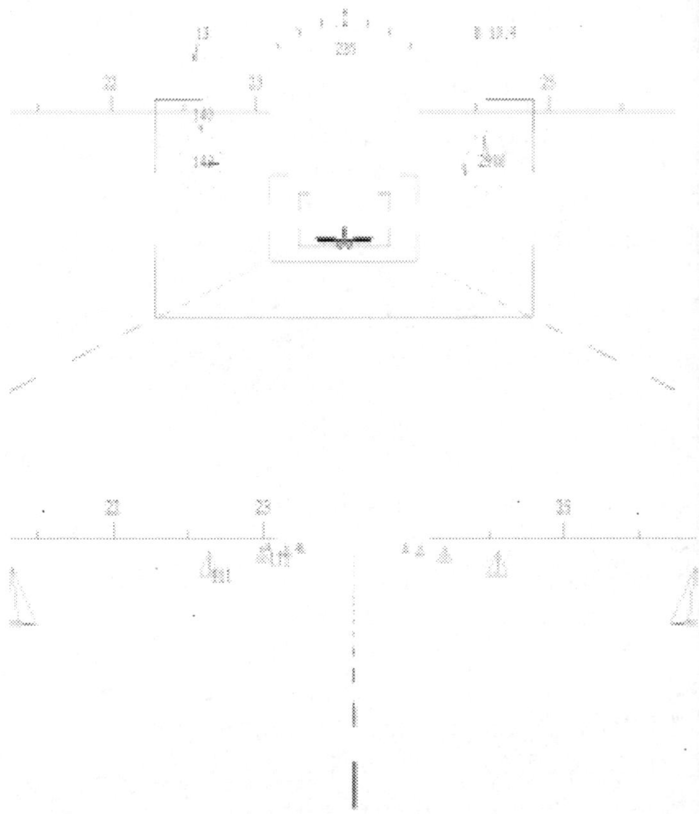

Figure 3. 3D HUD symbology used by Ververs (1998): (top) as pathway in the sky, (bottom) as runway overlay for taxi.

Attentional Tunneling

The high degree of superiority of the tunnel, whether positioned head up or head down, invites the question of whether it imposes any cost. The most prominent manifestation of such a cost, which has been observed in other head-up display studies, is that of **cognitive capture** or **attentional tunneling**. That is, the compellingness of the head-up display symbology can invite the unwanted and total capture of the pilot's attention to that symbology, at the expense of monitoring for and detecting **very unexpected** events in the far domain. Such an observation was initially made by Fischer, Haines, and Price (1980), who noted that pilots flying a final approach with a HUD were less likely to observe an aircraft taxiing onto the active runway, than when the instrumentation was presented head down. Wickens and Long (1995), documented the same phenomenon in a better controlled experiment, and although not all experiments replicate the statistically reliable HUD cost here, others have noted trends for longer latency detection of unexpected events (Martin-Emerson & Wickens, 1997; Lauber, Bray, Harrison, Hemingway, & Scott, 1982). Furthermore, the aggregate statistical trend revealed by a meta analysis of the data reveals a cost at the head-up location (Fadden et al., 1998).

Further scrutiny and experimental evaluations appear to illuminate conditions that modulate this attentional tunneling phenomenon. First, it appears to be manifest only when events are truly surprising (i.e., like a runway incursion, for which the pilot is totally unprepared). Second, it appears to be mitigated by the use of symbology that is conformal with imagery in the far domain. This is particularly true of the 3D runway overlay, that appears to act to "fuse" attention between the near and far domain (Wickens & Long, 1995; Lasswell & Wickens, 1995; Ververs, 1998). Third, attentional tunneling appears to be **enhanced** by the use of the 3D tunnel during flight (Ververs, 1998). While this finding would appear to contradict the second finding (conformal symbology reduces tunneling), the contradiction is resolved when careful analysis of the two displays is undertaken: the conformal runway symbology provides a direct linkage or overlay between the two domains [Foyle et al. (1995) and McCann & Foyle (1996) refer to this as "scene linking"], hence apparently facilitating the divided attention between both. In contrast, the tunnel provides a very **compelling,** ego-referenced source of guidance that does not have a visible counterpart in the far domain. Hence the tunnel appears to capture attention and preclude the detection of the unexpected, low salience events, much in the same way that attention is captured by the head-up location of non-conformal symbology.

In conclusion, it appears that tunnels are quite effective for guidance, and under most circumstances, their location in a HUD configuration will support the effective distribution of attention between the guidance and other display or far domain events. However the possible dangers of tunnel HUDs in inviting attentional capture and disrupting detection of the very unexpected event cannot be overlooked. Of course a second concern, raised by the 3D tunnel at either location, is that it provides a "keyhole" view of the world (Figure 1a) that cannot readily support more global hazard awareness.

Global Hazard Awareness: The Tethered Display

One alternative to the keyhole characteristics of the tunnel display whether head up or head down is to greatly expand the geometric field of view, as shown in Figure 1b. However, such expansion severely distorts the perceived location of hazards, and hence does not serve global hazard awareness well (Wickens & Prevett, 1995).

As an alternative, we have examined the "tethered" perspective shown in Figure 1c (Wickens, Liang, Prevett, & Olmos, 1996; Olmos, Liang, & Wickens, 1997a; Olmos et al., 1997b; Hickox & Wickens, 1997; Merwin, O'Brien, & Wickens, 1997; Boyer & Wickens, 1994). In this section, we ask how this perspective compares with both the immersed tunnel perspective and the 2D (coplanar) display, in both guidance and global hazard awareness. In most of these studies, we have generally found that such a perspective does not quite support the same level of performance in flight path guidance as the immersed display, although it fairs well when compared with the standard co-planar format. In particular, the tethered viewpoint appears to offer advantages when it is used for pilotage (i.e., navigation with reference to ground landmarks) in VMC (Olmos et al., 1997a; Hickox & Wickens, 1997). Such an advantage makes sense, given that the tethered view offers the pilot both a wide depiction of ground space, and a viewing perspective that is fairly congruent with the pilot's direct view of the forward terrain (Hickox & Wickens, 1997). Also, not surprisingly, the advantage of the 3D tethered perspective over the coplanar view appears to grow at lower flight levels when the pilot's view of the terrain is more forward looking (i.e., three dimensional; Hickox & Wickens, 1997; Schreiber, Wickens, Renner, Alton, & Hickox, 1998).

Many of our efforts in the research cited above have focused on the ability of the tethered display to depict global hazards, such as traffic (Merwin et al., 1997; Olmos et al., 1997a, 1997b; May, Campbell, & Wickens, 1996; Wickens, Miller, & Tham, 1996), terrain (Wickens, Liang, Prevett, & Olmos, 1996; Olmos et al., 1997a, 1997b; Wickens & Prevett, 1995; May et al., 1996), or weather, (May et al., 1996; Boyer & Wickens, 1994; Merwin et al., 1997). Here, summarizing the results of many of these studies, as well as those of other investigators (particularly Ellis and his colleagues; see Ellis, 1993 for a review), we find that the biggest deficiency of the 3D tethered view is the **ambiguity** with which position in 3D space, is represented on a 2D viewing surface, an ambiguity that is eliminated with coplanar displays. Such ambiguity is most pronounced along the line of sight, orthogonal to the viewing surface. The loss of resolution (and resulting ambiguity) along the line of sight becomes particularly concerning, whenever small differences in space must be judged, as when projecting traffic conflicts (Merwin et al., 1997). On the other hand, the negative effects of ambiguity are mitigated, whenever the movement of aircraft (or position of hazards) is constrained to two axes of airspace (e.g., level flight; Merwin et al., 1997; Wickens, Miller, & Tham, 1996).

Conclusions

Three-dimensional perspectives are being considered for a variety of applications, as conformal HUD imagery, tunnel primary flight displays, enhanced terrain renderings to supplement GPWS, traffic and weather displays, or displays of FMS automation modes (Owen & Suiter, 1997). In considering the strengths and limitations of such displays, in either head-up or head-down location, the designer and human factors practitioner must carefully evaluate the weaknesses of each (keyhole, distortion, and compellingness of the immersed perspective; ambiguity of the tethered perspective), as well as the task factors that may modulate the negative impact of these weaknesses. These factors must be contrasted with the visual scanning imposed by standard planar or co-planar head down displays. It is also the case that a number of these weaknesses may be partially offset by introducing various cognitive "remediations" such as visual momentum or artificial frameworks, which we do not address in this paper (but see Olmos et al., 1997b; Ellis, 1993). Thus effective "pilot centered" cockpit design cannot rely exclusively on the 3D graphics technology offered by the computer engineer, but must receive the input from the human factors community to inform as to the strengths and weaknesses of this technology, and the remediations for its weaknesses.

Acknowledgments

The authors of this paper acknowledge the contributions of several students at University of Illinois, whose work is cited, and contributed to the ideas expressed here. We also acknowledge the support of several funding agencies, supporting the work referenced here; including NASA Ames research Center (Grant NAG 2-886 Dave Foyle monitor; Grant NAG 2-996 Vern Battiste and Anna Trujillo technical monitors; Grant NAG 2-308 Vern Battiste and Sandra Hart; Technical monitors); Logicon Services Inc. (Grant F41624-94-D-6000 Larry Hettinger Technical monitor); The Federal Aviation Agency (Grant FAA DTFA 01-91-C-0045 Larry Cole, Technical monitor; Grant FAA DTFA 95-G-049 Dennis Beringer Technical monitor)

References

Adams, J. J. (1982). Simulator study of a pictorial display for general aviation instrument flight (NASA TP-1963). Hampton, VA: NASA Langley Research Center.

Boyer, B. S., & Wickens, C. D. (1994). 3D weather displays for aircraft cockpits (Technical Report ARL-94-11/NASA-94-4). Savoy, IL: University of Illinois, Aviation Res. Lab.

Ellis, S. R. (1993). Pictorial communication: Pictures and the synthetic universe. In S. R. Ellis (Ed.), Pictorial communication in virtual and real environments (2nd ed., pp. 22-40). Bristol, PA: Taylor & Francis Inc.

Fadden, S. (1998). Supporting pilot performance across task and display transitions. Unpublished Ph.D. Dissertation. Urbana, IL: University of Illinois.

Fadden, S., Ververs, P., & Wickens, C. D. (1998). Costs and benefits of head-up display use: A meta-analytic approach. Proceedings of the 42nd Annual Meeting of the Human Factors & Ergonomics Society. Santa Monica, CA: Human Factors and Ergonomics Society.

Fadden, S., & Wickens, C. D. (1997). Improving traffic awareness with a head-up flight path highway display. In R. Jensen & L. Rakavan (Eds.), Proceedings of the 9th International Symposium on Aviation Psychology. Columbus, OH: Dept. of Aerospace Engineering, Applied Mechanics, and Aviation, Ohio State University.

Fischer, E., Haines, R. F., & Price, T. A. (1980). Cognitive issues in head-up display (NASA Technical Paper 1711). Moffett Field, CA: NASA Ames Research Center.

Flohr, E., & Huisman, H. (1997). Perspective primary flight displays in the 4D ATM environment. Proceedings of the 9th International Symposium on Aviation Psychology. Columbus, OH: Ohio State University.

Foyle, D. C., McCann, R. S., & Shelden, S. (1995). Attentional issues with superimposed symbology: Formats for scene-linked displays. In R. S. Jensen (Ed.), Proceedings of the 8th International Symposium on Aviation Psychology. Columbus, OH: The Ohio State University.

Grunwald, A. J. (1984). Tunnel display for four-dimensional fixed-wing aircraft approaches. Journal of Guidance, Control, and Dynamics, 7(3), 369-377.

Haskell, I. D., & Wickens, C. D. (1993). Two- and three-dimensional displays for aviation: A theoretical and empirical comparison. The International Journal of Aviation Psychology, 3(2), 87-109.

Hickox, J. C., & Wickens, C. D. (1997). 3D electronic maps, design implications for the effects of elevation angle disparity, complexity, and feature type. Proceedings of the 41st Annual Meeting of the Human Factors & Human Factors Society. Santa Monica, CA: Human Factors & Ergonomics Society.

Jensen, R. S. (1981). Prediction and quickening in perspective displays for curved landing approaches. Human Factors, 23, 355-363.

Lasswell, J. W., & Wickens, C. D. (1995). The effects of display location and dimensionality on taxi-way navigation (Technical Report ARL-95-5/NASA-95-2). Savoy, IL: University of Illinois, Aviation Research Lab.

Lauber, J. K., Bray, R. S., Harrison, R. L., Hemingway, J. C., & Scott, B. C. (1982). An operational evaluation of head-up displays for civil transport operations NASA/FAA phase III final report. NASA Technical Paper 1815. Moffett Field, CA: NASA Ames Research Center.

Martin-Emerson, R., & Wickens, C. D. (1997). Superimposition, symbology, visual attention, and the head-up display. Human Factor, 39(4), 581-601.

May, P. A., Campbell, M., & Wickens, C. D. (1996). Perspective displays for air traffic control: Display of terrain and weather. Air Traffic Control Quarterly, 3(1), 1-17.

May Ververs, P. (1998). Conformal flight path symbology for head-up displays: Defining the distribution of visual attention in three-dimensional space. Unpublished Ph.D. Dissertation. Urbana, IL: University of Illinois.

McCann, R. S., & Foyle, D. C. (1996). Scene-linked symbology to improve situation awareness. In AGARD Conference Proceedings #575: Aerospace Medical Panel Conference of Situation Awareness (pp. 16:1 - 16:11).

McGreevy, M. W., & Ellis, S. R. (1986). The effect of perspective geometry on judged direction in spatial information instruments. Human Factors, 28(4), 439-456.

Merwin, D., O'Brien J. V., & Wickens, C. D. (1997). Perspective and coplanar representation of air traffic: Implications for conflict and weather avoidance. Proceedings of the 9th International Symposium on Aviation Psychology. Columbus, OH: Dept. of Aerospace Engineering, Applied Mechanics, and Aviation, Ohio State University.

Nataupsky, M., & Crittendon, L. (1988). Stereo 3-D and non-stereo presentations of a computer-generated pictorial primary flight display with pathway augmentation. Proceedings of the AIAA/IEEE 8th Digital Avionics Systems Conference. San Jose.

Olmos, O., Liang, C- C., & Wickens, C. D. (1997a). Electronic map evaluation in simulated visual meteorological conditions. International Journal of Aviation Psychology, 7(1), 37-66.

Olmos, O., Wickens, C. D., & Chudy, A. (1997b). Tactical displays for combat awareness: An examination of dimensionality and frame of reference concepts, and the application of cognitive engineering. Proceedings of the 9th International Symposium on Aviation Psychology. Columbus, OH: Dept. of Aerospace Engineering, Applied Mechanics, and Aviation, Ohio State University.

Owen, G. L., & Suiter, J. (1997). 3D FMS map. In R. Jensen & L. Rakavan (Eds.), Proceedings of the 9th International Symposium on Aviation Psychology. Columbus, OH: Dept. of Aerospace Engineering, Applied Mechanics, and Aviation, Ohio State University.

Regal, D., & Whittington, D. (1995). Guidance symbology for curved flightpaths. In R. S. Jensen (Ed.), Proceedings of the 8th International Symposium on Aviation Psychology (pp. 74-79). Columbus, OH: The Ohio State University.

Reising, J. M., Liggett, K. K., Solz, T. J., & Hartsock, D. C. (1995). A comparison of two head-up display formats used to fly curved instrument approaches. Proceedings of the 39th Annual Meeting of the Human Factors and Ergonomics Society (pp. 1-5). Santa Monica, CA: Human Factors and Ergonomics Society.

Schreiber, B. T., Wickens, C. D., Renner, G. J., Alton, J., & Hickox, J. C. (1998, in press). Navigational checking using 3D maps: The influence of elevation angle, azimuth, and foreshortening. Human Factors.

Theunissen, E. (1996). Influence of error gain and position prediction on tracking performance and control activity with perspective flight path displays. Air Traffic Control Quarterly, 3(2), 95-116.

Ververs, P. M. (1998). Conformal flight path symbology for head-up displays: Defining the distribution of visual attention in three-dimensional space. Unpublished Ph.D. Thesis. University of Illinois.

Ververs, P. M., & Wickens, C. D. (1998, in press). Head-up displays: Effects of clutter, symbology intensity, and display location on pilot performance. The International Journal of Aviation Psychology.

Weintraub, D. J., & Ensing, M. (1992). Human factors issues in head-up display design: The book of HUD (State of the Art Report). Wright-Patterson AFB, Dayton, OH: Crew System Ergonomics Information Analysis Center.

Wickens, C. D. (1997). Attentional issues in head-up displays. In D. Harris (Ed.), Engineering psychology and cognitive ergonomics: Transportation systems, Vol. 1 (pp. 3-21). Aldershot, UK: Ashgate Publishing Ltd.

Wickens, C. D. (1998, in press). Frames of reference for navigation. In D. Gopher & A. Koriat (Eds.), Attention and performance, Vol. 16. Orlando, FL: Academic Press.

Wickens, C. D., & Carswell, C. M. (1995). The proximity compatibility principle: Its psychological foundation and relevance to display design. Human Factors, 37(3), 473-494.

Wickens, C. D., Haskell, I., & Harte, K. (1988). Perspective flight path displays: Application of ergonomically driven display design principles (pp. 506-509), Vol. I. Proceedings of the IEEE International Conference on Systems, Man, and Cybernetics. New York: IEEE.

Wickens, C. D., Liang, C- C, Prevett, T., & Olmos, O. (1996). Electronic maps for terminal area navigation: Effects of frame of reference and dimensionality. International Journal of Aviation Psychology, 6(3), 241-271.

Wickens, C. D., & Long, J. (1995). Object- vs. space-based models of visual attention: Implications for the design of head-up displays. Journal of Experimental Psychology: Applied, 1(3), 179-194.

Wickens, C. D., Miller, S., & Tham, M. (1996). The implications of data link for representing pilot request information on 2D and 3D air traffic control displays. The International Journal of Industrial Ergonomics, 18, 283-293.

Wickens, C. D., & Prevett, T. (1995). Exploring the dimensions of egocentricity in aircraft navigation displays. Journal of Experimental Psychology: Applied, 1(2), 110-135.

Wickens, C. D., Todd, S., & Seidler, K. (1989). Three-dimensional displays: Perception, implementation, and applications (Technical Report ARL-89-11/CSERIAC-89-1). Savoy, IL: University of Illinois, Aviation Research Laboratory. (Also CSERIAC SOAR 89-001, Crew System Ergonomics Information Analysis Center, Wright-Patterson AFB, OH, December.)

PILOTS' SPATIAL MENTAL MODELS FOR MEMORY OF HEADING AND ALTITUDE

Susanne Delzell, San Jose State University, Moffett Field, California

Walter W. Johnson, NASA Ames Research Center, Moffett Field, California

Min-Ju Liao, National Research Council, Moffett Field, California

Abstract

A study was conducted that examined areas of a Cockpit Display of Traffic (CDTI) committed to memory during a task that required monitoring of aircraft separation. The data showed that pilots tended to recall aircraft primarily as a function of spatial range and bearing, and not as a function of temporal range.

Introduction

Researchers are currently developing design guidelines for cockpit displays of traffic information (CDTIs) to support cockpit-based separation from other aircraft. Although air-transport pilots have some experience monitoring traffic and potential conflicts on Traffic Collision Avoidance Display (TCAS), displays, self-separation is a new task. Therefore, the information needed by the pilots' in order to accomplish this task, and how they will use that information, are not well understood.

A cognitively based approach to this problem is to describe the relevant information, and how a pilot, in terms of a mental model uses it. A mental model is a cognitive construct that reflects a combination of predispositions (biases) and situation specific learning (see Tversky, 1991, for a discussion of spatial mental models). A description of this model can aid the development of good displays in two ways. First, it allows a designer to match the information and format of the display to the mental model when the mental model is appropriate. Second, it allows a designer to create a display that compensates where the mental model is incomplete or incorrect.

One method of gathering information about a mental model is to examine the contents of long-term memory (LTM). The information represented in LTM is determined (at least in part) by what a person attends to, and how permanently it is encoded. This, in turn, is determined by predispositions and situation specific factors. This study examined potential spatio-temporal determinants of the mental model that guide attention and retention of information when pilots are evaluating a self-separation task. Specifically, it focused on how the mental model was related to the x-y positions on the display.

Among the most critical elements of traffic information are distance of other aircraft to Ownship in space (spatial proximity), the absolute angular displacement of an aircraft relative to Ownship path (relative bearing), and the time it would take an aircraft to reach Ownship if that aircraft immediately turned onto a collision course (temporal proximity). These measures form the bases for three corresponding mental models that may guide a pilot's self-separation activity.

0-7803-5086-3 /98/$10.00

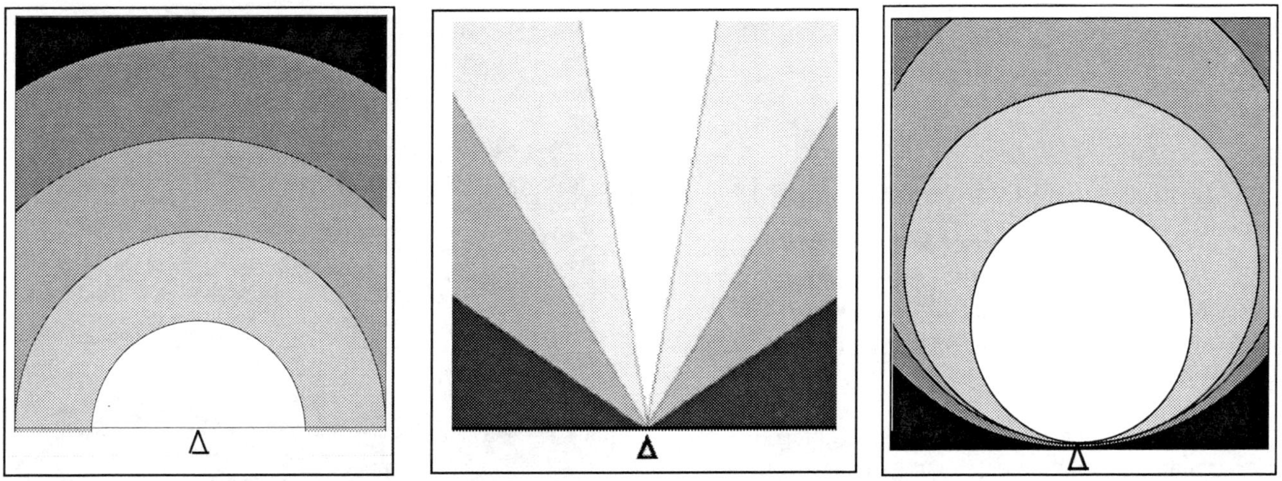

Figure 1. Recall predictions of three mental models. Left panel shows recall predictions based on spatial proximity (darker regions have poorer recall). The middle panel shows predictions based on relative bearing. The right panel shows predictions based on temporal proximity.

The predictions of these models are illustrated in Figure 1. The Spatial Proximity Model simply predicts that aircraft closer to Ownship will be better attended and encoded, and therefore better remembered. The Relative Bearing Model predicts that aircraft with smaller relative bearings will be remembered better. The Temporal Proximity Model predicts that aircraft that could reach Ownship in less time will be better remembered. For an airspace in which other aircraft are flying at approximately the same speed as Ownship, this corresponds to a set of nested centered in front of Ownship.

These three models represent different levels of sophistication with respect to traffic dynamics. The Spatial Proximity model is least sophisticated in that it only takes current separations into account. The Temporal Proximity model is more sophisticated, parsing the x-y space in terms of time needed for loss of separation. The Relative Bearing model reflects differences in closure rates (i.e., the fact that aircraft in front of Ownship tend to have higher temporal proximities than do aircraft to the side of Ownship).

In this experiment pilots were presented with a set of traffic scenarios on a CDTI and asked to evaluate their freedom to maneuver. The purpose of this evaluation task was to encourage the pilots to evaluate the display with respect to separation. After viewing and evaluating the scenario the pilots were interrupted to displace the contents of short-term memory. Following this, pilots were asked to recall as much information as possible about the traffic aircraft. These recall data were then used to determine which model or combination of models pilots likely use.

Methods

Participants

Twelve right-handed male participants with normal or corrected-to-normal vision were paid for their participation. All participants were active airline pilots experienced with glass cockpit instrumentation and with TCAS.

Apparatus

Scenarios were run on a 200MH Pentium Pro computer with an Omni Comp 300 SX OpenGL 3d accelerator graphics card. The monitor was 17 inches diagonal with 1600 horizontal by 1200 vertical resolution. Mouse and keyboard were standard. The size of the CDTI image on the monitor was 7" x 7". The

0-7803-5086-3 /98/$10.00 ©1998 IEEE

display was static over the period of monitoring time given.

Design

All pilots viewed 24 scenarios, with 12 aircraft presented on the CDTI during each scenario, for a total of 288 aircraft. The locations and headings of the aircraft were randomly generated subject to the following constraints:

- The 288 aircraft located by randomly assigning six aircraft that fit within the parameters of each of the 48 predefined and approximately equally sized x-y bins that covered the surface of the display.
- Within each bin, two of the six aircraft were located at each of the three altitudes: 31,000, 33,000 and 35,000 feet.
- For each pair of aircraft at each altitude within each bin, one aircraft was randomly assigned a heading that would eventually cross the path of Ownship, and one aircraft assigned a heading that would not cross the path of Ownship.
- The locations and headings of the aircraft were selected such that their future and past trajectories never resulted in them passing within 5 NM of Ownship, or within 2.5 NM of each other.
- No aircraft populated the area in front of Ownship for a ten-mile range (to avoid focusing attention on this area).
- The locations of the twelve aircraft were selected such that the altitude tags never overlapped.

Three altitudes were used in order to place a sufficient number of aircraft in each scenario without these aircraft being on conflicting trajectories. Furthermore, all combinations of heading (toward and away from Ownship path) and altitude were included within each x-y bin to ensure that the effect of location could be evaluated with minimal confounding effects of altitude and heading.

Description of the CDTI

Figure 2 shows the CDTI used in the study. At the bottom of the CDTI there is a filled triangular symbol with a short line segment extending out from it. The point of this triangle corresponds to the position of the Ownship, and the line segment is the predicted path for the next 30 seconds. The other aircraft are depicted by similar, but unfilled, triangular symbols with short lines oriented in their direction of travel. The traffic symbols also have associated altitude tags.

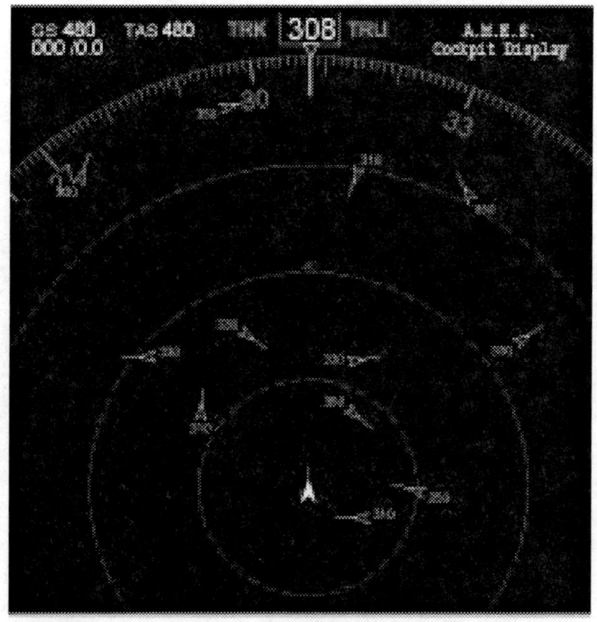

Figure 2. Example CDTI Scenario.

In addition to the aircraft, there are several other structural elements on the display. There is a compass arc at the top of the display, which shows Ownship direction. No aircraft were ever displayed beyond this compass arc. There is also a set of range rings on the display, which are circular arcs with the Ownship at the center. The range was set at 80 miles, with scale markers at 20-mile intervals. Speed was set at 480 knots for all aircraft, including Ownship and Ownship altitude was 33,000 feet. The depicted airspace extended 80 NM in front, 20NM behind, and 42.5 NM to the sides of Ownship.

Procedure and Task Description

It was explained to the pilots that each trial presented a different traffic scenario. They were told that their primary task was to evaluate how much freedom they had to safely change their heading without losing required separation from other aircraft (defined at 2.5 miles). The recall task was described as incidental to the primary task. It was emphasized that pilots should not memorize information from the display, but simply focus on the primary task as they would in a real flight situation, attending to information based on importance to the task. During each trial there were three separate stages which are described below.

CDTI Presentation and Inspection Stage:

At the beginning of this stage the message “Ready to Begin Next Trial” appeared and remained on the screen until a mouse button was pressed. At this time a CDTI similar to that depicted in Figure 2 was displayed and remained on the screen for one minute. During this one-minute interval the pilots were told to evaluate how much freedom they had to safely change their heading at all three altitudes (assuming a change of altitude) and to combine their assessments into one evaluation, which they would be asked for at the end of the trial.

Interruption Task Stage:

At the end of the first stage, the CDTI was replaced by a “interruption” task display (Figure 3). The purpose of this task was to interrupt visual and verbal short-term memory for the scenario and thus to simulate the effects of the various tasks that a pilot may do between sampling from the CDTI. Performance on the memory task would then be a measure of the traffic knowledge encoded in long-term memory. The interruption task required pilots to evaluate true/false statements about magnitude readings from four graphical “Engine Gauges” (e.g. “Engine 4 is greater than Engine 3 and greater than Engine 2”). Pilots were told that it was very important to be accurate, and they were given feedback (a beep) each time they gave an incorrect answer.

CDTI Maneuver Evaluation and Recall Stage:

After pilots responded to the interruption task, the third stage of the trial was presented. During this stage pilots rated their

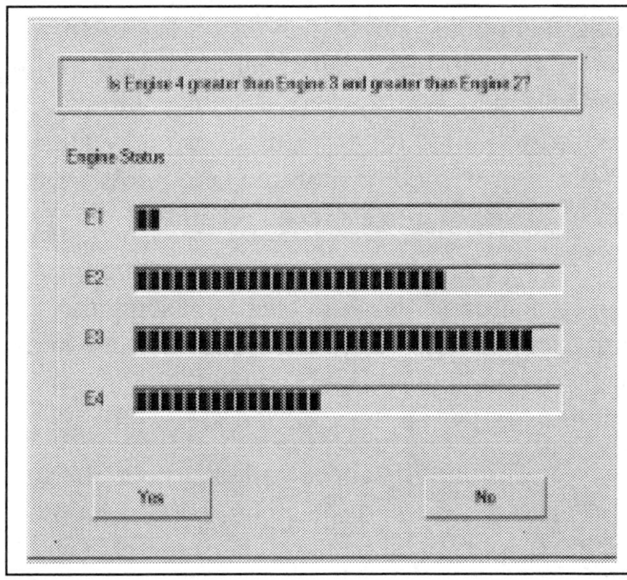

Figure 3. Engine display used for interruption task.

freedom to maneuver and recalled the information they remembered from the display.

When rating their freedom to maneuver, pilots were asked to use a cursor to select one of five possible ratings from "Very Low" to "Very High" as a measure of their overall freedom to maneuver. After the selection was made they went on to the memory recall task.

During the memory recall task pilots were asked to recall the headings and altitudes of as many aircraft as possible. The pilots were told not to guess at random, but that they could guess if they were fairly confident of their memory. The memory task began with a display of the x-y positions of the aircraft from the previous scenario. However, in this display the triangular symbols were replaced by small circles, and all altitude tags were removed. To the left of the display several more response controls and buttons were shown and used to collect recall responses using the following techniques:

Selection: The mouse and cursor could be used to select the symbol for any aircraft. This caused an additional circle to appear around the symbol, along with a line extending from that circle. Pilots then proceeded to any of the four sub-tasks (listed below) which could be performed in any order. Recalls could be revised/modified for any aircraft as many times as the pilot wished on any trial, and pilots were never forced to make a recall response.

Heading Recall: Recalled heading for a selected aircraft was indicated by placing the cursor on the line extending from the circle and dragging the line so that it aligned with the remembered ground track for that aircraft.

Altitude Recall: Recalled altitude for an aircraft was indicated selecting 31000, 33000 or 35000 from a menu on the right of the display.

Enter/Revise Another Aircraft: After pilots had entered altitude and/or heading information for a selected aircraft they could select a new aircraft or re-select a previously chosen aircraft.

Done Entering Information: When pilots decided they had entered all of the information they could remember for a scenario, they were told to click on a button labeled "Done Entering Information". When they confirmed that they were sure they were finished, the next trial began.

Results

Two dependent measures were analyzed using analyses of variance (ANOVAs) and regression analyses. Altitude responses were scored as correct if the correct altitude was chosen from the menu. Heading was partitioned into eight equal bins and scored as correct if the chosen aircraft heading was within 22.5 degrees of the actual heading. All aircraft headings and altitudes not responded to were scored as incorrect.

Analyses of Variance:

Prior to conducting the analyses of variance the display was partitioned into16 regions (Figure 4, middle panel). Each region was 21 NM wide by 21 NM long, with the 16 regions covering an area bounded by – 9 NM and +75 NM on the vertical ('Y') display axis, and –42 NM and +42 NM on the horizontal ('X') display axis. Within this area Ownship was located at the (0,0) coordinate.

Two-way fully within-subject ANOVAs were conducted on both the altitude and the heading recall probability data. Since the focus of this analysis was not on effects of display laterality, the data were collapsed across the display midline (e.g. combining regions 1 and 13, 5 and 9, . .), thereby reducing the 16 regions to eight regions. These eight regions were then divided into two levels of an X-Location factor and four levels of a Y-Location factor.

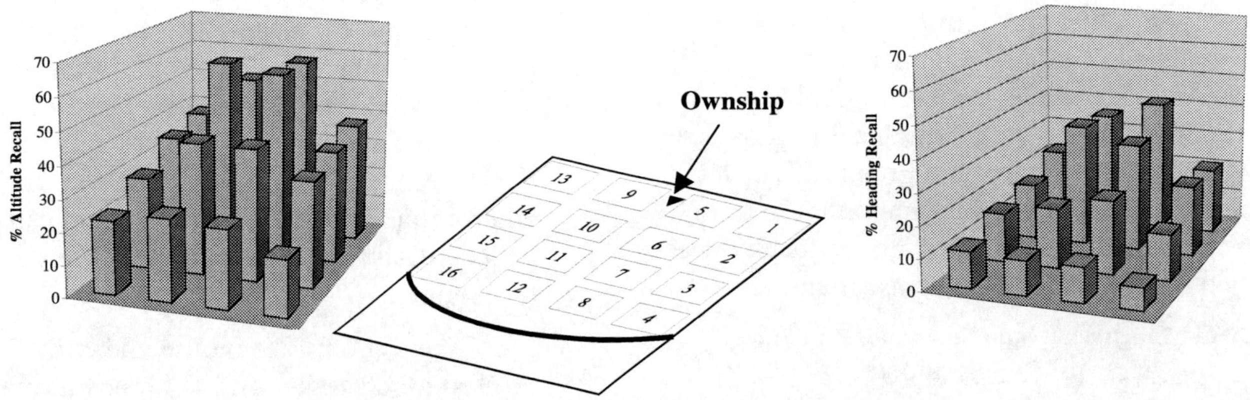

Figure 4. Percent recall for altitudes (left panel) and heading (right panel) as a function of display region. Depiction of display surface in center panel illustrates how graphs are related to display

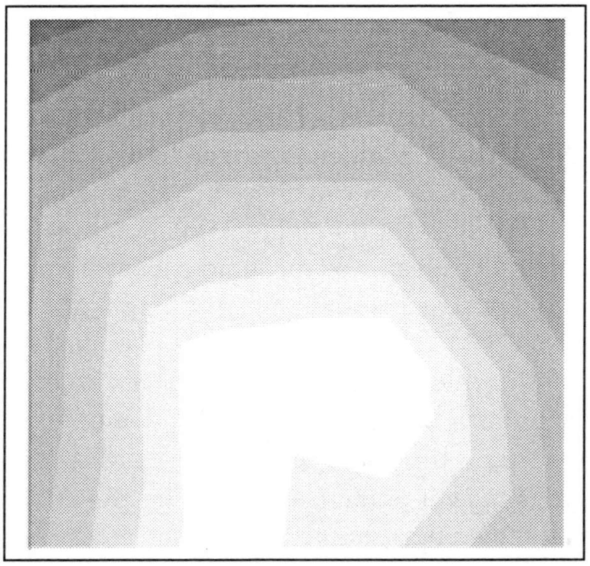

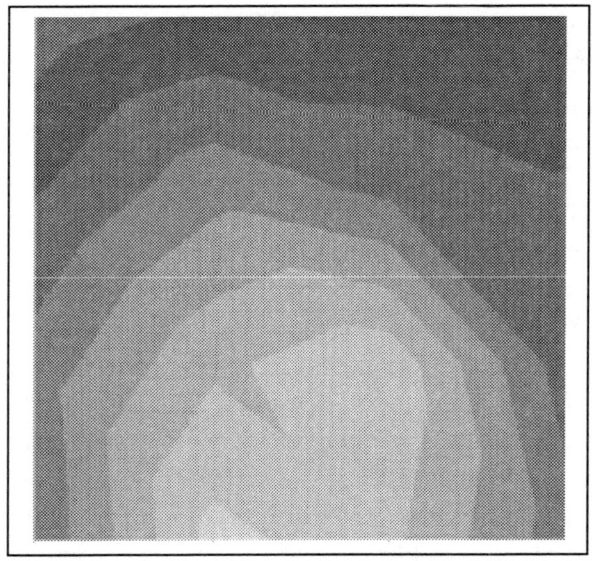

Figure 5. Probability contour plots for altitude recall (left panel) and heading recall (right panel). Ownship is located at the center bottom of graphs. Darker bands show poorer recall.

Both the altitude recall and heading recall ANOVAs yielded statistically significant ($p < .05$) main effects of X-Location, Y-Location, and a significant interaction of X-Location with Y-Location. The three-dimensional bar charts in Figure 4 show the distribution of the basic altitude and heading recall probabilities for the uncollapsed 16 regions. These charts show that best recall performance is found closest to Ownship for both variables. They also show that altitude was recalled better than heading.

Figure 5 shows best fitting contour plots for these same data. In both cases the iso-memory contours appear to be semi-circular ellipses with Ownship at the center and clearly illustrate how recall drops with increasing distance from Ownship. The elliptical form also shows that memory decreases less quickly along bearings closer to 0 (straight ahead), particularly for memory for altitude. However, memory does not appear to be especially sensitive to how quickly an aircraft could come into conflict with Ownship. Thus, data for altitude appears to reflect elements of the Spatial Proximity and the Relative Bearing Model, but not the Temporal Proximity model.

Regression Analyses:

The purpose of the regression analysis was to obtain a more quantitative test of the three models, and their corresponding measures, outlined in the Introduction. The Spatial Proximity model predicts that recall will be inversely related to the distance of an aircraft from Ownship. The Relative Bearing model predicts that recall will be inversely related to the relative bearing of an aircraft. The Temporal Proximity model predicts that recall will be inversely related to the time needed for an aircraft to reach Ownship. Prior to conducting the regression analysis, the probability of recalling each of the aircraft within each scenario was calculated by averaging across responses for all twelve participants. Furthermore, a large ($r = 0.45$) intercorrelation between the Spatial Proximity and Relative Bearing predictors (occurring because distance represented in front of ownship was greater than out to the sides) was removed by using only the data from aircraft within a 42.5 NM range. Significant intercorrelations remained between the Spatial Proximity and the Temporal Proximity predictors ($r = 0.21$), and between the Relative Bearing and the Temporal Proximity ($r = 0.50$).

For the heading variable, Relative Bearing accounted for 7% of the variability; Spatial Proximity accounted for 20% of the variability, while they jointly accounted for 25% of the variability. For the altitude variable, Relative Bearing accounted for 23% of the variability, Spatial Proximity accounted for 22% of the variability, while they jointly accounted for 40% of the variability. In all cases, and despite the intercorrelations with the other two predictors, Temporal Proximity accounted for less than 3% of the variability and was not statistically significant. The failure of Temporal Proximity as a predictor indicates that recall was not better for aircraft that were capable of quickly reaching Ownship than those that would take longer. Instead, pilots recalled aircraft better that were spatially close to Ownship and that were closer to straight-ahead.

Discussion

The mental model implied by the regression analysis of the recall data is primarily based on a combination of Spatial Proximity and Relative Bearing. The iso-memory contours are showing somewhat better memory for aircraft to the front of Ownship, than to the side. Notwithstanding this asymmetry, the near-circular form indicates that the effect of spatial proximity is dominant.

This can be contrasted with the forms that could be expected if the Temporal Proximity variable, or the Relative Bearing variable, dominated (Figure 1). The elliptical form does suggest, however, that the pilots' memories were not totally insensitive to the dynamics of the traffic situation. It shows that memories were at least somewhat better for the traffic in front of Ownship than to the sides of Ownship. The traffic in front includes the temporally most proximate aircraft, meaning, that they have the fastest potential closure rate (if they turned onto a conflicting course with Ownship). Still, a large emphasis is given in memory to aircraft that appear to be of small practical consequence (i.e. those directly to the sides and behind Ownship).

There may be several reasons for these findings. First, pilots were asked to evaluate their freedom to choose alternative headings, not just to scan for threat along their current heading. The task was meant to encourage a comprehensive encoding and evaluation of traffic. That is, the maneuver evaluation task required a pilot to compare the trajectories of a large number of other aircraft with that of Ownship. If the pilots considered a ±180-degree range of potential course changes equally, then our findings would not be surprising (i.e. memory would be as good in all directions from Ownship). However, the debriefing comments of the pilots argue against this interpretation. Eleven of the pilots stated that they only evaluated course changes of less than 30 degrees, and the twelfth pilot stated that he did not go beyond 40 degrees. Thus their statements indicated that they did not evaluate the more extreme maneuvers.

A second possibility is that the effects are task artifacts related to task demands. The memory task was designed as an incidental recall task, and not one on which the pilots should focus their attention or efforts. However, they clearly knew that it was of interest, and they would most likely have wanted to do as well as they could on this task. Therefore, they could have adopted a strategy to enhance recall, and the results could reflect that strategy and not one conducive to determining the basic memory model. However, again the debriefing comments of the pilots argue against this interpretation. All of the pilots indicated that they did not attempt to remember aircraft in order to meet the demands of the recall task. Most of them insisted that they remembered aircraft based on their performance of the primary task. This does not rule out the possibility, however, that task demands impacted their behavior at a subconscious level.

A third possibility is that the format of the display, or the nature of the task itself, may have influenced the salience of individual aircraft. In particular, since the CDTI is primarily an ego-centered display (in contrast to an air traffic controller's perspective-free display); it may have compelled attention to nearby aircraft, or to those closer to the immediate path. Research shows that memory for spatial information is affected by the observer's viewpoint, and that an ego-centered viewpoint holds a special perceptual status (Franklin, Tversky, and Coon, 1992). Thus these effects may have occurred despite a pilot's conscious intent. Comments by two of the pilots support this interpretation. These pilots said that they often "couldn't help remembering" aircraft behind them despite knowing that these aircraft could not possibly reach them.

The results of this study may be compared with studies of the memory of air traffic controllers in tasks requiring the monitoring of aircraft separation (e.g. Endsley and Rogers, 1996; Gronlund, 1997; Mogford, 1997). These studies also examined memory for aircraft information (e.g. recalled heading, altitude, and x-y location), sometimes relating it to whether aircraft were in conflict. These investigators focused upon potential relationships between memory and air traffic control performance. However, the researchers all noted that it is unclear what part of the information used to perform air traffic control is reflected in recall.

This caveat is certainly true for the present study. However, unlike the above studies, the point of the present study was to determine how spatial location affected the pilots' attention and encoding of information (although the relative sensitivity of these two processes to spatial location was not explored). Mental model structure was not related to the role of stored information in managing aircraft separation. Instead, this study used recall to uncover the spatial determinants of what is attended and encoded. This is based upon the premise, elaborated by Endsley (1995), that situational awareness, due to the degree to which it is attended and encoded should be reflected in recall.

The results indicate that pilots tend to pay undue attention to irrelevant aircraft when they are close to Ownship. This suggests that display designers need to find ways to focus pilots' attention on the most relevant aircraft.

References

Endsley, M. R., & Rodgers, M. D. (1996). Attention Distribution and Situational Awareness in Air Traffic Control. In Proceedings of the Human Factors and Ergonomics Society 40th Annual Meeting. Philadelphia, PA: Human Factors and Ergonomics Society (pp.82-85).

Endsley, M. R. (1995). Measurement of situational awareness in dynamic systems. Human Factors, 37, 65-84.

Franklin, N., Tversky, B., and Coon, V. (1992). Switching points of view in spatial mental models. Memory & Cognition, 20, 507-518.

Gronlund, S. D., Dougherty, M. R. P., Ohrt, D. D., Thompson, G. L. & Bleckley, K. M. (1997). The Role of Memory in Air Traffic Control. (Report No. DOT/FAA/AM-97/22) Washington, DC: Federal Aviation Administration.

Mogford, R. H. (1997). Mental Models and Situational Awareness in Air Traffic Control. The International Journal of Aviation Psychology, 7, 331-341.

Tversky, B. (1991). Spatial mental models. In G. Bower (Ed.), The psychology of learning and motivation: Advances in research and theory, Vol. 27, pp. 109-145. San Diego: Academic Press.

ED SITUATION AWARENESS IN THE FLIGHT DECK–ATC SYSTEM

Mica R. Endsley, SA Technologies, Marietta, GA

R. John Hansman, MIT International Center for Air Transportation, Cambridge, MA

Todd C. Farley, MIT International Center for Air Transportation, Cambridge, MA

Introduction

New technologies and operational concept changes have been proposed for implementation in the National Airspace System (NAS). These changes include improved datalink (CPDLC) technologies for providing improved weather, traffic, Flight Object (FO) and navigation information to the pilot and controller, and new forms of automation for both the flight deck and air traffic management system (FAA, 1997). In addition, the way business is conducted in the NAS is under consideration. Increases in the discretion provided to pilots (and dispatchers in commercial airlines) are being contemplated in an effort to increase system capacity and flexibility.

New concepts of operation (e.g., Collaborative Decision Making and Free Flight) allow for more control to be given to the cockpit or airline with correspondingly greater monitoring responsibilities on the ground. In addition, new technologies and displays make possible much greater information flow between the ground and the cockpit and also dramatic changes in the type of information provided. Designing to support these changes suggests two integrally linked questions:

(1) What display technologies and information are needed to support desired changes in operational responsibilities?

(2) How will the changes in information availability influence the negotiation process between the cockpit and the ground?

Each of these proposed changes (both in technology and operational concept) will have a marked impact on the performance, workload and Situation Awareness (SA) of both pilots and controllers. Typically such changes are evaluated independently in terms of the effects of the proposed change on either pilot performance or ATC performance. It is proposed here, however, that in order to fully understand the effects of such changes, the joint pilot/controller system must be considered. In particular, it will be shown that not only do changes in one part of the system affect the operators of that system, they may also induce shifts in the need for a shared understanding of the system between pilots and controllers (shared SA), and may induce shifts in the locus of control between these two agents as the degree of shared SA itself changes.

The model in Figure 1 is presented as a framework for understanding the changes induced on the overall system by changes to one part of the system. Historically the controller possessed a high degree of control for directing the separation activities of aircraft. With this role, there was little need (or opportunity) for shared situation awareness between pilots and controllers as it was assumed that pilots would obey controller commands. Only the controller needed to have SA regarding this information. For activities performed primarily by the pilot (e.g. systems operations) it is assumed only the pilot needs to have this information, not the controller.

0-7803-5086-3 /98/$10.00

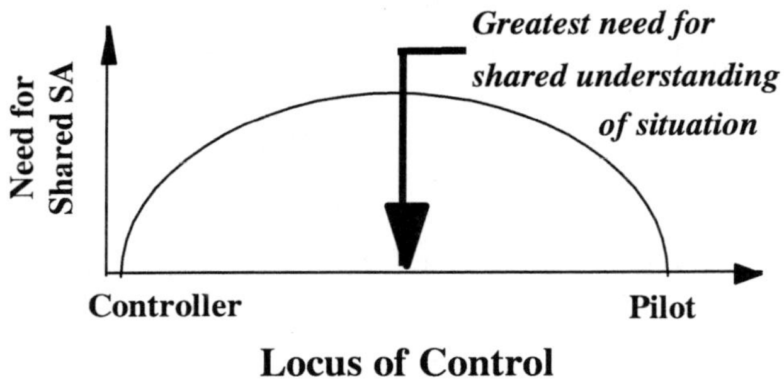

Figure 1. Shared SA and Locus of Control

As more information becomes available in the cockpit, the locus of control shifts to a middle ground. An example of this was the implementation of the TCAS collision avoidance system, where controllers had to deal with pilots interjecting unexpected actions (to the controllers) in response to TCAS alerts. In addition, TCAS was observed to change the negotiation process between controllers and pilots. With new operational concepts, such as an increase in self separation promised by some free flight advocates, it can be seen that the need for a shared common picture becomes much greater. Differences between controller SA and pilot SA may have substantial implications, resulting in problems such as increased negotiation or lack of immediate compliance to ATC commands by pilots. Therefore if pilots and controllers are going to have increasingly shared responsibilities for tasks (such as navigation, sequencing or separation activities) it will be increasingly important that they share a common picture of the environment.

Analysis of Shared SA Requirements

An analysis of shared goals and shared SA requirements between pilots and controllers in the NAS was conducted to examine this issue further. This was done by comparing an SA requirements analysis that was conducted for en route air traffic controllers (Endsley & Rodgers, 1994) with a similar analysis recently conducted for commercial airline pilots (Endsley, Farley, Jones, Midkiff, & Hansman, 1998). These two sets of goals and requirements were examined to form an understanding of (1) how these requirements overlap forming requirements for shared SA between pilots and controllers, (2) the mechanisms available to each party for acquiring SA on these factors in today's system, and (3) the implications of proposed changes to the system for shared SA and locus of control in the NAS of the future.

Example of Results

The pilot and controller goal hierarchies were analyzed and structured in order to identify areas of common or conflicting interest between pilots and controllers. Figures 2a and 2b depict the high-level goals of pilots and controllers, respectively. At these higher levels, the goal structures were highly parallel, and there was considerable overlap between the two. For example, both shared the need to avoid conflicts such as traffic and hazardous weather as well as the need to handle perturbations such as weather and emergencies.

In examining the detailed pilot and controller SA requirements, considerable overlap was present. At the same time there were also many differences, some of which were quite subtle, even for goals which at least superficially appeared to be the same.

For instance, pilots and controllers have a common goal to *maintain aircraft separation*. To the pilot, conformance to ATC clearances and instructions has been the major subgoal associated with achieving aircraft separation. Assuring that aircraft are conforming to clearances, on the other hand, is seen by controllers as a separate subgoal; and one which has been found to be neglected, particularly as workload increases (Endsley & Rodgers, 1998). In addition, if controllers do detect non-conformance to a clearance, they undertake to ascertain the reason for the non-conformance, focusing on determining the validity of the data they are receiving. Even if it is determined that non-conformance exists, the controllers may only undertake to correct the aircraft if it poses a particular problem.

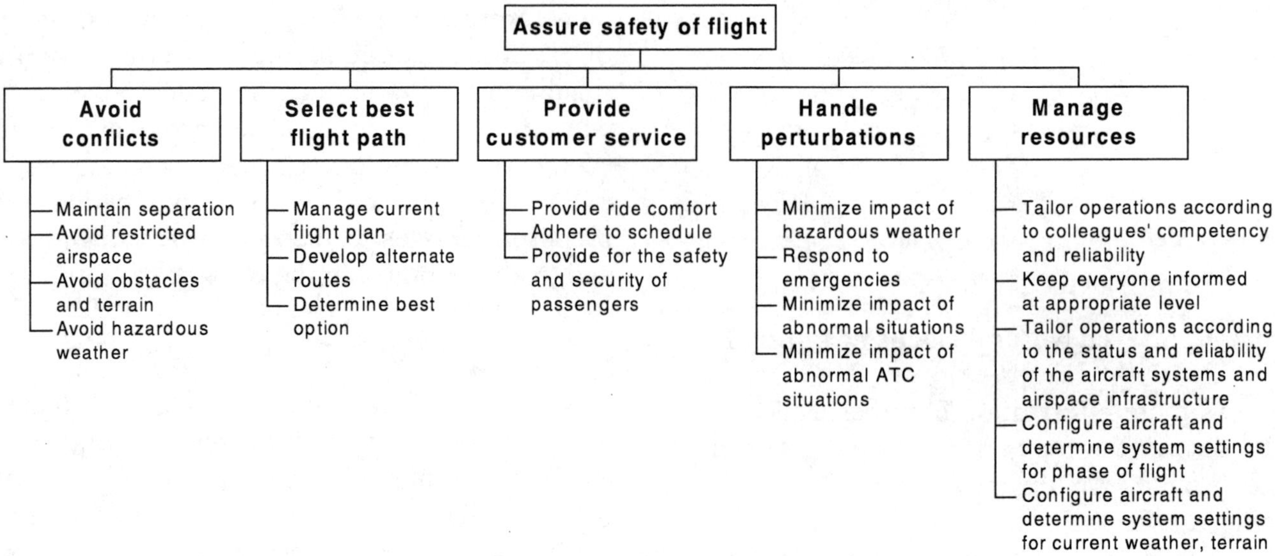

Figure 2a. Commercial Airline Pilot Top-Level Goal Hierarchy

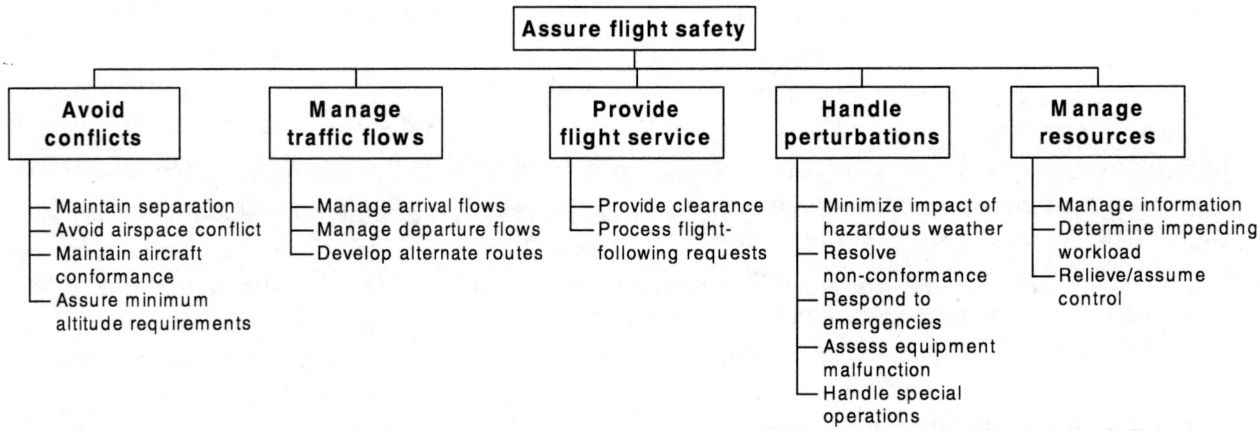

Figure 2b. En Route Air Traffic Controller Top-Level Goal Hierarchy

Pilots on the other hand, did not represent these types of concerns in their goal breakouts. They focused on ascertaining any deviation between the ATC clearance and their current and projected flight parameters (in common with the controllers). To do this they needed several pieces of information that did not show up in the controllers' breakout, including time available to meet the clearance and the magnitude and rate of change required to meet the clearance. This represents an "information gap" between ATC and pilots. Controllers rarely communicate time available for meeting a clearance (although sometimes they may in imprecise terms convey a sense of "urgency"). Pilots are left to figure out these higher-level SA requirements on their own. Suffering from a similar information gap, controllers may often have poor information on aircraft capabilities, weather and winds that are needed for making their own assessment of whether an aircraft will be able to conform to a clearance as expected.

Some interesting differences are also seen in the subgoals associated with determining whether aircraft are separated. While this subgoal clearly shows up for both ATC and the aircraft pilot, they appear to meet it in very different ways. The controllers very distinctly break their decisions up into a vertical separation and horizontal separation question, with SA requirements following these divisions. Furthermore, ATC compares aircraft separations to very specific limits, as appropriate for the particular airspace. No such distinctions were present for the aircraft pilots, however. They appeared to make a single decision regarding separation, and did not appear to use any fixed limits in assessing that separation.

Finally, if it is determined that some action is needed to resolve an aircraft conflict, pilots and controllers also appear to approach this subgoal differently. While many of the decisions appear to be similar (e.g., which aircraft to maneuver, time available, which type of maneuver to use), ATC appears to adopt a much more strategic focus to its decisions while the aircraft pilot's are far more tactical. For instance, ATC includes a specific consideration of the impact of a particular maneuver on air traffic and the projected number of changes that will be necessary to bring the aircraft back to its intended flight path and destination. They also specifically consider the capabilities of the aircraft and the pilots and how many previous actions (changes to clearance) had been given to a particular aircraft. They seek to optimize the flow of the traffic overall in making their decisions. The pilots' breakouts did not reveal these type of strategic considerations, but instead relied on more immediate tactical considerations, ATC clearances and the rules built into TCAS when deciding which aircraft would maneuver and when.

Many of these differences are natural outcomes of the current division of responsibility between ATC and the cockpit, and of the information available to each. Even so, certain conflicts which exist today between these two decision makers can be traced directly back to the subtle differences in each party's interests and the different information that is available to them.

One common example is re-route negotiation, wherein the pilot typically seeks to optimize his own route while the controller typically seeks to optimize the routes of many. The supporting information available to each is often different and sometimes exclusive. Given the potential for new technologies such as datalink to share information between the air and the ground, several issues arise:

(1) How will the sharing of information between pilots and controllers affect their ability to recognize routing conflicts and opportunities?
(2) Will it increase or decrease the amount of negotiation?
(3) Will negotiations become more collaborative or more contentious?

Simulator-Based Experiment

Based on the comparative SA requirements analysis, a part-task simulator experiment was conducted to investigate the effect of sharing weather and traffic information (via air-ground datalink) between pilots and controllers in negotiating route amendments. Weather and traffic were chosen as the key conflict categories because pilots and controllers reported in the SA requirements analyses (see Figures 2a & b) that weather and traffic are often the reason for re-routing and typically impose constraints on the available alternatives. Of particular interest in this experiment were indications of each subject's recognition of the other's constraints, his anticipation of the other's needs and/or preferences, his willingness to comply/cooperate, and his persistence in pursuing an alternate solution.

Experimental Design

The experiment paired an air transport pilot subject with an en route air traffic controller subject in a real-time simulated air traffic environment under present-day air traffic control procedures. Test scenarios featured traffic and weather elements specifically designed to create testable responses, a performance-based measure of situation awareness (Pritchett, Hansman, & Johnson,

1996). Testable response scenarios incorporate a hazard element (e.g., an intruder aircraft, a weather cell) that requires the subject to take action, provided s/he is aware of the situation. An appropriate action taken by the subject indicates situation awareness; inaction indicates a lack of situation awareness. Subjects interacted within the simulation environment to resolve the traffic and weather conflicts. The availability of shared traffic and weather displays was manipulated as the independent variable as shown in Table 1.

Table 1. Test Matrix

Configuration	Traffic information	Weather information
Without datalink	Controller only	Pilot only
With datalink	Shared	Shared

Because traffic displays are standard air traffic control equipment, the controller was always provided with traffic information, even in the non-datalinked configuration. Similarly, because weather radar is standard equipment in jet aircraft, the pilot was always provided with weather information, even in the non-datalinked configuration.

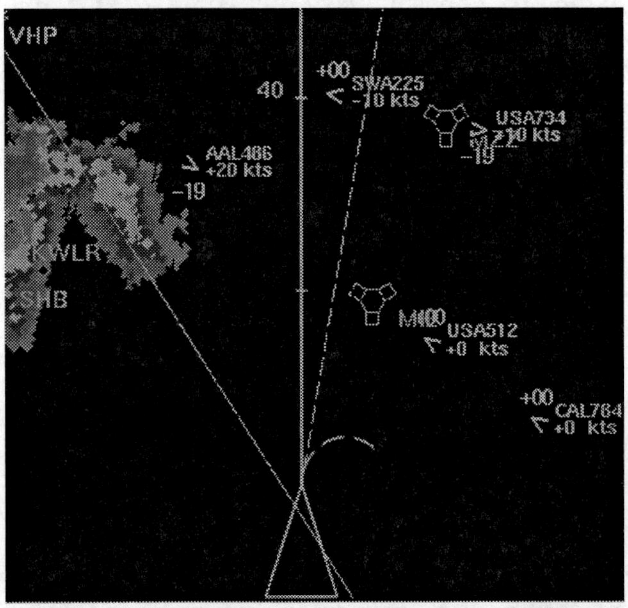

Figure 3. Cockpit map display with weather and traffic information

Figures 3 and 4 show the cockpit and ATC displays provided to the test subjects in the datalinked configuration, featuring both weather and traffic information on each. Weather information was depicted by six-color NEXRAD reflectivity images. Air traffic information was depicted on the controller's plan view display in a format consistent with current en route ATC equipment and included aircraft position, call sign, track, altitude, and ground speed. The pilot's cockpit display of traffic information (CDTI) depicted the position, call sign, track, relative altitude, and relative ground speed of all aircraft within 40 miles and 2600 feet in altitude.

Each pilot-controller subject pair performed three test scenarios two times, once with shared traffic and weather displays (i.e., "with datalink") and once without (i.e., "without datalink"). All scenarios took place in a high-level sector in Indianapolis Center airspace. Each scenario ran approximately ten minutes and featured between 12 and 18 aircraft transitioning the sector in the presence of convective weather activity. Those aircraft not piloted by the subject pilot were controlled by a pseudo-pilot, who also interacted with the subject controller and subject pilot via radio

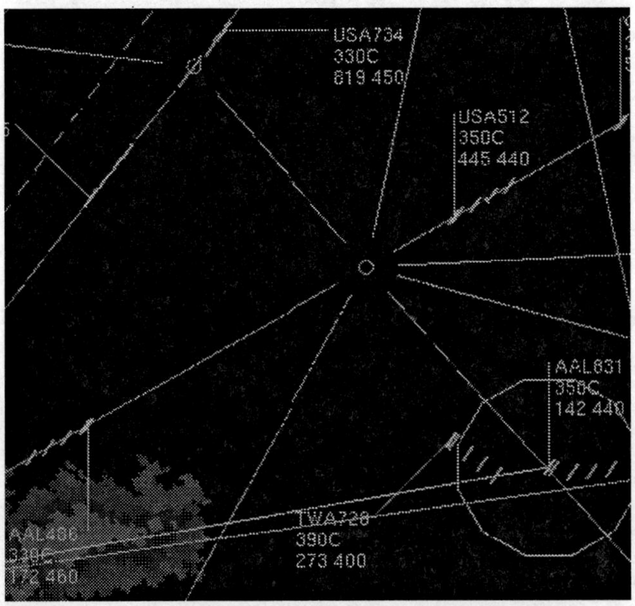

Figure 4. En route ATC plan view display with traffic and weather information

0-7803-5086-3 /98/$10.00 ©1998 IEEE

communication. Certain elements of each repeated scenario were changed (e.g., aircraft call signs, trajectories of non-factor traffic, etc.) in order to make the second run appear unique.

Pilot and controller situation awareness was measured using the testable response method. Radio communications were recorded and coded using a methodology adapted from Foushee, Lauber, Baetge, & Acomb (1986). Workload measurements were taken based on NASA–TLX (Hart and Staveland, 1988).

In order to observe pilot-controller interaction in a real-time, complex workload environment, MIT's distributed, interactive, multi-agent simulation facility was used (Amonlirdviman, et. al., 1998). The facility includes a part-task advanced cockpit simulator, a part-task en route controller workstation, a multi-aircraft pseudo-pilot station, and live voice communications between them, creating a real-time interactive air traffic environment.

Preliminary Results

Two pilot-controller teams have performed the experiment to date. Both controller subjects were Full Performance Level (FPL) ATC Specialists with an average of 10.5 years of experience, currently working the radar position at an Air Route Traffic Control Center (ARTCC). Both pilot subjects were jet transport pilots with an average of 12,850 hours.

Due to the small sample size, it is emphasized that the results presented here are preliminary. Good scientific judgment should be exercised before attempting to apply these results generally.

Situation Awareness

Each test scenario included one weather-related testable response condition and one traffic-related testable response condition. Both the pilot and controller were monitored for their awareness of each testable response condition.

Figure 5 summarizes their situation awareness with respect to weather. Pilots, having the benefit of the weather display for all test scenarios, demonstrated awareness of all of the weather-related testable response conditions. Controllers, without the benefit of a weather display in the non-datalinked configuration, demonstrated awareness of only half of the weather-related testable response conditions. When provided a weather display, however, controllers demonstrated awareness of all of the weather-related testable response conditions.

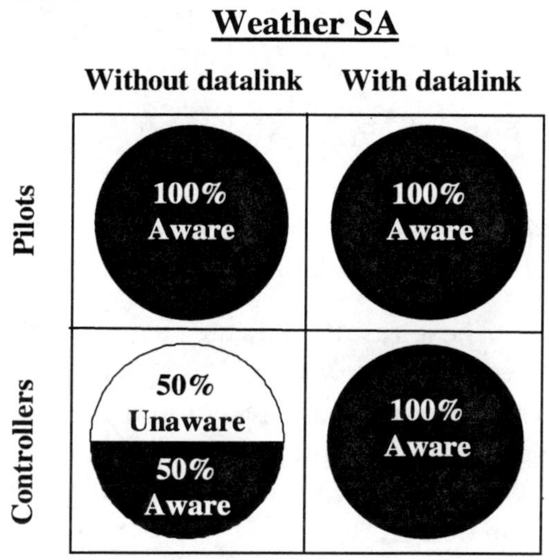

Figure 5. Pilot and controller awareness of weather-related testable response conditions.

Figure 6 summarizes pilot and controller situation awareness with respect to traffic. Pilots, without the benefit of a traffic display in the non-datalinked configuration, did not demonstrate awareness of any of the traffic-related testable response conditions. In many cases, the controllers recognized the traffic conflict before it became a significant threat to the pilot and would either advise the pilot of the traffic or vector him accordingly. In such cases, the pilot's testable response result was labeled "ambiguous". Controllers, having the benefit of their plan view traffic display for all test scenarios, demonstrated a high level of awareness of the traffic-related testable response conditions. In some cases, controller-issued clearance amendments inadvertently resolved the traffic-related testable response conditions before they arose; such cases were labeled "ambiguous" with respect to controller situation awareness.

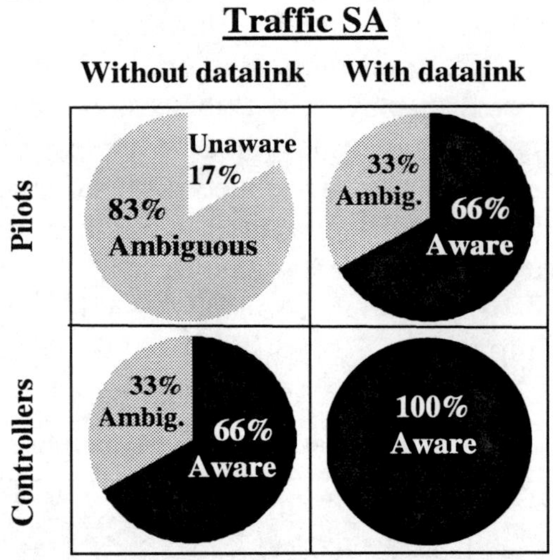

Figure 6. Pilot and controller awareness of traffic-related testable response conditions.

These results indicate that controller SA with respect to weather improves with the addition of a weather overlay to their plan view display. Similarly, the results suggest that pilot SA with respect to traffic improves with the addition of a CDTI.

Pilot and controller workload was measured using NASA–TLX. In general, workload was not affected in any systemic way, either individually or in a team sense.

In the twelve test scenarios completed thus far in the study, two operational errors[1] were observed, both occurring in the non-datalinked configuration. It is important to note that the test scenarios were challenging by design. Controllers were operating an air traffic sector other than their usual "home" sector and did not have the benefit of a conflict alert function or a D-side controller to assist them. However, the fact that both operational errors occurred in the non-datalinked environment does suggest that shared information may be beneficial in helping the controller maintain awareness with regard to separation issues.

[1] An operational error is defined as lateral separation of less than five miles and vertical separation of less than 1000 feet.

Communication and Negotiation

All radio communication was recorded, coded by category and topic, and analyzed. Figure 7 illustrates how the transactions conducted over the voice channel changed with the introduction of the datalink. As shown at the left, the number of transactions between the pilot and controller decreased when the datalink was introduced. However, the number of transactions for negotiating re-route clearances increased slightly, and the number of other transactions (including traffic advisories, ride reports, etc.) decreased.

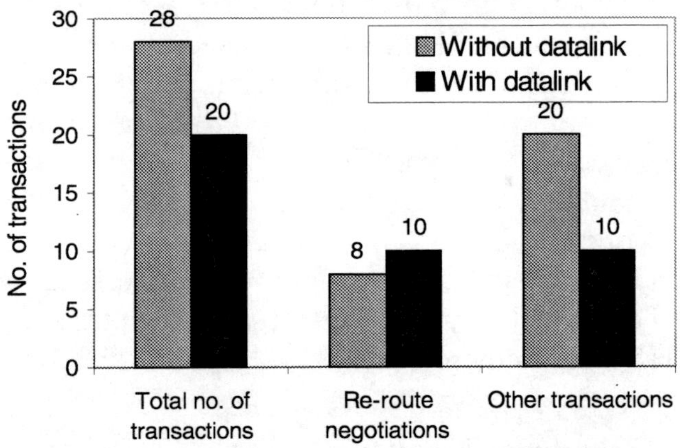

Figure 7. Voice communication transactions by topic

Figure 8 illustrates how the character of radio communications changed with the introduction of the datalink. The total number of transmissions, defined as "mic on" to "mic off", increased slightly (from 87 transmissions to 93) with the sharing of information via datalink. In analyzing these transmissions in more detail, several trends were observed. The number of pilot requests went up, consistent with the increase in re-route negotiation noted earlier. When the weather and traffic information was shared between the pilot and controller, there were four instances in which the pilot or controller voluntarily suggested a specific route amendment to the other party. This behavior was not evident in the non-datalinked configuration. Finally, controllers were much more proactive in providing weather advisories to pilots when they had the weather information overlay.

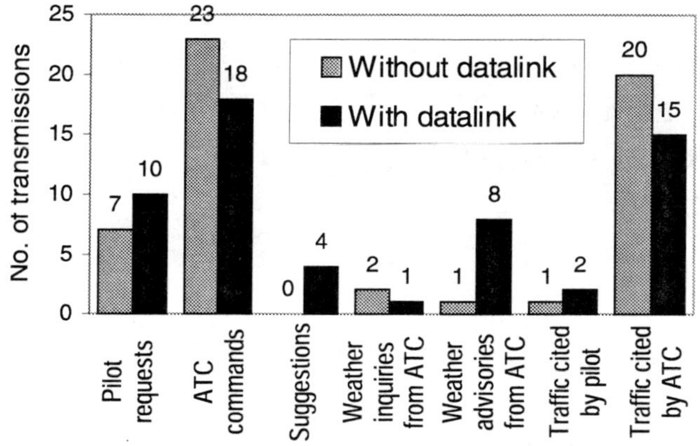

Figure 8. Voice communication transmissions by category

Conclusions

The comparison of pilots' and controllers' goals and SA information requirements revealed many parallels at the higher levels. At the more detailed, lower levels, however, subtle underlying differences were identified that lend insight into their decisions.

The preliminary experimental results indicate that situation awareness was improved when information was shared between the cockpit and the ground. Pilots' traffic situation awareness improved when traffic information was displayed; controllers' weather situation awareness improved when weather information was displayed. Operational errors were reduced in the process.

Pilots and controllers also exhibited behavioral changes. When information was shared, there were fewer requests for information, and controllers provided more frequent weather advisories to pilots. Pilots and controllers were more likely to suggest preferred route amendment alternatives to their counterparts, and re-route negotiation increased in the process.

Based on these preliminary results, it appears that shared information between flight crews and air traffic controllers may be beneficial in terms of improved situation awareness and the behaviors it seems to foster.

Acknowledgments

This work is supported by the National Aeronautics and Space Administration/Ames Research Center under grant NAG 2-716 and by TASC as part of the FAA Center of Excellence in Operations Research.

References

Amonlirdviman, K., Farley, T.C., Hansman, R.J., Ladik, J.F., & Sherer, D.Z. (1998). "A Distributed Simulation Facility to Support Human Factors Research in Advanced Air Transportation Technology", in Proceedings of the SISO Simulation Interoperability Workshop. Orlando, FL.

Endsley, M.R., Farley, T.C., Jones, W.M., Midkiff, A.H., & Hansman, R.J. (1998). Situation awareness information requirements for commercial airline pilots (ICAT-98-1). Cambridge, MA: MIT International Center for Air Transportation.

Endsley, M.R., & Rodgers, M.D. (1994). Situation awareness information requirements for en route air traffic control (DOT/FAA/AM-94/27). Washington, D.C.: Federal Aviation Administration Office of Aviation Medicine.

Endsley, M.R., & Rodgers, M.D. (1998). Distribution of attention, situation awareness, and workload in a passive air traffic control task: Implications for operational errors and automation. Air Traffic Control Quarterly, 6(1), 21-44.

FAA (1997) National Airspace System Architecture V 3.0 (draft)

Foushee, H.C., Lauber, J.K., Baetge, M.M., & Acomb, D.B. (1986). "Crew Factors in Flight Operations: III. The operational Significance of Exposure to Short-Haul Air Transport Operations". NASA Technical Memorandum 88322.

Hart, S.G. & Staveland, L.E. (1988). Development of NASA-TLX (Task Load Index): Results of empirical and theoretical research. In: Human Mental Workload (P.A. Hancock and N. Meshkati (Eds.)), pp. 139-183. North-Holland: Elsevier Science.

Pritchett, A., Hansman, R.J., & Johnson, E.N (1995). "Use of Testable Responses for Performance-Based Measurement of Situation Awareness" in (D. Garland and M. Endsley, eds.) Proceedings of the International Conference on Experimental Analysis and Measurement of Situation Awareness. Daytona Beach, FL: Embry-Riddle University Press.

SIMULTANEOUS DESIGN OF COCKPIT DISPLAY OF TRAFFIC INFORMATION AND AIR TRAFFIC MANAGEMENT PROCEDURES

A.R. Pritchett, Georgia Institute of Technology

Abstract

This paper examines two general considerations for designers of CDTI: the information required for intended applications, and the mechanisms capable for providing the pilot with that information. Methods for defining information requirements are summarized, and a case study using control theory analysis is given. The information available from procedures and from CDTI are compared. Finally, suggestions are given for the design of CDTI and air traffic procedures.

Introduction

The installation of Cockpit Display of Traffic Information (CDTI) in aircraft cockpits may enable new Air Traffic Management (ATM) operations. For example, with the availability of CDTI in the cockpit, pilots may be able to maintain a longitudinal separation behind another aircraft as specified by the controller. Other possible new operations include: passing other aircraft, both horizontally and vertically; sequencing into arrival streams; and maintenance of relative position within a 'platoon' of aircraft.

From the cockpit designer's point-of-view, enabling these new operations requires consideration of two general questions:

- What information does the pilot require to perform his or her part in these operations?
- From what sources, and in what formats, should the pilot receive this information?

This paper will address these two questions. The first section will highlight how numerical analysis of aircraft trajectories can provide a structured method of identifying minimum information requirements. As an example, analysis of a specific case study, maintaining a specified miles in trail, will be discussed. Based on these results, the capability of automatic control theory analysis tools will be discussed, as will its limitations due to modeling assumptions.

The second section will discuss the different mechanisms by which information about the other aircraft may be given to the pilot. While much attention has been paid to the ability of CDTI to present such information, it will also be suggested that air traffic control procedures – the proscribed set of tasks required for a particular ATM operation -- are a common source of traffic information for pilots, both through the communications that they establish and through the trajectories they pre-specify for all pilots. The type and reliability of information available through procedures will be detailed and compared with the information typically given by CDTI.

The final section of the paper will discuss the need for simultaneous analysis of CDTI information content and procedures. This simultaneous analysis is important during design of CDTI. This analysis may also be

0-7803-5086-3 /98/$10.00

valuable in analyzing changes to air traffic management procedures to identify information implicit in current procedures which may be changed or no longer transmitted to the pilot.

Methods of Determining Information Requirements

Several methods exist for ascertaining the information requirements for CDTI implementations. Each have strengths and weaknesses, highlighting the value in combining these methods for comprehensive results. Three of the most common include historical experiences, expert opinions, and experimentation with representative prototypes. The above three analysis methods are frequently used in an ad hoc manner during research and design. A variety of more structured analysis methods have also been documented. In the human-machine systems domain, these methods have relied upon normative models such as the Operator-Function Model [1], and more descriptive methods such as Ecological Task Analysis [2].

The range of available methods leads designers to another ad hoc process: method and model selections. Which model is appropriate for which situation depends strongly on the nature of the task. For determining information requirements for CDTI implementation, control theory analysis can serve as a valuable baseline.

Overview of Control Theory Analysis for Air Traffic Management Tasks

The applicability of control theory analysis stems from the manner in which it intrinsically incorporates the dynamic and temporal behavior of the complete air traffic management system.

The dynamic nature of the air traffic management system has several implications. Most apparent is the extent to which the relative positions of aircraft can change. An important element of the dynamic behavior is the large 'control effect' that the pilots, individually and collectively have over the system; not only are pilots reacting to situations, they are creating and modifying them. As such, any modeling of the environment needs to account for the coordinated evolution of system state and controller actions, rather than examining one-by-one discrete reactions to otherwise static conditions.

Timing of reactions is another important factor in many ATM tasks. The timing of pilot reactions can place profound constraints on their subsequent decision options; for example, the later the reaction to a loss-of-separation event, the greater the severity of a minimally acceptable resolution maneuver. Time-based modeling is required to understand these relationships.

A related effect is the need for continuous monitoring and frequent updates to command decisions. Unlike interaction with modal systems, ATM tasks are not well approximated by discrete models which typically assume a decision is made, and completed. A continuous-time model is instead needed to describing evolving system behavior.

Central to control theory analysis techniques is the numerical "Open Loop" model of the system's dynamic behavior through time. For modeling ATM operations, this dynamic model would comprise equations defining aircraft trajectories in response to pilot or autopilot controls.

This model of open-loop behavior is then linked with a controller, creating the 'Closed-Loop' model shown in Figure 1. This controller is assumed to access to some set of information; based on this information, it generates control commands, which, completing the circle, then modifies the subsequent dynamic behavior. The controller may be a pilot, a remote operator or some form of automation.

A controller may be based on several different algorithms. The simplest is a proportional controller which generates a control input equal to the error between its current and target states, multiplied by a controller gain. More complex controllers can be developed. Some use more sophisticated

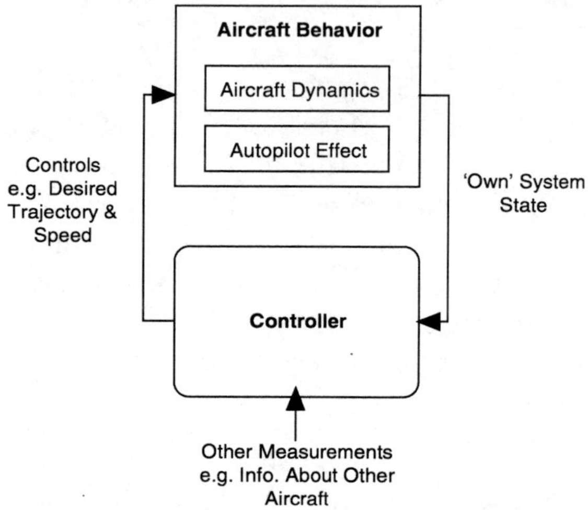

Figure 1. Closed Loop Control System

algorithms, while others attempt to estimate system states which they can not directly sense.

The dynamics of the closed loop control system can be simulated, providing close approximations to the continuous behavior. This encapsulates the continuous interplay between control and resulting behavior.

Control theory analysis can provide three items: general insight into the behavior of the task in general; specific identification of the upper bounds on performance achievable when any controller, be it human or automatic, attempts to control some aspect of the system based on a limited information set; and insight into how the behavior of one aircraft affects neighboring entities.

Case Study: Control Theory Analysis of Maintaining In-Trail Separation

The task of maintaining a specified in-trail separation requires pilots to space themselves longitudinally along an airway. Like many other ATM tasks for which CDTI implementation can be motivated, this task has several characteristics that make control theory analysis a useful early design tool.

Active control is required in several situations. Aircraft need to initially acquire their place in line or their specified spacing may be changed, requiring a re-acquisition of a new spacing. Pilots also need to monitor and respond to actions by other aircraft, such as instances where the preceding aircraft reduces speed in an arrival sequence.

For this analysis, a basic aircraft model was programmed for both the 'lead' aircraft and the 'own' aircraft following behind. The aircraft models use autopilot target velocity as an input. The autopilot, to capture this target velocity, commands an engine throttle setting which, through a lag time due to engine spool up, drives velocity and longitudinal position through thrust. The lead aircraft, at time '0', receives a command to slow from 600 fps (350 knots) to 425 fps (200 knots); the response of the own aircraft was then recorded.

The own aircraft was to maintain an in-trail spacing of 30000 feet (about 5 nm). A variety of simple controllers were simulated, each with different information sets available about the lead aircraft. These controllers calculated a target velocity for the own aircraft, which was then acted upon by its autopilot. The following sections indicate the achievable performance of these controllers .

Controller With Knowledge of Position Only

In this case, the own aircraft's controller is presumed to know only 'x', the distance the other aircraft is ahead. Based on this information, the controller calculates a target velocity according to the formula:

$$V_{Target} = V_{Own} + K_x(x_{Actual} - x_{Required})$$

Kx is the 'controller gain', describing how much compensation is added to the target velocity for errors in position.

The achieved error in position is shown in Figure 2. An oscillatory response was found which no values of the controller gain Kx can eliminate; instead the controller gains affect the magnitude and frequency of the oscillations. The throttle setting of the own aircraft is shown in Figure 3, which shows stop-to-stop overcompensation by the controller. These results highlight the need for the controller – in actual operations the pilot – to consider more than raw position information.

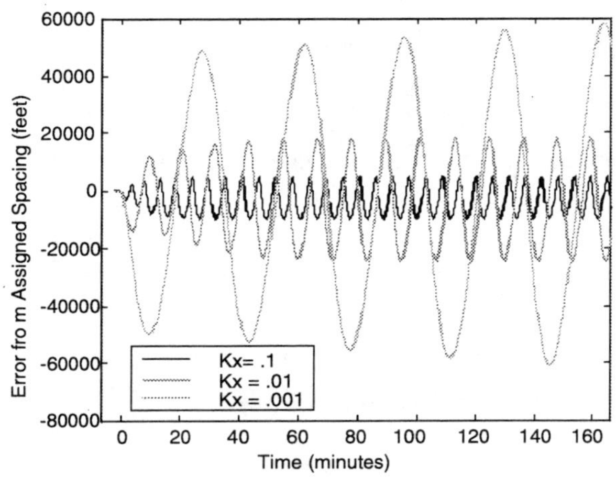

Figure 2. Error in In-Trail Separation, With Control Based on Knowledge of Lead Aircraft Position Only, For Different Control Gains

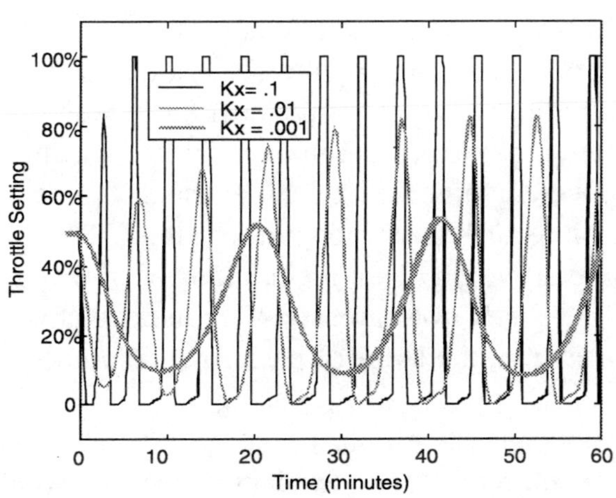

Figure 3. Own Aircraft Throttle Settings, With Control Based on Knowledge of Lead Aircraft Position Only, For Different Control Gains

Control With Knowledge of Position and Lead Aircraft Velocity

With knowledge of the lead aircraft's velocity, the target velocity can be set as:

$$V_{Target} = V_{Own} + K_x(x_{Actual} - x_{Required}) + K_v(V_{Lead} - V_{Own})$$

Kv is an additional controller gain which compensates for differences in velocity between the leading and following aircraft.

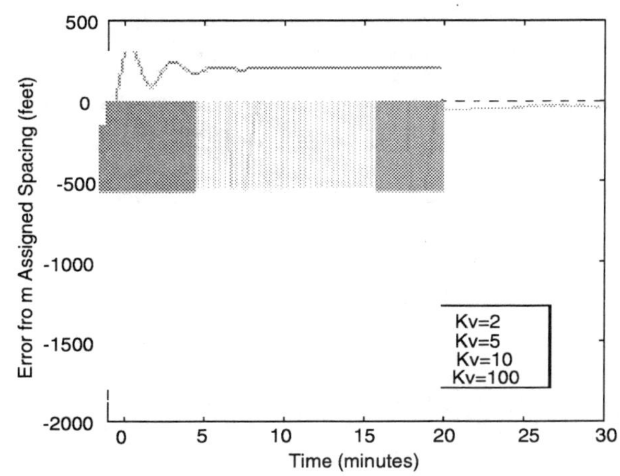

Figure 4. Error in In-Trail Separation, With Control Based on Knowledge of Lead Aircraft Position and Velocity, For Different Control Gains

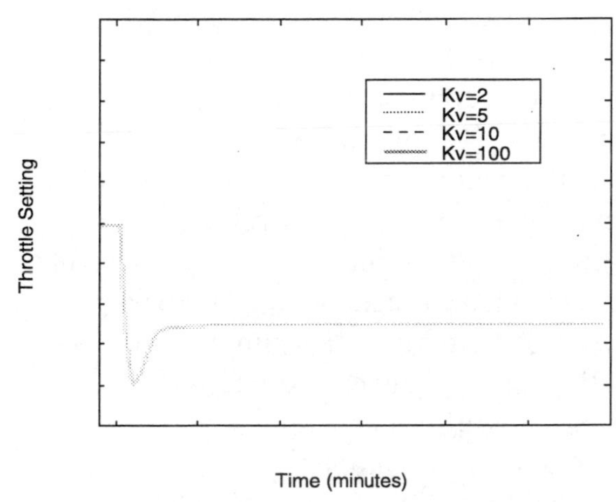

Figure 5. Own Aircraft Throttle Settings, With Control Based on Knowledge of Lead Aircraft Position and Velocity, For Different Control Gains

The achieved error in position is shown in Figure 4. The peak error is much smaller, and, depending on Kv, the desired spacing can be quickly attained. The throttle commands of the own aircraft are shown in Figure 5.

In this case, the contribution of the velocity information to the controller is sufficiently significant that reasonable changes in Kx were not found to cause noticeable changes in the system behavior.

Control With Knowledge of Position, Lead Aircraft Velocity, and Lead Aircraft Target Velocity

The target velocity can be further modified such that it depends primarily on the lead aircraft target velocity, with corrections for errors in position and velocity, as follows:

$$V_{Target} = V_{TargetLead} + K_x(x_{Actual} - x_{Required}) + K_v(V_{Lead} - V_{Own})$$

With perfect knowledge of target velocity, the achieved error in position is zero, as shown in Figure 6. Of course, in this analysis the two aircraft had identical responses to a change in target velocity; with different aircraft with different dynamic properties, some temporary error may arise due to one aircraft responding quicker than the other.

Care should be paid, however, to the effect that errors in the target velocity information have on the ultimate in-trail spacing. This controller depends so heavily on the velocity target of the lead aircraft that any error in that value causes steady-state errors in the spacing; the position and velocity error correction terms are not sufficiently powerful to override the inaccuracies effect. Human pilots are typically better than this simple controller at canceling steady-state errors, but to do so requires resolving an apparent conflict between two pieces of information.

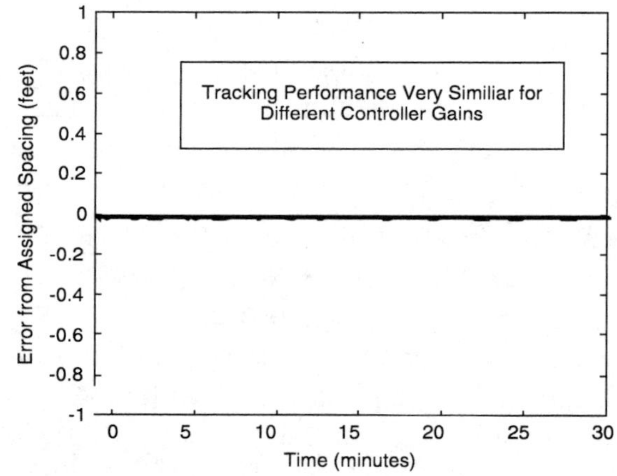

Figure 6. Error in Longitudinal Separation, With Control Based on Accurate Knowledge of Lead Aircraft Position, Velocity and Target Velocity

Summary of Two-Aircraft Cases

To summarize the previous three controller examples, three different controllers were tested, each of which used a different information set about the lead aircraft. The first controller used only information about the relative position of the lead aircraft, and was not able to react in a stable manner to a change in the lead aircraft's velocity.

The second controller attempts to both minimize the difference in position and in velocity between the lead and following aircraft, and has a stable response. Its performance can be evaluated by the maximum error in position and by the settling time (time required to converge on the desired separation). Different pairs of maximum error and settling time were attained with each of the different selections of Kv, as shown in Figure 7. The ideal pair would be a maximum error of 0 and a settling time of 0; however, with its available information, the locus of performance points achievable by a controller using only current position and velocity information can not reach this point, requiring instead a trade-off between the two performance measures.

When perfect knowledge of the other aircraft's target velocity is known, the controller's performance is shown in Figure 7 by the point at the origin. However, this controller depends strongly on the accuracy of

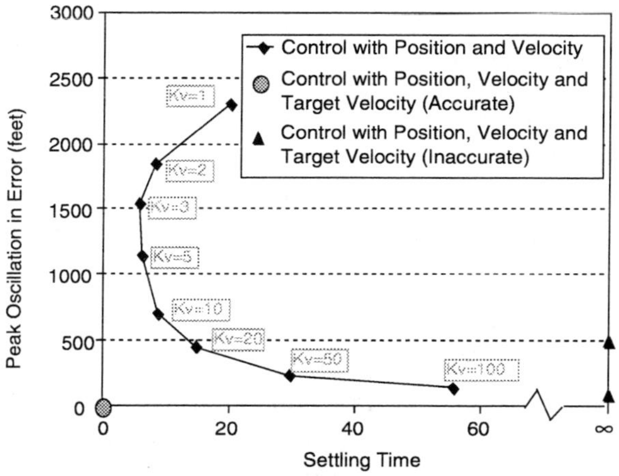

Figure 7. Comparison of Peak Oscillation and Settling Time for Different Controllers

the target information; if this information is inaccurate, then an error which never settles out can be created, indicated by the triangles in Figure 7 on the right side of the graph.

Extrapolation to Multiple Aircraft

The analysis described to this point examined only one 'own' aircraft's reaction to a deceleration by a 'lead' aircraft. This analysis was extended to a string of 10 aircraft, each which was to maintain the same in-trail separation within its place in line. The lead aircraft then decelerated in the same manner as in the previous analyses; all of the following aircraft used a position and velocity controller, with Kv set to a value of 5.

The reactions along the chain of aircraft were found to grow in magnitude and in settling time along the line. This effect is shown in Figure 8. While the separation between aircraft 1 and 2 is the same found earlier, the subsequent aircraft then start to react to the one shown before; their reactions grow in magnitude from an error of 1000 feet for aircraft 2 to an error of 4000 feet for aircraft 10. The responses of the later aircraft are also delayed as the 'information' about the deceleration ripples down the line. In addition, the behavior of the later aircraft is no longer smooth, but starts showing some oscillations in their response.

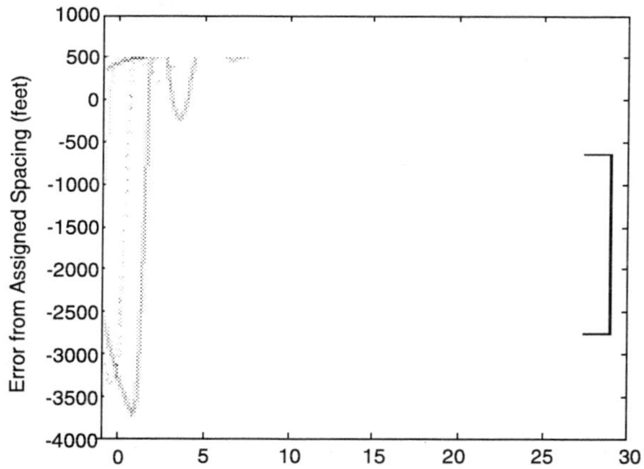

Figure 8. Error in Longitudinal Separation, Within a Line of 10 Aircraft (Aircraft #1 is Lead Aircraft)

Capabilities, Limitations and Further Extensions of Control Theory Analysis

Many fundamental assumptions implicit in control theory models, placing limitations on their resulting analysis. First, any model of a controller can not exactly specify what a human operator would do, only what they could do; this can require examining a large number of controller behaviors to examine the range of possible system behaviors. Second, just as discrete models can not easily capture the continuous trajectory behavior of the aircraft, traditional control theories are not usually well adapted to incorporating discrete decisions.

Within these limitations, control theory analysis can serve as an initial baseline, identifying the upper bound on the achievable performance of any controller. While it has the disadvantage that it does not provide a description of expected human behavior, it can identify design solutions that are not worth pursuing early in development. It also has the advantage of using established numerical techniques and modeling tools that can test possible control strategies – with their correlated available information sets – in a matter of hours. The results of control theory analysis can then highlight experimental issues for human-in-the-loop evaluations.

The case study in this paper only used very simple controllers. Such an analysis can be extended by using a number of more sophisticated control algorithms, such as:

- Controllers that can perform negotiation with other aircraft.
- Controllers that can guard against steady-state errors and adapt to inaccuracies in the available information.
- Controllers that estimate missing information, simulating a pilot's ability to infer rate information from memory of transitions from recent to current states.
- Controllers that only operate at discrete times, simulating pilots who can only attend to the task occasionally, or control using information that is updated at intervals.

Sources of Information: Procedures

CDTI is not the first mechanism for presenting traffic information in the cockpit. While it can provide the most compelling, most persistent, and most informative presentation of traffic information, much valuable information is, and has been, available both implicitly and explicitly from ATM procedures.

This section will discuss the types of information available from procedures. Then, the next section will propose methods of coordinating the information content of CDTI and procedures and guarding against unanticipated effects.

In this paper, the term *operations* has been used to describe the desired outcome within a traffic flow, such as having aircraft follow each other with a given in-trail separation. The term *procedure* refers more specifically to the proscribed actions required of each person in the ATM system to execute an operation. Because several different procedures may be possible to perform the same operation, regulatory bodies mandate one procedure. In doing so, they create a shared set of knowledge and expectations between pilots, and between pilots and controllers. Procedures thereby create a framework through which information can be passed.

The segment of the LARKS One Standard Terminal Arrival Route, shown in Figure 9, illustrates some of the information content in procedures. Even without any outside knowledge of other aircraft, when flying inbound from the LARKS intersection towards the FQF VOR (navigational station) pilots are given an expectation of what speeds they – and the neighboring aircraft – will fly, and what altitudes they will be at throughout the arrival. These expectations may be increased when pilots are familiar with the *modus operandi* of a particular airport; the author noted in jumpseat observation flights cases where pilots anticipated changes in the traffic flow with comments such as "the controllers at this airport always slow us at this intersection",

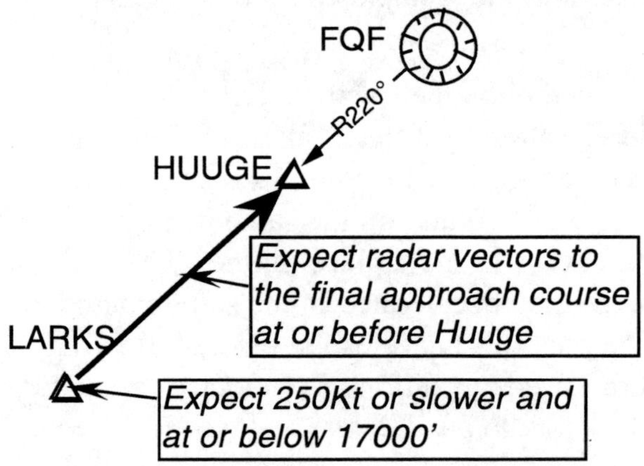

Figure 9. Schematic of LARKS One Arrival to Denver International Airport

information attainable to them only through repeated experience.

Procedures also provide information through the communications they require. This information may be explicit, such as conditions where the controller tells the pilot directly relevant information about other aircraft. For example, a pilot may be instructed "You are following a 767 heavy, 3 mile final", thereby

0-7803-5086-3 /98/$10.00 ©1998 IEEE

telling the pilot not only where the traffic is, but also allowing the pilot to anticipate what spacing the controller will steer him or her towards to meet wake vortex separations.

The information in procedural communications may also be implicit. For example, in surveys pilots reported that hearing position reports and instructions given to other aircraft help them not only with traffic separation but also with anticipating ATM commands for sequencing and spacing. [3]

The traffic information available from CDTI and from procedures differs in two ways. First, the procedural information tends to be *normative*, meaning it presents what the other aircraft ought to be doing; CDTI presents *descriptive* information, illustrating what the other aircraft are actually doing. Second, CDTI provides a *persistent* source of information, which pilots can refer to when desired. Information from procedures, on the other hand, may not be available when pilots can attend to it, limiting its effectiveness [4].

Simultaneous Design of CDTI and Procedures

The fact the CDTI and ATM procedures both provide traffic information suggests that they should be considered simultaneously during design and planning. Two design goals are possible: creating positive interactions between CDTI and procedures; and guarding against unanticipated interactions.

Creating Positive Interactions

The availability of information from two sources has the potential to add safety and efficiency to ATM operations. The two sources provide redundancy, and allow may allow for pilots to receive the information through multiple modalities, such as visual displays and (aural) voice communications.

These two information sources can potentially complement each other because of the different types of information they can provide. By combining a display of descriptive information with knowledge of normative conditions, pilots can both anticipate situations, and monitor for deviations from procedures that might impact their operations.

Guarding Against Unanticipated Interactions

Unanticipated interactions may arise between the information from procedures and from CDTI for three reasons. The first may arise when CDTI information suggests pilot actions not included in the procedures. The second arises from conflicting information, such as the case study's condition where the target information was not matched by the actual lead aircraft velocity.

A third effect may be found when procedures are changed. In prior operations, a CDTI system may have been judged sufficient, but with transition to more flexible procedures which does not pre-specify the actions of neighboring traffic, it may be discovered that pilots had been supplementing display content with procedural content now not available.

Further Research

The ability of control theory analysis warrants further study. It was proposed that this analysis can identify the minimum information required for ATM operations, and its trade-off with performance. Comparison with experimental results could examine how the results of this analysis may also serve as a predictive measure of pilot behavior. Further analysis may provide more detailed insight.

This paper also theorized that ATM procedures also provide traffic information to pilots, in addition to CDTI implementations. An experiment is under development to test this theory and to compare the manner in which traffic information from these two sources is used and trusted by pilots.

Acknowledgements

This work was supported by NASA grant NAG 2-1178, monitored by Dr. Walter Johnson of the NASA Ames Research Center.

References

[1] Mitchell, C.M. "GT-MSOCC: A Domain for Research on Human-Computer Interaction and Decision Aiding in Supervisory Control Systems", *IEEE Transactions on Systems, Man and Cybernetics*, 17(4), 1987

[2] Vicente, K.J. & Ransmussen, J. "Ecological Interface Design: Theoretical Foundations", *IEEE Transactions on Systems, Man and Cybernetics*, 22(3), 1992

[3] Pritchett, A.R. & Hansman, R.J. "Variations Among Pilots From Different Flight Operations in Party Line Information Requirements for Situation Awareness", *Air Traffic Control Quarterly*, 4(1), 1997

[4] Midkiff, A.H. & Hansman, R.J. "Identification of Important "Party Line" Information Elements and Implications for Situational Awareness in the Datalink Environment", *Air Traffic Control Quarterly*, 1(1), 1993

EXPERIMENTAL STUDY OF HELMET-MOUNTED DISPLAY SYMBOLOGY FOR TERRAIN AVOIDANCE DURING LOW-LEVEL MANEUVERS

Jonathan B. Taylor, MIT, Cambridge, MA
James K. Kuchar, MIT, Cambridge, MA

Abstract

A simulation study was performed to evaluate Helmet-Mounted Display (HMD) symbology for fighter aircraft Ground Collision Avoidance Systems (GCAS). Several candidate visual alert symbologies were tested, including an iconic cue, an aircraft-fixed guidance cue, a head-fixed guidance cue, and a head-fixed guidance cue with a pitch ladder. Subjects were given audio and visual side-tasks and responded to alerts in different aircraft attitudes. Shorter reaction times were observed when a head-fixed visual alert was given in addition to the standard audio alert. No significant differences in recovery performance (measured by response time and altitude loss) were observed between display formats. However, the display format had a significant effect on subject head motion. With head-fixed guidance cues, pilots were able to perform the recovery maneuver with their heads off axis; with aircraft-fixed or no guidance cues, pilots consistently turned their heads to face forward before maneuvering the aircraft.

Introduction

Controlled Flight Into Terrain (CFIT) is the second largest category of tactical aircraft Class A mishaps in the U.S. Air Force[1] [1]. CFIT has cost the Air Force an average of $180 million per year over the last decade, and has claimed 160 lives [2]. Though civil air transportation CFIT accidents occur at the rate of approximately 5 aircraft per year [3], Navy, Marine Corps, and Air Force aircraft have averaged 8 to 8.7 CFIT losses per year over the last decade [2,4].

Ground Collision Avoidance Systems (GCAS) were developed to combat the CFIT problem by providing pilots with time-critical warnings of an impending ground collision. Since their recent implementation in many military aircraft, the CFIT accident rate has declined. Despite the success of GCAS, problems remain which hinder the complete elimination of CFIT accidents. GCAS reliability is being enhanced by implementing the ability to scan ahead of the aircraft for terrain hazards. The development of accurate Global Positioning System (GPS)/Inertial Navigation System (INS) avionics coupled with Digital Terrain Systems (DTS) is making GCAS more reliable and accurate. However, the potential for improvements still exist regarding the pilot interface.

The ability to alert the pilot to a terrain hazard and provide the information that enables him or her to react quickly and accurately is a difficult challenge in modern fighter aircraft. Military missions often involve deliberate low-level flight in an information-intensive environment. These factors lead to high pilot workload, and dangerous situations can arise quickly and unexpectedly. A total of 44% of F-16 Class A mishaps from 1980-1989 occurred during combat maneuvering, and task oversaturation was cited as a contributor in over 20% of the accidents [5,6]. Additionally, channelized attention (where a pilot becomes so focused on a particular task that flying the

[1] involving loss of life, permanent disability, total aircraft loss, or over $1 million in damage

0-7803-5086-3 /98/$10.00 ©1998 IEEE

aircraft becomes secondary) was deemed a definite contributor in over 60% of the F-16 mishaps [5].

Current GCAS alerts consist of an audio alerting message and limited, iconic visual cues. Alerts may go unnoticed due to an information or work overload, and the pilot may react more slowly to alerts that are detected. A proposed method for addressing this problem is through the use of Helmet-Mounted Display (HMD) technology. Unlike other visual display formats, HMDs can display visual information in the pilot's field of view, regardless of his or her head position. In a fighter aircraft where pilots spend a great deal of time looking off-boresight, such a system could be an effective means of enhancing GCAS. This paper documents a simulation study of alert presentation issues for an HMD-enhanced GCAS using prototype alert and guidance displays.

Experimental Design

In the simulation study, pilots flew a T-38 simulator through a series of terrain encounters while performing visual and aural side tasks. In order to ensure consistent aircraft attitudes for testing alert responses, an autopilot flew the simulator through a series of maneuvers close to the ground. At certain predefined conditions during the testing, a simulated GCAS alert was given, consisting of a voice alert and one of several sets of visual symbologies. When the alert was given, control of the aircraft was transferred to the subject, who then flew a standard recovery maneuver. Aircraft state variables as well as the pilot's stick inputs and head position were continuously recorded.

Simulator Facility

The study was performed using a fixed-base T-38 simulator developed by The Charles Stark Draper Laboratory, using an HMD coupled to a head tracker. All of the visual graphics were presented to the pilot using two 1.3 in Liquid Crystal Displays in the HMD at a 480 x 240 color pixel resolution, corresponding to a 50° x 40° field of view. The HMD also featured earphones for the transmission of audio signals.

A pre-rendered scene centered at Randolph AFB in San Antonio, TX was chosen as a basis for the experiment because of its flat, featureless terrain. The terrain was textured green, and created the illusion of flat, grass-covered rolling hills. All buildings, runways, roads, clouds, and other structures were eliminated from the scene. The sky was colored pale blue, similar to a clear midday. A well-defined horizon was visible with no haze or fog effects. An opaque polygon was added to represent the inside of the cockpit, and obscured the pilot's view at angles greater than 45 degrees below the horizontal.

The only flight information symbology the pilot received at all times was a flight path vector. This symbol (using a conventional military icon) was displayed along the aircraft's nose and moved with regard to the aircraft angle of attack. Sideslip motion was not incorporated in the flight path vector. No head-down cockpit instrumentation was displayed in the scene.

A Logitech head tracker was used to determine pilot head orientation and operated via ultrasonic signals. The ultrasonic signals were received by a sensor mounted on the top of the HMD which resolved head angles at a rate of 50 Hz.

Experimental Protocol

During an experimental run, the autopilot flew the aircraft through prerecorded trajectories (flight profiles) while the pilot accomplished audio and visual side tasks. Four flight profiles were used, each taking from 80 to 155 sec. Each profile involved a series of pitch and roll changes of the aircraft and ended in specified pitch, roll, and altitude conditions as shown in Table 1. These profiles were designed to end in aircraft states for which a recovery maneuver would have to begin within 2 sec in order to provide at least 500 ft separation from terrain.

Table 1. Flight Profile End States

Profile	Pitch	Roll	Altitude
1	-10°	-30°	900 ft
2	-10°	60°	1000 ft
3	-20°	30°	1300 ft
4	-20°	-60°	1900 ft

At the end of each profile, when the specified attitude and altitude conditions had been reached, a GCAS alert was issued in one of five different formats. The pilot then was given control of the aircraft and flew a recovery maneuver. After the recovery, control was returned to the autopilot, and another profile and recovery was flown. The four profiles were each flown in random order using a given alert format. The order of the five alert formats was then counterbalanced to reduce learning effects.

Visual Side Task

The visual side task required the pilot to designate targets using a targeting circle affixed to the center of the pilot's field of view. Twelve targets (six 'friendly' red circles and six 'enemy' red squares) were constantly displayed around the aircraft. To divert the pilot's attention away from the front of the aircraft, the targets appeared in two areas to either side of the aircraft. Pilots designated targets by positioning their head so the targets fell within a targeting circle and depressing the top thumb button on the stick in the T-38 cockpit. When a target was designated by the pilot or if it left the defined area, a new target appeared randomly to the side of the aircraft.

Audio Side Task

The audio task required the pilot to listen to a series of callsigns and respond when the pilot heard his or her own callsign. The pilot received a constant stream of callsigns through the headphones on the HMD. The callsigns were recorded at an even rate in a male voice without inflection. Each was generated in random order at random intervals between 1.5 and 3 sec and had a duration of approximately 0.5 sec. The pilot was instructed to respond to their own callsign, 'Falcon 3', by pressing the side thumb stick button within one second of hearing it. Six other incorrect callsigns were used to increase the difficulty of the task: 'Falcon 1,' 'Falcon 4,' 'Eagle 1,' 'Eagle 3,' 'Farmer 2,' and 'Farmer 3.'

Alert Display Formats

Five alert display formats were used, summarized in Table 2. All displays included an audio alert which consisted of a voice message, 'Pull up! Pull up!,' spoken urgently in a woman's voice and repeated once after a 1.75 sec delay. Each audio alert message lasted approximately 0.75 sec.

One display format (A) did not include a visual component. The other four formats included some form of visual display in addition to the audio message. The visual alert display only appeared once the GCAS alert was triggered, and the alert symbology was removed once the subject had reached a steady wings-level climbing attitude.

The second alert format (X) included the audio warning and a break-X icon displayed in the center of the pilot's field of view. The break-X consisted of two intersecting lines forming an X, corresponding to commonly-used Head-Up Display symbology.

Table 2. Alert Display Formats

Display Format	Audio Alert	Visual Alert
A	Yes	None
X	Yes	Head-fixed break-X
AF	Yes	Aircraft-fixed guidance cue
HF	Yes	Head-fixed guidance cue
HFP	Yes	Head-fixed guidance cue with pitch ladder

The last three formats included a prototype guidance cue to aid the pilot in rolling the aircraft level and in performing the pull-up maneuver. This guidance cue, shown in Fig. 1, was integrated with the aircraft flight path vector and consisted of three components. First, an arrow extended from the flight path vector to indicate the direction of nose travel when the stick was pulled back. The second part of the symbology were two roll bars that extended from the flight path vector circle to indicate the direction to which the pilot should roll the aircraft. The bars remained fixed with respect to the horizon at angles of ±10° from vertical. As the aircraft rolled, the bars would remain fixed relative to the horizon, but would move with the flight path vector and pull arrow when the aircraft was pitched or yawed. The third part of the symbology was a pitch box positioned 20° above the horizon which functioned as a target pitch command.

In the AF display format, this set of guidance symbology was drawn in the aircraft-fixed reference frame, and could only be seen when the pilot was looking towards the nose of the aircraft. However, when an alert occurred, the targeting circle used for the visual side task was removed, providing an implicit visual cue regardless of the pilot's head position.

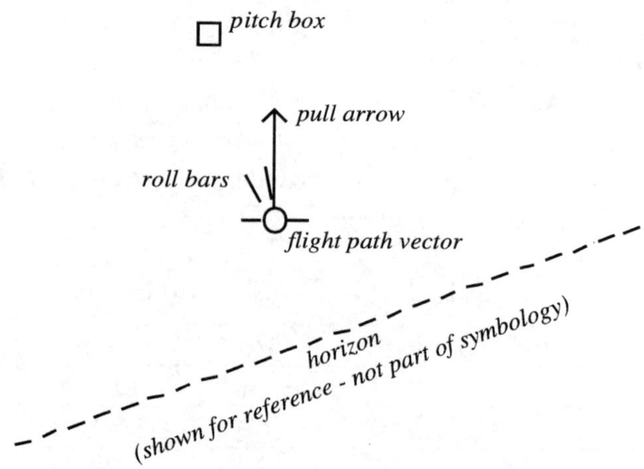

Figure 1. Alert Guidance Symbology
(aircraft shown pitched up and rolled right)

In the HF display format, the same guidance symbology was drawn in the head-fixed reference frame and always remained centered in the pilot's field of view. The guidance symbology orientation operated with respect to the aircraft body, regardless of the pilot's head position. Therefore, a mismatch between the symbology and the outside view would occur unless the pilot's head was centered forward. For example, if the pilot was looking directly out the right side of the aircraft and commanded a roll to the right, the symbology would show the roll marks and pitch box (if visible) rotating to the left, but the horizon the pilot viewed would be rising. Also, the aircraft-fixed flight path vector was not removed when this symbology was present. So, if a pilot was looking towards the nose of the aircraft, two flight path vectors were visible: one which moved with the aircraft and one with the attached symbology which moved with the pilot's head.

The final alert format (HFP) was identical to the head-fixed guidance cue (HF) except for the addition of a pitch ladder. The pitch ladder (Fig. 2), like the flight path vector, pull arrow, roll marks, and pitch box, was drawn in the head-fixed reference frame, but operated with respect to the aircraft body.

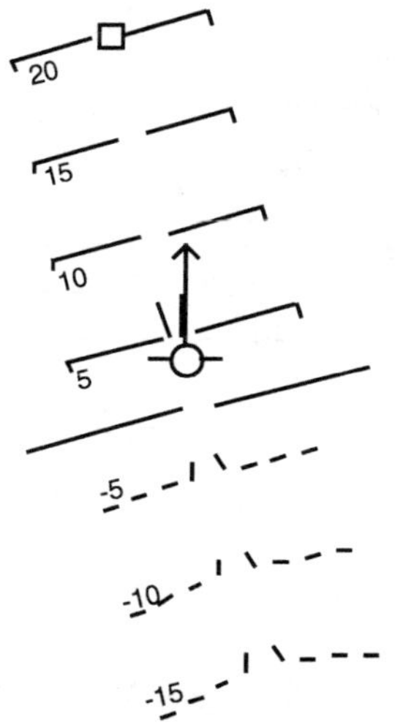

Figure 2. Head-Fixed Alert Guidance Symbology With Pitch Ladder

Results

Data were collected from 13 subjects in this study. Subjects ranged in age from 20 to 63, averaging 39, and ranged in powered flight experience from 77 to 3000 hrs, averaging 675 hrs. Subjects also varied in pilot ratings. All possessed a minimum of a Private Pilot License, 4 subjects had civilian instrument ratings, and 6 subjects had military flying experience, 3 of which had jet experience.

One focus of the study was the pilot's ability to recover from a terrain threat while looking off axis at the time of the GCAS alert. Accordingly, the data that are presented only include those cases in which the pilot was not viewing the aircraft's nose[2] when the GCAS alert was issued.

Reaction Time

Reaction time was defined as the time from when the GCAS alert was issued until the pilot moved the stick away from the neutral position. Average reaction time over all flight profiles for each display configuration is shown in Fig. 3.

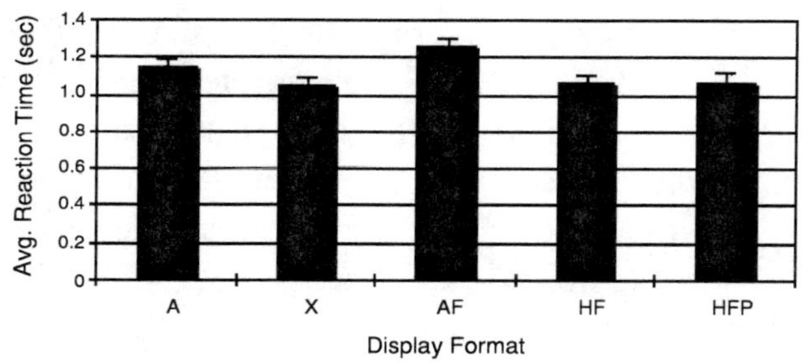

Figure 3. Average Reaction Time
(Bars show 1σ of estimate of mean)

Average reaction time was approximately 0.08 sec faster for display formats using head-fixed symbology (X, HF, and HFP) than for display formats with aircraft-fixed or no symbology (AF, A). These differences were statistically significant for the AF vs. X, HF, and HFP displays ($p < 0.01$). However, the differences between display format A and X, HF, or HFP were not statistically significant ($p > 0.05$). These results compare favorably with prior research evaluating the benefit of including visual alert information simultaneously with aural alerts [7].

Recovery Performance

Several metrics were used to examine the performance of the subjects in terms of their ability to recover and escape from the terrain threat. These metrics included: altitude loss following the alert; pitch and roll histories; time until positive pitch was attained; time until roll angle was reduced below 15°; and magnitude of roll angle overshoot during the recovery. Using these metrics, no statistically significant differences were found between display formats, suggesting that the pilots were able to perform the required roll and pitch maneuvers at a consistent level of performance regardless of alert format or guidance cue.

Pilot Head Motion

Statistically significant differences between displays was observed, however, in terms of pilot head motion following an alert.

[2] Corresponding to head yaw angles greater than 30° or head pitch angles greater than 25°

Fig. 4 shows the average head centering time for each display configuration. Head centering time was defined as the time from when the alert was issued until the pilot's head moved within the head yaw and pitch limits necessary to view the nose of the aircraft (±30° yaw, ±25° pitch).

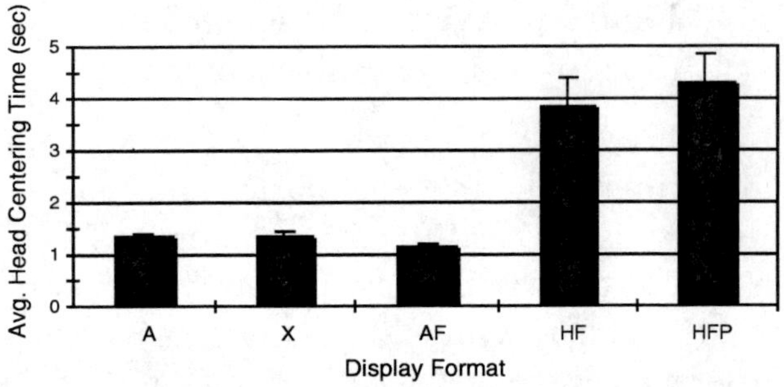

Fig. 4. Average Head Centering Time
(Bars show 1σ of estimate of mean)

As shown in Fig. 4, head centering times were significantly larger for the head-fixed guidance cue cases (HF and HFP), with $p < 0.01$. In other words, pilots tended to move their heads to the front more quickly with the A, X, and AF displays, and to keep their heads off-axis significantly longer with the HF and HFP displays. No statistically significant difference was observed between the HF and HFP display cases.

Additional insight can be gained by examining the time histories of pilot head motion for a few sample cases. Figure 5 shows the head yaw angle histories for all subjects in flight profile #1 (-10° pitch, -30° roll) for the AF and HF cases. In the plots, the alert was issued at time t = 0. The scatter in pilot head positions at times less than zero is a result of the visual side task which generally drove the pilots to keep their heads off-axis.

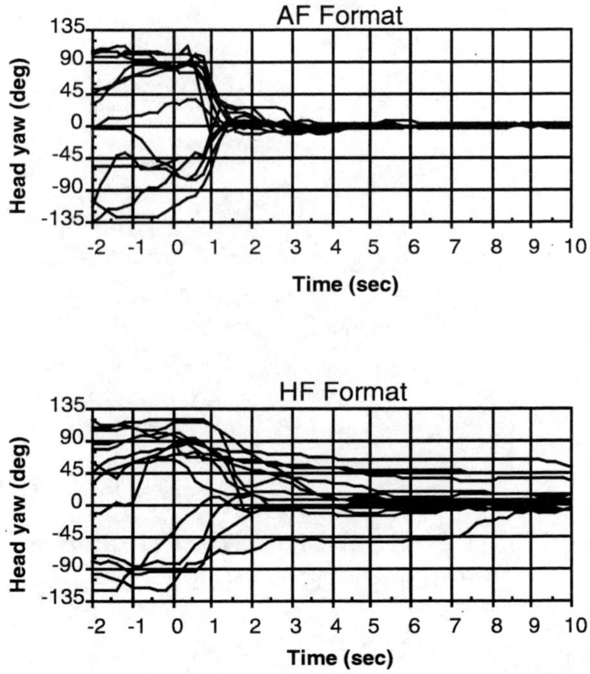

Fig. 5. Representative Head Yaw Histories
(Alert issued at t=0, Flight profile #1)

For brevity, only the plots for the AF and HF cases are shown. Plots for the A and X display formats were generally similar to the plot for AF shown. Likewise, the plot for HFP was similar to that of HF shown. A complete presentation and discussion of all conditions is available from Ref. 8.

These plots show significant differences in pilot head behavior between display configurations. In the A, X, and AF display cases, pilots centered their heads within 2 seconds following the alert. In the head-fixed guidance cue cases (HF and HFP), head behavior was more variable. In addition, the pilots often continued to keep their heads off-axis in the HF and HFP cases even well after the terrain threat had been resolved (approximately 4 sec after the alert was issued). Similar results were observed with respect to pilot head pitch angle and for the four different flight profiles.

Figure 6 shows plots of aircraft pitch angle vs. pilot head yaw angle for the AF and HF cases. As shown, in the AF case, pilots centered their heads rapidly as the pitch angle was increased, with the majority of the pull-up being performed with the head facing forward. In contrast, in the HF case, much of the pull-up maneuver was performed with the pilot's head significantly off-axis. Results for the A and X cases were similar to the AF case shown; results for the HFP case were likewise similar to the HF case shown.

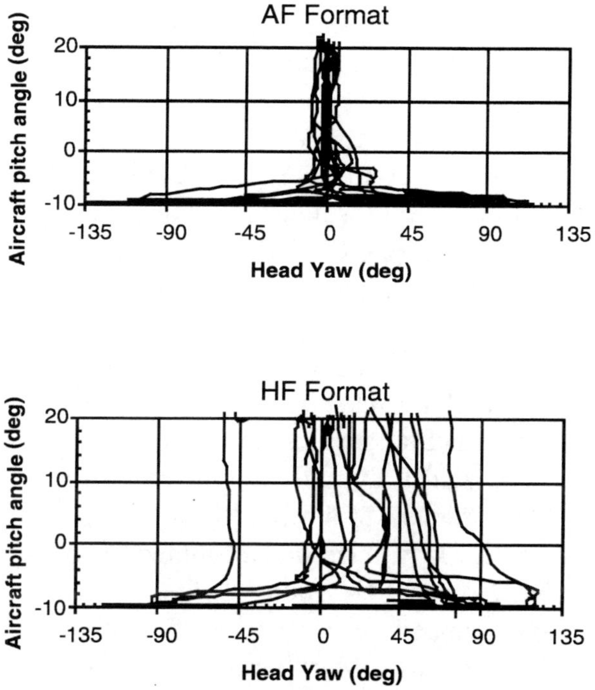

Figure 6. Aircraft Pitch vs. Head Yaw
(Alert issued at t=0, flight profile #1)

Subjective Ratings

The Analytical Hierarchy Process was used to obtain overall display preferences for each subject in terms of percentages [9]. Fig. 7 shows the resultant display preference for the group of subjects.

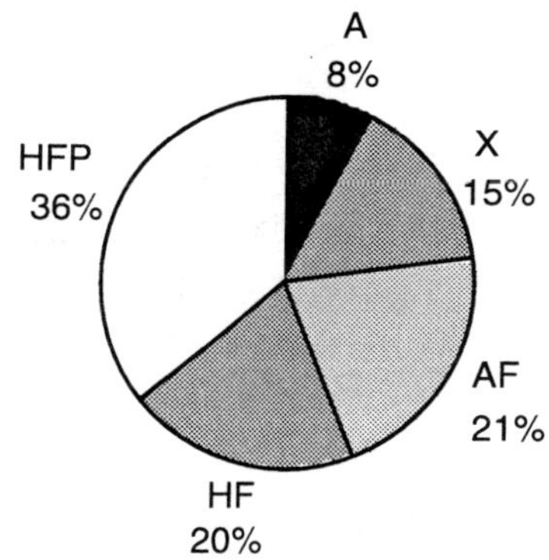

Figure 7. Overall Display Preference

The HFP display was preferred the most overall, followed by the AF and HF displays at nearly equal preferences, and then the X display and the A display. However, subjects largely differed in their individual preferences. Some subjects favored a particular display much more than the other displays. Other subjects tended to prefer particular categories of displays, such as head-fixed information (X, HF, and HFP), or guidance information (AF, HF, and HFP).

Conclusions

There appears to be no significant differences in GCAS recovery performance among the various display configurations tested. The aircraft pitch, roll, and altitude histories are nearly identical for each display configuration at each profile condition.

The major differences found in this study involve the pilot's head motion. Subject head behavior indicates a tendency to fixate on guidance and state symbology. When provided with aircraft-fixed guidance cues, pilots immediately turned their heads to look at the symbology, and continued to closely monitor the symbology during the recovery. When given head-fixed guidance cues, the pilots performed recovery maneuvers at the same level of performance without looking forward. This may imply that head-fixed symbology may enable pilots to perform off-axis tasks (e.g., target monitoring) while performing the pull-up maneuver. Further study is required, however, to determine whether this is actually the case.

Another implication relates to potential effects on the pilot's vestibular system. This study did not reproduce vestibular cues that would be felt while flying an actual aircraft. A GCAS alert and recovery can involve large head and aircraft angular rates. Differences in head motion relative to aircraft motion due to the different display configurations can lead to differences in pilot vestibular cues. Of particular importance are Coriolis or cross-coupling effects in which pilots can feel disturbing false vestibular cues when a sudden head motion is made orthogonal to a large aircraft angular rate [10]. These vestibular effects, which can lead to disorientation and vertigo, must be studied in more detail.

Acknowledgments

This research was performed at The Charles Stark Draper Laboratory, Inc., under Contract 27900. The authors would like to acknowledge Ed Bergmann, Tom McNamara, Tina Saunders, John Danis, Ed Bachelder, Dave Hauger, Wade Hampton, Linda Leonard, and Mario Santarelli from Draper Laboratory. The authors also thank Capt. Steve Jacobson, USAF, Dr. Mike Borowski of the Naval Safety Center, and DJ Atkins of the Air Force Safety Center. Special appreciation goes to the volunteer subject pilots.

References

1. Shah, D., *Ground Collision Warning System Performance Criteria for High Maneuverability Aircraft*, Wright-Patterson AFB, OH, Dec, 1988.
2. Statistical data from DJ Atkins, Air Force Safety Center, Kirtland AFB, NM.
3. O'Brien, J. and W. Edmunds Jr., "New Aircraft Technology and Associated Flight Crew Training", FSF 45th IASS & IFA 22nd International Conference, Long Beach, CA, 1992.
4. Statistical data from Dr. Mike Borowsky, Naval Safety Center, Norfolk, VA.
5. Holland, D., and J. Freeman, "A Ten-Year Overview of USAF F-16 Mishap Attributes from 1980-89", Human Factors and Ergonomics Society 39th Annual Meeting, San Diego, CA, 1995.
6. Hewitt, C., Henley, A., and J. Boyes, "A Ground and Obstacle Collision Avoidance Technique (GOCAT)", IEEE National Aerospace and Electronics Conference, 1991.
7. Boucek, G., Berson, B., Po-Chedley, D., and J. Hendrickson, "Aircraft Alerting Systems Standardization Study", 4th AIAA/IEEE Digital Avionics Systems Conference, St. Louis, MO, 1981.
8. Taylor, J. B., "Helmet-Mounted Display Symbology for Ground Collision Avoidance in Fighter Aircraft", CSDL-T-1324, Charles Stark Draper Laboratory, Cambridge, MA, September, 1998.
9. Yang, L. and R. J. Hansman, "Application of the Analytic Hierarchy Process for Making Subjective Comparisons Between Multiple Automation/Display Options", 6th IFAC/IFIP/IFORS/IEA Symposium, Cambridge, MA, June 27-29, 1995.
10. Boff, K., and J. Lincoln, *Engineering Data Compendium: Human Perception and Performance*, AAMRL, Wright-Patterson AFB, OH, 1988.

ADVANCED FLIGHT DECK FOR NEXT GENERATION AIRCRAFT

Robert A. Faerber, Rockwell Collins, Cedar Rapids, Iowa.
Timothy J. Etherington, Rockwell Collins, Cedar Rapids, Iowa.

Abstract

Alternative Flight Management System (FMS) human interface and Advanced Primary Flight Display (PFD) concepts are needed as the complexities of understanding and operating increased functionality on equipment continue to burden and confuse an already inundated pilot. In 1997 Rockwell Collins, Inc. designed an advanced FMS display/interface and enhanced an existing 3D PFD that significantly reduced human interface problems between users of these systems. Two independent human factors experiments were conducted in 1997. The first investigated the Advanced FMS Interface display format versus a conventional Control Display Unit (CDU) type FMS. The second examined an advanced 3D PFD, an adaptation of the NASA 3D pathways display, compared to a standard commercial PFD format. Results from both experiments revealed statistically significant findings at the 95th level of confidence favoring these advanced flight deck designs. The Advanced FMS Interface display format produced quicker reaction times and lower levels of mental workload than the conventional CDU. Likewise the advanced 3D PFD format resulted in less tracking error in both lateral and vertical axis, lower levels of mental workload, and increased situational awareness. These statistically significant findings lead us to believe that this type of advanced flight deck design has great potential for significantly reducing and/or eliminating human interface problems between users and the systems they interface with while at the same time reducing pilot mental workload and increasing situational awareness. This report will detail these findings and provide recommendations to cockpit designers of the next generation aircraft.

Advanced FMS Human Interface

The FMS has been scrutinized by leading aviation researchers, pilots, and safety officials for its opaque interface and difficulty of use during nonstandard flight operations. Indeed a leading factor in a fatal airline accident in South America (Strauch, 1997 and Dornheim, 1996) was directly associated with FMS mode awareness problems, non-distinction of different navigational aids in the FMS database, and poor pilot situation awareness. Although the interface between the pilot and the FMS has improved and changes continue to be made the FMS continues to be plagued with human interface problems.

Since 1996 Rockwell Collins Advanced Technology Center has aggressively attacked these human interface problems by designing, developing, and testing an alternative FMS prototype (see Figure 1 for an example of the graphical interface). Our objective has been simple and clear: design an advanced FMS interface that will significantly reduce and/or eliminate human interface problems between users and the FMS.

Progress

An in-depth literature review, user survey, and task analysis were initially completed enabling us to rapidly prototype three alternative FMS candidates for review and evaluation. A final candidate was then selected from these three as the most promising for reducing and/or eliminating the FMS man-machine interface problems. This Advanced FMS Interface prototype was then rapidly designed, subjected to iterative design modifications, and human factors evaluations.

0-7803-5086-3 /98/$10.00

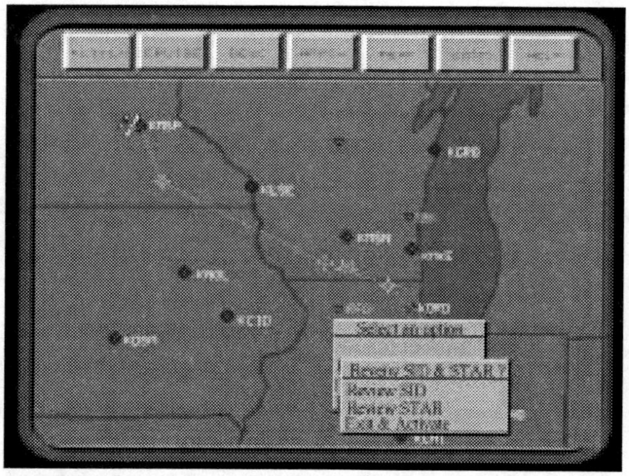

Figure 1 – Entering a SID and STAR into your Flight Plan

This prototype continues to be developed and plans for further human factors evaluations of these extended FMS functions are planned for the Fall of 1998.

Literature Review and Task Analysis

The review of literature (Sarter and Woods, 1994; Degani and Kirlik, 1995; Funk et. al., 1996; Sarter and Woods, 1992; Wickens, 1992; Tenney, Rogers, and Pew, 1995; and many others) has led to the formation of several key human factors and human interface issues that were addressed and applied in the design of the Advanced FMS Interface prototype display.

The major findings from these investigations point toward several human interface issues associated with the FMS. The most prominent are:

1) Knowledge gaps in understanding the functional structure of the FMS
2) Opaque interface
3) Interface too complicated
4) Database errors

To further investigate these findings from the literature review a task analysis was conducted on a representative CDU type FMS. The analysis was completed by utilizing a reference guide for the FMS, an available FMS simulation tool, and consultation with FMS engineers.

The task analysis included a Cognitive Task Analysis to highlight not only the necessary physical tasks or operations required to complete a desired goal, but to indicate the necessary cognitive operations required as well.

The results of the task analysis revealed that certain FMS operations are very lengthy, complicated, ambiguous, and often provide a low degree of feedback to the pilot. This data revealed that the level of cognitive detail coupled with the number of physical procedures is quite high for certain phases of flight, such as the landing phase. These results are commensurate with the human factors research literature previously reviewed.

Human Factors Analysis of FMS Prototype in a Part Task Simulator

Early evaluation of the Advanced FMS Interface was conducted to ascertain whether this new approach in conjunction with the entire flight deck display environment would indeed meet the objectives for this project. Formal human factors testing was initially limited to a relatively small number of FMS functions. This constraint was a result of difficulties integrating the full FMS functionality into the part task simulator.

Method

Participants. Six subjects participated. All subjects were pilots with a mean flying time of 3,250 hours.

Display Formats. Two different display formats were used, the Advanced FMS Interface utilizing a graphical user interface and a conventional CDU type FMS utilizing textual information augmented with color coding cues.

Scenario. The test scenario presented was a simulated flight from Minneapolis-St. Paul International Airport to Chicago O'Hare International Airport.

Experimental Design. The experiment was organized as a 2x5x6 repeated measures design.

0-7803-5086-3 /98/$10.00 ©1998 IEEE

Independent Variable. Display formats were used as factors. The *a priori* hypothesis asserted reaction times and pilot workload ratings would be lower for the Advanced FMS Interface display format than for the textual CDU.

Dependent Variables. Three dependent measures were taken: 1) reaction time to complete a given task, 2) pilot's subjective workload rating for each FMS task, and 3) subjective data taken after completion of all required tasks.

Procedure

Each experimental session included a briefing to the subjects of the nature of the experiment, followed by a brief demonstration/training session of all five FMS operations to each subject. The subjects were then given the opportunity to practice each of the five FMS operations a minimum of five times to allow them to become familiar with the FMS operation and the input device, a touchpad for the Advanced FMS Interface and the keypad for the CDU type FMS. Subjects performed all five FMS operations on the Advanced FMS Interface prototype first, which was integrated in InCS with minimal flight simulation followed by a very short break then completion of the same five FMS procedures on the CDU type FMS incorporated in a hardware in the loop simulation.

Results

Statistically significant findings at the 95th level of confidence using a t-test support our hypothesis that the Advanced FMS Interface display format would produce quicker reaction times and lower levels of mental workload than the CDU type FMS. An ANOVA also revealed a statistically significant main effect for display formats ($F_{1,23}= 4.99$, $p<0.03$). Furthermore, there were statistically significant findings at the 98th level of confidence using a t-test for the three subjective measures of Usability, Learnability, and Overall Attitude, favoring the Advanced FMS Human Interface over the conventional CDU interface (Faerber, R.A. 1998).

These preliminary results were very positive and allowed us to proceed with researching and investigating alternative human interfaces to the next generation FMS.

Expanded FMS Functionality and Human Interface

Additional FMS Functions

During 1998 the Advanced FMS Human Interface prototype has continued to explore alternative interface strategies and incorporate additional FMS features. The FMS functionality has evolved into incorporating icons for rapidly manipulating the display format, graphically editing the flight plan, selecting alternative menu formats, quickly and efficiently completing a hold function using an alternative interface, and rapidly manipulating one's flight path using drag and drop operations (see Figure 2).

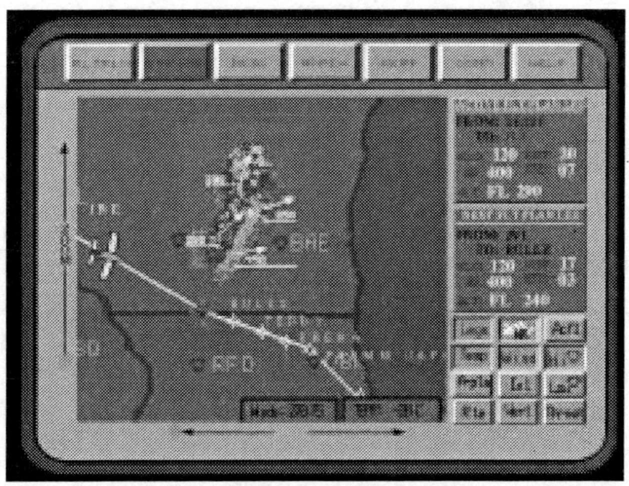

Figure 2 - Example of selecting icons to rapidly manipulate display format.

Multi-Modal Input Devices

Given that the FMS prototype does not utilize a CDU to enter data, alternative cursor control devices (CCDs) have been researched and investigated to find the most viable candidate. A variety of CCDs have been thoroughly reviewed via an extensive literature review. These CCD candidates ranged from conventional hand held controlled devices, such as trackballs, touchpads, and joysticks to more non-conventional CCDs, such as head and eye tracked devices as well as voice recognition technologies.

Because operation of the FMS must be able to be accomplished in a dynamic flight deck environment, we initially constrained our evaluations to hand held CCDs. These devices were investigated in simulated turbulent environments for gross and fine selection tasks that mimic certain Advanced FMS Human Interface operations.

Very preliminary evaluations suggest that a large trackball (diameter greater than three inches) produces the quickest reaction time, high accuracy levels, and high levels of user acceptance.

Coupled with a hand held CCD has been the integration and preliminary evaluation of a voice recognition system for entry of discrete data, such as numerical values. We hypothesize that this muti-modal input device (hand held CCD for specific selection tasks and voice for entry of data) will enhance the ability of the user to effectively and efficiently interface with the Advanced FMS Human Interface prototype.

Results from preliminary evaluations reveal that entry of discrete data via a voice recognition device is very promising. This multi-modal input device is planned for a thorough evaluation over the next year to identify the feasibility of using a multi-modal input device in an actual aircraft environment.

Results from human factors evaluations of these augmented features and functions are not available at this time. However, numerous comments from professional pilots and FMS experts agree that these enhancements are very positive and have significantly contributed to improving the human interface to the FMS.

Advanced Primary Flight Display

Initial electronic primary flight displays (PFDs) mimicked the analog equivalent instruments they replaced. These displays are not optimal from a human interface perspective since much of the information required to fly a navigation path requires integration of second and third order effects depending on whether raw data is used or the flight director is used. Research has been conducted in this area for over 20 years with very little change in the "modern" flight deck. There are many reasons for this large gap between theoretical and applied science. The conservative nature of the industry and the lack of cost effective hardware rank high on the list.

Display hardware that provides the graphical capability to realize new complex symbolic displays has been expensive, even by avionics standards. Navigation systems that precisely track position and compute the trajectories have been expensive or non-existent but as air traffic grows, an added emphasis on safety through human centered design is being stressed. Within 12 years, if real safety improvements are not found, a major airline accident will occur every week due to increased traffic growth alone. Additional ways are being sought to increase the capacity of the air traffic control system safely. Current PFDs are not optimal from a human centered design perspective. Although they can be used to fly curved approaches and conduct operations in the envisioned free flight environment, the increased workload and limited situational awareness aid they provide will render them inadequate if not uncertifiable for these operations.

Progress

A study of current research in at least five major display efforts was conducted and a reference list was compiled. This study was used to determine the current best 3-D/4-D

PFD concepts. Simple prototypes were constructed and the most promising aspects were combined and integrated into the Advanced Display Concepts (ADC) lab. An adaptation of the NASA pathways display was chosen for full implementation in InCS. A comparison study was conducted to determine if this advanced PFD was better than today's standard symbology. This study yielded statistically significant results to support the enhanced PFD.

Literature Review and Prototyping

The NASA pursuit tracking display was developed for the Vertical/Short Takeoff and Landing (V/STOL) Systems Research Aircraft (VSRA) program (Merrick, January 1990, November 1993). The display consists of a basic headup display (HUD) with a runway outline and pursuit aircraft symbology. The pursuit aircraft is a stick airplane that will change perspective as lateral and vertical tracking errors develop. The pursuit aircraft flies a perfect track ten seconds in the future. In this manner, the pursuit aircraft allows some anticipation of future required control movements. As such it works well for complex curved paths. In addition, the VSRA work developed a two circle method (Merrick, April 1995) to acquire any course from any arbitrary heading without excessive maneuvering.

The Grunwald tunnel display (Grunwald, March-April 1996) is an improvement over the NASA Dorighi tunnel display. The basic display is a square tunnel display with the edges defined as dashed lines going into the 3-D perspective space. Two square outlines are drawn, one two seconds ahead and the other six seconds ahead. The one at six seconds is semi-transparent with a smaller (one-fourth) cutout window placed at the desired path position. A predictor symbol (predicted aircraft position in six seconds), shaped like a cross and sized to fit the small cut-out window, is used for guidance. This is a very effective display for curved path segments and manages the curved-to-straight and straight-to-curved transitions with little tracking error.

The DELPHINS display work (Theunissen, 1996; Theunissen, 1997) is a set of PFD and MFD displays for 3-D navigation. The symbology is extensive and limited flight testing has been conducted. Some of the tunnel designs allow for less clutter than many of the specialized displays developed to date. The displays allow full integration of earth referenced navigation and terrain symbology behind the path symbology

A pathways display was developed at NASA Langley to support the high speed civil transport (HSCT) program. This has a unique billboard concept that may be beneficial on many current display applications. The display combines many HUD aspects including the standard flight path symbology with pathway symbology defined by goal posts and four angle brackets. The pursuit airplane from the NASA Ames research is used as the guidance symbol. The display incorporates fairly standard airspeed, altitude, and vertical speed tapes.

A Star Field display was developed on the Oz Project at the Naval Aerospace Medical Research Laboratory by Dave Stills, et al. The display makes extensive use of peripheral vision research. The display is highly integrated and presents a pictorial representation of all aircraft control parameters. Course, heading, and track errors can be perceived within a few degrees, and altitude may be easily monitored within 10 feet. The display presents a 360 degree integrated view of the aircraft navigational problem. The primary display elements are five rows of streaming light points (stars) that get smaller and dimmer in the distance. While on course and altitude, all stars are streaming away from the viewer in a straight-on perspective. Any slight relative motion of the star fields are easily discernible.

Advanced 3-D PFD Testing

The advanced 3-D PFD candidate was tested in a fixed base part-task simulator to determine if the new display format would indeed meet the objectives. Formal human factors testing was conducted over a period of four days. Six subjects completed a series of approaches on the Advanced PFD prototype and then the subjects completed the same procedures on a simulated air transport PFD.

Method

Participants. Six subjects participated. All subjects were pilots with a mean flying time of 2330 hours. All subjects except one were familiar with electronic flight displays and flight director formats. None of the subjects were familiar with pathways displays or the content of the advanced PFD. None of the subjects were familiar with headup displays or flight path vector symbology, which makes up a large part of the normal symbology on the advanced display. The sixth subject was naïve to instrument flying and was added to gauge ease of use and learnability of the advanced display.

Data Recording. Qualitative data was collected using questionnaire techniques. The situation awareness test (Endsley, M.R.1988) required the pilot to determine the distance to the runway, how far from localizer intercept, and to place the aircraft in proper position and orientation on the vertical and plan views of each approach chart. The workload assessment (Lysaght, et al, 1989) is a simple evaluation of mental effort, time load and stress level. Quantitative data was collected and plotted using simulation routines. The data was stored on disk and data reduction and plotting were performed off-line at the end of the tests. The data was used to determine tracking errors and to monitor other parameters.

Display Formats. Two different display formats were used. The first format was the advanced 3-D PFD presented on the outboard display in the simulator. The second format was the standard Boeing 747-400 PFD in the same location.

Scenario. The test scenario presented and thoroughly explained to all subjects was a simulated approach, with radar vectors, to Minneapolis-St. Paul International Airport runway 11R starting approximately 13 miles out. Two different levels of wind were used in the test, a no wind condition and a second with 15 knots of crosswind with 10 knots of variable gusts, lateral and vertical. Surface weather was indefinite ceiling, RVR 2000 feet in fog. Each display format was flown for each approach and each wind condition. Additionally, the advanced display format was flown using the curved approach procedure in a no wind condition to gather subjective data. All runs were terminated at threshold crossing

Experimental Design. The experiment was organized as a 2x2x6 repeated measures design.

Independent Variable. Display formats were used as factors. The *a priori* hypothesis asserted reduced tracking errors in both the lateral and vertical axis and pilot workload ratings would be lower for the advanced 3-D PFD format than for the standard PFD.

Dependent Variables. Three dependent measures were taken: 1) tracking errors, 2) pilot's subjective workload rating for each run, and 3) subjective data taken after completion of each run.

Procedure

Each experimental session included a briefing to the subjects of the nature of the experiment, followed by a brief demonstration/training session for each display. The subjects were then given the opportunity to practice using each display until they were comfortable with the presentation of all symbology. This practice session allowed for the pilots' behavior to stabilize prior to the formal test procedure. Subjects alternately performed the approach with each display, first with no wind and then with winds that simulated approximately 12 knots of crosswind with vertical and lateral gusts of 10 knots. Data recording began immediately and continued until threshold crossing. The situational awareness questionnaire was administered about one minute into the run before localizer capture. The workload assessment was completed at the end of each run. An additional run was completed for the complex curved approach path. Each experimental session lasted approximately two hours.

Results

Statistically significant findings at the 95th level of confidence using a *t*-test support the a priori hypothesis that the advanced 3-D PFD display format produced less tracking error in both the lateral and vertical axis. The display is easier to fly, resulting in less workload and increased situational awareness.

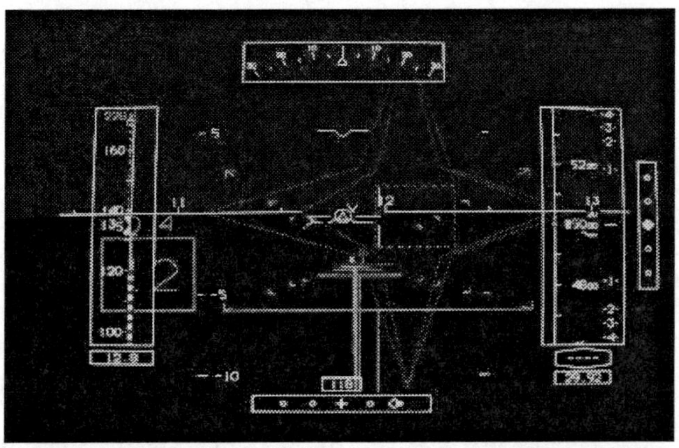

glideslope and localizer capture and down to the 50 foot decision height. Errors recorded over this narrow range eliminated the bias due to differing capture strategies and guidance calculations. This also eliminated the most difficult part of the procedure and if all the data were included the results would be even more significant in favor of the advanced PFD. All charted data is a mean measure of all six subjects.

Some better performance for vertical tracking was noted for the windy conditions for both displays. This inconsistency appears to be due to continuing adaptation (not enough practice time) to the flying qualities of the simulator. The difficulties in the simulator should have shown up as increased workload for both display formats and should not have caused errors in the collected data. The fact that the anomaly exists is either attributed to training time or some other unknown effect. If an additional study is conducted the wind and no wind conditions could be varied and this could then confirm or disprove the training effect theory. Additionally, the pilots were familiar with the standard display symbology and should have required less training time for that particular format. The relatively little variance in the raw data (not detailed here) for the pathways symbology supports the theory that the display is significantly easier to fly.

Although this report cannot contain all the plotted data, some significant trends were obvious. The pathways display produced significantly better tracks during all conditions. There was no overshoot on localizer intercept, even for the complex approach. Some of the overshoot, using the standard PFD, can be attributed to the flight director guidance calculations and not error in pilot tracking, yet this is still system error. The human tracking performance in this respect will be better, using the advanced PFD, than even an autopilot track using standard guidance, especially in significant crosswinds. For the pathways display, no significant altitude deviations were recorded and little tendency was noted in vertical oscillations due to the pilot getting out of sync with the autothrottle. For the standard display, these anomalies were easily noted by the test operator during the runs and were confirmed by examining the recording for the standard PFD runs. The author believes this is due to the ease of use of the flight path vector symbology and the fact that the normal diverging trends are easily understood from the display presentation.

Summary

Although these two studies were completed independently of each other, the extensions to the overall cockpit have significant ramifications for future flight deck designs. Increased workload and automated navigation procedures have changed the flight deck task from a controller to a managerial one.

Increased airspace complexity, increasing noise abatement procedures, and future curved approaches to accommodate differing traffic approach speeds will create the need for different flight deck approaches. Calls for increased flight safety and the need to reduce controlled flight into terrain also point to the need for a different flight display paridigm. These problems may be solved with the use of more visual realistic flight displays. The time has come to introduce new methods of navigation and aircraft control. These methods will increasingly be visual and 3-D/4-D based.

REFERENCES

1 Abbott, T.S., A Comparison of Two Control Display Unit Concepts and Their Effects on Flight Management System Training. NASA Technical Memorandum 4744, 1997.

2. Degani, Asaf and Kirlik, Alex, Models in Human-Automation Interaction: Inital Observations about a Modeling Approach. *Proceedings of the IEEE International Conference on Systems, Man, and Cybernetics.* (pp. 1-8), 1995.

3. Dornheim, Michael A., Recovered FMS Memory Puts New Spin on Cali Accident. *Aviation Week & Space Technology* (pp. 58-61), 1996.

4. Endsley, M.R., Situation Awareness Global Assessment Technique (SAGAT), In *Proceedings of the National Aerospace and Electronic Conference* (NAECON) (pp. 789-795), 1988.

5. Faerber, R.A., Advanced Flight Management System Human Interface. In *Proceedings of the International Conference on Human-Computer Interaction in Aeronautics (HCI-Aero)* (pp. 21-27), 1998.

6. Funk, K., Chou, C.D., & Madhavan, D., Studies of Cockpit Task Managment Errors. *The International Journal of Aviation Psychology,* (6(4), pp. 307-320), 1996.

7. Grunwald, A.J., "Improved Tunnel Display for Curved Trajectory Following: Experimental Evaluation," Journal of Guidance, Control, and Dynamics, Vol. 19, N0. 2, March-April, 1996, pp. 378-384.

8. Lysaght, R.J. et.al., Operator Workload: Comprehensive Review and Evaluation of Operator Workload Methodologies. Technical Report 851, United States Army Research Institute for the Behavioral and Social Sciences, 1989.

9. Merrick, V.K., Farris, G.G. and Vanagas, A.A, "A Head Up Display for Application to V/STOL Aircraft Approach and Landing", NASA TM-102216, January 1990.

10. Merrick, V.K., "Some VTOL Head-Up Display Drive-Law Problems and Solutions," NASA TM-104027, November 1993.

11. Merrick, V.K. and Jeske, J.J, "Flightpath Synthesis and HUD Scaling for V/STOL Terminal Area Operations," NASA. April, 1995.

12. Norman, D.A., The 'problem' with automation: inappropriate feedback and interaction, not ' over-automation'. *Phil. Trans. R. Soc. Lond.* (B 327, pp. 585-593), 1990.

13. Sarter, N.B., and Woods, D.D., Pilot Interaction With Cockpit Automation: Operational Experiences with the Flight Management System. *The International Journal of Aviation Psychology* (Volume 2, pp. 303-321), 1992.

14. Sarter, N.B., and Woods, D.D., Pilot Interaction With Automation II: An Experimental Study of Pilot's Models and Awareness of the Flight Management System. *The International Journal of Aviation Psychology* (Volume 4, pp. 1-28), 1994.

15. Strauch, Barry, Automation and Decision Making - Lessons Learned from the Cali Accident. *Proceedings from the Human Factors and Ergonomic Society 41st Annual Meeting.* (pp. 195-199), 1997.

16. Tenney, Yvette J., Rogers, William H., and Pew, Richard W., Pilot Opinions on High Level Flight Deck Automation Issues: Toward the Development of a Design Philosophy. NASA - CR - 4669, 1995.

17. Theunissen, E.R, "Integrated Design of a Man-Machine Interface for 4-D Navigation," Delft University Press, 1997.

18. Theunissen, E.R, "Towards the Integration of Perspective Displays into the Flightdeck," Delft University of Technology, January 1996.

19. Wickens, C.D., and Mireille, Raby, Strategic Workload Management and Decision Biases in Aviation. *International Journal of Aviation Psychology.* (Volume 3, pp. 211-240), 1994.

PROJECTED FLIGHT PATH DISPLAYS AND CONTROLLED FLIGHT INTO TERRAIN ACCIDENTS

Barry Strauch, National Transportation Safety Board, Washington, DC[1]

Abstract

In the early 1980's aircraft were introduced with automated "glass cockpit" displays that present computer processed airplane systems and navigation information. Today such displays have become standard equipment on nearly all recently manufactured air transport turbojets, irrespective of the airframe manufacturer or flight management computer designer. Among their numerous capabilities, these displays can portray navigation information in a readily interpretable manner and project the airplane's future lateral flight path with reference to critical navigation aids and areas of hazardous weather. As a result, pilots of these aircraft can quickly process navigation information and maintain a considerably enhanced awareness of their airplane's position as compared to pilots of analog-equipped aircraft. However, even with these capabilities several glass cockpit equipped aircraft have been involved in controlled flight into terrain accidents, in which the airplanes struck terrain following the pilots' failure to recognize their proximity to that terrain. Despite the considerable superiority of the presentation of position information in glass cockpit aircraft relative to older aircraft, these accidents demonstrate that under certain conditions pilots of such aircraft can still lose their position awareness. This paper examines three controlled flight into terrain accidents involving airplanes equipped with glass cockpit projected flight path displays, to determine commonalties among the errors that the pilots committed and limitations in the displays that may have contributed to those errors. Suggestions for research into the circumstances under which computer generated flight path displays can detract from position awareness will also be discussed.

Introduction

In the early 1980s with the advent of the Boeing 757 and 767, a new way of presenting critical information to pilots was introduced. These airplanes feature what has become known as "glass cockpit" displays, typically cathode ray tube (CRT) screens that present information generated by aircraft flight management computers. Today, all new air transport aircraft, most large regional aircraft, and some older aircraft that are being retrofitted employ this technology to present information to pilots.

Glass cockpit displays have primarily affected pilots' systems monitoring and navigation tasks. Navigation, which is ordinarily a complex cognitive task, can create large demands on pilots' limited processing resources (e.g., Aretz, 1991). These displays reduce pilots' cognitive

[1] The views expressed in this paper are those of the author and not necessarily those of the National Transportation Safety Board.

0-7803-5086-3 /98/$10.00 ©1998 IEEE

navigation efforts by presenting the airplane's current position and predicted flight path. Information for this display is generated from navigation data pilots have entered into the aircraft flight computers, and internally- and externally-based position information. Pilots can select this or other displays, although most opt for predicted flight path on at least one of the two navigation displays.

In addition, the navigation displays can also show navigation aids, airports, or other critical data pilots have requested, as well as overlay weather-radar data onto the predicted flight route. This latter feature allows pilots to quickly recognize the location of adverse weather relative to the predicted flight path, and to estimate the need for and the extent of deviation from that flight path. In addition, although vertical flight paths are not presented, the displays show the expected location of the airplane on the predicted flight path, at the conclusion of a steady climb or descent.

Pilots can also select flight routes stored in the aircraft data base and direct the aircraft computers to fly one or more of the selected routes. If choosing to fly along stored routes, the airplane will remain on a proper and safe flight path, so long as minimum altitude is maintained. Pilots control the altitude flown, although the airplane flight computers may indicate a preferred altitude that is within safe terrain clearance margins.

Controlled Flight Into Terrain

Controlled flight into terrain (CFIT) accidents result from the loss of pilot situation awareness relative to location and/or proximity to terrain. Situation awareness refers to comprehension of information a machine is giving them, in this case navigation information, as well as peoples' ability to project to the future from that information, e.g., the location of the airplane along the flight path (Endsley, 1995).

CFIT accidents occur when pilots either do know that terrain is ahead of or below them, or they fail to recognize their proximity to terrain in sufficient time to prevent an accident. This type of accident is among the major causes of contemporary commercial aircraft accidents (Khatwa & Roelen, 1996). It can occur in any airspace, but usually takes place in airspace that lacks air traffic control radar, in which pilots are not in contact with air traffic control, or where there are communication errors between controllers and pilots. Otherwise, controllers can usually use radar information to identify aircraft that are off course or close to terrain and alert the pilots to potential danger.

In analog instrumented aircraft, CFIT accidents can be attributed, in part, to pilot navigation errors. These are more likely to occur in analog-instrumented than in glass cockpit aircraft because of the considerably greater cognitive effort required to navigate in the older aircraft. In fact, only three CFIT accidents involving glass cockpit aircraft have occured in regular line flight[2] since their introduction in 1982, out of a total of

[2] On February 14, 1990, an Indian Airlines A-320 was involved in a CFIT accident while on approach to Bangalore, India. The airplane was destroyed and 90 of the 144 passengers and crew on board were killed in the accident. However, because the captain of the flight was undergoing initial operating experience, with a company check airman performing the functions of first officer, and instructional aspects of the flight were involved in the cause of the accident, it was not included in this paper.

156 such accidents studied by Khatwa & Roelen.[3]

While this number is small, the accidents tell us much about the use and misuse of glass cockpit flight displays and how, despite their enhancement of pilot situation awareness, features of these displays could actually militate against situation awareness. A close examination of each of these accidents can reveal much about potential errors in pilots' reliance on and use of glass cockpit instrumentation and on display limitations that may not have been apparent before the accidents.

Accident 1

On January 20, 1992, an Airbus A-320 operated by the French airline, Air Inter, on a flight from Lyon, struck terrain while in a descent for a landing in Strasbourg, France. All but nine of the 96 passengers and crew were killed in the accident (Commission of Investigation, 1993). The pilots had expected a circling approach, in which the airplane was to fly over the airport and then turn back and land in a direction different from that of the approach. However, after planning and preparing for the circling approach, air traffic control informed the pilots that only a straight in approach was immediately available. The pilots accepted a clearance for a straight in approach, but had to begin preparations for the new approach, in a shorter period of time than had been available for the initial approach. The attempt to complete all necessary tasks for the new approach in the time remaining led to a period of intense pilot workload.

Further, the new approach, unlike the initially intended one, was a "non-precision" approach. That is, the only information available to the pilots to direct them to the runway was lateral but not vertical guidance. Consequently, they selected a feature of the flight management computer that enabled them to enter a descent angle into the autopilot, so that the airplane would generate a vertical flight path that mimics a precision approach with vertical guidance. However, although the crew had entered the correct data for the descent angle and commanded the aircraft to fly that vertical path, they inadvertently selected a rate of descent and not a flight path angle, an error which both pilots failed to notice. The airplane began to descend at a high rate, which the crew had also failed to notice. The airplane struck a hill while turning towards the airport.

Accident 2

On July 31, 1992, an Airbus A-310 operated by Thai Airways International, on a flight from Bangkok, Thailand, struck a mountain while on approach to Katmandu, Nepal. The aircraft was destroyed and all 99 passengers and 14 crew members on board were killed in the accident (His Majesty's Government of Nepal, 1993). The pilots had begun an initial descent into Katmandu, but because of difficulties configuring the flaps they were unable to continue the approach. Although they quickly corrected what was a minor problem, they could not resume the approach from the point at which they had resolved the problem with the flaps. Because the approach into Katmandu required a steep descent angle, airplanes had to be properly configured from the beginning to be in a position to execute the approach. Therefore, the crew requested

---

[3] About 40% of the sample of accidents in the Khatwa & Roelen study included twin engine turboprop aircraft, as well as some single engine aircraft operated on-demand. All aircraft were operated for hire. About 23% involved jet aircraft and 35% turboprop aircraft used in scheduled passenger service.

permission to proceed to a navigation aid some distance from the airport, to enable them to begin a second attempt to approach Katmandu.

However, the pilots failed to notice that they had continued flying north past Katmandu, rather than turning back to the desired navigation aid south of their position in order to initiate the approach. This error occured during a period of intense workload in which they attempted to diagnose and correct the flap problem, locate the desired navigation aid, and communicate in English with the Nepalese air traffic controllers, a foreign language to both the pilots and to the controllers. The area north of the city, noted for its high Himalayan terrain, contained no navigation aids; all navigable airspace was to the south. As a result, the navigation display showed what was essentially a blank screen that was devoid of navigation aids.

Because the pilots were unaware that they were north of the city, despite the fact that several non-digital cockpit displays showed the actual heading of the aircraft relative to the city, they were unable to locate the navigation aid that was behind them, nor were they able to determine the cause of perceived difficulties with their navigation displays and aircraft flight computers. The report of the accident investigation found that both pilots interacted with the flight management computers in attempts to correct the perceived malfunction. In point of fact, the system had worked as designed. The airplane struck the 11,600 foot level of a 16,000 foot mountain, at an airspeed of 240 knots.

Accident 3

On December 20, 1995, an American Airlines Boeing 757, on a flight from Miami, Florida, crashed while on descent for an approach to Cali, Colombia. All but four of the 163 passengers and crew were killed in the accident (Aeronautica Civil of the Republic of Columbia, 1996). The flight had been established on an approach similar to that anticipated by the pilots of the Strasbourg accident airplane. They were to overfly Cali, circle and land in the direction opposite to that of the initial approach. About five minutes before the accident the crew was offered the opportunity to land straight in to Cali, a direct and more efficient route. They accepted the offer and initiated a descent that continued to the accident.

However, this change in the planned approach necessitated reentering into the flight computers data for the new approach. In so doing, this crew was unable to locate a navigation aid critical to the approach. Neither pilot recognized that the navigation aid had been deleted from the pilot-entered flight plan through an earlier action by the captain. Further, as the airplane continued to Cali it passed that navigation aid. Neither pilot recognized that it was now behind them, and that, unless specifically requested, the Boeing 757, as the A-310, shows navigation information forward of the airplane only.

Finally, the pilots commanded the flight management computers to fly to what they believed to be a navigation aid close to Cali. However, in their attempts to locate this navigation aid, neither recognized that what was actually entered into the flight management computer was a navigation aid located 130 miles away, near Bogota, that shared the same abbreviator on the navigation chart as the intended one. Nonetheless, the turn to Bogota was readily observable on the display of the predicted flight path. About a minute into it, the pilots recognized that they were turning away from Cali and commanded a turn back to the intended destination. During this turn, the airplane struck a mountain at the 8,900 foot level, about 33 miles northeast of Cali.

Lessons of the Accidents

The three accidents share many characteristics, despite the fact that they involved different airplanes, operated by crews of different airlines, in different parts of the world. In each of the accidents, the pilots made similar errors in misinterpreting or not perceiving navigation information, or in not understanding logic of the flight management computers, and ultimately in losing situation awareness regarding their proximity to terrain.

One element common to the accidents is that each involved a last minute change to the intended approach. These changes required the pilots to rapidly enter data into the airplane flight management computers and review the new flight paths, in less time than they available during the first preparation for the approaches. The need to make these changes within the reduced time intervals led to rapid increases in crew workload and the increased workload led to errors by the three crews.

Awareness Of Flight Computer Logic

Neither the crew of the Katmandu nor of the Cali accident airplane understood why they were unable to locate navigation aids critical to the execution of their respective approaches, despite the fact that the flight management computers worked as designed. Their persistent and ultimately unsuccessful attempts to locate them demonstrates what Sarter and Woods (1995) have described as the opaque nature of the automation interface that operators use to control automated systems. That is, because of the design of the system logic, its information feedback and other factors, aspects of the state and behavior of automated systems are obscure to the operators. Consequently, their ability to understand why the automated system is performing the way it is or how it will likely perform subsequently is restricted.

As seen in these two accidents, none of those pilots understood why the flight management computers were providing the navigation information they were. Rather, they mistakenly believed that they had either entered incorrect data or given incorrect commands to the system. Their repeated interaction with the system manifested their belief that with further effort they would provide the correct combination of data to obtain the desired navigation information, or determine why they had not obtained the desired information. Moreover, consistent with the findings of Hart and Bortolussi (1984), their errors themselves increased the levels of their workload, leading to further degraded performance. The initial errors in attempting to locate the critical navigation aids only heightened crew workload and further decreased the likelihood that they could determine the reasons for what they perceived to be errors.

Information Not Presented

Each crew also committed errors in navigating, errors that would otherwise have been unlikely given the unique features of glass cockpit aircraft flight displays previously discussed. The possible reasons for these errors reveal much about limitations of even readily interpretable displays.

Flight path information is presented in two dimensions while actual flight paths are three dimensional. The Strasbourg and Cali accidents, in which both airplanes were descending at the time they struck terrain, illustrate that during periods of high workload, pilots may lose sight of the vertical flight paths while focussing exclusively on the lateral. In routine flight they can rely with confidence on the lateral navigation performance of the airplane and attend to other tasks. As Adams et al. (1995) noted, the management of complex competing tasks, such as occurred in these accidents, requires pilots to prioritize the tasks and attend only to those most in need of attention. However, in so doing, other critical tasks could be omitted.

For example, in the Air Inter accident the crew believed that the vertical flight path was being effectively managed by the aircraft computer system. They therefore focused on the two dimensional navigation information displayed, to the detriment of attending to information not displayed, three dimensional navigation along the vertical flight path. Similarly, in the Cali accident the airplane continued to descend from the beginning of the initial approach to Cali to the accident, without crew action to modify or arrest the descent. Taken together, these accidents illustrate, as Klein (1996) noted, that under stress salient cues will become prominent. As applied to these accidents, the salience of the lateral information shown on the navigation display may have actually hindered the pilots' ability to attend to equally critical but absent information, the vertical flight path.

The overlay of weather radar presentations of hazardous weather onto projected flight paths on glass cockpit aircraft also reduces the cognitive workload of its users when trying to determine the effects of weather on flight paths. However, terrain is not portrayed on the glass cockpit flight displays, a factor of little import in airspace with air traffic control radar available, in areas without high terrain, or at those times when pilots are not encountering high workloads.[4] However, because the displays are the primary, and for most flight phases the only source of navigation information,[5] pilots encountering high workload may lose sight of the presence of nearby terrain if they have not yet begun an approach, as in the Cali accident, or if they are using charts that do not graphically portray terrain, as in the Katmandu accident. By contrast terrain information is shown on navigation charts, although its interpretability has been enhanced on new charts.[6]

Further, in most flight phases this absence is of little significance because of the factors noted, and because many of the routes flown are selected from among those stored in the aircraft data base. As these routes are determined by the civil aviation authorities of the governments that control the airspace, pilots are assured that they meet acceptable standards of terrain clearance, so long as the displayed flight route being flown is the stored route. If the route being flown was generated by the

---

[4] New generation of "enhanced ground proximity warning systems" present terrain information on navigation displays. Images are based on internal terrain data and derived aircraft position. At this time, major US-based and some overseas airlines have committed to installing these systems.

[5] Pilots of glass cockpit aircraft are only required to use approach charts and can use navigation displays for other navigation information.

[6] In fact, at the time of the Cali accident the major producer of navigation charts, the Jeppesen-Sanderson Company, was in the process of graphically portraying terrain information on their published approach charts to enable pilots to quickly recognize the location of hazardous terrain.

pilots and not from one stored in the data base, there is no assurance of terrain clearance. Yet, the glass cockpit display of terrain information does not indicate whether the presented routes are stored. The flight path display appears the same regardless.

The pilots of both the Cali and the Katmandu accident aircraft encountered an additional limitation of digital navigation displays, only forward flight path information is presented. The Thai Airlines crew did not note, until just before impact, that they were headed north of the city rather than south where the critical navigation aid was located. Likewise, the American Airlines pilots failed to recognize that the critical navigation aid they were seeking was behind them. By contrast, conventional navigation charts display all area navaids on the printed page. Once their location has been determined pilots can readily find the names and relative positions of nearby navigation aids. Again, in routine flight this is not a problem but in high workload conditions the absence of navigation information aft of an aircraft's position can hinder pilots' ability to gain situation awareness relative to lateral position.

Finally, the Katmandu accident illustrates a feature of the navigation displays that would not likely have been foreseen by the designers. That is, the absence of navigation data would be interpreted by flight crews as a fault in the system, rather than as an accurate portrayal of the airspace. Here the pilots misinterpreted the lack of displayed navigation information, caused by the absence of navigation information within the forward range of the display, as a system fault. Airlines do not train pilots for such an event and until this accident it is doubtful that pilots in line flight had seen a navigation display devoid of navigation information while over land. When the Katmandu accident crew encountered it, they misinterpreted it in a manner consistent with their training and their aircraft experience.

Discussion

Few accidents are caused by one or two errors and it is hoped that this discussion has touched upon the complexity of events generally involved in accident causation. In point of fact, the use and misuse of advanced flight displays should not be viewed without examining the use of the supporting flight management computers, in the context of the accident scenario. Considerable attention has been paid to this topic (e.g., Parasuraman and Riley, 1997) and little can be added to that literature.

In addition, several of the limitations noted herein have been or are being examined in the laboratory. For example, Kuchar & Hansman (1993) studied the absence of a display of terrain information. They found that when pilots recognized that were unable to depend on air traffic control for terrain avoidance, as is true in much of the airspace outside the US, the presentation of terrain information on flight displays enhanced pilot situation awareness regarding their proximity to terrain.

In addition, Haskell and Wickens (1993) examined the absence of vertical flight path displays. They found that pilot performance was generally superior when three-dimensional flight displays were used rather than two-dimensional. It is clear from the research that regulators, who ultimately determine the information that must be presented to pilots, as well as manufacturers and display designers, should address the limitations in current electronic flight displays to present to pilots a more complete portrayal of an airplane's flight path, and the environment in which the flight path occurs.

The accidents and these studies suggest that considerable work remains to be done to determine the "ideal" display of information to pilots. Although the display of navigation information has brought many benefits, these accidents demonstrate that under certain conditions limitations in the display of information can have serious adverse results. Consequently, it is recommended that any research into this subject consider the effects of high workload and stress on the ability of pilots to obtain information from the displays, to determine if the results are consistent with the intent of the designers.

References

Adams, M.J., Tenny, Y.J., & Pew, R.J. (1995). Situation awareness and the cognitive management of complex systems. *Human Factors, 37*, 85-104.

Aeronautica Civil of the Republic of Colombia. (1996). *Aircraft accident report: Controlled flight into terrain, American Airlines flight 965, Boeing 757-223, N651AA, near Cali, Colombia, December 20, 1995*. Santafe de Bogota, Colombia: Author.

Aretz, A.J. (1991). The design of electronic map displays. *Human Factors, 33*, 85-102.

Commission of Investigation, (1993). *Official report of Commission of Investigation into the accident which occurred on January 20, 1992, (between Bas-Rhin), of the Airbus A320 registered F-GGED operated by Air Inter.* Paris, France. Author.

Endsley, M. R. (1995). Toward a theory of situation awareness. *Human Factors, 37*, 32-64.

Hart, S. G., & Bortolussi, MR. (1984). Pilot errors as a source of workload. *Human Factors, 26*, 545-556.

Haskell, I.D., & Wickens, C.D. (1993). Two- and three-dimensional displays for aviation: A theoretical and empirical comparison. *International Journal of Aviation Psychology, 3*, 87-110.

His Majesty's Government of Nepal. (1993). *Aviation accident report, Thai Airways International Ltd. Airbus Industrie A310-304, HS-TID, near Kathmandu, Nepal, 23 NM NNE, 31 July 1992.* Kathmandu, Nepal. The Director General Department of Civil Aviation.

Khatwa, R., & Roelen, A.L.C. (1996). An analysis of controlled flight into terrain (CFIT) accidents of commercial operators, 1988 through 1994. *Flight Safety Foundation Digest,* April-May 1996, 1-45.

Klein, G. (1996). The effects of acute stressors on decision making. In J.E. Driskell and E. Salas (Eds.), *Stress and human performance.* Mahwah, New Jersey: Lawrence Erlbaum Associates.

Kuchar, J. K., & Hansman, R. J. (1993). An Exploratory study of plan-view terrain displays for air carrier operations. *International Journal of Aviation Psychology, 3*, 39-54.

Parasuraman, R., & Riley, V. (1997). Humans and automation: Use, misuse, disuse, abuse. *Human Factors, 39*, 230-253.

Sarter, N.B., & Woods, D.D. (1995). How in the world did we ever get into that mode? Mode error and awareness in supervisory control. *Human Factors, 37*, 5-19.

A DISPLAY CONCEPT FOR STAYING AHEAD OF THE AIRPLANE

Eric N. Johnson, Lockheed Martin Aeronautical Systems, Marietta, Georgia

David C. Hansen, Lockheed Martin Aeronautical Systems, Marietta, Georgia

Abstract

A display concept for depicting past, present, and future situation data in a consistent manner is presented. Anticipated advantages include better planning and improved operator workload management. Issues associated with the concept are discussed.

Introduction

For generations, flight crews have been trained to "stay ahead of the airplane". That is, they have been trained to always be prepared for their next task and to make plans and decisions prior to the latest available moment. To accomplish this, the highest level of situation awareness is a prerequisite: the crew must be able to predict the future situation and be cognizant of it s implications. Conventionally, this requires an expert. The expert can use this ability to smooth-over high workload periods, and can constantly reevaluate their decisions based on new information prior to fully implementing decisions.

Advances in airborne processing capabilities and datalink communications may enable a new type of human interface to be used in the future flight deck and air traffic control facilities: one in which the current system state and predicted states can presented to the operator in a cohesive and intuitive manner. In the concept presented here, the conventional notion of the horizontal situation display with flight plan, traffic, and weather information is extended by an ability to depict a predicted situation at a future time, based on the current situation, flight plans, and weather forecasts. A number of design issues associated with such a concept are discussed (in particular the accuracy of predicted information) in the context of enabling the human operator to make complex decisions with enhanced safety.

Background

Current Practice

Airspace users rely on a variety of sources for information about the past, present and future situation. Today, there is considerable variety in presentation of this information, depending on the source. Table 1 is a sample of these data sources. This situation is illustrated by looking at the ways a pilot gets

Table 1. Typical Flight Deck Information Sources

Information	Past Data	Current Situation	Future Situation
Traffic	Not Generally Available to Flight Crew	ATC (voice), TCAS (situation display)	Flight Plan (paper, situation display), ATC (voice)
Weather	Weather Maps (paper)	Radar (situation display), ATC (voice), Pilot Reports (voice)	Forecasts (paper)

0-7803-5086-3 /98/$10.00

information about past, present, and future traffic and weather information during a typical flight.

Enabling Technologies

A variety of technologies are becoming available that are allowing more of the information described in the previous section to be available to flight crews. This potential has lead many to propose changes to the National Airspace System (NAS) to take advantage of these technologies:

Datalink: Satellite Communication, Mode-S and other technologies are making it possible to reliably communicate a considerable bandwidth of digital data between aircraft and ground facilities, and even between aircraft.

Multi-Function Displays and Control: Reconfigurable Displays and Controls have come into common use in aircraft. These devices include reliable large-format color displays, alphanumeric keypads, function keys, and cursor control devices.

Airborne and Ground-Based Processing: Advances in computer processing are enabling more and more processing of data in the timely manner required for many aerospace applications. Considerably more data fusion, data prediction, alerting, and automation is potentially possible than is utilized today.

These technologies are enabling many to consider an overhaul of the ATC system to improve both safety and through-put. One current trend appears to be toward greater flight deck responsibility in flight path selection, often referred to as Free Flight. The use of predicted data may help operators with the task of evaluating options in a Free Flight environment.

Even without any new flight deck responsibilities, if flight crews can be given more of the information necessary to evaluate their progress, then it seems likely that safety will be improved due to the extra checks being made on decisions. This must be accomplished without causing their workload to increase to a point of diminishing returns. Arguably that point has already been reached in the terminal area, where new flight deck information might only be of benefit if it inherently decreases workload.

The Predictive Situation Display

A primary purpose for situation information is to allow flight crews and controllers to make flight plan decisions. For example, the purpose of weather information is to allow flight crews and controllers to make plans that cause all aircraft to miss undesirable weather. In making these decisions, the current situation and forecasts are used to predict the outcome of flight path options.

The predictive situation display is proposed as a way to apply the enabling technologies discussed above to help operators with flight planning and re-planning tasks. In this concept, a “look-ahead” time is selected to be used for generating a predicted situation display. Where a conventional display shows the current state of traffic and other hazards (along with, say, lines depicting flight plans in some designs), this display could also depict the predicted position of traffic and other hazards for that future time. How this look-ahead time is selected is discussed in a later section.

A notional example of this concept is depicted in Figure 1. In this implementation, the display shows current states in a de-emphasized manner (in this case as dotted lines) and predicted future states in a more conventional manner. This allows the user to easily visualize the negative outcome of a possible flight plan. Some form of strategic alerting system might have just given a warning that this was the predicted outcome, but here is a way to depict the outcome in a more useful manner.

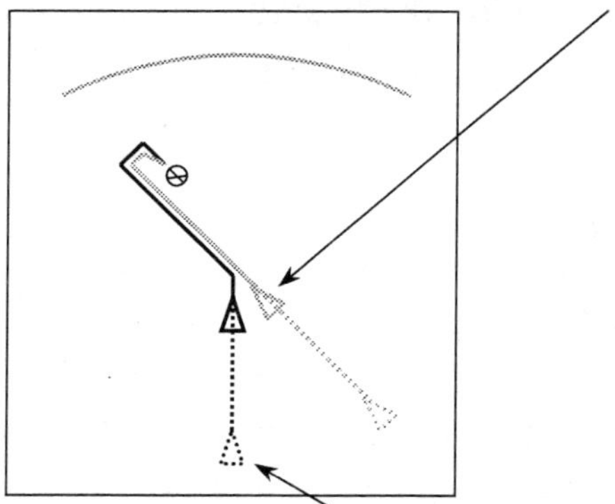

Figure 1. Depicting Predicted Situation for a Future Time, Notional Horizontal Situation Display (in Example, a Predicted Traffic Conflict is Depicted)

If the predicted information is adequate, this pre-planning can be arbitrarily far in the future. That is, the look-ahead time can rise and fall as the integrity of predicted information changes. If the predicted information changes dramatically, then the new outcome is presented and the re-planning can begin. The use of predicted information will discussed more in a later section.

The predictive situation display is proposed for showing the current and predicted-future state of:

- Ownship Navigation
- Other Traffic
- Weather (reports and forecast)
- Airspace (special use)
- Threats (for military use)

The use of this preview technique to depict these classes of information will be discussed below.

The advantages of this approach for displaying information are three-fold:

(1) Consistent portrayal of current, forecast, and past situation information: This is identified as a positive way to apply the previously mentioned enabling technologies to improve the flight crew's ability to do planning tasks, both in terms of accuracy (correctness) and frequency.

(2) Workload management: The capability to selectively plan ahead is recognized as a technique to reduce workload at times of peak workload, such at approach and landing. By planning for the approach while en-route, the workload in the approach is potentially reduced. In addition, the increased workload at times when the crew might otherwise become complacent or bored is a potential secondary benefit.

(3) Error detection and predictive alerting: Giving the flight crew the opportunity to selectively evaluate flight plans provides them the opportunity to catch their own errors or the errors of others in the system. Also, this can give them warning of impending alerts before they are actually triggered, potentially improving response (time and accuracy) if the alert cannot be avoided all together.

On the Use of Predicted Information

Fundamental to the usefulness of predicted information are accuracy and uncertainty. Predicting the location of an aircraft that is following a flight plan tens of minutes in the future can be quite accurate. Indeed, it might be predicted with nearly as much accuracy as knowledge of it's current location. However, between now and that future time that aircraft can *potentially* make major or minor changes to it's flight plan. It may have an emergency, or weather may not be as expected. This reduces the accuracy of predicted data.

Figure 2 is an illustration of the uncertainty of predicted data. The more that is known about the system the more accurate our predictions are expected to be. The more predictable the agents in the system are, the more accurate predictions will be. Irregardless, the accuracy of predicted data will generally get worse farther into the future – eventually becoming useless for planning purposes at some point. Because one cannot know the accuracy of predicted data until the future becomes the present, one can instead estimate uncertainty in the predicted data. This is a measure of how accurate the data is expected to be. In a highly dynamic environment, the uncertainty can be very high. In figure 2, uncertainty is low for past and current data, and generally increases for future data.

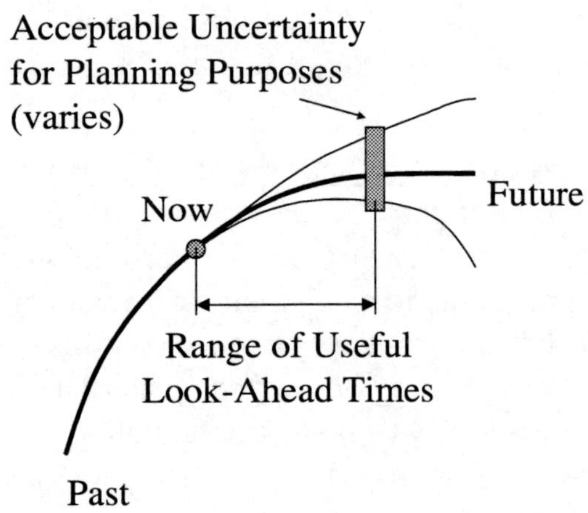

Figure 2. Accuracy of Predicted Data Generally Gets Worse Farther in the Future

With the concept of the predictive situation display presented here, where one selects a look-ahead time which an be adjusted between the current time (to see current data) to a future time (to see predicted data), we allow the flight crew to work at the time horizon that makes sense given the prevailing conditions. In the worst case situation, they would be working with data that is in the very near term, even current, because the situation has become highly dynamic and chaotic. In other conditions, they can plan far ahead, and then start looking for problems in the plan they and others have devised.

To give the flight crew some feedback about uncertainty, uncertainty estimates might be displayed. Another possibility is limiting the look-ahead time. A third option is to rely on the trained human pilot, perhaps the best qualified element in the system to evaluate uncertainty.

Control of the Look-Ahead Time

Issues associated with the selection of the look-ahead (or look-behind) time used to generate the predictive situation displays are discussed in this section. Overall, it is proposed that the flight crew select the preview time per predictive situation display manually. In this way, an individual crew member can plan as far ahead as appropriate for managing their workload or as procedures dictate.

Look-Behind

The predictive situation display concept may lend itself to depicting recorded data from the past as well as predicted information about the future in a single seamless display. This would be useful, for example, to look at trends in recorded weather data.

Time Slew

The preview time can be run at faster or slower than real time when going through recorded or predicted data. This is recognized as a useful way to show trends in the information. This would be particularly useful with weather and possibly with other information sources as well.

Changes to Predicted Information

Periodically, there will be a change to predicted data that can potentially change the desired flight path. This must be depicted in some manner. However, this may be as simple as a message that the weather forecast has changed – giving the crew the indication that they should re-check their plans. If they are currently viewing predictions for an affected future time, then that change can be depicted on the display immediately (just as, say, a weather radar changes in real-time on many current situation displays).

Depiction of Flight Plan Consequences

In this section, several display issues associated with the use of the predictive situation display for the evaluation of flight plans are discussed by information type. This discussion includes traffic separation, weather, airspace status, and terrain clearance.

Traffic Separation Information

Depicting the flight plans of other aircraft as lines and curves on a display is not adequate for the task of evaluating flight plans with regard to traffic separation. This is due to the time-dependence of trajectories. This is true no matter where a proposed set of flight plans originated, be it the aircrew, or an air traffic controller.

One desirable design element is to indicate on the display the location of a predicted separation problem area. The aircrew task, if they chose to develop a resolution themselves, would be to revise their flight plan to ensure that they do not enter the separation problem area at all, or at least enter it at a different time (arrive before or after other traffic). Such a display is depicted in Figure 3.

The second and complementing element that the aircrew would ideally have is a way to visualize their solution space, that is the types of changes that can be made to their flight profile to resolve the conflict. The preview concept used in concert with the separation problem zone is shown in Figure 4. In this case the preview time has been selected to be just before the separation problem is predicted to go into effect. Here, the aircrew can check to see if they can go above or below the other traffic; they might go left or right slightly -- but a conflict might still occur at a new time; or they could pull in behind or in-front-of the other traffic – perhaps the preferred option in this case.

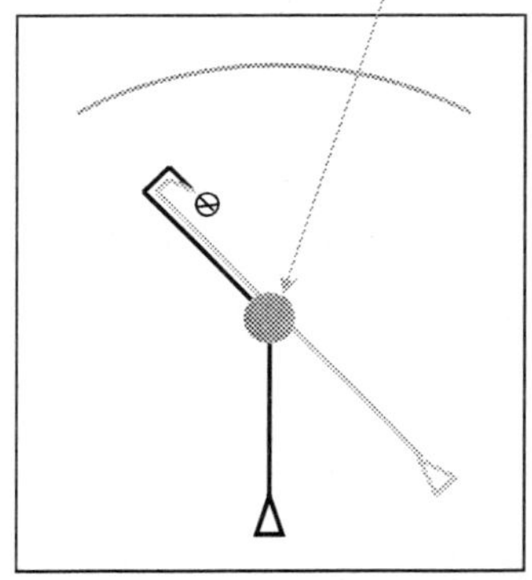

Figure 3. Traffic Conflict Information: Location of Conflicts are Depicted On a Notional Display, Enabling Aircrew to Visualize a Future Conflict

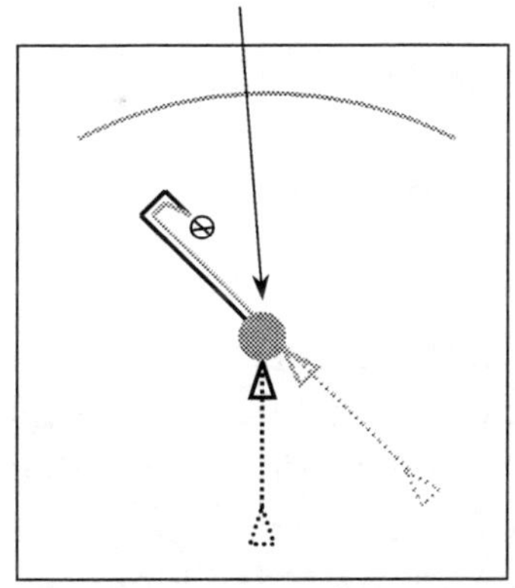

Figure 4. Traffic Conflict Information: Location of Conflicts are Depicted On a Notional Display, a Preview Time Selected Near Conflict Time

This raises the issue as to what flight plan the predicted data should be based on. For the purposes of evaluating options, it would be desirable to base predictions on other than the active flight plan, particularly for short-term or tactical decisions.

Weather Information

The depiction of on-board generated weather radar data and weather reports from various flight services and dispatch sources is another primary driver of flight planning. Figure 5 depicts a notional weather situation display which combines on-board derived weather information and weather reports on a single display. The on-board derived data may be from a radar, the off-board data may be from a ground-based radar. Pilot reports and other information might also be depicted graphically when available.

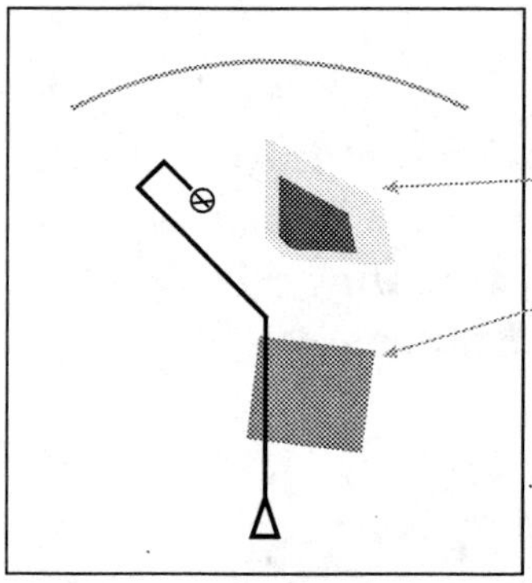

Figure 5. Nominal Weather Information: Weather Radar Data and Weather Reports Depicted

However, the aircrew would also like to be able to evaluate flight profiles (including go/no-go and divert decisions) based on weather forecasts. The preview concept can potentially be used to seamlessly switch between reports and forecasts. This is illustrated in Figure 6. Here, the display preview time has been moved into the future, and weather forecasts are depicted on the displays.

As discussed earlier, the flight crew might also like to see recorded weather information over the last few hours to see trends. This is recognized as a good reason to allow the predictive situation display to also be a recorded situation display depending on the look-ahead/look-behind time selected.

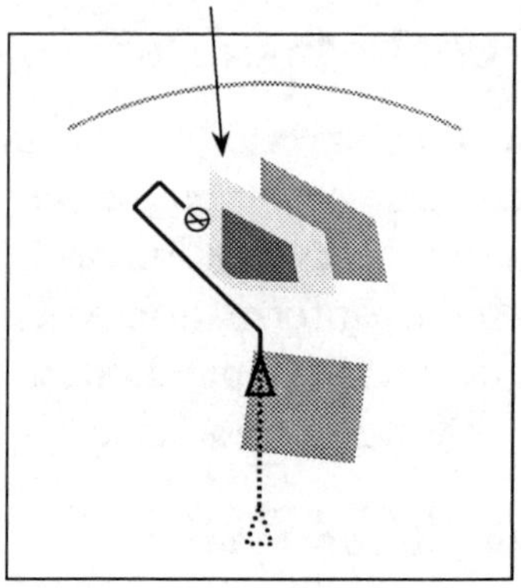

Figure 6. Weather Forecast Information: When Displaying Situation for a Future Time, Weather is Based on Forecasts Rather Than Weather Radar or Weather Reports

Airspace Status

The depiction of airspace status, both in terms of geometry and in terms of schedule, is similar to the problem of depicting certain types of weather information. Effectively, an area of known thunderstorms is a moving/evolving restricted airspace.

For example, a Military Operations Area could be shown as active for the scheduled time interval. A conflict alert similar to that used for aircraft traffic conflicts might also be depicted if the predicted aircraft path will conflict with the restricted airspace during a scheduled time.

It is worth noting that this concept might be expanded into a practical way to do air traffic control, where aircraft are assigned time-dependant airspace.

Terrain Clearance

A terrain database is recognized as an effective information source for evaluating flight plans with regard to terrain clearance. Automatically generated flight profiles can be generated with terrain clearance. Aircrew modified flight plans can be evaluated with regard to terrain clearance by automation. In any case, the final active flight plan can be depicted against the terrain database to the aircrew so that they might alter it based on more or less conservative criteria as appropriate.

Altitude vs. Time Display Format Concept

In order to visualize future traffic conflicts and hazards in three dimensions, some method of depicting four-dimensional data on two-dimensional displays needs to be employed. Perspective displays conformal to the external view from the aircraft may have an advantage during a very-near-term tactical situation, but some other vantage-point is required for mid- or long-term situations.

One concept that lends itself to this function is to complement conventional horizontal situation displays with an altitude vs. time display. This is similar to a profile display (showing altitude vs. down-range distance, or altitude vs. distance ahead of the aircraft along its heading or track). Figure 7 shows a notional version of such a display on the right, along with a notional horizontal situation display on the left. Common symbology can be used between the displays.

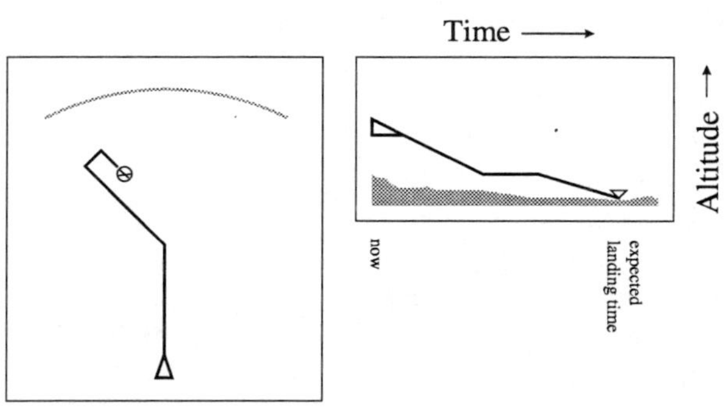

Figure 7. Altitude vs. Time Display Complementing a Notional Horizontal Situation Display

The altitude vs. time display can be used to depict terrain in much the same way as a profile display; that is, the terrain height of the area that the ownship is predicted to be over at a future time can be shown on the display. A required time of arrival, an element of a 4-D clearance that some have proposed, or a runway-use time slot can potentially be easily visualized since time is an axis on the display.

The use of these displays as predictive situation displays is shown in Figure 8. Here, the look-ahead time is shown graphically on the altitude vs. time display.

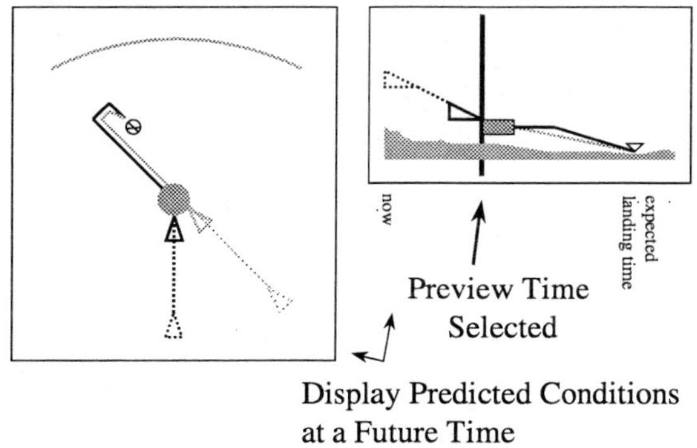

Figure 8. Notional Horizontal and Altitude vs. Time Predictive Situation Displays

In a conventional profile display where down-range distance is shown on the horizontal axis rather than time, angles of the flight path on the display are proportional to predicted aircraft flight path angle. In this altitude vs. time display, angles of the flight path on the display are proportional to predicted aircraft vertical speed.

Relationship to Farther-Term Advanced Interface Technologies

The concepts presented in this paper coupled with some more imaginative interface technologies being explored by researchers, yields some exciting potential applications. One approach would be to apply Three Dimensional (3D) display techniques. The hardware requirements of such a system could be limited to a head tracker, which would be used in conjunction with a conventional display to provide feedback on the viewpoint of the observer. The system would maintain the image in an apparent fixed location in space and movement of the head would create a parallax shift between objects in the display. Juxtaposition, or the eclipsing of one object by another, is a powerful visual cue in developing a perception of depth. The parallax feature would allow the operator to selectively optimize his vantage to observe either juxtaposition or relative distance. Stereo optic cues could be added to complement the parallax by use of shutter glasses that switch in phase with the display to ensure that each eye sees the displayed scene from its own perspective.

Another means by which a pilot could develop a future path for the aircraft would be to allow him to fly the aircraft in the future (predicted) world. In this scenario the pilot would require not only a 3D strategic display, as was described previously, but he would also require a tactical display with sufficient field of view to allow navigation of hazards. The Head Mounted Display (HMD) is an example of such a display since it has both the excellent field of view and the capability of full or partial stereo field overlap. To generate a path with this system, the pilot could actually fly in a future world in standard, accelerated, or decelerated time depending on the complexity of the task and the time available to develop the new path. There would be advantages of such a system for a stealthy aircraft negotiating rough terrain in hostile territory while avoiding fixed and moving threats and remaining cognizant of ownship radar vulnerabilities. In a commercial application, free flight may include similarly complex tasks that might benefit from such an approach.

Summary and Future Work

A concept for expanding upon the notion of a situation display to seamlessly include recorded past and predicted future situation information was presented. It was motivated by advances in airborne processing capabilities and datalink communications. It was arrived at as a way to apply these technologies to improve the ability of flight crews to evaluate decisions and manage workload. Several design issues associated with such a concept were discussed, including:

- Methods for Dealing with the accuracy and uncertainty of predicted information
- Selection and control of look-ahead/look-behind times
- Ways of depicting several specific types of information.

A low-fidelity simulator is planned to be utilized to accomplish an initial evaluation of this concept. Ongoing work includes a prototype of the predictive situation display with traffic and weather information. Techniques for look-ahead time control and handling large changes to predicted data are among the first to be examined. In addition, the implications of giving flight crews access to predicted data will be examined. Among these, better workload management, reduced error rate, and improved response to alerting systems are hypothesized.

SPATIAL TERRAIN DISPLAYS: PROMISES AND POTENTIAL PITFALLS

Eric Theunissen, Delft University of Technology, Delft, The Netherlands.

Abstract

One of the conditions for a controlled flight into terrain to occur is that the pilot is unaware of the terrain until it is too late. The technology is available to provide the pilot with an artificial presentation of the surrounding terrain. The way this technology will be implemented determines whether it actually contributes to an increase in safety. This paper addresses the potential of terrain data presentation to increase safety, the dangers associated with unintended use, and the likelihood of unjustified, presentation induced confidence and confirmation bias.

Introduction

In the last 20 years, more than half of the fatalities in western built jet airliners have been attributable to controlled flight into terrain (CFIT) (Aerospace, 1995). Research into the integration of terrain contour data on the navigation display (Kuchar and Hansman, 1993a) demonstrated that this increases the likelihood of detection of trajectories which result in a CFIT. Research into pilot avoidance maneuvers with so-called cockpit displays of traffic information (CDTI) has demonstrated that when pilots are presented with a plan-view of the situation in which altitude is presented at discrete locations, most of the escape maneuvers will be performed in the horizontal dimension (Ellis et al., 1987). Kuchar and Hansman (1993b) compared three different terrain alerting display formats and report that with the plan view display 80% of the time lateral maneuvers were executed versus 30% of the time with a perspective display and 5% of the time with a vertical profile display. From these studies it can be concluded that an emphasis on a certain spatial dimension causes a bias in the decisions, and certainly not always yields the best solution. Research has also demonstrated that the spatially integrated presentation of traffic will allow pilots to better determine the available options as compared to a plan-view presentation (Ellis et al., 1987). Based on these findings it can be argued that spatially integrated presentation of terrain data in the appropriate frame of reference is likely to have advantages relative to a color coded plan view of the terrain profile. There are many different ways to display spatially integrated data, and the frame of reference determines to a large extent for which applications the presentation is best suited. Therefore, the selection of the frame of reference should be based on task dictated information requirements. Often there are conflicts between guidance and awareness related design requirements. Furthermore, the level of realism conveyed by certain presentations can increase confidence in the conveyed information to a level which cannot be justified from the integrity of the data. Such unjustified confidence can result in a confirmation bias. Unintended use, combined with a confirmation bias and an error in the data can result in CFIT. This may seem an unlikely combination, but most combinations of events which resulted in a CFIT seemed unlikely. The rest of this paper will discuss new developments in the area of spatial navigation displays, address their

0-7803-5086-3 /98/$10.00 ©1998 IEEE

potential to increase safety, and identify a number of pitfalls which must be prevented.

The potential to increase safety

In the 1992 crash near Strasbourg of an Airbus A320 from Air Inter, a wrong Mode Control Panel entry resulted in a vertical speed 3300 ft/min instead of the desired flightpath angle of -3.3 degrees, thus yielding a forcing function error[1]. In the 1995 crash near Cali of an American Airlines Boeing 757, an erroneous flight management system (FMS) entry resulted in a forcing function error. After detection, pilots were not aware of dangerous terrain when taking corrective action, and only detected the terrain when it was too late. In the 1996 crash of an Air Force CT-43A, one of the factors was a too low MDA published by the Croation authorities. The absence of a working glideslope yielding a too large position error is one of the factors which resulted in the 1997 Korean Air crash on Guam. A software error in the Minimum Safe Altitude Warning (MSAW) system prevented timely detection by air traffic control (ATC). Confusing instructions from ATC, resulting in an erroneous forcing function, are likely to have played a role in the 1997 crash of a Garuda A300 near Medan. The limited visibility prevented a timely detection of the terrain.

To identify possibilities to reduce CFIT, an analysis of the different combinations of events which result in a CFIT is needed. Fig. 1 shows an overview of the combinations of events that are required for a CFIT to occur. Note that at the top of the tree an AND gate requires a forcing function error, a fixed object presented in the trajectory described by the forcing function, and no timely detection to occur. If one of these requirements is not met, a CFIT does not occur. One requirement for CFIT is that the pilot is unaware of the danger until it is too late. Thus, it is correct to conclude that a correct awareness of the surrounding terrain with a sufficient time horizon should eliminate CFIT.

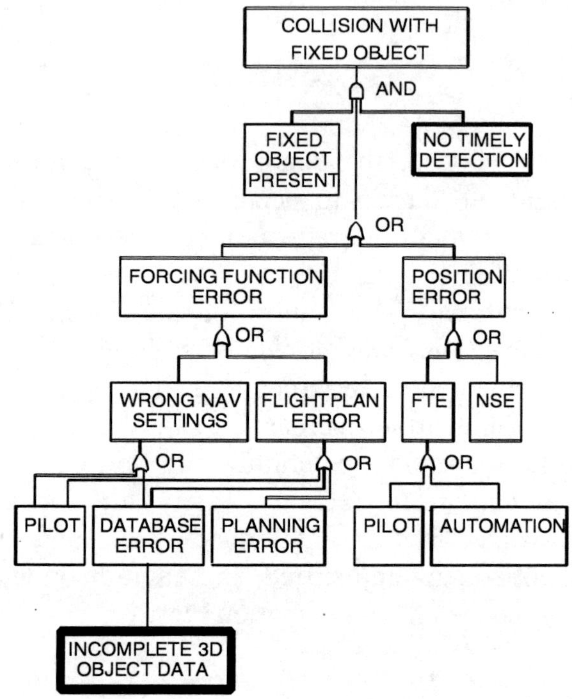

Figure 1. Overview of the combinations of events that are required for a CFIT to occur.

Terrain depiction

The presentation of terrain data can serve two different goals:

1. To increase the awareness of pilots regarding the surrounding terrain and support the timely detection of situations which get them too close to the terrain.
2. To use the terrain data for navigation and guidance purposes.

A direct visualization of the forcing function resulting from the mode control panel

[1] The forcing function refers to the data which describes the trajectory to be tracked. In case the data represents a trajectory which is different from the intended onr, this is refered to as a forcing function error.

(MCP) and FMS in combination with a depiction of the terrain increases the likelihood that a trajectory into the terrain is detected in time. The predictive capability allows pilots to see future results of their own actions during manual control or the automatic flight control system (AFCS) during supervisory control.

When using the data for guidance, the pilot determines the future trajectory based on available space indicated by the display. This concept originated in the military community. Way et al. (1984) describe the evaluation of pictorial formats which include a presentation of terrain obstacles. Young and Siferd (1990) describe an integrated terrain access and retrieval system (ITARS), in which terrain data '*is retrieved, decompressed, and distributed to avionic subsystems for enhanced navigation, terrain following, terrain avoidance, threat avoidance, sensor blending, and to any other system that requires spatial data*'. ITARS generates three simultaneous situation displays, a plan view format, a perspective view format, and a Head-Up display format. The clearance plane format is similar to the concept described by Kuchar and Hansman (1993b) as ownship relative altitude coding. Synthetic vision has also been proposed in the context of the high speed civil transport (HSCT) program to provide the pilots with a view of the runway without having to use a moveable nose as in the Concorde. In the context of the autonomous airplane, providing a synthetic view of the outside world is proposed to allow pilots VFR alike operations in IFR conditions. In Germany, synthetic vision has been pioneered by Institute of Flight Dynamics and Flight Control of Munich University of Technology. M ller and Sachs (1994) describe the architecture of a system using terrain elevation data, digital maps, and digital feature analysis data (DFAD). Actual flight tests of this system combining perspective guidance symbology with terrain data depiction were first performed in 1994 (Sachs et al., 1995).

Database integrity

It is unrealistic to assume that terrain databases are completely error-free. The problem of assuring sufficient integrity of a terrain database can be compared with that of a navigation database. Neither the makers of navigation systems or aircraft guarantee the accuracy of data in the navigation database, nor are the data checked as part of the flight management system certification process. The impact of a database error on overall safety depends on the operational procedure. When the 3-D terrain data is only used to increase the likelihood of timely detection of dangerous terrain, and other independent data is used as a reference for navigation, an error in the terrain database can cause a false alarm or a missed detection. The likelihood of a CFIT is the likelihood of an error in the flightplan which vectors the pilot into terrain multiplied by the likelihood of an error in the terrain database which causes a missed detection. Since with current displays the likelihood of detecting terrain in bad visibility is smaller than the likelihood of an error in the terrain database, displaying terrain data to increase terrain awareness can yield a potential increase in safety. However, when the 3-D terrain data is directly used for navigation in a sense that pilots use the image of the terrain to determine their future trajectory, an error in the database which goes undetected may cause a CFIT. The presentation of terrain data certainly has the potential to increase safety by improving terrain awareness, but for guidance purposes the data integrity requirements are substantially higher. The method of terrain data presentation has a significant influence on the likelihood of unintended use. *A system which aims at increasing terrain awareness but is not based on data with sufficient integrity for navigation purposes, should not provoke the pilot to use it for guidance.*

Realism, level of detail, unintended use, and confidence

With today's technology, a whole range of implementations, varying from planar color coded contour displays to almost photorealistic renderings of the surrounding environment are possible. With respect to a synthetic vision system (SVS), Jenkins and Kaiser (1994) state that '*whether synthetic imagery should recreate contact displays or provide optimal flight guidance cues*' is a fundamental design issue which still has to be resolved. Corwin et al (1993) discuss Synthetic Terrain Imagery which is based on the latter approach. They state their goal as: '*The intent of STI for HMD is to provide the pilot with the minimum necessary visual cues, not a virtual world*'.

For terrain awareness, a display does not need to provide flight guidance cues. To increase terrain awareness, one might consider a terrain display which shows a spatially integrated presentation of the terrain data. Fig. 2 shows an example of a guidance display format which includes a terrain depiction. It uses color coding to convey altitude cues, and texture to provide additional depth cues. The display system used consists of commercial-off-the-shelf components and costs less than 3k$.

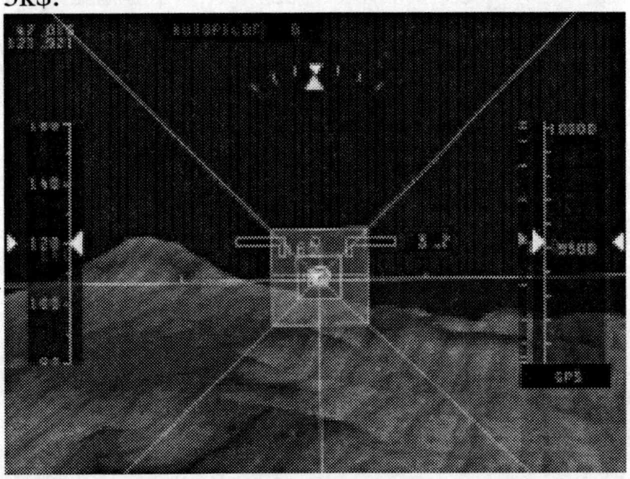

Figure 2. Example of a guidance display format which includes a terrain depiction.

Unfortunately, integrating terrain data for awareness purposes in a display meant for guidance introduces new problems. Earlier in this paper it was already indicated that the unintended use of terrain data for guidance and control should be prevented. It is certainly not unthinkable that a spatially integrated presentation of terrain data provokes unintended use for guidance purposes. The problem becomes even worse when people disregard the data which is intended to be used. The underlying problem is that the way data is presented influences the confidence people have in the correctness of the information they obtain from this data. The higher the fidelity of the depiction of the terrain data, the likelier it is that pilots will believe it and disregard information which tells them something is wrong. As stated by Wickens (personal communication), '*the more realistic it looks, I believe, the less likely it will be for pilots to retain some healthy mistrust in less than fully reliable data bases*'. In a study into key parameters for augmented vision systems, Greig et al. (1997) report that '*The transition from a well- defined SVS image to a blurred EVS image tended to reduce overall confidence in the display*'. This stresses the problem that confidence is influenced by image quality and perceived realism.

Thus, besides the fact that certain presentations might provoke unintended use, the likelihood increases that data contradicting the information obtained from the terrain depiction is disregarded. *From a safety point of view, any presentation induced confirmation bias is simply unacceptable.*

How to prevent pitfalls

The previous discussion illustrated that just simulating the outside world as seen through the window is not likely to be the best solution to provide terrain awareness with a

0-7803-5086-3 /98/$10.00 ©1998 IEEE

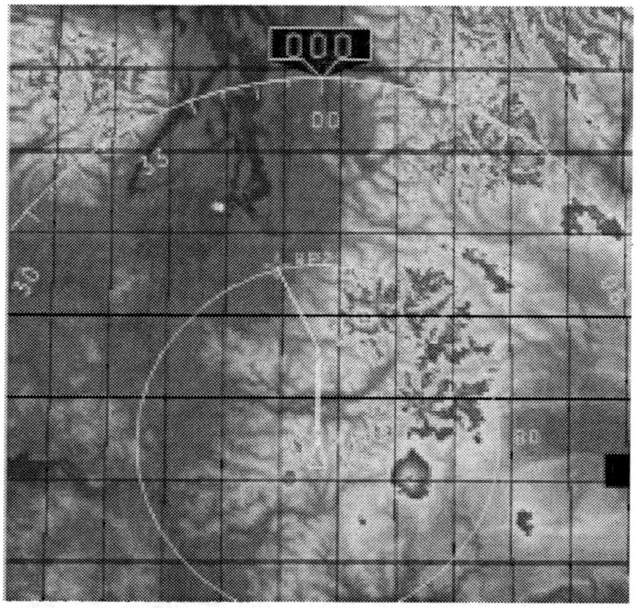

Figure 3: Example of a plan view terrain display.

With today's electronic flight instrument (EFIS) displays, data for aircraft control is presented on the primary flight display (PFD), and data for navigation on the navigation display (ND). Display concepts which include an egocentered presentation of the future trajectory on the PFD integrate guidance requirements by providing the pilot with information about constraints that can be used for control. An integration of terrain data on the PFD as for example presented in Fig. 2 necessitates an egocentric field of view. For integration of terrain data on the conventional ND, a plan view would be required. The limited field of view of an egocentric frame of reference always imposes restrictions on the amount of visible data required for terrain awareness.

When using a plan view of the terrain, similar to maps, color coding can be used to indicate different terrain altitudes (Fig. 3).

To minimize cognitive processing, objects indicating obstacles and/or threats should be presented in such a way that the presentation elicits spontaneous recognition. Kuchar and Hansman (1993b) compared the presentation of terrain contour data relative to mean sea level with a presentation of terrain contour data relative to ownship.

They report that the depiction relative to ownship yielded a significantly lower response time in the task of determining whether the route does not conflict with the terrain, and that pilots preferred the ownship-relative display.

In certain situations, it might be necessary for the pilot to focus his attention on a specific object, for example in case the object poses a potential hazard. Attributes such as color, intensity, blinking, and magnification can be used to emphasize such an object. Since the attention of pilots is influenced by their expectations and motivation, features must be used that are strong enough to attract their attention regardless of a certain confirmation bias.

Clearly, integrating terrain data on the display used for guidance and control, with its inherent restrictions imposed by the limited field of view, is not the best solution to provide terrain awareness. Several studies have demonstrated that with respect to awareness, a well selected exocentric frame of reference is superior to an egocentric frame of reference. The advantage of an exocentric perspective display is that the additional spatial dimension allows terrain altitude to be directly represented, and color coding can be used to provide redundancy. Fig. 4 shows an example of an exocentric view.

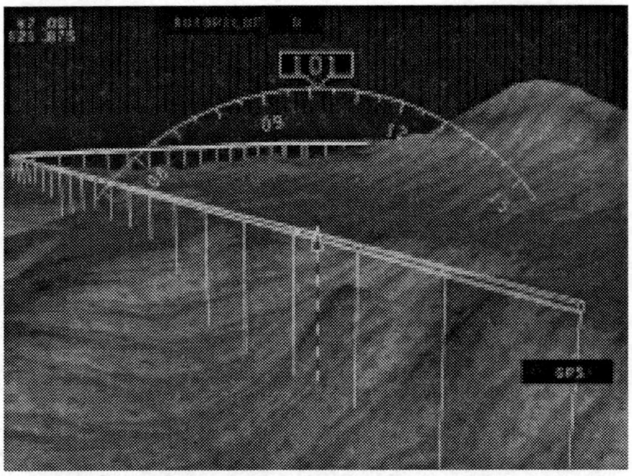

Figure 4. Example of an exocentric terrain display.

Wickens and May (1994) compared the ability of air traffic controllers and pilots to determine the aircraft state relative to terrain with a planar color coded representation versus a perspective view. They report that air traffic controllers were less accurate in determining whether the trajectory was safe with the perspective display. Furthermore, they report that little difference was found in the ability to vector a plane away from the mountains, and that with a flight planning task, the 2-D display resulted in a better accuracy but the 3-D display yielded a more efficient route in terms of altitude changes. The superiority of the plan-view in the terrain conflict detection task may have been caused by a less than optimal selection of the viewpoint in the exocentric frame of reference. Guidance with an exocentric display is not impossible. Olmos et al. (1997) describe an exocentric display for guidance purposes. To provide the pilot with sufficient guidance cues, a track-up format was proposed and several additional elements were integrated into the display. A predictor extension line showed the future heading, a color coded altitude drop line provided information about altitude deviations, and an attitude indicator provided information about pitch angle. The projected ground track, however, did not follow the terrain profile. In contrast, the exocentric display shown in Fig. 4 is not necessarily either a track-up or north-up format. To show the clearance along the planned route, the altitude poles of the flightpath follow the terrain profile, and no specific elements to support the guidance task have been integrated into the display. Furthermore, it is assumed that a separate display providing the information needed to performance the guidance and control task is available.

Technical feasibility

Based on the requirement of generating an artificial depiction of the environment, certain similarities with visual systems for flight simulators exists. However, there are important differences in objectives which influence the representation of the outside world. With the visual systems used in flight simulators, there is a continuous strive towards an increase in realism. This is typically achieved through implementation of all kind of details. With flight simulators the strive for maximum visual realism is driven by the assumption that the higher the fidelity of the visual scene, the better the training and transfer will be. To what extent this assumption holds is not known. As indicated by Owen (1996), '*realism and fidelity have received more attention than they deserve, given how little is known about what specific information is needed to support the simulation of particular tasks*'. With terrain data depiction for awareness purposes, the question regarding the level of detail of the representation must be addressed. As indicated previously, an increased level of visual realism tends to increase confidence in the correctness of the information conveyed by the presentation. This leads to the question: '*To what abstraction level can the representation be reduced while still allowing spontaneous recognition?*'. In the same context, another question which should be asked is: '*Does the use of realism enhancement techniques provide task relevant information or is it just more data?*'. The fact that we do not always know whether a certain item might convey tasks relevant information should not be a justification to include it. Textures increase realism by increasing the level of detail per polygon. A high level of realism is typically achieved by using phototextures. The use of textures is useful for increasing the spatial frequency per polygon, and in such a way can provide additional cues to allow estimates of relative position to be made and ambiguity to be resolved. However, use of phototextures also is likely to suggest the availability and correctness of information to a level of detail which just is not there. Wickens (1984) claims that '*there is emerging evidence that texture can be represented by a relatively small number of spatial features and may, in fact, be perceived analytically in terms of these spatial features, because the high degree of redundancy in texture allows much of its information contents to be captured by just a few levels of spatial frequency*'. Owen (1990) discusses the existence of an optimal texture density for task performance. He hypothesizes on the reason for the existence of an optimal density but indicates that '*the finding of optimal texture density may temper the current drive toward greater detail realism*'.

The components needed to generate a real-time presentation of terrain combined with flight guidance symbology are:

1. Powerful graphics processor.
2. Accurate 3-D position.
3. Database with terrain data.

New ground proximity warning systems such as the EGPWS rely on a massive onboard database in which a terrain altitude map is stored. Although this data may certainly not be used for navigation and guidance, it could be used to provide a more advanced depiction of the future flightpath relative to the terrain. Graphic processors with sufficient performance to render a realistic terrain image in real-time cost less than 30$. A compact prototype system based on commercial off-the-shelf components has been build for less than 3k$. Clearly, computer technology is not a bottleneck anymore.

Conclusions

A correct awareness with a sufficient time horizon of the surrounding terrain should eliminate CFIT. The integrity of today's terrain databases is far too low to allow the use of terrain data for navigation and guidance purposes. Since confidence is influenced by image quality and perceived realism, certain presentations of terrain data might provoke unintended use. With increased visual realism, also the likelihood increases that data contradicting the information obtained from the terrain depiction is disregarded. From a safety standpoint, any presentation induced confirmation bias is simply unacceptable. Given the problems with the data integrity, the objective of a terrain display can be summarized as:

1. Provide terrain awareness.
2. Increase the likelihood of detection of conflicts between the future flightpath and the terrain.
3. Avoid the use of the terrain awareness display as a guidance and navigation device.

Integration of terrain data into the primary flight display increases the likelihood that terrain data is used for guidance. The risk of unintended use for guidance can be reduced by selecting a planar view or exocentric frame of reference with appropriate design parameters. Computer technology is not a bottleneck anymore.

Acknowledgments

The research into spatially integrated data presentation for navigation, guidance, and control is performed in the context of the Delft Program for Hybridized Instrumentation and Navigation Systems phase II (DELPHINS II), which is being sponsored by the Dutch Technology Foundation STW.

References

Aerospace (February 1995). 'Controlled flight into terrain', pp. 16-19.

Corwin, W.H., Probert, A., and Royer, R. (1993). 'Synthetic Terrain Imagery for Helmet-Mounted Display', Proceedings of the 7th International Symposium on Aviation Psychology, pp. 108-114, Columbus, OH.

Ellis, S.R., McGreevy, M.W., Hitchcock, R.J. (1987). 'Perspective Traffic Display Format and Airline Pilot Traffic Avoidance', Human Factors, Vol. 29, No. 4, pp. 371-382.

Greig, I., Shaw, G. and Lumsden, B. (1997). 'Augmented Vision Systems for the All Weather Autonomous Landing Guidance of Fixed Wing Aircraft: An Investigation of the Key Parameters by Piloted Flight Simulation'. Proceedings of the 10th CEAS Conference, Amsterdam, The Netherlands.

Jenkins, J.P., and Kaiser, M. (1994). 'A NASA View of Synthetic Vision Systems', Proceedings of Synthetic Vision Systems for Civil Aviation, ISBN 1 857681851, Royal Aeronautical Society, London, UK.

Kuchar, J.J. and Hansman, R.J. (1993a). 'An Exploratory Study of Plan-View Terrain Displays for Air Carrier Operations', The International Journal of Aviation Psychology, 3(1), pp. 39-54.

Kuchar, J.J. and Hansman, R.J. (1993b). 'Part-Task Simulator Evaluations of Advanced Terrain Displays', SAE Aerotech '93 paper 932570, Costa Mesa, CA.

M ller, H. and Sachs, G. (1994). 'Synthetic Vision for Enhancing Poor Visibility Operations', IEEE AES Magazine, Vol. 9, No. 3, pp. 27-33.

Olmos, O., Liang, C., and Wickens, C.D. (1997). 'Electronic Map Evaluation in Simulated Visual Meteorological Conditions', The International Jounral of Aviation Psychology, 7(1), pp. 37-66.

Owen, D.H. (1990). 'Perception & control of changes in self-motion: A functional approach to the study of information and skill' In: R. Warren and A.H. Wertheim (Eds.), The Perception and Control of Self Motion. Hillsdale, N.J.

Owen, D.H. (1996). 'Redundant Sources of Visual Information for Active Control: Additive or Confirmatory?'. Proceedings of the 1996 IMAGE Conference, June 23-28, Scottsdale, AZ.

Way, T.C., Hornsby, M.E., Gilmour, J.D., Edwards, R.E., and Hobbs, R.E. (1984). 'Pictorial Format Display Evaluation', AFWAL-TR-84-3036, Wright Patterson AFB, OH.

Wickens, C.D. (1984). Engineering Psychology and Human Performance, Charles E. Merrill Publishing Company, Columbus, OH.

Wickens, C.D. and May, P. (1994). Terrain Representation for Air Traffic Control: A Comparison of Perspective with Plan View Displays. ARL-94-10/FAA-94-2.

Young, R. and Siferd, J. (1990). 'Integration of terrain elevation and feature data for tomorrow's cockpits', Proceedings of the 9th Digital Avionics Systems Conference, pp. 525-528.

SYNTHETIC VISION FLIGHT TESTS
FOR CURVED APPROACH AND LANDING

G. Sachs, Technische Universität München, München, Germany
K. Dobler, Technische Universität München, München, Germany
P. Hermle, Virtual Prototypes Inc., München, Germany

Abstract

A computer generated synthetic vision system was developed for improving poor visibility guidance in flight close to the ground. The system comprises a 3-dimensional outside world image, integrated guidance information including novel features (perspective flight path display) and high precision navigation as basic elements. A comprehensive flight test program consisting of five test series was set up for testing for a wide range of guidance applications of synthetic vision, including approach and landing. Demanding control tasks were specified and investigated. The results of the flight tests show that the synthetic vision system enabled the pilot to precisely control the aircraft and to accurately follow the specified trajectories. All guidance performance goals were achieved.

Introduction

Improvements for guidance and control of aircraft are offered by computer generated synthetic vision which provides the pilot with a 3-dimensional information of position, orientation and the future flight path of the aircraft. This is not possible with current instrumentation which basically provide information of current aircraft position and heading.

Recent papers report on the progress which has been achieved with visual aids based on computer generated synthetic vision (Refs. 1-4). The work covers theoretical issues and experimental investigations including flight tests.

It is the purpose of this paper to present flight test results which show how the guidance of aircraft can be improved with the use of synthetic vision in demanding control tasks.

Computer Generated Synthetic Vision Concept

The basic concept of computer generated synthetic vision as applied in the flight tests consists of the following elements (Fig. 1):

– 3-dimensional pictorial presentation of outside world
  A realistic 3-dimensional pictorial presentation of the outside world is generated by a graphics computer in real-time with an adequate update rate (30 images per sec.). A stored terrain database is used which contains all relevant information on elevation and on features of the terrain.

– Integrated guidance information
  Computer generated guidance information comprising novel visual elements is combined with the 3-dimensional pictorial presentation of the outside world and presented to the pilot as an integral part of the synthetic vision image. The novel guidance elements include a perspective flight path display for curved trajectory following (guidance tunnel).

0-7803-5086-3 /98/$10.00 ©1998 IEEE

– Precision navigation
  A precision navigation system provides the synthetic vision computer with position and attitude data. This is necessary for generating an image according to the actual field of view of the pilot. In the flight tests, a high precision navigation system was applied using differential global positioning and inertial sensor navigation data.

Two groups of data are used for generating the outside world image:

– terrain elevation data
– terrain feature analysis data

The terrain elevation data are referenced to a grid structure the elements of which have a size of 3" × 3" (or 1" × 1"), Fig. 2. A grid element represents an area of about 90 m × 60 m (or 30 m × 20 m) at the geographical latitude of the areas where the flight test took place.

For describing terrain feature data, three data groups were applied (Fig. 3):

– point features (buildings, bridges, power line pylons, etc.)
– linear features (roads, railways, rivers, etc.)
– areal features (cities, forests, waters, etc.)

For airport areas, a special treatment of terrain elevation and features is applied. This concerns the runway and other airport elements (taxi ways, apron, tower, terminal etc.) which are significant for the approach and landing task of the pilot. The special treatment yields a precise modeling as regards location, elevation, dimensions, shape etc.

The synthetic vision image which is presented to the pilot on a display shows the terrain segment which is in his momentary field of view. This section is determined by computational algorithms using the viewing pyramid (Fig. 4). A distance dependent resolution is introduced with regard to the displayed segment to support the real-time capability of the image generation software. This basically means that terrain segment elements further away are presented with less resolution than those which are nearer to the pilot.

Flight Tests Results

A comprehensive flight test series was performed aiming at a wide range of guidance applications of synthetic vision. Demanding control tasks were specified and investigated. The test program consisted of five flight test series:

– Precision approach and landing flight tests
  Braunschweig airport (Germany), October 10-14, 1994
– Low level flight tests in highly curved, narrow river valley
  Altmühl river (Germany), December 12-16, 1994
– Curved and steep approaches in mountainous area
  Lugano airport (Switzerland), July 31 - August 4, 1995
– Curved/steep/short approaches and low level and terrain following flights in mountainous area
  Offenburg/Schwarzwald (Germany), March 18-22, 1996
– Curved/steep approaches and curved trajectory following flights in mountainous area
  Freiburg/Schwarzwald (Germany), July 7-10, 1997

An overview of the flight test areas is provided by Fig. 5 which shows the locations of the test courses in Germany and Switzerland. Also shown is the place of the ground reference station for the Wide-Area-DGPS mode which was applied in the non-terminal flight tests (flights in river valleys and mountainous areas). A Local-Area-DGPS mode was used for approach and landing flight tests, with the ground reference station installed at the airport.

The vehicle used in the flight test program is a twin engine Dornier 128 which is operated by the Institute of Flight Guidance and Control of the Technische Universität

Braunschweig as a research aircraft (Fig. 6). The aircraft is equipped with a high precision navigation system. The high navigation performance is achieved by coupling differential global positioning and inertial sensor systems to yield an integrated precision navigation system, Ref. 5. In addition, computer and filter algorithms including error modeling are applied. Thus, it is possible to achieve a high precision for static as well as dynamic behavior.

Results of precision approach and landing flight tests at Braunschweig airport are presented in Figs. 7 and 8. The tests consisted of flights beginning at a height of about 350 m above ground and ending at touch down. The control task was to follow a flight path based on an ILS glide slope. In addition to the synthetic vision image visual information, the pilot was also provided with acoustic information which consists of the indication of the height above ground (400 ft, ... 10 ft, 5 ft, 4 ft, 3 ft, 2 ft).

A special feature of the guidance symbology for the touch down task is a height bar with marks which provide a precise indication of the vertical distance between the aircraft and the ground. The height bar length presented in the display decreases in accordance with the vertical motion of the aircraft approaching the ground. The height bar is shown not until reaching the final part of the approach trajectory immediately before touch down. The introduction of the height bar has proven as very useful for the pilot.

Flight test results for altitude histories presented in Fig. 7 show that the pilot precisely controlled the aircraft and held it on the specified glide path. The results also show that smooth touch downs are achieved. Results for the lateral motion are presented in Fig. 8. There are only minor deviations from the specified trajectory. They are further reduced in the final phase at the touch down condition.

Results from approach flight tests at Freiburg airport are shown in Figs. 9-11. Fig. 9 presents a vertical and horizontal view on the approach track. The pilot basically had to perform a curved and steep approach where the section between points A and B is of particular concern. This section is considered in the following.

The flight test results presented in Figs. 10 and 11 show vertical and horizontal deviations from the specified trajectory. The specified trajectory was presented to the pilot with the use of a guidance tunnel the dimensions of which are also indicated in Figs. 10 and 11. The histories of the deviations show that there are only minor excursions beyond the limits set by the width and height of the guidance tunnel. As a result, the pilots were able to accurately control the aircraft and hold it on the specified trajectory.

Conclusions

A computer generated synthetic vision system was developed for improving poor visibility guidance in flight close to the ground and evaluated in a comprehensive flight test program.

The synthetic vision system consists of the following basic elements:

- 3-dimensional outside world image
- Integrated guidance information comprising novel features like a perspective flight path display for curved trajectory following (guidance tunnel)
- Precision navigation

Several flight test series were performed aiming at a wide range of guidance applications of synthetic vision. Demanding control tasks were specified and investigated, including precision approach and landing flight tests and curved/steep/short approaches in mountainous area. The results of the flight tests show that the synthetic vision system enabled the pilot to precisely control the aircraft and to accurately follow the specified trajectories. All guidance performance goals were achieved.

0-7803-5086-3 /98/$10.00 ©1998 IEEE

References

1 Rathmann, U.: Künstliche Sicht. CCG-Kurs LR 4.05, "Moderne Unterstützungssysteme für Piloten", Carl-Cranz-Gesellschaft, Oberpfaffenhofen, 1992.

2 Möller, H., Sachs, G.: Synthetic Vision for Enhancing Poor Visibility Flight Operations. IEEE Aerospace and Electronics Systems Magazine, Vol. 9, pp. 27-33, 1994.

3 Theunissen, E.: Integrated Design of a Man-Machine Interface for 4-D Navigation. PhD Dissertation, TU Delft, The Netherlands, 1997.

4 Sachs, G., Dobler, K., Hermle, P.: Flight Testing Synthetic Vision for Precise Guidance Close to the Ground", AIAA Guidance, Navigation, and Control Conference Proceedings, pp. 1210-1219, 1997.

5 Vieweg, S., Schänzer, G.: Präzise Flugnavigation durch Integration von Satelliten-Navigationssystemen mit Inertialsensoren. DGLR-Jahrbuch 1992 I, pp. 171-177, 1992.

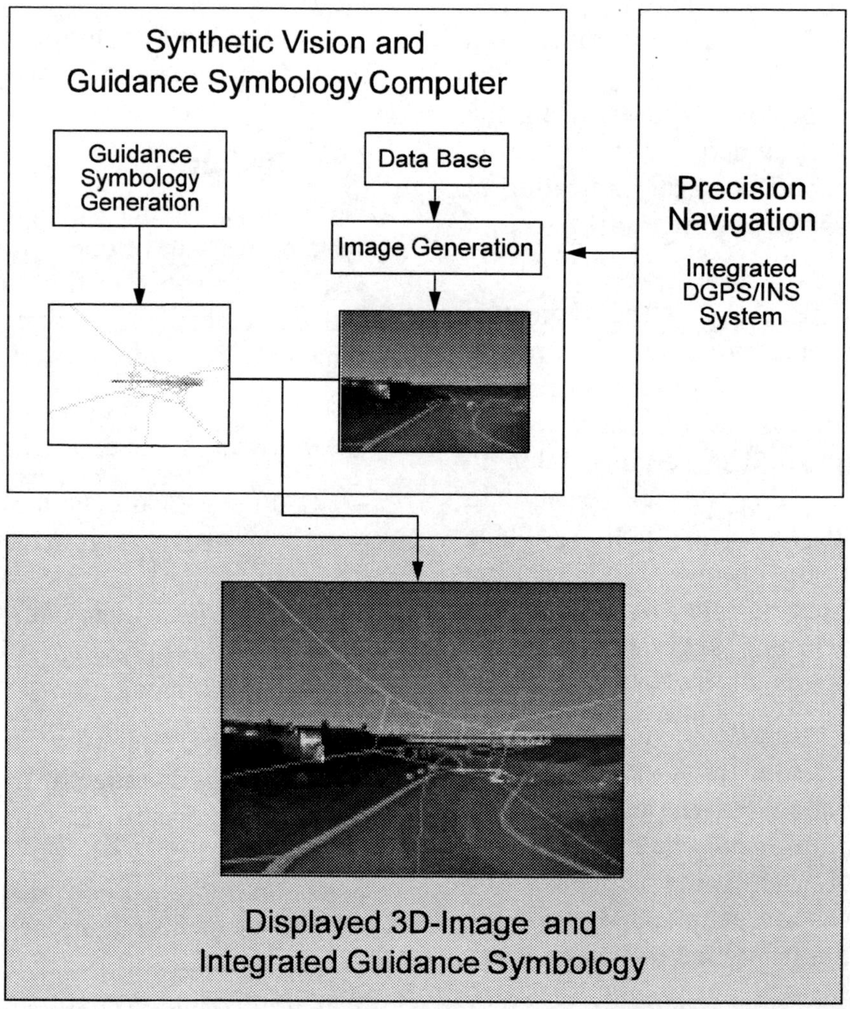

Fig. 1 Computer generated synthetic vision with integrated presentation of outside world image and guidance information (perspective flight path display)

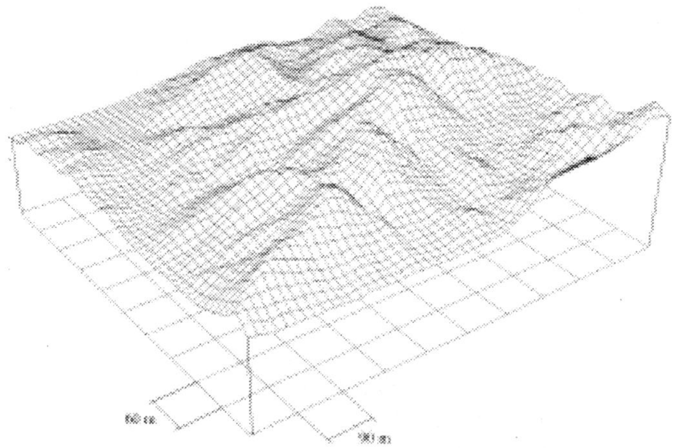

Fig. 2 Terrain elevation data

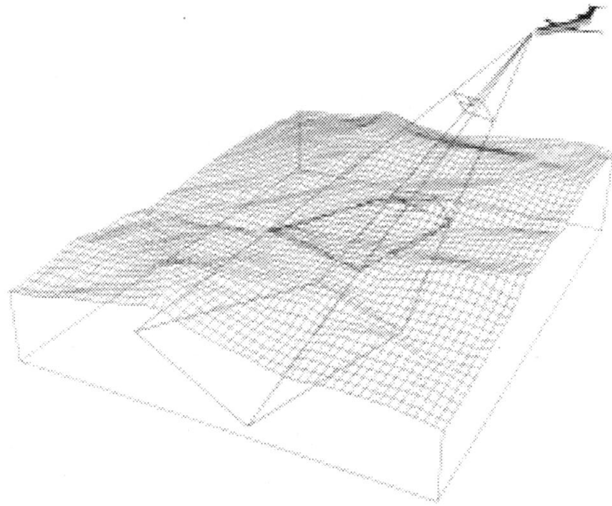

Fig. 4 Viewing pyramid for deteriming terrain segment displayed to the pilot

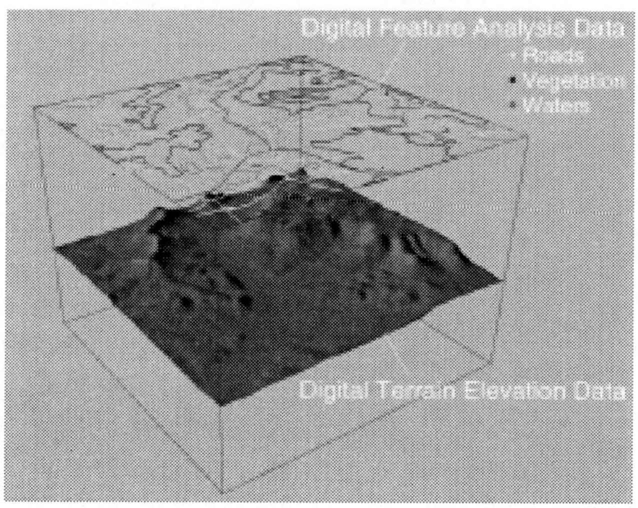

Fig. 3 Combining terrain feature analysis and elevation data

Fig. 5 Flight test areas and ground reference station for Wide-Area-DGPS

0-7803-5086-3 /98/$10.00 ©1998 IEEE

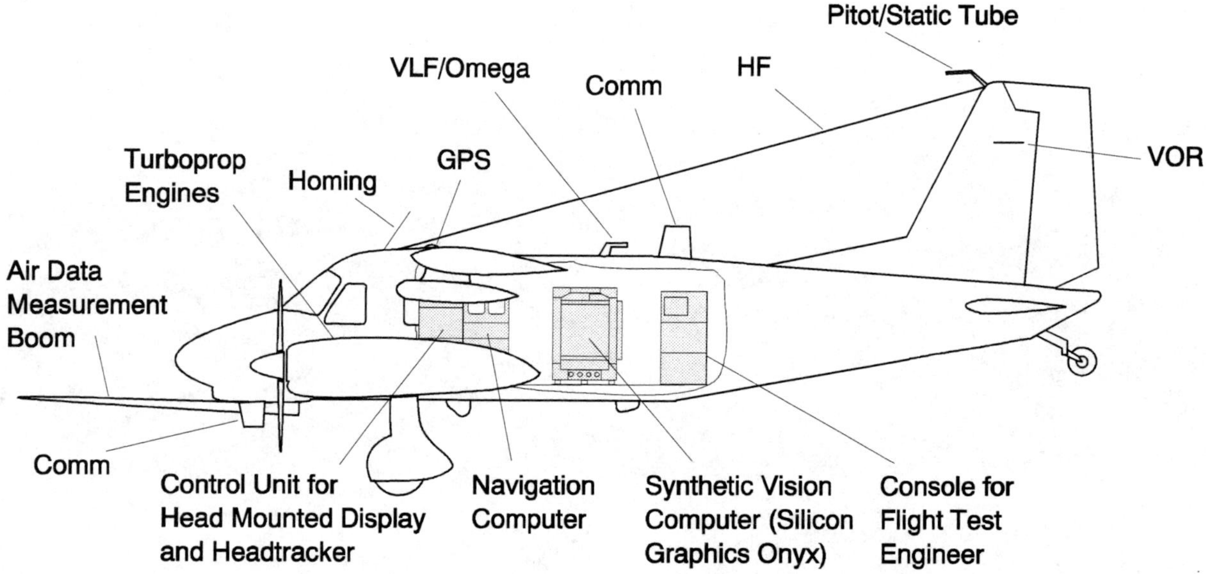

Fig. 6 Aircraft with test equipment

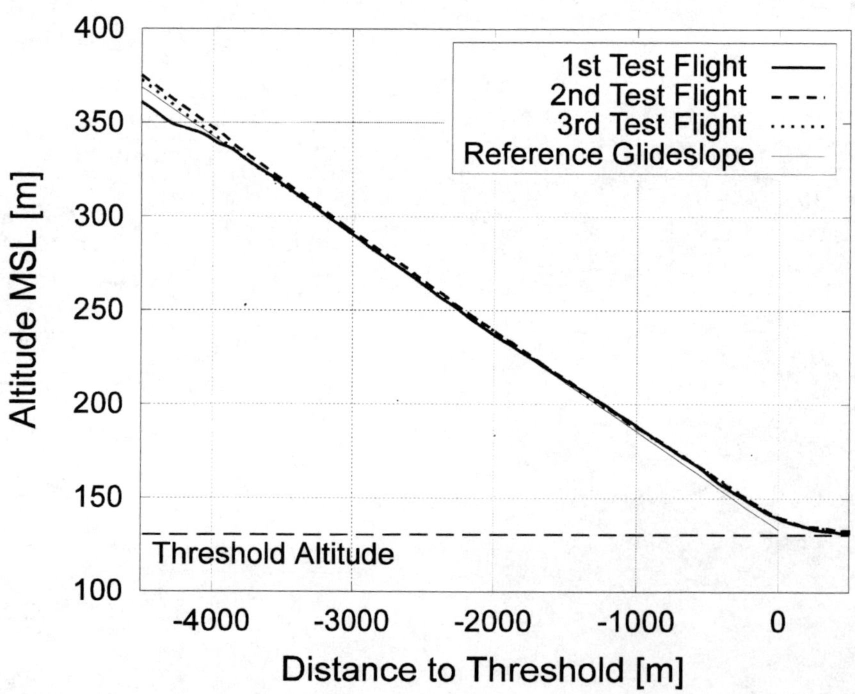

Fig. 7 Altitude history of precision approach and landing flight tests at Braunschweig airport

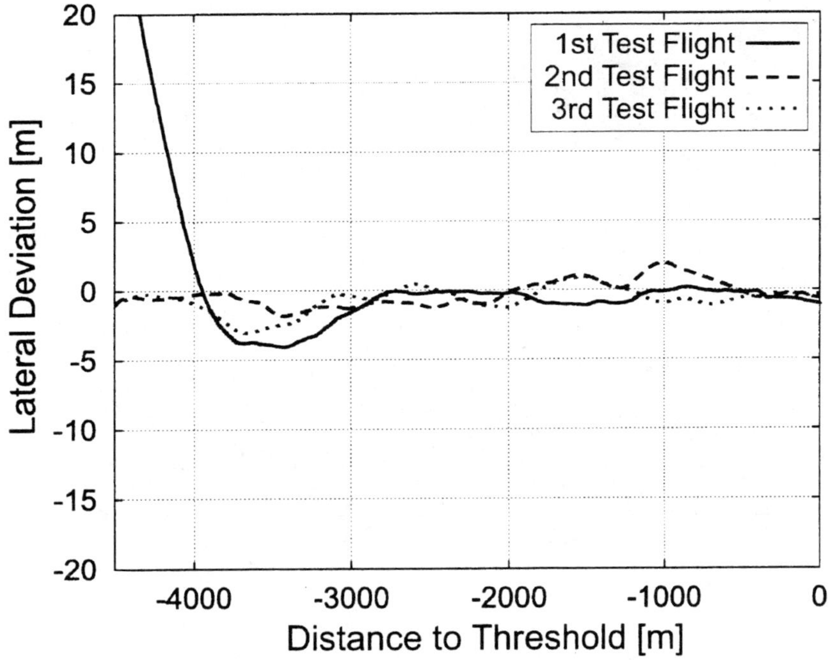

Fig. 8 Lateral deviations during precision approach and landing fligh tests at Braunschweig airport

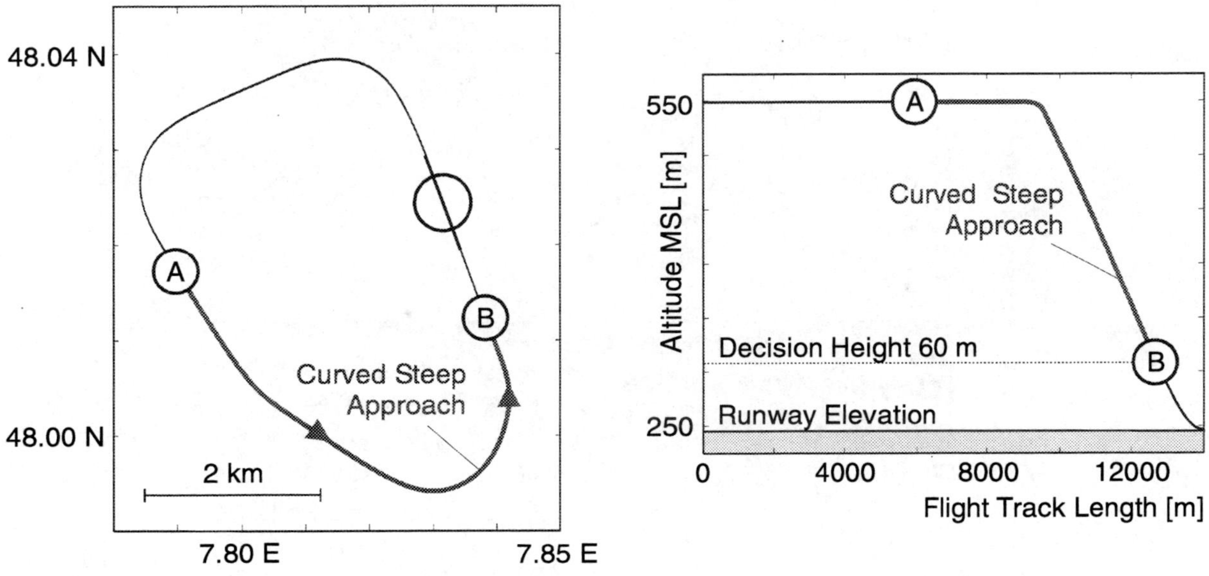

Fig. 9 Approach test course at Freiburg airport

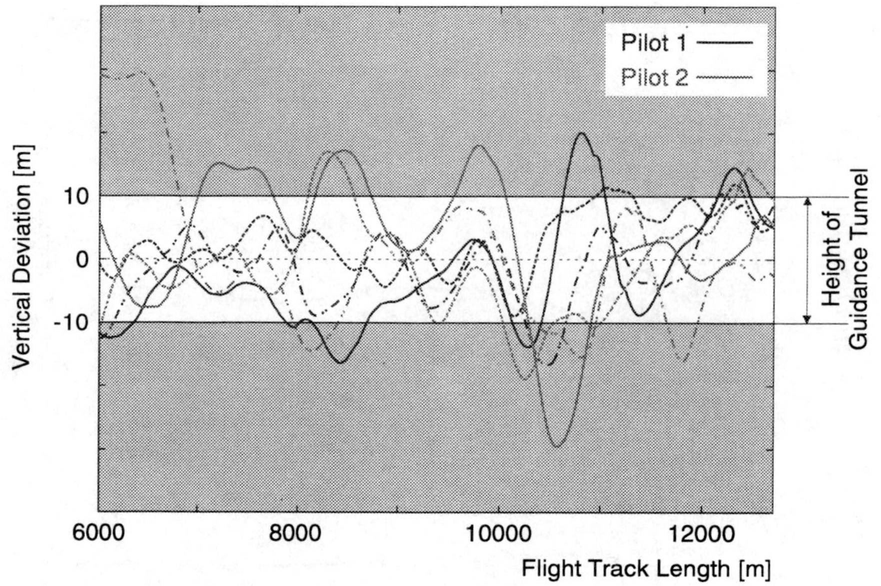

Fig. 10 Vertical deviations from nominal trajectory of Freiburg airport approach flight tests

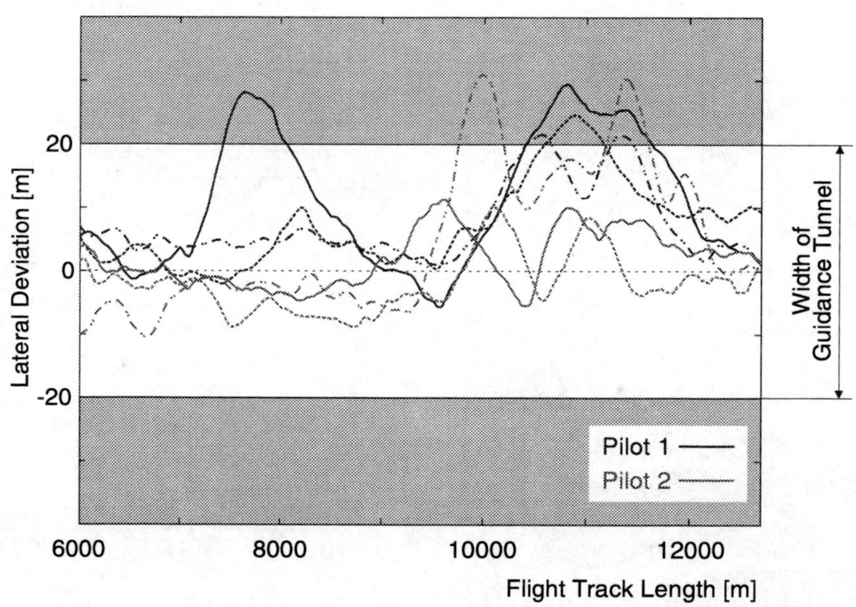

Fig. 11 Lateral deviations from nominal trajectory of Freiburg airport approach flight tests

GPS USER-INTERFACE DESIGN PROBLEMS

Kevin W. Williams, FAA Civil Aeromedical Institute, Oklahoma City, OK

Introduction

This paper is a review of human factors problems associated with the user-interface design of a set of Global Positioning System (GPS) receivers, certified for use in aircraft for instrument non-precision approaches. No GPS products will be mentioned by name, since the aim of the paper is not to criticize a particular GPS manufacturer. Instead, the paper will focus on design problems associated with the interfaces and specific inconsistencies across the set of interfaces that could cause confusion or errors during operation. Some specific problems to be addressed involve the layout and design of knobs and buttons; control labeling inconsistencies across units; the placement and use of warnings; feedback, or the lack thereof; and the integration of specific flying tasks while using the receivers. Recommendations for solving some of the problems are provided, as well as suggestions to the FAA, GPS manufacturers, and pilots regarding the future development and use of these products.

Data Collection

Most of the human factors problems reported in this paper were obtained from interviews with subject matter experts from the Federal Aviation Administration's (FAA's) Technical Programs Division, Flight Procedure Standards Branch (AFS-420). Additionally, data were collected through FAA internal memoranda (i.e., personal communications, S. Jackson, August, 1997; S. Jackson, February 18, 1998; S. Jackson, May 22, 1998, S. Winter, September 25, 1996), as well as from an FAA technical report (Winter & Jackson, 1996). Finally, data were taken from observation logs from a recently conducted operational test of a GPS Wide Area Augmentation System (WAAS), and from personal observation from the WAAS test.

Human Factors Guidelines

Initially, the FAA Aircraft Certification Human Factors and Operations Checklist for Standalone GPS Receivers (FAA, 1995) was the primary human factors reference source for GPS receiver design. The human factors and operations checklist is hereafter referred to as "the checklist." The checklist is a useful source of human factors design guidelines and includes references to several other commonly used guideline documents from both the military and civilian sectors.

Design Problems

The following design problems are not intended to be an exhaustive list. Many problems could not be included because of time and space constraints. Also, with the exception of complexity, the problems are not presented in order of importance since the importance of any particular problem cannot usually be determined without looking at how often it occurs and its effect on pilot workload and performance. Further research is required to make those determinations.

Complexity

Probably the most significant feature of GPS units, as far as the potential for user errors is concerned, is the sheer complexity involved in their operation. It is not that the manufacturers of these devices purposely designed units to be difficult to operate. Instead, the primary reason for their complexity is that they allow the performance of a large

0-7803-5086-3 /98/$10.00 ©1998 IEEE

number of tasks using a limited number of controls and a small display area.

One indicator of the complexity of a GPS unit is the size of its instruction manual. Manuals of between 100 and 300 pages are common. In addition to the large number of tasks that can be performed by a GPS unit, there is a relatively small area for the display and the controls -- the buttons, switches, and knobs -- needed to access GPS functionality. For this reason, most of the controls serve to activate multiple functions.

Taken together, the large number of accessible functions and limited number of controls will lead to the inadvertent activation of unwanted functions. When this happens, the pilot may not be familiar enough with the display configuration to correctly recover from the mistake without accessing the operations manual. The end result is confusion for the pilot, increased head-down time, and possible airspace blunders. It is unlikely that anyone would be familiar with the entire operations manual for a particular device, and it is also unlikely that a pilot would be able to know what to do in every situation without referring to the manual. One possible fix for this problem is to require an "undo" button/function on the unit. This would allow a pilot to return to the previous display configuration and proceed with the intended operation.

Knobs and Dials

The following problems have been noted for more than one GPS unit currently used for instrument approaches.

Activation Feedback

The checklist suggests that all knobs and buttons should provide adequate activation feedback to the user. Although this suggestion is part of the checklist's bench test section, it is important that activation feedback be sufficient during flight, when noise and vibration are present. At least one of the units on the market provided no tactile or auditory feedback indicating that a button had been activated. There was no noticeable tactile or auditory feedback when pressing the button, and visual feedback occurred only after a delay, when the screen display changed. A lack of feedback can lead to multiple button presses and the activation of displays and functions that were not intended.

Button Placement

The checklist indicates that the risk of inadvertent activation or deactivation of GPS functions should be minimized. An example of a violation of this principle is shown in Figure 1 and concerns the placement of 3 buttons.

The figure portrays a portion of the control interface of one of the more popular GPS units on the market. Three buttons are shown and located at the bottom of the display.

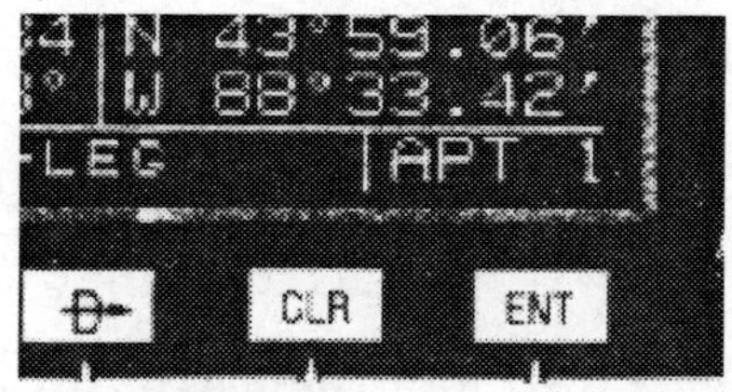

Figure 1: Example of poor button placement

They include the "direct-to" (capital D with an arrow), "CLR" (clear), and "ENT" (enter) buttons. While performing an approach, if the pilot elects to execute a missed approach, the procedure for this GPS unit (and most others) is to push the "direct-to" button and then the "enter" button. The placement of the clear (CLR) button between the direct-to and enter buttons increases the likelihood of accidentally pressing the clear (CLR) button during the execution of the missed approach procedure. In so doing, the display changes to one that is not expected, leading to confusion about how to recover. Furthermore, recovery may require extensive reprogramming of the GPS unit. This would certainly add to the already high degree of stress involved in executing an unplanned missed approach. Maintaining a specified minimum altitude and course, while trying to reprogram the GPS unit, could create a visually and mentally overloading situation. Repositioning the receiver buttons is warranted, though again, one possible solution to this problem would be an "undo" button/function.

Knob Issues

Many of the GPS units reviewed required the use of a rotary-type knob to select the information required for an operation. One example is in the selection of airport identification codes. These codes consist of three or four alphanumeric characters of the form F28, KSTL, OKC, etc., that identify various airports, VORs, and other navigational waypoints. Some receivers do not allow the input to "wrap around" from A, back to Z, using a single counterclockwise turn. Instead, to go from A to Z, the user must dial forward (clockwise) through the entire list. Also, if the users inadvertently dial past the intended character, they must continue to dial forward through the list until again reaching the correct character. This arrangement leads to significantly increased head-down time while using the receiver, a problem that has been mentioned in recent GPS studies (Williams, 1998; Wreggit & Marsh, 1998).

A second major issue concerning the use of knobs is that some receivers allow the knob to be in one of two physical positions, either out or in. Pulling the knob out enables different functionality than when the knob is pushed in. Using the knob in the alternate position (pulled out) is not required to operate the receiver, and the pilot may not be aware that it functions in this manner. A serious problem with this arrangement is that it is difficult to know whether the knob is out or in without first testing it. None of the units provide an indication of the knob's position, such as a warning light, on the display. As one might be in the habit of performing a certain function with the knob at a particular point in a flight (for example, during the set up for an approach), it is unlikely that the correct knob position will be ascertained. The unit could fail to respond in the expected manner, leading to confusion and requiring a correction. Using a procedural checklist would avoid this type of error; however, referencing a checklist would significantly increase the head-down time of the pilot.

Labeling

Annunciator Variations

Some GPS units have an associated annunciator panel to provide indications of GPS status and satellite availability to the pilot. The annunciator panel, if used, is not necessarily co-located with the GPS unit and depends on room available on the cockpit instrument panel. If the annunciator panel is not located near the GPS unit, or the navigational instrument used during an instrument approach, there is a possibility that the pilot will miss critical information displayed on the various instruments.

Also, annunciator panels can be made by a third party manufacturer, so that the panel for a particular unit can vary from airplane to airplane. The labeling of GPS modes varied across the units reviewed. Figure 2 shows two of the several different GPS annunciator panels reviewed.

In particular, notice the third box of each annunciator panel in Figure 2, labeled unit A and unit B respectively. TSO-C129a (FAA, 1996), Par (a)(3)(xi)(2), states that, "The equipment shall provide the capability for accomplishment of holding patterns and procedure turns. Activation of this function shall at least: a. Change automatic waypoint sequencing to manual. ... d. Permit the pilot to readily return to automatic waypoint sequencing at any time prior to the designated fix ('TO' waypoint) and continue with the existing flight plan."

For unit A, the annunciator button for manual interruption of autosequencing, for purposes of holding and procedure turns is called GPS SEQ, and includes two modes labeled AUTO and HOLD. Unit B labels their annunciator button GPS CRS, with the two modes labeled OBS and LEG. A third brand, not shown in the figure, does not have an external annunciator button and automatically

0-7803-5086-3 /98/$10.00 ©1998 IEEE

enters the "hold" function when outbound on an approach and enters an "auto" sequence function when inbound. As will be discussed below, there can be confusion over the labeling and operation of a panel and receiver if a pilot is required to suspend and restart autosequencing.

Button Labels

The number and variety of buttons included on each GPS display is different for every unit on the market. In addition, buttons that perform the same type of task on different units can have different labels. Table 1 presents several popular brands of GPS units, along with a listing of the buttons and their labels that are present on each of the units.

Table 1: Button Labels for GPS Units

Unit	Number	Labels
Unit A	7 buttons	CRSR (2), MSG, ALT, D->, CLR, ENT
Unit B	10 buttons	CRSR, MSG, D->, CLR, ENT, NRST, SET, RTE, WPT, NAV
Unit C	9 buttons	MSG, D->, ENT, WPT, NAV, FPL, CALC, AUX, APT/VOR
Unit D	10 buttons	MSG, EMG, NAV, DB, FPL, SYS, D->, SEL, INFO, ENT
Unit E	6 buttons	CRSR (2), D->, ALRT, MSG, ACK

In addition to the buttons listed, most of these units have one or two ganged knobs. A ganged knob consists of an outer, shorter knob, surrounding an inner, taller knob. As was mentioned earlier, the inner knob can be pulled out or pushed in to provide additional functionality for some of the units.

Keep in mind that all of the units reviewed provide essentially the same set of functions but require accessing specific functions in different ways. Some functions that are accessed through a knob on one unit may require a button press on a different unit. While most units have an "enter" (ENT) button to initiate a selection, Unit E in Table 1, for example, uses a button labeled ACK for this task. This lack of consistency and operational vagaries make it very difficult for a pilot to transition from one type of unit to another.

Procedural Problems

Selecting an Alternate Airport

The selection of an alternate airport requires many steps, since approaches can only be selected from the active flight plan and only one approach can be stored at one time. Approaches cannot be stored in inactive flight plans. The selection of an approach to an alternate is done after the decision to go to the alternate airport has been made. If the alternate airport is fairly close to the primary airport, very little time is available to accomplish the program change, as well as other required tasks. Many GPS units allow setting up a second flight plan. This feature could be used to establish a route to an alternate airport but not the approach itself. If the procedure for selecting a second flight plan is not often used, it might be difficult for a pilot to recall the details, should the situation warrant. Minimizing workload in this situation requires the pilot to be well rehearsed in the procedure beforehand. This advice, of course, holds for many of the problems discussed in this paper. The pilot could also program additional waypoints after the destination, if the receiver allows. However, as was determined during WAAS receiver testing (Winter & Jackson, 1996), some pilots do not take the time to input a complete flight plan into their GPS receiver, much less a flight plan plus additional waypoints to serve as an alternate destination.

Automatic vs. Manual Waypoint Sequencing

One of the most often cited problems occurring during the operational testing of these systems, to date, involved either placing the receiver in a mode where it automatically sequences from one waypoint to the next during the approach or in a non-sequencing mode. The non-sequencing mode is required for many overlay-type approaches (and some stand-alone approaches) in which a procedure turn, or hold-in-lieu-of procedure turn, is accomplished before establishing the aircraft on the final approach. In recently completed operational tests of these receivers (Winter & Jackson, 1996), it was reported that subject-pilots frequently forgot to take the GPS receiver out of the "hold" function after completing the procedure turn. Pilots were often unable to proceed to the next approach fix because the receiver was still in the "hold" function, and the pilot was often unable to determine the problem. Usually, the safety pilot had to prompt the subject-pilots about the failure, so as to prevent them from flying about aimlessly. Pilots were also sometimes unable to suspend autosequencing because the unit had already sequenced to the following waypoint before the pilot had taken the required actions.

Other testing found that the procedure for suspending and re-establishing automatic waypoint sequencing differed from unit to unit. For many of them, a button or switch on the annunciator panel was used to suspend and renew sequencing; however, for one unit, suspension and re-establishment of waypoint sequencing was primarily automatic. While this generally decreased pilot workload, if the pilot was required to suspend waypoint sequencing with this unit, the complex procedure was as follows:

- Press the direct-to button;
- Turn outer knob one click counterclockwise;
- Turn inner knob until proper course is selected;
- Press direct-to button again.

To then re-establish waypoint sequencing, the pilot was required to press the direct-to button twice. In addition to being non-intuitive, this procedure demonstrates the use of a button to accomplish a function for which it was not labeled and not originally designed.

Inconsistent Functionality

One final human factors problem to be discussed concerns differences in the way each GPS receiver functioned, depending on how it was installed in the aircraft. One example is found with the auto-slewing function for receivers connected to a horizontal situation indicator (HSI) used during a Distance Measuring Equipment (DME) Arc approach. GPS receivers can be installed in such a way that the HSI will automatically slew from one heading to the next during the execution of an approach procedure. A different installation with the same GPS unit will result in a configuration requiring the pilot to manually dial in new headings on the HSI while executing the same approach procedure. In this configuration, the GPS unit will present a message to the pilot each time a new heading should be selected. One possible scenario for pilots who rent aircraft is a situation in which they are given an identical airplane, with identical GPS equipment, but the instruments behave differently during the approach because of differences in the way the equipment was installed. In this situation, it may be difficult to determine if the aircraft has auto-slewing capability until after it is airborne.

A second example concerns the selection and de-selection of the GPS unit during an approach. Some receivers automatically de-select the GPS unit when an Instrument Landing System (ILS) frequency is dialed in to the navigation radio. Under this condition, the ILS display is receiving its information solely from the ILS ground station. For other receivers, the GPS signal is not de-selected unless it is done manually by pushing a button on the annunciator panel. Whether the de-selection is automatic or manual is determined solely by the installation of the GPS unit.

A situation can arise where the pilot believes that the GPS has been de-selected when it has not. This will result in a case in which the glide slope indicator on the ILS display is controlled by the ILS signal, but the course indicator on the same display is controlled from the GPS.

Conclusions

This paper was not intended to be overly critical of GPS manufacturers or currently used GPS equipment. However, it is important, given the problems stated in this paper, that efforts are made to discover the frequency of occurrence of these problems and what their effects are on pilot workload and performance. As was stated in the introduction, the primary problem with most receivers is that they have a large number of available functions, but a limited number of controls for activating those functions. Also problematic is the manner in which GPS functionality is implemented. Different procedures are required for the same functions for every receiver on the market – even some units made by the same manufacturer.

It is doubtful that easy solutions exist for all of the problems mentioned in this paper. However, opportunities for GPS improvements are presented below for the FAA, GPS manufacturers, and the end users – the pilots.

FAA

One suggested solution to the problem of receiver complexity is to reduce the number of different kinds of GPS approaches that receivers need to accommodate. The elimination of approaches (overlays and GPS approaches containing procedure turns) that require suspension of automatic sequencing of waypoints would eliminate the need for the extra functionality required to accomplish this task. Most of the approaches requiring suspension of automatic sequencing are overlay approaches. Overlay approaches are approaches that were previously established VOR or NDB approaches that have been redefined by the FAA as GPS approaches. Steve Jackson, of the FAA Flight Standards Division, has suggested that, "The GPS 'T' approach must be established as the standard to maximize GPS receiver capabilities" (personal communication, February 18, 1998). Operationally, a "T" approach would eliminate the need to suspend waypoint sequencing, and would likely reduce pilot workload during a high-workload portion of flight.

Manufacturers

Designing an "undo" function on all of these receivers would keep the pilot from becoming lost after making an entry error and reduce control inputs during critical phases of flight. Whether standards are imposed by the FAA, or volunteered by manufacturers, some standardization of button labels, annunciator panels, and displays is needed. At a minimum, a core set of GPS functions should be performed in essentially the same way for every unit on the market. John Steuernagle, of the Airplane Owners and Pilots Association Air Safety Foundation, has recommended that standardized procedures be designed for the following set of GPS functions (J. Steurenagle, personal communication, February 3, 1998):

- Selecting a waypoint
- Establishing a course to or from a waypoint
- Selecting and activating an approach
- Transitioning to a missed approach procedure
- Interruption of autosequencing.

Based on findings stated earlier, two items that could be added to this list are the re-activation of automatic waypoint sequencing, and the selection and activation of an approach to an alternate airport.

Pilots

The Aeronautical Information Manual (FAA, 1998, p. 1-1-49) recommends that, before using any particular receiver for instrument flight, pilots should practice GPS approaches under visual meteorological conditions until thoroughly proficient with all

0-7803-5086-3 /98/$10.00 ©1998 IEEE

aspects of their equipment (receiver and installation). The pilot should practice:

- Utilizing the receiver autonomous integrity monitoring (RAIM) prediction function
- Inserting a Standard Instrument Departure (SID) into the flight plan, including setting terminal Course Direction Indicator (CDI) sensitivity, if required, and the conditions under which terminal RAIM is available for departure (some receivers are not SID or STAR capable)
- Programming the destination airport
- Programming and flying the overlay approaches (especially procedure turns and arc)
- Changing to another approach after selecting an approach
- Programming and flying "direct" missed approaches
- Programming and flying "routed" missed approaches
- Entering, flying and exiting holding patterns, particularly on overlay approaches with a second waypoint in the holding pattern
- Programming and flying a "route" from a holding pattern
- Programming and flying an approach with radar vectors to the intermediate segment
- Indication of the actions required for RAIM failure both before and after the Final Approach Waypoint (FAWP)
- Programming a radial and distance from a VOR (often used in departure instructions).

In addition, Steve Winter, of the FAA Flight Standards Division recommends familiarization with the following additional procedures (S. Winter, personal communication, September 25, 1996):

- Recovering from flying past a waypoint where holding was intended, after failing to place the receiver in the hold mode;
- Adding another waypoint prior to the Initial Approach Fix (IAF) waypoint after entering the approach procedure data into the flight plan;
- Rejoining the course between two waypoints after being cleared and proceeding directly to another waypoint.

Pilots should never assume that familiarity with one GPS unit will facilitate learning to use another unit. During the course of the operational tests mentioned in this paper, there were several instances of pilots having difficulty transitioning to the units used in the test, despite familiarity with their own GPS units. Pilots should make certain that they are comfortable and proficient with the unit that is to be used for that flight. Also, pilots should not assume that a familiar type of GPS unit will interact with all avionics displays as expected, due to possible differences in installation procedures.

Finally, if a pilot finds himself/herself totally lost and unfamiliar with what is seen on the GPS display, they should have a backup procedure ready to implement. If all else fails, pilots should be prepared at any time during the flight to simply turn the unit off. They shouldn't follow their GPS unit into the ground.

References

FAA (1995). *FAA aircraft certification human factors and operations checklist for standalone GPS receivers (TSO C129 Class A).* DOT/FAA/AAR-95/3. U.S. Department of Transportation, Federal Aviation Administration, Washington, DC.

FAA (1996). *TSO-C129a, Airborne supplemental navigation equipment using the global positioning system (GPS).* U.S. Department of Transportation, Federal Aviation Administration, Aircraft Certification Service, Washington, DC.

FAA (1998). *Aeronautical information manual: Official guide to basic flight information and ATC procedures.* U.S. Department of Transportation, Federal Aviation Administration, Washington, DC.

Williams, K.W. (1998). *GPS design considerations: Displaying nearest airport information.* U.S. Department of Transportation, Federal Aviation Administration, Office of Aviation Medicine, Washington, DC. DOT/FAA/AM-98/12.

Winter, S. & Jackson, S. (1996). *GPS Issues* U.S. Department of Transportation,

Federal Aviation Administration, Standards Development Branch, Oklahoma City, OK. DOT/FAA/AFS-450.

Wreggit, S.S. & Marsh, D.K., II. (1998). *Cockpit integration of GPS: Initial assessment – menu formats and procedures.* U.S. Department of Transportation, Federal Aviation Administration, Office of Aviation Medicine, Washington, DC. DOT/FAA/AM-98/09.

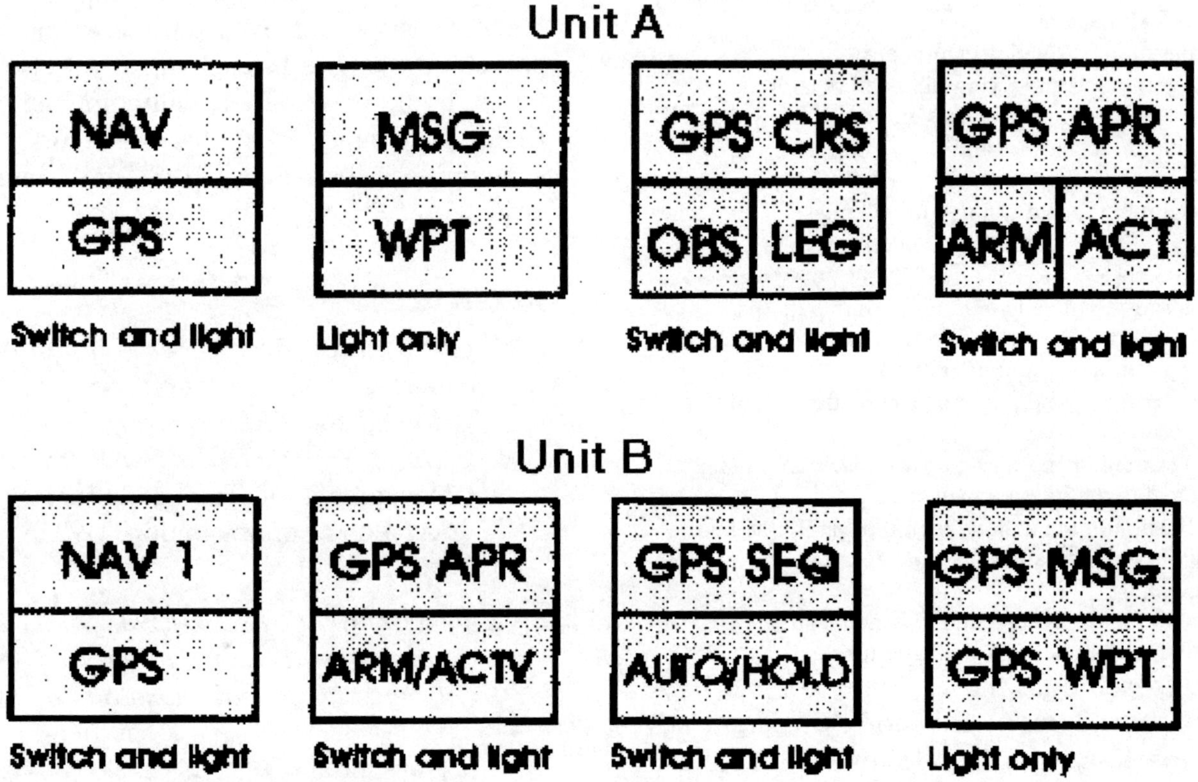

Figure 2: Two examples of annunciator panels

AN INTERNATIONAL USABILITY SURVEY OF GPS AVIONICS EQUIPMENT

Kurt M. Joseph, FAA Civil Aeromedical Institute, Oklahoma City, OK, USA
Dieter W. Jahns, SynerTech Associates, Bellingham, WA, USA
Michael D. Nendick, The University of Newcastle, Callaghan, New South Wales, Australia
Ross St. George, New Zealand Civil Aviation Authority, Lower Hutt, New Zealand

Abstract

The rapid introduction of Global Positioning System (GPS) receivers for airborne navigation has outpaced the capacity of international aviation authorities to resolve human factors issues that concern safe and efficient use of such devices. Current certification technical standards appear to have had little influence on standardizing receiver architectures, interfaces and operating manuals despite evidence from research simulation and flight tests that lack of standardization may undermine safety. The present research used factor-analytic techniques to reduce 308 pilots' ratings from a 163-item survey to 24 factors suitable for identifying human factors issues related to GPS receiver displays and controls, operating procedures, navigation performance, training, and other topical areas. Multivariate analysis of variance revealed that GPS experience and receiver type influenced pilot ratings for several of these factors. The results are consistent with previous research, and their application to certification procedures and standards is discussed.

Introduction

The accelerated development and introduction of Global Positioning System (GPS) receivers for use in airborne navigation has outpaced the capacity of international aviation authorities to fully implement regulations and guidance for the safe and efficient use of such devices (Nendick & St. George, 1996a). Technical Standard Order (TSO) C129 A1 is the document currently used to certify Instrument Flight Rules (IFR) GPS receivers for installation in the United States and Canada, and is accepted as the certification standard elsewhere, including Australia and New Zealand. However, it appears to have had little influence on standardizing receiver architectures, interfaces, and operating manuals (Heron, Krolak, & Coyle, 1997). Recent flight simulation research performed at the Federal Aviation Administration Civil Aeromedical Institute (FAA/CAMI) illustrates that some features of GPS-receiver interfaces can compromise their effective use and perhaps undermine safety (Williams, 1998; Wreggit & Marsh, 1998). Furthermore, FAA flight tests using brand-name receivers have revealed several human factors issues related to pilot interaction with receiver interfaces (Williams, in press; Winter & Jackson, 1996). These reports underscore the need for GPS-receiver standards that exceed the criteria specified in TSO C129 A1. Human factors issues associated with receiver architectures, interfaces, and operating manuals need to be addressed specifically.

At least two human factors references directly support TSO C129 A1. McAnulty's (1994) review of human factors principles and guidelines for the design of controls and displays for standalone GPS and Long Range Navigation (LORAN) receivers is germane to regulatory requirements. In addition, the FAA has issued an aircraft certification human factors and operations checklist for standalone GPS receivers (Huntley, Turner, Donovan, & Madigan, 1995). It includes both a bench and flight test, which were designed to assist certification personnel and manufacturers in evaluating receiver interface characteristics of GPS receivers to be certified in accordance with TSO C129 A1. The bench test assists in evaluations of GPS receivers that do not require aircraft installation and a test flight. The flight test helps in evaluating GPS-receiver design characteristics and functions during actual flight conditions. Both checklist tests focus on GPS-receiver controls, displays and operating characteristics. Together with the guidelines proposed by McAnulty (1994), the checklist tests represent important progress in resolving significant human factors issues associated with GPS receivers and their certification. However, evidence that considerable work remains on these issues is manifest in GPS manufacturer and certification personnel requests for guidance on numerous human factors issues for which they have no baseline data.

The research reported here is intended to complement existing human factors data by providing baseline measures of pilot perceptions of, and experiences with, GPS receivers. This research is based on the 163-item GPS User Survey 97 (Nendick, 1997), which was administered to 1,880 pilots in the United States (US). Nendick (1994) administered an earlier version of this survey to 227 New Zealand pilots, and is currently administering a further-modified, 1998 version to 4,000 Australian and New Zealand pilots. In the US, the survey represents the first systematic attempt to quantify a large sample of pilot perceptions of GPS receivers. Together with the survey data from New Zealand and that forthcoming from Australia, the US data provides a unique source of information that can be aggregated with existing human factors data (e.g., usability tests) to properly support the tasks of pilots, certification personnel, and GPS-receiver manufacturers. One of the many challenges for the FAA as it moves toward greater dependence on GPS-based navigation is to supplement TSO C129 A1 and future technical standard orders with human factors specifications. These should ensure usability through some basic standardization guidelines without prohibiting GPS-receiver manufacturers from adding new features to their GPS devices.

Method

Administration of the GPS User Survey was part of a larger effort to explore free-flight enabling technologies, like GPS, and their impact on pilot performance of tasks related to communication, navigation, and surveillance functions. During the course of its modification for use in the US, the original 125-item survey developed by Nendick (1994) was expanded to include 163 questions. These questions were categorized into the following topic areas: Receiver Controls and Displays, Operating Logic, Receiver Functions, Receiver Operations, Operating Procedures, Navigation Performance, Pilot Attitudes, and Training. Seven-point, Likert-type rating scales were used to measure pilots' responses to survey questions. Several open-ended questions were included within the survey to gather qualitative data on particular issues that couldn't be addressed using the question and rating format. Finally, several demographic questions were inserted to determine the extent to which the sample of pilots used in this research represented the General Aviation (GA) pilot population.

Table 1. Demographics for Pilot Samples

FAA Region	Total Hours			GPS Hours		
	Mean	Med	SD	Mean	Med	SD
AAL	1883	988	2919	414	240	621
ASW	1999	957	2887	210	40	448
Both	**1917**	**983**	**2905**	**354**	**200**	**582**

The 35-minute survey was administered to pilots in two different FAA Regions. Pilots had to have a current medical certificate, be at least a private pilot, and have a minimum of five hours of GPS flight time to be eligible for participation in the survey. Eighty-eight pilots from the FAA Southwest Region (ASW) completed the survey at FAA/CAMI in December 1997, and it was mailed to 1,800 FAA Alaskan Region (AAL) pilots in April 1998. Approximately 12% (N=220) of the pilots in the latter region returned a completed survey. Although this return rate is quite low, it is consistent with that obtained for other surveys of AAL Region pilots (e.g., see Driskill, Weissmuller, Quebe, Hand, & Hunter, 1997). The average age of pilots in the AAL and ASW Regions was 49 (SD=10) and 41 (SD=14), respectively. Table 1 shows the average, median, and standard deviation for total hours and GPS experience (in hours) for each pilot sample. Pilots from both regions were comparable in age and total hours. A comparison of GPS experience revealed that AAL Region pilots had nearly twice as many GPS hours than pilots in the ASW Region. The 50th percentile for GPS experience was 200 hours.

Table 2. Pilot License as a Function of Region

FAA Region	License Type (Actual and Observed in %)					
	Private		Commercial		Air Transport	
	Act.	Obs.	Act.	Obs.	Act.	Obs.
AAL	49.7	89.1	29.2	1.8	21.1	5.0
ASW	46.0	37.5	25.9	38.6	28.1	19.3

Table 2 compares the actual distribution of licenses within each region (see 1996 FAA Statistical Handbook of Aviation) with the distribution obtained for AAL and ASW Region samples. Observed percentages for each region sum to less than 100% because a small percentage of pilots did not indicate which license they held. Although the AAL Region sample is skewed more heavily, analysis of the license distribution in both regions indicates that private pilots comprised a very large majority of all survey respondents. Given the skewed distribution and low return rate, the results reported in the next section would be most representative of GA pilots who hold a private pilot license.

Results

Estimates of internal consistency (i.e., coefficient alpha) were calculated for each of the eight survey topic areas. In general, the following standardized item alphas indicate that pilot ratings for most of the survey topic areas were reliable: Receiver Controls and Displays (.92), Operating Logic (.90), Receiver Functions (.89), Receiver Operations (.91), Operating Procedures (.60), Navigation Performance (.70), Pilot Attitudes (.81), and Training (.56). The relatively low reliability estimates for Operating Procedures and Training may be due to the smaller number of items in these topic areas, and the large variety of GPS receivers, each of which has a different operating manual.

Underlying factor structures were extracted by entering pilot ratings from each of the survey topic areas into separate exploratory principal components factor analyses that utilized varimax rotation to obtain factor solutions. A multivariate analysis of variance (MANOVA) was then utilized to determine if differences in mean factor ratings were a function of FAA region (AAL and ASW), GPS experience (below the 50th percentile and above the 50th percentile), and GPS-receiver type (hand-held and panel-mount). Pilots with 200 hours or less ranked below the 50th percentile for GPS experience, whereas pilots with more than 200 hours ranked above the 50th percentile.

Pilot ratings were pooled across FAA region after preliminary analyses revealed differences that were attributable solely to the availability of ground-based navigational aids within each region. For example, regions located in the contiguous US have more very high, omni-directional range (VOR) navigational aids than the AAL Region. Consequently, pilots in the AAL Region were less likely to use a VOR navigational aid as a backup to GPS than were pilots in the ASW Region. Pilots in the AAL Region also were less likely to use a backup means of navigation because this region has fewer navigational aids than do regions located in the contiguous US. Hence, the following results are based on the exploratory factor analyses of each survey topic area, and the MANOVA, which included GPS experience and GPS-receiver type as independent variables, and mean factor ratings within each survey topic area as dependent variables.

Receiver Controls and Displays

Consistent with the findings of Nendick (1994), the following four factors emerged from the factor analysis of this topic area.

- Control Dimensions (button and/or knob operation, labeling, and layout)
- Accessories and Installation (power supply, antenna, receiver size and shape, fit into aircraft)
- Display Readability (readability of symbols and day/night illumination)
- Display Messages (warning indications)

Generally, higher ratings on these factors indicate a favorable view of the particular receiver that was rated. MANOVA tests of pilot ratings revealed that the main effect of GPS-receiver type was significant for both the Accessories and Installation and the Display Readability factors. Relative to hand-held receivers (M=5.6, SD=1.0), the accessories and installation associated with panel-mount receivers (M=6.0, SD=0.8) were assigned significantly higher ratings by pilots [F(1,184)=6.8, p<.01]. Panel-mount receivers (M=5.4, SD=1.1) also received significantly higher ratings on the Display Readability factor than did hand-held receivers [M=6.0, SD=1.0; F(1,184)=11.2, p<.01]. Despite these differences, all pilots gave fairly high ratings to the four factors that characterized receiver controls and displays.

Operating Logic

Two factors emerged from the factor analysis of this topic area.

- Programming Demands (route programming and review, waypoint entry, standardization)
- Cognitive Demands (practice required for proficiency, reliance on memory, operating complexity)

None of the interaction and main effect tests involving these factors were significant.

Receiver Functions

Two factors emerged from the factor analysis of this survey topic area.

- Navigation Feedback (desired track, distance to go, crosstrack error and CDI, and groundspeed)
- Receiver Integrity and Emergency Information (RAIM warning, satellite status, nearest airport, and present position)

None of the interaction and main effect tests involving these factors were significant.

Receiver Operations

Two factors emerged from the factor analysis of this survey topic area.

- In-flight Operation (one-handed operation, use in turbulence, and night and day use)
- Data Entry and Modification (creating and modifying route, pre-flight and in-flight data entry, undoing errors)

In general, higher ratings on these factors indicate a favorable impression of the particular receiver that was rated. There was a significant effect of GPS experience for the In-flight Operation factor [F(1,184)=4.0, p<.05]. Pilots with more GPS experience assigned significantly higher ratings (M=5.7, SD=1.1) for in-flight operation of receivers than pilots with less GPS experience (M=5.3, SD=1.1). Once again, the ratings for these factors were fairly high overall.

Operating Procedures

Four factors emerged from the factor analysis of this survey topic area.

- Validation of Receiver Data (in-flight cross-checks of information using other data sources)
- Operational Errors (incorrect data input, misreading the display, and mode errors)
- Navigation Contingencies (navigation backups for GPS, use of other navigational methods with GPS)
- Incidents (incidents resulting from misreading the display or incorrect input of data)

A significant main effect test for the Operational Errors factor revealed that the amount of GPS experience influenced pilot ratings [F(1,184)=5.0, p<.03]. Pilots with less GPS experience assigned significantly higher ratings (M=2.8, SD=1.0) for this factor than did pilots with more GPS experience (M=2.5, SD=1.0). This effect suggests that pilots with less experience report that they make more errors during their interaction with GPS receivers.

Navigation Performance

Four factors emerged from the factor analysis of this survey topic area.

- Awareness Issues (changes in situation awareness, lookout, and instrument scan)
- Workload Issues (changes in mental and physical workload, head-down time, and chart use)
- Course Tracking (tracking direct and to waypoints, and accuracy of tracking)
- Decision Making (navigation decisions, flying in marginal conditions, and flying near controlled airspace)

The main effect of GPS experience was significant for the Course Tracking factor [F(1,184)=4.8, p<.03]. Pilots with more GPS experience assigned significantly higher ratings (M=5.9, SD=0.9) for this factor than did pilots with less GPS experience (M=5.5, SD=1.0). This finding suggests that pilots perceive improvements in GPS course tracking as concomitant with gains in GPS experience.

Pilot Attitudes

Four factors emerged from the factor analysis of this survey topic area. The first two factors are consistent with the findings of Nendick (1994).

- Confidence in GPS (use for IFR navigation and non-precision approaches, and accuracy and reliability of GPS device)
- User Confidence (use of basic and complex GPS functions)
- Dependence on GPS (complacency, reliance on GPS for VFR and IFR)
- Use of Database (affordability and use of current data)

The main effect of GPS experience was significant for the User Confidence and Dependence on GPS factors. [F(1,184)=5.6, p<.02; F(1,184)=4.5, p<.04, respectively]. Pilots with more GPS experience assigned significantly higher ratings (M=6.0, SD=0.8) for the User Confidence factor than did pilots with less GPS experience (M=5.6, SD=1.0). The average ratings for this factor were very high. Compared to those with less experience (M=3.9, SD=1.2), pilots with more experience also assigned significantly higher ratings (M=4.3, SD=1.3) for the Dependence on GPS factor. Overall, the ratings indicated that pilots reported increased dependence on GPS as they gain more experience with it. The main effect test of GPS-receiver type for the Dependence on GPS factor approached the nominal significance level criterion of .05 [F(1,184)=2.9, p<.09]. Pilots assigned higher ratings to panel-mount receivers (M=4.4, SD=1.2) than they did to hand-held receivers (M=3.9, SD=1.2).

Training

Two factors emerged from the factor analysis of this survey topic area.

- Knowledge and Experience Required (technical training, knowledge level and human factors training needed)
- Knowledge and Experience Attained (user knowledge, training received, user manual)

The main effect of GPS-receiver type was significant for the Knowledge and Experience Required factor [F(1,184)=10.1, p<.01]. Pilots assigned higher ratings to panel-mount receivers (M=4.5, SD=1.0) than they did to hand-held receivers (M=4.0, SD=1.1). This finding indicates pilots perceive that panel-mount receivers require more training to operate proficiently. Finally, the main effect test of GPS experience for the Knowledge and Experience Attained factor approached the nominal significance level criterion of .05 [F(1,184)=3.0, p<.09]. Pilots with more GPS experience assigned higher ratings (M=4.4,

SD=1.1) than did pilots with less GPS experience (M=4.2, SD=1.0).

Discussion

The present research has produced several meaningful results. First, MANOVA tests indicated that GPS experience and GPS-receiver type did not interact to affect pilot factor ratings within each of the survey topic areas. Thus, the following discussion assumes that GPS experience and receiver type combine in an additive fashion to influence pilot perceptions of GPS receivers. Significant differences in pilot ratings as a function of GPS-receiver type suggested that panel-mount receivers were better than hand-held receivers on factors such as display readability, power supply, antenna, size and shape, and fit into aircraft. Pilot ratings also revealed that panel-mount receivers require more knowledge and training to operate proficiently. Pilots who used panel-mount receivers rated themselves as more reliant on GPS and more complacent than pilots who used hand-held receivers. In part, this finding may be due to the fact that panel-mount receivers provide continuous monitoring of GPS signal integrity, whereas hand-held receivers do not.

Significant differences in pilot ratings as a result of GPS experience were distributed across several survey topic areas. Pilot ratings suggested that with more experience, in-flight operation of GPS receivers became easier, fewer operational errors were made, course tracking improved, there were gains in knowledge of the receiver, and increased confidence in using basic and complex functions. One possible negative consequence associated with more GPS experience is that pilots perceive themselves to be more reliant on GPS and more complacent than pilots with less experience.

Consistent with ratings of the New Zealand pilots surveyed by Nendick (1994), the results reported here provide evidence of a discrepancy between pilot ratings of GPS-receiver design and pilot performance in GPS-receiver evaluations (e.g., see Wreggit & Marsh, 1998). A plausible explanation for this discrepancy is presented by Clarke (1994) and Nendick and St. George (1996b), who have warned of the pitfalls related to pilot perceptions of the operational simplicity of GPS. On one hand, pilots may be captivated by the inherent simplicity and minimal training required to execute frequently used basic receiver functions. On the other hand, as Winter and Jackson (1995) have demonstrated, pilots who have not acquired a significant amount of knowledge and training can become overwhelmed when they go beyond basic receiver functions to complex ones that are required for difficult navigational tasks. Examples of subtle human factors issues associated with complex receiver functions will likely be revealed during flight tests using GPS Wide Area Augmentation System (WAAS) terminal approaches, and in future tests of other difficult navigational tasks.

Another meaningful finding resulted from the exploratory factor analysis of each survey topic area. Separate sets of factors were identified that could be used as a basis for supplementing existing certification guidelines for GPS receivers. In combination, certification personnel, manufacturers and human factors specialists could examine the specific factors within each topic area for design issues relevant to certification. Such an effort would result in coverage of traditional human factors issues related to receiver displays and controls, as well as other topical issues, such as those related to operating procedures, navigation performance and training.

This approach to identifying human factors issues in certification has at least two advantages over the current approach. The first advantage is that certification personnel and manufacturers would collaborate with human factors specialists in examining issues and developing materials. Just as it is for GPS receivers, end-user participation is critical for the acceptance and use of certification test materials. Second, standardization of tests, checklists, and other certification material will go a long way to ensure that GPS receivers have standardized architectures, interfaces and operating manuals. Current practices place considerable onus on certification personnel to gather information and develop their own supplemental materials for certification tests. Consequently, the subtle variation in certification materials and procedures impose greater demands on certification personnel, and perhaps make the task of GPS-receiver standardization more difficult. Adopting a systematized approach for developing human factors standards in support of certification personnel and GPS manufacturers may ease this difficulty.

Acknowledgment

The authors gratefully acknowledge the voluntary participation of the 308 pilots who completed survey forms. Thanks also to Ms. Tara Bergsten, Ms. Janine King and Ms. Debbie Perry, of OMNI Corporation, for their support during survey distribution, and Dr. David Hunter, FAA Office of Aviation Medicine, for his administrative support of the survey. Finally, the authors appreciate the assistance of Mr. John Hallinan, FAA Alaskan Region Flight 2000 Program Manager, who organized survey-related activities in Alaska.

References

Clarke, B. (1994). *Aviator's guide to GPS*. New York, NY: TAB Books.

Driskill, W.E, Weissmuller, J.J., Quebe, J.C., Hand, D.K., & Hunter, D. (1997). The use of weather information in aeronautical decision-making: II. U.S. Department of Transportation, Federal Aviation Administration, Office of Aviation Medicine, Washington, D.C. NTIS # DOT/FAA/AM-97/23.

FAA (No date). *FAA Statistical Handbook of Aviation*, pp. 7-4/7-9 [Online]. Available: http://api.hq.faa.gov/handbook/1996/toc96.htm [1998, July 28].

Heron, R., Krolak, W. & Coyle, S. (1997). *A human factors approach to the use of GPS receivers.* Vancouver, BC: Heron Ergonomics, Incorporated.

Huntley, Jr., M.S., Turner, J.W., Donovan, C.S., & Madigan, E. (1995). *FAA aircraft certification human factors and operations checklist for standalone GPS receivers (TSO C129 Class A).* Technical Report DOT/FAA/AAR-95/03. Washington, DC: Federal Aviation Administration.

McAnulty, D.M. (1994). *Guidelines for the design of GPS and LORAN receiver controls and displays.* Technical Report DOT/FAA/RD-95/01. Washington, DC: Federal Aviation Administration.

Nendick, M.D. (1994). *Global Positioning System (GPS): Human factors aspects for general aviation pilots.* Unpublished masters thesis. Palmerston North, New Zealand: Massey University.

Nendick, M.D. (1997). *GPS User Survey 97.* Callaghan, NSW, Australia: The University of Newcastle.

Nendick, M.D. & St. George, R. (1996a). General aviation pilots and GPS: Some results from a New Zealand study. *Airways*, *9*, 12-16.

Nendick, M.D. & St. George, R. (1995). Human factors aspects of global positioning systems (GPS) equipment: A study with New Zealand pilots. In R. Jensen (Ed.) *Proceedings of the Eighth International Symposium on Aviation Psychology*, pp. 152-57. Columbus, OH: The Ohio State University.

Williams, K.W. (1998). *GPS design considerations: Displaying nearest airport information.* U.S. Department of Transportation, Federal Aviation Administration, Office of Aviation Medicine,

Washington, D.C. NTIS # DOT/FAA/AM-98/12.

Williams, K.W. (in press). *GPS user-interface design problems*. U.S. Department of Transportation, Federal Aviation Administration, Office of Aviation Medicine, Washington, D.C. NTIS # DOT/FAA/AM-98.

Winter, S. & Jackson, S. (1996). *GPS issues*. Technical Report DOT/FAA/AFS450. Oklahoma City, OK: Federal Aviation Administration.

Wreggit, S. & Marsh III, D. (1998). *Cockpit integration of GPS: Initial assessment-menu, formats and procedures*. U.S. Department of Transportation, Federal Aviation Administration, Office of Aviation Medicine, Washington, D.C. NTIS # DOT/FAA/AM-98/09.

THE 4D-DISPLAY

Harro von Viebahn[1], VDO Luftfahrtgeräte Werk GmbH, 60439 Frankfurt, Germany

Abstract

A perspective primary flight display - the 4D-Display - and an advanced navigation display format were jointly developed by VDO-L and the Darmstadt University of Technology. The goal of the new display formats is to improve the pilot's situation awareness - especially terrain and obstacle awareness - which in turn considerably reduces the risk of a CFIT event. The Displays also incorporate means for an accurate spatial and on-ground guidance. Therefore, a great potential for future applications is offered.

Several flight test campaigns were performed in order to evaluate the display concept and to prove the acceptance of the pilots. The flight tests comprised approaches, demonstrations of the taxi and the inherent ground proximity warning function and low-level terrain following tasks.

The concept was highly appreciated by pilots. It could be demonstrated that the synthetic vision technology offers a great potential for new applications such as curved approaches or approaches in difficult terrain and low visibility. In particular it was stated that that the 4D-Display is an appropriate means to avoid CFIT occurrences.

In order to provide the required graphics processing power for complex perspective displays VDO-L develops highly integrated, high performance graphics modules which meet the requirements of prospective applications.

[1] e-mail: viebahn@vdol.de

Introduction

Recent advances in technologies such as communication, data links, aviation data bases, and accurate navigation will increase the aircraft's operational capabilities. Consequently, new scenarios like free-flight, curvlinear approaches and newly designed approaches in difficult terrain are being discussed within the aviation community. These anticipated changes will also affect the tasks and information requirements of the cockpit crew. For example, CDTI is one of the approaches to cope with the changing operational environment in the cockpit. Generally speaking, an enhanced situational awareness information is required for the crew in order to operate in a safe and economically efficient way [FHT] [Cha96] [Cha97] [Bro].

In addition, requirements for future military air transport comprise low-level terrain following capabilities in poor visibility conditions and complex curved approaches utilising satellite and terrain based navigation systems. These flight procedures will demand a comprehensive visual support to put the crew in a position to successfully fulfil the mission even in a threatening environment [Lev].

In order to offer a solution VDO-Luftfahrt and the Darmstadt University of Technology have jointly pursued an approach which enables the pilot to perceive his spatial situation in relation to terrain, obstacles and other relevant fixed or moving objects. In

0-7803-5086-3 /98/$10.00 ©1998 IEEE

comparison to sensor Enhanced Vision Systems (EVS) the Synthetic Vision System (SVS) is completely independent of meteorological and visual conditions.

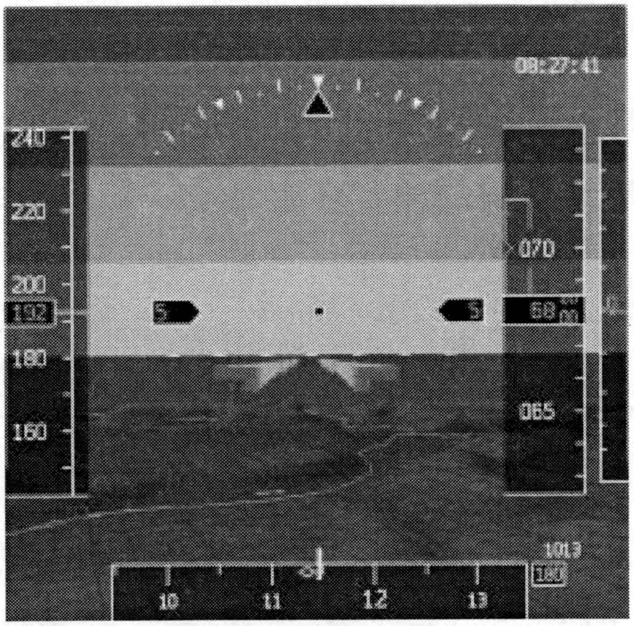

Figure 1. The 4D-Display – PFD Format

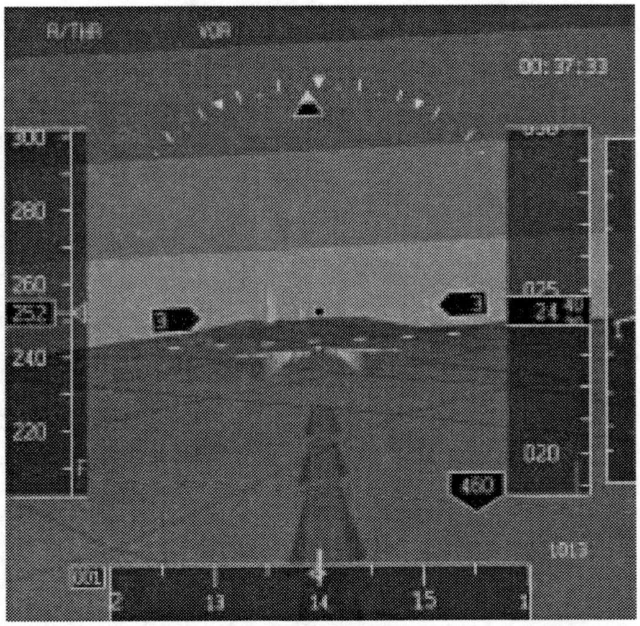

Figure . The 4D-Display – Approaching Terrain 2

Display Formats

The stylised three-dimensional (3D) view out of the cockpit window is the central source of information in the '4D-Display' format (Figure 1). A databases is used to generate the image of the surrounding terrain including the most significant objects like rivers or prominent obstacles (Figure 2) [Schi]. This artificial image of the outside world is extended by virtual elements. A spatial flight path predictor – representing the fourth dimension - and the predetermined flight path are presented on the display and enable the pilot to perceive the flight situation intuitively. Additionally, in a two-dimensional (2D) overlay, analogue indications provide exact values for speed, heading and altitude. A more detailed description of these display formats and the navigation display format is given in [Kub].

Proof Of Concept

Simulator Based Evaluation

A series of simulator experiments was carried out with 18 airline pilots to evaluate the 4D-Display. Each subject flew six flight tasks. Five of them were approaches to an airport with increasingly difficult conditions. Each task was flown twice, once using the 4D-Display and once using a conventional instrumentation. All relevant parameters were recorded on-line and later evaluated in detail. In addition, a questionnaire was answered after each task and after the pilots had finished the simulation session. As an important result it was found that the new concept was well accepted by the pilots. Even pilots without glass-cockpit experience quickly adapted to the 4D-Display.

When using the 4D-Display, the deviation from the demanded flight path was significantly reduced compared to the runs using the conventional instrumentation. Also the pilots' control input activities decreased.

0-7803-5086-3 /98/$10.00 ©1998 IEEE

After the simulation trials further development of the display formats was performed with continuos feedback from airline pilots in order to meet operational requirements. For instance, the pitch indication and the perception of the pitch rate was significantly improved by dividing the sky shading area into 10° segments.

Flight Tests

Test Aircraft

Three flight test campaigns were conducted using the flying testbed ATTAS (VFW 614) operated by the DLR[2]. The left crew station was equipped with a sidestick (fly-by-wire) and a commercial LCD screen mounted in front of the instrument panel (Figure 3). The LCD showed both display formats, the 4D-Display (PFD) and the Navigation Display (ND). On the right seat, a DLR safety pilot supervised the actions of the evaluation pilot.An observer seat in the rear cabin was also equipped with a LCD mirroring the test pilot's instruments. For precise navigation, an integrated INS, GPS, D-GPS system was installed onboard the aircraft [Schi]. All flights were performed manually without engaging autopilot functions.

Figure 3. DLR's ATTAS Cockpit

Approaches

During the first campaign eight airline pilots flew several flight tasks, mainly ILS- and non precision like approaches. Because of certification constraints a go-around had to be initiated at 500 ft AGL. The new display presentation for landings and taxi-operations could be monitored on the observer's LCD.

For spatial guidance, a channel like symbol was depicted on the PFD and ND representing glideslope and localizer information. While approaching a small airfield without any ground installations in a hilly terrain a `virtual` ILS-like channel was generated, which led the aircraft to the runway threshold. Additionally, terrain advisory information was provided by colouring the relevant areas in the ND as well as in the PFD.

Terrain following

During the second campaign, airforce transport pilots manually flew low-level missions along a preplanned channel with the 4D- and the Navigation-Display as primary instruments and the only source of visual information. The channel followed the contour of the mountainous terrain, taking into account the aircraft's performance (Figure 3). For safety reasons the missions were performed with an altitude offset, whereas the display formats represented a flight approximately 150 ft above ground. Pilots were also requested to leave and reenter the channel from different directions.

The trials showed that overshooting behind terrain peaks was minimized due to the easy cognition of the prospective flight path given by the channel. Analysis of the recorded flight data showed, that the horizontal deviation mainly stays within the limits of the channel. The mean of the horizontal path deviation was about zero indicating that the pilots kept track during the terrain following mission. The same was found for the vertical deviation except that the mean was shifted slightly above the commanded slope. The combination of the

[2] Deutsches Zentrum für Luft- und Raumfahrt

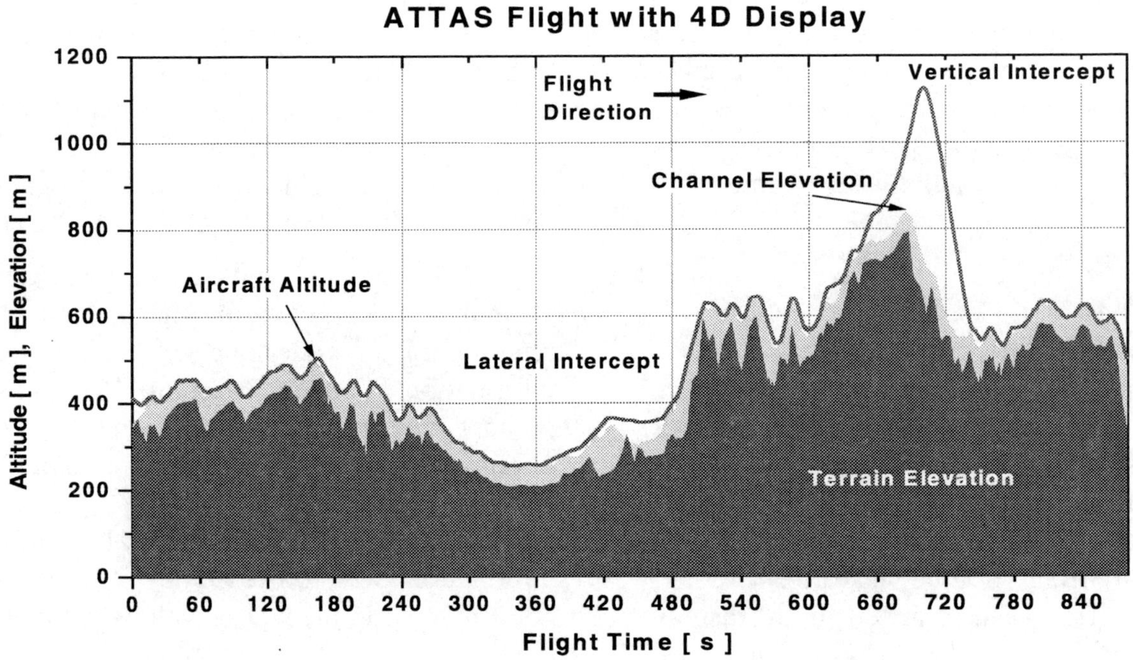

Figure 3. Terrain Following Profile and Aircraft Track

flight path predictor and the channel provides a sensitive, easily recognized indication when falling below the channel bottom thus avoiding dangerous terrain approaches.

The Third Campaign

After incorporation of improvements a third flight test campaign was conducted in August 97 which aimed at the evaluation and the demonstration of the capabilities of the Synthetic Vision System. 22 pilots – mainly experienced airline pilots - flew 25 sorties from Frankfurt's Rhein-Main-Airport. The 1,5 hour flight pattern comprised climb-out, enroute navigation, terrain approaches and warnings, ILS-approaches and demonstrations of the taxi functions in the busy environment of an international airport.

Results

The 4D-Display was very well accepted by nearly all participants. Pilots` assessment of the display concept yielded an overall improvement of situation awareness (Figure 4). In particular, it was stated that the displays are an appropriate means to eliminate CFIT events. Furthermore, 65% of the pilots think that this kind of representing information will reduce the workload in the cockpit. Comments taken from the questionnaires which the pilots filled in after the flight with the new displays were as follows (examples):

- very good representation
- considerable improvement of situation awareness
- easy anticipation of events or actions required
- a huge step forward for approaches in difficult terrain
- 4D-Display is a great progress
- filled with enthusiasm

- stunning how easy to fly and without problems
- excellent taxi guidance

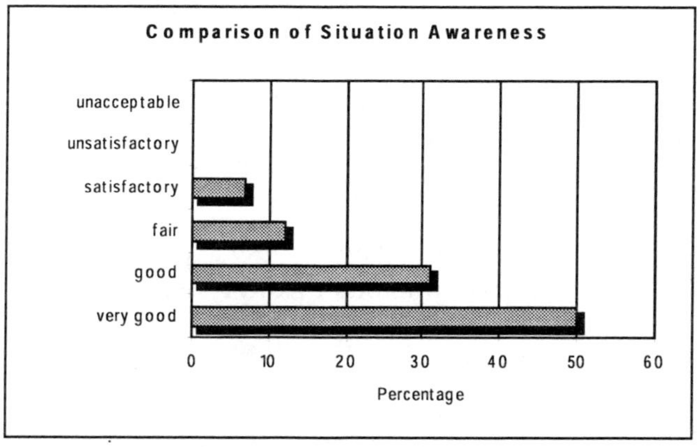

Figure 4. How do you rate your situation awareness with the 4D-Display compared to conventional displays?

These and earlier experiments showed that the perspective symbology enables the pilot to manually fly even spatially curved and critical low profiles following the terrain while reducing pilot's workload and improving the tracking of the demanded flight path at the same time. It could be demonstrated that the synthetic vision technology offers a great potential for new applications such as approaches in difficult terrain and low visibility.

Technology Approach

Today's cockpit displays are driven by Symbol Generation Units (SGU). The symbology presented to the pilot on a conventional PFD comprises the computer-graphical reproduction of the electromechanical predecessors like the ADI or the altimeter. From the computer graphics' point of view, these symbols are two-dimensional, flat graphical elements requiring only a few parameters to define their shape and dynamic behaviour.

With the demand for the presentation of a synthetic vision on large displays the requirements in terms of computational performance, data throughput for high resolution and real time graphics processing have increased tremendously. The former SGU changes its profile towards a high performance, embedded 3D graphics processing and generation unit. Based on the general requirements of modular avionics architectures additional restrictions concerning unit size, power consumption and heat dissipation were introduced which the designer has to cope with. So an evolutionary step in the development was made to accomplish the transition from conventional 2D capabilities of an SGU to the 3D perspective image generation technology.

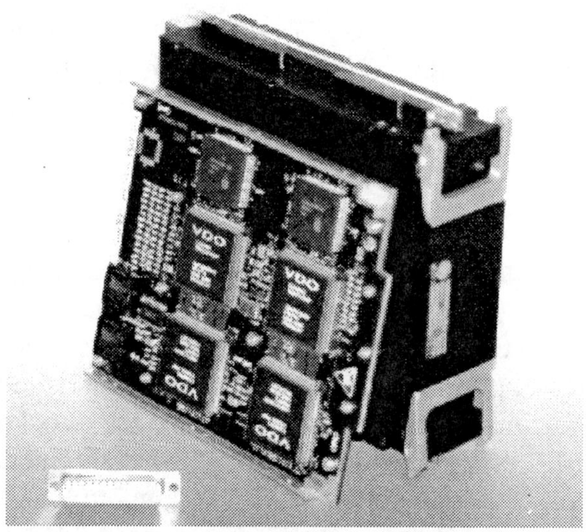

Figure 5. First Generation 3D Graphics Generator

In order to provide the required graphics processing power for complex perspective displays VDO-L developed the first generation of a highly integrated, high performance graphics module which meets the requirements of prospective applications. It is designed for an established modular avionics architecture and provides full 2D and 3D capabilities including transformations, shading, hidden surface removal, view clipping and anti-aliasing. The module (Figure 5) provides high resolution, coloured synthetic image generation through

- a scaleable multi-processor architecture
- a scaleable graphics architecture for rasterization, anti-aliasing of 2D- and 3D-graphic primitives and hidden surface removal
- a multi tasking high order language (HOL) graphics processing software
- a HOL application interface providing full programmability to the user
- digital encoded transmission of videosignals on high-speed fibre optical links

With this graphics module for avionics application the feasibility and performance was demonstrated successfully. Display formats like the 4D Display are no longer a fiction, they can be generated on board of an aircraft.

In addition to a powerful graphics generator an on board data base and data base server process is an essential component of a Synthetic Vision System. The data base elements have to be processed and rasterized in real-time for presentation on the cockpit displays. For example, using a terrain model with a grid resolution of 100m up to 25.000 quadrilaterals have to be processed and generated per frame just for the terrain representation on the display of a transport aircraft. With a frame or parameter update rate of 30 Hz this leads to more than 750.000 polygons per second. Such a detailed presentation of terrain is useful for flight phases like approach and landing but not necessarily required during cruise at high altitude levels. Nevertheless, the graphics generator has to provide maximum performance to cope with the worst case in terms of performance.

Currently the second generation of an integrated 3D graphics computer including mass data processing and data base management is developed. This development comprises

- new graphics functions
- performance increase
- safety and feedback features for flight critical applications
- further reduction of number of components through high-level ASIC integration

Furthermore the housing and the interfaces are compliant to ARINC standards for civil transport aircraft, thus enabling the integration into a existing avionics system for the next step of validation in an operational environment.

References

[Aga] AGARD *Glass Cockpit Operational Effectiveness*, Flight Vehicle Integration Panel, Working Group 21, AR 349, April 1996

[Bel90] Below, Ch., Kling, H., v. Viebahn, H.: *Experimentalprogramm zur räumlichen Flugführungsdarstellung*, DGON Symposium ‘Auswirkung neuer Technologien auf die Sicherheit im Luftverkehr’, Verlag TÜV Rheinland 1990

[Bel95] Below,Ch., Hammer, M. v. Viebahn, H.: *4D Flight Guidance Displays*, Synthetic Vision for Vehicle Guidance and Control, SPIE Volume 2463, Orlando April 1995

[Bro] Brooks, P.: *Highways In The Sky* Aerospace International, May 1998-07-22

[Car] Carbaugh, D.; Cooper,S.: *Avoiding Controlled Flight Into Terrain*, Airliner, Apr-Jun 1996

[Cha96] Chang, C.C.; Livack, G.S.; McDaniel, J.I.: *Emerging cockpit technologies for free-flight: Situational Awareness for Safety, Automatic Dependent Surveillance-Broadcast, air-to-air data link, and weatherlink.*

[FHT] FAA Human Factors Team: *The Interfaces Between Flightcrews and Modern Flight Deck Systems*, FAA, Washington, 1996

[Cha97] Chang, C.C.; Livack, G.S.; McDaniel, J.I.: *The safety case for enhanced cockpit situational awareness*

[Gru] Grunewald, A.: *Predictor Symbology in Computer Generated Pictorial Displays*, NASA-CR-169441, Israel Inst. of Tech., 1981

[Kub] Kubbat, W.J.; Lenhart, P.M.; v. Viebahn, H.: *4D Flight Guidance Displays, A Gate to Gate Solution,* 17th DASC paper #148, Seattle, 1998

[Lev] Levedag, S.; Kühn, T.; v. Viebahn, H.: *Moderne Sichtunterstützung zukünftiger Flugführungssysteme in Bodennähe*, DGLR, 4. Avionik Workshop, Neubiberg, 9. Nov. 1994

[Sch] Schiefele, J.; May, L.; Raabe, H.; Pfister, J.; Schmalz, C.; Dörr, K.-U-; Kubbat, W.: *Safety Relevant Navigation and Certifiable Databases for 3D Synthetic Vision Systems,* 17th DASC paper #147, Seattle, 1998

[Vie] v.Viebahn, H.: *Situation Awareness - A challenge for an avionics manufacturer*, Workshop 'Human Factors / Future Combat Aircraft', Munich, 1994

[Wic] Wickens, C., Haskell, I., Harte K.: *Perspective Flight Path Displays*, ARL-89-2/Boeing 89-1, Institute of Aviation, University of Illinois, 1989

[Wie] Wiener, Earl L., Nagel, David C.: *Human Factors in Aviation*, Academic Press, 1988

* * *

4D FLIGHT GUIDANCE DISPLAYS, A GATE TO GATE SOLUTION

Wolfgang J. Kubbat, Darmstadt University of Technology, Darmstadt, Germany

Peter M. Lenhart, Darmstadt University of Technology, Darmstadt, Germany

Harro von Viebahn, VDO-Luftfahrtgeraete Werk, Frankfurt, Germany

Abstract

Present state of the art flight deck instrumentation does a big deal to enhance safety and efficiency of today's aircraft operations. But a dramatic change in air traffic scenarios has already begun. The challenge to increase air traffic capacity and at the same time improve safety and efficiency significantly will only be met, if future cockpit instruments will guarantee the pilots spatial situation awareness in any phase of flight. The only way to achieve this is synthetic vision.

Commercial transport aircraft operate without exception under instrument flight rules (IFR). Therefore regarding flight deck instrumentation a key to assure spatial awareness is the visualization of the spatial environment in a natural way no matter what the actual visibility conditions are. Since computational performance increased very fast at the end of the eighties and this progress is still going on this is now possible by the means of synthetic vision. Consequently an experimental synthetic vision system (SVS) was jointly developed by the Darmstadt University of Technology and the VDO Luftfahrtgeräte Werk and underwent extensive flight tests in august 1997.

A geographical database, a precision navigation system and the 4D flight guidance displays driven by a symbol generator, capable to render 3D images, form the three pillars of the synthetic vision system.

The database will be discussed in #147 'Safety Relevant Navigation and Certifiable Databases for 3D Synthetic Vision Systems', hardware issues and the flight test campaign will be discussed in #169 'The 4D Display'. This paper will focus on the human machine interface (HMI).

The HMI currently consists of two perspective displays complementing one another. The Primary Flight Display (PFD) incorporates a synthetic insight out view, while the Navigation Display (ND) is characterized by a synthetic 'birds eye view'. By providing not only flight guidance but also taxi and traffic information the two displays support the crew from gate to gate.

Since most information is presented in a graphical way, it can intuitively be seized by the pilots. Especially in phases of high workload, the highly pre-processed information and its redundant presentation will significantly contribute to assure pilots situation awareness both in the air and on the ground.

Reasons to act

Increase in air traffic

With an annual increase of 5 - 8 % in air traffic (**Figure 1**). it is obvious, that ground measures alone will not be sufficient. Appropriate onboard measures, i.e. datalink, preprocessing and visualization in the cockpit are needed to integrate the cockpit crew in the ATC process (rather than being only managed by the ground). This however means, that the cockpit crew has to get a complete view of their situation relative to other traffic, the geographic situation and possible dangers. How else could

0-7803-5086-3 /98/$10.00 ©1998 IEEE

this be done than via three dimensional depictions?

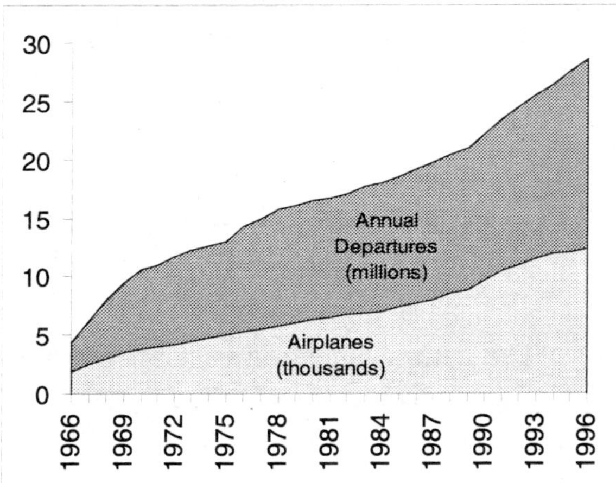

Figure 1. Jet airplanes in service and annual departures [1]

The (congested) airspace situation

Regarding accident rates and accident locations (**Figure 2**) some conclusions can be drawn:

- **a.** The maxima in accident rates occur in areas where in the near future no break through in ground based measures (ATC) can be expected. If we do not counteract this situation with onboard measures we might ran into an intolerable situation.
- **b.** The absolute accident maxima however are due to dense traffic in well controlled areas. Again: We must incorporate the cockpit crew into the air traffic control process via an appropriate human machine interface (apart from other measures).

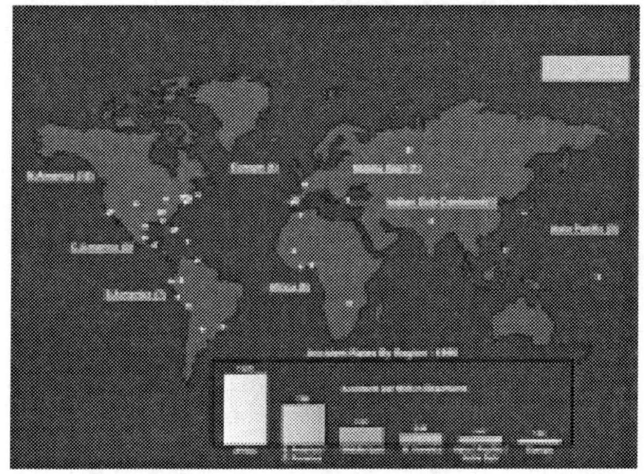

Figure 2. Accident locations - 1996 [2]

New modes of operation

Free flight, curved approaches or autonomous taxiing are technically feasible. They can however not be realized without an appropriate system to pilot interface. This must be three dimensional, allowing an instant and intuitive perception of the situation depicted.

Economical pressure

Nowadays delays in air traffic cause some billions of dollar losses for airliners and passengers. Increased IFR separation distances, increased distances in take off sequence, taxiing speed 50% less, no intersecting taxi traffic, increased taxiing distances (up to 150%), no flight operations at Frankfurt Rhein-Main-Airport for instance if visibility is below 125 m are the result. If we can fly and taxi under IFR conditions as we do under VFR, we could save this losses. However, this needs a solution: Synthetic vision.

Stagnation in accident rates

With the stagnation in accident rates (**Figure 3**) and the steady increasing traffic, we might drift into a situation of an information policy driven psychological collapse in air traffic.

[1] Source: Airplane Safety Engineering, Boeing Commercial Airplane Group

[2] Source: Mc Donnell Douglas

0-7803-5086-3 /98/$10.00 ©1998 IEEE

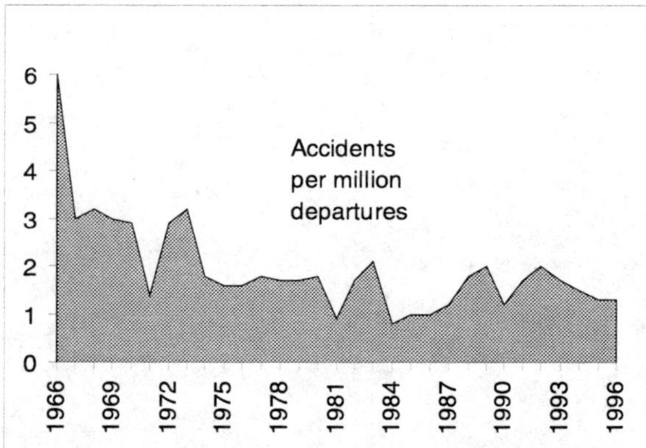

Figure 3. Hull loss accidents (worldwide commercial jet fleet 1966 - 1996) [3]

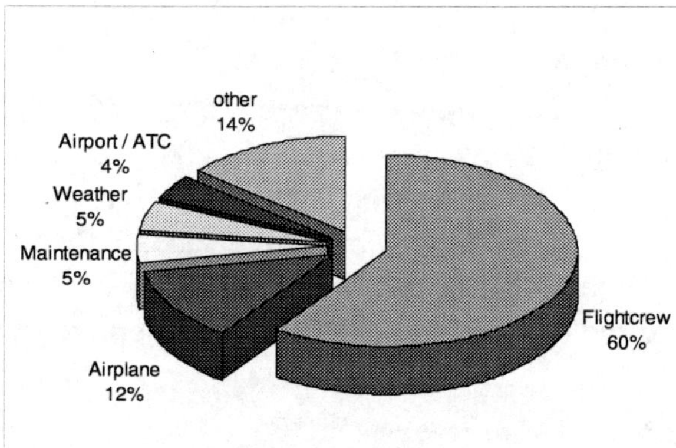

Figure 4. Primary Cause Factors of Accidents (Worldwide Commercial Jet Fleet 1986 - 1995) [4]

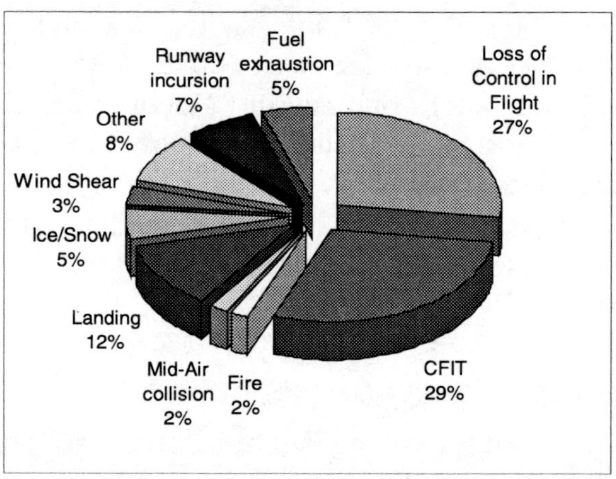

Figure 5. Fatal accidents by type of accident (1991 - 1995) [5]

Controlled flight into terrain (CFIT) continues to dominate accident statistics (**Figure 4 - 5**). 30 % of hull losses and 50 % of fatalities are due to CFIT. The described situation is even worse, if other than the big aircraft are considered (**Table 1**).

The main reason for CFIT is loss of spatial situation awareness by the flight crew. Hence, if the pilots spatial situation awareness can be improved significantly a dramatic reduction of fatal accidents can be expected.

The central question is: **Can nowadays flight deck technology cope with these challenges?**

Table 1. Relative CFIT Risk [6]

Type of operation	CFIT risk	Relative CFIT risk
Part 121 Airline	0.1 aircraft lost per million flights inside / outside North America	1
Part 135 Regional	2.0 aircraft lost per million flights	20
Part 91 Business Jet	2.2 aircraft lost per million flights inside / outside North America	22
Part 91 Turbo-Prop	4.3 aircraft lost per million flights	43

[3] Source: Airplane Safety Engineering, Boeing Commercial Airplane Group
[4] Source: Airplane Safety Engineering, Boeing Commercial Airplane Group
[5] Source: Airplane Safety Engineering, Boeing Commercial Airplane Group
[6] Source: D. Carbaugh, S. Cooper, 'Avoiding CFIT', Airliner Apr-Jun 1996

Flight deck technology

Figure 6. Ju 52 cockpit

Figure 7. Boeing B 747-200 cockpit

Figure 8. Boeing 777

A saturation of flight deck technology can be observed (**Figure 6 - 8**). To cope with the challenges of tomorrow's air traffic, flight decks must provide an enhanced situation awareness to the crew enabling them

- to move gate to gate with few support from the ground
- to land in adverse weather even on poorly equipped airports
- to "see" obstacles and terrain even during poor visibility
- to monitor surrounding traffic
- to cope with further automation

Thus, enhanced situation awareness can only be achieved with synthetic vision.

Synthetic vision system

The three pillars of a synthetic vision system

Any operational synthetic vision system (SVS) will be based on three core elements (**Figure 9**). First, their is the human machine interface, the 4D flight guidance displays which are conform to the principles of a human centered design and which provide a synthetic vision under all meteorological conditions.

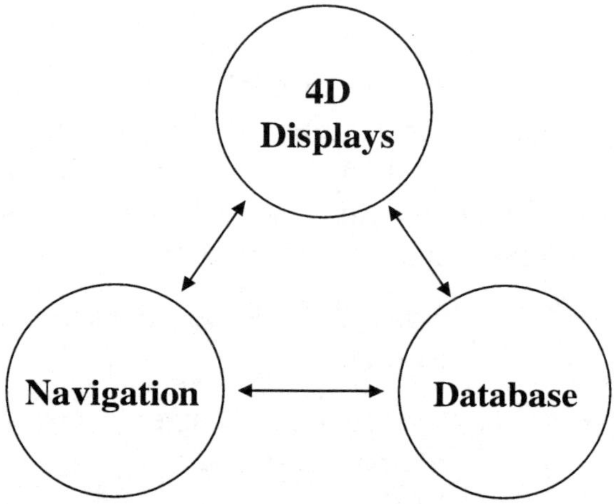

Figure 9. The three pillars of SVS

Second, their has to be an accurate navigation that enables the matching process of database information and reality. For reliability and accuracy reasons the navigation has to be based on several independent sources like GPS, INS and so on.

Third, a certified database has to provide the navigational and terrain features presented on the 4D displays.

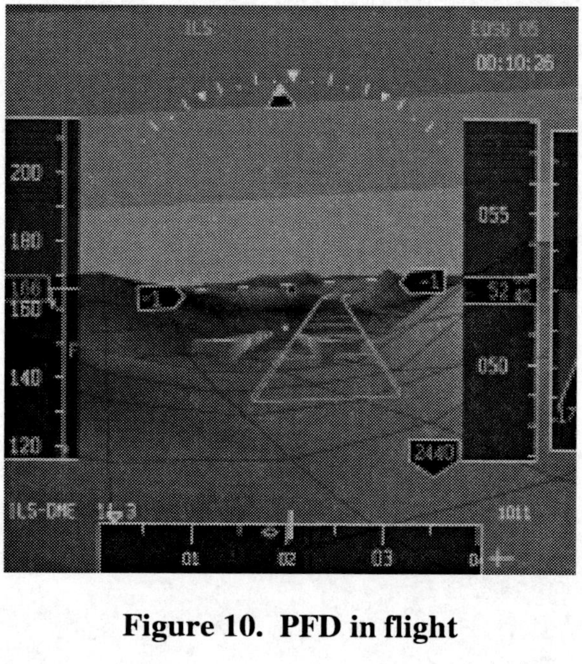

Figure 10. PFD in flight

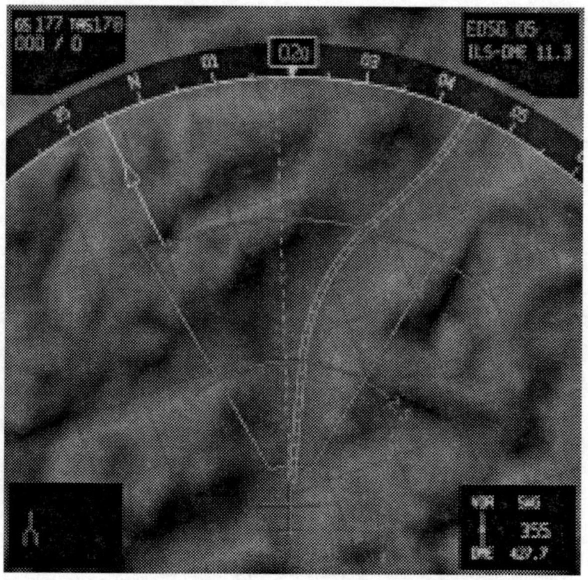

Figure 11. ND in flight

The human machine interface

General features

The 4D Flight Guidance Displays proposed here, are based on a contact analogue synthetic vision system. On the one hand there is an „inside out view“ of the actual situation called the Primary Flight Display (PFD) (**Figure 10**), on the other a parallel projection including the same features called Navigation Display (ND) or moving map display (**Figure 11**).

The artificial images of the outside world on PFD and ND are extended by virtual elements. For example a spatial flight path predictor and the predetermined flight path are visualized on the PFD (**Figure 10**) and permit the pilot to intuitively understand his actual situation and react with foresight.

This 3D view is superimposed by conventional 2D symbology like Basic T. They provide exact values for speed, heading and altitude. To avoid the different elements of the display from concealing each other, some of them like the tape scales are made transparent and are placed on the edge of the display.

Primary Flight Display

In the design of the PFD much attention has been paid to redundancy of information. There is for example the altitude information during approach phase: The first clue to the present altitude is given by the size in which the grid appears which is rendered on the surface of the terrain. This allows to estimate the altitude above the ground at a first glance. When flying a precision approach (**Figure 12**) the glide slope channel provides the required altitude very precisely in a purely graphical form. Shortly before touchdown the shadow of the flight path predictor gets into the field of view giving another very accurate altitude reference. In addition to these graphical hints a digital readout is integrated into the altitude scale. Another example for redundant information is the speed indication. The exact value can be obtained from the scale with its digital readout. As an intuitively perceivable presentation the visual flow field (especially the grid) gives a sense of the actual speed. Furthermore, the predictor changes its length and color according to the speed of the aircraft, thus giving a continuous feedback.

This redundancy of information helps to prevent the pilots from misinterpreting the presented data.

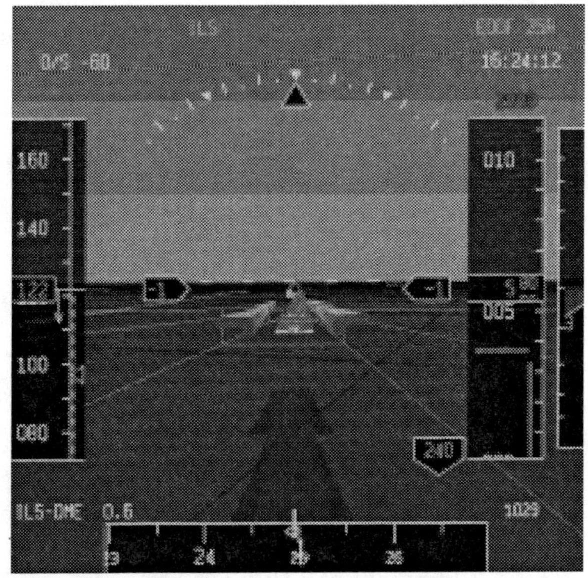

Figure 12. PFD on approach

Navigation Display

In order to enable the pilots to develop a complete imagination about terrain, obstacles and adjacent air traffic, the PFD had to be complemented by a ND (**Figure 11**).

The ND format is computed from the same databases as the PFD, containing the necessary information about the terrain underneath the aircraft. Though using parallel projection it provides again a three-dimensional impression (relief-picture).

The new display combines a conventional navigation display with a map and virtual elements such as navigational aids, air traffic routes, commanded or preplanned flight path, taxi paths and ground proximity warning symbology.

ND zooming is possible, allowing the pilot to observe in detail an area of his choice. The field of view may be selected with the pilots own aircraft at the center or the lower edge of the display to give the pilot an impression of the terrain and traffic surrounding him.

One example of the integrity of the PFD and ND is the feature of the PFD viewing area, appearing in the ND. Thereby the objects

0-7803-5086-3 /98/$10.00 ©1998 IEEE

visible on the PFD can clearly be associated with the ones on the ND.

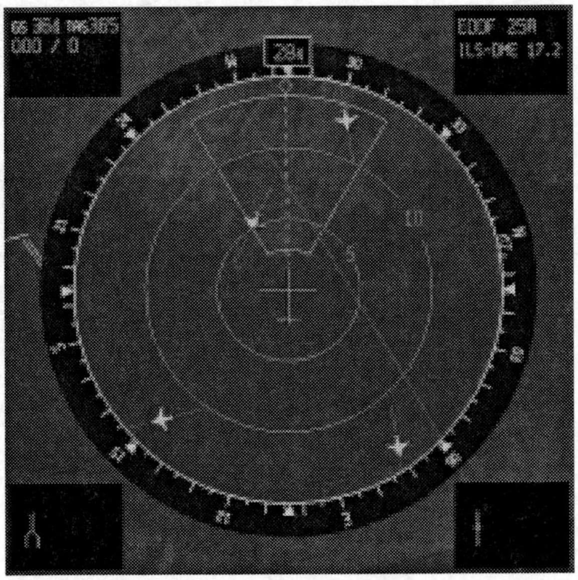

Figure 13. ND with traffic information

Gate to gate functionality

In contrast to the conventional PFD / ND combination, which can be found in any modern airliner and which provides only the information for flying the aircraft and for navigation, synthetic vision displays assist the crew from gate to gate in all phases of operation.

The combination of synthetic vision with the virtual elements predictor and approach corridor (**Figure 12**) allows autonomous precision approaches and landings in adverse weather, even on poorly equipped airports. Terrain and obstacles are always visualized and thus avoidance of them is assured even in low visibility conditions.

If traffic information is provided by data link or TCAS equipment, the crew can monitor position and movements of surrounding traffic both on PFD and ND (**Figures 13 -15**).

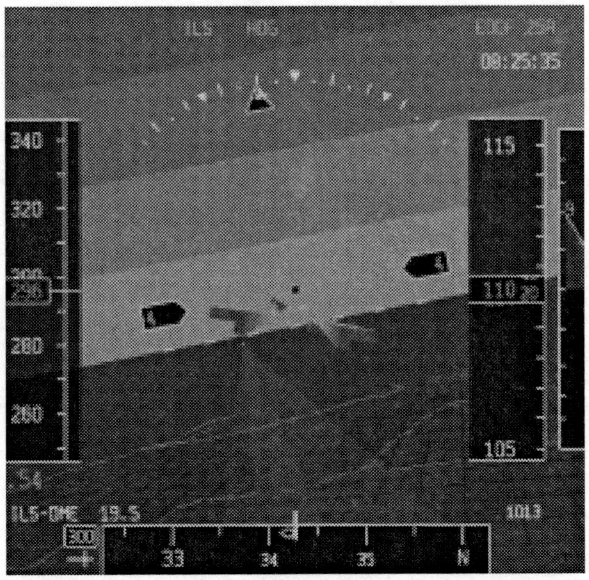

Figure 14. PFD with TCAS information

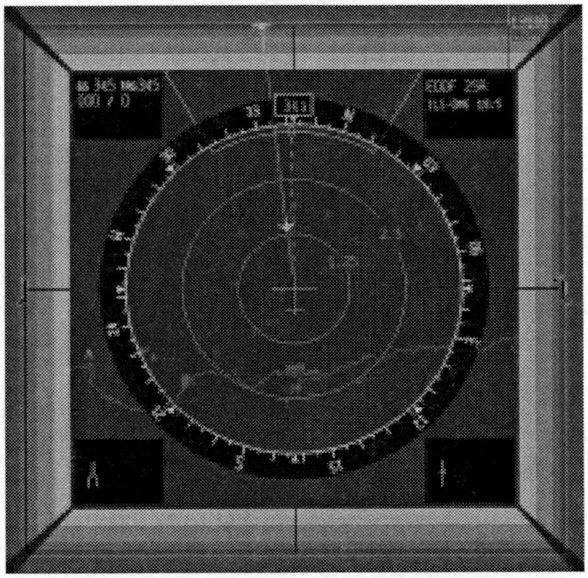

Figure 15. ND with TCAS information

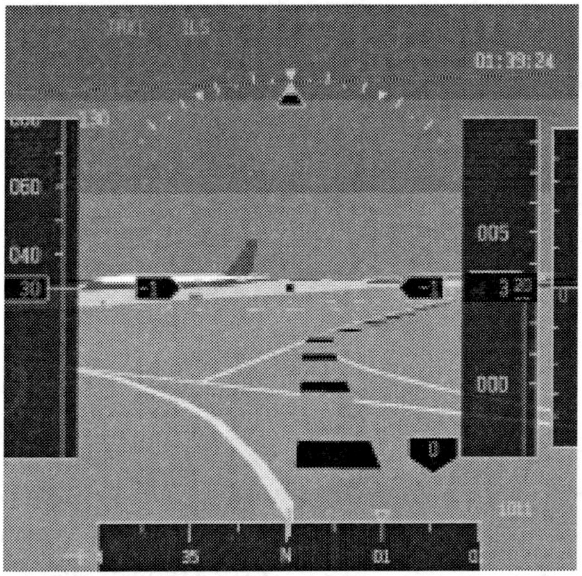

Figure 16. PFD in taxi guidance mode

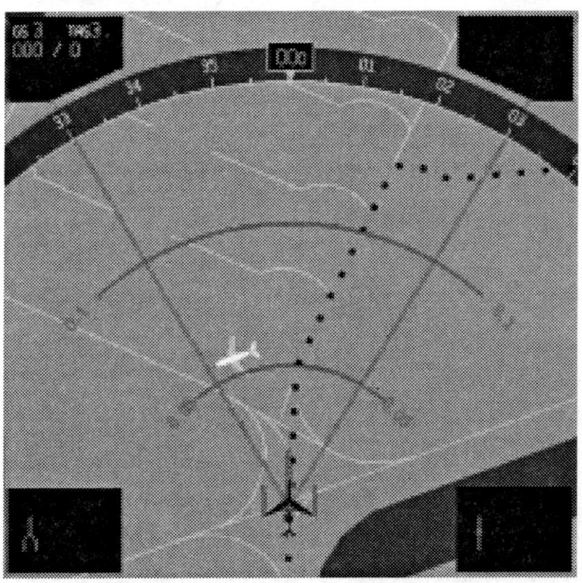

Figure 17. ND in taxi guidance mode

But for a gate to gate solution, taxi guidance is as important as inflight guidance. Both PFD and ND support advanced surface movement guidance and control (**Figures 16 and 17**). Since the database contains a detailed representation of the airport, taxi ways can be selected by the pilots or may be given by ATC. These selected taxi ways are highlighted and contain also information about the required taxi speed. In order to ease the steering task a taxi predictor including a steering angle indicator for the front wheel is integrated into the ND. In addition for large aircraft there is a predictor for the movement of the wing tips.

Arrangements have been made to include co-operative moving or non-moving obstacles in the display. Prerequisite is digital data transmission between aircraft, ground vehicles and ATC, which is expected to be available in near future.

Assuming appropriate taxi procedures and airport infrastructure, taxi with few support from ground control, relying on synthetic vision, even under low visibility may be possible. Delays caused by low ground visibility could be reduced. A first rough cost estimation for such a scenario showed that the investment for the SVS will be amortized in two years.

Conclusions

Synthetic vision is a powerful contribution to the solution of today's problems in air traffic, particularly safety, capacity and efficiency.

It consists of three pillars: 3D image generation, accurate and reliable navigation and reliable real time on board databases. The first two pillars are regarded as 'close to producebility' while the last one is feasible but needs further work.

.IMINARY FLIGHT TESTS OF PSEUDO-ATTITUDE USING SINGLE ANTENNA GPS SENSING

Richard P. Kornfeld, Department of Aeronautics and Astronautics, MIT, Cambridge, MA

R. John Hansman, Department of Aeronautics and Astronautics, MIT, Cambridge, MA

John J. Deyst , Department of Aeronautics and Astronautics, MIT, Cambridge, MA

Abstract

This paper describes a novel method for aircraft control based on pseudo-attitude. The pseudo-attitude information consists of flight path angle and pseudo-roll angle about the aircraft velocity vector axis and is completely observable from single antenna GPS velocity measurements. Pseudo-attitude information is presented on a pseudo-attitude display which shows flight path angle and pseudo-roll angle instead of the traditional pitch and roll information, thus yielding direct indications of the actual flight path state. Flight tests in a Piper Arrow aircraft showed that pseudo-attitude is equivalent in performance to conventional attitude, with no subjective or objective differences between them. Possible applications of pseudo-attitude include its use as a backup attitude system, in guidance display or flight control schemes or for fault diagnostics in redundant systems.

Introduction

This paper presents a method for aircraft control which is based on the use of pseudo-attitude and a novel pseudo-attitude display. It relies solely on the measurements of a single antenna GPS receiver. In recent years, GPS has found broad acceptance as a navigation aid and is integrated in many new aircraft cockpits as well as retrofitted into many existing aircraft. Thus pseudo-attitude control makes use of a readily available sensor and provides either an additional level of attitude redundancy, at minimal additional cost, or an equivalent level of redundancy at lower cost by reducing a level of hardware redundancy.

In the past, several GPS attitude determination methods have been proposed and successfully tested. They rely on the use of multiple antennas, separated by known baselines, to measure the carrier phase differences of the GPS signals between antennas. After resolving the integer ambiguity inherent in the measurements, the attitude can be calculated with high accuracy [1]. The use of a number of antennas incurs both the cost of the system as well as the cost of the installation. The method presented in this paper infers attitude from the aircraft trajectory, measured by a single antenna GPS receiver, under the assumption of coordinated flight using a simple point mass aircraft model.

Pseudo-Attitude

Overview

The display of pseudo-attitude provides attitude information which is completely observable from single antenna GPS velocity information. High quality GPS velocity information is available from Doppler frequency shift measurements of the GPS carrier tracking loop.

Pseudo-attitude is a GPS observable surrogate for traditional aircraft attitude and is referenced to the aircraft velocity vector. Pseudo-attitude consists of flight path angle γ relative to the local horizontal, substituting for traditional aircraft pitch angle, and a pseudo-roll angle $\tilde{\phi}$ about the aircraft velocity vector axis, substituting for traditional roll angle. The pseudo-roll angle $\tilde{\phi}$ is chosen as the effective bank angle which corresponds to the observed

0-7803-5086-3 /98/$10.00 ©1998 IEEE

lateral rate of change of the velocity vector. Pseudo-attitude, unlike traditional attitude, provides a direct indication of the flight path. It will be shown, however, that for coordinated flight the pseudo-roll angle closely corresponds to the traditional roll angle.

Figure 1 shows the velocity vector axis, $\mathbf{v}_g$, used for pseudo-attitude and the body axes, (x_b y_b z_b), used for the definition of the traditional attitude.

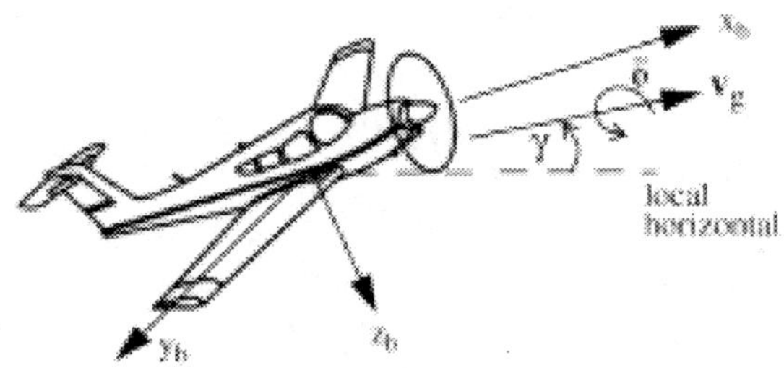

Figure 1: Attitude and Pseudo-Attitude Definition

The flight path angle γ is determined from the horizontal and vertical velocity components as

$$\gamma = \mathrm{atan}\left(\frac{-v_d}{\sqrt{v_n^2 + v_e^2}}\right)$$

where $\mathbf{v}_g = (v_n, v_e, v_d)$ denotes the GPS velocity measurements in a north, east and down reference frame.

The pseudo-roll $\tilde{\phi}$ is determined from the known aircraft acceleration **a** and the gravitational acceleration **g** as shown in Figure 2. First, a pseudo-lift acceleration vector **l** is defined as the vector difference of $\mathbf{a}_n$ and $\mathbf{g}_n$, the components of **a** and **g** perpendicular to the aircraft velocity vector $\mathbf{v}_g$, respectively. The pseudo-roll $\tilde{\phi}$ is then determined as the complementary angle between the lift vector **l** and a local horizontal reference .

The force diagram in Figure 2 closely resembles the force diagram of an aircraft flying a coordinated turn. The only difference is that in the diagram of Figure 2 the velocity vector axis is used to resolve the forces instead of the aircraft body x-axes (a coordinated turn has by definition no sideforce in the body axes frame). However, for most flight conditions the two axes are closely aligned. Consequently, the pseudo-roll angle corresponds closely to the traditional roll angle for coordinated flight.

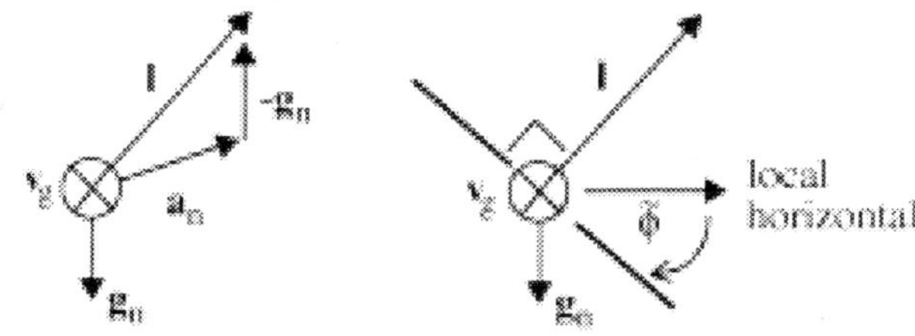

Figure 2: Determination of Pseudo-Roll

Both the aircraft velocity v_g and acceleration a are derived from GPS measurements, and pseudo-attitude is therefore completely observable from GPS velocity measurements. A detailed description of the pseudo-attitude synthesis is given in reference [2].

Pseudo-Attitude Display

Pseudo-attitude is displayed in a manner similar to the traditional attitude display. Figure 3 shows the comparison between traditional and pseudo-attitude displays. The left display shows the traditional artificial horizon in the case when the aircraft is in level flight and has a pitch angle of 5 deg and a roll angle of 20 deg. The right display shows the pseudo-attitude of the aircraft in the same flight condition. The roll representation matches the conventional display, but the pitch information is replaced with the flight path angle. As shown in Figure 3, because the aircraft is in a level turn the aircraft symbol lies on the horizon. In normal flight conditions, the difference between pitch angle and flight path angle is due to the angle of attack.

The minimum GPS velocity update rate necessary is determined by the minimum attitude display update rate required to obtain adequate aircraft controllability and pilot acceptance. Preliminary evaluations of different attitude display update rates, performed on the MIT Part Task Flight Simulator [3], suggest a desirable update rate of 6 Hz or more for conventional aircraft control, and a minimum update rate of 2 Hz. A GPS output and attitude display update rate of 10 Hz is used for this research.

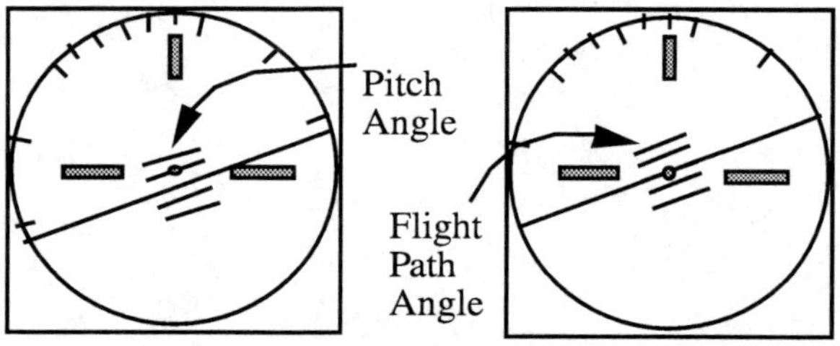

Figure 3: Attitude and Pseudo-Attitude Display

Properties of Pseudo-Attitude

Pseudo-attitude, unlike traditional attitude, provides a direct indication of the aircraft trajectory. Flight path angle gives a direct indication of the flight path state. By way of contrast, pitch angle only gives an indirect indication of flight path. In the lateral direction, pseudo-roll angle typically corresponds closely to traditional roll angle for coordinated flight. [2] gives a first order analysis of the difference between roll and pseudo-roll in these flight conditions.

During an uncoordinated maneuver, such as during a severe slip or yaw maneuver, or in severe wind shear condition, pseudo-roll angle may differ considerably from the aircraft bank angle. In these instances pseudo-roll indication can be used to maintain a straight ground track. Because the pseudo-roll angle correlates with the rate of change of the velocity vector, the pseudo-roll indication induces a self-correcting pilot control input which could be used in approach conditions with wind shear or in a forced slip (e.g. a twin-engine aircraft with a failed engine). In these conditions the pilot will tend to bank the airplane to correct for any indicated pseudo-roll angle and thereby maintain a straight ground track.

The use of pseudo-attitude is most appropriate in linear or quasi-linear flight regimes where the angle between the aircraft body x-axis and the velocity vector is small. In stall or post-stall flight conditions this assumption may no longer be valid.

Implementation

Aircraft velocity and acceleration are necessary to synthesize pseudo-attitude and both are observable from single antenna GPS measurements. In the implementation described in this paper, a Novatel GPS receiver, operated in single-point mode, is used as the primary velocity source. The velocity output is provided to a Kalman filter which estimates acceleration. The velocity and acceleration information is the input to the attitude synthesis algorithm which, together with the Kalman filter, is implemented in an onboard computer. The calculated attitude is then displayed on an electronic attitude display. Figure 4 shows a simplified block diagram of the attitude determination system.

Separate, but identical, Kalman filters are implemented for the north, east and down directions. Each Kalman filter uses a triple integrator plant as the process model with velocity, acceleration and jerk states. The jerk state is necessary because it corresponds to aircraft roll rate which typically is correlated over time. A detailed description of the Kalman filter is given in reference [2].

Since the velocity solution of the GPS receiver itself is likely to be based on a Kalman filter, the cascade of two filters may yield slightly suboptimal solutions. In an alternate

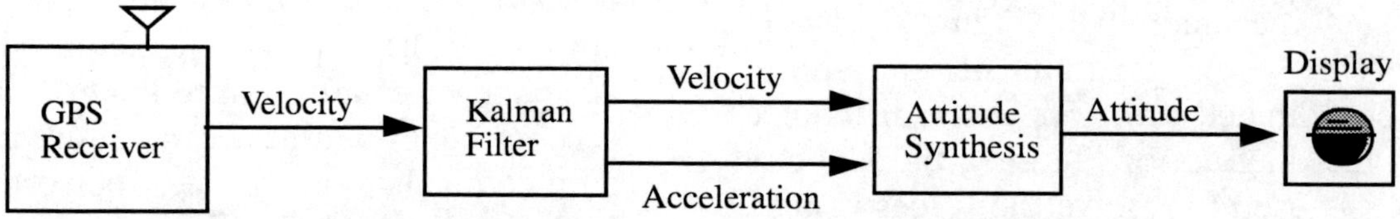

Figure 4: Simplified Block Diagram of the Attitude Determination System

approach, the external Kalman filter may provide the complete navigation solution using the receiver's raw measurements only. For reasons of rapid prototyping the latter approach was not pursued in this research.

Flight Test Setup

To demonstrate single antenna GPS based pseudo-attitude aircraft control and to compare it with traditional attitude a series of flight tests was undertaken. In addition, pilot evaluations of the usability of a pseudo-attitude display were made. The tests were conducted in an instrumented single-engine four-seat Piper Arrow aircraft shown in Figure 5.

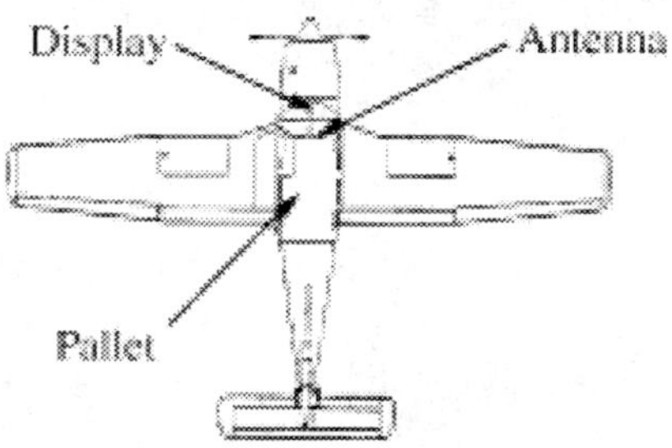

Figure 5: Flight Test Configuration

The attitude synthesis algorithm and the Kalman filter were implemented on a laptop computer connected to the flight test instrumentation as shown in Figure 6. The instrumentation consisted of a Novatel GPS receiver as the primary GPS velocity source, an INS/GPS unit (Rockwell Migits) as an attitude reference and an air data unit providing data for post-flight analysis. The Novatel GPS receiver was selected because its velocity update rate of 10 Hz met the display update rate requirements. Information from the INS/GPS and the air data systems were not used in the computation of pseudo-attitude but only provided reference data for post-flight evaluation of the GPS attitude system.

The Novatel GPS velocity data and the INS/GPS attitude data were transmitted at an update rate of 10 Hz to a Pentium based Laptop over standard RS-232 serial links. The air data was transmitted at a 2 Hz update rate. The laptop executed the attitude synthesis algorithm and stored the acquired data for post flight analysis. The pseudo-attitude was displayed on the laptop screen, at an update rate of 10 Hz. Optionally, the reference attitude provided by the INS/GPS unit could be displayed. In a preliminary flight test the Kalman filter was tuned to best satisfy pilot acceptance.

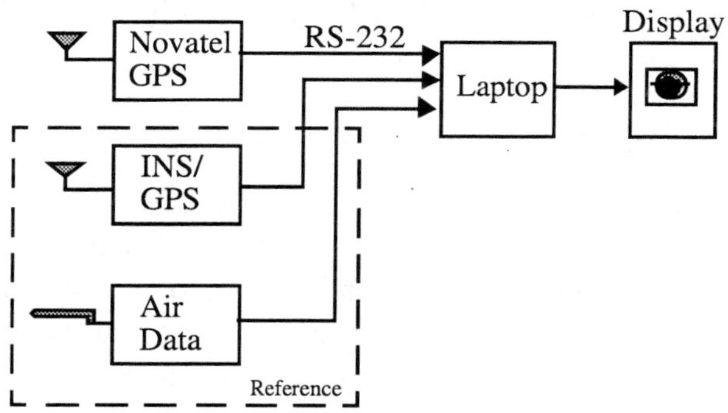

Figure 6: Flight Test Instrumentation

The flight test instrumentation was fixed to a test pallet which in turn was mounted to the aircraft structure. The Novatel GPS and the Migits GPS antenna were mounted on the inside surface of the wind shield close to the aircraft center of gravity. Satellite visibility during flight test typically fluctuated between 5 and 9 satellites.

In the initial flight test the implementation of pseudo-attitude synthesis was evaluated by comparing it to the reference pitch and roll attitude measured by the INS/GPS unit. In subsequent flight tests IFR-rated pilot subjects evaluated the usability of pseudo-attitude.

The flight test protocol included a pre-flight briefing explaining the principle of pseudo-attitude synthesis and the flight test procedures. The entire flight test was performed under simulated IFR conditions (hood). At the outset of the flight test, the pilot subjects flew the pseudo-attitude display for two minutes to familiarize themselves with the display.

After familiarization, the pilot subjects performed a sequence of coordinated flight maneuvers consisting of straight and level flight, two shallow and two steep turns, a 500 ft climb and a 500 ft descent during a 360 deg turn, using the traditional panel mounted mechanical attitude indicator. Cooper-Harper subject ratings of the traditional attitude display were obtained which served as a baseline for the comparison. The traditional attitude indicator was then covered and the pilot subjects repeated the procedure, this time using the synthesized pseudo-attitude display shown on the laptop screen held in front of them, and Cooper-Harper ratings were obtained.

The pilot subjects also performed an uncoordinated slip maneuver with full rudder deflection while maintaining a straight flight path, using the traditional attitude indicator. Cooper-Harper ratings were obtained serving as a baseline for the comparison. After covering the traditional attitude indicator, the pilots repeated the maneuver using the synthesized pseudo-attitude display and Cooper-Harper ratings were obtained.

Flight Test Results

Results from the flight tests are shown below, including a comparison of single antenna GPS based pseudo-attitude and traditional attitude measured by the INS/GPS unit. In addition, the pilot subject evaluations of the pseudo-attitude display are presented.

Comparison of Pseudo-Attitude and INS/GPS Reference Attitude

Comparisons of pseudo-attitude and roll and pitch angles measured by the INS/GPS unit are shown in Figure 7. The data was taken in condition of moderate turbulence and wind magnitude of 28 kts.

Figure 7a) shows the comparison of pseudo-roll and reference roll for straight and level flight. Due to the moderate turbulence, level flight was difficult to achieve. Figure 7b) shows the comparison for shallow turns and Figure 7c) for steep turns. As can be seen, the synthesized roll angle corresponds closely to the true roll angle.

A slight overshoot behavior of pseudo-roll can be seen at instances where the roll performs an abrupt change. The observable lag is approximately 0.5 seconds or less which corresponds to the Kalman Filter time constant. A lower delay time and better overshoot behavior may be achievable with a different filter tuning at the expense of more noise in the estimate.

Figure 7d) shows the comparison of synthesized flight path angle and reference pitch angle during straight and level flight in the presence of moderate turbulence. Figure 7e) shows the comparison for a 500 ft climb and Figure 7f) for a 500 ft descent during a 360 deg turn.

Figure 8: Attitude vs. Pseudo-Attitude

The plotted flight path angle largely follows the pitch angle with an offset of approximately 5 deg. A high frequency content is observable in the pitch angle which is not present in the flight path angle. This is due to the fact that flight path angle is less sensitive to disturbances from atmospheric turbulence.

Figure 8 shows a comparison of the traditional attitude and pseudo-attitude displays during a level turn of approximately 30 deg bank angle in one of the flight tests. As can be seen, the pseudo-roll indication on the laptop screen (on the right) corresponds closely to the traditional roll indication on the panel mounted artificial horizon (on the left).

Subjective Evaluation of Pilot Usability of Pseudo-Attitude

The evaluation of pilot usability of the pseudo-attitude system is based on flight tests performed by six IFR-rated pilots. Cooper-

Harper subjective evaluations of the pseudo-attitude and the traditional attitude indicator for the different flight maneuvers are shown in Table 1. The Cooper - Harper scale ranges from 1 to 10, where 1 is the highest rating.

Table 1: Cooper-Harper Subjective Evaluation of Pilot Usability

Pilots	Coordinated Flight		Slip	
	Attitude Indicator	Pseudo-Attitude Display	Attitude Indicator	Pseudo-Attitude Display
A	2	2	-	-
B	3	3	5	3
C	3	roll 3, fpa 6	3	3
D	3	4	2	3
E	3	3	4	3
F	2	2	2	1

The Cooper-Harper scale ranges from 1 to 10, where 1 is the highest rating.

Based on the six pilot evaluations, no statistically significant difference exists between the two displays. The examination of the individual ratings reveals that in the case of coordinated flight the pseudo-attitude performed equivalently to the traditional attitude indicator for the majority of pilots. It should be noted that for the evaluation of pseudo-attitude the pilot subjects had an unusual instrument scan pattern due to the position of the laptop screen held in front of them.

One subject (pilot C) objected to some of the display features in the pitch ladder (insufficient space between bars on the pitch ladder) and as a consequence voluntarily separated the Cooper-Harper ratings for flight path angle and pseudo-roll.

For the slip maneuver, the pseudo-attitude display shows a more consistent rating with four out of five pilots giving it an equal or better rating. This may be attributed to the fact that in order to accomplish the maneuver the pilot simply has to track zero pseudo-roll angle in order to fly a straight flight path.

Finally, two pilots flew an ILS approach solely with reference to the pseudo-attitude display and reported no difficulty.

Implications of Pseudo-Attitude

A primary application of pseudo-attitude may be its use as a backup attitude indicator in a panel mounted or handheld receiver configuration. Since most current attitude reference systems are inertial sensor based the use of GPS based pseudo-attitude provides functionally independent attitude information. Because no alignment or specific aircraft information is required, the implementation as a stand-alone hand held GPS receiver is particularly appealing. In addition, pseudo-attitude may be used as an independent attitude information source in redundancy management schemes and may serve as a cross-reference or tiebreaker in single- or dual-redundant primary attitude determination systems, respectively.

The availability of pseudo-attitude enables a number of flight control and guidance display applications to be implemented solely based on single antenna GPS information. These applications traditionally have relied on the availability of attitude sensing instrumentation and radio-navigation aids or inertial navigation systems (INS) for the necessary attitude, position and velocity reference. Consequently, the potential use of some of the applications have been limited to business-size or larger aircraft only. The use of pseudo-attitude as the horizon reference enables these applications to be driven by single antenna GPS information only. This has significant system integration and cost advantages and consequently allows for their implementation in General Aviation (GA) aircraft as well.

Guidance displays such as flight director and tunnel-in-the-sky displays may be implemented using single antenna GPS based position, velocity and pseudo-attitude information. Similarly, pseudo-attitude based autopilot modes such as ground track and altitude capture and hold or approach capture and hold can be realized using GPS based position, velocity and pseudo-attitude information. A GPS based autopilot is particularly suited for the control of small-scale, expendable UAVs where weight and power are critical.

The use of the GPS based pseudo-attitude display as a primary attitude source is currently premature. GPS integrity and availability have first to meet the current standards for flight critical instrumentation. These issues are currently under investigation in programs such as WAAS [4] and LAAS.

Conclusions

A novel method for aircraft control based on pseudo-attitude was successfully demonstrated. Pseudo-attitude is synthesized using single antenna GPS velocity measurements and consists of flight path angle and pseudo-roll angle about the velocity vector axis. It was shown to provide adequate attitude information with no subjective or substantial objective differences compared to the conventional attitude display.

The significance of pseudo-attitude lies in the fact that it makes use of a navigation aid which has found wide use in today's aircraft fleet, to provide an additional source of attitude information at minimal additional cost. Possible applications include its use as a backup attitude indicator in a panel mounted or hand held GPS receiver. Additional applications include the use of pseudo-attitude in flight director and perspective flight path displays or in autopilots for aircraft or UAVs.

Finally, the single antenna GPS based attitude information may be used in fault detection and isolation schemes for inertial sensor based attitude determination systems and may reduce costs by eliminating some redundant hardware, while maintaining high system integrity.

Acknowledgment

This research has been supported by a gift from Rockwell Collins which also provided the Migits INS/GPS unit.

References

[1] Cohen, C. E., *Attitude Determination*, GPS: Theory and Application, ed. Parkinson and Spilker, AIAA Progress in Astronautics and Aeronautics Vol. 163.

[2] Kornfeld, R. P., Hansman R. J., and Deyst J. J., *Single Antenna GPS Based Aircraft Attitude Determination*, ION, Proceedings of National Technical Meeting, Long Beach, CA, January 21-23 1998.

[3] Henderson, R. O., *A Study of GPS Based Attitude Indicators and Instrument Update Rates*, AIAA Mid-Atlantic Region I Student Conference, Old Dominion University, April 1997.

[4] Enge, P. K., *Wide Area Augmentation System*, GPS: Theory and Application, ed. Parkinson and Spilker, AIAA Progress in Astronautics and Aeronautics Vol. 163.

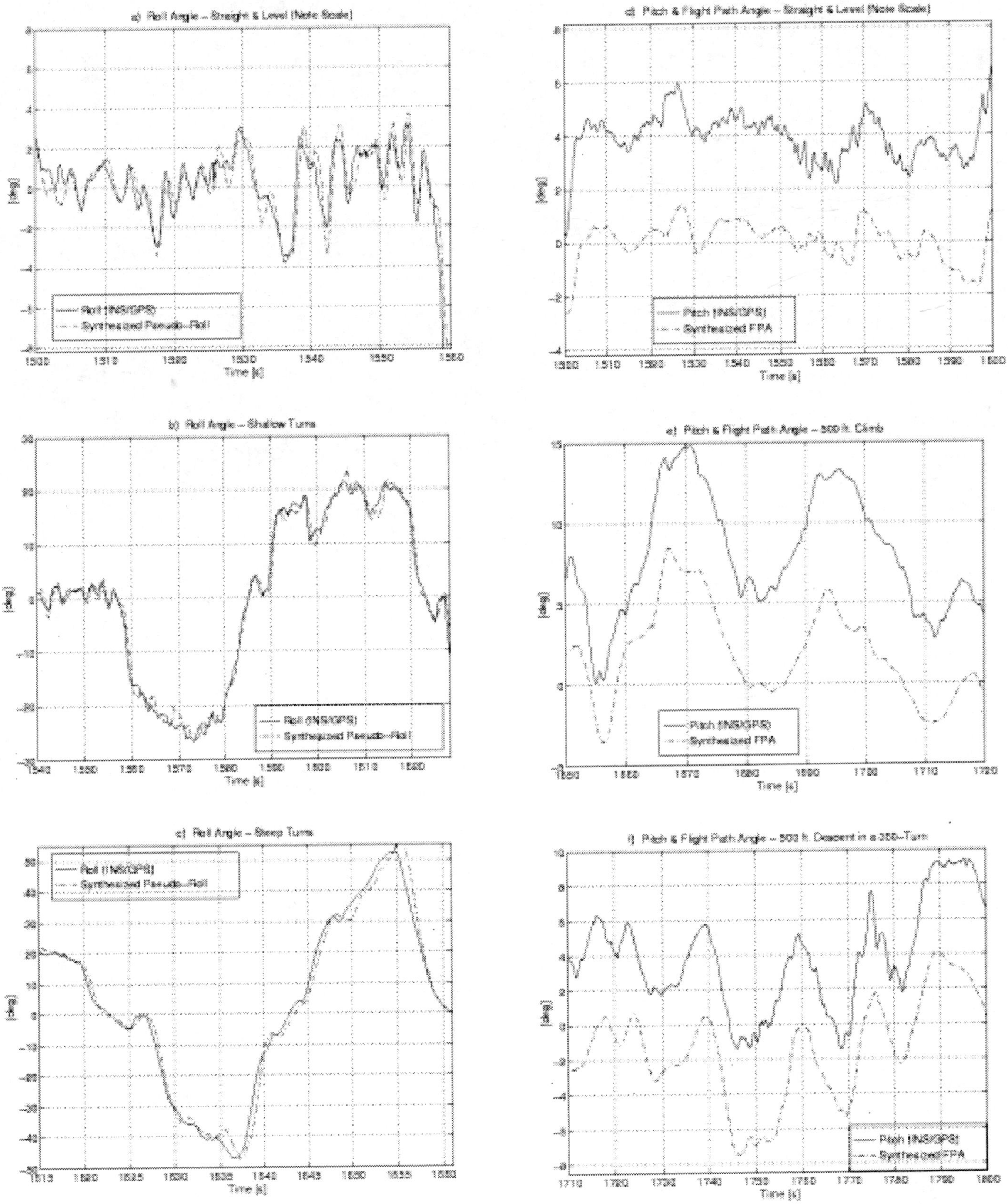

Figure 7: Comparison of Pseudo-Attitude and INS/GPS Reference Attitude

17th D A S C

The AIAA/IEEE/SAE Digital Avionics Systems Conference

Bellevue, WA
Oct. 31 – Nov. 7, 1998

Proceedings

Volume II

IEEE Catalog Number 98CH36267
ISBN 0-7803-5086-3 (Softbound Edition)
ISBN 0-7803-5087-1 (Casebound Edition)
ISBN 0-7803-5088-X (Microfiche Edition)

Prepress Production By:

ISOGEN International Corp.
2760 N. Academy Blvd, Suite 301
Colorado Springs, CO 80917
719 572-1706
sgml@isogen.com

Printed in USA by:

Lucent Technologies
Customer Information Center
2833 N. Franklin Road
Indianapolis, IN 80917

17th DASC
Presentations

Track F - Air Traffic Mgmt Systems

Session F1: Air Traffic Management

	Title	Presenter
1	Vertical Path Trajectory Prediction for Next Generation ATM	Mr. Anthony Warren
2	Advances in Flight Data Acquisition and Management Systems	Mr. Tom McDade
3	CNS/ATM Aircraft Customization Task (Not Published)	Mr. Arnold Oldach
4	USAF Initiatives for Global Air Traffic Management and Navigation Safety	Mr. Leo La Forge
5	A Simplified Aeronautical Telecommunications Network (ATN) Avionics Router	Mr. T. Signore
6	Avionics Architecture For Air Force Mobility Command Aircraft To Meet Cns/Atm And GATM Requirements	Mr. Donald Happel
7	A Pseudo Ramp Manager Workstation for the Laboratory Development of Airline-ATC Collaborative Arrival Planning Tools	Ms. Susan Dorsky

Session F2: Surface Movement 1: Display & Datalink

	Title	Presenter
1	Airport Surface Movement Technologies - Atlanta Demonstration Overview	Ms. Denise R. Jones
2	Description and Flight Test of a Rollout and Turnoff (ROTO) Head-Up Display (HUD) Guidance System	Mr. Richard Hueschen
3	Field Evaluation Of T-NASA: Taxi Navigation And Situation Awareness System	Mr. Anthony D. Andre Ph.D.
4	Airport Surface Operations Data Link Communications and DGPS	Mr. Steve Koczo
5	Controller-Pilot Data Link Statistics from NASA's 1997 Atlanta Flight Test	Dr. James Rankin

Session F3: Surface Movement 2: Surveillance

	Title	Presenter
1	Runway Incursion Reduction Program (RIRP) Surveillance System - NASA / FAA Atlanta Demonstration	Mr. Vincent Capezzuto
2	Development of Airport Surface Surveillance Performance Requirements	Mr. Rick Cassell
3	Analysis of ADS-B, ASDE-3 and Multilateration Surveillance Performance -- NASA Atlanta Demonstration	Mr. Carl Evers
4	Application of ADS-B for Airport Surface Surveillance	Mr. Dan Hicok
5	Comparison of A-SMGCS Requirements with Observed Performance of an Integrated Airport CNS System	Mr. Steven Young
6	What's Next for LVLASO: Status of Plans for a Year 2000 Flight Test and Demonstration (Not Published)	Mr. Wayne H. Bryant

Session F4: Communication, Navigation, Surveillance

	Title	Presenter
1	Bancroft's Algorithm for Solving Passive Multilateration Equations	Dr. Michael Geyer
2	An Integral Flight Director and Surveillance System for Helicopters in Metropolitan Service	Prof. Chin Lin
3	Data Processing Methods For Autonomous On-Board Surveillance And Collision Avoidance	Mr. Juan Besada Portas
4	Implementation of a Low-Cost SSR/ADS-B Aircraft Receiver Decoder	Mr. Alex Smith
5	Loop Technology (LOT) as an Alternative Surface Surveillance System	Mr. Vern Edwards
6	Broadcast Data Link Range Performance as a Function of Aircraft Size-Experimental Results (Not Published)	Robert Strain

Session F5: CNS/ATM Human Factors

	Title	Presenter
1	How Data Link Communication Might Affect Controller Workload in a Terminal Option	Dr. O. Veronika Prinzo PhD
2	Simulation Study of Vocoder Communication In Air Traffic Control	Dr. Earl S. Stein
3	Survey of In-Flight Replanning Performed on the Flight Deck	Mr. James K. Kuchar
4	Introducing New Technology to the Air Traffic Controller: Implications for Skill Acquisition and Training	Mr. Alfred L. Smith Jr.
5	Human Factors Issues in a Future Air Traffic Management System	Mr. Philip J. Smith
6	Man-In-The-Loop Part Of A Study Looking At A Free Flight Concept	Jacco Hoekstra

17th DASC
Presentations

Track G Aircraft Systems

Session G1: InFlight Entertainment

	Title	Presenter
1	Total Integrated Management (TIM) (Not Published)	Mr. Joseph R. Winston
2	Availability, Reliability, and Maintainability Prediction Model	Mr. James M. Hansen
3	Reality in Certification of IFE and Telephone Equipment	Mr. Ritch L. Triplett
4	A New Approach to Data Communications Utilizing the North American Terrestrial System (Not Published)	Mr. Mike O'Meara
5	Real-World Implementations of Intra-Aircraft Infrared Data Communications Networks	Mr. Patrick Potega
6	In-Flight Entertainment - Getting From Wishlist To Reality	Mr. Donald B. Lee
7	High Speed Networking of Multimedia Data for Passenger Entertainment (Not Published)	Mr. Greg Henrikson
8	Packet-based Networks	Mr. Steven W. Russert

Session G2: Aircraft & Satellite Avionics

	Title	Presenter
1	Airborne Reception of Data and Direct Broadcast TV using a Phased Array Antenna (Not Published)	Mr. David C. Vacanti
2	Guidance Characteristics of GNSS Landing Systems	Dr. Alex Stratton
3	Managing Aircraft Airworthiness Through Information Technology (Not Published)	Mr. William E. Larsen
4	Pro Line 21 Advanced Avionics System Architecture	Mr. Timothy Rayl
5	Techniques for Improved Reception of 1090 MHz ADS-B Signals	Mr. William Harman
6	Enhanced Ground Proximity Warning System (Not Published)	Mr. Barry C. Breen

Session G3: Commercial-Off-The-Shelf & Open Systems

	Title	Presenter
1	COTS Based Open Systems for Military Avionics	Mr. John Paul
2	Reducing Avionics Software Cost Through Component Based Product Line Development	Mr. David Sharp
3	Introduction to the Electronic Power Specification Standardization Activity	Mr. Sergio Navarro
4	Open System Design for CNI Avionics	Dr. George Mitschang
5	The Application of Commercial Processing Technologies to the Airborne Military Environment	Mr. Hugh S. Perry
6	Telecom Technologies (Not Published)	Brian Graber

Session G4: Unmanned Airborne Vehicles

	Title	Presenter
1	Redundant Control Systems for UAV's (Not Published)	Graham Gyattt
2	Flight Testing Perseus-B (Not Published)	Mr. Tom Clancy
3	Flight Testing a Large, Autonomous, Unmanned Aircraft (Or Global Hawk Begins to Soar) (Not Published)	Robert Ettinger
4	Flight Testing DarkStar (Not Published)	Mr. John Straub
5	X-36 Tailess Agility Aircraft Subsystm Integration (Not Published)	Willard J Harris
6	X-36 Tailless Agility Aircraft Subsystems Integration (Not Published)	Mr. Bill Harris
7	X-38 Avionics System (Not Published)	Christopher Nagy

Track H - Spacecraft Systems

Session H1: Commercial & Defense Space Systems

	Title	Presenter
1	IRIDIUM Low Earth Orbit Satellite Constellation	Mr. Peter Lemme
2	ATM Mobile Switch Requirements, Issues, and Concerns (Not Published)	Mr. Richard H. Paine
3	The 'Take-Off' of MOTS Avionics System for Space/Launch Vehicle Applications	Joe Cecchini
4	Rapid Development of avionic systems	Randy Black
5	Space-Based Wind-Sensing Lidar Design (Not Published)	Dr. Daniel Novoseller
6	Multifunctional Structures: A New Concept For Spacecraft Design Integrating Electronics, Structure And Thermal Control (Not Published)	Mr. Dave Barnett

17th DASC
Presentations

Track H - Spacecraft Systems (Continued)

Session H2: Space Exploration

1	Deep Space One Integration and Test Challenges: Getting to the Launch Pad in the Faster, Better, Cheaper World	Ms. Paula J. Pingree
2	The Mars Microprobe Mission Advanced Micro-Avionics for Exploration of the Martian Surface (Not Published)	Mr Randel Blue
3	From the Sun to Pluto	Ms. Karla B. Clark
4	NASA/JPL Mars Surveyor Program: New Challenges in a New Era (Not Published)	Mr. Robert L. Bunker
5	Interferometer Real Time Control for the Space Interferometry Mission (Not Published)	Mr. Charles E. Bell

Session H3: Microelectronics for Space

1	Ultra Low Power Rad Hard 12 Bit A/D Converter for Space-Based EO Sensors	Sven Nystrom
2	Integrating PWA Design and Analysis Using a Unix-Based Durability Toolset (Not Published)	Mr. Mostafa Rassaian
3	Amecom Direct Chip Attach Project	Tracey Clay
4	Digital Control of the MIDEX Spacecraft Power System (Not Published)	Ms. Karen Castell
5	Issues To Address In Use of Composite Materials for Electronic Packaging	Mr. Gary Trembley

Session H4: Space System Elements

1	X2000: Avionics for A Multi-Mission Spacecraft	Mr. Savio Chau
2	Applications For A Spacecraft Avionics Functional Model	Mr. Joseph F. Smith
3	A Table-Driven Control Method To Meet Continuous, Near-Real-Time Observation Requirements For The Solar X-Ray Imager	Mr. Kevin Shawn Wallace
4	The TRWIS III Hyperspectral Imager: Instrument Performance and Remote Sensing Applications	Dr. Stephanie Sandor-Leahy
5	Fault Protection Design of the Quikscat and Seawinds Instruments	Mr. Matthew B. Bennett
6	Development of a Low Cost Data Acquisition System for the Space Shuttle Solid Rocket Booster Program	Mr. Kevin Crawford

Track I - Automotive Systems

Session I1: Ground Vehicle Electronics

1	Software Process Improvement in an Automotive Electronics Organization	Mr. Gregory McHugh
2	Automotive System Design: Today and Tomorrow	Dr. Peter Hofmann
3	Specification and Testing of Automotive Powertrain Control System Software Using CACSD Tools	Mr. Steve Toeppe
4	Active Control of Vehicle Dynamics	Mr. Mike Fodor
5	The Next Generation Automotive Electrical Power System Architecture: Issues and Challenges.	Dr. John P. Miller PE, PhD
6	Automotive & Aerospace Circuit Fault Analysis	Mr. Craig Siegel
7	Requirements For A Real-Time Local Area Network Architecture In Land Combat Vehicles	Mr. Paul Richardson

Session I2: Intelligent Transportation System (1)- Vehicle Electronics

1	The Network Vehicle - A Glimpse into the Future of Mobile Multi-Media	Huan-Wun Yen
2	A Demonstration Project for the ITS Data Bus Prototype	Mr. Philip Spelt Ph.D.
3	Signal Processing and Waveform Generation in the Side Zone Automotive Radar	Mr. John C. Reed
4	Adapting Radar and Tracking Technology to an On-Board Automotive Collision Warning System	Dr. Edward Jocoy
5	Use of Map Data Information in an On-board Intersection Violation Detection System	Mr. John Pierowicz
6	SWIFT Project Results	Mr. Larry Senn

17th DASC Presentations

Track I - Automotive Systems (Continued)

Session I3: Intelligent Transportation System (2) - Info Management

Session I4: Hybrid Vehicles

Track J Information Systems

Session J1: Information System Application

Session J2: Information Integration

F. System Design Track – Air Traffic Management Systems

The ATM track is designed to cover recent developments in technologies and procedures for new air traffic management systems as we move into the 21st Century. The 5 sessions are designed to cover ATC provider plans for research, development, and the introduction of new tools and systems to improve aircraft safety and efficiency in the current ATC system and in a free flight environment.

Track Chair: Alex Smith, Rannoch Corp

Track F - Air Traffic Management Systems

Session F1: Air Traffic Management

Chairs: Steve Hill, Hanscomb AFB; Vu Duong, Eurocontrol Experimental Center

1	Vertical Path Trajectory Prediction for Next Generation ATM	Mr. Anthony Warren
2	Advances in Flight Data Acquisition and Management Systems	Mr. Tom McDade
3	CNS/ATM Aircraft Customization Task (Not Published)	Mr. Arnold Oldach
4	USAF Initiatives for Global Air Traffic Management and Navigation Safety	Mr. Leo La Forge
5	A Simplified Aeronautical Telecommunications Network (ATN) Avionics Router	Mr. T. Signore
6	Avionics Architecture For Air Force Mobility Command Aircraft To Meet Cns/Atm And GATM Requirements	Mr. Donald Happel
7	A Pseudo Ramp Manager Workstation for the Laboratory Development of Airline-ATC Collaborative Arrival Planning Tools	Ms. Susan Dorsky

Session F2: Surface Movement 1: Display & Datalink

Chair: Denise Jones, NASA Langley Research Center

1	Airport Surface Movement Technologies - Atlanta Demonstration Overview	Ms. Denise R. Jones
2	Description and Flight Test of a Rollout and Turnoff (ROTO) Head-Up Display (HUD) Guidance System	Mr. Richard Hueschen
3	Field Evaluation Of T-NASA: Taxi Navigation And Situation Awareness System	Mr. Anthony D. Andre Ph.D.
4	Airport Surface Operations Data Link Communications and DGPS	Mr. Steve Koczo
5	Controller-Pilot Data Link Statistics from NASA's 1997 Atlanta Flight Test	Dr. James Rankin

Session F3: Surface Movement 2: Surveillance

Chair: Steve Young, NASA Langley Research Center

1	Runway Incursion Reduction Program (RIRP) Surveillance System - NASA / FAA Atlanta Demonstration	Mr. Vincent Capezzuto
2	Development of Airport Surface Surveillance Performance Requirements	Mr. Rick Cassell
3	Analysis of ADS-B, ASDE-3 and Multilateration Surveillance Performance -- NASA Atlanta Demonstration	Mr. Carl Evers
4	Application of ADS-B for Airport Surface Surveillance	Mr. Dan Hicok
5	Comparison of A-SMGCS Requirements with Observed Performance of an Integrated Airport CNS System	Mr. Steven Young
6	What's Next for LVLASO: Status of Plans for a Year 2000 Flight Test and Demonstration (Not Published)	Mr. Wayne H. Bryant

Session F4: Communication, Navigation, Surveillance

Chairs: Jim Dieudonne, MITRE; Rick Cassell, Rannoch Corp.

1	Bancroft's Algorithm for Solving Passive Multilateration Equations	Dr. Michael Geyer
2	An Integral Flight Director and Surveillance System for Helicopters in Metropolitan Service	Prof. Chin Lin
3	Data Processing Methods For Autonomous On-Board Surveillance And Collision Avoidance	Mr. Juan Besada Portas
4	Implementation of a Low-Cost SSR/ADS-B Aircraft Receiver Decoder	Mr. Alex Smith
5	Loop Technology (LOT) as an Alternative Surface Surveillance System	Mr. Vern Edwards
6	Broadcast Data Link Range Performance as a Function of Aircraft Size-Experimental Results (Not Published)	Robert Strain

Session F5: CNS/ATM Human Factors

Chairs: James Kuchar, MIT

1	How Data Link Communication Might Affect Controller Workload in a Terminal Option	Dr. O. Veronika Prinzo PhD
2	Simulation Study of Vocoder Communication In Air Traffic Control	Dr. Earl S. Stein
3	Survey of In-Flight Replanning Performed on the Flight Deck	Mr. James K. Kuchar
4	Introducing New Technology to the Air Traffic Controller: Implications for Skill Acquisition and Training	Mr. Alfred L. Smith Jr.
5	Human Factors Issues in a Future Air Traffic Management System	Mr. Philip J. Smith
6	Man-In-The-Loop Part Of A Study Looking At A Free Flight Concept	Jacco Hoekstra

VERTICAL PATH TRAJECTORY PREDICTION FOR NEXT GENERATION ATM

Anthony W Warren, Boeing Company, Seattle, WA
Yaghoob S Ebrahimi, Boeing Company, Seattle, WA

Abstract

Current path prediction methodology during climb and descent phases of flight have much larger path uncertainty errors than during the cruise phase of flight. These uncertainty errors are a major concern for development of advanced ATM tools such as arrival and departure managers for regional flow planning and for medium term conflict detection in congested terminal and transition airspace. This paper describes a trajectory synthesis methodology and presents error sensitivity results for trajectory predictions. The trajectory synthesis method assumes the availability of high resolution, rapidly updated wind and temperature forecast data and enhanced ADS downlink of aircraft states and path intent. Trajectory sensitivity for a Boeing 737 aircraft climb profile is used to quantify system requirements and anticipated trajectory precision using this methodology.

Introduction

Many advanced concepts for Air Traffic Management explicitly utilize trajectory synthesis as an underlying element for air traffic decision support. For example, the URET conflict probe for En-Route ATM and the CTAS arrival management support tools explicitly use trajectory prediction methods to detect airplane conflicts and arrival flow problems. These systems are designed as augmentations to existing ATC operations, i.e. existing airplane systems and ground based infrastructure are used to develop trajectory predictions. However, existing methods of wind forecasting, radar surveillance, and flight planning are often not well suited to the task of predicting aircraft flight trajectories with 10 to 30 minutes lookahead, as needed for these applications. Trajectory synthesis during climb and descent phases is even more difficult, due to unknowns in aircraft performance and imprecision in estimating aircraft along-track velocity. This paper explores the issue of next generation ATM requirements for trajectory predictions, i.e. how much accuracy is needed for reliable conflict probes and regional flow management, and how could such requirements be met by appropriate airborne and ground equipage, and by further development of trajectory synthesis methodology.

Trajectory synthesis for climb and descent phases is more difficult than during cruise due to additional uncertainties such as airplane performance and reliability of a priori climb / descent profiles. The performance of the CTAS system for arrival scheduling [1] has been the subject of various studies and field testing which has established benchmarks for prediction uncertainty using existing technology. For example, [1] shows that for descent predictions, the ability of non-FMS aircraft to maintain a reference speed target may be on the order of 15-20 knots one sigma, and the radar tracker along-track ground speed error and the wind forecast error for the CTAS system may be of comparable magnitude. These errors are substantially higher than is desirable for decision support tools such as medium term conflict probe and meter fix scheduling.

Earlier methods for trajectory prediction in climb and descent segments use simplified

0-7803-5086-3 /98/$10.00

methods such as lookup tables of airspeed and vertical rate for specific aircraft types. These prediction methods typically yield errors on the order of one-three minutes in predicting time-of-arrival at a specific flight level or at a critical waypoint fix [2], and may not be adequate for scheduling aircraft or medium term conflict assurance during congested traffic periods. The heart of more recent methods is to tightly control velocity estimation errors due to sources such as aircraft surveillance, FMS speed scheduling, and wind forecasting. Although desired error levels for flow scheduling and conflict probe may be beyond that achievable with today's systems, they are potentially feasible for next generation ATC system architectures. This paper shows the feasibility of obtaining greatly improved trajectory forecasts, provided that the underlying CNS subsystems and ATM decision support are suitably enhanced for this purpose.

In this paper we assume performance requirements on desired trajectory uncertainty based on past studies [3] and on trajectory sensitivity studies summarized in the next section. We then examine a potential trajectory synthesis methodology which may be capable of achieving this level of performance for appropriately equipped (FMS /CNS) aircraft. This involves estimating aircraft 'equivalent weight' for enhanced prediction of aircraft climb and descent profiles.

Requirements for Medium Term Trajectory Predictions

Horizontal Plane Prediction Requirements

In our previous Monte-Carlo studies on conflict probe [3], we found that good horizontal probe performance could be directly related to along-track prediction error uncertainty. Typically, good conflict detection performance is obtained when one sigma prediction error is less than 2.0 nm. For a 20 min lookahead period, this means that the along track velocity error (delta_v) should satisfy delta_v * delta_time < 2 nm. Setting delta_time equal to 1/3 hr above yields delta_v < 6 knots rms, i.e. it is desirable that total velocity errors not exceed 6 knots. Similarly, the maximum sustained acceleration error that is acceptable in a vertical transition is such that delta accel * delta_time^2 /2 < 2 nm. Setting delta_time equal to 1/3 hr and translating this requirement into millig's yields delta_accel < 0.5 millig's (one sigma). These velocity and acceleration error requirements are the starting point for our trajectory synthesis method, i.e. horizontal error sources which could exceed these levels are to be mitigated, if possible, with technology elements which can reduce the horizontal prediction error components for climb and descent segments to no more than 2 nm rms.

Vertical Plane Prediction Requirements

For good vertical conflict detection performance, it is desirable to have no more than a 1000 foot altitude prediction error (one flight level for IFR traffic) at a two sigma or 95% accuracy level. This criterion is an aggressive goal for next generation systems, but is important for reliability of medium term conflict alerting and trajectory intervention, i.e. it is desirable to intervene only when an airspace conflict is highly likely in both the vertical and horizontal dimensions. Current trajectory estimation systems (manual or automated) are much less reliable in predicting vertical versus horizontal conflicts, which may result in excessive interventions such as level-offs in congested, extended terminal airspace. We therefore recommend precision in vertical prediction of 500 ft rms or better for climb and descent segments.

The above vertical criterion can be directly translated into an acceleration error criteria for FMS aircraft flying airspeed schedules in descent and climb segments. We illustrate this process and derive acceleration error requirements for a Boeing 737-400 aircraft modeled using an aircraft performance evaluation tool called INFLT. INFLT models trajectory flight paths by integrating simplified vertical plane, point mass equations-of-motion using nominal thrust, drag, and fuel flow tables. This code is similar to the trajectory methodology used in FMS aircraft and is comparable in modeling detail to the vertical trajectory synthesis method used for CTAS [4].

Current FMS aircraft use airspeed regulation in specific flight segments to implement climb and descent transitions. A nominal climb profile for the 737-400 aircraft is shown in Figure 1 below. Following retraction of the flaps and gear at about 1500 feet, the FMS typically schedules four flight segments to climb to cruise altitude:

*Segment 1: 250 knots CAS speed restriction
*Segment 2: Speed transition at 500 fpm Rate-of-Climb (ROC)
*Segment 3: 280 knots CAS climb speed
*Segment 4: 0.74 Mach climb to cruise altitude

This transition sequence is the recommended climb path for minimum fuel burn, and assures that airspeed stays within the flight envelope.

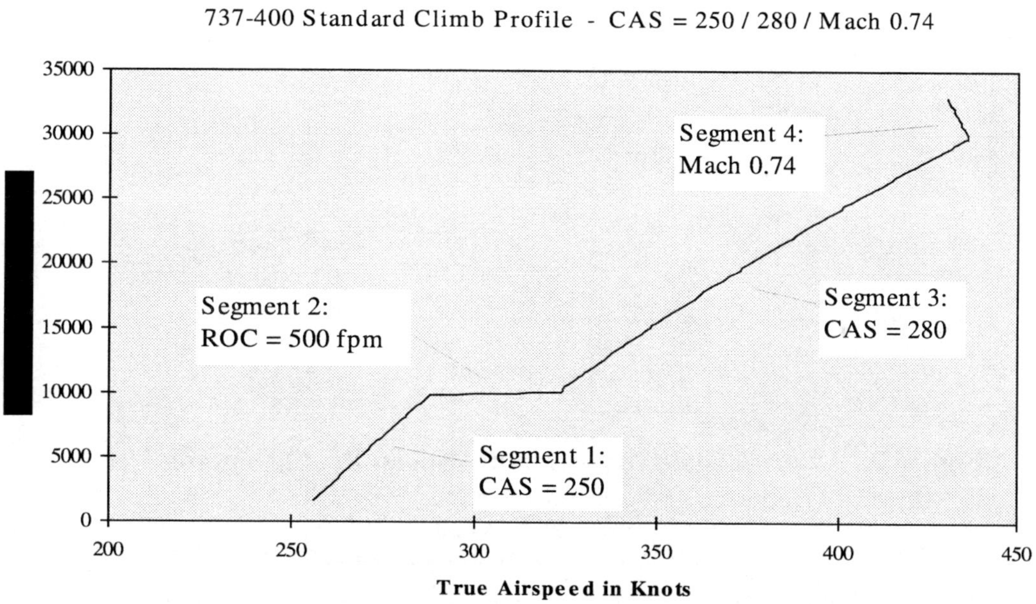

Figure 1. Nominal Climb Transition for 737-400 Aircraft

The sensitivity of the 737 aircraft to energy rate errors in estimating aircraft thrust, drag and weight was studied by examining the effect of small percentage differences in these variables on trajectory predictions. Figures 2 and 3 show prediction results for a 5% difference in modeled thrust profile. In this case, a take-off weight of 115,000 lbs was assumed for both trajectories, and sensitivity to aircraft thrust uncertainty was examined by assuming one aircraft flies maximum climb thrust and the other flies at 95% of maximum thrust. Figure 2 shows that the difference in the ground speed profiles grows with prediction time and exceeds 10 knots during the 280 CAS segment. However, the effect on the along track predictions is within acceptable limits (~ one nm error). In other words, this level of thrust uncertainty is acceptable for <u>horizontal</u> trajectory predictions. By contrast, Figure 3 shows that the altitude divergence is quite pronounced in this case, and the maximum altitude difference is about 2000 feet and delta time to Top-of-Climb is about 2.0 minutes. The difference in energy rate = net thrust minus drag divided by weight for these two trajectories is about 8% during the initial CAS=250 climb segment. This means that for a maximum deviation of 500 ft rms error in vertical profile, the initial energy rate error should not exceed 2% rms or about 0.25 millig's in horizontal acceleration. This requirement is about a factor of two more precise than that estimated above for horizontal prediction accuracy. Other sensitivity studies to errors in estimating energy rate yielded similar results.

In summary, for medium term conflict detection performance it is desirable to estimate horizontal velocity errors within 6 knots rms, and horizontal acceleration errors within 0.25 millig's or about 2% of initial climb levels. The latter requirement is necessary to achieve good vertical predictions with ~ 20 minute lookahead.

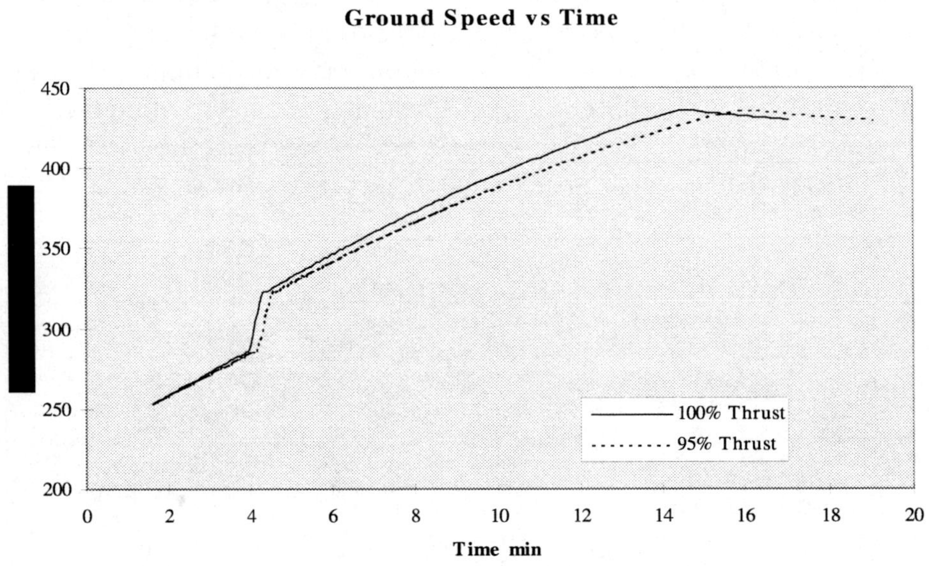

Figure 2. Thrust Sensitivity - Ground Speed Profile Versus Time

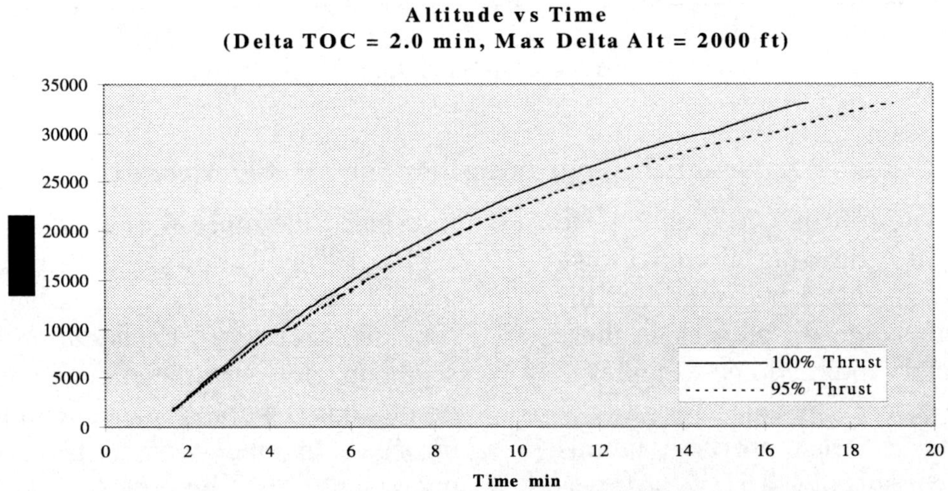

Figure 3. Thrust Sensitivity - Altitude Profile Versus Time

Technical Approach for Enhanced Trajectory Synthesis

We here assume that the most critical need for enhanced trajectory precision is in the terminal and extended terminal area of congested hub airports. Our technical approach assumes that aircraft equipped to support enhanced trajectory predictions will be given enhanced ATM procedures such as more efficient routings and arrival scheduling, and reduced tactical interventions to justify enhanced equipage. Specifically, we assume the use of GPS or baro-inertial based RNAV or FMS systems for precise path following, and the use of an air-ground datalink for broadcasting flight intent and aircraft specific parameters such as weight and intended speed schedule.

Given the above, the technical approach which we investigated for enhanced path predictability is based on three supporting technical concepts:

- Downlink of trajectory intent parameters
- The use of onboard and ground based intent monitoring for trajectory quality assurance
- Calibration of aircraft acceleration profile using onboard vertical rate measurements.

The downlinking of intent information is to assure reliability of path intent, which is a problem with current conflict probe and arrival management systems. Trajectory quality assurance includes position and intent conformance monitoring and overall estimation of trajectory prediction quality as a function of airborne equipage level, forecasting quality and other factors. Acceleration profile calibration is a means to achieve the accuracy requirements specified in the previous section. The rest of this paper focuses on the concept of 'Equivalent Weight' estimation to perform this function.

Equivalent Weight Profile Matching Concept

An earlier section established a requirement for acceleration profile accuracy of roughly 2% rms, relative to initial climb acceleration values. This level of accuracy is probably not achievable with current operational methods of estimating aircraft performance, which use look-up tables for thrust and drag, and estimated aircraft weight. Unknown factors such as age degradation of engine performance, off-nominal thrust due to environmental conditions and airline policy, and errors in estimating passenger weight can easily lead to rms errors on the order of 5 to 10%. In fact, the CTAS system is reported to have acceleration errors of this magnitude, due to uncertainty in estimating thrust, drag and weight [1].

The operational concept investigated in this paper for achieving more accurate acceleration profiles is to measure vertical rate accurately onboard the aircraft and then to match the estimated energy rate term (T - D) / W to the measured vertical rate during constant CAS or Mach speed segments. Given vertical rate measurements hdot', trajectory profile matching may be achieved by estimating equivalent weight We to match the following well known trajectory equation [5]:

$$\text{hdot}' \sim [(T - D(We)) / We] * Va / (1 + G * Va / g)$$

where T denotes net thrust, D(W) denotes drag as an implicit function of weight W, Va denotes airspeed, G denotes a partial derivative of speed with respect to altitude, and g denotes a sea-level gravitation constant.

Once We is accurately estimated, updated climb or descent trajectories can be synthesized, replacing actual weight with equivalent weight in the equations of motion. The estimation process for We can be implemented in the FMS, or by ADS-B broadcast of vertical rate to an ATC facility.

Equivalent Weight Estimation Accuracy

A least squares analysis was performed to quantify the percentage accuracy achievable in estimating equivalent weight. For this study we assumed an a priori error in estimating the energy rate term (T - D) / W of 10% rms. The nominal trajectory for the sensitivity studies was a 737-400 climb trajectory with takeoff weight = 120,000 lb. Sensitivity to two measurement parameters was analyzed, i.e. the accuracy of the vertical rate measurement, and measurement update time.

Vertical rate rms errors assumed the use of a GPS sensor and 3 bounding conditions depending on GPS sensor accuracy and local vertical wind. The best conditions occur when the GPS signal is free of SA noise, i.e SA is turned off, and there is only light vertical wind. In this case, both GPS velocity error and wind error are in the range of 0.1 to 0.3 m/s and the anticipated rms error is on the order of 0.5 knots. Substantially less accuracy is obtained when SA noise is broadcast or when encountering moderate vertical winds. In this case, an rms error on the order of 2.0 knots can be anticipated. Although very large vertical rate errors are occasionally anticipated, a 95% bounding case for normal operations would be about twice that of the moderate turbulence case, i.e. with an rms sensor error of 4.0 knots. Parameter sensitivity to these bounding cases is examined below.

Two cases were examined for measurement update time. The best case used one second updates, corresponding to onboard processing of available GPS measurements, or equivalently to ADS-B broadcasting at a one second update rate. The worst case examined use of a nominal five second update rate. Least squares accuracy performance for the one second updates and varying levels of sensor accuracy is shown in Figure 4. Figure 4 shows that with one second updates, most of the value of data smoothing is achieved with 20 seconds of vertical rate processing, and equivalent weight accuracy on the order of 1 to 2% rms is achievable, even with sensor rate errors on the order of 2 to 4 knots rms. Results with 5 second updates shows similar trends, with somewhat degraded performance. These results show that equivalent weight estimation is a feasible method for achieving the desired (~ 2% rms) acceleration accuracy.

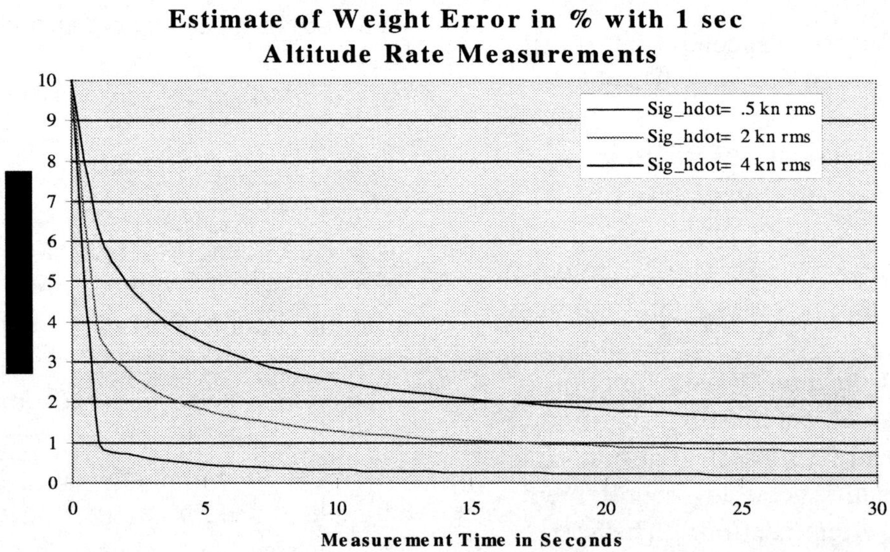

Figure 4. Equivalent Weight Estimation with One Second GPS Rate Data

Equivalent Weight Profile Study

The effectiveness of equivalent weight profile matching was evaluated using INFLT to model trajectory profiles. The 95% thrust climb profile for the 737-400 illustrated earlier was used as a Truth model for the equivalent weight study, i.e. we assumed that the actual profile flown is a 737-400 with take-off weight Wo = 115 klb and climb thrust = 0.95 * max climb thrust. Trajectory prediction accuracy was evaluated assuming equivalent weight matching at 3500 ft altitude to adjust for the error in thrust using max climb thrust as the nominal thrust level for trajectory predictions. Sensitivity to perfect weight matching, and to an estimation error of 1500 lbs (~ 1.2% weight error) were evaluated in this study.

Equivalent weight for profile generation was obtained by solving the following energy rate matching equation for We:

$$(To - D(We)) / We = (0.95 * To - Do) / Wo$$

where To = Max thrust at 3500 ft ~ 24.37 klb
Do = Nominal drag at 3500 ft ~ 7.08 klb
Wo = Nominal takeoff weight ~ 115 klb
We = Equivalent weight at takeoff .

Solving the above equations for We yields We ~ 121.5 klb. This calculation assumes perfect weight estimation. We also examined the case when We = 123 klb, i.e. with a 1.5 klb error in equivalent weight estimation.

Trajectory profiles for the truth model and with perfect weight estimation were nearly identical, validating the weight matching concept. Trajectory profile comparisons for We = 123 klb are shown in Figures 5 and 6. The ground speed profile shown in Figure 5 virtually matches that of the truth trajectory. This is to be compared with Figure 2, which shows ground speed sensitivity to an uncompensated thrust error. Figure 6 shows the altitude profile with equivalent weight compensation versus the truth model profile. This figure is to be compared with Figure 3. This sensitivity study shows the potential value in compensating for unknown energy rate errors, i.e. vertical prediction accuracy may be dramatically improved.

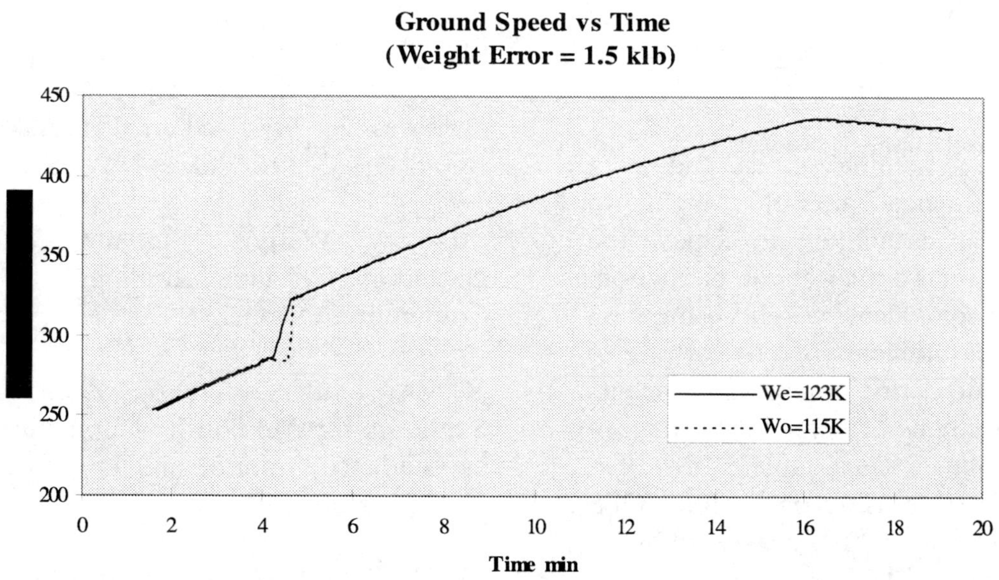

Figure 5. Equivalent Weight and Truth Model Ground Speed Profiles

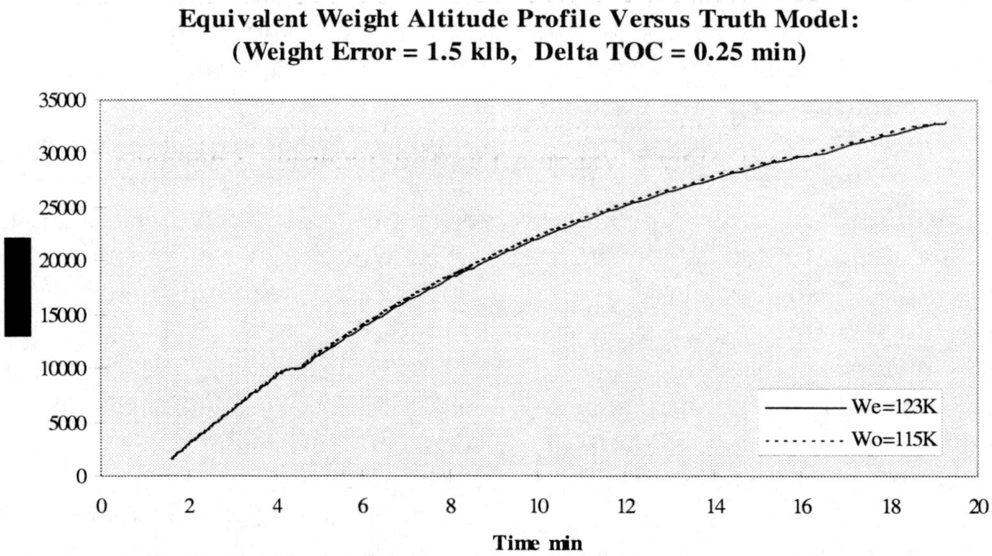

Figure 6. Equivalent Weight and Truth Model Altitude Profiles

Conclusions

Next generation ATM applications such as medium term conflict detection and traffic flow management may require enhanced trajectory prediction capability. Our studies show the need for controlling velocity and acceleration error sources in trajectory synthesis to 6 knots rms and 0.25 millig's. Energy rate uncertainty strongly affects vertical profile predictions, but is generally not a strong factor in horizontal path predictions.

The use of nominal thrust and drag profiles and estimated aircraft weight to compute energy rate is unlikely to achieve the desired accuracy for vertical flight prediction. The concept of equivalent weight estimation using onboard measurements of vertical rate to calibrate energy rate during climbs and descents is proposed as a means of achieving enhanced trajectory predictions. Our studies show that this concept is feasible using one second GPS measurements and a smoothing interval of about 20 seconds to stabilize the estimation process. Implementation of this concept would require downlinking equivalent weight to an ATC facility, or the use of ADS-B to broadcast vertical rate and other aircraft state parameters.

References

[1] C.G. Hunter et al., "CTAS Error Sensitivity, Fuel Efficiency and Throughput Benefits Analysis," TM96150-2, Seagull Technology Inc., July 1996.

[2] J.N. Beers et al., "The Application of Trajectory Prediction Algorithms for Planning Purposes in the Netherlands ATC System," AGARD-AG-301, Vol. 2, May 1990.

[3] A.W. Warren, "Medium Term Conflict Detection for Free Routing," 16th DASC Conference, October 1997.

[4] R. Slattery and Y. Zhao, "Trajectory Synthesis for Air Traffic Automation," Journal of Guidance, Control and Dynamics, Vol. 20, No. 2, March 1997.

[5] A. Benoit and S. Swierstra, "Generation of Aircraft Trajectories for On-Line Operations," AGARD-AG-301, Vol. 2, May 1990.

0-7803-5086-3 /98/$10.00 ©1998 IEEE

ADVANCES IN FLIGHT DATA ACQUISITION AND MANAGEMENT SYSTEMS

Thomas M McDade, AlliedSignal Air Transport and Regional, Redmond, Washington

Abstract

This paper describes the advances in the AlliedSignal Flight Data Acquisition and Management Systems (FDAMS) product line. AlliedSignal's advanced FDAMS integrates the functionality of the ARINC 717 Digital Flight Data Acquisition Unit (DFDAU), Aircraft Condition Monitoring System (ACMS), PCMCIA Interface , Smart Access Recorder (SAR), Digital ACMS Recorder (DAR) and Quick Access Recorder (QAR) into a single Line Replaceable Unit for ARINC 700 series aircraft. The mandatory DFDAU function supports both the latest 4X and 1X retrofit data frames. The FDAMS provides a user-configurable, single Hardware and System/Application software solution for multiple aircraft types and customer configurations.

Introduction

In the past, customer specific ACMS applications required extensive development time and effort. These fixed software applications are obsolete. AlliedSignal has simplified the process by developing a database driven ACMS application for reports, data recording, system reconfiguration and growth. Integration of single function LRU's into a common unit, and access to DFDR data via integrated PCMCIA card reduce cost and increase the user's flexibility and control over the system. 16 resident DFDAU frame formats support the latest 4x and 1x retrofit frames allowing for installation of the same hardware and software in multiple aircraft types and configurations.

Unique ACMS databases are developed for each customer/aircraft type utilizing an advanced Windows 95 based reconfiguration tool. The Airborne Database Reconfiguration Tool (ADRT), uses a Graphical User Interface (GUI) to define all of the Alpha Callup Parameters, 8 SAR channels, DAR frame formats, selection of QAR recording, ACARS options and Memory Partitioning. ADRT features VADAR, Visual Algorithm Definition for Avionics Reconfiguration. VADAR is used to develop SAR and DAR trigger logic, ACMS report logic, data collections, output formats and output rules.

Advances in AlliedSignal FDAMS with ACMS allow an entire ACMS system to be developed, exported and uploaded as databases without costly and untimely software modifications.

Aircraft Condition Monitoring Systems (ACMS)

The early generation ACMS applications were developed in the "C" programming language requiring extensive time to develop, test and modify. **These systems offered limited reconfigurability and customer specific requirements were often difficult to support quickly and economically**. As airline customers gained ACMS experience they frequently requested modifications to their systems. Functional growth in ACMS often took months or years from concept to implementation. AlliedSignal's development of VADAR has changed ACMS programming dramatically. Report trigger logic and formats once hard coded in software are now presented visually and easily modified and uploaded as a database. Airborne database

0-7803-5086-3 /98/$10.00 ©1998 IEEE

loading is accomplished by either a PCMCIA Card or an ARINC 615 device in less than 5 minutes.

VADAR allows any user to easily develop or modify ACMS functions such as aircraft exceedence monitors, aircraft performance reports, Flight Operations Quality Assurance (FOQA) monitors and data recording triggers without knowledge of programming languages. Requests from sub-system manufacturers to monitor components, changes to exceedence limits, new FOQA monitors or data recording triggers to support maintenance requests are now implemented in a timely matter. Up to 79 user-definable VADAR reports are provided with the standard unit. After product delivery, the ACMS users can implement additional reports and recording triggers specific to their operations. Since the entire ACMS database is reconfigurable, the ACMS user's can also modify and/or delete any of the standard programming.

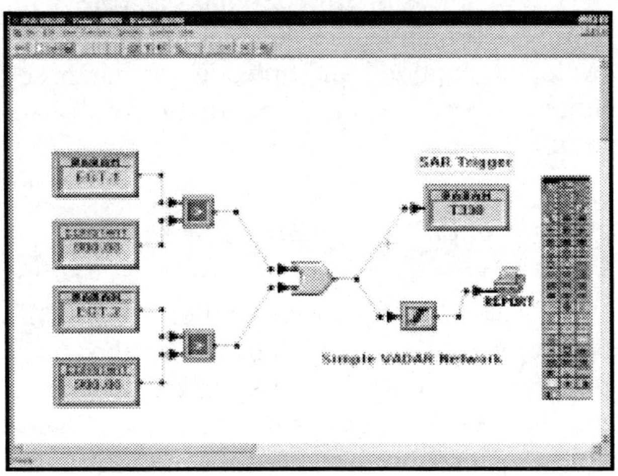

Figure 1 – Simple VADAR Network

Years of ACMS experience with multiple aircraft types and airline customers has allowed AlliedSignal to develop in VADAR a Pre-programmed set of Standard ACMS Reports. These reports provide a broad foundation to new and experienced ACMS users. The standard set of ACMS reports provided in the basic FDAMS ACMS Database is shown below:

- Engine Stable Cruise Performance Report
- Engine Takeoff Report
- Engine Gas Path Advisory Report
- Engine Mechanical Advisory Report
- End of Flight Report

These reports allow each ACMS user to monitor fleet performance accompanied with event driven Report data collections and DAR/SAR Recordings to assist in the improvement of aircraft maintenance, reliability and safety.

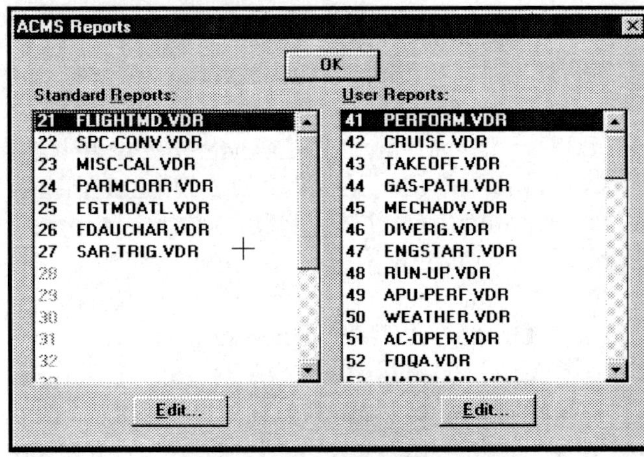

Figure 2 - ACMS Reports

Unscheduled engine maintenance, APU problems, hard landings and degraded or degrading sub-system performance is expensive and effects aircraft dispatchability and safety. Programs to objectively monitor flight operations and pilot performance for the purpose of training and safety are supported by aviation authorities and have been implemented at several airlines. AlliedSignal has developed via ADRT and VADAR, 2 optional ACMS packages, Advanced Maintenance reports and FOQA reports. These packages are intended for those airline customers who wish to implement advanced ACMS capabilities beyond the basic engine trending and engine event monitoring provided by the five basic ACMS reports. The cost of installing the FDAMS system can be offset if even one unscheduled engine removal is avoided. Delays associated with a malfunctioning APU often affect dispatchability. Hard landing reports can provide information relative to the extent of inspection after a hard landing

0-7803-5086-3 /98/$10.00 ©1998 IEEE

incident. AlliedSignal developed the Advanced Maintenance reports based on years of experience with numerous ACMS users. The Advanced Maintenance reports option consists of the following ACMS reports implemented with VADAR:

- Engine Divergence Report
- Abnormal Engine Start Report
- Aircraft Performance Report
- Engine Ground Run-up Report
- APU Performance Report
- Hard Landing Report
- Turbulence Report
- Overweight Report
- Weather Report
- Normal Engine Start Report
- Approach Configuration Report

Airlines using these maintenance reports can react to events quickly, performing inspections or scheduling maintenance at the first opportunity.

Regulatory agencies have initiated Flight Operations Quality Assurance (FOQA) programs in an effort to improve pilot training and flight safety. These programs are typically based on post-processing FDR data and are limited to the data contained in that data stream. Several airlines have independently initiated their own FOQA programs, and for these customers, AlliedSignal has developed a FOQA report as an option. The advantage of the AlliedSignal system is that all aircraft parameters available in the FDAMS are available for use in the FOQA report. This ACMS option consists of a standard but modifiable VADAR report which currently features over 85 event triggers, snapshot data collection and FOQA statistics developed with inputs from a combination airline customers.

The FOQA report when used in conjunction with an Airborne Communications Addressing and Reporting System (ACARS) provide an airline with event driven solution for improved operational safety. Updates to the FOQA report such as modification to existing triggers or additional triggers is easily accomplished with VADAR. Alternative ACMS systems with fixed FOQA algorithms do not offer the flexibility of the AlliedSignal ACMS system. If ACARS is not available to the user, then the FOQA statistics and supporting recorded events can be downloaded either via the PCMCIA Card or ARINC 615 device.

VADAR also allows the user to define the logic for triggering both the ARINC 573 DAR recorder and the 8 SAR channels available in the FDAMS. SAR channel recording is an extremely useful tool for monitoring aircraft data and the ability to easily redefine the triggering logic provides an airline with a powerful tool for maintenance and safety. For example, a FDAMS user recently combined VADAR triggering and SAR recording to diagnose an intermittent engine stall problem, and another airline and the associated engine manufacturer recently used this functionality to quickly diagnose an engine anomaly that only became apparent when the airplane was put into use at that particular airline.

The ACMS system provides for up to 79 user defined VADAR reports. With a little familiarization, any user can start developing VADAR reports or SAR/DAR triggers.

Some basic steps involved in creating a VADAR report are:

1. Use the widget pallet for selection of the parameter inputs, comparators, etc and implement a logical network.
2. If developing an SAR trigger, connect to a parameter output and select an open trigger parameter (T001 - T500).
3. For report content, select from the built-in parameter list to set up the data collections desired for the report.

4. If developing a new Report, connect to the report widget, define the report format, and select the output rule for the report (printer, storage, ACARS etc).

Each widget in the tool pallet also features online help to assist the user in the understanding of the widget's function, inputs and outputs. Figure 3 below is the Stable Cruise report and demonstrates some of the visual characteristics of VADAR. Note how simple it is to define the most complicated ACMS logic in such a simple graphical format.

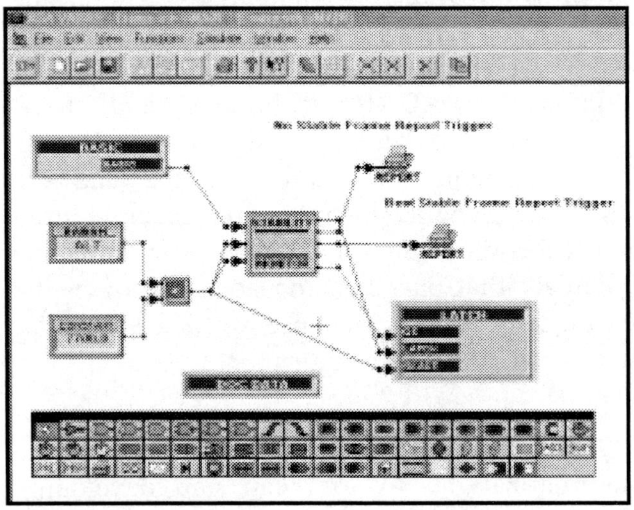

Figure 3 - VADAR Stable Cruise Report

Recent VADAR Enhancements

Several powerful advancements to VADAR include the stability widget, an advanced data collection widget, multiple report formats, new report formatting, string generators and extractors and a comments widget.

The stability widget, shown below, encapsulates functionality of the complex stable frame algorithm and performs data collection and averaging for such. As shown in Figure 4, all aspects of the previously hard coded stability parameters are now reconfigurable within this widget.

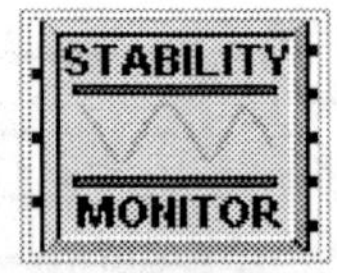

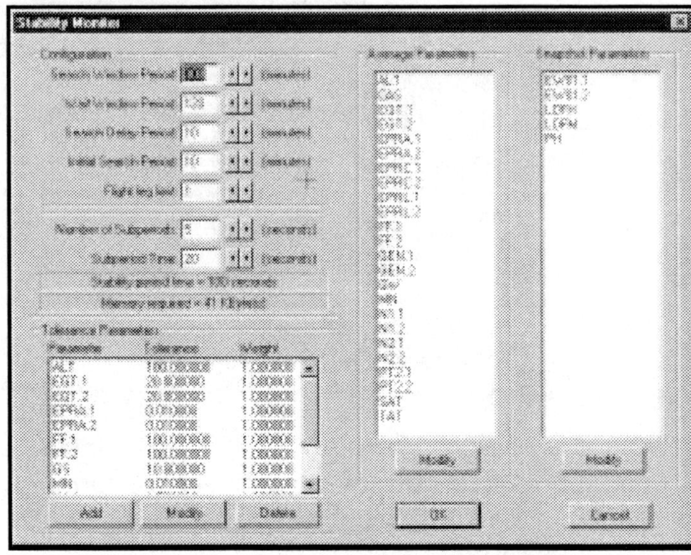

Figure 4 - Stability Monitor Definition

Advancements in the data collection capabilities have improved the users ability to collect and store data in the VADAR reports. The Data Collector Widget (shown below) is used to collect historical and real time values of set of aircraft parameters for subsequent output on an ACMS Report. The average value of the time series data collection is also available for output on the ACMS Report. A set of up to 128 aircraft parameters may be specified for collection via the internal parameter drop-down box of the Data Collector Widget. When the input Boolean enable recording command is TRUE, a snapshot of the current pre-event history is made, and then the output ready/busy pin goes to busy, and remains busy until data is collected for the specified post event time. Refer to figure 5 for a view of the "insides" of the Data Collector Widget.

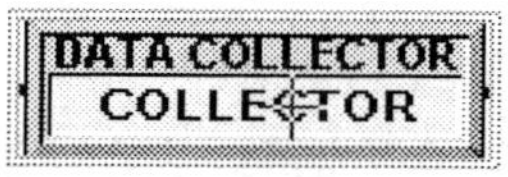

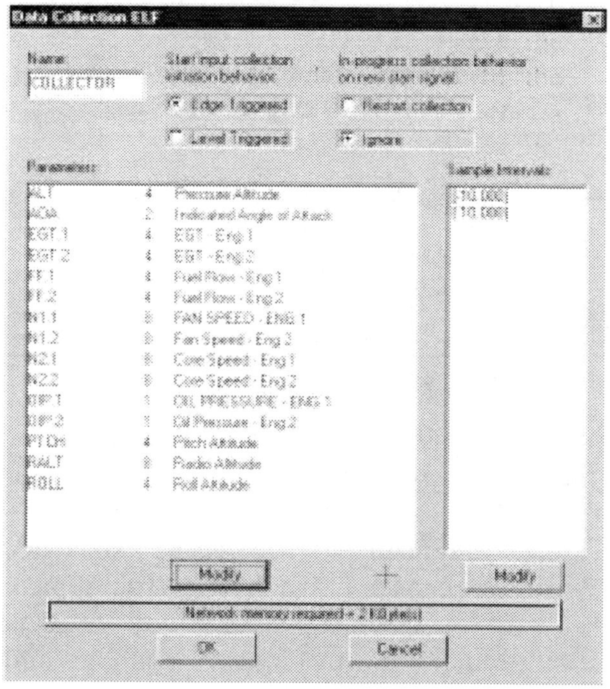

Figure 5 - Data Collection Definition

Multiple report formats are now selectable from within the Report Generator widget. An unlimited number of Report formats can now be defined for the same VADAR network (i.e., VADAR Report). As an example, an ACMS user could specify 5 report formats for an engine exceedence report, each with a potentially different output rule:

1. Primary engine data for the left engine printed at the end of flight
2. Secondary engine data for the left engine dumped to a PCMCIA Card
3. Primary engine data for the right engine printed at the end of flight
4. Secondary engine data for the right engine dumped to a PCMCIA Card
5. Exceedance Summary Data Downlinked via ACARS Immediately

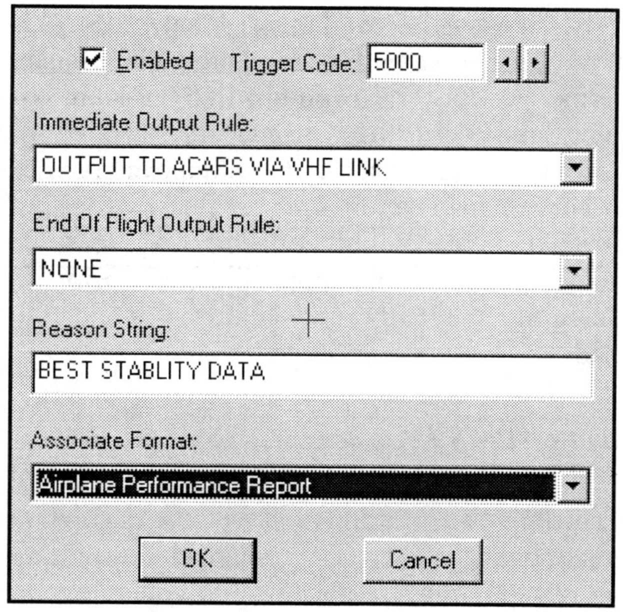

Figure 6 - Report Attributes

The editor used to define report formats has also been improved and integrated with the data collection widget as shown in figure 7.

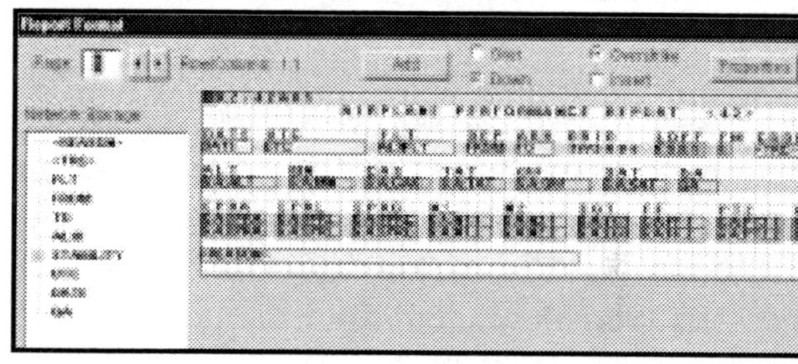

Figure 7 - Report Format Editor

String manipulation widgets have also been added to support character data in reports. The String Generator Widget provides the generator of a character string up to 15 characters as a function of an integer input. This Widget has been useful in the creation of Aircraft Tail Numbers as a function of the Aircraft Identification plug. The String Extractor Widget can be used to extract any character within a Character string.

The comment widget has been expanded from 15 characters to an unlimited character string complete with font and size selections. This improvement allows for clear documentation inside VADAR Networks.

Testing of new networks is accomplished using the simulation tool. Inputs to the network are simulated to validate the logic and verify the format of the report. Testing and verification of the new report prior to aircraft installation, ensures that the desired parameters are being collected and that the trigger logic is operating properly prior to changing the airborne database application load.

Smart ACMS Recorder (SAR)

The SAR function is a powerful recording function that allows the airline to create dedicated recording channels to monitor aircraft data triggered by specific system events for maintenance, system evaluation, trend analysis and troubleshooting purposes. The ACMS database reconfiguration tool (ADRT) provides the ability to easily reconfigure each SAR channel. Individual SAR triggers are developed in VADAR and can be easily added or modified as required. The recorded SAR Data can then be extracted by the PCMCIA Card or an ARINC 615 device. The extracted data can then be displayed in graphical, tabular, world plot, and/or cockpit animation form via the AlliedSignal Readout Tool.

The SAR function currently provides the capability to record 127 parameters on each of 8 channels. Refer to figure 8 for a view of the top level of the SAR editor.

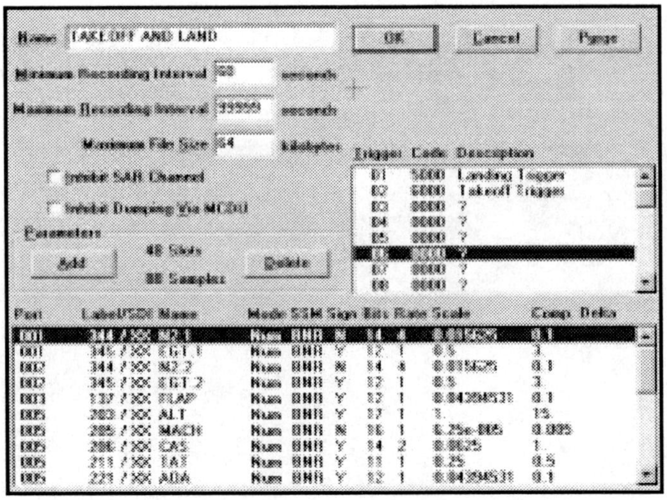

Figure 8 - SAR Channel Definition

SAR files can be stored internally or output to floppy disks or the PCMCIA card. The PCMCIA card offers the distinct advantage of 110Mb removable storage capacity standard, with higher capacity available. The typical SAR user can record weeks worth of data before removing the card for ground processing.

Digital ACMS Recorder (DAR)

The ACMS provides an ARINC 573 output for those airlines utilizing DAR recorders. The DAR frame format is easily defined using the reconfiguration tool ADRT and supports recording of 1X, 2X, 4X, 6X and 8x recording speeds. DAR triggering is defined in VADAR and assigned in the DAR editor. The integrated PCMCIA interface allows the user to output the DAR data stream to the PCMCIA card, eliminating the need for a separate DAR recorder. Figure 9 shows the section of the DAR editor which selects the DAR output control.

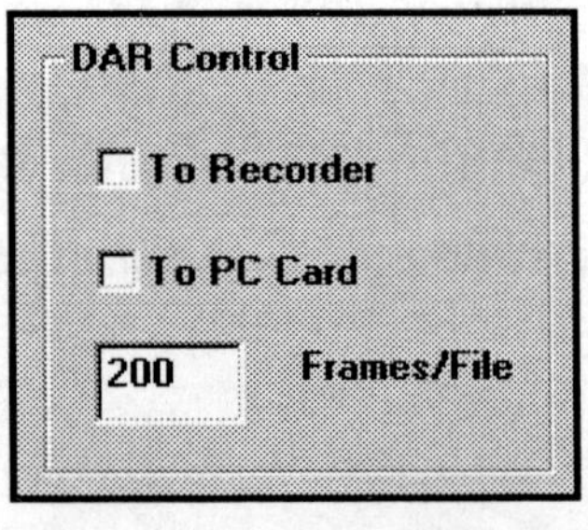

Figure 9 - Dar Control

Quick Access Recorder (QAR)

The ARINC 573 FDR data stream can also be output to the PCMCIA card by selecting a switch in the reconfiguration tool. The integrated PCMCIA interface allows the user to output the QAR data stream to the PCMCIA card, eliminating the need for a separate QAR recorder. Figure 10 shows the section of the DAR editor which selects the Programmable QAR output control.

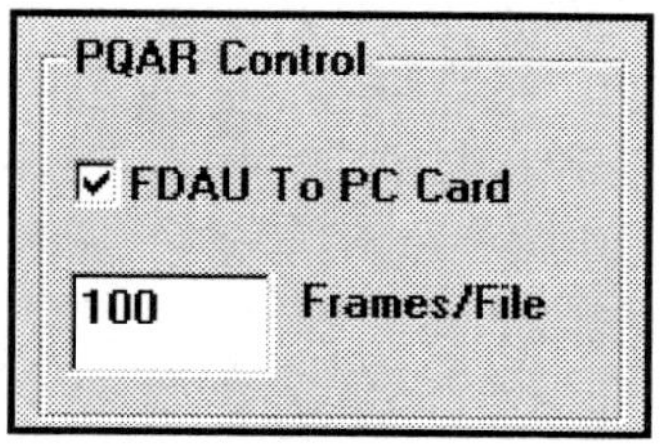

Figure 10 - PQAR Control

The storage capacity of the standard 110Mb PCMCIA Flash ATA card is dependant on the recorder data rate and selection of recorder options. The table 3 describes the PCMCIA storage capacity:

Table 3 - PCMCIA Storage

Data Rate (words/sec)	Capacity (Hours)
64	212
128	106
256	53

Integration of the recorder functions reduces cost, save weight, increase system reliability and offers airlines an expanded choice of recording options.

PCMCIA Data Loader and Storage

The PCMCIA card interface introduces a powerful advancement over older FDAU, Data management and ACMS systems. Flash ATA cards are used for data storage and dumping. The standard ATA card provided with the AlliedSignal FDAMS system contains 110Mb of data storage. Larger PCMCIA cards are available for airline customers who desire greater online storage capacity. The data stored on the PCMCIA card is easily read by ground processing equipment because the information is stored in basic DOS file formats. Additional cards of the same size and larger are commercially available

The PCMCIA interface also functions as a data loader utilizing a PCMCIA flash memory card. Typical load time of the FDAMS system/application or database is 5 minutes. Users will typically upload changes to the ACMS databases with a PCMCIA load card.

Aircraft Turbulence Report Option

The FAA, NOAA, United Airlines and AlliedSignal are currently developing a program to quantify and report in-flight turbulence in an effort to reduce turbulence related injuries. Reports downlinked via ACARS offer an efficient, objective measurement of turbulence while reducing reliance on ATC/pilot communications. When fully implemented, participating airlines will be provided with an another valuable advancement for improving weather forecasting, passenger service and safety.

Summary

The AlliedSignal advanced FDAMS with ACMS now provides an integrated solution for digital flight data acquisition, aircraft condition monitoring, and DAR/SAR/QAR aircraft data recording. Integration of the PCMCIA card interface provides for a convenient method of data storage, data retrieval and loading of ACMS system/application software and ACMS databases. The database driven ACMS application software eliminates the need for costly software modifications. Advancements in the industry's most powerful ACMS reconfiguration tool, VADAR, allow airlines ever increasing control and flexibility in the reconfiguration of the functionality of their onboard monitoring and recording systems.

0-7803-5086-3 /98/$10.00 ©1998 IEEE

With the advancements introduced by the FDAMS with ACMS, AlliedSignal now provides a customized ACMS for each individual airline customer. This customization is based on previously developed "off-the-shelf" databases, which significantly reduces the ACMS implementation cost and cycle-time.

List of acronyms

ACARS	Aircraft Communications Addressing and Reporting System
ACMS	Aircraft Condition Monitoring System
ADRT	Airborne Database Reconfiguration Tool
APU	Auxiliary Power Unit
ATC	Air Traffic Control
DAR	Digital ACMS Recorder
DFDAU	Digital Flight Data Acquisition Unit
DFDR	Digital Flight Data Recorder
FAA	Federal Aviation Administration
FDAMS	Flight Data Acquisition and Management System
FOQA	Flight Operations Quality Assurance
PCMCIA	Personnel Computer Memory Card International Association
NOAA	National Oceanic and Atmospheric Administration
SAR	Smart ACMS Recorder
QAR	Quick Access Recorder
VADAR	Visual Algorithm Definition for Avionics Reconfiguration

AN OVERVIEW OF CIVIL AIRSPACE CHANGES IMPACTING INTEROPERABILITY WITH MILITARY AIRCRAFT

Leo G. La Forge, Rockwell Collins, Cedar Rapids, Iowa

Abstract

Globally, airspace densities continue to increase at a rate which some fear could result in congestion levels approaching a gridlock condition. In response to this situation, Civil Aviation Authorities (CAA's) are transitioning to an air traffic management system that embraces the Future Air Navigation System (FANS) concepts originated by ICAO. This emerging system will automate most of the routine voice communications of the current system with data links that will provide communication between the aircraft and the air traffic controllers/computers. In the cockpit, aircrews will be required to assume an increased level of responsibility for safely maintaining separation assurance from ground and airborne hazards and obstacles. To ensure peacetime interoperability with the civilian sector, aspects of this new system (CNS-ATM) will ultimately be imposed on military aircraft either by direct mandate or through relegating non-compliant aircraft to sub-optimal routes.

Faced with the situation described above, the United States Air Force (USAF) has defined a number of initiatives to ensure that they will maintain access to global airspace. This paper will focus on two of these, Global Air Traffic Management (GATM) and Navigation Safety (NAV SAFETY). A regionally based overview of the anticipated changes in civil airspace requirements that will potentially impact military avionics equipage will be provided.

GLOBAL ACCESS, NAVIGATION, SAFETY

Recognizing the value in adapting to the emerging Free Flight environment, the U.S. Air Force has defined and initiated a broad set of initiatives called Global Access, Navigation, Safety or (GANS). Key elements of GANS include: Global Positioning System (GPS 2000), Joint Precision Approach and Landing System (JPALS), Air Traffic Control and Landing System (ATCALS), a Deployable ATCALS (DATCALS), Navigation Warfare (NAVWAR), Navigation Safety (Nav Safety), Global Air Traffic Management (GATM) and Cockpit Modernization. The purpose of GANS is to establish a framework that will enable the Air Force to maintain and/or enhance the current level of operational capability of their existing fleet of aircraft in the areas noted above.

The following provides a brief description of the two GANS elements (Nav Safety and GATM) that are most impacted by changes in the emerging CNS/ATM airspace environment.

Nav Safety

The USAF recognizes the advances made in safety enhancing avionics in the commercial market and has elected to incorporate some of these into their passenger carrying transports. These products include Terrain Awareness and Warning Systems (TAWS), Traffic Collision Avoidance Systems (TCAS II) and Flight Data Recorders.

0-7803-5086-3 /98/$10.00

GATM

The objective of GATM is to allow the USAF to continue to have unrestricted peacetime access to the evolving global air traffic environment. The operational concepts and requirements for GATM are derived directly from those developed for civilian airspace users. Following this approach, the military will remain fully interoperable within the civil airspace and will be positioned to reap the operational benefits associated with the ICAO Future Air Navigation System (FANS) concepts.

Cockpit Modernization

The introduction of GATM and Nav Safety capability into military aircraft will require a significant improvement in the Human Computer Interface (HCI) over that provided by the existing electro-mechanical and early generation electronic flight instruments. In addition to seeking Reliability and Maintainability (R&M) improvements over these older instruments, the shift to new technology products such as Liquid Crystal Displays (LCD's) will permit additional functionality to be accommodated in an efficient manner.

Evolving Civil Airspace Requirements

The management of civil airspace is evolving on regional basis with each geographical area developing rules and procedures intended to optimize operations in that region. Unlike an airline which can tailor their aircraft's equipage to satisfy the requirements of a particular region, the USAF has the requirement to go anywhere, at anytime, and therefore must consider the impact of compliance with requirements on a global basis. The section summarizes the emerging requirements of the primary operational airspace regions.

Oceanic

Operations in today's oceanic environment are based upon procedures that utilize HF voice communicated position reports to confirm compliance with previously filed flight plans. The infrequent reporting and inherent difficulties using HF voice communications associated with this manual control system requires that large separation standards be employed to maintain an acceptable level of safety.

North Atlantic Region

For reference, predominate traffic flows in the NAT region are reflected in Figure 1.

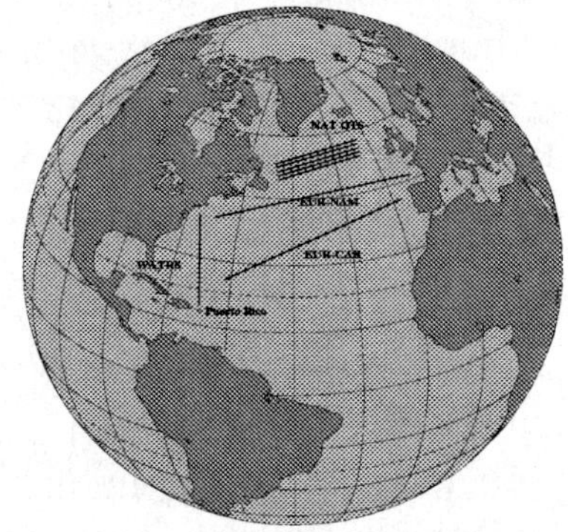

Figure 1 Major North Atlantic Traffic Flows [1]

In the ICAO North Atlantic (NAT) region, basic aircraft separations are 60 nm laterally, 2000 foot vertically, and a time of 15 minutes longitudinally. Longitudinal separations may be reduced to 10 minutes using constant mach techniques. Operations conducted in the NAT region between Flight Levels (FL) 285 and 420 now require that an aircraft be equipped to perform to Minimum Navigation Performance Specifications (MNPS). The MNPS requirements roughly equate to a Required Navigation Performance (RNP) of 12.6 nautical miles.

The recent implementation of Reduced Vertical Separation Minimum (RVSM) airspace on the North Atlantic Organized Track System (NATOTS) reduced the vertical separation to 1000 feet for those aircraft operating between FL330 and FL370. RVSM requirements will be extended to cover the full intended range of altitudes between FL290 and

0-7803-5086-3 /98/$10.00 ©1998 IEEE

FL410 by the year 2000. Non-equipped aircraft will be excluded from the airspace above FL280 in this region.

Pacific Regions

Airspace over the Pacific Ocean is divided into three primary regions. These are referred to as the North Pacific (NOPAC), Central Pacific (CENPAC) and South Pacific (SOPAC) regions. Predominate traffic flows in the Pacific regions are reflected in figure 2.

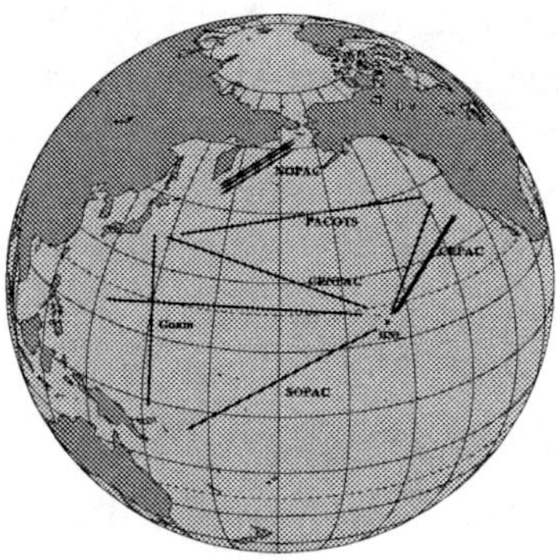

Figure 2 Major Pacific Ocean Traffic Flows [1]

The Pacific Organized Track System (PACOTS) routes aircraft traffic on flexible tracks (updated twice daily) between North America and Hawaii to Asia and Australia. The current separation standards on these routes are 100 miles laterally, 15 minutes in-trail longitudinally and 2000 feet vertically. Longitudinal time separations can be reduced to 10 minutes using constant Mach number techniques.

The NOPAC and Central-East Pacific (CEPAC) Composite Route Systems depicted in Figure 2 utilize fixed tracks which require separations of 50 nautical miles lateral and 1000 feet vertical on adjacent tracks. Required Navigation Performance of 10 nautical miles (RNP-10) was implemented on 23 April 1998 for all NOPAC Route and those ATS and Flexible Routes (PACOTS) which cross, join or diverge form the NOPAC within the Anchorage FIR, at and above FL280. RNP-10 implementation on CEPAC routes is expected in the near future.

Implementation of RNP-10 for CENPAC region (PACOTS only) is scheduled for 3 December 1998 and is expected to apply to all dense Pacific airspace by 2000.

RNP-10 was also implemented in the Tasman Sea area, in accordance with the ICAO Asia/Pacific Air Navigation Planning and Implementation Regional Group (APANPIRG) Regional CNS/ATM implementation plan. ATC in this region will attempt to accommodate aircraft not meeting the RNP-10 criteria, but RNP-10 compliant aircraft will be given priority over non-compliant aircraft.

In March 1996, the first dedicated "CNS" routes between Los Angeles and Sydney, Australia and between Los Angels and Auckland, New Zealand were opened in the SOPAC region. The functional requirements to operate on these routes include:

Controller Pilot Data Link Communication (CPDLC) via Inmarsat Aero-H SATCOM
Required Navigation Performance of 4 nautical miles (RNP-4)
Flight Plan Upload Capability to support Dynamic Aircraft Route Planning (DARP) operations
Required Time of Arrival (RTA) capability of 30 seconds
Automatic Dependent Surveillance (ADS)

The combined implementation of this group of functions is referred to as FANS-1 by Boeing or FANS-A by Airbus Industries. Per the FAA Strategic Plan for Oceanic Operations, aircraft operating in dense areas of the Pacific airspace will be required to be equipped with this level of "CNS" capability in the 2003 to 2005 timeframe. The objective is to permit reductions in lateral and longitudinal separations to 30 nm.

The economic incentive for users to implement FANS 1 functionality is largely obtained from the Dynamic Aircraft Routing Program (DARP). Under DARP, an aircraft can revise its route of flight while enroute in order to take advantage of a more efficient trajectory that is based upon current weather conditions. In the initial implementation of

DARP, Air Traffic Services will supply an updated routing for enroute aircraft. Subsequent implementation phases of DARP are planned to support user-defined routings and revisions.

Since 1994, the use of TCAS has been evaluated in the Pacific to facilitate in-trail climb (ITC) and descent (ITD) procedures. These procedures apply to an aircraft flying behind and 2000 feet above / below another aircraft operating on the same route. The trailing aircraft may request a path that crosses the altitude of the lead aircraft with as little as 15 nm longitudinal separation. The FAA plans to fully implement these procedures in the Oakland and Anchorage FIR's.

Once fully validated and operational in the North Atlantic, RVSM requirements are expected to be imposed on dense Pacific routes. The planned implementation date for RVSM in the Pacific falls in the 2000 timeframe.

Emerging requirements and functional capabilities in the oceanic environment are summarized in figure 3.

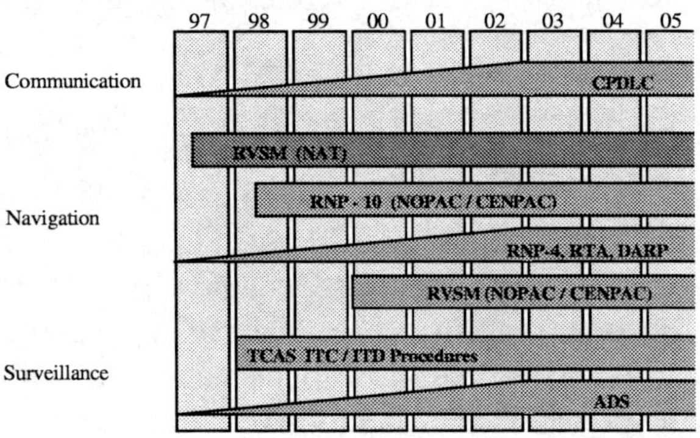

Figure 3 Combined Oceanic Requirements Timeline

European Civil Airspace requirements

Faced with an already highly populated air traffic environment, which is likely to double in density by the year 2015, the European Civil Aviation Conference (ECAC) nations are taking the lead from a regional perspective in imposing requirements for the carriage of new avionics equipage. Although not specifically applicable to military airspace users, it is generally believed that there will be significant negative operational impact to those non-equipped aircraft that normally operate as General Air Traffic (GAT).

Current European Initiatives and Schedules

BRNAV

Europe continues to move forward with plans to require area navigation (RNAV) capability for those civil aircraft operating on ATS routes. The first of two planned phases was implemented on 23 April 1998. It requires a basic level of area navigation equipment (BRNAV) which is essentially equivalent to a Required Navigation Performance level of 5 nautical miles (RNP-5). The certification requirements for BRNAV are defined in JAA document AMJ 20X-2 and more recently in FAA Advisory Circular AC-90-96.

The guidance provided in the JAA and FAA documents allows for multiple methods of achieving the required navigation performance associated with the BRNAV requirement The navigation systems / sensors discussed in the advisory circular include: INS (with time limitations), LORAN C (in areas of suitable coverage) and VOR / DME / ADF. Compliance with AMJ 20X-2 can be achieved by those systems that have been certified per FAA TSO-C129 provided that Receiver Autonomous Integrity Monitoring (RAIM) capability is included in the GPS based system.

RVSM

It is planned to extend the requirements for RVSM, now partially in effect over the North Atlantic, to the European Continent between FL290 and FL410. Trials are scheduled to begin in November 2000 with implementation of RVSM requirements effective in November 2001.

VHF 8.33 kHz Spacing

Eight European States including Austria, Belgium, France, Germany, Luxembourg, the Netherlands, Switzerland and

0-7803-5086-3 /98/$10.00 ©1998 IEEE

the United Kingdom have elected to implement 8.33 kHz channel spacing in late 1999. Operations conducted above FL245 (FL195 in France) will be impacted by this change. NATO has agreed that military aircraft which are frequent users of this airspace should be equipped with 8.33 kHz capable VHF radios.

FM Immunity

The ICAO ANNEX 10 requirements for on-board avionics to provide immunity to interference caused by FM broadcast stations are applicable to ILS Localizer, VOR and VHF communication receivers. With the exception of Germany, most countries are excluding the FM immunity requirement for VHF communications. Effective since 1-1-98 in some countries, restrictions are now being published that preclude the use of unprotected ILS receivers at locations having known interference problems. Other European nations, such as the UK, are allowing the FM immunity requirement for ILS to be deferred until 2001 to give operators time to equip with a multi-mode receiver (MMR).

Mode S

Mode S is secondary radar system that provides two basic functional capabilities: surveillance and datalink. Selective addressing allows direct communication with a specific aircraft. The data link function supports bi-directional air - ground uplink and downlink messages. Eurocontrol intends to mandate the carriage of at least a Level 2 Mode-S transponder with Downlink of Aircraft Parameters (DAP) capability for all aircraft operating IFR effective in January of 2003. This requirement will be extended to VFR operations effective January 2005.

ACAS II

The ECAC States have established implementation deadlines for carriage of an approved Airborne Collision Avoidance System. The ACAS II system is equivalent to the U.S. defined TCAS with Change 7.0 implemented. The Change 7.0 modifications are expected to provide the reduction in Traffic and Resolution advisories believed to be necessary to operate in dense European airspace. These requirements will be phased in beginning in January 2000 for heavier civil fixed wing turbine powered aircraft. These requirements will be extended to all turbine powered aircraft over 5700 Kg (or maximum seating >19) by January 2005. There is mounting political pressure within in ECAC States to extend the ACAS requirement to military transports as well.

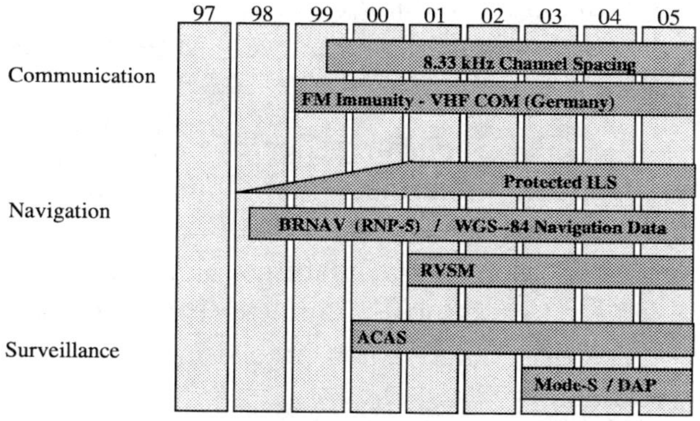

Figure 4 Current European CNS/ATM Initiatives

To implement the future concepts described in the EATMS OCD will require that users equip their aircraft with a number of additional enabling technologies. Those new functions and technologies that have a potential impact on an aircraft's CNS/ATM avionics architecture are described below:

PRNAV:

The second phase of precision area navigation (P-RNAV) requirements will be the equivalent to RNP-1 as defined in the RTCA MASPS. No firm implementation date for the implementation of PRNAV has been established. Due to the issues associated with equipping aircraft to meet these requirements the earliest date for PRNAV implementation would be in the year 2005.

GNSS / EGNOS:

Europe, like the U.S., has plans to deploy a Space Based Augmentation System (SBAS) in the 2000 timeframe to overcome the intentional degradation of accuracy associated with civilian use of the GPS. The European

GNSS-1 system will augment the existing GPS/GLONASS constellation with a system called EGNOS, which will provide an overlaid navigation augmentation system. EGNOS will provide coverage over the Atlantic and Indian Oceans, Europe, Africa, Western Asia and South America. The EGNOS system is expected seamlessly interface between its areas of coverage and those areas covered by the U.S. [FAA] deployed WAAS. The Europeans have also agreed to cooperate with Japan to maintain interoperability and compatibility with the planned Japanese Multifunction Transport Satellite (MTSAT).

EGNOS is being deployed to support aeronautical, marine and land mobile applications. The system will provide the accuracy, integrity, availability and continuity of service to support Enroute (RNP 1) and CAT I precision and non-precision approach operations.

Surveillance / Datalink:

Although self-separation of air traffic plays a strategic role in the future EATMS operational concept, EuroControl has not yet defined an implementation plan for an ECAC-wide surveillance system to support this concept. One of the fundamental issues being debated is the selection of a datalink standard on which to base an ADS-B system. A number of competing datalink technologies are now, or will soon be, assessed within the European Community. Mode-S and a self-organizing TDMA (S-TDMA, also referred to as VDL Mode 4) based data links are currently viewed as the favored approaches. With a planned "air-ground integration" implementation in the 2005 - 2012 timeframe, it is unlikely that any consensus will be reached or firm decisions made prior to the year 2000.

A number of datalink mediums are being considered to support digital controller – pilot communications (i.e. CPDLC) including VDL Mode 2, TDMA (VDL Mode 3) and commercial SATCOM. VDL Mode 2 is emerging as the most likely near-term standard for VHF datalink communication.

CONUS Airspace

The planned transition to modernize the CONUS airspace environment is documented in the Federal Aviation Administration's National Airspace System (NAS) Architecture, current version 3.0. From a CNS/ATM perspective, the operational capabilities defined in the NAS Architecture are traceable to the RTCA Task Force 3 report on Free Flight.

National Airspace System Architecture

The FAA's NAS Architecture calls for the transition to space-based navigation and landing aids. This transition will occur over a decade beginning in 2005 and would result in the decommissioning of approximately 50 percent of the ground based navigation stations in use today.

Central to the FAA's planned transition to space-based navigation is the deployment of the Wide Area Augmentation System (WAAS) which will provide Category 1 precision approach capability and limited primary means navigation for enroute applications. The WAAS is planned for deployment throughout the NAS in 2003. Also planned beginning in the 2003 time frame is a ground-based precision approach capability to Category II and III minimums. This capability requires the deployment of a Local Area Augmentation System (LAAS) at those airports that have the need for this level of capability.

The proposed concepts for surveillance will apply to all positively controlled airspace and employ both Primary and Secondary Surveillance Radar's, as well as, space-based dependent surveillance systems commonly referred to as ADS-B. The planned ADS-B capability will provide enhanced enroute coverage and will serve as the primary reference for surface surveillance. The version 3.0 NAS calls for an initial air-air capability in 2000 with air-ground integration in the 2006 timeframe. Mode-S is the favored data link for the ADS-B function in the CONUS region.

Similarly to other geographic airspace regions, plans are being developed to transition communications from voice-only to a more efficient blend of voice and data link communication methods. One of the major change drivers for the NAS is that there is an insufficient number of 25 kHz spaced VHF channels to support any significant expansion of services. The favored approach to resolve this issue in the CONUS region is to transition to digital voice and data communications using Time Division Multiple Access (TDMA) techniques employed by VDL Mode 3. The FAA refers to this new generation radio communication network as NEXCOM.

In the interim, the FAA plans to expand the system's CPDLC capability beyond the Oceanic regions to enroute operations in the 2002 timeframe. This capability will be provided via secondary service providers using VDL Mode 2. It is worth noting that within the NAS, it is planned to maintain UHF voice communication to support military air traffic control.

The planned introduction of these described capabilities in the near to mid term are reflected in Figure 5.

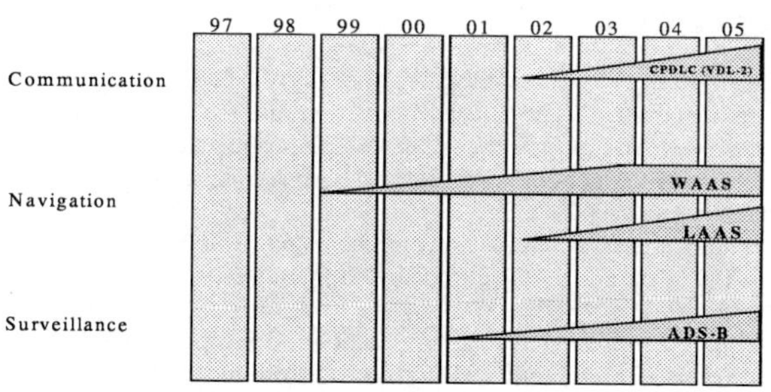

Figure 5: NAS 3.0 Introduction of Capability

Mandates for Carriage of Safety Related Equipment

Effective 30 December 1993 the FAA required per FAR 121.356 that civil aircraft operating in U.S. airspace and which are capable of carrying more than 30 passengers be equipped with a TCAS II system and the appropriate level Mode S transponder. TAWS

The FAA is currently developing plans for requiring a Terrain Awareness and Warning System (TAWS) on U.S.-registered aircraft. The FAA will propose a new section to Part 91 that will state that all U.S. registered, turbine-powered aircraft with more than six passenger-seats will be required to operate with an approved TSO-C151 TAWS. Parts 121 and 135 will also be changed to be compatible with Part 91.

Per the FAA, effective Dec 2000 all new aircraft will require TAWS and effective Dec 2004 all affected aircraft must have TAWS installed. This is estimated to affect 18000 U.S. registered aircraft. The FAA would also like this to be submitted to ICAO as part of Annex 6, Part 1. The FAA's goal is for TAWS become a requirement mandated by all CAA's.

Summary and Timing of Applicable Civil CNS/ATM Requirements

A consolidation of the emerging Communication, Navigation and Surveillance functional requirements and growth provisions are provided in the following tables.

Table 1. Combined Navigation Requirements

Navigation Requirement	Region	Effectivity	Date	Reference Documents
RVSM	NAT PAC Europe	FL290-FL410	1997 2000 2000	FAR part 91 Annex G FAA Order 8400.10
RNP-10	NOPAC Tasman Sea CENPAC	FL310-FL390 All Altitudes	1998 1998 1998	FAA Order 8400.12
BRNAV (RNP-5)	Europe	ATS Routes	1998	JAA AMJ-20X2 FAA AC 90-96 European AIC's
Protected ILS	Europe	As Identified	2001	ICAO Annex 10
RNP-4	PAC	CNS Routes	2003	RTCA DO-236 Strategic Plan For Oceanic Airspace
30 Second RTA	PAC	CNS Routes	2003	ARINC 702A RTCA Task Force III Report on Free Flight
Flight Plan Upload	PAC	CNS Routes	2003	ARINC 702A Strategic Plan For Oceanic Airspace
GPS Landing	CONUS		2003	RTCA DO-229

WAAS/LAAS				
PRNAV (RNP-1)	Europe		2005+	RTCA DO-236 EuroControl ATM Strategy for 2000+
Decommission NAVAID's	CONUS		2005 - 2010	NAS 3.0

Table 2 Combined Communication Requirements

Communication Requirement	Region	Effectivity	Date	Reference Documents
8.33 kHz Spacing	Europe	Above FL245 FL195 in France	1999	ICAO Annex 10
FM Immunity	Europe	As Identified	1999	ICAO Annex 10
CPDLC / AFN	PAC	CNS Routes	2003	FANS-1 ATS SR&O RTCA DO-219 Strategic Plan For Oceanic Airspace
ATN	Europe CONUS		2005	CNS/ATM SARPS EuroControl ATM Strategy for 2000+
VDL Mode 2	CONUS Europe		2002 (IOC)	ICAO Annex 10
VDL Mode 3	CONUS		2006	VDL-3 MOPS / SARPS
VDL Mode 4	Europe		2010	VDL-4 SARPS

Table 3 Combined Surveillance / Safety Requirements

	Region	Effectivity	Date	Reference Documents
TCAS II, v7.0 (ACAS II)	Europe Nav Safety	> 15,000 Kg or >30 passengers	2000	Specimen AIC - ACAS ARINC 735 RTCA DO-185A
Mode S w/ DAP	Europe	All IFR traffic	2003	Specimen AIC – Intro of SSR Mode S ARINC 718(A) ICAO Manual On Mode S Specific Services
ADS (-A)	PAC	CNS Routes	1998	ARINC 745 RTCA DO-212 FANS-1 ATS SR&O Strategic Plan For Oceanic Airspace
EGPWS	CONUS USAF Nav Safety Rqmt		2003 2001 - 2005	ARINC 723-1 RTCA DO-161A
Flight Recorder	USAF Nav Safety Rqmt		2001 - 2005	ARINC 747
ADS-B	CONUS Europe		2006	RTCA DO-242

The USAF operates a wide variety of aircraft types that fulfill an equally diverse range of performance and mission requirements. A number of studies have been, or are being, performed to assess the operational profile of a particular platform in order to determine which of the emerging civil airspace requirements are applicable to that aircraft type.

For example, the USAF has determined that prior to 2003, the following requirements must be met for a large transport aircraft in order to achieve the desired level of interoperability in civil airspace:

- RVSM
- RNP 10, BRNAV, 4 with growth to RNP 1
- 30 Second Required Time of Arrival Function
- ILS FM Immunity
- MLS with growth for GPS (WAAS/LLAS) approaches
- 8.33 kHz Channel Spacing
- Satcom Voice and Data
- VHF Datalink with growth to VDL Modes 2 and 3
- Growth for ATN compatibility
- ADS –A
- Growth to ADS-B
- CPDLC / AFN
- EGPWS
- Mode S
- TCAS II

Clearly, this level of CNS/ATM functionality will place the USAF transports on par with those aircraft operated by the most sophisticated airline operators. The challenge ahead will be to determine the specific needs and cost-effective avionics changes necessary for the balance of the fleet including bombers, tactical and special mission aircraft.

References

[1] "Air Mobility Command CNS/ATM Study: CNS/ATM Interoperability Report", The Mitre Corporation, December 1996

[2] "Strategic Plan For Oceanic Airspace Enhancements and Separation Reductions", Federal Aviation Administration, January 1997

[3] "The Impact of CNS/ATM Developments on Military Aviation", NATO - Committee For European Airspace Co-ordination, 21 March 1997

[4] "A Joint government / Industry Operational Concept for the Evolution of Free Flight", RTCA Select Committee, August 1997

[5] "Requirements Update", USAF Electronic Systems Center, September 1997

[6] "EATMS Operational Concept Document (OCD) ed. 1.0", EuroControl, March 1997

[7] "NAS Architecture Version 3.0 - Proposed Baseline", Federal Aviation Administration Briefing, October 1997

[8] FAA Web Site on Implementation of RNP-10, http://www.faa.gov/ats/ato/RNP.htm

[9] EuroControl Navigation Web Page, http://brnav.ecacnav.com/

[10] International Aircraft Owners and Pilots Association Web Page, AOPA News, http://www.aopa.ch/

[11] International Aircraft Owners and Pilots Association Web Page, AOPA News, http://www.aopa.ch/xnews.htm#fmimm
[12] Implementation of RNP-10 – Tasman Sea Area, AIP Supplement – H6/98, Australia, 23 April 1998
[13] Capstone Requirements Document, USAF 003-97 "Global Air Traffic Management" , United States Air Force, Draft 10/97
[14] Operational Requirements Document, AMC 007-97-I/II-A, "C-5A/B Global Air Traffic Management Program" United States Air Force, 6 February 98

A SIMPLIFIED AERONAUTICAL TELECOMMUNICATION NETWORK (ATN) AVIONICS ARCHITECTURE

T.L. Signore, Center for Advanced Aviation System Development, The MITRE Corp., McLean, Virginia

Introduction

The Aeronautical Telecommunication Network (ATN) is a set of standards [1] developed by the International Civil Aviation Organization which specifies the operation of a global network to support civil aviation communications. The specification requires the use of the Open Systems Interconnection (OSI) standards. This is in contrast to the Internet which uses the Transmission Control and Internet Protocol (TCP/IP) services developed by the Internet Engineering Task Force. The most important aspect of the ATN infrastructure is the accommodation of multiple civil aviation air/ground links. These links include a satellite system run by Inmarsat, very high frequency digital links (VDL2) operated by ARINC and SITA, Secondary Surveillance Links (Mode S) and VHF digital links (VDL3) operated by civil aviation authorities. Aircraft are expected to have multiple links installed for the following three reasons.

First, the extent of link availability will vary with location. Multiple line-of sight systems using VHF frequencies are operated by commercial companies and government agencies. There is no one VHF system available worldwide over all land areas. Satellite links supplement the VHF links in many areas where coverage is not available.

Second, the bandwidth of individual links is limited. Simultaneous use of multiple links will increase throughput and availability.

Third, governments that plan to operate data links as part of the ATN intend to restrict the use of these links to air traffic control and flight information services. An aircraft operator wishing to transfer both air traffic control and airline operational data would require two air/ground links under these circumstances.

The multiplicity of links available, the need to use them cost effectively, the need to segregate traffic by links, and the need to account for the mobility of the aircraft distinguish the ATN design from any existing network.

Clearly, the ATN specifications were designed for sophisticated and complex topologies. However, a class of aircraft operators that may have need of only one air/ground link dedicated to air traffic control purposes may exist. This paper explores some of the network and transport layer simplifications possible to the ATN under this circumstance.

The Present Requirements[1]

The present ATN specification requires the most general and the most capable avionics installation. One that can manage multiple air/ground links employing a router to arbitrate which link to use for each message. The arbitration is done by means of priority and quality of service (QoS) information included in each network packet presented to the ATN router for transfer. The ATN specification also provides rich support to end-systems in the aircraft, requiring OSI transport and complete upper layer functionality.

Figure 1 illustrates the most general avionics architecture that the ATN can employ—that is multiple air/ground data links and multiple end-systems connected to an ATN router. Any avionics end-system can send data

[1] The definitive statement of ATN requirements appears in reference [1], while reference [2] provides an excellent introduction to all aspects of ATN planned operation.

0-7803-5086-3 /98/$10.00 ©1998 IEEE

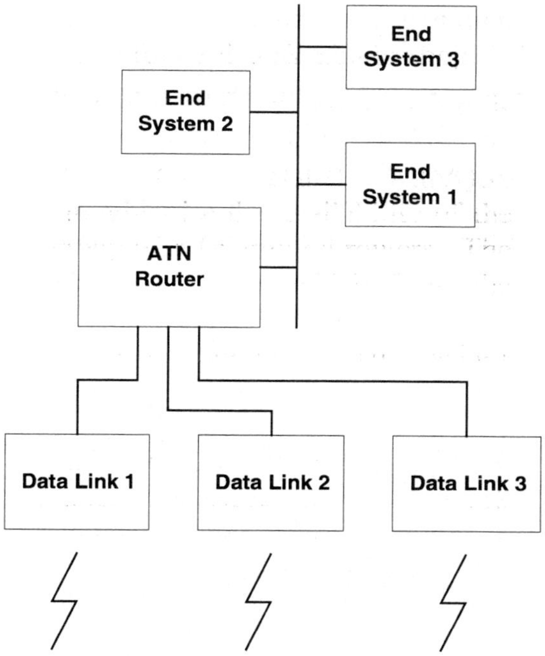

Figure 1. ATN Avionics Architecture

through any air/ground link and any air/ground link can be used by a ground end-system to reach any avionics end-system.

The airborne ATN router must have a peer router on the ground with which user data is exchanged. In general, it is expected that each air/ground link will have at least one ATN ground router associated with it through which the airborne router can communicate. The ATN support for this architecture has the following requirements.

IDRP to Manage Multiple Links

The choice of air/ground data link used for a message is dynamic and depends on the ground connectivity of the data link, the priority/QoS that the data link will support, and the priority/QoS specified by the end-system. For example, it may be that in one country the VDL2 link has ground connectivity to both that country's air traffic control network and the airline operation's center for the aircraft. In another country the same link, VDL2, only has connectivity to the airline operation's center. Thus, in order to send messages the operating characteristics of the data link—the ground connectivity, and priority/QoS which the data link can support—must be known dynamically for each installed link. It is function of the Inter-Domain Routing Protocol (IDRP) to exchange this information between the aircraft router and the ground router.

Router Address Exchange to Initiate System

The IDRP [3] protocol is a connection-oriented protocol which opens a "session" with each peer. A necessary requirement for this session is the OSI network address, or more precisely the Network Service Access Point (NSAP), of the remote peer router. In a mobile environment it is impossible to know *a priori* (particularly for ground ATN routers) the NSAP of the peer router. Therefore, the ATN has defined a dynamic mechanism to exchange ATN router NSAPs between local and remote routers. Whenever an air/ground link detects a new aircraft, it is required that the local ATN router be notified by the link. This notification is termed a "join event." The router, upon receipt, sends its NSAP address to the remote router using the link address provided in the notification.

Compression of Data Packets

As the bandwidth of the air/ground links is assumed to be limited, compression techniques are mandated by the ATN. The router performs the compression techniques before the data is transferred to the air/ground data link for transmittal. Multiple compression types are allowed; the one to use on a particular link must be negotiated between the two peer routers. Only after this occurs can data (including IDRP data) be sent between aircraft and ground. This negotiation (and one form of compression) requires the use of ISO8208 between the ATN router and the link avionics. ISO8208 defines the use of switched virtual circuits between Data Terminal Equipment (DTE) and Data Circuit Termination Equipment (DCE).

Dynamic Reassignment of Application Program Location

An application's location within the avionics is not standardized. That is, the address of the application (and consequently the address of the end-system in which the application resides) is not known *a priori*. This allows the aircraft operator the most flexibility in designing, installing, and operating the avionics components of the ATN. This flexibility allows application addresses to change while in flight or to differ from aircraft to aircraft. To support this, a program is needed which dynamically determines the address to use for a particular ground/aircraft application interaction. This program is denoted the Context Management (CM) application[2]. Essentially, the CM application allows databases on the ground to be updated dynamically with the required information for each ATN-equipped aircraft. Ground applications, wishing to contact an aircraft, would use this database to determine the applications installed in the aircraft and the addresses of the applications.

Synchronized (TP4) Transport Delivery

Each end-system is required to have a complete OSI transport layer level 4 (TP4) installation. All the ATN defined applications assumes a reliable transport layer which uses a sliding window buffer allocation scheme. This process requires explicit feedback as to which buffers can be cleared and/or resent.

Simplifications in a Single Link Environment

If an installation has only one data link, the following simplifications are possible.

IDRP Need Not Be Used in the Avionics

When there is only one air/ground data link, the router has no choice as to the means for data transfer. Information about ground connectivity is pointless, as the assumption is that the one link can reach all necessary ground processes. (That is, there is complete ground connectivity and delivery of a packet to any ground entry point is sufficient to have the packet delivered.) In a single link installation, the need for link QoS information is extremely diminished as the QoS exists mainly to choose the air/ground link. Information about the priority values which the link serves also becomes useless as an aircraft will not (or should not) have applications installed which use a priority which cannot be met by the one air/ground link. It could be argued that the QoS, connectivity, and priority might change dynamically with the link. However, it makes no sense to track this information if there is no choice as to what link to use. If it is very important for a domain to exclude certain types of aircraft originated traffic, it can do so on the ground. The lack of importance of dynamic information when a sole link is available means that IDRP is unnecessary. The IDRP code is approximately 19000 lines of C in the MERIT Gated implementation. Removal of IDRP is thus a significant simplification.

The designers of ATN recognized that the benefits of IDRP relative to its complexity might not make it suitable for all situations. For this reason an option is provided which allows for IDRP non-use. This option should be exercised in a single data link avionics environment.

Router Address Exchange Is Not Required

Without IDRP the main reason for NSAP address exchange is removed. It is still necessary to handle two problems that arise from the lack of NSAP exchange. The first occurs in the avionics and is very minor. One purpose of the remote NSAP is to use it as the next hop entry in the routing tables to distinguish alternate paths. Since there is only one path to the ground, there are no alternates. In this situation the avionics can have a fixed route entry specifying a default route to the ground. The entry would specify the link layer

[2] CM functionality is not limited to this purpose.

address and/or (radio) link interface to which the packet is to be forwarded as the next hop.

A solution to the second problem is crucial. The ground IDRP protocol uses the aircraft NSAP (and related path/QoS information) to advertise the availability of the aircraft to a group of ATN ground routers. The group is limited to routers which form the vector between the first ATN router and the home domain of the aircraft. This vector advertisement must still occur for it is this advertisement which allows mobile communication. Any ground process wishing to contact an aircraft constructs an OSI network packet containing the data to be sent and a destination NSAP address identifying the aircraft. This packet is eventually forwarded to one of the ATN routers forming the vector of updates. (The packet must intersect the vector because the home domain forms a superset of the aircraft's domain.) At this point the route to the aircraft is known and the packet can be forwarded as required. Thus, an aircraft NSAP is essential to provide mobile routing; yet it is proposed that the aircraft not provide this information. The solution is for the ground ATN router to derive the NSAP based on information it has. The router can combine the link address (already known from the join event) with a link service provider identification to form a unique aircraft NSAP or to form a key to look-up the NSAP. Thus, the ground can generate the airborne NSAP solely from link information. With this approach the aircraft need not perform any (network) initiation when using its sole link.

Further Simplifications

The following simplifications apply to any ATN installation regardless of the number of air/ground links.

Compression, While Useful, Can be done by the Link Layer

Router compression can be eliminated, assuming that compression (if desired) will be performed by the link. Compression by the link is the paradigm that is used today in most implementations. For example, personal computer (PC) modems support a form of Lemple-Ziv compression on the data sent. This approach also allows the link to choose a form of compression that optimizes its operation. Elimination of router compression also means that the ISO8208 protocol can be eliminated from the avionics. This is a significant advantage as the ISO8208 protocol is complex (3600 lines of C code in the BSDI 1.1 Unix Operating System). There is a concomitant reduction of link overhead as the ISO8208 Call Request, Call Accept, Data, Receive Ready, and Reset packets are not sent.

Dynamic Reassignment of Application Program Location Can Be Restricted

The generality afforded the aircraft operator in changing end-system addresses has resulted in a need to provide a dynamic application name to NSAP address cross-reference capability. There are installations that do not require a dynamic mapping approach. For example, when a network device fails, bringing on-line new equipment that is configured with the address of the failed device is transparent to network users. Another type of installation, which does not require any dynamic mapping, occurs when there is only one end-system, with no redundant equipment. In both cases, the application-to-NSAP mapping and the applications available for the installation are fixed for the life of the installation. The cross-reference information need only be recorded once in ground servers, *viz.* DNS or X500, to be available to network users. The information is available any time and in any location without depending on the aircraft to provide dynamically the information via the CM application. In such situations parts of the CM application need not be installed.

TP4 is not needed for Certain Classes of Installation

Connection-oriented transport code (TP4) represents a considerable increase in size and complexity over connectionless transport code, (CLTP). This is because CLTP does not provide a reliable delivery service or any form of flow control. The differences in complexity of the two transport designs are evident in their code sizes. The OSI code provided with the BDSI 1.1 Unix Operating System requires approximately 4500 lines of code C code for the TP4 protocol and 300 lines of code for the CLTP protocol. Elimination of TP4 would simplify significantly the ATN installation. This approach should be considered for two reasons. The first is that the proposed air traffic control services are all of a request/reply operation with a requirement that the controller or pilot intervene if the response is not received within a specified time. In essence, the pilot and controller are providing the connection transport function. Second, many of the proposed data links (VDL3 and Mode S) are of sufficient reliability that a TP4 operation may provide only marginal benefits. A CLTP only installation should be allowed for certain combinations of application and data links.

PC/Internet Connection Model Now Applicable

With the elimination of compression as a router requirement, combined with the elimination of IDRP and router NSAP exchange, the need for an ATN router in the avionics is abrogated. It is thus possible for an aircraft with one end-system supporting all air/ground applications to have a default route to the ground for all network traffic. In this situation the avionics design to support data link closely resembles that of a home PC connecting to an internet service provider. Except that the PC would have a permanent ATN (internet) address assigned to it and the service provider would use this address to support ATN mobile networking.

Conclusion

It would seem to be advantageous for countries that desire to provide air traffic services via mobile networking to profile a set of ATN options that allow aircraft to operate without an ATN router. Such a profile should also allow for simplification of the OSI transport layer by substituting a connectionless transport for a connection-oriented transport. These changes would be applicable only to aircraft that have one air/ground data link installed and consequently do not need the power of an ATN router in the aircraft. A large majority of such aircraft would be start-up commercial carriers and general aviation aircraft.

References

1. International Civil Aviation Organization, Proposed New Text, Annex 10, Volume III, Part 1, Chapter 3, Aeronautical Telecommunication Network, Technical Requirements, ATNP/2-WP/4, Appendix A, 14 November 1996.
2. ATN Workshop, 16-19 December 1997, Luxembourg, Proceedings, Volumes 1 and 2, McCallum Whyman Associates Ltd., Alresford, England.
3. ISO/IEC 10747:1994, Protocol for Exchange of Inter-Domain Routing Information among Intermediate Systems to Support Forwarding of ISO/IEC 8473 PDUs.

0-7803-5086-3 /98/$10.00 ©1998 IEEE

PROPOSED AVIONICS ARCHITECTURE FOR AIR FORCE AIR MOBILITY COMMAND AIRCRAFT TO MEET CNS/ATM AND GATM REQUIREMENTS

Donald A. Happel, Rockwell Collins, Cedar Rapids, Iowa

Abstract

Air Force Air Mobility Command (AMC) aircraft must meet military requirements for transport missions anywhere in the world while also addressing emerging Civil Aviation Authority requirements for Communication, Navigation and Surveillance (CNS) for operation in the Global Air Traffic Management (GATM) environment. An architecture designed to meet these requirements must accommodate existing avionics to the maximum extent possible to minimize costs and must also incorporate new avionics that take advantage of Commercial Off The Shelf (COTS)/Non Developmental Items (NDI) and open system architectures to be able to incorporate the growth necessitated by these emerging requirements.

This paper describes a generic architecture designed to meet these requirements for a tanker/transport platform. The architecture is designed so that future upgrades for these evolving requirements could be met with minimal impact to the aircraft. The implementation of this architecture is primarily a systems engineering effort in managing the human machine interface necessary to provide a seamless integration of GATM and navigation safety equipment including an Enhanced Ground Proximity Warning System (EGPWS), a Communication Management Function (CMF), a Multi-Mode Receiver (MMR), a Traffic Collision and Avoidance System (TCAS), new communications radios, a new Displays Subsystem and existing Flight Management Systems that are modified, as necessary, to support Required Navigation Performance (RNP) objectives.

Introduction

Technology advances are making possible new standards for Air Traffic Management (ATM). Increasing airspace densities are adding to pressure to reduce separation minimas and to better manage many Air Traffic Control (ATC) functions through the use of digital data links in lieu of voice communication.

Military air forces recognize the benefits of this advancing technology and require access to civil-controlled airspace to accomplish their worldwide missions. This paper describes an architecture for military aircraft that meets these evolving technologies and that provides growth to allow smooth transitions to a free flight environment for these aircraft.

Requirements and Objectives [1]

The growth in air traffic is resulting in major changes to the air traffic system. New technologies are being mandated in the area of communication, navigation and surveillance that aircraft must incorporate in order to continue to operate in the environment being defined by civil aviation authorities. Congested traffic means aircraft must better know where they are, know where other traffic is relative to them and be able to communicate their position and intent with both ATC and surrounding aircraft.

FY1999-2003 Defense Planning Guidance states that "components will support, as necessary, Federal Aviation Administration

0-7803-5086-3 /98/$10.00 ©1998 IEEE

efforts to determine equipment requirements needed for the Department to comply with the new International Civil Aviation Organization airspace architectures. Components also could structure their programs to accommodate investments necessary to comply with the pending decisions on Departmental compliance with the new Global Air Traffic Management (GATM) regime." This document identifies the capabilities identified in the GATM Capstone Requirements Document (CRD) (draft, 17 Oct 97) for integration into AMC aircraft.

To meet equipage requirements to ensure global airspace access, AMC aircraft will need to upgrade existing CNS equipment with state-of-the-art highly reliable, and fully integrated systems. COTS/NDI equipment that is compliant with civil standards (such as FAA technical standard orders) will be used as much as practical with minor missionization to meet military-unique requirements.

System Performance Requirements

The GATM architecture for AMC aircraft must meet (or comply with the intent of) FAA or other appropriate civil technical standards. Cockpit design must also meet the requirements of the Air Force cockpit endorsement process outlined in AFI 11-206. GATM equipment must be cockpit selectable, including the ability to turn it off. NVIS-B night vision goggles will be utilized in performing missions at night and low light conditions. GATM interface with current communication, navigation, and surveillance equipment not planned for replacement is imperative for program cost effectiveness and GATM equipment must be compatible with current aircraft electrical systems.

Communication Requirements

Increased navigation accuracies and reduced aircraft distances require position reporting to be accomplished using data link capabilities for prompt and accurate reporting to controlling agencies. Reduced separation standards, a primary objective of the CNS/ATM system, will also require a more responsive capability for backup voice communications with ATC in the event of emergencies, conflict resolution, contingency operations, weather avoidance, etc. At a minimum, data link capabilities must provide communication for ATC (Controller-Pilot Data Link Communication (CPDLC) application), flight service (flight plans, weather, etc.), and surveillance (Cockpit Display of Traffic Information (CDTI), Automatic Dependent Surveillance (ADS) A application) types of services. While providing data link capability, AMC aircraft systems must also provide/maintain UHF, VHF, HF, and SATCOM voice capability.

Communication system components include dual independent data link transceivers. A dual Communication Management Unit (CMU), or equivalent functionality, is required. Data link requirements will be mandated by region with satellite communications (SATCOM) and/or high frequency data link (HFDL) required for Beyond Line Of Sight (BLOS) and very high frequency (VHF) (planned US standard) and/or Mode S (possible future European standard) required for line-of-sight pilot-to-controller communications. The current third-party HF voice system used for transoceanic communication does not meet the future requirement for direct pilot-to-controller voice communication in the reduced separation environment. System components must have a growth path for easy upgradeability to meet future requirements. An integrated satellite communication system that meets ICAO/FAA requirements for oceanic ATC data link and direct pilot-to-controller voice communication is required.

VHF-AM frequency congestion in Europe has driven a requirement for 8.33 kHz channel spacing. VHF radios must meet 8.33 kHz channel spacing requirements and be upgradeable to meet future requirements for AVPAC (VDL Mode 2) and TDMA or other

0-7803-5086-3 /98/$10.00 ©1998 IEEE

technology that is adopted to resolve VHF-AM frequency congestion and add VHF digital link capabilities. Capability to communicate on the current 25kHz-spaced VHF frequencies must be retained (e.g., 125.025MHz, 125.050mhz, 125.075mhz, etc.).

Navigation Requirements

In the new ATM environment, navigation standards will be defined by specifying RNP. RNP requires an aircraft to be within a specific number of NM of its cleared position (cross track and along track) 95 percent of the time during the duration of the flight. In order to operate effectively in this airspace, and in all airspace as ground based navigation aids are decommissioned, aircraft navigation systems will require a satellite navigation capability such as GPS. To meet military-unique requirements, the integrated GPS system equipment must be Precise Positioning Service (PPS, Y-code) equipment which complies (as a minimum) with Air Force policy (26 Mar 97 AF/XO message "Implementation of AF Navigation and Safety Master Plan and Policy Clarification for GPWS, ADF, and GPS Navigation Systems"). Systems must also be readily upgradeable to incorporate NAVWAR solution required by FY06.

For GPS to be used as a supplemental means of navigation, as a primary means of navigation in oceanic/remote airspace, as the primary/sole means of navigation, to include nonprecision approach, it must meet applicable TSO requirements. For GPS/GPS augmentation systems to be used in the terminal area (RNP 1 and less) for nonprecision/precision approach capability, they must comply with applicable RTCA/FAA requirements which are now under development. Precision approach solutions must be compatible with the Joint Precision and Landing System (JPALS) solution as it relates to aircraft missions.

An FMS integrated multisensor and/or dual GPS navigation system will be required to operate in an RNP-4 or less environment. To ensure a low probability of losing navigation capability, dual FMS and dual navigation sensors will be required. GPS systems/sensors must have a growth path for easy upgradeability to meet future requirements. The current radio navigation aid systems (VHF omni-directional range, instrument landing system (ILS), distance measuring equipment, tactical air navigation, etc.) are required for at least the near-term (far-term plan is to decommission ground based navigation aids). An enhanced GPS system is required.

Potential FM radio interference with onboard ILS receivers in Europe has driven the requirement for an ILS receiver with greater immunity to FM interference (protected ILS). ILS receivers for aircraft that operate in Europe must meet these immunity requirements as outlined in ICAO Annex 10.

Both the UK and the Netherlands plan to install MLS at major international airports that USAF aircraft transit. To maintain compatibility with ground-based navigation aids in these two and possible other European countries, dual MLS installation on aircraft is required.

Surveillance/Safety Requirements

The new traffic environment will rely more heavily on aircraft systems to ensure safe separation of aircraft than does the current air traffic system. Two main functions for the aircraft surveillance element are required: an Airborne Collision Avoidance System (ACAS) that is interoperable with civil systems and an ADS system for automatic position reporting. The ACAS system currently used in the civil aviation community is the traffic alert and collision avoidance system (TCAS). There are two ADS functions. The first is ADS-A (addressed) used primarily for oceanic reporting when an aircraft is communicating with a specific controller. The second is ADS-B (broadcast) in which aircraft information is periodically broadcast to all line-of-sight users (other aircraft and ATC). ADS-B information will be displayed via Cockpit Display of Traffic

0-7803-5086-3 /98/$10.00 ©1998 IEEE

Information (CDTI) systems to greatly improve pilot situational awareness. CDTI and ADS-B are key to the implementation of free flight.

The main components of a surveillance system include, but are not limited to: a TCAS system, Mode-S transponder, displays, and a system to perform the ADS-A function. TCAS II (version 7 software or later), and Mode-S is required. A growth capability for ADS-B must be provided. Components must also have a growth path for future upgrade.

Avionics to enhance aircraft safety include an EGPWS. The EGPWS will provide aural and visual warnings for unsafe conditions using a digital terrain database and navigation data. The audio and visual warnings must be consistent with TSO-92c warnings and provide protection for excessive descent rates, excessive terrain closure rates, unsafe terrain clearance, unsafe landing flap and landing gear positions, altitude loss after takeoff and deviation below glideslope. A warning must be provided when landing short of the runway based upon proximity to the runway and radar altitude. Alerts will be provided for excessive roll attitudes and cautions and warnings provided for reactive wind shear during takeoff and approach. Audio output from the EGPWS must be selectively inhibited.

Avionics Crew Control

One of the most important features of an avionics architecture is the design of the man-machine interface. Crew interaction with aircraft avionics is increasing in importance as the avionics takes on a significantly larger role in the cockpit. The architecture described in this paper utilizes an innovative approach to the man-machine interface using eyes up displays for much of the crew/avionics interaction. This architecture expands on the capabilities of AMC aircraft by integrating new display and control technology into those cockpits.

Displays Interface

Crew interface with aircraft avionics is normally via control panels, knobs, switches and, more recently, using a Control and Display Unit (CDU). The CDU has been a means of centralizing a variety of crew control requirements and provides coordinated control of diverse functionality. However, with the transition from voice to data for a significant portion of a pilot's routine ATC interface, a heads down CDU can be a stumbling block to efficient communication.

The Rockwell Collins architecture reduces head down time by transferring a significant portion of the flight crew avionics interface to eyes up displays. Existing electro-mechanical displays and older electronic flight displays are replaced by Liquid Crystal Display (LCD) displays that provide functionality in an efficient manner.

In a typical configuration, the pilot and copilot each have a Multi-Functional Display (MFD) and a Primary Flight Display (PFD). For those configurations that require replacement of engine instruments, an Engine Indication and Caution Advisory System (EICAS) display is located to be visible to both the pilot and copilot. Systems that require it will have an EICAS and/or an MFD at the navigator station as well. A portion of the EICAS display is dedicated to engine instrument data and a portion to Caution Warning and Advisory (CWA) data.

In this architecture, all controls are naturally and easily accessible on the pedestal, side panel and glare shield. All mode and status annunciations that affect current flight safety, integrity or navigation and guidance parameters are in the primary field of view. Only non-immediate annunciations or alerts will be shown on the FMS CDU.

The architecture maximizes "eyes up" indications. Alerts and messages are prioritized on the MFD where they can be read, acknowledged and acted upon without the pilot's having to look down at a CDU.

0-7803-5086-3 /98/$10.00 ©1998 IEEE

MFD Displays

The MFD will typically display both heading up maps for both terminal and enroute flight operations and north up maps for flight planning and look ahead features. FMS generated flight plans are displayed on the MFD and can be edited in an eyes up manner.

The MFD also provides the data link interface for the CPDLC interface. Uplinked flight plan data can be displayed either textually or as a map prior to insertion into the flight plan. On some configurations, interface to the SATCOM will be provided on the MFD.

Radar, TCAS and EGPWS data can be displayed on either MFD as stand alone or as an overlay. Declutter capability is provided to ensure the crew can see priority data.

PFD Displays

The PFD presents complete ADI, HSI and air data information in the "Standard T" arrangement that is equal in size to existing electro-mechanical instruments. In addition to standard ADI display data, the ADI provides a display of thrust, roll and pitch axis mode annunciators for the currently active flight control or fuel performance mode. The crew can replace the conventional HSI display or a TCAS overlay with an arc HSI, a center map or an arc map.

Additional data displayed on the PFD includes: digital readouts for barometric and radar altitude (a linear moving tape provided for baro altitude); selected barometric pressure; altitude pre-select (flashes when the altitude is achieved); vertical speed (moving pointer on a fixed scale); digital vertical speed with a pointer indicating direction; vertical speed trend line; vertical speed advisory (magenta pointer indicating vertical speed required to reach next altitude constraint); TCAS resolution advisories (red and green speed regions); Indicated Air Speed (IAS both digitally and as a moving tape); visual cues for stall speed and maximum operating speed situations; mach; IAS and mach target speed; speed trend vector; and Angle Of Attack (AOA presented on a stationary -15° to +30° scale with a moving pointer).

TCAS Displays

The TCAS display will be visible from anywhere in the cockpit, either as a dedicated large-format TCAS display on the MFD or an overlay on the PFD navigational display. TCAS can be repeated on the EICAS display below the engine instruments and alongside the CWA display.

Architecture Features

The major subsystems of navigation, communication, displays and surveillance must coordinate functionality to successfully meet CNS/ATM requirements. Figure 1 illustrates how functions are allocated between these subsystems.

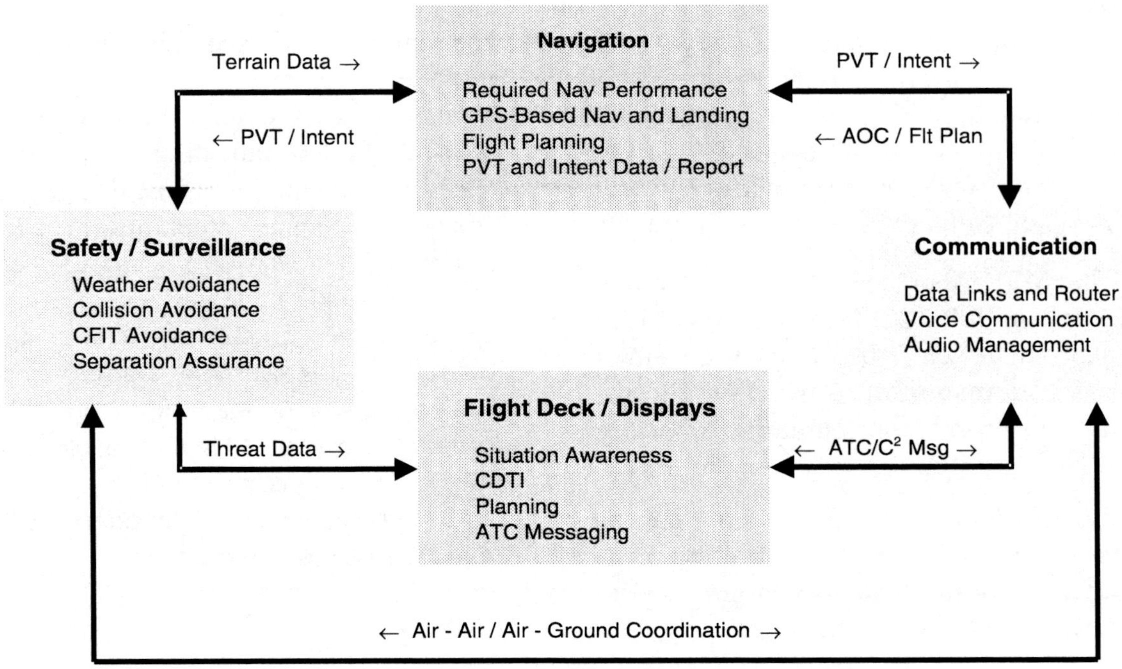

Figure 1: AMC Aircraft Solution Must Meet GATM Requirements

The system architecture for AMC aircraft must also utilize existing aircraft avionics, integrate COTS/NDI solutions as practical and utilize open system architectures to ensure growth for requirements still being formulated. MIL-STD-1553B databus architecture is standard on many AMC aircraft. ARINC standards define the protocol for many capabilities that support open system architecture concepts. This results in a mix of avionics equipment that must be integrated into these aircraft as the "best" architecture for that aircraft.

If the aircraft is equipped with a Malfunction Detection, Analysis and Recording System (MADARS), the MADARS is also interfaced via MIL-STD-1553B. The MADARS accepts status data from aircraft avionics units for maintenance purposes and must be interfaced with new capabilities. The I/O for most COTS equipment is over an ARINC bus. Figure 2 illustrates the allocation of major functionality to the bus architecture.

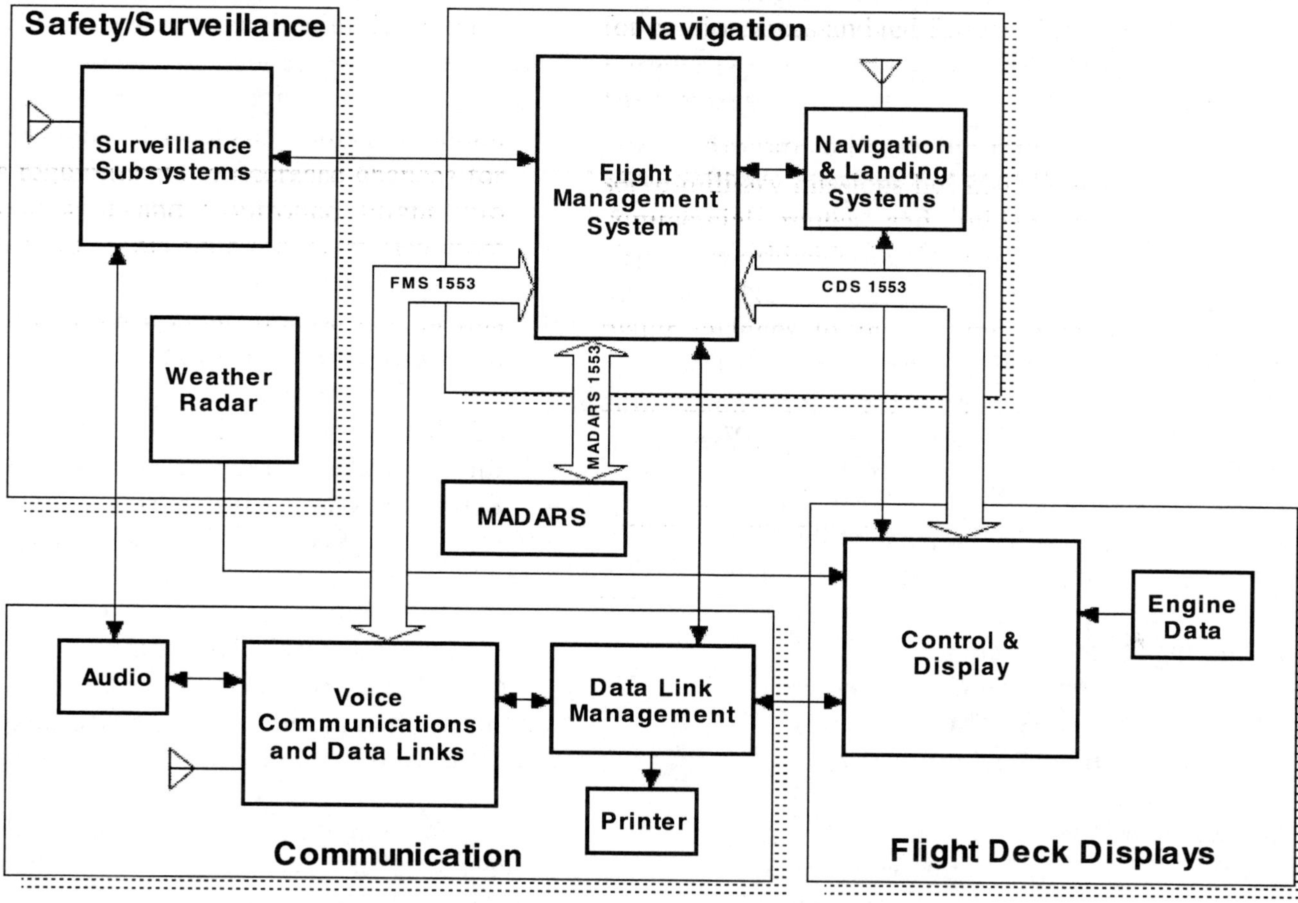

Figure 2: AMC Aircraft Bus Architecture

Flight Deck Display Subsystem

The pilot and copilot displays are driven independently. If a PFD or EICAS display senses loss of data from its primary source, it automatically switches to the secondary source displaying the same data. If a PFD display fails, the PFD display is automatically switched to that crew member's MFD. The crew can manually switch back to the MFD display if desired. If an EICAS fails, either crew member can display EICAS data on their MFD.

Communication Subsystem

Control of the communication avionics utilizes open architecture designs to ensure growth for future requirements while maintaining compatibility with existing avionics capabilities. The proposed communication architecture integrates existing communication radios with newer communication equipment to provide Beyond Line Of Sight (BLOS) voice and data link capabilities.

Dual ARC-210 radios controlled on the 1553 databus by the bus controller CDU provide 8.33 kHz channel spacing for VHF communications while also providing embedded COMSEC, HAVEQUICK I/II, SINCGARS, UHF AM/FM and UHF DAMA and Non-DAMA SATCOM. The ARC-210 radios will meet ICAO Annex 10 FM immunity requirements.

HF communications is provided by the ARC-190. Control of the HF is via the CDU on the 1553 databus. Control is provided for the ALE, AJ, SELCAL and HFDL, communications modes. In addition provisions

0-7803-5086-3 /98/$10.00 ©1998 IEEE

for messaging and input required for user programmable functions required for HF ALE, AJ, SELCAL and HFDL functions are provided.

SATCOM capability is provided with the SAT-2000 AERO-I SATCOM. SATCOM operations are controlled on the MFD.

The CMF will host existing Future Air Navigation Systems (FANS-1/A) requirements including ATS Facilities Notification (ATN), Automatic Dependent Surveillance (ADS) and CPDLC. The communication management function interfaces directly to the subnetworks via ARINC 429 to provide HF, VHF, SATCOM and Mode S data link routing. An ARINC 607 Aircraft Personality Module is supported. All data link control is provided on an MFD thereby eliminating nearly all head down time for this capability.

The communication architecture provides near term GATM capabilities. The hardware supported by this architecture provides the infrastructure to support future Aeronautical Telecomunications Network (ATN) requirements. As these requirements are established, they will be implemented in this architecture via new software loads. Operational Flight Program (OFP) loads will be via the standard ARINC 615 protocol or the MIL-STD-2217 protocol. The OFPs for each of these LRUs is loaded using the data loader under the control of a CDU thereby centralizing the loading function.

Navigation Subsystem

Present and future RNP requirements will necessitate the use of satellite based navigation sensors. European air facilities require MLS and support a need for ILS that is protected against FM interference (protected ILS). Future GPS based landing systems will require Local Area Augmentation System (LAAS) and Wide Area Augmentation System (WAAS) capabilities. The navigation subsystem architecture supports these requirements and provides growth for future GPS Landing System requirements.

The proposed navigation subsystem architecture provides sensors that support both near term GATM requirements and growth capability for long range requirements. Basic navigation sensors include triple Inertial Navigation Sensors (INSs), dual Central Air Data Computers (CADCs) and at least dual GPS receivers. The GPS receiver, which meets P/Y code military requirements, must also provide the Receiver Autonomous Integrity Monitor (RAIM) function. In all configurations, the pilot and copilot can independently select from INS1/GPS, INS2/GPS, INS1, INS2, or GPS only as a navigation source for their guidance computations. A third INS serves as a hot spare. Should an INS fail, the third INS can be switched to replace the failed unit. The selected solution drives the associated pilot/copilot displays and can be selected for coupling to the autopilot. Data based airspace requirements to fly RNP airspace are contained on data loader cartridges.

A Multi-Mode Receiver (MMR) provides VOR/ILS, MLS and Marker Beacon capability and has growth capacity for GPS. Dual TACAN receivers round out the navigation sensor complement. TACAN is used as an independent navigation source for the displays and autopilot and as an update source for the FMS integrated navigation solution.

Control for the navigation sensors is provided on CDUs and includes INS alignment, GPS initialization, and VOR/ILS, MMR and TACAN tuning. Tuning by identifier is provided using the navigation database resident on data loader cartridges.

When WAAS becomes a reality, a commercial GPS receiver could be installed in the MMR. As future GPS requirements progress to GRAM/SAASM, LAAS, NAVWAR and JPALS, this capability could also be housed within the MMR by adding a military GPS. This architecture design isolates the flight critical functions required for landing into one LRU. A data interface from the MMR to the FMS will allow use of MMR GPS data by the flight management navigation function.

Surveillance/Safety Subsystem

The surveillance and safety function addresses requirements to decrease chances for mid-air collisions and Controlled Flight Into Terrain (CFIT). This function must also meet evolving requirements in the ATM system. Separation assurance is an integral part of this system. The system defined to meet separation assurance is the ADS system. Initially, separation assurance will be provided by transmitting addressed data (ADS-A) with growth capability to support broadcasting ADS data (ADS-B).

A Traffic Collision and Avoidance System (TCAS) provides the crew with an awareness of surrounding traffic in a pre-defined airspace and provides aural and visual warnings and vertical avoidance maneuvers to maintain safe separation. A commercial standard TCAS II system incorporating Change 7.0 with a Mode S transponder and an IFF is necessary.

Control of the TCAS, IFF and Mode S would also be provided using a CDU. The FMS then provides position, track, velocity and intent data to the TCAS/Mode S and the TCAS provides resolution and traffic advisory data to the displays.

The EGPWS provides standard ground proximity warnings and includes altitude awareness callouts, an excessive bank angle alert, reactive windshear detecting and alerting, a terrain clearance floor function and a terrain awareness alerting warning. The EGPWS also provides a display output which can be viewed on either the MFD or the PFD.

EGPWS display output provides the pilots with visual detection of flight paths that could lead into conflict with precipitous terrain. The display output relates the present and projected aircraft position to the built-in terrain data base to provide timely alerts synchronized with aural alert messages. EGPWS data is displayed as green to signify no danger, yellow for caution areas and red for warning areas.

Summary

Aircraft utilized by military services serve military missions but many must also fly commercial routes and utilize commercial airports worldwide in the execution of those missions. New technologies are bringing about major changes to the air traffic system that military aircraft must adapt to or face elimination from major portions of civil-controlled airspace.

This paper presents an architecture suitable for military transport aircraft to provide these aircraft with an avionics suite that meets current civil requirements and has growth capability to meet future requirements. A feature of this architecture is the use of eyes up displays for the control of many new ATC requirements providing control of these functions without changing the primary field of view of flight deck crew members.

GATM related avionics that could be integrated into this architecture include dual CMFs, dual ARC-210 radios, SATCOM, ARC-190 HF, EGPWS, Mode S/TCAS II/IFF, and new cockpit displays. This configuration can meet current GATM requirements and provide the needed growth path to ensure these aircraft can remain flying in civil airspace as GATM requirements evolve.

REFERENCES

[1] Operational Requirements Document, AMC 007-97-I/II-A, "C-5A/B Global Air Traffic Management Program" United States Air Force, 6 February 98

[2] Capstone Requirements Document, USAF 003-97 "Global Air Traffic Management", United States Air Force, Draft 10/97

[3] AFI 11-206 Air Force Instruction, General Flight Rules

A PSEUDO RAMP MANAGER WORKSTATION FOR THE LABORATORY DEVELOPMENT OF AIRLINE-ATC COLLABORATIVE ARRIVAL PLANNING TOOLS

Susan Dorsky, Seagull Technology, Los Gatos, CA

George Hunter, Seagull Technology, Los Gatos, CA

Richard Zelenka, NASA-Ames Research Center, Moffett Field, CA

Cheryl Quinn, NASA-Ames Research Center, Moffett Field, CA

Santosh Mathan, Sterling Software, Redwood City, CA

Abstract

Significant changes to air traffic management and the relationship between the air traffic control (ATC) service provider and the system user are expected as the government/industry "free-flight" initiative begins to take hold. Under the free-flight paradigm, airspace users would select their own path and speed in real-time, with air traffic control imposing restrictions only when necessary. Shared decision making and collaboration between system users and service providers would be a central tenet, allowing increased flexibility of air traffic management operations and allowing system users to operate in their preferred manner. Automation decision support tools will play a critical role in assisting both airspace users and service providers in operating in the free-flight environment. Essential to the development of such new automation aids are laboratory simulations of such tools in the free-flight environment. In this work, an airline airport ramp manager workstation is described. The workstation was developed to support the laboratory simulation and development of collaborative airline-ATC aircraft arrival planning tools. Airline airport ramp towers at major "hub" airports orchestrate the arrival and departure of dozens of aircraft in rapid succession. Future decision support tools are being proposed that would allow the airline user to influence aircraft arrival characteristics for maximum operational benefit. Such characteristics include arrival sequencing and scheduling. The airline airport ramp manager workstation developed mocks essential aspects of airline systems commonly used at airport ramp towers to enable realistic simulations of new airline ramp tower decision support tools. The workstation presents a "Gantt chart" type display of arriving and departing aircraft sorted by airport gate, as typically used by "hub and spoke" air carriers in their airport ramp towers. The workstation employs a modern object-oriented design and is highly adaptable toward use in a variety of decision support tool simulations.

Introduction

Shared decision making and collaboration between air carrier system users and air traffic control (ATC) service providers is a critical foundation to enable the government/industry "free-flight" initiative [1]. Under free-flight, airspace users would select their own path and speed in real-time, with air traffic control imposing restrictions only when necessary, thus allowing increased flexibility of air traffic management operations and allow system users to operate in their preferred manner. Automation decision support tools will play a critical role in assisting both airspace users and service providers in operating in the free-flight environment. Essential to the development of such new automation aids are laboratory simu-

0-7803-5086-3 /98/$10.00 ©1998 IEEE

lations of such tools in the free-flight environment.

In support of the development of new ATC-airline decision support tools, the Pseudo Ramp Manager (PRM) workstation was developed. The PRM is used during laboratory simulations of proposed new collaborative airline-ATC decision support tools by the airline participant, and provides the critical information required by an airline when managing airport ramp hub operations. Commonly imposed airborne delays applied to arriving aircraft on a first-come, first-serve basis do not allow the system users to prioritize for late arrivals or to economically optimize their arrival sequence. Airline preferences are most important at the airline's "hub" operation, where off-schedule arrival aircraft are common. The airline seeks to minimize such schedule violations as they cause airport ramp difficulties and result in airline operational inefficiencies. Service provider / system user collaborative arrival planning decision support tools are being developed to allow such airline operating priorities.

The first use of the PRM will be in direct support of decision support tools developed as derivatives of the NASA/FAA Center/TRACON Automation System (CTAS), specifically those under the Collaborative Arrival Planning (CAP) component of CTAS. CTAS is a suite of decision support tools that provide computer generated advisories for both en-route and terminal area controllers to manage and control arrival traffic more efficiently [2]. CTAS CAP will allow the user to request and influence intra-airline arrival characteristics from ATC via CTAS, without negatively impacting ATC operations. The tactical CAP tool will assist and improve the handling of individual aircraft arrival preferences, allowing the user to better manage off-schedule aircraft or aircraft facing possible airport diversion. The strategic CAP tool will alter CTAS arrival sequence within a carrier's fleet given an arrival preference order or relative priority ranking. CAP tools will be developed through high-fidelity laboratory simulations that will require constant participation of airline and ATC personnel. The PRM is used to locate and identify problem "outlier" aircraft – those aircraft that are candidates for CAP tool assistance. The airline participant then formulates schedule requests for outlier aircraft and communicates the requests to CTAS using the PRM. The subsequent schedule improvements are then monitored using the PRM.

Pseudo Ramp Manager (PRM) Workstation

This section describes the requirements, functionality, design and development of the PRM workstation.

Requirements and Laboratory Use

The existing CTAS laboratory located at NASA Ames Research Center can simulate the entire CTAS with either live or recorded air traffic data. The primary requirement of the Pseudo Ramp Manager (PRM) workstation is to facilitate airline participation during CTAS simulations. As such, it emulates some of the tools found in the ramp tower of a major hub airline. It accepts as input airline connectivity data for the aircraft, cabin crew, flight attendants and passengers. It displays these data to show the status of arrival banks expected over the course of a day.

The display is interactive so that the airline participant can manage the arrival/departure traffic "banks." The PRM includes a generic version of existing ramp tower display systems of arrival flights, connections, and gates. The airline representative (e.g., airline ramp tower manager) will use the PRM during simulations to identify "outlier" aircraft and thus candidates for CAP tool assistance. Figure 1 illustrates the high-level CTAS architecture with the PRM in place. During field trials of the CAP tools, the PRM will be supplanted by the host airline's actual ramp management systems and displays.

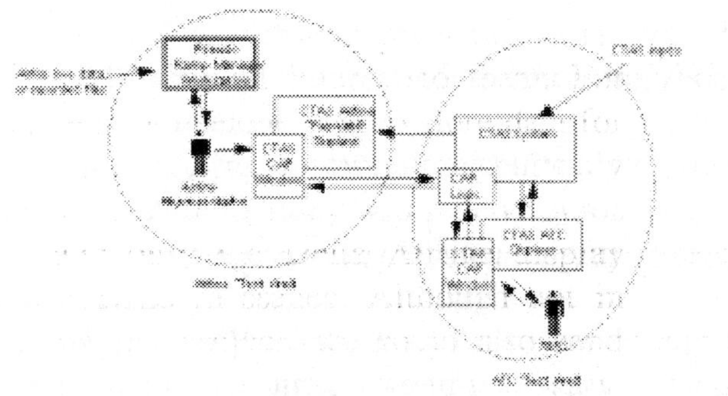

Figure 1. CAP Tool Development Laboratory Configuration with PRM.

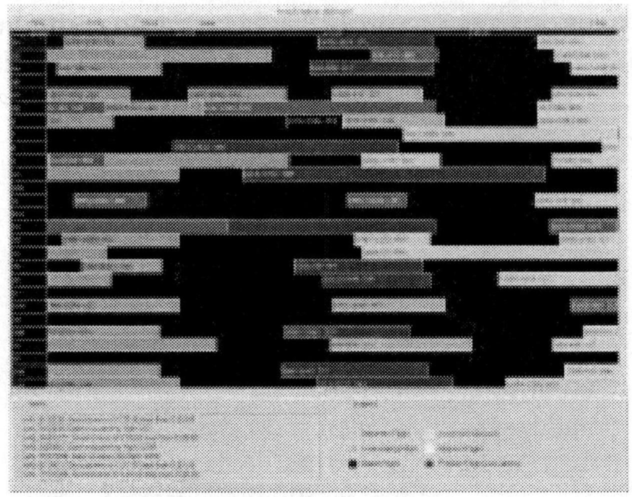

Figure 2. Main PRM Display of Aircraft at Airport Gates.

Functionality

The Pseudo Ramp Manager workstation employs a hockey puck or Gantt chart type of representation to depict the status of arrival banks expected over the course of a day. The PRM is capable of supporting the following tasks:

- monitor multiple arrival banks
- monitor and resolve problem alerts
- view information associated with individual flights
- update flight information for individual flights
- view gate characteristics
- modify gate characteristics
- make gate assignments
- find flights matching specific criteria
- record data entered by the user during a simulation

Figure 2 depicts the main PRM display from which the above tasks may be performed. Airport gates are shown along the left vertical axis, time is shown along the top horizontal axis and the aircraft are shown relative to both gates and time. The puck size depicts the time interval an aircraft is expected to occupy a gate and its color conveys the arrival/departure status and the alert status of the aircraft.

Individual dialog windows may be accessed from the menu bar on top of the main display or from context sensitive menus on the display itself. The user may view individual gate and flight information through some of these dialogs. The user may alter various gate and flight characteristics through the use of other dialogs. For example, upon receiving current arrival information from CTAS, the user can alter the on time for the flight. Also, if there is a conflict at a gate, the user can change the gate assignment for one of the conflicting flights. Figure 3 displays a dialog from which the user may update information associated with an individual flight. The user may want to close gates for maintenance or note a change in its APU status. Figure 4 displays the dialog s/he would use to edit these and other gate characteristics.

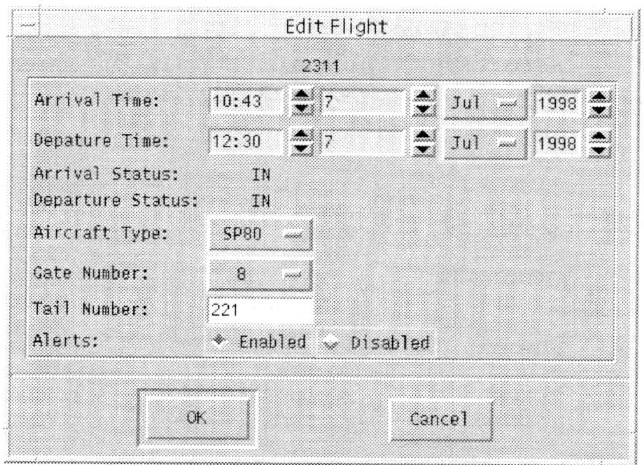

Figure 3 Modify Flight Dialog

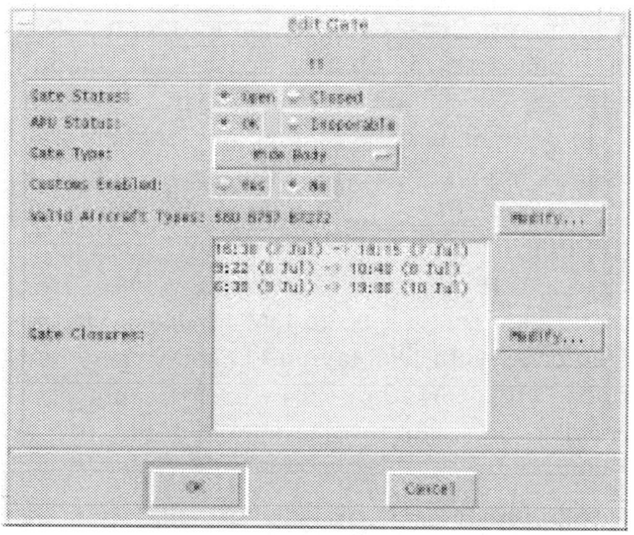

Figure 4 Modify Gate Dialog

The PRM may run in a "fast-time" mode, whereby the ASCII airline data files are used to sequentially update the PRM displays at a user-selected rate. The user may set the rate to receive data separately from the rate at which the display is updated. The PRM is capable of indicating to the user, through a color change, that the user has changed some data. It also allows the user to manually turn-off the display of system alerts. More details regarding usage are found in the PRM User's Guide [3].

Design

PRM was designed to be general enough to receive additional or modified data, data sources and graphics in future upgrades. Three separate processes comprise PRM: Data Acquisition (DA), Data Management (DM) and Display. These were designed so that any one of the processes can be replaced or modified without significantly affecting the other processes. This provides extensibility and flexibility.

This design also supports the Model / View / Controller (MVC) project framework. This framework decouples the application from the presentation and the control mechanism. We may design any display that connects to the DM as long as it uses the same interface. We may design any method of obtaining the data as long as it passes the data along to the data storage module using the defined interface.

The DA process (also referred to as module), receives data from outside sources. The data may come from ASCII data files or it may come through sockets using the TCP/IP protocol. The Factory design method was used with the module. This method allows the user to "define an interface for creating an object, but lets the subclasses decide which class to instantiate."[4] For example, the DA knows it must establish some method to receive messages from the outside. When the software was written and compiled, DA had no information regarding these connections and messages but when PRM is run, DA reads the configuration files that describe the connections and determines how to proceed. Therefore, by changing the configuration files we can control the DA input and need not alter DA itself. Also by using a general interface, the airline that is sending the data may format their messages in any way. Once DA has received the message, it converts it to the form used internally throughout PRM.

The DM process acts as the server to the other two processes. It receives messages from both DA and the Display. It uses the Observer design method, better known as Publish-Subscribe. This method "defines a one-to-many dependency between objects so that when one object changes state, all its dependents are notified and updated automatically."[4] We used

0-7803-5086-3 /98/$10.00 ©1998 IEEE

this method to allow different processes to receive messages only of interest to them. For example, we may use one DM to store data for multiple airlines during a simulation, but only the American Airline display will receive AAL messages and only the Delta Airline display will receive DAL messages. Although not in the scope of this project, we could also send information back to the airline when necessary. The DM developed for PRM stores the gate and flight data that come from the airline separately from the gate and flight data modified by the user of the display. This design method allows for different displays to view the same data. When one user changes data on one display, the change will be shown on another display as long as it is subscribing to the same data. We could also create a display that subscribes to the same data but displays it in a different format or we could decide to have one display show information regarding one group of gates and flights and another display showing a different group.

The Display process receives messages for DM, manipulates a portion of the data, computes additional data and stores them in a format that is applicable to the display. There are three parts to the Display module; the Communication Monitor (CommMonitor), the Local Data Storage (LDS) and the Graphical User Interface (GUI). Here we also decouple the receiving and sending of messages, the local data storage and the display. This module uses POSIX threads to prevent the simultaneous access to the data. The CommMonitor sends and receives messages to/from the DM. Once it decodes the message and prepares the data, it locks the LDS and creates or updates gate and flight data. It then unlocks LDS and notifies the display that new data have arrived and it should update itself. Then the display locks LDS to retrieve the data it needs and unlocks it when it is finished. The display also accesses LDS when the window is scrolled and when the user changes gate or flight data. After the user changes data, the altered flight or gate is sent to the CommMonitor which prepares the data into the correct message format and sends it back to DM. Figure 5 illustrates the three process PRM architecture.

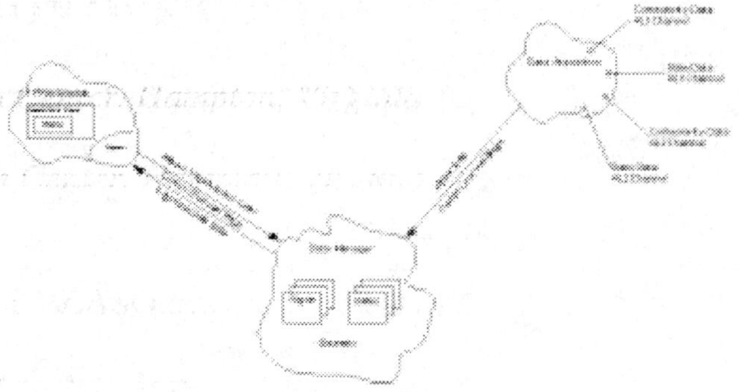

Figure 5 PRM Architecture

The Data Communications component of PRM is not a separate process nor module but is used by all of the modules. It contains the code that handles the messaging between processes. PRM defines its message structure as Communications Channel Messages (CommChannelMsg). When DA receives a message, it creates a CommChannelMsg from it and then sends it on to DM. When DM communicates with Display and Display communicates with DM they also send CommChannelMsgs back and forth as shown in Figure 5.

Development

In order to test and validate the PRM workstation, as part of the Object Oriented Analysis (OOA) part of the project, we developed Use Cases to verify each task. Use Cases describe who the users of the workstation are, how the user interacts with the workstation and how data move through the processes. Interaction diagrams were drawn that offer a picture of the Use Cases. During the design phase of the project, Object Oriented models were drawn which show the interaction of classes.

The PRM processes run on a SUN Ultra 2 computer with the Solaris 2.5.1 operating system. Seagull used the C++ programming language for its object-oriented features (Reference [5] discusses the importance of object oriented software) and to be compatible with software in the CTAS simulation laboratory.

The display was programmed with a combination of C++, Tcl/Tk scripting language and toolkit and OpenGL graphics.

Conclusions

An airline airport ramp manager workstation was developed for use in the laboratory development and simulation of new "free-flight" airline-ATC decision support tools. Such proposed tools would allow the airline user to influence aircraft arrival characteristics for maximum operational benefit.

The ramp manager workstation developed mocks essential aspects of airline systems commonly used at airport ramp towers. The workstation presents a "Gantt chart" type display of arriving and departing aircraft sorted by airport gate, as typically used by "hub and spoke" air carriers in their airport ramp towers.

The workstation employs modern object-oriented design and is highly adaptable toward use in a variety of decision support tool simulations.

References

[1] "Report of the RTCA Board of Directors' Select Committee on Free Flight", RTCA, Inc., Washington, D.C., January 18, 1995.

[2] Denery, D, Erzberger, H., "The Center/TRACON Automation System: Simulation and Field Testing", *Proceedings of the Advanced Workshop on Air Traffic Management*, Oct. 1995, Capri, Italy.

[3] Dorsky, S., Dow D., Henthorn R., Kondrasuk, G., "PRM User's Guide", Seagull Technology Technical Report, 98167-02 (June 1998)

[4] Gamma, E., Helm, R., Johnson, R., Vlissides, J., "Design Patterns, Elements of Reusable Object-Oriented Software", Addison-Wesley ISBN 0-201-63361-2 (1995)

[5] Dorsky, S., "Pseudo Airline Operations Control Workstation, Phase I", Seagull Technology Technical Report, 971167-01 (December 1997)

AIRPORT SURFACE MOVEMENT TECHNOLOGIES - ATLANTA DEMONSTRATION OVERVIEW

Denise R. Jones, NASA Langley Research Center, Hampton, Virginia

Steven D. Young, NASA Langley Research Center, Hampton, Virginia

Abstract

A flight demonstration was conducted in August 1997 at the Hartsfield Atlanta (ATL) International Airport as part of low visibility landing and surface operations (LVLASO) research activities. This research was aimed at investigating technology to improve the safety and efficiency of aircraft movements on the surface during the operational phases of roll-out, turnoff, and taxi in any weather condition down to a runway visual range of 300 feet.

The system tested at ATL was composed of airborne and ground-based components that were integrated to provide both the flight crew and controllers with supplemental information to enable safe, expedient surface operations. Experimental displays were installed on a Boeing 757-200 research aircraft in both head-up and head-down formats. On the ground, an integrated system maintained surveillance of the airport surface and a controller interface provided routing and control instructions.

While at ATL, the research aircraft performed a series of flight and taxi operations to show the validity of the operational concept at a major airport facility, to validate simulation findings, and to assess each of the individual technologies' performance in an airport environment. The concept was demonstrated to over 100 visitors from the Federal Aviation Administration (FAA) and the aviation community. This paper gives an overview of the LVLASO system and ATL test activities.

Introduction

NASA's Terminal Area Productivity (TAP) program is aimed at developing requirements for terminal area operations and technologies that will safely enable the same, or higher, capacity at the major airports in Visual Meteorological Conditions (VMC) and Instrument Meteorological Conditions (IMC). TAP research activities have been decomposed into four sub-elements: air traffic management, reduced separation operations, aircraft-ATC integration, and low visibility landing and surface operations. A flight test was conducted in 1997 as part of ongoing research under the LVLASO sub-element of TAP, in conjunction with the FAA and several industry partners.

In general, the LVLASO research is aimed at investigating technology to improve the safety and efficiency of aircraft movements on the surface during the operational phases of roll-out, turnoff, inbound taxi, and outbound taxi. This investigation becomes critical with the growing demands for air travel, the increasing number of reported surface incidents (287 in 1996) and fatal accidents (5 since 1990), and the economic, environmental, and geographic infeasibility of constructing new airports and/or runways. The goal of this research, which began in 1993, is to investigate technology to make better use of existing runways and ideally, enable safe VMC capacities (i.e. flow rates) on the surface in weather conditions down to a visibility of 300'.

0-7803-5086-3 /98/$10.00 ©1998 IEEE

Specifically, the objectives of the ATL flight test were to demonstrate a prototype system that has the potential to meet the LVLASO goal; validate selected simulation findings and the operational concept at a major airport facility; and assess the performance and suitability of the prototype as compared to (a) the operational requirements of an Advanced Surface Movement Guidance and Control System (A-SMGCS) [1], as well as (b) the requirements of NASA's conceptual system.

This testing was based on several pieces of prior and related work including "lessons-learned" in flight simulation studies at NASA's Langley [2] and Ames [3][4][5] Research Centers; a flight test performed at the FAA Technical Center in 1995 [6]; and two draft requirements documents [1] [7].

System Description

The surface operations system tested at ATL consisted of both ground and flight components that were integrated via three digital datalinks as well as the normal voice channels. The flight system provided the crew with enhanced guidance and situational awareness information through the use of a head-up display (HUD) and a head-down liquid-crystal display (LCD). These displays were integrated with onboard sensors and datalinks that provided the necessary input data as well as providing aircraft state data to the ground components. The displays were designed to function based on the phase of flight. The Roll-Out Turn-Off (ROTO) display symbologies and functions were used during high-speed roll-out and runway exit. The Taxiway Navigation and Situational Awareness (T-NASA) [3] display symbologies and functions were engaged during taxi. Regardless of the phase of flight, the information presented on the displays was intended to supplement missing visual cues in low visibility situations or at night, and to reinforce any available visual cues that may have an uncertainty associated with them (e.g. traffic positions, path to follow, etc.).

Similarly, ground components of the system provided the controller with supplemental information about traffic (e.g. position, identity, and intent), as well as a means for communicating with the flight crew over a digital link, in parallel with the normal voice channel. As with the flight crew, the information provided was meant to supplement missing visual cues and to reinforce uncertainties associated with the visual cues that were available.

Flight System

Figure 1 depicts the architecture for the experimental flight system employed at ATL. As mentioned previously, the testing was conducted using a Boeing 757-200 (B-757) research aircraft. Modifications to the flight deck included installation of three hardware devices (figure 2).

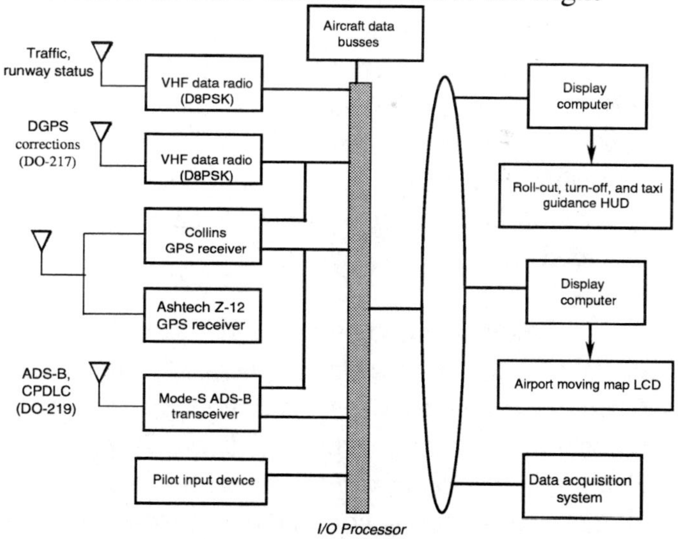

Figure 1. Flight System.

A *Head-Up Display* device was mounted in front of the left seat position and was used to display ROTO and taxi guidance

0-7803-5086-3 /98/$10.00 ©1998 IEEE

symbology. This specific HUD was manufactured by Flight Dynamics, Inc. and was capable of projecting a holographic image based on a raster-type graphics input. The field of view was 30 degrees horizontal by 24 degrees vertical.

A *Liquid-Crystal Display* device was mounted under the glare shield (left of center) and was used to render the raster moving map symbologies. This LCD was manufactured by Rockwell International. It was sunlight readable and provided a 1024x768 pixel resolution, an 8"x6" viewing area, and a 65 degree horizontal viewing angle which allowed for viewing by both crew members.

Figure 2. Flight Deck Layout

A *Pilot Input Device (PID)* was mounted on the center aisle stand and allowed the pilots to control the experimental displays.

Aft of the flight deck, pallet workstations contained the necessary on-board systems required for data acquisition/recording, power, flight management, audio/video recording/ telemetry, datalink, and display generation. Hardware aft of the flight deck included:

Dedicated computers were used to generate the experimental displays. A *Silicon Graphics Indigo2 Extreme computer* generated the map LCD display while a *Silicon Graphics Personal Iris (PI) computer* generated the HUD images. These computers supported a SCRAMNET (described below) interface that allowed them to communicate with each other and also the I/O subsystem. The software system design is described in [8].

Two identical *VHF data radios* were provided by Rockwell International. One radio was responsible for receiving DGPS corrections. The other radio was responsible for receiving traffic and runway status information provided by the ground surveillance system. The radios employed the Differentially encoded 8-Phase Key Shifting (D8PSK) modulation waveform and adhered to the RTCA standard protocol DO-217 [9].

An *Extended Mode-S transponder unit* was provided by Rockwell International. This unit contained a Mode-S radio, a GPS receiver, and an air datalink processor. The unit provided GPS position reports to the ground surveillance system. These reports adhered to Automatic Dependent Surveillance Broadcast (ADS-B) specifications [10]. This unit also supported the bi-directional Controller-Pilot Datalink (CPDLC) which adhered to the RTCA standard protocol DO-219 [11].

An *I/O processor* was responsible for reformatting data received by the experimental datalinks and providing it to the display computers. This processor also relayed data to be downlinked to the test controller at the ground site via the Mode-S transceiver. Finally, the processor integrated DGPS and Inertial Reference Unit (IRU) position data ensuring a continuous position update on the experimental displays. This also allowed for intermittent outages of the DGPS system and convergence to an accurate position when DGPS data was valid. A description of the algorithms employed for DGPS/IRU

0-7803-5086-3 /98/$10.00 ©1998 IEEE

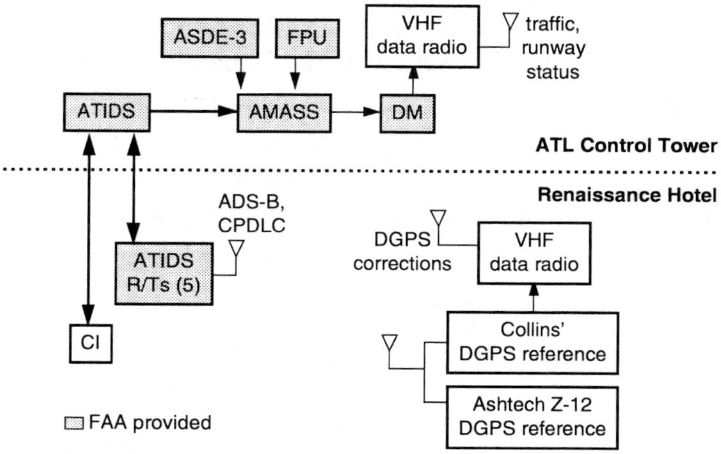

Figure 3. Ground System.

integration is given in [12].

The *Data Acquisition System (DAS)* stored digital data that was time stamped using the GPS time reference.

A *SCRAMNET I/O network* is a ring network that allows nodes to communicate via virtual shared memory blocks. For this testing, the four nodes on the SCRAMNET were the DAS, I/O processor, Indigo2, and PI.

A *video recording system* logged the following images: tail perspective, nose perspective, flight deck activity, scan-converted HUD display, scan-converted map display, and a view from near the pilot's eyepoint.

An *audio management system* allowed researchers to communicate from any seat position with (1) each other, (2) the flight deck, and (3) the ground. All audio received in the flight deck (by both the pilot and co-pilot) as well as voice transmissions to ground locations were recorded on the video recorders.

A *telemetry system* was capable of sending two of the six available video images to the ground simultaneously for viewing by visitors and ground participants during the testing.

An independent *GPS survey system* was employed using an Ashtech Z-12 receiver. This system recorded GPS data and, along with data stored at the ground site, allowed for post-processing that resulted in nominal 5cm accurate position data. This data was used to evaluate the accuracy of the experimental real-time position determining system.

Ground-Based System

The ground subsystem is illustrated in figure 3. It provided surveillance of the airport surface and enabled the transfer of required information among the functions implemented on the ground and the B-757 research aircraft.

The surveillance system consisted of four primary elements (ASDE-3, AMASS, FPU, and ATIDS). The first three are already part of the NAS and are used to provide controllers with supplemental traffic information in real-time such that safe separations can be maintained for surface movements. The fourth, ATIDS, is an FAA research and development project that is primarily aimed at providing identity information to controllers. The four elements were integrated in an attempt to provide full coverage of the airport surface, to provide identity information to both pilots and controllers, and to collect data so that multipath mitigation algorithms can be developed. Requirements for a surveillance function are listed in [1].

The *Airport Surface Detection Equipment (ASDE-3)* [13] captured position data (range and azimuth) at a 1 Hz rate for all aircraft or vehicles operating on the airport surface movement area. ASDE-3 does not require any equipage on aircraft or vehicles and is capable of detecting targets with a cross section as small as three meters. It operates in the Ku-band (15.7 - 16.2 GHz) and has a range specified to be 24,000 feet in all directions on the surface and up to 200' above the surface.

0-7803-5086-3 /98/$10.00 ©1998 IEEE

ASDE-3 and its associated display is scheduled for deployment at 36 airports. At the time of the testing, the ASDE-3 display was available and operational in the ATL tower cab although it was not fully commissioned.

Although ASDE-3 is a high performance radar system, it does have certain limitations. ASDE-3 has a 500' "cone-of-silence" area encircling the antenna in which targets are not visible. Further, there can be other coverage gaps with particular ASDE-3 installations as it is a line-of-sight radar. Also, ASDE-3 is susceptible to multi-path reports. This can result in a false target being reported and possibly displayed. Finally, ASDE-3 does not report target identity information. It is because of these issues the other systems described below were integrated with ASDE-3 for this testing to maximize coverage, minimize multi-paths, and provide identification which is required in [1].

The *Airport Surface Target Identification System (ATIDS)* captured position and identity data for aircraft and ground vehicles with operating ADS-B and Mode-S transponders by performing a multilateration function [14] through the use of five fixed receiver/ transmitters (R/Ts). In addition, ATIDS captured the ADS-B transmissions emanating from the research aircraft and also acquired data from the FPU (described below). The ATIDS update rate was specified to be 1 Hz. The coverage area for the ATL ATIDS was specified to be only on the north side of the airport out to 500' beyond the approach end of the runways and up to 500' above the surface.

The *Airport Movement Area Safety System (AMASS)*, as configured at ATL, provided the following: (a) tracking of ASDE-3 targets; (b) data fusion of ATIDS target data with ASDE-3 track data, and (c) safety logic to detect occupied runways and alert controllers and the test pilots. AMASS is an enhancement to the ASDE-3 radar that provides automatic alerts and warnings (visual and aural) to the controller [15][16]. AMASS is being designed to overlay information on the ASDE-3 display; however, for this testing, an independent AMASS display was used. AMASS was designed to track up to 200 targets.

For this testing, AMASS was also responsible for passing target information and runway status to a *Datalink Manager (DM)*. The *DM* converted this data into the protocol required by the D8PSK transmitter for transmission to the B-757. The DM was designed to be able to support multiple transmitter types simultaneously such that aircraft/vehicles with different receivers could acquire the traffic broadcast (if a reciprocal transmitter were connected to the DM). This enables alternate datalinks to be utilized.

A *Flight Plan Unit (FPU)* provided a transparent interface to the ARTS-IIIA system database. This allowed ATIDS to extract the Mode-A code, the aircraft call sign, and the aircraft type from the database, in real-time, and associate this information with specific Mode-S transmissions received. All retrieved information was forwarded to AMASS for use by the fusion function. This resulting fused surveillance data was provided to both the test ground controller and the B-757's flight crew enabling them to have the same "picture" of the airport surface traffic at any point in time. This is a requirement specified in [1].

An independent *GPS ground station* was implemented to provide differential corrections. It consisted of two GPS receivers and a VHF data radio. These components were identical to those used onboard the research aircraft. One of the GPS receivers was an Ashtech Z-12 that was responsible for storing data for post-processing to obtain high accuracy "truth" position data. The other was the Rockwell-Collins GPS receiver that operated in conjunction with the D8PSK radio transmitter to fully implement the RTCA DO-217 specification [9].

Display Symbologies

Moving Map LCD

The map LCD (figure 4) onboard the B-757 provided both crew members with:

- depiction of the airport layout
- depiction of current position and heading of the B-757
- depiction of current position of other traffic on the movement area
- display of ATC instructions including the taxi route
- display of runway status (indicated occupied runways)

This map display format was part of the T-NASA system that has undergone human factors testing in several simulation studies [3][4][5]. In addition to the input data received from the datalinks and the DGPS/IRU system

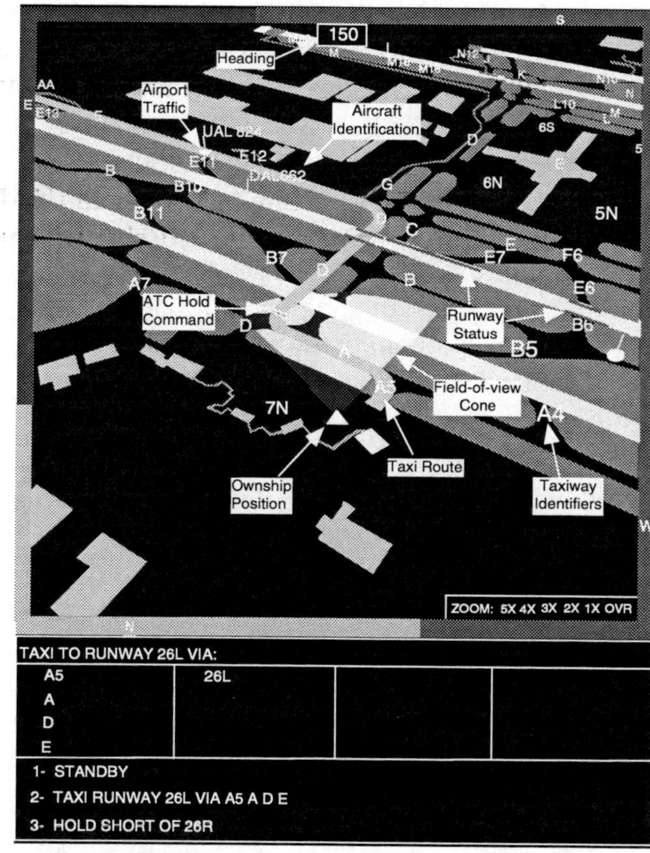

Figure 4. Moving Map LCD Symbologies.

onboard, an accurate airport database was also required. This database was provided by Jeppesen-Sanderson and included all runway/ taxiway edges and centerlines as well as hold-short lines. These were all required to be accurate to one foot (0.3m).

The flight crew interacted with the electronic map through the PID. The crew was able to select from six zoom levels, one of which was an overview of the entire airport. The airport overview zoom level was north up while all other zoom levels were track up. The crew also had the choice to display symbols for other traffic and, if displayed, show traffic identification labels, if desired. The capability also existed to scroll through the list of ATC instructions displayed in the lower portion of the map LCD.

In addition to rendering the display, the moving map computer generated downlink messages that were relayed to the test controller at the ground site. For example, if the B-757 deviated from the route issued by ATC, a message was sent to the test controller alerting him of this deviation. Similarly, if the B-757 got back on its approved path, a "taxi route resolved" message was sent to the test controller.

Roll-Out, Turn-Off, and Taxi Guidance HUD

On the HUD, from final approach until the B-757 had safely exited the runway, the roll-out and turn-off (ROTO) symbologies were enabled. Specifically, while in the landing phase, the ROTO system displayed symbology similar to that found on commercial HUD systems designed to provide landing guidance. Once the aircraft landed and the nose strut was compressed, the symbology transitioned from the in-flight symbology to the roll-out and turn-off guidance symbology (figure 5). While rolling out, the symbologies were presented to reinforce available visual cues that may be obscured due to visibility or darkness (i.e. runway edges and runway remaining markers) and to provide a deceleration profile to follow that would minimize runway occupancy time to the chosen exit. The goal was to maintain VMC roll-out turn-off times in IMC conditions or at night. After turning off of the runway, the pilot decelerated the aircraft to taxi speed, or to a stop, depending on controller instructions received.

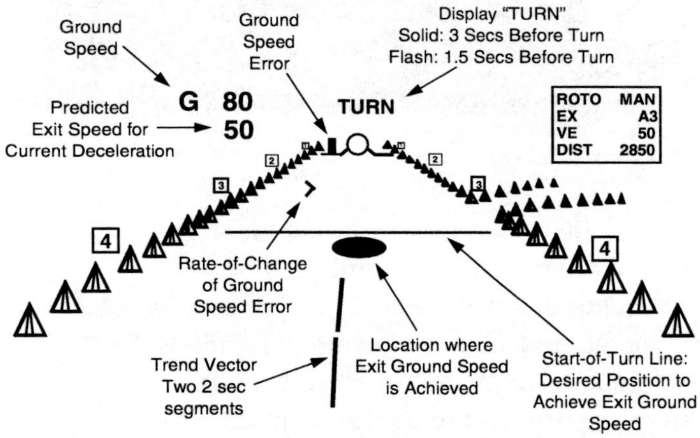

Figure 5. ROTO HUD Symbologies.

Once the taxi instructions were given by ATL control and were also delivered verbally by the test controller, the HUD symbology transitioned from the ROTO mode to the taxi mode (which was also part of the T-NASA system). The taxi symbols, which pertained to the approved taxi route, are shown in figure 6 and included:

- taxiway centerline markings

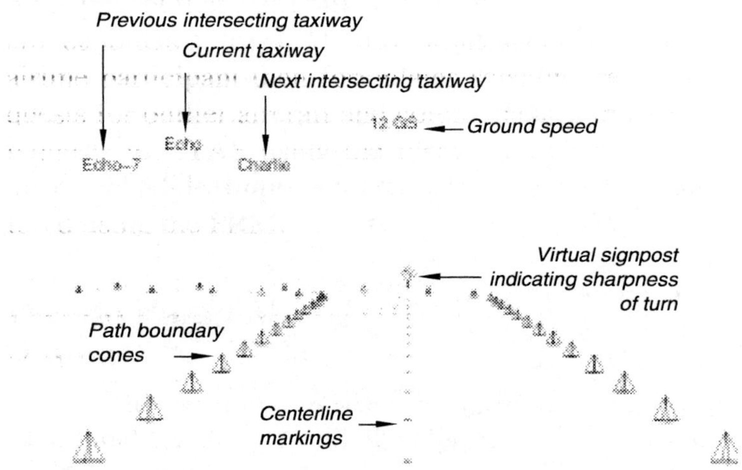

- path boundary cones
- indications of location and angle of turns
- ground speed
- previous, current, and next taxiway identifiers

All HUD symbols were displayed relative to the pilot's eye reference point such that they overlaid the outside scene.

Figure 6. Taxi HUD Symbologies.

Controller Interface

A *Controller Interface (CI)* (figure 7) allowed a test controller located at a ground test site to mimic ATC verbal instructions in parallel,

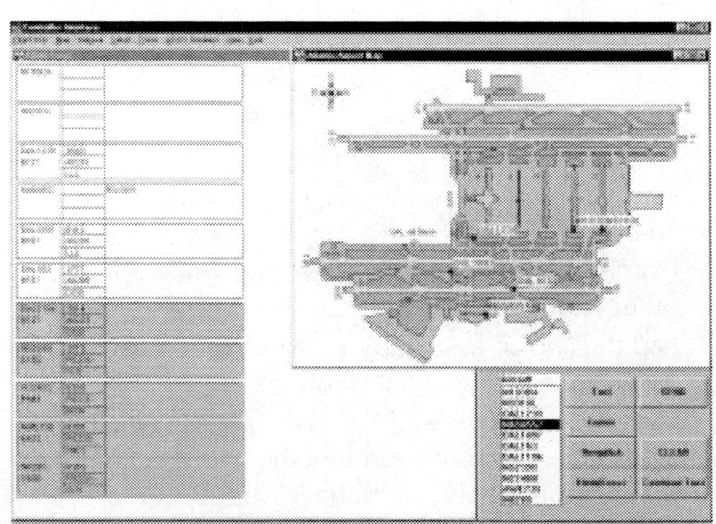

Figure 7. Controller Interface Display.

and then transmit these instructions digitally for display in the flight deck of the B-757. Two-way communications with the research aircraft were implemented using Mode-S Specific Services [17]. These adhered to the RTCA standard DO-219 [11]. The CI provided:

- electronic flight strips updated in real-time
- continuous display of surface traffic positions and identification on an airport map
- controller instruction capture and datalink to the B-757 via voice recognition or touchscreen
- alerts of route deviation by the B-757
- runway exit taken by the B-757

The CI was developed by St. Cloud State University [18].

Flight Test Operations

The deployment to ATL occurred during August 1997. Flight test runs were conducted using NASA test pilots and commercial B-757 captains as subjects. Demonstrations were conducted for over 100 visitors from the aviation community. These demonstrations included a briefing, an opportunity to view a flight test from the ground site, and a tour of the B-757.

All flight test runs were enacted with the following guideline: the operation shall follow, as close as possible, a routine flight operation from "gate-to-gate". The only difference, operationally, would be the additional tools provided to both the test pilots and the test controller that would show the potential for improving the safety and efficiency of the surface operation.

The crew of the B-757 was instructed to maintain radio contact as needed with the ATL ATC during the testing. Because the CI was at the prototype stage, a test controller was used. This controller was located at the ground site (not in the tower cab) and monitored ATL ATC communications. Any instructions designated for the B-757 were sent electronically to the aircraft, in parallel, using the CI.

The crew was also instructed to utilize the HUD and map LCD while maneuvering the B-757 on an as-needed basis. The HUD was to be used by the captain for supplemental guidance cues and enhanced situational awareness during landing, roll-out, turnoff, and taxi. The map LCD was to be used primarily by the first officer for situational awareness which could then be relayed to the captain if necessary. The captain could refer to the map LCD occasionally if desired. During test runs, the flight crew could manipulate the map LCD using the PID as desired. Specific details on how to use the LVLASO display system were provided as part of each pilots' training procedure prior to the flight experiment.

All flight test runs began in the ramp area located at the Fixed Base Operator (FBO) just north of runway 8L/26R. At initiation of a run, the B-757 was in position to begin taxi and the responsible flight deck crew member called for taxi instructions from ATL ATC. These instructions were sent to the aircraft by the test controller in parallel with the verbal ATC commands. The captain then taxied to the designated departure runway. After taking the runway, the B-757 would either (1) takeoff/circle/land or (2) taxi down the runway depending on the test run. Once clear of the runway, the B-757 verbally received a taxi instruction from ATC. Again, this instruction was sent to the B-757 by the test controller in parallel via datalink. After the crew verbally acknowledged receipt of the instruction, the captain taxied back to the FBO ramp area following the designated path. While taxiing, the captain was instructed to taxi at a normal taxi rate or higher if he felt safety was not being compromised.

Runs that required a landing tested both

the ROTO and T-NASA systems. If a takeoff was not required, only the taxi guidance system onboard (T-NASA) was evaluated.

Tests runs were conducted predominantly at night as this more closely represents a "low visibility" condition. A total of 53 test runs were successfully completed which resulted in 1378 minutes (almost 23 hours) of audio, video, and digital data. The average run time was 26 minutes.

Conclusion

This testing has successfully demonstrated and validated the potential for using technology and a holistic systems approach for improving the safety and efficiency of airport surface operations. By providing supplemental guidance and situational awareness information to both pilots and controllers, safety margins can increase since there is more confidence in the understanding of the current state of the airport surface. In poor visibility, at night, or at unfamiliar airports, this supplemental information becomes critical, particularly if VMC flow rates are expected to be maintained safely.

This paper was meant to be an overview of the testing that occurred at ATL in August 1997 as part of the TAP LVLASO program. Results and findings were not addressed here but are documented in [19]. Several other papers presented at the 17th Digital Avionics Systems Conference discuss the details and analysis of the various subsystems employed at ATL.

References

1. "Draft Manual of Advanced Surface Movement Guidance and Control Systems (A-SMGCS)", 16th Meeting of the International Civil Aviation Organization's All Weather Operations Panel, Montreal, Canada, June 1997.
2. Batson, V., Harris, R., and Hunt, P., "Navigating the Airport Surface: Electronic vs. Paper Maps", Proceedings of the 13th Digital Avionics System Conference, October, 1994.
3. Foyle, D., "Taxiway Navigation and Situation Awareness (T-NASA) System", Ninth International Symposium on Aviation Psychology, Ohio State University, April, 1997.
4. McCann, R., Andre, A., Begault, D., Foyle, D., and Wenzel, E., "Enhancing Taxi Performance Under Low Visibility: Are Moving Maps Enough?", Proceedings of the 41st Annual Meeting of the Human Factors and Ergonomic Society, pp. 37-41, 1997.
5. Foyle, D., Andre, A., McCann, R., Wenzel, E., Begault, D., and Battiste, V., "Taxiway Navigation and Situation Awareness (T-NASA) System: Problem, Design Philosophy and Description of an Integrated Display Suite for Low Visibility Airport Surface Operations", SAE Transactions: Journal of Aerospace, v105, pp. 1411-1418, 1996.
6. Jones, D., Young, S., Wills, R., Smith, K., Shipman, F., Bryant, W., and Eckhardt, D., "Flight Demonstration of Integrated Airport Surface Technologies for Increased Capacity and Safety", NASA TM-1998-206930, February, 1998.
7. Cassell, R. and Smith, A., "Development of Required Navigation Performance Requirements for Airport Surface Movement Guidance and Control", Proceedings of the 14th Digital Avionics Systems Conference, November, 1995.
8. Beskenis, S., Green. D., Hyer, P., and Johnson, E., "Integrated Display System for Low Visibility Landing and Surface Operations", NASA/CR-1998-208446, July, 1998.
9. "Minimum Aviation System Performance Standards, DGNSS Instrument Approach System: Special Category I (SCAT-I)", Document No. RTCA/DO-217, RTCA, August 27, 1993.
10. "Minimum Aviation Performance Standards for Automatic Dependent Surveillance Broadcast (ADS-B)", Draft Version 6.0, RTCA Special Committee 186, August 28, 1997.
11. "Minimum Operational Performance Standards for ATC Two-Way Datalink Communications", Document No. RTCA/DO-219, RTCA, August 27, 1993.
12. Plyler, V., Slominski, C., and Wolverton, D., "The LVLASO I/O Concentrator Software Description, Version 3.5", NASA/CR-1998-206907, January, 1998.
13. Go, G., and Ianniello, J., "Third Generation Airport Surface Detection Equipment Design", Proceedings of NAECON '94, May, 1994.
14. Friedland, B.; and Hutton, M., "New Algorithms for

Converting LORAN Time Differences to Position", Navigation: Journal of the Institute of Navigation, Vol. 20, No. 2, Summer, 1973.

15. Watnick, M., and Ianniello, J., "Airport Movement Area Safety System", Proceedings of the 11th Digital Avionics Systems Conference, October, 1992.
16. Ianniello, J., and Kruczek, R., "Airport Surface Collision Warning System Implementation", Proceedings of the 12th Digital Avionics Systems Conference, October, 1993.
17. "Minimum Operational Performance Standards for Air Traffic Control Radar Beacon System/Mode Select (ATCRBS/Mode-S) Airborne Equipment", Document No. RTCA/DO-181A, January 14, 1992.
18. Mattson, P., and Rankin, J., "Research and Design of an ATCT Ground Controller CPDLC Workstation", Proceedings of the 42nd Annual Air Traffic Control Association Conference, September, 1997.
19. Young, S., and Jones, D., "Flight Demonstration of Integrated Airport Surface Movement Technologies", NASA TM-1998-206283, January, 1998.

DESCRIPTION AND FLIGHT TEST OF A ROLLOUT AND TURNOFF HEAD-UP DISPLAY GUIDANCE SYSTEM

Richard M. Hueschen, NASA Langley Research Center, Hampton, VA

Walter W. Hankins, NASA Langley Research Center, Hampton, VA

L. Keith Barker, NASA Langley Research Center, Hampton, VA

Summary

An experimental rollout and turnoff (ROTO) guidance system installed on a B757-200 research aircraft was flight demonstrated at the Hartsfield-Atlanta International Airport (ATL) in August 1997 as part of NASA's Low Visibility Landing and Surface Operations (LVLASO) program. The system uses a Head-Up Display (HUD) to provide deceleration guidance along the runway and steering guidance to exit. Seven pilots made test landings on runway 8L/26R during both day and night clear weather visibility conditions.

Results of these tests show that the ROTO HUD system provided a consistent runway occupancy time (ROT) for a given exit. Pilots commented that the system was easy to use and indicated that they would like to see it installed on commercial transport aircraft.

Introduction

Every year airplane traffic increases, challenging airport capacity. In response, NASA's Terminal Area Productivity (TAP) program is developing technology to help airports improve capacity, especially in bad weather. The goal of the LVLASO portion of TAP is to develop technology to enable Visual Meteorological Conditions (VMC) capacity for landing and surface operations (rollout and turnoff, taxi-to-the-gate, taxi-to-takeoff, and takeoff) down to a visibility of 300 ft. The ROTO sub-element of LVLASO further focuses on technology that will maintain clear-weather runway occupancy time (ROT – *the time that an aircraft spends on the runway*) under these visibility conditions.

Today, there is no specific on-board system to give pilots deceleration guidance or steering guidance for turning the aircraft off the runway after landing. Pilots perform this operation much like an automobile driver decelerates to turnoff from one highway onto another. In low visibility weather conditions, the rollout and turnoff task becomes more difficult and ROT increases significantly.

NASA Langley Research Center (LaRC) sponsored research efforts to develop navigation, guidance, and automatic control systems for the rollout and turnoff operation during the period 1974 to 1982 (refs. 1, 2, 3, 4). These efforts were motivated by a new landing navigation system under development at that time (the Microwave Landing System or MLS) and a proposed buried-wire navigation system.

In the 1980's the Global Positioning System (GPS), a satellite navigation system, was under development. Flight test experiments beginning in 1990 using differential GPS (DGPS) navigation showed it would provide both precise lateral and longitudinal aircraft position information without the problems associated with MLS and the buried-wire navigation. Beginning in 1994 NASA LaRC sponsored research to develop a

0-7803-5086-3 /98/$10.00 ©1998 IEEE

rollout and turnoff system using DGPS to provide guidance for both manual and automatic control of a transport aircraft (refs. 5, 6). These efforts resulted in system designs and simulation results for a MD-80 and MD-11 aircraft and preliminary recommendations for Head-Up Display (HUD) guidance symbology for manual (pilot) control and monitoring of automatic control.

In 1996 LaRC began to develop manual HUD guidance for the rollout and turnoff operation using a motion-based cockpit simulation. These efforts culminated in the ROTO experimental system described in this paper.

ROTO System Description

Overview

Figure 1 is a block diagram of the ROTO system. The system included a Silicon Graphics Personal Iris to execute the ROTO software, a ten-positioned switch (figs. 2 and 3) for pilot input of the selected exit, and a Flight Dynamics HUD (projector and combiner glass) for presentation of ROTO guidance and information. The I/O processor blended Inertial Reference Unit and differential GPS (DGPS) inputs to provide smoothed DGPS aircraft position at a 25 Hz update rate. Basic aircraft sensor data and the smoothed DGPS inputs of aircraft position were processed along with a high-resolution database of the local airport's runway geometry to:

- Specify a particular runway exit as a goal for turnoff;
- Calculate a deceleration profile (from the aircraft's touchdown point to the start of the turn for the selected exit) for guidance;
- Extract certain static parameters for display (e.g., exit location relative to nominal touchdown location, nominal safe exit speed);
- Produce symbology to outline the runway and its exits;
- Generate symbology to display state variables for approach, landing, and rollout and turnoff.

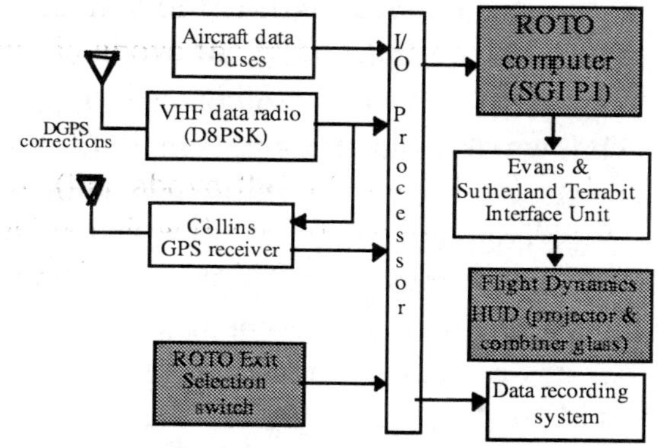

Figure 1. ROTO System

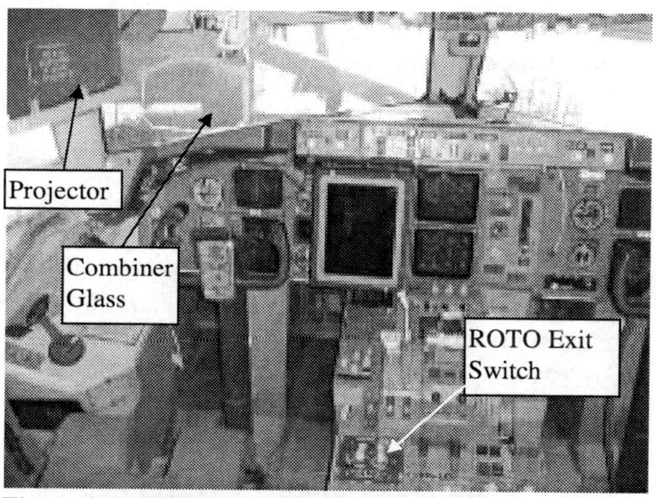

Figure 2. ROTO System Cockpit Hardware

Figure 3. ROTO Exit Switch

The HUD mounted above the left seat of the research aircraft (fig. 2) had a field of view of 30 degrees horizontal by 24 degrees vertical.

Guidance Algorithms

Two algorithms, Alg1 and Alg2, which provide deceleration guidance relative to a desired ground speed profile (computed at main gear touchdown as a function of distance-from-touchdown to the start of the turn) were flight tested. Alg1 computes a ground speed profile to achieve a desired exit speed assuming constant deceleration. Alg2 computes a profile based on a deceleration that changes non-linearly with distance. For Alg2, the desired deceleration starts at a low value, builds to a peak value, and then decreases as the exit is approached. Other guidance algorithms have been or are being developed using a cockpit simulator.

HUD Symbology

The ROTO Landing Symbology generated during approach and landing appears on the HUD when (1) the pilot selects a valid Instrument Landing System (ILS) frequency and (2) a valid ILS signal is received during intercept of the runway localizer. Upon nose gear compression after landing, the HUD display transitions to the ROTO Ground Symbology which supplies guidance and situational awareness for rollout and turnoff.

ROTO Landing Symbology

The ROTO Landing Symbology was overlaid on a basic set of in-flight situational and navigation symbology, patterned after a Flight Dynamics HUD. This basic symbology (unlabeled elements shown in fig. 4) includes pitch and roll attitude scales, a pitch reference symbol, pitch and roll flight director bars, a magnetic heading scale, an inertial flight path symbol, and localizer and glideslope deviation scales. Also displayed are digital readouts of the ground speed , calibrated airspeed, radio altitude, barometric altitude, and vertical speed.

The overlaid ROTO symbology is labeled in figure 4. The ROTO Box located in the upper right corner of the display provides static information about the selected exit. It shows the runway exit selected (A3 in fig. 4), the nominal safe exit speed (50 knots), the distance in feet from the expected touch down point to the exit (2850 ft.), and whether or not the ROTO system is engaged in the manual (MAN) mode or Automatic (AUTO) mode. In manual mode the pilot remains in full control after touchdown with guidance supplied by the system. In the automatic mode, braking and steering after landing are performed automatically according to the same guidance that would be given the pilot in the manual mode. The automatic mode is currently being developed and was not part of the ATL testing (Note that the AUTO position on the ROTO Exit switch currently commands the ROTO system to select an exit and does not arm the system for automatic control).

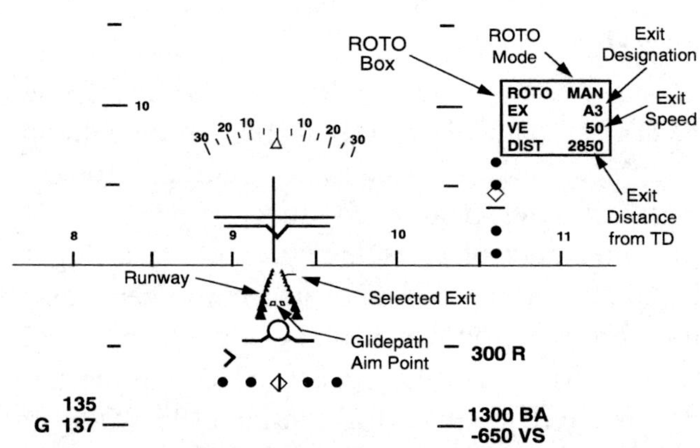

Figure 4. ROTO Landing Symbology Overlaid On Landing Symbology Based a Current HUD System

The remaining overlaid ROTO symbology includes outlines of the runway and selected exit and glidepath aim point markers. The runway and exit are outlined using small pyramid-like symbols called plops which are spaced 100 feet apart, remain fixed to the runway, and change in size within the real world scene as the aircraft maneuvers toward the airport. The glidepath aim point markers overlay the standard runway aim point markers which are located approximately 1000 feet from the runway threshold.

A temporary display is available if the Exit Switch (fig. 3) is placed in the RWY

0-7803-5086-3 /98/$10.00 ©1998 IEEE

position. In this position, a plan view of the landing runway showing available runway exits is momentarily displayed for five seconds (fig. 5). When the planview is active, switching to any of manual positions 1 through 8 will show the corresponding exit for that switch position as the selected exit and switching to AUTO will show the exit selected by ROTO. Five seconds after switching to a manual position or AUTO, the display will revert to the landing symbology (fig. 4).

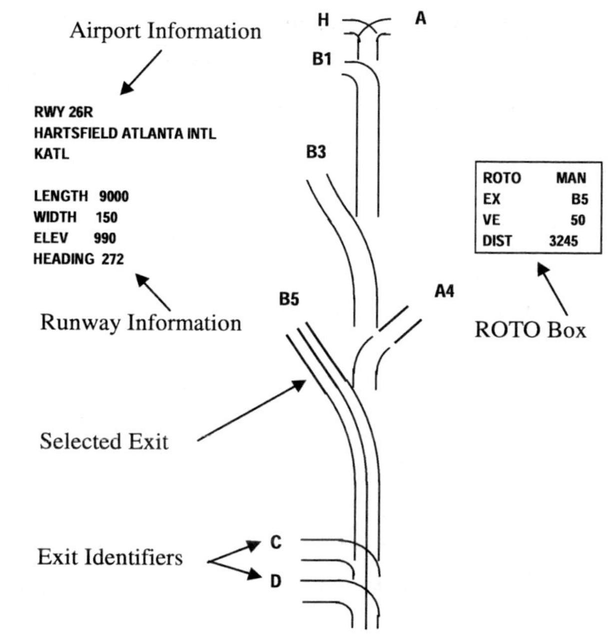

Figure 5. ROTO Runway Planview

ROTO Ground Symbology

At nose wheel touchdown, the HUD transitions from the Landing Symbology to the Ground Symbology (fig. 6) to provide guidance for rollout and turnoff. The plops, the ROTO box, and the flight path symbol remain. The rest of the Landing Symbology is removed. The primary Ground Symbology guidance symbols are the ground speed error bar, the "carrot", and the "football". The ground speed error bar, located on the left horizontal line (wing) of the flight path symbol, indicates the difference between the desired ground speed for the aircraft and its current ground speed. When the bar is above the wing, the aircraft ground speed is greater than the desired. The "carrot" symbol which shows the rate-of-change of the speed error is also referenced to the left horizontal line of the flight path symbol. If the "carrot" is above the reference line, the ground speed error bar will move upward (positive change). For example, if the speed error is positive (bar above the reference line) and the "carrot" is below the reference line, the bar will be decreasing towards the reference line.

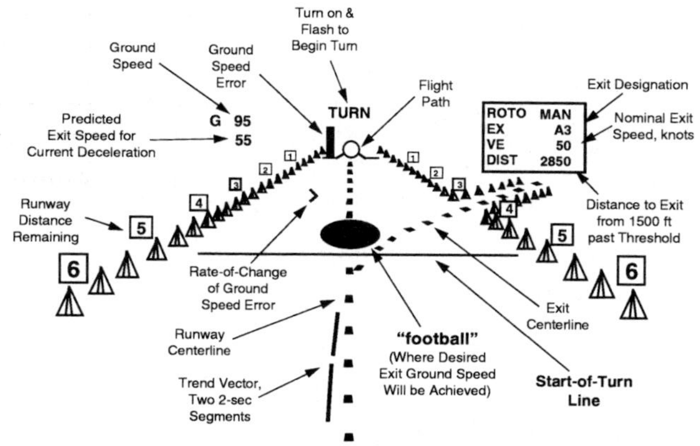

Figure. 6. ROTO Ground Symbology Display

Two additional symbols provide guidance for decelerating to the nominal exit ground speed (to be described later). One is simply a line drawn across the runway to locate the start of the turn onto the selected exit. The other, referred to as the "football", shows where along the runway the nominal exit speed is predicted to occur if the aircraft's current deceleration level is maintained. Increasing the deceleration level moves the "football" closer to the airplane. If the pilot centers the "football" at the start-of-turn line, the airplane is predicted to arrive at the start-of-the-turn with the nominal turnoff ground speed. The position of the football is computed using inputs of current position, ground speed, and deceleration and smoothed by a first order filter with a time constant of one second.

Among other ground symbology is a trend vector consisting of two segments showing the predicted track of the airplane. The end of the first segment is the predicted position in two seconds and the second segment the predicted position in four seconds. The main purpose of the trend vector is to help the pilot track the exit centerline during turnoff. The trend vector is computed using the current position, ground speed, and yaw rate and smoothed with a first-order filter with a time constant of 1.5 seconds.

Runway distance remaining boxes are drawn on each side of the runway indicating, in "thousands of feet", the distance to the end of the runway.

The word "TURN" is displayed above the flight path symbol three seconds from the start of the turn to alert the pilot that the turn onto the selected exit is imminent. It begins to flash at 1.5 seconds, continues for five seconds, and is removed.

A digital display of the ground speed is shown to the left and slightly above the flight path symbol. Predicted speed at the start of the turn is computed using current position, speed, and deceleration and smoothed by a first order filter with a time constant of 1.5 seconds. It is digitally displayed just below ground speed.

If the Exit Switch is in the AUTO position and the ROTO algorithms determine that the nominal exit speed cannot be achieved without exceeding acceptable passenger comfort deceleration levels, the word "NEXT" is displayed above the flight path symbol, the next available exit is automatically selected, and appropriate information for the new exit is displayed in the ROTO box. If the next available runway exit is the last exit, the word "LAST" is displayed. However, if the pilot has set the switch to any of positions 1 through 8 (manual exit selections), no automatic switch to the next available exit will occur nor will the word "TURN" be displayed. In this situation, the pilot may choose to manually select the next exit.

Exit Selection Logic

The ROTO system contains logic for automatic exit selection which operates only when the Exit Switch is in the "AUTO" position. There are two situations in which the logic is used. One is during approach and the other is after landing to reselect an exit if turnoff on the exit selected during approach cannot be achieved within acceptable deceleration limits.

During approach, an exit is automatically selected based on two assumptions: (1) the aircraft will touchdown at a point 1500 feet past the runway threshold, and (2) the aircraft will decelerate to the exit at a constant level not to exceed a midlevel value of 6.5 ft/sec^2 (based on passenger comfort) arriving there at the nominal speed specified for that exit in the database.

After landing, the system will automatically select the next available exit if the required aircraft deceleration to achieve the maximum exit speed exceeds 10 ft/sec^2 for one second. The next available exit is chosen using the second assumption for approach exit selection above and the current position and speed of the aircraft.

The numbers used in exit selection are nominal and can be changed to meet prevailing air safety regulations and the particular needs and desires of the user.

Runway Database

Specific runway information is stored in the database to compute deceleration guidance, select exits, draw plops for the selected exit, and to obtain data for the ROTO box. The information stored for each exit is (1) the nominal exit ground speed, (2) the maximum exit ground speed, (3) the angular turn of the exit (e. g. 30, 45 or 90 degrees), (4) the distance from the runway threshold to the start of the turn in feet, and (5) a deceleration buffer distance (used to compute the ground speed profile from the TD location to a point at a small distance before the start of the turn).

The nominal exit ground speed is the primary speed for the start-of-the-turn chosen such that lateral acceleration during turnoff is within acceptable passenger comfort limits. The maximum exit speed represents the speed at which lateral acceleration approaches unacceptable passenger comfort limits.

Flight Test

The ROTO system was flight tested and demonstrated at the Hartsfield-Atlanta International Airport during August 1997 as part of a larger LVLASO flight test demonstration which included testing of an airport surface guidance and navigation system (ref. 7). Three NASA research pilots and four airline pilots flew the system.

All ROTO test runs were initiated upon selection of the ILS frequency for approach and landing to the desired runway. All landings were done with the B757 autoland mode engaged and with the right-seat pilot in command. At nose wheel touchdown, the autoland system was disengaged by the right-seat pilot and the left-seat pilot took control of the aircraft performing the rollout and turnoff maneuvers using the ROTO HUD guidance.

The subject pilot's task was to set the throttles to idle reverse at nose wheel touchdown and, when needed, apply reverse thrust and brake to maintain the speed error bar relatively small. The pilots were instructed, that if a speed error existed, a positive error was preferred over a negative error.

Test Runs

A total of 37 test runs were conducted. Of those, 29 used high-speed exits (50 knots), seven used 15-knot exits and one used an eight-knot exit, 22 were conducted at night, four were training runs, seven were demonstration runs for visitors from the aviation community, nine were conducted using guidance algorithm Alg1 and 28 with Alg2.

Recorded Data

Digital sensor data were recorded onboard the aircraft and were timestamped using the GPS time reference. These included inertial data, air-data, control positions, and selected parameters computed by the ROTO algorithms (e.g., commanded speed and speed error). Video data of the HUD symbology as well as the view out of the front cockpit windshield from a camera mounted near the HUD combiner glass were recorded along with audio communications. Pilots filled out questionnaires after completion of their test runs.

Test Results

Runway Occupancy Time

Runway occupancy time (ROT) was computed for each run, using the recorded DGPS position data, as the difference between when the aircraft crossed the runway threshold and when the entire aircraft was clear of the edge of the runway. The bulk of the limited number of test runs used exits A6, B7, and B11 on runway 8L and exits A4, B1, B3, and B5 on runway 26R. Fourteen runs were conducted using exit B3, which is the second high-speed exit on 26R. Four of these runs were training runs. The remaining ten were used to compute the mean (51.3 sec) and standard deviation (2.0 sec) of ROT for this exit.

An independent LVLASO experimental system called DROMS (Dynamic Runway Occupancy Measurement System) had been in place for some time recording aircraft transponder position data at ATL for computation of ROT statistics on commercial traffic. The mean and standard deviation for 53 DROMS landings to exit B3 for B757 aircraft were 50.0 sec and 4.1 sec, respectively.

Note that ROTO and DROMS mean times for this exit are essentially the same while the DROMS standard deviation is twice as large. The number of ROTO test runs to other exits (six or less) is admittedly low for statistical computations. However, the comparisons of ROTO and DROMS ROT statistics for other exits showed the same trend as for exit B3 above; i.e., similar mean times, with DROMS standard deviations a factor of 1.5 to 1.6 larger.

Preliminary simulator results indicate that pilots follow the ROTO guidance equally well in low visibility or clear-weather conditions. The ROTO/DROMS comparisons above showed good clear-weather, real world comparability. Thus, cautious inference suggests that clear-weather runway capacity (or ROT) will be maintained in actual low visibility conditions using the ROTO guidance. Further, the lower standard deviation of the ROTO ROT data suggests a potential to more accurately predict ROT.

Pilot Comments

All four airline pilots indicated the HUD transitions from approach guidance to runway deceleration guidance to exit turnoff guidance were seemless and natural. They were also unanimous in finding the guidance acceptable and easy to follow after a few training runs.

Pilot Performance

The ground speed error plots of figures 7 and 8 show how closely pilots followed the ROTO guidance (i.e., the speed error bar on the HUD indicating aircraft ground speed minus command as a function of distance along the runway). Figure 7 shows the ground speed error for the ten runs used in the ROT calculations above (all but two used algorithm Alg2 - r54, r56) to exit B3 on runway 26R. The plots show the ground speed error during the time this guidance is displayed to the pilot (i.e., from nose gear touchdown to the start-of-the-turn). Note that the error is less than 10 knots.

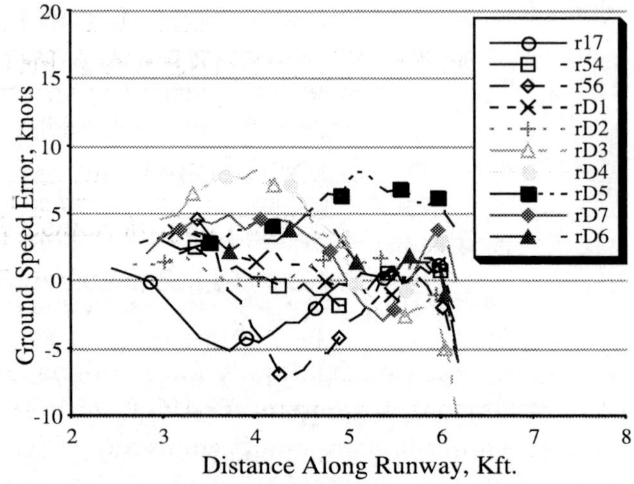

Figure 7. Speed Error vs Distance for Exit B3 on ATL 26R

Figure 8 shows the ground speed error for test runs made to exit B5 on runway 26R (1st high-speed exit). All but one run (r53) used guidance algorithm Alg2. Note that for all but two runs (r8, r53) the error remains below 10 knots. For both of those runs the nose gear letdown time was six seconds preventing the pilot from starting to control the error until the aircraft was 3000 feet down the runway. However, these larger errors did not result in any control problems. The pilot was still able to reduce the speed error to an acceptable level at the start-of-the-turn without exceeding deceleration limits specified in the ROTO guidance algorithms.

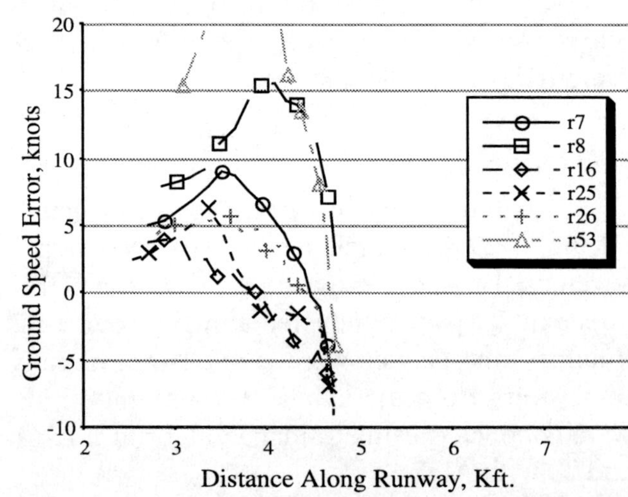

Figure 8. Speed Error vs Distance for Exit B5 on ATL 26R

Concluding Remarks

The ROTO system provided guidance that resulted in mean ROT values essentially the same as those for current clear-weather operations and standard deviation ROT values a factor of 1.5 to 2.0 less (depending on exit comparison). Assuming the pilots can follow the guidance in low visibility conditions, these ROT values imply that clear-weather runway capacity could be maintained in low visibility conditions. All the pilots liked the system and found the guidance easy to follow after a couple of training runs. In general, the pilots were able to control the aircraft ground speed along the runway to within 10 knots of commanded ground speed from the ROTO guidance.

References

1. Pines, S., Schmidt, S. F., and Mann, F. : "Simulation, Guidance, and Navigation of the B737 for Rollout and Turnoff Using MLS Measurements", NASA CR-144959, December 1975.

2. Pines, S., Schmidt, S. F., and Mann, F. : "Automated Landing, Rollout, and Turnoff Using MLS and Magnetic Cable Sensors", NASA CR-2907, October 1977.

3. Pines, S.; and Hueschen, R. M.: "Guidance and Navigation for Automatic Landing, Rollout, and Turnoff Using MLS and Magnetic Cable Sensors", Presented at the AIAA Guidance and Control Conference, Palo Alto, California. August 7-9, 1978.

4. Pines, Samuel: "Terminal Area Automatic Navigation, Guidance and Control Research Using the Microwave Landing System (MLS), Part I - Automatic Rollout, Turnoff, and Taxi", NASA CR-3451, August 1981.

5. Goldthorpe, Steve H; Kernik, Alan C.; McBee, Larry S.; and Preston, Orv W.: "Guidance and Control Requirements for High-Speed Rollout and Turnoff (ROTO)", NASA CR-195026, January 1995.

6. Goldthorpe, S. H.; Dangaran, R. D.; Dwyer, J. P.; McBee, L. S.; Norman, R. M.; Shannon, J. H.; and Summers L. G.: "Guidance and Control Design for High-Speed Rollout and Turnoff (ROTO)", NASA CR-201602, August 1996.

7. Young, S. D. and Jones, D. R.: "Flight Demonstration of Integrated Airport Surface Movement Technologies", NASA TM-1998-206283, January, 1998.

FIELD EVALUATION OF T-NASA: TAXI NAVIGATION AND SITUATION AWARENESS SYSTEM

Anthony D. Andre, San Jose State Foundation, San Jose, CA

Becky L. Hooey, Monterey Technologies, Inc., Los Gatos, CA

David C. Foyle, NASA Ames Research Center, Moffett Field, CA

Robert S. McCann, San Jose State Foundation, San Jose, CA

Abstract

This paper reports the results of a field evaluation of an advanced taxi navigation and situation awareness (T-NASA) system, aimed at improving the efficiency of aircraft ground taxi operations under low-visibility conditions. T-NASA consists of two main components: 1) a panel-mounted electronic taxi map display and 2) a heads-up scene-linked display (HUD). These components were installed in NASA's B-757 research aircraft and flight tested at Atlanta's Hartsfield International Airport. The results clearly demonstrated both the feasibility and effectiveness of the T-NASA system towards improving the efficiency of airport taxi operations. In addition, as a direct result of the evaluation, improvements were made to the design and procedures of the T-NASA system.

Introduction

Over the last five years, researchers at the NASA Ames Research Center have been involved in the development of an advanced taxi navigation and situation awareness system, called T-NASA. T-NASA was developed as part of the Low Visibility Landing and Surface Operations (LVLASO) element of NASA's Terminal Area Productivity (TAP) program, aimed at developing technology and procedures for improving the capacity, efficiency and safety of aircraft surface movements during the roll-out, turn-off and taxi in/out phases. More specifically, the goal of the program is to enable safe VMC surface movement capacities in IMC conditions down to 300' runway visual range (RVR). T-NASA attempts to accomplish this goal by providing the pilot (and ultimately the controller) with advanced navigation displays that provide information on ownship position and ground speed, traffic and other obstacles, and the path of the cleared taxi route.

Much has been previously published concerning the objectives of the LVLASO program, the development of T-NASA, and the results of various experiments and simulations aimed at determining the utility and usability of the T-NASA system.[1,2,3] However, while these studies provided valuable data towards the utility of T-NASA relative to the taxi displays currently available to pilots (i.e., paper taxi charts), they were conducted under conditions and constraints that did not allow for an accurate assessment of the impact of T-NASA

---

[1] Andre, A.D. (1995). Information requirements for low-visibility taxi operations: What pilots say. *Proceedings of the 8th International Symposium on Aviation Psychology*, pp. 484-488, Columbus, OH.

[2] Foyle, D., Andre, A., McCann, R., Wenzel, E., Begault, D., and Battiste, V., "Taxiway Navigation and Situation Awareness (T-NASA) System: Problem, Design Philosophy and Description of an Integrated Display Suite for Low Visibility Airport Surface Operations", SAE Transactions: Journal of Aerospace, v105, pp. 1411-1418, 1996.

[3] McCann, R.S., Foyle, D.C., Andre, A.D. and Battiste, V. (1996). Advanced navigation aids in the flight deck: Effects on ground taxi performance under low-visibility conditions. SAE Transactions: Journal of Aerospace, 105, 1419-1430.

0-7803-5086-3 /98/$10.00 ©1998 IEEE

relative to the operational goals of the LVLASO program.

Accordingly, the objectives of this flight test were: 1) to demonstrate that the T-NASA system, and the required supporting technology infrastructure, can be implemented at a major airport facility, 2) to validate the utility of the T-NASA system in the context of normal handling conditions at a major airport facility, and 3) to determine necessary changes in the design or use of the T-NASA system.

T-NASA Description

As evaluated here, T-NASA consists of two main components: 1) a panel-mounted electronic moving map display and 2) a HUD.

Taxi Map

The taxi map is a secondary display intended to provide pilots with navigation and situation awareness information in a heads-down moving map format. It is not designed to support inner-loop control of the aircraft. The taxi map used in the Atlanta Flight Tests, seen in Figure 1, had the following features:

View of airport layout. The taxi map provided two viewing modes of the airport layout: 1) a track-up, perspective mode that rotated around the fixed aircraft symbol, and 2) an overview mode, which presented a north-up fixed view of the entire airport. Using the pilot input device (PID), pilots could select either view. Also, within the track-up view, pilot's could select from five zoom levels (5x, 4x, 3x, 2x, 1x) which each provided increasingly greater levels of visual detail.

Ownship position. The position of the NASA B-757 was updated on the taxi map at a rate of ~25 Hz. It was marked with a white triangle centered 1/3 up from the bottom of the

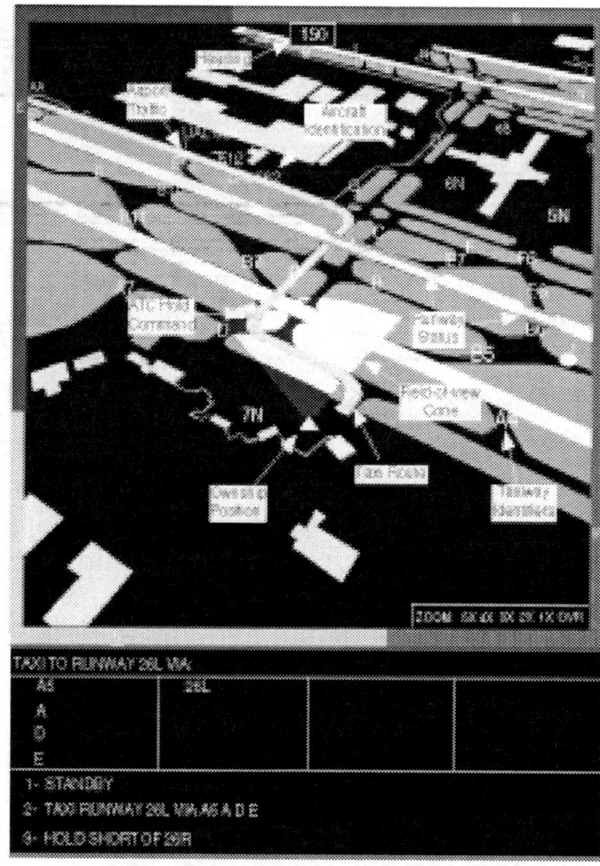

Figure 1. Taxi Map Display.

map. This icon was highlighted by a white arc or "wedge" in front of the triangle icon that highlighted the area of the map most important to the pilot; that is the area directly in front of the aircraft's path.

Airport traffic. Other aircraft were identified as white circular symbols and were labeled with their aircraft call numbers extending out from the aircraft symbol on a virtual flag pole. Pilots had the option to toggle aircraft labels on and off. Airport ground traffic was updated at ~1 Hz.

Hold bars. The taxi map displayed two types of hold bars. Hold instructions sent from ground control (GC) were designated using a red bar surrounded by a yellow border. The second type of hold bar was presented when another aircraft was taxiing on a runway. In this case, the map presented red hold bars across each taxiway that intersected the occupied runway.

Route guidance. The taxi map also presented route guidance information. The cleared path was painted magenta to be consistent with the existing cockpit symbology.

Clearances and messages. A text box underneath the map presented the GC instructions including the taxi route. Small pointers (<<, >>) on either side of the taxiway or runway name, indicated the aircraft's current position.

Heading bars. A moving frame rotated around the perimeter of the map representing the four compass directions. Each direction was represented as a uniquely colored bar.

Taxi HUD

The HUD displayed airport surface information and was intended to increase taxi speed and route navigation accuracy in low visibility conditions. The HUD was intended as a primary ground navigation display to be used in conjunction with the taxi map. The taxi guidance information was generated by the LVLASO software system. All information on the HUD was updated at ~25 Hz. The taxi HUD provided three dimensional scene-linked symbology to the left seat crew member during taxi. The HUD symbology is conformal in that it directly overlays elements that exist in the environment, but may be difficult to see under low-visibility conditions. The HUD symbology used in the Atlanta flight tests, presented in Figure 2, had the following features:

Taxi route. The cleared route was displayed in the form of a series of virtual cones located along both edges of the cleared taxiway, and a series of small squares arranged along the taxiway centerline. Virtual road signs indicated the location and degree of turns.

Location information. The HUD also displayed information regarding the current

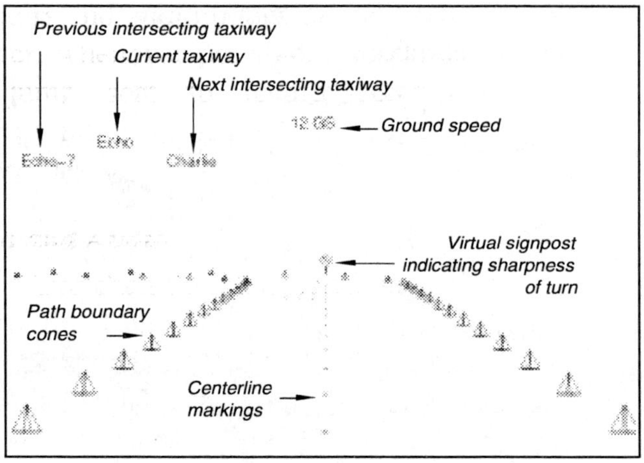

Figure 2. Taxi HUD Display.

taxiway/runway, the next taxi that intersects the current route, and the taxiway recently passed. This information was displayed in the upper left corner of the HUD (see Figure 2).

Ground speed. The HUD displayed ground speed in the form of a digital read out in the top-middle portion of the HUD.

Flight Test Design

The flight test involved two phases: The first phase served to validate the accurate operation of all systems involved and to refine the data collection and observation methods. The second phase represented the actual evaluation phase, in which both commercial airline pilots and NASA test pilots participated. We report here only the method and results of the evaluation phase (for a more detailed account, see Young and Jones, 1998).[4]

Pilot Selection And Training

Four B757 captains from four different commercial airlines were recruited to participate in the evaluation phase of the Atlanta flight test. All captains reported being

4 Young, S., and Jones, D. (1998, January). Flight Demonstration of Integrated Airport Surface Movement Technologies, NASA TM-1998-206283.

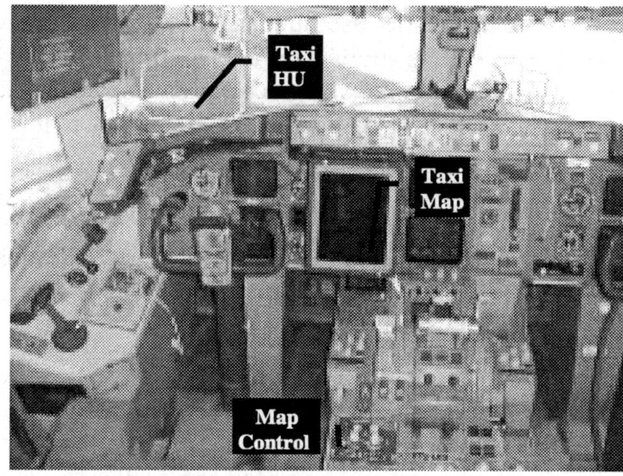

Figure 3. NASA B-757 Flight Deck

very experienced in taxiing an aircraft, but differed on their level of experience with taxiing in low-visibility conditions from not very experienced to very experienced. All pilots had previously flown into or out of the Atlanta Hartsfield Airport and were at least moderately familiar with the airport layout.

Prior to the Atlanta flight tests, the commercial pilots were provided with training materials for both the taxi map and taxi HUD components of T-NASA. Approximately one month prior to the Atlanta flight test, simulator training with the T-NASA system was conducted at NASA Langley Research Center.

Cockpit Configuration

The T-NASA cockpit configuration is shown in Figure 3. The LCD taxi map was mounted on the left side of the console between the primary flight displays and the EICAS displays. The map control device was situated such that it could only be reached by the captain or the co-pilot seated directly behind the center console. This control panel allowed the pilot to change the map field of view, show/hide traffic and show/hide traffic labels. The HUD was mounted in front of the captain (left seat).

The crew consisted of one commercial pilot and two NASA test pilots. The commercial pilot, the captain in the left seat, was responsible for the aircraft during landing and taxiing. One NASA pilot, sitting in the middle seat, acted as first officer and interacted with the captain, ATC, and GC to request clearances for taxi, take-off and landing. The second NASA pilot, in the right seat, remained isolated from experimental procedures. His task was to ensure the safety of the aircraft and crew.

Flight Test Trials

All 28 flight test trials were conducted during dusk and night hours (from 20:00 to 00:30 EDT). Each commercial pilot participated in seven trials which spread across two nights of testing. These trials were either taxi-only trials (including a high speed taxi) or taxi plus flight. The T-NASA display configurations that were tested were: 1) Jeppesen chart only, 2) taxi map only, and 3) taxi map + HUD. Table 1 depicts the number of trials that each pilot completed with each display configuration.

Table 1. Summary of Trials

Displays	# Trials
Jeppesen Chart (taxi only)	1
Map (taxi only)	1
Map + HUD (taxi only)	1
Map + HUD (taxi and flight)	4

Pilot Instructions

Before starting the taxi trials the pilots were briefed on the desired procedures, performance goals and design/implementation qualifications for the T-NASA system. Pilots were encouraged to use the taxi map on an occasional basis as a cross-check for resolving navigation or situation awareness issues. They were reminded that it was important that they not fixate too long on either the taxi map or the taxi HUD at the exclusion of the outside view.

Flight Test Procedure

Each flight test began in the ramp area just north of Runway 8L/26R, located at the Mercury Air Center FBO, which supported the test aircraft during testing. To begin each trial, the first officer called for taxi instructions from GC. On receipt of the instructions, the captain taxied the aircraft to the designated runway. The aircraft either completed a high speed taxi and exited the runway, or took off, circled, and landed. After roll-out and turn-off from the runway, the captain taxied the aircraft back to the ramp area.

The same route was used for the Jeppesen chart- and the taxi map-only trials to allow for a direct comparison between these conditions. This route involved a long taxi distance from the north side ramp to the south side of the airport and back to the north side ramp after a high speed taxi.

Data Collection Methods

Surveys

After each trial, the captain completed a brief 'post-trial survey' specific to the display configuration that was just tested. They were asked to reflect only on the trial that was just completed. Upon completion of the last flight test, both the captain and first officer were given a 'final survey' regarding the entire T-NASA system.

Debriefings

Each night, after the last trial was completed, an interview and debriefing session was conducted to allow the captain and first officer to express any observations, concerns, or suggestions about the T-NASA system.

Observations

T-NASA researchers observed and recorded crew behavior and communications related to the use of the T-NASA system. Researchers observed monitors located either in the hotel control operations room or from monitors and audio tracks in the aircraft cabin. Further, whenever possible, a researcher sat in the jump seat to record communications between pilots and observations of the out-the-window scene.

Video and Audio

For each trial, six cameras were used to record the following views: tail perspective, nose perspective, flight deck activity, scan-converted HUD, scan-converted taxi map, and a view from as close to the pilot's eye point as possible. Also captured on audio tape were communications within the cockpit, among the pilots, ATC, and GC.

Performance Data

A number of objective performance parameters were recorded for each flight test. Of particular interest to the T-NASA team were navigation errors, ground speed and the zoom level selections.

Results

Post-Trial Survey Results

Survey data illustrate how the commercial pilots felt that terminal area productivity may be improved by T-NASA technology. Specifically, pilots reported that, compared to the Jeppesen chart alone, the taxi map and taxi HUD reduced total taxi time, increased the safety of taxiing, were less distracting and reduced mental navigational workload while taxiing.

Reduced taxi time. Commercial pilots in both the taxi map and the taxi map + HUD conditions subjectively reported decreased taxi times over having no navigation aid at all.

Those pilots that reported a taxi time advantage were asked to select from a list of options that may have contributed to this positive result. Pilots responded that T-NASA technology reduced taxi times because less time was required to plan routes, taxi speeds were increased, less time was spent at confusing intersections, and, to a lesser extent, fewer stops were required while taxiing. Other reasons cited were that the taxi map + HUD combination allowed for greater situation awareness when changing tasks (i.e. clearing the runway and converting to taxi), and afforded the pilot greater confidence in his position on the airport surface.

Increased safety. Not surprisingly, all pilots commented that both the taxi map and the taxi map + HUD improved safety over having no navigation aid at all. However, only 75% of the pilots reported that the Jeppesen chart increased safety over having no map at all.

Increased situation awareness. All pilots reported that the Jeppesen chart, taxi map, and taxi map + HUD increased SA over having no navigation aid at all. In addition, pilots were asked to rate how easy it was to identify and locate their cleared taxi route on a scale from 1 (very easy) to 5 (very difficult). Pilots rated having more difficulty locating the cleared taxi route with the Jeppesen chart (M = 4.0) than with the taxi map (M = 1.25).

Reduced navigational workload. All pilots agreed that both the taxi map and the taxi map + HUD reduced mental navigational workload. Only 50% of the pilots reported reduced workload in the Jeppesen chart condition over having no navigation aid at all.

Final Survey Results

All four commercial pilots agreed unanimously on the following statements regarding the T-NASA system:

•T-NASA is beneficial relative to taxi information currently available.

•T-NASA aids the pilot in safely and accurately navigating the aircraft.

•T-NASA is very beneficial towards overall taxi efficiency (taxi speed, planning time, navigation awareness, communication needs).

•T-NASA will increase airport terminal area capacity in low visibility/IMC conditions.

•T-NASA increased situation awareness during taxi operations.

•T-NASA decreased cognitive navigation workload during taxi.

•T-NASA improved communications with Ground Control.

•T-NASA improved communications with flight crew.

•T-NASA allowed more time to look out the window, and required less time consulting the Jeppesen chart.

Taxi Performance

Taxi errors. An error was committed in one of the four trials using only the Jeppesen chart. The pilot taxied past the instructed taxiway turn and instead turned on the following taxiway. No errors were committed using the T-NASA system.

Taxi speed. Average taxi speeds were compared for each pilot's two south-side, taxi-only trials which were conducted with the Jeppesen chart alone and the taxi map. These two trials are directly comparable as they were closely matched for route length and number of turns. Unfortunately, it was not possible to include a matched route to allow a direct comparison with the taxi map + HUD trials. A moving velocity was calculated using velocity values greater than or equal to 6 Kts (values below 6 Kts were not recorded reliably). Three of the four pilots increased their taxi speeds with the taxi map over the Jeppesen chart. The mean taxi speed across all pilots was higher when taxiing with the taxi map (16.3 Kts) than when taxiing with the Jeppesen chart (14.6 Kts).

Map zoom usage. Prior to each trial, the pilots were asked to select the zoom level that they wanted to start taxiing with. This provided an estimate of the preferred taxi level, at least for the initial taxi out of the ramp phase. The 4x view was chosen most often (on 40% of the trials), followed by the 3x view (30%).

During the trials, all four pilots adopted a consistent strategy of zooming the taxi map in close when approaching turns and hold bars (5x, 4x, 3x), while zooming the map further out (1x, OVR) when planning the route, and while holding short. It was also determined that all five zoom levels were not necessary and that four zoom levels may be more desirable.

Usability of T-NASA Features

Taxi Map. Pilots were asked how easy it was to identify and quickly understand the symbols and information presented in the taxi map. They rated each feature on a scale from 1 (very difficult) to 5 (very easy). All map features were rated 3.0 (neutral) or higher. Compass Headings, Clearance Text, and Text Messages received the lowest ratings.

Taxi HUD. Pilots were asked how easy it was to identify and quickly understand the taxi HUD features. They rated each feature on a scale from 1 (very difficult) to 5 (very easy). All taxi HUD features were rated a 3.0 (neutral) or higher. While the edge cones, center line, and ground speed were easily understood, three problem areas were noted: The triad which depicts the previous, current, and next taxiway received ratings just slightly above neutral. Also the symbology that presents information about upcoming turns (distance to turn and degree of turn) received only neutral ratings. Finally, pilots noted problems with the HUD symbology during turns. In effect, the HUD showed no symbology during 90 degree turns since the forward view of the pilots is not aligned with the taxiway/runway.

Pilots' Comments

T-NASA system:

- "I can't wait until this system is installed in real-world aircraft."
- "I want to see this in the industry as soon as possible."
- "I'd like to go back to my company and say we ought to proactively endorse the development of this technology."
- "Once you get used to T-NASA it would be hard to live without it."
- "In Cat II and III it would really help. I can think of a couple of accidents it would have prevented."
- "This research represents a quantum step in air operations safety and technology, on the same level as the development of GPWS and TCAS."

Jeppesen chart:

- "I felt lost without the use of the LCD taxi map."
- "Mike-Juliet - Dixie - was a difficult corner to navigate. The Jeppesen chart was confusing."

Taxi map:

- "I'm really getting spoiled by this [taxi map] display."
- "I'm really developing an affinity for this display; it's value increases as the out-the-window visibility decreases."
- "Of everything I see here, what I like the most is the map with the magenta line showing my route."
- "I'm going to have a hard time taxiing an airplane without the map."
- "The map allows you to accelerate

with confidence and taxi at higher speeds".

Taxi HUD:

- "I was constantly processing the information on the HUD."
- "It [the HUD] improves situation awareness when changing tasks."

Design Implications

Beyond providing data to validate the utility of T-NASA, several important design issues were raised from the pilot's comments and researchers observations. The design suggestions listed below have already been studied and incorporated into a revised version of the T-NASA system.

Taxi Map

- Add ground speed indicator.
- Add aircraft type to traffic data tags.
- When hold bars are present on the taxi map, the portion of the route beyond the hold bar should be colored yellow.
- A taxi ATIS view should be added to the taxi map, providing static and dynamic information from the Jeppesen chart and ATIS (winds, closed runways, RVR, radio frequencies, etc.).

Taxi HUD

- Hold-short information should be provided on the taxi HUD.
- Eliminate the previous intersection ID on the taxi HUD. Place the current taxiway ID below the next intersection ID.
- Provide better turn guidance symbology in the taxi HUD.

Conclusions

The Atlanta field evaluation clearly demonstrated both the feasibility and effectiveness of the T-NASA system towards improving the efficiency of airport surface operations. The performance data, pilot surveys and observational data showed that the T-NASA system results in the following:

- reduced total taxi time
- reduced mental navigation workload
- increased situation awareness
- improved communications
- increased taxi efficiency
- increased safety

In addition, as a direct result of the field evaluation, improvements were made to the design and procedures for both the taxi map and the taxi HUD. These design and procedural changes have had a positive impact on the utility and usability of T-NASA.

We acknowledge that the ultimate goal of T-NASA—to increase taxi throughput under low-visibility conditions, was not demonstrated (due to safety precautions). Notwithstanding, based on the positive results obtained here in clear-weather night conditions, we are even more confident that T-NASA will safely enable VMC capacities in IMC conditions.

Acknowledgments

The authors wish to thank Denise Jones, Steve Young and David Green, in addition to the entire NASA LaRC flight test crew for their expertise and dedication to this effort. We are also grateful to the airline pilots and NASA test pilots who participated in this field evaluation. Questions about T-NASA should be directed to Dr. David Foyle, T-NASA Lead, at: dfoyle@mail.arc.nasa.gov

Airport Surface Operations
Data Link Communications and DGPS

Steve Koczo, Rockwell Collins, Cedar Rapids, IA[*]

Abstract

NASA, FAA and a number of industry partners conducted flight tests and demonstrations of airport capacity and safety enhancement technologies at the Hartsfield Atlanta International Airport (ATL) during August, 1997. As part of this team effort, Rockwell Collins provided avionics for the NASA Low-Visibility Landing and Surface Operations (LVLASO) system that was deployed on NASA's Boeing 757 Research Aircraft. Among the systems supplied and supported were differential (D)GPS, VHF data links for DGPS, traffic and runway status information uplink broadcast, Mode-S for ADS-B and Controller Pilot Data Link Communications (CPDLC), and a LCD head-down taxi display. This paper describes how these avionics elements were used in the LVLASO system and provides test results on data link and DGPS performance.

1.0 Introduction

The current air traffic environment in airport terminal areas experiences substantial delays when weather conditions deteriorate to Instrument Meteorological Conditions (IMC). Expected future increases in air traffic will put additional pressures on the National Airspace System (NAS) and will further compound the high costs associated with airport delays. To address this problem, NASA has embarked on a program to address Terminal Area Productivity (TAP). The goals of the TAP program are to provide increased efficiencies in air traffic during approach, landing, and surface operations in low-visibility conditions. The ultimate goal is to achieve efficiencies of terminal area flight operations commensurate with Visual Meteorological Conditions (VMC) at current or improved levels of safety.

During the last year, research activity culminated in the development, flight test and demonstration of a prototype Low-Visibility Landing and Surface Operations (LVLASO) system. A NASA led industry team and the FAA developed the LVLASO flight test system, integrating airport surface surveillance, aeronautical data link, DGPS navigation, automation systems, and controller and flight deck displays and interfaces. The LVLASO system supported both controllers and flight crews with guidance, control and situational awareness information to achieve improved efficiencies and safety of surface operations in IMC weather conditions. The LVLASO system was tested and demonstrated to a wide range of aviation industry members at the Hartsfield Atlanta International Airport during August, 1997 [1].

2.0 LVLASO Demonstration System

The system demonstrated at Atlanta consisted of several subsystems that were integrated to provide the intended capabilities of increased situational awareness and guidance information to air traffic controllers and pilots in conducting efficient and safe surface operations. Both ground sub-systems and airborne sub-systems provided services to meet the overall operational goals.

2.1 LVLASO Ground System Architecture

The LVLASO ground system, shown in Figure 1, consisted of the following sub-systems:

1. Surface surveillance sub-system
   a) Airport Surface Detection Equipment (ASDE-3) surface radar (skin-paint radar, counterpart to primary radar used in enroute surveillance).
   b) Airport Target Identification System (ATIDS)
      - Multilateration surveillance on transponder transmissions (signal received by multiple sites, position calculated from time-of-arrival of signals at each site)

0-7803-5086-3 /98/$10.00 ©1998 IEEE

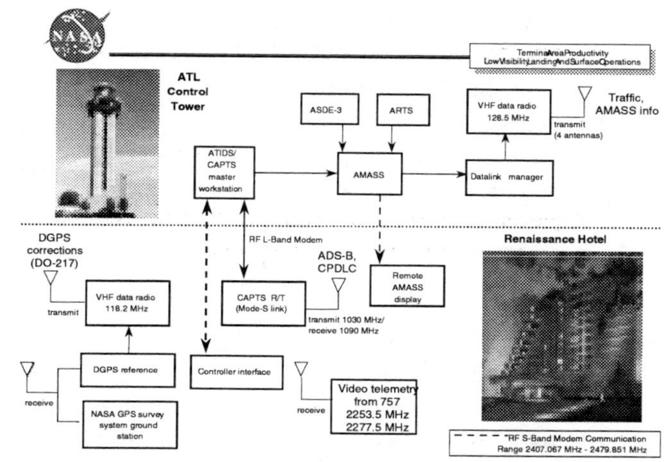

Figure 1 LVLASO Ground System

- Automatic Dependent Surveillance broadcast (ADS-B) (GPS position reports using Mode-S extended squitters)

c) Airport Movement Area Safety System (AMASS)

- Automation system that provides runway incursion warnings
- Data fusion of surveillance reports from ASDE-3, ATIDS and ARTS (Automated Terminal Radar System)

2. VHF traffic data link (broadcast uplink of traffic information and runway holdbars)
3. Differential GPS (DGPS) base station and VHF DGPS data link (broadcast uplink of DGPS correction information)
4. Controller Pilot Data Link Communications (CPDLC); data link messages between controllers and flight crews for routine communications in place of voice.

The physical location of the ground systems was as follows: 1) surveillance systems and VHF traffic data link were located at the Atlanta control tower; 2) the DGPS base station, VHF DGPS data link, the Controller Interface, and one of the five ATIDS receiver/transmitters (ground portion of Mode-S link) were located at the Renaissance Hotel located immediately to the North of the airport.

A room at the Renaissance Hotel was set up to serve as a pseudo ATC tower cab, with a test controller relaying the actual controller voice communications to the NASA 757 research aircraft (which served as the test vehicle) via CPDLC data link messages. A remote AMASS display was provided via modem link to provide the surface traffic display that controllers typically see.

2.2 Airborne Systems

The NASA 757 experimental system architecture is shown in Figure 2. LVLASO aircraft sub-systems were as follows:

1. Data Links
   a) VHF traffic broadcast data link receiver
   b) VHF DGPS broadcast data link receiver
   c) CPDLC via Mode-S link
   d) ADS-B downlink broadcast using Mode-S extended squitters
2. Airborne GPS receiver (provided DGPS position accuracies using DGPS correction inputs from the VHF data link)
3. Displays
   a) Airport moving map LCD head-down display (HDD)
   b) Roll-out, turn-off and taxi guidance head-up display (HUD)
4. Pilot input device (allowed pilot to select display modes and zoom levels; also served as data link acknowledgment to CPDLC messages).

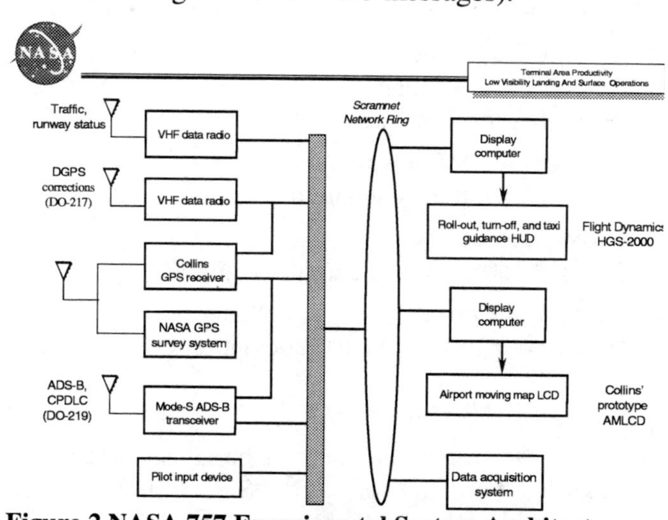

Figure 2 NASA 757 Experimental System Architecture

3.0 LVLASO Sub-System Description and Performance Results

This section describes each of the sub-systems supported by Rockwell Collins in more detail and, when available, provides performance results and observations based on data collected during preliminary sub-system tests and also during the actual LVLASO flight tests and industry demonstrations.

3.1 DGPS Sub-System

A Collins DGPS base station provided the DGPS correction information, which was broadcast to the test aircraft via VHF data link. Figure 3 shows the DGPS base station and VHF data link configuration. A Collins GNR-4000 GPS receiver onboard the NASA 757 provided DGPS position reports for use by the aircraft systems. NASA used a GPS survey system on the ground and onboard the aircraft using Ashtech Z-12 receivers to record position data for post-processing that resulted in nominal 5 cm accurate position data for experimental system accuracy evaluation.

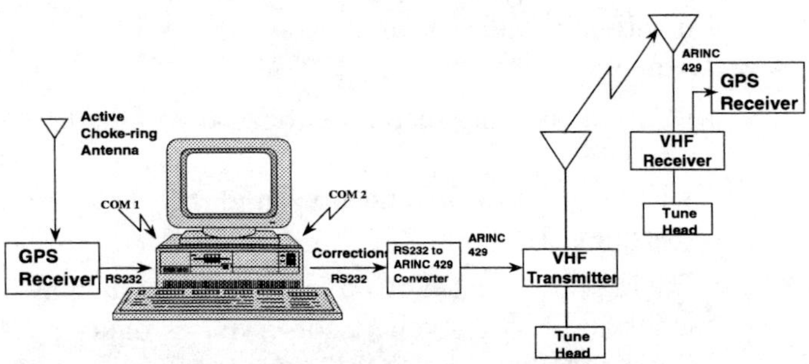

Figure 3 DGPS Base Station and VHF Data Link

DGPS performance results indicated the following performance for a 30 minute flight test run (other test runs provided similar results):

Horizontal RMS position error
(mean = 0.728 m, standard deviation = 0.485 m)

Cross track error
(mean = 0.056 m, standard deviation =0.494 m)

Along track error
(mean = -0.142 m, standard deviation =0.706 m)

3.2 VHF Data Links

Two identical pairs of broadcast VHF data links were utilized for 1) DGPS corrections uplink and 2) traffic and AMASS runway holdbar information uplink. The VHF data links were prototype radios using 31.5 kbps, D8PSK modulation similar to those planned for SCAT-I data link (RTCA DO-217, Appendix F) [2].

3.2.1 VHF DGPS Data Link Coverage Tests (Van Taxi Tests)

In preparation of formal LVLASO flight tests and demonstrations, VHF data link coverage tests were conducted on two separate occasions at ATL by taxiing a van in the ramp areas and on runways / taxiways. The purpose of these initial tests was to determine siting locations and expected airport surface coverage for the two VHF data links and their respective DGPS and traffic applications. Transmitters were located atop the Control Tower (using 128.5 MHz) and the Renaissance Hotel (118.2 MHz). Tests were conducted using horizontally and vertically polarized dipole antennas.

Coverage tests yielded excellent message reception rates of ~99.75% averaged over all tests with essentially no difference between horizontally and vertically polarized dipole antennas. Coverage was also excellent for both the Control Tower and Renaissance Hotel transmit sites. Figures 4 and 5 illustrate sample coverage tracks for ramp taxi and runway / taxiway taxi operations, respectively, for the Control Tower site. Coverage tracks were recorded using GPS position at a 1 per second rate. Lost or corrupted messages are highlighted.

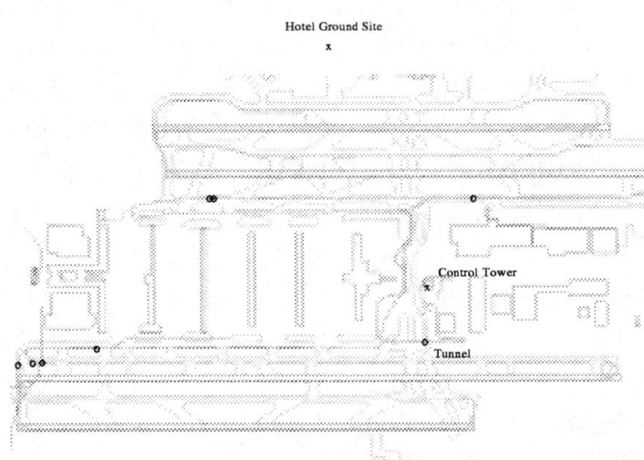

Figure 4 Coverage Test of Runways & Taxiways (Control Tower using Vertical Polarization)

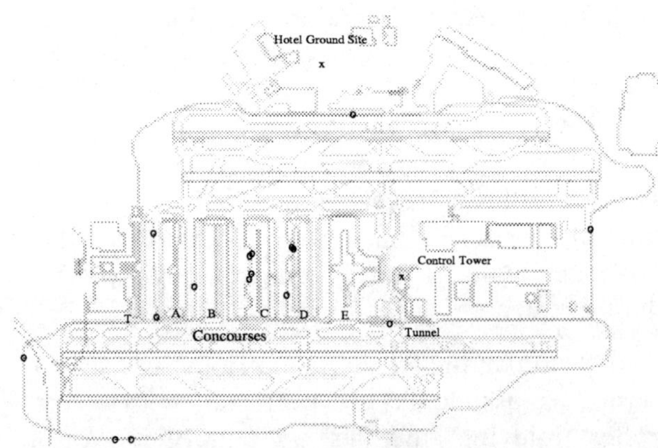

Figure 5 Coverage Test of Ramp Area (Control Tower using Vertical Polarization)

A few lost or corrupted messages were observed, primarily near the West walls of the concourses, where severe line-of-sight (LOS) blockages were intentionally sought out by taxiing as close as possible to the concourse buildings or very near large aircraft in the gate areas. Some messages were also lost along the loop road in areas where the road was ~50 ft below grade. Only a few messages were garbled in the movement area due to multipath.

Figures 6 and 7 show antenna placement at the Control Tower and the Renaissance Hotel, respectively. Since the control tower has four balconies (NW, NE, SE, SW corners), it was decided to place a dipole antenna at each balcony to avoid any potential blockage effects to LOS by the tower. Figure 6 shows one of the vertical dipole antennas pointing to the NW. The Renaissance Hotel is visible to the North of the airport (visible just to the left of the dipole antenna). Figure 7 shows a vertical dipole pointed to the South from the hotel. During the actual LVLASO test with the NASA 757, a horizontally polarized turnstile antenna was used and was located closer to the SE corner of the hotel roof.

The initial van taxi tests were conducted for the DGPS VHF data link application in September, 1996. A follow-on test in April 1997 confirmed the results of the earlier test but also provided additional information concerning the VHF traffic / AMASS holdbar data link. When transmitting DGPS, only ~ 1/4 to 1/3 of the maximum message length of the data link was used. For traffic and AMASS holdbar information during high traffic times, use of full-length messages was anticipated, thus a test was conducted to determine coverage for full-length messages. As expected, message reception rate for long traffic messages did decrease somewhat to 97.75 % in regions of severe LOS blockage. These tests confirmed that the link would work reliably for the LVLASO test. A more detailed description of the earlier van DGPS VHF data link coverage test is found in [3].

Based on the above van taxi coverage tests, it was decided to 1) transmit DGPS from the Renaissance Hotel site using at 118.2 MHz and using horizontal polarization, which concurs with ongoing ICAO and RTCA standardization activity for DGPS data link [2], i.e., DGPS data link in the aeronautical navigation band (108 to 118 MHz)

Figure 6 Control Tower Antenna Installation

Figure 7 Renaissance Hotel Antenna Installation

uses horizontally polarized signals-in-space, and 2) transmit VHF traffic / AMASS holdbars from the control tower at 128.5 MHz, since the LVLASO surveillance sub-system was also located there.

Based on the above frequency band selections, the NASA 757 VOR antenna was used for receiving VHF DGPS data link, while a top-mounted VHF communications blade antenna was used for the VHF traffic data link.

3.2.2 VHF Traffic and AMASS Holdbar Data Link

Traffic reports were uplinked once per second. An AMASS runway status message was also sent each second indicating whether or not runway / taxiway intersections were "hot", i.e., not to be crossed due to an active runway. These runway holdbars were displayed on the HDD in the NASA 757 flight deck for runway incursion situational awareness.

Traffic information and AMASS runway status were output by AMASS each one second surveillance scan. Traffic data was collected and formatted by the Data Link Manager (developed by PMEI Inc.) for transfer to the VHF data link transmitter. The Data Link Manager took as many targets as were available (up to 15) to build a single transmit message. For the worst case traffic loads of 40 to 50 aircraft observed during the Atlanta LVLASO tests, up to 3 maximum length messages were transmitted per second. This was well within the capabilities of the data link. A 31.5 kbps D8PSK VHF broadcast data link should be able to support ~240 aircraft using the message encoding used during the LVLASO flight test.

The VHF traffic data link performed reliably throughout the LVLASO flight tests and industry demonstration. Latency between actual position and the displayed position report of traffic on the HDD was ~ 2 seconds. This latency was primarily noticeable for high-speed aircraft during take-off or landing and did not appear to be a significant issue in providing reliable situational awareness.

3.2.3 VHF DGPS Data Link (NASA 757 Tests)

Data link performance was characterized for the VHF DGPS data link onboard the NASA 757 aircraft by recording DGPS position and data link message status outputs from the GNR-4000 GPS sensor. Signal strength outputs based on internal receiver Automatic Gain Control (AGC) information from the VHF data link receiver were also recorded. Three states of message status were recorded; 1) no message received, 2) message received but CRC failed, and 3) message received and CRC passed successfully indicating a correctly received message.

Figures 8 through 12 are representative VHF DGPS data link performance plots that depict the path traversed by the NASA 757 during flight-test runs. Message failure events are indicated by larger "squares" in the flight path (recall that messages were sent at a one per second rate).

From Figures 8 to 10 it is apparent that our goal of providing reliable signal coverage was met throughout all the flight tests for both VHF data links with one notable exception. The DGPS data link experienced some degraded performance in the NW corner of the terminal area beyond 5 NMI range when the NASA 757 was in flight. The cause for this degraded performance was directly attributable to the antenna placement of the DGPS ground station at the hotel transmit site.

While antenna placement at the hotel was ideal for surface operations, it was not optimal for terminal area operations as can be seen in Figure 7.

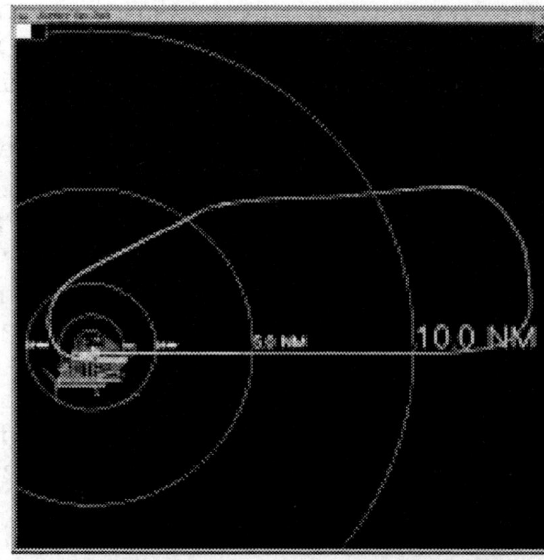

Figure 8 Flight Test Scenario # 1 - Takeoff on 26L, Downwind Leg, Landing on 26R

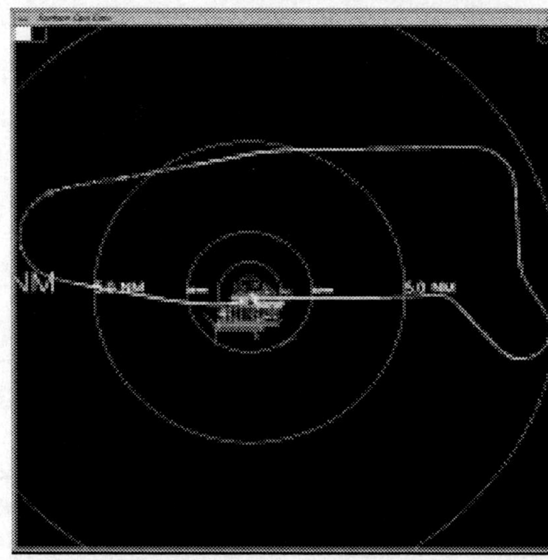

Figure 9 Flight Test Scenario # 2 - Takeoff on 26L, Downwind Leg, Landing on 26R

From Figure 7 it is apparent that the additional ~30 ft structure seen to the North (left) of the antenna on the hotel roof provided some signal blockage, primarily toward the NW direction.

Figures 8 and 9 illustrate two flight test runs where the NASA 757 performed a flight, with takeoff on runway 26L and a loop to the NW and subsequent downwind leg, and then turning South to intercept the Localizer for approach and landing on runway 26R. In Figure 8, the aircraft made an immediate turn toward the downwind leg, and thus the signal level remained strong throughout the West portion of the flight path. Even during the downwind leg that extended ~ 13 NMI beyond the location of the DGPS base station and VHF

Transmitter site (Renaissance Hotel), the signal remained quite strong and no message failures occurred. In Figure 9, the NASA 757 did a more gradual climb and turn and went ~8 NMI to the WNW before turning downwind. The signal level dropped severely and some messages were garbled or lost. Signal level improved substantially on the downwind leg. Some signal degradation occurred during the South leg and Localizer capture portion of the flight, but not as severe. The degradation in the NW corner of the flight path was the direct result of signal blockage due to antenna siting of the DGPS data link as indicated above.

Figure 10 illustrates the scenario when the NASA 757 conducted a takeoff on 8R and flew a downwind leg to return on 8L. The effect of antenna blockage is very evident to the WNW in the pattern, where numerous messages were either garbled or lost entirely beyond 5 NMI range. The test run illustrated in Figure 10 represents the worst case results that were observed throughout the tests. Once the antenna blockage is no longer a factor, signal strength is very good throughout the flight with no additional messages failures.

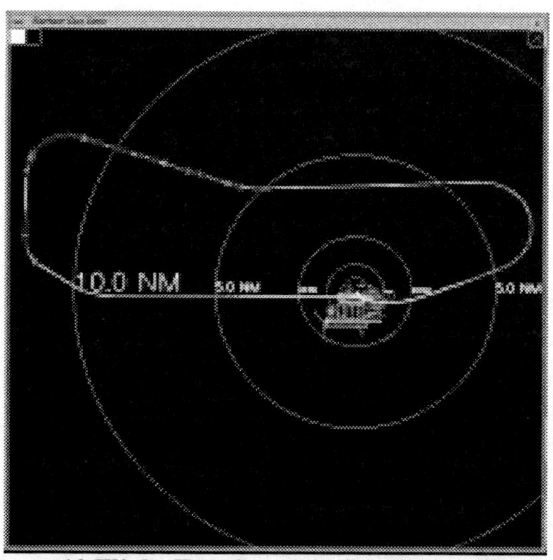

Figure 10 Flight Test Scenario # 3 - Takeoff on 8R, Downwind Leg, Landing on 8L

Figures 11 and 12 provide flight test data on scenarios that did not include flight. The aircraft was based at a ramp to the North of the airport. In Figure 11 the NASA 757 taxied from the ramp via taxiways Alpha, Dixie, and Echo for a simulated takeoff on 26L. The aircraft then performed a high-speed ROTO with exit on Echo 3 and return taxi to the ramp via Echo, Charlie and Alpha. As expected,

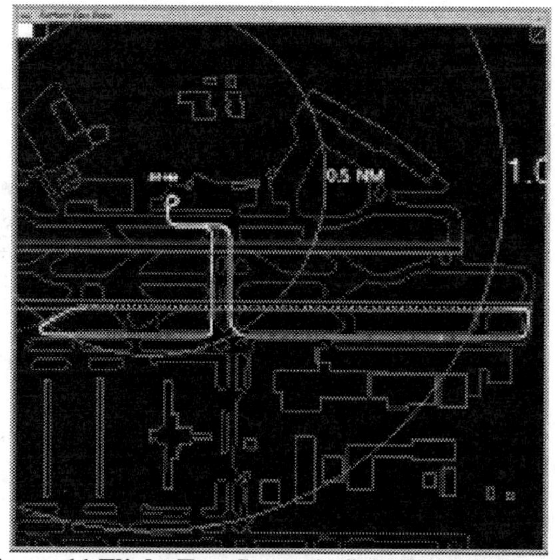

Figure 11 Flight Test Scenario # 4 - Taxi-Only Scenario with High-Speed ROTO

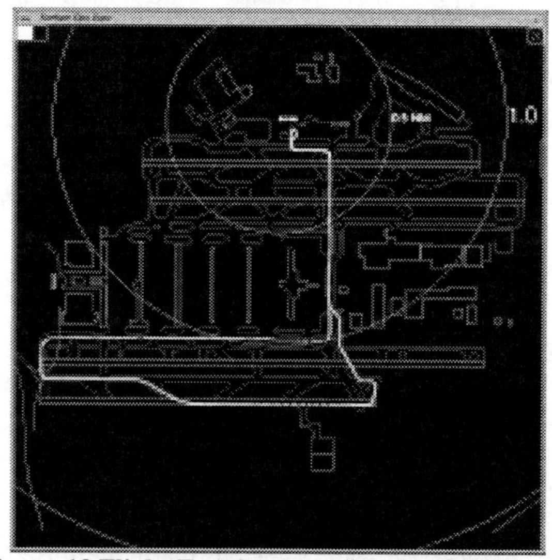

Figure 12 Flight Test Scenario # 5 - Taxi-Only Scenario with High-Speed ROTO

the signal level was very strong throughout the entire scenario. The few messages that were actually lost are attributed to multipath as a result of the large Delta hangars on the SE corner of the North runway area.

Figure 12 illustrates a taxi scenario to the South half of the airport. Again signal levels were strong except in a few areas along the taxiway directly South of the terminal buildings, where some signal blockages were observed. Only a couple messages were not received correctly in the flight test scenario.

Table 1 summarizes the message reception performance of the VHF DGPS data link for flight test runs with the NASA 757. Data collected for both taxi and airborne tests resulted in ~99.8 % messages reception probability (27 errors in 15792 messages for taxi, 76 errors in 37681 messages for airborne tests). Taxi message failures occurred only as single error events. For airborne flight tests one triple error event and 3 double error events occurred, with all other message failures being single events. Message failures in the NW corner were excluded since they were clearly due to LOS effects at the hotel site. Even when accounting for message failures in the NW corner (with suboptimal antenna siting), the message reception probability was 99.2%.

Probability of correctly receiving a single message per number of attempts	Taxi only tests (15792 messages)	Airborne tests (37681 messages)	Total all flight tests (53473 messages)
Single attempt	99.83%	99.80%	99.81%
Two attempts	100%	99.992	99.994%
Three attempts	100%	99.997%	99.998%

Table 1 Message Reception Probability vs Number of Transmission Attempts (Boeing 757 tests)

The few message losses that did occur were in the vicinity of the following regions: 1) NASA 757 ramp area (A5) / Taxiway A intersection (multipath), 2) in front of the Delta Hangar to the Southeast of runway 8R/26L (multipath), 3) on the South half of the airport when the line-of-sight was blocked by terminal buildings (particularly Terminal E), 4) for a brief instant at the time the NASA 757 rotated when executing some of the takeoffs on runway 8R (multipath and/or aircraft antenna null), 5) a few occasions in the ENE corner during flight, and 6) the NW corner due to line-of-sight blockage effects by the hotel.

3.3 Mode-S Link

The Mode-S link was utilized for Controller Pilot Data Link Communications (CPDLC) and also provided ADS-B extended squitters for surveillance. The Mode-S link consisted of ATIDS R/Ts (uplink on 1030 MHz) on the ground and the Mode-S transponder and associated Airborne Data Link Processor (ADLP) onboard the NASA 757 aircraft. The Mode-S transponder / ADLP provided the Mode-S Specific Services (MSSS) protocols needed for addressed CPDLC communications and broadcast ADS-B.

3.3.1 CPDLC Data Link

CPDLC in the LVLASO test was accomplished by using a test controller who repeated actual ATC voice instructions to the NASA 757 via the Controller Interface as data link messages. The data link messages were encoded using RTCA DO-219 message encoding using existing message definitions whenever possible, although additional messages were developed to include surface operations messages. The DO-219 encoded uplink messages were sent by the Controller Interface to ATIDS (see Figure 1) via modem for further encoding to MSSS protocol and transmission via 1030 MHz. The Controller Interface was developed by St. Cloud State University and ATIDS was developed by Cardion, Inc. Uplink messages were received by the NASA 757 Mode-S transponder / ADLP and forwarded to the I/O network and flight computer for processing and subsequent display on the HDD to the flight crew. Messages were acknowledged by the flight crew using the pilot interface device (PID), which encoded a downlink message via the Mode-S transponder / ADLP using DO-219 and MSSS protocols (1090 MHz downlink).

Note: Message retry protocols were implemented in the event a data link message collided with another transmission on the Mode-S link, which is entirely possible due to the random access protocol used by the Mode-S link. In addition, controller messages were highlighted to the controller when acknowledgments occurred. Failure to receive an acknowledgment was thus immediately evident to the ATC controller.

Table 2 summarizes the CPDLC messages used in the LVLASO flight test.

Two methods of repeating and converting ATC voice messages to data link messages were demonstrated by the test controller using either aural repetition of actual ATC messages and using a Verbex voice recognition unit to digitize the message, or a touch screen input to enter the data link message. In both cases, the digital message was encoded using DO-219. The Controller Interface is further described in [4].

3.3.2 ADS-B

The NASA 757 aircraft reliably transmitted ADS-B extended squitters at a 0.5 second update rate. ADS-B squitters provided aircraft position with DGPS accuracies. ATIDS surveillance of

ADS-B reports was superior to multilateration surveillance, since reception of the signal by just one ATIDS R/T allows tracking of the aircraft (versus reception by multiple R/Ts needed for multilateration).

LVLASO Uplink Messages	
Element ID	**Message**
117	Contact [icaoname][frequency]
120	Monitor [icaoname][frequency]
200	Hold Short [position]
212	Taxi [runway] Via [route]
219	Taxi [ramp] Via [route]
220	Cross [position] [without delay]
221	Continue Taxi
223	Taxi Into Position and Hold
224	Cleared For Takeoff
LVLASO Downlink Messages	
1	Roger
3	Unable
202	Taxiway Deviation
203	Turned-off on Taxiway [#]
204	Taxiway Deviation Resolved

Table 2 LVLASO CPDLC Messages

Figures 13 and 14 illustrate sample plots for ADS-B surveillance of the NASA 757 aircraft that were recorded. Figure 13 represents ADS-B surveillance for a taxi test to the South half of the airport (similar to the Figure 12 scenario). From Figure 13, obvious LOS outages were observed whenever the NASA 757 taxied near a terminal building. The outages were simply explained by the fact that the 5 ATIDS R/Ts were deployed only on the North half of the airport and thus did not provide LOS to the regions blocked in the South half. The problem would be easily corrected by deployment of ATIDS R/Ts to include the South half of the airport. With the exception of the blockage regions, ADS-B surface surveillance on the NASA 757 was reliable.

Figure 14 shows a flight scenario and again demonstrates good ADS-B surveillance with the exception of some outages in the NE corner of the flight pattern.

4.0 Summary

This paper describes results of flight test experiments of a prototype airport surface operations system. Focus of this paper is on the communications, navigation and surveillance (CNS) subsystems including VHF data link, DGPS and Mode-S ADS-B, which were demonstrated with promising performance results. These sub-systems are expected to play an important role in the overall National Airspace System in the future and should allow for significant improvements in surface operations. A surface operations system, such as LVLASO, provides improved situational awareness to both controllers and flight crews with attendant benefits of increased safety and improved operational performance, particularly in low-visibility conditions.

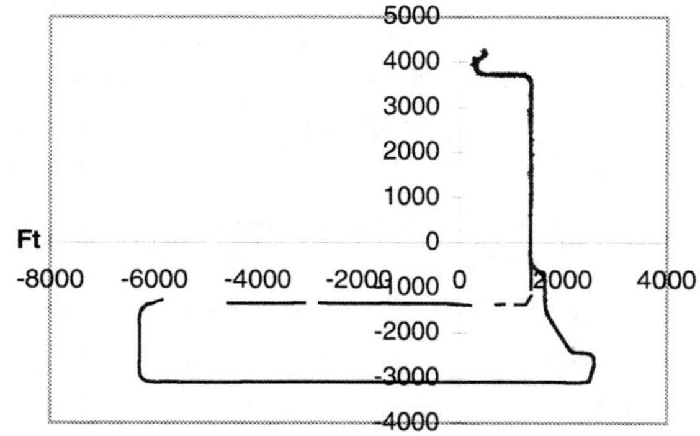

Figure 13 NASA 757 ADS-B Surveillance Coverage - Taxi Scenario

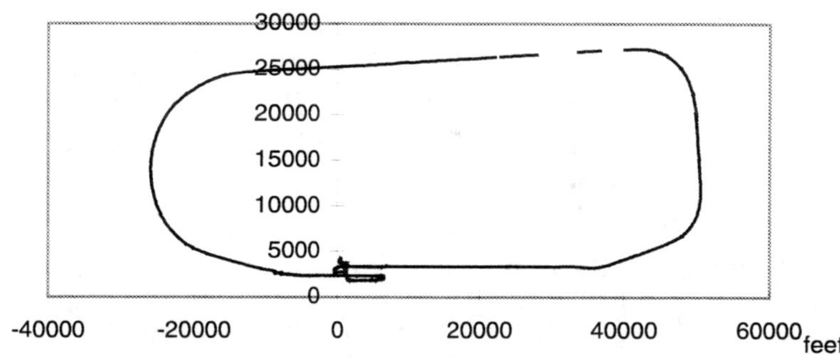

Figure 14 NASA 757 ADS-B Surveillance Coverage-Flight Scenario

References

[1] Young, S.D., and Jones, D.R.: "Flight Demonstration of Integrated Airport Surface Movement Technologies", NASA TM-1998-206283, January, 1998.

[2] RTCA DO-217, "Minimum Aviation System Performance Standards, DGNSS Instrument Approach System: Special Category I (SCAT-I)", August 27, 1993.

[3] S. Koczo, "Airport Surface Communications Performance of a VHF DGPS Data Link Using D8PSK Modulation", GNSSP WG-D/4, February, 1997.

[4] Rankin, J., and Mattson, P.,"Controller Interface for Controller-Pilot Data Link Communications", Proceedings of the 16th Digital Avionics Systems Conference, October, 1997.

CONTROLLER-PILOT DATA LINK STATISTICS FROM NASA'S 1997 ATLANTA FLIGHT TEST[1]

James M. Rankin, Avionics Engineering Center, Ohio University, Athens, OH [2]

Pat Mattson, St. Cloud State University, St. Cloud, MN

Abstract

Controller-Pilot communications at NASA's Low Visibility Landing and Surface Operations (LVLASO) flight test (Atlanta, GA 1997) used a Mode-S datalink to reinforce normal VHF radio communications. The Controller-Pilot Datalink Communications (CPDLC) channel followed a modified version of the RTCA DO-219 standard to uplink taxi routes and hold clearances to NASA's 757 research aircraft. A Controller Interface (CI) workstation encoded the air traffic controllers' instructions into the DO-219 format. The CI also used electronic flight strips and a graphical map to increase the controllers' situational awareness.

This paper investigates the statistical success of the CPDLC channel. "Lost" messages that did not appear on the display are analyzed for the fault source. Round trip time between the CI and the 757 displays are presented. Finally, the voice recognition accuracy statistics are examined.

Introduction

In August 1997, NASA sponsored a flight demonstration at Hartsfield-Atlanta International Airport. The flight demonstration highlighted research activities in NASA's Low Visibility Landing and Surface Operations (LVLASO) project. LVLASO, part of the Terminal Area Productivity (TAP) program, is focused on improving throughput and safety for ground operations during poor visibility.

To do this, the flight crew must have accurate knowledge of where the airplane is located on the airport, where other aircraft are located, and the assigned taxi route. This paper focuses on the third item, informing the pilot of the assigned taxi route. The taxi route is generated by the Controller Interface.

The Controller Interface (CI) provides a tool for Air Traffic Controllers to communicate with aircraft using Controller-Pilot Data Link Communications (CPDLC). The CI encodes common phraseology used by Air Traffic Controllers and flight crews into binary patterns to reduce message size and increase bandwidth. The coding schemes are defined in RTCA's DO-219 standard for ATC Two-Way Data-Link Communications. [1]

This paper reports the performance of the CI and CPDLC during the Atlanta Flight Demonstration.

---

[1] This research project was funded by NASA-Langley Research Center, Hampton, VA.

[2] Dr. Rankin was previously with St. Cloud State University, MN.

0-7803-5086-3 /98/$10.00 ©1998 IEEE

Test Configuration

In the LVLASO tests, the Controller Interface was the input and monitoring device for a Test Controller. The Test Controller listened to the voice instructions given by the Ground or Tower controller and repeated them into the voice recognition system.

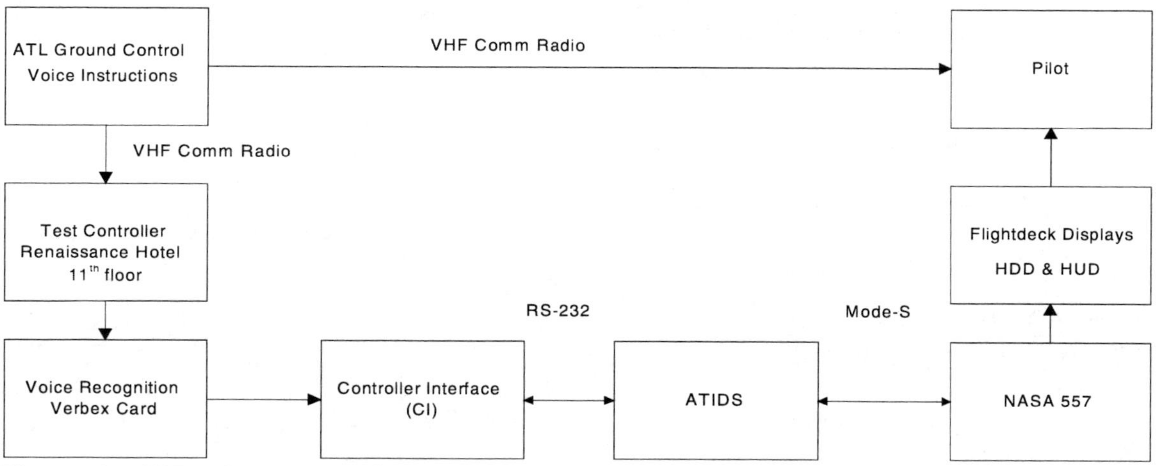

Figure 1. CPDLC Test Configuration

The NASA 757 used the CPDLC uplink to provide the flight crew with a graphical display of the taxi route and any Hold Short instructions. The 757 had a Heads-Down Display (HDD) that overlaid the commanded taxi route over the airport map. The crew could monitor their progress as the map was continually updated to show their current position. Other airport traffic was also shown on the display to enhance safety.

A Heads-Up Display (HUD) was installed in front of the pilot's windscreen. The HUD displayed the taxiroute by drawing taxiway symbology only for the commanded route. For example, as an aircraft arrives at an intersection with three possible exits, the HUD would only highlight the taxiway designated by ATC.

As shown in Figure 1, ATC instructions were relayed to the flight crew through the normal VHF radio. To enforce the voice traffic, the ATC instructions were relayed to the crew through the CPDLC channel. CPDLC messages were relayed between the CI and the NASA 757 via the Airport Traffic Identification System (ATIDS).

ATIDS provided position and ID of aircraft equipped with MODE-A/C/S transponders. ATIDS determined aircraft position based on multilateration; i.e., the time difference for a transponder transmission to be received at remote Receiver/Transmitters (R/Ts). In Atlanta, ATIDS had five Mode-S R/Ts arranged around the North side of Atlanta-Hartsfield airport. ATIDS also provided a two-way datalink to aircraft equipped with Mode-S transponders.

The CI was located in the "Test Control Room" on the 11th floor of the Renaissance Hotel that overlooks the Atlanta Hartsfield airport. The ATIDS system was in the ATC tower.

Controller-Pilot Data Link Communications (CPDLC)

Most of the DO-219 messages are for enroute or terminal area operations. For the LVLASO experiment, new CPDLC messages were created to handle surface movement. Table 1 lists the Uplink message set used in the experiment. Table 2 lists the Downlink message set. To identify messages that were defined for the LVLASO demonstration, numbers greater or equal to 200 were used. Numbers below 200 signify messages that are defined in DO-219.

Table 1. CPDLC Uplink Messages

No.	Message Description
1	STANDBY
2	REQUEST DENIED
3	ROGER
117	CONTACT [name][frequency]
120	MONITOR [name][frequency]
153	ALTIMETER [altimeter]
200	HOLD SHORT OF [position]
212	TAXI RUNWAY [runway] VIA [route]
219	TAXI RAMP [ramp] VIA [taxiroute]
220	CROSS [position] [without delay]
221	CONTINUE TAXI
222	UNAVAILABLE TAXIWAYS [taxiways]
223	TAXI INTO POSITION AND HOLD
224	CLEARED FOR TAKEOFF

The TAXI RUNWAY and TAXI RAMP messages added the most complexity of all the CPDLC messages. All taxi routes were created on the fly according to the actual route that was assigned by the Air Traffic Controller. There were no canned routes used in the Atlanta demonstration. The use of dynamic routes required the voice recognition system and the DO-219 encoding to be very flexible. It also required the display software on the 757 to be very intelligent. The display software had to decide if the assigned route was logical. For example, taxiways Alpha and Bravo are parallel to each and do not intersect. If an assigned route listed Alpha and then Bravo, the display computer would generate an error.

The CI also received CPDLC downlink messages from the 757. After leaving the runway, a CPDLC message was automatically generated to alert the controller of the aircraft's position. Automatic "TAXIWAY DEVIATION" and "TAXIWAY DEVIATION RESOLVED" messages were downlink anytime the aircraft deviated/returned from the approved taxi route.

More details of the DO-219 encoding scheme plus example encoding are provided in [2].

Table 2. CPDLC Downlink Messages

No.	Message Description
1	UNABLE
3	ROGER
201	REQUEST TAXI CLEARANCE
202	TAXI DEVIATION
203	TURNED-OFF ON TAXIWAY [taxiways]
204	TAXI DEVIATION RESOLVED

CI System Description

The Controller Interface was hosted on a P5-166 computer. A Verbex VAT31 voice recognition system was used to recognize the Controller's instructions. A 17" capacitive touch screen provided backup data entry. A TrueTime Model 560-5500 GPS-PC card provided accurate time information.

The CI displayed information to the controller in two main windows (see Figure 2). The Flight Strip Window displayed pertinent information about the aircraft and its flight plan. Flight strip information was provided from the ARTS database. The ARTS information included: flight identification number (e.g. NW742), aircraft type, Mode-C transponder code, and proposed departure or arrival time.

The right side of the flight strip displayed messages that had been sent to or received from the aircraft. The message text color indicated the status of a message.

An Airport Map window displayed the current position of all aircraft being tracked by ATIDS. The map in the window had a South up orientation. This orientation was necessary because the test controller was located in the

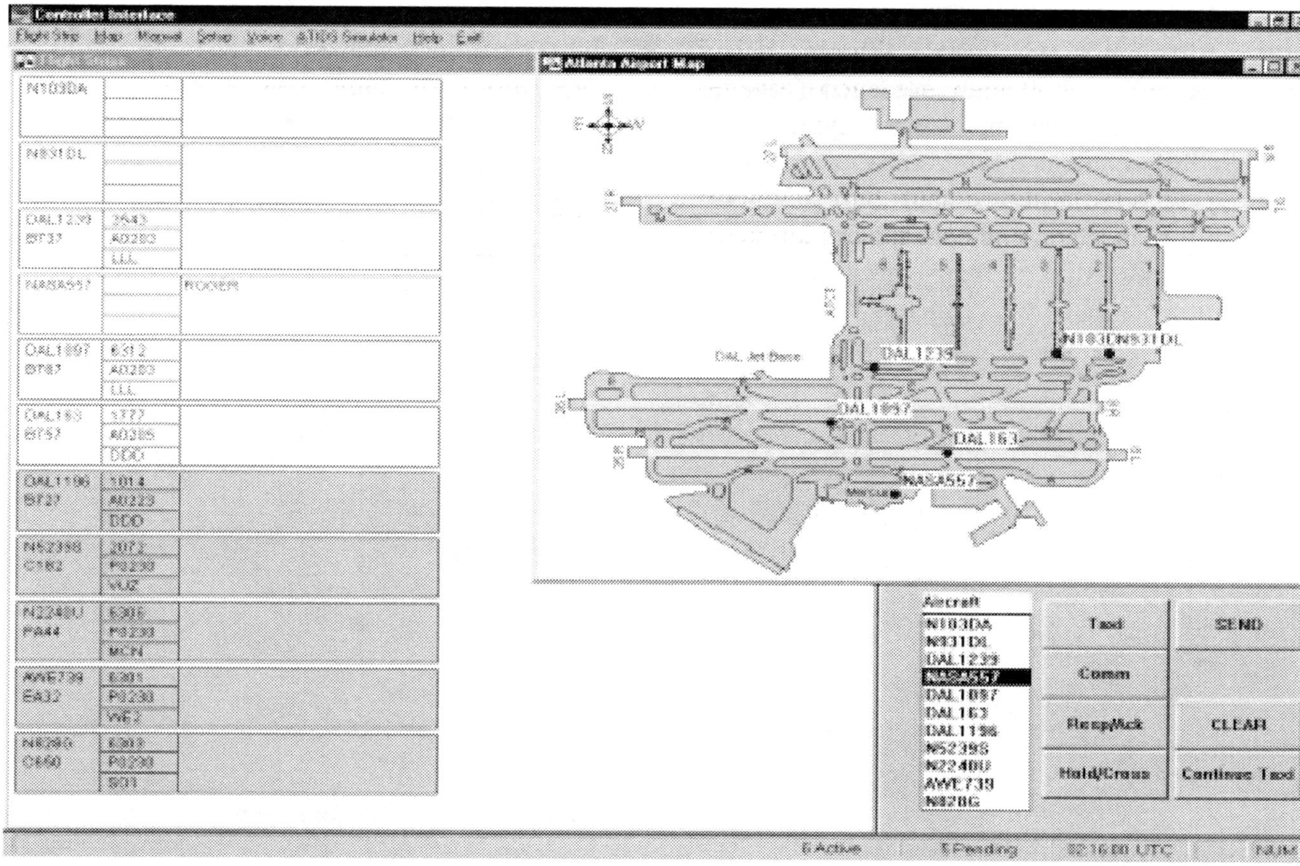

Figure 2. Controller Interface display Window

Renaissance Hotel that overlooked the Atlanta Hartsfield to the South.

CPDLC Datalink

CPDLC messages between the CI and the 757 were routed through several connections. An uplink message was sent from the CI to the ATIDS Master Station via a RF-modem. From the ATIDS Master Station, the uplink was sent via RF-modem to one of the 5 Mode-S R/Ts located on the North side of Hartsfield airport. A Mode-S datalink connected the ATIDS R/T to the Mode-S transceiver on the 757. Once a message arrived at the 757, it was transferred to the display computer. CPDLC downlinks passed through the same systems on the route back to the CI.

There were several different acknowledgement protocols defined for the CPDLC uplink messages. Since each uplink message was assigned a sequential number that was referenced by the acknowledgement message, it was possible to track the progress of a message.

CPDLC Uplink Statistics

Figure 3 shows the number of CPDLC uplink messages transmitted during 37 runs performed from August 18 through August 28, 1997. There were 470 uplinks transmitted from the CI during this period. Of the 470 messages sent, 433 (92%) were acknowledged by the ATIDS Master Station and 432 (92%) were acknowledged by the 757 display computer.

Caution must be used when interpreting these results. It is tempting to assume that only 1 message was lost between the ATIDS system and the 757. Actually, there were messages that were received and acknowledged by the 757 that were not acknowledged by ATIDS. In addition, a few messages appeared on the aircraft's map displays even though an acknowledgement was not received from the 757. (A cockpit camera and video telemetry allowed the test controller to monitor CPDLC messages, taxi routes, and hold short bars on the pilot's HDD.) Due to the experimental status of some systems, extensive protocol checking was not performed. The experimental systems processed messages even when handshaking checks were not satisfied.

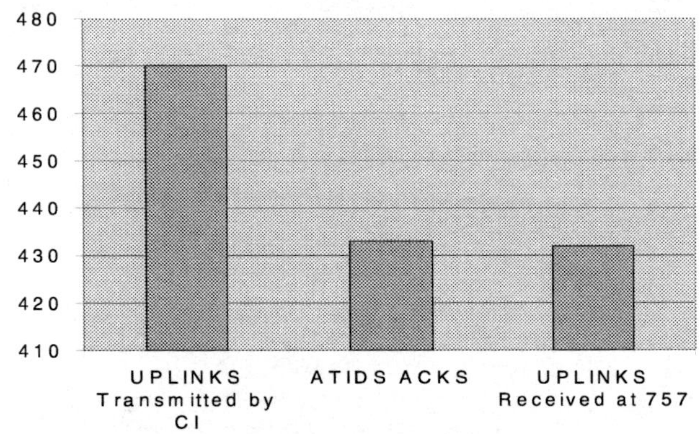

Figure 3. CPDLC Uplink Statistics

The two main reasons that 37 (8%) messages were lost between the CI and ATIDS were: 1) a retransmit capability was not programmed into the CI and 2) the ATIDS Status message was not monitored before the CI transmitted. If a message was not acknowledged by the ATIDS, it was left to the test controller to repeat/retransmit the message.

The ATIDS Status message relayed the ATIDS health to the CI. Messages transmitted when ATIDS was off-line were recorded as lost messages when they should not have been transmitted.

It should be noted that several messages were lost between the ATIDS and the 757 when the South side of the Atlanta airport was used. The five ATIDS R/Ts were concentrated on the North side. Holes in datalink coverage on the South side were caused by blockage of the R/Ts by buildings. Five of the 37 runs in this report were to the South side resulting in 7 lost messages.

CPDLC Downlink Statistics

There were 534 CPDLC downlink messages received by the CI as shown in Figure 4. This number differs from the number of uplinks due to the automatically generated messages such as Taxiway Deviation, Taxiway Deviation Resolved, and Turned-Off on Taxiway.

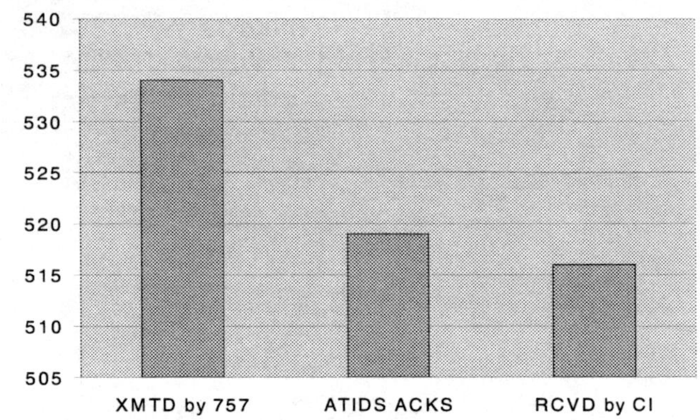

Figure 4. CPDLC Downlink Statistics

There were 519 (97%) CPDLC downlinks received successfully by the ATIDS. The CI received 516 (97%) of the CPDLC downlinks. A missed downlink was determined by examining the downlink message counter that was embedded in each message. A skipped message number meant a lost downlink message.

CPDLC Roundtrip Time

The latency time calculated from when an instruction was verbally issued by the Controller until it appeared on the map is an important statistic. At the LVLASO demonstration, this statistic was difficult to evaluate. The test controller(s) in the test control room listened to ATC's actual instruction via VHF radio. After determining the content of the instruction, the command was repeated by the test controller for the voice recognition system. This technique often added several seconds delay. Accuracy was stressed over speed to prevent the display of misleading information to the flight crew.

One possible measure of the datalink latency time is to determine the roundtrip time from the transmission of a CPDLC uplink to the receipt of the corresponding downlink acknowledgement. This statistic is possible because the NASA display computer on the 757 automatically generated downlink acknowledgments for each uplink message.

Figure 5 provides a histogram of the roundtrip time for a CPDLC message. The statistics are calculated on 128 messages over 10 runs. The average round trip time was 1.158 seconds. 95% of the messages made the round trip in less than 2.5 seconds.

The longest times are associated with taxi route messages. Once a taxi route message was received at the aircraft's display computer; the software was required to evaluate the taxi route. If the taxi route did not consist of contiguous segments, did not start from the aircraft's present position, or did not end at the taxi destination, then the software replied with an UNABLE. Because all the taxi routes in the Atlanta test were evaluated in real-time (not "canned"), the more complex routes needed more processing time.

VOICE RECOGNITION

The Verbex voice recognition system used on the Controller Interface was a speaker dependent system. A grammar file was created based on the phrases listed in Tables 1 and 2. Each test controller then trained the voice recognition system for those phrases.

A typical training session took approximately 20 minutes.

The grammar file listed all taxiway, runway, ramp, phonetic alphabet, ATC frequencies, and altimeter setting possibilities. A complete set of "canned" phrases for each taxi route was not defined. All taxi routes and hold short instructions were determined dynamically. The controller could give taxi instructions with up to 6 segments and the grammar file would recognize the route.

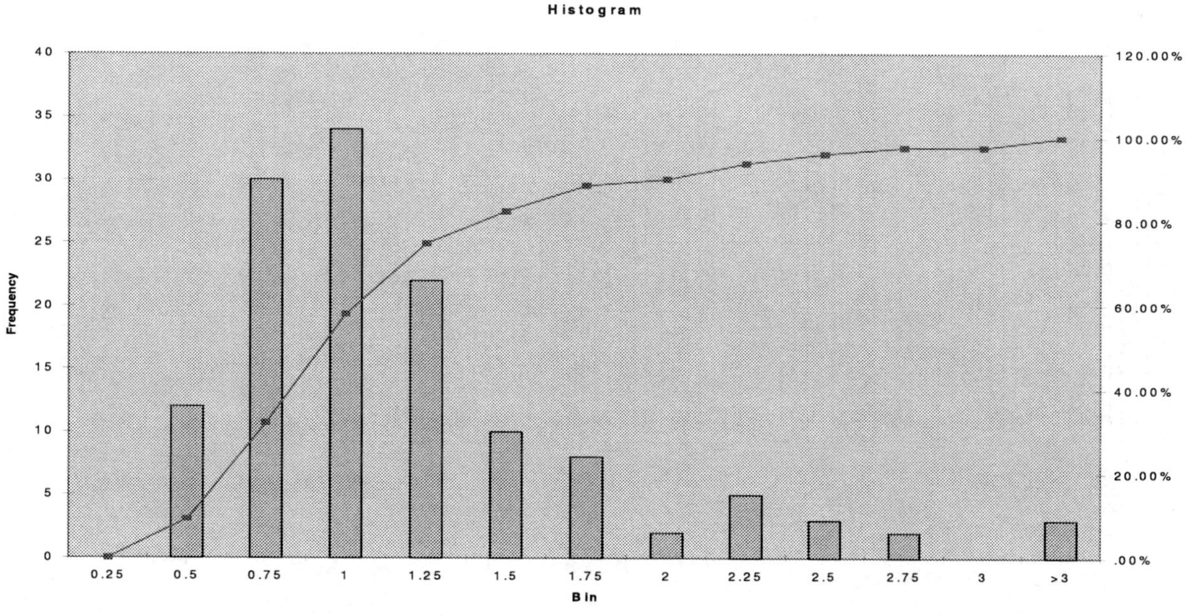

Figure 5. Histogram of round trip times for CPDLC messages

Initial results during the first two weeks of testing showed that 14% (27 of 194) of the voice recognition tries were not successful. Most of these errors (20) were caused because the system could not match the voice input to one of the trained phrases. These results required the filtering parameter in the grammar file to be adjusted.

If the filtering parameter is set too tight, then the words have to be enunciated exactly as they were trained. If an utterance does not match a trained word, then no recognition occurs. On the other hand, if the filter is too loose, then the voice recognition system will match any sound to one of the words that was trained. This reduces the number of Not Recognized errors, but the number of Recognition Errors grows. A proper filter coefficient was found that optimized the voice recognition performance.

The proper filtering was a concern during the last week of the demonstration when media, industry, and government representatives where invited to view the LVLASO system. During the flights, 15-25 people were in the Renaissance hotel room that was used as the test control room. At times, the ambient noise in the room was quite loud. The voice recognition system continued to perform well even in these conditions.

In addition to the filter parameter, the grammar file was tailored to the Atlanta Hartsfield airport. Taxiways and taxiway segments that did not exist at the airport were removed. For example, the taxiway descriptor Charlie was occasionally misunderstood as taxiway segment Charlie Eight that did not exist. By removing Charlie Eight from the possible grammar phrases, the accuracy of the voice recognition system was improved.

Figure 6 presents the voice recognition statistics for all runs from August 18 through August 28. The last filter parameter and grammar file changes were made just before

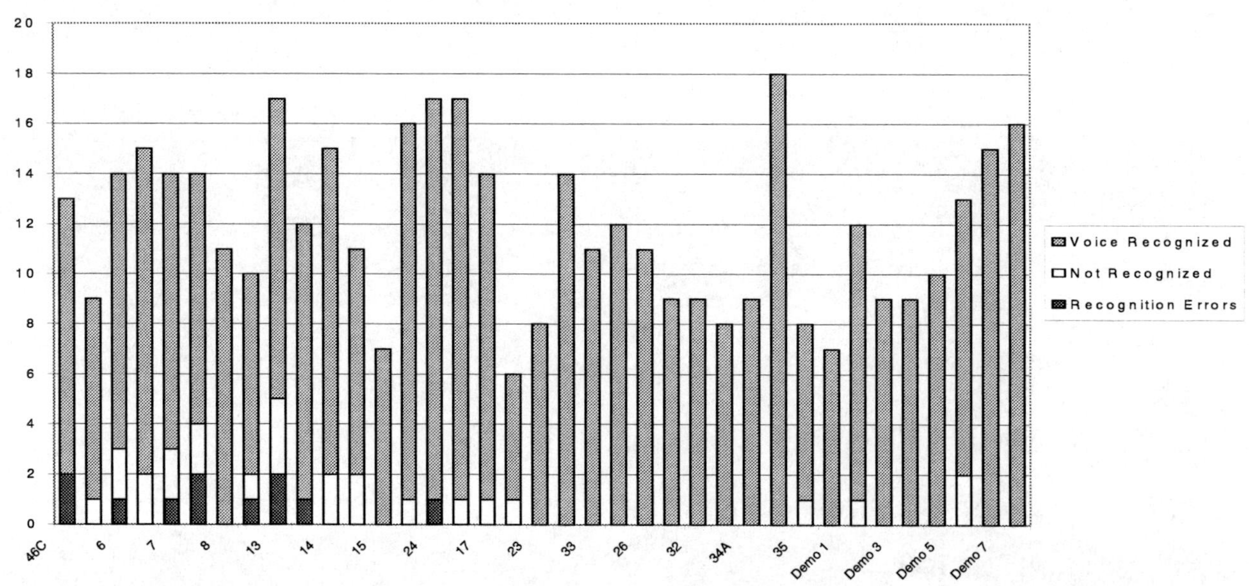

Figure 6. Voice recognition statistics

run 24. After the changes were made, the voice recognition system performed with 97% accuracy.

Conclusions

The Controller Interface was successful at providing the NASA 757 with taxiroute information via Controller-Pilot Datalink Communications. Many of the datalink errors that were observed could be eliminated with a more robust communications protocol.

Datalink latency does not appear to be a problem. ATC instructions can typically arrive at the flight deck in less than 1 second. The impact of the datalink latency may depend on meteorological conditions. In VFR conditions, the CPDLC datalink may not be needed to supplement voice communications. In IFR conditions, an extra half-second for the datalink to reinforce the routing will probably be insignificant.

The voice recognition system can work at a normal rate of speech and with a high ambient noise level. The main drawbacks with the voice recognition system as implemented are the requirements to use standard phraseology and to have individual voice files.

Air Traffic Controllers as well as other visitors during the public demonstrations thought that CPDLC could enhance surface operations during low visibility conditions. The use of color displays in the tower cab did not receive unanimous support, but there were many that thought that color maps and improved traffic displays would be useful. [3]

Summary

The Controller Interface was researched and designed as part of NASA's Low Visibility Landing and Surface Operations (LVLASO) research project. The initial intent of the CI was to allow a test controller to easily generate CPDLC Uplink messages to the NASA 757 aircraft.

After a successful flight demonstration at Atlanta, research in CPDLC implementation and controller workstations continues. Additional research is needed in robust communication protocols, security, and integrity issues.

Acknowledgements

The authors would like to thank the St. Cloud State University students that participated in the development of the Controller Interface. Electrical Engineering students were CarrieAnn Kowalke, Scott Boeshart, Chee-Keong Lai, and A.K. Muhammad Masud. Craig Berthiaume and Tsun-Chih Wei were from the Department of Aviation.

Carrie Kowalke assisted with the development while working as a NASA-Langley Summer Scholar in 1996. Electrical Engineering students Ross Weber and Robert Wimmer also assisted the CI and LVLASO project as NASA-Langley Summer Scholars in 1997.

References

1. "Minimum Operational Performance Standards for ATC Two-Way Data Link Communications", RTCA DO-219.
2. Rankin, J., and Mattson, P., "Controller Interface for Controller-Pilot Data Link Communications", Proceedings of the 16th DASC, October 1997.
3. Young, S.D., and Jones, D.R., "Flight Demonstration of Integrated Airport Surface Movement Technologies", NASA/TM-1998-206283, January 1998.

RUNWAY INCURSION REDUCTION PROGRAM (RIRP) SURVEILLANCE SYSTEM, NASA/FAA ATLANTA DEMONSTRATION

Vincent Capezzuto, Federal Aviation Administration, Washington, D.C.
David Olster, Department of Transportation, Cambridge, MA
Michael Curry, Trios Associates, Inc., Greenbelt, MD
Steven L. Pendergast, QuesTech, Inc., San Diego, CA

Background

Runway Incursions (RI's) in the United States have experienced a 16% average yearly increase from 1993 to 1997.[1] This fact, coupled with recent airport surface accidents in Atlanta, Detroit, Los Angeles and St. Louis, ranks RI's as a top priority in the Federal Aviation Administration (FAA) and is listed on the National Transportation Safety Board's (NTSB) "10 Most Wanted" safety improvements.[2]

It is generally accepted that increased situational awareness among all users and improved surveillance of the airport will be effective in reducing RI's that could lead to accidents. To this end, seamless surveillance coverage, identification of aircraft on the surface and information sharing with airport users have become a priority for the FAA. This priority is reflected in the FAA's "Concept of Operations for the National Airspace System (NAS) in 2005"[3] and the NAS Architecture.[4]

The FAA, the National Aeronautics and Space Administration (NASA), the Department of Transportation (DOT), and an industry team that included Trios Associates, Inc. and QuesTech Inc., conducted a demonstration of a seamless airport surface surveillance system and surveillance information distribution at Atlanta's Hartsfield International Airport (ATL) in August of 1997. The system combined surface surveillance data from two sources: 1) the Airport Surface Detection Equipment – Model 3 (ASDE-3) and 2) a prototype Airport Target Identification System (ATIDS), and received terminal flight information and position data from the Automated Radar Terminal System Model IIIA (ARTS-IIIA). The ASDE-3 provided primary radar surveillance and ATIDS provided both multilateration and Automatic Dependent Surveillance - Broadcast (ADS-B) for cooperative targets. This information was fed to a prototype Airport Movement Area Safety System (AMASS). The prototype AMASS was originally designed to receive ASDE-3 and ARTS-IIIA data. These two data sources were applied to safety logic to provide air traffic controllers with conflict alert advisories on arrival and surface targets. The prototype AMASS was enhanced by software and hardware changes to perform multiple sensor data fusion, which provided a complete track and identification picture of the airport movement area. This surface surveillance system, combined with enhanced data links and displays, demonstrated the capability to maintain situational awareness in the tower and cockpit.

This paper describes the performance of this surveillance system, as well as the impact of the system to current FAA airport services.

Site Coordination

A demonstration of this scale would not have occurred without the able participation of the local and regional personnel of ATL and

[1] "1997 Aviation Safety Statistical Handbook," Office of System Safety, FAA; 1998.
[2] "Most Wanted Transportation Safety Improvements," NTSB; May 5, 1998.
[3] Air Traffic Services Concept of Operations; FAA; April 1997.
[4] NAS Architecture, V. 3.0, DRAFT; FAA; December 1997.

the FAA's Southern Region. The coordination among national and local authorities facilitated the demonstration through airport access, frequency authorization, and Air Traffic and airline coordination.

Surveillance System Description

Figure 1 shows a block diagram of the entire demonstration ground system, with FAA systems highlighted. The centerpiece of the demonstration surveillance system was the prototype AMASS. Production AMASS systems are currently scheduled to be deployed at every ASDE-3 site starting in 1999. A production ASDE-3, already in place at ATL, and a prototype ATIDS system provided the surveillance of the airport surface.

The multiple surveillance sub-systems were fed into the prototype AMASS to be "fused" into an optimal report. Data fusion allows for the integration of two or more sensors to provide a position and identification report which is of higher integrity than any one sensor would be alone. This enhanced report in turn could improve the safety logic calculation within the AMASS. The improved surveillance and safety information was then distributed via data links to several airport surface users.

The function of each segment of the surveillance system is detailed below:

- ASDE-3: Provided raw digitized radar video.
- ATIDS: Provided position and flight identification information via multilateration and ADS-B.
- ARTS IIIA: Provided flight and position information for arrivals and departures.
- AIU: The ARTS Interface Unit provided arrival ID and position data from the ARTS IIIA to the AMASS.
- FPU: The Flight Plan Unit provided flight plan information from the ARTS IIIA to ATIDS.
- AMASS: Converted ASDE-3 video to target tracks. Combined all surveillance data to determine target position and identification using data fusion algorithms. Also provided safety alert logic to display warnings and advisories to controllers. Provided output information to other users.
- Data Link Manager: The Data Link Manager provided traffic and hold bar information from AMASS to be uplinked to the test aircraft.
- Remote AMASS: Utilized data link system to transfer limited AMASS data to remote location.

The target position, identification, and safety alert output of the AMASS was provided to NASA via the Data Link Manager for display in the cockpit.[5]

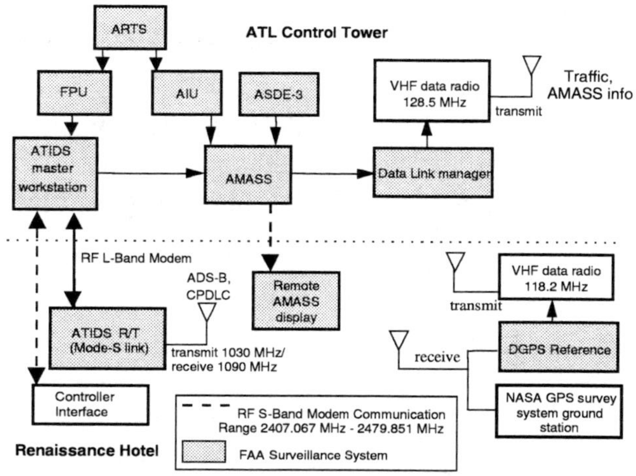

Figure 1 - Ground System Diagram

Fusion Performance

The use of multiple sensors in the surveillance system required the use of data fusion algorithms to provide an optimal position and identification solution. Three alternative data fusion architectures were considered:

- Central Fusion Architecture: Utilized position data from ASDE-3/AMASS and untracked ATIDS data.
- Federated Fusion Architecture: "Averaged in" ATIDS and ASDE-3/AMASS position and velocity data in a weighted fashion.
- Best Attribute Fusion Architecture: Associated ATIDS identification data with a fusion track, whose position was derived from ASDE-3 position data.

Consideration of data taken during previous testing of ATIDS and ASDE-3 suggested that the Federated Fusion Architecture would provide the optimal surveillance solution. The Federated approach performed tracking on ATIDS data, then associated the two sensor tracks to maintain the

[5] "Flight Demonstration of Integrated Airport Surface Movement Technologies," Young, S., Jones, D., NASA; 1998.

best overall tracking and identification solution for each target. Moreover, the Federated approach is easily adaptable to more and varying sensors, an important feature to the RIRP.

Figure 2 shows generation of fusion tracks under the Federated approach. Characterization of the sensors was performed to determine covariance estimates to form weighted matrices. AMASS and ATIDS target positions were extrapolated to fixed polling times. A fusion track state estimate was then formed using the weighted matrices associated with each of the two contributing sensor reports.

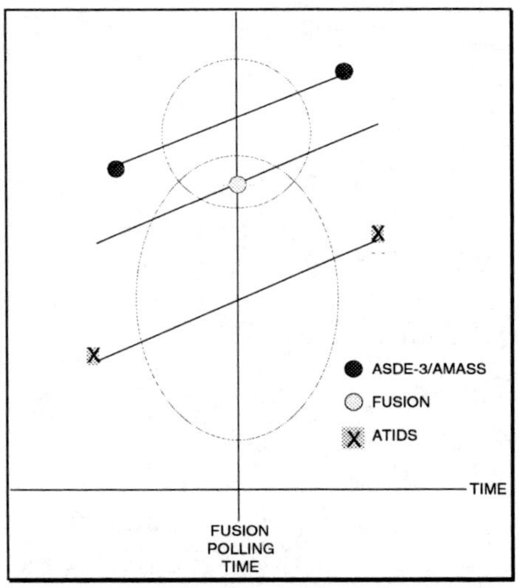

Figure 2- Federated Fusion Track Generation

The performance of the demonstrated surveillance system was post-evaluated in five areas: Detection Coverage, Track Continuity, Sensor/Fusion Error Analysis, Track Split/Merge Analysis, and Target Identification. The data for this report was taken over the five test runs in which all segments of the system were functional (Test Nos. 18, 24 27 34b and 35).[6]

Detection Coverage

Detection Coverage is characterized by the probability of detecting a target on the movement area. Detection coverage for the fusion output is not a simple combination of the sensors' coverage detection. This is due to filtering by fusion processing that eliminated "low-confidence" ATIDS target reports. However, the fusion output should produce coverage greater than any one sensor.

The increase in detection coverage during the demonstration was determined by comparing the amount of target positions detected by the ASDE-3/AMASS with the amount of fusion target positions produced for the same time frame. The area observed for this measurement was limited to the area for which the ASDE-3/AMASS was allowed to perform surveillance; i.e., fusion was not given credit for conducting surveillance outside of the surface area. Also, since the ATIDS multilateration system was optimized for the north half of the airport, no reports on the south side were considered. With these two factors in mind, the fusion increased surveillance reports by an average of 13.8% over the prototype ASDE-3/AMASS during the five test runs surveyed.

The areas on the airport surface where coverage was increased was consistent throughout the five test runs. Two main areas existed: 1) On Taxiway E in front of the Delta hangar, and 2) on the west side Taxiway F near the entrance to Ramps T and A (See Figure 3). These two areas are known to lack ASDE-3/ AMASS surveillance in Atlanta. The fusion algorithms and surveillance system, therefore, showed the ability to increase detection coverage during this demonstration.

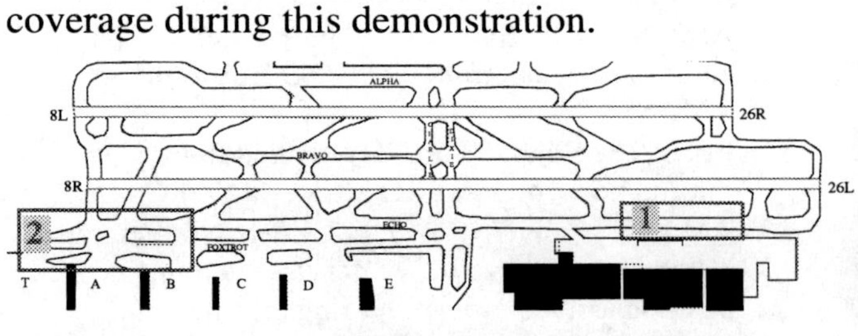

Figure 3- Areas of ASDE-3 Loss of Surveillance

[6] "Flight Demonstration of Integrated Airport Surface Movement Technologies," Young, S., Jones, D., NASA; 1998.

0-7803-5086-3 /98/$10.00 ©1998 IEEE

Track Continuity

The Federated Fusion approach introduces the concept of the "track locker," a process used to maintain target continuity when coverage lapses are found in sensor reports. This concept is useful when all sensor coverage is lost. In this situation, the unsupported fusion track is placed in the track locker. The last known fusion track state (position and velocity) is retained until its extrapolated area of uncertainty exceeds a threshold or predicts an exit off the surface movement area. This typically occurs for a period longer than the current single-sensor drop-track time (four seconds). In addition, any new tracks formed within an area consistent with possible track locker positions are evaluated for re-association. The track locker process allows for the retention of identification that would have been lost without fusion, as well as the elimination of reacquisition time, both of which improve surveillance.

An example from the demonstration is shown in Figure 4. During this test run, the ASDE-3/AMASS produced two separate tracks and ATIDS produced five separate tracks. The test aircraft also taxied through the "cone of silence" of the ASDE-3, shown by the circle drawn around the (0,0) point on the graph. Through this run, however, the fusion algorithm produced only one track. The system maintained location and identification of the test aircraft throughout the run, even with loss of detection from one system and multiple track formations from both surveillance systems.

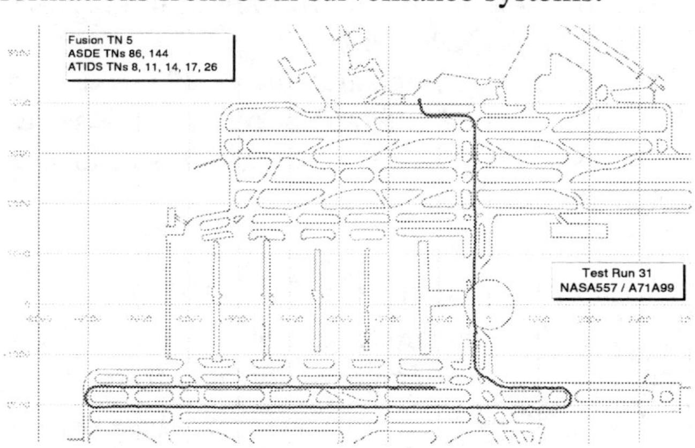

Figure 4- Track Continuity Example

A measurement was taken to quantify this improvement in surveillance. The increase in track continuity was measured by 1) the decrease in number of tracks and 2) the increase in track life. The decrease in number of tracks measurement quantifies the number of dropped targets which the fusion algorithms overcame to maintain surveillance. Because reacquisition of dropped tracks takes two seconds, reducing the number of dropped targets leads to an increase in the amount of time that the system will have surveillance. Over the five test runs, for tracks having multiple sensor support at some time during the test period, the fusion processing maintained an average of 36% fewer track segments than the ASDE-3/AMASS. Assuming a two second reacquisition time for the ASDE-3/AMASS, an average increase in track life of 1% was measured over the five test runs. The fusion algorithms, therefore, demonstrated the ability to increase track continuity during this test activity.

Sensor/Fusion Error Analysis

The sensors' respective accuracies impact the ability of the fusion algorithms to form high-integrity track reports. This impact, in turn, affects the ability of AMASS to accurately predict potential accident situations. Estimations of the relative accuracies of the surveillance systems and the fusion processing were made, and are shown in Table 1:

Table 1- Estimated Sensor/Fusion Relative Errors

Data Source	Std. Dev.
ASDE-3/AMASS	3.73 ft.
ATIDS/ADS-B Raw	7.50 ft.
ATIDS/Multilateration Raw	15.92 ft.
ATIDS Tracked	10.60 ft.
Fusion Tracked (Actual)	4.64 ft.
Optimum Federated Fusion (Projected)	3.52 ft.

The relative errors of ASDE-3/AMASS and ATIDS position reports were derived from a scatter analysis in reference to the fusion track position. The estimated error for the fusion position was derived from the variant difference from a polynomial fit of the track. In other words, the fusion track was line-fitted, then compared against the actual fusion positions to determine the error.

Accurate time-tagging of position information is critical to fusion performance. During the evaluation, a delay in the time tagging of the ATIDS ADS-B position reports was apparent. A comparison of the ATIDS ADS-B position reports and the ASDE-3/AMASS position for an aircraft that was accelerating indicated a time delay of up to 810 ms, which produced an error in the x position of –83 feet (the aircraft was accelerating in the positive x direction). This tendency of ATIDS to lag behind the ASDE-3/AMASS position may be explained by the variable delay in the transmission of the differentially-corrected Global Positioning System (DGPS) position by the aircraft.

The Mode S ADS-B transponder has a process time to determine the differentially-corrected GPS position of the aircraft. One prototype Mode S ADS-B transponder used in the evaluation had a process time of 220 ms, according to the manufacturer. The position will then be transmitted during the next random squitter. However, this position may be rebroadcast during the second squitter of that particular second (assuming a 2 Hz squitter rate for the transponder), making the apparent age of the differentially corrected position even older. Depending upon the timing of the random squitter, this position may be broadcast a third time. An example is shown in Figure 5:

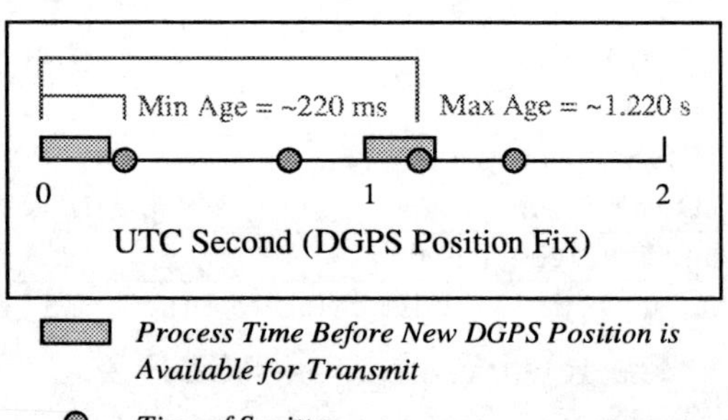

Figure 5- Example of GPS Position Age

In this example, the differentially-corrected GPS position fix received at Universal Time Coordinated (UTC) time "0" is transmitted three times, with a minimum age of 220 ms and a maximum age of 1.22 s. This delay can cause apparent errors in DGPS position reporting relative to a target in motion.

Track Split/Merge Analysis

Degradation in the performance of the individual sensors could produce occurrences where two targets are merged into one fusion track, or where a single target could produce multiple fusion tracks. These occurrences are considered degradations in surveillance.

During the five test runs studied, approximately eight incorrect track splits were experienced over the total test time of 2.68 hours (for a rate of 3 incorrect splits per hour). 35 incorrect track merges were experienced during the 2.68 hours of the five test runs (for a rate of 13 incorrect merges/hr).

Both of these degradations need to be minimized to allow optimum surveillance on the airport surface. Further refinement of the fusion processing should lead to an acceptable level of track merges and splits.

Target Identification

Identification of targets on the movement area was provided to the fusion algorithms by the ARTS IIIA through ATIDS. The fusion process had the ability to maintain identification of a target track, even if there was a loss of ATIDS surveillance on that target. During this demonstration, 7,297 position reports retained identification after loss of ATIDS surveillance, for an identification increase of 6.3%.

Fusion Performance Conclusion

The fusion algorithms showed the potential to allow improved detection coverage, track continuity, and identification in the surface surveillance system used in this

0-7803-5086-3 /98/$10.00 ©1998 IEEE

demonstration. Further development should yield algorithms which meet the surveillance and identification needs of the airport surface.

Impact on Current NAS

The RIRP team used the Atlanta demonstration opportunity to investigate the effects that enabling transponders on the airport surface would have on other operational systems. A summary of these effects follows.

Impact on Mode Select (Mode S) Radar

Historically, aircraft transponders have not been enabled on the surface due to the proximity to the terminal radars. With the introduction of Mode S transponders, which are designed to reduce transmissions, it may be possible to enable transponders to perform secondary surveillance. This section outlines a preliminary study on the effects on the Mode S radar from transponders on the surface.

Signal Interference

The operation of transponders on the surface increases the amount of 1090 MHz signals close to the Mode S sensor. This introduces potential interference with the sensor's ability to track aircraft. However, the Mode S sensor has the ability to re-interrogate aircraft several times in the same beam dwell time. Re-interrogation can overcome an increase in interference without dropping surveillance on a target.

No Mode S radar target drops were observed during the demonstration in Atlanta. Since most testing was conducted in low-traffic time periods (to reduce interference with normal airport operations), additional analysis was conducted to investigate the possibility that surveillance would be adversely affected in a worst case environment.

Using the Poisson probability model[7], the ATIDS system was placed in a high-capacity environment, with transponders enabled on the surface. The ATIDS system had virtually no effect on Mode S sensor surveillance by the end of the beam dwell time. This analysis corroborates with two previous studies performed separately by the Joint Spectrum Center[8] and MIT Lincoln Laboratory[9], which found that squitters on the airport surface do not cause the Mode S sensor to lose surveillance on targets.

False Targets

False Targets occur when signals from an aircraft reflect off a structure and are processed as a separate aircraft. The Mode S sensor and Mode S transponder have features which try to prevent false targets from occurring. The sensor utilizes selective addressing to detect two tracks with the same identification. Also, side lobe suppression is used to inhibit aircraft not in the main beam from replying to interrogations. The Mode S transponder is designed to respond only to discreet interrogations while on the ground. The transponder uses an aircraft's weight-on-wheels switch to determine whether or not it is on the surface.

Despite these precautions, false targets were observed on the ARTS IIIA scopes during the Atlanta demonstration. The false targets were consistent, appearing approximately three miles north of the airport (see Figure 6).

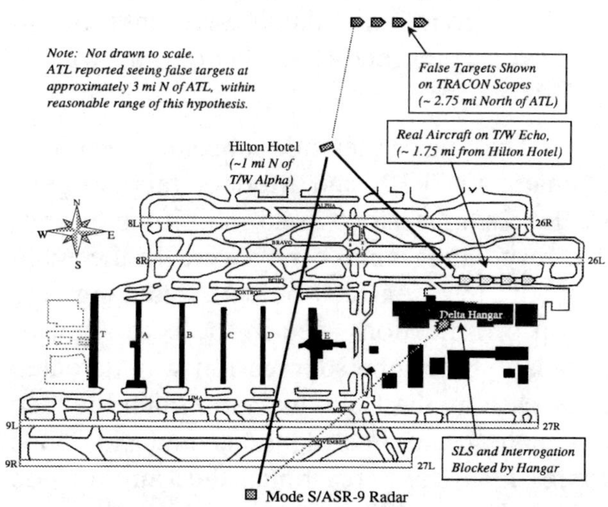

Figure 6- False Targets Seen During Demonstration

---

[7] Probability and Random Processes for Electrical Engineering,; Leon-Garcia, Alberto; Sep, 1993, pp. 106-108.

[8] "Mode S/GPS Squitter Interference Analysis," JSC, July 1997

[9] "GPS-Squitter Interference Analysis," MIT/LL, 1995.

0-7803-5086-3 /98/$10.00 ©1998 IEEE

Because of the location of the false targets, the reflective body was most likely the Hilton hotel located north of the airport. The real positions of the aircraft were most likely Taxiway E, in front of the Delta hangar. The total distance of a reflection from Taxiway E to the Hilton hotel to the Mode S sensor is approximately 2.75 miles, within range of the three mile estimation of the controllers who observed the false targets.

The precautions within the Mode S sensor and transponder against false targets were defeated by three factors: the "non-compliant transponder fix" in the sensor, blockage from the Delta hangar, and incorrectly operating or missing weight-on-wheels switches in the aircraft. Combinations of these three factors contributed to the display of false targets.

The "non-compliant transponder fix" is a modification to the Mode S sensor which forces the sensor to act as an older Air Traffic Control Radar Beacon System (ATCRBS) sensor during part of the surveillance cycle. Mode S transponders should not be affected by this, since they are designed not to respond to ATCRBS interrogations while on the ground. This precaution is defeated, however, if the Mode S transponders do not have the weight-on-wheels switch correctly set. Approximately 6% of the aircraft in Atlanta were observed to not have a correctly set weight-on-wheels switch.

The sensor uses side lobe suppression to eliminate ATCRBS and Mode S false targets. Side lobe suppression would inhibit the aircraft from responding unless they were in the main beam of the sensor, thereby reducing the amount of reflections seen in the bore sight of the radar. Side lobe suppression was defeated in this case by the Delta hangar, which blocked both main beam interrogations and side lobe suppression from reaching the aircraft on Taxiway E.

The false targets observed in Atlanta may be mitigated by the removal of the non-compliant transponder fix (currently planned by the FAA), and the addition of dynamic false target reflector files in the Mode S (also currently being developed by the FAA). The system demonstrated did not have any significant adverse effects on the Mode S radar.

Impact on TCAS

The Traffic Alert and Collision Avoidance System (TCAS) was not intended to participate in anti-collision activities with aircraft that are on the ground. TCAS uses information contained in the Mode S squitter and the reported altitude of ATCRBS aircraft to determine whether or not aircraft are on the surface. Most manufacturers also utilize the reported altitude of Mode S aircraft as a secondary filter. However, this feature is not required of TCAS manufacturers.

As previously mentioned, approximately 6% of aircraft on the surface of Atlanta did not correctly set their weight-on-wheels indicator. This incorrect setting caused extraneous TCAS traffic alerts for some aircraft arriving into ATL. It was determined after the demonstration that one TCAS manufacturer does not look at the reported altitude of Mode S aircraft, causing these extraneous alerts.

The effects on TCAS need to be investigated further to determine the full impact. These impacts will also have to be considered in light of the new version of TCAS (Change 7) being deployed in the near future.

Impact on DME

Because the Distance Measuring Equipment deployed throughout the NAS use frequencies near the surveillance operating frequencies, concerns were raised over the possibility that enabling transponders on the surface near navigational aids might interfere with their operation. The DME located on the surface of ATL operates on the frequencies of 1140 MHz and 1203 MHz.

No degradation of the DME was observed during this demonstration. Because of the distance between the frequencies of the transponders and the DME, the difference in the spacing of the pulses in the two systems, and the certification requirements of aircraft transponders, no degradation is expected.

Impact on NAS Summary

It was demonstrated that the surveillance system did not impair the current NAS system from maintaining surveillance on aircraft. However, operational concerns were raised and need further investigation. Specifically, the mitigation of false targets and the extent of the impact on TCAS need to be addressed. No effect was observed on the DME.

Conclusions/Future Work

The surveillance system tested in Atlanta demonstrated the ability to provide enhanced and seamless surveillance, identification of targets, and information sharing with users of the airport. Fusion algorithms allowed this enhancement of surveillance. Specifically, detection coverage, track continuity and target identification improved as a result of the system. It was observed that the surveillance system did not impair the current NAS system from maintaining surveillance on aircraft.

As a continuation of the R&D efforts in Atlanta, the RIRP team will be moving to Dallas-Ft. Worth International Airport (DFW) to conduct further test activities. DFW was chosen because of its high number of operations, complexity of the airport surface (with seven runways and three control towers), current equipment in place, and an invitation from local personnel to conduct test activities at the site.

The goal of the RIRP system remains to obtain seamless surface surveillance coverage and increased situational awareness in a robust system. This can be accomplished with the use of a system similar to that used during the Atlanta demonstration, with several new components, features, and increased reliability/usability. Figure 7 shows a diagram of the systems to be demonstrated in DFW.

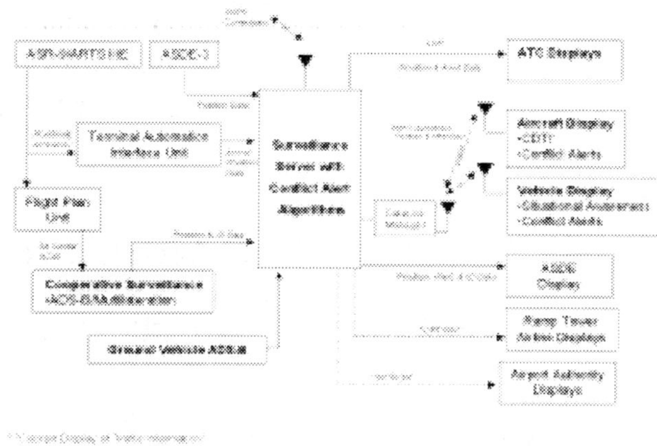

Figure 7- DFW System Block Diagram

In addition to the services demonstrated in Atlanta, the DFW system will provide aircraft identification in safety alert messages presented to the controllers and pilot/vehicle operators. DFW will also serve as a test bed for validating the initial Air Traffic requirements being developed for surface surveillance. A more rigorous analysis of system performance, which was unable to be produced in the time frame of the Atlanta demonstration, will also be conducted on the DFW system. Certification and Air Traffic procedure issues for surface surveillance will also be addressed.

The DFW system is also the architecture proposed to participate in Flight 2000 activities. Specifically, this architecture is foreseen to perform surveillance services at Anchorage, AK airport.

The future of NAS operations on the airport surface requires a surveillance system that provides seamless coverage, identification and information sharing among all users. Continued development of the surveillance system demonstrated in Atlanta will yield a system that meets the needs of all users of the airport surface.

DEVELOPMENT OF AIRPORT SURFACE SURVEILLANCE PERFORMANCE REQUIREMENTS

Rick Cassell, Rannoch Corporation, Alexandria, Virginia

Carl Evers, Rannoch Corporation, Alexandria, Virginia

Abstract

There are a number of new technologies being assessed for performing surveillance of aircraft and vehicles on the airport surface. The reason for increased interest in such technologies is to address the problem of runway incursions and the associated risk of collisions. Technologies are under development to assist in conflict detection and collision avoidance on the airport surface. Activities are underway to develop a set of performance standards by which the adequacy of these technologies can be judged. This paper presents a summary of recent efforts by the International Civil Aviation Organization (ICAO) and RTCA to develop surface surveillance performance requirements. This includes proposed requirements for four key parameters - Integrity, Continuity, Accuracy and Availability. The paper will also describe the methodologies being used to develop the requirements.

Introduction

Airport surface operations are receiving increasing attention from air traffic management both in the U.S. and abroad. The reasons for this are increasing traffic levels and the increased need for assuring safety. The safety concern has arisen due to an increase in the number of runway incursions at U.S. airports [Ref. 1]. A key element of the needs of air traffic management is providing adequate surveillance information for both aircraft and vehicles. Standards are under development by the International Civil Aviation Organization (ICAO) All Weather Operations Panel for Advanced Surface Movement Guidance and Control (A-SMGCS) [Ref. 2], the Airport Surface Navigation and Surveillance subgroup of RTCA Special Committee 159, and RTCA SC-186 for Automatic Dependent Surveillance - Broadcast (ADS-B) [Ref. 3]. In support of these activities, the FAA, and NASA's Low Visibility Landing and Surface Operations Program (LVLASO), Rannoch has been developing the performance requirements for surface movement surveillance. The only existing standards for surface surveillance are related to radar systems (i.e., ASDE-3). A generic set of standards is needed that can be applied to any technology, such as ADS-B. This paper presents a summary of recent efforts by ICAO and RTCA to develop surface surveillance performance requirements.

This paper contains the understanding and views of the authors only, and does not necessarily reflect the official position or views of NASA or the FAA.

Description of Surface Operations

Airport surface operations include several distinct phases that can have different surveillance requirements. These phases are illustrated in Figure 1. The distinguishing characteristics of these phases are the safety margins between aircraft and objects, and the

0-7803-5086-3 /98/$10.00

separation maintained between aircraft. Basically these margins are larger on the

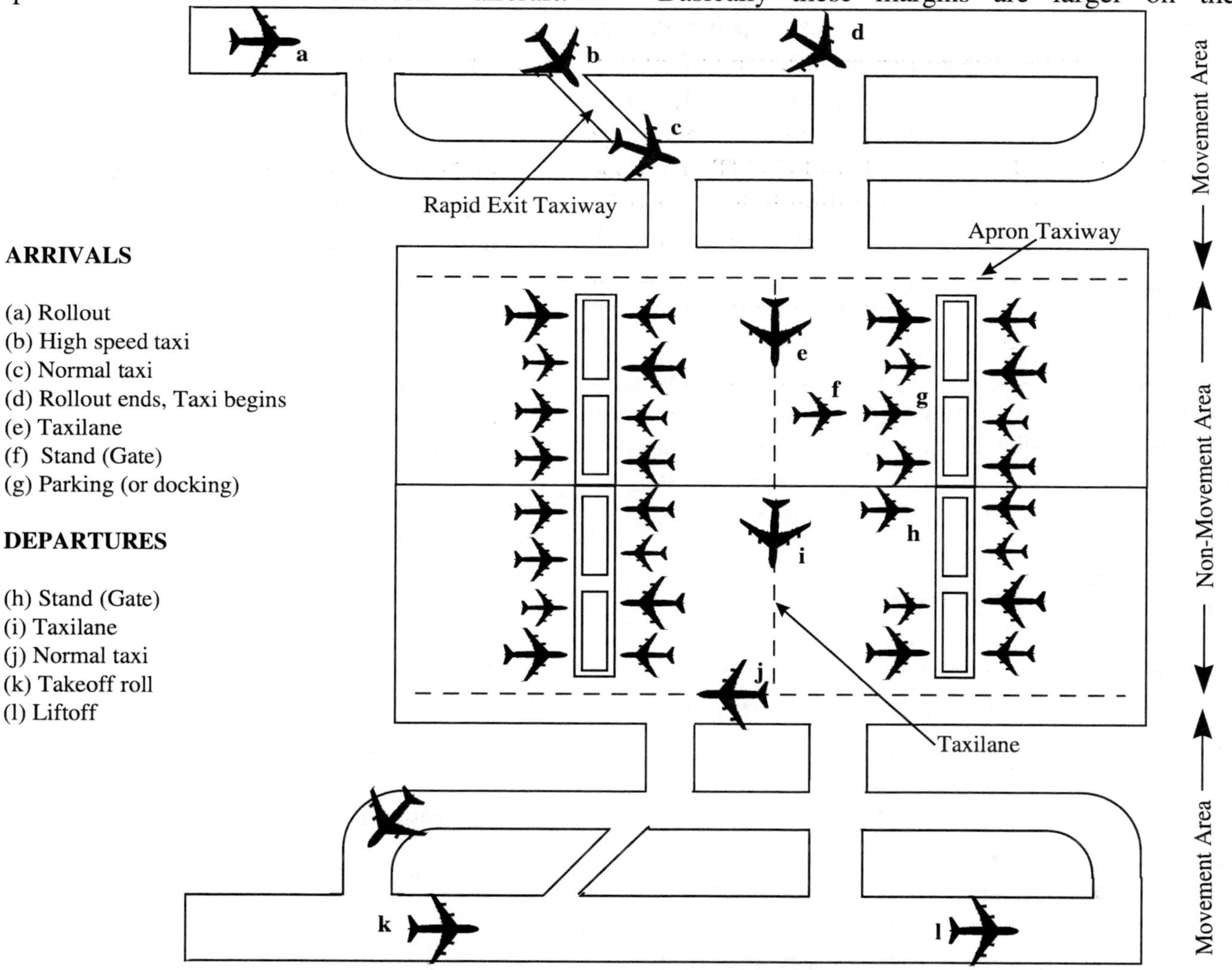

Figure 1. Airport Surface Phases of Operation

runways and primary taxiways than on the apron and gate areas.

The other key aspect of surface operations is visibility condition - how it affects the pilot's ability to taxi (and avoid other aircraft) and the controller's ability to observe and control traffic. ICAO has defined four distinct visibility conditions as follows [Ref. 4]:

- Visibility Condition 1.

Visibility sufficient for the pilot to taxi and to avoid collision with other traffic on taxiways and at intersections by visual reference, and for personnel of control units to exercise control over all traffic on the basis of visual surveillance.

- Visibility Condition 2.

Visibility sufficient for the pilot to taxi and to avoid collision with other traffic on taxiways and at intersections by visual reference, but insufficient for personnel of control units to exercise control over all traffic on the basis of visual surveillance.

- Visibility Condition 3.

Visibility sufficient for the pilot to taxi but insufficient for the pilot to avoid collision with other traffic on taxiways and at intersections by visual reference with other traffic, and insufficient for personnel of control units to exercise control over all traffic on the basis of visual surveillance. For taxiing this is normally taken as less than 400 meters visibility but more than 75 meters.

- Visibility Condition 4.

Visibility insufficient for the pilot to taxi by visual guidance only. This is normally considered to be 75 meters visibility or less.

Visibility conditions have a significant impact on surface surveillance requirements. Visibility governs the reliance of air traffic control on surveillance information. Generally, lower visibility places more demanding requirements on the surveillance function.

System Level Requirements

Target Level of Safety (TLS)

Target Level of Safety, or TLS, is the probability of an accident (fatal or hull loss) during aircraft movement on the aerodrome. The total mission TLS (including all phases: taxi-out, takeoff, climb cruise, descent, approach, landing and taxi-in) established by ICAO is targeted to be 1.0×10^{-7} [Ref. 2]. The final approach and landing phase was allocated 1.0×10^{-8}. Similarly, the other phases of flight have allocations that use only a small portion of the overall TLS. Therefore, the taxi phase should be allocated a comparable portion. An analysis of NTSB (National Transportation Safety Board) and worldwide fatal taxi accident rates revealed them to be 6.2×10^{-8} and 9.0×10^{-8} per operation, respectively. Based on the above considerations, the surface movement TLS is 1.0×10^{-8} (per operation). This is a margin of 6-9 over the historical accident rate and is in line with the allocations of TLS to the other phases of flight. It should be noted that this TLS applies to all visibility conditions for surface operations.

Surveillance Function risk

Function risk is the probability of an accident caused by a failure of a specific function. A-SMGCS has four functions - Guidance, Surveillance, Control and Routing. Surveillance is allocated a risk of 3.0×10^{-9} per operation, with the remainder of the TLS allocated to the other functions, as shown in Figure 2. The risk assigned to surveillance is subsequently allocated to all of the elements included in the surveillance function.

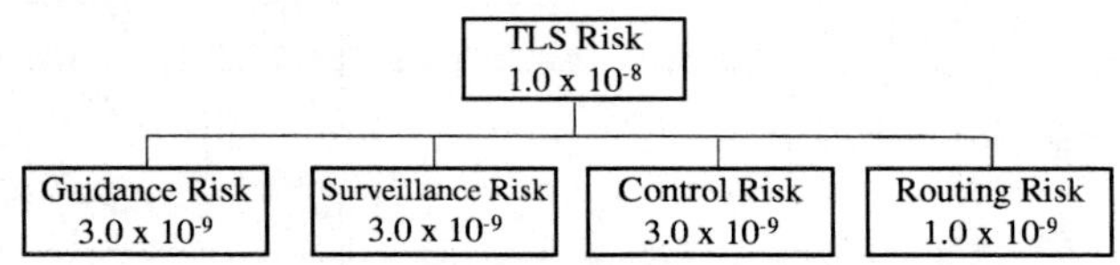

Figure 2. TLS Allocations

Surveillance performance requirements

The following is a list of the key parameters that encompass the performance of the surveillance function for surface movement.

- Coverage
- Covered speed (of vehicles)
- Position accuracy
- Update rate
- Update success rate
- Latency
- Integrity
- Continuity
- Availability
- Traffic loading

Each one of these parameters is described along with the proposed performance requirement. Also included is a brief synopsis of the rationale for the requirement.

0-7803-5086-3 /98/$10.00 ©1998 IEEE

Coverage area

Coverage area is that area of the airport in which surveillance is provided. The minimum region is the airport movement area. The movement area is defined in the U.S. as "the runways, taxiways, and other areas of an airport/heliport which are utilised for taxiing/hover taxiing, air taxiing, takeoff, and landing areas of aircraft, exclusive of loading ramps and parking areas." ICAO defines the movement area to include the apron area. The A-SMGCS coverage as defined by ICAO is intended to include those apron and ramp areas where aircraft are operating.

Altitude coverage

The minimum altitude coverage required is 500 ft. This allows for coverage of missed approaches and helicopter operations.

Approach coverage

The minimum approach coverage (for each runway direction) is 10 nautical miles, which can be partially satisfied by an approach radar. The purpose of this is to allow for the transition between approach and airport surveillance. Typically the surface surveillance sensor needs to provide coverage 2-3 miles beyond the airport boundary to ensure overlap with the approach surveillance radar.

Covered speed (of aircraft/vehicles)

The speeds of aircraft and ground vehicles which the surveillance function must accommodate are as follows:

• 0 to 250 knots for aircraft on final approach, missed approach and runways. This value is based upon the maximum speeds of aircraft at a distance of 10 NM.

• 0 to 80 knots for aircraft on runway exits. High-speed taxi normally occurs with the aircraft at a speed between 30 and 50 knots on rapid (high speed) exits. The maximum speed is 80 knots. This is based on measurements of operational speeds.

• 0 to 50 knots for aircraft on straight taxiways and 20 knots in curves. Normal/apron taxiway speeds range from 10 - 50 knots for straight sections. This is based upon a study that found 95 percent of aircraft taxi speeds to be less than 30 knots, with the maximum found to be 50 knots [Ref. 5]. On curves, the average speed has been found to be 10 knots, with the maximum being 20 knots.

• 0 to 80 knots for vehicles on the movement area. This accounts for the maximum speeds of emergency vehicles.

Position accuracy

Surveillance position accuracy is the difference between a target's measured position and its true position. It is specified in terms of the 95 percent horizontal and vertical accuracy performance. The requirements are:

- Longitudinal position accuracy: 10 meters
- Lateral position accuracy - runways and taxiways: 10 meters
- Lateral position accuracy - apron and gate regions: 3 meters
- Vertical position accuracy: 20 meters

The longitudinal position accuracy requirement is based on the effect of the surveillance accuracy on the ability to detect loss of separation and potential traffic conflicts or runway incursions. Two scenarios were analyzed - a runway incursion where the aircraft crosses the hold line, and the loss of longitudinal separation between two aircraft. The runway incursion scenario is designed to determine the warning time required of the surveillance system to the potential incursion, and to prevent the aircraft from proceeding onto the runway (see Figure 3). The geometry depicted is for an airport where large aircraft are operating, and the hold line is 75 m from the runway center line.

Based on this scenario and a sensitivity analysis of the effect of accuracy, it was determined that 10 meters would allow time (with some margin) for detection of the incursion and stopping the aircraft prior to entering the runway. This is based on the pilot being provided with conflict detection information directly. Little improvement is seen with the accuracy better than 10 meters.

Where the air traffic controller must be given the alert and issue instructions to the pilot, all accuracy values manifest an excessive time delay, resulting in an inability to prevent the aircraft from entering the runway. However, in general, surveillance accuracy better than 10 meters can result in significant improvements in system performance (in the form of shorter warning times) and allow more time for reaction to avert a conflict.

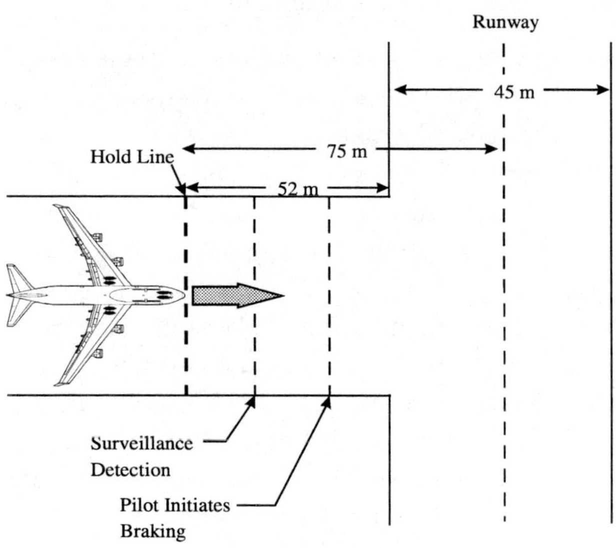

Figure 3. Runway Incursion Detection Scenario

A similar analysis was conducted for separation between two in-trail aircraft. This scenario was found to be slightly less demanding on accuracy in order to maintain the required separation. The 3 meter requirement for the apron and gate regions is based on the smaller safety margins between aircraft and objects in those areas of the airport.

Velocity accuracy

Surveillance velocity accuracy is the difference between a target's measured velocity (speed and heading) and its true velocity. It is specified in terms of the 95 percent lateral and vertical speed and heading performance. The speed requirement is 1 knot. The direction requirement has yet to be determined.

Update rate

Update rate is the maximum time interval at which surveillance reports must be updated. The requirement is one second, and is based on the same analysis as for position accuracy. Using the same scenario described there, sensitivity analyses determined that aircraft stopping distance is relative to update rate. Generally there is a direct linear relationship between update rate, warning time and stopping distance. That is, the faster the update rate, the longer the warning time, which results in shorter stopping distance. Maximum benefits are obtained when the requirement is set to one second.

Update success rate

Update success rate is the minimum probability of receiving a surveillance report on each target. The requirement is 98 per cent. A minimum is required to assure that the integrity of surveillance is maintained. It is only when several reports in a row are missed that integrity, and thus safety, could be compromised. A 2 percent chance of a single missed target report results in a probability of 8.0×10^{-6} of missing three consecutive reports. Therefore, 98 percent represents a reasonable minimum value, that will not result in a compromise of integrity.

Latency

Latency is the maximum time between when the target position is determined and its transmission to or use by the control function. The latency requirement is one second. Similar

0-7803-5086-3 /98/$10.00 ©1998 IEEE

to update rate, any delays in the time from position determination until its use in detecting loss of separation or conflicts, has a direct effect on the probability of successfully resolving the conflict. One second was chosen as a reasonable upper bound on this value.

Reference point accuracy

Reference point accuracy is the longitudinal accuracy of the location on the aircraft or vehicle to which the system will be referring in a surveillance position report. The requirement is a maximum of 3 meters. This is based on an accuracy requirement of 10 m, and an assumption that an additional 3 m of uncertainty in the reference point location adds minimal error to overall position determination. Assuming these two errors are independent then the overall position error would be determined on a root sum square basis and would equal $(10^2 + 3^2)^{1/2}$ or 10.4 m.

Integrity

Surveillance integrity relates to the trust which can be placed in the correctness of the surveillance information. Integrity includes the ability of the surveillance function to provide timely and valid warnings to the user when the system must not be used for the intended operation. Surveillance integrity risk is the probability of an undetected failure which results in incorrect surveillance information potentially leading to a loss of separation or other hazardous condition. The integrity requirements are shown in Table 1. The allocations are derived from the overall surveillance function risk of $3.0 x 10^{-8}$. Factored into the allocations is the controller's ability to detect erroneous information. The requirements reflect the fact that greater integrity is required of the surveillance information under lower visibility operations, where the controller has less capability to independently verify the information visually.

Table 1. Surveillance Integrity

	Visibility Condition		
	1	**2,3**	**4**
Integrity Risk (per hour)	2.0×10^{-3}	2.0×10^{-5}	2.0×10^{-6}
Time to Alert (seconds)	10	10	2

Continuity

Surveillance continuity is the capability of the surveillance function to perform the surveillance function without non-scheduled interruption during the intended surface operation. Surveillance continuity risk is the probability that the system will be interrupted and not provide surveillance information over the period of the intended operation. The requirements are shown in Table 2. As with integrity, these allocations are derived from the overall surveillance function risk of $3.0 x 10^{-8}$. Continuity requirements are not as stringent as integrity because when there is a loss of surveillance information the controller is aware of it and can take appropriate action (e.g. revert to visual procedures).

Table 2. Surveillance Continuity

	Visibility Condition		
	1	**2,3**	**4**
Continuity Risk (Per Hour)	2.0×10^{-2}	2.0×10^{-3}	2.0×10^{-3}

Availability

Surveillance availability is an indication of the ability of the surveillance function to provide usable service within the specified coverage area, and is defined as the portion of time during which the information is available for use by air traffic control and/or the aircraft flight crew. Surveillance availability is specified in terms of the probability of the surveillance function being available at the beginning of the intended operation. The requirement is 0.999, and applies under all visibility conditions.

Traffic loading

Traffic loading is specified in terms of the number of aircraft/vehicles to be covered by the surveillance function. The requirements are defined as the maximum number of aircraft and vehicle targets in the following regions:

- Aircraft on movement area: 100
- Aircraft on apron area: 100
- Ground vehicles on movement and apron areas: To be determined

The determination of the maximum number of aircraft on the movement area is based on the assumed peak traffic at a large airport with an optimum configuration for traffic capacity. The airport analyzed was Dallas Fort Worth [Ref. 6]. Another way of analyzing the peak amount of aircraft on the movement area is to simply calculate the maximum number of aircraft that could be accommodated on the taxiways waiting for departure or arriving. Based on this analysis, it was found that both Atlanta and Dallas-Fort Worth could accommodate about 150 aircraft. This represents the upper extreme, therefore the requirement can be set somewhat less than this. The value of 100 accounts for projected peak traffic, while adding some margin for increased taxi-out delays.

Conclusions

This paper has presented an overview of the surface surveillance requirements that have been developed by ICAO and RTCA to support A-SMGCS. While some of these requirements have undergone validation, additional analyses and testing is required to complete the validation effort. Some testing has been accomplished as part of NASA's LVLASO demonstration program. The results of these tests are reported in References 7 and 8. It is expected that the test programs underway and planned, both within the U.S. and abroad, will complete this validation process.

Acknowledgments

The authors gratefully acknowledge the contributions of the NASA LVLASO team, including Steve Young and Denise Jones, for their guidance and contributions to the research presented in this paper. We also wish to acknowledge Greg Joyner of the FAA for his leadership and guidance of the ICAO A-SMGCCS activities.

References

1. "FAA Aviation Safety Statistical Handbook - Runway Incursions," 1998.
2. "Draft Manual of Advanced Surface Movement Guidance and Control Systems (A-SMGCS)," 16th Meeting of the ICAO All Weather Operations Panel, Montreal, Canada, June 1997.
3. "Minimum Aviation System Performance Standard for Automatic Dependent Surveillance Broadcast (ADS-B)," RTCA/DO-242, February 19, 1998.
4. "Operational Requirements for A-SMGCS," Report of the 16th Meeting of the ICAO All Weather Operations Panel, Montreal, Canada, June 1997.
5. "Taxiway Deviation Study," British Airport Authority Planning Services, November 1987
6. "Dallas Fort Worth International Short-Term Capacity Enhancement Plan," 1994.
7. Evers, C., Cassell, R., and Lee, D., "Analysis of ADS-B, ASDE-3 and Multilateration Surveillance Performance - NASA Atlanta Demonstration," 17th DASC, November 1998.
8. Young, S. and Jones, D., "Flight Demonstration of Integrated Airport Surface Movement Technologies," NASA TM-1998-206283, January 1998.

LOOP TECHNOLOGY (LOT) AS AN ALTERNATIVE SURFACE SURVEILLANCE SYSTEM

Vern Edwards, Federal Aviation Administration (FAA), Washington, D.C.
Carl Evers, Rannoch Corporation, Alexandria, VA

Abstract

This paper provides an overview and preliminary results for demonstration of a prototype distributed loop-based system called LOT. This paper contains only the understanding and views of the authors and is not intended to reflect the official position of the FAA. The FAA completed the first phase of a prototype system installation in 1997 at Long Beach Airport (LGB) in California. This prototype applies inductive loops, a mature technology, to the airport surface surveillance application. The project also involves a technology transfer of neural network signal processing technology from Department of Defense to form a non-cooperative surface movement sensor system. LOT has the potential to be used in a standalone mode or as a supplemental sensor input to an Airport Surface Detection Equipment (ASDE) radar surface surveillance system.

Introduction

Most automation efforts to date have rightly concentrated on surface safety and incursion problems at higher level airports, where traffic is more dense and complex. The Airport Surface Detection Equipment-3 (ASDE-3) has been implemented to provide surface surveillance at these higher level airports. However, an alternative surveillance system is needed to address airports where ASDE-type equipment will not be installed or the monitoring requirement is too localized to warrant an ASDE-type system. The system employed should offer more flexibility in application (i.e., limited application vs. full airport), be scalable, and allow for incremental implementation to reduce or manage the cost. A non-cooperative surveillance system is preferred, as no new aircraft equipage is required to provide surveillance.

The FAA is evaluating LOT for its surface surveillance application. The project is currently comprised of two phases – a technical assessment and an operational assessment. The first phase, conducted in 1997, investigated the technical feasibility and performance of a distributed inductive loop-based system to provide surface surveillance. Evaluations were performed LGB in California and at Hyde Field in Maryland. LOT demonstrated the capability to detect aircraft and ground vehicles when they taxi over an inductive loop. LOT goes one step beyond the typical highway loop implementation by using the inductive signature to classify the aircraft or ground vehicle. Detection and classification information is used to track traffic in the airport loop coverage area. The second phase began in 1998, and the LOT system installed at LGB received enhancements of an approach radar interface, safety logic and a ground controller display. Evaluations are planned for late 1998.

While LOT can provide full runway/taxiway coverage, a more cost-effective implementation is a limited system installation that addresses the immediate and localized safety needs of the airport. Figure 1 illustrates four LOT applications, which are further discussed in the following sections. The flexibility exists to install inductive loops to support any one or combination of these applications at any location(s) on the airport.

0-7803-5086-3 /98/$10.00 ©1998 IEEE

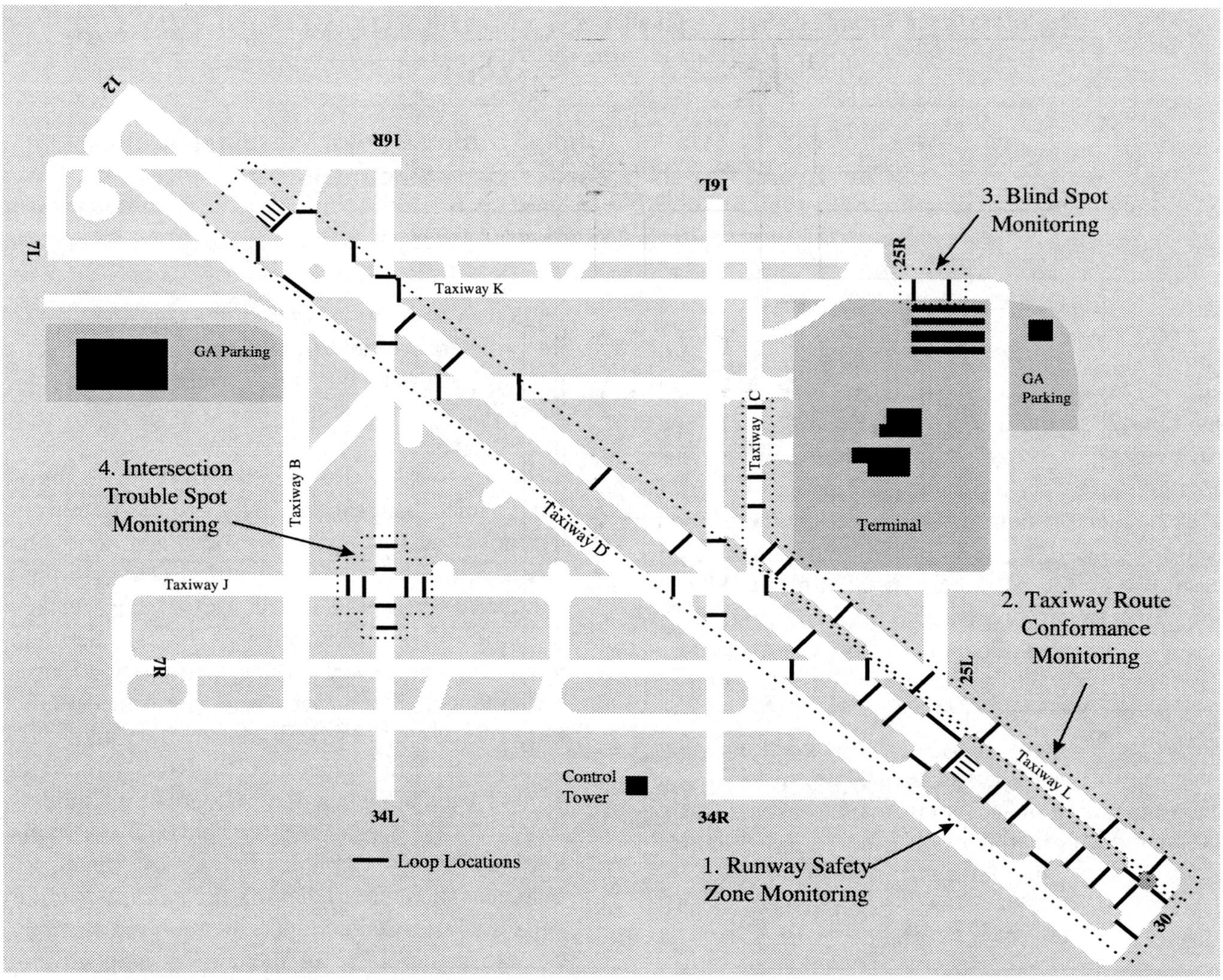

Figure 1. LOT Conceptual Applications

Operational Concept

Air traffic controllers can use LOT to aid in minimizing the risk of incursion incidents and improve situational awareness. In addition to providing an operational display of surface traffic where loops are installed, LOT can provide the following capabilities (See Figure 1): (1) runway safety zone violation, (2) standard taxi route conformance monitoring, (3) blind spot monitoring, and (4) monitoring of troublesome intersections. The system will be used by local and ground controllers to augment visual observations of aircraft and vehicle movement, and may be used by the controllers under the following conditions:

- Visibility is less than the most distant point in the active movement area where loops are installed.
- Traffic is operating in blind spots with respect to the tower.
- Verification of position of aircraft or vehicles at specific update points is needed.
- In the controller's judgment, the systems used will assist in the performance of their duties at any time.
- There are system-generated aural and visual indications of violation of runway threshold or an unsafe condition on runway.
- There are system-generated aural and visual indications of non-conformance with standard taxi route operation.

0-7803-5086-3 /98/$10.00 ©1998 IEEE

The LOT-derived position and identification information will be used to assist the controllers with the following activities:

- Formulating clearances and control instructions to aircraft on the movement area.
- Determining when the runway is clear of aircraft and vehicles prior to a landing or departure.
- Positioning aircraft and vehicles.
- Determining the exact location of aircraft, or their relationships to other aircraft on the movement area.
- Monitoring compliance with control instructions.
- Confirming pilot-reported positions.
- Providing directional taxi information (excluding heading) on pilot request.
- Identifying the weight class of aircraft to assist in maintaining proper departure separation standards.
- Determining when an aircraft has lifted off or touched down.

LOT Overview

System Architecture

The LOT system uses Commercial-Off-The-Shelf (COTS) inductive loop technology and a computer workstation to provide surface surveillance and surface traffic safety processing for aircraft and ground vehicles operating on runways and taxiways where inductive loops are installed. The LOT system consists of the following major elements:

Surveillance Subsystem

The surveillance subsystem forms the heart of a loop-based surveillance system. It consists of the sensor (loop wire), detection component and tracking component. This subsystem provides track reports to the safety subsystem for display and processing by the safety logic.

Inductive Loops

Inductive loops provide the sensing element for the Loop Detection Component (LDC). They can be located over the whole runway/taxiway surface or in strategic locations (i.e., troublesome intersections, blind spots). Inductive loops are implemented on the airport surface in rectangular saw cuts, which span most of the width of the runway and/or taxiway to be monitored (see Figure 2). In the demonstration system installed at LGB, the taxiway loops are 10'x65' and the runway loops are 10'x150'. The loops are spaced to achieve a nominal update rate of once every four seconds at typical taxi speeds. A loop consists of three turns of encapsulated stranded wire, which is laid down in a saw-cut groove. A backer rod is installed in two-foot sections spaced two inches apart on top of the loop wire. The saw cut is then covered with epoxy sealant to ensure against water seepage and prevent cracking of the surface where installed. A key factor in the overall performance of inductive loops is the quality of the installation and keeping the loop wires from moving.

Loop Detection Component (LDC)

Each inductive loop has an associated LDC. The LDC is located within the controller cabinet, which is usually within 500 - 1000 ft. of the loop itself. Each loop is connected to the LDC via a lead-in cable. In the LGB installation, each controller cabinet houses up to four LDCs. The LDC provides detection of aircraft and ground vehicles. The LDC also extracts signatures and performs classification and single loop velocity estimates.

The LDC drives a current through the loop wire at the loop resonant frequency and the current creates an electromagnetic field around the loop wire. The loops are laid out such that the shortest side or length is normal to the direction of movement of the aircraft or vehicle across the loop. The metallic structure of the aircraft/vehicle cuts (see Figure 3) through the electromagnetic field and sets up eddy currents, which tend to lower the inductance of the loop in proportion to the aircraft underside surface area and height above the loop [Ref. 1]. As the aircraft/vehicle continues to move over the loop, the result is a varying inductance change, which produces a shape-based signal "signature" or pattern output (see Figure 4).

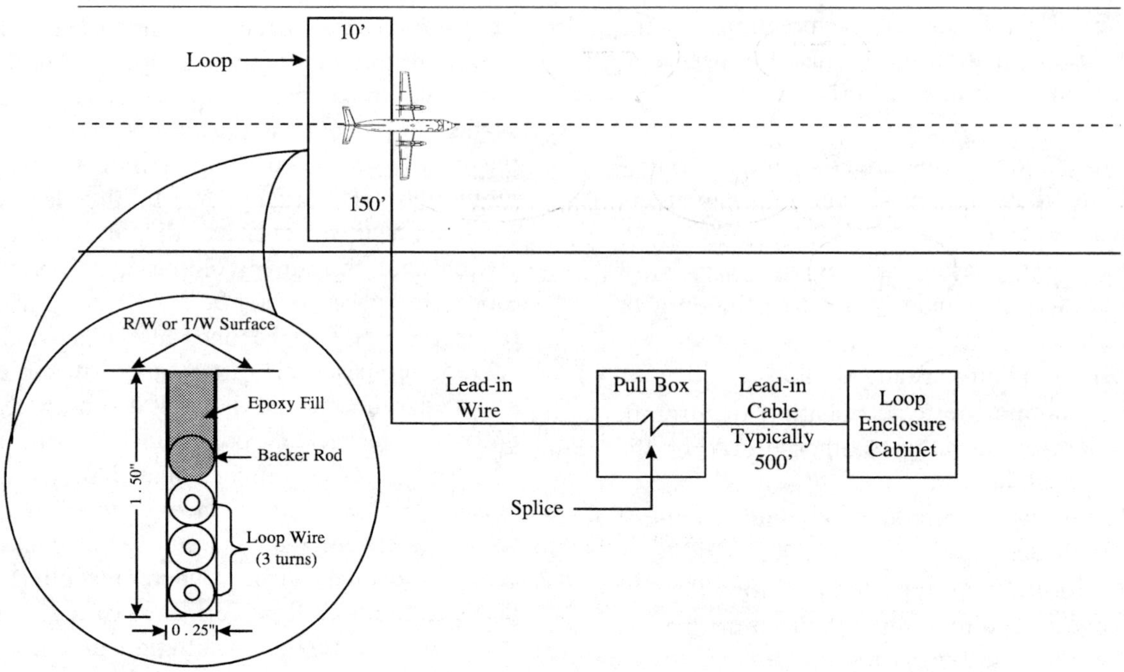

Figure 2. Physical Layout of Loop System Airfield Components with Loop Wire Cutaway View

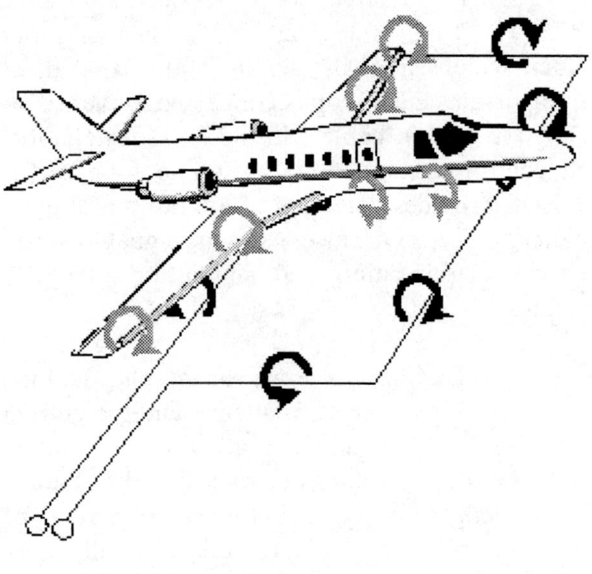

Figure 3. Aircraft Cutting Loop Flux Lines

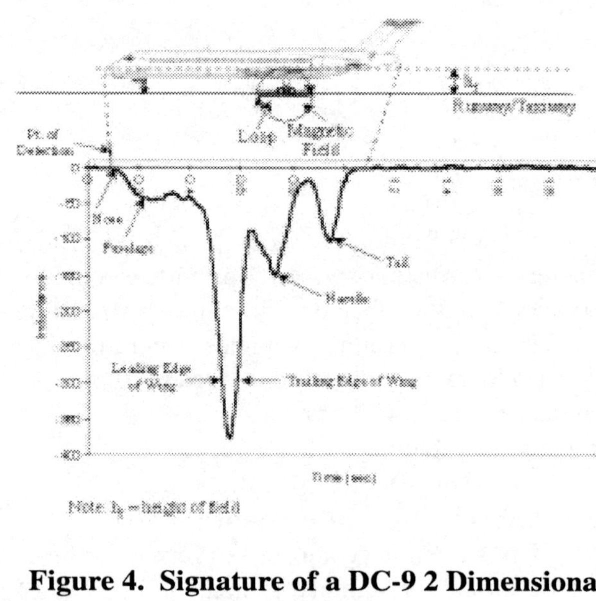

Figure 4. Signature of a DC-9 2 Dimensional Model

This signature is unique for each aircraft type. When the time-varying signal exceeds the loop detector detection threshold, detection is declared. The detection will not be dropped until the signal drops below the threshold and remains there for a parameter-determined length of time. Small ground vehicles can also exhibit a distinguishing signature (double hump), which can be discriminated from aircraft signatures (see Figure 5).

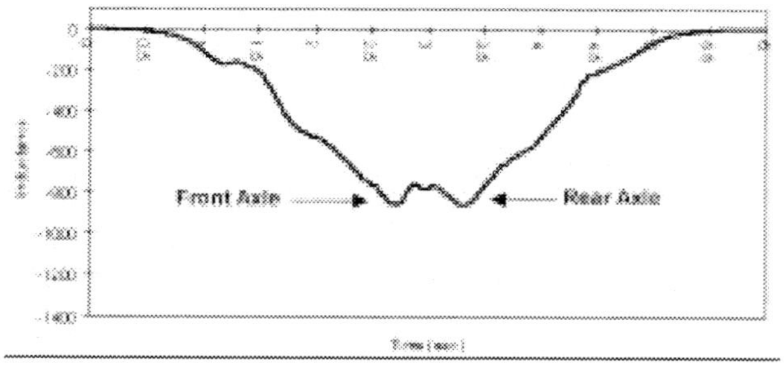

Figure 5. The Distinguishing Double-Hump Signature of a Small Ground Vehicle

Tracking Component (TC)

The TC is a computer workstation that performs loop-to-loop correlation and tracking function position, velocity, and heading based upon kinematics. It also performs data handoff between arrival surveillance data and the loop data from the airport surface.

Safety Subsystem

The safety subsystem provides for processing of track reports from the surveillance subsystem by the safety logic for detection of runway occupancy and runway safety zone violation.

Controller Display

The display provides the controller with a situational display of traffic movement in the inductive loop coverage area(s). Aural and visual indications are provided in the event that there is a violation detected by the safety logic (e.g., violation of runway safety zone, entry into blind spot area, etc.).

Neural Network Classification

The demonstration prototype LOT installed at LGB has a neural network classifier associated with each loop detector component. Although a certain amount of information and tracking can be done from detection alone, classification by neural network processing forms the essence for a loop-based surface system and its ability to track aircraft and vehicles. Classification provides information, which can be used to help accurately correlate a detection update with an established track.

As mentioned under *Loop Detection Component*, each aircraft type produces a unique signature. An aircraft's unique signature is based on its distinguishing characteristics or "features" (e.g., location of nose wheel engine placement, shape of wing, etc.). Just as the controller has been trained to recognize different aircraft type by visual observation of certain features, a properly trained neural network classifier can achieve similar results by emulating a similar process. An air traffic controller is trained by his supervisor to recognize and classify different aircraft types by visual observation of their features. Being able to identify the type of aircraft gives the controller much more information about the flight characteristics of that aircraft.

The neural network is trained, Figure 6, to the population set of aircraft and vehicle signatures appearing at the airport. Development of the signature training set requires collecting aircraft and ground vehicle signatures or patterns and visually truthing them by type. Once this training is completed, subsequent signature inputs for which it has been trained can be classified with high confidence. Presently, we are only attempting to classify three major classes: aircraft, ground vehicle, and unknown. The aircraft class is subdivided into the sub-classes of small (representing GA type aircraft), large (representing air carrier type aircraft), and heavy (representing the larger air carrier type aircraft). Planned enhancements to the neural network classifier may enable classification of many aircraft down to the specific type (e.g., B737, MD-11, DC-9, etc.). Preliminary testing indicates that an enhanced classification capability may enable the system to perform velocity estimates with surveillance from a single loop.

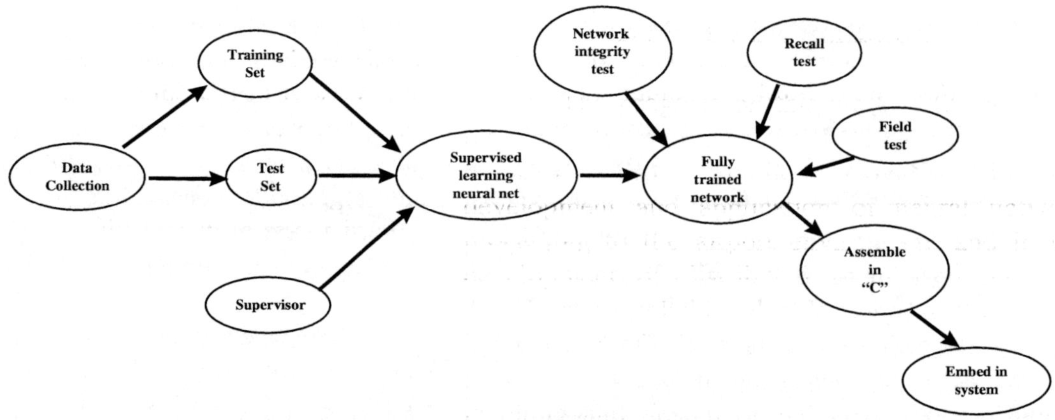

Figure 6. Neural Network Training Process

Preliminary LOT Evaluation Results

LOT evaluations [Ref. 4, 5, 6] were performed at Hyde Field and LGB. RTCA defines guidelines and recommended requirements for surface movement sensors [Ref. 7]. A qualitative assessment with respect to key RTCA-recommended requirements (in italics) is provided below. Future testing will better quantify performance results.

Probability of Detection > .9995

Detection performance testing with ground vehicles and a small aircraft was performed. A range of speeds up to 60 mph was tested. Loop crossing points from the centerline out to the runway or taxiway edge were tested, and various angles of crossing were tried. Test results indicate that there is a high probability of detection when the detector is set to provide sufficient sensitivity. During initial testing, missed detections were experienced for slow moving ground vehicles and small aircraft. However, using a detector set to provide improved sensitivity resolved the problem.

False Detection Reports < 1 per 8 hours

Erroneous detections were not experienced during any of the testing. The loops, being wires, act as antennas, yet all noise was well below the threshold of detection, thus having no impact on detector performance. Ground vehicles driven over the lead-in wire in the grass areas did not cause false detections. It was found that aircraft could be detected before entering a loop, if the detector sensitivity is set too high. Failure to set the sensitivity at the proper setting could potentially result in a false detection of a hold line crossing.

Coverage of Runway and Taxiway Centerlines where Surveillance is Required

Surveillance for the full width of the runway or taxiway was demonstrated. This provides the capability to detect ground vehicles, which sometimes travel along the edges of taxiways and runways. Reliable coverage is provided in regions where the loops are installed, thus demonstrating the capability to provide localized surveillance to address trouble spots. Accordingly, using a localized approach allows LOT to be implemented in a cost-effective manner. System scalability will be demonstrated in Phase II with the addition of more loops to expand the LOT coverage area.

Position Accuracy Within ±7.5 Meters

LOT determines position based on initial detection of an aircraft or ground vehicle. Ideally, this detection occurs as soon as the aircraft enters the loop just over the hold line. Initial detection occurs when the inductance measured by the detector exceeds a detection threshold. Detection typically occurs soon after the aircraft's nose or ground vehicle's front bumper crosses the wires at the loop entry point. Preliminary indications are that the ±7.5 meters can be achieved. Testing with small aircraft and ground vehicles results in accuracy performance better than ±2 meters.

Position Reported Within One Second of Passage Through Sensing Area

LOT provides an indication of detection prior to the aircraft transversal of the loop or sensor area, thus exceeding the requirement. A demanding application for latency is safety zone violation detection. Detection should occur soon after the aircraft proceeds from a stop at the hold line and the nose enters the runway safety zone. Initial testing has shown that detection will occur within four seconds for aircraft. This time is expected to improve to less than two seconds when the detector is re-configured for improved sensitivity in Phase II.

Classification to Four Weight Classes - Probability of Correct Classification > .9

Testing has been performed to demonstrate classification by aircraft type (e.g., B-737, MD-80) as opposed to weight class (i.e., heavy, large, small). Classification by type is a more demanding application. Figure 7 provides sample signatures. Both the heavy and large aircraft tend to have a peak inductance change associated with the wings. The level segment in front of the peak is associated with the fuselage. The gently sloping segment after the peak is associated with the fuselage and the tail. Unlike the large and heavy aircraft, small aircraft signatures tend not have a level segment in front of the peak associated with the wing. However, in many cases a segment associated with the small aircraft tail is identifiable. Large ground vehicles will give a single peak with no identifiable leading or trailing segments. Small ground vehicles will give a double peak signature. Initial testing indicates that many large and heavy aircraft can be classified by type. However, small aircraft classification by type may be difficult. The small aircraft signatures tend to be similar to each other, thus making it difficult to discriminate. However, small aircraft signatures are distinct from signatures of other aircraft classes. LOT can differentiate between ground vehicles and aircraft. While aircraft type information is desirable, it is not essential for an operational system. The goal for the LOT project is to classify traffic in one of the following categories: heavy aircraft, large aircraft, small aircraft, and ground vehicles. Neural network logic has been modified to perform classification by category as opposed to type. Classification performance testing will be performed in Phase II.

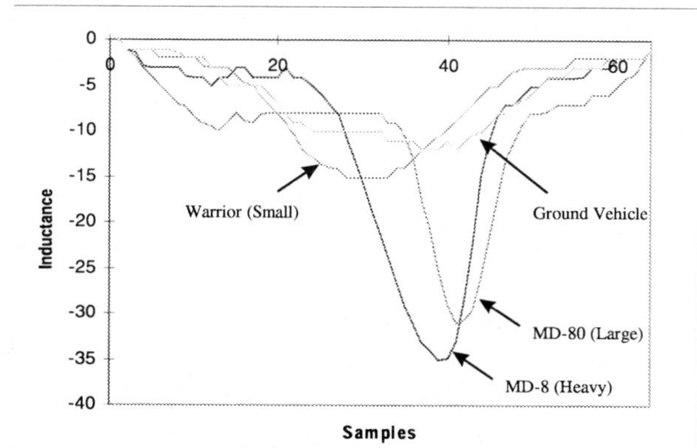

Figure 7. Aircraft and Ground Vehicle Signatures

Provide Velocity Measured for Speeds up to 20 Meters per Second

Velocity was measured for speeds up to 50 meters/sec. The primary means for estimating velocity is the two-loop method. An estimate of velocity is performed each time that an aircraft crosses any two consecutive loops. The speed is calculated using time between loop pairs and distance traveled equal to the separation between loops. Typical loop separations are 300 to 500 feet. Heading is determined from the loop crossing sequence. A second method for estimating velocity is the single-loop method, which is applied for taxi scenarios (e.g., aircraft stops between loops then moves again) where the two-loop method can not be used. Distance traveled equals aircraft length, and time equals single loop crossing time. Aircraft length is obtained via a database lookup, which is performed upon classification.

Velocity Accuracy Within ±2 Meters per Second for Velocities Exceeding 5 Meters per Second

Velocity is estimated using either the single-loop method or the two-loop method. The single-loop method has demonstrated velocity accuracy performance better than 5 meters/sec. for small aircraft. However, accurate classification to aircraft type is critical to achieving a high degree of accuracy. An assessment of the two-loop method has not been performed. Two factors that influence the accuracy of the two-loop method are variability in point of detection and vehicle dynamics. Vehicle dynamics will have the biggest influence on accuracy. For instance, an aircraft that changes acceleration between loops will introduce errors in the velocity estimate.

Indicated Direction is Correct Within ±89 Degrees

Reliable direction determination was demonstrated. The system determines direction of travel when an aircraft crosses two consecutive loops. The heading is assumed to be parallel to the pavement centerline in the direction of travel. Classification information is used to confirm that any two consecutive crossings are from the same aircraft. Taxiing with frequent stops is common for many airports. The system maintains knowledge of an aircraft's presence in the zone between loops, which is used to determine direction when the aircraft crosses the next loop. Testing was performed to determine if a single loop could be used to provide direction determination. Several loop shapes were tried, but single-loop direction determination was not demonstrated.

Conclusions

The initial evaluations have demonstrated that LOT can provide reliable aircraft and ground vehicle detection. Airport surface surveillance requires large loop sizes which places demands on detector sensitivity performance. Sufficient sensitivity is required to ensure reliable detection and classification. Initial indications are that the system can achieve the RTCA recommended requirements. The overall viability of LOT as an alternative airport surface surveillance will be demonstrated in Phase II. In particular, classification and detection of runway safety zone violations will be demonstrated. System performance will be evaluated against RTCA and International Civil Aviation Organization (ICAO) Advanced Surface Movement Guidance and Control Systems draft requirements [Ref. 8].

Acknowledgements

The progress on this work is the result of the dedicated work of the LOT team members. Orincon Corp. provided expertise in the development and application of neural network processing to the airport environment and in the development of distributed loop-based tracking. 3M Corp. provided invaluable technical support for independent off-site evaluations. Pete Mills of the FHWA provided insights and guidance for the planning and design of inductive loops. Others acknowledged are Volpe Center, TASC, ATC, Rannoch, Edwards & Kelcey, and JIL. A special thanks goes to Long Beach Airport for being the host for this prototype demonstrations system and their operations personnel who provided invaluable support during the installation and testing.

References

1. Federal Highway Administration (FHWA), *Traffic Detector Handbook*, Second Edition, National Technical Information Service (NTIS) Report No. FHWA, July 1990.
2. W.T. Illingworth, *A Practical Guide to Neural Nets*, Addison-Wesley Publishing, March 1991, ISBN 0-201-56376-0.
3. Advisory Group for Aerospace Research & Development (AGARD) Neuilly sur Seine, France, AGARD Lecture Series 179, *Artificial Neural Network Approaches in Guidance and Control*, September 1991, NTIS AD-A244 247.
4. *Inductive Loop Technology Off-Site Test Report*, FAA Report No. AND410LOT-OFF/0598, May 19, 1998.
5. Petrescu, J., *Loop Incursion Prevention Lab Ground Vehicle Analysis Test Report*, DOT Report Number OCR 98-4139-U-0091, March 1998.
6. *Draft Inductive Loop Technology On Site Test Report*, AND-410-LOT-OST-5-98, May 1998.
7. *Guidance and Recommended Requirements for Airport Surface Movement Sensors*, Document No. RTCA/DO-221, April 1994.
8. *Draft Manual of Advanced Surface Movement Guidance and Control Systems*, AWOP/16-WP/724, Draft 4, June 1997.

APPLICATION OF ADS-B FOR AIRPORT SURFACE SURVEILLANCE

Daniel Storm Hicok, Rannoch Corporation, Alexandria, Virginia

Derrick Lee, Rannoch Corporation, Alexandria, Virginia

Abstract

Automatic Dependent Surveillance Broadcast (ADS-B) is a function on an aircraft or ground vehicle that periodically broadcasts its state vector (horizontal and vertical position, horizontal and vertical velocity) and other information [1]. The broadcast ADS-B message provides surveillance information to other users, principally Air Traffic Control (ATC) and aircraft/vehicle operators. The applications for ADS-B include ATC display of traffic, runway incursion detection and alerting, and Cockpit Display of Traffic Information (CDTI). The Federal Aviation Administration (FAA) and the National Aeronautics and Space Administration (NASA) are investigating the suitability of this technology to support these applications in the airport surface environment. NASA has recently tested ADS-B using 1090 MHz data transmission in an airport surface environment as part of the Low Visibility Landing and Surface Operations (LVLASO) program. This paper presents results of the ADS-B testing and an assessment of how well 1090 MHz ADS-B performs with respect to surveillance system requirements established by the International Civil Aviation Organization (ICAO) and RTCA [2, 3]. In addition, data collected from vehicle to vehicle tests is analyzed to determine how well on-board avionics can utilize ADS-B data.

Introduction

In the United States, demand for air travel continues to increase, however, many airports are unable to accommodate this increase by building new runways. The FAA and NASA are investigating ways to increase airport efficiency and safety as air travel continues to increase over the next several decades. To realize these goals, methods must be developed to provide highly accurate surveillance data in all weather.

ADS-B is a technique for obtaining surveillance data. An ADS-B equipped aircraft or airport vehicle determines its own position and velocity (and possibly other information such as planned route) and broadcasts it periodically. Any properly equipped user within range can receive and use this information. Broadcasts are performed automatically without external stimulus (i.e., interrogation). ADS-B is known as dependent and cooperative surveillance; dependent because it relies on the availability of onboard navigation and transmission equipment, and cooperative because information can only be obtained from properly equipped aircraft.

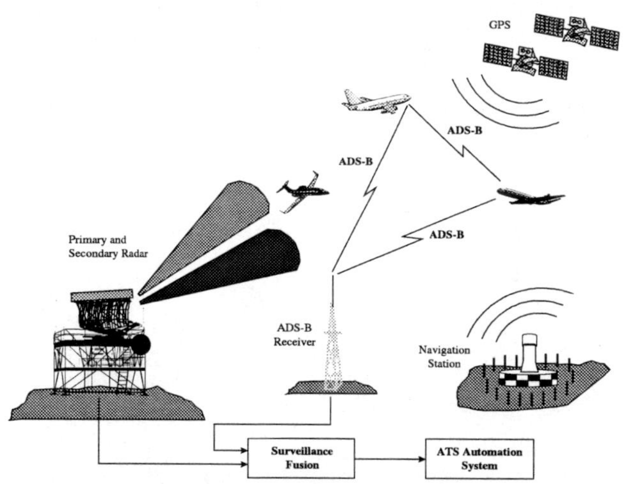

Figure 1. ADS-B Concept

The near term benefits of ADS-B in the airport surface environment include improved surveillance data for use by ATC and improved navigation and situational awareness tools in the cockpit. Airports can employ a system of ground stations that receive the ADS-B reports of all properly equipped traffic for use in ATC displays. Other aircraft can receive the ADS-B reports and use them in a Cockpit Display of Traffic Information (CDTI) for increased situational awareness and in collision avoidance logic.

0-7803-5086-3 /98/$10.00 ©1998 IEEE

If ADS-B is to be used in these applications, an implementation should be developed that meets the standards for an Advanced Surface Movement Guidance and Control System (A-SMGCS) as established by ICAO and RTCA. NASA recently conducted flight experiments with a 1090 MHz ADS-B equipped Boeing 757. This paper addresses this implementation of ADS-B and how well it performed with respect to A-SMGCS standards. This paper contains only the understanding and views of the authors and is not intended to reflect the official position or view of the FAA or NASA.

Test Description

The LVLASO program completed a series of flight experiments at Hartsfield International Airport in Atlanta in 1997. The program is focused on providing technology and operating procedures for achieving clear-weather capacity in instrument-weather conditions at airports while also improving safety [4]. Technologies under investigation are advanced flight deck displays and surveillance systems. The NASA 757 was equipped with a CDTI, which provided the flight crew with enhanced guidance and situational awareness information. The surveillance sensors tested were primary radar using ASDE-3/AMASS and multilateration and ADS-B using an Airport Surface Target Identification System (ATIDS). Each sensor was evaluated as a stand-alone system.

ATIDS is a system consisting of ground stations (known as receiver/transmitters or RTs) and a central processing station (known as a Master Work Station). The RTs receive and decode Secondary Surveillance Radar (SSR) transmissions and relay them to the Master Work Station, which can use the time of arrival to calculate a multilateration position of the aircraft. The RTs can also receive and decode ADS-B messages and relay them to the Master Work Station. This single system provides two sources of surveillance information. ATIDS in the LVLASO testing used 5 RTs to cover the north movement area. The location of each RT is indicated by a circle in Figure 2 with the number that was assigned to each.

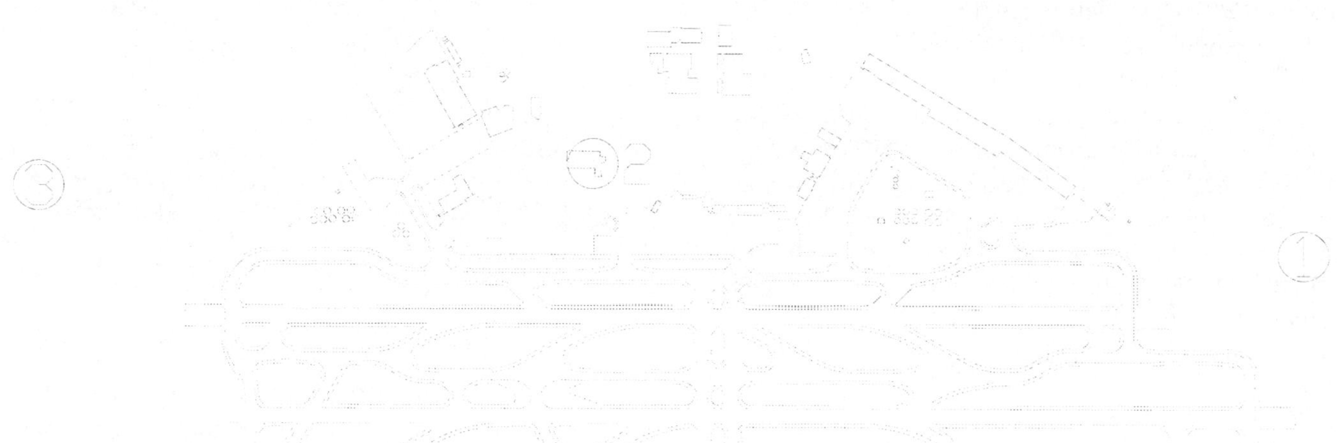

Figure 2. Ground Station Arrangement

0-7803-5086-3 /98/$10.00 ©1998 IEEE

The performance of an ADS-B ground system is a function of the location and number of receivers. The results presented here are dependent upon the configuration of the ATIDS receivers as installed in Atlanta. The receivers were sited to optimize their ability to meet the multilateration function. They are arranged to provide the best geometry for determining position. If the receivers were arranged to provide the best ADS-B coverage, it is likely that the ADS-B performance would be improved and the number of receivers required would be less. Therefore, the results presented here must be viewed in the context of an ADS-B system that doubles as a multilateration system.

The LVLASO test used local area differential global positioning service (GPS) as the navigation sensor. A base station broadcast GPS corrections by VHF datalink to the 757. Local area differential GPS typically has lateral position errors of several meters or less. The position determined by the navigation system was then broadcast.

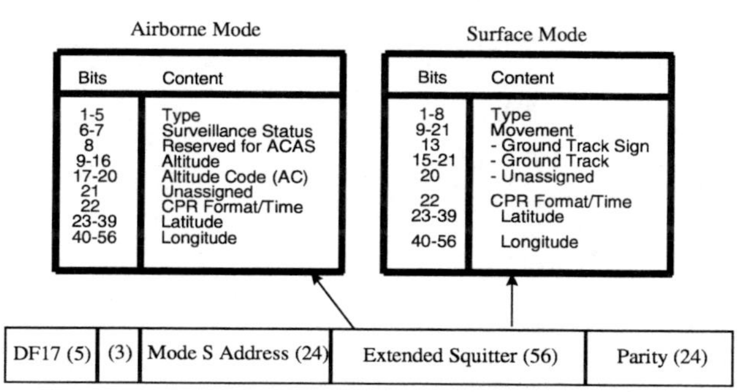

Figure 3. ADS-B Message Using DF 17

The NASA 757 was equipped with a GPS receiver and a 1090 MHz Mode S extended squitter transponder. This was a transponder capable of transmitting a 112 bit message which included the Mode S address and a 56 bit ADS-B message. This 112 bit message was formatted using downlink format 17 (DF 17) [5]. Aircraft position was calculated once per second and the most recently computed position was transmitted nominally twice per second. Two different ADS-B messages were transmitted, depending on whether the aircraft was airborne or on the airport surface as shown in Figure 3. ADS-B transmissions alternated between the top and bottom mounted antennas when airborne. ADS-B transmissions were radiated only from the top mounted antenna when the aircraft was on the ground.

Data Collection

1090 MHz ADS-B data was logged at the ATIDS Master Work Station for a series of NASA 757 runs. The runs included departures, arrivals and movement area taxiing. Extracts from the log files of several runs were used to create a master file for the north side movement area, which was used to evaluate coverage and update rate. This file was assembled to include a single pass on each of the runways and taxiways on the north movement area. The master file included unprocessed 1090 MHz message data and position processed data. The data collection was performed during low traffic periods, thus degradations in ADS-B surveillance performance due to garbling were minimized.

Ground System Performance

ICAO and RTCA [2, 3] have established tentative performance standards for airport surface surveillance systems. The data collected during the LVLASO trials was analyzed and the results are presented and compared to those standards.

Surveillance Coverage

A-SMGCS requirements state that an airport surface surveillance system must provide coverage of the entire movement area as a minimum. To determine if this requirement was met, the master file of ADS-B position reports was plotted on a map of the movement area to determine if coverage gaps existed. First this was done for each RT individually. A position was plotted for each ADS-B update that was received error free (all 112 bits correctly decoded). Statistics were compiled which describe the percentage of messages received error free relative to the total number of transmissions for the entire movement area and for each of six regions of the movement area. These statistics are presented in Table 1.

Table 1. ADS-B Ground System Reception Statistics

	Region 1	Region 2	Region 3	Region 4	Region 5	Region 6	All Regions
RT 0 (Delta Hangar)	52.4%	73.4%	65.5%	36.4%	57.5%	64.6%	54.8%
RT 1 (Ford Plant)	88.4%	58.3%	50.6%	80.2%	58.3%	78.2%	69.2%
RT 2 (Renaissance Hotel)	69.2%	42.9%	8.6%	68.8%	80.7%	54.4%	59.5%
RT 3 (FAA Region Bldg)	65.7%	91.0%	70.7%	47.3%	48.7%	62.5%	59.9%
RT 4 (Concourse C)	80.8%	54.6%	4.1%	56.4%	58.9%	53.6%	52.6%
At Least 1 RT of 5	98.9%	98.0%	95.0%	95.6%	98.1%	99.6%	96.5%

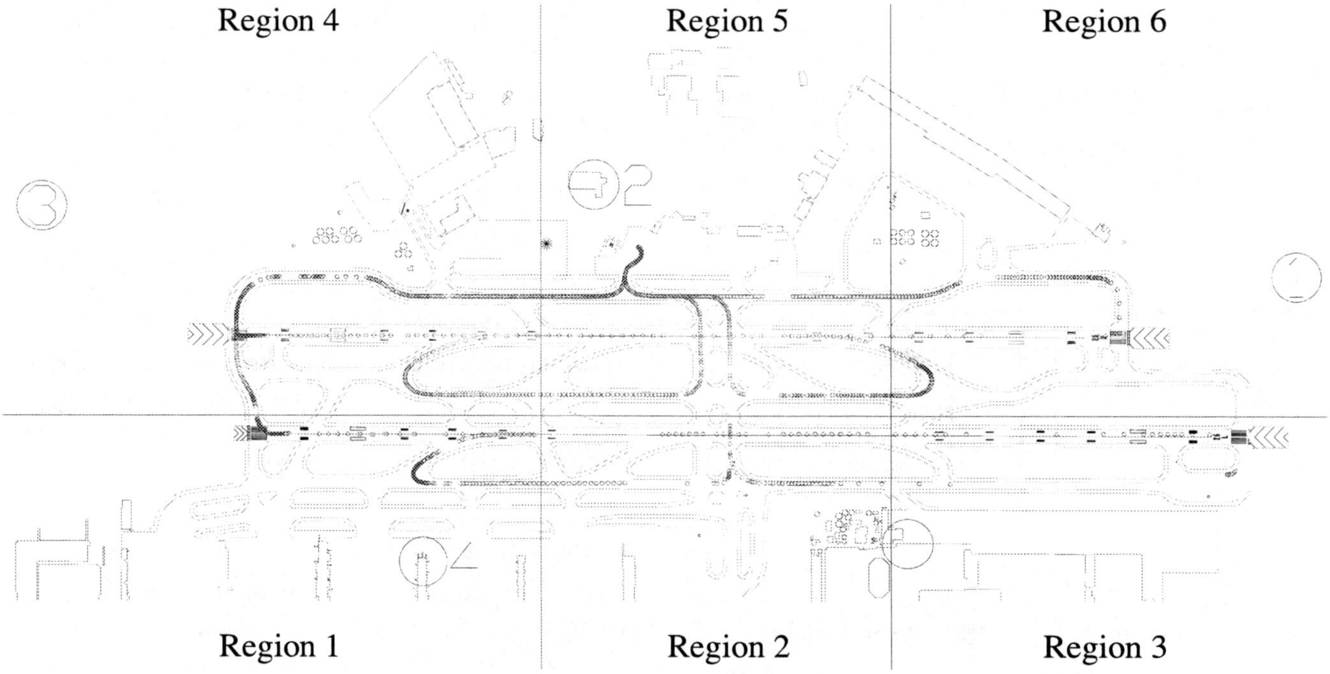

Figure 4. Error-free ADS-B Reception by RT 2

Figure 4 shows the ADS-B reports received error free by RT 2. This is typical of the reception performance of all the RTs. The statistics in Table 1 demonstrate that coverage is very good in some areas (80% - region 5) and almost non-existent in others (8% - region 3). The very low reception in region 3 was determined to be caused by multipath from the large Delta hangars that are located adjacent to this taxiway. For the entire movement area, RT 2 received approximately 60% of transmissions error free.

The coverage by single RTs demonstrated the limitations of 1090 MHz ADS-B. The broadcasts are susceptible to corruption from multipath. In other cases, line of sight between the aircraft and the RT was not maintained and the transmissions were not received. It is clear that no single receiver will be capable of providing complete movement area coverage. A network of carefully placed receivers is necessary to provide complete coverage.

Next, the performance of the system of receivers was analyzed. Positions reports were plotted where at least one of the five RTs received an error free transmission. The statistics are shown in the last row of Table 1. Approximately 97% of all transmissions were received by at least one of the RTs. In the areas where transmissions were missed, more runs were analyzed to determine if there were areas that consistently had coverage problems. It was concluded that the missed updates were randomly distributed throughout the movement area. Even though 100 percent of transmissions were not received, the entire movement area was covered by the system of 5 RTs.

Bit error analysis was conducted to determine if error correction would significantly improve reception performance. For each RT, the master file was examined to determine what percentage of total transmissions had been received with no errors, what percentage had no message bits in error but had failed parity (error in parity bits), and what percentage had 1, 2, or 3+ bits in error. The remainder either had too many errors to be recognized as from the NASA 757 or were not received at all. The RTs generally received approximately 60% with no error and about 6% with 2 bits or less in error. Further investigation is required to determine the performance improvement offered by error correction, but if messages with 2 or less bits in error are recoverable, this would increase the reception by about 6% for each RT.

During testing, a van was driven on the movement area to investigate 1090 MHz ADS-B coverage performance for ground vehicles. The reception performance of the van ADS-B transmissions was similar to the results obtained for the 757 (97.5%). Based on these results, ADS-B is expected to provide equivalent coverage of ground vehicles and aircraft.

The ADS-B equipped van was also driven through the ramp areas to evaluate the coverage in these regions. Large coverage gaps existed in several of the ramp areas. Reception percentage varied from 50% (Ramp 1) to 96% (Ramp 5). Overall, the ramp areas had 69% coverage, well below the 97% coverage of the movement area.

The LVLASO test confirmed that 1090 MHz ADS-B can be compliant with the A-SMGCS requirement to have complete movement area coverage of aircraft and vehicles as long as a network of carefully sited ground stations is used. To determine if 1090 MHz ADS-B will be suitable for ramp area surveillance, more testing will be required.

Surveillance Approach Coverage

The A-SMGCS requirements state that the surveillance system must provide coverage of aircraft on approach sufficient for transition from the terminal radar to the surface surveillance system. Approach coverage performance was assessed, even though the system was optimized for surface surveillance. RT 3 was the only RT aimed toward the approach path. This RT had the longest range of reception on the approach path, with a first detection 17 NM from the airport. However, there was a gap in coverage between 13.3 NM and 7.4 NM. At approximately 4.8 NM, all RTs started consistently receiving error free messages. Five miles is adequate range to provide seamless operations with the approach radar. With the addition of antennas sited for the approach region, it is likely that greater range will be achieved, allowing for better transition from the terminal radar.

Surveillance Position Accuracy

The ADS-B position reports used the Compact Position Reporting (CPR) format as a means of encoding latitude and longitude values in a 34-bit message [5]. This formatting scheme divides the globe into approximately square discrete grid coordinates. The true position is mapped to the closest grid position. To achieve globally unambiguous decoding, two different grid spacings are defined, one for odd seconds and one for even seconds. By comparing the differences in the odd and even second readings, a globally unambiguous solution may be obtained. Because this method is discrete, the positions reported exhibit discontinuities between updates. Figure 5 shows and example of an actual position and its corresponding reported position on each grid. The reported position will alternate from one second to the next as it is reported on different grids. The maximum error introduced by this encoding technique is half the grid size, or 0.75 meters.

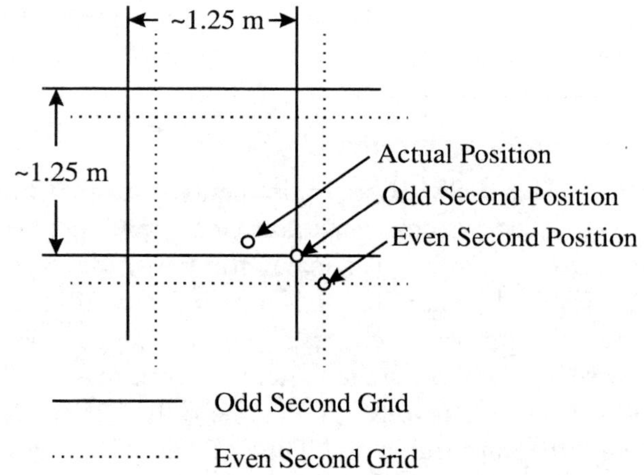

Figure 5. Example of CPR Encoded Position

ADS-B data collected during 757 taxi-only runs was analyzed to assess accuracy performance. The truth source was provided by an Ashtech Z-12 GPS, which achieves centimeter level accuracy. Straight segments of the taxi runs were used in the accuracy analysis. Cross track, along track, and total horizontal error were analyzed. Table 2 lists the results of the accuracy analysis. The 95% errors are less than 1.5 meters. This includes the error introduced by the CPR encoding.

	Mean	Std. Dev.	95%
Along Track Error (m)	0.0	0.5	±0.9
Cross Track Error (m)	-0.1	0.6	±1.4
Total Horizontal Error (m)	0.7	0.4	±1.42

Table 2. ADS-B Position Accuracy

The accuracy demonstrated by the ADS-B system easily meets the A-SMGCS requirement of 10 meters for longitudinal and lateral position accuracy in the movement area. The requirement for 3 meter accuracy in the ramp areas was not evaluated.

Update Rate

For purposes of this analysis, the definition of a position update was established to be any error-free ADS-B report rather than a new or unique position update. This is an important distinction because the GPS position was updated in the aircraft once per second, but position information was transmitted pseudo-randomly at a nominal rate of twice per second. This typically resulted in the same position being transmitted twice (and sometimes three times) consecutively. This definition was chosen because it better reflects the ability of the system to receive ADS-B information independent of the aircraft equipage. The performance of the system is not penalized for the position calculation rate of an individual aircraft. Also, it is expected that the calculation rate will be higher than 1 Hz in an operational system, so using this method better reflects the performance potential of this system.

The ADS-B unit used in the testing transmitted the Mode S extended squitter pseudo-randomly at a nominal rate of once every 0.5 seconds. To quantify the amount of missed updates, the actual time difference between valid (error-free) ADS-B updates was calculated for each update and statistics were compiled on this data. It was calculated that the probability of receiving an update within the nominal update interval plus the jitter (0.6 seconds) was 96.3%. There was a 98% probability of receiving an update within 0.95 seconds. This exceeded the A-SMGCS requirement of a 98% probability of receiving an update within 1 second.

Other Requirements

Identification. An airport surveillance system is required to present the identity of a target. ADS-B will satisfy this requirement by transmitting the flight number of the aircraft.

Surveillance Altitude Coverage. The A-SMGCS requirements state that coverage must be provided to at least 500 ft above the airport surface. The LVLASO test demonstrated that ADS-B can easily meet this requirement. The network of receivers consistently received transmissions up to several thousand feet.

Covered Speeds. A-SMGCS requirements state that a wide range of aircraft speeds must be accommodated. ADS-B easily meets these requirements.

In addition to the above requirements, other A-SMGCS requirements were not evaluated but are expected to be easily met. These include the accuracy of velocity data (less than 1 knot), the reference point accuracy (better than 3 m 95%), and probability of false targets. Other requirements were not evaluated and require more testing such as the ability to operate in all topography, the ability to enable incursion detection, the ability to detect deviations from assigned route, maximum traffic loading, latency, and surveillance integrity, continuity and availability. ADS-B will not comply with the requirement to detect unauthorized targets. ADS-B ground systems will only detect properly equipped aircraft and vehicles.

Vehicle to Vehicle Performance

In addition to testing the ADS-B ground system for compliance with surveillance system standards, several tests were performed to determine how well ADS-B could be used in vehicle to vehicle applications such as situational awareness and collision avoidance. This is an extension of the airborne application using direct aircraft to aircraft ADS-B transmissions to obtain traffic information. The alternative approach, which was also tested, is for a vehicle to obtain traffic information from a TIS (Traffic Information System). The TIS transmits surveillance data collected from the ground system to aircraft. In either implementation, the goal is to improve pilot situational awareness by a visual display of traffic information in the cockpit.

To determine the feasibility of this function on the airport surface, several vehicle to vehicle scenarios were tested. A Rannoch test van was equipped with an ADS-B receiver and antenna mast with an adjustable height (2.5 - 15 meters). An ATIDS RT inside the van provided the ADS-B reception and decoding. The van was used to simulate a general aviation or small commuter aircraft with a low antenna height (2.5 meters) and an air carrier aircraft by raising the antenna to 6 meters.

The first scenario, arrival monitoring, involved an aircraft at a runway/taxiway intersection waiting to cross while an aircraft is on final approach to that runway. To test this, the van was parked near the runway while the 757 was on approach. The test van antenna was positioned at a height of 3 meters. Line of sight was maintained throughout the approach. Reception was lost while the aircraft was banking when turning on to the final approach. During the straight segment of the approach, there were 6 time periods when gaps in updates exceeded 4.8 seconds (update rate of airport surveillance radar). The largest time gap was 10.8 seconds. These gaps were experienced mostly when the 757 was on the last five miles of the approach.

The second scenario, runway occupancy monitoring, was similar, involving an aircraft holding short of a runway while another aircraft takes off on that runway. The holding aircraft must be able to see the aircraft on the active runway on the cockpit display. To test this scenario, the van was parked while the NASA 757 landed or departed on one of the runways. Line of sight was maintained throughout the departure and during taxiing. There were significant drops in ADS-B updates during taxiing and departure. Several antenna heights and parking locations were tested. All site/antenna height configurations experienced significant gaps in coverage.

The third scenario was aircraft in-trail spacing monitoring. For taxi operations in low visibility, a following aircraft must be able to receive the ADS-B position reports of the aircraft in front of it. To test this, the van followed the NASA 757 while it taxied on the movement area. The van maintained a spacing of approximately 200 meters behind the 757. When the van lined up behind the 757 and both were traveling in a straight line, reception was lost. As soon as the 757 initiated a turn, reception was reestablished. Additional testing confirmed that there was a blind spot behind the 757 where ADS-B transmissions were not received.

All three scenarios confirmed that ADS-B has significant technical issues for direct vehicle to vehicle transmissions in the airport surface environment.

Conclusions

Although 1090 MHz ADS-B is susceptible to multipath and requires direct line of sight, a system of carefully placed ground stations can overcome these difficulties and provide highly accurate surveillance data of airport surface traffic. ADS-B is capable of meeting the accuracy, coverage and update rate requirements of an A-

SMGCS when local area differential GPS is the source of position data. To meet the update rate requirements, a transmission rate of at least twice per second is required. To provide complete movement area coverage, fewer receivers, placed in optimum locations for ADS-B, may meet the requirements.

For vehicle to vehicle ADS-B, there are significant difficulties associated with obtaining reliable vehicle to vehicle reception on the airport surface. More testing is required to determine the limitations of vehicle to vehicle use and possible ways of improving performance for this application.

Acknowledgments

The authors gratefully acknowledge the contributions of the NASA LVLASO team, including Steve Young and Denise Jones, for their guidance and contributions to the research presented in this paper.

References

[1] *Minimum Aviation System Performance Standards for Automatic Dependent Surveillance Broadcast (ADS-B)*, RTCA Document No. DO-242, February 1998.

[2] *Draft Manual of Advanced Surface Movement Guidance and Control Systems (A-SMGCS)*, ICAO All Weather Operations Panel, June 1997.

[3] *RTCA Report On The Role of the Global Navigation Satellite System (GNSS) in Supporting Airport Surface Operations,* Draft 2, RTCA Special Committee 159, July 1998.

[4] *Flight Testing of an Airport Surface Guidance, Navigation, and Control System*, Steve Young and Denise Jones, NASA Langley Research Center, January 1998.

[5] *International Standards and Recommended Practices for Aeronautical Telecommunications*, Annex 10, Volume 4, ICAO, July 1995.

COMPARISON OF A-SMGCS REQUIREMENTS WITH OBSERVED PERFORMANCE OF AN INTEGRATED AIRPORT CNS SYSTEM

Steven D. Young, NASA-Langley Research Center

Abstract

The International Civil Aviation Organization (ICAO) has recently drafted a reference document describing the operational requirements for Advanced Surface Movement Guidance and Control Systems (A-SMGCS) [1]. During the summer of 1997, NASA, the FAA, industry, and academia partners demonstrated a holistic system approach that has the potential to meet many of the proposed A-SMGCS requirements. An assessment of the field tested system and data resulting from the field testing is presented to determine its compliance with A-SMGCS requirements. In those areas where compliance was not demonstrated, a recommendation is presented suggesting further research or a modification of the system architecture.

Introduction

Advanced Surface Movement Guidance and Control Systems (A-SMGCS) is the term used by ICAO to describe a modular system that is needed in the future to support safe, orderly, and expeditious movement of aircraft and vehicles on the airport surface under all circumstances with respect to visibility, traffic density, and complexity of the airport layout. Operational requirements have been written that are technology-independent and provide guidelines for the analysis and development of implementation-specific requirements. The need for A-SMGCS has arisen due to the rising number of surface incidents, the increasing complexity of airports, the increasing number of operations, and the desire to maintain capacity in all weather conditions.

Concurrent with ICAO's development of these requirements, focused research at NASA has resulted in a holistic systems approach to supporting low visibility airport surface operations while maintaining capacity. Milestones of this research have included several flight simulation studies and two flight test activities. The most recent flight testing occurred during the summer of 1997 at the Hartsfield-Atlanta International Airport (ATL). NASA, the FAA, industry, and academia partners demonstrated, at ATL, a prototype system that not only met many of the goals of the research program, but also showed its potential for meeting the proposed A-SMGCS requirements. The prototype communication, navigation, and surveillance (CNS) system consisted of both ground-based and airborne components with digital data links and radio communications integrating the two.

In the following sections, an assessment of this field tested system and data resulting from the field testing is presented, along with the A-SMGCS requirements. Compliance is shown where appropriate. In the final section of the paper, for those areas where compliance was not demonstrated, a recommendation is presented suggesting further research or justifying a modification of the system design or the proposed requirement.

A-SMGCS Requirements

The current SMGCS design and implementation is based on procedures and enhanced lighting that is not always capable of supporting desired capacity levels under low visibility conditions [1]. A-SMGCS requirements have been drafted such that

0-7803-5086-3 /98/$10.00 ©1998 IEEE

regardless of the visibility conditions, aircraft operations can be safely supported at high levels of capacity. The requirements were written to be technology-independent and to allow airport operators to implement specific A-SMGCS functions depending on local circumstances or needs.

The A-SMGCS operational concept supports the expected growth in airport operations while minimizing delay and pilot/controller workload. It supports the sharing of information among the users (e.g. pilots, controllers, vehicle drivers) and supports provisions for improved guidance via both visual aids and electronic means. Complex traffic flows will require an A-SMGCS to function as a surface management system with responsibilities to provide for the planning of surface movements and to be a part of the global CNS/ATM concept which supports "gate-to-gate" operations.

In order to support safe efficient gate-to-gate operations, A-SMGCS must provide suitable facilities for surveillance, routing, guidance, and control of all surface movements (figure 1). The requirements for these four primary functions will depend on the visibility condition (VC), traffic density (TD), and airport layout (AL). The most stringent of these are: VC 4 (visibility less than 75m); TD heavy (greater than 25 take-offs or landings per runway per hour); and AL complex (an airport with more than one runway).

Surveillance

The A-SMGCS definition of surveillance is a function that captures identification and positional information on aircraft, vehicles, and objects within a specified airport area. Current SMGCS surveillance procedures are based on the "see and be seen" principal to maintain separation on the airport surface. Advanced methods of surveillance are

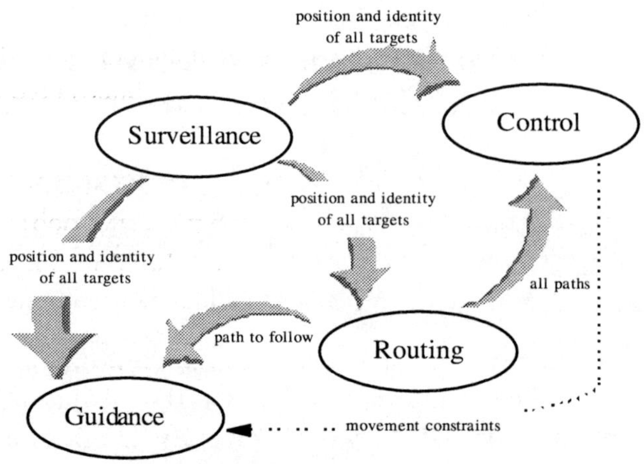

Figure 1. A-SMGCS Functional Dependencies.

required to ensure safe separations during periods of high density and low visibility. In these conditions, "see and be seen" procedures must be supplemented. Specific A-SMGCS requirements for the surveillance function are[1]:

S.1 Capable of providing accurate position information on all surface movements.

S.2 Capable of identifying and labelling all surface movements.

S.3 Cope with both moving and static aircraft and vehicles.

S.4 Capable of updating data required by the guidance and control functions.

S.5 Immune to adverse weather and topographical conditions.

S.6 Provide for health monitoring of surveillance equipment.

S.7 Provide control authorities with surveillance data.

S.8 Provided to an altitude to cover missed approaches and helicopter operations.

S.9 Provided to cover aircraft on approach out to an appropriate distance.

S.10 Support a seamless transition to terminal area surveillance.

[1] For brevity, the author has paraphrased some of the A-SMGCS requirements.

0-7803-5086-3 /98/$10.00 ©1998 IEEE

S.11 Detect runway incursions and incursions into unauthorized areas.

S.12 Detect deviations from assigned route.

S.13 Support the control function in detecting conflicts.

Routing

The routing function of A-SMGCS is defined as the planning and assignment of a route to individual aircraft and vehicles to provide safe, expeditious and efficient movement from its current position to its intended position. Currently taxi routes are provided via radio communications. On request, ATC provides flight crews with the route to follow using a standard phraseology. The route is then read-back by the pilot to confirm receipt and understanding of the radio transmission. Routes are generated by ATC based on the activity that can be seen from the tower cab and a mental picture of airport traffic and airport layout. Specific A-SMGCS requirements for the routing function are:

R.1 Enable a route to be designated for all aircraft and vehicles.

R.2 Allow for a change of destination at any time.

R.3 Allow for a change of route to the same destination at any time.

R.4 Capable of meeting the needs of dense traffic patterns at complex airports.

R.5 Shall not constrain the pilot's choice of runway exit after landing.

R.6 Support minimum taxi distances for any operational configuration.

R.7 Interact with the control function to minimize conflicts at intersections.

R.8 Responsive to operational changes (e.g. closed taxiways or temporary hazards).

R.9 Support standard terminology or symbology.

R.10 Provide routes in a timely manner on request from all users.

R.11 Provide a means of validating routes.

R.12 Provide the control authority with recommended routes for designation.

R.13 In automatic mode, designate and assign routes.

R.14 In automatic mode, support manual intervention by the control authority.

Guidance

The guidance function of A-SMGCS is defined as the necessary advisory information, provided in a continuous unambiguous reliable manner, such that pilots and vehicle operators can steer their aircraft or vehicle along the assigned route while maintaining an appropriate velocity. In today's environment, guidance is provided to flight crews via visual aids during airport surface operations. Pilots navigate on the surface by referring to painted markings (e.g. centerlines), signage, and in-pavement lights. Airport charts are used for a more global reference. Occasionally, progressive taxi instructions are issued by ATC to aid pilots. A-SMGCS requirements for the guidance function are:

G.1 Provide guidance for all possible route assignments.

G.2 Provide clear, unambiguous, indications to pilots and vehicle drivers such that they can follow their assigned route.

G.3 Enable pilots and vehicle drivers to maintain awareness of their position along the assigned route.

G.4 Capable of accepting a change of route at any time.

G.5 Capable of indicating restricted areas of the airport surface.

G.6 Allow for monitoring of the operational status of all guidance aids.

0-7803-5086-3 /98/$10.00 ©1998 IEEE

Control

The control function of A-SMGCS is defined as the application of measures to prevent collisions, runway incursions, and to ensure safe, expeditious and efficient movement on the airport surface. The control function is typically performed by allocating responsibilities to both pilots and controllers. The pilot performs "see and avoid", while the controller issues movement constraints (e.g. hold-short instructions) to reduce the likelihood of conflict. A-SMGCS requirements for the control function are:

C.1 Capable of supporting the maximum movement rate at any time.

C.2 Capable of supporting the planning of requested movements for up to one hour.

C.3 Detect conflicts and provide resolutions on designated routes in a timely manner.

C.4 Provide for required separation minima and generate alerts when separation falls below the minima.

C.5 Provide alerts for intrusions to runways/taxiways and activate protection devices (e.g. alarms).

C.6 Provide alerts for intrusions to critical or sensitive areas (e.g. radio navigation aids).

C.7 Provide alerts for intrusions to emergency areas.

C.8 Keep pilots, vehicle drivers, and controllers in the decision loop.

C.9 Support prioritizing control activities.

C.10 Sequencing of aircraft to ensure minimum delay and maximum utilization of airport capacity.

C.11 Capable of segregating support and maintainence vehicle operations.

C.12 Separate movements from obstacles, secure areas, and restricted areas.

C.13 Provide alerts when an aircraft or vehicle is predicted to enter a restricted area.

C.14 Provide deviation alerts.

System Requirements

Also described in [1] are several system-level requirements of A-SMGCS. In general, they allude to the operational environment envisioned for A-SMGCS. For example, the A-SMGCS should support all aircraft and vehicles types as well as all weather and visibility conditions. The system should be implementable at any airport. The system should support strategic planning of operations in advance. Human-machine interface requirements are also described in general.

LVLASO[2] CNS System Description

The LVLASO CNS system has been specifically designed by NASA to support safe efficient airport operations in low visibility conditions. Further, it has primarily been targeted for commercial transport application and large airport facilities. This goal is very close to the operational environment that poses the most stringent requirements on an A-SMGCS implementation (see page 2). Because this research compliments the goals of A-SMGCS, and a prototype system has been fielded, it becomes possible to assess compliance and to assess the validity of the proposed requirements against the field tested system.

The system implemented at ATL (figure 2) consisted of several subsystems that were integrated into a CNS system [2]. The system included:

- A ground-based surveillance system that provided surveillance data to ATC and to any equipped aircraft/vehicle via data link.

[2] LVLASO is the Low Visibility Landing and Surface Operations focused research program.

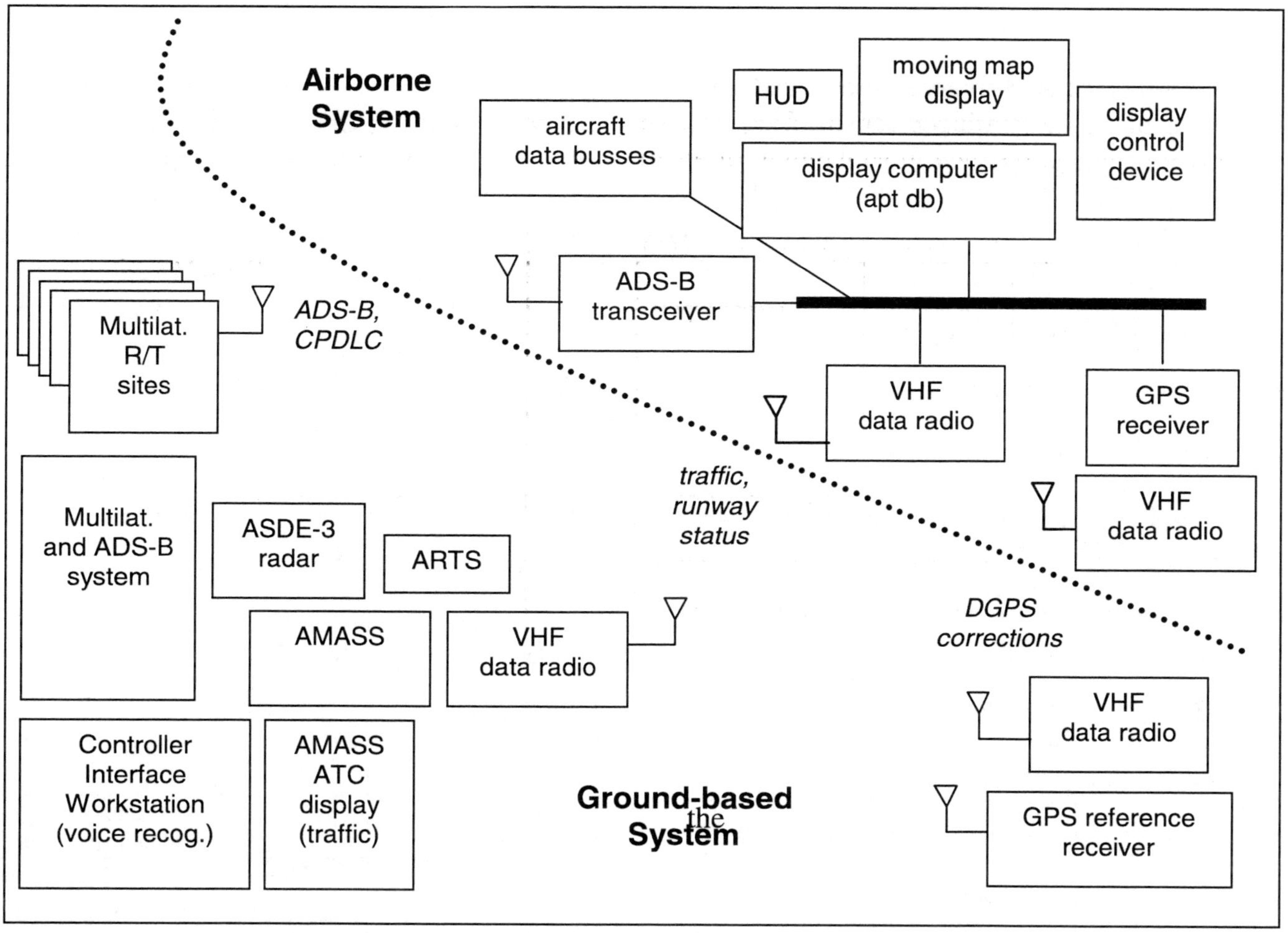

Figure 2. LVLASO CNS System Architecture.

- Flight deck displays that utilized DGPS and an onboard airport database to provide supplemental guidance, surveillance, routing, and control information to the flight crew during surface operations.
- Controller displays that provided ATC a supplemental means of surveillance as well as alerts of route deviation, incursions, and conflicts.
- Controller-pilot data link communications (CPDLC) that allowed ATC to issue instructions via an independent link in parallel with the normal voice channel, and to be alerted of pilot deviations.

Compliance Tables

The information contained in the following tables is intended to describe how the LVLASO CNS system design complies with A-SMGCS requirements as they are currently drafted. Although the airborne portion of the system was only implemented on a single aircraft at ATL, the compliance tables assume the envisioned system were implemented on all aircraft/vehicles. Also, the compliance tables are generated for the most stringent A-SMGCS case: low visibility (less than 75m); heavy traffic density; and a complex airport.

In terms of surveillance, many operational requirements of A-SMGCS were demonstrated during the ATL trials. Adverse weather (S.5) is not expected to affect the performance of ADS-B or multilateration. Although radar performance has been shown to deteriorate in periods of heavy rain, the ASDE-3 radar was specifically designed to minimize weather effects. Transitioning to terminal radar surveillance (S.10) was not demonstrated at ATL. However, it is a design requirement for the FAA's surface surveillance system and will be implemented in the next prototype.

Table 1. Surveillance Requirements

Req't	Comply[3]	Method/Note
S.1	Yes	Fusion of radar, multilateration, and ADS-B reports available via data link broadcast
S.2	Yes	Identification requires transponder
S.3	Yes	Surveillance is independent of aircraft/vehicle state of motion
S.4	Yes	Provided to both ATC and flight deck displays to support "see and avoid"
S.5	*	System not tested in adverse weather
S.6	Yes	Alerts of system failures are generated
S.7	Yes	Surveillance data depicted on multiple controller displays
S.8	Yes	Radar covers up to 500', multilateration to 300', ADS-B limit not measured but >> 500'
S.9	Yes	Only radar is marginal
S.10	*	Not tested
S.11	Yes	AMASS functionality
S.12	Yes	Performed onboard with deviation alert data linked to ATC via CPDLC
S.13	Yes	AMASS functionality

Selected performance requirements for the surveillance system are shown in Table 2 along with the observed performance achieved at ATL.

The surveillance system employed at ATL was not designed to cover the ramp areas of the airport. To fully implement A-SMGCS, this would need to be addressed. Radar is not recommended for ramp coverage; however, either multilateration (with appropriate siting) and/or ADS-B (which requires transponder equipage) can provide this coverage. Approach coverage was not demonstrated; however, this can be achieved by integrating with the approach radar as the two coverage areas were shown to overlap at ATL. During the period of testing, no more than 47 aircraft/vehicles were observed on the movement area at any one time. It is anticipated that more than 100 targets can be accommodated by a surveillance system architecture as employed at ATL.

Table 2. Surveillance Performance

Requirement	A-SMGCS req't	LVLASO observed
Surface coverage area	Movement area, ramp	Movement area
Altitude coverage	500' AGL	1000' (mlat.) >1000' (ADS-B) 500' (radar)
Approach coverage	10 nm	coverage overlaps with apprch. radar
A/C on movement area	100	47 (max. observed)
Horizontal accuracy (95%)	10 m	6.9 m (mlat.) 1.4 m (ADS-B) 6.4 m (radar)
Update rate (at 98% update success rate)	1 Hz	0.25 Hz (mlat.) 1.05 Hz (ADS-B) 0.95 Hz (radar)
Latency	1 sec	0.4 sec (mlat.) 0.1 sec (radar)

Finally, the latency listed is that observed by the ATC user. The latency observed in the flight deck was as high as two seconds due to the processing algorithms employed onboard. It is expected that the one second latency requirement can be achieved in the flight deck using a more efficient algorithm for displaying traffic updates.

In terms of routing, many operational requirements of A-SMGCS were demonstrated during the ATL trials. Generating routes that are minimum distance (R.6) is an automation function that can be implemented within the LVLASO system infrastructure either onboard or at the controller workstation. This requires

[3] "Yes" indicates compliance has been demonstrated in the field, "No" indicates the architecture will not support this requirement, and "*" indicates the architecture can support the requirement, although it was not necessarily implemented or validated in the field.

real-time surveillance data, desired destination, and an airport database. All of these are available at either location in the system.

Table 3. Routing Requirements

Req't	Comply	Method/Note
R.1	Yes	Voice + CPDLC
R.2	Yes	Voice + CPDLC
R.3	Yes	Voice + CPDLC
R.4	Yes	CPDLC will reduce the amount of voice congestion
R.5	Yes	Arrival taxi routes requested after runway exit
R.6	*	Not tested
R.7	Yes	Routes for all aircraft can be depicted on controller display; route deviation alerts generated
R.8	*	Not tested
R.9	Yes	ATC/pilot uses standard phraseology; routes depicted to pilots similar to enroute ND
R.10	Yes	Voice + CPDLC
R.11	Yes	On receipt via CPDLC, airborne system checks route for validity
R.12	*	Not tested
R.13	*	Not tested
R.14	*	Not tested

Responding to operational changes (R.8) is supported but was not tested/validated at ATL. Providing ATC with recommended routes (R.12) can be implemented within LVLASO's controller workstation. These can be standard routes that are stored and provided on request, or these can be determined dynamically based on current conditions and specified constraints (e.g. closed taxiways, traffic bottlenecks, or gate assignment). This was not demonstrated at ATL but is being considered for implementation and evaluation. Automatic modes of providing route instructions (R.13 and R.14) were not implemented or demonstrated at ATL. The LVLASO CNS system architecture could support this concept.

The performance requirement for the routing system that is relevant to the ATL LVLASO system is listed in Table 4. The other routing performance requirements listed in [1] relate to automatically generating routes which was not done at ATL.

Table 4. Routing Performance

Requirement	A-SMGCS req't	LVLASO observed
Time to transmit route to aircraft	1 sec	0.5 sec

Table 5. Guidance Requirements

Req't	Comply	Method/Note
G.1	Yes	Visual aids + ATC verbal + HUD + map display + DGPS
G.2	Yes	HUD + map display + CPDLC
G.3	Yes	HUD + map display + DGPS
G.4	Yes	Voice + CPDLC
G.5	Yes	Map display
G.6	Yes	Alerts of guidance display system failures were generated

As shown in Table 5, the operational requirements for the guidance function were met at ATL. The relevant A-SMGCS performance requirement for the guidance function is shown in Table 6 with the observed performance achieved at ATL. Compliance with accuracy requirements suggested for the gate areas cannot be shown as gate operations were not done during the ATL testing.

Table 6. Guidance Performance

Requirement (Airport Code E)	A-SMGCS req't	LVLASO observed
Horizontal accuracy (taxiway)	2.2 m (95%)	1.3 m (95%)

While the DGPS/INS navigation system employed on the B-757 performed well at ATL, it remains unclear whether the A-SMGCS requirements for low visibility gate operations can be achieved with this technology (i.e. DGPS). An alternate means of guidance may be necessary in these areas if high movement rates are expected to be maintained in low visibility.

Table 7. Control Requirements

Req't	Comply	Method/Note
C.1	*	Not tested
C.2	*	Not tested
C.3	*	Not tested
C.4	*	Not tested
C.5	Yes	AMASS functionality
C.6	*	Not tested
C.7	*	Not tested
C.8	Yes	Information is advisory only
C.9	Yes	ATC discretion
C.10	*	Not tested
C.11	*	Not tested
C.12	*	Not tested
C.13	*	Not tested
C.14	Yes	Onboard system detected deviations and alerted ATC

Many of the A-SMGCS operational requirements for the control function were not specifically tested at ATL (Table 7). The controller tools tested at ATL provided supplemental means for ATC to maintain awareness of traffic positions and to communicate with pilots in all weather conditions. These augmentations should allow ATC to support a higher movement rate in any visibility (C.1). In the LVLASO CNS system concept, resolving conflicts (C.3) and assuring separation (C.4) remains the responsibility of ATC as is currently done. Many of the automation functions suggested by A-SMGCS to aid controllers (C.10 to C.13) were not implemented at ATL.

The LVLASO control function was designed to minimize the impact on normal ATC procedures. Surveillance data is displayed to be used *as needed* to supplement visual acquisition. Voice recognition is used to capture instructions and forward them to pilots to reinforce the verbal radio transmissions and avoid misunderstandings/miscommunications. This approach does not inhibit the architecture from supporting the automation functions suggested by A-SMGCS if they are deemed necessary.

Summary

The LVLASO CNS system concept is based primarily on two theories:

- Current procedures, visual aids, radio communications, and human performance enable safe airport surface operations at near peak flow rates in conditions of good visibility.
- In conditions of poor visibility, safe airport surface operations can be maintained at near peak flow rates if radio communications and the limited visual cues are supplemented, or reinforced, using technological approaches to surveillance, routing, guidance, and control functions. The application of these technologies can be accomplished in such a way to minimize the impact on current procedures both in the flight deck and in the ATC tower. Information provided by the new technologies is done so in a natural manner that permits it to be used as needed.

The primary difference between the LVLASO system approach and A-SMGCS is in the area of automation. The LVLASO approach does not suggest automating any function at this time over and above what current procedures require. While A-SMGCS requirements for automation (e.g. automatic route generation/transmission, planning for the control of surface movements in advance, and generating minimum distance routes) will most likely improve efficiency, it is not clear that they will improve capacity over and above what current clear weather procedures allow.

One example of automation that the ATL system did support was the automatic alerting of ATC of a route deviation. This is

0-7803-5086-3 /98/$10.00 ©1998 IEEE

necessary to support current operational procedures in the tower. If visual conditions do not allow ATC to observe a deviation from a the tower cab, an indication must be provided to ATC to avoid a potential conflict.

The ATL system did not address required A-SMGCS functionality (surveillance, guidance, routing, and control) in the ramp area. This is being addressed as part of the ongoing research. Surveillance approaches being considered are ADS-B, multilateration, and ground loops. Additions to the airport geographic database along with enhanced DGPS/INS may support the guidance function. Routing and control augmentations will be addressed with respect to current ramp control procedures/systems.

Finally, required levels of integrity, continuity, and availability were not achieved by all components of the field-tested system. This is directly related to the maturity of the specific technologies employed. Components that adhered to established standards, or were certified, (e.g. the data links, the DGPS subsystem, and the flight deck displays) rarely failed or were unavailable (much less than 1% of the total test time) [2]. On the other hand, components/functions that were less mature (e.g. multilateration, surveillance sensor data fusion, and voice recognition), require further work to ensure adequate levels of availability, continuity, and integrity for full deployment of this concept.

Conclusion

This paper has shown that even in the most stringent conditions of visibility, traffic density, and airport complexity, the LVLASO CNS system architecture can provide for the primary A-SMGCS functions without requiring enhanced visual aids (e.g. lighting) or significant procedural changes. It does require additional equipage both in the tower, and in any aircraft and/or vehicles operating on the airport surface (GPS, data link, and display augmentations).

Future work will consider A-SMGCS-required automation functions and ramp operations that have not been addressed to date by the LVLASO research team.

References

[1] *Draft Manual of Advanced Surface Movement Guidance and Control Systems (A-SMGCS)*, 16th Meeting of the International Civil Aviation Organization's All Weather Operations Panel, Montreal, Canada, June 1997.

[2] Young, S., and Jones, D., *Flight Testing of an Airport Surface Movement Guidance, Navigation, and Control System*, Proceedings of the Institute of Navigation's National Technical Meeting, January 21-23, 1998.

SOLVING PASSIVE MULTILATERATION EQUATIONS USING BANCROFT'S ALGORITHM

Michael Geyer, US DOT Volpe Center, Cambridge, MA
Anastasios Daskalakis, US DOT Volpe Center, Cambridge, MA

Abstract

This paper applies Bancroft's algorithm, which was developed for GPS, to the solution of aircraft multilateration surveillance equations. Results developed are equally applicable to two-dimensional hyperbolic surveillance and navigation. Bancroft's algorithm provides a closed-form position solution with a time tag.

Nature of Multilateration Equations

A passive multilateration surveillance system measures the times of arrival (TOAs) of the same aircraft transponder emission, usually termed a "squitter," at three or more ground sites. It uses the TOAs to calculate the aircraft horizontal position. Aircraft altitude is obtained by other means, such as decoding Mode C messages embedded in the squitter.

Horizontal position can be computed from the minimum of three TOA measurements using several approaches. Often, differential TOAs (DTOAs) are formed, yielding two equations in two horizontal position variables and eliminating the unknown squitter Time of Emission (TOE). When an initial position estimate is available, the two equations can be linearized and a solution readily obtained. Another approach is to apply numerical iteration to the nonlinear (hyperbolic) DTOA equations.

The validity of the position solution obtained from either approach can be checked by computing DTOAs for the position solution and comparing them to the measured DTOAs. However, because the governing equations are essentially quadratic, they can have two solutions. Linear and numerical methods may unknowingly miss the correct solution, requiring that a new initial position estimate be selected and that the solution calculation be repeated (possibly multiple times).

This paper applies Bancroft's algorithm to calculate aircraft position from TOA measurements of a single squitter. The algorithm solves simultaneously the three nonlinear governing equations, and has several advantages over most other techniques: (1) it involves readily-programmed closed-form expressions for both possible solutions (correct and extraneous); (2) it yields an estimate for the squitter TOE which can be used for time tagging the associated position estimate; (3) for most aircraft locations of operational interest, the calculated TOEs can be used to identify the extraneous solution without additional calculations; (4) the regions of the horizontal plane where unresolvable ambiguities occur are known a priori, providing guidance for ground station siting; and (5) the analytic nature of the technique facilitates correlating solutions derived from a sequence of squitters.

Description of Bancroft's Algorithm

Bancroft's algorithm was developed for application to GPS (Refs. 1 and 2). It is applied here to situations involving two spatial dimensions. Bancroft's algorithm makes use of the Lorenz inner product for time-space vectors, which is defined as follows. Let **u** and **v** denote vectors of the form:

0-7803-5086-3 /98/$10.00 ©1998 IEEE

$$\mathbf{u}=\begin{bmatrix}x_u\\y_u\\ct_u\end{bmatrix},\quad \mathbf{v}=\begin{bmatrix}x_v\\y_v\\ct_v\end{bmatrix} \qquad \textbf{Eq. 1}$$

where x and y are spatial coordinates, c is the speed of light, and t is time. A vector can represent either an aircraft (in which case the time component is the squitter TOE), or a ground station (in which case the time component is the squitter TOA). The Lorenz inner product of **u** and **v** is defined to be

$$\langle \mathbf{u},\mathbf{v}\rangle = x_u x_v + y_u y_v - c^2 t_u t_v \qquad \textbf{Eq. 2}$$

Turning to the multilateration problem explicitly, the analysis is formulated in two spatial dimensions — the plane containing the ground stations. The three stations are located at (x_i, y_i), and the associated squitter TOAs are t_i, where i=1,2,3. This information is combined in the state vectors

$$\mathbf{s}_i=\begin{bmatrix}x_i\\y_i\\ct_i\end{bmatrix} \qquad \textbf{Eq. 3}$$

The aircraft is located at (x_a, y_a) and has squitter TOE t_a; its state vector is

$$\mathbf{s}_a=\begin{bmatrix}x_a\\y_a\\ct_a\end{bmatrix} \qquad \textbf{Eq. 4}$$

The nine ground-station scalar quantities are known. The three aircraft quantities are unknown and are to be found.

Each TOA measurement satisfies an equation of the form

$$(x_a - x_i)^2 + (y_a - y_i)^2 = c^2(t_i - t_a)^2 \qquad \textbf{Eq. 5}$$

which is equivalent to

$$2(x_i x_a + y_i y_a - c^2 t_i t_a) = x_a^2 + y_a^2 - c^2 t_a^2 + x_i^2 + y_i^2 - c^2 t_i^2 \qquad \textbf{Eq. 6}$$

or, in vector-matrix form,

$$2\mathbf{A}\mathbf{s}_a = \lambda \mathbf{1} + \mathbf{b} \qquad \textbf{Eq. 7}$$

where

$$\mathbf{A}=\begin{bmatrix}x_1 & y_1 & -ct_1\\x_2 & y_2 & -ct_2\\x_3 & y_3 & -ct_3\end{bmatrix} \qquad \textbf{Eq. 8}$$

$$\lambda = \langle \mathbf{s}_a, \mathbf{s}_a\rangle = x_a^2 + y_a^2 - c^2 t_a^2 \qquad \textbf{Eq. 9}$$

$$\mathbf{1}=\begin{bmatrix}1\\1\\1\end{bmatrix} \qquad \textbf{Eq. 10}$$

$$\mathbf{b}=\begin{bmatrix}\langle \mathbf{s}_1,\mathbf{s}_1\rangle\\\langle \mathbf{s}_2,\mathbf{s}_2\rangle\\\langle \mathbf{s}_3,\mathbf{s}_3\rangle\end{bmatrix}=\begin{bmatrix}x_1^2+y_1^2-c^2t_1^2\\x_2^2+y_2^2-c^2t_2^2\\x_3^2+y_3^2-c^2t_3^2\end{bmatrix} \qquad \textbf{Eq. 11}$$

Observe that Eq. 7 relates $\mathbf{s}_a$ to its Lorenzian norm λ.

When **A** is nonsingular, $\mathbf{s}_a$ is given by

$$\mathbf{s}_a = \tfrac{1}{2}\lambda \mathbf{A}^{-1}\mathbf{1} + \tfrac{1}{2}\mathbf{A}^{-1}\mathbf{b} \qquad \textbf{Eq. 12}$$

or

$$\mathbf{s}_a = \lambda \mathbf{d} + \mathbf{e} \qquad \textbf{Eq. 13}$$

where

$$\mathbf{d} = \tfrac{1}{2}\mathbf{A}^{-1}\mathbf{1} = \begin{bmatrix}x_d & y_d & ct_d\end{bmatrix}^T$$
$$\mathbf{e} = \tfrac{1}{2}\mathbf{A}^{-1}\mathbf{b} = \begin{bmatrix}x_e & y_e & ct_e\end{bmatrix}^T \qquad \textbf{Eq. 14}$$

Taking the Lorenzian norm of both sides of Eq. 13 results in a quadratic equation in λ

$$\lambda = \langle \mathbf{d},\mathbf{d}\rangle \lambda^2 + 2\langle \mathbf{d},\mathbf{e}\rangle \lambda + \langle \mathbf{e},\mathbf{e}\rangle \qquad \textbf{Eq. 15}$$

or

$$\alpha\lambda^2 + \beta\lambda + \gamma = 0 \qquad \textbf{Eq. 16}$$

where

0-7803-5086-3 /98/$10.00 ©1998 IEEE

$$\alpha = \langle \mathbf{d},\mathbf{d} \rangle = x_d^2 + y_d^2 - c^2 t_d^2$$
$$\beta = 2\langle \mathbf{d},\mathbf{e} \rangle - 1 = 2x_d x_e + 2y_d y_e - 2c^2 t_d t_e - 1$$
$$\gamma = \langle \mathbf{e},\mathbf{e} \rangle = x_e^2 + y_e^2 - c^2 t_e^2$$

When $\alpha \neq 0$ there are two solutions to Eq. 16

$$\lambda_+ = \frac{1}{2\alpha}\left(-\beta + \sqrt{\beta^2 - 4\alpha\gamma}\right)$$
$$\lambda_- = \frac{1}{2\alpha}\left(-\beta - \sqrt{\beta^2 - 4\alpha\gamma}\right) \qquad \textbf{Eq. 17}$$

and two solutions for the aircraft state

$$\hat{\mathbf{s}}_{a+} = \lambda_+ \mathbf{d} + \mathbf{e} = \begin{bmatrix} \hat{x}_{a+} \\ \hat{y}_{a+} \\ c\hat{t}_{a+} \end{bmatrix}$$
$$\hat{\mathbf{s}}_{a-} = \lambda_- \mathbf{d} + \mathbf{e} = \begin{bmatrix} \hat{x}_{a-} \\ \hat{y}_{a-} \\ c\hat{t}_{a-} \end{bmatrix} \qquad \textbf{Eq. 18}$$

In Eq. 18, the circumflex symbol is used to emphasize that these are actually candidate solutions. One is correct (equals $\mathbf{s}_a$); the other is extraneous. That is, an ambiguity exists.

There are pathological but feasible cases for which $\alpha = 0$. Then Eq. 16 has one solution

$$\lambda = -\frac{\gamma}{\beta} \qquad \textbf{Eq. 19}$$

Here there is only one solution for the aircraft state

$$\hat{\mathbf{s}}_a = \lambda \mathbf{d} + \mathbf{e} \qquad \textbf{Eq. 20}$$

which is equal to $\mathbf{s}_a$.

Analysis of Solutions

Behavior of "Plus" and "Minus" Solutions

To facilitate explanation of the behavior of the solutions found by Bancroft's method, the plane containing the three ground stations is partitioned into two regions (Figure 1):

- The Unfavorable Region (UR), comprising the three wedges bounded by the baseline extensions
- The Favorable Region (FR), comprising the rest of the plane.

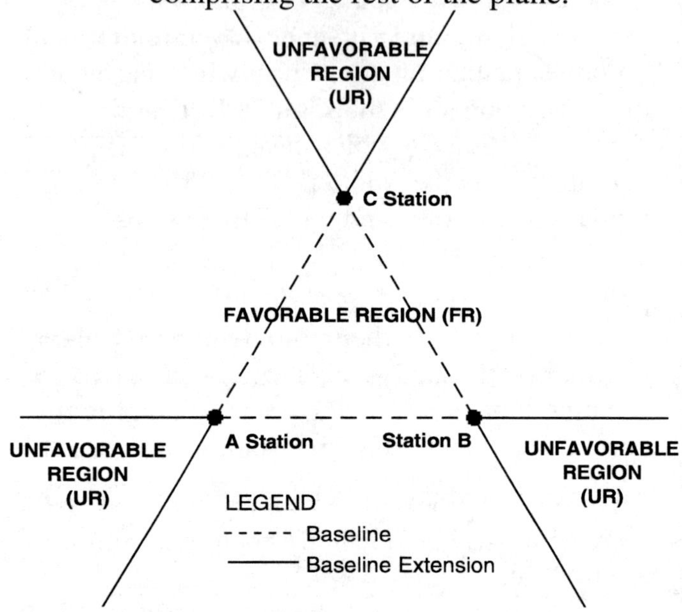

Figure 1 Regions for Bancroft Algorithm "Plus" and "Minus" Solutions

A multilateration system would be designed so that its operational area would be inside the triangle formed by the baselines connecting the stations. Such a system could also be used, with some accuracy degradation, for aircraft outside the triangle but near the baselines. It would not be intended for surveilling aircraft which are in the UR or in the FR but "very close" to the ground stations.

After simulating thousands of varied aircraft locations, the following Observations were formulated:

1□ The "minus" position solution $(\hat{x}_{a-}, \hat{y}_{a-})$ is always in the FR

2□ The "minus" TOE solution $\hat{t}_{a-}$ is always less than all TOAs

3□ The "plus" position solution $(\hat{x}_{a+}, \hat{y}_{a+})$ can be in the FR or UR

4□ The "plus" TOE solution $\hat{t}_{a+}$ can be

0-7803-5086-3 /98/$10.00 ©1998 IEEE

either less or greater than all TOAs.

5□ When the aircraft is in the FR, the "minus" solution $\hat{\mathbf{s}}_{a-}$ is correct

6□ When the aircraft is in the UR, the "plus" solution $\hat{\mathbf{s}}_{a+}$ is correct.

Figure 2 illustrates the validity of Observations 5 and 6. A grid of aircraft locations was selected and Bancroft's algorithm exercised on the corresponding TOAs. When the "plus" solution was correct, a "+" was placed at the grid point; when the "minus" solution was correct, a "o" was placed at the grid point. The two solutions are equal along the FR-UR boundaries, because the expression under the radicals in Eq. 17 is zero on these boundaries.

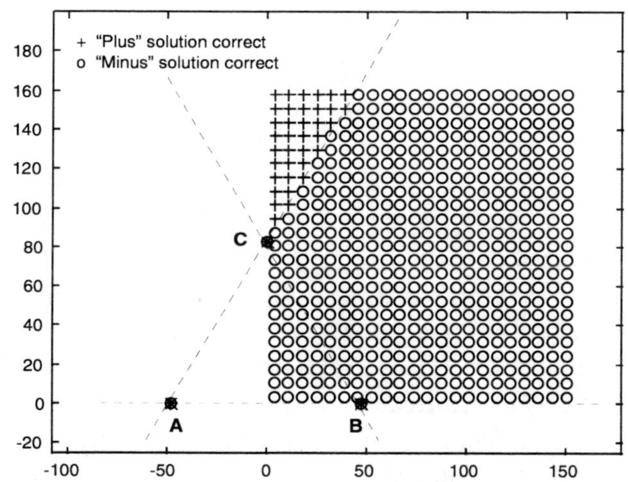

Figure 2 Illustrating Regions where "Plus" and "Minus" Solutions Are Correct

When an aircraft is being tracked reliably, Observations 5 and 6 provide a basis for correlating solutions derived from a sequence of squitters. They are particularly useful when an aircraft crosses a boundary between the FR and UR. However, they are not sufficient for resolving the solution ambiguity for a single squitter. Next it is shown how Observations 1-4 are used to resolve solution ambiguities for aircraft located in most of the FR.

Solutions Categorized by Hyperbola Branches

The solutions to Bancroft's algorithm can also be classified "geometrically," by forming two DTOAs and examining the intersections of the associated hyperbolas. Analytically, the DTOA equations are formed by taking the square-root of both sides of Eq. 5. Differencing two such equations eliminates the squitter TOE t_a and forms a hyperbolic equation with a

DTOA on the right-hand side.

Each hyperbola has two branches, one corresponding to the DTOA with the correct sign and the other to the DTOA with its sign reversed. Figures 3 and 4 shows hyperbolas for selected DTOAs together with solutions from Bancroft's algorithm. Two distinct forms of intersections occur.

Different Branches (Figure 3) — Each solution is formed by the intersection of hyperbola branches which are distinct from those which form the other solution. One solution corresponds to the correct DTOAs, and the other to their sign-reversed versions.

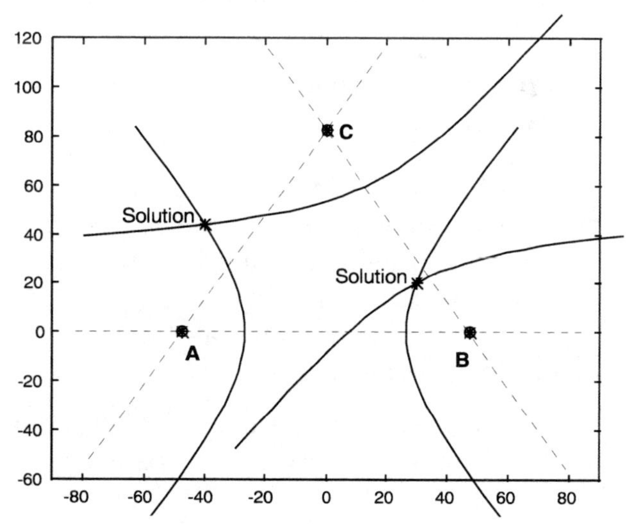

Figure 3 Separate Branches Solutions

0-7803-5086-3 /98/$10.00 ©1998 IEEE

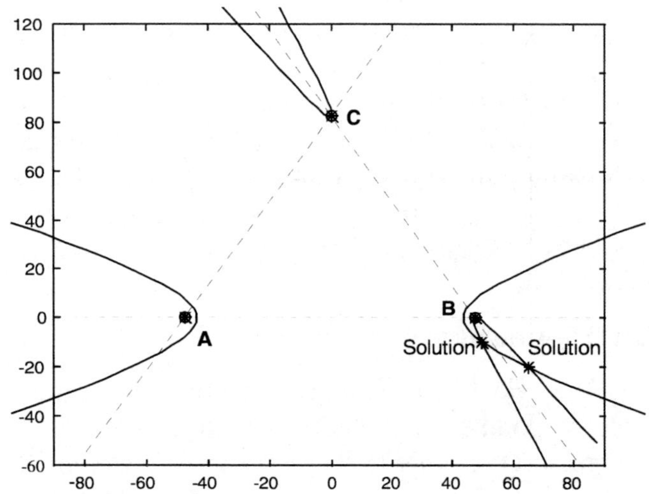

Figure 4 Same Branches Solutions

Same Branches (Figure 4) — The two solutions are formed by the intersection of the same branches of the hyperbolas, each of which corresponds to a correct DTOA.

Behavior of "Different Branches" Solutions

"Different branches" ambiguities can be identified by either of two Indicators, which occur simultaneously:

(a) *"Plus" position solution is in the FR*

(b) *"Plus" TOE is later than the three TOAs.*

It follows from either Indicator that the "minus" solution is correct: in the case of (a), based on Observations 1, 3 and 5; and in the case of (b), based on causality. These Indicators are satisfied for aircraft locations "well inside" the FR.

Figure 5 presents examples of "different branches" solutions, using a case where the aircraft is well inside the FR. The initial squitter occurs at position 0, which is 30 nmi due north of the coordinate system origin. The aircraft then proceeds east at a constant speed, moving 10 nmi between squitters over a 100 nmi track.

Based on the example construction, the "minus" solution is correct for all 11 squitters. Note that the initial "plus" solution is near the "minus" solution, but is rejected based on

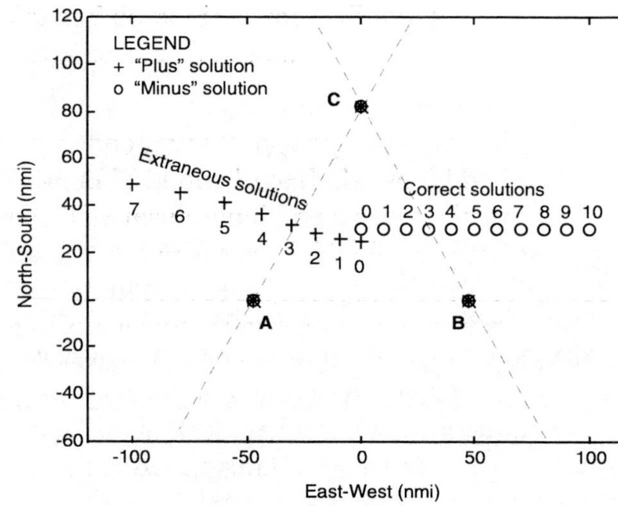

Figure 5 Example of Position Solutions when "Different Branches" Occur

Indicator (a). The "plus" solution then travels to the northwest, accelerating as it proceeds. The last three solutions are off the plot area.

Figure 6 shows the TOE solutions corresponding to Figure 5. The ground station TOAs are also shown. A sequence of actual TOEs would increase linearly with time. However, for clarity of presentation, the common trend is removed from both the TOEs and the TOAs — i.e., all TOEs are simulated as occurring at zero. Also, TOEs and TOAs are shown in units of nautical miles rather than seconds. Figure 6 demonstrates that the "plus" TOEs satisfy Indicator (b).

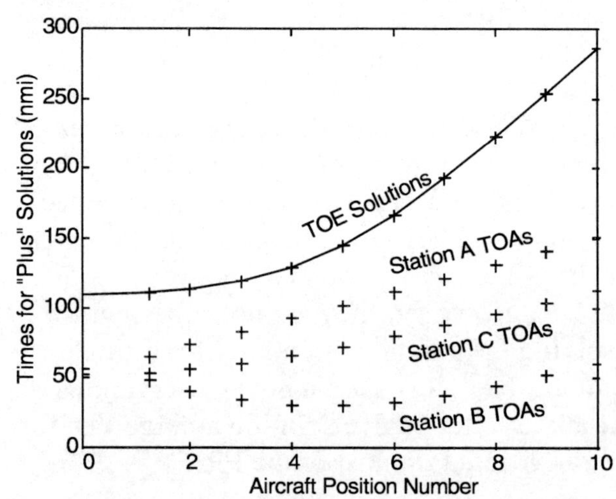

(a) "Plus" Solution

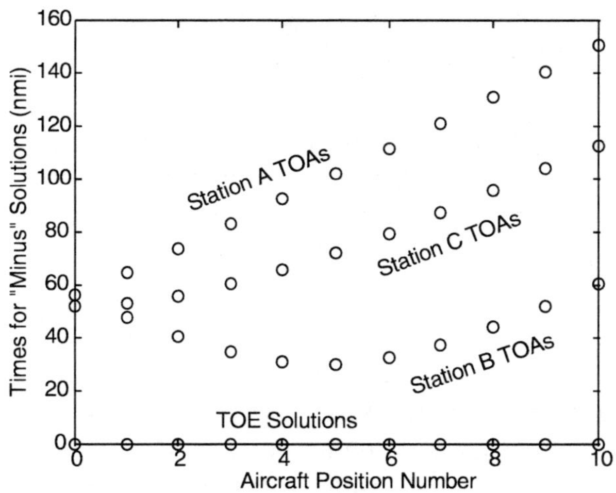

(b) “Minus” Solution

Figure 6 TOE Solutions for Figure 5

Behavior of “Same Branches” Solutions

“Same branches” ambiguities arise when Indicators (a) and (b) are not true. More directly, they are revealed by the following Indicators, which occur simultaneously:

(c) *“Plus” position solution is in the UR*

(d) *“Plus” TOE solution is earlier than all three TOAs.*

When these Indicators are true, either the “plus” or “minus” solution may be correct. The ambiguity cannot be resolved without additional data “Same branches” ambiguities occur for aircraft within or near the UR

Figures 7 (position solutions) and 8 (TOE solutions) illustrate the behavior of the Bancroft algorithm solutions for an aircraft that moves from the FR into the UR. The aircraft begins 20 nmi south of the origin and travels due east over a distance of 100 nmi, squittering every 10 nmi. The first six squitter positions are in the FR, while the last five are in the UR.

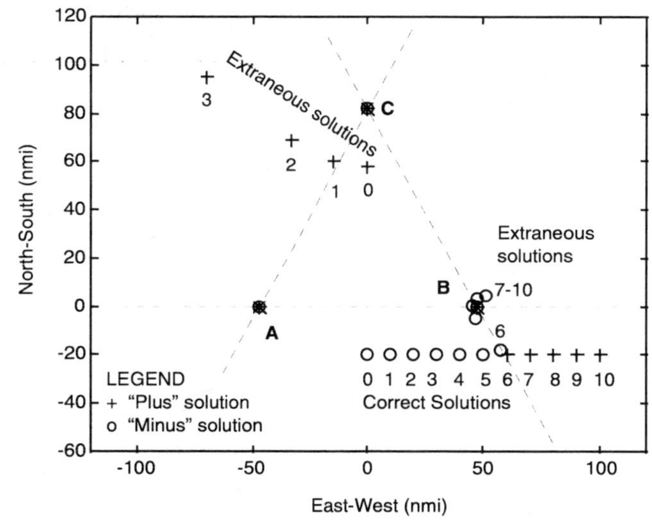

Figure 7 Example of “Different Branches” and “Same Branches” Ambiguities

Squitters 0 through 4 result in obvious “different branches” ambiguities. The “minus” solution can be selected based on Indicator (a) or (b).

For squitter 5, the “plus” position solution jumps to a location in the UR which is southeast of the plot area satisfying Indicator (c), and the “plus” TOE solution is earlier than all TOAs, satisfying Indicator (d). While it is clear from the construction of the example that the “minus” solution is correct, the Bancroft algorithm ambiguity cannot be resolved.

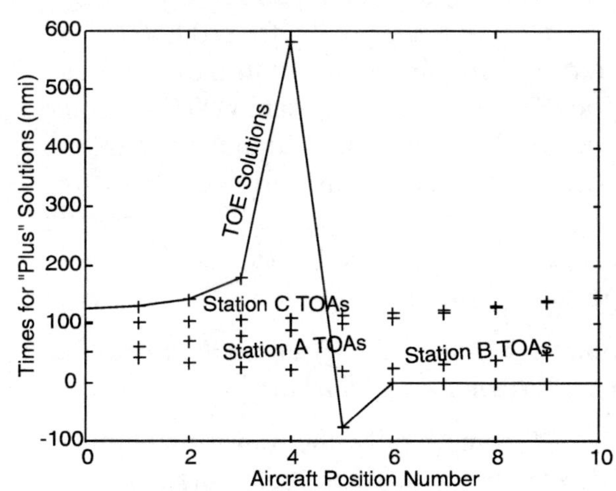

(a) “Plus” Solution

0-7803-5086-3 /98/$10.00 ©1998 IEEE

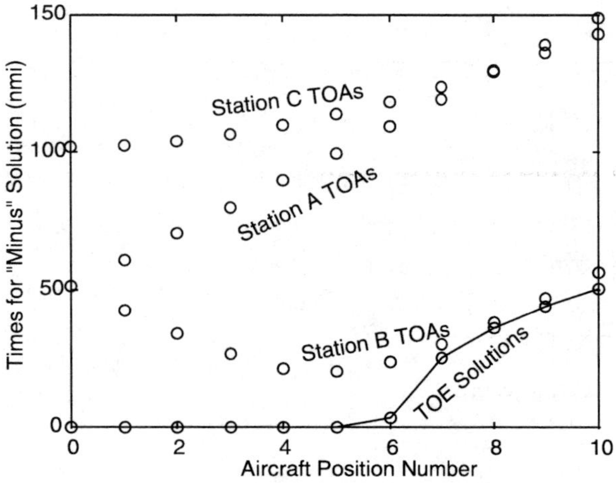

(a) "Minus" Solution

Figure 8 TOE Solutions for Figure 7

Indicators (c) and (d) are also true for squitter 6. However, in this case the "plus" solution is correct. Squitters 7 through 10 also result in "same branches" ambiguities. The extraneous "minus" position solutions cluster around the nearest station, and the "minus" TOEs are slightly less than the smallest TOA.

A thought experiment illustrates the fact that "same branches" ambiguities occur because certain DTOA pairs (with proper signs) can be caused by two different aircraft locations. Consider an aircraft that is in the UR due north of station C, and assume the stations form an equilateral triangle with side L. The equal DTOAs for station pairs A-C and B-C can vary, in distance units, from 0.866 L (aircraft infinitely far north) to L (aircraft at station C). However, the same DTOAs could be caused by an aircraft within the triangle, close to but due south of station C. It follows that no algorithm can resolve "same branches" ambiguities using only TOAs from a single squitter.

Boundary Between "Different Branches" and "Same Branches" Solutions

Geometrically, the transition between "different branches" and "same branches" solutions occurs when a pair of branches, one from each hyperbola, are asymptotically parallel and do not intersect. Figure 9 illustrates such a situation.

If either or both of the DTOAs associated with the Figure 9 hyperbolas were to decrease, at least one hyperbola would be less curved. The result would be a second intersection to the northwest of the plot area, and two "different branches" type solutions would occur. Conversely, if either or both of the DTOAs were to increase, at least one hyperbola would become more curved. A second solution to the southeast of the plot area would occur; the two solutions would be "same branches" type.

One-solution situations are pathological; the second solution can be thought of as being at infinity. These cases correspond to $\alpha = 0$ in Eq. 16.

The locus of all single-solution points can be formed either analytically by setting $\alpha = 0$ in Eq. 16 or geometrically from two hyperbolas by increasing one DTOA and decreasing the other, while maintaining parallel

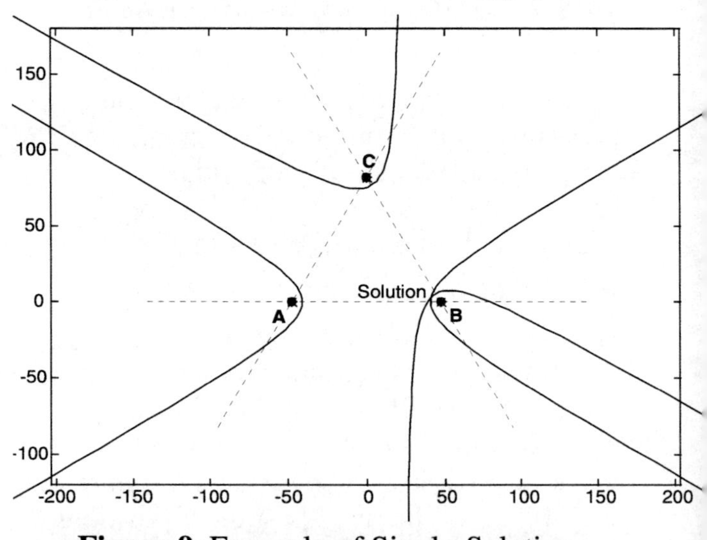

Figure 9 Example of Single-Solution

asymptotes. This locus forms a boundary between the regions where "different branches" and "same branches" ambiguities occur. Figure 10 is the result when the process is performed for three stations forming an equilateral triangle.

0-7803-5086-3 /98/$10.00 ©1998 IEEE

The "same branches" region includes the UR plus a bordering strip in the FR. If the common side length is L, the strip width is 0.0722 L (viz., 7.22% of L) at its narrowest point near a station, and 0.0776 L when measured along the baselines joining the stations. At distances ½L and L from a station, the strip width is 0.190 L and 0.205 L, respectively.

Acknowledgments

Gratitude is acknowledged to the Federal Aviation Administration, Surveillance Systems Engineering Group (AND-402), for supporting this research.

References

1. Bancroft, S., "An Algebraic Solution of the GPS Pseudorange Equations," *IEEE Transactions on Aerospace and Electronic Systems*, AES-21, November 1985, pp 56-59.
2. Chaffee, J., and Kovach, K., "Bancroft's Algorithm is Global Nonlinear Least Squares," *Proceedings of Institute of Navigation GPS-96*, September 1996, pp 431-7.

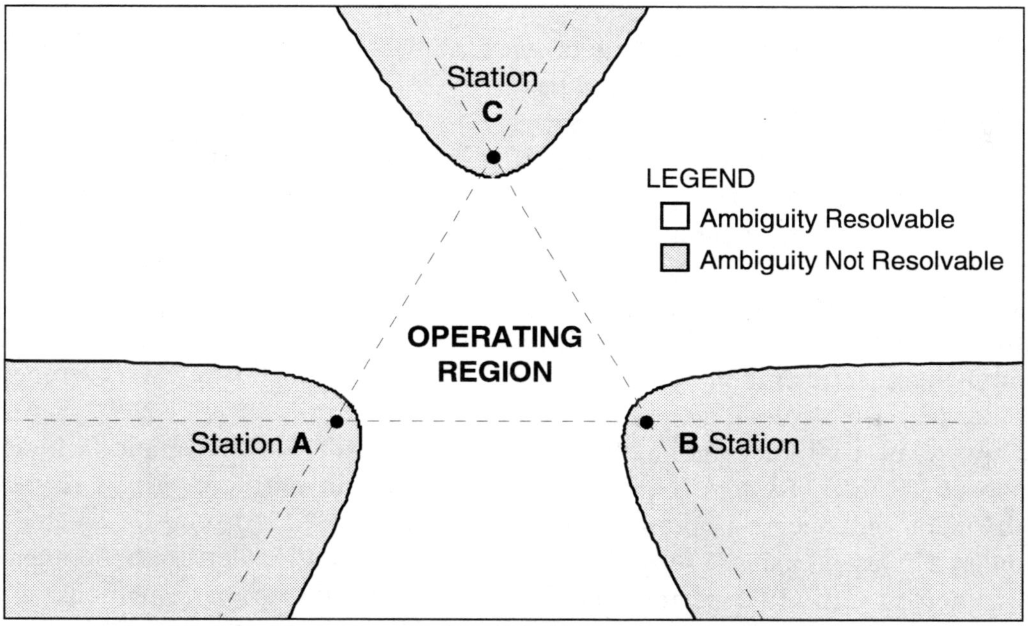

Figure 10 Three-Station Multilateration Ambiguity Regions

17TH DASC PAPER
AN INTEGRAL FLIGHT DIRECTOR AND SURVEILLANCE SYSTEM FOR HELICOPTERS IN METROPOLITAN SERVICE

(Conceptual Design and Preliminary Test)

Chin E. Lin S.J. Hong K.L. Chiu C.Y. Tsai
Department of Aeronautics and Astronautics
National Cheng Kung University
Tainan, Taiwan, R. O. China

Abstract

High density helicopter flights in metropolitan area arises conflict threat and challenges flight safety due to the absence of strict ATC surveillance. An integral system for flight director and surveillance (FDS) for helicopter operations in low altitude flights is proposed with feasibility study. The proposed FDS combines GIS with GPS data based on ADS concept to meet helicopter operation with collision avoidance. Both pilot and ground ATC are aware of conflict situation from the proposed FDS. System configuration with operation concept is presented with simulated tests using remote-piloted helicopters to verify FDS performance.

Introduction

Helicopter to be used as a general transport in Taiwan will increase air traffic control (ATC) load and challenge flight safety. Method to offer an effective means for helicopter ATC is most desired to reduce risk of flight collision in low altitudes. Within the scope of CNS/ATM developments, GPS data as well as Geodetic Information System (GIS) are applied together to receive geographic background to pronounce air traffic conditions to all aircraft. An Integral Flight Director and Surveillance System (FDS) proposes a idea for both airborne flight director, as an avionics, and ground surveillance, as ground ATC facilities.

The main loop of FDS contains automatic dependence surveillance system for local area broadcasting (ADS-LB), a communication network, GIS mapping-transform system, and ATC surveillance and collision avoidance. A metropolitan surveillance GIS display is designed for ground ATC with complete presence of flight conditions. ATC functions of separation, terrain and warnings are implemented for operation. In addition to ATC, an airborne flight director with dynamic GIS display is created to assist pilot of traffic alert for helicopters.

In this paper, methods to develop traffic collision avoidance for the proposed FDS are discussed by definitions of advisories [1,2] with computer software implementation on both the airborne segments and ground segment.

Two remote-piloted helicopters are used for simulated tests to examine GPS accuracy and FDS performance with feasibility [3].

Traffic Alert and Conflict Resolution for Helicopters

The helicopter flights are permitted as general transport under VFR in Taipei FIR. All light aircraft operating in VFR

0-7803-5086-3 /98/$10.00 ©1998 IEEE

except authorized flights shall follow the established VFR corridors and flight rules. Pilots flying in VFR corridors shall follow altitude, speed, operation constraints and voice report procedures [1].

Since helicopters may be used as a prompt transport among special industrial sites, this demand stimulates the increase of helicopter flights near metropolitan areas. Traffic alert and conflict resolution become important to helicopter flights.

Threat Advisory and Conflict Resolution

The basic concept for collision avoidance arises that the pilot should be alerted to a threat in time for his remedial action to operate the aircraft to escape by changing its speed or altitude. Indeed, if possible, ATC controller should also be aware of such situations before and after a traffic alert happens. The development of FDS should follow ICAO standards. FDS functions include: (a) threat detection, (b) generation of Proximity Advisory (PA), Traffic Advisory (TA), and Resolution Advisory (RA), (c) coordination to ground ATC station, (d) effective system performance. Different levels of advisories represent different threat of aircraft approach in terms of collision time [1,2].

Threat detection: Conflict detection and hazard criteria are based on the relative motion of multiple moving aircraft for either fix-wing or rotor-wing. Time and position of aircraft to the point of closest approach (PCA) in 3-D are calculated.

Protected Zone: Consider pilot's sensuous reaction to delay helicopter maneuver, a protected zone, called "bubble", is created for each aircraft. On the screen, the bubble can constantly remind the pilot of threats. This also includes the instant intrusion of other helicopters with proper vertical separation. A required vertical separation should be over 5 times of main rotor diameter, when one helicopter is right below or above another. For example of Bell-212, the vertical separation should be 240 feet. The bubble is then defined as a column of 2,050 feet radius with 240 feet height above and below. This also concerns with the limit of miss distance in 3-D space and provides enough vertical separation when PCA is accounted [2].

Generation of Advisories

Conflict resolution comprises two actions: (a) to alert the pilot to potential conflicts including terrain; (b) to present the conflict resolution from ATC. When threats or potential threats are detected, the proximate traffic within certain range and certain altitude, if altitude is reported, should be displayed and distinguished by color or symbol. Recommend maneuver is sent to the pilot as part of resolution. PA, TA, and RA are displayed in terms of estimated collision time. Both pilot and ATC receive all the same information.

The requirements of helicopter to identify a threat include 3-D position; relative velocity vector; and sensitivity state of helicopter.

Sensitivity state: According to the aircraft performance, the sensitivity state is defined. (1) S=1, a "STANDBY" mode in which TA and RA are inhibited; (2) S=2, a "TA only" mode in which RA is inhibited; (3) S=3-7, further states that enable to issue TA and RA to provide alert time.

Alter time: Referring ICAO standards of ACAS, a nominal RA before closest approach varies from 15 to 35 seconds, depending on the relative velocity vector of two aircraft. TA is generally issued around 5 to 20 seconds in advance to RA. Alert time depends on the sensitivity state. Considering Bell-212 at cruise speed 80-100 knots, S=3, TA=5 and RA= 15 seconds for example.

Resolution constraints: Ignoring the strategic maneuvers, the tactical resolution analysis is focused on safety constraint. Three maneuver patterns are taken into account as Altitude Maneuver to climb or descend 300 feet, which is above minimum vertical separation, Heading Maneuver to make a right turn in a 500 feet radius circle to make-ways, and Speed Change Maneuvers to accelerate or decelerate 10 knots, above "dead zone".

Coordination

The FDS coordination of Collision Avoidance logic is built into the ground ATC station with priority control to decide the resolution as shown in Fig. 1. For compatibility, a set of resolutions should be decided for all helicopters in a multi-aircraft conflict event simultaneously, and the resolution maneuver or traffic warning could be performed after confirmation between pilot and ATC. This implies that the coordination of airborne to ground uses the same configuration.

About priority control, assume that A, B, C and D aircraft are in a multiple conflict event. The Collision Avoidance logic issues one suite of resolutions for A, B, C and D in which the resolution between A and B is concerned first, then others. Three maneuvers are applied to let the highest priority aircraft "escape" from conflict threat. In operation, any fixed-wing aircraft can always get the highest priority.

FDS Architecture

In Fig. 1, the proposed FDS is mainly composed of three segments: (1) the airborne segment for all participating helicopters, (2) the ground segment for local ATC ground station, and (3) the software segment for performance requirements. Through radio transceiver and radio modem, each helicopter down-links its airborne data to the local ATC ground center. ADS-LB is defined following the ADS definition for local area application only. According to the airborne data, the ground station sends ATC commands, including collision avoidance resolutions, to the helicopters within its safety bubble, or broadcasts aircraft's relative position in time sharing mode. Aircraft's positions are projected on the GIS maps of both the airborne segment and the ground segment. Coordination is also displayed on each segment as well, with symbols and abbreviation. Such information in GIS background will clearly assist pilots or controllers to be aware of environments. The GIS supports necessary background of local map for flight director. Each local area is defined as a 25 miles by 25 miles area. All participating helicopters and ground station will generate the same GIS database during ATC monitoring.

Airborne Segment

The airborne segment is an avionics for helicopters. The main feature of the airborne segment is shown in Fig. 2-a. All fixed wing aircraft are not concerned.

Ground Segment

The ground segment is an ATC station to perform helicopter surveillance and control within the defined local area. The main feature of the ground segment is shown in Fig. 2.b.

Software Segment

The software segment contains GIS access, threat advisories generation logic, conflict resolution logic, and display generation logic. The proposed FDS chooses MapInfo as the core software for it's compatible with Microsoft Windows 95 and Office 97. Under such environment, the desktop mapping with Graphic User Interface (GUI) can easily be handled, and its MapBasic language can be used to integrate MapInfo functionality into applications using Visual Basic. MapBasic, Microsoft Excel and Visual Basic are the main software of concerns. The main function of the MapBasic is to receive GPS data and GIS map database. At the same time, this geo-information can be shown on the cells of Excel. While the program is running, a loop is created to check the flight conditions to send ATC commands to the cells on Excel. Visual Basic processes the geo-information and ATC commands in broadcasting using its format as:

GPS data	Encode	ATC command

Fig. 3 shows the coordination flow of software.

The software segment converts the collision avoidance advisories and algorithm into execution software as shown in Fig. 4. During the flight, both airborne segment and ground segment are running the same software program to generate conflict warnings onto displays.

Communication

Radio modem is used to down-link GPS data, and two-way VHF radio transceiver is used to exchange flight data and ATC commands in the FDS. Their effective range covers about 25 miles. It determines the defined local area of 25 miles by 25 miles in width. The ADS-LB is performed to have access airborne data down-link to ground ATC.

System Performance Test

The concept of FDS is composed of software engineering and hardware facilities. Most part of FDS is promoted by software integration to improve its function capability. Since helicopter test is very expensive, the FDS test used two remote-piloted helicopters for simulation, as shown in Fig. 5. Three vans are used, one plays as the ground segment, and the others operate as airborne segments. The remote control helicopters transmit flight data to these three vans.

The simulated test verifies the FDS with the following performance. (1) Data transmission for both radio modem and radio transceiver is capable deliver data within the local area. (2) As airborne GPS is calibrated initially, the new GPS data can be regarded as a shift from the calibrated position. It maintains repeatable error under 100 feet. For a helicopter of 80 knots, 100 feet error means about 1 second flight. Multipath exists during the simulated tests since the vans are running on the streets, but can be ignored while in real flights. (3) The FDS provides increased awareness to both ATC controllers and helicopter pilots. The ground station up-links the ATC command and flight director information to the airborne segments. The FDS provides the terrain with helicopter data on the moving map monitor via radio data-link. (4) The simulated tests examine the validity of GIS and GPS integration. Both airborne segments and ground segment are capable to display necessary flight information. (5) Collision avoidance logic is supported to alert to both pilots and ATC for threats.

A simulated test for four helicopters in demonstrated as shown in Fig. 6. The ground ATC station monitors helicopter flight in the local area. Airborne A and

Airborne B are approaching to cause a conflict warning. RA is firstly announced. At time t, both airborne segments and ground segment alert the conflict. Scenario assumes that Airborne A obtains higher priority, Airborne B has to make ways to make a 500 feet radius circular flight. (R5r) is pronounced on the airborne segment of Airborne B. Airborne A receives a (M) director to get the highest priority of flight. Airborne B receives a (R5r) director to make a right turn. At time t+4, Airborne A is "escaped" from collision. At time t+5, the alarm of Airborne A is dismissed. The participating helicopters C and D are not affected at this very moment, but vertical separation is pronounced by ATC to Airborne C to climb 300 feet and leave more vertical space. (□H+3) is pronounced to Airborne C. Since Airborne A is making a 500 feet circular flight, Airborne D may intrude its airspace by a RA. A reducing speed of 10 knots as (□S-1) is pronounced. In this figure, the solid line means the past trajectory and the dashed line indicates the expected trajectory. Airborne A and -B are two remote-piloted helicopters.

In Fig. 6, there are five monitor displays. The central one shows the ground ATC station to watch all four helicopters in this local area. Each corner shows the airborne display with conflict warnings after RA being issued. Each helicopter shows its flight direction with heading with intruders into RA. The flight paths of all helicopters in this local area are marked and extrapolated to show their performance. In Fig. 6, darker circle (green) means TA=5 sec., lighter circle (blue) means RA=15 sec., "star" means each sample point of helicopter position. Each window of Fig. 6 shows the concerned relative positions of intruders, such as A and B; while helicopter is flying without conflict threat will show a clear one, such as C and D.

Conclusion

A configuration of an integrated flight director and surveillance system (FDS) is proposed in this paper based on CNS/ATM concept. The FDS contains GIS mapping with GPS data and ATC collision avoidance with early-warning and conflict resolution. The proposed FDS integrates the airborne avionics to direct pilot flight and the ground ATC facility to maintain helicopter separation in low altitude. Both hardware and software are implemented for the FDS with simulated tests.

Further studies on cellular phone for communication, wider multiple local area operation, traffic alert, conflict resolution are concerned.

The proposed FDS is a feasibility study for helicopter applications. Conceptual design of the proposed FDS is demonstrated. The simulated test results encourage the motivation for advanced studies.

References

[1] T. Willianson, N. A. Spencer, "Development and Operation of the Traffic Alert and Collision Avoidance System (TCAS)", Proc. of IEEE, Vol.

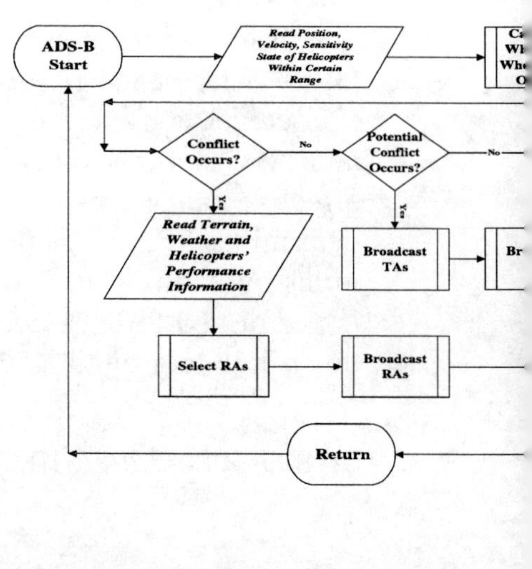

77, No. 11, Nov. 1989, pp.1735-1744.

[2] J. E. Rogers, "Terminal Area Surveillance System", IEEE Int'l Radar Conf., 1995, pp. 501-503, 0-7803-2120-0-95.

[3] J. Y. Gilkey, R. C. Galijan, "The Army GPS Truth Data Acquisition, Recording, and Display System at the White Sands Missile Range", IEEE Int'l Radar Conf., 1994, pp. 134-144, 0-7803-1435-2/94.

Fig. 1. FDS Concept and Operation.

Fig. 5. Simulated Tests.

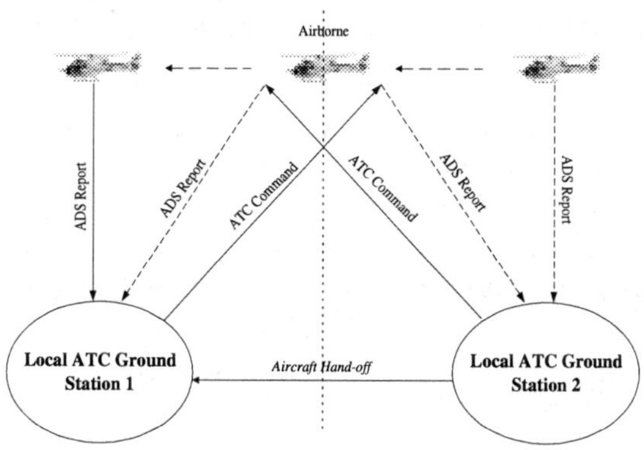

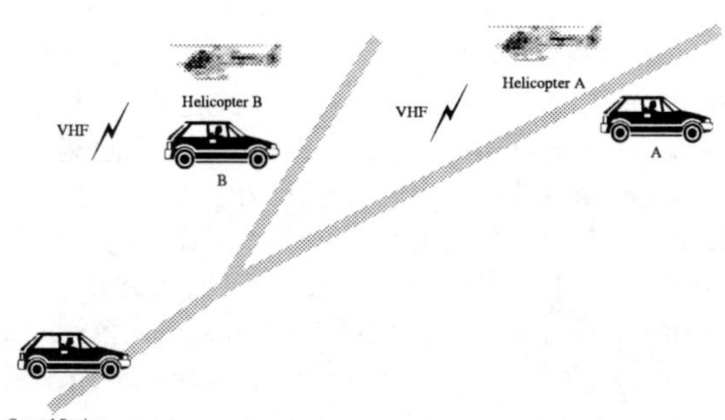

Fig. 3. Software coordination Flow.

Fig. 4. Flow Chart of Software Segment.

Fig. 2. The Proposed FDS Airborne

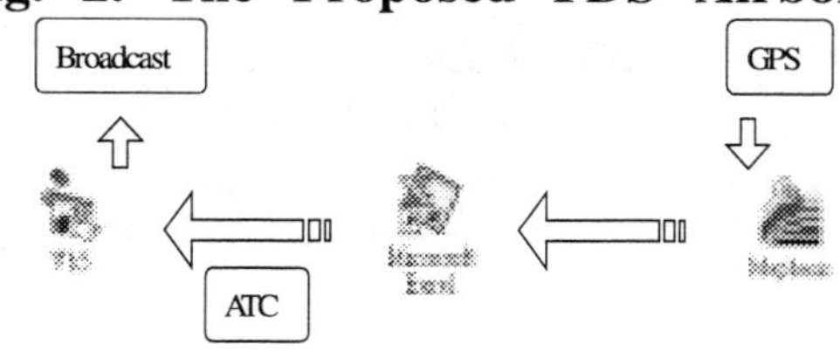

Segment and Ground Segment.

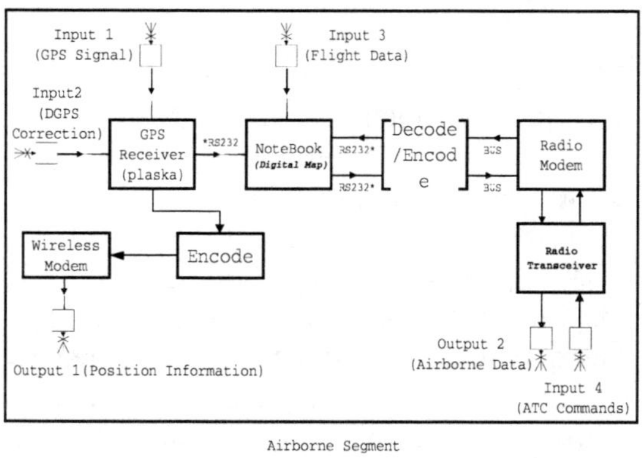

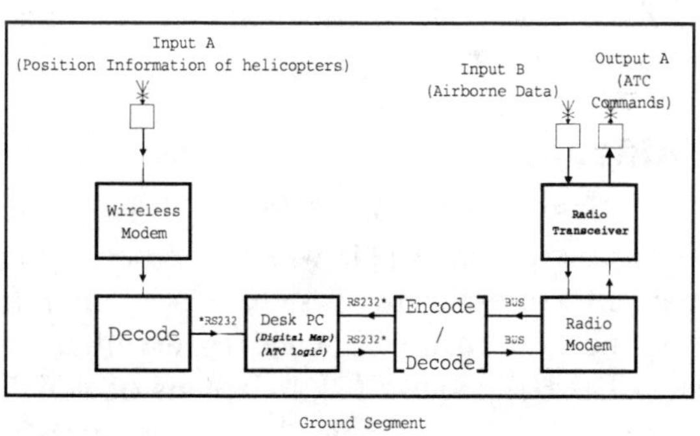

DATA PROCESSING METHODS FOR AUTONOMOUS ON-BOARD SURVEILLANCE AND COLLISION AVOIDANCE[1]

Juan A. Besada Portas, Universidad Politécnica de Madrid, Madrid, Spain.
Jesús García Herrero, Universidad Politécnica de Madrid, Madrid, Spain.
Francisco J. Jiménez Rodríguez, Universidad Politécnica de Madrid, Madrid, Spain.
José R. Casar Corredera, Universidad Politécnica de Madrid, Madrid, Spain.

Abstract

In this contribution, an ADS-B based airborne co-operative surveillance function, being the basis for an autonomous collision prediction function, is proposed. The main capabilities of the integrated data processing architecture provided will be detailed. This includes track formation, message-to-track association, coordinate transformation, position/velocity filtering, and conflict detection based on the concept of miss distance. Results illustrating the effectiveness of the system, based on rigorous modelling of navigation errors and communication channels, are presented.

1. Introduction

The integration of Global Navigation Satellite Systems (GNSS) [1], which are becoming operational as sole-means systems for aircraft navigation, and Automatic Dependent Surveillance-Broadcast (ADS-B) avionics [2], by means of which equipped aircraft can automatically broadcast their navigation information, will be the basis for future on-board safety systems. Within areas with radar coverage, future on-board anti-collision systems are expected to act as redundant safety systems with respect to ATC ground facilities, but they will be the only means for providing the pilot with anticipated potential collision warnings in non-radar coverage situations, both in oceanic or continental areas. Several configurations of airborne collision avoidance systems are currently being proposed, although standardisation of such a system is mandatory for it to become operational.

Currently, there is a system for airborne conflict detection (TCAS II) which is able to detect intruder aircraft closing ours, but cannot distinguish effectively aircraft in dangerous relative trajectories and those not so hazardous, because it only uses range measurements. Other limitation of that system (due to the unavailability of azimuth measurements) is the necessity of vertical manoeuvres to avoid intruders.

The second section of this contribution is devoted to the main reference frames to be used in the airborne safety and surveillance functions developed. The design of the conflict alert scheme requires a reliable navigation error model of the message data source, which is described in section 3. Section 4 will analyse the different data-links enabling ADS-B functionality, focusing on those channel effects that should be taken into account in the design of the algorithms described in section 5. Finally, in section 6, the performance of the system will be evaluated on exemplifying scenarios.

2. Coordinate Definitions and Transformations

The considered reference frames (see Figure 1) are three: 1) Navigation Frame, composed of longitude (λ), geodetic latitude (ϕ) and height (h); 2) Earth Centred – Earth Fixed (ECEF) frame; and 3) Local to Aircraft Cartesian Frame (LACF), whose axes point North (X) and West (Y).

[1] This work has been financed by CICYT.

0-7803-5086-3 /98/$10.00 ©1998 IEEE

Depending on ADS-B avionics, messages including 3D-velocity information (groundspeed, heading and vertical rate) can be sent.

Figure 1: Coordinate Frames.

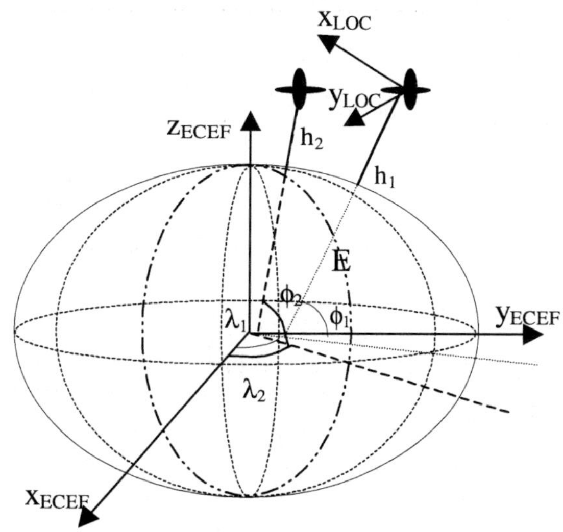

Tracking, and miss distance evaluation, is performed in LACF. For position determination, two coordinate transformations are applied, the first one for both aircraft (i=1,2), from navigation solution frame to ECEF frame:

$$\begin{bmatrix} x_{ECEF,i} \\ y_{ECEF,i} \\ z_{ECEF,i} \end{bmatrix} = \begin{bmatrix} \dfrac{a\cos\lambda_i}{\sqrt{1+\left(1-e^2\right)\tan^2\phi_i}} + h_i\cos\lambda_i\cos\phi_i \\ \dfrac{a\sin\lambda_i}{\sqrt{1+\left(1-e^2\right)\tan^2\phi_i}} + h_i\sin\lambda_i\cos\phi_i \\ \dfrac{a\left(1-e^2\right)\sin\phi_i}{\sqrt{1-e^2\sin^2\phi_i}} + h_i\sin\phi_i \end{bmatrix}$$

where a is the mayor axis of the Earth ellipsoid and e is its eccentricity (according to WGS-84). Then, in order to obtain the relative position estimates of aircraft 2 in aircraft 1 (ours) LACF, a translation plus a rotation is to be performed:

$$\begin{bmatrix} x_{LACF} \\ y_{LACF} \end{bmatrix} = F(\lambda_1,\phi_1)\left(\begin{bmatrix} x_{ECEF,2} \\ y_{ECEF,2} \\ z_{ECEF,2} \end{bmatrix} - \begin{bmatrix} x_{ECEF,1} \\ y_{ECEF,1} \\ z_{ECEF,1} \end{bmatrix}\right)$$

being $F(\lambda,\phi)$ the rotation/projection matrix:

$$F(\lambda,\phi) = \begin{bmatrix} -\sin\phi\cos\lambda & -\sin\phi\sin\lambda & \cos\phi \\ \sin\lambda & -\cos\lambda & 0 \end{bmatrix}$$

The relative velocity of both aircraft is the one with sense in LACF. Approximately:

$$\begin{bmatrix} v_{xLACF} \\ v_{yLACF} \end{bmatrix} = \begin{bmatrix} V_2 \cos\alpha_2 \\ V_2 \sin\alpha_2 \end{bmatrix} - \begin{bmatrix} V_1 \cos\alpha_1 \\ V_1 \sin\alpha_1 \end{bmatrix}$$

where V_i is the groundspeed and α_i is the heading of i-th aircraft, obtained from the navigator (for our aircraft) or from the ADS-B reports (for aircraft 2).

Measurements from both navigators should be referred to the same instant, which is achievable if all GNSS receivers are synchronised with UTC time.

3. Navigation Error Models

Location error expressed in LACF ($\Delta\vec{x}_{LACF}$), can be modelled as:

$$G(\lambda,\phi) = [F(\lambda,\phi) \quad \vec{0}]$$

$$\Delta\vec{x}_{LACF} = G(\lambda_1,\phi_1)(H^T H)^{-1} H^T \Delta\vec{\rho}$$

where H is the Nx4 projection matrix from ECEF coordinates to pseudo-range directions [1], and $\Delta\vec{\rho}$ is a Nx1 vector containing pseudo-range errors (N is the number of satellites used to get the navigation solution). To measure velocity GNSS receivers estimate Doppler shift of the signal carrier, performing a least-squares fitting of the data. It can be shown [1] that their velocity error ($\Delta\vec{v}_{LACF}$) is:

$$\Delta\vec{v}_{LACF} = G(\lambda_1,\phi_1)(HH^T)^{-1} H^T \Delta\vec{f}$$

where $\Delta\vec{f}$ is a vector with the errors in Doppler shift estimation (translated to velocity units). As satellite constellation and aircraft position are continuously changing, H changes accordingly, and so do position and velocity error statistics.

In a GNSS navigator, there are two different kinds of errors for pseudorange estimation: 1) Bias terms: due to Selective Availability, ionospheric and tropospheric effects, multipath, ephemeris prediction, etc. 2) Random terms: Receiver noise. For velocity estimators, the dominant term is that of receiver noise.

The preceding equations can be used in the characterisation of aircraft 1 biases and variances, but this is not the case for aircraft 2. The problem is the uncertainty in aircraft 1 about the satellites aircraft 2 selected to measure its position. In fact, if both aircraft are using the same satellites, most of the bias terms cancel when comparing both positions [3]. A conservative assumption, which is that of non-correlated errors between navigators, will be adopted. This assumption can deal correctly with encounters between aircraft with navigators of different types (GPS, WADGPS, and GLONASS)[1].

The concept of Dilution of Precision (DOP), relating errors in each coordinate with that in raw observations, is applied when dealing with GNSS errors:

$$XDOP = \frac{\sigma_x}{\sigma_\rho} \qquad VXDOP = \frac{\sigma_{Vx}}{\sigma_f}$$

where σ_ρ is the assumed pseudo-range RMS error, σ_x is the position RMS error in X (LACF) coordinate, σ_f is the assumed radial velocity RMS error to each satellite, σ_{Vx} is the velocity RMS error in X (LACF) coordinate, and XDOP and VXDOP are the Dilution of Precision both for X position and X velocity, respectively. Similar definitions hold for Y and Z local coordinates. It can be shown that XDOP and VXDOP are equal (and the same equality holds for Y), and can be obtained as:

$$\begin{bmatrix} XDOP_1^2 & XYDOP_1 \\ XYDOP_1 & YDOP_1^2 \end{bmatrix} = G(\lambda_1,\phi_1)(H^T H)^{-1} G^T(\lambda_1,\phi_1)$$

Most GPS receivers can directly give an estimation of XDOP and YDOP. Otherwise (and knowing satellite time-variable positions) they must be calculated. Satellite constellation should have been designed in order to make DOP not too dependent of receiver position, or of receiver type. So, other navigators DOP is in the order of ours:

$$\xi DOP_2 = (1+\delta)\xi DOP_1$$

$$\xi\psi DOP_2 = (1+\delta)^2 \xi\psi DOP_1$$

where ξ and ψ are two LACF coordinates, and δ is a constant, set to a conservative value (e.g. 0.3), assuming aircraft 2 navigator to be worse than ours. The assumed RMS navigator errors are [1]:

	$\sigma_{\rho b}$	$\sigma_{\rho r}$	σ_f
GPS	32.0 m	2 m	0.05 m/s
WADGPS	3 m	3 m	0.05 m/s

where $\sigma_{\rho b}$ accounts for the bias terms and $\sigma_{\rho r}$ for the random errors in pseudorange observable, while σ_f models the (random) errors in radial velocity.

4. Communication channels and associated quantification errors

Several communication means can be used to support ADS-B, being taken into consideration by several standardisation entities the following:

- Mode S Squitter [4].
- VHF Data Link, specially its VDL-Mode 4 version [5].

The broadcast channel based on Mode S squitter consists on transponders transmitting randomly (without need of Mode S radar interrogation, and with a time interval between consecutive reports of 0.4-0.6 seconds). Its main problem is its probability of losing messages, as far as they can collide with other replies from very close aircraft. The problem is increased because squitter frequency is shared with some other aeronautical systems, such as TCAS-II, Mode S replies, or SSR replies. In [4] some figures show the probability of losing an ADS-B report because of a collision, in terms of the number of aircraft near the receiver. For air-to-air applications, this problem is not too severe.

The 112 bit proposed Squitter message [4] has 56 bits devoted to ADS-B, with the following fields and associated quantification errors (LSB):

FIELD	LSB	AMBIGUITY RANGE
Latitude	0.0023'	10°
Longitude	0.0023'	10°
Barometric Height	25-100 ft	0-127kft
Navigator type	-	GPS/DGPS
Turn Indicator	-	TURN/NO

The GNSS system currently standardised is used to give measurements for integer UTC seconds, and so the same information will be sent, in most occasions, twice. It makes necessary to have a discarding logic (not to filter twice the same measurement) selecting only the new information.

VHF datalink appears as a possible alternative to the Squitter concept. One of the mayor features of VDL Mode IV is ADS-B implementation. Its media access control consists on TDMA with channel allocation by polling (ground-controlled mode) or contention resolution (autonomous mode). So, reports can be lost in autonomous mode. The slot allocation procedure makes report rate not completely constant in autonomous mode, but report collision is not likely to occur with the replies of nearby aircraft.

The VDL Mode-4 message for ADS-B would be User Sync (or Polled User Sync), whose fields, related to ADS-B functionality (and their range and quantification steps), are [5]:

FIELD	LSB	AMBIGUITY RANGE
Latitude[2]	0.016'	Total
Longitude	0.032'	Total
Baro. Height	16 ft	0-49kft
Speed	2 knots	0-4095 knots
Heading	.2°	0-359.8
Navigator type	-	GPS/DGPS

Quantification error is modelled as a uniform distribution of width that of the quantification step (LSB) in the tables. So, its bias is assumed to be zero, and its variance, over each of the measurements, is:

where ξ is the measurement dimension. Those variances must be translated to LACF coordinates, assuming the earth to be locally spherical, with local radius E. As X axis points Local North and Y axis points Local West, the quantification variances can

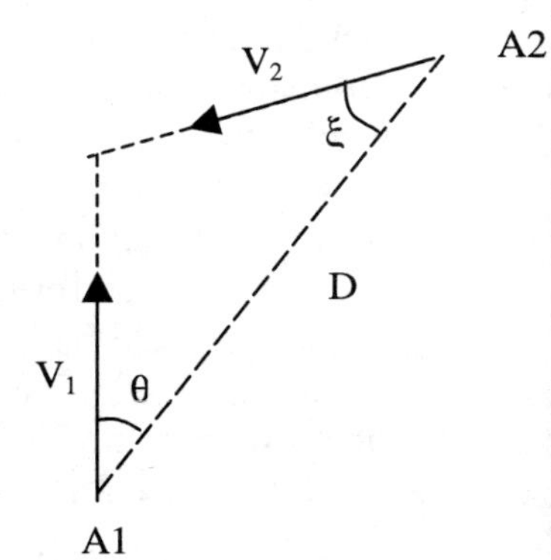

be assumed to be independent an equal to:

[2] In en-route applications, low resolution (with no ambiguities) will be used.

For velocity information, the variance transformations are as follows:

5. Data Processing Architecture and Algorithms

The proposed tracking and collision detection algorithms should be able of processing position and velocity. The architecture will be independent of the communication system used, but some of its functions will be redundant, unnecessary or inapplicable for a determined channel.

Fist of all, the potential ambiguity in longitude and latitude measurements (due to the limited length of the ADS-B messages) should be resolved. Ambiguous latitude and longitude should be translated to unambiguous values, taking into account the central reference of the ambiguity sector in which aircraft 1 is located. Besides, if our aircraft is near a border between two ambiguous zones, the ambiguity resolving algorithm should add or subtract the width of the ambiguity sector to those reports with a spatial difference (with respect to aircraft 1) greater than half an ambiguity sector.

Secondly, a selection of those reports with flight height similar to that of our aircraft should be performed. Tracking those aircraft in the flight levels just above and below ours helps to avoid anti-collision system induced collisions. Nevertheless, all processing will be performed for each horizontal layer. Also, a geographical selection of the reports from surrounding emitters in the horizontal plane should be performed. With GPS-Squitter, this selection could be left implicit to the operating range of the squitter. Otherwise, for the calculation of an interest region, an encounter geometry with an expedite aircraft, at each azimuth angle (angle between own trajectory and line of sight between both aircraft) is evaluated (see Figure 2).

Figure 2: Encounter Geometry.

So, a maximum range can be calculated, considering reaction times, minimum miss distances, and times to establish a consistent track. For each received report, the approximated azimuth angle (θ_{aprox}) and distance (D_{aprox}) to aircraft 2 are estimated as:

where b is the minimum Earth radius.

From the horizontal encounter geometry:

where K is a constant relating position error RMS (σ_H) and minimum allowed separation (s); $HDOP_i$ is the Horizontal Dilution of Precision ( ) of i-th aircraft; L_i is the i-th aircraft maximum extent (including safety margins); V_2 is maximum velocity; T_{track} is the time necessary to have a reliable track; T_{react} is the pilot reaction time to an advisory; T_{marg} is an additional safety margin time; and a_{min} is the minimum acceleration of the aircraft. The result, if we assume T_{track}=10 s, T_{react}=10 s, T_{marg}=5 s, L_1=100 m, L_2=100 m, a_{min}=1 m/s^2, $HDOP_1$=2, V_1=300 m/s, V_2=320 m/s, GPS receivers, K=5, and $\sigma_{QX}=\sigma_{QY}$=10 m, can be seen in the figure 3.

Figure 3: Example Geographic gate.

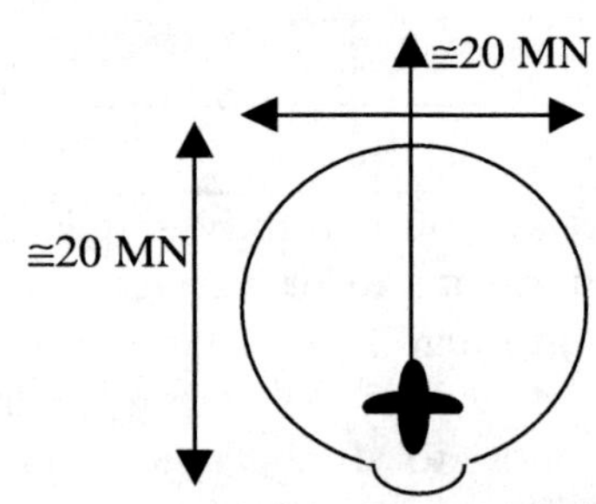

For those reports falling within the geographical gate ($D_{aprox}<D(\theta_{aprox})$), the Coordinate transformation to aircraft 1 LACF is applied (both to the measures and measurement uncertainties).

Then, the track to report association logic is applied. The report identification field is compared

0-7803-5086-3 /98/$10.00 ©1998 IEEE

with that of any established or tentative track, and they are associated only if both codes are identical.

Those reports not associated to any track are used for track initiation, and a tentative track for them is created, with the transformed position (and velocity if available) in the report. If the reports contain velocity information, a track can be initiated with a single report, but if they only contain position information, two reports are necessary. For position plus velocity reports, the initiation algorithm copies the transformed measurements:

$$\begin{bmatrix} \xi_f \\ V_{\xi f} \end{bmatrix} = \begin{bmatrix} \xi_{LACF} \\ V_{\xi LACF} \end{bmatrix}$$

where ξ is X or Y coordinate, and f subindex stands for filtered estimate.

To initiate filtered covariances, it is important to distinguish them from the bias terms. So, we only have to model the random terms from both navigators and quantification error. The variances, to be used in initiation and filtering, are:

$$\sigma_\xi^2 = \left(\xi DOP_1^2 + \xi DOP_2^2\right)\sigma_{pr}^2 + \sigma_{Q\xi}^2$$
$$\sigma_{V\xi}^2 = \left(\xi DOP_1^2 + \xi DOP_2^2\right)\sigma_f^2 + \sigma_{QV\xi}^2$$
$$\sigma_{\xi V\xi} = 0$$

where, once again, ξ can be X or Y. So, for position and velocity reports, the initiation covariance matrix will be:

$$P_{\xi f} = \begin{bmatrix} \sigma_\xi^2 & 0 \\ 0 & \sigma_{V\xi}^2 \end{bmatrix}$$

The initial estimates and covariances, with two consecutive positions initiation, will be:

$$\begin{bmatrix} \xi_f \\ V_{\xi f} \end{bmatrix} = \begin{bmatrix} \xi(t) \\ \dfrac{\xi(t) - \xi(t-T)}{T} \end{bmatrix} \qquad P_{\xi f} = \begin{bmatrix} \sigma_\xi^2 & \dfrac{\sigma_\xi^2}{T} \\ \dfrac{\sigma_\xi^2}{T} & \dfrac{2\sigma_\xi^2}{T^2} \end{bmatrix}$$

being T the time interval between the (equal identification) position reports.

The proposed data filtering scheme consists on an uncoupled Kalman filter, with different measurement models depending on the presence or absence of velocity information:

- Prediction step:

$$\Phi = \begin{bmatrix} 1 & T \\ 0 & 1 \end{bmatrix} \qquad Q = \left(\sigma_{a1}^2 + \sigma_{a2}^2\right)\begin{bmatrix} \dfrac{T^4}{4} & \dfrac{T^3}{2} \\ \dfrac{T^3}{2} & T^2 \end{bmatrix}$$

$$\begin{bmatrix} \xi_p \\ V_{\xi p} \end{bmatrix} = \Phi \begin{bmatrix} \xi_f \\ V_{\xi f} \end{bmatrix} \quad ; \qquad P_{\xi p} = \Phi P_{\xi f} \Phi^T + Q$$

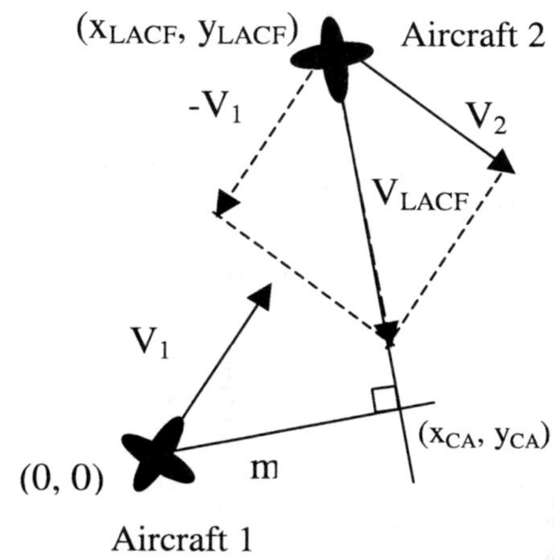

where T is the extrapolation time (between last filtering and current measurement instant), and σ_{ai} (i=1,2) is the plant noise variance for the i-th aircraft, accounting for aircraft manoeuvres. If we know any of the aircraft is not manoeuvring, σ_{ai} can be set to zero, but otherwise a reasonable value will be 3 m/s^2. So, the use of a turn indicator can be very effective.

- Filtering step (there are two filtering equations sets, depending on the availability of velocity):

1. No velocity available:

$$H = \begin{bmatrix} 1 & 0 \end{bmatrix} \qquad R = \sigma_\xi^2$$
$$K = P_{\xi f} H^T \left(H P_{\xi f} H^T + R\right)^{-1}$$
$$\begin{bmatrix} \xi_f \\ V_{\xi f} \end{bmatrix} = \begin{bmatrix} \xi_p \\ V_{\xi p} \end{bmatrix} + K\left(\xi_{LACF} - \xi_p\right)$$
$$P_{\xi f} = (I - KH) P_{\xi p}$$

2. Velocity available:

$$H = \begin{bmatrix} 1 & 0 \\ 0 & 1 \end{bmatrix} \qquad R = \begin{bmatrix} \sigma_\xi^2 & 0 \\ 0 & \sigma_{V\xi}^2 \end{bmatrix}$$

$$K = P_{\xi f} H^T \left(H P_{\xi f} H^T + R \right)^{-1}$$

$$\begin{bmatrix} \xi_f \\ V_{\xi f} \end{bmatrix} = \begin{bmatrix} \xi_p \\ V_{\xi p} \end{bmatrix} + K \left(\begin{bmatrix} \xi_{LACF} \\ V_{\xi LACF} \end{bmatrix} - \begin{bmatrix} \xi_p \\ V_{\xi p} \end{bmatrix} \right)$$

$$P_{\xi f} = (I - KH) P_{\xi p}$$

with I the 2x2 identity matrix.

The parameter T_{track}, which is used to select the interest area around aircraft 1 (ours), will depend on the number of plots necessary to obtain reliable position and velocity estimators.

A track cancellation procedure based on a number of allowed missed reports (N_m) and extrapolated position of the target is also included in the system. The extrapolation is performed each $T_{dismiss}$ seconds (e.g. 10) for every track, and those whose extrapolated position falls out of the interest region, and have more than N_m missed reports (e.g. 2), are dismissed.

A potential conflict prediction algorithm, based on the concept of miss distance [6] (see Figure 4), is then applied.

Figure 4: Miss distance.

The predicted miss distance (m) is estimated on the basis of filtered position and velocity in LACF. From those estimations, and from geometrical reasoning, m can be calculated as:

$$k = \frac{v_{Yf}}{v_{Xf}} \quad ; \quad x_{CA} = \frac{k^2 x_f - k\, y_f}{1 + k^2}$$

$$y_{CA} = \frac{y_f - k x_f}{1 + k^2} \quad ; \quad m = \sqrt{x_{CA}^2 + y_{CA}^2}$$

where k is the slope of the line passing through the current relative position of aircraft 2 and with the direction of the relative speed. Another important parameter is the time to close approach (t_{CA}). This is the time from present to the instant at which both aircraft are to be separated by the miss distance. From the previous figure:

$$t_{CA} = \frac{x_{ca} - x_f}{v_{Xf}} = -\left(\frac{x_f}{v_{Xf}} + \frac{y_f v_{Yf}}{v_{Xf}^2} \right)$$

Once those two quantities have been calculated, the conflict detection scheme tries to show potential conflicts involving our aircraft and those around it. First of all, a taxonomy (ordered in terms of likelihood and risk) of the possible cases of conflicts was developed, based on the potential intents of both pilots:

1. No pilot will perform a manoeuvre.
2. The other pilot is manoeuvring.
3. Our pilot is manoeuvring or to manoeuvre.
4. Both are manoeuvring.

The worst manoeuvres, in terms of conflicts, are those with a large projection over the direction of close approach. We can (possibly) know about our pilot present or future intent, and also, if we are using GPS-Squitter for ADS-B, about the other pilot's present intent (but not future), through the turn indicator.

The proposed approach for conflict avoidance will be to give negative directives (not to turn right or left) before giving positive directives (turn right or left). Potential conflicts will be analysed with every aircraft in the vicinity region, and then the directives for all should be combined, taking into account the previous taxonomy. This contribution will not deal with avoidance algorithms, and the combination of directives, but with methods for detecting all potential conflicts in the previous taxonomy, which could be used by an avoidance system. The conflict detection scheme comprises the following steps:

1. Acceleration projection over the Close Approach Line of Sight (CALOS, this between both aircraft at t_{CA} from present). This projection will be performed assuming maximum lateral and longitudinal accelerations (whose values depend on the conflict case we are studying):

$$a_{pLon,i} = \frac{a_{Lon} x_{CA} \cos\alpha_i + a_{Lon} y_{CA} \sin\alpha_i}{m}$$

$$a_{pLat,i} = \frac{a_{Lat} x_{CA} \sin\alpha_i - a_{Lat} y_{CA} \cos\alpha_i}{m}$$

$$a_{pTot,i} = \left| a_{pLon,i} \right| + \left| a_{pLat,i} \right|$$

where a_{Lon} and a_{Lat} are the longitudinal and lateral maximum accelerations assumed for each conflict case (manoeuvring aircraft, $a_{Lon} = 1.5\ m/s^2$, $a_{Lat} = 6\ m/s^2$, non-manoeuvring aircraft $a_{Lon} = a_{Lat} = 0$), $a_{pLon,i}$

0-7803-5086-3 /98/$10.00 ©1998 IEEE

and $a_{pLat,i}$ are the projections of those accelerations over the CALOS.

2. Calculation of the worst case miss distance (m_a) for each conflict case. We will assume (although it is not true), that, in a manoeuvre scenario, both aircraft will maintain constant (and equal to those just calculated values) approaching accelerations, from the time of last track update to the time of closest approach. For each manoeuvring assumption, in worst case:

$$m_a = m - \frac{1}{2}\left(a_{ipTot,1} + a_{ipTot,2}\right)t_{CA}^2$$

3. Comparison of m_a with a threshold, fixed for a level of probability of being less than a certain safety margin:

$$m_a > L_1 + L_2 + K'\sigma_{ma}$$

where K' is a constant related to the probability of having two aircraft nearer than $L_1 + L_2$, and σ_{ma} is the RMS in the estimation of m_a. Two different approaches could be used for the estimation of σ_{ma}. The first one relies on the linear first order approximation of equations relating m_a with the tracking terms. The adopted alternative approximation is based on the following assumptions:

- The errors in the estimated t_{CA} and acceleration projections have little effect in the overall error.
- The error in x_{CA} and y_{CA} comes from two effects: the linear extrapolation (which could be evaluated through covariance extrapolation of a Kalman filter, with σ_a=0), and the bias terms related with GNSS.

For the calculation of the bias terms, a CALOS Dilution of Precision (CDOP), is first evaluated as:

$$\vec{r}_{CALOS} = \begin{bmatrix} \frac{x_{CA}}{m} & \frac{y_{CA}}{m} \end{bmatrix}^T$$

$$CDOP^2 = \vec{r}_{CALOS}^{\,T} \begin{bmatrix} XDOP^2 & XYDOP \\ XYDOP & YDOP^2 \end{bmatrix} \vec{r}_{CALOS}$$

while the linear extrapolation error is:

$$\sigma_{Xp}^2 = [1 \quad 0]\Phi(t_{CA})P_{fX}\Phi(t_{CA})^T[1 \quad 0]^T$$

$$\sigma_{Yp}^2 = [1 \quad 0]\Phi(t_{CA})P_{fY}\Phi(t_{CA})^T[1 \quad 0]^T$$

$$\sigma_{pCALOS}^2 = \vec{r}_{CALOS}^{\,T} \begin{bmatrix} \sigma_{XpCALOS}^2 & 0 \\ 0 & \sigma_{YpCALOS}^2 \end{bmatrix} \vec{r}_{CALOS}$$

Finally, σ_{ma} can be assumed to be:

$$\sigma_{ma}^2 = CDOP^2\sigma_{pb}^2 + \sigma_{pCALOS}^2$$

4. Only those conflicts whose time to be in conflictive situation is equal or less than the total time to react should be informed to the pilot. Otherwise, targets flying along parallel paths would rise lots of false conflicts. The time to be in conflictive situation (t_{cs}) is:

$$a = v_{Xf}^2 + v_{Yf}^2 \qquad b = 2x_f v_{Xf} + 2y_f v_{Yf}$$

$$s_{av} = K\sigma_{ma} + L_1 + L_2 \qquad c = x_f^2 + y_f^2 - s_{av}^2$$

$$t_{cs} = \frac{-b + \sqrt{b^2 - 4ac}}{2a}$$

where s_{av} is the minimum allowed distance, while the total time to react is calculated as:

$$T_{reac\_tot} = T_{marg} + T_{react} + T + \sqrt{\frac{2s_{av}}{a_{min}}}$$

No directive is given before a potential conflict between two targets in linear movement can appear.

6. Results and Conclusions

All the experimental results in this paper were obtained by means of Monte Carlo simulation. The first experiment shows the capability of the system to distinguish hazardous situations. The scenario consists of two crossing targets (both with a speed of 300 m/s), which pass at a distance of M metres one from the other. The probability of detecting a conflict for the first case in the taxonomy is calculated, for both system configurations (STDMA drawn in continuous line, and Mode S squitter as dotted line, navigating with GPS).

Figure 5: Probability of detecting a conflict.

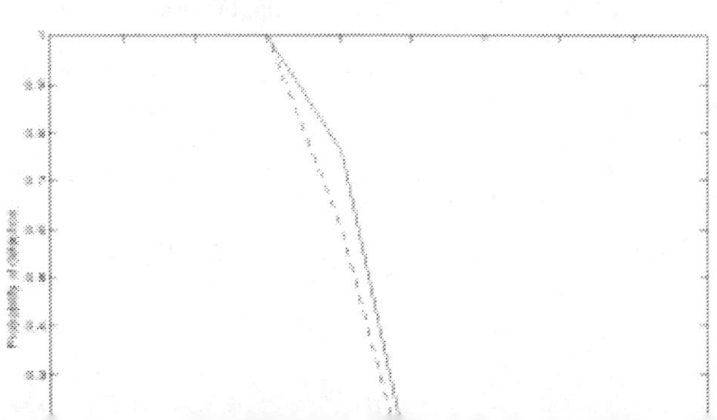

The mean time left for manoeuvre avoidance is evaluated for both system configurations, for a scenario with a real encounter with an expedite aircraft, depending on θ (see Figure 2).

Figure 6: Time left for manoeuvre avoidance.

The main conclusions of our study are:

- The proposed system can safely detect conflicts, even without WADGPS receivers.
- Both STDMA and GPS-Squitter seem to be good elections for this application. GPS-Squitter is based on a much more mature technology, that of TCAS and Mode S. The reduction of report period (possible with STDMA) could be used to increase performance when resolving a conflictive situation.
- Some improvements to the system could be:

1. Include a superset of the fields of STDMA bursts and squitters, including detailed turning information.
2. Make the GPS system give the position results sent by ADS-B with higher periodicity.

References.

[1]Understanding GPS. Principles and applications. Elliot D. Kaplan. Artech House Publishers. 1996.

[2] A Visionary Look at Aviation Surveillance Systems. George Donohue. IEEE AES Systems Magazine. October 1995.

[3]GPS Relative Accuracy for Collision Avoidance. Marc-Philippe Rudel, Jonathan Baldwin. Institute of Navigation Technical Meeting. 1997.

[4]GPS-Squitter: System Concept, Performance, and Development Program. Vincent A. Orlando, George H. Knittel, and Robert E. Boisvert. Lincoln Laboratory Journal. 1994.

[5]SARPS VDL Mode-4. Aeronautical Mobile Communications Panel (ACMP). Fourth Meeting. Montreal 1996.

[6]TCAS: Maneuvering Aircraft in the Horizontal Plane. Douglas W. Burgess, Sylvia I. Altman, and M. Loren Wood. Lincoln Laboratory Journal. 1994.

IMPLEMENTATION OF A LOW-COST SSR/ADS-B AIRCRAFT RECEIVER DECODER (SY-100)

Bennett Cohen, Rannoch Corporation, Alexandria, Virginia

Alex Smith, Rannoch Corporation, Alexandria, Virginia

Abstract

For many aviation research and development programs it is necessary to acquire accurate aircraft position and identification information. Many previous analyses in terminal areas have used radar surveillance data recorded from terminal radar systems. These analyses have required conversion and post processing of radar tapes from air traffic control. Problems with this type of analysis include a low data update rate (once per 4.8 seconds in the terminal area), sufficient coverage at low levels (below the radar cut-off), data latency (it can be up to 2 seconds old), and the intensive task of manually sorting and parsing data from the tapes. To assist with several research and development programs Rannoch has developed and implemented a real time stand alone radar receiver decoder system. The system decodes all radar data on the 1090 MHz and 1030 MHz (downlink and uplink) including Mode S, TCAS, ATCRBS, ADS-B, and IFF. In a typical terminal area the system decodes real time aircraft radar data at up to 10 Hz from each aircraft. This paper describes the application of the system to assist with several Air Traffic Control (ATC) research programs including wake vortex research. The equipment set up and integration with a wake vortex sensor suite is described and results are presented from field trials at JFK International Airport. This paper contains only the understanding and views of the authors and is not intended to reflect the official position or view of the US Government, the FAA or NASA.

Introduction

The Volpe Center, NASA and the FAA have been conducting wake vortex research at JFK, Memphis, DFW and other airports for a number of years. The purpose of their research is to characterize the behavior of wake vortices and to ultimately be able to predict their movement so that ATC will be able to warn aircraft before a wake vortex encounter. A key element in detecting and characterizing wake vortex behavior is determining what type of aircraft generated a particular set of vortices. In the past this data was either collected by an observer, (who had to be very familiar with different aircraft types), or by obtaining FAA radar data and determining which aircraft were the aircraft of interest and matching them up with the data from the wake vortex sensors. These methods proved to have numerous drawbacks. An observer was costly to employ around the clock, and occasionally made mistakes in identification and time recording. At the end of data collection, all of the observer's data had to be integrated by hand with the data from the wake vortex sensors.

Aircraft data from ATC radar required that prior arrangements be made with the FAA. Afterwards, this radar data required considerable effort to separate out the aircraft of interest from the rest of the aircraft and to match the radar data with the corresponding sensor data. Because airport surveillance radar rotates once every 4.8 seconds, the aircraft tracks are sometimes difficult to follow. As aircraft get close to the radar antenna they occasionally go below the coverage of the antenna, making it difficult to track the aircraft close in to the airport. Because of these shortcomings, Rannoch Corporation, under contract from the Volpe Center, integrated an SSR decoder to automatically identify landing aircraft models and types as well as provide other information.

0-7803-5086-3 /98/$10.00 ©1998 IEEE

Why Mode S

Rannoch set out to build aircraft identification equipment that would be low cost, passive, stand alone, and automated. We wanted to build equipment that would cost a fraction of a TCAS unit, that didn't interfere with any ATC or other equipment, and that would operate continuously in an unattended environment. Additionally, the equipment had to comply with the guidelines established in the ICAO Mode S SARPS, the RTCA Mode S MOPS and the Draft FAA / Industry ADS-B requirements. With these requirements we decided that the best approach would be to develop a device that listened to the Mode S transponder transmissions from aircraft. Mode S transponders are required equipment on all passenger carrying aircraft with more than 30 seats. This requirement includes foreign carriers that operate into the U.S. Many aircraft with 10 to 30 seats are also equipped with Mode S transponders, although this is not a requirement. Many business jets carry Mode S transponders as well.

Evolution of SSR

In the early days of radar, targets were identified by directing megawatts of RF energy through a narrow rotating beam of directed energy. This RF energy would be reflected off the target and some of the energy would bounce back to the transmitting radar antenna. A drawback of this method, called primary radar, is that the return RF energy from the target diminished as an inverse function of the square of the distance to the target. For example, the returned energy from a target that is twice as far away than another target is only one quarter of that of the closer target. Then in the 1950's a method was developed that extended the useful range of radar by eliminating the problem of the inverse square law. Aircraft began carrying transponders, radio transmitters that replied to the radar interrogation with their own RF energy as opposed to reflected radio energy. This is called secondary surveillance radar (SSR). The earliest methods of SSR enabled a target aircraft to reply to the interrogating radar with one of 64 (2^6) possible pulse codes. Further refinement of the technique enabled target aircraft to reply with up to 4096 (2^{12}) possible unique pulse codes used for discrete identification. This was called Mode A. Mode C enabled the ground based interrogator to ask for the pressure altitude of the aircraft from an encoding altimeter.

In the 1980's the Mode S transponder format was established. It allows for the transmission of many more pulses and therefore, much more information. The Mode S format actually consists of a subset of 25 possible uplink and downlink formats. The "short" downlink formats contain 56 pulsed bits of information, while the "long" formats contain 112 pulsed bits. Figure 1 attempts to show the relative length of the Mode A, C, and S short and long reply formats. Note that the Mode S long format has been truncated due to printing size considerations. The Mode S long formats are actually 120 microseconds long.

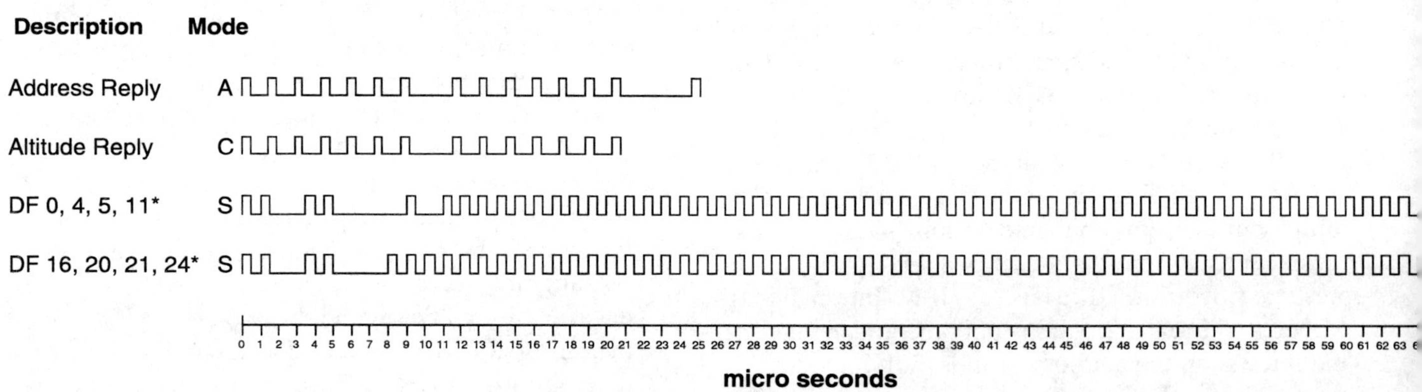

Figure 1. 1090 MHz Formats

At present only formats 0,4,5,11,16,20, 21 and 24 are used in aviation. The Rannoch decoder is capable of decoding all 25 formats. One format, DF 17, will be the first format used for the much

0-7803-5086-3 /98/$10.00 ©1998 IEEE

touted ADS-B (Automatic Dependent Surveillance-Broadcast) system. The format will contain information on aircraft latitude, longitude and altitude. This position data can be from GPS or other navigation systems. Aircraft will transmit DF-17s once per second without being interrogated. This is referred to as a "squitter" and will enable both aircraft and ground based controllers to receive very accurate position information. The Rannoch decoder was used during surface operation trials of the ADS-B system in July, 1997 at Atlanta Hartsfield International Airport.

How it Works

The Rannoch decoder is comprised of an antenna, a 1090 MHz receiver, an analog to digital converter and decoder, and Pentium processor and software as shown in Figure 2.

Figure 2. Decoder, Receiver & Processor

The antenna used for the wake vortex research project at JFK was a 4 pole yagi directional antenna (see Figure 3). The directional antenna allows for adjusting the reception of the decoder to the area of interest, which in this case is the final approach course. It also eliminates reception from other aircraft on the surface or operating to a parallel runway. A variety of different antennas including directional, omni-directional and sector antennas can be used with the decoder to achieve various coverage patterns.

Figure 3. Yagi Directional Antenna Used To Monitor Approaches to One Runway

The 1090 MHz receiver is a small (less than 2 lb.) unit that can either be attached to the antenna for maximum effectiveness or can reside next to the decoder processor. It is also capable of receiving 1030 MHz for monitoring ground and TCAS interrogations. The receiver receives DC power through the coax cable and can be switched remotely to operate at either 1030 MHz or 1090 MHz. The receiver was designed to operate with the modulation and bandwidth of the uplink and downlink formats. Our first prototype (shown in Figure 4) achieved a sensitivity of -90 dBm.

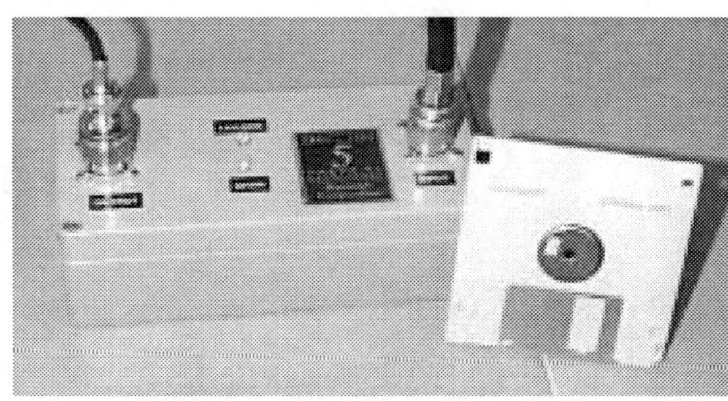

Figure 4. Mast Mountable SSR Receiver To Maximize Range

The analog to digital converter and decoder have been designed into a range of hardware architectures, from the PC ISA bus on the low end to the VME bus for commercial applications. The decoder circuitry is re-configurable at any time. The software extracts the encoded Mode S address, converts the address to a U.S. registration number (N number) and does error detection and correction. It also extracts the altitude, which is pressure altitude reported in either 25' or 100' resolution. Then, the N number is looked up into the database of all registered US aircraft. The output of the decoder consists of a computer time stamp, Mode S address, downlink format, N number, owner/operator, make, model, serial number and altitude (see Figure 5). The software is easily modifiable to include or exclude any fields from the database. The data is sent to a monitor, the hard drive and a serial port in real time. The software also enables a signal strength threshold to be set which excludes the reception of transponder transmissions which fall below a specified level.

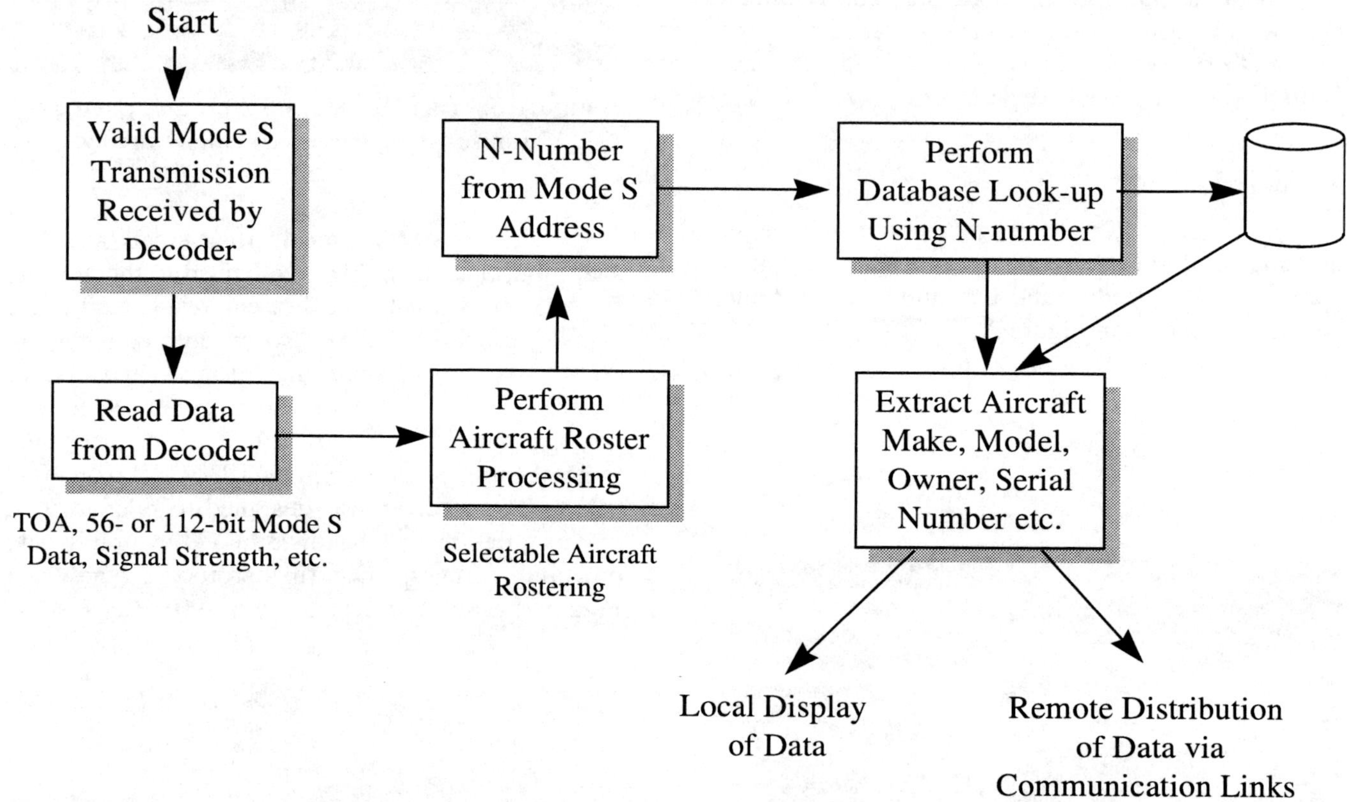

Figure 5. Real Time Decode & Look Up Process

Deployment Considerations

The purpose of the decoder at the JFK wake vortex sensor site was to identify individual aircraft as they passed over the wake vortex sensors. These sensors are located just inside the middle marker for runway 31R, about 3800 feet from the touchdown zone. A directional yagi antenna was used to focus on aircraft on final approach to 31R and ignore those on the ground and on parallel approach to runway 31L. The signal strength threshold was set to a level where signals were received for about three to thirteen seconds for each aircraft as it approached the sensor area. The variation in reception duration time is due to a number of factors including differences in transponder output power, antenna gain and ground speed. The signal strength threshold was set to a level that did not exclude any aircraft. On this model thresholds were selectable, via graphical user interface, from -90 dBm to -20 dBm.

Decoder Output

The following excerpt from the decoder output file shows typical data for an aircraft as it approaches the wake vortex sensor area. The first field of data is a time stamp from the computer's internal clock. The time is displayed in hours:minutes:seconds:hundredths. The next field is the Mode S address expressed as a six digit hexadecimal number. The third field is the downlink format. Notice in the following example that DF 0's are most prevalent. They are responses to TCAS interrogations. DF 4s, 11, and 20s also appear in this example. The next field is the U.S. registration number, or "N" number, which is derived from the Mode S address. The N number is used to look up into a database of US registered aircraft. The next five fields, owner/operator, manufacturer, model, serial number, and weight class all come from the database. The last field is pressure altitude which is encoded in downlink formats 0, 4, 16 and 20.

In the following typical data sample (Figure 6) from a United Airlines Boeing 777, twenty-five transponder transmissions were decoded over a period of about ten seconds. During that time the aircraft's pressure altitude decreased from 250 feet to 150 feet.

```
09:45:45:66 AAB001 11 788UA  UNITED AIR LINES INC  BOEING  777-222  26942  CLASS 3
09:45:46:87 AAB001 00 788UA  UNITED AIR LINES INC  BOEING  777-222  26942  CLASS 3  250
09:45:47:04 AAB001 00 788UA  UNITED AIR LINES INC  BOEING  777-222  26942  CLASS 3  225
09:45:47:70 AAB001 11 788UA  UNITED AIR LINES INC  BOEING  777-222  26942  CLASS 3
09:45:48:08 AAB001 00 788UA  UNITED AIR LINES INC  BOEING  777-222  26942  CLASS 3  225
09:45:48:30 AAB001 00 788UA  UNITED AIR LINES INC  BOEING  777-222  26942  CLASS 3  225
09:45:48:52 AAB001 00 788UA  UNITED AIR LINES INC  BOEING  777-222  26942  CLASS 3  225
09:45:49:07 AAB001 00 788UA  UNITED AIR LINES INC  BOEING  777-222  26942  CLASS 3  200
09:45:49:29 AAB001 00 788UA  UNITED AIR LINES INC  BOEING  777-222  26942  CLASS 3  200
09:45:49:45 AAB001 11 788UA  UNITED AIR LINES INC  BOEING  777-222  26942  CLASS 3
09:45:49:62 AAB001 04 788UA  UNITED AIR LINES INC  BOEING  777-222  26942  CLASS 3  200
09:45:50:06 AAB001 00 788UA  UNITED AIR LINES INC  BOEING  777-222  26942  CLASS 3  200
09:45:50:28 AAB001 00 788UA  UNITED AIR LINES INC  BOEING  777-222  26942  CLASS 3  200
09:45:50:50 AAB001 11 788UA  UNITED AIR LINES INC  BOEING  777-222  26942  CLASS 3
09:45:50:94 AAB001 00 788UA  UNITED AIR LINES INC  BOEING  777-222  26942  CLASS 3  175
09:45:51:10 AAB001 00 788UA  UNITED AIR LINES INC  BOEING  777-222  26942  CLASS 3  175
09:45:51:43 AAB001 20 788UA  UNITED AIR LINES INC  BOEING  777-222  26942  CLASS 3  175
09:45:51:60 AAB001 20 788UA  UNITED AIR LINES INC  BOEING  777-222  26942  CLASS 3  175
09:45:52:09 AAB001 00 788UA  UNITED AIR LINES INC  BOEING  777-222  26942  CLASS 3  175
09:45:52:31 AAB001 00 788UA  UNITED AIR LINES INC  BOEING  777-222  26942  CLASS 3  175
09:45:52:53 AAB001 00 788UA  UNITED AIR LINES INC  BOEING  777-222  26942  CLASS 3  175
09:45:54:07 AAB001 04 788UA  UNITED AIR LINES INC  BOEING  777-222  26942  CLASS 3  150
09:45:54:23 AAB001 04 788UA  UNITED AIR LINES INC  BOEING  777-222  26942  CLASS 3  150
09:45:55:06 AAB001 00 788UA  UNITED AIR LINES INC  BOEING  777-222  26942  CLASS 3  150
09:45:55:28 AAB001 00 788UA  UNITED AIR LINES INC  BOEING  777-222  26942  CLASS 3  150
```

Figure 6. Sample Decoder Data Real Time Output

Data Interpretation and Analysis

Descent Profiles

Descent profiles can be generated easily from the decoder data. Figure 7 shows the descent profiles for a SAAB 340 and an A320 Airbus. The SAAB was two minutes in trail of the Airbus on a visual approach to JFK's 31R. The graph clearly shows that the SAAB was above the flight path of the preceding Airbus. Perhaps the pilot of the SAAB was trying to stay above the wake vortices of the Airbus, or perhaps he was just making a longer landing to avoid a lengthy taxi to his gate.

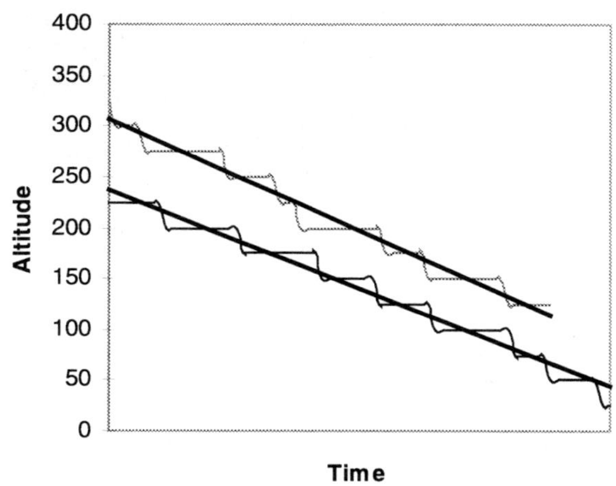

Figure 7. Aircraft Descent Profile Monitoring

Aircraft Type Breakdown

The decoder data can also be used to generate a breakdown of the type of aircraft that landed on a particular runway. The following pie chart (Figure 8) shows a breakdown of the aircraft types that landed on JFK's 31R in a 24 hour period. It is interesting to note that over 40% of the aircraft were twin turboprop commuters. Analyzing data over a longer period of time would provide statistics on aircraft separation in IFR/VFR, and could be used to build a database of pairwise separations (i.e., achieved separation between each aircraft type and weight class).

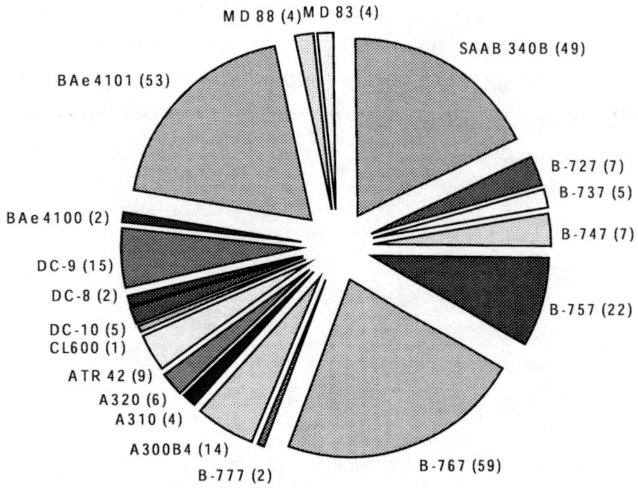

Figure 8. Aircraft Types Arriving at JFK

Ascent Profiles

The following graph of a DC-9 ascent profile (Figure 9) was generated from data collected at Rannoch's headquarters in Alexandria, Virginia. Aircraft arriving and departing Washington National Airport (DCA) are an excellent source of transponder transmissions in a terminal environment. The graph shows the DC-9 ascending at 2000 feet per minute.

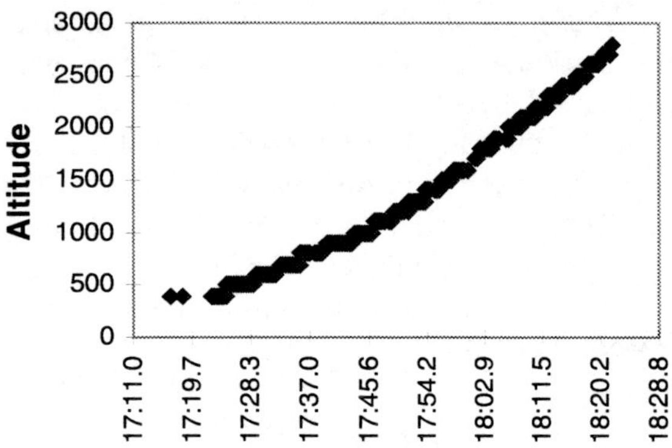

Figure 9. DC-9 Departure

Downlink Formats

The decoder receives and decodes all Mode S downlink formats, as well as Mode A, Mode C, and IFF. Each Mode S downlink format has a particular purpose. DF 0's are responses to TCAS interrogations. DF 4's are responses to ground based interrogations. DF 11's are "squittered" by Mode S transponders at a nominal rate of 1 Hz. Figure 10 shows the cumulative number of transmissions of DF 0's, 4's, and 11's for the DC-9 shown in the previous graph. Note that the DF 0's are transmitted much more frequently than the DF 4's or 11's. This is the expected response in a busy terminal area where there are many TCAS interrogations. Also note that the DF 11's are transmitted at the rate of about once per second. For this particular aircraft there are approximately 10 times as many TCAS replies as there are squitters or replies to ground interrogations.

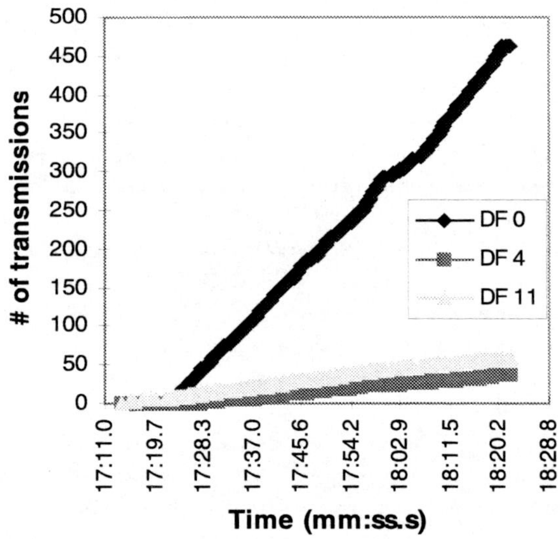

Figure 10. Cumulative DF0s, 11s, and 4s.

DF 4's, Response to Ground Interrogations

Figure 11 shows a more detailed view of the accumulation of DF-4 transmissions. A pattern of DF 4 replies can be seen every 4.8 seconds, which is the rotation rate of airport surveillance radar. Somewhat less obvious is a second pattern of replies, 180 degrees out of phase with the first pattern and also approximately every 4.8 seconds. Perhaps these two patterns of replies can be attributed to multiple interrogations during the testing of DCA's new ASR-9 prior to the decommissioning of the old ASR-7.

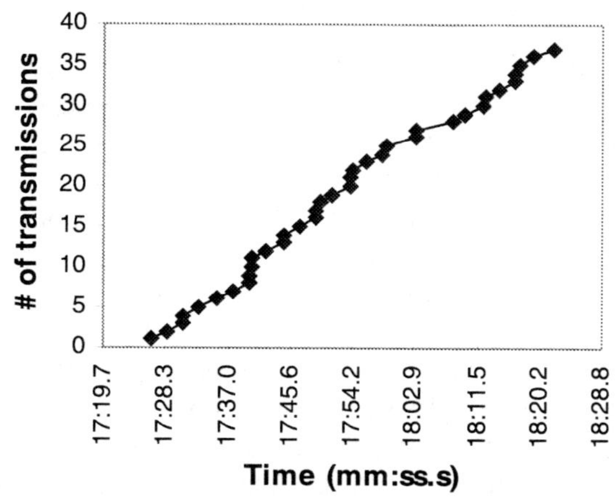

Figure 11. Response to Ground Interrogations

DF 16's, TCAS Resolution Advisory

Downlink format 16's are transmissions which are used by Traffic Alert and Collision Avoidance System (TCAS) units to communicate between aircraft. Figure 12 clearly depicts the beginning and end of a TCAS Resolution Advisory (RA) transmitted by a B-737 aircraft approaching Washington National Airport. TCAS RAs are instructions to the pilot to either descend or climb based on a TCAS encounter. We cannot tell, from the limited data in this instance, if the RA was a false alarm or not.

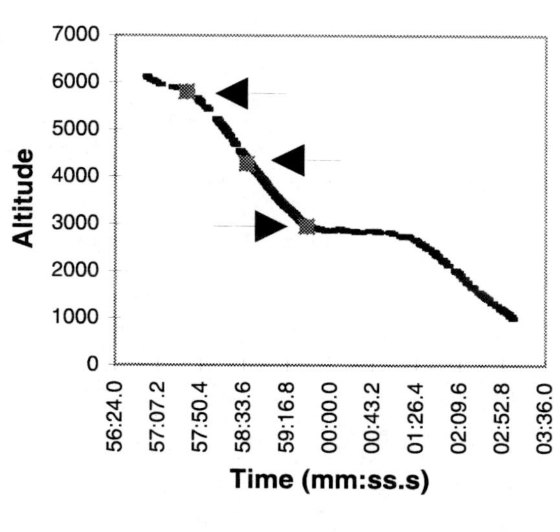

Figure 12. Aircraft Descending with DF 16s

Limitations

The ability of the decoder to accurately identify an aircraft is dependent upon the accuracy of the database. As new aircraft are put into service, and as aircraft registrations occasionally change, the database becomes less reliable. Fortunately, most airliners do not change ownership that often and therefore their registration numbers are not likely to change. The database is updated and made available on a quarterly basis. The decoder is unable to determine the registration number of foreign aircraft because there is no direct correlation between a non-U.S. registered aircraft's Mode S address and its registration number. However, some countries have made databases of their registered aircraft and corresponding Mode S addresses available. Additionally, the decoder can be used at an airport to determine a particular foreign aircraft's Mode S address and this can be associated in the database with the aircraft's registration number which can be entered manually.

Results of Deployment

For the wake vortex project the decoder operated continuously from May 13 to May 27, 1998. During one twenty-five hour period on May 21-22, the decoder logged approximately 2 Megabytes worth of data. The size of the file on any given day depends primarily on how much runway 31R was used for arrivals. On days when the wind did not favor 31R there is no data because there were no aircraft that flew over the sensors. Virtually all of the decodes yielded useful data in terms of a time when the aircraft crossed the wake vortex sensor area.

Applications of the Technology

One possible application for the decoder is for the monitoring of aircraft which exceed noise restrictions. Many airports have restrictions on the level of noise that can be generated on takeoff or landing. Large airports typically have very sophisticated noise monitoring systems which tap directly into the airport surveillance radar to record flight paths. Smaller airports however, can not afford to spend as much money on this sophisticated equipment and their needs could be met with a much less expensive system that could correlate a noise event with a particular arrival or departure.

Another application for the decoder is landing fee collection. Landing fees generate millions of dollars of revenue for airport operators. Fees that are not collected are lost revenue. Currently airport operators use methods such as reviewing fuel receipts to determine the number of arrivals and type of aircraft that a given airline operated into an airport in a given month. The decoder could easily provide a list of aircraft operations sorted by airline on a monthly basis to an airport operator.

As a component of an ADS-B ground station, the decoder will be able to decode advanced downlink formats, such as DF-17, so that an Air Traffic Control display can be generated. Finally, as a component of a multilateration system, several decoders can be used to calculate time of arrival of transponder transmissions, then share this data with a central processor which will use triangulation to determine the precise location and track of an aircraft.

Conclusions

One of the main goals during the development of the decoder was to make it less expensive than other equipment that is based on the same technology, such as TCAS. As the system was custom designed, we were able to achieve our cost goals for production equipment. Other comparable systems are usually based on modified commercial equipment, such as TCAS, meaning that there is unused or inapplicable functionality which the user has to pay for. In terms of performance, the customized elements of the system determine the range and decode rate, offering an optimum solution. During our trials we demonstrated the ability of our system to decode Mode A, C, S, and IFF transmissions. The device has been operationally deployed in support of several air traffic control research programs at major airports across the U.S. Observed data rates are an order of magnitude higher than standard airport surveillance radar. The data is suitable for graphing ascent and descent profiles and detailed information is available on U.S. registered Mode S equipped aircraft. The receiver decoder technology can be used by airport operators for noise monitoring, and landing fee collection. Air Traffic applications include ADS-B ground stations and receivers for multilateration systems. This type of technology is a more practical way to conduct research at airports. Rather than requesting radar data tapes and parsing data manually data can be collected automatically and processed in real time.

LOOP TECHNOLOGY (LOT) AS AN ALTERNATIVE SURFACE SURVEILLANCE SYSTEM

Vern Edwards, Federal Aviation Administration (FAA), Washington, D.C.
Carl Evers. Rannoch Corporation, Alexandria, VA

Abstract

This paper provides an overview and preliminary results for demonstration of a prototype distributed loop-based system called LOT. This paper contains only the understanding and views of the authors and is not intended to reflect the official position of the FAA. The FAA completed the first phase of a prototype system installation in 1997 at Long Beach Airport (LGB) in California. This prototype applies inductive loops, a mature technology, to the airport surface surveillance application. The project also involves a technology transfer of neural network signal processing technology from Department of Defense to form a non-cooperative surface movement sensor system. LOT has the potential to be used in a standalone mode or as a supplemental sensor input to an Airport Surface Detection Equipment (ASDE) radar surface surveillance system.

Introduction

Most automation efforts to date have rightly concentrated on surface safety and incursion problems at higher level airports, where traffic is more dense and complex. The Airport Surface Detection Equipment-3 (ASDE-3) has been implemented to provide surface surveillance at these higher level airports. However, an alternative surveillance system is needed to address airports where ASDE-type equipment will not be installed or the monitoring requirement is too localized to warrant an ASDE-type system. The system employed should offer more flexibility in application (i.e., limited application vs. full airport), be scalable, and allow for incremental implementation to reduce or manage the cost. A non-cooperative surveillance system is preferred, as no new aircraft equipage is required to provide surveillance.

The FAA is evaluating LOT for its surface surveillance application. The project is currently comprised of two phases – a technical assessment and an operational assessment. The first phase, conducted in 1997, investigated the technical feasibility and performance of a distributed inductive loop-based system to provide surface surveillance. Evaluations were performed LGB in California and at Hyde Field in Maryland. LOT demonstrated the capability to detect aircraft and ground vehicles when they taxi over an inductive loop. LOT goes one step beyond the typical highway loop implementation by using the inductive signature to classify the aircraft or ground vehicle. Detection and classification information is used to track traffic in the airport loop coverage area. The second phase began in 1998, and the LOT system installed at LGB received enhancements of an approach radar interface, safety logic and a ground controller display. Evaluations are planned for late 1998.

While LOT can provide full runway/taxiway coverage, a more cost-effective implementation is a limited system installation that addresses the immediate and localized safety needs of the airport. Figure 1 illustrates four LOT applications, which are further discussed in the following sections. The flexibility exists to install inductive loops to support any one or combination of these applications at any location(s) on the airport.

0-7803-5086-3 /98/$10.00 ©1998 IEEE

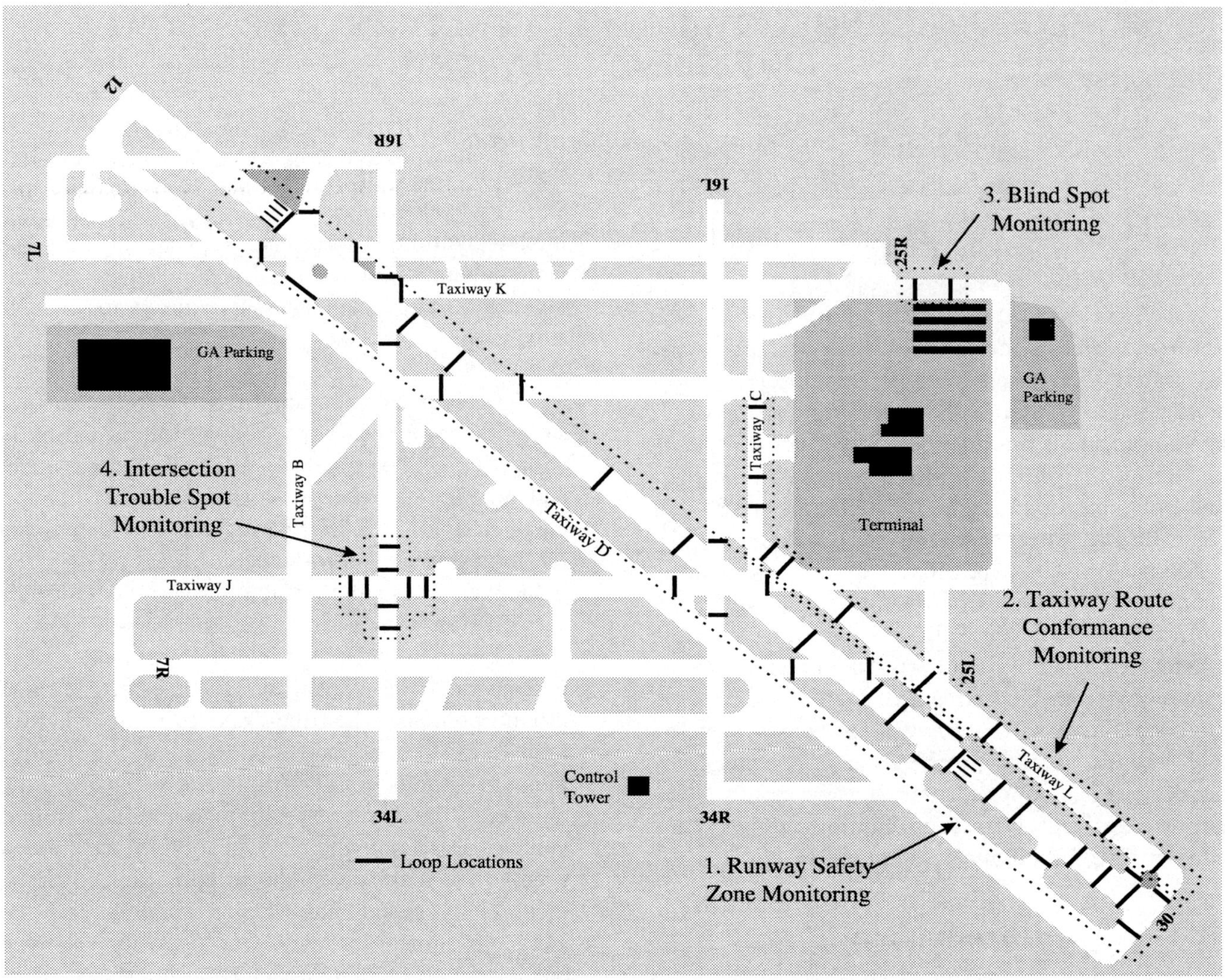

Figure 1. LOT Conceptual Applications

Operational Concept

Air traffic controllers can use LOT to aid in minimizing the risk of incursion incidents and improve situational awareness. In addition to providing an operational display of surface traffic where loops are installed, LOT can provide the following capabilities (See Figure 1): (1) runway safety zone violation, (2) standard taxi route conformance monitoring, (3) blind spot monitoring, and (4) monitoring of troublesome intersections. The system will be used by local and ground controllers to augment visual observations of aircraft and vehicle movement, and may be used by the controllers under the following conditions:

- Visibility is less than the most distant point in the active movement area where loops are installed.
- Traffic is operating in blind spots with respect to the tower.
- Verification of position of aircraft or vehicles at specific update points is needed.
- In the controller's judgment, the systems used will assist in the performance of their duties at any time.
- There are system-generated aural and visual indications of violation of runway threshold or an unsafe condition on runway.
- There are system-generated aural and visual indications of non-conformance with standard taxi route operation.

0-7803-5086-3 /98/$10.00 ©1998 IEEE

The LOT-derived position and identification information will be used to assist the controllers with the following activities:

- Formulating clearances and control instructions to aircraft on the movement area.
- Determining when the runway is clear of aircraft and vehicles prior to a landing or departure.
- Positioning aircraft and vehicles.
- Determining the exact location of aircraft, or their relationships to other aircraft on the movement area.
- Monitoring compliance with control instructions.
- Confirming pilot-reported positions.
- Providing directional taxi information (excluding heading) on pilot request.
- Identifying the weight class of aircraft to assist in maintaining proper departure separation standards.
- Determining when an aircraft has lifted off or touched down.

LOT Overview

System Architecture

The LOT system uses Commercial-Off-The-Shelf (COTS) inductive loop technology and a computer workstation to provide surface surveillance and surface traffic safety processing for aircraft and ground vehicles operating on runways and taxiways where inductive loops are installed. The LOT system consists of the following major elements:

Surveillance Subsystem

The surveillance subsystem forms the heart of a loop-based surveillance system. It consists of the sensor (loop wire), detection component and tracking component. This subsystem provides track reports to the safety subsystem for display and processing by the safety logic.

Inductive Loops

Inductive loops provide the sensing element for the Loop Detection Component (LDC). They can be located over the whole runway/taxiway surface or in strategic locations (i.e., troublesome intersections, blind spots). Inductive loops are implemented on the airport surface in rectangular saw cuts, which span most of the width of the runway and/or taxiway to be monitored (see Figure 2). In the demonstration system installed at LGB, the taxiway loops are 10'x65' and the runway loops are 10'x150'. The loops are spaced to achieve a nominal update rate of once every four seconds at typical taxi speeds. A loop consists of three turns of encapsulated stranded wire, which is laid down in a saw-cut groove. A backer rod is installed in two-foot sections spaced two inches apart on top of the loop wire. The saw cut is then covered with epoxy sealant to ensure against water seepage and prevent cracking of the surface where installed. A key factor in the overall performance of inductive loops is the quality of the installation and keeping the loop wires from moving.

Loop Detection Component (LDC)

Each inductive loop has an associated LDC. The LDC is located within the controller cabinet, which is usually within 500 - 1000 ft. of the loop itself. Each loop is connected to the LDC via a lead-in cable. In the LGB installation, each controller cabinet houses up to four LDCs. The LDC provides detection of aircraft and ground vehicles. The LDC also extracts signatures and performs classification and single loop velocity estimates.

The LDC drives a current through the loop wire at the loop resonant frequency and the current creates an electromagnetic field around the loop wire. The loops are laid out such that the shortest side or length is normal to the direction of movement of the aircraft or vehicle across the loop. The metallic structure of the aircraft/vehicle cuts (see Figure 3) through the electromagnetic field and sets up eddy currents, which tend to lower the inductance of the loop in proportion to the aircraft underside surface area and height above the loop [Ref. 1]. As the aircraft/vehicle continues to move over the loop, the result is a varying inductance change, which produces a shape-based signal "signature" or pattern output (see Figure 4).

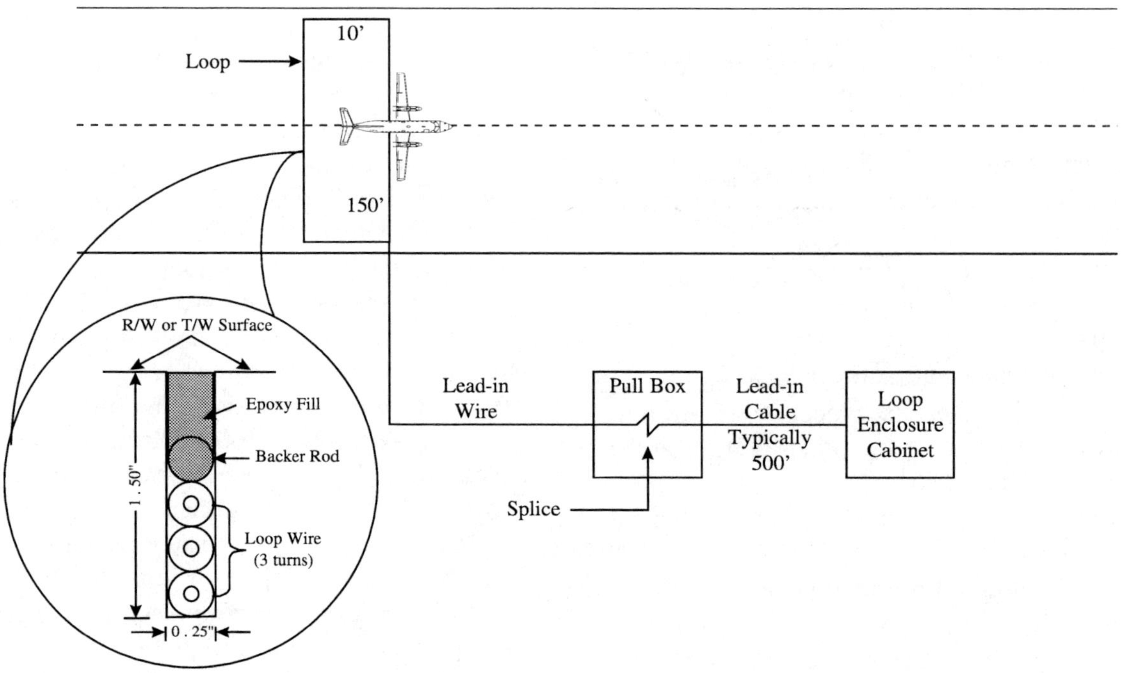

Figure 2. Physical Layout of Loop System Airfield Components with Loop Wire Cutaway View

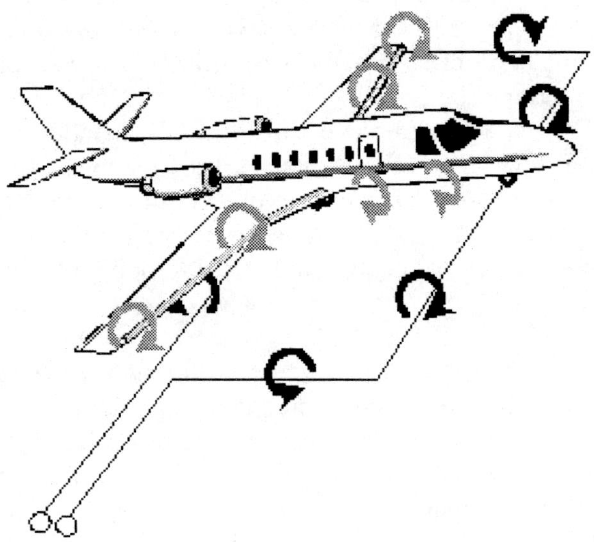

Figure 3. Aircraft Cutting Loop Flux Lines

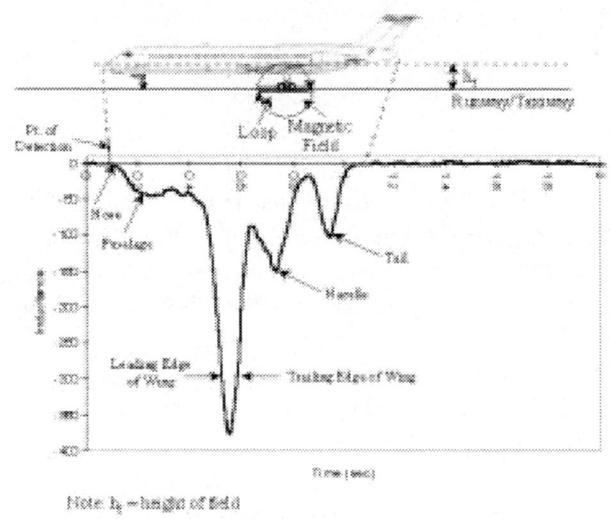

Figure 4. Signature of a DC-9 2 Dimensional Model

This signature is unique for each aircraft type. When the time-varying signal exceeds the loop detector detection threshold, detection is declared. The detection will not be dropped until the signal drops below the threshold and remains there for a parameter-determined length of time. Small ground vehicles can also exhibit a distinguishing signature (double hump), which can be discriminated from aircraft signatures (see Figure 5).

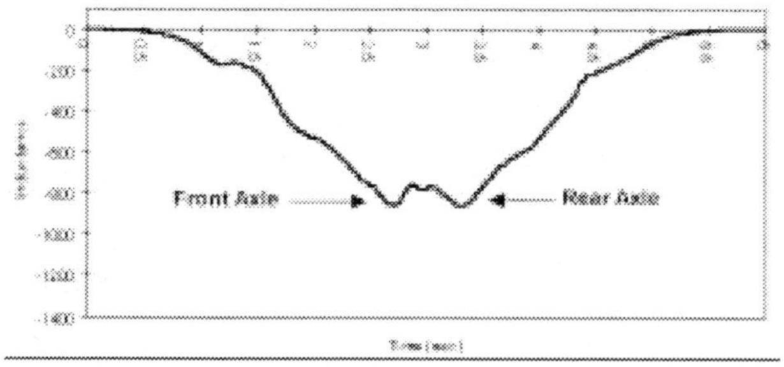

Figure 5. The Distinguishing Double-Hump Signature of a Small Ground Vehicle

Tracking Component (TC)

The TC is a computer workstation that performs loop-to-loop correlation and tracking function position, velocity, and heading based upon kinematics. It also performs data handoff between arrival surveillance data and the loop data from the airport surface.

Safety Subsystem

The safety subsystem provides for processing of track reports from the surveillance subsystem by the safety logic for detection of runway occupancy and runway safety zone violation.

Controller Display

The display provides the controller with a situational display of traffic movement in the inductive loop coverage area(s). Aural and visual indications are provided in the event that there is a violation detected by the safety logic (e.g., violation of runway safety zone, entry into blind spot area, etc.).

Neural Network Classification

The demonstration prototype LOT installed at LGB has a neural network classifier associated with each loop detector component. Although a certain amount of information and tracking can be done from detection alone, classification by neural network processing forms the essence for a loop-based surface system and its ability to track aircraft and vehicles. Classification provides information, which can be used to help accurately correlate a detection update with an established track.

As mentioned under *Loop Detection Component*, each aircraft type produces a unique signature. An aircraft's unique signature is based on its distinguishing characteristics or "features" (e.g., location of nose wheel engine placement, shape of wing, etc.). Just as the controller has been trained to recognize different aircraft type by visual observation of certain features, a properly trained neural network classifier can achieve similar results by emulating a similar process. An air traffic controller is trained by his supervisor to recognize and classify different aircraft types by visual observation of their features. Being able to identify the type of aircraft gives the controller much more information about the flight characteristics of that aircraft.

The neural network is trained, Figure 6, to the population set of aircraft and vehicle signatures appearing at the airport. Development of the signature training set requires collecting aircraft and ground vehicle signatures or patterns and visually truthing them by type. Once this training is completed, subsequent signature inputs for which it has been trained can be classified with high confidence. Presently, we are only attempting to classify three major classes: aircraft, ground vehicle, and unknown. The aircraft class is subdivided into the sub-classes of small (representing GA type aircraft), large (representing air carrier type aircraft), and heavy (representing the larger air carrier type aircraft). Planned enhancements to the neural network classifier may enable classification of many aircraft down to the specific type (e.g., B737, MD-11, DC-9, etc.). Preliminary testing indicates that an enhanced classification capability may enable the system to perform velocity estimates with surveillance from a single loop.

0-7803-5086-3 /98/$10.00 ©1998 IEEE

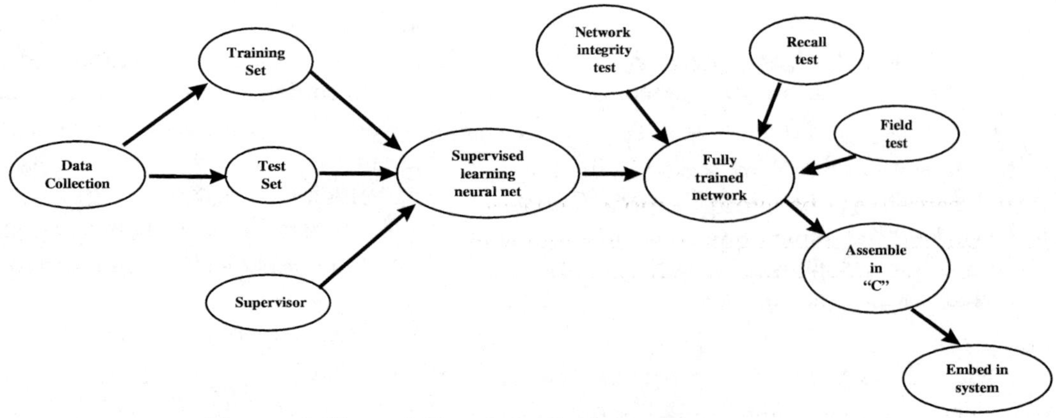

Figure 6. Neural Network Training Process

Preliminary LOT Evaluation Results

LOT evaluations [Ref. 4, 5, 6] were performed at Hyde Field and LGB. RTCA defines guidelines and recommended requirements for surface movement sensors [Ref. 7]. A qualitative assessment with respect to key RTCA-recommended requirements (in italics) is provided below. Future testing will better quantify performance results.

Probability of Detection > .9995

Detection performance testing with ground vehicles and a small aircraft was performed. A range of speeds up to 60 mph was tested. Loop crossing points from the centerline out to the runway or taxiway edge were tested, and various angles of crossing were tried. Test results indicate that there is a high probability of detection when the detector is set to provide sufficient sensitivity. During initial testing, missed detections were experienced for slow moving ground vehicles and small aircraft. However, using a detector set to provide improved sensitivity resolved the problem.

False Detection Reports < 1 per 8 hours

Erroneous detections were not experienced during any of the testing. The loops, being wires, act as antennas, yet all noise was well below the threshold of detection, thus having no impact on detector performance. Ground vehicles driven over the lead-in wire in the grass areas did not cause false detections. It was found that aircraft could be detected before entering a loop, if the detector sensitivity is set too high. Failure to set the sensitivity at the proper setting could potentially result in a false detection of a hold line crossing.

Coverage of Runway and Taxiway Centerlines where Surveillance is Required

Surveillance for the full width of the runway or taxiway was demonstrated. This provides the capability to detect ground vehicles, which sometimes travel along the edges of taxiways and runways. Reliable coverage is provided in regions where the loops are installed, thus demonstrating the capability to provide localized surveillance to address trouble spots. Accordingly, using a localized approach allows LOT to be implemented in a cost-effective manner. System scalability will be demonstrated in Phase II with the addition of more loops to expand the LOT coverage area.

Position Accuracy Within ±7.5 Meters

LOT determines position based on initial detection of an aircraft or ground vehicle. Ideally, this detection occurs as soon as the aircraft enters the loop just over the hold line. Initial detection occurs when the inductance measured by the detector exceeds a detection threshold. Detection typically occurs soon after the aircraft's nose or ground vehicle's front bumper crosses the wires at the loop entry point. Preliminary indications are that the ±7.5 meters can be achieved. Testing with small aircraft and ground vehicles results in accuracy performance better than ±2 meters.

Position Reported Within One Second of Passage Through Sensing Area

LOT provides an indication of detection prior to the aircraft transversal of the loop or sensor area, thus exceeding the requirement. A demanding application for latency is safety zone violation detection. Detection should occur soon after the aircraft proceeds from a stop at the hold line and the nose enters the runway safety zone. Initial testing has shown that detection will occur within four seconds for aircraft. This time is expected to improve to less than two seconds when the detector is re-configured for improved sensitivity in Phase II.

Classification to Four Weight Classes - Probability of Correct Classification > .9

Testing has been performed to demonstrate classification by aircraft type (e.g., B-737, MD-80) as opposed to weight class (i.e., heavy, large, small). Classification by type is a more demanding application. Figure 7 provides sample signatures. Both the heavy and large aircraft tend to have a peak inductance change associated with the wings. The level segment in front of the peak is associated with the fuselage. The gently sloping segment after the peak is associated with the fuselage and the tail. Unlike the large and heavy aircraft, small aircraft signatures tend not have a level segment in front of the peak associated with the wing. However, in many cases a segment associated with the small aircraft tail is identifiable. Large ground vehicles will give a single peak with no identifiable leading or trailing segments. Small ground vehicles will give a double peak signature. Initial testing indicates that many large and heavy aircraft can be classified by type. However, small aircraft classification by type may be difficult. The small aircraft signatures tend to be similar to each other, thus making it difficult to discriminate. However, small aircraft signatures are distinct from signatures of other aircraft classes. LOT can differentiate between ground vehicles and aircraft. While aircraft type information is desirable, it is not essential for an operational system. The goal for the LOT project is to classify traffic in one of the following categories: heavy aircraft, large aircraft, small aircraft, and ground vehicles. Neural network logic has been modified to perform classification by category as opposed to type. Classification performance testing will be performed in Phase II.

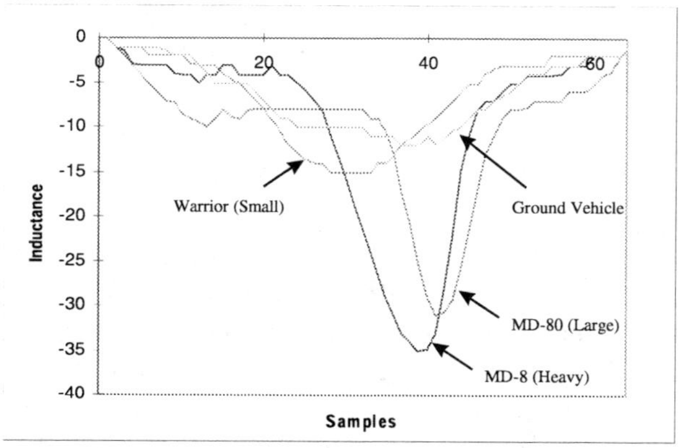

Figure 7. Aircraft and Ground Vehicle Signatures

Provide Velocity Measured for Speeds up to 20 Meters per Second

Velocity was measured for speeds up to 50 meters/sec. The primary means for estimating velocity is the two-loop method. An estimate of velocity is performed each time that an aircraft crosses any two consecutive loops. The speed is calculated using time between loop pairs and distance traveled equal to the separation between loops. Typical loop separations are 300 to 500 feet. Heading is determined from the loop crossing sequence. A second method for estimating velocity is the single-loop method, which is applied for taxi scenarios (e.g., aircraft stops between loops then moves again) where the two-loop method can not be used. Distance traveled equals aircraft length, and time equals single loop crossing time. Aircraft length is obtained via a database lookup, which is performed upon classification.

Velocity Accuracy Within ±2 Meters per Second for Velocities Exceeding 5 Meters per Second

Velocity is estimated using either the single-loop method or the two-loop method. The single-loop method has demonstrated velocity accuracy performance better than 5 meters/sec. for small aircraft. However, accurate classification to aircraft type is critical to achieving a high degree of accuracy. An assessment of the two-loop method has not been performed. Two factors that influence the accuracy of the two-loop method are variability in point of detection and vehicle dynamics. Vehicle dynamics will have the biggest influence on accuracy. For instance, an aircraft that changes acceleration between loops will introduce errors in the velocity estimate.

Indicated Direction is Correct Within ±89 Degrees

Reliable direction determination was demonstrated. The system determines direction of travel when an aircraft crosses two consecutive loops. The heading is assumed to be parallel to the pavement centerline in the direction of travel. Classification information is used to confirm that any two consecutive crossings are from the same aircraft. Taxiing with frequent stops is common for many airports. The system maintains knowledge of an aircraft's presence in the zone between loops, which is used to determine direction when the aircraft crosses the next loop. Testing was performed to determine if a single loop could be used to provide direction determination. Several loop shapes were tried, but single-loop direction determination was not demonstrated.

Conclusions

The initial evaluations have demonstrated that LOT can provide reliable aircraft and ground vehicle detection. Airport surface surveillance requires large loop sizes which places demands on detector sensitivity performance. Sufficient sensitivity is required to ensure reliable detection and classification. Initial indications are that the system can achieve the RTCA recommended requirements. The overall viability of LOT as an alternative airport surface surveillance will be demonstrated in Phase II. In particular, classification and detection of runway safety zone violations will be demonstrated. System performance will be evaluated against RTCA and International Civil Aviation Organization (ICAO) Advanced Surface Movement Guidance and Control Systems draft requirements [Ref. 8].

Acknowledgements

The progress on this work is the result of the dedicated work of the LOT team members. Orincon Corp. provided expertise in the development and application of neural network processing to the airport environment and in the development of distributed loop-based tracking. 3M Corp. provided invaluable technical support for independent off-site evaluations. Pete Mills of the FHWA provided insights and guidance for the planning and design of inductive loops. Others acknowledged are Volpe Center, TASC, ATC, Rannoch, and JIL. A special thanks goes to Long Beach Airport for being the host for this prototype demonstrations system and their operations personnel who provided invaluable support during the installation and testing.

References

1. Federal Highway Administration (FHWA), *Traffic Detector Handbook*, Second Edition, National Technical Information Service (NTIS) Report No. FHWA, July 1990.
2. W.T. Illingworth, *A Practical Guide to Neural Nets*, Addison-Wesley Publishing, March 1991, ISBN 0-201-56376-0.
3. Advisory Group for Aerospace Research & Development (AGARD) Neuilly sur Seine, France, AGARD Lecture Series 179, *Artificial Neural Network Approaches in Guidance and Control*, September 1991, Reproduce by the NTIS AD-A244 247
4. *Inductive Loop Technology Off-Site Test Report*, FAA Report No. AND410LOT-OFF/0598, May 19, 1998.
5. Petrescu, J., *Loop Incursion Prevention Lab Ground Vehicle Analysis Test Report*, DOT Report Number OCR 98-4139-U-0091, March 1998.
6. *Draft Inductive Loop Technology On Site Test Report*, AND-410-LOT-OST-5-98, May 1998.
7. *Guidance and Recommended Requirements for Airport Surface Movement Sensors*, Document No. RTCA/DO-221, April 1994, Prepared by: SC-178.
8. *Draft Manual of Advanced Surface Movement Guidance and Control Systems*, AWOP/16-

WP/724, Draft 4, June 1997.

How Data Link Communication Might Affect Controller Workload On A Terminal Option

O. Veronika Prinzo, Civil Aeromedical Institute, Oklahoma City, Oklahoma

Introduction

The Aeronautical Data Link System (ADLS) is being developed as part of the Federal Aviation Administration's modernization plan to improve safety, reduce delays, and increase the efficiency of resource utilization,. Terminal Data Link (TDL), an ADLS application, will provide five services to support Controller/Pilot Data Link Communication (Automatic Dependent Surveillance Panel, 1996).

These services will require controller use of a keyboard-input module to communicate with pilots. At terminal radar approach control (TRACON) facilities, the keyboard-input module includes an ABC-style alphanumeric keyboard and a trackball. As shown in Figure 1, the keys are arranged in accordance with the serial presentation of the English alphabet.

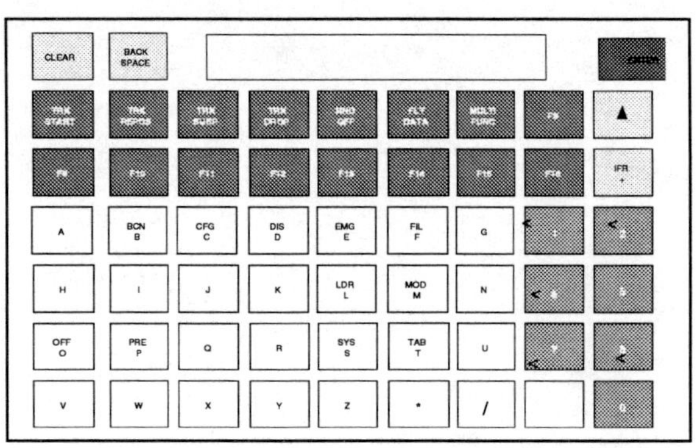

Figure 1. Example of an ABC-style Keyboard

As a controller composes a message, it appears in a preview area on the planned view display (PVD) and is cleared after the keyboard [ENTER] key is pressed. Other information presented on the PVD includes 1) targets, 2) tabular lists, 3) the system data area and, 4) a video map. In addition to the current visually presented information, CPDLC information could be presented visually in menus, status and history lists, and full data block displays on the PVD or an adjacent monitor.

There are several assumptions related to the introduction of menu displays. First, they should facilitate the use of the data link services to perform required functions. Second, they should reduce controller workload associated with data link communications.

As shown in Figure 2, in the Controller Instruction menu, the "A" indicates an altitude, "H" a heading, and "S" a speed. The first entry is listed next to the number "1." The "A090" indicates an altitude of 9000 feet. To send this instruction to a pilot, the controller would press the [F16] key followed by the number [1], place the cursor on the position symbol portion of the aircraft data tag, and press [ENTER].

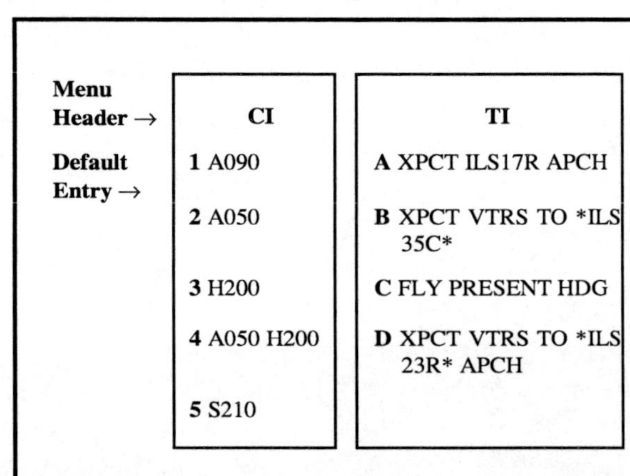

Figure 2. Controller Instruction and Terminal Information Menus

Terminal Information menu items consist of non-critical information that provides pilots with advance planning information such as, "expect ILS 17 right approach." This information could appear in the TI menu as "*A XPCT ILS17R APCH*."

There are several potential favorable outcomes and advantages for controller-use of CPDLC. First, by using data link to transmit non-safety related and routine information, there would be greater availability and access to voice-radio frequencies for more timely, flight-critical information. Second, controllers could reallocate the time they use to formulate and construct messages to strategic planning and related activities.

There are several potential unfavorable outcomes and disadvantages for controller-use of CPDLC The inherent delays associated with data link may be unacceptable for some TRACON operations. In a study conducted by the Data Link Benefits Study Team (DLBST, 1996), the total transaction time for data link messages ranged from 8- to more than 200-seconds. A closer examination of the overall frequency distribution of end-to-end transfer delay times revealed that most transactions were completed between 10-16-seconds. Completing transactions in under 10-seconds was a rare event (> 250 of the 9,000 messages).

The heads-away time associated with message selection may result in the controller not detecting a potential conflict or pop-up. The overlay of the data link menus onto the PVD may make it difficult for the controller to identify an AID. Providing pilots with multiple communication pathways may increase controller workload.[1] Communication problems often occur when controllers experience an overload brought on, in part, by heavy traffic and radio frequency congestion (cf. Morrison & Wright, 1989; Morrow, Lee, & Rodvold, 1993).[2]

Observational and anonymous reports submitted by pilots and controllers to the Aviation Safety Reporting System consistently state that air traffic controllers speak too fast (cf. MITRE Report, 1990). Similarly, voice quality changes were reported to change a function of cognitive, psychological, physiological or emotional stress (cf. Brenner et al., 1985, 1987; Griffin & Williams, 1987; Williams & Stevens, 1972). Hesitation pauses is associated with complexity of general planning and task difficulty (Eisler, 1968).

In light of those findings, Prinzo, Lieberman, and Pickett (1998) performed an analysis to determine if changes in workload could be ascertained from changes in controller's speech. Significant increases in pause duration and frequency were interpreted to indicate controllers generally used a more "thinking" mode when handling a light traffic load -- shifting to more "automatic" responses as load increased.

The costs and benefits of data linked communications to controllers has not been empirically determined. Speculations can be offered; however, comparisons between traditional voice versus data linked communication using a common set of definitions are preferred. A simulation study was performed to examine how CPDLC might affect communication practices and controller workload.

---

[1] At a 1997 meeting held at NASA-Ames, pilots told attendees they would transmit the same data link message over ACARS, FANS1, and HF radio if they did not receive a prompt reply. The oceanic controller had to respond to the same message three times.

[2] A common workload index (or overload) includes the number of aircraft under control, traffic volume and complexity, and duration and type of radio communication (Costa, 1993; Porterfield, 1997).

Method

Participants

Eight full-performance level (FPL) air traffic control specialists from a level 5 TRACON facility participated in this study. Collectively they had 12.6 mean years of terminal experience (SD = 3.8) with 9.8 mean years (SD =3.4) at the full performance level.

Equipment

The CAMI TRACON simulation laboratory includes two file servers, two controller workstations, two ghost pilot workstations and one remote position workstation. The laboratory provides full audio and video recording capability.

Materials

Scenario Development The TRACON facility provided HOST computer flight plan data for a 24-hour time interval. Those data were reformatted and two, 35-minute time intervals, one representing light- and the other heavy- arrival traffic, and one, 15-minute time interval were simulated.

Subjective Measure of Workload. Hart & Staveland, (1988) developed The NASA Task Load Index (TLX). It is a widely used and accepted multi-dimensional subjective rating procedure and index of perceived workload.

As shown in Figure 3, for the computerized version, all dimensions appeared on a single computer screen. The controller moved an indicator along each dimension and a value from 0-100 was recorded at the point where the mouse button was pressed. The values associated with each construct were coded and stored in an ASCIII text file.

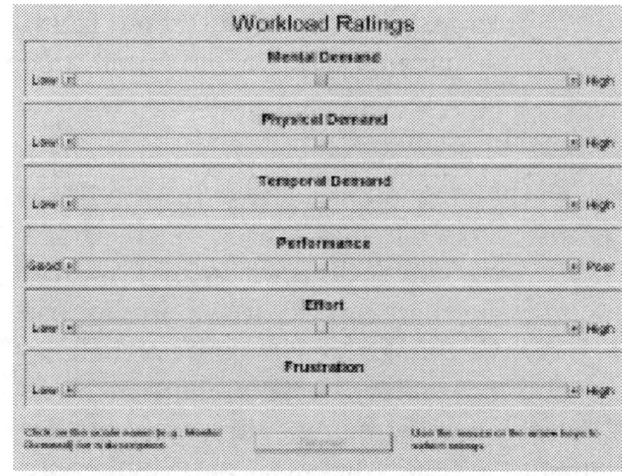

Figure 3. Computer Display of the NASA Task Load Index Version 1

Procedure

Following completion of the required administrative paperwork, the controller received 4-hours of familiarization training on the simulator and instruction on CPDLC using the ARTS IIIA keyboard. During that time, the trainer observed the controller demonstrating skill and dexterity with the keyboard-input module. Afterwards, the controller completed the first 30-minute scenario, provided TLX workload ratings, received a 15-minute break, and so on until each scenario was completed twice, once using voice-only, and again using voice+data link. Order of scenario presentation was counterbalanced across participants. All voice communications were recorded onto DAT tape and time stamped. Radar display and ARTS IIIA keyboard and trackball entries were recorded onto VHS tape.

Objectively Derived Measures of Workload

There were three classes of objectively derived measures of workload: aircraft counts, acoustic and physical measures, and communication measures. Workload was defined as the number of active aircraft on frequency when the controller initiated a transmission. A pause was defined as a unit of time devoid of hand movement on or about the ARTS IIIA keyboard or a moment of silence in an utterance greater than 25 ms.. Communication measures included number of restarts, adjusted time on frequency (subtracting pause duration from time on frequency), and total number of topics and syllables in a message.

Results

Subjective Workload Measures

A Mode of Communication (voice-only, voice+data link) by Traffic Density (heavy, light) repeated measures analysis of variance (ANOVA) was performed on participant TLX ratings. The mean subjective ratings for each dimension are presented in the main body of Table 1. For all analyses, alpha was set = .05.

There were significant main effects of Mode of Communication on the temporal demand $F(1,7) = 5.63$ and frustration $F(1,7) = 7.77$ dimensions. Physical demand $F(1,7) = 5.54$ approached significance, $p = .051$. Neither the Traffic Density nor Mode of Communication by Traffic Density interaction were significant (all p's > .08). As shown in the last column of Table 2, compared to voice-only, participants rated voice+data link as a higher workload environment.

Objectively Derived Measures of Workload

Four hundred and seventy-three (473) controller-generated transmissions were analyzed. The objectively derived measures of workload are presented in Table 2. A Mode of Communication (voice-only, voice+data link) by Traffic Density (heavy, light-to-moderate) repeated measures analysis of variance was performed for each dependent variable.

There were significant main effects of Traffic Density on the aircraft $F(1,7) = 355.98$ and adjusted time on frequency $F(1,7) = 11.946$. None of the Mode of Communication by Traffic Density interactions was significant. There were more aircraft on frequency during the heavy- than light-traffic density simulation and controllers spent more time on frequency during the light- than heavy traffic simulation.

Table 1. Mean Ratings on the NASA TLX by Mode of Communication and Traffic Density.

	NASA TLX Dimensions						
Simulation	Mental	Physical	Temporal	Performance	Effort	Frustration	Mean
Mode							
Voice+DL	35.06	33.19	32.50	29.00	36.50	31.75	33.00
Voice-only	23.38	18.31	17.75	18.00	24.31	13.38	19.19
Traffic							
Heavy	36.68	32.87	32.62	22.88	38.50	27.25	31.80
Light	21.75	18.63	17.63	24.13	22.32	17.88	20.39
Mode by Traffic							
Voice+DL Heavy	38.62	38.25	38.62	21.00	45.25	36.00	36.29
Voice+DL Light	31.50	28.12	26.38	37.00	27.75	27.50	29.71
Voice-only Heavy	34.75	27.50	26.62	24.75	31.75	18.50	27.31
Voice-only Light	12.00	9.13	8.88	11.25	16.88	8.25	11.06

There were significant main effects of Mode of Communications on pause frequency $\underline{F}(1,7) = 7.88$, pause duration, $\underline{F}(1,7) = 13.39$, number of restarts, $\underline{F}(1,7) = 16.12$, number of topics $\underline{F}(1,7) = 12.35$ and syllables $\underline{F}(1,7) = 6.57$ in a message. The adjusted time on frequency measure approached significance, $\underline{F}(1,7) = 5.191$, $\underline{p} = .057$. There were more frequent and longer pauses when controllers communicated with aircraft during the voice+data link than voice-only simulations.

Controllers restarted their messages more often and generally spent more time on frequency (adjusted for pause duration) during the voice+data link simulations. Pairwise comparisons revealed that the restarts occurred primarily for messages transmitted by data link. Pairwise comparisons performed on adjusted time on frequency revealed controllers spent more time on frequency during the voice+data link, light-traffic density simulation. Traffic density did not influence the amount of time the controller spent on a frequency during the voice-only simulations.

Controllers sent more topics per message during the voice-only simulations and their messages were longer. An examination of the transmissions revealed that 82% of the transmissions made during the voice+data link heavy-traffic density simulation were transmitted by data link and all of them contained two topics. Of the remaining 18% transmitted by voice, only 25% contained two topics. For the voice-only heavy-traffic density simulation, 64% of the transmissions contained 2 topics and an additional 34% contained 3-4 topics. Controllers demonstrated greater variability in the number of topics transmitted by voice than data link. It is not surprising that more syllables were present in voice-only transmissions given that the transmissions contained more topics.

Table 2. Means for Each Objective Measure of Workload Presented by Simulation

	Objective Measures of Workload						
Simulation	Number of Aircraft	Pause Frequency	Pause Duration	Number of Restarts	Adjusted Time on Frequency	Number of Topics	Number of Syllables
Mode							
Voice+DL	9.48	.12	.72	.11	3.50	2.21	17.86
Voice-only	9.73	.03	.03	.01	2.82	2.56	20.10
Traffic							
Heavy	12.85	.08	.38	.07	3.02	2.32	18.32
Light	6.35	.07	.36	.05	3.30	2.45	19.64
Mode by Traffic							
Voice+DL Heavy	12.78	.11	.72	.11	3.31	2.17	17.32
Voice-only Heavy	12.93	.05	.05	.02	2.73	2.47	19.31
Voice +DL Light	6.19	.12	.72	.10	3.68	2.24	18.39
Voice-only Light	6.52	.02	.01	.00	2.91	2.64	20.90

Discussion

Today's operational communications involves the use of voice-radio communications by controllers to provide pilots with timely messages pertaining to the safe and efficient passage of passengers and cargo to their destinations. As CPDLC is transitioned into the operational environment, the manner by which air traffic services are provided and the workload experienced by controllers will change.

Of the objectively derived measures of workload, pause frequency and pause duration, and the number of restarts, topics, and syllables in a transmission all varied as a function of mode of communication. Mode of communication did not alter the number of aircraft provided with air traffic services. This is not surprising since controllers provide aircraft with radar separation minima and speed restrictions as specified in FAA Order 7110.65 and Letters of Agreement. Mode of communication could alter these established practices.

Adjusting for pause duration, controllers still spent more time on frequency in the voice+data link than voice-only environment. The video recordings were examined to identify whether the longer time on frequency was attributable to message selection/construction tasks alone or other factors. Generally, it took controllers only about 2 seconds to select and construct their data link messages. The remainder of the time, the controller's hand was away from the ARTS IIIA keyboard either resting on or rotating the trackball. The message remained in a buffer until a particular event occurred. When controllers detected it, they transmitted the message to the aircraft.

Thus, the amount of time that lapsed from when the data link message was selected/constructed to when the [ENTER] key was depressed (i.e., the remainder of their adjusted time on frequency) could be spent performing other related tasks. Therefore, the adjusted time on frequency reflected the time controllers spent selecting/constructing data link messages monitoring the occurrence of a triggering event, and sending the message.

For example, once an aircraft was near a particular location, the controller would select/construct a data link heading or altitude instruction and place it in the buffer. Then, while scanning and monitoring the radar display for the "triggering" event to occur, controllers would communicate with other aircraft by voice radio. Those savings of time would become available to other cognitive processes; ones which could activate the sequences of mental operations (i.e., strategies) that controllers utilize to optimize their performance.

The longer and more frequent pauses in the voice+data link than voice-only simulation also suggest that controllers had more time and resources available for strategic processes and planning activities (Ericsson & Simon, 1980). Pauses only occurred for data link transmissions. Pause duration could reflect cognitive activity associated with strategizing during message construction (Prinzo et al., 1998). Research performed on sentence production indicates that hesitations/pauses generally result from a) breathing processes, b) pragmatic reasons, and c) linguistic planning (Carroll, 1986). It is reasonable to assume that breathing was not responsible for the pauses associated with data link transmissions. Since controllers selected pre-formatted message elements from the CI and TI menus, the activities associated with pragmatic reasons, such as adding emphasis, do not provide a tenable explanation.

The more plausible explanation is that linguistic planning, associated with the selection/construction of the data link message, was responsible. Participants usually pause when they are heavily involved in planning a new element (Bower & Clapper, 1989). It could be argued that the more frequent and longer pauses in data link messages were attributable to the controllers' unfamiliarity with selecting/constructing and transmitting message elements. However, that explanation is not supported by the data. Of the 472 transmissions analyzed, 431 did not contain pauses greater than .25 ms.

Generally, more syllables per message were transmitted during the light-traffic density simulations regardless of mode of communications. However, during the heavy-traffic simulations, more syllables per message were sent by voice-only. An examination of the number of topics included in a message revealed more were transmitted during the voice-only simulations in spite of varying traffic densities. Interestingly, there also were fewer transmissions in the voice-only simulations. The fewer transmissions made during the voice-only simulations reflected an economy in radio frequency usage to "send more less often." The opposite strategy emerged in the voice+data link environment where " send less more often" occurred far more frequently.

The subjective workload measures revealed that controllers perceived their workload to be greater in the voice+data link environment. When providing radar services in an environment that required both modes of communication, controllers perceived greater physical and temporal demands and increased levels of frustration. These findings are best explained by the objective workload measures.

An examination of those data revealed more restarts only during the voice+data link-radio simulations. Of the 36 message restarts, 33 involved the use of data link for the TOC. A closer examination of those data revealed that the time pressure that might be associated with traffic density was not a factor; 53% of the restarts involved heavy traffic and 40% involved light traffic. However, it might be the case that the 11-second delay in receiving a pilot acknowledgement to a CI message added to the perceived temporal demands experienced by the controllers.

It may be that the disparity in TOC procedures involving voice-radio or data link is responsible for the restarts. In the voice-radio environment, pilots typically close a transaction by saying "Good day," "See ya," or they read back the numbers in the new radio frequency. In data link TOC, there is no pilot acknowledgement. The lack of feed back may have prompted controllers to re-send the TOC to the aircraft. If uncertain as to whether they had completed the TOC, controllers might begin, press the clear button, and a few seconds later, re-enter the data link TOC. The perceived increase in frustration thus might be attributed to the lack of auditory or visual feed back in the TOC procedure.

Removing restarted transmissions from the analyses revealed that controllers still transmitted more messages to target aircraft when both modes of communication were available. This only occurred during the light-traffic density simulations. This result is not surprising since controllers frequently used the ARTS IIIA keyboard to communicate with pilots. The shift in communicating from a voice-only environment to one requiring both voice and manually entering messages could be responsible for the perceived change in physical demands. It is uncertain whether the individual or joint effects of increased keyboarding, providing air traffic services in an environment that supported a mixed mode of communications, and/or the inherent delays associated with data link transmissions were responsible for the perceived increase in workload.

References

Automatic Dependent Surveillance Panel (ADSP). (March, 1996). *Draft ICAO Manual of Air Traffic Services (ATS) Data Link Applications Version 0.3.*

Bower, G. H., & Clapper, J. P. (1989). Experimental methods in cognitive science. In M. P. Posner (Ed.), *Foundations of Cognitive Science* (pp. 245-300). Cambridge: MIT Press.

Brenner, M., & Shipp, T. (1987). Voice stress analysis. In J. R Constuck (Ed.), *Mental State Estimation* (pp. 363-376). Washington, DC: NASA Conference Publication, 2504.

Brenner, M., Shipp, T., Doherty, E. T., & Morrisey, P. (1985). Voice measures of psychological stress-Laboratory and field data. In I. R. Titze & R. C Scherer (Eds.), *Vocal fold physiology, biomechanics, acoustics, and phonatory control* (pp. 239-248). Denver: The Denver Center for the Performing Arts.

Carroll, D. W. (1986). *Psychology of Language*. Monterey CA: Brooks/Cole Publishing Co.

Costa, G. (1993). Evaluation of workload in air traffic controllers. *Ergonomics, 36*, 1111-1120.

Data Link Benefits Study Team (1996). *Benefits of Controller-Pilot Data Link ATC Communications in Terminal Airspace.* Final Report. DOT/FAA/CT-96-3. Federal Aviation Administration Technical Center.

Eisler, F. G. (1968). *Psycholinguistics: Experiments in Spontaneous Speech*. London: Academic Press.

Ericksson, K. A., & Simon, H. A. (1980). Verbal reports as data. *Psychological Review, 87,* 215-251.

Griffin, G. R., & Williams, C. E. (1987). The effects of different levels of task complexity on three vocal measures. *Aviation, Space, and Environmental Medicine*, *58*, 1165-70.

Hart, S. G., & Staveland, L. E. (1988). Development of NASA-TLX (Task Load Index): Results of empirical and theoretical research. In P. A. Hancock & N. Meshkati (Eds.), *Human Mental Workload* (pp. 239-250). Amsterdam: North-Holland.

Morrison, R.,& Wright, R. H. (April, 1989). ATC Control and communications problems: An overview of recent ASRS data. In R. S. Jensen (Ed.), *Proceedings of the Fifth International Symposium of Aviation Psychology, Volume 2* (pp. 902-907). Columbus, OH: The Ohio State University.

Morrow, D. G., Lee, A., & Rodvold, M. (1993). Analysis of problems in routine controller-pilot communication. *International Journal Aviation Psychology, 4*, pp. 285-302.

Porterfield, D. H. (1997). Evaluating controller communication time as a measure of workload. *International Journal of Aviation Psychology, 7*, 171-182.

Prinzo, Lieberman, and Pickett (1998). *An acoustic analysis of ATC communication*. Washington, DC: Federal Aviation Administration, Office of Aviation Medicine Technical Report.

The MITRE Corporation. (1990). *Study of pilot readback, controller hearback, communication errors*. (Number WP-90W00348). McLean, VA.

Williams, C. E., & Stevens, K. N. (1972). Emotions and speech: Some acoustical correlates. *The Journal of the Acoustical Society of America*, *52*, 1238-1250.

SIMULATION STUDY OF VOCODER COMMUNICATION IN AIR TRAFFIC CONTROL

Earl S. Stein, Ph.D., FAA William J Hughes Technical Center

Randy L. Sollenberger, Ph.D., FAA William J Hughes Technical Center

Atlantic City International Airport, New Jersey 08405

Abstract

Human factors researchers at the Federal Aviation Administration (FAA) William J. Hughes Technical Center conducted a real time person-in-the-loop evaluation comparing two vocoder systems and traditional analog radio communications. Vocoders digitize and compress speech signals to use less communication spectrum. Active Air Traffic Control Specialists (ATCSs) from FAA Terminal Radar Approach CONtrol (TRACON) facilities worked simulated traffic under moderate and high levels of task load. ATCSs used all three systems over a 1-week period. Results indicated that the controllers performed well when using the two vocoders and the analog communications system. They rated one vocoder as less intelligible and acceptable than the other but preferred analog radio. The more highly rated vocoder was a close second to analog radio in terms of intelligibility. Researchers concluded that vocoders are a potential solution to frequency conjestion.

Introduction

Air traffic control (ATC) is one of the most complex command and control systems in the world. It represents a delicate balance of people, machines, weather, and airspace. Air traffic controllers working in this environment must understand everything that is occurring in real-time and be able to anticipate what will happen in the near-term future. This is a key aspect of situational awareness (Endsley, 1988)[i]. Communication between responsible individuals is essential. Accurate and timely information flow makes air traffic management possible (National Research Council, 1998)[ii]. Unfortunately, the radio channels available for aviation are becoming very crowded. This is especially true around busy airports where rapid communication is critical.

With the growing complexity of today's ATC system, there is a critical need to increase communications capacity. The current bandwidth in aviation is 25 KHz. Controllers and pilots communicate using analog radio. However, in this system, everyone can hear everyone else and must wait their respective turns to make a transmission. Vocoder digital communication technology offers a potential solution to this problem by reducing the bandwidth necessary for each transmission. A reduction in this bandwidth will allow the addition of more channels and reduce radio congestion.

0-7803-5086-3 /98/$10.00

Both pilots and controllers have to understand vocoder communication and be willing to accept the transition from analog radio. This study investigates vocoder human factors issues using a real-time simulation under realistic ATC conditions.

Background Literature

Sanders and McCormick (1993)[iii] reviewed several aspects of a speech transmission system that can affect intelligibility. These include frequency distortion, filtering, amplitude distortion, and modification of the time scale. These concerns have stimulated commercial evaluation techniques. Dynastat (1995)[iv], for example, created a measure referred to as the Diagnostic Rhyme Test (DRT). This is one of the better known evaluation tools. However, it requires trained listeners who are usually not skilled in the job domain of typical operators of the system. In contrast, Crowe (1988)[v] used a subjective evaluation technique with untrained listeners. This technique was based on a five-point scale for rating the listening quality of speech communications for vocoder usage with the Aeronautical Satellite Service.

In another vocoder study, Dehel, Grable, and Child (1989)[vi] used an objective evaluation for vocoder assessment including an open word test, the DRT, and a number recognition test. Child, Cleve, and Grable (1989)[vii] had ATCSs use a five-point subjective scale to rate both intelligibility and acceptability of vocoder equipment for ATC communications. For most vocoders tested, the intelligibility rated approximately 5% higher than the acceptability. This suggested that participants could make independent subjective evaluations of both intelligibilty and acceptabilty.

Troll (1989)[viii] evaluated vocoder communications for satellite digital communication links using readback tests for whole and split sentences. Troll used actual ATCSs and a five-point scale to rate the intelligibility of vocoder communications. He established a relative comparison between different vocoder models with application to vocoder usage for controller/pilot communications by satellite links. The study found that controller memory became important as the length of the clearance increased. However, the method worked very well for split tests in which the controllers read back only key parts of a clearance such as a radio frequency or call sign.

The performance of vocoders can be expressed as a function of a host of independent variables characterizing each transmission. The present study evaluated the performance of 4.8 kbps vocoder voice messages in the airport terminal environment. It focused on the impact of two vocoders and how they compared to traditional analog voice communication.

Method

Participants

Sixteen ATCSs from 13 Level 5 TRACONs volunteered. All participants were full performance level (FPL) controllers. They ranged in age from 32 to 52 years old (Mean = 38.94, *SD* = 4.88) and ranged in experience from 8 to 34 years of active service (Mean = 17.06, *SD* = 6.69).

Simulation Environment

Researchers conducted the simulation in the Research Development and Human Factors Laboratory (RDHFL) at the William J. Hughes Technical Center. The simulation equipment consisted of state-of-the-art controller workstations with large high-resolution displays, a voice communication system, networked computer resources, and simulation software. A voice communication link to another room allowed controllers to issue commands to trained simulation pilots. The simulation pilots moved the aircraft radar targets using simple keyboard commands and communicated with the controllers using proper ATC phraseology.

Figure 1 illustrates the overall setup and organization of the simulation pilots, ATCSs, and observer. In each of the independent sessions, one simulation pilot (denoted as A1 or B1) operated all aircraft. Three other pilots communicated with the controllers. Each of these pilots transmitted with a different aircraft background noise and responded to controller clearances of the appropriate aircraft type. One pilot (denoted as A2 or B2) transmitted with a helicopter background noise, a second pilot (denoted as A3 or B3) transmitted with a propeller aircraft background noise, and the third pilot (denoted as A4 or B4) transmitted with a jet aircraft background noise. In addition to readbacks, the simulation pilots provided initial contact communications and replied to traffic advisories. A subject matter expert observed the controllers and conducted performance ratings.

Procedure

The main independent variable used in the simulation was the type of communication equipment. Each participant controlled different traffic scenarios using either vocoder A, vocoder B, or the analog radio simulator. The analog radio simulator was the control condition of the experiment, serving as the standard of comparison for the vocoders. The second independent variable was the level of ATCS taskload that the researchers varied by designing traffic scenarios with either a medium or high volume of traffic.

Participant controllers worked traffic in a generic TRACON sector that was developed and validated in a previous human factors simulation study (Guttman, Stein, & Gromelski, 1995)[ix]. Guttman et al. designed this airspace with several advantages relative to modeling an actual sector in simulations. It provides a realistic TRACON environment and is easy for controllers to learn. It consists of easily remembered fix names and simplified operating procedures.

An automated data collection system, developed at the RDHFL, produced a large set of system effectiveness measures for ATC simulation research (Buckley, DeBaryshe, Hitchner, & Kohn, 1983[x]; Stein & Buckley, 1992)[xi]. In addition to these objective performance measures, a Supervisory ATCS (SATCS) observed controllers and made over-the-shoulder ratings of performance. The SATCS used an observation form specifically designed for ATCS performance evaluation research (Sollenberger, Stein, & Gromelski, 1997)[xii]. Finally,

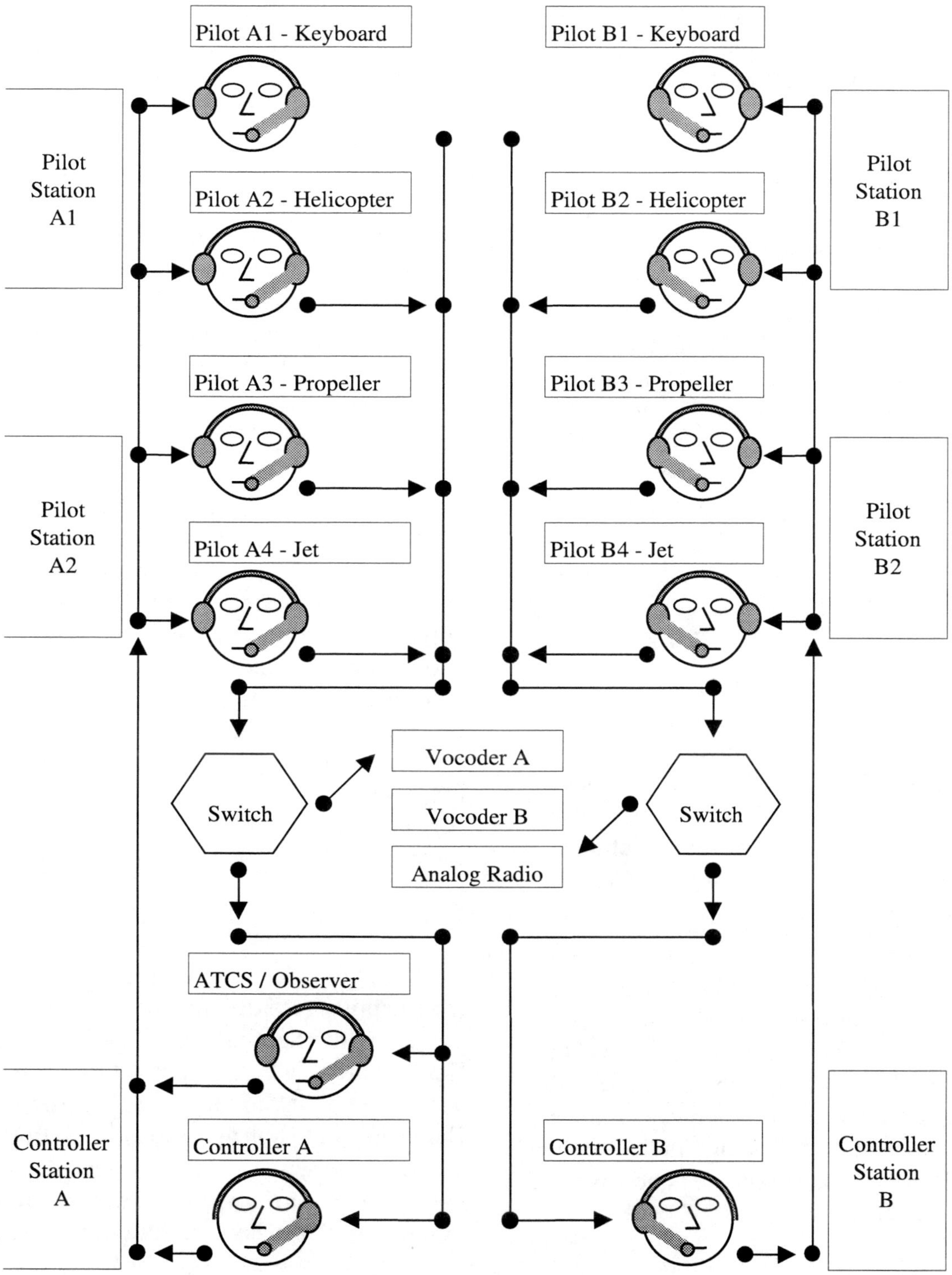

Figure 1. Simulation Setup and Organization of Controllers, Observers, and Simulation Pilots

controllers provided intelligibility and acceptability ratings for the vocoders and analog radio simulator after each scenario.

The researchers assigned ATCSs to one of three groups (denoted A, B, or C). The groups used each of the three communication systems and handled a different set of four traffic scenarios with each system. Each set of scenarios consisted of two medium and two high taskload scenarios. An important feature of the experimental design is that each controller worked each scenario only once. If controllers repeated the scenarios using different communication systems, the scenarios may have been easier to perform due to familiarity with the traffic problems.

Results

This type of simulation study generates a great deal of data that exceeds the space available here. These results are a subset of the data more fully described in the complete technical report by Sollenberger, La Due, Carver, and Heinze (1997)[xiii].

The results indicated that the vocoders did not affect controller workload measured with the air traffic workload input technique (ATWIT) (Figure 2). In general, overall intelligibility and acceptability ratings were highest for analog radio, only slightly lower for vocoder B, and lowest for vocoder A (see Figure 3 and 4 for details).

The results indicated an interaction between the communication equipment and aircraft background noises for both intelligibility and acceptability ratings. For jet and propeller background noises, intelligibility and acceptability were the lowest for vocoder A, but there were no significant differences between analog radio and vocoder B. For helicopter background noise, intelligibility, and acceptability were the highest for analog radio, but there were no significant differences between vocoder A and vocoder B.

Controller taskload did not affect intelligibility (Figure 3) and acceptability (Figure 4) ratings but had very strong effects on the other dependent measures. Safety, capacity, and efficiency indicators showed that controllers committed more errors, completed more flights, and issued more clearances in high taskload scenarios. Observer and controller self-performance ratings were generally lower in high taskload scenarios.

Discussion

The intelligibility and acceptability results of the simulation agreed with the findings of a previous field study where controllers rated audiotapes (La Due, Sollenberger, Belanger & Heinze, 1997)[xiv]. Both studies suggest that vocoder B is comparable to analog radio and vocoder A is less intelligible and

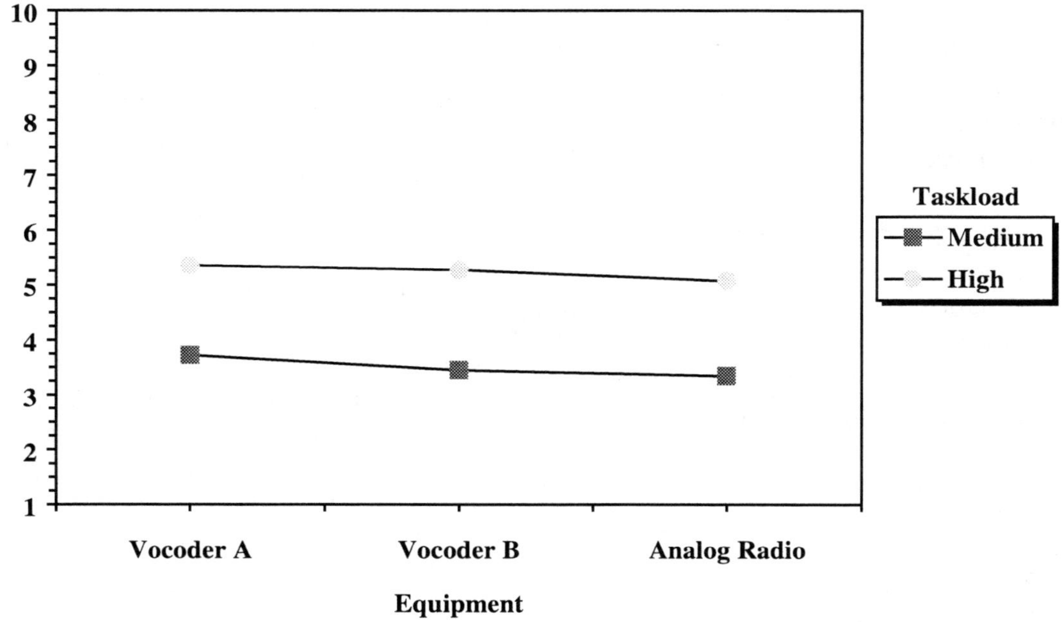

Figure 2. Mean Air Traffic Workload Input Technique (ATWIT) Ratings as a Function of Communication Equipment and Controller Taskload

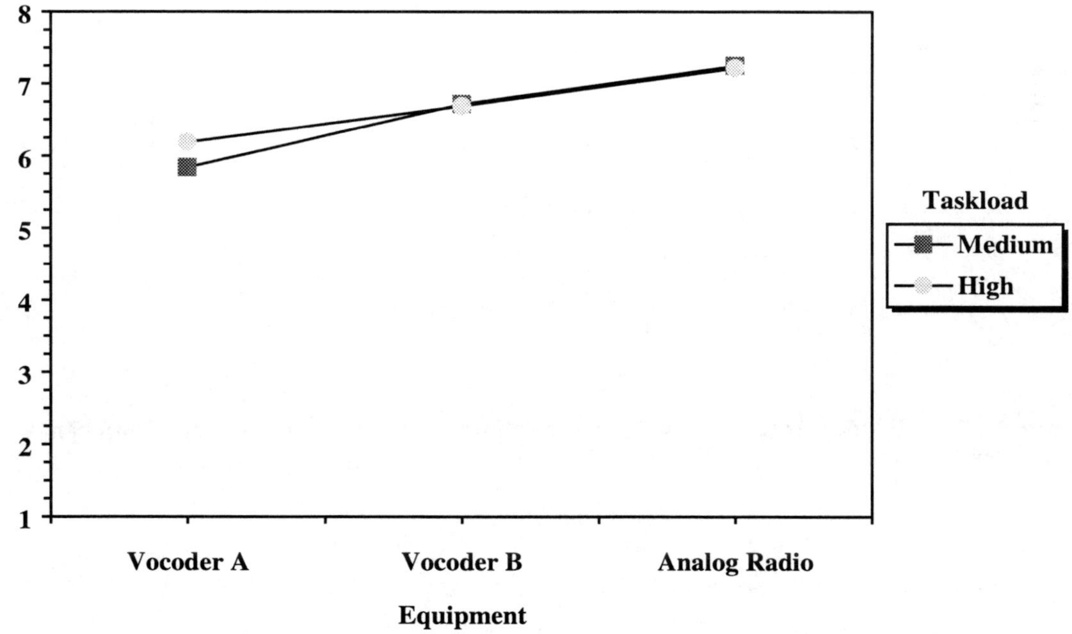

Figure 3. Mean Intelligibility Ratings for all Transmissions as a Function of Communication Equipment and Controller Taskload

acceptable. Although the researchers collected a large number of objective ATC performance measures and other subjective ratings, there were no other differences between the three communication systems. These results suggest that even the least preferred vocoder did not have substantial detrimental effects on ATCS performance. This study further demonstrated that controllers could make direct subjective evaluations on intelligibility and acceptability. These evaluations can separate systems on these dimensions.

This study examined a limited set of factors that could potentially influence the effectiveness of vocoders. Given time limitations, researchers did not evaluate the impact of regional or foreign accents. Also, controllers expressed some concern that vocoders could mask emotional expression in emergency transmissions. This may or may not have influenced controller interpretation of pilot communications.

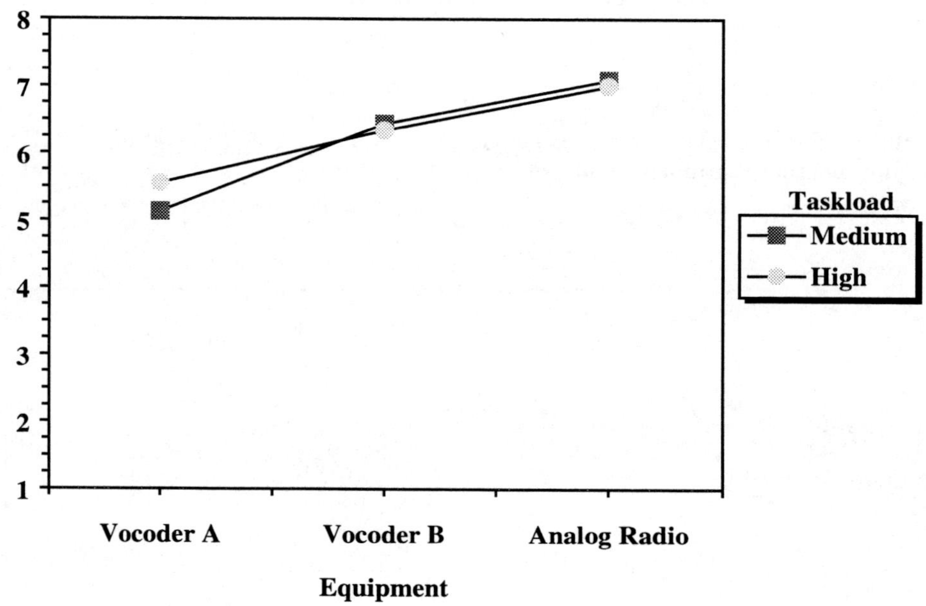

Figure 4. Overall Mean Acceptability Ratings as a Function of Communication Equipment and Controller Taskload

References

i Endsley, M. (1988). Design and evaluation for situational awareness enhancement. Proceedings of the Human Factors Society 32nd Annual Meeting. Santa Monica, CA: Human Factors Society.

ii National Research Council. (1998). The future of air traffic control. Washington, DC: National Academy Press.

iii Sanders, M. S. & McCormick, E. J. (1993). Human factors in engineering and design. McGraw-Hill.Book Company.

iv Dynastat. (1995). Speech intelligibility evaluation at Dynastat. Corporate Publication.

v Crowe, D. P. (1988). Selection of voice CODEC for the Aeronautical Satellite Service (AEEC subjective test report). Ipswich, UK: British Telecom Research Labs.

vi Dehel, T., Grable, M., & Child, J. (1989). Phase II testing and evaluation of low data rate voice CODEC equipment (DOT/FAA/CT-TN89/49). Atlantic City International Airport, NJ: DOT/FAA Technical Center.

vii Child, J., Cleve, R., & Grable, M. (1989). Evaluation of low data rate CODECS for air traffic control applications (DOT/FAA/CT-TN89/13). Atlantic City, NJ: DOT/FAA Technical Center.

viii Troll, N. L. (1989). Assessment of voice coders for ATC/pilot voice communications via satellite digital communication channels (CAA89004). London: Civil Aviation Authority.

ix Guttman, J., Stein, E. S., & Gromelski, S. (1995). The influence of generic airspace on air traffic controller performance (DOT/FAA/CT-TN95/38). Atlantic City, NJ: DOT/FAA Technical Center.

x Buckley, E. P., DeBaryshe, B. D., Hitchner, N., & Kohn, P. (1983). Methods and measurements in real-time air traffic control system simulation (DOT/FAA/CT-TN83/26). Atlantic City, NJ: DOT/FAA Technical Center.

xiStein, E. S., & Buckley, E. P. (1992). Simulation variables. Unpublished manuscript. Atlantic City: WJH Technical Center.

xii Sollenberger, R. L., Stein, E. S., & Gromelski, S. (1997). The development and evaluation of a behaviorally based rating form for the assessment of air traffic controller performance (DOT/FAA/CT-TN96/16). Atlantic City International Airport, NJ: DOT/FAA William J. Hughes Technical Center.

xiii Sollenberger, R. L., La Due, J., Carver, B., & Heinze, A. (1997). Human factors of vocoders air traffic control (ATC) environments phase II: ATC simulation. (DOT/FAA/CT-TN97/25). Atlantic City International Airport, NJ: DOTA/FAA William J Hughes Technical Center.

xiv La Due, J., Sollenberger, R. L., Belanger, B. & Heinze, A. (1997). Human factors evaluation of vocoders for ATC environments: Phase 1 field evaluation (DOT/FAA/CT-TN97/11). Atlantic City International Airport, NJ: DOT/FAA William J. Hughes Technical Center.

SURVEY OF IN-FLIGHT REPLANNING PERFORMED ON THE FLIGHT DECK

Deborah S. Hyams, MIT, Cambridge, MA
Terence P. Fan, MIT, Cambridge, MA
James K. Kuchar, MIT, Cambridge, MA

Abstract

A two-part survey of pilots is summarized in which the effects of pilot reports on decision making behavior was examined. The survey involved a questionnaire on the world wide web and a more focused series of personal interviews. Replanning decisions in response to turbulence and weather were examined.

Introduction

Air carrier flight plans are the result of a careful balance between flight schedule, environment (winds, weather, and traffic congestion), and aircraft performance, and they are designed to optimize the route so as to deliver the aircraft on time and with minimum fuel burn. However, these flight plans are based on forecast conditions and ideal flight patterns, and it is not uncommon for pilots to encounter unforeseen circumstances such as inclement weather, traffic conflicts, or a change in winds which may require a replanning of the original route.

When these in-flight replans are required, pilots, together with Air Traffic Control (ATC) and the Airline Operations Center (AOC), decide upon an alternate course. To make an informed decision, pilots need access to relevant data such as weather location and intensity and turbulence.

This paper describes two studies which solicited current preferences and decision-making behaviors. The focus was on determining the effect of pilot reports (or lack thereof) on decision-making.

Prior Research

Some in-flight replanning issues have been raised and explored in previous studies [1-11]. Several of these studies focused on eliciting pilot behavior and preferences for replanning[1-3] and on developing prototype expert systems to act as decision aids in the cockpit or for traffic management [5-8]. Other studies have focused on modeling the replanning process and on outlining the major research issues [8-10].

Still other studies have focused solely on the amount and types of information available to pilots and how it affects their decision-making process. These studies looked at "Party Line Information" (PLI) that pilots receive by overhearing communications addressed to other aircraft on shared VHF voice frequencies. This information includes elements such as pilot reports, traffic avoidance, and weather deviations. One finding which was common to all of the PLI studies was the perceived importance of weather information reported by other pilots. According to one study, pilots appear to use PLI as a supplemental information source[2]. This study performed a survey of active commercial pilots and asked them to rate different PLI elements in various stages of flight. For the cruise portion of flight, ride reports and weather information received the highest importance ratings. In fact, across all phases of flight, weather situation information and ride reports were rated first and third in

0-7803-5086-3 /98/$10.00 ©1998 IEEE

terms of importance, respectively. According to further results of the study, weather information and ride reports were two of the most accurate and available PLI elements. Also, in a following simulation study it was found that turbulence and weather deviations are extremely effective in inducing action responses from pilots.

Another survey of pilot perceptions of PLI information emphasized the pilots' need for specific traffic and weather information[4]. Still other research concentrated on the methods used by pilots to make decisions in high-risk situations[3]. According to this research, the majority of high risk situations reported by pilots were weather related. This research also highlighted the need to understand a pilot's information needs when designing decision aids. Understanding these needs will allow for the improved design of decision aid systems. If pilot reports and weather scenarios have a large influence on the in-flight replanning decisions then it is important to understand their influence so that it can be incorporated into a decision aid.

What was not clear from previous work, however, was the degree to which pilot reports induce or inhibit replanning decisions. Accordingly, this paper presents preliminary results of two survey efforts designed to investigate how turbulence and weather-related pilot reports affect pilot decision-making.

In-flight Replanning Survey

To better understand how pilot report information is used in the replanning process on the flight deck, two surveys of active pilots were conducted. The first survey was written in HyperText Markup Language (HTML) and was conducted via the world wide web. Users could fill out the survey at their leisure using a web browser, and their responses were delivered via anonymous email to the experimenters. Additionally, the online format allowed for efficient distribution to pilots around the world. The second, "follow-on" study was a more focused version of the online study and further explored pilots' preferences and decision-making behavior through personal interviews.

Subject Selection

A hypertext link to the online survey was posted at several popular aviation websites, including a general-interest web site that produces weekly newsletters to aviation enthusiasts (Avweb[14]) and a site devoted to users of Flight Management Systems (Bluecoat[15]). Because of the inability to control and verify the background of respondents over the internet, several open-ended questions on the survey were used to screen out non-pilots.

Over a three-week period, from mid-January to early February, 1998, a total of 309 valid survey responses were received from the online study. These surveys came from a variety of pilots with different levels of flight experience. The respondents were grouped into four categories according to the highest level of flight rating obtained: Private, Instrument, Commercial, and Air Transport Pilot. Among the 309 respondents, 91 (29%) were Air Transport Pilots. Because the focus of this effort was on airline in-flight replanning, the data discussed in this paper include only the responses from the Air Transport Pilots. These pilots had an average of 9,678 total flight hours with an average age of 44 years. A complete summary of the survey responses for all pilot types is available in Reference 16.

At this point the follow-on study has been completed by four pilots. The results of the follow-on study can not be statistically significant with such a small sample size, however, they can provide preliminary insight as to the use of pilot reports in in-flight replanning.

Ride Quality Pilot Reports

In the first part of the survey, pilots were presented with scenarios involving the prospect of turbulent weather. In each scenario, pilots were told that there was a region approximately 20 minutes ahead[1] along their route of flight in which moderate chop (turbulence) had been reported earlier[2]. Moderate chop is a state of continuous, rapid turbulence, which may be dangerous to unbelted passengers but is not dangerous for flight.

The scenarios were designed to test the following parameters: presence of a lead aircraft, turbulence on own aircraft, pilot report of moderate chop from a lead aircraft, and an altitude change performed by a lead aircraft in response to turbulence. These scenarios are shown in Table 1. Only scenarios 1-5 were presented in the online survey. However, all 8 scenarios were presented in the follow-on study.

In each scenario, respondents were asked to choose one of three options: proceed using the current flight plan, change the flight plan, or request more information. In addition, there was space for pilots to specify the type of information they would request and to give general comments with regard to the scenario.

In each of the first three scenarios, there were clearly-preferred courses of action that received a majority of support by the respondents (Fig. 1). Whether this preferred course of action was to continue with the current plan, replan, or to request more information depended on the specific scenario, as discussed below.

[1] The distances were given in miles as well as minutes in the follow-on study.

[2] The follow-on study stated that turbulence was forecast as opposed to reported.

Table 1: Turbulence Study Scenarios

Scenario	Turbulence on own A/C	Lead A/C present	Moderate chop on lead A/C	Altitude change by lead A/C
1				
2		X		
3		X	X	
4		X	X	X
5	X	X		
6	X			
7	X	X	X	
8	X	X	X	X

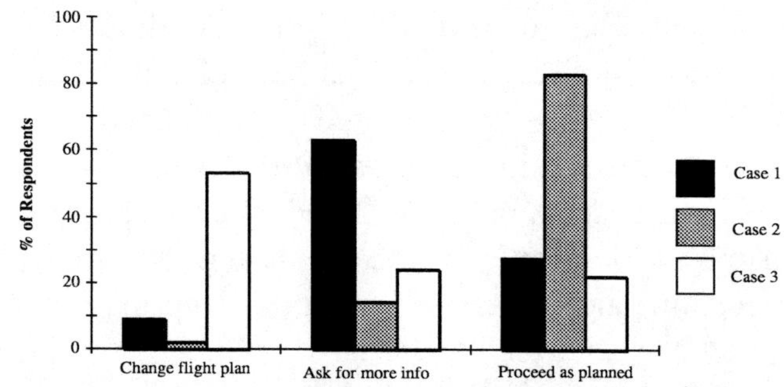

Fig. 1. Replanning Decisions: Scenarios 1-3

The results of the scenarios are reported below. Only results which are statistically significant at the 1% level are reported ($p < 0.01$).

Scenario 1 was a control case, in which the subject's aircraft was the only aircraft in the area. Pilot decision-making was therefore based solely on the prior report of turbulence 20 minutes ahead. In scenario 1, 63% of the pilots surveyed online indicated that they would request more information before deciding whether to continue with the current flight plan. In the follow-on study, all of the pilots also chose to ask for more information. This likely results from the ambiguous baseline condition in which the recency and scope of the turbulence report was not explicitly mentioned to the subject.

In scenario 2, the pilots heard via radio that another aircraft was flying approximately 5 to 10 minutes ahead of them on the same route and at the same flight altitude. However, this aircraft had not made a turbulence report. This scenario was designed to determine whether a lead aircraft could have an effect on pilot decision-making. In this scenario, pilot preference shifted significantly such that 83% indicated that they would continue on course. In the follow-on survey, one of the pilots chose to continue on course while the other three chose to ask for more information. From the online survey, it appears that the presence of a lead aircraft can have a significant influence on the pilots' replanning decisions to proceed as planned. Additionally, from talking to the pilots who completed the follow-on study, the information that they would request would be a ride report from the lead aircraft. This implies that the pilots are reassured by the presence of another aircraft, presumably because they expect the lead aircraft to report turbulence when and if any is encountered. Thus, the lead aircraft acts as a surrogate ride quality sensor.

In scenario 3, the pilots heard a report that the lead airplane was experiencing moderate chop. Comparing scenario 3 with scenarios 1 and 2 allows evaluation of the effect of a pilot report from a lead aircraft. As in scenario 2, scenario 3 resulted in a significant change in pilot preference. In this case, slightly more than half (53%) indicated that they would request a change in flight plan. The remaining pilot preference was split approximately equally between requesting more information and proceeding as planned. In the follow-on survey, two pilots choose to change the flight plan while the other two asked for more information. This scenario shows that a pilot report of turbulence from a lead aircraft shifts pilot preference toward changing the flight plan. However, approximately 25% of the pilots indicated that they would request more information, suggesting that pilot reports, though useful, do not provide all of the information that is desired.

In scenario 4, the lead aircraft, in addition to the pilot report of moderate chop, also requested a change in altitude in order to improve ride quality. In this case, the additional impact of a lead aircraft that changes its flight plan can be determined. In this scenario, the fraction of pilots requesting a change of plan was slightly less than in scenario 3 (46%) for the online survey, as shown in Fig. 2. In the follow-on survey, two pilots changed course and two asked for more information, the same as in scenario 3. Apparently, a request for deviation due to turbulence does not carry much additional information over a turbulence report alone, and therefore does not significantly change a decision to reroute. It may, however, factor into the type of rerouting that is requested.

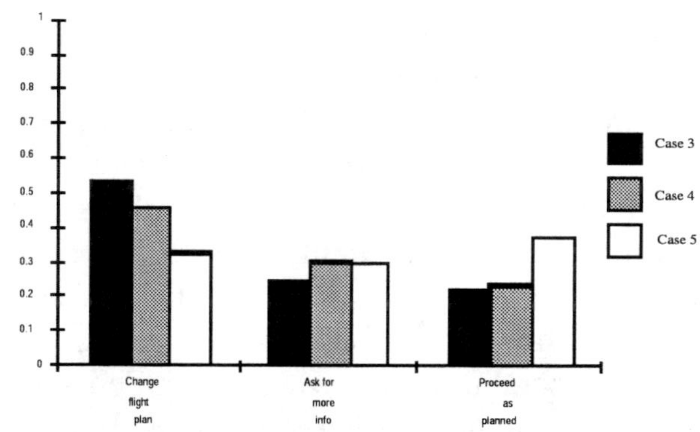

Fig. 2. Replanning Decisions: Scenarios 3-5

Finally, in scenario 5, the lead airplane was proceeding straight and had not reported any turbulence, but the subject's airplane was experiencing moderate chop. This allowed for the exploration of the effect of actual (as opposed to predicted) turbulence on decision-making. A somewhat surprising result was obtained in response to scenario 5 (Fig.2). Fewer pilots (33%) indicated that they would request a change in their flight plan in this case than had indicated they would do so in scenarios 3 (53%) or 4 (45%). The results of the follow-on study were less clear, mainly due to the small sample size. In the follow-on study one pilot indicated that he would proceed as planned, another chose to ask for more information, and the other two indicated that they would change course immediately. From the online survey it appears that subjects were less willing to change the flight plan while they were experiencing turbulence than when the subject's ride was smooth. The reason for this apparent contradiction is likely due to the presence of the lead aircraft. Namely, a lead aircraft that is not reporting turbulence is interpreted to indicate that there is no turbulence at its location. Thus, in scenario 5, the subjects likely decided that the turbulence would abate in the near future, and thus a change of altitude might not be warranted. In contrast, although the subject's ride was smooth in scenarios 3 and 4, the prospect of turbulence ahead, due to the pilot report, was sufficient to lead many subjects to request a change in flight plan.

The remaining three scenarios were not presented in the online survey; however, they were examined in the follow-on study. Scenario 6 is the baseline condition for all of the cases in which turbulence was acting on the pilot's own aircraft. In this scenario all of the pilots chose to ask for more information. As in scenario one, this likely results from the ambiguous baseline condition in which turbulence is forecast but no updates have been presented.

In scenarios in which the lead aircraft reports turbulence (scenario 7) and then requests a change in altitude (scenario 8), three out of the four pilots changed course immediately. The fourth pilot requested more information. This suggests that the presence of turbulence on their own aircraft causes pilots to give more weight to the reports of the aircraft ahead of them, especially when the report confirms what the pilot is experiencing.

In addition, several subjects commented on additional information they would have liked to have had in the scenarios. This additional information included: alternate altitudes or flight paths with smoother air (mentioned by 13 pilots); type of aircraft reporting the turbulence (10 pilots); spatial extent of the problem, or how long turbulent conditions would last (9 pilots); and the ride quality of other aircraft in the area (7 pilots). Other factors that would also be considered by the air transport pilots (as expressed in the additional comments for this question) include cabin concerns (i.e., whether meals are being served, mentioned by 5 pilots), the recency of the pilot report (3 pilots), fuel (mentioned twice) and penalty of diversion (mentioned twice). In the general comments section, diversion penalties, weather conditions, wind profile, airborne traffic, and cabin concerns (mentioned once or twice each) were also listed as other factors influencing the replanning decisions.

The effect of turbulence on the pilots' own aircraft appears to make the pilot more cautious. As suggested in the online study, it is as if the pilot report is a surrogate weather information source. The information from the follow-on study further adds that if the pilot's own aircraft is in calm air then turbulence reports from a lead aircraft are taken more skeptically. However, when there is turbulence acting on their own aircraft, the pilots react immediately to the reports of the lead aircraft.

Weather Case Scenarios

The second part of the follow-on study was a study of the impact of pilot reports on replanning choices around a weather system. In the study, pilots were presented with a weather radar display and three route options (Fig. 3). The choice of Route A was to remain on their current path at 31,000 ft (FL310), which led directly into the weather, Route B was a climb of 4000 ft to FL350, and the third choice, Route C, was a deviation to the North around the weather system at FL 310. Figure 3 is an example of a test scenario in which a report of moderate chop is being given by a lead aircraft on route A (as shown by the diamond symbol).

Figure 3: Example test scenario

Fourteen separate scenarios were tested, and the order was counterbalanced with each pilot in order to compensate for learning effects throughout the study. The scenarios went through the following situations on each route: absence of a lead aircraft, presence of a lead aircraft, lead aircraft reporting light chop, and a lead aircraft reporting moderate chop (Table 2).

The scenarios were presented with a light weather picture which showed an entirely green cell. Additionally, the scenarios involving Route A were repeated with a slightly worse weather display reading in order to explore the effect of harsher weather on the replanning process. This weather cell included some areas of yellow along the route of flight.

Table 2: Test Matrix for Weather Study

	Route A	Route B	Route C
No lead			
Lead; no report			
Lead; light			
Lead; moderate			

Analytical Hierarchy Process

For each scenario in the weather radar case study, the pilot was asked to use the Analytical Hierarchy Process (AHP) to rate the route choices that were presented. The Analytical Hierarchy Process is used to obtain subjective preferences of multiple options. It provides a means of evaluating multiple design options using weighted ranking scales [17]. The process breaks up multiple options into a series of paired comparisons which are then recombined to produce an overall weighted ranking. Most importantly, AHP retains information about the relative size of the intervals between the rankings. So it is possible to know *how much* better one option is than another. The final result of AHP is a pie chart in which the dominance of an option is given by its area in the chart; the larger the area, the more dominant the option.

Weather Scenario Preliminary Results

For most scenarios, route C was the most preferable option with routes A and B taking up smaller slivers of the pie chart as in Figure 4. What these results mean besides the fact that Route C was the most preferable option of all three, however, is that route B was the second most preferable, and was preferred about three times as much as route A.

While Figure 4 shows the results for most of the scenarios, it is especially interesting to look at the results of the scenarios which involve a lead aircraft on route C. The results for the case in which a lead aircraft is present on route C without any pilot reports is shown in Figure 5.

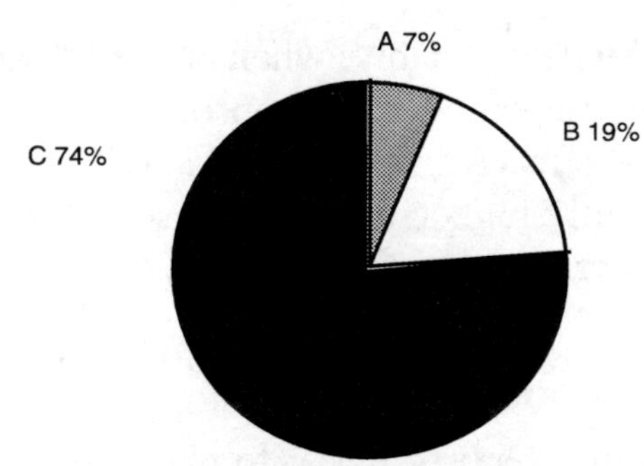

Figure 4: Dominance chart for most light and all moderate weather scenarios.

0-7803-5086-3 /98/$10.00 ©1998 IEEE

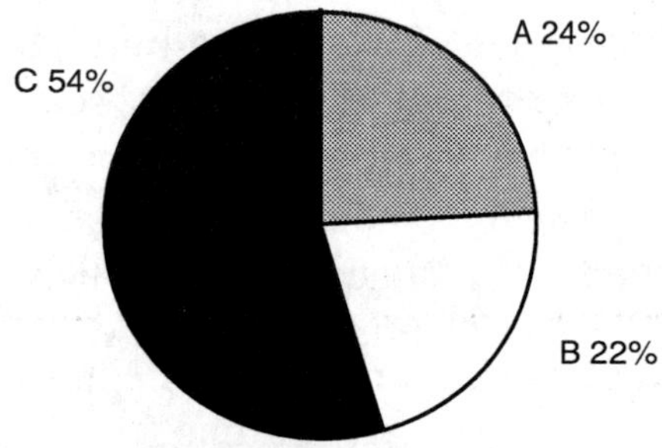

Figure 5: Dominance chart: lead aircraft route C

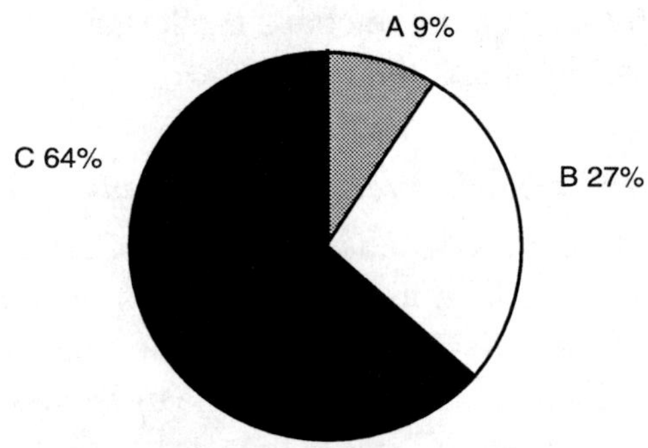

Figure 6: Dominance chart: moderate chop reported on route C

Although it seems strange that Route A would have such a high preference rating when the lead aircraft had chosen to fly on Route C, one pilot explained that since the lead aircraft was close to the weather system and not reporting any turbulence it was reasonable to assume that the ride through the weather system would probably be acceptable.

Understandably, when the lead aircraft was reporting moderate chop on Route C, the preference for Routes A and B fell to levels only slightly higher than the ones held without a lead aircraft, as shown in Figure 6.

Conclusions

Pilot reports appear to play a significant role in decision-making related to turbulence. A pilot report of turbulence from a leading aircraft along the route of flight alters the following-pilot's preference from "request more information" to "change flight plan". The lack of a pilot report from another aircraft in an area of previously-reported turbulence likewise significantly alters pilot preference from "request more information" to "proceed as planned". An aircraft that issues a pilot report for turbulence and also requests a change in altitude does not significantly alter following-pilot preferences to change course over that which occurs when the pilot report alone occurs, suggesting that the deviation request does not provide much additional information over the ride report itself.

The results from the weather scenarios also support the conclusions drawn in the turbulence scenario. Pilot reports from other aircraft have a strong influence on which route a pilot is willing to take. If there is a report of turbulence within the weather system then the pilot is not willing to follow the route into the system. However, if a pilot is near a weather system with a lead aircraft not reporting a lot of turbulence, then the pilot may be willing to follow the shorter route.

Acknowledgments

The authors are appreciative of the support provided by the NASA Langley Research Center, including Sally Johnson, Mark Ballin, and Terry Abbott. We are also appreciative of the collaboration with Bill Corwin at Honeywell, and Geoff Gosling at the University of California, Berkeley. The pilot respondents to the survey also deserve thanks for their time and input.

This research was supported by the National Center of Excellence for Aviation Operations Research, under Federal Aviation Administration Research Grant Number 96-C-001. This document has not been reviewed by the Federal Aviation Administration (FAA). Any opinions expressed herein do not necessarily reflect those of the FAA or the U.S. Department of Transportation.

References

1 Patrick, N. J. M., "Decision-Aiding and Optimization for Vertical Navigation of Long-Haul Aircraft", PhD Thesis, MIT, Cambridge, MA, August, 1996.

2 Midkiff and Hansman, "Identification of Important Party Line Informational Elements and the Implications for Situational Awareness in the Datalink Environment", Report ASL-92-2, MIT, Cambridge, MA, May 7, 1992.

3 Dershowitz, A., "The Effect of Options on Pilot Decision Making in the Presence of Risk", PhD Thesis, MIT, Cambridge, MA, October, 1997.

4 Pritchett, A., "Variations in Party Line Information Requirements for Flight Crew Situation Awareness In the Datalink Environment" , Report ASL-94-5, MIT, Cambridge, MA, May 18, 1994.

5 Prevot, Gerlach, Ruckdeschel, Wittig, and Onken, "Evaluation of Intelligent On-Board Pilot Assistance in In-flight Field Trials", 6th IFAC Symposium, Cambridge, MA, June 27-29, 1995.

6 Robinson, Davis, and Isaacson, "Fuzzy Reasoning-Based Sequencing of Arrival Aircraft in the Terminal Area", AIAA GNC Conference, New Orleans, LA, Aug, 1997.

7 Nguyen, T., and D. Ward, "A Neural-Network Based Inference Engine for a General Aviation Pilot Advisor", AIAA Aerospace Sciences Meeting, Reno, NV, January 6-10, 1997.

8 Rudolph, Homoki, and Sexton, "Diverter: Decision Aiding for In-flight Diversion", NASA CR-182070, 1991.

9 Abbott, T., "Functional Categories for Future Flight Deck Designs", NASA-TM-109005, Hampton, VA, August, 1993.

10 Rogers, Corwin, Riley, Quarry, and Dwyer, "Airborne-Based Conflict Probe", NASA Contractor Report, February, 1998.

11 NASA Aviation Safety Investment Strategy Team (ASIST), "Weather Investment Recommendations Summary", NASA Langley Research Center, April 15, 1997.

12 Pritchett, A., "Pilot Non-Conformance to Alerting System Commands During Closely Spaced Parallel Approaches", PhD Thesis, MIT, Cambridge, MA, January, 1997.

13 DeCelles, "The Delayed GPWS Response Syndrome", Aviation Research & Education Foundation, Herndon, VA. July, 1991.

14 www.avweb.com

15 www.NeoSoft.com/~sky/BLUECOAT

16 Fan, T., Hyams, D., and J. Kuchar, "Preliminary Study of In-flight Replanning Performed on the Flight Deck", NEXTOR Report RR-98-4, MIT, Cambridge, MA, April 30, 1998.

17 Yang and Hansman, "Application of the Analytical Hierarchy Process for Making Subjective Comparisons Between Multiple Automation/Display Options", 6th IFAC Symposium, Cambridge, MA. June 17-29, 1995.

INTRODUCING NEW TECHNOLOGY TO THE AIR TRAFFIC CONTROLLER: IMPLICATIONS FOR SKILL ACQUISITION AND TRAINING

Alfred L. Smith, Jr., Federal Aviation Adminstration, Washington, DC[1]

Albert Douglas, Jr., Federal Aviation Adminstration, Washington, DC

Introduction

This paper addresses the challenge the Federal Aviation Administration (FAA) faces in introducing and having accepted new technology within the Air Traffic Services community, specifically among Air Traffic Controllers (en route and terminal). This topic is extremely important since it is related to current efforts to replace aging equipment and introduce the latest in automation and other technologies. Over the past decade, air traffic control (ATC), has received criticism in the areas of safety, National Airspace (NAS) modernization, and computer technology (GAO, 1989).

The FAA has committed itself to an aggressive campaign to modernize the NAS by the year 2005. The end result of this modernization should be enhanced safety, produced by new technology, advanced automation, and availability of more commercial flights. For the ATC community, this agency commitment should mean getting reliable equipment and software that reduces workload and makes the management of air traffic safer and more efficient. Such advances, however, will not come without their share of problems. Not the least of these is acceptance of new technology by veteran controllers. Further, new technology requires new skills. People may be resistant both to the systems themselves and to the need to learn new skills. In conjunction with the training and skill acquisition for operation and use of new systems, those controllers involved in the development and prototyping process need to be provided with additional learning opportunities. These learning experiences should be used to develop an enhanced understanding of the systems design process. With this training, their role in providing feedback during the development process can be instrumental in producing technology that will be useable – and used. If the development of new tools and systems and replacement of old equipment is going to be successful and occur according to schedule, therefore, there must be new training programs. These programs must be designed not only to address the acquisition of novel skill sets, including skills needed to design systems, but also to take into account the potential underlying resistance present in the workforce.

Technology Acceptance

According to Webster's (1984), technology is defined as, 'The application of science especially to industrial or commercial objectives." Automation has a number of varied meanings but the definition adopted for

[1] The views expressed in this paper are those of the authors and do not necessarily reflect the position of the Federal Aviation Administration.

0-7803-5086-3 /98/$10.00 ©1998 IEEE

this paper is "a device or system that accomplishes (partially or fully) a function that was previously carried out (partially or fully) by a human operator (National Research Council, 1998, p.13)." For most of us non-scientists, technology means computer technology, both hardware and software. In ATC new technology means advanced forms of automation. According to Hancock (1995), technology generally speaking, is associated with things, objects, and systems that are new, modern, and exciting. Technology is supposed to represent ingenuity by which mankind benefits. It is the vehicle for achieving our industrial or commercial objectives. For ATC these industrial and commercial objectives are improved safety and efficiency; availability of the technology, and the desire to support the controller (National Research Council, 1998).

Frequently, efforts to bring on-line new technology to support the controller have met with difficulty. This difficulty has been experienced in the design and prototyping process as well as in the field. Part of the challenge has been to have the technology under consideration viewed as a product that can help enhance the way air traffic control task are executed. Effort has been made to convince personnel that revolutionary changes are for the better. In many respects attitudes that are expressed by some in the air traffic community, at times, give lip service to wanting to embrace technology but not really to seeing its full potential realized. There is a tendency to want to stay with what the controller knows (e.g., the screen looks the same or use of switches rather than buttons). Perhaps for some, the attitude can be characterized, as, "Yes, 'they' need to use new technology, but not me." Thus, some controllers are voicing concerns about the unreliability and lack of safety associated with the "old" equipment in use today, yet efforts to replace or introduce new equipment (i.e., new technology) are met with a general lack of willingness to accept it. In the design process, there tends to be little acceptance of any proposed new technology without major clamoring for redesign. This need for redesign really may be reflective of individual preferences but is being touted as related to general human factors' issues, which sometimes are valid. Often, however, the request for redesign can be interpreted as just a means to put off implementation of new technology.

By its very nature, technology and the products that result from it mean a different way of doing things. The implication of new technology, innovation and change, may mean to some controllers less and less reliance on the various proven techniques, methods, information and decision-making that have been established across time. This can be very unsettling. Personnel may be resentful. They may be intimidated by new approaches. They may be worried about whether they can "learn new tricks." They may be resistant both to the new technology being presented and to the need for the acquisition of new skills.

Clearly, acceptance of technology is related to an employee's attitudes or belief system (Gattiker, 1988). Very often, either people are all for it and embrace (accept) it when it is introduced, or they are against it and resist its introduction. Few people are ambivalent or take the middle ground. Attitudes are shaped by the individual's belief system based on prior experiences. They will be accepting of new technology if it fits in with their belief system. In the air traffic arena, that belief system may include beliefs about themselves- their capabilities, their skills, their likes and dislikes, as well as their beliefs about how their job should be preformed, how technology should assist them, and about computer technology in general. If the person has prior successful experience with automation, feels that the technology was helpful in the past, or did not have difficulty mastering its use, he/she is likely to be more willing to accept new technology

The literature posits a number of reasons why a person would accept or resist new technology. Martinko, Henry, and Zmud (1996) propose that how a person responds to technology is the result of a combination of factors that act to influence whether or not it is accepted or resisted. These factors are classified as external and internal influences.

External influences may include the characteristics of the system itself, e.g., how it looks, ease of use, etc. Attitudes exhibited by co-workers, how easily others use the system, and positive support from management are other examples of external influences, as is the learning/working environment. Being in an environment in which they can ask questions, voice concerns, and admit discomfort can due a lot to bring about acceptance of new situations. Internal influences may include attributions that result from past interactions with the technology and expectations about future performance. Fears related to staff reductions may also influence acceptance of new technology (Korunka, Weiss, and Zauchner, 1997).

In other words, how a person responds to technology is the result of a combination of factors, which interact to influence whether or not technology is accepted or resisted. When technology is resisted behaviors such as low levels of use, lack of use or by dysfunctional, e.g. harmful, use may be observed. Such behavioral reactions well documented in the literature. Still other behaviors, e. g., employee complaints, may be emitted as an indication of resistance (Korunka, et al., (1997).

New Skills

Technological advances will require changes in the way the job is performed. New information provided from new systems will be strategic rather than tactical in nature. The job of the controller will require different decision-making and planning skills from those that are currently needed. Thus, it is logical to assume that the skills needed by controllers to operate in future environments will differ from what they are today and will require new strategies in training.

Current systems provide the controller with information with which he or she makes decisions. The results of those decisions are expressed in instructions or advisories to pilots. Information includes aircraft locations, verbal reports from pilots, information from Traffic Management, supervisors, and other controllers. Under the present system, automation provides some of the required information to the ATCS [Air Traffic Control Specialist], but the controller makes all the decisions and performs most of the operations required to implement and communicate those decisions (Manning, et al. 1992, p.1)." As can be seen by this description of the process, automation plays an important role in providing necessary information, however the ATCS exerts system control, analyzing information by which decisions are made and instructions are given. Essentially, the controller sees himself/herself as in full control.

In the future, there is the very real possibility that the role of the controller will change significantly from what is observed today. The National Research Council (1997) states that a new generation of automated tools has the potential to replace the decision-making and planning activities of the controller. Due to the simplicity of these decision-making processes, however, the controller would still be needed to monitor the automation and intervene when the automation failed. Thus, the controller may need different cognitive skills dealing more with crisis management, and more technical skills in terms of making the equipment work correctly, and interpreting equipment problems, as well as the skills to operate new equipment and systems.

0-7803-5086-3 /98/$10.00 ©1998 IEEE

Reducing Resistance through User Involvement

The introduction of new technology in the workplace can be an unsettling proposition and without proper organizational interventions, can result in negative outcomes. Misconceptions and fears related to the introduction of technology can influence technology acceptance. One of the ways to eliminate resistance to new technology is to involve the controller (user) in the design process. Involving the user in systems design produces realistic expectations, which reduces resistance (Gattiker, 1988). The question is, will involvement by itself ensure that resistance to technology is minimized?

The FAA has tried to lessen the negative effects of new technology and increase acceptance of automation by involving users (usually union representatives) in the total process of development, including the requirements generation. Historically, controller involvement has not produced the positive outcomes one would expect from involving the user (Small, 1994). Management realizes that controller involvement is critical in the design process and ultimate product success, i.e., a product that gets accepted by the user. Webb (1996) states that "bad products, products that are unsuccessful or not well accepted by the user population, are not designed around user input (p. 76). The issue in Air Traffic is not the involvement of controllers in the process, but how best to achieve the desired outcome, that is, a product that is accepted and used. For this desired outcome to be realized training for controllers involved in the design process will have to occur.

If controllers (as users) are to make a positive contribution by involvement in the design process their contribution and expectations need to be defined. Damodaran (1996) argues that effective user involvement is not just feedback on technical issues, because many of the technical issues are what she calls socio-technical in nature. She suggests that, for the user to make a real contribution, the organizational context has to be carefully structured. She goes on to say that users need to be educated (e.g. site visits of similar systems, demonstrations and discussion sessions, etc.) and develop an understanding of the principles underlying the processes that they are involved. Further, she states that, "the strategy for participatory design must include the infrastructure (resources and capability) to promote understanding by all of the participants. The strategy must also include the clear allocation of responsibilities to key stakeholders and clarity regarding the respective roles of the different players. Users should be viewed as providers of key information. They should also be allowed to exert effective influence. In order to be able to this, then, according to Damodaran, users should be exposed to training in system design methods. Through such training, controllers may develop an appreciation of the design process, changing some of the pre-conceived notions about the system under consideration.

Other research has shown that when you involve the user in the development process and they are not appropriately trained you do not get the desired result. Heinbokel, Sonnentag, Frese, Stolte and Brodbeck (1996) in a study of commercial software development projects examined the effects of the user-centered approach through measures of user participation and user orientation. They found that both user participation and user orientation were negatively related to process and product quality, as well as to overall project success. According to Heinbokel et al., (1996) user participation or user orientation led to a number of problems. These problems included more sophisticated ideas about how the software should work during the course of the development process. This led to interruptions in the design process. Software developers found it difficult to incorporate changes late in development. Second, users feared job loss or a decline in working conditions and, as a result, they did not participate constructively. Third, the users were considered unpredictable. The behavior being best described as 'a child in a candy store' i.e., seeing something somewhere else, deciding that they must have it, and demanding changes and affecting schedule by doing so. Finally, user orientation as the authors put it may lead to higher levels of aspiration that reduce team effectiveness.

Webb (1996) believes that user involvement has obvious advantages but can have serious disadvantages also. He identifies some of the disadvantages as the lack of compatibility between the design environment that is dynamic, creative and innovative and the individual variance that users bring to that environment. Further he goes on to state that there is difficulty associated with the assessment of users. He argues that in creative environments, user centered design approaches are inappropriate. Webb states, " The user is not a designer and studies have shown that user designs are generally inferior to those of interface professionals (p.82)."

The conclusion to be drawn from the literature reviewed here is user involvement is necessary for development of products that will have usefulness. However, without appropriate training for participating controllers regarding the technology, the principles that guide its development and about their role in the process the success of the outcomes will be substantially reduced.

Training Programs

It is obvious from the discussion about that training for controllers involved in the design process is critical for program success. Educational training for these individuals should include information on the principles that underlie the given systems' development and operation. Controllers need to be allowed to view similar systems and see them perform in operation. They also need to take part in discussion sessions with systems designers that explain why and how things work, why some modifications are feasible while others may be impractical and/or impossible. In addition, training for potential users involved in the development phase must also recognize and address those issues, discussed above, that relate to apprehension and resistance to the new technologies and the need for new skills. Thus, the benefit of this design training is twofold. First, the controller will develop an understanding of the principles underlying the process for the technology in which they are involved. Secondly, the training should influence controllers' acceptance of the technology by removing or reducing their fears, concerns, misgivings, false beliefs, etc.

The people selected to conduct this type of training need both knowledge about the technology and special skills to work with people who lack that knowledge, may have little interest in the knowledge, and may be resistant to the training. The trainers should have an understanding or an awareness of the potential causes surrounding resistance to technology and change. With such an awareness, they should be able to incorporate information that will provide reassurances about the technology, e.g., that it will be learnable and usable, will not impact jobs in a negative manner, etc. Under this scenario, the trainers are not only providers of information about the subject matter but also instillers of confidence and reassurance.

Training for the end-users once the technology has been accomplished does not necessarily need to cover aspects about the design and the principles that govern the operation and functioning of technological systems in depth. There should be, however, at least some background information available to those users, albeit likely a very small number, who embrace technology and want to learn more about the systems than how to operate. At a minimum sufficient information should be included in the training for all controllers to take the "mystery" out of the technology. This part of the training should provide the opportunity to identify and clear up any misconceptions or preconceived notions about how systems do and do not work.

It is important that the skills need to work with new systems be fully and accurately identified. The training program must be long enough to provide the end-users sufficient practice of the necessary skills to establish a comfort level. Further, controllers should receive honest feedback during training so that they know how well they are doing and what further practice they may need. The environment should be one in which they can feel comfortable in asking questions and getting further assistance as needed. Once they are on the job, they should be able to ask for additional training help until they feel they have achieved mastery over the system. The availability of OJT assistance could do a lot to alleviate apprehension about skill acquisition.

Although involving knowledgeable controllers in the design phase will help to decrease resistance to new systems, it is improbable that this will completely eliminate end-user resistance. Thus, as with the training for controllers involved in design, training for end-users also must recognize and address the factors that contribute to the lack of acceptance of new technology. Further, in addition to being able to deal with those factors that inhibit user acceptance and eliminate or reduce them, the trainers must be fully knowledgeable about how to operate the new systems. Imagine the unimaginable: a reluctant or resistant controller being trained by some who cannot answer all the questions asked, demonstrate mastery of the technology, or allay the fears, trepidation, or intimidation of the user-trainee.

Conclusion

Industry usually retools its employees through comprehensive training programs to ensure that the necessary new skills are acquired and the competencies necessary to execute a new way of doing business are present. Training controllers in such a way as to maximize acceptance of new technology should have several aims. The first aim of a training program is to train people for development. The focus is to make those controllers who become involved in the development and design process knowledgeable about the techology. In this way, their expertise as controllers can best influence technology that controllers in the field will be willing to use. To get the most from this user involvement in the early phases, this training will also have to focus on ameliorating those factors that build resistance in users. The second aim of the training program is to prepare end-users. Again, one focus of the training must address the kinds of issues that create resistance in users. Further, the program must provide the controllers with the skills necessary to use the technology. Thus trainers must not only have expertise in the technology, but also expertise in conveying knowledge. At the same time, they must be able to address the issues that will be result in greater user acceptance.

0-7803-5086-3 /98/$10.00

References

Charytonowicz, J. (1997). Computer in shaping of human habitat. In Salnendy, G., Smith, M. J., & Koubek, R. J., (Eds), Design of computing systems: Cognitive considerations. Proceedings of the Seventh International Conference on Human-computer Interaction (HCI International '97) Vol. 1. Amsterdam, The Nethelands: Elsevier Science.

Damodaran, L. (1996). User involvement in the systems design process – a practial guide for users. Behaviour and Information Technology, 15, 363-377.

Gattiken, U.E. (1988*).* Technology adaptation: a typology for strategic human resource management, 7, 345-359.

Government Accounting Office. (1986). Air traffic control: FAA's interim actions to reduce mid-air collision*s.* (Report to the Chairman, Subcommittee on Transportation and Related Agencies, Committee on appropriations, U. S. Senate/RCED-89-149).

Government Accounting Office. (1986). Air traffic control: Computer capacity shortfalls may impair flight safety. (Report to the Chairman, Subcommittee on Transportation and Related Agencies, Committee on appropriations, U. S. Senate/RCED-89-63).

Government Accounting Office. (1986). Air traffic control: FAA's implementation of modernization projects in the field. (Report to the Chairman, Subcommittee on Transportation and Related Agencies, Committee on appropriations, U. S. Senate/RCED-89-92).

Hancock, P. A. (1995) Teleology for technology. In Parasuraman, R. & Moulona, M. (Eds). Automation and human performance: Theory and applications. Hillsdale, NJ: Erlbaum

Heinbokel, T., Sonnentag, S., Frese, m. Stolte, W., &Brodbeck, F. C. (1996). Don't underestimate the problems of user centeredness in software development projects- there are many! Behavior and information technology, 15, 226-236.

Korunka, C., Weiss, A., & Zauchner, S. (1997). An interview study of 'continouse' implementaions of information technology. Behavior and information technology, 16, *3-16.*

Manning, C. & Broach, D. (1992). Identifying ability requirements for operators of future automated air traffic control systems. (Report no. DOT/FAA/AM-92-26). Washington, DC: Federal Aviation Administration.

Martinko, M. J., Henry, J. W., & Zmud, R. W. (1996) An attributional explanation of individual resistance o the introduction of information technologies in the workplace! Behavior and information technology, 15, 313-330.

National Research Council (1997). Flight to the future: Human factors in air traffic control. Washington, DC.: National Academy Press

National Research Council (1998). Future of air traffic control: Human operators and automation. Washington, DC: National Academy Press

Small, D. W. (1994) Lessons learned: Human factors in the AAS procurement. (Report No. MP94W0000088). McLean, VA: The MITRE Corp.

Webb, B. R. (1996) The role of users in interactive systems design: When computers are theatre, do we want the audience to write the script? Behavior and information technology, 15, 76-83.

Webster's II New Riverside University Dictionary. (1984). Boston: Houghton-Mifflin.

HUMAN FACTORS ISSUES IN A FUTURE AIR TRAFFIC MANAGEMENT SYSTEM

Philip J. Smith, Institute for Ergonomics, The Ohio State University, Columbus, Ohio
Charles E. Billings, Institute for Ergonomics, The Ohio State University, Columbus, Ohio
David D. Woods, Institute for Ergonomics, The Ohio State University, Columbus, Ohio
C. Elaine McCoy, Department of Aviation, Ohio University, Athens, Ohio

The Shape of the Future Air Traffic Management System

Air carriers have indicated a need for a less constrained, more flexible air traffic management (ATM) system. In response to this need, RTCA (1995) advanced proposals for a "free flight" system in which the airlines will be able to pursue their objectives more efficiently. The RTCA proposal assumes a higher level of automation to assist air traffic service providers to detect and resolve short-term tactical conflicts, reducing the need to impose strategic constraints on the airspace system. It also implies that human operator roles and responsibilities in the future system may be markedly different from those in today's air traffic management system.

Figure 1 encapsulates some of the differences between the present and future system. It suggests that major changes in management style will be required. There will be a need for additional intent information, since flight path routings will normally come from pilots or airline operations centers (AOCs). The controller will normally be a monitor except when short-term constraints or hazards to air traffic are detected; new technology will assist the controller, or airspace manager. Enhanced traffic displays will also be needed to aid the pilot in his increased decision-making role. Note that closed-loop controller direction of air traffic is replaced by a "softer", indirect loop in which controllers, or air traffic managers, must rely on pilot/AOC intent information and on situation displays to "close" the control and surveillance loop.

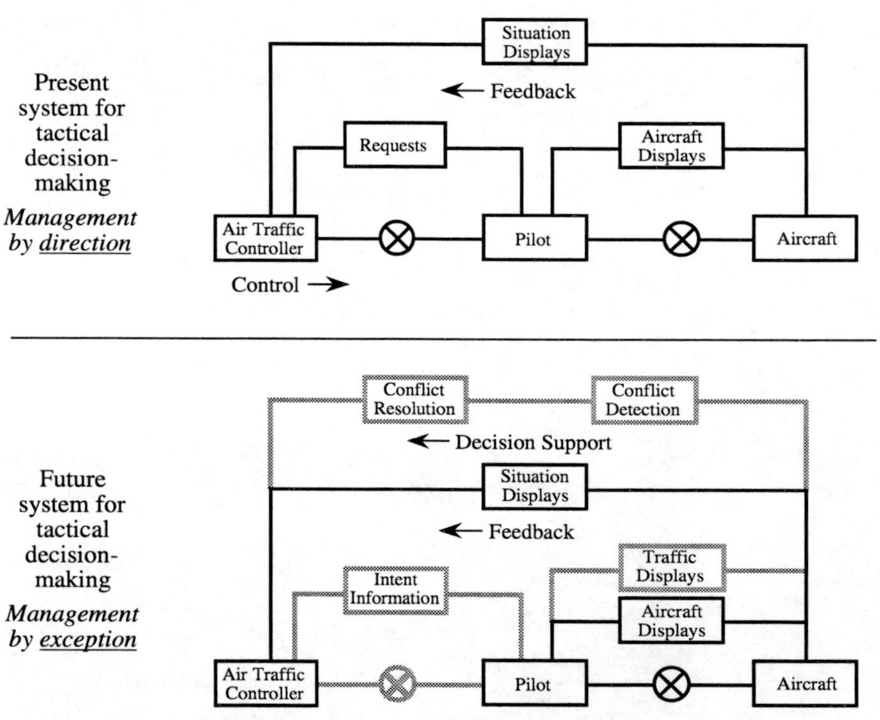

0-7803-5086-3 /98/$10.00 ©1998 IEEE

Figure 1: Comparison of present and future ATM System control loops (Smith et al., 1997)

Human Factors Implications of these Changes

Figure 2 points out that the construct illustrated in Figure 1 looks simple only because it does not take account of the enormous amount of information that is actually required for decision-making in the air traffic control system. A great deal of data (some of which will not be available to computers in the foreseeable future) determines what is an effective or ineffective decision in this complex distributed system.

Figure 3 indicates that there is also a great deal of variability in the system, some of it under control of human managers and operators, but much of it not controllable (weather, etc.). The aviation system is highly dynamic and data describing the system changes constantly, sometimes in unanticipated ways. Coping with unanticipated events, in particular, is something computers do not do well. This fact emphasizes the system's dependence on human operators who can cope with such variance, regardless of whether pilots or controllers are making the decisions.

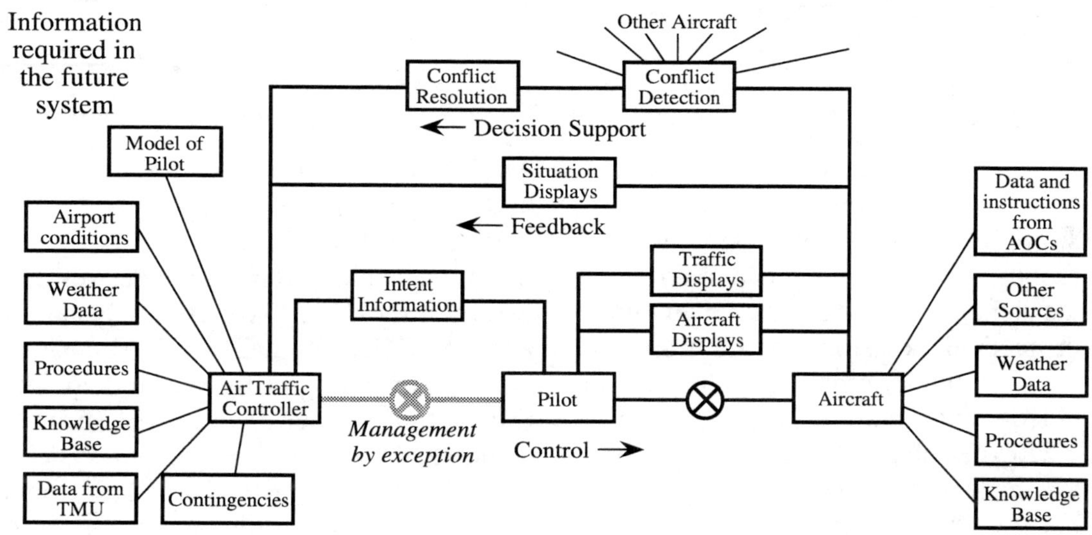

Figure 2: Information Required for Decision-Making in a Future System

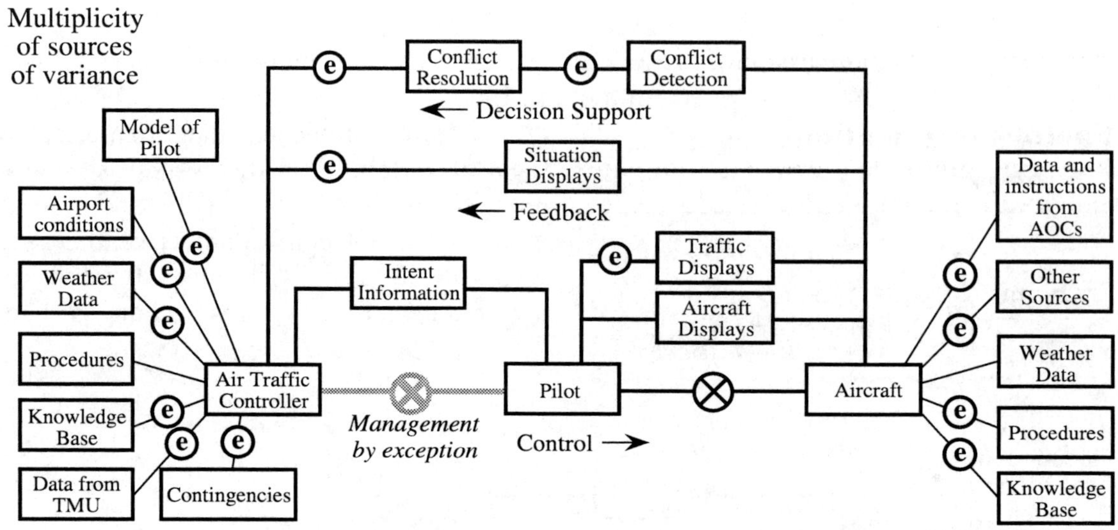

Figure 3: Sources of Variance in Air Traffic Management Data
("e" represents a source of variability or error in the process)

If decision-making is to be more widely distributed among controllers, traffic managers, pilots and dispatchers or AOCs, the information necessary for tactical decision-making must be more widely distributed than has been the case in the past. Effective management of ATM information will be absolutely critical to the success of a future system based on free flight concepts.

Redistributing decision-making authority is comparatively easy: it involves only the formulation of new policies and procedures for system management. Providing real-time, relevant information to support such decision-making, in forms that support uniformly safe decisions, is much harder in an information-bound system because of the enormous quantity of rapidly-changing data from which such information must be extracted. This difficult challenge must be the focus in developing such a future ATM system.

Further, though the future system is to emphasize tactical management, to allow users more flexibility, it is not clear from whom or what agency will come the strategic decision-making that will still be required to permit the system to function. Air traffic management involves the allocation of finite commodities (airspace and airport facilities) across time, among users. In today's system, there is competition among participants for these commodities, which are often insufficient to meet the desires of all users at certain times. FAA's ATC System Command Center (SCC) and Traffic Management Units (TMUs) presently fulfill the strategic management roles, often in coordination with airline AOCs. If a future air traffic management system is to rely more heavily on tactical air traffic management, another infrastructure must be provided to function as the system's strategic management arm and to act as a "referee" among participants.

Our studies (see references) have supported the thesis that attempts to increase operator flexibility in flight path management are often accompanied by a requirement to exchange appreciably more information among system participants. In the absence of additional information, operators find themselves unable to formulate good decision strategies and thus to effectively exercise the additional authority available to them. This was less of a problem in a system that was based on management by direction; Controllers and Traffic Management Units had all the information necessary to exercise system control safely, though not necessarily enough to exercise that control flexibly, in a manner that adequately met user objectives.

In a system based on more distributed control, however, the conventional system model is clearly inadequate. What is needed is a more collaborative approach to decision-making, based on more effective sharing of information among system participants (both human and machine). This is suggested in Figure 4, which illustrates the existing distributed control loop, backed by improved technology and an infrastructure which makes information more easily accessible to system decision-makers as they need it to make effective management and control decisions.

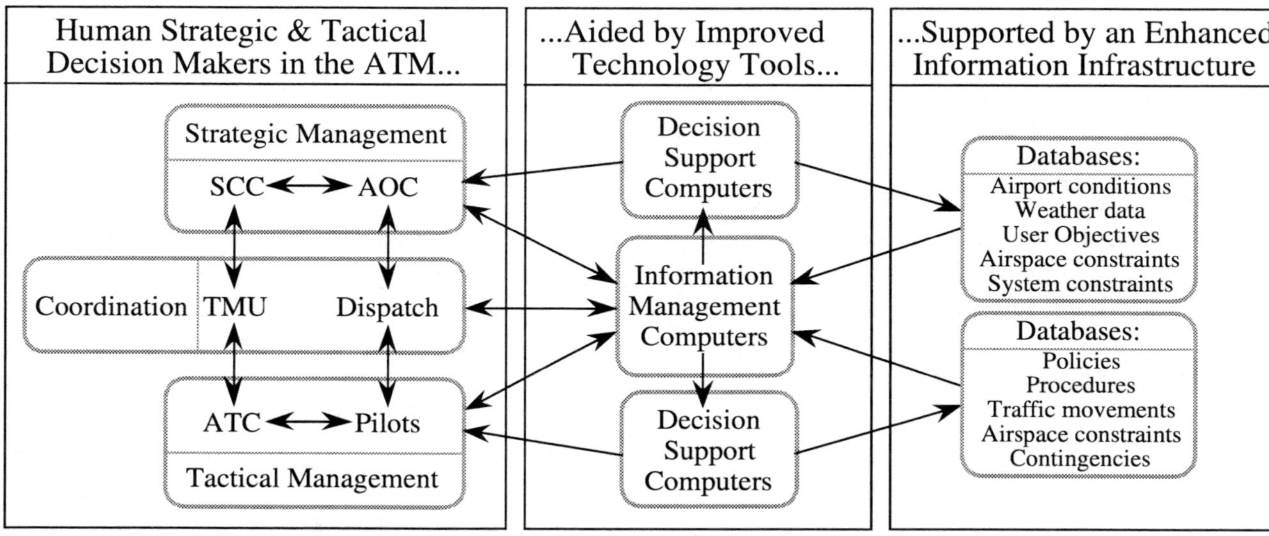

Figure 4: A model of technology-enabled cooperative problem-solving in a future ATM system

In figure 4, we have shown a control loop involving humans, because we are not convinced that strategic management and tactical control can be effectively implemented by machines in the complex air transportation context. We believe that the future system will require far more capable information management tools to assist and support human managers and operators, and that with such assistance in finding and utilizing information, skilled humans will be able to accomplish the tasks required in the future system, while preserving the ability to react to unplanned disturbances and anomalies in the system.

Examples Illustrating the Need for Information Exchange

As Figure 4 indicates, there are many different individuals who need to exchange information and coordinate their activities. This information exchange is needed to support both strategic and tactical management and control decisions.

Information exchange to support strategic decision-making

The concept of free flight is meant to extend to strategic as well as tactical decision-making. In both cases, the goal is to give the users more flexibility in making decisions or choices based on their business concerns, subject to constraints in the system that must be considered to assure safety and efficient overall use of system capacity. One of the implications of these constraints, however, is that there will be days when a certain route or segment of that route will not be able accommodate a particular flight at all, or will not be able to accommodate that flight under enroute free flight rules, because of weather, traffic congestion or airport restrictions.

One implication is that, as the locus of control for pre-flight planning of routes is further shifted to AOCs, such restrictions need to be communicated to dispatchers so that they can consider the restrictions during their pre-flight planning. Furthermore, the dispatchers making these decisions will need flight planning tools that can take these constraints into consideration. In essence, in a free flight environment, dispatchers need to know the state and intentions of the ATM system. They will need ATC forecasts just as they need weather forecasts.

This view was expressed by a controller who participated in one of our studies:

Controller: "Traffic management would determine what routes would be inappropriate for free flight based on predicted weather and traffic, and would program the computer. The dispatcher would send in the flight plan two hours ahead of time and the computer would analyze it. Then the dispatcher would get an acknowledgment that that flight can't go today free flight. Then he could do better flight planning on it. The pilots would know before they leave that they're not flying free flight."

Such a view of the strategic aspects of free flight thus raises a number of important research and development questions:

1. Can tools be developed to help identify such routes or route segments? (This includes communication tools to improve input from regional traffic management centers as well as tools that automatically look at collected data.)

2. Can tools be developed to help disseminate such restrictions to AOCs so that they can plan more efficiently and effectively?

3. What tools or procedures would reduce the number of situations where such restrictions would be necessary?

4. If some segments of a flight are under free flight rules while others are not, how will transitions be handled once a flight is enroute?

Tactical decision-making—sharing intent information

Just as dispatchers need to know the intentions of the ATM system to make decisions when developing flight plans, pilots and controllers will need to know the intentions of the other aircraft in their immediate vicinity in order to decide whether a particular maneuver is acceptable. This sharing of intent information will be critical if enroute free flight is to be safely be supported with acceptably small separation distances.

This view was articulated by a number of controllers in studies that we conducted. In referring to a hypothetical scenario in which there was a loss of separation due to inadequate coordination between a flight crew and a controller, observations like the following were made:

Controller: "The overload was caused by the fact that he was surprised. Something happened that he wasn't planning on. An airplane's moving on its own without telling me about it. That starts the whole workload ball building up. What was a very manageable situation all of a sudden now gets out of hand."

Controller: "Today if this happens, I've got a surprise, I can handle it because I know what everybody else is doing. They're doing what they're supposed to be doing. But now there's a whole lot of question marks after all these call signs. In fact, that's how I would probably identify the guys who are in free flight. I'd have the computer put a question mark after his call sign because I'm not positive what he's going to be doing."

Thus, this need to share intent information relevant to tactical decision making points to a number of additional questions that must be addressed:

1. How can intent information be communicated and displayed in an efficient and effective fashion? Can intentions be adequately expressed in a form that can be "understood" by computerized conflict probe systems?

2. How will coordination be achieved to ensure timely, effective use of intent information? If a flight crew broadcasts a new intention, how (and by whom) will its acceptability be determined (based on a first-come-first served paradigm, based on some established "rules of the road", or based on discussion and negotiation between all of the flights involved)?

3. How will the controllers who are monitoring a situation determine when they need to intervene? How much detail must be provided by flights in communicating intentions, so that such decisions can be made effectively?

Maintaining adequate situation awareness

Knowing the current stated intentions of other aircraft provides only part of the picture that must be considered in making tactical decisions. To be effective, the decision-maker must have full awareness of the situation (including weather, the intentions of all aircraft, the actual behaviors of all aircraft, likely changes in the intentions of all aircraft, and available alternatives). Since it is not likely that computers will be able to reason sufficiently about such complex factors, flight crews will have to have maintain this full awareness, as will the controller who is monitoring the situation. Any recommendations or warnings provided by the computer will be based on an incomplete model of the situation, and hence can be treated as only one source of information to support decision–making by the crew and the controller.

This view was reinforced by an example provided by another of the controllers in our studies:

Controller: "Some of those things, I just don't see the computer being able to predict or really help me with. You have huge thunderstorms, and every airplane in the sky is diving for a hole. I don't know how we communicate that to the computer, and how it's going to help me do the conflict prediction through this hole."

The research questions motivated by this requirement include:

1. What are the circumstances where enroute free flight is feasible in the sense that flight crews and controllers can access and integrate all of the information necessary to make decisions?

2. What are the different classes of information needed? How should such information be exchanged and displayed to ensure adequate situation awareness among managers, controllers and operators?

Conclusions

The first theme of this paper is that, as the locus of control is shifted, access to the relevant information must also be shifted. At a strategic level, as AOCs are given increased authority to file flight plans, they will need access to much more information about likely constraints due to traffic bottlenecks and weather. Similarly, at a tactical level, while enroute, pilots and controllers will need full access to the information necessary to maintain situation awareness.

A second theme is that in some situations, decision-making requires reasoning about uncertainty and consideration of data that is not easily communicated to a computer. Both of these factors imply that humans will continue to have the responsibility for maintaining full situation awareness and for integrating all of the relevant information to make decisions. Computer systems with capabilities such as conflict detection and resolution will simply be some of many sources of information that the people in the system must consider.

A third theme is that (consistent with discussions by various organizations about concepts like "dynamic density") there will be circumstances where the flexibility offered under free flight (during pre-flight planning or while enroute) will have to be curtailed. The participants in our studies generally agreed that there were situations where enroute free flight was feasible:

Controller: "There's so many times I've got one aircraft that's overtaking another and I want to say: 'Offset a couple miles and go around him or descend 300 feet and pass him.' I'll tell him the intent is to stay level. And then I'll broadcast to the other aircraft: 'He's going to come by you and all that.' Then, that takes my complete concentration away from those aircraft so that I can work on more. ... This is one way of taking some of my workload and shifting it to the cockpit."

There are also, however, situations which require levels of coordination that are difficult or impossible without a more centralized locus of control. Thus, one of the challenges is to clearly delineate such situations, and to develop procedures for ensuring the safe, timely transition of control from AOCs to traffic managers (for strategic concerns) and from flight crews to controllers (for tactical concerns).

Finally, it is important to recognize that strategic and tactical decisions are not in actuality independent. Dispatchers and traffic managers will have to know how the impacts of their strategic choices will affect tactical decision-making. Similarly, flight crews and controllers will need a better understanding of how their tactical decisions influence the overall strategic picture. Thus, as suggested in Figure 4, dispatchers, traffic managers, flight crews and controllers will need a much better understanding of how various system components and decisions interact, and will need to share better real-time information to make use of this knowledge.

Acknowledgments

This work was funded by the Advanced Air Transportation Technologies program at NASA Ames Research Center under Grant Number NAG2-995. We would like to express our appreciation to Judith Orasanu and to the controllers, dispatchers, pilots and traffic managers who assisted us in completing the work.

References

Denning, R, Smith, PJ, McCoy, E, Orasanu, J, Billings, C, Van Horn, A, Rodvold, M (1996): Initial Experiences with the Expanded National Route Program. *Proc. Human Factors and Ergonomics Society,* 40th Annual Meeting, 1996, 98-101.

RTCA (1995). *Report of the RTCA Board of Directors' Select Committee on Free Flight.* Washington, DC: RTCA, Inc., January 18, 1995.

Smith, PJ, McCoy, CE, Layton, C (1997): Brittleness in the Design of Cooperative Problem-Solving Systems: The Effects on User Performance. *IEEE Transactions on Systems, Man, and Cybernetics–Part A: Systems and Humans* 27(3) 360-371.

Smith, P, Billings, C, Woods, D, McCoy, E, Sarter, N, Denning, R, Dekker, S (1997): Can Automation Enable a Cooperative Future ATM System? *Proc. Ninth International Symposium on Aviation Psychology (Vol. I)*, 1481-1485.

Smith, PJ, McCoy, E, Denning, R, Obradovich, JH (1997): Cooperative Problem-Solving in the Air Traffic Management System. *Proc. of 1997 IEEE Symposium on Systems, Man and Cybernetics*, 3137-3141.

Smith, PJ, McCoy, E, Orasanu, J, Billings, C, Denning, R, Rodvold, M, Gee, T, Van Horn, A (1997): Control by Permission: A Case Study of Cooperative Problem-Solving in the Interactions of Airline Dispatchers with ATCSCC. *Air Traffic Control Quarterly* 4(4) 229-247.

Smith, PJ, Woods, D, McCoy, CE, Billings, C, Sarter, N, Denning, R, Dekker, S (1998): Using Forecasts of Future Incidents to Evaluate Future ATM System Designs. *Air Traffic Control Quarterly* 6(1) 71-86.

MAN-IN-THE-LOOP PART OF A STUDY LOOKING AT A FREE FLIGHT CONCEPT

J.M. Hoekstra, R.N.H.W. van Gent, R.C.J. Ruigrok,
National Aerospace Laboratory NLR, Amsterdam, Netherlands

ABSTRACT

The paper describes the Man-in-the-Loop part of a study looking at a Free Flight concept with detection and resolution of conflicts during cruise flight by means of airborne systems. The overall study included fast-time simulations to define a base-line concept, a safety analysis, and a Man-in-the-Loop simulator experiment to investigate human factors issues. Part of the human factors study was the integration of traffic information, conflict detection and conflict resolution advisories in the displays. Based on the results of the three sub-studies, the feasibility of the Free Flight concept as defined in this study, could not be refuted.

1. Introduction

1.1 Self Separation

Free Flight is an Air Traffic Management concept, which aims at a safe and more efficient airspace usage by self-optimization. The Free Flight allows the operators more freedom to select their path and speed in real-time (Ref.1). Two main elements are associated with Free Flight:

- Direct Routing
- Airborne Separation

In this paper the term Free Flight refers to the concept including both elements, not just direct routing.

The two elements are closely related because the traffic situation resulting from direct routing might be very hard to monitor from the ground. This is illustrated by the figures below.

With present day ATC, traffic is organized in a way that the air traffic controller can easily see where problems might occur. Figure 1A shows this in a schematic way.

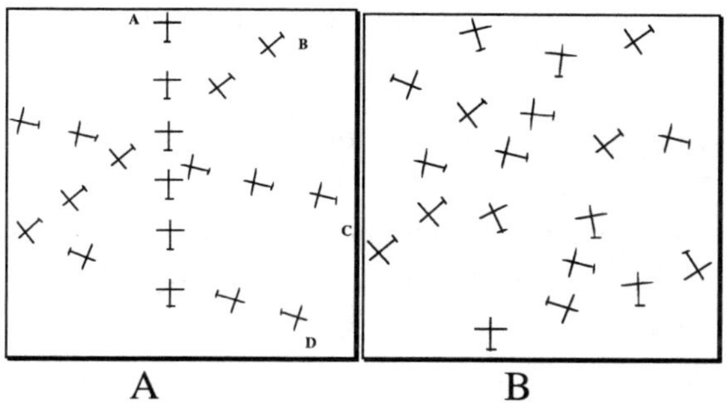

Figure 1 Schematic representation of how airways aid organization of traffic

Using the airways A, B, C and D and by imposing altitude restrictions to the different airways a large amount of traffic can be controlled by a single air traffic controller. These airways and altitude restrictions aid the controller in maintaining a mental model of the situation. However these airways and altitude restrictions deny the aircraft the possibility of flying their optimal direct route. If each aircraft would fly their optimal route, the picture would look more like Figure 1B.

Figure 1B shows a much less structured global overview and it is expected that ensuring separation of all traffic could become problematic for a single controller. This is because the controller has to identify and resolve for each aircraft the potential conflicts.

0-7803-5086-3 /98/$10.00

He or she then has to solve all problems without creating new ones. Additionally he has to do all of this within a short time span.

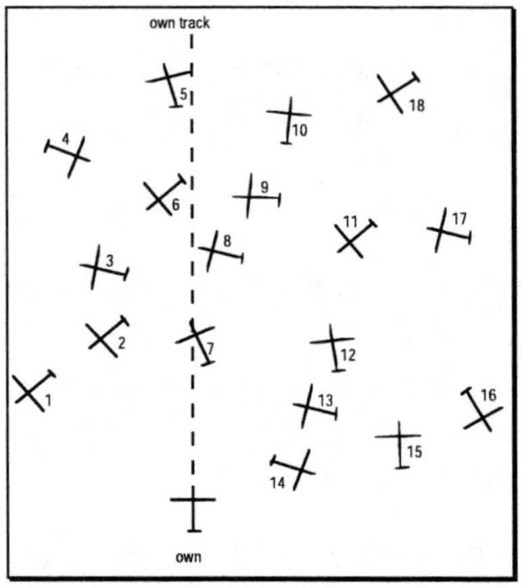

Figure 2 The Airborne perspective

The same picture shown from the perspective of an aircraft seems less problematic.

Figure 2 shows the problem as seen from one aircraft named "own". The first assumption is that all aircraft can "see" all other aircraft. This means that all aircraft are properly equipped (e.g. ADS-B and EFIS traffic display).

1.2 Goal of study

NASA has requested the NLR to study the human factors of self-separation. Specifically the probably increased workload was to be addressed as well as MMI issues like display design and alerting. For this study a simulator experiment has been carried out. However due to the lack of definition of how Free Flight was to be implemented first the concept had to be designed in more detail (Ref. 2). In the conceptual design phase of the project the rules-of-the-sky were defined and a safety study was performed to evaluate the chosen set of rules-of-the-sky. (Ref. 7)

One of the assumptions for the human factors study was that the crew was flying in a dedicated Free Flight area with only aircraft fully equipped with an Airborne Separation Assurance System (ASAS) and a Cockpit Display of Traffic Information (CDTI). The study described here was limited to cruise flight and does not cover transition issues, which occur when descending into controlled airspace.

2. Task analysis

2.1 Controlled vs. Free Flight

In Figure 3 in the main tasks when flying in controlled airspace and in the Free Flight situation are shown.

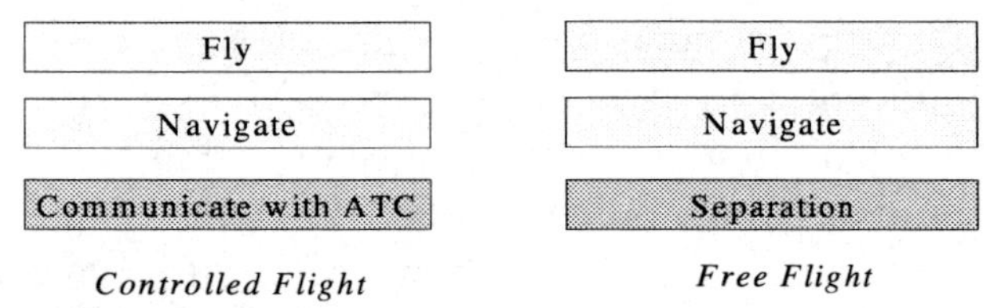

Figure 3 Comparison main tasks

When comparing the Free Flight situation with the controlled flight situation it is apparent there is mainly a change of tasks and not an extra task. The separation task mainly replaces the task of communication with ATC.

The communication task comprises of the following subtasks.

Table 1 Communication subtasks

Communicate with ATC
Reports
Requests
Monitoring for ATC calls
Responding to ATC calls
Monitoring traffic situation via R/T
Inter-traffic R/T

The workload associated with the sub-tasks varies depending on flight phase and the

ATC situation. During the arrival and approach and in areas with no or poor radar coverage the subtask 'Monitoring traffic situation' will be more demanding.

The tasks associated with Airborne Separation are listed in table 2.

Table 2 Separation subtasks

Separation
Monitoring traffic situation on display
Conflict detection
Conflict resolution
Inter-traffic R/T

In this list the Monitoring and Conflict detection tasks will be most demanding since the other sub-tasks are rarely executed. Typical frequency of conflicts per hour per aircraft is low. In a study by DERA (UK) (Ref. 3) a frequency of 881 conflicts per hour for 1400 aircraft with a separation of 5 nautical miles has been found for an average Western-European traffic density. This means 1762 aircraft per hour are involved in a predicted conflict and have to take action. For one aircraft this means an average of one conflict per 48 minutes. This was confirmed by NLR off-line traffic simulations indicating a maximum of one conflict per 20 minutes for traffic densities equal to three times the average Western-European density.

In the next sections the man-machine interface changes to aid the pilots in this new task will be discussed per subtask.

2.2 Traffic Monitoring

The purpose of the task of traffic monitoring is to increase the situational awareness. The mental picture of the traffic situation is used for several purposes: monitor long term dangerous situation, which might lead to too high traffic densities and conflicts, to know the maneuvering space and to correlate R/T information (e.g. turbulence reports) to position.

For monitoring the traffic situation the following data are necessary:

Table 3 Traffic data

Data for traffic monitoring
Position (latitude, longitude)
Altitude
Vertical speed
Ground speed
Ground track

These data are available through ADS-B (Ref.4).

Figure 4 Traffic symbol for navigation display

To display these data the TCAS symbology is not sufficient since it does not indicate track. Adding a track line will clutter the display more than using an arrow shaped symbol and therefore the symbology shown in Figure 4 has been adopted. It was shown on the navigation display instead of using a separate display for traffic. The symbol indicates only the sign of the vertical speed by using an arrow next to the traffic symbol. The altitude and ground speed is shown digitally above the traffic symbol.

2.3 Conflict detection

The purpose of the task of conflict detection is to predict an intrusion of the protected zone (Ref. 1) in the near future. The dimensions of the protected zone, shown in the figure below, are chosen to be equal to today's separation minima (with vertically 1000 ft for all levels).

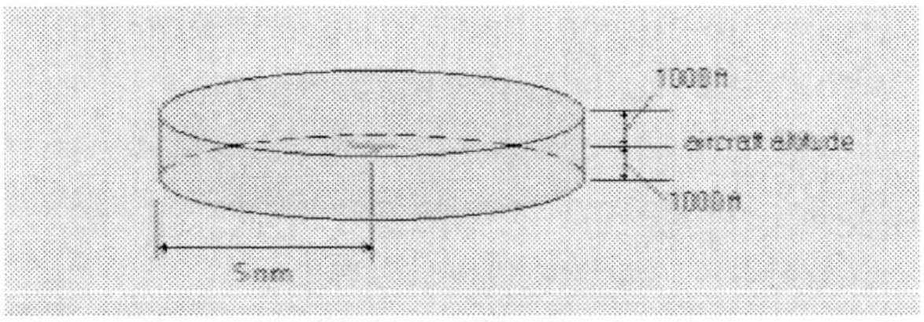

Figure 5 Protected zone dimensions

The actual computation of whether the current position and speed leads to a conflict in the near future is a task better fit to a computer than to a human. This task is therefore allocated to the conflict detection module, using the data listed in Table 3.

For this computation the near future is determined by the so-called look-ahead time. This look-ahead time should be as long as possible to allow more time to maneuver. On the other hand the longer the look-ahead time, the more false alarms will be generated due to the deteriorating quality of the predicted position of the traffic. Off-line traffic simulations indicated a look-ahead time of 5 minutes to be the optimum.

For the crew two alerting levels were defined: a predicted conflict between 5 and 3 minutes will generate a standard conflict alert and a conflict between 3 and 0 minutes will generate an urgent conflict alert. The conflict alert consists of a sound and the conflict symbology on the navigation display and a light next to the warning/caution lighted buttons.

The symbology was chosen to indicate the geometry of the conflict, since the conflict geometry forms the basis for resolution algorithm used (see next section).

2.4 Conflict resolution

From off-line traffic simulations using several traffic densities one resolution algorithm was chosen (Ref. 2). This algorithm is referred to as the modified voltage potential algorithm (Ref. 6). The term voltage potential is somewhat misleading since the actual resolution algorithm does not use 'fields' but is a very pragmatic way to solve a conflict. Figure 6 illustrates the method.

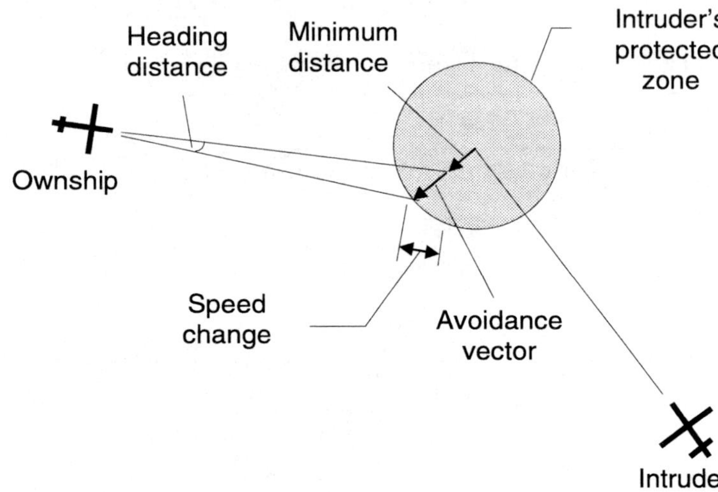

Figure 6 Resolution algorithm

The minimum distance position forms the starting point for the algorithm. When a conflict is predicted for two aircraft, the ownship will be inside the predicted protected zone of the intruder. To solve the conflict the predicted position of the ownship should be moved to the edge of the protected zone. The quickest way out of the protected zone is in the direction of connecting line of the intruder's predicted position to the ownship's predicted position. The vector to the edge of the protected zone forms the so-called avoidance vector, which in this case can be translated in a heading change and a speed change. For the intruder the avoidance vector will by definition have the other sign (since the role of ownship and intruder is reversed). Though both aircraft initially assume the other aircraft will not maneuver, this method is based on co-operative maneuvering. Advantages of this method are:

- **Effective**: solves conflicts
- **Efficient**: in terms of fuel and time
- **Fair**: distributes effort & costs among participants equally
- **Transparent**: Resolution understandable

- **Fail safe**: one aircraft can solve conflict

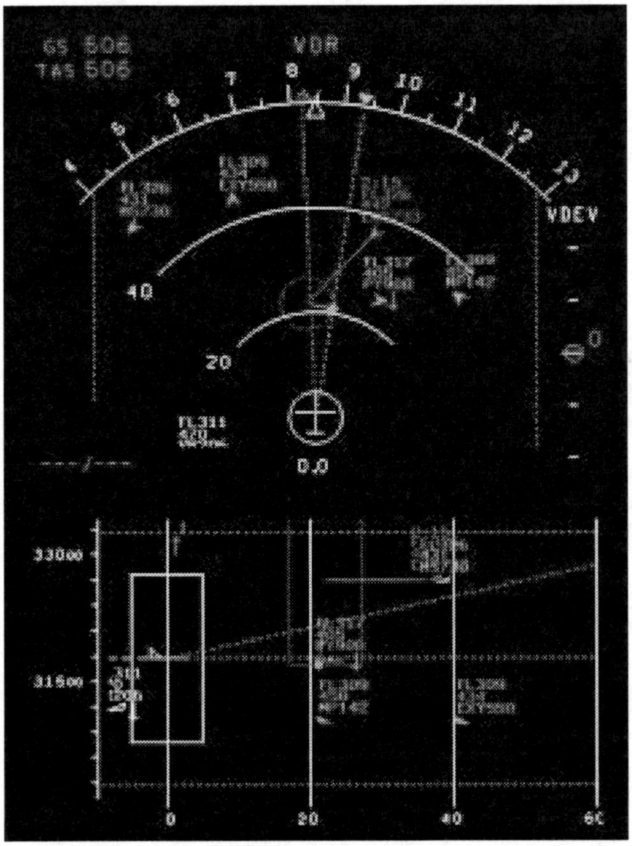

Figure 7 Conflict & Resolution symbology on navigation display

The symbology on the display is shown in the figure above. A vertical display was added to aid the pilots in understanding all three dimensions of the conflict geometry.

To aid the pilots in interpreting the display, the traffic symbols are color coded to indicate the range rate: white traffic is moving away, while cyan aircraft are getting closer. The red and amber symbology indicates the two alerting levels of the conflict. An aircraft will stay red or amber for a certain time to enable the pilots to identify the interesting aircraft even when the conflict was only detected during a very short period.

The extra display parameters controlled by the pilots were vertical range (also used for clipping traffic of the horizontal display) and declutter switches per line of the label.

Because of the cylindrical shape of the protected zone there are two options available: the horizontal and the vertical projection of the resolution vector. The resolution vector is translated to magenta bugs on the heading scale (indicates advised track), the vertical speed scale on the PFD and the speed scale on the PFD. The advised track & path lines are also shown in magenta on the navigation display.

The execution of the resolution can be performed manually by setting the advised values in the mode control panel. In the experiment also the possibility of automatic execution of the resolution by pressing a ‘load resolution’ button on the mode control panel. This option was offered separate and combined for the horizontal and vertical maneuver. This results in three options for resolution execution: manual, separate and combined. (see also ref. 8)

2.5 Inter-traffic R/T

To enhance situational awareness the call sign of the traffic is shown in the label of the traffic symbol on the navigation display. The main purpose is to enable the crew to communicate with an aircraft via the inter-traffic frequency. It also facilitates crew communication, because it serves as a unique reference to the traffic symbol.

3. Simulator Experiment

3.1 Experimental design

The independent variables for the experiment were:

- Traffic density: single, double and triple
- Resolution execution: manual, separate and combined

The resulting set of nine runs has been carried out under nominal conditions but also under non-nominal conditions. The non-nominal conditions were planned to explore several issues concerning fail-safe and man-machine interaction. The non-nominal situations were:

- Failure of ownship ASAS
- Failure of intruder ASAS
- Delaying factors in resolution

Of each category three variants were used resulting in a total of nine non-nominal situations. The place of the non-nominal situation in the experiment was varied between the crews.

The resulting experiment matrix consists of two times a 3 x 3 matrix, so a total of eighteen runs per crew. Eight crews, consisting of airline pilots, participated in the experiment (average experience 5500 hours).

3.2 Traffic Scenarios

The same program as used in the off-line simulation generated the traffic environment: the so-called traffic & experiment manager (TEM). The TEM simulated and recorded a traffic pattern using up to eighteen airports and en-route entry points. These recorded traffic scenarios contained traffic densities equal to one, two and three times the average Western European traffic density. On top of that extra aircraft were added to the scenario to ensure even more predicted conflicts for the crew in the simulator. The result was that crews would experience nine times as much conflicts in the triple density as in today's European airspace. Every flight lasted for twenty minutes, in the triple density they would typically encounter three predicted conflicts. About 300 aircraft were simulated in the experiment area simultaneously.

3.3 Configuration

The configuration (see Figure 8) consisted of three connected elements:

1. The research flight simulator with the subject crew.
2. The AIRSIM desktop simulator to take over an intruder in non-nominal cases.
3. The traffic & experiment manager (TEM) to generate the traffic scenarios

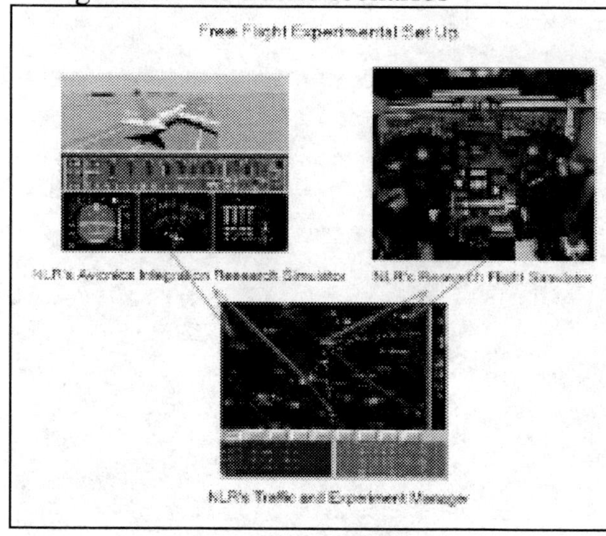

Figure 8 Experiment configuration

The research flight simulator consists of a full-glass transport cockpit with FMS, visual system and moving base. It was configured as a Boeing 747-400 cockpit for this experiment.

The AIRSIM desktop simulator is a simplified version of the host flight simulation program, fitted with EFIS, out-of-the-window view and a software mode control panel operated with the mouse.

3.4 Data gathering

Four types of data were gathered during the experiment:

- **Simulation data** recording: crew actions, aircraft parameters, traffic situation parameters etc.
- **Physiological measurements**: eye tracking, pupil diameter, respiration rate and heart rate variability.
- **Questionnaires** with questions concerning MMI, workload, safety and acceptability
- **Video recordings** of cockpit instruments and traffic situation.

A selection of the results is given in the next section.

4. Results

4.1 Workload

At this moment for workload only the questionnaire data is available. The workload has been rated on a Rating Scale of Mental Effort (RSME) of 0 –150. The figure below shows the effect of resolution execution on workload, the largest effect observed.

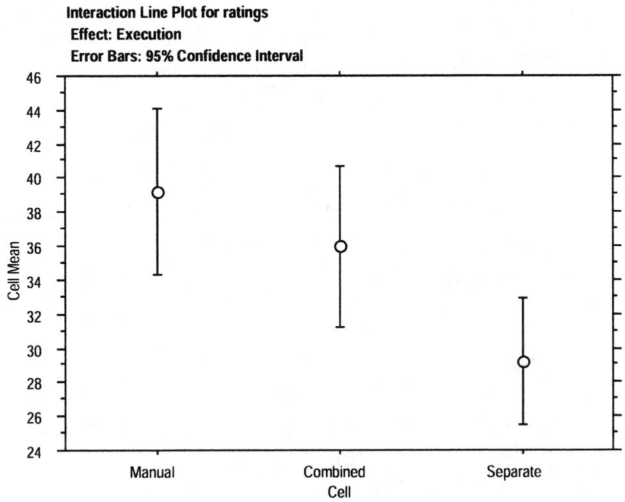

Figure 9 Subjective workload on a scale of 0-150

4.2 Acceptability

The subjective acceptability was rated using a scale of 1 to 5:

1 = Completely Unacceptable
2 = Unacceptable
3 = Acceptable
4 = Favorable
5 = Perfect in every way

The acceptability ratings for the flights show an effect of traffic density (see Figure 10).

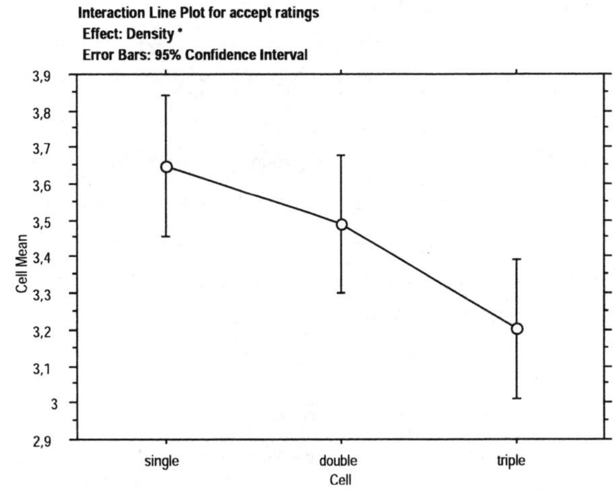

Figure 10 Acceptability ratings for three traffic densities

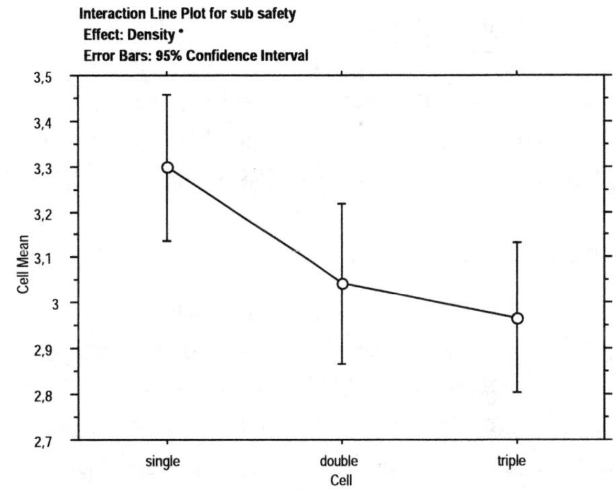

Figure 11 Subjective safety perception as function of traffic density

The subjective safety rating also use a scale of 1-5 to compare the impression of safety to today's controlled flights:

1 = ATC much safer
2 = ATC safer
3 = same as ATC
4 = FF safer than ATC
5 = FF much safer than ATC

The largest effect on subjective safety seen was traffic density (see Figure 11).

4.3 Eye tracking

The eye-tracking data showed that both the pilot flying and the pilot not flying spent 47 percent of the time watching the navigation display with the traffic. Even though the conversations took place just as in normal cruise flight and the workload was rated normal, the moving traffic pattern apparently provides an interesting picture.

5. Discussion

The expected increase in workload was not confirmed by the subjective workload measurements. The workload ratings show an average of 29 on the RSME scale for separate resolution loading. The RSME scale has been used in other experiments were the base-line rating for (ATC controlled) cruise flight was 27. This means the experienced workload under Free Flight conditions with the ASAS system as described is comparable the workload in controlled flight during cruise, even when the traffic density is high.

Even when flying in high-density traffic after 79 percent of the flights the crews rated Free Flight with this ASAS system as acceptable or better. The safety was regarded the same as today or better in 71 percent of the flights. This includes non-nominal events like ASAS failures or other aircraft not obeying the rules and even an emergency descent through the protected zone.

As a result of remarks by the subject pilots, after this experiment a predictive ASAS has been developed, showing the forbidden track angles, vertical speeds and speeds, which might lead to a short-term conflict. This will be tested in the next man-in-the-loop experiment.

6. Conclusion

None of the three sub-studies, conceptual design, safety analysis and man-in-the-loop simulations could refute the feasibility of the Free Flight concept.

7. References

1) RTCA, Report of the Radio Technical Commission for Aeronautics (RTCA) Board of Directors' Select Committee on Free Flight).
2) Hoekstra, Van Gent, Ruigrok, Conceptual Design of Free Flight with Airborne Separation Assurance, NLR, Paper AIAA-98-4239
3) Magill, Trajectory Predictability and Frequency of Conflict Avoiding Action, DERA (UK),Paper 34, CEAS Free Flight Symposium 1997
4) Website: http://iimorrow.com/caa_ads/adsb-info/mode_s/mode_s_presentation
5) Eby, Martin S. , A Self-Organizational Approach for Resolving Air Traffic Conflicts, The Lincoln Laboratory Journal Vol. 7, Nr. 2, 1994
6) Hoekstra, Van Gent, Ruigrok., Off-line Validation of a Conceptual Design for Free Flight in Cruise. NLR CR 1997.
7) Daams, J. et al., Safety evaluation of an initial Free Flight scenario with TOPAZ Draft version, NLR, 1997.
8) http://www.nlr.nl/public/hosted-sites/freeflight/ffdisp.html (leave '-' in!)

ABBREVIATIONS

ASAS Airborne Separation Assurance System
AIRSIM Avionics Integration Research SIMulator
ADS-B Automatic Dependent Surveillance - Broadcast
EFIS Electronic Flight Instrument System
FMS Flight Management System
MMI Man Machine Interface
NASA National Aeronautics and Space Administration
NLR Nationaal Lucht- en Ruimtevaartlaboratorium
R/T Radio Telephony
RSME Rating Scale for Mental Effort
RTCA Radio Technical Commission for Aeronautics
TEM Traffic and Experiment Manager

G. System Design Track - Aircraft Systems

The Aircraft Systems Track concentrates its focus on four areas of avionics development that are being pursued: "Open Systems" avionics development; In-flight Entertainment (IFE) development; Satellite / Aircraft operations development; and Unmanned Air-vehicle development. "Open Systems" involves the utilization of Commercial Off The Shelf (COTS) hardware and software in what has previously been custom applications. IFE involves what will be the heaviest, most-power consuming, most expensive and most "profitable" avionics system on any transport. Aircraft / Satellite operations are revolutionizing Aircraft CNI and this session will cover some of these changes. Unmanned Air Vehicles are being considered for more and more missions and have greatly modified aircraft design; this session will describe some of these.

Track Chairs: Erv Ulbrich, Boeing; Steve Grothe, Honeywell

Track G Aircraft Systems

Session G1: InFlight Entertainment

Chairs: Ana Christofferson and Craig Depner, Boeing

1	Total Integrated Management (TIM) (Not Published)	Mr. Joseph R. Winston
2	Availability, Reliability, and Maintainability Prediction Model	Mr. James M. Hansen
3	Reality in Certification of IFE and Telephone Equipment	Mr. Ritch L. Triplett
4	A New Approach to Data Communications Utilizing the North American Terrestrial System (Not Published)	Mr. Mike O'Meara
5	Real-World Implementations of Intra-Aircraft Infrared Data Communications Networks	Mr. Patrick Potega
6	In-Flight Entertainment - Getting From Wishlist To Reality	Mr. Donald B. Lee
7	High Speed Networking of Multimedia Data for Passenger Entertainment (Not Published)	Mr. Greg Henrikson
8	Packet-based Networks	Mr. Steven W. Russert

Session G2: Aircraft & Satellite Avionics

Chairs:George Cobley, Rockwell Collins; Orville Nyhus, Honeywell

1	Airborne Reception of Data and Direct Broadcast TV using a Phased Array Antenna (Not Published)	Mr. David C. Vacanti
2	Guidance Characteristics of GNSS Landing Systems	Dr. Alex Stratton
3	Managing Aircraft Airworthiness Through Information Technology (Not Published)	Mr. William E. Larsen
4	Pro Line 21 Advanced Avionics System Architecture	Mr. Timothy Rayl
5	Techniques for Improved Reception of 1090 MHz ADS-B Signals	Mr. William Harman
6	Enhanced Ground Proximity Warning System (Not Published)	Mr. Barry C. Breen

Track G Aircraft Systems (Continued)

Session G3: Commercial-Off-The-Shelf & Open Systems

Chairs: Glen Logan, Open Systems Joint Task Force;
Greg Solowey, GEC-Marconi Hazeltine

1	COTS Based Open Systems for Military Avionics	Mr. John Paul
2	Reducing Avionics Software Cost Through Component Based Product Line Development	Mr. David Sharp
3	Introduction to the Electronic Power Specification Standardization Activity	Mr. Sergio Navarro
4	Open System Design for CNI Avionics	Dr. George Mitschang
5	The Application of Commercial Processing Technologies to the Airborne Military Environment	Mr. Hugh S. Perry
6	Telecom Technologies (Not Published)	Brian Graber

Session G4: Unmanned Airborne Vehicles

Chairs:

1	Redundant Control Systems for UAV's (Not Published)	Graham Gyattt
2	Flight Testing Perseus-B (Not Published)	Mr. Tom Clancy
3	Flight Testing a Large, Autonomous, Unmanned Aircraft (Or Global Hawk Begins to Soar) (Not Published)	Robert Ettinger
4	Flight Testing DarkStar (Not Published)	Mr. John Straub
5	X-36 Tailess Agility Aircraft Subsystm Integration (Not Published)	Willard J Harris
6	X-36 Tailless Agility Aircraft Subsystems Integration (Not Published)	Mr. Bill Harris
7	X-38 Avionics System (Not Published)	Christopher Nagy

IFE AVAILABILITY PREDICTION MODEL

By James M. Hansen, Scitor Corp., Colorado Springs, Co

Introduction

This paper will provide the reader with information about a software model designed to predict availability of an In-Flight Entertainment (IFE) system. This paper will discuss the structure of the model, concept of operations, and future opportunities.

Description of an In-Flight Entertainment System

We will first define what comprises an In-Flight Entertainment (IFE) system. The purpose of the IFE system is to provide entertainment to passengers on long flights via video and audio programs. The video usually contains movies, news programs, short stories, and television programs. The audio has a selection of multiple channels with a variety of programs. Future systems will contain on-line shopping, games, gambling, and other personal computer related activities.

The hardware components necessary to deliver video and audio entertainment to passengers includes a video display, a handset controller, audio jacks for sound, computer process to handle data conversions and data requests, several high volume distribution boxes, and content delivery and storage units.

Model Structure and Processes

The IFE Availability Prediction Model has been designed to assess IFE system availability for a fleet of aircraft using a predefined flight schedule. This assessment is accomplished by simulating IFE unit failures and the time and resources required restoring the IFE system based on a specified maintenance and logistics support system. The impact of each failure of a Line Replaceable Unit (LRU) is obtained from a Failure Mode and Effects Analysis (FMEA) of the IFE system configuration to determine the resulting impact on seat availability. An overview of the Prediction Model is depicted in Figure 1. The model can be viewed as consisting of the following major elements:

1) Data Collection – This involves significant research and analysis of the data used to populate the model with the existing IFE system. This includes obtaining information on system design and impact of failed units.

2) Model Inputs - The basic inputs to the model are IFE configuration data and LRU Mean Time Between Failures (MTBF) and Mean Time to Restore (MTTR) data, airline schedules, definition of the maintenance and logistical support system and concept.

3) Fleet Activity Model – Use of the airline schedules to determine the duration of flights and the ground time available for maintenance.

4) LRU Failure Model - Generates LRU failures based on flight duration and LRU MTBF and maintains the status of LRU failures on each airplane. The probability of an LRU failure is 1 - probability of success = $1 - e^{-t / MTBF}$ where {t} is the flight duration.

5) Airport/Maintenance Facility Process - Models the process of restoring IFE systems when an airplane is at an Airport with maintenance and spare LRUs available. As depicted in Figure 1, the process is accomplished by performing any required maintenance on a first in first out (FIFO) basis by allocating all required

0-7803-5086-3 /98/$10.00 ©1998 IEEE

maintainers and spare LRUs to the first arriving airplane and then the next arriving plane is allocated remaining resources if any overlap exists. The model maintains a log of resources (maintainers available by skill level, quantity of spares remaining). Spare LRU re-stocking is simulated using a LRU replenishment time based on LRU type and Airport location.

6) Generate Model Outputs - Computes and maintains LRU failure data, resource utilization (maintainers, LRU spares), and seat availability and degradation each flight segment. Output reports are generated using the incremental outputs that are computed on arrivals and departures.

7) Report Customization – Customer specific reports based on weekly and monthly data points provide trending data for overall system performance.

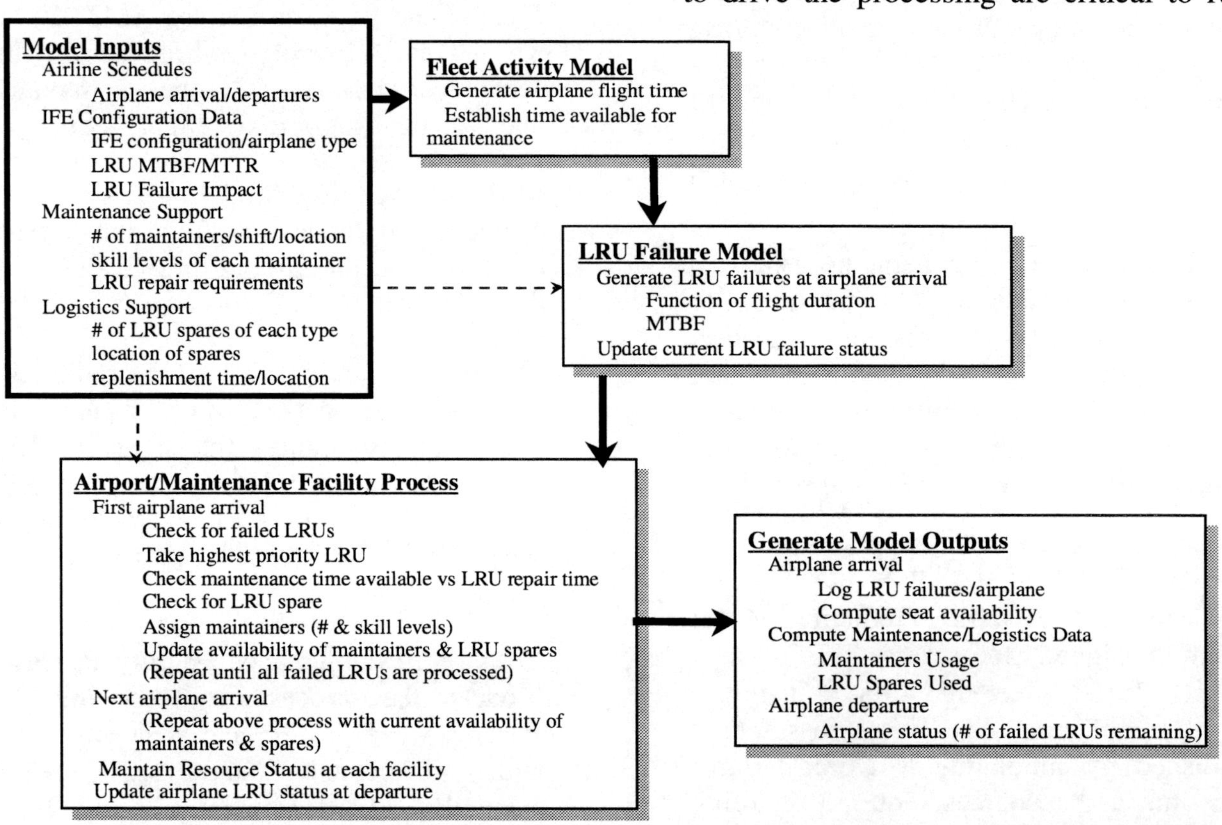

Figure 1: Overview of IFE Seat Availability Model

Concept of Operations

The concept of operations for the IFE Availability Prediction Model describes the following three processes:

1) Generation and preparation of model inputs.

2) Model operation.

3) Model report output generation and analysis.

Generation and Preparation of Model Inputs

The generation and preparation of data to drive the processing are critical to running the model and obtaining valuable results. The model has been structured to input data into various tables. IFE configuration data, LRU

MTBF (Mean Time Between Failures), and MTTR (Mean Time to Replace) data, airline flight schedules, definition of the maintenance and logistics are the basic set of input data required. Seating configuration for each airplane type is also required. The IFE configuration and LRU functions forms the basis for determining the impact of each LRU failure and corresponding impact on seat availability. The IFE configuration is used to establish the number of unique LRUs and the quantity of each LRU type on an airplane. MTBF and MTTR data for each LRU, plus the impact of LRU failures are the basic inputs required for the model. The MTTR includes the definition of the maintainer requirements (number and skill levels and time). The overall process to generate IFE LRU inputs for the model is depicted in Figure 2.

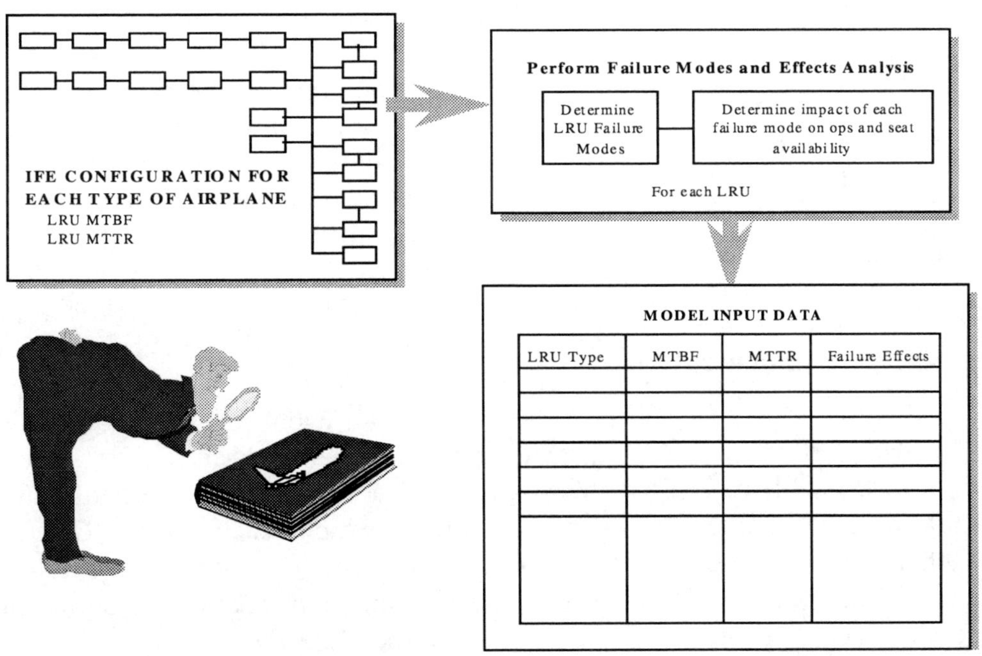

Figure 2: Overview of Process to Generate IFE LRU Model Inputs

Model Operations

Model operations will be described by first discussing what functions the model was designed to perform. Next, the process flow will be examined, along with a breakout of the functional flow. The final topic will discuss how the airport and maintenance processes were implemented.

Model Scope

This model has been designed to assess IFE system performance in terms of seat availability, system degradation, spares usage, and maintenance hours for IFE systems. Given a particular IFE system configuration and reliability, the model provides a way of evaluating alternative logistics support concepts. Evaluation measures consist of seat availability, spares usage and man-hours used by maintainers at each line station.

The model is based on standard computational methods for simulating failures and LRU restorals. Additional features of the IFE Seat Availability Model are the use of an airline schedule to model the operational characteristics and the use of seat availability and system degradation to evaluate the performance of the IFE system. The table driven design allows for flexibility and expansion to accommodate variations in fleet size, system configuration, line stations, and maintenance philosophy.

The IFE availability prediction model is based on the following concepts:

1) The IFE system is operating (fully powered on) from scheduled departure time to arrival time.

2) IFE LRU failures are based on constant failure rates; i.e., infant mortality failures and end of life failures due to mechanical

wear-out are not included. An LRU failure is independent of other LRU's failures; i.e., a failure in one LRU will not cause another LRU to fail.

3) Airline maintenance concept is that no maintenance (remove/replacement of LRUs) is performed while the airplane is in flight. Maintenance is only performed at Airports with qualified maintainers and with spare LRUs. No swapping of failed LRUs with operational LRUs within an airplane will be performed to fix a failure.

4) Allowable maintenance time at an Airport is assumed to be the time span from airplane arrival to departure minus passenger unloading and loading time. Airplane departure times will not be delayed to allow for IFE LRU replacements.

5) The model has been developed to enable seat availability to be defined to conform to an airline and IFE supplier contract specified definition of seat availability. Seat availability can be defined as any functional outage to a seat, which results in that seat or seats being unavailable. Seat availability is computed for each flight segment.

Airline schedule for the fleet of airplanes to be used in the model is established using the published schedule and assigning airplane tail numbers to each flight segment. The key input is the association of tail numbers to each flight. The model tracks each airplane using the tail number as the identifier and routes the airplane to each Airport using the airline schedule.

The maintenance and logistics support system and concept are specified in the model by defining the availability of LRU spares, number of maintainers of each skill level, LRU replenishment strategy and replenishment cycle times for each Airport facility. The model can also be run assuming there are excessive resources available at each designated Airport. The resulting model outputs will show the number of spares consumed at each Airport and the number of hours of maintainer time used.

Model Flow

The model simulates a fleet of airplanes, each operating an onboard IFE system over a scheduled span of time. To obtain statistically valid outputs from the model, the span of time to be simulated, and the number of times the model is run are dependent on LRU MTBF data. The cumulative flight hours of a single run of the model should nominally be 10 times the largest MTBF. The LRU failure and seat availability output data should be averaged over a minimum of 4 runs. For most IFE configurations, the six-month period used exceeds the 10 time the largest MTBF rule.

The arrival event of the model triggers the calculation of LRU failures for a particular flight segment. Failures are simulated for each IFE LRU on the arriving airplane using Monte Carlo random numbers to determine if an LRU has failed during the flight segment. The probability of an LRU failure is 1 - probability of success = $1 - e^{-t / MTBF}$ where **t** is the segment flight duration.

The arrival event of the model also triggers the maintenance cycle. The start of this cycle is a replacement identification process during which each failed LRU is examined to determine if priority, available spares, ground time and manning resources are sufficient at the line station to initiate a replacement. The highest priority failed LRU is assigned available resources first; then the next highest priority failed LRU is assigned the remaining resources and so on. If ground time is not long enough to complete the LRU replacement, the failed LRU is not restored; in no case will departure time be delayed for LRU replacement. Replacement status, spares, maintenance time and resource usage are calculated and logged. LRU replenishment computations are made prior to each arrival event. The seat availability is calculated at the time of arrival and as the aircraft departs. The difference between these two calculated values determines the system degradation.

The model treats departure of an airplane as a second event of the airplane / line station encounter and determines the departure status (departure seat availability) based on the replacements accomplished during ground maintenance. Full seat availability may not exist on departure if all failed LRUs have not been replaced.

The model works on a first in first out basis using flight arrival times. The model is serial event driven, the first airplane in a series of airplanes is assigned all necessary resources, the start time of the next maintenance cycle is determined by arrival time and spares availability and worker utilization of previous airplanes repair actions. Each subsequent arrival and departure at an Airport per the flight schedule input data is processed in a like manner.

LRU failures generated from the previous flight segment are assumed to have failed for the duration of the flight segment for computing the IFE seat availability for the previous flight segment. The Model computes seat availability assuming that an LRU failure that impacts any functionality at the seat makes the seat unavailable. The model calculates arrival status of an aircraft (arrival seat availability) using the simulated LRU failures with resulting impact on seat availability and the seat availability of the airplane at the start of the flight segment (the computed seat availability at departure time at previous Airport).

Model Logic Flow

The IFE Availability Prediction Model logic flow depicted in Figure 3, illustrates the processing that occurs as the model steps through the flight schedule. This high level flow identifies all the major processing that occurs during the model's operation.

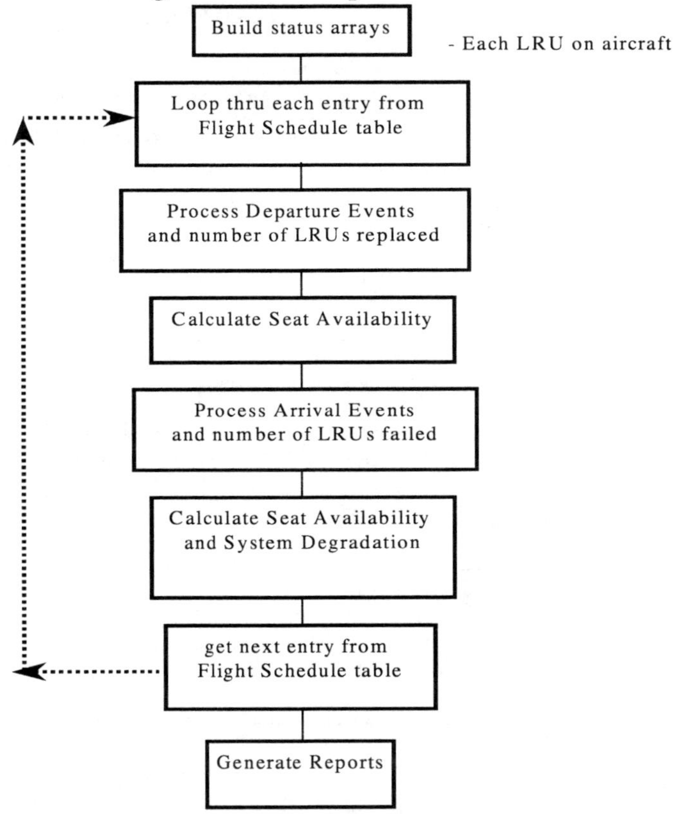

Figure 3: Process Flow of Model Operation

Airport / Maintenance Process

The IFE Availability Prediction Model processes the following activities during an Airport event. As the aircraft arrives at a given Airport, the simulation determines if any LRUs have failed. The resulting seat availability impacts are determined and recorded. The model then applies the available resources (workers and spares) to repair the prioritized LRUs. If the repair can not be completed, the configuration of the aircraft is logged, and the failure will persist until it can be repaired. This process is outlined in Figure 4.

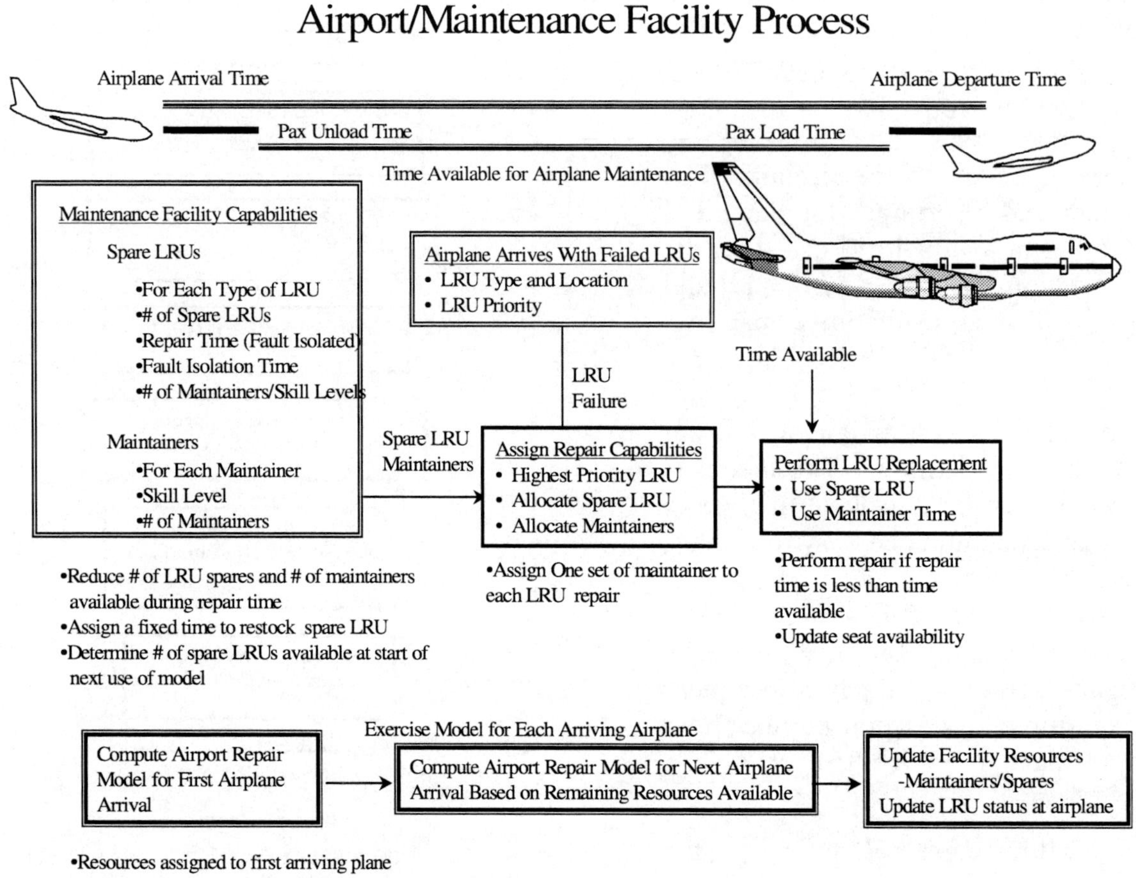

Figure 4: Airport/Maintenance Facility Process

Model Report Output Generation and Analysis

Model execution results from the logged data can be presented in several forms including:

1) by Airport and Line Station

2) by Aircraft Configuration Type

3) by Aircraft Configuration Type and Passenger Seating Configuration

4) by the Airline Fleet

5) by Service Class

Several different reports can be processed from the stored configuration and history data allowing reports oriented to specific criteria such as spares levels, manning levels, or locations.

The basic output of the model consists of tabular and graphical reports that are used in an analysis process. These reports provide insight into seat availability, system degradation, ground time sufficiency, spares, and manning resource usage, and fleet configuration data.

The current set of reports can be divided into three categories:

Baseline configuration data:

The baseline configuration reports provide insight into the data for each of the aircraft configurations, the flight and ground times, and the line stations used for sparing. This data provides a basis for evaluating the remaining reports and graphs.

Spare Usage graphs:

The spare usage graphs provide information into the volume and types of spares that are used at each line station and by which type of aircraft. This data is valuable when assessing resource allocations and distributions.

Seat Availability and System Degradation graphs:

The seat availability and degradation reports provide insight into how individual aircraft types are performing as well as how the overall fleet is doing. This data is useful when determining if the IFE system is meeting customer expectations and if the vendor is providing the desired level of operations.

These reports are designed to provide a basic set of data for analysis and decision making.

Future

This model was designed to be flexible and adaptable. It can be easily maintained and enhanced to meet the changing requirements and needs of engineering users. This was accomplished via a modular framework and clearly defined data processing sections. The goal of this tool is to provide the ability to predict over a long period of time the maintenance requirements for schedule driven systems. This tool is ideal for systems with hardware units exhibiting constant failure rates and known repair procedures. This tool could be adapted to handle the needs of any schedule driven systems. Some of the more obvious ones are buses, trains, and ships. We are currently looking at different ways to put this model to use.

0-7803-5086-3 /98/$10.00 ©1998 IEEE

REALITY IN CERTIFICATION OF IFE AND TELEPHONE EQUIPMENT

Ana M. Christofferson, Senior Specialist Engineer, DER, Boeing Cabin Systems, Seattle WA
Ritch L Triplett, Senior Engineer, DER, Boeing Cabin Systems - Rockwell IFE, Seattle WA

This paper provides an overview of the process followed by Boeing to certify In-Flight Entertainment (IFE) and Telecommunication Systems. It is not intended to provide an in-depth description of the certification process and/or applicable Federal Aviation Regulations.

The Certification Process
The Boeing Company, as an airframe manufacturer and system integrator, shall certify each airplane prior to its delivery. In seeking certification, airplane design shall comply with airworthiness design standards per Federal Aviation Regulations (FAR part 25, transport category aircraft, and FAR part 36, noise). Post delivery continued airworthiness, maintenance and airline operations are maintained by the airlines and shall comply with FAR parts 121 and 91.

There are four categories of certification processes which apply to airplanes produced by the Boeing Company. These processes are:

New airplane type certification
Process of certifying a new type model (for example, 777-200)

Type certification of major or minor changes to an existing model
Process followed for an amendment to an existing type certificate(for example, certifying a new SATCOM system on a 777-200)

Production airplane certification
Every commercial airplane delivered by Boeing requires a Certificate of Airworthiness. A Standard Certificate of Airworthiness is used for U.S. airlines. An Export Certificate of Airworthiness is used for airplanes delivered to foreign airlines.

Foreign certification
Many foreign countries require each Boeing airplane delivered be certified in that country or that the FAA certification be validated by that country's foreign regulatory authority. This process can vary from immediate approval of the FAA certification or a up to a full three year effort. Bilateral Airworthiness Agreements exist between the USA and 24 foreign countries.

The flowchart below provides a general overview of the certification process milestones and interdependencies.

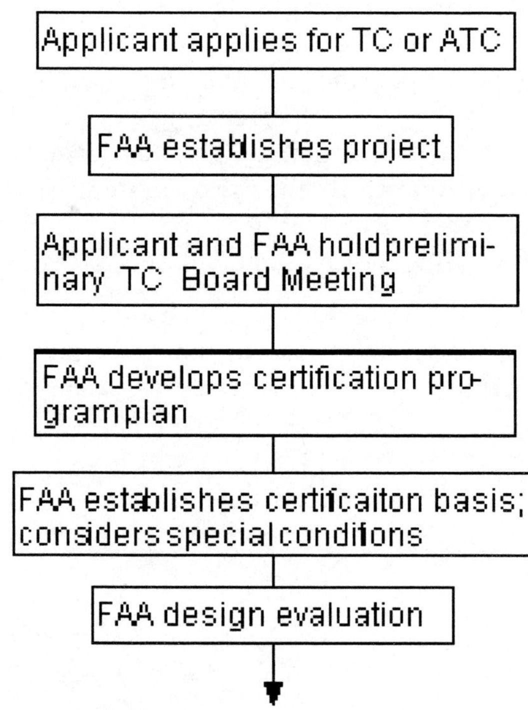

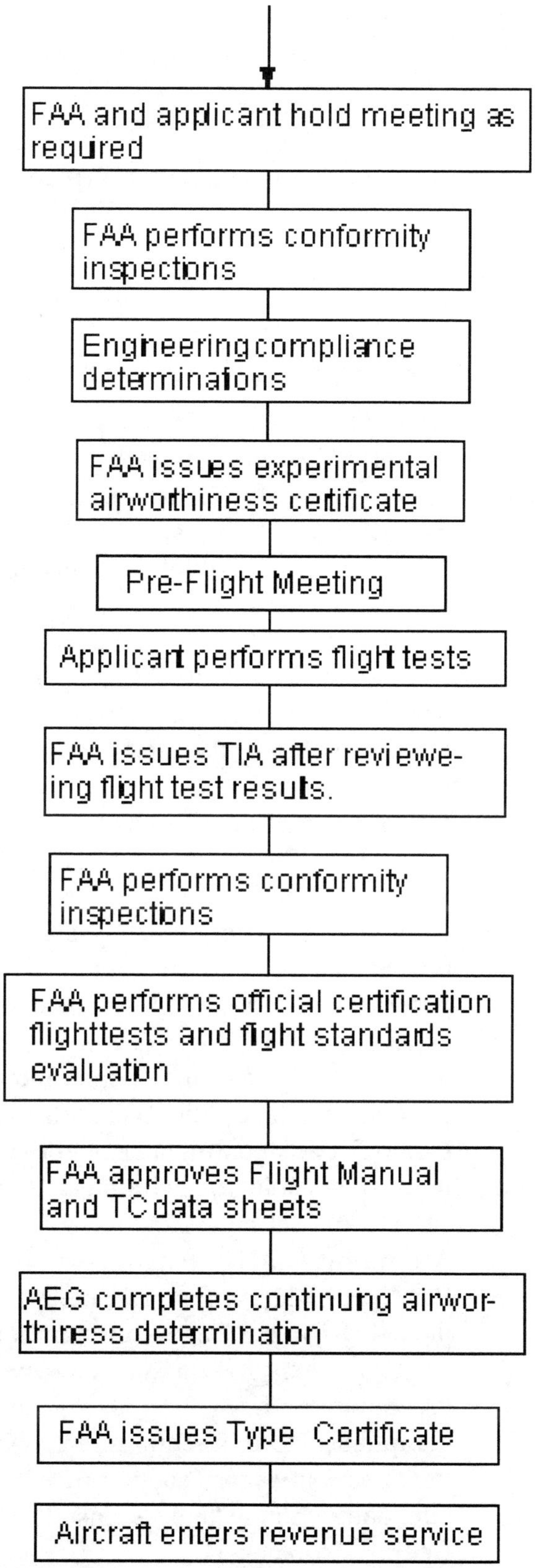

The certification process for type certification and configuration control during the design and manufacturing stages of an airplane program requires that all elements of the airplane be maintained in compliance with appropriate FARs. The manpower requirements necessary to assure compliance exceeds the resources of the FAA. A procedure was established to utilize the expertise within an airframe manufacturer's design organization to ensure that the design complies with appropriate regulations. This procedure calls for the use of Designated Engineering Representatives (DER) to represent the FAA in compliance control. FAR Part 183 describes the requirements for designating private persons to act as representatives of the Federal Aviation Administrator. Chapter 5 of FAA Order 8110.4 "Type Certification" covers Designated Engineering Representatives authority and responsibility to the FAA. FAA Order 8110.37 "DER Guidance Handbook" describes policies and procedures for use by the FAA Regional offices in administering the DER system. The DERs are selected for their demonstrated expertise in specific technologies.

The Boeing DER Role

DERs are individuals appointed by the FAA to act on their behalf to ensure that the design and manufacturing of the airplane follows configuration control practices and certification processes are maintained in compliance with the appropriate FARs.

The Boeing Co. organizes engineering groups according to the various areas of expertise. There is a DER per area of expertise that provides overview and approval of engineering data used to show compliance for specific FARs. For example, The Electromagnetic Compatibility (EMC) organization reviews and approves Qualification Test Data pertinent to DO-160C/D sections 18, 19, 20, 21, 22 and 23. This group is also responsible for conducting airplane testing (Electromagnetic Interference (EMI) Survey). However, the Payloads and Stress Engineering Organization reviews and approves data pertinent to DO-160C/D sections 7 and 8, or would coordinate delethalization testing of seat mounted equipment. The Cabin System (CS) DER has overall responsibility of the certification of the Cabin Systems equipment. The CS DER is mostly responsible for ensuring that the equipment design is suitable for airplane installation, that it performs it operates correctly when properly installed and that it does not interfere with airplane required systems and equipment (FAR Parts 25.1301, 25.1309 and 25.1431). The CS DER also reviews and approves engineering data for which they have been delegated authority to do so, and verifies that all other pertinent DERs have received, reviewed and approved pertinent data that they are responsible for. The CS DER will also coordinate the final flight test demonstration with the FAA and coordinate any AFM revision that may result from the FAA system evaluation.

Certification of Cabin Systems Equipment

A Cabin System is a set of avionics components that automate various crew and passenger functions, including safety, service, entertainment and telecommunications. Cabin Systems functions can be grouped in three categories:

Cabin Services System (CSS)
Provides basic functions including: cabin interphone, passenger address, passenger service and control of cabin lighting (747 and 777 only). CSS is basic on all Boeing airplane models.

In-flight Entertainment System (IFES)
Includes video, audio, games, etc. IFES is extremely volatile and competitive and is available in various configurations, including broadcast mode, interactive and video-on-demand.

Aeronautical Telecommunication System (ATS)
Includes telephony and facsimile services. ATS is available in various configurations: stand-alone, integrated with the IFES or a combination of both.

Usually each airplane customer has a unique Cabin Systems configuration. As a result, Cabin Entertainment and Telecommunication Systems are not basic to an airplane model. The certification of this equipment is accomplished following the *Amendment to Type Certificate* process. The certification of Cabin Systems is regulated by various parts of CFR Title 14, primarily in Parts 25 and 121. Applicable requirements are dependent upon the Cabin System functions performed and the level of integration with other airplane systems and equipment.

Major Cabin Systems Functions are certified to the level of criticality assigned by the FAA. The following is a list of Cabin System Functions and their associated level of criticality:

Passenger Address System
Address by CFR Title 14, Parts 25.1423 and 121.318. This system is present for the purpose of making announcements to the passengers with regard to safety preparedness and in actual emergency situations. CFR Title 14 specifies various pertinent requirements, including response times, intelligibility, and cabin attendant access to input devices.

Cabin Interphone System
Per CFR Title 14, Part 121.319, this system is an integral component of the Crewmember Interphone System, required on airplanes carrying more than 19 passengers for the purposes of conducting private conversations between the flight deck, cabin crew members, and ground personnel. CFR Title 14 requires that the system have signaling/alerting devices, and it must be able to distinguish emergency calls from routine calls.

Attendant Call System
Per CFR Title 14, Part 121 means must be provided for passengers to summon cabin crew members for assistance. Operational procedures may be used to show compliance to this requirement. To avoid the necessity for such procedures, an attendant call system is installed on all Boeing airplanes.

Other Required Systems/Functions
CFR Title 14 specifies requirements for additional systems/functions in the

airplane cabin, such as cabin lighting, information signs, lavatory smoke detectors, etc. Depending on the implementation of systems on different airplane models, some of these systems/functions may be fully or partially integrated into a Cabin System. If they are, the specifics of the pertinent CFR Title 14 requirements must be addressed.

Miscellaneous Non-required Systems

A majority of Cabin System equipment falls into this category(individual reading lights, video/audio system, telephones).

Certification requirements for non-required systems are the same as for required systems with regard to form, for example, flammability, smoke, toxicity and vibration. Equipment design shall be suitable for airplane application and operate correctly when properly installed (per CFR Title 14, Part 25.1301 and 25.1309). Additionally, non-required systems must not interfere with required systems and equipment (per CFR Title 14, Part 25.1431). Advisory Circular 25-10 (AC 25-10) provides the primary guidance for certification of non-required systems.

The Boeing SCSRD

Various industry standards, ARINC 628 for IFE or ARINC 746 for Telecommunications, govern the system architecture and internal/external interfaces of Cabin Systems equipment. However, Boeing currently utilizes the Standard Cabin System Requirements Document (SCSRD) document D6-36440 to specify Cabin Systems requirements. The SCSRD specifies the operational and functional requirements for the Cabin Services System, In-Flight Entertainment System and the Aeronautical Telecommunication System.

The SCSRD also defines installation requirements for Cabin System equipment, including mounting locations, weight limitations, electrical power capacity, cooling and wiring. It should be noted that the SCSRD provides the Boeing implementation of the standards and requirements defined per ARINC 628 and/or other applicable industry standards.

The SCSRD also defines applicable qualification requirements and deliverables in support of certification. Examples of deliverables that IFE and ATS suppliers provide in support of certification include hardware for laboratory testing, LRU technical specification data, system descriptions, interconnect diagrams, software documentation, development plans and schedules.

Engineering/DER Activities in the Certification of Cabin Equipment

The following is an outline of milestones and activities that take place from vendor selection to certification and delivery of Cabin System Equipment:

Airplane Implementation (Cabin Systems equipment suppliers are selected)

Initial Technical Coordination Meeting (Boeing-Customer-Vendor meeting. Configuration is defined: quantities, locations, part numbers, etc. Schedules are reviewed/planned, deliverables defined.).

Follow-on coordination meetings and telecons. (On-going activity through out the program cycle, normally 8-12 months. Issues addressed include: review of system definition, design, development plan, etc. Changes due to new requirements or problems found during testing are also addressed.)

Boeing Engineering Installation Design (Includes equipment installation, wire design, power, cooling, etc. Designed by Boeing Engineering per technical data provided by equipment supplier. Involves coordination with other Boeing organizations and their respective DERs.

DER reviews and approves equipment installation via 8110-3 Top Drawing Approval.

Participate in developmental laboratory testing at supplier facility.

Review and Approval of Qualification Test Plan and Procedure

Request for Test Article and Test Setup Conformity Inspection.

- 8120-10 form sent to Boeing DMIR office, requesting conformity of Test Article and Test Set Up.
- Boeing or Vendor QA completes Statement of Conformity Form (8130-9).

- FAA Inspector or DMIR completes 8100-1 FAA Conformity Inspection record

Environmental Qualification Testing is completed. Qualification Test Report and 8130-9 form is submitted to Boeing for Engineering/DER review.

Conduct integrated laboratory testing at Boeing facility, including evaluation of interfaces to other airplanes systems (ACARS, SATCOM).

Conduct preliminary EMI lab evaluation, as required.

Coordination with FAA:

- Letter of Definition
- Certification Plan
- Agreed on Method of Certification
- Coordination of any technical issues pertinent to certification.

Review and approval of Qualification Test Report, and other deliverables pertinent to certification (software documentation, etc.). Approval is via FAA 8110-3 form.

DER coordinate and schedule pertinent airplane testing to support certification:

- Engineering Functional Ground Test
- EMI Ground Test
- Engineering Flight Test
- Certification Flight Test

Coordination meeting or teleconference with the FAA to review the Certification Plan, proposed testing and any additional requirements.

Production functional tests are completed. Engineering ground (both functional and EMI) and flight test are completed to verify readiness for certification.

The FAA issues TIA (Test Inspection Authorization).

Airplane Conformity Inspections are completed.

FAA performs official certification flight tests and flight standards evaluation. Usually Cabin System Certification flights are concurrent with First of Model (FOM) tests.

FAA approves flight manual and TC data sheet/Flight Test Report.

FAA issues Type Certificate or Amended Type Certificate.

Special Considerations during the Certification Process

Equipment Environmental Qualification

Qualification testing is completed to ensure the cabin system equipment operates safely in an airplane environment. Qualification Test Data is also used to substantiate compliance to applicable FARs. Below is a list of qualification test requirements included in the SCSRD, chapter 5, sections 5.1 through 5.6:

Thermal	Altitude
Altitude	Humidity
Waterproofness	Fluids Susceptibility
Salt Spray	Sand and Dust
Fungus Resistance	Vibration
Shock	Acceleration
Explosion Proofness	Flammability
Power Input	Magnetic Effects
Conducted Susceptibility	Induced Signal Susceptibility
Audio Frequency Susceptibility	RF Conducted Susceptibility
RF Radiated Susceptibility	Lightning Induced Susceptibility
Induced Spikes Susceptibility	Interference Voltages on Signal Lines
Short Duration Interference	Conducted Emissions
RF Emissions	Structural Substantiation and Abuse Loads

Software Considerations

Software installed in most Cabin System equipment is considered level E (software whose anomalous behavior, as shown by the system safety assessment process, would have no effect on the aircraft operational capability or crew workload) per RTCA/DO-178B. According to Paragraph 2.2.2.(e) of DO-178B, once the software has been designated as level

E by the certification authority, no additional s/w documentation submittals are necessary. However, proper software configuration control practices are not exempt per guidelines of FARs 21.31(a), 21.123(b)(c) and 25.1301(b). Software associated with some of the Cabin Services functions (Passenger Address, Cabin Interphone, etc.) is considered level D per RTCA DO-178B and required additional documentation submittal in support of certification (Software Accomplishment Summary, Configuration Index Document, etc.).

Laboratory Testing
Integration testing is completed to validate the various Cabin System configurations. In addition to specific system functionality (video-on-demand, or ground-to-air fax transmission) testing is used to validate interfaces to other airplane systems. Laboratory testing is primarily engineering testing and is not used for certification credit. However, laboratory testing provides the DERs the opportunity to evaluate the system before it is installed on the airplane. Also, certain unique failure conditions can only be tested in a laboratory environment, for example delethalization testing (16g HIC) for equipment mounted in a seat back.

Top Drawing Approval and the Hold Process
DERs approve the equipment and wiring installation (Top Drawing approval) via FAA Form 8110-3. Due to the logistics of obtaining DER signatures for all specialties, DERs are asked to sign these forms up to three months before delivery. During the time between a DER signing the form and delivery of the airplane, certification issues can arise. The "hold" procedure enables a DER to control "ticketing" of an airplane after signing an 8110-3 form until an unsatisfactory condition on the airplane is corrected.

Document Review and Approval
In addition to the top drawings, DERs review and approve via 8110-3 forms the following documents:

- Environmental Qualification Test Plan
- Environmental Qualification Test Procedures
- Environmental Qualification Test Report
- Statements of Similarity
- Software Documentation (Accomplishment Summary, Configuration Index Document, Version Description Document) for level D software.

Airplane Testing
System functionality and non-interference to other airplane systems must be validated by airplane test per AC 25-10 paragraphs m and n, and direction from the FAA. DERs are responsible for coordinating with the Boeing Flight Test Organization and originating the required paper work to schedule airplane testing. First time IFE and ATS installations undergo several airplane level tests. Testing includes Engineering Functional Evaluation Ground Test (4-8 hours), Engineering EMI Evaluation Ground Test (8+ hours), Engineering Power Quality Ground Test (10+ hours), Engineering Flight Test (1-3 hours) and a Certification Flight Test (1-2 hours). Certification Flight Tests are witnessed by the FAA. Occasionally Certification Flight Tests are delegated to Engineering and Pilot DERs. A brief description of each of the airplane type testing is provided below:

- **Engineering Functional Evaluation Ground Test**
  Allows for the engineering staff to exercise IFE and ATS systems on the airplane. Verifies airplane interfaces are operational.
- **Engineering EMI Evaluation Ground Test**
  Cabin Systems equipment and wiring are evaluated for emissions in all bands used by airplane navigation and communication systems. Cabin Systems equipment is also subjected to and evaluated on airplane system interference such as power quality and radio frequency susceptibility.
- **Engineering Power Quality Ground Test**
  Cabin System equipment is evaluated for worst case voltage distortion levels.
- **Engineering Flight Test**
  Allows for engineering staff to exercise the system while in the air and ensures system is ready for demonstration to the FAA for certification.
- **Certification Flight Test**
  Present the Cabin System to the FAA for demonstration of operation and non-interference. Includes evaluation of flight deck systems to ensure non-interference that could be attributed to Cabin Systems.

IFE and ATS System Configurations
Boeing delivers various Cabin System IFE and ATS system configurations to allow for future growth, and accommodate different levels of capabilities and limitations. The certification effort associated with each of these configurations is different and may require additional coordination with the FAA. The CS configurations can be grouped in four categories: Partial Provisions, Full Provisions, Limited System Operation and Full System Operation. The following is a brief description of these categories and the associated certification effort:

- **Partial Provisions**
  Airplane installation provisions are provided to accommodate future installation and activation of an IFE/ATS system. These provisions can include equipment locations, mounting provisions, airplane wiring (less electrical power) and cooling. Occasionally an LRU is installed as part of the partial provisions (for example, an antenna). If circuit breakers are provided with the partial provisions, power lines are capped and stowed at the circuit breaker.
- **Full Provisions**
  Full provisions are available for previously certified systems. Full Provisions can include all mentioned in Partial Provisions with the addition of electrical power (power need not be capped and stowed at the circuit breaker).
- **Limited System Operation**
  An active system is installed but only a subset of its operational capability is available. This level of certification allows for the system to be demonstrated to the FAA, when the system is not fully operational. It establishes a certification baseline and potentially reduces the certification requirements as additional system functionality is incorporated. Certification of less than a fully operational system shall be coordinated with the FAA prior to the certification flight test. Any further system enhancement shall be evaluated on a case by case basis to determine the certification effort that is required.
- **Full System Operation**
  The IFE and ATS systems are fully operational and functional for demonstration to the FAA.

ADVANCED INFRARED (AIR): INTRA-CABIN WIRELESS DATA NETWORK

Patrick H. Potega, AirPower™, West Hills, CA

Background

The viability of infrared (Ir) data communications on commercial aircraft has, until now, been impractical. Previous efforts by Sennheiser, for example, to provide wireless audio throughout the cabin, were unsuccessful. Passenger and airline expectations were not met, primarily because of attempts to implement immature infrared technologies. One of the persistent problems with infrared technologies, in general, has been that the promises of performance have not always been backed-up with adequate hardware.

This paper explores a new generation of wireless data LAN hardware that is robust, reliable, fast, non-directional, light tolerant, and that has triple the range of "yesterday's" infrared. Advanced Infrared (AIr) delivers what airlines and passengers expect from wireless devices and systems. Capabilities are on par with short-range wireless RF, but without the inherent risks of radio waves. Being light waves, infrared is totally safe and benign. This has already been validated by previous infrared programs within commercial aviation.

Connectors and Cables

No one would disagree that cables and connectors are the bane of airframe manufacturers, airlines, and vendors. The penalties are obvious: weight, mass (congested wire troughs), and reliability. Two of these penalties translate to an airline's bottom line as fuel expenditures (to haul the weight) and maintenance costs to repair or replace damaged cables or connectors. Vendors are burdened with the cost of designing and building wire runs and providing connectors. Cabling and connectors can easily account for 30% of the price of a vendor's product.

Cutting through all of this, both for fleet operators and vendors, is the cost of certification and testing. Every time a cable is lengthened or shortened, a cost is incurred because of certification.

Airframe manufacturers are faced with trying to accommodate hundreds of vendors vying for space to run their cables. While standardization is an always-hoped-for expectation, fully-implemented standards are not yet a reality. Airframe manufacturers suffer the consequences of odd-ball connectors, interface boxes, and cables.

PED Connectivity

As the mobile computing world converges with in-flight entertainment (IFE), cabin equipment interfaces will focus on connectors that passengers access. Portable Electronic Devices (PEDs) that passengers bring aboard afford an opportunity for airlines to add significant value to the flight experience. Today's multi-media notebook computers are much better devices on which to view personal video than are the small-view screens the airlines are providing at each seat. Why would passengers want to watch a first-run movie on the small screen of a personal video viewer, when they come aboard carrying a 14.4" screen on the laptop?

Just provide a set of video- and audio-

0-7803-5086-3 /98/$10.00 ©1998 IEEE

out connectors, and a laptop-equipped passenger can have an in-flight movie experience that approaches home entertainment quality.

Watch Out For "Bluetooth"

But passenger connectivity, whether it be for video, in-seat power, or e-mail/fax data to an in-flight phone's handset, brings with it a potential antenna effect. Cable grids throughout the cabin become antenna arrays when PEDs are connected to them. EMI and RF harmonics from PEDs can amplify and transmute into dangerous signals or noise that could disrupt critical Nav and ILS equipment.

The aviation world is full of ironies. Just a few years ago, this world was buzzing with unconfirmed reports that laptop computers were "glitching" cockpit instrumentation and communications. Nothing was ever substantiated, but it has taken awhile for all of us to get emotionally comfortable with laptop use during flight. One of the positive after-effects of that controversy was that the mobile computing industry became obsessively scrupulous about emissions and radiation levels of its equipment.

Just when those old emotional fears are subsiding, a new laptop RF-communications platform specification has been announced. Based on cellular technology, the "Bluetooth" specification integrates the Tx and Rx hardware of a spread-spectrum 2.4 GHz cellphone into a laptop. The implications are obvious, especially since 2.4 GHz is being proposed as the RF bandwidth of an ARINC-approved wireless LAN that operates throughout the airport. But, more importantly, do we provide inadvertent antenna arrays—as cables throughout the cabin—to which passengers carrying Bluetooth devices can connect?

Benefits of a Wireless Infrared Network

Infrared is nothing more than a sophisticated, hi-tech semaphore. Its roots are from the early days of aviation when flashing signal lights were used to send code signals to aircraft about to land on carriers. With Ir, nanometer-range-specific light-emitting diodes to send digital data packets. Basically, it's beaming light. Infrared is a distant cousin to fiber-optics. . .it's light transmission without the "fiber." Advanced Infrared (AIr) is the antithesis of fiber-optics, in that it creates light diffusion, rather than columniation.

Because it's based on light transmission, Advanced Infrared is a "zero-weight" data communications technology. Not only does it eliminate heavy and failure-prone cables, it is intrinsically non-directional. It moves data from point to point where running cables would be impractical or expensive. AIr can take data from a seat to an overhead, for example, without having to run a cable from an armrest to the floor, under the floor, up the wall and to an overhead bin. It can send that data to a passenger several seat columns over, or to a cabin crew member who is passing down the aisle, without a file server. Advanced Infrared creates data connectivity where cable runs could never exist in a practical way.

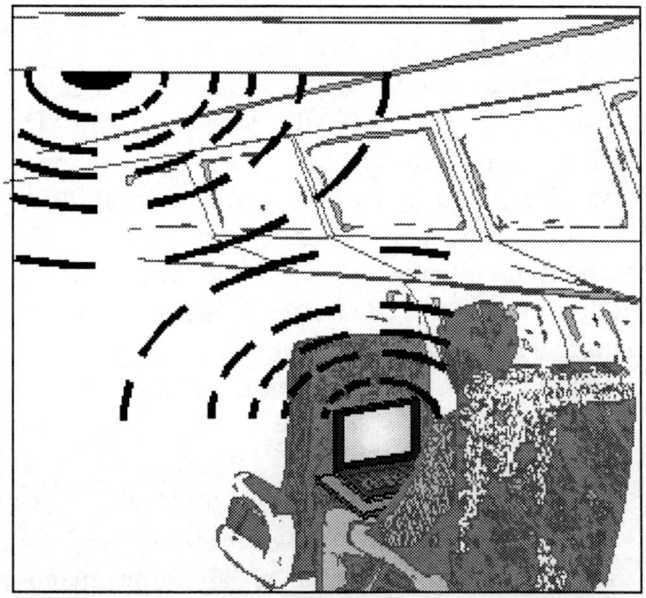

Figure 1. Overhead- or ceiling-mounted infrared LAN access points service clusters of 12-20 seats with non-directional Ir.

Based on today's fuel-burn-to-haul-weight statistics, a wireless infrared data LAN can deliver a significant ROI to an airline. For example, one airline indicates that it costs over $65K per year to haul about 100 pounds (based on a Boeing 777). Eliminating cables and connectors translates to an immediate economic benefit, which goes directly to the bottom line as reduced operating costs. Narrow-bodies are the major beneficiaries of Advanced Infrared (AIr), as wireless opens the door to IFE.

Providing data connectors at passenger seats is complex. Aside from which style of connector interface is required for a given application, issues of breakage and wear bring high replacement and maintenance costs. Passenger tolerance for unreliable or damaged connectors is traditionally quite low, so aggressive and vigilant maintenance schedules are mandatory.

"It costs over $65K per year to haul about 100 pounds. . . . Eliminating cables and connectors translates to an immediate economic benefit."

Infrared is, by definition, a "connector-less" data solution. Passenger use or abuse can damage or degrade the reliability of the data interface. Maintenance is addressed in the same way as any light bulb-replacement schedule.[1]

A Certification Home Run

The installation of an infrared wireless network is relatively non-invasive. Depending on the location of transceivers deployed throughout the cabin, most mechanical modifications are to ceiling panels, or to the utility light area of overhead bins. The only wiring needed to a LAN access point transceiver would be for low-voltage power. Since passenger seats are not impacted, certification costs of cables and connectors to each seat are eliminated. Installation uniformity across an aircraft type simplifies certification and testing procedures, as well as lowers cost.

Standard Shared-Resource for IFE Vendors

Unlike data cables, which can only be parsed for shared data with complex hardware, Advanced Infrared uses data packet address headers and other methods of sorting data. For example, e-mail packets are kept separate from faxes by packet headers and Ir stack identifiers. As a data infrastructure model, AIr allows for simultaneous co-existence of multiple vendors' data. As AIr's bandwidth increases to 16 Mbps, then 32 Mbps, the cabin-wide Ir LAN will be able to accommodate more and more server-based applications. This will dramatically decrease cabling. It will become a universal LAN infrastructure that can be cooperatively shared by a number of vendors without major modifications to the Advanced Infrared installation. This LAN model will lower IFE equipment costs by eliminating repeated purchases of dedicated cable runs.

RF and EMI Can't "Attach"

An AIr network cannot act as an antenna. RF and EMI, whether generated by equipment at the network's head-end, or by passengers' PEDs, cannot "attach" to the infrared light beams. Thus, any noisy or radiating devices are kept localized and isolated. Harmonic propagation created by devices attached to cables is avoided. This benefit of AIr delivers an extra layer of safety in an environment where numerous unknown passenger devices can pose risk.

---

[1] Infrared diodes have a finite life expectancy. The industry rates the useable life of a typical diode at 40,000 hours of usage. AIr diodes are rated the same, but the power management feature extends that by as much as 10 to 20%.

0-7803-5086-3 /98/$10.00 ©1998 IEEE

Out With the Old, In With the New

The old problems with infrared have been eliminated. Table 1 compares "yesterday's" infrared performance characteristics with those of Advanced Infrared (AIr):

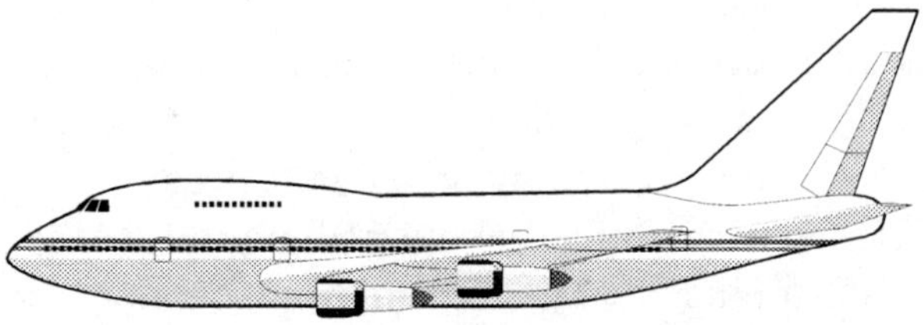

Table 1. Performance Comparison

	Pre-AIr Infrared	**Advanced Infrared (AIr)**
Range	1 meter (3 feet)	10+ meters (30+ feet)
Data Speed	4 Mbps	4 Mbps, but already capable of 16 Mbps[2]
Alignment Angle	30-degree cone[3]	120-degree cone[4]
Device Support	One Tx to one Rx (point-to-point)	Multi-point (number of simultaneous devices is only limited by available bandwidth)
Network-able	No[5]	Yes (true LAN)
Multi-point	No	Yes
Data-Loss Fault Tolerant	No	Yes (reliably "recovers" lost data packets)
Collision Avoidance	No	Yes (user devices don't interfere with each other)
Roaming	No	Yes
Auto Configuration	No (usually requires user configuration)	Yes (auto-ON and "Plug n' Play")
Spontaneous Networking	No	Yes ("*ad hoc*" network connections)
Ambient Light	Poor response	Very tolerant
Energy Management	Limited	Four levels
Standardized Platform	Yes (IrDA)	Yes (IrDA and SMBus)[6]

As can be readily seen, Advanced Infrared (AIr) affords a significant breakthrough in wireless LAN performance. The stats are even more impressive when one sees the technology in operation. For example, "Spontaneous Networking" is a server-less technology that allows multiple Ir-enabled devices to interconnect in a peer-to-peer model. This happens by simply being an AIr-enabled device in the environment, with no device reconfiguration. Thus, multiple passengers can join each other in an *ad hoc* video game session, simply by opening a game application. Passengers can join or drop out at will, without disrupting other players' activities.

Roaming is another key feature. A

[2] There are solid indicators that 32 Mbps is achievable.

[3] The typical 30-degree Ir cone has 100% reliability but only for data transfer in approximately 15 degrees of that beam.

[4] While AIr transmitters emit light in a 120-degree conical pattern, the functionality of AIr is actually that of diffuse infrared. Receivers read Ir data that has been redirected through bouncing off nearby surfaces, so that AIr's performance is comparable to a true diffuse Ir system.

[5] There are access points available that allow single-user devices to beam data into a hardwired network.

[6] Infrared Data Association (IrDA) specifications have been recognized globally for over five years, and 95+% of all laptops have IrDA-compliant infrared installed. System Management Bus (SMBus) is a recognized global standard for device power management, and AIr is seeking official compliance recognition to this standard.

passenger can cruise the aisles of the plane wearing an AIr-enabled wireless headset (it's battery-powered). As the headset passes through the LAN access points (embedded transceivers) strategically placed throughout the cabin, the available audio signals are seamlessly passed to the peripatetic headset. A real-world application for roaming is for hand-held computers carried by cabin crew members. Food and drink orders can be instantaneously passed from roaming cabin personnel to a server in the galley, thus saving countless steps.

Data-loss fault tolerance is one of the crown jewels of this new infrared LAN. Transceivers allow for data packet latency and also monitor lost packets. A packet can arrive at a transceiver up to 20 seconds late and still be reinserted in its proper sequence in the data stream. This also enables a transceiver to accept indirect signals that are bounced off a cabin wall, thus eliminating the need for line-of-sight transmission.

In AIr demonstrations, the access point and a client transceiver are pointed away from each other and participants are invited to wander back and forth between the two devices. Even in this situation, where signals are being blocked by foot traffic while the two devices are pointed *away* from each other, AIr still maintains a functional and uninterrupted stream of data.

Tolerance to ambient light has been a concern with infrared since its inception. Pre-AIr devices could not be used near unshaded windows. To make a point during Advanced Infrared demonstrations, a transceiver is positioned so that the full light from an overhead projector[7] is passing between it and the AIr access point. The infrared signal behaves flawlessly, so there will be no need to lower cabin window shades.

Compatibility with Legacy Equipment

Advanced Infrared requires special hardware to achieve the functionality indicated in Table 1. AIr hardware is scheduled for installation in laptops and other computer-type devices. However, because Advanced Infrared is IrDA-compliant, the wireless Ir LAN on an aircraft can be accessed by any IrDA device. More than 95% of all laptops come equipped with IrDA infrared ports, so passenger attach rates and utilization will be high.

Furthermore, as a recognized infrared technology, AIr LANs will also be prevalent in office environments and even in homes. Passengers will, thus, be familiar with the aircraft's LAN network.

Real-World, Practical AIr Implementations

The AirPower division of Lifestyle Technologies® manufactures a sophisticated In-Seat Power Supply System (ISPSS) for commercial and corporate aviation. When product design started in 1996, we set certain goals and criteria. Among them were to create an in-seat power supply system that has:

- Extremely low total system weight
- Reliability and safety through high levels of fault tolerance, and advanced warnings of potential failures
- Low Total Cost of Ownership (TCO) achieved by:
  - Minimal investment in installation time
  - Self-monitoring maintenance
  - Simplicity—eliminating complexity or risk of failure
- Ease of use for both airline and passengers
- Compatibility with other cabin sub-systems, as well as with PEDs
- Expandability

[7] The light spectrum of an overhead projector bulb is balanced sunlight at 3200-degrees Kelvin. Essentially, it is bright daylight as experienced at high noon.

Low system weight is difficult to achieve with power supplies. While ounces here and there can be saved with conscientious component selection and good circuit board design, shaving away pounds had to occur outside the box. Power cables are what they are —power delivery requires copper conductors— but data cables are another matter entirely. That's where Advanced Infrared provided a significant weight-savings breakthrough. We were able to eliminate all data wiring between the seat-based power supply and the head-end Master Control Unit (MCU). The wireless AIr LAN is the perfect data connectivity solution for our advanced AirPower "Power Monitor."

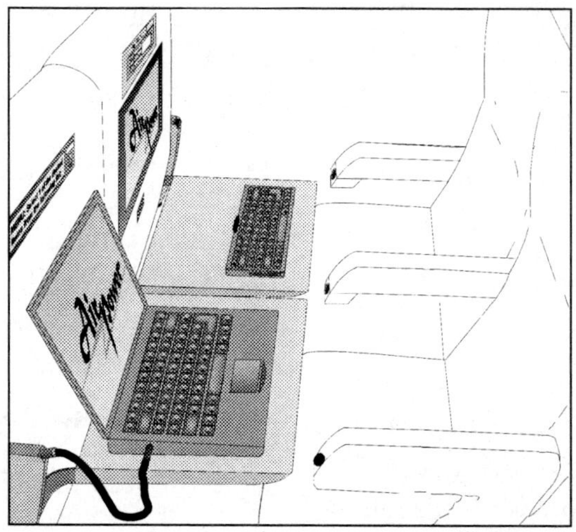

The Power Monitor performs real-time diagnostics of every seat-based power module, logs power consumption load schedules, red flags questionable power modules that may need servicing, tracks power module and connector life cycles related to MTTFs, and even reallocates power from seat to seat, as demand dictates. The total weight of this sophisticated Power Monitor sub-system is only the weight of the server, and the few feet of data cable required to connect to the first AIr access point. We recommend that airlines integrate Power Monitor into an existing server or, absent that, a 4-6 pound laptop makes a good Cabin Service Unit (CSU). The CSU can also double as an MCU. The total installed weight of Power Monitor can be as little as five pounds, which includes data connectivity to every passenger seat.

Thus, AirPower's objectives were achieved by integrating infrared into our cabin-wide power network. Low system weight, as well as advanced power diagnostics and monitoring, were achieved with a wireless LAN.

The Perfect Cigar

The shape of the aircraft cabin was put to good advantage for AirPower's data communications. The semi-circular cross section of a cabin interior, in conjunction with expanses of hard surfaces, prove an ideal environment for "caroming" light beams. We rely on AIr access points' ability to intercept bounced data light beams. The cigar shape of the cabin has proven to be an inadvertent and free benefit that enhances infrared communications. This has helped us eliminate further data cable runs around the seat area, and data signals are converted to light beams within a foot or so of AirPower's embedded power supply.

AirPower Installation

AirPower can be installed at every seat on the aircraft. We're not restricted to First and Business Class, primarily through the power conservation achieved with Power Monitor. The only way we could have ever gone beyond 120 seats without incurring huge weight penalties from additional cable runs, was by utilizing AIr's "zero-weight" wireless data LAN. We have done prototype installations of over 220 seats, and have been able to stay within the weight parameters of competitors' systems, while adding the extra benefits of AirPower's sophisticated power management. Typical installations can be completed in less than eight hours on a 757, and that includes bringing power to every seat in all classes.

Simplicity is achieved by having to run only power lines throughout the cabin interior. No data lines are laid down, so thousands of feet of cables (and hundreds of connectors) instantly disappear. Some may argue that we have not decreased total system complexity by trading off the inherent reliability of cables and connectors for computer chips and diodes. In actuality, the failure rates of infrared devices is virtually zero, based on the life expectancy of everything from TV remotes, to years of real-world infrared usage in laptops.

Further Cost Amortization

Since we view the wireless AIr LAN as a shared resource, the installation cost of an Advanced Infrared network is amortized over the number of other vendor systems that utilize the network. Data telephony is a natural, especially at the low 9600-baud rates currently used for e-mail and faxes. Cabin services is another area ideal for low-bandwidth applications. Transactional processing, for such activities as duty-free shopping, also falls into the shared-resource class of wireless applications. Thus, AirPower's use of infrared is not an aberration among a host of cable-bound cabin technologies. It is merely the first of a number of mainstream data activities that will find the AIr LAN the ideal platform for intra-cabin data communications.

More Bandwidth, More Participants

As Advanced Infrared's bandwidth climbs toward 32 Mbps, the more data-intensive in-flight applications will use the wireless network to send data throughout the cabin. Audio, and even full-motion (30 frames per second) video, will become a reality.

An alternative approach to waiting for Ir bandwidth to increase as single-diode modules is to simply add more Ir circuits. One installation model specifies an access point configured as a small hemispherical half-globe at strategic locations along the centerline of the cabin ceiling. Because AIr has a range of 30 feet, only 8-12 such 3" diameter hemispheres are required to service an entire cabin. Each access point covers approximately 12-20 seats.

To increase bandwidth, each access point can have additional diodes and circuitry. Separation of data streams can be achieved by shifting the nanometer range of the diode sets so that there is no contention. Changing the light-sensitivity range of diodes isn't always practical, but diodes can also be lensed so that they are directional, pointing only at specific seat clusters. This would result in a segmented LAN, with each segment tied to a specific function, such as video or audio.

Can A Single Server Be Far Behind?

Once the AIr LAN becomes a co-operative network, utilized by a number of vendors, discrete functions and applications could be integrated on fewer file servers. Given space and weight limitations that multiple, dedicated file servers create, it seems only logical that vendors consider consolidating their applications on a single piece of hardware.

Today, this seems more a dream than a possible reality, given the highly competitive environment among IFE vendors. However, profits drive any business, and the cost of designing, building, and installing sophisticated head-end hardware is a major economic hit for even the largest IFE system manufacturers.

Admittedly, there are individual needs for specialized hardware that supplements a generic file server, but these could be handled with standardized card racks, switches, and routers.

B.Y.O.D. (Bring Your Own Data)

As AirPower looks to the future, the company sees a passenger's travel experience revolving around smaller mobile computing devices. The movement toward sub-notebooks and hand-helds is already well established. This migration to smaller computing devices is not totally voluntary. There are pressures—some subtle, others blatant—that are contributing:

1. Concerns over in-flight battery charging. Charging laptop batteries on commercial aircraft should not be allowed for several reasons. While battery chemistries are getting *potentially* safer, battery charging will never be a risk-free activity.
2. Few air travelers need the robust computing power and myriad bells-and-whistles of today's high-end notebook computers. Except for business flyers who travel with multi-media presentations, most "road warriors" monitor e-mail and generate spreadsheets or word documents.
1. Airlines are enforcing new luggage restrictions. Laptops with 13" screens or larger, consume a sizable amount of briefcase or luggage space. If battery charging is banned on passenger aircraft (as it should be), travelers might opt to carry extra battery packs. While AirPower's In-Seat Power Supply System (ISPSS) can detect and prevent battery charging while providing primary power to a laptop, power consumption of notebooks still makes spare batteries a wise choice.
2. Laptop theft has reached epidemic proportions.

Data-Only Travel

AirPower's crystal ball sees a day coming when passengers won't need to bring a laptop aboard. Just as the need for bas to power a laptop has become superfluous with the advent of in-seat power so, too, will the deployment of personal viewing screens at every seat (*á la* Virgin Airways) eliminate the need for a laptop viewing screen. It's already on the plane!

As Advanced Infrared grows in bandwidth, passengers can use the cabin environment as a "Kinko's In the Sky." Physical drives on a file server can be logically partitioned so that each passenger has a private work area. Ir data packets can be encrypted to ensure privacy.

All that's missing is a suitable input device. That's where AirPower's "Keyosk Flyer" wireless and cordless keyboard comes in. It's a compact keyboard that has integrated data read/write hardware. Our prototypes use an Imation LS-120 drive. The diskettes hold 120 MBytes of data—enough storage to bring your own operating system, applications, and files.

Prior to landing, travelers write back to the LS-120 the data they uploaded to the file server, reformat the logical drive partition to remove any data, and deplane.

Figure 2. The Airpower "Keyosk Flyer" keyboard not only is wireless and cordless, but it has built-in removable-media data storage. Passengers can travel with an LS-120 diskette that holds 120 MB of data (enough for an operating system, applications and files). The in-seat screen (here TNCI's 11.3" model) and an infrared port are all that's needed. Data is stored on a local hard drive at the seat, so data security is preserved. Flight attendants hand out the keyboards, as a courtesy, or as rentals.

For more information on Advanced Infrared (AIr), contact: Patrick Potega, AirPower, 7021 Vicky Avenue, M/S AI-44, West Hills, CA 91307-2314. Phone: (818) 712-9129. E-mail: *lifetech@mcimail.com*. Web address: *www.airpower-in-seat.com.*

IN-FLIGHT ENTERTAINMENT - GETTING FROM WISHLIST TO REAL

Author: Donald B. Lee, Senior Specialist Engineer
Boeing Cabin Systems Enabling Technology
PO Box 3707 M/C OT/06, Seattle, WA. 98124
donald.b.lee@boeing.com

Introduction

Much of what we see in the consumer electronics stores will eventually find it's way into airplanes in the form of In-Flight Entertainment (IFE). This consumer electronic equipment is often referred to as "the leading edge of technology". Unfortunately, that same equipment that is the "leading edge" of consumer electronics, often starts to feel like the "bleeding edge" when you try to integrate it into IFE systems for today's modern aircraft. Unlike Avionics systems, the technologies behind IFE equipment are not usually mature before they are integrated into the aircraft. What may work well in a home environment, may not work well at all in the aircraft environment. Often this problem results in last minute redesign in an effort to make the equipment function as intended on the aircraft.

Both the FAA and the aircraft manufacturer impose many requirements on the suppliers and products of IFE. Systems need to be designed for the aircraft environment. Many variations on In-Flight Entertainment Systems have forced the creation of new standards for integration into commercial aircraft. Power, cooling, and EMC issues are becoming more of a concern as we move to full aircraft implementations. Beyond the technical evaluations and lab/aircraft testing, one key component is customer satisfaction.

This paper will cover some of the methods and tools we use to get IFE from wishlist to reality.

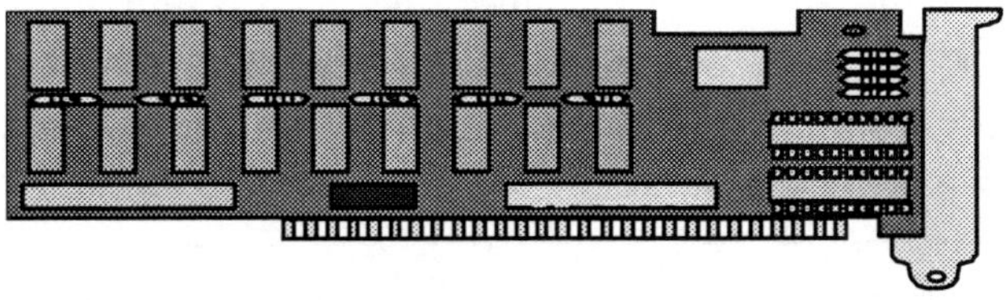

IFE - Commercial Electronics Origins

The In-Flight Entertainment (IFE) industry is unique among all the other suppliers of aircraft equipment and systems, because it is the passenger who drives both the need and the features of the system.

In Avionics systems it is the airline, government regulator, and the aircraft manufacturer that determine what systems are necessary for basic operation of the aircraft. These systems are transparent to the majority of the flying public. Example: A Flight Management System upgrade may help the pilot and airline operations greatly, but the passengers may not know about it or even care about it. All they are interested in is getting to their destination quickly and safely.

Passengers do not see the aircraft avionics, but they do see the cabin amenities. The typical passenger profile has changed over the years. As more families are flying and as more companies are trying to contain travel costs, passengers are looking for the most bang for the buck. If airline "A" has great service while airline "B" has great service along with In-Flight Entertainment, assuming ticket costs are equal; most passengers will choose airline "B".

0-7803-5086-3 /98/$10.00 ©1998 IEEE

Thus, in order to remain competitive in the market, increase passenger loads, and to differentiate themselves from competitors, airlines have been forced to embrace In-Flight Entertainment (IFE) on their aircraft.

Technology Enablers

In-Flight Entertainment functionality and systems are driven by technologies developed in the consumer electronics industry. IFE suppliers are leveraging these technologies to bring entertainment and home office functions to the individual passenger seat.

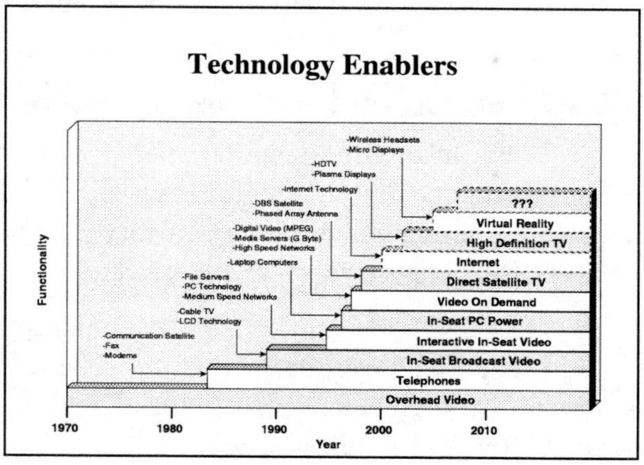

It all started in 1969 with PAN AM and the first broadcast audio system. That same year Air France retrofitted the first overhead video system. In 1984 Telephones, FAX, and Satellite communication systems came on-board. In 1989 the first In-Seat broadcast video system was installed. In the early 1990's came the first interactive In-Seat video system. Pushing the technology even farther, in 1997 airlines saw Video On Demand (VOD), and Direct Satellite TV come on the scene.

What will the future hold for In-Flight Entertainment? Some of the ideas currently being hashed out include full-cabin Audio/Video On Demand (AVOD), Internet and Office In The Sky, High Definition TV, Virtual Reality, and many others that I can't mention at this time. One thing is for sure - future IFE will definitely be exciting and a challenge for IFE suppliers, Airlines, and Aircraft Manufacturers alike.

IFE - Systems Design for the Aircraft

System Design Challenges - Business

In-Flight Entertainment is not the primary business of the major suppliers. Rather, IFE is typically a product line extension of their existing business. These relatively small business units are locked in fierce competition for market share. Low price and the promise of state-of-the-art functionality on a short flow are the primary competing factors. Staffing and capitol assets are minimized until a contract is won. Cost management has then become a priority, this has resulted in the suppliers failing to perform to their committed schedule on any major development to date - retrofit or forward production. In retrofit, the projects are simply delayed to later aircraft, for new production, the airframe manufacturer has been forced to bridge this gap and to maintain committed airplane schedules and deliver an IFE system with at least minimum functionality.

System Design Challenges - Technology

The aircraft environment is very different from the consumer electronics environment where IFE originates. Aircraft and system Electromagnetic Interference (EMI), power quality, and current harmonics continue to be a problem. Aircraft temperature changes, air pressure changes, and vibration are much more severe than what equipment would see in the home or hotel. Couple that with the fact that as system complexity increases, so does the need to use higher levels of integration than are possible with MIL Spec. Parts. Thus IFE suppliers are forced into the need to use commercial grade electronics where densities are higher and functionality/capabilities are greater.

This takes us to what I call the

0-7803-5086-3 /98/$10.00 ©1998 IEEE

"bleeding edge" of technology. Designing and packaging commercial grade electronics in such a way that it performs it's intended function on the aircraft, and meets all government and aircraft manufacturer requirements. This is the challenge for both IFE designer and Aircraft manufacturer. It involves knowledge of the target aircraft and state-of-the-art electronics and system design.

> *"We are trying to put functionality in our aircraft that is way advanced over the current state of consumer electronics. Thus, we end up being way ahead of the proper time to introduce it into the aircraft."*
> *(Del Silva, Associate Technical Fellow, Boeing)*

System Design Challenges - Issues

The many variations on In-Flight Entertainment systems have forced the creation of new standards for integration into commercial aircraft. ARINC 628 and various other committees have addressed many of the concerns by the creation of standards for installation and IFE system architecture. This is a good start, but there is still much more to do as more new suppliers come onto the scene with ever more complex systems.

IFE - The Qualification Mine Field

There are many roadblocks to successful implementation of an IFE system. Successful DO-160 Lab Testing on a single unit does not guarantee problem free installations on the aircraft. Because of this, aircraft manufacturers have created their unique requirements for entertainment systems based largely on actual installation and operation experience. The challenge was to get together all these requirements and tie them together under one document.

The Boeing Answer - D6-36440 SCSRD

At Boeing this document is called D6-36440 or the Standard Cabin Systems Requirements Document (SCSRD). The document was created to help suppliers navigate through the documentation submittal nightmare required by Boeing and the FAA. Below is a simple chart that illustrates what areas the Boeing SCSRD covers.

Cabin Systems Structured Analysis
The Integrated Cabin: SCSRD Requirements Coverage

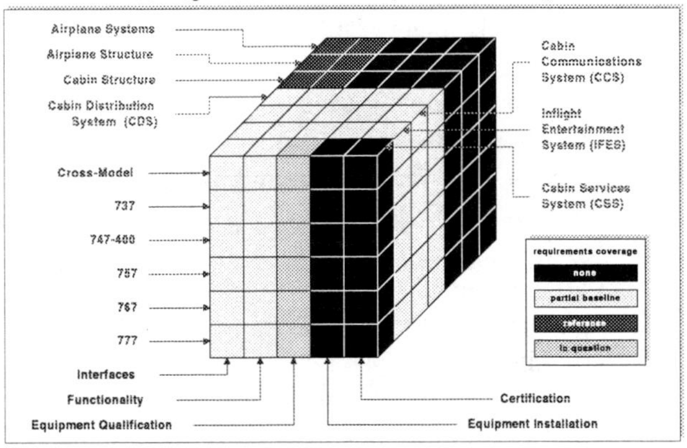

The SCSRD has become a partial Boeing baseline for all models in the areas of equipment qualification and installation of In-Flight Entertainment systems. The SCSRD also provides a partial baseline for Interfaces, Functionality, and the Cabin Distribution System as it pertains to Boeing aircraft. The document provides a common location for system test requirements and points back to the source documents.

Required Submittals for all IFE Systems

The are several required data submittals that Boeing needs from every supplier of In-Flight Entertainment equipment. The data, called Technical Data or T/D for short, is used to support installation of the IFE supplier's equipment in Boeing aircraft One of the most important aspects of this information addresses qualification of Line Replaceable Units (LRU).

The Qualification requirements specify the environment and test conditions in which the Cabin System Equipment must survive and operate.

All Technical Data must be written in English or provided with an English translation. Each document shall be prepared in accordance with good engineering form and practice.

Required Data Submittals - Systems

- System Development Plan
- System Requirements Document
- Plan to Support Certification
- Qualification Basis and Document Summary
- Hardware & Software Configuration Mgt. Plan
- Quality Assurance Plan
- System Integration Test Plan
- Operations Manual
- Component Part Number List
- System Interconnect Diagram
- System Install & Arrangement Diagram
- System Integration Test Procedure & Report
- Weight Analysis
- Service Provider Access Approval
- Safety Analysis
- Plan for Software Aspects of Certification
- Software Verification Plan
- Software Configuration Index
- Software Accomplishment Summary
- Environmental Stress Screening Program Plan
- Structural Substantiation Plan
- Seat Functional Test

Required Data Submittals - LRUs

- Outline Drawing
- Schematic & Block Diagram
- Acceptance Test Procedure (ATP)
- Specification Control Drawing (SCD)
- Environmental Stress Screening Program Plan
- Environmental Qualification Test Procedure
- Environmental Qualification Test Report
- Electrical Qualification Test Procedure
- Electrical Qualification Test Report
- Fire Properties Test Plan
- Fire Properties Test Report
- Structural Substantiation Data Package
- Static Test Report
- Component Maintenance Manual
- Safety Analysis

Required Data Submittals - Software

- Plan for Software Aspects of Certification
- Software Verification Plan
- Software Configuration Index
- Software Accomplishment Summary

The three technical data submittal types that give suppliers the most trouble are: Outline Drawings, Environmental Qualification Test Procedures and Reports, and Electrical Qualification Test Procedures and Reports.

The Outline Drawing is important because it is the document that is used by the most groups at Boeing. It contains the information we need to do the installation design, thus it is usually the first drawing through the Boeing system.

Outline Drawing Data

- Dimensions of Unit
- Placement and Content of the Label
- Center of Gravity
- Connector Part Number and Placement
- Mating Connector Identification
- Mounting Hold Down Identification and Location
- Max Weight of Unit (Hoisting Prov's if Required)
- Location and Content of Static Sensitive Stickers
- Environmental Characteristics (Temp, Vib, etc.)
- Location/Dimensions of Ventilation/Cooling Parts
- Finish & Color (Equip in passenger cabin area)
- Location/Access of Tuning/Adjustment Controls
- Sufficient Info for Materials Analysis and Cert.
- Equipment Power Dissipation in Watts for Operational & Idle Modes
- Equipment Cooling Requirements (Fans, Venting)

Additional Data for Cables and Wires

- Minimum Bend Radius
- Wire/Bundle Diameters
- Wire Lengths
- Recommended Mounting
- Connector Envelope Dimensions

Additional Item For Projectors

- Projector Throw Distance (Projectors Only)

Additional Data for Installation Kits

- Detailed Installation & Installation Drawings for Non-Standard Component Parts

The Environmental Qualification Test Procedure and Report contain the tests we use to determine if the proposed new equipment can survive the aircraft environment.

A commercial aircraft can be one of the most inhospitable places on earth for electronic equipment. Depending on where the equipment is located, it can expect temperature changes well beyond that of commercial electronics. If the proposed IFE can not operate in the full aircraft operating temperature range, a cooling system design will be required.

Vibration was a major cause of file server moving mass storage (Hard Drive) failures in the early IFE systems.

Humidity proved to be a challenge for much of the early IFE equipment and systems on long flights. As the humidity would go down, the electrostatic discharge problem was increased causing random seat entertainment failures. Many of the first passengers using these early systems were not happy campers...

Environmental QTP/QTR Data

- Test Setup Conformity (SCSRD 5.2.2.1.3.2)
- Performance Standards (SCSRD 5.2.2.1.1 g)
- Temperature, Altitude (SCSRD 5.2.2.2.1)
  - DO-160 P.4 CAT A1
  - Temperature 4.5
  - Alt Test 4.6.1 +15K Alt.
  - Decom 4.6.2 +45.1K Alt.
  - Overpress 4.6.3
- Temp Var (SCSRD 5.2.2.2.3)
  - DO-160 P.5 CAT C
- Humidity (SCSRD 5.2.2.2.4)
  - DO-160 P.6 CAT A
- Operational Shocks (SCSRD 5.2.2.2.8)
  - DO-160 P.7
- Vibration (SCSRD 5.2.2.2.7)
  - DO-160 P.8 CAT C & C'
- Impact & Implosion (SCSRD 5.2.2.2.10)
  - UL 1418 Paras 13.2 & 14.2
- Liquid Tolerance (SCSRD 5.2.2.2.6)
- LRU Surface Temp (SCSRD 5.2.2.2.2)
- Explosion Proofness (SCSRD 5.2.2.2.9)

The Electrical/EMI Qualification Test Procedure and Report contain the tests we use to determine if the proposed new equipment can survive on aircraft power, operate safely, and not interfere with or be impaired by other aircraft electronic systems.

Electrical/EMI QTP/QTR Data

- Test Setup Conformity (SCSRD 5.2.2.1.3.2)
- Performance Standards (SCSRD 5.2.2.1.1 g)
- Magnetic Effect (SCSRD 5.2.2.3.2)
- Power Input (SCSRD 5.2.2.3.1)
  - DO-160 P. 16 CAT A
  - D6-44588 P. 3.4.5.1, 6, 7
- Voltage Spike (SCSRD 5.2.2.3.3)
  - DO-160 P. 17 CAT A
- Audio Frequency Susceptibility (SCSRD 5.2.2.3.4)
  - DO-160 P. 18 CAT Z
- Induced Signal Susceptibility (SCSRD 5.2.2.3.5)
  - DO-160 P.19 CAT Z
- RF Conducted Susceptibility (SCSRD 5.2.2.3.6.1)
  - DO-160 P.20 CAT T
- RF Radiated Susceptibility (SCSRD 5.2.2.3.6.2)
  - DO-160 P.20 CAT T
- RF Emissions (SCSRD 5.2.2.3.7)
  - DO-160 P.21 CAT Z
- Lightning Induced Transient Sus (SCSRD 5.2.2.3.8)
  - D6-16050-4 Sect 7.4.4 Level L1. (PA Supt. Only)
- X-RAY Radiation (SCSRD 5.2.2.3.10)
- Grounding & Bonding (SCSRD 5.2.2.3.11)
  - D6-44588 P. 3.5.4
- Electrostatic Discharge (SCSRD 5.2.2.3.9.1)
- Max Power Demand (SCSRD 5.2.2.3.1)
- Current Harmonics (SCSRD 5.2.2.3.12)
- Power Factor (SCSRD 5.2.2.3.2)

The area on the Electrical Qualification submittals that generate the most discussion and debate (read: wailing and gnashing of teeth) is the proper way to setup the IFE equipment for electromagnetic testing.

At Boeing we require all equipment that is to be located in a monument (Purser Work Station or Galley) or is to be located in a seat must be tested in the floated configuration. Thus Boeing says suspend the equipment above the ground plane during testing and DO-160 says bond the equipment to the ground plane during testing in the EMI chamber.

What you have to realize is that the DO-160 specification was created to test aircraft avionics that were to be mounted in either the electronics bay on a well bonded rack, or in the cockpit. Both of these locations provide a good ground directly to the IFE equipment case through the rack or mounting hardware. This is not the case with equipment mounted in a monument or in a seat. Because the Purser Work Station and Galley walls are not conductive, we must rely on the wires for our ground return. The same is true of the seats. The seat track does not provide a good mechanical ground, so again we must rely on the wires for our ground. Since the goal of testing is to verify the equipment will operate in the aircraft environment, we must adjust the test setup to match the aircraft installation as much as possible.

The "Floating" IFE Test Setup

As you can see from the illustration below, the test is configured much like the DO-160 configuration with three major differences:

1. All equipment and wires are suspended above the ground plane by 5cm (~2in).
2. LISN are used on the HI side of power input only.
3. A 1-meter minimum chassis ground wire is used.

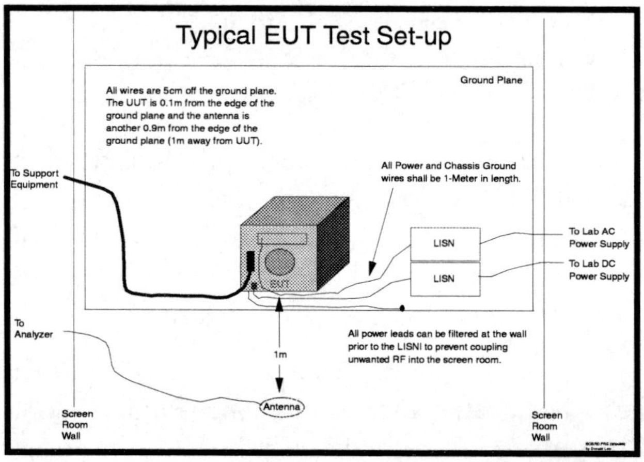

Composite Radiated Emissions Limit Line

The next tangent that Boeing takes from the DO-160 document is the Radio Frequency Emissions limit line. We use the DO-160 section 21 Category Z limit line everyone else uses with two additions: We add an HF interference notch of -10db below the specified curve in the frequency band from 2MHz through 30MHz, and a VHF interference notch of -10db in the frequency band from 100MHz through 150MHz.

To illustrate this, see the test photo below taken in the Boeing EMC lab from a COTS 10BaseT Ethernet test.

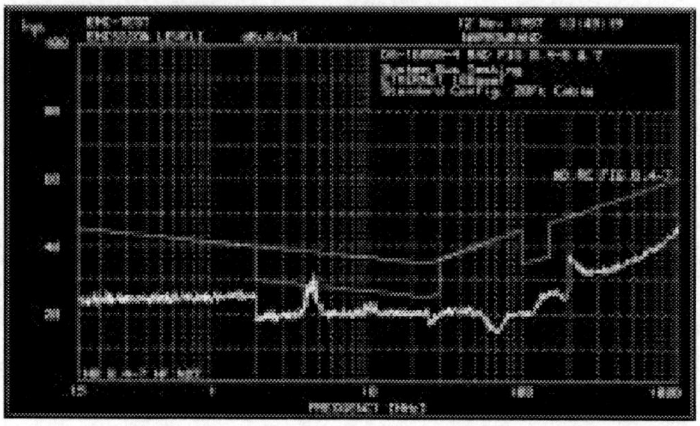

This test was taken before DO-160D was in effect, so the limit line was a modified DO-160C. The DO-160D limit line has the Boeing and other notches carved into it making it even more fun to try and pass.

The limit line below is the composite Boeing and DO-160D narrow band limit called out in revision B of the Boeing SCSRD. This will be the new requirement for radiated emissions testing for all IFE.

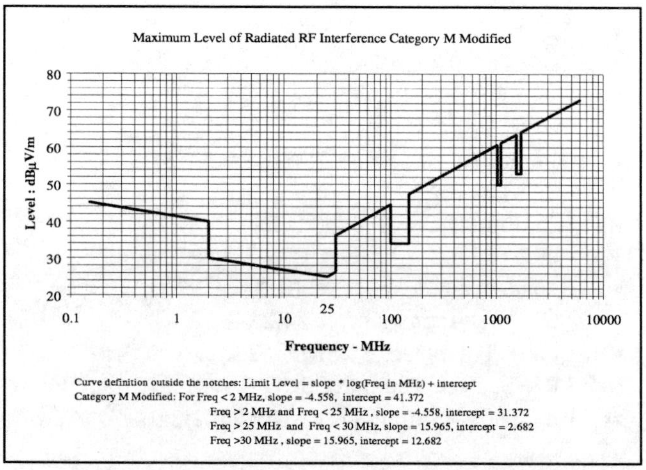

Sometimes an overage at a particular frequency may not be a problem on the aircraft due to IFE location or quantity. When we determine what emissions are a problem, we look to the aircraft frequency allocation chart below.

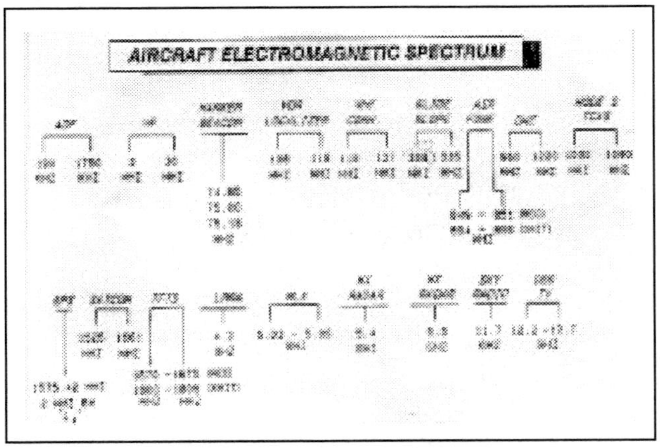

An aircraft is covered with antennas, each one used to perform a specific system function. The antennas that are most affected by IFE are the HF and VHF communication antennas. Since the HF and VHF are used for voice communication, we take great care to make sure that the entertainment system does not mess up voice communication on and off the aircraft. A "bad" emission in the lab at a frequency that is not in a navigation or communication radio's reception band may not be a problem. These are looked at on a case-by-case basis and depend on installation and other details.

IFE - The Processes for Success

Airline use of IFE is based on fleet decisions. The IFE marketplace is highly price and feature competitive with a relatively short product useful life. With this in mind, Boeing Cabin Systems has tried many business interface process structures before we came up with one that works fairly well for all parties. These methods breakdown into three process types that we will call Buyer Furnished Equipment (BFE), Seller Furnished Equipment (SFE), and our current process Cabin Systems Equipment (CSE).

Buyer Furnished Equipment (BFE) Process

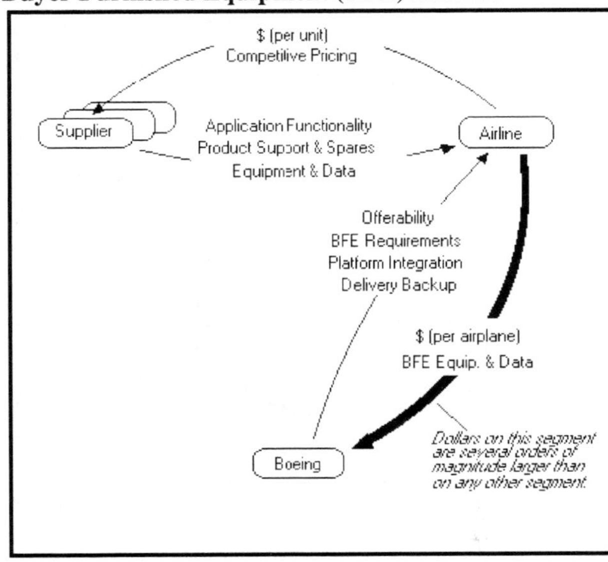

The BFE process was not equipped to handle the high level of system complexity, large scale integration, and multiple supplier coordination required.

Seller Furnished Equipment (SFE) Process

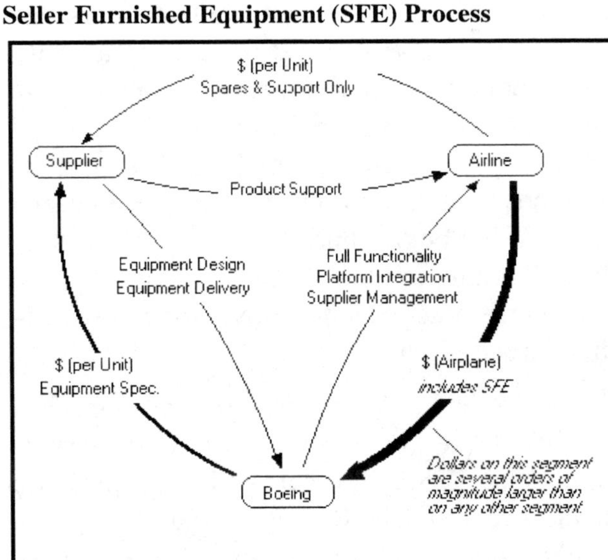

The SFE process could handle BFE deficiencies, but was unable to address airline fleet issues or airline desire for unique customization.

Cabin Systems Equipment (CSE) Process

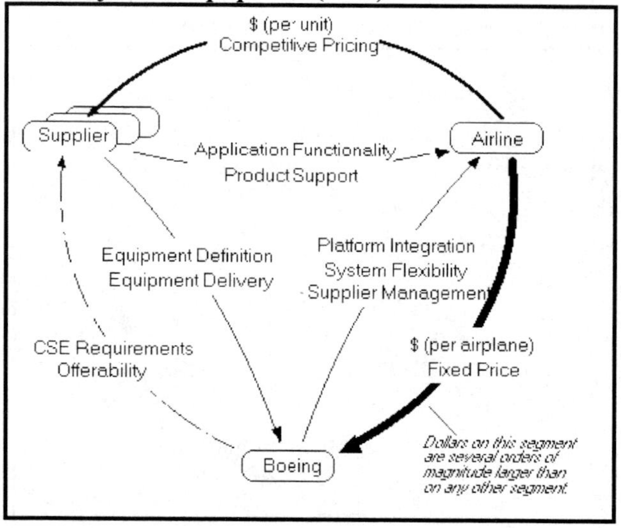

The CSE process bridged the gap with elements of both processes. Additional refinements and broader use within the cabin marketplace should improve effectiveness.

Retrofit First - The new way to do it

Boeing's new Retrofit First process came about because we needed a way to control the explosion of IFE systems and suppliers that wanted to get onto Boeing production aircraft. The production environment is much more rigid than the retrofit environment. We needed a way to give the airlines the latest IFE, and the suppliers a chance to install their systems and work out the bugs while staying out of the way of the Boeing production line. The only way to accomplish this was to move to a "Retrofit First" process.

To hedge their bets, several suppliers are turning to the Boeing Technical Service Agreement (TSA) to get help in areas we understand best: The Boeing Airplanes. We are excited about this, because everybody wins in this arrangement. The airline and IFE supplier get insights and help so they can apply the retrofit work to Boeing Offerability, and Boeing engineers get insights into the problems encountered during the system design and installation phases of the IFE project.

IFE - Criteria For Success

As the In-Flight Entertainment systems grow in complexity, we need to define what successful design is. This criteria is different for each party. For the Airline, the passengers like what they see and want to fly on their airplanes. For the IFE supplier, success is a system that works first time and continues to work without problems after being in-service for a few months. For the aircraft manufacturer, success is the airline is pleased with their new airplane and they want to order more.

IFE - Customer Satisfaction

Beyond the technical evaluations, lab and aircraft testing, one key component is customer satisfaction. To rate this you have to understand who your customer is. To be sure the airline is the customer, but who drives their requests. You guessed it, the flying passenger.

The typical passenger evaluates an In-Flight Entertainment system installed in their seat like any other piece of entertainment gear they might have in their home. If it is a stereo, it had better work as well as it does at home. Along that same line, if it looks like a TV it had better work as well as the TV at home. Don't expect them to marvel at the level of integration effort that put it in the seat, they only care if it works or not.

IFE - Conclusions

In closing, I would like to say that the next ten years look like they will be a very exciting time for the In-Flight Entertainment industry.

Catch the excitement of

Cabin Systems

Coming to an airplane near you

0-7803-5086-3 /98/$10.00 ©1998 IEEE

AIRPLANE CABIN NETWORK CONVERGENCE

Steven W. Russert, Boeing Applied Research & Technology
PO Box 3707, MC 7L-49, Seattle, WA 98124
steven.w.russert@boeing.com

Network Convergence

Advances in hardware and software have enabled effective transmission of voice and video over packet-based networks. The ability to combine voice, data and video on a single network can benefit both airframe manufacturers and airlines.

Most airplane cabins today have three networks; telephone, video, and data. The development of Digital Signal Processors (DSPs), more efficient voice and video encoder/decoders (Codecs), and high-speed network technologies makes it practical to combine the transport of all three of these types of information over a single network. While manufacturers and airlines can save money by exploiting this technology, the real opportunity lies in providing new services to the flying public.

This paper examines benefits and issues associated with converging cabin networks, and argues that creation of an open environment for development and deployment of services can prove worthwhile for all involved in IFE.

Network Integration: Technology Drivers

DSPs

DSPs (digital signal processors) are not new, even to desktop computers. DSPs have been used in traditional modem and sound boards. Now, however, low-cost, high-performance, multipurpose DSPs capable of running several different tasks concurrently are available. General-purpose DSPs with sophisticated operating systems, algorithm libraries, and friendly programming interfaces make sophisticated audio processing inexpensive and practical for compact, low-power devices. Using DSPs, one piece of hardware can perform multiple functions such as fax, modem, sound, MIDI (Musical Instrument Digital Interface), and video processing. Some desktop PCs (Personal Computers) already have DSPs on their motherboards.

The increasing power and ease of programming of these DSPs makes it practical to include advanced telephony functions on seat-back computers. Computationally demanding speech coder/decoders (Codecs) that would consume most of the power of a Pentium processor can be offloaded to a DSP, enhancing perceived voice quality. The DSP can be used to not only encode and decode voice signals, but can also be used for background noise reduction, especially useful in the aircraft cabin.

Audio Codecs

Selection of a speech codec is an important part of designing a converged voice and data network infrastructure. Codec performance is a function of bit rate, complexity, delay, and audio bandwidth.

Bit Rate

It is desirable that the codec bit rate be as low as possible within the constraints of meeting other requirements. One way to reduce the average bit rate of the codec is to use discontinuous transmission and combine it with a silence compression scheme. In such an arrangement, the codec only transmits its full

0-7803-5086-3 /98/$10.00 ©1998 IEEE

bit rate when there is active speech. During non-speech periods (silence or background noise) only a keep-alive signal is sent. Since non-speech periods are typically greater than 50% of the time in a two-way conversation, this reduces the average bit rate of the codec by over 50%.

Complexity

In circuit switched and wireless telephony, the codecs are most often implemented on special purpose DSPs. The speed of these devices is riding the same silicon curve as general-purpose processors. Nevertheless, for a DSP or a general-purpose processor implementation, MIPS (Millions of Instructions per Second) is the most important complexity attribute. Obviously lower complexity is better because it uses fewer resources, be they in the network or on terminal equipment, such as a PC. For the application of IP (Internet Protocol) telephony, it is desirable that the codec should use only a small fraction of the processing power of a seatback PC that is acting as a user's terminal.

Delay

In circuit switched telephony, a long distance call of 3000 miles in length has a minimum network delay of about 16 ms, just due to the speed of light, and the actual delay is higher. However, the perception of delay can be kept fairly well under control by the use of echo cancellers if the delay is below 200 ms. If the delay is between 200 and 400 ms, the user will perceive the delay even if there are no echoes, simply because the conversational dynamics are impaired – you can't interrupt the talker at the other end. If the delay is greater than 400 ms, the system must be used as if it were half-duplex, i.e. Push-to-Talk. Using IPv4 (version 4 of the Internet Protocol, the predominant Internet network protocol today), the present Internet has one-way delays that often exceed 400 ms. Geostationary satellites have a delay of about 500 ms due to the 55,000 mile round-trip distance.

All modern low bit rate speech coders process a block (or frame) of samples at a time, rather than a single sample at a time. Consequently the codec adds its own component of delay to the system. Typical blocks are 10 to 30 ms in length. In addition to the block length, the codec may also "look ahead" to help measure certain speech parameters. This "look ahead" also adds a delay. Finally, the processing of the speech signal at the encoder and the decoder must take place. The time it takes for this processing is a function of both the block size and the complexity. For a given complexity, a smaller block size means a lower delay.

Audio Bandwidth

In circuit switched telephony and digital wireless telephony, the speech signal is bandpass filtered and then sampled at 8000 Hz. The typical low frequency cut off is around 200 Hz and the high frequency cut off is around 3200 Hz. These cut off frequencies might be slightly higher or lower depending on the implementation and the application. In any case, this is often referred to as 3 kHz bandwidth or telephone bandwidth.

The importance of the audio bandwidth is that if the higher frequencies of a speech signal can be preserved, the reproduced speech sounds more like the speaker is right in the room with you. However, when a higher sampling rate is used it means there is more information being collected (because the sampling rate must be higher) and, in turn, this means that more information must be described. The consequence is that wider bandwidth codecs usually have a higher bit rate.

If a wideband codec is connected with the circuit switched telephone network, the network will filter out the higher frequencies and their benefit will be lost. Since there are a number of VOIP (Voice Over IP) applications that involve connecting with the circuit switched network, it is probably a waste of bit rate and bandwidth to use a wideband codec when interconnecting with the circuit switched network. However, the use of a high bandwidth codec for onboard telephony (where LAN bandwidth is available) will provide a more intelligible conversation.

Video Encoding Schemes

New techniques for compressing and transmitting video streams facilitate the use of packet networks for video transport. The H.261 ISO/IEC (International Standards Organization) standard provides for interactive video at multiples of 64 kilobits per second (kbps). Motion Picture Experts Group MPEG-1 and MPEG-2 are encoding techniques for high-quality video transport over higher-speed networks. MPEG-1 (ISO standard 93a) uses about 1.2 million bps (Mbps) for a 24 frame-per-second video stream at up to 768x576 resolution. MPEG is suitable for asymmetrical processing, where more computing power is used to efficiently compress the video stream, and less computing power can decode the stream for playback. The MPEG stream consists of I-frames, which contain all the information needed to define a video image, P-frames, which use the previous image and define changes to it, forming a new image, and B-frames, which define an image based on both previous and subsequent images. MPEG provides 16-bit audio coding at the same sampling frequencies as Compact Disk (CD) audio; 44.1 kHz and 48kHz, as well as at 32kHz.

MPEG-2 is an extension of MPEG-1 designed to provide higher resolution video, using bandwidths from 2 to 80 Mbps. MPEG-2 defines multiple profiles, with provision for support of interlaced video and other advanced features, including those to support HDTV. MPEG-1 audio coding supplies up to five full bandwidth channels (left, right, center and two surround channels), plus an additional low frequency enhancement channel, and/or up to seven commentary or multilingual channels.

High-speed networks

The advent of technologies like switched 100Mbs Ethernet, gigabit Ethernet, and RFC 1394 "FireWire" at up to 400Mbs means that the LAN bandwidth used by a telephone conversation is not significant in terms of network capacity. Properly engineered LANs and campus networks can accommodate IP telephony data volumes in addition to digital video distribution, web browsing and other applications. For voice transport over lightly loaded high-speed switch-based LANs, we can choose a voice coder that is simple, has low delay, and imposes little load on the end system, as even pulse code modulation (PCM) voice at 64kbps consumes a very small percentage of available network bandwidth. High-speed networks can also support multiple video streams, facilitating implementation of Video on Demand (VOD) service.

The challenge in transporting interactive voice over packet networks is delay, as discussed above. Delay in the network consists of transmission time, propagation time, and queuing. The time required to send a given number of bits is determined by the speed of the network, specifically the speed of the slowest link in the path. Propagation delay is determined primarily by the length of the path. Variation in delay (delay jitter) is caused by queuing. A short voice packet arriving at the head of the queue may have to wait until a longer data packet, already being transmitted, is finished. Multiple queues or other priority schemes cannot eliminate this delay variation. Higher network speeds reduce the delay variation, since less time is required to send a packet of information. An ATM (Asynchronous Transfer Mode) network may be used to provide a Constant Bit Rate virtual circuit (VC). This CBR VC emulates a dedicated circuit, providing the dedicated bandwidth and predictable delay of a dedicated circuit, but also the inefficiencies of sending idle bits when no traffic is presented by the endpoint. ATM also specifies Variable Bit

Rate VCs, which can provide acceptable performance for voice if delay variation in the network is tightly controlled.

Network Integration: Business Drivers

Cost

An airplane is cheaper to design and build if the network design and integration tasks can be done once for a group of planes, rather than on a plane-by-plane basis. Less complexity in assembly also means lower cost (especially if assembly can be done in one operation, rather than several). One of the seminal books on lean manufacturing[1] points out that an automobile with all available options can be built more cheaply on a production line where all cars have those options, than an auto with no options on a production line that builds with all combinations of options.

Eliminating redundant network infrastructures reduces the weight of cabling and components that must be carried aloft, reducing fuel consumption and increasing payload and range. Less power is required to operate network components and therefore less heat is generated, both of which result in fuel savings. Since space is at a premium in the airplane cabin, elimination of extra cabling and components adds to the space available for people and their baggage. Fewer networks also mean fewer problems with EFI (Electromechanical Frequency Interference).

In the Internet world, cost savings from IP telephony fall into two categories: toll call arbitrage and infrastructure efficiency. Arbitrage allows retail users to benefit from the different ways in which network use is charged for in current voice and data networks. Two end users can make local calls to their respective Internet Service Providers (ISPs) and conduct a long-distance telephone call for the connect time charged by the ISPs (often a monthly unlimited-use fee). Infrastructure efficiency gains will allow telephone carriers and corporations to save on infrastructure costs and pass those savings on to customers in order to compete.

New applications

The integration of voice and data enables new applications. Enhancement of voice mail is one area of interest. Voice mail systems have traditionally been run from the "station side" of the Private Branch Exchange (PBX) or telephone switch. This approach requires per-PBX integration effort and isolates the voice mail servers from each other. When the voice calls are carried on the data network, the voice mail server is just another node on the data network, and can communicate with other voice mail servers as well as e-mail servers (for integrated "universal in-box" functionality). Another application area is conferencing. Desktop conferencing is an extension of IP telephony, with whiteboard, application sharing and video added to regular phone call functionality. IP telephony may also enable new types of groupware, software that helps people work together more effectively, even when they are not sitting in the same building or office. IP telephony facilitates integration of telephony and web browsing, where a customer can peruse a catalog on a web site, then simply click a button to talk to a company representative about technical details of the product depicted. The company rep can have the web page the customer is viewing brought up automatically along with the voice connection, to increase the efficiency of sales/support staff.

[1] "The Machine that changed the world" Womack, JP;Jones,DT, Roos, D Copyright 1990, Harper Collins ISBN# 0892563508

Aircraft-specific drivers

Flying Public Expectations

The commercial aircraft is presented to the flying public as a high-technology environment. The availability of computing, telephony and Internet access everywhere from the office to the home, the school, and the public library builds an expectation that one can remain connected even in-flight.

Standards-based Networks

Data networks based on commercial Local Area Network (LAN) standards such as 10base-T, 100base-T, FDDI, and ATM are being installed on planes to take advantage of COTS hardware and software. Use of these standard networking technologies simplifies the integration of COTS applications, including IP Telephony.

Computing power on-board

Moore's law is making it possible to put compact and powerful computers in seatbacks. New portable computer technology provides low power requirements and low heat generation. Portable computer technology also provides answers to some of the environmental problems encountered onboard aircraft, like temperature variation, power fluctuation and vibration.

Windows CE

The Windows CE (according to Microsoft, CE doesn't stand for anything) operating system from Microsoft offers unique benefits for airplane cabin systems. Microsoft® Windows® CE is a modular operating system, made up of software building blocks that can be assembled in many different ways to create new devices. This means that if you want to build a device, you can pick and choose only those pieces you need and keep the size of your software consistent with the size of your device.

Windows CE uses a subset of the Microsoft Win32® Application Programming Interface (API) that is used on Windows-based desktop and server computers. Developers can use the same development tools, such as the Microsoft Visual C++® or Microsoft Visual Basic® development system, and communications protocols to create applications for Windows CE-powered devices.

One of the more important benefits of CE is its instant-on capability. This capability not only allows a seatback system to be available immediately without boot-up time, but with minimal battery backup can also isolate the system from power interruptions.

Airline Competition

Passenger services are an important way for airlines to differentiate their service from that of their competitors. Airplane fleets with standards-based network infrastructure and IFE systems that provide flexibility and easy integration of new applications will enable airlines to stay competitive in a rapidly changing market.

Benefits

Cost

Integration of voice and data networking in the airplane can save money for the airplane manufacturer as well as the operator. Reducing the number of components and network cabling systems on the airplane reduces weight, space requirements, power consumption, heat production, EFI problems and overall complexity. These reductions translate to increased payload and reduced operating costs. The manufacturer can save money by reducing the design and integration effort spent on each individual airplane, and the operator can save money on fuel, increased payload and reduced maintenance.

Service

Seat-to-seat

Telephony can be an attractive function within the airplane as well as air-to-ground. Many teams of people wish to discuss upcoming or just completed meetings. The current environment requires jousting with flight attendants attempting to move food and beverage carts. Seat-to-seat calling can provide a better working environment, since each party can have some space to spread out materials at their seat.

On-board conferencing

The obvious extension to seat-to-seat calling is on-board conferencing. H.323 clients (like Microsoft NetMeeting or Lucent MMCX) can be used on seatback computers and one server device can function as both an interface to the off-board facilities and a PBX for onboard conferencing. With the addition of a headset with active noise cancellation, the experience can be superior even to conversing with someone in the next seat, due to the reduction of ambient noise. This capability can turn dead time into productive time for travelling teams of salespeople or engineers. And who knows, perhaps on-board conferencing can reduce the churning caused by tour leaders trying to trade seats after boarding to bring their group together!

Flexibility

Technology is developing rapidly. New ways of using computers to deliver multimedia content to users are being introduced constantly. Airplanes with an integrated cabin network and computing infrastructure that uses standard software interfaces can be upgraded more quickly and easily.

Integrated off-board capacity

The converged network infrastructure on the airplane can facilitate use of a single air-to-ground link technology to carry both voice and data. This system may be a new infrastructure or enhancement of existing systems.

Issues

Technical Issues

Handset / interface

Combination telephone handset/remote control units are popular today. The advantages of noise cancellation noted above can only be achieved with a stereo headset. Cost may limit such devices to first or business class. Also, any unsecured i/o (input/output) device will inevitably be sat on by passengers, yanked on by the cord attachment, etc. – so it has to be tough. For control, touch screens would make for a nice interface, but add significant cost.

Embedded Computing Vs Customer Computers

Making several hundred networked multimedia computers work reliably is a challenge. Adding other computers of unknown configuration to the network is a real problem. DHCP (Dynamic Host Configuration Protocol) servers can provide IP address assignment as well as gateway and DNS (Domain Name Service) server addresses, if the computer being connected to the network has the necessary network adapter, has the TCP/IP (Transmission Control Protocol/Internet Protocol) protocol installed, and is properly configured for DHCP. The biggest challenge is that many users (who do not use DHCP assignment of these parameters) would have to manually reconfigure their network settings at the end of the flight. As laptop owners who work in multiple network environments know, the multiple profiles provided by Windows 95 today do not yield independent networking profiles. One solution for such ad-hoc networking would be to provide a dial-up connection capability onboard. The customer could then create a dial-up networking configuration for onboard use. This dial-up

0-7803-5086-3 /98/$10.00 ©1998 IEEE

type connection would actually be to a modem on the plane, so that the bandwidth of the air-to-ground link could be conserved. Providing 110vac power to laptop computers creates another set of penalties in weight and space, as well as additional EFI and power harmonics challenges. Promise for 'nomadic' computing comes from efforts underway to create an environment in which a user's profile is available on whatever device they use to access the network. Then the seatback computer could present the user's personal desktop profile. This, of course, requires the onboard network to be integrated with the Internet. Another technology that may prove useful is Jini™ from Sun Microsystems. Jini would provide a way for devices to register themselves on a network automatically, and for devices to discover and make use of the services of other devices on the network.

Market Issues

Existing IFE Systems

The proprietary IFE (Inflight Entertainment) systems already installed provide a defined set of entertainment options at a known cost. The original IFE vendor provides the upgrade path, retaining some of the infrastructure. Many narrow body airplanes have no IFE installed.

Who Supplies Networks?

Boeing wants to make it simpler to design and build an airplane, however Boeing does not want to compete with our suppliers. The inclusion of a standard network in the airplane is attractive, but only if it meets the needs of the customer airline and their selected IFE vendor.

Conclusion

While travelers usually desire entertainment, the business traveler increasingly needs to remain productive during flight. Creating a business-capable computing and networking environment on the airplane can produce new opportunities to provide services that differentiate an air travel experience in the eyes of the business traveler. A flexible, standards-based computing and networking infrastructure will facilitate implementing new entertainment options. The marriage of telephony and computing, the use of standard networks, and conformance to industry-standard APIs will allow the airplane IFE system to ride the Internet/web technology wave. Choose your paradigm now: locked-in customers via proprietary technologies and interfaces, or growth that takes advantage of the creative work being done throughout the world. The future most popular and profitable application for airplane cabin computing and networking may be under development right now; but completely outside of the airplane and IFE industries. The computing industry has shown that companies that take advantage of open interfaces prosper, while those that stay with closed systems wither. Do you want to offer only the applications developed by people in one IFE vendor's organization, or your choice of the applications being developed by thousands of developers around the world?

The airplane cabin presents many challenges that we must overcome to extend the reach of the Internet and the business computing environment there, but the effort promises great rewards.

GUIDANCE CHARACTERISTICS OF GNSS LANDING SYSTEMS

D. Alexander Stratton, Rockwell Collins, Inc., Cedar Rapids, IA

Abstract

Technical and operational concepts for precision landing have evolved around the signal in space characteristics of the Instrument Landing System (ILS). For the Global Navigation Satellite System (GNSS) Landing System (GLS), many of these characteristics will be dependent on computational processes inside the airborne receiver. Development of GLS standards creates an opportunity to improve on inherent ILS limitations while retaining useful features. This paper compares proposed GLS guidance algorithms including an approach where nominal ILS features are reproduced as well as alternatives proposed to eliminate features seen as undesirable. Simulated deviation outputs are derived for various maneuvers, including orbits, on-path, and near-path approach trajectories. The potential impact of alternative approaches on aircraft performance during capture, flare, and rollout is shown. Additional considerations addressed include the impact of changes on field verification and flight inspection and the applicability of these concepts to the Multi-Mode Receiver (MMR).

Introduction

Precision-landing systems, including ILS, GLS, and the Microwave Landing System (MLS), provide guidance in the form of vertical and lateral deviations from an intended approach path. Significant characteristics of ILS guidance arise from the signal in space (SIS) created by two antenna arrays [1]. A glide-slope array located near the touchdown zone provides a UHF signal for vertical guidance, while a localizer array typically located off the runway stop end provides a VHF signal for lateral guidance. Each signal contains 90 and 150 Hz amplitude modulations, and the antenna patterns are adjusted so that the modulation depths are equivalent along the final-approach path. Angular displacements from the final-approach path produce proportional changes in the depth of the individual modulations, so that a Difference in Depth of Modulation (DDM) is sensed by an airborne ILS receiver. The proportionality between angular displacement and DDM, called Displacement Sensitivity, is maintained within specifications by adjustments of the antenna patterns [2]. The receiver's DDM outputs can be displayed "raw" for manually piloted approaches and processed by flight-control systems to produce vertical and lateral steering commands for manual and automatically controlled landings.

For GLS systems, the characteristics described above are independent of the SIS; instead, they are produced by algorithms inside the GLS receiver. The development of industry standards including the Local-Area Augmentation System (LAAS) [3] and the Wide-Area Augmentation System (WAAS) [4] has spawned numerous discussions concerning the formulation of these algorithms. A natural approach to defining them is to reproduce the SIS characteristics of ILS – referred to herein as ILS Look-Alike. On the other hand, a variety of alternative schemes have been proposed to avoid introducing real and perceived ILS weaknesses into GLS. This paper compares several proposed algorithms in simulated maneuvers to quantify the differences between them. Of concern are vertical guidance during glide-slope capture and flare, and lateral guidance during rollout. Other consid-erations include the impact on field measurement. The ILS Look-Alike formulation is seen to provide the highest degree of compatibility with current operational Category III landing systems. Compatibility is of special value given the desire

0-7803-5086-3 /98/$10.00 ©1998 IEEE

to efficiently introduce GLS technology into the anticipated multi-mode landing environment.

ILS Guidance Characteristics

In addition to the characteristics discussed above, two other ILS characteristics will be referred to in later development of GLS algorithms: the conical glide-slope antenna pattern, and the dependency of localizer width on runway length. The antenna pattern of the image-array glide-slope is generated using the terrain about the antenna as a reflecting surface. Because the reflecting surface creates a "mirror image" of the antenna phase center under the reflecting surface, surfaces of constant DDM are conical. Because the glide-slope array is installed several hundred feet laterally offset from the runway centerline, the final approach path (i.e., the contour of zero DDM on centerline) is a hyperbola in space. The hyperbolic characteristic becomes most significant for heights within 200 ft of touchdown.

The localizer array is typically located several hundred feet past the stop end of the runway. The array is aligned with the projected runway centerline, providing a dominant 90 Hz modulation on the right course and 150 Hz modulation on the left course. For typical runways, the displacement sensitivity of the localizer is adjusted to provide a sensitivity of 0.00145 DDM/m at the runway threshold. Consistent sensitivity during the final stages of the approach is advantageous, but it also changes the angular width of the localizer, which affects localizer capture. On long runways, the angular width is held to a minimum value of 1.5° per 0.155 DDM, resulting in lower sensitivity than 0.00145 DDM/m at the threshold. On short runways, the width is held to a maximum value of 3° per 0.155 DDM, which results in a heightened sensitivity at the threshold.

Proposed GLS Guidance Characteristics

Airborne LAAS and WAAS equipment produces deviations by first determining position using Differential GNSS techniques. The LAAS VHF data broadcast will provide differential corrections and a Final-Approach Segment (FAS) data block (described below) that defines the final-approach path [3]. WAAS equipment will receive differential corrections from WAAS satellite broadcasts, and will access an on-board data base containing the FAS [4]. The deviation outputs are then calculated using an algorithm such as the ones developed in this section.

Final-Approach Segment Definition

All of the proposed deviation formulations use a standard FAS data block, as defined in Table 1 and illustrated in Figure 1. Example values of Table 1 are used in the simulations described later.

The algorithms below use a Cartesian coordinate system originating the RDP and including vertical, cross-track, and longitudinal unit vectors ($\mathbf{u}_{vert}$, $\mathbf{u}_{lat}$, and $\mathbf{u}_{rw}$, respectively). Local vertical for the approach is defined as normal to the WGS 84 ellipsoid at the RDP. The local level plane for the approach is defined as a plane perpendicular to the local vertical passing through the RDP (i.e. tangent to the ellipsoid at the RDP). The vectors $\mathbf{u}_{lat}$ and $\mathbf{u}_{rw}$ lie in the local level plane, with $\mathbf{u}_{rw}$ pointing toward the FPAP. The final approach path is defined as a line with an angle (defined by GPA) relative to the local level plane passing above the RDP at a height defined by DCH. The Glide-Path Intercept Point (GPIP) is the point where the final approach path intercepts the local level plane. The GPIP may actually be above or below the runway surface depending on the curvature of the runway.

Proposed GLS Vertical Guidance Definitions

ILS Look-Alike Vertical Deviations

The ILS Look-Alike formulation incorporates the glide-slope characteristics described above, including the conical offset glide-slope. The vertical deviations are stated with respect to a GLS Elevation Reference Point (GERP) as illustrated in Figure 1. The GERP is laterally offset from the GPIP by a fixed value 150 m.

Table 1. Final-Approach Segment Parameters

Data	Definition	Example Value
Runway Datum Point (RDP)	Latitude (WGS-84) Longitude (WGS-84) Height above WGS-84 ellipsoid	N39° 27.83290' W74° 35.43748' – 12.2 m
Flight Path Alignment Point (FPAP)	Latitude (WGS-84) Longitude	N39°27.05615' W74°33.56691'
Datum Crossing Height (DCH)	Desired height of the FAS at the RDP	15.2 m
Glide Path Angle (GPA)	Desired descent angle of the FAS	3°

For the sake of convenience (as well as for future applications), a *rectilinear* deviation (i.e., a deviation in distance) is defined as well as an angular deviation. The rectilinear vertical deviation (d_v in Figure 2) is defined as the distance of the computed airplane Guidance Control Point (GCP) position from the surface of a vertically-oriented circular cone with its apex at the GERP and with an angle defined by the GPA. The angular vertical deviation (α_v in Figure 2) is defined as the angle whose sine is the ratio of the rectilinear vertical deviation and the distance from the aircraft to the GERP (D_v in Figure 2).

The angular vertical deviation may be computed using the relationship:

$$\alpha_v = \tan^{-1}\left\{\frac{[\mathbf{u}_{vert}\cdot\Delta\mathbf{r}]}{\sqrt{[\mathbf{u}_{lat}\cdot\Delta\mathbf{r}]^2+[\mathbf{u}_{rw}\cdot\Delta\mathbf{r}]^2}}\right\} - \text{GPA} \quad (1)$$

where $\Delta\mathbf{r}$ is the vector pointing from the GERP to the aircraft GCP. The angular vertical deviation is expressed in terms of DDM via:

$$\text{Vertical Output} = \frac{0.175\alpha_V}{0.25\text{GPA}} \quad (2)$$

(Note that 0.175 DDM equals 150 μA). The width is scaled to 25% of glide-path angle, consistent with Annex 10 MLS characteristics, which are slightly inconsistent with Annex 10's 24% scaling for ILS. Digital equipment will provide linear scaling beyond the 0.175 DDM full-scale deflection. However, the angular vertical deviations are flagged invalid when the bearing to the GERP is more than ± 90° from the FAS bearing (i.e., the airplane is beyond the GPIP).

Conical With Apex at GPIP

An alternative to the ILS Look-Alike approach, referred to herein as Conical-GPIP, is to define the GERP at the GPIP (i.e., zero GERP Offset). Other than the zero offset (which removes the flare characteristic), the description of Conical-GPIP is the same.

Planar Vertical Deviations

The Planar alternative abandons the conical nature of the glide-slope entirely. In the planar formu-lation, the GPIP is defined as above, but the angular vertical deviation is computed via:

$$\alpha_v = \tan^{-1}\left\{\frac{[\mathbf{u}_{vert}\cdot\Delta\mathbf{r}]}{|\mathbf{u}_{rw}\cdot\Delta\mathbf{r}|}\right\} - \text{GPA} \quad (3)$$

Note that only the along-track and vertical components of $\Delta\mathbf{r}$ affect the computation. Output scaling is identical to ILS Look-Alike.

Hyperbolic Vertical Deviations

A fourth alternative preserves the conical glide-slope and its flare characteristic, but uses a more complex reference surface, a hyperboloid of two sheets, centered above the GPIP. Using this method, the angular vertical deviation may be computed via:

$$\alpha_v = \tan^{-1}\left\{\frac{[\mathbf{u}_{vert}\cdot\Delta\mathbf{r}]}{\sqrt{[\mathbf{u}_{lat}\cdot\Delta\mathbf{r}]^2+[\mathbf{u}_{rw}\cdot\Delta\mathbf{r}]^2}}\right\} - \alpha_{ref} \quad (4)$$

where $\Delta\mathbf{r}$ is a vector pointing from the GPIP to the GCP, and α_{ref} is a position-dependent reference path angle, which may be defined by:

$$\alpha_{ref} = \tan^{-1}\left\{\tan(GPA)\sqrt{\frac{D^2_{GERP}}{[\mathbf{u}_{lat}\cdot\Delta\mathbf{r}]^2+[\mathbf{u}_{rw}\cdot\Delta\mathbf{r}]^2}+1}\right\}$$

where D_{GERP} is the lateral GERP offset of a conical glide-slope providing an equivalent flare characteristic along the procedure centerline. Output scaling is the same as ILS Look-Alike (along the centerline).

Two Proposed GLS Lateral Deviation Definitions

ILS Look-Alike Lateral Deviations

The ILS Look-Alike formulation incorporates the ILS localizer characteristics described above, including the variable width. The deviations are referenced to a GLS Azimuth Reference Point (GARP). The GARP is defined to be beyond the FPAP along the procedure centerline by a fixed GARP Offset value of 305 m (approximately 1000 ft). The rectilinear lateral deviation (d_{lat} in Figure 3) is defined as the distance of the measured airplane Guidance Control Point (GCP) position from the lateral deviation reference plane (positive as shown). The angular lateral deviation (α_{lat} in Figure 3) is defined as the angle whose tangent is the ratio of d_{lat} and the horizontal distance from the GCP to the GARP (D_{lat} in Figure 3). The angular lateral deviation, α_{lat}, can be computed by the relationship:

$$\alpha_{lat} = \tan^{-1}\left\{\frac{\mathbf{u}_{lat}\cdot\Delta\mathbf{r}_l}{|\mathbf{u}_{rw}\cdot\Delta\mathbf{r}_l|}\right\} \quad (6)$$

where $\Delta\mathbf{r}_l$ is the vector pointing from the GARP to the aircraft GCP. The angular lateral deviation output is expressed in terms of DDM or µA, determined via:

$$\text{Lateral Output} = k\,\alpha_{lat} \quad (7)$$

The scale factor, k, is derived from Table 2 using the distance from RDP to GARP, D_G. The table values are chosen to limit the total course width (i.e., the angular difference between –0.155 DDM and 0.155 DDM) to less than 6 deg. (Note that 0.155 DDM equals 150 µA.) For typical runways, the scale factor is adjusted to maintain 0.00145 DDM/m sensitivity at the RDP. The equipment may present valid lateral guidance whenever availability criteria are met to take advantage of expanded lateral coverage vs. ILS.

Table 2 GLS Lateral Deviation Output Sensitivity

D_G	**Scale Factor k (DDM/deg)**
< 2040 m	$\frac{0.155}{3}=\frac{31}{600}$
between 2040 m and 4082 m	$0.155\cot^{-1}\left(\frac{106.9}{D_G}\right)$
> 4082 m	$\frac{0.155}{1.5}=\frac{31}{1200}$

Fixed-Width Lateral Deviation

The alternative Fixed-Width formulation is designed to eliminate the width variations as well as to eliminate the region of high sensitivity associated with flight over the localizer antenna. The Fixed-Width approach can be derived from the ILS look-alike approach, by fixing the GARP at a distance of 10,000 ft from the RDP (i.e., D_G = 10,000 ft regardless of the FPAP), which from Table 2 results in a fixed ±2° width. The same equations may be used to compute the angular deviation unless the computed GCP position is within 10,000 ft of the GCP. Because GLS is planned for use for rollout guidance on runways longer than 10,000 ft, use of angular deviations would produce unacceptably high sensitivity when passing over the GARP. To prevent this occurrence, the angular deviations are "linearized" (i.e., the rectilinear deviation is output in place of angular deviation) using a fixed

0-7803-5086-3 /98/$10.00 ©1998 IEEE

scaling of 0.00145 DDM/m when within 10,000 ft of the GARP:

$$\begin{matrix}\text{Lateral} \\ \text{Output}\end{matrix} = \begin{cases} k\alpha_{lat} & D_{lat} \geq 10{,}000\text{ft} \\ (0.00145)d_{lat} & D_{lat} < 10{,}000\text{ft} \end{cases} \quad (8)$$

Comparison of Proposed Definitions

Glide-Slope Characteristics

Glide-slope capture characteristics have emerged from discussions as a potential problem for planar deviations. Figure 4 compares the capture altitudes of the proposed GLS algorithms. The result is obtained by flying a simulated orbit about the FPAP at a distance of approximately 6.6 nm (corresponding to 5 nm. from the RDP when on course). When the bearing is more than ± 4° of centerline, the altitude differences are more than 4 m, (and the LAAS specified 95% system accuracy is 4 m). Therefore, differences in glide-slope capture characteristics are important if glide-slope is captured before localizer. This will be technically possible for GLS, so the lower capture altitude of the planar deviations may be a valid concern. Note the slight asymmetry of the ILS Look-Alike vs. the Conical-GPIP and Hyperboloid. This asymmetry is an unintended feature of the ILS, but it creates no known operational concerns.

Whether the flare characteristic provided by the ILS is needed has been an important issue. The low-altitude characteristics of the GLS algorithms are compared in Figure 5. The result is obtained by flying a straight-in approach intercepting the GPIP with a 3° glide-path angle. A recent study [5] on the integration of GLS with an existing air-transport flight-control system inadvertently re-discovered the value of the flare characteristic. The Monte Carlo analysis compared the ILS and GLS touchdown footprints using JAR-AWO certifi-cation criteria for a certified CAT-IIIb auto-land system. The initial results exhibited a mean offset (i.e., bias) between the two footprints. Further investigation showed that the simulated GLS, which used a planar vertical deviation charac-teristic, was causing the system to land short of the ILS by a significant amount as compared to the JAR-AWO acceptance criteria. Whether or not a particular auto-land system is affected by the ILS flare depends on the system's particular gain scheduling characteristics (e.g., how radio altimeter is blended in). Justification for elimination of the flare characteristic would appear to involve significant analyses of the installed systems. This is the key factor behind inclusion of the lateral offset in the ILS Look-Alikc approach.

Field Verification Considerations

Field evaluation of GLS is needed to ensure correct promulgation of the FAS data and to validate that the LAAS supports the intended operation. To perform these evaluations it is necessary to separate known design characteristics from actual errors. A basic tool for performing ILS field evaluations is an optical theodolite, which is situated near the glide-slope mast. By adjusting the angular orientation of theodolite to track an instrumented aircraft on the approach, instrument inherently generates a conical measure of "truth" for glide-slope evaluations. Therefore, the ILS Look-Alike approach is directly measurable through current field-evaluation methods (i.e., where it is currently sited [6]).

The impact of the alternative approaches is illustrated by simulating the proposed algorithms for a 3° approach with a 1° offset (to simulate flight technical error). The results of this simulation, as shown in Figure 6, indicate that a theodolite would observe approximately 20-40 uA raw error during field evaluations if the receiver implements Planar, Conical-GPIP, or Hyperboloid methods. Equip-ment which uses these methods can be used effectively only by a flight-reference system capable of three-dimensional positioning (e.g., a DGPS reference system or an automatic flight-inspection system). While such systems are used in more developed countries, it is nonetheless clear that field-evaluation specialists and flight-inspection system developers will need to understand the form of the GLS guidance algorithms implemented by their equipment in order to effectively evaluate GLS in the field. The ILS Look-Alike approach has an inherent compatibility with existing systems.

Comparison of Roll-Out Characteristics

Current Category IIIb auto-land systems provide a rollout function using ILS lateral guidance. These systems must regulate lateral displacement using internally computed estimates of rectilinear deviation. An example of such an estimate is:

$$\hat{d}_{lat} = \hat{s}\alpha_{lat}\hat{D}_{lat} = \alpha_{lat}(D_{lat} + e_D)(s + e_s) \qquad (9)$$

where s represents actual scaling from DDM to radians, e_D represents a bias in the determination of distance to the localizer, and e_s represents an error in scaling from DDM to radians. Figure 7 compares a hypothetical estimated d_{lat} computed during a simulated rollout with a constant lateral displacement of 10 m from the runway centerline. Using the ILS Look-Alike method, a 10% error is incurred due to simulated e_D and e_s. However, the Constant-Width method would produce significant under-estimate of the 10-meter error. It is clear that systems using the Constant-Width method will require significant analysis of rollout characteristics of the installed system.

Mitigating this issue is the expectation that next-generation auto-landing systems will be able to take advantage of the rectilinear deviation outputs directly provided by GLS to by-pass the estimation process. It also should be noted that the fixed width and consistent sensitivity of the Fixed-Width approach are operational advantages for Category I systems, for which rollout guidance is a non-issue.

Application to the MMR

The Multi-Mode Receiver (MMR) combines ILS, MLS, and GLS precision-landing modes in a single Line-Replaceable Unit. Rockwell Collins produces a series of modular ARINC 755 and 756 MMR equipment for transport-category aircraft. The MMR is intended to support display and auto-land systems from a common set of digital or analog interfaces. Use of the ILS Look-Alike method will enable greater autopilot and display-system commonality between landing modes. This is of particular significance for retrofit applications, where re-design of existing systems is prohibitive.

Conclusions

Given the long evolution of precision-landing technology, perhaps it is unsurprising that even eliminating so-called "warts" of the ILS might involve significant work. All of the schemes presented here are of similar receiver complexity, so equipment cost is not a consideration. Therefore, simplification of aircraft integration and certification is the over-riding factor behind the advocacy of the ILS Look-Alike approach. Future auto-landing systems will benefit from continuing the evolutionary development path, especially given the extended lifetime that the ILS will likely enjoy.

Acknowledgements

The concepts presented here were developed through discussions that included Daryl McCall, Gary McGraw, and Joel Wichgers of Collins, and Tim Murphy of Boeing. This work was supported by the Navigation Systems Engineering Department of Rockwell Collins.

References

[1] Kayton and Fried, Ed., *Avionics Navigation Systems*, John Wiley and Sons, 1969.

[2] *International Standards and Recommended Practices, Aeronautical Telecommunications, Annex 10 to the Convention on International Civil Aviation*, Vol. I, International Civil Aviation Organization, July 1996.

[3] RTCA Paper No. 037-98/SC159-778, *Minimum Aviation System Performance*

Standards for the Local Area Augmentation System, RTCA Inc., April 1998.

[4] RTCA DO-229a, *Minimum Operational Performance Standards for the Wide-Area Augmentation System*, RTCA Inc., July 1998.

[5] G. A. McGraw and J. A. Finley, "Analysis and Testing of GNSS Landing System Sensor Error Requirements," *Proceedings of Inst. of Navigation ION GPS-97*, Kansas City, MO, pp 1593-1602, Sept. 16-19, 1997.

[6] Order 8200.1A, *United States Standard Flight Inspection Manual*, U.S. Dept. of Transportation, Federal Aviation Administration, May 1996.

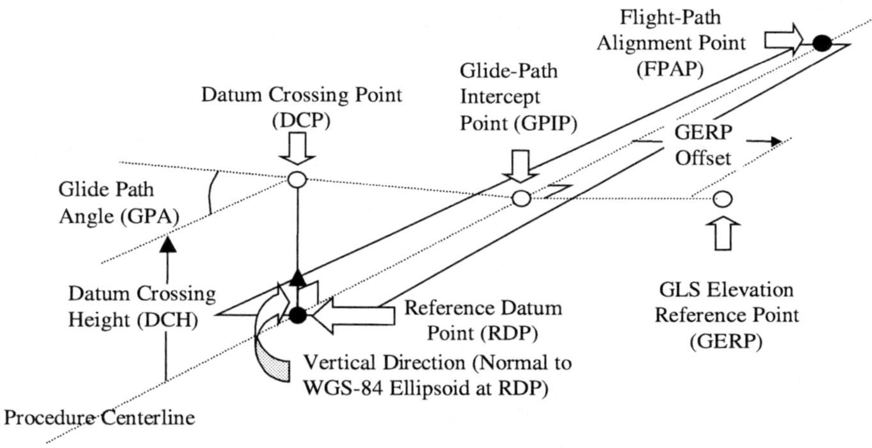

Figure 1. GLS Final-Approach Segment Definition

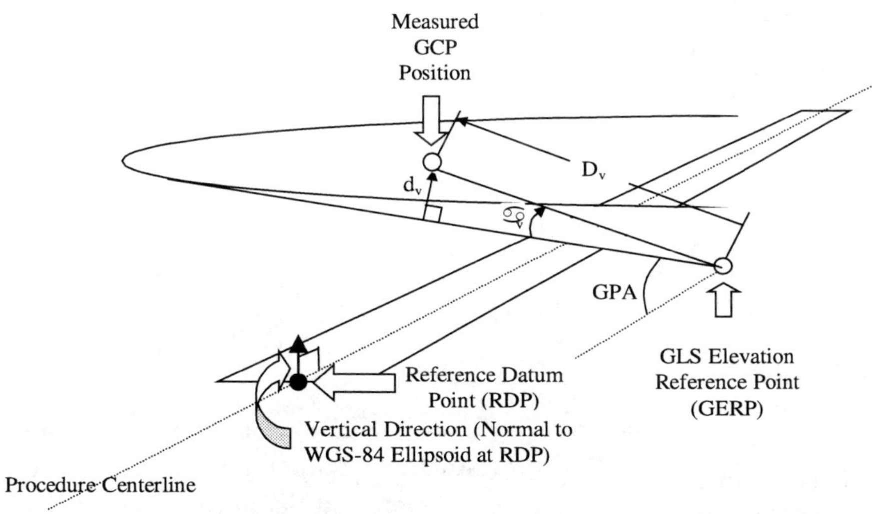

Figure 2. ILS Look-Alike Vertical Deviation Definition

0-7803-5086-3 /98/$10.00 ©1998 IEEE

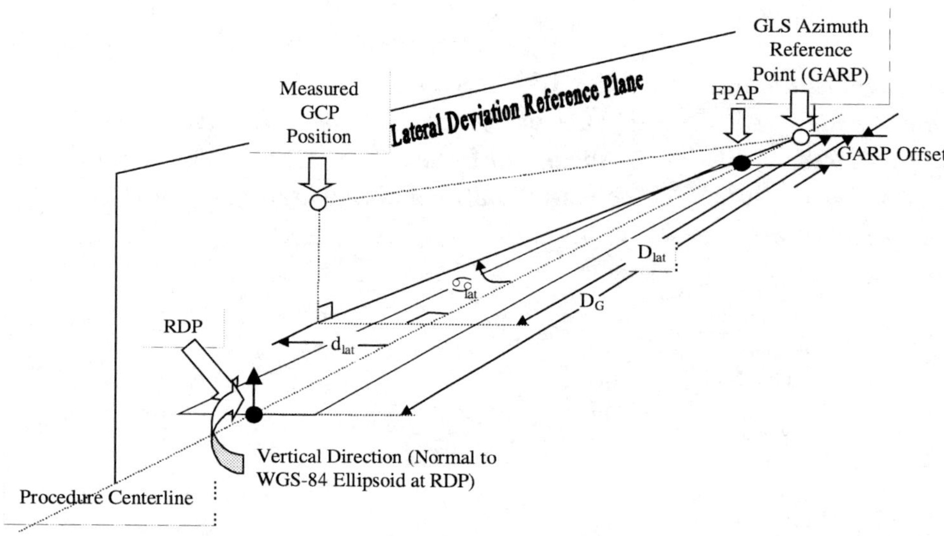

Figure 3. ILS Look-Alike Lateral Deviations

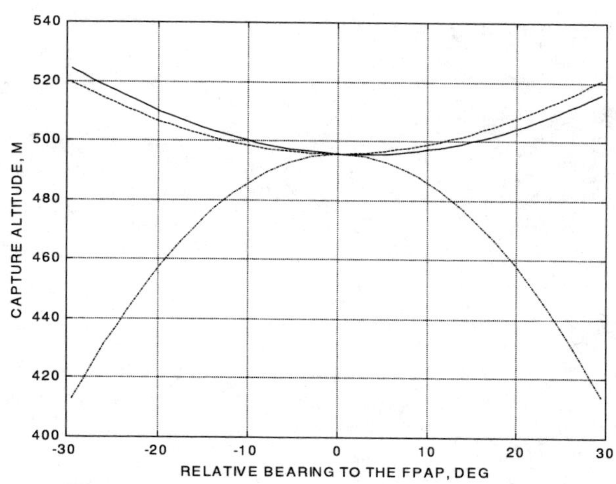

Figure 4. Comparison of Capture Altitude. Comparison of altitude for zero vertical deviation of ILS Look-Alike (solid), Planar (dot-dashed), Conical-GPIP (dashed), and hyperboloid (matches dashed) at approximately 5 nm. from the threshold.

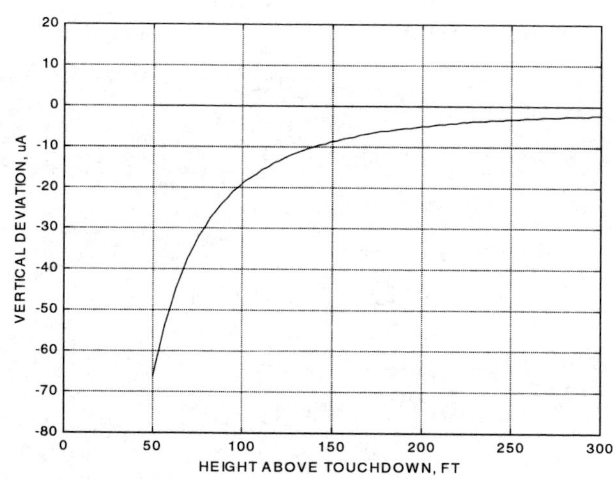

Figure 5. ILS Look-Alike Flare Characteristics. Comparison of ILS Look-Alike and Hyperboloid (solid) vs. Planar and Conical-GPIP (dot-dashed) vertical deviations on a straight-in approach to the GPIP with 3° glide-path angle.

0-7803-5086-3 /98/$10.00 ©1998 IEEE

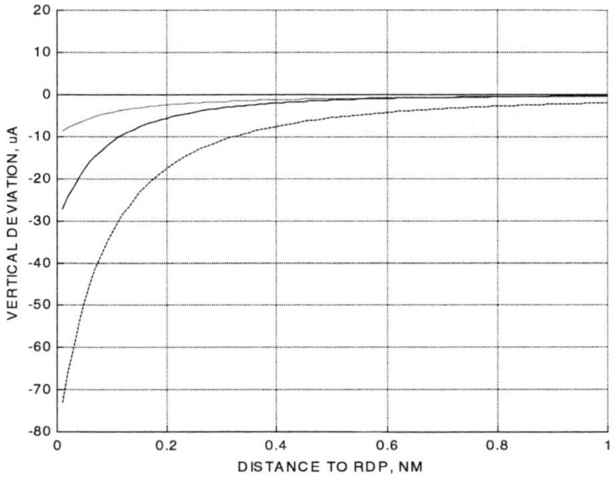

Figure 6. Comparison of Vertical Deviations Along Laterally-Offset Path. Comparison of ILS Look-Alike (solid), Planar (dot-dashed), Conical GPIP (dotted), and Hyperboloid (dashed) vertical deviations on a 1° offset approach to the GPIP with 3° glide-path angle.

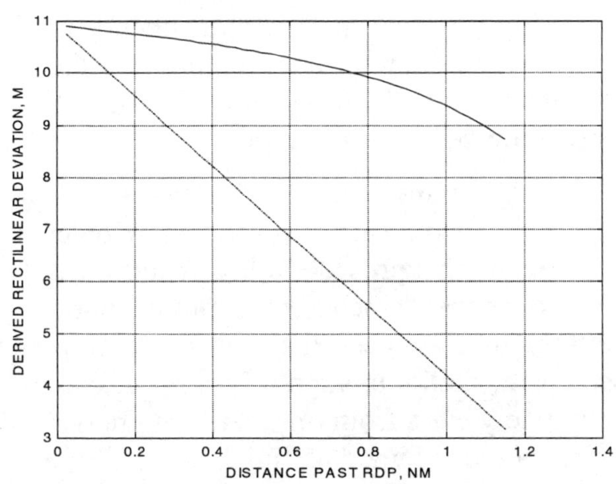

Figure 7. Comparison of Rollout Characteristics. Comparison of auto-land system's estimate of rectilinear deviation using ILS Look-Alike (solid) and Constant-Width (dot-dashed), along with actual rectilinear deviation (dashed) for a simulated rollout 10-m offset from centerline.

ADVANCED AVIONICS SYSTEM ARCHITECTURE

Gordon R. Fabian and Timothy R. Rayl, Advanced Systems, Rockwell Collins, Cedar Rapids, Iowa

Abstract

The Collins Advanced Avionics System Architecture uses an increased amount of integration to provide an unprecedented level of functional capability, fault tolerance and flexibility, as well as on-board diagnostic aids and other features designed to reduce crew work load and enhance the availability of system functions. Large format liquid crystal flight displays (LCD) and cursor control devices with voice activation integrate the display and control of system functions. This lessens crew work load while reducing the number of system displays and controls, providing a more intuitive cockpit. Functional processing is accomplished in an Integrated Processing Center (IPC) using standardized modules and Virtual Machine/partitioned processing and field-loadable software. This provides a flexible, cost effective, reusable architecture with inherent growth capability and true software mobility. The digital communication network achieves total connectivity among all subsystems. The connectivity is accomplished with data concentrators for legacy equipment and high speed digital Local Area Network Hubs for internal IPC data, IPC to IPC communication, and interface to other high data rate users. This significantly reduces system latencies, cross cockpit wiring and individual I/O connections.

Introduction

Important changes are taking place that are bringing new levels of capability and versatility to aircraft operations. The evolution of air traffic management has begun. Issues such as RVSM, FANS, and other developments leading up to free flight are placing new demands on aircraft and avionics manufacturers and operators alike.

Providing pilots with the flight-critical information they need–when they need it–is an increasingly important element of designing the flight deck around the operational requirements of the future. Critical issues such as information management, improving flight safety by reducing pilot workload, and enhancing situational awareness are fundamental design objectives of an advanced avionics system designed for next generation aircraft.

A flexible and scalable architecture needed to support a broad spectrum of capability and performance must include all of the elements and interfaces required for the implementation of a total avionics solution for a broad range of aircraft types.

All of this capability implies an information explosion requiring an avionics architecture that provides a high bandwidth, high connectivity means of communication between the avionics subsystems. Pushing this need is the ability to supply data for graphics on a timely basis from on-board data bases and from the ground via data link.

While this increase in capability is occurring, a need exists to improve the avionics dispatchability, reliability and maintainability while reducing the physical burdens of size, weight, wire count, etc. The architecture also must have the capability to be quickly customized to accommodate upgrades and new capabilities. In addition, the market is demanding life cycle cost reductions for the basic functions of avionics and expects continuing upgrades and easy additions to the system functionality. There also exists a desire for the use of standards and open architectures.

0-7803-5086-3 /98/$10.00 ©1998 IEEE

ADVANCED AVIONICS SYSTEM ARCHITECTURE

System Overview

A fundamental objective of the Collins Advanced Avionics System Architecture is to achieve a system with the flexibility to be cost-effectively applied across a broad market range and accommodate significant future functional change and growth.

This system architecture will support applications in the corporate jet, regional airline, commercial air transport, and military markets. These markets include the minimum-display-compliment, asymmetrical cockpit of the corporate entry jet to the multi-display cockpits of the corporate heavy jet, 30 - 100 passenger regional aircraft, large commercial air transport aircraft, and military transports and helicopters. This will allow the system to leverage all of the products and capabilities among the different markets.

Our system architecture definition is driven by several key attributes necessary to achieve the stated objective. These attributes include reduction of installed system cost, modularity of system elements to provide desired flexibility, maximizing the reusability of hardware and software system elements and system concepts, and improvement of system availability.

Functionality

A typical system implementation using this architecture, such as the Collins Pro Line 21 System, can be designed to support a broad range of aircraft application and functionality. This system includes all the elements for a total avionics system solution. The span of functionality includes traditional avionics functions such as sensors, navigation, communication, guidance, displays, and maintenance. Further, the system architecture is defined so that it is capable of being extended to support other functions such as aircraft utility system control and fly-by-wire.

Collins Advanced Avionics System Architecture integrates and displays aircraft system data and enables the crew to manage aircraft systems, communications, flight path, hazard avoidance and maintenance functions. The architecture supports multiple Adaptive Flight Displays (AFD) for various cockpit instrument panel configurations. The flexibility of the architecture allows these displays to be supplemented by one or more Head-up Displays (HUD).

Typical control panels are minimized with the use of display annunciations and Cursor Control Devices (CCD) and/or voice activation.

The Collins Advanced Avionics System Architecture encompasses innovations that greatly improve the flight deck and give the crew the ability to manage information, to control system state, and to improve safety. These include:

- Total data connectivity, which allows all functions to have access to all data, providing effective information management during normal, abnormal and emergency operations
- Use of a CCD and/or voice activation to interact with adaptive flight displays
- 3-Dimensional display, graphics rendering capability for lateral and vertical flight profile information, including 3D terrain for increased awareness of ground hazards
- Our system architecture also supports needs anticipated in the areas of functional growth and change. Changes will result from flight advancements such as evolving CNS/ATM concepts and safety improvements brought about by simplified cockpit operation and enhanced awareness, which will be

provided by improved display technologies

As such, our system architecture is also defined to anticipate significant functional growth and an increase in coupling between functions which have traditionally been stand-alone in nature.

In order to provide the required functional capability, our system architecture accommodates a number of attributes, which typically characterize avionics applications. These attributes include items such as criticality, determinism, real-time, and redundancy for availability. Specifically, the system hosts a mix of non-essential, essential, and critical functions within a common environment. The system also provides bounded latencies to support deterministic, real-time, periodic applications. Furthermore, the system supports a variety of event-driven applications in conjunction with the real-time applications.

The functionality required also drives a broad mix of data communication and bandwidth requirements, which our system accommodates. This includes requirements for periodic broadcast data, message passing, large data blocks or "file transfers," and ATN support.

Key Concepts

In order to meet the system objectives, several key concepts have been incorporated into our system architecture:

- Increased sharing of system resources through physical and logical integration of functions within a common processing environment
- Creation of a common processing resource that is structured to provide application software independence from the underlying platform hardware specifics
- Use of data concentration and bussing technology to minimize the physical wiring terminations and J boxes required in the aircraft while accommodating legacy interfaces
- Use of asynchronous communications and bussing technology to maximize data availability throughout the system providing increased system flexibility and fault tolerance
- System redundancy so that most single failures will not impact operation
- Use of common, standardized system "building blocks" to provide the capability to construct systems that address a broad range of aircraft applications (system scalability) with high levels of reuse

Architecture Components

The core of the Collins Advanced Avionics System Architecture is the Integrated Processing Center (IPC). Multiple IPCs perform application processing in the system as well as Input / Output data concentration from peripheral subsystems and devices.

Throughout the system, wherever significant integration of function or concentration is used, increased redundancy of system elements is employed to mitigate the consequences of single point faults. Figure 1 shows a simplified block diagram of one of our Pro Line 21 implementations that uses the Collins Advanced Avionics System Architecture.

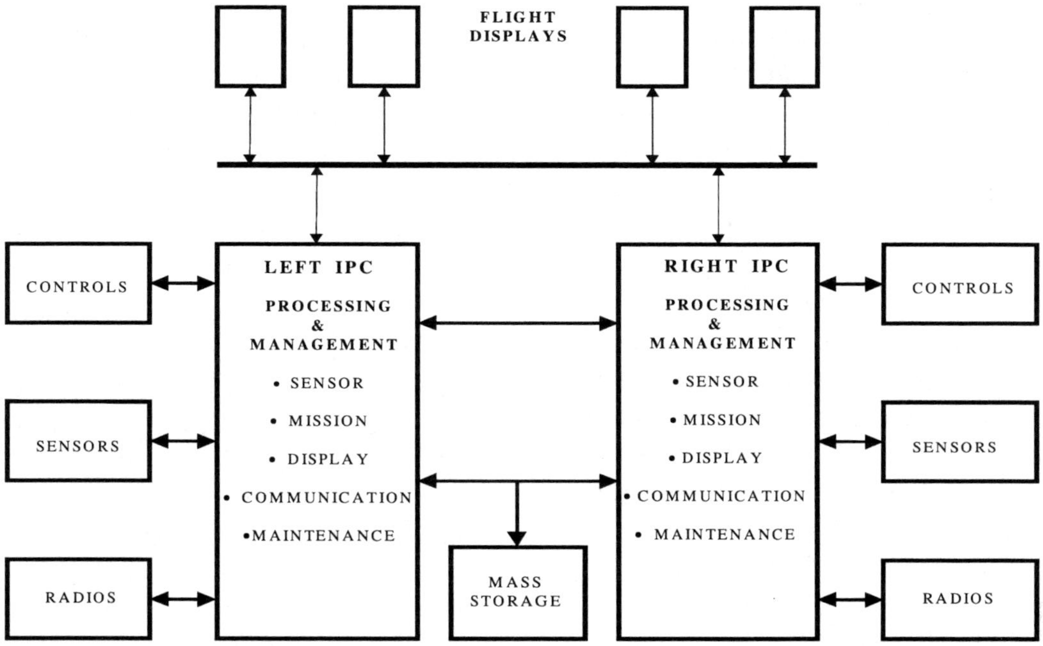

Figure 1 Advanced Avionics System Architecture

Typical Pro Line 21 system configurations will have two or more IPCs. However, the architecture is scalable to a single IPC. For the typical system solution, multiple IPCs are used to achieve the computational channel redundancy necessary to support function availability requirements, system fault tolerance requirements, physical segregation requirements, or meet physical installation constraints. The Pro Line 21 system's operation is based on multiple, similar, asynchronous computation channels provided by the IPCs. Each IPC hosts one or more computation channels, which provide the physical hardware platform and logical application software environment for hosting of system application functions.

Figure 2 shows the physical hardware platform for an IPC computation channel. It consists of a dual Local Area Network Hub, one or more Common Computing Resources (CCR), dual Power Supplies and Environmental Control.

Communications

Communication is key to this architecture. All functions must have the ability to access all of the aircraft data, which must be provided in a timely manner. In addition, the communications function must be able to communicate with existing, legacy aircraft equipment.

Several generic architectural topologies were investigated in order to identify a suitable avionics topology that meets flexibility, scalability and cost requirements.

The basic topologies distilled to multi-drop and point-to-point Hub and Spoke. The multi-drop topology was deemed unfavorable due to its installation complexity, need to synchronize the users for access, and need to employ extensive measures against a user fault that fails the entire multi-drop system. The advantages of Hub & Spoke were the ease of transition from current products, the ability for simultaneous communication between users and the lower cost and complexity of aircraft installation.

A Local Area Network (LAN) implements the Hub and Spoke topology. To provide availability and dispatchability, a dual implementation of the LAN is used. The LAN is used for inter- and intra- IPC communication and with other high data rate users as a replacement for the present ARINC 429 standard, which is extremely limited in bandwidth and label availability. Also used is an open standards bus, which supports the following LAN characteristics:

- Ability to support critical avionics data
- Full duplex operation
- Very high composite bandwidth with concurrent path capability
- Low latency with adequate determinism
- Allows bandwidth growth as computing power continues to grow

An Input / Output Concentrator (IOC) is used to provide the interface to legacy ARINC 429 type equipment. The IOC concentrates, normalizes and schedules the 429 data, then does a protocol conversion for receiving and transmitting data on the Local Area Network. As with the LAN, each IPC contains two completely independent IOC functions for increased availability and dispatchability.

To further reduce aircraft wire lengths, production breaks, etc., the same IOC components can be packaged in a unit remote from the IPC to provide local concentration.

Processing

One of the guiding principles for processing hardware and software is to utilize open, standardized industry specifications where possible, with avionics-grade implementations. This approach minimizes design time and maximizes the availability of tools and experienced engineers.

The Common Processing Resource (CPR) is a standard embedded processor design that is used in multiple applications. The prime interface to the CPR is an industry standard PCI bus. Optional CPR interfaces such as an additional PCI bus, an ISA bus, and/or an SCSI bus, may be populated for very high bandwidth communication with display functions, medium-speed interfaces for data such as discretes or aural warnings, and a high-speed interface for peripheral devices such as a mass storage devices.

The Common Computing Resource (CCR) module used in the Collins Advanced Avionics System is a high performance, general purpose computing platform. Each CCR module is a combination of hardware and software that is capable of hosting multiple applications at different criticality levels. Two CCRs in each IPC provide the processing throughput and functional redundancy required for the baseline system. Additional CCRs can be used for increased functionality.

The CCR has a combination of hardware and layered system software that provides hardware independence from the processing platform. This hardware independence enables application software to be insulated from changes in the hardware used in the underlying platform. Hardware independence is achieved by providing a "Virtual Machine" for the application software to execute on.

The CCR consists of a Common Processing Resource (CPR) which is coupled to a LAN bus node to support a redundant LAN connection to the system LAN hubs. The CPR

PCI bus provides a high-speed interface to the LAN node from the processor.

Other applications of the CPR include combining the CPR with display graphics generation and control hardware to provide flight display functions, and combining the CPR with non-digital bus interface hardware, analog to digital/digital to analog converters, to provide a data acquisition function.

Software Partitioning

The Real Time Operating System (RTOS) is a layered system that provides a multiprocessing, partitioned environment for application software. The partitioned environment eases system configuration and promotes cost savings through reuse and incremental certification.

In addition to providing independence for the processing platform, the CCR also uses a message-passing method of communication that provides independence from the bus technology and network topology. This insulates application software from changes in the underlying network and network devices.

In the Collins Advanced Avionics System Architecture, each application executes within its own "partitioned environment," which limits access to system resources. As far as the application software is concerned, each application is running in its own "Virtual Machine" and therefore in its own partition. This is the basis of both functionality and critical partitioning. Any attempt to violate another partition will trap to the kernel Operating System and the appropriate recovery action will be taken.

The memory, processor time and network resources required by each partition is defined at configuration time, using information provided by the supplier of a partition. Partitions are created at initialization time, and the partitioned memory use and network resources are validated and then allocated. At run-time, the RTOS enforces the partitioning of these resources.

Each partition does not know (and cannot know) that it is sharing the processor with other partitions. As specified in ARINC 653 (APEX), partitions are time-sliced by a fixed, preset schedule within a major, cyclic frame. Partitions do not have priorities. The RTOS also provides an interface for inter-partition communication. A partition does not know the location of the source or the destination partition for any inter-partition communication message that it sends or receives.

Again, to increase availability and dispatchability, dual power supplies and distribution systems are used in the IPC. Power line transient protection is provided along with a load shed capability to minimize power requirements during emergency battery operation. The power supplies are designed to be fail-independent so that failures occurring within either power supply do not affect the other power supply. Each power supply in this architecture normally supports only a portion of the load in the IPC. This allows the internal components of the power supply to be operated at greatly reduced levels during normal operation, thus improving power supply reliability. However, each power supply has the capability to support the IPC's full load requirements. This improves the fault tolerance of the system.

Environmental Control

The IPC Environmental Control Module function provides forced air cooling to the IPC and will provide heat for extreme low temperature operation.

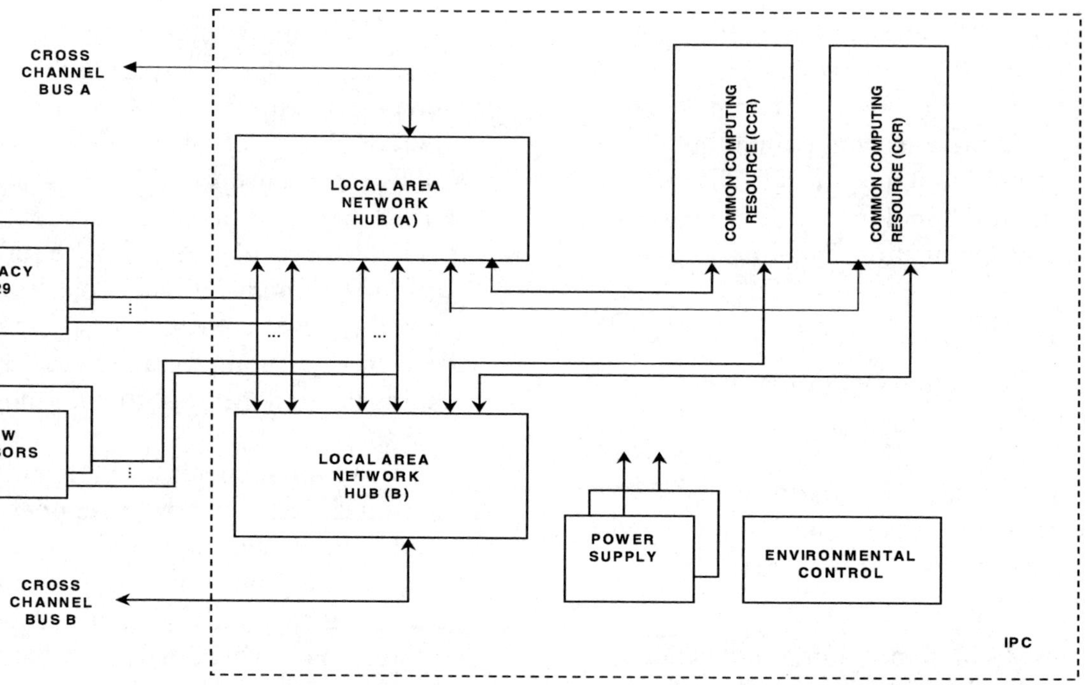

Figure 2 Integrated processing Center

Life Cycle Cost

This architecture provides for reduced Life Cycle Costs:

- Commonality and reusability reduces non-recurring cost and thus acquisition costs.
- On-aircraft software download of individual functions (partitions) allows software updates to be accomplished without affecting unrelated functional software or removing the equipment from the aircraft.
- Total data connectivity provides increased maintenance diagnostic capability, reducing the number of No Fault Founds.
- The partitioned environment allows one function to be changed without impacting others.
- Intuitive flight deck design with improved user interfaces reduces pilot training.
- Defined break points, such as the LAN and processor interfaces, allow updates to occur without the need to redesign the total system.
- Integration of functions naturally increases reliability, as there are fewer things to fail.
- Fleet commonality is enhanced due to the architecture's scalability, which extends the system's capability and redundancy requirements for aircraft of all sizes, from the minimum-display-compliment, asymmetrical cockpit of the corporate entry jet to the multi-display cockpits of the corporate heavy jet, 30 - 100 passenger regional aircraft and large commercial air transport aircraft.

- By taking advantage of the architecture's commonality, flexibility and scalability, additional cost can be saved by integrating non-avionics functions, such as Aircraft Utilities Systems Control, into the avionics system.

Summary

Collins Advanced Avionics System Architecture features a new, flexible architecture designed to support a broad spectrum of capability and performance. The highly scalable system includes all of the elements and interfaces required for the implementation of a total avionics solution for business, regional airline, commercial air transport and military aircraft.

New levels of avionics hardware and software independence allow the Advanced Avionics System Architecture to be quickly customized to accommodate upgrades and new capabilities. Reliance on Real Time Operating Systems and Application Programming Interfaces means no fundamental–or costly–changes in basic system architecture are required to take advantage of on-aircraft field-loadable software improvements.

TECHNIQUES FOR IMPROVED RECEPTION OF 1090 MHZ ADS-B SIGNALS *

W. Harman, J. Gertz, A. Kaminsky

Background

The recent development of ADS-B (Automatic Dependent Surveillance-Broadcast) is based on the use of the Mode S transponders now carried by all air carrier and commuter aircraft. ADS-B aircraft broadcast aircraft positions, identity, and other information via semi-random Mode S transponder squitters. Other aircraft or ground facilities receive the squitters and the associated position and status.

Squitter reception includes the detection of the Mode S 1090 MHz waveform preamble, declaration of the bit and confidence values, error detection, and (if necessary) error correction. The current techniques for squitter reception are based upon methods developed for use in Mode S narrow-beam interrogators and for ACAS. In both of these applications, the rate of Mode A/C fruit that is stronger than the Mode S waveform is relatively low, nominally less than 4,000 fruit per second.

Extended squitter applications now include long range (up to 100 nmi) air-air surveillance in support of free flight. This type of surveillance is sometimes referred to as Cockpit Display of Traffic Information (CDTI). In high density environments, it is possible to operate with fruit rates of 40,000 fruit per second and higher. Operation of extended squitter in very high Mode A/C fruit environments has led to the need to re-evaluate squitter reception techniques to determine if improved performance is achievable.

* This work is being performed under Air Force Contract No. F19628-95-C-0002 and sponsored by the Federal Aviation Administration. Opinions, interpretations, conclusions, and recommendations are those of the author and are not necessarily endorsed by the United States Government.

The purpose of this paper is to provide a summary of work in progress to investigate improved squitter reception techniques. Elements of improved squitter reception being investigated include (1) the use of amplitude to improve bit and confidence declaration accuracy, (2) more capable error correction algorithms, and (3) more selective preamble detection approaches.

Current Squitter Reception Techniques

Extended squitter uses the 112-bit Mode S waveform shown in Figure 1. This waveform encodes data using pulse position modulation (PPM). A chip in the leading half of the bit position represents a binary ONE. A chip in the trailing half bit position represents a binary ZERO. Reception of the reply begins with the detection of the four-pulse preamble. At preamble detection, a dynamic minimum triggering level (DMTL) is set 6 dB below the level of the preamble. Any signal received below this DMTL will not be seen by the reply processor. This eliminates the effect of low level Mode A/C and Mode S fruit on the reception process.

Once all of the bits have been received, error detection is performed using the 24-bit CRC contained in the reply. If no error is detected, the reply is passed on to surveillance processing. If an error is detected (indicated by a non-zero error syndrome), an error correction technique is applied. The current error correction algorithm can correct the errors caused by one stronger overlapping Mode A/C reply.

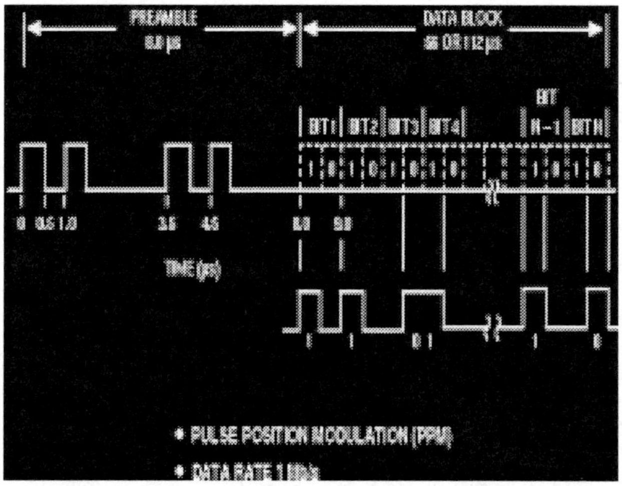

Figure 1. Mode S Reply Waveform.

Bit values are declared by comparing the amplitudes of the centers of the two chips; the chip with the greater amplitude is declared the bit value. This amplitude comparison technique limits bit errors caused by fruit of lower level than the squitter being received.

Confidence in the bit declaration is declared by observing if signals are above threshold in both chips. If only one chip is above threshold, it is declared to be high confidence since there is no evidence of overlapping fruit. If both chips have amplitude above threshold, low confidence is declared since some form of interference had to be present to cause this condition.

When a Mode A/C reply interferes with a Mode S reply, some of the Mode S bits may be declared in error. The Mode S error correction algorithm attempts to correct these errors by locating a string of 24 consecutive bits such that the low confidence bits in the window cover the bits in the corresponding error syndrome. If such a window is found, all low confidence bits matching the syndrome are flipped, and the reply is declared correct.

Use of Amplitude in Bit Declaration

The current technique of declaring a bit based upon the higher of the two chips will generate bit errors in cases of higher level overlapping ATCRBS fruit. The use of amplitude to correlate the received pulse with the preamble pulse level will improve bit declaration accuracy. Two techniques have been investigated. One is a very simple approach that uses only the amplitude measured at the center of each chip. The second is a somewhat more complex approach that takes advantage of four amplitude samples per chip. Note that current implementations already used such sampling to establish bit confidence. Each of these techniques is described in the following paragraphs.

An improvement in the declaration of Mode S data bits can be achieved if the actual amplitudes of the center samples of the '1' and '0' chips can be measured for each data position, rather than just a comparison of which chip sample is greater. All Mode S pulses, including those of the preamble, have approximately the same level (within 1 or 2 dB). Thus if the preamble level is measured, the expected level of each data pulse will be known. Then if both center samples of a data position are above threshold, but only one is within a 3 dB band centered at the preamble level, it would be reasonable to assume that the corresponding chip is the correct Mode S pulse location.

Figure 2 illustrates the new data and confidence declaration algorithms that result when the actual sample amplitudes can be measured, and both samples are above threshold. A bit is high confidence when 1 and only 1 of the two samples correlate with the preamble level. The correlating sample, rather than the larger sample, is declared to be the true data value. If both samples, or neither sample, correlates with the preamble, a low confidence bit is declared. In this case, the larger sample is selected as the bit value, as in the current technique.

With the center amplitude technique, a low confidence bit is declared only when the ATCRBS reply is within 3 dB of the Mode S reply. ATCRBS replies of either lower or higher power in general cause neither errors nor low confidence bits. Thus the region of concern

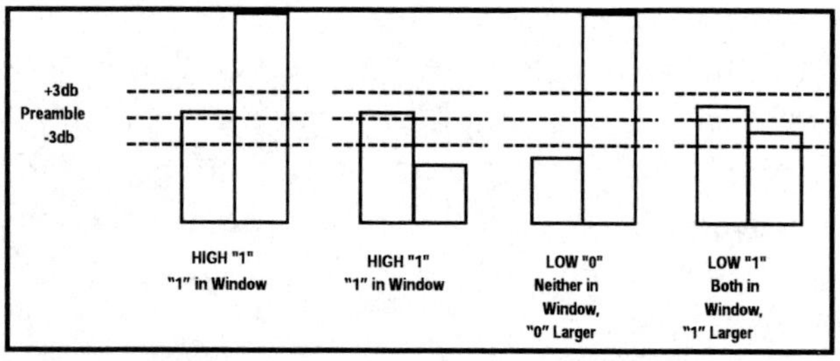

Figure 2. Center Amplitude Bit and Confidence Declaration.

for ATCRBS aircraft is reduced from all the aircraft at closer range to just those at approximately co-range. This is particularly significant when the system is attempting to listen to far-range Mode S aircraft.

The above amplitude data declaration approach can be improved if all 8 samples (4 per chip) that are taken for each Mode S bit position are utilized in the decision process. In particular, much more complex overlap situations can be resolved. For example, Figure 3 presents samples that might be observed when 2 ATCRBS interfering pulses are present, one in the '1' data chip and the other in the blank '0' data chip. The overlap of the Mode S pulse and the ATCRBS pulse, because of small frequency differences (nominal allowable range is 2 MHz), produce 4 different sample amplitudes, while the ATCRBS pulse by itself produces all samples at the same level. Even though the '0' center sample is the only center sample in the window, the full set of samples suggests that the correct data value is a '1'.

The algorithm developed for utilizing all 8 samples is a form of pattern matching. Each sample is defined with one of four values:

0: below threshold (-6 dB from preamble)

1: above threshold, but below the 3db preamble window

2: within the +/- 3db preamble window

3: above the 3db preamble window

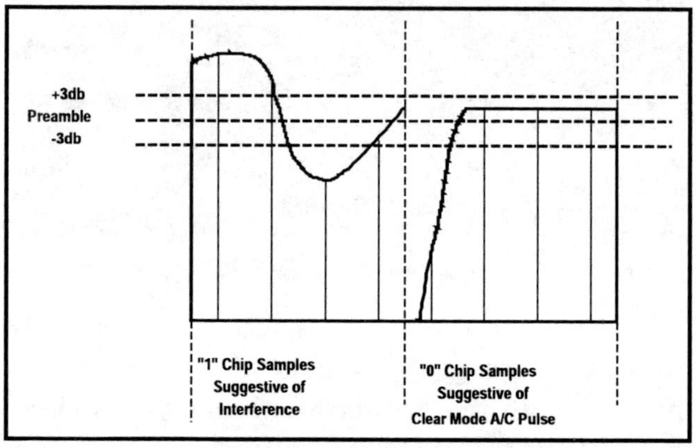

Figure 3. Multi-Amplitude Bit and Confidence Declaration Example. Only "0" Center Sample in Window, But High Confidence "1" Declared.

Since there are 8 samples, with 4 possible values each, a Mode S data position can have 4^8 = 65536 (64K) different sample patterns. Two 1-bit tables, each stored in a 64K x 1 ROM, are defined over the set of patterns: the first declaring the bit position to be a '1' or '0', the second high or low confidence. Once the pattern existing for a given bit is determined, two table lookups supply the proper declaration.

These tables are generated by running millions of simulations of Mode S replies in 40,000 fruit per second environments. For each bit of each trial, the pattern and the correct Mode S bit value are noted. A sample result could be, for example, 5876 examples of pattern 16453 were encountered, of which 5477 occurred when the Mode S bit was a '1'. The table values are then defined as follows, where the percents are sample occurrences when the bit was a '1':

H1: 90% or more

L1: 50% - 90%

L0: 10% - 50%

H0: 10% or fewer

The above 8 sample approach requires lookup tables of size 64K, adding significant cost to the hardware implementation of the decoder. A variation of this method that only requires tables of size 256, called the 4-4 approach, has been designed to reduce this expense.

The 4-4 method forms two estimates of the bit data and confidence values, one using the odd samples (1-3-5-7) and the other using the even samples (2-4-6-8); the final decision is then a combination of the individual estimates. Since each set includes samples in both chip positions, pattern matching is still possible, although the fineness of the pattern variation is cut in half. Since only 4 samples are in each set, and each sample is quantized to the same 4 levels as above, 4^4 = 256 patterns are possible for each set.

To counteract the loss of resolution, and to aid in the combining operation, 3 levels of confidence (high, medium, and low) are defined for each pattern. Following the simulation generation scheme described above, the table values are defined as follows:

H1: 90% or more

M1: 70% - 90%

L1: 50% - 70%

L0: 30% - 50%

M0: 10% - 30%

H0: 10% or fewer

Once the values and confidences are determined for each set of samples, odd and even, the composite values actually declared for the bit are found according to the following table:

Odd (1357) \ Even (2468)	H1	M1	L1	H0	M0	L0
H1	H1	H1	H1	L0	H1*	H1
M1	H1	H1	L1	H0*	L0	L1
L1	H1	L1	L1	H0	L0	L0
H0	L0	H0*	H0	H0	H0	H0
M0	H1*	L0	L0	H0	H0	L0
L0	H1	L1	L0	H0	L0	L0

* alternative is Low confidence; correct decision under study

Brute Force Error Correction Technique

If the bit declaration algorithm has performed its function properly, all errors in Mode S data values will reside in bits declared low confidence. If this is true, a simple approach to error correction is to try all possible combination of low confidence bits, and accept the set that matches the error syndrome (provided only one success is discovered). For obvious reasons, this method has been named the Brute Force Technique. It is applicable to any method of data and confidence declaration, with or without amplitude, and is applied after the other error correction techniques have failed.

Clearly, for processing time and error bounding reasons, the maximum number of low confidence bits to process must be limited. The number of cases to consider is given by 2^n if n low confidence bits exist for a reply; this grows exponentially with n (32 at n=5, 4096 at n=12). The undetected error rate is proportional to the number of cases, and thus also grows exponentially with n. Fortunately, the Hamming distance of 6 for the Mode S parity code implies that undetected errors are essentially zero if n<=5 is enforced. For this reason, a value of n=5 has been used chosen.

Improved Preamble Detection Algorithm

When a heavy Mode A/C fruit environment exists, squitter preambles are often lost due to the existence of earlier apparent preambles produced by Mode A/C reply pulses. Since preamble detection can be suppressed for up to 130 microseconds after each preamble located, a Mode S squitter starting during the false preamble dead time will be lost.

The PPM encoding used in the Mode S downlink data block requires that there be a pulse in either the '0' or '1' chip of each data block position. This is a pattern that is difficult to generate with ATCRBS fruit replies. Ideally, the improved algorithm could attempt to find pulses in all 56 (or 112) data positions. However, ATCRBS fruit interference could make it difficult to accurately find all the pulses – simulation and experiment have shown that high fruit rate environments often destroy one or more data pulses in a Mode S downlink message. Trying to qualify all the data pulse positions would also entail quite a bit of processing time. A compromise is to check for the pulses in the first 5 bits of the data block – the Mode S downlink "data format" (DF) field.

Figure 4 illustrates the results of the improved preamble detection algorithm as a function of fruit rate. The current algorithm, at 40K fruit per second (the target CDTI case), finds only 57% of the squitter preambles. The suggested method, of using 9 pulse preamble detections, increases this to 93% of the squitter preambles.

Simulation

The acceptance rate for alternative reply declaration and error correction technique combinations is shown in Table 1 for the case of 40 K fruit per second uniformly distributed between -5 and +10 dB relative to the squitter. The rows correspond to the degree of amplitude information employed in bit declaration. The first column gives the performance with no error correction, the second column the performance for the current Mode S error correction technique, and the third column gives the performance with the addition of the Brute Force technique (limited to 5 low confidence bits).

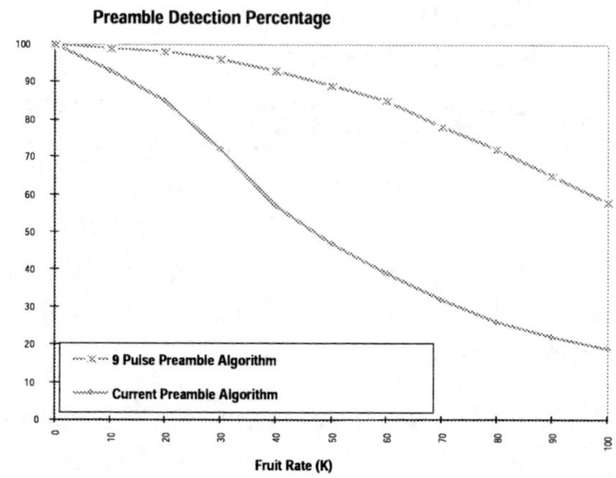

Figure 4. Preamble Detection Improvement with New Technique.

Table 1. Algorithm Performance Comparison

40K Fruit/Second (-5 dB To +10 dB)
Percent Acceptance Rate

	Error Correction Algorithm Employed		
Amplitude Information	**None**	**Conservative Mode S**	**+ Brute Force @5**
None (Current)	**3**	**8**	**18**
Center Sample	**13**	**23**	**36**
Multiple Samples	**40**	**52**	**63**

Measurements

The simulation work developing improved reception techniques is now being followed by a measurement program to validate the conclusions from the simulation. A 1090 MHz receiver is being used to receive signals that happen to exist at the time of the measurements, having been transmitted by aircraft flying in the area. The receiver has high sensitivity for receiving weak signals, which is appropriate for the final application of these techniques. Initially the receiver was used in a ground-based installation in the Boston area, and subsequently it has been installed on an aircraft for airborne reception, flying to the New York metropolitan area.

Plans are being made for flights to other areas, including the Los Angeles Basin.

The main objective of these measurements is to validate the simulation. Additional objectives include: (1) validate a proposal to reduce receiver bandwidth from 8 MHz (which is used by TCAS* receivers) to 4 MHz, and (2) assess possible limitations caused by man-made noise.

Configuration of Measurements

The configuration of the ground-based installation is illustrated in Figure 5. The receiving antenna, which is mounted on a tower, is a DME* antenna. This antenna provides omnidirectional reception in azimuth and some directivity in elevation, with maximum gain (7 dB) near the horizon.

The receiver was designed with the intention of having the maximum sensitivity that can be practically achieved in operational avionics. It uses a low-noise preamplifier, having a noise figure of 1.0 dB. The noise figure of the full receiver (including front end filter, limiter, cables, and all other effects) is 2.5 dB referred to the antenna port. This results in a system noise power of -105 dBm referred to the antenna port, when the system bandwidth (3 dB) is 4 MHz. For an aircraft installation, the same receiver is used without the long cable and its associated preamplifier. In this configuration, the system noise figure is also about 2.5 dB.

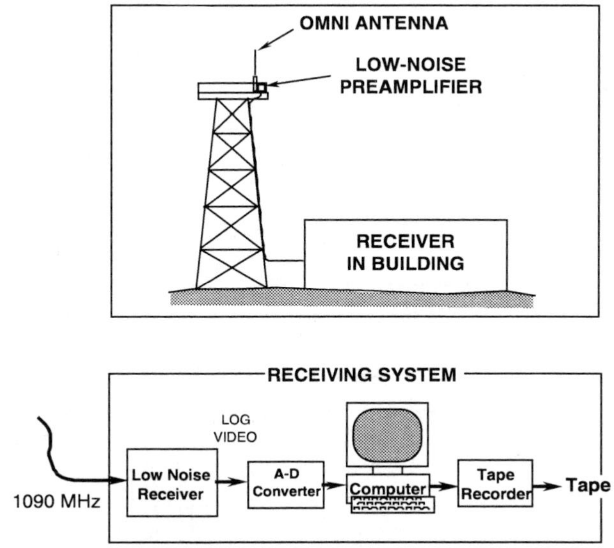

Figure 5. Receiver Ground Based Installation.

The receiver includes a mixer to convert to an intermediate frequency of 60 MHz, a bandpass filter, which can be changed to test different bandwidths (8 MHz and 4 MHz for example), and a log-video amplifier. The log-video signal is sampled at the rate 8 samples per μs and digitized by a 12 bit A/D converter. The receiving system was carefully calibrated in power level such that the recorded amplitudes can be associated with the absolute power levels received.

The nature of the data recorded by this system is shown in Figures 6 and 7. The data plotted in Figure 6 includes an ATCRBS reply, as marked. Note that this is a relatively weak reply, which would not have been detected by current 1090 MHz receivers. In TCAS, for example, the nominal receiver sensitivity is -74 dBm, whereas this reception is -88 dBm. Note also that the recording process samples everything, even the noise between pulses. Similarly a Mode S reply is shown in Figure 7.

---

* Traffic Alert and Collision Avoidance System (TCAS) is an airborne system for avoiding mid-air collisions.

* Distance Measuring Equipment (DME) uses an omnidirectional antenna.

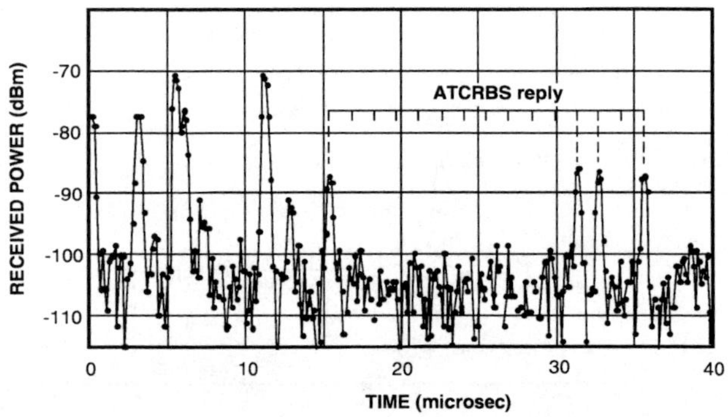

Figure 6. Received Waveform, an ATCRBS Reply.

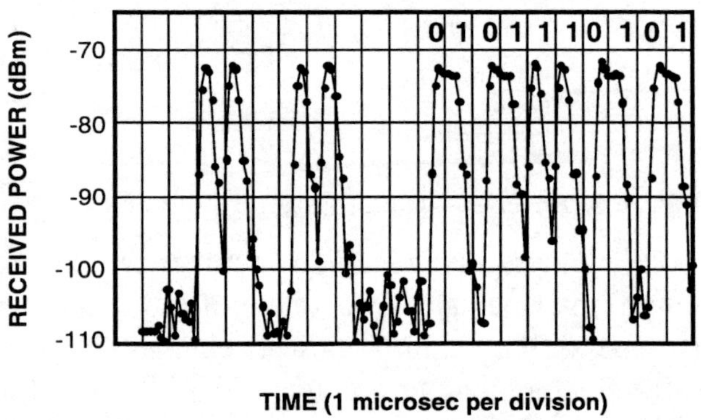

Figure 7. Received Waveform, Mode S Reply.

Measurement Results

While assessing reception performance, it is of interest to measure the fruit environment at this receiving station. Figure 8 gives the cumulative distribution of ATCRBS and Mode S fruit rates, as a function of received power level. Each point is the fruit rate for receptions at that power or stronger. The total fruit rate is seen to be 11,000 ATCRBS replies per second, which is a substantial rate. The results also indicate that the received power levels are primarily between -75 and -85 dBm. This appears to be a result of the range distribution of the aircraft in view.

Before making comparison between measured Mode S reception performance and that seen in simulation, it was necessary to modify the simulation in several ways. The original simulation runs were focused on reception of Extended Squitters, whereas these are not present in the measurement environment. Accordingly, the simulation was modified to focus on short squitters, which are received at a substantial rate in the measurements. Another change applies to bandwidth. Originally, the simulation was developed for a bandwidth of 8 MHz, which has previously been the standard for Mode A, C, and S reception. When these measurements were done with 4 MHz bandwidth, this caused an increase in the pulse risetimes. A corresponding change was made to the simulation to apply to the 4 MHz case.

Probability of correct reception is given in Figure 9 as a function of received signal power. This figure shows a comparison between simulated results and results from measurements. It should be noted that the processing is simulated in both cases. Measured waveforms were used in one case and simulated waveforms in the other. The lower plot applies to the simulation only, in which the presence of a desired signal is known.

In the upper plot, the presence of desired signals is inferred from the received waveform. This inference can be done effectively, but not perfectly as seen in these plots.

The comparison shows that the measurements yield results that are approximately the same as the results from simulation. These are preliminary comparisons, now being repeated and carried out for more cases.

Bandwidth reduction to 4 MHz was also investigated via measurements. The original proposal to reduce bandwidth was based on the

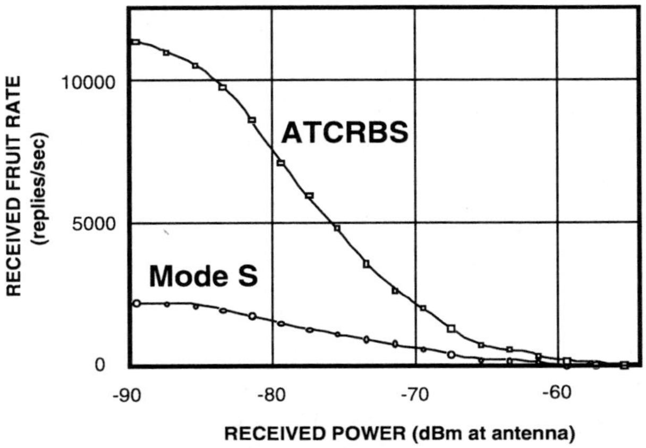

Figure 8. Fruit Environment, Measured at Lexington, MA.

following reasoning. Whereas Mode A, C transmission standards allow the carrier frequency to deviate by as much as +/-3 MHz, for Mode S the deviation is +/-1 MHz. This allows a bandwidth reduction for a receiver that is intended only for Mode S. Furthermore, for reception of Extended Squitters, it is not necessary to perform accurate time-of-arrival measurements and therefore the bandwidth can also be reduced for that reason.

Measurements have been made using both 6 MHz and 4 MHz bandwidths. Preliminary comparisons have been made between these cases, yielding an indication that 4 MHz bandwidth appears to be satisfactory. Mode S signals are still detectable, and the probability of correct reception is approximately the same as for 6 MHz.

An objective of the measurement program was to gain experience with effects such as man-made noise that might limit performance when using a high-sensitivity receiver. In fact, the first data collection indicated a significant problem of interference, in a form that appeared to be a very high noise level. Using a spectrum analyzer, the noise was identified to be a signal transmitted by a newly installed cellular phone tower about a mile away. This transmits a very strong signal at about 960 MHz. The cellular interference was eliminated by addition of a more effective front-end filter centered on 1090 MHz. Subsequently, man-made interference of much lower levels has been observed in some of the data collections. Further investigations are on-going.

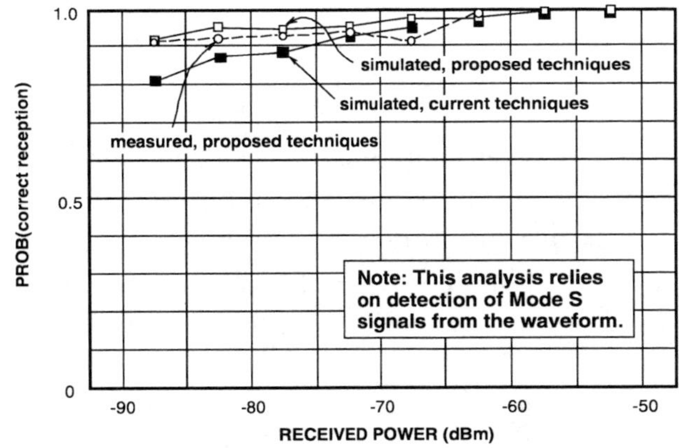

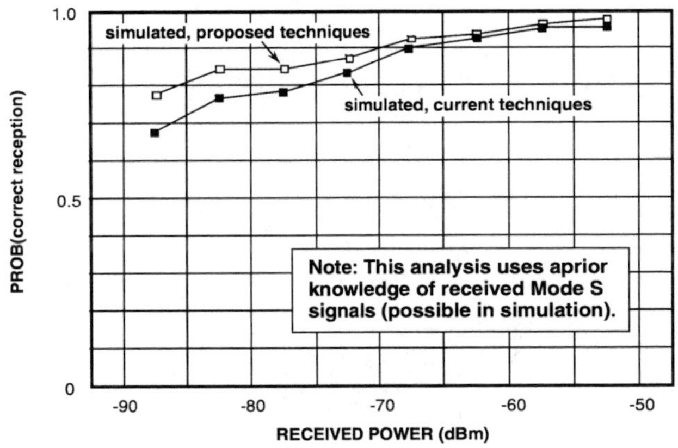

Figure 9. Performance Comparisons.

Airborne measurements are also being made with this system. In November 1997, the receiver was installed in a Convair 580 aircraft for data acquisition in the Boston and New York areas. Analysis of the data indicates significant differences in the fruit environments relative to the ground-based environment described above. A major change expected is that the aircraft antenna gain (approximately 0 dB) is much lower than the gain of the ground based antenna (approximately 7 dB). Therefore, the fruit rate above a given power level would be lower airborne, and this effect was seen to be true in the airborne data. Another expectation was that New York would have a higher interference environment than Boston, and this also was seen to be true in the recorded data. Performance assessments using the airborne data are in progress. Plans are being made for additional airborne measurement in other locations, including the Los Angeles Basin.

Summary

Several techniques have been identified to improve Mode S reception in a high interference environment. Initial assessments through simulation indicate that these techniques would yield significant improvements in reception reliability. Measurements are also being conducted to validate the simulation results. Results to date tend to indicate that the simulation is valid for predicting Mode S reception performance. More comparisons of this kind are in progress. The measurements have also yielded useful results relating to receiver bandwidth and man-made noise. Further progress is expected as more data is analyzed. When the simulation validation is complete, the simulation tool will be useful for predicting performance in very high interference environments.

0-7803-5086-3 /98/$10.00 ©1998 IEEE

COTS BASED OPEN SYSTEMS FOR MILITARY AVIONICS

John T. Paul, Naval Air Warfare Center, China Lake, CA

Abstract

The Department of Defense (DoD) has many acquisition programs that are aggressively implementing open architecture principles in new avionics systems. Since "open" is an unclear attribute, projects eventually give in to a point solution that has no flexibility to cost effectively keep up with rapid changes in technology.

The Open Systems Development Initiative (OSDI) project at Naval Air Warfare Center – Weapons Division (NAWC-WD), China Lake, CA, utilized Commercial Off The Shelf (COTS) products to study the feasibility of building an open system that has plug-and-play capabilities. Lessons learned from the AV-8B Open Systems Core Avionics Requirements (OSCAR) and the F/A-18 Advanced Mission Computers and Displays (AMC&D) programs clearly indicated that understanding the underlying interfaces is crucial to keeping the system as open as possible to take advantage of the rapid changes in technology. A matrix of Key Open Standard Interfaces (KOSI), called the KOSI matrix, was developed and an applicable standard was identified for each interface. A list of non-conforming interfaces was also identified and the use of extensions or wrappers was investigated in an attempt to comply with standards. Standardization, rather then optimization of such interfaces, was considered to be more beneficial. It became evident that, with the exception of ruggedization, there is no difference in the use of COTS products for either commercial or military systems. Performing a KOSI analysis helped identify the key interfaces and standards, thus enabling the OSDI system to be scalable, portable and interoperable.

A good KOSI matrix provides a vehicle for clear communication and helps systems integration and technology insertion to be less painful than what it is today. It helps reduce time-to-market and provides guidance to systems engineers and vendors to keep the system open.

Background

The AV-8B OSCAR and F/A-18 AMC&D programs have agreed to aggressive schedules to take advantage of new "cutting edge" technology as soon as possible. With a majority of the work performed by teams external to NAWC-WD, it was considered essential that in-house expertise be developed to understand the complexity of building an open system for the first time. A team of engineers with expertise in avionics systems, program domain, object oriented software development and tools was established for the OSDI project.

OSDI System Architecture

The goal of the system design was to use COTS hardware that communicated across a widely accepted backplane protocol. Contractors supporting the OSCAR and AMC&D programs had already performed trade studies for selection of processor, memory and backplane. To be compatible with systems being developed for these programs, their architecture was adopted for the OSDI project.

The system currently consists of a target computer utilizing a commercial chassis with a Versa Module Europe (VME) backplane

0-7803-5086-3 /98/$10.00 ©1998 IEEE

hosting a PowerPC (PPC) based Single Board Computer (SBC) providing 2 MIL-STD-1553 channels on a mezzanine. A mini Operational Flight Program (OFP), described in the OSDI Software Architecture section, executes on the SBC to provide mission processing.

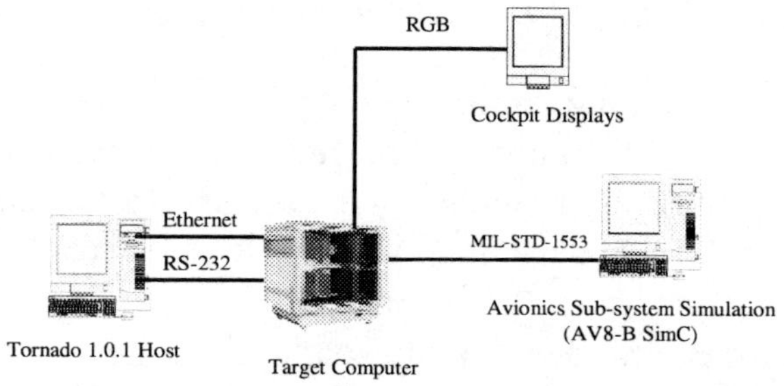

Figure 1. System Architecture

It is proposed to add another PPC based SBC to the system to implement a distributed architecture and move the display processing to the second SBC and provide an RGB output as shown in Figure 1. For the purpose of identifying interfaces, this paper considers the system to contain both the SBCs.

A Windows NT desktop computer provided the host platform for all development and integration through RS-232 and ethernet connections to the target computer. The following tools were installed on the host computer

- Tornado 1.0.1 provides the tools required to integrate the VxWorks OS/Kernel with vendor specific board support packages
- Microsoft Visual C++ provides the desktop development environment for the application software
- Green Hills Multi provides the tools for debugging the application software in a near real time environment
- Rational Rose/C++ provides an object oriented based design tool for the application program
- Microsoft Office 97 provides the documentation tools

The target computer communicated via the 1553 channels with the AV-8B SimC Simulator, which provided models of the various avionics sub-systems representative of an AV-8B Harrier aircraft.

OSDI Technical Architecture

In the target computer, the architecture was divided into three layers as shown in Figure 2.

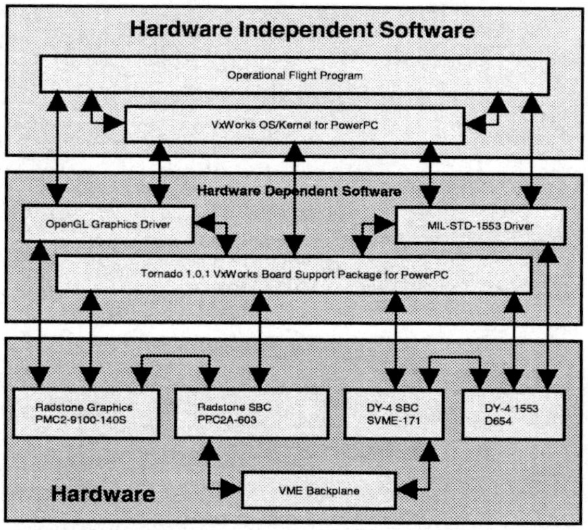

Figure 2. Technical Architecture

The hardware independent software layer consisted of the VxWorks OS/Kernel for PPC and the OFP itself. The hardware dependent software layer consisted of the board support packages and drivers. The Hardware layer consisted of the backplane, PPC SBCs and 1553/Graphics mezzanine cards. Adopting a layered architecture allowed us to isolate the OFP from any dependence on hardware. Table 1 lists various components, and their sources, to highlight the fact that COTS components from different vendors come together at several interfaces in the target computer.

Table 1. System Component List

System Component	Vendor
VME Backplane	Third party
DY-4 SBC SVME-171	DY-4
DY-4 1553 D654	DY-4

System Component	Vendor
Radstone SBC PPC2A-603	Radstone
VxWorks OS/Kernel for PowerPC	Wind River
VxWorks BSP for SVME-171	DY-4
VxWorks BSP for PPC2A-603	Radstone
MIL-STD-1553 driver	Third party through DY-4
OpenGL Graphics Driver	Radstone
Operational Flight Program	OSDI Team

OSDI Software Architecture

Object Oriented (OO) software development is now the de-facto methodology for major software systems. Design patterns are gaining a lot of visibility in their application

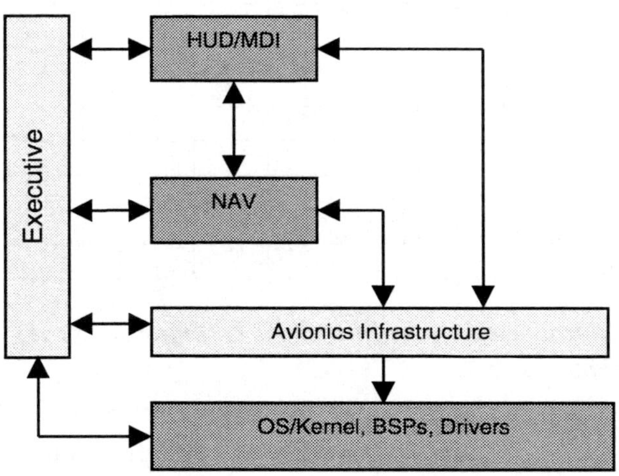

Figure 3. Software Architecture

to domain engineering. Based on research of design patterns, a Relaxed Layered Architecture, was adopted for the software. This allowed a layer to have direct access to services provided by any of the lower level layers. Figure 3 shows the software architecture of the target computer. The avionics infrastructure layer provided all the 1553/Graphics channel data for input and output, to and from the target computer. The Executive controlled the execution sequence of the various parts of the OFP. A Navigation (NAV) module provided basic aircraft position data for display on the Head Up Display (HUD) or the Multipurpose Display Indicator (MDI).

Standards and Conformance

Standards allow systems and their components to interact with each other through established channels of communication. Conformance to standards is not always a yes/no answer. Although conformance to standards solves many portability problems, efficiency, ease of implementation and historical practice often tend to reduce the level of conformance to standards. Generally, there are four levels of conformance.

- Strictly conforming
- Conforming
- National Body conforming
- Conforming with extensions

The OSDI project attempted to identify a standard for every critical interface in the system, and assign a required level of conformance to each.

Interfaces

For the target computer to be truly open, it was determined that a clear definition

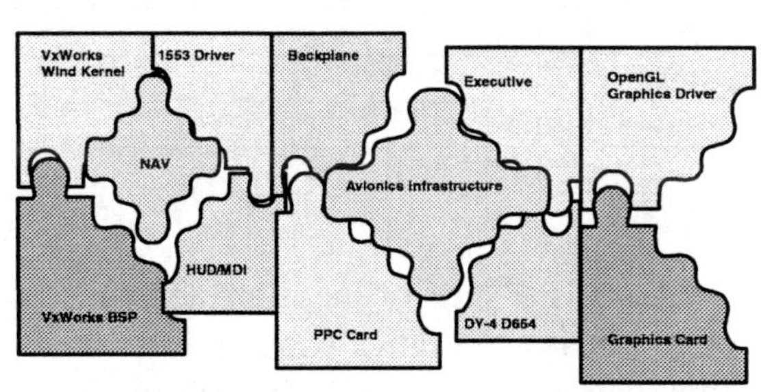

Figure 4. The Interface Challenge

of all the underlying interfaces must be

established. It was essential that an interface existing between any of the system components be assessed for compliance with a standard. As illustrated in Figure 4, bringing these components together to form a perfect fit is like putting together a puzzle with pieces that do not fit together. Herein lies the biggest challenge of building an open system. The following steps were used to generate the KOSI matrix:

Step 1: Identifying interfaces

Expanding the components listed in Table 1 to include those from Figure 3, a matrix of all the system components was developed as shown in Table 2. Based on the layered architecture approach, interfaces are listed from the top-most layers to the bottom layer. An interaction between any two components was indicated.

Table 2. Interface Matrix (Step 1)

	Executive	HUD/MDI	NAV	Avionics Interface	VxWorks OS/Kernel	OpenGL Driver	1553 Driver	VxWorks BSP for PPC	Backplane	DY-4 SVME-171	DY-4 1553 D654	Radstone PPC2A-603	Radstone Graphics PMC2
Executive		**X**	**X**	**X**	**X**								
HUD/MDI	X		**X**	**X**		**X**							
NAV	X	X		**X**									
Avionics Infrastructure	X	X	X				**X**						
VxWorks OS/Kernel	X					**X**	**X**	**X**					
OpenGL Driver		X			X								**X**
1553 Driver				X	X						**X**		
VxWirks BSP for PPC					X	X				**X**	**X**	**X**	**X**
Backplane										**X**		**X**	
DY-4 SVME-171								X	X		**X**		
DY-4 1553 D654							X	X		X			
Radstone PPC2a-603								X	X				**X**
Radstone Graphics PMC2						X		X				X	

Only one-half of the interfaces need to be looked at as the matrix has a duplicate of each interface. A total of 22 interfaces (in bold) were identified and considered for standardization.

Step 2: Identifying Key Open Interfaces

An open system is characterized by its ability to accept changes to individual components without requiring a major re-development effort. Any interface that provided communication between scalable, portable or interoperable components was considered to be critical in keeping the system open. All such interfaces were kept in the matrix as key open interfaces. Others were not considered as key open interfaces and hence were deleted from the matrix. Table 3 identifies the key open interfaces for the OSDI target computer.

Table 3. Key Interface Matrix (Step 2)

	Executive	HUD/MDI	NAV	Avionics Interface	VxWorks OS/Kernel	OpenGL Driver	1553 Driver	VxWorks BSP for PPC	Backplane	DY-4 SVME-171	DY-4 1553 D654	Radstone PPC2A-603	Radstone Graphics PMC2
Executive		X	X	X	X								
HUD/MDI			X	X		X							
NAV				X									
Avionics Infrastructure							X						
VxWorks OS/Kernel						X	X	X					
OpenGL Driver													
1553 Driver													
VxWirks BSP for PPC													
Backplane										X		X	
DY-4 SVME-171											X		
DY-4 1553 D654													
Radstone PPC2a-603													X
Radstone Graphics PMC2													

Step 3:Key Open Standard Interface (KOSI) Matrix

Each key open interface was then examined to see if there is any standard that governs its definition and use. Wherever possible, standards that were commercially accepted and available were used. Key open interfaces that have no standards available were identified for standardization. Members of the OSDI project team have been working with national standards committees and with the Open Systems Joint Task Force (OSJTF) to influence the standards development to meet the needs of the DoD. Table 4 identifies standards that were adopted for the OSDI project and identifies those interfaces that constitute the KOSI matrix.

The types of standardization that were observed are

- National Body
- Company Proprietary
- Implementation Specific

Table 4. The KOSI Matrix (Step 3)

	Executive	HUD/MDI	NAV	Avionics Interface	VxWorks OS/Kernel	OpenGL Driver	1553 Driver	VxWorks BSP for PPC	Backplane	DY-4 SVME-171	DY-4 1553 D654	Radstone PPC2A-603	Radstone Graphics PMC2
Executive		X	X	X	POSIX								
HUD/MDI			X	X		OpenGL							
NAV				X									
Avionics Infrastructure							None Available						
VxWorks OS/Kernel						POSIX	POSIX	POSIX					
OpenGL Driver													
1553 Driver													
VxWirks BSP for PPC													
Backplane										VME		VME	
DY-4 SVME-171											MaxPack		
DY-4 1553 D654													
Radstone PPC2a-603													PMC
Radstone Graphics PMC2													
Note: X = Implementation Specific													

The 1553 driver was an example of an interface that was vendor specific with no applicable standards. Drivers obtained from DY-4 and Radstone were compared and their interfaces were found to be very different. It was immediately recognized that some standardization of this interface was required. An object was created to hide the implementation details, and provide higher-level functions (initialize channel, create chain, add message to chain etc.) to the avionics infrastructure layer. In this case, standardization versus optimization was considered to be more beneficial to the project.

MaxPack is a mezzanine interface standard developed by DY-4 for its products. Nevertheless, PCI Mezzanine Card (PMC) is gaining industry wide acceptance very rapidly, as a standard for mezzanine interface. Many vendors, including DY-4, now provide PMC mezzanine cards.

In the application software, there were a few implementation specific interfaces. It's important to note that implementation specific interfaces are key interfaces and need to be controlled within the developer's organization. Parties from both sides of the interface must mutually agree upon any changes to these interfaces.

Open Systems Application Programming Interface (OSAPI)

Considering the wide range acceptability that open systems technology is gaining within the DoD, it should be recognized that a standard Application Programming Interface (API) is becoming a necessity rather than a goal. Team members from the OSDI project participated in a joint Navy & industry wide effort to identify standard services that a Navy open architecture system would need. This API was to be flexible enough to handle legacy avionics interfaces (MIL-STD-1553) as well as newer high bandwidth interfaces like Fibre Channel and PCI.

Open System Technical Architecture

The Open Systems Mission Computer Application Programming Interface (OSMC API) team consisted of members from the Navy, Boeing, General Dynamics Information Systems (GDIS), and Smith Industries to define the services that the API was required to provide for the AV-8B OSCAR and F/A-18 AMC&D programs. The services were documented in an Interface Requirements Specification (IRS) for the OSMC, which later was used as the baseline for the OSCAR program. The AMC&D program is now using the same baseline to define extensions specific to the program. Figure 5 is an example of an Open Systems API (OSAPI) based on the OSMC API.

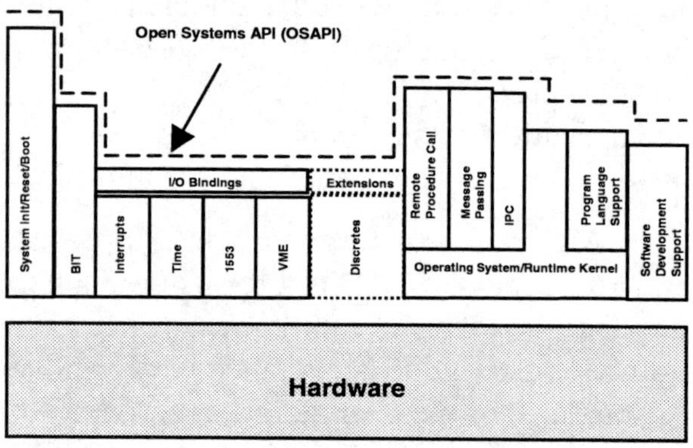

Figure 5. The OSAPI Definition

Generic API Services

The services that the OSAPI provides can be grouped by function, and categorized into the following list.

- System Initialization & Boot Loader
- Built-In Test (BIT)
- Device Drivers
  - Interrupts
  - Timers
  - Backplane (VME)
  - Legacy I/O (1553, 1760)
  - High speed I/O (Fibre Channel etc)
  - Discrete I/O
  - Bindings
- Operating System (OS)/Kernel
- Programming Language Runtime Support
- Software Development Interface

A recommendation was made to the OSJTF to facilitate the distribution of this OSAPI specification through their on-going efforts with open systems technology vendors. A copy of the OSMC API specification can be obtained from NAVAIR PMA-209, Patuxent River, MD.

DoD expects that the industry will voluntarily embrace this OSAPI specification, and contribute their time and money to enhancing the performance of the hardware, and related drivers, instead of developing new versions of the same interface.

Summary

The OSDI project provided the much-required hands-on expertise within NAWC-WD to intelligently work with external teams developing system components for the OSCAR and AMC&D projects. Standardization of key open interfaces is a challenge that must be met in the very initial stages of system analysis. Both industry and DoD participation is essential in any effort to standardize interfaces that bring commercial technology into the military environment. When a technology achieves industry standardization, the costs and risks of using it go down, and the number of vendors and users goes up. Standardization is the primary key to cost savings and to successfully build COTS based systems for the military.

Systems based on a KOSI matrix tend to be open indefinitely, as vendors develop faster and better technology that meet the established interface specifications. The much-hyped terms like portability, scalability and interoperability can now become a reality in military avionics systems. Systems adopting these standard interfaces have a drastically reduced time-to-market allowing the military to capitalize on rapidly changing technology. Changes are expected, but any changes to the KOSI matrix must be religiously controlled to avoid any surprises to system integrators and vendors providing the system components.

A system that does not have a KOSI matrix established is bound to end up, sooner or later, as a point solution. Such a system requires a major redevelopment effort to enhance its capability. In the past, when programs seemed to have unlimited budgets, and rigorous fiscal/personnel controls were not in place or practiced, we could still get a workable product. However, this required additional funds to correct problems that could have been avoided. But now, times are changing and a product at unlimited cost is no longer acceptable and perhaps unachievable.

Developing a KOSI matrix provides a solid definition of the system interfaces and becomes a valuable tool for system engineers as well as component vendors. It guarantees that components designed to comply with the matrix will indeed interface correctly. Additionally it provides an established guideline for system upgrades.

Acknowledgements

The author wishes to express his appreciation to the following members of the team for their invaluable contributions to the OSDI project: Dale Christensen, Chris Ricketts, Debra Borden, Dave Breitigam, Fernando Escober, John Veazey, Jeff Fisher, Lee Lucas, and Linda Shelton.

The author wishes to acknowledge the contributions of the members of the Navy and industry team who jointly defined the OSMC API.

References

Jones, Derek, *Verifying claims of Open Systems Conformance,* Knowledge Software Ltd., Farnborough, Hants GU14 9RZ, UK. Presented at Sunscope Conference 1994.

Buschmann, Frank, et. al., *Pattern-Oriented Software Architecture: A System of Patterns*, John Wiley & Sons, West Sussex, England, 1996. ISBN 0-471-95869-7.

Gamma, Erich, et. al., *Design Patterns: Elements of Reusable Object-oriented Software*, Addison Wesley Longman, Inc., 1994. ISBN 0-201-63361-2.

Acronym List

AMC&D	Advanced Mission Computer & Displays
API	Application Programming Interface
BIT	Built-In Test
COTS	Commercial Off The Shelf
DoD	Department of Defense
GDIS	General Dynamics Information Systems
HUD	Head Up Display
I/O	Input/Output
IRS	Interface Requirements Specification
KOSI	Key Open Standard Interface
MDI	Multi-purpose Display Indicator
NAV	Navigation
NAWC-WD	Naval Air Warfare Center – Weapons Division
OFP	Operational Flight Program
OSCAR	Open Systems Core Avionics Requirements
PCI	Peripheral Component Interface
OSJTF	Open Systems Joint Task Force
OO	Object Oriented
OS	Operating System
OSAPI	Open Systems Application Programming Interface
OSDI	Open Systems Development Initiative
OSMC	Open Systems Mission Computer
PMC	PCI Mezzanine Card
PPC	PowerPC
RGB	Red Green Blue
SBC	Single Board Computer
VME	Versa Module Europe

REDUCING AVIONICS SOFTWARE COST THROUGH COMPONENT BASED PRODUCT LINE DEVELOPMENT

David C. Sharp, The Boeing Company, St. Louis, Missouri

Introduction

In 1995, an initiative was launched at McDonnell Douglas (now merged with The Boeing Company) to assess the potential for reuse of operational flight program (OFP) software across multiple fighter aircraft platforms, and to define and demonstrate a supporting system architecture based upon open commercial hardware, software, standards and practices.[1] This initiative produced a set of reusable object oriented navigation software, which was flight tested on several different aircraft platforms hosted on different hardware configurations (both MIPS® and PowerPC® based).

The following year, this validation of the product line approach led to broader application of the techniques to the tactical aircraft mission processing domain. This paper describes key aspects of the component based logical architecture developed therein.

Software Project Goals

As illustrated in Figure 1, two primary software project level goals were defined:

1. Development of a reusable architecture framework for the product line.
2. Development of reusable application components for the product line.

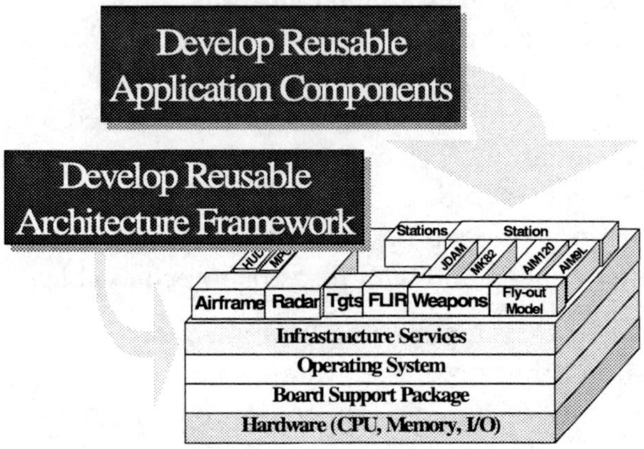

Figure 1: Software Project Goals

The project follows a two-pronged strategy: establish an architecture foundation that defines the overall system structure and component environment, and then create application components that reside therein. The combination of these two techniques provides the basis for an affordable approach to product line avionics software. This strategy follows Johnson's approach of decomposing reusable software systems into a framework and a component library.[2]

Architecture Goals

Given the driving project goals, there are two key goals of the software architecture itself:

1. Contain change.
2. Maximize reusability.

[1] Winter, Don C., "Modular, Reusable Flight Software For Production Aircraft", *15th AIAA/IEEE Digital Avionics Systems Conference* Proceedings, October, 1996, p. 401-406.

[2] Johnson, Ralph, *How to Develop Frameworks*, Conference on Object-Oriented Programming, Systems, Languages, and Applications, October 1997.

0-7803-5086-3 /98/$10.00 ©1998 IEEE

Naturally, there are many types of change that appear in a complex avionics mission computing product line:

1. Differences in avionics subsystems (e.g. radar, cockpit controls and displays, INS)
2. Differences in mission computing hardware – both in terms of number of processors and in terms of the exact microprocessor chosen.
3. Differences in system requirements.

In terms of the changes that need to be anticipated during design, system differences within the product line are similar to traditional changes to a single system. Therefore, in a product line project, there are two types of differences that must be contained:

1. Temporal differences – the evolution of a system over the course of time, and
2. Simultaneous differences – the differences between multiple products in the product line that must be concurrently supported.

The established adage of "encapsulate change" can be somewhat broadened to "encapsulate differences" when in a product line context.

Maximizing reusability relates to minimizing cost in a reuse-driven affordability process. One corollary to this goal is that the developed software must be granular enough that specific aircraft products can select the exact software elements applicable to their system. This requires elements with the ability to be plugged together to create a deliverable system.

Note that these architecture goals are a direct result of the original goal to develop software applicable to a full product line.

Patterns

Patterns have been widely used in the object-oriented community to capture best practices. They document the forces and issues providing the answer to the frequently asked question of *why* software was designed a certain way. Documenting the tradeoffs made while arriving at a solution guides future development of the produced artifact and reuse of the techniques in other areas. Understanding the consequences of a design choice is especially key in those cases where multiple potential solutions exist whose selection is dependent on specifics of the problem context. It is in these situations that patterns prove most valuable.

There are several different classification techniques, or taxonomies, of patterns in the literature. Gamma et al. divide their design patterns into three categories[3]:

1. Creational – patterns used in flexibly constructing objects
2. Structural – patterns outlining class or object composition
3. Behavioral – patterns describing how objects interact and distribute responsibility

Schmidt[4] and others classify patterns as:

1. Tactical – patterns that resolve localized design forces in an application
2. Strategic – patterns that address application-wide issues and thus have broad impact

and also as:

1. Domain Independent – patterns that cross application domains
2. Domain Dependent – patterns applicable to a specific application domain

Buschmann et al. break patterns into the following divisions[5]:

1. Architectural – patterns providing the overall skeleton of a system
2. Design – patterns dealing with localized design choices
3. Idioms – patterns of usage within a particular language

---

[3] Gamma et al, *Design Patterns: Elements of Reusable Object-Oriented Software*, Addison-Wesley, 1995, p. 10.

[4] Schmidt, Douglas C., "A Family of Design Patterns for Application-Level Gateways", *Theory and Practice of Object Systems*, Wiley & Sons, Vol. 2, No. 1, December 1996.

[5] Buschmann et al., *Pattern-Oriented Software Architecture: A System of Patterns*, Wiley & Sons, 1996, p. xiv.

Some of the numerous relationships between these classifications are:

- Design patterns are often tactical
- Architectural patterns are often strategic
- Strategic patterns are often domain dependent

Given this context, the development process relies heavily on leveraging domain independent patterns developed in other organizations and growing domain specific patterns within the home organization, particularly strategic ones. This combined set of patterns then forms a shared expertise base, which guides high level architecture development as well as individual development across the full organization.

Top Level Architecture

Two architectural patterns underlie the top level software architecture: the *Layers* pattern[6] and the *Model-View-Controller* pattern[7]. Using these patterns helps *generate* the overall architecture for the system[8].

Layers Pattern

The *Layers* pattern describes the widely used method of basing software systems on layers of abstraction and insulation. Low layers in the system provide hardware and operating system level facilities. Higher layers provide application functionality. The low layers form the foundation that insulates the application from specifics of the hardware configuration. This is an essential characteristic of a product line software system that is targeted to varying hardware configurations.

A key characteristic of layered architectures is that all dependencies flow downward. This helps avoid cyclic dependencies, and avoids low level abstractions from depending on high-level application considerations. Jacobson et al. mention layered architectures as a central aspect of reusable software systems.[9]

Model-View-Controller Pattern

The Model-View-Controller (MVC) pattern was developed within the Smalltalk community as an architecture for graphical user interface based systems. Its architecture is shown in Figure 2, and is itself a layered architecture.

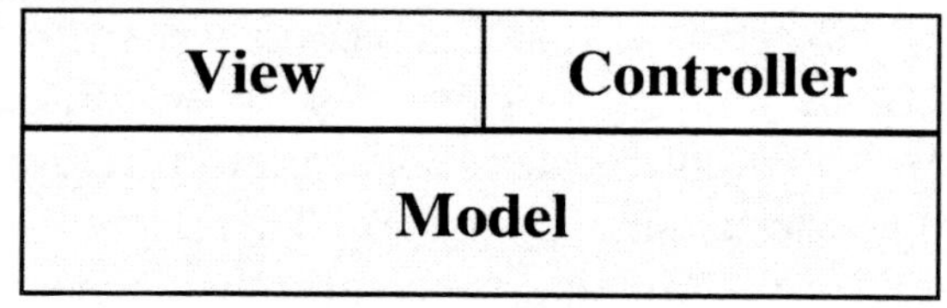

Figure 2: Model-View-Controller Architecture

One of the important characteristics of the MVC architectural pattern is the separation of the domain model from the user interface utilized to present domain information to users and allow them to manipulate it. This allows multiple views (i.e. displays) of the same information. It also separates the relative stability of domain models from the relative instability of user interfaces, resulting in better change encapsulation.

Top Level Architecture

The top level software architecture derives from applying the two aforementioned architectural patterns. In the case of the Layers pattern, it is applied recursively in hierarchical fashion. In all cases, a *relaxed* version of the Layers architecture is applied where a higher level layer may bypass intermediary layers and access low layers directly where necessary.

Figure 3 shows the top level architecture. The Model layer represents the core domain application as defined in the MVC architecture. The Operator layer contains both the View and Controller portions of the MVC

[6] Ibid., p. 31-52.

[7] Ibid., p. 125-144.

[8] Beck, Kent, and Johnson, Ralph, "Patterns Generate Architectures", European Conference on Object-Oriented Programming (ECOOP) Proceedings, 1994.

[9] Jacobson et al, *Software Reuse: Architecture Process and Organization for Business Success*, Addison-Wesley, 1997, p. 170-212.

pattern, sometimes referred to as a *Document-View* variant of the MVC architecture.[10]

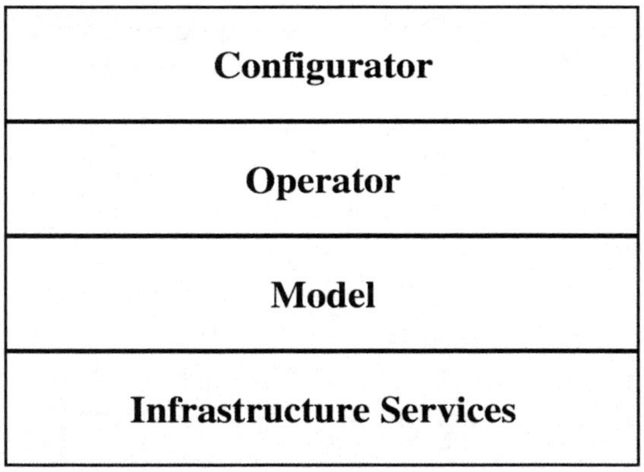

Figure 3: Top Level Layered Architecture

To these MVC based application layers, a foundational Infrastructure layer is added to provide a CORBA based run-time system and insulate the application from hardware configuration specifics. At the top, a Configurator layer is added which represents a notional "software system circuit board" and is responsible for instantiating all of the objects and establishing their relationships. The Configurator layer is responsible for taking all of the applicable software elements and plugging them together into a functional system.

Especially at the architectural level, the boundaries between software elements should foster change encapsulation and separate independent concerns. This inspection reveals the following benefits of the top-level architecture:

- The *Infrastructure Services-Model* boundary isolates hardware, run-time, and operating system issues from the application.
- The *Model-Operator* boundary isolates the core application from user interface volatility.
- The *Operator-Configurator* boundary isolates the software elements from their integration into a complete system.

Model Architecture

As stated earlier, the Layers pattern is applied recursively. One example of this is within the top level Model layer. Figure 4 illustrates this second level layered architecture.

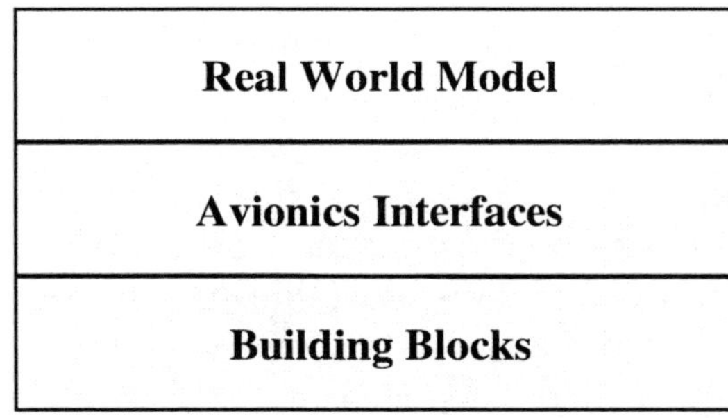

Figure 4: Model Layered Architecture

In this case, the Building Blocks layer represents general-purpose objects like vectors and matrices. Avionics Interfaces represents the mission computing interface to other avionics subsystems. The Real World Model contains the software representation of the problem domain. Examples of elements in the Real World Model are weapons, stations, targets, the ownship airframe, waypoints, and routes.

Applying the boundary inspection process as before at this level reveals the following benefits:

- The *Building Blocks-Avionics Interfaces* boundary separates more general-purpose domain independent elements from domain specific elements.
- The *Avionics Interfaces-Real-World Model* boundary forms a key separation isolating product line elements from the specifics of a single avionics system on a particular aircraft. This boundary is directly related to the original goal of containing avionics system specifics.

Summarizing, combining repeated application of these architecture patterns with domain specific considerations develops the overall system structure and leverages proven

---

[10] Buschmann, p. 140.

architecture experience within the product line software architecture.

Operational Flight Program Component Overview

Once the overall logical architecture structure is established, issues concerning software element development within this architecture surface. An important consideration is the granularity of the software elements. If an extremely fine granularity is chosen, the canonical software elements can be defined as being individual objects. This results in a multitude of elements that must be configured and integrated at the Configurator level. Near the other extreme, a single software element can be created for each of the layers. This fails, however to provide the configurability required by individual aircraft projects both in terms of independently selecting software elements applicable to their system and in allowing any required customizations to those elements. As Roberts and Johnson note in their Fine-Grained Objects pattern,[11] making software elements more reusable tends to make them finer-grained. The opposing force is that systems with more elements are more difficult to understand.

In the subject software system, the term Component was adopted to represent a software element somewhat in the middle of the granularity continuum. Given the sometimes ambiguous usage of the term, and the need to clearly specify its meaning within the project, a set of defining patterns was developed. Because of their pervasive usage within the system, they are classified as strategic patterns. These patterns can be simplistically stated as providing software elements that are as large as possible while still being reusable within the product line.

These patterns form a small language whose names can be used within the organization to easily communicate and guide development. There are potentially many different types of components that fit in this language, including:

1. Component – the basic component definition that applies to all of the more specific types
2. Configurable Component – components that allow customization by users
3. Distributable Component – components that are distributable when in a multi-processing hardware configuration
4. Real Time Component – components that provide real time quality of service information for scheduling purposes[12]

a) Active Component – components that have specific requirements under which they must run

b) Passive Component – components that execute only as a service to other components

This paper focuses on the foundational Component pattern itself. The Configurable Component pattern can be referenced in the literature and is not included here due to space constraints.[13]

Component Pattern

Roughly following the pattern outline of Gamma et al.[14], the following sections define the Component pattern.

Intent

Define software entities which are similar to objects but which can be composed

---

[11] Roberts, Don, and Johnson, Ralph, "Evolving Frameworks: A Pattern Language for Developing Object-Oriented Frameworks", *How to Develop Frameworks*, Conference on Object-Oriented Programming, Systems, Languages, and Applications, October 1997.

[12] Harrison, Timothy, Levine, David, and Schmidt, Douglas, "The Design and Performance of a Real-time CORBA Event Service", Conference on Object-Oriented Programming, Systems, Languages, and Applications Proceedings, October 1997, p. 184-200.

[13] Sharp, David C., "Reducing Avionics Software Cost Through Component Based Product Line Development", Software Technology Conference, April 1998.

[14] Gamma et al., p. 6-8.

of multiple smaller objects that together provide a set of services to a client.

Motivation

In multithreaded, multiprocessor applications, objects must deal with a number of complexities. Among these complexities are concurrent threads, shared resources, and multiple address spaces. In addition, large object oriented applications can easily comprise thousands or tens of thousands of objects. Components provide a practical way to group closely coupled objects that together provide a set of services to other objects. Grouping objects in this way allows for localization of certain concerns and provides a higher level of abstraction when considering software collaboration.

Applicability

Use the Component pattern when

- closely coupled objects provide a set of cohesive services to clients,
- a group of objects is reusable as a single entity, and
- especially when collaborating objects share similar synchronization requirements or thread priorities (e.g. processing rate)

Structure

Figure 5 shows the Component pattern structure using the Unified Modeling Language (UML).

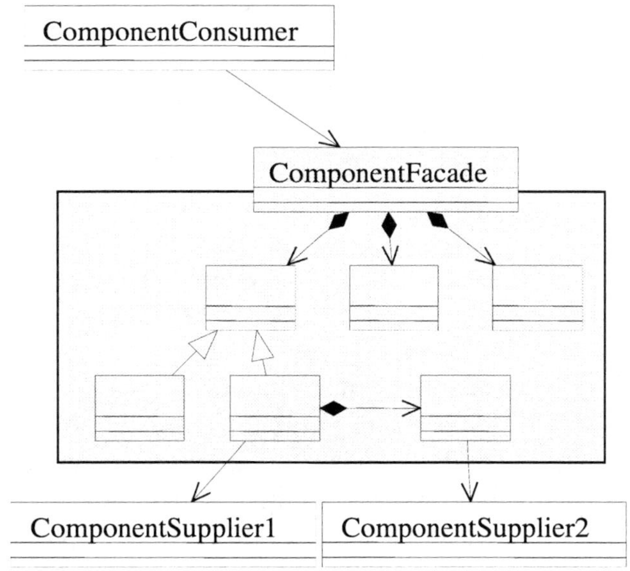

Figure 5: Component Pattern Structure

Participants

- **ComponentConsumer**
  - uses the component via the facade
- **ComponentFacade**
  - provides the public interface for the contained objects

Collaborations

- Consumers communicate with the component via the façade, which forwards them to the appropriate contained objects as applicable.
- Consumers do not access internal objects directly without first going through the facade.

Consequences

The Component pattern has the following benefits and drawbacks:

1. *Encapsulates change.* Promotes weak coupling between components by defining a clear interface via the component façade. The façade actually provides an interface to potentially many enclosed objects.
2. *Defines distribution boundaries.* Components form the potentially distributable software entities at a level higher than individual objects. Note that components are not always distributable, but all

distributable items are components. Objects within components are guaranteed to reside within the same address space, fully supporting pointer sharing.

3. *Localizes concurrency control.* Localizes concurrency control issues to the component facades so that most objects do not have to be concerned with thread locking mechanisms.
4. *Identifies pluggable entities.* The pluggable software entities, which developers may choose to incorporate or not incorporate into a specific executable, are represented by components. While some components are foundational and essential, the majority is selectable and pluggable.
5. *Enhances reusability.* Makes it easier to configure resulting executables by reducing the number of software entities that must be considered.
6. *Reduces complexity.* Similar to the reason for *Enhances reusability*, Component reduces complexity by reducing the amount of detail that component users must consider.

Implementation

Consider the following issues when implementing the Component pattern:

1. *Allowed calls.* No external component is allowed to call member functions of objects internal to the component without first going through the facade. Doing otherwise defeats the goal of localizing concurrency control and providing a single public interface for the whole component. Objects within the component are allowed to invoke member functions on external component façades. Summarizing, *calls in* go through the façade. *Calls out* have no such restrictions.
2. *Contained objects.* Components can contain references to internal objects. If a façade returns a reference to another object for direct use by a consumer, then some other mechanism must be provided for serialization, such as:
   a. the returned object must manage its own concurrency mechanism, etc, and therefore takes on the responsibilities of a component facade, or
   b. the component façade must provide a request/release locking mechanism for access to the reference to ensure that access is serialized.
3. *Façade usage.* Objects within the component should not call public methods in the Façade. Since the façade normally contains the concurrency locking mechanism, components are typically not reentrant. If an internal object calls a façade member function, it would potentially block indefinitely waiting for the current thread (its own) to release the lock. For this reason, the façade should not contain algorithms or other functionality in public member functions that objects internal to the component would want to invoke. While not strictly prohibited, making facades reentrant requires recursive locking which is more performance intensive.

Related Patterns

This pattern is a specialization of the Façade pattern.[15] Façade pattern issues apply here as well.

[15] Ibid., p. 185-193.

Conclusion

This paper focuses on two important characteristics of the software architecture:

- Layered Architecture
- Medium Grained Architecture

It is important to note that both of these characteristics derive directly from the original goals of encapsulating change and maximizing reuse. The layering aspect especially relates to developing an application that is independent of the hardware and avionics specifics of a single aircraft. The medium granularity of the components provides a system that can be deployed on varying numbers of processors in single processor or distributed hardware configurations. Both of these goals directly relate to the original decision of developing a product line software system.

Patterns form a very powerful technique for reusing and spreading *expertise*. The pattern driven architecture described herein leverages previous industry experience to address the original goal of defining a software architecture applicable to a fighter aircraft mission processing product line. The Component pattern language defines the reusable software elements and provides the complete development community with understanding of the underlying rationale, tradeoffs, and implementation concerns involved. The resulting component library embodies reusable application elements available for specific aircraft products. All of these elements, including:

- Architecture,
- Components,
- Patterns, and
- Developer expertise

provide reusable artifacts, which are exploited to reduce life cycle avionics software cost.

Acknowledgements

The concepts described herein were developed within the stimulating confines of the project Software Core Architecture Team. Their contributions, and especially the many insights provided by founding member Bryan Doerr, are gratefully acknowledged. Our work also reflects an extremely fruitful collaboration with Dr. Douglas C. Schmidt and his colleagues at Washington University in St. Louis, Missouri.

Introduction to the Electronic Power Specification Standardization Activity

Sergio Navarro, TRW, Dayton, OH
Carlos Gonzalez, Raytheon Systems Company, El Segundo, CA
Marvin Soraya, AFRL/IFSC, Dayton, OH

Abstract

The Electronic Power Specification Standardization (EPSS) activity is consistent with DOD's Open System Joint Task Force (OS-JTF) strategy to reduce systems acquisition and lifecycle costs through judicious use of commercial resources (standards, products, processes, and services). As a result, the deliverables from this activity are commercial industry-based guidelines/standards for use by the government and industry to enhance use of commercial resources. The EPSS activity focuses on standardization of elements relevant to electronic power distribution systems. Therefore this paper will describe: 1) the rationale for partitioning an electronic power distribution system into primary / secondary power distribution and power supply systems, 2) the various on going activities in electronic power distribution standardization as they relate to EPSS, and 3) the progress to date and plans of the present focus of the EPSS activity: a) IEEE P1515 Working Group, and b) EPSS Power Supply System Consortium.

Introduction

The EPSS activity is sponsored by the Open System Joint Task Force (OS-JTF) from the Office of the Secretary of Defense. OS-JTF's strategy is to reduce DOD systems acquisition and lifecycle costs through sensible use of commercial resources. Use of commercial resources implies [1] use of commercial standards, products, processes, practices, and/or services either commercial or industrial grade ("off-the-shelf" or customized). The activity is facilitated by the EPSS Core Team which consists of personnel from TRW, Raytheon Systems Company and AFRL / IFSC. Recently, the EPSS activity has gained the support of the Power Sources Manufacturer's Association (PSMA). The broader participating EPSS Working Group (the integrated product team (IPT)) consists of about 100 personnel from the military, commercial industry, and academia. In working with the IEEE organization, it was suggested that the EPSS standardization process should begin with guidelines[1] or recommended practices and, following their sufficient use, migrate to formal standards. Hence, throughout this paper, the term "standard" implies guideline, recommended practice, or formal standard.

In concert with industry, the EPSS activity focuses on developing standards for elements relevant to electronic power distribution systems. There are many models to represent what is meant by electronic power distribution systems; the representative EPSS model is shown in Figure 1. There are three main elements to this model: 1) primary power distribution system, 2) secondary power distribution system, and 3) power supply system. Partitioning in this manner was done not so much from a technical standpoint, but rather from a market standpoint. For instance, an industry exists which develops and produces primary / secondary power distribution systems (kW to MW high voltage power converters, power controllers, etc.); on the other hand, another industry exists which develops and produces power supply systems (5W to 1kW low voltage power converters, rack power controllers, etc.). The systems developed by these two industries may schematically look similar, but the components, practices, and processes used may vary greatly due to the differences in power and voltage levels. Consequently, it was deemed inappropriate to attempt to develop a standard that would try encompassing primary / secondary power distribution systems and power supply systems.

A more general overview of EPSS activity and goals is given in reference [2].

Ongoing Standardization Activities

Standards for electronic power distribution exist today; for instance, DO160 is a commercial airlines environmental conditions and test procedures standard and IEC-1204 is a computer power supply standard. So why are more standards needed? The answer lies in the NEEDS of the commercial electronic power distribution system customers. Commercial customers are beginning to want [3] increased modularity with standard building blocks and interfaces. As the computer industry has shown, the use of standards is a responsive approach that

[1] The terms guideline, recommended practice, and standard can be characterized in the following ways: guideline uses terms like "may do this", recommended practice uses terms like "should do this", standard uses terms like "must do this".

0-7803-5086-3 /98/$10.00

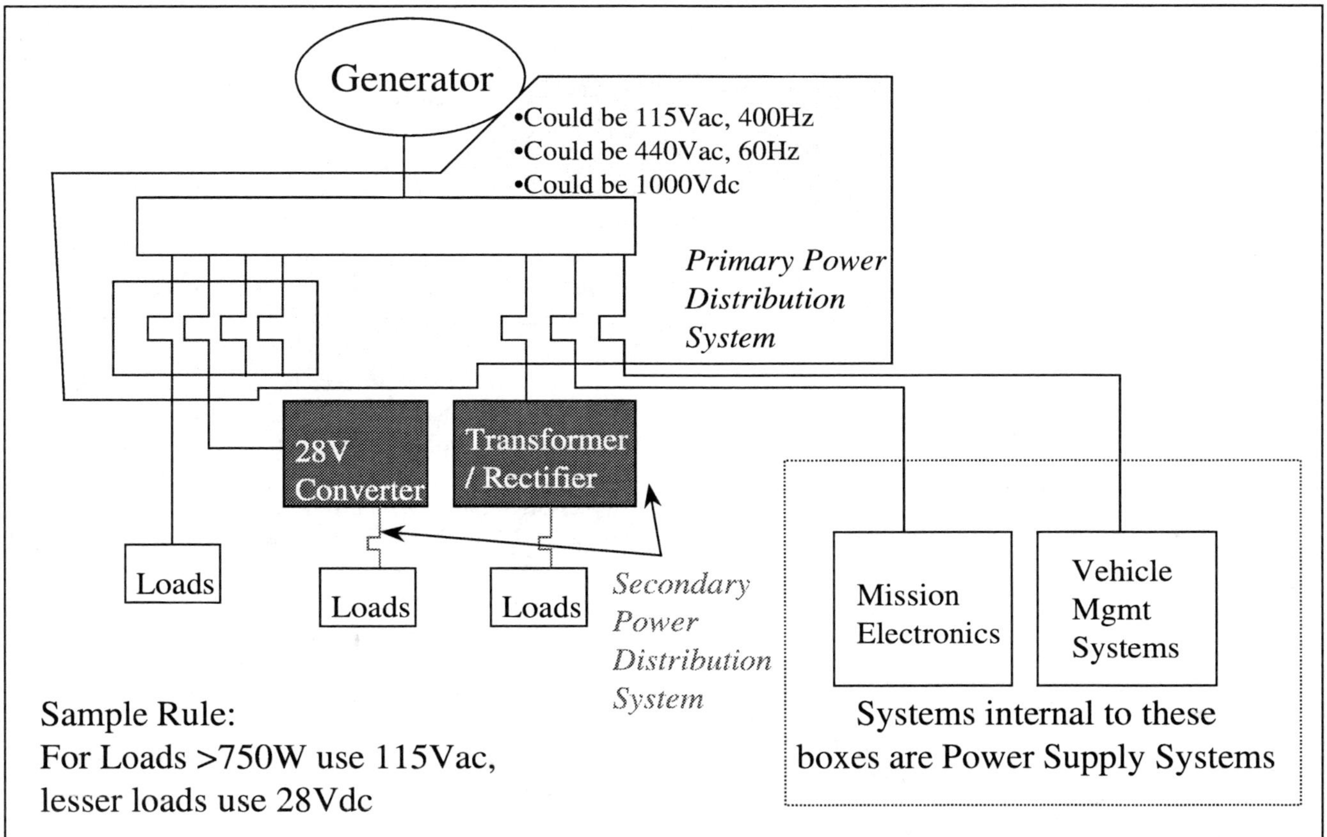

Figure 1 Electronic Power Distribution System Model

benefits both the customers and producers of modular products.

Below is a listing of the ongoing electronic power specification standardization activities with a brief description of each.

IEEE 1156.2 This is a standard for environmental specifications for computer systems. It contains minimum environmental withstanding conditions applicable to computer systems and their associated components. It defines such parameters as: atmospheric stress, corrosion, drop heights, electrical stress, EMI, ESD, fire resistance, flammability, humidity, level of qualification, mechanical stress, shock, sinusoidal vibration, thermal shock, thermal testing, vibration, etc. This standard has been approved for publication.

IEEE P1156.3 This will be a standard for Power Supply Specifications for Computer Systems. This standard is to create a generic power supply specification to support IEEE bus standards. This standardization activity is being performed because existing computer bus standards only provide a brief, incomplete set of specifications for a power supply and often none at all. This is an ongoing activity.

SAE AE-7 This is a standardization activity which is dealing with primary power distribution standards. The ultimate standard will define primary power distribution voltage, power quality, etc. This is an ongoing activity.

IEEE P1515 This is a standardization activity for electronic power distribution system specification standardization. The purpose of this standard will NOT be to create a generic specification, but rather to create a standard "specification language" which both customers and product developers use to unambiguously communicate specifications. Consequently, this standard will define a set of parameters that can unambiguously characterize the elements of power distribution systems; this will include description of system parameters and their applied test conditions, and test methods. This is an ongoing activity; it is described in more detail in the following section.

Power Supply System Consortium The deliverable intends to be a "building codes" standard for power supply system development; it will enable government and military contractors to reference and sensibly apply commercial resources. These "building codes" are essentially rules which guide the government and the military contractor in the development of power supply system products, processes, and practices to enable intelligent use of commercial resources. The purpose of these rules is to help the government to better utilize what commercial industry has to offer. This effort is a part of the EPSS activity; it is described in more detail below.

IEEE P1515 Progress and Plans

The IEEE P1515 Working Group was established at APEC '98 with the objective to develop the specification language for the core parameters that are used to characterize electronic power distribution systems elements (power supplies, energy storage devices, power switching devices, etc.). These parameters are used to describe AC/DC and DC/DC systems elements, and include single phase and three phase systems. The power levels extended to 20KW, with voltage levels up to 600V, and frequencies up to 400Hz.

The specification language consists of three sections. They are: 1) Definition, 2) Test Method, and 3) Test Condition. The Definition Section consists of the meaning of the parameter along with references to its location of measurement and its units of measure. Test Method is a description of the means by which to conduct the test for the parameter. In Test Method there is diagram of the test setup, an identification of the kinds of equipment, the loads for the tests, the stimulus description, and any other circumstances required to properly test the parameter. A description of the characteristics of the test equipment and the loads is defined in the Definitions section of the Recommended Practice. Test Condition identifies the kind of data that needs to be collected under given operating conditions. An example of a specification language is shown below. The use of this language on a product leads to consistent specification and performance testing by manufacturers. The result is an unambiguous understanding of the electronic power distribution system elements by the user, leading to customer satisfaction with minimal user confidence testing.

Based on an informal industry survey conducted in 1997, the Core Team activity led to the identification of 91 parameters (see Table 1 Electronic Power Distribution System Parameters); these 91 parameters are forming the basis of the P1515 Specification Language. Some parameters are more critical than others extending from such fields as Electromagnetic Interference (EMI) to voltage regulation, to environmental, fault characteristics, and simple mechanical parameters (pin outs, footprint, etc.). An abridged version of the P1515 Preliminary Draft Recommended Practice that included the language for 10 parameters was developed and used as a starting point at APEC '98. The feedback from the reviewers highlighted several areas for the EPSS Core Team's activity to address and enhance. In the APEC meeting the EPSS Core Team began to gain volunteers to join the (P1515) Working Group and identify the parameters they wanted to support in developing the specification language. Gaining volunteers has been an ongoing effort; it has resulted in achieving representation from: Boeing, Advanced Conversion Technologies, ELDEC, Power Paragon Inc., Raytheon Systems Company, Rockwell Collins, Sundstrand Aerospace, and VICOR. The support from the volunteers (as of May 98) has claimed 64 of the 91 parameters. To report the status of P1515 Working Group activities, IEEE has established a web area for the Group at address: http//grouper.ieee.org/groups/1515.

Specification Language Example
Dynamic Load Regulation: Example Definition
The transient response of the output voltage of an electronic power distribution system element due to a step change in the load current.
Dynamic Load Regulation: Example Test Method
For a load increase to the output voltage, V_o, use the test setup described in Figure 2. The waveform generator should produce a waveform as described on Figure 3. Figure 4 provides a definition for the various UUT (unit under test) measurement parameters; for example, output voltage, V_o, response time, dynamic regulation (excursion), recovery time, and final value of output, V_o, after the load transient.

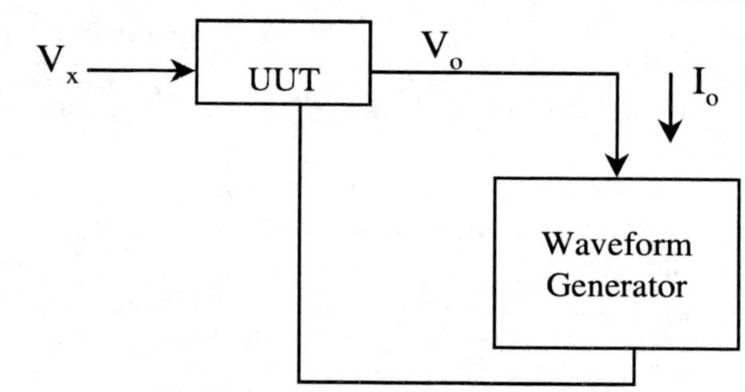

V_x = Input voltage adjusted to nominal value and measured with DMM
Vo = Output voltage measured with Oscilloscope
Io = Output load current measured with Current Probe

Figure 2 Test Setup to Generate Output Load Step Increase in Dynamic Load Regulation Test

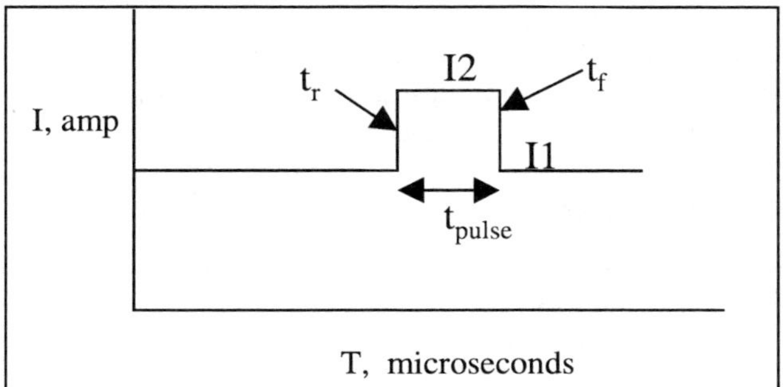

I1= average operational value of Io for UUT
I2= adjusted value of Io to maximum specified in specification
t_r = the time required for the Io current step to rise from 10 to 90 per cent
t_f = the time required for the Io current step to decrease from 90 to 10 per cent
t_{pulse} = minimum time to allow the UUT output voltage,Vo, to recover to within the specified accuracy limits

Figure 3 Load step increase waveform for dynamic load regulation

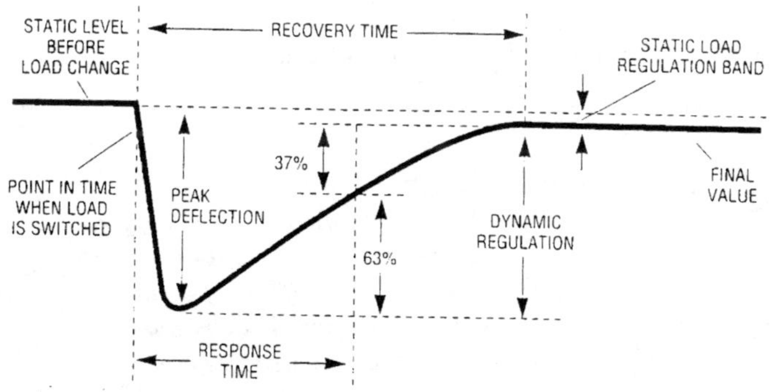

Figure 4 Typical Vo deflection due to step load increase to UUT

Example Test Condition: Dynamic Load Regulation for step load increase

Ambient Temp Range/	neg 40c	plus 25c	plus 50c
Parameters			
response time (sec)			
recovery time (sec)			
dynamic regulation (volts)			
Vo final value (volts)			
Vx = Vnom input voltage			
Io = Imax load current			
The load includes any line impedance in series with it.			
Record measurements of parameters at all three temperatures			

The P1515 Working Group has developed a process to follow for achieving the publication of the IEEE P1515 Recommended Practice. The plan is:

1. Incorporate each volunteer's first input into a first draft Recommended Practice for review at a meeting to be held in concurrence with NAECON (July 1998) in Dayton, OH. All consensus feedback will be incorporated into a subsequent draft of the Practice.
2. The Group will submit their updated specification language to incorporate into the second draft of the Practice for review, comment and feedback at the third P1515 Working Group meeting to be held in concurrence with INTELEC (October 1998) in San Francisco, CA.
3. Each volunteer will update their second draft and submit for distribution and consensus by the Working Group via the Internet.
4. IEEE P1515 Working Group to establish balloting group to review second draft; 75% balloting group approval required.
5. Submit the Recommended Practice to the IEEE Standards Board for final approval.

The P1515 Working Group will support the publication of the Recommended Practice and work to maintain buy-in from such organizations as the Power Sources Manufacturer's Association (PSMA) and electronic power distribution Automatic Test Equipment (ATE) community. The Group will maintain a constant association with these organizations to promote the Recommended Practice and help develop a process to insure its use in the industry. It is anticipated that the P1515 Recommended Practice will be interpreted and recommended for modification by individual companies and technical groups. The P1515 Working Group will collect and review the comments with IEEE to consider their incorporation into a revised version of the Recommended Practice. The Working Group will continue to attend or organize meetings or industry conferences and present the findings, highlights, and strategies to continue the Recommended Practice's penetration and utilization by developers and manufacturers until it has migrated from a IEEE Recommended Practice to a formal IEEE Standard.

Power Supply System Consortium Progress and Plans

The purpose of the Power Supply System Consortium is to develop a set of commercial industry based "building codes" standards for use by government/military industry in development of cost effective power supply systems. The standard intends to enable government/military industry to sensibly use commercial resources (products, process, practices, and services). Unlike the P1515 standard, the Power Supply System Consortium standard is mainly aimed at government contractors. However, the benefactors of the standard are both the government and the commercial industry because by use of this standard the government contractor will be able to lower power supply system costs with increased availability while the commercial industry will be able to increase its government market. In April of this year the first Power Supply System Consortium Workshop was held and a plan established for achieving better use of commercial resources. The following companies were represented at the workshop: Boeing Company, GEC-Marconi, Lockheed Martin, Northrop Grumman, Raytheon Systems Company, Smiths Industries, and TRW; results of the workshop are summarized below.

The workshop attendees constructively expressed many concerns about using commercial resources, but acknowledged the necessity of learning how to intelligently use commercial resources. It was also acknowledged that the development of a set of "consensus" based building codes is an effective method of enabling intelligent and practical use of commercial resources. It was stated that the building codes or rules could be realized only if such development is in concert with the commercial industry. It is this notion that made acquiring PSMA support so important for EPSS standardization activity. Accordingly, to guide the development of power supply system building codes, a consortium mission statement or objective was derived.

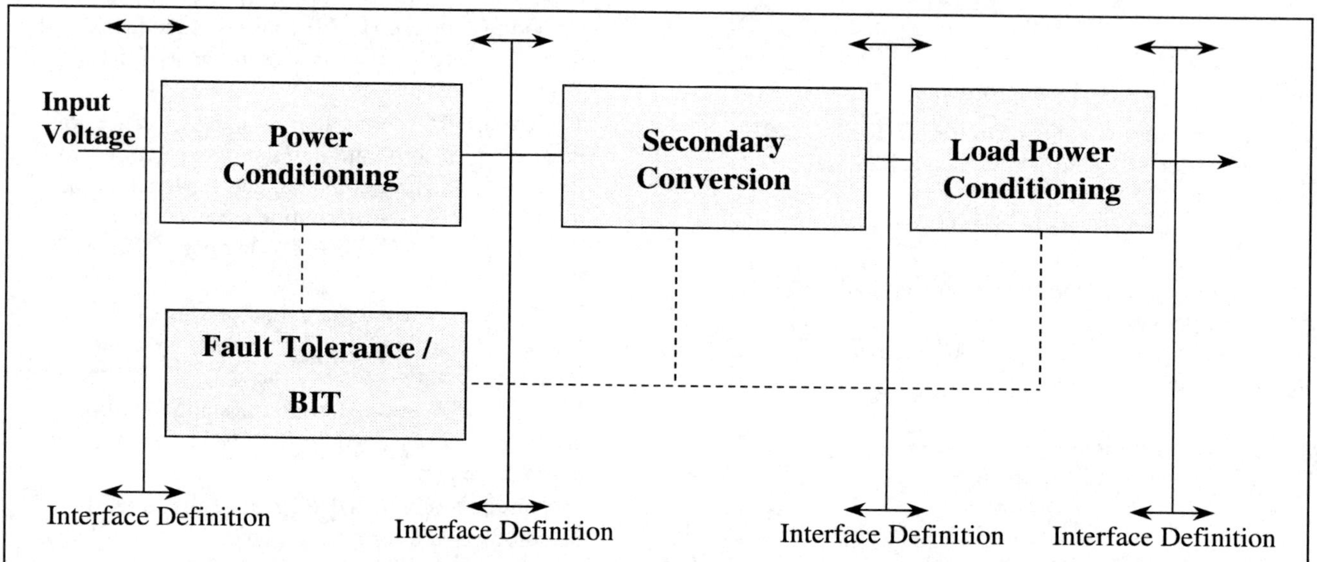

Figure 5 Power Supply System Electrical Performance Interface Definition

The consortium objective is to,

> *Facilitate the identification and resolution of technical and business case issues necessary to better harmonize the efforts of commercial and government power supply developers, manufacturers, integrators, and users*

Benefits of standard building codes for power supply systems can only be achieved if the proper mixture of structure and creativity is maintained. It is for this reason that it was decided that the Power Supply System Consortium building codes would only govern interfaces. This approach is also consistent with the OS-JTF's open systems approach and the emerging commercial power supply industry desire for modular systems with standard interfaces. The electrical performance interfaces to be defined are shown in Figure 5.

The "Load Power Conditioning to Loads" interface is somewhat unique because future trends of the loads (using electronics) need to be understood and forecasted to derive a long lived standard. Building codes for other than electrical/electronic interfaces will also be developed, for example:

- Mechanical interfaces: form factor, pin outs, thermal control, etc.

- Environmental interfaces: temperature range, shock, flammability, etc.
- System Effectiveness interfaces: Reliability, configuration management, quality, supportability, fault tolerance, Built-in-Test, etc.

Focusing on interfaces also minimizes the impact on the standard due to technology evolution.

During the April Consortium workshop, the effort to develop responsive building codes or rules was started. It began with government/military contractors (generally based on their own experience) defining their problems/impediments to utilizing commercial resources. These problems/impediments were identified and categorized, and are summarized below:

1. Lack and / or level of testing by commercial industry
2. Lack of reliability, maintainability, supportability, and / or interchangeability confidence in commercial resources
3. Lack of sufficient configuration management
4. Unavailability of desired documentation
5. Insufficient/questionable electrical, mechanical and environmental performance of commercial / industrial products
6. Lack of data sharing regarding available commercial resources
7. Present business practices

The Power Supply System Consortium has developed a plan to address the above impediments in deriving building codes for power supply system development; the results of which will be published in a later paper. The plan is summarized below:

1. Expand Consortium membership:
   - Actively involve Commercial, Telecommunication, and Industrial power supply vendors (Large, medium, and small)
   - Actively involve DoD (AF, Army, Navy, Marines, NASA)
   - Establish the EPSS Consortium as a recognized body
   - Further PSMA involvement in realizing the Consortium's objective
2. Support the completion of IEEE P1515 Recommended Practice....Standard data sheet (s) may be developed using P1515
3. Determine (and resolve) the hurdles of using commercial / industrial items in military power supply systems
   - "Items" include practices, components, products, and processes
   - Evaluate the cost benefit
4. Establish an industry-led coordination group for the development and maintenance of "power supply system" de facto building codes.
5. Develop a set of recommendations for DoD to modify procurement policies affecting power supplies to make it possible to use more commercial resources; resultant may be that the modified procurement policies will have applicability to more than just power supplies.

Summary

The government, military industry, and the commercial industry recognize the EPSS activities. There are many electronic power distribution system standardization activities; it would benefit all activities to coordinate efforts and share lessons learned. The IEEE P1515 specification language activity is acknowledged as an imperative first step; larger and more active industry participation is needed to achieve not only a timely standard but a standard that will be recognized and used by industry. The Power Supply System Consortium activity is just beginning, but would very much like industry's participation and opinion in developing responsive building codes to enable intelligent use of commercial resources for government/military power supply systems. More information on EPSS activities can be found on the WEB at http://www.acq.osd.mil/osjtf/standards3.html, or in the near future from a link from the PSMA WEB page.

References

1. COTS in the Changing Defense Environment, by Greg Saunders, 28-29 May 1998, COTS Conference 1998, Keynote Address
2. Electronic Power Specification Standardization, by S. Navarro, C. Gonzalez, M. Soraya, LtCol G. Logan, October 1997, http://www.acq.osd.mil/osjtf/standards3.html, Technical Paper
3. Power Technology Roadmap, by Don Staffiere, Jim Sarjeant, 15-19 February 1998, Applied Power Electronics Conference and Exposition (APEC), page 3-8

Table 1 Electronic Power Distribution System Parameters

Parameter	Parameter	Parameter	Parameter
AC Distortion	**Voltage Modulation**	**Efficiency**	**Frequency**
AC Distortion Factor	Voltage Modulation Amplitude	Power Dissipation	Frequency Modulation
AC/DC Distortion Spectrum	Voltage Operating Range	**Isolation/Grounding**	Frequency Stability
Harmonic Distortion	**Regulation**	Isolation Resistance	**Fault Characteristics**
Crest Factor	Dynamic Line Regulation	Isolation, Input-Output	Load Unbalance
Power Factor	Dynamic Load Regulation	Output Isolation	Output Load Characteristics
DC Distortion	Line Regulation	Elect. Bondg/Resistance to gnd	Fault Time
DC Distortion Factor	Load Regulation	Ground Isolation	Recovery Time
Noise	Point of Regulation	**Supportability**	Over Voltage Latch
Voltage Ripple	Current Regulation	Redundancy/Fault Tolerance	Output Over/Undervoltage
Current Ripple	Temperature Regulation	Maintainability	Input Over/Undervoltage
Spikes	Feeder Drop	**Reliability**	Overcurrent Protection
Single Spectral Line Amplitude	**Transients**	MTBF	Rupture Current
EMI	Transient,Voltage, Frequency	**Environmentals**	Rupture Capacity
Common Mode Output Current	Settling Time	Humidity	Input Protection
Differential Fault Current	Step Input	Operating Altitude	Corona
Harmonic Current	Overshoot/Undershoot	Salt Fog	Overload
Input Current Balance	Hold Up Time	Storage Altitude	Short Circuit
Inrush Current	**Load Voltage**	Vibration	Short Circuit/Overload Recovery
Emissions and Susceptability	Load Impedance	Thermal Shock	**Mechanical**
AC Voltage	Load Power	Operating Temperature Range	Pin Out
AC Voltage, DC Component	Load Voltage	Over Temperature	Footprint
AC Voltage, Peak	**Source Voltage**	Steady State	Height
AC Voltage, Steady State	AC Power	**On/Off Control**	Mounting
Voltage Phase Difference	DC Power	Cntrl Turn On/Turn Off Response	Connection Types
Voltage Unbalance	Source Impedance	Turn On to Turn off Hysteresis	Connection Locations
Phase Sequence	Input Capacitive Loading	Remote On/Off Control	
DC Voltage	Source Voltage		
DC Steady State Voltage Limit			
DC Voltage, Steady State			

OPEN SYSTEM DESIGN FOR COMMUNICATIONS NAVIGATION AND IDENTIFICATION (CNI) AVIONICS

Dr. George W. Mitschang, GEC Marconi Hazeltine, Wayne, NJ

Introduction

Saying one is using open system design approaches to build systems has a pleasing ring and quickly gains acceptance among government program sponsors. The PC model is usually noted as the premiere example of open systems. Much excellent work has been sponsored in further defining open system approaches for the next generation of systems. However, putting open system design into practice when one deals with legacy systems and interface, form/fit, and performance requirements is a significant challenge. Add to this, a typical company's own internal processes for system, hardware, and software design, which typically are aimed at optimizing single function performance, and the challenge is increased. For the future programmable CNI systems' market, this paper examines some of these issues, identifies additional business considerations that come into play, and briefly describes work underway at GEC Marconi Hazeltine.

Defining Open Systems

The Joint Task Force on Open Systems (JTF-OS) established a framework for viewing an open systems approach to system design using three types of architectures: operational, technical and system.

Before I discuss these, however, let me offer a general definition of architecture taken from the software world. An architecture is: "the structure of the components of a program/system, their interrelationships, and the principles and guidelines governing their design and evolution over time". These last few are important because they impose on the architecture the concept of future change and they force the architecture to identify interfaces in such a way that the impact of likely change is minimized (reference 1).

In the JTF-OS framework, the operational architecture provides the user with a means to define his operational requirements in performance terms. The system designer can see the big picture of how his system fits and the critical performance needs that are important to the user.

The technical architecture is the real "rubber meets the road" place where open systems moves from concepts and approaches to real defined elements of what is meant by open system, at least from the view of the creator of the technical architecture. The functional building blocks of the systems are defined as well as the interfaces between functions. General definitions of a software architecture may also be imposed to facilitate code transportability (isolate code from hardware implementation) or reuse for other system applications. An acquisition authority may also choose to impose form and fit requirements at this level to take advantage of commercial parts, boards and technology advance. A developer may also impose design constraints to facilitate design reuse of a company investment funded development.

At the system architecture level, all the system designer has to do is meet operational architecture performance within all the design constraints imposed by the technical architecture!

One does not have to look far to identify the building block or architectural elements that

0-7803-5086-3 /98/$10.00 ©1998 IEEE

comprise an open system approach. Those elements that are available through standards bodies can accommodate many digital-data transfer and board-control functions because of their similarity with applications in personal computers. As processors get faster and additional programmable elements such as Field Programmable Gate Arrays (FPGAs) enter the market; functions that previously could only be accomplished with custom hardware are able to move into software and firmware.

Reference Models

The process of defining open systems for new applications has benefited tremendously from work sponsored by DOD in the last 2 - 5 years (See references 2 - 5 for examples). The architectural categories noted above were products of that work and assisted in providing a common language for discussion and a forum for defining open system standards.

Standards

The avionics industry has established standards for form factors and interconnects that are widely used in commercial aircraft and were designed primarily for building single function systems. The form factors made possible standardized sizes for equipment and encouraged a wider variety of suppliers to produce cards and chassis. It also made it easier to supply power and control from the aircraft manufacturer's perspective.

Recently, government and industry radio standards activity has been addressed at defining standards for programmable radios. The Multi-Mode Information Transfer Systems (MMITS) forum and the Joint Tactical Radio System (JTRS) inspired Programmable Modular Communications Systems (PMCS) (reference 6) applied much of the JTF-OS open system work to programmable radios. They both defined reference models that would form the basis for more specific future standards definitions.

Interfaces

Much of the emphasis in the standards area derives from the approach applied to modern personal computers. The interfaces which define the boundaries into which computer cards are inserted, makes possible the multi-vender development of cards and applications. Standardized bus interfaces such as PCI and cPCI, define physical, electrical and data exchange protocols. These definitions enable modular, functional elements like modems, display drivers, and parallel port interfaces to be developed by many venders and used interchangeably in systems.

Can the same type of standards and interface definitions be applied to programmable radio systems for the avionics market?

One of the practical challenges of standards definition for programmable radios is establishing the appropriate functional level of modularity to specify a standard interface. In the JTF-OS tutorial (reference 2) this is referred to as the "atomic level" of the system architecture. Interfaces above the atomic level are controlled by open system standards; interfaces below the level are at the supplier's discretion. If one defines a system "atom" at too high a level, interchangeability of modules within the system will be at a high level (i.e., many functions) and likely to be expensive and limited. Also insertion of technology may be impaired, because replacement of larger, expensive multi-function elements may be required to replace a single technology element. If the system "atom" is too small, meaningful functional elements may not be defined at all and the problem comes down to building with component level elements that address items like pin configurations and virtually no benefit for an application like programmable radios is derived.

0-7803-5086-3 /98/$10.00 ©1998 IEEE

For the avionics market today, the dilemma for the company making development investment decisions is selecting from a number of possible standards, applying reference models, and incorporating legacy waveforms into a programmable and open system.

Constraints

If one were developing only new systems for new applications, applying standard and interface definitions would be relatively straightforward. The practical problem though, is that these standards must permit existing form and fit for retrofit applications; they must survive in harsher environments than were originally intended; and they must meet performance requirements from legacy systems that were able to optimize system elements for one and only one function.

Retrofit

For many applications, programmable CNI systems are required to fit within the form factor of available space and with existing interfaces. This may limit standard approaches that have their own form factors. In these cases using layered software development for the applications code make possible easier rehost of the code onto modules with unique processors that fit within the available space. Also programmable systems, using modern, denser technology may replace multiple single function units with smaller more compacts, multiple function replacements. This may free additional space for installing enhanced or additional capability, thus improving the overall capability of the platform.

Military Environment

CNI systems for military aircraft must operate in harsher environments than standard commercial approaches may permit. In these instances the system designer may still be able to use commercial components and standard modules, but would be required to condition the environment that these elements see by using specialized chassis and/or enclosures. See reference 7 for further information on this and other issues associated with the use of COTS in military systems.

Legacy Performance

With retrofit and new applications, interoperability with legacy systems is usually a requirement. The performance requirements of these systems usually is specified in terms applicable to single function (i.e., federated) equipment. The system designer needs to identify the legacy requirements that are directly aimed at interoperability and remove those that are terminal specific. Close coordination with the users and acquisition agency is required so that the designer arrives at the minimum set required for interoperability. For secure modes of legacy operation, entirely new sets of issues are involved due to required NSA certification that are beyond the scope of this paper. Suffice it to say that development of programmable crypto, using standard interfaces, and replacing obsolete units will be required. When accomplished, further reduction in size, weight and lifecycle support costs will be realized.

Future Requirements

A major benefit of programmable CNI systems is the potential for installing new capabilities through new software, making use of the programmable, functional modules already in the system. The significant challenge for the system designer is to select those open system standards that facilitate this approach. Staying close to standards that are state-of-the-art in the commercial arena means that technology advances will be easier to incorporate. Functional modules of the proper "atomic level" also permit easier insertion of modern technology. The critical interfaces that are an important part of the open systems definition, also make insertion easier.

Summary of Systems Approach

In summary, the performance based operational requirements constrained by design requirements for open systems and legacy interoperability, form factor, and harsher environments lead to the specific system architecture.

The system designer must also identify trends in technology and future operational needs, and attempt to build in reserve processing and growth to accommodate these.

Business Investment and Risk

Shrinking DoD development budgets force system developers today to invest in product development. Without specific requirements, the developers risk spending their scarce investment dollars on systems for which there may not be firm markets. By using open system standards in these internally funded efforts the hope is that acquisition agencies will be flexible in how they specify requirements so that multiple developers using open standards can compete. A developer doesn't want to be placed in a situation where he has selected one open standard and an acquisition agency has selected another.

Protection of Proprietary Information

An open systems approach does not require a developer to tell everything about a particular module that is being used in a system. If the interface requirements are properly defined, through standard hardware interfaces and software application program interfaces (API), then innovative use of devices and techniques that a developer incorporates into his system can be protected. For complex radio waveforms implemented on programmable systems, protecting proprietary techniques while at the same time maintaining an open systems approach is a real challenge.

For example, consider a programmable modem with an exceptional range of performance developed under company investment. For another vender to make use of that modem, i.e., write code for it, significant information about its internal workings would be required. The technical data released could compromise an innovative approach and the investment that funded it.

Investment

For the Joint Tactical Radio System markets, the government wants to maximize its own investment by design reuse that it funds and applying commercial technology developed in accordance with open system standards. A developer could select a generic approach that would apply across many JTRS markets or domains, or he/she could only go after one domain, such as the fighter airborne domain.

In the former approach, the developer is in a position to enter any of the domains with essentially the same core of SOFTWARE and system architecture. The specific procurement may require small additional investment or could even be funded to adapt the general approach to the specific application.

In the second approach, domain specific requirements may force choices that result in more expensive and restrictive solutions. Pursuing other domains may indeed be possible, but the recurring per unit cost is likely to be higher because airborne environmental requirements would require higher cost parts.

My view is to develop a generic approach, adopt it to a benign environment to demonstrate its capability and then adapt it to different environments as required. In a very general way, this is how the government develops technology – a DARPA funded concept or technology demonstration followed by transition to a weapons systems acquisition. The difference here is that a generic radio product, rather than technology, is the first step.

0-7803-5086-3 /98/$10.00 ©1998 IEEE

The transition is adapting that product for specific radio applications across land, sea, and air.

Per Unit Payback

The heritage of many military contractors was based on developing hardware and software in the development phase of a program and selling the hardware in the production phase of a program where 70% or more of the acquisition funds are spent. An open system approach encourages and makes greater use of COTS elements and reduces the manufacturing content of a production buy. In open systems, software, running on someone else's hardware, could be the end product for a military system. Now add the notion of investment funding to produce that software and selling it as a license. There may or may not be any hardware content in a production buy for a particular vender. Military contractors must consider return-on-investment in ways similar to commercial companies and procuring activities must be open to new ways of evaluating cost for procurements.

Design Approach Changes

COTS and open system usage requires a new way of thinking about system solutions. In the past, performance requirements have generally been the driving force behind the system design. One way to think of an open systems approach is that additional requirements have been introduced at the system. These requirements are not derived from operational needs, but from the "business" needs of the user and acquisition authority. They ultimately have lower cost as their underlying source. In addition, requirements for "growth or future capabilities", which by themselves are difficult to quantify in a high level system document, can be better defined with the interfaces and standards of an open system approach. This definition facilitates insertion of modern technology, or inserting new functional modules with improved performance.

System Partitioning Based on Available Elements

The system designer generally derives system functions from the operational requirements and partitions those functions in ways that satisfy the design constraints (imposed by open system requirements, for example). If the open system guidelines have been done properly, COTS products may be available to fill some or all of the <u>functional</u> modules of the system. Or, as noted above, developers of custom modules (such as are required as part of an avionics CNI system), use the open system functional guidelines as physical interface definitions for physical elements they build.

An example of this might be a transceiver module whose function is up- and down-conversion between low level RF and baseband digital samples. The open system guideline would define this element as a functional entity with controlled interfaces. A developer/supplier would build physical modules complying with the interface definition and incorporating the performance within the module that would be allocated to this function through a system analysis of a complete communication system. Another supplier, using the same interface standard, could develop a module for higher frequency or wider bandwidth for future needs.

Design Compromise for Multiple Applications

The design approach suggested at the end of the previous section requires broader architectural decisions. For example, to maximize the use of a COTS general-purpose computer board in a radio system, one might group together radio functions appropriate to digital message processing and implement them on that board. One might also implement on that board functions, such as signal processing,

0-7803-5086-3 /98/$10.00 ©1998 IEEE

which normally would be done in special-purpose components, but can be run acceptably in a processor that achieves results through "brute force" processing. The savings in non-recurring engineering (NRE) cost and the application to multiple communications types makes this extremely attractive. In so doing, however, the environment that the board operates in, must be assured by other system elements (e.g., chassis).

DoD's initiative to eliminate MILSPECs opens the door for more extensive use of commercial parts. The acquisition authority must specify the real operational environment of the system as opposed to the test environments that were used in most specifications. COTS suppliers are motivated to produce a part that is reliable in as wide a range of applications as economically feasible. He can sell more and he doesn't have to repair as many! These same elements can be used in all but the harshest military environments through proper conditioning of the environment by the surrounding system elements.

To summarize this notion, the question system designers might ask themselves changes from "What do I have to build to meet the requirements?" to "How can I meet the system requirements with these open system standards and building blocks?"

Make / Buy Shift

Using an open systems approach that encourages and makes greater use of COTS elements is in itself a make / buy decision. If one believes in the "question change" posed in the previous paragraph, than the open systems approach places much more emphasis on the buy side of the make / buy decision. Of course, this presumes that the module being procured complies with the open system standard selected for the system, or alternatively, the open system definition is broad enough to not severely limit module buys.

Rapid Prototyping

One of the significant system design benefits of stipulating open systems in a system requirement is the ability of the developer to prototype risky aspects of the system using commercially available products. Even if the final system must have special purpose hardware due to harsh environments, by requiring standardized interfaces, COTS boards, chassis, and diagnostic tools can be used to verify design concepts and make system changes long before expensive development commitments are made. Verification can be done rapidly and at very low cost and with customer involvement, a low risk development will result that the customer believes in.

Example of a Programmable CNI Design Approach

GEC Marconi Hazeltine is now developing a generic programmable radio for the JTRS market. Building on our strengths in wideband datalinks and using the recent guidance documents from PMCS, we have defined an architecture that we believe can be applied across multiple JTRS domains. The interesting aspect of open system approaches and flexible programmable systems is that the actual system block diagram looks almost exactly like the system functional block diagram. Today's programmable logic and high-speed digital signal processing and general-purpose processing chips, permit hardware modules to be allocated complex functions. For a communications system, one can isolate RF receive and transmit, signal processing modulation and demodulation, message manipulation, cryptographic processing and user interfaces all in individual hardware modules and interconnect them with standard busses and protocols. Our design uses COTS boards along with custom designs that adhere to standard board interfaces. COTS operating systems provide a rich software development environment and a wide selection

0-7803-5086-3 /98/$10.00 ©1998 IEEE

of off-the-shelf software for hardware drivers and interfaces.

Summary

Practical application of open system approach for a specific product like a programmable radio is a challenge for the developer. More specific interpretation of open system for these radios is required if multiple developers are to invest to develop competitive products. The current plan for the new JTRS program is to form government and industry groups to provide this interpretation. Commercial companies have proven that useful standards can be developed in competitive markets. It will be up to military suppliers now to follow those examples for the future military radio market.

References:

1. Software Architecture: an Executive Summary; Technical Report CMU/SEI-96-TR-003; ESC-TR-96-003; February 1996; Paul C. Clements and Linda M. Northrop.
2. Open Systems Tutorial; Open System - Joint Task Force; Office of the Undersecretary of Defense (Acquisition and Technology); first presented at DASC, 26 – 30 Oct 1997.
3. DoD Technical Architecture Framework for Information Management, Version 3.0, 30 April 1996.
4. DoD Joint Technical Architecture, Version 2.0 Second Draft, 31 October 1997.
5. Defense Information Infrastructure (DII) Common Operating Environment (COE) Integration and Runtime Specification (I&RTS), Version 3.0, July 1997; prepared for: Joint Interoperability and Engineering Organization, Defense Information Systems Agency.
6. Programmable Modular Communications System (PMCS) Guidance Document – Revision 2 31 July 1997
7. COTSCON '98 Proceedings; May 28 and 29, 1998; sponsored by Military and Aerospace Electronics.

0-7803-5086-3 /98/$10.00 ©1998 IEEE

THE APPLICATION OF COMMERCIAL PROCESSING TECHNOLOGIES TO THE AIRBORNE MILITARY ENVIRONMENT

Hugh S. Perry, General Dynamics Information Systems, Minneapolis, MN

Introduction

Airborne, military processing for multiple tactical and surveillance aircraft applications has been one of General Dynamics Information Systems' (GDIS) core technologies and business areas for more than 20 years. These computers have invariably used unique processor designs, without employing open systems architectures or commercially available interface standards. The change from these unique, point designs to the use of commercially available processor hardware and software solutions, as the result of Secretary of Defense Perry's 1994 directive, has caused a dramatic impact on the way we do business. This paper addresses the approaches that GDIS is continuing to adopt to meet this new environment, and the lessons we have learned as the result of these experiences.

The range of platforms that are currently in need of upgraded mission processing subsystems is expanding, and includes the AV-8B, F-15, F-18 aircraft, as well as other fixed and rotary wing platforms. This paper discusses how we are using commercial devices in the military environment, specifically how they are now being applied to the Ground Proximity Warning System (GPWS) for the UK Harrier II, and the mission computer for the US AV-8B aircraft. This tactical operational environment requires increasing processor performance, while providing compatibility with legacy standards, for example MIL-STD-1553.

This paper outlines the lessons learned to date as these upgrade processor programs transition through the integration, qualification and performance verification phases of development, and into early production.

A Paradigm Shift in the Way We Do Business

Secretary of Defense Dr. Perry in 1994 directed that more affordable methods be employed within DoD that would significantly reduce the cost of doing business. This included rethinking the use of traditional military standards and procurement processes using a cost benefit analysis, and to more closely align DoD with the trends in the commercial industry.

Dr. Perry, the Department of Defense, and industry recognized that the collective buying power of military systems relative to the commercial electronic business base is diminishing. With this small leverage, DoD can no longer influence the marketplace to any large extent.

Subsequent to Dr. Perry's memorandum, the avionics industry began providing avionics mission management and processing products based on open system architectures and commercial-off-the shelf (COTS) technologies. The AV-8B Open Systems Core Avionics Requirement (OSCAR) program is one of the early and significant programs that have emerged as the result of the "Perry Directive." This program provides a replacement for the Navy standard AN/AYK-14 family of computers based on open system architectures and COTS technologies.

0-7803-5086-3 /98/$10.00 ©1998 IEEE

Impact of the Use of COTS

As a result of this directive, the perception has arisen that all that needs to be undertaken is to take commercially available, ruggedized, processing and interface modules, place them in a standard chassis, and we have our COTS equipment that will meet the needs of our tactical aircraft. This has not been our experience.

The reality that exists is that there is the need for an integrator who fully understands the requirements of the aircraft system, and will ensure that all the COTS and ruggedized-off-the-shelf (ROTS) components are indeed compatible with each other, and will meet these requirements. Where additional hardware and software functions need to be provided to meet the aircraft requirements, modifications and additions to the available devices need to be implemented by the integrator. In addition, the equipment will need to be employed and supported in operational service over a period of perhaps 20 years, while accommodating the fact that individual technologies will roll every one to three years, with little or no coordination between them. This is indeed a challenge.

Evaluation of COTS Technologies

It is most important to be aware of where the commercial marketplace is applying its research and development funds so as to ensure that avionic processor designs can fully leverage these commercial technologies. The military equipment suppliers have to recognize that a rapidly shrinking proportion of microcircuit manufacturing capacity is now being applied to full military-environment devices.

GDIS selected the Motorola-IBM-Apple PowerPC line of products in mid 1994 as the processing engine, and began an internal research and development (IRAD) project to design an advanced common processor (ACP) based on this highly capable product family. A significant influence in this decision was the availability of established Ada software development systems for the PowerPC. There have subsequently been further software development systems developed for the PowerPC microprocessor, including applications using C and C++.

The ACP product development has now progressed through several design iterations. The initial version, shown in Figure 1, is based on the PowerPC 604 microprocessor. One major objective of this IRAD project was to design a processing system using no custom logic. The processor module includes a standard PCI (peripheral component interface) module local bus, PCI to Ethernet interface, PCI to VME 64 interface, 8 Mbytes of DRAM, 1 Mbyte of Flash and 256 Kbytes of secondary cache memories. This logic fits on one side of a 6 inch x 9 inch AN/AYK-14 compatible module. The second or backside of the module has provisions for multiple input/output (I/O) interfaces, and currently incorporates two redundant MIL-STD-1553 interfaces.

Compatibility with the legacy AN/AYK-14 computer was maintained to support several flight demonstrations in tactical military aircraft during 1996. This design also allows the user to bridge into a legacy system bus using field programmable gate array (FPGA) technology which provides a slave 32-bit high performance processor capability for existing legacy systems.

This module was inserted into the existing AN/AYK-14 chassis during these successful flight tests. Flight demonstrations of this processor demonstrated COTS technologies in the AV-8B and F-18 tactical environments, and provided an empirical data base for future incorporation of COTS technology into demanding applications.

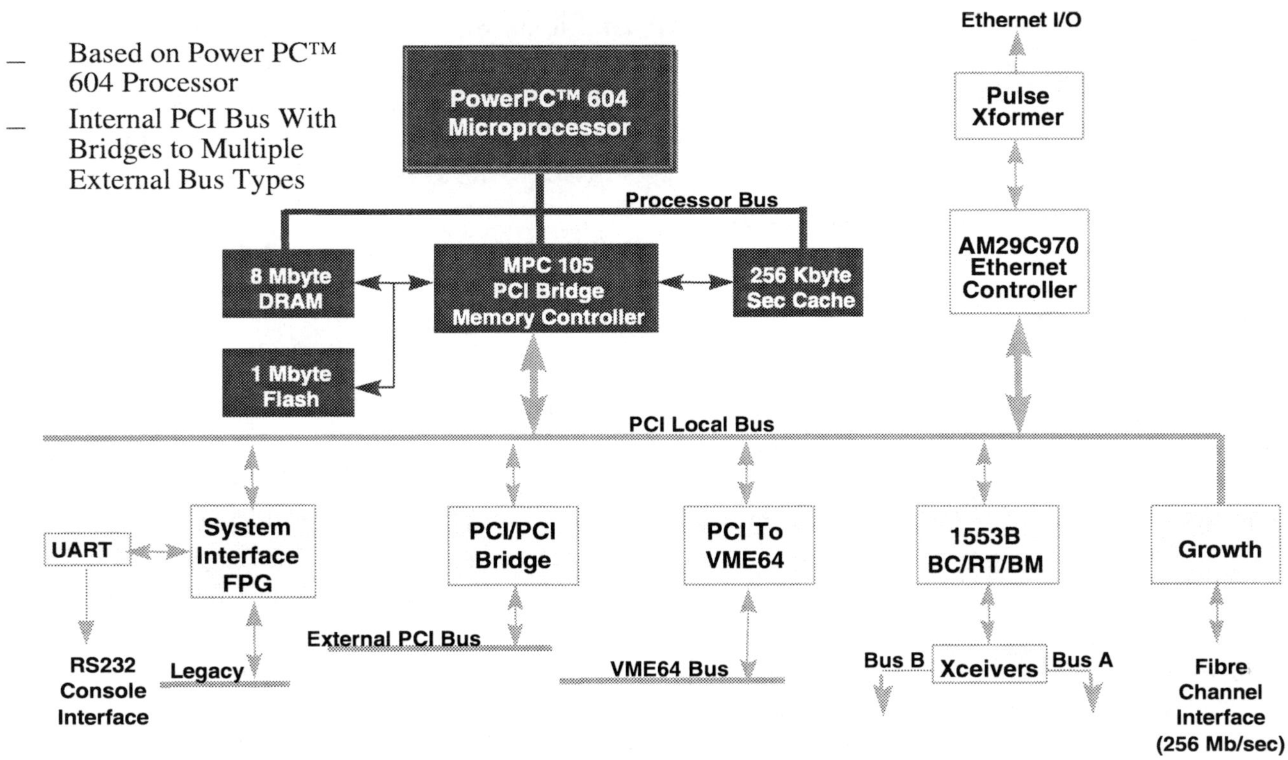

Figure 1. The Advanced Common Processor (ACP) Design

The implementation of this architecture for the aircraft demonstrations is shown in Figure 2.

Figure 2. ACP Design Flown on AV-8B and F-18 Aircraft in 1996

Technology Transition to Real World Programs

GPWS Program

Our first application of this technology was for the GPWS program. GPWS provides the pilot advisory information on terrain situation awareness relative to the aircraft flight path, and advises via a voice output, and the head up display, of impending terrain impact. The GPWS employs the Terrain Processing Module (TPM) to perform the system functions. The module design in Figure 1 was adapted to the specific needs of the TPM by increasing Flash memory for the digital terrain environment data base, and a voice output capability. It utilized the PowerPC 603e processor for reduced power consumption. The TPM has now completed design verification and environmental qualification testing, and has demonstrated and verified performance requirements over the specified operating environments.

OSCAR/Harrier Program

In late 1996, following the successful demonstration of the ACP on the AV-8B and F-18 aircraft, the development of the AV-8B OSCAR program began.

The OSCAR processing subsystem is based on a commercially available Common Processor Module from DY-4 Systems Inc. Processing, interface and specialized functions are configured on standard IEEE 1101.2 6U VME 64 form factor modules. The baseline OSCAR subsystem consists of a chassis with a VME backplane including other interfaces, a Common Processing Module (CPM), a Utility Module (UTM), and a Power Supply Module (PSM). Significant growth capability is provided with up to eight (8) additional 6U VME modules, and additional power conversion capability.

The first application of this growth capability is for the Terrain Processing Module (TPM) for the UK Harrier GPWS. However, the design will accommodate modules which provide additional capabilities, including additional processing capability, input/output functions (e.g. MIL-STD-1553, fibre channel interfaces), Cockpit Display Processing, Digital Map Display Processing, and Sensor (radar, sonar, infra-red, electro-optical, etc.) Preprocessing. A simplified diagram of the subsystem is shown in Figure 3, and the installation of the modules in the chassis is shown in Figure 4.

Chassis Backplane (CBP)

The CBP is a chassis with interconnecting wiring that is physically similar to an existing AN/AYK-14 chassis, that is an ATR chassis with forced air cooling ducts, rack mounting and front panel located external connectors. The CBP has required careful thermal design since the size of the VME modules is slightly wider than that of 6 inch x 9 inch modules, thus reducing the amount of air flow across the finned surfaces for a given pressure drop. The thermal and physical characteristics of the CBP are required to be the same as the AN/AYK-14 computer it replaces, so that the design has minimal impact on the existing aircraft installation.

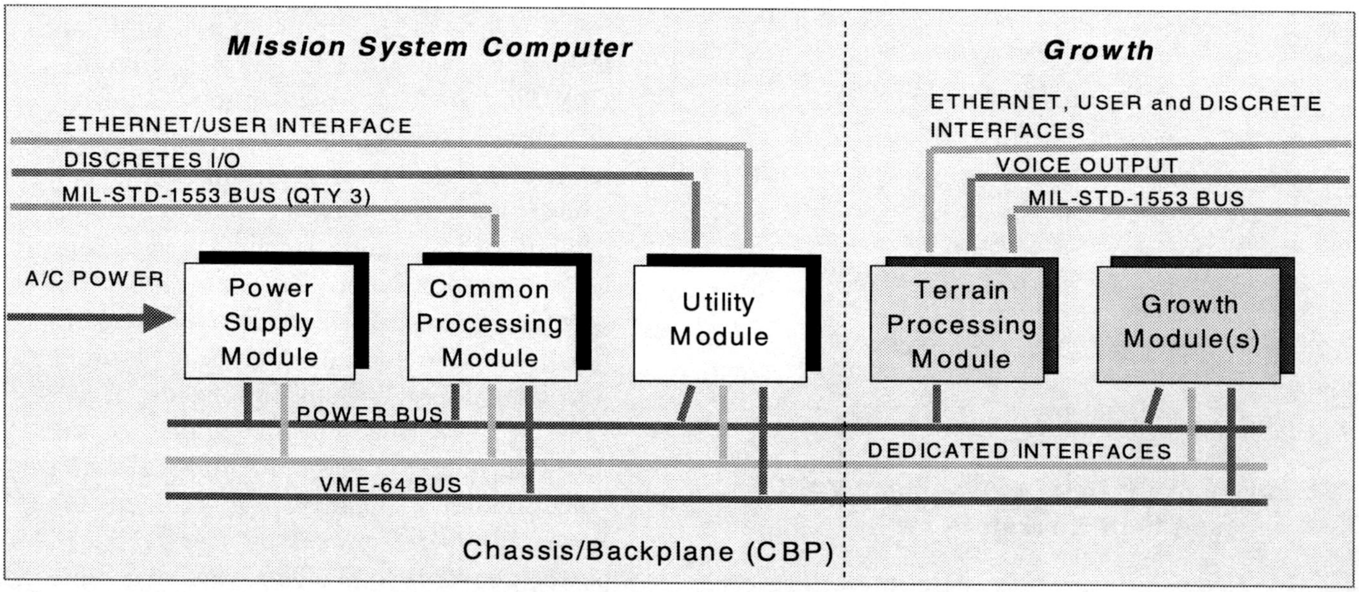

Figure 3. Simplified OSCAR/Harrier II Subsystem Diagram.

Common Processor Module (CPM)

The CPM is a PowerPC 603e single board computer with MAXPack ™ interfaces. The CPM characteristics include:

- High performance processor with growth path
- 16 MB DRAM
- 8 MB Nonvolatile FLASH memory, directly executable
- Timers: Watchdog timer, Operating System timer, 4 - 16-bit timers
- Three MIL-STD-1553A/B interfaces, with BC/RT capability, direct and transformer coupled outputs
- IEEE 1 101.2 6U VME form factor
- VME-64 backplane interface, system controller capable
- Ethernet development interface

Utility Module (UTM)

The UTM provides input/output discretes compatible with existing aircraft interfaces, non-volatile memory, and an Ethernet hub for the modules in the subsystem.

Power Supply Module (PSM)

The PSM provides regulated power from the aircraft supplies which complies with the VME bus requirements, provides growth to accommodate lower rail voltages for future device families, and allows growth to power a chassis that is fully populated with VME modules.

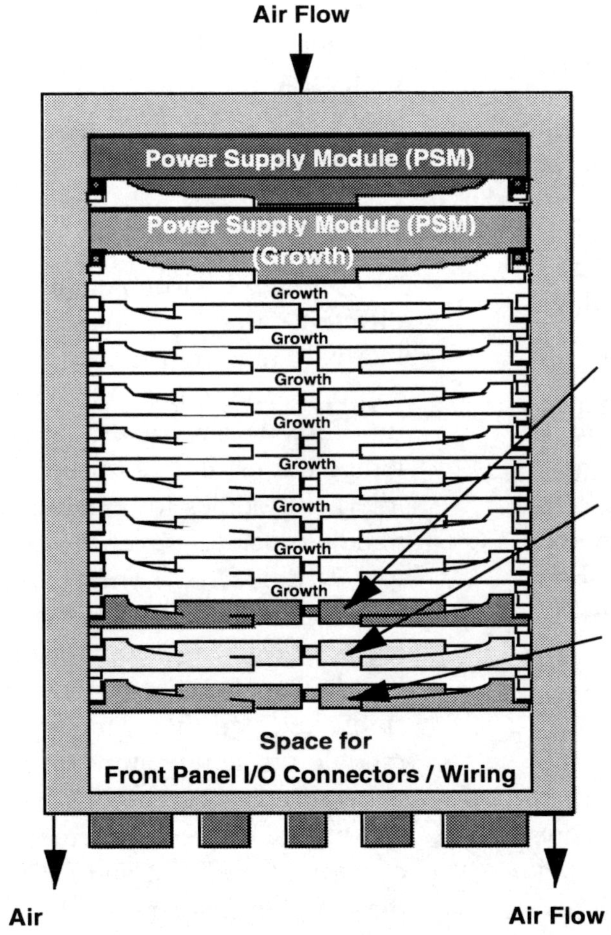

Figure 4. OSCAR Chassis with TPM Installed

0-7803-5086-3 /98/$10.00 ©1998 IEEE

Program Summary

In summary, the roles played by the processing subsystem integrator are to put technical, program and warranty "wrappers" around the COTS/ROTS products, so as to provide the integrated services the aircraft primes need.

To translate these needs into our experience on the OSCAR/Harrier programs from a technical standpoint, the following specific activities needed to be performed:

- Provide application-specific hardware interfaces between COTS processing and other elements and the Prime Contractor's platform
- Provide unique designs of modules employing commercial/industrial components when board-level modules are not available to perform the specified functions
- Provide the application-specific mechanical (chassis), thermal and environmental conditioning needed to allow the COTS elements to operate in the platform environment
- Provide the conditioned power environment needed for the COTS elements, using the power provided by the platform. This is usually defined by MIL-STD-704.
- Provide the software environment that interfaces with the platform, and adapts the software functions and facilities provided by our software vendors to meet these needs. These include the Operating System, Built-in-Test functions, special input/output interfaces, and application program services.

Some Lessons Learned to Date

Requirements Definition

The definition of the customer's performance requirements could be expected to match those provided by the COTS vendors. While this is true to an extent, the aircraft requirement is still a stringent set of performance and environmental needs. The result is that a detailed set of performance specifications is still required, and that the full requirements analysis process must be undertaken.

The result of these detailed requirements is that the OSCAR subsystem design requires additional interfaces and functions to adapt the COTS design for our application. The Utility Module is a case in point. It provides the platform-unique discrete and test interfaces for the processor. Provisions for growth in the number and type of these interfaces have been included such that the module may be used in other future applications.

The lesson we have learned here is that the full requirements analysis process needs to be undertaken, and that we should expect to need to add both hardware and software functions to tailor the COTS design to meet the requirements of the platform. This is indeed a significant element of the "value added" by the integrating organization.

Thermal Requirements

The environmental control systems of different aircraft impose significantly differing thermal requirements on the design. For example, the initial OSCAR design was based on forced air flows, pressure drops and temperatures that allowed the design of the internal modules to be required to operate with a rail temperature over the range - 40 degrees C to + 71 degrees C. These requirements allowed the use of both industrial and military temperature range components, particularly in the Power Supply Module (PSM). Further study revealed that a higher upper temperature module rail temperature for the AV-8B aircraft is required under certain extreme operational conditions. This requirement drove a redesign of parts of the PSM and UTM to accommodate these higher temperatures.

The lesson that has been learned from this experience is that we may need to provide more than one version of a module design, in these

0-7803-5086-3 /98/$10.00 ©1998 IEEE

cases the PSM and UTM, to accommodate the needs of different applications. This will avoid the need to supply the highest cost military design unit to applications that have less stringent requirements.

"Uprating" of Components

A significant effort has been undertaken to investigate the environmental uprating of COTS elements such that they may employed outside their specified temperature range. As a part of this effort, support has been provided by the Computer-Aided Life-Cycle Engineering (CALCE) Electronic Packaging Research Center (EPRC), University of Maryland, to participate in the research activities on plastic encapsulated microcircuits (PEMs), interconnect and reliability.

Each PEM device used in the system has been subjected to a thorough analysis to determine whether the die and packaging techniques used will indeed withstand the environment. Close customer coordination is needed to satisfy their potential concerns relating to long term reliability. Where the PEMs are shown to be unacceptable, a military grade component needs to be used.

As a result, the OSCAR design has moved to using military grade components where the thermal conditions require them. This is a conservative approach, which may be able to be modified in future, as the uprating of COTS becomes a more accepted procedure. The TPM has solved this issue by undertaking a one-time production buy of components, the rationale being that the parts are sourced from the same vendor and production lot, and it is this set of parts that have been used for qualification and accelerated life testing.

Equipment Maintainability

Several critical requirements contribute to the ease or otherwise that the equipment may be maintained. Two such parameters are mean time between failures (MTBF) and built-in-test (BIT) coverage. There are a number of ways of estimating subsystem MTBF; the classical approach is to use the guidelines established in MIL-STD-217, other approaches use vendors' reliability data, and operational life experience data. The MIL standard approach results in a conservative failure rate.

The aircraft prime contractors then often place additional "factors" based on their experience, this further reduces the MTBF estimate. The result is that in order to achieve the specified failure rate; full military components need to be used. Short development timescales do not permit extensive testing of industrial or commercial parts prior to their inclusion in the baseline design to qualify them for use in the application. An example of this issue has been with the DC to DC converters in the PSM. The effect is that the cost of the unit has had to be increased by the use of military parts.

Our experience with the Commercial Ball Grid Array device (CBGA) (MPC 105) used on the TPM as the PCI bridge, left no option but to solve the problem of attaching the CBGA reliably to the printed wiring board (PWB). No other package style is available, and so extensive testing of the normal and recommended method of attaching the device resulted in an unacceptably high failure rate after thermal cycling. It was fortunate that program schedules permitted the design and testing of a patented 'interposer' device, using compliant leads, that is placed between the MPC 105 and the PWB. It permits thermal stresses to be taken by the pins in the compliant leads, and avoids cracking the soldered joints of the ball grid array device itself. This approach has successfully converted the CBGA into a ruggedized off-the-shelf (ROTS) compliant lead grid array (CLGA) device.

The OSCAR program may be one of the first programs to provide a performance warranty to our customer. A short turn around on faulty units in-service would be provided. This approach would place the responsibility onto the integrator to provide a reliable subsystem, and should help to reduce the concerns the prime contractors have over the failure rates of individual subsystems.

Going Forward from Here

The OSCAR/Harrier II program is perhaps the first program at GDIS that has applied COTS processing to avionic systems, and a number of technical issues have had to be resolved within the timescale of the program. The solutions employed may not always be the most cost effective solutions, but have been necessary to meet the customer requirements in the time available.

It is clear that a parallel technical activity is recommended in the organization to resolve many of these technical challenges before they need to be addressed by the program. The results of these activities will go a long way to establishing and building a data base of technical expertise. It will include a detailed knowledge of our suppliers' capabilities to provide PEMs that will satisfactorily perform in avionic environments.

The question of parts obsolescence is a very real issue when commercial technologies are used in our equipment that we have to guarantee operational life and availability over 10 to 20 years. Experience already gained from OSCAR and Harrier II tells us that many parts will not be available when we come to procure more parts in 1 to 3 years. Apart from adopting a lifetime buy policy, which does attract many additional parts control challenges, we need to be prepared for such technology developments, and have solutions available before the problem actually occurs.

To achieve this goal, it is suggested that teams be established within the organization whose responsibility is to address the total technical (systems, hardware and software) approach, the procurement approach to components and subsystems (modules, mezzanine cards, hybrid devices, etc.), and the processes that need to be employed. This activity has started, but was not yet sufficiently mature to influence the OSCAR/Harrier designs to date. However, the importance and effectiveness of these teams will become more critical as new and improved modules are developed to grow the functions of the OSCAR family of processing solutions.

Summary and Conclusions

The OSCAR/Harrier II program has broken new ground in terms of introducing COTS processing technologies into military avionic applications. Our customers are requiring performance definitions significantly in excess of those required in the past for point solution designs. While they are attempting to tailor requirements from old military specifications to the "real" operational environment, their requirements are still challenging.

While the solutions provided to our customers are employing commercial standards and off-the-shelf products, a major role to be played by the integrating organization is to become the "trusted subsystem integrator." The organization will "put wrappers" around the commercial technologies so as to meet the customers' needs.

Many technical challenges are presenting themselves as we use non-military technology, and there is an increasing need to establish technical, manufacturing and business centers of expertise to resolve these challenges before they become a major problem to a specific program, and result in expensive solutions.

However, these programs have demonstrated that the 1994 directive to supply COTS-based equipment to military avionic systems can indeed be implemented at reduced cost and improved system performance.

RCC-975

H. System Design Track - Spacecraft Systems

The Space Systems Track focuses on the unique forms of travel by different vehicle configurations. The sessions in this track cover the needs of future space missions in terms of avionics elements and enabling microelectronics technology. Applications for commercial, civil, and defense space systems will be discussed. The requirements of specific future space missions will also be covered.

Track Chairs: Mike Jahan, Jet Propulsion Labs; David Barnett, Lockheed Martin Astronautics

Track H - Spacecraft Systems

Session H1: Commercial & Defense Space Systems

Chairs: Hendrik Gelderloos, Honeywell Space Systems;
Dave Barnett, Lockheed Martin Astronautics

1	IRIDIUM Low Earth Orbit Satellite Constellation	Mr. Peter Lemme
2	ATM Mobile Switch Requirements, Issues, and Concerns (Not Published)	Mr. Richard H. Paine
3	The 'Take-Off' of MOTS Avionics System for Space/Launch Vehicle Applications	Joe Cecchini
4	Rapid Development of avionic systems	Randy Black
5	Space-Based Wind-Sensing Lidar Design (Not Published)	Dr. Daniel Novoseller
6	Multifunctional Structures: A New Concept For Spacecraft Design Integrating Electronics, Structure And Thermal Control (Not Published)	Mr. Dave Barnett

Session H2: Space Exploration

Chairs:

1	Deep Space One Integration and Test Challenges: Getting to the Launch Pad in the Faster, Better, Cheaper World	Ms. Paula J. Pingree
2	The Mars Microprobe Mission Advanced Micro-Avionics for Exploration of the Martian Surface (Not Published)	Mr Randel Blue
3	From the Sun to Pluto	Ms. Karla B. Clark
4	NASA/JPL Mars Surveyor Program: New Challenges in a New Era (Not Published)	Mr. Robert L. Bunker
5	Interferometer Real Time Control for the Space Interferometry Mission (Not Published)	Mr. Charles E. Bell

Session H3: Microelectronics for Space

Chairs:

1	Ultra Low Power Rad Hard 12 Bit A/D Converter for Space-Based EO Sensors	Sven Nystrom
2	Integrating PWA Design and Analysis Using a Unix-Based Durability Toolset (Not Published)	Mr. Mostafa Rassaian
3	Amecom Direct Chip Attach Project	Tracey Clay
4	Digital Control of the MIDEX Spacecraft Power System (Not Published)	Ms. Karen Castell
5	Issues To Address In Use of Composite Materials for Electronic Packaging	Mr. Gary Trembley

Session H4: Space System Elements

Chairs:

	Title	Author
1	X2000: Avionics for A Multi-Mission Spacecraft	Mr. Savio Chau
2	Applications For A Spacecraft Avionics Functional Model	Mr. Joseph F. Smith
3	A Table-Driven Control Method To Meet Continuous, Near-Real-Time Observation Requirements For The Solar X-Ray Imager	Mr. Kevin Shawn Wallace
4	The TRWIS III Hyperspectral Imager: Instrument Performance and Remote Sensing Applications	Dr. Stephanie Sandor-Leahy
5	Fault Protection Design of the Quikscat and Seawinds Instruments	Mr. Matthew B. Bennett
6	Development of a Low Cost Data Acquisition System for the Space Shuttle Solid Rocket Booster Program	Mr. Kevin Crawford

IRIDIUM® AERONAUTICAL SATELLITE COMMUNICATIONS

Peter W. Lemme, ASK; Kirkland, Washington

Simon M. Glenister, Iridium LLC; Washington, District of Columbia

Alan W. Miller, Motorola; Chandler, Arizona

Overview

The ever-increasing demand to stay in touch, to be able to communicate anywhere and anytime, has created a market for Low Earth Orbit (LEO) satellite communications services such as the Iridium System. The Iridium satellite communications network is being developed to support the needs of the passenger, the cabin crew and the flight crew for:

- Aeronautical Public Correspondence (APC)
- Aeronautical Administrative Communications (AAC)
- Aeronautical Operational Control (AOC)
- Air Traffic Services (ATS).

Iridium Communications

Iridium LLC owns a network of satellites that provides a connection between mobile subscribers and gateway earth stations. The gateways provide connections to the public switched telephone network (PSTN), as well as to Aeronautical Gateway Subsystems (AGS). AGSs provide connections to packet data networks, and are used to initiate ground to air "safety" voice calls.

Iridium Satellite

Iridium satellites are long, slender, triangular in shape, approximately four meters high, and weigh approximately 700 kg (see Figure 1).

The main mission antennas communicate with subscriber units through tightly focused antenna beams forming a continuous pattern on the earth's surface. The main mission antenna subsystem of each satellite includes three phased array antennas, each containing an array of transmit/receive modules. Collectively, the 48 beams produced by each satellite combine to cover a circular area with a diameter of approximately 4,450 kilometers (see Figure 2).

As the satellite beam footprint moves over the ground, the subscriber signal is switched from one beam to the next or from one satellite to the next in a hand-off process. As the satellites approach the poles, their footprints converge and the beams overlap. Outer beams are then turned off to eliminate this overlap and conserve spacecraft power.

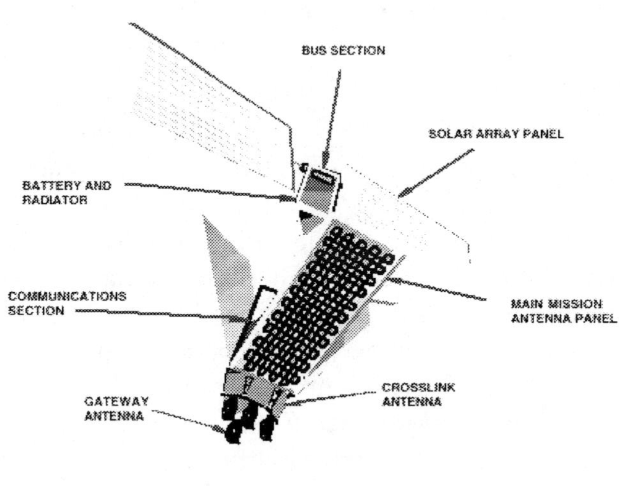

Figure 1. Iridium Satellite

Iridium is a registered trademark and service mark of Iridium LLC

0-7803-5086-3 /98/$10.00 ©1998 IEEE

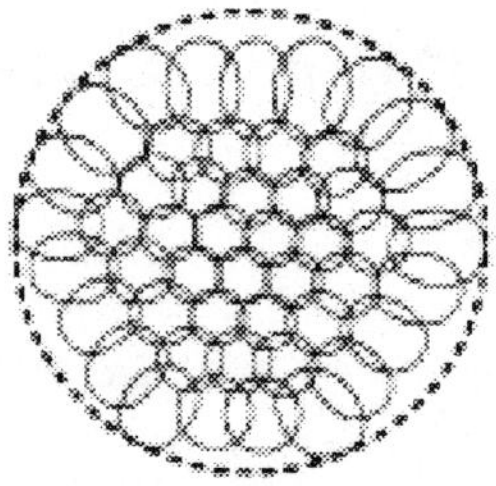

Figure 2. Iridium Spot Beam Map

Each satellite has four satellite-to-satellite crosslinks to relay digital information around the globe. These crosslink antennas point to the nearest spacecraft in the same plane (fore and aft) and in the two adjacent co-rotating planes. "Feeder link" antennas relay information to the terrestrial gateways and the System Control Segment earth stations. Table 1 lists the available frequency bands for each link. The spacecraft payload is the dominant element in the architecture, with high-speed digital switching handling complex telephony routing.

Table 1. Available Frequency Bands

L-Band up/down-link	1616 – 1626.5 MHz
K-Band down-link	19.4 -19.6 GHz
K-Band up-link	29.1 - 29.3 GHz
Crosslink	23.18 – 23.38 GHz

Iridium Satellite Constellation

Sixty-six operational satellites are configured in six (near-polar) orbital planes, with 11 satellites in each plane, covering all areas of the Earth. The satellites are phased appropriately in co-rotating planes up one side of the earth, across the poles, and down the other side of the earth. The first and last planes rotate in opposite direction, creating a "seam". Co-rotating planes are separated by 31.6 degrees, and the seam planes are 22 degrees apart. The orbits have a mean altitude of 780 kilometers and each satellite will circle the Earth once every 100 minutes.

Inter-Satellite Networking

The Iridium architecture incorporates certain characteristics, such as call hand-off, that allow the space segment communications link with subscriber equipment to be seamlessly transferred from satellite to satellite as the satellites move over the area where the subscriber is located. The crosslink antennas permit satellites in the constellation to communicate with one another. This inter-satellite networking capability is a significant distinguishing feature of the Iridium System and provides a number of benefits. Inter-satellite networking will afford a subscriber worldwide access to the Iridium System (to make or receive calls), irrespective of gateway location, by routing a call from satellite to satellite until it is connected to the gateway that is most appropriate to the destination of the particular call. This feature also permits the Iridium System to route calls around gateways or satellites as required, in the event of a temporary outage. Because crosslinks are used, the gateway earth stations need only process the information designated to them. Crosslinks allow for reduced ground infrastructure requirements, enabling least cost routing and full global coverage (including the oceans and polar regions).

System Control

Motorola operates the system control segment: monitoring, managing and controlling the satellites. The Satellite and Network Operations Center (SNOC) is located in Landsdowne, Virginia, USA; the Backup Operations Center (BOC) is located in Italy; and the Tracking, Telemetry and Control (TT&C) stations are located in northern Canada, Hawaii, and Iceland. These facilities will manage the performance and status of the individual satellites. The SNOC manages the system control segment by assigning earth terminals to satellites. The SNOC also manages the network by developing and distributing routing tables for use by the satellites and gateways and directing traffic through the network.

Gateways

Gateway earth stations provide call-processing services, such as subscriber validation and interconnection between the world's PSTNs and the Iridium System. Iridium calls are connected to the PSTN through an International Switching Center. Gateways communicate with the space segment via gateway link antennas on the satellites and ground-based antennas, or earth terminals, at each terrestrial gateway facility. Each gateway typically includes three or four antennas (with geographic diversity to overcome local weather effects if necessary), a controller to manage communications with the constellation, an operations center to perform local network management, a paging message origination controller, and a switch that connects the gateway to the local PSTN. Each gateway will also include a subscriber database used in call-processing activities, such as subscriber validation. The switch and the associated subscriber databases are based on GSM (Global System for Mobile Communications) technology and support the mobility management protocols required for inter-operation with other GSM-based systems. Gateways will also generate call detail records used in billing.

Aeronautical Gateway Subsystem

For aeronautical services, an additional subsystem, known as the Aeronautical Gateway Subsystem (AGS), participates in all safety voice and all packet data communications. The AGS is responsible for the terrestrial end of the Air/Ground data link protocol. The AGS also provides connectivity to terrestrial users for safety voice services, and provides security features to screen incoming safety voice call requests. The AGS is generally co-located with a gateway and is directly wired to the output of the gateway Siemens GSM-D900 circuit switch.

Initially an AGS will be deployed at each of three gateways (Europe, US, Asia), providing geographic diversity. The possibility of deploying AGSs at additional gateways is being considered.

GSM

The Iridium System utilizes proven GSM (Global System for Mobile Communication) technology. Use of this standard GSM architecture will allow service enhancements that are developed for GSM on the ground to be uploaded to the satellites via the ground stations and offered to all Iridium subscribers.

Mobile communications systems are becoming increasingly prevalent and are influencing communications patterns. People traveling globally are more and more accustomed to being able to communicate through "roaming". "Roaming" is the term given to the ability to travel and communicate through different cellular service providers while using the same phone and phone number with billing to a single account number. With the Iridium service, aircraft can become another roaming environment. GSM compatibility facilitates this option. Through the use of a SIM (subscriber identity module), the Iridium System uniquely identifies each subscriber for call routing, billing and account verification. By using a SIM card, any GSM subscriber with Iridium service authorization can effectively customize that phone to operate over the Iridium System. With the SIM card in place, calls made to the subscriber's mobile number will automatically be received at the seat phone. The SIM card thus provides a more comfortable and convenient method for passengers to bill and pay for a call.

Passengers who do not use a SIM card can be provided a prepaid, Iridium stored value card, that can be offered as an amenity, or passengers can pay via traditional methods (e.g. using credit cards).

All Iridium terminals, including airborne terminals, are registered for service with one of the gateways. This is known as the

"Home Gateway". Aeronautical users are registered in the Home Gateway as determined by their country of origin. As Iridium users roam around the globe, the Iridium System and other affected GSM networks maintain a record of their current location. The Iridium System automatically maintains this record as long as the mobile unit SIM card is reachable. When this location is serviced by a gateway other than the Home Gateway, that gateway is notified of their presence. This second gateway is known as the "Visited Gateway". For example, an Italian citizen travelling in the United States will have the Italian Gateway as his Home Gateway and the North American Gateway as his Visited Gateway. The Home and Visited Gateway may be identical. Both the originating and terminating customers may have Home and Visited Gateways; therefore a total of four gateways may be involved in portions of an Iridium circuit.

For Iridium voice calls, the gateways exchange signaling information to establish the billing and routing for the call and make any necessary connections to the public telephone network. When the call is answered, the connection is switched so that the satellite constellation is used to the maximum extent possible. For calls between Iridium terminals, this means that the gateways participate only in the call setup and tear down, and are not involved in the actual call transaction itself. For calls between an Iridium terminal and a non-Iridium terminal, a gateway is always involved.

Iridium Aeronautical Installation

Iridium airborne terminal equipment is being developed utilizing the same core components that are used in hand held land mobile terminals. These modules allow a cost effective means to provide the Iridium communications service to all aviation users.

Iridium airborne terminal equipment for air transport aircraft is being developed to industry characteristic ARINC 761 and will support multiple, simultaneous channels of communication. Further product variants are also being developed for use in military, government, and corporate aircraft. Iridium equipment could even be developed for use in light, general aviation aircraft, possibly using a portable terminal.

"Safety of Flight" Communications

The Iridium Service is being developed to support the requirements for communications necessary for the safe operation of aircraft. These voice and data communications typically provide a link between the flight crew/aircraft systems and ATC/dispatch/maintenance personnel or automation.

Qualification

The Iridium aeronautical team is actively supporting the preparation of the ICAO SARPs (Standards and Recommended Practices), the RTCA MASPS (Minimum Aeronautical System Performance Standards), and the RTCA MOPS (Minimum Operational Performance Standards) to apply to "next generation" satellite service for aeronautical safety application. The Iridium System is planned to comply with these standards as well as the FCC's Type Acceptance criteria.

Priority, Precedence and Preemption

Iridium aeronautical services will share the Iridium infrastructure with non-aeronautical communications services. To ensure that safety-related aeronautical services, normally ATS and AOC, obtain timely access to resources, the Iridium Aeronautical System includes provision for Priority, Precedence and Preemption (PPP) of system resources.

Functionality for PPP is required in the avionics (AESs); the gateways (GESs) and the space vehicles (SVs) to support the identification and tracking of safety related communications.

The implementation of the SV priority call setup and handoff scheme includes a call release mechanism in the event that no channels are available during call setup or handoff (intra- or inter-satellite). This call release mechanism is invoked by the SV for support of calls designated as aeronautical safety calls by the avionics or the gateways. The SV does not limit the number of aeronautical safety calls. The entire resources of the satellite are available for aeronautical safety calls.

Aeronautical safety traffic has the highest protection from load management of any user service. The SV includes a design implementation of multiple levels of high priority preemptive calls over the regular non-safety traffic. When an acquisition request is received at the SV, the request is accepted or rejected based on the Acquisition Class. Once accepted, a traffic channel assignment, consisting of a frequency access, an uplink time slot, and a downlink time slot, will be made based on the priority level. This priority level will be maintained throughout the duration of the call. In the event that an unused traffic channel is not available for a high priority acquisition or handoff request, the SV will demand a confirmed call release from a lower priority user to provide a traffic channel for the higher priority call. Immediate preemption adds no more than 250 milliseconds to the call setup delay that would have been experienced if preemption were not required.[2]

To ensure that the priority of the call is maintained by the satellite subnetwork for the duration of the call, the priority code is included in the call overhead messaging between the SVs during each Inter-SV Handoff. The high priority aeronautical safety priority calls are queued for handoff before those calls of lower priority. Thus high priority aeronautical calls "move to the head of the line" for all handoffs.

Performance

Iridium aeronautical services operate over an Iridium communications circuit. There are two major random components that contribute to transfer delay:

- call setup time
- effective channel throughput

Of these components, the call setup time is the dominant factor. In the worst case call setup times may be 17-21 seconds at the 95th percentile.[3]

Once a circuit is established, voice communications are subject to latency. Contributions to latency include voice coding/decoding, propagation delays through the ether, and processing delays introduced by any crosslinks that might be utilized. In any case, latency is expected to be between 270 and 390 ms.[4]

For data communications, the effective full duplex channel throughput ranges between 2400 bps and 600 bps, depending on the interference and multipath environment. Using the standard ICAO message length of 128 octets, or 1024 bits, the contribution to transfer delay due to effective channel throughput is between 427 ms and 1.7 seconds. Thus, effective throughput is only a small contributor to Iridium aeronautical transfer delay.

Availability

An individual Iridium satellite provides coverage to a particular subscriber for no more than about nine minutes (11 satellites/100 minutes). The most catastrophic failure mode of a satellite would create a hole in the coverage, causing a communications outage greatest at the equator, and decreasing in size progressively towards the polar regions. Adjacent satellite coverage can be expanded to minimize the affected area, filling the hole at about 45 degrees latitude. In addition, each of the 6 orbital planes will have a spare satellite

[2] ICAO AMCP, Working Group A, Working Paper 539

[3] ICAO AMCP, Working Group A, Working Paper 539

[4] ICAO AMCP, Working Group A, Working paper 403

that can be maneuvered into position in a timely manner.

Once a satellite outage has been detected, the affected area can be predicted with accuracy, and this information can be made available to users to allow preparation and planning. As the Earth rotates while the satellites orbit, the area affected by an outage is continuously changing.

Dual Dissimilar

Iridium communications can complement other communications services such that the combination of the services are highly reliable, and may not be affected by a common mode failure. Examples of systems that may be complemented include VHF, HF and even Inmarsat Satellite Communications.

Voice

Iridium communications are being designed to provide a direct voice link between the flight crew and air traffic control, airline flight dispatch, maintenance, medical or other operational staff. Voice communications are essential for timely or interactive communications, or to recover from the failure of data link communications. The Iridium System is planned to also provide communications between two or more aircraft, and between an aircraft and a land mobile Iridium terminal.

The AGS provides some translations to facilitate telephone number discovery. However, the development of an industry specification to support real-time telephone number discovery using data link (to benefit any communications service) is not available and must be undertaken to achieve truly "immediate" direct communications between pilot and controller.

Flight Crew Initiated

The flight crew selects (through a control/display unit) a telephone number to reach a ground party. For Air Traffic Control, a special six digit code has been assigned to each Flight Information Region (FIR). The AGS will translate this code to an unlisted telephone number and will connect the flight crew to a pre-determined (usually supervisory) station at the selected Air Traffic Control Facility. The six digit code is published on enroute air navigation charts once approved and published by an individual state in their Aeronautical Information Publication (AIP).

Many airlines will pre-program a directory of phone numbers to allow speed dialing operational centers (e.g. dispatch, maintenance). These centers can then conference other parties into the call, as needed. Although not technically needed, some airlines will restrict the flight crew from calling any phone number not previously programmed into the directory. This is to ensure the system is not used inappropriately.

Calls to the Flight Deck

A caller to the flight deck must first be authenticated by calling an AGS and providing the necessary security codes. Once authenticated, the caller identifies the aircraft to be called, and the priority of the call. The AGS sets up the call, and connects the calling party.

The calling party must be able to identify the aircraft in such a way such that the correct telephone number can be determined. The AGS may be able to utilize the ICAO 24 bit codes assigned to each aircraft to determine the proper telephone number. Flight crews are encouraged to provide the eight digit octal representation of the ICAO address of their aircraft in their filed flight plan to facilitate timely call setup.

Packet Mode Data

ACARS (Aircraft Communications, Addressing and Reporting System) provides a packet data interface to exchange information between the flight crew/aircraft systems with airline/ATC systems and personnel. The industry is beginning to transition to other packet data networks. The Iridium System is planned to support all of the packet data networks.

0-7803-5086-3 /98/$10.00 ©1998 IEEE

Examples of packet data applications include:

- Movement messages
- Engine trend monitoring
- Equipment fault reporting
- Flight Information Services
- Automatic Dependant Surveillance
- Direct Controller/Pilot Communications

ACARS

The Iridium System is planned to support both the existing ACARS packet data interface ("satellite category A") to a default Data Link Service Provider (DSP) and a new protocol ("satellite category B") allowing simultaneous communications to multiple DSPs.

ATN

The Aeronautical Telecommunications Network (ATN) is planned to be supported consistent with ICAO SARPS.

X.25

The Iridium System is planned to support a direct X.25 connection between a host computer and applicable aircraft systems.

Non-Safety Communications

The Iridium System is planned to be integrated with industry standard ARINC 746 cabin telecommunications units (CTU). The Iridium System will complement any terrestrial passenger communications system currently installed. Just as in its hand-held service, the Iridium System may be used in a least cost routing configuration. If terrestrial service is available and is lower cost, the call will be routed on the terrestrial service (NATS, TFTS). If there is congestion on the terrestrial system or it is unavailable, the call will be routed to the Iridium System. Iridium aeronautical service is planned to support basic telephony, fax, packet data service, modem data messages, and packet data services.

Voice

Placing Calls from an Aircraft

Placing calls from an aircraft conforms to the dialing procedures of the installed cabin telephony system. Passengers who are not using an Iridium System compatible SIM card can either charge their communications by using a standard credit or telephone billing card, or they may have their charges "deducted" electronically from an Iridium prepaid card. The prepaid card is an amenity that can be sold or given to passengers in advance.

Receiving Calls on an Aircraft

Market research has shown a strong desire for reliable, simple ground to air calling. This function is also likely to generate a high number of air to ground calls in response to ground to air communications. When a passenger uses a SIM card to register on the aircraft, a ground caller just dials the subscriber's regular mobile number and will not even be aware that the call is routed to an aircraft.

If a SIM card reader is not available on the aircraft, the passenger may still have the option to register manually. This involves entering a 10 to 24 digit registration code (usually a phone number) and identifying a credit card for billing authority. A caller then dials a special Iridium aero access number, and then provides the 10 to 24 digit registration code of the desired called (registered) party, as well as a call-back number. The registered party is notified and given the option to take the call immediately, or call back.

Circuit Mode Data

Facsimile Transmissions

Facsimile is a store and forward feature, operating at 2.4 kbps. An aircraft may be equipped with facsimile (fax) machines or passengers may be able to use their laptop computer with fax software. Passengers sending faxes from the aircraft will follow the same procedures as those used to place an air-to-ground call.

A passenger may receive faxes while in flight after receiving notification that a fax has been sent to an Iridium "mailbox". Anyone sending a fax to a passenger will simply dial the passenger's fax number and will not be aware that the fax will ultimately be received in the air.

Modem Data Messages

The Iridium aeronautical service will support PC data transmissions at 2.4 kbps. Passengers with laptops and modem connections can plug into the modem jack at the seat (if one is provided) and initiate a transmission in the same manner as they would place a phone call.

Packet Mode Data

Value added services such as on-board shopping and information retrieval can be supported via packet data communications using the Iridium aeronautical service. Passengers may perform credit card transactions (e.g., rent cars, make hotel reservations), order goods and services, and access news and information sources.

Robust Terrestrial Communications

Occasionally, all terrestrial communication links to an ATS facility become unavailable. This creates a challenge to maintain safe aircraft separation. Iridium equipment can provide an immediate backup communications link to provide direct aircraft communications, as well as a link to other facilities.

Search and Rescue

Aircraft incident and accident locations may be anywhere, and are typically very remote. The Iridium System can provide the surviving passengers and crew with a communications link to search and rescue, perhaps using an Iridium battery power hand-held terminal. Search and rescue operations could use Iridium services to coordinate their activities, without regard to their location. The ability to both make and receive calls would be particularly useful.

Pager

The Iridium System includes a global messaging service that delivers numeric or alphanumeric messages to paging units similar in form to the familiar pagers widely used today. For paging transmission, spatial diversity is used to achieve high message character accuracy in the presence of fading and adverse pager unit orientation. The flexible message format accommodates variable-length messages as required by several different pager unit types. Tests have demonstrated that pages may be reliably received in an aircraft and special aeronautical paging zones have been established to enable a paging customer to advise the system of his or her approximate air routing, such that pages can be received throughout the flight. In the "you-page-me-I'll-call-you" paradigm of individuals who carry a pager and cellular phone the aircraft page is expected to generate a call by the passenger.

SelCal

The Iridium pager is a passive, receiving device and will require no aircraft interface. It is possible that the Iridium pager could be used to replace HF SelCal. SelCal codes are reaching capacity, and retrofit will be costly. Iridium pagers are simple to retrofit into any aircraft, and can provide short messaging in addition to simple notification.

Conclusion

The Iridium Communication System is being designed to support the ever increasing demand within the aeronautical community to be in contact, serving the needs of passengers, crew members and aircraft systems.

Reference: Raymond J. Leopold and Ann Miller, "The IRIDIUM Communication System," IEEE Potentials, April 1993.

THE TAKE-OFF OF MOTS AVIONIC NAVIGATION SYSTEMS FOR SPACE/LAUNCH VEHICLE APPLICATIONS

Joseph Cecchini, Honeywell Inc, Clearwater, Florida

ABSTRACT

Application of a Military-off-the-Shelf (MOTS) navigation system. The Honeywell H-764G, is being qualified as a primary navigation system for new generation low cost launch vehicles.

Current commercial market needs are driving leading national aerospace companies to closely examine the new performance and cost advantages of high volume military-off-the-shelf products. Leveraging from existing military applications and heritage, Honeywell is opening the door with a smart, low-cost, Space Integrated GPS/INS (SIGI) system as shown in Figure 1 for

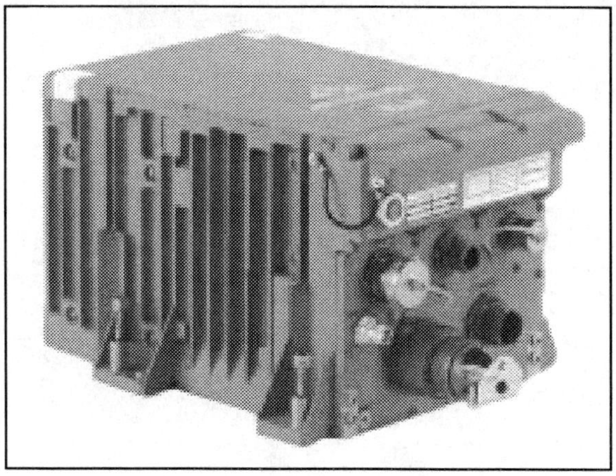

Figure 1. New SIGI Design

commercial launch vehicles.

Important differences between military and commercial space launch environments need to be considered. While the military environments. are very demanding, launch vehicle navigation systems also require thorough testing and additional environmental tests. This paper addresses the new environmental conditions and tests of Honeywell current navigation systems used for military applications but which can be used for space and launch vehicle missions. The bulleted items which follow list the major categories of test and the inherited synergism between military and space missions.

For extension of the MOTS system to meet space requirements, several new space and launch vehicle environmental tests had to be verified. These new environments included those listed below:

- Low vacuum pressures of space
- Rapid De-pressurization/ Re-pressurization
- Differential Pressure
- Corona discharge
- Thermal temperature
- Vibration
- Shock

This paper also details the factors and considerations which went into the development of the test plans and procedures used to qualify Honeywell's MOTS product for space and launch vehicle navigation.

Through close co-operation between Honeywell's Space Systems division and the Sensor and Guidance Products division, joint teamwork provided the resources and expertise to qualify a SIGI to support space missions.

Shorter time to develop a low cost MOTS for space has forced navigation system suppliers like Honeywell to lead the effort in fielding MOTS products with more extensive test and verification.

By thorough test and evaluation using calibrated test equipment and controlled conditions, new use of MOTS systems can be realized to meet the low cost needs of today's current and tomorrow's future space and launch vehicles.

0-7803-5086-3 /98/$10.00 ©1998 IEEE

The extension of military navigation products to launch vehicle and space application faces several challenges unique to space flight environments. Many important lessons were learned as a result of this effort. Thise include design enhancements, test and verification methods for application of a MOTS navigation system in the harsh environments of space and launch vehicles. A detailed test plan and procedure was developed and reviewed. Properly controlled environmental tests were defined and performed to verify system performance and quality both during and after exposure to the new requirements of space and launch vehicles. A standard H-764G Embedded GPS/INS, SN 2397, was pulled from the production line at Honeywell's Sensor and Guidance Operation in Clearwater, Florida and used as the test article. This paper details the environmental tests developed to qualify the MOTS system for space application and reports the results from the testing which have been performed to date.

A detailed test evaluation of the unit under test, UUT, was performed both during and after exposure to the multiple. Functional navigation runs used before, during and after exposure to the different test environments, served to quickly validate proper system operation of the navigation system. Honeywell's navigation system tightly couples GPS satellite pseudo range data with ring laser gyros and accelerometers sensors in a Kalman Filter. Navigation measurements and repeatability are monitored and compared with external truth conditions while the test article is placed at a known location and orientation.

Space Environmental Issues

One of the major issues involves the mechanical enhancements necessary for application in low pressure of space. Low vacuum pressure presents additional corona plasma concerns. Another important area for test verification is thermal temperature extremes and sudden rate of temperature change. The heat dissipation of an electronic navigation system in un-pressurized micro-g space environments, presents additional challenges.

Mechanical Enhancements

Previous space applications such as the Hercules program tested on a shuttle flight in late 1992, demonstrated the proper operation of Honeywell's attitude reference unit (ARU) while operating in space. However, this space mission was for a short duration in a pressurized shuttle cabin compartment. To meet the larger needs of commercial launch vehicle and space flight, Honeywell initiated an alternate approach to extend the use of their current military guidance and navigation systems out to new space applications. Starting with the Embedded GPS/INS (EGI) navigation system, the H-764G, Honeywell made several design modifications and enhancements. The differential pressure between the interior and exterior of the chassis is a function of the altitude. In order to prevent structural distortion due to differential pressure, two pressure relief valves were incorporated into the chassis of the SIGI as shown in Figure 2. The original EGI design planned for aircraft application had a structural rigidity to withstand maximum pressure differential of 5 lb/in^2. The original EGI navigation system designed and built for military aircraft applications is shown in Figure 2A.

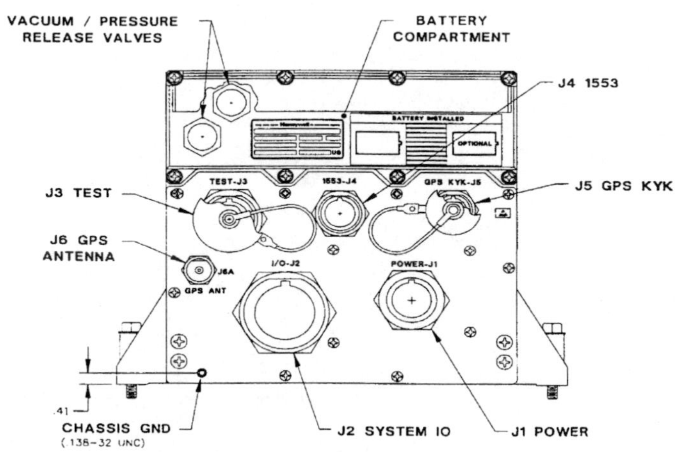

Figure 2. Thermal Vacuum Profile

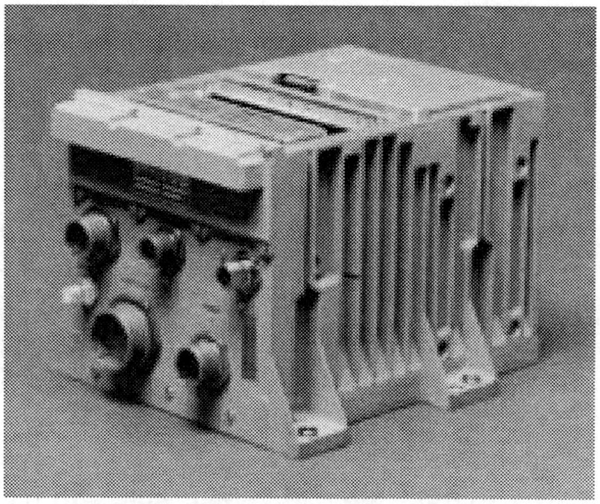

Figure 2A. Old EGI Design

Space Vacuum Environment

Rapid de-pressurization and re-pressurization tests were performed. A profile of this is shown in Figure 3 (Rapid Depress/Re-Pressurization). The SIGI was tested and verified at these pressures. The results of the test verified proper operation of the pressure relief valves with no signs of structural distortion.

Exposure to and operation in unpressurized space vacuum presented several new challenges which needed to be addressed.

These included:

- Rapid de-pressurization and re-pressurization
- Corona free navigation operation at all altitudes
- High Vacuum >300k feet Cold soak
- Thermal vacuum

Corona Discharge

Another extremely important issue for launch vehicle applications, include operation at altitudes between 50k to 140k feet. It is at these altitudes where corona. plasma of high voltage signals can occur. At ground level air is an excellent insulator but at higher altitudes where air is thinner, electro arching of high voltage can damage other exposed electrical components and circuitry. For commercial launch vehicle applications operating at these altitudes, a verification test was determined as necessary and developed. Figure 4 (Corona plot) details the pressure and respective altitude at which the SIGI was tested and verified.

In order to ensure corona free operation, the SIGI test article was mounted in vacuum chamber (See Figure 5) and commanded to navigation mode while operating with "live sky" GPS satellite signals. System operation was monitored at all times and recorded to a save file using Honeywell's test equipment operating the CDATS test software. A high frequency antenna sensor was also placed in the test chamber to monitor any RF emisions from corona plasma that might occur. This antenna sensed any and all RF emissions and fed the measurements back to Honeywell's HP 70000 Spectrum Analyzer test equipment which displayed the detection level and provided a record of the highest voltage levels detected.

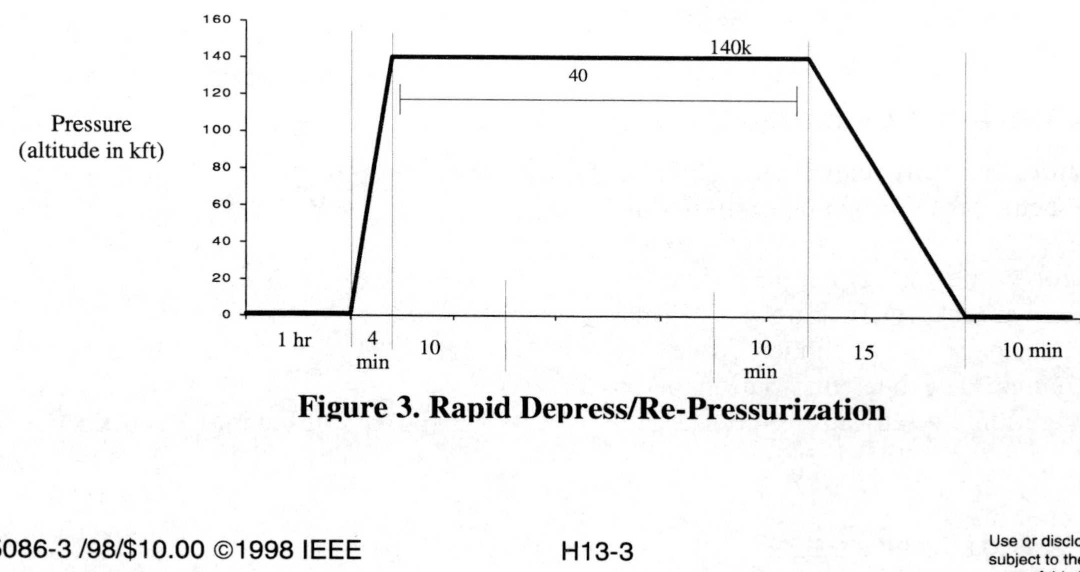

Figure 3. Rapid Depress/Re-Pressurization

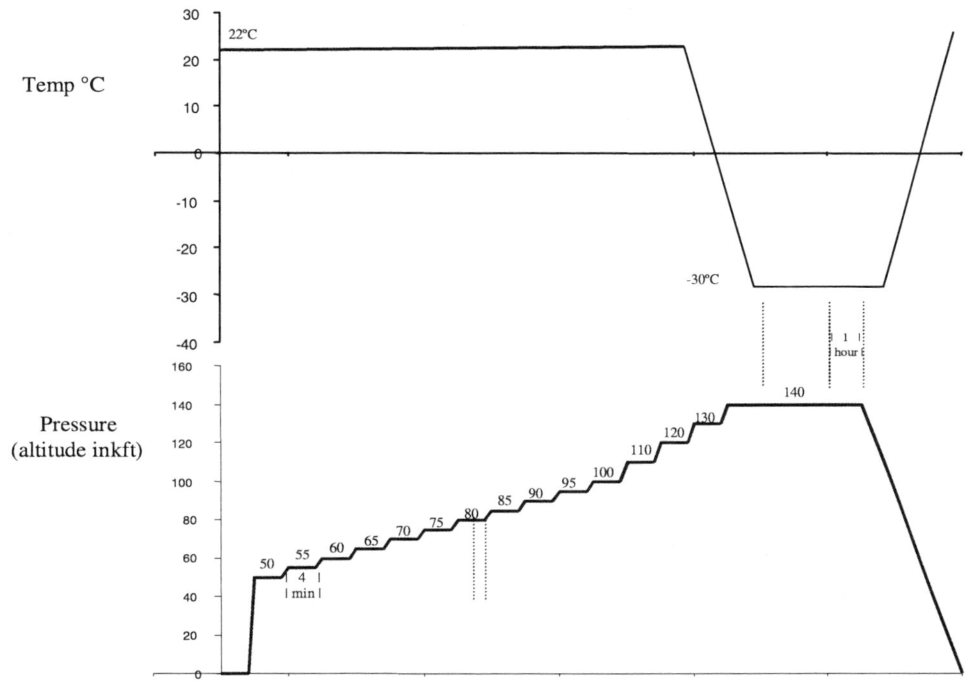

Figure 4. Corona Plot

Figure 5. Vacuum Chamber

Starting at a test chamber pressure equivalent to 50,000 feet, the SIGI performance and the corona discharge sensor were monitored for a period of four minutes dwell at the specified altitude. The pressure was then increased in five thousand foot increments with another dwell for four minutes. This test at higher and higher altitude pressures was repeated with the same dwell at each altitude. At 80,000 feet, the test article was intentionally powered off and back on. This was done to verify that the high voltage pulse used to start Honeywell's Ring Laser Gyros lasing would not induce any corona. This step was also run to verify a mission requirement to initiate power of the navigation system while in flight at the corona senstive altitudes. No corona high-voltage arcing was detected. The test article's navigation performance remained stable and accurate at each of the test altitudes /pressures. Once the equivalent pressure level of 100,000 feet altitude was reached, the pressure step increments were increased to 10,000 feet. The same dwell time of four minutes was used at each altitude as the pressure was increased from 100,000 feet to 140,000 feet. Again no sign of corona was observed. This test was conducted over the equivalent pressure range from 50,000 feet to 140,000 feet where corona discharge can occur.

Cold Soak/Temperature

At the conclusion of the corona test shown in Figure 4, the unit was powered off and the ambient temperature was taken to -30°C for cold soak duration of more than 12 hours. The unit was then monitored with external sensors to verify internal temperature prior to the power being restarted the next day. This power on restart from a cold start was performed to simulate extended stay in orbit of space vehicle without power. Once power was re-applied, the navigation system restarted and a functional navigation run was made which did verify proper operation. A detailed thermal profile was measured and recorded using internal temperature sensors. The thermal characteristics were used to verify the minimum and maximum temperatures at which the SIGI operates.

Thermal/Vacuum Test

Thermal vacuum tests were used to verify navigation system over full range of temperature space environments. Thermal/ vacuum tests were performed on the SIGI operating in a navigation mode using "live sky" GPS satellite constellation data. The thermal vacuum profile is shown in Figure 6. Vacuum chamber pressure levels of ~300,000 feet (5xE-5 Torr) and mounting interface temperatures of (–54°C) to (+60°C) were used.

Temperature consideration plays an important role in thermal operation of the unit. There is no heat transfer by convection in an unpressurized compartment. This plays an important role in determining the maximum operating temperature. Internal components were monitored while thermal vacuum tests were being performed. Testing showed internal temperature 20°C greater than interface mounting temperature. The primary method of heat transfer was conduction through the mounting feet on the test unit to the heat sink. Radiation was a secondary source of heat transfer but at greatly diminished levels compared to the conduction. The test was run to determine temperature differences and further supported the importance of vehicle barbecue roll maneuver used to stabilize temperature within satellites/launch vehicle.

Standard thermal test required three cycles of maximum high/low temperatures. This temperature cycling was done to support the multiple mission re-use of the SIGI on RLV applications. The unit is designed to be powered off for extended periods while in space. Testing was performed to validate the SIGI powers up and properly operates after extended periods at low temperatures.

Verification of proper moding and navigation after environmental was performed to validate the design.

Structural Strength

In addition to potential mission re-use, the physical and mechanical requirements for space launch systems are very demanding. In order to verify the structural strength and integrity of Honeywell's SIGI, multiple vibration and shock

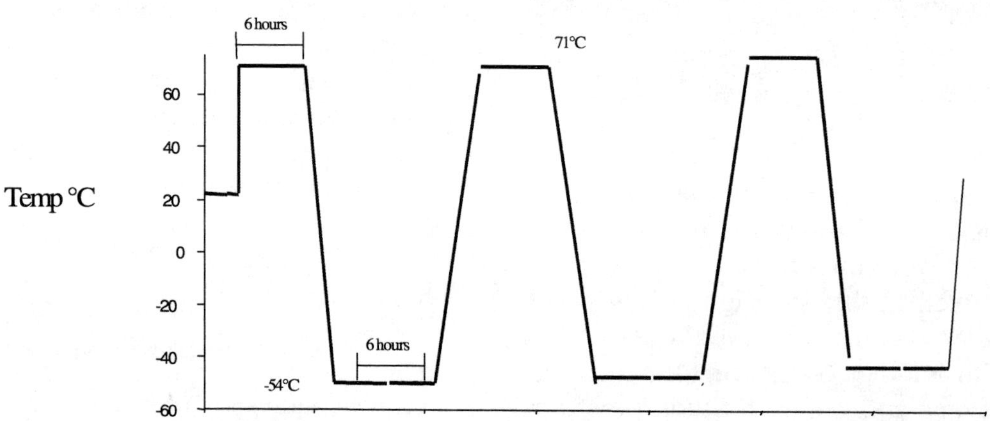

Figure 6. Thermal Profile

tests were performed. The vibration test included both random vibration and sine sweep vibrations. (See Figures 7 and 8) Additional shock tests were conducted off site at one of the users launch test facilities. The shock tests were performed with dual barrel pneumatic pressures to achieve 7500g-shock level as shown in Figure 9. This test was done to measure pyro separation shock levels that occur on the launch vehicles at stage separations.

Vibration

Random Vibration

The standard technique used to evaluate random vibration responses consists of monitoring performance prior to, during and after vibration excitation. The SIGI was mounted on a vibration test fixture in preparation for the acceptance and qualification random vibration test of 9.54 grms and 19.087 grms for durations of 225 seconds for each axis. Prior to each exposure, a functional navigation run was performed and recorded using GPS satellite constellation. Performance of the test article was monitored before, during and after the random vibration exposure. Figure 8 defines the random vibration performance profile. The endurance level vibrations are shown in Figure 9. This profile was provided by a commercial launch vehicle customer familiar with actual flight vibration environments.

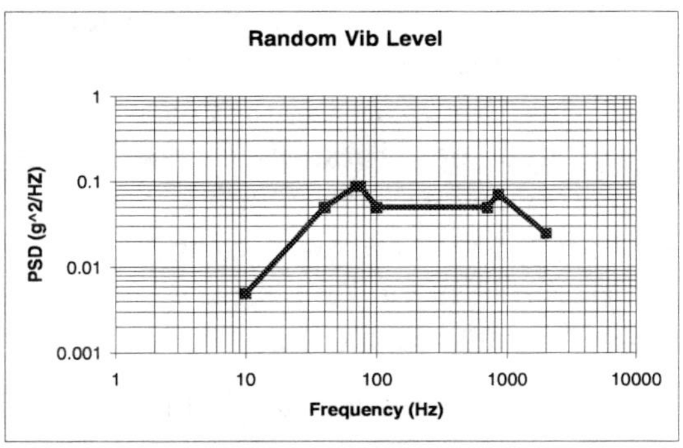

area =	91.03
grms =	9.54

Figure 7. Performance Level Random Vibration

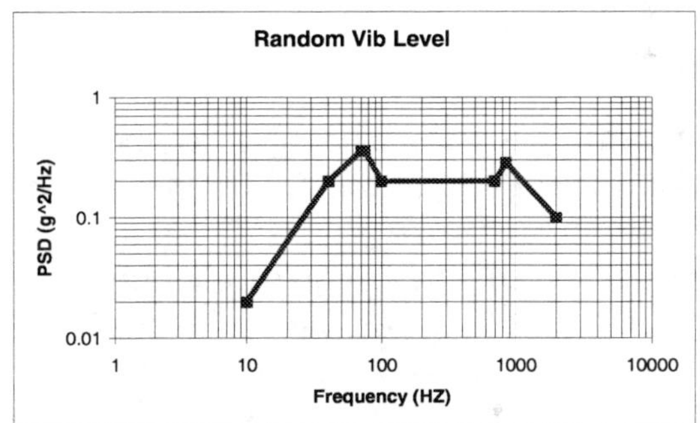

area =	364.11
grms =	19.08

Figure 8. Endurance Level Random Vibration

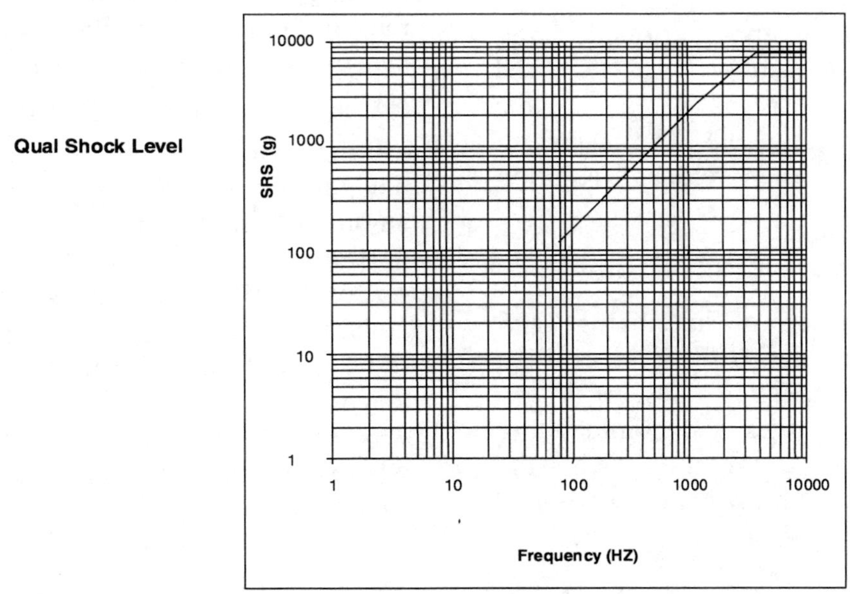

Figure 9. Shock Level

Random Vibration

Honeywell performed both performance and endurance level vibration testing. At endurance level vibration, loss of GPS lock to the satellites was seen. After endurance level vibration stopped, re-acquisition of the satellites quickly occurred. The endurance level vibration simulated the entire life cycle which the unit was expected to see. During "Y" axis testing at the endurance level, loss of GPS lock occurred but no re-acquisition occurred after vibration stopped. A problem review board (PRB) was consulted and troubeshooting identified suspect subassembly. The GPS supplier was contacted and corrective action was taken to securely bond oscillator lead wires to the GPS board. After corrective actions were taken, proper navigation resumed.

An intermittent loss of power to the unit was seen during the vibration. Additional attempts to re-power the unit were unsuccessful. Physical examination of the power supply module revealed sheared pins on the connector. The PRB convened and recommended to use a CTS connector manufacturer. The power supply connector replacement was made and re-test of unit was performed. The remainder of the vibration tests were run successfully.

Test Monitor Equipment

Test monitor operation was performed using CDATS test program running on PC and monitoring I/O messages over the MIL-STD-1553 interface. The test program displays the position and velocity solution determined by the SIGI.

Functional Navigation Runs

Typical operation consists of initializing GPS Receiver, commanding the GPS receiver to a navigation mode, initializing system, gyro compassing for four minutes and commanding system to navigate. Prior to vibration testing the system was placed into the navigation mode and navigation performance verified.

Vibration Test Log

Test equipment performance plots were printed during testing. Normal operation was seen. A test snapshot of the system performance was recorded approximately 100 seconds after vibration started. No loss of GPS satellite links occurred during the performance level.

The geometric dilution of precision GDOP and other DOPs were quickly lost after the endurance vibration began. These quality of performance values are computed as a function of the constellation geometry. Loss of one or several of the satellites causes an immediate reduction of the quality.

The SIGI's GPS receiver stayed on lock with one satellite, SVID 25, the entire duration of the endurance vibration. The duration of the vibration was for 3.7 minutes as shown in Figure 9. Once the vibration stopped, satellite reacquisition was performed quickly within one minute. Internal self-test was examined and no LATCH BITS, built in Test Bits, were reported. As part of the detailed test procedure, functional

NAV run was performed after the endurance vibration for a five minute period. For test monitor purposes, accelerometer sensors were mounted internally to the GPS receiver, the processor and the inertial electronic circuit card assemblies. This was done to not only monitor the vibration level of the system/chassis but to also provide detailed measurement of response acceleration seen at the card level.

Similar to an actual flight mission scenario, the navigation system undergoes the greatest amount of shock and vibration during the early portions of the mission and at the stage separation phase. The expected levels similar to endurance level vibrations are followed by less rigorous but sustained flight environment. For this reason, the endurance level vibration test was followed by another performance level test where system navigation and accuracy are closely monitored.

Sine Sweep Vibration

Another critical test to the evaluation of launch navigation test consists of SINE Sweep Vibration. Honeywell performed sine vibration testing at 8g and 13g levels. Additional 30g sine vibration tests were performed at the test facility of one of the primes. Navigation performance runs were checked while built-in self tests were performed during the vibration exposures. Each axis was vibrated over frequency range from 450-550 Hz with a 60 second sweep rate up and 60 second sweep rate down. Diagnostic tests were run at reduced levels to determine any resonant frequencies. No resonant frequencies were found. Functional NAV runs were performed prior to and after each axis vibration. No loss of GPS lock occured during the 8g or 13g Sine sweep tests. Functional navigation performance runs were made that verified nominal navigation operation.

Shock Testing

Following successful random vibration testing, shock level testing was performed on the test article. Working in cooperation with one of the launch system primes, Honeywell was invited to perform the shock testing at their environmental test facility. The test series were set up and performed. Test data taken during and after the shock test indicate proper navigation operation prior to and after the shock exposure. Multiple shocks were performed on each axis. The navigation system performance was verified as nominal after the shock exposures.

Summary

Honeywell's SIGI has been subjected to the environmental tests that are representative of launch vehicle and space applications. The SIGI test unit was verified to perform under acceptance test level environments and to perform after exposure to survival endurance level environments

Functional and performance tests run after the test unit was exposed to acceptance and endurance level environments, verified proper operation. As a result of the extensive testing performed at launch vehicle levels, Honeywell's engineering staff was able to determine the limitations of the existing design. In turn, the engineers were able to make recommendations and design enhancements. The unit was then subjected to another complete battery of tests. Honeywell's SIGI navigation performance was nominal throughout the testing. The difficult space environments were provided by the primes. Honeywell is currently under contract supporting both launch vehicle and space programs with its Military Off-the-Shelf (MOTS) SIGI. By continuing to improve existing MOTS products, Honeywell is leading the technology advancements of affordable space navigation systems into the 21st century.

Abbreviations

MOTS	Military-Off-the-Shelf
GPS	Global Positioning System
INS	Inertial Navigation System
LV	Launch Vehicle
RF	Radio Frequency
RLV	Re-usable Launch Vehicle

0-7803-5086-3 /98/$10.00 ©1998 IEEE

RAPID DEVELOPMENT OF AVIONICS SYSTEMS

Randall H. Black HONEYWELL Space Systems Division Glendale, AZ

Abstract

As hardware development techniques have improved, the time to market in avionics systems has become increasingly dependent upon systems analysis and software development. At the same time the lion's share of life cycle costs have migrated from development to integration and maintenance. In order to remain competitive in today's government and private sectors, avionics system developers must employ a scheme which reduces systems analysis, software development, and overall integration efforts. Such a scheme includes an effective process, a flexible architecture, and the tools necessary to greatly reduce life cycle costs. This paper presents a spiral development process with an associated architecture and tool template designed to reduce time to market. Specific uses of this template on the International Space Station and the Boeing 777 avionics system are described along with the next generation implementation.

Introduction

As the digital aerospace industry has matured, life cycle costs have decreased unevenly. While hardware development methods have improved significantly, system and software development efforts have been less successful. Specific phases of the product life cycle that entail a disproportionate amount of effort are: *system modelling and analysis*; *software development*; *hardware-software integration*; and *system maintenance*. In order to remain competitive in today's government and private sectors, system developers must employ a scheme which reduces these phases of the overall product life cycle. It is essential that such a scheme include an effective process, a flexible architecture, and the tools necessary to reduce life cycle costs. One such scheme has been successfully utilized on the Boeing 777 avionics controller and the International Space Station command and data handling system. These efforts have provided the foundation for a flexible scheme which may be tailored to the specific needs of a variety of development efforts.

Rapid Prototyping Spiral Development Process

A key element in improving system development time is a process which allows critical design decisions to be made early in the life-cycle. A spiral development process, particularly one which emphasizes modelling and prototyping, promotes early definition and validation of design decisions. A rapid prototyping spiral development process includes three major loops: *Virtual Development; Rapid Prototyping;* and *Flight Development.* Figure 1 illustrates this process.

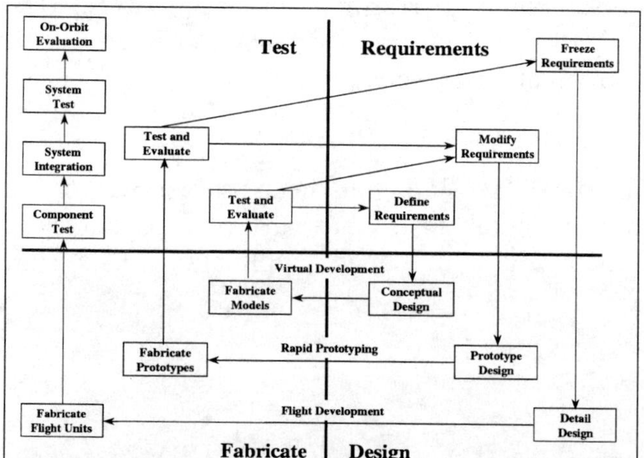

Figure 1 – Rapid Prototyping Spiral Development Process

0-7803-5086-3 /98/$10.00 ©1998 IEEE

Each of the three major loops of the spiral begins in the Requirements quadrant. In the virtual development loop, the requirements are initially defined, followed by the creation of a conceptual design of the target system. This design is translated into models using one of the available modelling tools such as MatLab or MATRIXx. The entire system is modelled including hardware components, software components, dynamics, environment, control algorithms, etc. The system is tested and validated in model form. Alternate architectures can be explored and trade studies conducted using the model system. The process either iterates within the Virtual Development loop or transitions to the Rapid Prototyping loop of the spiral.

In the rapid prototyping loop, the requirements are modified as needed. Based upon the models which were verified in the virtual development loop, the prototype components are designed and fabricated. The prototype system is tested and verified. Once again the process may iterate or it may proceed to the flight development loop of the spiral.

In the flight development loop, the requirements are frozen and the design finalized. The flight components are then fabricated, integrated, tested, and flown.

Throughout the spiral development process, changes may be evaluated using the system model rather than consuming scarce prototype or flight components. Likewise, maintenance problems can be rapidly isolated and potential solutions tested in the virtual environment.

Using this scheme allows the system design to be validated during the Virtual Development loop of the spiral, prior to expending time and resources on prototype or flight products. During the Rapid Prototyping and Flight Development loops of the spiral, the system is verified to function as predicted in the virtual environment. The virtual environment excels at proving the functional capabilities of the system, while the real-time interactions are proven in the prototyping environment.

Rapid Development Architecture

A flexible architecture which supports modelling, prototyping, and validating the proposed system is critical to realizing the potential life cycle cost savings inherent in the spiral development process. To accomplish these goals, a rapid development architecture requires three major components: a *Modelling System*; a *Simulation System*; and a *Target System*. Figure 2 illustrates the interaction between each of these components.

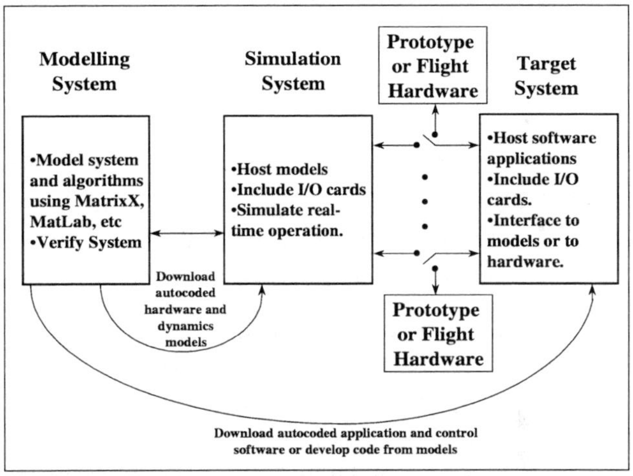

Figure 2 – Rapid Development Architecture

Modelling System

The modelling system is used to support the virtual development loop of the spiral development process. Using a tool such as MatLab or MATRIXx, a model of the system is developed, including the environment, sensors, actuators, control algorithms, etc. With the models in place, the system may be tested and the functions validated. Trade studies can be conducted rapidly by modifying the models and evaluating the results. Use of this scheme allows functional and interface requirements to be solidified and validated early in the life cycle. This is particularly useful for the software developer who normally must wait for the initial hardware design to be complete

0-7803-5086-3 /98/$10.00 ©1998 IEEE

before ascertaining detailed requirements relating to hardware drivers, test values, etc.

The simulations created in the modelling system are used throughout the development process to verify that the prototype and flight systems behave as predicted by the models. Since the overall system has been validated in the virtual environment, consistency between the prototype/flight and virtual systems produces a high degree of confidence in the functionality of the final product.

Simulation System

The simulation system consists of processing capability and I/O cards. Once the virtual system has been validated, the hardware, dynamics, and environmental models are downloaded into a processor in the simulation system. This may be accomplished either by autocoding the models or by creating original code based upon the model algorithms.

With the models downloaded into the simulation system, a virtual model of the hardware and environment is available for use in software development and hardware-software integration. This allows the testing of software implementation prior to the availability of prototype or flight hardware. The simulation system also consists of I/O cards that have the same type of interfaces as those used by the physical hardware components. This allows the developers to use the simulations and hardware components interchangeably and with equal confidence. Use of the prototype and flight hardware allows real-time interactions to be tested and validated.

In addition to the models and I/O cards, a third component is required for the simulation system. In simulation system the hardware and environment models must interface to the outside world at real-time rates. An interface between the I/O cards and the models is necessary to simulate real-time functions, including the capability to provide data at cyclic rates. This real-time I/O capability is generated using the processing capabilities of the simulation system.

Target System

The target system consists of the system processor and I/O cards. Following the validation of the virtual system, the control algorithms and application software models are downloaded into the target system. This may be accomplished either by autocoding the models or by creating original code based upon the model algorithms. The target system interfaces to either the simulation system or to the physical hardware devices using the I/O cards in the target system. These I/O cards may be COTS, prototype, or flight cards, but they must have the same interface capabilities that the target processor will use in the flight system to communicate to the hardware components.

As the prototype or flight hardware components become available, they replace the models in the simulations system, while the models of the environment may still be used to check out the overall product.

Benefits

Since the simulation system models the real-world hardware and environment, the target system may interface to the simulation system, the physical hardware devices, or a combination of the two. Models used in the virtual system may be autocoded into the simulation system to validate the software and system designs throughout the prototype and initial integration efforts. The use of this type of rapid development architecture reduces the life cycle costs in a number of key areas:

- Systems analysis using a model of the system simplifies trade studies and allows early validation of the system architecture.
- Validation of the system architecture allows early settling of functional and interface requirements.
- The software development and hardware-software integration efforts occur in parallel with the detailed design, parts procurement, and fabrication of the prototype and flight components.

- Simulations are created once and reused throughout the life cycle.
- Errors detected in-flight may be duplicated using the virtual system, allowing potential solutions to be easily developed and tested in that environment.

Real World Examples

The rapid development architecture has been used successfully on a diverse range of projects. Included are the Boeing 777 avionics system and the International Space Station command and data handling system. While the specific architecture on these two projects varied significantly, the development scheme remained constant.

Boeing 777

The avionics system on the Boeing 777 was designed by Honeywell using an integrated modular approach which included control processor modules and I/O modules. These modules, combined with a high reliability ARINC 653 backplane data bus, comprise the Integrated Modular Avionics (IMA) system. The control processor modules use an Advanced Micro Devices Am29050 microprocessor. IMA I/O cards support ARINC 429, ARINC 629, analog, and digital interfaces.

Figure 3 illustrates the rapid development architecture used in the development of the Boeing 777 avionics controller.

Modelling was performed in MATRIXx on Sun or DEC workstations. The simulation system consists of a VME-based workbench called a Validation Facility (VALFAC). The processor on the VALFAC is an R3000 with appropriate I/O interfaces and a simulation language interpreter used for test scripts. As a modular system, all or part of the IMA could be developed and integrated as the target system.

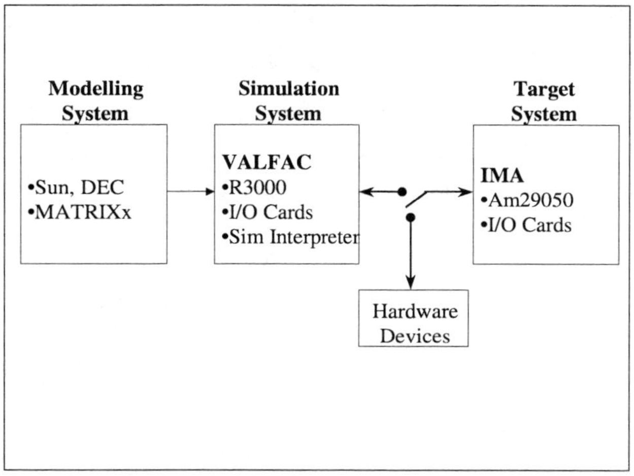

Figure 3 – Boeing 777 Avionics Controller Rapid Development Architecture

The software on the Boeing 777 avionics controller consisted of over 600,000 source lines of code (SLOC). The rapid development approach, along with the use of other tools supporting the IMA, resulted in a significant reduction in the software development effort. The average development time for the Boeing 777 avionics controller software was 1.8 hours per SLOC compared to an industry average[1] of 4.2 hours per SLOC. Honeywell continues to use this scheme to produce avionics controller software. The average development time for these software efforts of less than 35,000 SLOCs is 1.1 hours per SLOC.

International Space Station

The International Space Station command and data handling system is comprised of a number of Multiplexer-DeMultiplexers (MDMs) connected in a distributed processing scheme. The MDMs consist of an Intel 80386SX microprocessor and associated I/O cards, including MIL-STD-1553, analog, digital, solenoid driver interface, and serial interconnections.

Figure 4 illustrates the rapid development architecture used in the

[1] Per the SLIM® tool database produced by Quantitative Software Management®(QSM, Inc McLean, VA)

development of the International Space Station command and data handling system.

The modelling and simulations portion of this implementation of the rapid prototyping architecture are combined into an MDM Application Test Environment (MATE). Modelling is performed in MATRIXx on a Sun SPARCstation using a toolset produced by Integrated Systems, Inc. (ISI). Simulation is performed using multiple Baja R4700 processors and appropriate I/O interfaces. The MATE also includes a data logging processor for collection and post-processing of data points. Both the MDM and its functional equivalent unit have been used as the target system in this scheme.

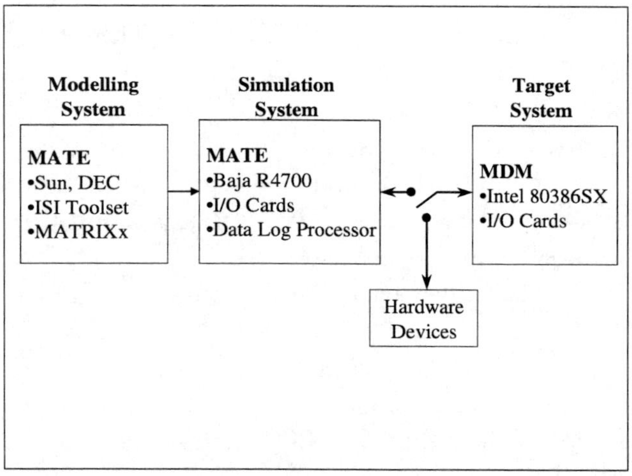

Figure 4 – International Space Station Command and Data Handling Rapid Development Architecture

Application software on the International Space Station is being developed by a number of companies throughout the world. No numbers have been collected to assess the savings being realized by this scheme. However, more than 60 MATE systems are in use throughout the United States and in Russia, Japan, Italy, etc. MATEs have been linked together using reflected memory to create the Software Verification Facility and NASA's Johnson Space Center which tests the entire distributed command and data handling application software functions.

Next Generation Architecture

While the VALFAC- and MATE-based rapid development architectures have proven to be highly effective, the technology is aging rapidly, and the size of the systems (each of which require multiple racks of electronics) is prohibitive for many environments. Based on improving technology and lessons learned from previous systems, Honeywell has produced the next generation rapid development architecture, dubbed the Software Development System (SDS).

Significantly smaller than its predecessors, the SDS consists of a Windows NT system, a VME chassis, and a target system. While its predecessors were designed with a specific target system in mind, the SDS has been designed to be flexible enough to support a variety of target systems and applications. Figure 5 illustrates the components of the SDS.

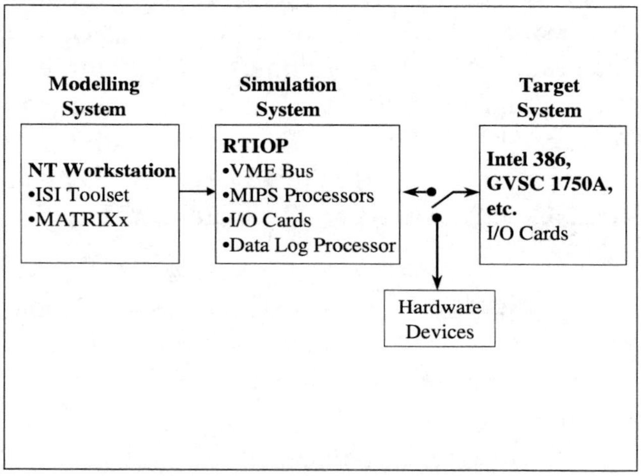

Figure 5 – Software Development System Architecture

The SDS *modelling system* consists of an NT Workstation running the ISI toolset. ISI tools center around the MATRIXx modelling scheme and include utilities to support virtual system definition, test, and validation. Once models of operational environment, vehicle components, and vehicle dynamics are created, they may be modified to allow evaluation of alternate architectures, or for use on future designs.

The SDS *simulation system*, the Real-Time I/O Processor (RTIOP), consists of a single VME chassis containing two MIPS processors, a data logging processor, and a number of I/O cards. One MIPS processor runs the autocoded models of the hardware, environment, and dynamics. The second MIPS processor interfaces between the models and the I/O cards to simulate real-time data rates. An optional data logging processor card uses a Motorola 68040 processor to store data points for post-processing. Honeywell has developed real-time simulations in support of analog, digital, SCSI, serial, and Mil-Std-1553 I/O protocols. The RTIOP uses a VME backplane allowing COTS I/O cards to be installed into the system easily. This capability permits the number and type of I/O cards to be changed with little or no impact to the SDS architecture and tools.

The SDS architecture is flexible enough to allow the *target system* to be whatever processor is used in the system under development. SDS has been verified using the Intel 386SX processor and the GVSC 1750A processor. However, the choice of target systems is largely independent of the remainder of the SDS architecture. If autocoding is used to develop software applications, a compiler which runs on an NT host and generates code for the target processor is required. Additionally, the target system must include the I/O cards which match those used by the physical hardware components. Other than autocode and I/O card interfaces, the target system is independent of the remainder of the SDS.

As the next generation rapid prototyping architecture, SDS provides the best of previous architectures as well as additions derived from lessons learned and user inputs. Some of the features which have been found to enhance the efficiency of the rapid development environment, and which are a part of the SDS design, include:

- Use of a powerful toolset such as ISI simplifies the development and validation of a system in a virtual environment.
- Models and simulations which are developed may be reused by modifying specific parameters to match the system under development.
- Use of a common backplane such as VME in the simulation system allows the use of COTS I/O cards. This supports the rapid development of a simulation environment for use in the testing and integration of software and interfaces.
- Creating software for the simulation system which supports common I/O protocols and simulates real-time operations allows the target system to use models and hardware interchangeably.
- The independence of the target processor allows the developer the flexibility to use the same development environment for multiple systems. This decreases the learning curve inherent in typical development systems which are closely tied to a specific target processor.

Conclusion

In today's highly competitive aerospace industry, both government contracts and the commercial market require a product which can be produced quickly and efficiently. In order to meet the demands of this environment, a design scheme is required which includes an efficient process, a flexible development architecture, and the tools necessary to reduce life cycle costs. One such scheme includes a spiral development process which emphasizes modelling and prototyping, a rapid development architecture such as the Honeywell Software Development System, and powerful COTS modelling tools. This scheme has been demonstrated to be highly effective in reducing the overall life cycle costs, particularly in the areas of systems analysis, software development, and hardware/software integration.

DEEP SPACE ONE INTEGRATION & TEST CHALLENGES: GETTING TO THE LAUNCH PAD IN THE FASTER, BETTER, CHEAPER WORLD

Paula J. Pingree, JPL, Pasadena,CA

Abstract

This paper describes the integration and test challenges of verifying and validating the Avionics hardware and flight software which have been experienced in meeting the New Millennium Program Deep Space One (DS1) project's faster, better, cheaper requirements.

This paper gives a high level overview of the development and application of the two flight system testbeds (DS1 Testbed and Hotbench), the testbed activities that have supported the DS1 Spacecraft Assembly, Test and Launch Operations (ATLO), and the testing performed to prepare for and support post-launch mission operations.

In the "Faster, Better, Cheaper" environment of the New Millennium Program, DS1 Integration & Test (I&T) has defined new methods and decision criteria to meet our requirements and goals while assessing and minimizing the risk in the paths we have taken. Our successes and failures are largely yet to be seen as we approach our July 1 launch date. *[NOTE: Since submittal of the abstract for this paper, the DS1 Launch date was changed to October 15, 1998.]*

This paper describes the challenges that have been faced and some that have been overcome during the DS1 I&T phase. It will present the risks that have been accepted in our attempts to test completely the DS1 Avionics system in preparation for launch and mission operations. Our early lessons learned will also be described.

Overview

New Millennium Program/DS1 Project

The New Millennium Program (NMP) is comprised of a series of advanced technology validation missions defined and implemented as an integrated program for NASA by JPL. Deep Space One (DS1) is the first NMP deep space mission. Like all NMP missions, the main objective of DS1 is to space-validate a suite of advanced technologies - the payload for these missions. The validation of these technologies gives promise to enabling 21st century space science missions with low life-cycle mission cost. There are twelve such technologies for DS1, two of the most prominent being Ion Propulsion and Autonomous Navigation. A mission profile was chosen which serves as an appropriate test track for the selected complement of advanced technologies. The baseline mission profile includes a near-Earth asteroid encounter with the possibility of an extended mission providing a comet encounter. The nominal launch date for the DS1 spacecraft is October 15, 1998.

The Launch Date Slip

A JPL-internal Risk Review of the DS1 Project was held in March 1998 which led to the decision to postpone the launch of DS1 until October. As a result of the slip in the launch date, May 1, 1998 was identified as the scheduled goal for the completion of several key tasks involving the DS1 Avionics Team including:

0-7803-5086-3 /98/$10.00 ©1998 IEEE

1. DS1 Project Budget Replan
2. Release of the DS1 I&T Test Plan
3. Delivery of the Launch Load Flight Software (FSW) version to the DS1 Testbed
4. Integrated Electronics Module (IEM) hardware stability - both Flight (FLT) and Engineering Model (EM) versions

All goals were met by May 1, 1998 with the exception of achieving IEM hardware stability. This area is addressed further in the section titled "Avionics Hardware I&T." Also, as a result of the Launch date slip, the FSW capability was descoped to include only that capability which is essential to launch, achieve attitude stability and begin thrusting with the Ion Propulsion System (IPS). Deferred FSW capabilities, primarily related to new technology validation include, the Low Power Electronics/Power Activation & Switching Module (LPE/PASM), data filtering for the Diagnostic Sensor Electronics Unit (DSEU), the Multifunctional Structure (MFS), and the Remote Agent Experiment (RAX).

Testbed Development

Background

The DS1 Testbed and the Hotbench, together, are an integral part of the overall spacecraft I&T effort. The Testbeds are primarily electronics and software laboratories operating nearly interchangeably and equally sharing the load of the test schedule effort. The DS1 Testbed was delivered by JPL's industry partner, Spectrum Astro, Inc. (SAI), to JPL in April 1997, which kicked off the start of DS1 Avionics I&T, just 14 months prior to the original launch date.

In August of 1997, the spacecraft "Partial Bus" was delivered from SAI to JPL which began JPL's Assembly, Test and Launch Operations (ATLO) effort. The Partial Bus was delivered with EM versions of the IEM boards. A second testbed, called Hotbench, was used at SAI to complete flight board testing prior to delivery of those boards to the spacecraft. Hotbench was delivered to JPL at the end of January 1998, when Flight Board acceptance testing was complete.

Description of Testbed Architecture

The heart of the DS1 Avionics System is the Integrated Electronics Module (IEM). The DS1 Testbed and Hotbench IEMs each contain a VME backplane and twelve integrated avionics boards. The VME master and slot one controller in the system is the radiation-hardened RISC6000 (RAD6K) processor which hosts the spacecraft Flight Software (FSW). The other avionics boards integrated in the IEM are:

1) BTF - Bus Transfer Function. Contains a 1750A processor in which the bus controller software for the 1553B interfaces resides.
2) ACE - Attitude Control Electronics. Contains unique interfaces for the Inertial Measurement Unit (IMU), Sun Sensor Assembly (SSA) and another DS1 new technology, MICAS - Miniature Integrated Camera and Spectrometer.
3) UDL - Uplink/Downlink Board. Provides the command and telemetry interface for the spacecraft.
4) GPB - General Purpose Board. Provides various timing functions including the watch-dog timer registers for the RAD6K and BTF processors.
5) GDE - Gimbal Drive Electronics. Provides the 1553B interface to the solar array gimbals and the gimbal assembly of the ion engine.
6) PCE - Propulsion Control Electronics.
7) PDE - Propulsion Drive Electronics. Together with the PCE, provides the control and drive interface for the reaction control system (i.e., hydrazine thrusters).

8) PCD - Pulse-Code Modulated, Digital board. Collects and formats the spacecraft digital signals.
9) PCA - Pulse-Code Modulated, Analog board. Collects and formats the spacecraft analog signals.
10) IPC - IEM Power Converter. Provides power conversion for the VME backplane, IMU and several IEM boards.
11) EEPROM - Contains FSW loads and parameters, Navigation files, Fault Protection scripts and stored sequences.

The DS1 Testbed and Hotbench systems are each comprised of the IEM, the Electrical Ground Support Equipment (EGSE) Signal Rack, three Sun Workstations, an NT computer and an external power supply for the IPC. Figure 1, below, illustrates the block diagram components and their interfaces.

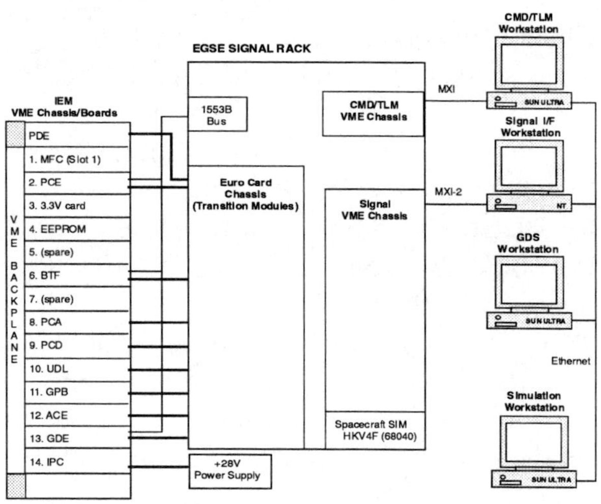

Figure 1. DS1 Testbed/Hotbench Block Diagram

Testbed engineers execute their tests of the FSW via the three Sun workstations. The JPL Multimission Ground Data System (GDS) operates on one workstation. From that workstation spacecraft commands are issued and telemetry is monitored. The telemetry displays are the same as those used in ATLO for spacecraft testing and are the same as those that will be used by the Mission Operations team during flight operations.

The EGSE Host machine provides the command and telemetry interface between the GDS workstation and the EGSE Signal Rack hardware. The third Sun workstation is used to run the Simulation software which closes the loop with the IEM hardware by simulating the external hardware and system dynamics.

Testbed Applications and Test Scope

The key responsibilities of the DS1 Testbed Team are to 1) ensure the health and safety of personnel and hardware, 2) understand and follow hardware handling and test constraints, 3) perform electrical integration and test of avionics hardware and electrical interface verification of "external" hardware, 4) perform verification and validation of flight software, 5) perform System-level tests and provide Spacecraft I&T support, 6) support flight operations training, 7) provide anomaly resolution services.

The test scope of the DS1 Testbed/Hotbench includes testing of the FSW functionality per Level 3 requirements, nominal interface operation between the IEM hardware, the EGSE and the Simulation, commands and command constraints, telemetry and visibility, sequence validation, and FSW verification. These are the capabilities of the testbed environment.

The Testbeds are limited by the lack of external hardware, and rely heavily on the simulation models and dynamics to perform closed-loop testing. Whenever possible, real hardware is substituted for simulation models in order to increase the fidelity of a test. A semi-permanent resident in Hotbench is the DCIU (Digital Control Interface Unit), the 1553 interface to the Ion Propulsion System, and its simulators for the Xenon Feed System (XFS) and Power Processing Unit (PPU). The

MICAS breadboard is regularly available in the DS1 Testbed to perform imaging and test its interface to FSW components such as the "Photo Opportunity" process of Autonomous Navigation.

Testbed Activities

Avionics Hardware I&T

Upon delivery of the DS1 Testbed, the first level of delivered FSW, the hardware managers, were used to test individual board interfaces in an open-loop configuration. Although this testing was far from flight-like, it provided a learning platform for the Testbed team members to familiarize themselves with the hardware, software and system operations.

Approximately 8 months later, the FSW was sufficiently developed to begin closed-loop testing. This testing uncovered several IEM board problems requiring modification to both the EM and FLT models. Every Avionics board in the IEM required some level of modification after delivery and integration into the testbed systems. Lacking Flight Spare hardware, this level of rework resulted in a "musical boards" scenario of trying to keep the spacecraft and both Testbed's operational, if possible. Despite a very focused effort, as of the writing of this paper, IEM board stability has not yet been achieved. Many modifications have been required due to hardware design flaws, but several EM boards remain to be reworked to remedy workmanship problems.

Other external hardware that has been tested in the DS1 Testbed include the EM Stellar Reference Unit, the breadboard LPE, the Flight Spare Inertial Measurement Unit and the EM Solar Array Drive Assembly.

FSW Verification & Validation

The first FSW version containing fully-integrated Attitude Control Subsystem (ACS), Navigation (NAV) and Fault Protection (FP) functionality was delivered to the testbeds on February 27, 1998. A FSW version without ACS, NAV & FP integrated together was tested for approximately three weeks prior to this integrated delivery. This "preview" allowed for the integrated version to be delivered from the Testbed to ATLO on March 3, 1998, in time for scheduled Spacecraft Systems tests. Full FSW Launch +30 day capability was delivered to ATLO on May 4, 1998. FSW Verification and Validation has continued to progress well in both the spacecraft and testbed systems.

Spacecraft/ATLO Support

The DS1 Avionics Team supports spacecraft test activities by providing procedures, supporting test execution, performing anomaly resolution, etc. The DS1 Testbed Team, an Avionics sub-team, tests each FSW version and delivers it to the spacecraft. As indicated in the previous section, new FSW versions do not spend much time in the Testbed before delivery to ATLO. The DS1 Avionics team has gained quick confidence in their product and processes by supporting these deliveries. In contrast to a common strategy on other flight projects, not all tests that are executed on the spacecraft are first executed on the Testbed. This is due to the short ATLO schedule and the Project's desire to get as much spacecraft test time as possible with the latest and greatest version of FSW.

The DS1 Testbed is also used to verify all IEM hardware prior to installation on the spacecraft, whether it be the first delivery of that item or a board that has been modified.

Flight Operations Preparation

As of the writing of this paper, approximately three months prior to launch, the Testbed team has had little direct involvement in the preparation for flight operations. The Testbed team is not available to support this activity due to the aggressive schedule to complete FSW system-level testing. This is an accepted risk to the DS1 Project. The DS1 Testbed Team will transition from the Avionics Team to the Flight Engineering Team at Launch. Three members of the DS1 Testbed Team will go on to support mission operations after Launch +30 days.

Test Methods and Reporting

The DS1 I&T Test Plan

In response to the Risk Review Board's indication that the Testbed Test Plan was not adequate, the Avionics Team spent approximately 4-6 weeks developing a comprehensive test plan for each area of the FSW functionality. The DS1 I&T Test Plan, in final form, was released on May 1, 1998. Cognizant engineers from the DS1 Avionics Team provided test case definitions and estimated hours to perform each test. The DS1 I&T Test Plan covers all areas of FSW capability including ACS, NAV, Data Management, Fault Protection, IPS, MICAS, Telecom, Thermal Control, DSEU and Power. The Test Plan also includes test cases for Mission Activity sequences and End-to-End Information System (EEIS) demonstrations.

The DS1 I&T Test Plan was converted to database format and is actively used by the test engineers to track the testbed system time utilized and the percentage complete of each test case that is executed. On a weekly basis, the Testbed Testing Summary is generated which charts the actual test hours and "Earned Value" against the test plan estimates. This method of tracking our testing progress is described in more detail in the "Test Case Database" section, below.

Test Procedures

From each of the Test Cases defined in the DS1 I&T Test Plan, test procedures are developed by Testbed engineers working in each of the test areas. The required inputs and success criteria of each test case are defined in the test procedure. Based on ground commands and telemetry monitoring, the test procedures are executed to verify the expected results. Non-real-time analysis is also an important part of claiming success in a test.

Test Reporting

Daily Log Entries

A daily log database originally created for Cassini ATLO was modified for DS1 to allow electronic input of daily test activity via the Web. This database is actively used to chronicle the DS1 Testbed and Hotbench tests. Individual test engineers write their daily log entries identifying the test goals, accomplishments and problems encountered. In the daily status meeting the Daily Logs are reviewed, status is reported, and actions are assigned. The daily logs can be queried via the Web which allows for quick and easy recall to the details of every test report. This has been a critical tool for reporting the test activity of the DS1 Avionics Team.

Problem Reports

The DS1 Testbed Problem Report (PR) database is an internal project database contained on a PC web server. The DS1 I&T homepage provides links to enter, search/modify, and print problem reports from the database. It is the philosophy of DS1 I&T to report all known problems found in test and it is the individual test engineer's responsibility to do so. A weekly PR meeting is held to review new PRs and to assess open PR status and due dates for action. A smaller group periodically reviews all PRs submitted for closure and updates their status to closed, if appropriate.

It is recognized and accepted by DS1 that using a project-internal problem reporting system, as opposed to JPL's institutional PR system, poses a risk due to potential lack of visibility and rigor in the assessment of each problem. The internal PR database system was first used successfully on the Mars Pathfinder project and was adapted from it's example for DS1.

As of July 29, 1998, a total of 496 Problem Reports have been written in the DS1 Testbed and Hotbench. Of those, 27 are open, 30 are marked for re-test, 6 are deferred to post-launch, 92 are submitted for closure and 341 are closed.

Test Case Database (Tracking our "Earned Value")

Accompanying the release of the DS1 I&T Test Plan was the Test Case Database. Created in Excel, each test case area (e.g., ACS, Navigation, Fault Protection, etc.) was translated into a worksheet with columns added to track the Actual Hours spent on the testbed system for each test case, the test engineers estimation of the percent complete of each test and the calculated "Earned Value" of each test based on the equation:

Earned Value = Estimated Hours x Percent Complete

From the database, a weekly summary of Earned Value and Actual Hours is compiled and presented in chart form as illustrated in Figures 2 & 3 provided at the end of the paper.

Figure 2 is the combined result for all areas of testing. It shows an estimated goal of 1600 test hours and the allocated plan across five months in order to achieve the goal. In most test areas a "times-two" factor was applied to the original estimate of hours and is represented in the curve. The Combined Testing chart shows the Testbed Earned Value to be on track and the actual test hours spent around 1.5x the original estimate.

Figure 3 shows the test summary for ACS. After several weeks of tracking the progress in this area it was observed that the original 2x estimate was inadequate. For several weeks, in order to complete tests with full earned value, nearly 3x the original estimate of hours was required. The Earned Value of the ACS tests were also falling short of the goal during this period.

In response to this indication, in mid-June, the ACS Test Lead organized a focused team of ACS test engineers combining members of the Testbed, Simulation and FSW development teams. Within two weeks, the actual hours spent and Earned Values converged toward the curve of the goal. This is a very good example of how to use an implemented tracking system, recognize a problem area, make adjustments to correct the problem, and see a quick result.

Test Scheduling & Priorities

In May 1998, Meeting Maker schedules were implemented to register requests for DS1 Testbed and Hotbench system time. DS1 Team members can view the current testbed schedules from their office PC or Mac. In the daily status meeting, the DS1 Testbed and Hotbench two-week schedules are quickly reviewed and short-term test schedule priorities are established. This system was slow to catch on, but is continuing to grow in its use and value.

Current Status (as of July 29, 1998)

Since delivery of the Launch Load FSW version to the DS1 Testbed on May 1, there have been five (5) subsequent releases to ATLO which have primarily fixed FSW bugs found in test. Another FSW version is scheduled to be delivered to the spacecraft on August 3, prior to its shipment to the Cape on August 14. The "final" FSW version is scheduled for delivery to the spacecraft on August 24.

DS1 Testbed/Hotbench verification and validation of the FSW capabilities is scheduled to be complete by September 21, 1998. The DS1 I&T Test Plan is being re-planned to more accurately identify and schedule the remaining required tests and will be released on August 3. Both testbeds will be utilized extensively for launch and operations readiness testing in the final weeks before launch. Indications from our test tracking summary are that we will meet that FSW testing goal and be ready to support the October 15th launch.

Avionics I&T Lessons Learned

Colocation improves communication and enhances test productivity. The Mars Pathfinder Project advocated this all along. The DS1 Avionics Team and the Testbeds were not colocated until May 1998, where in just weeks the improvement in team interaction was observable.

Partnering with industry does not preclude careful review of deliverable designs, acceptance test procedures and test results. It was originally planned by the DS1 Avionics Team to carefully review the Spectrum Astro deliverable products of the IEM hardware including board-level schematics, acceptance test procedures and test results to validate the hardware designs. As time did not permit, and with a philosophy of partnering with industry allowing for built-in trust to the work performed, this review was not completed by the DS1 Avionics team. This proved in hindsight to be regrettable. By the time the system-related hardware problems were beginning to be discovered in closed-loop system-level tests, DS1 was 5 months from the original launch date.

A lesson still to be learned is how to do Faster, Better, Cheaper projects without burning out the team. On the "Faster, Better, Cheaper" DS1 project, many individuals consistently work at a high level of intensity, working long hours over several consecutive weeks or months.

Conclusion

The DS1 Project has many dedicated people who under the stress of what remains to be completed have already achieved many significant accomplishments of schedule, budget and technical merit. The DS1 Avionics Team, working in a very dynamic hardware and software environment, continues to meet the testing challenges of a "Faster, Better, Cheaper" project. Defining and implementing new methods for reporting and tracking test progress has helped the DS1 Avionics Team maintain confidence and remain on track for the October 15 launch date.

References

"Deep Space One Basebody Control: Validating the Data Path From Computer to Thruster," Paula J. Pingree, Ching F. Leang, and Ralph R. Basilio, 1997 DASC.

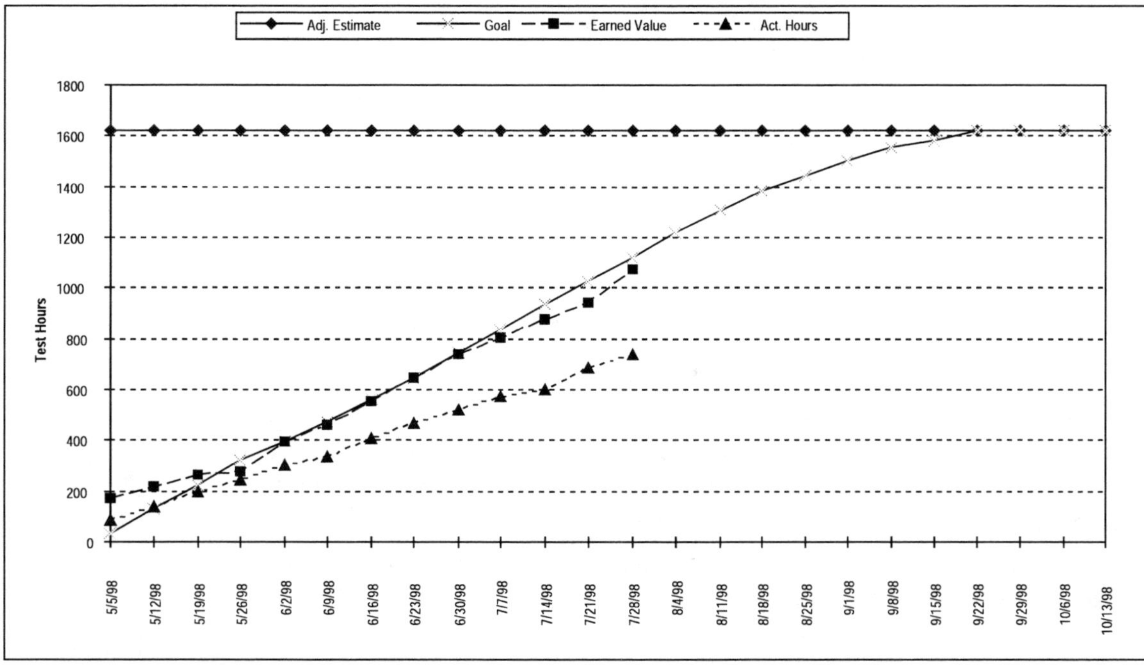

Combined Testing	Date	Adj. Estimate	Goal	Earned Value	Act. Hours	% Complete
	7/28/98	1640	1124	1071.82	738.5	65%

Figure 2. Test Database Summary Chart for Combined Testing

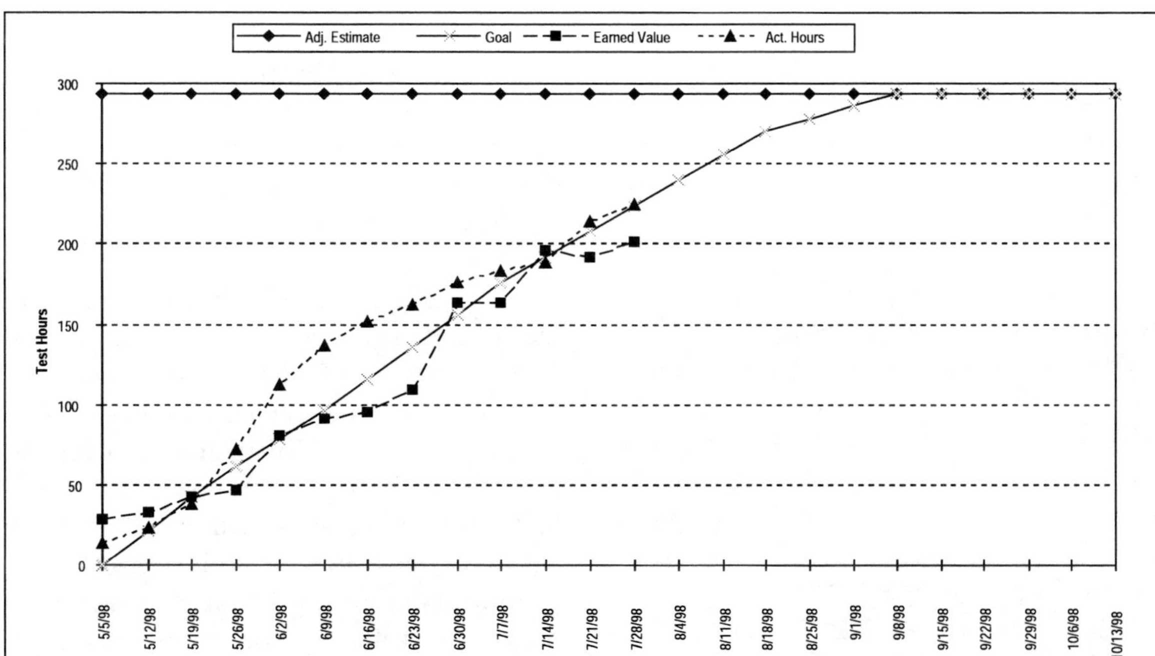

ACS Testing	Date	Adj. Estimate	Goal	Earned Value	Act. Hours	% Complete
	7/28/98	302	224	201.3	224.5	67%

Figure 3. Test Database Summary Chart for ACS Testing

From the Sun to Pluto

by Karla B. Clark
Jet Propulsion Laboratory
California Institute of Technology

Abstract

From the Sun to Pluto, all by way of Jupiter. The Outer Planets/Solar Probe (OP/SP) Project first three missions will all visit Jupiter before reaching their destinations of the Sun, the Jupiter moon Europa and the Pluto/Charon system. Using a unique sun shield/high gain antenna Solar Probe will twice encounter the Sun at a distance of three solar radii. Solar Probe will make local measurements of the birth of the solar wind and be able to image surface features as small as 60 kilometers across. The Europa Orbiter mission will send a small spacecraft into the radiation intensive environment around Jupiter and its moon Europa to determine if a water ocean exists below the moon's icy surface. Pluto is the last of the known planets in our solar system left to be locally explored. The Pluto/Kuiper Express will complete this reconnaissance by sending a spacecraft on its long journey to the outer reaches of the solar system.

These technically challenging missions will all require significant technological advances to make them achievable at significantly lower costs when compared to similar previous missions. A core piece of technology which will be common amongst the three missions will be the avionics system being developed under NASA's Advanced Deep Space Systems Development Program (ADSSDP). One of the first deliveries from ADSSDP is the X2000 avionics architecture which includes the main engineering computing system, the attitude control sensors, the power electronics and the core software used to operate the spacecraft in ground test and in flight. The avionics hardware technology utilizes integration of functions among a set of multichip modules with standard interfaces to achieve lower production costs, power and mass. Additionally, the packaging density improvements allow the reduction of the shielding mass required for the survival of the Europa Orbiter electronics within the intense Jupiter radiation belts.

X2000 INTRODUCTION

NASA's Advanced Deep Space Systems Development Program (ADSSDP) at the Jet Propulsion Laboratory is developing advanced technology components and systems to enable low-cost, yet ambitious, missions to the outer planets, Mars, and other destinations in the early 21st century. One of the program's products will be a series of engineering development multimission spacecraft systems which we refer to as the X2000 Buses. The X2000 engineering model spacecraft system are to qualify the design and to enable flight units to be procured at a low recurring cost.

The qualification of the First Delivery of the X2000 Bus is currently planned for February 2001. Some of the software development, however, is expected to continue beyond that time. The First Delivery Bus is being designed to perform both the proposed Europa Orbiter mission and the Pluto/Kuiper Express mission with a common build-to-print spacecraft avionics bus. Mission-unique elements, such as science, probes, propulsion modules, and radiation shielding, are being designed modularly with standard, common interfaces to minimize cost.

The avionics, software, and engineering sensors in the First Delivery are also being designed to fulfill the needs of other missions,

0-7803-5086-3 /98/$10.00 ©1998 IEEE

including New Millennium Deep Space 4/Champollion, Solar Probe, and select spacecraft in the Mars Program mission set. X2000 components and systems would also be available for Discovery missions and Earth orbiting scientific and commercial spacecraft as high-performance components with very low mass and low recurring cost.

OP/SP Mission Set Summary

Europa

Jupiter's moon Europa fascinated scientists after the images from the Voyager Mission indicated that the surface of Europa was unusually smooth and lacked visible craters, suggesting that it was very young. Combined with information about its bulk composition, which indicated it had a veneer of water ice, and the knowledge that Europa experienced strong, heat-inducing tides, this finding led to the tantalizing suggestion that a water ocean might be present below the moon's surface. The data were of insufficient resolution to allow much more than theoretical speculation, and the Galileo observations were awaited with eagerness.

The Galileo images did not disappoint. In a June 1996 image, strong evidence appeared for surface cracking into ice floes, reinforcing a Voyager interpretation. Then, the close Europa flybys found the first direct evidence of cryo-volcanism on a Jovian moon. These discoveries were followed by evidence of what appear to be icebergs now apparently frozen into place, but which seem to have been floating on some substrate that is difficult to conceive of as anything but liquid. But while increasingly compelling, there was as yet no unequivocal determination of the existence of a global ocean on Europa or its size.

About the time of these discoveries, the Jet Propulsion Laboratory began advanced studies of

Figure 1: Europa Orbiter Spacecraft
Error! Not a valid link.

a mission called the Europa Orbiter which would determine if an ocean is present and the thickness of the overlaying ice layer. If a thin ice layer could be confirmed, future spacecraft could be designed to penetrate the ice and look beneath its surface. Data from Galileo and IUE indicate that materials other than ice, including sulphur, are present on the surface of Europa. Over time, these materials could conceivably be churned into the water/ice below. It is this material (probably including organics), the presence of water, and the tidally driven internal heat source that intrigue scientists who speculate on the possibility of life in the oceans of Europa.

The current design for the Europa Orbiter (see Figure 1) would take about 3 to 4 years to reach the Jovian system and an additional 1½ to 2 years to reach orbit around Europa. The mission consists of one month in orbit around the moon taking data and relaying the data back to Earth. The mission duration is driven primarily by the intense total ionizing dose radiation levels present in the Jovian system. Even with the short lifetime, the mission is scientifically engaging and rich and would lay the groundwork for future missions to Europa and other Jovian moons such as Io.

For the Europa mission, the X2000 First Delivery avionics bus will be interfaced with a 750 kg wet mass dual-mode bipropellant propulsion module. This will provide for attitude control, Trajectory Correction Maneuvers (TCM's), Jupiter Orbit Insertion (JOI), and Europa Orbit Insertion.

One proposed Europa science suite features a radar sounder to remotely determine the depth of the surface ice and determine if a liquid water ocean exists beneath it. Visible and thermal imaging is also included to map the surface and determine composition and structure. Accurate tracking of the spacecraft orbit, in conjunction with a laser altimeter, will be used to determine the tidal flexing of Europa and provide key information on the internal structure and nature of possible subsurface oceans.

Pluto

Pluto is the only planet in our Solar System which has not been explored by a spacecraft. Pluto and its large moon Charon form a binary system that has an orbit varying dramatically in distance from the Sun. Currently, Pluto's orbit is within the orbit of Neptune, but 124 years hence it will be outside Neptune's orbit at an aphelion of 49 AU. This variation in distance will cause Pluto's atmosphere to sublimate (or condense) so that it will become increasingly more difficult to detect or utilize for science or aerobraking.

Even with the long flight times of 9 to 10 years, the Pluto/Kuiper Express mission is designed to reach Pluto/Charon while there is still

Figure 2: Pluto/Kuiper Express Spacecraft
Error! Not a valid link.

a significant atmosphere. It is unknown when, or even if, the atmosphere will "collapse". It may just gradually diminish. There are conflicting opinions, all based on the same very limited data. Characterization of the Pluto/Charon system will help answer questions about this unusual binary system and will contribute to our understanding of the formation of our Solar System.

After a fast reconnaissance flyby of the binary system, the trajectory will be altered to fly by a member of a group of objects referred to as the Kuiper Disk Objects. Kuiper Objects, predicted in the late 1940s by Edgeworth and Kuiper, were only discovered in the early 1990s yet now there are several dozen of these objects catalogued. These small objects form a disk around our solar system and are believed to be remnants of the solar system's formation and to be the primary source of short period comets. By studying at least one of these objects, scientists hope to learn more about the possible origin of the volatiles that form the Earth's atmosphere and oceans.

Figure 3: Solar Probe Encounter Configuration

Error! Not a valid link.

The extreme distance from the Sun, long lifetime, and fast flyby speed make this a very challenging engineering mission, as well as an exciting science mission. The Pluto/Kuiper Express spacecraft is depicted in Figure 2.

For Pluto/Kuiper Express, the X2000 First Delivery bus will be mated with a 35 kg wet mass hydrazine monopropellant propulsion module. The propulsion module will provide for attitude control, TCMs, a probe deflection maneuver at Pluto, and retargeting to Kuiper disk objects. The propulsion module is also designed to structurally support an experimental microspacecraft probe that would be separated from the main spacecraft and targeted to conduct atmospheric science.

Solar Probe

Solar Probe is an exploratory mission to our star which gives us life and whose effects on the earth and solar system are profound. We are only beginning to understand the relationship between the sun, its atmosphere (the corona), and the solar effects on the earth. The recent observatory missions (YOHKOH and SOHO) have given us new data to answer old questions and create new questions that only the Solar Probe can answer. The mission is designed to take scientific instruments into the solar atmosphere to within 3 solar radii (2.1 Gm) of the Sun's atmosphere where they will make measurements to determine what causes the heating of coronal particles (to well over a million degrees), as well as what are the sources and acceleration mechanisms in the solar winds. The low-altitude passes of the Solar Probe spacecraft over the polar regions will allow imaging that has here-to-fore been impossible and at perspectives that will never be attained from near-Earth observatories.

This close approach to the Sun requires that several technical challenges be undertaken. Materials in the exposed portions of the spacecraft must survive the extreme temperatures during the encounter with the Sun. An integral heat shield and high-gain antenna design has been developed to shield the spacecraft from the extreme heating while providing high-rate, real-time communication with the Earth during the critical solar encounter period (see Figure 3).

This innovative concept is accomplished through the unique trajectory and a new application of high temperature materials technologies. The trajectory uses a Jupiter gravity assist maneuver to provide the unique quadrature geometry at perihelion that enables the shield/antenna combination.

A complex set of power sources, including multiple solar arrays and batteries, is required to survive the extremes in solar radiance between the dim cold of a Jupiter

Figure 4: Europa Orbiter Spacecraft BusError! Not a valid link.

gravity assist flyby and the intense heating of a close encounter with the Sun.

Because of the unique shield geometry and thermal requirements of the mission, the X2000 First Delivery bus cannot be used “as is”. The core avionics will be similar, if not identical to hose used on Pluto/Kuiper Express but physically there has to be some tailoring. The slices will be mounted on custom designed panels designed specially for Solar Probe. Most of the core software from the First Delivery bus will also be utilized.

The propulsion requirements for Solar Probe are very similar to Pluto Express; therefore, Solar Probe may use the same tankage and components at a recurring cost. The Propulsion Module integration structure for Solar Probe will be different from Pluto Express, however, due to the unique shield geometry and thermal constraints.

X2000 First Delivery Bus

A concept of the First Delivery spacecraft bus configured for the Europa Orbiter is shown in Figure 4.

Microelectronics

To accommodate the diverse requirements of the multiple missions identified above, the architecture of the X2000 First Delivery microelectronics must be scaleable. This requirement has a number of implications, one of which is the need for standard interfaces. This will allow modules to be added or removed from the system to meet the requirements (performance, mass, power, volume) of a specific mission.

The microelectronics are a highly integrated stack of standard sized (10 x 10 cm) Multi-Chip Modules (MCMs) and Chip-On-Board (COB) subassemblies. Both of these technologies utilize unpackaged bare microchip die directly bonded onto a standard size substrate. The substrate contains all the printed circuit traces required for electrically connecting the die together. At the periphery of each of these subassemblies are all of the electrical terminals required to interface the subassembly with other elements of the spacecraft. These subassemblies are often referred to as “slices.”

The slices are stacked on top of one another using a new packaging technology that allows different combinations of MCMs and COBs, or even conventional PC boards, to be assembled together in any order, with the electrical terminals on the periphery in contact with those of the adjacent slices (Figure 5).

There are standard terminal locations for data buses, power buses, and other signals, so that these can be reconfigured in a plug and play fashion. To change between single-string and dual-string configurations requires only the removal or addition of slices. In the same manner, more memory or processors can be added, or individual slices can be upgraded to the next generation simply by replacing those slices with the new ones, leaving the remainder of the slices in the stack unchanged.

The performance of the system must not be affected by the addition or removal of modules. As the number of modules increase, the system should not degrade. This is achieved with a multi-master bus, avoiding the performance bottleneck that might occur with a dedicated master. Each module in this architecture is individually coupled to the bus and has its own resources.

In the X2000 architecture, the role of the flight computers is flexible and scaleable, allowing the number of processors to be increased or decreased depending on mission requirements. It also allows the system to be more fault tolerant, so that if one computer fails, another can take over its tasks.

Figure 6 shows a typical configuration which could be utilized for the Europa and Pluto missions. Although there are 39 total slices in this configuration, they are built up from only 10 different types of slices that must be designed and qualified. The radiation shielding is also modular and tailored to the requirements of each mission.

Figure 6: Dual-String Micro-Electronics Stacks

Error! Not a valid link.

Software and Autonomy

From the software's point of view, the microelectronics stack resembles a local area network, with multiple processors running standard operating systems and languages that communicate over high bandwidth interconnections. Each processor (flight computers and microcontrollers) will run a commercial real-time network-aware operating system. The software will be distributed, multi-threaded, and object oriented.

Telecommunications

The communication system for each spacecraft consists of a high gain antenna, solid state power amplifier and Space Transponding Modem. The communication is performed via an X-band link between the spacecraft and the Deep Space Network (DSN). The unique trajectory of the Solar Probe will allow real-time data transfer during first sun encounter. The other two missions, and the second pass of the Solar Probe, will utilize the on-board memory for data storage until the data can be relayed back to Earth.

Sensors and Attitude Control

The temperature sensors are plugged into the spacecraft using a standard I²C low-to-

Figure 5: X2000 Micro-Electronics Packaging

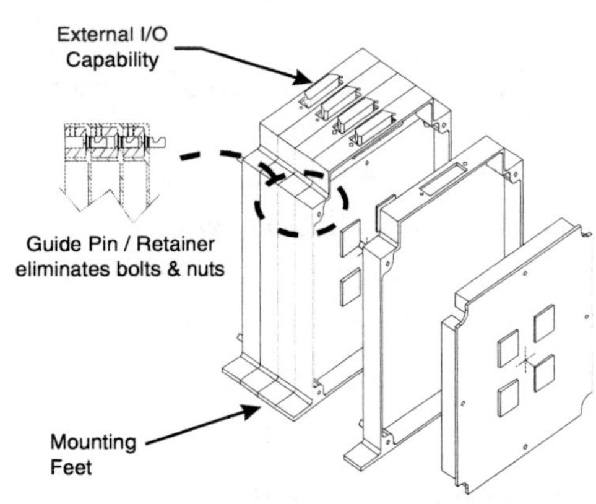

moderate data rate sensor bus. The Inertial Measurement Units (IMUs) and the star tracker plug into the 1394 bus, provided for all the spacecraft components. The specific attitude control sensors for these missions have not yet been chosen. Obviously, for the Europa raditation environment, non-recurring development will be necessary to achieve low mass solutions to this tough requirement. Pluto/Kuiper Express and Solar Probe may be able to use commercially available low mass sensors.

Bus Structural and Thermal Design

The baseline bus design for Europa Orbiter and Pluto/Kuiper Express is shown in Figure 4. Six rectangular composite honeycomb panels attach to a lightweight titanium frame to form a stiff cubical structure to which components are mounted internally and externally. The panels are modular and can be independently removed for access and service. The exact configuration of the avionics on each panel will depend on the numbers of each slice required to perform the mission. Due to volume constraints formed by maintaining the majority of the spacecraft within the umbra of the high gain antenna (or in the shade) during closest approach to the sun, Solar probe utilizes a 4 panel configuration with science instrument filling gaps between the panels.

The internal portion of the bus is maintained at a temperature range of -30 to 0ºC. Heat is obtained both from the operating electronics and thermal communication with the attached propulsion module, which in the case of Pluto Express and Europa Orbiter is warmed with waste heat from an attached Radioisotope Power Source (RPS). Radiative heat loss from the bus is controlled by thermal blankets, and thermostatically controlled louvers on one of the bus panels correct for variations in the heat loads inside the bus.

Propulsion Modules

The baseline propulsion subsystem for Europa Orbiter is a dual mode bi-propellant system designed to approximately 2600 m/s of Delta V capability. The Pluto/Kuiper Express and Solar Probe missions only required small amounts of Delta V and thus are using monopropellant systems. Small milli-newton thrusters are being developed to help reduce the amount of propellant required for Europa Orbiter and Pluto/Kuiper Express while still meeting the accurate pointing requirements.

Power Sources

All three Outer Planet/Solar Probe missions are being designed to be powered by RPS. The avionics bus is designed to include a battery and charge control electronics. The battery is used for brief periods where power consumption must exceed the output of the RPS, mainly during propulsion maneuvers and communication to earth.

MISSION DATA SYSTEM

In past JPL deep space missions, there have been four distinct data systems: the testbed system, the flight software on-board the spacecraft, the uplink system, and the telemetry and archiving system (the latter two often called the Ground Data System, but they are in fact two distinct streams of subsystems that cannot communicate electronically). The future JPL software system instead embodies a new concept of a single Mission Data System (MDS) which is a collection of end-to-end services, and is used throughout the life cycle of the mission, through design, development, integration and test, operations, and archiving. This MDS is being pioneered by X2000 First Delivery but will continue to develop well past the intitial launches of the Outer Planets/Solar probe missions. Mission Unque software will be written using the X2000 core capbailities to form the mission unique Data System.

End-to-end services means that a subsystem (e.g., a spacecraft battery) provides device driver flight software, tactical goal-oriented flight software (e.g., charge battery), strategic goal-oriented flight software (e.g., achieve and maintain maximum charge prior to orbit insertion), planning models for the planner (same planner on-board as on-ground as for system engineering scenarios), performance analysis based on telemetry, and commanding. This is, in fact, no increase in scope for the project: because all these tasks already need to be done for each subsystem. The difference is that now a subsystem is responsible, and empowered, to provide a complete plug-and-play package.

CONCLUSIONS

The Outer Planets/Solar Probe missions are three very scientifically diverse missions going to the extremes of the solar system. One will brave the extreme heat of the sun, while another explores the vast coldness of Pluto's orbit while the third will encounter deadly radiation doses in the Galilean System. The avionics for these three distinct spacecraft will be alike in every way possible to reduce the recurring costs for entire project. Sate-of-the-art avionics are required to meet the performance and mass requirements for the set of missions. The X2000 First Delivery avionic set is pushing the state of the art in achieving a highly integrated, yet modular and scaleable, spacecraft architecture that will have high performance capabilities, yet very low mass.

The software for the avionics is being written to accommodate all the Outer Planets/ Solar Probe missions as well as form the core for all future JPL missions. The MDS presents daunting challenges in management and in technical execution, requiring considerable sophistication. It is probably not cost effective to replace all the legacy systems that have evolved over decades. It is also not reasonable to expect that the technology developed for the First Delivery will not become obsolete with time. Therefore, the Mission Data System will include legacy code whose implementation may be obsolete but whose function is still vital, new state-of-the-art code, and code that is yet to be imagined which will be implemented with whatever comes after.

ACKNOWLEDGMENTS

Many people on the X2000 and Outer Planets/Solar Probe Project Teams contributed to the work described herein and provided assistance to this publication.

The work described in this paper was performed at the Jet Propulsion Laboratory, California Institute of Technology, under contract with the National Aeronautics and Space Administration.

Author Biography:

Karla B. Clark is the Program Engineer for the Outer Planets/Solar Probe Project. She has worked on flight spacecraft since joining JPL 11 years ago. She has been involved with such programs as Mars Observer, TOPEX, Mars Global Surveryor and most recently Cassini. Karla was also a Group Supervisor for the Power Systems and Electronics Group in the Avionics Equipment Section for 3 ½ years.

ULTRA-LOW-POWER, RADIATION-HARDENED 12-BIT ANALOG-TO-DIGITAL CONVERTER FOR SPACE-BASED ELECTRO-OPTICAL SENSORS

Sven C. Nystrom, Todd R. Smith, William R. Peterson, and David L. LeFevre, Raytheon Systems Company, El Segundo, California

Daryl T. Butcher, Larry Gipson, and Martin M. Spanish, Technology Applications Group, La Habra, California

Abstract

We have achieved a breakthrough in analog-to-digital converter (ADC) technology by demonstrating a robust, self-correcting 12-bit ADC ASIC architecture with about six times lower power dissipation than existing devices. The prototype ADC has noise and average differential nonlinearities so low that we believe it has the resolution required for 14 bits. The measured ADC power dissipation is 100 mW at 6 megasamples per second (MSPS)—six times lower than that of the best commercially available radiation-hardened ADC, which has only 11 effective bits of resolution. Four sample ADCs fabricated in a commercial, non-radiation-hardened process worked perfectly at total dose levels up to 50 krad(Si), a level sufficient for many space programs. The productized ADC is expected to be extremely radiation hardened: >300 krad(Si).

Introduction

We have developed an ADC ASIC that will dramatically improve the performance of analog signal processing (ASP) electronics and ultimately reduce the cost of space-based electro-optical (EO) sensors. Such sensors have special signal processing and electronics requirements; key parameters are power dissipation, speed, total-dose radiation hardness, and resolution. The ADC requires

- Ultra-low power dissipation: <20 mW/MSPS
- High speed: >5 MSPS
- Radiation hardness: 50–300 krad(Si)
- Resolution: 12 or more effective bits

Commercially available ADCs cannot meet those requirements. Space-based EO sensors frequently require hundreds of ADCs to process the analog signals. Not only do such large numbers of ADCs dissipate a great deal of power, but that power is also extremely difficult to extract—especially if the ADCs are located on the gimbal of a scanning sensor.

We have achieved a breakthrough in radiation-hardened ADC technology for space-based EO sensors, providing a figure-of-merit performance 10× greater than that of the best commercially available ADCs. Our technology brings a dramatic reduction in ASP power dissipation, with tremendous advantages for the system that, combined, allow use of a smaller cooler and/or radiator. The payoff: significant reductions in sensor weight and, therefore, cost.

ADC Development Strategy

To develop an ultra-low-power radiation-hardened ADC, we designed a 12-bit prototype chip, using radiation-hardening design techniques; but we manufactured the prototype in an inexpensive, non-radiation-hardened, quick-turnaround 1.2-μm bulk CMOS process. The design, fabrication, and test cycle was iterated several times to achieve

0-7803-5086-3 /98/$10.00 ©1998 IEEE

an optimized proof of design. In a staggered parallel fashion, the design was converted to a radiation-hardened version. Sample ADCs were fabricated in parallel to obtain immediate feedback on radiation hardness and to uncover process transition issues.

We chose a novel, fully differential, pipelined algorithmic ADC architecture that tolerates process variations and is highly immune to noise. It allows easy extension of the initial 12-bit design to 14 bits.

Certain common limitations in ADC resolution are caused by errors due to variations in capacitors, unwanted charge injection, and accuracy of the comparator circuitry. Our ADC architecture contains digital correction circuitry that remedies many such errors. The result: an extremely robust design that is immune to statistical variations in device parameters introduced by any CMOS process.

Performance Results

ADC performance is measured over a range of codes. A *code* is the smallest step—or *least significant bit* (LSB)—within the dynamic range of the ADC. A 12-bit ADC has 2^{12} codes. Two key ADC performance parameters are

- Differential nonlinearity (DNL)—the linearity of each ADC code
- Integral nonlinearity (INL)—the running sum of the DNL results over the dynamic range of the ADC

As Figure 1 shows, the average DNL measured on a 12-bit sample ADC is excellent (about 0.3 LSB), but a few spurious DNL spikes exceed 1.0 LSB. The spikes are caused by an INL anomaly associated with the capacitor layout, which affects amplifier gains. The graphed INL results over dynamic range have a triangular shape, with smaller embedded triangles. Careful evaluation of the results plotted in the figure reveals that the INL triangles correspond to the code locations of the DNL spikes.

In a space-based EO sensor, the INL of the ADC is added to the nonlinearity of the focal plane array (FPA), which is eventually calibrated out in the digital signal processor. The ADC INL, ±0.1% of full scale, is still much less than the nonlinearity of the FPA, besides meeting the requirements for EO space

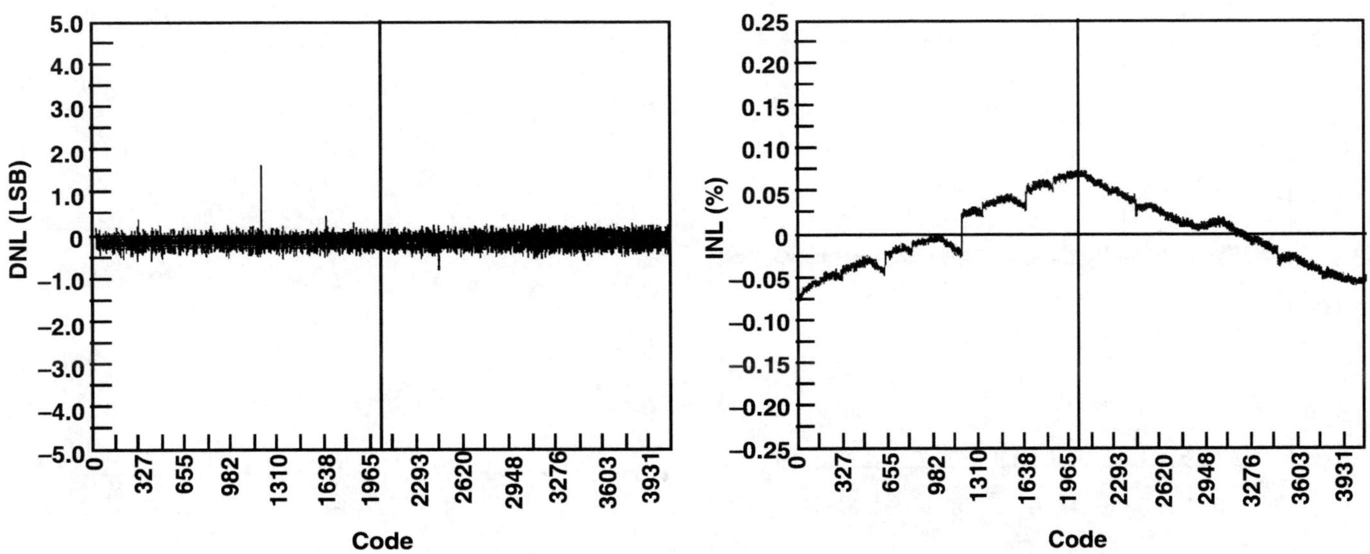

a. Measured DNL

b. Measured INL

Figure 1. Typical DNL and INL Measurements on 12-Bit Sample ADC Exhibit Excellent Average DNL of About 0.3 LSB and INL Less Than ±0.1% of Full Scale.

0-7803-5086-3 /98/$10.00 ©1998 IEEE

sensor applications. However, the INL performance correlates with the DNL spikes that exceed 1.0 LSB. As a result, the ADC resolution is less than 12 bits over portions of the dynamic range—a problem we can easily correct on the next design iteration simply by optimizing the capacitor layout to resolve the associated INL anomaly.

To help understand DNL and INL performance, we conducted a behavioral simulation of the ADC architecture and algorithm. The simulation revealed how variations in the circuit parameters affect ADC performance. Typical simulation results, graphed in Figure 2, demonstrate that the measured ADC performance is replicated by the behavioral simulation. We analyzed the simulated DNL and INL performance of the ADC as circuit parameters were systematically modified. By evaluating the effects of errors in specific circuit parameters, we identified the cause of the correlated DNL and INL anomalies. Two matched capacitors, which determine amplifier gains, had been laid out differently. As a result, stray capacitance was systematically inserted into one of the capacitors. The layout problem has been corrected in the latest design iteration and new sample chips will be fabricated.

Because the DNL and noise performance of the 12-bit sample ADCs was equivalent to what should be expected of a 14-bit ADC, we added two additional stages to the design. The resulting 14-bit ADC was also fabricated in a non-radiation-hardened CMOS process. Initial test results, plotted in Figure 3, indicate that the 14-bit ADC performs as well as the 12-bit. A unique feature adds flexibility to the design, allowing the 14-bit ADC to operate as a 12-bit device: the two unused stages are powered down to prevent them from contributing to the total power dissipation.

Twelve 12-bit sample ADCs were packaged and characterized for average and peak DNL, INL, and power dissipation. Some samples were also tested for thermal stability, and some for total dose hardness. Of the twelve, one was bad, two were marginal, and the remaining nine were good. That represents a yield of 75%—an excellent result for the first design iteration of a prototype ADC.

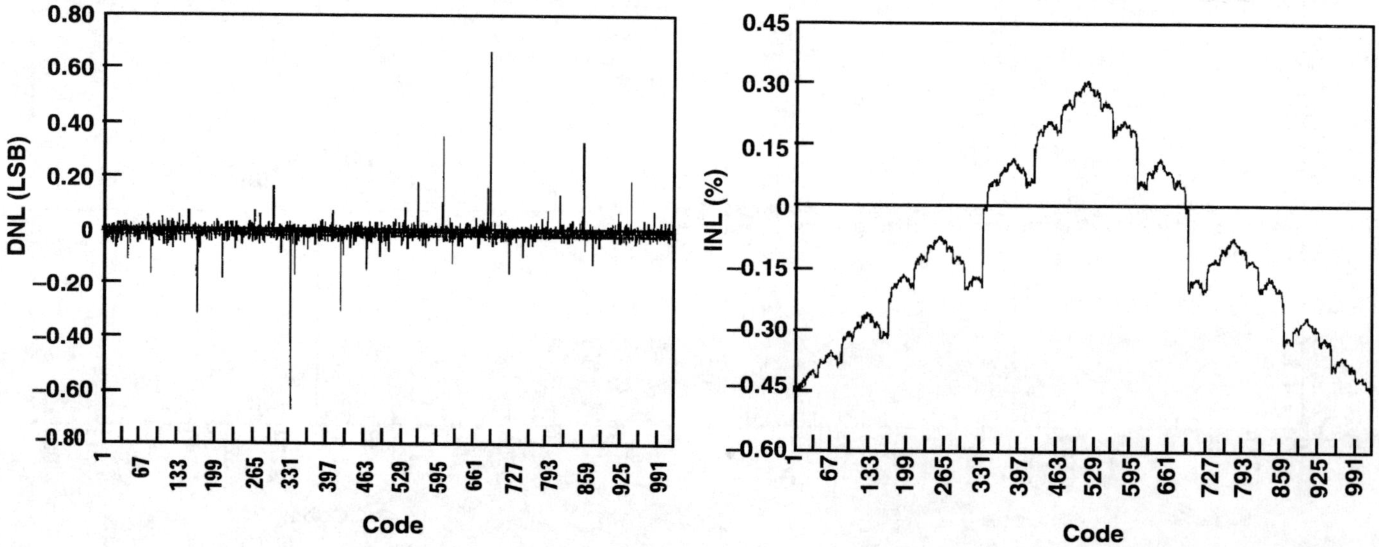

a. Simulated DNL

b. Simulated INL

Figure 2. Behaviorial Simulation of 12-Bit Sample ADC Exactly Replicates Measured DNL and INL Performance.

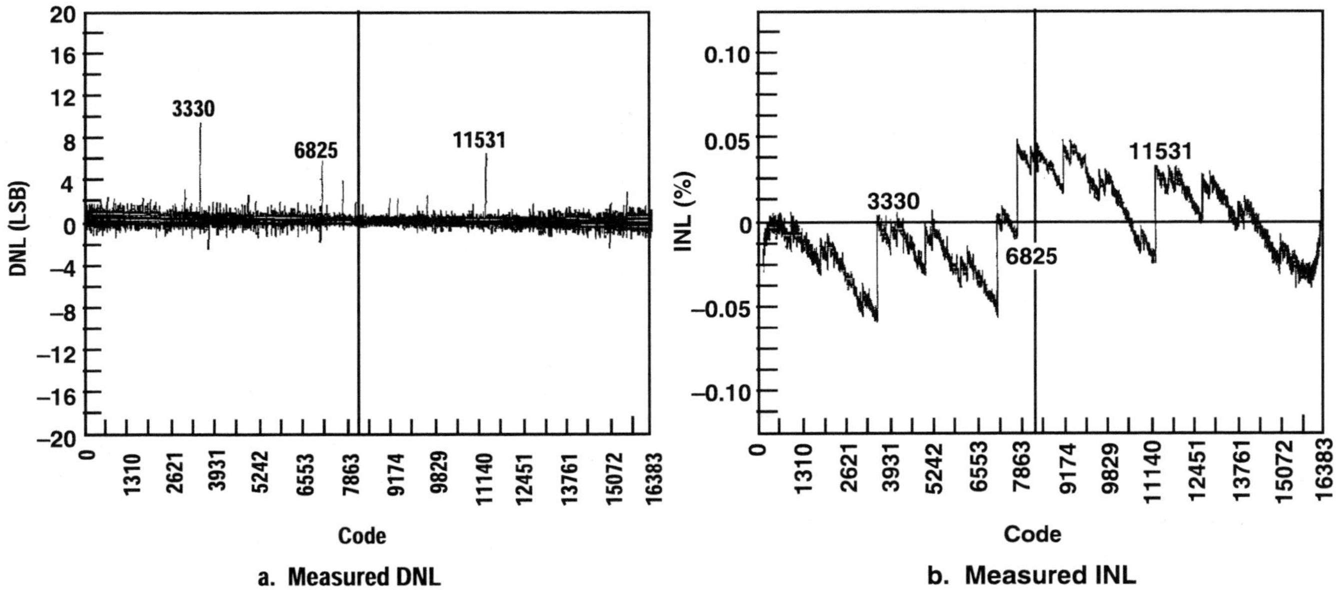

Figure 3. Typical DNL and INL Measurements on 14-Bit Sample ADC Exhibit Excellent Average DNL of About 0.9 LSB and INL Less Than ±0.1% of Full Scale.

Thermal stability is critical during calibration of a space sensor. Therefore, we tested four samples at higher temperatures to determine the drift in gain and offset over temperature. The results, summarized in Table 1, were excellent, meeting the stringent temperature stability requirements for radiometric calibration.

Radiation tests, described below, were conducted on four ADCs (serial numbers 003, 005, 007, and 008).

Radiation Tests

Four of the twelve sample 12-bit ADCs were tested in a total-dose ionizing environment provided by a gamma source. DNL and INL were measured at 25-krad(Si) increments of total dose levels. The ADCs received worst-case dc biases during the dose, and data were taken within 10 minutes after irradiation.

The sample ADCs were processed in a commercial foundry and thus not intended to be radiation hardened. Nevertheless, they proved to be unusually tolerant of total-dose ionization, as Figure 4 shows. The chips exhibited minimal performance degradation up to 50 krad(Si)—adequate for some space sensor applications. Performance significantly degraded at 100 krad(Si), though the ADC was still operational at 10 bits. Such high radiation tolerance is achieved through digital error-correction circuitry that also remedies many of the physical device effects induced by total-dose ionization. Further, future 12-bit and 14-bit ADC designs will be fabricated in a radiation-hardened mixed-signal CMOS process that incorporates channel stops to prevent surface leakage.

Future Development Plans

We plan to correct the INL anomaly (which causes the spurious DNL spikes, described above) by revising the layout to optimize capacitor matching. We will also proceed with four design enhancements, summarized in Table 2, to further reduce power dissipation.

Table 1. Summary of Test Data on 12-Bit Sample ADCs Demonstrates Yield of 75%.

Device Serial Number	Functional	DNL: Avg DNL (LSB) (>95% of Codes)	DNL: Peak DNL (LSB) (<1% of Codes)	Peak INL (LSB)	Power Dissipation (mW)	Thermal Tests (ΔT = 70°C): Gain Drift (%)	Thermal Tests (ΔT = 70°C): Offset Drift (LSB)	Total-Dose Radiation Tests INL/DNL Degradation (Rad (Si))
001	Yes	<±0.4	+1.2/–1.1	±4	95	N/A	N/A	N/A
002	Yes	<±0.4	+1.5/–0.6	±4	95	N/A	N/A	N/A
003	Yes	<±0.4	+3/–1.4	±5	95	N/A	N/A	25K to 50K
004 (control)	Yes	<±0.4	+1/–0.9	±4	95	0.2	0.04	N/A (control)
005	Yes	<±0.3	+1.6/–1.2	±4.1	96	N/A	N/A	25K to 50K
006	Yes	<±0.5	+2.5/–0.7	±4	96	0.18	0.12	N/A
007	Yes	<±0.3	+1.4/–1.4	±4.8	100	N/A	N/A	50K to 100K
008	Yes	<±0.3	+2.5/–1.7	±3.5	94	N/A	N/A	25K to 50K
009	No	N/A	N/A	N/A	89	N/A	N/A	N/A
010	Yes	<±0.4	+3.5/–0.5	±4	95	0.15	0.1	N/A
011	Yes	<±0.4	+3.5/–0.6	±4	97	0.18	0.1	N/A
012	No	N/A	N/A	N/A	100	N/A	N/A	N/A

Test conditions:
- +5-V bias
- 1-MSPS sampling rate
- Tested using TMT ASL-1000 automated tester

Key: ☐ Good ☐ Marginal ▩ Poor

Table 2. Future design enhancements to further reduce power dissipation.

Design Enhancement	Expected Power Reduction
Scale design to 0.6-μm process	1.4×
Scale amplifier capacitor	2.0×
Convert to current-mode design	1.5×
Use cryogenic operation	2.0×

First, we will scale the prototype ADC to operate at 3.3-V biases (instead of 5 V) and to use a 0.6-μm CMOS process that should reduce power dissipation by 1.4×.

Second, because power dissipation is dominated by the capacitance in the amplifier circuits, we will reduce the amplifier capacitance in successive stages to decrease power dissipation by an additional factor of 2. Reducing amplifier capacitance increases noise, but, by the third stage, the signal-to-noise ratio is high enough that increased noise is not an issue.

Third, we will convert some preamplifiers to current-mode designs that will allow further power savings and enable future designs to be fabricated in a 0.3-μm process.

Fourth, we will develop a cryogenic ADC—a critical component for space sensors with a very complex cryocooler/vacuum interface between the FPA and the ASP. Locating ADCs on or near the FPA greatly reduces the complexity of the FPA/ASP interface: the number of wires is reduced and the interface becomes digital, thereby increasing noise immunity. Normally, such a system architecture is not possible unless ADC power is reduced to less than about 5 mW/MSPS—a very low value, but one that we believe we can achieve. Ultimately, we plan to replace the FPA/ASP wire interface with optical fibers.

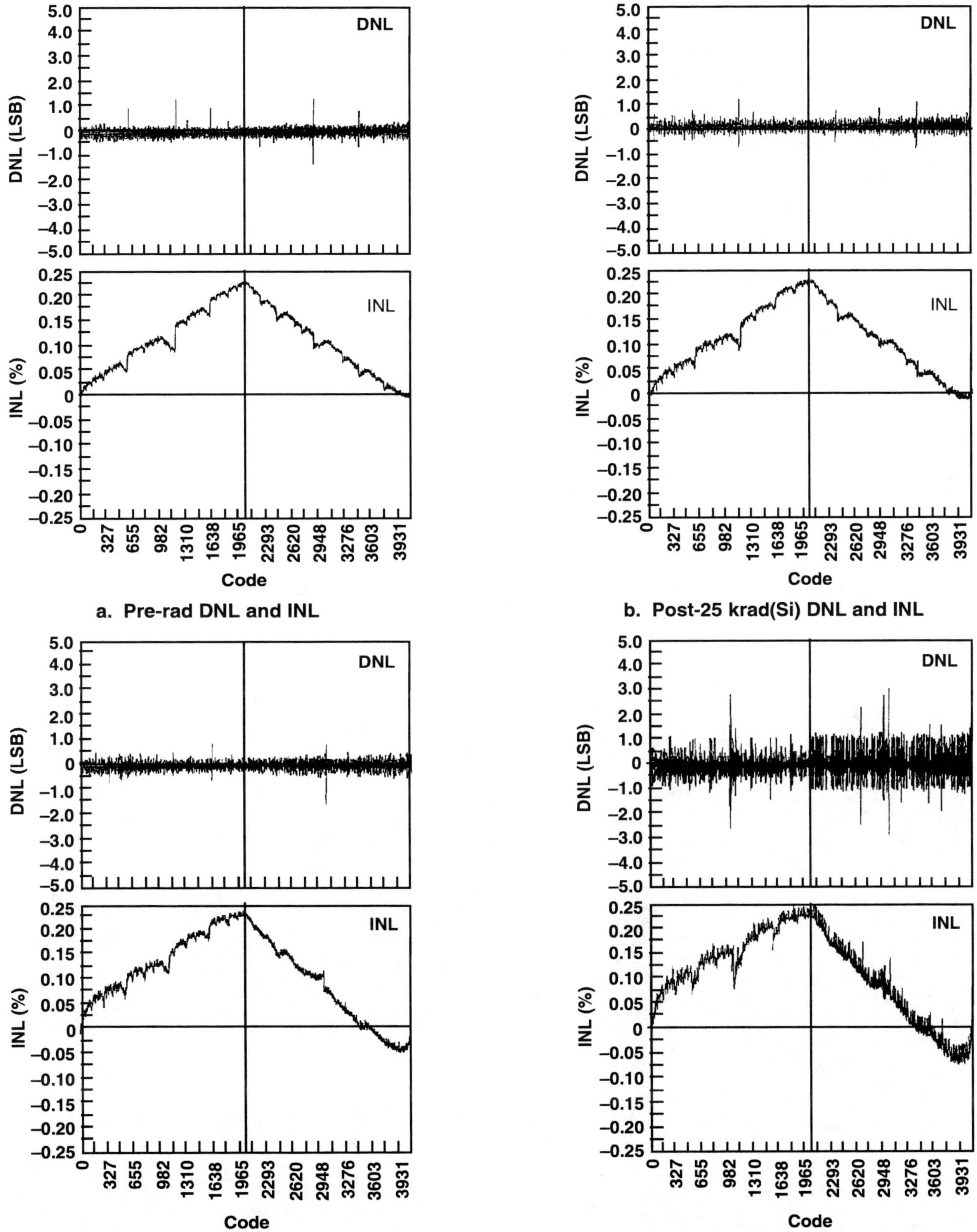

a. Pre-rad DNL and INL

b. Post-25 krad(Si) DNL and INL

c. Post-50 krad(Si) DNL and INL

d. Post-100 krad(Si) DNL and INL

Figure 4. Radiation Test Results on 12-Bit Sample ADCs Exhibit Excellent Performance Up to 50 krad(Si) on Non-radiation-hardened Chip.

0-7803-5086-3 /98/$10.00 ©1998 IEEE

Acknowledgments

The authors wish to thank Raytheon test engineers Gilberto R. Medrano and Wing H. Fong for their support in the characterization of the ADC, as well as Raytheon analog design engineers Malcolm W. Jahr and Mark V. Martin for their technical support.

LITTON AMECOM DIRECT CHIP ATTACH PROJECT

Tracey K. Clay - Litton Amecom; College Park, MD
Craig R. Auletti - Litton Amecom; College Park, MD

Introduction

Direct Chip Attach (DCA) is the next logical progression for the electronics industry. DCA, by definition, is the process of attaching the microelectronic die directly to the substrate rather than installing the die inside an electronic package. The benefits are miniaturization, weight reduction, elimination of packaged parts screening, and better electrical performance. Depending on the package configuration and the required I/O, area reductions can range between four and eight times smaller than standard SMT packages and the weight reductions are even better in most cases. DCA also provides the user with greater design flexibility and eliminates the layer of NRE costs associated with multichip module (MCM) and chip scale (CSP) packaging. DCA can be categorized into two approaches: Chip-On-Board (wire bonding) and Flip Chip.

Chip-On-Board (COB) Technology

Chip-On-Board is the process of attaching the die to the substrate with the bond pads facing upward, and wire bonding the die pads to corresponding pads on the board surface. This process has been utilized in packaged devices for many years on ceramic substrates. Recently, Litton Amecom has developed in-house design and assembly processes for COB on organic substrates. Reliability validation was completed in February 1998 and the assembly of flight hardware was initiated in March 1998. Some of the details of the assembly and testing will be presented in later paragraphs and appendices.

Flip Chip Technology

Flip Chip is the process of bumping the pads on die and attaching the die to substrates with the bumped pads facing downward on the board. Flip Chip technology was initially developed by IBM and has been around for about thirty years (C4 Process). Litton Amecom has evaluated several flip chip manufacturing techniques, one of which was electrically monitored real-time during environmental testing (reliability validation). For space and military applications, flip chip is preferred over COB because the die pads and passivation are encapsulated with an underfill material and there are no exposed wires that could be damaged during the handling associated with integration and testing. Another plus is the fact that the flip chip process is more compatible with the standard SMT reflow and manual processes (particularly in the area of high pressure solvent cleaning).

The Development Program Objective

The objective of this development program, was to develop the design and assembly processes for one or both of the direct chip attach techniques and demonstrate the space flight worthiness of the assembled hardware. The preferred DCA technique which was the main focus early in the project, was Flip Chip because of the handling issues associated with COB technology. The satellite hardware that was targeted for the first space flight hardware application of DCA technology was two identical memory modules (>40GB total). Due to program requirements, the size of the assembly was restricted to a 13.4" x 9.2" assembly consisting of two CCAs laminated to an aluminum heatsink. Preliminary area studies indicated the necessary circuitry could not be

0-7803-5086-3 /98/$10.00 ©1998 IEEE

incorporated into the volume constraint using standard SMT devices, hence, a DCA alternative had to be implemented to meet the mission objective. As one would expect, the mere size of the module presented significant assembly equipment and reliability issues for both DCA techniques.

The infrastructure for COB is fairly well established. Practically all die that are commercially marketed are designed specifically for wire bonding. Several equipment manufacturers have developed automatic die attach machines and automatic wire bonders which have been utilized successfully for many years. Processes for manufacturing and utilizing die bond adhesives, gold wire, and encapsulants have all been developed and are well established. The key issues identified for the Litton program, were handling and rework for the proposed space application where the one-of-a-kind assemblies undergo an extensive integration and testing cycle. Chip-on-board was considered the "fall-back" position since there had already been some studies and tests conducted prior to the start of the Litton Amecom project. However, none of the studies incorporated the combination of standard surface mount devices with COB on standard CCA substrate materials (particularly substrates as large as 13" x 9").

The greater challenge was the development of the Flip Chip design and assembly processes. Today's users of Flip Chip are typically high volume OEMs that have the luxury of spreading the cost of wafer processing over several thousand units. The die are processed (tested and bumped) at the wafer level because of the economies of scale that can be achieved. The challenge that Litton Amecom has undertaken, with some limited success, is single die testing and bumping which is a necessity for military and space hardware (typically procured in small quantities). The initial focus was to determine the viability of using flip chip technology for flight hardware given the predetermined hardware and schedule constraints of the program.

Printed Wiring Board Issues

The transition from MCM-C (ceramic substrates) to MCM-L (laminate substrates) is limited by the printed wiring board (pwb) material physical properties. The materials that were considered for this program were Epoxy/Kevlar™, Polyimide/Glass, and Thermount®, a non-woven aramid epoxy. The Kevlar material was eliminated because of the relatively large hole sizes required for adequate pwb plated-through-hole reliability. The large hole sizes greatly handicap the pwb designer's ability to save pwb surface area. Thermount® was considered for DCA because of its relatively low in-plane CTE of 6-9 ppm/°C, low dielectric constant (Dk) with epoxy resins (3.9), excellent dimensional stability (0.02%), and smooth surfaces; however, in mid 1996, none of the Thermount® pwb fabricators were capable of achieving the density (0.003" lines & spaces) required for standard die flip chip attachment. The only acceptable alternative for high reliability applications was polyimide/glass. At least one printed wiring board supplier had already demonstrated they could fabricate these challenging designs in prototype quantities. Several other pwb fabricators purported they could also fabricate polyimide/glass pwbs with the same density.

For flip chip, assuring the pad geometry on the board matched the bond pad layout of the die was a major concern. Since typically the die in question are wire bonded in packages, there is a minimal amount of concern in the die manufacturing industry to standardize on die pad layouts or to fix the pad positions (no artwork changes) during die shrink efforts. As a result, the pwb pad geometry could be designed for one revision of the die pad layout and another revision could be physically received because the earlier revision die is out-of-stock. The Litton Amecom approach to resolving this issue was to procure the full compliment of the die that could be conceivably needed for the program and include in the purchase orders the revisions of each die pad artwork. Then all design efforts revolved around the artwork specified in the purchase order. As soon as the die were received, the locations of the die pads were physically verified to match the pwb pad design before the completion of the pwb fabrication process. Using this methodology, there were no problems with die to board mismatches. While this issue is diminished for COB pwb pad layouts, it is still of great concern given the fact that wire lengths should be minimized to reduce failures resulting from vibration.

Known Good Die (KGD) Testing Issues

Known Good Die testing is a major concern for the direct chip attach industry because conventionally all reliability testing for standard SMT components is done after the die is packaged. Several methodologies have been used including slave boards, wafer test fixtures, individual die test fixtures, and Known Good Lot testing where a percentage of the die on a wafer is packaged and the test results of the packaged die are considered to reflect the unpackaged die of the same wafer lot. The approaches utilized for this project were: 1) Full parametrics at room temperature and hot (125°C) and 2) Full parametrics at room temperature and functional testing at 125°C. All testing was done at the wafer level.

The Flip Chip Development Activity

The schedule constraints dictated the development of a space flight proven process for flip chip technology in eighteen months. After resolving the major printed wiring board issues, the issue of placement was addressed.

Several flip chip aligner/bonder equipment manufacturers were queried regarding placement accuracy and maximum board size. Only one was identified to have an off-the-shelf system for handling the anticipated maximum board sizes. Equally important as machine accuracy, is the ability for the aligner/bonder to have a heater built into the placement head so the attach process can be performed simultaneously with placement. During Litton Amecom's previous experience with memory stacks (20 mil pitch Flip Chip), it was noted the die tended to drift during reflow. Performing the attach while the chip is held in place ensures a higher attachment yield. In the final analysis, the equipment cost was significant enough for contract assembly houses to be researched and Hycomp, Inc. in Marlborough, MA was contracted to do all of the flip chip assembly including die bumping.

Bumping Die For Flip Chip

Typically, die are bumped in wafer form (before scribing) which means a program requirement of 20 die could result in the purchase of several hundred to get the die in wafer form. Wafer bumping is achieved by using either a vapor deposition or electroplating process to accurately bump die with any one or combination of several materials (e.g. Sn-Pb, indium, gold, nickel, etc.). Wafer bumping the die with solder or indium allows the circuit card assembly (CCA) manufacturer to easily incorporate the flip chip process within the standard SMT process. Wafer bumping distributes the NRE, fixturing, and processing cost over several hundred die rather than a single die; however, due to the die quantities involved for the flight program, only single die bumping was considered viable for this project. The only relatively economical method of single die bumping was the "stud bump" method. Stud bumping is the process of using a wire bonder to make a gold ball on each pad of the die and tearing away the wire to leave a single ball bond on each pad. Although Litton Amecom experimented with gold stud bumping die for several months to develop the technique, it was decided that Hycomp, Inc. would perform all of the stud bumpingusing an automatic wire bonder for process control purposes.

Flip Chip Attach

If the die can be bumped at the wafer level, there are several options to the attach method; however, since gold stud bumping was deemed the most feasible alternative for low volume production applications, soldering as an attach method was not an option. Soldering gold stud bumped die to a tin-lead plated substrate would result in solder joint embrittlement. Given these constraints, non-conventional attach methods were pursued:

1. Z-Axis anisotropic conductive adhesives, which are materials loaded with metal particles that when compressed and thermally cured create an electrical path between the die pad and the pwb pad. The metal loading is such that there is no electrical path between adjacent conductors.
2. Conductive Polymer attach, in which a conductive polymer is deposited on the bumps or pads only and cured to create an electrical path. An underfill material is used for mechanical strength.
3. Non-Conductive Adhesive attach describes the process of compressing gold bumps to the gold pads of the pwb through a non-conductive adhesive that is cured to hold the die and connection intact. The interconnection is entirely mechanical and is analogous to pin/socket connector technology.

Z-axis anisotropic conductive adhesives were considered the preferred method because it had been successfully implemented for the repair of ribbon connectors for laptop computer displays and it was purported to offer excellent reworkability. The "Z-axis" materials are offered in sheet and liquid forms. Only two alternatives were given serious consideration because many of the materials were still in the development stage at the time of the investigation. Only one of the materials (sheet adhesive) was available that could withstand the processing temperatures of the heatsink-to-board lamination process. Exceeding the working temperature of the material once the electrical connections have been established causes opens that are not reworkable. After several trials with the only available material, it was determined that the metal particle density was not adequate for flip chip applications. It was also noted that the thickness of the material is critical for assuring proper electrical conductivity and underfilling. The process window proved to be too small for the stud bumping process given the fact that slight variations in ball locations and sizes could cause random disconnects.

In theory, the conductive polymer approach also offered great promise with regard to reworkability assuming the interconnects could withstand the test process prior to underfilling. In practice, the unprotected interconnects proved to be too fragile. In some cases where we were able to underfill and confirm the interconnects were still intact, opens occurred during subsequent assembly processing. Eight out of eight flip chip die using the conductive polymer failed either before or during the lamination process; therefore, no reliability testing was completed.

The non-conductive adhesive approach samples; however, were successfully assembled and reliability tested in two locations on a single CCA. Two out of two successfully passed thermal cycling and vibration testing as described later in this paper. Although the attach methodology was successfully tested, there is no plausible method for rework attach errors. The board and die are damaged beyond repair if a die has to be removed and/or replaced. For this reason and the fact that the sample size was inadequate, the decision was made not to use flip chip for flight hardware until further developments and testing are completed.

Flip Chip Underfill Materials

An underfill material is used for flip chip devices to provide a barrier against water absorption and contamination. More importantly, the underfill provides a buffer for the CTE mismatch between the board material and the chip material. Inducing air bubbles during the underfill process must be avoided to prevent air entrapment that may facilitate/aggravate CTE mismatch stresses. No in-house process development was initiated for the underfill operation due to the fact that Hycomp, Inc. had an established methodology for their conductive polymer attach process. In the cases of the Z-Axis adhesive attach and the non-conductive adhesive attach processes, the adhesive functions as the underfill material.

The Chip -On-Board Development Activity

Litton Amecom has had more that twenty years of wirebonding experience; however, none of the experience had included wire bonding to an organic laminate substrate of this size (13" x 9"). Equipment specifications were reviewed given the fact that the substrates for typical wirebonding applications are equal to or smaller than 4" x 4". The wire bonder was modified to increase the throat depth of the machine from 4" to 5.8". A special heated workstage and associated tooling was designed and fabricated to maintain the predetermined substrate temperature during wire bonding. A second fixture was fabricated to heat the heatsink of the laminated assembly which would in turn serve as the conduction heater for the attached substrate. The standard cleaning (UV Ozone), die placement, and die attach equipments were utilized for their respective processes.

Consideration for rework was incorporated into the pwb footprint design to allow for a maximum of three rework cycles on the wedge end of the wire bond (on the pwb pad). Compound bonding was included in the reliability evaluation to qualify rework at the ball end of the wire bond. Additionally, pull tests and cross-section analysis were performed to assure the reliability of the compound bonds.

The encapsulation that was selected was a "glob top" approach for which the ball end of the bond at the gold to aluminum interface was encapsulated while the wedge end was not. This approach protects the gold to aluminum interface while maintaining the appropriate stress relief in the wire. If the die and wires are totally encapsulated, failures could occur during thermal cycling due to the CTE mismatches. Given the same stress relief consideration, the CCAs were parylene coated (vapor deposition) as opposed to the standard industry spray coatings. The parylene coating coats the wire without creating a web between the wire and the substrate material underneath.

From the inception of the project, handling for COB was considered a major issue. Handling fixtures were designed and fabricated to provide mechanical protection for the CCA during the single pwb processing, the double-sided laminated assembly processing, transport between the CCA manufacturing facility and the box integration facility, and box level testing. Covers for each die type was also developed to facilitate the lamination process and subsequent rework cycles. In spite of all of the fixturing , several wires were damaged during the course of handling especially during the removal of the lamination covers. Awareness and training were critical for minimizing wire handling damage.

Reliability Validation

The substance of this project was the fact that Reliability/ Environmental Testing was performed to qualify the DCA hardware produced. The following tests were used to validate the reliability of the test items. Five validation (Concept) CCAs were subjected to random vibration: 10 Grms (or G's) over a frequency range of 20 to 2000 hertz for three minutes in each axes (X,Y, & Z), and Thermal Cycling: 500 cycles; -20°C to 85°C. Two of the five Concept CCAs were then subjected to 85/85 Temperature/Humidity bias testing (500 hours).

Both Flip Chip and COB were incorporated into the Concept CCA design (see Figure 1 for a picture of the laminated assembly and Appendix A for the process flow). The reliability of the Concept CCAs was established through the exposure of the test item to the above environmental tests. Periodically during each exposure, the acceptability of each component on the test board was assessed through the use of one (or more) of the following inspection techniques:

1. Functional testing: SRAMs (37 I/O, 0.289" x 0.344" die) were functionally tested using test vectors generated by an on-board FPGA to determine if the devices continued to perform electrically as anticipated. The health status of the devices could be queried with an off-board computer. Any device which failed to respond to the stimuli in the anticipated fashion resulted in the lighting of an on-board LED. Each "failure" was recorded and the functional testing was resumed with the confirmed failures masked. The testing was completely autonomous during normal operation; identification of failed devices/interconnects required manual intervention/interrogation.
2. Visual Inspection: The attach points for the balance of the components that were not functionally tested were visually inspected.

Final Results Of The Concept (Validation) CCAs

All five Concept assemblies (each having two circuit cards laminated to a heat sink) completed the vibration testing and the thermal cycling with no failures assignable to the test environment. Both assemblies that were subjected to the 85/85 Temperature/Humidity bias testing failed before reaching the prescribed 500 hours of exposure. Appendix B summarizes the results of each individual test vehicle. Based on the data provided by the testing, the chip-on-board process was qualified for flight hardware.

Flight Hardware Status

Following reliability confirmation utilizing the Concept CCAs, one ETU Memory CCA has been assembled and integrated into a Flight Chassis as of May '98. Two flight assemblies have been started through the manufacturing processes and are scheduled to be integrated into the flight chassis in August 1998. The memory cards (13" x 9") are designed with chip-on-board technology.

APPENDIX B
DIRECT CHIP ATTACH TEST SUMMARY (AS OF 2/16/98)

Below is a table that summarizes the results of the electrical testing during and after the environmental exposures.

LOT NO.					COMMENTS
	A -SIDE S/N	B-SIDE S/N	FLIP CHIPS PASSED THRU	WIRE BONDS PASSED THRU	
I[1]	N/A	N/A	N/A	T/C & VIBE	HANDLING DAMAGE TO U3 AFTER LAMINATION; ASSEMBLY FAILED DURING TEMP/HUMIDITY BIAS
II	2	8	FAILED	T/C & VIBE	ASSEMBLY FAILED DURING TEMP/HUMIDITY BIAS
III[2]	1	6	FAILED	T/C	BOTH SIDES FAILED (U3) AT VIBE; EXPERIENCED ANOMALOUS FAILURES OF SIDE B (U3) THROUGHOUT TESTING AFTER STEP 37
IV[3]	7	4	T/C & VIBE	T/C & VIBE	SIDE A U3 (WIRE BONDS) DAMAGED AT LAMINATION;
V	5	3	FAILED	T/C & VIBE	SERIAL NUMBER 3 CLEANED IN ENSOLV PRIOR TO WIRE BONDING.

T/C = 500 THERMAL CYCLES
VIBE = ALL THREE AXES

The following footnotes provide additional details regarding the deviations from the flow depicted in the Test and Evaluation Plan for the Chip on Board (COB) Concept Development Boards. It should be assumed, as in the case of the Lot V assemblies, that the entire assembly processing and testing was completed as planned unless otherwise noted in the comments in the chart above and/or in the footnotes below.

[1] The Lot I assemblies served as the pilot assembly for the test circuit/Actel programming and for COB testing. It did not have any assembled flip chips (steps 10 through 23 in flow chart omitted). At step 30, the wire at pad 30 was incorrectly bonded in all four locations (U3 and U6 on both boards). All four die were reworked by removing the wire and installing a new wire in the correct location. One wire was damaged on side A by an engineer during removal of the lamination die covers (step 34). Two additional wires on the same side A were inadvertently removed by the engineer just prior parylene coating.

[2] The Lot III sub-assemblies (Concept Board serial numbers 1 and 6) experienced die failures at step 30 (just after wire bonding). The die at location U6 was removed and replaced on both serial numbers.

[3] The Lot IV Concept assembly had the only working flip chip die after step 34

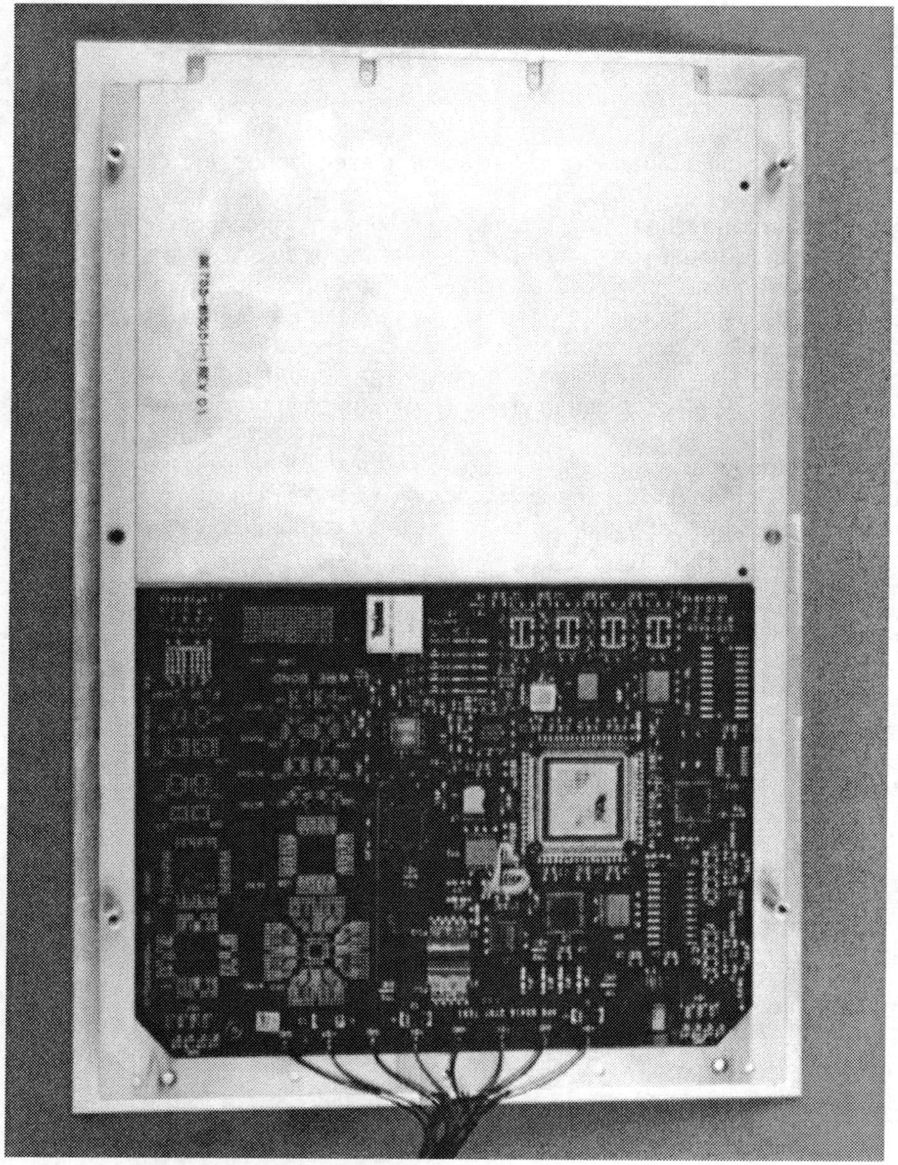

Figure 1

Figure 2

ISSUES TO ADDRESS IN USE OF COMPOSITE MATERIALS FOR ELECTRONIC PACKAGING

Gary C. Krumweide, Composite Optics, Incorporated, San Diego, CA
Gary A. Tremblay, Composite Optics, Incorporated, San Diego, CA

Abstract

There is an ever growing trend towards greatly reducing the size and mass associated with aerospace avionics and electronics, especially the structural components (i.e. heatsinks, thermal cores, enclosures, racks or decks). For several years effort have centered on substituting lesser density materials for, the most often used, aluminum alloy materials. Materials like graphite/cyanate, Carbon-Carbon and Albermet™ have been considered viable alternatives to aluminum for these structures. But with the higher packing density of the electronic and with increase in power requirements there becomes even greater need to dissipate heat more efficiently.

Together these trends necessitate a better understanding of the issues to address when proposing new material candidate to satisfy future electronic packaging requirements.

The understanding of these issues is more crucial when the trend is to integrate the electronics into the structures; as currently identified by the term "multi-functional structures" (MFS). Several spacecraft components will be space tested using this concept, MFS, before the turn of the century.

Composite Optics, Incorporated (COI) has provided several state-of-the-art lightweight, high-performance electronic enclosure (and heatsinks) to aerospace customers using graphite fiber reinforced polymer matrix composites (GFR/PMC) materials. Examples of flight qualified electronic enclosures which demonstrate the capabilities of advanced composite technology to reduce mass for satellite and aircraft applications will be discussed. For the most part these applications represent a "substitution" of a lower density material (i.e. GRF/PMC) for aluminum in an existing enclosure design. A discussion of the issues (i.e. EMI, grounding, radiation, cost, etc.) associated with the usage of GFR/PMC for these enclosures will be presented in this paper, along with the methods employed by COI to satisfy their current and future customers requirements.

1.0 Introduction

In today's atmosphere of increasing technology capabilities and changes and limited economic resources, it is a requirement to be truly "Better, Faster, Cheaper" to survive. This has brought the low-cost electronic packaging programs to the forefront to get as much of the latest technology in orbit in the shortest timeframe from the design station to the launch pad. It is inherently necessary to increase the payload-to-structure mass ratio for future missions to meet higher mission requirements within program budget constraints.

The requirement for electronic packaging to maintain suitable thermal management while achieving improved mechanical performances and/or lighter weight configurations has drawn increasing focus during the past 5 years. This focus has extended beyond the usual space applications to aircraft avionics. During approximately the same period, the advanced high thermal conductivity carbon fiber technology was progressing beyond the development phase and making its way to flight applications. But its

0-7803-5086-3 /98/$10.00 ©1998 IEEE

flight insertion was limited because it was being utilized in hardware applications where performance was paramount and cost and schedule issues were second tier. Mechanical performance and efficient thermal management was occasionally compromised by the geometry requirements or selected processing of the candidate hardware. To merge the application needs and technology benefits and to make significant advancement in the application of these high conductivity composite materials, a different approach was necessary. It had to have consistent performance results, meet all mechanical, thermal, and EMI requirements, and be affordable through all phases of development and production procurement.

Given the need for lightweight, high performance electronic packaging for various science, military, and commercial satellite applications, COI has set out to address several challenging issues in the production of electronic enclosures. The feasibility of producing lightweight, high performance electronic enclosures affordably can be attributed to recent technology advancements in materials, innovative design/fabrication approaches, plating, and radiation shielding. These technology advances and several supporting programs (SBIRs, BAAs, CRADs, internal IRADs) have allowed COI to provide a full service development/ design/ fabrication/ testing capability to the aerospace community that provides an alternative solution to the current State-of-the-Practice (SOP) in electronic packaging to address the following issues:

1. How to achieve lightweight enclosures at less than 50% of the current metal baseline technology weight that also addresses all thermal management issues.

2. How to address EMI and grounding requirements on a composite enclosure.

3. How to maintain acceptable levels of radiation shielding without significantly deteriorating the weight savings of a composite enclosure.

4. How to address the low through-the-thickness (K^3) thermal conductivity characteristic of polymer matrix composite materials in an enclosure design to meet thermal management requirements

5. How to reduce NRE cost and schedule issues

6. How to reduce recurring costs of enclosure to an affordable level compared to current SOP technology

7. How to address wear surfaces and hard mounting and bolted interfaces.

In order to illustrate how a specific issue (above) is being addressed for a nominal space (and airborne avionics) environment enclosure, a typical electronic enclosure will be presented. Design and manufacturing approaches that address the performance requirements and these issues will be discussed.

2.0 Technical Discussion

The above issues and solutions will be discussed separately in order to understand their differences and interactions.

Affordable Manufacturing Approach (SNAPSAT™)

SNAPSAT™ is an acronym for "Short Notice Accelerated Production SATellite". It is a proven design and manufacturing technology for advanced composite materials that provides affordable productivity by reducing shop setup and recurring touch labor. It is incorporates the best mechanical, thermal, and economic attributes of different material systems into a structure design that utilizes flatstock details with self locating features. The manufacture of these individual details and the final assembly is amenable to automated processes with minimal tooling. Details and advantages of this technology are described as follows:

Design Approach

1. Minimal Shaped Details: It is a design approach with all details being precise flatstock laminate patterns with self locating mortise and tenon (slot and tab). Simple molded details such as angle clips are used at corners as required for structural strength. This allows for utilizing off-the-shelf stock laminates or an early start on material procurement and lay up prior to final design release.
2. Assembly Level Design: Hardware design can be maintained on the top assembly or subassembly level when using any of the CAD systems. Flat pattern details can be isolated and arranged on a flat pattern file for CNC trim. Only angle shaped details need to be mono-detail dimensioned. This reduces design effort.
3. Simple Integration of Advanced Materials: The latest technology materials or combinations of materials within an assembly can be inserted without processing development. Details can be engineered within a subassembly to incorporate high cost, high performance (mechanical or thermal conductivity) materials only where necessary to meet program requirements.

Nonrecurring

1. **Minimal Tooling:** The flatstock approach utilizes minimal number and complexity of tooling by using self locating flatstock details. This reduces the NRE cost and schedule for prototype and first unit flight productions. Fixed geometry tooling is not cost effective for short run programs and it does not allow for simple configuration modifications to match the program changes without significant cost and schedule impacts. Sometimes this is a force-fit situation. With SNAPSAT, configuration can be significantly changed shortly before drawing release without affecting program cost or schedule.
2. Simplified Tooling: Shop orders for detail trim and assembly can be simplified to pictures and short instructions since all of the critical processes are standard generic procedures.

Hardware Manufacturing

1. Detail Trimming: Precise trimming of flatstock details can be completed on a CNC controlled waterjet or router. Multi-layer of the same detail can be stacked and commonly trimmed to yield multi sets of details in a single pass. without compromising the level of accuracy.
2. "Group Part Processing": A shipset of flat details can be processed (handled, prepared for bonding, piece part identified and stored) for assembly as a single laminate group of details if the details are maintained connect to each other via small tabs. This eliminates the processing cost concerns associated with increased part count of the flatstock approach.
3. Automated Assembly: Precisely located details can be subassembled and/or assembled without dry-fitting and bonded using automated bonding techniques (wicking adhesives, robotic adhesive dispensing, adhesive silk screening, etc.).

Material Selection

Material selections for each application is based on different criteria and performance requirements. Ultimately, the materials that are selected are critical to the program's success and invariably have to have a proven heritage in space.

The typical electronic enclosure has performance requirements that deemed it necessary to select a combination of composite materials to meet the thermal management and stiffness issues and to be affordable. A high thermal conductivity carbon fiber such as Amoco's K1100 or K800 or Mitsubishi's K13X series prepregged with a epoxy or cyanate ester resin system in a unidirectional tape format are the normal candidates. See Table 1 Utilizing these high "K" materials in either a pseudo-isotropic or highly biased format provides the thermal management for the structure for conductive, convective or radiative cooling..

All other items that do not interact with or contribute to the thermal management utilize a low cost, medium modulus, high strength composite fiber material such as T300, M40, M46, or AS4 to maintain affordability.

Metal fittings and card rails utilize aluminum for wearability, thermal conductivity, lightweight, and low cost issues. Bonded surfaces are prepared for bonding with an epoxy primer. Details are bonded using a room temperature structural / conductive adhesive such as Hysol's 9395or Emerson Cuming's EC-285.

Table 1

Material as	Type	Density (g/cm^3)	Thermal Conductivity 1-dir (W/M-K)
Aluminum (Al)	2024	2.7	170
P120/CE	Isotropic	1.75	190
P120/CE	Unidirectional	1.75	380
K13C2U/CE	Isotropic	1.75	160
K13C2U/CE	Unidirectional	1.75	300
K800/CE	Isotropic	1.75	183
K800/CE	Unidirectional	1.75	330
K1100X/CE	Isotropic	1.75	300
K1100X/CE	Unidirectional	1.75	650

3.0 Electronic Enclosure Description/Design Consideration

The typical enclosure application is originally baselined to use the SOP aluminum plate approach until mission established technical requirements dictated a change to an advanced composite technology. At this point, COI and the systems integrator must closely work together to concurrently designed these structures using the flatstock design approach and with attention to the manufacturing details to make it affordable.

Enclosure Configuration

The typical satellite electronic enclosure (Figure 1) utilizes a double-walled configuration with a series of high conductivity ribs between the walls (see Figure 2).

0-7803-5086-3 /98/$10.00 ©1998 IEEE

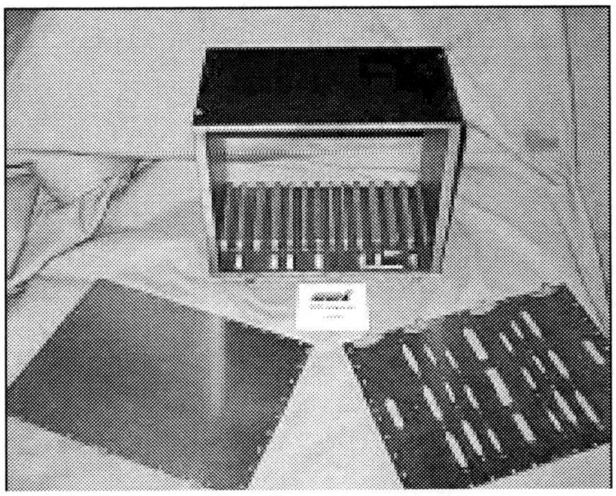

Figure 1
Typical Satellite Electronics Enclosure

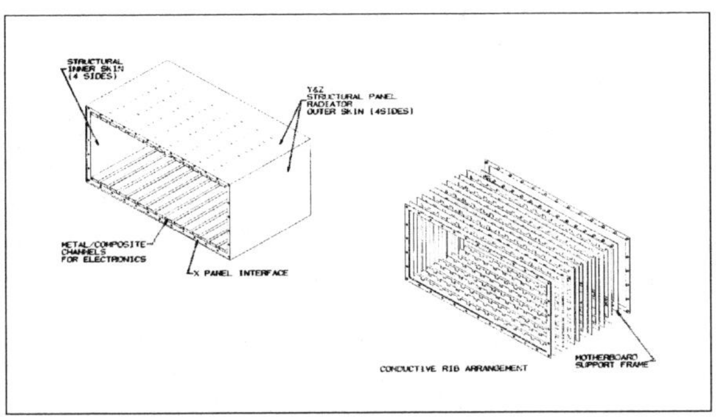

Figure 2

The ribs details have tab features that protrude the inner skin wall to achieve a direct contact with the electronic card rail and provide a high conductivity path to the conductive ribs. These ribs provide a conductive thermal path to the metal mounting fittings or to the outer skins for radiative cooling. Typically, the high thermal conductivity GFR/PMC are selected for all conductive ribs in the equipment panels and M46J.954-3 Gr/CE is used for all structural skin and ribs in the structural frame, decks and equipment panels. The K800 or K1100 ribs can be fabricated in a biased lay up to provide maximum stiffness (approx. 40 Msi; 4 times aluminum) and thermal conductivity (approx. 300 to 500 W/mK; 2 to 3 times aluminum) in the detail longitudinal direction at 65% the density of aluminum or as a pseudo-isotropic configuration. Typical material thicknesses are 0.030" for the structural outer and inner skin members, 0.030" for the conductive ribs.

To address EMI performance requirements, after the composite details are assembled, the enclosure is electroless plated with a 0.0003" thick nickel layer on the inner skin and rib surfaces (both sides). This metal plated surface can also serve as a grounding plane since the electrical conductivity of the advanced composite fiber is not sufficient or easily accessible to meet a 2.5 or 5 milli-ohm requirement. A localized bus bare is also sometimes imbedded in between the skins to connect the card rails to the mounting pads for grounding.

Incorporating a radiation shielding material into the enclosure is as easy as substituting COI's lightweight, hybrid composite material system for the structural outer skin of the enclosure. This material is 30% to 40% lighter than its equivalent aluminum material and has an equivalent shielding characteristic (see Figure 3).

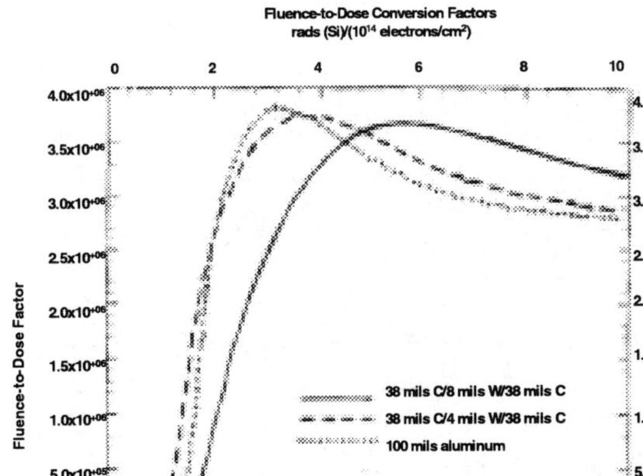

Figure 3

0-7803-5086-3 /98/$10.00 ©1998 IEEE

Other approaches to radiation shielding is to use localized spot shielding on the enclosure, electronic chip shielding, or a strategic arrangement of the electronics to shield the most vulnerable components with other components.

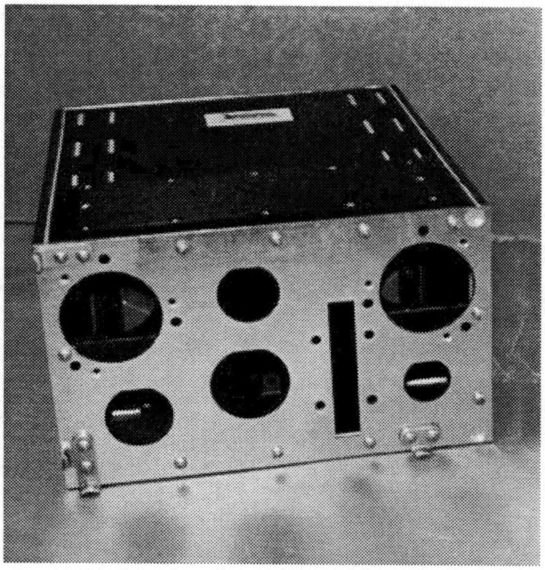

Figure 4

For typical airborne avionics, the configuration approach is very similar except that thermal management utilizes convective cooling of the ribs or closely spaced fins with an air flow (see Figure 4). These ribs are directly in contact with the electronic card mounting rails and are configured to direct the heat as efficiently as possible from the rails to the air flow.

4.0 Performance Requirements

Performance requirements for satellite enclosures can be divided into two broad categories: 1) launched/flight phase; and 2) on-orbit operations requirements. The launch / flight phase requirements are well defined by the selected launch/aircraft vehicle, while the on-orbit/flight specifications are set by the mission/science requirements. The enclosure discussed in this paper addressed typical requirements for small satellite missions. A summary of performance requirements are presented in Tables 2A and 2B.

Table 2A

	14 Card Config. Requirements	14 Card Config. Actual	8 Card Config. Requirements	8 Card Config. Actual
Launch/Flight Requirements:				
Application	Space	Space	Airborne	Airborne
Launch/Flight Loads: a) Lateral b) Longitudinal c) Shock	142g 142g		20g 20g	
Temperature Exposure	-45°C/+60°C	Compliant	-62°C/+95°C	Compliant
Volume (in.)	15.3x10.0x7.6	Compliant	9.1x5.5x10.7	Compliant
Enclosure Wt (lb.):Alum. Composite	11.7 < 4.66	4.40 (<38% Al)	5.8 <4.0	3.80(<66%Al)
Fully Integrated Wt.(lb.)	26.7	Compliant	18	Compliant
Mechanical Stiffness (Hz) (fully loaded)	>600 >800	>900	>600	>800

Table 2B

Operations Requirements:				
Thermal Performance a) Max.Power b) Rail V Temp.Rail-to-Base c) Baseplate Power Density d) Lower Rail Temp e) Upper Rail Temp.	 49 W Total <V5°C 1.0 W/in^2 75.2°C 85°C	 Compliant Compliant 1.4 W/in^2 74.5°C 80.1°C	 205 W Total 89°C 89°C	 Compliant 88°C 88°C
EMI Shielding	>20dB@10 GHz	Compliant	>45 dB@25-750MHz	Compliant
Grounding-top to base	<25 Milli-Ohm	Compliant	<2.5 Milli-Ohm	Compliant
Dimensional Tolerance (in) a) Rails b) Volume & Features c) Flatness-Baseplate	 ±0.005" ±0.010" 0.010"	 Compliant Compliant Compliant	 ±0.005" ±0.010" 0.010"	 Compliant Compliant Compliant

5.0 Manufacturing Approach

COI's SNAPSAT™ technology is based on a simple approach of utilizing flatstock laminate for all details (except corner joints as necessary). These details are designed to be self-locating using mortise-and-tenon (tab and slot) features for geometric accuracy and mechanical joint strength. This allows for the easiest implementation of automated material fabrication (commodity sheet stock), detail trim (CNC waterjet) and subassembly/ final assembly (automated bonding prep and adhesive dispensing) efforts. Some of these procedures/ processes are presently available at COI and were used on these programs. Figure 5 shows the simple trim pattern for the flat structural and high conductivity details that efficiently utilizes all of the flatstock laminate material.

Work-in-process testing of the laminate assures quality performance of the processed material prior to assembly. The simplified assembly tooling that maintains critical feature accuracy and is easily adjustable to late changes is shown in Figure 6.

6.0 Program Performance Parametrics and Targets

Results from the several electronic packaging programs executed at COI have provided a wealth of information on effectively implementing this technology. Experience and lessons learned in the design, analysis, and tooling technical approach have guided COI to establish achievable targets for program performance in the near term and far term as follows:

0-7803-5086-3 /98/$10.00 ©1998 IEEE

Parameter	Current Performance	Near Term Performance	Far Term Performance
Design/Analysis NRE Timeframe	16 to 20 weeks	12 to 15 weeks	8 to 12 weeks
Prototype Fabrication Timeframe	8 to 10 weeks	6 to 8 weeks	5 to 6 weeks
Production Fabrication Timeframe	1 unit/2-1/2 weeks	As Required	As Required
Unit Cost Ratio (vs. Aluminum) a) Low Qty Production b) High Qty Production	 2.5X to 3.0X 2.0X to 2.5X	 2.2X to 2.7X 1.8X to 2.3X	 2.0X to 2.4X 1.6X to 2.0X
Unit Labor/Mat'l Cost Ratio a) Low Qty Production b) High Qty Production	 75% / 25% 65% / 35%	 70% /30% 60% / 40%	 70% / 30% 55% / 45%

The far term and high quantity production targets are the author's projections based on a full implementation of SNAPSAT™ automation capabilities and a full production scenario. In select applications, these projections may be considered conservative.

7.0 Conclusions

COI has designed, analyzed, and demonstrated a simplified, cost- effective method for the production of satellite and airborne avionic electronic packaging. This process produces an all-composite electronic enclosure that is lightweight and fully compliant with all mechanical, thermal, electrical, and radiation shielding requirements, thus providing substantial improvement over aluminum designs. The fabrication technology that has been developed produces significant savings in production time and expense over previous composite processes. Presently it is an affordable technology and will be competitive with aluminum electronic enclosures in expense and speed of production on a "Best Value" basis with more heritage and full exploitation of the SNAPSAT technology. It is also applicable to a wide variety of structures.

BIOGRAPHY

GARY A. TREMBLAY

Mr. Tremblay has over 20 years of advanced composite experience as a designer, program manager, and product manager. He has been intimately involved with the application of advanced composite materials in aerospace hardware at Composite Optics, Incorporated (COI), San Diego, CA since 1981. He has served as program manager for critical hardware and/or primary structure on notable programs such as the Hubble Space Telescope, ACTS, Telstar, FORTÉ, in addition to several comsat reflectors such as SATCOM, GSTAR, Superbird, DBS, and the SSMIS radiometer. He served as Product Manager of Spacecraft Structures at COI for five years with the responsibility of managing all of the technology and business associated with bus structures, solar array substrates, yokes/booms and equipment panels. In 1995, he was promoted to the position of Business Area Manager of Electronic Packaging, a new product technology at COI and is responsible for all business and technology aspects of advancing the application of advanced, high conductivity composite materials to electronic hardware.

A TABLE-DRIVEN CONTROL METHOD TO MEET CONTINUOUS, NEAR-REAL-TIME OBSERVATION REQUIREMENTS FOR THE SOLAR X-RAY IMAGER

Kevin Shawn Wallace, NASA - Marshall Space Flight Center, Huntsville Alabama

Terry A. Brown, NASA - Marshall Space Flight Center, Huntsville, Alabama

Kathleen A. Freestone, NASA - Marshall Space Flight Center, Huntsville, Alabama

Abstract

The design of the Solar X-Ray Imager (SXI) for the Geostationary Operational Environmental Satellite (GOES) presented an unusual scenario for controlling the sequence of solar observations. The SXI is an operational instrument, designed not primarily for scientific research, but for providing operational data used by the National Oceanic and Atmospheric Administration (NOAA) to forecast the near-term space weather. To this end, a sequence of images selected to cover the full dynamic range of the Sun will be executed routinely. As the dynamics of the Sun have differing temporal cadences, the frequency of each type of image will vary. These images must be routinely received at the forecast center in near real-time, 24-hours a day, with minimum interruptions. While these requirements clearly lead to a 'routine patrol' of images, the parameters for each do not form a static set. The dynamics of the Sun change with the 11-year solar cycle; the performance of the imaging system will vary with age and on-orbit conditions; and forecasting techniques may change with time. An additional complication is the highly restricted commanding window and slow commanding rate.

To fulfill these requirements, the SXI is designed to utilize a table-driven approach. Sequences are defined using structured loops, with nested repetitions and delays. These sequences reference combinations of imaging parameters that in turn reference tables of parameters than can be loaded by ground commands. (See Figure 1.) Multiple sequences can be built and stored for later execution. The result is an approach that provides a flexible, yet autonomous SXI capable of meeting arbitrary forecasting requirements and operating within the commanding constraints.

Operating Background

The SXI includes a focal plane assembly, a twelve-position broadband filter wheel, and a grazing incidence x-ray mirror. The focal plane assembly consists of a microchannel plate (MCP) used as the x-ray detecting element, a phosphor-coated optical taper, and a 512x512 charge coupled device (CCD). Incident x-rays striking the MCP create an electron avalanche that then strikes the phosphor deposited on the face of the optical coupler, and emits visible light detectable by the CCD. The phosphor is an yttrium oxysulphide selected to match the peak sensitivity of the CCD. Both the MCP and the phosphor are supplied with programmable high voltages that are used to control the gain. As such, the physical parameters that define an image are integration time, filter position, and two high voltage gains. It is also necessary to synchronize the image integration to the stepping of the Solar Array Drive Assembly (SADA) that provides a period of quasi-stability every 3.38 seconds.

0-7803-5086-3 /98/$10.00 ©1998 IEEE

The SXI is designed to acquire solar images in a foreground/background method. The foreground imaging, known as Routine Patrol consists of the images required to meet the primary operational objectives of the SXI. Because of the limited graduation of dynamic range and spectral bandwidth, only a finite number of image sets are operationally useful at a given time, although as the solar cycle evolves, the image set will change as well. The useful cadence of these images is determined both by the speed of evolution of solar phenomenon, and the approximately 40 seconds required to downlink a full-frame image at the 100kbps downlink rate. As a result, there will be periods of time when Routine Patrol will not be executing images. As originally envisioned, a Flare Patrol would operate as the background imaging sequence whenever Routine Patrol is not acquiring or downlinking images. Flare Patrol is not required to meet the operational SXI requirements, and would have been geared toward both scientific research and the development of algorithms to automatically detect and monitor solar flares at higher cadences, by limiting the field of view to the area surrounding the flare. However, the development of a detection algorithm that would be simple, effective, and relatively flexible proved elusive within the constraints (primarily schedule) upon the SXI design. For this reason, this bonus feature is not implemented. When Routine Patrol is not executing or downlinking an image, the SXI downlinks housekeeping and ancillary data.

Image Acquisition

Routine Patrol consists of a cyclical repetition of a set of images over a period of minutes or hours, with the capability to extend over months. The requirement is to execute a series of images, each with a different frequency. The primary example used for constructing the control concept is capability to execute a series of five images every ten minutes, an additional set of two images every hour, and an additional image every six hours. Thus, at the sixth hour, all eight images are acquired. An additional feature is the ability to add or delete images at each of these cadences, and to add an additional image cadence – perhaps one image every 24 hours – by modification to the existing table, rather than a complete reconstruction. The SXI Control Concept meets these objectives, and can be extended to arbitrary sequences using the rules of the table construction.

However, before the temporal structure of the images can be constructed, it is first necessary to define the exposure setting to be used for each image. Solar image acquisition is controlled via four table types stored within the SXI in both Random Access Memory (RAM) and Read Only Memory (ROM). These tables are known as the Exposure Setting Table (EST), Parameter Tables (PTs), Imaging Control Tables (ICTs), and Storage Count Tables (SCTs). Each image is defined by an entry in the EST that consists of a series of indexes into the PTs for each parameter. These exposures are then linked together by the ICT to form a sequence that behaves similarly to the rules of nested do-loops. (Figure 1.) The number of completed repetitions of the image at each entry (or ID) in the ICT are stored in a Storage Count Table (SCT).

The basic ICT structure was developed to allow imaging of the Sun. This control concept has been extended to other operations. The SXI can execute an ICT as Routine Patrol, Flare Patrol, MCP Startup, or XRP Oscillation Diagnostic. By exploiting the commonality between imaging and the other SXI operations, the amount of software required is significantly reduced. In each case, the rules for constructing the ICTs are the same. However, the SXI software will execute the ICT differently for each operation, resulting in slight variations in how the hardware is enabled, and how the data is downlinked. There is a default ICT for each operation, and each may be redefined by ground command.

Parameter Tables (PTs)

The Parameter Tables (PTs) contain the actual imaging parameters needed to take an image, such as filter position, integration time, or gain. Each PT consists of a series of entries that begins in ROM and continues into RAM. RAM entries in each table can be modified or added, up to the limit for each table. It is only the PTs that are specific to the particular SXI detector design. By selecting the appropriate set of PTs, the SXI Control Concept can be extended to any type of instrument.

Exposure Setting Table (EST)

A single Exposure Setting Table (EST) exists that contains multiple Setting Index entries. Each Setting Index references a set of Parameter Table index numbers and defines a single image. The first 800 entries are resident in ROM. Up to 223 additional entries can be uploaded into RAM via ground commands, accounting for a maximum of 1023 EST entries.

Storage Count Table (SCT)

A Storage Count Table (SCT) consists of a single numerical entry for each ID of the active ICT. Each table will contain up to 255 entries. One table is dedicated to Routine Patrol, one to Flare Patrol, and one to MCP Startup. The fourth table is a general purpose table that can be used for Routine Patrol, Flare Patrol, MCP Startup or XRP Oscillation Diagnostic.

When a table is made active using the appropriate command, two bits in the command are used to select either the general purpose SCT or the default SCT that has been reset. As there is no default SCT for XRP Oscillation Diagnostics, only the general purpose SCT can be used. Therefore, when the command is executed to make a certain ICT active for XRP Oscillation Diagnostic, the general purpose SCT is not reset before it is used. The general purpose SCT can only be reset by ground command.

All SCTs are set to zero during power-up initialization or software reset. An SCT will not be reset between images if the ICT is halted and then restarted without making a new ICT active. This allows execution to resume without reloading the Storage Counts.

Imaging Control Table (ICT)

An Imaging Control Table (ICT) is used to link entries from the Exposure Setting Table into continuous, repetitive sequences. There are six fields for each line (or ID) of the ICT. (See Table 1.) Multiple ICTs exist, covering a wide range of operational objectives, solar cycle activity, and seasonal or lifetime variations in detector response. Only one ICT table is active for each type of ICT. Any existing ICT can be designated the ICT for Routine Patrol. As implemented for SXI, 40 ICTs will exist in ROM, and up to 23 additional ICTs can be created in RAM by ground command, allowing a maximum of 63 ICTs, each with a maximum of 255 entries. The key feature of the ICT is that although the entries in the ICT are labeled with a sequential structure, the looping structure is independent of this labeling, being governed instead by the Transfer Fields. This allows the addition of images that are nested deeply within the structure to be included by appending entries, rather than requiring restructuring of the table to include the images as sequential entries in the ICT.

Repetitions Field (FV 1)

The Repetitions Field indicates how many times this particular ID should be executed. The number of repetitions executed is stored in a Storage Count Table. Each time a given ID is executed, the Storage Count is incremented by one.

Exposure Setting Index Field (FV 2)

This field contains the index into the EST that will define the image to be executed.

Delay Fields (FV 3 and 4)

The Delay Field is a time delay in minutes from the start of execution of one image referenced by an ID to the start of execution of the next ID, either Jump-To or Loop-Back, depending upon the status of the storage count. The Loop-Back Delay is used when the storage count is less than the number of repetitions. The Jump-To Delay is used when the storage count is equal to the number of repetitions. Splitting the delay into these two types not only simplifies conceptualization and construction of the ICT, it reduces the number of entries required to construct complex tables with multiple loops.

Transfer Fields (FV 5 and 6)

There are two Transfer Fields that contain the ID of the next image to be executed. The Transfer Field can not reference a different ICT table. The Loop-Back ID defines the nested inner loop, while the Jump-To ID defines the outer loop. When the value of the Storage Count is equal to the value of the Repetitions Field, the software resets the Storage Count to zero, and transfers execution to the ID indicated by the Jump-To ID entry. Otherwise, the software transfers execution to the ID indicated by the Loop-Back ID, and does not reset the Storage Count.

Special Features Fields

The SXI has been designed to include the capability to apply special properties to each image. These include the selection of one or more areas on the CCD, referred to as windows, that can be downlinked in lieu of the entire image; whether the image should be processed with a linear or a logarithmic transfer function; whether the filter-wheel should be returned to the radiation shield between images; whether the image data should be discarded; whether the ICT should be halted following the image; whether and how the image should be synchronized to the solar array oscillation; and whether the image should be omitted when ground command sets the SXI to a protective mode during solar flares that could damage the detector if longer, high-gain images are made. Each of these features is assigned to one or more bits in the Special Features Field. (See Table 2.)

There are two types of Special Features Fields: the Image Special Features Field (ISFF) and the Global Special Features Field (GSFF). The ISFF is extracted from a Parameter Table for each entry in the EST. The GSFF is a variable set by ground command. The Resultant Special Features Field (RSFF) is the logical OR of the ISFF with the GSFF. This allows certain special parameters to be set either for certain images or globally. Each bit of the Special Features Field is defined such that a "0" indicates the normal/default mode and a "1" indicates the override/special mode. By defining each bit in this manner, the GSFF can be used to activate the override condition but can not deactivate an override condition that is defined in the ISFF.

Flow Control

An ICT always starts executing on a one-minute boundary, defined as zero seconds past the minute. This boundary is selected due to the nominal one image per minute requirement for SXI, and the relatively long 40 seconds downlink time for a full-frame SXI image. Images with non-zero values in the appropriate Delay Field will be initiated on a subsequent one-minute boundary. The restriction to one-minute boundaries allows easy prediction of when a particular image within the sequence will be executed, even when the downlink and integration time may vary. The gaps between consecutive images would also allow time for the background task

Flare Patrol to become active, if Flare Patrol had been implemented. A delay time of zero is interpreted to mean 'as soon as possible,' and the actual start time for the image will depend only upon the time for on-board processing. Predictability for an image sequence with delays of zero is possible, but more difficult.

In order to alter the flow of the ICT, a command is provided to set the Storage Count of any ID within an SCT to any valid value. There is also a command to reset all of the Storage Counts in the specified table to zero. The SCT is normally reset for Routine Patrol when a new ICT is made active and run. Modifying the SCTs was originally designed to allow synchronization of certain images to ground milestones. However, the satellite's operations time-line usually interrupts the SXI imaging every six hours for housekeeping operations. As such, the ICTs will normally be reset at that time, and have, in general, been constructed to operate within those bounds. Modifications to the SCT can also be used to reduce the number of repetitions that will be executed by an existing ICT.

A bit in the Special Features Field can be used to 'Execute and Halt.' This allows an ICT to be autonomously terminated. If the active ICT is not changed, and the SXI is commanded to resume imaging, the ICT will begin with the next entry indicated by the Transfer Field of the image that caused the halt.

Efficient Table Operations

The key to the effectiveness of the SXI Control Concept is to provide a large set of template tables that cover various scenarios, then to utilize the indirection and tiered table structure to an operational advantage. The SXI tables span both ROM and RAM. Tables and entries resident in ROM are unalterable, but ICTs can be copied to RAM for modification and execution.

The SXI will be launched with 40 ICTs, and 800 EST entries loaded into ROM. Each of the PTs is also between 50% and 75% filled in ROM. Each of these parameters, exposures, and sequences is chosen to cover the operational range of imaging – including the changes that will be required with the evolution of the solar cycle, as well as seasonal variations in detector response. In the case of the ICTs, most contain multiple sequences that loop within a limited section of the table, allowing more sequences than permitted by the limit of 63 uniquely numbered ICTs. This limit is imposed by bit-stuffing the commands defining ICT operations, in order to minimize the number of commands sent to the SXI for nominal operations. A command to set the current ID of an ICT to any valid ID allows the 'mini' ICT to be started from within the main ICT with a minimum of commanding.

RAM Pointers

One powerful technique utilizes a ROM ICT that indexes an EST entry in RAM. At startup the RAM entry in the EST need not exist, so long as it is created prior to executing the ICT. This allows the structure of the ICT to be utilized, but the actual images of the patrol to be defined by loading the values into the RAM EST. A similar technique allows an EST resident in either ROM or RAM to use pointers to RAM PTs. The RAM PTs can be loaded in orbit, and modified as required to adjust the SXI response.

Since the selection of usable combinations of parameters is limited, multiple ICTs can point to the same EST entries; and multiple EST entries can utilize the same PT entries in different combinations. Thus, by changing the value of one parameter in a PT, the executed images in dozens of EST entries in many different ICTs are simultaneously altered. As an example, the gain of the detector is strongly affected by temperatures that will change seasonally. To account for these changes, the programmed value will be altered periodically, resulting in less variation in the actual voltage, and hence less variation in the gain of similar images made throughout the year. This can be accomplished by having the ICTs and EST entries for the desired patrol point to a set of 'favorite' gain voltages used to image the various features of the Sun. Throughout the year, these favorites can be modified by a few volts at a time, in order to level off the temperature effects.

Shadow RAM

A further sophistication of pointers is allowed by implementing a RAM shadow of the most important ROM parameters. At power-up or software reset, the SXI will copy the first five ICTs, the first twenty EST entries, and the first seven entries of each PT from ROM into RAM. The ROM ICTs and ESTs can reference the shadow values in RAM and execute without modification. When minor adjustments are required, and as time in the commanding sequence can be scheduled, the shadow locations in RAM can be modified. As an example of this feature, the SXI contains several filters that are redundant. By running the image patrols with shadow RAM entries for the filter-location, the default from ROM can be used without modification, so long as the filter remains undamaged. If, however, the filter ruptures, develops pin-holes, or is otherwise suspect, the RAM location can be modified to contain the redundant filter, and effectively make this correction for all EST entries that utilize this construction for selecting the filter.

Other Implementations Of ICTs

In addition to Routine Patrol, there are three other implementations of ICTs used to provide SXI control capability. These are MCP Startup, XRP Oscillation Diagnostic, and Flare Patrol.

MCP Startup

The microchannel plate requires a specified, controlled startup procedure. This involves the gradual increase of both high voltage power supplies (HVPS). At each step of the procedure, the voltage is maintained for a specified amount of time, while currents on the MCP are monitored. The MCP Startup procedure is similar to an imaging sequence, with the exception that the HVPS voltages are applied to the MCP for an extended duration.

An ICT for MCP Startup is constructed exactly the same as for Routine Patrol. The primary difference is that the software will enable the hardware into the extended integration mode that does not deactivate the HVPS between images. The SXI software uses a different format for the downlinked data, but continues to downlink images, unless the image data is discarded by the Global Special Feature Field. There are two ICTs for MCP Startup resident in ROM, and other tables can be constructed by ground command. The active ICT for MCP Startup can be changed from the default value to any other existing ICT. The MCP Startup procedure is then begun by commanding the SXI to begin the MCP Startup.

XRP Oscillation Diagnostic

The High Accuracy Sun Sensor (HASS) located inside the SXI is used to provide pointing knowledge during images. The SXI is mounted to the X-ray Positioner (XRP) that is in turn mounted to the solar-array yoke, and driven by the Solar Array Drive Assembly (SADA). The pointing of the SXI is thus a function of the motion of both the XRP and the SADA. A special mode, XRP Oscillation Diagnostic (XOD), has been provided to collect HASS data nearly continuously, in order to observe the dynamic oscillations experienced by the SXI from both of these sources. While executing XOD, SXI collects repetitive HASS measurements every 16 milliseconds. Because of the similarity in hardware of this diagnostic to imaging, an ICT is used to pass parameters to the software to control how the diagnostic is executed.

The SXI software downlinks HASS data in lieu of image data for any ICT that is executed as XOD. There is a default ICT resident in ROM for XOD, and other tables can be constructed by command. The default ICT collects data for 6.2 seconds, a period slightly less than twice the period of stepping of the SADA. A gap of approximately 58 measurements occurs each time the SXI downlinks the previous data set. By ground command, the active ICT for XOD can be changed from the default value to any other existing ICT. The XOD is then begun by commanding the SXI to begin XRP Oscillation Diagnostic.

Flare Patrol

The rules governing the Flare Patrol ICT are the same as for Routine Patrol. For the current SXI, it is possible to define an ICT and make it active for Flare Patrol. However, the commands to execute Flare Patrol are disabled.

Summary And Recommendations

The SXI Control Concept has proven effective at meeting the challenges set for flexibility and performance. The ICT and EST concept can be applied directly to other designs, with different types and quantities of parameter tables used to reflect the configuration of a different instrument. However, there are several areas where a new design utilizing the these concepts can be improved.

Ideally, there should be a Storage Count Table (SCT) for each ICT. The restriction to share SCTs between ICTs has not hampered the SXI, but might be a constraint on future designs utilizing this concept.

Ideally, there should be the capability to preserve table modifications by transfer to non-volatile memory. Due to various design constraints that dictated a rigorous management of board-space for design components, the SXI does not include non-volatile memory. During orbital eclipse season, the SXI will lose power fairly frequently over the span of a few weeks, and will thus loose the table modification resident in RAM. It is an operationally intensive effort to reload the values, and the SXI must operate from ROM tables with few, if any, modifications until sufficient commanding time is found to reload the tables with optimum values. A design that includes non-volatile memory for storing table modifications would easily alleviate this issue.

Acknowledgments

Contributions to the development of the SXI Control Concept, both direct and indirect, are the result of the close cooperation of the authors with NOAA and the entire SXI team. Citations of credit for the development of the concepts in this document are given particularly to:

Richard N. Grubb
Chief Engineer
NOAA Space Environment Center

Dr. Patricia Bornmann
Responsible Scientist
NOAA, Space Environment Center

Dr. John Davis
Chief, Solar Physics Branch
NASA, MSFC

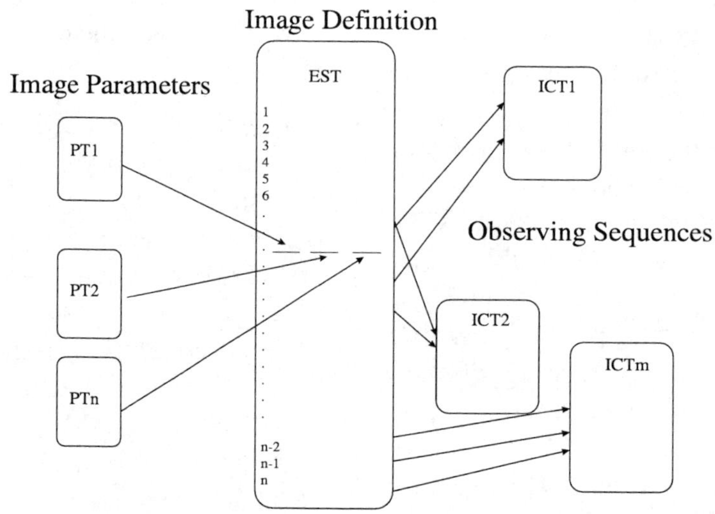

Figure 1: SXI Control Concept Table Structure

	FV=1	FV=2	FV=3	FV=4	FV=5	FV=6
ID	**Reps**	**EST Index**	**Loop-Back Delay**	**Jump-To Delay**	**Loop-Back ID**	**Jump-To ID**
1	R1	ES1	LD1	JD1	LB1	JT1
2	R2	ES2	LD2	JD2	LB2	JT2
m-1	Rm-1	ESm-1	LDm-1	JDm-1	LBm-1	JTm-1
m	Rm	ESm	LDm	JDm	LBm	JTm

Storage Count
SC1
SC2
SCm-1
SCm

Table 1: ICT Structure

	Description	**#Bits**	**Notes**
LSB	Window(s) Selection	6	One bit for each window 1=yes; 0 = no
	Flare Test Image	1	Not Implemented
	Linear/Logarithmic Output	1	Selects the gain channel for the video
	Execute Image Then Halt	1	For automatically halting the ICT
	Return To Radiation Shield	1	Sets the filter-wheel to the radiation shield between images
	Disable Line Advance	1	For disabling shifting of the CCD to track the Sun
	Disable SADA Synchronization	1	
	Discard Image Data	2	Image data is not downlinked after image execution
	Calculate SADA Offset	1	
MSB	Exposure Override	1	If this bit is set to 1 and the internal detector protect enabled flag is set to 1, the image will be omitted

Table 2: Special Features Field

THE TRWIS III AIRBORNE HYPERSPECTRAL IMAGING SYSTEM

Stephanie R. Sandor-Leahy, TRW Space and Electronics Group, Redondo Beach, California

Abstract

The TRW Imaging Spectrometer III (TRWIS III) airborne hyperspectral imager is a pushbroom sensor that collects imagery in 384 contiguous spectral channels covering the 400nm to 2450nm wavelength range. The instrument was completed in 1996 and underwent its first flight season in the second half of that year. Since then, TRWIS III has been used to gather data for applications including remote classification of desert vegetation, forest species, and man-made materials. Along with in-flight operational demonstrations, the instrument has undergone laboratory performance validation and radiometric calibration. This paper will describe the TRWIS III instrument and its current applications to remote sensing.

Introduction

Hyperspectral imaging spectrometers produce data with high spectral resolution, typically in the range of 5 to 10 nm, giving processors the ability to detect subtle spectral features and differentiate objects from a background. This powerful capability is important for remote sensing applications (e.g. geologic typing and surveying, agricultural monitoring and optimization, environmental damage assessment, forestry surveys, etc.) and for military applications (camouflage detection, identification friend or foe, target typing, mine detection, etc.).

The TRWIS III instrument was designed as a research grade sensor capable of producing high fidelity hyperspectral data in the 400 to 2500 nm wavelength range. The top-level performance requirements were for an instrument with extremely high signal to noise ratio (SNR), co-registered spectral channels taken simultaneously, accurate location knowledge, and accurate radiometric calibration. Another design requirement was that the instrument could be adapted to a wide variety of aircraft platforms. The TRWIS III instrument has met these design criteria and generated an extensive database of hyperspectral imagery.

TRWIS III System Description

The TRWIS III system consists of a sensor head, containing a pair of co-boresighted grating spectrometers, two electronics racks, a digital data recorder, and a color video camera. The Visible/Near Infrared (VNIR) spectrometer covers the wavelength range from approximately 400 nm to 1000 nm in 5 nm wide bands, and the Shortwave Infrared (SWIR) spectrometer operates from approximately 900 nm to 2500 nm in 6.25 wide bands.

Each spectrometer consists of a set of refractive foreoptics that image the scene onto a slit. Light passing through the slit is dispersed perpendicular to the slit by a flat grating and then imaged onto a two-dimensional focal plane array (FPA). One dimension of the array, the dimension along the slit, provides spatial scene information. The second dimension of the array, along which the light from any given point in the slit has been dispersed, provides spectral information. An image is generated by moving the instrument across a scene in a pushbroom fashion, perpendicularly to the instrument's slits, and recording frames of spectral and spatial information detected by the VNIR and SWIR FPA's. The image cubes generated in this manner contain 256 pixels in

0-7803-5086-3 /98/$10.00 ©1998 IEEE

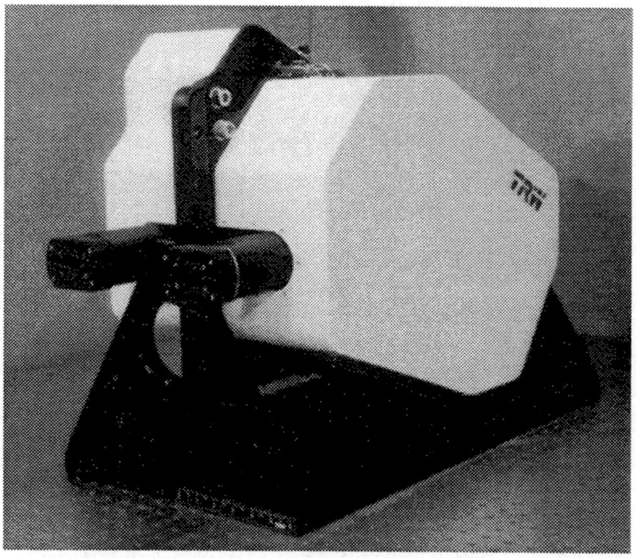

Figure 1. TRWIS III Instrument Sensor Head

the cross-track direction and 384 spectral bands per pixel.

TRWIS III was designed with a variable frame rate, which allows for imaging at altitudes from 600m to 12km. The 0.9mrad instantaneous field of view (IFOV) of the instrument results in spatial resolution ranging from 0.5m to 11m. The system includes on-board navigation equipment, including a Global Positioning System (GPS) with a differential correction receiver and an inertial navigation sensor (INS). Navigation data from these systems is used in TRWIS III data post-processing for image rectification and geo-location. The TRWIS III sensor head configuration is shown in Figure 1.

System Electronics

A custom electronics data recorder interface board has been designed and built. The board connects TRWIS III to a high speed, high volume digital tape recorder (an Ampex DIS 160i unit). The interface allows formatted image data and calibration files to be written directly to digital tape, bypassing intermediate computer storage media. One Ampex tape can store up to 165 Gbytes of digital data; therefore, the length of flight lines is no longer limited by storage considerations. The interface also can play back recorded data. This capability allows the user to review recorded data at any time after acquisition. The recorded scenes can be played back at any of the three available frame rates and displayed in a waterfall fashion as side-by-side, three-band VNIR and SWIR images.

The interface electronics allow data to be recorded at average rates of 16 Mbytes/second and can also be used separately from TRWIS III, as a stand-alone unit, to record and play back data from the Ampex. The interface contains 3 differential output video ports and one single-ended output video port. In the TRWIS III system one of these differential video ports is connected to a frame grabber board allowing in-flight, real-time viewing of data as it is being recorded. The interface board also contains one unused input port which is configured so that in a future system upgrade navigation data can be imbedded directly in each image frame rather than written to computer hard drive. This upgrade would automatically synchronize navigation data (position and attitude) with each frame of an image.

Calibration System

TRWIS III has In-Flight Calibration Sources (IFCS's) for both the VNIR and SWIR spectrometers. These sources have replaced the original IFCS's [1]. An IFCS unit consists of two quartz tungsten halogen lamps (6 Watt bulbs running at 1 Amp) mounted inside a blackened, light-tight assembly that attaches to the housing of the spectrometer foreoptics. Each assembly also contains two panels, either of which can be moved in front of the spectrometer's entrance aperture. The inner panel is made from a 99% reflectance standard and the outer panel from a doped reflectance standard (Erbium Oxide in the VNIR case and Holmium Oxide for the SWIR).

0-7803-5086-3 /98/$10.00 ©1998 IEEE

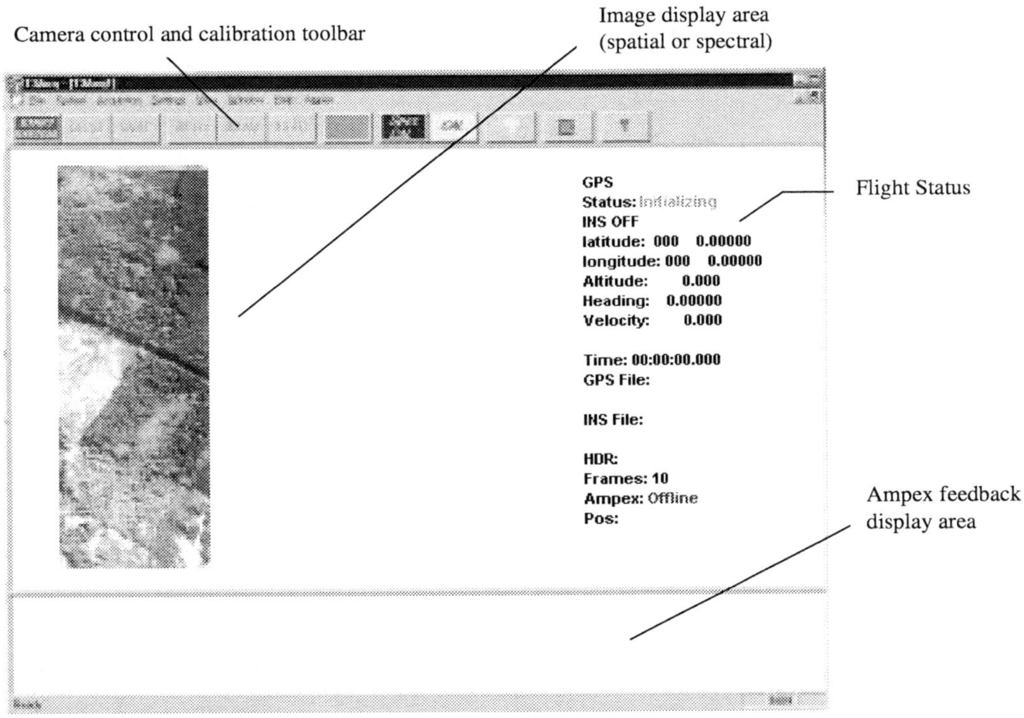

Figure 2. TRWIS III Data Acquisition Software User Interface.

In-flight calibration consists of closing the calibration assembly cover, recording frames of data with the lamps off, then turning the bulbs on, and recording the instrument's response to the illuminated reflectance standard. This procedure is followed once before and after each flight line. The calibration process is completely automatic and is controlled by one click of the mouse button from the system operator's control panel (see Data Acquisition Software). This calibration technique allows the entire throughput of the optical system as well as the FPA response to be measured in-flight.

In addition to an in-flight radiance calibration of the instrument, the calibration system also provides data for spectral calibration. Since the new calibration system now includes a panel of material from a doped reflectance standard, an instantaneous measurement of the instrument's cross track spectral error can be made. That is, the spectrum of the doped material can be used to determine the wavelength center for each pixel in the FPA.

Data Acquisition Software

Since the instrument was first flown, the in-flight data acquisition software has been upgraded. The new TRWIS III Acquisition Program (T3ACQ) is an operator-friendly, Windows-based program that controls acquisition of GPS, inertial navigation sensor, and spectral image data. The software is easy to use and allows the operator to rapidly control instrument operations. T3ACQ has a wide variety of options that can be used to provide the operator with real-time feed back on system performance and status.

By running the acquisition software, the operator calls up an interface that includes a number of menu options, a status display, and an image display. Menu options are grouped by functionality, and a tool bar contains the most frequently used options (see Figure 2).

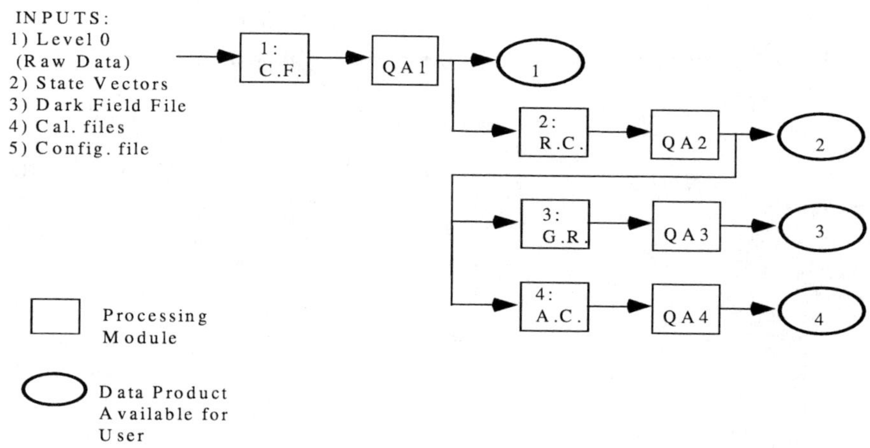

Figure 3. TRWIS III Data Calibration and Correction High-Level Process Flow.

Instrument operating commands are entered using a mouse to click buttons or sliders that control instrument parameters. The operator can initialize the system and select camera parameters such as frame rate, VNIR iris settings, gains and offsets. The interface allows these parameters to be changed at any time during flight.

Once the system is initialized, the operator can select from a variety of data display or record options. Spectral data can be recorded to the Ampex and a spatial waterfall image displayed in real time using operator-selected band numbers for blue, green and red. Also VNIR and SWIR imagery can be displayed side-by-side in real time as spatial, monochrome, waterfall images. There is an option to record data to available memory rather than to the Ampex, and for diagnostics there is an option that only updates the waterfall display and does not record data. Note that the software can display either spectral or spatial data so that the operator has the flexibility to view real-time data directly from the focal planes.

Data Correction and Calibration

A ground-based software system has been developed for formatting, correcting and radiometrically calibrating TRWIS III data. This data processing system was designed to be modular as shown in Figure 3. The system was also designed to be as automated as possible and to keep a log file, generated as a data set is processed, which lists each processing step applied to the data. These steps include cube formation (C.F.), radiometric calibration (R.C.), georectification (G.R.), and atmospheric correction (A.C.). The quality assurance (QA) steps are currently designed to be a combination of manual and automatic checks which look for artifacts not corrected by the processing modules. The software is configured so that cube formatting, radiometric correction, and the associated automatic QA steps can be performed on platforms including SPARC, SGI, and PC workstations. There are four data products generated by the system - these products are labeled 1 - 4 on Figure 3 and are listed in Table 1. The products are numbered for clarity and do not correspond to any standard definitions of processing levels for remote sensing data.

Table 1. Available TRWIS III Data Products

Products	Definition
1	Data converted to 16 bits with an added descriptive header/footer
2	Product 1 with gain and bias radiometrically corrected (note: automated QA2 steps include identification of bad pixels and replacement using surrounding pixel values)
3	Processed 3 band resampled (geo-rectified) and terrain corrected image with reference back to product 2 for all 384 bands
4	Product 2 with atmospheric attenuation effects removed (i.e. data available in reflectance units)

The software system is controlled by a master script file that configures the modules and initiates the processing by prompting theuser for a series of input files (i.e. raw data, lab calibration, in-flight calibration, and navigation files). Once the system has been initialized, the cube generation module takes the raw data file, formats it into 16-bit data of a selected interleave type - typically Band Interleaved by Line - and attaches a 256 byte descriptive header and a footer to the data. The header contains image acquisition details (e.g. number of image lines, number of image bands, image latitude/longitude). The output of the cube generation module is a raw data cube that can be read by off-the-shelf software packages.

The cube calibration module takes as input a formatted raw data cube that has been processed by the cube generation module and applies gain and offset corrections to the data. The offsets are calculated using in-flight dark files. These dark files consist of frames of data taken with the calibration cover closed before and after an image acquisition (see Section 2.2). The cube calibration module averages the frames of data to derive an offset correction then subtracts this offset from each frame of image data. The gain correction is then applied to each frame of the dark-subtracted data. In general, the gain corrections are derived from laboratory measurements of pixel responsivity. The laboratory gain file can be modified using the in-flight calibration data. Each image data frame is multiplied by the gain corrections to yield a radiometrically corrected image cube in units of radiance. Once again, the radiometrically corrected data, which is the output of this module, can be read by off-the-shelf software.

The other two processing steps that can be applied to the data are georectification and atmospheric correction. The georectification software used in this system was developed specifically for this project by the TRIFID Corporation, a company that specializes in geo-correction of imagery collected from satellite and airborne platforms. An interface was written between the TRW and TRIFID software so that the TRIFID software can take TRW-generated imagery and navigation data and automatically geo-locate and rectify the image data. Note that internal to the TRIFID software are a number of quality assurance checks which are performed automatically. The output of the TRIFID software are a 3-band georectified and geolocated image and a file listing the spatial corrections applied to each pixel. This file can be used to automatically correct any or all of the remaining 381 spectral bands of the image.

Atmospheric water vapor levels can be retrieved from the TRWIS III radiance data. The method was developed using ATREM [2], which was modified to work with TRWIS III imagery. The basis of the retrieval is a curve fit

Figure 4. High resolution TRWIS III imagery. The spatial resolution of the images is approximately 1m. The images were generated using 3 bands from the VNIR to produce "true color" images.

of calculated spectra in the 1.14um and 0.94um water vapor band absorption regions to observed spectra. The derived water vaporvalues are then used to remove the effects of water vapor absorption features throughout the measured spectral range.

The atmospheric correction module takes as input a calibrated TRWIS III image (in radiance units) and a text file which contains a description of conditions at the time of image acquisition and some instrument parameters (e.g. latitude/longitude, wavelength centers, etc.). The module verifies the wavelength center.

TRWIS III Data Collections

Since its first flight season in 1996, the instrument has been used for a number of remote sensing applications. TRWIS III images have been acquired over a wide variety of scenes, including forests, deserts, oceans, urban areas, agricultural fields, and military scenes (see Figure 4). The instrument has been flown under many different environmental conditions and at altitudes ranging from 1,500 to 28,000 feet [3].

In many instances, ground truth data has been collected concurrently with instrument overflights. Ground measurements have been

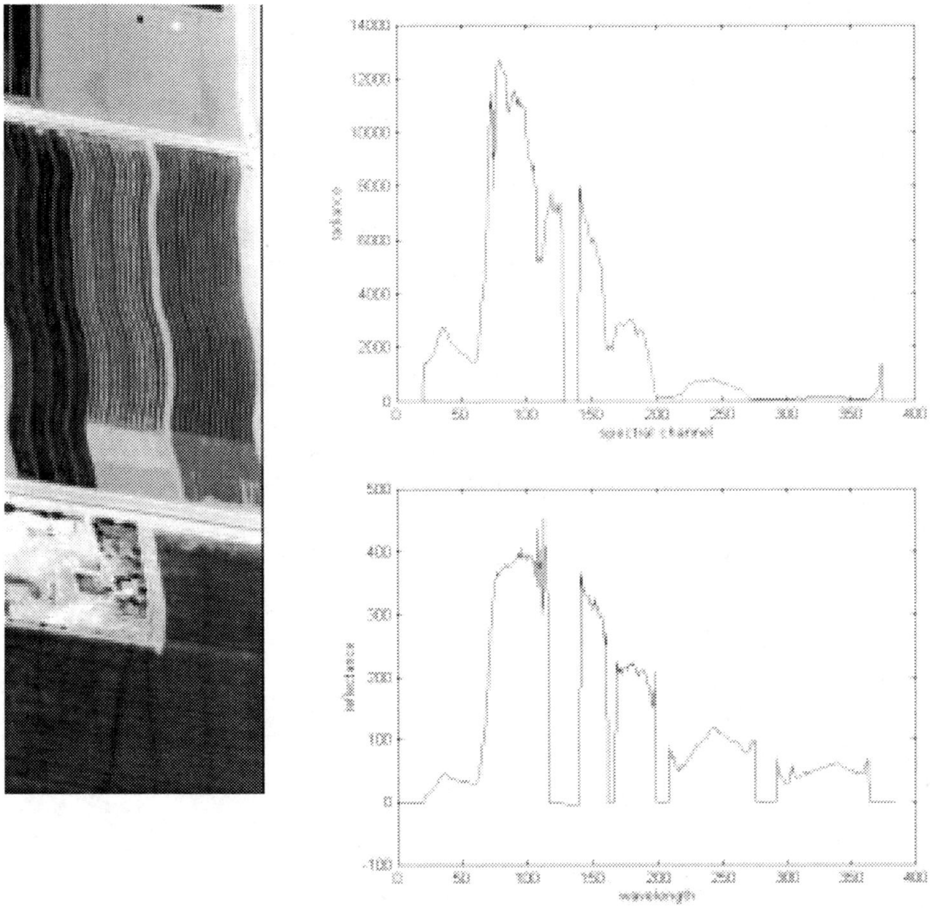

Figure 5. TRWIS III image over agricultural fields (left). Radiance (upper right) and reflectance (lower right) spectra from the center field. The radiance units are 0.01 watts/m^2-sr-um. Reflectance has been multiplied by 1000.

taken using a field spectrometer with a spectral range from 350nm to 2500nm. Both spectral signatures and GPS data have been gathered for targets of interest (e.g. vegetation, soil, white and dark calibration panels) in many TRWIS III scenes. This ground truth data has been archived along with the hyperspectral imagery, forming an extensive database of spectral information.

Typically, the instrument has been used in applications that require remote detection and discrimination of land cover types. A current, on-going project requires that TRWIS III data be used to differentiate multiple varieties of the same crop (see Figure 5). A series of flights have taken place over agricultural fields at various geographical locations across the United States. Between 5 to 10 GBytes of TRWIS III imagery has already been acquired for this project. The images are currently undergoing radiometric calibration, atmospheric correction, and analysis. Preliminary results indicate that varietal differences can be detected in the hyperspectral images, and a thorough analysis of the flight and ground data is currently underway.

Conclusion

The TRWIS III hyperspectral imager has been operating successfully since 1996.

Thus far the instrument has collected imagery over a wide variety of scenes, such as forests, deserts, oceans and military targets. Since the early phases of its development, the instrument has undergone significant upgrades including - but not limited to - a new in-flight calibration system and new data acquisition software. The upgrades were designed and implemented by the TRWIS III engineering team after many flight hours of experience operating the system. In conjunction with instrument maintenance and upgrade activities, a data processing system has been developed. The system is highly automated, except for specific quality assurance steps that are performed manually. There are four data products available from the processing system, including atmospherically corrected and georectified products.

Over its last several flight seasons the TRWIS III instrument has produced large quantities of high-quality hyperspectral imagery. TRWIS III is a commercially available system, operating in the 400nm to 2450nm wavelength range, that is capable of producing data with high signal to noise ratios, spectral and spatial resolutions, and radiometric accuracy.

Acknowledgment

Although this paper has one author, all aspects of the TRWIS III instrument program have been a true team effort. The author would like to thank the following people, who have designed, built, operated, and modified the TRWIS III instrument: Debra Beiso, Miguel Figueroa, Mark Folkman, Darrell Gleichauf, Ted Hedman, Peter Jarecke, and Sveinn Thordarson.

References

[1] M. A. Folkman, R. K. DeLong, C. T. Willoughby, D. A. Gleichauf, S. Thordarson, M. Figueroa, W. Procino, "TRWIS III: An Aircraft-Based Hyperspectral Imager", *Proc. Optical Science, Engineering, and Instrumentation Symposium*, **SPIE 2819-15**, 1996.

[2] B. Gao and A. F. H. Goetz , "Column Atmospheric Water Vapor and Vegetation Liquid Water Retrievals From Airborne Imaging Spectrometer Data", *J. of Geophysical Research*, Vol. 95 No. D4, pp. 3549-3564, 1990.

[3] J. Pearlman, M. A. Folkman, P. Jarecke, S. Sandor-Leahy, D. Xu, "The TRWIS III Hyperspectral Airborne Imager: Capabilities and Directions", *Proc. ISSSR,* December, 1997.

FAULT PROTECTION DESIGN OF THE QUIKSCAT AND SEAWINDS INSTRUMENTS

Matthew B. Bennett, Jet Propulsion Laboratory, California Institute of Technology, Pasadena, CA
Joseph F. Smith, Jet Propulsion Laboratory, California Institute of Technology, Pasadena, CA
William B. Wilkinson, Federal Data Corporation, Pasadena, CA

Abstract

The QuikScat and SeaWinds instruments are radar scatterometer instruments that will be used to measure ocean surface winds. The QuikScat instrument will be launched on a dedicated spacecraft in November 1998, and the SeaWinds instrument will be launched on the Japanese ADEOS-II spacecraft in the summer of 2000.

The instrument is designed to continuously operate in a wind observation mode for nearly the entire three year mission. However, a number of fault and external conditions can occur that will interrupt the instrument's continuous wind observations. These types of faults include the failures in the radar unit's TWTA, communication errors with the spacecraft, communication errors between the instrument's three subsystems, software errors in the computer subsystem, and possible effects of cosmic ray or solar induced single event upsets in the instrument's computers.

In general, the philosophy of the instrument's autonomous fault response is to perform different levels of resets in order to clear a fault that is causing a particular type of problem. In general, the instrument attempts to recover from the fault in a manner that will allow the instrument to resume normal operations without ground intervention. However, if the fault does not clear with a reasonable level of effort by the autonomous algorithms in the instrument, the instrument places itself into a safe standby mode and waits for ground interaction. In no case does the instrument attempt to recover from faults by switching redundant units. The switching of redundant units is to only be performed under command and control from the ground.

This paper describes the fault protection mechanisms that have been designed into the spacecraft, in order to react to certain faults and failures in the instrument. In addition, it explains how these mechanisms escalate their response when a fault is not cleared by their initial response. Also, this write-up describes the actions that the spacecraft will take on behalf of the instrument in the case of a spacecraft failure that will require the shutdown of the instrument.

Background

The QuikScat and SeaWinds instruments are scatterometer microwave radar instruments which are designed to provide a measurement of ocean surface wind velocity over 90% of the earth's ice free oceans every two days. The instruments will be operated continuously to provide global, long term wind vector data for studies of ocean circulation, climate, air-sea interactions, and for weather prediction. Present plans are for two SeaWinds instruments to be flown in the next two years. The first instrument (SeaWinds on QuikScat) will fly on the dedicated QuikScat spacecraft with a scheduled launch date of 1 November 1998. This instrument is a replacement for the successful NSCAT instrument which was lost due to a spacecraft failure on 30 June 1997. A drawing of the QuikScat spacecraft appears in Figure 1, below. The second SeaWinds instrument (SeaWinds on ADEOS) will be launched with four other instruments on the Japanese ADEOS-II spacecraft in the summer of 2000. A drawing of the ADEOS-II spacecraft appears in Figure 2, below.

0-7803-5086-3 /98/$10.00 ©1998 IEEE

FIGURE 1

The instruments consist of three separate subsystems (see Figure 3 at the end of this paper). The Scatterometer Electronics Subsystem (SES) is the radar transmitter, receiver, traveling wave tube amplifier (TWTA), and radar signal processing and control electronics. The SeaWinds Antenna Subsystem (SAS) is the radar antenna, spin motor, and motor control and monitoring electronics. The Command and Data Subsystem (CDS) is the command, control and data handling center for the instrument, and the command, telemetry, and power interface with the spacecraft. All subsystems have a backup redundant unit.

The differences between the QuikScat & SeaWinds versions of the instrument include the interfaces to the spacecraft for command, telemetry, power, and switching, and the addition of survival heaters for QuikScat.

General Fault Protection Approach

The instrument and the spacecraft incorporate a variety of fault protection features with the primary objectives of: (i) avoiding permanent damage or degradation to the instrument, or (ii) avoiding disruption of normal operations, to the instrument itself, or to the spacecraft. Efforts are made to minimize the loss of science data, although in no case is autonomous action taken to swap to redundant units.

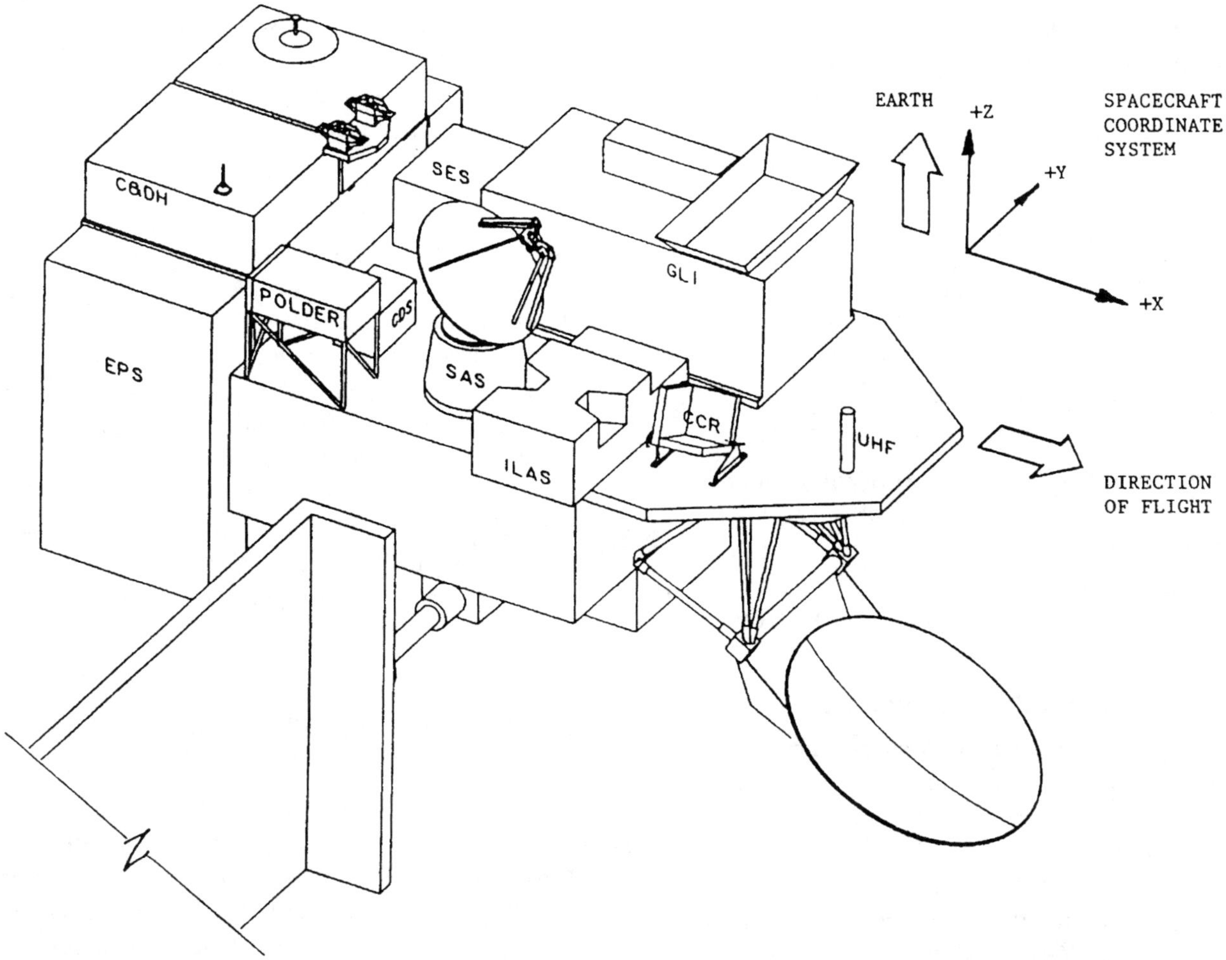

ADEOS-II SPACECRAFT
FIGURE 2

Hardware Provided Fault Protection

Hardware fault protection is used for faults that require immediate action to prevent permanent damage or degradation to the instrument.

Hardware Under-Voltage Trips

Each subsystem element (SES electronics, TWTA, CDS, SAS) has hardware under-voltage trip and reset features. Each subsystem power supply resets and prevents operation when its input voltage falls below specified trip levels. Also, the QuikScat spacecraft shuts down the instrument electronics if the spacecraft bus voltage drops below the undervoltage trip point. The spacecraft performs the shutdown by removing power from the payload electronics bus and automatically supplying it to the instrument survival heaters.

TWTA Hardware Safing

The TWTA incorporates internal hardware fault protection features to remove high-voltages and prevent RF transmission in response to anomalous conditions of under-voltage and over-current. The TWTA responds to these anomalous conditions by entering a reset mode that removes the high voltages for a cathode warm-up period (approximately 210 seconds). For undervoltage conditions, the warm-up period is not started until the input voltage is returned to normal levels.

0-7803-5086-3 /98/$10.00 ©1998 IEEE

SAS Hardware Spin Inhibit

The SAS antenna is mechanically caged during launch. The SAS electronics inhibits the continuous application of power to the drive motor when in the caged configuration.

Memory Integrity Protection

The CDS RAM memory is protected from single bit errors by an error detection & correction code. If a single bit error is detected during a read of a memory location, then the hardware automatically corrects the data, and rewrites it to the proper memory location.

To insure the integrity of SES parameters in its memory, the CDS automatically resends operating parameters to the SES at the beginning of each packet cycle. Additional protection is gained by the SES processor executing its software from radiation hard PROM.

Internal Computer Fault Monitoring

The CDS and SES computers provide internal fault protection for watchdog time-outs. In addition, the CDS has an algorithm for responding to missed CDS interrupts missed. Finally,. the CDS employs a two level reset response mechanism that attempts to restore the instrument's operating mode with a "soft" reset" if possible, before resorting to a full "hard" reset.

Watchdog Time-Out

Both the CDS and SES computers have internal watchdog timers. The watchdog timers countdown from a starting value of 500 milliseconds. A timer is reset to its starting value periodically during nominal computer processing. A fault occurs if a watchdog timer counts down to zero. A fault implies that the computer has experienced a failure that prevented it from resetting the timer. An example of such a fault is an infinite loop caused by an error in program memory.

In response to a watchdog timeout, the corresponding subsystem resets itself and attempts to continue normal operations. The CDS will perform a "soft" reset in response to a watchdog timeout. After it is complete, the CDS will continue operations in the current mode with the current set of variable parameters. The SES will perform a system reset in response to a watchdog timeout. After it is complete, the CDS will (per normal operations) reload the set of parameters and continue operations in the current mode.

Missed Interrupt Faults

The CDS services interrupts using a polling mechanism. The interrupts nominally occur in a deterministic order. If the CDS detects that an interrupt is asserted out of order, it is counted as a missed interrupt. If the count during a packet is less than the threshold, then the CDS assumes that the cause is a power on transient. In this case, the CDS resets the SES to resync it to the CDS. If the number of missed interrupts is equal to the threshold or greater, then the CDS assumes there is CDS software problem, and performs a "soft" reset.

CDS Two-level Reset Processing

The CDS flight software implements two levels of resets: a "soft" reset and a "hard" reset.

The purpose of the "soft" reset is to provide a graceful recovery from a fault by returning to the operating mode and state before the reset event. The flight software code is reloaded from PROM. Critical variables needed for restoring the instrument operating state are retained. Also restored are ultra-critical variables that provide error diagnostic telemetry. Variable are restored using a 2 out of 3 priority voting scheme. If all three copies disagree, then default values are loaded. The events that can cause a "soft" reset are a "soft" reset command from the spacecraft, a missed interrupt fault, a multiple SES data loss fault, a multiple SAS data loss fault, or a CDS watchdog time-out.

The purpose of the "hard" reset is to provide a last resort autonomous recovery from a fault. In this case, no attempt is made to restore the instrument operating mode. The instrument is initialized to standby mode. If the TWTA was on when the "hard" reset occurred, the TWTA will still be powered when the instrument goes to standby mode. However, the TWTA is safe because the instrument is in standby mode, and the SES will not be transmitting. Software is reloaded from PROM and only the ultra-critical variables are restored. The events that can cause a "hard" reset are a CDS hardware power on reset, a CDS unrecoverable operating system fault, and 3 CDS "soft" resets within the 5 minutes.

CDS Provided Fault Protection

The CDS software provides for the detection of errors, the recovery from errors, and the safing of the instrument, in response to the anomalous conditions described below.

TWTA Initialization/Reset Faults

The CDS periodically monitors the SES engineering telemetry status word for the presence of TWTA fault protection trips. The CDS immediately turns off the TWTA and transitions into standby mode as the recovery response to persistent trips. Intermittent trips are handled by the TWTA internally. It resets itself back to its nominal power on state after 210 seconds. A TWTA fault protection trip is caused by either an undervoltage or overcurrent within the TWTA. Occasional trips can be caused on-orbit by spurious radiation hits.

CDS performs the following processing to filter out intermittent trips. Upon detection of a trip, the CDS increments a trip counter. The CDS increments the counter as long as a trip condition continues. If trip conditions are not present for a time, then the CDS decrements the trip counter during this period. The CDS continues this process of counting up the trip counter when a trip is present, and counting down this trip duration counter when no trip is present. If the trip counter ever reaches the persistence threshold, then the CDS executes the recovery response.

Low TWTA Drive Power Faults

The CDS monitors the TWTA drive power during transmissions in order to protect the TWTA from the adverse effects of low drive power. If the TWTA drive power persists below the fault threshold, then the CDS disables the TWTA grid in all instrument modes to protect the TWTA from receiving low drive power. The CDS then transitions to the receive only mode to allow the ground to safely diagnose the SES without it transmitting, and with the TWTA left powered.

SES Supplemental Heater

The SES supplemental heater is used to heat the SES when the TWTA is unpowered. During changes to modes of operation that power on the TWTA, the CDS automatically turns the SES supplemental heater off. This protects against the ground forgetting to unpower the heater. The SES supplemental heater may cause an SES over temperature fault if it is left on while the TWTA is powered.

Missing Equator Crossing Update

The spacecraft sends an equator crossing signal to the instrument once per orbit when it crosses the ascending node. The CDS maintains an orbit step counter which is reset when it receives this signal. CDS uses the orbit step counter to index into tables of radar parameters that are orbit position dependent. These parameters are doppler and range coefficients used to compute the transmit frequency and receive gate delay.

The CDS continues operations as normal if the orbit step counter overflows and rolls over before CDS receives an equator crossing signal. When the orbit step counter rolls over, it points to the first entry in the doppler and range coefficient tables. CDS continues to operate using these doppler and range table entries. CDS also continues to increment the orbit step counter. If an equator crossing command is still not received upon the counter reaching the second orbit step, then the CDS reports a missing equator crossing error in telemetry, and continues operations as normal.

If the equator crossing signal fails permanently, then the CDS has an alternate mechanism for maintaining the orbit step. The instrument can be commanded to estimate the orbit period and respond to equator crossing synchronization commands from the spacecraft. These commands will be sent by the operations team as often as necessary to maintain the required orbit step accuracy.

Subsystem Communication Errors

The CDS monitors its communications with the SES and SAS for outages. The CDS resets the other subsystem at the first indication of an outage. If these outages do not clear within a defined period of time, then the CDS performs a "soft" reset. This is done because it is assumed that if the problem can not be resolved in the other subsystem, then the problem could be in the CDS flight software.

SES Data Errors

The CDS detects an SES data error when the SES sends an incorrect amount of digital telemetry to the CDS within the prescribed time period. The SES should return a deterministic amount of data to the CDS .for each command and engineering telemetry request. Data not received due to transmission anomalies or parity errors are dropped by the CDS hardware. When too few data are received, the CDS cannot determine which data were not received. The CDS treats the data received as if it is in the expected order and puts that data in telemetry. However, it marks the remaining positions in telemetry (where data should have gone) with error codes.

Multiple SES Data Error Faults

A multiple SES data error fault is defined as at least 8 different SES data errors during a packet. At the first occurrence of this fault, the CDS resets the SES the end of the present packet. In the case that the SES may have experienced a POR or a watchdog time-out, the CDS waits for the fifth packet, and checks again for the fault. If the fault persists in the fifth packet, the CDS assumes the fault lies within itself, and initiates a "soft" reset.

SAS Data Errors

The CDS detects an SAS data error when the SAS sends an incorrect amount of antenna position data to the CDS within the prescribed time period. SAS data errors are handled like SES data errors with the following addition. The CDS calculates the antenna position based on the last SAS antenna position data and reports it in telemetry. The CDS uses the antenna position to command the SES doppler shift and receive gate delay. Upon returning from a POR, the CDS initializes the antenna position to zero until a valid input is received from the SAS.

Multiple SAS Data Error Faults

A multiple SAS data error fault is defined as at least 8 different SES data errors during a packet. In the case that the SAS may have experienced a POR, the CDS waits for the fourth packet, and checks again for the fault. If the fault persists after the waiting period, the CDS assumes the fault lies within itself, and it initiates a "soft" reset.

SES Single Event Upset Faults

The SES may experience single event upsets (SEUs) when the spacecraft flies over the South Atlantic Anomaly. The CDS has a countdown timer to reset the SES periodically for automatic recovery from these potential SEUs. The countdown timer is restarted at each ascending node crossing. The default value for the countdown duration is the time it takes the spacecraft to cover 75% of an orbit. This duration results in an SES reset once per orbit over the South Pole.

Spacecraft Provided Fault Protection

The ADEOS-II and QuikScat spacecraft provide fault protection to the instrument for spacecraft faults and for monitoring selected analog measurements.

Spacecraft Faults

QuikScat Undervoltage Response

The QuikScat spacecraft removes power from the payload bus when the spacecraft bus experiences an undervoltage fault. Powering off the payload bus turns off all instrument electronics and replacement heaters. The spacecraft applies power to the instrument survival heaters through a separate power interface.

SeaWinds Light Load Mode Response

The ADEOS-II spacecraft will command the instrument to thermal safe mode in response to anomalous spacecraft conditions (e.g., spacecraft bus under-voltage, loss of attitude control). The instrument thermal safe mode powers all instrument electronics off and all replacement heaters on.

Spacecraft Response to Instrument Faults

The spacecraft monitors selected instrument analog measurements. If the value of a measurement goes beyond a pre-determined set of limits, the spacecraft will make a response. The response of the spacecraft to an out-of-limits condition is to place the instrument in standby mode, and place the subsystem at issue into a thermal safe condition.

Analog Measurement Monitors

The spacecraft monitors instrument passive analog temperature measurements. The spacecraft responds to all out of bounds conditions (a temperature too high or too low) by first commanding the instrument into the standby mode. The spacecraft then powers off the subsystem with the out of bounds condition, and turns on its replacement heater. If the subsystem with the out of bounds condition is the CDS, then the entire instrument is commanded to thermal safe mode.

SeaWinds Spin Rate Monitor

In addition to temperature measurements, the ADEOS-II spacecraft monitors the SAS antenna spin rate. The spacecraft responds to an out of bounds spin rate condition by commanding the instrument into standby mode, powering off the SAS electronics, and turning on its replacement heater.

Conclusion

A spacecraft instrument fault protection design has been implemented with the following key features:

⇒ Fault responses escalate the scope of the response if their first attempts to recover are unsuccessful.

⇒ The approach taken for a fault response depends on whether the instrument health is at risk.

⇒ A fault response is chosen that maximizes the safety of the instrument with a minimal concern for science data loss if the instrument health is at risk.

⇒ A fault response attempts to restore the instrument's operating mode, parameters, and nominal science data collection if the instrument health is not at risk.

⇒ First attempts at recovery are those with the smallest impact to operations if the instrument health is not at risk.

⇒ Instrument safing and discontinuation of nominal operations is taken as a last resort if the instrument health is not at risk.

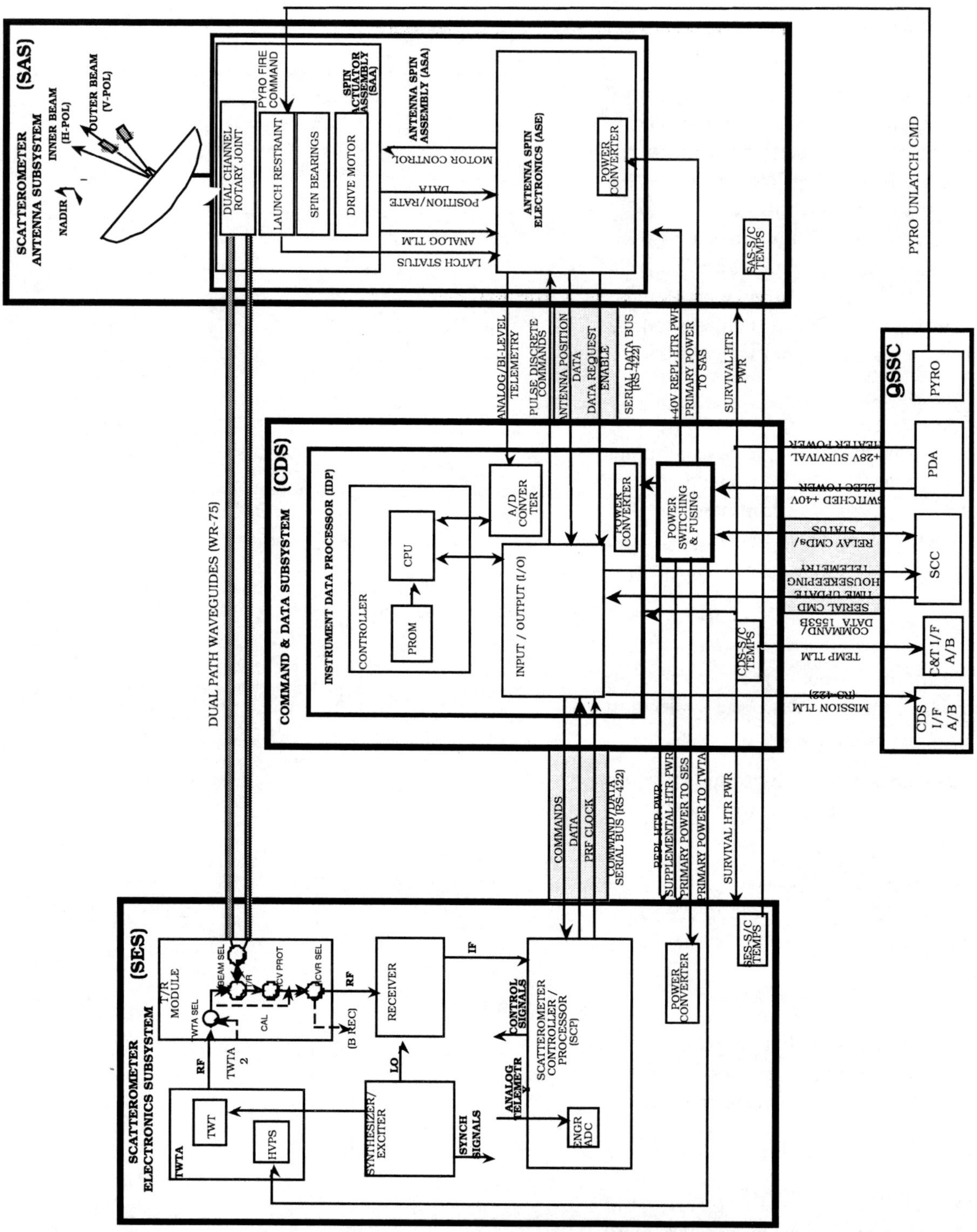

Figure 3. QuikScat Scatterometer Block Diagram

DEVELOPMENT OF A LOW COST DATA ACQUISITION SYSTEM FOR THE SPACE SHUTTLE SOLID ROCKET BOOSTER PROGRAM

Kevin Crawford, Computers and Data Systems Division, Astrionics Laboratory, NASA Marshall Space Flight Center, Huntsville, Alabama

David Pinkleton, Reusable Space Systems, Boeing North American Inc., Huntsville, Alabama

Background

The space shuttle solid rocket booster (SRB) is made up of a segmented solid rocket motor and booster assemblies (see Figure 1). These components are combined to form the solid rocket booster flight configuration.

During the first two minutes of ascent, the SRBs provide 5.8 million pounds of thrust, 70 percent of the vehicle's total thrust. The SRBs separate from the vehicle at an altitude of approximately 28 miles, while the shuttle is travelling at about 3,100 mph. The SRBs parachute into the Atlantic Ocean and are recovered for re-use.

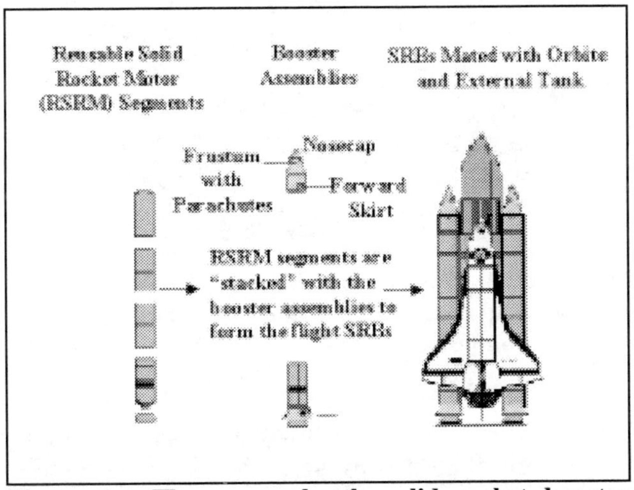

Figure 1. The space shuttle solid rocket booster consists of a solid rocket motor and major subassemblies: the forward assembly and aft assembly. These components are combined with the orbiter and external tank to form the shuttle flight vehicle.

The shuttle has been flying for 17 years. NASA plans to continue flying the shuttle until 2030. To meet this requirement, NASA has undertaken a program to upgrade various shuttle components. Among these is the proposed removal of spray-on foam insulation (thermal protection system – TPS) now applied to the SRBs to insulate SRB components. The TPS removal upgrade will save weight and contribute to improved SRB and shuttle vehicle performance. The program will also save the time now spent in applying and removing the TPS. NASA Marshall Space Flight Center (MSFC) wants to verify thermal, vibration and acoustic models that justify TPS removal.

NASA also seeks to refine the data used to predict SRB flight performance in relation to flight environments.

In a related issue, MSFC wants to understand SRB water impact and splashdown loads, which have caused damage to the forward skirts in previous missions.

A developmental flight instrumentation (DFI) system collected data from the SRBs' first four flights in the early 1980's, and the first three flights during the shuttle return-to-flight phase after the Challenger accident. However, the DFI data collected are of low fidelity and do not correlate well with cases of observed structural damage. The DFI data also is not of the fidelity required to confirm thermal models related to the proposed removal of the TPS.

System Requirements

In support of the Upgrades Program, in the mid-1990s, MSFC began evaluating options for collecting SRB data. The DFI system was evaluated for reuse, but the cost to re-fly it was prohibitive, primarily because of the system's dependence on external power, and the requirement to run cables through the SRB systems tunnel. The MSFC Structures and Dynamics (S&D) Laboratory, in association with the MSFC SRB Project Office, established a set of requirements for a compact, stand alone data acquisition system:

- Compact, light weight, stand-alone (internal power, data storage, time coding)
- Software-programmable
- Minimum sample rate 100 Hz with a goal of 5 kHz
- Capable of surviving harsh ascent, descent and splashdown environments
- Capable of accommodating multiple data inputs:
    - Thermocouples, resistant thermocouple devices (RTDs)
    - Heating gauges
    - Strain gauges
    - Pressure transducers
    - Accelerometers.

Development Approach

MSFC performed a study to determine the cost effectiveness of developing and manufacturing the system in-house as opposed to contracting an outside source. The study recommended an in-house effort, and predicted a three-year concept-to-flight development time. The study also recommended taking maximum advantage of commercial off-the-shelf (COTS) components.

MSFC consulted with its Shuttle Integration contractor, Boeing North American (BNA), who was in the process of conducting remote data acquisition trials on military and civil transportation platforms. BNA performed an analysis of available data acquisition systems. The SoMat Model 2100 Field Computer was selected after passing preliminary outgassing and functional testing.

The formal program to develop an Enhanced Data Acquisition System (EDAS) was initiated in February 1997. MSFC undertook further system analyses:

- determining the location of the data acquisition unit in the SRB
- identifying EDAS/SRB integration requirements
- identifying requirements for the EDAS enclosure, connectors, cables, and cable routing.

EDAS Enclosure

The EDAS enclosure, shown in Figure 2, is made of aluminum. It was designed and machined at MSFC. The size of the enclosure was based on two major considerations: space availability in the SRB aft skirt (where EDAS was originally to be installed); and the size of SoMat "stack" components. Enclosure weight including the SoMat unit is 13 lbs.

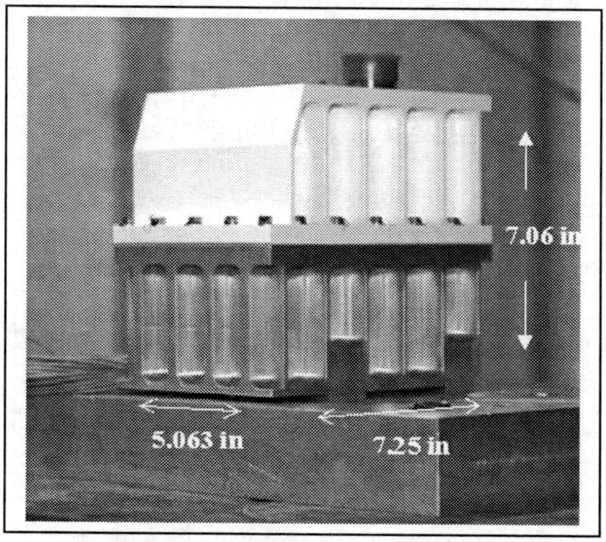

Figure 2. EDAS enclosure (containing SoMat and related components) on vibration table during qualification testing at MSFC. Dimensions noted.

The MSFC Structures and Dynamics Laboratory conducted an analysis to determine the optimum enclosure size. MSFC generated a rapid prototype model of the enclosure and used it for a fit check on an aft skirt to ensure the enclosure size was accurate.

As a result of qualification testing, MSFC decided to mount the EDAS inside the forward skirt. This would distance the EDAS from the direct shock of water impact. The forward skirt also provided more space for mounting the EDAS and associated cables and connectors.

SoMat Model 2100 Field Computer

The SoMat Model 2100 Field Computer consists of a processor, memory board, a power/communications module, a DC-DC converter, and as many analog transducer modules (channels) as are required by the user. The EDAS SoMat units contained six channels each. Figure 3 provides a cutaway view of the SoMat 2100 Field Computer. Figure 4 shows a SoMat unit sitting outside an EDAS enclosure.

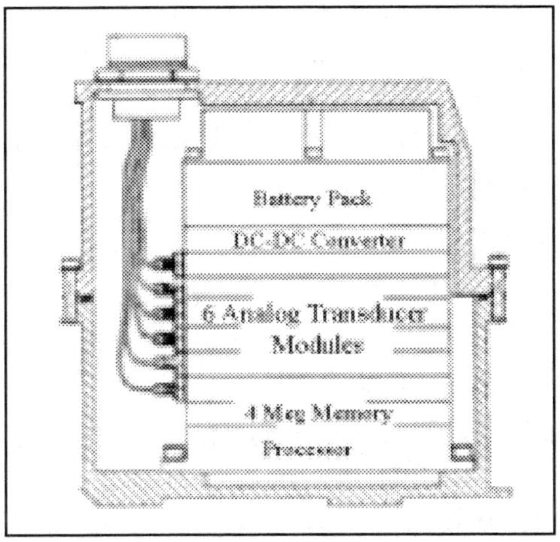

Figure 3. Cutaway diagram of SoMat unit inside enclosure (crosshatched).SoMat Enhancements

Wires/Connectors

To fit the EDAS enclosure, it was necessary to replace the straight SoMat (Lemo) connectors with 90-degree (Lemo) connectors. MSFC technicians strengthened the 30-guage wire/connector interface by adding heat-shrink tubing.

Momentary Inertial Switch (G-Switch)

The standard SoMat unit uses three 9-volt batteries in series. A DC-DC converter provides a range of 18-36 volts. The battery configuration provides 3.5 hours of operating time. SoMat provides a "sleep mode" that can be programmed to activate the system at a given date and time. The contingencies involved in spaceflight, however, mean that shuttle flight hardware can be stored for up to six months prior to use. MSFC identified requirements for a momentary inertial switch.

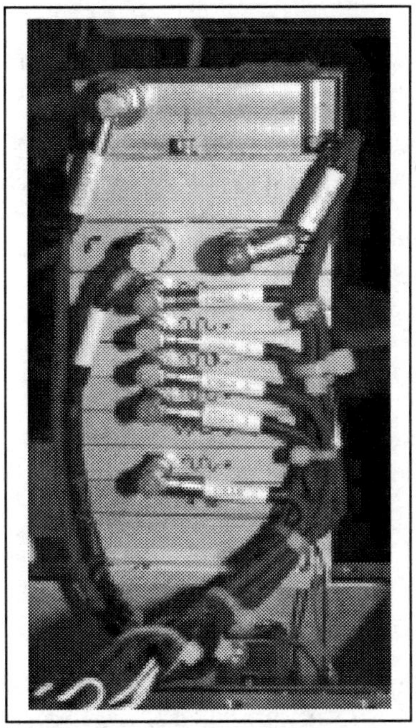

Figure 4. SoMat stack prior to being placed into the enclosure.

Because of the limited area provided by the EDAS enclosure, the switch was required to be smaller than any existing product. MSFC also required that the switch close and complete the power circuit upon experiencing .75 units of gravitational force (Gs) in the x direction only.

Inertial Switch, Inc. modified an existing momentary inertial switch (G-switch) model and provided a prototype to MSFC. See item 6. under EDAS Qualification Testing for a description of problems/corrections related to switch vibration testing.

Arming Command

MSFC requested that SoMat add an arming command feature to the Model 2100 Field Computer software. On the current configuration, power is applied to the unit at all times via the internal batteries, and the unit is activated when the user initiates the start command. With the arming command, the SoMat unit does not collect data until power is applied. The arming command is now a standard feature of the Model 2100 software.

Sensors

All EDAS sensors were procured from commercial sources. The sensors were selected and installed by MSFC and the SRB prime contractor, USBI, especially for EDAS.

Sensor types and associated measurements are provided below:

- thermocouple - interior air temperature
- platinum resistant thermal device PRT - interior skin temperature
- calorimeter - exterior skin heat flux
- pressure transducer - interior pressure
- accelerometer - vibration
- strain gauge - strain loads.

EDAS Qualification Testing

The MSFC S&D Laboratory performed vibration testing on an EDAS unit to qualify it for flight. The focus of the testing program was to qualify the *enclosure* for flight. As described below, a few problems with the SoMat unit, installed inside the enclosure, were encountered and corrected.

When qualification testing began, plans called for EDAS to be installed in the SRB aft skirt. The aft skirt impacts the water at splashdown. As a result of qualification (vibration) testing, the decision was made to install EDAS in the forward skirt, insulating the unit from the shock of water impact.

The qualification unit was tested in x, y and z orientations for the equivalent of 20 mission cycles of liftoff, ascent, and water impact vibrations. The magnitude of vibration was three times the nominal frequencies as currently defined by early 1980s developmental flight instrumentation (DFI) data. The three-sigma factor of safety is standard for the space shuttle program. The worst-case vibration applied during testing was 200 Gs, simulating water impact.

The following problems were encountered and corrected during vibration testing:

1. The captive screws that hold the SoMat modules together backed out, causing broken bus pins. The problem was corrected by applying Locktite™ to the screw threads.
2. The screws on the SoMat memory board also backed out. This condition was corrected by applying Locktite™.
3. Vibration of the SoMat DC-DC converter (supplied by Datel) caused the connectors between the converter and the circuit board to break. SoMat uses two vendors for DC-DC converters. SoMat provided a unit with a converter supplied by Calex. The Calex unit passed all vibration testing.
4. SoMat DC-DC converter modules, jumpers and fuses backed off, causing loss of power. This problem was corrected by bonding them with a silicon-based adhesive.
5. The 3.5-volt lithium battery on the SoMat power/com module became detached. The problem was corrected by bonding the battery to the printed circuit board with a silicon-based adhesive.
6. During its first vibration test, the internal components of the prototype G-switch were forced out through the potting material located at the opening used during switch assembly to insert the wiring. This problem was corrected by drilling a hole through the side of the switch housing, and a pin inserted to further prevent movement of the switch wiring.

EDAS Configuration/Operation

EDAS consists of a SoMat Model 2100 Field Computer, enclosure, G-switch, relay, internal cables and connectors, and a 55-pin J-1 connector. See Figure 5 for an EDAS block diagram.

An original goal of the EDAS program was to achieve a sample rate of 5 HKz. The SoMat unit provided a sample rate of 1500 sample per second.

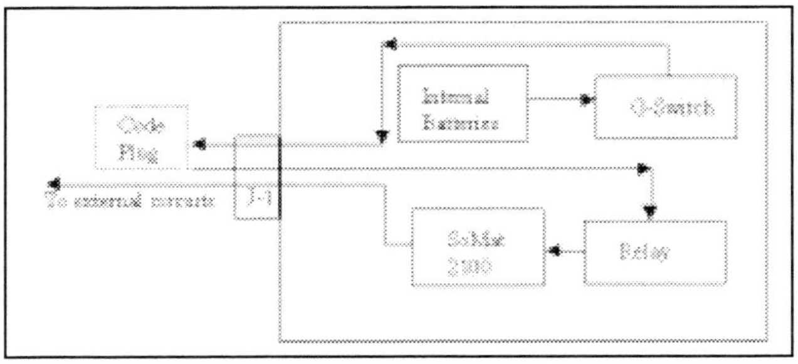

Figure 5. EDAS block diagram.

Additional Components

The following components were used in addition to the EDAS units.

- Laptop computer for communications, commands, calibration and data download
- Thermocouple conditioners (Manta Corporation) to provide reference junction
- Code plug (jumper mode).

WINTCS software, provided as part of the standard SoMat configuration provides the interface between the SoMat unit and the command/data retrieval computer.

The WINTCS software is used to calibrate/configure the SoMat unit. Each channel is calibrated separately.

MSFC developed an EDAS unit tester to provide the capability to functionally test an EDAS unit. The unit tester can be operated on either internal, 27-volt battery power, or external DC power.

The unit tester provides a jumper mode to allow the switching of an EDAS between internal or external power. This allows the calibration of an EDAS unit without using the EDAS internal (battery) power.

Using the unit tester power source or the EDAS batteries, the unit tester simulates the EDAS code plug, tripping the circuit between the g-switch and the relay. This activates the g-switch to allow for functional testing of the EDAS unit.

EDAS Installation

United Space Alliance (USA), the shuttle processing contractor at Kennedy Space Center, installs, calibrates and arms the EDAS units inside the SRB forward. Each EDAS unit is mounted to the SRB structure by four bolts, one of which serves as ground. USBI installs the sensors and cables linking the sensors to the EDAS units.

Prior to the closing of the forward skirt doors, the code plug (jumper wire) is installed, completing the circuit between the G-switch and the latching circuit. The EDAS units can then sit until the shuttle vehicle is ready for flight.

EDAS First Flight

Four EDAS units were installed in the forward skirt on the right hand booster for shuttle mission STS-91. The flight configuration is shown in Figure 6. The launch occurred on June 2, 1998. The mission ended with the orbiter landing at Kennedy Space Center on June 12.

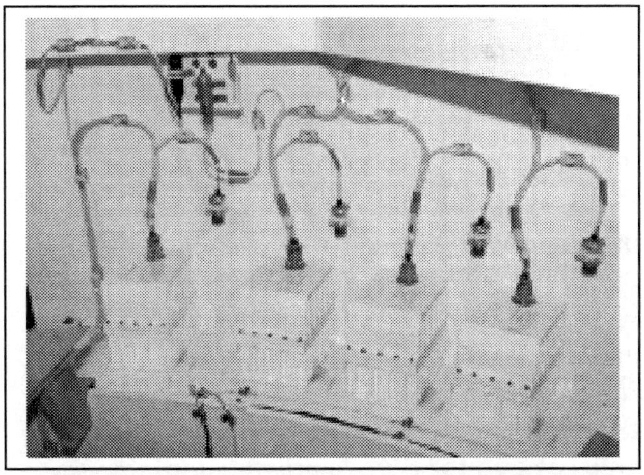

Figure 6. The four EDAS flight units as installed in the right-hand SRB forward skirt for shuttle mission STS-91.

Data were collected on 21 of the 24 EDAS channels. The two accelerometers each required a filter module. One strain gauge malfunctioned during calibration during flight processing at KSC. The decision was made to proceed without the sensor.

Postflight Inspection/Data Download

After recovery of the SRBs and upon post-flight inspection, USBI technicians noted no damage to or displacement of the EDAS units. USBI removed the EDAS units and provided them to the MSFC team for inspection, data download and refurbishment.

Close inspection by EDAS technicians also revealed no physical damage to the EDAS units. After downloading the data, it was concluded that all required data was captured and recorded with no anomalies.

Figure 7 shows a nominal shuttle mission profile, and the major events of interest to MSFC SRB engineers: liftoff, ascent, SRB separation, SRB descent, main parachute deployment, and splashdown (water impact). Sample plots of interior air temperature, exterior skin heat flux, and vibration data as recorded by individual EDAS channels are provided and correlated to the ascent/descent profile.

EDAS Refurbishment

During refurbishment of an EDAS, the SoMat unit remains in the enclosure. After the data is downloaded, the unit is functionally checked via the unit tester and the SoMat-provided WINTCS software. The top half of the enclosure is then removed and interior is visually inspected. The batteries are replaced, and the top half of the enclosure is re-installed, followed by the final functional test. The EDAS is then ready to be stored by USBI in an area secured for flight hardware until the next flight.

Planned EDAS Enhancements

MSFC is investigating two EDAS enhancements: a higher sample rate of up to 5

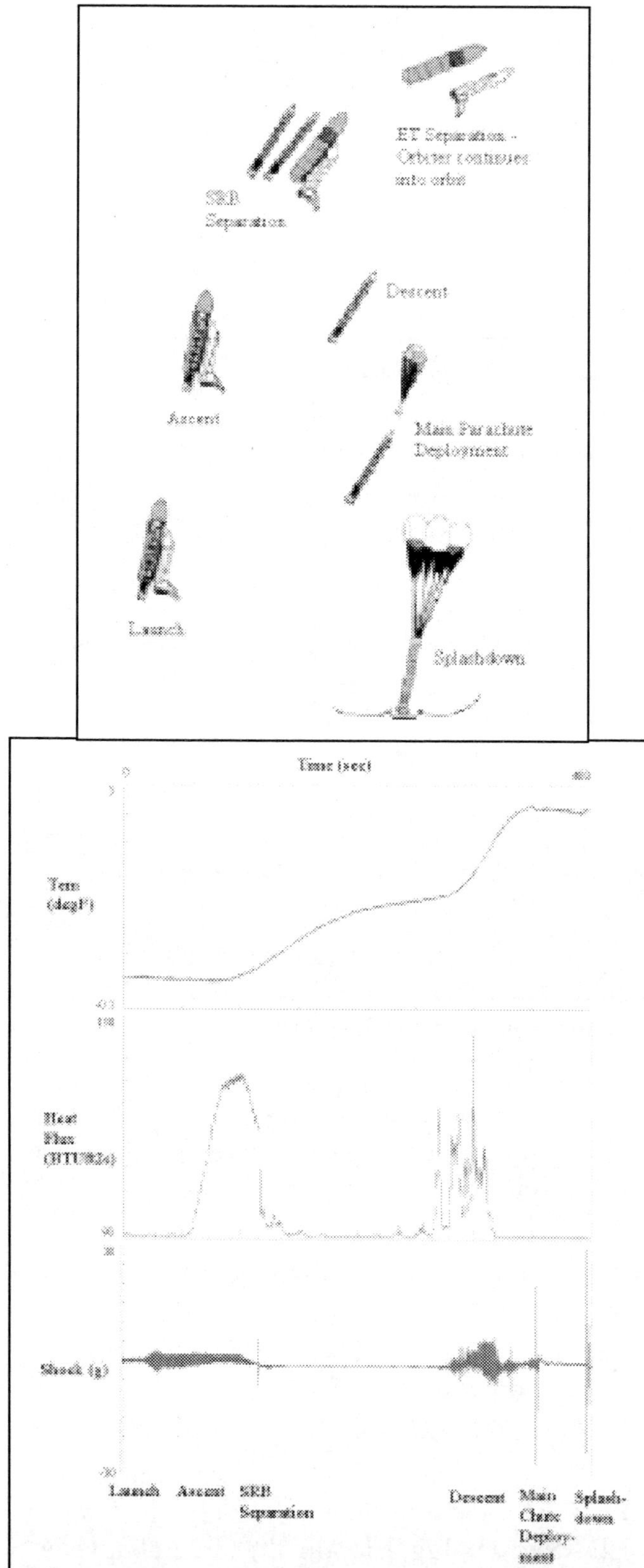

Figure 7. EDAS STS-91 interior temperature, exterior skin heat flux, and vehicle vibration data correlated to a nominal shuttle ascent/descent profile.

kHz, and a programmable version of a SoMat or Somat-like field computer. The higher sample rate would allow the capturing of higher frequencies and the refinement models of component flight environment models. The programmable EDAS computer would allow the capability to use software rather than hardware jumpers on the SoMat modules.

Future Applications

The current EDAS configuration is now scheduled to fly on shuttle missions scheduled for October 1998 and January and May 1999. The decision to re-fly was based on the EDASs' performance during STS-91.

The MSFC shuttle external tank and reusable solid rocket motor project offices have begin investigating the EDAS concept for collecting flight environmental data on their respective hardware.

H. System Design Track - Automotive Systems

Wayne Devereux

Since the introduction of electronic engine controllers in the late 1970's into production cars and light trucks electronic growth has continued unabated. This exponential growth in electronic content is fostering the electrification of the automobile. The electronic content of today's ground vehicles is expected to more than double by 2005 due to increased "infotronics" and the application electrically powered ancillaries. The Automotive Electronics Track is focused on the electronics and sensor systems being employed and envisioned in existing and future automotive and ground vehicles, hybrid vehicles, and Intelligent Transportation Systems (ITS) for commercial and military, fleet vehicles, Commercial Vehicle Operators (CVO) and even future Unmanned Ground Vehicles (UGVs). This Track shall reach across the converging engineering practices and technologies which have the highest synergy with the space and aerospace communities.

Presenters will introduce the cutting edge automotive electronics systems to provide all attendees an excellent cross-industry view of where the automotive electronics industry is, and is going.

Track Chairs: Wayne Devereux, Veridian; Paul Skalny, National Automotive Center

Track I - Automotive Systems

Session I1: Ground Vehicle Electronics

Chair: John M. Miller

1	Software Process Improvement in an Automotive Electronics Organization	Mr. Gregory McHugh
2	Automotive System Design: Today and Tomorrow	Dr. Peter Hofmann
3	Specification and Testing of Automotive Powertrain Control System Software Using CACSD Tools	Mr. Steve Toeppe
4	Active Control of Vehicle Dynamics	Mr. Mike Fodor
5	The Next Generation Automotive Electrical Power System Architecture: Issues and Challenges.	Dr. John P. Miller PE, PhD
6	Automotive & Aerospace Circuit Fault Analysis	Mr. Craig Siegel
7	Requirements For A Real-Time Local Area Network Architecture In Land Combat Vehicles	Mr. Paul Richardson

Session I2: Intelligent Transportation System (1)- Vehicle Electronics

Chairs: Allan Kirson, Motorola, Inc; Dick Bishop, ITS America

1	The Network Vehicle - A Glimpse into the Future of Mobile Multi-Media	Huan-Wun Yen
2	A Demonstration Project for the ITS Data Bus Prototype	Mr. Philip Spelt Ph.D.
3	Signal Processing and Waveform Generation in the Side Zone Automotive Radar	Mr. John C. Reed
4	Adapting Radar and Tracking Technology to an On-Board Automotive Collision Warning System	Dr. Edward Jocoy
5	Use of Map Data Information in an On-board Intersection Violation Detection System	Mr. John Pierowicz
6	SWIFT Project Results	Mr. Larry Senn

Track I - Automotive Systems (Continued)

Session I3: Intelligent Transportation System (2)- Info Management

Chairs: Dick Bishop, ITS America; Allan Kirson, Motorola, Inc.

1	The In-Vehicle Information System Development Center at ORNL, An In-Vehicle Information System for ITS Information Management	Mr. Philip Spelt Ph.D.
2	Decision Making for Road Departure Warning Systems (Not Published)	Dr. Thomas Pilutti
3	Human Factors in Intelligent Transportation Systems in the Vehicle Environment (Not Published)	Mr. Gene Farber
4	Reconfigurable Field Research Vehicle For Human Factors Experiments	Mr. R. Wayne Devereux
5	The Carnegie Mellon TruckSim, The Advanced Human Factors Research and Driver Training Research Facility	Mr. Richard Grace
6	The Drowsey Driver Report from Carniege Mellon	Mr. Richard Grace
7	Determining and Controlling Emergent Behavior in Intelligent Transportation Systems	Mr. James L. Overholt

Session I4: Hybrid Vehicles

Chairs: Henry Oman, Consulting Engr; Steve Cortese, Lockheed Martin Control Systems

1	Energy Power Management in Hybrid Electric Vehicles	Mr. William Morchin
2	Solar Energy for Electrical Vehicles: Systems Analysis	Mr. Henry Oman
3	On-Board Energy and Power Management on Electric Vehicles: Effect of Battery Type	Mr. Henry Oman
4	HybriDriveTM Hybrid Electric Drive System (Not Published)	Mr. Tim Grewe
5	Optimization of Complex Powertrain Systems for Fuel Economy and Emissions (Not Published)	Ilya Kolmanovsky
6	Design of a Low Production Cost Hybrid HMMWV	Mr. Jeff Armfield
7	Fundamental Analysis of a Terminal Voltage Based Battery State Regulation Technique	Mr. Herman L. N. Wiegman

SOFTWARE PROCESS IMPROVEMENT IN AN AUTOMOTIVE ELECTRONICS ORGANIZATION

Gregory C. McHugh, Visteon Automotive Systems, Dearborn, MI

Introduction

The software content of automotive electronic modules has been increasing rapidly in recent years. Projections indicate that automotive electronics will follow the trend seen in other areas of the electronics industry (e.g., avionics, telecommunications, and consumer electronics) where the software engineering capability of the organization will be a key factor in successful product development. As automotive electronics organizations recognize that a world-class product development capability is dependent on a world-class software engineering capability, software process improvement initiatives have become a key component in their long-term strategic plans.

This paper describes software process improvement efforts that have taken place within Visteon Automotive Systems over the last six years in response to this growth in software content in automotive electronics.

Automotive Software Overview

It has been over 20 years since the initial introduction of software into automotive electronics. Electronic engine control systems at Ford Motor Company in the late 1970's included 2-4 kilobytes of assembly code in the microcontroller. By the mid 1980's the code size for a powertrain control system had grown to 50-60 kilobytes and by the mid 1990's the code had grown to 90-110 kilobytes. Current powertrain control systems have over 200 kilobytes of code.

In recent years the software code size in the overall vehicle has been doubling every 3-4 years. A 1995 Lincoln Continental contained 250 kilobytes of code in 15 microcontrollers for the implementation of basic vehicle functions (not including the audio system and cellular phone) and the code size will be about 500 kilobytes for the same class of vehicle in 1999.

Software growth is now accelerating with the introduction of high-end audio systems and navigation, guidance, and communication systems using cellular and GPS technologies. The development of open architecture standards for connecting consumer electronics devices into the vehicle electronics system will lead to another significant increase in vehicle software content. Microsoft has entered the automotive electronics arena with the introduction of the Auto PC. Automotive electronics is following the trend seen in avionics, telecommunications, and consumer electronics where the growth of software content in the product has required electronics organizations to focus on software engineering as a core product development capability.

Organization Overview

Visteon Automotive Systems was created in 1997 as a global Ford Motor Company enterprise with a focus on the design, development, and delivery of integrated systems to automotive manufacturers on a worldwide basis. Visteon has 82,000 employees located in 21 countries. 1997 global sales revenues for Visteon Automotive Systems were $17.2 billion.

0-7803-5086-3 /98/$10.00 ©1998 IEEE

Visteon Automotive Systems is organized into seven divisions:

- Chassis Systems
- Climate Control Systems
- Electronic Systems
- Exterior Systems
- Glass Systems
- Interior Systems
- Powertrain Control Systems

Within these divisions are 24 Strategic Business Units (SBU's), with each SBU accountable for its own profit and loss.

This paper describes the software process improvement effort within the automotive electronics development activities of five Strategic Business Units in the Electronic and Powertrain Control Systems divisions of Visteon. A centralized Software Engineering Process Group (SEPG) coordinates software process improvement efforts across these various software development activities within Visteon.

Software Product Overview

The software process improvement activities described in this paper are focused on the following product areas within Visteon:

- Audio
- Body
- Climate Control
- Driver Information
- Electric Vehicle
- Powertrain[1]
- Safety
- Security
- Steering and Suspension

Within these product areas, there are about 200 personnel working on the development and testing of embedded product software. These personnel support over 200 specific software products with an average code size of 24 kilobytes. 85% of the currently supported code is written in the C language and 15% in assembly language.

Software Process Improvement

The remainder of this paper describes the following aspects of the software process improvement activities at Visteon:

- Process Improvement Strategies
- Software Engineering Process Group
- Industry Standards
- Examples of Best Practices
- Future Plans

Process Improvement Strategies

The current software process improvement efforts at Visteon can be directly traced to a software improvement activity that was initiated in 1992 within the Electronics Division of Ford Motor Company. During the last six years there has been a consistent focus on improving the software development capability within the automotive electronics activities at Ford even as these activities have been reorganized several times into the current Visteon organization.

The initial strategy outlined in 1992 focused on the development of a standard software process that could be used by all product areas. The goal was to have this standard process represent best practices captured from within the organization and from outside of Ford.

Much of the effort in the first few years of the process improvement program was focused on getting participation and buy-in by all of the software development activities which are dispersed throughout the product development groups within the organization. By 1995, all of the software activities within the Electronics Division were participating in the software process improvement activities

[1] The activities within the Powertrain Control Systems Division described in this paper did not include engine and transmission control software development, these activities were part of a separate software process improvement effort.

0-7803-5086-3 /98/$10.00 ©1998 IEEE

and were being supported by a centralized process improvement group.

In 1993, Ford Electronics Division started working with the Software Engineering Institute (SEI) at Carnegie Mellon University to provide training and support for software process improvement. The Software Capability Maturity Model (CMM) from the SEI has been used as the basis for software process improvement at Visteon. The product areas outlined above were externally assessed to be at CMM Level 2 in May of 1998 and there is a goal to reach CMM Level 3 for these product areas in 1999.

Software Engineering Process Group

As noted above, a decision was made early in the software process improvement efforts to focus on the development of a standard software process across product areas. A centralized Software Engineering Process Group (SEPG) supports this strategy by developing, maintaining, and improving this standard software process. The software product supervisors within the software development activities are the owners of this standard software process and they approve all changes to the process proposed by the SEPG. Overall goals for process improvement (e.g., reaching CMM Level 3) are shared objectives of the product development organizations and the SEPG.

Industry Standards

As a global supplier of automotive electronics to automotive manufacturers and to after-market sales, Visteon must meet a worldwide set of industry and customer standards, including:

QS-9000 / ISO 9000

Automotive suppliers are required to meet QS-9000 quality standards, a U.S. automotive industry version of ISO 9000 that consolidates the quality requirements of Ford, General Motors, and Chrysler into a single standard. Visteon has been certified for both QS-9000 and ISO 9001.

TickIT / ISO 9000-3

TickIT certification is becoming a requirement for sales to the government in the U.K. TickIT auditing uses the ISO 9000-3 guidelines for the application of ISO 9001 to the development, supply, and maintenance of software. TickIT replaces the generic requirements of ISO 9001 with specific software quality management and software quality assurance standards. Visteon is currently evaluating the requirements for TickIT certification.

Ford Standards

Ford Motor Company has defined a set of world-wide vehicle software development standards that go beyond the QS-9000 quality requirements. These standards apply to all suppliers of vehicle electronics for Ford vehicles and Visteon complies with these standards.

MISRA

The Motor Industry Software Reliability Association (MISRA) in the U.K. has released two sets of guidelines for automotive software:

- Development Guidelines for Vehicle Based Software, Nov. 1994
- Guidelines for the Use of the C Language in Vehicle Based Software, April 1998

The Ford standards mentioned above are consistent with the MISRA software development guidelines and Visteon is currently incorporating the MISRA C language guidelines into our coding standards and process.

Examples of Best Practices

The "best practice" software process improvements made at Visteon that are generally applicable to most software development organizations include:

Peer Reviews

The use of technical peer reviews of software development work products is probably the most effective quality improvement practice that has been added to the Visteon software process. Visteon followed the practice of many organizations and included an implementation of Peer Reviews in our process at an early stage rather than waiting until they are required by the CMM at Level 3.

Introduction of Peer Reviews into the Visteon software process had an immediate impact on the quality of the software work products, 88 reviews were conducted in the first 4 months of implementation and 255 major defects were removed from work products using this method. During the first seven months, 321 reviews were conducted and 750 major defects were corrected. On average, 2.3 major defects are being removed in each Peer Review. Major defects are defined in the Visteon Peer Review process as an error in a software work product that could cause a functional failure of the product or could be detected in operation.

The lack of wide-spread implementation of Peer Reviews is a surprising fact based on the proven effectiveness of the method. The definitive paper on "Software Inspections", a formalized technical peer review process, was published by Michael Fagan of IBM in 1976. The peer review has been proven over the last 20 years to be a highly effective process improvement in many organizations but this procedure is still not widely implemented across software development organizations.

It should be noted that the technical peer review has also been applied effectively as a quality improvement process for hardware design work products.

Risk Management

Another example of a CMM Level 3 process requirement that was implemented early as part of the CMM Level 2 process improvement effort is the definition of a risk management process for software projects. The Software Risk Management process used at Visteon defines a set of activities that are undertaken to identify, communicate, and resolve software risks. The goal is to identify and manage any risks that could cause the software project to fail to meet its objectives for schedule, cost, functionality, quality, reliability, real-time performance, or use of critical target system resources.

Personal Software Process (PSP)

Visteon has been using the SEI Software CMM as the basis for organizational software process improvement for the last five years. A method for applying the CMM process improvement principles to individual work has been developed by Watts Humphrey of the SEI in the form of the Personal Software Process (PSP). Two years ago Visteon set a goal of providing PSP training and follow-on coaching to all software personnel and we expect to reach that goal within the next year. Visteon is also involved in a pilot of the Team Software Process (TSP), also developed by Watts Humphrey at the SEI. The TSP extends and refines the CMM and PSP methods to guide engineers in their work on development and maintenance teams.

C Coding Standards

As noted in the Software Product Overview section above, 85% of Visteon software is written in the C language with the remainder written in assembly language. C is generally the only high-level language that is available for the types of embedded microcontrollers used in automotive electronics and we do not expect to see any significant shift away from C in the near future. Given that we are dependent on C as a high-level language, we are implementing a coding standard that defines a use of the C language that is appropriate for high reliability applications. The MISRA "Guidelines for the Use of the C Language in Vehicle Based Software", noted above in the section on industry standards, describes the automotive industry guideline for the use of C in high reliability applications that is being implemented at Visteon.

Future Plans

Software engineering initiatives planned for the next year at Visteon include:

Software Reuse

Most software organizations have a goal of increasing the level of software reuse. Increasing the level of reuse of software work products has the potential to allow the organization to deliver more complex software products while at the same time improving quality, reducing development time, and improving efficiency and productivity. Experience has shown that a software reuse strategy must be planned and managed. Visteon is implementing a software reuse initiative using a strategy that includes:

- The development of product line architectures within each product area to facilitate the reuse of designs from project to project
- The implementation of a "software bookshelf" that contains software components that have been certified to be of high quality
- Dedicated "software design champion" personnel that support the reuse initiative
- The reorganization of several product areas into "core" sections (with the responsibility for the development of reusable software components) and "application" sections (with the responsibility for delivering integrated software applications to the customer)

The early results of using this strategy in some of the product areas is encouraging with a significant increase in reuse of software development work products on new projects.

Training and Education

The software engineering staff at Visteon has doubled in size in the last few years mostly through the recruitment and hiring of personnel with experience in embedded software development. The need for additional personnel is expected to continue for the foreseeable future but we do not expect to find a significant number of experienced personnel available for hire in the current job market. Visteon is currently expanding our training program for software engineers to provide embedded software development training for new hires without embedded experience, new college graduates who are not usually trained in embedded development, and current employees in other engineering fields who are interested in becoming trained in embedded software development.

Visteon has two employees who have completed the Masters of Software Engineering program at Carnegie Mellon University on a full-time basis under an educational fellowship program and we expect to have additional personnel take advantage of this program. Personnel attending the Masters of Software Engineering program typically return to a position that includes responsibility for the training and mentoring of less experienced personnel.

Overseas Software Development

Visteon currently has three worldwide software development activities, two located in our engineering centers in the United States and the U.K., and a third dedicated software development activity in India. The Software Engineering Process Group that supports the current process improvement initiatives in the United States at the Dearborn engineering center is being expanded to support all of the software development activities worldwide. Growth of software development capability overseas is expected to be significant in the future in order to support local engineering of regional products and to take advantage of the availability of software engineering expertise that is in short supply in the U.S. and Europe.

Summary

This paper has provided an overview of software process improvement activities at Visteon Automotive Systems over the last six years. The motivation for this long-term process improvement activity is the continued rapid growth in the software content of the automotive electronics products produced at Visteon.

As part of this process improvement effort, an organization of about 200 software personnel working in nine different product areas has implemented a standardized software process that was assessed at SEI CMM Level 2 in May of 1998. Future plans include reaching SEI CMM Level 3 in 1999.

AUTOMOTIVE SYSTEM DESIGN: TODAY AND TOMORROW

Florian Wohlgemuth, Daimler-Benz Research, Stuttgart, Germany
Peter Hofmann, Daimler-Benz Research, Stuttgart, Germany

Abstract

No other technology enhanced dramatically the functionality of cars in the last years as the information- and communication-technologies. Electronic based systems provide the driver with a maximum of security, comfort, information and entertainment.

Such electronic systems are specified by the car manufacturer and produced by the supplier. A model-based specification allows a detailed information interchange between these partners and limits mis-understandings.

Today, models of such systems are developed by a structured analysis process. The desired function for one control unit is hierarchically divided, until the behaviour can be described by a state machine. Tomorrows electronic car systems provide enhanced functionality, are networked and interact with each other and external services. Designing such systems requires a powerful way to structure, describe and reuse the components of the car and the surrounding environment.

Introduction

In the past years, the number of electronic components of a car grew continuously, enhancing the functionality dramatically. They control the engine, guarantee active safety and enhance interior comfort.

These functions are realized by electronic control units (ECU), small processors with limited storage capacity and processing power compared to their desktop-models, but with better operational heat and virbration characteristics and shock-resistance. They monitor the environment through sensors and react by actuators usually within a certain period of time. Such systems are called embedded real-time systems.

Today, the ECUs of a car are interconnected by a controller area network (CAN) to share information and allow interaction. Communication facilities allow us to drift away form the present assignment of one functionality per ECU and to scatter the functionality to different locations. Multiple functions can be realized on one ECU or distributed over many ECUs.

The development of a new ECU is a joint effort of the car manufacturer and the supplier. The car manufacturer writes a specification of the desired functionality for which the software and hardware is delivered by the supplier. This specification is often a plain text file. To be cost-effective, the supplier decides usually to use standard hardware components and to implement nearly all of the desired functionality in software.

To eliminate communication problems with plain language documents, such as misinterpretations and contradictions, a model-based specification is going to be established. The tool used for building such models is introduced in the next section.

The development of electronic control units is part of the development of a whole car. Other development steps need information about the ECU-functionality and even about the ECU-design, such as cabling, diagnostics, documentation and so on. The coupling or integration of the activities respectively the tools is an ongoing challenge.

0-7803-5086-3 /98/$10.00 ©1998 IEEE

Approach Today

Natural language specifications are usually inaccurate, misleading, ambigue and sometimes contradictory. This error-source can be reduced by writing the specification in a formal language, yielding a model, where several aspects can be mathematically proven to be correct. Complicated language expressions however, should be hidden to the user. Acceptance grows, if the formal model of the specification is presented by visual means to the engineer, which are natural in his working domain.

Statecharts [Harel 87] are such a visual language with a formal foundation. The belonging development method suggests to decompose the functionality into smaller functions until their behavior can be described with an automata [Harel 1990]. As these automata still tend to be complex, Harel introduced with his Statechart formalism hierarchical and parallel states in order to reduce the number of states and transitions.

Statemate, a tool from i-Logix Inc., implements the Statechart formalism. It allows the simulation of the model, even if parts of the model aren't completely specified. Thus, the developer can check parts of his model at a very early point of the development process.

To discuss the model with colleagues or with a supplier which are not familiar to the concepts of the statechart formalism, Statemate introduced panels. With these an engineer can easily built a GUI and connect it to the model. During simulation of the model, the panels generate the stimuli and display the reaction. This way, the complete system, including sensors and actors, can be simulated.

The design entities of the model can be annotated with explanatory text. These annotations can be extracted automatically from the model into a structured (i.e. SGML) document.

Finally, Statemate allows the generation of C-code out of the model. This generated code does not meet the time and space restrictions of the ECU target hardware. But for rapid prototyping purposes, this code can be run on a special universal computer which can be mounted in the car. The proper interaction of the model with the surrounding sensors and actuators and the correct realisation of the required functionality can be tested with this environment.

A similiar development environment exists with MatrixX for the area of continuous applications.

Challenges

The ongoing development of new functionality in the car makes the software for the ECUs more complicated. The total complexity of the system car is further increased by the intergration of many functions onto one ECU and vice versa, the distribution of one functionality on many ECUs.

The car can be seen as a distributed system and is itself a node in a distributed system because of the growing number of telematic applications. The future ability of car functions to communicate with external services is responsible for further inreasing the complexity. Furthermore new electronic devices for entertainment and information (CarPC) urge to be intergrated in the car. Figure 1 shows these new aspects of mobility.

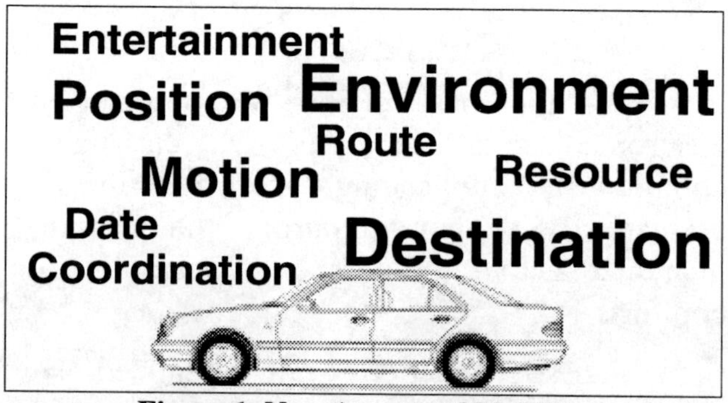

Figure 1. New Aspects of Mobility

Last but not least, it should be possible to extend the functionality realised by the software in the ECU after the car is sold to the customer. This is possible, because most of the

0-7803-5086-3 /98/$10.00

necessary sensors and actors are already built in today, and new functionality will be realised in software only.

All these new applications should be specified by a formal model as stated above. Thus, each model is growing and the total number of models is growing, too. There are many forces trying to enlarge the models in various directions (Fig. 2):

- horizontally, by adding new functionality or coupling it with other models.
- vertically, by adding information from the analysis or aspects of the operating system.

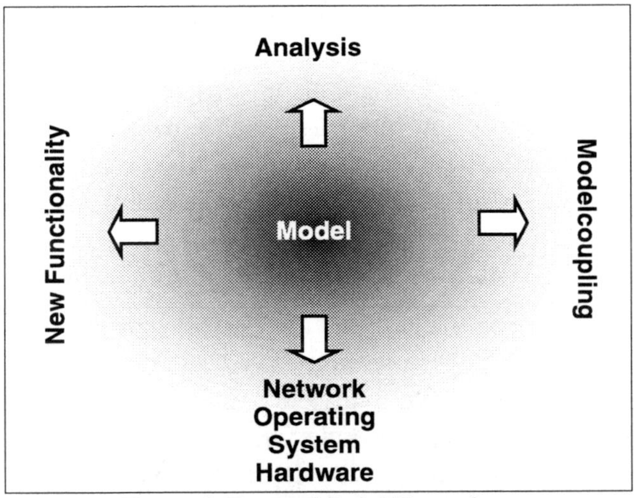

Figure 2. Future Model Extensions

This quadratic growth imposes problems to todays methods and tools applied in the automotive sector. An approach which relies on structured analysis produces a decoupled function hierarchy but a tight coupling by the data. This is a major obstacle to the scalability.

As mentioned in the last section, all these new applications need to be described properly. The design cycles however, get shorter.

Tomorrows way

Object-orientation is the appropriate way in the traditional IT-area to design large, complex data-processing systems. The successful practice in this area encourages us to transfer this technology to the modeling of embedded real-time systems. Applied well, this technique leads to structures grouped around durable aspects of the product: the information.

In contrast to methods following the Structured Analysis approach, which divide the design space in data-structures and functions, object-orientation groups functions around data-structures they use. The definition of the data-structures constitutes together with the data modifing functions a class. As the data-structures are hidden and can only be accessed or altered though the functions of a class, we achieve a interaction of classes through well-defined interfaces.

Beside the reuse of existing classes, the ability to adapt classes is another key feature of object-orientation. Specialization and generalisation of classes are possible by inheritance (Fig. 3). Class B inheriting from another class A possesses all data-structures and functions of A. By adding or redefining functions, B specializes class A. On the other hand is class A more abstract than class B and can be used for another specialization. Generalisation of classes can be actively pursued to reach a new class which potentially fits in a number of different areas.

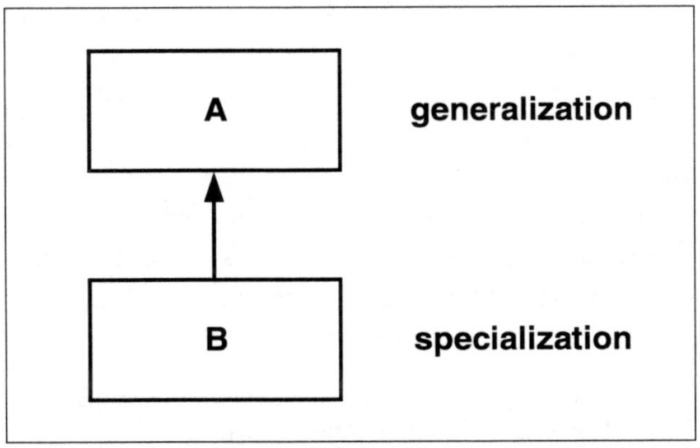

Figure 3. Inheritance

Inheritance permits the redefinition of a class' function. How can we prevent a redefinition damaging the overall behavior of the class? The solution is to introduce aspects of a function, which can't be changed through redefinition. Pre- and post-conditions of a function, are such static aspects. A precondition can only be weakened, to ensure continued usability for other classes using the inherited version of the class. On the other hand can postconditions only be stronger than their precursors, thus guaranteeing the results of the inherited version.

The concept of Design-By-Contract [Meyer 1997] extends the interpretation of these assertions. Both are part of an imaginary contract between the supplier and the client of a class. The client ensures that the precondition is met, whereas the supplier guarantees the results specified in the postconditions. This way the semantics of a function can be formally described.

Back to object-oriented concepts. At run-time, classes are instanciated by filling the attributes with values. These instances are called objects. To guarantee at compile-time that a called function exists, strong static typing is needed. It also helps to find the implementation of a function fast along the inheritance chain. [Meyer 1997].

An object occupies memory to store the information. At some point in time the object may become superfluous and the requested memory can be freed. Manual disposing of objects is a common source of problems in the traditional IT. Automated memory management with garbage collection helps in this field. In the area of time critical applications however, a garbage collection at arbitrary time with undetermined length is a major obstacle. Modern garbage collectors work incrementally with sufficiently short cycles and support multiple threads [Magnusson 1995].

Real-Time and Object-Orientation

Some aspects of the target domain require adaptions and extensions to the object-oriented paradigm. These are summed up below and explained in detail in the following sections:

- guaranteed processing times
- hardware-interfaces
- states
- distributed objects
- priorities
- integration with operating systems

Time contracts

The timely completion of functions is crucial for the correct interaction with the surrounding environment. When the maximum response time of a specific function gets known during the design process, it should be attached to the function. [Kim 1994] suggests to attach this deadline information to the function declaration. We extend this view and regard this information to be part of the postconditions as in the principle of Design-by-Contract. This way, the mathematical specification of the behaviour of a function gets more complete. Potential clients can rely on the maximum execution time. Furthermore the execution time can roughly be checked by summarizing the ensured deadlines of the called subfunctions.

However, the extension with time information increases the hardware dependency. On sufficiently fast hardware configurations, the code will execute correctly with respect to its specification, on a slower target it will not. The failed postcondition leads directly to the function consuming too much CPU time which can be modified accordingly. If a further efficiency improvement is not feasible, the target hardware must be re-evaluated. So, as the model should cover the complete system, this hardware dependeny is desired.

Connecting the hardware

Static typing allows no random type casts and pointer arithmetic, thus making direct access to the hardware impossible. Integrating specialized I/O-hardware into the systems requires access to the hardware. The solution is, to wrap the desired functionality by a class and implement the hardware-related parts in a more hardware-oriented but still platform independet language, such as C.

States

Many applications in the car are inherently state-based. These state-properties should be described by a state machine. To avoid two separated concepts, automatas should be embedded in the object-oriented paradigm. One possible way is to associate with each class a statechart [Douglass 1997]. An object can thus be seen as a machine, reacting to events. Inheritance however makes this technique complicated, multiple inheritance is actually unsupported. Furthermore, as a state automata can principally receive any event and ignore it, the interface of a class gets blurred.

Distributed Objects

As mentioned already in the first paragraphs, tomorrows cars contain a distributed system and are nodes of a larger system by themselves.

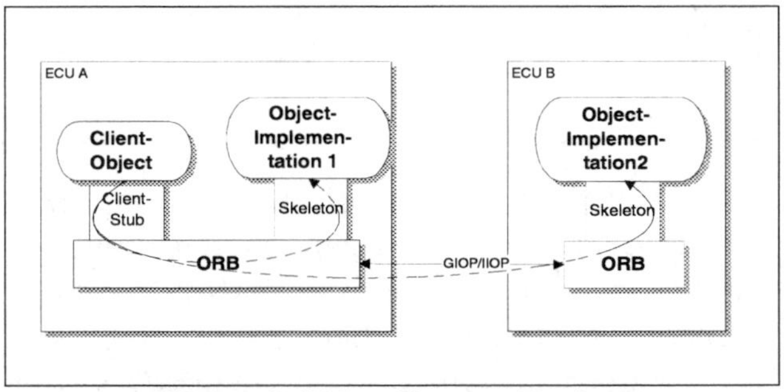

Figure 4. Object Request Broker

As the final location of a piece of code gets more and more irrelevant, it is desirable to describe the system "car" as independent as possible of its actual distribution on the ECUs. CORBA, a standard born in the classic IT, provides mechanisms to hide the distribution (Fig. 4). Messages are sent in a network-independent format to the Object Request Broker (ORB), which in turn transmits the message to the receiver, if necessary over the network. For embedded systems are tailored versions necessary, which conform to the time and space restrictions.

Furthermore a smooth integration into the development environment would make the distribution more transparent. It is sufficient, to mark a class as active, that is, objects of that class have their own thread of execution.

Priorities

One ECU controls multiple aspects simultaneously. This is done with the usual time-slicing techniques from multi-tasking operating systems. The tasks have different priorities according to their relative importance or to their time requirements.

Determining the correct priority for an active object should be formalized and tool-supported. A possible way is, to compare each active object and decide which one gets preferred in case of simultaneous activation. The priorities are then formally derived from this pair-wise comparison [Awad 1996]. Combining priorities with inheritance needs further clarification.

Operating System

To achieve a seamless development, interfacing with the operating system should also happen through classes.

Unified Modeling Language

The Unified Modeling Language (UML) is an attempt to unify the look and feel of the various representations accompanying different methods.

Because of its universal approach, many diagram types have been included. For automotive purposes only a subset is needed and UML's close relation to C++ must be reconsidered.

Among others, it contains three important diagrammatic notations:

- interaction diagrams
- class structure diagrams
- state-transition diagrams

As UML is only the modeling language, its successful usage – especially in the automotive sector – needs to be clarified. A first attempt is described in the next section.

Object-oriented System-Design

The proper application of object-oriented technology requires new skills from the engineers. Training the technology isn't sufficient, a method providing a guideline would do more. The following paragraphs summarize a first suggestion.

The first step in the development of an ECU functionality is to identify the participating classes resembling the desired functionality. The electromechanical parts of the environment to be controlled give a first orientation. Sensors and actors are good candidates for example. Their interaction during a certain control-scenario is then described by an interaction diagram (or message sequence diagram), where already known maximum response times can be denoted. Figure 5 depicts the interaction when a folding top is opened.

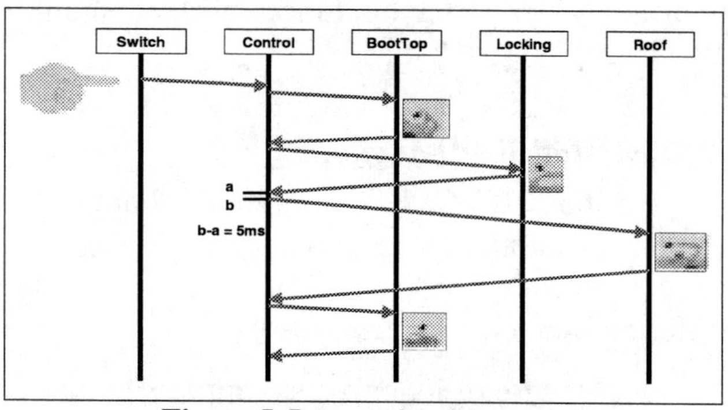

Figure 5. Interaction Diagram

The communication described in the interaction diagram is used to build the class structure. In the class strcuture diagram further relations between classes, like inheritance, are layed out. The behaviour of each class can be described in a statechart, a form of state diagram. The interaction diagram can be used again to identify the events sent to the class.

This is in short terms a description how to get quick to a working system. Reuse of classes with this scheme will take place only by chance. Producing actively classes intended for reuse in say, one model range or even spanning several model ranges, needs further activities such as abstraction and generalisation. The goal must be to achieve and maintain a class-framework for the electronic functionality of a car.

Development Environment

One goal of development environments should be to relieve the user from repeating tasks. Instead he should concentrate on the major task: the creative design of the system. This quality greatly depends on the founding concepts for object-orientation. Strict rules allow more checking at compile-time and reduce error-sources at run-time.

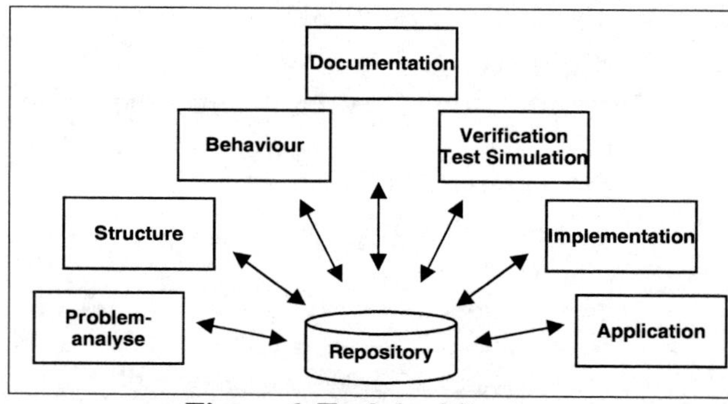

Figure 6. Tool-Architecture

With respect of the enourmous requirements toward a seamless development environment we propose a repository based approach. A number of handy tools (e.g. one for each diagram-type described above) access the information stored in a central database (the repository) and offer editing facilities within a certain visualisation concept (Fig. 6). So the tools offer different views of the system under development and allow the editing within. The consistency between those different views is maintained by the common database.

Such a toolbox of dedicated, small and easy-to-use tools offers the ability to group the needed tools around the development process. As every tool modifies the information in the database, the order of their deployment is irrelevant. Furthermore we gain reversability: results from a tool dedicated to a late development step can always be viewed by a "early" tool.

Beyond these aspects this tool-architecture allows an easy extension with domain-specific tools. They represent the information of the repository in a way, that is common to engineer working in this domain. Modifications using such a domain-specific tool need to be translated back to object-model used in the repository. The "basic" tools remain thus capable to display and modify the results. Specialisations remain visible and have no effects on the underlying object-oriented paradigm.

Summary

After giving a short overview, how the ECUs of a car are designed today, we showed that these techniques won't be sufficient to master tomorrrows challenges. Today, electronic control units are developped using commercial tools like Statemate, MatrixX, Matlab and some others. There are two important problems. First, these tools are very large and don't support a proper data-interchange with other tools of the toolchain. Second, we need small and very flexible tools, well integrated in the development process. Further we need more methodological support to handle complex design of reusable components.

Similar problems forced the classic IT to switch to the object-oriented programming paradigm. The advantages of this approach were shown. Modifications and extensions to the object-orientation necessary to suit well for the area of embedded real-time systems were discussed. Their further clarification and tailoring for the automotive sector is our goal of research for the next years.

References

[Awad 1996] M. Awad, J. Kuusela, J. Ziegler: Object-Oriented Technology for Real-Time Systems: a Practical Approach Using OMT and Fusion; Prentice-Hall; 1996

[Douglass 1997] Bruce Powel Douglass: Real-Time UML: Developing efficient Objects for Embedded Systems; Addison Wesley; Reading, Massachusetts; 1997

[Harel 1987] David Harel: Statecharts: A Visual Formalism for Complex Systems; in: Science of Computer Programming, Vol. 8; Elsevier Science Publishers; Amsterdam; 1987

[Harel 1990] David Harel, Hagi Lachover, Amnon Naamad, Amir Pnueli, Michal Politi, Rivi Sherman, Aharon Shtull-Trauring, Mark Trakhtenbrot: Statemate: A Working Environment for the Development of Complex Reactive Systems; in: IEEE Transactions on Software Engineering, Vol 16; 1990

[Kim 1994] Kane Kim, Hermann Kopetz: A Real-Time Object Model RTO.k and an Experimental Investigation of Its Potentials; Proc. COMPSAC '94 (IEEE Computer Society's 1994 Int'l Computer Software & ApplicationsConf.), Nov.1994, Taipei

[Magnusson 95] Boris Magnusson, Roger Henriksson: Garbage Collection for Hard Real-Time Systems; in: In Proceedings of the International Workshop of Object-Orientation in Operating Systems (IWOOOS'95), Lund, Sweden, 1995

[Meyer 1997] B. Meyer: Object-Oriented Software Construction (2nd ed.); Prentice-Hall; 1997

0-7803-5086-3 /98/$10.00 ©1998 IEEE

SPECIFICATION AND TESTING OF AUTOMOTIVE POWERTRAIN CONTROL SYSTEM SOFTWARE USING CACSD TOOLS

Steve Toeppe, Ford Research Laboratory, Ford Motor Company, Dearborn, MI

Scott Ranville, Ford Research Laboratory, Ford Motor Company, Dearborn, MI

Ken Butts, Ford Research Laboratory, Ford Motor Company, Dearborn, MI

Introduction

A trial evaluation of commercially available Computer-Aided Control Systems Design (CACSD) tools has been successfully completed. In this study, engineers responsible for the production powertrain control system used the CACSD tools to develop subsystem features that comprise approximately 60 percent of a complete engine control algorithm. Powertrain control algorithms are a rich combination of supervisory logic that determines the operating mode of the powertrain system and the associated transformations from controller input to output that are appropriate for that mode. In this paper, we present 1) the characteristics of an executable software specification methodology that captures this rich behavior and 2) automated procedures for verifying the software realization of the specified behavior.

CACSD tools have long been used in the automotive industry to develop simulation models [11, 25, 26], validate control law designs [9, 10], to accelerate the development process [18], and verify module implementation [16, 17]. Recent efforts have focused on extending CACSD tools into the realm of executable software specifications wherein the behaviors of the specification model and the software implementation are intended to be the same [3, 21]. That is, the specification models are complete in the sense that no additional functionality (e.g. diagnostics, failure-mode-effects management, or reset logic) needs to be added to the specification model during software development.

The costs of preparing a complete executable specification are recouped in several ways. One important way is through expected output generation for unit and structural testing [8]. A key reference for unit testing is found in [2]. This book introduces a large variety of testing techniques including control flow, domain, data flow, and state-machine testing. Another good overview paper is found in [26]. A detailed definition of an important test coverage class called Modified Condition / Decision Coverage (MCDC) is presented in [4]. This paper also describes how to determine what test cases satisfy this coverage class. MCDC coverage is called for in military specification RTCA/DO-178A [23]. Data flow coverage is another important testing class. References can be found in [7] and [14]. A comparison of the effectiveness of different testing techniques is presented in [1].

We describe the characteristics of the executable software specification and present an example in the **Executable Software Specification** section. A structural testing methodology to verify the realization of the software specification is presented in the section called **Structural Testing**. We conclude with the **Future Work** section.

Executable Software Specification

In this section we provide motivation for the executable software specification,

0-7803-5086-3 /98/$10.00 ©1998 IEEE

describe its characteristics, and present an example.

Purpose

The executable software specification model has four primary purposes:

- to allow the control law designer to validate the design prior to committing software engineering resource to implementation. This can be accomplished via simulation, systems analysis, and rapid prototyping.
- to completely and unambiguously describe the desired functional behavior of the software. It is intended that software design can emanate from this specification. The software design may include architectural analysis, data type definition, and module packaging.
- to generate expected outputs from input test stimuli to support software verification processes.
- to completely and graphically describe the behavior of the control law to the systems development community. In the automotive industry, this community includes several hundreds of vehicle calibrators, advanced developers, and researchers. It is important to note that the systems developers are assured that there is no functional difference between the specification and the realization.

Characteristics

The executable specification is used to describe the continuous-state and discrete-state behaviors of the control algorithm during all modes of operation. Typically, these modes include normal operation, reset, failure detection, and failure-mode-effects-management. The air charge estimation, exhaust gas recirculation, spark, knock control, fuel, and the diagnostic executive features are among those algorithms that have been modeled. To date, this modeling method has not been used to capture real-time-operating system, low level input / output driver or communication driver behavior.

In order to be consistent with structured requirements methodologies [13], the transformation of controller inputs to outputs and the activation of these transformations are strongly separated in the specification model. This separation was not possible in many commercial CACSD tools until recently. For instance, The MathWorks introduced the finite-state-machine modeling tool called Stateflow™[1] in June of 1997 [24]. It is important to note that the separation between transformation and activation behaviors was an important characteristic of the previously employed ladder logic specifications. Consequently, the various Ford engineering organizations have readily endorsed the new modeling method.

To ensure that specification models are prepared consistently, Ford has developed a modeling style guideline that restricts the control designer to a constrained set of tool features. Thus, the large training investment associated with the deployment of the CACSD based development process is used effectively. The guideline also facilitates the generation of software coding guidelines and tools for process automation such as automated user documentation generation, automated test harness and test vector generation, tool-to-tool translators and (potentially) automated code generation. The specification models are peer reviewed for functional quality and compliance to style guidelines prior to delivery to the software and calibration engineers.

Because the functional behavior of the software realization is to match that of the model, the set of simulation semantics of the CACSD tool serves as a formal definition of the specification model. The specification can be considered unambiguous because the user can gain clarification by 1) referring to the style guidelines for usage scenarios and 2) simulating the specification model. Modern

[1] MATLAB and Simulink are registered trademarks and Stateflow is a trademark of The MathWorks, Inc.

CACSD tools are open and it is possible to query and modify almost all of the characteristics of the specification model through command scripts. This capability is very important for those that wish to 'tune' the commercial tools to meet specific needs of their organization. Process automation and third party tool integration efforts also rely on the open nature of the CACSD tool suite.

Preliminary work has shown that, under restrictions consistent with our modeling style, automated procedures may be made available to formally verify the discrete-state behavior of the specification model with symbolic model checking tools [22]. Links to hybrid systems analysis and synthesis methods [5,6] are also being explored.

In our case, the executable specification models must be consistent with a significant base of legacy software, designs and architectures. Specific file organization, function prototypes, data definitions, calibration definitions, coding styles and naming conventions are characteristics of this legacy software base. The legacy design and architecture focus on specific tasking models and input/output interfaces. The modeling methodology ensures that the executable specification models can be easily integrated into this well-defined legacy architecture.

The tasking architecture is taken into account by organizing the hierarchical specification models into execution contexts that are run at a single rate. Each task and asynchronous event source has a corresponding execution context. This approach is contradictory to the normal approach used with most CACSD tools. However, in order to fully leverage the software testing and code generation automation potential, we chose to have a close correlation with the software tasking architecture. It is also a very practical approach for Ford's usage scenario where the tasking architecture and execution rates are already known. Finally, a correlation between select specification model hierarchies and C function call prototypes is required to support integration with legacy code and to facilitate tool automation.

Example Using MATLAB® / Simulink® / Stateflow™

Ford has worked cooperatively with The MathWorks since mid-1994 on the concept of executable software specifications with special focus on the automotive powertrain domain. In this section we show how The MathWorks tool suite, in particular MATLAB®, Simulink®, and Stateflow™, can be used to create executable software specifications.

The usage of the "function-call" triggers is intended to provide simulation behavior that provides a functional equivalent to a "C" function call. When a "function-call" is issued, the Stateflow™ chart issuing the call relinquishes control to the subsystem associated with the "function-call". It is possible for the called subsystem to issue subsequent "function-calls". A situation analogous to a "C" function call tree occurs where control is returned to the calling subsystem when the called subsystem completes its operation. The Simulink® simulation engine executes the sequence of function calls and the associated subsystem actions in zero simulation time. Hence, the start time is an accurate simulation, but there is not yet any support of execution delays. The close correlation between the "function-call" hierarchy and the "C" call hierarchy permits automated code generation and automated unit testing to be supported.

The following snippets are taken from the executable specification model of the idle speed set point calculation. Figure 1 is a Simulink® data flow diagram representing the behavior of the algorithm that 1) performs the idle speed specific failure-mode-effects-management of the ambient temperature estimate and 2) adjusts the desired idle speed setpoint according to the state of the air conditioning system. Each data flow diagram has a Stateflow™ control specification that manages the controller execution by passing control to another Stateflow™ control specification located at a lower level data flow diagram or by issuing a "function-call" trigger to transformation processes. Transformation processes are typically Simulink® block diagrams packaged into a "function-call" triggered subsystem, but Stateflow™ charts may be employed if the transformation is of a discrete-state nature. Returning to Figure 1, the *N_D_ctl* Stateflow™ chart is activated by another Stateflow™ control chart located at a higher level in the model hierarchy. The activating "function-call" trigger is called *trig_amb* in this instance.

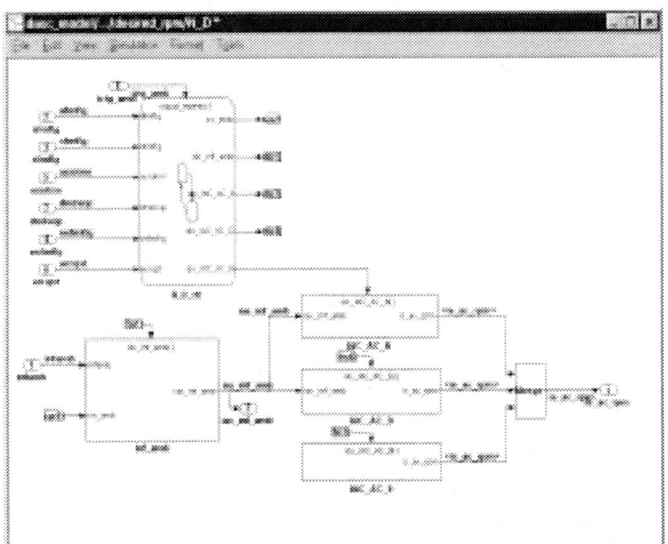

Figure 1. Data flow Diagram Using Simulink®

The transformation processes *inf_amb, IAC_AC_N, IAC_AC_D,* and *IAC_AC_0* shown in Figure 1 are controlled by "functional-call" trigger outputs from the *N_D_ctl* Stateflow™ chart. They are all Simulink® block diagrams packaged into "function-call" triggered subsystems. *IAC_AC_N, IAC_AC_D, and IAC_AC_0* are three different algorithms that calculate the same data variable (*ia_ac_rpm*) depending on the mode of the powertrain system. Note the Simulink® merge block is used to indicate that the three subsystems are actually writing to the same data variable.

Figure 2 is the expansion of the *inf_amb* transformation process. Via a selector switch setting determined in the *N_D_ctl* Stateflow™ chart, this process assigns either the *inf_amb* data value or a failure-mode-effects substitution value to the *iac_inf_amb* data value. The process executes once upon the receipt of the *do_inf_amb* "function-call" trigger. The sample time for the unit delay block is set to "-1" to specify event driven update characteristics.

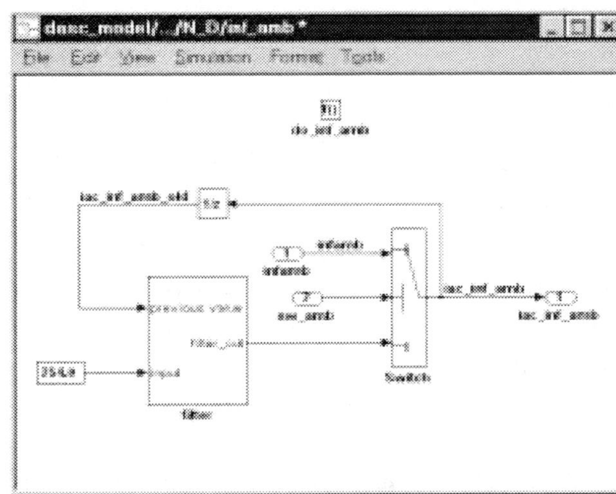

Figure 2. Transformation Process Using Simulink®

Stateflow™ charts are largely based on the statechart formalism developed by David Harel [12]. However, the Stateflow™ semantics add flow chart type behavior to the formalism. A Stateflow™ chart may contain pure statechart (finite-state-machine) behavior, pure flow chart behavior, or a mixture of the two behaviors. This extension has been very important in powertrain controller specification where a great deal of combinational logic (modeled as flow chart behavior) is found. An example of mixed statechart and flow chart behaviors is shown in the *ac_on* state of the *N_D_ctl* Stateflow™ chart (Figure 3). The reader can refer to the Stateflow™ User's Guide [24] to gain an understanding of the behavior of this chart.

0-7803-5086-3 /98/$10.00 ©1998 IEEE

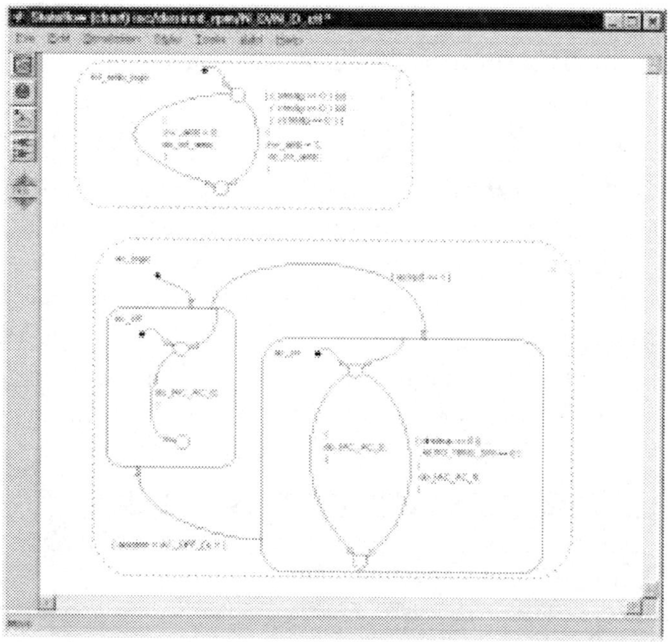

Figure 3. Control Using A Stateflow™ Chart

The MathWorks has introduced the concept of priority associated with different rate operations. The concepts of preemption, latency and duration are not supported in the standard tools. However, the mechanisms provided within the "S-Function" permit the introduction of customized blocks for delays, latency and coarse preemption. Research efforts are underway to develop automated measurement and estimation methods to parameterize these delay and latency blocks. The delay and latency blocks will be incorporated into the specification model after algorithms have been determined to be functionally sound but have not been proven to be supportable with the current software architecture and processing resources. Thus, this enhanced simulation model will be run to determine if the algorithms can be realized on the production target processor in a robust manner.

Structural Testing

The automated testing methods employed are a hybrid approach using a combination of white box structural testing and black box functional testing. Initial results are very promising and we plan to continue developing this hybrid testing approach. Specialized tools have been developed that permit the hybrid approach to unit testing to be conducted in a highly automated environment within a CACSD tool suite. This allows full leveraging of the flexible development, simulation and analysis environment.

The model is used as an executable specification that provides the idealized expected results associated with any given test vector [8]. Test vectors can be obtained from structural characteristics of the model, structural characteristics of the code, functional test scenarios and constrained random number generation.

Example Using MATLAB® / Simulink® / Stateflow™

Test coverage is provided from two points of view. Stateflow™ provides coverage metrics that attempt to indicate executable specification coverage that is somewhat weaker than Modified Condition/Decision Coverage (MCDC) coverage where each conditional within a predicate equation is verified as independently affecting the outcome of a decision. The C Unit is also instrumented with one of the many commercially available test coverage tools that provide coverage metrics that will detect MCDC criteria. Currently, all testing is performed on the host engineering workstation. We plan to support target based testing in the future.

The software test tool user interface (Figure 4) permits the software engineer to easily set up a particular test to compare a selected unit within the specification model to a corresponding C unit. A typical definition of a C unit is a C function. However, the tool is also being used for the "low level" integration testing of several functions residing in a larger module. The primary restriction placed upon the specification model is that the C function interface under test must correspond to the Simulink® subsystem interface. Each variable name must match and the function prototype must be known. The GUI permits selection of the data dictionary, generation of the harness, generation of various support files, establishment of various plots, scopes and other post analysis methods. A separate data dictionary GUI and manager is provided to handle all object definitions including data, vectors, arrays, communication message definitions and other miscellaneous objects.

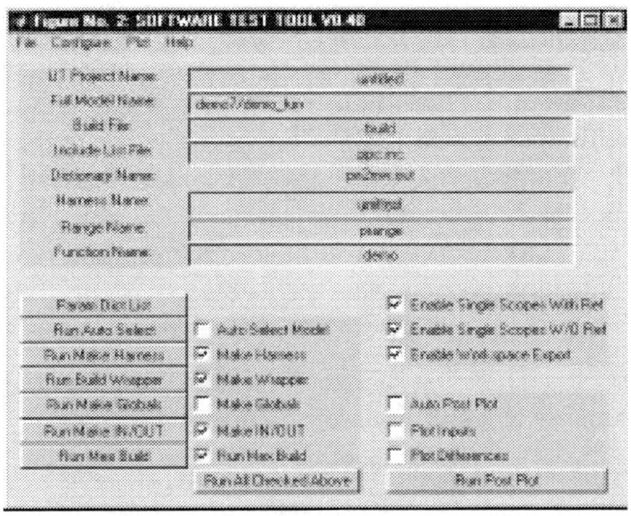

Figure 4. Software Test User Interface

Automated test harness generation permits an appropriate specification model subsystem to be used as a reference for establishing expected values for "C" unit testing. The specification model subsystem and the "C" unit are combined into a Simulink® harness to directly accommodate the test to be performed. MATLAB® "m-script" and "handle graphics" are used to implement the test tool GUI and test harness generation. The test harness generator will automatically interrogate the specification model to determine the parameter interfaces. Additional information about the parameters is obtained from the data dictionary that holds all relevant information about the data. The data characteristics are then used in the automatic generation of a "S-Function" wrapper around the "C" unit under test. The wrapper permits all required Simulink® signals to be mapped to the parameters or calling interfaces of the unit under test. Proper data type casting is supported. If fixed-point math is used, the proper value adjustments are made to map from Simulink® "doubles" to the fixed-point integer notation. The tool supports Simulink® vectors that represent arrays, structures and groupings of signals.

The test harness (Figure 5) is made up of a signal generator *(SIGNAL)*, the reference model (*MODEL*), the C unit under test (*CUNIT*), a sequencer to order the test execution (*CALLER*) and the error detector (*COMPARE*). The test harness will exercise both the C function and the corresponding model using the same set of input variables. The CALLER Stateflow™ chart controls the order of execution. The outputs are then compared and differences between the actual and expected value suggest that further review is needed.

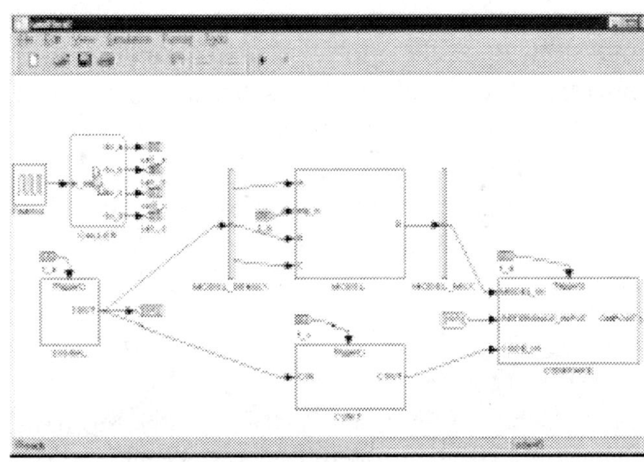

Figure 5. Software Test Harness

The actual and expected results can be analyzed in a number of ways. The test harness generator automatically establishes individual scopes for each output variable. The ability to cross-reference input signals is also provided to aid in the isolation of the source of specific errors. Differences between the signals are also calculated so that several types of thresholding algorithms can be used to automatically detect a pass or fail condition. Post plots of all signals are also available. The typical usage of the tool involves simulating 100 to 1000 test points and then analyzing the results of the error detector. Numerous subtle errors have been detected during testing and error detection productivity has increased by 400 percent when compared to current verification practices.

The stimulus currently being used is a pseudo random generator that is constrained to work within the operating ranges of the data parameters. Boolean and other discrete valued input types produce random discrete values. The random number generation uses a different seed for each of the variables. The overall distribution for each parameter is uniform over the number of test points. Of course, random numbers do not guarantee a statistically significant coverage criteria. Our experience for complex application models has shown 80 to 90% branch or decision coverage. However, it is possible to insert manually developed signal patterns to achieve 100% branch coverage. More rigorous levels of coverage are desired and consequently more advanced test vector generation technology is in development. The test harness generator has been designed to support the introduction of any source of test vectors, thus permitting seamless tool upgrades and easy transition between complementary tools.

Future Work

The Ford powertrain development community has endorsed the CACSD based modeling methodology. The formal nature of the models and the open nature of the CACSD tool suites admit continued analysis, automation, and integration opportunities. We have alluded to some of these opportunities (e.g. formal verification, latency-enhanced simulations, and automated code generation) in the previous sections. In the following, we conclude with the future opportunities that are most directly related to structural analysis and testing technologies.

Ford is currently working with tool vendors to develop and enhance automatic unit test and software integration test vector generation to provide various levels of control and data flow coverage. In the main, Ford will use the Modified Condition/Decision Coverage control flow test coverage criteria while stronger and weaker forms will be used as needed. The automated test vector generation tools will provide a minimal set of MCDC test vectors. A minimal set of test vectors that provides the desired levels of coverage and rigor is highly desirable to prevent redundant test vectors that will identify a single error numerous times. MCDC coverage suggests that the number of test cases required for any predicate equation of n terms is bounded by $[(n+1)$ to $2n]$. Therefore, a test vector generator that blindly generates 2^n vectors is not an acceptable solution for MCDC coverage.

The test vector generator will also provide test stimulus for the following cases:

- Equation Input / Output Stressing [2]
- Predicate Equation Boundary Value Testing [2]
- Loop Testing [2, 14]
- Data Flow Analysis [2, 7]
- State Machine Traversal Vectors [2]
- Fixed-Point Math Equation Stressing

All of the specification criteria required for the test vector analysis and generation is available in the specification diagrams, data dictionary and additional software design elaboration details. Thus, the algorithms and theory currently available within the unit testing community could be readily applied to our environment. For example, let us consider the standard control flow graph representation.

The Stateflow™ chart shown in Figure 6 can be represented in a control flow graph as shown in Figure 7. The new concept of waiting / sequencing is shown with the bolded lines. The bolded lines indicate where the Stateflow™ chart is waiting for the next iteration or step in processing. We believe data flow or definition-usage diagrams [2, 7] can also be extracted into a standard form to admit further analysis.

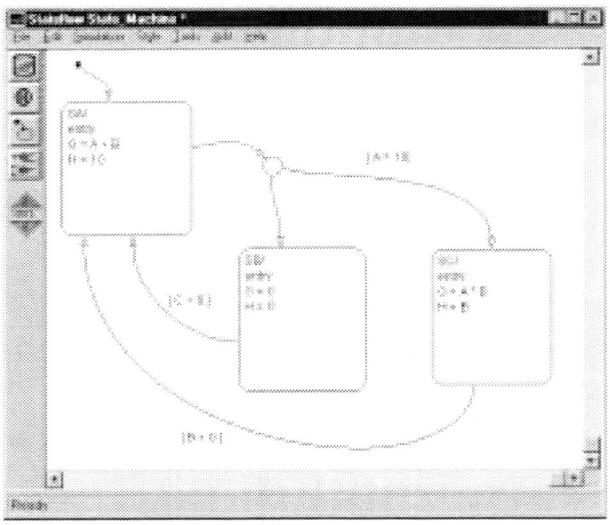

Figure 6. Typical Stateflow™ Chart

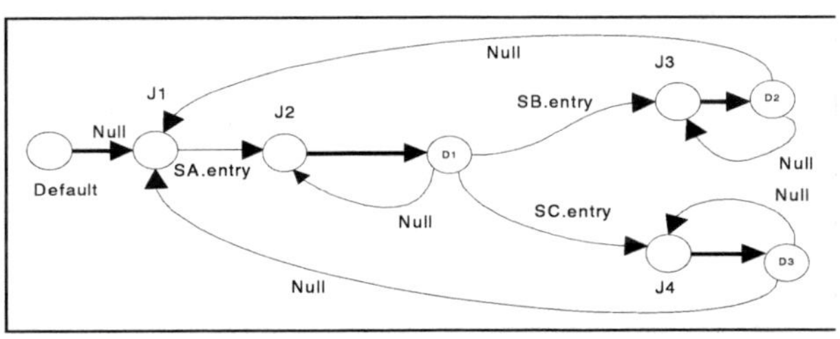

Figure 7. Control Flow Graph Corresponding to Figure 6

We also believe software integration testing will use data flow connectivity information to establish end-to-end test vectors for an integrated application. Just as in the unit testing case, numerous complexities in analysis can be more easily addressed if the control and data flow representations are in a standard form. The entire application will be integrated together in the CACSD environment using a large-scale model manager and automated wiring scripts. Appropriate points of correlation will be established between the integrated software and the large-scale model. Test vectors will then be applied to both the integrated model and the software.

Additional information can be derived from the control flow graph to establish a specific operating state or path of execution. The control and data flow representations are also the foundation for the next generation of automatic code generation technology that employs integrated application wide optimizations. The test vectors will also be used in the future to establish specific paths of execution for automated timing measurements on the target or through cycle accurate microprocessor simulators. The timing measurements will be used to populate scheduling analysis tools to determine if a given tasking architecture is schedulable.

Acknowledgements

We would like to acknowledge our visionary and enthusiastic colleagues at Ford Advanced Vehicle Technology (Paul Smith, Dennis Raymond, Craig Stephens, Pat Menter, Dan Pinkos, Chris Davey, Barb Mattison, Nada Shaya, Joe Suffredini, Bob Baskins, Kevin Rzemien, Rajeev Ramanan ...) who have made significant contributions to the modeling and testing methodologies.

References:

1. Basil and Selby, "Comparing the Effectiveness of Software Testing Strategies," IEEE Transactions on Software Engineering, Vol. SE-13 No. 12, Dec. 1987.
2. Beizer, B., **Software Testing Techniques**, Second Edition, International Thomson Computer Press, 1990.
3. Butts, K., "An application of integrated CASE/CACSD to automotive powertrain systems," Computer-Aided Control System Design, 1996., Proceedings of the 1996 IEEE International Symposium on , Page(s): 339-345
4. Chilenski and Miller, "Applicability of modified condition/decision coverage to software testing," Software Engineering Journal, September 1994.
5. Chutinan, A., and Krogh, B. H., "Computing approximating automata for a class of linear hybrid systems," To appear in Hybrid Systems V, Lecture Notes in Computer Science. Springer-Verlag, 1998.
6. Chutinan, A., and Krogh, B. H., "Computing polyhedral approximations to dynamic flow pipes," submitted to the 37th Conference on Decision and Control: Invited Session on Synthesis and Verification of Controllers for Hybrid Systems, 1998.
7. Clarke, Podgurski, Richardson, and Zeil, "A Formal Evaluation of Data Flow Path Selection Criteria," IEEE Transactions on Software Engineering, Vol. 15 No. 11, November 1989.

8. Cohen, E., and White, L., "A Domain Strategy for Computer Program Testing," IEEE Transactions on Software Engineering, Vol. SE-6, No. 3, May 1980.

9. DePoyster, M.R.; Hoying, J.F.; Majeed, K.N., "Rapid prototyping of chassis control systems," Computer-Aided Control System Design, 1996, Proceedings of the 1996 IEEE International Symposium on , Page(s): 141-145

10. Dorey, R. E., Maclay, D., "Rapid Prototyping for the Development of Powertrain Control Systems," Computer-Aided Control System Design, 1996, Proceedings of the 1996 IEEE International Symposium, Page(s): 135-140.

11. Godbole, D., and Karahan, S., "Automotive Powertrain Modeling, Simulation and Control Using Integrated System's CASE Tools," Society of Automotive Engineers #940180, International Congress & Exposition, 1994.

12. Harel, D., "Statecharts: A Visual Formalism for Complex Systems," Science of Computer Programming 8, 1987. Pages 231 – 274.

13. Hatley, D., and Pirbhai, I., **Strategies for Real-Time System Specification**, Dorset House Publishing, 1988, 1987

14. Huang, J., "Detection of Data Flow Anomaly Through Program Instrumentation," IEEE Transactions on Software Engineering, Vol. SE-5 No. 3, May 1979.

15. Jung-Ho Kim.; Byeong Yong Jeong; Dong-Il Dan Cho; Hoyoun Kim "AUTOTOOL, a PC-based object-oriented automotive powertrain simulation tool," Intelligent Transportation System, 1997. ITSC '97., IEEE Conference on , Page(s): 753 -758.

16. Kendall, I. R., Jones, R. P., and Thomas, S. M., "Simulation as a Means of Achieving "The Impossible": An Investigation into the use of Simulation in the Development of Electronic Control Systems at Jaguar Cars," to be presented at Simulation '98 – Innovation through Simulation, IEE, University of York, U.K., September, 1998.

17. Krohm, H., and Gheorghiu, V., "Hardware-in-the-Loop simulation for an elcetronic clutch management system," Society of Automotive Engineers, 950420, SP-1080, February 1995.

18. Moskwa, J.J., Anthony, J.W., Babbitt, G.B. and Rubin, Z.J., "Synthesis of Software and Hardware in PCRL for Powertrain Design and Development," IEEE Control Systems Magazine, Special Issue on Powertrain Control, 1998.

19. Niinomi, T., Krogh, B.H., and Cury, J.E.R., "Computing exact automata representations for a class of hybrid systems," Proceedings of the 35th IEEE Conference on Decision and Control, December, 1996.

20. Niinomi, T., Krogh, B.H., and Cury, J.E.R., "Refinements of approximating automata for synthesis of supervisory controllers for hybrid systems," in **Hybrid Systems III**, Springer-Verlag, 1996.

21. Ohata, A., "Executable Software Specification," presented at the Global Automotive Engineering Seminar, Troy, MI, June 1998.

22. Rausch, M., Chutinan, A., and Krogh, B. H., "Verification of StateCharts and Hybrid Dynamic Systems," Ford Research Laboratory Seminar, April 1998.

23. "Software Considerations in Airborne Systems and Equipment Certification," Military Specification RTCA/DO-178A. (Now DO-178B, also known as EUROCAE ED 12B in Europe).

24. **Stateflow Users Guide**, Version 1, The MathWorks Inc., 1997.

25. Stefanopoulou, A. G., Cook, J. A., Freudenberg, J. S., Grizzle, J. W., Haghgooie, M., and Szpak, P., "Modeling and Control of a Spark Ignition Engine with Variable Cam Timing," Proceedings of the 1995 American Control Conference, Seattle WA, June 1995.

26. Weeks, R., and Moskwa, J., "Automotive Engine Modeling for Real-Time Control using MATLAB/SIMULINK," Society of Automotive Engineers #950417, International Congress & Exposition, 1995.

27. White, L., "Software Testing and Verification," Advances in Computer, Vol. 26, 1987.

ACTIVE CONTROL OF VEHICLE DYNAMICS

Michael Fodor, Ford Motor Company, Dearborn, Michigan

John Yester, Ford Motor Company, Dearborn, Michigan

Davor Hrovat, Ford Motor Company, Dearborn, Michigan

Introduction

Digital electronic controls have been applied to various automobile subsystems for nearly two decades. Initially, they were used for powertrain controls, improving fuel economy and reducing emissions. More recently, digital controls have been used to improve the dynamic response of the vehicle to braking, accelerating, and steering inputs from the driver. Systems such as anti-lock brakes and traction control are industry mainstays with active yaw control currently being introduced throughout the luxury car brands. Active suspensions and four-wheel steering have also been added to automotive products with less market success. Because such systems have a potential to improve the safety and ease of use of vehicles, their use is expected to increase with accompanying increases in system complexity and subsystem interaction.

The purpose of this paper is to introduce the topic of vehicle dynamic controls, giving a sense of the diversity of problems associated with this field. This is accomplished by defining a generic framework that applies to all vehicle dynamic control systems and by highlighting three vehicle subsystems that can be cast in this framework, active suspensions, traction control, and active yaw control. The level of dynamic improvement produced by each of these subsystems will be shown. Some comments about future trends in this field will be presented, followed by a discussion of development tools and procedures.

Motivation and Framework for Dynamic Controls

Modern automobiles perform well when operated under typical conditions such as on clean, dry, and smooth roads at moderate speeds. Under these conditions passive mechanical systems are sufficient to produce predictable and pleasing dynamic response of the vehicle to steering, accelerating, and braking inputs. However, as operating conditions become less typical, as is the case with operating the vehicle on slippery or rough roads at speeds too high for conditions, the dynamic response of the vehicle to operator inputs can become unpredictable and unsettling to the driver, increasing driver stress and possibly reducing safety. Thus, the primary motivation for active vehicle dynamic controls is to increase the range of conditions under which the vehicle behaves predictably. Additionally, active controls can also help to enhance vehicle comfort and response under ideal conditions.

A generalized framework for active vehicle dynamic controls can be defined as shown in Figure 1. Here, driver inputs are applied to both the vehicle (as is normally done with the passive vehicle system) and to a reference model. The vehicle response is then compared to the model response, and the difference is used by the controller and its actuators to force the vehicle response toward the idealized model response. This conceptual model can be applied to individual subsystems

0-7803-5086-3 /98/$10.00 ©1998 IEEE

as well as interacting combinations of subsystems.

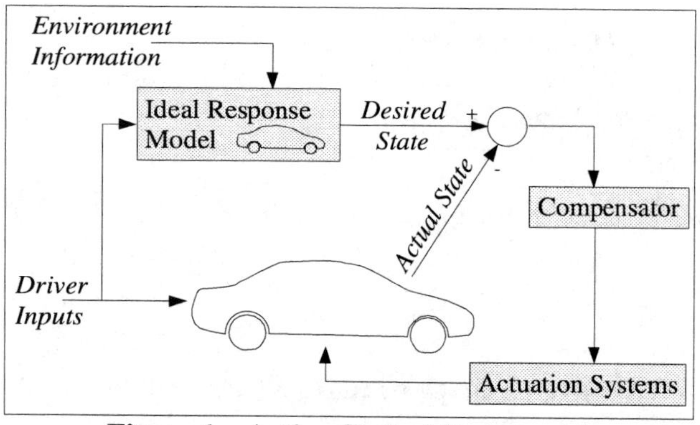

Figure 1 – Active Control Framework

Actuation means commonly available in automotive systems include control of powertrain output torque, brake torque, and suspension forces. Less commonly available are control of rear wheel steering and side-to-side and front-to-rear transfer of powertrain torque. Control of suspension forces is discussed next.

Active Suspensions

The purpose of an automotive suspension is to isolate the passenger from the road roughness while providing proper vehicle handling. Suspension designs must strike a balance between reducing accelerations of the vehicle body (sprung mass) which cause passenger discomfort and reducing the tire's hopping motion as it traverses a rough driving surface which reduces handling capability. In passive suspensions with statically tuned components, this tradeoff between ride and handling results in stiffly sprung sports cars and poorly handling but plush riding luxury cars. If a suitable means for actively altering the suspension tuning is available, the ride/handling tradeoff can be changed as driving conditions change. This adaptability as well as the potential of active safety improvements provide the impetus for the use of active suspensions.

Actively controlled suspensions have in practice been divided into two classes, active and semiactive. Semiactive systems, like completely passive systems, only dissipate energy, having no source of power for adding energy to the overall vehicle system. They typically consist of a tunable damper operating in mechanical parallel with a static spring. Active suspensions contain power sources and can alter their dynamic characteristics within a given frequency range. They typically consist of a force-producing device such as an electro-hydraulic servo system or fully electrical actuator.

Active suspension control methods have been the subject of a large amount of academic and industrial research. The quarter car model (one wheel's suspension and its associated sprung mass portion) is easily linearized, lending the active suspension problem to a large variety of control theories (see Ref. [1]). One of the most revealing methods of active suspension control study has been the application of linear optimal control theory to the problem. This method allows the tradeoff between sprung mass acceleration and wheelhop (ride versus handling) to be studied explicitly in the context of full and partial state feedback.

Some results of the application of linear optimal controls to the suspension problem are shown in Figure 2 which presents the tradeoff between wheelhop and sprung mass acceleration, both of which are normalized using the magnitude of the vertical road surface velocity (the fundamental system disturbance), as seen by a traversing tire, and vehicle speed. System operation moving toward the origin of the plot is preferable as this reduces both the wheelhop and the sprung mass acceleration. The passive suspension curve in Figure 2 is produced by sweeping through a range of suspension damping values for a given fixed suspension stiffness. A typical tuning point, marked on the Figure 2 passive suspension curve, represents the best damping value for the given stiffness, and any adjustment from this value results in an increase to both sprung mass acceleration and wheelhop. The optimal control solution to the suspension problem calls for weights to be placed on wheelhop and sprung mass acceleration. As these weights are changed, the resulting optimal control gains produce the suspension behavior described by the Figure 2 curve for the active case. Here, a more favorable performance tradeoff can be realized by adapting to different road and speed conditions.

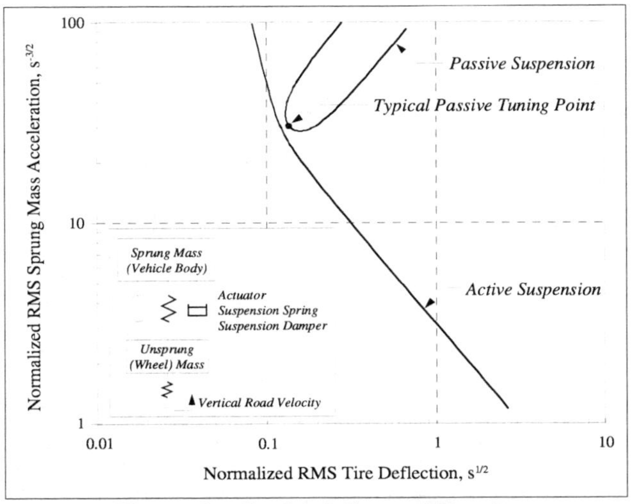

Figure 2 – Optimal Control Results

Active suspension controls begin to fit the dynamic control framework depicted in Figure 1 when the controller gains are scheduled as a function of operating conditions. For a given handling (steering) condition, the allowable magnitude of wheelhop is a fixed number. Ideally, a vehicle would fully utilize this wheelhop constraint as road roughness changed. This is accomplished by increasing the ratio of wheelhop to road vertical velocity as road roughness decreases (moving to the right on the x-axis of Figure 2), allowing for a reduction in sprung mass acceleration. Thus, by monitoring the magnitude of wheelhop, the active suspension controls can schedule gains to maintain an optimal tradeoff between ride and handling as operating conditions change.

Production applications of active suspensions have been introduced to the market slowly and sparingly, primarily due to the high cost of actuators and the limited benefits realized so far. These tradeoffs have forced manufacturers to use either semiactive systems or limited capability active systems. As technological solutions to the actuator problems surface, active suspensions will become more widespread, especially as their inherent active safety potentials fully materialize.

Traction Control

Vehicle motion is caused almost entirely by forces generated at the tire/road interfaces, and these forces provide the largest non-linearity and uncertainty in the vehicle dynamic control problem. Figure 3 depicts typical tire forces generated as a function of longitudinal tire slip for an arbitrary steering angle and road surface friction. Slip is a measure of the relative velocity between the contact patch of the tire and the road surface. At low slip, the absolute longitudinal tire force is low as compared to the lateral. This relationship reverses at higher slips, and at very high slips the longitudinal force passes its maximum value and begins to decrease. High slips occur when wheels are locked during braking or are over-spinning during acceleration. It is in this high slip range that braking, accelerating, and steering all suffer.

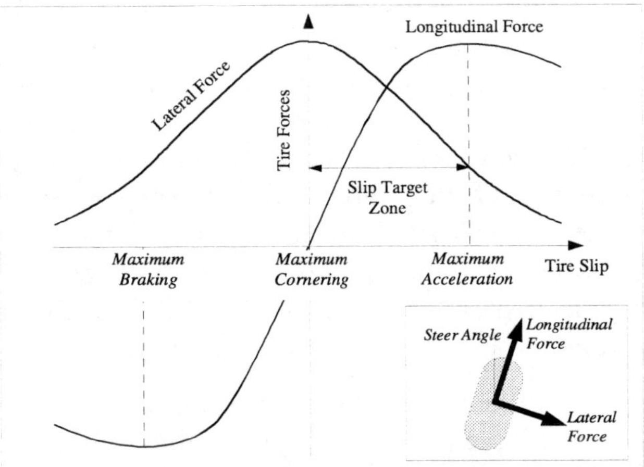

Figure 3 – Tire Characteristic

The first system to actively control tire slip was ABS. Its purpose is to eliminate high-slip operation, improving vehicle steerability and reducing stopping distances. It is out of ABS work that traction control arose, being the analog of ABS in the accelerating mode. The purpose of a traction control (TC) system is to manage vehicle acceleration and steerability by targeting an appropriate driven wheel slip, based on driver control inputs, and meeting this target using closed-loop control and the available actuators. Figure 4 depicts the generalized TC structure with engine torque and brake torque as the control actuation.

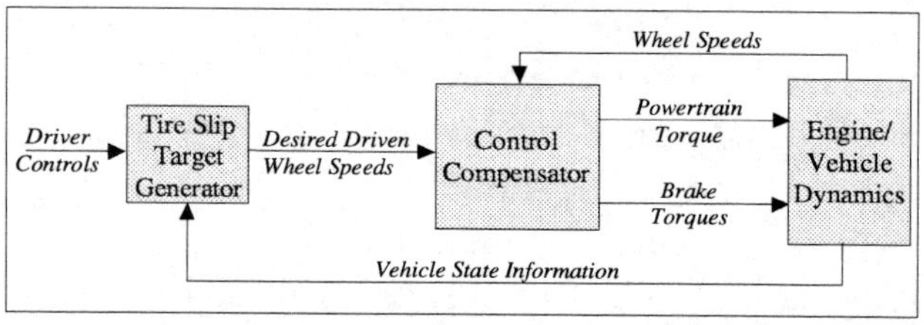

Figure 4 – Traction Control Structure

Wheel slip targets are based on vehicle speed, driver steering intent, accelerator pedal position, and estimated road surface friction. In the absence of a steering angle sensor, yaw rate is used to determine the driver's steering intent. In the absence of a yaw rate sensor, yaw rate is deduced using the difference between the non-driven wheel speeds. When the vehicle is cornering (large steering intent as compared to the vehicle's longitudinal speed) low tire slips are targeted to produce the necessary higher lateral tire forces. When large accelerator pedal positions are present and the driver's steering intent is low, larger slip levels are targeted to produce the larger longitudinal forces needed for better acceleration. Slip targets are further adjusted according vehicle speed and estimated surface friction in an effort to adjust to the changes in tire frictional characteristics that occur as speed and surfaces change.

In general, closed-loop tire slip control is a multiple input multiple output problem. For vehicles with an open differential, the torques at the driven wheels are equal when brake torques are zero. For driving surfaces with uniform friction, powertrain-only actuation is sufficient for tire slip control. In this case, tire slip can be controlled by modifying the powertrain output torque using engine spark advance, cylinder cutting, cylinder air/fuel ratio, transmission shifting, or electronic throttle. Currently spark/fuel control is the most widely used torque control method due to its low cost. Because each of these actuation methods operates with its own bandwidth and, in some cases pure delay, the mixing of actuation systems must be done carefully. To complicate the problem, some of these actuation methods can effect engine emissions and combustion noise quality.

If the friction coefficient for the tire/road interfaces is different for the driven wheels (a "split-mu" condition), some brake torque is necessary to maintain the desired slip for each tire. Brake torques at each wheel are controlled by pumping fluid into or releasing fluid from each brake caliper. Typically, this process utilizes two valves at each caliper and tends to be very non-linear. Brake compensator tuning requires attention to this non-linear brake system behavior. Controlling brakes improperly can lead to unnecessarily large transmission loads and increased system noise. Such behavior is characterized by aggressive brake application and the excitation of high frequency driveline oscillatory modes. Poor brake system performance can also be exacerbated by poor powertrain torque management. Designing the brake and powertrain controls so that they cooperate well and interact smoothly under all possible road conditions is one of the more challenging aspects of traction control design. The quality of slip control is a major differentiator between production systems since it has a significant influence over the vehicle's noise, vibration, and harshness during a TC event.

Because the overall vehicle system dynamics change with changing road friction, it is advantageous to schedule control system gains as a function of friction (mu). Mu is determined by estimating the powertrain output torque and observing the rate of change of wheel speeds. Figure 5 shows the closed loop control system tracking a desired wheel speed (wheel speed and slip are connected through vehicle speed) as the vehicle transitions from snow to ice and back to snow. A mu-based control gain is also plotted in addition to estimated mu.

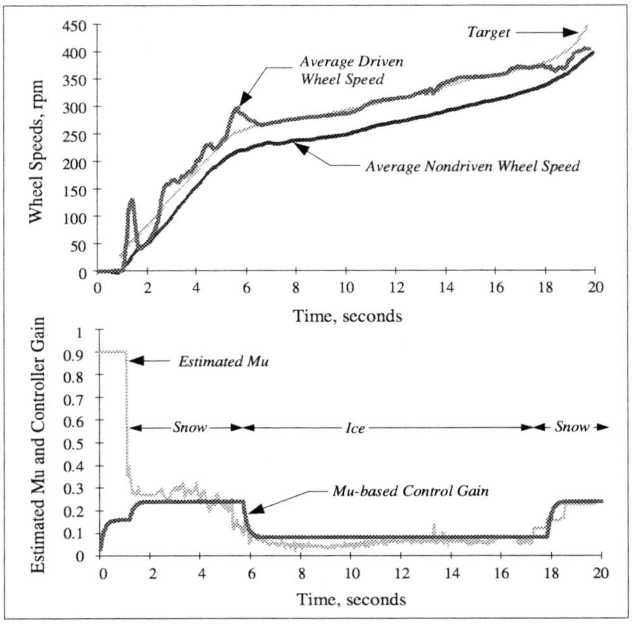

Figure 5 – TC Operation With Gain Scheduling

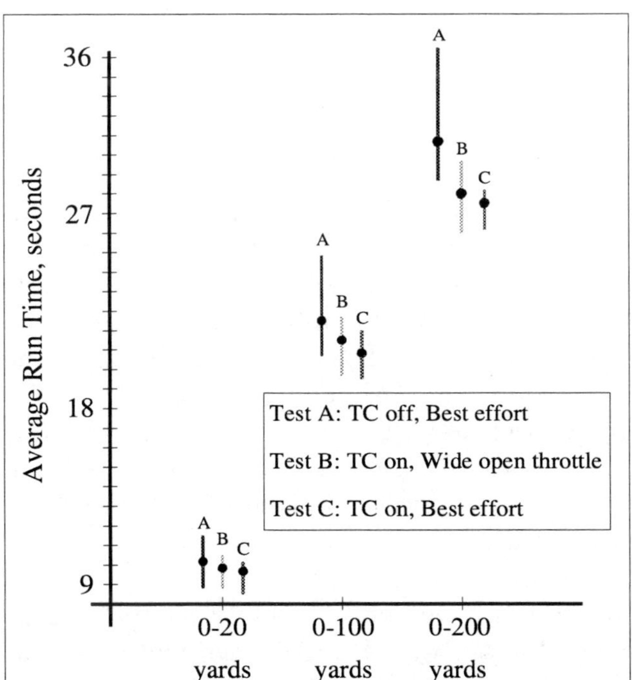

Figure 6 – Acceleration Performance on Ice

For straight-line acceleration, data collected for several drivers driving with and without TC show that with TC an inexperienced, perhaps less agile driver will perform at least as well if not better than test drivers experienced in driving on slippery surfaces without TC. Timings conducted for these tests are shown in Figure 6. The difference in acceleration between best effort with and without TC for the 20 yard run relates to roughly 1 car length, a significant difference when crossing an icy intersection. Similar improvements in vehicle steerability and stability are also evident with traction control engaged.

Yaw Control

Because traction control systems typically lack steering wheel angle, yaw rate, and lateral acceleration sensors, they cannot determine that a vehicle is steering improperly. However, by extending a TC system with these sensors and by controlling wheel slips individually, closed loop control of a vehicle's steering response to driver inputs can be accomplished. Active yaw control requires two major functions, producing a desired yaw rate target and meeting this target through closed-loop control. This functionality fits well with the Figure 1 vehicle dynamic control framework.

Yaw rate target generation, like tire slip target generation in traction control, depends on road/tire friction, vehicle speed, and driver steering intent. High-mu vehicle operation at moderate speeds produces a well-behaved relationship between steering wheel angle and yaw rate that can be predicted either through an empirically generated map or approximated using a relatively simple mathematic function. Usually this relationship is considered ideal and is used as the yaw rate target with possible modifications for mu. As mu decreases the vehicle's ability to corner decreases as well, limiting the yaw rate that the vehicle can sustain while still maintaining a proper attitude with respect to its velocity direction. Thus, some modification for surface friction coefficient is usually reflected in a yaw controller's yaw target. One possible yaw rate target can be given by:

$$\dot{\psi} = \frac{v\delta}{l}$$

where δ is the angle of the front wheels with respect to the longitudinal axis, v is the vehicle speed, and l is the vehicle's wheelbase. Possible friction-based upper and lower limits on this yaw rate target could be given by

$$\dot{\psi}_{\max/\min} = \pm\frac{\mu g}{v}$$

where μ is the coefficient of friction and g is acceleration due to gravity. Surface friction can be estimated as in traction control, as well as by using the measured lateral vehicle acceleration. For closed loop control, the yaw target is compared to the measured yaw rate, and the difference is used to mechanically alter the vehicle's yaw moment as described next.

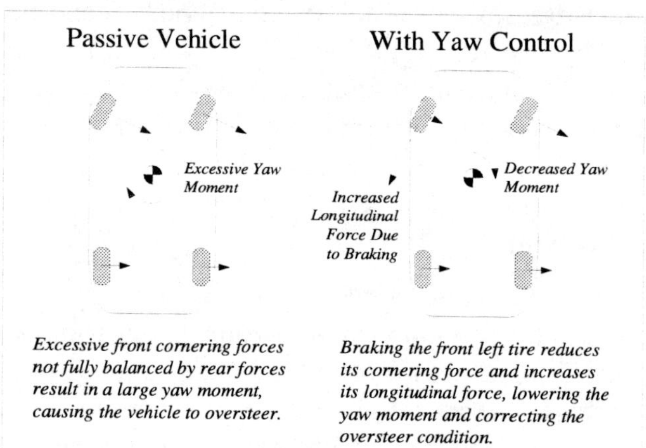

Figure 7 – Yaw Control Using Differential Braking

In present yaw control systems, actuation is achieved through active brake control. Vehicles turn due to a yaw moment induced by tire forces. To alter a vehicle's yaw moment, brakes can be applied to one side of the vehicle or the other (see Figure 7). The resulting change in tire slip alters the tire forces that combine to produce yaw torque. When applying this differential braking, increasing the absolute tire slip raises the tire's absolute longitudinal force while decreasing it's lateral force (see Figure 3). Therefore, if active brake control is available on front and rear tires (in some cases only the driven wheel brakes have active control) differential braking may be applied to either the front or rear tires, depending upon the driving circumstances, to produce the combination of lateral and longitudinal forces which are most favorable to adjusting yaw torque while maintaining overall vehicle lateral adhesion (cornering ability).

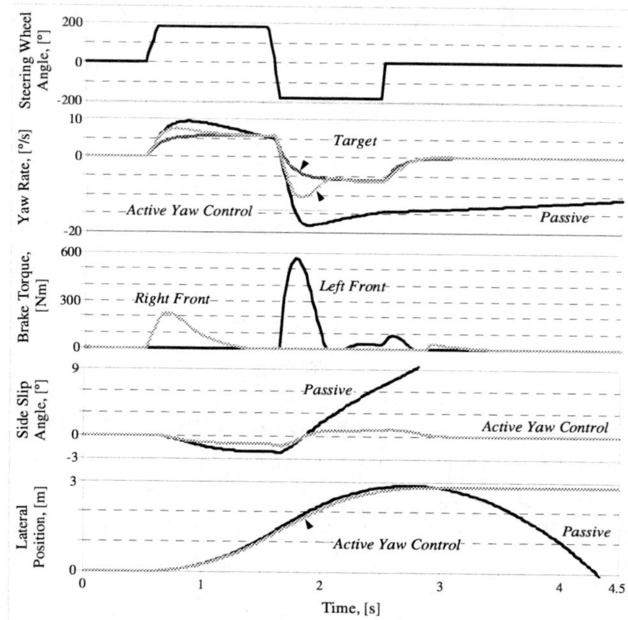

Figure 8 – Yaw Control During a Lane Change

Figure 8 depicts several key states for two vehicles coasting through a lane change on snow, one with and one without active yaw control. When the initial steering input is applied (to the left), the vehicle's yaw rate overshoots the desired. To correct this, the right front brake is applied to reduce the yaw rate. When the steering input is reversed, the vehicle's yaw rate again attempts to exceed the desired and the left front brake is applied for correction. Slip angle, the angle measured between the vehicle's velocity vector and its longitudinal axis, is also shown and serves as an indicator of system attitude control. Under normal conditions slip angle is usually limited to around +/- 5 degrees. The smaller slip angles and cleaner transition to the second lane for the active yaw control case are indicators of the overall driveability improvement brought by yaw control. A full-motion video of the Figure 8 data is included as the file *Active Yaw Demo.avi* on the 17th DASC proceedings CD-ROM.

Other design issues associated with active yaw control include integrating its functionality with traction control and dealing with the low-cost sensor sets typically associated with automotive systems.

Future Vehicle Dynamic Controls

Future developments in vehicle dynamic controls will begin with general improvements to the already existing systems. Better and more mature hardware will be accompanied by algorithms that more effectively utilize the hardware to produce systems that gradually migrate toward optimal.

These improvements will be followed by the addition of new actuation systems, perhaps including more advanced four-wheel drive, augmented two-wheel steering, four-wheel steering, and systems that provide a separate driving torque at each wheel (some of these systems have already been introduced in small quantities). These new systems may allow vehicle dynamic control to become less of an intervention technology that engages only in extreme conditions (such as current implementations of ABS, TC, and yaw control), providing continuous adjustment of vehicle response to driver inputs.

With the addition of more actuation systems, increased interaction between these systems will follow. Rather than a collection of subsystems, each with its own separate computational engine, future dynamic control systems may be comprised of a single computer operating on a unified set of desired system states, commanding the output of a collection of actuators. As with aircraft, such a unified control system will be capable of detecting failing sensors and actuators and will operate gracefully in the presence of such failures, providing the best possible control despite hardware degradation. Finally, vehicle dynamic controls may become integrated with intelligent vehicle systems to provide functions such as active obstacle avoidance, fully automatic steering and braking, crash mitigation, and the like.

Development Methods/Tools

With the introduction of digital electronic controls in the automotive industry, traditional hardware-based trial and error engineering has given way to computer aided design. Most of the tools and methods being used and introduced are familiar to controls practitioners in the aerospace controls field. One primary difference may be that with automotive systems there are less development-related safety issues than in the aerospace industry. This circumstance allows for rapid design iteration and begins to blur the lines between algorithm development and system testing. The development process can be divided into five basic steps: modeling, desktop simulation, man and hardware in the loop simulation, in-field testing, and iteration.

Various levels of system modeling can be conducted in the process of controls development. Initially, the assembly of very simple models that capture only the basic states of system operation are preferred. Such models allow designers to understand the basic raw functionality of the plant being controlled without being hampered by extraneous dynamics that only cloud the big picture. These simple models are easier to validate against actual hardware and provide a framework for designing the basic overall control structure. Simple models are later appended with previously unmodeled dynamics to insure the robustness of initial control designs and to help avoid "surprises" when testing begins. The use of first principles models is usually preferred to using numerical identification techniques. One of the main benefits of modeling is to provide the designer with as much understanding of the physical system behavior as possible.

Desktop simulation has become increasingly more powerful in recent years. Using control system design tools, a designer can quickly analyze the performance of a system under a variety of control schemes, arriving at a preliminary design. Current tools can efficiently convert high-level designs into machine code for use in real-time testing and development.

Although still in their early stages, man-in-the-loop and hardware-in-the-loop design are becoming more common in the automotive industry. These methods allow testing of control concepts in the presence of the full dynamic response of subsystems (humans included) that can be difficult to model. Man-in-the-loop development usually takes place in a driving simulator and provides insight into the human factors aspects of the man/machine interface.

In-field testing provides designers with the full realization of dynamic system response to real-world inputs. It is here that hopefully all previously unmodeled system dynamics surface and the need for further design iteration can be identified. Hardware testing has been simplified by the emergence of portable development controllers that are easily reprogrammed, allowing for fast controller tuning and algorithm changes in the field.

Perhaps the most important advancement in development tools has been the ability to combine "desktop" simulation and testing together in the field. Many automotive control applications can be simulated on a portable computer allowing the controls developer to be both the test engineer and the algorithm designer. System development times can be reduced significantly by combining the development steps as closely as possible in both time and space. Future improvements in this area may see in-field testing and desktop simulation combined with hardware/man-in-the-loop simulation and testing.

Conclusion

The overview presented here only begins to address some of the basic design aspects of three systems which are either commonly available as products or have been extensively researched. The depth of design considerations in this field is considerable. As each of these considerations is mastered by the engineering community, vehicle dynamic controls will continue to deliver safer, more pleasing products to consumers at greater value. Ultimately, the influence of these systems on automobiles will approach the influence that aircraft controls have had in their industry. Active control of vehicle dynamics has become a rich field of study and innovation for the automotive industry and will become increasingly more critical to the marketability of automotive products in the future.

Acknowledgements

The authors would like to recognize the contributions of Jahan Asgari, Ming Kuang, and Craig Simonds to the collective body of knowledge needed to produce this paper. They are all members of the Interactive Systems Modeling and Controls project within the Vehicle Electronic Systems Department at Ford Research Laboratories.

References

[1] Hrovat, D. "Survey of Advanced Suspension Developments and Related Optimal Control Applications", *Automatica*, Vol. 33, No. 10, pp. 1781-1817, 1997.

[2] Mack, J., et al. Eds. *ABS/TCS and Brake Technology Developments*, SAE '94, No. SP-1018.

[3] van Zanten, A.T., et al. "VDC Systems Development and Perspective", SAE '98, No. 980235.

THE NEXT GENERATION AUTOMOTIVE ELECTRICAL POWER SYSTEM ARCHITECTURE: ISSUES AND CHALLENGES.

John M. Miller, Ford Motor Co./Scientific Research Laboratory, Dearborn, MI.
Paul R. Nicastri, Ford Motor Co./Scientific Research Laboratory, Dearborn, MI.

Abstract

The electrical and electronic content of contemporary passenger cars and light trucks now exceeds 20% of the vehicle cost. Semiconductor content alone is growing at 16%/yr, primarily in embedded microcontrollers, safety and security systems and entertainment systems. As the automobile becomes more electrified, the demand for installed capacity to support future electronic and ancilliary functions is expected to exceed the historical 4%/yr. Today, the automobile electrical system consumes approximately 1.2 kW to support essential loads. With electrified ancillaries, like power assisted steering, the demand will grow to 3 kW and higher in the near term (10 to 15 years). Meeting this predicted demand in a low voltage system is one issue facing automotive electrical system design today. The challenge lies in how to meet such future demand at affordable cost and uncompromised quality. This paper summarizes candidate electrical architectures that are considered viable alternatives to today's system.

Introduction

The automotive industry today is faced with electrical/electronic system complexity concerns and electrical power demands that are stretching the capabilities of present on-board power supplies. The pervasive application of electronics for customer comfort, convenience and safety features such as electronic automatic climate control, entertainment systems, antilock brakes (ABS) and traction control (TC) systems, not only contribute to exponential growth in wiring harness size but to power demand as well. A typical luxury class vehicle today has 2.5 km of wire in the harness, a subset of some 350 connectors (250 different types) and nearly 1500 cut leads (i.e., # of circuits). A 1950's automobile contained a mere 45m of wiring. The introduction of multiplex communications has helped reduce the bulk of the electrical harness for most body electrical functions such as window, mirror and seat controls. Multiplex communications facilitates the architectural move from point-to-point load control to remote power switching. For the most part, electronic content is growing at a rapid pace.

After the introduction of electronic engine controls and emissions controls in the 1980's the application of electronics to the automobile was growing at roughly 26%/yr. During the 1990's this rate has dropped to 21%. Of the total semiconductor production (some 10^6 m^2 of silicon per year) the automotive industry consumes only 2.7%. Today's high silicon consumers are the telecommunications, computer and consumer electronics industry. The bulk of silicon used in the automotive industry today goes into information processing, computational and signal processing. There is little precedent for power electronics - the use of information to process power. The best illustrations of power electronics in an automobile are in ignition systems, alternator regulator, climate control blower motor speed control and most recently in electrically assisted steering systems.

Cost is an over arching concern in automotive products. Technologies introduced are typically mature and provide the customer demanded function at minimal cost in a demanding environment. The thermal environment for each of the three major partitions of the modern automobile are underhood at 150 °C (and increasing to 180 °C),

0-7803-5086-3 /98/$10.00

passenger compartment at 85°C and trunk (including package tray and in other enclosed body areas at 115°C. Work in progress in electronics reliability, specifically, component uprating will help to redefine the application of electronics in the automotive environment [1].

Standardization of the electrical system is of paramount importance, especially when contemplating increasing the system voltage. The automobile is an open architecture system for both communications and power supply. When the industry changed from 6V to 12V batteries in 1955 the changeover was accomplished within two years for all but a couple of manufacturers [2]. The major subsystems impacted then were few, namely generator, battery, ignition, lighting, horn, clock and cranking motor. Today, the number of affected subsystems is enormous in comparison. The consequences of changing the electrical system today will impact nearly all of the component and full system supply base. That is why automotive OEM's in North America, Europe and Asia are participating in the development of global standards for the next generation electrical system. The Consortium for Advanced Automotive Electrical/Electronic Components and Systems organized and coordinated by MIT [2] now comprises some 20 companies with the goal of establishing the 42V specification as the new voltage standard.

Open wiring systems are the hallmark of automotive electrical systems. Accessible fuse panels, power distribution boxes and junction boxes are facilitated by distribution voltage levels that are considered safe. The Society of Automotive Engineers (SAE) in their guideline J2232 [3], based on the work of interested engineers who formed the SAE Dual/Higher Voltage Committee in 1988, recommended 60 Vdc as the maximum safe voltage with dynamic overvoltages, including periodic ripple, to 65 Vdc for open wire systems. In Europe both DIN and VDE have comparable criteria setting 60 Vdc as the upper limit. When both static and dynamic overvoltages are taken into account the maximum power supply voltage should be less than 50 Vdc [3]. The proposed 42V/14V dual voltage system adheres to these guidelines and further, recognizes the importance of maintaining backward compatibility with existing component infrastructure.

Why are we considering a change? For two reasons: first, the present vehicle power supply is becoming overburdened by the demand of higher power features and functions and second, today's system is too inefficient to meet the demands of more stringent fuel economy and emissions regulations on the horizon [4]. This paper will discuss the issues that must be addressed in proposing a change to the vehicle electrical system and what challenges we face when doing so.

Automotive Near Term (CY2010)

Electrical and electronic growth in automotive systems today are fueled by the introduction of new telematics and infotronics systems. By telematics we mean communications between the vehicle and other vehicles and the external world. By infotronics we mean information systems within the vehicle such as autoPC for centralization of the functions needed for entertainment, CD changer, navigation aides and others. Improvements in technologies supporting fuel economy and emissions include:

- Electric assist steering
- Electric transmission controls
- Electric Brakes
- Electro-mechanical engine valves
- Beltless engine concepts

Some future features will be implemented because the customer demands it, some because they become regulated, and others because they offer the manufacturer an advantage in meeting legislated criteria such as fuel economy or emissions.

The single distribution voltage, point to

point wiring, and energy storage of today's electrical charging system will likely become inadequate to meet the demands of any combination of the above list of functions. It is expected that as the vehicle electrical load reaches a level between 150 - 200 amps, the present 14V power supply system will become impractical. **Figure 1** provides a reasonable projection of the average power demand over time, but the actual demand on any particular vehicle will likely be more of a stairstep, as additional high power loads are implemented.

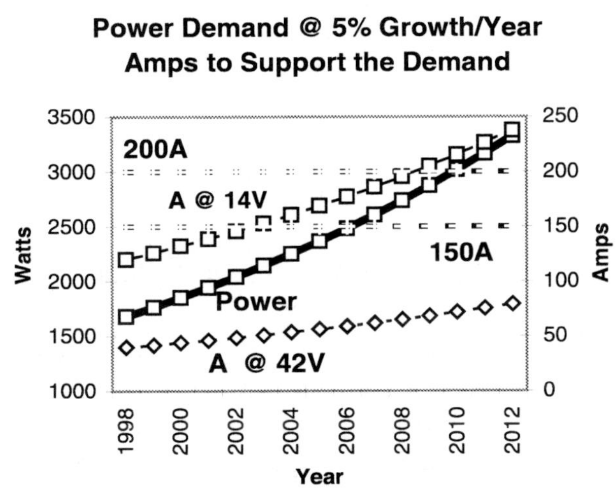

Figure 1. Projected increase in electrical load.

An illustration of the present automobile electrical system architecture is shown in **Figure 2**.

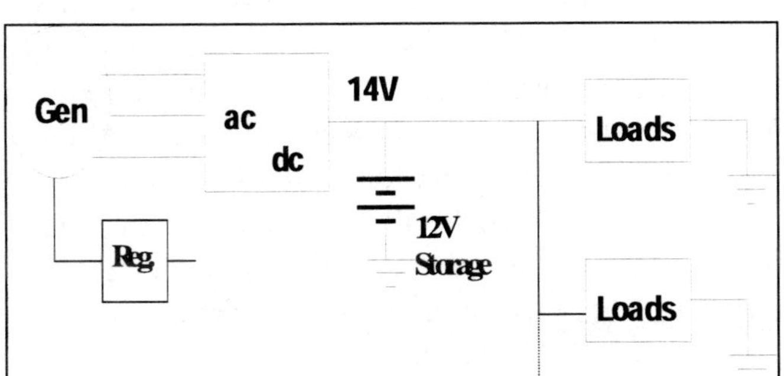

Figure 2. Present electrical system architecture.

The charging system comprises an alternator with rectifier and regulator for maintaining the charge on a lead-acid storage battery and a distribution system which includes fusing and grounding for wire and component protection.

The proposed dual voltage electrical system architecture is shown in **Figure 3**. In this proposed system storage is necessary at each voltage level using lead-acid batteries to maintain bus regulation during load switching.

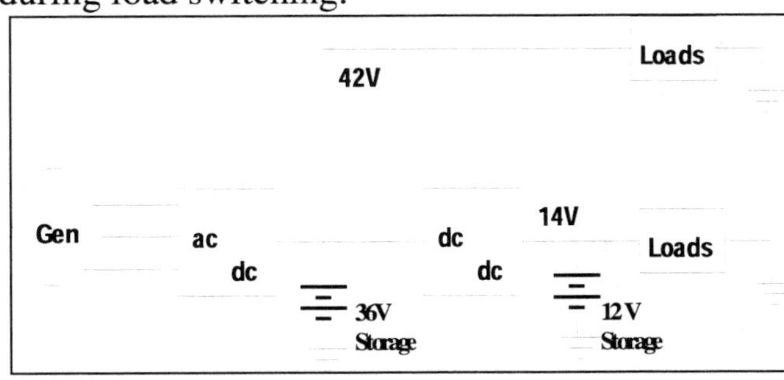

Figure 3. Proposed 42V/14V system architecture.

In the automotive near term, lead acid will remain the technology of choice for energy storage primarily because of cost and availability. This is not to say that technology advances such as seen now by the development of spiral wound and thin metal foil (TMF) lead acid systems will not be used.

Automotive Far Term (2010+)

At and beyond the near term, and as customer demand for features drives the introduction of more electrified ancillaries, the average electrical load for conventional passenger car and light trucks will approach 3 kW (215 A) and higher. The base electrical load on today's luxury vehicle is nearly 1.2 kW. When future electrical content is added to this base load the demand will indeed be 3 kW average. New features expected in and beyond the near term are highlighted in **Table 1** by season. Features such as electromechanical engine valve actuators present a power demand that tracks engine operating speed, hence, driver usage. Vehicle usage patterns are captured in drive cycles

Table 1. Electrical Loads circa 2010

Type of Load	Peak Pwr (W)	Summer Avg. (W)	Winter Avg. (W)	Average (W)
Electromechanical Engine Valve Actuator	2400	800	800	800
Engine Cooling Fan(s)	800	300	-	-
Engine Coolant Pump	300	300	30	-
Electric Assist Power Steering	1000	100	100	100
Heated Glass (front)	2500	-	250	-
Electrically Heated Catalytic Converter	3000	60	120	-
Electric Active Suspension (4 wheels)	12000	360	360	350
Infotronics & telematics	100	100	100	100
Total	-	2020	1760	-

Practical Electrical System Architectures

The dual voltage architecture described in **Figure 3** forms the core of next generation electrical system architectures. Alternatives to this DC/DC converter architecture include dual winding alternators and a single, 42V distribution bus as shown in **Figure 4 a & b**. The dual stator architecture can be realized by an electric machine with two independent sets of 3-phase windings - a 6 phase machine [6]. The high phase order electric machine can be designed to have either high mutual coupling between the independent 3-phase groups or little mutual coupling between 3-phase groups. Mutual coupling enhances regulation of both outputs from a common field winding but permits transients in one output to be coupled into the second output. The case of no mutual coupling minimizes the transient coupling issue but exacerbates the regulation.

Electrical storage potential shown in the architectures below is the open circuit voltage of the traditional electrochemical cell stack. The bus voltage specified is the nominal, under charge, regulated voltage. The difference between bus regulated voltage and battery potential accounts for battery internal resistance of the current collectors, electrolytic concentration gradient, and polarization potential

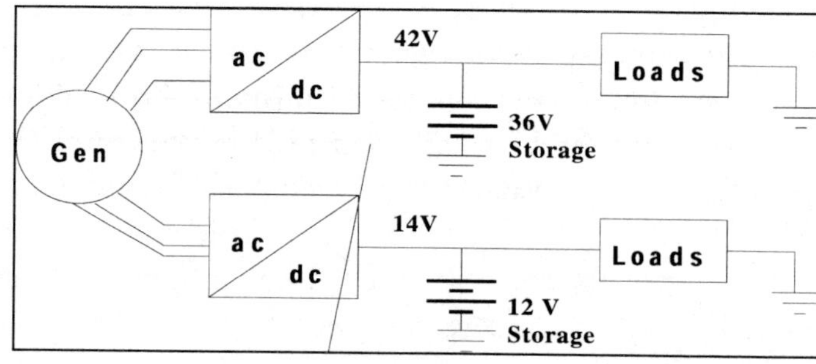

4 a) Dual winding alternator architecture

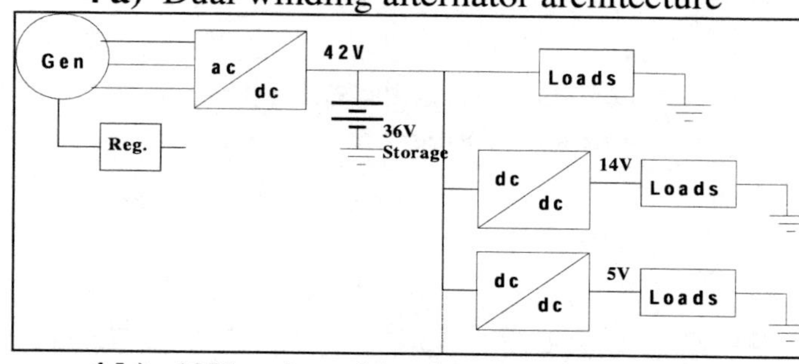

4 b) 42V point of load converter architecture

Figure 4. a) dual stator 42V/14V architecture, b) multiple voltage, distributed converter architecture.

In addition to the power system architectures shown in **Figures 3** and **4** it is possible to distribute using high frequency AC (HFAC). In this case, single phase AC is distributed via conventional 2 wire cable, coaxial cable or flat lead [7]. There are several benefits of using HFAC, most notable, ease of voltage level changing using simple toroidal transformers. For example, if the power main in **Figure 4b** was converted to a 25 kHz, 33V_{rms} distribution bus then the DC/DC converters shown would become interface transformers followed by an AC/DC rectifier stages feeding the load. An issue with AC distribution is the difficulty of maintaining bus voltage regulation in the presence of random load scheduling (e.g., customer usage patterns).

Architecture Selection

Selection of an appropriate power architecture is guided by choice of attributes and

0-7803-5086-3 /98/$10.00 ©1998 IEEE

appropriate metrics. In the automotive industry the attributes used are cost, weight, efficiency (or measure of losses) and reliability (FIT, λt or R/1000 - all interchangeable). Each architectural component is characterized in these four attributes. Metrics are used to convert attributes to cost or fuel economy measures. For example, the metric for weight is 50 kg = \$80 = 100 W (losses) = 0.4 mpg (0.17 km/l). Full vehicle architectures are compared using a CAE tool developed by Techlogix, Inc. called MAESTrO (Multi-attribute Automotive Electrical System Trade-off Optimization), [8]. A pairwise Pareto optimization is used as the final architecture comparator. Wire length and component placement appearing in the input schematic insure that electrical losses are accounted for in the final assessment. In fact, the CAE tool will size wire harness conductors based on voltage drop constraints and temperature.

Architecture Evaluation

Once a practical architecture has been identified using static simulation such as MAESTrO it is important to do further dynamic evaluation using either transient simulation (SABER) or using a hardware simulator. The dual voltage architecture described in this paper is currently undergoing hardware evaluation in our Research Laboratory, Architecture Lab shown in **Figure 5**.

Figure 5. Architecture Evaluation Lab.

The baseline architecture is a brass board evaluator consisting of the full complement of electrical and electronic components and harness from a 1998 Lincoln Continental vehicle. This board (2m x 4m perforated sheet metal painted yellow) called a "yellowboard" forms the ground plane/chassis return for the vehicle electrical system. Electrical performance is nearly identical to the in-vehicle situation. Some differences occur due to 2-dimensional rather than 3-dimensional placement and metal shielding effects.

Hardware performance is controlled and recorded using a central control station for sending drive cycle vehicle speed information to the alternator drive stand (standalone unit in **Figure 6**), electrical load switching to the yellowboard via a multiplex link to the vehicle Standard Corporate Protocol (SCP) network and data collection via a National Instruments data acquisition unit. The graphical user interface (GUI) for this interface is shown in **Figure 7.** In the GUI, the upper left hand corner does configuration control, the center top is for drive cycle selection. The lower portion of the GUI is used for data display, alternator drive stand rpm, with acquired signal display in the center and vehicle control on the lower right.

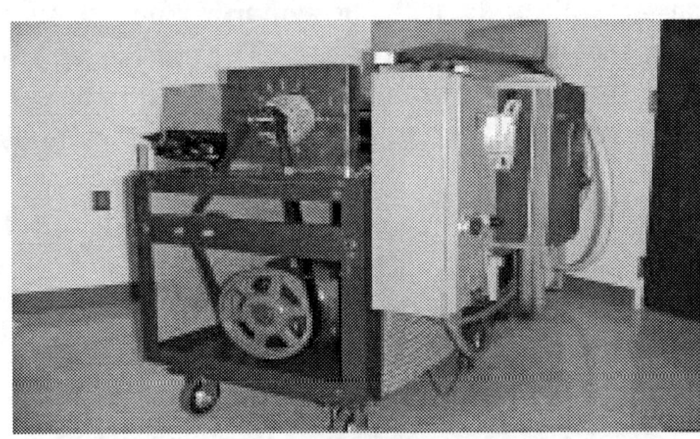

Figure 6. Alternator drive stand for hardware simulator.

During operation, full control over the vehicle hardware simulator and alternator drive stand are performed from the central control station. Dynamic display of various signals acquired during test are selected from the lower control text block in **Figure 7**.

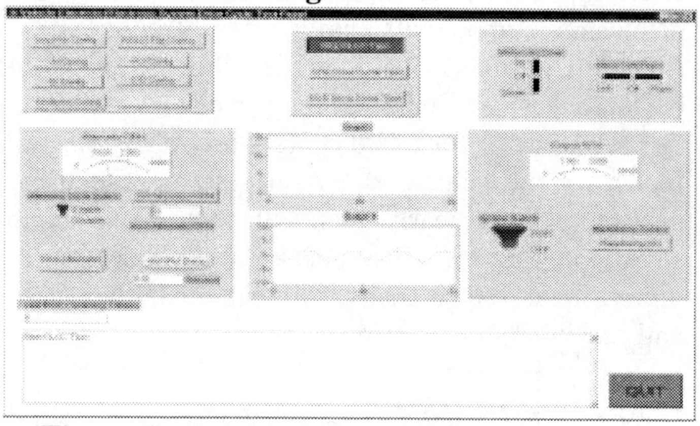

Figure 7. Graphical User Interface to vehicle hardware simulator

Architecture Issues

Electrical power system architecture optimization can be visualized by considering three aspects of the automobile as a closed system with the constraint that fuel economy is maximized. Customer usage is captured in standardized drive schedules such as the Federal Urban Drive Cycle (FUDs), Federal Test Procedure (FTP-75) or other specialized and sometimes proprietary drive cycles. The architecture itself may consist of technologies that are still on the technology horizon and as such represent future option content. The third aspect of architecture optimization is the inclusion of customer usage patterns to define the functions duty cycle (i.e., ratio of ON-time to drive cycle period).

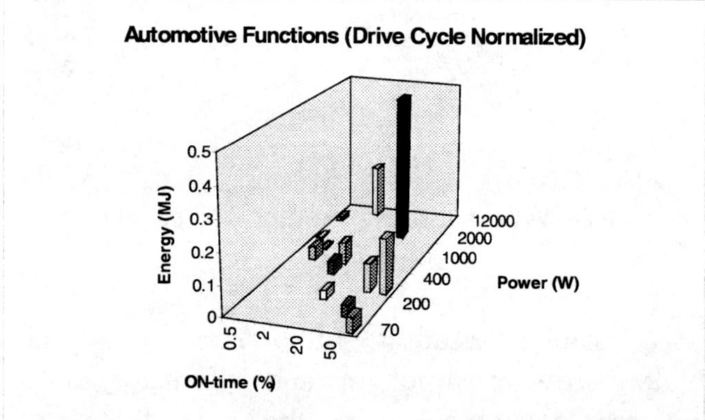

Figure 8. Technology horizon of E/E functions

How the vehicle is operated in urban or rural driving is captured in standardized drive schedules. How the customer uses the various selectable options, the customer usage pattern though generally not known, is estimated as the percent ON-time. For the purpose of vehicle architecture design, each feature or function is assigned a duty cycle.

Figure 8 is a representation of the automotive technology horizon. The x-axis represents customer determined ON-time of the function, eg, duty cycle. The y-axis is the power required by the function during the time it is selected by the customer. For example, during engine cranking the starter motor draws approximately 3.5 kW from the battery for about 3 seconds. Some features, once selected remain on at their peak power until turned off. Other features, once selected have predetermined ON-times. An illustration of this would be windshield and backlite heaters used to demist or deice the glass that remain ON for approximately 10 minutes. The total ON-time of the function divided by the drive cycle period (approx. 740 s for the FTP75 cycle) results in the net duty cycle for that function.

The full complement of electrical and electronic functions, when plotted as the ON-time power and drive cycle normalized energy space shown in **Figure** 8, represents the total energy required from the vehicle power supply. As content grows in the future the energy needed grows in direct proportion. The summation of energy consumed by all the installed functions, averaged over the drive cycle period, represents the average power demand for the vehicle. This accumulation must equal the rated capacity of the vehicle charging system in order to realize a charge sustaining strategy. By charge sustaining, we mean that the battery state of charge at the completion of the drive cycle (mission) will equal the initial state of charge. In other words, the customer will not experience a dead battery if this strategy is followed.

Today the peak capability is 1.7 kW and rising at 4%/yr. Considering the future loads shown in **Table 1**, it is conceiveable that the automotive ancillary power system demand will plateau out in the range of 3 to 6 kW (the high estimate if electric drive air conditioning is used).

The issues presented to the next generation electrical system can be summarized:

- The charging system must be charge sustaining over standardized drive cycles, using realistic customer usage profiles for load scheduling. Since it is not possible to characterize all possible drive scenarios, the charging system can be further classified as deficit charge at idle or net charge at idle.
- Voltage regulation. A goal of the Consortium is to standardize on the nominal bus voltage of 42V and the regulation limits. Elimination of high energy transients (load dump) from the distribution bus is essential to guarantee power quality.
- Protection against reverse battery installation and maybe multiple voltage jump starts have been and continue to be critical specifications to an automobile electrical system specification.
- Fusing and circuit protection is another issue presented by a higher voltage. Today's 32V rated automotive fuses may not be adequate for 42V applications.

Architecture Challenges

Having stated the issues facing architectural change, and these are significant, we propose the following as the salient challenge. Moving to a higher distribution voltage, while accepting the need for storage at each voltage level, requires that the total on-board electrical energy storage will be higher than the current architecture. The implications to this are considerable since lead-acid storage technology is essentially at fixed energy density. Weight will be added. This means not only direct cost of the component, but the cost of weight, must be considered. The vehicle infrastructure of maintaining the state of charge and providing appropriate seasonal regulation to the batteries all adds cost. For the new architecture to be affordable, the cost to implement a dual bus system must be offset by savings in the remainder of the vehicle. This is the architectural challenge now facing automotive OEM's.

Conclusions

The automotive industry is entering a transition phase in which vehicle manufacturers recognize the need for higher distribution voltage to meet the challenge of future electrical loads. A number of OEM's and suppliers are collaborating, in a pre-competitive sense, on what voltage level will meet the demands of higher electrical content in the future yet be affordable to the customer. This collaboration is under the auspices of the MIT Industry Consortium. The authors offer this summary statement regarding the goal of this collaboration: *Technical experts of automotive OEM's and suppliers, as members of the Consortium on Advanced Automotive Electrical/Electronic Components and Systems, have agreed that when it becomes necessary, the vehicle nominal voltage should be increased from 14V to 42V. This group agrees that a voltage range needs to be established and will work to develop a specification leading to an industry standard.*

Acknowledgement

The authors wish to acknowledge Henry Huang, Shahram Zarei, Mitch McConnell and Mark Lippman and others in Ford Vehicle Electronic

Systems Department for their contributions to this program.

References

[1] M.B. Wright, D. Humphrey, F. P. McCluskey, "Uprating Electronic Components for Use Outside Their Temperature Specification Limits," IEEE Transactions on Components, Packaging, and Manufacturing Technology, Part A, Vol. 20, No. 2, pgs. 251, 256, June 1997

[2] J.M.Miller, D. Goel, D. Kaminski, H.P. Schöner, T. Jahns, "Making the Case for a Next Generation Automotive Electrical System," International Congress on Transportation Electronics, Convergence 1998, Hyatt-Regency Hotel, Dearborn, MI. Oct 19-21, 1998

[3] SAE J2232, "Vehicle System Voltage - Initial recommendations," SAE Information Report, SAE Handbook, 400 Commonwealth Drive, Warrendale, PA. 15096-0001, June, 1992.

[4] DOE/S-0124, "Comprehensive National Energy Strategy," National Energy Policy Plan Pursuant to Section 801 of the DOE Organization Act, Washington, DC, April, 1998.

[5] J. G. Kassakian, H.-C. Wolf, J. M. Miller, C.J. Hurton, "Automotive electrical systems circa 2005," IEEE Spectrum, pgs. 22-27, Aug. 1996.

[6] J.M.Miller, P.J. McCleer, L.J. Giacoletto, "A Proposal for Electric Power Plant for Future ICE Powered Vehicles," IEEE Workshop on Power Electronics in Transportation - Symposium on Vehicle Electrical Systems Architecture, Hyatt-Regency Hotel, Dearborn, MI., Oct 24-25, 1996.

[7] M. Kokes, H.-P. Schöner, A. Steimel, "Modeling the Series Resonant Converter with State Vectors for an AC-Power Distribution System in Automotive Applications," Electrical Engineering 79, pgs. 465-472, 1996.

[8] K.K.Afridi, "MAESTrO - A Software Tool for the Design and Evaluation of Advanced Automotive Electrical Power Systems," IEEE Workshop on Power Electronics in Transportation-Symposium on Vehicle Electrical Systems Architecture, Hyatt-Regency Hotel, Dearborn, MI., Oct. 24-25, 1996.

AUTOMOTIVE AND AEROSPACE CIRCUIT FAULT ANALYSIS

Craig Siegel, Analogy, Inc., Beaverton, Oregon
Subba Somanchi, Analogy, Inc., Beaverton, Oregon

Abstract

This paper introduces a methodology for applying simulation techniques to the development of test programs for analog, mixed-signal and mixed-technology circuits and systems. Using the Saber simulator and its Testify test manager, the normal and fault behavior of a complete electro-hydraulic system is analyzed to obtain information used in generating a test program for the system. This information can also be used in reliability and safety studies.

Introduction

Electrical circuits and systems designed for automotive and aerospace applications often interface to electromechanical and hydraulic systems. Therefore, complete functional verification and fault testing of these circuits and systems requires observation of not only the electrical components, but also analysis of how the electronics affects the mixed-technology system. This presents a problem for the designer as well as the test engineer since most simulation tools operate efficiently in either the electrical domain or the mechanical/hydraulic domain – but not both.

Simulation can provide valuable information to design and test engineers regarding the operation of these circuits, information that may be difficult or very time consuming to obtain on the test bench. This paper focuses on the use of simulation in the fault analysis of mixed-technology systems. The Saber® mixed-signal simulator is used for these analyses because of its unique ability to efficiently and accurately simulate both electrical and mechanical/hydraulic systems [1]. Saber's Testify™ test manager is used for fault insertion and analysis and InSpecs™ is used for stress and statistical analysis.

Since Saber can simulate both individual circuit components and complete systems, this paper will use the terms "system" and "Unit Under Test" to refer to both circuits and systems under investigation.

The Need for Complete Systems Analysis

Simulation of mixed-signal (containing analog and digital components) and mixed-technology (containing electrical and non-electrical components) systems has traditionally been accomplished through a "divide and conquer" methodology. That is, those portions of the UUT (Unit Under Test) containing only one technology are simulated in isolation without regard for their impact on the other portions of the system, or the impact of other portions on the system on the UUT. This can lead to erroneous results due mainly to 1) not considering the loading effects of the non-electrical components, and 2) not having accurate signals in systems depending on feedback from devices of other technologies.

Therefore to achieve accurate simulation results of mixed-technology systems, it is necessary to simulate the entire system as a single unit, incorporating all the different technologies of the system in the simulation model. This becomes especially important in the analysis of faults where the test engineer needs to analyze the operation of the UUT in the presence of faulty components.

0-7803-5086-3 /98/$10.00 ©1998 IEEE

While there are many tools available and techniques established for the analysis of digital UUTs, there is a lack of adequate tools for the analysis of analog, mixed-signal and mixed-technology UUTs. It is not surprising, therefore, that the quality of TPSs and diagnostic routines for these UUTs often are of significantly lower quality than their digital counterparts [2]. This paper illustrates a methodology using unique simulation tools that can significantly increase the quality and effectiveness of these TPSs and diagnostic routines.

An Electro-Hydraulic Example

A simplified electro-hydraulic braking system is used to illustrate a methodology for performing fault analysis on mixed-technology UUTs. The techniques described here can be applied to any mix of technologies. The electro-hydraulic braking system is shown in Figure 1.

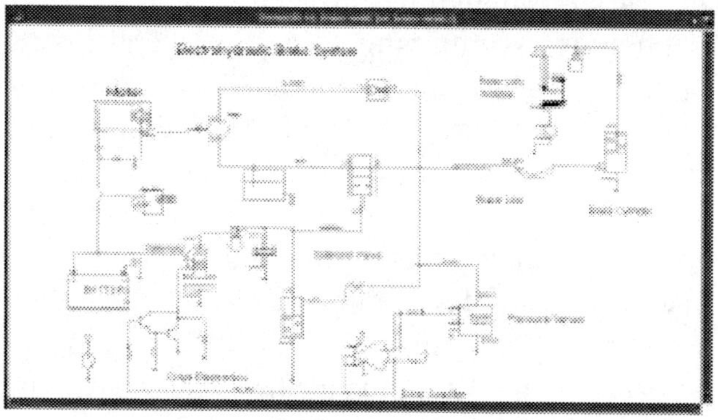

Figure 1. Electro-hydraulic Brake System Schematic

The brake system is composed of electronic, electromechanical and hydraulic components A battery powered DC motor and hydraulic pump provide pressure. A check valve, in series with a solenoid-controlled two-way valve establishes the controlled braking pressure at the midpoint. The actual operating pressure is adjustable with solenoid current as the valve spool and solenoid are rigidly connected.

An outer pressure control loop uses an electronic sensor to measure brake pressure. This is compared with a commanded reference voltage which would come from a brake pedal position sensor. The difference signal drives the base of a dual-transistor amplifier which controls solenoid current..

The brake assembly is modeled as a single acting cylinder with spring return, plus attached load mass and a mechanical hard stop or travel limit. Pressure is applied to the brake through a long rigid line and short flexible hose.

In this example we simulate 1.0 second of activity. The activation switch is turned on at 10 msec and the pump pressure responds quickly. The pressure at the wheel rises after a delay of 75 msec but remains low initially. Once in contact with the rotor, the pressure builds to near its maximum of 10 MPa after 200 msec.

At 400 msec the reference voltage signal (3.5V) is set to 1V. The brake pressure falls, settling at approximately 3MPa after considerable ringing. After reduced pressure operation for 300 msec, the reference signal is raised to 3V and the pressure rises to 7.6MPa, again after considerable ringing. Results of a nominal (unfaulted) Saber simulation are shown in Figure 2.

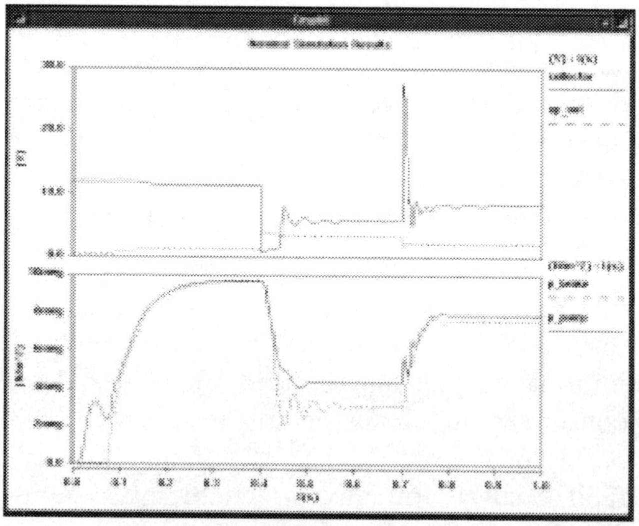

Figure 2. Nominal Simulation Waveforms

Mixed-Technology Test Methodology

Using the Saber simulator and its associated test and analysis tools, the test engineer can gain valuable insight into the operation of the UUT both under nominal (nonfault) and faulted conditions before proceeding to bench testing or running a Test Program Set (TPS) on the ATE (Automatic Test Equipment). The Testify test manager automates the fault simulation and analysis process. It's primary use is the in the development of Test Program Sets and diagnostic tests to detect failing components in a UUT [2]. Testify automates the following operations:

- Development of a set of tests to detect faults in the UUT
- Analysis of the UUT under normal operating conditions
- Insertion and simulation of hard faults in UUT electronics
- Insertion and simulation of parametric faults in any UUT components
- Analysis of the tests' ability to detect the presence of the inserted faults by measuring results of fault simulations against nominal results.

Using the information from Testify, the test engineer can develop a TPS or diagnostic routine for the UUT that can then be verified on the bench or ATE.

The mixed-technology test methodology consists of the following steps.

- Create a schematic of the UUT.
- Specify a set of tests or diagnostics.
- Determine the nominal operation and operational limits.
- Insert and simulate electrical hard faults.
- Insert and simulate parametric faults.
- Perform stress analysis.

CREATE A SCHEMATIC OF THE UUT

The schematic is used as the input to Saber. Existing schematics of the design can be used since Testify creates fault-capable models from existing schematic symbols. So normally there is no need to create additional fault-capable models to perform the analysis. The schematic can be created in Saber's SaberSketch or another schematic environment compatible with Saber. The schematic shown in Figure 1 was created in SaberSketch.

SPECIFY A SET OF TESTS OR DIAGNOSTICS

A set of tests is created to exercise the UUT and detect the presence of a faulted component. The simulation commands specified for the simulator are very similar in function to the tests that will be performed on the bench or on the ATE. ATE stimulus commands correspond to source inputs and ATE measurements correspond to measurement commands automatically applied to simulation results. Testify provides graphical menus and forms to guide the test engineer through the process of creating the simulation commands. Figure 3 shows the setup of a transient simulation to measure the voltage at the output of the error amplifier. In this example we will run a set of 13 tests, including measurements of node voltages drive circuitry voltages, pump and brake shoe pressure, solenoid armature position, and shaft rotation overshoot.

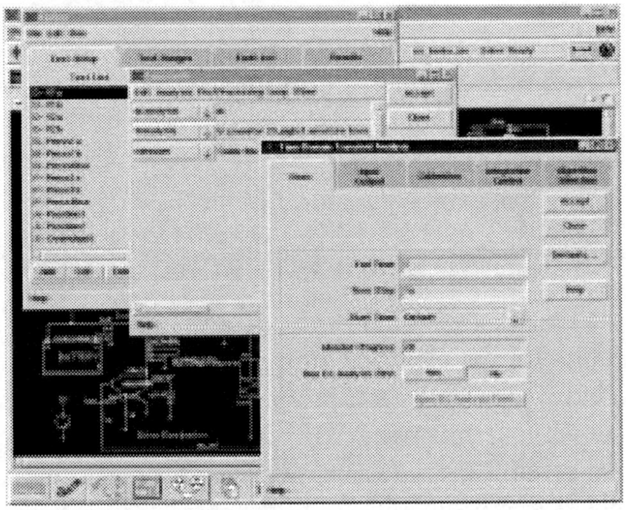

Figure 3. Testify Test Setup Menus

DETERMINE NOMINAL OPERATION AND OPERATIONAL LIMITS

The first step in the analysis of the UUT is to determine its normal or nominal operation. Valuable insight regarding UUT operation can be gained in this step, especially in cases where operational specifications or data for the UUT is lacking.

Next the operational limits, or the acceptable range of operation must be established. These are the upper and lower limits of measured operational parameters against which faulted behavior will be compared to determine if a test was able to detect the presence of a fault. These limits can be established in Testify by running a Monte Carlo analysis, or by simply inputting operational or test specifications. The results of the operational limits simulations are shown in Figure 4.

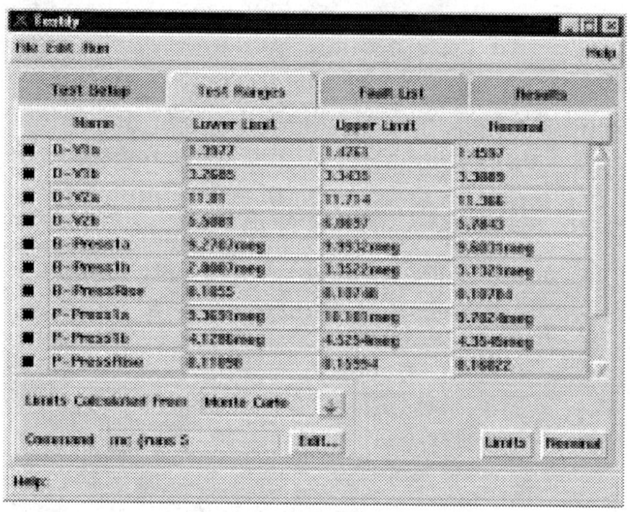

Figure 4. Operational Limits Results

INSERT AND SIMULATE HARD FAULTS

Testify creates a default set of hard faults for the electrical components of the UUT. These hard faults are pin-pin shorts, pin-supply shorts and pin opens. If additional hard faults are required, the test engineer can specify them using Testify menus. In this step of the process, the test engineer can select which faults are to be activated by Testify. He can select all possible faults, a subset of faults, or individual faults.

Testify will then control simulations in the following manner. It will activate the first fault, run the required simulations, take the specified measurements, and compare the results to the operational limits. If the measurement falls outside the limits, then aberrant behavior has been observed due to the presence of the fault and the fault is determined to be "detected" by that test. All specified tests are run for the fault and "detect" nondetect" status recorded.

Testify then resets the fault and activates the next one, continuing until all faults have been inserted and simulated and measurements taken. Note that Testify makes use of the behavioral capabilities of Saber's MAST® modeling language to insert and activate the faults [3,4]. This means that there is minimal impact on simulation times. Normal simulation times were impacted by only 5% due to the presence of the fault-capable models in the UUT netlist.

A portion of the results of the Testify hard-fault analysis is shown in Figure 5. A summary report is shown in Table 1 at the end of this paper. This output can now be examined to determine which tests must be run to detect and diagnose electrical faults in the system. One test – brake pressure after the stimulus level (reference voltage) has been lowered – detected the presence of all electrical faults. A sequence of tests can now be developed to isolate individual faults down to acceptable ambiguity groups.

INSERT AND SIMULATE PARAMETRIC FAULTS

To test the mechanical and hydraulic elements in the UUT, we use Testify's parametric fault insertion. This allows us to "push" component operating parameters outside of their acceptable range. In this example we insert parametric faults simulating stuck valves and solenoids, partially plugged lines and orifices and insufficient pump discharge. A Portion of these results is shown in Figure 6.

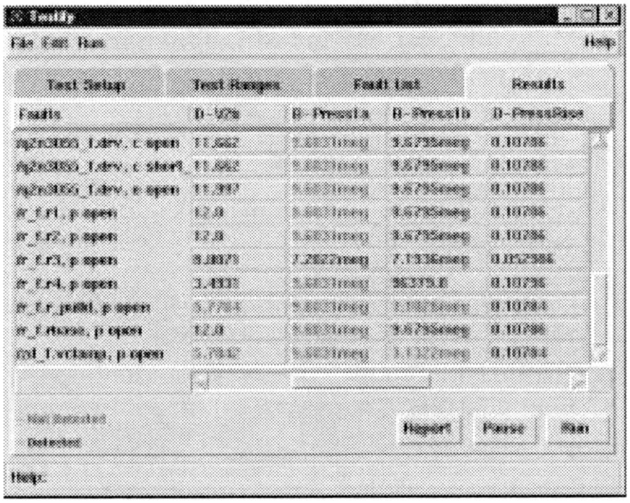

Figure 5. Hard Fault Analysis Results

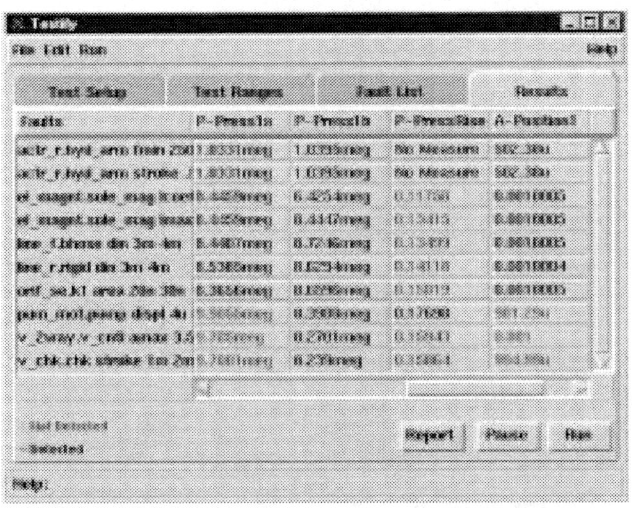

Figure 6. Parametric Mechanical/Hydraulic Fault Results

For illustration purposes we inserted parametric faults only in non-electrical components. However we could have inserted parametric faults into electrical components as well. Also, the parametric tests could have been run in conjunction with the hard faults in one simulation step.

PERFORM STRESS ANALYSIS

The final step in the Mixed-Technology Test Methodology is to run stress tests on the faulted simulation results. Currently this is a separate step in Testify, but in the future this will be more tightly integrated into the Testify process. The stress test can be performed on electrical and non-electrical faults.

This test will tell the test engineer if any components were overstressed or driven out of their safe operating areas due to the presence of the inserted fault. This information can be extremely valuable in determining if tests are destructive. It can also be used for safety and reliability analysis. Figure 7 and Figure 8 show the output of two stress tests for the braking system. In the first stress report, the Q2n2222a (Stage1) transistor in the drive circuit is overstressed due to the presence of a partially blocked orifice in the solenoid valve. In the second stress report, the same transistor is severely overstressed if the collector of Q2n3055 (Stage2) transistor is left open.

Stress Analysis: Clogged Solenoid Valve Orifice						
Instance	Name	Derated Value	Actual Value	At	Stress Ratio (%)	Bar Chart 0 - 100%
q2n2222a stage1	tj max	175	797	700m	515	xxxxxxxxxx
q2n2222a stage1	pd Max	500m	1.54	700m	308	xxxxxxxxxx
q2n2222a stage1	tj avg	175	267	---	161	xxxxxxxxxx
actr_e bcyl	ftens max	-235m	-291m	61.9m	124	xxxxxxxxxx
el_magnt sole_mag	ftens max	-223	-235	445m	105	xxxxxxxxxx
stop_t b_pad	ftens max	-19.1k	-19.8k	400m	103	xxxxxxxxxx
q2n2222a stage1	Pd avg	500m	481m	---	96.1	----------
actr_r hyd_arm	ftens max	-2.19m	-1.48m	89.4m	67.8	-------
q2n2222a stage1	vce max	40	26.8	701m	67	-------
q2n3055 drv	pd max	100	64.7	708m	64.7	------
q2n3055 drv	vce max	50	27.7	701m	55.3	------
zd vclamp	pd max	240	102	701m	42.4	----

Figure 7: Clogged Solenoid Valve Stress Report

Stress Analysis: Q2n3055 Collector Open						
Instance	Name	Derated Value	Actual Value	At	Stress Ratio (%)	Bar Chart 0 - 100%
q2n2222a stage1	tj max	175	2.8k	700m	1850	xxxxxxxxxx
q2n2222a stage1	pd max	500m	5.54	700m	1110	xxxxxxxxxx
q2n2222a stage1	tj avg	175	876	---	567	xxxxxxxxxx
q2n2222a stage1	pd avg	500m	1.7	---	339	xxxxxxxxxx
stop_t b_pad	ftens max	-19.1k	-19.2k	412m	101	xxxxxxxxxx
actr_r hyd_arm	ftens max	-2.19m	-2.2m	79.6m	100	xxxxxxxxxx
actr_e bcyl	ftens max	-235m	-235m	65.6m	100	xxxxxxxxxx
q2n2222a stage1	vce max	40	23.2	701m	58.1	------
pum_mot pump	torq max	20	7.78	412m	38.9	----
q2n2222a stage1	vcb max	75	22.4	701m	29.9	---
q2n2222a stage1	ic avg	800m	151m	---	18.9	--
q2n3055 drv	vce max	50	6.02	88m	12	-
el_magnt sole_mag	ftens max	-223	-643m	412m	0.288	.

Figure 8. Drive Transistor Open Stress Report

Conclusion

Using Testify and the Saber Simulator, the test engineer now has a tool to help him develop Test Program Sets and diagnostic routines for UUTs that contain a mix of technologies. This allows him to analyze a complete system, taking into account all the interactions between the electrical and non-electrical portions of the UUT.

This paper has illustrated a methodology using a unique set of simulation tools that can significantly increase the test engineer's knowledge of how these UUTs will behave both under normal operating conditions and in the presence of faults. Using Testify results from hard and parametric fault analysis, the test engineer has valuable information to help generate better TPSs and diagnostic routines. It can also provide information for reliability and safety studies.

References

[1] M. Donnelly, C. Siegel and D. Witt, "Design Analysis of an Electronically-Controlled Hydraulic Braking System Using the Saber Simulator, SAE International, 1994.

[2] Majernik et al, "Using Simulation to Improve Fault Coverage of Analog and Mixed-Signal Test Program Sets", AUTOTESTCON, 1997

[3] "MAST Reference Manual", Analogy, Inc., Beaverton, OR, 1991

[4] H.A. Mantooth and M. Figenbaum, "Modeling With an Analog Hardware Description Language", Kluwer Academic Publishing, 1995

Table 1. Fault Analysis Report

Testify Results - Short Version (Measurement Data Not Listed)	Test Volt Drive1a	Test Volt Drive1b	Test Volt Drive2a	Test Volt Drive2b	Test Pressure Brake1a	Test Pressure Brake1b	Test PressRise Brake	Test Pressure Pump1a	Test Pressure Pump1b	Test PressRise Pump	Test Position Arm1	Test Position Arm2	Test Ovrshoot Shaft
Electrical Faults													
/d_f.freew, p open	N	N	N	N	N	N	D	N	N	D	N	N	N
/d_f.freew, p short_n	D	D	D	D	D	D	D	D	D	D	D	D	D
/op1_f.err_amp, inm open	D	D	N	D	N	D	D	N	D	D	N	D	D
/op1_f.err_amp, inm short_vcc	D	D	D	D	N	D	D	N	D	D	N	D	D
/op1_f.err_amp, inm short_vee	X	D	D	D	D	D	D	D	D	D	D	D	D
/op1_f.err_amp, inp open	D	D	N	D	N	D	D	N	D	N	N	D	D
/op1_f.err_amp, inp short_vcc	X	D	N	D	N	D	D	N	D	N	N	D	D
/op1_f.err_amp, inp short_vee	D	D	D	D	N	D	D	N	D	D	N	D	D
/op1_f.err_amp, out open	D	D	D	D	N	D	D	N	D	D	N	D	D
/op1_f.err_amp, out short_vcc	X	D	D	D	N	D	D	N	D	D	N	D	D
/op1_f.err_amp, out short_vee	D	D	D	D	N	D	D	N	D	D	N	D	D
/q2n2222a_f.stage1, b open	N	D	D	D	N	D	D	N	D	D	N	D	D
/q2n2222a_f.stage1, b short_e	N	D	D	D	N	D	D	N	D	D	N	D	D
/q2n2222a_f.stage1, c open	N	D	D	D	N	D	D	N	D	D	N	D	D
/q2n2222a_f.stage1, c short_e	X	D	D	D	N	D	D	N	D	D	N	D	D
/q2n2222a_f.stage1, e open	N	D	D	D	N	D	D	N	D	D	N	D	D
/q2n3055_f.drv, b open	N	D	D	D	N	D	D	N	D	D	N	D	D
/q2n3055_f.drv, b short_e	N	D	D	D	N	D	D	N	D	D	N	D	D
/q2n3055_f.drv, c open	N	D	D	D	N	D	D	N	D	D	N	D	D
/q2n3055_f.drv, c short_e	X	D	D	D	N	D	D	N	D	D	N	D	D
/q2n3055_f.drv, e open	N	D	D	D	N	D	D	N	D	D	N	D	D
/r_fr1, p open	D	D	D	D	N	D	D	N	D	D	N	D	D
/r_fr2, p open	X	D	D	D	N	D	D	N	D	D	N	D	D
/r_fr3, p open	D	D	D	D	D	D	D	D	D	D	D	D	D
/r_fr4, p open	D	D	D	D	N	D	D	N	D	D	N	D	D
/r_fr_pulld, p open	N	N	N	N	N	N	D	N	N	D	N	N	N
/r_frbase, p open	N	D	D	D	N	D	D	N	D	D	N	D	D
/zd_f.vclamp, p open	N	N	N	N	N	N	D	N	N	D	N	N	N
/zd_f.vclamp, p short_n	X	N	N	N	N	N	D	N	N	D	N	N	N
Hydraulic/Mechanical Faults													
actr_r.hyd_arm fmin 250 300	D	D	D	D	D	D	D	D	D	D	D	D	D
actr_r.hyd_arm stroke .5m 1m		D	D	D	D	D	D	D	D	D	D	D	D
el_magnt.sole_mag lcoef 800 700	D	D	D	D	D	D	D	D	D	N	D	D	D
el_magnt.sole_mag lmax 120 150	D	D	D	D	D	D	D	D	D	N	D	D	D
line_f.bhose din 3m 4m	D	D	D	D	D	D	D	D	D	N	D	D	D
line_r.rigid din 3m 4m	D	D	D	D	D	D	D	D	D	N	D	D	D
orif_se.k1 area 20n 30n	D	D	D	D	D	D	D	D	D	N	D	D	D
pum_mot.pump displ 4u 5u	D	D	D	D	N	D	D	N	D	D	N	D	D
v_2way.v_cntl amax 3.5u 5u	N	D	D	D	N	D	N	N	D	N	N	D	D
v_chk.chk stroke 1m 2m	N	D	D	D	N	D	N	N	D	N	N	D	D

CVLAN: A REQUIREMENTS DEFINITION FOR A REAL-TIME LOCAL AREA NETWORK IN LAND COMBAT VEHICLES

Paul C. Richardson, U.S. Army TARDEC, Warren, MI

Introduction

Data communications in land combat vehicles involves the transmission of messages that contain device control signals and real-time information. The delivery of these messages must not only be reliable, but they must occur within specified time-constraints. The CVLAN (combat vehicle local area network) architecture is a definition of services and functions for network communications in land combat vehicles. The focus of the CVLAN is twofold: (1) Develop a robust LAN architecture that can guarantee all message time constraints under nominal conditions and that can adapt to variations in network loading induced by system and network faults. (2) Leverage apriori knowledge of system characteristics to reduce uncertainty during critical time segments. The architecture is defined as a hierarchical protocol layering that specifies basic functions and services provided. Each layer provides a set of services to the layer above it. Within the service provider, a set of functions implements the services transparently to the service requester.

The LAN in a combat vehicle is generally connected to a WAN via one or more FM radios that act as bridges or routers. We define the boundaries of the CVLAN as the computing resources internal to the vehicle. The delivery and time constraints of messages will not be considered beyond the input/output buffers of the radios. This delivery of real-time messages over a WAN is beyond the scope of this paper. The boundary between the internal vehicle CVLAN and the larger, external WAN are shown in figure 1.

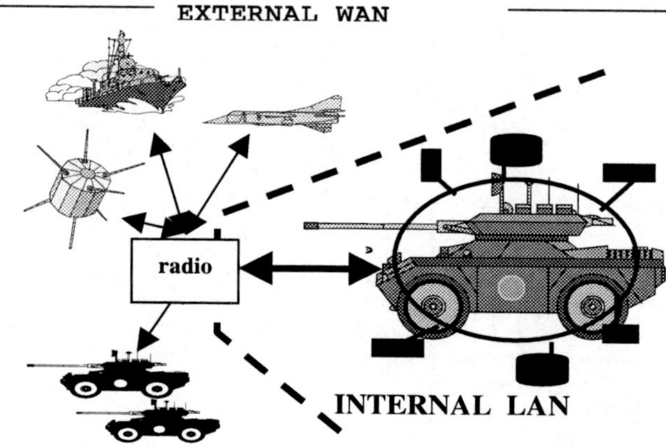

Figure 1: Partitioning of vehicle LAN and External WAN

In the remainder of this paper we present the background and motivation for this effort, then define the requirements for the LLC and MAC sub-layers.

Background & Motivation

The first generation of LAN based Combat Vehicles has reached production. The motivation for introducing a LAN in these vehicles was to reduce the number of dedicated data links through the use of shared media communications. The growing number of dedicated data links had significantly increased maintenance cost estimates. The extensive shielding required for transmission media in these vehicles resulted in considerable space and weight claim. These cost drivers made a LAN based architecture an attractive alternative to the use of dedicated data links.

The LAN architecture in these vehicles utilizes static message scheduling, static addressing, and centralized media access

0-7803-5086-3 /98/$10.00 ©1998 IEEE

control [15]. All messages are synchronous (transmitted periodically). There is no effort to guarantee delivery of any one message. The arrival of the next iteration of a message is assumed to compensate for any lost or corrupt messages. Lost or corrupt messages are assumed to occur infrequently, and are ignored or discarded as they occur. The strength of this approach is that it is deterministic and relatively easily verified. Because combat vehicles perform life-critical operations in hostile, non-deterministic environments, deterministic LAN behavior and verification of correct LAN operation are of paramount importance. The drawback of this static approach is that it limits the systems ability to respond to various system or network faults, since it precludes variations in network loading.

A primary goal of the CVLAN is to provide the capability to adapt to variations in network loading. A number of events, discussed below, can occur during combat operations that can cause variations in network loading. These are fault induced, transient and permanent variations that can possibly overload network resources.

- A media or component fault can cause a twofold variation in network loading. Initially, there is a temporary surge in message traffic as the system responds to the fault. After reconfiguration, there may be a permanent change in network loading and available bandwidth.
- Asynchronous real-time messages, or real-time messages that arrive at irregular intervals, require the designer to make some estimate about message inter-arrival rates [4]. This estimate must weigh efficiency against a worst case arrival scenario. A pessimistic estimate may cover every occurrence of an asynchronous message, but may be very inefficient. An optimistic estimate will be more efficient, but can fail in a high stress situation. In a combat operation, unforeseen events may cause asynchronous messaage arrivals to exceed thir estimated arrival rate, potentially overloading network resources.
- Corrupt or lost data necessitates retransmission of data. This results in cause unpredictable, transient surges in network traffic.

Any of the above occurrences could cause network capacity to be exceeded. In this situation, some messages will unavoidably be late. In a system where the untimely arrival of a message can result in serious injury or death, it is important to have the ability to ensure that critical real-time messages are delivered within their time constraints.

The second focus of the CVLAN is to utilize apriori knowledge of system characteristics to simplify the architecture and remove uncertainty from critical time segments. The transmission of critical, real-time messages in a hostile and non-deterministic environment necessitates a fault tolerant and deterministic LAN architecture. The consideration of fault tolerance and determinism adds significant complexity to the LAN architecture definition. However, known system characteristics can be used to mitigate these complexities, and to remove uncertainty from time critical segments. The CVLAN has a fixed message set, which allows us to allocate memory and initialize control information during system start-up. An isolated network with a fixed number of hardware components and no intermediate switches allows us to define a simplified communications protocol (ie: reduced layers). In terms of network utilization and criticality of messages, a combat operation is the worst case operational state for the system. For the purpose of guaranteeing time constraints of real-time messages, this is the only state we need consider [14]. These observations and simplifications allow us to reduce uncertainty and processing overhead during time critical processing.

There are several combat vehicles utilizing LAN architectures in the development phase. The recent success of commercial LAN architectures with respect to scalability, reliability, and maintenance has led to efforts to increase the efficiency and robustness of the shared media communications in these systems. The CVLAN architecture utilizes decentralized media access control, dynamic message scheduling, and logical addressing to provide a more robust architecture in response to the concerns of combat vehicle developers.

Protocol Definition

The architecture is based on the IEEE802 reference model [8], shown in figure 2. This architecture consists of a logical link sub-layer (LLC), media access control sub-layer (MAC), and a physical layer (PL). In this paper, we assume that a physical layer is present and provides a reliable bit stream. We define requirements for the LLC and MAC sub-layers.

Looking from the top down (figure 1), the user interface for message transmissions is defined as service access points (SAPs) in the LLC layer (in this context, a user is any task that takes part in LAN message transmissions). The LLC provides user tasks with services and functions to ensure the reliable and timely delivery of messages. The primary role of the MAC layer in a real-time LAN is to furnish the LLC layer with a deterministic and orderly means of transmitting data frames on shared media. The PL consists of the physical elements that provide a reliable bit stream to the MAC layer.

LLC Services and Functions

The LLC sub-layer provides the user with a set of services for the transmission of real-time messages. The services allow user tasks to initialize communications channels, to send and receive messages, and provides user tasks with the status of transmission requests. The service request format includes explicit

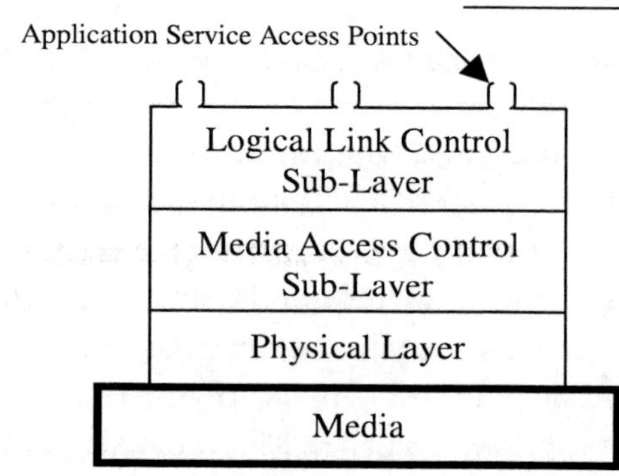

Figure 2 : IEEE 802 Reference Model

parameters for message time constraints and importance. The LLC layer must provide basic functions to ensure reliable and timely delivery of messages. The functions defined for this layer are framing, message addressing, message scheduling, flow control, and error control. Figure 3 depicts a functional block diagram for the CVLAN LLC. Arrows indicate the flow of data and control for outgoing and incoming messages.

Framing

Messages are divided into frames to prevent arbitrarily long data transfers from starving other messages. A frame consists of data and control information, transmitted as a unit on the network media. Control information includes addressing, error detection, data length, and priority. Data is the user defined frame payload.

For the CVLAN, we use frame boundaries to preempt message transmissions. This is done allow a high priority message to proceed ahead of lower priority message that had arrived earlier. Framing also facilitates flow control and error control, since a frame retransmission can be significantly shorter than a message retransmission. Message control parameters are propagated down to individual frames.

Addressing

Message addressing involves routing messages from their source to their correct destination. For each message, user tasks must provide control information that can be used by the LLC to route messages and provide notification of arrival to the destination tasks.

In addressing messages, we have selected logical addressing over physical addressing. Physical addressing maps addresses to specific hardware components. Logical addressing maps addresses to logical identifiers. The use of logical addressing permits the broadcast, multicast, or singlecast of any message independent of its physical location [2], facilitating system fault management. If a component fails, its tasks can be relocated to other resources, and its messages can still be transmitted to the same logical address. Physical addressing, by definition is hardware dependent.

Logical addressing of messages requires the MAC protocol to support logical frame addressing. Figure 4 depicts hardware independent addressing through the use of logical addressing. The logical address of the MAC frame can be mapped to message identification tables in the LLC layer. The tables then reference control information and data for individual messages. Communications channels can be modified for tasks that have been relocated to different hardware platforms by modifying the message identification table and references to control information and data.

The use of logical addressing with a fixed message set allows us to remove overhead processing and reduce uncertainty in the critical time segment of a message. The critical time segment of a message, t_{cs}, is defined as the time between the arrival for a message send request at its source, t_a and the time, t_r, that it is received at its destination.

$$t_{cs} \equiv t_r - t_a \qquad (1\text{-}1)$$

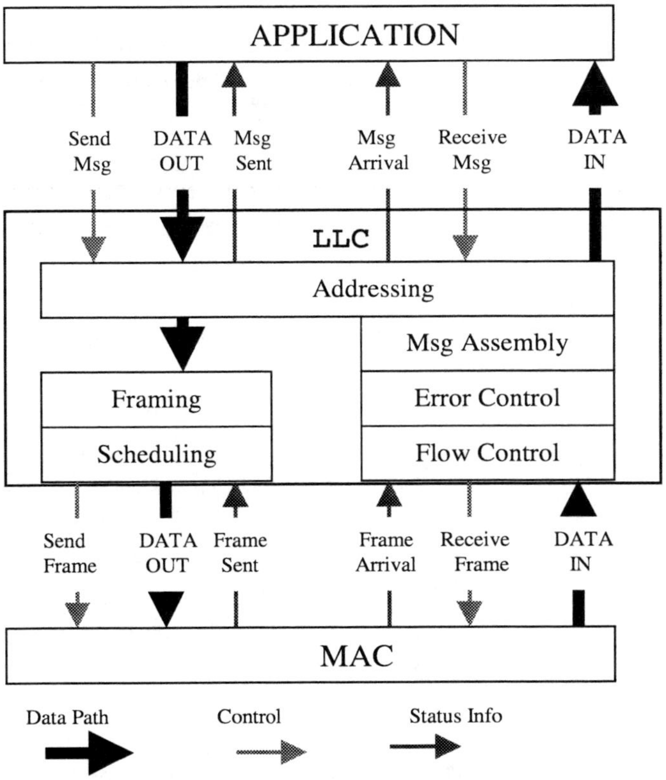

Figure 3: LLC Functional Block Diagram

To illustrate this point, consider the transfer time of a message, $t_{transfer}$, which can generally be divided into four states, each with an associated time: setup, awaiting transmission, transmission, and network delays.

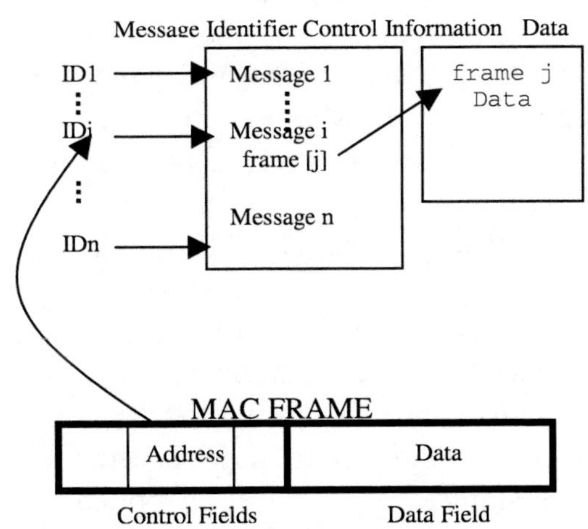

Figure 4: Logical Addressing

The setup state consists of allocation of memory for message storage, preparation of user data for transmission, and initialization of control information. Wait time refers to the delay caused by contention for shared resources. Transmission time refers to the actual generation of electrical signals, and is proportional to the message length. Delay time is the time the signal is delayed in traversing the physical network from its source to its destination. Using this convention, we can describe message transfer time as:

$$t_{transfer} = t_{setup} + t_{wait} + t_{transmit} + t_{delay} \qquad (1\text{-}2a)$$

Using conventional addressing means, we have the entire message transfer time occurring between the message arrival and its delivery.

$$t_r - t_a = t_{setup} + t_{wait} + t_{transmit} + t_{delay} \qquad (1\text{-}2b)$$

Message transmit time and network delays are relatively small when compared to set-up time and wait time, and can be determined. Network delays generally consists of signal propagation delay and node delays. Transmit time is dependent on the message length and network bit rate. Wait time for messages can vary widely depending on the message-scheduling algorithm, the media access control, and the arrival rate of messages. Set-up time involves allocating storage space for a message and initializing its control information. It is difficult to bound setup time independent of the system, since message setup depends on the allocation of system resources.

Using logical addressing and leveraging apriori information regarding the sources, destinations, and lengths of the messages, we can perform the assignment of control information and allocation of storage space for messages outside of the critical time segment. This is significant, because it reduces the time of transmission and removes uncertainty in transmission time from the time critical path. As a result of this, equation 1-2 simplifies to:

$$t_{cs} \equiv t_r - t_a = t_{wait} + t_{transmit} + t_{delay} \qquad (1\text{-}3a)$$

Since transmit and delay times are relatively small, and known, we have:

$$t_{cs} \equiv t_r - t_a = t_{wait} + c \qquad (1\text{-}3b)$$

Thus we have reduced the critical time segment for a real-time message transfer to the wait time for shared resource contention plus a constant, and the variation in the critical time segment to just the wait time.

Scheduling

Message scheduling in the CVLAN must guarantee that all message time constraints are met under nominal conditions, and must ensure that critical messages are delivered within their time constraints under fault conditions, if at all possible. Recall that during combat operations, system and network faults can cause the network demand to exceed network capacity. In this situation, the scheduler must consider both the timeliness and importance of messages in order to ensure that critical messages are delivered in a timely manner, perhaps at the cost of a non-critical message arriving late.

The CVLAN will use benefit based dynamic scheduling with dynamic message prioritization. Dynamic scheduling permits message transmissions to be altered based on changing network conditions. The dynamic prioritization of messages allows us to reflect the relative values of messages as they change with time. Benefit based scheduling allows us to consider both the timeliness and importance of message transmissions and accounts for late message arrivals in the event of a network overload[11].

Message scheduling takes place when a scheduling event occurs. These events could be *message sent, message received, frame sent*, or *frame received*. During scheduling, priorities are assigned to pending messages based on their state information. Message state information includes individual time constraints, message history, and messages importance. The priorities of messages are propagated to their frames. A lower priority message can be preempted at a frame boundary to allow the transmission of a higher priority message. The specifics of the scheduling algorithm and the metric used to measure its performance are beyond the scope of this paper and will be addressed in a separate effort.

Flow Control

Flow control is concerned with the reliable delivery of frames to the receiving node. A frame that is not delivered to its destination is "lost". Frames are most often lost when memory allocated for frame storage is overflowed because the input into the storage area exceeds the output. Frames can also be lost when control information has become corrupted, and the frame cannot be identified. Non-real time flow control algorithms, such as sliding window protocols [8], introduce timing uncertainties that make them unsuitable for real-time LANs. The CVLAN LLC layer must provide a service to retransmit frames in the event they are lost. The service must consider the time constraints of lost frames.

The CVLAN facilitates flow control by reducing the likelihood of an overflow. In an isolated LAN, with no intermediate switching or storage of frames, the only place an overflow can occur is at the receiving node, either in the MAC receive buffer or the LLC receive buffer. Recall that a fixed message set allows us to pre-allocate message storage space in the LLC during system initialization. In this way we can eliminate overflow in the LLC sub-layer.

Error Control

Error Control is a technique for detecting and managing bit errors in frames caused by transmission impairments. A bit error is the inversion of a logical bit (either 1 to 0 or 0 to 1)and can occur either singly or in-groups. Transmission impairments are caused by a number of phenomena, among them noise, attenuation, and delay distortion.

The CRC (Cyclic Redundancy Check) is an efficient and accepted means to detect errors [8]. Once detected, the frame must be either fixed, or discarded and retransmitted. The overhead and time cost of present error-correction algorithms makes them unsuitable for real time systems. The alternative, frame retransmission, is undesirable yet tractable, if the number of errors can be kept low. The use of appropriate signal encoding techniques and shielding can make the occurrence of bit errors rare. For the safety critical messages in a combat vehicle, retransmission's of lost or erroneous frames must be guaranteed by the LLC. Consider for example, the dramatic negative consequences that may arise from an erroneous retransmission of a "shoot main gun" message.

MAC Services and Functions

The primary role of the MAC sub-layer is to provide system wide arbitration for use of the shared media. In a combat vehicle LAN, key attributes for the MAC protocol are deterministic media access times and fault tolerance.

The MAC must provide support for a bound channel access time (BCAT) for frames requesting media access. The BCAT is defined as the elapsed time between the arrival of a frame transmission request from the LLC sub-layer and the granting of media access to the MAC sub-layer. For real-time systems, a BCAT is required to determine the overall frame transmission time. This in turn is used to determine the ordering of message transmissions and the feasibility of schedules in the LLC sub-layer.

The MAC sub-layer services allows the LLC to send and receive frames and provides status of frame transmissions to the LLC. On receipt of a 'send frame' request from the LLC, the MAC algorithm begins contention for access to the media. Once media access is obtained, the frame is transferred to the physical layer and transmitted to the receiving node. When a frame is received from the physical layer, the MAC compares the address frame with its assigned logical addresses. If there is a match, the MAC must notify the LLC that a frame has arrived, and initiate a frame transfer to the LLC buffers. The MAC layer must also provide support for error control, flow control, and logical addressing.

Media arbitration can generally be classified as centralized or decentralized. In centralized arbitration, there is a central point of control, or "bus master" that governs access to the media. In decentralized, or distributed arbitration, control is regulated by arbitration among peer nodes.

The strength of centralized control is its inherent determinism when combined with static scheduling. In this strategy, the bus master allocates media bandwidth based on a fixed schedule, allowing each node to transmit at a predetermined time. There are several weaknesses in this approach. The bus master is a single point of failure, a failure in the bus master causes a general network failure. The centralized control/static scheduling approach precludes variation in network loads. Earlier we described scenarios in combat operations that can cause fluctuations in network loading. Centralized control utilizes bandwidth less efficiently than distributed control. A two-way frame exchange between the master and the transmitting node is required, where the master must grant permission for the node to transmit, then the node can begin transmission.

In decentralized or distributed MAC protocols, there is no central point of control, media contention is arbitrated among peer nodes. The distributed approach generally does not have a single point of failure and is able to accommodate variations in network loads. This approach also make more efficient use of network bandwidth, since it does not require the two-way transmission found in centralized protocols.

There are several MAC protocols that utilize distributed media arbitration, including CSMA-CD[8], CSMA-CA[15], and token passing protocols[16,17]. CSMA-CD cannot accommodate variation in network deterministically, and hence will not be considered for the CVLAN. CSMA-CA has limitations on bandwidth and geographic span that preclude its consideration from CVLAN. The token passing approach can be utilized deterministically, does not have the bandwidth/geographical span limitation found in CSMA-CA, and has the benefits of the decentralized approach discussed above. This is the approach we will utilize in the CVLAN.

Token passing protocols utilize a special control frame, or token, to control access to the media. The token is transmitted between nodes. If a node wishes to transmit, it must first capture the token. Only the node that holds the token can transmit. Once the node has completed its transmission, it "frees" the token for subsequent capture.

There are two general variations on token passing protocols: priority driven protocols (PDP) or time sliced (TS) protocols[18]. In PDP protocols, frames are assigned priorities, and transmitted based on those priorities. The priorities are generally encoded on the token. In TS protocols, each node is assigned a time slice for frame transmission. A free token is transmitted unidirectionally between neighboring nodes. On receiving a free token, each node transmits frames for its assigned time slice, then generates a free token for its downstream neighbor. The thorough examination of token passing protocols is beyond the scope of this effort, but will be examined in follow on work.

Summary

This effort provides the framework from which to design and implement a real-time LAN architecture for combat vehicles. The focus of this effort was to define a real-time LAN architecture capable of adapting to fault induced variations in network loading, and still meet the time constraints of critical messages. We specified benefit based, dynamic scheduling as the means by which this can be accomplished. In order to facilitate meeting time constraints, we proposed a means to reduce uncertainty and processing time in a message's critical time segment through logical addressing and the use of apriori information.

There is currently an implementation effort underway for this architecture. The first part of this effort is the design and implementation of a logical link sub-layer based on the requirements described herein. This effort will focus on developing a benefit based scheduling algorithm and metric to measure scheduler performance. Another significant aspect of this effort is the integration of the LLC functions in such a way that message set-up processing can be removed from the critical time segment of a message. In the second part we will analyze the performance of this sub-layer on the two variations of token passing protocols. The goal of this effort will be to assess the suitability of either protocol for the types real-time message transmissions found in combat vehicles.

The realization of this architecture will provide a means for the developers of combat vehicles, or any other LAN based system that operates in a harsh, non-deterministic environment, to attain a more maintainable system that is able to adapt to fault induced variations in network loading.

REFERENCE LITERATURE

[1] Scheduling Algorithms for Multiprogramming in a Hard Real Time Environment, C. Liu, J. Layland, ACM 20(1):46-61

[2] Real Time Computing Systems: The Next Generation, J. Stankovic, IEEE Real Time Systems, Feb 1988

[3] A Software Overview of HARTS: A Distributed Real-Time System, Kang G. Shin, IEEE Software, vol9, no5, pp 58-68, Sept 1992

[4] Scheduling Real-Time Processes Using Fuzzy Logic, F.R. Brown, B.A. Abbott, Utah State Univeersity

[5] The Spring Kernel: A New Paradigm Ror Real-Time Systems, John Stankovic, Krithi Ramamritham, University of Massachusetts

[6] Real Time Systems Design and Analysis, Phillip Laplante, IEEE Press 1997

[7] Communication and Computer Networks, Michael Woodward, IEEE Computer Society Press, 1994

[8] Data and Computer Communications, 5th Edition; William Stallings; Macmillan, 1996

[9] Computer Organization and Architecture, William Stallings, Prentice Hall, 1996

[10] Computer Networks, Andrew Tanenbaum; Prentice Hall, 1990

[11] Eliminating the 'Hard'/'Soft' Real-Time Dichotomy, D. Jenson, Embedded System Conference, Apr 94

[12] A New Metric For Real Time Scheduling, Fon Brown, Department of Electrical and Computer Engineering, Utah State University

[13] A LAN Architecture for Communications In Distributed Real Time Systems, K. Arvind, J Stankovic, IEEE Real Rime Systems, Vol 3, No. 2, May 1991

[14] An Introduction to Real-Time Data Communications in Land Combat Vehicles, Paul Richardson, U.S. Army TARDEC, 1997

[15] CAN Specification 2.0, 1991 Phillips Semiconductors

[16] ANSI Standard X3T9.5 (FDDI: Token Ring MAC), May 1987

[17] IEEE Standard 802.5-1989, A Token Ring Access Method and Physical Layer Specification, IEEE, New York, 1989

[18]Real Time Performance of Two Token Ring Protocols, Kamat and Zhao, Advances In Real Time Systems, Prentice Hall, 1996

[19] M1 Series Tank Program, System Segment Design Document for Bus Operations (Limited Dissemination)

THE NETWORK VEHICLE – A GLIMPSE INTO THE FUTURE OF MOBILE MULTI-MEDIA

R. Lind, R. Schumacher, R. Reger, R. Olney, H. Yen and M. Laur

Delphi Delco Electronics Systems, Kokomo, Indiana

R. Freeman

IBM Corporation, White Plains, New York

ABSTRACT

The Network Vehicle is the Delphi Automotive Systems' vision for the future convergence of computers, the communications infrastructure, and the automobile. It features many advanced functions such as: satellite video, Internet access, virtual navigation, remote vehicle diagnostics and control, games, mobile office, automotive web site, and customized real-time stock quotes and sports scores. These features are enabled by an integrated planar antenna that is capable of multiple satellite reception, a client-server network architecture, and unique human-vehicle-interfaces such as color reconfigurable head up and head down displays, steering wheel controls, voice recognition, text-to-speech, and large touch screen active matrix liquid crystal displays (LCD's). The software applications are written in Java, using Application Programming Interfaces (API's) to reduce the complexity and cost of the source code.

INTRODUCTION

The Network Vehicle, a new technology initiative by Delphi Delco Electronics Systems and its partners (IBM, Netscape Communications, and Sun Microsystems) is aimed at offering more productivity tools, convenience, safety, and entertainment to millions of commuters who spend hours each day cruising the roads or stuck in traffic. It is designed to demonstrate what technologies and software can do for the vehicle of the future. The Network Vehicle, pictured in Figure 1, made its debut at COMDEX '97 and has since been all over the world appearing in demos, conferences, and technology shows, including SAE Congress '98, and CeBit '98.

Figure 1. The Network Vehicle

SYSTEM OVERVIEW

The Network Vehicle is created by integrating existing hardware and software technologies including voice recognition, wireless communications, global positioning via satellite, head-up displays, Java™ technology, microprocessors, Web access, and

0-7803-5086-3 /98/$10.00 ©1998 IEEE

other Internet/intranet features. The voice recognition technology allows drivers and passengers to verbally request and listen to e-mail messages, locate a restaurant or hotel, ask for navigation help or for specific music or sports scores, and use voice-activated telephone services, all done safely without interfering with driving. Drivers can use a head-up display projected through the windshield to navigate to their destination or check vehicle functions without taking their eyes off the road.

Passengers using individual terminals next to their seats can do even more, including interacting with the Internet, watching television or playing games. Existing services such as theft deterrent technologies and emergency services could also be integrated. These "smart" features are enabled mainly because of the real-time data-streaming capabilities over a wireless network using Java-based technology. The Network Vehicle also has an integrated cellular phone, Netscape Communicator software for Web browsing and e-mail, a removable personal digital assistant (PDA) and docking station. There is also a vehicle web site that provides driver support for a wide array of customized capabilities such as enabling remote monitoring and control of vehicle systems in emergency situations.

Onboard And Off-Board Communications Networks

The key to the functions of the Network Vehicle is its ability to communicate efficiently both onboard and off-board of the vehicle. Figure 2 depicts the onboard network architecture of the vehicle.

At the heart of the onboard system is a network computer that links up a number of sub-networks, including a Class II bus (for engine, head-up and head-down displays, door lock and door actuator, RFID, and cell phone, etc.), an Ethernet (for passenger computers and displays), and a mobile media link (for speakers and CD players). Since the subnetworks all have different clock speed and functionality, gateways are used to translate the messages from one sub-network to another so as to ensure harmonious operations.

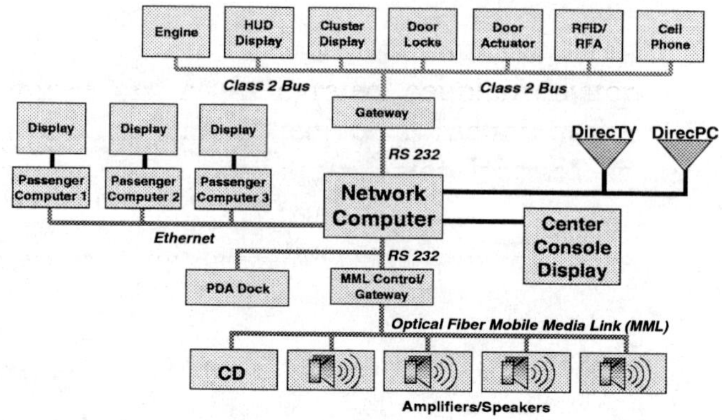

Figure 2. Onboard Network Architecture

The off-board network is shown in Figure 3. The high bandwidth requirements for on-demand audio and video functions are fulfilled with DirecPC and DirecTV satellite broadcasts. A wireless modem provides the uplink out of the vehicle directly to Internet service providers. The downlink return path from the Internet to the Network Vehicle can come through either the satellite (400 kbps), or through the wireless modem. The downlink satellite video and data signals are collected by the Network Vehicle's roof-mounted antenna and processed by their respective receivers.

Figure 3. Off-Board Network Architecture

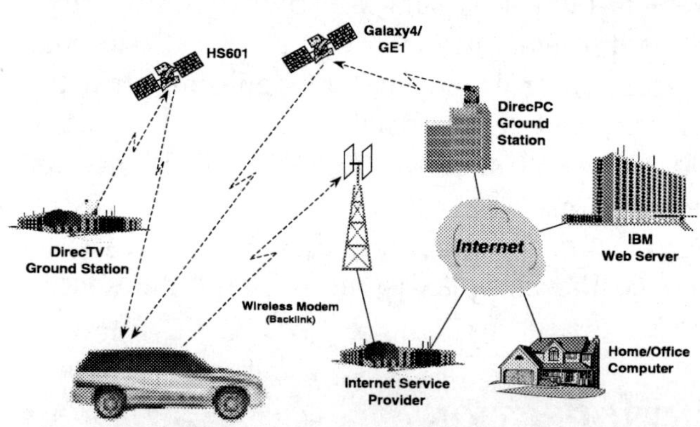

In order to take full advantages of the resources and services available to the drivers and passengers, the vehicle must have robust external communications paths. It is also vitally important that the vehicle configuration includes a firewall to ensure the integrity of the vehicle operation. This means that regardless of what happens to the multimedia applications that are running in the vehicle, the powertrain system control, the chassis control, brakes, airbags, and other critical functions must maintain their integrity. In the Network Vehicle, this isolation is implemented through multiple gateways as shown in Figure 2.

KEY TECHNOLOGIES AND SYSTEM COMPONENTS

The key technologies and system components behind the Network Vehicle are advanced speech recognition software, mobile media link (high-speed fiber optic data bus), reconfigurable head-up display, high-bandwidth communications that connect the vehicle with the outside world, and a suite of automotive computer software. These elements are described below.

Speech Recognition and Text-to-Speech System

IBM has modified its advanced speech recognition and text-to-speech system, ViaVoice™, for the automotive environment. ViaVoice allows the driver to access virtually all the vehicle's features through voice commands and enables the vehicle to talk back using synthesized speech. For example, the driver can: execute vehicle system commands such as lock doors, play CD, and change radio station, request travel directions and traffic updates from the Web or other sources, check e-mail and voicemail, request news, sports, and stock information. The speech recognition system can understand most drivers instantly, with no system training required, and it has been tuned to offer optimal performance even in a potentially noisy vehicle environment.

Driver and Center Console Displays

The Network Vehicle is equipped with three displays for the driver: the head-down display (HDD), the head-up display (HUD), and the center console display.

The HDD system displays standard graphics for an instrument panel: road speed, engine speed, engine status, door lock/ajar, and fuel level. It also displays the functions for the steering wheel buttons since these functions vary with the entertainment mode that is selected. The HUD projects a virtual image through the windshield that gives information to drivers without requiring them to take their eyes off the road. It displays road speed, engine status, waiting e-mail indication, navigation information, microphone on/off, and a text message area for giving feedback to the driver.

The center console's touch-screen LCD serves as a user interface for controlling nearly all of the Network Vehicle's multimedia functions, (office, navigation, entertainment, and information). When in entertainment mode, the display is configured as a radio- or CD-style faceplate whose buttons and controls are activated by touch (see Figure 4). It is reconfigured to display e-mail, navigation maps, Web browser, cellular phone faceplate (shown in Figure 5), and more. Voice-activated commands can be used with all of the center console functions and, in some cases, is combined with text-to-speech input to minimize driver distraction.

Figure 4. Audio on Demand Faceplate

Figure 5. Cellular Phone Faceplate Display

Passenger Displays

The Network Vehicle has color LCD touch panels for the back-seat and front-seat passengers. As with the center console display, the passenger displays serve as user interfaces for controlling nearly all of the functions on the Network Vehicle. Unlike the center console display, however, the passenger displays can show video from DirecTV or DVD players, and they allow touch-screen access to all functions.

Planar Satellite Antenna

The antenna technology used in the Network Vehicle is the Continuous Transverse Stub (CTS) array. It is chosen because of its simplicity, planar construction, and potential for being an inexpensive product. Figure 6 shows the cross sectional view of a typical CTS antenna, realized as an array of broad continuous transverse radiating stubs, finite in height, extending from the upper conductive plate of an open parallel-plate transmission-line structure.

As a receiving antenna, the induced longitudinal current components in the parallel plate structure are interrupted by the transverse-oriented stubs and excite propagating waves in the parallel-plate structure. This simple architecture allows for a complex two-dimensional planar array to be realized as an "extrusion" of a one-dimensional (constant cross-section) geometry. This has the effect of replacing a conventional "$NxN=N^2$" element structure (of discrete radiators, couplers, etc.) with a less complex "monolithic" array comprised of "*N*" integrated coupler/radiator features. The simple "tee" cross-section of the integrated CTS coupler/radiator forms an inherently low-"Q" (non-resonant) element which exhibits significant advantages (as compared to slot or patch radiators) in terms of wide-angle scanning capability, polarization purity, bandwidth, and dimensional insensitivity.

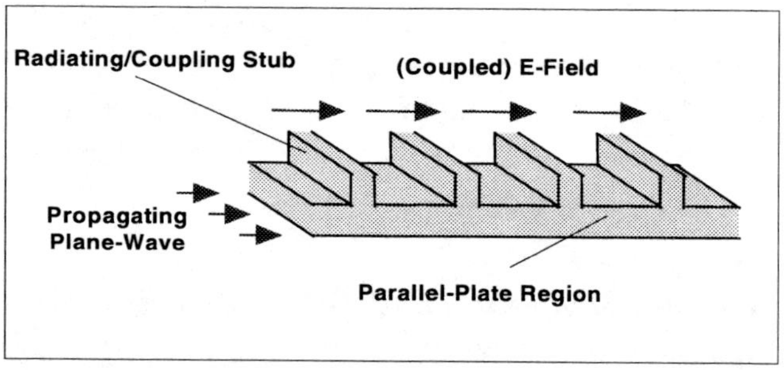

Figure 6. CTS Antenna Cross Section

Customer Application Web Site

The driver and passengers of the Network Vehicle can take advantage of Internet services provided by automotive-oriented ISPs (Internet Service Providers). The customer service site developed for the Network Vehicle showcases the types of features that could be offered when intelligent vehicle systems are connected to the Internet.

Because the Network Vehicle's systems are accessible via its command and control application, which also has secure access to the Internet, many scenarios are possible. Functions like parking lights and door locks on the Network Vehicle can be controlled remotely, providing safety and convenience. An example of such a web page is shown in Figure 7.

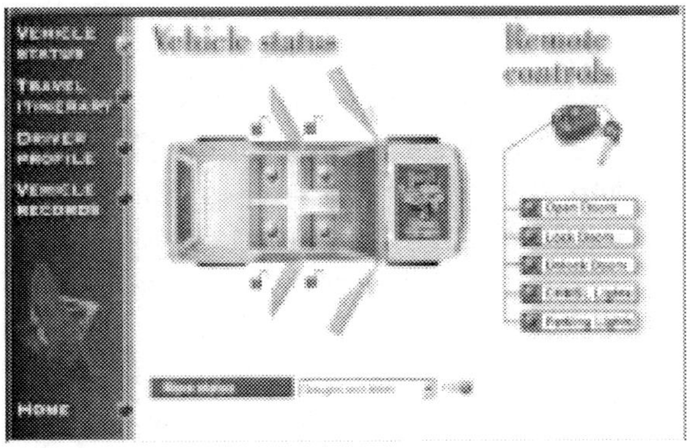

Figure 7. Monitor and Control of Vehicle Status via Internet

Under voice control of the driver, the Network Vehicle can upload vehicle status according to a profile set up in advance. An application monitors the diagnostic information available from components in the engine compartment. Should the diagnostics indicate that the latest reading from engine sensors falls outside a normal range, the application can notify the customer service Web site, where the severity of the reading can be determined. According to this determination, the control on the Network Vehicle could be used by the service representative to alert the driver; the alert could be done using text-to-speech and head-up display, and then features of the application, e-mail, and office could be used to arrange a service appointment. Figure 8 illustrates the sensor data that are being monitored through the vehicle web site.

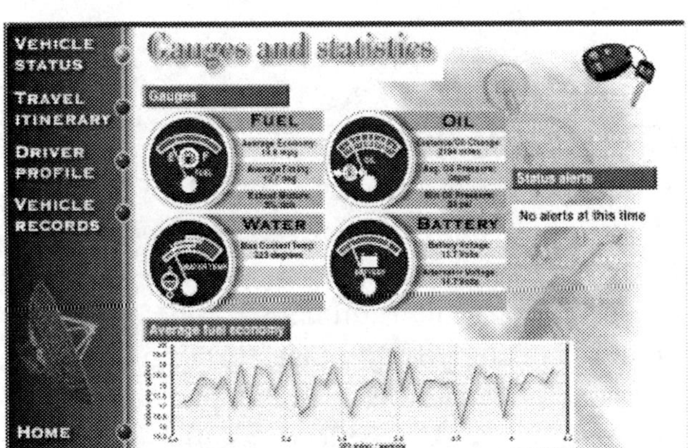

Figure 8. Remote Monitoring of Vehicle Gauges via a Web Site

Other features demonstrated take advantage of Internet service providers' ability to communicate with the Network Vehicle. Trip plans created by vehicle owners or concierge services can be stored at the Internet service provider. By accessing global positioning satellite (GPS) data directly from the vehicle, the driver and passengers can obtain maps, route guidance, fuel, lodging, and restaurant information while traveling. Owners of the Network Vehicle can also use their Web browser to set up a profile, including preferences like radio stations, personalized audio content, service records, and emergency service numbers.

Personal Digital Assistant Docking

An important role of the Network Vehicle is providing office functions to mobile users, using the same data the users have in their office desktop computers. This function is mainly provided through vehicle applications, Internet connectivity, and advanced, voice-based user interfaces. It is important for the Network Vehicle to access and synchronize personal and business data with PDAs, which is becoming increasingly important in our mobile, connected world.

A slot built into the center console of the Network Vehicle accepts the IBM WorkPad™ and uses the Network Vehicle's computer to add speech recognition and text-to-speech functions. This enables the driver to listen to schedules and to update files and other office data stored in the WorkPad. So the driver could, for example, issue a voice command to read calendar entries from the WorkPad or dictate a to-do list while driving. As PDA and smart card technology advances, the Network Vehicle's PDA integration could support the use of PDA-type devices for the secure transport and access of personal, financial, and business information as people move from one network access point to another.

0-7803-5086-3 /98/$10.00 ©1998 IEEE

Automotive Computer Software

Virtual machine-based, object-oriented technologies were applied to the Network Vehicle's development because they support the special requirements posed by the vehicle. Smalltalk, a well proven object-oriented environment and the latest Java™ technologies were chosen to develop the Network Vehicle's software.

A command and control application, written in Smalltalk and running on the vehicle's main processor, monitors vehicle status, controls devices such as cell phone and microphone, supplies driver information via head-up display, manages off-vehicle information flow, and controls vehicle software, such as the voice recognition and various convenience passenger applications. This application coordinates with the voice recognition, the head-up display, and the touch screen to provide feedback and control of the various systems, helping to keep the driver's eyes on the road. For driver convenience and safety, this application also interfaces via the Internet to a customer service provider, who can remotely monitor the vehicle status, unlock doors, turn on exterior lighting, etc.

Graphical user interfaces for Network Vehicle's center console and passenger displays are developed with Java to provide access to various in-vehicle and off-vehicle applications, such as navigation, audio/video entertainment, and office. An Internet server application was developed with VisualAge® for Java to provide communication between Internet-based customer applications and the vehicle systems. These applications support a message-based protocol communicating through the Internet to the in-vehicle command and control application, which in turn interfaces to the vehicle systems and the user interface application. Java applets were also developed for the customer site in order to provide animated vehicle status and control graphics.

Media BUS and In-Vehicle Electronics

The Network Vehicle contains a number of audio and video components that are configured and controlled via a high-speed (100 Mbps) digital fiber-optic mobile media link (MML). Audio and video devices like CD players can send their output signals via the multiplexed, fiber-optic bus to drive other components like amplifiers and displays. A control signal is also carried on the MML bus to configure and control the devices - configure the components, switch signals, and adjust variables such as volume, fade, and balance. The MML gives the application full use of the vehicle's multimedia systems to develop functions and user interfaces tailored to driver and passenger use.

Combining a command and control application with the MML in the Network Vehicle made possible features such as context-sensitive volume reduction during the use of the phone and the text-to-speech feature. It also allows "faceplate-style" user interface to integrate Internet radio with the sound system, and common and consistent user interfaces to define, review, and play the audio and video content preferred by the driver and passengers. The vehicle also features a satellite receiver system, which provides DirecTV and DirecPC access. DirecPC offers high-speed (440 kbps) Internet data access while DirecTV offers over 175 channels of digital entertainment programming.

FEATURES AND FUNCTIONS OF THE NETWORK VEHICLE

The main features of the Network Vehicle are manifested in the following three areas: (1) Driving Aids, (2) Business Tools, and (3) News and Entertainment. The Network Vehicle is designed such that, through software upgrades the customer can be provided with new features and functions as they become available.

Driving Aids - Get Where You're Going Efficiently

The Network Vehicle can access a simulated integrated global positioning system and hence advanced navigation capabilities via Internet. Therefore, it literally knows where it is and where it's going. It can display the appropriate maps or simply provide route directions on the head-up display, command console or as spoken instructions. Drivers could control the navigation system using commands like "destination" to select a new destination or "route guidance" to get turn-by-turn instructions. An example of such a display is shown in Figure 9.

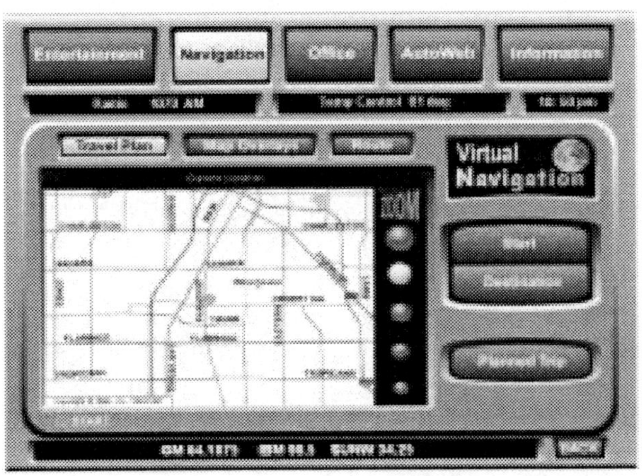

Figure 9. Example of a Navigation Function Display

The ability to monitor key vehicle systems means that the Network Vehicle can detect and avoid problems before they happen. If key vehicle systems show signs of failing, a warning light appears on the head-up display, or the vehicle tells the driver what the problem is. In the future, a technician or a vehicle manufacturer could take a look under the hood remotely to identify and diagnose problems, determine whether you can keep driving, or recommend obtaining service.

If you think you left the Network Vehicle's doors unlocked or the lights on, just visit the vehicle web site from your home or office computer to check, then fix the problem remotely. The Network Vehicle's future remote control capabilities not only could allow you to actually start the vehicle in the parking lot from the office, but also to set the heater or air conditioner from your office computer, so that the vehicle is comfortable and ready to roll when you are. You could even plan trips on the vehicle web site, then download them to your vehicle. Figure 9 shows an example page of the vehicle web site where remote control parameters can be set.

Business Tools – You're Never Out of Touch

The Network Vehicle is on the Web, with a DirecPC satellite link beaming the Internet to the vehicle at 400 kbps, more than 15 times the speed of a standard 28.8 kbps modem. Web sites can be viewed on the color reconfigurable displays at each of the three passenger seats, and the Network Vehicle can download selected information to the driver such as maps for navigation.

The Network Vehicle is an extension of your office, seamlessly offering access to everything from e-mail and address book to the critical files you need. The driver can be alerted to new e-mail via the head-up display and can ask the vehicle to read e-mail. With voice commands, a to-do list can be called up from the docked Workpad. E-mails and appointments can be downloaded to the docked Workpad PDA, for review after leaving the vehicle. In the future, the Network Vehicle will be able to communicate pages and play voice mail. The driver will be able to use voice commands to call up a personal address book, then dictate e-mail messages which the vehicle can turn into text and send. Passengers will be able to view faxes on their LCD displays.

The Network Vehicle's cell phone is fully integrated with other systems, so a simple voice command tells the phone to dial any number. The Network Vehicle uses its own in-vehicle audio system as a speakerphone, allowing hands-free conversations.

News and Entertainment - Getting There is More Than Half the Fun

Via the DirecTV satellite link, passengers will be able to view hundreds of broadcast and cable channels. Each passenger can watch a different channel at the same time, with headsets to keep the noise level to a minimum. Using the Network Vehicle web site, you can preprogram your favorite channels and shows. In the future, the Network Vehicle will include a DVD player so you'll be able to watch the latest movies on disc.

The Network Vehicle offers AM/FM radio, a CD player and audio from the Internet, all accessible with voice commands. From the Network Vehicle web site, one can program personal radio presets. Or, take advantage of various capabilities on the Internet, such as downloadable audio books, or audio-on-demand for an infinite choice of music selections or radio stations worldwide that are now broadcasting over the Web. Each passenger can also use an LCD display to watch movies, surf the Net or play the latest computer games. Using Netscape Communicator, the Network Vehicle will allow you to speak "Read Stocks" and hear the current value of each stock in your portfolio. In the future, the vehicle will read you the latest news from Web sites, or provide customized news from the vehicle web site.

CONCLUSION

The Network Vehicle has proven that by properly integrating several existing functions and technologies one can get a good glimpse of what the future of mobile multimedia holds. The range of applications enabled by the availability of wideband communications to the car is enormous. Reconfigurable displays and upgradable software will allow vehicle owners to take advantages of the latest applications and services. However, to make this vision a reality, a great deal of improvement and development is still needed. For example, an affordable and robust mobile satellite tracking antenna system is yet to be realized. It is clear though, the integration of computer, communications, and software technologies with vehicles will have a dramatic impact on the future of automotive industry.

A DEMONSTRATION PROJECT FOR THE ITS DATA BUS PROTOTYPE

Philip F. Spelt, Oak Ridge National Laboratory ,Oak Ridge, Tennessee

Allan M. Kirson, Motorola, Inc., Northbrook, IL

Introduction

With the continuing evolution of products for the new "Information Age," a variety of information sources are finding their way into road vehicles, helping to bridge the gap between the highway and the driver. These devices can provide information on route guidance, safety and hazard alerts, and traffic congestion advisories in real time. The devices and the information they provide should be tailored to driver needs, the traffic and environmental conditions and other situation parameters affecting the driver's travel. Safe installation and operation are a function of the extent to which devices can be easily installed and obtain vehicle information required for device functioning, while not corrupting existing vehicle data and functions. A dual-bus architecture, shown in Figure 1, has been designed as the basis for a family of Society of Automotive Engineers (SAE) standards, in which an ITS Data Bus (IDB) is connected to the vehicle multiplex bus through a gateway.

When implemented, this architecture permits electronic device manufacturers to build a single automotive version of their product which can plug into the IDB in any vehicle. In addition, the IDB is independent of all auto systems, with two exceptions: DC power, and the ability to exchange limited types and amounts of data with the vehicle through the gateway. The gateway, under control of the auto company, acts as a "firewall", allowing only authorized message traffic to pass between the auto's multiplex bus and the IDB devices. This "firewall" has significant safety benefits, as it prevents ITS devices (both OEM and after-market) from interfering with basic functions of the vehicle. The family of SAE standards which define the IDB was submitted for committee ballot this past summer.

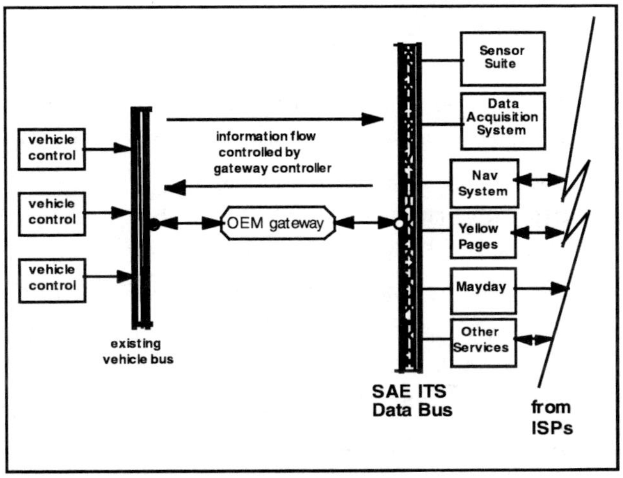

Figure 1. IDB dual bus architecture.

IDB Demo98

This paper presents a "snapshot" of an evolving project to demonstrate the installation of the IDB in production vehicles. The demonstration shows the operation of a variety of electronics components from different manufacturers. The goal of Demo98 is to demonstrate several of the functional requirements and capabilities of the IDB. These include "hot plug-and-play" of devices, interoperability of a device in different vehicles, and the information-sharing capacity of devices connected to the IDB.

0-7803-5086-3 /98/$10.00 ©1998 IEEE

The demonstration has been scheduled for showing at a number of automotive and consumer electronics shows beginning in the fall of 1998 and extending into the first part of 1999. The capability to be demonstrated is critical to the ITS goals of interoperability and safety. Interoperability of devices refers to the ability of the user to plug devices into any IDB-equipped vehilce and use them; safety of operation is enhanced by permitting the user's own device, with which the driver is already familiar, to be transferred into any vehicle being operated. These characteristics are inherent in any IDB-equipped vehicle.

Achievement of this project's goals required modification, integration, testing, and installation of hardware; and development and testing of new software components. Most of the displays will be stationary, although the vehicles can be demonstrated moving if conditions permit. Therefore, in the scenarios described later in the paper, the results of a vehicle trip, rather than the actual moving vehicle, will be demonstrated. Moreover, Demo98 has not implemented the full capabilities of the IDB standard developed by the SAE Committee. For example, the gateways only provide the specific services required by the Demo98 scenarios, rather than the full set of services specified in the SAE standard. In addition, only device messages specific to Demo98 are implemented by the participating companies, rather than those which would serve the full range of services a device manufacturer would wish to provide in a fully-developed commercial product.

IDB Demo98 Participants

Demo98 is being conducted by a consortium of private industry, under the combined direction of the SAE and the Oak Ridge National Laboratory (ORNL). Private industry participants donated three vehicles, a variety of electronics systems, and the labor required to make device-specific modifications. ORNL built the benchtop system and developed the Demo98 conformity test to permit evaluation of the functionality of each of the devices, as well as their interoperability in the IDB environment.

Companies participating in Demo98[1] include the United States' "Big Three" automobile manufacturers: Chrysler (Concorde LXi), General Motors (Cadillac SLS), and Ford (Lincoln Continental). A variety of devices has been contributed by a number of companies, shown in Table 1 below. In addition to their material contributions, several of the participating companies have contributed funds to support the development and presentations of Demo98.

Demo98 Scenarios

Four demonstration scenarios have been defined. Each lasts less than five minutes, and includes each of the three vehicles. Thus, a complete demonstration of the IDB capabilities, including audience questions, should take less than twenty minutes.

The Plug-and-Play Scenario is the most fundamental and general of the four scenarios developed for Demo98. It is intended to demonstrate the IDB characteristics of "hot plug-and-play" and the as can reasonably be shown with the devices provided by the participants. Thus, as the title implies, the hot plug-and-play capabilities of the IDB protocol are shown by inserting and removing devices while the IDB is alive and functioning. This demonstrates the inherent reconfigurability designed into the IDB protocol, by showing that current devices, which may have depended on another no longer available device, will still function on the IDB. For example, a pager would send the phone number from an incoming call to the cell phone to enable the driver to make a voice-activated call. However,

[1] As of August 1, 1998. Because this is a "living" demonstration, additional participants will be joining as time passes.

when the phone is removed from the IDB, the pager simply announces the incoming call to the driver through the wav-player. Similarly, in those vehicles in which the gateway can control the audio mute function, an incoming phone call mutes the radio audio and uses the vehicle speaker system to provide hands-free phone functions. Thus, this scenario uses at least some functions from all the IDB-compliant devices, although the primary emphasis is on the IDB capabilities.

Table 1. IDB Devices

Company	Device
Shell Oil Company	RFID tag
Motorola, Inc.	GPS, AMPS phone, pager
AMP, Inc.	Wiring and connectors
Molex, Inc.	Connectors
Cellport Labs, Inc.	Web server, WAW gateway
SanpOn Diagnostics	OBD2 data
Kenwood	CD changer
Audiovox	Security system
Safety Warning Fndtn.	Safety warning system
Strategis Consulting	HP Palmtop, WAV player, software
Oak Ridge Nat'l Lab	System integration, conformity test, etc.

The remaining scenarios are designed to provide a demonstration of not only the functionality of the IDB, but also of many of the devices plugged into the IDB. For example, the radio-frequency identification (RFID) tag provides a key function in both the Rental Car and the Traveling Salesman scenarios. The Rental Car scenario presents the vehicle at a service station fuel pump, where it interacts with the pump through the RFID tag. After a welcome from the pump, the fuel purchase is accomplished. Using vehicle information passed through the IDB gateway, the RFID can establish the type and quantity of fuel needed, and provide pricing information to the driver. Vehicle diagnostic data can be transmitted to the service station across the webserver. Any recommended service is communicated back to the driver, followed by the final fuel purchase transaction information.

In the Traveling Salesman Scenario, a number of devices serve the needs of a traveler while on the trip. The vehicle provides latitude and longitude positioning of the vehicle, fuel level, odometer reading, etc. In a (simulated) moving vehicle, the GPS updates the palmtop (PDA) along the route, permitting easier reporting of expenses on the trip.

The other scenario, the Stranded Motorist, uses a variety of devices on the IDB to help the driver of a disabled vehicle obtain the appropriate help. Following the breakdown, a user-initiated vehicle diagnostics scan is performed and reported, along with the GPS-determined vehicle location, to an independent service provider (ISP) via the web server. Communication back to the vehicle is provided through the cell phone and the web server. An approaching service vehicle, with the appropriate trasnponder, reads the RFID tag and receives the most recent vehicle diagnostics scan.

Building IDB Demo98

One purpose of IDB Demo98 is to create an initial proof-of-concept IDB which will demonstrate the viability of both the concept and its ease of implementation. Accomplishing this task required a benchtop prototype, the development of a conformity test for the demonstration, and installation of an IDB and gateways in each of the vehicles.

The Benchtop Physical System

The benchtop IDB serves as both a proof-of-concept prototype and a test facility for compliance of devices created for Demo98. It was built and remains functional in the In-Vehicle Information Systems R & D facilities at Oak Ridge National Laboratory. As stated earlier, a primary purpose of Demo98 it so develop a system which permits evaluation of the IDB concept and functioning of the IDB protocol. As such, not all features of the IDB standard are implemented. The benchtop system permits only two (key-off and key-on) of the four power modes specified; due to parts availability, the actual IDB connectors are not used in Demo98; and only a limited set of the ultimate message set is used.

The benchtop IDB consists of the appropriate twisted-pair (TP) backbone wiring for signals (two pairs TP) and power (one pair TP), together with necessary device trunks of similar configuration. The benchtop system is configured for a full complement of sixteen devices, the maximum permitted by the IDB standard. In this laboratory setup, one of these connections is allocated to the reference node, described below.

The power supply consists of a 12-volt battery connected to a standard automobile alternator belt-driven by an AC electric motor. The motor is, in turn, powered through a 24-hour multi-cycle timer, which permits emulation of a variety of driving conditions -- multiple short trips typical of urban driving, or a sustained period typical of a cross-country trip. This arrangement permits evaluation of the current draw of an IDB with a full complement of devices, necessary to be sure that the IDB complies with vehicle power draw limitations. As Demo98 evolves into a full benchtop implementation of the IDB, such capability will be useful for emulating the full range of power modes.

Benchtop Software

Software for the benchtop system, as well as for the Demo98 vehicle installations, consists of a set of drivers created by Strategis Consulting. These drivers provide device and gateway manufacturers with the capability of sending and receiving messages on the IDB, by adding or removing the IDB message header material. The manufacturers are responsible for creating the necessary software to translate information back and forth between their devices and the message content as received from the IDB. Some of these messages are general IDB messages, others may be specific to a particular device, or set of a manufacturer's devices.

The Demo98 reference node consists of software provided by Strategis Consulting, running on a Hewlet-Packard 620LX palmtop device under Windows CE. The reference node provides the ability to send and receive all messages used in Demo98, to and from all devices installed on the IDB. As such, it serves as the prototype for a full conformance test, to be developed as part of the SAE IDB standards family. Additional testing capacity is provided by Demo98 read and write programs, which send or receive multiple copies (hundreds or thousands at a time) of an IDB message as rapidly as they can be inserted onto the bus. The read program checks the messages for errors, thus providing an initial check of bus capacity and performance.

The Conformity Test

The conformity test is conducted using the devices and software described in the previous section, with the addition of a PC connected to the IDB in "snoop mode" -- that is, it monitors bus activity without consuming one of the allotted sixteen connections. The 300MHz Pentium II PC includes a CyberResearch 40 MHz data acquisition board with a 64K buffer, which provides digital oscilloscope functions at a fast enough rate to monitor the bit rise and fall times. There is also a programmable protocol analyzer package which provides information about message integrity.

The conformity test for a particular device consists of performing the series of tests listed below and monitoring the following IDB characteristics:

signal voltages and timing
bus activity:
-- collisions
-- token passing
-- devices present (especially "device")
-- devices active (especially "device")
-- device-specific messages

These activities are monitored for each device as it is tested in each of the following steps:

1. Monitor voltages and bus activity with IGN off. Record any anomalies.

2. Connect "device" with IGN off, record voltages and any bus activity.

3. Turn on IGN (voltage nominal), monitor and record bus activity as above.

4. Generate interrupts, monitor and record bus activity as above.

5. Raise voltage to Vmax; re-test

6. Simulate fully loaded bus; re-test

7. Lower voltage to Vmin; re-test

8. Simulate fully loaded bus; re-test

9. Return voltage to nominal and test power-down sequence, monitor and record bus activity as above.

This sequence of tests is performed for each device incorporated into the IDB Demo98, in both stand-alone and group mode. Results of a device's test are shared with the device manufacturer upon completion of the test program for that device.

Vehicle Installation

Vehicles were received from the manufacturers with the gateways installed. ORNL installed the backbone of the IDB, as well as audio connectors into the vehicles' audio systems. The gateways were evaluated by inserting the reference node into the vehicle IDB and performing the appropriate tests. Then, benchtop-certified devices were installed in the vehicles, and performance was evaluated in that environment, with the engine both off and running. Essentially the same series of tests was performed as was used for benchtop certification. As expected, once devices passed the benchtop test, there were no problems with performance in the vehicles.

Testing IDB Demo98

Once the three vehicles were prepared, it was necessary to evaluate the entire Demo98 process. This procedure required pitching a 30-foot by 60-foot tent, in which the vehicles and supporting booth infrastructure could be set up. This was done in late September, 1998, at ORNL.

The scenarios had previously been scripted and several professional spokes-persons were trained in conducting them. The Demo98 team were concerned that the scripts as created, coupled with the planned physical booth layout, would create problems with the movement of people through the booth. The ORNL test demonstration permitted a fine tuning of the scenario demonstration procedures, as well as evaluating the two alternative vehicle-parking arrangements: for showings in which the booth has a full back wall, the three vehicles are parked facing the front of the booth, with sufficient room between them to permit having doors open and still allow movement between them; for booths with 360-degree access, the middle vehicle faces toward one long side, and the two side vehicles face the other side. This arrangement mitigates against the appearance that the booth faces in a particular direction.

Acknowledgments

The work described in this paper is supported by CRADA No. ORNL98-0516 with the Society of Automotive Engineers, Warrendale, Pa., sponsored by the Laboratory Technology Research Program, Office of Energy Research, U.S. Department of Energy, under contract DE-AC05-96OR22464 with Oak Ridge National Laboratory, managed by Lockheed Martin Energy Research Corporation.

SIGNAL PROCESSING AND WAVEFORM GENERATION

IN THE SIDE ZONE AUTOMOTIVE RADAR

John C. Reed, H E Microwave, Tucson, AZ

Abstract

Recently, a "Side Zone" automotive radar, also known as the Side Detection System (SDS), has been introduced [1, 2]. The radar has a significant amount of signal processing aimed at tracking targets, estimating their ground speed, and rejecting clutter. The introductory papers give an overview of the radar including performance requirements, requirements analysis, architecture design, signal processing, block diagrams, and test result summaries. This paper is primarily concerned with, and expands on, the signal processing techniques used in the SDS radar. Also, the method of waveform generation is described in detail, as it is intricately related to the radar signal processing.

Introduction

The Side Detection System uses a low cost "sub-miniature" radar for remote sensing, and is designed to assist the driver of a car or truck (radar platform) in making a lane change maneuver. The radar field of view is constructed so as to cover the space which is least visible to the platform driver through mirrors and head turning. The system is used in conjunction with both inside and outside rear view mirrors by the platform driver to improve awareness of other vehicles or objects in the platform's adjacent lane.

"Clutter Rejection" Requirements

In addition to providing visibility into so called "blind spots" as may exist on various car and truck platforms, the system concept demanded that detected objects be discriminated between potentially hazardous (target) or clearly non-hazardous (clutter) objects.

The discrimination requirement was motivated by field tests on systems which did not have the discrimination feature. The tests showed that a significant percentage of drivers found the system to be annoying when the platform was driven next to the roadside. When the platform is next to the roadside, the detection zone of the radar spills well past the edge of the road, putting off-road objects into the radar's field of view (fig. 1). The system would indicate the presence of objects (alarm) that were next to the roadside such as trees and telephone poles, even though consideration of a lane change off the road is precluded. The result was that the drivers received alarms at undesired times when no potential lane change was possible.

In order to reduce the number of extraneous alarms, criteria for classification were determined and applied to the situation [2]. The criteria for classification was to alarm on any object (typically a vehicle) within either of the platform's adjacent lanes that was moving, either faster or slower, *in the same direction* as the platform. This criteria leads to alarms being generated on vehicles traveling

0-7803-5086-3 /98/$10.00

along side the platform, and not generating alarms on objects which are stationary such as trees and parked cars. Oncoming traffic is also rejected. By application of this capability, the over-all number of SDS alarms to the driver is greatly reduced during a typical drive experience. The time of alarms is then limited to those times when the alarm is of potential benefit.

Radar Performance Requirements Summary

Coverage Requirements

Figure 1 summarizes the area of coverage required of the radar. There a coordinate system is defined which facilitates the signal processing discussion.

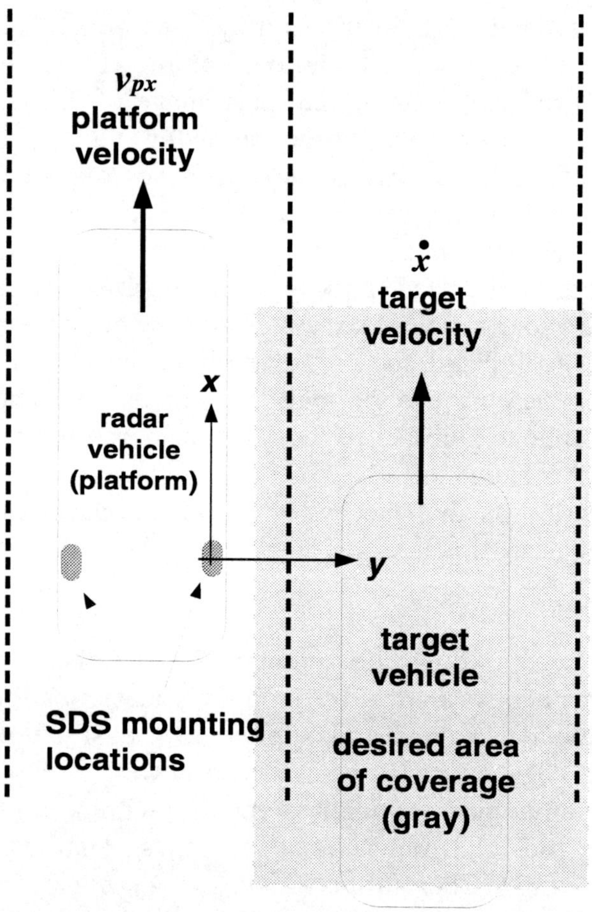

Fig. 1. The desired area of coverage for a right-side side zone automotive radar is depicted in gray. A coordinate system is defined so that the platform velocity defines the positive *x* direction.

Discrimination (Classification) Function

The discrimination criteria is stated mathematically as [2] :

$$\text{alarm when } \hat{\dot{x}} \geq -v_{px} + \epsilon . \qquad (1)$$

where

$\hat{\dot{x}}$ = estimated x - directed target velocity with respect to the radar

v_{px} = platform velocity with respect to the ground

ϵ = an error term

The platform velocity is known via the speedometer in the platform, thus *the essence of the radar function is to estimate the x-directed target velocity with respect to the radar.* Once this is accomplished, the simple classification criteria (eq.1) may be applied for each target.

Targets and Clutter

There exists a great variety of target and clutter objects. Targets whose relative *x*-directed velocity with respect to the platform is less than 40 m.p.h. , while having non-zero ground speed in the same direction as the platform, must be detected. Bicycles fit the criteria, and system sensitivity must support the detection of bicycles.

Clutter objects vary greatly in nature. Some objects, such as guard rails, are highly distributed in physical size. There are numerous "discrete" type clutter object examples : trees, light posts, mailboxes, etc.

Radar Environment

The radar environment is that of highly dense clutter in a typical scenario. Also, both target and clutter objects are quite close to the radar by traditional standards, thus the SDS works in a unique radar environment. The environment may at times be very rich in target objects also. The impact of the radar environment is that there will occur many detections in any typical radar cycle, all of which have to be sorted and assigned to either existing target tracks or used for new track generation.

Response Time

The SDS must detect and analyze the trajectory of objects and make a classification decision within 100 msec of the object's coming into the radar's field of view.

Cost

Since the intended application of the design is for commercial and private vehicles, unit production cost is of paramount importance. Cost considerations overwhelm other considerations, and are the primary design driving factors.

System Architecture Design

As mentioned above, the primary function of the SDS radar is to detect objects within the required detection zone and estimate their *x*-directed velocity. The design seeks to accomplish the task in the smallest possible cost structure.

Design Approach

The design approach is to use very wide azimuth beam antennas (i.e. very low azimuth resolution), and signal processing to make up the functionality difference between these and narrow beam (expensive) antennas. The required coverage zone on each side is split into two, rather than a multiplicity of, azimuth sectors, as shown in fig. 2.

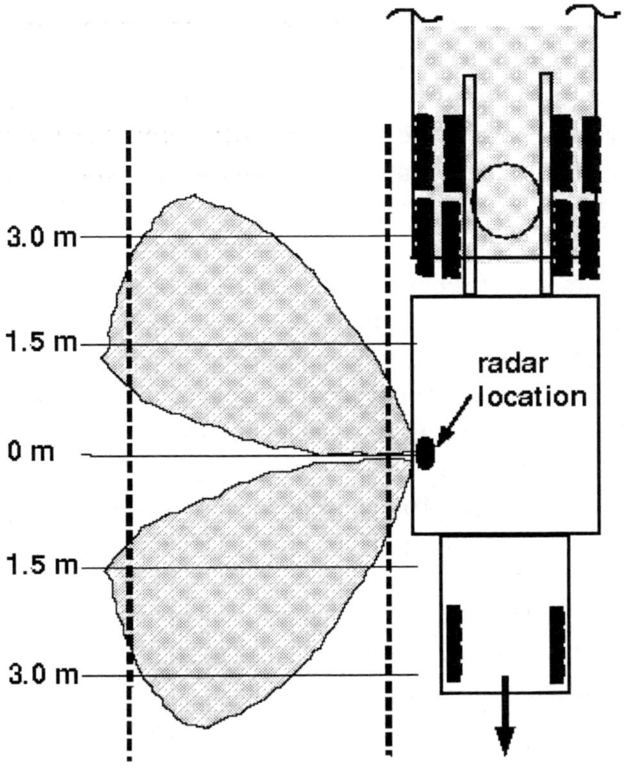

Fig. 2. Actual detection zones used in the SDS for a heavy truck application. The SDS sequentially switches between these zones by switching its transceiver channels between two antenna pairs.

The signal processing scheme is to use a Kalman filter to form target "tracks", where the target detections and measured parameters, range and range rate, are sequentially analyzed. The target's temporal trajectory in range and range rate space is analyzed as the measurements become available, and the target "state" (position and *x*-directed velocity) is continually updated. Once there are a sufficient number of detections and measurements on a target track, the estimate of the target *x* velocity is applied to the discrimination criteria, and the classification decision is made. If there are not enough measurements within a given time limit (100 msec) on a target , no alarm is given.

Required System Resources

Since narrow beam antennas are precluded due to cost, the system design carefully considered what capabilities were needed in the radar to provide the discrimination feature.

Ambiguities in range and range rate information cannot be tolerated in this application, therefore it was decided to use a pulse Doppler type of waveform. As discussed below, low cost methods were used to generate the waveform, and signal processing capability found in cost effective Digital Signal Processing (DSP) integrated circuits was relied upon for data reduction.

The range resolution used is slightly under 1 meter, and the range rate (Doppler) resolution is ~ 2.0 meters/second.

A relatively sophisticated tracking processor is used to estimate the *x* velocity of the target. This estimate is built through use of a tracking filter and the use of assigning a track to each potential target. The tracker not only estimates the target velocity, but provides mechanisms where the multiplicity of detections / measurements on each radar cycle (due to the target and clutter rich environment) are sorted and individual "hits" from the various objects are assigned to appropriate tracks.

Waveform Generation and Radar Signal Processing

Range Resolution Drives the Waveform Generation Method

Chirp (linear FM) waveforms were considered along with narrow pulse waveforms during the design process. Using long duration wideband chirp waveforms was selected to achieve high range resolution, high energy on target, and allow a low cost data sampling method. By using "stretch" processing (explained below), the sample rate of the system A/D may be significantly reduced, saving costs.

Overall description

A frequency vs. time diagram of the SDS pulse Doppler waveform is shown in figure 3. It is essentially a pulse train, as seen in classical pulse - Doppler waveforms, where each pulse is a wideband (200 MHz) chirp.

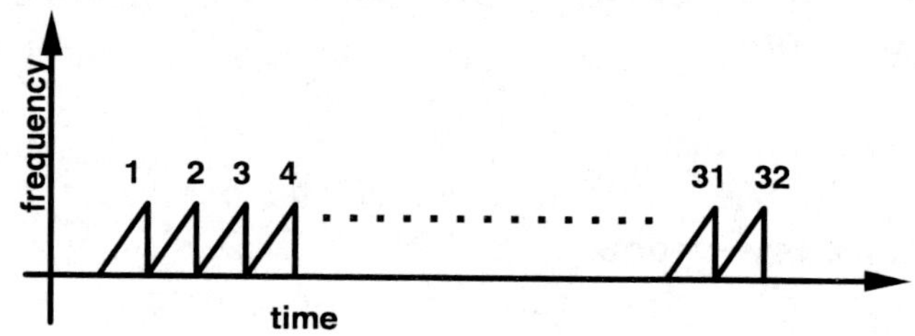

Fig. 3. SDS Waveform frequency vs. Time diagram. Each ramp represents a linear FM pulse, where there are 32 pulses in the pulse train processed simultaneously

The chirp signals are generated directly at the carrier frequency of 24 Ghz by a Voltage Controlled Oscillator. This oscillator signal also serves as a Local Oscillator for the homodyne receiver.

The expanded pulse duration is ~ 100 usec; the compressed pulse is ~ 5 nsec. The expanded pulse train is shown in figure 3. The "stretch" signal processing concept mixes a replica of the transmit waveform with the return signals from scattering objects (targets or clutter), delayed by their respective times of flight, during the chirp time. The result is a baseband signal where the frequency of the signal is proportional to the range of the target.

Radar Signal Processing

By sampling the baseband signal during the chirp time, and performing an FFT on the data, (termed a "ranging" FFT by virtue of the application), pulse compression is achieved. The FFT may be thought of as implementing a bank of contiguous bandpass filters which separate the signal returns from different objects at different ranges. In this way, a bank of 16 range bins are formed. There are 32 data samples taken during each chirp, resulting in a data sampling rate of approximately 100 Khz.

The progressive phase shift of the complex FFT result for each range bin is then analyzed by another FFT across the 32 pulses to determine the target Doppler frequency, if any. The two FFTs provide unambiguous range and range rate (Doppler) information at the end of the pulse train for all the detected objects in the detection zone (beam position) being processed.

Figure 4 shows an example of the output of the two dimensional FFT for one dwell in the rear pointing zone (ref. fig. 2) when the radar is viewing multiple targets of greatly differing amplitudes.

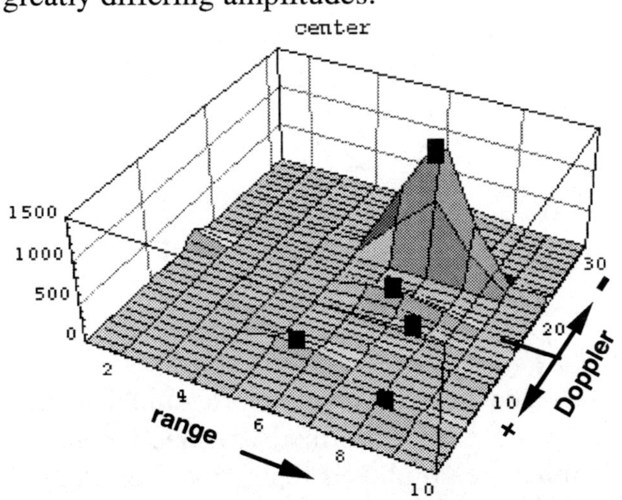

Fig. 4. Graphical result of the SDS two dimensional FFT on baseband data in a multi-target environment. Detected objects, and their location in the range-Doppler space are marked by the heavy dark lines. The dynamic range of the target amplitudes tend to disguise all but one of the targets in this view.

Doppler Effects

Depending on the range rate parameter of the target, different Doppler frequency shifts will be observed. The waveform design process was especially cognizant of the need for fully de-coupled range / Doppler information, thus the range bins described above were separated in frequency with adequate spacing such that the highest Doppler shift target would not "slip" a range bin due to its Doppler. In other words, the frequency span of the range bins must be high enough to support the highest expected Doppler shift. This critical consideration is satisfied by providing for an adequately high frequency sweep rate of the carrier during the chirp.

In the SDS, the sweep rate of the chirp is 200 MHz / 100 usec, resulting in a baseband frequency of 13.3 Khz per meter of range. Targets spaced by more than 3/4 meter in range will result in baseband signals spaced by more than 10 Khz. This was arranged to support a ± 5 Khz Doppler coverage without experiencing range bin errors due to Doppler effects. The SDS uses a carrier at 24.125 Ghz hence the Doppler coverage represents approximately ± 30 meters / second of range rate coverage.

Note that even though there is only one down convert mixer in the hardware, complete I/Q receiver capability is provided by effecting the target returns in baseband as *bandpass signals*, whose complex envelopes are preserved as manifested in the ranging FFT complex result.

Fig. 5. SDS Signal Processing Flow Diagram. The 2D FFT and data extraction processors work on individual channel data. The other processors work on data from both channels.

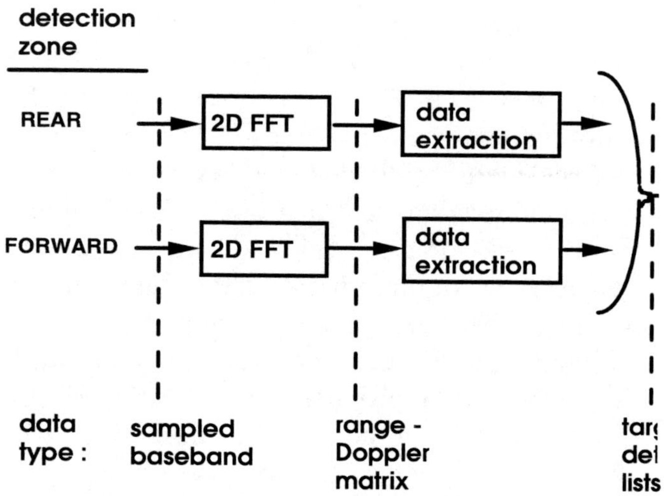

Radar Data Processing

Executive Processor

The "executive" software cycles the radar through its sequences of processes in each beam position, then executes processes that use data from both beams. The entire process is then repeated. A salient point is that the timing of various processes is dependent on the target load in the radar, i.e. the input size of the data sets. For large amounts of objects in the radar's field of view, certain processes take more execution time, hence the radar's cycle time is load dependent. In the heaviest of clutter environments, total cycle time can approach 28 msec, whereas average cycle times in the field are usually ~ 21 msec. Figure 5 shows the signal processing flow.

Data Extraction

This processor scans the range - Doppler (range rate) matrix of data, implements the detection function, and estimates the range / range-rate parameters of every detected target. The fundamental concept of the method used in "data extraction" is a simplified digital version of the optimum target parameter estimator[3], which in effect correlates the data with a replica of the time derivative of the transmit (compressed pulse) waveform. Although the technique used here is a crude approximation to the rigorous equation, it is fast and effective. The transmit waveform derivative model is first scanned along each Doppler bin in range for detection, then orthogonally scanned within a target bearing range bin to estimate target Doppler.

The output of "data extraction" is a list of the detections which occurred. Each detection, or "target report" has amplitude, range, range rate, azimuth sector, and a time stamp associated with it; all of these parameters are stored in a report vector. Table 1 gives the list of detections from the range - Doppler matrix shown in figure 3. The detection locations in the range - Doppler space are marked in the diagram .

Table 1. Example detection list, as generated by processing the data in fig. 3.

range (m)	range rate (m/sec)	Amplitude
5.40	22.34	70
4.94	5.03	70
3.41	17.10	100
4.30	0.06	190
4.42	-11.25	1650

"Ghost"

The "Ghost" processor combines the detection reports from the two detection zones, and thins the combined report list according to redundant hit criteria. Reports which are very close together in range and range rate are grouped, and the largest amplitude report within each group is used for further processing. The others in the group are discarded.

Tracking - Estimator

The primary purpose of the "tracking - estimator" processor is to estimate target *x*-directed velocities from a sequence of target

0-7803-5086-3 /98/$10.00 ©1998 IEEE

detection reports. Each object will, as it passes through the radar detection space, generate a sequence of reports, usually one report on each radar cycle. As there are usually a multiplicity of objects being observed (detected) by the radar simultaneously, an important required function of the "tracking - estimator" is to intelligently sort the stream of incoming target reports into data sequences which accurately reflect the actual sources of those reports. This sorting allows the radar to "track" (periodically update the position and velocity of an object) multiple targets.

In the general (and typical) case, there are more than one objects being "tracked" by the radar at any given time. In heavy clutter environments, such as driving next to parked cars, the target report list output by the "ghost" processor may have as many as six report vectors on each radar cycle. In this scenario, there are numerous objects to "track", thus the need to sort the various reports as they become available, and "associate" each report with the appropriate "track". Reports which are not associated with (do not belong to) an existing "track" are used to initiate new tracks.

A "track" is an estimate of the position and velocity , {*x-dot, x, y*} of an object. A track history is the position and velocity trajectory of an object, as it is observed by the radar as the object moves through the radar field of view. A radar measurement data history associated with an object is the position trajectory of the target in range - Doppler space as the object moves through the radar field of view. The track history position data is used to predict the next expected range - Doppler coordinates to be measured by the radar at any given time.

This entire tracking and estimating process is cast in the framework of a Kalman filter system "state" estimator [4]. The "system" is the radar and target, the "system state" is the position and velocity of the target in x - y coordinates, reference fig. 1.

System Dynamics

The system model in SDS is the target position and velocity with respect to the radar. The only available measurements are range and range rate. Salient assumptions were made in the filter design regarding constraints on target dynamics in order to simplify the required computations, allowing the embedded computing element to be of practical size and cost.

1. The target y-directed velocity (across lane) is assumed to be zero.

2.All target accelerations are assumed to be zero.

These assumptions are workable when considering the system sampling rate of approximately 45 Hz. The assumptions also reduce the only matrix inversion computation to a 2 X 2 matrix inversion, which significantly relieves processing loads.

"Extended" Non-linear Kalman Filter Equations

The state vector and measurement vectors for the SDS tracker are

$$\vec{x} = \begin{bmatrix} \dot{x} \\ x \\ y \end{bmatrix} = \begin{bmatrix} x_1 \\ x_2 \\ x_3 \end{bmatrix} \quad \vec{z} = \begin{bmatrix} r \\ \dot{r} \end{bmatrix} .$$

The state equations of motion are

$$\dot{\vec{x}} = \mathbf{F} \bullet \vec{x} \quad \text{where } \mathbf{F} = \begin{bmatrix} 0 & 0 & 0 \\ 1 & 0 & 0 \\ 0 & 0 & 0 \end{bmatrix} . \qquad (2)$$

The measurement equation is

$$\vec{z} = \mathbf{h}(\vec{x}) + \vec{v} = \begin{bmatrix} \sqrt{x_2^2 + x_3^2} \\ \dfrac{x_1 x_2}{\sqrt{x_2^2 + x_3^2}} \end{bmatrix} + \vec{v} \qquad (3)$$

where $\mathbf{h}(\vec{x})$ is a non-linear function of the state vector and $\vec{v}$ is a normal random variable vector with covariance **R** which characterizes the noise accompanying the measurement. The state transition matrix, which predicts the state vector at a given time from the last known state vector is

$$F = \begin{bmatrix} 1 & 0 & 0 \\ \Delta t & 1 & 0 \\ 0 & 0 & 1 \end{bmatrix}$$

where

$$\hat{\vec{x}}_k(-) = F \bullet \hat{\vec{x}}_{k-1}(+) \qquad (4)$$

where

$\hat{\vec{x}}_{k-1}(+)$ = the state vector estimate after a measurement at time $t = k\text{-}1$,

$\hat{\vec{x}}_k(-)$ = the state vector estimate immediately prior to a measurement at time $t = k$,

Δt = the time difference between the current and previous measurement. By time stamping the data, and computing Δt for each measurement, the variable cycle time of the radar is essentially invisible to the trackers, and non-uniform sampling periods are handled in this way.

As explained below, the "predictor" processor sorts through the new incomming data and associates the appropriate measurement vector with a track. Once the measurement data is determined to be associated, the state vector is updated by the measurement per the following:

$$\hat{\vec{x}}_k(+) = \hat{\vec{x}}_k(-) + \mathbf{K}_k \left[\vec{z}_k - \mathbf{h}_k\left(\hat{\vec{x}}_k(-)\right)\right]. \qquad (5)$$

where

$$\mathbf{K}_k = \mathbf{P}_k(-)\mathbf{H}_k^T \left[\mathbf{H}_k \mathbf{P}_k(-)\mathbf{H}_k^T + \mathbf{R}_k\right]^{-1} \qquad (6)$$

$$\mathbf{P}_k(-) = F\mathbf{P}_{k-1}(+)F^T \qquad (7)$$

$$\mathbf{P}_k(+) = \left[\mathbf{I} - \mathbf{K}_k \mathbf{H}_k\right] P_k(-) \qquad (8)$$

$$\mathbf{H}_k = \mathbf{H}_k\left(\hat{\vec{x}}(-)\right) = \begin{bmatrix} 0 & \dfrac{x_2}{\sqrt{x_2^2 + x_3^2}} & \dfrac{x_3}{\sqrt{x_2^2 + x_3^2}} \\ \dfrac{x_2}{\sqrt{x_2^2 + x_3^2}} & \dfrac{x_1\left(1 - \frac{x_2^2}{x_2^2 + x_3^2}\right)}{\sqrt{x_2^2 + x_3^2}} & \dfrac{-x_1 x_2 x_3}{\sqrt{x_2^2 + x_3^2}} \end{bmatrix} \qquad (9)$$

and the matrix **P** is the error covariance matrix. After the system state vector estimate is updated by new data with eq. (5), the error covariance matrix is updated also with eq. (8). A "track", or track file, physically is then the stored variables associated with equations (5-9), $\hat{\vec{x}}_k(+)$, $\mathbf{P}_k(+)$, and a time of last update stamp. A track file is generated and maintained for each object being tracked through the radar's field of view.

Alarm Criteria

If a track has been generated and data associated with it for three of four radar cycles, the track is considered "valid", and the x-dot estimate is passed to the discriminator for alarm evaluation. Up until this point, there is no attempt to determine whether the track is on a clutter object or a target of interest. The only clutter vs. target decision is in the discriminator.

Predictor

When data become available, the function of sorting the data and allocating the correct reports to the appropriate tracks is based on predicting the state vector from previous data and the known time lapse, eq (4). From this result, the expected measurement vector is computed from eq. (3). This result is then used to compare to the incoming measurement data to "find" which report belongs to a track, i.e. to "associate" a report with a track. In essence, the target report which is closest to the predicted report is selected to update the track, eqns. (5-9). All reports are tested for association with all existing tracks on every radar cycle.

In order to capture the correct target report, a window in range - Doppler space is set up around the predicted range - Doppler data point . If there are no measurements within this window, a "missed hit" is recorded, and the track remains without update until the next radar cycle is completed, and new data is available to test for association to the track.

Initialization

A track file, or "track", comes into existence by virtue of target reports which are not associated with existing tracks. Such reports generate track file data by means of geometrical considerations, given the known detection patterns, fig. 2. Special consideration is given to the initial estimates of what values the track file, especially the initial state vector, should have. Even if the initial state estimate has significant errors, subsequent input data rapidly improves the state vector accuracy, which in fact is whole point of the Kalman filter.

Discriminator

The "discriminator" is merely the simple calculation given by eqn. (1). Error terms exist to account for inaccuracies in the platform speed data and in the target velocity estimate.

Field Test Results

Numerous road tests have been performed using a population of approximately 30 radars and 10 vehicles with good results. Commercial truck drivers, using a right side only system, have found the system performance to be very good in filling their low visibility spots. At the same time, the drivers found the false alarm rate (due exclusively to clutter) to be pleasingly low.

Target Tracking

Field tests specifically showed the tracking Kalman filter approach to be quite effective. No anomalies were observed through out the test period, approximately 1 1/2 years, which could be associated with a Tracking - Estimator process weakness. Even in situations where the tracking - estimator was stressed by very poor *initial* estimates of the state vector, a correct alarm decision was always reached by the required decision time. In those cases where the filter could not rapidly converge to an accurate estimate of target velocity, the estimate error was biased such that a correct alarm decision was made anyway.

Clutter(False) Alarm Rate

Crude estimates of the effectiveness of clutter rejection were made during field testing, rigorous estimates being precluded by the massive amounts of data required to do so. It is estimated that the clutter reduction factor is approximately 10^4, where 1 of every ten thousand clutter objects that came into the radar's field of view generated a false alarm.

References

[1] J. C. Reed, "Side zone automotive radar", *Proc. 1997 National Radar Conference*, Syracuse, NY, May, 1997, pp. 186 - 190.

[2] J. C. Reed, "Side zone automotive radar criteria for target classification", *Proc. Intelligent Vehicles '95 Symposium*, Detroit, MI, Sept. 1995, pp. 361 - 363.

[3] A. D. Whalen, Detection Of Signals In Noise, Academic Press, Orlando, 1971, Sec. 10.7.

[4] A. Gelb, ed., Applied Optimal Estimation, M.I.T. Press, Cambridge, MA, Ch 4.

ADAPTING RADAR AND TRACKING TECHNOLOGY TO AN ON-BOARD AUTOMOTIVE COLLISION WARNING SYSTEM

Edward H. Jocoy, Consultant, Calspan Corporation, Buffalo, New York
James R. Knight, Consultant, Columbia, Maryland

Abstract

Calspan Corporation is currently developing an on-vehicle threat detection system for intersection collision avoidance (ICA) as part of its ICA program with the National Highway Traffic Safety Administration.

This paper briefly reviews the system design and describes recent efforts that include the development of a multitarget tracker/Kalman filter and collision avoidance (C/A) algorithm. A current headway radar was mounted on the Calspan Instrumented Vehicle (CIV) and in-traffic data were recorded. Warning functions, developed through simulation, were applied to the recorded data (non-real-time signal processing) and key Kalman filter/tracker parameters were varied. In addition, real-time performance in traffic is demonstrated.

This research demonstrates the feasibility of preventing many intersection collisions.

Introduction

Intersection collisions constitute approximately twenty-six percent of all accidents in the United States. This translates into 1.7 million crashes annually. Because of their complexity and demands on the perceptual and decision making abilities of the driver, intersections present an increased risk of collisions between automobiles. A program to develop systems to prevent these collisions is being sponsored by the National Highway Traffic Safety Administration Office of Crash Avoidance Research (NHTSA OCAR). The Intersection Collision Avoidance Using IVHS Countermeasures Program is a five year effort that proceeded from an analysis of accident data to quantify the problem, to the fabrication and testing of a candidate countermeasure system. Previous papers and reports have described the use of accident data to determine intersection crash characteristics[1] in 4 crash scenarios. Table 1 summarizes the percentages of accidents associated with each scenario. Previous papers have also described a candidate threat detection system[2] to be applied to intersection crashes.

Table 1.
Intersection Collision Scenarios

Scen-ario	Characteristics	Percen-tage
1	Parallel Path - Left Turn Across Path	23.8
2	Perpendicular Path - Inadequate Gap	30.2
3	Perpendicular Path - Violation of TCD	43.9
4	Premature Intersection Entry	2.1
	Total	100.0

Many variants of the 4 crash scenarios exist. The ICA collision avoidance system would prevent or reduce the severity of intersection crashes. Further discussion of the scenarios can be found in the Task 5 Interim Report[3] for this program, which also details the components of the Intersection Collision Avoidance Countermeasure. This paper will

---

[1] J.A. Pierowicz, "Development of an In-Vehicle Intersection Collision Countermeasure," SPIE Paper No. 2902-34, *SPIE Proceedings, Transportation Sensors and Controls: Collision Avoidance, Traffic Management, and ITS,* Volume 2902, pp. 284-292, Boston, 1996.

[2] E.H. Jocoy, J.A. Pierowicz, "Threat Detection System for Intersection Collision Avoidance," SAE Paper No. 980851, International Congress and Exposition, Detroit, MI, Feb. 1998.

[3] J.A. Pierowicz, E.H. Jocoy, M.M. Lloyd, O. Kelly, A. Bittner, *Intersection Collision Avoidance Using IVHS Countermeasures, Task 5: Design of Testbed System,* Calspan SRL Corporation, Buffalo, 1996.

0-7803-5086-3 /98/$10.00

focus on the continuing development of the tracker/Kalman filter and collision warning algorithm. System performance in Scenario 2 will be addressed. The system consists of a single radar assembly that will monitor vehicle traffic along the approaching lanes of traffic (a 3-radar system is under development). A metric of gap time based on predicted time of arrival at the intersection will be described. The use of the gap time metric to provide a warning to the driver will be illustrated. Both post processing of recorded data and in-traffic, real-time performance will be used to evaluate the system.

Threat Detection System

Figure 1 shows some of the basic components of the ICA threat detection/tracking/collision warning system. As indicated, the intersection collision avoidance system utilizes a Graphical Information System (GIS) and the Global Positioning System (GPS) to complement the on-board sensors. The GIS consists of a digital map database which contains information regarding roadways and traffic control devices (TCD) within a given area.

The GPS/GIS allows the countermeasure to identify the roadway on which it is traveling and the approach of an intersection. The intersection countermeasure utilizes the GIS map datafile to determine the configuration of the approaching intersection and angles of adjoining roadways. This feature allows the threat detection system to scan the intersection's highest risk areas during the target vehicles' approach to or traversal of the intersection.

Signal processing operates on targets detected within threat regions. The "size" of the intersection provided by the GIS will be used in developing a metric called gap time which utilizes the time predicted by the tracker for vehicles to enter and exit the intersection. The tracker uses the target's range, range rate and heading to develop tracks. The collision warning algorithm develops the gap time metric and issues warnings to the subject (radar) vehicle if the gap time is insufficient for the radar vehicle to execute a maneuver such as crossing an intersection. Other information such as the radar vehicle's distance to the intersection, speed and intended turn actions are also used (it takes longer for the radar vehicle to cross the intersection than to make a right turn into the intersection).

While the GPS/GIS system is absolutely required for an operational system, this paper will concentrate on the radar, tracker and warning systems. Intersection dimensions will be assumed or actually measured. Figure 2 illustrates 3 scenarios and the distances to be computed using range, range rate and bearing from a 3-radar system currently implemented on the Calspan Instrumented Vehicle. Three independent trackers/Kalman filters are implemented for the 3 trackers. Only in the collision warning system are the warnings that are developed from each sensor/tracker merged. In this paper samples of data and results of parameter variations from one radar/tracker will be given. A discussion of the tracker follows.

Tracker/Collision Avoidance System

Figure 3 is a block diagram of the tracking process and is generally typical of a multitarget tracker[4]. In this tracker detections give birth to a "candidate" track. Each candidate track can be promoted to a track which is maintained until deleted. Detections from the radar are used to update the tracks at the end of each processing interval. "Association" logic tries to associate the various detections (from multiple targets) with the track representing the appropriate target. The Kalman filter predicts a new state (e.g., position) of each track or candidate track based on previous updates. A "gate" is placed around the predicted positions and logic is used to determine if the detection is within the gate and hence is associated with the track. This process becomes complex when there are multiple closely-spaced tracks, each of which may claim a detection, or when more than one detection falls within a gate, or no

[4] Multitarget Tracking with Radar Applications, Blackman, Samuel, Artech House, 1986.

0-7803-5086-3 /98/$10.00 ©1998 IEEE

detections fall within a gate. Most gates are generalized in the sense they gate in more than one dimension; e.g., position, speed, acceleration, etc. The gate currently used in the ICA tracker is a "maximum likelihood gate".

As implied by Figure 3, the association logic and principal stages in track life for the ICA tracker are:

- If a detection does not fall within the gate of an existing candidate track or track, a new candidate track is formed.
- If a detection does not associate with any candidate track, the candidate track is deleted.
- If a candidate track does associate with a detection, it is promoted to a track.
- If a track does not associate with any detection for some specified number of updates it is deleted.
- If a track is deleted, the predicted times to the intersection can be held for a specified time (typical selections are 0-5 sec). This helps radars with limited angular coverage provide warnings when the target is no longer observed. The feature would not be necessary with a full coverage system.
- A track is deleted if it has a speed that is too negative (e.g., 10 fps) indicating an "opening" target. (Targets with closing ranges generate "positive" speeds in our system.) A track is also deleted if its speed is unreasonably large (for a car) or its (x,y) position unreasonably large with respect to an intersection and its roadways.

Kalman Filter

Figure 4 shows the block diagram of the Kalman filter used in the ICA tracker. The Kalman filter is discussed in many references[5]. The state variables selected for the Kalman filter are position (x,y), speed (S), and acceleration ($\dot{S}$). The measurement vector is range (R), range rate ($\dot{R}$) and bearing (Θ). A typical "North and East" coordinate system is centered at the ICA vehicle. A plan view would indicate North as x, East as y and bearing as the pointing angle of the antenna (see Figure 2). The Kalman filter implemented is an "extended" Kalman filter[5] which accommodates the non-linear measurement matrix relating target state and radar measurements. The inputs to the Kalman filter are radar updates processed by the association logic. The output is an updated estimate of the target state vector. From this state vector of position (distance to center of intersection) speed and acceleration, the predicted times to enter and exit the intersection are computed (recall, the dimensions of the intersection, with respect to its center, are obtained from the GPS/GIS system). All variables are defined in Figure 4. Variables with carets are predictions; a superscript "+" indicates the variable is post update ("a posterior") while a "-" indicates the variable is pre update ("a priori")[5]. The subscript k indicates the iteration or update number. The $\tilde{Z}_k$ variable is the error between the input measurements and the predicted measurements. It is called the residual or innovations vector. The R matrix contains the range, range rate and angle measurement error values and is typically diagonal if there is no correlation between range, range rate and angle (as is assumed in this case). The diagonal elements are the variances of estimated errors in range, range rate and angle. It is not an updated quantity in our application. The Q matrix contains the "plant" noise. The Q matrix is diagonal in our application and the diagonals are the variances of the anticipated state errors of the x position, y position, speed, and acceleration. A 4x4 matrix (since our state vector has 4 components), it is not an updated quantity in our application. Simulation trials of a simulated target indicated that effects of the variances ${\sigma_R}^2$ and ${\sigma_\Theta}^2$ of the R matrix and ${\sigma_x}^2$ and ${\sigma_y}^2$ of the Q matrix were small. However, using larger values of ${\sigma_{\dot{R}}}^2$ error of the R matrix and ${\sigma_S}^2$ and ${\sigma_{\dot{S}}}^2$ of

[5] Kalman Filtering Theory and Practice, Grewal, M.S., Andrews, A.P., Prentice Hall, 1993.

the Q matrix seem to help reduce the splitting of tracks that frequently occurs. For example, in simulation trials (i.e., using simulated traffic) of ICA tracker, varying the $\sigma_{\ddot{S}}^2$ term in the Q (i.e., the Q(4,4) term) matrix with (1) a decelerating target and (2) a constant velocity target resulted in a change in the number of tracks and RMS error between state and true acceleration as shown in Table 2. (Note that Table 2 uses standard deviations (SD) instead of variances.) For a small Q(4,4), the single target decelerating at 5 ft/sec/sec generated 3 tracks with acceleration estimates having rather large RMS values. Since the acceleration is a state variable used to predict the time that the target enters and exits the intersection and subsequent warnings, one can expect errors in the prediction times and the start and stop of the warnings. (See Figures 6 and 7 for examples of predicted times and warnings from in-traffic data). As Q(4,4) is increased to 1.0, the number of tracks decreases as does the RMS acceleration error implying better prediction. For a constant velocity target, little change is seen as Q(4,4) is varied. Clearly, there are compromises to be made in the selection of the Kalman filter input variables. An in-depth evaluation has not yet been completed, but selected values have performed quite well in the many simulation and real-time evaluations.

Table 2.
Varying Kalman Filter SD of Accel. Q(4,4)

Q(4,4) fps/s	Tgt. Accel. fps/s	No. of Tks.	RMS of Accel. Error (fps/s) Trk. 1	Trk. 2	Trk. 3
0.1	-5	3	13.7	12.7	2.8
1.0	-5	1	1.8	--	--
0.1	0	1	1.6	--	--
1.0	0	1	2.3	--	--

Experimental Results

In this section, some sample results from an in-traffic experiment are presented. Both real time audio and video as well as digital data were recorded (the video with audible warnings occurring in real time will be shown at the presentation). The scenario is No. 2 (see Figure 2) with the Calspan Instrumented Vehicle located at the edge of a two-lane highway (one lane each way) as if waiting to cross the highway. The radar is pointed left and in this case is not scanning. Figure 5 shows the slant ranges from the radar to each target over a 40 sec time interval. Initially, there is a cluster of cars approaching the "intersection". (Note: there is not an actual intersection in this experiment but the collision avoidance system computes warnings as if there was.) Two substantial gaps exist during which the ICA vehicle could cross the "intersection". In the 35 to 40 sec time period a truck passes preceded and followed by a gap. Finally a single car passes. The truck and single car were isolated targets. It is clear from their tracks that the tracker generated more than one track for each target. This phenomenon occurs as the target nears the radar and "fills the beam". Apparently the radar isolates more than one reflection point. Simulation indicates this to be the case, although the phenomenon is being studied further. A merging of tracks may be in order. However, the track splitting at near range does not adversely affect the warnings which are derived from predicted times (see Figure 6). The tracker processes the range, range rate and bearing angle obtained from the radar to estimate the "state" of the target with 4 state variables of position (x,y) speed and acceleration. The state is appropriately integrated in the direction of the roadway to yield the time predicted for the target to enter and exit the intersection. Figure 6 shows these predicted times as a function of real time along with a defining sketch. Although the plot is congested, note that each track produces two essentially parallel sets of common symbols: one line indicating when the target will enter the intersection, the other indicating when the target will exit the intersection. The symbols are consistent with those of Figure 5. The gaps when it would be safe for the ICA vehicle to cross are clearly apparent. Also shown in Figure 6 are the times when the ICA vehicle would occupy the intersection should it choose to cross. This time is based on an assumed acceleration from stand still. Of course if the ICA vehicle began to enter the intersection the time to cross would continually decrease. When a target and the ICA vehicle are predicted to occupy the intersection simultaneously, warnings occur. Warnings developed from the predicted times in Figure 6 are shown in Figure 7. A composite warning for all targets is shown at the bottom of the figure. The gaps when it would be safe to cross are indicated.

Summary

As part of its Intersection Collision Avoidance (ICA) program for the National Highway Traffic Safety Administration Office of Crash Avoidance Research, Calspan has implemented and is testing an on-board collision warning system for intersection traffic scenarios. To accomplish this, Calspan has adapted radar and tracking technology as well as GPS/GIS and computer technology to an automotive application. This paper has reviewed in some detail the tracker/Kalman filter used in the system and presented some results of post-processed data collected with a single radar on the Calspan Instrumented Vehicle (CIV). The CIV is currently being instrumented with 3 radars and in-traffic experiments are continuing.

Acknowledgments

This work is being performed for the Office of Crash Avoidance Research, National Highway Traffic Safety Administration. The authors would like to thank Mr. Art Carter and Dr. August Burgett of OCAR for their support and direction. The authors would also like to thank Messrs. Chris Nowak and Bert Pirson for their support in the acquisition and preparation of the in-traffic radar and video data and Ms. Molly Johnston for processing and preparing much of the data presented herein.

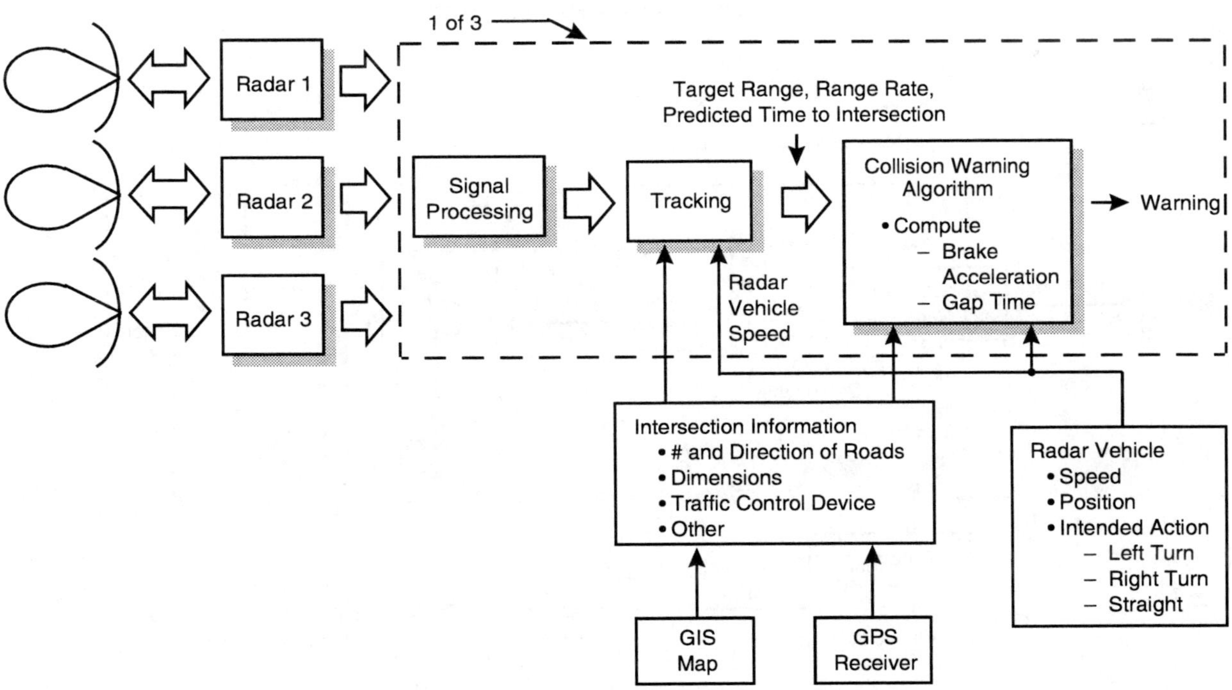

Figure 1. ICA Threat Detection System Showing Detailed Use of Intersection Information

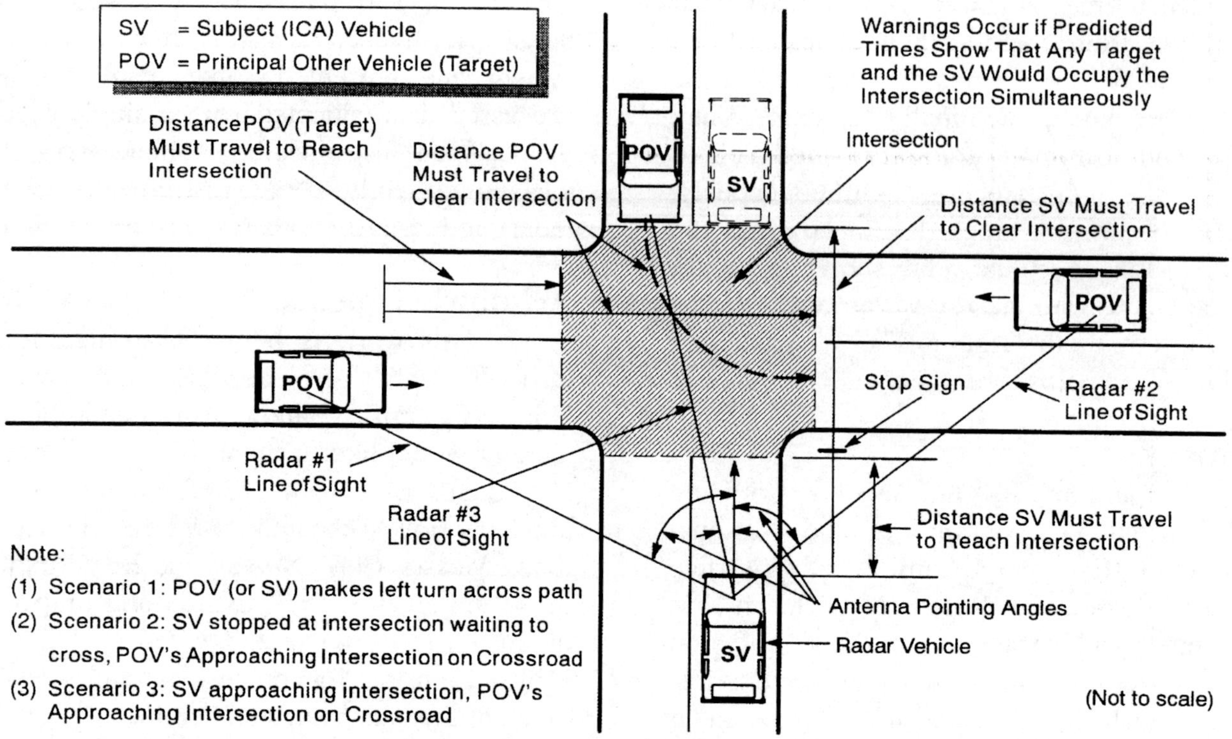

Figure 2. Scenarios 1, 2, and 3 Showing Distances Used to Compute Predicted Times and Warnings

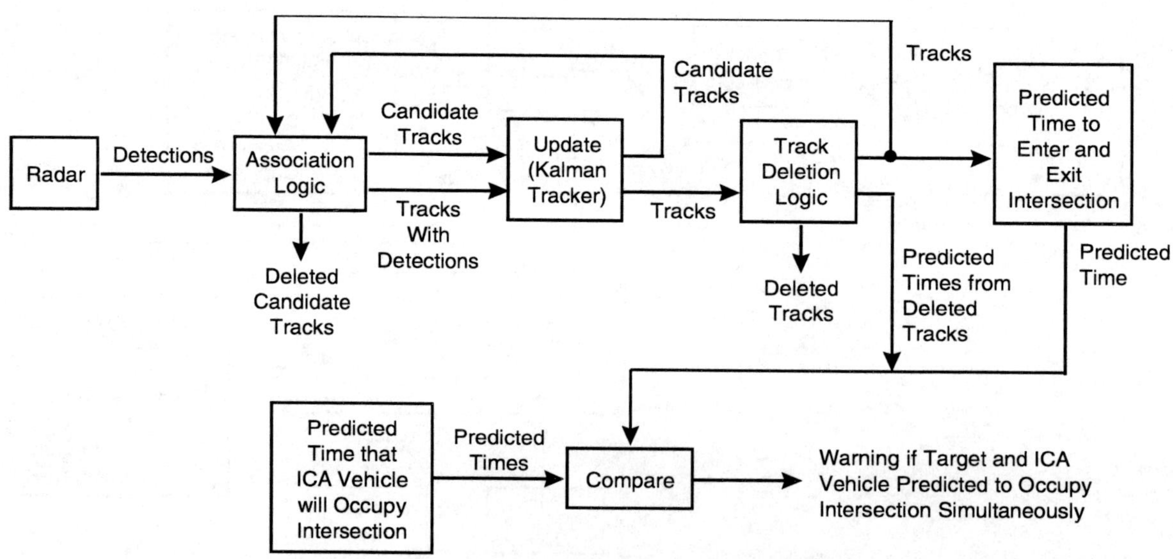

Figure 3. Collision Avoidance System Block Diagram Showing Tracker Logic

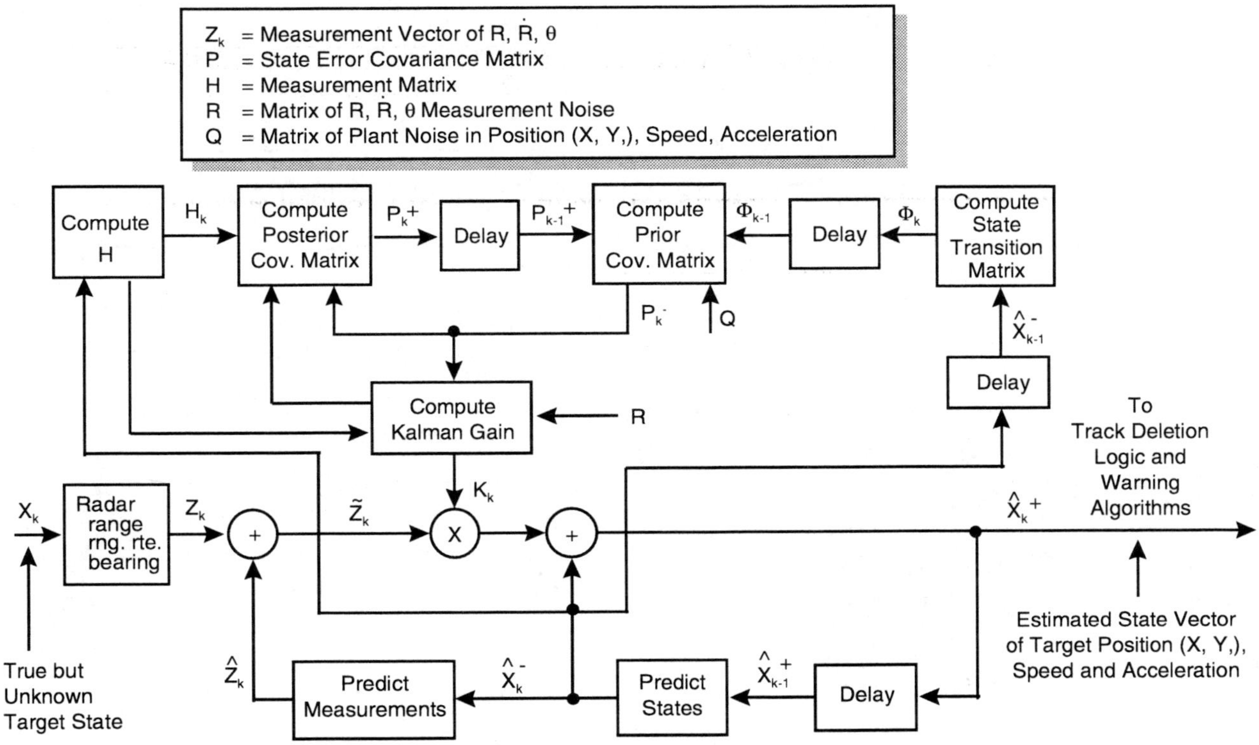

Figure 4. Kalman Filter for ICA Tracker

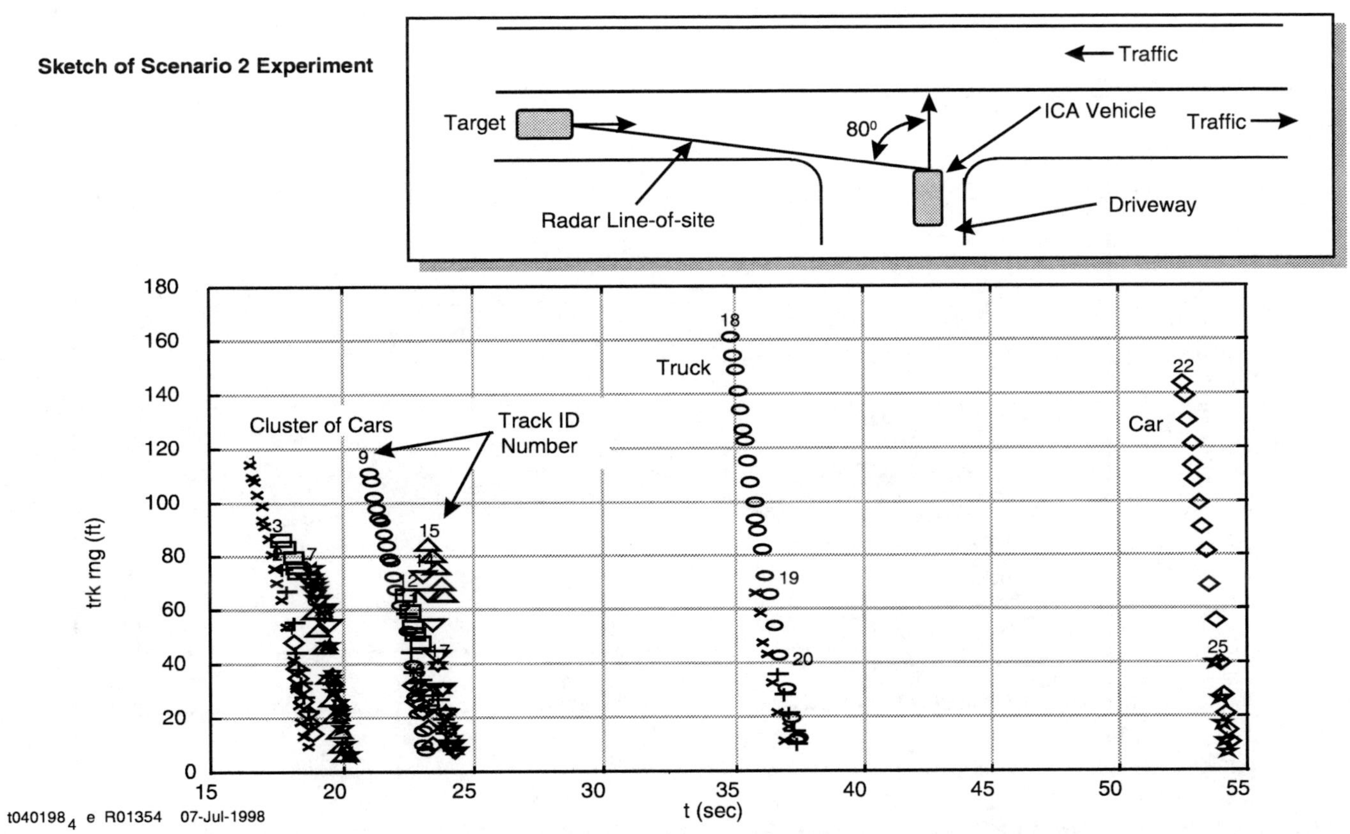

Figure 5. Track Derived Range for Scenario 2 Experiment

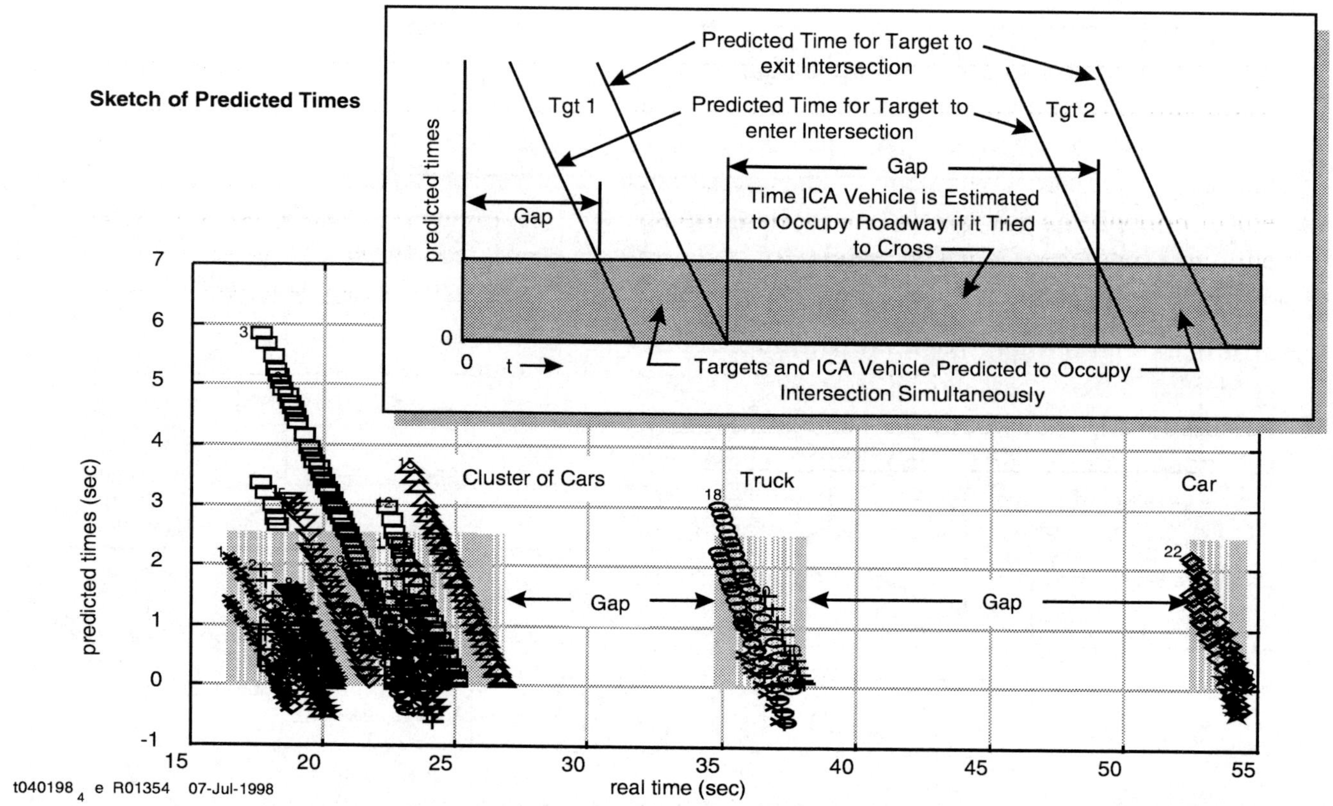

Figure 6. Predicted Time to Enter and Exit Intersection, Scenario 2 Experiment

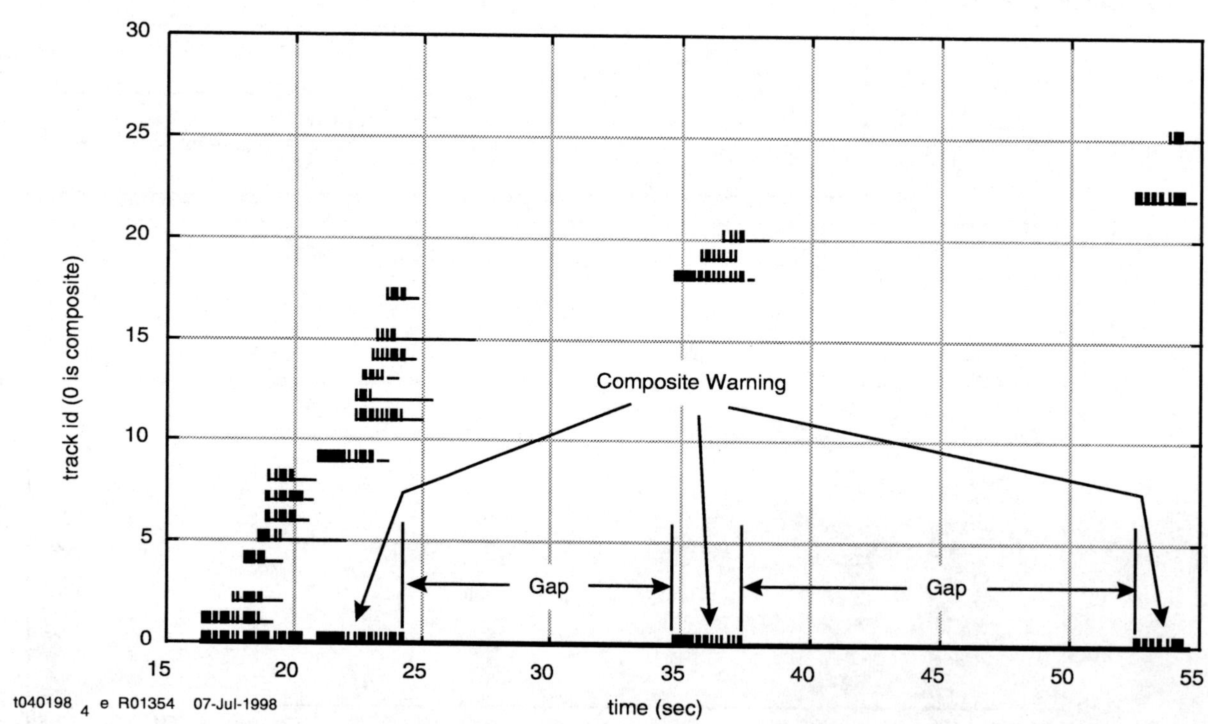

Figure 7. Binary (0-1) Warning Function for Target (Track ID) and Composite for All Targets

USE OF MAP DATA INFORMATION IN AN ON-BOARD INTERSECTION VIOLATION DETECTION SYSTEM

John A. Pierowicz, Principal Research Engineer, Calspan Corporation, Buffalo, NY

Abstract

Collisions within intersections constitute approximately twenty-six percent of all Police-reported crashes each year. Of these crashes, 47 percent occur at intersections controlled by stop signs. About 75% of intersection crashes may be attributed to driver errors, such as failure to observe the traffic control or driver inattention. Calspan is developing an on-board countermeasure system to prevent collisions at intersections. This effort is part of the Intersection Collision Avoidance Using IVHS Countermeasures program being sponsored by the National Highway Traffic Safety Administrations' Office of Crash Avoidance Research (NHTSA OCAR). A feature of this countermeasure is the detection of the intersection the vehicle is approaching, and the determination of the traffic control device at that intersection. This information is derived from an on-board geographical information system (GIS) in conjunction with vehicle position data acquired from global positioning system (GPS).

The use of this data, combined with other vehicle state information, can be used to determine the potential for driver violation of the intersection. This paper will describe the use of information from the GIS along with other vehicle parameters to provide a warning of intersection violation to the driver. Examples of the application of this technique will be provided.

Introduction

Intersection collisions account for approximately twenty-six percent of all police-reported collisions reported annually. Efforts to understand the intersection collision problem focus on the events and circumstances surrounding the crash in an attempt to identify opportunities for crash intervention. Through an in-depth clinical analysis of an intersection collision data sample, patterns in the crash sequence were identified[1]. This lead to the classification of four distinctively unique intersection collision scenarios. The distribution of the intersection crash population is shown in Table 1.

Table 1.
Intersection Collision Scenarios

Scen-ario	Characteristics	Percent
1	Parallel Path - Left Turn Across Path	23.8
2	Perpendicular Path - Inadequate Gap	30.2
3	Perpendicular Path - Violation of TCD	43.9
4	Premature Intersection Entry	2.1
	Total	100.0

Causal analysis of the data set concluded that nearly 75.0% of intersection collisions were due to "driver error", including *Driver Inattention* (28.7%), *Faulty Perception* (33.9%), and *Vision Impaired/Obstructed* (11.1%). This information directed the development of an Intersection Collision Avoidance (ICA) countermeasure that will mitigate causal factors by warning drivers of potential errors and pre-empting driver control of the vehicle to avoid a collision.

Of the collision scenarios, Scenarios 1 and 2 require the use of a system to detect

[1] J.A. Pierowicz, E. Bollman, L. Parada, *"Intersection Collision Avoidance Using IVHS Countermeasures, Task 1 Report: Analysis of Accident Data"*, Calspan Corporation, Buffalo, 1994.

0-7803-5086-3 /98/$10.00 ©1998 IEEE

approaching vehicles, and to provide distance and speed information to the driver. The countermeasure used in this program utilizes a commercially available headway radar system to provide data to a Calspan-designed tracker. Scenario 3 is a special case that involves violation of the traffic control by the countermeasure-equipped vehicle. It is these crashes for which this system is designed. For example, information about the traffic control device (TCD) at the intersection (available from an on-board geographic information system (GIS) map) and the distance of the vehicle to the intersection (available from the map and an on-board global positioning system (DGPS) system) can be used to prevent the vehicle from violating the TCD without intervention by the on-board radar. Many variants of the scenarios exist, including violations of the TCD by the principal other (target) vehicle (POV). Further discussion of the scenarios can be found in the Task 5 Interim Report[2] for this program, which also details the components of the Intersection Collision Avoidance Countermeasure.

Intersection Violation Detection

The detection of potential intersection violations may be accomplished through the use of on-board vehicle equipment. Primary in this equipment is Global Positioning System (GPS) and electronic maps. The GPS system has been gaining in utility for automotive navigation for years. The advent of low-cost GPS receivers, and the availability of map datafiles has allowed this system to be used for route guidance, MAYDAY, and fleet tracking purposes. Expanding these tools to collision avoidance applications requires little modification.

The key to this application is the ability to determine vehicle position with respect to the intersection. Armed wit h this information, it is possible to calculate the potential for a vehicle to stop prior to the intersection. Figure 1 illustrates the algorithm for intersection violation detection utilized in this application. The vehicle state functions of position, velocity, and heading are provided by the GPS functions. The use of differentially-corrected GPS increases the accuracy of the position determination, and allows for a reduction in the number of false alarms generated by the detection system.

Information regarding the type of traffic control device at the intersection the vehicle is approaching is required for the collision avoidance function. This information is provided by the GIS system. This data is not available in current map datafiles. Consultation with map vendors indicate that this data is available, but not included in present files due to lack of applications. Calspan constructed a separate datafile with traffic control data within a test area in suburban Buffalo for testing purposes. This datafile relates a traffic control with specific roadway segments and vehicle heading information.

The requirement to stop at an intersection is dependent upon the traffic control at that intersection. If the intersection is controlled by a stop sign, the requirement is always to stop prior to entering the intersection. If the intersection is controlled by a phased signal, the requirement to stop is dependent upon signal phasing. The current application of the countermeasure is limited to intersections controlled by stop signs. This capability may be expanded intersections controlled by phased signals in the future.

The detection of a potential intersection violation is facilitated by the monitoring of a metric termed a_p. This metric is the braking effort required to prevent vehicle entry into the intersection. This metric is a function of driver time delays, vehicle distance to intersection and vehicle speed. The magnitude of this metric is compared to normative data on driver deceleration at intersections. This data was collected previously in the Intersection Collision Avoidance Program. The exceedance of this threshold is utilized to trigger warnings to the driver.

[2] J.A. Pierowicz, E.H. Jocoy, M.M. Lloyd, O. Kelly, A. Bittner, *"Intersection Collision Avoidance Using IVHS Countermeasures, Task 5: Design of Testbed System"*, Calspan SRL Corporation Buffalo, 1996.

0-7803-5086-3 /98/$10.00 ©1998 IEEE

Examples of a_p for a normal intersection approach, and one where a violation of the intersection occurred is shown in figure 2 and 3. As may be observed by comparing these figures, during the approach resulting in a violation the magnitude of a_p increases significantly until the warning threshold is exceeded. The exceedance of this warning threshold may be used to activate a warning to the driver, or to activate a vehicle control system to initiate braking prior to driver intervention. Testing of the a_p with driver data from test series has indicated that this metric can detect potential intersection violations, with a low false alarm rate.

Use of Map Data

The implementation of the a_p metric in the countermeasure requires that vehicle position be known, and that this position data be related to upcoming intersections. An on-board map datafile is used to provide this information. The application utilizes position data derived from the GPS, updated at 10Hz, to locate the vehicle on a specific roadway segment. The map datafile decomposes roadways down to individual segments consisting of nodes and line segments. These nodes and segments are assigned various properties. Some properties of these segments are position, length, ID number, and adjoining roadway segments. This property allows intersections to be differentiated from other roadways. To this set of properties, Calspan has added a field for traffic control device at the intersection node.

With the vehicle located on a specific roadway segment, vehicle heading data is used to determine the node it is approaching. An association algorithm within the software determines if the node is an intersection. If the node is an intersection, other data such as the adjoining roadway segments and traffic control for each roadway is determined. Other map features are used within the Intersection Countermeasure. For example, the incident angle to each adjoining roadway segment is calculated and used to dynamically point the vehicles' radars to detect other vehicles on collision paths.

Driver Information

Figure 4 illustrates the information that is derived from the map datafile. Contained within this screen is vehicle position, speed, vehicle heading, current road vehicle is traveling on, next intersection (with roadway names), and distance to intersection. A graphical depiction of the traffic control at the next intersection is provided at different times in the approach sequence, depending upon trigger mode selected. As may be observed the amount of data present on this screen is too much for the driver to use in an intersection environment. The selection of what information to present to the driver will be finalized after testing of the system with a set of subjects.

Currently, the data may be provided to the driver in two modes, an advisory mode, or a warning mode. The advisory mode is illustrated in Figure 4. The warning of traffic control is presented to the driver 300 ft prior to intersection entry. This distance allows the driver to react to the intersection without using extreme control actions, such as severe braking. The advisory mode may be useful to older drivers, or drivers who are unfamiliar with the area in which they are driving.

The warning mode of operation utilizes the a_p metric previously described to send traffic control information to the driver. In this mode of operation, the system would only provide traffic control information to the driver when the system determines that there is a potential for intersection violation. At other times, with the driver stopping prior to intersection entry, the system would not provide any traffic control information. This mode of system operation is the baseline mode of operation for the current Intersection Collision Avoidance System.

System Testing Plans

As part of the Intersection Collision Avoidance Using ITS Countermeasures Program, Calspan will test the efficiency of this

0-7803-5086-3 /98/$10.00 ©1998 IEEE

system to prevent driver violation of intersections. As a surrogate measure of system performance Calspan will conduct a series of driver experiments with and without the system. A subset of drivers will be requested to drive a route within the test area. The manner in which the driver controls the vehicle, such as braking distance to intersection, and level of braking, will be recorded by an on-board data acquisition system. A second set of drivers will be requested to duplicate the route tests, with the warnings system activated. Changes in driver behavior, such as earlier barking, lower levels of braking, will be used to determine the efficiency of the system. In addition, comments from the drivers with regard to acceptability of the system will be recorded and used in the evaluation.

Summary

Calspan, as part of the Intersection Collision Avoidance Using ITS Countermeasures Program has developed an intersection violation warning system that utilizes GPS and data from an on-board map datafile to determine the potential for intersection violation by the equipped vehicle. Calspan will perform testing to determine system performance and driver acceptability.

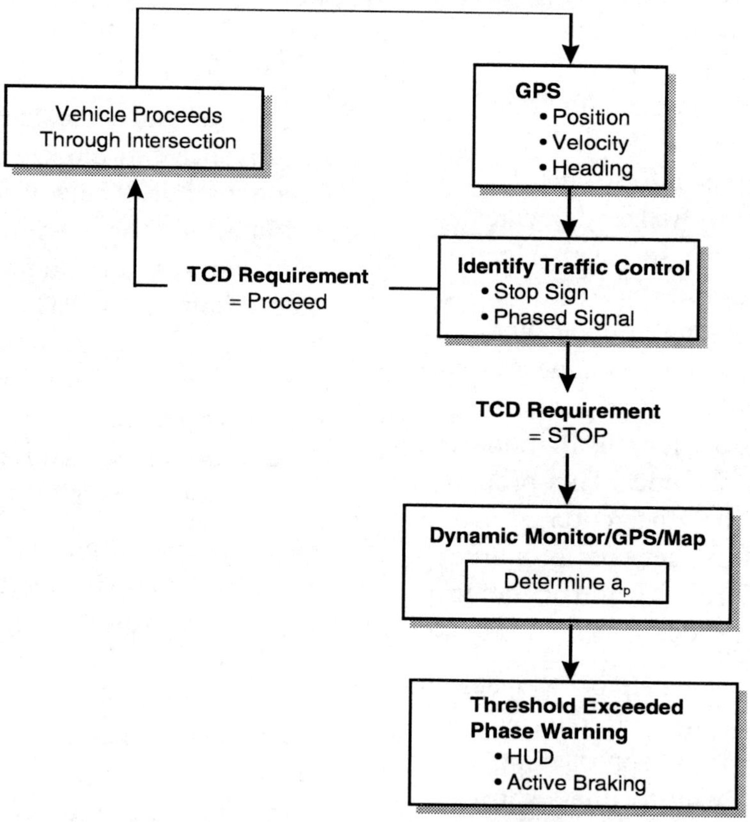

Figure 1. Intersection Violation Detection Algorithm

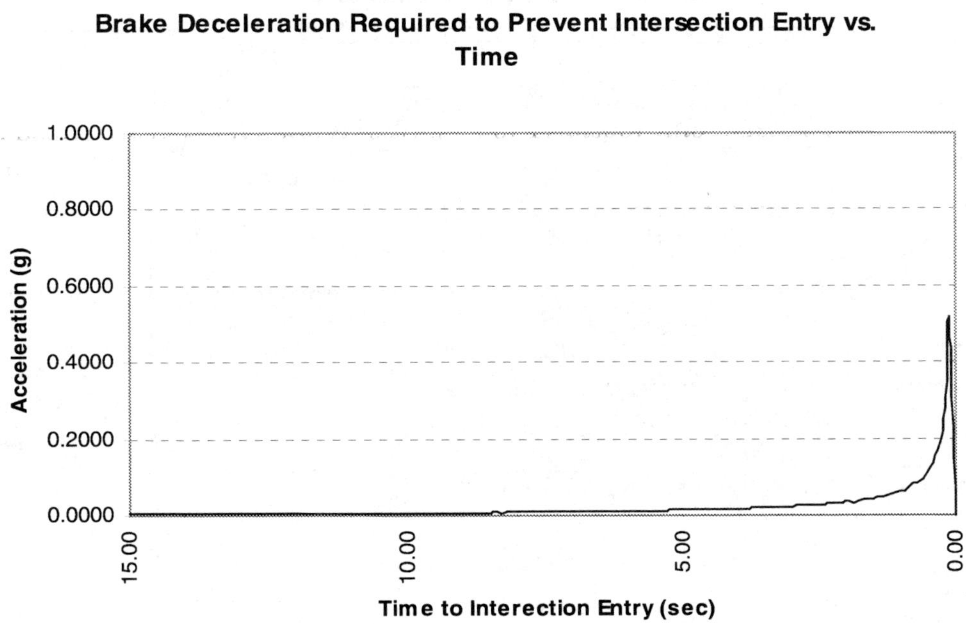

Figure 2. Illustration of a_p Metric for Typical Intersection Approach

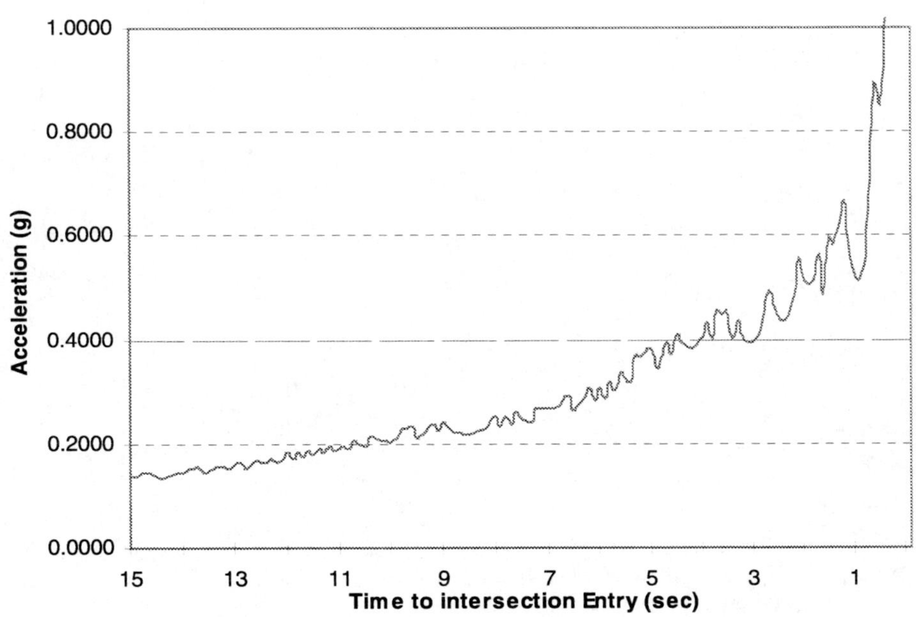

Figure 3. Illustration of a_p Metric for Intersection Violation

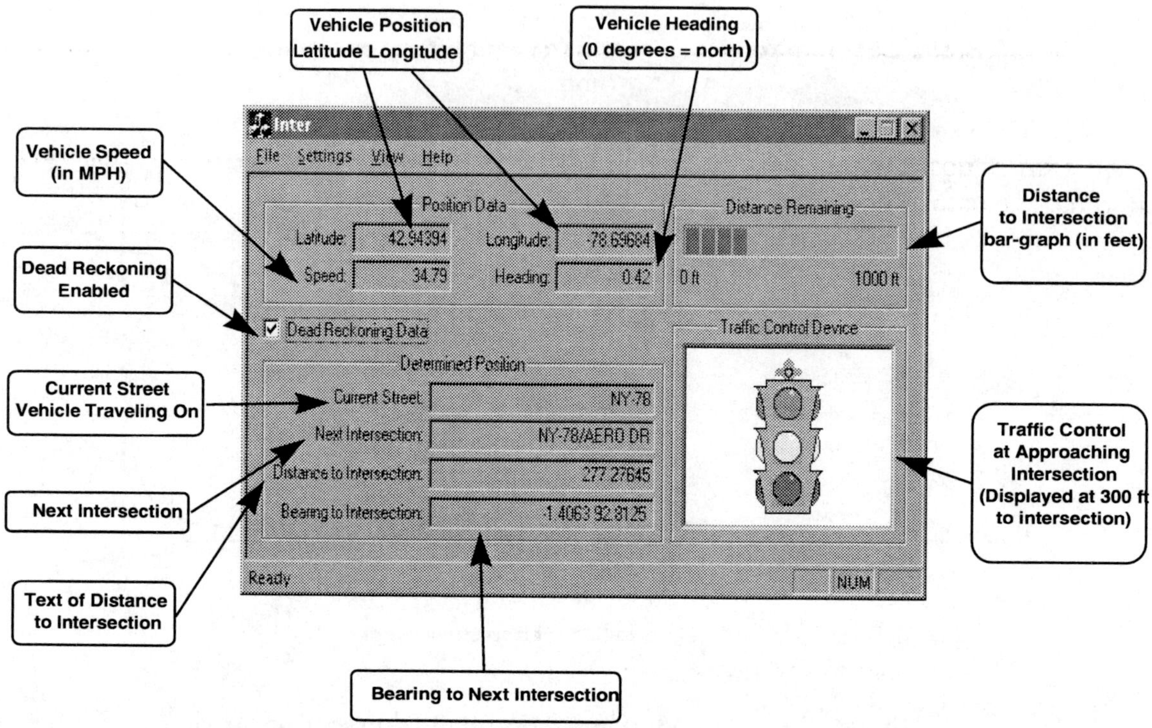

Figure 4. Illustration of Data Screen

SWIFT ATIS
CONSUMER REACTIONS AND MARKETABILITY

Larry L. Senn, Washington State Department of Transportation, Seattle, Washington

Introduction

The Seattle Wide-area Information For Travelers (SWIFT) Project successfully delivered traffic and transit information to 670 test participants in 1997 and 1998. This made SWIFT one of the largest tests of Advanced Traveler Information Systems (ATIS) in the United States to date. The usefulness of traveler information and what potential users thought about the delivery systems and devices provide valuable lessons for commercial deployment and insight into the needs and desires of the marketplace.

The systems developed to obtain and distribute the traveler information remain in place and are being used for services provided by Seiko Communications and others through the Seattle SmartTrek Intelligent Transportation Systems (ITS) deployment project.

An in depth evaluation of the SWIFT project was conducted by SAIC that included components of market studies including consumer acceptance, deployment costs, and details of the system architecture.

The findings in this paper came from those studies and my personal experience as the SWIFT project manager for the Washington State Department of Transportation. These studies are not in final form as of this writing, so no specific references are included in this paper.

Some of the most significant findings of SWIFT are:

- there is a market for traveler information,
- most users reported using the information, and
- a subscription service would likely be profitable.

System Components

The SWIFT system collected data from existing sources and used internet links to supply information derived from the sources to Seiko Communications, an FM subcarrier provider that operates a paging network in Seattle and other cities. The FM link then broadcast the information to users though three types of devices: a pager watch, an in-dash navigation unit, and a laptop computer. Each device offered unique capabilities.

Data Sources and Information

The Seattle area is fortunate to have a state of the art freeway control system which utilizes traffic loop sensors to collect traffic information for the purpose of operating a ramp metering system. This data, however, is stored in an esoteric manner not uncommon to custom built systems. Extracting the data and converting it into ready to transmit freeway speeds was done by researchers at the University of Washington.

King County Metro Transit (Metro), the local transit provider, also had a vehicle location system that tracked all of the coaches running regular routes. The system was intended to provide an additional level of security and to monitor real time performance of the transit system. Locations are expressed as a distance along the route, which reflects the nature of the "signpost" location system, as opposed to GPS which would locate the coach with respect to a grid. Again the University of Washington team took the raw data, converted locations into latitude and longitude, which were packaged for

0-7803-5086-3 /98/$10.00 ©1998 IEEE

transmission along with the route number, run number, and direction.

SWIFT used a variation of the International Traveler Information Interchange Standard (ITIS) message formats. These were originally developed for the lower speed RDS subcarrier in Europe. The systems developed by the University of Washington are still in place and can provide outputs to other ATIS systems as they are deployed. At present Washington State Department of Transportation is carrying the costs and the information is available without charge to a commercial provider of traveler information.

Traffic incident and other anecdotal information is regularly collected by Metro Networks, a commercial provider of traffic information to radio stations in most metropolitan areas in the U. S. A traffic workstation, developed by Etak Inc., was used to input the incidents and formatted it for transmission, along with user specific messages for test participants that had pager watches.

The FM Subcarrier

Messages were forwarded to Seiko Communication for transmission over their High Speed Data System (HSDS) FM subcarrier network. This network was already established to provide paging services through Seiko's MessageWatch, a commercially deployed wrist watch pager. The subcarrier sends data over a sideband to the regular FM radio entertainment channels and has a raw data rate of 19,000 bits per second.

For their pagers, Seiko uses two techniques to overcome potential message loss due to multipath or low signal strength. Each market area uses multiple stations to increase coverage, and each message is sent three times from each station over a six minute interval, consequently multiple messages in a short time period would queue and cause delay in message receipt. The receiver in the MessageWatch is frequency agile and searches for and locks onto the strongest signal. Each watch has a unique address and assigned time slots, and the receiver is turned off when not receiving to conserve the battery. The watch is synchronized to the time based multiplexed signal through time updates, which means every watch displays GPS time. Limitations in the watch receiver required that traffic messages be sent as individual pages in SWIFT.

Other SWIFT devices received "asynchronous", or broadcast information. These messages in the ITIS format were broadcast in the manner that regular radio works, allowing a single message to be received by all devices. This minimizes bandwidth and makes a system that could be scaled up to support an infinite number of users. Although ITIS messages were broadcast from all HSDS stations, each message was sent only once from each station, but the entire set of traveler information was repeated on a regular interval so that a recently activated receiver could be updated in less than a minute.

Only seven percent of the SWIFT test participants reported problems with reception. An occasional missed message did not appear to reduce the usefulness of the system, but users that experienced chronic reception problems at home or at work found the system to be far less valuable.

User Reactions

Test participants used a SWIFT device for about a year. During this time information was collected through a series of 4 mail back surveys and 13 focus groups. Throughout the test it became obvious that most of the participants thought of the SWIFT system and devices as something of a beta test rather than a finished product. Even so, over half of the participants said they were willing to pay $5 to $10 a month for a SWIFT like service ... if many of the improvements listed in this paper were made.

Usefulness of Traveler Information

One of the primary goals of SWIFT was to determine the usefulness of the information to users. Nearly 80% of the test group reported using a device at least once a week, and 70% indicated that they modified their commute route once a week as the result of traveler information. Participants seemed to like the added source of information, particularly with respect to traffic problems, even though SWIFT information was ranked a close second to the radio for obtaining traveler information.

Users suggested providing information for more arterials. Seattle has only one freeway, so this information is important when considering an alternate route. Participants also reported, on the average, to be only "somewhat satisfied" with the timeliness and accuracy of messages. SWIFT messages were generally broadcast within a minute of being entered, but it also takes time to identify and classify incidents before the message goes out. Improving the accuracy and timeliness of incident and congestion information on a vast network of freeways and arterials remains a challenge to more successful ATIS deployments.

One user tested the veracity of traffic incident messages by deliberately driving into the area, only to get stuck in a major traffic jam two days later. Another participant reported trying to avoid congestion reported on the freeway by taking a “collector distributor” only to find that the incident was on their alternate route.

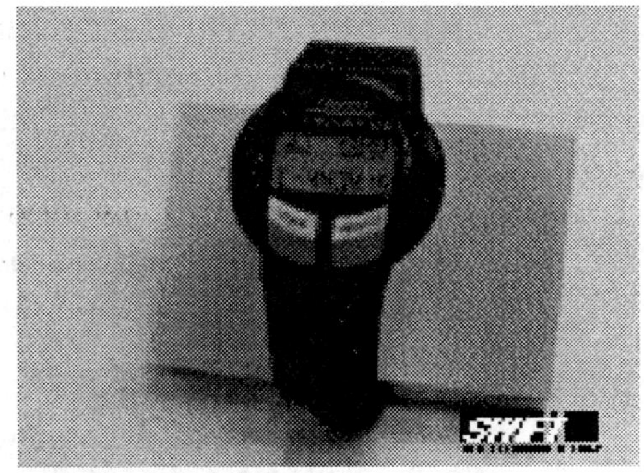

Convenience, Seiko MessageWatch

Easily the most popular device fielded by the SWIFT team was the Seiko MessageWatch. The somewhat bulky wrist watch pager encountered far fewer problems than the other SWIFT devices, partly due to it already being commercially deployed.

It was the most graphic example of people wanting convenience when it comes to information. The innocuous pager watch was easily carried by almost every one of the 470 participants. It was even more endearing that the pager watch provided sports scores, lotto numbers, and financial market information to amaze friends with. It also automatically reset to daylight savings time.

Even though traffic messages were a bit cryptic, due to a limited display that could not show every letter in the alphabet, users quickly learned how to read them. Users submitted a travel profile to describe their commute, and the profile was used to screen out unwanted road segments and times.

Delco Navigation Unit

Delco Electronics Inc. produced an in dash navigation device with an embedded HSDS receiver. The navigation unit used GPS and a location database to give the driver direction and distance to the selected destination. Traffic incidents were received in ITIS format, converted to phonetic text, run through a text to voice generator, and announced through the car radio speakers.

Problems encountered with the Delco units included one glitch that caused the HSDS receiver to drop off line, and another that caused incorrect reporting of incident directions. Both problems required software updates. Problems persisted with the paging function were moot because the car was not where people wanted to get pages. These things undoubtedly led participants to view the navigation device as less than a finished product. Most spoke well of the navigation unit, but when their units were removed many said they would miss the CD player radio, which it was paired with, more than the unit itself.

Although users rated the importance of voice announcements lower than the navigation features, it is an approach to keeping the drivers attention on the road. Users agreed that the voice was both safe and did not distract from driving. However, accessing a destination from the database, which required numerous button actuations as well as visual attention, was a human factors concern. Such attention demanding functions may need to be locked out when the vehicle is in motion.

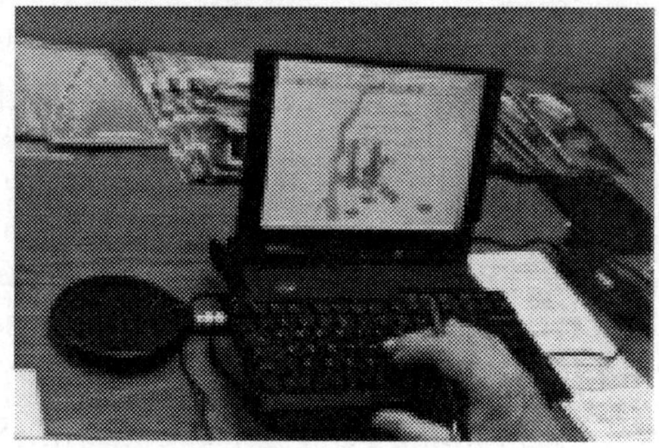

Laptop Computer, Map Graphics

The third device was a standard IBM compatible color laptop computer connected to an HSDS wireless modem. The laptops were the only device to display freeway speeds and bus locations.

All the traveler information was presented on a map based display. Speeds were represented through colored dots along the freeway , where green indicated speeds over 45 mph, yellow 30 to 45 and red under 30. A separate set of dots represented the HOV lanes. Incidents were displayed as icons on the map, and clicking on an icon would display a window with details about the problem. Busses were shown as icons, complete with route number, and clicking on them gave details about that trip.

The intent was to make a totally portable traveler information package, but limited battery resources in the PC and its weight and bulk made it difficult to lug around. Most of the participants in the laptop group were selected because they rode transit at times. The weight and difficulty of setting everything up on a bus proved too difficult for users. As a result, most used SWIFT laptops at one, or both, ends of their commute prior to departing. One participant mounted a laptop on her dashboard until an officer suggested she remove it.

It took about a minute to set up the hardware, boot the software, and get information. This made the information less than immediately available. Some of the PCs had an "instant on" feature that greatly shortened the process.

The laptop also gave the most information about the quality of reception through a “signal meter”. Curiously, this “signal meter” was the only use of differential GPS in SWIFT, as the “meter” represented a percentage of the number of differential signals in a fixed time period. Although the improved feedback may have contributed to the opinion that the laptop had the worst reception compared to other SWIFT devices; the opinions may also reflect reception test results which indicated the PC receiver’s higher gain made it more susceptible to multipath interference. Users also occasionally encountered problems in establishing communications between the laptop and the HSDS modem.

Test participants indicated a low confidence that all busses were being reported. This appeared to be supported by the transit “veracity” tests that showed up to 30% of the busses were not reported by SWIFT. Metro has already taken steps to improve the completeness of ATIS transit information.

Profit - Marketability

SWIFT's Deployment Cost study examined the life cycle costs and potential revenues associated with the system. The study estimated that it would take 4 1/2 years to recover the $7.9 million it would cost to deploy and operate SWIFT in a market. After cost recovery profits of $2.3 million per year could be shared between the partners providing the service.

Summary

There are many lessons to be learned from the SWIFT test of ATIS systems. The most basic is that potential customers don’t want to fuss a lot to get the information, and that they want it on demand. This was supported by the popularity of the MessageWatch compared to other devices. Use of the message watch only involved reviewing the messages and time they were sent and was also useful in providing information both before and during travel.

Another lesson is that the device must be easily used in the intended environment. This was supported by the inability of the laptop devices to conveniently and promptly provide enroute information to bus and auto users.

SWIFT showed that there is a desire for improved traveler information and a potential customer base. The biggest hurdle for traffic information appears to be improving the quality and the number of routes covered. SWIFT participants universally asked for more roads to be covered and for better reporting regarding the impact of an incident on congestion. Even the MessageWatch users reported they ignored the lowest of three levels of congestion indicated in traffic messages.

Collection of traffic performance information has traditionally been problematic. The objective is to give the commuter the information they need in the quickest and easiest manner. The trick is to know what is important to the traveler. Automated systems eventually coming on-line may serve to bridge this information gap, but for the current services to fill the void it will take a major increase in the effort currently employed and improvements in systems that handle the information.

SWIFT showed that traveler information is a viable business. National standards need to be developed for message formats and signal compatibility for widespread deployment to occur. A broadcast system, as employed in SWIFT for the navigation units and PCs, could potentially serve a vast number of customers using minimal transmission resources.

AN IN-VEHICLE INFORMATION SYSTEM FOR ITS INFORMATION MANAGEMENT

Philip F. Spelt, Oak Ridge National Laboratory, Oak Ridge, Tennessee

Daniel R. Tufano, Sr., Oak Ridge National Laboratory, Oak Ridge, Tennessee

Introduction

This paper presents an In-Vehicle Information System (IVIS) research and development (R & D) facility to investigate management of messages from Intelligent Transportation System (ITS) Services. These services include navigation and route guidance (N & RG) systems, real-time traffic management, and non-ITS devices such as cell phones and pagers, as well as other information to be presented to the driver. The introduction into road vehicles of multiple information sources related to the ITS raises issues concerned with driver distraction and cognitive overload, as well as problems which stem from dealing with multiple device interfaces. The IVIS serves as the interface between the driver and all the information sources, including both input from and information display to the driver. However, introduction of an IVIS raises a number of issues which relate to things such as proprietary messages, message prioritization across devices from different manufacturers, and safe access to the vehicle manufacturer's proprietary data bus.

The IVIS Platform is designed to support three important activities: A) identify emerging research issues to addressed by the introduction of additional ITS-related devices, as well as items such as cell phones, into vehicles; B) educate the driving public and the transportation community about the benefits of using integrated, human factors-guided IVISs, and obtaining feedback from users about such a system; C) provide a research system for the iterative and cumulative prototyping of IVIS system candidates, as well as testing devices themselves in an integrated setting.

One goal of the IVIS project has been to develop a fully-integrated IVIS which provides a single complete interface between the driver and the various aftermarket subsystems installed in the vehicle. An IVIS also provides information management to mitigate against the potential sensory and information overload from multiple separate systems. Using modern digital technology, IVIS tailors information both to the driver's needs and to the driving environment, while conveying the information in a manner which enhances its quality, utility, and meaning. Furthermore, it must perform this function in a way which greatly assists rather than interferes with the already difficult and demanding task of driving. An IVIS dynamically 1) **filters** out all information except what the driver needs or wants; 2) **fuses** or merges related information from different sources; 3) **displays** information in a clear manner that adapts to ambient conditions, operational conditions, and driver preferences; and 4) **synchronizes** the display of information to give the driver and vehicle enough time and distance to respond to it, keeping the information displayed only for as long as it applies. In addition, this system must account for driver characteristics (such as hearing deficiencies in presenting acoustic signals and messages), driver goals (e.g., never show tourist attractions), and trip goals (destination, route preferences, etc.). A laptop PC-based demonstration of the benefits of these information management techniques will be shown at the DASC conference.

0-7803-5086-3 /98/$10.00 ©1998 IEEE

IVIS Development Platform

Oak Ridge National Laboratory (ORNL) has created an In-Vehicle Information Development Center which contains a number of interrelated facilities. An important aspect of the Center is the IVIS emulation system, which serves as a development platform on which both the logic of information management and the techniques for integrating actual subsystems can be developed and evaluated. The IVIS Development Platform, shown in the figure below, comprises a number of PCs (six at present) on an ethernet-based LAN, as illustrated in Figure 1. In this system, various PCs emulate components of a possible IVIS. The IVIS PC provides for the development and testing of the rule base which performs the information management and display. The scenario generator creates the simulated trip for the network. The trip is a sequence of events and associated timing either from an actual trip, or from a fictitious trip created to produce the types of situations needed to test the functioning of the logic and display routines of the IVIS.

The other PCs in the LAN emulate the functions of various ITS-related devices which might be installed in a vehicle. In its simplest operation, the development platform is driven by output from the scenario generator, which provides the event signals for a trip. These signals are received by the other PCs in the LAN, which then emulate the functions of the device(s) they represent by sending appropriate messages onto the LAN. Those messages destined for the driver are collected by the IVIS PC and processed for display, using any of a variety of information management functions, some developed in this facility, others developed by other researchers.

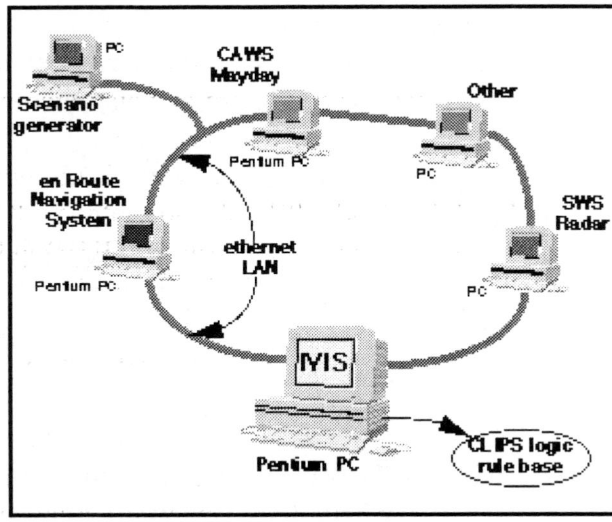

Figure 1. The IVIS LAN

For example, as the simulated trip proceeds, the scenario generator might broadcast on the LAN that the vehicle is approaching Main and Elm streets. The N&RG subsystem would then send a message to "turn left at Main & Elm", while the Safety Warning System (SWS) might be programmed to announce that a fire truck is approaching along Main from the left. These messages, along with others produced by any of the other emulated devices, are then processed for display by the IVIS PC. Using this system, both the logic of information management and the physical display of messages through the visual and/or auditory channel can be investigated and evaluated prior to installing and testing such an information management system in a more realistic setting, such as a driving simulator or a real vehicle.

Specific IVIS Platform activities

There are six ongoing activities or functions in the IVIS Development Center.

1. Develop integration techniques for devices. Each class of device (e.g., navigation system, cell phone, GPS) presents its own set of characteristics and requirements which must be accommodated in a properly-designed IVIS. The IVIS Platform permits ORNL to design and assess the effectiveness of subsystem integration techniques.

2. **Develop information management algorithms** and other driver interface techniques for IVIS. Specific algorithms for information management and presentation must embody the principles and guidelines developed by other FHWA and NHTSA efforts, as well as in private research. Such algorithms contribute to the prototype interfaces which are described in point three.

3. **Demonstrate prototype interfaces** that exemplify the human factors state-of-the-art in driver interface design and information management. This function serves to enlighten the transportation community and other visitors about the progress in IVIS development, and to solicit their input.

4. **Illustrate the difference between interfaces designed with and without use of human factors principles and guidelines.** By showing the same virtual trip with and without an IVIS in place, the benefits of appropriate application of human factors research results and guidelines can be graphically demonstrated to both users and members of the transportation community. This methodology permits demonstration of the degree to which unmanaged information can distract from the primary task of driving.

5. **Develop a test procedure for the ITS Data Bus**. The IDB is a bus protocol being developed by an ITS committee of the SAE to provide "plug-and-play" capabilities for ATIS subsystems into a vehicle. Once the IDB standard is developed, there will be need for a standardized test procedure to evaluate candidate ATIS devices. This test procedure will be developed using the IVIS platform.

6. **Serve as a User Facility** for industry partners to integrate their products into an IVIS environment, by applying the techniques developed in part 1 above to their specific products. Several companies have indicated that they need a facility in which to investigate methods for integrating their particular hardware/software products into an IVIS environment, as well as for exploring the effectiveness of the human factors aspects of their subsystems.

Evaluating Actual Devices

As IVIS development progresses, it is important to evaluate the performance of the IVIS as it interacts with actual devices. To accomplish this, the units can be inserted into the LAN, either as substitutes for one of the PCs, or using the PC as an interface to the LAN, as shown in Figure 2. For example, a radar detector embodying the SWS message delivery capability has been integrated into the development platform through a two-way RS232 port into one of the PCs. This data port permits the SWS unit to send its messages into the PC, which, in turn, transforms the messages into a form to be put onto the LAN, where the IVIS PC obtains them for processing and display. The two-way RS-232 port permits the detector to be triggered by the scenario generator through the LAN. For full feasibility testing, the detector can also be triggered by a reduced-power radar transmitter situated near the detector in the IVIS Development Center. As IVIS development proceeds, more and more actual subsystems will be integrated in the emulation network, using the SWS radar detector integration as a model.

Industry participation, by both vehicle and device manufacturers, has been received partly through an outreach task during the IVIS project, and partly in conjunction with the ITS Data Bus Demo98 project (the subject of another paper in this session). This enables the

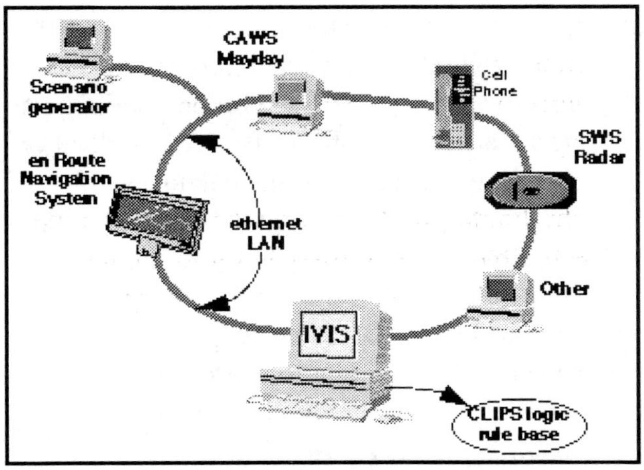

Figure 2. The IVIS LAN with devices.

the IVIS Development Center to receive input on industry perspectives on IVIS development, and provides an opportunity for cooperative development of standards and techniques for integrating the various products. In the past few years, industry representatives, along with FHWA and NHTSA personnel, have attended periodic workshops and working group meetings at which progress on the system was reviewed, suggestions for improvements were made, and non-proprietary information was exchanged. Some of the subsystems and integration work are also provided by manufacturers.

The Driving Platform

A low fidelity, non-motion-, computer-based driving simulator, produced by Systems Technology, Inc., is integrated into the facility. This platform is PC-based, and is integrated into a vehicle shell, with a set of driver controls (steering wheel, etc.), providing a realistic driving environment. The system permits high-level programming of events (e.g., other traffic, road curvature) and measures driver performance (e.g., root mean squared steering error), capturing relevant performance measures required to evaluate the impact of presenting ITS-based information to the driver. The integration of the simulator with the IVIS LAN is shown in Figure 3.

The fidelity of the driving platform is enhanced through the introduction of additional equipment, such as the "driving buck," as well as other features. For example, a touchscreen capable LCD head-down display is installed near the center console of the buck. A head-up display (HUD) is also simulated by using a PC-driven projector to display HUD information on the projection screen for the driving simulator. In addition, technologies from the DASCAR (Data Acquisition System for Crash Avoidance Research) Project that was designed and developed at ORNL for NHTSA, are utilized in the driving platform environment to enhance the data acquisition capabilities.

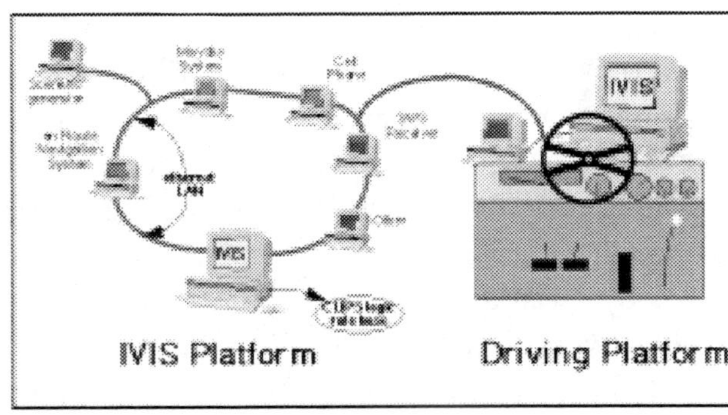

Figure 3. The dual IVIS platform.

The ORNL Research Vehicle

ORNL has dedicated $1M of internal funds for a two year (FY98 & 99) project to develop and demonstrate a capability to conduct experimental human factors research on the effects of advanced displays and information systems on driving behavior. The project focuses on mature driver issues concerning technology such as HUDs, Multi-Modal Displays (audio & visual), and Adaptive Cruise Control (ACC). The motivations for this project are: to leverage and complement the products of the IVIS Center, to provide a test bed for the future evaluation of an IVIS prototype, and to anticipate the research and development needs of the Intelligent Vehicle Initiative (IVI). There are two aspects to this capability: a laboratory-based driving platform and an instrumented vehicle.

The driving platform being used by this project is the one described above. This simulator will be used to conduct research in areas which may be too dangerous to do in the instrumented vehicle on the road; e.g., potential collisions caused by display-induced driver distraction. Techniques which are deemed safe after being tested in the simulator will be transferred to the instrumented vehicle.

The instrumented vehicle will have some components of the Data Acquisition System for Crash Avoidance Research (DASCAR), which ORNL developed for NHTSA. A set of digital video cameras will be installed to record driver actions and environmental occurrences. Vehicle parameters (e.g., speed, pedal force, etc.) will be monitored by a suite of sensors. Driver physiological variables (e.g., heart rate) will be sensed by wrist and finger worn devices linked to a belt worn computer, which will send these data to the main data acquisition system in the trunk, via a wireless data link. In addition to the data acquisition system, the vehicle will also have advanced displays and vehicle control and information systems. The displays will be a HUD, an audio display, and a flat panel LCD (same model as in the laboratory driving platform, with higher brightness). A navigation system and ACC will also be installed. The ACC will perform three functions: providing headway and time to collision data to the data acquisition system, provide headway and time to collision data to the driver display systems, and perform adaptive longitudinal control of the vehicle. Finally, a lane tracking system will has been procured and installed. Given these capabilities, the ORNL instrumented vehicle will approximate a baseline version of a first generation IVI light vehicle.

When the IDB and the associated gateway have been developed and tested in the IDB Demo project (see the companion paper in this session), they will be installed in the ORNL research vehicle. At this point, the data acquisition system (DAS) will be modified to operate as a node on the IDB. This will allow the DAS to get real-time data from the information systems and correlate their messages to the driver with other data events. This architecture will also support the installation of an IVIS prototype, when it is developed.

The Planned IVIS User Center

There are plans to designate the IVIS Development Center as a formal Department of Energy (DOE) User Facility, to be used by private sector developers of devices which will be installed into road vehicles to provide Advanced Traveler Information Services (ATIS) information. Such a designation facilitates use of the Center by outside organizations and researchers.

Several U.S. and international standards are being developed to facilitate the safe installation and use of electronics devices in cars at any time during the vehicle's lifecycle. In the future, consumers will be able to customize their cars, just as they do their personal computers today. These standards include an ITS Data Bus, which will be used to connect non-mission-critical devices, such as pagers, cell phones, navigation systems, GPS receivers, etc., to the car. Therefore, it will be carrying much of the message traffic that is being considered for analysis in the IVIS program. Other standards efforts include the standardization of message sets (e.g., mayday messages), and a message prioritization scheme to be used in displaying ATIS information to drivers. Early standardization and testing of these message sets and the ITS Data Bus will allow device manufacturers to introduce a wide variety of products quickly. This brings with it the need to resolve those issues which the IVIS research is intended to address. A coordinated development program will ensure that these efforts are in step from the beginning. Any standards development program must include a test phase, during which the standards are tested for simplicity, ambiguity, performance, etc. This testing is sometimes performed by one of the standards development committee participants, with the good favor of his or her employer, but this cannot always be guaranteed. The IVIS Center provides an ideal environment in which to test the IDB, since it can be used in a "real" simulation, with heavy message loads from the various devices either emulated or installed in the IVIS LAN.

The National Transportation Research Center

In addition to the DOE User Facility designation, discussions are also under way to create an IVIS Development Center in the newly designated National Transportation Research Center (NTRC). The NTRC was motivated by a desire to expand existing transportation R&D at the Oak Ridge National Laboratory and the University of Tennessee, Knoxville, by creating an entity that is responsive to the needs of industry as well as those of the government. The NTRC represents a partnership between the Department of Energy (DOE) and its management and operating contractor, Lockheed Martin Energy Research Corporation, The University of Tennessee (UT), and The Development Corporation of Knox County. Each party brings important capabilities and assets to the partnership. DOE brings the ORNL with its scientific and technical capabilities plus an existing funding base of nearly $80 million in transportation R&D from multiple sponsors. In addition, DOE provides access to the National Prototyping Center and the Oak Ridge Centers for Manufacturing Technology (ORCMT) located at the Oak Ridge Y-12 facility. UT brings nationally recognized R&D programs in transportation logistics and advanced vehicle technologies with a base funding approaching $20 million.

An important key to making the NTRC possible is the Development Corporation, which initiated the idea and is providing the site for the construction of a new R&D complex midway between Oak Ridge and UT. The building will provide approximately 85,000 sq. ft. of space with a net usable area of around 65,000 sq. ft. Roughly 20,000–25,000 sq. ft. will be assigned as conventional office space with the complement being used as special purpose laboratories and user facilities. The Center will house approximately 150 personnel and will have about 13 individual R&D laboratories or facilities. Approximately 80–90 of the staff will be from ORNL with another 30–40 faculty, research associates, and students from UT. The remaining space will be reserved for private enterprise users, collaborators, and visiting scientists and engineers. The Center will also contain approximately 4,000 sq. ft. of flexible space for small conferences, multiple classrooms, or to host public functions associated with the Center.

A long range goal of the NTRC is to become a nationally recognized and respected research, development, testing, and evaluation facility that can respond to the transportation needs of both the public and private sectors. To achieve this goal, a number of collective competencies will be emphasized:

- Engine, Propulsion, and Power Systems
- Information/Decision Support Systems
- Materials, Structures, and Mechanical Systems
- Vehicles, Materials Handling, and Packaging Systems
- Safety, Security, and Human Factors
- Systems Operation and Control

The Human Factors Engineering Laboratory will include capabilities for visualization, simulation, ergonomics, human perceptions associated with traffic, vehicle operations, etc., and will include appropriate

0-7803-5086-3 /98/$10.00

space for the IVIS Center and associated offices for staff.

Conclusions

Given the issues and knowledge gaps associated with the installation of multiple devices in vehicles, it is hard to predict with any level of confidence how system integrators will integrate these multiple systems. The problem of having multiple devices installed in vehicles has gained worldwide recognition, and work is in progress on the development of standards and guidelines which address the problems described in this paper. This paper has presented a R & D Platform and associated facility in which research on these problems is being conducted. The combination of a "desktop" LAN, a benchtop IDB, a driving simulator, and an instrumented research vehicle provide researchers with the opportunity to evaluate designs for in-vehicle devices across the spectrum of research settings, from laboratory to on-the-road vehicles. Those interested in exploring the research opportunities at the Oak Ridge National Laboratory IVIS Development Center should contact the authors.

Acknowledgements

IVIS research was sponsored by the Department of Transportation, Federal Highway Administration Contract No. 1883E020 A1, under Department of Energy Contract No. DE-AC05-96OR22464 with Lockheed Martin Energy Research Corp.

Research on the Research Vehicle is sponsored by the Laboratory Directed Research and Development Program of Oak Ridge National Laboratory, managed by Lockheed Martin Energy Research Corp. for the U. S. Department of Energy under Contract No. DE-AC05-96OR22464.

RECONFIGURABLE FIELD RESEARCH VEHICLE FOR HUMAN FACTORS EXPERIMENTS

R. Wayne Devereux, Veridian, Veda Operations, Lexington Park, Maryland

Jennifer Wilkinson, Veridian, Veda Operations, Lexington Park, Maryland

Introduction

Future transportation vehicles and highways may present drivers with information overload, causing distractions, high workload and fatigue that could impair the driver's ability to operate the vehicle safely. The human factors research community requires a tool to evaluate subjects under controlled experimental conditions and to determine drivers' responses to future yet-to-be-defined vehicle and highway information. This paper discusses the system architecture used in a reconfigurable research tool developed to perform computer controlled human factors experiments on the highways in the field. This design uses the vehicle's reconfigurable instrument panel, displays and audio messages to stimulate the driver while a data acquisition system records driver and vehicle responses. The experiments focus in the areas of intelligent vehicle display type, including a Head Up Display (HUD); placement and symbology; route navigation; in-vehicle signing; and, audio messaging.

The Human Factors Field Research Vehicle or Real Simulator (RealSim) architecture uses 5 parallel computers to control experiments. This tool allows researchers to evaluate various means of presenting information and assessing driver's responses while performing controlled field experiments. The experimenter can physically relocate and orient different sized graphical displays within the instrument panel and also vary the symbology, color, location and duration of images and audio sounds presented through software programs. The experimenter can control data acquisition and archiving and monitor the experiment in real time.

The system architecture is guided by the functional requirements and design specifications [1, 2, 3]. The primary goals of this design are:

- ***Reconfigurable Open Architecture***
- ***Advanced Driver Visual and Audio Stimulus***
- ***Dynamic Multipurpose Data Acquisition***
- ***Capability for Driver Input and Control***
- ***Controlled Experiment Planning/Execution***
- ***Planned Upgrade and Growth Path***

The full scale development phase of this research vehicle was completed and delivered to the Federal Highway Administration (FHWA). The research vehicle, off-vehicle development stations, and vehicle development workshop are operated by FHWA at the Turner-Fairbanks Highway Research Center in McLean, Virginia.

System Architecture

The RealSim system is designed with Commercial-Off-The-Shelf (COTS) equipment in a modular open architecture using parallel processing and real-time multi-process Operating System software. This design is flexible and adaptable to experimental differences and is expandable to support undefined future program requirements. The RealSim in-vehicle design uses 5 onboard Industry Standard Architecture (ISA) single board 80486/80586 Personal Computers (PCs) and other ISA support boards. These systems are housed in 3 industrial 19 inch rack computer chassis. Two of the three chassis have 20 slot 10/10 split backplanes, while the third chassis maintains a non-split 20 slot backplane, providing ample slots for growth.

0-7803-5086-3 /98/$10.00 ©1998 IEEE

The RealSim system architecture includes an on-board Experimenter Station System (ESS) to control the experiments and present information to the driver and the experimenter on independent displays. It also includes a reconfigurable Data Acquisition System (DAS) which acquires analog and digital vehicle response data, driver performance data, and objective and subjective operator workload.

The research vehicle also has video cameras, microphones, and recorders for acquiring video and audio data of the driver's performance and out-of-vehicle data. These systems are integrated using Ethernet protocol, custom Shared Memory Boards, and custom real-time control software developed for this project. This modular architecture is adaptable to new applications involving controlled experimental research, such as training simulators, and complex data acquisition and control systems.

The RealSim in-vehicle component layout is shown in Figure 1.

The complete RealSim architecture consists of 6 major in-vehicle subsystems and 4 major off-vehicle subsystems, listed in Table 1.

Table 1
RealSim Systems

IN-VEHICLE SYSTEMS

- ***In-vehicle Display/Control System (IDCS)***
- ***Navigation and Map System (NMS)***
- ***Experimenter Station System (ESS)***
- ***Data Acquisition System (DAS)***
- ***Video and Audio Acquisition System (VAS)***
- ***Research Vehicle System (RVS)***

OFF-VEHICLE SYSTEMS

- ***Off-vehicle Experiment Development Station***
- ***Off-vehicle Graphic Development Station***
- ***Off-vehicle Video Playback System***
- ***Research Vehicle Garage and Workshop***

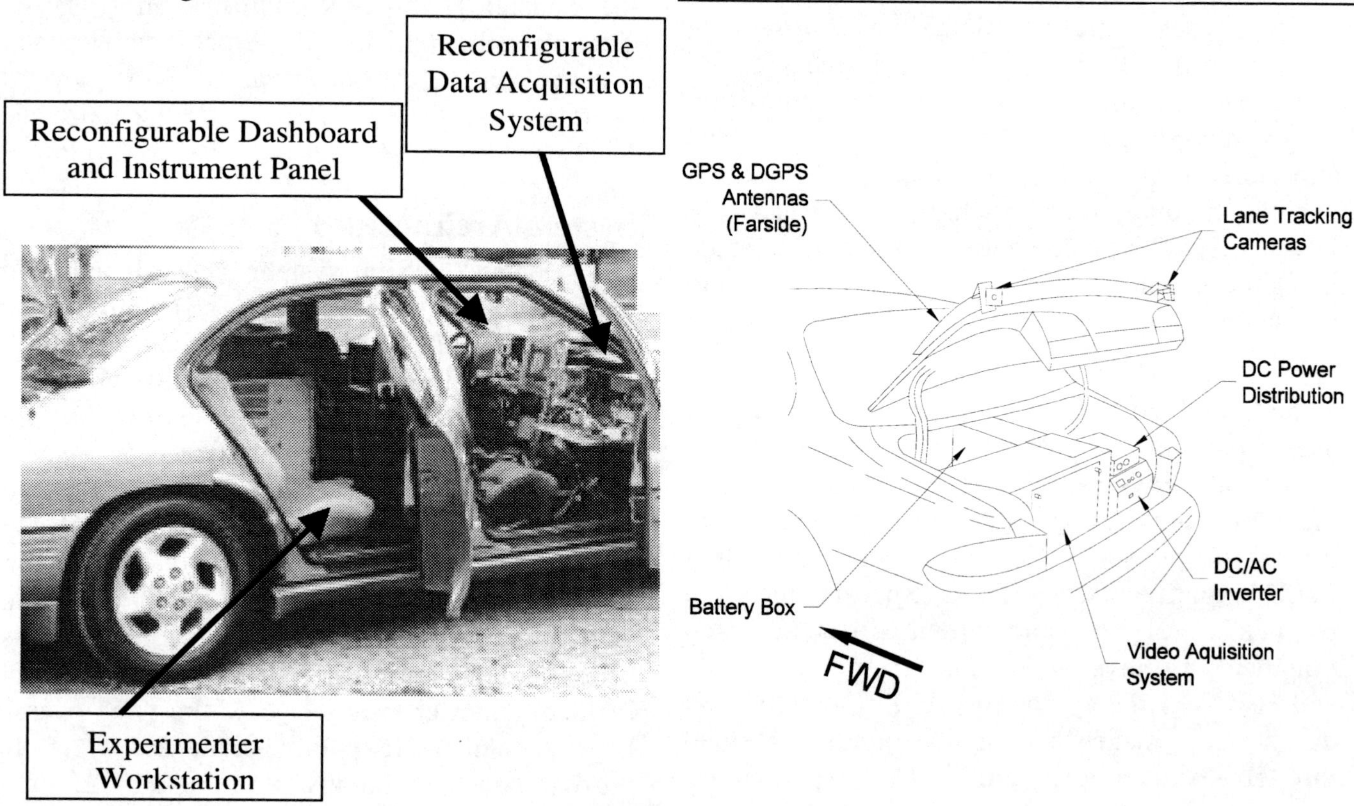

Figure 1
Vehicle Component Locations

Using the RealSim system architecture, a human factors researcher is able to configure the vehicle's instrument panel display type, placement and symbology on any of 7 computer driven Liquid Crystal Displays (LCDs). The experimenter will also be able to generate any audio message using a computerized sound board. All timing (start, duration, stop, etc.) and graphical details (placement, symbology, color, etc.) of each display and audio signal (sound, amplitude, etc.), as well as details of each DAS component (types of input, numbers of channels, sample and archive rate, file name, etc.), are under event-driven control of the Experimenter's Station System.

With the RealSim system, an experimenter "defines" a scenario for the system using a near-natural language. For example: If speed is greater than 55 mph, then begin DAS data archiving of parameters stored in setup file name "dataarch.01", then display "Slow Down" on display #3 and the HUD while announcing "Excessive Speed" over the audio speaker. During this experiment data is collected on the driver's response to the information presented.

Another experiment could be conducted concerning route planning. For example: If the vehicle's positional data are within a given location, the on-board computer could give warning to the driver of an upcoming turn. The system can display a blinking right turn arrow or other messages while measuring and recording the driver's responses.

In-Vehicle Display and Control System (IDCS)

The In-vehicle Display and Control System (IDCS) design is highly reconfigurable using software and physical display placement. The drivers view of the IDCS is shown in Figure 2. The design uses a unique PC based computer graphics generation software and special graphic display buffer boards. The graphical system consists of 2 graphics processor boards, 6 graphics buffer boards, and custom software to drive up to 6 vehicle displays using either RS-170 video or VGA signals. These include 4 IP displays, 1 HUD, and 1 In-vehicle Sign display (for displaying highway sign symbology). Color LCD displays are used for the drivers primary displays and a moving map display. Using custom fluorescent backlighting, brightness levels are achieved in excess of 400 nits in each display. This brightness level affords the most cost-effective sunlight readability offered by today's technology. Each of these displays are packaged in custom enclosures for mounting and thermal cooling inside the dashboard.

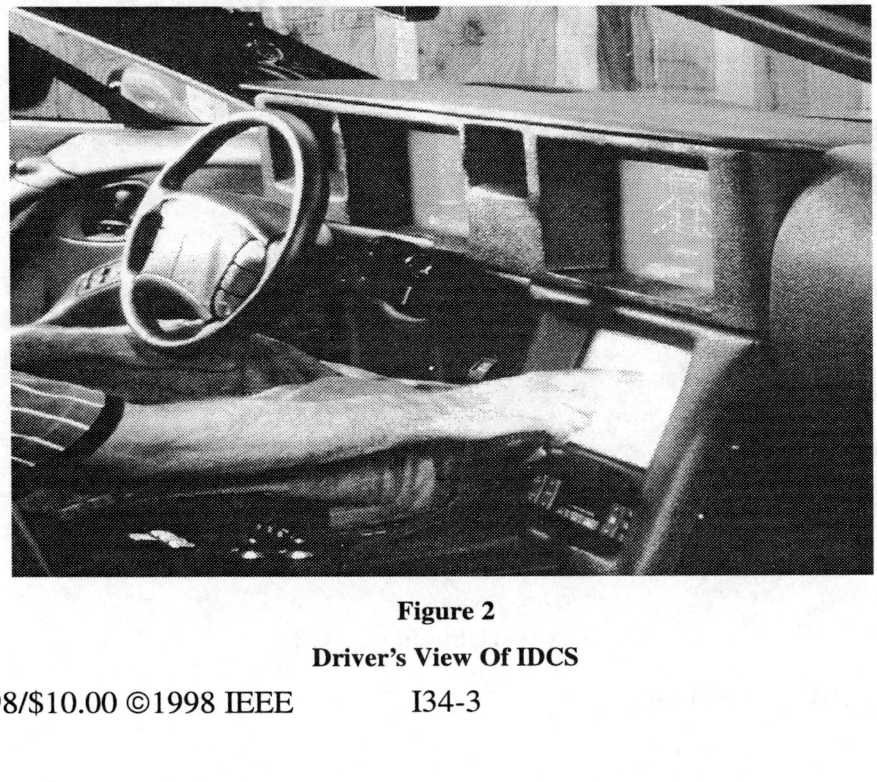

Figure 2

Driver's View Of IDCS

Real-time data from the DAS and ESS are used to drive the displays with actual data. Any of the IDCS displays are able to present basic vehicle performance, status, warning and routing information in any graphical or symbolic manor desired by the experimenter. Also, the driver may control some display functions using control switches to drive the DAS data.

Although the RealSim has an in-vehicle moving map display, it is not part of the IDCS. The moving map display is part of the Navigation and Map System (NMS) and is driven by a Global Positioning System (GPS) receiver and a Differential GPS receiver. The GPS/DGPS data are used by a Windows 95 based geographical mapping software to position the vehicle on the moving map display. A large sunlight readable LCD map display is driven from the VGA output and is located in the dashboard within reach of the driver. The display has an Infrared (IR) Touch Screen overlay for control of functions such as map, traffic, or route planning information. A Gyro-Compass is used to update and maintain precise position and direction information even during moments when no GPS signal is received.

By editing spreadsheet files before hand, the experimenter can display different routing and geographic information to the driver on the map. This provides the experimenter a method to evaluate driver's responses to events such as changing a driver's planned route due to a simulated traffic jam or accident. Each new routing screen can be triggered and displayed by the ESS's event-driven protocol as a function of time, distance, position, or experimenter's input.

Experimenters Station System (ESS)

The in-vehicle Experimenter's Station System provides a workstation environment for the experimenter in the rear seat of the research vehicle, shown in Figure 3. The ESS design uses a multi-processor real-time Operating System to run custom Graphical User Interface (GUI) and Action/Event Processor software. The ESS includes a Sound board and speakers for driver audio stimulus. The ESS interfaces to the experimenter's keyboard and display, the only keyboard in the vehicle. This keyboard and display are interfaced to all 5 onboard PC keyboards and displays through a switch located on a control panel at the experimenter's side.

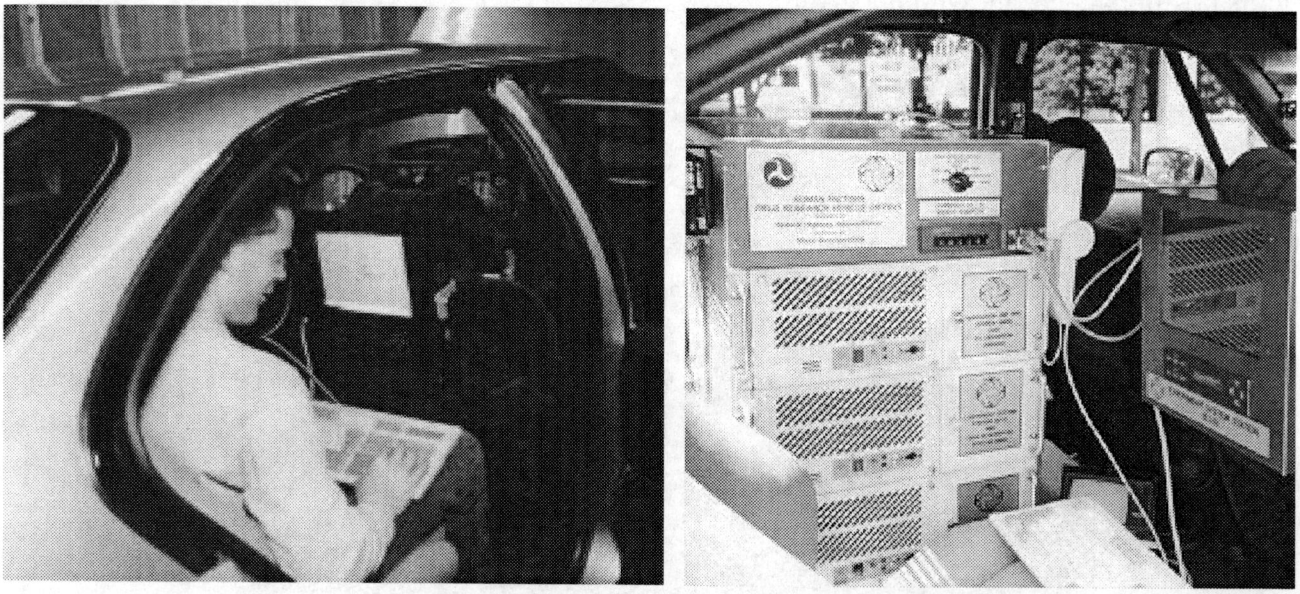

Figure 3
ESS Work Station Layout

The ESS software has 2 separate operating modes, Experiment Development and Trial Run-time. The ESS Experiment Development software is used by the experimenter off-vehicle to create and develop the scenario and debug the experiment program arguments in the laboratory. The ESS Trial Run-time software is used in the rear-seat of the research vehicle to control the entire suite of RealSim computers, displays, sensors and recorders. The experimenter uses it to run the experiment and to acquire and archive scenario data. These experiments are event-driven and computer controlled. This method of controlled event-driven experimentation affords the most rigor possible under field conditions.

Using the ESS Development Screen and the custom near-natural language, the experimenter programs experiment scenarios into the ESS Event List. This list causes scenario based event-driven actions to be generated. The experimenter can test the Event List by stepping through it using programmed data to provide precise experiment debugging functions. The three basic elements of an RealSim ESS experiment include; Conditional Events, User Defined Variables, and Data Acquisition System Setup. Examples of a Conditional Event and a user Defined Variable are shown in Figures 4a and 4b. The basic elements of an ESS Action and Equation are shown in Figures 5 and 6.

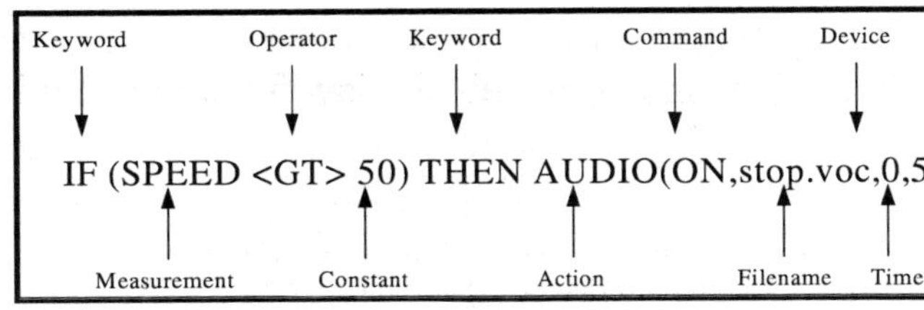

Figure 4a

Example of a Conditional Event

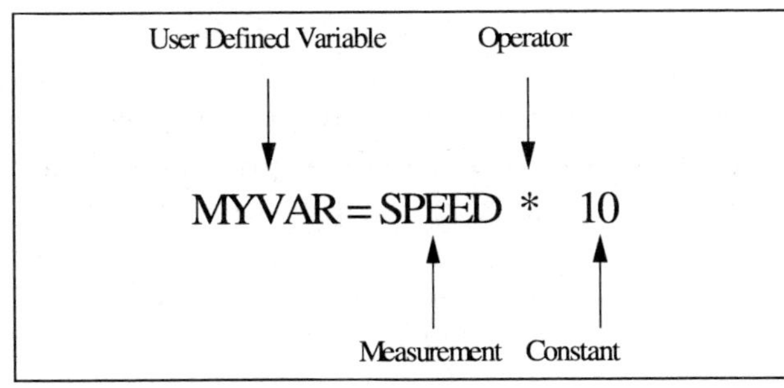

Figure 4b

Example of a User Defined Variable

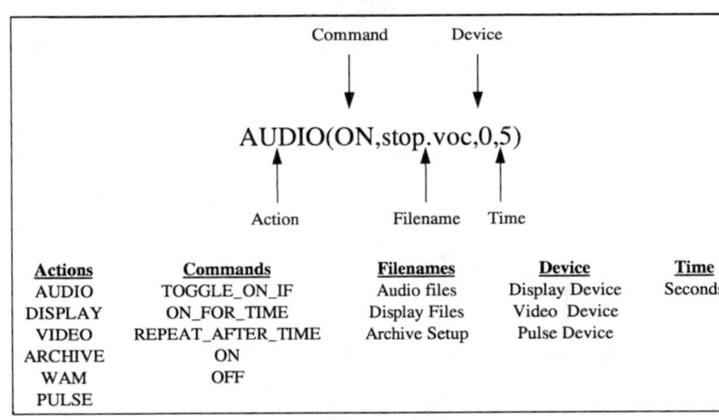

Figure 5

Elements Of An ESS Action

Keywords	Measurements	Equation Operators		Constants & User Defined Vars.	Actions
If	ACCEL_POS	**	SQRT	MYVAR	AUDIO
THEN	BREAK_POS	SIN	MAX	50.0	DISPLAY
	STEERING_WHEEL_ANGEL	COS	LOG	.	VIDEO
	CRUISE_CONTROL	TAN	LOG10	.	ARCHIVE
	HAZARD_FLASHER	ASIN	<GT>	.	WAM
	FUEL_LEVEL	ACOS	<LT>	.	PULSE
	HEADLIGHT_DIMMER	ATAN	<EQ>		
	HEADLIGHTS	ABS	<GE>		
	HORN	ALOG	<LE>		
	OIL_LEVEL_WARNING	ALOG10	<OR>		
	SEAT_BELT	EXP	<AND>		
	TURN_SIG_LEFT	+			
	TURN_SIG_RIGHT	-			
	WINDSHIELD_WIPER	*			
	ACCELLERATION_LNG	/			
	ACCELLERATION_LAT				
	HEADING				

Figure 6
Elements Of An ESS Equation

Using the ESS Run-Time Screen, the user creates the Trial specific file, prior to collecting data. Each Trial is unique to a subject number and is performed real-time. During these Trials, the Event List and Action Processor cause changing graphics to be presented on the displays and audio messages to be generated from the sound board. During the Trial, the Event List also controls the archiving of the required response data. Also during Run-time, real-time data can be reviewed using the Quick Look screen. An electronic Experimenter's Log is also available through the ESS GUI to document occurrences throughout the Trial.

The fundamental elements of a Trial are based upon the defined Experiment. These Trial elements are subject demographics, MS-DOS ASCII formatted archive files, and an Experiment Log file. Each subject's demographic file includes gender, age, visual acuity, driving experience, and 5 programmable open fields. Each formatted Trial archive file includes; a demographic header, variable name list, and time stamped data. Each Experiment Log file includes time, position (lat/long), and text that is input by the experimenter during the Run-time Trial.

Data Acquisition System (DAS)

The in-vehicle Data Acquisition System design uses COTS boards and custom software. The DAS includes an Analog to Digital board, a Digital Input/Output board, a Counter board, and a removable 2.1 Gigabyte SCSI Hard Drive. An external custom interface circuit is used to interface between the DAS and the research vehicle's onboard Engine Computer Module's (ECM) data link to acquire real-time vehicle and engine performance data.

The DAS data includes vehicle performance data (such as steering wheel position, RPM, speed, distance, etc.) and also driver response button data used to evaluate driver workload. The DAS system provides complete data acquisition of 16 analog channels and almost 50 digital input channels with sample rates up to 400 Hz. The analog inputs provide vehicle and driver response data used to measure driver response times. The vehicle analog signals include accelerator pedal and steering wheel positions, brake force and pressure, tri-axial accelerometer rate, fuel level, coolant temperature and pressure, etc. The driver response analog signals are from 3 slide potentiometers on a custom Drivers Response Panel mounted in the center console.

The vehicle discrete inputs include dashboard turn signals, high beam, seatbelt, hood/trunk/door adjar, over temperature, and other normal status indications. The driver response interrupt driven discrete signals include 4 momentary push buttons and 2 multi-position rotary switches from the Driver Response Panel and 8 additional momentary push buttons mounted on the steering wheel.

The DAS digital inputs are from a counter board which decodes the pulsed output representing vehicle speed and from an interface with the vehicle's computer over the ECM data link. For short intervals of time, the DAS controls the ECM to acquire vehicle performance data such as speed, fuel level, oil pressure, etc. Additionally, the DAS acquires an RS-232 digital input from an in-vehicle 916.5 MHz Radio Frequency (RF) transceiver dedicated to communication with an off-vehicle computer controlled RF transceiver. With this interface, the DAS acquires an RF transmitted signal which the experimenter can use to simulate off-vehicle electronic sign posts and other future highway "events." This allows an off-vehicle computer to control onboard actions such as initiating an in-vehicle display to show a "Stop" sign or announcing an "Accident Ahead" voice file.

For additional capability, 8 digital output channels and 2 analog output channels are provided under ESS software control to provide a variety of different stimulus (flashers, lights, buzzers, sound, etc.) to the driver.

The DAS's analog and digital data acquisition boards are setup and controlled using unique reconfigurable and separate acquisition and archiving software modules. While developing the experiment, the experimenter has complete control of what data are acquired and at what sampling rates. As new data are acquired, the DAS time stamps each data set with actual DAS time of acquisition. The DAS time is initialized with GPS time to assure accuracy. Once the data are time stamped the DAS writes the data into the Shared Memory board's Current Value Table (CVT) with the acquired time, as shown in Figure 7. The CVT is constantly updated with the most recent value of the acquired data and the time of acquisition. The data maintained in the CVT are used to drive the ESS Event and Action processes, the IDCS process and the data Archive process.

Prior to running the experiment, the experimenter must also create the required Archive setup files for all data that must be stored during the entire experiment. The setup files specify each set of parameters and the desired archiving rate. During the experiment, the data is archived to the computer's removable 2.1 GB hard drive. These parameters are stored whenever the ESS defines the Archive setup file to be opened. The ESS and DAS allow multiple Archive files to be created and opened concurrently. This capability allows archiving of the same or different data at different archive rates. These files can be archived at the same time or at different times. The archived data and ESS's Experimenter Log file are MS-DOS ASCII files that are compatible with statistical processing software packages such as SPSS.

Video and Audio Acquisition System (VAS)

The in-vehicle Video and Audio Acquisition System (VAS) design uses COTS video recording equipment. The VAS includes 6 black and white CCD video cameras, 2 audio preamplifiers and microphones, and 6 Hi-8 mm video/audio tape recorders (VTR) to tape and record all video and audio signals. An external IR LED is mounted to the bumper to illuminate the roadway and reduce shadowing effects. The VAS embeds time and position information onto the video tape using GPS data acquired from the NMS through an external RS-232 Broadcaster interfaced to the Reader and Translators.

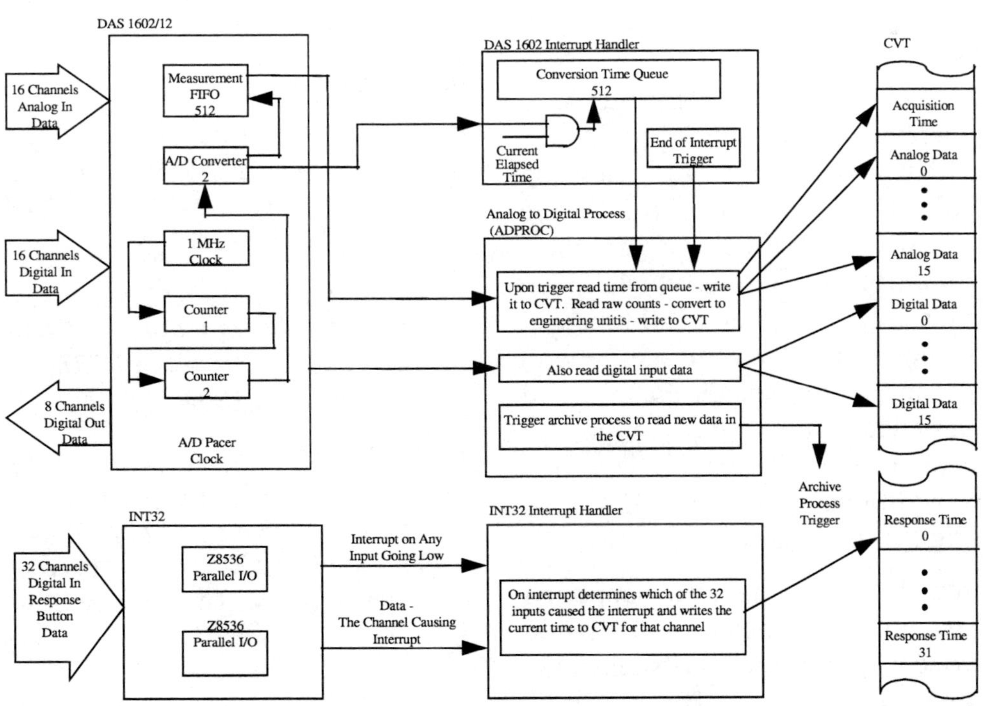

Figure 7

Acquisition Processing

The 6 VTR recorders are controlled by the ESS's Event List during a Run-time Trial through an RS-232 interface for Record and Stop control only. The VAS records the overall driver and vehicle responses using reconfigurable cameras locations that are generally dedicated to:

- ***2 cameras acquire driver head and eye movement***
- ***2 cameras acquire driver's actions and traffic patterns***
- ***2 cameras acquire vehicle lateral placement and lane tracking***

Two lapel microphones, one for the driver and one for the experimenter, are used to record verbal communications during the experiment trial to aid in post experiment data analysis. The two microphones are recorded separately on the VTR's respective right and left audio channels. Each of the 6 VTRs have the voice recorded on both stereo channels to ease the post-test processing of the video data.

Off-Vehicle Systems

Three additional PC chassis are dedicated as off-vehicle development stations; the ESS Development Station, the IDCS Development Station, and the Software Maintenance Development Station. These stations provide the experimenter the platforms to develop experiments and sounds using the ESS software and developing new graphical and map symbols and displays. Additionally, an off-vehicle video playback system is used to read the GPS time and position information embedded in the video's VITC. The time and position data are overlaid onto each frame of video and re-recorded onto a new tape. This step offers the experimenter the ability to edit additional comments onto the dubbed tape using the video titler.

The research vehicle is also supported with a work area in a 4.5 meter by 12 meter transportable wooden garage built to shelter the vehicle in a controlled climate. The garage is used to integrate the vehicle equipment, prepare the vehicle for new experiments, support upgrades and repairs, and provide a workspace for housing the off-vehicle development stations.

Summary

The RealSim's open architecture is modular and flexible for adapting to present and future human factors research experimental requirements. Using this architecture as a research tool, the future designers can evaluate optimum methods of presenting new information to the drivers of tomorrow's vehicles. New research data is being collected and analyzed by FHWA which will aid in understanding the requirements, specifications, and designs for safe human interaction with future vehicles and highways.

Acknowledgment

This work was sponsored by FHWA under contract number DTFH61-93-C-00093. The authors gratefully acknowledge the support provided by the FHWA Turner-Fairbanks Highway Research Center, Human Factors Program, in the development of this vehicle.

References

1. *Human Factors Field Research Vehicle Functional Requirements Document*, Veda Incorporated, Report 76661-94U/P070282, April 15, 1994
2. *Human Factors Field Research Vehicle Design Specification Document*, Veda Incorporated, Report 77823-94U/P070282, November 8, 1994
3. R. W. Devereux, *Systems Architecture of a Reconfigurable Field Research Vehicle for Human Factors Experiments*, Society of Automotive Engineers (SAE) 1995 Future Transportation Technology Conference

THE CARNEGIE MELLON TRUCKSIM: A TOOL TO IMPROVE DRIVING SAFETY

Richard Grace, Carnegie Mellon Research Institute, Pittsburgh, PA

Alberto Guzman, Carnegie Mellon Research Institute, Pittsburgh, PA

James Staszewski, Carnegie Mellon Research Institute, Pittsburgh, PA

Bethany A. Peters, Westminster College, New Wilmington, PA

Melissa Mallis, University of Pennsylvania, Philadelphia, PA

David F. Dinges, University of Pennsylvania, Philadelphia, PA

Abstract

Carnegie Mellon Driving Research Center is presently involved in the design and development of an Advanced Human Factors Research and Driving Training Research Facility. The facility has been designed to address human factors issues and driver training issues. Human factors interests include developing countermeasures for fatigue and driver/vehicle interface issues. Driver training issues include validating the usefulness of simulators for driver training, developing effective curricula and investigating simulator fidelity needed for effective training. A key component of the facility is the Carnegie Mellon TruckSim that will be capable of simulating a variety of commercial and emergency vehicles using interchangeable cabs mounted to a common motion platform. TruckSim's modular configuration will allow for rapid and cost effective design of experiments and training scenarios. A first research program to evaluate fatigue countermeasures is presented as an example.

Introduction

The Carnegie Mellon Driving Research Center (DRC), working together with ISIM Corporation, has designed and developed TruckSim, a state-of-the art driving simulator specifically designed for human centered research. TruckSim has been designed to provide a realistic driving experience. A key design criteria is a balanced perceptual environment with realistic visual, motion/vibration and sound cues.

TruckSim, as its name suggests, is currently configured as a Class 8 truck with a Freightliner FLD cab. The configuration was chosen to advance DRC's ongoing research in the heavy truck area. However, TruckSim has been designed to be reconfigurable in order to simulate a variety of vehicles including passenger cars, buses and off-road vehicles. Each cab/body to be used on the simulator is equipped with a dedicated sound system plus sensor; and actuators associated with the driver controls and displays. A truck cab/body is bolted to the motion base and integrated to the simulation control station through a set of quick connect plugs.

TruckSim also includes a user friendly experimenter console. Using the console, an experimenter can rapidly and cost effectively implement complex experimental designs. The easy to use graphical user interface allows the experimenter to precisely control the interaction of the simulated vehicle with over 100 additional vehicles or pedestrians. A behavioral scientist with little or no programming experience can really use this

0-7803-5086-3 /98/$10.00 ©1998 IEEE

intuitive interface.

TruckSim can be used for a wide variety of research programs.

Important classes of research problems that can be studied in simulation are:

Driver Training

- Commercial driver training
  - Novice driver training
  - Driver retraining (continuing education)
- Young driver training
- Older driver training/evaluation

Driver Workload

- Visual allocation/distraction/demand
- Manual demand
- Cognitive load assessment
- Information prioritization

Infrastructure

- Road signing/traffic advisory
- Automated highway/Intelligent Transportation System
- In-vehicle signing

Driver Impairment

- Drowsiness detection and warning
- Drug/alcohol effects
- Age effects
- Rehabilitation testing

Vehicle Design

- Display design
- Control Design
- Warning/emergency evaluation
- Equipment design
- Interface optimization

Simulator Fidelity

- Simulator sickness

All of these listed areas are of potential interest and may be addressed in the future using the TruckSim. To address a wide range of research topics the TruckSim is available to the research community on an hourly-fee basis.

Simulator Technology

There is no question that simulation is being used effectively today. But with certain technical advances and more emphasis on behavioral factors, simulation offers the potential for making even greater contributions at a significantly lower cost than previously possible. In the past, the importance of the relationship between how a simulator is used and its effectiveness has not been fully appreciated. As a consequence, simulators are often not as versatile or cost-effective as they could be. The contributions that behavioral science and human factors engineering has made to the design and use of simulators is underrepresented.

In the past, ground-vehicle simulators were not flexible enough to allow for rapid reconfiguration necessary to efficiently evaluate new transportation technology or cost effective training. As the rate of new systems development increased over the years, it became important to be able to rapidly simulate and test these systems. The CMRI team has formulated the design for a facility that provides cost-effective and rapidly configurable operator-in-the-loop simulation for use in basic and applied research.

Simulator Design

The TruckSim design is based on the ISIM Mark II simulator. Additions/upgrades of the Mark II design include improved visual resolution an enhanced four degree of freedom motion base and the addition of Eaton programmable transmission simulator, and the addition of the experimenter's console. See figure 1 for the simulator configuration.

- Insert Figure 1 here -

Figure 1: Carnegie Mellon TruckSim ©

The simulator's design includes:

Visual System

- Angular field of view - 180 degrees horizontal, 33 degrees vertical
- Forward display channels - 3 channels, 1000 x 800 pixel resolution
- Brightness of forward channel projectors - 300 lumens
- Rear display channels type - 2 channels, 600 x 480 pixel resolution, flat panel
- Minimum screen drawing rate - 30 Hz
- Anti-aliasing - implemented in hardware
- Projector position - inside of screens
- Number of moving vehicles per channel - 100

Motion System

- Motion platform degrees of freedom - 4 - roll, pitch, surge and heave
- Roll angular amplitude and acceleration - ±6 degrees, 1 G maximum acceleration
- Pitch angular amplitude and acceleration - ±6 degrees, 1 G maximum acceleration
- Surge linear amplitude and acceleration - 0.1 meters, 0.5 G maximum acceleration
- Heave linear amplitude and acceleration - 0.1 meters, 0.5 G maximum acceleration
- Small amplitude bandwidth - 15 Hz
- Maximum latency of motion system - 50 milliseconds
- Motion simulation capabilities - road surface irregularities (i.e., potholes, rumble strips, uneven road surface), off road surfaces (i.e., curb, soft shoulder, driving off road surface), tactile response for multiple tire types

Audio System

- Number of sound channels - 4
- Sound simulation capabilities - multiple tire type sounds, velocity dependent wind noise, transmission grinding cue, additional sounds can be added

Other Features

- Transmission simulation type capabilities - Eaton / gate lockout, manual grinding feedback
- Steering wheel bandwidth - 20 Hz
- Operator capabilities - scenario editing, playback, scoring
- Vehicle dynamics model - operator adjustable, including 3D tire model
- Road data base - configurable data base including urban, rural, mountains, highway scenes

Traffic Control

- Number of traffic objects - over 100 objects (vehicles, pedestrians, etc.)
- Automated intelligent traffic control - control of traffic density, aggressiveness, circulatory and Swarm traffic
- Scenario controlled traffic - Specific actions of any of the traffic objects can be specified using a windows based GUI.
- Operator controlled traffic - The operator can also take control of any traffic object.

TruckSim's visual system provides a forward field of view and two sideview mirror views. The forward field of view is implemented with three LCD projectors each providing 1200 x 1000 pixel resolution within each 60° of view. The sideview mirrors are implemented in LCD displays. The visual data base consists of a 21 mile 4-6 lane highway loop, a 9 block urban setting, and a rural setting. The driver can move seamlessly throughout the data base.

The motion system includes a four degree of freedom motion platform. The motion platform includes independent actuators for roll, pitch, surge and heave. Independent actuation was used in order to enhance the high frequency performance so important in achieving good road feel. The motion platform is capable of simulating road surface irregularities such as potholes, rumble strips, uneven road surfaces, curbs and driving off

0-7803-5086-3 /98/$10.00 ©1998 IEEE

road.

A vital part of TruckSim's excellent road feel is the implementation of a true three dimensional tire model. The road/tire interacts at a 1000 Hz rate allowing inclusion of fine road irregularities such as pot holes and rumble strips.

The system architecture is shown in Figure 2. All simulator control and the operator interface are contained in a dual Pentium Pro PC, four Real 3D 1000 image generators are employed to generate the visual scene, a Pentium PC is employed for the experimenter's console. All sensors and actuators are integrated directly into the dual Pentium Pro. Communication among the systems is accomplished using a 100 baseT ethernet.

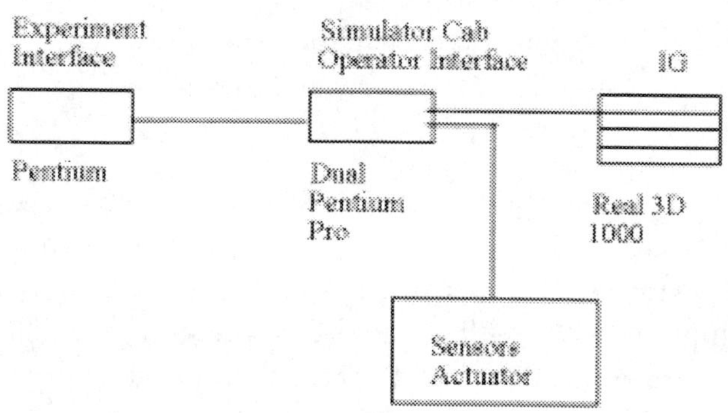

Figure 2: System Configuration.

- ***Multiple Vehicle Simulation.*** The simulator accepts a variety of vehicle bodies on the motion base. In addition, the modular software design allows for rapidly changing the vehicle dynamics software without disturbing the remainder of the system.

- ***Simple Experimental Design and Execution.*** A number of features have been included to facilitate experimental setup. A scenario builder allows the experimenter to define driving scenarios using a point and click interface. Data associated with these scenarios have been selected for collection and analysis using this package. In addition, methods for time synchronizing auxiliary data collection devices such as video tapes, physiological data and event strobing have been provided by the TruckSim.

The proper design of these components is largely based on expert front-end analysis. Since this process required considerable specialized effort, the CMRI team was successful because of their relevant area of expertise. Physical requirements for simulators cannot be considered in isolation because they interact with the way a device is used. The team utilized their depth of knowledge concerning the nature of these use factors and their interactions with equipment factors as a basis for simulator design within the facility.

Research Example: Drowsy Driver Study

An area currently under investigation involves detecting drowsiness and warning or alerting drivers of heavy vehicles in order to prevent costly accidents due to fatigue. An investigation of fatigue lends itself to simulator research. It is possible to simulate the conditions under which a driver would be likely to be drowsy without posing a hazard on the actual roadways. Specific details of one ongoing experiment are described below.

Background

This experiment has been designed to stress the effect of feedback to drivers experiencing drowsiness while driving the TruckSim. The feedback to be considered is:

- Information feedback
- An audible alarm
- Alerting Stimuli: olfactory stimulus and auditory stimulus

The objectives of this study are:

- To determine if the above feedback promotes arousal of the driver.
- To determine the driver's behavioral response to the feedback related to rest breaks and safe driving practices.

Proposed Experiment

The study goal is to test the effects on drivers' drowsiness (as defined by PERCLOS) and driving behaviors of a drowsy-driver detection system that includes a PERCLOS detector and algorithm coupled with a drowsiness/alertness gauge containing an audio and visual alarm, which in turn is coupled to one of two alerting stimuli: (A) a peppermint or menthol olfactory stimulus (specific type to be determined); and (B) an auditory alert command that the driver take countermeasure action immediately. The alerting stimuli will be delivered only after drowsiness has been presented for a sustained period of time (to be determined). The study will be implemented and performed using a total of 16 drivers.

All 16 subjects will undertake a 6-hr nighttime driving scenario on two separate occasions (serving as their own controls and counterbalancing for order conditions). These two drives will be performed using the Carnegie Mellon Research Institute truck simulator (CMRI TruckSim), and will occur in the morning (starting at approximately 0600hr to 0800hr) after drivers have completed a standard out-and-back, short-haul overnight drive for the company. This will ensure that drivers are performing in the simulator after a minimum of 12 hr awake, and near their circadian time for vulnerability to drowsiness. This approach is expected to produce drowsiness episodes in most, if not all drivers. However, since the magnitude and duration of these drowsiness episodes are not under full experimental control, and given the sample size of n = 16 drivers, it is imperative that there be a minimum sampling period of at least 3 hr during the treatment nighttime drive, to ensure that there is an adequate time base in which to capture drowsiness episodes and record driver responses (drowsiness and behaviors) to the two types of alerting stimuli.

Alerting stimuli are not considered at the same level as visual feedback on drowsiness. Rather, there is a progression from alerting stimuli. There is a very high potential for driver dissatisfaction with a system in which alerts are always paired immediately with detection of drowsiness, given the likely problems of false alarms, habituation to auditory and/or olfactory stimulation, and driver annoyance with repeated alerts. Therefore it is simply not realistic to expect that the final drowsy-driver on-line technology will function by attempting to arouse a driver every time drowsiness is detected. It is far more likely that, like other warning displays in a motor vehicle, the first acceptable mode for feedback regarding drowsiness detection will be a visual signal informing the driver that drowsiness is detected. Its purpose is to inform and prompt a self-directed change in driver behavior. Only after such self-initiated change is not undertaken, and drowsiness continues for a period of time deemed increasingly riskier, would one expect another level of feedback—namely that of arousal induction by alerting stimuli.

There is a logical progression from visual information feedback that drowsiness is detected, to alerting stimuli when drowsiness is sustained (i.e., an alerting stimulus only occurs after alarm feedback that unacceptably high levels of drowsiness has been present for some period of time). Thus, alerting stimuli will be contingent on a high level of drowsiness being sustained for (too) long a period, which will be predetermined, but is expected to be on the order of 5 to 10 minutes. This will permit assessment of the main effects on dependent variables of (1) visual feedback versus no visual feedback (n = 16); and (2) of the effects of an alerting auditory stimulus versus an alerting olfactory stimulus.

The specific conditions of the study are as follows:

Half of the subjects will undergo the control condition first and the intervention condition second, and half will undergo the intervention condition first and the control condition second.

In the intervention condition, all subjects will experience a visual gauge plus an auditory alarm for drowsiness with progression to an alerting stimulus.

Half of the subjects (8 drivers) will be exposed to the auditory alerting stimulus ("Take corrective action") on the outbound half of their drive, and the olfactory stimulus on the inbound half of their drive. Of these eight, half (n = 4) will experience this condition during the first nighttime drive and the remaining half (n = 4) will experience it during the second nighttime drive.

The other half of the subjects (8 drivers) will be exposed to the olfactory stimulus on the outbound half of their drive, and the auditory alerting stimulus on the inbound half of their drive. Of these eight, half (n = 4) will experience this condition during the first nighttime drive and the remaining half (n = 4) will experience it during the second nighttime drive.

Simulated nighttime driving scenario

The overnight, out-and-back drives (with and without drowsiness alarms/alerting stimuli) will be performed using the TruckSim. It is a high-fidelity ISIM Mark II Mobile Driving Simulator with the following features:

1. 21-mile looped freeway scenario with 2 and 3 lanes in each direction; 180° visual field of view; high resolution; programmable traffic controlled in direction, density and aggression.

2. full sized Freightliner FLD cab

3. 4 degrees-of-freedom motion platform (surge; heave; acceleration; shoulder rumble strips; road surface irregularities; etc.)

4. moving truck sound and vibration

5. vehicle instrument lights, and high and low beam exterior lights

6. interstate scenario includes multiple exits, rest area, flat test area, and runaway ramps

Drowsiness detection

An "on-line" drowsiness detection system based on the duration of eyelid closure of 80% or more (ignoring normal blinks) will be employed throughout both control and intervention driving conditions, using the basic observational criteria of PERCLOS[1]. The automated on-line PERCLOS system being developed by CMRI will be used. The cumulative total duration of time that eyelids are judged to be closed will serve as the primary biobehavioral drowsiness dependent variable for comparison between the control and intervention conditions to determine whether drowsiness feedback alarms and/or alerting stimuli reduce a driver's drowsiness. For the intervention conditions only (i.e., feedback visual alarms and alerting olfactory or auditory stimuli), eye closure drowsiness data will also form the basis for determining delivery of a drowsiness alarm.

Drowsiness alarms

Drowsy-driving feedback alarms will be audio and visual. Determination of the cumulative duration of time that the eyes are judged to be closed in order to trigger a drowsiness alarm will be based on analyses performed on data from the validation study of PERCLOS at the University of Pennsylvania.[2] Coherence between PERCLOS and vigilance lapses was greater for 20-minute periods of

[1] Weirwille, W.W. (1994). "Overview of Research on Driver Drowsiness Definition and Driver Drowsiness Detection." 14th International Technical Conference on Enhanced Safety of Vehicles (ESV), Munich, Germany.
[2] Dinges, D.F., Mallis, M.M., Maislin, G., and Powell, J.W. (1998). "Evaluation of Techniques for Ocular Measurement as an Index of Fatigue and as the Basis for Alertness Management, Draft Final Report," submitted to US DOT.

testing than for 1-minute, 2-minute, 4-minute, 5-minute, and 10-minute periods. A moving-window period between 1-minute and 20-minutes will be used as the basis for delivering a drowsiness alarm to drivers, when a threshold level of drowsiness is detected (as determined by the automated CMRI PERCLOS algorithm).

Alerting Stimuli

Two of the more promising alerting stimuli will be delivered to subjects only after drowsiness has been present for a sustained period of time (to be determined). This drowsiness detection is based on preliminary studies and published literature reviewed to date. We anticipate these will consist of a peppermint [or menthol] odor and an auditory warning - "Take corrective action".

Rest break behaviors

Drivers will be permitted to determine the timing and duration of their rest breaks, with the exception that the rest break at the midpoint of the nighttime drive (i.e., at the turn-around point) will be permitted to last no more than 15 minutes and cannot include napping. A careful record will be kept of their voluntary rest breaks (inter-break driving interval, break timing, break duration, break activities) during both the control and intervention nighttime drives. A comparison of these activities between the two drives will provide the data needed to determine whether drowsiness alarms-alerting stimuli alter a driver's tendency to take breaks and the nature of the breaks.

Unsafe driving behaviors

Throughout the two nighttime drives, three types of unsafe driving behaviors will be recorded from the TruckSim: (1) time spent speeding; (2) off-roadway partial departures not involving crashes (i.e., into the breakdown lanes); and (3) crashes (i.e., complete off-roadway departures; collisions with moving and/or stationary objects).

Statistical Analysis

By using subjects as their own controls and counterbalancing the order of nighttime drive conditions, the effects of drowsiness feedback (i.e., alarms and alerting stimuli) can be made on all 16 subjects. A test of repeated assessment will be undertaken first, however, in the event that mere repeated exposure to the nighttime drive has marked effects on primary dependent variables independent of (or in interaction with) the intervention condition. If this is the case, statistical comparisons will be made between the 8 subjects who had the control condition first and the 8 who had the intervention condition first. Since there will be random assignment to order, this between-subject alternative analysis will ensure that the experiment yields meaningful comparisons even if the repeated exposure to the nighttime driving scenario has an unexpected effect.

The results of this experiment will tell us the effectiveness of an automated, on-line, PERCLOS-based, drowsiness-detection system with feedback in reducing driver drowsiness on a nighttime drive with time pressure. Within this broad question the study will also determine whether the nature of the feedback (i.e., alarm vs. alerting stimuli) influences the effectiveness and use of the system by drivers. Below this level, the project will also provide information on whether or not the visual feedback alarm and alerting stimulus (i.e., auditory alert or olfactory stimulus) will optimally reduce drowsiness and lead to safer driving behaviors. Consequently, in addition to resolving whether such a system actually reduces the duration and severity of drowsiness episodes while driving, the study will determine whether the deployment of such a system encourages drivers to engage in additional rest breaks to further promote alertness, or whether it encourages continuous driving with even fewer breaks. Similarly it will tell us whether unsafe driving practices are increased or decreased by the system. These latter questions are the most frequently cited issues of contention when the deployment of drowsy-driving technologies are discussed. This will be the first project to directly measure these outcomes and provide quantifiable data on these controversial issues.

The use of the TruckSim for the above experiment will provide a wealth of information without posing a threat to the driver or others on the roadways. From this study, information will be gained about drowsiness detection systems and feedback in reducing driver drowsiness.

Conclusions

Used properly, ground vehicle simulators can be an extremely effective tool for improving driver safety. The value of using simulators for training and human factor research has been clearly shown in aviation where simulators have been used for 30 years. The challenge in front of us is to apply simulation in an effective and economic manner to the ground transportation industry.

Realistic driving simulators are available today at a reasonable cost, and the cost is expected to fall sharply over the next few years. To achieve widespread and effective use of simulation, we must proceed quickly with a combination of applied and basic research initiatives focused on building the tools needed for widespread use of simulation.

Applied research programs can provide the rapid assessment of simulator effectiveness for training and other purposes. These efforts will provide a quantification of the cost and benefits for employing simulators for widespread training. These efforts can also quantify how well lessons are learned in a simulator (whether the lessons are driving skills or the value of a heads-up display) and how well these lessons transfer to the real world of the highway.

Basic research can, over a period of time, provide a more in-depth understanding of simulators and their effective application. Driving simulator technology is being engineered, even though developers are in unanimous agreement that they are proceeding without knowledge that should be guiding their activities. Driving simulators are currently being used for research and training, even though doubt exists about whether the products of these efforts will generalize to the natural environments targeted.

A DROWSY DRIVER DETECTION SYSTEM FOR HEAVY VEHICLES

Richard Grace, Carnegie Mellon Research Institute, Pittsburgh, PA

Vicky E. Byrne, Carnegie Mellon Research Institute, Pittsburgh, PA

Damian M. Bierman, Carnegie Mellon Research Institute, Pittsburgh, PA

Jean-Michel Legrand, Carnegie Mellon Research Institute, Pittsburgh, PA

David Gricourt, ESIGELEC, France

Robert K. Davis, Carnegie Mellon Research Institute, Pittsburgh, PA

James J. Staszewski, Carnegie Mellon Research Institute, Pittsburgh, PA

Brian Carnahan, University of Pittsburgh, Pittsburgh, PA

Abstract

Driver drowsiness/fatigue is an important cause of combination-unit truck crashes. Drowsy driver detection methods can form the basis of a system to potentially reduce accidents related to drowsy driving. We report on efforts performed at the Carnegie Mellon Driving Research Center to develop such in-vehicle driver monitoring systems. Commercial motor vehicle truck drivers were studied in actual fleet operations. The drivers operated vehicles that were equipped to measure vehicle performance and driver psychophysiological data. Based on this work, two drowsiness detection methods are being considered. The first is a video-based system that measures PERCLOS, a scientifically supported measure of drowsiness associated with slow eye closure. The second detection method is based on a model to estimate PERCLOS based on vehicle performance data. A non-parametric (neural network) model was used to estimate PERCLOS using measures associated with lane keeping, steering wheel movements and lateral acceleration of the vehicle.

Introduction to the Drowsy Driver Problem

There are approximately 1.6 million truck tractors and 3.6 million trailers used in the motor carrier industry today. Advanced technology offers a significant potential to improve the productivity of the industry and help reduce the approximately 200,000 crashes in which combination-unit trucks are involved each year.

Driver drowsiness/fatigue is an important cause of combination-unit truck crashes. Recent analyses of the problem estimate that 1,200 deaths and 76,000 injuries can be attributed to fatigue related crashes annually.[1] The cost of these crashes is an estimated $12.4 billion per year and $2,060 per vehicle over the lifetime of a combination-unit truck.[2] This toll motivates efforts to identify and implement effective countermeasures for driver fatigue.

---

[1] Rau, P. S. (1996). NHTSA's drowsy driver research program fact sheet. Washington, D. C.: National Highway Traffic Safety Administration.

[2] Wang, J. S., Knipling, R. R., and Blinceo, L. J. (1996). Motor vehicle crash involvements: a multi-dimensional problem size assessment. Sixth Annual Meeting of the American Intelligent Traffic System, Houston, TX April 14-18, 1996. Washington, D. C.: National Highway.

We report on efforts performed at the Carnegie Mellon Research Institute (CMRI) Driving Research Center (DRC) to develop such countermeasures. The objective of this work is to devise and develop an in-vehicle system for Commercial Motor Vehicles designed to continuously monitor drivers for drowsiness. If it is possible to detect when a driver is drowsy and may fall asleep behind the wheel, it may be possible to implement accident prevention countermeasures.

Psychophysiological measures (e.g., devices to detect slow eye closure) and driver performance (e.g., steering wheel movements, lateral lane position, longitudinal speed, lateral and/or longitudinal acceleration, braking) are both classes of measures that could provide information to contribute to drowsiness detection. Vehicle based systems and psychophysiological systems can be applied separately or in combination to enhance system robustness. Both a video-based system to measure eye closure and driver performance-based detection methods are currently under consideration at the Carnegie Mellon DRC.

The development of a special camera system intended to obtain psychophysiological measure of eye closure is the main focus of this paper. First, there will be a discussion of the foundation of this eye-closure-based drowsiness detection system work, followed by detailed account of the Perclos camera development. Next, the other work based on driver performance measures will briefly be mentioned in order to provide insight into an alternative approach to drowsiness detection. A summary of the progress made at the DRC will be included along with future work intended to move closer to the DRC goal of developing an in-vehicle system for Commercial Motor Vehicles to continuously monitor drivers for drowsiness.

Fundamental Principles of a Video-Based Drowsiness Detection System

A special video-based system is under development at CMRI that measures eye closure, more specifically, PERCLOS. PERCLOS is defined as the proportion of time the eyes are closed 80% or more for a specified time interval.[3] Recent results from efforts at the DRC[4] and the University of Pennsylvania[5] have shown that ocular measures of drowsiness, in particular PERCLOS, can be effective in identifying lapses in attention due to drowsiness and that PERCLOS can be measured directly using video processing techniques. Significant progress has been made in fielding a device that directly measures PERCLOS in real time.

Field Study

Data collected from the DRC field study (see footnote 4) involving truck drivers driving their normal evening/early morning hour shift over a two week period provided a foundation for the work discussed here. Duty periods covering a 7 pm to 7 am time interval were selected to maximize the frequency with which drowsy episodes were likely to occur and to observe drivers when they were likely to pose the greatest threat to safety.

On-road data collection involved professional commercial motor vehicle drivers driving vehicles identical to those used by their firm, but specially instrumented to collect vehicle performance and psychophysiological

---

[3] Weirwille, W.W. (1994). "Overview of Research on Driver Drowsiness Definition and Driver Drowsiness Detection," 14th International Technical Conference on Enhanced Safety of Vehicles (ESV), Munich, Germany, May 23-26.

[4] Grace, R., Staszewski, J.J., Byrne,V.E., Davis R.K., Bierman, D. M., Jarrett, J., Swecker, G.W. (1998) "Develop, Test, and Evaluate a Drowsy Driver Detection and Warning System for Commercial Motor Vehicles, Draft Final Report," submitted to US DOT.

[5] Dinges, D. F., Mallis, M.M., Maislin, G., Powell, J.W. (1998) "Evaluation of Techniques for Ocular Measurement as an Index of Fatigue and as the Basis for Alertness Management, Draft Final Report," submitted to US DOT.

measures as unobtrusively as possible. Instrumentation consisted of a video camera and recording system for capturing drivers' faces and sensors for collecting drivers' head positions. Also recorded were steering wheel movements, the vehicles' lateral acceleration, and the vehicles' position relative to roadway lane boundaries.

Approximately 500 hours of recorded video images were collected. The video tapes were monochromatic and through the use of 880 nanometer light source, a pronounced retinal reflection resulted making the eyes appear to glow.

Specially trained coders were employed to manually generate ocular measures of drowsiness from the video tape records using previously developed techniques for measuring PERCLOS (see footnote 3). The coder's task involved estimating the degree to which a driver's eyes were open or closed. From this data, PERCLOS was extracted. The reliability of the manual coders was assessed to be sufficient.

Based on the manual coding of PERCLOS, the field study was successful in capturing several drowsy episodes from overnight drivers. Two of the seven participating drivers showed frequent and severe bouts of drowsiness. Figures 1 and 2 show a typical night of observed drowsiness for these two drivers. Each plot spans the portion of the night's run where the bout was observed (time is indicated on the horizontal axis), and the peaks and troughs of the plot itself represent the proportion of time PERCLOS was 80% or greater within each minute. Therefore, a value of 0.5 on the plot indicates that the driver's eyes were scored at least 80% closed for half of that paritcular minute.

The manual coding provided information about the number, duration, and severity of drowsy episodes amongst the drivers, however it is impractical and costly to consider placing a human in the cab of a truck to monitor drivers. Nevertheless, the information gained from the manual coders served as a baseline of comparison for automated systems currently under development.

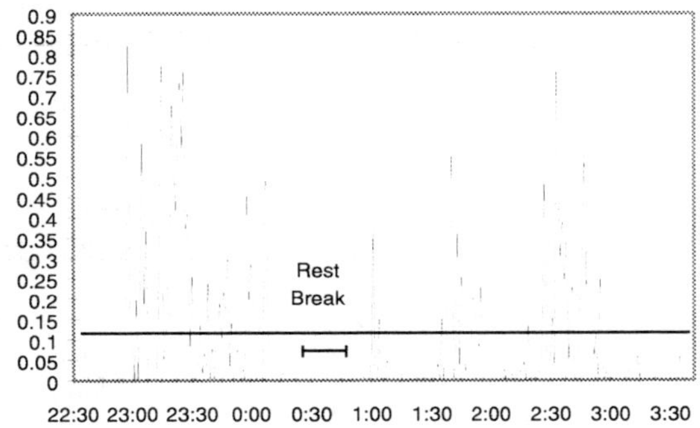

Figure 1: One night of observed drowsiness for Driver A.

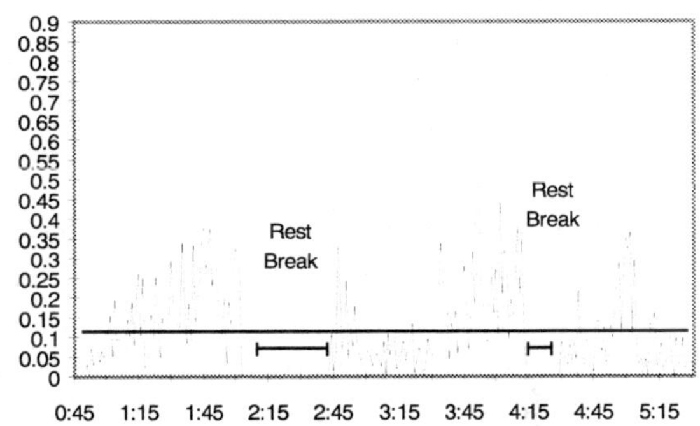

Figure 2: One night of observed drowsiness for Driver B.

Semi-automated Eye Tracking System

As a move towards more economically efficient measurement of eyelid closure, a semi-automated eye tracking system was developed. Some of the eye tracking procedures developed in this semi-automated system provided the necessary groundwork for the fully automated PERCLOS camera that was to follow.

The system tracked both eyes directly from a digitized video image, and independently calculated the percentage eye closure for each eye based on a pixel count. The system provided an eye closure value based on the eye that was most open.

The most important result to come out of testing the semi-automated eye tracking system was the system's ability to track the eyes with minimal operator intervention at a maximum rate of 6 frames per second. Through the course of testing, the system correlated very highly with manually coded data, and almost perfectly with itself on repeated trials.

However, the most significant problem with the semi-automated eye tracking system was its tendency to lose the eye under various circumstances (thereby necessitating operator intervention). Specifically, large-scale head movements would often take the eyes outside of the algorithm's search space, rendering them lost. Competing bright spots near the eyes (due either to additional infra-red reflections other than the eye itself, or in some cases, ambient light) would sometimes cause the system to fail. In particular, images of drivers with eyeglasses could not be processed by the system because of the additional reflections created by the eyeglasses. The PERCLOS camera was able to eliminate many of these spurious reflections via a subtraction process (described later). Large-scale head movements did not present a problem to the PERCLOS Camera because of its altogether different method of frame handling (described later).

Also, significant speed improvements were realized with the PERCLOS camera, as it processed information via dedicated hardware, an utilized more refined data handling techniques.

Through the development, testing, and analysis of the semi-automated eye tracking system, several key elements were decided upon as critical to the success of the PERCLOS camera. First and foremost, the PERCLOS camera must be able to accurately measure eye closure in real time and for all drivers. In addition, the camera must sit stably in the cab of the truck, and yet be inconspicuous enough so as not to be a distraction or to hinder the driver's vision in any way. Finally, the camera must possess sufficient memory to be able to store an entire run's worth of data.

PERCLOS Camera

Overview

DRC has designed a prototype of the PERCLOS camera for measuring eye closure in a heavy truck environment. The prototype eye tracking system was designed using charge-coupled device (CCD) imaging technology and a dedicated PC/104 with PCI Bus computer platform. The system is currently capable of measuring driver eye closure under low-light conditions, and calculating the resulting PERCLOS values in real time.

Prototype

The PERCLOS camera works by exploiting a basic physiological property of the human eye: That the retina reflects different amounts of infra-red light at different frequencies. At the 850 nanometer wavelength (nm), the retina reflects 90-percent of the incident light. However, at 950nm, the retina reflects only 40-percent of the incident light. In order for the camera to work, it requires two equally illuminated images of the driver's face—one created with 850nm light, and one created with 950nm light. The only difference in the two images is the intensity of the retinal reflections, so when the second image is subtracted from the first, only the retinal reflections remain. Once isolated, the retinal image sizes are measured, and the results are used to calculate PERCLOS. This process is repeated ten times per second.

In order to get two identical images with different sources of light, the system utilizes two separate cameras, both focused on the same point, yet situated at a 90-degree angle to one another. The image is passed through a beam splitter which reflects the image onto the lenses of both cameras. In order to isolate the correct wavelengths of light, one camera is outfitted with an 850nm filter, and one with a 950nm filter. The result is two cameras which receive identical signals at the same point in time, but which filter only the specific wavelengths of light necessary for the subtraction process.

Technical Components

Listed below is a list of the required instrumentation and computer hardware for the camera prototype:

Instrumentation box components

- 2 CCD cameras
- 2 filters (850 and 950 nm)
- 1 beam splitter
- a fan to reduce heat from the infra-red source
- a board built to synchronize the 2 cameras and control the current of the LEDs
- 1 LED ring with 850nm and 950 nm LEDs

Hardware components

- 1 timing controller
- 2 frame grabbers
- a PC104 w/ PCI Bus and a PCMCIA hard disk to store data

A diagram of the PERCLOS Camera, its instrumentation and hardware, is given below (Figures 3 and 4):

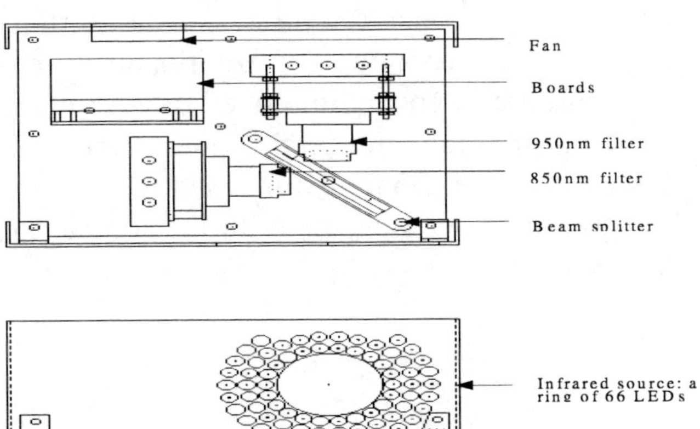

Figure 3: Instrumentation of PERCLOS Camera

Figure 4: Hardware (PC/104 Stack) for PERCLOS Camera

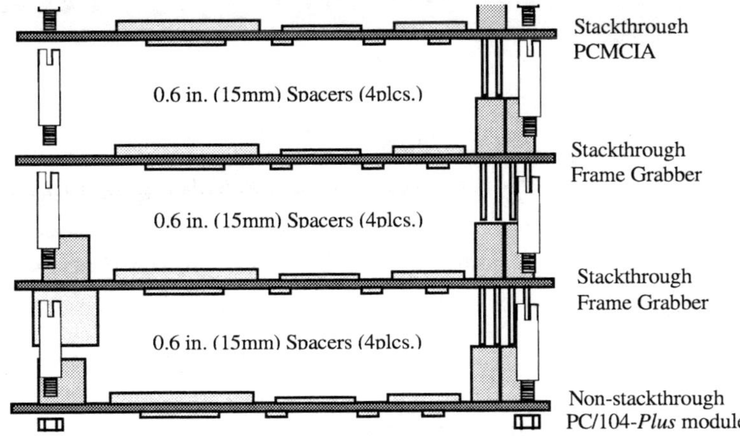

Because the camera is being designed to fit in the cab of a commercial motor vehicle, there are several important considerations to make with respect to its portability:

- Size and weight - System must be small enough and light enough to fit unobtrusively in cab.
- Power consumption - System must have sufficiently low power consumption. Since power consumption is directly proportional to heat generation, reduced power consumption may be required to support operation in high ambient temperatures.
- Shock and vibration - System must be able to withstand a wide range of motion-induced stress for an extended period of time.
- Temperature and humidity - System must be able to withstand large fluctuations in ambient temperature and humidity.
- Electrostatic and electromagnetic interference - System must be protected from possible interference emitted from radio/radar signals in the vicinity, as well as from other electrical devices in the truck.
- Power supply irregularities - System must be able to withstand possible fluctuations in its power current caused by irregularities in vehicle power sources (e.g. starting of the engine)
- Maintenance and repair - System must have a high mean time between failure (MBTF) rate.

Knowledge gained from the laboratory semi-automated eye tracking system was used to design and construct this rugged version of the automated PERCLOS camera system for field-testing.

The System's Algorithm

The process by which the PERCLOS camera reads in video images and processes the images to generate eye closure values involves eight steps:

1. Initialization - Prior to calling any program routines, the system must be initialized. Among the processes carried out during initialization are: 1) Determining which frame grabber is which and assigning them to their corresponding PCI bus slots, and 2) Setting the correct image size, thereby saving memory and bandwidth for later use.

2. Frame Grabbing - The frame grabbing process stores in memory the images captured by each camera. It operates on a simple loop which 1) grabs a frame from the video stream, and 2) processes the information in it.

3. Image Synchronization - The synchronization routine ensures that the frames being processed from each camera line up perfectly. This is necessary in order to guarantee that the correct images are being subtracted from one another.

4. Subtraction - The subtraction process is the principal routine in the system. Essentially, it is a very simple, computationally-inexpensive routine, which subtracts each value in the 950-image buffer from each value in the 850-image buffer, and stores the result in a subtraction buffer.

5. Bright-Spot Segregation - The bright-spot segregation routine clusters all regions in the subtraction buffer which contain non-zero pixel values. The routine first gathers all the bright pixels contained within a certain area, and then passes the values contained in each cluster to the extraction routine which determines whether the cluster is an eye, or noise.

6. Eye Extraction - The eye extraction routine discriminates the eye from surrounding noise. It does this by summing the pixel values of each cluster, and choosing the two largest clusters above a certain threshold to make the eye/noise distinction. While the summing process is computationally inexpensive and quite fast, it isn't 100% accurate. For this reason, an additional process is carried out which calculates the correlation between each candidate cluster and an "ideal" eye. The two highest scoring clusters are chosen as the eyes, and all other clusters are discarded.

7. Eye Measurement - The eye measurement routine calculates the height (i.e. degree of openness) of each eye, and chooses the larger of the two values for storage.

8. Data Storage - The data storage routine stores the time and the calculated height of each observation in a data file on the PCMCIA hard disk.

Screen captures illustrating the 850nm image, the 950nm image, and the subtracted image are shown below (Figures 5, 6, 7):

Challenges Met with Current PERCLOS Camera

The PERCLOS camera prototype solved many of the problems encountered with the semi-automated eye tracking system. Most significantly, it can run in real time without operator intervention. Additionally, eye recognition was greatly improved by correlating shape information with bright spot information. Also, the glare from drivers wearing glasses was eliminated completely with the subtraction process.

Figure 5: The 850nm image

Figure 6: The 950nm image

Figure 7: The subtracted image

Perclos Camera Placement into Heavy Trucks as a Detection System

The field test device is designed to operate under harsh conditions of vibration and electrical noise found in the environment of a typical tractor trailer cab. The device has recently been completed and has been installed for initial field-testing.

The initial field-testing will involve data collection during truck runs with selected drivers, laboratory analysis of the collected data, and making software refinements to optimize the accuracy of the PERCLOS measurements.

Future Directions in PERCLOS Camera Development

While the prototype of the PERCLOS Camera satisfies the requirement of a real time eye tracking system, there are changes that could be made to improve the system's operation and efficiency. Specifically, there are four changes that are scheduled to be made to the next generation PERCLOS Camera eye tracking system:

1. There will be a shift from a CCD image sensor to a complementary metal oxide semiconductor (CMOS) image sensor. The CMOS will not only provide greater sensitivity in the near infrared light spectrum, but it will also provide improved timing control, eliminating the problems with synchronization experienced in the prototype.

2. The PC104 microcomputer will be replaced with a microcontroller. This will eliminate the need for the frame grabbers as it will make the unit entirely self contained.

3. In order to allow more radiant energy while the image is captured, the system will pulse the LEDs. The pulsing will provide a higher signal-to-noise ratio, which will ultimately improve the system's ability to extract the eyes, and lower the system's total power consumption in the process. This will also serve to reduce the amount of infra-red light going to the eye.

4. The final change will involve redesigning the optics. A single lens will be placed in front of the filtering mechanism which will allow for the reduction in size of the optical set up.

An Alternative Approach to Drowsiness Detection – The Driver Performance-Based Model

Another approach to drowsiness detection that could be used in place of, or in conjunction with, the psychophysiological measure of eye closure involves the estimation of PERCLOS based on measurements of driver performance. Examples of driver performance measures include: Steering wheel movements, lateral lane position information, longitudinal speed, lateral and/or longitudinal acceleration, and vehicle braking. The categories of driver performance variables used in the CMRI DRC study (see footnote 4) included: Steering-related measures; lane-related measures; accelerometer-related measures; and head-position-related measures. Altogether, these measures provide a complete record of the driver's performance throughout the course of a run..

Results obtained from linear regression techniques indicated that seven variables were accounting for the majority of the variability

found in the PERCLOS measure. Therefore, it was these driver performance variables that were initially considered viable inputs to a performance-based drowsiness detection device. Specifically, the standard deviation of lane position, 4-steering wheel related measures, and 2-accelerometer related measures were isolated as the variables most contributing to the prediction of PERCLOS. It was this reduced set of variables that was then picked out for use with the neural networks. The neural networks are currently being developed and tested in order to determine the combination of driver performance variables that best model PERCLOS.

One potential drowsiness detection system might include both the video-based and driver performance- based systems to provide enhanced drowsiness detection specific to the habits and characteristics of individual drivers. Each approach provides its own set of costs and benefits, which should also be considered in future development.

Conclusions

The Carnegie Mellon DRC has put forth a considerable effort toward developing a video-based system to detect drowsiness in heavy vehicle truck operators. The work is firmly grounded in a field study also conducted at the DRC.

The current video-based system, the PERCLOS Camera, successfully measures eye closure in real time under nighttime driving conditions.

This ongoing research project is currently devoted to gathering field test data with the prototype system, improving and fine-tuning the system's performance, and further developing driver performance-based models of drowsiness detection.

Beyond the current efforts, the DRC also plans to begin exploring the various alerting/warning systems that could be triggered when drowsiness is detected.

DETERMINING AND CONTROLLING EMERGENT BEHAVIOR IN INTELLIGENT TRANSPORTATION SYSTEMS

James L. Overholt, National Automotive Center, U.S. Army TACOM, Warren, MI

Wendy K. Krill, National Automotive Center, U.S. Army TACOM, Warren, MI

Ka Chi Cheok, Oakland University, Rochester, MI

Abstract

Intelligent Transportation Systems (ITS) are the integration of current and emerging technologies in fields such as information processing, communications, and electronics applied to solving surface transportation problems. One specific use of ITS technologies seeks to incorporate a variety of on-board sensors to assist the driver in the efficient and safe operation of a vehicle.

The interaction of a series of independent ITS entities amongst themselves and/or with their environment, can lead to the development (or emergence) of highly complex behaviors or properties. In most cases these behaviors will be unanticipated by the original system designer. The potential for systems instability is a definite concern. Intelligent transportation systems have the potential for generating these undesired effects.

This paper will examine the current methods used to predict and control the emergence of unpredictable behaviors in complex driving assistance systems, especially concentrating on minimizing the overall effects of this phenomenon on the performance, stability and safety of ground vehicles.

Introduction

Intelligent Transportation Systems are a collection of technologies, systems, and transportation management concepts that collectively aim to make surface transportation more efficient and safer. It is a "system of systems," envisioned to evolve over the next 20 years. For the purposes of bringing focus to collaborative government and private research efforts, ITS has been bounded and defined in terms of 29 user services, under 6 umbrella categories;

- travel and traffic management
- public transportation management
- electronic payment
- commercial vehicle operation
- emergency management
- advanced vehicle safety systems

Under the *advanced vehicle safety systems* category the user service titles include;

- longitudinal collision avoidance
- lateral collision avoidance
- intersection collision avoidance
- vision enhancement for crash avoidance
- safety readiness
- pre-crash restraint deployment
- automated vehicle operation

In all of these service titles, "sensed" information is critical in the determination of a course of action for the driver and for the automated controls. When conflicting information is sensed, a potential instability occurs if the system is not designed to handle competing resources.

Multi-Objective Systems

Hierarchical systems contain multiple objectives that need to be satisfied in order for the overall system to perform its designed function. These tasks will each have an associated priority, which is used to properly

0-7803-5086-3 /98/$10.00 ©1998 IEEE

place the emphasis of the system control. The control policy created for this type of hierarchical structure must strive to accomplish all of the tasks according to the order dictated by the relative priorities. In actuality, the priority listing of each task is not fixed but will change based on factors such as the current state(s) of the plant and the magnitude of the control (called a *varying or "dynamic" priority hierarchy*).

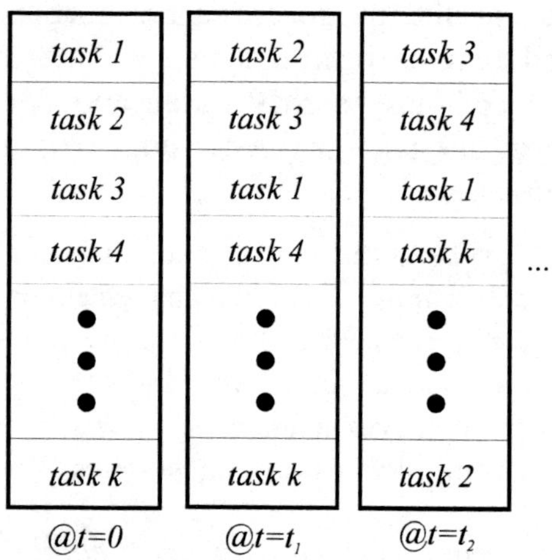

Figure 1: Task Priority Listings

In addition, these priority changes will occur with very little (if any) forewarning (**Figure 1**). Perhaps the biggest concern is that with the shifting priorities of the system, if a lower priority task is accomplished it may result in the system becoming unstable. Hierarchical systems control policies need to incorporate a structure, which tries to anticipate and handle priority changes in the multiple tasks of the system (**Figure 2**).

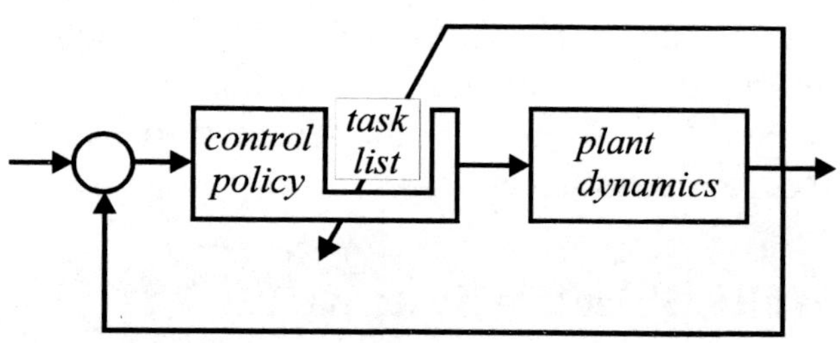

Figure 2: Varying Priority Control

All advanced vehicle safety systems are multi-objective systems that have the same operating concerns as those mentioned above.

Emergent Behavior

Getting artificial mechanisms to correctly choose between two or more conflicting situations, on a consistent basis, is a monumental task for even the simplest of processes. Many of the approaches put forth by the intelligent control community advocate the use of training and adapting "primitive" structures to assist in the solution of the conflict resolution problem. In turn, complex process behaviors come about by the simultaneous application of these primitive entities for the control of much more complicated problems. The resulting *unpredictable* output is sometimes referred to as the system *emergent behavior*.

I Robot by Isaac Asimov, Edward Lorenz's discovery of chaotic dynamics phenomena and John Conway's "Life" simulations all present examples of systems that are simply described but result in emergent behaviors. Asimov presented a possible *science fiction* framework for robots and artificial intelligence of the future with his *Three Laws of Robotics*. Lorenz's computer model for predicting weather conditions was found to be highly susceptible to small changes in the system initial conditions, resulting in vastly different outcomes. Finally, Conroy developed a set of simplistic, rigid laws governing the life and death of blips on a computer screen, all for the purpose of imitating bacterial life in a Petri dish. In all of these cases, each individual set forth very simple system descriptions and processes that led to complex, *unpredictable* results.

Inexact Descriptions

Complex control policies for large-scale nonlinear systems could be developed through the implementation of simplistic rule sets. The rules would take into consideration the states of the system, and if applicable the surrounding environment. Precision of the final results would not be a necessary condition of the output. However, only those systems with a

0-7803-5086-3 /98/$10.00 ©1998 IEEE

less then exact goal would be suitable for this approach. For example, driving a car from point A to point B can be considered an inexact task. A rule should be simplistic and take into account only one or two states of the system, e.g. *if the road is* ***veering left*** *and the car is* ***going fast*** *then* ***slow down slightly*** *and* ***turn the steering wheel left***.

Contextual-based Reinforcement

If a simplistic rulebase framework is to be developed for the control of complex nonlinear systems, it will need to incorporate a mechanism to resolve the frequent conflicts that will occur. Since the rules are developed to handle very general cases, a *contextual based resolution mechanism* is required to differentiate between specific situations.

Human beings operate under a very similar framework. Every day we deal with and resolve conflicting stimuli in order to function in the real world. Our day is spent striving to accomplish a never-ending series of tasks using a set of complex adaptive rules. When these rules are in conflict, an internal resolution mechanism is triggered and some final choice is made (this resolution capability develops through teaching, experience and heredity). If the choice is "correct", that choice is *reinforced* through some external reward system. If the choice is "incorrect", we (hopefully) learn not to make a similar selection should the same situation ever occur again (i.e. *negative reinforcement*).

Behavioral Control Approaches

Although the control literature is rife with behavioral approaches, the most popular is the *subsumption architecture* created by Rodney Brooks, of MIT.

Another that has recently been gaining acceptance is the development of the *threshold fuzzy systems* methodology by Jim Overholt, of Oakland University.

Multi-Layer Approach

In his seminal paper, "*A Robust Layered Control System For A Mobile Robot*", Rodney Brooks proposes a layered control approach where the control strategies are built upon fundamentally simple control behaviors (e.g. avoid objects). As more complicated tasks are required more layers to the control structure are added. The advantage of this approach is that the simple, foundational layers need not change once they have been tuned and incorporated. More and more layers can be added to accomplish a variety of complex tasks. While the system is running, all of the layers are processing information without the benefit of any *central control*. All control processes are trying their best to accomplish their particular tasks. When conflicts occur, the system will default to a lower layer process. Brooks called this de-centralized, layer-based control the *subsumption architecture*. Layers run asynchronously, monitoring their input sensors and sending control actions to their respective output signals. There is no "handshaking" between the layers. Outputs can be inhibited (by the outputs of other layers) and inputs to the layers can be suppressed. This is the mechanism in which higher level layers subsume the role of lower levels. Brooks' *subsumption architecture* (**Figure 3**) allows simple insect-like mobile robots to maneuver and navigate in an instinctual way.

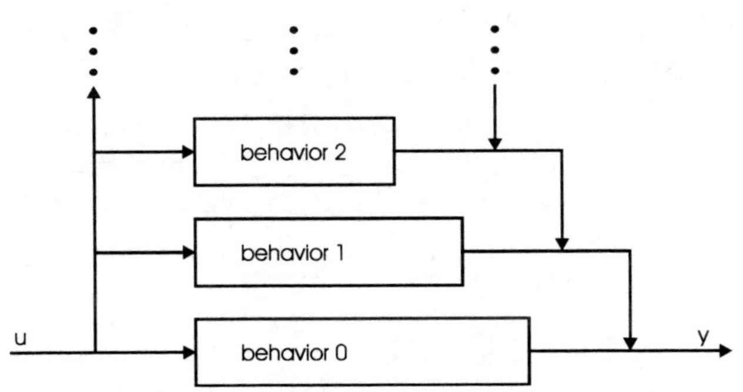

Figure 3: Subsumption Architecture

Limit Cycle Avoidance

Brooks' method requires that the users choose the fundamental layers and determine the hierarchy of the system tasks. For a simple mobile robot this is straightforward, however for other complex multi-objective control systems this is not an easy task. What is needed is a system that will allow the designer to place his emphasis and resources on the development of the behaviorist rulebases. The determination of the "layering" or "dominance structure" should be automatically determined based on the requirements of the system. In addition, some amount of planning and memory should be incorporated so that limit cycles can be avoided. This is fundamentally where the idea of *threshold fuzzy systems* differs from the *subsumption architecture* approach.

Threshold Fuzzy Systems

A *threshold fuzzy system* will include all the standard components of a Mamdani fuzzy system with the additional feature of a *behaviorist fuzzy rulebase* (BFR) as well as the inclusion of a *rule dominance mechanism* (RDM). Behaviorist fuzzy rules are a set of simply structured rules used for the control of a single task. The RDM is used to moderate and referee conflicts between the conflicting rules of the BFR.

The idea for *threshold fuzzy systems* is based on developing nonlinear control schemes where a designer could use their intuitive knowledge of the system's input-output relationships for the control design process. **Figure 4** shows the components of a TFS.

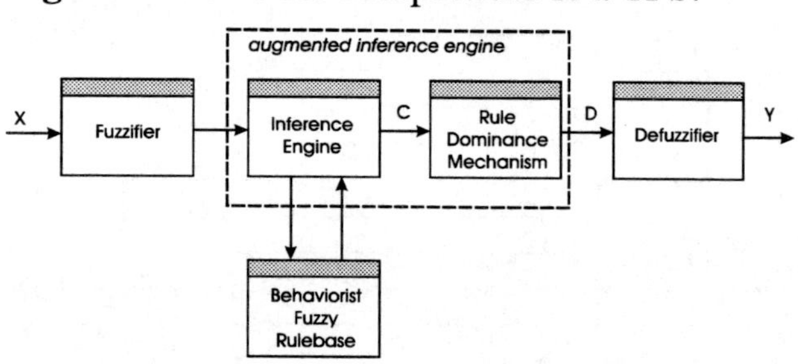

Figure 4: Threshold Fuzzy System

RDM Definition

The inclusion of a BFR into the standard Mamdani fuzzy control scheme is not sufficient to control all of the system responses of multi-objective plants. The addition of an RDM acts to modify the contribution of a fuzzy rule if a conflicting rule's contribution has reached a predetermined level.

The modification of a conflicting rule is done at the point where the output of the rule is about to be used in some sort of aggregation scheme. **Figure 5** shows a geometric version of the RDM functionality.

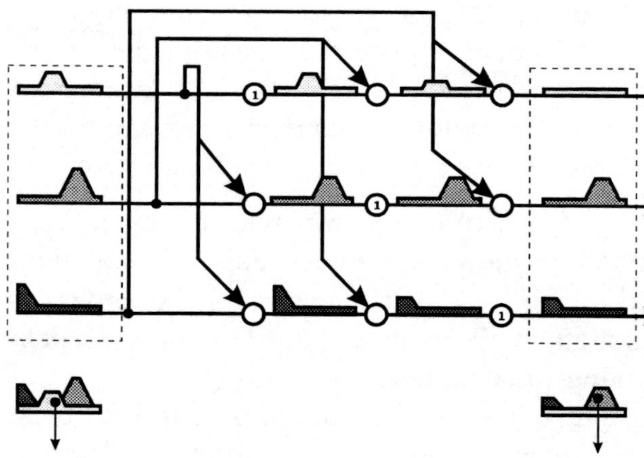

Figure 5: Mutually Conflicting 3-Rule System with an RDM

Examining the illustration shows that the output shifted to the right for the 'dominated' set of contributions. The contribution of rule #2 was affected the least. The contribution of rule #1 was entirely eliminated by the domination of rule #3. Rule #3 was reduced solely by the contribution of rule #2. The end result shows that the system was modified in favor of rule #2.

Simple Threshold Fuzzy System

Given the *conflicting behaviorist fuzzy rules*,

Rule 1: If x_1 is A then y is R, and

Rule 2: If x_2 is B then y is S,

than a simple *behaviorist fuzzy system* and a *rule dominance mechanism* is represented in

Figure 6. Each *node* in the RDM network represents a dominating pair from the BFR. The lower, left node (with the *dominance function* $\mathbf{F}_{2,1}$ in the middle) represents the dominance of *Rule 1* over *Rule 2*. The upper, right node (with the *dominance function* $\mathbf{F}_{1,2}$ in the middle) represents the dominance of *Rule 2* over *Rule 1*.

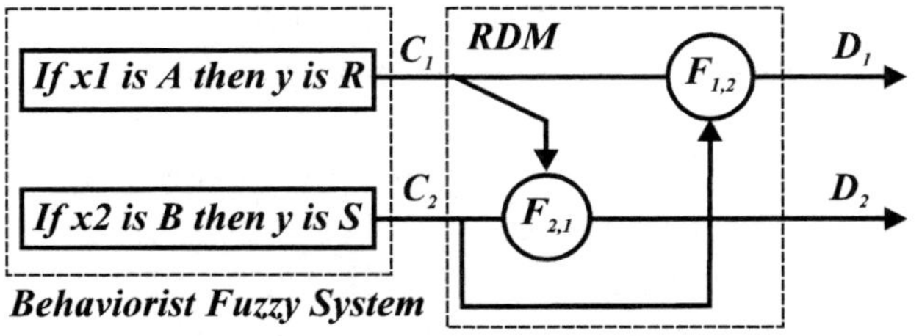

Figure 6: Simple 2-Rule RDM

The inputs to the dominance function are the contribution of the rule being dominated and the contribution of the rule doing the dominating. In general, the dominance function $\mathbf{F}_{m,n}$ implies that Rule 'm' is being dominated by Rule 'n'.

The dominance functions must be found so that the TFS will behave in the desired manner. This is accomplished through a supervised training scenario.

Supervised Training of a TFS

The idea of finding an optimal set of dominance functions will depend on the "expert" training data that can be used for he task. The construction of the TFS architecture is straightforward and in some sense automatic (especially when the BFR has been designed). The basic training scheme uses a "backpropagation-like" search methodology to find the optimal set of function parameters. A "least-squares" error function is used to determine the point at which training is complete. It is this neural network synergism that places the TFS into the neuro-fuzzy taxonomy.

Besides the gradient search method, a *genetic learning algorithm* could also be employed to determine the near optimal dominance function parameters.

Whatever method is used, the data has to be "rich enough" in order to capture a wide variety of the different scenarios that the system is meant to handle.

Conclusions

Intelligent Transportation Systems are essentially multi-objective, hierarchical systems. In order to assure robustness in the performance of these systems, built in behavioral conflict resolution mechanisms must be included.

The more popular methods espouse the use of behavioral techniques and neural network or genetic algorithm –based training for the determination of optimal system parameters.

References

1. Mamdani, E. H. "Applications of fuzzy algorithms for control of simple dynamic plant." *Proc. IEE*, 121 (12), 1585-1588.
2. Wang, L. X. *A Course In Fuzzy Systems And Control*. Prentice-Hall, NJ (1997).
3. Brooks, R. A. "A robust layered control system for a mobile robot." *IEEE J. of Robotics Automat.* (March 1986), 14-23.
4. Anderson, T. L. & Donath, M. "Animal behavior as a paradigm for developing robot autonomy." *Robotics and Autonomous Systems* 6 (1990), 145-168.
5. Asimov, I. I Robot, (1950).
6. *Chaotic Dynamics*. Prentice-Hall (1989).
7. Gardner, M. "Mathematical Recreations: John Conway's Life.", Scientific American, Oct. (1970).
8. Overholt, J.L., Cheok, K.C., Smid, G.E., "Control of Multi-Objective Plants Using Threshold Fuzzy Systems", proceedings of the World Automation Congress 98.
9. IMPLEMENTATION of the NATIONAL INTELLIGENT TRANSPORTATION SYSTEMS PROGRAM, A Report to Congress, 1994-1995.

Energy Management in Hybrid Electric Vehicles

William C. Morchin, President, Electric Bicycle Company
1345 East Main Street, Auburn, WA . Phone (253) 939-6722

Abstract

A hybrid electric vehicle is propelled with stored energy from a battery or flywheel, plus energy produced by burning fuel in an engine. The cost of the energy consumed, as well as the quantity of air pollutants released, can be reduced by optimizing (1) the ratings of the battery and engine, and (2) the power output that will be delivered by each source under the expected driving conditions. Life-cycle cost can be minimized by running the engine at a constant speed and power, and by avoiding deep discharges of the battery.

An on-board "energy manager," which contains an embedded computer, can track the energy content of the battery and optimize the load division between the battery and engine when given a travel time by the driver over a specified route. It can command the engine to deliver more power whenever the alternative is a life-shortening deep discharge of the battery. The vehicle designer needs to perform a system-engineering analysis to optimize the ratings of the engine and battery. For this he needs to understand the power required to move the vehicle at desired speeds over hills on the expected routes, and through headwinds that vary from day to day.

In this supporting analysis we modeled the hybrid vehicle by using Lagrange's calculus. We applied the model to a battery-powered electric bicycle which travelled a 200-mile route over a variety of hills. An unexpected result from this model is the energy consumed while travelling over a hilly route. With a specified travel time, no energy could be saved by climbing hills slowly and going down them rapidly. We did not consider the propulsion efficiency in this calculation. The next step will be to incorporate into the model variations in efficiency of the engine and the propulsion motor as speed is varied.

Introduction

In a hybrid electric car the fuel-burning engine can run at a speed and output power which gives the highest efficiency. Transient power for start-ups and hill-climbing can be supplied by a battery. Prototype hybrid vehicles have achieved fuel economies of over 100 miles per gallon.

Modeling is the key route to high efficiency in a hybrid electric vehicle. In aerospace engineering we have developed computational fluid dynamic models of airplanes. A scaled version of the modeled airplane is built and tested in the wind tunnel. The mathematical model is adjusted to make it reproduce the wind-tunnel results. The designer can then input small changes in the engine performance, aerodynamic form, and weight of his airplane. The model will predict the airplane's new performance in terms of fuel consumption and flight time for pertinent flight routes. The old DC-10 airplane delivered 30 passenger miles per gallon of jet fuel. The new 777 airplane, which was configured with computational fluid dynamics, delivers up to 90 passenger miles per gallon.

The hybrid-vehicle variables that can be optimized by modeling are many. Components that can be optimized include

0-7803-5086-3 /98/$10.00 ©1998 IEEE

engine rating, motor rating, and battery energy capacity. Variables to be considered in minimizing life-cycle cost include the size and type of battery. Deeply discharging most batteries reduces their lifetime as measured in charge/discharge cycles. Adopting a bigger battery reduces the depth of discharge and lengthens battery life, but at the cost of fuel for hauling the heavier battery over hills.

On-board energy management, which is now possible with small high-capacity embedded computers, is a new tool that can make hybrid electric vehicles even better. The energy manager can perform simple functions, such as informing the driver when the battery's energy content is falling below a designated minimum. By using digital data that is now available the energy manager can perform even more sophisticated tasks. We have digital data bases that contain the elevations of highways at 0.1-mile intervals. After the driver designates the route and destination of a trip and the desired travel time, the energy manager could calculate the energy required in each segment of the travel. For example, there might be a mountain pass on the route. Upon arrival at the beginning of the climb to the pass the energy manager would (1) calculate the required energy, (2) determine the energy that will be generated by the constant-power engine, (3) get the battery's energy from its charge/discharge history, and (4) advise the driver on the maximum speed with which he can drive up to the pass without excessively discharging his battery.

We used a method based on Lagrange's calculus[i] to show that it is possible to determine, using an on-board computer, the minimum energy required to travel between designated points. The method solves, by using the calculated travel energy and specified trip time the individual speeds one should travel, with available power, over each road grade. This method determines the minimum energy by setting to zero the derivative of a created function, using the operator's desired travel time.

Formulation of the Minimization Technique

We determine the minimum energy use by first writing the equations of the energy used for propelling a ground vehicle. The four components of energy use are acceleration, friction, air drag, and hill climbing. Energy recoverable from deceleration and hill descent in this study is ignored. The equations for consumed energy **(E)** for air drag, climbing hills and friction are[ii], respectively:

For air drag

$$\mathbf{E_{drag} = (C_d DA/2)(V_g+V_a)^2 s_i} \qquad \textbf{Eq,1a}$$

for hills

$$\mathbf{E_{hill} = 9.81WGs_i} \qquad \textbf{Eq.1b}$$

for friction

$$\mathbf{E_{fric} = 9.81WC_r s_i} \qquad \textbf{Eq.1c}$$

The terms in these equations are: $\mathbf{C_d}$ is coefficient of aerodynamic drag, **D** is air density (gm/m^3), **A** is frontal area (m^2), $\mathbf{V_g}$ is ground speed (m/s), $\mathbf{V_a}$ is component of wind speed head-on to vehicle velocity (m/s), $\mathbf{s_i}$ is the distance traveled (m), **W** is the vehicle weight including riders (kg), **G** is the road surface grade (a number less than 1) $\mathbf{C_r}$ is the coefficient of rolling resistance (a number less than 1). These equations give the energy in joules (watt seconds) For

[i] Kaplan, W, Advanced Calculus, Addison Wesley, 1952.

[ii] Morchin, W.C., "Battery-Powered Electric Bicycles", IEEE Technical Applications Conference, NORTHCON 94, Seattle, WA, October 12-14, 1994, Conference Record pp 269-274,

purposes of brevity we let

$C_1 = (C_dDA/2)$, $C_2 = 9.81WG$,

And $C_3 = 9.81WC_r$

To illustrate the problem, we consider the minimization of **E = E(t,s)** where **g(t,s)=0** and **h(t,s)=0** are given and where **t** represents time and **s** distance. A subscript refers to a particular segment of time and distance.

$$g(t,s) = t_1 + t_2 + t_3 + \cdots + t_i - T = 0 \qquad \textbf{Eq.2}$$

$$h(t,s) = s_1 + s_2 + s_3 + \cdots + s_i - S = 0 \qquad \textbf{Eq.3}$$

The equations **g=0** and **h=0** describe two "surfaces" in space and the problem is thus one of minimizing **E(t,s)** as **(t,s)** varies on the curve of intersection of these "surfaces". At a minimum point, the derivative of **E** along the curve, i.e., the directional derivative along the tangent to the curve, must be 0. This directional derivative is the component of the vector **σE** along the tangent. It follows that **σE** must lie in a plane normal to the curve at the point. This plane also contains the vectors **σg** and **σh**. Hence (Kaplan section 1-5) there must exist scalars λ_1 and λ_2 such that

$$\sigma E - \lambda_1 \sigma g - \lambda_2 \sigma h = 0 \qquad \textbf{Eq.4}$$

at the critical point. This is equivalent to the two scalar equations:

$$\delta_t E - \lambda_1\, \delta_t g - \lambda_2\, \delta_t h = 0 \qquad \textbf{Eq.5}$$

$$\delta_s E - \lambda_1\, \delta_s g - \lambda_2\, \delta_s h = 0 \qquad \textbf{Eq.6}$$

where δ_t and δ_s are the partial derivatives with respect to **t** and **s**. These two equations, together with the equations **g(t,s) = 0** and **h(t,s) = 0** serve as four equations in the four unknowns: **t, s,** λ_1, and λ_2.

Using the previously defined terms for the energy of traveling up a grade, overcoming air drag and friction we get:

$$E_i = C_{1i}s_i + C_{2i}(s_i/t_i + V_{ai})^2 s_i + C_{3i}s_i \qquad \textbf{Eq.7}$$

where the subscript **i** is used to denote a portion, s_i, of the total traveled distance **S** that has unique physical characteristics from other portions of the traveled distance.

Working with each ith portion of the traveled distance in **equation 5** and performing the derivatives of **equation 7** we obtain:

$$-3C_{2i}S_i^3/t_i^3 - 4C_{2i}V_{ai}S_i^2/t_i^2 - \lambda_1 = 0 \qquad \textbf{Eq.8}$$

and repeating the process with **equations 6** and **7** we get:

$$C_{1i} + 3C_{2i}S_i^2/t_i^2 + 4C_{2i}V_{ai}S_i/t_i + C_{2i}V_{ai}^2 + C_{3i} - \lambda_2 = 0 \qquad \textbf{Eq.9}$$

To proceed with the solution we will simplify our work by assuming the wind speed denoted by V_{ai} is zero.

Equations (5), (6), (8), and (9) are the basis for determining the unknowns noted above. We note that the terms s_i are predetermined when one has knowledge of the route. However we will proceed for the purpose of arriving at a solution for the other unknowns. Starting with equations (5) and (8) , we will solve for t_2 , t_3 , • • • , t_i in terms of t_1:

$$-3C_{21}S_1^3/t_1^3 = \lambda_1 = -3C_{22}S_2^3/t_2^3 \qquad \textbf{Eq.10}$$

Taking note of the term C_{2i}, which equals $0.5C_dDA$, we will make a reasonable assumption that the term is constant for all the portions of the traveled distance. The assumption may not hold up well if there is a significant elevation change, such as from 0 to 10,000 feet, throughout the distance traveled. With this assumption we find that $t_2 = t_1 s_2/s_1$ and similarly $t_3 = t_1 s_3/s_2, \cdots, t_i = t_1 s_i/s_1$. Substituting these terms into **Equation (2)** we obtain:

$$t_1 + t_1 s_2/s_1 + t_1 s_3/s_1 + \cdots + t_1 s_i/s_1 = T \qquad \textbf{Eq.11}$$

Simplifying we obtain: $t_1[\, s_1/s_1 + (S-s_1)/s_1] = T$

so that then

$$t_1 = Ts_1/S, \quad t_2 = Ts_2/S, \quad t_3 = Ts_3/S, \quad t_i = Ts_i/S \qquad \textbf{Eq.12}$$

This is the solution we seek. We find that solving **equations (3)** and **(9)** results in the same solution for the t_i terms.

We have applied these results to a limited trip scenario to illustrate the important results of this technique. To construct the scenario we used the distribution of grades determined for each 0.1 mile interval along the Seattle-to-Portland Bicycle Classic 200 mile route[iii]. The distribution is shown in **Figure 1.** We simplify this to a table of values:

Table 1. Simplified Distribution of Segments of Distance Traveled

Grade, %	Portion of distance,%
0	86
2	7
3	4
4	3

We analytically determined the energy use, assuming 100 percent efficiency, of an electric-powered bicycle for a certain distance that was proportioned into segments according to Table 1. The analysis used a specified travel time to complete the specified route. The travel time was allocated to the travel segments according to the time algorithm of **Equation (12).** The outcome of the analysis showed that minimum energy is used for a user specified total travel time if a constant speed is applied no matter if going up various road grades or not. We illustrate this with **Figure 2**. The figure shows a certain percent change in energy use for preset changes in travel speed over the various road grades. We see that for the combination of speeds there was always more energy used than if a constant speed was used. We attempted to achieve less energy use by going slower for increased road grades. It was not possible to find other combinations of speed distributions that could better the constant speed use. We interpret this as meaning one uses more energy to overcome aerodynamic drag on the flat terrain (going faster to make up for lost time for going slower on the higher road grades) than overcoming gravity on the hills. This seems counter-intuitive. However, recall that this is for a 100% efficient system.

When we consider motor losses and presumably other mechanical losses we find that the above energy minimization algorithm of **Equation (12)** does not provide for minimum energy use. It is very difficult to analytically model motor efficiency in a general sense and therefore hard to incorporate into the mathematical models. . Motor efficiency depends upon the materials used, motor configurations, and the motor controllers. In addition the efficiency as a function of motor power is highly nonlinear. **Figure 3** shows an example of one such motor. It is the motor that the Zap Bicycle Company uses. We have incorporated this

[iii] Morchin, W.C. "Trip Modeling for Electric-Powered Bicycles", IEEE Technical Applications Conference, Northcon95, Seattle, WA, November 4-6, 1996, Conference Record pp 373-377

motor efficiency curve into our computer model for determining minimum energy use.

Figure 4 shows the results of using the motor efficiency characteristics. We found that about 1 ½ percent energy could be saved compared to the previously determined constant speed operation. We found that for this particular situation that reducing ones speed on the 4% grades by about 25%, on the 3% grades by about 11%, and on the 2% grades by about 7% along with increasing ones speed on flat terrain by about 2% lead to the best results.

Conclusion

We believe the results described here can be applied to any type of motor or engine propelled ground vehicle. However, our studies are limited to electric-powered bicycles. Much more energy can be saved for larger vehicles than bicycles. The present study should be extended to include a wide variety of terrain and the effects of system losses. The algorithms need to be developed to include these system losses. The mathematical difficulties introduced by the effects of head winds should be solved. And finally a hardware demonstration program is needed to validate and demonstrate the energy savings that can be made possible.

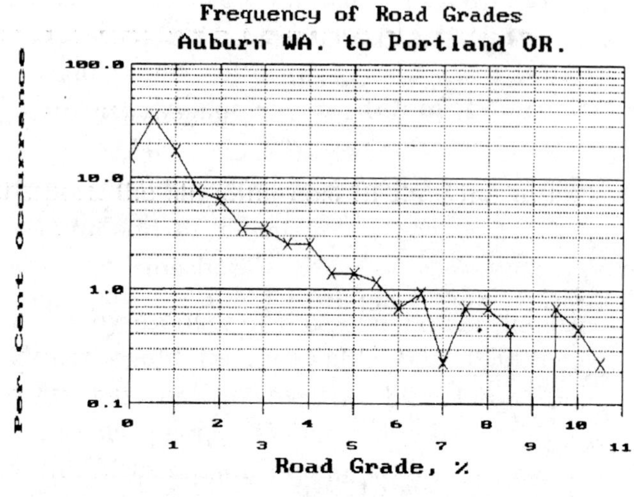

Figure 1. Distribution of frequency of occurrence for road grades on Seattle-to-Portland Bicycle Classic route. 180 mile distance from Auburn WA to Portland Or.

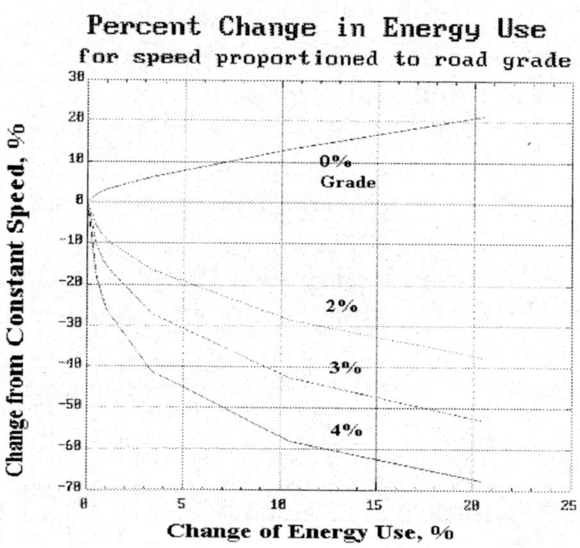

Figure 2. Illustration of percent change in energy use over hypothetical route with road grades of 0%, 2%, 3%, and 4%. Example conditions are for an electric-powered bicycle without pedal assistance and assuming 100% system efficiency. Weight of bicycle and rider is 142 kg, coefficient of aerodynamic drag is 1, frontal area is 0.42 m^2, air density is 1.209 kg/m^3, and coefficient of rolling resistance is 0.005.

Data shown are normalized to a bicycle ground speed of 10.4 m/s (23.3 mph).

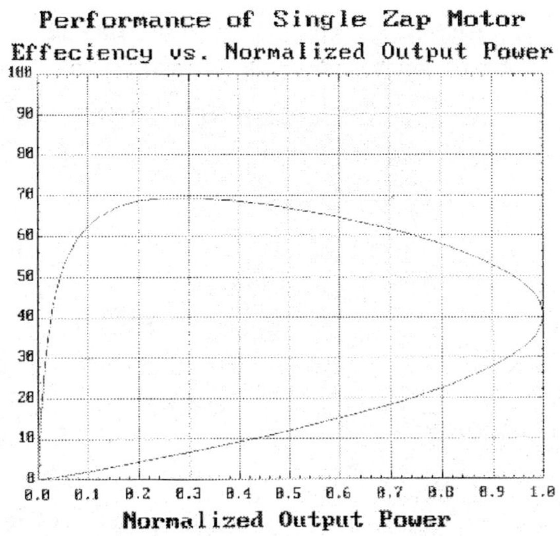

Figure 3. Example of efficiency variation of a permanent magnet field brushed motor with output motor power. Motor power shown is normalized to maximum motor output power. Example motor is that of ZAP Bicycle Co. on an electric-powered bicycle.

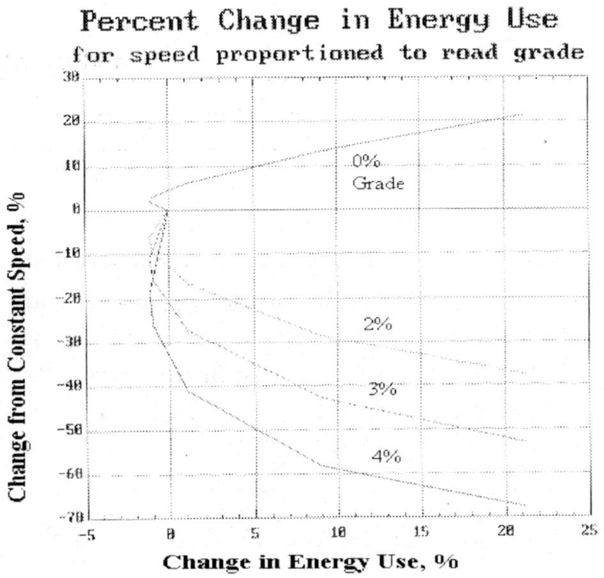

Figure 4. Illustration of energy usage when motor efficiency is used. Curves are calculated for the same conditions as for Figure 2 except motor efficiency is 67.1% for 0% grades, 67.5% for 2% grades, 64% for 3% grades, and 59% for 4% grades. We found that about 1 ½ percent energy could be saved, compared to that shown in Figure 1. We found that for this particular situation that reducing ones speed on the 4% grades by 25%, on the 3% grades by 11%, and on the 2% grades by about 7% along with increasing speed on flat terrain by about 2% lead to the best results.

SOLAR ENERGY FOR ELECTRICAL VEHICLES: SYSTEMS ANALYSIS

Henry Oman, Consulting Engineer, Seattle, Washington

Abstract

The Sun delivers to the Earth's surface during daylight hours around 85 milliwatts per sq cm (79 watts per sq ft) of energy on a clear day. This energy source has not been practical for vehicle propulsion because it may not be available when needed, and the vehicle may not be properly oriented for receiving it. Sunlight can be used to grow plants that can be converted into fuels, but the resulting fuel generally costs more to produce than does gasoline. Now available is a new tank for carrying solar energy--the zinc-air battery, which can deliver propulsion energy from sunlight with greater efficiency and less value-added expense. This battery generates electric energy by combining zinc from its anode with oxygen from air to form zincate. At the recharging station the zincate is replaced with zinc, and the zincate is sent to a electro-winnowing plant where the zinc is recovered.

The main cost of delivered solar power is the interest on the money invested in the solar-cell arrays. This interest, which persists even when the sun is not shining, has made solar power plants non-competitive for generating power for American electric utilities. In a tropical desert clouds rarely interrupt the intense sunlight, and the zinc output from a solar-powered electro-winnowing plant can be stockpiled. In a tropical desert a hectare-size solar-power plant with 30-percent efficient solar cells can in eight hours recover enough zinc to power 36,000 km (22,000 miles) of travel in lightweight 4-passenger Irmscher electric vehicles. The cost of the energy for a vehicle's 120 km travel range would be $2.88. If the desert processing plant is 5000 km from the user's port, the shipping cost of the zincate and zinc would add 17.5 cents to the cost of 120 km travel in the Irmscher.

A by-product of the solar-power plant in the desert could be renewable fuel that local residents can use in electric bicycles. One oriented solar panel, 10 meters by 10 meters in size, with 30-percent efficient solar cells, can in one day deliver enough energy for 14,700 km (9135 miles) of travel on bicycles. This by far exceeds the travel distance that could be obtained in one day by riding on an animal that is pastured on a 10-by-10 meter area.

Introduction

Electric cars powered by on-board solar-cell arrays, plus energy stored in batteries, have travelled great distances and also won races. However, on-board solar power is not practical for ordinary day-to-day automobile propulsion because sunshine is not available on cloudy days, nor during nights which can be long in the winter. Also, a vehicle's orientation during travel on roads and streets may not be the best for collection of solar radiation, and road-side buildings and trees cast shadows on the vehicle's solar panels.

Solar power not been practical for electric utilities either. The energy is free, but the interest on the invested cost of the solar-plant accumulates even when no power is being generated at night. Furthermore, a fuel-burning power plant must be running and ready to carry the solar-plant's load whenever clouds interrupt sunshine. Also, a utility's peak load comes in the early evening when the sun is near the horizon and light intensity is low. Batteries and pumped-hydro have been tested for supplying peak loads, but they can't compete with the

0-7803-5086-3 /98/$10.00 ©1998 IEEE

aeroderivative gas turbine, with which the independent power producers make electricity that costs only 3 cents per kWh. Some utilities now invite customers to make extra monthly contributions of money for building and operating solar and wind-power plants.

Solar energy has been used for propelling vehicles for centuries. For example, the bicyclist's propulsion power comes from third-hand value-added solar energy. During summer the no-cost sunlight is absorbed in a corn field to produce ensilage. During winter this product is fed to cows that produce milk. The milk is hauled to a value-adding creamery which delivers cheese that enables the bicyclist to travel 1000 miles with energy which costs him $9.20.

A new development, the zinc-air storage battery, can make energy from the sun practical for propelling electric vehicles. In the paragraphs that follow we evaluate the economics of using solar energy to electro-winnow the zincate from discharged batteries. The product is zinc-fuel for recharging the batteries. This electroforming can be done at a solar power plant located on a tropical desert where the intense sunlight is rarely dimmed by rain clouds. The principal cost is the interest on the solar-power plant investment. A small maintenance crew will be needed. The cost of waterborne shipping of the zinc and zincate between port cities is small.

Zinc-Air Powered Electric-Vehicles

The key to the new solar-power option is the zinc-air battery described by Jonathan Goldstein at the 31st *Intersociety Energy Conversion Engineering Conference* ([1]). This battery makes electric vehicles practical because it stores over seven times the watt-hours-per-kg that can be stored in common lead-acid batteries. The low weight of the zinc-air battery is a result of two factors. One is the weight of the negative electrode, which is zinc that weighs 7.13 grams per cc. Lead weighs 11.35 grams per cc. The second factor is the storage of oxygen in the from of lead oxide on the positive electrode. In the zinc-air battery the oxygen is obtained from the atmosphere.

Germany's Deutsche Post AG is testing 65 postal vans that cut the fuel cost in half by replacing gasoline energy with electric-utility energy. One van with a 150-kWh battery recently traveled on one charge from Chambray, France, across the Alps, to Turin, Italy, a distance of 152 miles. This trip included 93 miles of continuous climb.

Twice a week the German postal van goes to a "Refueling Station," where the oxidized zinc cassettes in the van's battery are replaced in 10-minutes (Figure 1). The cassettes are sent to an electro-winnowing plant in Bremen which services German postal-van fleets in Bremen, Cologne, Mainz, and Hamn, and also vehicles in Stockholm, Sweden.

Recharging a Zinc-Air Battery

During discharge of a zinc-air battery the zinc at the anode combines with oxygen from the air to produce zinc oxide:

$$2\ Zn + O_2 \rightarrow 2\ ZnO$$

This reaction theoretically delivers 1.65 volts per cell and 1350 watt-hours of energy per kg of zinc. Normal cell voltage is 1.15 volts, and present batteries deliver up to 200 watt-hours per kg at a 5-hour discharge rate.

The recharge of an electrical-vehicle battery consists of four steps:

1. The zinc oxide is removed from the current-collector frames.

2. The zinc oxide is dissolved in a potassium-hydroxide solution to form a zincate solution:

$$ZnO + 2KOH + H_2O \rightarrow K_2Zn(OH)_4$$

3. In electro-winnowing the zinc is recovered as metal:

$$K_2Zn(OH)_4 \rightarrow Zn + 2KOH + H_2O + 1/2\ O_2$$

4. The zinc is compacted onto the current-collector frame and is inserted into a separator bag.

The recharging energy, consumed in Step 3, can be supplied at a remote solar-energy power plant. The zincate solution can be shipped in a tanker. The recovered zinc-powder can be shipped as a slurry, or as a dry powder that can be blown in and out of the transporting vessel.

Cost of Solar Power Generation

A desert-area solar-power plant could power the electro-winnowing of zinc oxide from discharged zinc-air electric-vehicle batteries. For example, the ARCO solar-power plant, which was built in 1982 next to the Southern California Edison's Lugo substation, had 108 panels, each 9.75 m by 9.75 m (32 by 32-feet) in size. Each panel, with the then available 10-percent efficient cells, produced 9580 watts peak power. This panel, if kept oriented in equatorial sunshine, could produce 65 kWh per day. One hectare (2.47 acres) of desert with these panels could deliver 3400 kWh per day. A winnowing plant receiving this energy could keep 40 postal vans in service.

In the early 1980's the Sacramento Municipal Utility District built a 1.2 megawatt (MW) solar power plant at a cost of $12 million ([2]). The plant has no fuel cost, but the interest on the investment at today's 10 percent would be $1.2 million per year. The energy output being obtained at that time in large California solar power plants was 3 million kWh per year per megawatt of peak-power capacity. The interest of the investment would correspond to a cost of 40 cents per kWh of output power.

The efficiency with which solar cells can convert sunlight into electricity is approaching 30 percent ([3]). Thermophotovoltaic and AMTEC solar converters are approaching the efficiency of solar cells ([4]) and ([5]). These developments could introduce an opportunity for making desert areas even more valuable than farm lands for producing exportable products with energy.

Use of 30-percent efficient solar cells would triple the annual energy output of this solar-power plant. A further increase in output would be obtained at a tropical desert where the sunlight is nearly overhead all year long. These factors would reduce the cost of solar power at the plant site to perhaps 10 cents per kWh.

Floyd Wyczalek described Opel's Corsa Irmscher sedan that carries 14.4 kWh of energy in nickel-cadmium batteries which can deliver 6.8 kW of propulsion power. The Irmscher's mass is 1100 kg plus 160 kg payload, and the car offers 100 to 120 km of travel range on a battery charge ([6]). A hectare-size solar electro-winnowing plant with 30% efficient solar cells can in eight hours produce enough zinc to power 36,000 km (22,000 miles) of travel in these lightweight 4-passenger electric vehicles. If we assume a 50 percent battery-recharge efficiency, then the energy for driving the four-passenger Opel Irmscher over its 120-km range would cost $2.88.

Shipping Energy in Zinc Pellets

Electric-vehicle energy in the form of zinc would need to be shipped from a solar-powered electro-winnowing plant in the desert to a metropolis where the energy is used. The zinc oxide formed during battery discharge would be shipped back to the desert. These movements add cost. However, the price of gasoline also includes shipping cost. Petroleum energy is shipped from the Persian Gulf to the United States, and empty tankers carry water ballast on their return trips.

0-7803-5086-3 /98/$10.00 ©1998 IEEE

Another new rechargeable zinc-air battery was described at the 12th Annual Battery Conference on Applications and Advances. "Recharging" requires replacing the zinc-oxide-containing electrolyte and refilling a hopper on the battery with zinc pellets (Figure 2). The depleted electrolyte could be shipped in an ocean-going tanker to a port near the desert solar-powered zinc-extraction plant. The restored electrolyte could be returned in a tanker.

We can estimate the cost of shipping zinc-air fuel by assuming that the zinc and electrolyte deliver 200 watthours of energy per kg of mass. That energy content has been measured in laboratory tests. The 14.4 kWh required to propel the Opel Irmscher through its 120-km travel range would require a fuel mass, m_f:

$$m_f = \frac{14{,}400 \text{ Wh}}{200 \text{ wh/kg}} = 72 \text{ kg}$$

We assume that the depleted and regenerated electrolytes can be shipped in an ocean-going tanker, and that the zinc pellets recovered at the solar site can also be shipped back in a tanker. The cost of shipping petroleum in a modern tanker is 2.5 cents per 100 ton-miles. This corresponds to 1.67 cents per 100 metric-ton kilometers. If the port city is 1000 km from the solar site, then the Irmscher's 72 kg of fuel mass for traveling 120 km has to go 2000 km. The cost of shipping this 72 kg over the 2000 km round trip (c_f) becomes:

$$c_f = \frac{1.67 \text{ cents X } 72 \text{ kg X } 2000 \text{ km}}{100 \text{ ton-km X } 1000 \text{ kg/ton}} = 2.4 \text{ cents}$$

Alternatively the zinc could be shipped in a modern bulk carrier such as the Irena Dan. She can carry 81,040 tons of cargo which is loaded and unloaded with blowers. The cost of the 25-person crew plus fuel for her Sulzer propulsion engines corresponds 0.048 cents per ton mile. The shipping cost for the 2000-km round trip of zinc fuel, to and from the solar-reprocessing plant, would then correspond to 3.5 cents per 120 km of travel in the Irmscher electric car.

If the solar reprocessing plant is 5000 km from the user's port, this shipping cost would rise to 17.5 cents per 120 km of Irmscher travel. The more significant cost to the driver would be the taxes that are added to the fuel price by governments to support highway construction and maintenance.

Other Uses of Tropical Sunshine

A community that seeks prosperity must export products so that it can buy new technology and needed goods. Exportable food can be produced when sunshine illuminates irrigated farm land. Genetically engineered plants plus new agriculture technology are continually improving per-acre productivity. A marine farm in Thailand can grow seventy tons of dried green algae per acre per year. Examples of the energy productivity from agriculture areas are:

Product and Location	Tons per Acre-per-year of Land
Hybrid sorghum, Texas	16
Fast-growing alder trees, U.S.	25
Napier grass, Florida pond	50
Green algae, Thailand	70

The green algae, when dried, has a heating value of 7500 Btu per pound, and it can be converted into methane ([7]). This methane can fuel an aeroderivative gas turbine in a combined-cycle power plant that has a 50-percent efficiency. Then the fuel from one acre will support a continuous power output of 17.6 kW. At a wholesale rate of 3 cents per kWh the plant will earn $4600 per year from the algae produced on each acre.

Such options are not available in desert regions which have plentiful sunshine, but little water for supporting plant growth. Electric power could be generated with solar cells, and exported, but that requires an expensive transmission line which would be in service only when the sun shines. Sunlight can be concentrated to produce hydrogen from water with a complicated chemical process. On the other hand, electric power can electrolyze water with a very simple process. The hydrogen that is produced can be combined with nitrogen to make ammonia, which can be liquefied and shipped to farms for fertilizer.

Fuel for Local Transportation

A by-product of the solar plant could be renewable fuel for use by local residents in battery-powered electric bicycles. One ten-by-ten meter solar panel, equipped with 30-percent efficient solar cells, would in one day generate 195 kWh. An electric bicycle while travelling over moderately hilly terrain consumes 10.6 watthours per km ([8]). With 80-percent battery charging efficiency, one day's energy output from the ten-by-ten meter panel will propel electric bicycles a distance of 14,700 km (9135 miles). This far exceeds the daily travel distance that could be available by riding on an animal that lives on a 10-by-10 meter pasture.

Conclusion

A new renewable fuel for electric vehicles is zinc, which combines with oxygen from the atmosphere to form zincate as the vehicle's battery is discharged. The zinc is recovered from the zincate by electrochemical winnowing. This "recharging" operation could be done at a remote location, such as a solar power plant on a desert. This new battery technology, plus the development of solar cells that approach 30 percent in efficiency, opens new ways of exporting energy from tropical deserts. The solar energy would be converted to vehicle fuel at the solar array site. The energy-input received from the Sun costs nothing. The interest on the money invested in the solar power plant, plus shipping the used and renewed fuel between the solar plant and a 1000-km-distant port city can cost only $2.90 for every 120 km travelled by a four-passenger Irmscher car. The only remaining source of atmosphere pollution is the diesel-powered ship that hauls the depleted and renewed battery-fuel.

Furthermore, a single 10-meter by 10-meter solar panel with 30-percent solar cells can generate 195 kWh in one day. This energy can propel battery-powered electric bicycles a distance of 14,700 km (9135 miles).

Figure 1. Two times each week the electric-powered German postal van stops at a refueling station for exchanging the electric-fuel cassettes in its zinc-air batteries. The discharged cassettes are later regenerated.

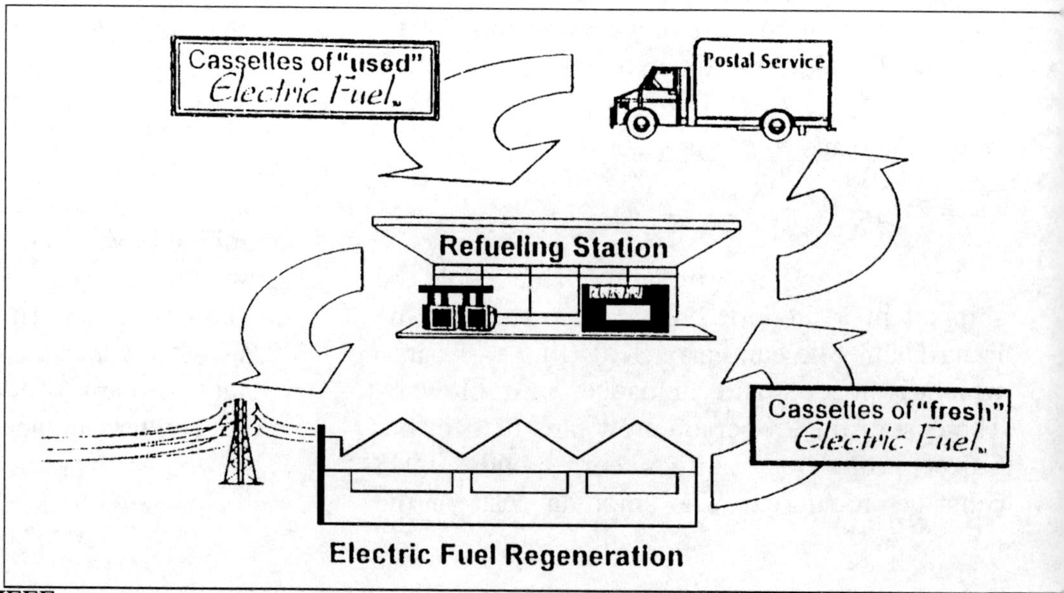

0-7803-5086-3 /98/$10.00 ©1998 IEEE

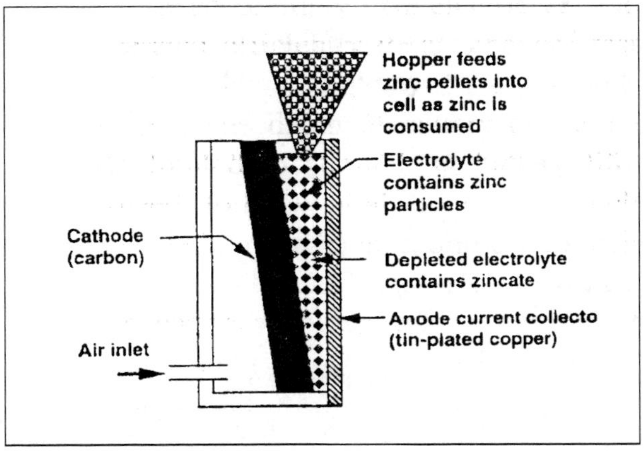

Figure 2. In another zinc-air battery the zinc from the hopper replaces the zinc consumed in the electrolyte.

---

1. Goldstein, J. R., "Field Test of the Electric Fuel Zinc-Air Refuelable Battery System for Electric Vehicles," *Proceedings of the 31st IECEC,* 1996, pp 1925-1929.

2. Anderson, M., and Spencer, R., "The SMUD Photovoltaic Power Plant," *Proceedings of the IEEE Power Engineering Summer Meeting, July, 15-20,* 1984, page 3.

3. Chmielewski, A. B., et al, "The New Millennium Power Technology," *Proceedings of the 31st IECEC,* 1996, pp 2193-2198.

4. Fatemi, N. S., et al, "High Efficiency Converters for Thermophotovoltaic Applications," *Proceedings of the 31st IECEC,* 1996, pp 2238-2242.

5. Van Hagen, N. S., et al, "Thermionic/AMTEC Cascade Converter Concept for High-Performance Space Power," *Proceedings of the 31st IECEC,* 1996 pp 629-634.

6. Wyczalek, F,, "Hybrid Electric Vehicles (EVS-13 Osaka)," *Conference Record, NORTHCON 96,* November 4-6, 1996, pp 409-412.

7. Enssani, E., "Liquid Hydrocarbons from Biomass Grown on Waste," *Proceedings of the 25th IECEC,* 1988, pp 1953-1957.

8. Morchin, W. E. "Trip Modelling for Electric Powered Vehicles," *Conference Record, NORTHCON 96,* November 4-6, 1996, pp 372-377.

ON-BOARD ENERGY AND POWER MANAGEMENT ON ELECTRIC VEHICLES: EFFECT OF BATTERY TYPE

Henry Oman, Consulting Engineer, Seattle, Washington

Abstract

The performance of a gasoline-powered vehicle is commonly quantified in terms of available acceleration, speed, fuel economy, travel range with a tankful of fuel, purchase cost, and life-cycle cost. A convenient gasoline-refueling infrastructure is in place today. The per-mile cost of energy for propelling an electric vehicle is a fraction of the per-mile cost of energy for a gasoline-propelled vehicle. However, an electric vehicle's energy has to be stored in a battery that must be recharged. No one battery can enable its electric vehicle to fully meet all of these performance requirements. For example, lithium batteries, which store the most energy per kg of weight, are being manufactured for laptop computers, where long life is not needed. The computer becomes obsolete before the battery wears out. The zinc-air battery offers long travel range with low weight, but it requires a completely new supporting infrastructure.

In a hybrid vehicle a fuel-burning engine's output power is supplemented with a battery. With an efficient Meijer version of the Stirling-cycle engine delivering constant power, a 100-miles per gallon fuel economy is feasible. A battery can deliver peak power and recover energy with dynamic braking. The optimum vehicle performance with a given battery can be obtained if the vehicle has an on-board energy manager which limits the life-reducing stresses on the battery and continually displays to the driver the remaining travel distance available with the energy in the battery. Today's best candidate batteries for electric vehicles are the lead-acid, nickel-cadmium, nickel-metal hydride, lithium-ion, and zinc-air types. In this report we describe their characteristics and limitations which are pertinent to the design of an on-board energy manager.

Introduction

Electric energy is a low-cost fuel for an automobile. A gallon of gasoline, when burned in a common 25-percent efficient Otto-cycle engine, will deliver about 9 kWh of propulsion energy. This same 9 kWh is offered for 27 cents by independent electric-power producers. They generate electric power from natural gas with aeroderivative gas turbines in combined-cycle power plants, the latest of which have over 60 percent efficiency. However, the 9 kWh per gallon of gasoline corresponds to 3534 watt-hours per kg of gasoline weight. Today the best watt-hours-per-kg performer in batteries is the lithium-ion electrochemistry. It can deliver around 100 watt-hours per kg of battery weight during a discharge.

Electric energy is not being commonly used today for propelling automobiles for these reasons: (1) The supporting infrastructure needed for supplying vehicles with electric energy is not in place, and (2) commercially available storage batteries are heavy, expensive, and have limited life. Engineering solutions to each of these obstacles are available, but the required financial and manufacturing capability is not yet available for various reasons. However, in the downtown areas of many cities the engine-powered vehicles are polluting the air to intolerable levels. Restrictions on the operation of engine-powered vehicles are inevitable, and the demand for electric vehicles is bound to grow.

In fuel-burning vehicles the rate of energy consumption has not been critical, so the driver selects his speed of travel and positions the

0-7803-5086-3 /98/$10.00 ©1998 IEEE

engine's throttle accordingly. An electric vehicle must be operated efficiently to get the greatest range from the energy content of the limited-weight battery. The depth of battery discharge needs to be limited to attain long battery life. The driver needs to know at all times how much energy he has left in his battery, and how much travel range he still has available. During charging the battery's voltage and current must be controlled, taking into account the battery's temperature and state-of charge.

On-board energy management is the key to efficient performance of an electric vehicle, but each battery electrochemistry requires unique energy management. Battery types are being evaluated, but no "best" electric-vehicle battery has yet been adopted. For example, a nickel-cadmium battery can easily deliver the desired vehicle acceleration. On the other hand, longer driving range on a single battery charge is available from a lithium-ion battery. However, this battery needs to be supplemented by ultra-capacitors if the acceleration available from nickel-cadmium batteries is required.

Route to 100 Miles-per-Gallon Hybrid

A hybrid electric vehicle can reduce urban air pollution, and also use fuel-energy efficiently. For example, a five-passenger hybrid car can deliver 100-miles-per-gallon if it is propelled by a Stirling engine and a nickel-cadmium battery. This hybrid combination avoids the efficiency limitation of today's Otto-cycle automobile engine that has to be designed to deliver high power and high speed, yet run most of the time at low power and low efficiency.

Ivo Kolin has recorded the efficiencies and fuel consumptions of 30 thermodynamic cycles that have been built into engines ([1]). On top of the list is the Meijer version of the Stirling cycle. It had twice the efficiency of the 25-percent efficient gasoline-burning Otto-cycle engine that powers cars today. Kolin's diagrams also explain why the Stirling-cycle engines have not been used in automobiles. The heat from the burning fuel must go through the metal of a heat exchanger. When you push the throttle to the floorboard the burner output gets hotter, but a few seconds elapse before the working fluid in the cylinders gets more energy.

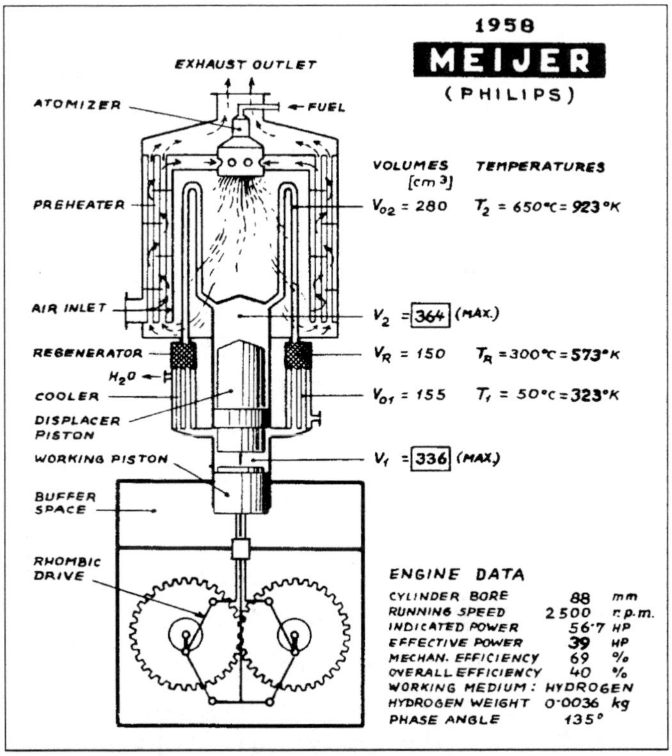

Figure 1. **Meijer's Stirling Engine**

Kolin's cross-section drawing shows that Meijer's engine is too complicated to build in a factory that makes the old cast-iron automobile engines (Figure 1). However, today we have numerically controlled metal-working machinery and robots that assemble components. Stirling-cycle machines are already in production for refrigeration and heat-pumping. The required high-temperature technology is already used in aircraft jet engines.

The Stirling engine's speed-of-response problem can now be easily solved with a hybrid propulsion system in which the engine runs at

[1]. Kolin, Ivo, "The Evolution of the Heat Engine," Moriya Press, 1998.

constant power and load peaks are supplied by a battery and a 95-percent-efficient permanent-magnet motor. The needed energy management functions are available in computer-based motor controllers.

Pertinent Battery Characteristics

Battery characteristics that affect the functions of the electric-vehicle's on-board energy manager are summarized in the paragraphs that follow.

Long Battery Life

In the early 1900's there were more electric cars than gasoline-powered cars on the roads. However, the lead-acid batteries of that day had short life in charge/discharge service. Owners of cars with dead batteries discovered that a new gasoline car would cost less than a new battery. In today's gasoline automobiles long life in lead-acid starting-batteries has not been needed. A battery that dies just before its guaranteed life expires is a desired commercial product. On the other hand, the lead-acid batteries used to supply uninterruptible power in telephone central-offices have lasted for decades.

Batteries are being used to supply auxiliary power in missiles, airplanes, and spacecraft. The needed qualities included deep discharges and long life in terms of cycles of charge and discharge. These batteries were carefully designed, manufactured with quality-controlled procedures, and tested for cycle-life in the specified temperature, acceleration, and vibration environments. Development of a battery that met new requirements took years of calendar time.

Sudhan S. Misra showed how this development time can now be reduced by modeling. He described flooded lead-acid batteries that had been installed in the 1980's for supplying uninterruptible (UPS) and emergency power ([2]). Premature failures had occurred after the batteries were placed in service. Recent tests showed that these failures could have been avoided by simple alternatives like a different specific gravity of the electrolyte, and a thicker glass-mat retainer between plates. With modeling, which is now possible, these failure mechanisms can be discovered during the battery design, rather than after the battery has been in service for five years.

Long battery life in spacecraft has been achieved with carefully designed and tested nickel-cadmium and nickel-hydrogen batteries. A qualification test showed that the lifetime of a new single-pressure-vessel nickel-hydrogen battery is 25,000 charge/discharge cycles. Such a battery, if installed in an automobile and recharged every night, could deliver to the owner and his descendants 68 years of service in their car.

Low Battery Weight

Battery weight is compared in terms of energy content in watt-hours per kg. For non-transportation applications battery weight has not been a significant consideration. The weights and costs of batteries for electric vehicles was summarized by Jasper Hardin in his plenary address at the *12th Annual Battery Conference on Applications and Advances*. His data is summarized in Table 1.

Table 1. Weights and costs of batteries that are applicable to electric vehicles.

Battery	Weight Watt-hours/kg	Cost $/kWh
Lead-acid	20 to 30	50 to 80
Nickel-cadmium	<50	70 to 100
Nickel-metal hydride	<60	800 to 1200
Lithium-ion	>100	300 to 600

2. Misra, Sudhan S., and associates, "Update on UPS Battery Life Characteristics," Proceedings of the *13th Annual Battery Conference on Applications and Advances*, 1997, pages 279-84.

0-7803-5086-3 /98/$10.00 ©1998 IEEE

The weights in Table 1 are based on the total energy that can be extracted from a fully charged battery. However, battery-life, when measured in cycles of charge/discharge, is affected by the depth of discharge. For example, we once tested an ordinary engine-starting battery for automobiles, discharging it until its voltage dropped 20 percent below its full-charge voltage. After 50 charge/discharge cycles the battery failed.

Available today are design techniques that can reduce the weight of traditional batteries. For example, today's common lead-acid battery contains a lot of lead that merely supports its lead electrodes and carries current. A cell manufactured by Optima Advanced Technologies has the active lead in two long thin strips that are rolled into cylinder shaped cells ([3]). The positive plate is continuously bonded to a lead disk at one end of the cylinder, and the negative plate is bonded to the disk at the other end. This design delivers 660 watts per kg and achieves an energy density of 29 watthours per kg.

Low Battery Cost

For some applications, performance is far more important than cost. For example, for hand-held telephones and laptop computers the key requirements are low weight and long operating time between battery chargings. Users happily pay an extra $100 for the best possible battery, which today is an assembly of lithium-ion cells. As a result, lithium-battery manufacturers are building big factories as fast as they can in order to supply this market.

Many of today's batteries are manufactured with traditional labor-intensive processes. More efficient processes are now available, but investors must support the construction of the large new factories that can build lower-cost batteries. The investors are not yet convinced that electric vehicles will become a sufficiently significant battery market.

[3]. Olsen, John B., and Sexton, Elizabeth D., "A High Power Spiral Wound Lead-Acid Battery for Hybrid Electric Vehicles," Proceedings of the *12th Annual Battery Conference on Applications and Advances*, 1996, pages 145-148.

On-Board Energy and Power Management--The Requirements

An electric vehicle's on-board energy and power manager can contribute to enhanced performance, increased travel range, and reduced life-cycle cost. The vehicle's battery is the key component that is to be managed. At this time, we have no standard battery for electric vehicles, so a system-engineering analyses must evaluate the possible alternatives for each battery type being considered. The leading battery types are lead-acid, nickel-cadmium, nickel-metal hydride, lithium-ion, and zinc-air. Their pertinent characteristics follow.

Lead-Acid Batteries

These batteries powered the electric cars in the early 1900's. Amateur electric-car builders still use lead-acid batteries, but their low energy content, generally under 30 watt-hours per kg, limits the car's travel range. Hence designers of electric-car are basing their designs on nickel-metal hydride and lithium-ion batteries. Also, today's lead-acid batteries have very short life when they are deeply discharged frequently.

Nickel-Cadmium Batteries

In San Francisco 60 trolley buses carry nickel-cadmium batteries for supplying propulsion power when they leave the trolley lines to enter the garage, or detour around road repairs. A New York State Consortium's low-floor demonstration bus required 46 kW maximum power, and 500 watts per kg at 23° C. The battery's life requirement was 600,000 charge/discharge cycles with 2-percent depth of discharges. A nickel-cadmium battery with a 6-year life would cost $18,000, compared to $7500 for a lead-acid battery with a 1.1-year life ([4]). The per-day life-cycle costs were $19.50 for lead-acid and $8.50 for the nickel-cadmium (Table 2). However, the nickel-cadmium battery's energy content is at best around 30 watt-hours per kg. This limits a vehicle's travel range if the battery is its only source of propulsion energy.

Table 2. Comparison of initial cost and life-cycle cost of batteries.

	Nickel Cadmium	Lead Acid
Initial Battery Cost	$18,000	$7500
Maintenance/Year	$110	$300
Estimated Life, years	6	1.1
Total Life Cycle Cost Per day	$8.50	$19.50
Battery Weight	660 kg	1320 kg

Nickel-Metal Hydride Batteries.

Ovonic is making nickel-metal hydride batteries for hybrid electric vehicles in ratings of 20, 30, and 60 ampere hours. The 60 ampere-hour battery has a power capability of 620 watts per kg and 1700 watts per liter. Its specific energy is 70 watt-hours per kg. The round-trip energy efficiency for the battery in a hybrid vehicle is between 82 and 88 percent. A typical self-discharge rate is one percent per day. A "Selectra" car travelled 250 miles on one charge. A Ford Taurus "range-extender" hybrid electric car, in which the battery supplements the engine, achieved a highway fuel economy of over 60 miles per gallon.

Lithium-Ion Batteries.

The lithium-ion cell's 100 watt-hours per kg and high efficiency offer great potential for electric vehicles. However, large cells have not yet met spacecraft requirements, such as lifetimes of 2500 charge/discharge cycles at 70 percent depth of discharge. Modeling indicates that these lifetimes can be met, and 15-ampere-hour cells have reached 6080 charge/discharge cycles at 10 percent depth of discharge. Intense development programs are under way, and Linnea Brush expected the price of lithium-ion batteries to drop by 25 percent per year. This price drop will rapidly expand their use. In 1997 these batteries were used in 60 percent of the laptop computers, 22 percent of hand-held telephones, and 12 percent of other portable devices.

Every cell in a lithium-ion battery must be monitored by the charge-manager. The charge-current of those cells in a battery that are fully-charged must be bypassed to avoid life-shortening charge unbalance.

Zinc-Air Batteries for Electric Vehicles.

A vehicle powered by a zinc-air battery is not burdened by having to carry its oxidizer, which is oxygen from the surrounding air. Achieved energy contents range up to 230 watt-hours/kg. The most advanced zinc-air battery is Electric Fuel Ltd's 150 kWh, 850-kg, unit that powers 12 German Deutsche Post 4.6-ton Mercedes-Benz vehicles. After travelling 300 km the van stops at a battery exchange station where its battery is replaced within a few minutes by a refueled battery. The discharged battery goes to a cassette handler that pulls the zinc cassettes from every cell for subsequent electro-winnowing into fresh cassettes at a regeneration plant. Three regeneration plants are in operation in Europe.

The present zinc-air battery is not applicable to a hybrid electric vehicle because it cannot be charged by the on-board engine-generator.

---

[4]. Disosway, Marc, "Development of High Power Nickel-Cadmium Batteries for Hybrid Vehicles," Proceedings of the *13th Annual Battery Conference on Applications and Advances*, 1997, pages 19-24.

0-7803-5086-3 /98/$10.00 ©1998 IEEE

Sensors and Processors for Battery Energy Management

The on-board energy-managing computer must perform many functions. For example, it needs to protect the propulsion motor from overloads during hill-climbing and sudden startups. It needs to inform the vehicle driver how much travel distance is available from his battery without making life-shortening deep discharges. For these functions the energy manager needs inputs from the battery monitors. In a hybrid vehicle the energy manager must also command the engine-driven generator to stop charging a battery when it is fully-charged.

Established and thoroughly tested sensors can track the energy content of a battery. They integrate the charge and discharge currents, and report the battery status to the energy manager. They can even track the self-discharge energy loss in an idle battery.

Battery charging controls can vary from simple to very complex. At one end of the scale is a charger for a nickel-hydrogen battery. During charging the battery's water is electrolyzed into hydrogen and oxygen. When the water is gone the hydrogen pressure reaches a peak, and no more electrochemical activity occurs.

At the other end of the scale is the lithium-ion battery. Full charge is best achieved by changing the charge voltage at various stages during charge. Charge current must likewise be limited. Charging must be stopped when full-charge is reached. The allowable charging voltage and current are different at different temperatures. If one cell of series-connected cells in a battery reaches full charge before the others do, then that cell will be overcharged and damaged before the remaining cells reach full charge.

Fortunately these most-complex problems have been solved for small batteries that are used in hand-held telephones and lap-top computers. An integrated circuit that is bonded to a lithium-ion battery was described by David K. Heacock ([5]). The circuit senses the voltage and temperature of the battery, and adjusts the charging voltage to achieve full charge and long battery life. The integrated circuit even tracks and remembers the self-discharge of the battery during standby, and reports it to the computer when the battery is placed into service. Heacock mentioned that the cost of this integrated circuit is around $2 each when purchased in 50,000 quantities.

Conclusion

A need for reducing automobile-generated air pollution in downtown areas has resulted in a battery-development program that is greater than any ever before undertaken. New systems-engineering and test-simulation tools are producing new storage-battery designs that deliver more power and energy per kg of weight. However, new-batteries need to be tested for life, measured in cycles of charge and discharge, when used in an electric-vehicle environment. At this time the most promising batteries are the lithium-ion and zinc-air types. The zinc-air battery requires a new type of supporting infrastructure which includes electroforming zincate into zinc. During charging of a lithium-ion battery the voltage must be adjusted whenever the state-of-charge and battery temperature change. Integrated circuits that perform these functions are already available. The key to long battery life, which reduces life-cycle cost, is an on-board energy manager which reduces battery stress by limiting peak power and deep discharges. The energy manager can supplement the battery power with ultracapacitors for peak power requirements, and suggest driving strategies to the vehicle driver for increasing his travel range in given conditions.

---

[5]. Heacock, David K., and Fundaro, Peter A., "Low Cost Battery Management, Widening the Use of Sophisticated Battery Management in Inexpensive Products," Proceedings of the *13th Annual Battery Conference on Applications and Advances*, 1997, pages 247-79.

DESIGN OF A LOW PRODUCTION COST HYBRID HMMWV

Jeffrey W. Hodgson, The University of Tennessee, Knoxville, TN

William R. Hamel, The University of Tennessee, Knoxville, TN

Craig B. Rutherford, The University of Tennessee, Knoxville, TN

Jeffrey S. Armfield, Oak Ridge National Laboratory, Oak Ridge, TN

Abstract

There is considerable interest within the military in improving the fuel efficiency of its vehicles. Hybrid powertrains for military applications offer the potential for improved fuel economy, enhanced stealth (silent operation) capability, and auxiliary field power generation.

While recent hybrid electric HMMWV conversions have shown impressive performance, the likely estimated production cost is high. In cooperation with the National Automotive Center (part of the United States Army Tank-automotive and Armaments Command, TACOM), Oak Ridge National Laboratory (ORNL) and The University of Tennessee (UT) have conducted a conceptual design for a low production cost hybrid electric powertrain for the HMMWV. The design takes maximum advantage of commercial off-the-shelf (COTS) technology. The parallel, electric assist configuration offers hybrid performance equivalent to the conventional base vehicle for limited time periods, silent electric operation, and limp-home capability in the event of electric drive system failure.

The design goals for the hybrid electric powertrain, the overall system design, and design decisions necessary to meet performance goals are reviewed. Performance simulation results show that the proposed powertrain design should meet or exceed the stated design goals.

Introduction

The United States Army in pursuing ways to reduce the fuel requirements of its assets has expressed interest in ways by which certain vehicles can be converted to hybrid electric operation. This paper deals with one vehicle in the military's inventory -- the High Mobility Multi-Purpose Wheeled Vehicle (HMMWV) shown in Figure 1.

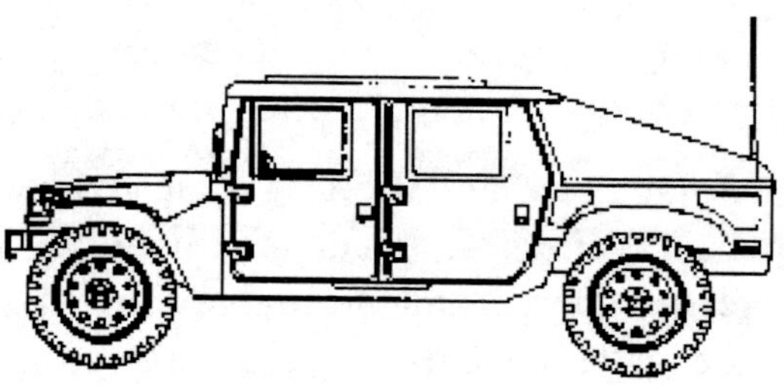

Figure 1. m114 HMMWV

The HMMWV is a highly flexible light wheeled vehicle used by all branches of the United States military. Its high mobility and the wide range of add-on shelters and vehicle configurations have allowed the HMMWV to replace a number of special purpose light to medium duty wheeled vehicles.

The engine used in the HMMWV is a 6.2 L diesel V8 with a peak power of 155 hp at 3600 RPM. This engine is coupled to a 3 speed electronically controlled automatic transmission and a 2-speed chain driven transfer case. With a 2500 lb payload, the HMMWV has a range of about 300 miles using 25 gallons of fuel -- a fuel economy of approximately 12 mpg. Increased fuel efficiency would result in both a greater range between refueling and a decreased level of tanker support required for this vehicle class.

Hybrid Electric Vehicles

The use of hybrid electric drives in motor vehicles is not a new concept. In a review article, Wouk [1] points out that a hybrid electric vehicle (HEV) was patented in 1905 and a commercially available HEV was on the road in 1910.

0-7803-5086-3 /98/$10.00 ©1998 IEEE

HEV Configurations

Three HEV configurations have appeared in practice: the series hybrid, the parallel hybrid, and the dual hybrid.

Series Configuration

In the series configuration, propulsive power to the vehicle's drive wheels is provided by one or more electric motors energized from a battery pack. The engine is used to drive a generator that supplies electricity to the battery pack. Figure 2 shows a series configuration schematic.

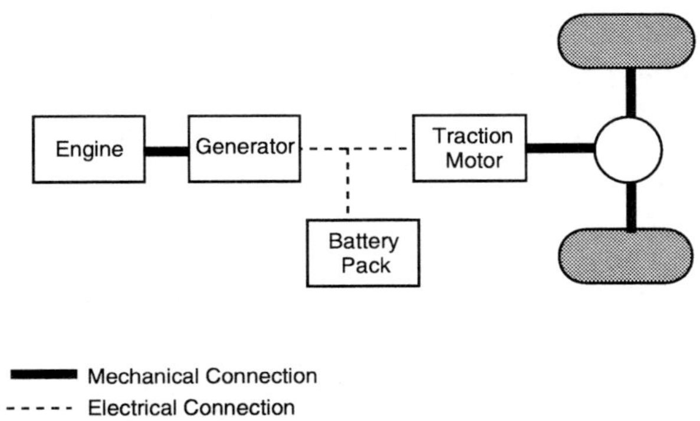

Figure 2. Series Configuration

In the series configuration the speed of the engine is decoupled from the vehicle speed. The series configuration requires that the electric motor provide all the vehicle's tractive power, which, in turn, imposes requirements on the battery pack. In general, the series configuration has fuel economy advantages under urban (stop and go) driving conditions since the engine (which typically has low efficiency under low load) can be turned off until the battery pack's state of charge (SOC) drops to the point where it requires recharging.

One characteristic of the series configuration that may be important in military applications is that if the electric drive becomes inoperative the vehicle is rendered immobile. On the other hand, pure electric operation gives the same vehicle performance as hybrid operation -- until the battery pack state of charge drops to the point where it can no longer deliver the required power to the motor.

Parallel Configuration

In the parallel configuration propulsive power to the drive wheels can be provided by the electric motor, the engine, or both. Figure 3 shows

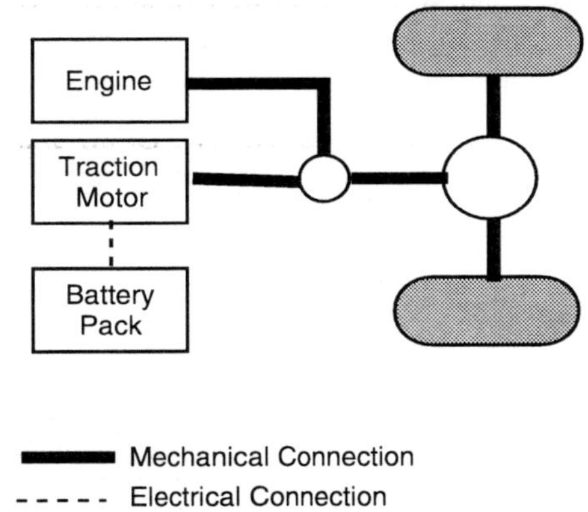

a parallel configuration schematic.

Figure 3. Parallel Configuration

The parallel configuration control strategy adopted can emphasize operation using the engine with the electric motor providing additional power when required (electric assist), or operation relying on the electric motors with the engine providing additional power when required. Since the maximum power requirements of the vehicle can be satisfied by operating both the motor and the engine, the motor can have a lower power rating than in the case of an equivalent series configured HEV. In general, the parallel configuration has better fuel economy than the series configuration under highway (high speed) conditions because the engine can be used to drive the vehicle directly -- thus avoiding some of the efficiency losses associated with the series configuration.

An advantage of the parallel configuration is that if the electric drive system becomes inoperative the vehicle can continue to function with reduced power through the redundancy of the conventional engine power transmission path ("limp home" capability).

Dual Configuration

The series and parallel configurations have fuel economy advantages and disadvantages relative to each other, which depend on vehicle operating conditions. Ideally, it would be desirable to have a powertrain that could switch between these two configurations as driving conditions change. A system is currently in production (the Toyota Prius) that allows this [2]. This configuration, termed the dual configuration, can operate as either a series or a parallel configuration and has the capability to provide a continuously variable transmission (CVT) between the engine and the drive wheels even during parallel operation. The system is shown in Figure 4.

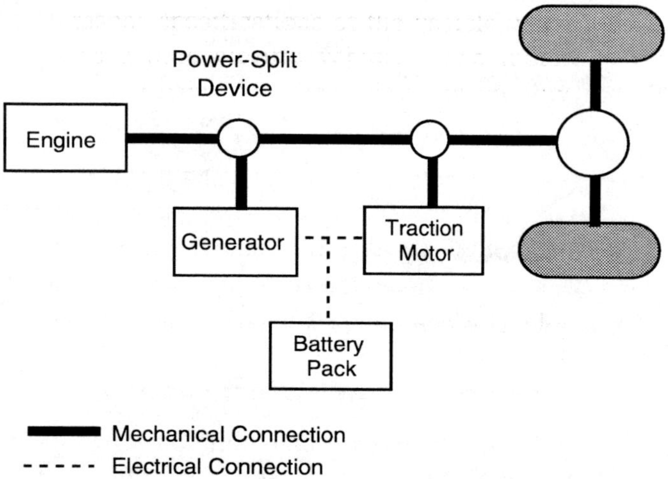

Figure 4. Dual Configuration

The power flow flexibility of the dual arrangement also results in the greatest control system complexity.

Cost Factors

The powertrain design and the cost of the components associated with the configuration selected influence the overall cost of a HEV. The primary cost factors are: (1) the type of electric motors, (2) the number of electric motors, (3) the type of batteries, and (4) vehicle integration.

There are many alternatives regarding electric drive motors. Light-weight and high efficiency motors normally used in HEVs are typically permanent magnet DC motors that incorporate pulse width modulated drive converters. At the power levels required for HEVs, these motors are produced in low quantities and are expensive compared with industrial electric motors of the same capacity. A limited number of production electric vehicles use less expensive (and less efficient) AC synchronous type motors with solid-state vector drives. The series configuration requires one generator and at least one traction motor, the parallel configuration requires at least one traction motor/generator, and the dual configuration requires at least one traction motor and one motor/generator. The designer has the freedom to incorporate multiple electric motors in parallel with any of these configurations, but since the unit cost of motor/converter combinations for this general class of motors is not a strong function of motor size, powertrain costs increase proportional to the number of motors.

There is a wide range of cost variation with battery storage alternatives. Lead acid batteries are manufactured on a large scale with common materials, and are by far the least expensive. Higher energy storage density alternatives such as nickel cadmium, nickel metal hydride and lithium ion are two to four times more expensive depending on the particular technology. Design tradeoffs between battery cost and weight are major cost drivers.

The final major cost factor is vehicle integration, which refers to the cost and complexity associated with incorporating the HEV drivetrain into the HMMWV chassis. Ideally, the HEV drivetrain should consist of retrofittable module(s) that can be installed in a standard HMMWV with minimal modifications to the chassis. Vehicle integration costs, and military deployment costs can be reduced by giving careful consideration to the electrical and mechanical vehicle modifications required.

Comparisons

Table 1 gives some qualitative comparisons of the three configurations with respect to key

Table 1. Configuration Comparisons

	Series	**Parallel**	**Dual**
Overall Cost	medium	low	High
Powertrain Complexity	low	medium	High
Control System Complexity	low	medium	High
Motor Size	largest	low	Lowest
Number of Electric Motors/ Generators	≥ 2	≥ 1	≥ 2
Battery Pack Power Demand	largest	low	Lowest

Powertrain Redundancy	no	yes	yes

system attributes. The term "redundancy" refers to the ability of the vehicle to propel itself on either of its propulsion systems (electric or conventional).

The increased operational flexibility obtained with the parallel and dual configurations comes with higher complexity, but they have the advantages of reduced battery pack size and robust backup capabilities.

Improved Fuel Economy

Hybrid electric vehicles have the potential to offer better fuel economy than conventional vehicles due to several factors: the inherently better efficiency of electric motors compared to engines, the ability to operate the engine only when necessary, the ability to operate the engine at its best efficiency, and the capability for recovering kinetic energy normally dissipated during braking.

Hybrid HMMWV Design Issues

As stated earlier, a hybrid electric HMMWV has the potential to increase fuel efficiency, operate with low noise and low thermal signatures, and to provide auxiliary power generation in the field.

The following discussion highlights baseline performance requirements, and hardware issues pertaining to the basic hybrid system design and components.

Baseline Performance Requirements

The HMMWV is a fielded equipment item and the military has developed explicit performance requirements for it. The baseline requirements (including appropriate hybrid vehicle considerations) related to drivetrain parameters are:

- Maximum Gross Vehicle Weight (GVW): 7700 lbs
- Fording depth: 30 inches
- Grade Operation: At its GVW the vehicle must be able to ascend a 60% grade at 2 mph and to stop and start on this grade. Also at its GVW the vehicle must be able to ascend a 5% grade at 55 mph.
- Acceleration: At its GVW the vehicle must be able to accelerate from 0 to 30 mph in 6 seconds or less, and from 0 to 50 mph in 30 seconds or less (both on a level hard surface road).
- Speed: At its GVW in high gear the top speed must be at least 60 mph and in low gear the top speed must be less than 2.5 mph.
- Range: As a hybrid the range must meet or exceed the standard 300-mile range. In the pure electric mode, the range must be at least 20 miles at 30 mph.
- While the vehicle must be capable of meeting the baseline vehicle performance requirements as a hybrid, its performance as a pure electric, or in any backup modes, may be appropriately de-rated.

Hardware Considerations

The military is giving greater consideration to the use of industrial grade COTS in fielded equipment to reduce costs and enhance maintainability. The role of COTS components in the low cost hybrid HMMWV design is critical. Manufacturers are continually announcing new developments in electric motors, generators, and batteries of all sorts, but at the present time many of these high performance components are in low-volume production, or at the prototype stage of development and are not commercially available. There are wide ranges of price-performance ratios for these components, which must be incorporated into the design optimization process. The following discussion summarizes some of these issues.

Engines

The standard HMMWV comes equipped with a 6.2 liter displacement diesel engine with a rating of 155 bhp at 3600 rpm. The torque curve is fairly flat at 235 $\pm$ 15 ft-lb over the speed range of 1200 to 3600 rpm. Any engine chosen for the hybrid version must use diesel fuel and must be able to provide equivalent drive wheel power in conjunction with the electric motor.

Motors

On the surface there appear to be many motors that might be available for use in this application, but closer examination of candidate motors suggests that permanent magnet DC motors have some advantages over other designs. These motors are lightweight, have high efficiency, and have been developed especially for hybrid electric drives.

Motor requirements are based primarily on the performance specifications of the vehicle when operating as a pure electric vehicle. The motor selection is driven by cost and availability considerations.

Battery Pack

The battery pack design specifications are dictated by the electric motor selection, the vehicle's pure electric range requirement, space and weight constraints, safety, recharging, and cost considerations.

Motor Requirements. The electric motor imposes certain requirements on the battery pack selection -- namely voltage and current capability. The electric motor requires a certain voltage supply range in order to operate satisfactorily and so the battery pack must be made up of a certain number of cells in series to generate the required pack voltage (this is often referred to as a *module*).

If a particular battery design cannot meet the motor's current requirements with one module, the designer may consider wiring two or more modules in parallel.

Range Requirement. Assuming that the motor's voltage and current requirements can be met, the designer must then examine how much *energy* storage the battery pack has to provide. This requirement is almost always dictated by the range that the vehicle must achieve under pure electric conditions.

Safety. There are safety considerations associated with battery use and these include exposure of personnel to battery chemicals (acids), fire, and battery explosion [3]. The designer must work with the battery supplier to make sure that the proper thermal environment is provided for the battery pack (provisions for cooling the battery pack) and that ventilation of the battery pack is provided -- particularly if hydrogen evolution could exist. Because of the high voltage levels required for motor operation, shock isolation and safety interlocking are key issues. The battery pack will be one of the heaviest vehicle components. Its structural design and vehicle mounting must be commensurate with the nominal and off-nominal shock and vibration requirements of the HMMWV.

Recharging. During regenerative braking, the kinetic energy of the moving vehicle is converted into electrical energy by driving the motor as a generator thus charging the battery pack. While it is likely that a battery pack designed to *deliver* large currents can also *accept* large regeneration currents (at least for brief periods of time), the control system of an electric vehicle has to incorporate some means of monitoring battery conditions (state of charge, temperature) to avoid overcharging the battery pack.

Transmission

From a systems integration standpoint, it is apparent that the simplest transmission to integrate into the overall system is the existing vehicle transmission. There are several considerations which must be made prior to making a choice of transmission, however. These include:

- Engine and motor input speeds
- Engine and motor input torques
- Required wheel torque at a given vehicle speed
- Transmission efficiency
- Size and weight
- Robustness of the component
- Component should be a COTS available item

Controls and Electronics

HEVs involve the coordinated operation of several power sources during the dynamic operation of the vehicle. Ideally, the mode switching of the system should be completely transparent to the driver such that the vehicle performs like a standard vehicle. These issues are increasingly more complex for the parallel and dual configurations compared to the series configuration. The most effective way to implement the hybrid vehicle control system including driver displays and controls is as a digital control system. There are a large number of established sources of COTS-type microcomputer systems, interface electronics, sensors and real-time operating system software suitable for the hybrid HMMWV.

Selected Hybrid HMMWV

Configuration

In light of the general design objective of producing a low production cost hybrid HMMWV which retains the performance characteristics of the standard vehicle, the configuration selected for more detailed study is a parallel, electric assist design. The schematic of this configuration is shown in Figure 5.

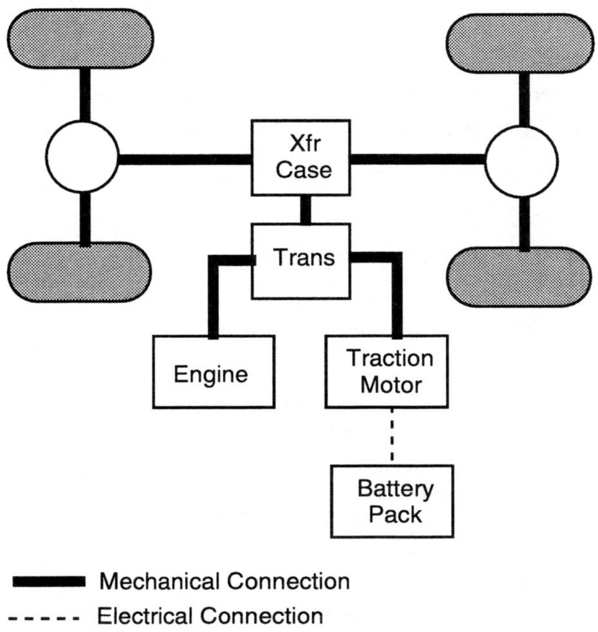

Figure 5. Selected Configuration

The features of this configuration that led to its selection as the candidate design are all primarily driven by cost considerations and are as follows:

- Only one electric motor/generator is necessary.
- No generator is used because the motor serves as the generator when required.
- Under the assumption that the conventional vehicle does not demand maximum power for extended periods of operation, the engine and the electric motor are sized such that together they can provide at least the same drive wheel power as the conventional engine.
- Lead acid batteries are used in the battery pack.

Components Selected for the Conceptual Design

Preliminary simulations of vehicle performance were performed as a first cut analysis of component options for the hybrid HMMWV. The following components were used:

- 1.9 L Volkswagen TDI engine.
- Unique Mobility SR-286 motor.
- Hawker Genesis 38 amp-hr sealed lead-acid batteries.
- GM 3L80 3-speed automatic transmission.

A comparison of the power rating of the standard engine, the proposed motor, and the proposed engine are shown in Figure 6. Note that the combined power of the proposed motor and proposed engine exceeds that of the standard engine at all speeds.

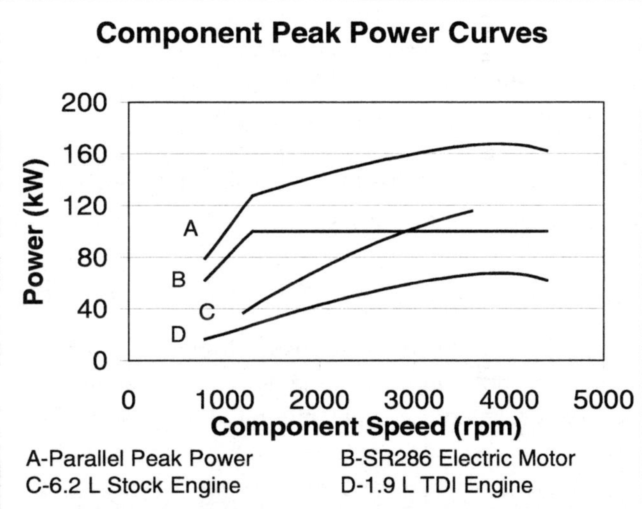

Figure 6. Component Power Curves

Control Strategy

The control system for the hybrid HMMWV prototype will be implemented with a COTS PC system, real-time operating system, and interface electronics that facilitate efficient development in a high level software environment such as the C++ language. Automotive and industrial sensors will be used. The architecture and software resulting from this approach can be readily converted in the future to a lower cost multi-electronic control unit architecture that is typical of production vehicles today.

The basic control strategy is similar to

other designs [4], and involves the creation of three basic driving modes that are driver selectable. The pure electric mode provides the quietest operation in which driver throttle commands are sent directly to the electric traction motor controller. The engine only mode involves total dependence for tractive power from the engine. This mode is primarily a back up mode for recovery in the event of an electric drive failure. The parallel mode allows the electric traction and engine drives to work in concert to provide a blend of total power, which meets instantaneous road power demands. The engine load conditions are monitored and used as the control key for supplementing the amount of electrical drive power. In the parallel mode the electric drive can also be placed in the charging mode in which a portion of the engine power is used to operate the motor as a generator to recharge the batteries when the SOC drops below the established minimum. Parallel assist and regeneration is controlled automatically by the vehicle control system. The control system provides automatic regenerative braking in which the braking system's hydraulic pressure is measured and used to smoothly blend electrical regenerative braking with the traction motor.

Design Simulation

Simulations were used for a number of functions including component sizing, transmission shift point selection, and integrated control system testing.

Simulation Tools

Simulations are used to insure that potential components are adequate to meet or exceed the gradeability, acceleration, speed, and range requirements. The initial calculations are based primarily on individual vehicle component performance and utilize MATLAB™. A more fully integrated vehicle systems approach is also under development in the SIMULINK™ environment using an ADVISOR™ based vehicle model.

Based on earlier work with the ADVISOR™ hybrid vehicle simulation program [5], a MATLAB™ simulation package has been developed to serve as a design tool for motor, engine, and battery pack sizing.

Preliminary studies were performed regarding range, top speed and gradeability. Parameters considered in this model include power transmission efficiency, aerodynamic drag [6], road grade, rolling resistance, and component inertia.

Acceleration

In the acceleration analysis, the vehicle performance was modeled using a transmission limited acceleration behavior [7]. At low transmission input shaft speeds, the torque input is limited by the design of the transmission. At higher speeds, the engine and electric motor provide less power than the transmission's capability and they become the acceleration-limiting factor.

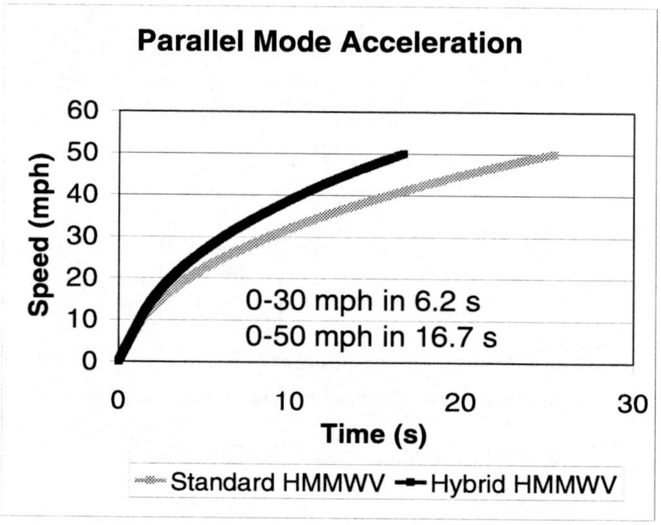

Figure 7. Projected Acceleration

As Figure 7 shows, the proposed hybrid configuration acceleration capability exceeds that of the stock vehicle.

As a verification of the validity of the existing simulation, the stock vehicle was modeled and compared to existing data on the production HMMWV. The simulated time for the 0-30 mph acceleration of the stock vehicle was approximately 10% greater than the standard vehicle time of 8 s, and the 0-50 mph simulated time was within 5% of the 24 s time achieved by the standard HMMWV. This discrepancy is most likely due to the transmission efficiency values utilized in the simulation.

Gradeability

Gradeability is an important consideration in any off-road high mobility vehicle. Since it is possible to be in a situation which would require traversing a steep grade in either the parallel, pure electric, or in the limp-home (engine only) mode, a simulation was conducted using the same software as the acceleration analysis for each of these situations.

Table 2 shows the speed at which the vehicle can climb a 60% grade. Simulation indicates that the minimum requirement (2 mph) can be met in all the operating modes. At 2 mph in the pure electric mode, the hybrid can travel 8.4 miles up the 60% grade before the state of charge reaches a critically low state.

Table 2. Hybrid HMMWV Gradeability

Operating Mode	Speed (mph)
Parallel (peak rating)	10.0
Parallel (continuous rating)	8.3
Pure Electric (peak rating)	6.0
Pure Electric (continuous rating)	4.5
Engine only	3.5
Standard HMMWV	7.0

The vehicle is also required to maintain a speed of 55 mph on a 5% grade while fully loaded. The hybrid vehicle will be capable of achieving this level of performance in all modes with a pure electric range of 16.5 miles under these conditions.

Pure Electric Range

The required pure electric range for this design is 20 miles at 30 mph.

Table 3. Vehicle Range at Selected Speeds

Vehicle Speed (mph)	Range (miles)	Energy for 20-Mile Range (kW-hr)
20	31.6	8.17
30	28.1	9.21
40	24.3	10.66
50	20.6	12.52

Table 3 shows the vehicle range at several speeds at a depth of pack discharge resulting in a 1.67 VDC/cell [8] which gives a pack voltage of 270.5 VDC. For these calculations, the vehicle was assumed to be on a level, hard surface at constant load. The last column of this table shows the energy requirements for a 20-mile range at each of the vehicle speeds evaluated.

Conclusion

The conceptual design of a low cost hybrid HMMWV is well underway and results at this point suggest that a parallel and charge-sustaining configuration will able to achieve performance, fuel economy, and cost objectives.

References

1. Wouk, Victor, “Hybrids: Then and Now”, IEEE Spectrum, July, 1995.
2. Toyota Motor Corporation, "Toyota Hybrid System", Toyota Press Information, Tokyo, Japan, 1997.
3. Juzkow, Marc W. and Steven T. Mayer, “Design Considerations for Lithium-ion Cells Part II: Safety and Abuse Testing”, PolyStar Corporation, 1997.
4. Tucker, Brian E., “The Development and Implementation of a Control System for a Hybrid Electric Vehicle,” MS Thesis, Mechanical Engineering, University of Tennessee, May 1997.
5. He, Xiaoling, “Hybrid Electric Vehicle Simulation and Evaluation,” MS Thesis, Mechanical Engineering, University of Tennessee, August 1997.
6. Munson, Bruce R., Theodore H. Okiishi, and Donald F. Young, Fundamentals of Fluid Mechanics, John Wiley & Sons, New York, 1994.
7. Gillespie, Thomas D., Fundamentals of Vehicle Dynamics, pgs. 79-123, Society of Automotive Engineers, Warrendale, PA, 1992.
8. Hawker Energy Products, Genesis Selection Guide, 1st edition, Hawker Energy Products Inc., 1997.

J. System Design Track - Information Systems

This new track developed from a recognition that avionics systems are increasingly directed to accessing, storing and providing the right information when required. This 'data-driven view' of avionics is assessed from how information systems are applied and how they integrated into the modern cockpit.

Track Chair: Thomas K. Brown, Draper Laboratory

Track J Information Systems

Session J1: Information System Application

Chair: Frank White, Draper Laboratory

Session J2: Information Integration

Chair: Thomas K. Brown, Draper Laboratory

AN APPLICATION OF COTS ETHERNET FOR AVIONICS

Brian Reynolds, Rockwell Collins, Inc. Cedar Rapids, Iowa

Abstract

To oversimplify, current aircraft digital buses such as ARINC-429 and ARINC 629 move too little data over too many wires in a manner that is too inflexible. In addition, in some cases they are also expensive to implement. The use of Commercial Off the Shelf (COTS) standards or components to address these problems is attractive, but brings on it's own issues. This paper examines one use of COTS, and some of the issues which have been encountered.

So What is COTS?

The acronym "COTS" is generally understood to mean "Commercial Off The Shelf," however there are two very different facets to the term. The first is what I will call "Commercial Off the Shelf – Standards" and the second is "Commercial Off the Shelf – Implementation," (abbreviated COTS-S and COTS-I respectively). In the commercial world this differentiation is unnecessary as the "user" of a COTS item is typically buying an implementation which is expected to comply with the standards. This works in the commercial environment, but does not work in our regulated environments where we are concerned with not only what something does; but also equally concerned with how it does it to ensure it does what it is supposed to do, and nothing else.

An example of COTS-S is the IEEE 802.3-ISO/EEC 8802.3 for Carrier Sense Multiple Access with Collision Detection (CSMA/CD). An example of COTS is when one buys 802.3 compliant parts from a supplier.

We do ourselves a great disservice when we, within our context, use the term COTS without differentiating between the standards and the implementation. The problem sets are very different and interfere with each other if it is not clear to which one we are referring. So let's agree to use the non-interchangeable terms (at least for the remainder of this paper): COTS-S, COTS-I and when one truly means both, COTS.

What is Attractive About COTS

In one word – Cost. By reusing something that is already widely used, it is expected that the total system life cycle cost can be lowered. Consider the following:

No requirements to develop

Use readily available development tools and environments

Special training is minimized

Buy what you need for rapid prototyping/Proof of Concept and requirements validation

Use standard tools and test equipment both for the manufacture and for the end user

Leverage commercial production rates with commensurate lower unit costs.

The term "COTS" can be usefully applied to the first 5, however the last item clearly brings up how the standard is implemented and has some very tough issues by itself. So a better way of looking at this is COTS Standards bring us the benefits of:

Mature standards

Widely available parts early in product development

0-7803-5086-3 /98/$10.00 ©1998 IEEE

Commercial available skills sets

A mature development and test environment

COTS-Implementation will be considered later in this paper.

OK – Let's Use COTS-Standards, But What Does That Mean?

Consider for this example, IEEE 802.3 Carrier sense, multiple access with collision detection (CSMA/CD)[1]. The Xerox Corporation originally developed Ethernet (which was a Xerox trademark until it was released into the public domain by Xerox; however Xerox is still the patent holder on the underlying technology) comprised of three principal areas: (See Figure 1 for both Ethernet and IEEE 802.3 frame formats)

The format of the data to be sent and received (the frame format)

The manner in which a transmitter gets access to the bus (Media Access and Control or MAC which includes interframe gap, preamble, and start of frame delimiter)

The electrical characteristics (signaling) of the actually transmitted signal (the Physical layer)

The first is the most interesting. Xerox defined five data fields:

Destination Address

Source Address

Type

Payload

Frame Check Sequence

So far so good. As Ethernet moved into wider acceptance, IEEE picked up the effort to create a standard. IEEE, however, works within a wider contest of the International Standards Organization (ISO), and ISO was developing its own communications standards which are now commonly referred to as Open Systems Interconnect (OSI). The design philosophy of OSI is that there are clearly defined layers in a communications environment, and each layer provides "services" to the layer above it through "Service Access Points." In addition, each layer can either perform simple error checking and discard on error (connectionless) or handle error detection and retries (as well as buffer management) as a connection oriented protocol. So when the 802.3 committee started up they were faced with a rather large dilemma. Ethernet was well established and was in fact a de facto (but not de jure) standard which had no concept of layers, service access points, or how to manage connection. Therefore Ethernet had to be "improved" so that it would be consistent with the other OSI layers. Their method for solving this was both creative and practical. They reused the Ethernet "Type" field. Ethernet had long established that the maximum number of bytes which could be carried as an Ethernet frame payload was 1500. Ethernet uses signaling characteristics to define the start and end of a frame, so it has no need for a "length" field. Instead the "Type" defines the protocol which is carried by the frame. For example, 0800 (hex) is defined by RFC[2]-1122 which signifies that an Internet Protocol (IP) Protocol Data Unit is being carried by the frame.

The 802.3 committee used the fact that all "types" were numerically larger than 1500.

[1] International Standard ISO/IEC 8802-3 - ANSI/IEEE Std 802.3 Carrier sense Multiple access with collision detection (CSMA/CD) access method and physical layer specifications.

[2] Request for Comment. The Requests for Comments (RFCs) are a series of notes, started in 1969, about the Internet (originally the ARPANET). The notes discuss many aspects of computing and computer communication focusing in networking protocols, procedures, programs, and concepts, but also including meeting notes, opinion, and sometimes humor.

The specification documents of the Internet protocol suite, as defined by the Internet Engineering Task Force (IETF) and its steering group the IESG), are published as RFCs.

Standard 1 defines the current set of RFC's which have reached "standard" status. See http://www.rfc-editor.org/ for a complete list of RFC's and their status.

So if the Type/Length field is equal to or less than 1500, the frame is in compliance with 802.3, and if bigger, it is not an 802.3 frame and hence not of interest to 802.3. 802.3 was then free to add an OSI Service Access Point fields as part of 802.2 Logical Link Control, as well as an LLC Control Frame so that the Link Layer could be either connection oriented, connectionless, or acknowledged connectionless. This was nice as far as it went, however Ethernet (the Xerox version with the "Type" field) was (and still is) very well established in the "real world." The Internet Engineering Task Force[3] defines a method for encapsulating an Ethernet frame over 802.3 by adding yet another field, the "SubNetwork Access Protocol" (SNAP) to the LLC header which (in addition to the Organizational Unit Identifier field of SNAP) added the Ethernet Type field back in so that upper layer protocols (typically the Internet Protocol at the Network layer) could figure out what to do with the payload. However RFC-1042 goes one step further. In addition to SNAP, it also defines the contents of the LLC header! So the Destination and Source Link Layer Service Access Points fields are fixed at "AA, AA," the Link Protocol is fixed at type 2 (connectionless) indicating that the next field will be interpreted as OUI and Ethernet Type.

What's the practical effect of this? No one uses 802.3 for definition of an Ethernet frame because through convention and precedence the LLC and SNAP headers provide no additional benefit over "straight" Ethernet. Windows can't even generate an 802.3 frame, while NT can be forced to generate an RFC-1042 compliant version of an 802.3 frame (that is with the fixed values shown in Figure 1C).

So the dilemma we find ourselves in is that we want to use a COTS standard because of all the reasons outline earlier; namely that IEEE 802.3 is available, but the "COTS world" doesn't use the 802.3 frame format! Simply put, the "standard" is not *The Standard*. So to reap the benefits of using COTS, we must, at least in this area, deviate from the Standard. It is worthwhile to recognize that in IEEE 802.3 revision "x" the foot note which, while acknowledging Ethernet as "other use," has been deleted[4] so that an Ethernet Frame is now allowed by IEEE 802.3. But in doing so there is now a naming problem. By bringing Ethernet into 802.3, how does one differentiate between the "old 802.3" – i.e. the Length frame and an Ethernet frame – i.e. the Type frame? We have chosen to use "802.3-Type Frame and "802.3-Length Frame" when it is necessary to differentiate between the two within an IEEE 802.3 specification context. The general term for an "802-3-Type frame" is Ethernet, and what is what is normally used.

The issue of 802.3-Length vs 802.3-Type is being debated within the Subnetwork Working Group of the ARINC Data Link committee. At the time of preparation of this paper the matter is unresolved, however my recommendation is that we stick with a straight -up Ethernet (e.g. 802.3 Type) frame like the rest of the world.

So a COTS Implementation Can Deviate from a COTS Standard?

When the Standard would require us to implement a function, feature, or service which is contrary to the reasons stated earlier on "What is Attractive About COTS" the answer is not only *Can*, but I would go one step further and say *Must* deviate from a "true" implementation. However any such deviation must be considered a proper subset of the standard in order to capitalize on COTS benefits.

Let's take another look inside of IEEE 802.3. This Standard is organized into several basic parts:

[3] In RFC-1042

[4] Clause 3.2.6 also is changed to allow the use of the Type/Length field to indicate the "MAC client protocol," thus bringing 802.3 into agreement with real world applications.

- MAC Service Specification
- MAC Frame Structure
- Media Access Control
- Layer Management
- Physical Signaling Service
- Physical Signaling and Attachment Unit Interface
- Specifications for Media Attach Units for:
- 10Base5, 10Base2, 10Broad36, 1Base5, 10BaseT, 10Base-F (several flavors). Additional revisions have added 100Base-T, 100Base-FO, 1000Base-FO, and full duplex operation.

The Service Specification is useful because it gives context and contains no actual requirements. The Frame Structure requires LLC which is presently not implemented in "real" Ethernet networks, so that gets "profiled out." Media Access Control[5] goes to the heart of how CSMA/CD works, so and as long as it does not require the use of LLC, it can be accepted as a requirement set. Layer Management is also required for interoperability as it defines the Management Information Base (MIB) objects used for network management. Physical Signaling Service, Attachment Unit Interface, and Media Attach Units all define connectors and signaling schemes. If you want your 10Base-T device to interoperate with another 10Base-T device, this is where it's specified.

In later revisions to this standard, a very useful notion from the OSI world has been brought into play - that is the Protocol Implementation Conformance Statement (PICS). The PICS is provided by the supplier of a device to show that they have actually complied with all of the applicable requirements contained in the specification. Generally limited to layer 1 (PHYsical Signaling), the PICS defines which parts of IEEE 802.3 are mandatory or conditional. Looking at the PICS for 10Base-T[6]; transmit and receive functions, signal levels, cable types and drive characteristics are all listed in the PICS.

Some Examples of "Profiling"

While I spent some time earlier discussing the problems of "LLC or No LLC," the actual frame is not covered by the PICS because it is considered a "higher level" than the Medium Attach Unit characteristics contained in the PICS. A good example of how industry has also "profiled" IEEE 802.3 is in the area of Collision Detection[7]

The title of 802.3 is CSMA/CD. However, some time in the past Ethernet moved away from coax (10Base2 and 10Base5) to telephone grade unshielded twisted pair, with one pair for transmit and another for receive. By using other services performed by a bridge (IEEE 802.1D), it is possible to limit a collision domain to just the link between a node and a bridge. Under these circumstances there will be no collisions and the link can operate in full duplex (simultaneously transmitting and receiving independently). HOWEVER, Clause 14 (which specified the behavior of twisted pair) retained the CSMA/CD methodology and requires (actually past tense now) that Collision Detection be mandatory. And CSMA/CD, when enabled, acutely precludes the use of the more efficient (no collision) Full Duplex! The supplier community responded by implementing the requirements for Collision Detection (allowing conformance to the PICS), but in addition they provided a means of disabling Collision Detection (allowing use of the faster and more efficient Full Duplex). Until 1998 an Ethernet Network Interface Card (NIC) operating in Full Duplex mode was not compliant with IEEE 802.3, even through the device was, "as shipped," capable of being compliant. With the adoption of IEEE 802.3x,

[5] Clause 4

[6] Clause 14

[7] Clause 14.10.4.5.5

Full Duplex Operation is now allowed, and the specification has caught up with the implementations.

A more straightforward example of profiling is the type of connector and cable to be used. The commercial world has a need to define the characteristics of the connector and cable to ensure operability. The connector chosen is the ISO/IEC 8877:1992 jack commonly called the RJ-45. A bigger version of the common telephone jack, the RJ-45 provides eight contacts, and is efficient and compact for office and closet use. Commercial repeaters and bridges will incorporate up to 128 Ethernet connections of the front of the switch unit, and the RJ-45 allows a "patch panel" approach to interconnecting end users with network service. But we have a different standard. Our units need to operate in a more hostile environment both physically and electrically. We cannot tolerate the problems which plastic connectors would give. Our wires come out the back of Line Replaceable Units using an ARINC specified connector which is suitable for an on-aircraft environment. So it is required that all of the PICS statements related to the Medium Dependent Interface (i.e. the connector) be changed to "Conditional." With the condition being "Mandatory if interconnection to IEEE 802.3 devices is implemented." The "profiling" allows us to do what we have to do, unless we need to interconnect with an IEEE 802.3 device. For that we would have to provide a RJ-45 connector.

Another example of profiling is what type of wire is actually used to interconnect IEEE 802.3 devices. The authors of IEEE 802.3 did not actually specify what cable is to be used. Instead they state:" These characteristics are generally met by 100 m of unshielded twisted pair cable composed of 0.5mm (24AWG) twisted pairs." So using a different cable type does not require profiling; but the cable must "perform" within the specifications of the link segment characteristic[8] if we expect the interface (transmitters and receivers) to work as specified. The supplier will provide devices which assume a compliant cable plant. Changing the cable plant for our needs may change the operating characteristics of the commercial devices, and if we want to use these devices we must live within the operating characteristics.

Similarly IEEE 802.3[9] defines "safety requirements" such as grounding, and "applications of telephony voltages." We would add the whole gamut of RTCA DO-160 requirements to that, but this is:

1) implementation specific and

2) a system requirement.

But How Can You Handle Implementation of COTS-S?

Up to now we looked at how a COTS specification can be "tailored" or "profiled" to fit within our requirements. However this is only part of the problem. Once the requirements definition (or Standards) is finished and accepted, the next required step is to actually build them into an acceptable implementation.

In the past we, as an industry, have been capable of using the unique requirements of the military to help solve our implementation problems. The military demanded (and got) high reliability, wide temperature, fixed (that is invariant) devices in production. As we all know, that was yesterday, and today is a different world. As a taxpayer I'm happy. As a systems designer I'm very unhappy. Commercial COTS-I parts are not designed for high reliability. The "fact" that they exhibit years and years of service is, in my opinion, a fortunate accident of good engineering rather than a conscious effort. In addition it cannot be considered even as a "fact," because there is no record keeping on how long components

[8] Clause 14.10.4.7.1

[9] Clause 14.10.4.5.15

actually last in the field. Commercial pressures will drive a vendor to supply components which have "acceptable" behavior. A company which produces devices with poor reliability and high failure rates will not stay in business because the market place will go elsewhere. But that gives us small comfort when our customers have come to expect MTBFs of 100,000 to 200,000 hours. And we, as good suppliers, provide reliability guarantees to our customers as if we were using high reliability components in our systems.

DO-160 requires operating temperatures which are not supported by commercial temperature range parts. The military is moving towards "extended" temperature ranges, but even these will not meet the requirements of DO-160. But consider how our products are used. An aircraft sits overnight in Helsinki and flies to Jeddah. Our customers expect our products to work over these temperature extremes from the time a maintenance worker starts the aircraft in the morning until it is shut down at the end of its operating day. Few COTS components can operate over these extremes unless they have assistance. DO-160 is a system/box requirement, and it is possible to provide a better temperature environment through good thermal management. Likewise for HIRF/EMI and lightning. They are challenges, but not do automatically disqualify an implementation at the piece part level.

What are the Real Issues with COTS Implementation?

There are three significant issues with COTS-I:

- Design methods and practices
- Commercial variability
- Supplier protection.

Our industry has a design model based, to a large extent, on the concept that complexity can be managed through process. First brought forward in RTCA DO-178A where process was applied to software aspects of certification, and later was improved[10] in RTCA DO-178B. RTCA DO-178 has a supplier insert formalized monitoring processes into software development. The rational is that if a process is properly monitored, the probability of the software meeting its stated requirements is greatly improved. These development and test processes brings formality to the software preparation, and by being applied to each step in that preparation the result is the implementation of highly complex functionality which has been incrementally examined through out its development to ensure that it does, and only does, what it is specified to do. Documents are created during each step to ensure that the process has been followed. It is a commonly held belief by many that an aircraft cannot be certified until the weight of paper exceeds the weight of the aircraft (I've never heard if that is dry or wet, but I think it's wet.) DO-178B greatly assists in this effort.

The regulatory authorities, having got software development "in shape," are now turning their attention to hardware, specifically Application Specific Integrated Circuits (ASIC). Their notion is that there is all this process control in software, but none when an ASIC is used to embed software into hardware. RTCA Special Committee SC-180 is working to develop ASIC development processes to hardware based on the DO-178 mindset.

What Does "Process" Mean in the Commercial World?

The effect of this is that we can expect the regulatory authorities to group all "highly complex devices" into a single category, independent of whether there are well known standards or not. In this mindset, an IEEE 802.3 compliant Media Access Control Device

[10]As part of the RTCA SC-167 committee I had a hand in this improvement so I'm perhaps a bit biased.

from AMD or Intel has the same lack of credence as a device which Collins might develop which is unique to how our LCD devices interface to a proprietary processor.

At this time there is no notion of being able to take credit for a device which implements a well known, well understood standard, which does a few things very well and very fast, and for which the production rate is literally thousands per day; as opposed to a unique set of requirements which have never existed until a specific proprietary need was defined, and for which perhaps the total production of the life of the product may be a less than few thousand. If we as an industry cannot find a way to take credit for commercial use, then a further discussion on the use of COTS implementations is moot.

OK - Say We Can Get The COTS-I Design Approved, What About Commercial Variability?

The second serious problem is control of commercial variability. An example would be where a supplier "improves" a COTS-I and we, as users, have no visibility into what was changed or why. A mask change for the convenience of the manufacture, for example, is typically not reflected in a part number change, so we, at the end of the supplier-user pipe, can't see the change or understand how it may or may not impact our use. This problem must be acceptably managed before COTS-I can become acceptable.

Supplier Protection?

And finally there is the problem of "man-rating." Many companies prohibit the use of their products under circumstances where people may be injured or killed without prior approval. This is widely, and properly, believed to limit the liability of a supplier for uses beyond their control, but for which they could be held liable. This is a business issue between corporations, and one that must be addressed by corporate consul.

Summary

This paper has presented the COTS domain as consisting of two separate and distinct entities:

- COTS-S or Standards
- COTS-I or implementation

By understanding which of these we are concerned with the nature of the strengths and weaknesses inherent in each can be brought to focus more easily addressed.

The use of COTS, either the standards or the implementation, can be likened to making lemonade. To make lemonade one takes lemons, which are inherently sour, and by adding water and a sufficient amount of sugar you get an acceptable drink. While it would be possible, given sufficient time and money, to genetically reengineer the lemon to produce more liquid which is sweet, it is generally recognized that this is not really viable from either a cost or schedule perspective.

COTS is like lemons, and implementing COTS solutions is like making lemonade. Don't redesign what is given, but add to the system what is necessary to get what you need. That is cost effective, and practical.

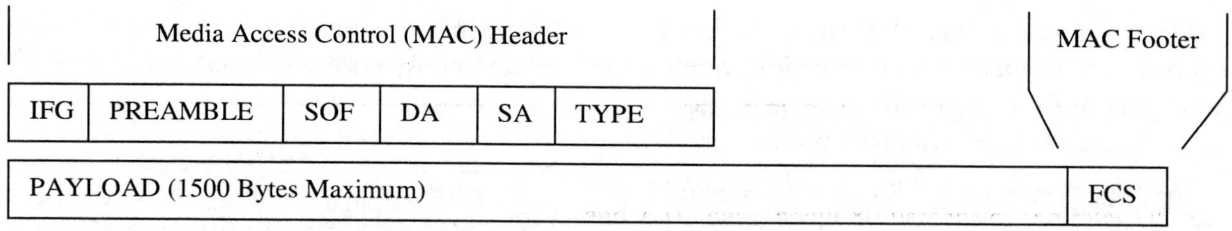

Figure 1-A. Original Ethernet Frame

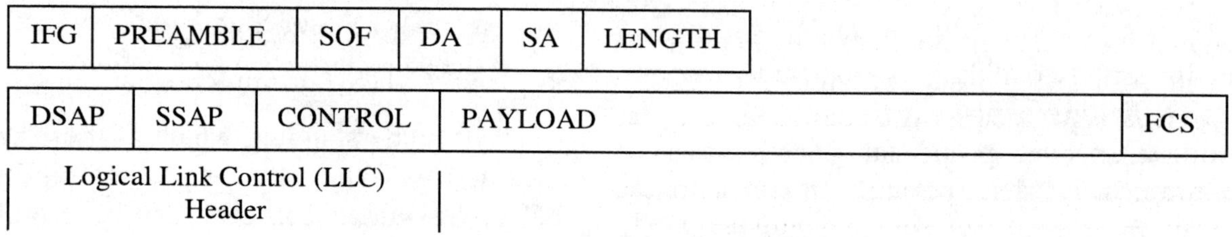

Figure 1-B. Format Specified By IEEE 802.3

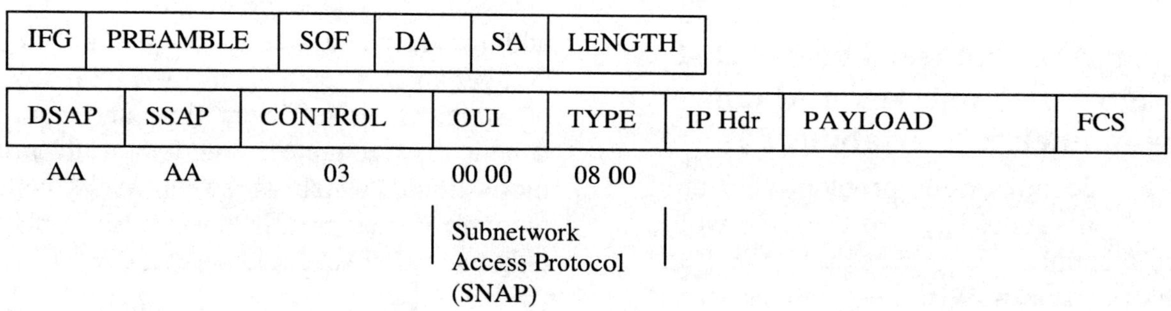

Figure 1-C. IP Datagram, Encapsulated in an IEEE 802.3 Frame (RFC-1042)

DA Destination Address
DSAP Destination Link Service Access Point
FCS Frame Check Sequence
IFG Interframe Gap
MAC Media Access Control
OUI Organizational Unit Identifier
SA Source Address
SSAP Source Link Service Access Point
SOF Start of Frame Delimiter

Figure 1. COTS Ethernet Frame Formats

THE REMOTE AIRCRAFT FLIGHT RECORDER AND ADVISORY TELEMETRY SYSTEM - RAFT (PATENT PENDING)

And It's Application to Unifying The Total Digital Avionics System

Sy Levine, Safety Intelligence Systems (SISTMS), Topanga, CA

1. Introduction/Abstract

RAFT is a worldwide, real-time aircraft remote monitoring and recording system that takes an aircraft's Digital Flight Data Recorder, DFDR, monitoring parameters out of an archival data base and plugs them into a safe, readily available, usable accident prevention system. RAFT combines the DFDR sensor data with the data from the Air Traffic Management/Control (ATM/C) system along with GPS/GLONASS, Map, Terrain and Weather information to actively anticipate and prevent accidents. It ends the information vacuum created by the aircraft and the ATM/C where presently each of them, acting independently, don't have sensors that directly measure the necessary parameters required to prevent a crash. By the sharing of the digital data, all of the necessary crash prevention parameters become visible and usable to actively anticipate and prevent accidents. Privileged non-safety related data is ciphered at the aircraft to insure air carrier confidentiality. In addition the global telemetry of the DFDR parameters allows aircraft monitored data to be simply and safely stored on the ground. Thus making it readily available for aircraft statistical analysis programs that enhance air carrier efficiency and safety. Also, in the advent of a crash, it provides a timely accurate global estimate of the downed aircraft's location for search, recovery and hopefully rescue operations. It establishes an aircraft global data super highway that uses high bandwidth satellite and ground internet communication links to supply the aircraft advisories necessary to enhance air space capacity, operational efficiency, security and substantially reduce accidents.

RAFT brings to aviation what the internet brought to data visibility and utilization. It unifies the National Airspace System (NAS) and fills the information vacuum that has been responsible for twenty years of a stagnant air carrier fatal accident rate. It makes all of the necessary safety data visible and readily available to the people who need to solve problems in a timely and cost effective manner before they become fatal accidents. RAFT also provides functional redundancy and simplifies the communication system. It is the only system capable of meeting the national goal of reducing the fatal accident rate by a factor of five in ten years and provides the necessary safety net that should be put in place prior to any transition to free flight.

2. Main Section/Description

2.1 BACKGROUND

More accidents can be attributed to aircraft data utilization than to stress and fatigue failures of the airframe. **The present federated airspace system is failing[1] to reduce the accident rate mainly because it presents the data too late, is prone to single thread failures, difficult to modify, costly and doesn't meet the needs of the new millennium.** On 9/26/97 there was a 234-fatality, Garuda Indonesia Airlines A-300 accident. The plane went into a mountain because the air traffic controller directed the pilot to make a left turn. The pilot misunderstood the instructions. The information was being displayed via the air traffic control (ATC) radar blip or dot. At the instant time of the crash the dot disappeared from the screen. A screen dot contains no information as to the direction\rotation of an aircraft. Yet the direction of the aircraft is a DFDR parameter provided by the inertial navigation system (INS). If that aircraft rotation data had been available in real time to the ATC in a simple display that depicted the actual aircraft correctly rotating, and as such would have provided the necessary information to prevent this accident.

In the 1960s there was a very successful infusion of technology into the NAS —placement of the INS into commercial aircraft. The initial INS were inserted on the B-707 by Pan American Airlines. In order to prevent a single-thread failure and provide redundancy of this critical function, two INS were installed. The FAA required that two INS units be functional prior to takeoff. To mitigate the risk of a delay, if one of the INS experienced a problem prior to take off, a third INS was installed on the B-747. In a similar fashion the autopilot, like the INS, is now functionally redundant. These units now have a very high functional availability. For enhanced safety it is desirable to eliminate other single-thread functional failures by having at least dual redundancy for fatal accident prevention.

An example of single-thread failure is the 228-fatality crash of the Korean Airlines B-747 Flight 801 that occurred on 8/6/97 at Agana Guam. At the time of the accident the ATC tracking radar was working, but the glide slope unit was inoperative. With the loss of the glide slope radar, the plane's altitude was low resulting in a crash into a hill. The altimeter on the aircraft was working. In this instance, if the aircraft's altitude data had been cooperatively shared with the ATC tracking radar data and used in an integrated display, it could have provided a backup for the malfunctioning glide slope unit. The aircraft's glide slope display can be made transparent to the failed microwave system so that the pilot would not require a different presentation.

Thus, by the cooperative sharing of the DFDR and ATC data in real-time, accidents like this can be avoided. Because of the major advancements in computers, memory and communications bandwidth and technology it is now possible to provide safety function redundancy at a fraction of the present federated system cost.

The present DFDR system is analogous to having a patient in intensive care being monitored. However few people look at the data until the patient dies or after release from the hospital. The DFDR is so important that in the advent of an ocean crash we risk lives and expend vast amounts of time and money searching for it. Even with these Herculean efforts the recovered DFDR's recorded data may be unreadable. After recovery of a readable DFDR, we utilize its data in a playback mode to perform a post-flight animation and analysis to determine the root cause of a crash. This is necessary to take a corrective action that will prevent the future reoccurrence of the accident.

The DFDR data is also routinely collected for post-flight statistical analysis. With six real-time programs that cooperatively share the DFDR and ATC data, about 80 percent of the fatal US air carrier crashes can be prevented. Post-flight analysis of non-crash data is important for flight safety and this work is presently being carried out in the Flight Operational Quality Assurance (FOQA) program. Yet the big pay back in flight safety does not come from post flight analysis but comes from using this data in real-time application programs that are targeted for accident prevention. These real time programs that share safety data will result in dramatic increases in air capacity, safety, security and operational efficiency.

The present safety data vacuum, created by not sharing all of the sensors required to solve safety related problems, has been the major cause of the lack of reducing the air carrier fatal accident rate in over twenty years[1]. With the growing use of air travel this translates to an ever increasing number of unnecessary and presently preventable air fatalities. By the sharing of the digital safety data, all of the necessary crash prevention parameters become visible and usable to actively anticipate and prevent accidents. The flight crew is responsible for control of the aircraft and ATC is responsible for the airspace and airport areas. The flight crew and the controllers are in a codependent symbiotic cooperative relationship for the safety and economic benefit of the public. It is this relationship that brought radars to both the pilot and the air traffic controllers utilization. Therefore the system should aid in minimizing misunderstanding and expedite communication to minimize fatalities. Thus air travel is a cooperative synergistic enterprise between the flight operation centers, pilots and the air controllers that requires precise communication for safe transportation. Simple shared safety advisory icons, warning of problems, only inserted into displays when there are potential and existing problems, will significantly improve safety and situation awareness while decreasing work loads. These safety icons/alarms would only come on during potential and existing problems and thus would not clutter or distract from the normal ATC displays. At present once alarms come on in a plane there is a lot of voice communication that takes place, between the flight crew and the traffic controllers, that can easily be misunderstood. It also puts stress on both the pilot and controller, depletes precious reaction time and increases work loads. The 1/25//90 Avionca crash, that had 73 fatalities, is just one case of where communication problems under stressful conditions have led to fatalities. The Avionca plane ran out of fuel while in a ATC directed holding pattern over JFK Airport. The crew reported to the ATC that they were running out of fuel but did not use the word emergency. The FAA listed the cause of this fatal crash as pilot error. This syntactical mistake is no reason for people to die since a simple emergency low fuel safety icon can automatically be displayed on the ATC monitor, similar to the low fuel warning light on automobiles, to alert the controller of the dangerous low fuel status. The low fuel warning light or oil pressure warning lights in an automobile doesn't increase the drivers workload but simply increases the situation awareness and prevents catastrophic failures. Once a controller receives a low fuel warning light he can then set the landing priorities to expedite a safe landing. The fuel supply is another one of the DFDR parameters whose data is presently locked in the aircraft DFDR and thus the ATC doesn't have any visibility other than with voice communication with the aircraft that the fuel is low. The problem was not the crews voice communication, pilot error or the air controller but the present unforgiving system that shares its safety data begrudgingly and has been willing to sacrifice innocent people for simple and normal human errors. The lack of automatically sharing of cooperative safety data did not permit the controller to have an unambiguous situation awareness of the pre-crash emergency. The 8/6/97 Korean B-747 crash is also another case of the failure single thread verbal communication system. The ATC could have continually verbally requested the aircraft's altitude from the flight crew and together with his radar range information successfully verbally guided the plane down to a safe landing. Also if the aircraft's DFDR altitude data going to the DFDR was readily and automatically available to the ATC, it in combination with the ground radar could have easily provided an automatic back up for the malfunctioning glide slope radar. In the future after RAFT is operational and a 6/8/95 ValuJet type fire occurs, the fire and smoke sensors added to the plane and recorded in the DFDR, will light icons and sound alerts not only on the flight decks displays but also on the air traffic controller's console. It is the controllers responsibility to make sure that a plane that is or may be experiencing a critical flight problem on the ground is not cleared for a takeoff and that a plane experiencing a critical flight problem in the sky is cleared for a safe landing in the shortest possible time as well as to alert the necessary ground support people to care for the passengers and flight crew so as to minimize further disasters. The present system is too verbally dependent in communication for the next millennium. Just as the internet is moving from just printed data to printed, graphics and acoustics; RAFT will permit the migration of the shared aviation system's safety information to move from essentially verbal communications to acoustic, graphic and printed. With the use of commonly shared visual icons as well as acoustic alerts the situation awareness of potential problems will be raised so as to prevent fatal accidents.

2.2 Detailed Description

RAFT is a worldwide real-time aircraft remote monitoring recording system that is used for enhancing airspace capacity, operational efficiency, passenger safety and security. It brings the digital flight data recorder information out of an archival database and plugs it into a real-time usable accident prevention system. It ends the information vacuum created between the aircraft and air traffic controller. Presently each is acting separately and lacks the sensors to directly measure the necessary parameters required to prevent aircraft mishaps. Combining these data sensors enhances the effective sensor suite so that many events can be anticipated. This event anticipation capability provides the visibility and time mandatory for the prevention of accidents. A best estimate of the location of a downed aircraft for timely search, rescue and retrieval operations may be provided by linking the global telemetry of the DFDR parameters to a ground processing and distribution station.

RAFT updates the federated system and unifies the communications approach so that the relevant data parameters are globally visible and readily available for timely and cost-effective problem resolution. It is a system engineering approach that potentially can eliminate or minimize the need for the costly and time intensive recovery of the aircraft's recorder. An alternate is to keep the existing FDR and to use RAFT as a redundant system that essentially eliminates the need to recover or post routine flight down load (FOQA) the recorder in all but a very small percentage of the crashes. Thus RAFT alleviates a broad spectrum of problems. As time progresses and RAFT proves its reliability over years of service the existing FDR will atrophy (similar to Omega and sextant star fixing navigation).

Twenty-four hours worth of DFDR data, for all of the US air carrier, taxi and cargo aircraft is not an insurmountable data quantity to be managed and can be contained on 32 gigabytes of disk space. This amount of data presently fits on only two personal computer (PC) disks. Because of the advances in information processing and communications technology, a modern airspace information system can be built that will effectively simplify the aircraft avionics and provide redundancy as well as add capability. One major advantage of this system is that it can provide routing, weather, map and topographical data while also providing redundancy.

RAFT remotes the DFDR, and its communications system will integrate the ATC, air carrier dispatch and the aircraft into an integrated system that will reduce the number of aircraft LRUs while providing redundancy. The two-way data super highway can handle the information of the DFDR data transmitted from the aircraft to the ground and by balancing the uplink and downlink transmissions, it can also provide the communications pipeline for safety advisories, weather, terrain/map and differential GPS corrections transmitted from the ground to the aircraft. It is a total global system approach to the problem that is aimed at reducing maintenance, enhancing operational efficiency, and increasing air space capacity and safety. The following paragraphs, figures[0], and descriptions illustrate the advantages of RAFT system.

Figure 1 depicts the RAFT Avionics System. It shows an aircraft consisting of a sensor multiplexer transceiver, also called the sensor multiplexer receiver and transmitter (SMART), that receives the inputs from the aircraft's performance and control sensors. The GPS and/or GLONASS navigation satellite data, if available, are also other inputs into the SMART. The RAFT transceiver antennae radiates to a communications satellite the sensor data compiled from SMART along with cargo bay, passenger compartment, and cockpit video information, and acoustic data. SMART also receives advisory data from the communications satellite, which is then shown on the Advisory Display System (ADS) panel located in the cockpit. Although Figure 1 depicts two antennae, it is possible to utilize only one antenna depending on the uplink and downlink frequency selected for the communications satellite. Although the figure shows separate LRUs for the display and data multiplexer functions, it is possible to utilize the existing LRUs aboard the aircraft, in multi-functional mode, for the RAFT displays.

Figure 2 depicts the worldwide communications link. It illustrates an aircraft communications with the nearest satellite and then the communications satellite link. The aircraft's data reception satellite then relays this data to other satellites in a line-of-sight communications data link until it reaches the closest satellite with an unobstructed data transmission path to the Central Ground Based Processing Station (CGBS). Communications is duplex, and thus the CGBS receives aircraft data and communicates with each aircraft or with all of the aircraft in the net simultaneously. Major safety data advisories in the form of cautions, warnings, and alerts are transmitted to the operating aircraft based on the ground processing of the information from the aircraft and ATC system. The ATC system consists of air traffic control radar and GPS surveillance data along with in-route weather information and map & topographical databases. These advisories are transmitted to the cockpit ADS as depicted in Figure 1.

Figure 3 depicts the CGBS. It shows the processed and stored data, aircraft simulators, aircraft advisory generators, display and control of the CGBS, and data transmission modules to the ATC, aircraft manufacturer, and air carriers. Because some of the data are air-carrier privileged, a number of the data parameters are ciphered at the aircraft's SMART so that only the air carrier has the encrypt keys. This is similar to the telecomputing banking and Internet charge card systems. The ground transmission of the data will utilize existing high band width fiber optic backbone communications links with capabilities of 45 to 155 million bits per second. These links are presently being used on the Internet.

Figure 4 depicts the Ground Based Distribution system. It shows the ground processor communicating with the map, topographic, and weather data base systems. It also depicts the aircraft manufacture and air carrier communications links and the ATC/M communication system. RAFT encompasses both the Terminal Radar Approach Control (TRACON) and the En-route ATC/M functions.

Some other examples of accident occurrences relating to data deficiencies in the present ATC/M system are airport runway/ground incursions. The salient example of this type of accident occurred on 3/27/77 at Tenerife in the Canary Islands. Two Boeing 747s (KLH and Pan American) crashed head on along a runway killing 583 people. Although this accident occurred some 20 years ago its root cause still exists. Two recent examples of runway/ground incursions include (1) the 2/1/91 LAX, Los Angeles, CA ground collision where a US Air Boeing 737 landed or a Skywest Fairchild Metroliner killing 34; and (2) the 12/3/90, Romulus, MI Northwest Boeing 727 and Northwest DC-9 runway incursion that killed 8. These runway/ground incursion killers can and will continue to occur unless RAFT is implemented to plug the hole in the information system. RAFT will alter the future ATC/M and CAS global displays. The current blip/point/dot plan position indicator (PPI) type radar displays, which are 1950 vintage carry over presentation technology, are incapable of showing the aircraft rotation. RAFT displays depict actual aircraft and shows the status of their brakes, landing gear, thrust, track, Euler angles; and a safe path or collision alert in a timely simple presentation. These new displays will decrease the flight deck and controller workload while increasing their situation awareness. The present blip/dot/point displays are ambiguous and as such require excessive interpretation and concentration to be utilized in future systems. They are incapable of meeting a low fatal-accident rate system criteria. Most of the current ATC and CAS physical displays can be programmed to RAFT compatible presentations. These will depict actual aircraft, terrain and map data on simple human friendly displays that will increase the situation awareness of both the flight crew and traffic controllers while decreasing their work-load. This physical representation of aircraft and terrain while minimizing the formerly ambiguous blip/point/dot displays is an excellent example of where the fusion of DFDR and ATC data simplifies human interface during routine and stressful operations.

Figure 3 and 4 combine to provide not only the ATC/M function but also a global weather and air turbulence reporting and advisory system for operational efficiency and safety. The most current weather maps are made available from government and flight operation center meteorology departments to the flight crew via the telemetry system. RAFT provides a global communication system so that routing decisions are jointly made by the flight crew and the flight operational centers based on the best available weather and airport status data. The ATC/M are automatically informed and can participate in these critical decisions. Planes that are experiencing weather data anomalies, such as clear air turbulence and lightning, report their findings to the central ground based weather system in order to update the data base. Decisions such as fuel remaining, distance to destination and aircraft location are automatically factored into the efficiency and control equation to assure safety. The ground processor performs the real time booking keeping effort to assure that an aircraft isn't following too closely into the clear air turbulence disturbance wake created by a preceding aircraft that was flying close to the same local trajectory. The computer keeps track of wake extinction coefficients by aircraft, derived analytically and experimentally, to assure that the proper time delay and safe separation distance are met. The weather data created in the flight deck is automatically time and position tagged. Since the system has all of the best data available, and a global wide bandwidth communication system, it can disseminate the information to the all of the people who need to solve the weather and routing problems in a timely manner. Thus optimal decisions for the routing of an aircraft can be made.

Figure 5, Tenerife et al, Raft Provides Automated Collision Avoidance Alerts, ATC/M and CAS Enhanced Capability Display, generically shows the situation of the 583-fatality head on Tenerife collision. It illustrates two aircraft on a runway orthogonal to each other. Both aircraft have their thrust on, brakes off, and are heading for a collision. Due to the inability of the ATC to see that brakes are off, thrust on, or the possibility of a collision, a fatal accident could occur. The ATC under these conditions, depends on voice communications with the pilots, which may be misinterpreted. RAFT because of its access to all of the parameters going to the flight recorder, can show the ATC the brakes, thrust, velocity, and heading of both of the planes in a simple graphic display as well as process the data to anticipate a collision. The map inputs to the system can provide the ATC with pictorial displays similar to those shown in Figure 5. The processor solves the estimated collision point long before it occurs and sends automated advisories/alerts that will enable the aircraft to take anti-collision maneuvers. Using RAFT the crash-avoidance advisories can be sent to the aircraft via manual as well as automated voice and ADS alerts. The exact colors, blinkers and human engineering crash prevention alerts will be optimized by working with air traffic controllers and pilots. The pictorial data will be made to work with existing monochromatic displays; however, color monitors provide more human performance enhancements, are more user friendly, more effective, and are the current display technology.

Figure 6, Tenerife, et al., No More, Raft Provides A Safe Trajectory Display ATC/M & CAS Enhanced Capability, depicts a safe takeoff condition for the previous example. Here the display will show a green safe path for takeoff for the plane that has thrust on and is moving. The red aircraft at the cross-runway condition is stopped with brakes on and engine thrust low. The digital processor computes a safe trajectory, which is depicted by a safe trajectory arrow. This simple ATC display pictorial is an example of the type of display RAFT is capable of providing. With the existing radar, or even future GPS non-cooperative system, these accidents will continue to occur since the ATC doesn't have access to the brake and thrust vectors that are recorded away in the DFDR. Using radar and/or GPS and differentiating the position vectors to get velocity and then differentiating again to get acceleration is both too noisy and time consuming to use for collision avoidance in the close encounter ground/runway incursions areas.

The conditions requiring a safe takeoff trajectory arrow are more complex than just the collision case. Clear air turbulence, weather, topographical and runway status, length, and icing conditions will be part of a safe-to-take-off simulation that takes into account the aircraft type, weight, etc. to arrive at a pilot advisory. The caution/warning safe-to-take-off advisories can be automatically sent to pilots on the ADS displays. Accidents such as the 3/22/92, Fokker F-28-MK4000, that crashed at the end of the runway at La Guardia Airport, Flushing, NY are too often attributed to pilot or controller error, but are in fact system breakdowns. The complex relationships required for take off under adverse conditions should be aided by a pre-takeoff computer simulation that advises the pilot of the probability of a successful take off on the ADS display. It is possible that the takeoff and landing safe algorithms will be time dependent since it may be important to alter the time between takeoffs to account for the air turbulence wakes generated by the preceding aircraft. Turbulence wake extinction coefficients could be used, or past history based on tests, may have to be used in the absence of active laser/microwave turbulence sensor data. The RAFT system aids the controller and prevents work over-load by providing the bookkeeping of the time dependent operations and simple aircraft animated real-time visual displays.

The Avianca, 1/25/90, Boeing 707 accident that killed 73 in Cove Neck, NY as a result of an aircraft running out of fuel after being put into a holding pattern, is an example of where data being sent to the DFDR should be used in real-time. The ATC would have knowledge of the remaining fuel with its estimated flight time capability and not solely depend on voice communications. A fuel-remaining icon can flash on the ATC's display; for example, to indicate when an aircraft has only 15 minutes of fuel remaining. The fuel caution icon could be only illuminated during the low remaining fuel conditions and thus would not clutter the display during routine operations. The low fuel icon displayed on the ATC terminal would be similar to the low fuel warning light in an automobile. This Avianca accident was also attributed to pilot/first officer error when in reality the pre-RAFT existing federated, non-cooperative safety system was the cause of these fatalities.

In a similar fashion the fire monitors can have their alarms displayed simultaneously in the cockpit and as a fire icon on the ATC's display. This remote alarm capability is similar to many of the fire alarm systems installed in private residences, businesses and government buildings. This dual alarm system would have provided an early warning, followed by mitigating actions which would have prevented the Value Jet, Miami, 5/11/96, 100-fatality accident.

The entire class of Controlled Flight into Terrain (CFIT) is still another example of where RAFT will substantially reduce accidents. This is accomplished by terrain map databases uplinked to the cockpit for synthetic vision as well as the terrain map information superimposed on the ATC in air and runway displays. The displays will also show the aircraft animation and not just ambiguous dots to show their rotations. These displays can be simpler than the present blip/point/dot rotation displays. RAFT permits these high fidelity displays by way of its cooperative safety data sharing capability.

In addition the present airborne X-Band, 3 cm wave length weather radar sensor, which is capable of seeing objects in all weather, day and night, must get restored to it's original multi-function mode. Even though CFIT has been responsible for a high percentage aviation fatalities (see TABLE 1), over the last thirty years the airborne radar has been allowed to atrophy from an anti-collision and weather capability to just a weather radar. This radar must be returned to its original multi-function mode of weather and surveillance. In the maritime industry it is the X-band radar that provides the anti-collision capability for both boats and ships. The low cost X-band airborne radar, implemented via a modified control and display system, can eliminate almost all of the CFIT. This day and night, all weather, radar will provide the visibility of terrain to essentially eliminate CFIT fatal accidents such as the 8/6/97 Korean Airlines, B-747 Agana, Guam, the 4/3/96 US Air Force, B-737, Dubroynik, Croatia, the 12/20/95 American Airlines, B-757 (Cali) Buga, Columbia and the 9/27/97 Garuda Indonesia Airlines, A-300 Buah Nabar, Indonesia.

The X-band radar supplemented with runway corner reflectors will also provide a much higher signal strength robust landing system than the GPS stand alone versions. The stand alone GPS landing systems are too easily prone to navigation data outages. This is due to the GPS navigation receivers being susceptible to only nano-volt electro-magnetic interference from intentional jamming and unintentional man made or natural electro-magnetic L-band radiation noise sources. The aircraft's radar can also provide the robustness needed for Cat. I, II and III landing systems and combined with GPS and INS in a complementary filter system it will provide the required redundancy and high operational availability. There are even compelling arguments for keeping the present micro-wave Instrument Landing System (ILS) and integrating it with the other sensors to add more redundancy into this critical function. This complementary filter system will minimize the GPS deficiencies (e.g.: A one watt jammer - the size of pack of cigarettes- will render the GPS navigation signals useless for 25 miles. Four watt GPS jammers that fit into the palm of your hand are now internationally available to render GPS useless for a 100 mile area.). Thus the X-band radar adds an all weather direct viewing rf visibility to the flight deck that enhances the crews visual perception to essentially eliminate CFIT. The information system would be supplemented with synthetic vision objects provided by map and terrain data bases. The X-band radar, which can see both fixed and mobile targets, can be used to automatically register the synthetic visual system as well as to eliminate synthetic vision problems that result from faulty data bases or mobile objects. Thus the radar and data base map systems complement each other to enhance the optical visual system.

Figure 7, the RAFT CAS Display presents another example of the benefits of RAFT. The system can provide the aircraft with a Collision Avoidance System (CAS) aircraft display or an enhanced Traffic Alert CAS (T-CAS) display that works in the close encounter environment. The present aircraft CAS only works when the planes are widely separated, moving at constant velocity, and statistically have few or no collisions. This enhanced T-CAS display (that can be put aboard the aircraft) has the necessary data parameters to anticipate and prevent crashes. This figure shows two planes on a collision course. With the sharing of cooperative data, such as the velocity and Euler angle vector data that come from the INS and go to the DFDR and the aircraft's present T-CAS transponder data, all of the necessary data parameters to anticipate and prevent a collision exist. In the close encounter environment, differentiating the existing aircraft CAS radar or GPS position vectors to derive velocity is too noisy and time consuming for a reliable collision avoidance capability. Thus RAFT, by its cooperative sharing of data, provides means toward the global prevention of many of the in-air collisions (e.g. Lockheed C141 Starlifter and German/Russian aircraft that collided over the southern coast of Africa on 9/13/97 killing 33; Boeing 747 and Ilyushin II-76TD collided over New Delhi, India killing 349). These fatalities can be avoided with RAFT CAS or a RAFT enhanced ATC or preferably both. Another advantage of the RAFT implementation is that it permits the CAS display in the flight deck and ATC to work both in the air and on the ground to prevent air, runway, and ground to air boundary incursions. It also will permit the ATC and aircraft to select and view, on demand, identical displays to prevent misunderstandings of intent.

Figure 7 also depicts a plane flying on a non-collision course that is experiencing two problems. One problem is that it only has one of it's landing gears down and the other is that it only has fifteen minutes of flight time fuel remaining. These are depicted with a flashing landing gear and low fuel remaining icons that pulsate so as to increase the flight crew and controllers situation awareness of the problems. The problem icons only come on when a problem exists so as not to increase operator work loads or unnecessarily clutter the displays during routine operations. In fact, like the warning icons in a car, these icons decrease work load by permitting the operator to concentrate on the main task and not so much on the inferences of the metered displays or misinterpretations of verbal communication.

Figure 8 depicts how RAFT provides an aircraft data super highway, similar to the Internet that respects an air carrier's privileged data. It shows air carriers having both privileged and safety related data. This data is transmitted to a global satellite communications link that provides two-way communications to the ground CGBS. Once the data is on the ground the data packets are sorted and distributed to the air carriers and ATC. Each air carrier receives its own data, including safety and privileged encrypted data. At their facility they can decipher their privileged data. The safety data on the other hand is cooperatively shared with the air traffic controller management systems and air carriers. This will fill the present safety data vacuum by providing the sensor data necessary to prevent most information deficient accidents. It provides the FOQA data to the air carriers over an Internet link on a timely basis at essentially no cost to the air carriers. RAFT substantially increases safety and increases air carrier operational efficiency. Once the data is at the air carriers the standard FOQA post-flight software tools can be utilized. In addition RAFT permits a paper-less inspection, maintenance log and Aircraft Safety Reporting System (ASRS). Thus, it corrects the deficiencies in the present flight recorders and provides a significant enhancement to air safety and operational efficiency by providing an aircraft data super highway. It unifies the communications approach so that relevant data parameters are globally visible and readily available to users.

Figure 9, Aircraft On-Board Layout, depicts the sharing of DFDR data with the ground system. It shows some of the most salient parameters being shared to greatly enhance air safety, security and efficiency. It also illustrates how RAFT will bring a much needed level of globally enhanced capability, redundancy and optimality to the existing system that is presently overly dependent on voice communication and inferences. The 1998 losses of ATC radar visibility of the President's plane over CONUS from the ATC/M system is still another reason for RAFT. It provides a satellite based ATC/M surveillance system that would be a redundant back-up to the existing ground based radar ATC systems. In the low air traffic density areas that extend beyond the radar horizon or don't have radar coverage it would provide a non-redundant ATC/M global safety net function. RAFT, by it's global coverage, can prevent many of the collisions that take place beyond ATC/M surveillance radar horizons. Thus RAFT provides significant enhancements to the present system that is prone to failures of omission, commission and misinterpretation as well as equipment outages.

Figure 10, shows a chronology of the communication costs per plane per average flight. The curve shows that the communication costs for a global system is being drastically reduced. This is a direct result of the technology advancements increasing the channels and bandwidth while reducing the cost of Low Earth Orbit (LEO) digital data communication satellites. By the year 2003 several low cost high band width LEO satellite communication systems will be operational and other low cost LEO satellite constellations will be in the process of getting to be globally fully operational. These LEO systems, projected to the year 2008 will bring down the average cost of the RAFT system to only nine dollars per flight. It is estimated that the safety benefit alone for RAFT at that time will represent a savings to the total of all air carriers of over $400 million per year. When the other operational efficiency benefits of RAFT are factored in, the savings will reach over a billion dollars per year.

Table 1 is a tabulation of the worldwide air carrier fatalities and fatal accidents between 1987 and 1996. This table[0] is a compilation of accidents sorted by causal type for all of the world's air carrier operators and by only the US operators. It tabulates the percent of fatalities and the percent of fatal accidents by accident type for the present system and compares these actual statistics with an estimate of what they would be if an operational RAFT system were in place. It shows that the RAFT system is more effective for the US operators than for the worldwide operators. This is because world wide air carriers presently experience more sabotage and highjacking than the US air carriers. Even though RAFT is not very effective in preventing sabotage and highjacking accidents, it can help through its video system by ensuring that the person checking in at the ticket counter is the same person that boards the plane. Other applications of the passenger compartment video is to monitor, record and hopefully discourage passengers from trying to enter the flight deck, seriously interfering with the function of the flight attendants, and endangering the aircraft. This is similar to the security video monitoring done in many businesses and government offices. A cargo area video sensor also serves as a backup fire and smoke detector as well as a detector of potentially dangerous cargo shifts. RAFT can reduce US fatal accidents by 78 percent.

3. CONCLUSION/EPILOGUE

RAFT is a world wide real-time remote aircraft flight recording telemetry system for enhanced air space capacity, passenger safety, security and operational efficiency. It utilizes existing state-of-the art communications, Internet, computer and software technology to unify the total avionics system. The DFDR sensor information, supplemented with video and radar data, is brought out of an archival database and into a real-time usable accident prevention system. In addition, it ends the information vacuum created by the aircraft and air traffic controller, where presently each acting independently, don't have the necessary measurement sensors that are required to prevent a crash. This information vacuum has compromised the safety net and is the major cause of the stagnant air carrier fatal accident rate. It has led to a situation where **currently air travel is over nine times more lethal than bus travel**[1]. By the cooperative combining of the aircraft and ground data sensors, and thus sharing the safety parameters in real time, the effective sensor suite is enhanced and the system can now anticipate many types of crashes. This crash anticipation capability provides the visibility and time necessary for the prevention of accidents. Furthermore, by the global transmission of telemetry of the DFDR parameters to a ground processing and distribution station, it provides a best estimate of a downed aircraft position for timely search and rescue operations. It also minimizes and eventually can eliminate the need for the costly and time-intensive recovery of the flight recorder. RAFT unifies the air space communications information system by providing an aircraft data super highway, similar to the Internet, to ensure that the relevant data parameters are globally visible and readily available to the people who need them in order to timely and optimally solve problems in a cost-effective manner prior to them becoming accidents. Furthermore, it optimizes the safety net and adds a level of redundancy to the present and planned suboptimal capacity and safety systems, which are prone to single thread failures. The system, which can be operational in five years, alleviates a broad spectrum of operational efficiency, air space capacity and air safety problems. RAFT provides the safety net that should be in place prior to any transition to free flight.

NOTES:

0. Due publication page limitations the figures and table referenced in the paper are omitted from the printed proceedings but were presented at the DASC conference.

1. Recent DOT Statistics show that air travel is over nine times more lethal than bus travel:

- US Air Carriers have 4.8 fatalities per 100 million miles traveled based on 5.9 billion vehicle miles.
- US Buses have 0.5 fatalities per 100 million miles traveled based on 6.4 billion vehicle miles.

(Buses were taken for the comparison statistics with carrier aircraft since both are classified as multi-passenger transportation carrier vehicles and their annual vehicle miles are equivalent.)

The average of the last four consecutive years of air travel has one of the highest average number of fatalities per mile traveled of any of the four consecutive year periods in the last twenty years. Also the air carrier fatal accident rate has remained essentially constant over the last twenty years. This constant fatal accident rate is in spite of the advances in pilot training do to the use of high fidelity flight simulators, aircraft materials due to enhanced fabrication methodology and superior metallurgy that has made them stronger and less subject to fatigue, enhanced avionics due to large scale integrated (LSI) semi-conductors that made the electronics smaller and more reliable, and improvements in engines and fuel that have made them more reliable. In the years between 1965 and 1970 there was a significant reduction in the fatal accident rate and fatalities due largely to improvements in jet engines that made them more reliable, microwaves that provided enhanced surveillance radar ATC/M and Instrument Landing Systems (ILS), and inertial navigation systems (INS) that reduced the aircraft's dead reckoning position errors. The radar based ATC also significantly enhanced the automated sharing of safety data between the plane and the ground monitoring system. Since the 1970's, there has not been a significant increase in the number of safety parameters that are automatically shared between the flight deck and the ATC. It has been this stagnation in avionics information that has directly caused the two decades of stagnation in the air carrier fatal accident rate. RAFT, which can be operational in five years, ends this information vacuum and thus reduces the fatal accident rate while making air travel more economical.

SAFETY RELEVANT NAVIGATION AND CERTIFIABLE DATABASES FOR 3D SYNTHETIC VISION SYSTEMS

J. Schiefele[1], L. May, J. Pfister, H. Raabe, Ch. Schmalz, K.-U. Dörr, W. Kubbat
Dept. Flight Mechanics and Control, Darmstadt University of Technology

Abstract

In Synthetic Vision Systems (SVS) a 3D flight guidance display depicts environmental data and mission critical information to pilots in order to increase situation awareness. With these systems Controlled Flight Into Terrain (CFIT) can be reduced and air traffic management for all phases of flight greatly enhanced.

Such a system consists of three necessary components:

1. Synthetic Vision Display (SVD)
2. High Precision Navigation
3. High Accurate Databases

While the Navigation System has to provide information about current position, attitudes, and sensor integrity, the databases provide information for terrain, obstacles, airports and navigation aids.

This papers shows that this two essential prerequisites for a SVD are at a very far developed stage and may be available as commercial equipment and procedures in the near future.

Introduction

Today's commercial transport aircraft cockpits usually contain an electronic Primary Flight Display (PFD) and Navigation Display (ND) in front of the pilot. These displays replace some of the traditional clock instruments such as artificial horizon, speed indicator, altitude indicator, and variometer.

In addition to conventional formats, a SVD depicts a 3D perspective image of terrain [Vie97] [Helm97] [Proc97/2]. Also virtual information such as a "tunnel in the sky" or a landing channel are integrated [Belo95]. A Synthetic Vision System consists of three key elements [Hamm94] (Figure 1).

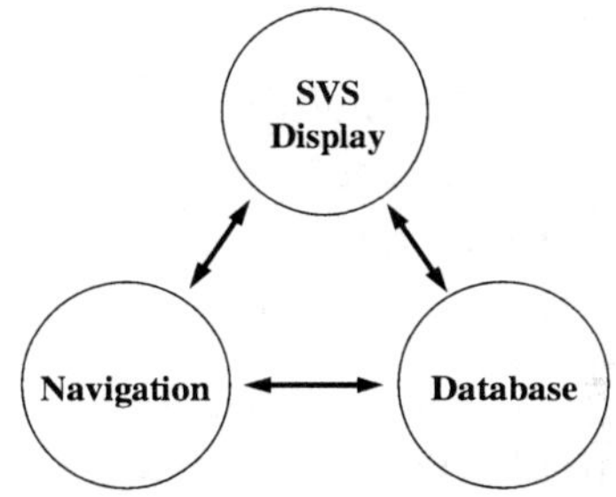

Figure 1. Key Elements for Synthetic Vision

A navigation unit must generate highly accurate and reliable position information [Pfist97]. The error must be in the sub-meter range. Otherwise, a flight approach with a SVS cannot be conducted because the SVD would not match reality. Finally, the database has to contain all information displayable to the pilot [Schi97].

Navigation System

Since 3D flight guidance displays require high precision positions and attitudes, the navigation unit has to satisfy them in terms of accuracy, reliability, and integrity. For navigation purposes integrity is defined as the „ability of a system to provide timely warnings to users when the system should not be used“ [Seeb89].

Requirements

Existing navigation systems use standard ground based Nav-Aids like ILS or DME combined with a standard inertial navigation.

[1] Contact: e-mail: schief@fmr.tu-darmstadt.de, phone: (+49) 6151-16-2890

0-7803-5086-3 /98/$10.00 ©1998 IEEE

These facilities cannot always provide the precision required by a SVS.

For that purpose usually a Global Navigation Satellite System (GNSS) is used. But in safety critical aviation applications any GNSS shows severe lacks for following requirements:

- Continuity: satellite masking can cause outages of GNSS.
- Availability: channel jamming or certain interference phenomena in some areas can also cause outages of GNSS.
- Integrity: multipath, cycle slips and incorrect satellite data can corrupt the received data and result in position errors up to 100m.

The developed navigation system uses sensors available on a modern aircraft, aided by GNSS. Even though GNSS has the above disadvantages, it is a very precise sensor. With respect to these problems a board-autonomous sensor, like an inertial navigation unit has to be integrated. A modern Kalman-Filter combines the precision of satellite navigation and the continuity and availability of board-autonomous sensors. That way a best estimate of the current position is available.

Sensor Fusion

The incorporated sensors are redundant to each other to allow failure detection and isolation (FDI). So the integrity of the sensors is guaranteed and the filter algorithm is prevented to use corrupted measurements. The measurements of dissimilar sensors have to pass a multilayer failure detection and isolation as depicted in Figure 2.

In a first step a „built-in-test" is executed which investigates the hardware components. If one sensor is defective, it will be excluded from the filter.

The delivered data is checked for plausibility by comparing the data with physical limitations. Data outlying these limits will not be processed in the filter algorithm.

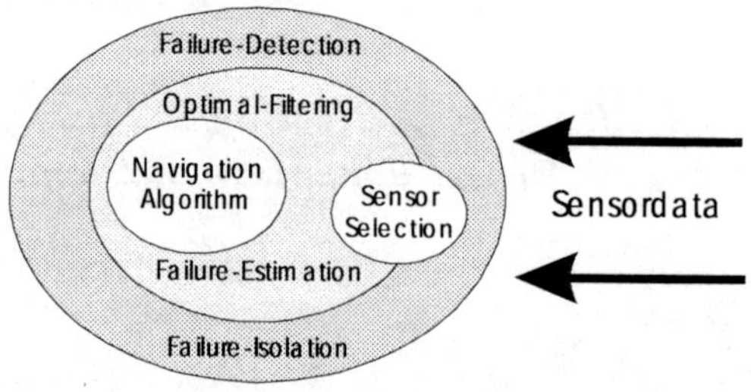

Figure 2. Structure of Sensor Fusion

In a third step all measured data is cross-checked. That way the different sensors are checked against each other using the redundancy of all sensors.

Finally the residuals of the filter are investigated using the history and the current sensor data and the values are combined with the filter estimates. After the FDI module identified any malfunctioning sensor the filter is able to reconfigure itself by adapting to the new sensor assembly. That way the filter is shielded from falsified data because this data would decrease the whole filter performance. After recovery the sensor is re-included.

The filter is realized by an extended Kalman-Filter linearized along the estimated path. It contains a set of three dimensional equations to model the sensor output and all dominant sensor errors.

Application ATTAS

The navigation concept described above was applied in the ATTAS of the German Aerospace Center DLR (Figure 3) and was demonstrated in combination with SVD in the Frankfurt/Main area 1997.

Figure 3. ATTAS (VFW 614)

The integrated navigation system consisted of following sensors:

- Honeywell H764 (inertial platform). This sensor provides a stationary signal characteristic and represents the core of the navigation system.
- D-GPS/NovAtel with kinematic carrier phase tracking. This is used as an additional sensor for position determination.
- Barometer. It is combined with the INU to stabilize the vertical channel.
- Radar-Altimeter. Used in low altitude in combination with database information. Thus high accuracy in the vertical channel could be reached during landing and take-off phases. The integrity check excludes the sensor from the Kalman-Filter above 2500 ft .

Figure 4. Navigation performance

The measurements have to pass the FDI and the filter module. The resulting error of the navigation solution is in the sub-meter range. Certain parts of the flight route were off the usual flight routes and in the vicinity of the ground.

In order to verify the navigation results a test vehicle was equipped with these measurement units. The system was tested at the apron area of Frankfurt/Main airport in May 98. As depicted in Figure 4, the navigation solution provides a very high performance in the sub meter range.

Database Issues

Today's use of databases in cockpits is limited to Flight Management Systems (FMS) or Terrain Warning and Alert Systems (TWAS). Currently, these devices use individual proprietary data in very low resolution displays [Proc97/1].

A SVD depends on a large amount of different data including the data currently available in a cockpit. This data will origin from different sources. Worldwide availability of data have to be ensured to allow operation in a commercial aviation environment.

Some of the basic information for SVD is derived from elevation data used for terrain depiction. To retain the pilots confidence in the SVD it is highly recommended that this data is as accurate as possible.

Different Data Sources

Depending on the SVS application different data sources have to be selected. The data sources needed for a SVS in a commercial transport aircraft can be found in Tables 1 and 2.

Table 1. Data Sources - Critical Data

Content	*Class*	*Availability*
NAV-Aids Airways ILS-Channel	Nav-Data	• worldwide in use
Obstacles Runways/ Thresholds Taxilines/ Parking Pos.	Cultural Data	• ICAO Annex 11/15 • for 12/97 required error smaller 1m

The data is divided in two principal logical groups. Flight critical data must be very reliable and highly accurate because it is used for primary means of navigation. In advisory data source inaccuracies can be tolerated if they cannot be removed from the data sources. They only provide additional information cues to a pilot. A pilot has to rely on flight critical data such as NAV-data for primary navigation.

Table 2. Data Sources - Advisory Data

Content	*Class*	*Availability*
Terrain	Terrain Data	• worldwide available • different sources • verification started
River/City Outlines Airport Buildings	Cultural Data	• partly available

All data can be classified in three classes:

1. NAV-Data
2. Cultural Data
3. Terrain Data.

These classes are derived from their geographical relation. NAV-Data contains all data included in today's Flight Management Systems (FMS). They are available worldwide and in daily use. Providers are Jeppesen and Swiss-Air. Cultural-Data are all man made objects such as obstacles (towers, houses, etc.), runways with attached thresholds, taxilines, and parking positions. All of these data sources have to be available in a high accuracy and reliability to be used for a SVS guided approach, take-off, or taxi phase [Belo95] of an aircraft. These sources were supposed to be available by the end of 1997. ICAO Annex 11 and Annex 15 require them to be provided in a very high accuracy (up to 1m). Terrain data is classified as advisory data because a commercial pilot is not supposed to navigate on the terrain. Terrain is only used to indicate a threat a pilot has to stay away from. The data is available worldwide in many different resolutions. Finally, rivers, roads, and city outlines in a raw resolution can be added for pilot information.

Data Creation

It is assumed that the needed data sources for SVS can be created at a high accuracy level.

NAV-Data Creation

Different investigations into today's navigation data-bases demonstrated that they still contain errors. The data is copied into this data-base from many different national Aeronautical Information Publications (AIP). A general problem is that dependent on the national authorities different reference systems are used. Examples are WGS84, NAD83, ED50 or even UTM. Therefore, a certain spherical coordinate in different reference systems does not describe the same location. It might introduce errors of several hundred meters depending on the global position. The provided accuracy and update frequency depend on national authorities and vary greatly.

It is assumed that in the near future all NAV-data sources will be greatly enhanced. The reason is that through the availability of GNSS in most aircraft database errors can be easily detected. This increases the pressure on national authorities significantly.

Terrain Creation

World wide terrain is available. At the moment there are many different satellites generating terrain models. For Example SPOT-Image, Landsat-TM, European Research Satellite (ERS)-I-II and Ikonos 1. The Shuttle Radar Topography Mission (SRTM) launching in September 1999 will generate a high resolution (30m) terrain model collected from 56° South to 60° North. In 2002 at the earliest, the processed 30m data will become limited distribution to authorized organization.

All digital elevation models (DEM's) are generated with different acquisition methods and independent sensors. The terrain generation method varies dependent on the provider. Therefore, these models can be used for verification purposes.

To get better statements about the terrain database error, a comparison with a high precision data source must be performed. Although such a source will usually cover only small parts of the database area, statistical evaluations are possible.

Figure 5. Terrain Data Comparison

In Figure 5 an overlay of a work data (Digital Chart of the World) set with high precision reference data (Airport Safety Model) is shown. In general it has been recognized that elevation errors of flat regions are very small, while significant errors occur when examining dynamic regions with reasonable mountains.

Airport Creation

Airport data is generally unavailable today. Even though the data for runways, thresholds, taxi-ways, taxi-lines, and parking positions were required to be generated by the end of 1997 only few countries full-fill the requirement.

High resolution images from all airports can be achieved through the satellites mentioned above. Satellite images can be georeferenced to WGS84 coordinates. From such images or an airport diagram the airport can be digitized and a 2D vector model generated. Finally, the model can be enhanced to a 3D database and stored in a hierarchical data format defined below. Figure 6 describes this process.

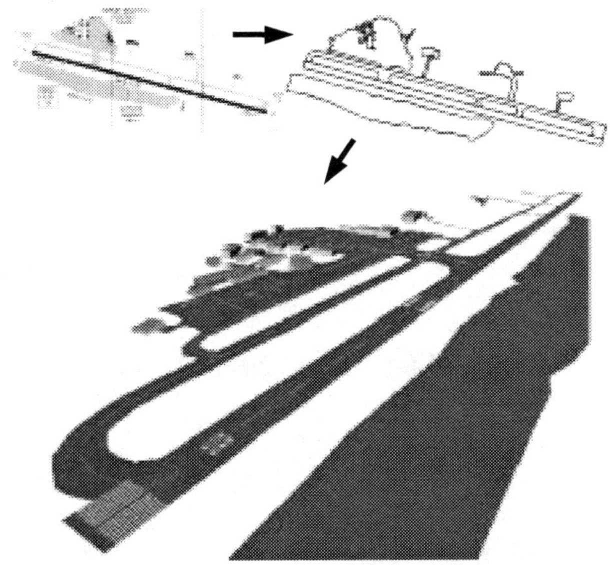

Figure 6. Airport Creation (Juneau, AK)

For the verification of the airport no complete local surveys are needed. Only some control points have to be measured by D-GPS survey. These control points can be correlated with the generated database.

Database Concept

While the previously described data generation processes ensures availability of data, data formats and contents used for most terrain, navigation, and cultural databases are not directly usable for SVS. For world-wide applications these databases are huge and cannot be taken into an aircraft. In addition, a SVS needs to render a 3D perspective image of the database in real-time at about 30 frames/sec [Belo95]. Today's graphics engines cannot load, store, and render most data-sets in real-time because of their storage formats and content density. In order to overcome these problems a three step process is proposed.

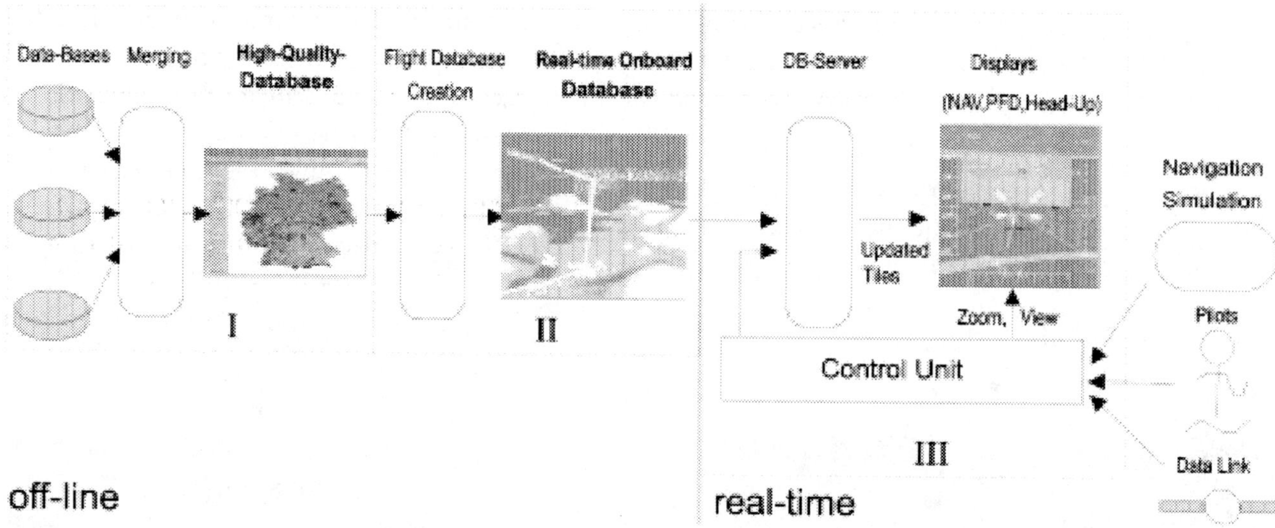

Figure 7. Three Step Data Processing

In the first step all needed data sources are integrated into a High Quality Database (HQ-DB). This HQ-DB contains all needed information for a world wide application. Selected data is then converted into the Real Time Onboard Database (RTO-DB) allowing real-time storage and rendering in an aircraft. The RTO-DB can be taken into an aircraft and the Database Sever (DB-Server) selects appropriate database regions (tiles) from the RTO-DB based on navigation input. It also processes short-term updates taken from a data-link and hands the final database to a SVD.

High Quality Database Creation

The HQ-DB is stored and handled with a commercial Geographical Information System (GIS). A GIS handles and merges large databases as they are required by world wide applications. With a GIS principal consistency checks on the database can be conducted. For instance, it can be prechecked whether or not airways intersect the terrain or a given airport is located under the terrain. To a certain extent it also helps to transfer data into WGS84 which is the selected reference system. General transformation is impossible without introducing an unpredictable error. The GIS tool is able to handle almost all data formats used in aviation applications.

Real Time Onboard Database Creation

The HQ-DB cannot be handled by a SVS due to rendering and size limitations. Therefore, a RTO-DB creation process was established.

The Inventor File Format [Wern95] was picked as real-time data storage format. It stores graphical data in a hierarchical format allowing real-time extraction. Objects with higher priority are stored first while unimportant features are moved to the back of the hierarchy. Within the format a certain order for storing additional feature information is declared. A certain object such as a runway can be identified with a geometrical and additionally logical representation. This is necessary because a SVD has to declutter selected information.

Such a database is still too large to be handled by a SVS in real-time rendering. Therefore, data reduction is applied to the terrain. This is the data source containing most polygons and amount of storage in the database. Level of Detail (LOD) generation creates terrain databases at different resolution and accuracy dependent on the distance to the aircraft.

Database Server

The RTO-DB can be taken into an aircraft. The DB-Server concept allows the system to handle the RTO-DB for real-time applications. Depending on navigation position the appropriate tiles are selected. Each tile currently consists of 12'x18' and 36 tiles are loaded at a time. This represents a region of about 100km x 100km in the continental US. The server will also include short term updates provided from a data link. This data could be NOTAM or weather information. For example, the information about a closed runway can be directly integrated and displayed in a warning color.

Applications

The developed concept is not limited to the usage of SVS. Instead, an ordinary simulator visual is supported. In difference to the SVS data sources satellite images are mapped on to the 3D terrain model. For take-off, landing, and taxiing the same 3D model of an airport is used as for the SVS. With this concept the differences between real in-flight data and simulator data vanishes. An example image from the simulator visual can be seen in Figure 8.

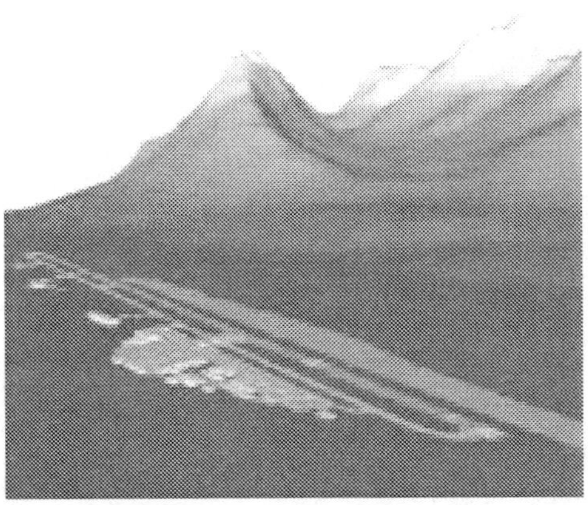

Figure 8. Simulator Visual (Juneau, AK)

In order to demonstrate the capabilities databases for Frankfurt/Main, Braunschweig, Hahn, and Juneau were generated. The quality of the RTO-DB for Frankfurt and Hahn were demonstrated in taxi- and in-flight tests [Vie98].

Certification of Database Processing

Up to now only FMS databases are used in flying avionics systems. Currently, their processing and handling is uncertified. Databases based TWAS will be incorporated shortly. Examples are Dassaults Ground Collisions Avoidance System (GCAS) and Allied Signals Enhanced Ground Proximity Warning System (EGPWS). For the generation of terrain and navigation databases certification standards and procedures are defined. RTCA-SC181/EUROCAE WG13 implemented DO-200A providing guidelines for the certification of database generation. The underlying concept is to certify the data processing but not the database itself. An typical process is depicted in Figure 9.

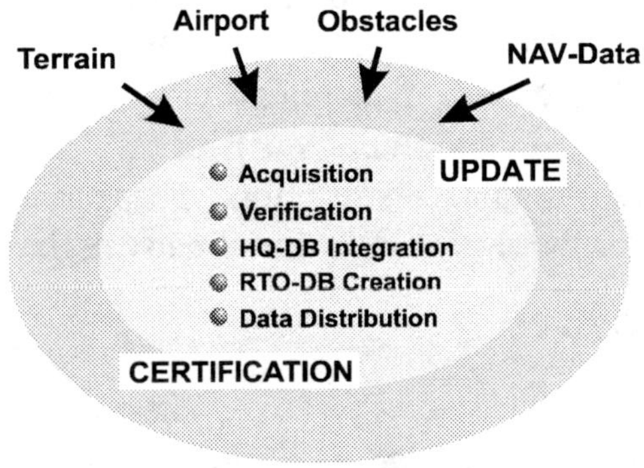

Figure 9. Database Creation Process

The end-user (e.g. an airline) is liable for the quality of the data. The certification of the software used in the processing are based on well established standards such as DO-178B, MIL-STD-498, DOD-STD-2167A and DIN-ISO 9000-3.

In addition, RTCA-SC193/EUROCAE WG44 develop industrial requirements for accuracy, resolution, and contents (terrain, obstacle, and airport data) for future aviation databases.

The SVS program in the NASA Aviation Safety Initiative will focus on enabling technologies for SVD. As a key component the development of a certifiable process for database generation is determined.

At the moment there are many different initiatives that will soon enable the integration of terrain, obstacle, and airport data into flying avionics systems. Several companies have already established contact with certification authorities for a DO-200A certification. The major problem for the future is not the certification of data processing but the availability of unclassified high accuracy data.

Conclusion

A Safety Relevant Navigation for a SVD can be achieved by integrating several commercially available sensors in an environment that allows Failure detection and filtering.

For high accurate and reliable databases some work has to be done. Existing technologies have to be combined into certifiable processes that ensure data integrity throughout the steps of acquisition, verification, real-time preparation, and distribution. The basic concepts for these processes are already defined.

References

[Seeb89] G. Seeber; „*Satellitengeodäsie: Grundlagen, Methoden u. Anwendungen*"; Berlin, New York, Verlag de Gruyter, 1989

[Schi97] J. Schiefele, L. May, H.Raabe; „*A concept for Creating In- and Pre-Flight Databases*"; SPIE; Vol. 3088; 1997

[Belo95] C. Below, H. von Viebahn, M. Hammer; „*4D Flight Guidance Displays*"; SPIE; Vol. 2463; 1995

[Vie97] H. von Viebahn, C. Below, M. Purpus; „*Flight Tests of the 4D Flight Guidance Displays*"; SPIE; Vol. 3088; 1997

[Vie98] H. von Viebahn; „*4D-Flight Guidance Displays*"; 2. Workshop on Synthetic Vision, NASA Langley; 1998

[Helm97] A. Helmetag, R. Kaufhold, M. Purpus; „*3D Flight Guidance Displays*", CEAS Symposium Free Flight; Amsterdam; 1997

[Hamm94] M. Hammer, M. Gross, R. Kaufhold; „*Natürliche Displays im Cockpit*", DGLR Fachausschuß Anthropotechnik, Braunschweig, 1994

[Proc97/1] P. Proctor; „*Major Airlines Embrace Enhanced GPWS*"; Aviation Week & Space Technologie; pp. 46-48; April 21; 1997

[Proc97/2] P. Proctor; „*Epic Avionics In Flight Test*"; Aviation Week & Space Technologie; p. 70; Sept. 22; 1997

[Wern95] J. Wernecke: „*The Inventor Mentor*"; Addison Wesley; 1995

[Pfist97] J. Pfister et al.; „A Concept for High Accuracy Navigation and Databases in Free Flight Synthetic Vision Systems"; CEAS Symposium Free Flight; Amsterdam; 1997

[May98] L. May, J. Schiefele, H.Raabe; „*Controlled Digital Elevation Data Decimation for Flight Applications*"; SPIE; Vol. 3364; 1998

[Raabe98] H.Raabe, L. May, J. Schiefele; „*Creating Flight Application Databases with Controlled Elevation Data Decimation*"; IMAGE Conference; Scottsdale, AZ; 1998

THE ROLE OF DOCUMENTATION IN A CONCURRENT ENGINEERING ENVIRONMENT

Lani Hajagos, Adobe Systems Incorporated, San Jose, CA

Abstract

Concurrent engineering is proving a valuable technique for reducing product development cycle time and costs. Engineering, purchasing, manufacturing, and customer service working together can identify problems early and avoid costly reworks.

Documentation is an important part of the product development process, and can benefit from being part of the concurrent engineering environment.

This paper examines the ways in which information can be shared among internal and external documents, and how that information can be coordinated with other types of product information in the concurrent engineering process.

The concurrent engineering process

In most product categories, the company which is first to market gets the largest market share. Therefore, organizations are re-engineering their product development and manufacturing processes in order to reduce product cycle time.

Concurrent engineering is a proven technique for reducing both time and expenses associated with product development. Many groups are involved in the development of a product:

- Marketing, which provides product requirements
- Engineering, which develops product specifications and designs
- Packaging, which designs the containers in which the product will be delivered
- Technical support, which must devise methods for supporting the product once it is in the field
- Manufacturing, which must figure out the best way to produce the product
- Purchasing, which creates bills of material for the product and determines how to source the parts
- Documentation, which creates documents describing how to install and maintain the product
- Sales and marketing, which produce product collateral

The previous process

In the "bad old days", information was passed from one group to another in a sequential fashion. Manufacturing and technical support rarely got to see the product requirements and specifications until the design was almost finalized. This meant that changes that they requested had major impacts on the design, resulting in delays and increased expenses.

How concurrent engineering works

Concurrent engineering is effective because it allow all concerned to be involved with and give input to all phases of product development.

For example, technical support may have valuable input to the marketing requirements document, based on their experience supporting other products. And manufacturing can provide valuable input to the engineering specifications, especially if the proposed design is expensive to produce.

Companies that have instituted concurrent engineering generally have some sort of a product data management (PDM) system into which the various engineering designs, BOMs, and manufacturing designs are stored and managed. The PDM system allows users in the various groups to access these document, and controls access to protect

0-7803-5086-3 /98/$10.00 ©1998 IEEE

against unauthorized changes. Some PDM systems also manage ECOs.

The role of documentation

Much of the information about the product exists as documents rather than design files. This includes the marketing requirements, engineering specifications, documentation plans, support plans, and marketing plans. The information is very interrelated, and has direct bearing on the design files and ECOs (which generally are also documents).

All too often, however, these documents are not controlled as part of the PDM system. This means that it is difficult to keep the documents up to date as the design changes.

Managing all the information

A comprehensive PDM system should incorporate documents as well as the design files. More importantly, the information in the documents should be linked so that individual requirements, so that as a requirement changes that change can be reflected in other documents.

For example, technical support may suggest a change to product requirements based on experience with other products. Based on this input, marketing updates the requirements. It's important that the other groups be notified of this modification so that they can modify their items as necessary. This could be done via a manual process: the product manager sends out an email notifying each group of the change. Or, if the information is linked among documents in the PDM system, this notification can happen automatically as the marketing requirements document is updated.

The point is that ideally, documents related to the product development process should be managed as part of the current engineering process, along with the design files and other engineering files.

Title: Multimedia Analysis and Archive System (MAAS) for Joint Analysis Center (JAC) Molesworth England in Support of the Bosnia PeaceKeeping Mission

By Michael S Gaydeski, General Dynamics-Information Systems, Bloomington, MN

Background

The armed forces of the early 21st century must possess knowledge dominance, both in their peacetime efforts to educate, train and modernize themselves, and in their efforts to deter aggression and engage in lethal and non-lethal operations. Knowledge dominance must build on the twin national strengths of high-quality people and information technology. Such dominance makes achievable the common goal expressed in Joint Vision 2010 for an American military that is "persuasive in peace, decisive in war, preeminent in any form of conflict."

Knowledge dominance requires the collection, processing, storage and dissemination of ideas, facts, data, sounds and images, both as raw information and in the form of products to which the human mind has provided additional value. This knowledge must be successfully managed so that its meaning is understood, and so that exactly the right knowledge sets are available to the right people at the right time. This knowledge must derive from ongoing and past military operations (institutional, synthetic and actual), and from whatever other sources contribute to informed warriors attaining dominant maneuver, precision engagement, force protection at all levels, and focused logistics. This knowledge must fuel operational learning, in respect to improved ways of deterring war or waging combat or non-combat operations, and in respect to achieving the greatest possible transfer of learning in education and training.

Information technologies provide the means to knowledge dominance through capturing past and present experience and making it available to current education, training and operations. However, realizing the full potential of these technologies requires careful melding of the sciences for assembling and making knowledge available in effective form with the best multimedia technology. Without thinking through the best way of applying technology to accomplish the effective and speedy transmission of just the right knowledge at just the right time and place in just the right fashion, the ocean of data in the information age can overwhelm as well as enable.

In the wake of the Cold War, the Army -- in partnership with other services, the Joint community, the National Archives and partners in the academic and business sectors -- developed a Virtual Research Library (VRL) under the Center for Army Lessons Learned (CALL), at Fort Leavenworth, Kansas. The Army recently re-conceptualized this library as a centerpiece of the University After Next (UAN). The UAN purpose is twofold. First, it educates and trains the Army's contribution to Joint Vision 2010, which is termed the Army After Next (AAN).

0-7803-5086-3 /98/$10.00 ©1998 IEEE

Second, it provides online experts and knowledge sets tailored to meet the requirements of leaders (Army and Joint) in the field as they prepare for and conduct operations. In split-based operations, conducted abroad but with direct support from out-of-theater, the UAN projects knowledge to forces wherever they are deployed.

The VRL is composed of raw sources in archives, products with value added in a library, and of a knowledge gateway that strives towards imposing the discipline of a research library on the chaos of the Internet. The contents and links are Joint as well as Army. The library is a multimedia system, aspiring to translate every known format in which knowledge is stored and communicated into digital objects available by way of a web browser. It includes tools for making this online knowledge interactive, for managing it as a record and according to the principles of library and archives science, for searching and retrieving it, for visualizing and mining it, for summarizing and parsing it, and for analysis. The VRL strives to protect and reduce the secrecy of information through application of technology to achievement of multilevel secrecy, and computer-assisted and fully automated declassification, reclassification, public release and sanitization.

The VRL at CALL constitutes the first step towards Army knowledge management. Working with the Joint Staff and Joint Warfighting Center, it is a linchpin for organizational learning by the American military. Indeed Harvard Business School and the Wall Street Journal agree that this Army model should be applied across the private sector, as does a consortium of academic and commercial leaders. It has had substantial impact on operations in Somalia, Rwanda, Haiti, Bosnia and the Persian Gulf, in supporting peacetime engagement in disaster and disturbance prevention and relief. The library has profoundly affected records management, the development of training and doctrine literature, and the use of technology to assist in the transfer of learning both at a distance and in an institutional setting. Due to its impact, the CALL VRL received awards from the Vice President of the United States and the Smithsonian Institution, and Director, CALL recently received the lead for development of the UAN. The UAN effort is closely associated with the new model community initiative at Fort Leavenworth, which strives in partnership with private industry and the pilot application of national reinvention authority to create the digital base operations, campus infrastructure and town of the future. The following Figure 1 shows the general information components of a system and how they need to flow for productive operations

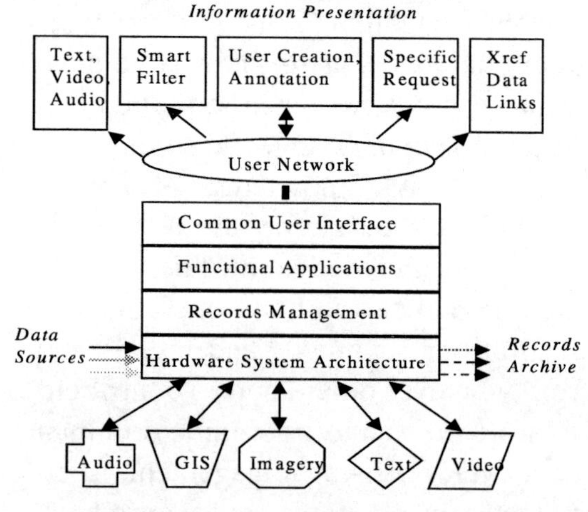

Figure 1. General Information System Components and Data Flow

MAAS for UAV

The process capturing mission video data to the dissemination of video mission information has multiple technology facets and challenges. The most significant challenges are the quantity of data, defining the value in the motion imagery and the dissemination of the products. Mission video data includes UAV video, other reconnaissance platform video, LANTIRN video and hand-held camera video. The technologies involved to transform the data to information include: analog imagery, digital imagery, information management, high performance computing and utilization of tremendous amounts of storage. The typical field site solution today is to retain the video mission data as analog video stored on VHS tapes. Organizations such as DARPA, NIMA, the National Technology Alliance (NTA), NSF and other organizations are funding programs to apply research to various segments or combination of segments of the capture of mission video data to the dissemination of mission video information (value added products). The commercial marketplace is developing video and multimedia products to address the needs of television video production, educational and business video productions and consumer video and multimedia applications.

The NTA with General Dynamics Information Systems (GD-IS) as the Systems Integrator, has been working on a complete solution from the perspective of the mission video analyst. This solution addresses issues from the capture of mission video data to the dissemination of the mission video information. The evolving proof-of-concept systems target key enablers for the integration of state-of-the-art commercial and government video applications and technologies meeting the needs of the video analyst with intuitive and natural interfaces to the system. The needs of the video analyst includes: the capture of the data, storage and archive of the data, the analysis of the mission video data, the organization of the analysis information and the retrieval of the data using the value added analysis information. Some subtle needs of the "mission video analyst" are highlighted by the fact they are mission analysts and not professional television video production people, and the analyst may be on a minimum tour of duty rotation with minimum time allotted for learning the analysis tools.

This paper defines the next generation of the proof-of-concept Multimedia Analysis and Archive System (MAAS-2) with the initial installation being in the Joint Analysis Center at Molesworth England in support of the Bosnia peace keeping effort. The paper describes the operational scenarios in use today at one of the premier European sites and the progression of the proof-of-concept video analyst systems NTA and GD-IS have done and are proposing to do.

The most sophisticated site/user of the UAV Bosnia mission video data is the Multinational Intelligence Coordination Cell, MICC, at the Joint Analysis Center, JAC, Molesworth England. This site was dominated by manual processes that require the analyst to focus on VHS video productions and editing techniques rather than exploitation and dissemination of information.

A synopsis of the mission video process is as follows:

1. The reconnaissance mission video VHS tape(s) are either captured on VHS tape(s) by the MICC from a broadcast over JBS (Joint Broadcast System) or VHS and Hi-8 video tapes received directly from the reconnaissance missions.
2. The videotapes are archived to the bookshelves.
3. The mission analysis is performed by a video image analyst, typically the day after the mission, at a work area consisting of a PC, a VCR and a video monitor. The analyst reviews the mission video and creates indexes to the mission VHS tape by creating "video sequences of interest metadata" entries into a database on the PC.
4. The dissemination of video analysis information based on RFIs, request for information, is by voice, hardcopy documents, digital image frames, digital image mosaics of video frames, hard copy images, digital video clips (very limited) or a VHS tape of analog video clips all distributed through physical media rather than taking advantage of electronic dissemination.
5. Ancillary information received from other sources is also obtained by the MICC and saved in the original format, i.e. paper and digital reports and open source paper and digital documents. This information may be appended to the RFI products if relevant and the analyst is aware of the information.

NTA and GD-IS video analyst proof-of-concept solutions: In 1996 the NTA and GD-IS teamed to evaluate the capability of existing technologies and commercial products to be integrated into an end-to-end proof-of-concept solution to automate the capture of mission video data and the electronic dissemination of mission video information. The Video Server Testbed, VSTB, implemented and demonstrated the automation of the capture of the mission video data to the electronic dissemination of the mission video information to the Center for Army Lessons Learned Database at Ft. Leavenworth. The end-to-end proof-of-concept solution automatically captured a real time UAV video feed, used the voice on the video to obtain the database metadata information and to parse the video stream into video clips, created mosaic image sequences from the video clips in real time, and presented the "analysis package" to the CALLDB server for inclusion in the UAV database at CALL. The analysis package included the database metadata, the video clip with voice and the mosaic image(s) of the video clip. The commercial technologies integrated to implement this proof-of-concept solution included high-and SUN workstations, television production video tools and available government technologies.

In the fall of 1997, NTA posed the challenge to GD-IS to provide a proof-of-concept system addressing the needs of the MICC video analysts at JAC Molesworth in the context of their process. The proof-of-concept system needed to support the analyst's transition from analog video analysis with VHS tape archive to digital video analysis with information management archives.

The proof-of-concept system also had to address minimizing the recurring price of subsequent copies of the system. This proof-of-concept system is called the Multimedia Analysis and Archive System (MAAS).

The MAAS system has addressed the challenges posed by the NTA of addressing the needs of the MICC video analysts, follows their existing process and minimizes the recurring price of the system. The intuitive user interface allows the analyst to produce their value added information products without being aware of the commercial video and information management tools integrated in the MAAS. To minimize the recurring price of copies of MAAS, MAAS uses commercial technology targeted to the small business and sophisticated consumers. The next generation MAAS, MAAS-2, will provide:

1. provide a WEB interface for the network connected users to query the analysis information contained in MAAS-2,
2. provide network connectivity to key networks within the JAC,
3. provide for more digital video storage and hierarchical storage management of the increase in storage,
4. improve the image quality by migrating to MPEG-2 video format
5. work with the video analysts to provide for improved analysts interfaces, i.e. a hand free environment with voice recognition, and
6. provide operational and feature enhancements identified in the course of using MAAS-2 for 6 to 9 months after installation.

Example of Application of the MAAS for User Product Output

This summary example encapsulates General Dynamics Information Systems (GD-IS) efforts on the MAAS which is to be connected to the Ft Leavenworth/CALL Digital Information Technology Testbed (DITT) and the Defense Advanced Research Project Agency (DARPA) Warfighter Visualization (WV) program. Figure 2 shows how the MAAS relates to the operational UAV community. The main objective of the WV effort is to design, develop and populate a Geo-Spatial Database (GDB) to allow precise, real-time registration of video data against controlled reference imagery for support applications needing precise video-based geo-location. Also, to develop rapid, efficient methods for populating a GDB like DITT with results of video data exploitation in precise, geographically based, coordinate systems.

The challenge is to use rapid, geo-controlled, 3D-scene reconstruction and automatic updates from imaging sources only. This is combined with the need to filter and smooth relatively noisy telemetry data position and attitude estimates, along with real-time registration measurements to achieve precise mensuration and object location determination.

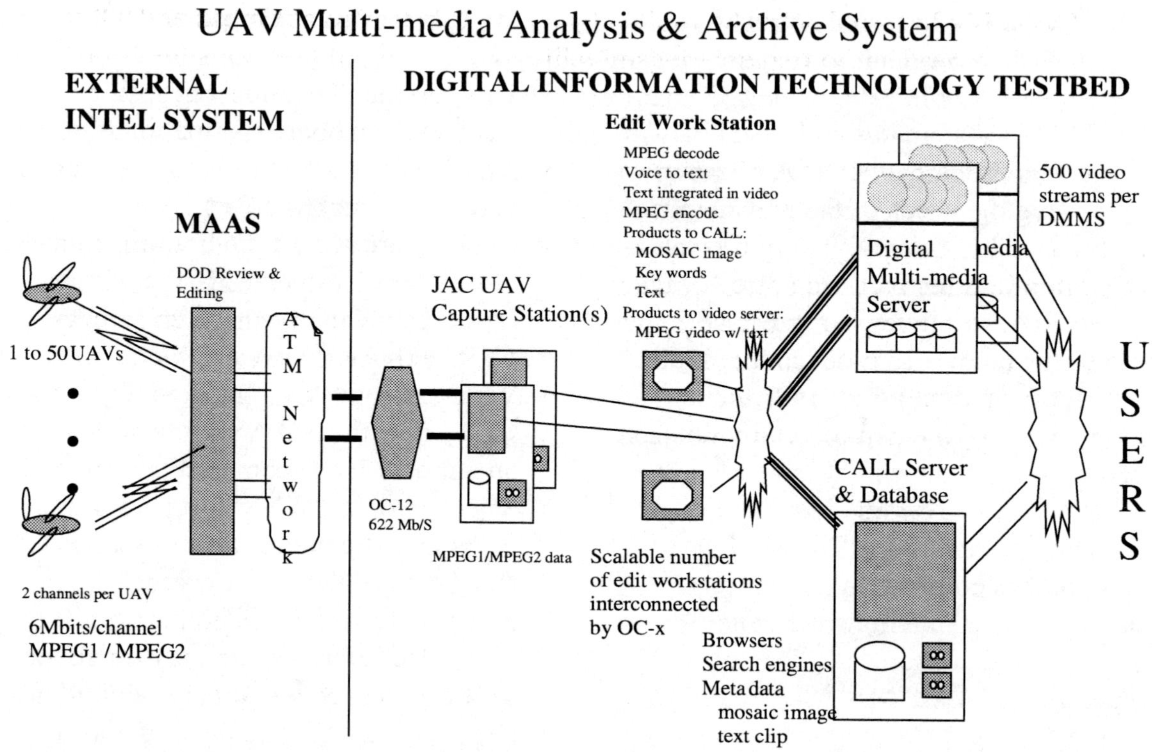

Figure 2. MAAS as a Part of the Intel Community

The Problem

The specific task that was explored was the combat assessment component of the targeting cycle. The combat assessment component is further broken down into additional elements. We focused on the requirement for rapid physical battle damage assessment and providing additional tools to help with the functional damage assessment and overall target status determination. After performing these functions, rapid dissemination of battle damage assessment information is required.

Our Approach

Our approach to meeting this challenge is to move the "system" to a 3D paradigm in which data sources would be acquired and stored in a system such as the Center for Army Lessons Learned/Digital Information Technology Testbed (CALL/DITT). Images, textual information, functional areas, etc.would be tied to 3D scene coordinates and retrievable through web interfaces (searches using relevant metadata such lat/long, time of day, and BE#) such as is done at JAC Molesworth for UAV video. This results in the products being available in 3D scene coordinates. A key component of the overall approach is to include a set of visualization tools to validate all of information sources and data representations.

A major aspect of this effort was to develop several algorithms for automatically comparing either the reference imagery with the new mission image, or the existing site model (constructed using reference imagery) with the newly acquired mission image. The intermediate and final results from these Change Detection Techniques are provided.

Automatic Change Detection Techniques

Output and intermediate results are made available from applying 6 different change detection techniques to some sample data. The area is Griffiss AFB in Rome, New York. The 6 techniques use various combinations of a Reference image (previous collection), a Mission image (current collection), and the 3D Site Model for the site. These are all stored in the DITT or MAAS. The techniques and comparisons are as follows:

1. Edge Verification - For each object in the site model attempt to determine whether the object is still present in the Mission image by looking at the edge energy in the mission image where the visible edges should be found.

2. Line Segment Comparison - Extract and compare the line segments from the Mission image and the Aligned Reference image. Line segments are compared in windows according by expected length and orientation.

3. Intensity Comparison - Compare the Mission intensities with the Aligned Photonormalized Reference Intensities. Group together regions of large differences and classify regions of significant difference.

4. ANS Comparison - Compute the ratio of the Mission and Aligned Reference intensity moments along with correlation as a feature vector (17 components) to supply to a trained Artificial Neural System Classifier (3 Layer Back Propagation Implementation).

5. Shadow Comparison - Compare the locations of shadows predicted for a given mission image (using sun position and site model objects) to

shadows detected for a given mission image.

6. Hausdorff Edge Comparison - Compare the line segments extracted via Hausdorff enhancement for a mission image against a predicted set of visible edges for a mission image.

In addition, a method for combining the results of the various change detection strategies was developed. All detectors produce measures of change (0-1) over all modeled objects. The outputs can be combined using a weighting or voting strategy. A voting strategy allows the user to set conservative thresholds for any individual change detection operator.

In a larger context, we also build a bridge between the 3D data and commercially available presentation packages. This was done by creating information in a www browser-type format. Finally we designed and constructed a prototype BDA product and made it available to users for evaluation.

The MAAS/DITT/WV helps solve major military and related commercial problems dealing with change detection. Output products managed by DITT and MAAS include: the image sources (UAV video), maps, DTED, 3D Model Supported Target Area Renderings, 3D Model Supported Exploitation Cultural Feature Enhancements, Change Detection Algorithm Results, Synthetic Target Area Views, and Fly Through Videos for Mission Planning/Targeting/BDA using Synthetic Views and Actual UAV Video Data.

Summary

The MAAS is integrated into the Ground Segment of the UAV videooperations. It is applicable to all ground and airborne applications where volumes of video and like imagery are part of the user operations. Characteristics of MAAS are:

- Networked PC / NT workstations & servers solution
- Scalable to multiple parallel mission video imagery encoding & capture streams
- Scalable to multiple parallel video imagery analyst workstations
- Scalable on-line storage and archive
- Capable of utilizing and integrating the UAV ESD information automatically into the video analysis database when the ESD information is available to the theater
- Proof-of-concept system installed and operational at the Multinational Intelligence Coordination Cell (MICC) at the Joint Analysis Center RAF Molesworth England--Video image analysts for UAV, P-3, C-130 and hand held camera video imagery
- Working with the Center for Army Lessons Learned (CALL) at Ft. Leavenworth to upgrade the CALL database to include multimedia information including UAV video imagery from the MICC

The Aftermarket Integration of Modern Electronic Information Systems in Classic Aircraft – Human Factors Design Considerations

Charles F. Adam, Federal Aviation Administration, Lakewood, California (USA)
Robert B. Barnes, Robert B. Barnes Associates, Inc., Scottsdale, Arizona (USA)

Abstract

Today, operators of Classic Aircraft are seeking the addition of new technologies, such as the integration of GPS/FMS, and even complete Electronic Information Systems (EIS), in order to meet the demands of a rapidly changing operational environment. The addition of such items to an already crowded flight deck poses problems that the original manufacturer never had to consider. The most obvious is the lack of adequate and appropriate space ("real estate") in which to install such upgrades.

Unfortunately, such additions in an already limited area, can lead to very different and quite unique crew interfaces due to the location chosen for the new display/control box (or boxes). This lack of space can also lead to the complex redesign, combining, and miniaturization of current systems simply to provide sufficient "real estate" to install the new box(es). This, in itself, may very well lead to increased crew workload which may not be immediately apparent to design engineers.

This paper examines the human factors issues involved with the integration of modern electronic information systems into the flight deck of classic air transport category aircraft through the Supplemental Type Certificate process.

Background

Our industry is faced with both a business opportunity and a human factors challenge. There are more than 6,500 transport category aircraft which were built prior to 1983 and are still in operation. More than 20% are Boeing 727s but there are also considerable numbers of Boeing 737-100/200, MDC DC-9, Boeing 747, DC-10, and L-1011 on this list. There are even some early models of other Boeing and Airbus Industrie aircraft built since 1983 which now qualify for potential flight deck upgrades.

The reason for such flight deck upgrades varies. Many operators want to extend the life of their fleets well into the next century by adding upgraded avionics systems and related life-extension modifications to meet the increasing complexity of the operating environment. But recently, there has also been a trend to mandate the installation of such upgrades as FMS, GPS, TCAS, and EGPWS as a result of some very unfortunate accidents.

Currently the most popular upgrades involve EHSI, GPS, and/or FMS. However, the issue of upgrade to Classics will become even more complicated with the introduction of the Future Air Navigation System (FANS), Automatic Dependent Surveillance (ADS), and the transition to total GPS (WAAS and LAAS) and Free Flight.

As more such upgrades are mandated, the flight deck has the potential to become a much more workload intensive environment. Mandated solutions become a bottom-up project for the operator (e.g. *"we need _______, so put it in."*). Gone is the top down integration of new additions to the flight deck based upon broad system safety and crew workload assessments. The emphasis is becoming *"put it in – somewhere"* and is driven by money constraints usually attached to FY budgets. No one appears

0-7803-5086-3 /98/$10.00 ©1998 IEEE

to be thinking of the various task combinations or permutations being placed in front of the flight crew as a result of an individual upgrade, especially when it is the third or fourth such individual upgrade to the same aircraft.

Three major human factors issues being seen in current STC applications for avionics upgrades to Classics are:

- Switchology – Combinations and Permutations
- Combination of Functions in One Display/Control Box
- Distribution of Functions Throughout the Flight Deck

Switchology

The addition of switches sometimes is required to meet availability and reliability requirements. However, pilots are trained to make yes-no decisions in an abnormal condition so as to reduce the probability of error. The combinations and permutations made possible due to the addition of avionics function on the flight deck goes away from goodness and increases the probability for error. For example, the Boeing 727 flight deck layout shown in Figure-1 is representative of the type of avionics upgrades that are being implemented in the Classics.

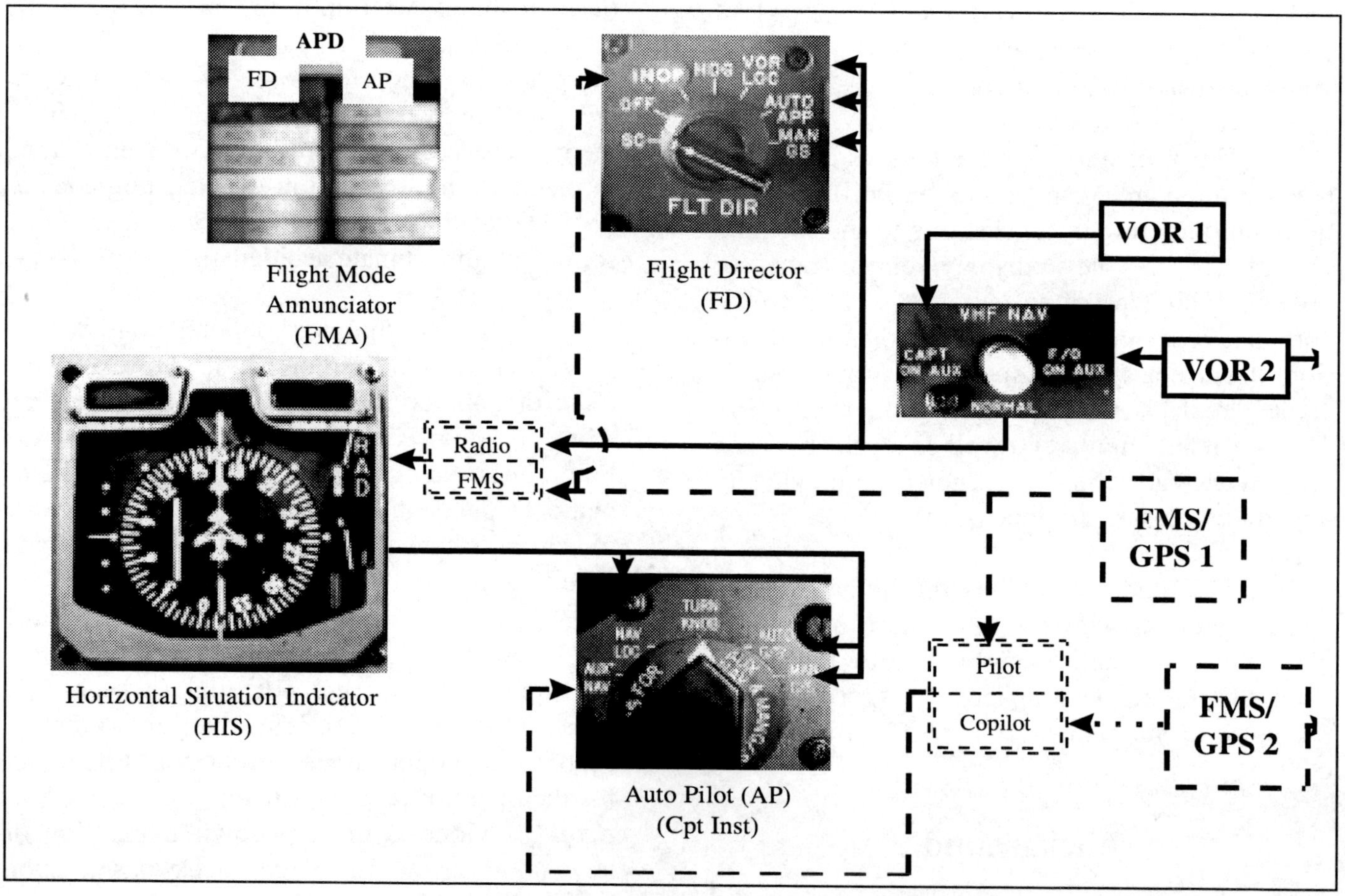

Figure-1. Boeing 727 Flight Deck Example

In Figure-1, The solid lines reflect the original production design while the dashed lines indicate merely the addition of FMS/GPS. The addition of other avionics capabilities, such as TCAS and/or GPWS, increases the complexity even more.

In the original design, either pilot could select the VOR/ILS or Heading Select as the sensor source for the Flight Director (FD) and it was annunciated on the Flight Mode Annunciator (FMA) of the respective Flight Director. The Autopilot (AP) commands were taken from the Captain's VHF NAV. During an abnormal situation (e.g. failure of the Captain's VHF NAV), the Copilot's VOR/ILS guidance information could be routed to the Captain's Horizontal Situation Indicator (HSI) and to the Autopilot through the VHF NAV switch. Any failure of the VOR/ILS signal caused the AP to disconnect and revert to the TURN KNOB Mode.

Some of these classic Boeing 727s, especially the over water and transcontinental models, were initially modified with long range navigation systems for use at "high altitudes" (but not in the terminal area). Such additions changed the overall functionality of the Flight Director, Autopilot, and Flight Mode Annunciator from the original design. Some of the HSIs were even modified to display INS on the HSI's face.

Installation of FMS/GPS in the Boeing 727 has now extended into the terminal area what was originally only a long range navigation solution for use at altitude. Unfortunately, replacing these older systems by simply "dropping in the FMS" has resulted many times in a disorganized and complex series of switching modes on this classic flight deck. Some examples of such switching problems seen in design submittals include:

1. Autopilot coupled to the FMS with the HSI displaying the VOR. This design created a situation in which there could be no tracking data displayed to the pilot. In addition, it was found that toggling the RADIO/FMS switch would not cause an Autopilot disconnect. Therefore, the pilot would not be forced to re-engage the AP and become aware of what nav mode was actually commanding the aircraft.

2. Copilot's VHF NAV data can be selected by the Captain but there are no provisions for FMS cross flight deck. The basic design always provided the Captain with the capability to select a cross-side sensor given a failure of the Captain's primary navigation source. This design philosophy should be implemented for installations of FMS/GPS especially as we migrate to lower approach minima when the Captain is flying the approach.

3. Copilot's FMS could be connected to the AP but none of the Copilot's other functions were integrated into the AP. This design once again created the possibility that the Captain, who is normally the only interface with the AP, would have no tracking data displayed.

Combination of Functions

Figure-1 illustrates that there are a number of functional combinations involving the VHF NAV 1 & 2 and the FMSs which can be displayed on just the Captain's flight instruments. Similar combinations are repeated on the Copilot's side as well.

Navigating by VOR and ADF in the basic aircraft design required some different displays in order to determine the crossing or interception of radials. This is not the case with current flight management or navigator systems. Today's flight deck designs have both pilots' FDs and HSIs displaying the aircraft commanded solution.

The retrofit of such designs to the Classics, however, has produced a number of different paths (or combinations) in which data can be displayed. In one FMS installation proposed for a Boeing 747-200, the navigation solution could have switched 3 times before ever being displayed on the HSI and it still would not reflect the proper source. In order to solve this problem, it was necessary for the integrator to disable the number of possible combinations for the FMS moding and then the switching integration was reduced to follow the basic aircraft design. This reduced the possible number of pilot induced failure modes dramatically. Although this might not have been the best solution, it was the most affordable solution without undergoing considerable redesign to bring it closer to today's flight deck design philosophy.

Distribution of Functions

Additional concerns with the integration of these systems involve switch mode annunciation and location. To change a control head face plate with a new "backlite" label is one of the easier tasks. Changing an OEM mode annunciator for either a single or limited number of aircraft installation is impossible. Unfortunately, the most economical solution is to add more switches and lights, then place them around the flight deck wherever there is space. In the Boeing 727 example described earlier, the FMS MESSAGE, RAIM, and APPROACH lights are found at the bottom of the instrument panel. The RADIO/FMS selector switches are "around" the HSIs. Labels on the AP and FD knobs are stick-on, some backlite. Labels in the HSIs remain RAD and in some cases INS with placards showing that "INS=FMS". The result? The flight crew must determine the displayed source by looking at lights in two different locations.

Another HF Issue

Flight crew training is also an increasing problem with these upgraded classics. The older aircraft similators used for training typically do not represent the highly modified configurations being flown on the line today. The wide number of aftermarket installations of displays, TCAS, Windshear, and lastly FMS makes it nearly impossible to configure simulators to match those aircraft flown by the smaller operators. As a result, training in non-normal procedures and overall flight crew evaluation to include the use of operational checklists can introduce the possibility of negative transfer which could adversely affect flight crew performance during an actual non-normal condition.

Today's Typical Challenge

The installation of an FMS/GPS can provide the crew with better situational awareness by displaying navigational data on a map display. Therefore, an operator decided to replace a mechanical HSI with an electronic EHSI. The integration challenge involved finding sufficient real estate for the EHSI control panel plus FMS/GPS displays. The solution was to consolidate control of the ADF, TAC, and IFF functions in the FMS CDU.

Next, the operator wanted to add TCAS. Space constraints required the replacement of the IVSI with an electronic TVSI so that the TCAS could provide both resolution advisory guidance and traffic information. However, in order to control this TCAS function, the FMS needed to be modified to select TRA and targets above and below the aircraft's flight path.

Most recently, the same operator decided to add EGPWS. This required the addition of controls to switch between TCAS, EGPWS, and Wx Radar all on the EHSI display.

What is the driving issue? Crowded cockpits and the need to add flight deck functionality force bottom-up solutions. If these functions must be added, who is responsible for the overall integration to ensure that the flight deck retains its basic design workload integrity? This example involved one customer but a different vendor for each additional installation. FAA Order 8110.4A (March 2, 1995) is quite clear when it states in Chapter 4, par 28, page 75 that the following note must be included in every multiple STC:

> *"Compatibility of this design change with previously approved modifications must be determined by the installer."*

FAR 25.1309, ARP 4754, and ARP 4761

FAR Part 25, Section 1309, defines the regulatory requirement for some form of a system safety analysis during certification. During 1996, the Society of Automotive Engineers (SAE) issued two Aerospace Recommended Practices (ARP) relating to the certification of highly integrated or complex aircraft systems. They are intended to provide guidance to applicants in achieving compliance with FAR. ARP 4754 reviews existing procedures and provides additional guidance for the integration of new electronic systems on the flight deck. ARP 4761 provides guidelines for conducting the safety assessment process on civil airborne systems and equipment. Both of these documents include references to human factors and/or operational considerations which are relevant for STC applications involving individual avionics additions and/or upgrades.

Until the publication of ARP 4754 and ARP 4761, the System Safety Assessment had been primarily an engineering evaluation and was rarely required for a Supplemental Type Certificate. Now, due to these two ARPs, which will soon be formally recognized when the new FAR 25.1309 and AC 25.1309-1B are published, FAA is requiring STC applicants to meet the objectives of a System Safety Assessment from an integration standpoint on all STC applications.

Summary

It is imperative for design engineers to remember that the addition of one more function to a classic flight deck may very well drive pilot workload to an untenable level through an increase in functional.combinations and/or the incompleteness of information displayed.

There is currently no methodology for combining the system safety assessment with a workload assessment to arrive at an appropriate evaluation for the affect of additional mode displays, switching functions, and distribution of essential components around the flight deck. Nor, is there any regulatory requirement which relates FAR 25.1309 to FAR 25.1523 for upgrades to Classics. This historically has been done by the OEM on new designs but a workload evaluation has not yet been required for the integration of advanced electronic information systems under a Supplemental Type Certificates.

Without the OEMs providing at least basic guidance to upgrade the FMA, FD, and AP to accept a new FMS, there will continue to be limited and inconsistent changes introduced through the STC process to these already complicated flight decks. Classic aircraft are bought, sold, and leased between operators on a regular basis which results in a mixed fleet even within a given series. Design support for these Classics is necessary in order to provide consistent flight deck designs throughout the fleet.

Recommendations

1. Since a retrofit to Classics is always based on operational economics, there needs to

be some rational approach which still meets the overall safety goals but enables the integrator to control the price of the new design. One such approach might be to evaluate the new design on the basis of its intended operational environment and demonstrate it to be "at least as safe as the original design." (ref 1)

2. Flight crew experience needs to be included early in the design process so that engineers and pilots can design for an acceptable workload for the new STC as it integrates with all previous STCs on the aircraft. Operational pilots can provide critical input to the engineering team by evaluating and commenting on component functionality, pilot-device interface, cockpit integration issues, plus compatibility with current and projected airspace operational requirements.

3. Involving qualified pilots working with mock-ups, bench-top demonstrators, and, if necessary, even full-motion simulators (if properly configured) early in the design process can effectively reduce total design cost while enabling design engineers to thoroughly evaluate all workload-related issues.

4. Changes to the crew interface due to the addition of new technology in Classic aircraft should involve the active participation of the OEM. There is no cost incentive nor do the aftermarket integrator and operator have the capabilities to consider the broad operational affects of crew workload during the current evolution to FMS/GPS for precision approaches.

Terms & Abbreviations

ADS	Automatic Dependent Surveillance System
AP	Autopilot
ARP	Aerospace Recommended Practice
Box	Systems Engineers tend to think of "boxes" as the LRUs down in the EE bay; while Flight Test Pilots tend to refer to avionics displays or controls as "boxes." The latter represents the usage in this paper.
EGPWS	Enhanced Ground Proximity Warning Systems
EIS	Electronic Information System
FANS	Future Air Navigation System
FD	Flight Director
FMA	Flight Mode Annunciator
FMS	Flight Management System
GPS	Global Positioning System
HSI	Horizontal Situation Indicator
LAAS	Local Area Augmentation System
OEM	Original Equipment Manufacturer
STC	Supplemental Type Certificate
TCAS	Traffic Collision Avoidance System
TVSI	Traffic and Vertical Speed Indicator
WAAS	Wide Area Augmentation System
Wx	Weather

References

1. Barnes, R.B. and Adam, C.F., "The Aftermarket Integration of Modern Electronic Information Systems in Classic Aircraft, Human Factors System Safety Considerations," 16th International System Safety Conference, Seattle, WA, 14-19 September 1998.

2. Kingsley-Jones, M., and Sheppard, I., "Ageing Airliner Census 1997," Flight International, Reed Business Information, UK, 9-15 July 1997, pp. 37-50.

3. Society of Automotive Engineers, ARP 4033: Pilot-System Integration, SAE International, Warrendale, PA, 1995.

4. Society of Automotive Engineers, ARP 4761: Guidelines and Methods for Conducting the Safety Assessment Process on Civil Airborne Systems and Equipment, SAE International, Warrendale, PA, 1996.

5. Society of Automotive Engineers, ARP 4754: Certification Considerations for Highly-Integrated or Complex Aircraft Systems, SAE International, Warrendale, PA, 1996.

6. U.S. Department of Transportation, Federal Aviation Administration, Order 8110.4A, Washington, D.C., 2 March 1995.

7. U.S. Department of Transportation, Federal Aviation Regulations, Washington, D.C., 1998.

Authors' Biographies

Charles F. Adam, Flight Test Pilot, Federal Aviation Administration, Aircraft Certification Service, Transport Airplane Directorate, 3960 Paramount Blvd., Lakewood, CA,90712, tel -- +(562) 627-5369, fax -- +(562) 627-5210, e-mail – Chip.Adam@faa.dot.gov.

Mr. Adam has participated as a test pilot on the certification of the MD-11 and as the focal point for two-person crew conversions of the Boeing 727 plus restricted category versions of the Lockheed C-130, P-3, and L-188 "Electra." He currently is the FAA test pilot for the integration of the Collins FMS 800 and FDS 255 in the KC-10 and C-12, the FMS 800 in the C-9 and E-4, and Litton 410 FMS in the T-43. He holds a Bachelor's Degree in Electrical Engineering, a Master's Degree in Business Administration, and is a graduate of the U.S. Navy Test Pilot School. He completed twenty-two years of military service with ten of those years as an Experimental Test Pilot in which his assignments included research, development, and operational flight testing.

Mr. Adam is a member of the Society of Experimental Test Pilots, the American Institute of Aeronautics and Astronautics, the Society of Automotive Engineers, and the SAE's G-10 and A-4 Committees for heads-up displays.

Robert B. Barnes, President, Robert B. Barnes Associates, Inc., 8711 E. Pinnacle Peak Road, #337, Scottsdale, AZ, 85255, USA, tel -- +(602) 585-5703, fax -- +(602) 585-5703, e-mail – RbarnesAZ@worldnet.att.net.

His company has specialized for 10 years in assisting technology-based companies enter the commercial marketplace. As a result, he is regularly involved with the commercialization of leading-edge systems. Frequently, he provides human factors-related project management assistance to aerospace companies. He holds a Bachelor's Degree in Aerospace Engineering and a Master's Degree in Educational Psychology plus a Professional Certificate in Aviation Safety. A former USAF instructor pilot and flying safety officer, he also holds a Certificate of Master Instructor in USAF Flying Training.

His recent aviation-related experience includes the development and direction of the Cockpit Workload and Crew Complement Compliance Plans for a two-pilot cockpit conversion of the Boeing 727; for a two-pilot conversion of the Lockheed 188 "Electra"; and for an automated flight deck management system to be used in two-pilot versions of the Boeing 727 and other large transport airplanes. His experience includes the certification of after-market Flight Management Systems for air transport aircraft.

Mr. Barnes is a member of the American Institute of Aeronautics and Astronautics, Society of Automotive Engineers, System Safety Society, Association of Aviation Psychologists, and the SAE's G-10 Committee on Aerospace Behavioral Engineering Technology.